D1233372

Second Edition

Brain Mapping

Second Edition

Brain Mapping

The Methods

Edited by

Arthur W. Toga

Laboratory of Neuro Imaging
Department of Neurology
Division of Brain Mapping
UCLA School of Medicine
Los Angeles, California

John C. Mazziotta

Ahmanson-Lovelace Brain Mapping Center
Department of Neurology, Radiological Sciences and
Medical and Molecular Pharmacology
UCLA School of Medicine
Los Angeles, California

ACADEMIC PRESS

An imprint of Elsevier Science

Amsterdam · Boston · London · New York · Oxford · Paris · San Diego · San Francisco
Singapore · Sydney · Tokyo

This book is printed on acid-free paper. 

Academic Press
An imprint of Elsevier Science.
525 B Street, Suite 1900, San Diego, California 92101-4495, USA
http://www.academicpress.com

Academic Press
84 Theobalds Road, London WC1X 8RR, UK
http://www.academicpress.com

Library of Congress Catalog Card Number: 2002016354

International Standard Book Number: 0-12-693019-8

PRINTED IN CHINA
02 03 04 05 06 07 RDC 9 8 7 6 5 4 3 2 1

Contents

10 Magnetoencephalographic Characterization of Dynamic Brain Activation: Basic Principles and Methods of Data Collection and Source Analysis

Matti Hämäläinen and Riitta Hari

11 Transcranial Magnetic Stimulation

Alvaro Pascual-Leone and Vincent Walsh

III Tomographic-Based Data Acquisition

12 High-Field Magnetic Resonance

Kamil Ugurbil

13 Functional MRI

Joseph B. Mandeville and Bruce R. Rosen

14 Magnetic Resonance Spectroscopic Imaging

Andrew A. Maudsley

15 Principles, Methods, and Applications of Diffusion Tensor Imaging

Susumu Mori

16 Neuroanatomical Micromagnetic Resonance Imaging

P. T. Narasimhan and Russell E. Jacobs

17 CT Angiography and CT Perfusion Imaging

M. H. Lev and R. G. Gonzalez

18 Imaging Brain Function with Positron Emission Tomography

Simon R. Cherry and Michael E. Phelps

19 SPECT Functional Brain Imaging

Michael D. Devous, Sr.

IV Postmortem

20 Postmortem Anatomy

Jacopo Annese and Arthur W. Toga

21 Quantitative Analysis of Cyto- and Receptor Architecture of the Human Brain

Karl Zilles, Axel Schleicher, Nicola Palomero-Gallagher, and Katrin Amunts

V Analysis

22 Statistics I: Experimental Design and Statistical Parametric Mapping

Karl J. Friston

23 Statistics II: Correlation of Brain Structure and Function

Roger P. Woods

24 Advanced Nonrigid Registration Algorithms for Image Fusion

Simon K. Warfield, Alexandre Guimond, Alexis Roche, Aditya Bharatha, Alida Tei, Florin Talos, Jan Rexilius, Juan Ruiz-Alzola, Carl-Fredrik Westin, Steven Haker, Sigurd Angenent, Allen Tannenbaum, Ferenc Jolesz, and Ron Kikinis

25 Combination of Transcranial Magnetic Stimulation and Brain Mapping

Tomáš Paus

26 Volume Visualization

Andreas Pommert, Ulf Tiede, and Karl Heinz Höhne

VI Databases and Atlases

27 The International Consortium for Brain Mapping: A Probabilistic Atlas and Reference System for the Human Brain

John Mazziotta for the International Consortium for Brain Mapping

28 Subpopulation Brain Atlases

Paul M. Thompson, Michael S. Mega, and Arthur W. Toga

VII Emerging Concepts

29 Radionuclide Imaging of Reporter Gene Expression

Gobalakrishnan Sundaresan and Sanjiv S. Gambhir

30 Mapping Gene Expression by MRI

Angelique Y. Louie, Joseph A. Duimstra, and Thomas J. Meade

31 Speculations about the Future

John C. Mazziotta and Arthur W. Toga

Contributors

Numbers in parentheses indicate the pages on which the authors' contributions begin.

Katrin Amunts (573)
Institute of Medicine, Research Center Juelich, D-52425 Juelich, Germany

Sigurd Angenent (661)
Harvard Medical School, Boston, Massachusetts 02115

Jacopo Annese (537)
Laboratory of Neuroimaging, Department of Neurology, UCLA School of Medicine, Los Angeles, California 90095-1769

Aditya Bharatha (661)
Harvard Medical School, Boston, Massachusetts 02115

Paul D. Cheney (189)
Smith Mental Retardation Research Center and Department of Physiology, University of Kansas Medical Center, Kansas City, Kansas 66160

Simon R. Cherry (485)
Department of Biomedical Engineering, University of California at Davis, Davis, California 95616

Lawrence B. Cohen (77)
Department of Cellular and Molecular Physiology, Yale University School of Medicine, New Haven, Connecticut 06520

Michael E. Dailey (49)
Department of Biological Sciences, University of Iowa, Iowa City, Iowa 52242

Michael D. Devous, Sr. (513)
Nuclear Medicine Center and Department of Radiology, The University of Texas Southwestern Medical Center, Dallas, Texas 75390-9061

Joseph A. Duimstra (819)
Division of Chemistry, Beckman Institute, California Institute of Technology, Pasadena, California 91125

Joseph P. Erinjeri (159)
Departments of Neurology and Neurological Surgery, Washington University School of Medicine, St. Louis, Missouri 63110

Chun X. Falk (77)
Department of Cellular and Molecular Physiology, Yale University School of Medicine, New Haven, Connecticut 06520 and RedShirtImaging, LLC, Fairfield, Connecticut 06432

Karl J. Friston (605)
The Wellcome Department of Cognitive Neurology, University College London, WC1N 3BG London, United Kingdom

Sanjiv S. Gambhir (799)
Crump Institute for Molecular Imaging, Department of Molecular and Medical Pharmacology, School of Medicine, University of California at Los Angeles, Los Angeles, California 90095-1770

Alan Gevins (175)
San Francisco Brain Research Institute and SAM Technology, San Francisco, California 94108

R. G. Gonzalez (427)
Department of Radiology, Massachusetts General Hospital and Harvard Medical School, Boston, Massachusetts 02114

Alexandre Guimond (661)
Harvard Medical School, Boston, Massachusetts 02115

Steven Haker (661)
Harvard Medical School, Boston, Massachusetts 02115

Matti Hämäläinen (227)
MGH/MIT/HMS Athinoula A. Martinos Center for Biomedical Imaging, Charlestown, Massachusetts 02129

Riitta Hari (227)
Brain Research Unit, Low Temperature Laboratory, Helsinki University of Technology, 02015-HUT, Espoo, Finland

Karl Heinz Höhne (707)
Institute of Mathematics and Computer Science in Medicine (IMDM), University Hospital Hamburg-Eppendorf, 20251 Hamburg, Germany

Russell E. Jacobs (399)
Beckman Institute, Division of Biology, California Institute of Technology, Pasadena, California 91125

Ferenc Jolesz (661)
Harvard Medical School, Boston, Massachusetts 02115
Ron Kikinis (661)
Harvard Medical School, Boston, Massachusetts 02115
M. H. Lev (427)
Department of Radiology, Massachusetts General Hospital and Harvard Medical School, Boston, Massachusetts 02114
Angelique Y. Louie (819)
Division of Biology, Beckman Institute, California Institute of Technology, Pasadena, California 91125
Joseph B. Mandeville (315)
Massachusetts General Hospital NMR Center and MGH/MIT/HMS Athinoula A. Martinos Center for Biomedical Imaging, Charlestown, Massachusetts 02129
Andrew A. Maudsley (351)
Department of Radiology, University of Miami School of Medicine, Miami, Florida 33136
John C. Mazziotta (3, 33, 727, 831)
Ahmanson-Lovelace Brain Mapping Center, Department of Neurology, Radiological Sciences and Medical and Molecular Pharmacology, UCLA School of Medicine, Los Angeles, California 90095
Thomas J. Meade (819)
Division of Biology, Beckman Institute, California Institute of Technology, Pasadena, California 91125
Michael S. Mega (757)
Laboratory of Neuro Imaging, Brain Mapping Division, and Alzheimer's Disease Center, Department of Neurology, UCLA School of Medicine, Los Angeles, California 90095-1769
Susumu Mori (379)
Department of Radiology, Johns Hopkins University School of Medicine, Baltimore, Maryland 21218
P. T. Narasimhan (399)
Beckman Institute, Division of Biology, California Institute of Technology, Pasadena, California 91125
Hellmuth Obrig (141)
Division of Neuroimaging, Department of Neurology, Neurologische Klinik, Charite, Schumannstraße 20-21, 10117 Berlin, Germany
Nicola Palomero-Gallagher (573)
Institute of Medicine, Research Center Juelich, D-52425 Juelich, Germany
Alvaro Pascual-Leone (255)
Laboratory for Magnetic Brain Stimulation, Beth Israel Deaconess Medical Center, Harvard Medical School, Boston, Massachusetts 02115
Tomáš Paus (691)
Montreal Neurological Institute, McGill University, Montreal, Canada H3A 2B4
Michael E. Phelps (485)
Department of Molecular and Medical Pharmacology, UCLA School of Medicine, Los Angeles, California 90095
Andreas Pommert (707)
Institute of Mathematics and Computer Science in Medicine (IMDM), University Hospital Hamburg-Eppendorf, 20251 Hamburg, Germany
Nader Pouratian (97)
Laboratory of Neuro Imaging, UCLA Department of Neurology, Los Angeles, California 90095-1769
Jan Rexilius (661)
Harvard Medical School, Boston, Massachusetts 02115
Alexis Roche (661)
Harvard Medical School, Boston, Massachusetts 02115
Bruce R. Rosen (315)
Massachusetts General Hospital NMR Center and MGH/MIT/HMS Athinoula A. Martinos Center for Biomedical Imaging, Charlestown, Massachusetts 02129
Juan Ruiz-Alzola (661)
Harvard Medical School, Boston, Massachusetts 02115
Axel Schleicher (573)
C. and O. Vogt Institute of Brain Research, Heinrich-Heine University Duesseldorf, Duesseldorf, Germany
Gobalakrishnan Sundaresan (799)
Crump Institute for Molecular Imaging, Department of Molecular and Medical Pharmacology, School of Medicine, University of California at Los Angeles, Los Angeles, California 90095-1770
Florin Talos (661)
Harvard Medical School, Boston, Massachusetts 02115
Allen Tannenbaum (661)
Harvard Medical School, Boston, Massachusetts 02115
Alida Tei (661)
Harvard Medical School, Boston, Massachusetts 02115
Paul M. Thompson (757)
Laboratory of Neuro Imaging, Brain Mapping Division, and Alzheimer's Disease Center, Department of Neurology, UCLA School of Medicine, Los Angeles, California 90095-1769
Ulf Tiede (707)
Institute of Mathematics and Computer Science in Medicine (IMDM), University Hospital Hamburg-Eppendorf, 20251 Hamburg, Germany
Arthur W. Toga (3, 97, 537, 757, 831)
Laboratory of Neuro Imaging, Division of Brain Mapping, Department of Neurology, UCLA School of Medicine, Los Angeles, California 90095
Kamil Ugurbil (291)
Center for Magnetic Resonance Research, University of Minnesota Medical School, Minneapolis, Minnesota 55455
Arno Villringer (141)
Division of Neuroimaging, Department of Neurology, Neurologische Klinik, Charite, Schumannstraße 20-21, 10117 Berlin, Germany
Matt Wachowiak (77)
Department of Cellular and Molecular Physiology, Yale University School of Medicine, New Haven, Connecticut 06520

Vincent Walsh (255)
Experimental Psychology, University of Oxford, Oxford OX1 3UD, United Kingdom

Simon K. Warfield (661)
Department of Radiology, Brigham and Women's Hospital and Harvard Medical School, Boston, Massachusetts 02115

Ling Wei (159)
Departments of Neurology and Neurological Surgery, Washington University School of Medicine, St. Louis, Missouri 63110

Carl-Fredrik Westin (661)
Harvard Medical School, Boston, Massachusetts 02115

Roger P. Woods (633)
Department of Neurology and Brain Mapping Center, UCLA School of Medicine, Los Angeles, California 90095

Thomas A. Woolsey (159)
Departments of Neurology and Neurological Surgery, Washington University School of Medicine, St. Louis, Missouri 63110

Karl Zilles (573)
Institute of Medicine, Research Center Juelich, D-52425 Juelich, Germany and C. and O. Vogt Institute of Brain Research, Heinrich-Heine University Duesseldorf, Duesseldorf, Germany

Michal R. Zochowski (77)
Department of Physics, University of Michigan, Ann Arbor, Michigan 48109

Preface

The need for this second edition of *Brain Mapping: The Methods* speaks volumes about the rapid pace of development in this field. Not only have significant developments and refinements of existing methods occurred in the past few years but so has the emergence of entirely new approaches to mapping the structure and function of the brain.

The first edition of this book is a hard act to follow. The book's comprehensive coverage of all the major methods used in brain mapping was extremely well received. It can be found in most every neuroimaging laboratory in the world. *Brain Mapping: The Methods* is utilized by the majority of graduate training programs focusing on brain mapping. However, a field such as brain mapping moves rapidly. The technological advances are relentless and the introduction of entirely new approaches occurs at an astounding rate. This second edition is both essential and timely.

There continues to be an almost exponential increase in the number of scientists and laboratories engaged in brain mapping. The seemingly overwhelming accumulation of experimental results may be partly influenced by the ease with which experiments can be performed using off the shelf instrumentation. This book continues its tradition of including a comprehensive survey of methods used to study the brain. Consideration of the complementary methodology, included in this text, will further empower investigators with the ability to acquire a deeper understanding of the complex structure/function relationships in the healthy and diseased brain.

The result of maintaining a comprehensive coverage was a considerably larger book. Thus, its value now extends from a complete treatise on the subject matter to an all-inclusive reference as well. For example, in the field of magnetic resonance imaging alone, what was once two chapters (structural and functional MRI) is now five chapters including all the newly developed methods using magnetic resonance imaging to unveil even more information about the biochemistry, physiology, angio- and myelo-architectural details of the brain.

Some of the chapters were included in the first edition but received updates, whereas others are completely new. We selected methods that provided new and exciting insights, had been applied by multiple laboratories and had been validated and published sufficiently to fully understand their inherent limitations and assumptions. We maintained coverage across a wide span of measurements from the molecular level to the macroscopic scale. We added stimulation techniques that can be used either independently or in concert with other brain mapping methods. There are several chapters that outline strategies for combining methods to achieve simultaneous, complementary measures of brain activity and circuitry.

The contributing authors are pioneers in the field. They are often the developers of the technique or the first to fully characterize it when applied to the brain. They are neuroscientists that have an in-depth understanding of the underlying bases of the technique and first hand experience using it to study the complexities of the brain.

The book is organized by section. The introductory section includes chapters outlining the concepts of mapping, its history, requirements and applications. The second section contains chapters describing both optical based approaches to measure and map brain function and electrophysiological methods to record and activate cortical brain regions. There is also a chapter on transcranial magnetic stimulation, a technique increasingly used to assist in mapping complex cortical circuits. The section on tomographic-based data acquisition includes a wide range of methods utilizing magnetic resonance imaging, recognizing the dramatic influence and diversity that this class of imaging has had on mapping brain structure and function. Methods to map the anatomy, angiography, white matter and biochemistry of brain are now mature and frequently applied in health and disease. Coupled with postmortem maps, these methods enable a comprehensive survey of brain. There is an analysis section that describes the wide array of strategies employed to yield the most information possible from single and multi-modality maps. The final section, as in the first edition of this book, attempts to see beyond the state-of-the-art and predict the realization of several exciting opportunities for mapping the brain.

Arthur W. Toga
John C. Mazziotta

Acknowledgments

The fact that we decided it was time to prepare another edition of this book says a great deal. Not only about the subject matter and the field but about the scientists, many of whom are authors in this volume. Taking time out from prolific scientific activities to write such outstanding review chapters is clearly appreciated by the editors and the readers. You all should be proud of your contribution. Collectively these chapters represent the sum of considerable achievement in the field. Those of you that worked with me before must be accustomed to all the emails and red ink, to those of you that are new to this experience, 'wasn't it worth it?'

Assembling all the little pieces and checking all the little details takes more effort than most of us are willing to admit. Without my staff at the Laboratory of Neuro Imaging, it would not have been completed. Special thanks go to Sandy Chow and Lidia Uce for their organizational skills and perseverance once again. As always, the work that is represented by chapters that I wrote is a group effort.

Financial support for the research described in several chapters came from a number of institutions, including our NCRR supported Resource (RR 13642), the National Library of Medicine (LM 05639) and the Human Brain Project funded jointly by the National Insitute of Mental Health, the National Institute of Drug Abuse, the National Cancer Institute and the National Institute of Neurological Disorders and Stroke (52176).

I think I have finally figured out there may not be any personal notes distinct from professional ones. I am only me. By now my family is not only used to my time away doing these projects but they expect it. For that I am more than grateful, I am dependent. To my wife Debbie and children, Nicholas, Elizabeth and Rebecca, here is further affirmation of that gratitude.

Arthur W. Toga

I second Arthur's comments about the motivation to create this second edition. In the short time since the original text was published so much has happened in neuroscience. Brain Mapping has taken on an ever increasing role in providing insights into the human brain's structure and function in health and disease.

We are deeply indebted to the fine scientists that have contributed to this second edition and applaud their efforts and responsiveness, given their already hectic schedules and demands.

I also wish to express thanks to those institutions, foundations and agencies that support Brain Mapping at UCLA. Of the federal agencies in the United States, special thanks goes to the Human Brain Project that supports the International Consortium for Brain Mapping (ICBM) and is funded by the National Institute of Mental Health, the National Institute of Drug Abuse, the National Cancer Institute and the National Institute for Neurological Disease and Stroke (MH/DA 52176) and the National Center for Research Resources (RR12169). In the private sector, I wish to acknowledge the generous and ongoing support of the Brain Mapping Medical Research Organization and Support Foundation, the Pierson-Lovelace Foundation, The Ahmanson Foundation, the Tamkin Foundation, the Jennifer Jones Simon Foundation, the Robson Family, the Capital Group Companies Charitable Foundation and the North Star Fund.

The Brain Mapping Center staff, Palma Piccioni, Laurie Carr, Leona Mattoni and Kami Longson were essential to complete my role in this project. Our colleagues at Academic Press, specifically, Jasna Markovac, Ph.D., Hilary Rowe and Emily Scheese, continue to provide guidance, encouragement and solid production values for the Brain Mapping series. My thanks to all these individuals.

To my daughters, Emily and Jennifer, who recently mentioned that even now when they return from college, they still see me doing "homework" nights, mornings and weekends, I say this text is a product of that homework. I appreciate your sense of humor, tolerance and spirit.

Finally, to my co-author, I suggest we write less now and play more golf. We're just as awful as we were five years ago.

John Mazziotta

I

Introduction

1

Introduction to Cartography of the Brain

Arthur W. Toga* **and John C. Mazziotta**

**Laboratory of Neuro Imaging, Division of Brain Mapping, Department of Neurology, UCLA School of Medicine, Los Angeles, California 90024*

Where the telescope ends, the microscope begins,
which of the two has the grander view?
—VICTOR HUGO, LES MISÉRABLES (1862)

Cartography is the ancient art–science of making maps. Even before recorded history man has always tried to communicate the concept of place and its character. Whether the cartographic topic is a description of a shoreline or the boundaries of the basal ganglia, the fundamental issues are the same. In the study of brain, understanding the structure of the entire organ is as important as identifying the genetic program for cell replication. In either scale, cartography seeks to guide our explorations and record our progress. A good map must represent reality as we know it.

The goal of this chapter is to provide an introduction and guide through a complicated set of intersecting fields, reflecting the depth and scope of disciplines needed to carry contemporary brain mapping into the future. It is clear that the present state of brain mapping has been profoundly influenced by efforts in such diverse fields as art, geography, neurobiology, mathematics, and computer graphics. The traditions and languages of these and other scientific disciplines provide us with the tools for description and exploration, but also provide the context for future discovery.

In this second edition of this text, we have attempted to remain true to the full history of this very modern endeavor while updating the remarkable advances that have occurred in the few short years since its first publication. We have also added sections describing the variety of maps newly created and put forth the notion that the brain map and the database are inseparable and mutually dependent.

I. Introduction to Cartography

The words map, chart, atlas all bring to mind the notion of geography. But that is too restrictive. Although maps are usually graphic in nature, they can represent not only the spatial domain of an object, but virtually our complete understanding of it. Thus, maps can include both the *where* and the *what* of an object. Throughout history man has produced maps of the things around him. However, never before have so many things succumbed to the cartographer's descriptive quest. While it is often claimed of any era, no period in the history of mapmaking rivals the present in sophistication and effort (Hall, 1993). It seems that more and more of scientific discovery is structured around an attempt to create a map. Cartography has become a way of organizing data, representing them, and communicating results.

A. History of Cartography

Knowledge, or at least belief, must be possessed before a map can be made. It is also a clear record of the science and culture of the time. The format and the content of maps depict not only the mapmaker's understanding, but, to a certain degree, the thinking and concerns of the time. The history and evolution of mapmaking (see Robinson and Petchenik, 1976; Wilford, 1982) is fascinating, exciting, and predictive of its potential. It is interesting to note that the most important maps of the past were focused on charting the land and sea. Later, the object of interest was increasingly larger in scope, including the depths of the seas, other planets, and even other solar systems. But most recently, the topic of choice has turned inward to the creation of a map of the human genome and, the topic of this book, the human brain.

Brain maps are corollaries to geographical maps. A brief account of mapmaking history not only provides a description of progress but also defines the disciplines that guide us even today. Many of the organizational principles that were developed to structure map data are still in practice. These include coordinate systems, hierarchical nomenclature, and multiple modalities.

By the 5th century B.C., Greek scholars realized that the earth was a globe that offered irresistible opportunities for symmetrical subdivision. The first attempt incorporated parallel lines around the sphere that were called "climata." The functional significance of these structural lines lay with the result that the length of the longest day was roughly equivalent for all regions within a defined zone. The Greek, Hipparchus (ca. 165–127 B.C.), developed a system similar to modern-day latitude and longitude, but instead of using regular intervals to separate the divisions, he first used ancient and prominent cities. The ability to recognize and refer to landmarks was as critical to the success of coordinate systems as it is today. Hipparchus later realized that a regular, evenly divided system would be better for providing a comprehensive coordinate system to locate all space. Ptolemy (ca. 127–151), however, is the undisputed father of modern geography because he devised a way to project the spherical earth onto a planar surface, something that was a prerequisite to the construction of accurate maps. [There are several groups in the brain mapping community who also subscribe to the value of planar or inflated representations of the somewhat spherical but topologically complex cerebral cortex (Van Essen *et al.,* 1997; Dale, *et al.,* 1999)]. His theories still had problems though, one of which was the placement of the earth at the center of the solar system. (He was also responsible for the placement of north at the top, solely because more was known about that hemisphere at the time.) By 1000–1200 A.D. grids were being emphasized and geographical and celestial coordinate spaces linked. These efforts mark the beginning of multidimensional and multimodal mapmaking. However, the grids of that time generally did not make allowances for the curvature of the earth. Mercatur (1512–1594) solved that problem with what became known as the Mercatur projection. Longitudinal lines were equivalent to cuts in an orange; sections could be flattened and stretched so that their points were elongated, resulting in a single large rectangle. Longitude became a series of parallel lines.

The same problems that vexed the early geographical mapmakers continue to challenge the developers of a brain map. How to represent the complexities of the object (earth or brain) sufficiently to guide exploration and accurately record progress and at the same time keep them understandable and usable. Brain (and other) maps must have appropriate presentation strategies to convey the pertinent information most obviously without obscuring other data, reference systems suitable for equating the map with the real object, and a product that is compatible with its intended use.

B. The Role of Instrumentation

The extraordinary pace of development of measurement devices over the past quarter century has revolutionized our ability to explore and chart everything around us. This leap from naked-eye observation to instrument-aided vision was one of the greatest advances in the history of human exploration. We can visualize and measure things at both extremes of scale, from angstroms to light-years. A map of the human genome is complete (Collins, 2001) and radio astronomists have created charts that probe the very beginning of the universe. Neither of these endeavors would be possible without the extraordinary sensitivity and variety of detectors. As will be noted in other chapters, there are usually trade-offs in time and space as detectors are often inefficient or highly specific in their collection of data. Currently, many detectors, measuring some aspect of brain structure or function, take advantage of electromagnetic emissions, absorptions, or reflections as they interact with the brain (Orrison *et al.,* 1995; Mazziotta *et al.,* 1995; Stark and Bradley, 1992; Seeram, 1994). By plotting several of these brain mapping instruments (see Fig. 1) relative to the electromagnetic wavelength used, it is clear that many possibilities exist for future detectors. But taking advantage of this information requires additional instrumentation beyond that of the detector. The volume of data that can be acquired, the speed with which it accumulates, and the manipulation of the source signals

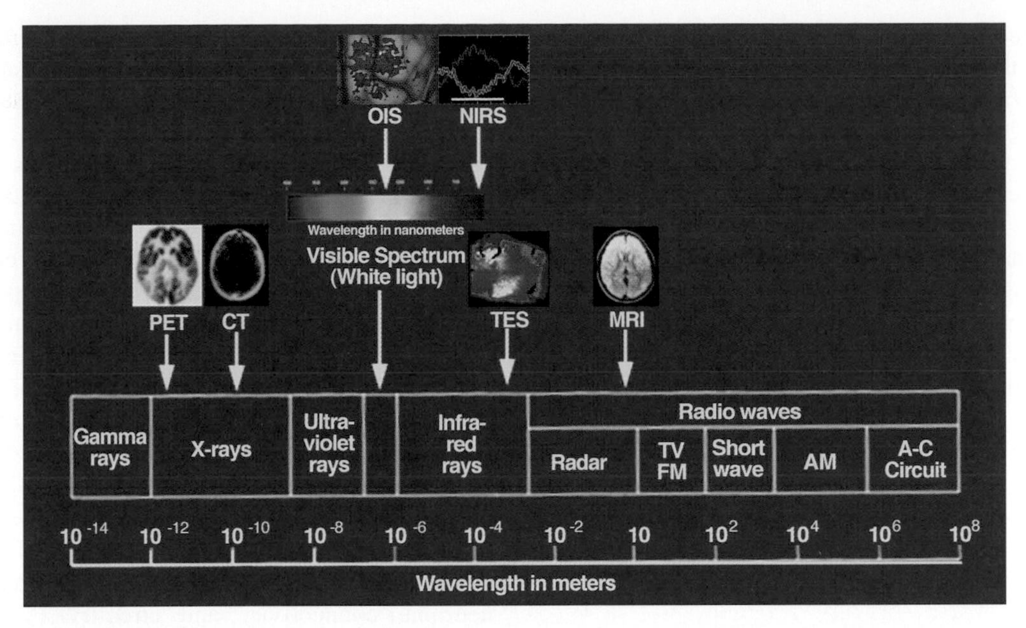

Figure 1 Electromagnetic spectrum. Many brain mapping methods take advantage of signals along the electromagnetic spectrum to derive the data describing brain structure or function. Positioning several methods graphically demonstrates the fact that there is potential room for many other detectors. In order to place these methods on a single graph, some liberties were taken. This graph shows the position of positron emission tomography (PET) as near 10^{-11} m. Positron annihilations produce gamma radiation in the 511 KeV range and using laws of physics, we can calculate the following (using Planck's constant), $\nu = E/h; f = E/2\pi h; f(\text{PET}) = 511 \times 10^3$ eV$/2\pi \times 4.135 \times 10^{-15}$ eV s $= 1.97 \times 10^{17}$ Hz, which results in a relative position along the graph. Magnetic resonance imaging (MRI) does not measure radio frequencies, but uses them to alter the spin of the nucleus. Computerized tomography (CT) utilizes X-rays as a penetrating beam of energy and optical intrinsic imaging (OIS), near-infrared spectroscopy (NIRS), and thermoencephalography (TES) take advantage of light measurements in the visible, near-infrared, and infrared bands, respectively.

needed for interpretation have mandated the application of numerical machines—computers.

C. Computers

In brain mapping, as in other cartographic efforts, computers are used to analyze, visualize, archive, and disseminate the information that constitutes the map. Every method described in this book utilizes digital processing for one or more of these tasks. In fact, many of the modern approaches used to study the brain are possible only with computers and several even acknowledge that dependence in their name or acronym. The revolution mentioned above was even more acicular when it comes to the inclusion of the computer. For example, the ability to examine the rapid neuronal or biochemical events of an activated region of brain often requires the sampling of data streams exceeding many millions of cycles per second. Conversion of these raw data into something comprehensible during the experiment has afforded the opportunity to alter the protocol in response to results in real time. Computerized representations can be constantly revised, each new version building upon the last. The available speed of computers permits measurement and plotting of change and rates of change, even when the time course is milliseconds. Computer-aided methods offer a smooth transition between measurement and interaction. The computerized map becomes an instrument that allows experiments to be conducted beyond that possible with the source data. Complex computational strategies enable the creation of statistical maps that describe multiple modalities, parameters, and populations of subjects representing a diverse array of characteristics. Simulations can be performed. Further, the results can be approximate, incorporating confidence limits based upon the amount of information contained in the collection.

D. Visualization

The most evident application of computers in mapping is in the field of visualization. Computerized visualization of data that have been converted into a cartographic image of a spatial domain best communicates the meaning of quantitative information. Some people can see in their mind's eye the beauty and structure in mathematical or statistical relationships, but most require a visual representation to best appreciate these dependencies. Visualization has played an important role in the history of mathematics and is an absolute necessity for cartography. Computerized visualization is used to plot data from a variety of sources to a representative map. Indeed, the brain maps described throughout this book are almost universally computerized renditions made incomplete when force-fit to printed form. The use of color, contours, and other visual cues to differentiate receptor densities, metabolic rates, electrical field potentials, and other attributes of structure or function produces multimodal, thematic, and holistic views of the brain. The product is not only informative but in many cases provocative, in that the overall patterns or relationships produced may tell a story entirely different from that told individually. Increasing the complexity and completeness of these computerized maps demands careful attention to their composition. The conscientious reader is directed to Tufte (1983 and 1990).

It is important to note that computerized visualization (graphic) techniques have been used not only to visualize already known phenomena but also, more interestingly, as participants in the process of solving problems not yet completely answered. The filtering effect of a colorized map can make sense of a confusing array of numbers. Manipulating the way data are visualized can improve the apparent signal-to-noise ratio. More and more, analytical and graphic-capable computers are used in a symbiotic way with multiple sensitive detectors as an effective instrument for brain mapping.

II. The Dimensions of a Brain Map

The difference between a map of a city and one of the planet is primarily detail. Although it is conceivable (and probably forthcoming) to incorporate all detailed maps (for a given subject) into a single database (Koslow and Huerta, 1997; Wong, 1998), present efforts typically focus on one point in the spatial continuum. This is certainly true for maps of the brain. For example, there are neuromaps of the whole brain (Toga *et al.,* 1995b; Damasio, 1995) and maps of small collections of neurons (Troyer *et al.,* 1994; German *et al.,* 1988). There are comprehensive maps of brain structure at a variety of spatial scales based upon 3D tomographic images (Damasio, 1995), anatomic specimens (Talairach *et al.,* 1967; Talairach and Tournoux, 1988; Ono *et al.,* 1990; Duvernoy, 1991) and different histologic preparations which reveal regional cytoarchitecture (Brodmann, 1909) and regional molecular content such as myelination patterns (Smith, 1907; Mai *et al.,* 1997), receptor binding sites (Geyer *et al.,* 1997), protein densities, and mRNA distributions. Other brain maps have concentrated on function, quantified by positron emission tomography (PET; Minoshima *et al.,* 1994), functional MRI (Le Bihan, 1996), or electrophysiology (Avoli *et al.,* 1991; Palovcik *et al.,* 1992). Additional maps have been developed to represent neuronal connectivity and circuitry (Van Essen and Maunsell, 1983), based on compilations of empirical evidence (Brodmann, 1909; Berger, 1929; Penfield and Boldrey, 1937). Each clearly has its place within a collective effort to map the brain, but unless certain precautions are taken (enabling common registration), they will have to remain as individual and independent efforts. Figure 2 illustrates a number of examples found in current neuroscience.

Figure 2 Maps of the visual system. To illustrate the variety of neuroscientific maps at different spatial and temporal scales, for different objects of study, and in method of presentation, this collage depicts numerous examples of maps of the visual system. Some are anatomical, others functional, and still others a combination of the two. Some maps incorporate the traditional components of cartography such as coordinate systems and a naming convention while others are merely visual representations of observed phenomena. The maps selected here for example demonstrate the diversity of approaches and applications. (**A**) Schematic of the visual system, dated 1083, that imposed a mathematical basis for the theory of vision and the theory that light radiated in straight lines from each point of an object. It was originally drawn by Alhazen, an Arabic scholar, and published in his *Kitab al-Manazir* (the numeric references are modern). (**B**) A drawing by Descartes published after his death in 1662, demonstrating the formation of images on the retina. (**C**) Introduction of the concept of cortical localization of function in 1810–1819. Franz Joseph Gall believed that these functions could be associated with cranial features—phrenology. Some of these features, he thought, included a sense of space and color. (**D**) Different elements of the retina were identified in 1866–1872, including the photoreceptors associated with scotopic and photopic vision. These 10 layers of the retina were drawn by Max Schultze. (**E**) Brodmann's 1909 numerical map of the human brain, based on a cytoarchitectonic study, recognized 52 discrete areas. The distinctions were based on cortical thickness, horizontal laminations, thickness of the laminae, type and number of cell types, cell density, and certain staining characteristics. (**F**) A map describing the major visual pathways and their terminations based on studies of 19th and 20th century scientists. **A–F** reprinted, with permission, from Finger (1994). (**G**) Map of cortical areas in the macaque. Current understanding of the layout of different visually related areas, wherein each is indicated by a different color-coded scheme. For example, occipital lobe—purple, blue, and reddish hues. Arial boundaries have been demarcated on this cortical map. Visual areas were identified in one hemisphere by architectonic criteria, pattern of interhemispheric connections, and transpositions of boundaries from a different brain, usually with a different type of representation. (**H**) Hierarchy of visual areas depicting 32 visual cortical areas, shaded according to same scheme as in F. At the bottom, RGC and LGN represent the primary source of visual inputs to the cortex. These areas are connected by 187 linkages, most of which are reciprocal pathways. **G** and **H** reprinted, with permission, from Felleman and Van Essen (1991). (**I**) Activation of visual cortex as mapped with *f*MRI. During photic stimulation, localized increases in signal were detected in the primary visual cortex. Reprinted, with permission, from Belliveau *et al.* (1991). (**J**) Visual cortex activation as mapped using PET. (**K**) Map of orientation selectivity in the striate cortex. Orientation preferences are distributed in continuous patches where they change linearly but slowly. This has been inferred by combining differential images [one positive, one negative image of cortex responding alternately to complementary (orthogonally oriented contours moving bidirectionally) stimuli]. Positive (dark) values indicate preferences for the test orientation, while negative (light) values indicate preferences for the orthogonal orientation. Reprinted, with permission, from Blasdel (1992). (**L**) Computer simulation of the formation of ocular dominance domains consistent with anatomical and physiological data. Each retinal projection is represented in a different color. The eye-specific firing was simulated by passing elongated "waves" of activity across each retina. Ocular columns arise during the ingrowth of fibers serving a simulated retina. Reprinted, with permission, from Montague *et al.* (1991). (**M**) Variability map encoding cortical variance in a population. (**N**) Asymmetry map illustrating right–left differences in sulcal lines. (**O**) Displacement map color coded in millimeters showing cortex displacement to map an individual to an average. (**P**) Tensor map indicating magnitude and direction in a population. (**Q**) Probability map showing the local likelihood of belonging to a given population. (**R**) Disease-specific model comparing the ventricular system and Alzheimer's patients with age-matched controls. (**S** and **T**) Maps of systems within the brain, such as the sulcal ribbons of **S** and the ventricles of **T**. These maps can encode variability, diffusion, probability, asymmetry, or other statistics. **Note the chronological evolution of the subject of the map as well as the improvement in sophistication**.

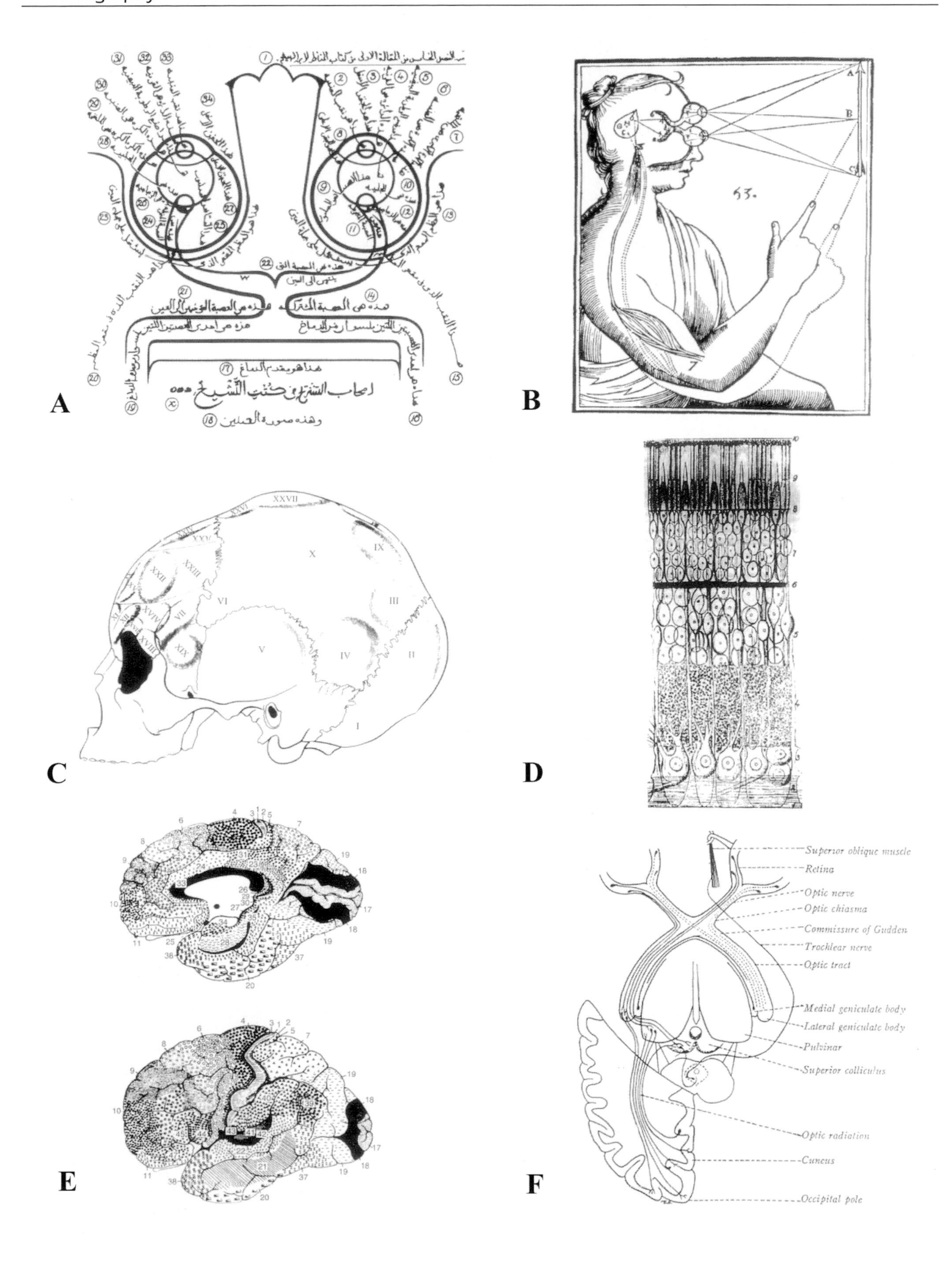
A
B
C
D
E
F
Superior oblique muscle
Retina
Optic nerve
Optic chiasma
Commissure of Gudden
Trochlear nerve
Optic tract
Medial geniculate body
Lateral geniculate body
Pulvinar
Superior colliculus
Optic radiation
Cuneus
Occipital pole

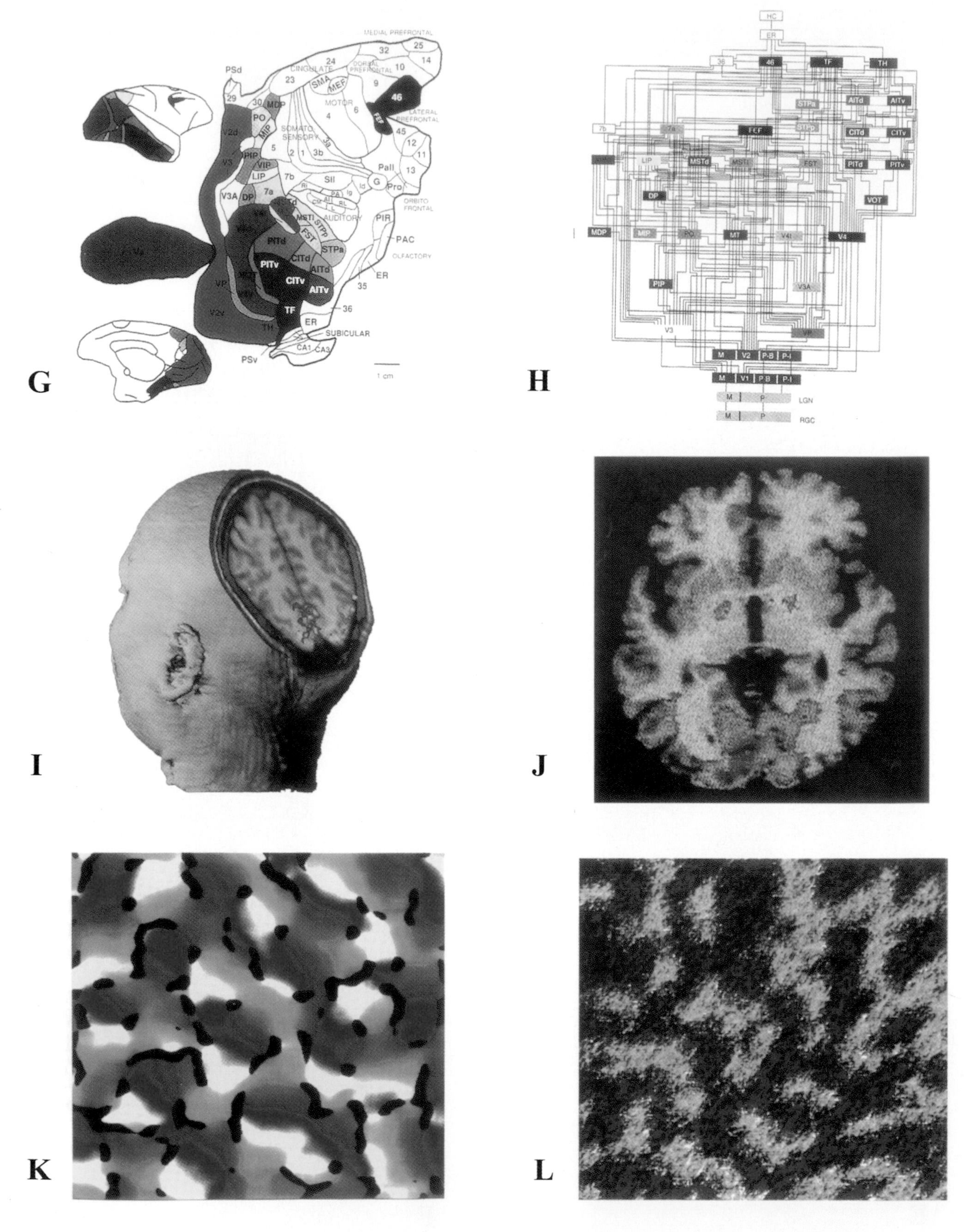

Figure 2 *Continued.*

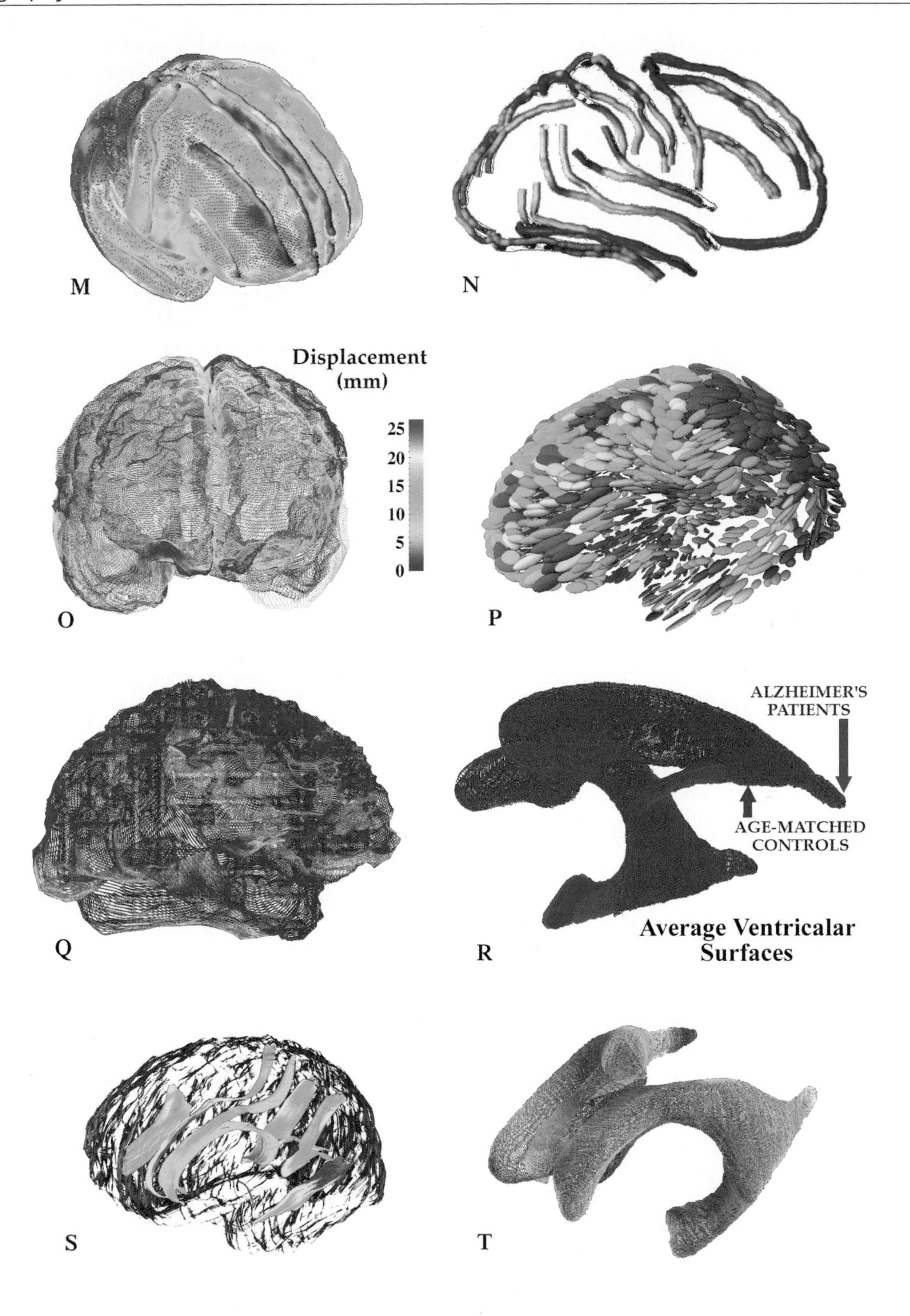

Figure 2 *Continued.*

A. Spatial Resolution

There are brain maps of structure derived from tomographic (Damasio, 1995), anatomic (Duvernoy, 1991), and histologic (DeArmond *et al.,* 1989) preparations. Each of these has different degrees of spatial resolution, none are compatible with the others, and they each were developed for different purposes. Brain maps such as these are most often applied as atlases, incorporating a coordinate system useful for stereotactic localization. In some maps, the intent is to generate a data set that helps in the interpretation of functional scans. Given the fact that there is neither a single representative brain nor a clear understanding of variation within an average, the notion of multimodal and multisubject maps deserves considerable attention later in this chapter and in other chapters of this volume. Other maps have concentrated on function, such as PET (Minoshima *et al.,* 1994); electrophysiology (Avoli *et al.,* 1991; Palovcik *et al.,* 1992); circuitry (Van Essen and Maunsell, 1983); and compilations of empirical evidence (Brodmann, 1909; Berger, 1929; Penfield and Boldrey, 1937). Historically, functional brain maps have captured a great deal of attention. Phrenological maps (Macklis and Macklis, 1992) were among the first to present spatial representations of brain function (albeit erroneously). The interest generated by these and related brain maps may be due to the apparent human need for gaining more insight into what it is that makes us who we are.

Generally, spatial resolution has been determined by the kind of detector used to collect the data. Increasing the field of view typically results in lower spatial resolution. Similarly, increasing the temporal resolution often reduces the spatial resolution. These trade-offs are due to the physical properties of both the instrumentation and the biology. They are discussed in detail in Chapter 2. Efforts to computerize the relationship between results from approaches tuned to one or the other end of the spatial/temporal scale ultimately will provide the best of all worlds, but currently these efforts are still under development. It should be noted, however, that computational and digital storage limitations are increasingly obviated by dramatic improvements in capacity, speed, and performance of computers and data storage.

B. Attribute Data

The dimensions of a map are not restricted to the spatial domain. The *what* of a map often goes far beyond structural observations and can include a myriad of attributes better characterizing each location. Most commonly, these include measures of the neuronal, metabolic, and vascular behavior of the anatomic regions. But as mapping strategies become more refined, additional information about the state of the subject, the methods used to collect the data, and related issues can be included with the map. The notion of the brain map and the database as a unified representation enables an integrated, computerized, as well as comprehensive, description of brain structure and function, along with the particulars of acquiring those data. Given the relentless revision and evolution of brain maps, comprehensive documentation of each version is critical to efficient and appropriate use of the information.

C. Time Series

A map of changes in the brain can be relatively slow, as with the development (Thompson *et al.,* 2000b; Sowell *et al.,* 1999), degeneration (Mega *et al.,* 2000b), evolution, or growth of tumors (Fetcho, 1992; DeClerk *et al.,* 1993; Haney *et al.,* 2001), or fast, as with electrophysiology (ERP and MEG) and blood flow (*f*MRI, ultrasound, laser Doppler, and OIS). Note that not all methods provide the same degree of sampling and that some methods are tomographic and others are not. The insight provided by time-series maps has been applied to ontogenic processes at the gross (Armstrong *et al.,* 1995) and cellular (Rance *et al.,* 1994) levels. They are also informative in studies that examine a graded response to input or even dose/response characteristics. The evolution of response magnitude is best characterized with a time-series map. Since the rate of change is rarely linear, a time-series map can help localize the contributing factors that influence the development of the response. Compelling examples of this can be seen with measures of cortical vascular and metabolic responses to stimulation using optical intrinsic signal imaging (Narayan, 1994a,b; Borg *et al.,* 1994; Haglund *et al.,* 1992; Grinvald *et al.,* 1991) *in vivo* and voltage- and calcium-sensitive dyes (Sugitani *et al.,* 1994; O'Donovan *et al.,* 1993) *in vitro.*

The time-series map itself is, in most cases, static. It can represent a rate of change or can be a series of maps, each representing a snapshot in time. Recently, we have warped annualized tissue loss in the hippocampus associated with Alzheimer's disease. There, regional, dynamic effects can be depicted in a specific map (Thompson *et al.,* 2001b). Electronic maps, on the other hand, published as computer disks or distributed via a network, can retain more information about the timing. The map itself can be an animation (Palovcik *et al.,* 1992) capable of playing back the sequence of events as measured. Increasingly, as the technology becomes more commonplace and the biological questions require it, such multimedia maps will find more utility. In multiple sclerosis, for example, serially acquired images can be "played back" to examine sites where future lesions will occur, allowing a better understanding of the natural course of pathophysiology in this disorder.

D. Derivatives and Probabilities

As was noted above the detectors used to collect anatomical and physiological brain data are often restricted in their modality and spatial/temporal field of view. The combination of several maps to derive a holistic representation is often informative. Maps that present correlative, multimodality (Chen *et al.,* 1990; Mazziotta *et al.,* 1995; Strother *et al.,* 1994), or other types of derived data can extend the map's usefulness beyond that available from the individual sources alone. This kind of derivative map is in contrast to the reductionistic approach necessary to collect the data.

The rapid development of informatics approaches to brain mapping has revolutionized our ability to integrate disparate maps. The notion of confidence limits attached to maps is relatively new (Mazziotta *et al.,* 1995). But the additional comprehension obtained by attaching probabilities or otherwise accommodating variability to the map data can be invaluable, especially when the map is intended to represent more than one observation. The value is even greater if the map represents a population (Evans *et al.,* 1993; Toga and Thompson, 1996, 1997; Thompson, *et al.,* 1997, 2000a). However, the methods necessary for a statistical framework to handle these new holistic measures of brain structure and function are still evolving.

The use of spatial normalization schemes has yet to completely accommodate the most variable brain structure, the cortex. The cortex is also the site of interest for most functional activation studies. Considerable normal variation in sulcal geometry has been found in primary motor, somatosensory, and auditory cortex (Missir *et al.,* 1989; Rademacher *et al.,* 1993); primary and association visual cortex (Stensaas *et al.,* 1974); frontal and prefrontal areas (Rajkowska and Goldman-Rakic, 1995); and lateral perisylvian cortex (Geschwind and Levitsky, 1968; Steinmetz *et al.,* 1989, 1990; Ono *et al.,* 1990). More recent 3D analyses of anatomic variability, in postmortem, *in vivo* normal and diseased populations, have found a highly heterogeneous pattern of anatomic variation (Thompson *et al.,* 1996a,b, 1998).

In view of the complex structural variability between individuals, a fixed brain atlas may fail to serve as a faithful representation of every brain. Since no two brains are the same, this presents a challenge for attempts to create standardized atlases. Even in the absence of any pathology, brain structures vary between individuals not only in shape and size, but also in their orientations relative to each other. Such normal variations have also complicated the goals of comparing functional and anatomic data from many subjects (Rademacher *et al.,* 1993; Roland and Zilles, 1994). Clark *et al.* (1991) presented initial encouraging results for a statistical model that assessed the probability that an individual PET was normal or similar to patients with Huntington's disease. Evans *et al.* (1993) developed large (>305) collections of MRI data and combined them into a common database resulting in a single average atlas. Thompson *et al.* (1998) and others (Pederson *et al.,* 1999; Narr *et al.,* 2000a,b; Thompson *et al.,* 2001a; Sowell *et al.,* 2000a,b, 2001) have created many population-based atlases that describe the morphology of healthy and diseased cohorts.

III. The Full Scope of Brain Mapping

A. Coordinates to Nomenclature

We use neuroanatomic labels to communicate, maps use coordinate space to quantitate. Equating the relationship between neuroanatomic labels and a Cartesian (or polar) coordinate system is one of the by-products of a map. The principles of most brain maps (and systems based upon them, e.g., stereotactic surgery) depend on a spatial structure defined by the Cartesian coordinate system. Descartes (1637) observed that it was possible to define locations of individual points in space by relating that point to three orthogonal intersecting planes. To define a location, three coordinates must be specified, each describing the distance from a plane of origin referenced to a reliable set of anatomical points. Figure 3 illustrates the value of coordinate systems for 3D targeting.

Sophisticated multimodality mapping and quantitative morphometric analyses have been accomplished through the use of the ubiquitous human brain coordinate system of Talairach and Tournoux (1988). The emergence of maps such as those of Tiede *et al.* (1993) and Steinmetz *et al.* (1989) has allowed the precise localization of targets deep to the surface of the brain and established the relationship between accepted neuroanatomic nomenclature and coordinates. Others label neuroanatomical structures for visual identification without the express purpose of stereotactic localization (e.g., Damasio, 1995; Duvernoy, 1991). Most modern efforts to create atlases are electronic and thus are easier to update with additional subdivisions, aliases, or even alternative coordinate systems. Despite these advances, most published maps remain disadvantaged by coarse spatial detail and low subject representation.

B. Templates

The use of previously established templates as aids for the delineation of anatomy (or functional anatomy) has been the holy grail of many a neuroscientist. The expectation is that appropriately developed maps should provide these. The idea is to use a standard set of outlines as the basis for identifying defined regions of the brain. There are several difficulties. First, there is the imprecise fit between a stylized template and a given brain data set. Second, the accuracy and detail present in the map containing the templates must fit the data. Nevertheless, several recent approaches have proven useful in this quest to use templates for automatic segmentation of a whole brain. These are best illustrated (and most successful due to minimal

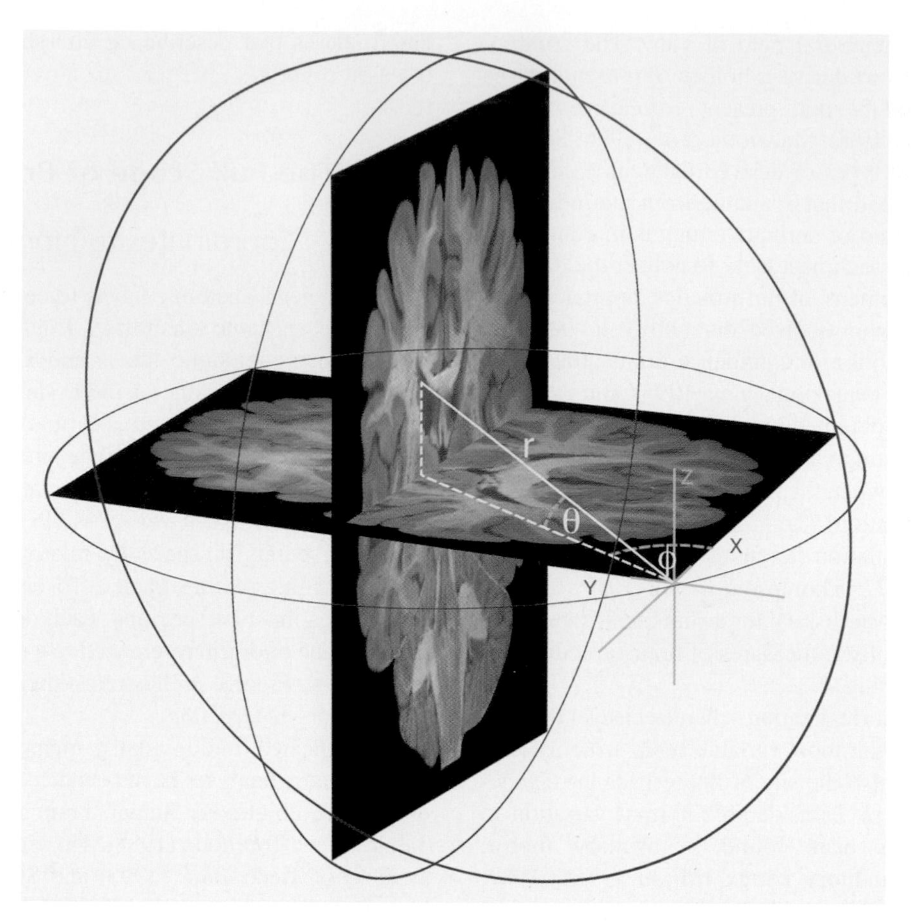

Figure 3 Coordinate referenced maps. A numerically indexed reference system greatly enhances the utility of a map. However, it imposes a quantitative requirement that certain data cannot meet. Brain data are usually referenced to anatomical landmarks and an adaptable coordinate system to accommodate the size and (some) shape variability across data sets. The use of coordinate systems (Cartesian and Polar) to identify any point within the 3D space occupied by the brain data is illustrated. Indexing is usually based on distance from a specific anatomic point. All stereotaxic approaches require 3D localizations. Only a single slice through the brain volume is shown for clarity. Traditionally, anatomically referenced systems are employed, and transformations to other systems are possible with defined relationships. The transformation to different scales or different points of origin are straightforward but may require some interpolation of the data.

morphologic variability) in atlases of nonhuman neuroanatomy, most notably the rat (e.g., Paxinos and Watson, 1986; Swanson, 1990). Electronic versions containing point and vector lists describing anatomic structures were developed by Bloom *et al.* (1990) and Swanson (1992), respectively. There also are publications in the human literature. Greitz *et al.* (1991) and Roland *et al.* (1994), among others, developed computerized brain atlases that can be adapted to fit individual anatomy. The objective in these atlases was to provide accurate anatomic delineations for functional studies, but the significantly greater variability in humans adds a distortion correction requirement that is difficult to implement and validate (see below). Numerous other approaches combined carefully delineated individuals or populations with spatial normalization to effectively label the anatomy of any individual placed in the atlas (Lancaster *et al.,* 1997; Schmahmann *et al.,* 1999; Toga *et al.,* 2001a; Thompson *et al.,* 2001b).

C. Anatomic Collections

Cytoarchitectural and chemoarchitectural collections of anatomy are most often used as the golden standard for structural delineations. The arsenal of histochemically differentiating stains has enabled anatomists to reliably describe subnuclei, laminae, and other subdivisions throughout the brain. However, collections themselves do not make a map. But they can provide the source material to build one. Such an approach has been attempted using the Yakovlev (1970) brain bank. As is discussed in detail in Chapter 20, Postmortem Anatomy, of this book, maps derived from these collections were unsatisfactory for several reasons. First, the resolution available using *in vivo* tomographic techniques is low; specifically, gray/white differentiation and identification of other features such as vascular anatomy are poor. Second, the data collected usually are not complete enough to be compared with other modalities that excel at one or another aspect of neuro- and related anatomies. Third, the data do not have the potential to be organized as a 3D volume and frequently are not in a digital format. Thus, the contribution of anatomic collections is best seen when used in combination with other brain mapping methods or if the collections were acquired using novel digital, 3D-compatible techniques. Morosan *et al.* (2001) and Binkofski *et al.* (2000) have developed several collections that provide detailed maps of the cyto- and chemoarchitectures equated to other maps of the same subject. Toga *et al.* (1994a,b, 1995a,b) also has created 3D surveys of the entire human brain at 100-μm resolution from postmortem material relative to MR and PET antemortem of the same subject.

D. Visualizations

As was noted above, visualization is an essential element in cartography and especially useful in a brain map. Visualization enables us to extract meaningful information from complex data sets. The multidimensional representation of brain lends itself to a variety of computerized visualization techniques concerned with modeling, manipulation, and display. If we can visualize a structure, we can map it. By employing the techniques of image processing, image synthesis, and computer graphics, we combine the utility of image and number, enabling us to statistically measure the visual representation. Morphometric measurements such as area, volume, and surface complexity can be determined from visible and digital representations. Further, such statistical analyses can be extended to compare and correlate a given view of the brain with data from other modalities and other subjects. The results can be rendered as independent visualizations or in relationship to brain anatomy, for example.

Regardless of its source, organizing data into volumes can make visualization easier. A complete collection of brain slices can be reconstructed into a volume, whether it comes from noninvasive or physically sectioned methods. Volume visualization has received considerable interest in recent years, resulting in several publications entirely devoted to it. Chapter 26 of this book provides comprehensive and excellent coverage of this topic. [Other sources include Friedhoff and Benzon (1989), Kaufman (1991), Robb (1985), Toga (1990b), and Udupa and Herman (1991). Related advanced topics in medical volume visualization can be found in Hohne *et al.* (1990) and Levoy *et al.* (1991).]

Visualizing measured quantities within a meaningful spatial framework is the intent of digital image display. Choosing the most appropriate method of presentation requires a diverse skill set that includes artistic, psychophysical, and statistical considerations. Since the goal of display is to describe, summarize, and, in some instances, interact with the data, great care is necessary in its design. For brain mapping purposes we must use an image composition that conveys several important characteristics about the data sets simultaneously. These include spatial, densitometric, correlative, and sometimes temporal information. Assuming 3D, the location and orientation of the reconstructed data and the shape, size, and relationship between substructures within the model describe the spatial features of the data set. The intensity or magnitude of a response is conveyed by the value of the pixel or voxel. These values can be transformed to represent a physiological measurement and/or pseudocolored to enhance their differences. Correlating multiple modalities assumes the data sets are in register following the necessary geometric transformations. Display of the correlations can be in the form of superpositioning multiple data sets, each with different color assignments (Mazziotta *et al.*, 1995), texture mapping one modality upon another (Payne and Toga, 1990), or a statistical representation of the relationship between data sets (Toga *et al.*, 1986, 2001a; Toga, 1993, 1994).

E. Transformations

In brain mapping, the need for control of image acquisition error, accurate morphometric measurements, and multisubject comparison has raised practical challenges. Combining images across modalities and subjects for inclusion in a common map requires positional and shape transformations to make them occupy the same coordinate space. A large literature focused on mathematical manipulations of the spatial representation of the brain data attests to the importance of this field (Toga, 1998). Cartographic transformations have long dealt with the problem of projecting a spherical data set (the earth) upon a flat surface (a map). However, since brain maps have, for the most part, adopted a volumetric format spatial transformations have concentrated on the multimodality and multisubject problem. [Note that several investigators have adopted alternate geometric representations, such as cortical flattening (Schwartz and Merker, 1986; Van Essen and Maunsell, 1980) and spherical and inflated topologies (Fischl *et al.*, 2001).] For example, accurate interpretation of PET or other views of functional anatomy can be improved by correlations with standardized templates. MRI can be used to identify the anatomy on an individual basis and stereotactic atlases can be used as a common reference coordinate system. The development of more generalized representations requires the ability to compare brains from different subjects and depends on the goodness of fit between scans and the stereotactic atlas. However, no single representation, used as an atlas, can prove accurate even within a homogeneous population of subjects. Thus, it is necessary to transform one brain to conform to another and to quantitate the degree of deformation necessary to make them coincident. Chapters 24 and 28 of this book deal with these and related issues in greater detail.

Image-warping algorithms, specifically designed to handle 3D neuroanatomic data (Bajcsy and Kovacic, 1989; Christensen *et al.*, 1993, 1996; Collins *et al.*, 1994, 1995; Thirion, 1995; Rabbitt *et al.*, 1995; Davatzikos, 1996; Thompson and Toga, 1996; Bro-Nielsen and Gramkow, 1996), can be used to transfer all the information in a 3D digital brain atlas onto the scan of any given subject, while respecting the intricate patterns of structural variation in their anatomy. These transformations must allow any segment of the atlas anatomy, however small, to grow, shrink, twist, and even rotate, to produce a transformation which represents and encodes local differences in topography from one individual to another. Such deformable atlases (Seitz *et al.*, 1990; Evans *et al.*, 1991; Miller *et al.*, 1993; Gee *et al.*, 1993; Christensen *et al.*, 1993; Sandor and Leahy, 1994, 1995; Rizzo *et al.*, 1995) can be used to carry 3D maps of functional and vascular territories into the coordinate system of different subjects, as well as information on different tissue types and the boundaries of cytoarchitectonic fields and their neurochemical composition.

Any successful warping transform for cross-subject registration of brain data must be high dimensional, in order to accommodate fine anatomic variations (Christensen *et al.*, 1996; Thompson and Toga, 1998). This warping is required to bring the atlas anatomy into structural correspondence with the target scan at a very local level. Another difficulty arises from the fact that the topology and connectivity of the deforming atlas have to be maintained under these complex transforms. This is difficult or simply impossible to achieve in traditional image-warping manipulations (Christensen *et al.*, 1995). Physical continuum models of the deformation address these difficulties by considering the deforming atlas image to be embedded in a three-dimensional deformable medium, which can be either an elastic material or a viscous fluid. The medium is subjected to certain distributed internal forces, which reconfigure the medium and eventually lead the image to match the target. These forces can be based mathematically on the local intensity patterns in the data sets, with local forces designed to match image regions of similar intensity.

F. Databases

Historically, as maps and databases evolved there was a distinction between them. Whereas a map is a collection of information; a representation of our understanding of an object, a database is designed with more interaction in mind. Its function is to organize and archive data records *and* provide an efficient and comprehensive query mechanism. Modern digital maps, however, have become databases. They have incorporated analyses and query functions. They combine information from different sources and establish relationships. Computerized brain atlases of several types have been developed, providing different degrees of database functionality.

One of the first database brain maps was developed by Bloom *et al.* (1990). They created an electronic version of atlas delineations from the Paxinos and Watson (1986) neuroanatomic atlas of the rat brain. These outlines were equated with coordinates and nomenclature so that the user could request information regarding structural groupings and systems. Since the system was based upon a hypercard (Apple Computer Corp.) database, the user could add information to this anatomic framework as an anatomy laboratory organizer. In a similar vein, Swanson (1992) provided a digital version of his anatomic delineations to his atlas of the rat brain. Cortical connectivity in the macaque monkey also has been organized as a database (Felleman and Van Essen, 1991). The most sophisticated attempt in the human brain mapping literature is BrainMap (Fox *et al.*, 1994; Fox and Lancaster, 1994). This database incorporates a true relational database structure intended to encapsulate data from diverse methods on brain activation. The anatomic framework is based upon the Talairach system and the database relates information about the activation task, the methods, the bibliography, and other pertinent data. However, it is a metadatabase comprising published results rather than primary data. These and other database brain maps have all reduced the data in some way as a matter of practical concern. Usually the image data are excluded and only boundary information is retained. Arya *et al.* (1993) created a prototype that did test the inclusion of source image data, but the database was never fully populated. More recently, an effort at Dartmouth University entitled the *f*MRI Data Center archived some *f*MRI data for cognitive science experiments published in the *Journal of Cognitive Neuroscience.* The journal's insistence on data deposition as a condition of publication created considerable controversy (Marshall, 2000; Aldhous, 2000) but sparked widespread dialogue about the critical issues of data ownership, human subject protection, data sharing, authorship, and security. Consortia that pool data also have emerged with database structures to facilitate the identification of and navigation through huge collections of thousands of well-characterized subjects. The ability to include potentially very large amounts of source image data is an import consideration for future implementations, but one requiring technological innovations in data compression, search systems, and visualization. The section at the end of this chapter entitled The Atlas Is the Database further discusses this increasingly important topic.

IV. Relationships to Other Biological Maps

Compared to maps of the brain, other efforts in biological cartography are simpler and more mature. These projects have already evolved into cooperatives involving international participation. They have established standards, languages, and technologies that facilitate the federation of their efforts and they have grappled with the informatics requirements and sociology of scientific cooperation. The Human Genome Project is an excellent example that provides lessons for brain mapping.

A. The Human Genome

The impact of efforts to map the human genome on the field of neuroscience is undisputed. The benefit on human lives of the identification of markers that predict the development of Huntington's disease, chromosome 4 (4p16.3) (Youngman *et al.,* 1989); muscular dystrophy, X chromosome (Hodgson and Bobrow, 1989; Chen *et al.,* 1990); and cystic fibrosis, chromosome 7 (7q31) (Wainwright *et al.,* 1985; White *et al.,* 1985); for example, are some of the most vivid testaments to the value of developing a comprehensive map. Genes that may confer a predisposition to common diseases such as Alzheimer's have also been localized to specific chromosomal regions. The idea to map the human genome was first articulated in 1980 (Botstein *et al.,* 1980) and the first map produced in 1987 (Donis-Keller *et al.,* 1987). Since then the degree of detail added has been almost exponential. However, the $3 billion effort known as the Human Genome Project has produced more than information about the very foundation of human life, it has clearly demonstrated the advantages of development of a complete biological map. Molecular biologists now publish within the structure of common reporting standards, utilizing electronically accessible databases, and can share information efficiently and beneficently.

Its success to date has been possible because of major contributions from many countries and the extensive sharing of information and resources. This coordination has been achieved largely by scientist-to-scientist interaction, facilitated by the Human Genome Organization, which has taken responsibility for some aspects of the management of the international chromosome workshops, in particular.

As the genome project has proceeded, progress along a broad range of technological fronts has been beneficial to other scientific endeavors. Among the most notable of these developments have been new types of genetic markers that can be assayed along with better experimental strategies and computational methods for assembling the results into physical maps. Second, mapping efforts are directed at both larger and smaller targets. At one end, methodological developments have generated low-resolution maps of the entire genome. At the other end of the scale, detailed mapping, sequencing, and the identification of genes within maps and sequences has greatly enriched the genome maps that are produced.

B. Cooperation

Federation is the cooperative creation of multiple, independent databases (for example, each serving a scientific subcommunity) with sufficient commonality of syntax and semantics so that any number of databases can be viewed simultaneously. BrainMap, a database of human functional neuroanatomy (Fox *et al.,* 1994), is designed to follow an idealized community database quite closely. Users query a repository of functional–anatomical associations derived from PET, *f*MRI, and ERPs to create meta-analyses that transcend paper, laboratory, or imaging modality. But, BrainMap cannot yet be federated with other databases, as they may operate on fundamentally different data objects and have no common syntax. On the other hand, the Human Brain Project already funds a number of projects (e.g., Mazziotta *et al.,* 1995, 2001a,b) having the expressed purpose of cooperation. Given the diversity of brain mapping techniques, scales, and foci proposed in the funded portfolio, considerable effort will be needed to avoid mutual exclusiveness among the projects (Kotter, 2001; Pechura and Martin, 1991). Several international workshops have focused on facilitating the necessary global efforts (e.g., the EU–US workshops in "Databasing the Brain," July 1–2, 2001; http://www.nesys.uio.no/workshop/). Meetings such as this along with numerous funding agency commitments to communication and cooperation will encourage federation of the emerging community databases.

V. Stereotaxy

The term stereotaxic, derived from Greek and Latin roots, refers to a three-dimensional system. Stereotactic adds the Latin connotation "to touch," thus referring to the interaction with the space occupied by the object, in this case the brain. Stereotaxy, as it pertains to brain mapping, concerns the coordinate system used to locate structures reliably in different individuals. As was noted earlier in this chapter, the history of cartography contributed a great deal toward the development of coordinate systems appropriate for accurate brain mapping. For example, where Hipparchus first used prominent cities as the landmarks for his earthly subdivisions, anatomically obvious points of reference are employed as the "origin" for brain systems.

A. History

Animal experimenters primarily used external landmarks such as the external auditory meatus, infraorbital ridge, upper incisors, and in some cases bony sutures (Lundsford, 1988). The dependence on cranial landmarks, however, was unsuitable for use in systems designed for humans. Therefore, Spiegel and Wycis (1952) developed a system that relied on internal brain landmarks identified with lateral X-ray films. Their "stereoencephalotome" was used to place lesions in the pallidum for Parkinson's patients. Later these investigators developed the first stereotactic atlas (Spiegel and Wycis, 1952) based upon the location of the pineal gland and posterior commissure from a single celloidin-embedded brain. Shrinkage was accounted for by scoring the brain prior to processing. They were also aware of residual variability and included a table of several subcortical structures. Variations on this original approach spawned several atlases of the human brain (Talairach *et al.*, 1957; Schaltenbrand and Bailey, 1959; Andrew and Watkins, 1969; Van Buren and Maccubin, 1962), some of which are still in common use today.

B. A Uniform Descriptive and Measurement System

Coordinate systems used in stereotaxy and all brain mapping endeavors are not absolute. They are, as described above, based upon anatomic landmarks. But which ones? Their selection is crucial to the accuracy of the resulting system. Since the development of the AC-PC line-based Talairach system, significant controversy remains regarding the appropriateness of these anatomic landmarks. Despite an elaborate scaling and transformation process, residual variability, even with relatively homogeneous subpopulations, remains (Evans *et al.*, 1992; Galaburda *et al.*, 1990). Thus, considerable extensions to this system utilizing local deformation are generally considered necessary to adequately accommodate variability in anatomic features (Toga and Thompson, 2001b; Woods *et al.*, 1999; Thompson *et al.*, 1996a,b).

Several inadequacies plague existing stereotactic coordinate systems and require improvement. One is the notion of symmetry. Human neuroanatomic atlases currently in common use assume hemispheric equivalence. However, there are numerous studies describing the differences between dominant and nondominant hemispheres, both cortically (Geshwind and Levitsky, 1968; Galaburda *et al.*, 1990; Wada *et al.*, 1975) and, possibly, subcortically (Orthner and Sendler, 1975). Similarly sexual dimorphism is not well accommodated in present systems. At the midsagittal plane of the brain, the massa intermedia is more often present (Rabl, 1958), and both the massa intermedia and anterior commissure (Allen and Gorski, 1986) are larger and the splenium of the corpus callosum may be more bulbous (De Lacoste-Utamsing and Holloway, 1981), in females than in males. Age, ethnicity, and other characteristics of the individual also clearly contribute to the degree of variability of the brain.

C. Atlases

An atlas is a special form of map that incorporates stereotaxic elements. Computerized stereotactic atlases are becoming increasingly important for the localization and segmentation of human brain anatomy and physiology derived from various tomographic modalities. These in turn can be used to compare the relationship between brains of different subjects. Atlases are the typical format for describing the relationship between a specified stereotactic coordinate system and neuroanatomic nomenclature. This provides the opportunity to use neuroanatomic labels to communicate while still having a coordinate system to quantitate, localize, and interact with brain space. Several stereotaxic approaches have been employed, as described above, the most notable being the proportional transformational system of Talairach *et al.* (1967; Talairach and Tournoux, 1988). This and similar systems, combined with computerized atlases (Bajcsy *et al.*, 1983; Bohm *et al.*, 1991, 1992; Evans *et al.*, 1991; Seitz *et al.*, 1990; Greitz *et al.*, 1991), allow the structural interpretation of functional studies. As discussed above, there is considerable research devoted to transforming brains to a common atlas. The use of more extreme transformations including local deformation (Benayoun *et al.*, 1994; Bajcsy & Kovacic, 1989; Miller *et al.*, 1993; Thompson *et al.*, 1996b, 2000a; Toga and Thompson, 1997) may improve the cortical fit (for review, see Toga, 1998), but in many cases certain sulcal geometries are absent in some individuals (Ono *et al.*, 1990). Since the first edition of this book was published, new strategies to accurately map the topologically complex and highly variable cortex have revolutionized population-based atlases.

Random vector field approaches have been employed to construct population-based atlases of the brain (Thompson and Toga, 1997). Briefly, given a 3D MR image of a new subject, a high-resolution parametric surface representation of the cerebral cortex is automatically extracted. The algorithm then calculates a set of high-dimensional volumetric maps, elastically deforming this surface into structural correspondence with other cortical surfaces, selected one by one from an anatomic image database. The family of volumetric warps so constructed encodes statistical properties of local anatomical variation across the cortical surface. Specialized strategies elastically deform the sulcal patterns of different subjects into structural correspondence, in a way which matches large networks of gyral and sulcal landmarks with their counterparts in the target brain.

Methods devoted to developing cortical parameterization and flattening algorithms optimally transform maps of cortical features onto a simpler, nonconvoluted surface such as a 2D plane (Van Essen and Maunsell, 1980; Carman *et al.*, 1995; Schwartz and Merker, 1986; Drury *et al.*, 1996), an ellipsoid (Dale and Sereno, 1993; Sereno *et al.*, 1996), or a sphere (Davatzikos, 1996; Thompson *et al.*, 1996a, 1997, 1998).

Cortical parameterization offers substantial advantages for mapping cortical topography and provides potentially clearer interpretations of how architectonic fields and functional loci are related in the cortex (Van Essen *et al.*, 1997). Parameterization and flattening techniques have also been applied to cortical models derived from the Visible Human data sets (Spitzer *et al.,* 1996), and the resulting templates have served as the foundation for developing related atlases of cortical regions (Drury and Van Essen, 1997). These cortical atlases serve as a structural framework upon which architectonic, functional, and electrophysiological data can be compared and integrated. Parametric models of the cortex also make comparisons of cortical anatomy more tractable in disease states (Thompson *et al.,* 1997) and even across species (Van Essen *et al.,* 1997). They can also help overcome problems caused by wide cross-subject variations in cortical geometry, by supporting nonlinear registration of cortically derived functional data and histologic brain maps localized at the cortex (Thompson *et al.,* 1996a,b; Davatzikos, 1996; Mega *et al.,* 1997).

Proponents of parametric techniques that significantly alter the cortical geometry (as in flattening or inflating) point to the improvements in examination of deep sulcal features. On the other hand, these geometric transformations necessarily impose an artificial topology to the cortical mantle and, depending on the underlying mathematical strategy, introduce "cuts" or other structural manipulations. Proponents of parametric techniques that retain the natural gyral and sulcal complexity argue that preservation of this biological topology is essential for accurate mapping. The price paid for retention of these material "folds" is incomplete visualization and frequently incomplete modeling of the full depth of the sulcal valley. Needless to say, each approach has its role to play and increasingly combinatorial approaches are applied where interior products include altered geometric representations.

VI. Nomenclature

The history of neuroanatomy and its underlying language is often punctuated with the names of anatomists and other neuroscientists. As these investigators made their discoveries, they had to create a lexicon to communicate their results. The importance of naming conventions for modern brain mapping lies not only with the need for investigators to communicate efficiently and with specificity; because of the computerized nature of brain maps a standardized nomenclature is critical. The development of neuroanatomic nomenclature has resulted in several different standards of terminology, each intending to solve inconsistencies in abbreviations, hierarchies, or species homologies. The development of a universal ontology that establishes an acceptable nomenclature and its relationship is essential to the creation of usable databases or metadatabases across laboratories and institutions.

A. Scale

Nomenclature has a spatial resolution also. The names and descriptions of lobes of the brain are obviously less precise than sulcal anatomy but sulci are more variable, and in some cases, not always present. Structural divisions in the midbrain and other subcortical regions continue to be refined with increasingly smaller parcellations based upon cytoarchitecture and chemoarchitecture. A hierarchical nomenclature system, properly structured, can accommodate multiple layers of finer and finer spatial scale. Nomenclature also must be extensible so that as structures are further subdivided or otherwise characterized, the naming system can accommodate the additional information. Multiple modalities at multiple spatial scales help establish relationships. Data from single subjects, premortem and postmortem, provide a unique view into the relationship between *in vivo* imaging and histologic assessment. Mega *et al.* (1997) scanned Alzheimer's patients in the terminal stages of their disease using both MRI and PET. These data were combined with postmortem 3D histological images showing the gross anatomy (Toga *et al.,* 1994b) and a Gallyas stain of neurofibrillary tangles. This multimodal, but single-subject, atlas of Alzheimer's disease relates the anatomic and histopathological underpinnings to *in vivo* metabolic and perfusion maps of this disease.

B. Aliases

It has not been possible to get anatomists to agree on a single system of neuroanatomical names. For this reason, the National Library of Medicine has created a metathesaurus to include databases of nomenclature (Tuttle *et al.,* 1990). Bowden and Martin (1995) created a brain hierarchy that related the most widely accepted English and Latin terms for neuroanatomic structures. NeuroNames, as their hierarchy is known, was designed to complement the international standard *Nomina Anatomica* for computer applications. One of the most important contributions of NeuroNames is the fact that it encourages aliases. It relates the terms from English and Latin nomenclatures, provides synonyms, relationships to other structures, species information, and bibliographic references. The use of aliases can greatly improve the cooperation among different scientists and, in some cases, can better describe a given location. Since one nomenclature system, or another (e.g., von Bonin and Bailey, 1947; Riley, 1943; Szabo and Cowan, 1984; Paxinos, 1990), may not completely label the entire brain, an alias approach that combines naming from multiple systems will prove more comprehensive. Issues such as mutual exclusivity and unambiguity are critically important for computerized brain maps, however (Bowden and Martin, 1995).

C. Nomenclature Maps

A map of neuroanatomic names has direct application for teaching as well as for a computerized interface to a database. Utilizing a graphical interface Toga *et al.* (1996) have

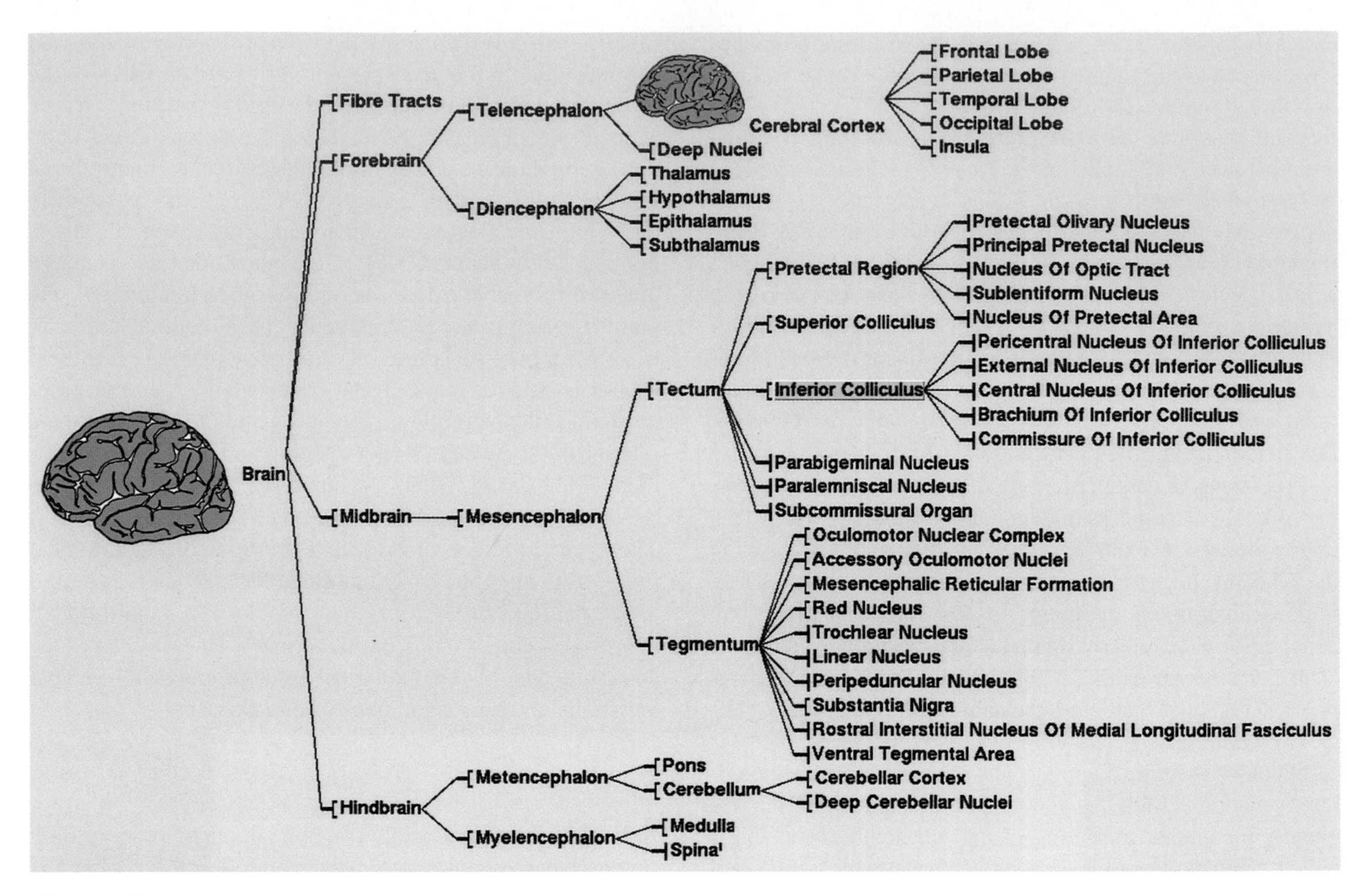

Figure 4 BrainTree. A hierarchical representation of the anatomical nomenclature serves to establish the relationship among structures and provides a map of the organization of the nervous system. As an electronic map, it provides an efficient interface into other maps that can be indexed with anatomical names. This implementation is interactive and provides a graphical display that changes as the user moves up and down the hierarchy. By activating a region (note the highlighted name, inferior colliculus), its subdivisions are listed as a subtree of the graph. These can be turned on or off. Only a small part of the entire program is shown for illustration. The program was developed by Colin Holmes of the Laboratory of Neuro Imaging (UCLA).

developed an approach that displays the relationship between structure names from the whole brain to small nuclei. The computer program, known as BrainTree, is an interactive and hierarchical presentation of nomenclature. It is a map of names (see Fig. 4). As this approach matures, it provides a ready-made interface to a coordinate system because all that is required is establishing the relationship between each voxel and the name of the corresponding structure. Whether these relationships are established as absolute or probabilistic depends on the application.

VII. Detection Devices

Brain mapping methods are numerous and varied, often complementing one another with differing degrees of invasiveness, accuracy, and object of measurement. Some chart anatomy, while others measure an aspect of physiology. Together, they describe brain structure and function. A brief review of some of these methods (follows) illustrates the assumptions and limitations of each and points to the need for multimodality surveys for comprehensive brain mapping.

A. Structure

The physical sectioning of postmortem tissue is the only measure of structure that permits enhancement of the anatomy by histology. However, its value in brain mapping is limited by its postmortem requirement and destructive data acquisition. Anatomic segmentations and descriptions of cytoarchitecture and chemoarchitecture provide some of the most detailed views of structure available today and these techniques remain valuable as a reference for other less demanding measures. Chapter 20, Postmortem Anatomy, covers these methods as applied to whole human brain preparations.

Magnetic resonance imaging, introduced in the early 1980s, provides the best *in vivo* structural images of the human brain in terms of spatial and contrast resolution. There is, depending upon the pulse sequences chosen, good contrast between gray and white matter and multiple scans on the same subject can produce isotropic surveys at ever increasing resolution of the entire brain. Further, variations in the pulse sequences can result in image characteristics that emphasize one or more aspects of the tissue, and by combining multiple scans more complete views of brain anatomy can be achieved. Chapter 16, Neuroanatomical Micromagnetic Resonance Imaging, describes MRI methods using high magnetic field strengths.

B. Diffusion Tensor Imaging

Diffusion tensor MR imaging (DTI) is a technique of value in the *in vivo* tracing of fiber pathways in the human brain and spinal cord (Conturo *et al.,* 1999; Stieltjes *et al.,* 2001; Basser and Pierpaoli, 1996; Pierpaoli *et al.,* 1996). The advent of DTI has already provided the opportunity to examine the location and directions of large portions of nerve fibers in the myelin sheets in human subjects, a topic that has languished for most of the past century (Flechsig, 1920). This method provides local measures of water diffusion and diffusion anisotropy by optimizing the MR signal to the random motion of water molecules. Since the brain is a compartmentalized structure, the motion of water within axons and along fiber tracks that run in the same direction can be visualized. As fiber bundles start to diverge (e.g., near cortical or nuclear targets), the unidirectional tensor for water molecule movement becomes more diffuse and the signal is lost. Thus, DTI is optimally effective for identifying large bundles of fibers running in parallel.

CT was developed more than 10 years before the advent of MRI and still is used in clinical practice because of its rapid image acquisition. It measures the differential absorption and scattering of X-radiation that is passed through the object of study. However, the spatial resolution and ability to clearly image soft tissues are inferior to those of MRI. Blood, bone, and teeth are well imaged with this method.

Magnetic resonance angiography images another aspect of brain structure—the vascular anatomy. The development of MRI pulse sequences that exaggerated the diminished signal from vascular compartments containing flowing blood resulted in images that demonstrated these vascular spaces (Bradley and Waluch, 1985; Ross *et al.,* 1989; Kaufman *et al.,* 1982). The result is a three-dimensional digitally sampled image of the medium and large-diameter arteries and veins. Since the data are three-dimensional and digital, it can be combined with other MRI data sets to produce composite images of both parenchymal and vascular neuroanatomy in a single synthetic image.

C. Computed Tomographic Angiograph (CTA)

The development of CTA, previously called helical or spiral CT, is the latest advance in our ability to perform imaging of the vascular system of the brain and neck in a fashion that is of particular interest to the brain mapping community because of its excellent spatial resolution (Preda *et al.,* 1998). Intermediate between MR angiography and conventional catheter angiography, CTA provides excellent three-dimensional anatomical views of both the intracranial and the cervical vascular systems. With this technique, high-speed CT imaging during intravenous administration of iodinated contrast is coupled with the movement of the patient bed through the scanner gantry. In one vascular transit time, the entire head is imaged. This strategy has been used to provide high-resolution images of the three-dimensional structure of the normal vascular anatomy of the brain as well as aneurysms, arteriovenous malformations, and other vascular abnormalities. Conventional angiography results in images that are two-dimensional projections of the blood vessel lumens. Often the neck of an aneurysm and its relation to its parent or daughter arteries can be hard to discern unless a number of different views are obtained at different angles. This exposes the patient to multiple doses of radiation and an increased burden of iodinated contrast material. Using CTA, a single administration of contrast allows for the reconstruction of the entire vascular complex. The data set can then be rotated and displayed from any angle, allowing the neurovascular anatomy to be closely examined, particularly if presurgical planning strategies are required. This detailed three-dimensional information occurs without the requirement for multiple radiation or contrast exposures for the subject.

D. Function

There are many approaches to measurement of functional changes in brain. While some methods directly monitor electrical events in neurons, others often take advantage of secondary effects of increased neuronal firing rates. Metabolic demand and the requisite changes in blood delivery are useful for localizing the sites and magnitude of brain activity. Several of the methods introduced below are the topic of chapters later in this book.

Magnetic resonance spectroscopy measures the concentration of specific compounds in the brain that may change regionally during specific behaviors. Such data can be produced on conventional clinical imaging instruments (Prichard and Rosen, 1994; Burt and Koutcher, 1984; Demaerel *et al.,* 1991; Gadian *et al.,* 1987; Petroff *et al.,* 1985a,b) and, thus, combined with other structural data sets. While proton spectra have been most widely applied, other isotopes that generate nuclear magnetic resonance spectra include carbon-13, sodium-23, fluorine-19, and phosphorus-31, albeit with lower signal to noise and typically requiring higher field strength instruments. Human brain studies with these methods have included the observation of increments in lactic concentrations during brain activation by behavioral tasks (Prichard *et al.,* 1991).

*f*MRI clearly has revolutionized functional activation studies by mapping changes in cerebral venous oxygen concentration that correlate with neuronal activity. Such approaches require fast two-dimensional brain imaging (Cohen and Weiskoft, 1991; Stehling *et al.*, 1991; Frahm *et al.*, 1992). Using echo-planar imaging, two-dimensional slice data can be acquired in 40 ms with an in-plane anatomical resolution of about 1 mm. While this technique was originally employed using exogenous contrast enhancement with gadolinium (Belliveau *et al.*,1991), it has been extended to measuring intrinsic signals that occur as a result of differential concentrations of deoxyhemoglobin in venous blood in brain regions that increase their neuronal activity (Ogawa *et al.*, 1992; Kwong *et al.*, 1992). Thus, functional maps of the human brain can be obtained without ionizing radiation or the administration of exogenous contrast material.

Developed in the mid-1970s, research applications for PET were originally focused on the functional anatomy of the brain (Mazziotta and Gilman, 1992; Phelps *et al.*, 1986). With PET, positron-emitting isotopes, produced at a local cyclotron, are then chemically integrated into radiopharmaceuticals with known biological behavior. Compounds labeled with oxygen-15, carbon-11, nitrogen-13, and fluorine-18 are injected intravenously or inhaled by subjects. This technique is, by its very nature, biochemical and physiological. Currently PET is used to measure cerebral metabolism, blood flow and volume, oxygen utilization, neurotransmitter synthesis, and receptor binding. The spatial resolution of PET is approximately 5 mm/voxel. Modern tomographs are able to sample the entire intracranial contents and studies can be repeated as frequently as every 10 min with the short half-lived isotopes (oxygen-15) until a maximum radiation dose is reached. PET studies of cerebral activation have provided new insights into the functional neuroanatomy of sensory, motor, language, and memory processes (Frackowiak and Friston, 1994; McIntosh *et al.*, 1994; Freund, 1990; Molchan *et al.*, 1994; Perani *et al.*, 1992; Rapoport, 1991). Further, PET has elucidated the metabolic, hemodynamic, and biochemical events that occur with epilepsy; cerebrovascular disease; degenerative disorders such as Alzheimer's, Parkinson's, and Huntington's disease; and psychiatric disorders such as affective diseases, obsessive compulsive disorder, and schizophrenia (Mazziotta and Gilman, 1992).

SPECT instruments were developed in the 1950s and 1960s and were immediately applied to studies of cerebral function (Lassen and Holm, 1992). Like PET, SPECT uses radiopharmaceuticals administered intravenously or by inhalation to evaluate function in the human brain. These radiopharmaceuticals incorporate isotopes including xenon-133, iodine-123, technesium-99m, and others that emit single-photon radiation, most typically in the form of gamma rays. The longer half-lives of these isotopes do not require their on-site production. Though less quantifiable than PET, SPECT provides estimates of cerebral perfusion, blood volume, and receptor distribution. SPECT techniques have a current spatial resolution of approximately 9 mm/voxel.

Electroencephalography (EEG), developed and optimized in the early part of the 20th century, combined with signal averaging and statistical methods is used in a wide variety of cortical mapping studies. Largely limited to surface dipoles, the signals reflect neuronal activity in the superficial layers of the cerebral cortex and the accompanying distortion by volume conductance within tissue and through the skull. Spatial resolution with the technique is determined by the density of electrode placements but typically is on the order of a few square centimeters at the cortical surface. Temporal resolution is exquisite and comparable to the time frames of neuronal events (e.g., milliseconds). The application of electrodes directly on the surface or into the depth of brain tissue (i.e., electrocorticography) increases spatial resolution because of the elimination of the distorting features of electrical conductance through the skull. Temporal resolution remains unchanged but sampling is limited to the exposed surface of the cortex. With depth electrodes the spatial resolution can be increased to a volume as small as 100 μm/ dimension with an attended reduction in sampling. The invasive nature of these latter two techniques limits their use to certain patients requiring surgery, however.

Magnetoencephalography (MEG) takes advantage of the fact that the weak electrical fields in the brain that are detected by EEG also induce a magnetic field that can be externally measured. The extremely low magnitude of these fields requires the use of supercooled devices in rooms that are isolated from the external magnetic and electrical environment. Since magnetic rather than electrical fields are detected using MEG, distortion caused by the effects of the skull are eliminated. Proponents stress that this lack of distortion allows for more accurate localization of electrical activity three-dimensionally in the brain. A comparison of EEG and MEG techniques in human subjects with implanted current sources (i.e., presurgical epilepsy patients with implanted electrodes) demonstrated a comparable spatial localization accuracy for the two methods (Cohen *et al.*, 1990). As with EEG, MEG has a temporal resolution that is in the millisecond time frame.

There are a variety of optical intrinsic imaging techniques. These include optical techniques that look at changes in potential-dependent dyes, fluorescent-ion concentration, NADH-dependent energy absorbents, and visible or infrared light reflection from the surface that can be related to blood flow (Grinvald, 1984; Cohen and Lesher, 1986; Kauer, 1988; Frostig *et al.*, 1990; Ts'o *et al.*, 1990). These approaches have recently been used for intraoperative measurements of functional activity and demonstrate the advantage of high temporal and spatial resolution (Masino *et al.,* 1993; Narayan *et al.,* 1994a,b; Toga *et al.,* 1995a). The obvious disadvantage of these approaches is that they require exposure of the cortical surface (or thinned bone in animal preparations) and thus limit their use to intraoperative evaluations of patients undergoing craniotomies.

As can be seen from this section, each brain mapping method has distinct advantages and disadvantages. The temporal and spatial resolution of each method determines what can be measured. These variables, added with other requirements such as repeatability, instrumentation, and invasiveness, determine when and where the method can be applied. Table 1 summarizes these variables. Combining the results from several imaging methods within a single framework can provide a comprehensive map of brain structure and function. Data combination can be performed within a given subject across modalities or by grouping subjects into a common average (Woods *et al.,* 1992, 1993; Greitz *et al.,* 1991; Collins *et al.,* 1994).

VIII. Brain Maps: Content and Format

The variety of maps devoted to the brain has resulted in a diverse array of products. In many cases the focus of the map dictates the format of the map and hence its construction. For example, maps devoted to anatomy must include boundaries that describe each structure and, therefore, usually include images of contours (sometimes superimposed over section images) or (digital) vector lists. Stereotactic maps include a coordinate system, often a Cartesian-based frame surrounding the images that comprise the map, and have traditionally been based on single individuals representing that population. Maps intended for referencing other digitally acquired data are themselves digital and compatible with applications that require spatial transformations. The need for comprehensive maps incorporating many of these requirements demands a computerized approach that is adaptable to a variety of applications (see Fig. 5).

Figure 5 Multimodality map. This 3D model illustrates the combination of data from multiple modalities. Anatomical data are used as the framework for mapping metabolic rates. All contributing data are placed within a common coordinate system. The map comprises a volume from which a specific plane can be extracted retaining the coordinate system. The volume can be resampled along any plane and presented at any resolution below that of the source data. This illustration shows several parallel cuts, approximately in the coronal plane, and the separation of sections from the volume. Left to right equals rostral to caudal. The first slice is MRI, the next is a Nissl-stained section, the middle slice is a blockface image, the next is a PET image, and last is a white matter stain. This multimodal composite comes from a single subject.

A. Volumes

Tomographically sampled brain data, whether anatomic or physiologic, are often reconstructed into a volume. It becomes a 3D map, suitable for subsequent resampling along a plane or boundary to derive the surface of a structure. The voxels that make up the volume can be manipulated to represent different physiologic parameters such as blood flow or glucose metabolism. Fewer voxels can be used to represent the volume for less demanding applications and the entire volume can be spatially transformed to fit specific coordinate systems. The volume as a map provides the most data about the brain but, depending on the resolution, can require significant storage space.

B. Explicit Geometry

Surfaces provide a complementary (to volumes) representation useful for mapping. Surfaces explicitly define the geometry of the structure exterior. As such they provide no information about the inside of an object. However, they do enable the calculation of morphometric statistics needed for quantitative comparisons of shape. While it is possible to calculate equivalent statistics using volume-based models, explicit geometry and graph theory makes these computations more efficient. Maps based upon explicit geometry have been used extensively in anatomic atlases (Zulegar and Stansbesand, 1977; DeArmond *et al.,* 1989) but the templates that describe the delineations are rarely available for extrapolation to planes not included in the original data set. Obviously, combining the volumetric and explicit geometry information into a single map provides the best of both worlds, but differences in data structure requirements and presentation strategies obligate careful design.

C. Statistical Representations

Coordinates, voxel values, and morphometrics alone can provide a numerical form of map. Tabular data describing the location, shape, and density of a measurement may provide as much map information as a visual presentation (at the expense of intuitive appreciation). Since all digital maps are already numerical, deriving these data is straightforward.

Table 1 Brain Mapping Methods Used in the Study of Human Health and Disease, along with the Types of Measurements They Provide and Some of the Clinical Situations in Which They May Be of Use

Method	Measurements provided	Disorders	Advantages	Limitations
X-ray computed tomograhy (CT)	1. Brain structure 2. Blood–brain barrier integrity	1. Acute/chronic hemorrhages 2. Acute trauma 3. General screening of anatomy 4. Focal or generalized atrophy 5. Hydrocephalus	1. Excellent bone imaging 2. 100% detection of hemorrhages 3. Short study time 4. Can scan patients with ancillary equipment 5. Can scan patients with metal/ electronic devices	1. Ionizing radiation 2. Poor contrast resolution
Magnetic resonance imaging (MR)	1. Brain structure 2. Brain and cervical vasculature 3. Relative cerebral perfusion 4. Chemical concentrations 5. Fiber tracts 6. Blood–brain barrier integrity	1. Acute ischemia 2. Neoplasms 3. Demyelinating disease 4. Epileptic foci 5. Degenerative disorders 6. Infections 7. Preoperative mapping	1. High spatial resolution 2. No ionizing radiation 3. High resolution 4. High gray–white contrast 5. No bone–generated artifact in posterior fossa 6. Can also perform chemical, functional, and angiographic imaging	1. Long study duration 2. Patients may be claustrophobic 3. Electronic devices contraindicated 4. Acute hemorrhages problematic 5. Relative measurements only
Positron emission tomography (PET)	1. Perfusion 2. Metabolism 3. Substrate extraction 4. Protein synthesis 5. Neurotransmitter integrity 6. Receptor binding 7. Blood–brain barrier integrity	1. Ischemic states 2. Degenerative disorders 3. Epilepsy 4. Movement disorders 5. Affective disorders 6. Neoplasms 7. Addictive states 8. Preoperative mapping	1. Can perform hemodynamic chemical, and functional imaging 2. Quantifiable results 3. Absolute physiologic variables can be determined 4. Uniform spatial resolution	1. Ionizing radiation 2. High initial costs 3. Long development time for new tracers 4. Limited access 5. Low temporal resolution
Single-photon-emission computed tomography (SPECT)	1. Perfusion 2. Neurotransmitter integrity 3. Receptor binding 4. Blood–brain barrier integrity	1. Ischemic states 2. Degenerative disorders 3. Epilepsy 4. Movement disorders	1. Can perform hemodynamic, chemical, and functional imaging 2. Widely available	1. Ionizing radiation 2. Relative measurements only 3. Nonuniform spatial resolution 4. Low temporal resolution
Xenon-enhanced computed tomography (XECT)	1. Perfusion	1. Ischemic states	1. Uses existing equipment	1. Ionizing radiation 2. High xenon concentrations have pharmacalogic effects
Spiral computed tomography (CT angiography, CTA)	1. Vascular anatomy 2. Boney anatomy	1. Vascular occlusive disease 2. Vascular and boney anatomy only	1. Provides high-resolution vascular images	1. Ionizing radiation 2. Vascular and boney anatomy only
Electroencephalography surface (EEG)	1. Electrophysiology	1. Epilepsy 2. Encephalopathies 3. Degenerative disorders 4. Preoperative mapping	1. No ionizing radiation 2. High temporal resolution 3. Widely available 4. Can identify epileptic foci	1. Low spatial resolution 2. Weighted toward measurements
Magnetoencephalography (MEG)	1. Electrophysiology	1. Epilepsy	1. No ionizing radiation 2. High temporal resolution 3. Can identify epileptic foci	1. Low spatial resolution
Transcranial, magnetic stimulation (TMS)	1. Focal brain activation	1. Preoperative mapping	1. No ionizing radiation 2. Potential for therapy 3. Can be linked to other imaging methods (PET fMRI)	1. Low spatial resolution 2. Has produced seizures in certain patient groups
Optical intrinsic signal imaging (OIS)	1. Integrated measure of blood volume, metabolism, and cell swelling	1. Intraoperative mapping	1. No ionizing radiation 2. High temporal resolution 3. High spatial resolution	1. Complex signal source 2. Invasive only (intraoperative)

Database and probabilistic approaches to mapping depend on statistical representations primarily. Furthermore, numerical forms are the most efficient for electronic communications and can be transformed subsequently to a visual representation. Publishing the locations of functional activation in Talairach coordinates suitable for databasing is a prime example of statistical representation. Mapping variance, biases, and other population-based statistics is possible using probabilistic approaches in which confidence limits can be set depending on the question being asked. Quantitating variance also permits determination of significance between groups, assuming sufficient power.

D. Multimodal Maps

Combining data from more than one modality or subject is an important development for brain mapping. The statistical requirements for this are explored elsewhere in this book, but how these data can be presented in a map deserves attention.

Multimodal maps are most often applied to the problem of anatomic mapping of functional data. Where a site of activation is located requires a study of the underlying structure of the same or representative individuals. Assuming the data are transformed to the same coordinate system, locations can be identified with statistical reference or visually with one modality mapped upon the other using a variety of visualization methods. The purpose of many atlases (Resnick *et al.,* 1993) is to establish this structure/function correspondence. As described above, the availability of anatomic templates is a sought after advantage to the interpretation of physiological studies.

A map that includes averages (Evans *et al.,* 1993; Holmes, *et al.,* 1998) or probabilistic or variance (Toga *et al.,* 2001a; Mazziotta *et al.,* 1991, 2001a,b) data requires collections across multiple subjects (or multiple observations of the same subject) and can, if required, be disassembled to portray the data of a single individual (observation). In addition, the map includes information about how that individual compares to the rest of the database. The value of comparing an individual to the entire (collective) map is in classification of that individual as belonging to the group that makes up the map or not. The implications are clear for clinical and experimental situations. Obtaining this, and related, information from a map requires careful consideration regarding the design of the map itself and the interface to it. It has already been established that the across-subject variance within the brain is itself variable depending on the coordinate system (Fox and Woldorff, 1994), the transformation method (Strother *et al.,* 1994), and the anatomic region (Rademacher *et al.,* 1993). Thus, the map must include an interactive capability to survey the relationship between data making up the map. Figure 6 describes a computerized prototype.

E. Dynamic Maps

In many ways, static representations of brain structure are ill suited to analyzing dynamic processes of brain development and disease. The intense interest in brain development and disease mandates the design of mathematical systems to track anatomical changes over time and map dynamic patterns of growth or degeneration.

F. Temporal Maps of Brain Structure

Current structural brain imaging investigations typically focus on the analysis of three-dimensional models of brain structure, derived from volumetric images acquired at a single time point from each subject in the study. However, serial scanning of human subjects, when combined with a powerful set of warping and analysis algorithms, can enable disease and growth processes to be tracked in their full spatial and temporal complexity.

One of the most promising applications of warping algorithms is their use in mapping dynamic differences (Gee *et al.,* 1993). Maps of anatomical *change* can be generated by warping scans acquired from the same subject over time (Thompson *et al.,* 1998; Thirion *et al.,* 1998). Serial scanning of human subjects (Fox *et al.,* 1996; Freeborough *et al.,* 1996; Thompson *et al.,* 1998) or experimental animals (Jacobs and Fraser, 1994) in a dynamic state of disease or development offers the potential to create four-dimensional models of brain structure. These models incorporate dynamic descriptors of how the brain changes during maturation or disease. For a range of patient populations, 4D models of the brain can be based on imaging and modeling its three-dimensional structure at a sequence of time points. In a changing morphology, warping algorithms enable one to model structural changes that occur over prolonged periods, such as developmental, aging, or disease processes, as well as structural changes that occur more rapidly, as in recovery following trauma or tumor growth. A four-dimensional approach can provide critical information on local patterns and rates of tissue growth, atrophy, shearing, and dilation that occur in the dynamically changing architecture of the brain (Toga *et al.,* 1996; Thompson *et al.,* 1998).

G. Mapping Growth Patterns in Four Dimensions

Algorithms to create four-dimensional quantitative maps of growth patterns in the developing human brain have been created (Thompson *et al.,* 1998; Thompson and Toga, 1998). Time series of high-resolution pediatric MRI scans were used to produce tensor maps of growth and spatially detailed information on local growth patterns, quantifying rates of tissue maturation, atrophy, shearing, and dilation in the dynamically changing brain architecture.

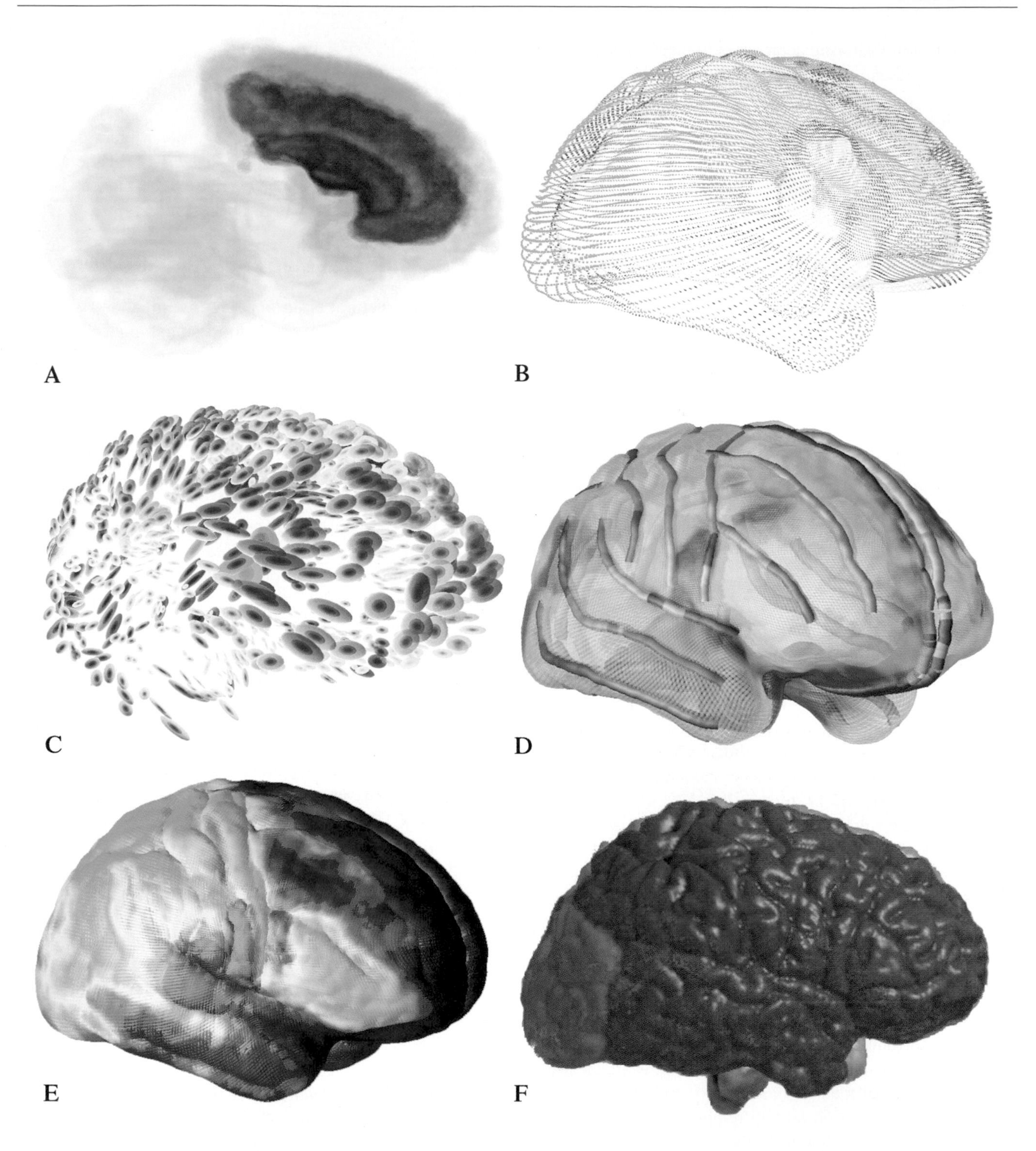

In order to map growth patterns both longitudinal and cross-sectional, experimental designs may have to be accommodated. In the Thompson *et al.* (2000b) study, young normal subjects (mean age 8.6 ± 3.1 years) were scanned at intervals ranging from 2 weeks to 4 years. While difference maps help to determine whether structural change has occurred in dementia (Freeborough *et al.,* 1996), these maps do not localize change, nor do they provide three-dimensional measures of dilation, contraction, or shearing of anatomic regions. To address this, parametric mesh models (Thompson *et al.,* 1996a,b, 1997, 1998) were created to represent a comprehensive set of deep sulcal, callosal, caudate, and ventricular surfaces at each time point. Parameterized cortical surface models were also automatically extracted

Figure 6 Statistical maps. (**A**) The inherent variability contained across a subpopulation of subjects offers the opportunity to characterize that population using a probability map. The map comprises a collection of data sets, each transformed to the same coordinate space. Anatomic structures are identified and delineated for each contributing data set and combined in the probabilistic map. The degree of consistency or variability for each structure can be measured and displayed in one of several ways. Part of the subvolume probabilistic atlas for the Alzheimer's disease (AD) population is shown. The atlas is constructed using the MRI data from 30 AD subjects randomly chosen from a large AD database. The focus is centered around the left frontal lobe with white matter (green color), gray matter (blue), and cerebrospinal fluid (red) clearly visible as fuzzy clouds. The color saturation indicates the chance that each voxel is part of the specified tissue type of the left frontal lobe. Higher probability values yield higher confidence that the 3D coordinate is part of the region of interest and conversely, lower color saturation shows decreased confidence associating the stereotactic location to the corresponding region. The atlas contains 180 probabilistic regions and is used for automated volumetric studies, as well as for statistical analysis of functional brain data for the AD and elderly populations. (**B**) Average cortical surface model from 15 male normal subjects. Parametric cortical models were extracted from T1-weighted magnetic resonance data using signal intensity information after the removal of extracortical tissue. A complex vector-valued flow field drives cortical surface anatomy from each subject into correspondence and retains variability information about the extent of the deformation in color. Hotter colors represent increased variability in cortical surface anatomy within the group. (**C**) Tensor map of cortex. Striking differences are found, even among normal human subjects, in the gyral patterns of the cerebral cortex. Tensor maps can be used to visualize these complex patterns of anatomical variation. In this map, color distinguishes regions of high variability (pink colors) from areas of low variability (blue). Ellipsoidal glyphs indicate the principal directions of variation—they are most elongated along directions in which there is greatest anatomic variation across subjects. Each glyph represents the covariance tensor of the vector fields that map individual subjects onto their group average anatomic representation. This map is based on a group of 20 elderly normal subjects. The resulting information can be leveraged to distinguish normal from abnormal anatomical variants using random tensor field algorithms. (**D**) 3D cortical surface and sulcal variability map. The surface model represents the average cortical surface from 28 normal subjects. The color indexes variability measured from homologous coordinate points from cortical surface models extracted from individual subjects in 3D stereotaxic space. The major cortical sulci are averaged and mapped onto the average cortical surface model and variability is indexed in different colors. Variability is calculated as the root mean square magnitude of displacement vectors required to map equivalent coordinate points from each anatomical model onto the group average. (**E**) The anatomic variation in gyral/sulcal cortical anatomy in a group of normal children (ages 6–16) is shown. (**F**) A representative picture of the subparcellation of the cerebral lobes in a child. In order to describe changes in tissue types (gray, white, and CSF) and volumetry during normal childhood and adolescence, we delineate the cerebral hemispheres into component lobes to assess which regions may most be affected by developmental processes.

from each of the mutually registered histogram-matched scans. The deformation field required to match the surface anatomy of one scan with the other was extended to the full volume using a continuum-mechanical model based on the Cauchy–Navier operator of linear elasticity (Thompson and Toga, 1998). Deformation processes recovered by the warping algorithm were then analyzed using vector field operators to produce a variety of *tensor maps*. These maps were designed to reflect the magnitude and principal directions of dilation or contraction, the rate of strain, and the local curl, divergence, and gradient of flow fields representing the growth processes recovered by the transformation.

In the near future, these brain mapping techniques will provide the ability to map growth and degeneration in their full spatial and temporal complexity. Despite logistic and technical challenges, these mapping approaches hold tremendous promise for representing, analyzing, and understanding the extremely complex dynamic processes that affect regional anatomy in the healthy and diseased brain.

H. The Atlas Is the Database

The integration of data from multiple modalities, subjects, and laboratories as well as scales and dimensions to produce a map of the brain brings to mind the capabilities of a database. A brain map is, after all, the result of a query. Brain maps are now virtually all computerized. They are calculated from a collection of data and increasingly include descriptions of multiple observations. Populations of subjects are included when variability measures are needed. They can be recalculated with a new set of data. While the resulting map may represent the sum of its individual contributions, clearly the brain map and the database from which it is derived are inseparable and dependent upon one another to be usable. The database without adequate structure such as a uniform coordinate system or ability to compare, contrast, sum, or categorize its contents will have limited value. Similarly, in modern neuroscience, a brain map that cannot assimilate disparate observations about structure and function or accommodate new data will be quickly outdated. Thus, in brain mapping, the atlas is the database.

IX. Summary

An introduction to cartography of the brain is, by its very intent, broad in its coverage. This chapter has described the emergence of cartographic strategies and their application to a better understanding of brain structure and function. Maps of the brain are different from maps of other objects, because they must accommodate so many diverse aspects of neuroscientific inquest. Therefore, there are many different versions of brain maps. The degree to which each is successful depends not only upon the available technology for acquiring the data, analyzing them, and taking advantage of them, but on how it is used with complete understanding of its underlying assumptions and limitations. Equally important is the faith garnered by these maps. The degree to which maps of the brain are developed and used by the scientific community depends on the collective belief that the data are accurate, reliable, representative, and above all, useful.

Acknowledgments

The authors thank colleagues in the Laboratory of Neuro Imaging and Brain Mapping Center for their assistance. Special thanks go to Andrew Lee for his digital artwork. This work was supported in part by the National Institutes of Health (RR13642 and RR00865), the National Library of Medicine (R01 LM05639), and the Human Brain Project (P20 MH52176) funded jointly by the National Institute of Mental Health, National Institute on Drug Abuse, National Cancer Institute, and National Institute for Neurologic Disease and Stroke.</ACK>

References

Aldhous, P. (2000). Prospect of data sharing gives brain mappers a headache. *Nature* **406**, 445–446.

Allen, L. S., and Gorski, R. A. (1986). Sex differences in the human massa intermedia. *Soc. Neurosci. Abstr.* **13,** 46.

Andrew, J., and Watkins, E., eds. (1969). "A Stereotaxic Atlas of the Human Thalamus and Adjacent Structures: A Variability Study." Williams & Wilkins, Baltimore.

Armstrong, E., Schleicher, Omran, H., Curtis, M., and Zilles, K. (1995). The ontogeny of human gyrification. *Cereb. Cortex* **1,** 56–63.

Arya, M., Cody, W., Faloutsos, C., Richardson, J., and Toga, A. W. (1993). QBISM: A prototype 3D medical image database system. *Data Eng.* **16,** 38–42.

Avoli, M., Hwa, G. C., Kostopoulos, G., Oliver, A., and Villemure, J.-G. (1991). Electrophysiological analysis of human neocortex *in vitro:* Experimental techniques and methodological approaches. *Can. J. Neurol. Sci.* **18,** 636–639.

Bajcsy, R., and Kovacic, S. (1989). Multi-resolution elastic matching. *Comput. Vision, Graph. Image Process* **46,** 1–21.

Bajcsy, R., Lieberson, R., and Reivich, M. (1983). A computerized system for the elastic matching of deformed radiographic images to idealized atlas images. *J. Comput. Assisted Tomogr.* **7,** 618–625.

Basser, P. J., and Pierpaoli, C. (1996). Microstructural and physiological features of tissues elucidated by quantitative-diffusion-tensor MRI. *J. Magn Reson.* **3,** 208–219.

Belliveau, J. W., Kennedy, D. N., McKinstry, R. C., Buchbinder, B. R., Weiskoff, R. M., Cohen, M. S., Vevea, J. M., Brady, T. J., and Rosen, B. R. (1991). Functional mapping of the human visual cortex by magnetic resonance imaging. *Science* **254,** 716–719.

Benayoun, S., Ayache, N., and Cohen, I. (1994). An adaptive model for 2D and 3D dense non-rigid motion computation. *INRIA Int. Rep.* No. 2297.

Berger, H. (1929). Uber das elektrenkephalogramm des menschen. *Arch. Psychiatrie Nervenkrankheiten* **87,** 527–580.

Binkofski, F., Amunts, K., Stephan, K. M., Posse, S., Schormann, T., Freund, H. J., Zilles, K., and Seitz, R. J. (2000). Broca's region subserves imagery of motion: A combined cytoarchitectonic and fMRI study. *Hum. Brain Mapp.* **11,** 273–285.

Blasdel, G. G. (1992). Differential imaging of ocular dominance and orientation selectivity in monkey striate cortex. *J. Neuroscience*, **12(8),** 3115–3138.

Bloom, F. E., Young, W. G., and Kim, Y. M. (1990). "Brain Browser Hypercard Application for the Macintosh." Academic Press, San Diego.

Bohm, C., Greitz, T., Seitz, R., and Eriksson, I. (1991). Specification and selection of regions of interest (ROIs) in a computerized brain atlas. *J. Cereb. Blood Flow Metab.* **11,** A64–A68.

Bohm, C., Greitz, T., and Thurfjell, L. (1992). The role of anatomic information in quantifying functional neuroimaging data. *J. Neural Transm. (Suppl.)* **37,** 67–78.

Borg, F. P., Salkauskus, A. G., and MacVicar, B. A. (1994). Mapping patterns of neuronal activity and seizure propagation by imaging intrinsic optical signals in the isolated whole brain of the guinea-pig. *Neuroscience* **58,** 461–480.

Botstein, D., White, R. L., Skolnick, M. H., and Davies, R. W. (1980). Construction of a genetic linkage map in man using restriction fragment length polymorphisms. *Am. J. Hum. Genet.* **32,** 314–331.

Bowden, D. M., and Martin, R. F. (1995). Neuronames brain hierarchy. *NeuroImage* **2,** 63–83.

Bradley, W. G., and Waluch, V. (1985). Blood flow: Magnetic resonance imaging. *Radiology* **154,** 443–450.

Brodmann, K. (1909). Vergleichende lokalisationslehre der grosshirnrinde in ihren prinzipien dargestellt auf grund des zellenbaues. *In* "Some Papers on the Cerebral Cortex." Barth, Leipzig. [Translated as "On the Comparative Localization of the Cortex," 1960, pp. 201–230. Thomas, Springfield, IL]

Bro-Nielsen, M., and Gramkow, C. (1996). Fast fluid registration of medical images. *In* "Visualization in Biomedical Computing, Hamburg, Germany, Lecture Notes in Computer Science" (K. H. Hohne and R. Kikinis, eds.), Vol. 1131, pp. 267–276. Springer-Verlag, Berlin.

Burt, C. I., and Koutcher, J. A. (1984). Multinuclear NMR studies of naturally occurring nuclei. *J. Nucl. Med.* **25,** 237–248.

Carman, G. J., Drury, H. A., and Van Essen, D. C. (1995). Computational methods for reconstructing and unfolding the cerebral cortex. *Cereb. Cortex* **5,** 506–517.

Chen, G. T., Pelizzari, C. A., and Levin, D. (1990). Image correlation in oncology. *In* "Important Advances in Oncology" (V. T. DeVita, S. Hellman, and S. A. Rosenberg, eds.). Lippincott, Philadelphia.

Christensen, G. E., Rabbitt, R. D., and Miller, M. I. (1993). A deformable neuroanatomy textbook based on viscous fluid mechanics. *In* "27th Annual Conference on Information Sciences and Systems," pp. 211–216.

Christensen, G. E., Rabbitt, R. D., Miller, M. I., Joshi, S. C., Grenander, U., Coogan, T. A., and Van Essen, D. C. (1995). Topological properties of smooth anatomic maps. *In* "Information Processing in Medical Imaging" (Y. Bizais, C. Carillot, and R. Di Paola, eds.), pp. 101–112. Kluwer Academic Publishers, Brest, France.

Christensen, G. E., Rabbitt, R. D., and Miller, M. I. (1996). Deformable templates using large deformation kinematics. *IEEE Trans. Image Process.* **5,** 1435–1447.

Clark, C. M., Ammann, W., Martin, W. R., Ty, P., and Hayden, M. R. (1991). The FDG/PET methodology for early detection of disease onset: A statistical model. *J. Cereb. Blood Flow Metab.* **11,** A96–102.

Cohen, C. B., and Lesher, S. (1986). Optical monitoring of membrane potential: Methods of multisite optical measurement. *Soc. Gen. Physiol. Ser.* **40,** 71–99.

Cohen, D., Cuffin, B. N., Yunokuchi, K., and Maniewski, R. (1990). MEG vs EEG localization test using implanted sources in the human brain. *Ann. Neurol.* **28,** 811–817.

Cohen, M. S., and Weisskoff, R. M. (1991). Ultra-fast imaging. *Magn. Reson. Imaging* **9,** 1–37.

Collins, D. L., Neelin, P., Peters, T. M., and Evans, A. C. (1994). Automatic 3D intersubject registration of MR volumetric data in standardized Talairach space. *J. Comp. Assisted Tomogr.* **18,** 192–205.

Collins, D. L., Holmes, C. J., Peters, T. M., and Evans, A. C. (1995). Automatic 3D model-based neuroanatomical segmentation. *Hum. Brain Mapp.* **3,** 190–208.

Collins, F. S. (2001). Contemplating the end of the beginning. *Genome Res.* **11,** 641–643.

Conturo, T., Lori, N. F., Cull, T. S., Akbudak, E., Snyder, A. Z., Shimony, J. S., McKinstry, R. C., Burton, H., and Raichle, M. E. (1999). Tracking neuronal fiber pathways in the living human brain. *Proc. Natl. Acad. Sci. USA* **96,** 10422–10427.

Dale, A. M., Fischl, B., and Sereno, M. I. (1999). Cortical surface-based analysis. I. Segmentation and surface reconstruction. *NeuroImage* **9,** 179–194.

Dale, A. M., and Sereno, M. I. (1993). Improved localization of critical activity by combining EEG and MEG with MRI cortical surface reconstruction—A linear approach. *J. Cognit. Neurosci.* **5,** 162–176.

Damasio, H. (1995). "Human Brain Anatomy in Computerized Images." Oxford Univ. Press, Oxford/New York.

Davatzikos, C. (1996). Spatial normalization of 3D brain images using deformable models. *J. Comp. Assisted Tomogr.* **20,** 656–665.

De Lacoste-Utamsing, C., and Holloway, R. L. (1981). Sexual dimorphism in the human corpus callosum. *Science* **216,** 1431–1432.

DeArmond, S. J., Fusco, M. M., and Dewey, M. M. (1989). "Structure of the Human Brain: A Photographic Atlas." Oxford Univ. Press, New York.

DeClerk, Y. A., Shimada, H., Gonzalez-Gomez, I., and Raffel, C. (1993). Tumoral invasion in the CNS. *J. Neuro-Oncol.* **18,** 111–121.

Demaerel, P., Johannik, K., Van Hecke, P., Van Ongeval, C., Verellen, S., Marchal, G., Wilms, G., Plets, C., Goffin, J., Van Calenbergh, F., *et al.* (1991). Localized ^{1}H NMR spectroscopy in fifty cases of newly diagnosed intracranial tumors. *J Comput. Assisted Tomogr.* **15,** 67–76.

Descartes (1637). "Discours de la Methode."

Donis-Keller, H., Green, P., Helms, C., Cartinhour, S., Weiffenbach, B., Stephens, K., Keith, T. P., Bowden, D. W., Smith, D. R., Lander, E. S., *et al.* (1987). A genetic linkage map of the human genome. *Cell* **51,** 319–337.

Drury, H. A., Van Essen, D. C., Joshi, S. C., and Miller, M. I. (1996). Analysis and comparison of areal partitioning schemes using two-dimensional fluid deformations. Poster presentation at the Second International Conference on Functional Mapping of the Human Brain, Boston, Massachusetts, June 17–21, 1996. *NeuroImage* **3,** S130.

Drury, H. A., and Van Essen, D. C. (1997). Analysis of functional specialization in human cerebral cortex using the visible man surface based atlas. *Hum. Brain Mapp.* **5,** 233–237.

Duvernoy, H. M. (1991). "The Human Brain." Springer-Verlag, New York.

Evans, A., Marret, S., Torrescorzo, J., Ku, S., and Collins, L. (1991). MRI–PET correlation in three dimensions using a volume of interest (VOI) atlas. *J. Cereb. Blood Flow Metab.* **11,** 169–178.

Evans, A. C., Collins, D. L., Mills, S. R., Brown, E. D., Kelly, R. L., and Peters, T. M. (1993). 3D statistical neuroanatomical models from 305 MRI volumes. *Proc. IEEE_Nucl. Sci. Symp. Imaging Conf.* 1813–1817.

Evans, A. C., Marrett, S., Neelin, P., Collins, L., Worsley, K., Dai, W., Milot, S., Meyer, E., and Bub, D. (1992). Anatomical mapping of functional activation in stereotactic coordinate space. *NeuroImage* **1,** 43–53.

Felleman, D. J., and Van Essen, D. C. (1991). Distributed hierarchical processing in the primate cerebral cortex. *Cereb. Cortex* **1,** 1–47.

Fetcho, J. R. (1992). The spinal system in early vertebrates and some of its evolutionary changes. *Brain Behav. Evol.* **40,** 82–97.

Finger, S. (1994). "Origins of Neuroscience: A History of Explorations into Brain Function." Oxford Univ. Press, New York.

Fischl, B., Liu, A., and Dale, A. M. (2001). Automated manifold surgery: Constructing geometrically accurate and topologically correct models of the human cerebral cortex. *IEEE Trans. Med. Imaging* **20,** 70–80.

Flechsig, P. (1920). "Anatomie des Menschlichen Gehirns und Ruckenmarks." Thieme, Leipzig.

Fox, N. C., Freeborough, P. A., and Rossor, M. N. (1996). Visualization and quantification of rates of cerebral atrophy in Alzheimer's disease. *Lancet* **348,** 94–97.

Fox, P. T., and Lancaster, J. L. (1994). Neuroscience on the Net. *Science* **266,** 994–996.

Fox, P. T., and Woldorff, M. G. (1994). Integrating human brain maps. *Curr. Opin. Biol.* **4,** 151–156.

Fox, P. T., Mikiten, S., Davis, G., and Lancaster, J. L. (1994). BrainMap: A database of human functional brain mapping. *In* "Functional Neuroimaging" (R. Thatcher, M. Hallett, T. Zeffiro, E. R. John, and M. Heurta, eds.). Academic Press, San Diego.

Frackowiak, R. S., and Friston, K. J. (1994). Functional neuroanatomy of the human brain: Positron emission tomography—A new neuroanatomical technique. *J. Anat.* **184,** 211–225.

Frahm, J., Gyngell, M. L., and Hanicke, W. (1992). Rapid scan techniques. *In* "Magnetic Resonance Imaging," pp. 165–203. Mosby, St. Louis.

Freeborough, P. A., Woods, R. P., and Fox, N. C. (1996). Accurate registration of serial 3D MR brain images and its application to visualizing change in neurodegenerative disorders. *J. Comput. Assisted Tomogr.* **20,** 1012–1022.

Freund, H. J. (1990). Premotor area and preparation of movement. *Rev. Neurol.* **146,** 543–547.

Friedhoff, R. M. (1989). *In* "Visualization: The Second Computer Revolution (R. M. Friedhoff and W. Benzon, eds.). Abrams, New York.

Frostig, R. D., Lieke, E. E., Ts'o, D. Y., and Grinvald, A. (1990). Cortical functional architecture and local coupling between neuronal activity and the microcirculation revealed by *in vivo* high-resolution optical imaging of intrinsic signals. *Proc. Natl. Acad. Sci. USA* **87,** 6082–6086.

Gadian, D., Frackowiak, R., Crockard, H., Proctor, E., Allen, K., and Williams, S. (1987). Acute cerebral ischemia: Concurrent changes in cerebral blood flow, energy metabolites, pH, and lactate measured with hydrogen clearance and ^{31}P and ^{1}H nuclear magnetic resonance spectroscopy. I. Methodology. *J. Cereb. Blood Flow Metab.* **7,** 199–206.

Galaburda, A. M., Rosen, G. D., and Sherman, G. F. (1990). Individual variability in cortical organization: Its relationship to brain laterality and implications to function. *Neuropsychologia* **28,** 529–546.

Gee, J. C., Reivich, M., and Bajcsy, R. (1993). Elastically deforming 3D atlas to match anatomical brain images. *J. Comput. Assisted Tomogr.* **17,** 225–236.

German, D. C., Walker, B. C., Manaye, K., Smith, W. K., Woodward, D. J., and North, A. J. (1988). The human locus coeruleus: Computer reconstruction of cellular distribution. *J. Neurosci.* **8,** 1776–1788.

Geschwind, N., and Levitsky, W. (1968). Human brain: Left–right asymmetries in temporal speech region. *Science* **161,** 186.

Geyer, S., Schleicher, A., and Zilles, K. (1997). The somatosensory cortex of man: Cytoarchitecture and regional distributions of receptor binding sites. *NeuroImage* **6,** 27–45.

Greitz, T., Bohm, C., Holte, S., and Eriksson, L. (1991). A computerized brain atlas: Construction, anatomical content, and application. *J. Comput. Assisted Tomogr.* **15,** 26–38.

Grinvald, A. (1984). Real-time optical imaging of neuronal activity. *Trends Neurosci.* **7,** 143–150.

Grinvald, A., Frostig, R. D., Siegel, R. M., and Bartfield, E. (1991). High-resolution optical imaging of functional brain architecture in the awake monkey. *Proc. Natl. Acad. Sci. USA* **88,** 11559–11563.

Haglund, M. M., Ojemann, G. A., and Hochman, D. W. (1992). Optical imaging of epileptiform and functional activity in human cerebral cortex. *Nature* **358,** 668–671.

Hall, S. S. (1993). "Mapping the Next Millennium." Random House, New York.

Haney, S., Thompson, P. M., Cloughesy, T. F., Alger, J. R., and Toga, A. W. (2001). Tracking tumor growth rates in patients with malignant gliomas: A test of two algorithms. *Am. J. Neuroradiol.* **22,** 73–82.

Hodgson, S. V., and Bobrow, M. (1989). Carrier detection and prenatal diagnosis in Duchenne and Becker muscular dystrophy. *Br. Med. Bull.* **45,** 719–724.

Hohne, K. H., Fuchs, H., and Pizer, S. M., eds. (1990). "3D Imaging in Medicine: Algorithms, Systems, Applications." Springer-Verlag, New York.

Holmes, C. J., Hoge, R., Collins, L., Woods, R., Toga, A. W., and Evans, A. C. (1998). Enhancement of magnetic resonance images using registration for signal averaging. *J. Comput. Assisted Tomogr.* **22,** 324–333.

Jacobs, R. E., and Fraser, S. E. (1994). Magnetic resonance microscopy of embryonic cell lineages and movements. *Science* **263,** 681–684.

Kauer, J. S. (1988). Real-time imaging of evoked activity in local circuits of the salamander olfactory bulb. *Nature* **331,** 166–168.

Kaufman, A., ed. (1991). "Volume Visualization." IEEE Comput. Soc., Washington, DC.

Kaufman, L., Crooks, L., and Sheldon, P. E. (1982). Evaluation of NMR imaging for detection and quantification of obstructions in vessels. *Invest. Radiol.* **77,** 554–560.

Koslow, S. H., and Huerta, M. F. (1997). "Neuroinformatics: An Overview of the Human Brain Project." Erlbaum, Hillsdale, NJ.

Kotter, R. (2001). Neuroscience databases: Tools for exploring brain structure–function relationships. *Philos. Trans. R. Soc. London B Biol. Sci.* **356,** 1111–1120.

Kwong, K. K., Belliveau, J. W., Chester, D. A., Goldberg, I. E., Weisskoff, R. M., Poncelet, B. P., Kennedy, D. N., Hoppel, B. E., Cohen, M. S., and Turner, R. (1992). Dynamic images of human brain activity during primary sensory stimulation. *Proc. Natl. Acad. Sci. USA* **89,** 5675–5679.

Lancaster, J. L., Rainey, L. H., Summerlin, J. L., Freitas, C. S., Fox, P. T., Evans, A. C., Toga, A. W., and Mazziotta, J. C. (1997). Automated labeling of the human brain: A preliminary report on the development and evaluation of a forward-transform method. *Hum. Brain Mapp.* **5,** 238–242.

Lassen, N. A., and Holm, S. (1992). Single photon emission computed topography. *In* "Clinical Brain Imaging: Principles and Applications" (J. C. Mazziotta and S. Gilman. eds.), pp. 108–134. Davis, Philadelphia.

Le Bihan, D. (1996). Functional MRI of the brain: Principles, applications and limitations. *Neuroradiology* **23,** 1–5.

Levoy, M. (1991). Methods for improving the efficiency and versatility of volume rendering. *Prog. Clin. Biol. Res.* **363,** 473–488.

Lundsford, L. D. (1988). "Modern Stereotactic Neurosurgery." Nijhoff, Boston.

Macklis, R. M., and Macklis, J. D. (1992). Historical and phrenologic reflections on the nonmotor functions of the cerebellum: Love under the tent? *Neurology* **42,** 928–932.

Mai, J., Assheuer, J., and Paxinos, G. (1997). "Atlas of the Human Brain." Academic Press, New York.

Marshall, E. (2000). A ruckus over releasing images of the human brain. *Science* **289**(5484), 1458–1459.

Masino, S. A., Kwon, M. C., Dory, Y., and Frostig, R. D. (1993). Characterization of functional organization within rat barrel cortex using intrinsic signal optical imaging through a thinned skull. *Proc. Natl. Acad. Sci. USA* **90,** 9998–10002.

Mazziotta, J. (1991). Clinical PET: A reality. *J. Nucl. Med.* **32,** 46N, 54N.

Mazziotta, J., Toga, A. W., Evans, A., Fox, P., Lancaster, J., Zilles, K., Woods, R., Paus, T., Simpson, G., Pike, B., Holmes, C., Collins, L., Thompson, P. M., MacDonald, D., Iacoboni, M., Schormann, T., Amunts, K., Palomero-Gallagher, N., Geyer, S., Parsons, L., Narr, K., Kabani, N., Le Goualher, G., Boomsma, D., Cannon, T., Kawashima, R., and Mazoyer, B. (2001a). A four-dimensional probabilistic atlas of the human brain. *J. of the Amer. Informatics Assoc.* **8,** 401–430.

Mazziotta, J., Toga, A. W., Evans, A., Fox, P., Lancaster, J., Zilles, K., Woods, R., Paus, T., Simpson, G., Pike, B., Holmes, C., Collins, L., Thompson, P., MacDonald, D., Iacoboni, M., Schormann, T., Amunts, K., Palomero-Gallagher, N., Geyer, S., Parsons, L., Narr, K., Kabani, N., Le Goualher, G., Boomsma, D., Cannon, T., Kawashima, R., and Mazoyer, B. (2001b). A probabilistic atlas and reference system for the human brain: International Consortium for Brain Mapping (ICBM). *Philos. Trans. R. Soc. London B* **356,** 1293–1322.

Mazziotta, J. C., and Gilman, S. (1992). "Clinical Brain Imaging: Principles and Applications." Davis, Philadelphia.

Mazziotta, J. C., Toga, A. W., Evans, A., Fox, P., and Lancaster, J. (1995). A probabilistic atlas of the human brain: Theory and rationale for its development. *NeuroImage* **2,** 89–101.

Mazziotta, J. C., Valentino, D., Grafton, S., Bookstein, F., Pelizzari, C., Chen, G., and Toga, A. W. (1991). Relating structure to function *in vivo* with tomographic imaging. *In* "Ciba Foundation Symposia," Vol. 163, "Exploring Brain Functional Anatomy with Positron Tomography," pp. 93–112. Wiley, Chichester.

McIntosh, A. R., Grady, O. L., Ungerleider, L. G., Haxby, J. V., Rapoport, S. I., and Horwitz, B. (1994). Network analysis of cortical visual pathways mapped to PET. *J. Neurosci.* **14,** 655–666.

Mega, M. S., Chen, S., Thompson, P. M., Woods, R. P., Karaca, T. J., Tiwari, A., Vinters, H., Small, G. W., and Toga, A. W. (1997). Mapping pathology to metabolism: Coregistration of stained whole brain sections to PET in Alzheimer's disease. *NeuroImage* **5,** 147–153.

Mega, M. S., Lee, L., Dinov, I. D., Mishkin, F., Toga, A. W., and Cummings, J. L. (2000a). Cerebral correlates of psychotic symptoms in Alzheimer's disease. *J. Neurosurg. Psychiatry* **68,** 1–4.

Mega, M. S., Fennema-Notestine, C., Dinov, I. D., Thompson, P. M., Archibald, S. L., Lindshield, C. J., Felix, J., Toga, A. W., and Jernigan, T. L. (2000b). Construction, testing, and validation of a sub-volume probabilistic human brain atlas for the elderly and demented populations. Sixth International Conference on Functional Mapping of the Human Brain, San Antonio, Texas. *NeuroImage* **11,** S597. [Abstract]

Miller, M. I., Christensen, G. E., Amit, Y., and Grenander, U. (1993). Mathematical textbook of deformable neuroanatomies. *Proc. Natl. Acad. Sci. USA* **90,** 11944–11948.

Minoshima, S., Koeppe, R. A., Frey, K. A., Ishihara, M., and Kuhl, D. E. (1994). Stereotactic brain atlas of positron emission tomography: Reference for indirect anatomical localization. *J. Nucl. Med.* **35,** 949–954.

Missir, O., Dutheil-Desclercs, C., Meder, J. F., Musolino, A., and Fredy, D. (1989). Central sulcus patterns at MRI. *J. Neuroradiol.* **16,** 133–144.

Molchan, S. E., Sunderland, T., McIntosh, A. R., Herscovitch, P., and Schreurs, B. G. (1994). A functional anatomical study of associative learning in humans. *Proc. Natl. Acad Sci. USA* **91,** 8122–8126.

Montague, P. R., Gally, J. A., and Edelman, G. M. (1991). Spatial signalling in the development and function of neural connections. *Cereb. Cortex* **1**(**3**), 199–220.

Morosan, P., Rademacher, J., Schleicher, A., Amunts, K., Schormann, T., and Zilles, K. (2001). Human primary auditory cortex: Cytoarchitectonic subdivisions and mapping into a spatial reference system. *NeuroImage* **13,** 684–701.

Narayan, S. M., Santori, E. M., and Toga, A. W. (1994a). Mapping functional activity in rodent cortex using optical intrinsic signals. *Cereb. Cortex* **4,** 195–204.

Narayan, S. M., Blood, A. J., and Toga, A. W. (1994b). Imaging optical reflectance in rodent barrel and forelimb sensory cortex. *NeuroImage* **1,** 181–190.

Narr, K. L., Thompson, P., Sharma, T., Moussai, J., Zoumalon, C., Rayman, J., and Toga, A. W. (2000a). 3D mapping of gyral shape and cortical surface asymmetries in schizophrenia: Gender effects. *Am. J. Psychiatry* **158,** 244–255.

Narr, K. L., Thompson, P., Sharma, T., Moussai, J., Cannestra, A. F., and Toga, A. W. (2000b). Mapping morphology of the corpus callosum in schizophrenia. *Cereb. Cortex* **10,** 40–49.

O'Donovan, M. J., Ho, S., Sholomenko, G., and Yee, W. (1993). Real-time imaging of neurons retrogradely and anterogradely labelled with calcium-sensitive dyes. *J. Neurosci. Methods* **46,** 91–106.

Ogawa, S., Tank, D. W., Menon, R., Ellerman, J. M., Kim, S. G., Merkle, H., and Ugurbil, K. (1992). Intrinsic signal changes accompanying sensory stimulation: Functional brain mapping with magnetic resonance imaging. *Proc. Natl. Acad. Sci. USA* **89,** 5951–5955.

Ono, M., Kubik, S., and Abernathey, D. (1990). "Atlas of the Cerebral Sulci." Thieme, Stuttgart.

Orrison, W. W., Levine, J. D., Sanders, J. A., and Hartshone, N. F. (1995). "Functional Brain Imaging." Mosby, St. Louis.

Orthner, H., and Sendler, W. (1975). Plannimetrische volumetrie an menschlichen gehirnen. *Fortschr. Neurol. Psychiat.* **43,** 191–209.

Palovcik, R. A., Reid, S. A., Principe, J. C., and Albuquerque, A. (1992). 3-D computer animation of electrophysiological responses. *J. Neurosci. Methods* **41,** 1–9.

Paxinos, G., ed. (1990). "The Human Nervous System." Academic Press, San Diego.

Paxinos, G., and Watson, C. (1986). "The Rat Brain in Stereotaxic Coordinates." Academic Press, San Diego.

Payne, B. A., and Toga, A. W. (1990). Surface mapping brain function on 3D models. *Comput. Graph. Appl.* **10,** 33–41.

Pechura, C. M., and Martin, J. B. (1991). "Mapping the Brain and Its Functions: Integrating Enabling Technologies into Neuroscience Research." Natl. Acad. Press, Washington, DC.

Pedersen, N. L., Miller, B. L., Wetherell, J. L., Vallo, J., Toga, A. W., Knutson, N., Mehringer, C. M., Small, G. W., and Gatz, M. (1999). Neuroimaging findings in twins discordant for Alzheimer's disease. *Dementia Geriatric Cognit. Disord.* **10,** 51–58.

Penfield, W., and Boldrey, E. (1937). Somatic motor and sensory representation in the cerebral cortex of man as studied by electrical stimulation. *Brain* **60,** 389–443.

Perani, D., Gilardi, M. C., Cappa, S. F., and Fazio, F. (1992). PET studies of cognitive function: A review. *J. Nucl. Biol. Med.* **36,** 329–336.

Petroff, O., Prichard, J., Behar, K., Alger, J., den Hollander, J., and Shulman, R. (1985a). Cerebral intracellular pH by ^{31}P nuclear magnetic resonance spectroscopy. *Neurology* **35,** 781–788.

Petroff, O., Prichard, J., Behar, K., Rothman, D., Alger, J., and Shulman, R. (1985b). Cerebral metabolism in hyper- and hypocarbia: ^{31}P and ^{1}H nuclear magnetic resonance studies. *Neurology* **35,** 1681–1685.

Phelps, M. E., Mazziotta, J. C., and Schelbert, H. R. (1986). "Positron Emission Tomography and Autoradiography." Raven Press, New York.

Pierpaoli, C., Jezzard, P., Basser, P. J., Barnett, A., and Ci Chiro, G. (1996). Diffusion tensor MR imaging of the human brain. *Radiology* **3,** 637–648.

Preda, K., Gaetani, P., and Radriguez y Baena, R. (1998). Sparro CTN geography and surgical correlations in the evaluation of intercraniel aneurysms. *Eur. Radiol.* **8,** 739–745.

Prichard, J., Rothman, D., Novotny, E., Petroff, O., Kuwabara, T., and Avison M. (1991). Lactate rise detected by ^{1}H NMR in human visual cortex during physiological stimulation. *Proc. Natl. Acad. Sci. USA* **88,** 5829–5831.

Prichard, J. W., and Rosen, B. R. (1994). Functional study of the brain by NMR. *J. Cereb. Blood Flow Metab.* **14,** 365–372.

Rabbitt, R. D., Weiss, J. A., Christensen, G. E., and Miller, M. I. (1995). Mapping of hyperelastic deformable templates using the finite element method. *Proc. SPIE* **2573,** 252–265.

Rabl, R. (1958). Strukturstudien an der massa intermedia des thalamus opticus. *J. Hirnforsch.* **4,** 78–112.

Rademacher, J., Caviness, V. S., Steinmetz, H., and Galaburda, A. M. (1993). Topographical variation of the human primary cortices: Implications for neuroimaging, brain mapping and neurobiology. *Cereb. Cortex* **3,** 313–329.

Rajkowska, G., and Goldman-Rakic, P. S. (1995). Cytoarchitectonic definition of prefrontal areas in the normal human cortex. II. Variability in locations of areas 9 and 46 and relationship to the Talairach coordinate system. *Cereb. Cortex* **5,** 323–337.

Rance, N. E., Young, W. S., and McMullen, N. T. (1994). Topography of neurons expressing luteinizing hormone-releasing hormone gene transcripts in the human brain. *J. Comp. Neurol.* **339,** 573.

Rapoport, S. I. (1991). PET in Alzheimer's disease in relation to disease pathogenesis: A critical review. *Cerebrovasc. Brain Metab. Rev.* **3,** 297–335.

Resnick, S. M., Karp, J. S., Turetsky, B., and Gur, R. E. (1993). Comparison of anatomically-based regional localization: Effects on PET–FDG quantitation. *J. Nucl. Med.* **34,** 2201–2207.

Riley, H. A. (1943). "An Atlas of the Basal Ganglia, Brain Stem and Spinal Cord." Williams & Wilkins, Baltimore.

Rizzo, G., Gilardi, M. C., Prinster, A., Grassi, F., Scotti, G., Cerutti, S., and Fazio, F. (1995). An elastic computerized brain atlas for the analysis of clinical PET/SPET data. *Eur. J. Nucl. Med.* **22,** 1313–1318.

Robb, R. A., ed. (1985). "Three Dimensional Biomedical Imaging." CRC Press, Boca Raton, FL.

Robinson, A. H., and Petchenik, B. B., eds. (1976). "The Nature of Maps: Essays towards Understanding Maps and Mapping." Univ. of Chicago Press, Chicago.

Roland, P. E., Graufelds, C. J., Wahlin, L., Ingelman, L., Andersson, M., Ledberg, A., Pedersen, J., Akerman, S., Dabringhaus, A., and Zilles, K. (1994). Human brain atlas: For high-resolution functional and anatomical mapping. *Hum. Brain Mapp.* **1,** 173–184.

Roland, P. E., and Zilles, K. (1994). Brain atlases—A new research tool. *Trends Neurosci.* **17,** 458–467.

Ross, J., Masaryk, T., Modic, M., Harik, S., Wiznitzer, M., and Selman, S. (1989). Magnetic resonance angiography of the extracranial carotid arteries and intracranial vessels: A review. *Neurology* **39,** 1369–1376.

Sandor, S. R., and Leahy, R. M. (1994). Matching deformable atlas models to pre-processed magnetic resonance brain images. *Proc. IEEE Conf. Image Process.* **3,** 686–690.

Sandor, S. R., and Leahy, R. M. (1995). Towards automated labeling of the cerebral cortex using a deformable atlas. *In* "Information Processing in Medical Imaging" (Y. Bizais, C. Barillot, and R. Di Paola, eds.), pp. 127–138. Kluwer Academic Publishers, Brest, France.

Schaltenbrand, G., and Bailey, P. (1959). "Introduction to Stereotactic Operations, with an Atlas of the Human Brain." Thieme, New York.

Schmahmann, J. D., Doyon, J., McDonald, D., Holmes, C., Lavoie, K., Hurwitz, A. S., Kabani, N., Toga, A. W., Evans, A., and Petrides, M. (1999). Three dimensional MRI atlas of the human cerebellum in proportional stereotaxic space. *NeuroImage* **10,** 233–260.

Schwartz, E. L., and Merker, B. (1986). Computer aided neuroanatomy: Differential geometry of cortical surfaces and an optimal flattening algorithm. *IEEE Comput. Graph.* **April,** 36–44.

Seeram (1994). "Computed Tomography Physical Principles: Clinical Applications and Quality." Saunders, Philadelphia.

Seitz, R. J., Bohm, C., Greitz, T., Roland, P. E., Eriksson, L., Blomqvist, G., Rosenqvist, G., and Nordell, B. (1990). Accuracy and precision of the computerized brain atlas programme for localization and quantification in positron emission tomography. *J. Cereb. Blood Flow Metab.* **10,** 443–457.

Sereno, M. I., Dale, A. M., Liu, A., and Tootel, R. G. H. (1996). A surface-based coordinate system for a canonical cortex. Proceedings of the Second International Conference on Human Brain Mapping, Boston. *NeuroImage* **3,** S252.

Smith, G. E. (1907). A new topographical survey of the human cerebral cortex, being an account of the distribution of the anatomically distinct cortical areas and their relationship to the cerebral sulci. *J. Anat.* **41,** 237–254.

Sowell, E. R., Levitt, J., Thompson, P. M., Holmes, C. J., Blanton, R. E., Kornsand, D. S., Caplan, R., McCracken, J., Asarnow, R., and Toga, A. W. (2000a). Brain abnormalities in early onset schizophrenia spectrum disorder observed with statistical parametric mapping of structural magnetic resonance images. *Am. J. Psychiatry* **157,** 1475–1484.

Sowell, E. R., Mattson, S. N., Thompson, P. M., Jernigan, T. L., Riley, E. P., and Toga, A. W. (2001). Mapping callosal morphology and cognitive correlates: Effects of heavy prenatal alcohol exposure. *Neurology* **57,** 235–244.

Sowell, E. R., Thompson, P. M., Holmes, C. J., Jernigan, T. L., and Toga, A. W. (1999). *In vivo* evidence for post adolescent brain maturation in frontal and striated regions. *Nat. Neurosci.* **2,** 859–861.

Sowell, E. R., Thompson, P. M., Mattson, S. N., Tessner, K. D., Jernigan, T. L., Riley, E. P., and Toga, A. W. (2000b). Voxel-based morphometric analyses of the brain in children and adolescents prenatally exposed to alcohol: Abnormalities in posterior temporo-parietal cortex. *NeuroReport* **12,** 515–523.

Spiegel, E. A., and Wycis, H. T., eds. (1952). "Monographs in Biological Medicine," Vols. 1 and 2, "Stereoencephalotomy." Grune & Stratton, New York.

Spitzer, V., Ackerman, M. J., Scherzinger, A. L., and Whitlock, D. (1996). The visible human male: A technical report. *J. Am. Med. Inform. Assoc.* **3,** 118–130. [http://www.nlm.nih.gov/extramural_research.dir/visible_human.html]

Stark, D. D., and Bradley, W. G. (1992). "Magnetic Resonance Imaging." Mosby, St. Louis.

Stehling, M. K., Turner, R., and Mansfield, P. (1991). Echo-planar imaging: Magnetic resonance imaging in a fraction of a second. *Science* **254,** 43–50.

Steinmetz, H., Furst, G., and Freund, H. J. (1989). Cerebral cortical localization: Application and validation of the proportional grid system in MR imaging. *J. Comput. Assisted Tomogr.* **13,** 10–19.

Steinmetz, H., Furst, G., and Freund, H. J. (1990). Variation of perisylvian and calcarine anatomic landmarks within stereotactic proportional coordinates. *Am. J. Neuroradiol.* **11,** 1123–1130.

Stensaas, S. S., Eddington, D. K., and Dobelle, W. H. (1974). The topography and variability of the primary visual cortex in man. *J. Neurosurg.* **40,** 747–755.

Stieltjes, B., Kaufmann, W. E., van Ziljl, P. C. M., Fredericksen, K., Pearlson, G. D., Solaiyappan, M., and Mori, S. (2001). Diffusion tensor imaging and axonal tracking in the human brain stem. *NeuroImage* **14,** 723–735.

Strother, S. C., Anderson, J. R., Xu, X. L., Liow, J. S., Bonar, D. C., and Rottenberg, D. A. (1994). Quantitative comparisons of image registration techniques on high-resolution MRI of the brain. *J. Comput. Assisted Tomogr.* **18,** 954–962.

Sugitani, M., Sugai, T., Tanifuji, M., Murase, K., and Onoda, N. (1994). Optical imaging of the *in vitro* guinea pig piriform cortex activity using a voltage sensitive dye. *Neurosci. Lett.* **165,** 215–218.

Swanson, L. W. (1990). "Brain Maps: Structure of the Rat Brain." Elsevier, Amsterdam/New York.

Swanson, L. W. (1992). Computer graphics file. Version 1.0. Elsevier, Amsterdam/New York.

Szabo, J., and Cowan, W. M. (1984). A stereotactic atlas of the brain of the cynomolgus monkey (*Macaca fascicularis*). *J. Comp. Neurol.* **222,** 265–300.

Talairach, J., and Tournoux, P. (1988). Principe et technique des etudes anatomiques. *In* "Co-planar Stereotaxic Atlas of the Human Brain—3-Dimensional Proportional System: An Approach to Cerebral Imaging" (M. Rayport, ed.), Vol. 3. Thieme, New York.

Talairach, J., David, M., Tournoux, P., Corredor, H., and Kvasina, T. (1957). "Atlas d'Anatomie Stereotaxique: Reperage Radiologique Indirect des Noyaux Gris Centraux des Regions Mesecephalo-sous-Optique et Hypothalamique de l'Homme." Masson, Paris.

Talairach, J., Szikla, G., Tournoux, P., Prosalentis, A., Bordas-Ferrier, M., Covello, L., Iacob, M., and Mempel, E. (1967). "Atlas d'Anatomie Stereotaxique du Telencephale." Masson, Paris.

Thirion, J. P. (1995). Fast non-rigid matching of medical images. INRIA Intern. Rep. 2547, INRIA Project Epidaure, France.

Thirion, J.-P., Prima, S., and Subsol, S. (1998). Statistical analysis of dissymmetry in volumetric medical images. INRIA Tech. Rep. RR-3178.

Thompson, P. M., and Toga, A. W. (1996). A surface-based technique for warping 3-dimensional images of the brain. *IEEE Trans. Med. Imaging* **15,** 1–16.

Thompson, P., Schwartz, C., Lin, R. T., Khan, A. A., and Toga, A. W. (1996a). 3D statistical analysis of sulcal variability in the human brain. *J. Neurosci.* **16,** 4261–4274.

Thompson, P. M., Schwartz, C., and Toga, A. W. (1996b). High-resolution random mesh algorithms for creating a probabilistic 3D surface atlas of the human brain. *NeuroImage* **3,** 19–34.

Thompson, P. M., and Toga, A. W. (1997). Detection, visualization and animation of abnormal anatomic structure with a deformable probabilistic brain atlas based on random vector field transformations. *Med. Image Anal.* **1,** 271–294.

Thompson, P. M., MacDonald, D., Mega, M. S., Holmes, C. J., Evans, A. C., and Toga, A. W. (1997). Detection and mapping of abnormal brain structure with a probabilistic atlas of cortical surfaces. *J. Comput. Assisted Tomogr.* **21,** 567–581.

Thompson, P. M., and Toga, A. W. (1998). Anatomically-driven strategies for high-dimensional brain image warping and pathology detection. *In* "Brain Warping" (A. W. Toga, ed.), pp. 311–336. Academic Press, San Diego.

Thompson, P. M., Moussai, J., Zohoori, S., Goldkorn, A., Khan, A. A., Mega, M. D., Small, G. W., Cummings, J. L., and Toga, A. W. (1998). Cortical variability and asymmetry in normal aging and Alzheimer's disease. *Cereb. Cortex* **8,** 492–509.

Thompson, P. M., Woods, R. P., Mega, M. S., and Toga, A. W. (2000a). Mathematical/computational challenges in creating deformable and probabilistic atlases of the human brain. *Hum. Brain Mapp.* **9,** 81–92.

Thompson, P. M., Giedd, J. N., Woods, R. P., MacDonald, D., Evans, A. C., and Toga, A. W. (2000b). Growth patterns in the developing human brain detected using continuum-mechanical tensor mapping. *Nature* **404,** 190–193.

Thompson, P. M., Mega, M. S., Vidal, C., Rapoport, J., and Toga, A. W. (2001a). Detecting disease-specific patterns of brain structure using cortical pattern matching and a population-based probabilistic brain atlas. *In* 17th International Conference on Information Processing in Medical Imaging, Davis, California, U.S.A., June 18–22, pp. 488–501.

Thompson, P. M., Mega, M. S., Woods, R. P., Zoumalan, C. I., Lindshield, C. J., Blanton, R. E., Moussai, J., Holmes, C. J., Cummings, J. L., and Toga, A. W. (2001b). Cortical change in Alzheimer's disease detected with a disease-specific population-based brain atlas. *Cereb. Cortex* **11,** 1–16.

Tiede, U., Bomans, M., Hohne, K. H., Pommert, A., Riemer, M., Schiemann, T., Schubert, R., and Lierse, W. (1993). A computerized three-dimensional atlas of the human skull and brain. *Am. J. Neuroradiol.* **14,** 551–559.

Toga, A. W. (1990a). The animation of 3D structure–function relationships. *Drug News Perspect.* **3,** 197–201.

Toga, A. W. (1990b). On computing the anatomy of the brain. *Pixel* **1,** 2.

Toga, A. W. (1990c). Images of brain. *BRI Bull.* **14,** 11–15.

Toga, A. W. (1990d). Visualizing the 3-D structure, function of brain. *Clin. Neuropharmacol.* **13,** 462–463.

Toga, A. W. (1993). Imagery of the brain. *In* "Visual Cues: Practical Data Visualization" (P. R. Keller and M. M. Keller, eds.). IEEE Press, New York.

Toga, A. W. (1994). Visualization and warping of multimodality brain imaging. *In* "Advances in Functional Neuroimaging: Technical Foundations" (R. W. Thatcher, M. Hallett, T. Zeffiro, E. R. John, and M. Huerta, eds.), Chap. 6, pp. 171–180. Academic Press, San Diego.

Toga, A. W. (1998). "Brain Warping." Academic Press, San Diego.

Toga, A. W., Santori, E. M., and Samie, M. (1986). Regional distribution of flunitrazepam binding constants: Visualizing Kd and Bmax by digital image analysis. *J. Neurosci.* **6,** 2747–2756.

Toga, A. W., Ambach, K. L., and Schluender, S. (1994a). High-resolution anatomy from *in situ* human brain. *NeuroImage* **1,** 334–344.

Toga, A. W., Ambach, K. L., Quinn, B. C., Hutchin, M., and Burton, J. S. (1994b). Post mortem anatomy from cryosectioned whole human brain. *J. Neurosci. Methods* **54,** 239–252.

Toga, A. W., Ambach, K., Quinn, B. C., and Mazziotta, J. C. (1995a). Postmortem cryosectioning as an anatomic reference for human brain mapping. *J. Cereb. Blood Flow Metab.* **15**(Suppl. 1), S600.

Toga, A. W., Mazziotta, J. C., and Woods, R. P. (1995b). A high resolution anatomic reference for PET activation studies. *NeuroImage* **2,** S83.

Toga, A. W., and Thompson, P. (1996). Visualization and mapping of anatomic abnormalities using a probabilistic brain atlas based on random fluid transformations. *Lect. Notes Comput. Sci.* **1131,** 383–392.

Toga, A. W., Thompson, P. M., Holmes, C. J., and Payne, B. A. (1996). Informatics and computational neuroanatomy. *J. Am. Med. Inform. Assoc.* **1996**(Suppl.), 299–303.

Toga, A. W., and Thompson, P. M. (1997). Measuring, mapping and modeling brain structure and function. *SPIE Med Imaging* **3033,** 104–115.

Toga, A. W., Thompson, P. M., Mega, M. S., Narr, K. L., and Blanton, R. L. (2001a). Probabilistic approaches for atlasing normal and disease-specific brain variability. *Anat. Embryol.,* **204**, 267–282.

Toga, A. W., and Thompson, P. M. (2001b). The role of image registration in brain mapping. *Image Vision Comput.* **19,** 3–24.

Troyer, T. W., Levin, J. E., and Jacobs, G. A. (1994). Construction and analysis of a database representing a neural map. *Microsc. Res. Tech.* **29,** 329–343.

Ts'o, D. Y., Frostig, R. D., Lieke, E. E., and Grinvald, A. (1990). Functional organization of primate visual cortex revealed by high resolution optical imaging. *Science* **249,** 417–420.

Tufte, E. R. (1983). "The Visual Display of Quantitative Information." Graphics Press, Cheshire, CT.

Tufte, E. R. (1990). "Envisioning Information." Graphics Press, Cheshire, CT.

Tuttle, M. S., Sheretz, D. D., Olson, N., Erlbaum, M., Sperzel, D., Fuller, L., and Nelson, S. (1990). Using Meta-1—The first version of the UMLS metathesaurus. *In* "Proceedings of the 14th Annual Symposium on Computer Applications in Medical Care," pp. 131–135. IEEE Comput. Soc., Los Alamitos, CA.

Udupa, J. K., and Herman, G. T., eds. (1991). "3D Imaging in Medicine." CRC Press, Boca Raton, FL.

Van Buren, J., and Maccubin, D. (1962). An outline atlas of human basal ganglia and estimation of anatomic variants. *J. Neurosurg.* **19,** 811–839.

Van Essen, D. C., and Maunsell, J. H. R. (1980). Two dimensional maps of the cerebral cortex. *J. Comp. Neurol.* **191,** 255–281.

Van Essen, D. C., and Maunsell, J. H. R. (1983). Hierarchical organization and functional streams in the visual cortex. *Trends Neurosci.* **6,** 370–375.

Van Essen, D. C., Drury, H. A., Joshi, S. C., and Miller, M. I. (1997). Comparisons between human and macaque using shape-based deformation algorithms applied to cortical flat maps. Third International Conference on Functional Mapping of the Human Brain, Copenhagen, May 19–23, 1997. *NeuroImage* **5,** S41.

von Bonin, G., and Bailey, P. (1947). :The Neocortex of *Macaca mulatta.*" Univ. of Illinois Press, Urbana. [Monographs in Medical Science]

Wada, J. A., Clarke, R., and Hamm, A. (1975). Cerebral hemispheric asymmetry in humans. *Arch. Neurol.* **32,** 239–246.

Wainwright, B. J., Scambler, P. J., Schmidtke, J., Watson, E. A., Law, H.-Y., Farrall, M., Cooke, H. J., Eiberg, H., and Williamson, R. (1985). Localization of the cystic fibrosis locus to human chromosome 7cen–q22. *Nature* **318,** 384–385.

White, R., Woodward, S., Nakamura, Y., Leppert, M., O'Connell, P., Hoff, M., Herbst, J., Lalouel, J. M., Dean, M., and Vande Woude, G. M. (1985). A closely linked genetic marker for cystic fibrosis. *Nature* **318,** 382–384.

Wilford, J. N. (1982). "The Mapmakers." Random House, New York.

Wong, S. T. C. (1998). "Medical Image Databases." Kluwer Academic, Dordrecht/Norwell, MA.

Woods, R. P., Cherry, S. R., and Mazziotta, J. C. (1992). Rapid automated algorithm for aligning and reslicing PET images. *J. Comput. Assisted Tomogr.* **16,** 620–633.

Woods, R. P., Dapretto, M., Sicotte, N. L., Toga, A. W., and Mazziotta, J. M. (1999). Creation and use of a Talairach-compatible atlas for accurate, automated nonlinear intersubject registration and analysis of functional imaging data. *Hum. Brain Mapp.* **8,** 73–79.

Woods, R. P., Mazziotta, J. C., and Cherry, S. R. (1993). MRI–PET registration with automated algorithm. *J. Comput. Assisted Tomogr.* **17,** 536–546.

Yakovlev, P. (1970). Whole brain serial histological sections. *In* "Neuropathology: Methods and Diagnosis" (C. G. Tedeschi, ed.), pp. 371–378. Little, Brown, Boston.

Youngman, S., Sarfarazi, M., Bucan, M., MacDonald, M., Smith, B., Zimmer, M., Gilliam, C., Frischouf, A. M., Wasmuth, J. J., and Gusella, J. F. (1989). A new DNA marker (D4S90) is located terminally on the short arm of chromosome 4, close to the Huntington's disease gene. *Genomics* **5,** 802–809.

Zulegar, S., and Stansbesand, J. (1977). "Atlas of the Central Nervous System in Sectional Planes: Selected Myelin Stained Sections of the Human Brain and Spinal Cord." Urban & Schwarzenberg, Baltimore.

2

Time and Space

John C. Mazziotta

Ahmanson–Lovelace Brain Mapping Center, Radiological Sciences, and Medical and Molecular Pharmacology, UCLA School of Medicine, Los Angeles, California 90024

I. Introduction

The nervous system is fertile ground for mapmaking. Ever since the earliest investigations of nervous system function, physical as well as theoretical depictions of brain function have been visualized using some form of mapping method. If one focuses on the structural aspects of brain anatomy, an important difference between geographical maps of the earth and neuroanatomical maps of the brain is quickly realized. That is, that while there is one single unique physical reality to the geographical organization of the earth, neuroanatomy must represent a variable physical reality that differs from individual to individual. Thus, human brain map atlases of structure and function require a representation that accounts for variance among individuals. Further, neuroscientists have yet to agree on a standard reference system and nomenclature to define brain location. This again differs from geographical maps for which the conventions of longitude, latitude, and altitude represent a universal standard.

Despite these differences among geographical and neuroanatomical maps, there are many similarities. Both earth and brain maps can represent a seemingly infinite number of variables. For earth maps these may take the form of population density, crime rate, transportation systems, temperature, environmental pollutants, etc. (Tufte, 1983, 1990). Similarly, brain maps can describe regional blood flow, metabolism, receptor density, fiber connections, etc. Both earth and brain maps can be depicted at different scales. For the earth this can range from global depictions to local neighborhoods. In the brain, maps can be equally global, regional, nuclear, cellular, or ultrastructural. Both earth and brain maps change over a wide time scale. The continental drifts that occur over hundreds of millions of years can be contrasted with meteorological maps that vary from minute to minute. Evolutionary changes in brain anatomy can similarly be contrasted with electrophysiological fluctuations that occur in the millisecond time range.

One can conclude this analogy by examining the types of data depiction schemes that have been used for the earth and are now, with ever-increasing regularity, being applied to the brain. Three-dimensional spherical models of the earth, such as globes, are commonly found in most elementary school classrooms. Flattening of this true representation of the earth induces aberrations that have led to many schemes for the projection and flattening of the three-dimensional surface onto a flat sheet. All of these approaches result in either errors or discontinuities in the representation. The complex gyral patterns of the human brain represent a similarly difficult task (Felleman and Van Essen, 1991; Sereno and Dale, 1992; Sereno *et al.,* 1994). The heterogeneous internal structure of the earth is rarely depicted in anything more complex than cross-sectional slices. The same can be said of brain maps. The challenge of producing a flattened brain map that depicts both cortical surface and internal nuclei is a formidable task, indeed.

The advent of digital representation provides a meaningful and highly flexible new tool to aid in the mapmaking process. With a true three-dimensional representation, the observer can select, scale, and view any orientation at will provided the appropriate data set is available for survey. Quantifiable information can then be retrieved by pointing to the appropriate location in space and opening a file that produces tabular data about a given region or a selected view of detailed structure at a higher degree of magnification.

Tools are currently available to analyze brain structure at a macroscopic level, particularly with the advent of *in vivo* imaging techniques such as X-ray computed tomography (CT) and magnetic resonance imaging (MRI). Microscopic anatomy has been of interest to investigators for more than a century. The same can be said, although for a shorter time frame, of the ultrastructural anatomy of the brain. Critical, though largely missing, components of these methods are the appropriate bridging technologies that can span the gamut from macroscopic to microscopic and microscopic to ultrastructural scales. Yet, in order to have a continuous, smooth scaling from the ultrastructural level to the macroscopic level such methods must be established. In this chapter, the types of maps of the nervous system that can be produced and the opportunities as well as constraints that currently exist for their development and ultimate use will be discussed. It should be realized, however, that the creation of formal digital and standardized approaches to mapping of the brain are less of a luxury and more of a requirement in the current neuroscientific era, as the explosion in neuroscientific information demands an organized approach to communication between neuroscientists that will increase both the efficiency and the quality of data analysis (Barinaga, 1993; Huerta *et al.,* 1993; Koshland, 1992; Mazziotta, 1984; Mazziotta and Gilman, 1992; Mazziotta and Koslow, 1987; Pechura and Martin, 1991b).

II. Critical Variables in Brain Mapping Techniques

While an infinite number of variables can be represented in maps, the two most obvious and immediately important are those of spatial organization and temporal pattern (Fig. 1 and Table 1). The resolution, both temporal and spatial, of measurements made with most brain mapping techniques is one factor that defines their applications. Spatial resolution determines the scale and, hence, the types of measurements that are attainable with the method (Fig. 2). By convention, high-resolution devices are capable of accurately defining smaller structures. With increasing spatial resolution one moves up the scale from macroscopic (e.g., CT, MRI, PET, SPECT) to microscopic (e.g., light microscopy, confocal microscopy, visible light, and near-infrared optical intrinsic signal imaging) to ultrastructural measurements (e.g., electron microscopy).

Similarly, temporal resolution is an important determining factor in the types of data that can be acquired with a given

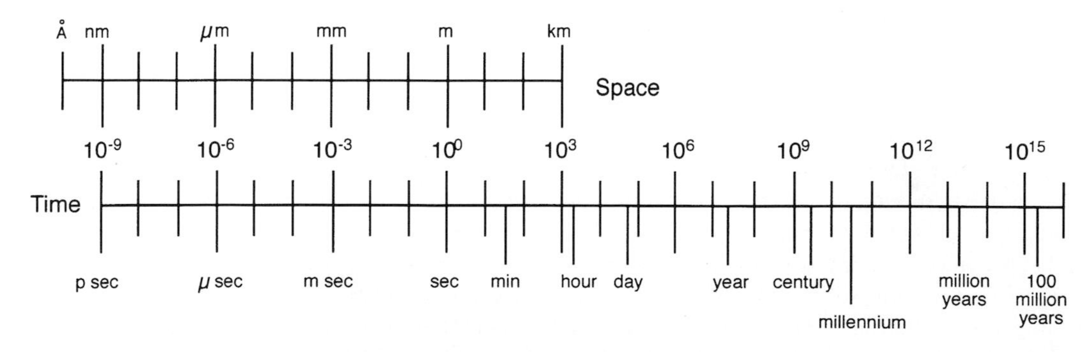

Figure 1 Log scale of time and space. Magnitudes relative to synaptic events and cellular processing would span from 10^{-7} to 10^{-2} meters in space and 10^{-4} to 10^{1} with regard to time. Molecular events can occur at a much smaller spatial scale and more rapid time frame. Evolutionary events take place in time frames as long as 10^{11} to 10^{15} s. Adapted, with permission, from Shephard (1990).

Table 1 Brain Mapping Techniques: Their Temporal and Spatial Resolutions and Sampling Volumes

Method	Energy source	Spatial resolution (mm)	Temporal resolution (s)	Risks	Constraints	Product
Structural MRI	Radio waves	1	N/A	None known	Immobilization, loud	Structure, vasculature, white matter
*f*MRI	Radio waves	4–5	4–10	None known	Immobilization, loud, cooperation	Relative CBF
EEG/MEG	Intrinsic	10	0.01	None known	Artifact, lack of unique localization	Electrophysiology of brain events
MR spectroscopy	Radio waves	10	10–100	None known	Immobilization, loud	Relative chemical concentrations
Optical/near-IR intrinsic signals	Near-IR or visible light	0.05	0.05	None known	some immobilization, surface > depth, limited FOV	Relative CBF, absolute values may be possible
Nuclear (PET and SPECT)	Radiation	5–10	60–1000	Ionizing radiation	Radiation limits immobilization	Physiology, absolute values
Transcranial magnetic stimulation	Magnetic fields	10	0.01	Low risk of seizures	Immobilization, loud	Electrophysiology, conduction times
Post mortem	N/A	0.01	N/A	N/A	Post mortem changes	Microarchitectures, chemoarchitectures

Note: N/A, not applicable

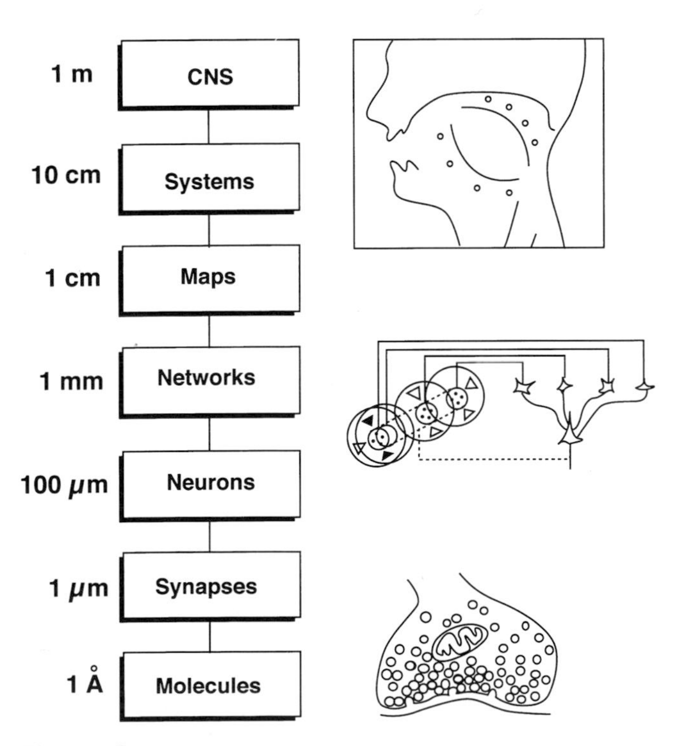

Figure 2 Brain mapping methods and their characteristics. A plot of spatial resolution and the types of neuronal systems they measure. Many methods overlap with regard to spatial characteristics but each is unique with regard to its full complement of attributes. Adapted from Churchland and Sejnowski (1988).

method and, hence, its applications in neuroscience research. On the temporal resolution spectrum, one could begin with the lowest forms of temporal resolution. Examples in this category would include the ability to examine fossilized remains and cranial casts, both within and between species, in order to evaluate the evolution of the macroscopic aspects of the nervous system in a time frame of thousands to millions of years (Jerison, 1989) (Fig. 1). Structural *in vivo* imaging over years with serial studies performed longitudinally in the same individual can provide insights into maturation, development, aging, and the structural impact of the natural course of diseases. Techniques that provide temporal resolution in the time frame of hours provide the opportunities to examine other types of events. For example, ligands labeled with radionuclides can be injected into subjects and these individuals can then be studied with PET or SPECT, with kinetic data being collected for many hours. Positron-emitting isotopes with half-lives of approximately 2 h (e.g., fluorine-18) are ideally suited to this type of experiment. The 2-h half-life of fluorine-18, when incorporated into an appropriate molecule, can provide insights into the uptake of the compound in the brain and the selective binding to specific receptor subtypes as well as the clearance of unbound compound from nonspecific sites in the brain.

Techniques that allow for measurements in the time frame of minutes include PET measurements of cerebral blood flow, volume, glucose, and oxygen metabolism. Techniques that provide accurate data in the time frame of seconds to hundreds of milliseconds include *f*MRI, optical intrinsic signal imaging, and tracer dye techniques. Finally, electrophysiological (e.g., EEG) and magnetoencephalographic (MEG) techniques would be those chosen to examine events in the millisecond time frame.

In addition to evaluating time and space factors in the domain of brain mapping, there are numerous other variables that are important to consider and which help to define the applicability of a given technique to a neurobiological question. Sampling volume is the critical variable of this type (Fig. 4). The *in vivo* brain imaging methods such as PET, MRI, CT, and SPECT provide global macroscopic views of the entire intracranial contents. At the other end of the spectrum would be depth electrodes used for intracellular recording that provide data from individual neurons. In general, sampling volume is inversely related to spatial resolution. This rule does not, however, hold for the relationship of sampling volume to temporal resolution. For example, *f*MRI, when used with echoplanar techniques, can sample the entire intracranial contents but still maintain a relatively high temporal resolution.

Sampling frequency is another key variable to assess. Some methods allow for only a single examination, a situation typical of postmortem or autoradiographic experiments. Other methods have a sampling frequency dictated by the need for clearance of an indicator substance (e.g., PET or SPECT radiopharmaceuticals, intravascular dyes, etc.), a factor that may also determine the upper limit of measurement that can be performed based on radiation or toxin exposure levels. Techniques that are entirely noninvasive are the most frequently repeatable. Included in this category would be *f*MRI, EEG, and MEG. Human experiments can be performed in animal models if brain size is adequate (e.g., PET studies in primates). Animal experiments can often not be performed in humans because of logistic or ethical issues.

The degree of invasiveness is an important factor in evaluating methods for brain mapping applications. As was noted above, completely noninvasive methods most minimally perturb normal brain function and, from the point of view of safety, have the widest range of applications because the risks of their use are so low. Measurements that involve the administration of substances that can produce toxic or allergic reactions as well as those that contain radioisotopes have added risk and more limited use because of exposure factors. Measurements that require direct access to brain tissue, such as optical intrinsic signal imaging and depth or surface electrophysiology, are reserved for clinical situations in which human subjects require surgery for specific neuropathological disorders or for animal models. These approaches, by their very nature, require physical manipulation of the brain tissue or its surrounding structures and, as such, have an inherently higher propensity to alter functional integrity. Human subjects evaluated with these techniques are, by definition, patients with cerebral abnormalities and, as such, are not examples of normal brain function. The most invasive methods involve the sacrifice of an animal. These terminal experiments provide valuable types of information but not serial, longitudinal measurements.

Other factors that should be kept in mind when evaluating brain mapping methods include whether there are any requirements for anesthesia or other medications that may alter brain function. For studies of postmortem tissue, one should consider whether a given method examines this tissue *in situ* or in excised specimens. Finally, the costs and complexity of establishing a given method for use in one's laboratory are additional considerations in defining the applications of specific techniques. In fact, these factors can often be the driving force in determining whether a specific laboratory can make use of a particular method.

III. The Concept of Resolution

Resolution can be simply defined in operational terms. Resolution is the ability to distinguish two entities as separate and distinct. This simple definition becomes more complex when applied to real-world situations, as will become obvious below in the discussion of spatial resolution. Nevertheless, the basic rules apply and can be consistently used when considering the concept of resolution.

A. Spatial Resolution

Using the operational definition, spatial resolution is the ability of a method to distinguish two separate objects that are positioned close to one another (Fig. 4). This depends on both the intrinsic resolving power of the instrument in question as well as the environment in which the objects exist and the display parameters for the resultant data set. For example, an individual with 20/20 vision should be able to resolve two average-size people walking side by side at a distance of 20 to 30 feet. However, under low light conditions, fog, or other adverse visual environmental factors, the perception may be that there is only one large individual. If conditions are very poor neither individual may be seen at all. Similarly, if an observer looks at two dots on a video monitor that can easily be distinguished when the monitor is properly adjusted, they may merge into one hazy cloud if the monitor is out of focus (Fig. 4). Thus, one sees that the concept of spatial resolution and detection can be separated into the intrinsic resolving power of the instrument and the environmental and display factors that affect contrast, focus, and other detection factors.

1. Intrinsic Resolution

A thorough understanding of intrinsic resolution is important. The intrinsic resolution of an imaging system is determined by the effective resolution of individual detectors in the system. One approach to its measurement is to use a sharply defined object with high contrast to the background and move it parallel to the face of a detector in the system. The result is called the line-spread function and it is a measure of the intrinsic resolution of the detector. The line-spread function is a Gaussian curve, the width of which at half its maximal height [full width half-maximum (FWHM)] is usually given as a measure of the intrinsic resolution of a detector. This type of measurement defines the limit of spatial resolution of a particular detection system.

2. Signal-to-Noise Factors

The signal-to-noise ratio (S/N) is a measure of error caused by statistical inaccuracies of data as well as instrument errors. A measure of the statistical accuracy of data from an imaging device is defined as the reproducibility of the measurement and can be expressed in terms of standard deviation or variance of a series of identical measurements. A common way to express S/N for an imaging system is in terms of the average standard deviation of a pixel value over a series of measurements from a uniform phantom signal (Hoffman and Phelps, 1986). These measurements are defined over a specific number of accumulated imaging sessions in a specified area in the center of the object of interest. Systems with the maximal S/N will provide the highest accuracy and most reproducible results. In general, such systems provide optimal image resolution for a given intrinsic resolution. Many factors contribute to S/N for imaging systems and these factors vary from imaging device to imaging device. As such, a discussion of the S/N characteristics of different imaging techniques can be found in the appropriate chapter of this text.

3. Image Resolution

Image resolution is affected by sampling, both linear and angular; by the fineness of the grid in the final image display (pixel dimension); and by the degree of spatial smoothing during or following the imaging process; as well as by the hardware design of the overall system (Hoffman and Phelps, 1986). Differences between system resolution and intrinsic resolution provide the measure of how well the entire imaging system is working as a unit (Hoffman and Phelps, 1986). The choice of the object used to identify image resolution can have a significant effect on the resultant answer. If a single object of high contrast is imaged, there will be minimal background activity to contribute to noise in the measurement (Fig. 4). In addition, minimal smoothing and filtering of the data, in a manner that maximizes the identification of single objects, will produce a higher spatial resolution value than one would actually realize in practice with a complex and heterogeneous neuroanatomical structure. The latter, being a distributed image typical of *in vivo* situations found in the nervous system, adds background noise to all points in the image. Thus, the signal-to-noise environment is much worse in a distributed source and noise reduction in the system can be achieved only by increasing the total number of measurements or by spatial averaging/smoothing of the data (Hoffman and Phelps, 1986). The greater the spatial averaging, the smoother and less noisy the appearance of the image. This occurs, however, at the expense of spatial resolution. When resolution is critical in noisy image sets, smoothing the image may provide boundary information and the calculation of quantitative information can then be performed on the original unsmoothed data set in a two-step process.

4. Axial Resolution

Axial resolution pertains to techniques that involve slabs or slices of tissue, irrespective of whether they are obtained *in vivo* and noninvasively, as would be the case with tomographic techniques such as CT, MRI, PET, or SPECT, or by physical sectioning of tissue the axial thickness of which is determined by the interval movements of the blade relative to the tissue block. Axial resolution in tomographic imaging contributes greatly to the overall three-dimensional spatial resolution since spatial averaging occurs in three dimensions in tomographic techniques as opposed to physical sectioning of the data (Hoffman *et al.*, 1979). A more complete discussion of this issue is provided under Partial Volume Effects.

5. Modulation Transfer Function

While the line-spread function can be used to determine detector spatial resolution, the modulation-transfer function (MTF) can be used to define system performance of an imaging device in terms of the ability to measure spatial frequencies. This is typically obtained by scanning phantoms consisting of alternating bars of high and low signal. The spatial frequency (cycle/distance) of a bar phantom with bar widths of N is 1 cycle/2N (i.e., the width of a pair of bars). Thus, a 5-mm bar phantom has a basic spatial frequency of 1 cycle/cm. A qualitative rule of thumb in imaging is that the resolution of a bar phantom requires a FWHM resolution approximately equal to 1.4–2.0 times the width of the bars (Hoffman and Phelps, 1986). For example, a system with a 10-mm image resolution should be able to resolve 5- to 7-mm bar phantoms.

The MTF is an imaging analog of the frequency response curve used to evaluate audio equipment. In audio equipment evaluations, pure tones of various frequencies serve as input for the component to be tested and the relative amplitude of the output signal is measured. The comparison between the amplitudes of the input and output signals as a function of different frequencies defines how much of each frequency will pass through the system to the listener. Similarly, the MTF of an imaging system measures the fraction of the amplitude of each spatial frequency that can be transferred from the object to the final image. The MTF can be calculated from the line-spread function by

$$\text{MTF}(\upsilon) = \int A(x)\cos(2\pi x)dx / \int A(x)dx, \quad (1)$$

where $A(x)$ is the magnitude of the line-spread function at a distance x from the origin of the line-spread function coordinate system, and υ is the spatial frequency in units of cycles/distance (Hoffman and Phelps, 1986). The limits of integration are usually over the full field of view of the imaging system but at least 10 times the FWHM. The result is a function that gives the fraction of signal amplitude the system will transfer to the image at each spatial frequency.

If the MTF is very low at a given spatial frequency, it is most likely that data from the system with that frequency are caused by statistical noise in the data set. The highest frequency one can hope to reliably measure is 1 (2 × sampling distance). In any imaging system, the overall MTF is the result of the MTFs of the components of the imaging chain, such as the MTF corresponding to the intrinsic resolution of a single detector, the MTF due to linear sampling, etc. Each step has an image degradation factor and any loss in the MTF for a given step can never be recovered (Hoffman and Phelps, 1986).

6. Partial Volume Effects

All imaging techniques are subject to partial volume effects. Simply defined, partial volume effects are three-dimensional averaging that takes place within the volume defined by the spatial resolution of the system. For tomographic techniques such as CT, MRI, PET, and SPECT, the term partial volume has come to represent the diminished contrast of an object that only partially occupies the thickness of the tomographic plane. Since most imaging devices today utilize voxels whose axial dimensions exceed their transverse dimensions, axial partial volume effects are more pronounced and, therefore, more widely recognized. Transverse partial volume effects also can be significant, particularly when the volume sampled by a detector in the transverse direction is large relative to the structure size (Hoffman *et al.,* 1979; Mazziotta *et al.,* 1981). Such a situation exists in *in vivo* human tomographic imaging, in EEG, and at the spatial limits of autoradiography (Lear, 1986; Lear *et al.,* 1983, 1984). Since the central nervous system consists of a composite of heterogeneous tissues, the three-dimensional structure size relative to the three-dimensional resolution of the imaging instrument provides an estimate of partial volume effects. That is to say, if a structure of interest is small relative to the voxel (three-dimensional volume element) of resolution of the imaging device, then the quantitative values for and the appearance of that object will be blurred due to averaging of the object with its surrounding structures. When the object of interest is large relative to the size of the resolution voxel of the imaging instrument then it is seen clearly and its quantitative values will be accurately measured to the limits of precision of the imaging device. When an object of interest has at least one dimension smaller than the width of the line-spread function of the imaging system (or approximately two times the FWHM), that object will only partially occupy the sensitive volume of the detectors viewing that dimension. As a result, there is an underestimation of the signal from the object. In general, the object must be approximately two times the FWHM of the imaging system in all dimensions in order to be accurately visualized and quantified (Hoffman and Phelps, 1986; Mazziotta *et al.,* 1981).

Another factor exists with regard to large objects. When the border of one object with another falls within the volume of a single resolution voxel, averaging will occur between the object of interest and its neighbor on the other side of the boundary. Once again, this is an opportunity for partial volume effects with resultant blurring of the object edges. A typical example would be the boundary of the ventricle with the caudate nucleus in human *in vivo* tomographic imaging. The smaller the voxel of resolution of the imaging device, the sharper such borders will be. The issue of orientation of objects relative to the position and placement of the resolution voxels of the imaging device is also important. If a given subject is scanned in two different positions relative to the imaging device in, for example, a PET instrument, different results may be obtained along borders because differing relationships and varying partial volume effects will occur in the boundary voxels along that border (Mazziotta *et al.,* 1981). Thus, for the most consistent and reproducible imaging results, the positioning of the object (e.g., subject) and the orientation of the imaging device (e.g., scanner gantry) should be consistent between imaging sessions to get the most accurate and, therefore, reproducible measurements across the entire sampling volume. For imaging devices with uniform and approximately Gaussian line-spread functions, the partial volume effects will simply scale with image resolution. Hoffman *et al.* (1979) have defined a term known as the recovery coefficient, originally defined for use with PET systems, that refers to the ratio of the apparent-to-true signal. For a bar phantom, the recovery coefficient can be calculated from the image line-spread function and the narrow dimension of the bar. The recovery coefficient is equal to the ratio of the area of the line-spread function overlapped by the bar to the total area of the line-spread function. Within the image plane one can assume the object always overlaps the center of the line-spread function. However, in the axial dimension the overlap can be in any part of the line-spread function and this factor must be considered in data interpretation. If the shape of the object is simple (i.e., cubic) the recovery coefficient of the object can be estimated as the product of the recovery coefficients of its three dimensions (Hoffman *et al.,* 1979). Thus, if the imaging system has an image resolution of 1 cm FWHM in all three dimensions, a 1-cm cube would have recovery coefficient of $(0.75)^3 = 0.42$. That is, only 42% of the signal from the object would be recorded in the final image.

In the brain, the shapes of various structures are not regular, making precise quantitation difficult in most instances. For structures smaller than two times the FWHM of the imaging system in each dimension, there will be spillover of activity from objects with high signals to those with low signals. This results in averaging and loss of spatial resolution (Mazziotta *et al.,* 1981).

7. Resolution Uniformity

Ideally, one would like all components of an imaging system to have equal resolution, thereby defining a single value for the entire system. In practice, this is often not true (Hoffman *et al.,* 1982). For example, in tomographic imaging, particularly emission tomography, such as PET and SPECT, spatial resolution is maximal in the center of the field of view and diminishes as a function of distance away from the center (Hoffman *et al.,* 1982). The discussion of resolution uniformity is unique to each type of imaging technique and, as such, the reader is referred to the specific techniques defined in other chapters of this text. Nevertheless, it suffices to say that image resolution cannot always be considered to be uniform and the investigator is well advised to know the spatial resolution at each point in the effective field of view of the imaging device in order to properly interpret data.

8. Object Movement

Loss of image resolution results from object movement in the field of view. This may result from gross movements of a subject's head in a tomographic imaging device or physiologic movement of the tissue under the field of view, as might occur with respiratory or cardiac variations in intrinsic signal imaging. Imaging approaches that employ postmortem tissue, such as autoradiography or light and electron microscopy, do not suffer from this artifact unless it is investigator induced.

For the situation in which the subject is voluntarily moving in the scanner, the primary remedies to the problem are a specially designed head restraint system (Mazziotta *et al.,* 1982) and very short scan times. In practice, no head restraint system short of rigid immobilization of the skull to the scanner, as is sometimes used in patients with stereotactic frames bolted to their skulls, can result in absolute immobilization of the subject. Many head restraint devices have been described in the literature but none of these absolutely limit head movement. The use of extremely short scan times minimizes the opportunity for movement during the interval in which image data are being collected. This has proven to be very effective for *in vivo* tomographic techniques such as CT, for which scan times have declined dramatically over the past 2 decades from minutes to seconds. A final approach to control for voluntary movement is to track the movement by an external sensing device and then correct for the movement or discard data that occurred during the movement. A serious problem with correcting data for movement was mentioned above, namely, nonuniformity of resolution. That is, if a subject moves from one position to another in an imaging device with nonuniform resolution, different values will be obtained for the same structure when they are imaged in a different location in the field of view. In PET devices this is due to the fact that image resolution deteriorates away from the center of the field of view. In MRI it can occur because of inhomogeneities in the magnetic field.

Respiratory and cardiac motion are also difficult problems. In the case of optical intrinsic signal imaging, monitoring the respiratory and cardiac cycle is typically employed (Toga *et al.,* 1995). Since this type of imaging occurs intraoperatively (in both humans and animal models), ventilation can be momentarily suspended during the course of this short measurement. In addition, gating with the cardiac cycle can help to minimize physiological motion artifacts. More recently it has become clear that physiologic motion contributes to artifacts that occur in *f*MRI data sets (Kwong *et al.,* 1992). Factors of importance in this situation include physical motion of the brain related to the cardiac and respiratory cycles as well as potential differences in oxygenation and cerebral blood volume that vary with these cycles.

B. Temporal Resolution

Analogous to the measurement of spatial resolution, temporal resolution is defined as the ability of a system to detect two events that occur in close temporal proximity. When considering temporal resolution for brain mapping techniques, two aspects should be considered: first, the temporal resolution of the measurement and second, the temporal resolution of the process being measured. To begin with the latter, the temporal resolution of biological processes spans a tremendous gamut (Fig. 1). At one end of the spectrum, there is the time frame of evolution that takes place over the course of millions of years and represents the slowest biological brain mapping process of interest. At the other end of the spectrum are molecular interactions that occur in the submillisecond time frame. Perhaps the most commonly discussed temporal events are those of action potentials that occur in a time frame of milliseconds as their electrochemical events unfold. These action potentials are propagated along fiber tracts in time frames that range from tenths to hundredths of milliseconds. If one examines physiologic, chemical, or hemodynamic events that occur as a consequence of electrophysiological events, then these must be considered at some "temporal distance" from the electrophysiological event of interest. An example is the measurement of cerebral blood flow change as an estimate of local changes in electrophysiological events. Nevertheless, indirect measurements of variables such as cerebral blood flow and volume, utilization of glucose and oxygen, or changes in substrate extraction are assumed to be both delayed and dispersed relative to the electrophysiological event that initiates them. In addition, the coupling between such events may not always be tightly linked either in magnitude or in timing. The relationship between these events, both spatially and temporally, is a current area of great interest in the field of brain mapping.

Turning to the speed of measuring biological events relevant to brain activity, a general rule may be offered. That is, the speed of the measurement must be faster than the speed of the biological event to obtain an accurate measure of it. For example, if a device that measures electrical potentials from the scalp has a response time that can be measured in microseconds, it should accurately depict the biological events that occur in a time frame approximately three orders of magnitude greater. Alternatively, if cerebral blood flow events can be measured with PET in a time frame that ranges from 30 s to 2 min, the measurements will be inaccurate or inappropriate for determining specific temporal information about blood flow changes that occur on the order of a few seconds.

Measurement time may also have a number of other characteristics. Certain methods require a short interval to actually acquire specific data about a measurement, such as glucose metabolism, but the method associated with that measurement requires certain biological conditions to be fulfilled before it can be made. For example, in the measurement of glucose metabolism, radiolabeled deoxyglucose is administered intravenously and then, after a period of 40 min, tissue is assayed (either by sacrifice of the animal and autoradiography or by *in vivo* PET imaging) in a time frame that is typically much more rapid than the time required for the steady-state conditions (i.e., 40 min) to be achieved (Sokoloff *et al.,* 1977).

Techniques can take advantage of gating or response-locked data collection strategies to obtain enough information to make a measurement by averaging over many trials. Examples of this approach include *f*MRI, optical intrinsic signal imaging, and electrophysiological evoked potentials. In this strategy, the stimulus-induced signal is small relative to the other biological information and the noise of the measurement. However, by averaging over many events, this signal remains constant while the noise and other biological events become randomized, leading to a progressive emergence of the stimulus-induced signal concomitant with a reduction in "contaminating" information (Thatcher, 1995). Numerous brain mapping instruments employ a gating feature that can be time-locked to a stimulus to record data in this fashion.

IV. Sampling

Brain mapping methods vary in both the volume and the frequency at which they sample data (Fig. 3). Both aspects are important in understanding the appropriate applications of the technique as well as its shortcomings for evaluating specific neurobiological events.

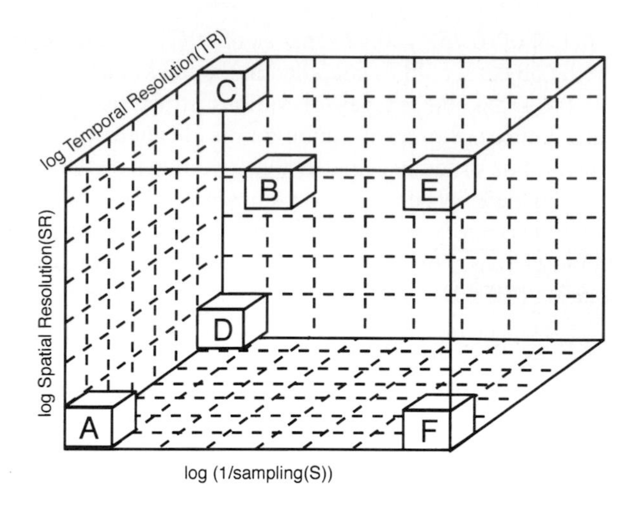

Position	SR	TR	S	Method
A	+	+	+	None Known
B	-	+	±	MEG
C	-	-	-	Kety-Schmidt
D	+	-	-	MRI, Postmortem
E	-	+		EEG
F	+	+		Unit Recordings

Figure 3 Three-dimensional plot depicting theoretical attributes of different techniques in terms of spatial and temporal resolution as well as sampling. Note that the axes are logarithmic and that the inverse of sampling is plotted on the *x* axis. The ideal technique would have exquisite spatial and temporal resolution as well as sampling the entire brain "A". No such technique is currently known and available for *in vivo* study of the human brain. In the table, a "+" indicates good performance by the method for a given attribute while a "–" indicates relatively poor performance. Note, for example, that position "F", identified with unit recordings of single cells or clusters of cells, produces data with exquisite spatial and temporal resolution but severely undersamples whole brain activity since it records only from a few cells. Contrast this with Kety–Schmidt arteriovenous difference measurements, which sample biochemical and physiological processes across the entire brain by examining arteriovenous differences in compounds and global blood flow. This method has complete sampling but poor spatial and temporal resolution. When considering brain mapping techniques, sampling should be considered along with spatial and temporal resolution.

A. Sampling Frequency

Some brain mapping methods can be used only to make a single measurement of biological information. This typically involves the sacrifice of an experimental animal or the examination of postmortem human tissue. The majority of brain mapping methods, however, are capable of repeated measurements. Such measurements can be divided into repeated independent measurements (e.g., serial CBF deter-

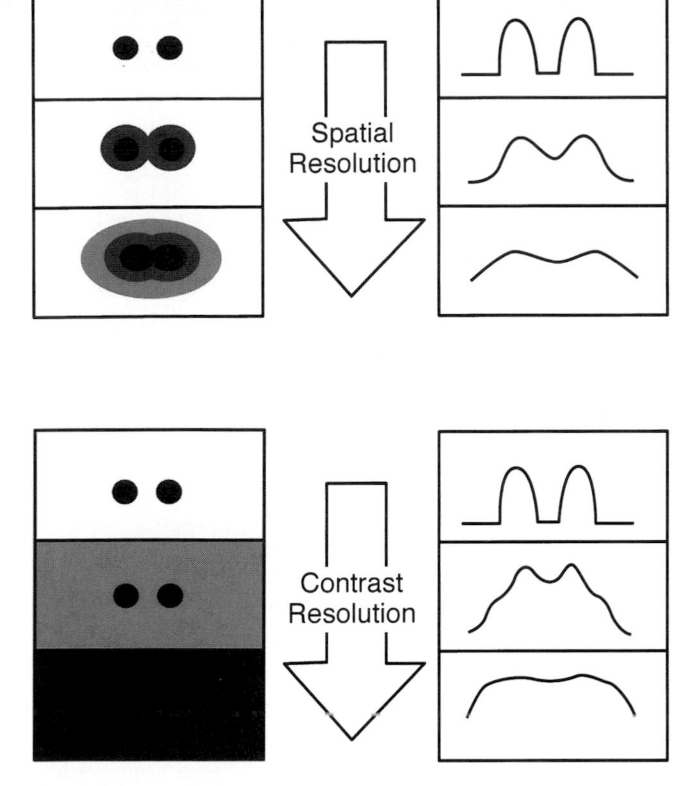

Figure 4 Spatial and contrast resolutions. On the left is the visual appearance of two sharply defined and closely spaced objects depicted as spatial resolution decreases (top) or contrast resolution decreases (bottom). On the right, a histogram through the center of these two objects is plotted at each spatial and contrast resolution. Note that as spatial resolution decreases, the distinctness of the boundary of the object falls and they ultimately merge into one broad object. Note that as contrast resolution decreases the ability to detect the object from the surrounding region is diminished.

minations), continuous measurement (e.g., EEG), single long-duration kinetic measurements (e.g., PET measurements of ligand binding), or single simultaneous measurements of multiple variables (e.g., double- or triple-labeled autoradiography).

One of the most common sampling frequency approaches is that of serial independent measurements. This would apply to determinations made with CT, MRI, *f*MRI, MRS, PET, SPECT, optical intrinsic signal imaging, and transcranial magnetic stimulation. Of interest with these methods is the total number of samples that can be acquired and the interval between such samples. The total number of these samples is typically determined by toxic limits to the subject or the ultimate tolerance of the subject to continued testing. With regard to toxin exposure, the upper limit of measurements is determined for radioisotope and contrast injection techniques by dose considerations, be they radiation or pharmacologic exposure. These upper limits are, in turn, determined by the sensitivity of the technique to acquire sufficient signal for a valid measurement relative to a unit dose of exposure. A goal of all techniques that have this limitation is to increase sensitivity so as to augment the number of trials available per subject. For example, in PET, oxygen-15-labeled water is used to measure cerebral blood flow. The total exposure allowed in normal human subjects varies in different locales but is in the range of 300–800 mCi per subject per year. Early scanners required as much as 100 mCi to make a single measurement, thereby limiting the total number of trials per subject to between 3 and 8 depending on the center. As instrument sensitivity increased, dose requirements decreased by an order of magnitude to 10 mCi per trial, in state-of-the-art instruments (Cherry *et al.,* 1991, 1992, 1993; Defrise *et al.,* 1990; Townsend *et al.,* 1991). Thus, even in the most restricted regulatory environment, up to 30 trials per subject could be performed on the basis of dose limitations alone. Similarly, initial studies using *f*MRI required the administration of gadolinium-labeled contrast material in order to measure blood volume changes associated with stimulus-induced neuronal activity (Belliveau *et al.,* 1991). Advances in the understanding of these MRI approaches resulted in the development of a method that can determine hemodynamic responses to stimuli that are based on intrinsic changes in venous blood oxygenation rather than having the requirement for exogenous contrast material (Kwong *et al.,* 1992; Ogawa *et al.,* 1992). This completely removes the requirement to limit the total number of trials per subject based on toxin exposure.

When exogenous agents are used to make a measurement, their influence must typically be eliminated from the biological system before the measurement can be repeated. For nonradioactive agents, this may require actual excretion from the body or, more commonly, redistribution of the agent into a different biological space so as not to confound future measurements with agents previously delivered. For radiopharmaceuticals, the physical decay of the radioisotope is sufficient to allow for a new administration of the radiopharmaceutical without having any background activity.

A simple estimate is that five physical half-lives of the radioisotope should elapse before a subsequent administration can be made, when no modification of the basic measurement technique is employed. Thus, for oxygen-15, the isotope most commonly used with PET to measure cerebral blood flow, the physical half-life is approximately 2 min. With this agent a CBF measurement can be made every 10 min in a normal subject (Fox *et al.,* 1988; Jones *et al.,* 1976; Mazziotta *et al.,* 1985). Practicalities of generating the isotope and creating a sterile dose for subject administration generally result in about a 15-min sampling interval between trials. Despite the fact that, as noted above, up to 30 10-mCi doses of oxygen-15-labeled compounds can be given to a single subject based on dosimetry rules, it is impractical to have subjects spend more than 3 or 4 h in a scanner having such measurements performed. Therefore, the practical limitation is still dictated by the relatively slow sampling frequency (i.e., 15 min) and leads to an upper limit of trials per subject of between 12 and 16 for a given session. Nevertheless, the ability to have multiple trials of the same task has significant statistical advantages in data analysis and in determining reproducibility of the results.

With regard to the tolerance of the subject, there are those methods that require surgical exposure of the brain to perform a given measurement. Examples in this category would include intraoperative electrophysiology or optical intrinsic signal imaging. In these situations, particularly relevant in humans, the amount of added intraoperative time to make such measurements must be balanced with the value of such data in achieving the intended clinical and surgical goals. Thus, despite the fact that these methods can measure events of neurobiological interest in the time frame of milliseconds, there is an upper limit to the total measurement time determined more by the invasiveness of the procedure then by the unit measurement duration itself.

Some techniques can make measurements continuously. The most obvious example is electroencephalography. Using this method, measurement of scalp electrical activity can be made continuously for hours or, in some cases, days at a time. This method has the interesting and rare feature of having very high temporal resolution but no absolute limit to the duration of the measurement period.

It is often of value to look at the kinetic events that occur in the course of a single measurement. An example of this situation is understanding the physiology of the blood flow response to a site of increased neuronal activity. If one has a rapid method for measurement, relative to the change in blood flow, the kinetics of the process can be observed. Such a situation would occur with the use of optical intrinsic signal imaging or ultrafast *f*MRI to examine CBF responses in animal models or humans in response to a stimulus-induced change in neuronal activity. By sampling a single behavioral event rapidly, the kinetics of the biological process can be observed (Huang and Phelps, 1986). Similarly, the kinetics of uptake, binding, and washout for a ligand–receptor interaction in the human brain, or in that of an animal model, can be determined using kinetic measurements with PET or SPECT.

Occasionally it is of interest to measure a number of processes simultaneously. This can be performed by combining different brain mapping techniques in the course of a single experimental intervention such as measuring evoked potential responses electrophysiologically during the determination of a CBF response with PET or *f*MRI. Alternately, multiple agents can be administered simultaneously and a single method used for their analysis. An example of this situation would be that of double- or triple-labeled autoradiography (Lear, 1986). In this situation, different radiopharmaceuticals are administered simultaneously to an experimental animal, each labeled with an isotope that differs greatly in its physical half-life (Lear *et al.*, 1983, 1984; Lear, 1986). By varying the isotope and the radiopharmaceutical, for example, the measurement of dopamine D_2 receptor–ligand binding, cerebral blood flow and glucose metabolism could be measured simultaneously in the same experiment. This could be accomplished by the co-injection of tritium-labeled spiperone (physical half-life approximately 11 years, target dopamine D_2 postsynaptic receptors), carbon-14-labeled antipyrine (physical half-life 5000 years, determination of cerebral blood flow), and fluorine-18-labeled deoxyglucose (physical half-life approximately 2 h, determination of cerebral glucose metabolism) followed by autoradiography and image subtraction techniques (Lear, 1986). This approach allows for the valid comparison between biological processes in the same animal without the confounding effects of experimental error that differs between subjects or differences in individual anatomy.

B. Sampling Volume

Ideally, one would like to know about a given biological process for the entire brain using a technique that has high spatial resolution. Nevertheless, it is generally found that methods with high spatial resolution have low sampling volumes (Fig. 3). For example, depth electrodes that measure electrophysiological events sample tissue in a range of less than 100 μm. While methods that assay global brain function, such as CT, MRI, PET, and SPECT, have spatial resolutions in the range of millimeters. While it is obvious that techniques with low spatial resolution provide less detailed data about a given biological process, observations or erroneous conclusions that occur from low-volume sampling of a given measurement may be more insidious. This is true because the area of interest is small relative to the potentially large network of brain sites that may be involved (Mesulam, 1990; Posner and Dehaene, 1994). A technique that measures only a small volume of tissue in a larger system requires a very well characterized hypothesis with regard to the site of interest compared to techniques with a larger sampling volume. The low-resolution high-sampling-volume technique will miss important events because even a signal of high magnitude but low spatial extent may be diluted and go unobserved using a method with coarse spatial resolution. Conversely, detailed examination of a local event with high spatial resolution and low volume sampling may lead to conclusions that would be far different if seen in the context of other participating sites of activity at centers in the brain not sampled in the small measurement volume. It is ideal to have a logical progression of the use of techniques from ones with large sampling volumes and lower spatial resolution to those with progressively higher spatial resolutions and lower sampling volumes that are focused on areas of interest identified by earlier steps in the survey. This is analogous to serially selecting objective lenses of increasing magnification when using light microscopy. In brain mapping, this may require switching to alternate techniques or making composite results with a combination of techniques applied either serially or simultaneously.

V. Sites Accessed

Methods vary with regard to the amount of tissue as well as the site in the nervous system that they can access. While postacquisition data processing may be able to recombine and reposition information acquired from a certain orientation into a more comprehensive view of the entire organ, sites of acquisition may impose an absolute constraint on the applications of a given technique. For example, there are a number of methods that are purely cortical. These include transcranial magnetic stimulation, optical intrinsic signal imaging, and scalp electroencephalography. Note that these techniques span the spatial range of resolution from microscopic (optical intrinsic signal imaging) to macroscopic (transcranial magnetic stimulation and surface EEG). Occasionally, techniques designed to assess one part of the brain can be modified and used for another. An example would be electrophysiology. Originally used intraoperatively during epilepsy and tumor surgery to locate behavioral functions and seizure foci in the cortex, they were modified and used as an implanted depth electrode technique to record seizures as well as to stimulate regions in the hippocampus and parahippocampal areas.

With few exceptions, most brain mapping techniques are not specifically designed to examine white matter. Instead, they are primarily tools to assess neuronal function in gray matter structures both cortical and subcortical. Exceptions to this rule include the evaluation of myelinization using postmortem tissue and specific stains, structural MRI, and PET or SPECT with radiopharmaceuticals that localize in white matter structures. In addition, certain electrophysiological techniques measure central conduction time (i.e., evoked potentials) as a means by which to estimate the macroscopic integrity of long fiber tracts in the central nervous system. Identifying neuronal targets by their fiber projection systems is performed by the injection of dyes into gray matter regions during life in animal models and sacrificing the animal at a later date to examine the ultimate site of transport of the compounds (Ralston, 1990). Modern techniques are attempting similar experiments using compounds that change local magnetic susceptibility and imaging performed with MRI methods (Jacobs and Frasier, 1994; Li *et al.,* 1999).

As was mentioned in the introduction to this chapter, a critical and much needed set of methods in the brain mapping armamentarium contains those that allow one to bridge large scales in spatial resolution or sampling volume. That is, techniques tend to cluster in the macroscopic, microscopic, or ultrastructural spatial domains. Few if any techniques can cross these boundaries and provide the necessary linking technologies to appropriately place and scale microscopic data into a macroscopic model. Some exceptions exist and are being exploited for their unique position in the spatial continuum of brain mapping methods.

Postmortem cryomacrotome studies serve as an example of a bridging technology that spans spatial domains (Quinn *et al.,* 1993; Rauschning, 1979; Toga *et al.,* 1994a,b; Van Leeuwen *et al.,* 1990). This technique allows for the assessment of whole human brain (and the brains of smaller species as well) anatomy from a structural point of view. Using these approaches, the entire head, including the brain, meninges, skull, and extracranial tissues, is sectioned after postmortem freezing (Toga *et al.,* 1994a,b). These sections can be as thin as 20 μm. Digital images are made of the entire block face and, currently, have a spatial resolution in the range of 100–200 μm. Higher magnifications can be achieved, with resolution as high as 30×30 μm per pixel by zooming the optics onto a small area of the block face. A series of these higher resolution images can be assembled as a patchwork tessellation that recreates the entire macroscopic surface of the tissue. New digital cameras have been announced that may increase the spatial resolution by a factor of 16 (4×4) in the plane of section. At these spatial resolutions, identification of the individual cells will be possible. Thus, with a single technique one can obtain microscopic data and place it in the appropriate macroscopic reference system. Since tissue sections are actually collected, they can be stained to provide information about cyto-, chemo-, and myeloarchitecture. Further, as atlases are built using these methods, the high-resolution cryomacrotome brain images can be stained with conventional or "landmark" stains (e.g., Nissl stain). With such an approach, an investigator who studies, for example, GABA receptors in the human hippocampus can incorporate data into an appropriate macroscopic atlas. In preparing the tissue, this investigator would process every *N*th section using one of the landmark stains that are part of the atlas. The investigator would then digitize the information from both the GABA receptor sections and the landmark-stained sections. Using alignment, registration, and warping tools, the investigator would then register the landmark-stained sections with the atlas and then use the same mathematical transformations to enter the GABA receptor information into the appropriate region of the atlas. Once referenced, the appropriate visualization of these new data could be performed in the macroscopic domain.

Another, less direct but *in vivo,* bridging technology is the marriage between PET or SPECT imaging and autoradiography. In both, an identical compound can be used to trace a specific process in the brain. For macroscopic *in vivo* imaging, this compound would be labeled with the appropriate positron- or single-photon-emitting radioisotope and external imaging would be obtained demonstrating the macroscopic behavior of the compound (Huang *et al.*, 1980; Phelps *et al.*, 1979). For the microscopic counterpart, the compound would be labeled with an appropriate autoradiographic radioisotope (Sokoloff *et al.*, 1977), the same experiment performed in an animal model, and autoradiography of the tissue performed to obtain the resultant images. Image sets obtained in this parallel fashion can be used to validate or refute animal models. For valid animal models, invasive or more detailed experiments can be performed using approaches that would be logistically impossible or unethical in human subjects.

MRI techniques are applicable in both the macroscopic world of human subjects (Prichard and Brass, 1992) and microscopically (Damasio *et al.*, 1991; Jacobs and Frasier, 1994). This is true for both structural and spectroscopic MRI and probably will be applied to all aspects of MR imaging within the capabilities of its resolution. Finally, electrophysiological techniques are equally applicable macroscopically in human subjects (e.g., surface EEG) and in local microscopic studies of animals (e.g., depth electrodes).

VI. Invasiveness

As was discussed in the section on the frequency of sampling, the degree of invasiveness and the practical/logistical aspects of performing a brain mapping measurement are important variables that should be discussed in the context of space and time. Obviously, the most invasive experiments require exposure or surgical manipulation of the nervous system itself, up to and including sacrifice of an animal involved in the measurement. Other techniques are designed to be employed in the purely postmortem setting. Intraoperative techniques that are utilized in human subjects add significant time to surgical procedures and, as such, their value with regard to patient care must be weighed against the added risk of increased exposure to anesthesia and intraoperative complications. Some intraoperative techniques such as awake questioning of the patient during cortical stimulation also add stress and anxiety in addition to prolongation of the procedure. Human intraoperative methods are, by definition, not performed in normal subjects and, therefore, do not provide pure information about normal brain function. This caveat is important to remember when comparing human data from normal subjects collected noninvasively with data from patients assessed intraoperatively who suffer from epilepsy, brain tumors, or vascular abnormalities. Since many of these pathologic states may have been long-standing, developmental or compensatory reorganization of the brain may have also taken place, adding further variability to the data and making it less representative of the normal condition. Further, both human and animal brain mapping experiments that involve intraoperative manipulation of the brain or its surrounding structures will perturb the underlying function to varying degrees. This also will add noise or variance to the data and make them less representative of the function of the organ in the natural state.

Methods that use exogenous compounds, whether they are radioactive or nonradioactive, add a degree of invasiveness because of the exposure of the subject or animal (and at times the investigator) to that agent. Nevertheless, these methods are typically applied in normal subjects and do not directly perturb brain function unless the mass of the administered agent is so large as to interfere with the natural chemical or physiological processes. Often the exogenously administered agent is not blood-borne but can be in the form of energy. This is true of some of the *in vivo* human techniques such as CT (X-irradiation) or MRI (magnetic fields and radio-irradiation). It is taken for granted that those techniques that use ionizing radiation are more invasive and have more restrictive boundary conditions for exposure in specific subject groups. Transcranial magnetic stimulation and, possibly, high-field *f*MRI techniques directly excite neuronal tissue in ways other than those that naturally occur under physiologic conditions. Care must be taken in this situation to determine what effects such perturbations produce and that their comparison with physiologic states may not always be valid.

A number of techniques require anesthesia. This is typically true of larger animal experiments that involve *in vivo* tomographic methods such as PET, SPECT, MRI, and CT. This is because of the immobilization requirement of these methods combined with ethical issues about producing muscular paralysis in an awake animal. Intraoperative techniques in human subjects and most animal models also require some degree of sedation although the patient is often awake during the actual measurement. Nevertheless, drugs are required for the preparatory phase of craniotomy, and their lasting effects, or required continued use at lower levels, undoubtedly distort the signal from the natural physiological one. Some techniques induce a certain degree of sedation or anesthesia by virtue of the method itself. An example is stable xenon CT blood flow measurements. As the inhaled concentration of stable xenon increases, progressive sedation and, ultimately, anesthesia can occur. In this case, the actual contrast agent used to make the blood flow measurement induces a perturbation in the system to be measured, in this case, anesthesia.

VII. Conclusions

As can be seen, a wide range of brain mapping techniques currently exist for use in human subjects and animal models. Each has its own unique advantages and disadvantages and all vary on the continuum with regard to spatial and temporal resolution as well as sampling frequency and volume. Special issues with regard to the sites that these methods can access, their degree of invasiveness, requirement for anesthesia, and repeatability all contribute to the selection of the appropriate approach for a given neurobiological situation. As has been stressed throughout this chapter, it is important to understand the limitations and constraints for each method so as to interpret the results appropriately and to use the resultant data to build reliable hypotheses that can be rigorously tested with further experimentation. The ideal brain mapping technique would have extremely high spatial and temporal resolution with the capacity to sample a large volume of the brain continuously. Its costs would be low as would its invasiveness, making it applicable in many settings, in human subjects as well as animal models. At present, no such method exists. Nevertheless, the combination of data sets acquired from many different techniques synthesized into an ever-growing atlas of brain structure and function provides the most unifying means by which to span the spectrum of all these variables. Current and developing tools as well as the increasing power and decreasing cost of digital approaches to the management of such data sets make this goal appear to be not just a possibility in the near future but an actual requirement as the volume of information and the need to standardize it across laboratories, experiments, and species continues to grow.

Developing tools and mapping the brain are important challenges for neuroscience. The increase in the quality and variety of techniques that provide input for brain mapping experiments is rapidly expanding. Simultaneously, the speed and memory capacity of computing devices are advancing at an ever-accelerating pace while cost is dropping, making powerful desktop manipulations of brain mapping data not only feasible but a reality in most laboratories. The appropriate mathematical and statistical models are being developed to provide advanced population-based probability atlases of the human brain and the databases to use them.

Once a proper framework for the organization and storage of neuroscientific data across spatial scales and temporal domains is available the results of every experiment and clinical examination involving the nervous system could ultimately have an appropriate place for future reference. This depends on the ingenuity and farsightedness of the creators of the reference system to provide an approach that is flexible, compatible with existing as well as future technologies, and presented in a manner that is acceptable, in both the technical and the sociological sense, to the neuroscience community at large. Such a system will be neither easy to create nor inexpensive. Nevertheless, when one examines the amount of data collected in both clinical and research settings today that becomes inaccessible soon after acquisition, one quickly realizes the economy of developing a system for storage and reference of this untapped and yet very costly information. Time and funds spent to organize and store these data that reflect the convenience of the investigators as well as the confidence and credibility ratings for the quality of each data set will provide a usable system that will stand the test of time.

Finally, one should return to the concept of using these systems not simply as libraries or databases but rather as rich sources of neuroscientific information upon which one can base hypothesis generation and test such theories against actual data that one need not personally collect. Similarly, the ability to rigorously correlate *in vivo* human data acquired tomographically or by other methods with postmortem tissue that is available for the myriad of immuno-, histo-, and biochemical stains provides a two-way system to develop a more thorough understanding of the microscopic anatomy of the brain that is driven by hypotheses generated from human *in vivo* experiments. Given the rapid growth and the amount of neuroscientific information, and the pace with which it increases, such organizational, storage, and conceptual systems should no longer be considered a luxury but rather a necessity.

Acknowledgments

I thank Arthur Toga, Ph.D., for his thoughtful comments in the review of this chapter, Andrew Lee for preparation of the graphic materials, and Laurie Carr for preparation of the manuscript itself. Partial support for this work was provided by a grant from the Human Brain Project (P20-MHDA52176), the National Institute of Mental Health, the National Institute for Drug Abuse, the National Cancer Institute, and the National Institute for Neurological Disease and Stroke. For generous support, the author also thanks the Brain Mapping Medical Research Organization, the Brain Mapping Support Foundation, the Pierson–Lovelace Foundation, The Ahmanson Foundation, the Tamkin Foundation, the Jennifer Jones-Simon Foundation, the Capital Group Companies Charitable Foundation, the Robson Family, the Northstar Fund, and the National Center for Research Resources, Grants RR12169 and RR08655.

References

Barinaga, M. (1993). Neuroscientists reach a critical mass in Washington. *Science* **262,** 1210–1211.

Belliveau, J., Kennedy, D., McKinstry, R., Buchbinder, B., Weisskoff, R., Cohen, M., Vevea, J., Brady, T., and Rosen, B. (1991). Functional mapping of the human visual cortex by magnetic resonance imaging. *Science* **254,** 716–719.

Cherry, S. R., Dahlbom, M., and Hoffman, E. J. (1991). 3D PET using a conventional multislice tomograph without septa. *J. Comput. Assisted Tomogr.* **15,** 655–668.

Cherry, S. R., Dahlbom, M., and Hoffman, E. J. (1992). Evaluation of a 3D reconstruction algorithm for multi-slice PET scanners. *Phys. Med. Biol.* **37,** 779–790.

Cherry, S. R., Woods, R. P., Hoffman, E. J., and Mazziotta, J. C. (1993). Improved detection of focal cerebral blood flow changes using three-dimensional positron emission tomography. *J. Cereb. Blood Flow Metab.* **13,** 630–638.

Churchland and Sejnowski (1988). Perspective on cognitive neuroscience. *Science*, Nov. 4; **242** (4879):741–745.

Damasio, H., Kuljis, R. O., Yuh, W., and Ehrhardt, J. (1991). Magnetic resonance imaging of human intracortical structures *in vivo*. *Cereb. Cortex* **1,** 374–349.

Defrise, M., Townsend, D.W., and Geissbuhler, A. (1990). Implementation of three-dimensional image reconstruction for multi-ring tomographs. *Phys. Med. Biol.* **35**, 1361–1372.

Felleman, D., and Van Essen, D. (1991). Distributed hierarchical processing in the primate cerebral cortex. *Cereb. Cortex* **1**, 1–47.

Fox, P. T., Mintun, M., Reiman, E., and Raichle, M. (1988). Enhanced detection of focal brain responses using intersubject averaging and change-distribution analysis of subtracted PET images. *J. Cereb. Blood Flow Metab.* **8,** 642–653.

Hoffman, E. J., and Phelps, M. E. (1986). *In* "Positron Emission Tomography and Autoradiography: Principles and Applications for the Brain and Heart" (M. E. Phelps, J. C. Mazziotta, and H. R. Schelbert, eds.), pp. 237–286. Raven Press, New York.

Hoffman, E. J., Huang, S.-C., and Phelps, M. E. (1979). Quantitation in positron emission computed tomography. 1. Effect of object size. *J. Comput. Assisted Tomogr.* **3,** 299–308.

Hoffman, E. J., Huang, S.-C., Plummer, D., and Phelps, M. E. (1982). Quantitation in positron emission computed tomography. 6. Effect of nonuniform resolution. *J. Comput. Assisted Tomogr.* **6,** 987–999.

Huang, S.-C., and Phelps, M. E. (1986). Principles of tracer kinetic modeling in positron emission tomography and autoradiography. *In* "Positron Emission Tomography and Autoradiography: Principles and Applications for the Brain and Heart" (M. E. Phelps, J. C. Mazziotta, and H. R. Schelbert, eds.), pp. 287–346. Raven Press, New York.

Huang, S.-C., Phelps, M. E., Hoffman, E. J., Sideris, K., Selin, C. E., and Kuhl, D. E. (1980). Non-invasive determination of regional cerebral metabolic rate of glucose in normal human subjects with 28-F-fluoro-2-deoxyglucose and emission computed tomography: Theory and results. *Am. J. Physiol.* **238,** E69–E82.

Huerta, M., Koslow, S., and Leshner, A. (1993). The Human Brain Project: An international resource. *Trends Neurosci.* **16,** 436–438.

Jacobs, R. E., and Frasier, S. E. (1994). Magnetic resonance microscopy of embryonic cell lineages and movements. *Science* **263,** 681–684.

Jerison, H. (1989). Brain size and the evolution of mind. *In* "Fifty-ninth James Arthur Lecture on the Evolution of the Human Brain." American Museum of Natural History, New York.

Jones, T., Chesler, D. A., and Ter-Pogossian, M. M. (1976). The continuous inhalation of oxygen-15 for assessing regional oxygen extraction in the brain of man. *Br. J. Radiol.* **49,** 339–343.

Koshland, D. (1992). The dimensions of the brain. *Science* **258,** 199.

Kwong, K. K., Belliveau, J. W., Chesler, D. A., Goldberg, I. E., Weisskoff, R. M., Poncelet, B. P., Kennedy, D. N., Hoppel, B. E., Cohen, M. S., and Turner, R. (1992). *Proc. Natl. Acad. Sci. USA* **89,** 5675–5679.

Lear, J. L. (1986). Principles of single and multiple radionuclide autoradiography. *In* "Positron Emission Tomography and Autoradiography: Principles and Applications for the Brain and Heart" (M. E. Phelps, J. C. Mazziotta, and H. R. Schelbert, eds.), pp. 197–235. Raven Press, New York.

Lear, J. L., Ackermann, R., Carson, R., Kameyama, M., Huang, S.-C., and Phelps, M. E. (1983). Evaluation of cerebral function using simultaneous multiple radionuclide autoradiography. *J. Cereb. Blood Flow Metab.* **3,** S97–S98.

Lear, J. L., Ackermann, R., Carson, R., Kameyama, M., and Phelps, M. E. (1984). Multiple-radionuclide autoradiography in evaluation of cerebral function. *J. Cereb. Blood Flow Metab.* **4,** 264–269.

Li, W.-H., Fraser, S. E., and Meade, T. J. (1999). A calcium-sensitive magnetic resonance imaging contrast agent. *J. Am. Chem. Soc.* **121,** 1413–1414.

Mazziotta, J. (1984). Physiologic neuroanatomy: New brain imaging methods present a challenge to an old discipline. *J. Cereb. Blood Flow Metab.* **4,** 481.

Mazziotta, J., and Gilman, S. (1992). "Clinical Brain Imaging: Principles and Applications." Davis, Philadelphia.

Mazziotta, J. C., and Koslow, S. H. (1987). Assessment of goals and obstacles in data acquisition and analysis for emission tomography: Report of a series of international workshops. *J. Cereb. Blood Flow Metab.* **7,** S1–S31.

Mazziotta, J. C., Huang, S. C., Phelps, M. E., Carson, R. E., MacDonald, N. S., and Mahoney, K. (1985). A noninvasive positron CT technique using oxygen-15 labeled water for the evaluation of neurobehavioral task batteries. *J. Cereb. Blood Flow Metab.* **5,** 70–78.

Mazziotta, J. C., Phelps, M. E., Plummer, D., and Kuhl, D. E. (1981). Quantitation in positron emission computed tomography. 5. Physical–anatomical factors. *J. Comput. Assisted Tomogr.* **5,** 734–743.

Mazziotta, J. C., Phelps, M. E., Meadors, K., Ricci, A., Winter, J., and Bentson, J. (1982). Anatomical localization schemes for use in positron computed tomography using a specially designed head holder. *J. Comput. Assisted Tomogr.* **6,** 848–853.

Mesulam, M.-M. (1990). Large-scale neurocognitive networks and distributed processing for attention, language and memory. *Ann. Neurol.* **28,** 597–613.

Ogawa, S., Tank, D. W., Menon, R., Ellerman, J. M., Kim, S. G., Merkle, H., and Ugurbil, K. (1992). Intrinsic signal changes accompanying sensory stimulation: Functional brain mapping with magnetic resonance imaging. *Proc. Natl. Acad. Sci.* **89,** 5951–5955.

Pechura, C. M., and Martin, J. B. (1991). "Mapping the Brain and Its Function." Natl. Acad. Press, Washington, DC.

Phelps, M. E., Huang, S. C., Hoffman, E. J., Selin, C. E., and Kuhl, D. E. (1979). Tomographic measurements of regional cerebral glucose metabolic rate in man with (18-F) fluorodeoxyglucose: Validation of method. *Ann. Neurol.* **6,** 371–388.

Posner, M., and Dehaene, S. (1994). Attentional networks. *Trends Neurosci.* **17,** 75–79.

Prichard, J., and Brass, L. (1992). New anatomical and functional imaging methods. *Ann. Neurol.* **32,** 395–400.

Quinn, B., Ambach, K. A., and Toga, A. W. (1993). Three-dimensional cryomacrotomy with integrated computer-based technology in neuropathology. *Lab. Invest.* **68,** 121A.

Ralston, H. J. (1990). Analysis of neuronal networks: A review of techniques for labeling axonal projections. *J. Electron Microsc. Tech.* **15,** 322–331.

Rauschning, W. (1979). Serial cryosectioning of human knee joint specimens for a study of functional anatomy. *Sci. Tools* **26,** 47–50.

Sereno, M. I., and Dale, A. M. (1992). A technique for reconstructing and flattening the cortical surface using MRI images. *Soc. Neurosci. Abstr.* **18,** 585.

Sereno, M. I., McDonald, C. T., and Allman, J. M. (1994). Analysis of retinotopic maps in extrastriate cortex. *Cereb. Cortex* **6,** 1047–3211.

Shephard (1990). "Synaptic Organization of the Brain." Oxford Univ. Press, London.

Sokoloff, L., Reivich, M., Kennedy, C., Des Rosiers, M., Patlak, C., Pettigrew, K., Sakurada, O., and Shinohara, J. (1977). The C-14-deoxyglucose method for the measurement of local cerebral glucose utilization: Theory, procedure, and normal values in the conscious and anesthetized albino rat. *J. Neurochem.* **28,** 879–916.

Thatcher, R. W. (1995). Tomographic electroencephalography/magnetoencephalography: Dynamics of human neural network switching. *J. Neuroimaging* **5,** 35–45.

Toga, A. W., Ambach, K. L., and Schluender, S. (1994a). High-resolution anatomy from *in situ* human brain. *NeuroImage* **1,** 334–344.

Toga, A. W., Ambach, K., Quinn, B., Hutchin, M., and Burton, J. S. (1994b). Postmortem anatomy from cryosectioned whole human brain. *J. Neurosci. Methods* **54,** 239–252.

Toga, A. W., Cannestra, A., and Black, K. (1995). The temporal/spatial evolution of optical signals in human cortex. Submitted for publication.

Townsend, D. W., Defrise, M., Geissbühler, A., Spinks, T., and Jones, T. (1991). Three-dimensional reconstruction for a multi-ring positron tomograph. *Prog. Clin. Biol. Res.* **363,** 139–154.

Tufte, E. R. (1983). "The Visual Display of Quantitative Information." Graphics Press, Cheshire, CT.

Tufte, E. R. (1990). "Envisioning Information." Graphics Press, Cheshire, CT.

Van Leeuwen, M. B. M., Deddens, A. J. H., Cerrits, P. O., and Hillen, B. (1990). A modified Mallory–Cason staining procedure for large cryosections. *Stain Tech.* **65,** 37–42.

II

Surface-Based Data Acquisition

3

Optical Imaging of Neural Structure and Physiology: Confocal Fluorescence Microscopy in Live Brain Slices

Michael E. Dailey

Department of Biological Sciences, University of Iowa, Iowa City, Iowa 52242

I. Introduction

A. Mapping Neural Organization at the Cellular and Synaptic Levels

Understanding brain function, both in the normal and in the diseased state, requires a thorough knowledge of the anatomical and physiological substrates. This necessitates a mapping of the organization and interconnection of populations of neurons, but must also include an intimate understanding of the physiological workings of the individual neural elements, both neurons and glia, since neural function ultimately depends on the organization of synaptic connections at the cellular and subcellular levels (Shepherd, 1998).

A major goal in neurobiology is to provide a detailed map of the anatomical connections between individual neurons and groups of neurons. This should include not only a "wiring diagram" of connections within and between brain regions, but also a qualitative and quantitative description of the subcellular distribution of synaptic contacts. Such a map certainly would prove useful for understanding neural circuitry at a network level. But even an exhaustive map of neuronal synaptic connectivity is not sufficient for figuring out how the anatomy subserves neural function. Information about the physiological properties of the synaptic contacts (e.g., strength, sign, and number) and their consequences for neuronal targets is also needed.

Even from an anatomical standpoint, a single *static* map of connections may not accurately reflect neural connectivity, especially during development and in diseased states in which physical connections may be changing. Such phenomena fall into the realm of modifiability, or *plasticity,* and there is now a major effort to understand how *structural* plasticity may subserve *functional* plasticity in the developing, adult, and diseased brain. Neural organization can thus be considered to be in a dynamic state, especially at the subcellular level, and we are in essence challenged with mapping an actively changing terrain! Such considerations highlight a need for time-resolved microanatomy as well as a conjoining of anatomical and physiological observations at the network, cellular, and subcellular levels.

Given that neural organization is in a dynamic state, it is important not only to generate a functional *map* of the brain, but also to elucidate the principles and mechanisms governing the development and plasticity of neural organization. It is anticipated that principles of neural organization will be elucidated by studying the process of construction of network connections during ontogeny. However, spatiotemporal changes in neural organization are at their highest during development, presenting developmental brain cartographers with a most challenging task.

B. The Problem: Cellular Diversity and Complexity of Neural Tissues

A major obstacle to mapping the anatomical substrates of neural function is the complexity of neural organization. Axonal and dendritic processes of neurons have very elaborate shapes, and each cell type, by definition, has unique morphological and physiological characteristics. Moreover, even within a rather homogeneous population of cells, it seems likely that no two cells have exactly the same morphology and pattern of connectivity. In the central nervous system (CNS), axonal processes often take complex paths to reach target regions and, once there, can ramify profusely. Axonal branches can also innervate multiple target regions. Likewise, dendritic branches are typically highly branched and can be recipient to tens of thousands of axonal synaptic contacts.

Over the past century, neuroanatomists have made tremendous strides toward mapping basic neuronal structure and connectivity, largely due to the extensive application of the Golgi technique (e.g., Ramón y Cajal, 1911). The Golgi stain generates a dense reaction product within most or all of the intracellular volume of individual neurons, and since it labels only a small percentage of cells in any given tissue volume, it provides a good method for examining the complex structure of individual neurons at the light microscope level. However, the Golgi technique is limited by unpredictable staining patterns such that the neuroanatomist has little control over the number and type of cells labeled. Furthermore, it is applicable only to postmortem tissue and is therefore of limited value for studies of a physiologic nature.

With the advent of techniques for injecting cells with a tracer dye while making electrophysiological recordings, it became possible to determine in detail the anatomical features of neurons following physiological characterization. This was an important step in efforts to correlate cellular physiology and morphology in brain tissue. Moreover, neuronal tracers such as horseradish peroxidase (HRP) afford the collection of three-dimensional (3D) information on the structure of neurons at both the light and the electron microscope (EM) level (e.g., Deitch *et al.,* 1991). HRP can be injected into tissue to label individual neurons or populations of neurons, but it is visible in the light microscope only after fixation and enzymatic reaction. Consequently, it is not a useful marker for assessing the structure of live cells. Fluorescent tracer dyes such as Lucifer yellow (LY), on the other hand, permit light and electron microscopic observations and, by virtue of their fluorescence, are visible in living cells. In the case of LY, staining for EM is accomplished by HRP immunohistochemistry using antibodies against the LY (Taghert *et al.,* 1982; Holt, 1989). A number of studies have used LY microinjection in conjunction with confocal imaging to examine the 3D structure of neurons in both living (Smith *et al.,* 1991, 1994; Turner *et al.,* 1991, 1993, 1994) and fixed (Belichenko *et al.,* 1992, 1994a,b; Belichenko and Dahlström, 1994a,b, 1995a,b; Trommald *et al.,* 1995) brain tissues. However, two major drawbacks to using LY are that it requires the tedious process of microinjection into single cells and, for EM-level analysis, the cytoplasm of the labeled cell is obliterated by the immunohistochemical reaction. Moreover, these methods of labeling do not directly reveal synaptic contacts and they have not been shown to be useful for directly imaging changes in cellular structure over time.

C. One Solution: Vital Fluorescent Labeling and 3D Confocal Imaging in Brain Slices

Much can be learned about the functional organization of the brain by correlating single-cell physiological analyses with static images of cellular morphology. However, functional mapping dictates that we integrate the structural and physiological features of individual cells into the larger context of tissue organization. This can be facilitated by simultaneously viewing the functional interrelationships and interactions of *many* cells within organized networks. Live brain slice preparations provide an outstanding opportunity to assess the dynamic structural and physiological features of cells, at high spatial resolution, within complex three-dimensional tissue environments.

In the past, it has been necessary to "reconstruct" neural structures and cellular relationships from the microscopic examination of several adjacent, relatively thin tissue sections. This is because conventional (both light and electron) microscopic techniques did not provide sufficiently high spatial resolution of cellular and subcellular structures in thick tissue specimens. Although it is still necessary in most cases to section whole brain for microscopic examination, the development and application of modern confocal microscopy (White *et al.,* 1987; Fine *et al.,* 1988; Lichtman, 1994; Dailey *et al.,* 1999) and, more recently, multiphoton microscopy (Denk *et al.,* 1990, 1994) have provided a means to examine structures at high spatial resolution in much thicker (>100 μm) tissue slices. The advantages afforded by confocal microscopy are derived from its ability to collect optical sections of a thick tissue specimen while rejecting light from out-of-focus components of the specimen (Wilson, 1990; Pawley, 1995). These features have made confocal imaging an indispensable tool for analysis of neural organization at the subcellular, cellular, and tissue levels of organization (Turner *et al.,* 1996). Indeed, the confocal microscope was initially conceived and developed with the goal of elucidating the neural organization of the brain (Minsky, 1961, 1988).

The successful application of modern optical techniques to elucidate neural organization has depended on the availability of suitable markers of cellular structure and physiology. Among the most widely used markers are the vital fluorescent membrane dyes for labeling cell surfaces and the fluorescent Ca^{2+} indicator dyes. More recently, it has been possible to express fluorescent proteins or fusion proteins in neurons and glia in order to visualize specific cellular structures such as synapses. This chapter discusses the use of these fluorescent probes and molecular markers in conjunction with high-resolution optical imaging with the confocal microscope to examine the dynamic anatomy and physiology of live brain tissues. It concentrates on the application of these methods to study neural organization at a cellular and subcellular level in tissue slices from developing rodents. However, several of these methods are more widely applicable to studies of normal and diseased adult brain tissues, both pre- and postmortem, as studies from other groups have shown (see references for a sample).

II. Live Brain Slice Preparation and Culture

Live brain slice preparations have been used extensively to study fundamental physiological properties of neurons and local neural circuits using both electrophysiological and optical approaches and, more recently, for examining dynamic morphological features of neurons and glia. The primary advantages of tissue slice preparations are their greater optical and physiological accessibility over *in vivo* conditions (Pozzo-Miller *et al.,* 1993) and the maintenance of structural and functional integrity of intrinsic synaptic connections vis-à-vis dissociated cell culture preparations. The major disadvantages are: (1) possible structural and physiological alterations (damage) to the tissues as a consequence of the isolation procedure; (2) loss of extrinsic neuronal connections such that the isolated tissues cease to exhibit normal functionality (i.e., the tissues are no longer integrated in the larger scheme of brain function); and (3) a finite period of time before tissue rundown (generally, a maximum of several hours). These drawbacks notwithstanding, the significant advantage that *in vitro* brain slice preparations hold for *optical accessibility* has driven the development of slice imaging methodology. Consequently, progress on the development of sensitive fluorescent-light microscopy, new fluorescent probes of cellular anatomy and physiology, and inexpensive computing capabilities has provided a powerful set of tools for investigating the organization and function of neural tissues. Moreover, good methods are now available for maintaining healthy, live brain slices *in vitro* for long periods of time. This section describes methods currently used for preparing, maintaining, labeling, and imaging live slices of developing rodent brain.

Isolated tissues derived from immature animals generally fare much better than adult tissues. In the case of the rat hippocampus, tissues taken from animals older than about 1 week of age generally do not remain healthy for longer than a few hours *in vitro.* It seems likely that this limitation is due, in part, to a developmental shift from anaerobic- to aerobic-based tissue metabolism. Moreover, developing brain tissues seem to suffer less stress under the hypoxic and hypoglycemic conditions that may occur during the tissue isolation procedures. Indeed, if brain tissues are isolated from developing animals, they can be maintained *in vitro* for long periods of time (weeks to months), extending well beyond the corresponding time point from which tissues can be maintained *ex vivo* when taken from older (>P7) animals. Consequently, many studies of live mammalian brain tissues have employed slice cultures derived from immature brain (see Gähwiler *et al.,* 1997).

There are two common methods for maintaining brain slices *in vitro* (Gähwiler *et al.,* 1997). One method is based on the so-called "roller tube" culture technique of Gähwiler (1984; Gähwiler *et al.,* 1991). This technique provides a convenient way of mounting tissue slices on glass coverslips for labeling, long-term culturing, and optical imaging. Another slice culture method (Stoppini *et al.,* 1991) in which slices are grown on porous filter membranes also has proven to be very suitable for long-term culture. Both culture methods (Fig. 1) involve rapidly removing the tissues of interest, then slicing the tissues with a tissue chopper (Stoelting, Chicago, IL) or a Vibratome at a thickness of 300 to 500 μm. In the case of roller tube cultures, the brain slices are then secured to alcohol-cleaned glass coverslips (11 × 22 mm) with a mixture of chicken plasma (10 μl; Cocalico) and bovine

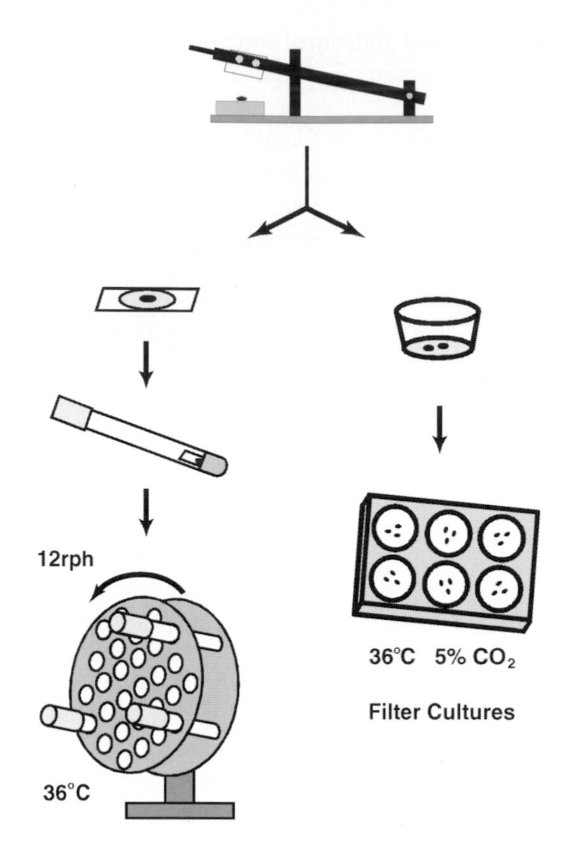

Figure 1 Preparation of brain tissue slices and slice cultures. Live tissues are isolated from neonatal rat and are sliced at a thickness of 300–400 μm using a manual tissue chopper. Tissue slices may be cultured by one of two methods (Gähwiler *et al.*, 1997): roller tube (left column) or filter membranes (right column). In the roller tube method, tissue slices are secured to rectangular glass coverslips using a plasma/thrombin clot. Coverslips with attached slices are inserted into culture tubes with 1 ml of Hepes-buffered growth medium, and culture tubes are placed in a roller drum and rotated at 12 rph to aerate the tissue. In the filter membrane method (Stoppini *et al.*, 1991), slices are placed on porous filter membrane inserts in six-well plates and cultured in bicarbonate-buffered growth medium in a CO_2 incubator. In both instances, cultures are maintained at 36°C, and growth media are exchanged two or three times a week.

thrombin (10 μl; Sigma, St. Louis, MO). Tissues are adherent within about 10 min, at which point the coverslips are placed in a test tube with 1 ml of Hepes-buffered culture medium containing 25% serum. The tubes are kept in a warm room (36°C) and are rotated at 12 rph in a roller drum tilted at 5° with respect to horizontal. This roller tube method provides constant gentle rolling to aerate the tissue slices, and viable slices with an organotypic tissue organization can be maintained *in vitro* for up to several weeks. However, the slices thin appreciably over time (down to about 100 μm thick within 1–2 weeks *in vitro*), and there can be a significant proliferation of glial cells (Dailey and Waite, 1999). The roller tube cultures are especially useful for high-resolution microscopy since the slices are adherent to—and can be viewed through—a stable glass coverslip (Terasaki and Dailey, 1995; Dailey, 1999).

Slice cultures grown on filter membranes (Stoppini *et al.*, 1991) provide a means for short- or long-term culturing of brain tissues without the need to physically rotate the tissues or attach them to coverglasses. Tissue slices are prepared as described above, then placed on cell culture inserts (Falcon 3090 or 3102) containing polyethylene terephthalate, track-etched porous membranes (1 μm pore size). The inserts are placed in six-well tissue culture plates, and culture medium (1 ml containing 50% MEM, 25% Hanks' balanced salt solution, 25% horse serum, 2 mM glutamine, and 0.044% $NaHCO_3$) is added to each well. The filter cultures are maintained in an incubator containing warmed (36°C), humidified air (5% CO_2). Translucent culture inserts are used to facilitate periodic inspection of the cultured tissues under a dissecting microscope.

Each tissue slice culture technique has its advantages and disadvantages. In terms of culture preparation, the filter cultures require less effort because there is no need for lengthy processing of coverslips. Moreover, once the slices are situated on the membranes, the tissues can be placed immediately into the incubator without concern for whether the slices are firmly attached to the culture substratum. Some investigators report that tissues cultured by the filter membrane technique show reduced gliosis in comparison to those cultured by the roller tube technique (del Rio *et al.*, 1991), suggesting that long-term cultured brain tissues may fare better on membranes. On the other hand, the roller tube cultures are more easily mounted for microscopic viewing, since the slices are securely attached directly to a piece of glass that can serve as the microscope coverslip (Terasaki and Dailey, 1995). For live tissue imaging, a more complicated scheme is required for mounting tissues in a way that permits on-stage perfusion of solutions (Dailey, 1999).

III. Labeling Neuronal and Glial Cells in Brain Tissue Slices

A. Visualizing Neural Structure with Fluorescent Membrane Dyes

To examine the structure of neurons in tissue slices, we have used a family of long carbon-chain, carbocyanine dyes (see Table 1), which incorporate into and diffuse laterally within plasma membranes of neurons (Honig and Hume, 1986; Honig, 1993) and other cells that come into contact with the dye. The rapid and complete surface labeling is especially useful for determining the morphology of cells such as neurons that have very long and elaborate branching processes. Consequently, membrane dyes have found widespread use as markers of axonal projections (Baker and Reese, 1993).

Table 1 Common Fluorescent Membrane Dyes for Assessing Neuronal Structure and Mapping Axonal Projections[a]

Name	Abbreviation	$Abs_{max}{}^{b}(\lambda)$	$EM_{max}{}^{c}(\lambda)$	Reference
$DiIC_{18}(3)$	DiI	550 nm	565 nm	Honig and Hume (1986)
$DiI\Delta^{9,12}C_{18}(3)$	Fast DiI	549 nm	563 nm	
$DiOC_{18}(3)$	DiO	484 nm	501 nm	Honig and Hume (1986)
$DiIC_{18}(5)$	DiD	644 nm	663 nm	Agmon *et al.* (1995)
4-Di-16-ASP	DiA	491 nm	613 nm	Mendelowitz *et al.* (1992)

[a]These dyes are available from Molecular Probes, Inc. (Eugene, OR). Data are from Molecular Probes' catalog and information sheets.

[b]Absorption maximum.

[c]Emission maximum.

A useful property of these membrane dyes is that they can label cells in both living and formaldehyde-fixed tissues. This is because formaldehyde fixation does not extensively cross-link lipids, so that fluorescent lipids can intercalate into the surface membrane and diffuse freely within the plane of the membrane. However, a form of DiI that is formaldehyde fixable is also available (Molecular Probes, Eugene, OR; Catalog No. D7000).

The diffusion rate of the most commonly used fluorescent lipid, DiI, is reported to be 6 mm per day in living tissue (Product Sheet MP282; Molecular Probes). In live brain slices, adequate levels of staining can be achieved in neuronal processes over 1 mm away from the labeling site within a few hours. We have successfully imaged both DiI- and DiO-labeled axons for several hours, but some workers have suggested that live cells labeled with DiO remain somewhat healthier during imaging than those stained with DiI. Perhaps this is because DiI may stain living cells more strongly than DiO (MP282; Molecular Probes).

Often it is most useful to label only a small percentage of the total number of cells within a tissue volume because even the best optical microscopes are unable to resolve the details of fine axonal and dendritic processes when all the tissue elements are stained with the same dye. In certain cases it is desirable to label a select subset of cells. For neural tissue, labeling a select population of cells can often be accomplished simply by varying the location of the dye application. Cell bodies of projection neurons can be back-labeled by injecting a tracer into target regions. Alternatively, axonal projections can be labeled by applying dye to the region of cell bodies, to dendrites, or along known axonal tracts. However, we have found that surface labeling with the membrane dyes (e.g., DiO) seems to be more efficient in the anterograde versus the retrograde direction along axons in living tissue (Dailey and Smith, 1993). This may be related to axonal transport or membrane trafficking patterns in neurons.

We have used several different methods for introducing membrane dyes into tissue slices. One approach involves pressure injection of a solution of dye. A stock solution (~0.5%) of dye is made in *N,N*-dimethylformamide, dimethyl sulfoxide, or vegetable oil and is injected through a glass micropipette (2–3 μm tip diameter) using a Picospritzer (General Valve). Small, localized injections can be made by presenting a series of brief pressure pulses (1 ms duration, 80 psi) to the back of the pipette.

Another membrane dye-labeling method works very well with relatively thin (50–100 μm) tissue slices, such as those carried as roller tube cultures for longer than 1 week. To label cells, the tip of a glass pipette is dipped into a saturated solution of DiI in ethanol, then inserted, and either removed after a time or broken off within the tissue. The dye solution dries onto the surface of the pipette, leaving a coating of dye crystals that contacts and labels cell membranes when inserted into the tissue. This method seems to produce labeling of fewer but more brightly stained cells with less granular background labeling of neighboring cells. Both the injection and the crystal insertion labeling protocols, when applied to roller tube cultures, benefit from the fact that the plasma clot holds the tissue slice in place during the labeling procedure.

There has been considerable interest in fluorescent markers whose excitation and emission spectra are in the red or near infrared. Although one potential drawback of the longer wavelength red light (vis-à-vis blue or green light) is that the spatial resolution is reduced somewhat, it is expected that optical imaging in tissues will be improved with red dyes because: (1) there is less background autofluorescence from tissues at the longer wavelengths; (2) biological tissues should diffract red light less, thereby improving the collection of light for image formation; and (3) the red light is of lower energy than UV, blue, and green light and therefore should produce less photodynamic damage (see later). There may not always be a dye available with suitable spectra, but in the case of the long-carbon-tail fluorescent membrane dyes, there is a longer-wavelength version of DiI, known as DiD, which has a five-carbon linking bridge (see Table 1). The excitation/emission maxima of DiD are 644/663 nm (in methanol), compared to 550/565 nm for the three-carbon bridge of the classic DiI (MP282; Molecular Probes). Thus, laser lines from the argon–krypton (647 nm) or helium–neon (633 nm) lasers can be used to excite DiD. A study by Agmon *et al.* (1995) indicated that DiD is, in fact, superior to DiI for examining axonal projections in brain slices by confocal microscopy.

A new and potentially powerful approach to obtain Golgi-like labeling of neuronal and glial cells with fluorescent lipids has been described (Gan *et al.,* 2000). This technique, referred to as "DiOlistics," is a modification of a biolistics approach to gene transfection (see later; Lo *et al.,* 1994). Tiny gold or tungsten particles (0.4–1.7 μm) coated with one or more fluorescent lipid dyes (such as DiO, DiI, or DiD) are propelled into brain tissues using a "gene gun" (Bio-Rad, Hercules, CA). Cells whose surfaces are contacted by dye-coated particles become labeled. In the case of neurons, the entire axonal and dendritic arbors, including synaptic spines, are labeled. The DiOlistics approach works in both live and paraformaldehyde-fixed brain tissues, including postmortem tissues derived from human brain (Gan *et al.,* 2000). In live tissues, cell labeling is very rapid, with complete labeling of complex dendritic arbors reportedly occurring within 5 min (apparent diffusion coefficient 10^7 cm^2/s). Labeling is severalfold slower in fixed tissues, consistent with many previous studies utilizing fluorescent lipophilic dyes. The number of labeled cells can be varied by altering parameters such as the density of particles in the "bullets," the pressure and distance at which the tissues are shot, and the number of shots fired at the tissues. Using particles coated with different combinations of various dyes with different fluorescent spectra, it is possible to label and distinguish many neurons within a small tissue region. For example, using combinations of three different lipophilic dyes, Gan *et al.* (2000) could distinguish individual cells labeled in one of seven different spectral patterns. Using three-dimensional multichannel confocal fluorescence imaging (see later), it is possible to identify processes from single cells even in densely labeled portions of complex brain tissues. This multicolor labeling methodology thus offers a new approach to mapping the complex organization and relationship of neuronal and glial cells in brain tissues.

Finally, it should be noted that fluorescent lipophilic dyes are useful for morphological studies of neural tissue at both the light and the electron microscope level. This is because the fluorescence emission can convert (i.e., photoconvert) diaminobenzidine (DAB) to an oxidized, electron-dense reaction product that is deposited locally near the fluorophore (Sandell and Masland, 1988). Thus, in the case of membrane dyes that label the plasmalemma, one can produce a nice outline of the cell surface that preserves the intracellular structure of labeled cells. This method could prove to be very useful for determining the ultrastructural characteristics, including the synaptic connectivity, of specific populations of neurons that have been selectively labeled with a fluorescent dye (e.g., von Bartheld *et al.,* 1990; Gan *et al.,* 1999). Moreover, because membrane dyes can be used as vital stains, it is now possible to observe cells in the living state with the light microscope and, subsequently, to examine the very same cells in the electron microscope. This should facilitate studies of neuronal structure and connectivity.

B. Immunofluorescent Labeling

Antibodies provide a very powerful means for labeling specific populations of cells and subcellular structures in neural tissues, although their use is usually limited to fixed specimens. Sometimes it is desirable to assess the 3D organization of immunostained structures, and optical imaging of immunolabeled thick brain slices provides a relatively convenient means for doing this, so long as the antibodies are able to penetrate the tissues adequately and clear optical sections of sufficient resolution can be collected. Immunohistochemical staining of thick brain slices has been shown to be feasible, and confocal microscopy affords collection of the 3D immunofluorescence data with high spatial resolution (e.g., Vincent *et al.,* 1991; Welsh *et al.,* 1991).

As an example, we have used antibodies against synaptic proteins in acutely isolated and in cultured brain slices (Dailey *et al.,* 1994; Qin *et al.,* 2001). One such antibody, generated against the protein synapsin-I (syn-I), has been shown to be specific to nerve terminals (DeCamilli *et al.,* 1983a,b) where it is associated with small synaptic vesicles (Navone *et al.,* 1984). Immunostaining of hippocampal slices (100–400 μm thick) with syn-I antiserum is performed after light fixation (2% formaldehyde for 10 min) and a rigorous extraction process (1% Triton X-100 for 24 to 72 h) (Dailey *et al.,* 1994). Penetration of the rather large antibody proteins into the thick tissues appears to be a major limitation. Therefore, we (1) perform the membrane permeabilization step on a rotating stage to provide constant mechanical agitation (~100 rpm) and (2) lengthen the primary and secondary antibody incubation times to several hours or overnight (4°C). This staining procedure provides a sufficient immunofluorescent signal to image small synaptic structures as much as 50 to 100 μm deep into a brain slice. The availability of a wide spectrum of fluorescent probes as well as confocal imaging systems with multiple laser lines permits double- and triple-label immunohistochemical analyses of brain tissues (Sergent, 1994; Wouterlood *et al.,* 1998). As indicated earlier, the use of longer wavelength fluorophores, such as Cy5 (650/667 nm) (Amersham Pharmacia Biotech, Inc., Piscataway, NJ) or members of the Alexa Fluor series [Alexa Fluor-633 (632/647 nm), Alexa

Fluor-647 (650/668 nm), or Alexa Fluor-660 (663/690 nm), from Molecular Probes], may improve the detectable fluorescence signal from deeper portions of the specimen.

Immunohistochemical staining also may be combined with staining by fluorescent membrane dyes (Elberger and Honig, 1990). These methods even permit immunohistochemistry in combination with time-lapse observations of live cells in brain slices. For example, O'Rourke *et al.* (1992) followed DiI-labeled migrating neuroblasts in slices of developing cerebral cortex by time-lapse confocal microscopy and then fixed the slices and immunohistochemically stained the tissues with an antiserum to reveal radial glial fibers. They first photoconverted the DiI to permanently mark the labeled cells with a stable, electron-dense DAB reaction product and then permeabilized the tissues for antibody staining.

C. Gene Transfection and Expression of Fluorescent Proteins

We have seen that fluorescent membrane dyes provide a means for assessing gross cellular morphology and that immunohistochemical methods afford more specific labeling of synaptic structures in fixed preparations. However, the complexity of neural tissues often makes it difficult to map the organization of synaptic structures in relation to cellular structure. One promising approach is to use molecular genetic tools to label synaptic structures within individual cells or defined populations of cells. The exploitation of a jellyfish green fluorescent protein (GFP) as a reporter of gene expression (Chalfie *et al.*, 1994), and more recently as a marker of protein distribution in cells, has opened the possibility of visualizing specific cell populations and cellular structures in both living and fixed brain tissues. Indeed, GFP and related fluorescent proteins (Tsien, 1998; Tsien and Prasher, 1998) are increasingly being used to mark neurons and glia in brain tissues (e.g., Lo *et al.*, 1994; Moriyoshi *et al.*, 1996; Zhuo *et al.*, 1997; Vasquez *et al.*, 1998; van den Pol and Ghosh, 1998; Chamberlin *et al.*, 1998.)

GFP expression in neurons and glia can be used in a variety of ways to facilitate mapping of brain anatomy and microstructure. Expression of soluble GFP fills the entire extent of neurons and glia, and this can be used to define cellular anatomy. Modification of the GFP to target it specifically to the plasma membrane (Moriyoshi *et al.*, 1996; Tamamaki *et al.*, 2000) may facilitate lateral movement in long processes and thus provide even better results for anatomical studies in neurons. Under control of cell-type- or region-specific regulatory genes, GFP can be used as a reporter system to label specific neuronal or glial cell populations (e.g., Zhuo *et al.*, 1997; Oliva *et al.*, 2000; Spergel *et al.*, 2001). Alternatively, GFP may be fused to proteins of interest containing specific cellular targeting sequences. For example, GFP fused to a synaptic protein results in specific labeling of pre- or postsynaptic structures in transfected neurons (e.g., Arnold and Clapham, 1999; Ahmari *et al.*, 2000). This offers the exciting possibility of mapping specific synaptic structures at the single-cell level within brain tissues.

A variety of approaches have been used to introduce foreign genes such as GFP and GFP-fusion proteins into cells and tissue slices *in vitro*. These include viral constructs for infection (Vasquez *et al.*, 1998) as well as nonviral transfection methodologies such as particle-mediated biolistics (Lo *et al.*, 1994), liposome-mediated transfection (Murphy and Messer, 2001), and single-cell electroporation (Haas *et al.*, 2001). For transfection experiments in brain slices (Marrs *et al.*, 2001; Qin *et al.*, 2001), we utilize the biolistics approach (Fig. 2) based on the methods of Lo *et al.* (1994). A Helios gene gun (Bio-Rad) is used following the manufacturer's instructions. Colloidal gold particles (1 μm diameter, 8 mg) are combined with 0.05 M spermidine (100 μl; Sigma), 1 M $CaCl_2$ (100 μl), and DNA (15–20 μg). Hippocampal slices are shot (2–3 mm, 70–80 psi) between 5 and 10 days in culture and then quickly returned to the incubator. Slices are fixed (4% formaldehyde in culture medium, 10 min, 4°C) 1–2 days later. In some cases, antibody staining is performed on tissues following fixation.

D. Ca^{2+}-Sensitive Fluorescent Probes for Studying Neuronal and Glial Physiology

There are now numerous probes for addressing questions of cellular physiology (Mason, 1993; Yuste *et al.*, 1999a). These include both fluorescent and nonfluorescent probes of intracellular calcium, magnesium, protons, sodium, zinc, and several other physiologically important molecules. Some of these probes are ratiometric and provide estimates of absolute concentrations of the molecular species of interest.

The use of voltage-sensitive dyes has played an important role in mapping patterns of neural activity and in the organization of synaptic connections in isolated tissue slices (e.g., Grinvald *et al.*, 1988). However, fluorescent calcium-sensitive dyes (e.g., fura-2 and fluo-3) (Grynkiewicz *et al.*, 1985; Minta *et al.*, 1989) have been used more widely to study the physiological properties of neural and glial cells as well as the physiological organization of developing and mature nervous system tissues. The fluorescent calcium probes are especially useful because large changes in intracellular calcium are associated with neural electrical activity (Ross, 1989), and it is clear that intracellular calcium is an important second messenger associated with a wide variety of neuronal functions (Kennedy, 1989; Ghosh and Greenberg, 1995). An early study by Yuste *et al.* (1992) showed the power of Ca^{2+} imaging in isolated brain tissue slices for determining features of cellular communication in the developing cerebral cortex. Several studies since have utilized Ca^{2+} imaging in semi-intact brain tissues to study

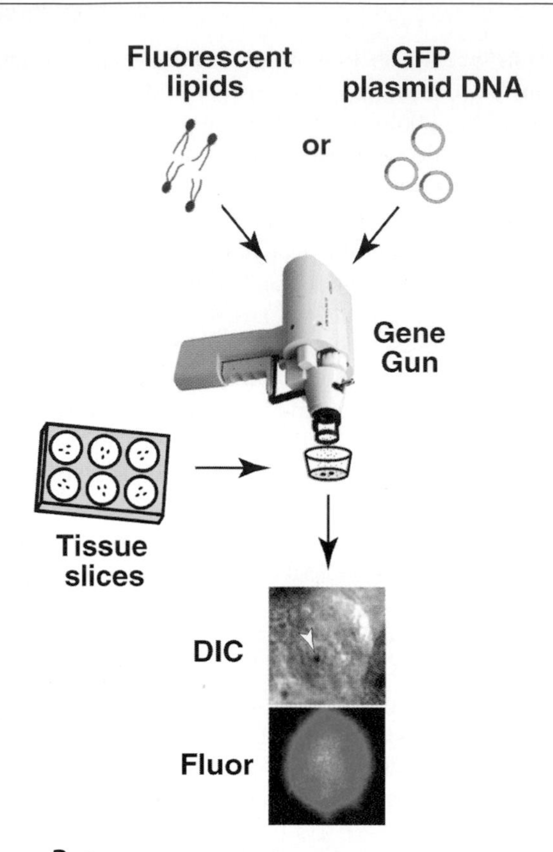

Figure 2 Fluorescent labeling of neurons and glia in brain tissues by gene gun-mediated particle bombardment (biolistics). Tiny (0.4–1.7 μm) gold or tungsten particles are coated either with fluorescent lipophilic dyes (e.g., DiI, DiO) or with plasmid DNA encoding GFP or GFP-fusion proteins. Particles carrying fluorescent dye or DNA are propelled into tissues using a blast of helium. In the case of dye labeling ("DiOlistics"; Gan *et al.*, 2000), cells are labeled when contacted by dye-coated particles. Labeling of axonal and dendritic arbors is very rapid (~5 min) in live tissues due to the diffusion of the lipophilic dye within the plane of the plasma membrane. Both living and fixed tissue preparations are amenable to DiOlistics labeling. Particles carrying various combinations of fluorescent dyes can be used to differentially label individual cells within tissues. For gene transfection (Lo *et al.*, 1994), particles bearing DNA must penetrate the cell and come to reside within or near the cell nucleus. Gene expression for 12 to 48 h is usually necessary to produce sufficient fluorescent signal. Particles bearing different plasmids [e.g., encoding green (GFP) or red (DsRed) fluorescent proteins] can be used to differentially label cells. A transmitted light (differential interference contrast, DIC) image and fluorescence image (Fluor) of the same cell are shown 1 day after being shot with plasmid DNA encoding the GFP. Note the DNA-bearing gold particle (arrowhead in DIC image) located within the cell.

the development and organization of neural circuits in brain (Dailey and Smith, 1994; Guerineau *et al.*, 1998) and retina (Wong *et al.*, 1998).

In addition to neurons, astrocytes (Cornell-Bell *et al.*, 1990; Dani *et al.*, 1992) and other glial cells (Jahromi *et al.*, 1992; Reist and Smith, 1992; Lev-Ram and Ellisman, 1995) in the nervous system also exhibit sizable intracellular fluctuations in calcium. Moreover, it is well documented that a variety of stimuli can induce waves of calcium activity within glial cell networks. Such *trans*-glial calcium signals have been proposed to have important roles in long-range cellular signaling within the brain (Charles *et al.*, 1991; Cornell-Bell and Finkbeiner, 1991; Smith, 1994; Charles, 1998). Thus, in generating a "map" of functional networks within the brain, it will be important to consider the organization of both neuronal and nonneuronal tissue components.

There are a variety of ways of introducing the physiological probes into neural tissue slices. Perhaps the most reliable is by direct intracellular injection. Microinjection allows one to select individual cells of interest as well as control the concentration of the intracellular dye. This approach has been used successfully to image intracellular calcium at high spatial and temporal resolution within neuronal cell bodies and processes as small as individual dendritic spines (Tank *et al.*, 1988; Müller and Connor, 1991; Guthrie *et al.*, 1991; Svoboda *et al.*, 1996; Yuste *et al.*, 1999b). However, one may want to label a *population* of identified neurons, or specific axonal projections and synaptic terminals, rather than single isolated cells. This can be achieved by localized superfusion (Regehr and Tank, 1991) or injection (O'Donovan *et al.*, 1993) of dyes into axonal tracts, leading to uptake and anterograde and retrograde transport of the dye. Such an approach is useful so long as the axons are bundled or spatially confined and the dye can be applied close to the site of interest since labeling is limited by intracellular diffusion.

Bulk loading of cells using membrane-permeant dyes provides less specific, more widespread labeling of neurons and glia in tissue slices. This approach makes use of dyes with acetoxymethyl (AM) groups linked to fluorophores by ester bonds. The AM–ester form of the dye is membrane permeant until the ester bonds are cleaved by endogenous esterases within the cells (Tsien, 1981). Once the ester bonds are cleaved, the dye molecule is trapped within the cell where it becomes a useful indicator of cellular physiology.

With regard to staining brain tissue, AM–ester loading works better on slices from embryonic tissues or in slices that have been maintained in culture for a period of time. For example, suitable fluo-3 AM loading of acutely prepared (<8-h-old) CNS tissue slices is difficult to achieve, but slices cultured for a week or two stain robustly. In fact, some success is achieved even after 12 to 24 h *in vitro* (Dailey and Smith, 1994). It is unclear why staining is enhanced in the cultured slices, but one possibility is that the tissue "loosens up" somewhat. Others have also found extreme difficulty staining acute CNS tissue slices with AM–ester dyes when the tissue was taken from rats older than 10 days of age (Kudo *et al.*, 1989; Yuste and Katz, 1991; O'Donovan *et al.*, 1993). Adams *et al.* (1994) suggested that the penetration of AM ester dyes into thick brain slices may be facilitated by cleaving some ester bonds prior to labeling.

IV. Imaging Methodology

A. Confocal Microscopy

The principle of confocal imaging is illustrated in Figure 3. A wide range of confocal imaging systems are now available. We used several different confocal imaging systems in making the observations described here, including three commercially available microscopes and a custom-built confocal microscope. The commercial confocals used were a Bio-Rad MRC-500 (Bio-Rad), modified as described previously (Smith *et al.,* 1990), a Noran Odyssey that was capable of collecting semiconfocal fluorescence images at video rate (30 frames/s), and a three-laser Leica TCS NT confocal system (Leica, Heidelberg, Germany). The custom-built microscope was a relatively low-cost, optical-bench style inverted microscope designed and built by Stephen Smith (Stanford University). For illumination, the microscope was equipped with a 25-mW argon-ion laser (Ion Laser Technology) and a 15-mW helium–neon laser (MWK Lasers). The laser beams were steered though a shutter, neutral density filter wheel, and excitation filter wheel and onto a neutral beam-splitting mirror that reflected 7% of the laser light to the specimen. The filter wheels were operated by stepper motors controlled by the host computer. The neutral beam-splitting mirror, which substituted for the dichroic mirror that is typically used in fluorescence microscopy, permitted single detector imaging of different fluorophores without the need to remove or adjust the beam-splitting mirror. Light was detected through an adjustable circular aperture in front of the detector, a gallium–arsenide photomultiplier tube. This microscope had a very high throughput and was thus well suited to long, time-lapse experiments. The microscope objective lenses we found to be most useful for the custom-built microscope were a dry Nikon 20×/0.75 Fluor, a Zeiss Plan-Neofluor 25×/0.8 (oil–water–glycerin), and an oil-immersion Olympus DApo 40×/1.3 UV objective. For the Leica system, we used a 20×/0.7 dry Plan Apo or a 63×/1.2 water Plan Apo (220 μm working distance) lens.

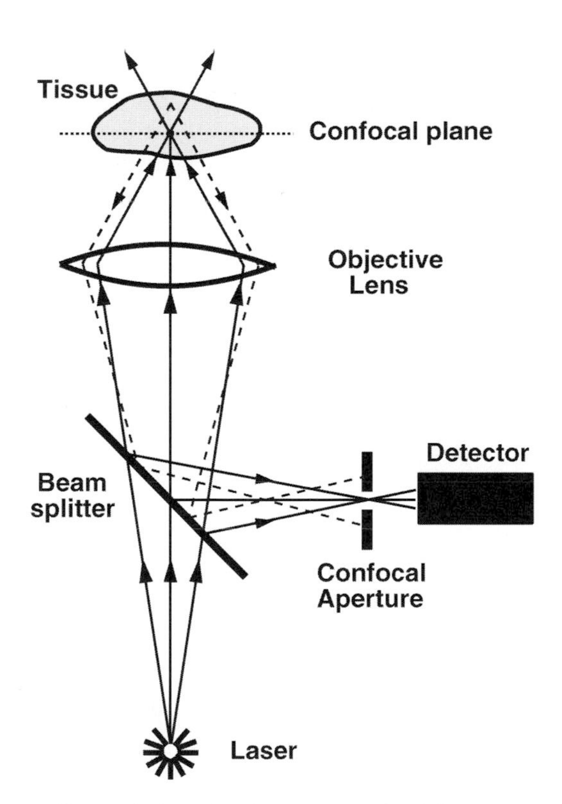

Figure 3 The principle of laser scanning confocal fluorescence microscopy. Laser light for fluorescence excitation is focused by a microscope objective lens to a diffraction-limited spot within a thick tissue specimen, such as a brain slice. Fluorescence emission is collected by the same objective lens and is directed via a dichroic or neutral beam-splitting mirror to a photomultiplier tube (detector). Light (dashed line) from regions of the specimen outside the confocal plane is rejected by the pinhole aperture in front of the detector (i.e., only light from a single, narrow focal plane is detected).

B. Three-Dimensional Imaging

An important feature of the confocal microscope is its ability to collect three-dimensional information on the structure of complex cells and cellular relationships at high spatial resolution (Lichtman, 1994). With regard to neural organization in the brain, the 3D information obtainable with the confocal microscope is helping to map the relationship of neurons within functional networks (Smith *et al.,* 1991) and localize synaptic structures at both the light (Hosokawa *et al.,* 1992, 1994; Belichenko and Dahlstrom, 1995b) and the EM (Deitch *et al.,* 1991) level of organization.

The question of how best to display the 3D image data is challenging, and the problem is compounded when 3D data are collected over time (generating 4D data sets; see later). For single time-point data sets, each of the individual optical sections along the axial (z) dimension can be displayed separately (e.g., Fig. 11A), or the axial stack of images can be combined to generate a pair of stereo images (e.g., Fig. 10). Stereo-pair images can be generated from any through-focus data set simply by shifting each successive image in the stack a small amount (i.e., by 1 pixel) before combining the images. For each of the stereo-pair image stacks, individual images are shifted in the opposite direction (i.e., shifted left for one stereo-pair image and right for the other). Generally it is best to recombine images using a maximum brightness operation (rather than a simple addition) to reduce the buildup of background noise. The disadvantage of recombining the image stack (either for stereo or for nonstereo viewing) is that adjacent image plans may contain a small amount of out-of-focus light from a given structure, and when images are recombined, this can reduce the sharpness and contrast of the structure. This problem can be effectively defeated in the confocal microscope by stopping down the pinhole aperture in order to reject the maximum amount of out-of-focus light and reduce flare. However, this will substantially decrease the detected signal and is most useful under conditions under which phototoxicity is not a concern and the illumination intensity can be turned up to compensate for the loss of signal.

C. Dynamic Imaging in Live Brain Slices

We have applied confocal imaging technology in conjunction with vital fluorescent labeling to examine dynamic changes in cellular structure and physiology in live brain slices. Some of the important considerations for live brain slice imaging are discussed next.

1. Maintaining Brain Slices on the Microscope Stage

Factors that seem to be critical for maintaining healthy brain slices on the microscope stage include temperature, pH, glucose levels, and oxygenation (Dailey, 1999). Slice physiologists have long known that oxygen deprivation can have severe effects on physiological properties such as synaptic activity, although CNS tissues from developing mammals seem to be fairly resistant to hypoxia (e.g., Dunwiddie, 1981) and hypoglycemia (Crépel *et al.*, 1992).

It is not always easy to assess the health of living tissue on the microscope stage, but in the case of dynamic processes such as cell division, migration, or axon extension, one would expect that the cells perform these activities at rates near that expected based on other methods of determination. Also, one should become suspicious if the rate of activity consistently declined or increased over time when imaged. For example, exposure of fluorescently labeled axons to high light levels can reduce the rate of extension or cause retraction. In contrast, high light levels can produce a long-lasting increase in the frequency of Ca^{2+} spikes in fluo-3-labeled astrocytes in cultured brain slices. In many cases, there will not be a useful benchmark for determining phototoxic effects, but consistent changes during imaging will serve to warn the concerned microscopist. It may be worth sacrificing a few well-labeled preparations to determine if different imaging protocols, such as lower light levels or longer time intervals between images, will significantly alter the biological activity under study.

The specific requirements for maintaining healthy tissues during imaging will dictate specimen chamber design (Dailey, 1999). Two important chamber considerations are whether to superfuse the tissue with bathing medium and whether to use an open or closed chamber. The closed chamber has the advantages of preventing evaporation during long experiments and stabilizing temperature fluctuations. We found that brain slices maintained in closed chambers (volume ~1 ml) with Hepes-buffered culture medium remain viable and vigorous for about 6 h, after which point the chamber medium acidifies and cell motility declines noticeably. However, when the old chamber medium is exchanged with new medium, the cells "jump to life" again. This crude method of periodic medium exchange has supported the continuous observation of DiI-labeled migrating neuroblasts in brain slices on the microscope stage for as long as 45 h (O'Rourke *et al.*, 1992). However, when using this approach, one runs the risk of mechanically disturbing the chamber or inducing a temperature change and thereby causing a jump in focus.

Continuous superfusion provides a more reliable method of medium exchange and introduction of experimental reagents. A variety of perfusion chambers with either open or closed configurations are available (Warner Instruments, Inc., Hamden, CT). Sometimes it is necessary to design and construct very sophisticated temperature and fluid level control systems (e.g., Delbridge *et al.*, 1990). Such chambers may permit a very rapid exchange of media, which is often necessary for physiological experiments requiring a fast exchange of reagents. Programmable automated perfusion systems that permit rapid switching between one of several perfusion channels are available (Warner Instruments or AutoMate Scientific, Inc., Oakland, CA). Some experimental conditions require only relatively simple, low-cost chambers and perfusion systems. We have used an inexpensive, custom-made perfusion system to continuously superfuse tissue slices on the microscope stage for many hours (Fig. 4). The tissue can remain healthy for 20 h or more when perfused (10 to 20 ml/h) with either the culture medium (Dailey

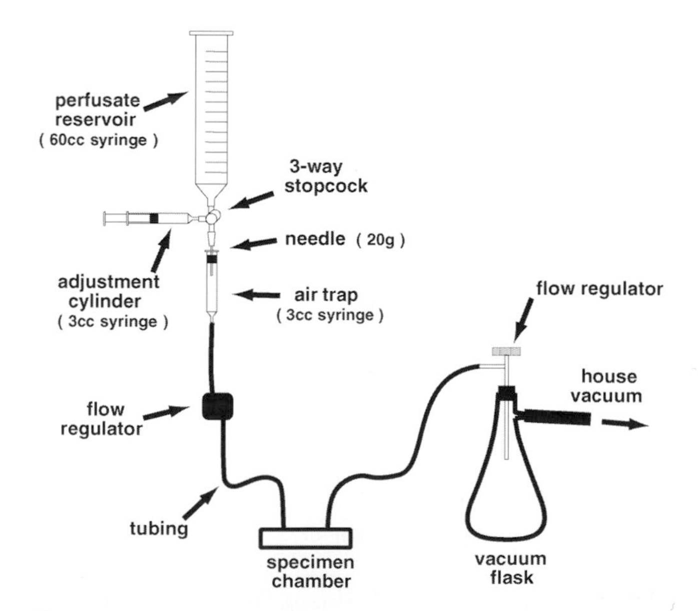

Figure 4 Perfusion system for maintaining live brain slices on the microscope stage. Perfusate (normal saline or growth medium) is delivered at ~10 ml per hour by gravity to one port in the specimen chamber and removed from another port by vacuum suction through a filter paper wick. The flow rate, which is monitored at the air trap, is controlled by constricting the tubing on the delivery side. The specimen chamber and microscope stage area are heated to ~35°C using forced air. For details of the specimen chamber, see Terasaki and Dailey (1995).

et al., 1994) or normal saline, both of which are buffered with 25 mM Hepes.

Specimen heating may be essential for some experiments, but this can induce an agonizing battle with focus stability (see later) as the chamber and stage components heat up. A sufficient period of preheating can help alleviate some of these problems. A relatively simple heating device can be constructed by modifying a hair dryer to blow warm air onto both the chamber and the stage (Dailey, 1999). It is also important to monitor the temperature of the perfusing medium very near the specimen. Low-cost microprocessor temperature controllers that reduce fluctuations in the heating/cooling cycle are available (Omega Engineering, Stamford, CT).

2. Depth of View

The goal of studies in tissue slices is to examine biological structure and physiology within a complex cellular environment that approximates that found *in situ.* In the case of live brain slices, it is generally desirable to image as far from cut tissue surfaces as possible to avoid artifacts associated with tissue damage. For example, the cut surfaces of developing brain tissues contain an abnormal cellular arrangement that includes a plethora of astrocytes and microglia, as well as a mat of growing neuronal processes. Although time-lapse imaging of these regions provides striking footage of glial cell movements, proliferation, and phagocytosis (Smith *et al.,* 1990; Dailey and Waite, 1999; Stence *et al.,* 2001), one has to be cautious when drawing conclusions about the normalcy of these events in relation to mature, intact brain.

With oil-immersion lenses, useful fluorescence images seem to be limited to a depth of 50 to 75 μm or so into the tissue. When imaging deep (>50 μm) within tissue, spatial resolution can suffer from several factors, including (1) weak staining of cells due to poor dye penetration; (2) light scatter by the tissue components; and (3) spherical aberration.

The first problem (weak staining) can be overcome if the dye is injected into the tissue with a minimum of disruption. Also, as noted previously, the loading of some dyes may be enhanced by culturing the tissue briefly. Conceivably, light scatter by the tissue can be minimized by using longer wavelength dyes. Imaging at longer wavelengths may also reduce phototoxic effects since the light is of lower energy. Finally, the problem of spherical aberration, which is exacerbated when imaging through various media with differing indices of refraction, is improved by using water-immersion objective lenses (see Pawley, 1995). Fortunately, microscope manufacturers have been responsive to the need for well-corrected water-immersion objective lenses with long working distance and high numerical aperture.

3. Signal-to-Noise and Spatiotemporal Resolution

When imaging any dynamic biological events, there is generally a trade-off between the signal-to-noise ratio ($R_{s/n}$) (which affects the spatial resolution) and the temporal resolution. The demands of high spatial resolution (high $R_{s/n}$) restrict temporal resolution. $R_{s/n}$ is proportional to the total number of photons collected. Thus, for scanning confocal imaging systems, higher $R_{s/n}$ can be achieved by: (a) intensifying the fluorescence staining; (b) increasing the incident illumination; (c) improving the collection efficiency of the system; or (d) increasing the absolute number of photons collected by lengthening the dwell time on a single-pixel, line, or frame-by-frame basis.

A primary concern when imaging live, fluorescently labeled cells is the photon collection efficiency of the optical system. This is especially critical when imaging dynamic processes, such as axon or dendrite extension, over long periods of time. Too much light can quickly halt growth. Systems with a higher collection efficiency will afford lower light levels, thus permitting more frequent sampling or experiments of longer duration before photodamage ensues.

A common means of improving the resolution is to average successive images, or frames. This can certainly increase the $R_{s/n}$, but if cell structure or physiology changes rapidly, frame averaging can "smear" the data in both the spatial and the temporal domains. It may be better to increase pixel dwell time to improve $R_{s/n}$, although this can also induce temporal distortion within single images such that the top and bottom portions of a scanned image are collected several seconds apart (see later discussion).

It should be noted that $R_{s/n}$ will also drop off as one focuses into the thick tissue specimen. Consequently, it is generally necessary to increase the incident (excitation) light to maintain a comparable spatial resolution when collecting three-dimensional data from thick tissue specimens such as brain slices. This problem can be compounded by weaker staining of structures located in deep portions of tissues.

4. Focus Drift

The thin optical sections produced by the confocal microscope reduce out-of-focus flare and improve resolution over standard wide-field optical imaging methods. However, with such a shallow depth of focus, even very small changes in the position of the microscope objective relative to the structures of interest within the specimen can create problems. This is particularly evident when imaging thin, tortuous structures such as axons or fine dendritic processes within live brain slices. A moving focal plane can, for example, give one the deceiving impression of axon or dendrite extension or retraction. This problem is compounded when imaging cells and cell processes that are actively moving within tissues.

One obvious approach is to image the cells in four dimensions (3D × time) (Fig. 5). Such "volume imaging" can keep structures in view in the face of minor tissue or stage movements; it is also helpful for monitoring axonal and dendritic

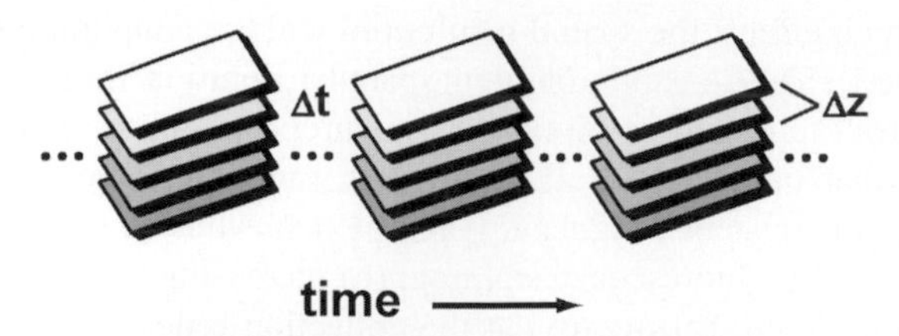

Figure 5 Protocol for time-resolved 3D confocal imaging of fluorescently labeled cells in thick brain slices. *z*-axis stacks of optical sections, spaced in the axial dimension by distance *z*, are collected at time intervals (*t*). The optical sections have a finite thickness that varies with the microscope objective and the detector pinhole configuration. Typical axial spacing of *z*-axis images is 0.8 to 2 μm. For following relatively slow processes such as cell migration and axon and dendrite extension, stacks of 5 to 20 images are collected at 5- to 10-min intervals.

processes that grow through one focal plane to another. Our strategy has been to image with the detector aperture (confocal pinhole) in an open configuration, which translates into a pinhole size roughly two to four times the Airy Disk. Although this reduces the axial resolution somewhat, it has the dual advantage of achieving a higher signal-to-noise ratio at a given illumination intensity as well as broadening the optical section. On the custom-built microscope, the open pinhole configuration gives an apparent optical section thickness of about 3 μm when using a 40×/1.3 objective. Thus, for each time point, images are collected at axial step intervals of 2 to 3 μm. In the case of the Leica TCS NT system, the axial steps are less than 1 μm when using a 63×/1.2 water lens. The guiding principle here is to space the image planes in the axial dimension to maximize the volume of tissue imaged but not to lose continuity between neighboring optical sections. When stacks of images are collected at 3- to 5-min intervals at power levels of ~50–75 μW (near the back aperture of the objective), DiI- or GFP-labeled cells do not appear to suffer phototoxic effects and can be imaged continuously for over 20 h (40× objective, zoom 2). Image stacks can be later recombined using a maximum brightness operation (Image-J; Scion Corp., Frederick, MD) so that segments of thin (<1 μm) axons and dendrites that course through the various *z*-axis images appear contiguous in a single image. Unfortunately, even when *z*-axis stacks of images are collected, tissue movements can be so severe as to necessitate a continuous "tweaking" of the focus to keep cells within the stack. Thus, it is helpful to store the recently collected images in such a way that they are quickly accessible and can be reviewed on the fly to make corrective focus adjustments. Alternatively, one can keep track of focus drift by simply marking in-focus features on an acetate sheet taped to the computer monitor. Corrective focus changes can be made as needed during the time-lapse experiment to keep the features within a given image plane.

Ideally, one would like an automated means of maintaining the desired plane of focus, especially for long time-lapse imaging sessions. Although there are several autofocus methods that work for simple specimens, imaging structures in tissues presents significant challenges because there is no single image plane within the specimen on which to calculate focus. One approach is to calculate focus from a stationary reference plane, such as the surface of the coverslip (Ziv and Smith, 1996). This may work as long as the tissue volume of interest maintains a fixed relationship to the coverslip, which is not always the case.

5. Photodynamic Damage

For observations on live, fluorescently labeled cells, the problem of photodamage is generally a limiting factor for achieving high spatial and temporal resolution. A trade-off exists between illumination intensity and resolution. Stronger incident illumination intensities enable faster data collection at a given spatial resolution.

In neural tissues, some of the most obvious signs of phototoxicity include a decrease in intracellular vesicle traffic, axon and dendrite retraction, blebbing of labeled processes and cell bodies, triangulation of dendritic branch points, and sustained rises in intracellular Ca^{2+}.

As discussed earlier, there are theoretical reasons for thinking that use of the longer wavelength dyes, such as DiD, will reduce photodamage. Indeed, it has been reported that DiD seems to cause less collateral damage to living erythrocytes than DiI (Bloom and Webb, 1984). This is consistent with the idea that the longer wavelength light used to excite DiD (i.e., 633 or 647 nm) is of lower energy than the excitation lines for DiI. However, our preliminary time-lapse observations suggest that axonal and dendritic elements in brain slices stained with DiD are *more* susceptible to photodamage than those stained with DiI. Thus, it is important to carefully assess each dye to determine its properties under the particular experimental conditions under which it will be used.

6. Data Management

Imaging tissues in three dimensions over time can produce very large data sets. For example, collection of 20 optical sections (512 × 512-pixel arrays) at 5-min intervals over a 10-h-long experiment generates 600 Mbytes of image data per channel. It is now feasible and often useful to collect two or more image channels simultaneously (e.g., two fluorescence and one transmitted light channels), generating gigabytes of image data per day. Fortunately, storage and retrieval of mass digital data are not nearly as problematic as they were just a few years ago. For example, recordable compact discs (CDs) offer a highly versatile, cost-effective, and nearly universal means of archiving large amounts of image data. Recordable digital video discs, with a severalfold larger storage capacity, may soon supplant CDs as a common medium of choice for archiving large data sets.

V. Application: Mapping Neural Structure and Physiology in Developing Brain Slices

We have applied the methods for fluorescence labeling and confocal imaging described earlier to study the organization and development of neural systems in isolated brain slices. From an anatomical standpoint, this has included three-dimensional mapping of axonal and dendritic branches, as well as synaptic structures. Dynamic structural changes associated with axonal and dendritic growth and synapse formation are also being examined in order to elucidate principles of developmental plasticity. From a physiological standpoint, we have investigated patterns of neuronal and glial activity in both acutely isolated and organotypically cultured brain slices from developing rodents. These examples serve to demonstrate the spatial and temporal resolution available for mapping neural structure and physiology in semi-intact mammalian brain tissues.

A. Organization and Growth of Axonal Fibers

As a model system, we have examined the organization and development of the hippocampal mossy fiber system. The mossy fibers are the axons of dentate granule cells that synaptically contact pyramidal neurons in area CA3 (Henze *et al.*, 2000). The organization of the mossy fiber projection was examined by staining with vital fluorescent membrane dyes. Injection of DiI or DiO into the dentate gyrus of live hippocampal slices labels within a few hours the full extent of the mossy fibers that project into area CA3. In slices taken from mature animals, one can see individual giant varicosities that are spaced along the length of mossy fibers (Fig. 6). These varicosities likely correspond to the giant presynaptic terminals that impinge on the CA3 pyramidal cell dendrites (Blackstad and Kjaerheim, 1961; Chicurel and Harris, 1992).

Confocal 3D reconstruction was used to examine the three-dimensional organization and development of mossy fiber axons and their giant terminals. Individual fibers were found to have serpentine pathways through the complex tissue environment, and individual synaptic varicosities could be clearly resolved (Fig. 7).

We next used time-lapse imaging to explore the dynamics of mossy fiber growth in developing hippocampal brain slices. Fibers in live tissue slices were labeled with DiI or DiO as described earlier, then imaged over a period of several hours. Single-focal-plane images were collected at 1- or 2-min intervals or, more frequently, stacks of five to seven images were collected at 5-min intervals. Based upon such 4D imaging movies, mossy fiber axons were found to extend within target regions at rates of about 10–30 μm/h (Dailey and Smith, 1993; Dailey *et al.*, 1994). Growth was often saltatory, showing phases of rapid growth that were interrupted by short quiescent periods, and axons reaching the edge of the fiber bundle frequently went through several rounds of retraction and redirection. In some cases, new branches sprouted laterally from the shafts of axons (Fig. 8).

B. Structure and Development of Neuronal Dendrites

Most excitatory synaptic connections in the brain are formed on dendritic projections called spines. We used fluorescence labeling and confocal imaging to address the development and dynamics of neuronal dendrite branches and spines (Dailey and Smith, 1996). To examine dendrites of pyramidal neurons in slices, fluorescent membrane dye is injected near the region of the basal dendrites (Fig. 9).

When dye is injected into slices taken from early postnatal rats, the elaborate organization of neuronal cell bodies and dendritic arbors is revealed. The three-dimensional organization of dendrites is best appreciated in stereo-pair images collected from a tissue volume at a relatively low magnification (Fig. 10). At higher magnification, confocal images reveal the fine microstructure of dendrites. Dendritic branches are studded with numerous filopodia and spine-like protrusions that extend into the surrounding tissue in all directions. These spiny protrusions are sites of synaptic termination by afferent axons.

Time-lapse imaging of fluorescently-labeled dendrites in developing tissue slices demonstrated that dendritic microstructure is quite dynamic (Dailey and Smith, 1996; Marrs *et al.*, 2001). Developing dendritic branches bear a combination of fleeting filopodia-like protrusions and relatively stable spine-like structures (Fig. 11). Since dendritic spines typically correspond to axonal synaptic input, the dynamic changes in spiny structures on developing dendrites may be a morphological correlate of synaptic plasticity. Based on time-lapse imaging, there is now substantial evidence that dynamic dendritic filopodia are precursors to more stable dendritic spines (Ziv and Smith, 1996; Friedman *et al.*, 2000; Marrs *et al.*, 2001). The extent to which dendritic spines and the synaptic structures associated with spines are dynamic in more mature tissues remains an open question (Okabe *et al.*, 1999).

In the future it will be of interest to examine directly the interaction and dynamic changes of pre- and postsynaptic components at identified sites of synaptic contact. This may be facilitated in experiments in which axonal and dendritic processes are labeled with different dyes to facilitate their identification (Fig. 12).

C. Organization of Neural Synapses

Mapping neuronal connections would be simpler if there were unique markers of the cells involved and of the synaptic contacts between each particular set of neurons. Some synaptic systems in the CNS do, in fact, have unique structural

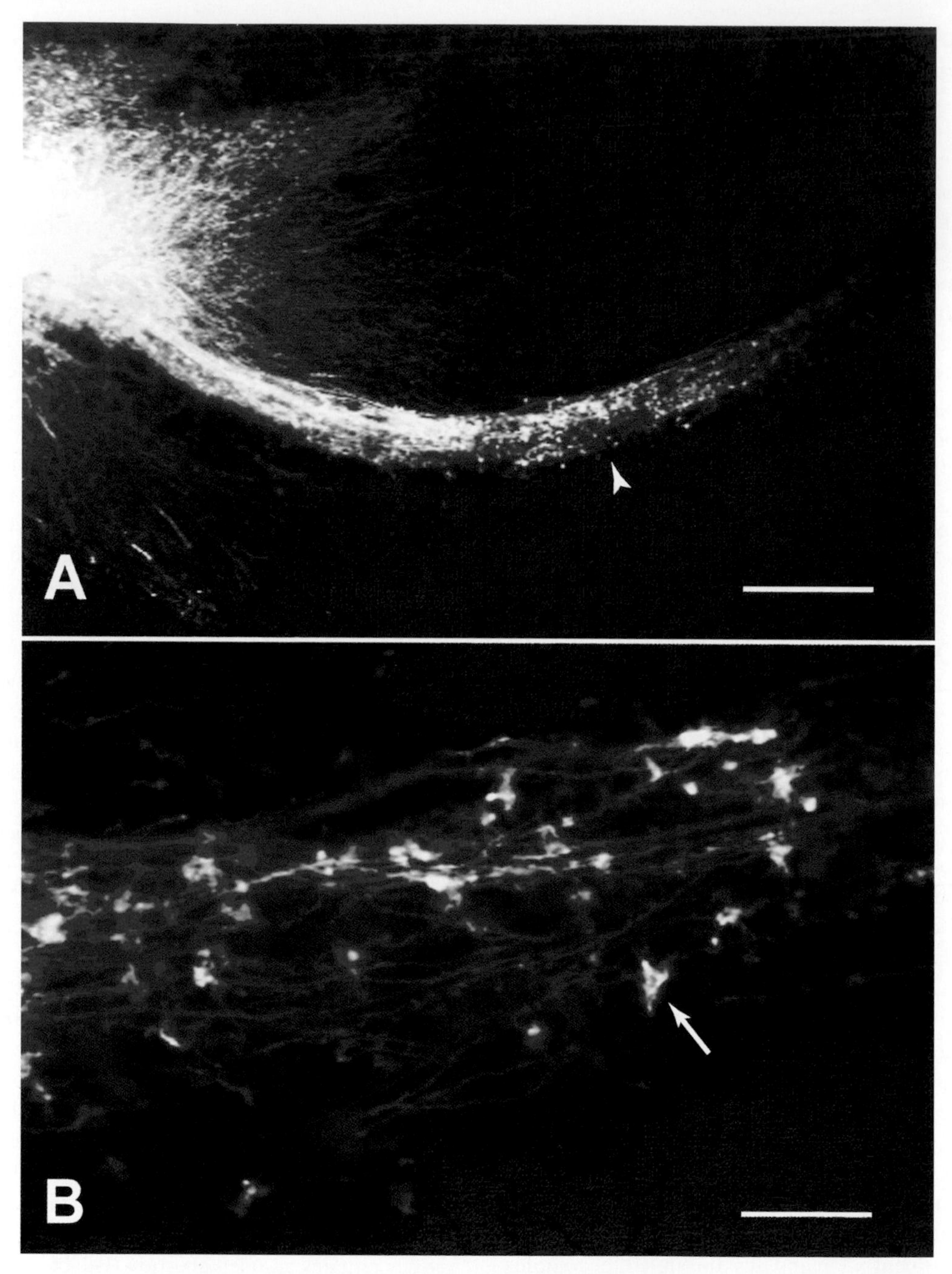

Figure 6 DiI-labeled mossy fibers in a live brain slice prepared from a 3-week-old rat. (**A**) Low-magnification view showing site of dye injection (*) near the dentate gyrus, which is the location of the granule cell bodies and the source of the mossy fibers. Note the tight bundle of labeled mossy fibers (arrowhead) that extend from the dentate gyrus into area CA3. Scale bar, 100 μm. (**B**) High-magnification view showing individual fibers and varicosities. The large, *en passant* varicosities (arrow) correspond to giant mossy fiber synaptic terminals. Scale bar, 25 μm.

features that facilitate their mapping. The mossy fiber synapses, for example, are characterized by giant (>2 μm) presynaptic terminals and large, complex postsynaptic spines that are readily identifiable even by light microscopy. We have examined the three-dimensional organization of mossy fiber terminals in brain slices using the antibody to syn-I, described earlier. Immunohistochemical staining of thick tissue slices isolated from formaldehyde-fixed rat brain reveals mossy fiber giant synaptic terminals (Fig. 13).

We also have used immunohistochemical techniques to assay the development and organization of mossy fiber synaptic contacts in brain slices cultured for 1–2 weeks

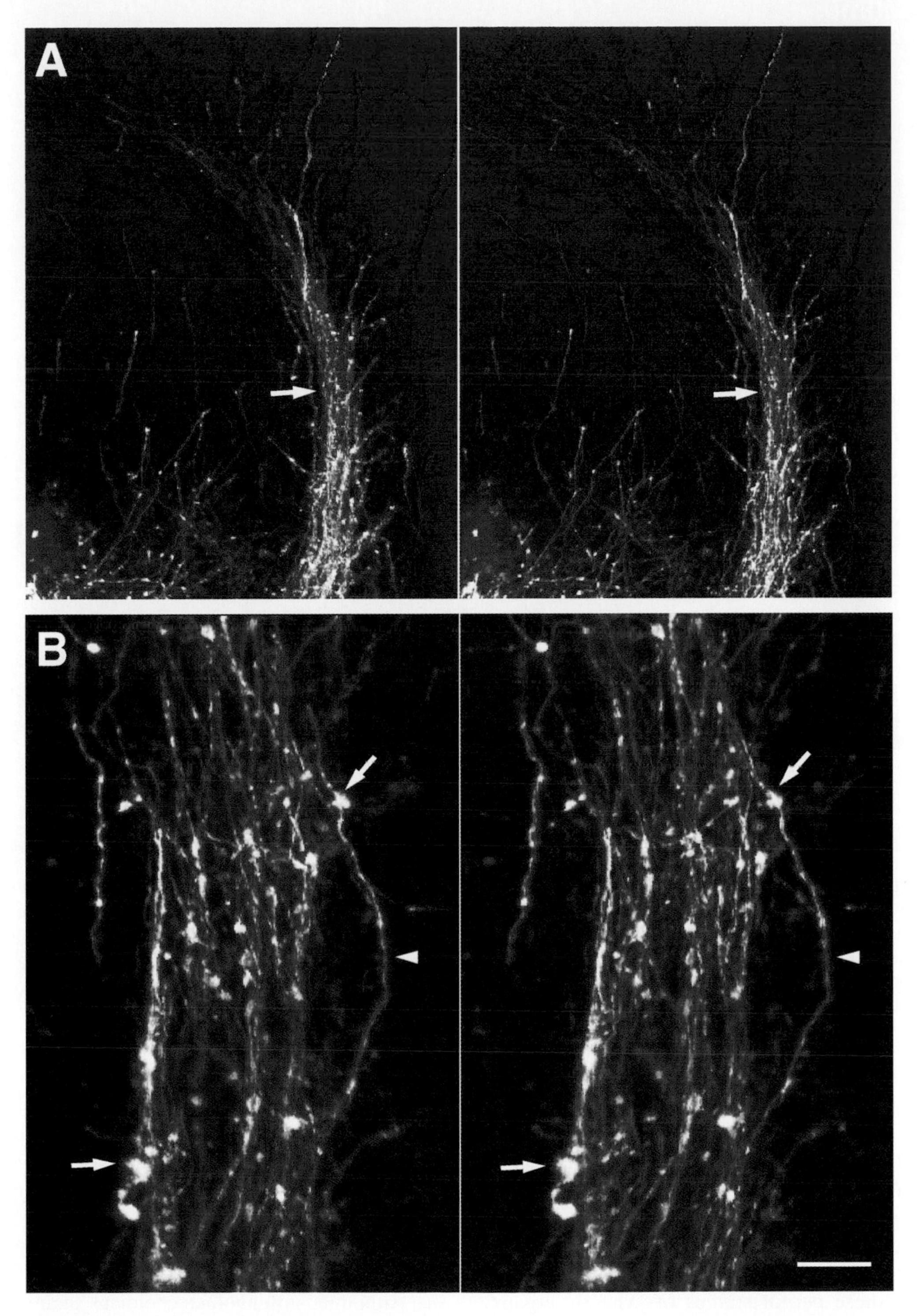

Figure 7 Three-dimensional organization of DiI-labeled mossy fibers in brain slices from developing rat. (**A**) Stereo-pair images of the mossy fiber bundle (arrow) in a slice from a P5 rat, at a time when the mossy fibers are just growing out and forming synaptic contacts with CA3 pyramidal neurons. Depth of view is 30 μm. (**B**) Stereo-pair images of mossy fibers in a live slice prepared from a P12 rat. Note the tortuous course of individual axons (arrowhead) within the mossy fiber bundle. The giant *en passant* varicosities (arrows), corresponding to synaptic terminals, can be seen along the length of the mossy fiber axons. Depth of view is 15 μm. Scale bar, 50 μm for A, 10 μm for B.

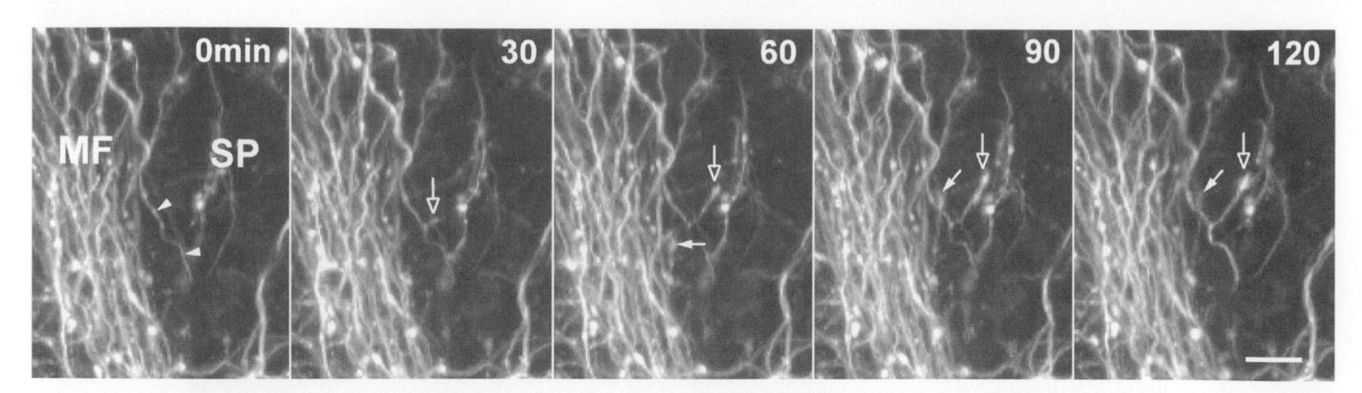

Figure 8 Time-lapse confocal imaging of growth of mossy fiber axons within the target region. The mossy fibers (MF) were labeled by injection of DiI into the dentate hilus. Note sprouting of a branchlet (open arrow) from the side of a mossy fiber axon (arrowheads) into the layer of pyramidal cell bodies (SP). The growing tip of another mossy fiber (filled arrow) can be seen to extend past the first fiber. These are extended-focus images composed of five optical sections collected at 2.5-μm axial step intervals. The slice (same as shown in Fig. 7A) was acutely prepared from a P5 rat. Elapsed time is shown in minutes. Scale bar, 20 μm.

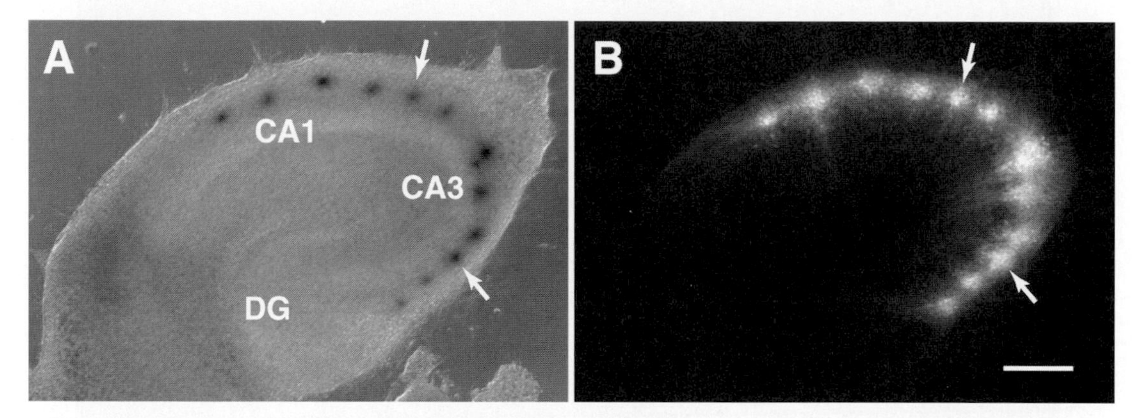

Figure 9 Vital staining of neuronal cell bodies and dendrites by injection of fluorescent membrane dye into isolated brain slices. (**A**) Transmitted light image showing organization of a live hippocampal slice (P3 rat). Black dots (arrow) in the stratum oriens of areas CA3 and CA1 correspond to sites where crystals of fluorescent dye were inserted. Dye inserted at these sites is picked up by basal dendrites of pyramidal cells, labeling cell bodies and the entire extent of the dendritic arbor (see Fig. 10). DG, dentate gyrus. (**B**) Fluorescence image, corresponding to A, showing sites of fluorescent dye (DiD) labeling (arrows). Scale bar, 250 μm.

(Dailey *et al.,* 1994). Three-dimensional imaging of syn-I-stained tissues revealed that the giant synapses maintain a stereotypical distribution along the apical dendrites of CA3 pyramidal neurons in cultured slices, although individual terminals appear slightly smaller and less complex than their *in vivo* counterparts (Fig. 13).

Antibodies against postsynaptic proteins also have been developed, and these permit studies of synaptic organization as well (Qin *et al.,* 2001). For example, we have used antibodies against a postsynaptic density protein, PSD95, which is a PDZ-domain scafold protein that serves to organize and link neurotransmitter receptors to the postsynaptic cytoskeleton (Cho *et al.,* 1992; Kistner *et al.,* 1993; Sheng and Sala, 2001). Immunohistochemistry using the anti-PSD95 antibody in hippocampal tissues highlights regional differences in the organization and size of synaptic structures (Fig. 14). The large and complex postsynaptic structures at mossy fiber synapses are especially evident in the stratum lucidum, where mossy fiber axons normally course, but they are also evident at lower density in the pyramidal cell body layer. In contrast to area CA3, immunostaining in area CA1 shows only small, typical postsynaptic structures.

Although immunohistochemical staining can provide information on the size and density of synaptic structures in brain tissues, they yield little information on the distribution of synaptic structures in relation to individual cells. To address questions of synaptic organization on individual cells, we utilized gene gun transfection and expression of a GFP–PSD95 fusion protein (Qin *et al.,* 2001). To label postsynaptic structures, we used a GW1 vector containing enhanced GFP fused in frame to the C-terminus of PSD95,

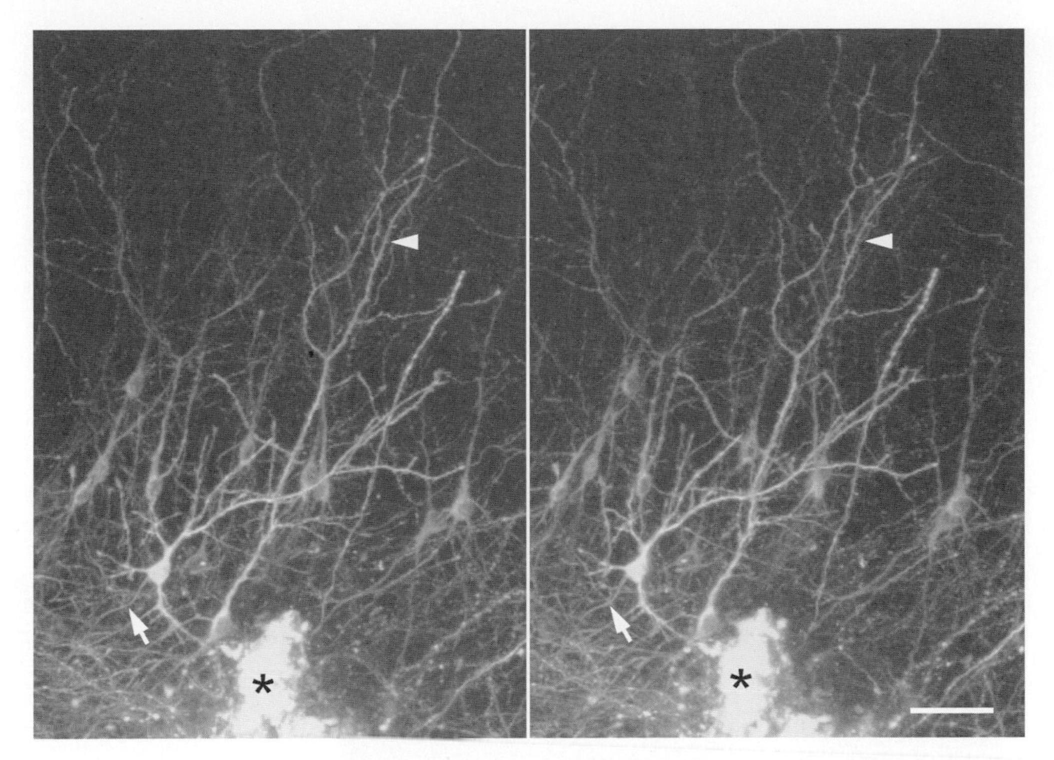

Figure 10 Stereo-pair images showing DiD-labeled CA1 pyramidal neurons near a site of dye crystal insertion (*). Note 3D organization of labeled cell bodies and apical dendrites (arrowhead), which branch and course throughout the thickness of the brain slice. In one case, the axon (arrow) can be seen emerging from the pyramidal cell body. Depth of view is 40 μm (20 images at 2-μm-step intervals). Slice is from a P6 rat and cultured for 5 days. Scale bar, 50 μm.

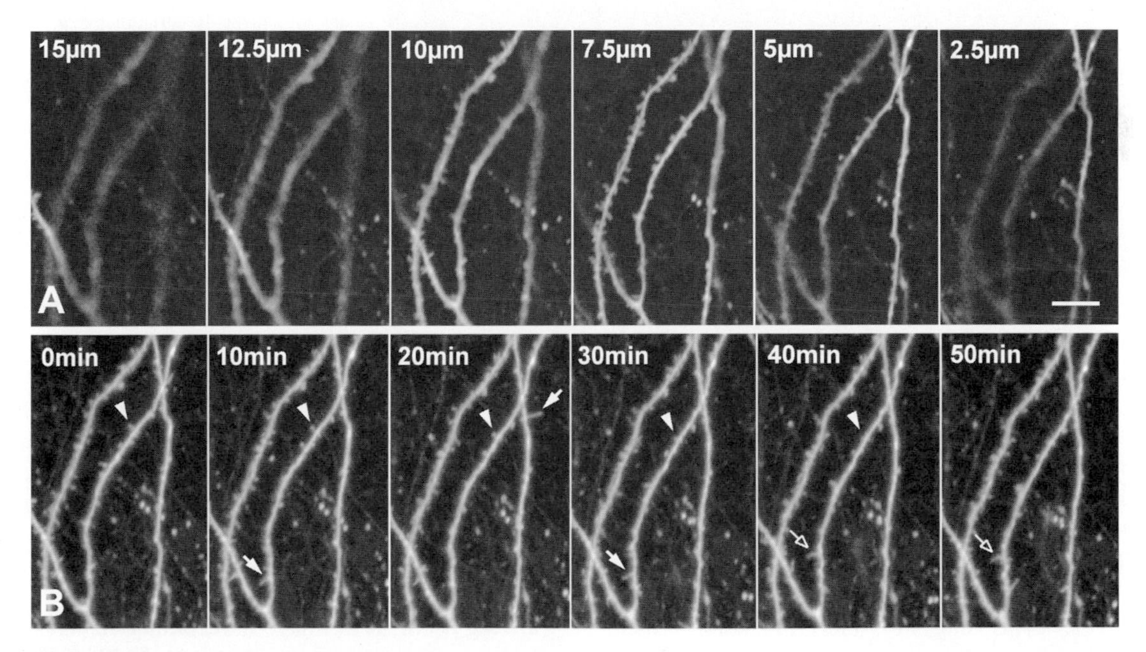

Figure 11 Imaging dendritic spine dynamics in a live brain slice from developing rat. (**A**) Through-focus series of a portion of a DiD-labeled dendrite from a CA1 pyramidal neuron in a live brain slice (P5, 7 days *in vitro*). The optical section depth is indicated in micrometers. These images were collected with the confocal pinhole aperture in the fully open configuration to maximize the signal detected. As a result, some out-of-focus flare is evident in the various optical sections. Scale bar, 25 μm. (**B**) Time-lapse sequence of same field as in A showing dynamics of dendritic spines. These are extended-focus images made by combining six optical sections collected at the depths indicated earlier. Note the shortening of a spine (arrowhead) and transient extension of filopodia-like protrusions (filled and open arrows). Such changes in dendritic structure may reflect plasticity in synaptic function. Time is shown in minutes.

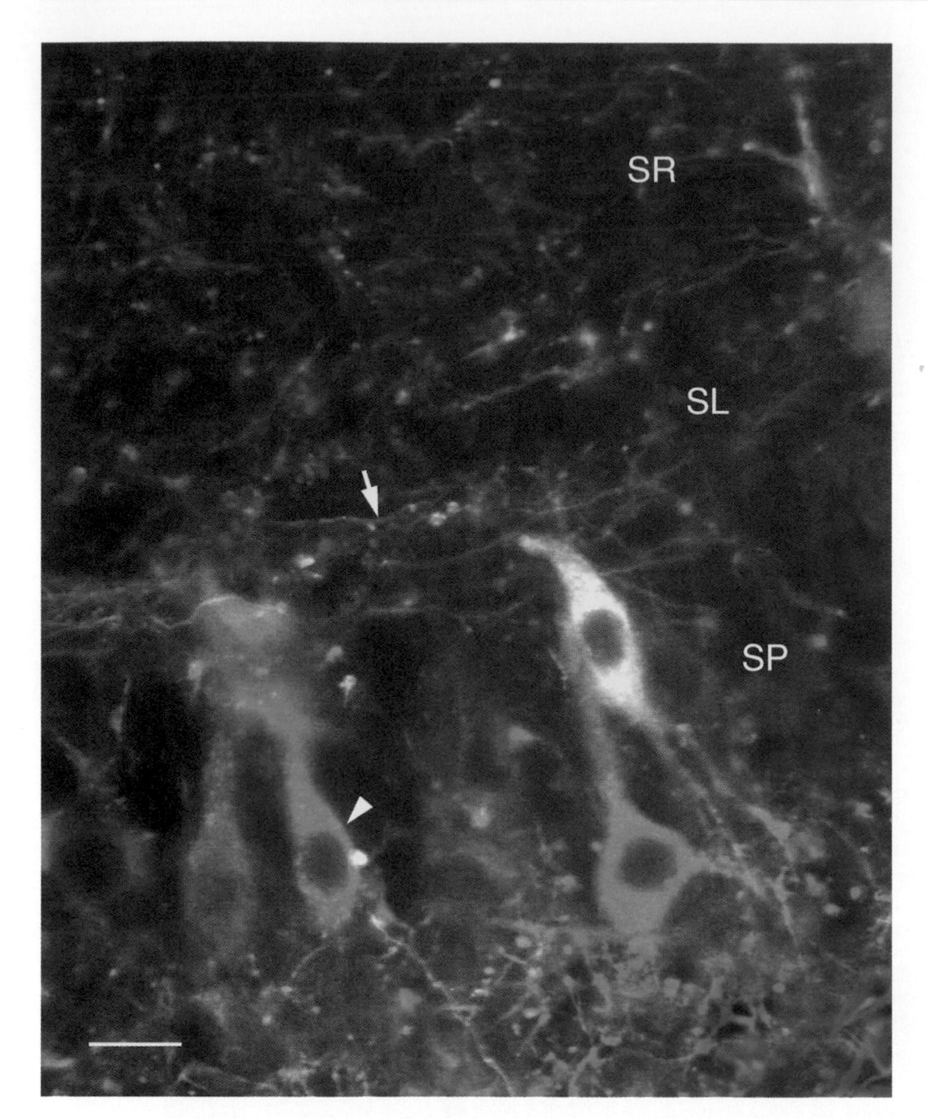

Figure 12 Double labeling showing mossy fibers (red) and CA3 pyramidal neurons (green) in a live brain slice (P4, 1 day *in vitro*). Mossy fibers, which normally course within the stratum lucidum (SL), were labeled by injection of DiI into the dentate hilus. Pyramidal neurons, whose cell bodies are located in the stratum pyramidale (SP), were labeled by injection of DiO into the stratum oriens of area CA3. Such double-labeling experiments can reveal the interaction of axonal projections and target neurons. SR, stratum radiatum. Scale bar, 25 μm.

under control of a human cytomegalovirus promoter (gift from D. Bredt, University of California at San Francisco).

Expression of the fusion protein, which targets normally to synapses, labels all synaptic sites in the transfected neurons. Confocal analysis and three-dimensional reconstruction of transfected tissues yield striking images of synaptic localization. Such data can reveal dramatic differences in the organization of synapses, on neighboring cells and even on different portions of the same cell, thus helping to define unique synaptic domains and identify distinct neuronal cell types (Fig. 15).

Notably, cells expressing GFP and GFP-fusion proteins can be examined in the living state, enabling studies on the dynamic formation and remodeling of synaptic structures (Okabe *et al.,* 1999; Marrs *et al.,* 2001). This general approach can be used to study other synaptic components, including signaling molecules (e.g., CaM kinase II) and neurotransmitter receptor subtypes (e.g., NMDA, AMPA, and metabotropic receptors). Moreover, it should be possible to express and distinguish more than one fluorescent protein in a single cell, allowing for "whole-cell mapping" of functionally distinct types of synapses.

D. Intracellular Ca2+ Transients in Neurons and Glia

Brain mapping encompasses both structure and physiology. Mapping multicellular patterns of neural activity in brain tissue at high spatial and temporal resolution continues to be an important goal in neurobiology. One method that is proving useful is to image intracellular changes in calcium. Such changes are a consequence of electrical activity in neurons (Ross, 1989; Sinha and Saggau, 1999; Smetters *et al.,* 1999), and glia also have been found to generate intracellular Ca^{2+} signals that reflect cellular physiology.

Changes in intracellular Ca^{2+} were assessed by imaging a fluorescent, Ca^{2+}-sensitive dye (fluo-3 or fluo-4). To image multicellular patterns of activity, it was necessary to use a

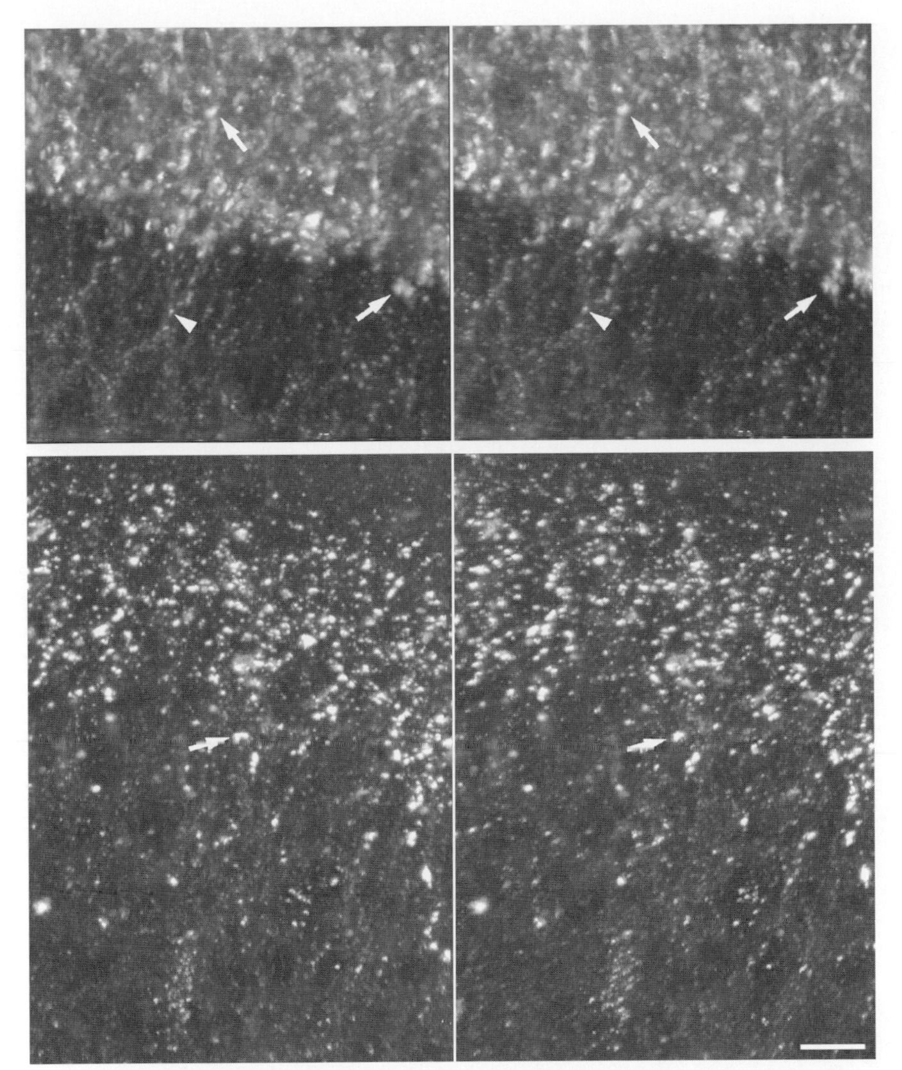

Figure 13 Three-dimensional organization of synaptic terminals in hippocampal brain slices revealed by synapsin-I (syn-I) immunostaining. Syn-I is a synaptic vesicle-associated protein found in virtually all CNS presynaptic terminals (DeCamilli *et al.,* 1983a,b). Indirect immunohistochemistry with anti-syn-I antibodies and a fluorescein secondary antibody was used to label synaptic terminals in these fixed and permeabilized brain slices. The large fluorescent punctae correspond to mossy fiber giant synaptic terminals in hippocampal area CA3. (**Top**) Stereo-pair images of a syn-I-stained slice from a P13 rat. Giant synaptic terminals (arrows) occupy a dense band along the apical aspect of the pyramidal cell body layer. Small synaptic boutons (arrowhead), corresponding to non-mossy-fiber terminals, are seen within the cell body layer. Depth of view is 30 μm (15 images at 2-μm axial step intervals). (**Bottom**) Stereo-pair images from an organotypic brain slice prepared from a P7 rat and cultured for 14 days. Giant mossy fiber synapses (arrow) develop and maintain a normal distribution throughout the thickness of the cultured slices. Depth of view is 40 μm. Scale bar, 20 μm.

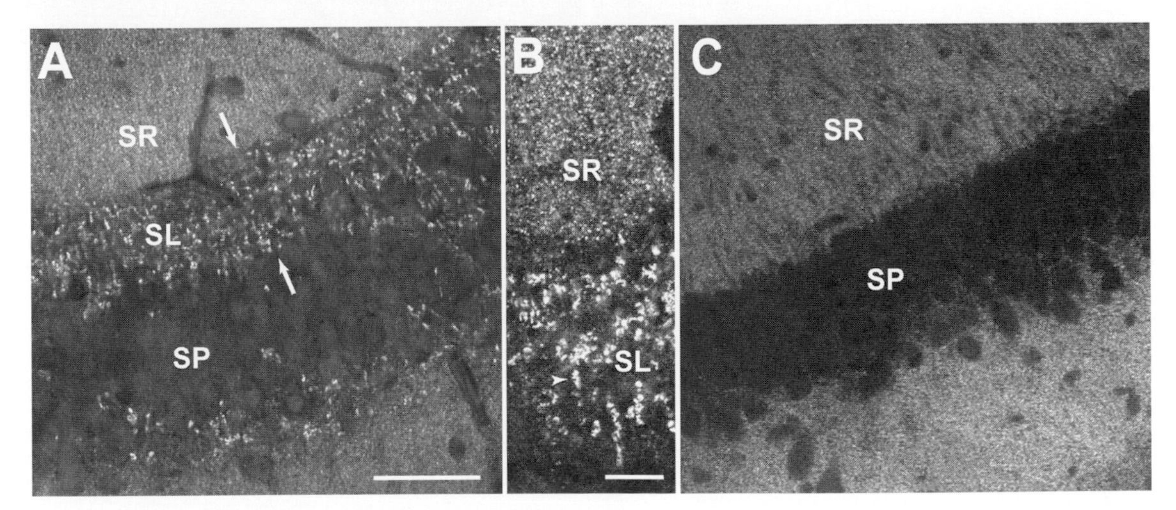

Figure 14 Use of antibodies to study the organization of postsynaptic structures in brain tissues. Hippocampal tissue slices (from P15 rat) were fixed and labeled with antibodies to PSD95, a prominent postsynaptic density scaffold protein at excitatory synapses. (**A**) Labeling in area CA3 reveals differences in staining patterns within the different strata. The stratum lucidum (SL) (between arrows), corresponding to the primary mossy fiber axon tract, contains many giant postsynaptic density (PSD) clusters. However, some giant clusters are also found along basal dendrites in the stratum pyramidale (SP). (**B**) Higher magnification view showing large postsynaptic clusters (arrowhead) in the SL. Only small synapses are evident in the stratum radiatum (SR). (**C**) In contrast to area CA3, no large synaptic clusters are apparent in area CA1. Scale bar, 50 μm for A and C, 20 μm for B.

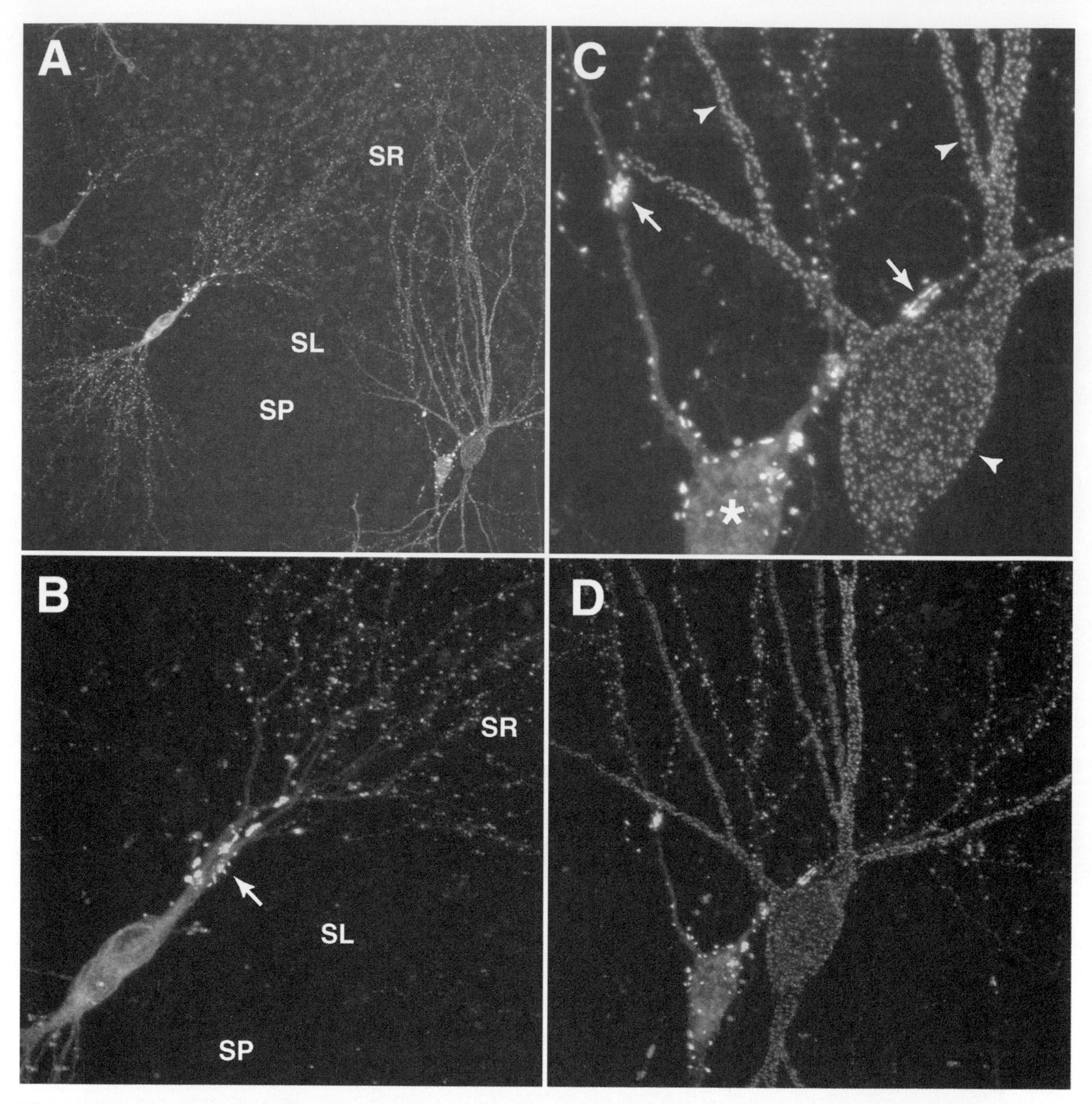

Figure 15 GFP labeling of synapses in transfected neurons in a hippocampal slice. Using a gene gun, neurons were transfected with a plasmid vector encoding a GFP–PSD95 fusion protein to label postsynaptic densities (PSDs). For details, see Qin *et al.* (2001). (**A**) Five neurons were transfected in this field of view from area CA3. (**B**) Higher magnification, stereo image of the cells above showing small postsynaptic structures on secondary dendrites in the SR and large postsynaptic structures on primary apical dendrites in the SL. The large PSD clusters (arrow) correspond to synaptic contacts of the mossy fibers. Use red–green stereo glasses (available from the author) to view depth in the image. (**C**) Higher magnification view of cells shown at lower right in A. Note the differences in morphology as well as patterns of GFP–PSD95 localization in these two adjacent neurons. The cell on the left (*) bears large GFP–PSD95 clusters (arrows) typical of mossy fiber synapses on a CA3 pyramidal neuron (cf. A and B). The cell on the right, which has a distinctive morphology with long, thick dendrites bearing few branches, is densely covered with small synapses but no large clusters. (**D**) Stereo image of cells shown in C. These images demonstrate the feasibility of mapping all morphological synaptic structures on individual neurons, as well as distinguishing neurons on the basis of synaptic organization.

bulk loading procedure (using the membrane-permeant, AM–ester forms of dyes described earlier) so that most or all of the cells in the tissue slice were labeled. This method provided suitable labeling of neurons and glia in live hippocampal slices prepared from developing rat. Time-lapse imaging showed that developing pyramidal neurons exhibit spontaneous Ca^{2+} transients that reflect neural synaptic activity (Dailey and Smith, 1994). Although most pyramidal neurons appear to be independently active, occasionally there are small groups of synchronously-active neurons (Fig. 16). Such patterns of activity may reflect electrical coupling between neurons during development. The significance of these activity patterns remains unclear, although similar patterns of activity are found in the developing neocortex and are thought to help set up the organization of chemical synaptic contacts between groups of neurons (Yuste *et al.*, 1992).

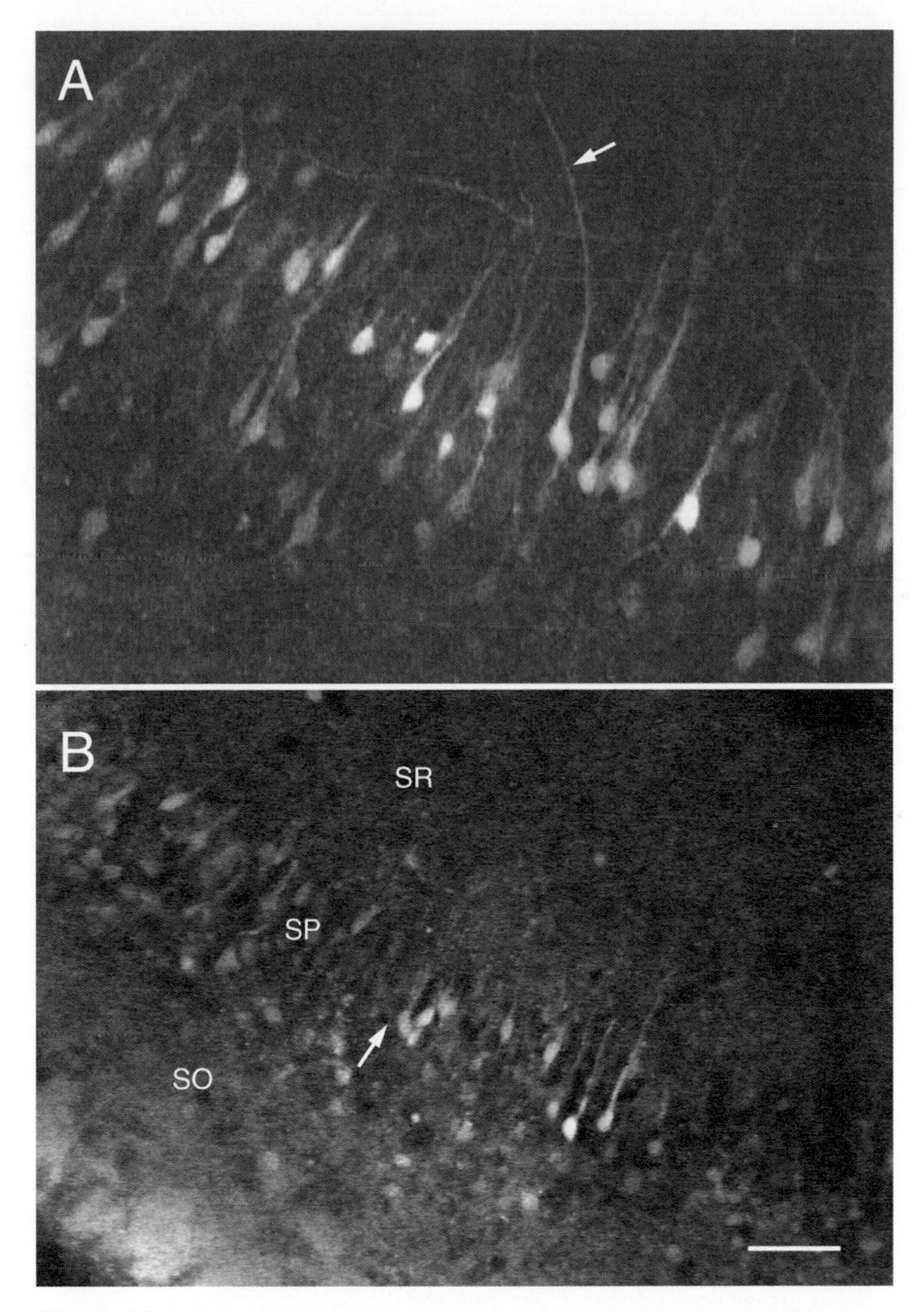

Figure 16 Patterns of spontaneous neural activity in brain slices from developing rat hippocampus. Neural activity is detected by a Ca^{2+}-sensitive fluorescent dye, fluo-3. An increase in fluorescence intensity corresponds to a rise in intracellular Ca^{2+}. Each image is a composite of three separate images taken at 12-s intervals and color coded such that different colored cells were active at different times. Both fields (A and B) show activity of pyramidal neurons in area CA1 of hippocampal slices from a P3 rat. Very few glial cells are evident at this developmental stage. (**A**) Note the high level of spontaneous Ca^{2+} activity in pyramidal cell bodies and apical dendrites. Activity of adjacent neurons is largely nonsynchronous. (**B**) Small groups of synchronously active neurons (red cells at arrow) are occasionally seen at this stage of development. This pattern of activity is likely due to the coupling of pyramidal neurons via gap junctions. Coupling between neurons is known to occur during development but diminishes as chemical synaptic connections are established. Scale bar, 25 μm for A, 50 μm for B.

A qualitatively different pattern of neuronal activity is found when hippocampal slices are maintained in culture for several weeks. When slow-scanning confocal images are collected, horizontal bands of high Ca^{2+} in neuropil and in neuronal cell bodies are seen (Fig. 17). Such patterns of neural activity probably reflect epileptiform-like activity, which is known from electrophysiological studies to develop in these cultured slices (McBain *et al.*, 1989).

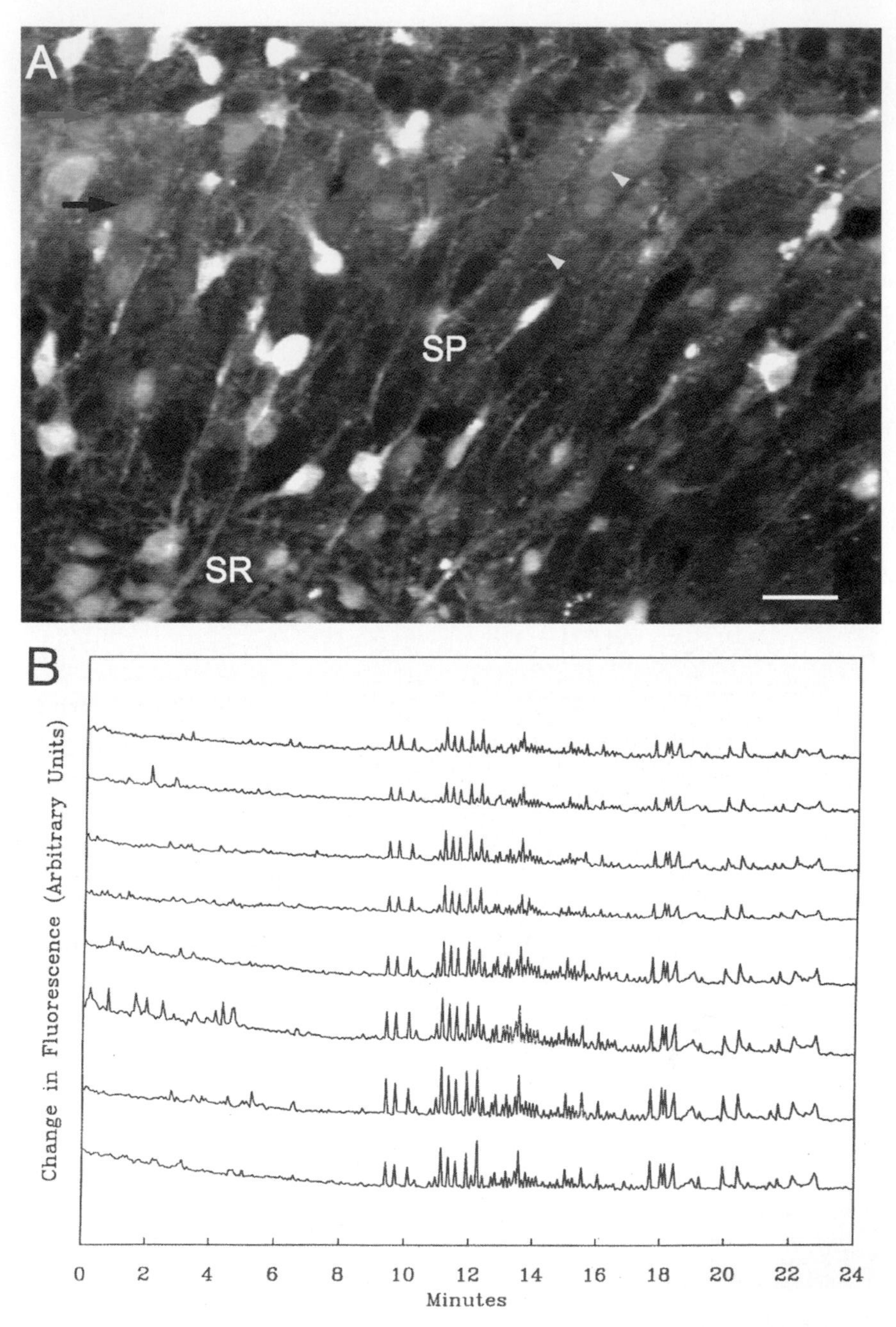

Figure 17 Patterns of spontaneous epileptiform activity in hippocampal slices detected by slow-scanning confocal imaging. (**A**) Composite image showing Ca^{2+} activity at three different time points (encoded blue–green–red, respectively) collected at 7-s intervals. At two of the three time points, a burst of synchronized activity was evident in many neurons (green and red horizontal stripes). Such patterns of activity are indicative of synchronized bursting, which is known from electrophysiological studies to occur in these brain slice cultures. The horizontal patterns are due to the slow scan rate (about 10 ms per horizontal line of resolution in this case) relative to the rate of Ca^{2+} rise in the cells. Note the high level of fluorescence near the onset of activity (arrows) followed by a gradual decline in fluorescence intensity as the scan collects data from portions of the field below. Scale bar, 25 μm. (**B**) Graphical display of spontaneous synchronized Ca^{2+} activity in neurons. Traces of fluorescence intensity over time are shown for eight neurons that were situated roughly in a horizontal line across the image field so that the time of sampling was identical for all these neurons. Images were collected at 7-s intervals. Initially, there was a low level of spontaneous activity. At about the 9-min time point, the neurons started to exhibit synchronized Ca^{2+} activity (upward spikes) that persisted for many minutes.

When the cultured brain slices are imaged at a higher time resolution with a fast scanning confocal microscope, the time course of the fast neuronal Ca^{2+} transients associated with epileptiform activity is better resolved. A comparable pattern of activity can be pharmacologically induced in cultured brain slices perfused with $GABA_A$-receptor antagonists, picrotoxin (100 μM) or gabazine (13 μM), which disinhibit brain slices (Fig. 18). These images serve to demonstrate the trade-off between high spatial and temporal resolution in scanning confocal imaging (see earlier discussion). Collection of image data at high rates generally reduces the spatial resolution because many fewer photons are collected at each point in space.

In addition to neuronal activity, imaging fluo-3-labeled brain slices with the confocal microscope reveals patterns of glial cell activity (Figs. 18C and 19). Previous studies revealed that transcellular waves of activity can pass through gap junctions connecting astrocytes (Cornell-Bell *et al.,* 1990). In brain slices, the intracellular glial Ca^{2+} activity has been linked to the local release of neurotransmitter during neural synaptic activity (Dani *et al.,* 1992). These observations suggest a dynamic, functional interplay between the neuronal and the glial networks in brain tissues (Smith, 1994; Charles, 1998; Araque *et al.,* 2001; Bezzi and Volterra, 2001).

VI. Conclusions and Future Prospects

The development of novel fluorescent probes of cellular structure and physiology (Mason, 1993; Tsien and Waggoner, 1995) has had a profound impact on studies of brain structure and function at the network, cellular, and subcellular levels. Coupled with the technical advances in high-resolution optical imaging, fluorescent markers provide a valuable set of

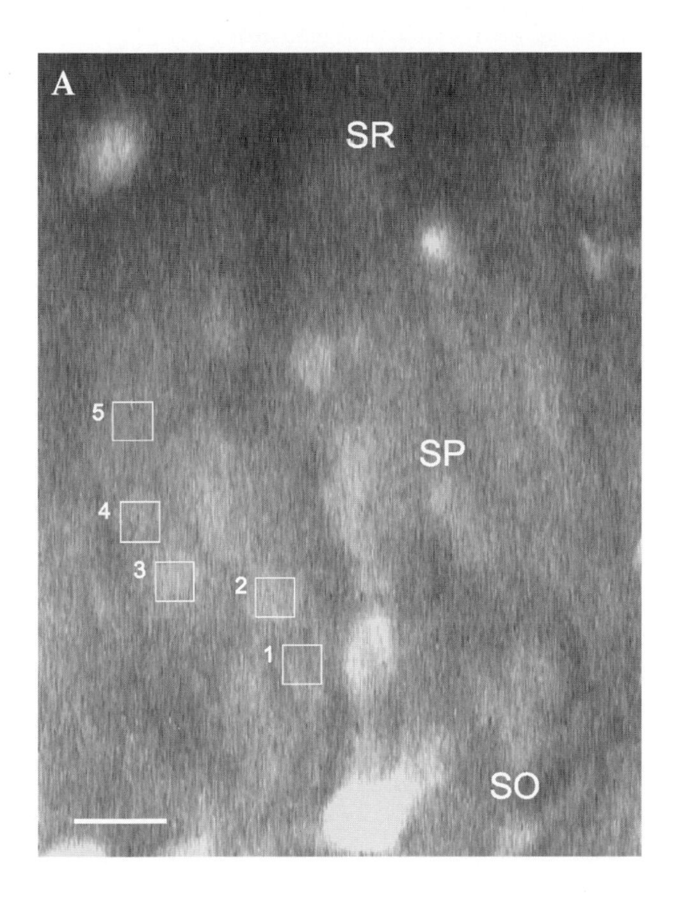

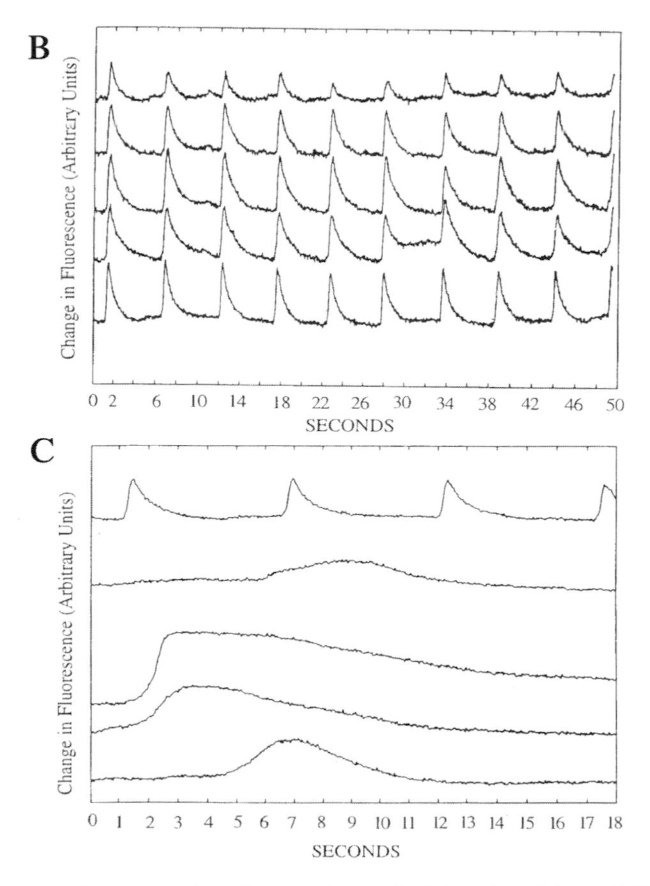

Figure 18 Patterns of picrotoxin-induced epileptiform activity in hippocampal slices detected by fast-scanning confocal imaging of intracellular calcium. The high time resolution was achieved using a confocal microscope (Noran Odyssey) that collects full-field images at video rate (30 Hz). (**A**) A single scan (nonaveraged) image showing the relatively low spatial resolution of fluo-3-labeled pyramidal neurons when imaged at high time resolution (compare with slow-scanned images in Figs. 16 and 17A). The slice is oriented such that the pyramidal cell body layer (SP) runs in a horizontal band across the middle of the field. Numbered boxes indicate the location of some of the pyramidal cell bodies, which were more clearly evident in averaged images. SR, stratum radiatum. SO, stratum oriens. Scale bar, 20 μm. (**B**) Plots of fluorescence intensity over time for the five boxed regions, corresponding to five pyramidal cell bodies, in A. Note how the high time resolution of the fast-scanning imaging reveals the repetitive, synchronized Ca^{2+} spikes in the neurons. (**C**) Plots of fluorescence intensity for a neuron (upper trace) and for four other nearby cells (not shown) that are most likely glia. Note the various patterns and slower time courses of Ca^{2+} transients in the nonneuronal cells that are resolved with the fast-scanning imaging.

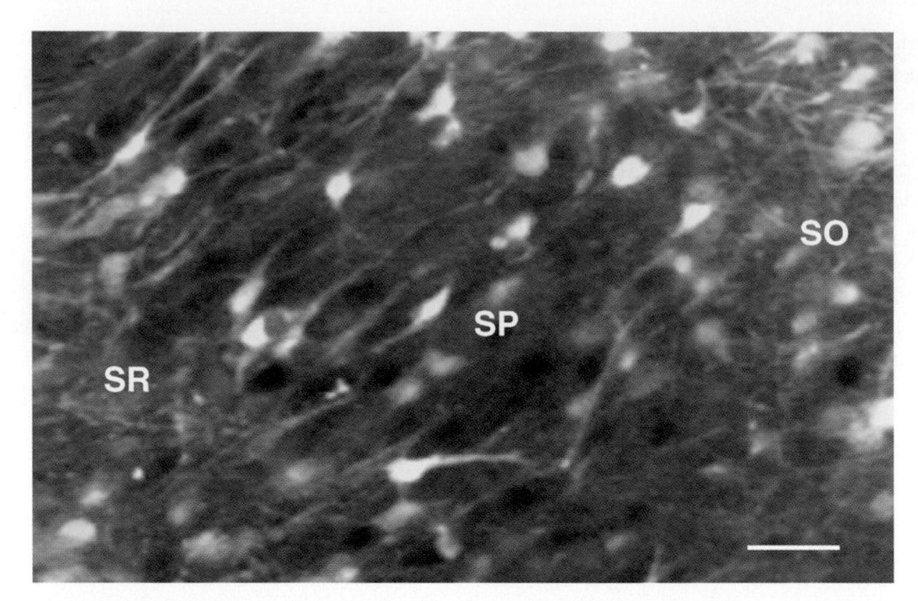

Figure 19 Glial cell Ca^{2+} transients in area CA3 of an organotypic hippocampal brain slice. This is a composite image showing Ca^{2+} activity at three different time points (encoded blue–green–red, respectively) collected at 7-s intervals. The layer of pyramidal cell bodies (SP) runs diagonal from upper left to lower right. Cell bodies and fine cellular processes of these nonneuronal cells, most probably astrocytes, are spontaneously active. Astrocyte activity is most prominent in the layers (SR and SP) adjacent to the pyramidal cell body layer, where there is a high density of excitatory synaptic contacts onto pyramidal cell dendrites. Intercellular waves of Ca^{2+} activity have been shown to propagate over long distances through such organized glial networks (see text). Scale bar, 50 μm.

tools for mapping the functional organization of the brain. The wide variety of fluorescent membrane dyes now available with varying spectral properties permits simultaneous labeling and discrimination of different populations of cells. This will continue to be useful to hodologists interested in more precisely identifying the organization and interrelationship of neural projections in brain tissues.

One area of great potential growth is in the development of new structural and functional probes that would be useful for identifying specific populations of neurons or for labeling functional synaptic contacts in live brain tissues. Ideal probes will permit functional analysis at high spatial and temporal resolution. The widespread use of GFP and similar proteins indicates that genetically engineered fluorescent probes should continue to play an important role in elucidating neural structure and function (see Yuste *et al.,* 1999a). New probes that are likely to participate significantly in this endeavor include calcium-sensitive (Miyawaki *et al.,* 1997), pH-sensitive (Miesenbock *et al.,* 1998), and voltage-sensitive (Siegel and Isacoff, 1997, 1999; Sakai *et al.,* 2001) fluorescent proteins, especially those that are targeted to synapses (Sankaranarayanan *et al.,* 2000). It seems likely that genetic probes whose expression can be restricted to certain brain regions or cell types will contribute significantly toward mapping the functional anatomy of the brain (Spergel *et al.,* 2001). Already, there are several transgenic mouse models that express reporters such as GFP in restricted subsets of neurons (van den Pol and Ghosh, 1998; Oliva *et al.,* 2000) or glia (Zhuo *et al.,* 1997).

Given the advancements in vital fluorescent probes and sensitive imaging techniques, it is now possible to map the 3D structure of single neurons and glial cells in live brain slices over a period of many hours. The ability to collect 2D and 3D image data sets from live neural tissue slices at high spatial resolution, over long periods of time and at relatively short time intervals, is revealing new information on the dynamics of neural structure in brain tissues. The time-resolved 3D imaging methods described here capture more of the dynamic events occurring within tissue and also provide the researcher with assurance that observed structural changes are not due to movement in and out of a focal plane. We can expect these vital fluorescence labeling and imaging methods to be applied more widely to studies of structural and functional neural organization in a variety of brain regions, especially with the use of genetic probes that can be targeted to specific brain regions and cell types. Moreover, it is anticipated that cell-type-specific probes will enable new strategies for investigating structural and functional relationships of neurons and glia.

Much can be done with the existing optical imaging technology. Nevertheless, future developments will undoubtedly continue to address constraints on high-resolution imaging deep (>50 μm) within tissues. Improvements are already being realized by using water-immersion lenses (to reduce spherical aberration) and longer wavelength dyes (to reduce light scatter by the tissue and minimize phototoxic effects). Newer optical techniques such as multiphoton imaging (Denk *et al.*, 1990, 1994) are being used more widely and are making significant contributions to mapping neural structure and function. Multiphoton imaging provides intrinsic three-dimensional resolution (Williams *et al.*, 1994) while confining fluorescence excitation to a single narrow focal plane, thus reducing the risk of photodynamic damage. With regard to imaging cellular and subcellular structure, multiphoton excitation yields a substantial improvement over conventional confocal for imaging fluorescently labeled cells at deeper (100–500 μm) levels in live tissues (Mainen *et al.*, 1999; Majewska *et al.*, 2000).

Another very exciting optical technique that should greatly facilitate the mapping of functional neural organization in brain slices is based on laser photostimulation (Farber and Grinvald, 1983). A modification of this technique (Callaway and Katz, 1993; Katz and Dalva, 1994) employs a scanned laser beam to focally release a caged (photoactivatable) form of the excitatory neurotransmitter, glutamate, thus stimulating nearby neurons. The development of new probes that can be uncaged with multiphoton excitation (Augustine, 2001; Matsuzaki *et al.*, 2001) should enable higher resolution studies in more intact tissues. In conjunction with electrophysiological recordings or physiological imaging, these methods should continue to provide opportunities for high-resolution mapping of neural circuits in excised brain slices.

At present, most of the work at the cellular and subcellular levels of resolution is being done in excised tissue slices. However, future work will continue to probe more of the outstanding questions in intact, functioning preparations (e.g., Dirnagl *et al.*, 1991; Them, 1993; Svoboda *et al.*, 1999; Lendvai *et al.*, 2000; Ilyin *et al.*, 2001). At any rate, it is clear that neuroscientists have a repertoire of powerful optical tools for dissecting and mapping the functional organization of brain tissues.

Acknowledgments

Some of the work described herein was performed by the author in the laboratory of Dr. Stephen J. Smith (Stanford University). More recent work in the author's laboratory at the University of Iowa was supported by grants from the NIH (NS37159) and Whitehall Foundation (98-6).

References

Adams, S., Bacskai, B., Deerinck, T., Ellisman, M., Heim, R., Lev-Ram, V., Makings, L., and Tsien, R. (1994). Recent progress in imaging Ca^{2+} and other signals. *In* "Society for Neuroscience Short Course III Syllabus," pp. 29–36. Soc. for Neurosci., Washington, DC.

Agmon, A., Yang, L. T., Jones, E. G., and O'Dowd, D. K. (1995). Topological precision in the thalamic projection to neonatal mouse barrel cortex. *J. Neurosci.* **15,** 549–561.

Ahmari, S., Buchanan, J., and Smith, S. J. (2000). Assembly of presynaptic active zones from cytoplasmic transport packets. *Nat. Neurosci.* **3,** 445–451.

Araque, A., Carmignoto, G., and Haydon, P. G. (2001). Dynamic signaling between astrocytes and neurons. *Annu. Rev. Physiol.* **63,** 795–813.

Arnold, D. B., and Clapham, D. E. (1999). Molecular determinants for subcellular localization of PSD95 with an interacting K^+ channel. *Neuron* **23,** 149–157.

Augustine, G. J. (2001). Illuminating the location of brain glutamate receptors. *Nat. Neurosci.* **4,** 1051–1052.

Baker, G. E., and Reese, B. E. (1993). Using confocal laser scanning microscopy to investigate the organization and development of neuronal projections labeled with DiI. *Methods Cell Biol.* **38,** 325–344.

Belichenko, P. V., and Dahlström, A. (1994a). Dual channel confocal laser microscopy of Lucifer yellow-microinjected human brain cells combined with Texas red immunofluorescence. *J. Neurosci. Methods* **52,** 111–118.

Belichenko, P. V., and Dahlström, A. (1994b). Micro-mapping of the human brain: 3-dimensional imaging of immunofluorescence and dendritic morphology using dual channel confocal laser scanning microscopy. *Hum. Brain Mapp.* **1,** 185–195.

Belichenko, P. V., and Dahlström, A. (1995a). Studies on the 3-dimensional architecture of dendritic spines and varicosities in human cortex by confocal laser scanning microscopy and Lucifer yellow microinjections. *J. Neurosci. Methods* **57,** 55–61.

Belichenko, P. V., and Dahlström, A. (1995b). Mapping of the human brain in normal and pathological situations: The single cell and fiber level, employing Lucifer yellow microinjection, carbocyanine dye tracing, immunofluorescence, and 3-D confocal laser scanning microscopy reconstruction. *Neurosci. Protocols* **50,** 1–30.

Belichenko, P. V., Dahlström, A., and Sourander, P. (1992). The application of confocal microscopy for the study of neuronal organization in human cortical areas after Lucifer yellow microinjection. *In* "Biotechnology Applications of Microinjection, Microscopic Imaging and Fluorescence" (P. H. Bach, C. H. Reynolds, J. M. Clark, P. L. Poole, and J. Mottley, eds.), pp. 29–36. Plenum, London.

Belichenko, P. V., Oldfors, A., Hagberg, B., and Dahlström, A. (1994a). Rett syndrome: 3-D confocal microscopy of cortical pyramidal dendrites and afferents. *NeuroReport* **5,** 1509–1513.

Belichenko, P. V., Sourander, P., Malmgren, K., Nordborg, C., von Essen, C., Rydenhag, B., Lindstrom, S., Hedstrom, A., Uvebrant, P., and Dahlström, A. (1994b). Dendritic morphology in epileptogenic cortex from TRPE patients, revealed by intracellular Lucifer yellow microinjection and confocal laser scanning microscopy. *Epilepsy Res.* **18,** 233–247.

Bezzi, P., and Volterra, A. (2001). A neuron–glia signalling network in the active brain. *Curr. Opin. Neurobiol.* **11,** 387–394.

Blackstad, T. W., and Kjaerheim, A[o]. (1961). Special axo-dendritic synapses in the hippocampal cortex: Electron and light microscopic studies on the layer of mossy fibers. *J. Comp. Neurol.* **117,** 133–157.

Bloom, J. A., and Webb, W. W. (1984). Photodamage to intact erythrocyte membranes at high laser intensities: Methods of assay and suppression. *J. Histochem. Cytochem.* **32,** 608–616.

Callaway, E. M., and Katz, L. C. (1993). Photostimulation using caged glutamate reveals functional circuitry in living brain slices. *Proc. Natl. Acad. Sci. USA* **90,** 7661–7665.

Chalfie, M., Tu, Y., Euskirchen, G., Ward, W. W., and Prasher, D. C. (1994). Green fluorescent protein as a marker for gene expression. *Science* **263,** 802–805.

Chamberlin, N. L., Du, B., de Lacalle S, and Saper, C. B. (1998). Recombinant adeno-associated virus vector: Use for transgene expression and anterograde tract tracing in the CNS. *Brain Res.* **793,** 169–175.

Charles, A. (1998). Intercellular calcium waves in glia. *Glia* **24,** 39–49.

Charles, A. C., Merrill, J. E., Dirksen, E. R., and Sanderson, M. J. (1991). Intercellular signaling in glial cells: Calcium waves and oscillations in response to mechanical stimulation and glutamate. *Neuron* **6,** 983–992.

Chicurel, M. E., and Harris, K. M. (1992). Three-dimensional analysis of the structure and composition of CA3 branched dendritic spines and their synaptic relationships with mossy fiber boutons in the rat hippocampus. *J. Comp. Neurol.* **325,** 169–182.

Cho, K. O., Hunt, C. A., and Kennedy, M. B. (1992). The rat brain postsynaptic density fraction contains a homolog of the Drosophila discs-large tumor suppressor protein. *Neuron* **9,** 929–942.

Cornell-Bell, A. H., and Finkbeiner, S. M. (1991). Ca^{2+} waves in astrocytes. *Cell Calcium* **12,** 185–204.

Cornell-Bell, A. H., Finkbeiner, S. M., Cooper, M. S., and Smith, S. J. (1990). Glutamate induces calcium waves in cultured astrocytes: Long-range glial signaling. *Science* **247,** 470–473.

Crépel, V., Krnjevic, K., and Ben-Ari, Y. (1992). Developmental and regional differences in the vulnerability of rat hippocampal slices to lack of glucose. *Neuroscience (New York)* **47,** 579–587.

Dailey, M. E. (1999). Maintaining live cells and tissue slices in the imaging setup. *In* "Imaging Living Cells" (R. Yuste, F. Lanni, and A. Konnerth, eds.), pp. 10.1–10.7. Cold Spring Harbor Laboratory Press, New York.

Dailey, M. E., and Smith, S. J. (1993). Confocal imaging of mossy fiber growth in live hippocampal slices. *Jpn. J. Physiol.* **43,** S183–S192.

Dailey, M. E., and Smith, S. J. (1994). Spontaneous Ca^{2+} transients in developing hippocampal pyramidal cells. *J. Neurobiol.* **25,** 243–251.

Dailey, M. E., and Smith, S. J. (1996). The dynamics of dendritic structure in developing hippocampal slices. *J. Neurosci.* **16,** 2983–2994.

Dailey, M. E., and Waite, M. (1999). Confocal imaging of microglial cell dynamics in hippocampal slice cultures. *Methods* **18,** 222–230.

Dailey, M. E., Buchanan, J., Bergles, D. E., and Smith, S. J. (1994). Mossy fiber growth and synaptogenesis in rat hippocampal slices *in vitro. J. Neurosci.* **14,** 1060–1078.

Dailey, M. E., Marrs, G., Satz, J., and Waite, M. (1999). Concepts in imaging and microscopy: Exploring biological structure and function with confocal microscopy. *Biol. Bull.* **197,** 115–122.

Dani, J. W., Chernjavski, A., and Smith, S. J (1992). Neuronal activity triggers Ca waves in hippocampal astrocyte networks. *Neuron* **8:**429–440.

DeCamilli, P., Cameron, R., and Greengard, P. (1983a). Synapsin I (Protein I), a nerve terminal-specific phosphoprotein. I. Its general distribution in synapses of the central and peripheral nervous system demonstrated by immunofluorescence in frozen and plastic sections. *J. Cell Biol.* **96,** 1337–1354.

DeCamilli, P., Harris, S. M., Huttner, W. B., and Greengard, P. (1983b). Synapsin I (Protein I), a nerve terminal-specific phosphoprotein. II. Its specific association with synaptic vesicles demonstrated by immunocytochemistry in agarose-embedded synaptosomes. *J. Cell Biol.* **96,** 1355–1373.

Deitch, J. S., Smith, K. L., Swann, J. W., and Turner, J. N. (1991). Ultrastructural investigation of neurons identified and localized using the confocal scanning laser microscope. *J. Electron Microsc. Tech.* **18,** 82–90.

Delbridge, L. M., Harris, P. J., Pringle, J. T., Dally, L. J., and Morgan, T. O. (1990). A superfusion bath for single-cell recording with high-precision optical depth control, temperature regulation, and rapid solution switching. *Pflügers Arch.* **416,** 94–97.

del Rio, J. A., Heimrich, B., Soriano, E., Schwegler, H., and Frotscher, M. (1991). Proliferation and differentiation of glial fibrillary acidic protein-immunoreactive glial cells in organotypic slice cultures of rat hippocampus. *Neuroscience* **43,** 335–347.

Denk, W., Strickler, J. H., and Webb, W. W. (1990). Two-photon laser scanning fluorescence microscopy. *Science* **248,** 73–76.

Denk, W., Delaney, K. R., Gelperin, A., Kleinfeld, D., Strowbridge, B. W., Tank, D. W., and Yuste, R. (1994). Anatomical and functional imaging of neurons using 2-photon laser scanning microscopy. *J. Neurosci. Methods* **54,** 151–162.

Dirnagl, U., Villringer, A., and Einhäupl, K. M. (1991). Imaging of intracellular pH in normal and ischemic rat brain neocortex using confocal laser scanning microscopy in vivo. *J. Cereb. Blood Flow Metab.* **11,** S206.

Dunwiddie, T. V. (1981). Age-related differences in the *in vitro* rat hippocampus: Development of inhibition and the effects of hypoxia. *Dev. Neurosci.* **4,** 165–175.

Elberger, A. J., and Honig, M. G. (1990). Double-labeling tissue containing the carbocyanine dye, DiI, for immunocytochemistry. *J. Histochem. Cytochem.* **38,** 735–739.

Farber, I. C., and Grinvald, A. (1983). Identification of presynaptic neurons by laser photostimulation. *Science* **222,** 1025–1027.

Fine, A., Amos, W. B., Durbin, R. M., and McNaughton, P. A. (1988). Confocal microscopy: Applications in neurobiology. *Trends Neurosci.* **11,** 346–351.

Friedman, H. V., Bresler, T., Garner, C. C., and Ziv, N. E. (2000). Assembly of new individual excitatory synapses: Time course and temporal order of synaptic molecule recruitment. *Neuron* **27,** 57–69.

Gähwiler, B. H. (1984). Development of the hippocampus *in vitro:* Cell types, synapses, and receptors. *Neuroscience (New York)* **11,** 751–760.

Gähwiler, B. H., Thompson, S. M., Audinat, E., and Robertson, R. T. (1991). Organotypic slice cultures of neural tissue. *In* "Culturing Nerve Cells" (G. Banker and K. Goslin, eds.), pp. 379–411. MIT Press, Cambridge, MA.

Gähwiler, B. H., Capogna, M., Debanne, D., McKinney, R. A., and Thompson, S. M. (1997). Organotypic slice cultures: A technique has come of age. *Trends Neurosci.* **20,** 471–477.

Gan, W.-B., Bishop, D., Turney, S. G., and Lichtman, J. W. (1999). Vital imaging and ultrastructural analysis of individual axon terminals labeled by iontophoretic application of lipophilic dye. *J. Neurosci. Methods* **93,** 13–20.

Gan, W.-B., Grutzendler, J., Wong, W. T., Wong, R. O. L., and Lichtman, J. W. (2000). Multicolor "DiOlistic" labeling of the nervous system using lipophilic dye combinations. *Neuron* **27,** 219–225.

Ghosh, A., and Greenberg, M. E. (1995). Calcium signaling in neurons: Molecular mechanisms and cellular consequences. *Science* **268,** 239–247.

Grinvald, A., Frostig, R. D., Lieke, E., and Hildesheim, R. (1988). Optical imaging of neuronal activity. *Annu. Rev. Physiol.* **68,** 1285–1366.

Grynkiewicz, G., Poenie, M., and Tsien, R. Y. (1985). A new generation of Ca^{2+} indicators with greatly improved fluorescence properties. *J. Biol. Chem.* **260,** 3440–3450.

Guerineau, N. C., Bonnefont, X., Stoeckel, L., and Mollard, P. (1998). Synchronized spontaneous Ca^{2+} transients in acute anterior pituitary slices. *J. Biol. Chem.* **273,** 10389–10395.

Guthrie, P. B., Segal, M., and Kater, S. B. (1991). Independent regulation of calcium revealed by imaging dendritic spines. *Nature (London)* **354,** 76–80.

Haas, K., Sin, W. C., Javaherian, A., Li, Z., and Cline, H. T. (2001). Single-cell electroporation for gene transfer in vivo. *Neuron* **29,** 583–591.

Henze, D. A., Urban, N. N., and Barrionuevo, G. (2000). The multifarious hippocampal mossy fiber pathway: A review. *Neuroscience* **98,** 407–427.

Holt, C. E. (1989). A single-cell analysis of early retinal ganglion cell differentiation in Xenopus: From soma to axon tip. *J. Neurosci.* **9,** 3123–3145.

Honig, M. G. (1993). DiI labelling. *Neurosci. Protocols* **50,** 1–20.

Honig, M. G., and Hume, R. I. (1986). Fluorescent carbocyanine dyes allow living neurons of identified origin to be studied in long-term cultures. *J. Cell Biol.* **103,** 171–187.

Hosokawa, T., Bliss, T. V., and Fine, A. (1992). Persistence of individual dendritic spines in living brain slices. *NeuroReport* **3,** 477–480.

Hosokawa, T., Bliss, T. V., and Fine, A. (1994). Quantitative three-dimensional confocal microscopy of synaptic structures in living brain tissue. *Microsc. Res. Tech.* **29,** 290–296.

Ilyin, S. E., Flynn, M. C., and Plata-Salaman, C. R. (2001). Fiber-optic monitoring coupled with confocal microscopy for imaging gene expression in vitro and in vivo. *J. Neurosci. Methods* **108,** 91–96.

Jahromi, B. S., Robitaille, R., and Charlton, M. P. (1992). Transmitter release increases intracellular calcium in perisynaptic Schwann cells in situ. *Neuron* **8,** 1069–1077.

Katz, L. C., and Dalva, M. B. (1994). Scanning laser photostimulation: A new approach for analyzing brain circuits. *J. Neurosci. Methods* **54,** 205–218.

Kennedy, M. B. (1989). Regulation of neuronal function by calcium. *Trends Neurosci.* **12,** 470–475.

Kistner, U., Wenzel, B. M., Veh, R. W., Cases-Langhoff, C., Garner, A. M., Appeltauer, U., Voss, B., Gundelfinger, E. D., and Garner, C. C. (1993). SAP90, a rat presynaptic protein related to the product of the *Drosophila* tumor suppressor gene dlg-A. *J. Biol. Chem.* **268,** 4580–4583.

Kudo, Y., Takeda, K., Hicks, T. P., Ogura, A., and Kawasaki, Y. (1989). A new device for monitoring concentrations of intracellular Ca^{2+} in CNS preparations and its application to the frog's spinal cord. *J. Neurosci. Methods* **30,** 161–168.

Lendvai, B., Stern, E. A., Chen, B., and Svoboda, K. (2000). Experience-dependent plasticity of dendritic spines in the developing rat barrel cortex *in vivo. Nature* **404,** 876–881.

Lev-Ram, V., and Ellisman, M. H. (1995). Axonal activation-induced calcium transients in myelinating Schwann cells: Sources and mechanisms. *J. Neurosci.* **15,** 2628–2637.

Lichtman, J. W. (1994) Confocal microscopy. *Sci. Am.* **271,** 40–45.

Lo, D. C., McAllister, A. K., and Katz, L. C. (1994). Neuronal transfection in brain slices using particle-mediated gene transfer. *Neuron* **13,** 1263–1268.

Mainen, Z. F., Maletic-Savatic, M., Shi, S. H., Hayashi, Y., Malinow, R., and Svoboda, K. (1999). Two-photon imaging in living brain slices. *Methods* **18,** 231–239.

Majewska, A., Yiu, G., and Yuste, R. (2000). A custom-made two-photon microscope and deconvolution system. *Pflügers Arch.–Eur. J. Physiol.* **441,** 398–408.

Marrs, G. S., Green, S. H., and Dailey, M. E. (2001). Rapid formation and remodeling of postsynaptic densities in developing dendrites. *Nat. Neurosci.* **4,** 1006–1013.

Mason, W. T., ed. (1993). "Fluorescent and Luminescent Probes for Biological Activity: A Practical Guide to Technology for Quantitative Real-Time Analysis." Academic Press, San Diego.

Matsuzaki, M., *et al.* (2001). Dendritic spine geometry is critical for AMPA receptor expression in hippocampal CA1 pyramidal neurons. *Nat. Neurosci.* **4,** 1086–1092.

McBain, C. J., Boden, P., and Hill, R. G. (1989). Rat hippocampal slices 'in vitro' display spontaneous epileptiform activity following long-term organotypic culture. *J. Neurosci. Methods* **27,** 35–49.

Mendelowitz, D., Yang, M., Andresen, M. C., and Kunze, D. L. (1992). Localization and retention in vitro of fluorescently labeled aortic baroreceptor terminals on neurons from the nucleus tractus solitarius. *Brain Res.* **581,** 339–343.

Miesenbock, G., DeAngelis, D. A., and Rothman, J. E. (1998). Visualizing secretion and synaptic transmission with pH-sensitive green fluorescent proteins. *Nature* **394,** 192–195.

Minsky, M. (1961). Microscopy apparatus. [U.S. Pat. No. 3,013,467]

Minsky, M. (1988). Memoir on inventing the confocal scanning microscope. *Scanning* **10,** 128–138.

Minta, A., Kao, J. P. Y., and Tsien, R. Y. (1989). Fluorescent indicators for cytosolic calcium based on rhodamine and fluorescein chromophores. *J. Biol. Chem.* **264,** 8171–8178.

Miyawaki, A., Llopis, J., Heim, R., McCaffery, J. M., Adams, J. A., Ikura, M., and Tsien, R. Y. (1997). Fluorescent indicators for Ca^{2+} based on green fluorescent proteins and calmodulin. *Nature* **388,** 882–887.

Moriyoshi, K., Richards, L. J., Akazawa, C., O'Leary, D. D., and Nakanishi, S. (1996). Labeling neural cells using adenoviral gene transfer of membrane-targeted GFP. *Neuron* **16,** 255–260.

Müller, W., and Connor, J. A. (1991). Dendritic spines as individual neuronal compartments for synaptic Ca^{2+} responses. *Nature (London)* **354,** 73–76.

Murphy, R. C., and Messer, A. (2001). Gene transfer methods for CNS organotypic cultures: A comparison of three nonviral methods. *Mol. Ther.* **3,** 113–121.

Navone, F., Greengard, P., and DeCamilli, P. (1984). Synapsin I in nerve terminals: Selective association with small synaptic vesicles. *Science* **226,** 1209–1211.

O'Donovan, M. J., Ho, S., Sholomenko, G., and Yee, W. (1993). Real-time imaging of neurons retrogradely and anterogradely labelled with calcium-sensitive dyes. *J. Neurosci. Methods* **46,** 91–106.

Okabe, S., Kim, H.-D., Miwa, A., Kuriu, T., and Okado, H. (1999). Continual remodeling of postsynaptic density and its regulation by synaptic activity. *Nat. Neurosci.* **2,** 804–811.

Oliva, A. A., Jr., Jiang, M., Lam, T., Smith, K. L., and Swann, J. W. (2000). Novel hippocampal interneuronal subtypes identified using transgenic mice that express green fluorescent protein in GABAergic interneurons. *J. Neurosci.* **20,** 3354–3368.

O'Rourke, N. A., Dailey, M. E., Smith, S. J., and McConnell, S. K. (1992). Diverse migratory pathways in the developing cerebral cortex. *Science* **258,** 299–302.

Pawley, J. B., ed. (1995). "Handbook of Biological Confocal Microscopy," 2nd edition. Plenum, New York.

Pozzo Miller, L. D., Petrozzino, J. J., Mahanty, N. K., and Connor, J. A. (1993). Optical imaging of cytosolic calcium, electrophysiology, and ultrastructure in pyramidal neurons of organotypic slice cultures from rat hippocampus. *NeuroImage* **1,** 109–120.

Qin, L., Marrs, G. S., McKim, R., and Dailey, M. E. (2001). Hippocampal mossy fibers induce assembly and clustering of PSD95-containing postsynaptic densities independent of glutamate receptor activation. *J. Comp. Neurol.* **440,** 284–298.

Ramón y Cajal, S. (1911). "Histologie du Système Nerveux de l'Homme et des Vertébrés." Maloine, Paris.

Regehr, W., and Tank, D. W. (1991). Selective fura-2 loading of presynaptic terminals and nerve cell processes by local perfusion in mammalian brain slice. *J. Neurosci. Methods* **37,** 111–119.

Reist, N. E., and Smith, S. J. (1992). Neurally evoked calcium transients in terminal Schwann cells at the neuromuscular junction. *Proc. Natl. Acad. Sci. USA* **89,** 7625–7629.

Ross, W. N. (1989). Changes in intracellular calcium during neuron activity. *Annu. Rev. Physiol.* **51,** 491–506.

Sakai, R., Repunte-Canonigo, V., Raj, C. D., and Knopfel, T. (2001). Design and characterization of a DNA-encoded, voltage-sensitive fluorescent protein. *Eur. J. Neurosci.* **13,** 2314–2318.

Sandell, J. H., and Masland, R. H. (1988). Photoconversion of some fluorescent markers to a diaminobenzidine product. *J. Histochem. Cytochem.* **36,** 555–559.

Sankaranarayanan, S., De Angelis, D., Rothman, J. E., and Ryan, T. A. (2000). The use of pHluorins for optical measurements of presynaptic activity. *Biophys. J.* **79,** 2199–2208.

Sergent, P. B. (1994). Double-label immunofluorescence with the laser scanning confocal microscope using cyanine dyes. *NeuroImage* **1,** 288–295.

Sheng, M., and Sala, C. (2001). PDZ domains and the organization of supramolecular complexes. *Annu. Rev. Neurosci.* **24,** 1–29.

Shepherd, G. M., ed. (1998). "The Synaptic Organization of the Brain," 4th edition. Oxford Univ. Press, New York.

Siegel, M. S., and Isacoff, E. Y. (1997). A genetically encoded optical probe of membrane voltage. *Neuron* **19,** 735–741.

Siegel, M. S., and Isacoff, E. Y. (1999). Green fluorescent proteins for measuring signal transduction: A voltage-sensor prototype. *In* "Imaging Living Cells" (R. Yuste, F. Lanni, and A. Konnerth, eds.), pp. 57.1–57.7. Cold Spring Harbor Laboratory Press, New York.

Sinha, S. R., and Saggau, P. (1999). Simultaneous optical recording of membrane potential and intracellular calcium from brain slices. *Methods* **18,** 204–214.

Smetters, D., Majewska, A., and Yuste, R. (1999). Detecting action potentials in neuronal populations with calcium imaging. *Methods* **18,** 215–221.

Smith, K. L., Turner, J. N., Szarowski, D. H., and Swann, J. W. (1991). Three-dimensional imaging of neurophysiologically characterized hippocampal neurons by confocal scanning laser microscopy. *Ann. N. Y. Acad. Sci.* **627,** 390–394.

Smith, K. L., Turner, J. N., Szarowski, D. H., and Swann, J. W. (1994). Localizing sites of intradendritic electrophysiological recordings by confocal light microscopy. *Microsc. Res. Tech.* **29,** 310–318.

Smith, S. J. (1994). Neural signalling: Neuromodulatory astrocytes. *Curr. Biol.* **4,** 807–810.

Smith, S. J., Cooper, M., and Waxman, A. (1990). Laser microscopy of subcellular structure in living neocortex: Can one see dendritic spines twitch? *Symp. Med. Hoechst* **23,** 49–71.

Spergel, D. J., Kruth, U., Shimshek, D. R., Sprengel, R., and Seeburg, P. H. (2001). Using reporter genes to label selected neuronal populations in transgenic mice for gene promoter, anatomical, and physiological studies. *Prog. Neurobiol.* **63,** 673–686.

Stence, N., Waite, M., and Dailey, M. E. (2001). Dynamics of microglial activation: A confocal time-lapse analysis in hippocampal slices. *Glia* **33,** 256–266.

Stoppini, L., Buchs, P.-A., and Muller, D. (1991). A simple method for organotypic cultures of nervous tissue. *J. Neurosci. Methods* **37,** 173–182.

Svoboda, K., Tank, D. W., and Denk, W. (1996) Direct measurement of coupling between dendritic spines and shafts. *Science* **272,** 716–719.

Svoboda, K., Tank, D. W., Stepnoski, R. A., and Denk, W. (1999). Two-photon imaging of neuronal function in the neocortex in vivo. *In* "Imaging Living Cells" (R. Yuste, F. Lanni, and A. Konnerth, eds.), pp. 22.1–22.11. Cold Spring Harbor Laboratory Press, New York.

Taghert, P. H., Bastiani, M. J., Ho, R. K., and Goodman, C. S. (1982). Guidance of pioneer growth cones: Filopodial contacts and coupling revealed with an antibody to Lucifer yellow. *Dev. Biol.* **94,** 391–399.

Tamamaki, N., Nakamura, K., Furuta, T., Asamoto, K., and Kaneko, T. (2000). Neurons in Golgi-stain-like images revealed by GFP-adenovirus infection in vivo. *Neurosci. Res. Suppl.* **38,** 231–236.

Tank, D. W., Sugimori, M., Connor, J. A., and Llinas, R. R. (1988). Spatially resolved calcium dynamics of mammalian Purkinje cells in cerebellar slice. *Science* **242,** 773–777.

Terasaki, M., and Dailey, M. E. (1995). Confocal microscopy of living cells. *In* "Handbook of Biological Confocal Microscopy" (J. B. Pawley, ed.), 2nd edition, pp. 327–346. Plenum, New York.

Them, A. (1993). Intracellular ion concentrations in the brain: Approaches towards in situ confocal imaging. *Adv. Exp. Med. Biol.* **333,** 145–175.

Trommald, M., Jensen, V., and Andersen, P. (1995). Analysis of dendritic spines in rat CA1 pyramidal cells intracellularly filled with a fluorescent dye. *J. Comp. Neurol.* **353,** 260–274.

Tsien, R. Y. (1981). A non-disruptive technique for loading calcium buffers and indicators into cells. *Nature (London)* **290,** 527–528.

Tsien, R. Y. (1998). The green fluorescent protein. *Annu. Rev. Biochem.* **67,** 509–544.

Tsien, R. Y., and Prasher, D. C. (1998). Molecular biology and mutation of GFP. *In* "GFP: Green Fluorescent Protein. Properties, Applications, and Protocols" (M. Chalfie and S. Kain, eds.), pp. 97–118. Wiley–Liss, New York.

Tsien, R. Y., and Waggoner, A. (1995). Fluorophores for confocal microscopy: Photophysics and photochemistry. *In* "Handbook of Biological Confocal Microscopy" (J. B. Pawley, ed.), 2nd edition, pp. 267–279. Plenum, New York.

Turner, J. N., Szarowski, D. H., Smith, K. L., Marko, M., Leith, A., and Swann, J. W. (1991). Confocal microscopy and three-dimensional reconstruction of electrophysiologically identified neurons in thick brain slices. *J. Electron Microsc. Tech.* **18,** 11–23.

Turner, J. N., Swann, J. W., Szarowski, D. H., Smith, K. L., Carpenter, D. O., and Fejtl, M. (1993). Three-dimensional confocal light microscopy of neurons: Fluorescent and reflection stains. *Methods Cell Biol.* **38,** 345–366.

Turner, J. N., Szarowski, D. H., Turner, T. J., Ancin, H., Lin, W. C., Roysam, B., and Holmes, T. J. (1994). Three-dimensional imaging and image analysis of hippocampal neurons: Confocal and digitally enhanced wide field microscopy. *Microsc. Res. Tech.* **29,** 269–278.

Turner, J. N., Swann, J. W., Szarowski, D. H., Smith, K. L., Shain, W., Carpenter, D. O., and Fejtl, M. (1996). Three-dimensional confocal light and electron microscopy of central nervous system tissue and neurons and glia in culture. *Int. Rev. Exp. Pathol.* **36,** 53–72.

van den Pol, A. N., and Ghosh, P. K. (1998). Selective neuronal expression of green fluorescent protein with cytomegalovirus promoter reveals entire neuronal arbor in transgenic mice. *J. Neurosci.* **18,** 10640–10651.

Vasquez, E. C., Johnson, R. F., Beltz, T. G., Haskell, R. E., Davidson, B. L., and Johnson, A. K. (1998). Replication-deficient adenovirus vector transfer of GFP reporter gene into supraoptic nucleus and subfornical organ neurons. *Exp. Neurol.* **154,** 353–365.

Vincent, S. L., Sorensen, I., and Benes, F. M. (1991). Localisation and high-resolution imaging of cortical neurotransmitter compartments using confocal laser scanning microscopy: GABA and glutamate interactions in rat cortex. *BioTechniques* **11,** 628–634.

von Bartheld, C. S., Cunningham, D. E., and Rubel, E. W. (1990). Neuronal tracing with DiI: Decalcification, cryosectioning, and photoconversion for light and electron microscopic analysis. *J. Histochem. Cytochem.* **38,** 725–733.

Welsh, M. G., Ding, J. M., Buggy, J., and Terracio, L. (1991). Application of confocal laser scanning microscopy to the deep pineal gland and other neural tissues. *Anat. Rec.* **231,** 473–481.

White, J. G., Amos, W. B., and Fordham, M. (1987). An evaluation of confocal versus conventional imaging of biological structures by fluorescence light microscopy. *J. Cell Biol.* **105,** 41–48.

Williams, R. M., Piston, D. W., and Webb, W. W. (1994). Two-photon molecular excitation provides intrinsic 3-dimensional resolution for laser-based microscopy and microphotochemistry. *FASEB J.* **8,** 804–813.

Wilson, T., ed. (1990) "Confocal Microscopy." Academic Press, San Diego.

Wong, W. T., Sanes, J. R., and Wong, R. O. L. (1998). Developmentally regulated spontaneous activity in the embryonic chick retina. *J. Neurosci.* **18,** 8839–8852.

Wouterlood, F. G., Van Denderen, J. C., Blijleven, N., Van Minnen, J., and Hartig., W. (1998). Two-laser dual-immunofluorescence confocal laser scanning microscopy using Cy2- and Cy5-conjugated secondary antibodies: Unequivocal detection of colocalization of neuronal markers. *Brain Res. Protocols* **2,** 149–159.

Yuste, R., and Katz, L. C. (1991). Control of postsynaptic Ca^{2+} influx in developing neocortex by excitatory and inhibitory neurotransmitters. *Neuron* **6,** 333–344.

Yuste, R., Peinado, A., and Katz, L. C. (1992). Neuronal domains in developing neocortex. *Science* **257,** 665–669.

Yuste, R., Lanni, F., and Konnerth, A., eds. (1999a). "Imaging Neurons: A Laboratory Manual." Cold Spring Harbor Laboratory Press, Cold Spring Harbor, NY.

Yuste, R., Majewska, A., Cash, S. S., and Denk, W. (1999b). Mechanisms of calcium influx into hippocampal spines: Heterogeneity among spines, coincidence detection by NMDA receptors, and optical quantal analysis. *J. Neurosci.* **19,** 1976–1987.

Zhuo, L., Sun, B., Zhang, C. L., Fine, A., Chiu, S. Y., and Messing, A. (1997). Live astrocytes visualized by green fluorescent protein in transgenic mice. *Dev. Biol.* **187,** 36–42.

Ziv, N., and Smith, S. J. (1996). Evidence for a role of dendritic filopodia in synaptogenesis and spine formation. *Neuron* **17,** 91–102.

4

Voltage and Calcium Imaging of Brain Activity: Examples from the Turtle and the Mouse

Matt Wachowiak*, Chun X. Falk*†, Lawrence B. Cohen*, and Michal R. Zochowski*‡1

**Department of Cellular and Molecular Physiology, Yale University School of Medicine, New Haven, Connecticut 06520;*
†RedShirtImaging, LLC, Fairfield, Connecticut 06432;
‡Center for Complex Systems, Warsaw School of Advanced Social Psychology, Warsaw, Poland

I. Why (and Why Not) Voltage and Calcium Imaging

An optical recording with a voltage- or calcium-sensitive dye has two advantages compared to intrinsic signal imaging and *f*MRI signals. First, the voltage or calcium signals are

[1]Present address: Department of Physics, University of Michigan, Ann Arbor, MI 48109.

relatively direct measures of neuron activity. Thus far, these signals have been uniquely interpretable as changes in membrane potential or intracellular calcium concentration. Second, these signals are fast (response time constants of <0.01 ms for voltage-sensitive dyes and <10 ms for calcium dyes). The speed of their responses means that they provide real-time information about brain activity.

On the other hand, voltage- or calcium-sensitive dye measurements are invasive. They are more invasive than *f*MRI measurements in that they often require opening the skull, and they are more invasive than both *f*MRI and intrinsic signal measurements in that they require the use of dyes (with the attendant possibility of pharmacological and/or phototoxic effects).

Optical recordings with voltage- or calcium-sensitive dyes have been used to study activity in processes of individual neurons, in monitoring spiking from many individual cell bodies, and in monitoring coherent (synchronous) activity in populations of neurons. In this chapter we discuss only measurements of population signals with examples from two *in vivo* preparations: the olfactory bulb of the turtle and of the mouse. In the two examples, the optical signals in response to odorants are small (fractional intensity changes, $\Delta I/I$, of between 10^{-4} and 3×10^{-2}); therefore, optimizing the signal-to-noise ratio in the measurements is important. This chapter begins with a general discussion of optical recording methods, including the choice of dyes, light sources, optics, cameras, and minimizing noise. The general approach to optimizing the signal-to-noise ratio is three pronged. First, test dyes to find the dye with the largest signal-to-noise ratio. Second, reduce the extraneous sources of noise (dark noise, vibrations, line frequency noise, animal movement, etc.). Third, maximize the number of photons measured to reduce the relative shot noise (noise arising from the statistical nature of photon emission and detection). The second part of the chapter provides a more detailed description of voltage-sensitive dye measurements in the turtle and calcium dye measurements in the mouse.

Different kinds of staining were used in the two preparations described in this chapter. For the voltage-sensitive dye, the olfactory bulb was superfused for 60 min in a solution of the dye. From previous experience we expected that this procedure would stain all of the neurons in the bulb and that the amount of dye, and thus the fluorescence intensity, would be large. For the calcium dye, we attempted to selectively stain the nerve terminals of the olfactory receptor neurons by introducing the dye into the nose, having it be taken up by the receptor neurons, and then waiting several days for the dye to be transported to the terminals in the olfactory bulb. We expected that the amount of dye reaching the bulb would be relatively small.

Two kinds of cameras were used in the experiments described here; both have frame rates faster than 1000 fps. One camera is a photodiode array with a total of 464 pixels and the second an 80×80-pixel CCD camera. Even though the spatial resolution of the two cameras differs rather dramatically, the most important difference is in the range of light intensities over which they provide an optimal signal-to-noise ratio. The CCD camera is optimal at low light levels and the photodiode array at high light levels (see below).

II. Signal Type

In both preparations described in this chapter, fluorescence changes were measured. However, with the voltage-sensitive dye and with many calcium indicators there are also changes in absorption that accompany activity (e.g., Gupta *et al.,* 1981). In *in vitro* slice preparations, Jin, et al. (2002) found that absorption signals have a larger signal-to-noise ratio. While transmitted light measurements are not easy in an *in vivo* preparation, larger signals might be obtained if they were used.

III. Dyes

A. Voltage-Sensitive Dyes

The voltage-sensitive dyes used here are membrane-bound chromophores whose absorption or fluorescence properties change in response to changes in membrane potential. In a model preparation, the giant axon from a squid, these optical signals are fast (Fig. 1), following membrane potential with a time constant of <10 μs (Loew *et al.,* 1985), and their size is linearly related to the size of the change in potential (e.g., Gupta *et al.,* 1981). Thus, these dyes provide a direct, fast, and linear measure of the change in membrane potential of the stained membranes.

Several voltage-sensitive dyes (e.g., Fig. 2) have been used to monitor changes in membrane potential in a variety of preparations. Figure 2 illustrates four different chromophores (the merocyanine dye, XVII, was used for the measurement illustrated in Fig. 1). For each chromophore, approximately 100 analogues have been synthesized in an attempt to optimize the signal-to-noise ratio that can be obtained in a variety of preparations. (This screening was made possible by synthetic efforts of three laboratories: Jeff Wang, Ravender Gupta, and Alan Waggoner, then at Amherst College; Rina Hildesheim and Amiram Grinvald at the Weizmann Institute; and Joe Wuskell and Leslie Loew at the University of Connecticut Health Center.) For each of the four chromophores illustrated in Fig. 2, there were 10 or 20 dyes that gave approximately the same signal size on squid axons (Gupta *et al.,* 1981). However, dyes that have nearly identical signal size on squid axons could have very different responses on other preparations, and thus tens of dyes usually have to be tested to obtain the largest possible signal. A common failing

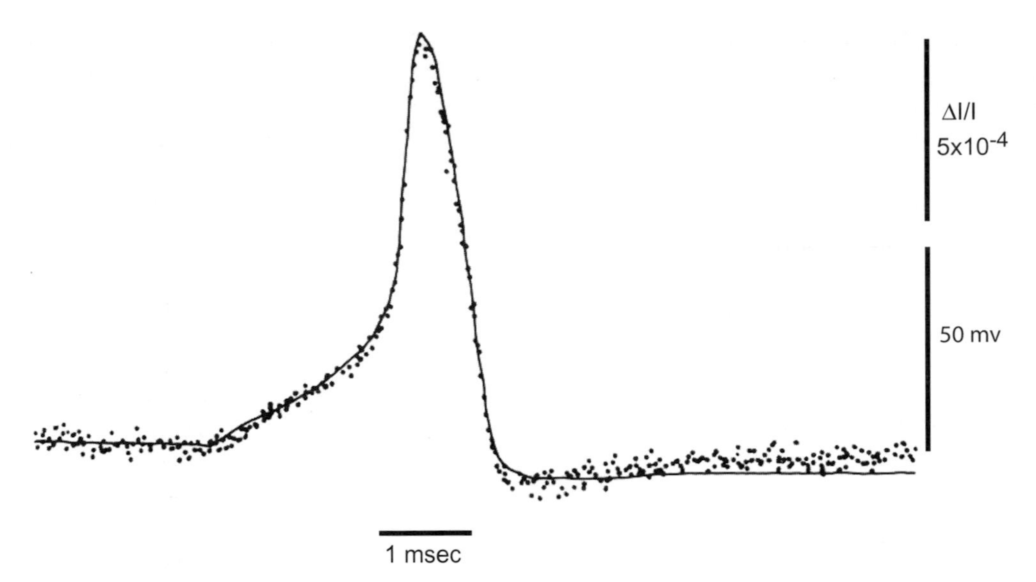

Figure 1 Changes in absorption (dots) of a giant axon stained with a merocyanine dye (XVII, Fig. 2), during a membrane action potential (smooth trace) recorded simultaneously. The change in absorption and the action potential had the same time course. In this and subsequent figures the size of the vertical line represents the stated value of the fractional change in intensity, $\Delta I/I$, or fluorescence. The response time constant of the light-measuring system was 35 μs; 32 sweeps were averaged. The dye is available from Nippon Kankoh-Shikiso Kenkyusho, Okayama, Japan.

was that the dye did not penetrate through connective tissue or along intercellular spaces to the membrane of interest.

The following rules of thumb seem to be useful. First, each of the chromophores is available with a fixed charge which is either a quaternary nitrogen (positive) or a sulfonate (negative). Generally the positive dyes have given larger signals when dye was applied extracellularly in vertebrate preparations. Second, each chromophore is available with carbon chains of several lengths. The more hydrophilic dyes (methyl or ethyl) work best if the dye has to penetrate through a compact tissue (vertebrate brain).

B. Calcium Dyes

Figure 3 shows the chemical structure of a calcium-sensitive dye, Calcium Green-1, together with a plot of the fluorescence spectrum as a function of the free calcium concentration. This dye signal reaches 50% of its maximum at a

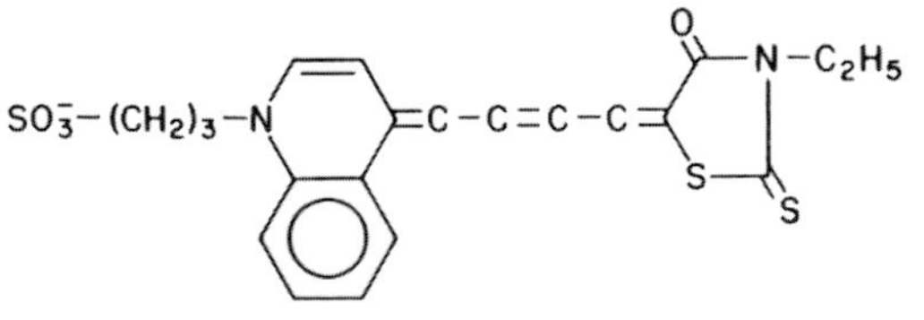

XVII, Merocyanine, Absorption, Birefringence

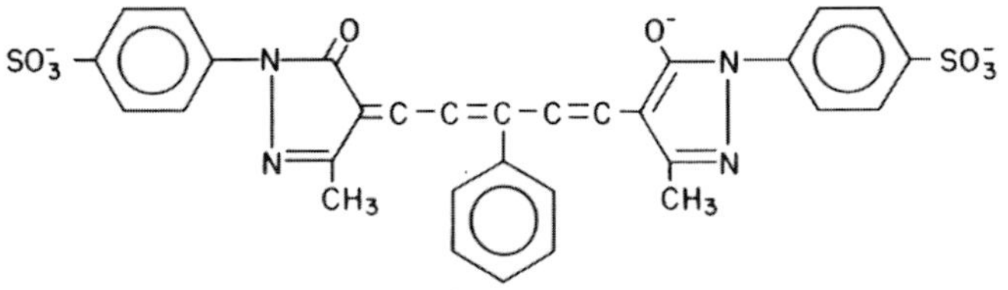

RH155, Oxonol, Absorption

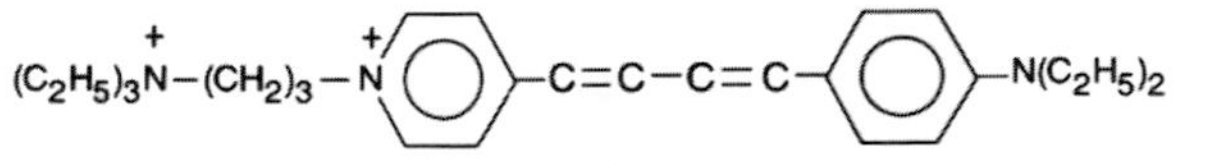

RH414, Styryl, Fluorescence

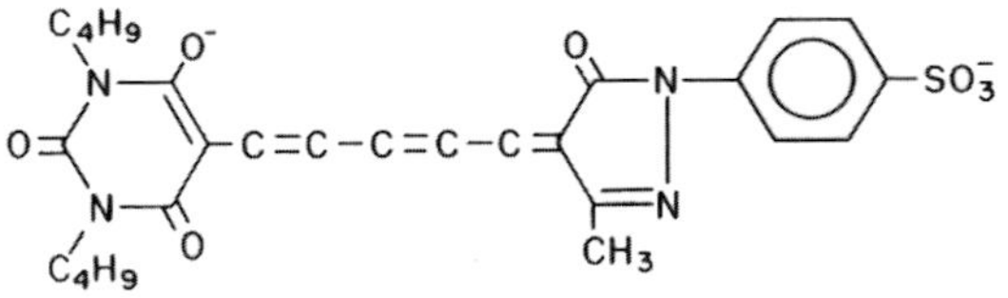

XXV, Oxonol, Fluorescence, Absorption

Figure 2 Examples of four different chromophores that have been used to monitor membrane potential. The merocyanine dye, XVII (WW375), and the oxonol dye, RH155, are commercially available as NK2495 and NK3041 from Nippon Kankoh-Shikiso Kenkyusho Co. Ltd., Okayama, Japan. The oxonol, XXV (WW781), and styryl, RH-414, are available commercially as dye R-1114 and T-1111 from Molecular Probes (Eugene, OR).

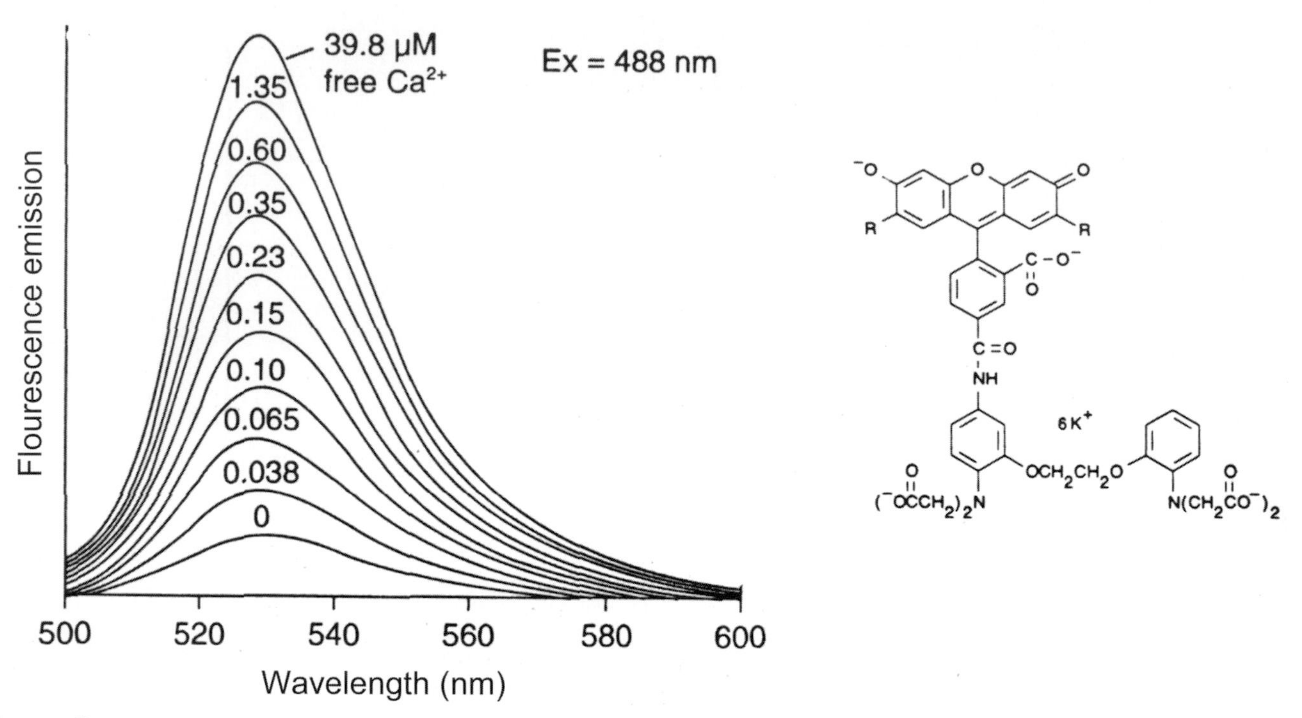

Figure 3 The chemical structure (right) and the emission spectra as a function of calcium concentration of Calcium Green-1. The conjugate with the 10-kDa dextran is available commercially from Molecular Probes as C-3713. Data were taken from the *Handbook of Fluorescent Probes and Research Chemicals,* 6th edition, Molecular Probes.

calcium concentration of about 0.2 μM. In contrast to the voltage-sensitive dyes, the calcium dyes are located intracellularly. The dye is presumed to be in the axoplasm and to report changes in the calcium concentration in the axoplasm although it is possible that some of the dye is in intracellular compartments. These dyes are slower to respond and, because they also act as buffers of calcium, the calcium signal in the presence of dye may substantially outlast the change in calcium concentration that would occur in the absence of the dye.

Figure 4 compares the time courses of a voltage-sensitive dye signal (using the styryl dye di-4-ANEPPS) and the calcium signal (using calcium crimson dextran) in an *in vitro* turtle olfactory bulb preparation. The calcium signal in response to an olfactory nerve shock was measured first using dye in olfactory receptor neuron nerve terminals that had been transported from the cell bodies in the nose. Following this measurement, the preparation was stained with the voltage-sensitive dye and the measurement repeated. While the calcium signal reached a peak only 20 ms after the voltage-sensitive dye signal, the calcium signal returned toward the baseline much more slowly.

Many calcium-sensitive dyes are commercially available. However, in order to obtain selective staining of the olfactory neuron nerve terminals in the olfactory bulb the dye must be transported from the receptor cell bodies in the nose to the terminals in the bulb. In our experience only dextran-conjugated dyes are transported to the nerve terminals after infusion into the nose.

Additional information about the choice of dyes is given in Sections X and XI below in which the *in vivo* experiments are described.

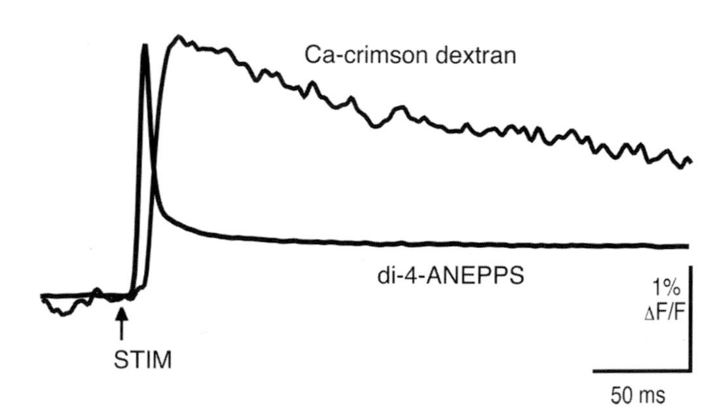

Figure 4 Calcium- and voltage-sensitive dye signals evoked by olfactory nerve shock in an *in vitro* turtle olfactory bulb preparation. The calcium crimson dextran signal was recorded first, then the olfactory bulb was stained with di-4-ANEPPS and the voltage-sensitive dye signal recorded using the same wavelength filters, stimulus intensity, and detectors. The calcium dye was perfused into the nose of the turtle together with 0.5% Triton X-100; the optical measurements were made several days later to allow time for transport of the dye to the nerve terminals in the olfactory bulb. Following the calcium dye measurement, the bulb was bathed in a solution of the voltage-sensitive dye, which stained all of the cell types in the preparation. Traces were digitally low-pass filtered at 200 Hz (no high-pass filtering).

IV. Amplitude of the Voltage or Calcium Change

Both of the signals discussed in this chapter are presented as a fractional intensity change ($\Delta F/F$). These signals give information about the time course of the potential or calcium concentration change but no direct information about the absolute magnitude. However, in some instances, approximate estimations can be obtained. For example, the size of the optical signal in response to a sensory stimulus can be compared to the size of the signal in response to an epileptic event (Orbach *et al.,* 1985). Another approach is the use of ratiometric measurements at two independent wavelengths (Gross *et al.,* 1994). However, to determine the amplitude of the voltage or calcium change from a ratio measurement, one must know the fraction of the fluorescence that results from dye in the expected location, i.e., bound to active vs inactive membranes for voltage-sensitive dyes or dye free in the axoplasm vs dye bound to protein or in intracellular compartments for calcium dyes. These requirements are only approximately met in special circumstances.

V. Noise in the Optical Measurements

A. Shot Noise

The limit of accuracy with which light can be measured is set by the shot noise arising from the statistical nature of photon emission and detection. The root-mean-square deviation in the number emitted is the square root of the average number emitted (I) over a long measuring period. Therefore, the signal in a light measurement will be proportional to I and the noise in that measurement will be proportional to the square root of I. Thus the signal-to-noise ratio (S/N) is proportional to the square root of the number of measured photons; more photons measured means a better signal-to-noise ratio. The S/N is also inversely proportional to the square root of the bandwidth of the photodetection system (Braddick, 1960). The basis for this square-root dependence on intensity is illustrated in Fig. 5. In Fig. 5A the result of using a random number table to distribute 20 photons into 20 time windows is shown. In Fig. 5B the same procedure was used to distribute 200 photons into the same 20 bins. Relative to the average light level there is more noise in the top trace (20 photons) than

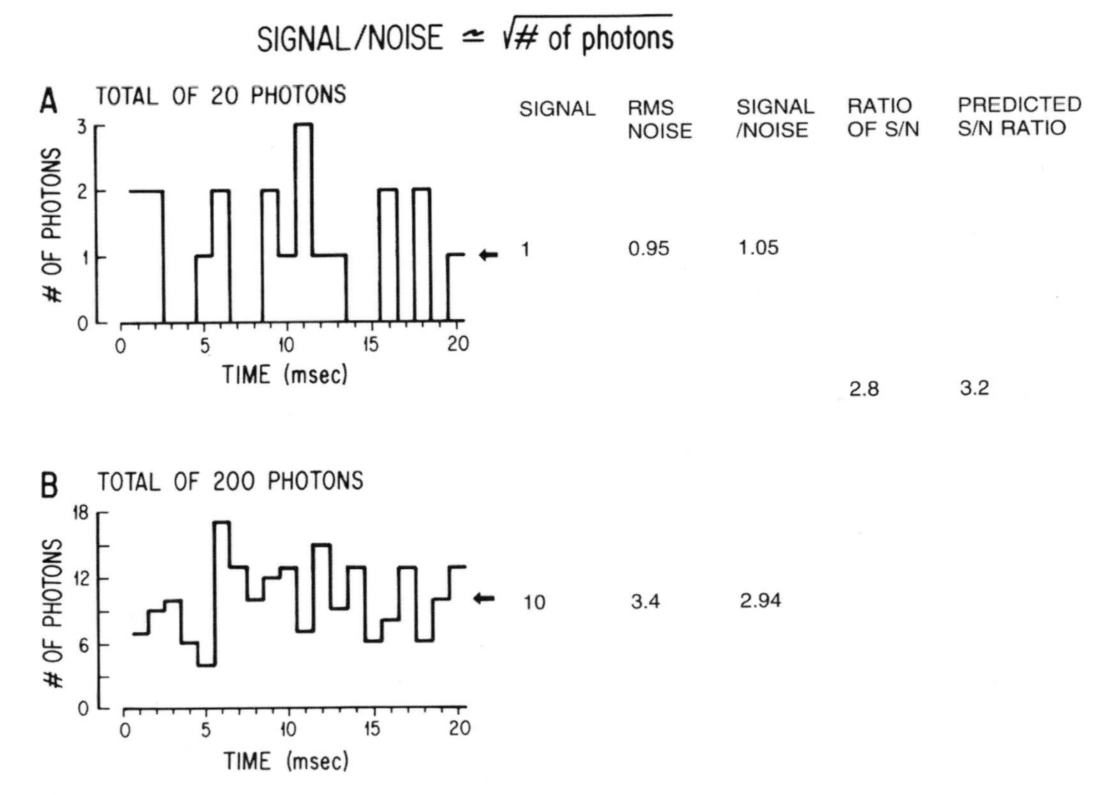

Figure 5 Plots of the results of using a table of random numbers to distribute 20 photons (**A**) or 200 photons (**B**) into 20 time bins. The result illustrates the fact that when more photons are measured the signal-to-noise ratio is improved. On the right, the signal-to-noise ratio is measured for the two results. The ratio of the two signal-to-noise ratios was 2.8. This is close to the ratio predicted by the relationship that the signal-to-noise ratio is proportional to the square root of the measured intensity.

in the bottom trace (200 photons). This square-root relationship is indicated by the green line in Fig. 6, which plots the light intensity divided by the noise in the measurement (S/N) versus the light intensity. At high light intensities this ratio is large and thus small changes in intensity can be detected. For example, at 10^{10} photons/ms a fractional intensity change of 0.1% can be measured with a signal-to-noise ratio of 100 in a single trial. On the other hand, at low intensities the ratio of intensity divided by noise is small and only large signals can be detected. For example, at 10^4 photons/ms the same fractional change of 0.1% can be measured with a signal-to-noise ratio of 1 only after averaging 100 trials. In a shot-noise-limited measurement improvement in the signal-to-noise ratio can be obtained only by (1) increasing the illumination intensity, (2) improving the light-gathering efficiency of the measuring system, or (3) reducing the bandwidth.

Figure 6 also compares the performance of two particular camera systems, a photodiode array (blue lines) and a cooled, back illuminated, CCD camera (red lines), with the shot-noise ideal (green line). The photodiode array approaches the shot-noise limitation over the range of intensities from about 3×10^6 to 10^{10} photons/ms. This is the range of intensities obtained in absorption measurements and fluorescence measurements on *in vitro* slices and intact brains in which all of the neurons are stained via direct application of the dye. On the other hand, the cooled CCD camera approaches the shot-noise limit over the range of intensities from about 3×10^2 to 5×10^6 photons/ms. This is the range of intensities obtained from fluorescence measurements on calcium dyes transported from the nose to nerve terminals in the olfactory bulb. The discussion that follows will indicate the aspects of the measurements and the characteristics of the two camera systems which cause them to deviate from the shot-noise ideal. The two camera systems illustrated in Fig. 6 are examples with outstanding dark noise, quantum efficiency, and saturation characteristics; cameras that do not have as good a performance would be dark-noise limited at higher light intensities, would be farther from ideal at all light levels, and/or would saturate at lower intensities.

B. Extraneous Noise

A second type of noise, termed extraneous or technical noise, is more apparent at high light intensities, at which the

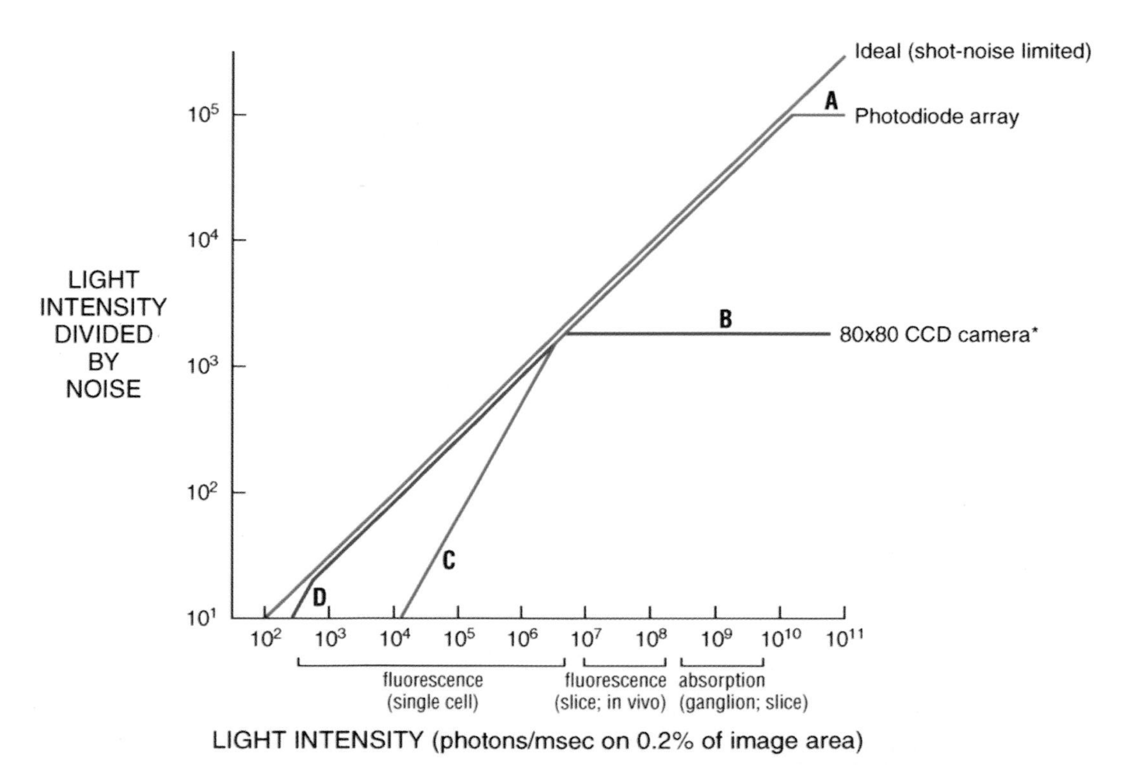

Figure 6 The ratio of light intensity divided by the noise in the measurement as a function of light intensity in photons/ms/0.2% of the object plane. The theoretical optimal signal-to-noise ratio (green line) is the shot-noise limit. Two camera systems are shown, a photodiode array with 464 pixels (blue lines) and a cooled, back-illuminated, 2-kHz frame rate, 80 × 80-pixel CCD camera (red lines). The photodiode array provides an optimal signal-to-noise ratio at higher intensities while the CCD camera is better at lower intensities. The approximate light intensity per detector in fluorescence measurements from a single neuron or from a transported dye is indicated on the left. The fluorescence intensity from a slice or an *in vivo* preparation after soaking with the dye is indicated in the middle. The signal-to-noise ratio for the photodiode array falls away from the ideal at high intensities (A) because of extraneous noise and at low intensities (C) because of dark noise. The lower dark noise of the cooled CCD allows it to function at the shot-noise limit at lower intensities until read noise dominates (D). The CCD camera saturates at intensities above 5×10^6 photons/ms/0.2%.

fractional shot noise and dark noise are low, thus the sensitivity to this kind of noise is high. One type of extraneous noise is caused by fluctuations in the output of the light source (see below). Two other sources of extraneous noise are vibrations and movement of the preparation. A number of precautions for reducing vibrational noise have been described (Salzberg *et al.,* 1977; London *et al.,* 1987). These include avoiding buildings near roads with heavy traffic, preferring basement rooms, and increasing the distance from large air conditioners.

The pneumatic isolation mounts on many vibration isolation tables are more efficient in reducing vertical vibrations than in reducing horizontal movements. One solution is air-filled soft rubber tubes (Newport Corp., Irvine, CA). Minus K Technology sells Biscuit bench-top vibration isolation tables with very low resonant frequencies. They provide outstanding vibration isolation in both vertical and horizontal directions. Nevertheless, it is difficult to reduce vibrational noise to less than 10^{-5} of the total light. For this reason the performance of the photodiode array system is shown reaching a ceiling in Fig. 6 (segment A, blue line).

C. Dark Noise

Dark noise will degrade the signal-to-noise ratio at low light levels. Because the CCD camera is cooled and the photosensitive area (and capacitance) is small, its dark noise is substantially lower than that of the photodiode array system. The excess dark noise in the photodiode array accounts for the fact that segment C in Fig. 6 is substantially to the right of segment D.

VI. Light Sources

Three kinds of light sources have been used. Tungsten filament lamps are a stable source, but their intensity is relatively low, particularly at wavelengths less than 480 nm. Arc lamps are somewhat less stable but can provide more intense illumination. Measurements made with laser illumination have been substantially noisier (Dainty, 1984).

It is not difficult to provide a power supply stable enough so that the output of a tungsten filament lamp fluctuates by less than 1 part in 10^5. In absorption measurements, in which the fractional changes in intensity are relatively small, only tungsten filament sources have been used. On the other hand, fluorescence measurements often have larger fractional changes that will better tolerate noisy light sources, and the measured intensities are lower, making improvements in the signal-to-noise ratio from brighter sources attractive. Opti-Quip, Inc. (Highland Mills, NY) provides 150-W xenon power supplies, lamp housings, and arc lamps with noise that is in the range 1–3 parts in 10^4. The 150-W bulb yielded three times more light at 520 ± 45 nm than a tungsten filament bulb. The extra intensity is especially useful for fluorescence measurements from single neurons or from weakly stained nerve terminals. If the dark noise is dominant, then the signal-to-noise ratio will improve linearly with intensity; if the shot noise is dominant it will improve as the square root of intensity.

VII. Optics

A. Numerical Aperture and Depth of Focus

The need to maximize the number of measured photons is a powerful factor affecting the choice of optical components. In epifluorescence, both the excitation light and the emitted light pass through the objective, and thus the intensity reaching the photodetector is proportional to the *fourth* power of numerical aperture (Inoue, 1986). Clearly, the numerical aperture (NA) of the objective is a crucial factor in increasing the number of measured photons. A Macroscope (RedShirtImaging, LLC, Fairfield, CT) provides a 0.4 NA at 4×, 0.5 NA objectives can be obtained at 10×, and 0.9 NA or higher at magnifications of 20× and above. In addition to determining the intensity reaching the photodetector, NA also determines the depth of focus. Salzberg *et al.* (1977) found that the effective depth of focus for a 0.4 NA objective lens was about 600 μm and Kleinfeld and Delaney (1996) found an effective depth of focus of 200 μm using 0.5 NA optics. In some instances (e.g., Zecevic *et al.,* 1989) a lower numerical aperture was accepted in order to have a larger depth of focus.

B. Light Scattering and Out-of-Focus Light

Light scattering can limit the spatial resolution of an optical measurement. The top of Fig. 7 shows that when no tissue is present, essentially all of the light (750 nm) from a small spot falls on one detector. The bottom illustrates the result when a 500-μm-thick slice of salamander olfactory bulb is present. The light from the small spot is spread to about 200 μm. Mammalian cortex appears to scatter more than the salamander bulb. Thus, light scattering will cause considerable blurring of signals in adult vertebrate preparations.

A second source of blurring is signal from regions that are out of focus. For example, if the active region is a cylinder (a column) perpendicular to the plane of focus, and the objective is focused at the middle of the cylinder, then the light from the in-focus plane will have the correct diameter at the image plane. However, the light from the regions above and below are out of focus and will have a diameter that is too large. The middle section of Fig. 7 illustrates the effect of moving the small spot of light 500 μm out of focus. The light from the small spot is spread to about 200 μm. Thus, in preparations with considerable scattering or with

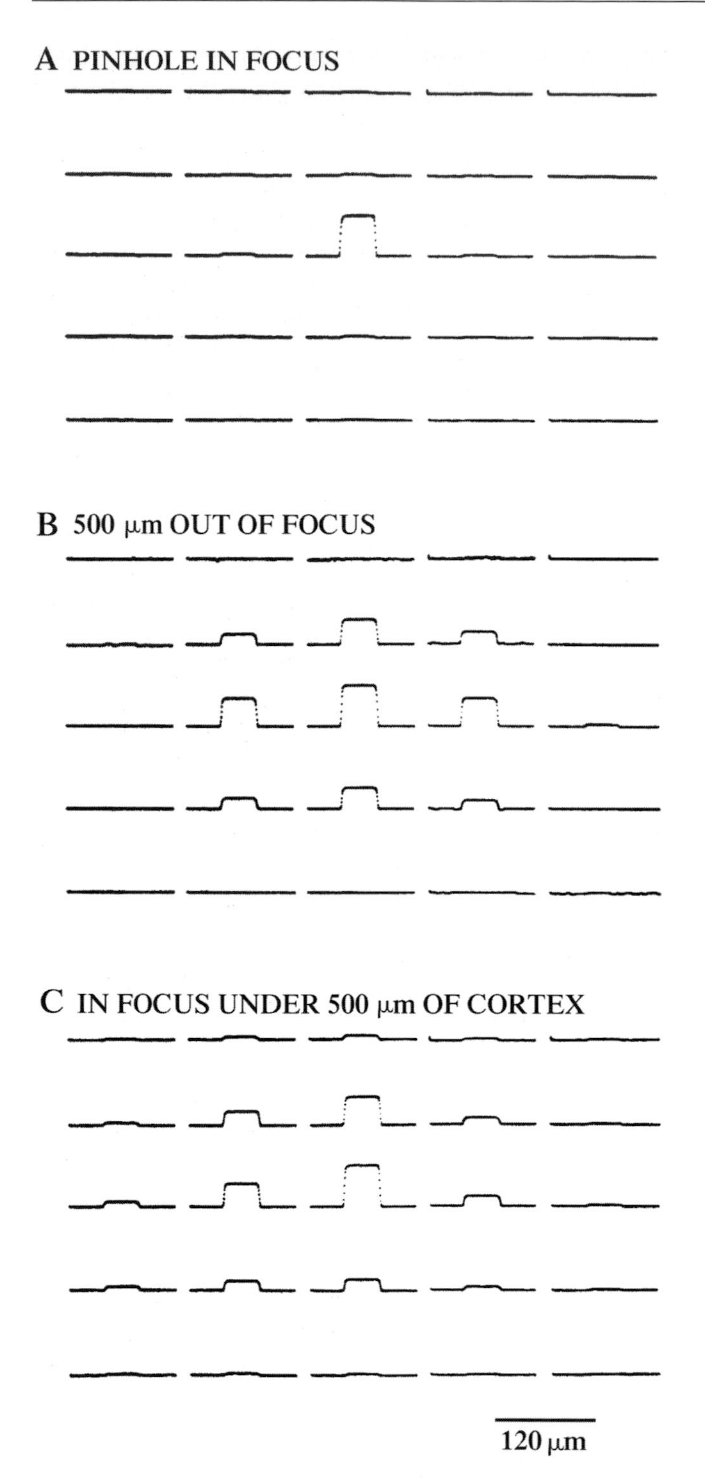

Figure 7 Effects of focus and scattering on the distribution of light from a point source onto the array. (**A**) A 40-μm pinhole in aluminum foil covered with saline was illuminated with light at 750 nm. The pinhole was in focus. More than 90% of the light fell on one detector. (**B**) The stage was moved downward by 500 μm. Light from the out-of-focus pinhole was now seen on several detectors. (**C**) The pinhole was in focus but covered by a 500-μm slice of salamander cortex. Again the light from the pinhole was spread over several detectors. A 10 × 0.4 NA objective was used. Kohler illumination was used before the pinhole was placed in the object plane. The recording gains were adjusted so that the largest signal in each of the three trials would be approximately the same size.

out-of-focus signals, the actual spatial resolution may be limited by the preparation and not by the number of pixels in the imaging device.

C. Confocal Microscopes

The confocal microscope (Petran and Hadravsky, 1966) substantially reduces both the scattered and the out-of-focus light that contributes to the image. A recent modification using two-photon excitation of the fluorophore further reduces out-of-focus fluorescence and photobleaching (Denk *et al.,* 1995). With both types of microscope one can obtain images from intact vertebrate preparations with much better spatial resolution than can be achieved with ordinary microscopy. However, at present the sensitivity of these microscopes is relatively poor and many milliseconds are required to record the image from a single very thin *x–y* plane. The kinds of problems that can be approached using a confocal microscope are limited by these factors.

VIII. Cameras

Because the signal-to-noise ratio in a shot-noise-limited measurement is proportional to the square root of the number of photons converted into photoelectrons (see above), quantum efficiency is important. Silicon photodiodes have quantum efficiencies approaching the ideal (1.0) at wavelengths at which most dyes absorb or emit light (500–900 nm). In contrast, only specially chosen vacuum photocathode devices (phototubes, photomultipliers, or image intensifiers) have a quantum efficiency as high as 0.2. Thus, in shot-noise-limited situations, a silicon diode may have a signal-to-noise ratio that is substantially larger. The discussion below considers only silicon diode systems that have frame rates of about 1 kHz or higher.

Two important considerations in choosing an imaging system are the requirements for spatial and temporal resolution. Because the signal-to-noise ratio in a shot-noise-limited measurement is proportional to the square root of the number of measured photons, increasing either temporal or spatial resolution will reduce the signal-to-noise ratio.

A. Parallel Readout Arrays

Photodiode arrays with 256–1020 elements are now in use in several laboratories (e.g., Iijima *et al.,* 1989; Zecevic *et al.,* 1989; Nakashima *et al.,* 1992; Hirota *et al.,* 1995). These arrays are designed for parallel readout; each detector is followed by its own amplifier whose output can be digitized at frame rates of >1 kHz. While the need to provide a separate amplifier for each diode element limits the number of pixels in parallel readout systems, it contributes to the very large (10^5) dynamic range that can be achieved. A discussion of amplifiers has been presented earlier (Wu and Cohen, 1993). A parallel readout array system is commercially available [NeuroPlex-II (464 pixels), RedShirtImaging, LLC].

B. Serial Readout Arrays

By using a serial readout, the number of amplifiers is greatly reduced. Furthermore, it is simpler to cool CCD chips to reduce the dark-current noise and read noise and, in addition, the much smaller pixel size reduces the diode capacitance, which also contributes to the much lower total dark noise of the CCD camera. However, because of saturation, currently available CCD cameras cannot be used with the higher intensities available in some neurobiological experiments. Saturation accounts for the bending over of the CCD camera performance at segment B in Fig. 6. A dynamic range of even 10^3 is not easily achieved with currently available CCD cameras. Thus, these cameras will not be optimal for fluorescence measurements in which the staining intensity is high. The light intensity would have to be reduced with a consequent decrease in signal-to-noise ratio. On the other hand, CCD cameras are close to ideal for measurements in which there are fewer than 10^7 photons/ms/0.2% of the image (Fig. 6). The intensity that is obtained from calcium dyes transported to olfactory receptor nerve terminals (Section XI) is within the range in which the CCD camera is close to ideal. Table 1 compares three CCD cameras with frame rates near 1 kHz.

In the following two sections we present examples of the use of voltage- and calcium-sensitive dyes to map spatial and temporal patterns of activity in the olfactory system. The examples illustrate how the different features of these dyes can be used to visualize different features of odor encoding and processing in the vertebrate olfactory bulb.

IX. Comparison of Local Field Potential and Voltage-Sensitive Dye Recording

We compared local field potential electrode recordings from two positions 2.3 mm apart on the surface of the turtle visual cortex and optical recordings from the same two locations. Figure 8 illustrates simultaneous local field potential and optical recordings. The top pair of superimposed traces are the local field potential recordings from the two sites. There is considerable overlap in these two recordings; the correlation coefficient calculated for the two traces was 0.9. The bottom pair of traces are the voltage-sensitive dye recordings from the two sites. There is less overlap in the optical recordings; the correlation coefficient calculated for the two traces was 0.6. Thus, in comparison with the optical recordings, the local field potential recordings appear to blur spatial differences.

The relative spatial resolution of the two methods was estimated by determining the distance on the cortex at which a pair of voltage-sensitive dye recordings would have a correlation coefficient equal to the local field potential recordings. In six trials from two preparations the correlation coefficient for the optical measurement was equal to that for the local field potential measurement when the two optical measurements were a distance apart that was 0.21 ± 0.05 (SEM) of the distance of the two local field potential electrodes. Thus, the optical measurement has a linear spatial resolution about 5 times better than the electrode measurement and a two-dimensional resolution about 25 times better.

X. Voltage-Sensitive Dye Recording in the Turtle Olfactory Bulb

Odor stimuli have long been known to induce stereotyped local field potential responses in the olfactory bulb consisting of sinusoidal oscillations of 5 to 80 Hz riding on

Table 1 Characteristics of Fast CCD Camera Systems (as Reported by the Manufacturer)

	Frame rate (fps)	Well size[a]	Read noise[b]	Back illum.[c]	Bits a-to-d	Pixels
MiCAM 01[d]	1333	—	—	No	12	92 × 64
Dalsa CA-D1-0128[e]	756	300,000	360e⁻	No	12	128 × 128
NeuroCCD-SM[f]	2000	300,000	20e⁻	Yes	14	80 × 80

[a]Number of electrons that can be stored at each pixel before saturation occurs.
[b]The rms read noise; at fast frame rates (1 kfps) the read noise will be the dominant dark noise.
[c]A back-illuminated camera will have a quantum efficiency (number of photoelectrons/photon) of about 0.9. A front-illuminated camera will have a quantum efficiency of <0.4.
[d]www.scimedia.co.jp
[e]www.dalsa.com
[f]www.redshirtimaging.com

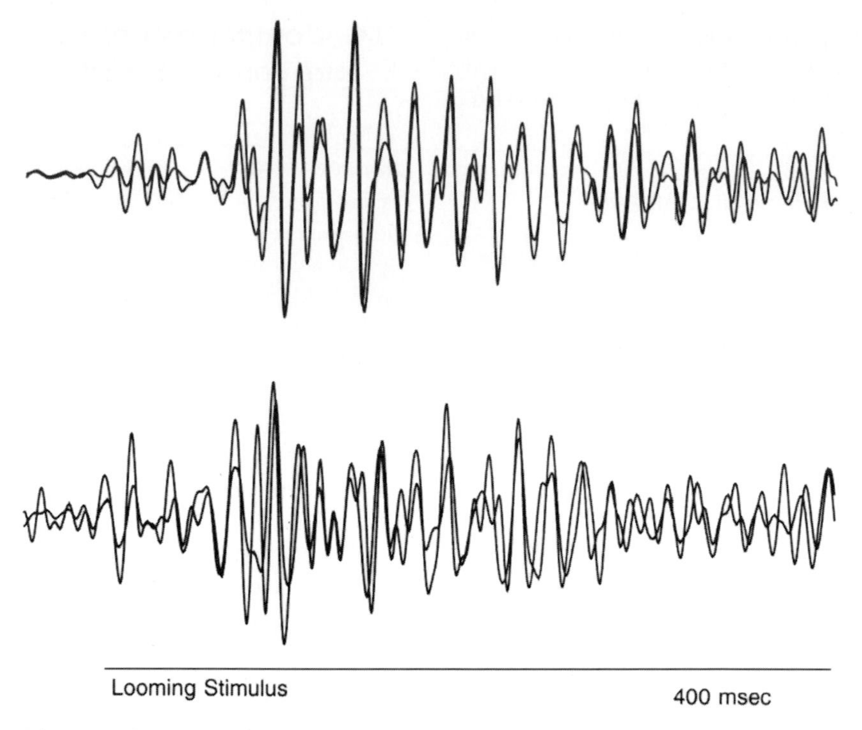

Figure 8 Comparison of voltage-sensitive dye and local field potential recordings. Simultaneous optical and local field potential recordings were made from two positions on the turtle visual cortex that were separated by 2.3 mm. The top pair of superimposed traces are the local field potential recordings from the two sites. The bottom pair of traces are the voltage-sensitive dye recordings from the two sites. There is much less overlap in the optical recordings. Thus, the voltage-sensitive dye recordings have better spatial resolution. Both sets of recordings were band-pass filtered (10–30 Hz) (J. Prechtl, L. B. Cohen, and D. Kleinfeld, unpublished results).

top of a slow "DC" signal. This kind of local field potential signal implies that a population of neurons is somehow synchronously active. Since its first discovery in the hedgehog (Adrian, 1942), odor-induced oscillations have been observed across phylogenetically distant species including locust, frog, turtle, rabbit, monkey, and human. We measured the voltage-sensitive dye signal that accompanies these oscillations in the box turtle. Because the optical measurements have a spatial resolution that is about 25 times better than local field potential measurements (Fig. 8), a more detailed visualization of the spatiotemporal characteristics of the oscillations was obtained.

A. Initial Dye Screening

In initial experiments, several voltage-sensitive dyes were screened using an *in vitro* preparation. The dyes were dissolved in turtle saline (see below); the olfactory bulbs were incubated in the dye solution for 60 min. The bulb was then imaged on the photodiode array (NeuroPlex) and the olfactory nerve was shocked via a suction electrode (Orbach and Cohen, 1983). Both the signal size and the penetration of the dye into the bulb were measured; the results are shown in Table 2. The styryl dye RH414 (Grinvald *et al.*, 1994), 0.01 to 0.2 mg/ml in saline, penetrated throughout the thickness of the bulb and had a relatively large signal. The dye staining appeared to be uniform in the different layers of the bulb, suggesting that the dye stains all cell types approximately equally.

Table 2 Results of Dye Screening on the Turtle Olfactory Bulb

Dye	$\Delta F/F$ (%)	Depth of staining (mm)	Structures R	n	X
JPW 1063[a]	0.4–0.9	200	Butyl	2	Et_3N^+
JPW 1113[a]	<0.4	200–300	Ethyl	3	SO_3
JPW 2005[a]	0.6–1.7	100–800	Ethyl	3	Et_3N^+
JPW 3031[a]	0.7–0.9	300–400	Propyl	3[b]	Me_2,EtOHN$^+$
RH 414[c]	1.9–2.4	600–800	Ethyl	3	Et_3N^+
RH 773[c]	1.1–1.6	600–800	Ethyl	3	Me_2,EtOHN$^+$
RH 795[c]	1	Uneven	Ethyl	3[b]	Me_2,EtOHN$^+$

[a]J. P. Wuskell and L. Loew, personal communication.
[b]2-OH
[c]A. Grinvald *et al.* (1994).

B. Methods

Three species of box turtle, *Terepene carolina, T. triunguis,* and *T. ornata,* 300–600 g, were used. The turtles were anesthetized by immersion in wet ice for 2 h. A craniotomy was performed over the olfactory bulb using a Dremel drill with a small round bit. To facilitate staining the dura and arachnoid mater were then carefully removed. A segment of polyethylene tubing was inserted into the outlet of the nasal cavity in the roof of the mouth and fixed in place by Krazy Glue and epoxy. Suction applied to this tube was used to draw odorant through the nose.

The turtle was then allowed to warm to room temperature; during this time the bulb was stained by superfusing a solution of the dye in turtle saline on the bulb for 60 min. The solution was changed every 10 min. To reduce movement artifacts during the optical recording, the animals were restrained. The tip of the nose was clamped to the recording apparatus with a piece of flexible plastic and the body was taped to the apparatus. These procedures, as well as those used for the mouse (see below), were approved by the Yale University and the Marine Biological Laboratory Animal Care and Use Committees.

Odor delivery was done with an olfactometer (Fig. 9, left, bottom) copied from Kauer and Moulton (1974) with minor modifications. The output of the odor from the applicator was monitored by measuring the CO_2 in the carrier gas with a CO_2 detector (Beckman medical gas analyzer, LB-2, Schiller Park, IL). The upper trace in Fig. 9, left, top, represents the command pulse sent to the solenoid pump controlling odor delivery; the lower trace is the CO_2 level detected by the gas analyzer. The concentration of odorant used in Fig. 10 is 10% cineole: 1 volume of cineole mixed with 9 volumes of clean air. The accuracy and stability of the flow dilution system over

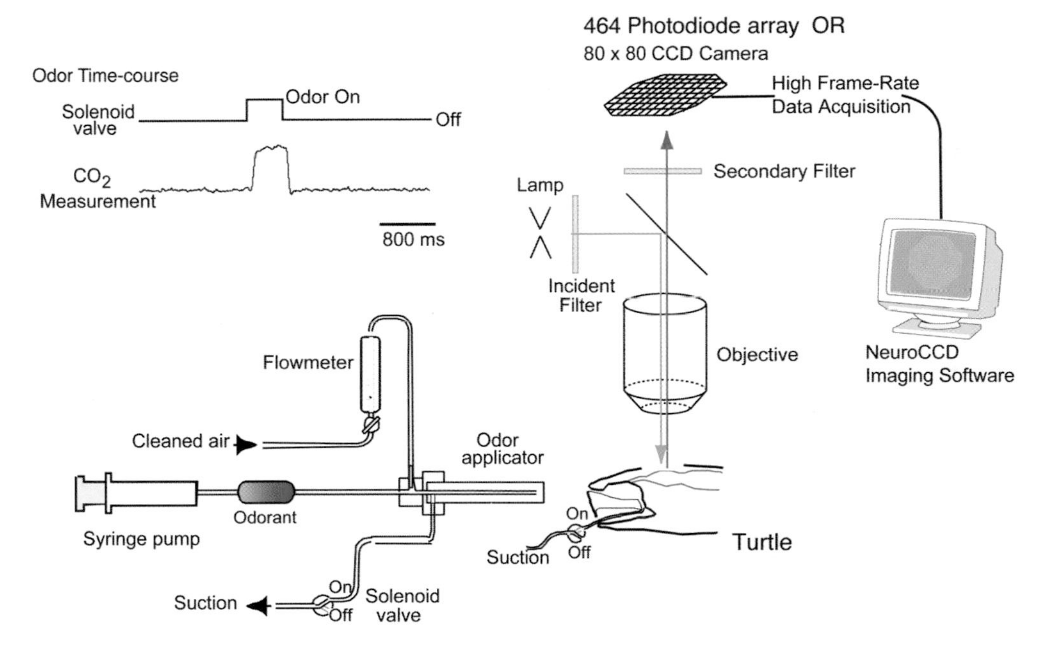

Figure 9 (**Left**) Top: Time course of the odor output from the olfactometer measured by monitoring the CO_2 in the carrier gas. The upper trace shows the time course of the command pulse delivered to the suction solenoid of the outer barrel of the odor applicator. The lower trace is the output of the CO_2 detector probe. There is a delay of about 100 ms between the command pulse and the arrival of the pulse at the CO_2 detector. The odor pulse is approximately square-shaped. Bottom: Schematic diagram of the olfactometer. Compressed air containing 1% CO_2 was used as the carrier gas. It was cleaned, desiccated, and then mixed with room air saturated with odorant vapor in the odor applicator. The flow rates of the air and the odorant vapor were controlled by a flowmeter and a syringe pump, respectively. The odor applicator had two barrels; the outer one was normally under suction to remove the odor. Turning off of the suction to the outer barrel released odorant from the end of the applicator. (**Right**) Schematic diagram of the optical imaging apparatus. The olfactory bulb was illuminated using a 100-W tungsten halogen lamp (voltage-sensitive dye) or a xenon arc lamp (calcium dye). The incident light passed through a heat filter and a band-pass interference filter (520 nm for voltage-sensitive dye and 480 nm for the calcium dye) and was reflected onto the preparation by a long-pass dichroic mirror (590 nm for the voltage-sensitive dye and 510 nm for the calcium dye). For the voltage-sensitive dye experiments the image of the preparation was formed by a 25-mm, 0.95 f camera lens onto a 464-element photodiode array after passing through a 610-nm long-pass secondary filter. For the calcium dye the image of the preparation was formed by a 10.5× or 14× objective lens onto a 80 × 80 CCD camera after passing through a 530-nm long-pass secondary filter. The secondary filter is needed to block reflected incident wavelengths that are transmitted by the dichroic mirror.

the range used was confirmed with a photoionization detector. Dedicated lines for each odorant avoided cross-contamination.

Optical imaging was carried out with optics (Fig. 9, right) optimized for light collection efficiency at low magnification. Because the fluorescence intensity in epifluorescence is proportional to the fourth power of the objective numerical aperture and conventional microscope optics have small numerical apertures at low magnifications, we used a 4× Macroscope (RedShirtImaging, LLC) based on a 25-mm focal length, 0.95 f, C-mount camera lens (with the C-mount end facing the preparation) (Salama, 1988; Ratzlaff and Grinvald, 1991). The intensity reaching the photodetector was 100 times larger with the Macroscope than with a conventional 4×, 0.16 NA, microscope lens.

Fluorescence was measured using a 464-element photodiode array camera (NeuroPlex) placed at the real inverted image formed by the Macroscope. The preparation was illuminated using a 100-W tungsten–halogen lamp. The excitation filter was 520 ± 45 nm. A 580-nm long-pass dichroic mirror was used to reflect the excitation light onto the preparation. The secondary filter was a RG610 long-pass filter. The band-pass filters in the amplifiers were set to 0.07–125 Hz. The data were recorded at a frame rate of 250 Hz.

C. Example Result

The recordings of voltage-sensitive dye responses to odorant stimulation from seven selected diodes are shown on the right of Fig. 10. The locations of these diodes are indicated by the numbered squares on the image of the olfactory bulb on the left. In rostral locations (detectors 1 and 2), there was a relatively long-lasting, 15-Hz oscillation with a long latency. On a diode from a middle location (detector 4) there was a relatively brief, short-latency oscillation and, on a diode from the caudal bulb (detector 7), the oscillation was of a lower frequency and long latency. In areas between two regions, the recorded oscillations were combinations of two signals. In addition, a DC signal, which appears as a single peak after high-pass filtering in Fig. 10, was observed over

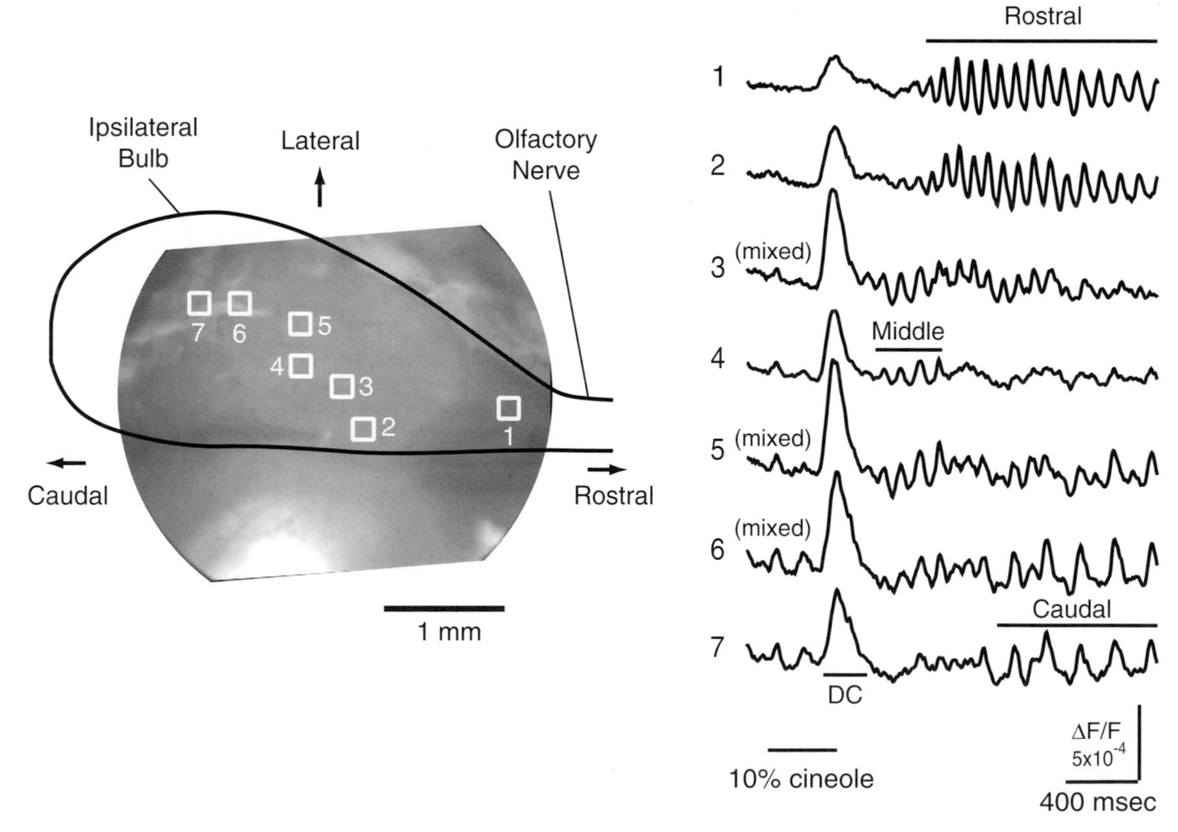

Figure 10 Simultaneous optical recordings from seven different areas of an olfactory bulb. An image of the olfactory bulb is shown on the left. Signals from seven selected pixels are shown on the right. The positions of these pixels are labeled with squares and numbers on the image of the bulb. All seven signals have a filtered version of the DC signal at the time indicated by the bar labeled "DC." The oscillation in the rostral region has a high frequency and relatively long latency and duration (detectors 1 and 2). The oscillation from the middle region has a high frequency and short latency and duration (detector 4). The oscillation from the caudal region has a lower frequency and the longest latency (detector 7). The signals from detectors between these regions (3, 5, and 6) appear to contain a mixture of two components. The horizontal line labeled "10% cineole" indicates the time of the command pulse to the odor solenoid. The data are filtered by high-pass digital RC (5 Hz) and low-pass Gaussian (30 Hz) filters.

most of the ipsilateral olfactory bulb. In addition to differences in frequency and latency, the three oscillations also had different shapes—the rostral and caudal oscillations had relatively sharp peaks while the middle oscillation was more sinusoidal. Thus, as a result of making a simultaneous measurement from many sites using a signal with improved spatial resolution, three independent oscillations were identified in the turtle olfactory bulb.

The noise in the above measurement of fluorescence from a bulk-stained vertebrate brain is consistent with expectations from a shot-noise-limited measurement (Section V). In the turtle experiments, the photocurrent on each detector was about 2×10^{-8} A (equivalent to 10^8 photons/ms). Because we digitally low-pass filtered the data at 30 Hz, the effective sample period was 33 ms and thus the number of photons/sample period was 3×10^9. The shot noise in this measurement should have then been about 2×10^{-5} of the resting intensity (Fig. 6). Consistent with this prediction, the noise in the measurements shown in Fig. 10 is less than 10^{-4} of the resting fluorescence.

XI. Calcium Dye Recording in the Mouse Olfactory Bulb

Voltage-sensitive dyes were effective in revealing spatiotemporal patterns of global activity in the olfactory bulb (Section X). In contrast, selective labeling of olfactory receptor neuron terminals with a calcium-sensitive dye was used to visualize the spatial patterns in the input to the bulb. While functional imaging methods such as *f*MRI and intrinsic imaging reveal patterns of glomerular activity (e.g., Xu *et al.*, 2000; Rubin and Katz, 1999), with these methods it is difficult to relate the signals to a particular class of neurons. Labeling neurons with voltage- or calcium-sensitive dyes via retrograde or anterograde transport allows selective monitoring of activity in defined neuronal populations (e.g., O'Donovan *et al.*, 1993; Tsau *et al.*, 1996; Kreitzer *et al.*, 2000). This approach was first developed for the olfactory system by Friedrich and Korsching (1997, 1998) in the zebrafish and later adapted for use in the turtle (Wachowiak *et al.*, 2002) and mouse (Wachowiak and Cohen, 2001).

Spatially organized patterns of neuronal activity have long been hypothesized to play an important role in odorant recognition (Adrian, 1953; Kauer, 1991). In the mouse, the majority of olfactory receptor neurons express 1 of ~1000 olfactory receptor proteins (Malnic *et al.*, 1999). All of the 10,000–20,000 neurons expressing the same receptor protein converge onto a few (one to three) glomeruli in stereotyped locations of the olfactory bulb (Vassar *et al.*, 1994). Thus, odorant-evoked activity of receptor neurons distributed across the olfactory epithelium is transformed into a spatially organized pattern of input to olfactory bulb glomeruli. We used calcium-sensitive dyes to image receptor neuron input to glomeruli in the dorsal olfactory bulb of the mouse.

A. Initial Dye Screening

We tested one calcium-sensitive dye, Calcium Green-1 (Molecular Probes, Eugene, OR), that was not conjugated to dextran. While it did appear to label receptor neurons in the olfactory epithelium, we did not detect labeling in the receptor neuron terminals in the olfactory bulb. Thus, our choice of calcium dyes was limited to those that can be obtained as dextran conjugates. We tried both Calcium Green-1 dextran and Fluo-4 dextran (Kreitzer *et al.*, 2000). In our hands, labeling was more reliable and the fluorescence signals were larger with Calcium Green-1 dextran. We tested both the 3000- and the 10,000-kDa dextran conjugates of Calcium Green-1. No clear difference was observed. We found that a concentration of 4% Calcium Green-1 dextran resulted in sufficiently bright labeling of olfactory bulb glomeruli. Concentrations of 2% or lower were considerably less bright. The Calcium Green-1 dextran solution was stored at 4°C for up to 2 weeks before use. Freeze–thawing this solution appeared to result in loading of receptor neurons but little or no transport of dye to the axon terminal. While loading and transport appeared to be successful with Fluo-4 dextran, labeling of glomeruli was less clear, presumably because of the low resting fluorescence of this probe (Kreitzer *et al.*, 2000). Also, odorant-evoked Fluo-4 dextran signals were smaller in amplitude and not as reliably detected across preparations.

B. Methods

Because dextran-conjugated dyes are membrane impermeant, loading olfactory receptor neurons with Calcium Green-1 dextran requires treatment with a permeabilizing agent. Friedrich and Korsching (1997) found that coapplication of Calcium Green-1 dextran with a dilute solution of Triton X-100 detergent was an effective method for loading zebrafish olfactory receptor neurons. Triton treatment appears to cause a transient loss of the receptor neuron cilia, which contain olfactory receptor proteins as well as the transduction machinery necessary for generating an electrical response to odorant stimulation. The cilia and molecular components necessary for odorant responsiveness appear to regenerate normally after 1–2 days (Friedrich and Korsching, 1997; Wachowiak and Cohen, 2001). However, the concentration of Triton used in the loading procedure is critical for obtaining adequate labeling with minimal damage to the olfactory epithelium and must be determined empirically for each new species used.

For loading mouse olfactory receptor neurons with Calcium Green-1 dextran, mice were anesthetized with ketamine (90 mg/kg, ip) and xylazine (10 mg/kg, ip) and 2 μl of 0.25% Triton X-100 in mouse Ringers was injected into the nasal cavity. After 60 s, 8 μl of 4% Calcium Green-1 dextran was injected at a rate of ~0.4 μl/min. Mice recovered from anesthesia and were held for 4–8 days before imaging. We found that this detergent "shock" procedure labeled more consistently and with less damage to the epithelium than application of Calcium Green-1 dextran dissolved in Triton. With higher Triton concentrations or more prolonged application, axon terminals in the olfactory bulb glomeruli were well labeled but the epithelium was severely damaged, as evaluated by the lack of an EOG response or by visual inspection. It was difficult to achieve homogeneous loading of olfactory receptor neurons projecting to glomeruli in all regions of the olfactory bulb. However, if the animal was placed on its back throughout the loading procedure, widespread labeling of most, if not all, dorsal glomeruli was achieved with a success rate of >70%.

For imaging, mice were anesthetized with pentobarbital (50 mg/kg, ip). A double trachaeotomy was performed so that an artificial sniff paradigm could be used for precise control of odorant access to the nasal cavity. This helped to ensure a rapid onset of the signal, which was important when multiple trials were averaged. The mice breathed freely through the lower trachaeotomy tube. The bone overlying the olfactory bulbs was thinned and Ringers solution and a coverslip was placed over the exposed area. The heart rate was maintained at 400–500 bpm by periodic injection of pentobarbital and was sometimes stabilized by directing a continuous stream of pure oxygen over the lower tracheotomy tube. The dorsal surface of one olfactory bulb was illuminated with 480 ± 25 nm light using a 150-W xenon arc lamp (Opti-Quip, Highland Mills, NY) and 515-nm long-pass dichroic mirror, and fluorescence emission above 530 nm was collected (Fig. 9, right). Images were acquired and digitized with an 80 × 80-pixel CCD camera (NeuroCCD; RedShirtImaging LLC) at 100–200 Hz and time-binned to a 25-Hz frame rate before being stored to disk. Fluorescence was imaged using a 10.5×, 0.2 NA objective (spatial resolution, 22 μm per pixel assuming no scattering or out-of-focus signals) or a 14×, 0.4 NA objective (16.5 μm per pixel resolution). The olfactometer used was the same as that described above for the turtle experiments.

While odorant-evoked signals were detected in single trials (e.g., Fig. 11C), we typically collected, then averaged, responses of two to four consecutive odorant presentations in order to improve the signal-to-noise ratio and to obtain a measure of trial-to-trial variability. We waited a minimum of 45 s between trials. Repeated presentations of the same odorant at this interstimulus interval evoked signal amplitudes that typically varied by less than 10%. The primary source of extrinsic noise was movement associated with respiration and heartbeat. The noise was largest in regions adjacent to major blood vessels, and so pixels overlying these regions were removed from the data set (omitted) prior to analysis. Occasional trials with widespread artifactual signals (primarily due to movement) were discarded. After averaging, data from each pixel were temporally filtered with a 1- to 2-Hz low-pass Gaussian and a 0.017-Hz high-pass digital RC filter (both filters have a low sharpness).

To correct for unequal labeling of glomeruli, the signal from each pixel was divided by its resting fluorescence obtained at the beginning of each trial. Because a significant part of the resting fluorescence arises from dye in axons, which appear not to experience an increase in calcium (Wachowiak and Cohen, 1999), this correction is only partially successful. To construct the spatial maps of input to the bulb, response amplitudes for each pixel were measured by subtracting the temporal average of a 400-ms time window just preceding the stimulus from a 400-ms temporal average centered around the peak of the response. For display, maps of response amplitudes were smoothed slightly by increasing the pixel dimensions from 80 × 80 to 160 × 160 and interpolating between pixels. Response maps were normalized to the maximum signal amplitude for that trial. No spatial filtering or background subtraction was performed.

Response amplitudes for individual glomeruli were measured from the maps in two ways: by averaging responses from 4 adjacent pixels in the center of the glomerulus and by measuring the amplitude of a one-dimensional Gaussian function fit to a profile (2–4 pixels wide) of the signal through a glomerulus (Meister and Bonhoeffer, 2001). The first method gives a measure of signal amplitude relative to the resting fluorescence, while the second method gives a measure of amplitude relative to the local background signal. The maps of signal amplitude are a measure of the activation of a population of 10,000 olfactory receptor neurons converging onto each glomerulus.

Strongly activated glomeruli could be easily counted by visual inspection. For counting glomeruli with smaller amplitude signals, the signal profile was fit to a Gaussian and evaluated according to criteria for size (half-width within 2 SD of the mean measured for 101 test glomeruli), amplitude (amplitude greater than five times the root mean square spatial noise, measured from adjacent nonactivated areas), and appearance in multiple trials.

C. Example Result

Calcium Green-1 dextran loading resulted in labeling of olfactory receptor axon terminals in the olfactory bulb (Figs. 11A and 11B). We imaged odorant responses from the dorsal olfactory bulb of anesthetized mice 4–8 days after loading. Previous experiments using this imaging method in the zebrafish and turtle olfactory bulbs showed that the Calcium Green-1 fluorescence increases reflect action potential-evoked calcium influx into receptor neuron axon terminals (Friedrich and Korsching, 1997; Wachowiak and Cohen, 1999). Odorant presentation evoked rapid (200- to 500-ms rise time) increases in fluorescence up to 6% $\Delta F/F$ (Fig. 11C). Spatial maps of the response amplitude measured from each pixel showed well-defined foci of fluorescence increases (Fig. 11D), often corresponding to individual glomeruli visible from the resting fluorescence (Fig. 11B). Odorant-evoked

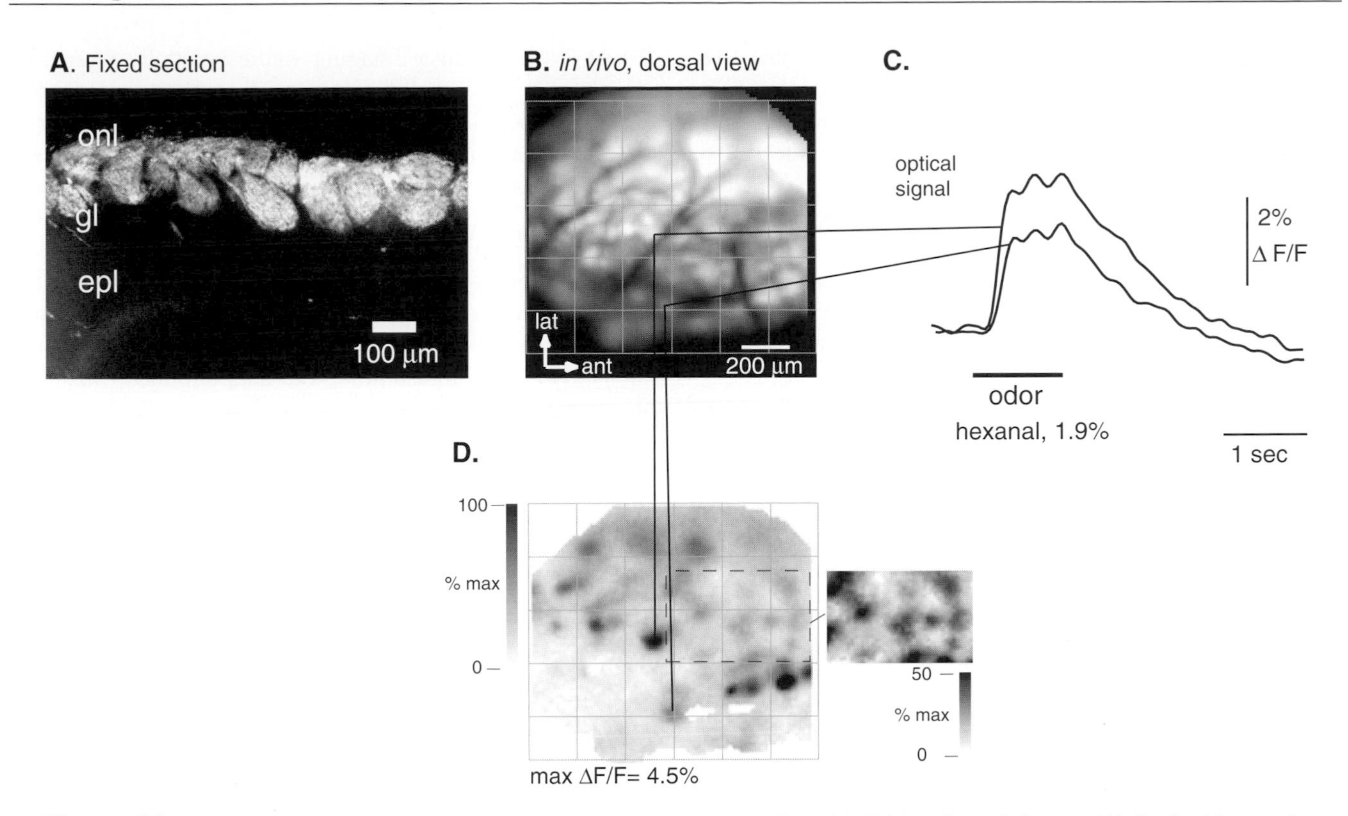

Figure 11 Imaging mouse olfactory receptor neuron activation after *in vivo* loading with Calcium Green-1 dextran. (**A**) Confocal image of a section through one olfactory bulb fixed 5 days after loading. Fixable rhodamine dextran (10 kDa) was used instead of Calcium Green-1 dextran to preserve labeling in fixed tissue. Glomeruli are strongly labeled. There is no evidence of transsynaptic labeling. onl, olfactory nerve layer; gl, glomerular layer; epl, external plexiform layer. (**B**) Resting Calcium Green-1 dextran fluorescence imaged *in vivo,* 7 days after loading. The image was contrast-enhanced to emphasize individual glomeruli. Blood vessels appear as dark lines. The saturated regions in the upper right are from olfactory nerve bundles that obscure underlying glomeruli. Lines originate from two glomeruli whose responses are shown in (C) and (D). (**C**) 1.9% hexanal evoked rapid (~200 ms rise time) increases in fluorescence in the two glomeruli indicated in (B) (lines). Each trace shows the optical signal measured from 1 pixel and from a single trial after band-pass filtering from 0.017 to 2 Hz. (**D**) Gray-scale map of the evoked signal for the trial shown in (C), showing foci of fluorescence increases. A region of the map normalized to 50% of the maximum signal (inset) shows additional smaller amplitude foci. Signal foci in the upper portions of the image are blurred due to out-of-focus glomeruli and light scattering from overlying axons.

signals showed a range of amplitudes in different glomeruli. Using an expanded gray scale (Fig. 11D, inset), it is clear that even smaller signals have glomerular localization. The spatial distribution and amplitude of the signals were consistent across repeated odorant presentations and were different for different odorants.

The noise in this measurement of fluorescence from transported dye is also consistent with expectations from a shot-noise-limited measurement (Section V). In these measurements made at a frame rate of 100 Hz, the number of photoelectrons per pixel per millisecond was approximately 2×10^4, much lower than in the measurements from the voltage-sensitive dye described above. Because we digitally low-pass filtered the data at 2 Hz, the effective sample period is 500 ms and thus the number of photons/sample period is 10^7. The shot noise in this measurement should then be approximately 3×10^{-4} of the resting intensity (Fig. 6). Consistent with this prediction, the noise in the measurements shown in Fig. 11C is less than 2×10^{-3} of the resting intensity. In preliminary measurements, we found that the shot noise and the noise from respiration and heart beat were similar in magnitude. For some detectors the noise was dominated by shot noise, for others it was dominated by the extraneous movement noise.

The spatial resolution shown in Fig. 11D is on the order of 20 μm, far better than might have been anticipated from the measurements illustrated in Fig. 7. However, both factors that could contribute to blurring are minimized in the measurements shown in Fig. 11. First, scattering will be minimal because the Calcium Green-1 dextran is only in the outer layer of the olfactory bulb. Second, out-of-focus signals will be nonexistent because the glomerular layer is only 100 μm thick.

In addition to the rapid, dye-dependent increase in fluorescence due to presynaptic calcium influx, odorants sometimes evoked a slower and longer lasting decrease in fluorescence, which we attributed to changes in the intrinsic optical properties of the olfactory bulb tissue (absorption or light scattering). This intrinsic signal was often apparent only at higher odorant concentrations (>1% of saturated vapor) and was strongest along major blood vessels (Fig. 12). Focal signals indicative of glomeruli were not apparent with this intrinsic signal. In freely breathing animals anesthetized with Nembutal, this signal had a slow onset and relatively small amplitude and so did not affect the spatial character or amplitude of the calcium signal. However, artificial respiration with pure oxygen caused an increase in the amplitude and speed of onset of the intrinsic signal such that the calcium signal was partially or completely obscured. A similar effect was seen in animals anesthetized with urethane (3 g/kg, ip). We have not tested other anesthetics.

Intrinsic imaging studies have reported fewer glomeruli activated by comparable concentrations of these same odorants (Belluscio and Katz, 2001; Meister and Bonhoeffer, 2001). These intrinsic imaging studies spatially filtered or thresholded the signals before counting glomeruli. To test whether either kind of processing might affect the number of detected glomeruli, Fig. 13B shows the response to 1.9% hexanal (from Fig. 13A) after smoothing with a Gaussian kernel and subtracting the smoothed data (Meister and Bonhoeffer, 2001), and Fig. 13C shows the same response after thresholding the data at 2 standard deviations above the mean (Belluscio and Katz, 2001). High-pass filtering preserves nearly all of the glomeruli visible in the unfiltered image. However, thresholding the data eliminates many of the glomeruli.

Five glomeruli were identifiable across animals based on their location and responses to one or two odorants; all five

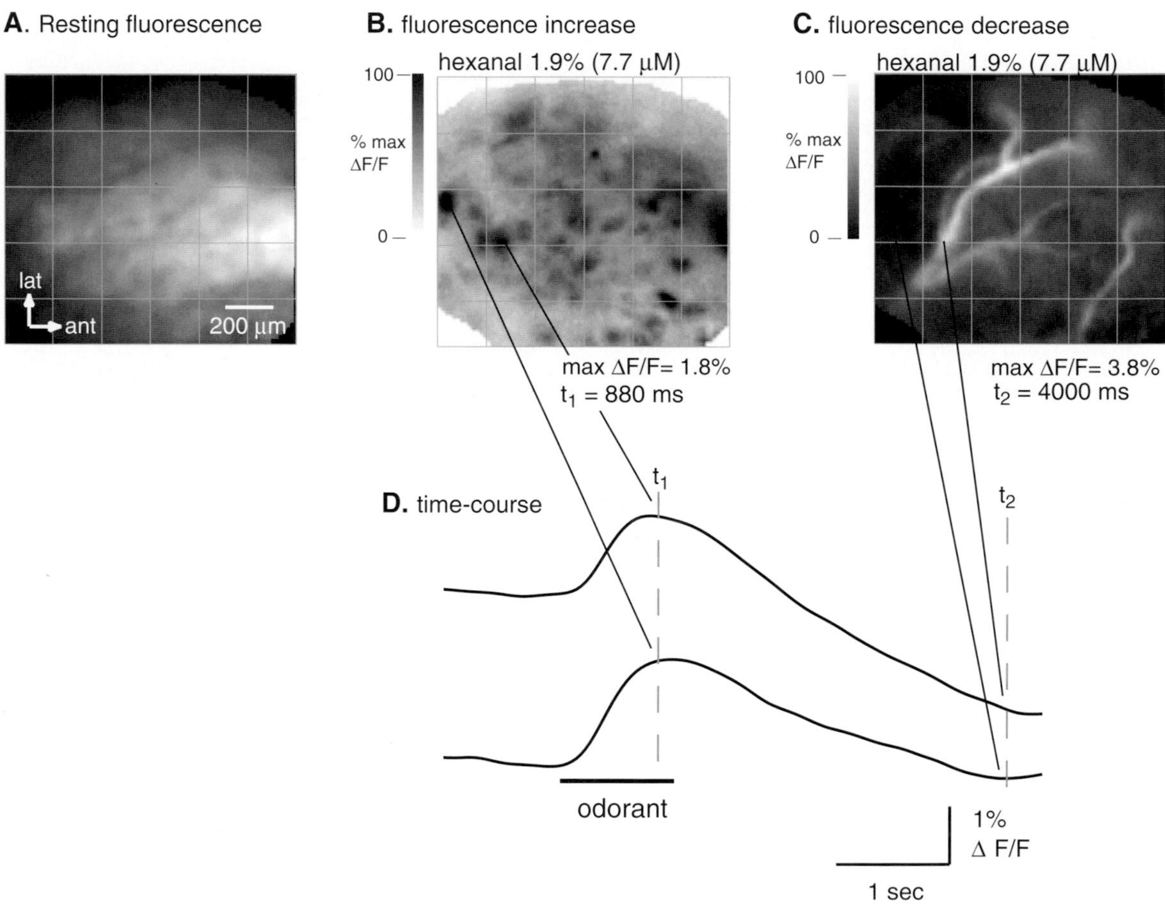

Figure 12 Odorant stimulation can evoke decreases in fluorescence that reflect changes in intrinsic optical signals. (**A**) Resting Calcium Green-1 dextran fluorescence imaged from the dorsal olfactory bulb at 14× magnification. This image was not contrast enhanced. This preparation was anesthetized with Nembutal and artificially ventilated with pure oxygen. (**B**) Response map of the fluorescence increase evoked by hexanal. The map was constructed as described in the text, with the time window for measuring response amplitude centered at t_1 (880 ms after odorant onset). In this map, darker shades represent larger increases in fluorescence. There were no decreases in fluorescence apparent at this latency. Hexanal evokes input to many widespread glomeruli. The time course of the response measured from the two glomeruli indicated by the lines is shown in (D). (**C**) Response map of the "fluorescence" decrease evoked by the same presentation of hexanal. The time window for measuring the response amplitude was centered at t_2 (4000 ms after odorant onset). In this map, lighter shades represent larger decreases in fluorescence, with black indicating no change. There were no increases in fluorescence apparent at this latency. The largest decreases in fluorescence occur along blood vessels. (**D**) Time course of the optical signal recorded from the two locations indicated by the lines in (B) and (C). Each trace is from a single pixel and is filtered from 0.017 to 1 Hz. The top trace, which is from a location near a major blood vessel, shows a strong decrease in fluorescence (t_2) following the initial fluorescence increase (t_1).

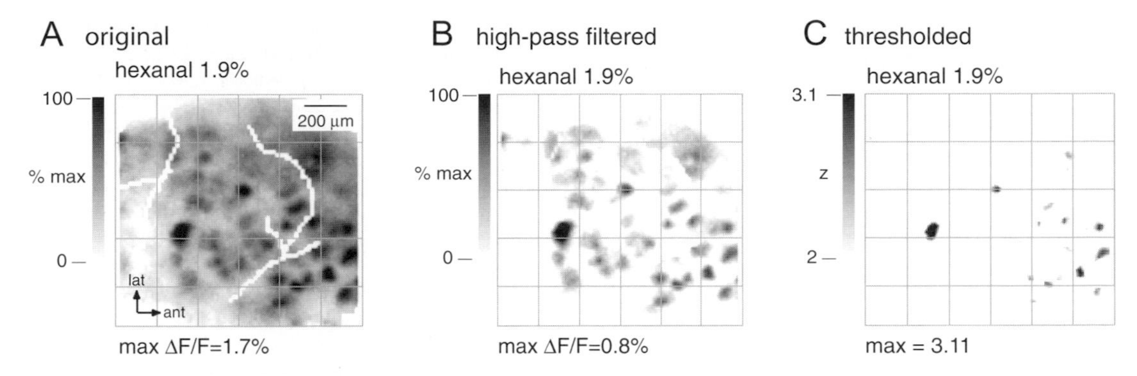

Figure 13 Response maps evoked by 1.9% hexanal (14× magnification) in an *in vivo* mouse olfactory bulb preparation. (**A**) The original data before spatial filtering or thresholding. (**B**) The same data as in (A) after smoothing with a Gaussian kernel of width $\sigma = 140$ μm and subtracting the smoothed from the original image. (**C**) The same data as in (A) after thresholding at 2 SD above the mean.

showed complex response specificities to odorants of different functional groups and molecular size. Maps of receptor neuron input were chemotopically organized at near-threshold concentrations but, at moderate concentrations, they involved many widely distributed glomeruli (Fig. 11D). These results suggest a high degree of complexity in odorant representations at the level of input to the olfactory bulb.

XII. Intrinsic Imaging and Fluorescence Signals from In Vivo Mammalian Brain

Measuring intrinsic imaging and/or fluorescence signals from directly stained *in vivo* mammalian preparations is more difficult than that described above for the mouse because the fractional intensity changes are smaller and the light intensity is larger and thus interference from movement noise is larger. Two methods for reducing the movement artifacts from the heartbeat are, together, quite effective in such measurements. First, a subtraction procedure is used in which two recordings are made but only one of the trials has a stimulus (Orbach *et al.,* 1985). Both recordings are triggered from the upstroke of the electrocardiogram so both should have similar heartbeat noise. Second, an air-tight chamber is fixed onto the skull surrounding the craniotomy (Blasdel and Salama, 1986). When this chamber is filled with silicon oil and closed, the movements due to heartbeat and respiration are substantially reduced.

XIII. Summary and Future Directions

In this chapter we have described methods for making optical measurements of neuron activity with the goal of optimizing the signal-to-noise ratio and thus optimizing the information that can be obtained from the measurement. While it is clear that attention to a large number of factors is required, many of these factors are fairly well understood.

Because the light-measuring apparatus is already reasonably optimized (see above), any improvement in the sensitivity of these optical measurements will need to come from the development of better dyes and/or investigating signals from additional optical properties of the dyes. The voltage-sensitive dyes in Fig. 2 and the vast majority of those synthesized are of the general class named polyenes (Hamer, 1964). While it is possible that improvements in signal size can be obtained with new polyene dyes [see Waggoner and Grinvald (1977) and Fromherz *et al.* (1991) for a discussion of maximum possible fractional changes in absorption and fluorescence], the signal size on squid axons has not increased in recent years (Gupta *et al.,* 1981; L. B. Cohen, A. Grinvald, K. Kamino, and B. M. Salzberg, unpublished results), and most improvements (e.g., Grinvald *et al.,* 1982; Momose-Sato *et al.,* 1995; Tsau *et al.,* 1996; Antic and Zecevic, 1995; Shoham *et al.,* 1999) have involved synthesizing analogues that work well in new applications or on specific preparations.

The best of the styryl and oxonol polyene dyes have fluorescence changes of 10–20%/100 mV in situations in which the staining is specific to the membrane whose potential is changing (Grinvald *et al.,* 1982; Loew *et al.,* 1992). Recently, Gonzalez and Tsien (1995) introduced a new scheme for generating voltage-sensitive signals using two chromophores and energy transfer. While these fractional changes were also in the range of 10%/100 mV, more recent results are about 30% (J. Gonzalez and R. Tsien, personal communication). However, because one of the chromophores must be hydrophobic and does not penetrate into brain tissue, it has not been possible to measure signals with a fast pair of dyes in intact tissues (T. Gonzalez and R. Tsien, personal communication; A. Obaid and B. M. Salzberg, personal communication).

An important new direction is the development of methods for neuron-type-specific staining. Three quite different approaches have been tried. First, the use of retrograde staining procedures has recently been investigated in the embryonic chick and lamprey spinal cords (Tsau *et al.*, 1996). An identified cell class (motoneurons) was selectively stained. While spike signals from individual neurons were sometimes measured in lamprey experiments, further efforts at optimizing this staining procedure are needed. The second approach is based on the use of cell-type-specific staining developed for fluorescein by Nirenberg and Cepko (1993). It might be possible to use similar techniques to selectively stain cells with voltage-sensitive or ion-sensitive dyes. Third, Siegel and Isacoff (1997) constructed a genetically encoded combination of a potassium channel and green fluorescent protein. When introduced into a frog oocyte, this molecule had a (relatively slow) voltage-dependent signal with a fractional fluorescence change of 5%. More recently, Sakai *et al.* (2001) and Ataka and Pieribone (2001) have developed similar constructs with very rapid kinetics. Neuron-type-specific staining would make it possible to determine the role of specific neuron types in generating the input–output function of a brain region.

Optical recordings already provide unique insights into brain activity and organization. Clearly, improvements in sensitivity or selectivity would make these methods more powerful.

Acknowledgments

The authors are indebted to their collaborators Vicencio Davila, Amiram Grinvald, Kohtaro Kamino, David Kleinfeld, Les Loew, Bill Ross, Guy Salama, Brian Salzberg, Alan Waggoner, Jian-young Wu, Joe Wuskell, and Dejan Zecevic for numerous discussions about optical methods. This work was supported by NIH Grant NS08437-DC05259, NSF Grant IBN-9812301, a Brown-Coxe fellowship from the Yale University School of Medicine, and an NRSA fellowship, DC 00378.

References

Adrian, E. D. (1942). Olfactory reactions in the brain of the hedgehog. *J. Physiol.* **100,** 459–473.

Adrian, E. (1953). Sensory messages and sensation: The response of the olfactory organ to different smells. *Acta Physiol Scand.* **29,** 5–14.

Antic, S., and Zecevic, D. (1995). Optical signals from neurons with internally applied voltage-sensitive dyes. *J. Neurosci.* **15,** 1392–1405.

Ataka, K., and Pieribone, V. A. (2001). A genetically-targetable fluorescent probe of channel gating with rapid kinetics. *Biophys. J.*, **82**, 509–516.

Belluscio, L., Katz, L. C. (2001). Symmetry, stereotypy, and topography of odorant representations in mouse olfactory bulbs. *J. Neurosci.* **21**, 2113–2122.

Blasdel, G. G., and Salama, G. (1986). Voltage-sensitive dyes reveal a modular organization in monkey striate cortex. *Nature* **321,** 579–585.

Braddick, H. J. J. (1960). Photoelectric photometry. *Rep. Prog. Phys.* **23,** 154–175.

Dainty, J. C. (1984). "Laser Speckle and Related Phenomena." Springer-Verlag, New York.

Denk, W., Piston, D. W., and Webb, W. (1995). Two-photon molecular excitation in laser-scanning microscopy. *In* "Handbook of Biological Confocal Microscopy" (J. W. Pawley, ed.), pp. 445–458. Plenum, New York.

Friedrich, R., and Korsching, S. (1997). Combinatorial and chemotopic odorant coding in the zebrafish olfactory bulb visualized by optical imaging. *Neuron* **18,** 737–752.

Friedrich, R. W., and Korsching, S. I. (1998). Chemotopic, combinatorial, and noncombinatorial odorant representations in the olfactory bulb revealed using a voltage-sensitive axon tracer. *J. Neurosci.* **18,** 9977–9988.

Fromherz, P., Dambacher, K. H., Ephardt, H., Lambacher, A., Muller, C. O., Neigl, T., Schaden, H., Schenk, O., and Vetter, T. (1991). Fluorescent dyes as probes of voltage transients in neuron membranes: Progress report. *Ber. Bunsenges. Phys. Chem.* **95,** 1333–1345.

Gonzalez, J. E., and Tsien, R. Y. (1995). Voltage sensing by fluorescence energy transfer in single cells. *Biophys. J.* **69,** 1272–1280.

Grinvald, A., Hildesheim, T., Farber, I. C., and Anglister, L. (1982). Improved fluorescent probes for the measurement of rapid changes in membrane potential. *Biophys. J.* **39,** 301–308.

Grinvald, A., Lieke, E. E., Frostig, R. D., and Hildesheim, R. (1994). Cortical point-spread function and long-range lateral interactions revealed by real-time optical imaging of macaque monkey primary visual cortex. *J. Neurosci.* **18,** 9977–9988.

Gross, E., Bedlack, R. S., Jr., and Loew, L. M. (1994). Dual-wavelength ratiometric fluorescence measurement of the membrane dipole potential. *Biophys. J.* **67,** 208–216.

Gupta, R. K., Salzberg, B. M., Grinvald, A., Cohen, L. B., Kamino, K., Lesher, S., Boyle, M. B., Waggoner, A. S., and Wang, C. H. (1981). Improvements in optical methods for measuring rapid changes in membrane potential. *J. Membr. Biol.* **58,** 123–137.

Hamer, F. M. (1964). "The Cyanine Dyes and Related Compounds." Wiley, New York.

Hirota, A., Sato, K., Momose-Sato, Y., Sakai, T., and Kamino, K. (1995). A new simultaneous 1020-site optical recording system for monitoring neural activity using voltage-sensitive dyes. *J. Neurosci. Methods* **56,** 187–194.

Iijima, T., Ichikawa, M., and Matsumoto, G. (1989). Optical monitoring of LTP and related phenomena. *Soc. Neurosci. Abstr.,* **15,** 398.

Inoue, S. (1986). "Video Microscopy," p. 128. Plenum, New York.

Jin, W. J., Zhang, R. J., and Wu, J. Y. (2002). Voltage-sensitive dye imaging of population neuronal activity in cortical tissue. *J. Neuroscience Methods* **115**, 13–27.

Kauer, J. (1991). Contributions of topography and parallel processing to odor coding in the vertebrate olfactory pathway. *Trends Neurosci.* **14,** 79–85.

Kauer, J. S., and Moulton, D. G. (1974). Responses of olfactory bulb neurones to odour stimulation of small nasal areas in the salamander. *J. Physiol. (London)* **243,** 717–737.

Kleinfeld, D., and Delaney, K. (1996). Distributed representation of vibrissa movement in the upper layers of somatosensory cortex revealed with voltage-sensitive dyes. *J. Comp. Neurol.* **375,** 89–109.

Kreitzer, A. C., Gee, K. R., Archer, E. A., and Regehr, W. G. (2000). Monitoring presynaptic calcium dynamics in projection fibers by in vivo loading of a novel calcium indicator. *Neuron* **27,** 25–32.

Lam, Y.-w., Cohen, L. B., Wachowiak, M., and Zochowski, M. R. (2000). Odors elicit three different oscillations in the turtle olfactory bulb. *J. Neurosci.* **20,** 749–762.

Loew, L. M., Cohen, L. B., Salzberg, B. M., Obaid, A. L., and Bezanilla, F. (1985). Charge-shift probes of membrane potential. Characterization of aminostyrylpyridinium dyes on the squid giant axon. *Biophys. J.* **47,** 71–77.

Loew, L. M., Cohen, L. B., Dix, J., Fluhler, E. N., Montana, V., Salama, G., and Wu, J. Y. (1992). A napthyl analog of the aminostyryl pyridinium class of potentiometric membrane dyes shows consistent sensitivity in a variety of tissue, cell, and model membrane preparations. *J. Membr. Biol.* **130,** 1–10.

London, J. A., Zecevic, D., and Cohen, L. B. (1987). Simultaneous optical recording of activity from many neurons during feeding in *Navanax*. *J. Neurosci.* **7,** 649–661.

Malnic, B., Hirono, J., Sato, T., and Buck, L. (1999). Combinatorial receptor codes for odors. *Cell* **96,** 713–723.

Meister, M., and Bonhoeffer, T. (2001). Tuning and topography in an odor map on the rat olfactory bulb. *J. Neurosci.* **21,** 1351–1360.

Momose-Sato, Y., Sato, K., Sakai, T., Hirota, A., Matsutani, K., and Kamino, K. (1995). Evaluation of optimal voltage-sensitive dyes for optical measurement of embryonic neural activity. *J. Membr. Biol.* **144,** 167–176.

Nakashima, M., Yamada, S., Shiono, S., Maeda, M., and Sato, F. (1992). 448-detector optical recording system: Development and application to *Aplysia* gill-withdrawal reflex. *IEEE Trans. Biomed. Eng.* **39,** 26–36.

Nirenberg, S., and Cepko, C. (1993). Targeted ablation of diverse cell classes in the nervous system *in vivo. J. Neurosci.* **13,** 3238–3251.

O'Donovan, M. J., Ho, S., Sholomenko, G., and Yee, W. (1993). Real-time imaging of neurons retrogradely and anterogradely labeled with calcium sensitive dyes. *J. Neurosci. Methods* **46,** 91–106.

Orbach, H. S., and Cohen, L. B. (1983). Optical monitoring of activity from many areas of the *in vitro* and *in vivo* salamander olfactory bulb: A new method for studying functional organization in the vertebrate central nervous system. *J. Neurosci.* **3,** 2251–2262.

Orbach, H. S., and Cohen, L. B. (1985). Optical mapping of electrical activity in rat somatosensory and visual cortex. *J. Neurosci.* **5**, 1886–1895.

Petran, M., and Hadravsky, M. (1966). Czechoslovakian patent 7720.

Ratzlaff, E. H., and Grinvald, A. (1991). A tandem-lens epifluorescence microscope: Hundred-fold brightness advantage for wide-field imaging. *J. Neurosci. Methods* **36,** 127–137.

Rubin, B., and Katz, L. (1999). Optical imaging of odorant representations in the mammalian olfactory bulb. *Neuron* **23,** 499–511.

Sakai, R., Repunte-Canonigo, V., Raj, C. D., Knopfel, T. (2001). Design and characterization of a DNA-encoded, voltage-sensitive flourescent protein. *Eur. J. Neurosci,* **13**, 2314–18.

Salama, G. (1988). Voltage-sensitive dyes and imaging techniques reveal new patterns of electrical activity in heart and cortex. *SPIE Proc.* **94,** 75–86.

Salzberg, B. M., Grinvald, A., Cohen, L. B., Davila, H. V., and Ross, W. N. (1977). Optical recording of neuronal activity in an invertebrate central nervous system: Simultaneous monitoring of several neurons. *J. Neurophysiol.* **40,** 1281–1291.

Shoham, D., Glaser, D. E., Arieli, A., Kenet, T., Wijnbergen, C., Toledo, Y., Hildesheim, R., and Grinvald, A. (1999). Imaging cortical dynamics at high spatial and temporal resolution with novel blue voltage-sensitive dyes. *Neuron* **24,** 791–802.

Siegel, M. S., and Isacoff, E. Y. (1997). A genetically encoded optical probe of membrane voltage. *Neuron* **19,** 735–741.

Tsau, Y., Wenner, P., O'Donovan, M. J., Cohen, L. B., Loew, L. M., and Wuskell, J. P. (1996). Dye screening and signal-to-noise ratio for retrogradely transported voltage-sensitive dyes. *J. Neurosci. Methods* **70,** 121–129.

Vassar, R., Chao, S., Sitcheran, R., Nunez, J., Vosshall, L., and Axel, R. (1994). Topographic organization of sensory projections to the olfactory bulb. *Cell* **79,** 981–991.

Wachowiak, M., and Cohen, L. B. (1999). Presynaptic inhibition of primary olfactory afferents mediated by different mechanisms in the lobster and turtle. *J. Neurosci.* **19,** 8808–8817.

Wachowiak, M., and Cohen, L. B. (2001). Representation of odorants by receptor neuron input to the mouse olfactory bulb. *Neuron,* **32**, 723–735.

Wachowiak, M., Cohen, L. B., and Zochowski, M. (2002). Distributed and concentration invariant spatial representations of odorants by receptor neuron input to the turtle olfactory bulb. *J. Neurophysiol.,* **87**, 1035–1045.

Waggoner, A. S., and Grinvald, A. (1977). Mechanisms of rapid optical changes of potential sensitive dyes. *Ann. N. Y. Acad. Sci.* **303,** 217–241.

Wu, J. Y., and Cohen, L. B. (1993). Fast multisite optical measurements of membrane potential. *In* "Fluorescent and Luminescent Probes for Biological Activity" (W. T. Mason, ed.), pp. 389–404. Academic Press, London.

Xu, F., Kida, I., Hyder, F., and Shulman, R (2000). Assessment and discrimination of odor stimuli in rat olfactory bulb by dynamic functional MRI. *Proc. Natl. Acad. Sci. USA* **97,** 10601–10606.

Zecevic, D., Wu, J. Y., Cohen, L. B., London, J. A., Hopp, H. P., and Falk, C. X. (1989). Hundreds of neurons in the *Aplysia* abdominal ganglion are active during the gill-withdrawal reflex. *J. Neurosci.,* **9,** 3681–3689.

5

Optical Imaging Based on Intrinsic Signals

Nader Pouratian and Arthur W. Toga

Laboratory of Neuro Imaging, UCLA Department of Neurology, Los Angeles, California 90095-1769

I. Introduction

Several functional brain mapping techniques have been developed over the past 3 decades which have revolutionized our ability to map activity in the living brain, including positron emission tomography (PET), functional magnetic resonance imaging (*f*MRI), optical imaging, and, more recently, near-infrared spectroscopy (NIRS) and transcranial magnetic stimulation (all are discussed in other chapters of this book). Each modality offers distinct information about functional brain activity and has certain advantages and limitations. In choosing a functional imaging modality for experiments, one should consider a modality's spatial and temporal resolution, the etiology of its brain mapping signal, the practicality of the imaging methodology, as well as the cost of implementation. In this chapter, the methodological details of optical imaging of intrinsic signals are explored, with special attention to the considerations listed above.

Optical imaging of intrinsic signals maps the brain by measuring intrinsic activity-related changes in tissue reflectance. Functional physiological changes, such as increases in blood volume, hemoglobin oxymetry changes, and light scattering changes, result in intrinsic tissue reflectance changes that are exploited to map functional brain activity. This offers a distinct advantage over extrinsic signal imaging, such as dye imaging, which may cause phototoxicity, especially in *in vivo* preparations, and thereby alter the normal physiology of the sample. It is unclear how normal physiology may be affected by the addition of dyes and radioisotopes or electrode insertion. By not requiring any contact with the tissue of interest whatsoever, optical imaging of intrinsic signals is ideally suited to studying chronic preparations, in which an investigator may wish to image a sample over a period of days, weeks, or months, and for intraoperative mapping of the human cortex during neurosurgery (Mazziotta *et al.,* 2000).

Although activity-related intrinsic optical changes in tissue reflectance associated with electrical activity or metabolism were first observed over 50 years ago (Hill and Keynes, 1949), it was not until the 1980s that these intrinsic optical changes were used to map cortical activity *in vivo* (Grinvald *et al.,* 1986). Since this initial report, intrinsic optical changes have been reported in rodents (Masino *et al.,* 1993; Narayan *et al.,* 1994), cats (Frostig *et al.,* 1990; Bonhoeffer and Grinvald, 1991), monkeys (Ts'o *et al.,* 1990; Grinvald *et al.,* 1991), and humans (Haglund *et al.,* 1992; Toga *et al.,* 1995a). The increasing popularity of optical imaging of intrinsic signals is largely because this technique offers both high spatial and high temporal resolution simultaneously. The spatial resolution of intrinsic imaging is unparalleled among *in vivo* imaging techniques (on the order of micrometers), making it ideal for studying the fine functional organization of sensory cortices as well as the physiology of neurovascular coupling at the level of the arteriole, venule, and even capillaries. Although the temporal resolution of optical imaging is not as great as with electrophysiological techniques, imaging is commonly performed at video frame rates (30 Hz). This is more than sufficient for imaging the slowly evolving perfusion-related responses, which peak 3 to 4 s after stimulus onset.

Because of these advantages, the number of studies using optical imaging of intrinsic signals has been growing rapidly (especially now that optical imaging systems are commercially available). This chapter provides a detailed methodology for investigators to design their own imaging system, with special attention to the limitations of certain approaches and different strategies that have been devised by various groups to overcome them. Understanding the various limitations and strategies will give investigators greater versatility in designing their systems and experiments and avoid making the commercially available optical imaging systems "black boxes" that merely produce functional maps.

This chapter surveys a wide array of optical imaging techniques and applications, including discussions of how optical spectroscopy has significantly advanced our understanding of intrinsic signal etiology, advantages and disadvantages of the different species used for optical imaging, different approaches to cortical immobilization, advances in detector technology, recent advances in both single-wavelength and spectroscopic analysis, baseline vasomotion and how it complicates data analysis, recent advances in optical imaging in humans, and integrating optical imaging with other functional imaging techniques to better understand the etiology of functional brain mapping signals.

II. Sources of Intrinsic Signals and Wavelength Dependency

Although investigators regularly draw conclusions about neuronal activity from functional imaging studies, most modern functional imaging techniques, including optical imaging of intrinsic signals, do not directly measure neuronal activity (Villringer and Dirnagl, 1995). Instead, they detect activity-related changes in perfusion and metabolism, such as increases in cerebral blood flow and changes in hemoglobin oxygenation. An understanding of the hemodynamic response and its relationship to electrophysiology and metabolism is therefore required to attach the appropriate significance to the results being reported. Moreover, it is critical to understand which aspects of the hemodynamic response underlie intrinsic signals.

It has been known for well over a century that neuronal activity causes local perfusion-related and metabolic changes (Roy and Sherrington, 1890). While electrophysiological changes occur on the order of milliseconds, perfusion-related changes occur on the order of seconds. The increase in blood flow supplies nutrients such as oxygen and glucose to metabolically active neuronal areas. The cascade of events includes regional vasodilatation (Ngai *et al.,* 1988), blood flow changes (Cox *et al.,* 1993; Lindauer *et al.,* 1993), blood volume increases (Frostig *et al.,* 1990; Belliveau *et al.,* 1991; Narayan *et al.,* 1995), and changes in relative hemoglobin concentrations (LeManna *et al.,* 1987; Kwong *et al.,* 1992; Malonek and Grinvald, 1996; Mayhew *et al.,* 1999; Nemoto *et al.,* 1999). Metabolic changes include increases in local oxygen consumption (Frostig *et al.,* 1990; Malonek and Grinvald, 1996; Vanzetta and Grinvald, 1999) and glucose utilization (Sokoloff *et al.,* 1977; Fox *et al.,* 1988). All of these electrophysiological and metabolic changes may contribute to intrinsic optical changes.

Several studies have demonstrated close spatial coupling of neuronal activity and perfusion-related mapping signals (see Villinger and Dirnagl, 1995, for review). Intrinsic signals, specifically, have been shown to be in good spatial agreement with electrophysiological activity in rodents (Masino *et al.,* 1993; Narayan *et al.,* 1994), cats (Grinvald *et al.,* 1986; Frostig *et al.,* 1990), nonhuman primates (Grinvald *et al.,* 1986; Frostig *et al.,* 1990), and humans (Haglund *et al.,* 1992; Toga *et al.,* 1995a; Pouratian *et al.,* 2000a). Interestingly, in most cases, optical signals seem to overspill regions of electrophysiological activity. This phenomenon of "spread" mostly has been observed in the rodent somatosensory cortex in response to whisker stimulation (Godde *et al.,* 1995; Chen-Bee and Frostig, 1996; Masino and Frostig, 1996). Narayan *et al.* (1995) reported that intrinsic optical and intravascular fluorescent dye maps overspilled regions of electrophysiologic activity (using single-unit recordings) by about 20%. Optical maps encompass not only the principal barrel but also adjacent barrels. This may be due to low-level neuronal activity which occurs in adjacent barrels in response to stimulation of adjacent, nonprincipal whiskers.

Three major components of optical signals have been identified: blood volume changes, hemoglobin oxymetry changes, and light scattering (Frostig *et al.,* 1990; Malonek and Grinvald, 1996; Mayhew *et al.,* 1999; Nemoto *et al.,* 1999).

A. The Blood Volume Component

The first component of the intrinsic signal originates from functional changes in blood volume related to either dilation of local vasculature or local capillary recruitment or both in cortically active areas. To determine the contribution of blood volume changes to intrinsic signals, Frostig and colleagues (1990) injected and imaged a fluorescent dye (Texas red dextran, MW 70,000; Molecular Probes) that was restricted to the intravascular compartment and compared the fluorescence signal with the intrinsic signal changes at a hemoglobin isobestic point (point of equal absorption of oxy- and deoxyhemoglobin), 570 nm. The signals were essentially identical (both showing increased absorption due to increased blood volume), leading the authors to conclude that the major component of the intrinsic signals measured at 570 nm originates primarily from blood volume changes. Narayan and colleagues (1995) subsequently compared functional intravascular fluorescent dye maps (once again using Texas red dextran) with intrinsic optical reflectance decreases at 610 and 850 nm, finding similar timing and localization of intrinsic signals and blood volume changes. Both studies found that the blood volume changes, however, extend beyond the electrophysiologically active cortex, suggesting that the majority of activity-dependent blood volume changes and related intrinsic signal changes are not specific to or finely regulated at the level of individual functional domains.

B. The Oxymetry Component

The second component of intrinsic signals arises from activity-dependent changes in hemoglobin oxygen saturation. Current theory holds that the first change to occur following neuronal activation is likely a brief burst of oxidative metabolism at the site of neuronal activity (Fox and Raichle, 1986; Frostig *et al.,* 1990; Vanzetta and Grinvald, 1999). This burst of oxidative metabolism is believed to produce a local increase in deoxyhemoglobin concentrations which is very tightly spatially coupled with electrophysiologically active neurons. Consistent with this theory, studies using oxygen-dependent phosphorescence quenching dyes demonstrated that oxygen tensions briefly decrease following functional activation (Vanzetta and Grinvald, 1999). These initial increases in oxidative metabolism and deoxyhemoglobin concentrations remain controversial since not all groups have observed them (Vanzetta and Grinvald, 1999; Kohl *et al.,* 2000; Lindauer *et al.,* 2001). Within seconds of activation, cerebral blood flow increases in excess of cerebral metabolic rate of oxygen (~6:1 according to Fox and Raichle 1986), causing a *decrease* in capillary and venous deoxyhemoglobin due to an "overperfusion" of blood rich in oxyhemoglobin (Ngai *et al.,* 1988; Frostig *et al.,* 1990; Narayan *et al.,* 1995). Intrinsic signals in the 600 to 630 nm range have a biphasic time course reminiscent of the changes expected in deoxyhemoglobin concentrations: an initial increase in absorption followed by a phase reversal and more prolonged decrease (Fig. 1). This is attributed to the fact that in this wavelength range, the absorption of light by oxyhemoglobin is negligible compared to that of deoxyhemoglobin. Although it is clear that other optically active processes, such as light scattering and changes in total hemoglobin, still contribute to the intrinsic signals at these wavelengths, it is commonly held that the major contributor to intrinsic signals in this wavelength range is the change in deoxyhemoglobin concentration (Frostig *et al.,* 1990; Nemoto *et al.,* 1999).

Imaging at wavelengths that are sensitive to oxygen extraction (i.e., 600–630 nm) may produce maps that are more spatially correlated with underlying neuronal activity than wavelengths that are influenced by blood volume changes (i.e., 550 nm) (Frostig *et al.,* 1990; Grinvald *et al.,* 1991; Hodge *et al.,* 1997; Vanzetta and Grinvald, 1999). This may be because fast changes in oxidative metabolism are more tightly coupled to electrical activity than the more delayed perfusion-related responses (Vanzetta and Grinvald, 1999). Imaging at 600–630 nm also offers the greatest signal-to-noise ratio and, when focused 2.0 mm beneath the cortical surface (using an imaging apparatus with a shallow depth of field), least emphasizes blood vessel artifacts (Hodge *et al.,* 1997).

C. Light Scattering Component

Light scattering changes that accompany cortical activation, which contribute to intrinsic signals, arise from ion and water flux; morphological changes (i.e., expansion or contraction) in vascular, cellular, and extracellular conformations (which may be intimately related to ion and water movement); and blood volume changes (Lou *et al.,* 1987;

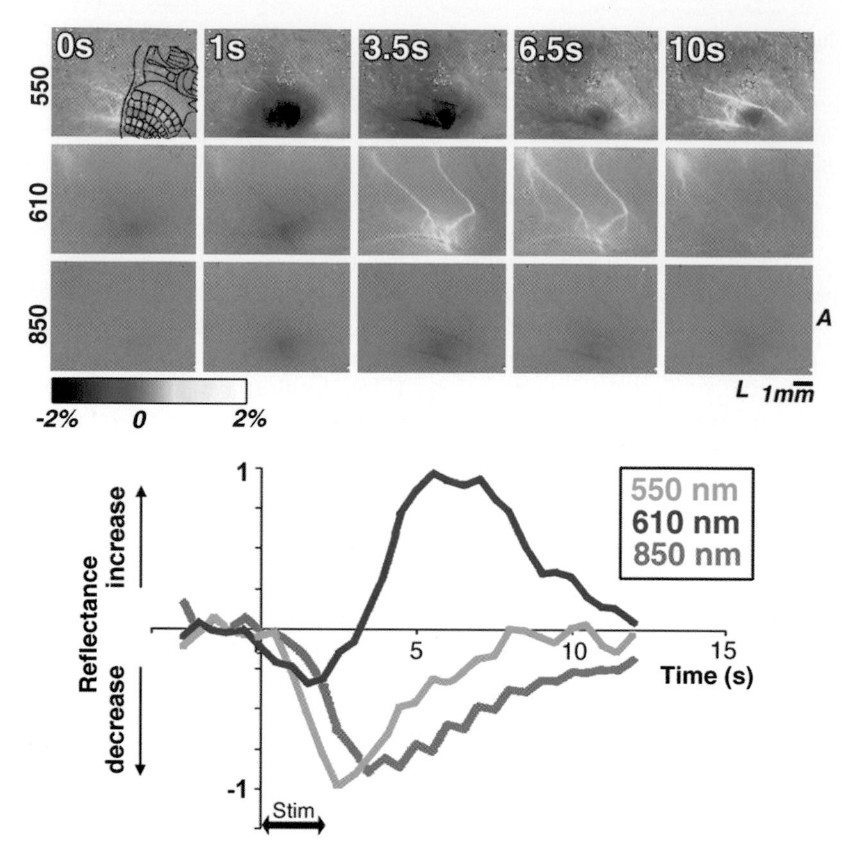

Figure 1 Distinct spatial/temporal patterns of functional intrinsic signals depending on wavelength. Optical responses to 2-s C1 whisker stimulation over the rodent somatosensory cortex are displayed at 550, 610, and 850 nm. 550 nm: The response at 550 nm is believed to represent changes in total hemoglobin (see Fig. 2) and is typically high intensity and monophasic. The monophasic time course is shown in the graph at the bottom. Overlaid on the 0 s image is a schematic representation of the rodent somatosensory barrel cortex for comparison. 610 nm: Responses at 610 nm are believed to emphasize changes in deoxyhemoglobin (see Fig. 2). Following stimulation, there is initially a focal increase in absorption (at 1 s, interpreted as an increase in deoxyhemoglobin and oxygen extraction) followed by a more widespread decrease in absorption (3.5–6.5 s, interpreted as a decrease in deoxyhemoglobin concentration). The second phase is related to the BOLD *f*MRI signal. The biphasic time course of optical responses at 610 nm is shown in the graph at the bottom . 850 nm: At 850 nm, neither isoform of hemoglobin absorbs much light. Instead, the signal is believed to originate from light scattering changes. This signal, like at 550 nm, is monophasic (see time course depicted on bottom graph) but is significantly less intense than the 550-nm response. Time courses were calculated by determining the average reflectance change within a statistically defined region of interest, using methods described by Blood *et al.* (1995). Time courses were normalized for comparison across wavelengths. The height of the graphs, therefore, does not reflect the magnitude of the reflectance changes. The optical signals and time courses were derived from averaging 12 trials in a single animal. A, anterior; L, lateral.

Narayan *et al.,* 1995). The light scattering component exists at all wavelengths but becomes a significant source of intrinsic signals at wavelengths with very small hemoglobin absorption (i.e., above 630 nm, see Fig. 2). In fact, in the near-infra red range (>750 nm), the light scattering component dominates the intrinsic signal. Although the specific etiology of light scattering changes cannot necessarily be defined in *in vivo* preparations, Narayan and colleagues (1995) demonstrated that functional light scattering changes at 850 nm correlated well, both spatially and temporally, with blood volume changes (as measured using intravascular fluorescent dyes). This is consistent with the fact that erythrocytes are the primary scatterers in whole blood; light scattering changes can be induced either by functional changes in the number of erythrocytes or by functional erythrocytic distortions (Zdrojkowski and Pisharoty, 1970). It should not, however, be assumed that all light scattering changes at 850 nm represent blood volume changes. For example, O'Farrell and colleagues (2000) showed that only the latter two of three phases of the triphasic 850-nm optical response to cortical spreading depression represented blood volume changes. The first phase may represent changes in the extracellular compartment or dendritic beading, which both may potentially change the light scattering properties of the tissue (O'Farrell *et al.,* 2000).

The light scattering signal introduces a potential confound to intrinsic signal mapping studies. Light scattering both blurs images and expands the apparent area of activity. In the cortex, however, the estimated error due to light scattering is smaller than 200 μm (Orbach and Cohen, 1983).

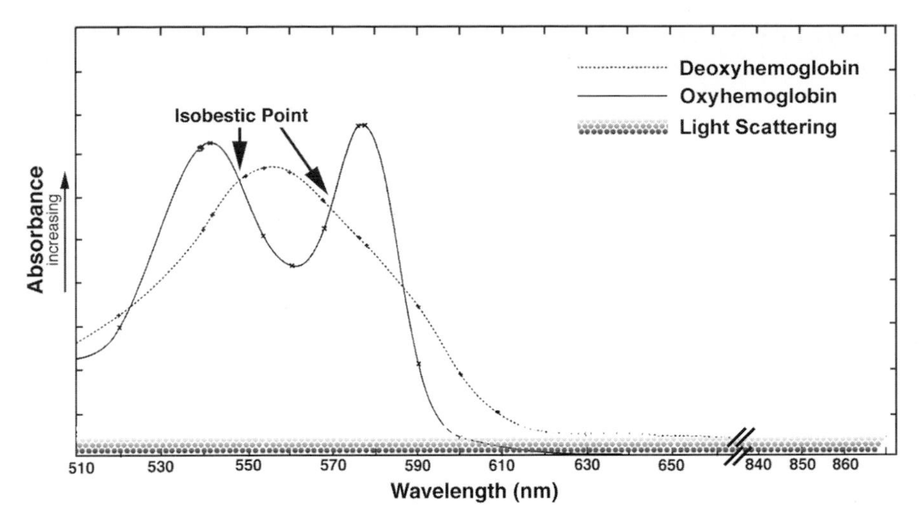

Figure 2 Hemoglobin absorption curves. The absorption curves for oxy- and deoxyhemoglobin are displayed to demonstrate the differential absorption of light by the primary moieties of hemoglobin at different wavelengths. This makes it possible to emphasize different physiological processes by imaging at specific wavelengths (see Section II.D.1). For example, imaging at 550 and 570 nm (the isobestic points) emphasizes changes in total hemoglobin because at these wavelengths the two major moieties of hemoglobin absorb equally. When imaging in the 600 to 630 nm range, on the other hand, oxyhemoglobin absorption is negligible so most of the intrinsic signal response is believed to represent changes in deoxyhemoglobin concentrations. At longer wavelengths both hemoglobin moieties have negligible absorbance and light scattering effects predominate. Light scattering is believed due to cellular swelling and changes in blood volume and blood flow, as well as many other unaccounted-for phenomena. Circles at the bottom of the graph represent light scattering occurring at all wavelengths, but its impact on the optical signal becomes significant only at longer wavelengths. The differential absorption of the major hemoglobin moieties is also essential for spectroscopic analysis of intrinsic signals (see Section II.D.2).

Recently, Nomura and colleagues introduced a new protocol for mapping light scattering changes *in vivo* without any contribution from hemoglobin absorption (Nomura *et al.,* 2000). This was accomplished by exchange transfusion with fluorocarbon (Green Cross, Osaka, Japan). Fluorocarbon is artificial blood with adequate oxygen-carrying capacity to maintain life for a number of days but without any absorption in the visible and near-infrared range. Although this approach allows imaging isolated light scattering changes *in vivo,* results using this model may not be extensible to intrinsic signals measured with whole blood intact since the oxygen-carrying capacity and solubility of fluorocarbon are significantly different from those of hemoglobin. These differences result in a doubling of cerebral blood flow and preferential flow increases to the cortex and cerebellum (Lee *et al.,* 1988).

Similarly, the light scattering signal has emerged as an extremely useful mapping signal in slices (Stepnoski *et al.,* 1991) and the isolated brain. The light scattering signals obtained in these *in vitro* preparations are, in a way, simpler to interpret than *in vivo* signals because they are not superimposed on signals arising from hemoglobin-related changes. However, the difficulty of ascribing a specific signal etiology to the light scattering signal still remains.

D. How Can the Different Components of the Intrinsic Signal Be Resolved?

The major chromophores that influence intrinsic signal changes are oxyhemoglobin and deoxyhemoglobin, each of which has a unique absorption spectrum (Fig. 2). Based on the differential absorption of light by these two chromophores, two approaches have been developed to determine the etiology of intrinsic signals. The first approach, herein referred to as "single-wavelength imaging," is to emphasize different physiological processes by imaging at specific wavelengths of light which either accentuate the differences between the chromophores (e.g., at 610 nm, deoxyhemoglobin absorbs significantly more than oxyhemoglobin) or minimize the difference between the chromophores (e.g., 570 nm is an isobestic point at which oxy- and deoxyhemoglobin absorb equally) (Fig. 2). The second approach, herein referred to as "spectroscopic imaging," is to image multiple wavelengths simultaneously and use post hoc spectroscopic analysis to determine the contribution of the different intrinsic signal components.

1. Single-Wavelength Imaging

The selection of wavelengths for single-wavelength imaging is primarily based on the absorption spectra of oxy- and deoxyhemoglobin. As would be expected, the time course and spatial involvement of optical responses at different wavelengths are unique (Fig. 2) (Hodge *et al.,* 1997; Nemoto *et al.,* 1999). For example, at 610 nm, the absorbance of oxyhemoglobin is negligible compared to that of deoxyhemoglobin (Fig. 2). Therefore, imaging at 610 nm is believed to emphasize changes in deoxyhemoglobin (Frostig *et al.,* 1990, Nemoto *et al.,* 1999). Consistent with blood oxygen level-dependent (BOLD) *f*MRI studies (Hu *et al.,* 1997; Logothetis *et al.,* 1999; Yacoub *et al.,* 1999) and oxygen-dependent phosphorescence-quenching dye studies (Vanzetta and Grinvald, 1999), optical imaging at 610 nm indicates an initial focal increase in deoxyhemoglobin (analogous to the "initial dip") followed by a more global decrease in deoxyhemoglobin (Fig. 1). In contrast, imaging at 550 or 570 nm, which are isobestic points, should emphasize changes in total hemoglobin, since deoxyhemoglobin and oxyhemoglobin absorb equally at these wavelengths. Accordingly, the time course and spatial extent of the response at these wavelengths is significantly different from that observed at 610 nm. Finally, imaging at 850 nm, which is near infrared and at which the absorption of both hemoglobin moieties drops dramatically, is dominated by light scattering changes. Studies comparing intrinsic and intravascular dye signals suggest that these light scattering changes correlate well with changes in cerebral blood volume and may not be directly influenced by hemoglobin levels and oxygenation (Fig. 2) (Narayan *et al.,* 1995; Cannestra *et al.,* 1998a; O'Farrell *et al.,* 2000).

The signal at any particular wavelength is multifactorial; it can only be said that certain components are *emphasized* at a particular wavelength. They cannot and should not be assumed to represent just those components described here. Also, different wavelengths of light penetrate the cortex to different degrees (longer wavelengths penetrate deeper into the cortex) and therefore one should be cautious about interpreting results across wavelengths since different cortical volumes may be sampled at different wavelengths (Mayhew *et al.,* 1999).

Although intrinsic signals at various wavelengths originate from different sources, they can all be used for functional mapping. These differences in intrinsic signals across wavelengths may be exploited depending on the scientific question at hand. To characterize the functional architecture of, or "map," the brain, it may be more appropriate to image at 610 nm. However, to characterize blood volume changes, it would be better to image at 550 or 570 nm.

2. Spectroscopic Analysis of Intrinsic Signals

Spectroscopic analysis of intrinsic signals has become increasingly popular because it allows a more specific determination of the etiology of intrinsic signals. Two approaches have been suggested. The first approach is to acquire simultaneous spectral information from many locations in the form of a spatiospectral image. In order to gain the spectral dimension, a spectrophotometer is used to spectrally decompose the image and at least one spatial dimension must be sacrificed. Still, under most circumstances, signals originating from cortex, draining veins, and feeding arterioles can be separated. In order to retain at least one spatial dimension from the imaged area, the optical imaging apparatus must be modified to have two image planes, instead of the one plane used in the conventional apparatus (see Section IV.C). In the first image plane, a narrow slit isolates a selected line across the cortical surface. This cortical slit (Fig. 3) is then projected onto a dispersing grating (in the spectrophotometer), whose grooves are aligned parallel to the slit, so that the image of the selected cortical band is dispersed into its spectral components along the orthogonal axis. Thus, in the second image plane, a two-dimensional image is produced, with one spatial dimension and one spectral dimension, showing the spectral information about multiple cortical points simultaneously (Fig. 3). The detector is placed in the second image plane to capture this spatiospectral image.

Alternatively, in order to retain two dimensions, a filter wheel can be used to acquire multiwavelength data *near* simultaneously. By interleaving images at different wavelengths throughout a single trial and by imaging at different wavelengths at the same time point across different trials, a four-dimensional volume set can be acquired (two spatial dimensions, time, and wavelengths), which can be used for subsequent spectroscopic analysis. The limitation of this approach is that the multiwavelength/spectral data being collected are not simultaneous, and it is well known that intrinsic signal responses can vary greatly across trials (Chen-Bee and Frostig, 1996; Masino and Frostig, 1996).

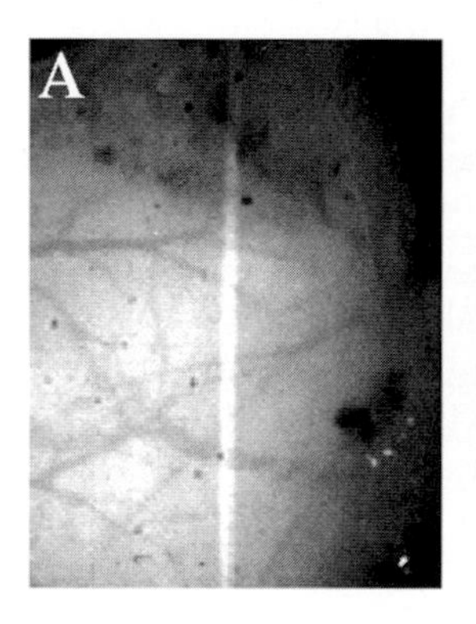

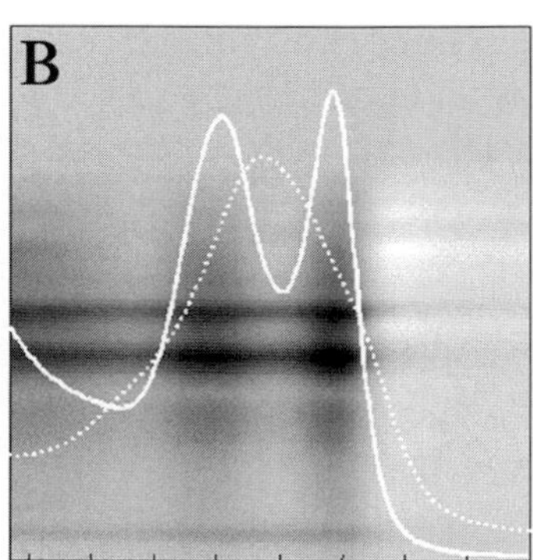

Figure 3 The "cortical band" and optical spectroscopy. (**A**) Image of the cortical surface with the position of the cortical slit from which spectroscopic data are collected. White light is transmitted through only the indicated portion of the image and directed into a spectrophotometer to create a spatiospectral image as shown on the right. (**B**) Image of the spatiospectral pattern obtained after the gratings in the spectrophotometer disperse the light transmitted through the cortical slit. This image is obtained in the second image plane where the camera detector is positioned. The image was obtained by subtracting the peak (~3.5 s after stimulus onset) of the averaged response (30 trials) following electrical whisker stimulation from the mean baseline (8 s) before stimulation. The y axis corresponds to the position down the slit, shown in (A). The x axis in the spatiospectral image corresponds to the wavelength. The "textbook" absorption spectra of oxy- and deoxyhemoglobin are superimposed on the image (oxy—solid line, deoxy—dotted line). The increases in absorption (darker vertical regions) are primarily due to an increase in the proportion of oxyhemoglobin. The predominant dark horizontal bands reflect the changes in blood volume in the middle cerebral artery. Adapted, with permission, from Mayhew *et al.* (2000).

This limitation can be statistically overcome by acquiring a large number of trials (i.e., high *N*) so that the variability of the response is accounted for.

Once spectral data have been obtained, a modified Beer–Lambert Law is used to extract the contribution of the different chromophores to the intrinsic signal. The spectral changes are fitted by the known spectra of oxyhemoglobin and deoxyhemoglobin as well as cytochromes and light-scattering components. The first such study applied a linear component analysis to extract the etiology of the intrinsic signals (Malonek and Grinvald, 1996), but this approach has been called into question recently. More sophisticated analyses are now available which incorporate terms such as wavelength-dependent path-length factors (Mayhew *et al.*, 1999; Nemoto *et al.*, 1999). (Details on methodology are provided in Section VI.B.)

E. The Time Course of Intrinsic Signals

As alluded to earlier, intrinsic signal changes at different wavelengths have different time courses. There are two major patterns of intrinsic signal time courses: monophasic (decreased reflectance) and biphasic (initially decreased reflectance followed by increased reflectance). In the visible spectrum up to ~590 nm, functional intrinsic signal changes monophasically decrease reflectance. Between 600 and 760 nm, a biphasic pattern is observed. Finally, at wavelengths greater than 760 nm, a monophasic pattern reminiscent of wavelengths less than 590 nm is observed.

For monophasic responses, responses generally appear approximately 1 s after stimulus onset, peak between 3 and 4 s after stimulus onset, and return to baseline by approximately 8 s (Nemoto *et al.*, 1999) (Fig. 1, bottom). For the biphasic responses, the initial response to stimulation is usually a little more rapid, appearing within 500 ms, peaking between 1.5 and 2 s after stimulus onset, reversing phases at about 3 s, peaking in the opposite polarity approximately 5 s after stimulus onset, and returning to baseline by approximately 10 s (Fig. 1, bottom). This time course is generally consistent with the time course of the "initial dip" in deoxyhemoglobin concentrations (Menon *et al.*, 1995; Malonek and Grinvald, 1996; Mayhew *et al.*, 1999; Nemoto *et al.*, 1999; Cannestra *et al.*, 2001).

Based on spectroscopic studies published, time courses of the individual components that contribute to intrinsic signal changes have also been characterized. Deoxyhemoglobin concentrations are believed to rise immediately following stimulus onset, peak between 1 and 2 s after stimulus onset, return to baseline 2–3 s after stimulus onset, decrease and peak between 4 and 6 s after stimulus onset, and finally return to baseline concentrations. Oxyhemoglobin concentrations on the other hand are slower to respond to stimulation, increasing approximately 1 s after stimulus onset, peaking at approximately 4 s after stimulus onset, and returning to baseline (monophasic) (Malonek and Grinvald, 1996; Mayhew *et al.*, 1999; Nemoto *et al.*, 1999).

III. Preparation of an Animal for Optical Imaging

A. Species

Optical imaging of intrinsic signals is done in a variety of species, including rodent (Narayan *et al.*, 1994; Masino and Frostig, 1996), cats (Grinvald *et al.*, 1986; Frostig *et al.*, 1990), nonhuman primates (Bonhoeffer and Grinvald, 1991), and humans (Haglund *et al.*, 1992; Toga *et al.*, 1995a). Each offers distinct advantages and drawbacks.

1. Rodents

The rodent somatosensory cortex is organized somatotopically as a ratunculus (Chapin and Lin, 1984), analogous to the homunculus. The posteromedial barrel subfield (PMBSF) of the rodent somatosensory cortex (Woolsey and Van der Loos, 1970) has cytoarchitecturally and functionally discrete groups of cells called "barrels," each representing one contralateral whisker (Woolsey and Van der Loos, 1970; Simons, 1978) (Fig. 4C). Anterior and medial to the PMBSF, there are sensory cortices representing the hindlimb and forelimb of the rodent. Electrophysiology (Chapin and Lin, 1984) and 2-deoxyglucose autoradiography (Durham and Woolsey, 1977; Kossut, 1988) confirm that functional somatotopy is tightly coupled with cytoarchitecture, making this system ideal for investigating the coupling of neuronal activity and perfusion-related and metabolic responses. Another advantage of the rodent model is the small intersubject anatomic variability in rodent (Toga *et al.*, 1995b). The rodent brain is also lissencephalic, thereby eliminating much of the geometric registration issues that complicate gyrencephalic studies. Because of the small size of rodents, imaging in these subjects is also amenable to a thin bone preparation and avoids the need for implantation of an imaging chamber, significantly simplifying imaging procedures (see Section III.C.2). Finally, and perhaps most importantly, the rodent model allows investigators to conduct population-based studies instead of basing measurements on a small number of subjects, as is often the case in studies involving cats, primates, or humans. This may be a critical point, considering that representations across populations may vary greatly (Chen-Bee and Frostig, 1996). Chen-Bee and Frostig (1996) have shown that functional whisker representations in rodents may be between 54.6% smaller and 50.6% larger than the average areal extent across the population. The major drawback to using the rodent model is that the extensibility of results to human subjects is unclear. If the goal of the research is to characterize neurophysiology to better understand the human condition, it is clearly preferable to use

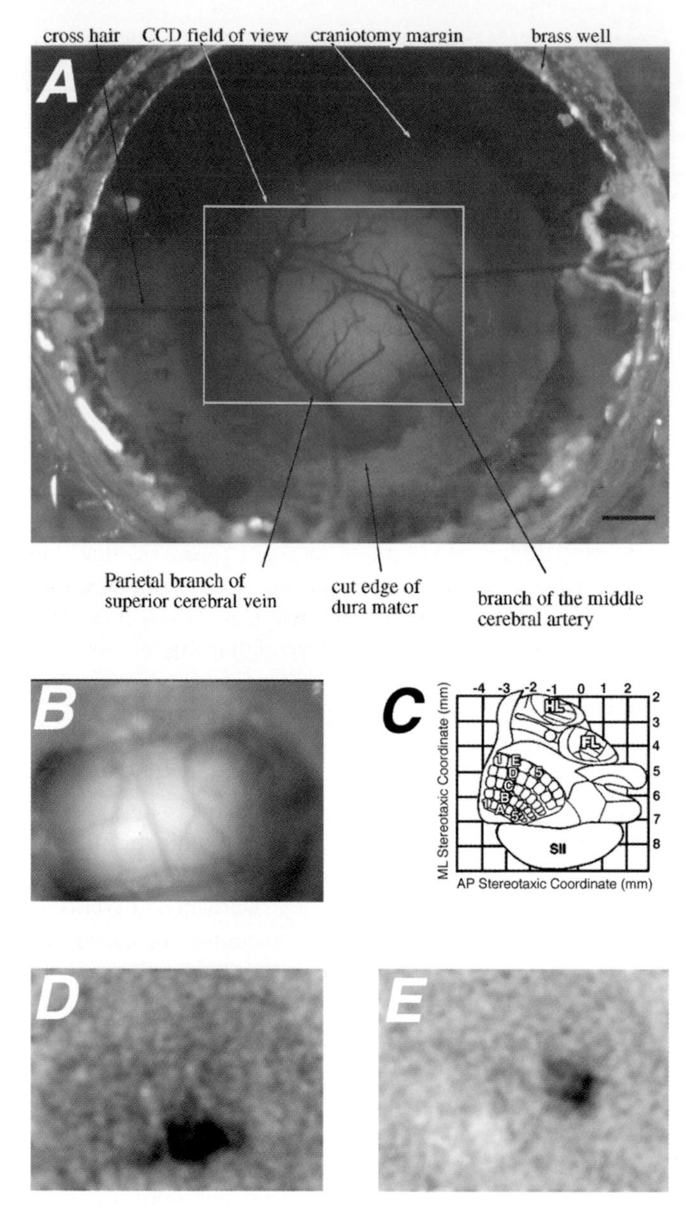

Figure 4 Cranial window for optical imaging. (**A**) Photograph of closed cranial window over the rodent somatosensory cortex, demonstrating a clear view of major cerebral vessels as labeled. Note that the closed cranial window design can be adapted to different species. Although a rodent model is portrayed here, similar systems are used in cats and primates. (**B**) Image of another cranial window overlying rodent somatosensory cortex as observed through a 610-nm filter. Once again, prominent venous vessels are observed at this wavelength, which emphasizes deoxyhemoglobin absorption. The image is slightly blurred because the imaging plane is focused below the cortical surface. (**C**) Stereotactic coordinates of the rodent somatosensory cortex. FL, forelimb; HL, hindlimb; SII, secondary somatosensory cortex. Other labels indicate the corresponding whiskers for each barrel found in the rodent somatosensory cortex. Notice that the area of somatosensory cortex responding to hindlimb stimulation is anterior and medial to the area that responds to whisker stimulation. (**D** and **E**) Activations observed at 610 nm corresponding to whisker and hindlimb stimulation of the rodent, respectively, using the cranial window preparation displayed in (B). Activations appear in appropriate locations and are consistent with the schematic stereotactic map displayed in (C).

subjects that are phylogenetically closer to humans, such as nonhuman primates.

2. Cats

Intrinsic imaging studies in cats have significantly furthered our understanding of both neurophysiology (Grinvald *et al.,* 1986; Frostig *et al.,* 1990; Malonek and Grinvald, 1996) and the organization of the visual cortex (Bonhoeffer and Grinvald, 1991, 1993a). The larger size of the brain relative to the rodent and its greater complexity are both significant advantages of using this subject over rodents. For example, rodents have very poorly developed visual cortices, which would have prohibited all of the major advances in visual cortex characterization achieved using intrinsic imaging (Bonhoeffer and Grinvald, 1991, 1993a,b; Grinvald *et al.,* 1991; Grinvald, 1992; Victor *et al.,* 1994; Shmuel and Grinvald, 2000). The higher complexity of the cat brain makes it more likely that results can be extrapolated to humans. Furthermore, the larger size of the brain has made it easier to isolate and characterize functional perfusion-related changes in different brain compartments (for example, see Malonek and Grinvald, 1996).

3. Nonhuman Primates

Primates offer the best opportunity to conduct studies that are extensible to human subjects without actually facing the challenges of intraoperative brain mapping. However, this very significant advantage comes at a cost. The care of primates is much more expensive and time consuming than that of rodents and cats. Because of the investment required in each subject, population studies of more than a few subjects are not practical. Furthermore, because of the thickness and density of the cranium and dura, imaging in primates requires craniotomy and removal of the dura. In many respects, these are small costs to pay to obtain physiological data that can be directly applied to humans.

4. Humans

Intraoperative mapping provides an unparalleled opportunity to examine the basic physiology and the organization of the functioning human brain. The opportunity exists to study questions that are impossible in other species, such as language organization and higher cognitive functions (Cannestra *et al.,* 2000; Pouratian *et al.,* 2000a). At the same time, intraoperative mapping poses unique challenges for the acquisition, analysis, and interpretation that are over and above those issues that are part of any other intact *in vivo* brain mapping. These include the operating room environment, time constraints, spatial resolution, status of the brain under anesthesia or performance during awake protocols, a dynamic cortical geometry, and other considerations. Imaging must be compatible with the intraoperative environment, most important of which are sterility and mobility. Another challenge faced by intraoperative imaging is that the site of imaging must be determined by clinical indications since subjects are primarily being admitted for treatment of intracranial pathologies. This may result in suboptimal imaging in many cases. Also, as anesthesia is dictated by clinical standards, the investigator cannot control anesthetic type or depth. It should be noted, also, that human experimentation requires institutional approval of human subjects research.

B. Anesthesia

Anesthesia has a strong effect on the coupling between cerebral blood flow and neuronal activity (see, for example, Buchweitz and Weiss, 1986). Therefore, anesthesia must be chosen with great care and levels must be monitored very closely. The level of anesthesia is especially important since a major component of intrinsic signals is related to hemoglobin oxymetry changes. Excessive anesthesia may adversely alter the ventilation and therefore relative hemoglobin concentrations of the subject. Shtoyerman and colleagues (2000) demonstrated that if anesthesia is well controlled, intrinsic signals under general anesthesia will have time courses similar to those observed in awake, behaving animals, although smaller in magnitude. Under most circumstances, experiments are conducted under general anesthesia. Barbiturates and gas anesthetics (e.g., isoflurane) are both popular and work well for optical imaging. This does not preclude the possibility that other anesthetics may be suitable for experiments. However, one should remember that the effect of anesthetics on neurovascular coupling may be species and/or strain dependent.

Anesthetic depth should be monitored continuously throughout experiments (especially those in which the effect of a certain manipulation is being ascertained) since anesthetic depth may affect the magnitude of the intrinsic signals. In general, anesthetic depth can be determined in all species by determining the response to a noxious stimulus. For example, in rodents, one can monitor depth by testing the toe-pinch reflex. The goal should always be to maintain a constant anesthetic depth for the duration of imaging.

There is increased interest in measuring intrinsic signals in awake, behaving subjects. As mentioned earlier, intrinsic signals in awake, behaving monkeys have been shown to have temporal profiles similar to those in general-anesthetized subjects, although with a larger magnitude (Shtoyerman *et al.*, 2000). Imaging responses in awake subjects offers the exciting opportunity to characterize neurophysiology without any possible pharmacological side effects. Conversely, such studies may help elucidate the pharmacological effects of certain drugs on cerebral perfusion. However, these experiments come at the expense of decreased investigator control of the subject. The most significant complication is the introduction of increased cortical movement and noise to the imaging procedure. However, the benefits of imaging awake, behaving subjects may far outweigh the costs since it affords the opportunity to image language and cognitive functions in humans, which may otherwise be impossible using intrinsic signal imaging (Cannestra *et al.*, 2000; Pouratian *et al.*, 2000a).

Optical intrinsic signals arise primarily from changes in cerebral blood flow, cerebral blood volume, hemoglobin oxygenation, and light scattering. It is therefore critical to monitor physiological parameters that may alter cerebral perfusion and metabolism independent of functional activity to ensure that any observed changes are not due to changes in systemic variables.

1. Ventilation and End-Tidal CO2

The ventilation rate and end-tidal CO_2 of the subject can and will have a direct impact on the hemoglobin oxygenation (see Section III.B). Under most circumstances, intubation or tracheotomy should be performed so that the investigator has complete control over respiration. Monitoring these parameters is also critical when a craniotomy has been performed since hypoventilation can result in cerebral edema and alter the normal neurophysiology of the brain.

2. Blood gases

In some cases, it may be appropriate to monitor arterial blood gases to ensure that the subject is adequately ventilated. Blood gas analysis provides measures of not only the oxygen and carbon dioxide content of the blood, but also its pH, all of which may affect cerebral perfusion. Blood gas analysis is particularly important in cases in which normal homeostatic mechanisms and reflexes may be severely impaired (e.g., deep anesthetic states and seizures).

3. Pulse Oxymetry

A pulse oxymeter allows noninvasive monitoring of the oxygen saturation and heart rate of the subject. Monitoring the heart rate is critical since a drop in heart rate will result in a decreased cardiac output (cardiac output = heart rate × stroke volume of left ventricle) and may affect cerebral perfusion.

4. Blood Pressure Monitoring

Similar to monitoring heart rate, it is important to ensure that blood pressures do not change since this may also impact cerebral blood flow and volume (Ferrari *et al.*, 1992).

5. Core Body Temperature

Finally, it is critical to monitor core body temperature since thermoregulation is one of the first homeostatic mechanisms to be compromised during anesthesia. Changes in core body temperature can significantly alter peripheral vascular resistance and systemic perfusion (because of the body's attempt to retain heat). Animals that are not adequately thermoregulated will frequently fail to produce any optical signal changes. A self-regulating heating blanket, which monitors the subject's core body temperature (usually rectal) and adjusts the temperature of the blanket appropriately, is highly recommended.

C Immobilization of Cortex

To take advantage of the superior spatial resolution of optical imaging, it is critical that images acquired before and during activation be in identical locations, since the former will be subtracted from the latter during analysis. The movement of the brain due to respiration and heartbeat presents a major obstacle. Five different strategies have been devised to overcome this challenge: the cranial window, the thin-skull preparation, stabilization using glass plates, synchronization with respiration and heartbeat, and postacquisition image registration. These approaches are not mutually exclusive and under several circumstances, more than one strategy can be used in a single experiment (e.g., cranial window and synchronization with respiration and heartbeat, Grinvald *et al.,* 1991; glass plate and postacquisition image registration, Haglund *et al.,* 1992).

1. The Cranial Window

The first method to stabilize the cortex employs an elaborate chamber system (a "cranial window") (adapted from Bonhoeffer and Grinvald, 1996). The chamber, composed of a circular, stainless steel ring, with an inlet and outlet valve to which tubing can be attached, is mounted on the skull with dental cement before the skull is opened (Fig. 4). The normal procedure is to make the trephination of the skull and to mount the chamber with dental cement before taking the piece of bone out of the skull. If trephination is carried out with a high-speed drill, heat can accumulate and may be particularly dangerous and harmful to the cortex. Excessive heat should be controlled for with constant irrigation during drilling. To perfect the seal between the chamber and the skull, dental wax is melted (with a microcauterizer) into the remaining gaps between the inside of the chamber and the skull. After the chamber is mounted, the bone and dura can be removed. Prior to excising the dura, large dural blood vessels should be occluded with thread or forceps to avoid contact between the exposed cortical surface and blood. Alternatively, the superficial dural vessels can be occluded prior to resection of deeper dural layers. Finally, the remaining cerebrospinal fluid or saline is removed from the cortex using small triangles made from cellulose fibers (Sugi, Escheburg, Germany).

The chamber is then sealed with a round coverslip and filled with silicon oil (e.g., Dow Corning 200, 50 cSt) or artificial cerebral spinal fluid via the inlet valve. It is critical that excessive pressure not be applied to the cortex. This is best achieved by having the fluid flow into the chamber from an upright syringe without the piston and adjusting the level of the syringe with respect to the chamber for precise regulation of the pressure of the fluid in the chamber. If the chamber is filled perfectly, i.e., without any air bubbles or cerebrospinal fluid droplets, this arrangement provides an ideal optical interface and, at the same time, stabilizes the brain.

In long-term experiments the stainless steel chamber just described has to be modified in several important ways. For chronic recordings, the inlets of the chamber can be closed with screws. Moreover, if the window of the chamber is large, it is important to have a metal lid that can be screwed into the chamber instead of the breakable coverglass. Finally, the chamber should be produced from titanium instead of stainless steel because titanium, although difficult to machine, is strong, light, and, above all, highly inert to bodily fluids. Even with implantation times over many months, no difficulties have been observed with a chamber made from this material (Shtoyerman *et al.,* 2000). In long-term experiments it is also of great importance that the chamber is mounted on the skull such that there is no danger of the chamber detaching even after long survival times. This is particularly problematic in young animals in which the bone is often still relatively soft. It has proven useful to clean and degrease the skull with ether and to place screws in the bone next to the chamber. These screws are then covered with the dental cement for mounting the chamber and thus help anchor the chamber firmly onto the skull.

Cortical edema, or herniations, due to hypoventilation or other reasons may be extremely traumatic when using a cranial window (especially when the exposed cortical area represents a large fraction of the brain surface area, such as in rodent preparations). There are several ways to deal with this problem, including injecting high-molecular-weight sugars (e.g., mannitol), lowering the position of the body, hyperventilation, applying 10–20 cm of hydrostatic pressures in a closed chamber for a limited period of time, or puncturing the cisterna magna.

2. Thin Skull Preparation

The thin skull preparation is a strategy used in rodents that takes advantage of the fact that the rodent skull is not very thick to begin with. Masino and colleagues (1993) demonstrated that if the rodent skull is uniformly thinned using either a scraping instrument or a dental drill over the cortical area of interest, intrinsic signal could be imaged through the intact skull. If a dental drill is used to thin the skull, care should be taken to thin slowly in order to (1) prevent cortical damage from excessive heat due to friction and (2) ensure that the skull is thinned uniformly throughout the area of interest. Heat damage can also be prevented by constant irrigation with saline during drilling. In order to increase the translucency of the skull for the duration of the imaging period, silicon oil should be applied to the skull. If the bone has been thinned adequately, arterioles and venules should be easily visualized in the field of view of the detector. The one difficulty with this preparation is that during the thinning procedure, epidural and subdural hematomas, which will interfere with signal detection, can be induced. If hematomas are observed, subjects should be excluded from studies. The utility of this approach has been confirmed by other studies as well (Cannestra *et al.,* 1998a; Polley *et al.,* 1999a,b; O'Farrell *et al.,* 2000).

3. Glass Plate

The use of a sterile glass plate atop the cortex immobilizes the cortex by applying pressure to the cortical surface to prevent its movement (Haglund *et al.,* 1992). This approach has been used only in human subjects. While this method may achieve immobilization, it is not considered ideal because it physically interferes with the cortex and therefore may alter normal functional blood volume and cellular swelling patterns.

4. Synchronization with Heartbeat and Respiration

Synchronization of image acquisition with respiration and heartbeat in subjects who are being ventilated is achieved by halting respiration after exhalation for a short time (less than 1 s). The respiration is then reinitiated when an appropriate trigger circuit detects the next heartbeat. If data acquisition is synchronized this way, the images will always be collected in the same phase of heartbeat and respiration, enabling cancellation of physiologically induced motion artifacts during data analysis. Synchronization between heartbeat, respiration, and data acquisition reduces the noise, even in a well-sealed chamber, by a factor of approximately 1.5 (Grinvald *et al.,* 1991; Toga *et al.,* 1995a).

In awake subjects or those not being ventilated, synchronizing image acquisition with respiration and heart rate requires monitoring of pneumographic and electrocardiographic waveforms. All trials (control and experimental) should begin at the same point in time during the respiration cycle, after which acquisition is synchronized to the cardiac cycle (500 ms post-R-wave). Experimental and control trials should be collected alternately on sequential respiratory expirations. Each experimental image will have a separate control image taken from either the preceding or subsequent expiration cycle. Since data acquisition occurs at similar time points during every respiration cycle, all images are collected with the brain in a similar position, minimizing the effect of periodic brain motion.

5. Postacquisition Image Registration

Under certain circumstances (e.g., intraoperatively), the investigator may be restricted from employing the previously described strategies for cortical immobilization. This is particularly true in the intraoperative environment, in which imaging time needs to be minimized and additional physical procedures for research purposes may not be authorized. Postacquisition image registration utilizes automated image registration (AIR) algorithms (Woods *et al.,* 1992). When acquiring images independent of pneumographic or electrocardiographic waveforms, the precise location of the cortex within the field of view varies from image to image. AIR can be used to realign all images in a series to a reference image either at the beginning or in the middle of the acquisition series. The realignment of images is intensity-based, using the extremely large difference in light intensities emanating from sulci and gyri to place the images into correspondence. Higher order warps may be necessary to achieve proper realignment if there is significant motion in the *z* direction (i.e., toward and away from the detector). Functional changes in cortical reflectance do not interfere with this realignment algorithm since functional reflectance changes, which are <1% in magnitude, are minimal compared to the difference in reflectance from sulci and gyri (gyral light intensity levels can be up to 2000 times that of sulci). Postacquisition image registration effectively compensates for movement between images and minimizes the need to interface with operating room monitors (Cannestra *et al.,* 2000, 2001; Pouratian *et al.,* 2000a).

IV. The Apparatus

A. The Camera

1. Photodiode Arrays

The first functional maps based on intrinsic signals were created using a 12 × 12 photodiode array (Grinvald *et al.,* 1986). Photodiode arrays generally offer a larger dynamic range (>17 bits) than other types of detectors and the capability for much higher temporal resolution (on the order of milliseconds). It is clear, however, that the low number of pixels in standard diode arrays limits their usefulness for high spatial resolution intrinsic imaging. While some may argue that these fast cameras with their millisecond time resolution may be of value for studying the time course of intrinsic signals (for example, with the imaging spectroscopy approach discussed earlier), the temporal resolution gained by using a photodiode array is not necessarily required (depending on the experiment) since the time course of perfusion-related signals is on the order of seconds. These devices may be particularly useful for optical imaging of voltage-sensitive dyes (as opposed to intrinsic signals) for monitoring extremely fast electrophysiological changes that occur within 100–200 ms.

2. Video Cameras

The first attempt to use video cameras to image cortical activity was made in 1974 by Schuette and collaborators (1974). More than 10 years later, Gross and colleagues digitally intensified video imaging with voltage-sensitive dyes to image voltage changes across single cell membranes (Gross *et al.,* 1986). Shortly afterward, Blasdel and Salama (1986) used a similar technology with an *in vivo* preparation to obtain spectacular images of the functional architecture of the macaque visual cortex. Compared to photodiode arrays, the increased spatial resolution of video cameras is at the expense of temporal resolution (Kauer, 1988). However, for intrinsic signal imaging, the extremely high temporal resolution of photodiode arrays is not a critical parameter. A more important limitation is the signal-to-noise ratio (SNR) of standard video cameras of approximately 100:1 to 1000:1.

3. Slow-Scan CCD Cameras

Slow-scan digital CCD (charge-coupled device) cameras offer very good SNR while retaining the advantages of high spatial resolution and moderate cost. Like video devices, CCDs have lower temporal resolution than photodiode arrays. However, as mentioned earlier, this is largely of no consequence since intrinsic signal time courses are on the order of seconds. Ts'o and colleagues (1990) were the first to use CCD cameras for intrinsic signal imaging to image the functional architecture of the visual cortex in the living brain.

Several important parameters which may influence image quality, SNR, and acquisition capabilities should be considered when comparing CCD cameras. They are discussed in detail below.

a. Shot noise. The magnitude of intrinsic signal changes is exceedingly small, on the order of 0.1%. In order to be able to ascribe statistical and biological significance to such measurements, one must be able to differentiate these intrinsic signal changes from stochastic fluctuations in photon emissions. That is, one has to ensure that small changes in reflectance are physiological in origin and not caused by the statistical fluctuations of the light-emitting process. The number of photons that can be attributed to statistical fluctuations equals the square root of the total number of photons emitted. Consequently, the number of photons needed to detect a signal change of 0.1% with a SNR of 10 is 100,000,000 (signal = 100,000, noise = 10,000, therefore SNR = 10). These calculations highlight the fact that light intensity and the well capacity of the CCD have to be chosen appropriately such that an adequate number of photons is accumulated during the experiment. Note that it is not necessary to accumulate this number of photons in a single frame (in fact, this would be impossible considering that the well capacity of most CCDs is on the order of 300,000 to 700,000). Rather, this number represents the total number of photons that should be collected over all trials; multiple trials can also be averaged to increase the SNR (by the square root of the number of trials).

b. Well capacity. The well capacity of a CCD denotes the total number of photons that can be accumulated on 1 pixel before there is charge overflow or saturation. Because of the previously mentioned considerations (see Shot noise above), it is important that well capacities be as large as possible. Well capacities generally range from 300,000 to 700,000; smaller well capacities can also be used but will require a greater number of trials to be able to detect similar small intrinsic signal changes. One way to increase the effective well capacity in existing chips is to combine the charge from several adjacent pixels using "on-chip binning." (This, however, results in a loss of spatial resolving power so the advantage of binning relative to increased spatial resolution must be considered.) Normally, on-chip binning is limited to 2×2 or 3×3.

c. Analog-to-digital converters. Although purchasing a CCD with a greater well capacity will theoretically allow an investigator to increase SNR (as explained above), this advantage is lost if the high-quality camera output is digitized using only an 8-bit analog-to-digital (AD) converter. Eight-bit AD converters have become an industry standard because they offer an economical means to digitize video signals. Although image enhancement strategies have been developed to allow continued use of such converters (i.e., differential video imaging, see section below), under most circumstances it is essential to incorporate a frame grabber with sufficient resolution to preserve the quality of the data obtained by the camera. To illustrate, imagine a CCD chip with a 500,000-electron-well capacity. Using an 8-bit AD converter would result in the equivalent of approximately 2000 electrons per level (256 levels total). Assuming baseline images are acquired in the middle of the dynamic range of the camera (~250,000), this would mean that each gray level of digitization would be approximately equal to a 0.7% intensity change. If the goal of intrinsic signal imaging is to detect signals as small as 0.1%, this is clearly insufficient. In contrast, if a 14-bit frame grabber were to be used, this would be equivalent to 30 electrons per level (16,384 levels total) or a 0.01% intensity change per level. Although the increased resolution comes at a price, it is a worthwhile investment since it would otherwise negate the significant investments made in the camera.

d. Advances in sensitivity and signal to noise. Recent advances in CCD camera technology have significantly increased the sensitivity and SNR of CCD chips. In general, back-thinned CCD chips are associated with greater sensitivity and quantal efficiency and have become standard in almost all CCD cameras. Intensified CCDs and electron-bombardment CCDs (EB-CCDs) have also recently been introduced into the market. In both types of cameras, a photocathode between the image and the CCD converts photons to electrons, acting to intensify the image and increase the quantal efficiency of the CCD. In EB-CCDs, the photocathode releases electrons (with variable gain) to be subsequently accelerated across a gap (via a high-voltage gradient) and "bombard" the CCD chip. The advantages of these new devices over a traditional slow-scan CCD are the additional gain and accompanying speed, allowing exposure times of as little as 1 ms while maintaining very good spatial resolution. These advantages are at a cost of diminished dynamic range. The increased gain, sensitivity, and SNR may be critical in certain applications, such as intraoperative imaging of human cortex in which image exposure needs to be minimized to reduce intraimage cortical movements.

e. Frame transfer. CCD cameras require that that during readout of the acquired image, illumination of the area containing the image information be avoided by all means. Some cameras use a mechanical shutter during the readout time to accomplish this. The mechanical shutter approach is problematic for optical imaging applications since the fastest readout times are on the order of 50 ms and therefore would require the shutter to be closed (and therefore not acquiring data for that period of time). Additionally, the large number of exposures in a single experiment and lifetime of a CCD would be incompatible with the limited lifetime of a mechanical shutter. Alternatively, many CCD cameras offer a "frame transfer mode." In frame transfer mode, half of the light-sensitive area of the CCD chip is covered with an opaque coating. After each exposure the accumulated charges from the illuminated area are shifted to the opacified area within approximately 1 ms. While the "transferred frame" is read out from the opacified portion of the CCD, the light-sensitive region of the CCD can begin to accumulate charges from the next exposure. Since this mode of operation allows acquisition of "back-to-back" images and avoids complications of a mechanical shutter with a limited lifetime, the frame transfer mode of operation is the recommended mode of operation.

4. Differential Video Imaging

As explained earlier, when a high-quality camera output is digitized using only 8 bits, the advantages of the camera are lost. Consequently, several image enhancement approaches have been developed to allow the use of an 8-bit frame grabber while retaining the details of the data offered from a higher resolution imaging device. Most of these techniques, however, use procedures that enhance the image only after its initial 8-bit digitization. These approaches in fact cannot detect intensity changes that are smaller than 1 part in 256 (~0.4%) since these intensity values would have been reduced to the same gray-scale value on initial digitization. Another common approach to image enhancement is to subtract a DC level from the data and amplify the resulting signal prior to its digitization. This approach is, however, applicable only to flat images with very low contrast.

An alternative approach for image enhancement is to process (i.e., subtract) images prior to digitization so that the difference images are digitized instead of original images. This approach uses analog differential subtraction of a stored "reference image" from the incoming video images. An apparatus which uses this alternative approach is commercially available (Optical Imaging, Inc., Germantown, NY). It uses analog circuitry to subtract a selected reference image from incoming camera images and then performs a preset analog amplification of the differential video signal before digitizing it using an 8-bit analog-to-digital converter. Although this approach is not as precise as using a higher resolution analog-to-digital converter and storing the original images for subsequent calculations, it offers exquisite sensitivity, is less expensive, requires less storage space, and has been used successfully by numerous groups. The noise can be further reduced, as with any approach, by trial averaging. The digital image of a given reference image and the corresponding enhanced sequence of images can later be combined.

Another significant advantage of the commercially available system is that an enhanced image is displayed in real time (video rate) on a monitor, thus providing important online feedback to the investigator. This feedback allows problems to be corrected at the earliest stages of the experiment rather than waiting until data have been analyzed offline. Because of the strongly amplified picture, minor optical changes that would later result in large artifacts are immediately noticed. These include moving bubbles of air or cerebrospinal fluid in the closed chamber, excessive noise due to imperfect stabilization of the cortex, or minute bleeding.

A major limitation of differential video imaging is that the original raw optical images are not stored and may therefore limit the versatility of the optical imaging apparatus. This prevents post hoc offline analysis of data using methods other than the subtraction–ratio approach. Such special analyses may be necessary if the investigator does not know *a priori* which image is to be used as the reference or control image (see Section VI.A.1) or if the investigator wishes to compare activations relative to a different control after the experiment has already been conducted. Although the original raw images can theoretically be back-calculated, the resulting images would be of low resolution (8-bit). For a review of different analytic approaches, see Section VI of this chapter.

B. Illumination

Illumination parameters are clearly critical for intrinsic optical imaging since it is the reflectance changes in this illumination pattern which determine the intrinsic signal changes. The wavelengths of the illuminating light depend on the components of the intrinsic signals that the investigator is attempting to emphasize (see discussion above). Apart from this consideration it is also important that the wavelength used provides sufficient penetration into the tissue and, when using a thin bone preparation, through the skull. Shorter wavelengths (e.g., 500 to low 600 nm range) do not penetrate tissue as well as longer wavelengths, especially in the near-infrared range (e.g., 700 and 800 nm range).

In order to relate the obtained activity maps to the anatomical landmarks, it is also useful to record pictures of the blood vessel patterns. These are best obtained with the brain illuminated by green light, which will highlight both arterial and venous vessels. A band-pass filter between 550 and 590 nm with a narrow range (+5 nm, e.g., 550 + 5 nm) is therefore very useful for obtaining the blood vessel picture. Moreover, it may be useful to increase the depth of field of the image by reducing the lens aperture in order to eliminate the blurring of cortical vasculature due to a curved cortical surface.

1. Lamps

A standard 100- or 150-W tungsten halogen lamp housing [e.g., Newport (Irvine, CA), Oriel (Stratford, CT)] with a focusing lens is suitable for illumination. The lamp should be powered by a high-quality regulated, voltage-stabilized power supply in order to achieve strong and stable illumination. It should provide adjustable DC output up to 15 V and 10 A (this can be controlled via serial port in many light sources allowing automated control of incident light levels). Ripple and slow fluctuations should be smaller than 1:1000. An excessive ripple can be improved by adding large capacitors in parallel with the output. The uniformity of the emission spectra is yet another important consideration in selecting an appropriate lamp. For example, arc lamps, which have characteristic distinct lines in their emission spectra, would be inappropriate for optical imaging since uniform lighting across wavelengths would be ideal.

2. Filters and Filter Wheels

Filters and filter wheels should be placed between the lamp housing and the light guides (Fig. 5). An adapter which can hold at least two filters should be fitted so that it connects to the lamp housing on one side and a dual- or triple-port light guide on the other. Alternatively, a high-speed filter wheel can be placed in this position. Most standard filter wheels (e.g., Sutter Instruments, Novato, CA) hold 10 filters and are computer controlled. Using a filter wheel allows acquisition of intrinsic signals at multiple wavelengths interleaved within the same trial. Incorporation of a filter wheel into the optical imaging apparatus will prevent the acquisition of truly back-to-back images since time must be allocated between successive images for the filter wheel to move (usually approximately 50 ms between adjacent filter positions). This, however, may not be an issue due to the relatively slow time course of perfusion-related responses being imaged by intrinsic signal imaging. Band-pass filters that would be useful for single wavelength imaging include: (1)

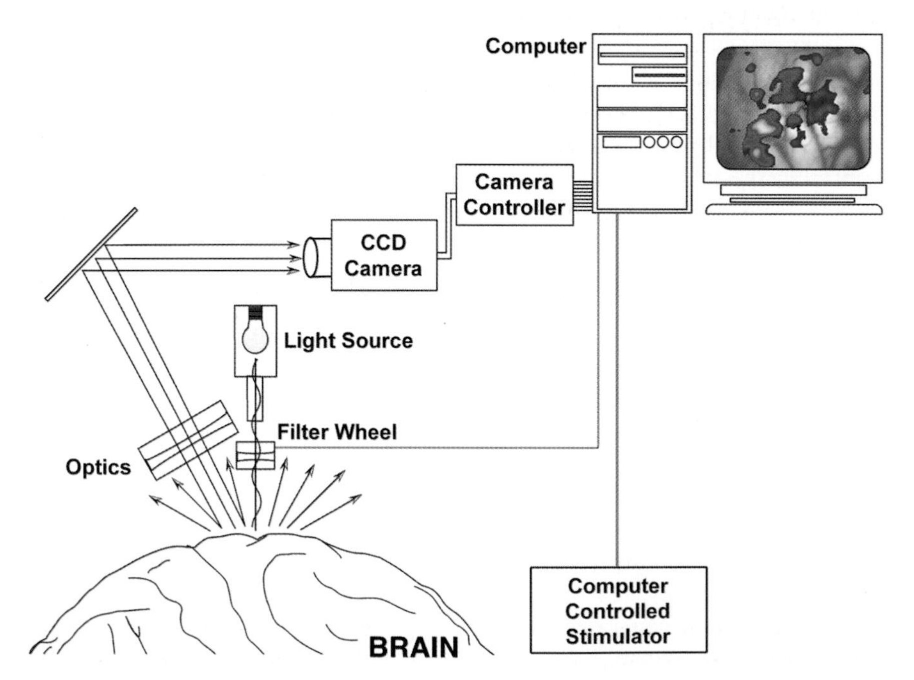

Figure 5 Schematic diagram of the optical imaging of intrinsic signals setup. A light source provides white light which is filtered by a computer-controlled filter wheel before illuminating the cortex. The reflected light is directed through appropriate optics and captured by a CCD camera (or other detector), which is synchronized with stimulation and the filter wheel by a computer-controlled camera controller. The camera controller contains an analog-to-digital converter that digitizes the image before sending the data to the computer to be stored on hard disk. Images are captured during rest and during stimulation. The two sets of images are then compared to detect at which pixels there are significant changes in reflectance. These changes in reflectance correlate with underlying electrophysiological changes.

green filter, 505, 550, or 570 nm—for blood vessel pictures and imaging total hemoglobin; (2) orange/red filter, 600, 605, 610, or 630 nm—for imaging hemoglobin oxymetry changes (the difference between oxy- and deoxyhemoglobin is maximized at 600 nm); (3) near-infrared filters, 730, 750, 790, 850 nm—for imaging functional light scattering changes; (4) heat filter KG2; (5) long-wavelength heat filter RG9. In general, it is best to use filters with a narrow band pass so that the investigator knows precisely what wavelengths are being imaged. However, using excessively narrow band-pass filters (i.e., ~1 nm) may limit the amount of light that reaches the imaging device and therefore limit the resolving power of the imaging setup. In general, investigators should attempt to use filters with a full width half-maximum of 10 nm. A 3 OD attenuator (×1000 attenuation) is also often useful in order to artificially produce a signal of 1000 to test the apparatus (see below).

3. Light Guides

Although theoretically epi-illumination *through the lens* should be ideal to achieve a uniform illumination, this mode of illumination does not work well for imaging the live brain because the brain is a curved surface and some parts of the brain absorb more light than others. Therefore, for *in vivo* imaging, using two or three flexible light guides has proven by far the most useful tool in providing evenly illuminated images of the cortex. It is important to confirm even illumination of the brain by looking at an online image from the imaging device. This is especially critical when imaging using near-infrared light that is not visible to the human eye. Liquid light guides provide a more uniform illumination than fiber optic guides. Adjustable lenses can be used at the front end of the light guides to focus the incident light on the cortex. Schott (Mainz, Germany) offers suitable light guides and small, adjustable lenses that attach at the preparation side.

4. Shutter

Some groups have found it advantageous, particularly when experimenting with higher light intensities, to have a shutter on the light source so that the cortex is illuminated only during data acquisition. Note that this shutter is not used during imaging since it would introduce vibration into the imaging experiment and would have a limited lifetime due to the excessive number of image exposures used during any given optical experiment. It is important to ensure that if a shutter is used, it introduces minimal vibration into the system. Good shutters can be obtained from a variety of sources [e.g., Uniblitz (Rochester, New York), Prontor (Germany)].

C. The Macroscope (Adapted from Bonhoeffer and Grinvald, 1996)

Photographic macrolenses have a large depth of field, resulting in large blood vessel artifacts in the functional maps. These artifacts often hampered the observation of subtle features in the recorded maps. To alleviate this problem, Ratzlaff and Grinvald (1991) constructed a "macroscope" tandem-lens arrangement with an exceedingly shallow depth of field.

This device is essentially a microscope, with a low magnification (around 0.5–10×), composed of two "front-to-front" high-numerical-aperture photographic lenses. By doing this, the macroscope provides an unusually high numerical aperture compared to commercial, low-magnification microscope objectives, resulting in a very shallow depth of field (e.g., 50 μm, nominal for two coupled 50-mm lenses with $f = 1:1.2$). Therefore, when focused 300–500 μm below the cortical surface, the surface vasculature is sufficiently blurred and artifacts from it virtually disappear (Ratzlaff and Grinvald, 1991).

1. Lenses

The macroscope can easily be built by connecting two camera lenses front to front. The magnification of this tandem-lens combination is given by $f1/f2$, where $f1$ is the focal length of the lens close to the camera and $f2$ is that of the lens close to the brain. To build a macroscope from conventional 35-mm camera lenses, items (1) and (2) listed below are needed, plus one or more of the combinations of lenses that are listed.

(1) C-mount to camera adapter (e.g., "Pentax" if Pentax lenses are used).

(2) Adapter for the tandem-lens arrangement. A solid ring with proper threads to connect the front part of each of the camera lenses. (This lens thread is usually used for standard camera filters.) To minimize vibration and to protect the lenses it is advantageous to add support.

(3) For a magnification of 1× (covering approximately 9 × 6 mm^2 with a standard camera), use two 50-mm Pentax lenses with $1/f$ of at least 1.2. Alternatively, a video lens with a shorter working distance (3 cm) but an even higher numerical aperture (0.9) can be used.

(4) For a magnification of 2.7× (covering approximately 3.3 × 2.2 mm^2), use one 50-mm and one 135-mm lens. Pentax offers a 135-mm lens with a numerical aperture of 1.8. If the 50- and 135-mm lenses are installed in the reverse order, then the tandem lens will cover a very large portion of the cortex (approximately 22 × 14 mm^2).

(5) A 2× standard camera extender provides flexibility for additional magnification or for demagnification.

(6) A zoom lens covering the range of 25–180 or 16–160 mm can be used as the top lens. A zoom lens has the advantage of allowing adjustment of the magnification without the replacement of lenses. This flexibility is, however, achieved at the cost of a lower aperture that causes a larger depth of field.

(7) For imaging human cortex during neurosurgery, a zoom lens with a larger working distance starting at approximately 10 cm is preferable to a tandem-lens combination.

In the tandem-lens combination, commercial home video CCD lenses may also be used as the lens next to the camera. Their use as the lens next to the cortex may be problematic whenever the working distance is important. The advantage of using the home video lenses is that the numerical aperture of home video CCD lenses is often larger than that of 35-mm camera lenses.

The camera can also be mounted on a conventional microscope or on an operative microscope, preferably one that offers a high numerical aperture and, consequently, a short working distance (5–7 cm). The final working distance, or distance from the first lens to the specimen, should be at least 3 cm in order to manipulate the specimen, including proper alignment of light guides and electrode placement. Numerical aperture, illumination, working distance, and precise mechanical stability should all be considered in the final design.

2. Camera Mount

The video camera should be rigidly mounted to a vibration-free support. The ideal arrangement is to mount the camera to an immobile support system so that the camera has no means of moving. This is especially important because many modern imaging devices have built in cooling devices that may vibrate the camera. By placing the camera on a firm support, this source of vibration can be eliminated. In such a system, the subject would be placed on an *xyz* translator so that it would be moved into the appropriate field of view of the camera and translated in the *z* direction to focus. The next best approach would be to allow the macroscope one direction of movement, the *z* direction, to enable it to focus. Still the subject would have to be placed on an *xy* translator to move it into the appropriate location relative to the field of view of the camera for imaging. Finally, the camera can be mounted onto an *xyz* translator, but this arrangement gives the camera the most means of inadvertent movements. For all arrangements, the *z* direction of movement should preferably have a coarse, large travel distance control as well as a fine focus control. Furthermore, it is advantageous to construct the camera holder such that it allows rotation of the camera around its optical axis as well as tilting it at any desired angle.

D. The Spectrophotometer

For spectroscopic analysis of optical intrinsic signals, it is necessary to include a spectrophotometer in the optical imaging apparatus. The details of how the spectrophotometer is integrated into the apparatus have been described earlier (Section II.D.2). Briefly, a modified macroscope setup which has two image planes is used, instead of the one plane used in the conventional apparatus. In the first image plane, a narrow slit isolates a selected line across the cortical surface which is projected onto a dispersing grating (in the spectrophotometer). The spectrally decomposed image is then projected onto the second image plane, where it is captured by the CCD. Spectrophotometers are commercially available (e.g., Roper Scientific, Trenton, NJ). Gratings should be appropriately chosen to disperse the reflected light in the visible spectrum.

V. Data Acquisition

A. Basic Experimental Setup

The basic experimental setup for intrinsic optical imaging experiments is shown in Fig. 5. First, the subject's head must be stabilized either by a stereotaxic frame (in animal subjects) or by a Mayfield apparatus (in humans). Once the cortical area of interest has been exposed, the brain is illuminated with filtered light using light guides. The camera should be positioned over the cortical exposure such that the area of interest is centered within the field of view of the camera. Images are taken of the cortex at rest and during activation. The camera controller digitizes the images and forwards the data to the data acquisition computer.

B. Timing and Duration of a Single Data Acquisition

As mentioned earlier, optical imaging of intrinsic signals offers the best combination of both spatial and temporal resolution compared to other brain mapping methods. It is advisable therefore to design experiments such that a time course can be reconstructed from the data. Since the time course of intrinsic signals are on the order of seconds, it is often sufficient for frames to be collected every 250–500 ms in order to capture the temporal profile of the response.

When should imaging begin relative to stimulus onset? Although blood volume changes do not occur until approximately 1 s after stimulus onset, hemoglobin oxymetry changes may begin within a few hundred milliseconds of neuronal activation (Malonek and Grinvald, 1996; Mayhew *et al.*, 1999; Nemoto *et al.*, 1999). It is therefore essential that image acquisition begin, at the very latest, simultaneous with stimulus onset. Analysis of optical images, however, often requires at least one baseline image (i.e., prior to stimulus onset) in order to determine percentage change in signal from baseline. Some methodological approaches require an even greater number of baseline frames (see for example Zheng *et al.*, 2001) in order to be able to estimate and "subtract out" baseline vascular oscillatory signals. Therefore, ideally it is advisable to collect approximately 8 s of baseline images prior to stimulus onset.

How long should images be collected after stimulus onset? Most of the components that contribute to intrinsic signals return to baseline between 8 and 12 s after stimulus onset. Therefore, in order to capture the entire temporal profile of the intrinsic signal response, one should image for approximately 15 s after stimulus onset. The intrinsic signal response, however, generally peaks within 5 s after stimulus onset so many groups image for only 5 s after stimulus onset. If the investigator is interested in only the peak mapping signal and not in response time course, this approach suffices and is actually advantageous because it reduces the volume of data. For experiments that aim to investigate physiology, the "return to baseline" phase of the intrinsic signal may be an important component that should be imaged. For example, Berwick *et al.* (2000) found that inhibiting neuronal nitric oxide synthase did not affect the early increase in deoxyhemoglobin, but significantly dampened the late functional increases in total hemoglobin and oxyhemoglobin. These later changes would not have been identified if imaging did not last longer than 5 s.

C. Interstimulus Interval

Intrinsic signals decay back to baseline in 12–15 s for a stimulus lasting 2 s. The interstimulus interval, or the time between successive stimuli, should therefore not be too short in order to avoid systematic errors in the resulting functional maps. It is also important to avoid short interstimulus intervals because of reports of "hemodynamic refractory periods," or reduced vascular response capacities with temporally close stimuli (Cannestra *et al.*, 1998b; Ances *et al.*, 2000). The magnitude of the optical response to a stimulus presented immediately after another temporally close stimulus is significantly reduced. However, choosing excessively long stimulus intervals in practice also results in lower quality maps since fewer images can be averaged in the same amount of time. Moreover, systematic errors can at least partly be avoided by randomizing the sequence of stimuli.

D. The Amount of Data

It is important to collect and average a sufficient number of trials to increase the signal-to-noise ratio of the imaging experiment. SNR increases with the square root of the number of trials. The number of trials used by different groups varies between 8 and 128. In most cases, however, 32 trials seems sufficient to achieve acceptable SNR. For a more precise methodology of determining the correct number of trials, see Section IV.A.3.

High-resolution optical imaging produces vast amounts of data. Ideally, to maximize map quality during offline analysis, one should store every single frame acquired. Normally, this is impractical. The amount of storage space needed for a typical experiment lasting 1 h with a data acquisition time of 23 s (8 s prestimulus, 15 s poststimulus) and an interstimulus interval of 37 s could amount to 30 Gbyte [23 s $\times$ 30 video frames/s $\times$ (768 $\times$ 576) pixels $\times$ 60 trials $\times$ 2 bytes/pixel ~ 40 Gbyte]. Twenty hours of data collection (which is not exceptional) would require ~1 Tbyte. Beyond the problem of storage space, one obviously would be faced with enormous amounts of time for data transfer and image analysis. These considerations necessitate a massive reduction of the amount of data. The first reduction of data occurs on the CCD chip itself with on-chip binning (e.g., 2 $\times$ 2 or 3 $\times$ 3 binning). Next, video frames can be accumulated into a single image for data analysis purposes, effectively reducing the temporal resolution of data acquisition. Moreover, data accumulated under identical stimulus conditions (i.e., same time relative to identical stimulus) are normally averaged 8–32 times, which again reduces the amount of data by this number. For those investigators who focus on data analysis techniques, preserving data from each trial may still be important.

E. Testing the Apparatus (Adapted from Bonhoeffer and Grinvald, 1996)

Two tests have proven particularly useful in testing the optical imaging apparatus before using it for experiments: (1) testing SNR and (2) testing whether data acquisition and stimulus presentation are properly timed.

(1) The following procedure is used to generate an artificial test signal that is comparable to a typical intrinsic signal as recorded from the living brain. A LED display of the number "8" (with seven individual LEDs) is connected to the data acquisition system such that for the different stimulus conditions the different segments of the LED display are switched on. The brightness of this LED is set to use the full dynamic range of the camera. This brightness is then attenuated by a factor of 1000 with a 3 OD filter. If, finally, this whole arrangement is then illuminated with red light so that its brightness is again almost at the saturation level of the camera, one has a device that produces modulation of 1 in 1000 on a relatively high absolute light intensity. The optical imaging apparatus can then be tested by acquiring data under these conditions and seeing whether this very weak modulation can be picked up by the system with the proper SNR.

(2) In order to test the consistency of data acquisition, data analysis, and stimulation, it is useful to run a test experiment in which a visual stimulator produces patterns that can easily be distinguished from each other. If the stimulator is controlled by the data acquisition program in the same manner as in a real experiment and if the camera is pointed directly onto the stimulator screen, one has a very simple testing procedure: data analysis of the pattern that the camera imaged from the screen will immediately reveal any inconsistencies in the stimulation, data acquisition, and data analysis procedures.

VI. Data Analysis for Mapping Functional Architecture

Intrinsic signal data can be collected in two ways: (1) single wavelength or (2) multiple wavelengths simultaneously using a spectrophotometer. The analysis of these different types of data is quite distinct and so is discussed separately in this section. One should keep in mind that intrinsic signal maps can and will differ depending on the method of analysis used. Therefore, it is critical to understand the assumptions and limitations of each approach, both when *selecting* a particular approach and when *interpreting* results. The methodology, assumptions, and limitations of different analytic methods are discussed in this section.

A. Analysis of Single-Wavelength Data

1. Ratio Analysis

In order to obtain activity maps from the cortex, images have to be acquired both while the cortex is stimulated and while the cortex is at rest. A "reference image" is then subtracted from and divided into all subsequent images on a pixel-by-pixel basis in order to determine the percentage change in reflectance at each pixel at each time point. The reference image is divided into the difference images in order to normalize for uneven illumination. The selection of the reference image is not trivial, has important implications, and can influence the appearance of maps. Three different approaches have been employed: (a) prestimulus cortical image, (b) cocktail blanks, and (c) poststimulus cortical image.

a. Prestimulus cortical image. Using a cortical image prior to stimulus onset as a reference image makes the fewest assumptions about the functional architecture and physiology of the brain (Fig. 6). The advantage of a picture of the inactive cortex is that no assumption is made about the complete set of stimuli that are required to activate the cortex uniformly (as is the case with cocktail blanks, see below). The disadvantage of using the blank picture is that it can cause very strong, activity-related blood vessel artifacts in the maps (Fig. 6). These vascular artifacts often overwhelm the fine mapping details that emerge when a cocktail blank is used for analysis instead. Other strategies than the cocktail blank approach have been developed to reduce these vascular artifacts, including using a poststimulus cortical image as a reference image [Chen-Bee *et al.,* 1996; see (c) below], using a macroscope (Ratzlaff and Grinvald, 1991; see Section IV.C), and focusing below the blood vessels so that they do not contribute significantly to the intrinsic signal maps (Hodge *et al.,* 1997). Alternatively, Chen-Bee and colleagues (2000) have recommended averaging only the first 1.5 s after stimulus onset when imaging in the low 600 nm range, which coincides with the period of fast oxygen consumption of active neurons but precedes the activation of large surface vessels, thereby minimizing the contribution of vascular artifacts to the final map (Chen-Bee *et al.,* 2000).

b. Cocktail blank. The cocktail blank is the sum of multiple individual cortical images under different stimulus conditions. It is intended to represent the uniformly activated cortex. The only difference between the single-activity map and the cocktail blank hypothetically is the orientation of the stimulus; it is argued that the picture is not confounded with an additional difference in overall activity (i.e., functional vascular activity). The disadvantage, however, is that assumptions are made about the functional architecture and neuronal interactions within the cortex of interest, resulting

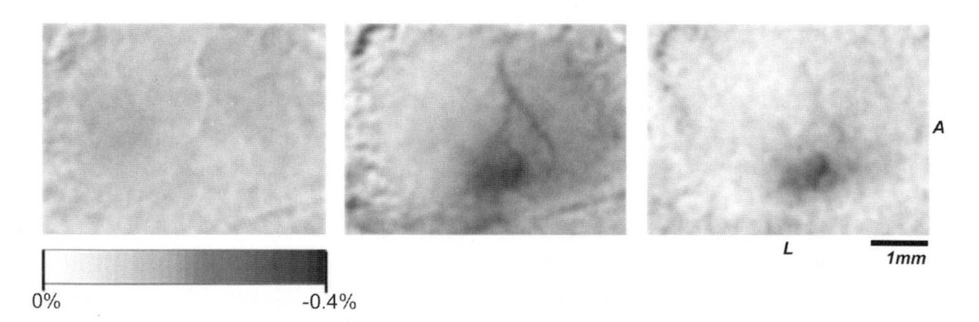

Figure 6 The effect of using different reference images (prestimulus vs poststimulus cortical image) on optical map. Optical reflectance changes over the right somatosensory cortex of rodent were imaged in response to left C1 whisker deflection at 610 nm. **Left:** This image shows no significant optical changes in a control frame prior to stimulus onset. **Middle:** Using a prestimulus cortical image as a reference image for ratio analysis can result in significant activity-related vascular artifacts in the optical map. Specifically, decreases in reflectance in draining veins are seen medial to the focus of optical activity overlying the location of the stimulated barrel. **Right:** When an average of a prestimulus (0.5 s prestimulus) and poststimulus (2.5 s poststimulus) cortical image is used as the reference image for ratio analysis, vascular artifacts can be minimized (Chen-Bee, *et al.*, 1996). This method of data analysis exploits the different time courses of stimulus-dependent *cortical* activity and stimulus-dependent *vessel* activity. The significant vascular artifacts seen in the middle image are greatly reduced by using this alternative approach, providing a much more focal representation of cortical activation. A, anterior; L, lateral.

in a circular argument. If the goal of intrinsic signal mapping is to characterize the functional organization of the cortex, then how can one create a cocktail blank, which requires assumptions about the underlying functional cortical architecture? When recording iso-orientation maps from the primary visual cortex of the cat, for instance, it is normally assumed that a complete set of all orientations activates the cortex evenly and therefore the stimulus set is used to calculate the cocktail blank. In some cases, when a cocktail blank composed of all the different presupposed stimuli for a given cortical area is divided by the unstimulated blank, an underlying structure emerges in the resulting map, suggesting that some characteristic of the stimuli used disproportionately activates some regions of the cortex. In fact, it was speculated by Bonhoeffer and Grinvald (1993a) that a clear pattern obtained when a cocktail blank was divided by a prestimulus reference might hint that spatial frequency maps may exist in cat area 18. Indeed such spatial frequency maps have been demonstrated using optical imaging. This example shows that a cocktail blank under some circumstances can be inadequate since it imposes a structure onto activity maps. It is therefore important to divide the cocktail blank by a prestimulus image to ensure that it does not contain any structural information within itself. Conversely, this procedure may be beneficial because it can give hints to the existence of additional stimulus attributes that are represented on the cortex in a clustered fashion.

c. Poststimulus cortical image. This method of data analysis exploits the different time courses of stimulus-dependent *cortical* activity and stimulus-dependent *vessel* activity in order to minimize the contribution of vascular artifacts to intrinsic signal maps (Chen-Bee *et al.*, 1996) (Fig. 6). Intrinsic signals from blood vessels tend to persist longer than those arising from the cortex. Therefore, cortical images greater than 2–3 s after stimulus onset provide a "map" of the vascular signal. If this vascular signal is used as part of the reference image, the intrinsic signals from the vessels can be minimized. To do this, the reference image used is an average of a prestimulus cortical image and an image after stimulus-dependent cortical activity has subsided but intrinsic vascular activity persists (usually 2–3 s after stimulus onset). Chen-Bee and colleagues (1996) have shown the success of this approach in reducing vascular components in intrinsic signal maps. This approach, however, assumes that the vascular and cortical intrinsic signals can be well separated in time, which is not always the case. Furthermore, this approach tends to underestimate intrinsic signal magnitude since the poststimulus image will inevitably contain intrinsic signals emanating from the area of interest even after the majority of the cortical signal has subsided.

Single-condition maps. Single-condition maps are calculated by taking the activity map obtained with one particular stimulus and dividing this image by the reference image. The resulting map then shows the activation that this particular stimulus causes.

e. Differential maps. In differential maps, to maximize the contrast, one activity map is divided by (or subtracted from) the activity map that is *likely* to give the complementary activation pattern. This approach makes further assumptions about the underlying cortical architecture (which may or may not be true) and therefore must be carefully considered. One example of a problem using differential maps is that a gray region in such maps (e.g., pixel value of ~128) can correspond to cortical regions that were not activated by either stimulus or alternatively to regions that were strongly activated by the two stimuli but with equal magnitude. Single-condition maps using the cocktail blank are free of these problems. These considerations are similar to those encountered when choosing between unipolar and bipolar electrodes for electrophysiological recordings, in that single-condition maps are analogous to unipolar electrodes and differential maps are analogous to bipolar electrodes.

Two different methods can be used to calculate differential maps. One possibility is to calculate the ratio between the maps A and B. The other possibility is to subtract the two maps and then divide the result by a general illumination function like the cocktail blank: $(A - B)$/blank. Whereas the first calculation provides information about the relative activity at each pixel in A vs B, the second calculation highlights the *difference* between the maps as the important entity. Although these two calculations may seem very different, they are equivalent under the assumption that the reflectance changes are exceedingly small relative to baseline (Bonhoeffer, 1995).

The assertion is that

$$\frac{V-H}{B} \cong \frac{V}{H} \tag{1}$$

where V is the image obtained with a vertical grating, H is the image obtained with a horizontal grating, and B is the blank image.

Since the changes of reflected light intensity are small compared to the absolute values of the reflected light intensity, one can write

$$V = I_0(1 + \Delta V) \text{ and } H = I_0 + \Delta H), \text{ with } \Delta V, \Delta H \approx 10^{-3}.$$

Since $B \approx I_0$, the left-hand side of the first equation can be rewritten as

$$\frac{I_0(1+\Delta V)-I_0(1+\Delta H)}{I_0} = \Delta V - \Delta H \tag{2}$$

and the right-hand side of the same equation can be written as

$$\frac{V}{H} = \frac{I_0(1+\Delta V)}{I_0(1+\Delta H)} = \frac{(1+\Delta V)}{(1+\Delta H)}. \tag{3}$$

Since $1/(1 + x) \approx 1 - x$ for $x < 1$ and since $\Delta H < 1$, one can further write

$$\frac{(1+\Delta V)}{(1+\Delta H)} \approx (1+\Delta V)(1-\Delta H). \quad (4)$$

Thus, the original equation can be rewritten as

$$\Delta V - \Delta H \approx 1 + \Delta V - \Delta H - \Delta V \Delta H.$$

For intrinsic signals, ΔV and ΔH are in the range of 10^{-3}. Therefore, $\Delta V \Delta H$ is approximately 10^{-6}, which is 1/1000 of the intrinsic signal change that is normally observed and therefore negligible, so it can be disregarded. Moreover, since the scaling of the activity maps is done with an arbitrary offset, the 1 on the right side of the equation can be disregarded. Consequently, under these circumstances, dividing images and subtracting images yield the same result.

2. Principal Component Analyses

Because of the exceedingly small magnitude of the intrinsic mapping signals in the background of relatively large magnitude global signals (e.g., blood volume changes, which are generally not specific to the activation paradigm), other analytic paradigms than the "ratio analysis" presented above have been implemented to extract the significant mapping components of intrinsic optical signals. In particular, several groups have focused on variations of principal component analyses to exclude the so-called global signals and noise from the final intrinsic signal map (Cannestra *et al.*, 1996; Gabbay *et al.*, 2000; Stetter *et al.*, 2000; Zheng *et al.*, 2001). Principal component analysis (PCA) assumes that the data are made up of a linear sum of signals, which can be decorrelated based on differences in variance. The mathematical procedure for PCA has been outlined in detail (Geladi *et al.*, 1989; Yap *et al.*, 1994). Briefly, given a set of images [X_t = 0 to N] from time points 0 to N, a single matrix [X_S] is constructed such that the columns represent a pixel intensity of a specific image and the rows represent the intensity time course of a single pixel from the image set. Matrix $[X_S]^T$ is transposed, and the covariance matrix [C] determined from the relation: $[C] = [X_S]\ [X_S]^T$. The eigenvalues (?^{i}) and eigenvectors (V?^{i}) are determined by the Jacobi transformations technique.

With high SNR (as is often the case in intrinsic signal imaging in rodents), the first principal component can be used as an accurate measure of the time course of the intrinsic signal response (Cannestra *et al.*, 1996) (Fig. 7). The eigenvector corresponding to the first principal component can then be used to reconstruct the data set without noise. In cases of low SNR, however, more complex approaches to

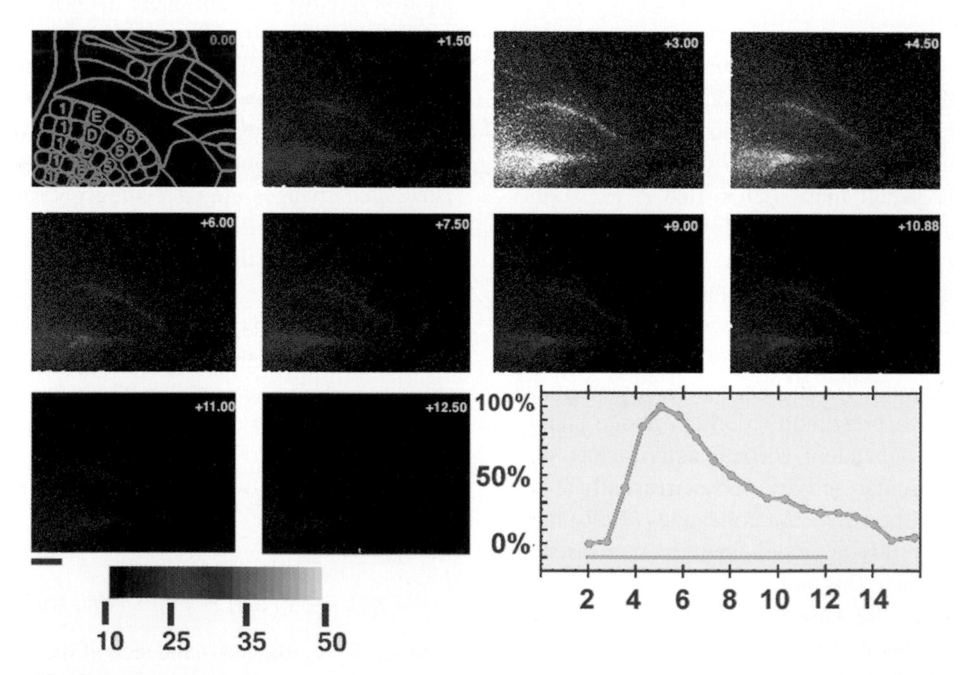

Figure 7 Principal component reconstruction of intrinsic signal responses at 850 nm over rodent somatosensory cortex. Whisker C1 on the contralateral face was stimulated through a 30° angle at 10 Hz for 10 s. With high SNR, the first principal component can be used as an accurate measure of the time course of the intrinsic signal response (Cannestra *et al.*, 1996). The images displayed here are a reconstruction of the data set using the eigenvector corresponding to the first principal component. Using principal component analysis can significantly reduce the contribution of noise to the final data reconstruction. **Bottom right:** Calculated time course of the first eigenvector, showing an initial peak intrinsic signal response approximately 4–5 s after stimulus onset and a return to baseline after stimulus termination. Scale bar is 1 mm. Color bar represents reflectance decrease × 10^{-4}. Adapted, with permission, from Cannestra *et al.* (1996).

PCA of intrinsic signal maps must be used. Gabbay and colleagues (2000) recommended that only those principal components whose time course correlates with the stimulus presentation sequence be used for the reconstruction of the final intrinsic signal map. Specifically, a contiguous range of principal components is chosen whose probability of being genuinely related to the stimulus is high. This range typically excludes the higher power principal components, which they ascribe to "vegetative processes," including circulation and respiration, and the lower power components, which are ascribed to background noise (Gabbay *et al.*, 2000).

PCA is limited by the fact that the only means of separating out signals is variance. If mapping signals and background noise have similar variances, PCA cannot separate these signals. Blind signal separation (BSS) builds upon PCA, attempting to separate out the underlying signal sources by making assumptions about the statistical structure of the underlying signal sources. One example of BSS is by extended spatial decorrelation (ESD) (Stetter *et al.*, 2000). ESD assumes that each signal component varies smoothly across space and that each component has zero cross-correlation functions with the other functions. Subsequent to PCA and selection of significant principal components, ESD can be used to extract the mapping signal, for better separation of mapping signals from global signals and movement artifacts (Stetter *et al.*, 2000).

3. Correlation Maps

In BOLD *f*MRI analysis, a common analysis is to determine the correlation of the measured signal with a predetermined hemodynamic response function (Bandettini *et al.*, 1993; Cohen, 1997). Although popular for BOLD *f*MRI analysis, this methodology is not commonly applied to intrinsic signal imaging. The major advantage of this technique over those presented above is that it can be used to generate statistical maps (based on the correlation coefficient), providing an estimate of the statistical reliability of the maps generated. These statistical maps, however, should be carefully considered since Pearson's test of correlation was designed to evaluate the correlation of independent samples. Consecutive time points in a physiological response are *not* independent; correlation analysis is therefore not ideal.

Mayhew and colleagues (1998) examined the correlation of intrinsic signals under red and green light illumination with three different hemodynamic models (with different lags). Not surprisingly, the maps revealed that responses to stimulation occurred with different lags at different spatial positions depending on wavelength. Therefore, in order to use correlation analysis effectively it is critical to know the timing of the response that is to be mapped and extracted. It can be extremely useful in separating out signals with time courses known to be different. Clearly, this analytic approach is limiting in that it requires assumptions to be made about the time course of intrinsic signals in advance. Although the time courses of intrinsic signals have been well characterized (see Section II.E), it should not be assumed that the time courses of intrinsic signals are uniform across different species and anesthetic states. Moreover, this technique cannot separate out signals of different etiologies with similar time courses. For example, if the mapping signal of interest has a time course similar to that of some global confounding signal, the two would be inseparable using this approach.

4. The Generalized Linear Model

The generalized linear model (GLM), like PCA, holds that any observed response is a linear sum of multiple individual underlying responses. In contrast to PCA but similar to correlation maps, when using GLM, the data are fitted to *predetermined* signal time courses instead of signals being extracted based on differences in variance (as is done with PCA). Briefly, GLM analysis can be represented mathematically as $X = G \cdot \beta + e$, where X is a vector representing the time course of a given pixel, G is the design matrix containing the series of predetermined response functions, β is a vector of coefficients representing the degree to which each response function contributes to the overall signal, and e is random noise. This equation is solved at each pixel using a least-squares approach.

Using GLM, one can determine how predetermined (or hypothetical) hemodynamic and metabolic responses may combine to contribute to the final observed intrinsic signal response at each pixel. Based on these calculations, maps can be generated which indicate to what degree the different predetermined response functions contribute to the signal at each pixel. Like correlation maps, GLM can also be used to generate maps of the statistical reliability of the results of the analysis. This method differs from correlation maps in that the time course of a pixel does not necessarily have to match a single predetermined function, but may be a linear sum of multiple predetermined functions.

Mayhew and colleagues (1998) demonstrated the application of GLM for analyzing intrinsic optical responses and discussed the limitations of this approach in detail. Briefly, they used three different response functions: one representing the hypothesized initial increase in deoxyhemoglobin peaking 2–3 s after stimulus onset (i.e., the first phase of the 600- to 630-nm response), one representing changes in total hemoglobin (peaking 3–4 s after stimulus onset), and one representing the second phase of the 600- to 630-nm response (peaking 8–10 s after stimulus onset). They further exploited the power of this technique by determining if adding other functions to the design matrix would reduce the residual signal after analyses. In particular, the effect of incorporating a response function for baseline vasomotion, or vascular oscillatory behavior, was assessed (see Section VI.C).

This approach has many of the same limitations as correlation maps, in that analysis must begin with a preconceived notion of temporal response profiles. Residual maps demonstrate that if inappropriate functions are used, the residual signal retains a significant amount of spatial and temporal information, suggesting the inadequacy and limited applicability of this approach.

5. Analysis Using Weak Models

The major limitation of correlation maps and GLM is that the investigator must make assumptions about the time course of the response. Recently, Zheng and colleagues (2001) have suggested using "weak model" constraints or modifications to two of the above analyses (BSS and GLM). The so-called weak models make three assumptions: (1) The response begins after stimulation. The temporal profile of the response, however, is not specified. Rather, the models hold only that the time course is flat prior to stimulation and changes after stimulation. (2) The spatial distributions of noise and the mapping signal are different but constant over time. (3) The signal is a linear combination of all signal sources (similar to PCA and GLM). The details of these approaches are not provided here as they are explained in detail elsewhere (Zheng *et al.*, 2001). These techniques have been shown to be more effective at isolating mapping signals compared to other approaches. The difficulty with these approaches, however, arises from the need to record extended baselines (~8 s) prior to stimulation in order to model the response appropriately. Zheng and colleagues (2001) have shown that the method is still applicable with shorter prestimulus baselines (1 s), although not as effective.

B. Analysis of Spectral Data

By integrating a spectrophotometer into the optical imaging apparatus, investigators can simultaneously collect spectral and spatial information regarding the intrinsic signal response (Malonek and Grinvald, 1996; Mayhew *et al.*, 1999, 2000; Kohl *et al.*, 2000; Lindauer *et al.*, 2001). The analysis of the spectral data, however, remains controversial and the approach taken significantly alters the outcome and final interpretation of results.

All spectroscopic analysis methods are based on the Beer–Lambert Law,

$$\Delta A_\lambda = \varepsilon_{\text{chromophore},\lambda} \cdot \Delta[\text{chromophore}] \cdot l,$$

where ΔA_λ is the change in absorbance at a given wavelength, $\varepsilon_{\text{chromophore},\lambda}$ is the absorption coefficient of a given chromophore at a given wavelength, $\Delta[\text{chromophore}]$ represents the change in concentration of that chromophore, and l is a non-wavelength-dependent term which is required for proper dimensionality. Assuming several chromophores contribute to the changes in absorption at a given wavelength, this expression can be modified:

$$\Delta A_\lambda = \varepsilon_{a,\lambda} \cdot \Delta[a] \cdot l + \varepsilon_{b,\lambda} \cdot \Delta[b] \cdot l + \varepsilon_{c,\lambda} \cdot \Delta[c] \cdot l.$$

Given three chromophores, absorbance changes must be measured using at least three different wavelengths so that this algebraic problem can be solved as a system of three equations. In many cases, more than three wavelengths are collected, and so a least-squares approach is used to fit the experimental data to the absorption spectra.

The basic Beer–Lambert relationship expressed above does not contain a wavelength-dependent path-length factor. Many groups argue that a path-length factor is necessary since different wavelengths of light have different scattering patterns and therefore travel different distances and have different path lengths (Mayhew *et al.*, 1999, 2000; Kohl *et al.*, 2000; Lindauer *et al.*, 2001). Considering this, the equations can be modified such that

$$\Delta A_\lambda = \text{DP}_\lambda(\varepsilon_{a,\lambda} \cdot \Delta[a] + \varepsilon_{b,\lambda} \cdot \Delta[b] + \varepsilon_{c,\lambda} \cdot \Delta[c]),$$

where DP_λ represents a wavelength-dependent path-length factor which depends not only on the wavelength of imaging but also on tissue properties (including assumptions of absolute concentrations of oxy-Hb and deoxy-Hb).

The major chromophores usually incorporated into spectroscopic analysis include oxyhemoglobin, deoxyhemoglobin, and cytochrome oxidase.

1. Linear Spectroscopic Analysis

The first spectroscopic study of intrinsic optical signals assumed that absorbance changes at a given wavelength were due to a linear sum of absorbance changes due to each chromophore (Malonek and Grinvald, 1996). In addition, a wavelength-independent factor, called LS, was included to represent a wavelength-independent light scattering component. This modification was based on previous optical imaging studies in blood-free slice preparations that showed that light scattering changes are independent of wavelength (Cohen and Keynes, 1971; Grinvald *et al.*, 1982).

Changes in absorbance were therefore estimated as

$$\Delta A_\lambda = \varepsilon_{a,\lambda} \cdot \Delta[a] + \varepsilon_{b,\lambda} \cdot \Delta[b] + \text{LS}.$$

A least-squares routine was used to fit the data. The authors state that using other wavelength-dependent models for the light scattering term did not affect the calculated chromophore time courses (Malonek and Grinvald, 1996).

Using this linear approach, Malonek and Grinvald demonstrated an initial increase in deoxyhemoglobin concentrations and a delayed increase in oxyhemoglobin concentrations (Malonek and Grinvald, 1996). Support for this approach has been provided by Shtoyerman and colleagues (2000), who showed that similar chromophore concentration changes and time courses are calculated regardless of wavelengths used for imaging and analysis. They suggest the robustness of the results across wavelengths validates linear spectrographic analysis. This approach, however, has been criticized because it is well known that tissue penetration as well as light scattering patterns is wavelength dependent.

Therefore, this approach may be an oversimplification and produce erroneous results. Kohl and colleagues (2000) posit, "Ignoring its wavelength dependence in the calculation of concentration changes has the same effect as calculating with wrong extinction spectra." This debate remains critical since the time course of chromophore changes depends critically on the analytic approach used (Mayhew *et al.,* 1999; Lindauer *et al.,* 2001).

2. Path-Length Scaling Spectroscopic Analysis

Nemoto and colleagues (1999) were the first group to include a path-length-dependent factor in the spectroscopic analysis. They include an additional path-length factor into the analysis (DP_λ above). The wavelength-dependent path length was assumed to be *constant* throughout the acquisition period and not to be affected by functional changes in the hemoglobin concentration and oxygenation state under physiologic conditions (Nomura *et al.,* 1997). They still incorporated a wavelength-independent term that was presumed to represent light scattering changes [using the same rationale proposed by Malonek and Grinvald (1996) above]. Using this approach, Nemoto and colleagues also detected an early increase in deoxyhemoglobin concentrations followed by a delayed increase in oxyhemoglobin concentrations (Nemoto *et al.,* 1999). This approach may also be limited by the fact that DP_λ may change with functional activation since the tissue properties change significantly with changes in blood flow and hemoglobin oxygenation.

Both Mayhew and colleagues (1999, 2000) and Dirnagl and collaborators (Kohl *et al.,* 2000; Lindauer *et al.,* 2001) have proposed using a Monte Carlo simulation of light transport in tissue to determine DP_λ (see Mayhew *et al.,* 1999; Kohl *et al.,* 2000; Lindauer *et al.,* 2001; for methodological and theoretical details regarding Monte Carlo simulations). According to these groups, the DP_λ factor changes dynamically with the brain as the relative concentrations of deoxyhemoglobin, oxyhemoglobin, and cytochrome oxidase change with functional activation. These groups use the same general equation as Nemoto and colleagues (1999) except that DP_λ is dynamic. Interestingly, this modification eliminates the calculated initial increase in deoxyhemoglobin concentrations. Lindauer and colleagues (2001) therefore argue that there is no evidence for the early increase in deoxyhemoglobin concentrations. Mayhew and collaborators (1999), however, take the analysis one step further by applying GLM to remove the 0.1-Hz baseline vasomotion from the signal (see discussion of baseline vasomotion in section below). After applying this manipulation, they demonstrate that an early increase in deoxyhemoglobin concentrations can be identified (Fig. 8). Mayhew and colleagues have also shown that when blood pressure is controlled for and increased to physiological levels using

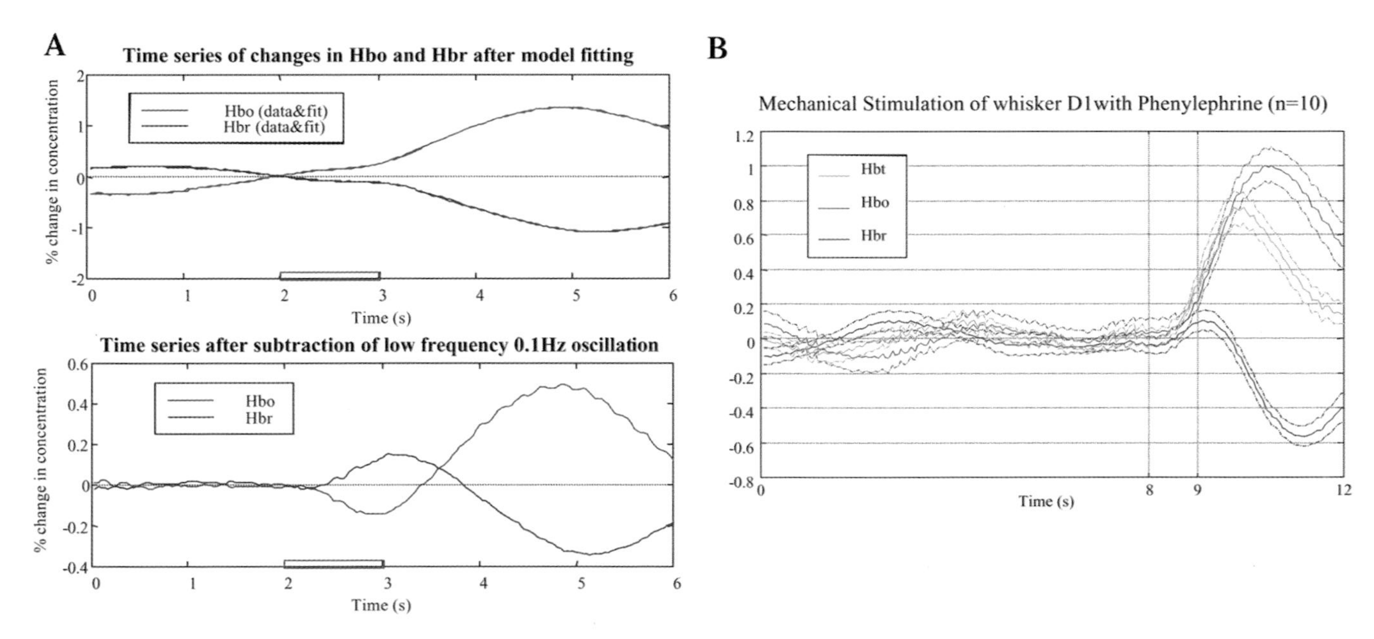

Figure 8 Spectroscopic analysis reveals initial increase in deoxyhemoglobin concentrations. (**A, top**) Initial spectroscopic analysis using path length scaling analysis reveals an decrease in deoxyhemoglobin concentrations concomitant with an increase in oxyhemoglobin concentrations in response to whisker stimulation over the rodent somatosensory cortex. (**A, bottom**) Subtraction of the 0.1-Hz underlying vasomotion signal, however, reveals that the initial increase in deoxyhemoglobin concentrations is present and was initially (top) not apparent due to the vasomotion signal complicating analysis. (**B**) Recent studies by Mayhew and colleagues (2000) suggest that the vasomotion signal can be minimized by increasing the systemic blood pressure of the rodent using phenylephrine. Spectroscopic analysis using path length scaling without subtraction of a baseline oscillatory vascular signal (which has been suppressed by increasing blood pressure) reveals an initial increase in deoxyhemoglobin concentrations. The differences observed between the graphs on the left highlight the need for investigators to carefully consider any analysis technique used to interpret results. Adapted, with permission, from Mayhew *et al.* (1999, 2000).

phenylephrine, the contribution of the background vasomotion is minimized and the early increase in deoxyhemoglobin concentrations can be identified even without applying GLM (Fig. 8). Although these groups argue that this approach better approximates true wavelength-specific path lengths, these calculations are based on simulations and are not measured values. It has been proposed that true path lengths for different relative concentrations of oxyhemoglobin and deoxyhemoglobin can be calculated for each wavelength to better estimate the true changes in chromophore concentrations. This approach, however, has not yet been applied, so its efficacy is unclear.

As should be clear from this discussion, the approach taken to the analysis of the spectral data can significantly alter the conclusions of the investigation. Therefore, investigators should critically appraise any analytical or mathematical approach used to evaluate spectroscopic data. It is important to acknowledge the assumptions and limitations underlying each technique.

C. Baseline Vasomotion

Several groups are increasingly noting baseline oscillatory behavior in intrinsic signals (see Mayhew *et al.*, 1996, for example). The oscillations are most probably vascular in origin and represent baseline vasomotion or alternating relaxation and constriction of neurovasculature. These baseline oscillations are approximately 0.1 Hz, observed at all wavelengths (although more prominent in the red and green range), and may be several times the magnitude of activity-related intrinsic optical responses (Mayhew *et al.*, 1996). The presence of these large baseline oscillations, if not accounted for, may therefore alter the interpretation or analysis of intrinsic maps.

Most groups deal with this oscillatory behavior by averaging several trials and thereby attempting to eliminate the regular background oscillations from the final map. This is possible since baseline oscillations are random over time but the functional response is locked to stimulus presentation. It is important, therefore, to collect a sufficient number of prestimulus images to make sure that the baseline oscillations are adequately suppressed by trial averaging. Although some argue that this approach does not adequately cancel the effect of these large oscillatory patterns, many groups have been able to produce functional intrinsic signal maps using this approach (see for example Polley *et al.*, 1999a).

Another approach to removing these slow oscillations is to use the so-called "first-frame analysis." Since the time course of these "vascular" oscillations (~10 s) is significantly slower than the duration of a trial (in some cases ~3–4 s), the vascular signal can be assumed to be represented as a fixed pattern in all the frames acquired during the trial: if the first data frame is taken prior to any stimulation, it will contain the spatial representation of the oscillatory signal, but no information about the pattern of the stimulus-evoked activity. Thus, to minimize slow noise, the first frame can be subtracted from all subsequent frames before any additional analysis.

Mayhew and colleagues have implemented several algorithms (presented above) to deal with and try to extract these oscillations from the intrinsic signal maps, including GLM and analyses using weak models (Mayhew *et al.*, 1998; Zheng *et al.*, 2001). These approaches attempt to model the slow intrinsic oscillations and subsequently extract the signal prior to any subsequent analysis. Details are provided in Sections VI.A.4 and VI.A.5.

D. Color Coding of Functional Maps

Color coding of functional maps can be extremely helpful in relaying a large quantity of information within a single functional map (Blasdel and Salama, 1986; Ts'o *et al.*, 1990; Bonhoeffer and Grinvald, 1991; Blasdel, 1992a,b). For instance, for comprehensive analysis of the organization of iso-orientation domains in visual cortex, a color-coded display in which the color at each pixel represents the angle of preferred grating orientation at each pixel is advantageous (Fig. 9B). Preferred orientations are determined by using vector sums of single-condition responses to different orientation gratings at each pixel. So-called "angle maps" can be produced to display only the angle of the resulting vector, with no information about the magnitude of the resulting vector (originally introduced by Blasdel and Salama, 1986). Alternatively, the magnitude of the resulting vector (which relates to the degree of orientation preference) can also be conveyed by using differences in color intensity. Note that a small-magnitude response can represent multiple situations. For example, a small-magnitude vector may represent either a poor response to a single orientation or strong but slightly unequal responses to gratings moving in opposite directions.

To overcome this ambiguity, one can calculate maps called "hue-lightness-saturation maps" (HLS maps) (Fig. 9B). This type of map simultaneously displays the three values of (1) preferred orientation, (2) overall response strength, and (3) tuning sharpness (Ts'o *et al.*, 1990; Bonhoeffer and Grinvald, 1993a). As with angle maps, color codes for orientation preference. Also, like angle maps, the overall response magnitude at each pixel (i.e., the response of one particular site summed over all orientations) is represented by color intensity. The final parameter displayed in these maps, which is different from angle maps, is color saturation, which reveals the quality of the orientation tuning (the worse the tuning the less saturated, i.e., the "whiter" the color). Consequently, if many different orientations activate a particular pixel (as was presented hypothetically at the end of the previous paragraph), the color of that pixel would be

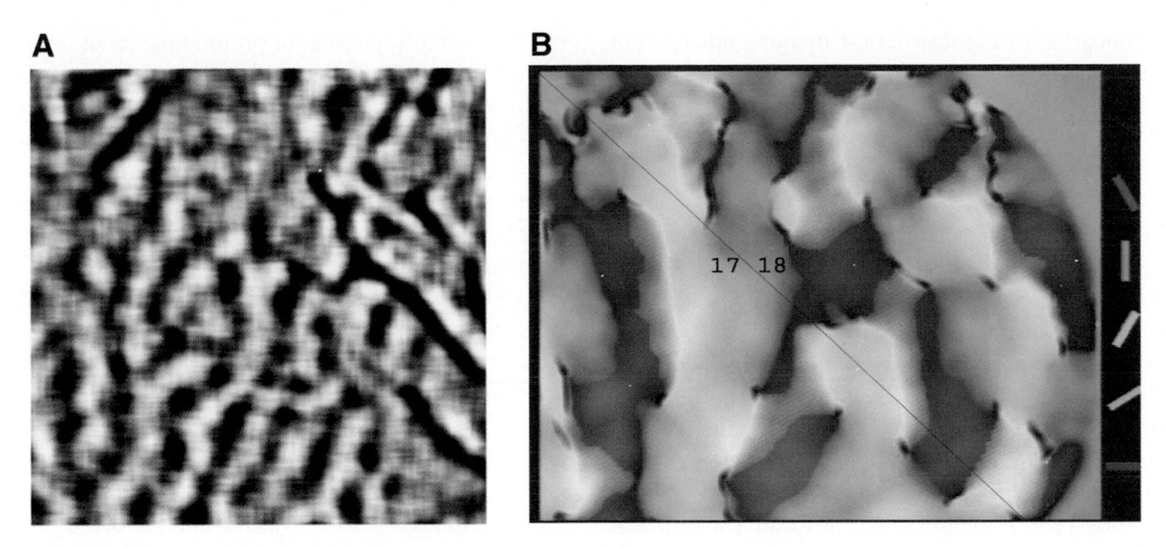

Figure 9 Optical intrinsic signal mapping of visual cortex. (**A**) Ocular dominance map, demonstrating distinct "strips" in a 1-cm^2 patch of cortex, activated by a stationary bar presented monocularly to the visual system of a rhesus monkey. (**B**) HLS map of orientation selectivity in areas 17 and 18 of the cat visual cortex. This image shows the preference of neurons to lines of different orientations, when presented to the retina. The legend on the right shows the relationship between the color of each pixel and orientation. The brightness of each pixel indicates the selectivity of each point in the map: dark indicates points in the map that are not particularly selective to any orientation, while bright points signify points in the map that are tuned specifically to a given orientation. (A) was adapted, with permission, from Macknik and Haglund (1999). (B) was provided by S. L. Macknik and G. G. Blasdel (Harvard Medical School).

less saturated, or "whiter." By conveying more information about the underlying maps which were summed to produce the orientation maps, these HLS maps eliminate much of the ambiguity present in the angle maps.

VII. Chronic Optical Imaging

In addition to its superior spatial *and* temporal resolution, one of the strengths of optical imaging of intrinsic signals is that it is relatively noninvasive, in that the cortex is not touched or damaged by the imaging procedures. Consequently, subjects can easily endure long-term optical imaging, both within the span of the day and over a period of several weeks. In order to make optical imaging over a period of weeks or months successful, several methodological modifications may be necessary.

A. Imaging through the Intact Skull and/or Dura

Several groups have now demonstrated successful intrinsic signal imaging through both intact dura and thinned bone (Blood *et al.,* 1995; Polley *et al.,* 1999a). Frostig and colleagues (1990) showed, in cats, that orientation columns in the visual cortex identified by imaging through the dura were identical to ones obtained directly from the exposed cortex after removing the dura.

Different wavelengths will penetrate the thin bone and dura to different degrees and so imaging may not be possible through the intact dura and skull in all subjects. Originally, imaging through the thin bone preparations was done using only near-infrared light because these longer wavelengths are better able to penetrate tissue than shorter wavelengths. However, thin bone preparations have also successfully been used for imaging between 500 and 700 nm (see for example, Polley *et al.,* 1999a). The quality and extent of bone thinning will largely determine the likelihood of successful imaging. Success will depend on the degree to which the skull is thinned (in rodents, the bone should be thinned to approximately 250 μm so that the branches of the major cerebral arteries and veins can be visualized using green light illumination) and whether any epidural or subdural hematomas were induced by the thinning procedure (if so, such subjects should be excluded from studies). Polley and colleagues (1999a) elegantly used a thin bone preparation in rodents to study experience-dependent modification of intrinsic optical maps of whisker representations in the somatosensory cortex. Subjects were imaged at three different time points, each 4 weeks apart, while the whiskers and the environments were manipulated to determine their effects on whisker representations. This study was clearly made possible by using the thin bone preparation that enabled the investigators to image the same subjects over a period of months (Polley *et al.,* 1999a).

When using a thin bone or intact dura preparation it is necessary to provide prophylaxis against infection. Investigators should make sure that antibiotics and analgesics used in the care of research subjects do not have vasoactive properties that may otherwise introduce a confounding factor in the interpretation of chronic optical imaging studies. For example, steroids and opioids are often recommended as analgesics. These drugs, however, may alter vascular responsiveness and so investigators should opt for other analgesics such as nonsteroidal anti-inflammatory drugs.

B. Chronic Optical Imaging in the Awake Monkey

Experiments in awake monkeys offer many advantages for the study of higher cognitive functions. Since such studies require very long preparatory periods and financial investment, which are devoted to training the animal, it is essential that the imaging should not be restricted to a single experiment and thus that chronic recordings are feasible.

The foremost problem with chronic recordings in primates is maintaining the cortical tissue in optimal condition for long periods of time. In order to achieve this and at the same time provide good optical access to the brain, Shtoyerman and colleagues (2000) examined the feasibility of implanting a transparent artificial dura made of silicon. After affixing chambers to the primate's skull with stainless steel screws, the skull was removed from the region circumscribed by the chamber. Once the cortex was not exerting any pressure on the dura (see Section III.C.1), the dura was opened and the cortex was *immediately* covered with the artificial dura (homemade). After the artificial dura was in place, the chamber was filled with an agarose and antibiotic solution. Agarose is necessary to prevent rapid regrowth of the natural dura. Finally, the chamber was sealed with a glass plate and a metal plate to protect the exposed cortex. Subjects were treated 7–14 days with antibiotics and regularly tested for infections. Regularly sampling the chamber for both bacterial and mycological infections is key since infections can spread quickly from within the chamber into the cortex. The authors report that the immediate covering of the cortex is crucial for prevention of infections and maintaining the cortex for long periods (Shtoyerman *et al.*, 2000). The advantages of this procedure are that the cortex remains visible for periods of greater than 1 year and the artificial dura is easily penetrable by electrodes (which is key for matching intrinsic signal maps with electrophysiological maps).

C. Reproducibility of Optical Maps

Defining the reproducibility of maps over time is critical for chronic imaging studies, especially those which aim to define the effect of an experimental manipulation (e.g., monocular deprivation or whisker plucking) on functional representations. In order to conclude that a particular treatment alters functional maps, one must first be confident that maps do not spontaneously change over time.

Having established a methodology for chronic imaging in primates, Shtoyerman and colleagues (2000) explored the stability and reproducibility of functional maps over a period of greater than 1 year. Ocular dominance columns followed over 1 year did not change significantly and appeared extremely reproducible. The variability of ocular dominance column maps over a period of 1 year was similar to the variability of ocular dominance column maps from a single imaging session or over a few consecutive days (Shtoyerman *et al.*, 2000) (Fig. 15).

The reproducibility of maps was also investigated in rodents by Masino and Frostig (1996). Unlike the results of Shtoyerman *et al.* (2000), they found that functional whisker representations in rodents may vary by as much as 48% over a period of months. Although the location of the whisker maps was consistent over months, they identified nonsystematic changes in size, shape, and response magnitude of the intrinsic signal map (Masino and Frostig, 1996). Despite the intrinsic variability in functional representations, the rodent model can still be used to investigate the effect of a manipulation on functional maps by demonstrating that a manipulation causes an even *greater* variability in functional maps. Polley and colleagues (1999a), in fact, showed that whisker plucking can change functional whisker representations in rodents by greater than 100%, which clearly cannot be attributed to intrinsic fluctuations in functional maps alone. Moreover, they showed that functional maps consistently expanded or contracted depending on the manipulation (Polley *et al.*, 1999a).

Many factors other than intrinsic cortical map changes may contribute to chronic variability in functional mapping signals. Most importantly, anesthetic depth will greatly influence the extent of intrinsic signals. Therefore, investigators should have a consistent method of inducing and assessing anesthetic depth to ensure that differences in maps are not due to anesthesia. Slight changes in the quality of the stimulus presented may also greatly influence the intrinsic signal map. With respect to visual stimuli, stimulus contrast, brightness, and speed can all be expected to impact the amplitude and possibly the extent of intrinsic signal maps. In rodent experiments using the barrel cortex, it is important to ensure that the degree of whisker stimulation (e.g., 15° vs 30°) and the distance of the whisker perturbator from the base of the whisker are consistent across experiments. It is critical that all parameters except for the experimental manipulation itself be kept constant in order to be able to ascribe changes in functional representation to the manipulation rather than other confounding factors.

The advantage of chronic imaging is the ability to image changes in functional representation, either developmental (Bonhoeffer, 1995) or plasticity related (Dinse *et al.,* 1997b; Antonini *et al.,* 1999; Polley *et al.,* 1999a; Pouratian *et al.,* 1999; Nguyen *et al.,* 2000). In some cases, these experience-dependent changes may become a confounding factor. It is important to consider what effect, if any, the chronic implantation of a chamber and the incision of the scalp may have on the behavior of the animal and therefore experience-dependent cortical representations. In some cases, it may be difficult to isolate the relative effects of the optical imaging preparation and the experimental manipulation on the functional map.

VIII. Optical Imaging of the Human Neocortex

Intrinsic signal imaging of the human neocortex is particularly attractive as an intraoperative mapping modality because it can rapidly assess the functional activity of a large area of exposed cortex with very high spatial resolution (micrometers). The high spatial resolution of optical imaging can be quite clinically useful. By mapping the functional organization of the exposed cortex, resection of pathological tissue can be maximized and iatrogenic damage to healthy cortex can be minimized. Studies using optical imaging in humans, however, have been limited because of their technical difficulty (Haglund *et al.,* 1993; Toga *et al.,* 1995a; Cannestra *et al.,* 1996, 1998a,b, 2000, 2001; Pouratian *et al.,* 2000a).

Haglund and colleagues were the first to observe optical signals in humans, reporting activity-related changes in cortical light reflectance during cortical stimulation, epileptiform afterdischarges, and cognitive tasks (Haglund *et al.,* 1992). Large optical signals were found in the sensory cortex during tongue movement and in Broca's and Wernicke's language areas during naming exercises. They also reported that surrounding afterdischarge activity, optical changes were of the opposite sign, possibly representing inhibitory surround. This was a landmark paper as it demonstrated the use of optical methods in humans, which up until this point had been employed only in monkeys, cats, and rodents. Since then, there have been reports describing the evolution of optical signals in the human cortex (Toga *et al.,* 1995a), the mapping of primary sensory and motor cortices (Cannestra *et al.,* 1998a), and the delineation of language cortices within (Cannestra *et al.,* 2000) and across languages (Pouratian *et al.,* 2000a) and comparing intraoperative intrinsic signals with preoperative BOLD *f*MRI signals (Cannestra *et al.,* 2001; Pouratian *et al.,* 2001).

A. Imaging during Neurosurgery

A major concern for intraoperative optical imaging studies is that the imaging system be compatible with the traditional operating room environment since the primary concern must always be the welfare and health of the subject (Fig. 10). Intraoperative optical imaging, fortunately, can be done with very little modification of the traditional operating room environment. The imaging device (e.g., CCD) can be mounted directly onto the operating microscope. The scope and mounted camera, under a sterile drape, can then be

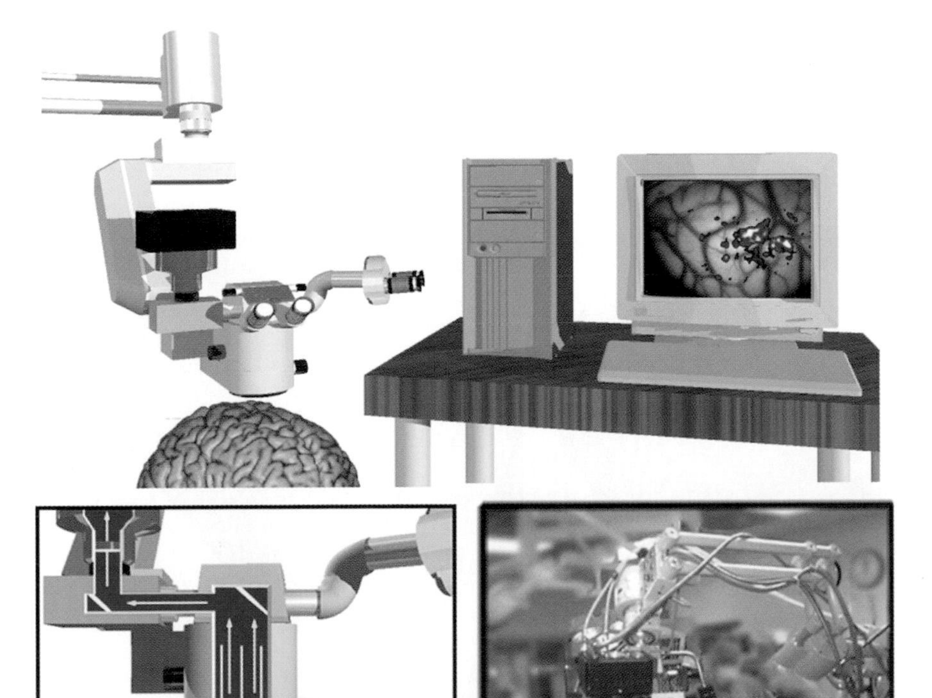

Figure 10 Schematic of intraoperative setup for optical imaging of intrinsic signals. The basic setup for optical imaging in the intraoperative environment is very similar to that seen in Fig. 5. The major modification is that the CCD camera (or other detector) must be mounted onto the operating microscope (top and bottom right). Illumination is provided through the operating microscope optics and the reflected light is directed back through the microscope optics and toward the mounted CCD (bottom left). **Bottom right:** Photograph of the CCD camera used by Toga and collaborators mounted onto the operating microscope immediately prior to optical imaging in the operating room.

moved into the sterile field over the cortical exposure. Sterilized circular polarizing and heat filters are placed under the main objective of the operating microscope to reduce glare artifacts from the cortical surface. Voltage-stabilized white light illumination is provided by the operating microscope light source, through a fiber optic illuminator. Reflected light is filtered through a band-pass transmission filter before being captured by the camera. In general, the temporal resolution of imaging ranges between 200 and 500 ms per frame (which is sufficient to image the slowly evolving perfusion-related response) with image exposure times between 50 and 200 ms. The image exposure time should be kept short in order to minimize intraimage brain movement.

The movement of the cortex due to cardiac and respiratory motion complicates intrinsic signal imaging intraoperatively. Imaging during awake procedures (in which the patient is performing cognitive tasks) is an even greater challenge because the subject's respiration rate is not under the direct control of the anesthesiologists.

1. Strategies for Reducing Movement Artifacts

Strategies for reducing cortical movement have previously been discussed in Section III.C. Relevant approaches for intraoperative imaging are briefly discussed here. Several strategies have emerged to minimize the effect of cortical movement during imaging, including imaging through a sterile glass plate which lies atop the cortex (Section III.C.3) (Haglund *et al.,* 1992), synchronizing image acquisition with respiration and heart rate (Section III.C.4) (Toga *et al.,* 1995a), and using postacquisition image registration (Section III.C.5) (Haglund *et al.,* 1992; Cannestra *et al.,* 2000; Pouratian *et al.,* 2000a). While each significantly improves the SNR of intrinsic signal image analysis, each also has some drawbacks that should be considered. The first report of intrinsic signals in humans used a glass plate to immobilize the cortex. However, it is unclear how this physical restraint of the cortex affects the physiology which underlies the intrinsic signal etiology. Synchronizing data acquisition with respiration and heart rate ensures that the brain is in the same position during each image acquisition. However, this strategy can limit temporal resolution and may be technically challenging since it requires interfacing with intraoperative anesthesiology monitoring equipment. Finally, postacquisition image registration alleviates the need to interface with operating room equipment and has no temporal resolution limitations, but the movement of the cortex during acquisition may introduce movement-related cortical reflectance changes that are independent of activity, even after image registration. For more detailed discussion, see Section III.C.

2. Clinical Utility of Intrinsic Signal Maps

Other than the technical challenges discussed, the major limitation of intraoperative optical imaging as a clinical tool is that optical imaging does not directly detect neuronal activity. Instead, optical signals arise from physiological processes (perfusion-related and metabolic) that are coupled to, yet whose temporal/spatial profiles are somewhat distinct from, neuronal activity. Although several groups have characterized the relationship between intrinsic signals and electrophysiological activity, a precise understanding of this relationship will be necessary in order to base clinical decisions on intrinsic signal maps. In a survey of 10 human subjects, Pouratian *et al.* found that 98% of sites deemed active by electrocortical stimulation mapping (ESM) demonstrated optical changes at 610 nm (Pouratian *et al.,* 2000c) (Fig. 11). [ESM is the current gold standard of intraoperative cortical

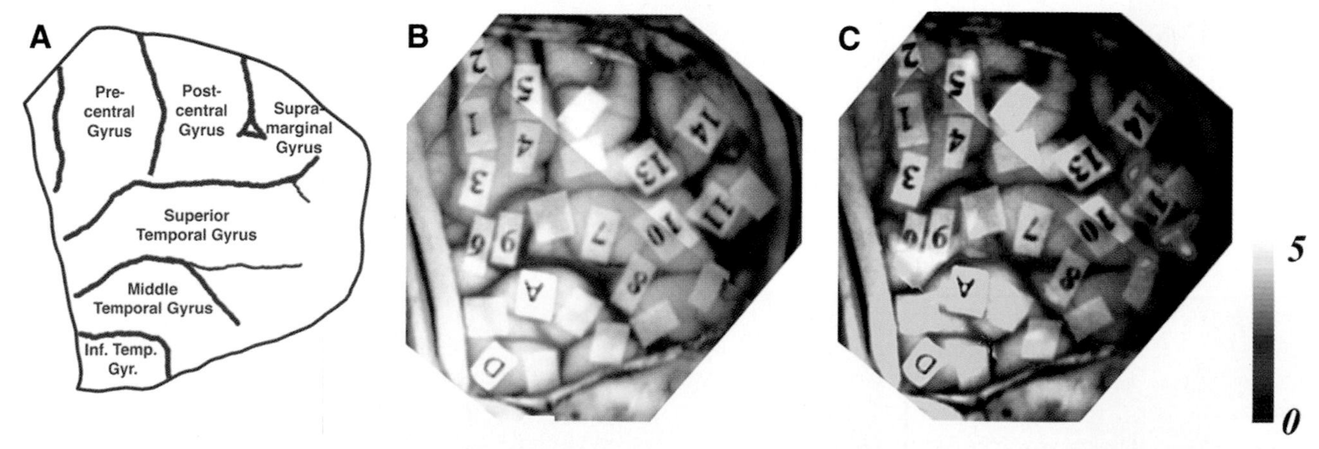

Figure 11 Correlation of optical signals with electrocortical stimulation maps intraoperatively. (**A**) Schematic indicating the gyral morphology displayed in (B) and (C). (**B**) Electrocortical stimulation map (ESM) showing areas essential for language processing tagged with numbers, areas not essential for language processing with blank tags, and areas which when stimulated induced afterdischarge activity. (**C**) Optical map (positive 610-nm signal) show very high spatial correlation with ESM map shown in (B). Note that all areas which were deemed active by ESM in (B) have optical activity within millimeters of the numbered tags and those areas with blank tags do not show adjacent optical activity. Color bar represents percentage reflectance change. Adapted, with permission, from Pouratian *et al.* (2000a).

mapping: a direct current is applied to the cortical surface via bipolar electrodes (4 to 20 mA for 2–7 s). Stimulation results in a movement if over motor cortex, a perception (over sensory or limbic areas), or task disruption in awake language testing.] In addition, ~25% of areas deemed *inactive* by ESM were also optically active, indicating that intrinsic signal maps extend beyond the regions indicated by ESM. This phenomenon of "spread" has also been observed in the rodent somatosensory cortex in response to whisker stimulation (Godde *et al.,* 1995; Chen-Bee and Frostig, 1996; Masino and Frostig, 1996). In humans, the spread may be due, in part, to the fact that optical imaging detects both essential and secondary (i.e., active but not necessary for completion of task) cortices while ESM detects only essential areas. Alternatively, this spread may represent an inexact colocalization of optical signals and electrophysiological activity due to imprecise physiological coupling of neuronal activity, metabolism, and perfusion. The significance of the spread of optical signals must be clarified before intrinsic signal maps can be used for clinical decision-making.

B. Optical Imaging through the Intact Human Skull—NIRS

Several studies have demonstrated that imaging in animals can be performed through the intact (thinned skull) (Frostig *et al.,* 1990; Blood *et al.,* 1995; Masino and Frostig, 1996). In these studies, the skull was thinned to a few hundred micrometers. Is optical imaging through the intact human skull, measuring 515 mm, at all realistic? Using longer wavelengths of light (near infrared) that penetrate the skull much more efficiently than visible light, several groups have been able to produce functional maps of the human cortex through the intact human skull. This method, known as near-infrared spectroscopy, can be considered a variant of intrinsic optical imaging since it also measures metabolism and perfusion-related changes in light absorbance. Imaging through the intact skull, however, introduces and necessitates significant methodological and analytical modification of intrinsic signal imaging techniques. Consequently, NIRS is discussed in detail in its own chapter (Chapter 6, this volume).

C. Applications

Intraoperative optical imaging offers the opportunity to answer questions about basic neurophysiology (Cannestra *et al.,* 1996, 1998b) and characterization of brain organization and relationships (Cannestra *et al.,* 1998a, 2000, 2001; Pouratian *et al.,* 2000a) and to develop a clinical tool that can help guide clinicians through procedures. Clinically, mapping the cortex prior to procedures can help delineate functional boundaries by which resections can proceed, ensuring that as much pathologic tissue as possible is resected and healthy tissue remains intact. Another potential application of intrinsic imaging in human neurosurgery is in the visualization of epileptic foci (if they are on the surface of the brain) with a precision much better than currently achieved with electrical recording (approximately 1 cm).

Although this chapter focuses on intrinsic signal imaging and only intrinsic signal changes have been reported in humans, optical signals can also be measured using dyes that are restricted to physiological compartments, such as intravascular dyes (Frostig *et al.,* 1990; Narayan *et al.,* 1995), and dyes that are sensitive to physiological events, such as oxygen-dependent phosphorescence quenching dyes (Vanzetta and Grinvald, 1999). Optical signals which are specific to pathological tissue, such as tumors, can also be detected by using tracers specific to tissue properties or even metabolic rates (Daghighian *et al.,* 1994), which may also improve neurosurgical planning. To date, the toxicity of such dyes has made this approach prohibitive in humans.

IX. Combining Optical Imaging with Other Techniques

Intermodality comparisons are useful because each imaging technique offers different information about the cortical organization, anatomy, and physiology of the cortex of interest. Specifically, intrinsic signal imaging can be combined with tracer injections for investigating anatomical and functional relationships. Alternatively, intrinsic signal imaging can be combined with other perfusion-related brain mapping modalities to characterize the etiology of perfusion-related neuroimaging signals, since each technique offers slightly different information about perfusion responses and holds different assumptions.

A. Targeting Tracer Injection into Selected Functional Domains

Tracers and markers are often injected into cortex to determine the anatomic connectivity between one cortical area with another. It is unclear whether the markers are injected into a single functional domain or into several neighboring regions with different functional properties (e.g., the border between two cortical areas with different orientation preferences). Maps obtained with optical imaging can be very helpful for systematically targeting tracer injections into discrete functional domains. By combining optical imaging with tracer injection techniques, both functional and anatomical data can be obtained from a single subject, enabling investigators to define functional-anatomical relationships with much higher precision.

The most critical step in combining *in vivo* optical imaging with histological observations is to precisely match the optically derived maps with the histological sections. This can be accomplished by matching the pattern of superficial blood vessels recorded *in vivo* using the optical imaging apparatus with reconstructions of the vascular patterns observed in histological sections. Alternatively, fiducial lesions can be recorded in the optical image and later identified and matched with the same lesions in the histological samples. In order to get good correspondence between the optically acquired maps and the histology, it is very important that the cortical tissue is not flattened or distorted in any other way before the histology is performed.

Optical imaging has been combined with histological investigation in the visual cortex of macaque monkey as well as in area 18 in the cat (Malach *et al.,* 1993; Kisvarday *et al.,* 1994). In one such study, an anterograde tracer, biocytin, was injected into monocular domains, binocular domains, and cortical areas corresponding to different orientations, as defined by functional intrinsic optical signal mapping. By using optical-map-directed injections, the authors were able to demonstrate that intrinsic anatomical connections tended to link monocular regions with same-eye ocular dominance columns (skipping columns corresponding to the other eye), binocular regions projected to other binocular regions but not to monocular regions, and cortical areas subserving a specific orientation preference connected with other areas with a similar orientation preference (Malach *et al.,* 1993). It becomes clear that studies of this sort can lead to a better understanding of the relationship between anatomy and function in the cerebral cortex at both the single-cell and the macroscopic level.

B. Combining Optical Imaging with Electrical Recordings or Stimulation

The chamber described in Section III.C.1 can be modified to be compatible with electrical recordings or stimulation. This chamber (originally described in *Brain Mapping: The Methods,* first edition, Chapter 3) consists of a chamber with a square coverglass that is much larger than the chamber diameter. The chamber is sealed by pressing the coverglass against an O ring. The special coverglass is fitted with a rubber gasket, flexible enough to be penetrated by a hypodermic needle. Both the coverglass (held by a thin metal frame) and the syringe (attached to a hydraulic manipulator used to penetrate the rubber gasket on the coverslip) are attached to an *xy* manipulator and can thereby be moved relative to the base of the cranial window. Once the syringe is in the vicinity of the recording site of interest, an electrode that resides within the needle is pushed out of the needle and advanced into the cortex, using the hydraulic manipulator.

This device is very useful for the electrical confirmation of optically obtained functional maps as well as for targeted microelectrode recordings. Shmuel and Grinvald (1996) used it to perform both perpendicular and nearly tangential penetrations studying the relationships between unit responses and optically imaged functional domains for orientation and direction. The same device is theoretically applicable for microstimulation during optical recording experiments.

Evoked potentials and electroencephalographs (EEG) can also be recorded simultaneously with intrinsic signal imaging without the modified cranial chamber. Using a thinned bone preparation, a burr hole can be drilled over the cortex of interest and an electrode can be advanced through it (Chen *et al.,* 2000; O'Farrell *et al.,* 2000). This method is feasible only for recording of evoked potentials (EPs) or continuous EEG of populations of neurons. EPs are often considered the most appropriate measurement of electrophysiological activity because, like intrinsic signal imaging, EPs measure net cortical responses over broad regions rather than activity of individual neurons (Fig. 12). Although single-unit recordings may reveal more detailed information about the activity of individual neurons, these recordings require craniotomy and administration of intravenous mannitol to reduce cerebral swelling, potentially interfering with chronic imaging experiments and optical signals arising from cellular swelling, respectively.

Intrinsic signal imaging of direct cortical stimulation has been done in both nonhuman primates (Haglund *et al.,* 1993) and humans (Haglund *et al.,* 1992). Haglund and colleagues found that while subthreshold stimulation did not produce any optical changes, the spatial extent and intensity of intrinsic signal response to suprathreshold stimulation depended on the duration and magnitude of bipolar cortical stimulation (Fig. 13). Further, they found that intrinsic signal changes did not activate nearby ocular dominance columns or orientation patches; stimulation-induced intrinsic signal changes were localized. Similar results were found in humans, in which graded responses to stimulation intensity were observed.

C. Combining Optical Imaging and fMRI

BOLD *f*MRI has become one of the most commonly used perfusion-based brain mapping modalities (Ogawa *et al.,* 1992). Although most believe the BOLD signal is due to cerebral blood flow (CBF) increases in excess of increases in the cerebral metabolic rate of oxygen, resulting in decreased capillary and venous deoxyhemoglobin concentrations (see Chapter 13 of this volume for more details), the exact etiology of the BOLD signal continues to be debated (Boxerman *et al.,* 1995; Menon *et al.,* 1995; Hess *et al.,* 2000; Kim *et al.,* 2000). Characterization of the etiology of BOLD is critical for the accurate interpretation and analysis of BOLD *f*MRI brain mapping studies.

Studies comparing BOLD with optical intrinsic signals are particularly advantageous for characterizing BOLD signals because the etiology of optical signals has been well scrutinized (Frostig *et al.,* 1990; Narayan *et al.,* 1995; Malonek and Grinvald, 1996; Nemoto *et al.,* 1999). Although many cite optical imaging studies to support the etiology of the BOLD signal, only a few intrasubject comparisons of optical intrinsic signals and BOLD *f*MRI exist (Hess *et al.,* 2000; Pouratian

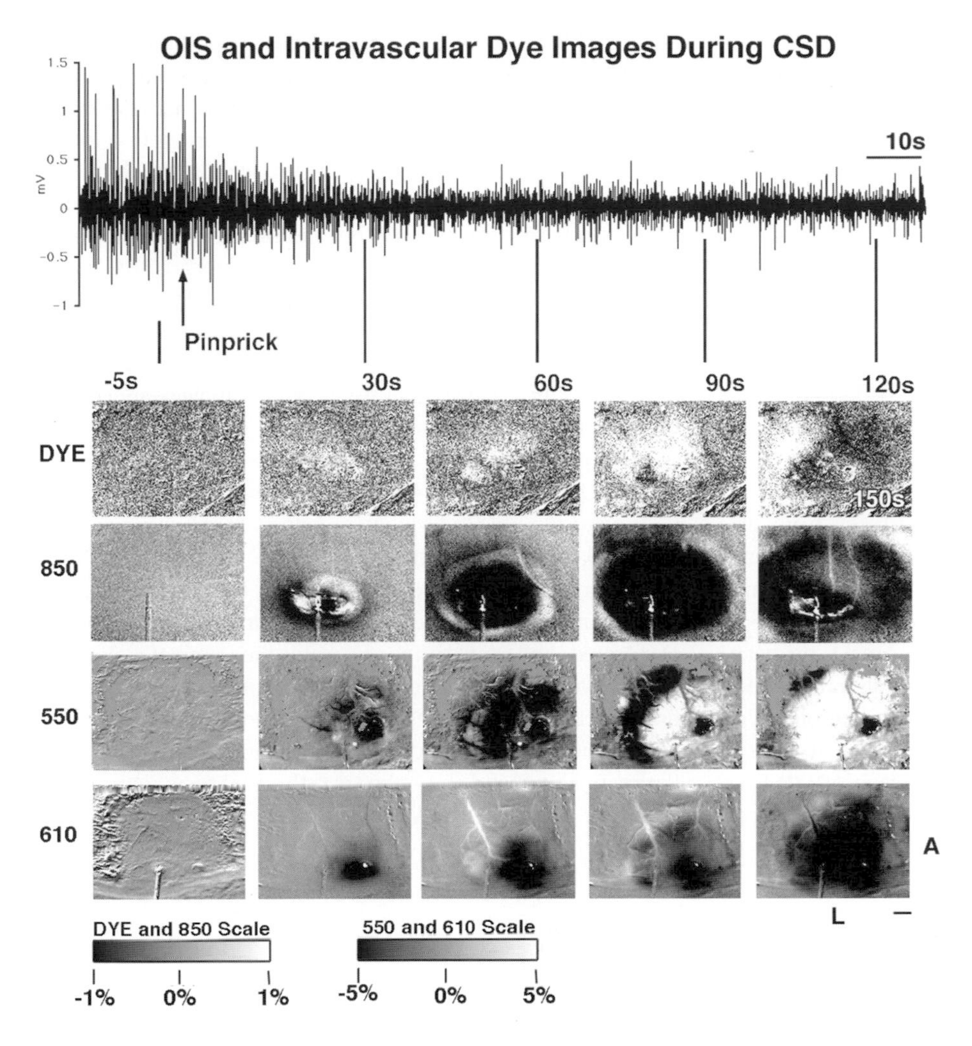

Figure 12 Triphasic optical intrinsic signal response to cortical pinprick-induced cortical spreading depression. **Top:** EEG demonstrating suppression of spontaneous cortical activity within 20–30 s of pinprick. **Bottom:** Intrinsic signal responses at different wavelengths and intravascular dye imaging. At least three distinct phases can be identified at all three wavelengths of intrinsic signal imaging. Intravascular dye imaging, on the other hand, indicates only two phases coincident with the second and third phase of the intrinsic signal responses, suggesting that these two phases are due to changes in blood volume. The first phase, on the other hand, is hypothesized to be due to light scattering changes. Interestingly, at 610 nm, different polarity signals are observed arising from the parenchyma and vasculature, suggesting that different physiological processes occur in these different compartments. This differentiation once again highlights the advantage of the high spatial resolution of optical imaging. Figure provided by Alyssa Ba.

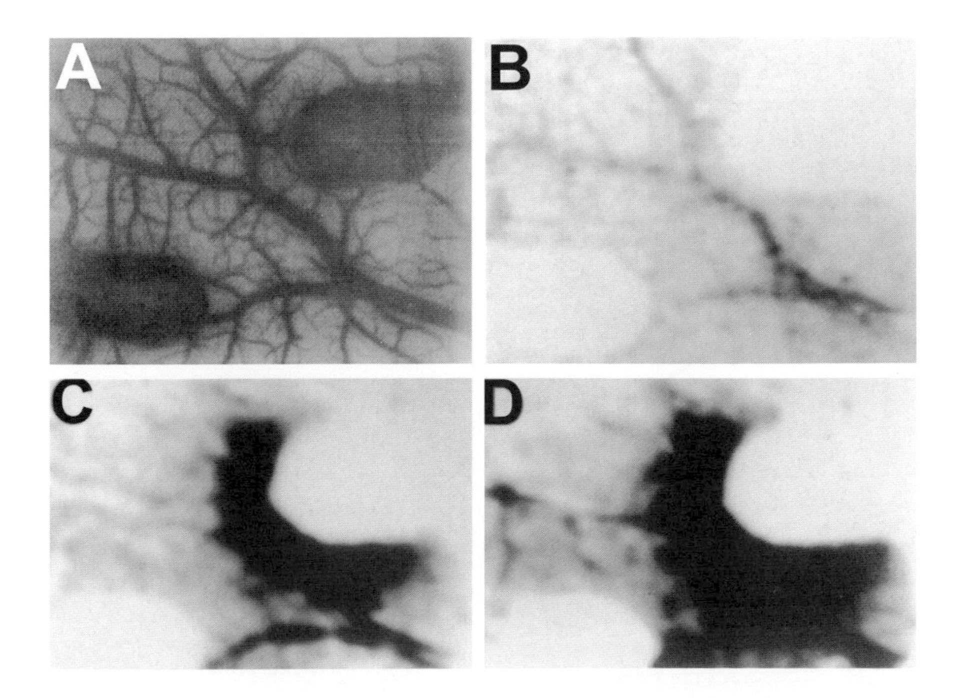

Figure 13 Optical imaging of bipolar cortical stimulation. (**A**) Raw image of primate cortex. The two darkened circles on the top right and bottom left indicate placement of the bipolar electrode. (**B**) Subthreshold stimulation of the cortex. (**C**) 1 mA stimulation of the cortex. (**D**) 1.5 mA stimulation of the cortex. Darkened areas in (C) and (D) indicate activity-related changes. Note that activity-related changes are limited to the area surrounding the electrodes. This study, however, does not preclude the possibility that ESM may cause activation of distant cortices. Adapted with permission from Haglund *et al.* (1993).

et al., 2000a; Cannestra *et al.,* 2001). Considering that many optical imaging studies draw conclusions and discuss the implication of such studies on functional magnetic resonance imaging (see for example Malonek and Grinvald, 1996; Vanzetta and Grinvald, 1999; Mayhew *et al.,* 2000), characterization of the relationship between the signals from these two modalities is critical.

Comparing *f*MRI signals, which are tomographic, with optical signals, which are surface signals, presents a challenge. In order to compare mapping signals from multiple modalities within a single subject, the different modalities must be warped into a common space (Fig. 14). The common space used is the three-dimensional (3D) cortical model (i.e., cortical extraction) of each subject's brain (generated from high-resolution T1-weighted MR scans, see middle of Fig. 14 for example). Before extracting cortical surfaces from the high-resolution T1-weighted MR scan, each scan is skull and cerebellum stripped and corrected for radiofrequency nonuniformity (Sled *et al.,* 1998). A 3D active surface algorithm is used to generate an external cortical surface mesh for each subject (MacDonald *et al.,* 1994). In order to align the *f*MRI activations with this cortical model, one must determine the transformation necessary to align high-resolution structural scans (which are coplanar with the *f*MRI scans of interest) with the T1-weighted MR scans used to create the cortical model. This transformation should be a rigid-body (i.e., six parameter—three rotations and three translations) transformation since two structural scans of the same subject are being aligned. The resulting transformation is then applied to the original *f*MRI data to line up the functional data with the 3D cortical model (see Pouratian *et al.,* 2001, and Cannestra *et al.,* 2001, for examples of this approach). Optical intrinsic signal (OIS) maps are projected onto the cortical model by matching sulcal landmarks on the cortical model with identical sulcal landmarks on the raw optical images. The accuracy of these projections is demonstrated in Fig. 15 (middle). It is clear from Fig. 15 that sulci in the optical images are continuous with sulci on cortical extractions, confirming the success of these projections.

The Ojemann group has adopted a similar methodology (Corina *et al.,* 2000). However, instead of relying on sulcal landmarks for registration, they use the position of the cortical blood vessels. Preoperatively, an MR angiogram (an MRI scan sequence which highlights surface vasculature) is performed and superimposed onto the subject's cortical extraction. After reflection of the dura and exposure of the brain surface, the arteries and branch points that have been superimposed onto the cortical extraction are identified on the cortical surface and subsequently used to register intraoperative maps with preoperative scans.

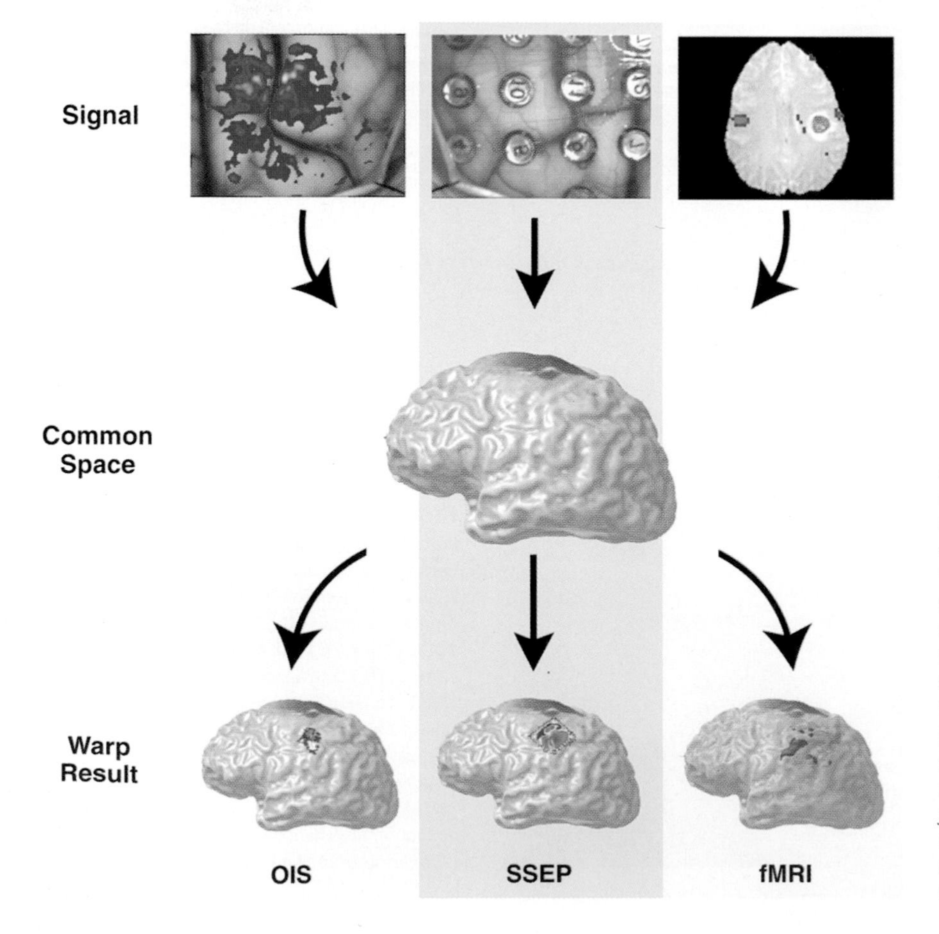

Figure 14 Intermodality comparisons. In order to compare mapping signals across modalities, the different modalities being compared must be warped into a common space. The common space is the cortical extraction of each subject's brain. Depending on the modality, different warping strategies are employed to warp the different mapping signal into a common space. While optical imaging is a surface imaging technique and requires projection of the data onto the cortical surface, *f*MRI is a tomographic technique and is warped into the common space using rigid body transformations to align the *f*MRI volume with the high-resolution MRI used to extract the cortical surface. The darkened area at the top of each cortical extraction demarcates tumor localization.

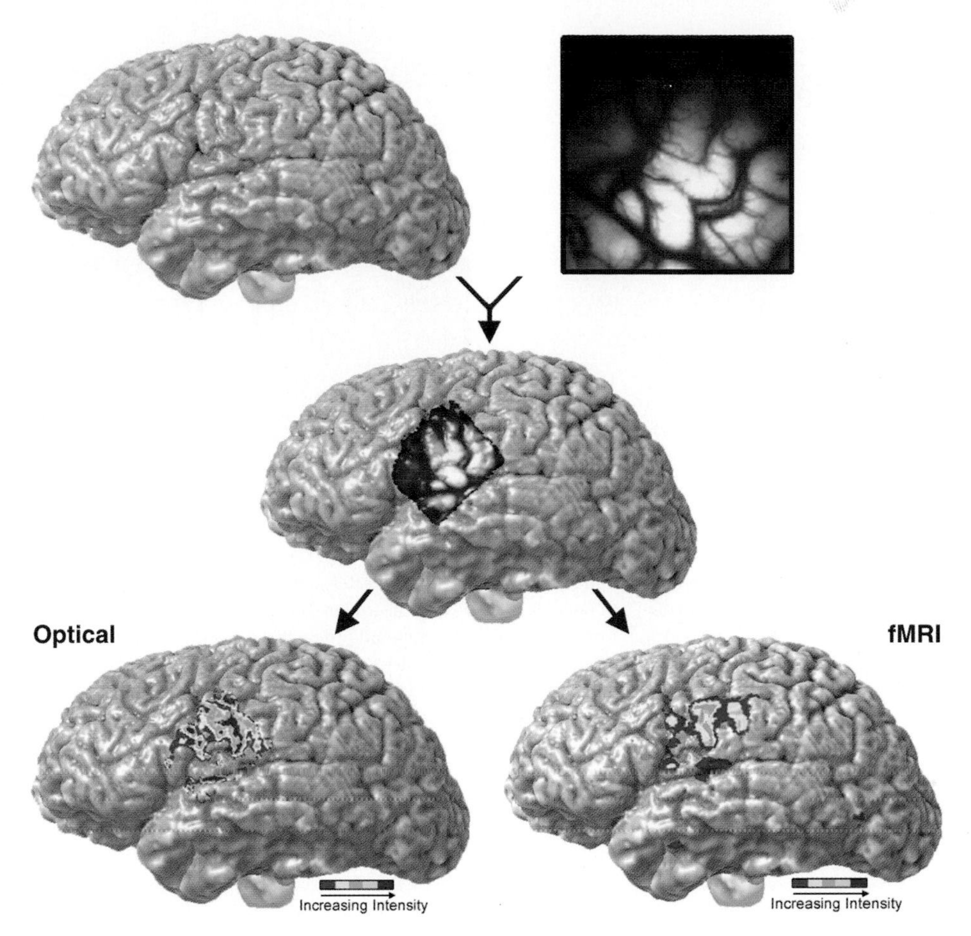

Figure 15 Strategy for comparing intrinsic signal imaging (a surface-based mapping technique) with functional magnetic resonance imaging (a tomographic technique). A 3D cortical surface extraction is used as the common space for comparing multiple modalities (top left, see Fig. 14). A photograph of the cortical surface imaged for intrinsic signal imaging (top right) is warped onto this surface by matching sulcal landmarks on the photograph and the 3D surface extraction. The accuracy of this warp is confirmed by the fact that the sulci in the optical image (projected photograph) are continuous with sulci on cortical extraction (middle). The warping parameters used to project the photograph onto the cortical surface are then applied to the intrinsic signal map to project the functional signal onto the cortical surface (bottom left). Other modalities, such as *f*MRI, can also be projected into the "common space," or the 3D surface extraction (bottom right), in order to compare across modalities.

Recently, Hess *et al.* (2000) imaged both BOLD signals and OIS in rodents, demonstrating that positive BOLD signals can occur even if deoxyhemoglobin levels are persistently elevated (as determined by decreased reflectance at 605 nm). It is unclear, however, whether such results can be generalized to human subjects, whose brains are gyrencephalic (vs lissencephalic in rodents) and whose vascular architecture differs significantly from rodents (Bär, 1981; Duvernoy *et al.,* 1981; Dirnagl *et al.,* 1991; Reichenbach *et al.,* 1997). In contrast to these findings, Cannestra and colleagues have demonstrated a spatial/temporal mismatch between the early negative 610-nm response [which Hess *et al.* (2000) and others attribute to increases in deoxyhemoglobin] and the BOLD signal in response to brief (2 s) somatosensory stimulation (Cannestra *et al.,* 2001). While the BOLD signal is centered on sulci, the negative 610-nm response is centered on gyri (Fig. 16). Furthermore, when the negative 610-nm OIS activation was used as an objective region of interest for analysis of BOLD signals, an initial dip was identified in the BOLD signal (Cannestra *et al.,* 2001). In a case study comparing *f*MRI and OIS activations, Pouratian and collaborators (2000a) found that the *positive* 610-nm response and the BOLD signal demonstrated much greater spatial correlation (than the negative 610-nm response and the BOLD signals). Despite the greater spatial correlation, the positive 610-nm response still demonstrated greater gyral activations than BOLD signals (Pouratian *et al.,* 2000a) (Fig. 16).

Recent technological advancements have led to the development of a low-cost combined optical imaging and functional MRI (3 T) system for animals (Paley *et al.,* 2001). The system allows concurrent MR imaging, optical spectroscopy, and interventional capabilities, for both physiological monitoring of the rodent and experimental manipulation. With the

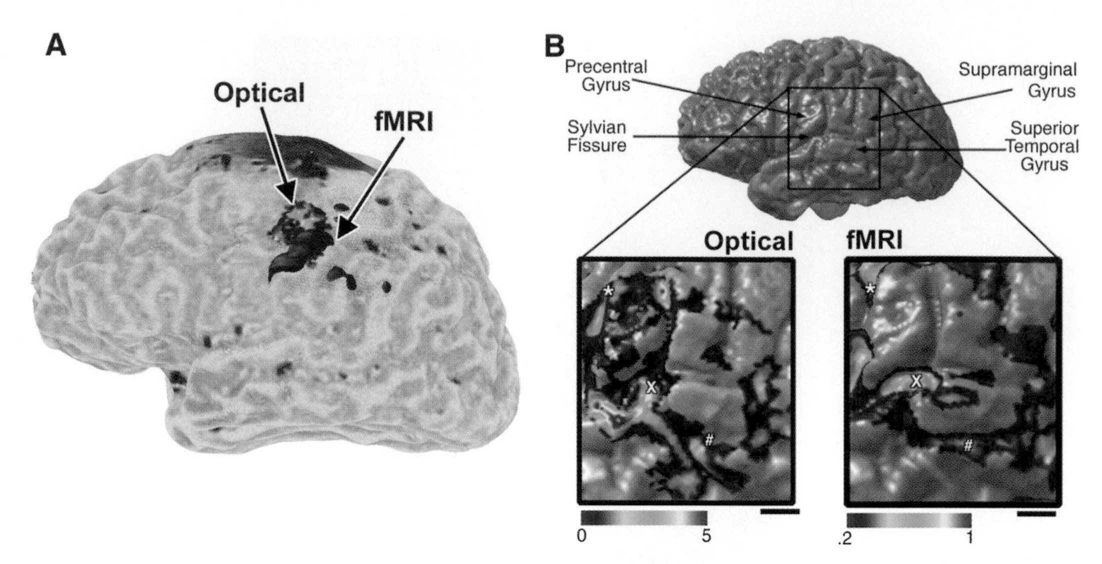

Figure 16 Comparison of 610-nm optical signals and BOLD *f*MRI. (**A**) Responses to brief (2 s) somatosensory stimuli (110-Hz finger vibration) were imaged using both optical imaging of intrinsic signals and BOLD *f*MRI. The initial decreased reflectance at 610 nm (labeled "Optical") was neither spatially nor temporally (data not shown) correlated with the BOLD *f*MRI signal (labeled "*f*MRI"). While the *f*MRI signal centered on the sulcus and extended deep within the sulcus (where the venous vasculature is located), the negative 610-nm response centered on gyri and better matched somatosensory evoked potential maps (data now shown). Adapted, with permission of Oxford University Press, from Cannestra *et al.* (2001). (**B**) The positive 610-nm response, or late increased 610-nm reflectance, is more spatially and temporally correlated with the BOLD signal. In a case study comparing *f*MRI and OIS activations, we found that the *positive* 610-nm response and the BOLD signal demonstrated much higher spatial correlation (compare bottom left optical signal with bottom right *f*MRI signal). Despite the higher spatial correlation, the positive 610-nm response still demonstrated greater gyral activations than BOLD signals. Adapted, with permission, from Pouratian *et al.* (2000a).

advent of interventional MR imaging devices for neurosurgery, intraoperative optical imaging can also be integrated with intraoperative MR (iMR) for better comparison of mapping signals within a single subject. The advantage of having concurrent intraoperative optical imaging and MRI is that both modalities can be used to image the brain after craniotomy and dural reflection so that the brain is imaged in the same space and (possibly) time. The iMR environment imposes imaging restrictions above and beyond that of the normal operating room environment in that investigators must be cautious of high magnetic field physically affecting the imaging device and interfering with the imaging system (and distorting image capture), which was not designed to be MR compatible. Nonetheless, Toga and colleagues have confirmed the feasibility of intraoperative optical imaging in the intraoperative MR environment (unpublished data).

D. Combining Optical Imaging and Micro-PET

Micro-PET offers the opportunity to image functional changes in the rodent brain with millimeter resolution (Kornblum *et al.,* 2000). A comparison of optical imaging and micro-PET data has yet to be published but with recent technological advancements, it is feasible and may offer a greater understanding of the etiology of optical signals and better elucidate the nature of perfusion-related responses. The specific advantage of micro-PET is that known tracers are used (e.g., [^{18}F]fluorodeoxyglucose) so that the etiology of the functional changes observed by micro-PET are unambiguous. Similar intermodality comparisons were done several years ago between *f*MRI and PET, serving to help clarify the etiology and significance of the *f*MRI signal (Ramsey *et al.,* 1996).

The challenges faced in comparing intrinsic signals and micro-PET are similar to those faced in comparing intrinsic signals with *f*MRI. Namely, while micro-PET is a tomographic technique, intrinsic signals are surface signals. Consequently, a common space must be identified in which to compare the signals. By analogy, it seems that the appropriate space would be the three-dimensional model of the subject's brain. Similar to the other techniques, fiducial markers must be used to align the maps appropriately after imaging. Unfortunately, micro-PET does not offer any anatomical information by which to perform image registration. Alternatively, radioactive sources can be placed during micro-PET imaging and identified on the optical image for subsequent registration.

X. Applications

Based on the specific advantages which optical imaging of intrinsic signals offers, including ease of implementation, high spatial and temporal resolution, relatively noninvasive imaging, the opportunity to image a single subject chronically, and the ability to determine the time course of underlying functional chromophore changes, four areas of application to which optical imaging has been applied have emerged. These include characterizing the functional architecture of sensory cortices, investigating patterns of functional perfusion and neurovascular physiology, imaging development and plasticity, and characterizing the dynamic profile of disease processes.

A. Characterization of Visual and Somatosensory Cortices

The high spatial resolution of optical imaging has made it an ideal modality for characterizing the fine functional architecture of sensory (both visual and somatosensory) cortices. Studies by Bonhoeffer and Grinvald (1991, 1993) perhaps best highlight the utility of this modality for characterizing the functional architecture of the visual cortex in cats. By imaging visual cortex responses to gratings of different orientations, they established "pinwheels" as an important structural element underlying the organization of orientation domains in the visual cortex (Bonhoeffer and Grinvald, 1991, 1993a) (Fig. 9B). Similarly, optical imaging of intrinsic signals has been used to demonstrate ocular dominance domains in the primary visual cortex of macaque monkeys (Ts'o *et al.*, 1990) (Fig. 9A). Methodological improvements made it possible to investigate more subtle features of cortical organization. For example, Ts'o and colleagues recently reporting using optical imaging in conjunction with single-unit electrophysiology and cytochrome oxidase (CO) histology to characterize the functional organization within the CO stripes of visual area V2 of primates with greater detail. They report previously unrecognized subcompartments within individual CO stripes that are specific for color, orientation, and retinal disparity (Ts'o *et al.*, 2001). The organization of V2 is reminiscent of V1, for which Victor and colleagues used optical imaging of intrinsic signals to identify orientation, spatial frequency, and color (Victor *et al.*, 1994).

Macknik and Haglund (1999) have also used optical imaging of intrinsic signals to characterize the primate visual cortex. Unlike some other optical imaging studies that seek only to characterize the functional architecture of the cortex, this study went on to ask: What is the stimulus for visual system activation, the physical visual scene or the perception of the visual scene? (Macknik and Haglund, 1999) That is, in the case of a visual illusion in which a physical stimulus is not perceived, does the visual system still respond to that physical stimulus? These optical imaging studies demonstrate that, for the visual illusions investigated, the visual system does not in fact respond to the entire physical visual stimulus, but rather the visual perception (Macknik and Haglund, 1999). These types of studies begin to demonstrate the versatility and the diverse array of questions that can be answered using optical imaging of intrinsic signals.

Optical imaging has similarly been applied in human subjects to characterize somatosensory and language cortices. Cannestra and colleagues stimulated the face, thumb, and index and middle fingers to obtain maps of cortical activation for each. While peak responses were distinct for each activation paradigm, nonpeak signals were more dispersed and overlapped (Cannestra *et al.*, 1998a). This is consistent with a report by Godde and colleagues who, using rodents, reported large and overlapping cortical representations with distinct foci in response to stimulation of different digits (Godde *et al.*, 1995). The organization of language cortices in humans has also been characterized both within (Cannestra *et al.*, 2000) and across languages (Pouratian *et al.*, 2000a). While Pouratian and colleagues showed that different languages in a bilingual subject can produce both overlapping and distinct areas of activation, Cannestra and collaborators showed that spatial and temporal activation patterns of both Broca's and Wernicke's area is (language) task dependent (e.g., a visual object naming task activates Broca's area different from an auditory responsive naming task) (Fig. 17).

B. Characterization of Functional Perfusion and Neurovascular Physiology

Intrinsic optical signals have multiple sources, including perfusion- and metabolism-related etiologies, making it a suitable modality for characterizing functional perfusion and neurovascular physiology. It is an especially powerful tool for answering these types of questions because of its superior spatial and temporal resolution. Optical imaging studies have been critical for describing the time course of functional perfusion-related changes, including changes in blood volume (Frostig *et al.*, 1990; Narayan *et al.*, 1995) and hemoglobin concentration changes (Frostig *et al.*, 1990; Malonek and Grinvald, 1996; Nemoto *et al.*, 1999; Mayhew *et al.*, 2000, 2001). Moreover, intrinsic signal imaging was used to describe baseline vasomotion and to compare the magnitude of these oscillatory changes with functional perfusion-related changes (Mayhew *et al.*, 1996). In addition to characterizing the time course of these functional signals, optical imaging has been used to determine the parameters which influence the magnitude of functional signals, including the absolute number of stimuli as well as the frequency of stimulation (Blood *et al.*, 1995; Polley *et al.*, 1999b). Several studies also have used optical imaging to determine how functional signals are affected by either temporally or spatially adjacent stimuli (Blood and Toga, 1998; Cannestra *et al.*, 1998b) (Fig. 18). Blood and colleagues demonstrated that simultaneous activation of nearby cortices can affect the magnitude of the functional signals emanating from the regions of interest (Blood and Toga, 1998) (Fig. 18A). Cannestra and colleagues

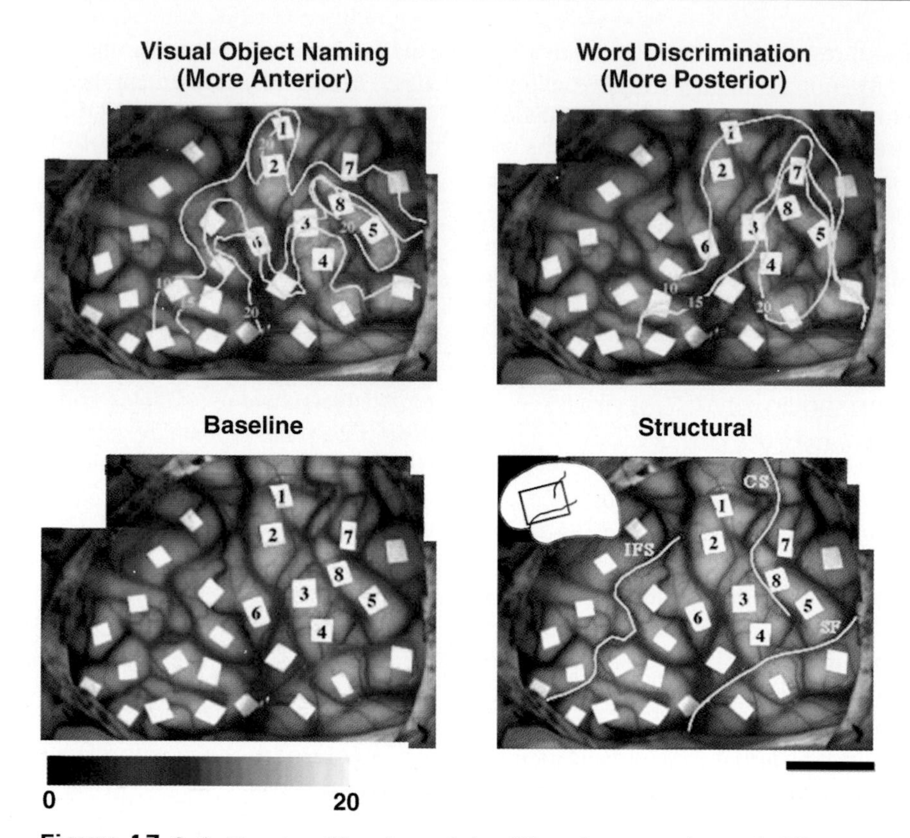

Figure 17 Optical imaging of Broca's area during different language tasks reveals different patterns of activation. Each language task requires differing degrees of semantic, syntactic, and lexical processing. It was hypothesized, therefore, that different language tasks (e.g., visual object naming vs word discrimination) would activate both Broca's and Wernicke's area differently. Activation patterns for two different language tasks are shown, visual object naming and word discrimination. While the first task activated more anterior regions of Broca's area, the latter task activated more posterior regions. In the study by Cannestra and colleagues (2000), the investigators were able to use optical imaging of intrinsic signals to define subdivisions of both Broca's and Wernicke's area and to describe different patterns of temporal activations within each region. **Bottom left:** Baseline image showing no optical activity. **Bottom right:** Identification of gyri of interest and position of exposure relative to a schematic brain diagram. Color bar denotes reflectance decrease at 610 nm $\times 10^{-4}$. Scale bar is 1 cm. Adapted, with permission, from Cannestra *et al.* (2000).

reported "neurovascular refractory periods," showing that functional responses are diminished in magnitude if they are preceded by another temporally close response (in the face of identical electrophysiological responses) (Cannestra *et al.*, 1998b) (Fig. 18B).

With the advent of spectroscopic intrinsic signal imaging, many studies have been published which aim to characterize the time course of functional changes in blood volume, different hemoglobin moieties, and light scattering (Malonek and Grinvald, 1996; Nemoto *et al.*, 1999; Kohl *et al.*, 2000; Mayhew *et al.*, 2000, 2001; Shtoyerman *et al.*, 2000; Lindauer *et al.*, 2001). These studies differ in their spectroscopic analysis techniques. In general, however, most studies suggest an early increase in deoxyhemoglobin concentrations upon cortical activation, followed by a delayed increase in total blood volume and increase in oxyhemoglobin concentrations [exceptions include Kohl *et al.* (2000) and Lindauer *et al.* (2001), which do not observe the initial increase in deoxyhemoglobin concentrations]. Clearly, the temporal profile of functional changes is still questioned and much work remains to be done to overcome the ambiguities and questions.

Recently, optical imaging has been combined with other imaging modalities in order to better describe functional perfusion-related changes and understand the etiology of the different brain mapping signals. Three recent reports have compared optical imaging with *f*MRI (see Section IX.C) (Hess *et al.*, 2000; Pouratian *et al.*, 2000a; Cannestra *et al.*, 2001) (Fig. 16). These types of studies are critical since each modality holds different assumptions and limitations, and intermodality comparisons may be one of the best ways of overcoming the limitations of each individual technique.

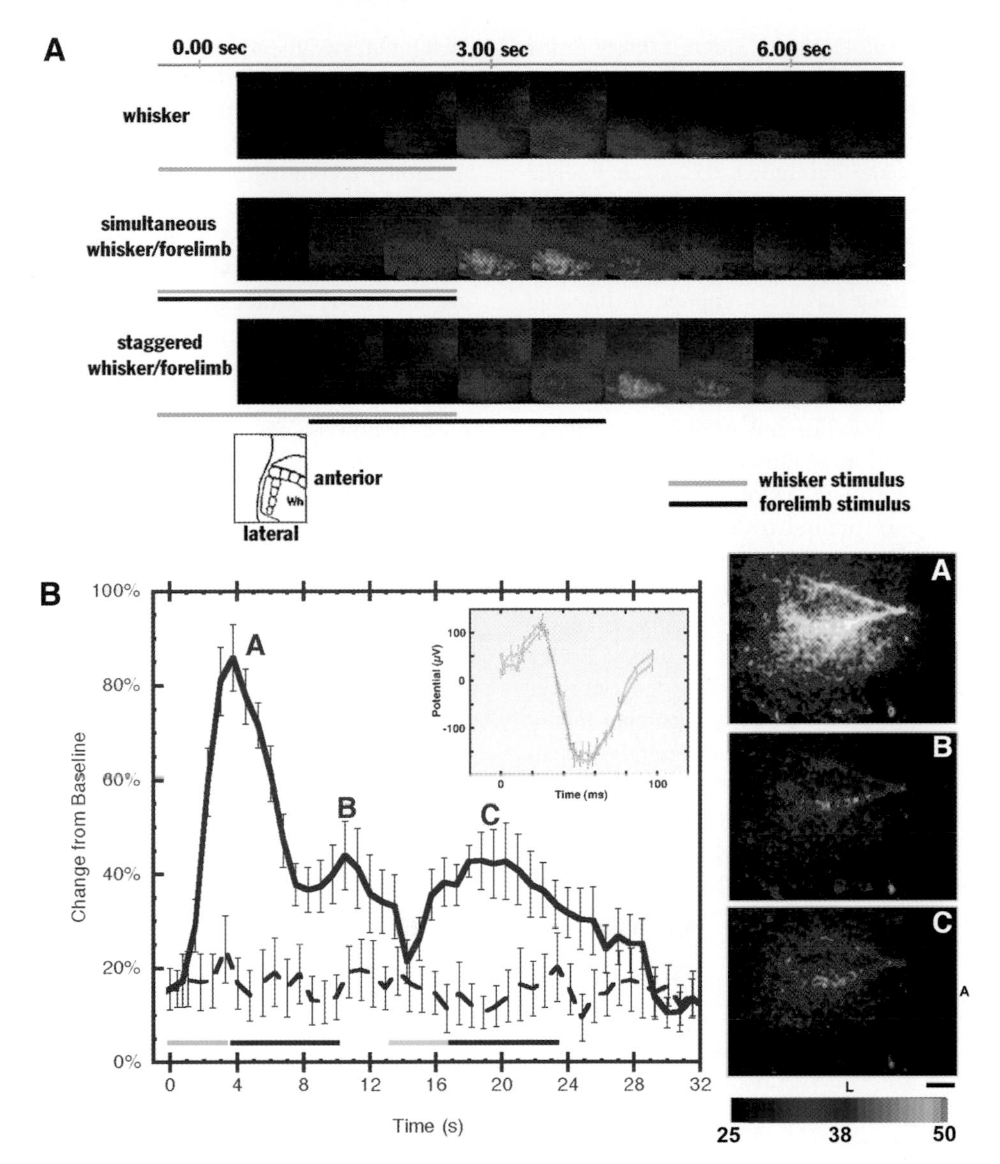

Figure 18 Modulation of optical responses to whisker stimulation by spatially (A) and temporally (B) adjacent stimuli. Optical imaging has been used extensively to characterize factors that modulate functional perfusion-related responses. (**A**) Top: 850-nm optical response to 3 s whisker stimulation. Middle: Optical response to simultaneous whisker and forelimb stimulation. By simultaneous stimulation of the forelimb, the perfusion-related response over the whisker somatosensory cortex is potentiated. Bottom: Optical response to whisker stimulus followed by forelimb stimulation (staggered condition). The response over the whisker somatosensory cortex is delayed and potentiated. Adapted, with permission, from Blood and Toga 1998. (**B**) Hemodynamic refractory periods. The time course of the optical response to two consecutive whisker stimulations (2 s each) with a 3-s interstimulus interval. The response to the second stimulus is nearly half that of the response to the first stimulus even though the evoked potentials (shown in the top right window) are unchanged for the two stimuli. Horizontal bars at bottom indicate stimulus timing. Dashed line represents baseline optical activity. Error bars are standard errors of the mean. Time courses were calculated using principal component analysis. Adapted, with permission, from Cannestra *et al.* (1998).

C. Imaging Development and Plasticity

More recently, several groups have taken advantage of the ability to use optical imaging for chronic imaging to characterize both developmental and plasticity-related changes in functional architecture and perfusion-related signals. The development of orientation domains has been characterized in ferret visual cortex, from the time that they are first observed (p32) to the time at which they are fully developed (p43) (Chapman *et al.,* 1996). The authors demonstrate that during this time (which is the peak of the critical period) the structure of the maps is remarkably stable (Chapman *et al.,* 1996). A similar analysis of the development of kitten primary visual cortex has been reported (Bonhoeffer, 1995). Pouratian and collaborators characterized developmental changes in intrinsic signals in the rodent "barrel" cortex, showing that while some intrinsic signals change throughout the developmental period (i.e., 850 nm), intrinsic signals at other wavelengths remain relatively stable and unchanged throughout development (i.e., 550 and 610 nm) (Pouratian *et al.*, 2000b).

Recently, optical imaging has been applied to imaging plasticity-related changes in functional signals (Prakash *et al.*, 1996; Dinse *et al.*, 1997a; Antonini *et al.*, 1999; Polley *et al.*, 1999a). While Prakash and collaborators demonstrated *rapid* representational changes in barrel cortex response after addition of either brain-derived neurotrophic factor or nerve growth factor, other groups have used optical imaging to demonstrate more long-term changes in functional representations secondary to physical manipulations of the subject (Dinse *et al.*, 1997a; Antonini *et al.*, 1999; Polley *et al.*, 1999a). The ability to image chronically was key to the success of such studies. Polley and colleagues, for example, showed that long-term whisker removal and differential housing of rodents can alter intrinsic signal responses in the barrel cortex differently in two distinct ways (Polley *et al.*, 1999a). While a single whisker's functional representation underwent expansion following innocuous removal of all neighboring whiskers, a contraction could also be induced, using the same manipulation, if the animal was given a brief opportunity to use its whiskers for active exploration of a different environment (Polley *et al.*, 1999a). Pouratian and collaborators have similarly investigated the effect of innocuous whisker removal on functional responses to stimulation of the spared whisker (Fig. 19). Optical responses at 850 nm were not potentiated relative to age-matched control until 2 weeks after initial whisker plucking, despite reports that electrophysiological and autoradiographic changes may occur within minutes to days (Kossut, 1998; Kelly *et al.*, 1999). Clearly, the chronic imaging ability of optical imaging combined with its superior spatial resolution makes this modality amenable to many more questions of plasticity-related changes in cortical architecture.

D. Imaging Disease

Optical imaging of intrinsic signals has recently been applied to imaging disease. Haglund and colleagues (1992) first reported imaging epileptiform afterdischarges following direct cortical stimulation of the human neocortex. They reported that activation patterns were dependent on the intensity and duration of the afterdischarge activity and often found a pattern of alternating increased and decreased reflection, which they hypothesized to represent a central area of activation surrounded by areas of cortical inhibition or decreased activity. More recently, Chen and colleagues

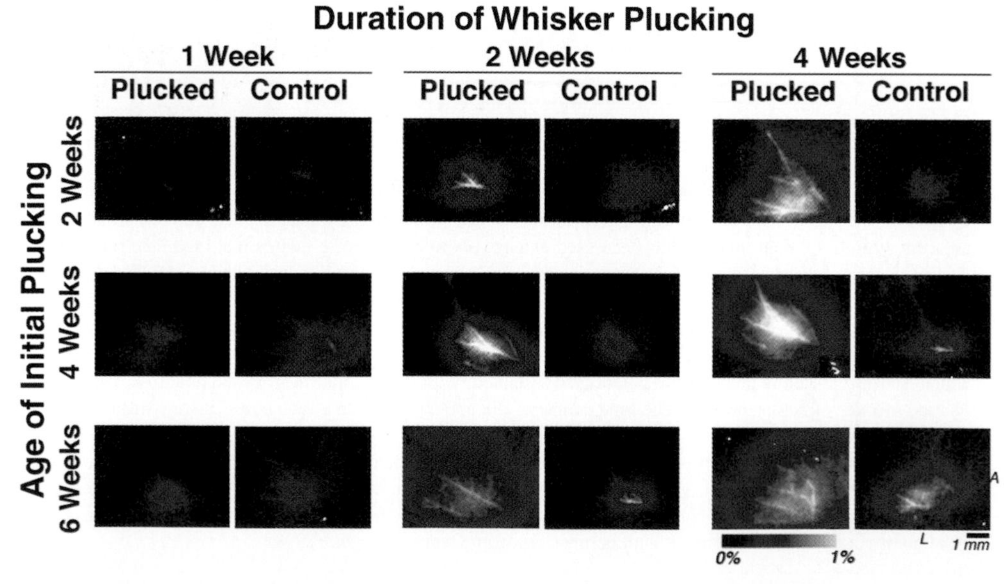

Figure 19 Plasticity-related changes in optical images. The effect of removing all but one whisker on functional optical responses was investigated in the rodent somatosensory cortex. Whisker removal induced more expansive and intense optical responses to spared whisker stimulation relative to age-matched controls. Images in columns labeled "Plucked" are representative responses from each experimental group while images in columns labeled "Control" are representative responses from age-matched controls. Images in the first column demonstrate that increases were not observed after 1 week of whisker plucking. When duration of whisker plucking was increased to 2 and 4 weeks (as represented by the second and third columns), optical responses were consistently larger and more intense in each group relative to age-matched controls. Color bar represents percentage reflectance decrease. A, anterior; L, lateral.

reported imaging seizure propagation in the rodent neocortex, induced by topical penicillin application (Chen *et al.,* 2000). Seizure-related cortical reflectance changes correlated well with EEG epileptiform discharges and often preceded initial EEG spikes, suggesting that OIS may provide sensitive cues for seizure detection. This may have important clinical implications. Many have hypothesized that optical imaging may be used intraoperatively to image epileptic foci and to delineate functional borders prior to neurosurgical intervention. These studies suggest that this may in fact be possible.

Optical imaging has also been used to study cortical spreading depression (CSD) (Yoon *et al.,* 1996; O'Farrell *et al.,* 2000). CSD is an important disease model for migraine (Lauritzen, 1994). Yoon and colleagues first described CSD *in vivo,* reporting the propagation of a nonuniform wavefront (Yoon *et al.,* 1996). In order to characterize CSD further, O'Farrell and collaborators imaged the CSD response to cortical pinprick for a longer duration and compared the intrinsic signal response to intravascular dye imaging results. They identified a triphasic response to CSD, with an initial highly uniform wavefront, which was not matched by intravascular dye signals, and two subsequent intrinsic signal phases (of alternating polarity), which were matched by intravascular dye signals (O'Farrell *et al.,* 2000) (Fig. 12).

These studies only begin to show the exciting application potential of optical imaging for imaging and characterizing functional disease progression in the neocortex. No doubt, this application will become increasingly important as optical imaging continues to be integrated into the clinical setting.

XI. Comparison of Intrinsic Optical Imaging with Other Imaging Techniques

A. *f*MRI and PET

*f*MRI and PET rely on signal sources similar to those of intrinsic optical imaging (Fox and Raichle, 1986; Ogawa *et al.,* 1990; Belliveau *et al.,* 1991). In some cases, they yield similar results, whereas in other cases they are complementary. As explained in Section II.B, the early increase in deoxyhemoglobin concentrations is far more localized to the site of electrical neuronal activity than blood flow and blood volume changes, which spread considerably beyond the electrically active cortex. Thus, it is not surprising that the resolution that can be attained in optical imaging when imaging between 600 and 630 nm (which is approximately 100 μm) is more than an order of magnitude better than that achieved with *f*MRI or PET, which normally image blood flow and volume changes (Cannestra *et al.,* 2001). Studies suggested that if *f*MRI could focus on the initial deoxygenation of the blood, higher resolution and more specific functional maps could be created (Frahm *et al.,* 1994; Cannestra *et al.,* 2001). Consistent with these hypotheses, Kim and colleagues recently used the initial increase in deoxygenation, or "initial dip," in the BOLD *f*MRI signal to map orientation columns with much higher spatial resolution than is traditionally available using *f*MRI (Kim *et al.,* 2000). However, the validity of initial dip imaging remains extremely controversial, with some groups doubting its validity (Kohl *et al.,* 2000; Lindauer *et al.,* 2001) and others adamantly supporting its existence and utility for high-resolution functional mapping (Vanzetta and Grinvald, 1999; Kim *et al.,* 2000).

Technical differences in *f*MRI, PET, and optical imaging instrumentation are also responsible for differences in spatial and temporal resolution. It is commonly accepted that optical imaging offers the best combination of both spatial and temporal resolution of brain function. Other factors, however, such as the possibility to measure structures deep within the brain and the absolute noninvasiveness, at least of *f*MRI, cannot be overestimated.

B. NIRS

NIRS, as mentioned earlier, may be considered a variant of intrinsic signal imaging. NIRS signals share a similar etiology with intrinsic signal etiology: perfusion- and metabolism-related changes (Villringer *et al.,* 1993; Obrig *et al.,* 2000). Similar to optical imaging, NIRS detects changes in the attenuation of light with functional activation and, using spectral decomposition techniques, estimates concentration changes of the different hemoglobin moieties. Although NIRS offers highly specific signals in terms of defining the etiology of the signal, it suffers from poor spatial resolution. Increasing the number of optodes has improved the spatial resolution of the technique. However, significant interpolation must still be used to create two-dimensional images of functional activation changes. For a more detailed look into the advantages and disadvantages of NIRS, see Chapter 6, this volume.

C. Laser Doppler Flowmetry (LDF)

LDF exploits the Doppler phenomenon to detect changes in flow. Unlike optical intrinsic signal imaging, which can image and detect functional changes over several millimeters of exposed cortex simultaneous with relatively high spatial resolution (~25 μm), LDF offers very poor spatial resolution with extremely high temporal resolution (milliseconds). The LDF probe must be placed over a certain area of interest and used to detect blood flow changes under that probe but offers no spatial resolution. Mayhew and colleagues (1998) pointed out that within the area imaged by LDF, there can be several different time courses and response patterns which LDF would not be able to discern. Therefore, the spatial resolution offered by LDF may not be adequate to capture the subtleties of the cascade of perfusion-related responses occurring within adjacent compartments in the cortex (i.e., capillaries vs venules). The increased temporal resolution of LDF is generally not very useful since the time course of perfusion-related changes in the brain is on the order of seconds.

The advantage of LDF over intrinsic signal imaging is that the nature of the signal is well understood. The signal is unambiguously due to changes in flow patterns within the volume of tissue being sampled. Furthermore, LDF probes can be adjusted to measure flow from different depths of tissue. Optical imaging depth of imaging can also be adjusted by raising or lowering the plane of focus. However, LDF still offers the opportunity to monitor flow changes occurring at greater depths than is available by intrinsic signal imaging.

Scanning laser Doppler flowmetry (SLDF) offers two-dimensional images of flow patterns. Although traditionally used for retinal imaging, recently this technology has been used to monitor flow changes on the surface of the rodent brain following functional activation (Nielsen *et al.*, 2000). This offers an exciting means of monitoring flow changes over a large spatial extent with relatively good spatial resolution. Unlike traditional LDF, however, depth flow recordings are not possible. The cost of increased spatial resolution is severely impaired temporal resolution: The temporal resolution of SLDF (on the order of several seconds) is limited and so dynamic imaging of blood flow changes with functional activation may not be possible.

D. Dye Imaging

One of the principal shortcomings of intrinsic imaging is its limited temporal resolution. Voltage-sensitive dyes are in this respect at least three orders of magnitude better. A fine temporal resolution is, of course, required for a detailed understanding of the flow of information and its processing at different cortical sites. This fine temporal resolution is, however, also the reason for some of the technical difficulties associated with voltage-sensitive dyes. It is, for instance, much easier to obtain a good SNR if signals are slow: a measurement of small slow signals with a rise time of a second rather than a millisecond will, for instance, yield a 33-fold improvement in SNR due to the square root relationship between the number of samples and the SNR. Moreover, the excellent temporal resolution with the current voltage-sensitive dyes is achieved at the cost of a limited spatial resolution. The spatial resolution with the current voltage-sensitive dyes is, however, not ultimately limited by the number of pixels of the photodetector or by the available computer and instrumentation capabilities, but rather by obtainable signal size and concomitant photodynamic damage. Without a significant improvement in the quality of currently available dyes, a 64×64-pixel array is close to the usable limit.

In addition to voltage-sensitive dyes, intravascular dyes (Texas red dextran, MW = 70,000; Frostig *et al.*, 1990; Narayan *et al.*, 1995), oxygen-sensitive phosphorescence quenching dyes (Vanzetta and Grinvald, 1999; Lindauer *et al.*, 2001), and molecule-specific dyes (e.g., DAF-2, a nitric oxide-sensitive dye) are available for imaging specific processes. These dyes all suffer from the same limitation as voltage-sensitive dyes, namely, the SNR using these dyes is significantly less than with intrinsic signal imaging and they can induce photodynamic damage. The advantage of using such dyes is that the investigator knows unambiguously the etiology of the signal being imaged, whereas with optical imaging and optical spectroscopy the exact nature of the signal being observed is not entirely clear.

In addition to the photodynamic damage associated with the use of any dye upon prolonged or intense illumination, the use of dyes has other difficulties, such as bleaching, limited depth of penetration into the cortex, and possible pharmacological side effects. While the extent of pharmacological side effects on cortical function has not been carefully evaluated, it is already clear that stained cortical cells still maintain their principal response properties.

XII. Conclusions and Outlook

Optical imaging based on intrinsic signals is unique among functional neuroimaging techniques in that it offers both high spatial (on the order of micrometers) and high temporal (on the order of milliseconds to seconds) resolution *simultaneously*. Moreover, no other functional neuroimaging technique offers comparable spatial resolution, allowing visualization of the fine structure of individual functional domains within cortical areas. Additionally, high spatial resolution maps can be acquired over relatively large cortical areas. The resolution of optical imaging makes it ideal for a variety of experimental investigations, including studying the functional organization of and functional relationships within the brain, characterizing neurovascular physiology, and examining pathophysiological changes due to certain disease processes.

Optical imaging of *intrinsic signals* is a particularly attractive methodology because of its relatively noninvasive nature. Although imaging intrinsic signals is invasive in the sense that the skull has to be thinned or opened in order to gain optical access to the brain, the brain itself is left completely untouched: no physical cortical contact is required, no dyes (which may be phototoxic) are necessary, and animal sacrifice is not necessary for acquiring results. In addition to providing data about the brain in its native state, the noninvasive nature of optical imaging makes this modality ideal for chronic investigations, such as development and plasticity, which may last up to several weeks or months.

One of the more exciting and challenging applications of optical imaging of intrinsic signals is studying the human cortex. This not only allows us to characterize the physiology and functional architecture of the human brain, but may also be clinically useful as an intraoperative functional brain mapping tool for neurosurgical guidance.

Recent advances in detector technology and microscope optics have increased the sensitivity, speed, and applications of optical imaging. Recent advances in spectroscopic techniques, for example, have made it possible to make highly specific determinations of changes in hemoglobin concentrations. No doubt, technological advances will continue to revolutionize this modality. We anticipate that by explicitly stating some methodological considerations involved with optical imaging and by introducing investigators to the different approaches used to analyze optical imaging data, we have increased interest in this modality and demonstrated its versatility.

Acknowledgments

The overall structure of this chapter has been adapted from a chapter by Tobias Bonhoeffer and Amiram Grinvald that appeared in the first edition of *Brain Mapping: The Methods*. Moreover, some of the content has been adapted from the original chapter. The authors are indebted to these authors for their original contribution. The authors also thank members of the UCLA Laboratory of Neuro Imaging for contributing figures to this chapter, including Andrew Cannestra, Michael Guiou, Sanjiv Narayan, Alyssa O'Farrell Ba, and Sameer Sheth. The authors also thank Sameer Sheth for his critical review of the manuscript. N.P. is supported, in part, by the Medical Scientist Training Program (GM08042) and a National Research Service Award (MH12773-01). Additional support was provided by research grants to A.W.T. (NIMH MH/NS52083).

References

Ances, B. M., Greenberg, J. H., and Detre, J. A. (2000). Effects of variations in interstimulus interval on activation-flow coupling response and somatosensory evoked potentials with forepaw stimulation in the rat. *J. Cereb. Blood Flow Metab.* **20,** 290–297.

Antonini, A., Fagiolini, M., and Stryker, M. P. (1999). Anatomical correlates of functional plasticity in mouse visual cortex. *J Neurosci.* **19,** 4388–4406.

Bandettini, P. A., Jesmanowicz, A., Wong, E. C., and Hyde, J. S. (1993). Processing strategies for time-course data sets in functional MRI of the human brain. *Magn. Reson. Med.* **30,** 161–173.

Bär, T. (1981). Distribution of radially penetrating arteries and veins in the neocortex of rat. In *Cereb. Microcirc. Metab.* pp. 1–8 (J.C. Navarro and E. Fritschka *eds.*) Raven Press, New York.

Belliveau, J. W., Cohen, M. S., Weisskoff, R. M., Buchbinder, B. R., and Rosen, B. R. (1991). Functional studies of the human brain using high-speed magnetic resonance imaging. *J. Neuroimaging* **1,** 36–41.

Berwick, J., Johnston, D., Zheng, Y., Coffey, P., Mayhew, J., and Whiteley, S. J. O. (2000). Spectroscopic analysis of the action of 7-nitroindazole. *Soc. Neurosci. Abstr.* **26,** 1737.

Blasdel, G. G. (1992a). Differential imaging of ocula dominance and orientation selectivity in monkey striate cortex. *J. Neurosci.* **12,** 3115–3138.

Blasdel, G. G. (1992b). Orientation selectivity, preference, and continuity in monkey striate cortex. *J. Neurosci.* **12,** 3139–3161.

Blasdel, G. G., and Salama, G. (1986). Voltage-sensitive dyes reveal a modular organization in monkey striate cortex. *Nature* **321,** 579–585.

Blood, A. J., Narayan, S. M., and Toga, A. W. (1995). Stimulus parameters influence characteristics of optical intrinsic signal responses in somatosensory cortex. *J. Cereb. Blood Flow Metab.* **15,** 1109–1121.

Blood, A. J., and Toga, A. W. (1998). Optical intrinsic signal imaging responses are modulated in rodent somatosensory cortex during simultaneous whisker and forelimb stimulation. *J. Cereb. Blood Flow Metab.* **18,** 968–977.

Bonhoeffer, T. (1995). Optical imaging of intrinsic signals as a tool to visualize the functional architecture of adult and developing visual cortex. *Arzneimittelforschung* **45,** 351–356.

Bonhoeffer, T., and Grinvald, A. (1991). Iso-orientation domains in cat visual cortex are arranged in pinwheel-like patterns. *Nature* **353,** 429–431.

Bonhoeffer, T., and Grinvald, A. (1993a). The layout of iso-orientation domains in area 18 of cat visual cortex: Optical imaging reveals a pinwheel-like organization. *J. Neurosci.* **13,** 4157–4180.

Bonhoeffer, T., and Grinvald, A. (1993b). Optical imaging of the functional architecture in cat visual cortex: The layout of direction and orientation domains. *Adv. Exp. Med. Biol.* **333,** 57–69.

Bonhoeffer, T., and Grinvald, A. (1996). Optical imaging based on intrinsic signals: The methodology. *In* "Brain Mapping: The Methods" (A. W. Toga and J. C. Mazziotta, eds.). Academic Press, San Diego.

Boxerman, J. L., Bandettini, P. A., Kwong, K. K., Baker, J. R., Davis, T. L., Rosen, B. R., and Weisskoff, R. M. (1995). The intravascular contribution to *f*MRI signal change: Monte Carlo modeling and diffusion-weighted studies in vivo. *Magn. Reson. Med.* **34,** 4–10.

Buchweitz, E., and Weiss, H. R. (1986). Effect of withdrawal from chronic naltrexone on regional cerebral oxygen consumption in the cat. *Brain Res.* **397,** 308–314.

Cannestra, A. F., Black, K. L., Martin, N. A., Cloughesy, T., Burton, J. S., Rubinstein, E., Woods, R. P., and Toga, A. W. (1998a). Topographical and temporal specificity of human intraoperative optical intrinsic signals. *NeuroReport* **9,** 2557–2563.

Cannestra, A. F., Blood, A. J., Black, K. L., and Toga, A. W. (1996). The evolution of optical signals in human and rodent cortex. *NeuroImage* **3,** 202–208.

Cannestra, A. F., Bookheimer, S. Y., Pouratian, N., O'Farrell, A. M., Sicotte, N. M., Martin, N. A., Becker, D., Rubino, G., and Toga, A. W. (2000). Temporal and topographical characterization of language cortices using intraoperative optical intrinsic signals. *NeuroImage* **12,** 41–54.

Cannestra, A. F., Pouratian, N., Bookheimer, S. Y., Martin, N. A., Becker, D., and Toga, A. W. (2001). Temporal spatial differences observed by functional MRI and human intraoperative optical imaging. *Cereb. Cortex* **11,** 773–782.

Cannestra, A. F., Pouratian, N., Shomer, M. H., and Toga, A. W. (1998b). Refractory periods observed by intrinsic signal and fluorescent dye imaging. *J. Neurophysiol.* **80,** 1522–1532.

Chapin, J. K., and Lin, C. S. (1984). Mapping the body representation in the SI cortex of anesthetized and awake rats. *J. Comp. Neurol.* **229,** 199–213.

Chapman, B., Stryker, M. P., and Bonhoeffer, T. (1996). Development of orientation preference maps in ferret primary visual cortex. *J. Neurosci.* **16,** 6443–6453.

Chen, J. W. Y., O'Farrell, A. M., and Toga, A. W. (2000). Optical intrinsic signal imaging in a rodent seizure model. *Neurology* **55,** 312–315.

Chen-Bee, C. H., and Frostig, R. D. (1996). Variability and interhemispheric asymmetry of single-whisker functional representations in rat barrel cortex. *J. Neurophysiol.* **76,** 884–894.

Chen-Bee, C. H., Kwon, M. C., Masino, S. A., and Frostig, R. D. (1996). Areal extent quantification of functional representations using intrinsic signal optical imaging. *J. Neurosci. Methods* **68,** 27–37.

Chen-Bee, C. H., Polley, D. B., Brett-Green, B., Prakash, N., Kwon, M. C., and Frostig, R. D. (2000). Visualizing and quantifying evoked cortical activity assessed with intrinsic signal imaging. *J. Neurosci. Methods* **97,** 157–173.

Cohen, L., and Keynes, R. D. (1971). Changes in light scattering associated with the action potential in crab nerves. *J. Physiol. (London)* **212,** 259–275.

Cohen, M. S. (1997). Parametric analysis of *f*MRI data using linear systems methods. *NeuroImage* **6,** 93–103.

Corina, D. P., Poliakov, A., Steury, K., Martin, R., Mulligan, K., Maravilla, K., Brinkly, J. F., and Ojemann, G. A. (2000). Correspondences between language cortex identified by cortical stimulation mapping and *f*MRI. *NeuroImage* **11,** S295.

Cox, S. B., Woolsey, T. A., and Rovainen, C. M. (1993). Localized dynamic changes in cortical blood flow with whisker stimulation corresponds to matched vascular and neuronal architecture of rat barrels. *J. Cereb. Blood Flow Metab.* **13,** 899–913.

Daghighian, F., Mazziotta, J. C., Hoffman, E. J., Shenderov, P., Eshaghian, B., Siegel, S., and Phelps, M. E. (1994). Intraoperative beta probe: A device for detecting tissue labeled with positron or electron emitting isotopes during surgery. *Med. Phys.* **21,** 153–157.

Dinse, H. R., Godde, B., Hilger, T., Haupt, S. S., Spengler, F., and Zepka, R. (1997a). Short-term functional plasticity of cortical and thalamic sensory representations and its implication for information processing. *Adv. Neurol.* **73,** 159–178.

Dinse, H. R., Reuter, G., Cords, S. M., Godde, B., Hilger, T., and Lenarz, T. (1997b). Optical imaging of cat auditory cortical organization after electrical stimulation of a multichannel cochlear implant: Differential effects of acute and chronic stimulation. *Am. J. Otol.* **18,** S17–18.

Dirnagl, U., Villringer, A., Gebhardt, R., Haberl, R. L., Schmiedek, P., and Einhäupl, K. M. (1991). Three-dimensional reconstruction of the rat brain cortical microcirculation in vivo. *J. Cereb. Blood Flow Metab.* **11,** 353–360.

Durham, D., and Woolsey, T. A. (1977). Barrels and columnar cortical organization: Evidence from 2-deoxyglucose (2-DG) experiments. *Brain Res.* **137,** 168–174.

Duvernoy, H. M., Delon, S., and Vannson, J. L. (1981). Cortical blood vessels of the human brain. *Brain Res. Bull.* **7,** 519–579.

Ferrari, M., Wilson, D. A., Hanley, D. F., and Traystman, R. J. (1992). Effects of graded hypotension on cerebral blood flow, blood volume, and mean transit time in dogs. *Am. J. Physiol.* **262,** H1908–1914.

Fox, P. T., and Raichle, M. E. (1986). Focal physiological uncoupling of cerebral blood flow and oxidative metabolism during somatosensory stimulation in human subjects. *Proc. Natl. Acad. Sci. USA* **83,** 1140–1144.

Fox, P. T., Raichle, M. E., Mintun, M. A., and Dence, C. (1988). Nonoxidative glucose consumption during focal physiologic neural activity. *Science* **241,** 462–464.

Frahm, J., Merboldt, K. D., Hanicke, W., Kleinschmidt, A., and Boecker, H. (1994). Brain or vein—Oxygenation or flow? On signal physiology in functional MRI of human brain activation. *NMR Biomed.* **7,** 45–53.

Frostig, R. D., Lieke, E. E., Ts'o, D. Y., and Grinvald, A. (1990). Cortical functional architecture and local coupling between neuronal activity and the microcirculation revealed by in vivo high-resolution optical imaging of intrinsic signals. *Proc. Natl. Acad. Sci. USA* **87,** 6082–6086.

Gabbay, M., Brennan, C., Kaplan, E., and Sirovich, L. (2000). A principal components-based method for the detection of neuronal activity maps: Application to optical imaging. *NeuroImage* **11,** 313–325.

Geladi, P., Iasksson, H., Lindqvist, L., Wold, S., and Esbensen, K. (1989). Principal component analysis of multivariate images. *Chem. Intell. Lab. Syst.* **5,** 209–220.

Godde, B., Hilger, T., von Seelen, W., Berkefeld, T., and Dinse, H. R. (1995). Optical imaging of rat somatosensory cortex reveals representational overlap as topographic principle. *NeuroReport* **7,** 24–28.

Grinvald, A. (1992). Optical imaging of architecture and function in the living brain sheds new light on cortical mechanisms underlying visual perception. *Brain Topogr.* **5,** 71–75.

Grinvald, A., Frostig, R. D., Siegel, R. M., and Bartfeld, E. (1991). High-resolution optical imaging of functional brain architecture in the awake monkey. *Proc. Natl. Acad. Sci. USA* **88,** 11559–11563.

Grinvald, A., Lieke, E., Frostig, R. D., Gilbert, C. D., and Wiesel, T. N. (1986). Functional architecture of cortex revealed by optical imaging of intrinsic signals. *Nature* **324,** 361–364.

Grinvald, A., Manker, A., and Segal, M. (1982). Visualization of the spread of electrical activity in rat hippocampal slices by voltage-sensitive optical probes. *J. Physiol. (London)* **333,** 269–291.

Gross, D., Loew, L. M., and Webb, W. W. (1986). Optical imaging of cell membrane potential changes induced by applied electric fields. *Biophys. J.* **50,** 339–348.

Haglund, M. M., Ojemann, G. A., and Blasdel, G. G. (1993). Optical imaging of bipolar cortical stimulation. *J. Neurosurg.* **78,** 785–793.

Haglund, M. M., Ojemann, G. A., and Hochman, D. W. (1992). Optical imaging of epileptiform and functional activity in human cerebral cortex. *Nature* **358,** 668–671.

Hess, A., Stiller, D., Kaulisch, T., Heil, P., and Scheich, H. (2000). New insights into the hemodynamic blood oxygenation level-dependent response through combination of functional magnetic resonance imaging and optical recording in gerbil barrel cortex. *J. Neurosci.* **20,** 3328–3338.

Hill, D. K., and Keynes, R. D. (1949). Opacity changes in stimulated nerve. *J. Physiol. (London)* **108,** 278–281.

Hodge, C. J., Jr., Stevens, R. T., Newman, H., Merola, J., and Chu, C. (1997). Identification of functioning cortex using cortical optical imaging. *Neurosurgery* **41,** 1137–1144; discussion 1144–1145.

Hu, X., Le, T. H., and Ugurbil, K. (1997). Evaluation of the early response in *f*MRI in individual subjects using short stimulus duration. *Magn. Reson. Med.* **37,** 877–884.

Kauer, J. S. (1988). Real-time imaging of evoked activity in local circuits of the salamander olfactory bulb. *Nature* **331,** 166–168.

Kelly, M. K., Carvell, G. E., Kodger, J. M., and Simons, D. J. (1999). Sensory loss by selected whisker removal produces immediate disinhibition in the somatosensory cortex of behaving rats. *J. Neurosci.* **19,** 9117–9125.

Kim, D. S., Duong, T. Q., and Kim, S. G. (2000). High-resolution mapping of iso-orientation columns by *f*MRI. *Nat. Neurosci.* **3,** 164–169.

Kisvarday, Z. F., Kim, D. S., Eysel, U. T., and Bonhoeffer, T. (1994). Relationship between lateral inhibitory connections and the topography of the orientation map in cat visual cortex. *Eur. J. Neurosci.* **6,** 1619–1632.

Kohl, M., Lindauer, U., Royl, G., Kuhl, M., Gold, L., Villringer, A., and Dirnagl, U. (2000). Physical model for the spectroscopic analysis of cortical intrinsic optical signals. *Phys. Med. Biol.* **45,** 3749–3764.

Kornblum, H. I., Araujo, D. M., Annala, A. J., Tatsukawa, K. J., Phelps, M. E., and Cherry, S. R. (2000). In vivo imaging of neuronal activation and plasticity in the rat brain by high resolution positron emission tomography (microPET). *Nat. Biotechnol.* **18,** 655–660.

Kossut, M. (1988). Modifications of the single cortical vibrissal column. *Acta Neurobiol. Exp. (Warsz)* **48,** 83–115.

Kossut, M. (1998). Experience-dependent changes in function and anatomy of adult barrel cortex. *Exp. Brain Res.* **123,** 110–116.

Kwong, K. K., Belliveau, J. W., Chesler, D. A., Goldberg, I. E., Weisskoff, R. M., Poncelet, B. P., Kennedy, D. N., Hoppel, B. E., Cohen, M. S., Turner, R., *et al.* (1992). Dynamic magnetic resonance imaging of human brain activity during primary sensory stimulation. *Proc. Natl. Acad. Sci. USA* **89,** 5675–5679.

Lauritzen, M. (1994). Pathophysiology of the migraine aura. The spreading depression theory. *Brain* **117,** 199–210.

Lee, P. A., Sylvia, A. L., and Piantadosi, C. A. (1988). Effect of fluorocarbon-for-blood exchange on regional cerebral blood flow in rats. *Am. J. Physiol.* **254,** H719–H726.

LeManna, J. C., Sick, T. J., Pirarsky, S. M., and Rosenthal, M. (1987). Detection of an oxidizable fraction of cytochrome oxidase in intact rat brain. *Am. Physiol. Soc.* **253,** C477–C483.

Lindauer, U., Royl, G., Leithner, C., Kuhl, M., Gold, L., Gethmann, J., Kohl-Bareis, M., Villringer, A., and Dirnagl, U. (2001). No evidence for early decrease in blood oxygenation in rat whisker cortex in response to functional activation. *NeuroImage* **13,** 988–1001.

Lindauer, U., Villringer, A., and Dirnagl, U. (1993). Characterization of CBF response to somatosensory stimulation: Model and influence of anesthetics. *Am. J. Physiol.* **264,** H1223–1228.

Logothetis, N. K., Guggenberger, H., Peled, S., and Pauls, J. (1999). Functional imaging of the monkey brain. *Nat. Neurosci.* **2,** 555–562.

Lou, H. C., Edvinsson, L., and MacKenzie, E. T. (1987). The concept of coupling blood flow to brain function: Revision required? *Ann. Neurol.* **22,** 289–297.

MacDonald, D., Avis, D., and Evans, A. C. (1994). Multiple surface identification and matching in magnetic resonance imaging. *Proc. SPIE* **2359,** 160–169.

Macknik, S. L., and Haglund, M. M. (1999). Optical images of visible and invisible percepts in the primary visual cortex of primates. *Proc. Natl. Acad. Sci. USA* **96,** 15208–15210.

Malach, R., Amir, Y., Harel, M., and Grinvald, A. (1993). Relationship between intrinsic connections and functional architecture revealed by optical imaging and in vivo targeted biocytin injections in primate striate cortex. *Proc. Natl. Acad. Sci. USA* **90,** 10469–10473.

Malonek, D., and Grinvald, A. (1996). Interactions between electrical activity and cortical microcirculation revealed by imaging spectroscopy: Implications for functional brain mapping. *Science* **272,** 551–554.

Masino, S. A., and Frostig, R. D. (1996). Quantitative long-term imaging of the functional representation of a whisker in rat barrel cortex. *Proc. Natl. Acad. Sci. USA* **93,** 4942–4947.

Masino, S. A., Kwon, M. C., Dory, Y., and Frostig, R. D. (1993). Characterization of functional organization within rat barrel cortex using intrinsic signal optical imaging through a thinned skull. *Proc. Natl. Acad. Sci. USA* **90,** 9998–10002.

Mayhew, J., Hu, D., Zheng, Y., Askew, S., Hou, Y., Berwick, J., Coffey, P. J., and Brown, N. (1998). An evaluation of linear model analysis techniques for processing images of microcirculation activity. *NeuroImage* **7,** 49–71.

Mayhew, J., Johnston, D., Berwick, J., Jones, M., Coffey, P., and Zheng, Y. (2000). Spectroscopic analysis of neural activity in brain: Increased oxygen consumption following activation of barrel cortex. *NeuroImage* **12,** 664–675.

Mayhew, J., Johnston, D., Martindale, J., Jones, M., Berwick, J., and Zheng, Y. (2001). Increased oxygen consumption following activation of brain: Theoretical footnotes using spectroscopic data from barrel cortex. *NeuroImage* **13,** 975–987.

Mayhew, J., Zheng, Y., Hou, Y., Vuksanovic, B., Berwick, J., Askew, S., and Coffey, P. (1999). Spectroscopic analysis of changes in remitted illumination: The response to increased neural activity in brain. *NeuroImage* **10,** 304–326.

Mayhew, J. E., Askew, S., Zheng, Y., Porrill, J., Westby, G. W., Redgrave, P., Rector, D. M., and Harper, R. M. (1996). Cerebral vasomotion: A 0.1-Hz oscillation in reflected light imaging of neural activity. *NeuroImage* **4,** 183–193.

Mazziotta, J. C., Toga, A. W., and Frackowiak, R. S. J., eds. (2000). "Brain Mapping: The Disorders." Academic Press, San Diego.

Menon, R. S., Ogawa, S., Hu, X., Strupp, J. P., Anderson, P., and Ugurbil, K. (1995). BOLD based functional MRI at 4 tesla includes a capillary bed contribution: Echo-planar imaging correlates with previous optical imaging using intrinsic signals. *Magn. Reson. Med.* **33,** 453–459.

Narayan, S. M., Esfahani, P., Blood, A. J., Sikkens, L., and Toga, A. W. (1995). Functional increases in cerebral blood volume over somatosensory cortex. *J. Cereb. Blood Flow Metab.* **15,** 754–765.

Narayan, S. M., Santori, E. M., and Toga, A. W. (1994). Mapping functional activity in rodent cortex using optical intrinsic signals. *Cereb. Cortex* **4,** 195–204.

Nemoto, M., Nomura, Y., Sato, C., Tamura, M., Houkin, K., Koyanagi, I., and Abe, H. (1999). Analysis of optical signals evoked by peripheral nerve stimulation in rat somatosensory cortex: Dynamic changes in hemoglobin concentration and oxygenation. *J. Cereb. Blood Flow Metab.* **19,** 246–259.

Ngai, A. C., Ko, K. R., Morii, S., and Winn, H. R. (1988). Effect of sciatic nerve stimulation on pial arterioles in rats. *Am. J. Physiol.* **254,** H133–139.

Nguyen, T. T., Yamamoto, T., Stevens, R. T., and Hodge, C. J., Jr. (2000). Reorganization of adult rat barrel cortex intrinsic signals following kainic acid induced central lesion. *Neurosci. Lett.* **288,** 5–8.

Nielsen, A. N., Fabricius, M., and Lauritzen, M. (2000). Scanning laser-Doppler flowmetry of rat cerebral circulation during cortical spreading depression. *J. Vasc. Res.* **37,** 513–522.

Nomura, Y., Fujii, F., Sato, C., Nemoto, M., and Tamura, M. (2000). Exchange transfusion with fluorocarbon for studying synaptically evoked optical signal in rat cortex. *Brain Res. Protocols* **5,** 10–15.

Nomura, Y., Hazeki, O., and Tamura, M. (1997). Relationship between time-resolved and non-time-resolved Beer–Lambert Law in turbid media. *Phys. Med. Biol.* **42,** 1009–1022.

Obrig, H., Wenzel, R., Kohl, M., Horst, S., Wobst, P., Steinbrink, J., Thomas, F., and Villringer, A. (2000). Near-infrared spectroscopy: Does it function in functional activation studies of the adult brain? *Int. J. Psychophysiol.* **35,** 125–142.

O'Farrell, A. M., Rex, D. E., Muthialu, A., Pouratian, N., Wong, G. K., Cannestra, A. F., Chen, J. W. Y., and Toga, A. W. (2000). Characterization of optical intrinsic signals and blood volume during cortical spreading depression. *NeuroReport* **11,** 2121–2125.

Ogawa, S., Lee, T. M., Kay, A. R., and Tank, D. W. (1990). Brain magnetic resonance imaging with contrast dependent on blood oxygenation. *Proc. Natl. Acad. Sci. USA* **87,** 9868–9872.

Ogawa, S., Tank, D. W., Menon, R., Ellermann, J. M., Kim, S. G., Merkle, H., and Ugurbil, K. (1992). Intrinsic signal changes accompanying sensory stimulation: Functional brain mapping with magnetic resonance imaging. *Proc. Natl. Acad. Sci. USA* **89,** 5951–5955.

Orbach, H. S., and Cohen, L. B. (1983). Optical monitoring of activity from many areas of the in vitro and in vivo salamander olfactory bulb: A new method for studying functional organization in the vertebrate central nervous system. *J. Neurosci.* **3,** 2251–2262.

Paley, M., Mayhew, J. E., Martindale, A. J., McGinley, J., Berwick, J., Coffey, P., Redgrave, P., Furness, P., Port, M., Ham, A., Zheng, Y., Jones, M., Whitby, E., van Beek, E. J., Wilkinson, I. D., Darwent, G., and Griffiths, P. D. (2001). Design and initial evaluation of a low-cost 3-tesla research system for combined optical and functional MR imaging with interventional capability. *J. Magn. Reson. Imaging* **13,** 87–92.

Polley, D. B., Chen-Bee, C. H., and Frostig, R. D. (1999a). Two directions of plasticity in the sensory-deprived adult cortex. *Neuron* **24,** 623–637.

Polley, D. B., Chen-Bee, C. H., and Frostig, R. D. (1999b). Varying the degree of single-whisker stimulation differentially affects phases of intrinsic signals in rat barrel cortex. *J. Neurophysiol.* **81,** 692–701.

Pouratian, N., Bookheimer, S. Y., O'Farrell, A. M., Sicotte, N. L., Cannestra, A. F., Becker, D., and Toga, A. W. (2000a). Optical imaging of bilingual cortical representations: Case report. *J. Neurosurg.* **93,** 686–691.

Pouratian, N., Kamrava, A., Nettar, K. D., O'Farrell, A. M., and Toga, A. W. (2000b). Developmental changes in perfusion-related responses in rodent somatosensory cortex. *Soc. Neurosci. Abstr.* **26,** 133.

Pouratian, N., Kamrava, A., O'Farrell, A. M., and Toga, A. W. (1999). Plasticity induced changes in functional perfusion. *NeuroImage* **9,** S288.

Pouratian, N., Martin, N. A., Cannestra, A. F., Becker, D., Bookheimer, S. Y., Sicotte, N. M., and Toga, A. W. (2000c). Intraoperative sensorimotor and language mapping using optical intrinsic signal imaging: Comparison with electrophysiologic techniques and *f*MRI in 40 patients. *In* "2000 American Association of Neurological Surgeons Annual Meeting," San Francisco, California.

Pouratian, N., Sicotte, N. L., Rex, D. E., Cannestra, A. F., Martin, N. A., Becker, D., and Toga, A. W. (2001). Spatial/temporal correlation of optical intrinsic signals and BOLD. *Magn. Reson. Med.* **47**, 760–776.

Prakash, N., Cohen-Cory, S., and Frostig, R. D. (1996). Rapid and opposite effects of BDNF and NGF on the functional organization of the adult cortex in vivo. *Nature* **381,** 702–706.

Ramsey, N. F., Kirkby, B. S., Van Gelderen, P., Berman, K. F., Duyn, J. H., Frank, J. A., Mattay, V. S., Van Horn, J. D., Esposito, G., Moonen, C. T., and Weinberger, D. R. (1996). Functional mapping of human sensorimotor cortex with 3D BOLD *f*MRI correlates highly with H_2(15)O PET RCBF. *J. Cereb. Blood Flow Metab.* **16,** 755–764.

Ratzlaff, E. H., and Grinvald, A. (1991). A tandem-lens epifluorescence macroscope: Hundred-fold brightness advantage for wide-field imaging. *J. Neurosci. Methods* **36,** 127–137.

Reichenbach, J., Venkatesan, R., Schillinger, D., Kido, D., and Haacke, E. (1997). Small vessels in the human brain: MR venography with deoxyhemoglobin as an intrinsic contrast agent. *Radiology* **204,** 272–277.

Roy, C. W., and Sherrington, C. S. (1890). On the regulation of the blood-supply of the brain. *J. Physiol. (London)* **11,** 85–108.

Schuette, W. H., Whitehouse, W. C., Lewis, D. V., O'Connor, M., and Van Buren, J. M. (1974). A television fluorometer for monitoring oxidative metabolism in intact tissue. *Med. Instrum.* **8,** 331–333.

Shmuel, A., and Grinvald, A. (1996). Functional organization for direction of motion and its relationship to orientation maps in cat area 18. *J. Neurosci.* **16,** 6945–6964.

Shmuel, A., and Grinvald, A. (2000). Coexistence of linear zones and pinwheels within orientation maps in cat visual cortex. *Proc. Natl. Acad. Sci. USA* **97,** 5568–5573.

Shtoyerman, E., Arieli, A., Slovin, H., Vanzetta, I., and Grinvald, A. (2000). Long-term optical imaging and spectroscopy reveal mechanisms underlying the intrinsic signal and stability of cortical maps in V1 of behaving monkeys. *J. Neurosci.* **20,** 8111–8121.

Simons, D. J. (1978). Response properties of vibrissa units in rat SI somatosensory neocortex. *J. Neurophysiol.* **41,** 798–820.

Sled, J. G., Zijdenbos, A. P., and Evans, A. C. (1998). A nonparametric method for automatic correction of intensity nonuniformity in MRI data. *IEEE Trans. Med. Imaging* **17,** 87–97.

Sokoloff, L., Reivich, M., Kennedy, C., Des Rosiers, M. H., Patlak, C. S., Pettigrew, K. D., Sakurada, O., and Shinohara, M. (1977). The [^{14}C]deoxyglucose method for the measurement of local cerebral glucose utilization: Theory, procedure, and normal values in the conscious and anesthetized albino rat. *J. Neurochem.* **28,** 897–916.

Stepnoski, R. A., LaPorta, A., Raccuia-Behling, F., Blonder, G. E., Slusher, R. E., and Kleinfeld, D. (1991). Noninvasive detection of changes in membrane potential in cultured neurons by light scattering. *Proc. Natl. Acad. Sci. USA* **88,** 9382–9386.

Stetter, M., Schiessl, I., Otto, T., Sengpiel, F., Hubener, M., Bonhoeffer, T., and Obermayer, K. (2000). Principal component analysis and blind separation of sources for optical imaging of intrinsic signals. *NeuroImage* **11,** 482–490.

Toga, A. W., Cannestra, A. F., and Black, K. L. (1995a). The temporal/spatial evolution of optical signals in human cortex. *Cereb. Cortex* **5,** 561–565.

Toga, A. W., Santori, E. M., Hazani, R., and Ambach, K. (1995b). A 3D digital map of rat brain. *Brain Res. Bull.* **38,** 77–85.

Ts'o, D. Y., Frostig, R. D., Lieke, E. E., and Grinvald, A. (1990). Functional organization of primate visual cortex revealed by high resolution optical imaging. *Science* **249,** 417–420.

Ts'o, D. Y., Roe, A. W., and Gilbert, C. D. (2001). A hierarchy of the functional organization for color, form and disparity in primate visual area V2. *Vision Res.* **41,** 1333–1349.

Vanzetta, I., and Grinvald, A. (1999). Increased cortical oxidative metabolism due to sensory stimulation: Implications for functional brain imaging. *Science* **286,** 1555–1558.

Victor, J. D., Purpura, K., Katz, E., and Mao, B. (1994). Population encoding of spatial frequency, orientation, and color in macaque V1. *J. Neurophysiol.* **72,** 2151–2166.

Villringer, A., and Dirnagl, U. (1995). Coupling of brain activity and cerebral blood flow: Basis of functional neuroimaging. *Cerebrovasc. Brain Metab. Rev.* **7,** 240–276.

Villringer, A., Planck, J., Hock, C., Schleinkofer, L., and Dirnagl, U. (1993). Near infrared spectroscopy (NIRS): A new tool to study hemodynamic changes during activation of brain function in human adults. *Neurosci. Lett.* **154,** 101–104.

Woods, R. P., Cherry, S. R., and Mazziotta, J. C. (1992). Rapid automated algorithm for aligning and reslicing PET images. *J. Comput. Assisted Tomogr.* **16,** 620–633.

Woolsey, T. A., and Van der Loos, H. (1970). The structural organization of layer IV in the somatosensory region (SI) of mouse cerebral cortex. The description of a cortical field composed of discrete cytoarchitectonic units. *Brain Res.* **17,** 205–242.

Yacoub, E., Le, T. H., Ugurbil, K., and Hu, X. (1999). Further evaluation of the initial negative response in functional magnetic resonance imaging. *Magn. Reson. Med.* **41,** 436–441.

Yap, J. T., Chen, C. T., Cooper, M., and Treffert, J. D. (1994). Knowledge-based factor analysis of multidimensional nuclear medicine image sequences. *Proc. SPIE* **2168,** 289–297.

Yoon, R. S., Tsang, P. W., Lenz, F. A., and Kwan, H. C. (1996). Characterization of cortical spreading depression by imaging of intrinsic optical signals. *NeuroReport* **7,** 2671–2674.

Zdrojkowski, R. J., and Pisharoty, N. R. (1970). Optical transmission and reflection by blood. *IEEE Trans. Biomed. Eng.* **17,** 122–128.

Zheng, Y., Johnston, D., Berwick, J., and Mayhew, J. (2001). Signal source separation in the analysis of neural activity in brain. *NeuroImage* **13,** 447–458.

6

Near-Infrared Spectroscopy and Imaging

Arno Villringer and Hellmuth Obrig

Division of Neuroimaging, Department of Neurology, Neurologische Klinik, Charité, Schumannstrasse 20-21, 10117 Berlin, Germany

I. Introduction

The preceding chapters illustrate the wealth of information on brain function to be obtained by optical methods in studies of the exposed brain. In this chapter, we describe approaches based on light in the near infrared (NIR) allowing one to similarly gather data on the human brain through the intact skull. By comparison with other noninvasive functional neuroimaging methods such as positron emission tomography (PET), functional magnetic resonance imaging (*f*MRI), and magnetoencephalography (MEG), NIR spectroscopy (NIRS) surpasses these techniques in (1) flexibility, (2) biochemical specificity, and (3) high sensitivity to detect small substance concentrations. (4) Also NIRS may extend beyond the imaging changes in hemodynamics, potentially allowing for the imaging of mitochondrial metabolism and even neuronal activity.

(1) Regarding flexibility, *f*MRI, PET, and MEG require large and bulky instruments interfering with clinical monitoring of patients. Also it is impossible to examine a number of normal/physiological situations such as standing or walking. Subjects who cannot fully collaborate (children, demented people, etc.) cannot be examined and studies at the patient's bedside are not feasible. In contrast optical methods assume only a connection of light, e.g., via fiber optics, requirements similar to those of EEG technology. The much higher degree of flexibility allows for studies on gait (Miyai *et al.*, 2001) and at the patient's bedside (von Pannwitz *et al.*, 1998), as well as in babies (Hintz *et al.*, 1999).

(2) A further shortcoming, especially of the widely used *f*MRI, is the lack of a clearly defined relationship of the

observed signal (e.g., the BOLD signal) to a quantifiable physiological and biochemical parameter. Optical methods offer higher biochemical specificity of the signal, and approaches for quantification of substance concentrations are more straightforward (see considerations of a modified Beer–Lambert Law below).

(3) Optical methods are sensitive to very low substance concentration by employing fluorescence methods. In theory, the sensitivity can go down to the level of single molecules (Harms *et al.,* 2001). The application to the noninvasive human studies may offer a potential similar to PET, whose obvious disadvantage is the necessity of radioactive tracers.

(4) Finally, existing technologies measure a "vascular signal," such as *f*MRI and H_2O-PET; an intracellular metabolic signal, such as FDG-PET; or a signal representing neuronal activity more directly, such as MEG or EEG. In producing images of different aspects of brain function temporal and spatial resolution greatly differ between techniques. High temporal resolution is the domain of electrophysiological methods (EEG/MEG) while excellent spatial resolution is achieved by techniques based on hemodynamics. Though the combination of the respective strengths has been investigated by combined approaches, technical requirements are complex and signal-to-noise ratio is lowered by such approaches. In principal, optical methods offer the option to monitor vascular, metabolic–cellular, and neuronal responses. Furthermore, in combined approaches with *f*MRI, PET, and MEG/EEG there is no interference with the biophysics of the other techniques.

II. Optical Window for Noninvasive Studies

In studies on the exposed brain surface the signal is dominated by light which is reflected from the brain surface. However, there is also some contribution from deeper structures. These contributions may be specifically highlighted by methods such as confocal laser scanning microscopy [see Chapter 1; also for *in vivo* studies of the rat brain there is work from our group (Villringer *et al.,* 1989; Dirnagl *et al.,* 1991)] and optical coherence tomography (Drexler *et al.,* 2001). The depth penetration of light, however, depends on its wavelength. In confocal microscopy at 488/514 nm, a penetration of about 250 μm into cerebral tissue can be achieved (Dirnagl *et al.,* 1991).

When using dual photon technology with an excitation at 830 nm (Kleinfeld *et al.,* 1998) the depth of penetration can be increased down to 600 μm below the brain surface. Depth penetration of light in the visible wavelength range is limited by the high absorption due to hemoglobin. Also beyond 950 nm light penetrates tissue poorly, since water is a strong absorber in this spectral range. The range between about 650 and 950 nm (which is part of the near-infrared[1] wavelength range) can be termed a biological "optical window," framed by hemoglobin and water absorption. At this light, we can even "see" through the skull of an adult human. Light will penetrate down to the cortex and thus allow for noninvasive studies of the human brain. This optical window is used for noninvasive near-infrared studies.

A. Transmission/Reflection Mode for Near-Infrared Studies

Near-infrared studies may be performed either in "transmission" or in "reflection" mode. When the diameter of the object is relatively small, e.g., a forearm (Hillman *et al.,* 2001), a breast (Franceschini *et al.,* 1997), or a neonate's head (Hintz *et al.,* 1999), transmission of near-infrared light through the "sample" is feasible. It should be noted that, due to pronounced scattering, the light does not travel along a straight line from sender to receiver, but rather the path of the photon is represented by the product of the geometrical distance (d) and a so-called differential path-length factor (DPF). This fact has to be considered when applying imaging algorithms as well as when performing concentration measurements (see below: modified Beer–Lambert Law).

Objects too large for transmission mode spectroscopy require the approach in reflection mode. Studies on the adult brain have been performed by connecting the light source with an optic fiber to the subject's head, and another fiber which connects to the receiver is placed at a distance of approximately 3–6 cm. A typical setup to investigate cerebral tissue in an adult human subject is given in Fig. 3. In this setup the sample volume corresponds to a flat semilunar volume beneath and between the two light fibers. In addition to extracerebral "contamination," signal can be ascribed to the cerebral cortex. Whether and how much deeper layers contribute to the observed signal is a matter of controversy (for more detail see Firbank, 1998). The sensitivity to the deeper layers can be increased by using time-resolved technology instead of purely intensity-based assessment of attenuation changes (Hemelt and Kang, 1999).

B. Absorption and Scattering and a Modified Beer–Lambert Law

The main interactions of light with tissue are absorption and scattering. For the simple model system of a cuvette, the effects of scattering and absorption on the path of "individual photons" is schematically illustrated in Fig. 2.

[1]The term "near infrared" indicates that this is the part of infrared light closest in wavelength ("near") to visible light.

For a quantitative assessment of the concentration of absorbing chromophores, the Beer–Lambert Law states that light attenuation (A) is proportional to the concentration of the absorbing molecule (c). The proportionality factor is termed specific extinction coefficient ε:

$$A = \varepsilon \times c \times d.$$

However, this relationship assumes infinitesimal concentrations and disregards scattering. This assumption clearly does not hold for spectroscopy in tissue. Scattering prolongs the path length of light as illustrated by photon 2 in Fig. 2 and thus the path length becomes longer than the distance between sender and receiver. In a typical transcranial study of the human brain, the mean path length of light is about six times as long as the distance between sender and receiver (Duncan *et al.,* 1995). In order to account for the longer path length, in a modified Beer–Lambert Law a DPF (B) is introduced. A second modification of the Beer–Lambert equation is necessary since light may be lost due to scatter, not reaching the detector (photon 1). The detector cannot differentiate between the loss due to absorption and that due to scatter. Therefore, in the modified Beer–Lambert Law a term, G, is introduced. This factor depends on the size of the detector and the geometry of the system. From these considerations, the modified Beer–Lambert Law is derived:

$$A = \varepsilon \times c \times d \times B + G.$$

Assuming constant B and G gives

$$\Delta A = \varepsilon \times \Delta c \times d \times B,$$

an equation frequently used for the assessment of concentration *changes.*

The terms B and G are "correction" terms accounting for scatter, which is assumed constant to allow for the determination of concentration changes. In principle, however, scatter by itself may also become a measurement parameter. In fact, there is evidence that light scatter corresponds to physiological processes in nervous tissue as outlined below.

III. Other Optical Parameters Relevant for Near-Infrared Studies

Usually, the energy (of the photon) which is taken up by an absorbing molecule (e.g., hemoglobin or water) is translated into thermal energy. However, depending on the molecule, after a certain delay emission of light at a longer wavelength may occur. When the delay is shorter than 10^{-8} s, the phenomenon is termed *fluorescence,* when it lies between 10^{-8} and 10^{-6} s it is termed *delayed fluorescence,* when it is longer than 10^{-6} s it is termed *phosphorescence.* Based on fluorescence and phosphorescence molecules can be detected at extremely low concentrations. Most fluorescent tracers are excited by visible light (see Chapter 1); however, tracers for near-infrared studies are being designed, thus permitting noninvasive fluorescence studies even in human subjects.

Another optical parameter represents a special case of light scattering. When light is scattered by moving particles the frequency slightly changes (Doppler shift). This frequency shift depends on the velocity of the moving particle, the direction of the movement, and the number of interactions with a moving particle. Thus the Doppler shift offers an opportunity to assess the movement of particles in tissue. This method has been implemented as laser-Doppler flowmetry (LDF) for the measurement of cerebral blood flow (Haberl *et al.,* 1989; Dirnagl *et al.,* 1989) on the brain surface and through thinned skull preparations in the animal. Near-infrared applications employing a correlation spectroscopy approach allow for noninvasive examinations in the animal (Cheung *et al.,* 2001).

IV. Technical Approaches for Near-Infrared Spectroscopy and Imaging

The systems which are available for near-infrared studies in humans differ with respect to the principle approach of data acquisition (time-resolved, continuous wave), the technical specifications (number of discrete wavelengths, continuous spectrum), and the number of "channels" used for data acquisition.

The simplest approach uses a *continuous wave* (cw) light source with either discrete wavelengths (typically between two and seven) (Cope and Delpy, 1988) or a light source emitting across the entire NIR spectrum (Matcher and Cooper, 1994; Heekeren *et al.,* 1999). The only optical parameter measured is attenuation. The light source may be a laser or a LED (for the discrete wavelength approach) or a simple halogen lamp (for the continuous spectrum approach). The advantage of the cw approach is its simplicity and flexibility; also a high signal-to-noise ratio is reached. A disadvantage is a strong contribution to the signal changes by superficial, extracerebral structures. The separation of deep and superficial layers can be approximated by multiple source-detector separations.

Another approach uses a pulsed light source typically with a pulse duration on the order of picoseconds (Nomura and Tamura, 1991) [*time resolved spectroscopy* (TRS)]. In addition to the assessment of total light intensity, this approach assesses the distribution of photon arrival times. Based on this additional parameter a multilayer depth resolution has been proposed (Steinbrink *et al.,* 2001). Also by assessment of the individual DPF the quantification of the concentration changes can be much improved.

Since time-resolved measurements require rather demanding technology, frequency-domain monitors have been developed to more easily assess the mean time of flight of photons. Instead of using a pulsed light source the intensity of the injected light is sinusoidally modulated at a high frequency (100–150 MHz). The reflected light will also show this modulation. The phase delay of the modulation is proportional to the mean time of flight [*phase-modulated spectroscopy* (PMS) or *frequency domain NIRS system*]. Measurements based on phase changes are more sensitive to deeper structures than intensity-based measurements (Hemelt and Kang, 1999); compared to a cw approach the depth resolution allows for a coarse differentiation between a superficial and a deeper layer, which may be sufficient to better detect truly cerebral changes. Furthermore, small changes in phase may be related to changes in light scattering (Gratton *et al.,* 1995).

All of the above-mentioned approaches may be performed over just a single site (one channel) or many sites. When a larger number of emitter–detector pairs is used, it is possible to apply various image algorithms (Heekeren *et al.,* 1999; Benaron *et al.,* 2000; Igawa *et al.,* 2001). An example for noninvasive functional optical imaging is given in Fig. 7 (Benaron *et al.,* 2000). The instrument used in this experiment was a TRS system which also allowed for some depth resolution of the optical signal. Figure 7 illustrates the spatial match of the optical signal (right part) to the *f*MRI signal (left part) in the same subject performing a motor (finger tapping) task.

V. Physiological Parameters of NIRS Measurements

Light which is reflected from or transmitted through nervous tissue is influenced by changes in optical properties of the illuminated tissue. This phenomenon was known long before modern imaging techniques had been developed (Hill and Keynes, 1949), and the generation of high-resolution functional maps of the brain is a cornerstone of invasive functional brain research by *optical intrinsic signals* (see Chapter 5) in animals (Grinvald, 1992) and human subjects (Haglund *et al.,* 1993; Cannestra *et al.,* 1998). To derive physiological parameters from such intrinsic optical signals, however, is a complicated task, since various physiological events which occur during brain activity are associated with different kinds of optical changes. Thus, in order to understand optical signals in terms of underlying physiological parameters, the different contributions to the intrinsic optical signals have to be disentangled (pioneered in the work of Malonek and Grinvald, 1996). In this section, we discuss which physiological parameters influence optical parameters in the tissue and how far the translation to noninvasive NIRS in the human is possible. An overview is given in Table 1.

Absorption changes in biological tissue are dominated by intrinsic chromophores. Based on their color (i.e., their absorption spectra) concentration changes of different compounds can be determined according to the above explained modified Beer–Lambert Law. For measurements assessing changes in optical properties only those endogenous chromophores which absorb light differentially depending on their functional state are relevant. For example, the absorption spectrum of hemoglobin depends on its oxygenation state, a fact well known in the visible part of the spectrum (venous vs arterial blood) and extending into the NIR (see Fig. 1). Another relevant chromophore is cytochrome *c* oxidase (Cyt-Ox), the terminal enzyme of oxidative phosphorylation whose absorption spectrum depends on its redox-state. Cyt-Ox measurements are possible with visible light and with light in the near-infrared wavelength range.

Table 1

Physical (optical) parameter	Physiological parameter measurable in optical methods on exposed brain surface	Physiological parameter measurable by transcranial near-infrared methods
Absorption	[Oxy-Hb] [Deoxy-Hb] [Total Hb]	[Oxy-Hb] [Deoxy-Hb] [Total Hb]
	Cyt-Ox redox state	Cyt-Ox redox state
Light scattering	Fast scattering signal Membrane potential	Fast scattering signal (?)
	Slow scattering signal Volume changes of cell and subcellular compartments	Slow scattering signal (?)
Fluorescence	Autofluorescence: NADH Exogenous fluorescent tracers (contrast agents)	Autofluorescence: No significant autofluorescence in NIR NIR fluorescent tracers (contrast agents)
Phosphorescence	Exogenous phosphorescent tracers (contrast agents)	Feasible in principal with near- infrared phosphorescent dyes
Doppler shift	Cerebral blood flow (laser-Doppler flowmetry), speckle imaging	Feasible transcranially in rodents (Cheung *et al.*, 2001), not clear whether feasible in humans

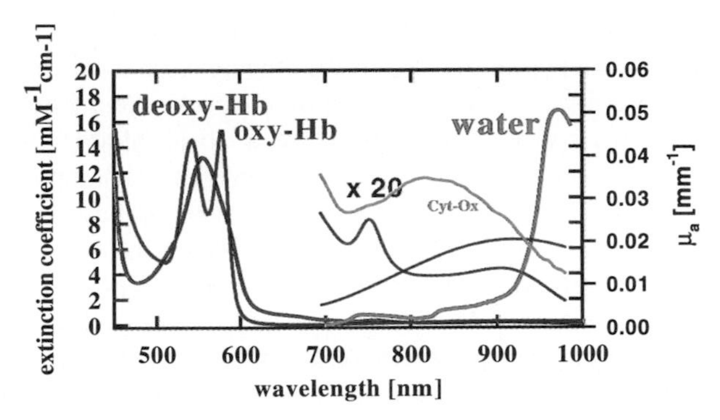

Figure 1 The extinction spectra of different biologically relevant chromophores from 450 to 1000 nm. Note that the global absorption between 650 and 950 nm is quite low compared to the high absorption of the hemoglobins below 650 nm and the high absorption of water beyond 950 nm. In between light absorption is low enough to allow penetration to a depth of some centimeters, thus making spectroscopic analysis of optical changes in the cerebral cortex feasible. The magnification (×20) of the oxy-Hb, deoxy-Hb, and Cyt-Ox spectra within this "spectroscopic window" illustrates the option to differentiate between them by their differential spectral properties.

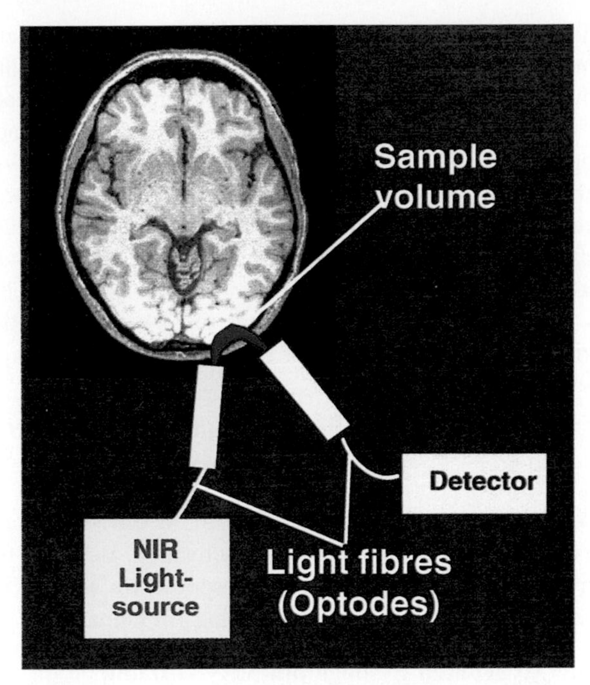

Figure 3 Sketch of NIRS setup in reflection mode. Light from the light source is guided to the head by a fiber-optic bundle, a so-called "optode." A second optode will collect the light which leaves the head at a distance of some centimeters. The probability that these photons have traveled a path in the banana-shaped sampling volume is much higher than the probability that the photons have traveled through deeper or more superficial layers of the head. Under the assumption of constant scatter the volume does not change over time and the changes in attenuation can be attributed to changes in chromophore concentration in this volume. The difficulty in exactly defining the shape and extent of the volume is reflected by the difficulty in quantifying the changes.

However, these measurements are not interchangeable, since measurements employing visible light mainly report on the Cyt-a/a3 redox state within Cyt-Ox, whereas measurements employing near-infrared light are dominated by another redox center of the enzyme, Cu_A. Many other absorbers are present in brain tissue, e.g., melanin or water; however, these absorbers show no oxygenation-dependent change in color and will not change absolute concentration over the course of the experiment. These chromophores limit penetration depth of the light into tissue by background absorption, but will not influence the changes in absorption due to functional activation or other cerebral processes.

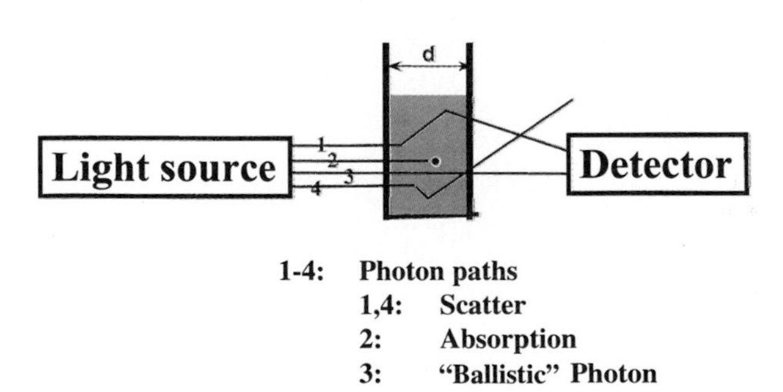

Figure 2 Cuvette model of the relation between scatter, absorption, and light attenuation. The different photons (1–4) demonstrate the different potential "fates" in a scattering and absorbing medium. Very few photons will directly reach the detector without scatter or absorption events (ballistic photon 3). Some photons will be absorbed by the chromophore (photon 2). Others will not reach the detector, since they are scattered out of the sampling volume (photon 4). Finally some photons will reach the detector but will have traveled a longer path than the geometrical distance (d) between light source and detector (photon 1). Assuming a constant scatter the ratio of photons reaching the detector is proportional to the number of type 2 photons. The fact of multiple scatter will introduce an enlargement of the sampling volume and the mean path length, accounted for by the differential path-length factor (DPF) in the modified Beer–Lambert Law.

Independent of concentration changes of light-absorbing molecules, brain activity is accompanied by changes in scatter. Changes in the scattering properties of the tissue have been described on two time scales. Fast changes in scatter are temporally linked to changes in membrane potential (Stepnoski *et al.,* 1991), while slow scatter changes probably reflect cellular or subcellular volume changes (edema) (see below).

The optical phenomena of fluorescence and phosphorescence can also be taken advantage of for optical studies. However, the only relevant endogenous fluorescent component is NADH, sensitive to UV light, which has a very low penetration depth into tissue. In the near infrared, no relevant autofluorescent is present in biological tissue. The main relevance of fluorescence lies in detecting contrast agents (*extrinsic signals;* see also Chapter 4). For such studies, fluorescent dyes are used which change their fluorescence depending on certain physiological/biochemical parameters. For example, dyes have been developed which are sensitive to ion concentrations (Ca, K, etc.), voltage across membranes, pH, etc. Linking a fluorescent tracer to antibodies allows for a mapping of receptors or other structures reacting with the respective antibody. Whereas so far most fluorescent dyes have been designed to be excited with visible light, recently, more and more dyes for the near-infrared wavelength range have been developed, potentially useful for noninvasive near-infrared studies (Bremer *et al.*, 2001). In the animal the phenomenon of phosphorescence has been employed for measuring oxygen concentration (Rumsey *et al.*, 1988; Lindauer *et al.*, 2001).

A. Assessment of Physiological Parameters and Applications

We will now describe the behavior of the different parameters in response to functional brain activation. Based on this characterization of a typical NIRS response we will discuss potential applications of the methodology in humans.

1. Oxygenated Hemoglobin (Oxy-Hb) and Deoxygenated Hemoglobin (Deoxy-Hb)

In studies on intact brain tissue (either on exposed cortex or noninvasively through the intact skull) concentration changes in oxy-Hb and deoxy-Hb dominate the optical changes evoked by brain activity. As outlined above, oxy-Hb and deoxy-Hb can be identified through their characteristic absorption spectra. The typical behavior of Δ[oxy-Hb] and Δ[deoxy-Hb] associated with brain activity consists of an increase in [oxy-Hb] and a decrease in [deoxy-Hb] with temporal kinetics similar to a parallel increase in local cerebral blood flow. These concentration changes reflect a transient hyperoxygenation caused by a focal blood flow increase larger than the increase in oxygen consumption (Fox *et al.*, 1988). This pattern has been observed on the exposed brain cortex (Frostig *et al.*, 1990; Malonek and Grinvald, 1996; Mayhew *et al.*, 2001; Lindauer *et al.*, 2001), but also noninvasively through the intact skull (Villringer *et al.*, 1993; Obrig *et al.*, 1996; Meek *et al.*, 1995; Colier *et al.*, 2001); an example is given in Fig. 4. In optical studies as well as *f*MRI studies, it has been reported that this hyperoxygenation is preceded by a transient hypo-oxygenation (Frostig *et al.*, 1990; Menon *et al.*, 1995; Malonek and Grinvald, 1996). However, it is still controversial whether this early hypo-oxygenation ("dip") is generally associated with brain activity or whether it occurs only rarely (Fransson *et al.*, 1998; Lindauer *et al.*, 2001), perhaps under specific experimental conditions such as hypoxia or hypocapnia (Kim and Dirnagl, preliminary results presented in personal communication).

2. Cytochrome *c* Oxidase Redox State

Another relevant chromophore changing its spectral absorbance in the NIR depending on its oxidation state is Cyt-Ox. The spectroscopic difference between its oxidized form and the reduced form can be used to detect metabolic changes in response to functional activation by noninvasive NIRS approaches (see Fig. 1). Studies on the exposed cortex

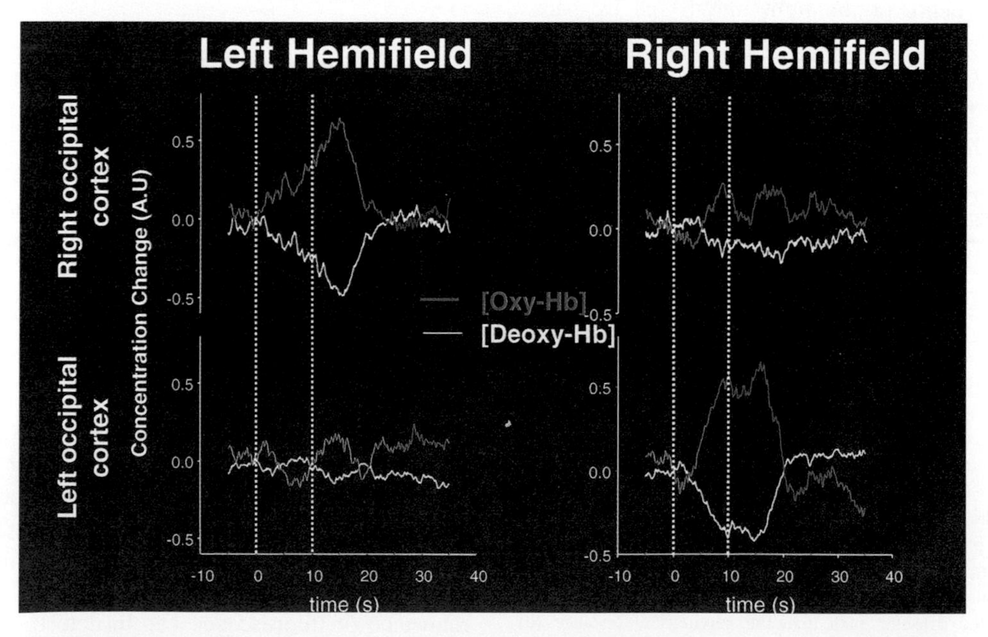

Figure 4 An example of a functional activation study with NIRS. The subjects underwent visual stimulation in either hemifield by a checkerboard paradigm for 10 s (stimulation period is marked by the broken lines). The two probe pairs were fixed over the right and left occipital regions (upper and lower traces). The results demonstrate that during stimulation of the contralateral hemisphere there is a "typical" NIRS response over the respective hemisphere consisting of an increase in [oxy-Hb] and a decrease in [deoxy-Hb]. The changes are not quantified but given in arbitrary units (AU).

of animals operating within the visible spectrum indicate that a transient oxidation of the heme a/a3 center occurs associated with increase brain activity (Lockwood *et al.*, 1984). This finding, however, has remained controversial. In the human we recently reported evidence for a similar change for the Cu_A center dominating the optical changes induced by Cyt-Ox in the NIR (Figs. 5a and 5b). During a checkerboard stimulation the absorption spectrum measured

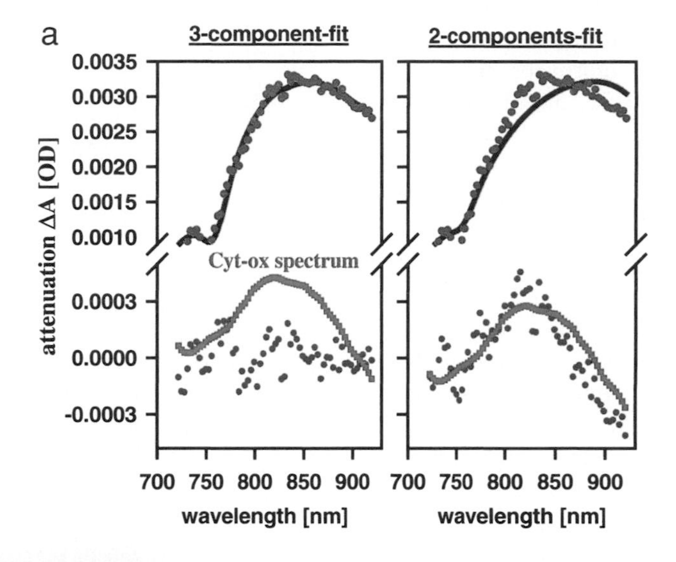

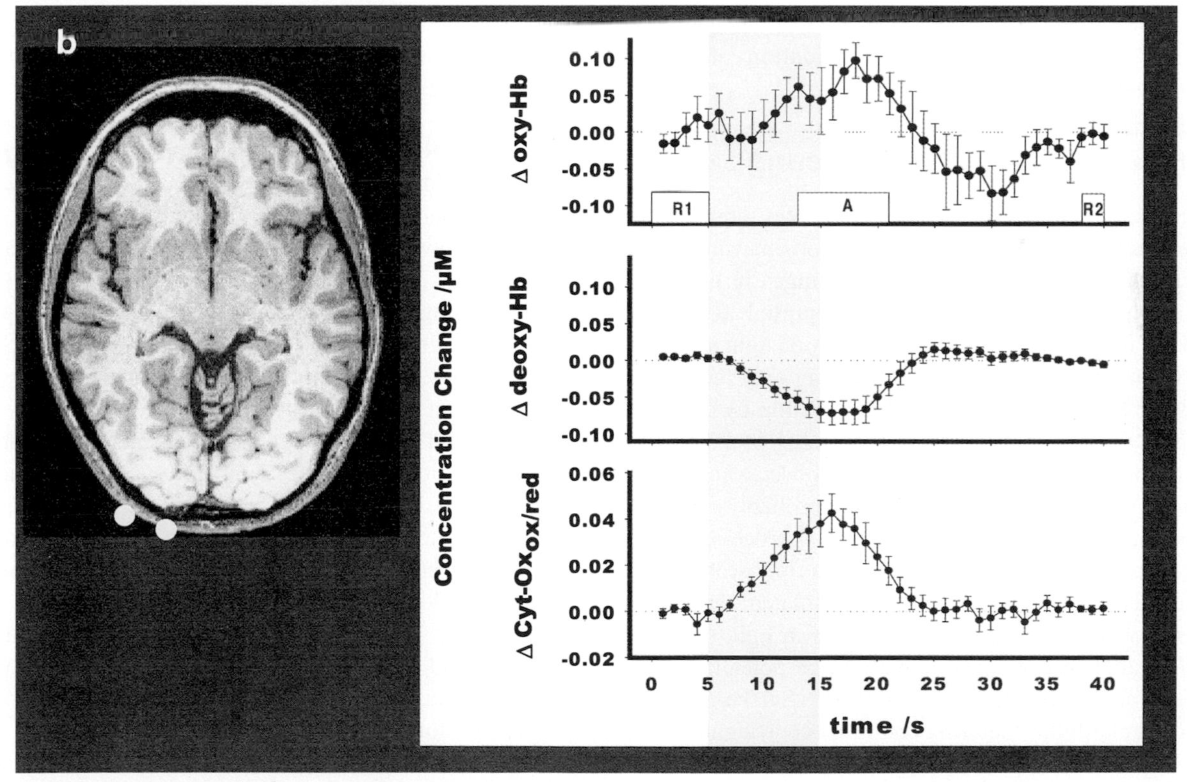

Figure 5 (**a**) A spectral analysis of a subtraction spectrum (spectrum "activation" minus spectrum "rest") is given (blue dots in upper row). The stimulation consisted of checkerboard stimulation, the measurement was performed over the occipital cortex [see (b)]. The black line in the upper right is the result of a two-component fitting procedure with the components "oxy-Hb" and "deoxy-Hb." Please note, the fitting seems not "perfect." The residuals of this fit are given below (red dots). These residuals seem not randomly distributed; rather the configuration seems to resemble the spectrum of Cyt-Ox in this wavelength range (green line). In the upper left, a three-component fit is given (black line) including Cyt-Ox together with oxy-Hb and deoxy-Hb. Note the improvement in the fit. (**b**) The changes in [oxy-Hb], [deoxy-Hb], and [Cyt-Ox] over the right occipital region in response to a checkerboard stimulation (10 s) using the three-component fit which is given in (a). The image on the left shows the positions of the optodes, the right-hand graphs show micromolar changes in the chromophores from a "baseline" set to 0. Upon stimulation an increased oxidation of Cyt-Ox is noted.

over the occiput of adult volunteers is altered. The changes, i.e., the difference spectrum between the resting state and the activated visual cortex, could not be explained by hemoglobin chromophores alone (Fig. 5a). Assuming changes only in oxy-Hb and deoxy-Hb the residuals of the fit showed the wavelength dependence of the absorption spectrum of Cyt-Ox. Including the latter into the fitting procedure (three-component fit) the fit significantly improved, and residuals showed near-random distribution (Fig. 5a). This does not prove a contribution of cytochrome oxidase redox shifts to the changes in optical properties in response to the stimulus; however, the opposite is also true for the simulations performed on a multilayer head model, demonstrating that the changes can theoretically be explained by cross talk between the parameters. The Monte Carlo simulation yielded erroneous Cyt-Ox changes on the order of magnitude that we find in response to visual stimulation (Uludag *et al.,* 2001). Currently we are examining differential cortical activation for stimuli which differ in their respective activation of blob (high Cyt-Ox concentration) and interblob (low Cyt-Ox concentration) areas in the visual cortex. The preliminary result is that some of the findings exclude the possibility of Cyt-Ox being a mere spectroscopic artifact. For example, when [oxy-Hb] and [deoxy-Hb] changes are of the same magnitude a differential oxidation of Cyt-Ox cannot be explained by cross talk.

B. Fast and Slow Light Scattering Signals

Whereas the above-mentioned molecules are detected as absorbers in cerebral tissue, other phenomena which occur during brain activation are related to changes in light scattering. Such scattering changes were reported as early as 1949 (Hill and Keynes, 1949). Two types of light scattering changes can be differentiated by their temporal characteristics: a slow and a fast scattering signal. While by definition these signals differ with respect to the timing of their occurrence (millisecond versus seconds after onset of nerve cell stimulation) and in their magnitude, they probably also differ with respect to the underlying physiological process. The *fast light scattering optical signals* occur on a time scale of milliseconds (Hill and Keynes, 1949). In isolated nerve cells, Stepnoski *et al.* have shown that changes in light scattering closely parallel the changes in membrane potential during action potentials (Stepnoski *et al.,* 1991). Salzberg's group reported similar light scattering changes occurring in the terminals of the neurohypophysis (Salzberg and Obaid, 1988). On the exposed cortex, it has been reported by Rector *et al.* that these signals can be detected (Rector *et al.,* 1997), and recently Rector and George presented images of the fast optical signals (Rector *et al.,* 2001). The translation to noninvasive studies in human adults has been pioneered by the group of Gabriele Gratton. Using a "frequency domain" near-infrared system, fast optical signals (EROS, or event-related optical signals) are reported to be detectable noninvasively through the intact scalp (Gratton *et al.,* 1995; Gratton and Fabiani, 2001) (see Fig. 6). Such an extension of noninvasive NIRS to studying electrophysiologically induced signals would be of supreme relevance to the investigation of neurovascular interactions and for functional imaging in general. However, so far no other group has successfully pursued this approach. Our group reported fast optical signals using an intensity-based near-infrared approach (Steinbrink *et al.,* 2000). The signal-to-noise ratio (SNR) in these studies is poor, preventing systematic exploration of the signals in larger subject groups.

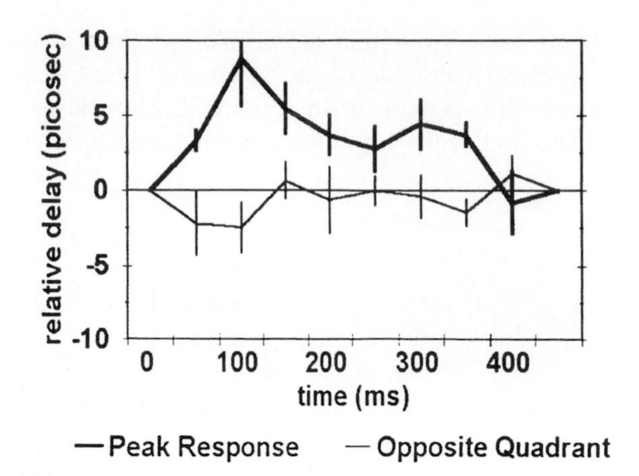

Figure 6 The so-termed EROS reported by the group of G. Gratton. Note that in response to the reversal of the visual stimulus (black/white grid) at 0 ms there is a substantial increase in mean time of flight (~10 ps), peaking at 100 ms, which is in agreement with the latency of the electrophysiologically well-established P100 peak in VEP recordings. This response is seen only when the quadrant of the visual field which projects on the cortical area underlying the probes is stimulated (solid vs thin line). Such fast optical signals have been reported in a number of studies by the group, but as yet have not been confirmed by other groups. (Adapted, with permission, from Gratton *et al.,* 1995.)

Slow light scattering signals have been reported in bloodless brain slices occurring on a time scale of seconds after the onset of stimulation (MacVicar and Hochman, 1991). Also, slow scattering signals have been observed in intact brain after separation from (the much stronger) hemoglobin signals (Malonek and Grinvald, 1996; Kohl *et al.,* 1998a). The precise origin of the signals is still unclear. Most favored candidates are changes in cell volume and/or extracellular volume, respectively (MacVicar and Hochman, 1991; Holthoff and Witte, 1996). However, cellular volume cannot be the sole determinant of this scattering signal since different types of brain stimulation, each leading to cell swelling/shrinkage of extracellular space, induced opposite changes in the intrinsic signal. It is thus likely that in addition to whole cell volume, other factors such as volume alterations of intracellular organelles are also important determinants of this slow scattering signal (Buchheim *et al.,* 1999; Haller *et al.,* 2001).

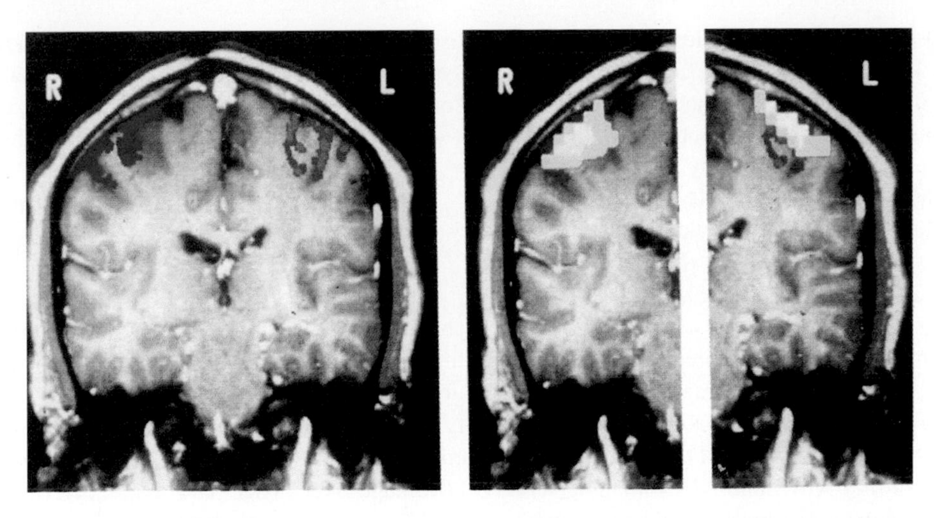

Figure 7 Response to a simple finger tapping task in a subject who was investigated by NIRS imaging and BOLD-contrast *f*MRI. The left presents the results of the *f*MRI study, showing the BOLD contrast in response to unilateral movement of the right (red pixels) and left (blue) hand. The right side shows the overlay of the *f*MRI data and the results from the optical imaging (yellow pixels for either side). Note the good spatial correlation between the two measurements. (Adapted, with permission, from Benaron *et al.*, 2000.)

VI. Near-Infrared Spectroscopy and Imaging: Applications

So far we have given an overview of the general potential of optical techniques in neurophysiological research. The core of the present chapter, however, is on noninvasive applications in humans. There is no doubt that the potential of optical methods may be limited by signal-to-noise ratios when a slab of ~1 cm highly scattering and absorbing tissue (i.e., the skull) is introduced, obscuring and distorting the image of the cerebral cortex. However, one should bear in mind that the first images presented in many techniques resembled blurred sketches of known locations of cortical activation. Our incentive to further noninvasive optical methods is based on the enormous wealth of physiological information which has been gathered from the invasive applications and the perspective of obtaining some of this information at the patient's bedside.

A. Functional Imaging of the Human Brain

From the above considerations it follows that near-infrared spectroscopy in principal allows for assessment/imaging of brain activity based on

- vascular signals (oxy-Hb, deoxy-Hb),
- metabolic signals (Cyt-Ox), and
- fast optical signals presumably related to electromagnetic neuronal activity.

1. Functional Imaging Based on Hemoglobin Oxygenation

As outlined above, *increased brain activity*[2] is associated with a local hyperoxygenation, i.e., an increase in [oxy-Hb] and a decrease in [deoxy-Hb]. Using NIRS, it has been shown that these changes can be measured noninvasively through the intact skull (Hoshi and Tamura, 1993; Kato *et al.*, 1993; Villringer *et al.*, 1993; Obrig *et al.*, 1996; Hock *et al.*, 1996). Taking advantage of this phenomenon, functional studies have been performed on a wide variety of different tasks and conditions, employing visual stimulation (Kato *et al.*, 1993; Meek *et al.*, 1995; Hock *et al.*, 1996; Colier *et al.*, 2001), motor tasks (Obrig *et al.*, 1996; Colier *et al.*, 2001), language paradigms (Sakatani *et al.*, 1998; Watanabe *et al.*, 1998), and other cognitive tasks (Villringer *et al.*, 1993; Hock *et al.*, 1995; Hoshi *et al.*, 2000). It has been demonstrated that the NIRS signal correlates well with CBF changes as derived from PET (Villringer and Chance,

[2]The term "increased brain activity" cannot be equated with "increased neuronal firing rate" or "increased excitatory activity" or any other electrophysiologically defined term. Rather this term is used similarly in PET and *f*MRI studies, which measure either metabolic or vascular responses to changes in brain activity. In these studies, any change in brain activity inducing an increase in local metabolism/blood flow is taken to indicate an "activation." It has been shown that in principle not only excitatory events associated with increased neuronal firing rate (Rees *et al.*, 2000; Heeger *et al.*, 2000), but also inhibition may be associated with increased metabolism and blood flow (Mathiesen *et al.*, 1998).

1997) and the BOLD signal in *f*MRI studies (Kleinschmidt *et al.*, 1996; Toronov *et al.*, 2001). Taking advantage of the flexibility of NIRS, studies in neonates and children have been performed (Isobe *et al.*, 2001; Hintz *et al.*, 2001; Bartocci *et al.*, 2000) and recently a study during walking has been presented (Miyai *et al.*, 2001). The bedside capability has been illustrated by studies in which hemoglobin oxygenation changes were monitored during epileptic seizures (Steinhoff *et al.*, 1996; Villringer *et al.*, 1994; von Pannwitz *et al.*, 1998).

A number of other studies have explored the feasibility of multiple-channel ("imaging," "topography") studies in human adults and babies (Benaron *et al.*, 2000; Hintz *et al.*, 2001; Isobe *et al.*, 2001; Miyai *et al.*, 2001). Based on these findings, we are positive that NIRS and NIRS imaging will claim a position for studies at the bedside and for clinical studies and those in neonates.

2. Functional Imaging Based on Cytochrome *c* Oxidase Redox State

Studies on the exposed cortex of rats have indicated that changes in brain activity are associated with alterations of the redox-state of cytochrome *c* oxidase (Lockwood *et al.*, 1984). In humans, there have also been attempts to assess the Cyt-Ox redox state noninvasively (Nollert *et al.*, 1995) and studies indicate that monitoring of the Cyt-Ox redox state may be a useful indicator of severe hypoxia (Nollert *et al.*, 1997), which may become relevant for brain monitoring in cardiac surgery. Such changes are mediated by insufficient oxygen supply, which cannot be assumed to occur under physiological conditions in the human. During functional brain activation, our group has provided evidence that a transient increase in Cyt-Ox oxidation occurs (Heekeren *et al.*, 1999). Such changes related rather to changes in mitochondrial membrane electrochemical potential shift provide a new means of assessment of brain activity. It should be noted that the issue of Cyt-Ox assessment is still controversial due to potential cross talk of the much larger hemoglobin signal, as we and others have emphasized (Sakamoto *et al.*, 2001; Uludag *et al.*, 2001). Currently it is our goal to define response patterns of all three chromophores which can be judged "real" in analogy to our previous work establishing a decrease in [deoxy-Hb] and an increase in [oxy-Hb] as the typical response pattern over an activated area.

3. Functional Imaging of Fast Light Scattering Events Associated with Brain Activity

As outlined above, a fast light scattering signal exists which traces activity of nervous tissue on a time scale of milliseconds. Using a PMS system, Gratton's group has reported a large number of studies measuring these fast optical signals (Gratton *et al.*, 1995; for an overview see Gratton and Fabiani, 2001). So far, other groups have not implemented these measurements. Our group has recently used a different approach employing an intensity-based system and showing a fast optical signal associated with somatosensory stimulation. So far, SNR remains poor and we have not seen a similar signal with other modalities.

B. Studies on Neurovascular Coupling in Health and Disease

As mentioned above, the typical pattern of qualitative changes associated with brain activity consists of an increase in [oxy-Hb] and a decrease in [deoxy-Hb]. This differentiation of oxy-Hb and deoxy-Hb and the possibility of quantifying the observed concentration changes (especially when employing TRS and PMS–NIRS approaches) is a major advantage over *f*MRI. The major determinant of the *f*MRI BOLD signal is also deoxy-Hb; however, based on *f*MRI measurements it seems not possible to draw any quantitative conclusions. Using NIRS, we and others have addressed issues of neurovascular coupling. We have shown that during "deactivation" of occipital regions associated with acoustically cued saccades known to decrease occipital cerebral blood flow (Paus *et al.*, 1995), [oxy-Hb] decreases and [deoxy-Hb] increases; i.e., it can be followed that during a transient decrease in brain activity the drop in blood flow exceeds the drop in oxygen consumption (Wenzel *et al.*, 2000). We have also studied issues of linearity of the response in [deoxy-Hb] and [oxy-Hb] to different durations of brain activity. Our results indicate a linear behavior when the duration of the stimulation exceeds several seconds (Wobst *et al.*, 2001); however, at shorter stimulation periods strong nonlinear effects come into play. Of course, in order to study neurovascular coupling, it is important to have a measure of both the neuronal activity *and* the vascular response. Here, optical methods provide two interesting approaches. First, the simplicity and flexibility of optical studies makes a combination with electrophysiological approaches such as EEG and MEG relatively simple (Onoe *et al.*, 1991; Igawa *et al.*, 2001), thus allowing for simultaneous acquisition of event-related (electromagnetic) potentials and vascular response (Obrig *et al.*). This approach allows one to study neurovascular coupling, in particular in longitudinal examinations. For example, it is controversial whether neurovascular coupling (in particular the interactions between changes in local blood flow/velocity/ volume versus oxygen consumption versus neuronal activity) is affected during prolonged neuronal activity. Using the combined EEG/NIRS approach during a prolonged visual stimulation we could demonstrate that the vascular response seems to behave in a rather linear relationship to the neuronal activity.

Another interesting combination is NIRS with magnetoencephalography. MEG offers the opportunity to measure slow DC potentials (Mackert *et al.*, 2001). The simultaneous measurement of changes in [oxy-Hb] and [deoxy-Hb] with

NIRS demonstrated a close relationship between the vascular changes and the magnetoelectrical DC signals (preliminary results). The temporal similarity is striking with respect to the vascular response lag of about 6 s. These data indicate that among the different aspects/correlates of neuronal activity which can be assessed with noninvasive methods, DC activity may be most closely related to the vascular response.

As mentioned above, the optical signal itself also contains a fast component which seems to be related rather directly to neuronal activity in a fashion similar to electrophysiological methods. The group of Gratton has put forward data which could allow for noninvasive measurement of neurovascular coupling just by the optical approach (Gratton *et al.,* 2001). Future studies will challenge the feasibility of this approach (e.g., issues of SNR and artifacts) and it must be investigated to see whether it excels the combination of NIRS with EEG/MEG.

Studies on neurovascular coupling are probably of particular relevance in pathological situations. A number of studies indicate that neurovascular coupling may be altered in several kinds of disorders of the nervous system. After global ischemia, there may be a dissociation between evoked neuronal activity, metabolism, and blood flow response (Schmitz *et al.,* 1998). In patients with carotid stenosis an inverse vascular response has been reported during functional brain activation (Röther *et al.,* 2002); a similar observation has been made in subjects with Alzheimer's disease (Hock *et al.,* 1997) (both studies, however, without a measure of neuronal activity). In patients with epilepsy, a pathological alteration of neurovascular coupling has been suggested interictally as well as during epileptic seizures. In these patients, neuroimaging studies which rely on the vascular response (*f*MRI, PET) as a "reporter of brain activity" can be correctly interpreted only when the alteration of neurovascular coupling is known. One approach may be the definition of a "coupling index," which establishes some kind of quantitative relationship between a parameter of neuronal activity and a parameter of the vascular response (Obrig *et al.*) which may be corrected for in the interpretation of the vascular studies; another approach may be to simultaneously measure an aspect of neuronal activity.

Knowledge of a pathological alteration of neurovascular coupling will be essential for the interpretation of results in functional neuroimaging methods when applied to pathology. In addition, disturbed coupling may play a pathophysiological role in certain neurological disorders. In the penumbra of ischemic stroke, disturbed vascular response to neuronal activity leads to the phenomenon of peri-infarct depolarization, which has been shown to induce a drop in hemoglobin oxygenation using NIRS in an animal model. Similarly, it has been suggested by studies of Jens Dreier that a disturbance of neurovascular coupling may be the pathophysiological mechanism behind delayed neuronal damage after subarachnoid hemorrhage (Dreier *et al.,* 2000). In epilepsy it is known that epileptic seizures may induce secondary brain damage. It is tempting to speculate that inadequate neurovascular coupling not meeting the excessive demand during a seizure may be an indicator of subsequent secondary brain damage. The clinical examples given above emphasize the need for measuring neurovascular coupling in clinical settings at the patient's bedside. The flexibility of noninvasive optical methods seems to meet this clinical need particularly well.

C. Other Applications of Near-Infrared Spectroscopy and Imaging

A number of other applications have been suggested for near-infrared spectroscopy and imaging. NIRS has been successfully used for perioperative monitoring during cardiac surgery, and interestingly, the Cyt-Ox signal seemed to be most useful in predicting poor "cerebral" outcome after the surgery (Nollert *et al.,* 1997). Other groups have suggested using NIRS during carotid surgery (Kirkpatrick *et al.,* 1998). One group has used a NIRS system for the detection of intracranial hemorrhage with encouraging results (Gopinath *et al.,* 1995). Monitoring hemoglobin saturation (Quaresima *et al.,* 2000) in patients in the intensive care unit seems another quite likely application for the method.

Whereas above we have outlined what kind of studies one may choose to pursue with NIRS, we will now discuss practical issues concerning NIRS measurements.

VII. Practical Aspects of NIRS Measurements

NIRS is a rather undemanding technique and can be easily applied at the bedside. Also it allows for coregistration with a number of other functional methods (*f*MRI, PET, EEG, MEG), with which there is little or no interference concerning the simultaneous acquisition of data. Nevertheless the design and execution of a NIRS study should take into account some basic considerations, which will allow for an easier judgment of the results acquired.

A. Study Environment

Apart from the general requirements so as to not distract the subject's attention, NIRS measurement should be preferably performed in a dimmed room and the optical probes may need extra light shielding to reduce ambient light. This requirement depends on the NIRS system used since some of the imaging systems apply an "encoding" of the different sources by modulating the incident light with a specific frequency.[3] Thereby the detector can differentiate light that stems from the source from ambient light, which does not exhibit this frequency. However, any light detector has a maximal capacity, which is optimally exploited when the reflected light predominantly originates from the light source. Changes in illumination should be avoided, and though the bedside measurement is technically possible, ambulatory patients should definitely be examined in a special room, the ward setting introducing unforeseen environmental artifacts.

The subject's positioning during the experiment is critical to reduce movement artifacts; like in many a functional method these are a major reason for offline data rejection. More specifically two different kinds of movement artifacts must be considered in NIRS. The movement of the subject may be tolerated as long as it does not critically interfere with the experimental procedure. In fact, as has been shown by a recent study on gait, NIRS does have the potential to monitor cerebral function during tasks difficult to study in the rather limited space of the magnet bore or scanner. In addition to the rather general origin of artifacts, the movement of the probes will introduce changes in reflectance, which cannot easily be rejected by the data analysis. If the probe is not firmly fixed to the head, a small movement may increase or decrease the proportion of light reaching the detector. But even if the probe is perfectly attached to the skin, a movement may lead to sheering forces thereby altering the interoptode distance. This, as has been demonstrated above, violates an essential assumption of the constant path length of the photons during the course of the measurement. Though displacement may be small one should consider the fact that a change in interoptode spacing of 1 cm alters the magnitude of the reflected light by roughly one order of magnitude. We conclude that a perfect fixation of the probes on the subject's head is the prerequisite of a good NIRS measurement. Trivial as the issue may seem, we stress this point since to our own experience such a perfect fixation is not quite as trivial to achieve and has as yet limited some of the interpretation of measurements in patients who may cooperate less well in part as a result of their illness (ICU patients, acute stroke patients, and most notably epileptic patients in the course of a seizure).

Unfortunately the optimum of a firm attachment of the probes may not coincide with the subject's optimal comfort. Again there is the general notion to not extend measurement time beyond a convenient duration. With more specific regard to NIRS, pressure on the skull will initially reduce skin blood flow, generally desirable, since extracranial contributions to absorption and systemic changes in blood pressure are judged a source of signal contamination. Longer lasting pressure, however, will lead to focal hyperemia, and tight bandages around the head may cause headache, which will induce both meningeal and systemic alterations of blood flow. The trade-off between convenience and probe stability should not be disregarded and sufficient care for this starting point of the measurement is advisory.

[3]In cw monitors this encoding frequency is not used to assess the change in phase of the reflected compared to the incident light. However, frequency domain monitors by nature of their operating with a modulated light source usually take advantage of this principle.

B. Probe Localization

Unless an imaging device is used the issue of probe localization will rely on skull rather than cortical topography. We generally apply the somewhat coarse topography of the 10–20 system. It is justified to take the midpoint between the two probes as the area which will most strongly contribute to the changes measured. However, as has been demonstrated by a layered head model, the assumption of a semilunar sampling volume, selectively "scratching" the cortical surface at the center of the pair of probes, is a gross oversimplification. The CSF must be considered to act like a light channel, thus extending the sampling volume in latitude and limiting its penetration depth. There are more demanding alternatives to the 10–20 system. Individual anatomical reference may be provided by CT or MRI, which is often available when neurological patients are examined; however, 3D reconstruction of the slices at the bedside may overcharge the examiner's skills. If an MRI scan is performed after the NIRS measurement one should not miss the option of marking the probe position by lipid-containing capsules (e.g., vitamin E), which allow for a post hoc correlation to individual cortical anatomy. More elaborate techniques with projection of the alleged sampling volume on a 3D reconstruction are appealing, though the pseudo-exactness reached will not necessarily solve (maybe improve, however) the question as to the exact biophysical definition of the volume sampled. Thus visualization of such kind may be reserved for well-controlled scientific studies.[4] Applying an imaging device reduces the problem of localization and focal changes can be judged as a physiological check for their intracerebral origin, since focal changes in skin blood flow in response to a stimulus do not seem plausible. Nonetheless, spatial resolution with respect to cortical

[4]The question of how much the use of demanding techniques to visualize results enhances the understanding of functional data may be doubted on a more general level. One should not forget that even the "classical" superposition of functional MRI measurements on the anatomical scan taken at the beginning of a measurement usually implies interpolating pixels and reaching an exactness of spatial resolution that sometimes demonstrates the physiology of the software much more than that of the subject studied.

topography should be considered rough, which is essential for the design of the experimental protocol.

C. Interoptode Distance (IOD)

The distance between the light emitter and the detector has been an issue of some debate. Generally by increasing interoptode distance the contribution of deeper layers increases. Noise, however, also increases due to a logarithmic decay in intensity with distance. The concept of a differential path-length factor is justified as long as IODs greater than 2.5 cm are used. Below this distance a clearly nonlinear behavior of the DPF has been demonstrated. IODs greater than 5 cm usually result in very low light intensities measured. Depending on the study's goal, this may be partly compensated for by a lower sampling rate. Generally, if the study design offers no or few options to check for the truly cortical origin of the signal changes a larger IOD and an area of the head with a high light penetration should be chosen (e.g., the front beyond the hairline). We typically use IODs of 3–4 cm over the hairy areas of the head, sometimes reducing the distance in subjects with dark skin or hair. It should be noted, though, that due to interindividual variance, and even by nature of the variable thickness of the extracerebral layers within a subject, there is no IOD which can be regarded as a safeguard for the truly cerebral origin of the oxygenation changes monitored. Again the experimental design should include some kind of control task or control measurement, which will help to differentiate between generalized or extracerebral changes and the cortical changes aimed at.

D. Sampling Rate

Technically NIRS has an exquisite temporal resolution, theoretically limited but by the velocity of light. The choice of the sampling rate thereby essentially depends on the estimation of a signal-to-noise ratio appropriate to the study's aim. We see no reason to miss potential temporal information by reducing sampling frequency, unless the system used applies inappropriately slow software, losing data during the storing process. The SNR dependent on the yield of light should rather be adjusted by offline analysis. However, physiological sources of "noise" must be considered. The first commercial monitors had maximal sampling frequencies around 1 Hz. The proximity to the heartbeat is unfortunate, producing potential aliasing in the data. Similar considerations apply to frequencies around the respiration rate at ~0.2 Hz and the spontaneous slow oscillations of 0.04–0.14 Hz, the latter interfering with (causing?) the lag in vascular response. Unless long procedures, such as surgery or long-term monitoring of ICU patients, are monitored we recommend sampling rates of 4–10 Hz. Thus heartbeat can be easily filtered, and potentially a procedure to assess individual DPF can be applied (Kohl *et al.*, 1998b).

E. Coregistration of Basic Physiological Parameters

By and large this option has been greatly neglected in the design of NIRS monitors. This is all the more surprising, since the routine implementation of a rough measure of skin blood flow seems inexpensive and will be a prime aide to differentiate between extra- and intracerebral changes monitored. A simple optical probe operating at two wavelengths will provide a reference of cranial but (rarely, N.B.) cerebral changes in blood oxygenation. External pulse oxymeters are the more refined option, sometimes complicating the data coregistration and unfortunately often showing a lag between the measurements, due to the devices being mostly designed for clinical use.[5] The best option to measure skin blood flow is an LDF device. The sampling volume may reach only the upper layers of the skin, but trends rather than magnitude will be helpful for offline detrending of the NIRS data. Electrocardiograms can be monitored for heart rate, but the additional information is of little use as long as the NIRS sampling rate is high enough. If the heart rate cannot be easily seen in the NIRS data (mostly in the oxygenated hemoglobin trace) this should raise skepticism as to whether the NIRS monitor is operating at all. Blood pressure measurements are desirable since pressure may be raised by complex tasks. The resulting increase in blood flow will induce task-related changes in the NIRS parameters, which cannot be differentiated from the truly cortical contribution, unless additional information on topography or depth resolution is available. Conventional blood pressure assessment with an arm cuff is of little help, since sampling intervals are on a time scale far beyond the NIRS sampling rate. There are alternative monitors, which provide continuous noninvasive assessment. The fact that the absolute values of these monitors are unreliable is of little relevance for our intent. Again the trend in systemic blood pressure is relevant for the correction of the NIRS results. Unfortunately this weakness of the continuous blood pressure monitors ended the production and distribution of such devices. Scientific, prototype kinds of devices can be acquired, though, or if one is lucky, one's institution may hold one of the older models.

[5] In addition to computational lags often difficult to deduce from the company's specifications, clinically used monitors have a tendency to "hatch" their data. Monitoring an ICU patient makes many a physiological parameter invasively monitored available. However, by the iron law of patient safety the connection to an external computer is often effectively prevented by the company's policy.

F. Experimental Protocols

It is our experience that the very general advice to maximally reduce complexity of the protocols, which will much enhance the power of the results deduced, should head an article on NIRS protocols. But there may be an even earlier discussion as to whether a NIRS study will be adequate to answer the issue in question. Of course, there is no valid rule which will help to definitely answer the methodological choice. Here are some of our present concepts of what we consider a question that can be appropriately answered by NIRS. The authors feel confident that these guidelines will loosen with the advance of NIRS technology and it is our hope that a number of the "rules" will turn out to be conservative by the evidence of studies challenging the limitations of the method.

1. NIRS Has an Excellent Temporal Resolution

This strength of the method is obvious compared to PET but also to *f*MRI, though the latter is steadily increasing its maximal sampling frequency. When the vascular response to a functional task is investigated the "effective" temporal resolution will primarily result from the latency of the vascular response. However, there are a number of open questions as to an early increase in [deoxy-Hb], as to successive activation in different cortical areas, or as to spontaneous oscillatory phenomena of cerebral blood flow and metabolism, which will definitely profit from high temporal resolution. In such studies it is of help to define a paradigm which is known to activate regions of the cerebral cortex, so as to reduce the effect of erroneous localization of the NIRS probes. By defining two comparatively demanding tasks hypothesized to differentially activate the area under investigation the time courses can be reliably compared by the method. Resting periods between the different tasks are somewhat mandatory in most block designs, though the brain imaging community has taken a strange kind of liking to the ill-defined state of the brain at rest, a state rarely met by the investigators. For single-trial designs a jitter of the interstimulus intervals and exact temporal monitoring of the execution of the task are recommended, much like studies using event-related BOLD-contrast *f*MRI. The undemanding experimental environment of NIRS examination is lending itself to these requirements. Interindividual comparison of different response latencies needs more powerful statistical approaches, considering that suboptimal localization and individual opaqueness of the subject's extracerebral layers influence the SNR. The analysis of response latency may strongly suffer from such experimental bias, when groups of patients are compared to "healthy" controls. Here qualitative rather than quantitative differences are to be looked for. This may apply for many functional techniques, but the power of the results of those techniques with a fuzzy spatial resolution might profit when we remember that the increase in noise induced by poor localization of the area activated may well lead to a bias of the magnitude and latency of a response. To sum up, questions as to the temporal characteristics of a vascular response seem adequate for the technology, as long as there is a rather solid knowledge as to where the response is expected topographically.

2. NIRS Has a Poor Spatial Resolution

This point is trivial when a single probe pair is used; however, the imaging systems provide but a blurry image of the brain's surface. Any structure beyond the cortex is invisible to the method, which may be termed an advantage compared to electrophysiological recordings. The method may be helpful to coarsely localize changes in oxygenation. This may in fact be sufficient, when spontaneous cortical processes are to be investigated. For example, a patient with seizures will not spend a day in the magnet but telemetry may be easily supplemented by a NIRS recording supplying information on the hemisphere hosting the epileptic focus. This can in turn help to better interpret the EEG recordings. Concerning physiological questions the study design may more reliably provide a check for localization than the attempt to unambiguously position the probes in the "right" position. A priori knowledge of the area activated is necessary. For example, a study on the motor response to a simple tapping task should include a "control task" presumably activating the motor cortex contralateral to the probe position. Thereby the response to the two tasks can be compared and a rough check as to the contribution of blood flow changes not related to the focal activation is possible. If imaging systems are used, the topographical information is mostly sufficient since a focal response cannot be attributed to extracerebral or systemic changes in hemodynamics. As yet the potential to differentiate between two foci of altered regional cerebral blood flow has not been thoroughly investigated. There are a number of factors contributing to the spatial resolution, which will potentially be reached by NIRS. The biophysical basis of the method is the highly scattering medium in which changes are assessed. At a depth of a few millimeters the ray of photons rather resembles a cloud, with a stochastic distribution of the direction of flight. The ability to localize an object in a highly scattering medium has been investigated in fish-tank models and methodologies can be compared by such an approach. *In vivo,* the knowledge of the exact optical properties (μa and μs) of the tissue are rarely known and will differ interindividually, thereby fundamentally limiting the method's spatial resolution. Other factors depend on more or less technical aspects of the imaging system. The number of probe pairs, their distribution, and the interprobe spacing define the data set on which an image is based. More complex imaging algorithms have been proposed. They rely on image reconstruction algorithms which respect that the light sampled by a detector stems from sources with different distances. For application in the adult head, as yet the SNR has limited such approaches. With a typical IOD of 3–4 cm the one but next source will not significantly contribute photons to a sampling probe in a distance of 6–8 cm. Applying time-resolved techniques for imaging and optimizing detector sensitivity the spatial resolution may well be increased by future systems, based on such image reconstruction algorithms. To sum up, spatial resolution is the major shortcoming of the method. This limits its power to answer questions of functional anatomy of the cerebral cortex in the adult. When there is some a priori knowledge as to the distribution of the areas under investigation, activation may be differentiated when the distance between the foci is on the order of some centimeters.

3. NIRS Has a High Parameter Specificity

The straightforward Lambert–Beer approach to assess concentration changes by NIRS has the advantage of reliably differentiating between oxy-Hb and deoxy-Hb changes. This is of special importance since deoxy-Hb changes are also the basis for BOLD-contrast *f*MRI. BOLD contrast is, however, susceptible to a number of other vascular parameters such as blood volume, and aspects of linearity are the issue of some controversy in the literature. Here NIRS offers the unique option to independently assess changes in cerebral oxygenation, and combined approaches have been demonstrated to not impose major demands on the study design (Kleinschmidt *et al.*, 1996). The strength of NIRS to contribute to the physiology of BOLD-contrast *f*MRI has been demonstrated and a number of issues are certainly a future avenue for NIRS's place in functional imaging. As soon as compounds of smaller concentrations are included there is the possibility of spectroscopic cross talk. This limits the reliability of the changes assessed in the redox state of cytochrome *c* oxidase. Currently the parameter must be considered "risky"; on the other hand the option of noninvasively monitoring cellular metabolism opens a perspective, especially when pathological conditions are to be studied. At present, a study design directed toward issues concerning the redox state of Cyt-Ox should respect that the analysis should include some check for the possibility of cross talk. While a definite exclusion of cross talk in advance is not possible, there is the option to apply different algorithms for the analysis, which are based on either assumption, i.e., assuming or not a contribution to the spectral changes assessed. The prerequisite for such analyses is high spectral resolution of the NIRS approach. The full spectrum approach is desirable, though algorithms with four and more discrete wavelengths may be analyzed similarly and the authors are confident that response patterns can be defined which will help to discern erroneous from real changes in the enzyme's redox-state. To sum up for the vascular parameters, oxy-Hb and deoxy-Hb, the method offers a high specificity as to the underlying compound. For cytochrome *c* oxidase redox changes, the analysis must consider the possibility of cross talk.

VIII. Problems and Perspectives

This book covers the methods for functional brain imaging. Assuming that all neuroimaging approaches are available and one could freely choose one's instrument, where is the role of near-infrared spectroscopy and imaging? Table 2 gives an overview of advantages and disadvantages of NIR methods and indicates most likely applications of the method.

The most significant disadvantages are the relatively poor spatial resolution (on the order of centimeters) and the limited depth penetration (probably mostly confined to the cortex). The most important advantages are the biochemical specificity of the signals and the flexibility of the approach.

Table 2

Advantage	Disadvantage	Research application	Potential clinical application
Good temporal resolution	Poor spatial resolution	Neurovascular coupling (linearity, individual determination of coupling index? etc.)	Assessment of disturbed neurovascular coupling in neurological disorders
Biochemically defined measurement parameters	Poor depth penetration	Functional imaging in babies, children	Monitoring of cerebral oxygenation in comatose patients in intensive care unit
Assessment of intra vascular and intracellular metabolic events and neuronal activity	Susceptibility to artifacts	Functional imaging in "natural" situation (walking, standing, etc.)	Monitoring of tissue at risk during surgery or after stroke (ischemic penumbra)
High flexibility (bedside examination)		Add-on information to *f*MRI and PET studies (Cyt-Ox, fast signals)	Detection of hematoma (epidural, subdural)
Extremely good sensitivity to small concentrations		Molecular imaging in the mouse brain	Molecular imaging in neurological disorders

Furthermore, the ability to detect extremely small substance concentrations should give the perspective of an alternative to PET. In rodent studies, cross-sectional optical CT for detection of fluorescent tracers (Ntziachristos and Weissleder, 2001) is feasible; in humans, it is questionable whether information about the entire brain can be obtained; however, the cortex should also be accessible to molecular studies.

References

Bartocci, M., Winberg, J., Ruggiero, C., Bergqvist, L. L., Serra, G., and Lagercrantz, H. (2000). Activation of olfactory cortex in newborn infants after odor stimulation: A functional near-infrared spectroscopy study. *Pediatr. Res.* **48,** 18–23.

Benaron, D. A., Hintz, S. R., Villringer, A., Boas, D., Kleinschmidt, A., Frahm, J., Hirth, C., Obrig, H., van, H. J., Kermit, E. L., Cheong, W. F., and Stevenson, D. K. (2000). Noninvasive functional imaging of human brain using light. *J. Cereb. Blood Flow Metab.* **20,** 469–477.

Bremer, C., Tung, C. H., and Weissleder, R. (2001). In vivo molecular target assessment of matrix metalloproteinase inhibition. *Nat. Med.* **7,** 743–748.

Buchheim, K., Schuchmann, S., Siegmund, H., Gabriel, H. J., Heinemann, U., and Meierkord, H. (1999). Intrinsic optical signal measurements reveal characteristic features during different forms of spontaneous neuronal hyperactivity associated with ECS shrinkage in vitro. *Eur. J. Neurosci.* **11,** 1877–1882.

Cannestra, A. F., Black, K. L., Martin, N. A., Cloughesy, T., Burton, J. S., Rubinstein, E., Woods, R. P., and Toga, A. W. (1998). Topographical and temporal specificity of human intraoperative optical intrinsic signals. *NeuroReport* **9,** 2557–2563.

Cheung, C., Culver, J. P., Takahashi, K., Greenberg, J. H., and Yodh, A. G. (2001). In vivo cerebrovascular measurement combining diffuse near-infrared absorption and correlation spectroscopies. *Phys. Med. Biol.* **46,** 2053–2065.

Colier, W. N., Quaresima, V., Wenzel, R., van der Sluijs, M. C., Oeseburg, B., Ferrari, M., and Villringer, A. (2001). Simultaneous near-infrared spectroscopy monitoring of left and right occipital areas reveals contra-lateral hemodynamic changes upon hemi-field paradigm. *Vision Res.* **41,** 97–102.

Cope, M., and Delpy, D. T. (1988). System for long-term measurement of cerebral blood and tissue oxygenation on newborn infants by near infrared transillumination. *Med. Biol. Eng. Comput.* **26,** 289–294.

Dirnagl, U., Kaplan, B., Jacewicz, M., and Pulsinelli, W. (1989). Continuous measurement of cerebral cortical blood flow by laser-Doppler flowmetry in a rat stroke model. *J. Cereb. Blood Flow Metab.* **9,** 589–596.

Dirnagl, U., Villringer, A., Gebhardt, R., Haberl, R. L., Schmiedek, P., and Einhaupl, K. M. (1991). Three-dimensional reconstruction of the rat brain cortical microcirculation in vivo. *J. Cereb. Blood Flow Metab.* **11,** 353–360.

Dreier, J. P., Ebert, N., Priller, J., Megow, D., Lindauer, U., Klee, R., Reuter, U., Imai, Y., Einhaupl, K. M., Victorov, I., and Dirnagl, U. (2000). Products of hemolysis in the subarachnoid space inducing spreading ischemia in the cortex and focal necrosis in rats: A model for delayed ischemic neurological deficits after subarachnoid hemorrhage? *J. Neurosurg.* **93,** 658–666.

Drexler, W., Morgner, U., Ghanta, R. K., Kartner, F. X., Schuman, J. S., and Fujimoto, J. G. (2001). Ultrahigh-resolution ophthalmic optical coherence tomography. *Nat. Med.* **7,** 502–507.

Duncan, A., Meek, J. H., Clemence, M., Elwell, C. E., Tyszczuk, L., Cope, M., and Delpy, D. T. (1995). Optical pathlength measurements on adult head, calf and forearm and the head of the newborn infant using phase resolved optical spectroscopy. *Phys. Med. Biol.* **40,** 295–304.

Firbank, M., Okada E., and Deply, D. T. (1998). A theoretical study of the signal contribution of regions of the adult head to near-infrared spectroscopy studies of visual evoked responses. *Neuroimage* **8,** 69–78.

Fox, P. T., Raichle, M. E., Mintun, M. A., and Dence, C. (1988). Nonoxidative glucose consumption during focal physiologic neural activity. *Science* **241,** 462–464.

Franceschini, M. A., Moesta, K. T., Fantini, S., Gaida, G., Gratton, E., Jess, H., Mantulin, W. W., Seeber, M., Schlag, P. M., and Kaschke, M. (1997). Frequency-domain techniques enhance optical mammography: Initial clinical results. *Proc. Natl. Acad. Sci. USA* **94,** 6468–6473.

Fransson, P., Kruger, G., Merboldt, K. D., and Frahm, J. (1998). Temporal characteristics of oxygenation-sensitive MRI responses to visual activation in humans. *Magn. Reson. Med.* **39,** 912–919.

Frostig, R. D., Lieke, E. E., Ts'o, D. Y., and Grinvald, A. (1990). Cortical functional architecture and local coupling between neuronal activity and the microcirculation revealed by in vivo high-resolution optical imaging of intrinsic signals. *Proc. Natl. Acad. Sci. USA* **87,** 6082–6086.

Gopinath, S. P., Robertson, C. S., Contant, C. F., Narayan, R. K., Grossman, R. G., and Chance, B. (1995). Early detection of delayed traumatic intracranial hematomas using near-infrared spectroscopy. *J. Neurosurg.* **83,** 438–444.

Gratton, G., Corballis, P. M., Cho, E., Fabiani, M., and Hood, D. C. (1995). Shades of gray matter: Noninvasive optical images of human brain responses during visual stimulation. *Psychophysiology* **32,** 505–509.

Gratton, G., and Fabiani, M. (2001). The event-related optical signal: A new tool for studying brain function. *Int. J. Psychophysiol.* **42,** 15–27.

Gratton, G., Goodman-Wood, M. R., and Fabiani, M. (2001). Comparison of neuronal and hemodynamic measures of the brain response to visual stimulation: An optical imaging study. *Hum. Brain Mapp.* **13,** 13–25.

Grinvald, A. (1992). Optical imaging of architecture and function in the living brain sheds new light on cortical mechanisms underlying visual perception. *Brain Topogr.* **5,** 71–75.

Haberl, R. L., Heizer, M. L., and Ellis, E. F. (1989). Laser-Doppler assessment of brain microcirculation: Effect of local alterations. *Am. J. Physiol.* **256,** H1255–H1260.

Haglund, M. M., Ojemann, G. A., and Blasdel, G. G. (1993). Optical imaging of bipolar cortical stimulation. *J. Neurosurg.* **78,** 785–793.

Haller, M., Mironov, S. L., and Richter, D. W. (2001). Intrinsic optical signals in respiratory brain stem regions of mice: Neurotransmitters, neuromodulators, and metabolic stress. *J. Neurophysiol.* **86,** 412–421.

Harms, G. S., Cognet, L., Lommerse, P. H., Blab, G. A., Kahr, H., Gamsjager, R., Spaink, H. P., Soldatov, N. M., Romanin, C., and Schmidt, T. (2001). Single-molecule imaging of L-type Ca(2+) channels in live cells. *Biophys. J.* **81,** 2639–2646.

Heeger, D. J., Huk, A. C., Geisler, W. S., and Albrecht, D. G. (2000). Spikes versus BOLD: What does neuroimaging tell us about neuronal activity? *Nat. Neurosci.* **3,** 631–633.

Heekeren, H. R., Kohl, M., Obrig, H., Wenzel, R., von Pannwitz, W., Matcher, S. J., Dirnagl, U., Cooper, C. E., and Villringer, A. (1999). Noninvasive assessment of changes in cytochrome-c oxidase oxidation in human subjects during visual stimulation. *J. Cereb. Blood Flow Metab.* **19,** 592–603.

Hemelt, M. W., and Kang, K. A. (1999). Determination of a biological absorber depth utilizing multiple source-detector separations and multiple frequency values of near-infrared time-resolved spectroscopy. *Biotechnol. Prog.* **15,** 622–629.

Hill, D. K., and Keynes, R. D. (1949). *J. Physiol.* **108,** 278–281.

Hillman, E. M., Hebden, J. C., Schweiger, M., Dehghani, H., Schmidt, F. E., Delpy, D. T., and Arridge, S. R. (2001). Time resolved optical tomography of the human forearm. *Phys. Med. Biol.* **46,** 1117–1130.

Hintz, S. R., Benaron, D. A., Siegel, A. M., Zourabian, A., Stevenson, D. K., and Boas, D. A. (2001). Bedside functional imaging of the premature infant brain during passive motor activation. *J. Perinat. Med.* **29,** 335–343.

Hintz, S. R., Cheong, W. F., van, H. J., Stevenson, D. K., and Benaron, D. A. (1999). Bedside imaging of intracranial hemorrhage in the neonate using light: Comparison with ultrasound, computed tomography, and magnetic resonance imaging. *Pediatr. Res.* **45,** 54–59.

Hock, C., Muller-Spahn, F., Schuh-Hofer, S., Hofmann, M., Dirnagl, U., and Villringer, A. (1995). Age dependency of changes in cerebral hemoglobin oxygenation during brain activation: A near-infrared spectroscopy study. *J. Cereb. Blood Flow Metab.* **15,** 1103–1108.

Hock, C., Villringer, K., Muller-Spahn, F., Hofmann, M., Schuh-Hofer, S., Heekeren, H., Wenzel, R., Dirnagl, U., and Villringer, A. (1996). Near infrared spectroscopy in the diagnosis of Alzheimer's disease. *Ann. N. Y. Acad. Sci.* **777,** 22–29.

Hock, C., Villringer, K., Muller-Spahn, F., Wenzel, R., Heekeren, H., Schuh-Hofer, S., Hofmann, M., Minoshima, S., Schwaiger, M., Dirnagl, U., and Villringer, A. (1997). Decrease in parietal cerebral hemoglobin oxygenation during performance of a verbal fluency task in patients with Alzheimer's disease monitored by means of near-infrared spectroscopy (NIRS)—Correlation with simultaneous rCBF–PET measurements. *Brain Res.* **755,** 293–303.

Holthoff, K., and Witte, O. W. (1996). Intrinsic optical signals in rat neocortical slices measured with near-infrared dark-field microscopy reveal changes in extracellular space. *J. Neurosci.* **16,** 2740–2749.

Hoshi, Y., Oda, I., Wada, Y., Ito, Y., Yutaka, Y., Oda, M., Ohta, K., Yamada, Y., and Mamoru, T. (2000). Visuospatial imagery is a fruitful strategy for the digit span backward task: A study with near-infrared optical tomography. *Brain Res. Cognit. Brain Res* **9,** 339–342.

Hoshi, Y., and Tamura, M. (1993). Dynamic multichannel near-infrared optical imaging of human brain activity. *J. Appl. Physiol.* **75,** 1842–1846.

Igawa, M., Atsumi, Y., Takahashi, K., Shiotsuka, S., Hirasawa, H., Yamamoto, R., Maki, A., Yamashita, Y., and Koizumi, H. (2001). Activation of visual cortex in REM sleep measured by 24-channel NIRS imaging. *Psychiatry Clin. Neurosci.* **55,** 187–188.

Isobe, K., Kusaka, T., Nagano, K., Okubo, K., Yasuda, S., Kondo, M., Itoh, S., and Onishi, S. (2001). Functional imaging of the brain in sedated newborn infants using near infrared topography during passive knee movement. *Neurosci. Lett.* **299,** 221–224.

Kato, T., Kamei, A., Takashima, S., and Ozaki, T. (1993). Human visual cortical function during photic stimulation monitoring by means of near-infrared spectroscopy. *J. Cereb. Blood Flow Metab.* **13,** 516–520.

Kirkpatrick, P. J., Lam, J., Al-Rawi, P., Smielewski, P., and Czosnyka, M. (1998). Defining thresholds for critical ischemia by using near-infrared spectroscopy in the adult brain. *J. Neurosurg.* **89,** 389–394.

Kleinfeld, D., Mitra, P. P., Helmchen, F., and Denk, W. (1998). Fluctuations and stimulus-induced changes in blood flow observed in individual capillaries in layers 2 through 4 of rat neocortex. *Proc. Natl. Acad. Sci. USA* **95,** 15741–15746.

Kleinschmidt, A., Obrig, H., Requardt, M., Merboldt, K. D., Dirnagl, U., Villringer, A., and Frahm, J. (1996). Simultaneous recording of cerebral blood oxygenation changes during human brain activation by magnetic resonance imaging and near-infrared spectroscopy. *J. Cereb. Blood Flow Metab.* **16,** 817–826.

Kohl, M., Lindauer, U., Dirnagl, U., and Villringer, A. (1998a). Separation of changes in light scattering and chromophore concentrations during cortical spreading depression in rats. *Opt. Lett.* **23,** 555–557.

Kohl, M., Nolte, C., Heekeren, H. R., Horst, S., Scholz, U., Obrig, H., and Villringer, A. (1998b). Determination of the wavelength dependence of the differential pathlength factor from near-infrared pulse signals. *Phys. Med. Biol.* **43,** 1771–1782.

Lindauer, U., Royl, G., Leithner, C., Kuhl, M., Gold, L., Gethmann, J., Kohl-Bareis, M., Villringer, A., and Dirnagl, U. (2001). No evidence for early decrease in blood oxygenation in rat whisker cortex in response to functional activation. *NeuroImage* **13,** 988–1001.

Lockwood, A. H., LaManna, J. C., Snyder, S., and Rosenthal, M. (1984). Effects of acetazolamide and electrical stimulation on cerebral oxidative metabolism as indicated by the cytochrome oxidase redox state. *Brain Res.* **308,** 9–14.

Mackert, B. M., Wubbeler, G., Leistner, S., Trahms, L., and Curio, G. (2001). Non-invasive single-trial monitoring of human movement-related brain activation based on DC-magnetoencephalography. *NeuroReport* **12,** 1689–1692.

MacVicar, B. A., and Hochman, D. (1991). Imaging of synaptically evoked intrinsic optical signals in hippocampal slices. *J. Neurosci.* **11,** 1458–1469.

Malonek, D., and Grinvald, A. (1996). Interactions between electrical activity and cortical microcirculation revealed by imaging spectroscopy: Implications for functional brain mapping. *Science* **272,** 551–554.

Matcher, S. J., and Cooper, C. E. (1994). Absolute quantification of deoxyhaemoglobin concentration in tissue near infrared spectroscopy. *Phys. Med. Biol.* **39,** 1295–1312.

Mathiesen, C., Caesar, K., Akgoren, N., and Lauritzen, M. (1998). Modification of activity-dependent increases of cerebral blood flow by excitatory synaptic activity and spikes in rat cerebellar cortex. *J. Physiol.* **512,** 555–566.

Mayhew, J., Johnston, D., Martindale, J., Jones, M., Berwick, J., and Zheng, Y. (2001). Increased oxygen consumption following activation of brain: Theoretical footnotes using spectroscopic data from barrel cortex. *NeuroImage* **13,** 975–987.

Meek, J. H., Elwell, C. E., Khan, M. J., Romaya, J., Wyatt, J. S., Delpy, D. T., and Zeki, S. (1995). Regional changes in cerebral haemodynamics as a result of a visual stimulus measured by near infrared spectroscopy. *Proc. R. Soc. London (Biol.)* **261,** 351–356.

Menon, R. S., Ogawa, S., Hu, X., Strupp, J. P., Anderson, P., and Ugurbil, K. (1995). BOLD based functional MRI at 4 tesla includes a capillary bed contribution: Echo-planar imaging correlates with previous optical imaging using intrinsic signals. *Magn. Reson. Med.* **33,** 453–459.

Miyai, I., Tanabe, H. C., Sase, I., Eda, H., Oda, I., Konishi, I., Tsunazawa, Y., Suzuki, T., Yanagida, T., and Kubota, K. (2001). Cortical mapping of gait in humans: A near-infrared spectroscopic topography study. *NeuroImage* **14,** 1186–1192.

Nollert, G., Mohnle, P., Tassaniprell, P., and Reichart, B. (1995). Determinants of cerebral oxygenation during cardiac surgery. *Circulation* **92,** 327–333.

Nollert, G., Shin, Nagashima, M., and Shum-Tim, D. (1997). Cerebral oxygenation during cardiopulmonary bypass in children. *J. Thoracic Cardiovasc. Surg.* **114,** 871–873.

Nomura, Y., and Tamura, M. (1991). Quantitative analysis of the hemoglobin oxygenation state of rat brain in vivo by picosecond time-resolved spectrophotometry. *J. Biochem. Tokyo* **109,** 455–461.

Ntziachristos, V., and Weissleder, R. (2001). Experimental three-dimensional fluorescence reconstruction of diffuse media by use of a normalized Born approximation. *Opt. Lett.* **26,** 893–895.

Obrig, H., Israel, H., Kohl-Bareis, M., Uludag K., Wenzel, R., Müller, B., and Villringer, A. (2002). Habituation of the visually evoked potential (VEP) and its vascular response: Implications for neurovascular coupling in the healthy adult. *Neuroimage*, in press.

Obrig, H., Hirth, C., Junge-Hulsing, J. G., Doge, C., Wolf, T., Dirnagl, U., and Villringer, A. (1996). Cerebral oxygenation changes in response to motor stimulation. *J. Appl. Physiol.* **81,** 1174–1183.

Onoe, H., Watanabe, Y., Tamura, M., and Hayaishi, O. (1991). REM sleep-associated hemoglobin oxygenation in the monkey forebrain studied using near-infrared spectrophotometry. *Neurosci. Lett.* **129,** 209–213.

Paus, T., Marrett, S., Worsley, K. J., and Evans, A. C. (1995). Extraretinal modulation of cerebral blood flow in the human visual cortex: Implications for saccadic suppression. *J. Neurophysiol.* **74,** 2179–2183.

Quaresima, V., Sacco, S., Totaro, R., and Ferrari, M. (2000). Noninvasive measurement of cerebral hemoglobin oxygen saturation using two near infrared spectroscopy approaches. *J. Biomed. Opt.* **5,** 201–205.

Rector, D. M., Poe, G. R., Kristensen, M. P., and Harper, R. M. (1997). Light scattering changes follow evoked potentials from hippocampal Schaeffer collateral stimulation. *J. Neurophysiol.* **78,** 1707–1713.

Rector, D. M., Rogers, R. F., Schwaber, J. S., Harper, R. M., and George, J. S. (2001). Scattered-light imaging in vivo tracks fast and slow processes of neurophysiological activation. *NeuroImage* **14,** 977–994.

Rees, G., Friston, K., and Koch, C. (2000). A direct quantitative relationship between the functional properties of human and macaque V5. *Nat. Neurosci.* **3,** 716–723.

Röther, J. Knab, R., Hamzei, F., Fiehler, J. R., Buchel, C., and Weiller, C. (2002). Negative dip in BOLD fMRI is caused by blood flow—oxygen consumption uncoupling in humans. *Neuroimage* **15**, 98–102.

Rumsey, W. L., Vanderkooi, J. M., and Wilson, D. F. (1988). Imaging of phosphorescence: A novel method for measuring oxygen distribution in perfused tissue. *Science* **241,** 1649–1651.

Sakamoto, T., Jonas, R. A., Stock, U. A., Hatsuoka, S., Cope, M., Springett, R. J., and Nollert, G. (2001). Utility and limitations of near-infrared spectroscopy during cardiopulmonary bypass in a piglet model. *Pediatr. Res.* **49,** 770–776.

Sakatani, K., Xie, Y., Lichty, W., Li, S., and Zuo, H. (1998). Language-activated cerebral blood oxygenation and hemodynamic changes of the left prefrontal cortex in poststroke aphasic patients: A near-infrared spectroscopy study. *Stroke* **29,** 1299–1304.

Salzberg, B. M., and Obaid, A. L. (1988). Optical studies of the secretory event at vertebrate nerve terminals. *J. Exp. Biol.* **139,** 195–231.

Schmitz, B., Bock, C., Hoehn-Berlage, M., Kerskens, C. M., Bottiger, B. W., and Hossmann, K. A. (1998). Recovery of the rodent brain after cardiac arrest: A functional MRI study. *Magn. Reson. Med.* **39,** 783–788.

Steinbrink, J., Kohl, M., Obrig, H., Curio, G., Syre, F., Thomas, F., Wabnitz, H., Rinneberg, H., and Villringer, A. (2000). Somatosensory evoked fast optical intensity changes detected non-invasively in the adult human head. *Neurosci. Lett.* **291,** 105–108.

Steinbrink, J., Wabnitz, H., Obrig, H., Villringer, A., and Rinneberg, H. (2001). Determining changes in NIR absorption using a layered model of the human head. *Phys. Med. Biol.* **46,** 879–896.

Steinhoff, B. J., Herrendorf, G., and Kurth, C. (1996). Ictal near infrared spectroscopy in temporal lobe epilepsy: A pilot study. *Seizure* **5,** 97–101.

Stepnoski, R. A., LaPorta, A., Raccuia-Behling, F., Blonder, G. E., Slusher, R. E., and Kleinfeld, D. (1991). Noninvasive detection of changes in membrane potential in cultured neurons by light scattering. *Proc. Natl. Acad. Sci. USA* **88,** 9382–9386.

Toronov, V., Webb, A., Choi, J. H., Wolf, M., Michalos, A., Gratton, E., and Hueber, D. (2001). Investigation of human brain hemodynamics by simultaneous near-infrared spectroscopy and functional magnetic resonance imaging. *Med. Phys.* **28,** 521–527.

Uludag, K., Kohl, M., Steinbrink, J., Obrig, H., and Villringer, A. (2001). Crosstalk in the Lambert–Beer calculation for near-infrared wavelengths estimated by Monte Carlo simulations. *J. Biomed. Opt.,* **7**, 51–59.

Villringer, A., and Chance, B. (1997). Non-invasive optical spectroscopy and imaging of human brain function. *Trends Neurosci.* **20,** 435–442.

Villringer, A., Haberl, R. L., Dirnagl, U., Anneser, F., Verst, M., and Einhaupl, K. M. (1989). Confocal laser microscopy to study microcirculation on the rat brain surface in vivo. *Brain Res.* **504,** 159–160.

Villringer, A., Planck, J., Hock, C., Schleinkofer, L., and Dirnagl, U. (1993). Near infrared spectroscopy (NIRS): A new tool to study hemodynamic changes during activation of brain function in human adults. *Neurosci. Lett.* **154,** 101–104.

Villringer, A., Planck, J., Stodieck, S., Bötzel, K., Schleinkofer, L., and Dirnagl, U. (1994). Noninvasive assessment of cerebral hemodynamics and tissue oxygenation during activation of brain cell function in human adults using near infrared spectroscopy. *Adv. Exp. Med. Biol.* **345,** 559–565.

von Pannwitz, W., Obrig, H., Heekeren, H., Müller, A., Kohl, M., Wolf, T., Wenzel, R., Dirnagl, U., and Villringer, A. (1998). Clinical approaches in near-infrared spectroscopy. *In* "Transcranial Cerebral Oximetry" (G. Litscher and G. Schwarz, eds.), pp. 166–183. Pabst Science, Berlin.

Watanabe, E., Maki, A., Kawaguchi, F., Takashiro, K., Yamashita, Y., Koizumi, H., and Mayanagi, Y. (1998). Non-invasive assessment of language dominance with near-infrared spectroscopic mapping. *Neurosci. Lett.* **256,** 49–52.

Wenzel, R., Wobst, P., Heekeren, H. H., Kwong, K. K., Brandt, S. A., Kohl, M., Obrig, H., Dirnagl, U., and Villringer, A. (2000). Saccadic suppression induces focal hypooxygenation in the occipital cortex. *J. Cereb. Blood Flow Metab.* **20,** 1103–1110.

Wobst, P., Wenzel, R., Kohl, M., Obrig, H., and Villringer, A. (2001). Linear aspects of changes in deoxygenated hemoglobin concentration and cytochrome oxidase oxidation during brain activation. *NeuroImage* 13, 520–530.

7

Dynamic Measurements of Local Cerebral Blood Flow: Examples from Rodent Whisker Barrel Cortex

Thomas A. Woolsey, Ling Wei, and Joseph P. Erinjeri

Departments of Neurology and Neurological Surgery, Washington University School of Medicine, St. Louis, Missouri 63110

I. Why Measure Local Cerebral Blood Flow?

William Harvey's experimental demonstration of the circulation of the blood in 17th century is the foundation of physiology (Harvey, 1978). Biological systems are dynamic. To fully appreciate functions in an organism and the mechanisms underlying these functions, their dynamics should be understood. The renewed interest in the dynamic behavior of the cerebral circulation as a means to detect changes in neuronal activity follows on a fundamental relationship first fully enunciated by Roy and Sherrington, namely that increased neuronal activity increases local cerebral blood flow (LCBF) (Kety and Schmidt, 1945; Posner and Raichle, 1994; Roy and Sherrington, 1890). Imaging modalities and studies based on this relationship are detailed in this book. To appreciate the information these techniques provide (and do not provide), it is necessary to know the extent to which changes in the vascular system reflect changes in the nervous system. The two are closely linked (for details, see Woolsey *et al.*, 1996).

Flow (Q) is movement of materials having the dimensions of volume per time. Increases in volume during the same time interval indicate increased flow; decreases in transit times through a system of fixed volume also indicate increased flow. Generally, system volume and time in the system change together such that neither alone can account fully for the extent to which flows change. Several methods can measure flow itself or be combined to take into account both volume and transit time and derive flow.

Many approaches devised to measure blood flow are based on physical measurements related to free diffusion and dilution of chemically inert tracers and well-developed theory to interpret steady-state or equilibrated flows (Kety, 1960). These are thoroughly reviewed by Heistad and Kontos (1983) and need not be recapitulated here. This chapter reviews techniques used to study the dynamic real-time changes in LCBF by direct observation of the brain surface in experimental animals. The animals were tested with natural sensory (whisker) stimulation, temporary systemic physiological changes, or altered local hemodynamics.

Just as Leeuwenhoek watched red blood cells (RBCs) in capillaries of the transparent membranes of the thin fins of the tadpole tail under the microscope (1688, in Landis, 1982), Rosenblum (1969, 1970) demonstrated that cortical flow could be watched directly in mice. Fluorescent compounds, microscopes, video cameras, and computers now make it possible to follow changes in the vasculature in exquisite detail with quantification. Many of these methods, most of which are implemented in our laboratories, are summarized in this chapter. They are illustrated from experiments in which the exposed cerebral cortex of the rat or mouse was observed directly by video microscopy. This overview is based on work published in detail elsewhere (Cox *et al.*, 1993; Dowling *et al.*, 1996; Erinjeri and Woolsey, 2002; Ido *et al.*, 2001; Liang *et al.*, 1995; Moskalenko *et al.*, 1996; Rovainen *et al.*, 1993; Wei *et al.*, 2001; Woolsey *et al.*, 1996).

II. Function and Structural Contexts

The diverse structure of the mammalian central nervous system is characteristic and specific for regions with different functions. In experimental animals and humans, electrical recordings identify the functions of these regions, many of which can be readily recognized structurally. A number of studies have demonstrated elegantly the correlation of function with finer underlying neuronal architecture (Hubel, 1982; Simons, 1985). The detailed vasculature of brain regions differs also, and shifts in vascular patterns correspond closely to shifts in brain architecture. Vessel patterns could be evaluated for their coincidence with and contribution to LCBF changes evoked by local neuronal activity. Figure 1 shows an example from mouse cortex stained for cell bodies and a corresponding section that demonstrates vascular patterns. Just as for cell bodies and other markers (e.g., Welker and Woolsey, 1974; Woolsey, 1990, 1993; Woolsey and Van der Loos, 1970), the pattern of the vessels varies by cortical region and layer (e.g., Cox *et al.*, 1993). Conveniently, blood flow to the cerebral cortex is through vessels that enter and leave from the pial surface (Fig. 1).

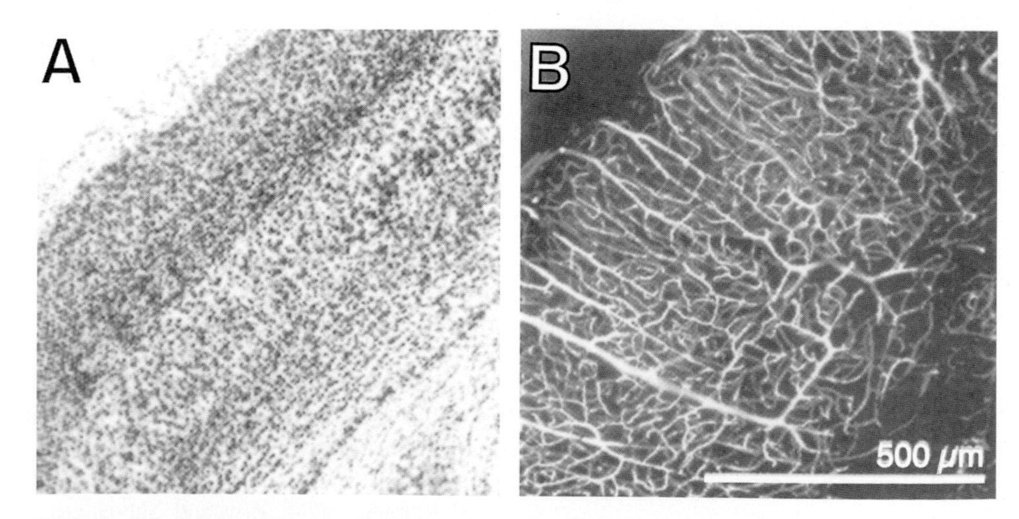

Figure 1 Perpendicular sections through the cerebral cortex of the mouse. (**A**) The layers of the cerebral cortex seen with a Nissl stain for cell bodies identify regions by cytoarchitectonic criteria. Structure correlates with recordings of evoked electrical activity and neuroanatomical connectivity and is used to detect boundaries between different functional areas postmortem. (**B**) A similar section from another mouse perfused with fluorescein in gelatin rendered in three dimensions from scans every 1 μm through a 200-μm section with a scanning laser microscope (Zeiss LSM confocal microscope). Blood vessel patterns correspond to different cortical layers and cytoarchitectonic regions and are distinctive for different functional brain regions.

There are pathways for lateral and collateral flow in the cortical parenchyma but, to a first approximation, flow in surface vessels indicates flow to parenchyma below. Within the brain parenchyma, capillary density correlates directly with metabolic capacity (Boero *et al.,* 1999; Borowsky and Collins, 1989; Klein *et al.,* 1986).

III. Global Tracers

Two classes of radioactive tracers have relatively wide use for studies of blood flow. Both permit a direct quantification and as such these methods constitute the "gold standard" for studies of flow in animals.

Radioactive occlusive microspheres are convenient flow markers (Heistad *et al.,* 1977). After injection, they are trapped in vessels of differing diameters, depending on the size of the beads. If the noncompressible beads are well mixed with blood then they distribute in proportion to flow. Beads labeled with different isotopes can be injected and flows at different injection times determined in the same animal (Marcus *et al.,* 1976). The animal is sacrificed, and the tissues are dissected and counted for radioactivity. Local flows can then be calculated. The method is limited by the number of isotopes that are practical to count, their radioactivity, and the dissection. This method can be used to test or calibrate other methods (Heiss and Traupe, 1981; Horton *et al.,* 1980; Muller *et al.,* 1996), to measure global changes in LCBF (Ido *et al.,* 2001), and for comparison with various histological preparations (Zhao *et al.,* 2001).

The second method, refined by Sokoloff and his colleagues in rat (Sakurada *et al.,* 1978) and mouse (Jay *et al.,* 1988), depends on the rapid, free diffusion of an intravascular marker. The markers, [^{14}C]iodoantipyrine (IAP) or ^{99}Tc-hexamethylpropyleneamine oxime (HM-PAO; Lear, 1988; Neirinckx *et al.,* 1987), are delivered to tissues in proportion to local flow, where they accumulate in the short term. Tagged markers are immobilized in brain by quick freezing after sacrifice. Sections cut on a freezing microtome and dried rapidly to minimize diffusion are used to expose X-ray film (high resolution for relatively low-energy ^{14}C-tagged IAP). With frequent arterial samples, a blood concentration of the isotope is determined throughout the experiment and, with autoradiographic standards, radioactivity is related to density. Standard equations are used to convert film density into blood flow in ml/100 g/min (Sakurada *et al.,* 1978). Strategies have been developed to bypass the difficult cannulations and blood sampling to facilitate the use of this technique in mouse models (Maeda *et al.,* 2000).

Autoradiograms of two sections from the rat brain are depicted in Fig. 2. The whiskers on the left face were stimulated manually at 3–5 Hz to strongly activate appropriate parts of the central trigeminal pathways (e.g., Durham and Woolsey, 1978; Ginsberg *et al.,* 1987; Jacquin *et al.,* 1993b). In the example shown, 10 min after the stimulus was begun, the animal was given IAP by a ramp iv injection, stimulation was continued for 30 s until decapitation, and the brain was quickly frozen (see Gross *et al.,* 1987). A series of 20-μm sections was used to expose X-ray film. The pattern of label, which resembles the white matter/gray matter differences in unstained sections, indicates the spatial resolution of the method.

From the autoradiogram, whisker stimulation increases blood flow in trigeminal brainstem nuclei on the same side and in the somatosensory ventroposteromedial nucleus of thalamus and somatosensory (barrel) cortex on the opposite side. The location of the activity in the main pathway and

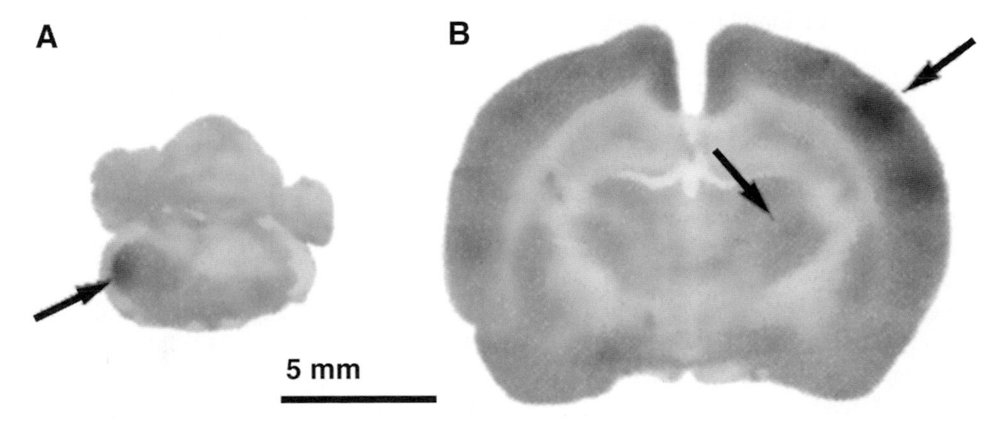

Figure 2 [^{14}C]Iodoantipyrine autoradiograms of transverse 20-μm sections at two levels from a rat brain. The whiskers on the left side of the face were stimulated continuously from 10 min before and during the 30-s injection until sacrifice. Increased density of the trigeminal pathway demonstrates localization and specificity of the LCBF increases with sensory activation of the trigeminal pathway. (**A**) Transverse section through the brain stem at the level of the spinal trigeminal nucleus, subnucleus interpolaris. (**B**) Transverse section through the ventroposterior nucleus of the thalamus and the somatosensory cortex. Arrows point to trigeminal structures with stimulus-evoked flow increases which correspond to appropriate structures identified by cytoarchitecture from adjacent Nissl-stained sections.

cortex is confirmed from adjacent sections stained for cell bodies. Calculated flows (in ml/g/min) on the active side in comparison to the side not stimulated are 1.83 ± 0.24 vs 1.34 ± 0.11 (mean ± SD; n = 5 rats each) in subnucleus interpolaris of the brain stem, 2.56 ± 0.34 vs 1.83 ± 0.24 in thalamus, and 4.75 ± 0.66 vs 2.44 ± 0.24 in layer IV of barrels. These images illustrate the principal characteristics of this method; results are quantifiable, highly localized, and comprehensive, regardless of location in the nervous system. The relative flows provoked by a defined stimulus can be assessed throughout a particular neural system but dynamic information is available only from multiple subjects. Other drawbacks include poor temporal and spatial resolution, labor intensiveness, and cost.

IV. Volatile Tracers

Inert, freely diffusible blood-borne tracers distribute to tissues and then wash out in relation to flow (Kety, 1960). For instance, *Xe, inhaled and monitored by external detectors, was a tracer used for early localization studies of evoked changes in cerebral blood flow in humans (Ingvar, 1975; Lassen *et al.*, 1977) and rodents (Linde *et al.*, 1999). Inhaled gasses are delivered to and cleared from the bloodstream relatively quickly and, within the limits of radiation exposure, can be used for repeated measurements of flow under different conditions.

H_2 has excellent diffusional properties (for reviews, see Moskalenko *et al.*, 1980; Young, 1980) and can be detected in tissues for polarography with platinum electrodes. The oxidation of H_2 to $2H^+$ catalyzed on Pt generates current in proportion to H_2 concentration and Ph_2 in the tissue. Tissue Ph_2 can be followed over time after the gas is inhaled (Fig. 3A) and consists of an accumulation and clearance of the tracer. The rate constant from the exponential disappearance of H_2 can be used to measure blood flow around the Pt electrode. H_2 measurements have been compared with radioactive microspheres (Heiss and Traupe, 1981). The volume of tissue measured depends in part on the configuration of the electrode. The method is good for steady-state conditions. A generalized flow stimulus, such as inhaled CO_2, tested with H_2 demonstrates increased flows (Fig. 3A). For studies of dynamic flows, this method is cumbersome and lacks temporal resolution.

Lübbers and colleagues (Stosseck *et al.*, 1974) modified the H_2 clearance method to permit continuous monitoring of flows. They passed a small direct current through one Pt electrode to generate H_2 by hydrolysis while recording H_2 currents from another Pt electrode nearby. For a good example of refinement and calibration of this method in kidney see Parekh *et al.* (1993). H_2 generation obviates the need to breathe the test gas repeatedly and can provide continuous recordings of Ph_2. The currents do not appreciably alter neuronal activity. A ratio is calculated from the CO_2-provoked changes in Ph_2 with inhaled H_2- and CO_2-stimulated changes in Ph_2 with electrolytically generated H_2 to determine flow changes with a steady local generation of H_2 (Fig. 3B). The system gives information on changes in flow but, because of electronic filtering and the time course of diffusion, the extent of rapid changes is damped.

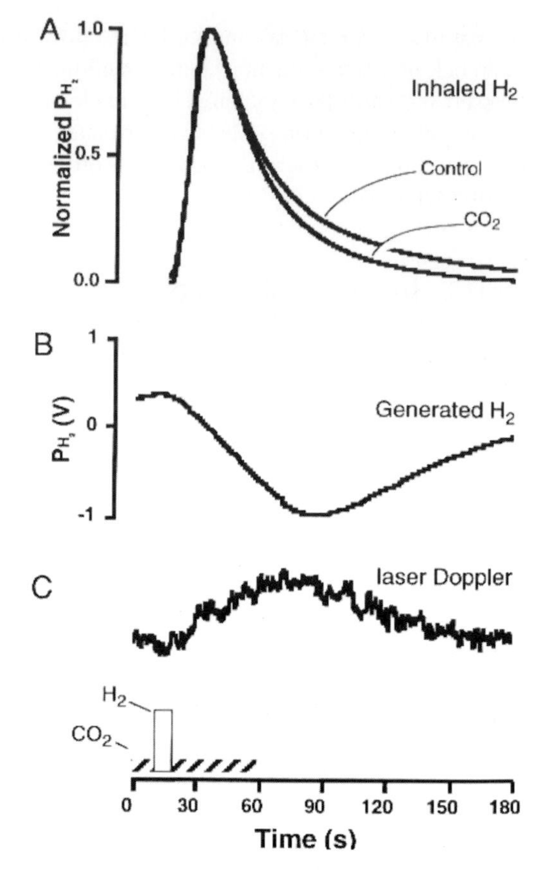

Figure 3 Polarography monitors Ph_2, which is proportional to the local $[H_2]$ at a Pt electrode. Inhaled H_2 is delivered by the blood to the tissues and disappears with a rate constant proportional to blood flow. (**A**) Records showing the appearance of inhaled H_2 (bottom, above the time scale) at a Pt electrode in the cerebral cortex. Inhaled 7.5% CO_2 (bottom, above the time scale), a powerful general stimulus of LCBF, shortens the H_2 clearance. (**B**) Two or more Pt electrodes in the brain can record LCBF continuously; one generates H_2 with small currents against a distant reference, and Ph_2 is recorded from the other(s). The increase in LCBF is provoked by inhaled 7.5% CO_2 in air. The ratio of changes in Ph_2 produced by inhaled CO_2 measured with inhaled and locally generated H_2 is used to calibrate electrodes and recording sites to estimate flow changes. The location of the recording electrodes is determined by postmortem histology. (**C**) A laser-Doppler recording taken in the same preparation shows the relative increase in LCBF in response to CO_2 with a time course similar to that recorded with the Pt electrode in (B).

Electrodes can be targeted to active cortex through the exposed dura in relation to vascular landmarks, electrophysiology, flow responses monitored by laser-Doppler flowmetry, and/or intrinsic optical signals. Positioning Pt electrodes is limited by the geometry of the fixed electrode arrays which must be navigated past surface vessels. Electrode locations are confirmed by postmortem histology. Flow responses monitored over time have been used to measure latencies to the onset of flow changes, direction of flow changes (increased or decreased), changes in different stimulus parameters, changes with stimulation of different parts of the periphery to activate different barrels and intracortical circuits, and changes in flow of different layers of cortex (Moskalenko *et al.*, 1994, 1996, 1998; Woolsey *et al.*, 1996). Localization of the responses in space can be improved with the configuration of the electrodes (smaller electrode surfaces and closer spacing between electrodes). A good demonstration of the resolution possible is the localization of clearance near single retinal vessels and across the retina (Yu *et al.*, 1991).

Infrared thermal encephalography (IR-TE; Rausch and Eysel, 1996) is an imaging technique which uses heat as a volatile tracer. The basis of this method is the transfer of thermal energy from circulating blood to cortical tissue depending on local perfusion. The thermoscanners used in IR-TE measure two-dimensional temperature fields. Thus, IR-TE images reflect LCBF over the cortical surface, unlike hydrogen clearance methods, which reflect LCBF at a certain point in or on the brain.

V. Doppler Flowmetry

The frequency shift in reflections from objects moving to or away from a source, the Doppler effect, can be used to estimate mean velocities and the magnitude of moving objects. Ultrasonic Doppler probes have been targeted by manipulators to monitor velocities in medium-sized surface vessels on the cortex of dogs and cats (Busija *et al.*, 1981). With optical laser-Doppler flowmetry, frequency shifts in coherent light in the near infrared to which tissue is relatively transparent can be used to estimate RBC velocity and volume of moving RBCs. Taken together, these parameters register flow. Fiber optic probes positioned over the brain record relative flow changes with different stimuli and can even be used through the thinned intact skull (Gerrits *et al.*, 1998). The sampling is limited to a $\leq 1\text{-mm}^3$ volume below the probe tip. Probes, like electrodes, can be advanced into tissue for readings from deep sutures. Temporal resolution is on the order of 200 ms, which is mainly determined by the design of the commercial instruments. Cranial windows permit targeting probes under direct vision. The advantages of this approach are ease of use and online display (Irikura *et al.*, 1994; Lindauer *et al.*, 1993). The laser can be scanned in two dimensions, allowing for a map of laser-Doppler parameters to be generated (Ances *et al.*, 1999; Wardell and Nilsson, 1996). Since light of different wavelength penetrates the cortex to differing degrees, flow can be inferred at different depths of cortex (Fabricius *et al.*, 1997). The disadvantages are uncertain absolute flow values, the sizes of volumes measured, and the fact that the visualization of cortex underneath the optical probes is obscured. Reproducibility of laser-Doppler signals can be improved by signal averaging (Ances *et al.*, 1998; Detre *et al.*, 1998). Flow by laser Doppler has been verified by direct comparisons with IAP (Dirnagl *et al.*, 1989; Fabricius and Lauritzen, 1996), H_2 clearance (Haberl *et al.*, 1989; Skarphedinsson *et al.*, 1988; Uranishi *et al.*, 1999), microspheres (Lindsberg *et al.*, 1989), and arterial diameter (Ngai *et al.*, 1995). Related techniques include optical Doppler tomography (Chen *et al.*, 1998), laser-Doppler anemometry (Seki *et al.*, 1996), and laser speckle (Dunn *et al.*, 2001).

VI. Video Microscopy

Direct observation of cerebral vessels requires magnification. Low-magnification systems with appropriate macro lenses and dissecting microscopes for modest magnification can be used to view changes over wide areas of exposed brains of larger animals, including humans (Feindel *et al.*, 1968; Ratzlaff and Grinvald, 1991). However, the compound microscope is necessary to resolve vessels and flow markers under 10 μm in diameter for velocity and flux measurements (Hudetz, 1997).

Automated techniques include the dual slit method, which determines the timing between hemoglobin (Hb) absorption profiles at different points on the same vessel (Ma *et al.*, 1974). Typically, the surface of the brain is exposed surgically under anesthesia and a simple chamber (cranial window) is constructed to maintain physiological conditions (temperature, pO_2, artificial cerebrospinal fluid, etc.) and covered with glass to provide good optics (Fig. 4; Cox *et al.*, 1993; Hudetz *et al.*, 1995; Morii *et al.*, 1986). Ports in the cranial window permit adjustment of intracranial pressure and changing of solutions (as to introduce drugs) and probes, including electrodes (Fujita *et al.*, 2000; Jallo *et al.*, 1997). Arterial and venous lines are used for injecting markers and for monitoring physiological parameters; the animals are ventilated mechanically and body temperature is thermostatically controlled. Standard physiological parameters such as end-tidal CO_2, EKG, and mean arterial pressure are recorded throughout with a computer interface. Arterial blood gasses can be sampled periodically. Long-working-distance microscope objectives (1–40×) provide satisfactory resolution and access to the cranial window.

For studies of the rodent cortex, whisker stimulation is varied in frequency, amplitude, and duration with accurate timing. Additional probes can be positioned in the field of view, such as a fiber optic probe for laser-Doppler flowmetry and Pt electrodes for polarography. Incident light of different frequencies (green light to enhance the contrast of

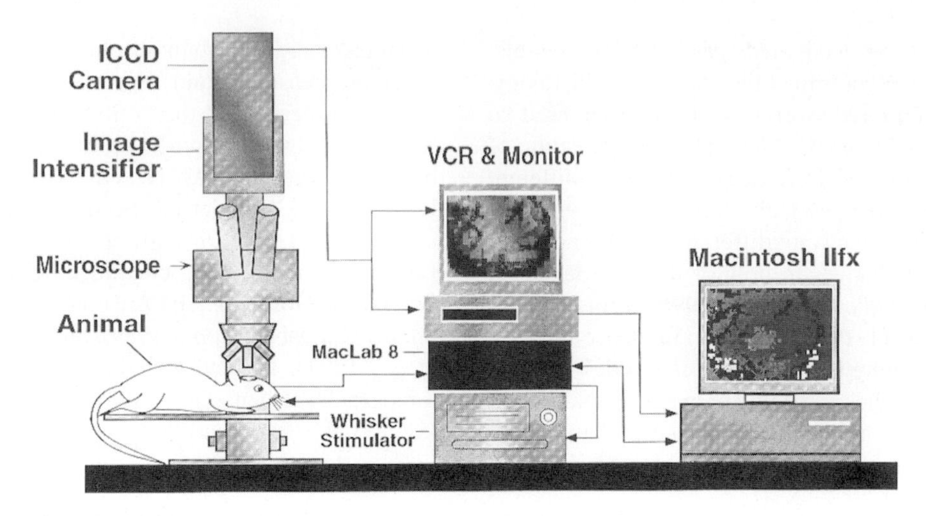

Figure 4 Setup for viewing vascular responses on the brain surface. The brain surface is exposed through a craniotomy and a chamber is installed with a coverslip (cranial window) through which the brain is illuminated at low light levels. Intensified images are digitized directly with a small computer and are recorded on videotape for later analysis. Simultaneous changes in physiological parameters (e.g., blood pressure, ECG, laser-Doppler flows, exhaled pCO_2) and peripheral hardware (stimulator) are recorded by and controlled from a computer.

Hb) or epi-illumination (at excitation frequencies for fluorescent probes) are used. The microscope is fitted with a CCD video camera. An image intensifier significantly reduces the light levels to which the brain is exposed to avoid photodamage. Images are digitized through an image-capture board in the computer for analysis "online" and recorded on videotape for later study and analysis.

The full extent of the exposed cortex can be mapped with respect to the image of the surface (e.g., Erinjeri and Woolsey, 2002). This indicates the locations of surface arteries, arterioles, veins, and venules and where they penetrate the cortex. These static maps can be related to deep structures if the brain is sectioned after the experiment in an orientation that is approximately parallel to the surface that was seen in the living brain. This step relates specific parts to identified surface vessels. Drawings of sections (camera lucida) and the photographic surface image are superimposed accurately with commercial software (Cox *et al.,* 1993).

Image information can be analyzed with analysis software from a number of suppliers. However, Wayne Rasband's magnificent public domain creation, NIH Image for Macintosh computers (available at http://rsb.info.nih.gov/nih-image/), and its multiplatform successor, ImageJ (available at http://rsb.info.nih.gov/ij/), is the package of choice for its versatility, flexibility, and "user friendliness." This software permits measurements of the vessel diameter, the length of filled vessels, the velocity of stroboscopically illuminated fluorescent beads, and the brightness of selected parts of the video field for dye transits. It also supports averaging, subtracting, and making ratios of the images or pixels for detecting intrinsic signals or to outline vessels filled with contrast. Macro programs written with this package can be used to estimate labeled RBC frequency and velocity and to automate repetitive procedures such as generating dye concentration integrals. In conjunction with other commercially available software for graphical and statistical analyses and for drawing, NIH Image and ImageJ provide the power for analysis similar to that of improved video equipment for capturing visual information.

VII. Localization of Activity Changes

Electrophysiological recording of evoked responses is the traditional way to determine the location of sensory cortical areas (e.g., Woolsey *et al.,* 1979). It takes considerable time to generate detailed "maps" point by point for accurate localization representation within a map (Merzenich *et al.,* 1983), even with surface electrodes (Woolsey, 1967). The fortuitous discovery of an "intrinsic" signal related to local brain activity that can be detected by processing images from sensitive cameras has the potential to "map" relatively large areas of cortex simultaneously (Grinvald *et al.,* 1986). The sources of these signals differ at different illumination and filter wavelengths and are postulated to arise from blood, blood vessels, neurons, and glia (Frostig *et al.,* 1990; Haglund *et al.,* 1992; Masino *et al.,* 1993; Narayan *et al.,* 1994). Because the intrinsic signal is weak, the camera and the image processor both need sensitivity and frame averaging, and these uniquely configured devices often require special purpose software and are marketed as stand-alone packages. The time to acquire images for averaging and subtraction can be long (ranges of 5–20 min are usual). Intrinsic signal mapping can be performed through the thinned skull (Frostig *et al.,* 1990) and in species as large as humans

(Haglund *et al.,* 1992) or as small as mice (Erinjeri and Woolsey, 2002; Prakash *et al.,* 2000).

Based on a strategy described by Peterson and Goldreich (1994), the video setup such as illustrated in Fig. 4 can be used to monitor an intrinsic signal. The fluorescent filter on the microscope transmits light from 520 to 560 nm within the range of absorption for total Hb and therefore RBC volume (Frostig *et al.,* 1990). Alternatively, filters in the 605 to 620 range can be used to detect light absorbed predominantly by deoxyhemoglobin related to oxygen delivery from capillaries to tissue (Erinjeri and Woolsey, 2002; Malonek *et al.,* 1997). Sensory stimulation leads to a change in reflectance that can be detected by averaging video frames with a video board in a Macintosh computer and NIH Image. For the flow techniques described in this chapter, the configuration of the hardware and software is unchanged. The localization of the signal is always appropriate by histology (Dowling *et al.,* 1996). At 540–560 nm, the principal component of the intrinsic signal recorded is a change in red cell mass, a parameter that changes in relation to flow. The analysis of dynamic aspects of flow with subsequent high magnification is greatly assisted by this localization step at low magnification. Since the stimulus images are compared to images without stimulation, movements of the high-contrast components of the system-surface arteries and veins-are detected as well (Figs. 5 and 6).

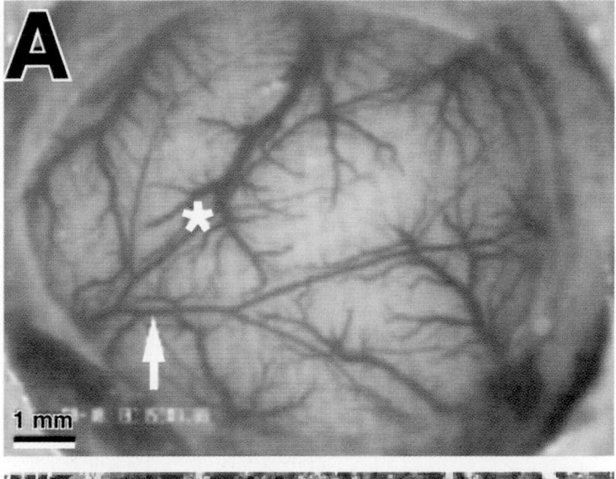

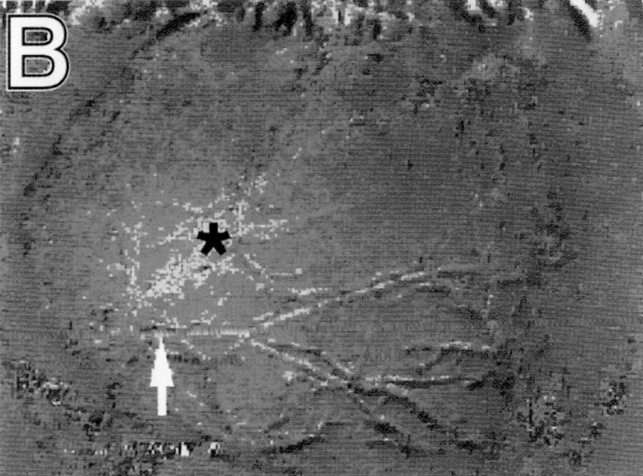

Figure 5 (**A**) The left cortex of a rat through a cranial window. The left middle cerebral artery (arrow points to a lateral branch) courses in a medio-posterior direction to supply most of the cortical surface. Veins drain in all directions to dural venous sinuses. The asterisk marks the locus of the center of the optical response shown in (B). Medial is up. The scale applies to both images. (**B**) Optical properties of active cortex are different from those of resting cortex. Hb has a strong signal at 520–560 nm and is detected in active cortex. This difference image is from an average over 500 ms (15 video frames) taken 7 s into stimulation of all the contralateral whiskers at 4 Hz (warmer colors indicate greater differences). With histological reconstructions, the green area is superimposed directly over the whisker barrels in layer IV. Blood vessel movements with the pulse cause useful vessel "shadows" in the difference images which are useful for relating intrinsic signals to the image shown in (A). Changes in vessel dimensions, local to the area of activity (to the left and slightly below the asterisk), can also be detected.

VIII. Diameter

Pial arterial diameter changes are the principal vessel "artifacts" that investigators of optical changes in active brain encountered and sought to "correct" (Blasdel and Salama, 1986; Ts'o *et al.,* 1990). Vasodilation is readily measured in surface vessels with diameters ≈30 μm without any special marking technique because of the contrast between Hb and brain. For instance, Ngai and colleagues (1988) demonstrated increases in the diameter of arteries supplying the hindlimb region of the rat somatic cortex with sciatic nerve stimulation, which they have correlated with laser-Doppler records (Ngai *et al.,* 1995) and velocity changes (Ngai and Winn, 1996). The measurements can be made on each frame from video records and are a good index of flow-related changes in real time. RBCs do not demonstrate the full diameter of a vessel because of the 3-μm layer of plasma next to the luminal wall with laminar flow (e.g., Rovainen *et al.,* 1993).

Colored plasma markers fill the vascular lumen and, when the plane of focus is through the vessel axis, "true" internal vessel diameter can be measured to ≈1 μm. Standard vital dyes, patent blue, and fluorescein provide contrast for such measurements. With image intensification, low light levels through epifluorescence illumination give excellent images of the vascular system after injection of fluoresceinated compounds that distribute with plasma. The image in Fig. 6 shows changes in surface vessels produced by apnea, a powerful vasodilator, after injections of fluoresceinated dextrans. Figure 6 demonstrates that the arteries, not the veins, actively dilate. In contrast to focal activation (as in Fig. 5), all of the arterial tree dilates. Frame-by-frame analysis of the diameter changes indicates the timing of the response, its magnitude, and its recovery. All parts of the arterial tree dilate at the same time during apnea and, after normalization, changes in all parts of the tree, including constriction, are similar. This method permits simultaneous assessment of different parts of

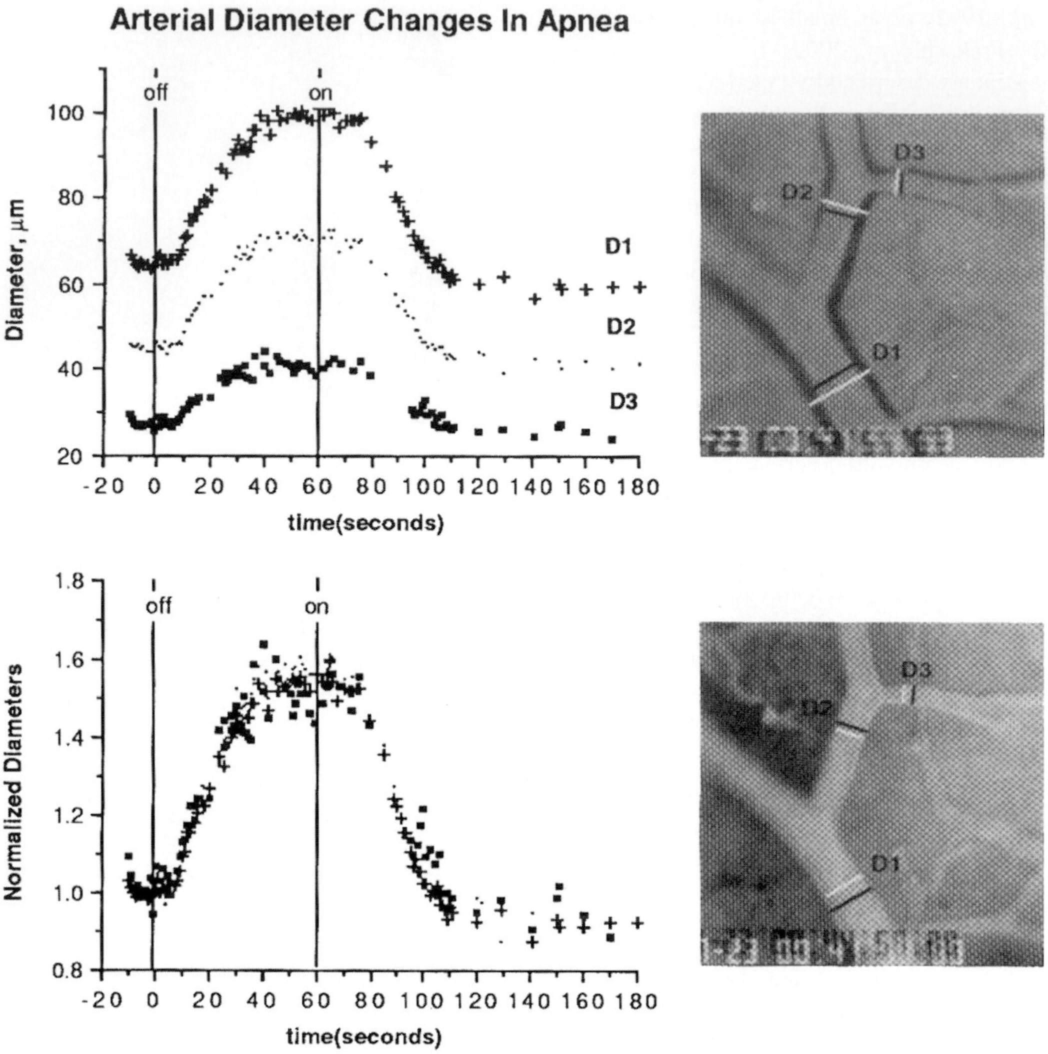

Figure 6 Intravascular plasma markers, such as fluorescein or fluoresceinated dextrans, demonstrate the internal diameter of surface vessels. Dynamic diameter changes are evident in video sequences, especially during fast replay. Difference images before (above right) and after (below right) 1 min of apnea show significant global increases and decreases in the diameter of arteries (shadows), but not in veins. Diameter measurements at three locales (D1, D2, D3) show similar timing of diameter changes throughout the arterial tree. Normalization reveals that the relative magnitudes of the changes are the same. The arterial tree constricts when breathing resumes as shown by the undershoots.

the surface vasculature for studies of localization. In this and similar images taken with sensory (Cox *et al.,* 1993; Liang *et al.,* 1995; Liu *et al.,* 1994) or pharmacological stimulation (Woolsey and Rovainen, 1991) surface arteries change, whereas surface veins do not.

IX. Intravascular Dyes

Angiographers use the timing of movement of radio-opaque dyes to estimate flow abnormalities (e.g., Lasjaunias and Berenstein, 1990). Transits have been measured in animals with carbon particles (Schiszler *et al.,* 2000; Tomita *et al.,* 1983) and dilution of RBCs with saline or dye (Eke, 1993). In the cranial window, this speed can be quantitated by following the time course of boluses of intravascular markers as they first appear in arteries and then capillaries and veins (e.g., Wei *et al.,* 2001). With image processing, small regions in the same field of view are selected over an arteriole, capillaries (operationally brain surface without obvious vessels or the whole field of view), and a venule. The preferred dyes and fluorescein and fluorescent dextrans are injected briskly into any part of the vascular system and monitored with epi-illumination or laser scanning confocal microscopy (Morris *et al.,* 1999). Injection into a large systemic vein is satisfactory for demonstration of the anatomy and for measurements before, during, and after a manipulation. Injection of small boluses (0.01 ml) of fluorescein or fluoresceinated dextrans via the ipsilateral external carotid artery provides brighter and faster signals because the concentrated dye reaches the brain more directly (Fig. 6; Wei *et al.,* 1995). The volume is small and the dye is quickly diluted in the general circulation so that signals from the second and third circulation times are weak and little dye accumulates with multiple injections.

The dye appears in the arterioles quickly after a bolus injection. Brightness in "tissue" pixels is due to plasma in capillaries and appears over a longer time course. Dye appears last in the venules, often with mixed laminar streaming and complex time courses. Relative flow changes can be accurately followed by comparing arteriovenous transit times (AVTTs) with stimulation to those determined before and after. The shape of the dye appearance curves depends on the mechanics of injection; the differences in the normalized integrals of arterial and venous dye brightness are insensitive to these variations and form the final basis for comparisons with this method (Fig. 7). Flows were slowed with a 1.6- and 7-fold increase in AVTTs in two veins draining different parts of cortex around an experimentally created area of ischemia (Wei *et al.,* 1998, 2001). The principal practical advantage of this method is that measurements can be made nearly automatically at relatively low magnification over many parts of a field of view from videotape recordings with computer macroroutines. Its principal disadvantage is that observations can be made only intermittently.

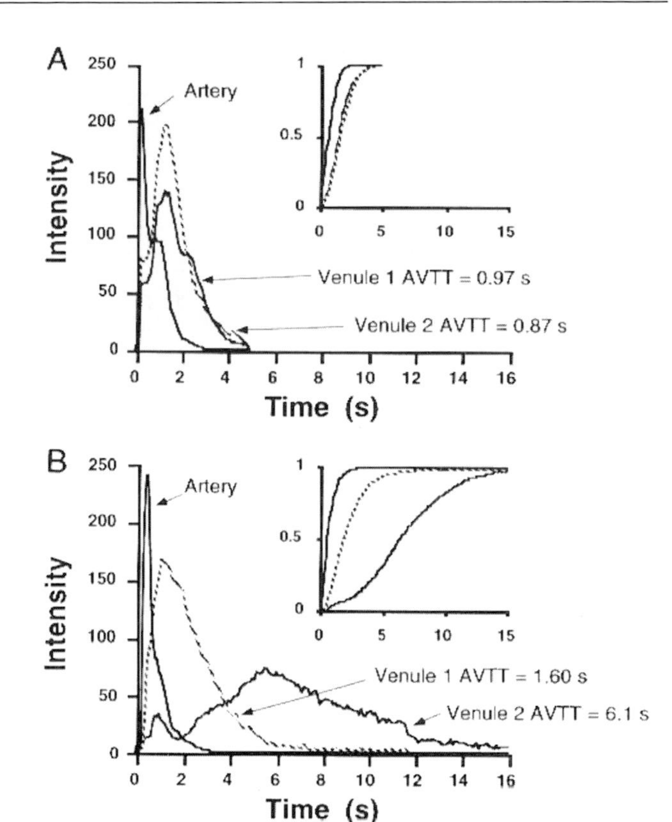

Figure 7 Arteriovenous transit times (AVTTs) of boluses of intravascular markers between identified arterial and venous pairs before and after ligation of cortical arterioles. (**A**) A small bolus of fluorescein injected into the ipsilateral external carotid artery promptly filled a cortical artery. Shortly afterward the dye appeared nearly simultaneously in two nearby surface venules. The inset shows these curves integrated and normalized. AVTTs were calculated from the areas between the arterial and the venular integrals and are similar for the two venules. (**B**) After ligation of arteriolar branches supplying the whisker cortex, there was no significant change in the arrival of dye in the nonischemic artery, but the time to the two venules was obviously delayed. (For details of the paradigm see Wei *et al.,* 1995, 2001). The inset shows shifts of the integrals to the right. The calculated AVTT to venule 2 inside the ischemic region is approximately four times those to venule 1 outside the obviously ischemic region. Increased AVTTs indicate decreased flow.

X. Intravascular Particles

In vessels ≥10 μm in diameter, the two principal components—plasma and RBCs—are partially separated in laminar flow so that the RBCs are concentrated in the vessel axis. Conversely, plasma, platelets, and white blood cells are more concentrated at the luminal wall. In laminar flow, velocity varies from the axis of the vessel, where it is fastest, to the luminal surface, where it is slowest (Rovainen *et al.,* 1993). IAP and H_2 monitor changes in whole blood but do not resolve the differences in timing and distribution of plasma and RBCs (see Fig. 9). However, with magnification, some markers are distinctly segregated in blood, a useful fact for detailed studies of mechanisms (e.g., Rovainen *et al.,* 1992).

Brightly labeled fluorescent latex beads of various sizes are available from commercial suppliers. Their properties for *in vivo* studies are excellent. Because the beads are rigid, they must be small enough to pass the capillary bed (<4 μm in diameter). They can be resolved at low light levels to minimize phototoxic effects. At video frame rates, individual beads can be tracked only in capillaries and small venules; at the higher velocities in larger vessels they appear as streaks, the lengths of which are proportional to velocity (Woolsey and Rovainen, 1991). With stroboscopic illumination (400 Hz), the swiftest beads are "stopped in their tracks" and are illuminated several times in a single video frame (Rovainen *et al.,* 1992). The distance between each consecutive image of the bead is the distance traveled between flashes, which permits measurement of velocity (Fig. 8A).

The beads are injected in trace amounts in small boluses to avoid confusion between multiple beads illuminated simultaneously in the same vessel (Fig. 8A). Their concentration decreases in time as they are cleared from the blood. Beads partition to plasma, where they can be seen moving between the hemoglobin of RBCs and the vessel wall. Fewer beads are swept up in the central flow through a vessel and are intercalated between RBCs. The speed of the

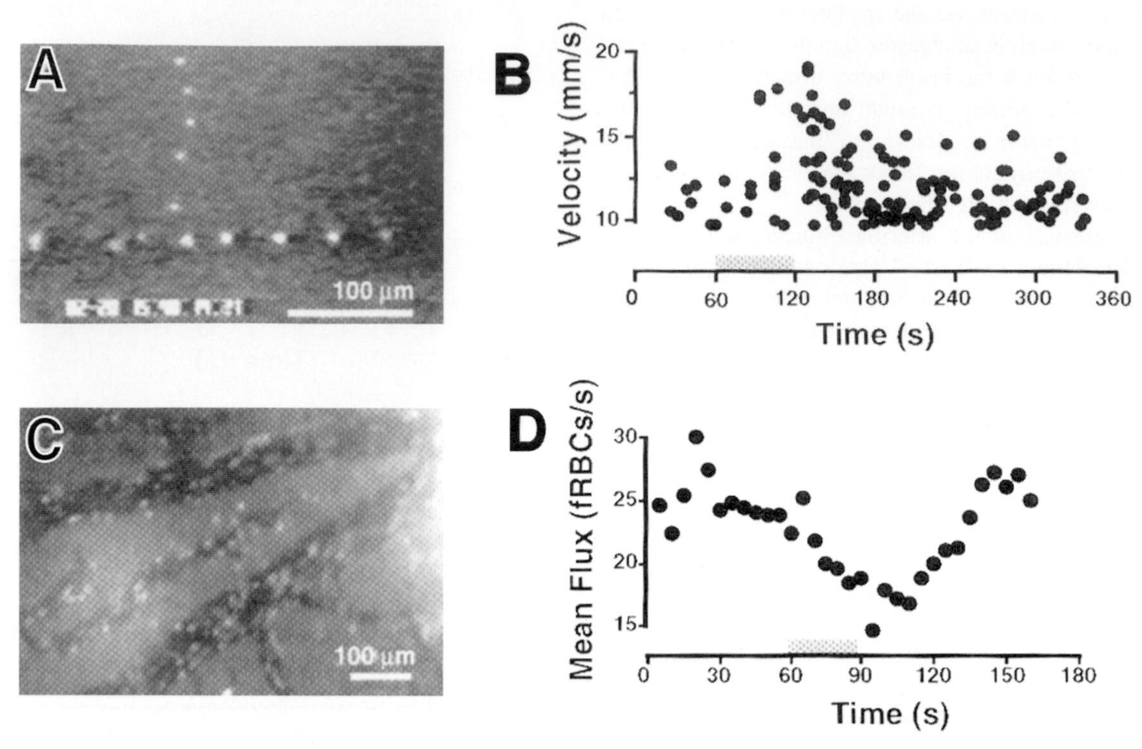

Figure 8 (**A**) Stroboscopic illumination of intensely fluorescent 3-μm beads permits estimates of flow velocities. Tracks for three beads are shown. One moves left to right through the parent arteriole, a second slower bead flows bottom to top into a small branch, and a third enters the frame from the left. Some intercalate between RBCs and approximate maximal velocities [upper envelope of velocities in (B)]. (**B**) Velocities increase with increased flows in activated cortex during whisker stimulation (gray bar at 60 s). This shows beads moving faster than 10 mm/s. The increase in velocity with stimulation of a row of whiskers was observed in an arteriole feeding the cortex over the appropriate barrel row (see Fig. 9). (**C**) RBCs labeled with fluorescein and/or other dyes are larger, more persistent markers. Fluorescently labeled RBCs (fRBCs) have $T_{1/2}$ of about an hour and can be used to measure fluxes in the same venule over extended periods of time. Note the lower magnification compared to (A). (**D**) The frequency of fRBCs in surface venules (RBC flux) is an index of flow, which, in this example, falls in cortex next to activated barrel cortex during stimulation of the appropriate whiskers.

fastest beads approximates the maximal velocity (V_{max}) of RBC flow and, given enough beads (observations), can be measured reliably. From V_{max} and vessel diameter, peak flow can be calculated for individual vessels. The principal drawback to this method is that at the trace amounts of beads required, the incidence and persistence of beads in the bloodstream are low, limiting determination of velocity time courses. The speeds of all beads can be used to calculate the harmonic mean velocity, which is used traditionally together with diameter to calculate nonpulsatile flow in a vessel. The observations can be used to determine peak shear stress at the luminal surface for modeling signal transduction by the endothelium (Fujii *et al.*, 1991; Rovainen *et al.*, 1993; Wang *et al.*, 1992). Bead velocity increases in appropriate vessels with sensory stimulation (Fig. 8B; Cox *et al.*, 1993). However, the repeated injections of beads and their relatively small size (requiring higher magnifications with a reduced field of view and having a narrow focal plane) prompted exploration of other methods for following dynamic flow changes over larger areas of the brain surface.

The high concentration of hemoglobin in RBCs gives them excellent contrast with the surrounding plasma and tissues. In capillaries, where RBCs move in single file, the velocity and fluxes of individual corpuscles can be measured automatically (Hudetz *et al.*, 1995). Special microscope hardware and software such as coincidence detectors (detecting single RBCs as they file by one, then the other of two nearby regions of a capillary) can measure RBC velocities efficiently in capillaries in the focal plane of the microscope (Hudetz *et al.*, 1992). Villringer *et al.* (1994) measured RBC flows in capillaries after injections of fluorescein with a laser-scanning confocal microscope in the single-line mode (≈512 Hz) to follow the movements of single red cells in capillaries of the cortex regardless of the orientation of the vessels. Two-dimensional laser-scanning microscopy (Seylaz *et al.*, 1999) and two-photon microscopy (Helmchen *et al.*, 2001; Kleinfeld *et al.*, 1998) allow imaging of capillary RBC movement 200–500μm below the cortical surface. Use of such approaches to measure flows in cortical capillaries at or just below the cortical surface is demanding for the tissue stability required (pulse and respiration) and restricts the area to be examined (now easily identified by the intrinsic signal).

In larger vessels (diameter >10 μm), individual unlabeled RBCs cannot be resolved by CCD cameras at standard video rates because of their speed and number. Labeling RBCs as tracers is desirable because of their "physiological " characteristics (Hb carriers, longer $T_{1/2}$, flexibility, and uniform distribution after adequate mixing) and larger size (for better visualization at the lower magnifications to see larger areas of cortex simultaneously—active vs not active). RBCs from donor rats that are labeled with fluorescein isothiocyanate (Sarelius and Duling, 1982) and other commercially available labels (PHK; Sigma) can be stored for a week before being resuspended in serum for injection (Liang *et al.,* 1995). Labeled RBCs are stable markers throughout an acute experiment. In trace quantities, fluorescently labeled RBCs can be counted at intermediate magnifications as they flow slowly in venules leaving the brain parenchyma and coursing on the surface. The number of labeled cells/time is termed flux and is a measure of RBC flow in the vessel being observed. Software to count these events automatically has been developed (Knuese *et al.,* 1994) and, as with the other video methods, analysis can be done readily from videotapes post facto. Stroboscopic illumination of fluorescent RBCs has been used to measure blood velocity in arteries similar to that described for fluorescent beads (Johnson, 1995). Using CCD cameras with ultrafast frame rates (up to 1000 frames/s), fluorescently labeled RBC can be tracked to measure velocity in arteries (Ishikawa *et al.,* 1998).

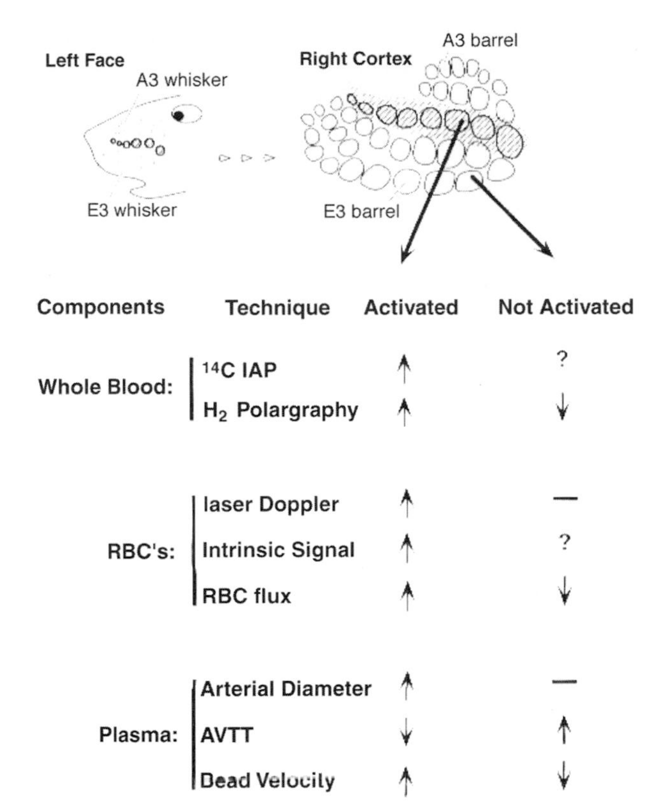

Figure 9 Stimulating a row of whiskers (the middle row C) activates the appropriate barrels (row C) and generally increases flow in the cortex of row C (shading). Different methods monitor different flow-related changes in different components of blood. Qualitative changes observed in rat barrel cortex with whisker stimulation are indicated for each technique. As shown in Fig. 8, flow may be reduced or unchanged in nearby regions that are not directly activated (not hatched). For some methods (?) the effects in cortex not directly activated have not yet been determined. (For details, see Woolsey *et al.,* 1996.)

XI. Localization of Flow Changes

Each of the approaches discussed has been used to study sensory-evoked flow changes in the rodent brain (Greenberg *et al.,* 1979) and changes in development (Foltz *et al.,* 1992; Rovainen *et al.,* 1992; Wang *et al.,* 1992) and in experimental pathology (Wei *et al.,* 1995). The whisker barrels of rats, mice, and hamsters are proven useful models of the columnar functional organization characteristic of the cerebral cortex (e.g., Jacquin *et al.,* 1993a; McCasland and Woolsey, 1988; Simons, 1995). Functional activation of neurons in obvious groups (a single barrel is principally activated by an appropriate single whisker) can be used to correlate flow changes to activation of the neuronal population that surface vessels supply. The general results are summarized in Fig. 9. Increased flows to an area with increased neural activity are observed with each method: increased IAP labeling, increased arterial diameter, decreased AVTT, increased H_2 washout, increased blood volume, increased RBC velocity, increased arterial velocity, and increased RBC flux. The extent to which these changes are sharply localized depends on the method employed. With the microscope, changes are observed in single vessels (it is possible to analyze all of the vessels in a microscopic field) at the brain surface. To measure parenchymal flows, autoradiography determines tissue perfusion to a resolution of ≈100 μm (see Bryan and Duckrow, 1995). Dynamic techniques having poorer spatial resolution could be improved by changing the configuration of H_2 electrodes (dependent on tip size and interelectrode spacing) and the diameter of a fiber-optic filament for laser-Doppler flowmetry (dependent on the hemisphere illuminated in the tissue) to follow time courses of parenchymal flow changes. Several (but not all) methods demonstrate decreased flow in areas not directly stimulated (Fig. 8D). Studies are under way in several laboratories to determine what factors, neural and/or hemodynamic, contribute to these decreases in flow. Figure 9 summarizes the findings qualitatively.

XII. Conclusions and Prospects

Exceptional opportunities now exist for understanding brain function from flow-based markers. Flow changes indicating changes in neuronal activity can be detected

Table 1 Summary of Techniques

Method	Craniotomy	Spatial resolution	Temporal resolution	Observations	Tissue	Measure
^{14}C IAP ^{99}Tc HM-PAO autoradiography	–	≈100 μm	30 s	Once	Whole brain	Flow
Occlusive beads	–	>1 mm	10 s	Several	Whole brain	Flow
Laser-Doppler flowmetry	+ (or thinned skull)	≈1 mm	200 ms	Continuously	At probe tip	"Flow," "velocity," "volume"
H_2 polarography	+ (or thinned skull)	>300 μm	<1 s	Continuously	At electrode tip	Flow
Intrinsic signal	+ (or thinned skull)	<1 μm	<33 ms	1.2–30/min Video rates (30–60/s)[a,b]	At or near brain surface	Hemoglobin volume[a]
Vessel diameters	+	<1 μm	17 ms	Video rates (30–60/s)[a,b]	At or near brain surface	Volume
AVTT	+	<1 μm	17 ms	4/min	At or near brain surface	Flow
Fluorescent beads	+	<1 μm	Strobe rates, e.g., 2.5 ms	Video rates (30–60/s)	At or near brain surface	Velocity
Fluorescent RBCs	+	<1 μm	17 ms (strobe 2.5 ms)	Video rates (30–60/s)	At or near brain surface	Flow, velocity, volume

[a]Detection rate, depth of view, and signal source depend on illumination λ, acquisition, and processing.

[b]Some cameras permit faster video rates (i.e., 1000/s).

noninvasively. It is imperative to have a firm handle on the basis of those changes. Table 1 summarizes the salient features of each method. The cranial window technique requires a craniotomy and is invasive up to the cortical surface. It is possible to install windows chronically to overcome the limitations imposed by anesthesia, and awake, sedated subjects can be secured under the microscope. Direct observation of single vessels is limited to comparatively small microscopic fields. Several of these techniques can be considered together to estimate local flow changes which, depending on assumptions, approach the values from the gold standard methods. A number of the video techniques have been used to determine the timing, extent, and stimulus parameters in activation of cortical blood flow changes with natural stimulation. As these and other factors are further understood, they will provide useful information for the design of activation and imaging sequences (e.g., Sereno *et al.,* 1995).

Acknowledgments

This work was supported by NIH Grants NS 28781 and NS 37372 and by an award from the Spastic Paralysis Foundation of the Illinois–Eastern Iowa District of the Kiwanis International.

References

Ances, B. M., Detre, J. A., Takahashi, K., and Greenberg, J. H. (1998). Transcranial laser Doppler mapping of activation flow coupling of the rat somatosensory cortex. *Neurosci. Lett.* **257,** 25–28.

Ances, B. M., Greenberg, J. H., and Detre, J. A. (1999). Laser Doppler imaging of activation-flow coupling in the rat somatosensory cortex. *NeuroImage* **10,** 716–723.

Blasdel, G. G., and Salama, G. (1986). Voltage-sensitive dyes reveal a modular organization in monkey striate cortex. *Nature* **321,** 579–585.

Boero, J. A., Ascher, J., Arregui, A., Rovainen, C., and Woolsey, T. A. (1999). Increased brain capillaries in chronic hypoxia. *J. Appl. Physiol.* **86,** 1211–1219.

Borowsky, I. W., and Collins, R. C. (1989). Metabolic anatomy of brain: A comparison of regional capillary density, glucose metabolism, and enzyme activities. *J. Comp. Neurol.* **288,** 401–413.

Bryan, R. M., and Duckrow, R. B. (1995). Radial columns in autoradiographs generated from tracer methods for measuring cerebral cortical blood flow. *Am. J. Physiol.* **269,** H583–H489.

Busija, D. W., Heistad, D. D., and Marcus, M. L. (1981). Continuous measurement of cerebral blood flow in anesthetized cats and dogs. *Am. J. Physiol.* **241,** H228–H234.

Chen, Z., Milner, T. E., Wang, X., Srinivas, S., and Nelson, J. S. (1998). Optical Doppler tomography: Imaging in vivo blood flow dynamics following pharmacological intervention and photodynamic therapy. *Photochem. Photobiol.* **67,** 56–60.

Cox, S. B., Woolsey, T. A., and Rovainen, C. M. (1993). Localized dynamic changes in cerebral blood flow in rat barrel cortex with whisker stimulation. *J. Cereb. Blood Flow Metab.* **13,** 899–913.

Detre, J. A., Ances, B. M., Takahashi, K., and Greenberg, J. H. (1998). Signal averaged laser Doppler measurements of activation-flow coupling in the rat forepaw somatosensory cortex. *Brain Res.* **796,** 91–98.

Dirnagl, U., Kaplan, B., Jacewicz, M., and Pulsinelli, W. (1989). Continuous measurement of cerebral cortical blood flow by laser-Doppler flowmetry in a rat stroke model. *J. Cereb. Blood Flow Metab.* **9,** 589–596.

Dowling, J. L., Henegar, M. M., Liu, D., Rovainen, C. M., and Woolsey, T. A. (1996). Rapid optical imaging of whisker responses in rat barrel cortex. *J. Neurosci. Methods* **66,** 113–122.

Dunn, A. K., Bolay, H., Moskowitz, M. A., and Boas, D. A. (2001). Dynamic imaging of cerebral blood flow using laser speckle. *J. Cereb. Blood Flow Metab.* **21,** 195–201.

Durham, D., and Woolsey, T. A. (1978). Acute whisker removal reduces neuronal activity in barrels of mouse SmI cortex. *J. Comp. Neurol.* **178,** 629–644.

Eke, A. (1993). Multiparametric imaging of microregional circulation over the brain cortex by videoreflectometry. *In* "Optical Imaging of Brain Function and Metabolism" (U. Dirnagl, A. Villringer, and K. Einhaupl, eds.), pp. 183–191. Plenum, New York.

Erinjeri, J. P., and Woolsey, T. A. (2002). Spatial integration of vascular changes with neural activity in mouse cortex. *J. Cereb. Blood Flow Metab.*, **22**, 353–360.

Fabricius, M., Akgoren, N., Dirnagl, U., and Lauritzen, M. (1997). Laminar analysis of cerebral blood flow in cortex of rats by laser-Doppler flowmetry: A pilot study. *J. Cereb. Blood Flow Metab.* **17,** 1326–1336.

Fabricius, M., and Lauritzen, M. (1996). Laser-Doppler evaluation of rat brain microcirculation: Comparison with the [^{14}C]iodoantipyrine method suggests discordance during cerebral blood flow increases. *J. Cereb. Blood Flow Metab.* **16,** 156–161.

Feindel, W., Hodge, C. P., and Yamamoto, Y. L. (1968). Epicerebral angiography by fluorescein during craniotomy. *Prog. Brain Res.* **30,** 471–477.

Foltz, G. D., Rovainen, C. M., and Woolsey, T. A. (1992). Developmental changes in identified cerebral arterioles and venules in individual mice at two ages. *Soc. Neurosci. Abstr.* **18,** 154.

Frostig, R. D., Lieke, E. E., Tso, D. Y., and Grinvald, A. (1990). Cortical functional architecture and local coupling between neuronal activity and the microcirculation revealed by in vivo high-resolution optical imaging of intrinsic signals. *Proc. Natl. Acad. Sci. USA* **87,** 6082–6086.

Fujii, K., Heistad, D. D., and Faraci, F. M. (1991). Flow-mediated dilatation of the basilar artery in vivo. *Circ. Res.* **69,** 697–705.

Fujita, H., Matsuura, T., Yamada, K., Inagaki, N., and Kanno, I. (2000). A sealed cranial window system for simultaneous recording of blood flow, and electrical and optical signals in the rat barrel cortex. *J. Neurosci. Methods* **99,** 71–78.

Gerrits, R. J., Stein, E. A., and Greene, A. S. (1998). Laser-Doppler flowmetry utilizing a thinned skull cranial window preparation and automated stimulation. *Brain Res. Protoc.* **3,** 14–21.

Ginsberg, M. D., Dietrich, W. D., and Busto, R. (1987). Coupled forebrain increases of local cerebral glucose utilization and blood flow during physiologic stimulation of a somatosensory pathway in the rat. *Neurology* **37,** 11–19.

Greenberg, J., Hand, P., Sylvestro, A., and Reivich, M. (1979). Localized metabolic–flow couple during functional activity. *Acta Neurol. Scand.* **60,** 12–13.

Grinvald, A., Lieke, E., Frostig, R. D., Gilbert, C. D., and Wiesel, T. N. (1986). Functional architecture of cortex revealed by optical imaging of intrinsic signals. *Nature* **324,** 361–364.

Gross, P. M., Sposito, N. M., Pettersen, S. E., Panton, D. G., and Fenstermach, J. D. (1987). Topography of capillary density, glucose metabolism, and microvascular function within the rat inferior colliculus. *J. Cereb. Blood Flow Metab.* **7,** 154–160.

Haberl, R. L., Heizer, M. L., Marmarou, A., and Ellis, E. F. (1989). Laser-Doppler assessment of brain microcirculation: Effect of systemic alterations. *Am. J. Physiol.* **256,** H1247–H1254.

Haglund, M. M., Ojemann, G. A., and Hochman, D. W. (1992). Optical imaging of epileptiform and functional activity in human cerebral cortex. *Nature* **358,** 668–671.

Harvey, W. (1978). "Circulation of the Blood." University Park Press, Baltimore.

Heiss, W. D., and Traupe, H. (1981). Comparison between hydrogen clearance and microsphere technique for rCBF measurement. *Stroke* **12,** 161–167.

Heistad, D., and Kontos, H. (1983). Cerebral circulation. *In* "Handbook of Physiology" (J. Shepherd, F. Abboud, and S. Geiger, eds.), Vol. 3, pp. 137–182. Am. Physiol. Soc., Bethesda, MD.

Heistad, D. D., Marcus, M. L., and Mueller, S. (1977). Measurement of cerebral blood flow with microspheres. *Arch. Neurol.* **34,** 657–659.

Helmchen, F., Fee, M. S., Tank, D. W., and Denk, W. (2001). A miniature head-mounted two-photon microscope: High-resolution brain imaging in freely moving animals. *Neuron* **31,** 903–912.

Horton, R. W., Pedley, T. A., Meldrum, B. S., and Chir, B. (1980). Regional cerebral blood flow in the rat as determined by particle distribution and by diffusible tracer. *Stroke* **11,** 39–44.

Hubel, D. H. (1982). Exploration of the primary visual cortex, 1955–78. *Nature* **299,** 515–524.

Hudetz, A. G. (1997). Blood flow in the cerebral capillary network: A review emphasizing observations with intravital microscopy. *Microcirculation* **4,** 233–252.

Hudetz, A. G., Feeher, G., Weigle, C. G. M., Knuese, D. E., and Kampine, J. P. (1995). Video microscopy of cerebrocortical capillary flow: Response to hypotension and intracranial hypertension. *Am. J. Physiol.* **268,** H2202–H2210.

Hudetz, A. G., Weigle, C. G. M., Fenoy, F. J., and Roman, R. J. (1992). Use of fluorescently labeled erythrocytes and digital cross-correlation for the measurement of flow velocity in the cerebral microcirculation. *Microvasc. Res.* **43,** 334–341.

Ido, Y., Chang, K., Woolsey, T. A., and Williamson, J. R. (2001). NADH: Sensor of blood flow need in brain, muscle, and other tissues. *FASEB J.* **15,** 1419–1421.

Ingvar, D. H. (1975). Patterns of brain activity revealed by measurements of regional cerebral blood flow. *In* "Brain Work: The Coupling of Function Metabolism and Blood Flow in the Brain" (D. H. Ingvar and N. A. Lassen, eds.), pp. 397–413. Munksgaard, Copenhagen.

Irikura, K., Maynard, K. I., and Moskowitz, M. A. (1994). Importance of nitric oxide synthase inhibition to the attenuated vascular responses induced by topical L-nitroarginine during vibrissal stimulation. *J. Cereb. Blood Flow Metab.* **14,** 45–48.

Ishikawa, M., Sekizuka, E., Shimizu, K., Yamaguchi, N., and Kawase, T. (1998). Measurement of RBC velocities in the rat pial arteries with an image-intensified high-speed video camera system. *Microvasc. Res.* **56,** 166–172.

Jacquin, M. F., McCasland, J. S., Henderson, T. A., Rhoades, R. W., and Woolsey, T. A. (1993a). 2-DG uptake patterns related to single vibrissae during exploratory behaviors. *J. Comp. Neurol.* **332,** 38–58.

Jacquin, M. F., Renehan, W. E., Rhoades, R. W., and Panneton, W. M. (1993b). Morphology and topography of identified primary afferents in trigeminal subnuclei principalis and oralis. *J. Neurophysiol.* **70,** 1911–1936.

Jallo, J., Saetzler, R., Mishke, C., Young, W. F., Vasthare, U., and Tuma, R. F. (1997). A chronic model to simultaneously measure intracranial pressure, cerebral blood flow, and study the pial microvasculature. *J. Neurosci. Methods* **75,** 155–160.

Jay, T. M., Lucignani, G., Crane, A. M., Jehle, J., and Sokoloff, L. (1988). Measurement of local cerebral blood flow with [^{14}C]iodoantipyrine in the mouse. *J. Cereb. Blood Flow Metab.* **8,** 121–129.

Johnson, P. C. (1995). *Biophoton. Int.* **2.**

Kety, S. S. (1960). Measurement of local blood flow by the exchange of an inert, diffusible substance. *Methods Med. Res.* **8,** 228–236.

Kety, S. S., and Schmidt, C. F. (1945). The determination of cerebral blood flow in man by the use of nitrous oxide in low concentrations. *Am. J. Physiol.* **143,** 53–66.

Klein, B., Kuschinsky, W., Schrock, H., and Vetterlein, F. (1986). Interdependency of local capillary density, blood flow and metabolism in rat brains. *Am. J. Physiol.* **251,** H1333–H1340.

Kleinfeld, D., Mitra, P. P., Helmchen, F., and Denk, W. (1998). Fluctuations and stimulus-induced changes in blood flow observed in individual capillaries in layers 2 through 4 of rat neocortex. *Proc. Natl. Acad. Sci. USA* **95,** 15741–15746.

Knuese, D. E., Feher, G., and Hudetz, A. G. (1994). Automated measurement of fluorescently labeled erythrocyte flux in cerebrocortical capillaries. *Microvasc. Res.* **47,** 392–400.

Landis, E. M. (1982). The capillary circulation. *In* "Circulation of the Blood, Men and Ideas" (A. P. Fishman and D. W. Richards, eds.), pp. 355–406. Am. Physiol. Soc., Bethesda, MD.

Lasjaunias, P. L., and Berenstein, A. (1990). "Surgical Neuroangiography," Vol. 3. Springer Verlag, Berlin.

Lassen, N. A., Roland, P. E., Larsen, B., Melamed, E., and Soh, K. (1977). Mapping of human cerebral functions: A study of the regional cerebral blood flow pattern during rest, its reproducibility and the activations seen during basic sensory and motor functions. *Acta Neurol. Scand. Suppl.* **64,** 262–263, 274–275.

Lear, J. L. (1988). Quantitative local cerebral blood flow measurements with technetium-99m HM-PAO: Evaluation using multiple radionuclide digital quantitative autoradiography. *J. Nucl. Med.* **29,** 1387–1392.

Liang, G. E., Thompson, B. P., Erinjeri, J. P., Liu, D., Rovainen, C. M., and Woolsey, T. A. (1995). RBC flow and vessel diameter changes with stimulation of rat whisker barrel cortex. *Microcirc. Soc. Abstr.* **2,** 96.

Lindauer, U., Villringer, A., and Dirnagl, U. (1993). Characterization of CBF response to somatosensory stimulation: Model and influence of anesthetics. *Am. J. Physiol.* **264,** H1223–H1228.

Linde, R., Schmalbruch, I. K., Paulson, O. B., and Madsen, P. L. (1999). The Kety–Schmidt technique for repeated measurements of global cerebral blood flow and metabolism in the conscious rat. *Acta Physiol. Scand.* **165,** 395–401.

Lindsberg, P. J., O'Neill, J. T., Paakkari, I. A., Hallenbeck, J. M., and Feuerstein, G. (1989). Validation of laser-Doppler flowmetry in measurement of spinal cord blood flow. *Am. J. Physiol.* **257,** H674–H680.

Liu, D., Dowling, J., Spence, M. E., Rovainen, C. M., and Woolsey, T. A. (1994). Blood flow responses in the rat barrel cortex during whisker stimulation. *Soc. Neurosci. Abstr.* **20,** 1422.

Ma, Y. P., Koo, A., Kwan, H. C., and Cheng, K. K. (1974). On-line measurement of the dynamic velocity of erythrocytes in the cerebral microvessels in the rat. *Microvasc. Res.* **8,** 1–13.

Maeda, K., Mies, G., Olah, L., and Hossmann, K. A. (2000). Quantitative measurement of local cerebral blood flow in the anesthetized mouse using intraperitoneal [^{14}C]iodoantipyrine injection and final arterial heart blood sampling. *J. Cereb. Blood Flow Metab.* **20,** 10–14.

Malonek, D., Dirnagl, U., Lindauer, U., Yamada, K., Kanno, I., and Grinvald, A. (1997). Vascular imprints of neuronal activity: Relationships between the dynamics of cortical blood flow, oxygenation, and volume changes following sensory stimulation. *Proc. Natl. Acad. Sci. USA* **94,** 14826–14831.

Marcus, M. L., Heistad, D. D., Ehrhardt, J. C., and Abboud, F. M. (1976). Total and regional cerebral blood flow measurement with 7-, 10-, 15-, 25-, and 50μm microspheres. *J. Appl. Physiol.* **40,** 501–507.

Masino, S. A., Kwon, M. C., Dory, Y., and Frostig, R. D. (1993). Characterization of functional organization within rat barrel cortex using intrinsic signal optical imaging through a thinned skull. *Proc. Natl. Acad. Sci. USA* **90,** 9998–10002.

McCasland, J. S., and Woolsey, T. A. (1988). High resolution 2DG mapping of functional cortical columns in mouse barrel cortex. *J. Comp. Neurol.* **278,** 555–569.

Merzenich, M. M., Kaas, J. H., Wall, J., Nelson, R. J., Sur, M., and Felleman, D. (1983). Topographic reorganization of somatosensory cortical areas 3b and 1 in adult monkeys following restricted deafferentation. *Neuroscience* **8,** 33–55.

Morii, S., Ngai, A., and Winn, H. (1986). Reactivity of rat pial arterioles and venules to adenosine and carbon dioxide: With detailed description of the closed cranial window technique in rats. *J. Cereb. Blood Flow Metab.* **6,** 34–41.

Morris, D. C., Zhang, Z., Davies, K., Fenstermacher, J., and Chopp, M. (1999). High resolution quantitation of microvascular plasma perfusion in non-ischemic and ischemic rat brain by laser-scanning confocal microscopy. *Brain Res. Protoc.* **4,** 185–191.

Moskalenko, Y. E., Dowling, J., Liu, D., Rovainen, C. M., Spence, M. E., and Woolsey, T. A. (1996). LCBF changes in rat somatosensory cortex during whisker stimulation: Dynamic H_2 clearance studies. *Int. J. Psychophysiol.* **21,** 45–59.

Moskalenko, Y. E., Rovainen, C. M., Woolsey, T. A., Wei, L., Liu, D., Spence, M. E., Semernia, V. N., Weinstein, G. B., and Malisheva, N. G. (1994). Comparison of local CBF measurements by H_2-clearance with H_2 inhalation and with transient electrochemical H_2 generation in brain tissue. *Sechenov. Physiol. J.* **80,** 119–125.

Moskalenko, Y. E., Weinstein, G. B., Demchencko, I. T., Kislyakov, Y. Y., and Krivchenko, A. I. (1980). "Biophysical Aspects of Cerebral Circulation." Pergamon, Oxford.

Moskalenko, Y. E., Woolsey, T. A., Rovainen, C., Weinstein, G. B., Liu, D., Semernya, V. N., and Mitrofanov, V. F. (1998). Blood flow dynamics in different layers of the somatosensory region of the cerebral cortex on the rat during mechanical stimulation of the vibrissae. *Neurosci. Behav. Physiol.* **28,** 459–467.

Muller, T. B., Jones, R. A., Haraldseth, O., Westby, J., and Unsgard, G. (1996). Comparison of MR perfusion imaging and microsphere measurements of regional cerebral blood flow in a rat model of middle cerebral artery occlusion. *Magn. Reson. Imaging* **14,** 1177–1183.

Narayan, S. M., Santori, E. M., and Toga, A. W. (1994). Mapping functional activity in rodent cortex using optical intrinsic signals. *Cereb. Cortex* **4,** 195–204.

Neirinckx, R. D., Canning, L. R., Piper, I. M., Nowotnik, D. P., Pickett, R. D., Holmes, R. A., Volkert, W. A., Forster, A. M., Weisner, P. S., and Marriott, J. A. (1987). Technetium-99m d,l-HM-PAO: A new radiopharmaceutical for SPECT imaging of regional cerebral blood perfusion. *J. Nucl. Med.* **28,** 191–202.

Ngai, A., Ko, K., Morii, S., and Winn, H. (1988). Effect of sciatic nerve stimulation on pial arterioles in rats. *Am. J. Physiol.* **254,** H133–H139.

Ngai, A. C., Meno, J. R., and Winn, H. R. (1995). Simultaneous measurements of pial arteriolar diameter and laser-Doppler flow during somatosensory stimulation. *J. Cereb. Blood Flow Metab.* **15,** 124–127.

Ngai, A. C., and Winn, H. R. (1996). Estimation of shear and flow rates in pial arterioles during somatosensory stimulation. *Am. J. Physiol.* **270,** H1712–H1717.

Parekh, N., Zou, A. P., Jungling, E., Endlich, K., Sadowski, J., and Steinhausen, M. (1993). Sex differences in control of renal outer medullary circulation in rats: Role of prostaglandins. *Am. J. Physiol.* **264,** F629–F636.

Peterson, B. E., and Goldreich, D. (1994). A new approach to optical imaging applied to rat barrel cortex. *J. Neurosci. Methods* **54,** 39–47.

Posner, M. I., and Raichle, M. E. (1994). "Images of Mind: Exploring the Brain's Activity." Freeman, New York.

Prakash, N., Vanderhaeghen, P., Cohen-Cory, S., Frisen, J., Flanagan, J. G., and Frostig, R. D. (2000). Malformation of the functional organization of somatosensory cortex in adult ephrin-A5 knock-out mice revealed by in vivo functional imaging. *J. Neurosci.* **20,** 5841–5847.

Ratzlaff, E. H., and Grinvald, A. (1991). A tandem-lens epifluorescence microscope: Hundred-fold brightness advantage for widefield imaging. *J. Neurosci. Methods* **36,** 127–137.

Rausch, M., and Eysel, U. T. (1996). Visualization of lCBF changes during cortical infarction using IR thermo-encephaloscopy. *NeuroReport* **7,** 2603–2606.

Rosenblum, W. I. (1969). Erythrocyte velocity and a velocity pulse in minute blood vessels on the surface of the mouse brain. *Circ. Res.* **24,** 887–892.

Rosenblum, W. I. (1970). Effects of blood pressure and blood viscosity on fluorescein transit time in the cerebral microcirculation in the mouse. *Circ. Res.* **27,** 825–833.

Rovainen, C. M., Wang, D. B., and Woolsey, T. A. (1992). Strobe epi-illumination of fluorescent beads indicates similar velocities and shear rates in brain arterioles of newborn and adult mice. *Microvasc. Circ.* **43,** 235–239.

Rovainen, C. M., Woolsey, T. A., Blocher, N. C., Wang, D. B., and Robinson, O. F. (1993). Blood flow in single surface arterioles and venules on the mouse somatosensory cortex measured with videomicroscopy, fluorescent dextrans, non-occluding fluorescent beads, and computer-assisted image analysis. *J. Cereb. Blood Flow Metab.* **13,** 359–371.

Roy, C. S., and Sherrington, C. S. (1890). On the regulation of the blood supply of the brain. *J. Physiol. (London)* **11,** 85–108.

Sakurada, O., Kennedy, C., Jehle, J., Brown, J. D., Carbin, G. L., and Sokoloff, L. (1978). Measurement of local cerebral blood flow with iodo[^{14}C]antipyrine. *Am. J. Physiol.* **234,** H59–H66.

Sarelius, I. H., and Duling, B. R. (1982). Direct measurement of microvessel hematocrit, red cell flux, velocity, and transit time. *Am. J. Physiol.* **243,** H1018–H1026.

Schiszler, I., Tomita, M., Fukuuchi, Y., Tanahashi, N., and Inoue, K. (2000). New optical method for analyzing cortical blood flow heterogeneity in small animals: Validation of the method. *Am. J. Physiol.* **279,** H1291–H1298.

Seki, J., Sasaki, Y., Oyama, T., and Yamamoto, J. (1996). Fiber-optic laser-Doppler anemometer microscope applied to the cerebral microcirculation in rats. *Biorheology* **33,** 463–470.

Sereno, M. I., Dale, A. M., Reppas, J. B., Kwong, K. K., Belliveau, J. W., Brady, T. J., Rosen, B. R., and Tootell, R. B. H. (1995). Borders of multiple visual areas in humans revealed by functional magnetic resonance imaging. *Science* **268,** 889–893.

Seylaz, J., Charbonne, R., Nanri, K., Von Euw, D., Borredon, J., Kacem, K., Meric, P., and Pinard, E. (1999). Dynamic in vivo measurement of erythrocyte velocity and flow in capillaries and of microvessel diameter in the rat brain by confocal laser microscopy. *J. Cereb. Blood Flow Metab.* **19,** 863–870.

Simons, D. J. (1985). Temporal and spatial integration in the rat SI vibrissa cortex. *J. Neurophysiol.* **54,** 615–635.

Simons, D. J. (1995). Neuronal integration in the somatosensory whisker/barrel cortex. *In* "Cerebral Cortex" (E. G. Jones and I. T. Diamond, eds.), Vol. 11, pp. 263–297. Plenum, New York.

Skarphedinsson, J. O., Harding, H., and Thoren, P. (1988). Repeated measurements of cerebral blood flow in rats. Comparisons between the hydrogen clearance method and laser Doppler flowmetry. *Acta Physiol. Scand.* **134,** 133–142.

Stosseck, K., Lübbers, D. W., and Cottin, N. (1974). Determination of local blood flow (microflow) by electrochemically generated hydrogen: Construction and application of the measuring probe. *Acta Physiol. Scand.* **134,** 133–142.

Tomita, M., Gotoh, F., Amano, T., Tanahashi, N., Kobari, M., Shinohara, T., and Mihara, B. (1983). Transfer function through regional cerebral cortex evaluated by a photoelectric method. *Am. J. Physiol.* **245,** H385–H398.

Ts'o, D. Y., Frostig, R. D., Lieke, E. E., and Grinvald, A. (1990). Functional organization of primate visual cortex revealed by high resolution optical imaging. *Science* **249,** 417–420.

Uranishi, R., Nakase, H., Sakaki, T., and Kempski, O. S. (1999). Evaluation of absolute cerebral blood flow by laser-Doppler scanning—Comparison with hydrogen clearance. *J. Vasc. Res.* **36,** 100–105.

Villringer, A., Them, A., Lindauer, U., Einhaupl, K., and Dirnagl, U. (1994). Capillary perfusion of the rat brain cortex: An in vivo confocal microscopy study. *Circ. Res.* **75,** 55–62.

Wang, D. B., Blocher, N. C., Spence, M. E., Rovainen, C. M., and Woolsey, T. A. (1992). Development and remodeling of cerebral blood vessels and their flow in postnatal mice observed with in vivo videomicroscopy. *J. Cereb. Blood Flow Metab.* **12,** 935–946.

Wardell, K., and Nilsson, G. E. (1996). Duplex laser Doppler perfusion imaging. *Microvasc. Res.* **52,** 171–182.

Wei, L., Craven, K., Erinjeri, J., Liang, G. E., Bereczki, D., Rovainen, C. M., Woolsey, T. A., and Fenstermacher, J. D. (1998). Local cerebral blood flow during the first hour following acute ligation of multiple arterioles in rat whisker barrel cortex. *Neurobiol. Dis.* **5,** 142–150.

Wei, L., J. P., E., Rovainen, C. M., and Woolsey, T. A. (2001). Collateral growth and angiogenesis around cortical stroke. *Stroke* **32,** 2179–2184.

Wei, L., Rovainen, C. M., and Woolsey, T. A. (1995). Ministrokes in rat barrel cortex. *Stroke* **36,** 1459–1462.

Welker, C., and Woolsey, T. A. (1974). Structure of layer IV in the somatosensory neocortex of the rat: Description and comparison with the mouse. *J. Comp. Neurol.* **158,** 437–454.

Woolsey, C. N., Erickson, T. C., and Gilson, W. E. (1979). Localization in somatic sensory and motor areas of human cerebral cortex as determined by direct recording of evoked potentials and electrical stimulation. *J. Neurosurg.* **51,** 476–506.

Woolsey, T. A. (1967). Somatosensory, auditory and visual cortical areas of the mouse. *Johns Hopkins Med. J.* **121,** 91–112.

Woolsey, T. A. (1990). Peripheral alteration and somatosensory development. *In* "Development of Sensory Systems in Mammals" (J. Coleman, ed.), pp. 465–520. Wiley, New York.

Woolsey, T. A. (1993). Glomerulos, barrels, columns and maps in cortex: An homage to Dr. Rafael Lorente de Nó. *In* "The Mammalian Cochlear Nuclei: Organization and Function" (M. A. Merchán, J. M. Juiz, D. A. Godfrey, and E. Mugnaini, eds.), pp. 479–501. Plenum, New York.

Woolsey, T. A., and Rovainen, C. M. (1991). Whisker barrels: A model for direct observation of changes in the cerebral microcirculation with neuronal activity. *In* "Brain Work and Mental Activity: Alfred Benzon Symposium" (N. A. Lassen, D. H. Ingvar, M. E. Raichle, and L. Friberg, eds.), Vol. 31, pp. 189–200. Munksgaard, Copenhagen.

Woolsey, T. A., Rovainen, C. M., Cox, S. B., Henegar, M. H., Liang, G. E., Liu, G., Moskalenko, Y. E., Sui, J., and Wei, L. (1996). Neuronal units linked to microvascular modules in cerebral cortex: Response elements for imaging the brain. *Cereb. Cortex* **6,** 647–660.

Woolsey, T. A., and Van der Loos, H. (1970). The structural organization of layer IV in the somatosensory region (SI) of mouse cerebral cortex: The description of a cortical field composed of discrete cytoarchitectonic units. *Brain Res.* **17,** 205–242.

Young, W. (1980). H_2 clearance measurement of blood flow: A review of technique and polarographic principles. *Stroke* **11,** 552–564.

Yu, D. Y., Alder, V. A., and Cringle, S. J. (1991). Measurement of blood flow in rat eyes by hydrogen clearance. *Am. J. Physiol.* **261,** H960–H968.

Zhao, W., Busto, R., Truettner, J., and Ginsberg, M. D. (2001). Simultaneous measurement of cerebral blood flow and mRNA signals: Pixel-based inter-modality correlational analysis. *J. Neurosci. Methods* **108,** 161–170.

8

Electrophysiological Imaging of Brain Function

Alan Gevins

San Francisco Brain Research Institute and SAM Technology, San Francisco, California 94108

I. Introduction

A central goal of functional brain mapping is to isolate local neuronal activity associated with sensory, motor, and cognitive functions or with disease processes. To be truly effective in this regard, an imaging modality needs both millimeter precision in localizing regions of activated tissue and subsecond temporal precision for characterizing changes in patterns of activation over time. Increasingly fine anatomical resolution is available from functional magnetic resonance imaging, but the length of the temporal sample required is still currently an order of magnitude too long to resolve the rapidly shifting neurophysiological processes of cognition and certain disease processes, including epileptic seizure onsets. The electroencephalogram (EEG), in contrast, has a temporal resolution as fine as the analog-to-digital sampling rate used to record it (typically in the 1- to 10-ms range). Furthermore, the high sensitivity of the EEG to changes in sensory, motor, and cognitive activity particularly has been recognized since Hans Berger (1929) reported a decrease in the amplitude of the dominant (a) rhythm of the EEG during mental arithmetic.

These features would seem to make the EEG an ideal complement to *f*MRI and PET in mapping brain function. Additionally, the fact that EEGs can be recorded from ambulatory subjects in natural environments means that EEGs can provide a unique source of information about brain function in the real world beyond the scanner's barrel. Sadly, the EEG is being somewhat neglected as attention is focused on three-dimensional localization of brain function using the newer brain mapping technologies. To those accustomed to viewing the crisp little dots of activation on *f*MRIs, the cortical lobular-level spatial detail provided by conventional EEG recordings seems coarse by comparison. Although the ability to infer the three-dimensional distribution of electrical sources in the brain from scalp EEG (or magnetoencephalogram, MEG) recordings has fundamental physical limits, the amount of spatial information that can be recovered from the scalp-recorded EEG is usually underappreciated. Indeed, the low level of spatial detail in conventional EEGs reflects the simple fact that only a small number of scalp sites (2, 3, 8, or 19) are sampled during most routine recordings; moreover, digital spatial signal-enhancing methods to correct the blurring due to conduction through the highly resistive skull have not often been applied. A number of laboratories have demonstrated the increase in detail possible by making EEG recordings with 64 to more than 100 electrodes, performing signal processing to deblur the data, and coregistering these data with structural and functional MRIs. Above and beyond such advances, it must be remembered that three-dimensional localization is but one of many possible reasons for measuring brain function. In many basic science and clinical applications, the aim is to *monitor,* rather than *localize,* brain function. In such instances, the EEG continues to be the method of choice because of its high sensitivity, unimposing recording conditions, and low cost.

This chapter briefly reviews methodological issues in using the EEG and stimulus-related brain electrical activity such as evoked or "event-related" potentials (ERPs) to image cognitive brain functions. We begin by briefly considering the neural basis of the EEG and its measurement at the scalp. We then consider innovations in EEG recording and analysis technology that improve spatial resolution and integration of EEG with MRI. Many of the basic issues covered here are directly extensible to magnetoencephalographic recordings as well; a summary of differences between EEG and MEG approaches is presented in Wikswo *et al.* (1993) and references cited therein. A more detailed discussion of conventional recording and analysis techniques in clinical and experimental electroencephalography can also be found elsewhere (Regan, 1989). Finally, use of EEG as a diagnostic instrument in clinical neurology is beyond the scope of this chapter; a comprehensive discussion of clinical EEG can be found in many places (Chiappa, 1983; Aminoff, 1986; Daly and Pedley, 1990).

II. The Electroencephalogram and Averaged Event-Related Potentials

A. Neuronal Generation and Transmission to the Scalp

The ongoing EEG is generated when currents flow either into (a current "sink") or out of (a current "source") a cell across charged neuronal membranes. The EEG does not derive from summated action potentials. Instead, the EEG recorded at the scalp is largely attributable to graded postsynaptic potentials (PSPs) of the cell body and large dendrites of vertically oriented pyramidal cells in cortical layers 3 to 5 (Lopes da Silva, 1991). These synaptic potentials are of much lower voltage than action potentials, but they also last much longer and involve a larger amount of surface area on cellular membranes and, as a result, the extracellular current flow produced by their generation has a relatively wide distribution. The columnar structure of the cerebral cortex facilitates a large degree of electrical summation rather than mutual cancellation. Thus, the extracellular EEG recorded from a distance represents the passive conduction of currents produced by summating activity over large neuronal aggregates. At the level of the scalp surface most of the recordable signals in the ongoing EEG presumably originate in cortical regions near the recording electrode. However, relatively large signals originating at more distal cortical sites might often make a significant contribution to the activity observed at a given scalp recording site, and signal averaging techniques can resolve even relatively small signals generated subcortically (e.g., auditory-evoked potentials generated in brain stem nuclei).

Scalp-recorded EEGs in the waking state in healthy adults normally range from several to about 75 μV but, in pathological states (such as epileptic seizures), can range up to 1 mV or more. Several factors determine the degree to which a cortical potential will be recordable at the scalp, including the amplitude of the signal at the cortex, the size of a region over which PSPs are occurring in a synchronous fashion, the proportion of cells in that region which are in synchrony, the location and orientation of the cells in relation to the scalp surface, and the amount of signal attenuation and spatial smearing produced by conduction through the intervening tissue layers of the dura, skull, and scalp. PSPs are thought to be synchronized by rhythmic discharges from thalamic nuclei (Lopes da Silva, 1991), with the degree of synchronization of the underlying cortical activity reflected in the amplitude of the EEG recorded at the scalp. For example, it has been estimated that if cortical activity is synchronous over an area of several square centimeters, it produces macropotentials that are evident at the scalp over a somewhat larger area and with reduced amplitude (Abraham and Ajmone-Marsan, 1958; Cooper *et al.,* 1965). In contrast, desynchronization of the EEG in a region of cortex presumably reflects increased mutual interaction of a subset of the population engaging in "cooperative activity," and such desynchronization is associated with decreased amplitude of the scalp-recorded signal in the α and β ranges (see below).

Many hypotheses have been brought forward concerning the role of EEG macropotentials in the processing and transmission of neural information. For example, John and colleagues (1973) proposed that the amplitude versus time distribution (i.e., waveshape) of low-frequency (<12 Hz) macropotentials (without reference to local spatial properties) encodes information concerning the interpretation given to stimuli based on past experience. This view is supported by reports that similar waveshapes of macropotentials from widespread cortical and subcortical structures are associated with similar memory retrieval processes. Another view focuses on the local spatial distribution of the amplitude of higher frequency macropotentials in specific structures. For example, Freeman (Freeman, 1978; Freeman and Baird, 1987) describes 40- to 80-Hz EEG bursts of the rabbit olfactory bulb which might represent a type of carrier wave whose spatial modulation encodes factors associated with both a stimulus and the internal state of the animal, and other investigators have described similar spatial modulation of high-frequency activity in the visual (Gray *et al.*, 1989) and sensorimotor (Murphy and Fetz, 1992) cortices of other species. Although much further research will be required to clarify these issues, current evidence suggests that the EEG can reflect important aspects of mass neuronal information processing.

B. Acquisition and Application of EEG Signals

To measure the EEG extracranially, electrodes are attached to the scalp with a conducting paste or liquid; each electrode is connected with an electrically "neutral" lead attached to the chest, chin, nose, or ear lobes (a "reference" montage) or with an "active" lead located over a different scalp area (a "bipolar" montage). Differential amplifiers are used to record voltage changes over time at each electrode; these signals are then digitized with 12 or more bits of precision and are sampled at a rate high enough to prevent aliasing of the signals of interest. EEGs are conventionally described as patterns of activity in four frequency ranges: δ (less than 4 Hz), θ (4–7 Hz), α (8–12 Hz), and β activity; β is sometimes subdivided into $\beta1$ (13–20 Hz) and $\beta2$ (21–35 Hz) bands. The origin of scalp-recorded potentials in the β, and especially the γ (above about 35 Hz), bands is ambiguous, arising either from cortical activity or from scalp muscle action potentials. The relationship of activity in all these bands to perception, cognition, and behavior is critically reviewed in Gevins and Schaffer (1980); in short, complex mental activity and sustained attention result in increased signal power in the lower frequency ranges (below α) and decreased signal power at the higher ranges (α and above). (Although there are a number of reports of increased γ-band activity recorded at the scalp during mental tasks, it has not been proven that such activity is not produced in whole or in part by scalp muscle action potentials.) The harmonic composition of the EEG is usually complex and only occasionally approaches a sinusoidal form. Like other brain function measurement modalities the EEG (and MEG) can be contaminated by physiological and instrumental artifacts, including those originating from such diverse sources as the subject's eye or head movements, heart beats, or poor electrode contacts, and care must be taken to correct for or eliminate such artifacts before further analyses are performed.

The spontaneous EEG is a sensitive and useful measure of task-related changes in neuronal activity. However, as with any functional neuroimaging modality, it is essential to pay careful attention to experimental design and to include appropriate control conditions to ensure that observed changes in the composition of the EEG are actually related to the experimental processes of interest and not some extraneous variable. In addition, brain signals are known to vary with age, handedness, gender, level of autonomic arousal, alertness and fatigue, habituation, and use of caffeine, nicotine, alcohol, and drugs. These factors should be taken into account when trying to measure physiological signs of mental activity in the brain. In the case of cognitive studies in particular, an important and obvious concern is the need to measure behavior during task performance to validate the assumption that the subject is performing what the experimenter assumes he or she is doing and to allow grouping of task trials with similar reaction time and accuracy characteristics. Another consideration is that "cognitive" tasks actually involve a variety of perceptual, motor, and cognitive activities. For example, in addition to recall of an item, a "memory" task can require watching or listening to a stimulus and activating a response switch. The interpretability of the outcome of an EEG study thus depends on the effectiveness of methodological controls designed to isolate and manipulate specific cognitive processes while eliminating or holding constant other task-related activities. When controls are inadequate, the error can be made of associating an EEG pattern with a certain abstract psychological construct (e.g., "spatial imagery"), when in fact it might actually be associated with a variety of other cognitive, perceptual, and motoric processes. Even when such factors are controlled, it is difficult to deduce the characteristics (order, time of onset, and duration) of the component operations of any given complex task. That is, as with *f*MRI and PET, sampling several seconds of EEG for analysis obscures possible correlates of the component cognitive operations and is based on the supposition that the component operations together constitute a distinguishable neural state. In most instances, there is no strong evidence supporting this supposition. This situation can be improved by analysis procedures utilizing shorter data samples time locked to an event (see next section) or by experimental procedures that isolate individual components of the cognitive tasks.

C. Event-Related Potentials

The electrical response of the brain to a specific stimulus is referred to as an evoked (or event-related) potential. Because of their low amplitude, especially in relation to the background EEG activity, such stimulus-related cerebral potentials usually cannot be identified in routine EEG recordings. With the use of computer signal-averaging techniques, however, the signal (or ERP)-to-noise (or background EEG activity) amplitude ratio can be increased so that the ERP is clearly delineated. For cognitive studies, typically 25–100 trials are averaged to form a clear ERP. Clinically, measurement of ERPs can provide a means of evaluating the functional integrity of the brain stem and sensory pathways. They permit subclinical lesions of these pathways to be detected and localized. ERPs may also help establish or support a neurological diagnosis, help follow the course of certain disorders, and help assess sensory function when behavioral testing cannot be undertaken (Chiappa, 1983).

As a research tool, ERPs can provide valuable information about the precise timing and cortical distribution of the neuroelectrical activity generated during cognitive processes. An averaged ERP waveform consists of a series of positive waves; a significant difference in latency, amplitude, duration, or topography of one or more of these waves between experimental conditions which differ in one specific cognitive factor is assumed to reflect the mass neural activity associated with that cognitive factor. Measurement of changes in the amplitude and timings of peaks in the series of ERP waves allows inferences to be made about the sequence and timing of task-associated processes, such as prestimulus preparation, encoding of stimulus features, operations such as matching or comparison of stimulus codes and memory codes, evaluation of the meaning of the stimulus, and response selection and execution (for general reviews of the electrophysiology of cognition, see Hillyard and Picton, 1987; Picton, 1992; Gevins and Cutillo, 1995).

The first reports of ERP events related not just to external (exogenous) factors, such as stimulus properties or motor activity, but to cognitive (endogenous) factors occurred in the 1960s when Gray Walter reported an ERP event related to expectation (Walter *et al.,* 1964). The "contingent negative variation" (CNV) is a low-frequency negative wave preceding an expected stimulus, often visible in the unaveraged EEG. The CNV was found to vary during stages of conditioned learning, with motivation, attention, distraction, and other cognitive variables. At almost the same time, Samuel Sutton, Victor Zubin, and E. Roy John discovered a positive wave peaking about 300 ms after infrequent, task-relevant stimuli (Sutton *et al.,* 1965). This "P300" or "P3" peak was considered cognitive because its scalp topography and behavior did not depend on physical stimulus properties per se, but on their interpretation or meaning to the subject in the context of the task. Many types of paradigms evoke P300, but the strongest factors in its elicitation are stimulus novelty and task relevance. An especially cogent finding was the elicitation of P300 by the absence of stimuli. This "missing stimulus ERP," obtained by averaging brain potentials registered to the time the missing stimuli should have occurred, lacks early exogenous peaks such as N100, but contains P300 and other endogenous waves.

The study of cognition using amplitude and timing measurements of ERP waves and peaks is built on experimental paradigms adapted from cognitive psychology. In the most general terms, it is assumed that a cognitive behavior can be viewed as a combination of sequential (and in some cases parallel) subprocesses in which information from stimuli is encoded, compared with memory and prior experience, and acted on. The division into component subprocesses may be somewhat arbitrary and depends on the particular cognitive model used, but it is assumed that these subprocesses can be systematically manipulated and that the electrical signals generated by these processes can be distinguished at the scalp (see also Gevins and Cutillo, 1995).

It is important to also keep in mind that the averaged ERP method assumes that the component subprocesses comprising a cognitive behavior do not vary much in timing from trial to trial. This assumption should be made with the understanding that there may be a certain amount of trial-to-trial variation in the timing of neurocognitive processes that affect both the latency and the amplitude of ERP components since they are interdependent. For this reason, measures of the integrated amplitude ("area") of an averaged wave are sometimes used, and "latency correction" of a particular peak prior to averaging can be done (Woody, 1967). Further, the average ERP waveform is assumed to be a linear composite of spatially and temporally overlapping "components" generated in different neural systems. This overlap makes the measurement of individual components in the composite waveform an important consideration. In some cases, relatively simple experimental or analytic methods may be effective in distinguishing components, as with the isolation of the N200 (task-relevant feature extraction) wave from the P200 by subtraction of the waveform to standard stimuli from the waveform to target stimuli. But it is only sometimes possible to distinguish overlapping components by such experimental manipulations. Principal components analysis (PCA) is often used as a means of objectively defining ERP components. The PCA method can be useful, but is limited by the fact that physiologically independent neural generators may produce evoked potential time series that are statistically correlated. Other practical problems have been described, such as the misallocation of variance among components and the arbitrariness of the rotation used (Wood and McCarthy, 1984; Dien, 1998). Alternatives to PCA have been developed which may be more effective in extracting small signals in cases of low signal-to-noise ratio, including comparative factor analysis (Rogers, 1991).

Finally, while in any experiment it is essential to control all factors not directly related to the intended experimental manipulation, it is worth repeating that special attention to experimental control is required in neurocognitive imaging studies since, as noted earlier, brain signals are sensitive to many sensory and motor as well as cognitive factors, and neurocognitive signals may be small in relation to other simultaneously occurring neural processes. In addition, to control for irrelevant stimulus, physiological state, and response-related factors by experimental design, *a posteriori* review of data sets to verify these controls is highly recommended. If small differences between experimental conditions are found, discarding outlier trials can eliminate such differences. Factors that can differ in final data sets despite *a priori* experimental control include physical stimulus properties, frequency and ordering of stimuli, task difficulty, automatization, and the timing, force, duration, hand, and nature of response movements (Gevins and Schaffer, 1980).

III. Improving the Spatial Resolution of the Electroencephalogram

A. High-Density Recordings

Recording from more electrodes is the first requirement for extracting more spatial detail from scalp-recorded EEGs. The 19-channel "10/20" montage of electrode placement (Jasper, 1958) commonly used in clinical EEG recordings has an interelectrode distance of about 6 cm on a typical adult head. This spacing may be sufficient for detecting signs of gross pathology or for differentiating the gross topography of ERP components, but is insufficient for resolving finer grained topographical details that may be of importance in studying cognition. By increasing the number of electrodes to over 100, average interelectrode distances of about 2.5 cm can be obtained on an average adult head. This is within the range of the typical cortex-to-scalp point spread function (i.e., the size of the scalp representation of a small, discrete cortical source) (Gevins *et al.*, 1990b). Although studies have been reported based on simultaneous sampling of as many as 256 scalp electrodes (Suarez *et al.*, 2000), it is questionable whether there is any useful improvement in spatial resolution beyond that obtained with 100 or so electrodes.

B. Spatial Sharpening and Anatomical Integration

For electrical (but not magnetic) recordings, the usefulness of such increased spatial sampling remains limited by the distortion of neuronal potentials as they are passively conducted through the highly resistive skull (Gevins *et al.*, 1991; van den Broek *et al.*, 1998). This distortion amounts to a spatial low-pass filtering, which causes a blurring of the potential distribution at the scalp. In recent years a number of spatial enhancement methods have been developed for reducing this distortion.

The simplest and most widely used of these methods is the spatial Laplacian operator, usually referred to as the Laplacian Derivation (LD). It is computed as the second derivative in space of the potential field at each electrode. The LD is proportional to the current entering and exiting the scalp at each electrode site (Nunez, 1981; Nunez and Pilgreen, 1991) and is independent of the location of the reference electrode used for recording. The LD is relatively insensitive to signals that are common to the local group of electrodes used in its computation and thus relatively more sensitive to high spatial frequency local cortical potentials. A simple method of computing the LD assumes that electrodes are equidistant and at right angles to each other, an approximation which is reasonable at only a few scalp locations, such as the vertex. A more accurate approach is based on measuring the actual three-dimensional position of the electrodes and using 3D spline functions to compute the LD over the actual shape of a subject's head (Le *et al.*, 1994). The main shortcoming of the LD is that it unrealistically assumes that the skull has the same thickness and conductivity everywhere, which limits the improvement in spatial detail that the method can achieve.

This shortcoming of the LD can be ameliorated by using a realistic model of each subject's head to locally correct the EEG potential field for distortion resulting from conduction to the scalp. One such method is called "finite element deblurring." It provides a computational estimate of the electrical potential field near the cortical surface by using a realistic mathematical model of volume conduction through the skull and scalp to downwardly project scalp-recorded signals (Gevins *et al.*, 1991; Le and Gevins, 1993; Gevins *et al.*, 1994a). Each subject's MRI is used to construct a realistic model of his or her head in the form of many small tetrahedral elements representing the tissues of scalp, skull, and brain. By assigning each tissue a conductivity value, the potential at all finite element vertices can be calculated using Poisson's equation. Given that the actual conductivity value of each of these finite elements is unknown, a constant value is used for the ratio of scalp to skull conductivity; the conductivity of each finite element is set by multiplying this constant by the local tissue thickness as determined from the MRI. Thus, even though true local conductivity is unknown, the procedure is well behaved with respect to this source of uncertainty, because it successfully accounts for relative conductivity variation due to regional differences in scalp and skull thickness.

The deblurring method has been shown to be reliable and more accurate than the LD (Gevins *et al.*, 1991, 1994a), an improvement that occurs at the expense of obtaining and processing each subject's structural MRI. Although deblurring can substantially improve the spatial detail provided by scalp-recorded EEGs, it does not usually provide additional information about the location of generating sources. Nevertheless, the improved spatial detail facilitates formation of more specific hypotheses about the distribution of active cortical areas during a cognitive task.[1]

Exploratory studies of deblurring and other high-resolution EEG techniques focused on the spatial enhancement of sensory ERPs, for which a great deal of *a priori* knowledge exists concerning their underlying neural generators (Gevins *et al.,* 1994a). These studies demonstrated that deblurred somatosensory responses isolated activity to the region of the central sulcus. Similar localization is obtained with movement-related potentials. For example, Fig. 1 illustrates the results of deblurring potentials locked in time to a button-press response made with the right hand. The major foci of activity occur in the contralateral side (left hemisphere) in the somatomotor region of the pre- and postcentral gyri. Demonstrations like these help verify the reasonableness of the approach. A better validation is obtained by comparison of the deblurred potentials with subdural grid recordings in epileptic patients undergoing evaluation for ablative surgery. To date these validation studies have produced a reasonable degree of agreement between the deblurred potentials and those measured directly at the cortical surface (Gevins *et al.,* 1994a).

Recent developments suggest that high-resolution EEG methods are useful tools in the experimental analysis of higher order brain functions; functional localization of cognitive processes inferred from spatially enhanced and anatomically registered neurophysiological measurements can be compared with the results of lesion studies and other neuroimaging approaches. As a complement to these approaches, the fine-grain temporal resolution of ERP measurements, in combination with improved topographic detail, adds valuable insights gained by characterizing both the regionalization of functions and the subsecond dynamics of their engagement. For example, spatial enhancement of EEGs related to component processes in reading has yielded results that are highly consistent with current knowledge of the functional neuroanatomy thought to be involved with visual pattern recognition and language functions (Gevins *et al.,* 1995a).

Modern EEG methods have also been used to study subsecond and multisecond distributed neural processes associated with working memory, the cognitive activity of creating a temporary internal representation of information during focused thought (Gevins *et al.,* 1996; McEvoy *et al.,* 1998). In task conditions that placed a high load on working memory functions, subjects were asked to decide if the stimulus on each trial matched either the verbal identity or the spatial location of a stimulus occurring three trials previously (about 13.5 s before). This required subjects to concentrate on maintaining a sequence of three letter names or three spatial locations concurrently; they had to update that sequence on each trial by remembering the most recent stimulus and could drop the stimulus from four trials back. In two corresponding

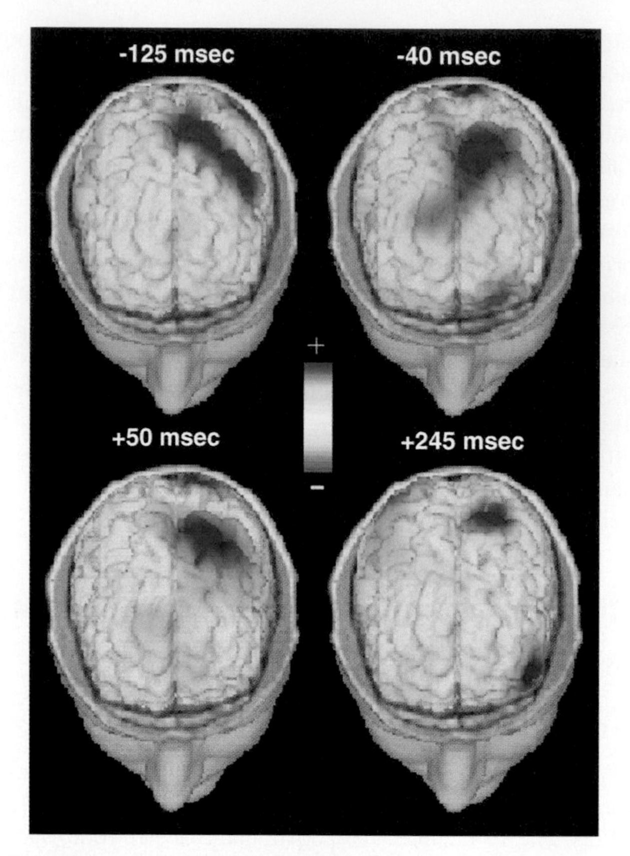

Figure 1 Deblurred movement-related potentials. Evoked potentials, time-locked to the onset of a button-press response made with the right middle finger, were recorded from a high-density (128-channel) electrode montage attached to the scalp. The blurring effects of the scalp and skull were mathematically removed using the deblurring method, and the resulting spatially sharpened data were projected onto the cortical surface, which was constructed from the subject's MRI. Activity at two instants prior to the button press (top) and two instants after the button press (bottom) are plotted. Both pre-and postresponse activation are strongly lateralized to the left hemisphere—the hemisphere contralateral to the hand used to make the response. The lateralized activation of the precentral gyrus prior to and immediately after a button-press response is shown. Approximately 250 ms after the response, the focus of activation moves to the postcentral gyrus.

[1]The issue of blurring of brain signals by the skull can largely be avoided by recording the magnetic rather than the electrical fields of the brain, because the skull has no effect on magnetic field topography. However, this transparency does not eliminate the need for utilizing a high density of sensors to accurately map the spatial topography of brain magnetic fields. Furthermore, the problems of localizing generator sources are as severe for MEG as they are for EEG (see text). Further, the cost of MEG technology is about 50 times greater than that required for EEG studies, and the associated infrastructure required to perform MEG studies is more complex and inflexible. Thus, for most laboratories, and for some applications (particularly those in which a subject's head cannot be immobilized, e.g., long-term monitoring or ambulatory recordings), MEG does not provide a viable alternative to EEG recordings.

control conditions, only the verbal identity or spatial location of the first stimulus had to be remembered. Both spatial and verbal working memory tasks produced highly localized momentary modulation of ERPs over prefrontal cortical areas relative to control conditions, with deblurred voltage maxima approximately over Brodmann's areas 9, 45, and 46 (Fig. 2). These brief (~50- and 200-ms) events occurred in parallel with a sustained ERP wave, maximal over the superior parietal lobe and the supramarginal gyrus, with a slight right-hemisphere predominance. It began ~200 ms after stimulus onset, returned to near baseline by ~600 ms poststimulus in control conditions, and was sustained up to ~1 s or longer in the working memory conditions. The subsecond ERP effects occurred in conjunction with multisecond changes in the ongoing EEG, of which the θ band power focused over midline frontal cortex is shown in Fig. 2 (Gevins *et al.*, 1997;

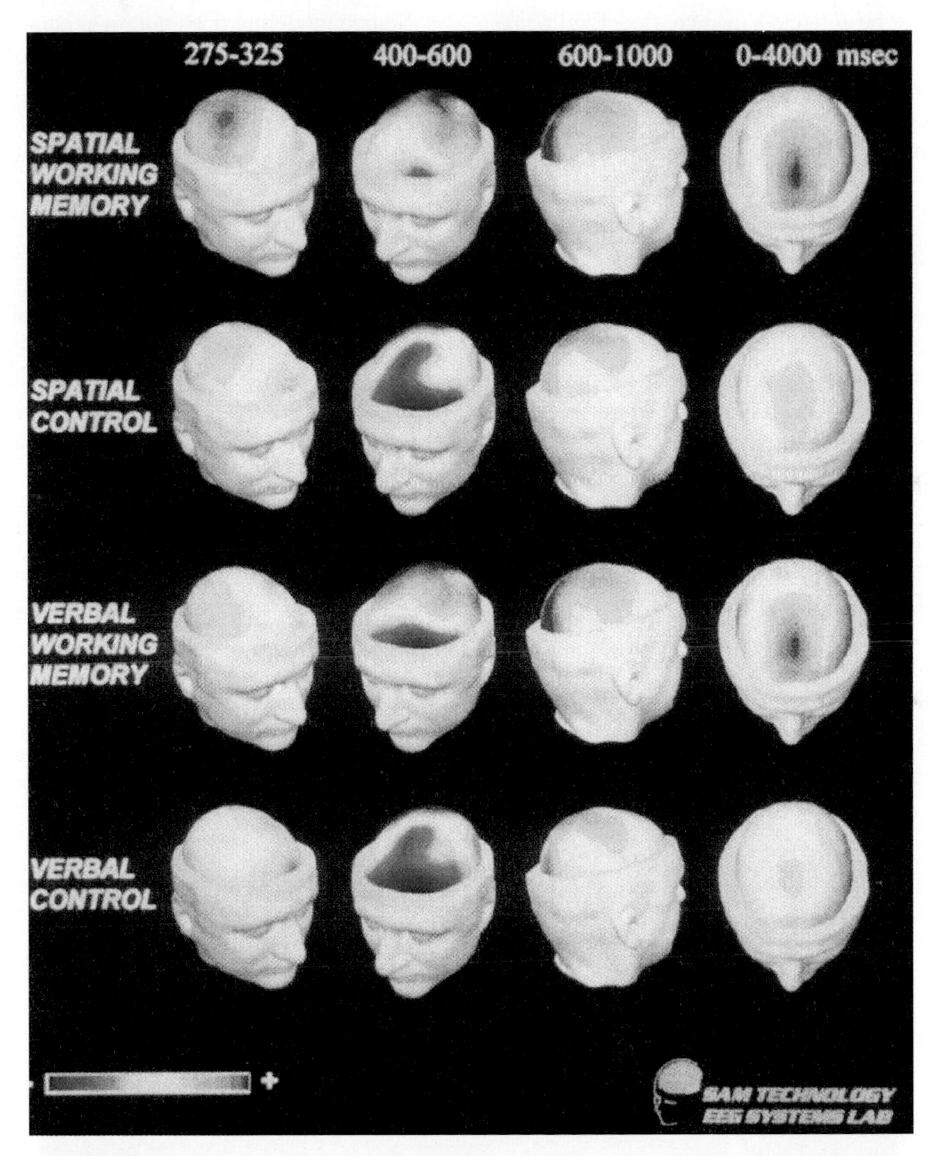

Figure 2 Deblurred event-related potentials and ongoing EEG related to sustained focused attention. High-resolution EEG methods have made it possible to simultaneously measure both subsecond phasic and multisecond tonic regional brain activity during performance of cognitive tasks. In this experiment a sequence of increased subsecond ERP peaks and waves was observed over frontal (first and second columns) and parietal (third column) cortices during a difficult working memory task, in comparison to control conditions with lower working memory requirements. These subsecond changes in the working memory tasks were accompanied by longer lasting (4-s) increases in ongoing EEG θ band power (rightmost column). These EEG findings suggest that various types of attention are associated with neural processes that have distinct time courses in distinct neuronal populations. Amplitude scale is constant across experimental conditions within each column; ERP scale is voltage, EEG scale is z-scored spectral power.

Smith *et al.,* 1999). These EEG findings may provide the first direct evidence in a single experiment supporting the idea that the various types of attention are associated with neural processes with distinct time courses in distinct neuronal populations. The increased θ band power may be a marker of the continuous focused attention required to perform the task and may reflect activity in the anterior cingulate gyrus (Gevins *et al.,* 1997) (Fig. 4). In contrast, the momentary attention required for scanning and updating the representations of working memory may be reflected in increased ERP peaks over lateralized regions of dorsolateral prefrontal cortex, while maintenance of a representation of the stimuli being remembered may be reflected in the parietally maximal ERP wave and other concomitant changes in the EEG (Gevins *et al.,* 1996, 1997).

C. Identifying the Generators of EEG

Neither the Laplacian Derivation nor the more advanced EEG spatial enhancement algorithms such as deblurring, nor MEG recordings, provide any conclusive three-dimensional information about where the source of a scalp-recorded signal lies in the brain. In some cases, such as when healthy subjects perform difficult cognitive tasks and strong signals are recorded over areas of association cortex (i.e., dorsolateral prefrontal, superior and inferior parietal, inferotemporal and lateral temporal), the hypothesis that EEG potentials are generated in these areas is the most plausible. However, counterexamples can always be presented. In addition to visual examination of the potential field distribution, "dipole modeling" provides another method for generating hypotheses concerning the neuroanatomical loci responsible for generating neuroelectric events measured at the scalp (Scherg and Von Cramon, 1985; Fender, 1987). Dipole modeling uses iterative numerical methods to fit a mathematical representation of a focal, dipolar current source, or collection of such sources, to an observed scalp-recorded EEG or MEG field.

Source modeling does not, in general, provide a unique or necessarily physically correct answer about where in the brain activity recorded at the scalp is generated. This is so because solving for the source of an EEG or MEG distribution recorded at the scalp is a mathematically ill-conditioned "inverse problem," which has no unique solution; additional information and/or assumptions are required in order to choose among candidate source models. Some of this *a priori* information is obvious—for example that the potentials must arise from the space occupied by the brain. Thus, anatomical data from structural MR images can usefully constrain inverse solutions (Spinelli *et al.,* 2000). In contrast, other assumptions border on presupposing unknown information (i.e., that the potentials arise only from the cortex or that the number of active cortical areas is known).

One simple, convenient, and potentially clinically useful approach for potentials elicited by simple sensory stimulation is to assume that the scalp potential pattern arises from a single point dipole source (e.g., Fig. 3). Although not

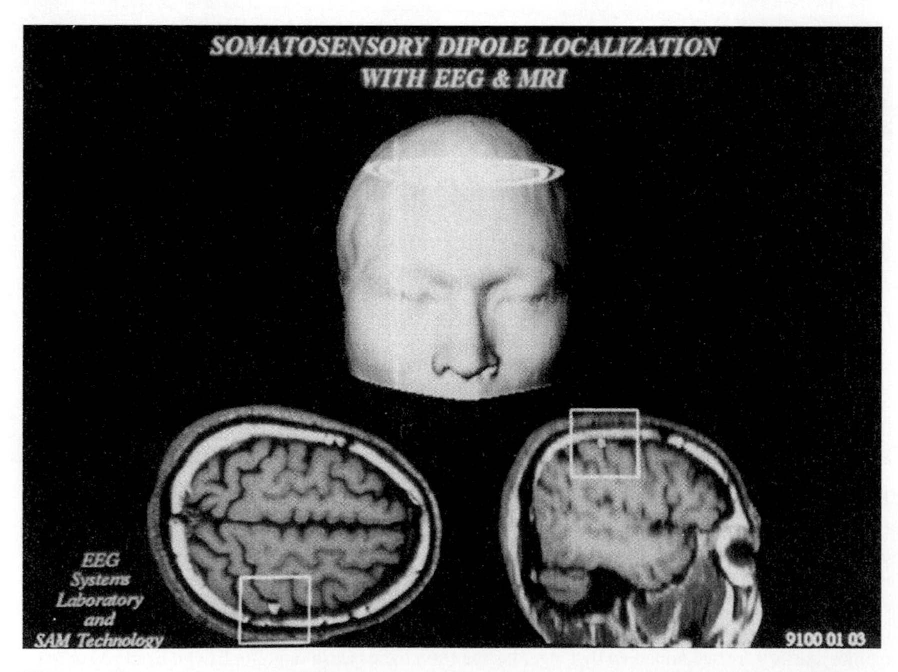

Figure 3 Localization of an EEG dipole model in the somatosensory cortex of the right hemisphere from scalp-recorded data evoked in response to transient electrical stimulation of the left index finger. This popular type of source generator localization modeling produces anatomically plausible results in the case of simple sensory stimulation.

anatomically or physiologically realistic, such simple models can sometimes be useful for locating the center of mass of primary sensory cortex and hence major functional landmarks such as the central sulcus. When justified by simple-voltage topography (e.g., Fig. 4), models of this sort can also be useful for generating initial hypotheses about the possible sources underlying other phenomena.

Most complex scalp-recorded neurophysiological phenomena are not well approximated by a single-dipole-source model. Efforts to obtain estimates of the strengths and 3D locations of the underlying neuronal generators when there are multiple, time-overlapped active sources are necessarily subject to widely recognized practical and theoretical difficulties (Miltner *et al.,* 1994). Efforts are under way to develop improved methods for source analysis for electrical phenomena that are likely to arise from multiple and/or distributed sources (Wang *et al.,* 1993; Gorodnitsky *et al.,* 1995; Tesche *et al.,* 1995; Grave de Peralta-Menendez and Gonzalez-Andino, 1998; Koles, 1998). Even so, regardless of the method used to formulate them, source generator hypotheses must ultimately be independently verified. In rare cases, this might be done in patient populations in the context of invasive recordings performed for clinical diagnostic purposes (Smith *et al.,* 1990; Halgren *et al.,* 1998). More commonly, another type of imaging modality, such as *f*MRI, has to be employed. Indeed, one promising approach to this issue is to use information about the cortical regions activated by a task as mapped by *f*MRI to constrain source models and to derive information about the spatiotemporal dynamics of those sources from ERP measurements (George *et al.,* 1995; Simpson *et al.,* 1995; Liu *et al.,* 1998; Mangun *et al.,* 1998; Sereno, 1998; Dale and Halgren, 2001).

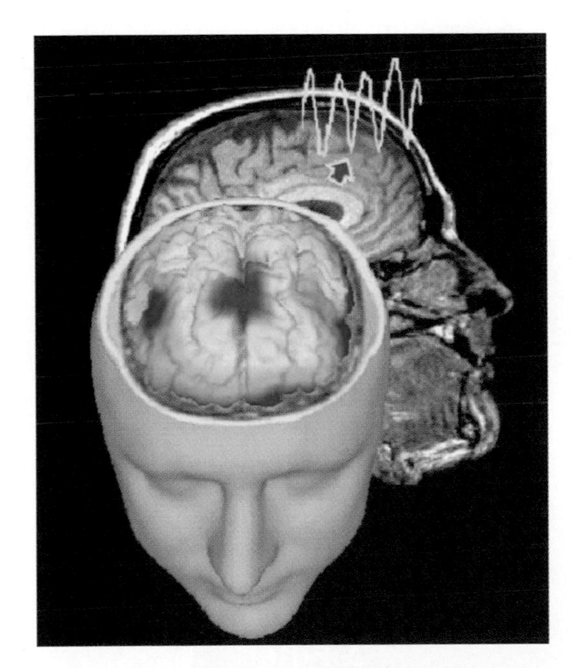

Figure 4 Deblurred frontal midline θ EEG activity and localization of corresponding source model in the region of the anterior cingulate cortex. Topographic data correspond to the difficult working memory task condition depicted in Fig. 2 (Gevins *et al.,* 1997). The data were processed with the deblurring method, and the spatially sharpened results were projected onto the cortical surface, which was constructed from the subject's MRI. The upward-oriented arrow superimposed on the midline sagittal image depicts the localization of a point dipole source model for these data.

Future progress in such multimodal integration efforts will require advances in knowledge regarding how blood flow measures are coupled to electromagnetic signals; at present their relationship is only poorly understood and not explicitly accounted for in such resulting source models. Although it appears that there is an approximately linear relationship between neuronal activity integrated over time and the amplitude of the hemodynamic response (Gratton *et al.,* 2001), this relationship does not necessarily imply a close relationship between patterns of increased blood flow and electromagnetic sources (Nunez and Silberstien, 2000). As noted above, whether synaptic activity is detectable at the scalp depends on the degree to which it is additive, which in turn requires that synapses be activated synchronously and that such synchronous activation occurs in a population of cells that has an architecture such that current flows summate rather than cancel and that has a geometric orientation that supports propagation to the scalp. Relative synchrony in activation does not necessarily imply a relative increase in metabolic activity, and metabolic intensity is not necessarily closely related to the presence or absence of laminar architecture or a suitable orientation. Because of these inherent ambiguities, source models based on coregistration of EEG (or MEG) with *f*MRI or PET measures should be viewed very cautiously.

IV. Analysis of Functional Networks

Independent of whether definitive knowledge of source configurations exists, changes in the spatial distribution of EEG phenomena can be used to characterize the neural dynamics of thought processes. Even the simplest cognitive tasks require the functional coordination of a large number of widely distributed specialized brain systems. A simple response to a sensory stimulus involves the coordination of sensory, association, and other areas that prepare for, register, and analyze the stimulus; the motor systems that prepare for and execute the response; and other distributed neuronal networks. These distributed networks serve to allocate and direct attentional resources to the stimulus, to relate the stimulus to internal representations of self and environment in order to decide what action to take, to initiate or inhibit the behavioral response, and to update internal representations after receiving feedback about the result of the action.

It is only in recent years that imaging studies of brain metabolism and blood flow have begun to examine functional networks in terms of statistical covariation between regions of interest (Nyberg *et al.,* 1996; McIntosh, 1998; Moeller *et al.,* 1999; Buchel and Friston, 2000). However, such approaches have been employed for many years in the analysis of patterns of EEG and ERP activity. In the ongoing EEG, hypotheses about functional interactions between cortical regions are sometimes drawn from measurements of statistical interrelationships between time series recorded at different sites. These can be quantified by various measures of spectral, waveshape, or information-theoretic similarity, including spectral coherence (Walter, 1963), correlation (Brazier and Casby, 1952; Livanov, 1977; Gevins *et al.,* 1981, 1983), covariance (Gevins *et al.,* 1987, 1989a,b), information measures (Callaway and Harris, 1974; Mars and Lopes da Silva, 1987), nonlinear regression (Lopes da Silva *et al.,* 1989), and multichannel time-varying autoregressive modeling (Gersch, 1987).

Some of the above methods can be used to characterize the spatiotemporal relationships of subsecond ERP components. Since the ERP waveform delineates the time course of event-related mass neural activity of a neuronal population, the coordination of two or more populations during task performance should be signaled by a consistent relationship between the morphology of the ERP waveforms emitted by these populations, with consistent time delay (Gevins and Bressler, 1988). If the relationships are linear, as they often appear to be, this coordinated activity might be measured by the lagged correlation or covariance between the ERPs, or segments of ERPs, from different regions (Gevins *et al.,* 1987, *et al.,* 1989a,b). One such measure of this type of process is referred to as an event-related covariance (ERC). (Of course, a significant covariance of this type is only a measure of statistical association and does not map the actual neuronal pathways of interaction between functionally related populations.) Studies of the neurogenesis of ERCs are still in their infancy (Bressler *et al.,* 1993; Gevins *et al.,* 1994b), and any interpretations of ERCs in terms of the underlying neural processes that generate them must thus be made very cautiously. (It is noted, however, that ERC results to date have been highly consistent with the known large-scale functional neuroanatomy of frontal, parietal, and temporal association cortices.) In the meanwhile, ERCs have provided fascinating glimpses of the complex, rapidly shifting distributed neuronal processes that underlie simple cognitive tasks.

The ERC technique has yielded its most interesting results as a tool for studying preparatory attentional networks, the changes in brain activity associated with readiness for an impending event or action. For example, subjects in one experiment (Gevins *et al.,* 1987, 1989a,b) performed a task that required graded finger pressure responses with either the right or the left hand proportional to visual numeric stimuli from 1 to 9. The hand to be used was cued 1 s before the stimulus. A 375-ms ERC analysis window spanned the interval preceding the stimulus number in order to measure how ERP patterns differed according to the hand subjects expected to use. Figure 5 shows right-hand preparatory ERCs for seven subjects for those trials for which the response (~0.5 to 1 s later) was subsequently either accurate or inaccurate. The set of subsequently accurate trials is characterized by covariances of the left prefrontal electrode with electrodes overlying the same motor, somatosensory, and parietal areas that were involved in actual response execution. (Simultaneous measurement of flexor digitori muscle

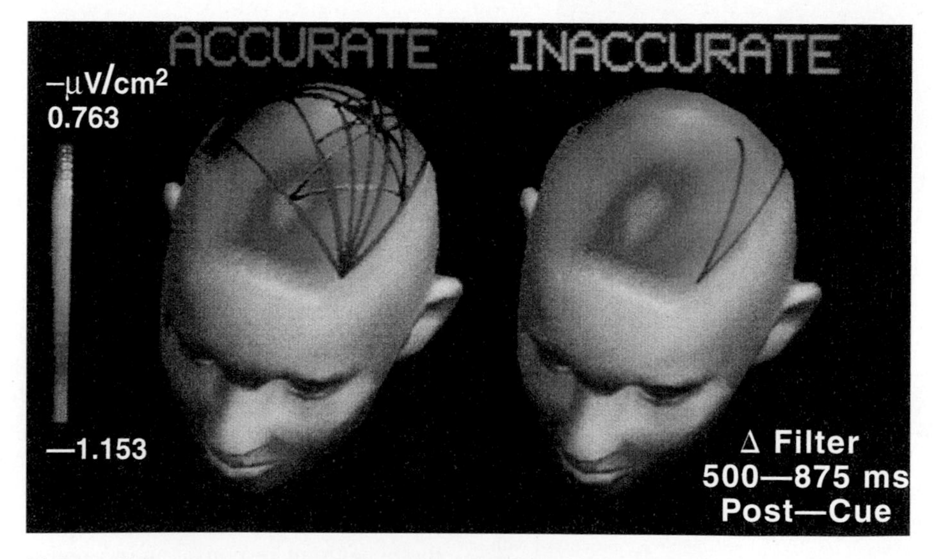

Figure 5 Preparatory event-related covariance (ERC) patterns preceding accurate and inaccurate responses (Gevins *et al.,* 1987). ERCs involving left frontal, midline precentral, and left central and parietal electrode sites are prominent in patterns preceding accurate responses (by 0.5 to 1 s) (left). The number and magnitude of ERCs are smaller preceding inaccurate responses (right).

activity showed that the finger that would subsequently respond was not active during the preparatory interval.) The preparatory patterns preceding inaccurate responses differed markedly from those preceding accurate responses, with fewer ERCs between the left frontal site and other electrodes. Such results suggest that one important role of frontal lobe integrative mechanisms is the anticipatory scheduling and coordination of the activation of those specialized brain regions that will participate in an upcoming cognitive event.

V. The EEG as a Monitoring (vs Imaging) Modality

Although the EEG has clear limitations with respect to its use as a method for fine-grain 3D anatomical localization of neurofunctional systems, it has clear advantages relative to other neuroimaging techniques as a method for tracking cognition and brain function over time. That is, the EEG provides a sensitive, unimposing, portable, and low-cost monitor of cognition for research, clinical assessment, and other applications (Gevins, 1998; Gevins *et al.*, 1998). The compactness of EEG technology also means that, unlike other functional neuroimaging modalities (which require massive machinery, large teams of technicians, and complete immobilization of the subject) EEGs can even be collected from an ambulatory subject who is literally wearing the entire recording apparatus (Gilliam *et al.*, 1999).

Such properties of the EEG have been demonstrated to have significant clinical utility. Continuous scalp EEG monitoring has long been an essential tool in the diagnostic evaluation of epilepsy (Thompson and Ebersole, 1999) and in the evaluation and treatment of sleep disorders (Carskadon and Rechtschaffen, 1989). It is also coming to play an increasingly important role in neuro-ICU monitoring (Vespa *et al.*, 1999) and in gauging level of awareness during anesthesia (John *et al.*, 2001; O'Connor *et al.*, 2001). The EEG as a monitoring modality has also demonstrated potential outside of the clinical realm as a means of tracking variations in alertness (Akerstedt *et al.*, 1982, 1987) or cognitive workload (Wilson *et al.*, 1994; Gevins *et al.*, 1995b; Smith *et al.*, 2001) in individuals engaged in complex and demanding naturalistic tasks.

The same factors that make the EEG useful as a continuous monitor of cognitive status also make it ideal for incorporation into a test of cognitive function administered repeatedly to track changes in neural function that characterize or accompany disease progression and treatment. Traditionally, behavioral measures have been almost exclusively relied upon as a means to detect cognitive impairment. However, compensatory efforts by subjects might mask real, functionally relevant, changes in brain state. Conversely, a low level of test performance may reflect motivational rather than ability factors. Since EEG measures can provide independent evidence of variations in alertness and attentiveness or mental effort, incorporating them into tests of cognitive function might lead to more sensitivity and less ambiguous clinical assessment tools.

A test that combines task-related behavioral and EEG measures is emerging out of research on the neurophysiology of working memory or the ability to control attention and sustain its focus on a particular active mental representation(s) in the face of distracting influences (Engle *et al.*, 1999). Animal studies (Goldman-Rakic, 1987, 1988) and human lesion and neuroimaging studies indicate that tasks requiring attention and working memory activate a functional network linking regions of medial and dorsolateral prefrontal cortex with posterior association cortices (Frisk and Milner, 1990; Posner and Peterson, 1990; Miller and Desimone, 1994; Smith and Jonides, 1995; Smith *et al.*, 1995; Owen *et al.*, 1996). As noted above (e.g., Figs. 2 and 4), high-resolution EEG studies have produced results consistent with this view and have helped to identify neurophysiological signals associated with the mental effort required to focus and sustain attention and hold information in working memory (Gevins *et al.*, 1996, 1997; McEvoy *et al.*, 1998).

Measures of individual differences in working memory capacity tend to be positively correlated with performance on psychometric tests of cognitive ability and other indices of scholastic aptitude (Carpenter *et al.*, 1990; Kyllonen and Christal, 1990). Consistent with such observations, EEG and ERP measures collected during working memory task performance have been found to be excellent predictors of individual differences in overall cognitive ability as measured by conventional neuropsychological tests, and regional differences in such measures can be used to discriminate individuals with relatively greater verbal ability from individuals with relatively greater nonverbal ability (Gevins and Smith, 2000). Thus, such signals have substantial face validity as potential indices of the integrity of cognitive function.

Under normal conditions the neurophysiological signals modulated by task-imposed variations in working memory demands tend to be very stable. In a recent study in which the test–retest reliability of task-related EEG spectral features was examined in well-practiced subjects, correlations of $r \geq 0.9$ were found between two test sessions with a 1-week lag (McEvoy *et al.*, 2000). Despite this apparent stability when other factors are held constant, there is also evidence indicating that such task-related neurophysiological signals are highly sensitive to stressors that can impose some form of mild transient cognitive impairment. For example, working memory-sensitive neurophysiological signals have been demonstrated to be sensitive to sleep loss (Gevins *et al.*, 1990a; Gevins and Smith, 1999), low doses of alcohol (Gevins and Smith, 1999; Gevins *et al.*, 2001; Ilan and Gevins, 2001), drowsiness associated with the common over-the-counter antihistamine diphenhydramine (Gevins *et al.,* 2001), marijuana (Nichols *et al.,* 2001), and cognitive changes associated with prescription medications frequently used to treat neurological (Chung *et al.,* 2002) and psychiatric (McEvoy *et al.,* 2001) disorders. From such results an assessment technique that incorporates both behavioral and neurophysiological measurements is beginning to emerge that may ultimately prove to have wide ranging clinical utility.

VI. Summary and Conclusions

The neurophysiology of mentation involves rapid coordination of processes in widely distributed cortical and subcortical areas. The electrical signals that accompany higher cognitive functions are subtle, are spatially complex, and change both in a tonic multisecond fashion and phasically in subsecond intervals in response to environmental demands and internal representations of environment and self. No one brain imaging technology is currently capable of providing both near-millimeter precision in localizing regions of activated tissue and subsecond temporal precision for characterizing changes in patterns of activation over time. However, by combining several technologies, it seems possible to achieve this fine degree of spatiotemporal resolution. Modern high-resolution EEG is especially well suited to monitoring rapidly changing regional patterns of neuronal activation accompanying purposive behaviors, while *f*MRI seems ideal for precisely determining their three-dimensional localization and distribution. It is a topic of current research to determine how to combine EEG and *f*MRI data from the same subjects doing the same tasks. Because of the relative low expense and unobtrusiveness of the technology required for EEG recordings relative to other means of neurofunctional assessment, combined with its high level of sensitivity to changes in neuronal activity, the EEG has long played an important role in contexts in which it is important to continuously monitor brain function. Its most promising future contributions are also likely to be closely related to this monitoring function.

Acknowledgments

We thank our many colleagues at the San Francisco Brain Research Institute (formerly called EEG Systems Laboratory) and SAM Technology, past and present, for their contributions to the work described here. This research was supported by grants from the Air Force Office of Scientific Research, the National Institute of Mental Health, the National Institute of Neurological Disorders and Stroke, the National Science Foundation, the National Aeronautics and Space Administration, the Air Force Research Laboratory, the Office of Naval Research, the National Institute of Alcoholism and Alcohol Abuse, the National Institute of Drug Abuse, the National Institute of Child Health and Human Development, and the National Institute of Aging of the United States federal government.

References

Abraham, K., and Ajmone-Marsan, C. (1958). Patterns of cortical discharges and their relation to routine scalp electroencephalography. *Electroencephalogr. Clin. Neurophysiol.* **10,** 447–452.

Akerstedt, T., Torsvall, L., *et al.* (1982). Sleepiness and shift work: Field studies. *Sleep* **5**(Suppl. 2), S95–106.

Akerstedt, T., Torsvall, L., *et al.* (1987). Sleepiness in shiftwork: A review with emphasis on continuous monitoring of EEG and EOG. *Chronobiol. Int.* **4,** 129–140.

Aminoff, M. J. (1986). "Electrodiagnosis in Clinical Neurology." Churchill Livingstone, New York.

Berger, H. (1929). Uber das Elektroenzephalogramm des Menschen. *Arch. Psychiatry* **87,** 527–570.

Brazier, M. A. B., and Casby, J. U. (1952). Crosscorrelation and autocorrelation studies of electroencephalographic potentials. *Electroencephalogr. Clin. Neurophysiol.* **4,** 201–211.

Bressler, S. L., Coppola, R., *et al.* (1993). Episodic multiregional cortical coherence at multiple frequencies during visual task performance. *Nature* **366,** 153–155.

Buchel, C., and Friston, K. (2000). Assessing interactions among neuronal systems using functional neuroimaging. *Neural Networks* **13,** 871–882.

Callaway, E., and Harris, P. (1974). Coupling between cortical potentials from different areas. *Science* **183,** 873–875.

Carpenter, P. A., Just, M. A., *et al.* (1990). What one intelligence test measures: A theoretical account of the processing in the Raven Progressive Matrices Test. *Psychol. Rev.* **97,** 404–431.

Carskadon, M. A., and Rechtschaffen, A. (1989). Monitoring and staging human sleep. *In* "Principles and Practice of Sleep Medicine" (M. H. Kryger, T. Roth, and W. C. Dement, eds.), pp. 943–960. Saunders, Philadelphia.

Chiappa, K. H. (1983). "Evoked Potentials in Clinical Medicine." Raven Press, New York.

Chung, S. S., McEvoy, L. K., *et al.* (2002). Neurophysiological assessment of cognitive dysfunction associated with phenytoin. *Clin. Neurophysiol.,* in press.

Cooper, R., Winter, A., *et al.* (1965). Comparison of subcortical, cortical, and scalp activity using chronically indwelling electrodes. *Electroencephalogr. Clin. Neurophysiol.* **18,** 217–230.

Dale, A. M., and Halgren, E. (2001). Spatiotemporal mapping of brain activity by integration of multiple imaging modalities. *Curr. Opin. Neurobiol.* **11,** 202–208.

Daly, D. D., and Pedley, T. A. (1990). "Current Practice of Clinical Electroencephalography," 2nd edition. Raven Press, New York.

Dien, J. (1998). Addressing misallocation of variance in principal components analysis of event-related potentials. *Brain Topogr.* **11,** 43–55.

Engle, R. W., Tuholski, S., *et al.* (1999). Individual differences in working memory capacity and what they tell us about controlled attention, general fluid intelligence and functions of the prefrontal cortex. *In* "Models of Working Memory" (A. Miyake and P. Shah, eds.), pp. 102–134. Cambridge Univ. Press, Cambridge, UK.

Fender, D. H. (1987). Source localization of brain electrical activity. *In* "Methods of Analysis of Brain Electrical and Magnetic Signals" (A. S. Gevins and A. Rémond, eds.), Vol. 1, pp. 355–403 Elsevier, Amsterdam.

Freeman, W. J. (1978). Spatial properties of an EEG event in the olfactory bulb and cortex. *Electroencephalogr. Clin. Neurophysiol.* **44,** 586–605.

Freeman, W. J., and Baird, B. (1987). Relation of olfactory EEG to behavior: Spatial analysis. *Behav. Neurosci.* **3,** 393–408.

Frisk, V., and Milner, B. (1990). The relationship of working memory to the immediate recall of stories following unilateral temporal or frontal lobectomy. *Neuropsychologia* **28,** 121–135.

George, J. S., Aine, C. J., *et al.* (1995). Mapping function in the human brain with magnetoencephalography, anatomical magnetic resonance imaging, and functional magnetic resonance imaging. *J. Clin. Neurophysiol.* **12,** 406–431.

Gersch, W. (1987). Non-stationary multichannel time series analysis. *In* "Methods of Analysis of Brain Electrical and Magnetic Signals" (A. S. Gevins and A. Rémond, eds.), Vol. 1, pp. 261–296. Elsevier, Amsterdam.

Gevins, A. (1998). The future of electroencephalography in assessing neurocognitive functioning. *Electroencephalogr. Clin. Neurophysiol.* **106,** 165–172.

Gevins, A., and Smith, M. E. (1999). Detecting transient cognitive impairment with EEG pattern recognition methods. *Aviat. Space Environ. Med.* **70,** 1018–1024.

Gevins, A., and Smith, M. E. (2000). Neurophysiological measures of working memory and individual differences in cognitive ability and cognitive style. *Cereb. Cortex* **10,** 829–839.

Gevins, A., Le, J., *et al.* (1994a). High resolution EEG: 124-channel recording, spatial deblurring and MRI integration methods. *Electroencephalogr. Clin. Neurophysiol.* **90,** 337–358.

Gevins, A., Smith, M. E., *et al.* (1997). High-resolution EEG mapping of cortical activation related to working memory: Effects of task difficulty, type of processing, and practice. *Cereb. Cortex* **7,** 374–385.

Gevins, A., Smith, M. E., *et al.* (1998). Monitoring working memory load during computer-based tasks with EEG pattern recognition methods. *Hum. Factors* **40,** 79–91.

Gevins, A., Smith, M. E., *et al.* (2001). Tracking the cognitive pharmacodynamics of psychoactive substances with combinations of behavioral and neurophysiological measures. *Neuropsychopharmacology,* in press.

Gevins, A. S., and Bressler, S. L. (1988). Functional topography of the human brain. *In* "Functional Brain Imaging" (G. Pfurtscheller, ed.), pp. 99–116. Huber, Toronto.

Gevins, A. S., and Cutillo, B. A. (1995). Neuroelectric measures of the mind. *In* "Neocortical Dynamics and Human EEG Rhythms" (P. Nunez, ed.), pp. 304–338. Oxford Univ. Press, New York.

Gevins, A. S., and Schaffer, R. E. (1980). A critical review of electroencephalographic EEG correlates of higher cortical functions. *CRC Crit. Rev. Bioeng.* **4,** 113–164.

Gevins, A. S., Bressler, S. L., *et al.* (1989a). Event-related covariances during a bimanual visuomotor task. Part I. Methods and analysis of stimulus and response-locked data. *Electroencephalogr. Clin. Neurophysiol.* **74,** 58–75.

Gevins, A. S., Bressler, S. L., *et al.* (1989b). Event-related covariances during a bimanual visuomotor task. Part II. Preparation and feedback. *Electroencephalogr. Clin. Neurophysiol.* **74,** 147–160.

Gevins, A. S., Bressler, S. L., *et al.* (1990a). Effects of prolonged mental work on functional brain topography. *Electroencephal. Clin. Neurophysiol.* **76,** 339–350.

Gevins, A. S., Brickett, P., *et al.* (1990b). Beyond topographic mapping: Towards functional–anatomical imaging with 124-channel EEGs and 3-D MRIs. *Brain Topogr.* **3,** 53–64.

Gevins, A. S., Cutillo, B. A., *et al.* (1994b). Subdural grid recordings of distributed neocortical networks involved with somatosensory discrimination. *Electroencephalogr. Clin. Neurophysiol.* **92,** 282–290.

Gevins, A. S., Cutillo, B. A., *et al.* (1995a). Regional modulation of high resolution evoked potentials during verbal and nonverbal matching tasks. *Electroencephalogr. Clin. Neurophysiol.* **94,** 129–147.

Gevins, A. S., Doyle, J. C., *et al.* (1981). Electrical potentials in human brain during cognition: New method reveals dynamic patterns of correlation. *Science* **213,** 918–922.

Gevins, A. S., Le, J., *et al.* (1991). Seeing through the skull: Advanced EEGs use MRIs to accurately measure cortical activity from the scalp. *Brain Topogr.* **4,** 125–131.

Gevins, A. S., Leong, H., *et al.* (1995b). Towards measurement of brain function in operational environments. *Biol. Psychol.* **40,** 169–186.

Gevins, A. S., Morgan, N. H., *et al.* (1987). Human neuroelectric patterns predict performance accuracy. *Science* **235,** 580–585.

Gevins, A. S., Schaffer, R. E., *et al.* (1983). Shadows of thought: Shifting lateralization of human brain electrical patterns during brief visuomotor task. *Science* **220,** 97–99.

Gevins, A. S., Smith, M. E., *et al.* (1996). High resolution evoked potential imaging of the cortical dynamics of human working memory. *Electroencephalogr. Clin. Neurophysiol.* **98,** 327–348.

Gilliam, F., Kuzniecky, R., *et al.* (1999). Ambulatory EEG monitoring. *J. Clin. Neurophysiol.* **16,** 111–115.

Goldman-Rakic, P. (1987). Circuitry of primate prefrontal cortex and regulation of behavior by representational memory. *In* "Handbook of Physiology: The Nervous System—Higher Functions of the Brain" (F. Plum and V. Mountcastle, eds.), Vol. 5, pp. 373–417. Am. Physiol. Soc., Bethesda, MD.

Goldman-Rakic, P. (1988). Topography of cognition: Parallel distributed networks in primate association cortex. *Annu. Rev. Neurosci.* **11,** 137–156.

Gorodnitsky, I. F., George, J. S., *et al.* (1995). Neuromagnetic source imaging with FOCUSS: A recursive weighted minimum norm algorithm. *Electroencephalogr. Clin. Neurophysiol.* **95,** 231–251.

Gratton, G., Goodman-Wood, M. R., *et al.* (2001). Comparison of neuronal and hemodynamic measures of the brain response to visual stimulation: An optical imaging study. *Hum. Brain Mapp.* **13,** 13–25.

Grave de Peralta-Menendez, R., and Gonzalez-Andino, S. L. (1998). A critical analysis of linear inverse solutions to the neuroelectric inverse problem. *IEEE Trans. Biomed. Eng.* **45,** 440–448.

Gray, C. M., Engel, A. K., *et al.* (1989). Oscillatory responses in cat visual cortex exhibit inter-columnar synchronization which reflects global stimulus properties. *Nature* **338,** 334–337.

Halgren, E., Marinkovic, K., *et al.* (1998). Generators of the late cognitive potentials in auditory and visual oddball tasks. *Electroencephalogr. Clin. Neurophysiol.* **106,** 156–164.

Hillyard, S. A., and Picton, T. W. (1987). Electrophysiology of cognition. *In* "Handbook of Physiology" (V. B. Mountcastle, ed.), Vol. 5, pp. 519–584. Am. Physiol. Soc., Bethesda, MD.

Ilan, A. B., and Gevins, A. (2001). Prolonged neurophysiological effects of cumulative wine drinking. *Alcohol,* **25**, 137–152.

Jasper, H. H. (1958). The ten–twenty electrode system of the International Federation. *Electroencephalogr. Clin. Neurophysiol.* **10,** 371–375.

John, E. R., Bartlett, F., *et al.* (1973). Neural readout from memory. *J. Neurophysiol.* **36,** 893–924.

John, E. R., Prichep, L. S., *et al.* (2001). Invariant reversible QEEG effects of anesthetics. *Consciousness Cognit.* **10,** 165–183.

Koles, Z. J. (1998). Trends in EEG source localization. *Electroencephalogr. Clin. Neurophysiol.* **106,** 127–137.

Kyllonen, P. C., and Christal, R. E. (1990). Reasoning ability is little more than working memory capacity?! *Intelligence* **14,** 389–433.

Le, J., and Gevins, A. S. (1993). Method to reduce blur distortion from EEGs using a realistic head model. *IEEE Trans. Biomed. Eng.* **40,** 517–528.

Le, J., Menon, V., *et al.* (1994). Local estimate of the surface Laplacian derivation on a realistically shaped scalp surface and its performance on noisy data. *Electroencephalogr. Clin. Neurophysiol.* **92,** 433–441.

Liu, A. K., Belliveau, J. W., *et al.* (1998). Spatiotemporal imaging of human brain activity using functional MRI constrained magnetoencephalography data: Monte Carlo simulations. *Proc. Natl. Acad. Sci. USA* **95,** 8945–8950.

Livanov, M. N. (1977). "Spatial Organization of Cerebral Processes." Wiley, New York.

Lopes da Silva, F. (1991). Neural mechanisms underlying brain waves: From neural membranes to networks. *Electroencephalogr. Clin. Neurophysiol.* **79,** 81–93.

Lopes da Silva, F., Pijn, J. P., *et al.* (1989). Interdependence of EEG signals: Linear vs. nonlinear associations and the significance of time delays and phase shifts. *Brain Topogr.* **2,** 9–18.

Mangun, G. R., Buonocore, M. H., *et al.* (1998). ERP and fMRI measures of visual spatial selective attention. *Hum. Brain Mapp.* **6,** 383–389.

Mars, N. J., and Lopes da Silva, F. H. (1987). EEG analysis methods based on information theory. *In* "Methods of Analysis of Brain Electrical and Magnetic Signals" (A. S. Gevins and A. Rémond, eds.), Vol. 1, pp. 297–307. Elsevier, Amsterdam.

McEvoy, L. K., Chellaramani, R., *et al.* (2001). Effects of a benzodiazepine on behavioral and neurophysiological measures of working memory. Presented at the Annual Meeting of the Society for Neuroscience, San Diego, CA.

McEvoy, L. K., Smith, M. E., *et al.* (1998). Dynamic cortical networks of verbal and spatial working memory: Effects of memory load and task practice. *Cereb. Cortex* **8,** 563–574.

McEvoy, L. K., Smith, M. E., *et al.* (2000). Test–retest reliability of cognitive EEG. *Clin. Neurophysiol.* **111,** 457–463.

McIntosh, A. R. (1998). Understanding neural interactions in learning and memory using functional neuroimaging. *Ann. N. Y. Acad. Sci.* **855,** 556–571.

Miller, E. K., and Desimone, R. (1994). Parallel neuronal mechanisms for short-term memory. *Science* **263,** 520–522.

Miltner, W., Braun, C., *et al.* (1994). A test of brain electrical source analysis (BESA): A simulation study. *Electroencephalogr. Clin. Neurophysiol.* **91,** 295–310.

Moeller, J. R., Nakamura, T., *et al.* (1999). Reproducibility of regional metabolic covariance patterns: Comparison of four populations. *J. Nucl. Med.* **40,** 1264–1269.

Murphy, V. N., and Fetz, E. E. (1992). Coherent 25–35 Hz oscillations in the sensorimotor cortex of awake behaving monkeys. *Proc. Natl. Acad. Sci.* **89,** 5670–5674.

Nichols, E. A., Ilan, A. B., *et al.* (2001). Effects of marijuana on neurophysiological correlates of working and intermediate-term memory. Presented at the Annual Meeting of the Society for Neuroscience, San Diego, CA.

Nunez, P. L. (1981). "Electric Fields in the Brain: The Neurophysics of EEG." Oxford Univ. Press, New York.

Nunez, P. L., and Pilgreen, K. L. (1991). The spline–Laplacian in clinical neurophysiology: A method to improve EEG spatial resolution. *J. Clin. Neurophysiol.* **8,** 397–413.

Nunez, P. L., and Silberstien, R. B. (2000). On the relationship of synaptic activity to macroscopic measurements: Does co-registration of EEG with fMRI make sense? *Brain Topogr.* **13,** 79–96.

Nyberg, L., McIntosh, A. R., *et al.* (1996). Network analysis of positron emission tomography regional cerebral blood flow data: Ensemble inhibition during episodic memory retrieval. *J. Neurosci.* **16,** 3753–3759.

O'Connor, M. F., Daves, S. M., *et al.* (2001). BIS monitoring to prevent awareness during general anesthesia. *Anesthesiology* **94,** 520–522.

Owen, A. M., Evans, A. C., *et al.* (1996). Evidence for a two-stage model of spatial working memory processing within the lateral frontal cortex: A positron emission tomography study. *Cereb. Cortex* **6,** 31–38.

Picton, T. W. (1992). The P300 wave of the human event-related potential. *J. Clin. Neurophysiol.* **9,** 456–479.

Posner, M. I., and Peterson, S. E. (1990). The attention system of the human brain. *Annu. Rev. Neurosci.* **13,** 25–42.

Regan, D. (1989). "Human Brain Electrophysiology." Elsevier, New York.

Rogers, L. (1991). Determination of the number and waveshapes of event related potential components using comparative factor analysis. *Int. J. Neurosci.* **56,** 219–246.

Scherg, M., and Von Cramon, D. (1985). Two bilateral sources of the late AEP as identified by a spatio-temporal dipole model. *Electroencephalogr. Clin. Neurophysiol.* **62,** 32–44.

Sereno, M. I. (1998). Brain mapping in animals and humans. *Curr. Opin. Neurobiol.* **8,** 188–194.

Simpson, G. V., Pflieger, M. E., *et al.* (1995). Dynamic neuroimaging of brain function. *J. Clin. Neurophysiol.* **12,** 432–449.

Smith, E. E., and Jonides, J. (1995). Working memory in humans: Neuropsychological evidence. *In* "The Cognitive Neurosciences" (M. Gazzaniga, ed.), pp. 1009–1020. MIT Press, Cambridge, MA.

Smith, E. E., Jonides, J., *et al.* (1995). Spatial versus object working memory: PET investigations. *J. Cognit. Neurosci.* **7,** 337–356.

Smith, M. E., Gevins, A., *et al.* (2001). Monitoring task load with multivariate EEG measures during complex forms of human computer interaction. *Hum. Factors,* **43,** 366–380.

Smith, M. E., Halgren, E., *et al.* (1990). The intracranial topography of the P3 event-related potential elicited during auditory oddball. *Electroencephalogr. Clin. Neurophysiol.* **76,** 235–248.

Smith, M. E., McEvoy, L. K., *et al.* (1999). Neurophysiological indices of strategy development and skill acquisition. *Brain Res. Cognit. Brain Res.* **7,** 389–404.

Spinelli, L., Andino, S. G., *et al.* (2000). Electromagnetic inverse solutions in anatomically constrained spherical head models. *Brain Topogr.* **13,** 115–125.

Suarez, E., Viegas, M. D., *et al.* (2000). Relating induced changes in EEG signals to orientation of visual stimuli using the ESI-256 machine. *Biomed. Sci. Instrument.* **36,** 33–38.

Sutton, S., Braren, M., *et al.* (1965). Evoked potential correlates of uncertainty. *Science* **150,** 1187–1188.

Tesche, C. D., Uusitalo, M. A., *et al.* (1995). Signal-space projections of MEG data characterize both distributed and well-localized neuronal sources. *Electroencephalogr. Clin. Neurophysiol.* **95,** 189–200.

Thompson, J. L., and Ebersole, J. S. (1999). Long-term inpatient audiovisual scalp EEG monitoring. *J. Clin. Neurophysiol.* **16,** 91–99.

van den Broek, S. P., Reindeers, F., *et al.* (1998). Volume conductions effects in EEG and MEG. *Electroencephalogr. Clin. Neurophysiol.* **106,** 522–534.

Vespa, P., Nenov, V., *et al.* (1999). Continuous EEG monitoring in the intensive care unit: Early findings and clinical efficacy. *J. Clin. Neurophysiol.* **16,** 1–13.

Walter, D. O. (1963). Spectral analysis for electroencephalograms: Mathematical determination of neurophysiological relationships from records of limited duration. *Exp. Neurol.* **8,** 155–181.

Walter, W. G., Cooper, R., *et al.* (1964). Slow potential changes in the human brain associated with expectancy, decision and intention. *Electroencephalogr. Clin. Neurophysiol. Suppl.* **26,** 123–130.

Wang, J. Z., Williamson, S. J., *et al.* (1993). Magnetic source imaging based on the minimum-norm least-squares inverse. *Brain Topogr.* **5,** 365–371.

Wikswo, J. P., Gevins, A. S., *et al.* (1993). The future of EEG and MEG. *Electroencephalogr. Clin. Neurophysiol.* **87,** 1–9.

Wilson, G. F., Fullenkamp, B. S., *et al.* (1994). Evoked potential, cardiac, blink, and respiration measures of pilot workload in air-to-ground missions. *Aviat. Space Environ. Med.* 100–105.

Wood, C. C., and McCarthy, G. (1984). Principal component analysis of event-related potentials: Simulation studies demonstrate misallocation of variance across components. *Electroencephalogr. Clin. Neurophysiol.* **59,** 249–260.

Woody, C. D. (1967). Characterization of an adaptive filter for the analysis of variable latency neuroelectric signals. *Med. Biol. Eng.* **5,** 539–553.

9

Electrophysiological Methods for Mapping Brain Motor and Sensory Circuits

Paul D. Cheney

Smith Mental Retardation Research Center and Department of Molecular and Integrative Physiology, University of Kansas Medical Center, Kansas City, Kansas 66160

I. Introduction and Historical Perspective

One of the most important developments in the past century of brain research is the recognition that specific aspects of the sensory and motor periphery are mapped in an orderly fashion to specific regions of the brain. Discovering the nature of the map and its orientation continues to be one of the most important challenges in neuroscience. But why is brain mapping important? Maps reveal the way the brain is organized and, in part, how it does its job. Several decades of electrophysiological brain mapping have shown that as more refined tools are applied to this problem, more detailed and specific maps of functional localization emerge. Those leading the movement in the early 19th Century against localization of function in the brain would indeed be astounded at the extent to which they were wrong (Schäfer, 1900).

Electrophysiological brain mapping began in the mid-1800s with the recognition by Fritsh and Hitzig (1870) that electrically stimulating specific parts of the cerebral cortex could evoke movements in the periphery. Moreover, the movements that were evoked were not random but had a certain order about them such that movements of adjacent body parts could be evoked by adjacent regions of cortical tissue. This began the serious experimental study of brain mapping that has continued to the present day (Table 1). Developments of particular significance in the history of electrophysiological brain mapping include the human cortical motor maps drawn by Penfield and Boldrey (1937) based on stimulation experiments in patients undergoing neurosurgical procedures. Also, the detailed map of the monkey motor cortex published by Woolsey in 1952 was particularly influential. The introduction by Adrian (1926) of techniques for recording the spike discharges of single axons led to the later refinement of microelectrode techniques in the 1950s for mapping sensory areas of the brain at the level of single neurons (Hubel, 1957). The development by Herbert Jasper in 1958 of techniques for recording from single neurons in awake animals and its application by Evarts to studies of the properties of neurons in relation to trained movements in monkeys revolutionized the study of the motor system. The awarding of the 1981 Nobel prize in Medicine or Physiology to David Hubel and Torsten Wiesel for their elegant electrophysiological studies of the organization of visual cortex brought well-deserved recognition of their contributions to understanding the representation of the visual periphery in the cerebral cortex. The introduction in 1967 of intracranial microstimulation by Hiroshi Asanuma and colleagues has proven to be a very informative and revealing tool for investigating transformations between sensory and motor maps in several regions of the brain (Asanuma and Sakata, 1967). Finally, the introduction by Merton and Morton in the 1980s of transcranial electrical and magnetic stimulation of the cortex has provided new noninvasive tools by which to explore cortical function and motor maps in humans (Merton and Morton, 1980).

Brain mapping efforts have focused most intensely on primary sensory and motor areas because their relationship

Table 1 Chronology of Significant Developments in Electrophysiological Brain Mapping

1700s–mid-1800s	A long period dominated by the phrenologists, who ascribed a wide range of functions including personality traits, weaknesses, etc., to the protrusions (gyri) of the brain.
1860s	Hughlings Jackson's observations on the patterns of activation of muscle groups during convulsions convince him of localization of function in the brain.
1870	Fritsch and Hitzig first discover electrical excitability of cortex in the dog and produce the first experimentally determined brain motor map.
1875	Ferrier produces the first motor map of the primate brain and transfers the centers to an outline of the human brain.
1926	Adrian first records the extracellular spike activity of a nerve fiber.
1937	Penfield and Boldrey report data showing the first experimentally determined motor map of the human brain. Penfield's work continued for many years, culiminating in 1950 in the now famous human motor homunculus figure that adorns virtually every textbook of neuroscience.
1952	C. Woolsey produces a detailed map of the monkey's cortex using 60-Hz sinusoidal current in trains of 2 s duration.
1957	Pioneering studies of Vernon Mountcastle and colleagues using microelectrode recording to map primary somatosensory cortex.
1957	Nobel-prize-winning studies of Hubel and Wiesel begin. These studies mapped the columnar organization of primary visual cortex in the monkey.
1958	Herbert Jasper in Montreal makes the first recordings from neurons in an awake monkey during behavior.
1965	Edward Evarts uses Jasper's technique to investigate the properties of neurons in awake monkeys during voluntary movement.
1967	Hiroshi Asanuma and co-workers introduce the technique of intracortical microstimulation for detailed mapping of cortical motor output.
1985	Transcranial magnetic stimulation is used to activate motor cortex in humans noninvasively.

to the periphery is clearest and easiest to understand. In general, the farther away from the periphery one goes, either on the sensory input side or on the motor output side, the more fuzzy the map becomes. However, it can be argued that all regions of the brain probably have a map of some type, but for many regions the map is based on subtle or abstract features whose relation to the environment is not readily apparent. In general, as one moves to areas of the brain farther from the periphery, more abstract properties are encoded by single neurons and the maps for these regions become more complex. For example, neurons in the hippocampus, an area long implicated in memory storage, have been mapped for specific locations or places in the environment. Some neurons discharge only when an animal occupies a specific place within its environment (Breese *et al.,* 1989; Muller and Kubie, 1989; Speakman, 1987).

The future of brain mapping holds in store many fascinating revelations because there is so much yet to be learned. Nevertheless, great progress has already been made in understanding the maps of primary sensory and motor areas of the brain using electrophysiological methods. Because mapping motor function of the brain poses unique problems for brain mapping that have inspired a rich variety of methodologies, the primary emphasis of this chapter will be on motor system rather than sensory system mapping. Maps based on electrophysiological approaches have been determined for somatosensory, visual, and auditory systems in a variety of species but detailed presentation of these maps for every sensory system is beyond the scope of this chapter. However, the methodology upon which these sensory system maps are based, namely unit recording with microelectrodes, has also been used extensively for motor system mapping and this approach will be discussed in a later section of this chapter.

II. Structural versus Functional Brain Maps

Brain mapping takes many forms, but maps of all types can be categorized as either structural or functional in nature. Structural mapping reveals aspects of the neural hardware, that is, the anatomical connectivity linking one part of the brain to another part and the principles that apply to the way in which these linkages are ordered. Methods for this type of mapping almost always consist of electrical stimulation of neurons and pathways. The measured output variable may take several different forms, including intracellular potentials, field potentials reflecting underlying synaptic inputs to an area of the brain, extracellularly recorded unit potentials, or electromyographic (EMG) activity. The distinguishing factor in structural mapping is that a part of the system is stimulated electrically or magnetically, and a response, usually in the form of some type of electrical activity, is recorded from another part of the neural circuit.

Unlike structural methods of brain mapping, functional mapping, as the name implies, reveals the spatial representation within the brain of some natural parameter such as touch receptors on the body surface, movement of a body part, the frequency of an auditory stimulus, or the location of a light stimulus in the visual field. Functional mapping involves the orderly localization of natural stimuli or movements to examine organization within a particular brain region.

III. Strengths of Electrophysiological Mapping Methods Compared to Other Brain Mapping Methods

Electrophysiological methods provide the greatest resolution for mapping the localization of function within the brain because such maps can use as their basis the most elementary unit of brain structure—the single neuron. This is true both of sensory maps, with which responses of single neurons to various modalities of sensory stimuli can be investigated, and for motor maps, with which the target muscles of individual output neurons can be determined using new techniques that will be described in later sections of this chapter. The properties of many single units in an area can then be pieced together to construct a more global map. Electrophysiological mapping methods, then, are ideally suited to answering detailed questions about the properties of individual neurons or discrete clusters of neurons and their relation to the larger map.

Electrophysiological mapping also has disadvantages. First, microelectrode recording and stimulation are invasive methods that can be applied in humans only under unique circumstances, for example, during surgeries for intractable epilepsy or tumor removal. However, the recently introduced methods of transcranial electrical and magnetic stimulation overcome this limitation. Second, brain mapping with these methods is a tedious and time-consuming process in which the properties of hundreds or even thousands of neurons (or sites) must be determined before a complete map can be assembled. Such recordings in a single animal generally require several hundred electrode penetrations and either months of work in chronically implanted animals or, typically, 24 or more hours in anesthetized animals.

In contrast, imaging methods of various types (*f*MRI, PET) can provide a complete picture of brain activity as a "snapshot" at one moment in time during a particular behavior of interest. However, the strengths of electrophysiological mapping, namely high spatial resolution down to the level of single neurons and time resolution in the millisecond range, are the weaknesses of the new brain imaging methods, although progress is being made to improve these aspects of imaging technology as discussed in other chapters. Time resolution with high channel-density EEG recording and magnetoencephalography (MEG) is excellent but spatial resolution can be problematic. Imaging techniques based on histological procedures such as 2-deoxyglucose mapping provide outstanding spatial resolution but time resolution is still a weakness. Nevertheless, the global picture of the whole brain in relation to a particular behavior or function provided by imaging techniques is very difficult to obtain with electrophysiological mapping procedures. Therefore, combining electrophysiological and imaging methods constitutes a very powerful approach. Brain imaging techniques provide a complete picture of the whole brain, identifying regions that are involved in a particular function and some basic features of the organizational map that may exist. The map can then be investigated further with electrophysiological analysis to examine more detailed features of organization.

An example of the detail possible with electrophysiological methods is the map in Fig. 1 of primary somatosensory cortex in a monkey taken from the elegant work of John Kaas and colleagues (Nelson *et al.,* 1980). In this experiment, systematic microelectrode penetrations were made at intervals of 250 μm throughout the postcentral gyrus in a monkey under anesthesia. With electrophysiological mapping, the area devoted to each body part can be carefully determined and often relatively sharp boundaries can be identified. The map shows that there are complete representations of the body surface in both area 3b and area 1. Not illustrated in the map is the fact that within each digit there is also an orderly representation in which the most distal ends of the digit maps in areas 3b and 1 are contiguous, that is, the maps in 3b and 1 are mirror images of each other. Maps such as these currently cannot be derived by any means other than electrophysiological approaches.

Microelectrodes for unit recording generally consist of sharpened rods of tungsten or platinum–iridium that have been insulated with glass, Epoxylite, or Parylene, leaving an uninsulated tip of about 5–12 μA. Such electrodes generally have an impedance of 0.7–1.5 MΩ, which is ideal for recording from neurons in cerebral cortex and many other brain areas. The technology for fabricating metal microelectrodes and the electronic instrumentation needed for unit recording are well documented and have been reviewed in detail elsewhere (Lemon, 1984).

IV. Contrasts between Sensory versus Motor System Mapping

Sensory system mapping consists of determining the spatial representation of a specific sensory parameter within a particular brain region of interest. This is normally done by recording from single neurons or neuronal clusters in repeated systematic electrode penetrations spaced at an appropriate density, usually 0.2–1.0 mm, while exploring the sensory space of interest. A great advantage from an experimental standpoint is the fact that the investigator has virtually complete control over all sensory stimulus parameters. This usually means that a neuron's receptive field and its modality specificity can be determined relatively quickly and with considerable certainty. Because of the relative ease with which sensory receptive fields can be identified, very complete maps, such as the one illustrated in Fig. 1, are possible. Most significantly, with sensory system mapping there is only one output variable—unit recording from the region of interest—and one input variable—the stimulus of interest applied throughout the space that should contain the neuron's receptive field.

The motor system presents unique difficulties for mapping. In contrast to sensory system mapping, a wide range of different input and output measures have been applied to motor system mapping. Many of these measures can be rigorously applied only in awake animals trained to make movements of interest. In fact, the next section of this chapter lists six output measures and seven input measures that have been or could be used in studies of motor system mapping (Table 2). Some of these procedures are directly analogous to sensory system mapping, but most are fundamentally different.

What is the motor system analog of the sensory system receptive field and how can this feature be determined? A receptive field is the sensory space of a neuron's adequate stimulus, that is, the naturally occurring stimulus modality to which the neuron is most responsive. The sensory receptive

Table 2 Methods for Mapping the Motor System of the Brain

Output (response) measures
Brain unit activity associated with movement
Evoked movements
Monosynaptic muscle spindle Ia reflex testing
EMG activation
Single-motor-unit activity
Intracellular recording from motoneurons
Input (stimulus) measures
Macrostimulation of the cortical surface
Transcranial magnetic stimulation
Transcranial electrical stimulation
Repetitive microstimulation
Single-pulse microstimulation (stimulus-triggered averaging of EMG)
Single-cell action potentials in spike-triggered averaging of EMG
Passive joint movements

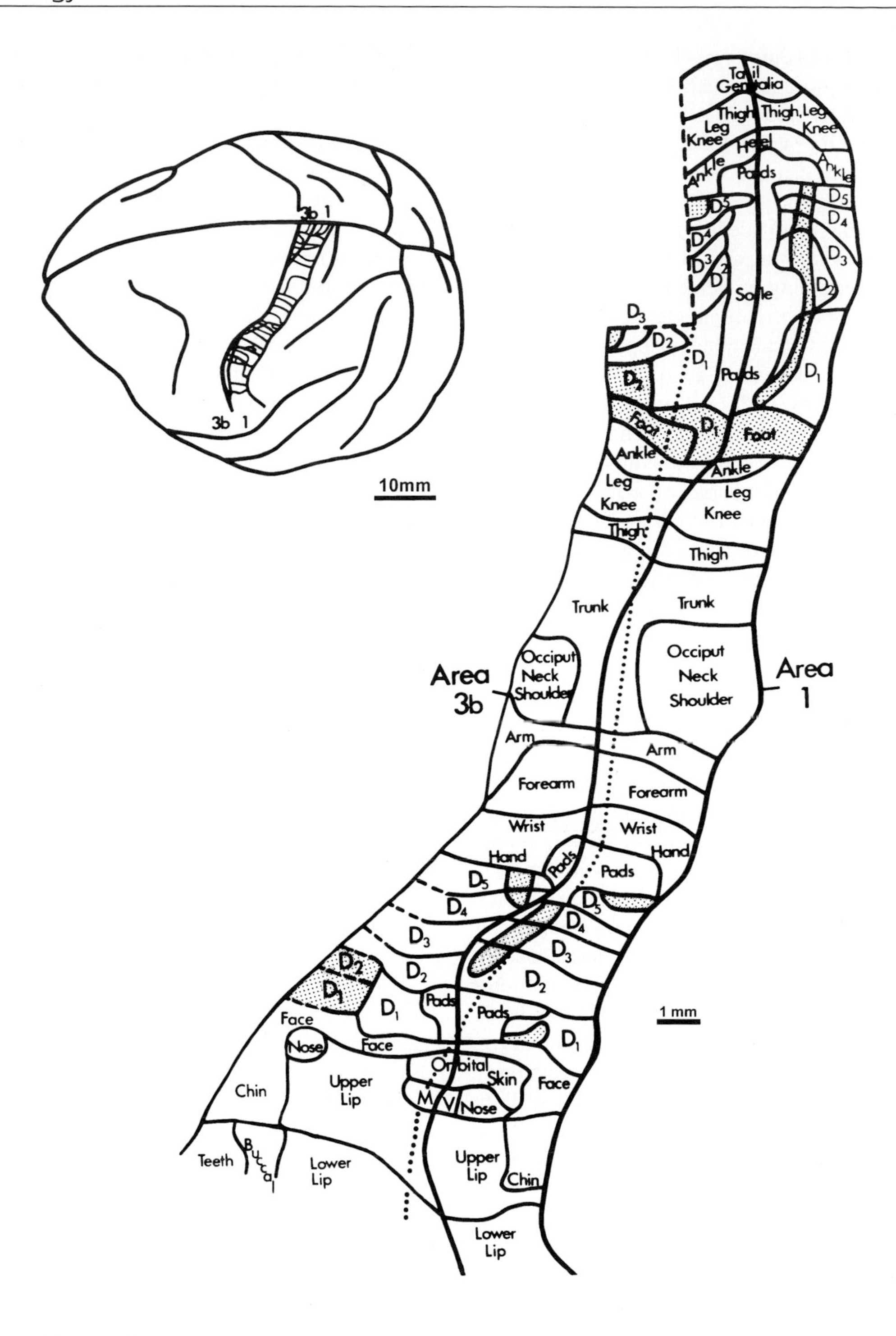

Figure 1 Map of body surface representation in areas 3b and 1 of primary somatosensory cortex in the cynomolgus monkey determined by systematic electrode penetrations at intervals of 250–350 μm and testing every 250 μm within an electrode penetration. Heavy lines mark the borders between 3b and 1. Shaded areas correspond to the hairy dorsum of the digit representation. The dotted line indicates the position of the central sulcus. The dashed line indicates regions of the representation that are contained within the medial wall of the hemisphere. Partial maps from several monkeys were combined to produce a complete body surface map. Note the highly orderly representation of body parts beginning most medially with the tail and genitalia and proceeding laterally to the foot, ankle, leg, thigh, trunk, neck, shoulder, arm, wrist, hand, digits, face, and, most, laterally the lips. (Reproduced, with permission, from Nelson *et al.,* 1980.)

fields of many individual neurons can be merged to yield an overall map of sensory space for the brain region of interest. The analogous parameter for neurons in motor regions of the brain is the motor field (Fetz and Cheney, 1979). Different subtypes of motor fields can be identified: movement fields, functional muscle fields, and structural muscle fields. Movement fields are determined by identifying all the movements which engage the activity of a particular neuron. It is important to note that determining movement fields is a form of functional mapping just as determining the receptive fields of sensory neurons constitutes functional mapping. An assumption is often made that underlying a neuron's movement field are anatomic connections with motoneurons of muscles involved in producing the movement that engaged the cell's activity. Although this is a reasonable assumption, establishing causality between the discharge of a neuron and the activity of a muscle with which it is correlated requires additional procedures that will be described in a later section of this chapter. Rigorous movement field mapping is difficult experimentally because it requires training animals to make isolated movements at many different joints. Mapping functional muscle fields also poses problems because it requires activation of muscles in isolation and in most cases this is difficult if not impossible.

Motor system mapping can also be based on the identification of structural muscle fields. Structural muscle fields are the target muscles of a single neuron, that is, the motor nuclei with which the neuron is synaptically linked. The linkage must be relatively direct, either monosynaptic or disynaptic, to produce the type of robust control over muscle activity needed of an output pathway. It is important to note that the structural muscle field is based on data about the cell's anatomic linkage with motoneurons, and in this sense it is fundamentally different from other types of motor fields. How can data about structural muscle fields be obtained? Neuroanatomic labeling techniques might be one possibility, but this technology has not advanced to the point that all the axonal terminations of single neurons injected into brain structures such as the cortex can be traced all the way to the spinal cord. However, some very elegant studies of axon terminal reconstruction after injection of tracers into corticospinal axons have been published (Lawrence *et al.*, 1985; Shinoda *et al.*, 1981). Figure 2 is an example showing the terminals of a single corticospinal axon branching at C7 to innervate at least four motor nuclei. In this case, Shinoda and colleagues were able to find sites of close contact between terminal boutons and proximal dendrites of motoneurons in three of the four motor nuclei. Lawrence *et al.* (1985), using similar methods, showed that each corticospinal axon generally makes only one synaptic contact with a given motoneuron. A variety of electrophysiological approaches also exists for determining structural muscle fields, and these will be discussed in a later section. However, parallel studies using both anatomical and electrophysiological methods are the most powerful approach to muscle field mapping.

V. Output Measures for Mapping Motor System Organization

Output measures are the end points upon which motor maps are based. Therefore, the properties of the output measure used will influence the character of the resulting map. Peripheral output measures for brain mapping can take several different forms (Table 2). The most commonly used measures for motor system mapping are evoked movements and EMG activation. Tabulating the topography of evoked movements from electrical stimulation of a brain structure is the simplest and quickest method for generating a motor map. No sophisticated recording devices are needed. Movements evoked from a given site are simply noted based on the body part involved in the movement. This technique must be combined with a stimulation method that will activate sufficient premotor and motoneurons to produce an overt movement that can be visually detected. Limitations of this technique include its lack of sensitivity, lack of information about specific muscles that are affected by a particular stimulus, and inability to detect subliminal or weak effects on motor nuclei.

Much greater sensitivity can be obtained by directly recording EMG activity from individual muscles. Evaluation of stimulation applied to a brain region of interest can be evaluated from single responses in raw EMG records. Alternatively, the EMG responses can be averaged over several stimulus trials to produce a more reliable and quantitative end-point measure. Recording EMG activity has many advantages for detailed mapping. Techniques for EMG activity have been well documented (e.g., Loeb and Gans, 1986; Park *et al.*, 2000).

The most common output measure in sensory system electrophysiological mapping is the recording of single neurons or clusters of neurons in relation to natural stimuli. This is one of the least common methods applied to motor system mapping because of the difficulty in training animals to make the large repertoire of active movements that would be required for a detailed and comprehensive map. However, this type of mapping has been done on a more limited scale. Schieber and Hibbard (1993) trained monkeys to move individual digits of the hand in isolation and determined the extent to which single motor cortex cells were related to isolated movements of each digit. They found that motor cortex cells were related to movements of multiple digits and that the territories containing cells related to different digits overlapped extensively.

A more common approach to motor system mapping using unit recording as the output measure is actually a form of sensory mapping that involves determining a neuron's responses to passive movements of different joints (Soso and Fetz, 1980; Lemon and Porter, 1976; Wong *et al.*, 1978). Many studies have established strong correlations between the joint involved in a motor cortex cell's active movement response and the location of its sensory receptive field

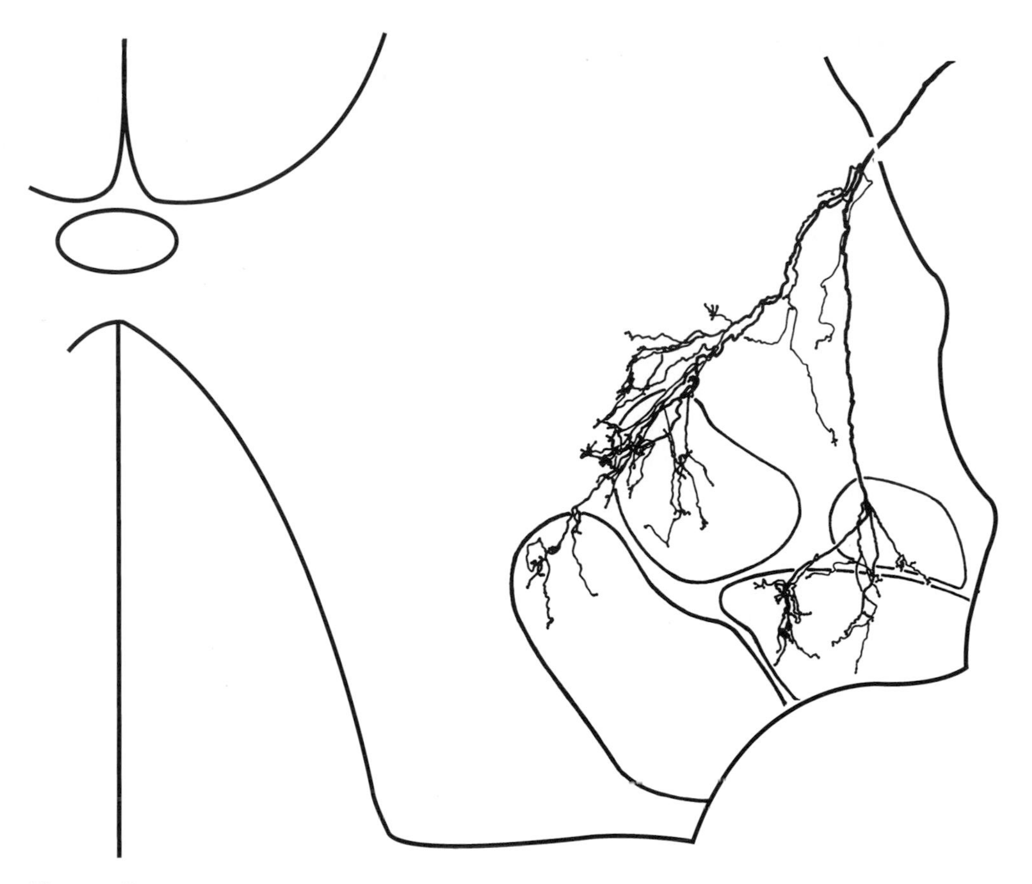

Figure 2 Transverse reconstruction of the terminal arborization of a hand area corticomotoneuronal axon in the C7 segment of the monkey spinal cord. This map was constructed from 12 serial transverse sections. The ulnar nerve motoneurons (upper two motor nuclei) and the radial nerve motoneurons (lower two motor nuclei) were labeled by retrograde transport of HRP. (Reproduced, with permission, from Shinoda *et al.*, 1981.)

(Rosen and Asanuma, 1972; Wong *et al.*, 1978). Therefore, maps based on passive joint movements provide an approximation of what the active movement map would look like (Murphy *et al.*, 1978).

One additional output measure for motor system mapping deserves mention. Preston and colleagues (1967) and later Asanuma *et al.* (1976) used monosynaptic reflex volleys recorded from muscle nerves in response to stimulation of muscle spindle Ia afferents as an output measure for detecting the sign and distribution of cortical influence on different muscle groups. This was a very informative approach capable of revealing not only excitatory effects but also inhibitory effects. This approach provided an initial tabulation of the net effect of corticospinal output on different muscle groups of the upper and lower extremity.

VI. Electrical Stimulation and Other Input Measures for Mapping Motor System Organization

Input (stimulus) measures for mapping the motor system vary in refinement and volume of tissue activated. Most input measures used for motor system mapping rely on some form of electrical stimulation to activate a collection of neurons in the vicinity of the stimulating electrode. The output response then reflects the organization of the collection of stimulated neurons rather than any single neuron. Nevertheless, despite its artificial nature, electrical stimulation of the brain has been and remains an important and meaningful method of mapping motor system organization.

The magnitude of the response obtained with electrical stimulation will depend on several factors, including stimulus current, stimulus duration, stimulus polarity, and electrode dimensions. Extracellular current density is the factor that determines whether a neuron or axon will be stimulated. The current density must achieve a certain threshold value to excite a neuronal element. Of course, the larger the electrode the greater the current that must be applied to produce an effective current density near the electrode. Currents of 1–2 mA are commonly applied with brain macroelectrodes (1–3 mm in diameter), while currents on the order of 2–5 μA are sufficient with microelectrodes (5–10 μm exposure). The voltage changes near a small electrode tip can be very large. Thus the use of stimulating electrodes with small tips (termed microstimulation) can produce greater localization of the stimulated elements, yielding greater map resolution. However, this added effectiveness near the electrode tip applies only for the first 50 μm from the electrode. At distances of several electrode tip radii away, the current will become dispersed to such an extent that electrode size will no longer make a difference (Rank, 1975, 1981).

The extent of effective current spread from a stimulating electrode is an important issue for interpretation and for defining boundaries within the map. Current spread from electrodes is given by the expression

$$I = a + kd^2,$$

where I is the effective current, that is, the current passed through the electrode tip necessary to excite an axon or soma to discharge an action potential; a is the minimum current necessary for excitation when the electrode is immediately adjacent to the axon or cell body; k is a proportionality constant which is dependent on the electrode characteristics and the nature of the tissue being stimulated; and d is the distance between the electrode and the neuron or axon. It is clear from this expression that the current required to activate a particular neuronal element increases in proportion to the square of the distance between the electrode and the element to be stimulated.

Rank (1975) has summarized the results of many experiments on electrical stimulation of axons and cell bodies. Figure 3 is a log–log plot of current strength against distance after current pulses from many different experiments have been normalized to 200 μs. The data shown are for a monopolar cathode at different distances from the recording electrode. Despite considerable variability, the data in Fig. 3 show that myelinated axons and their cell bodies have much lower thresholds than small CNS neurons and unmyelinated axons. For example, a 100-μA current pulse will excite a large myelinated axon (65 m/s conduction velocity) at a distance of 1.2 mm, but a smaller axon (25 m/s) lying only 500 μm from the electrode tip will not be activated. One can also conclude that all neurons within a 500-μm radius will be excited by this stimulus. At distances of 5–7 mm from the electrode, the effective currents are very high, 2–5 mA. Asanuma and Arnold (1975) reported that stimulation with six 200-μs pulses through a microelectrode with an exposed tip of 15–30 μm produced bubbles when the currents reached 60 μA. Therefore, currents higher than these can be delivered safely using only electrodes with large exposures. In any case, most commercially available stimulators would not be capable of the voltages required for delivering currents much greater than 100 μA through a microelectrode with an impedance of 0.7–1.5 M.

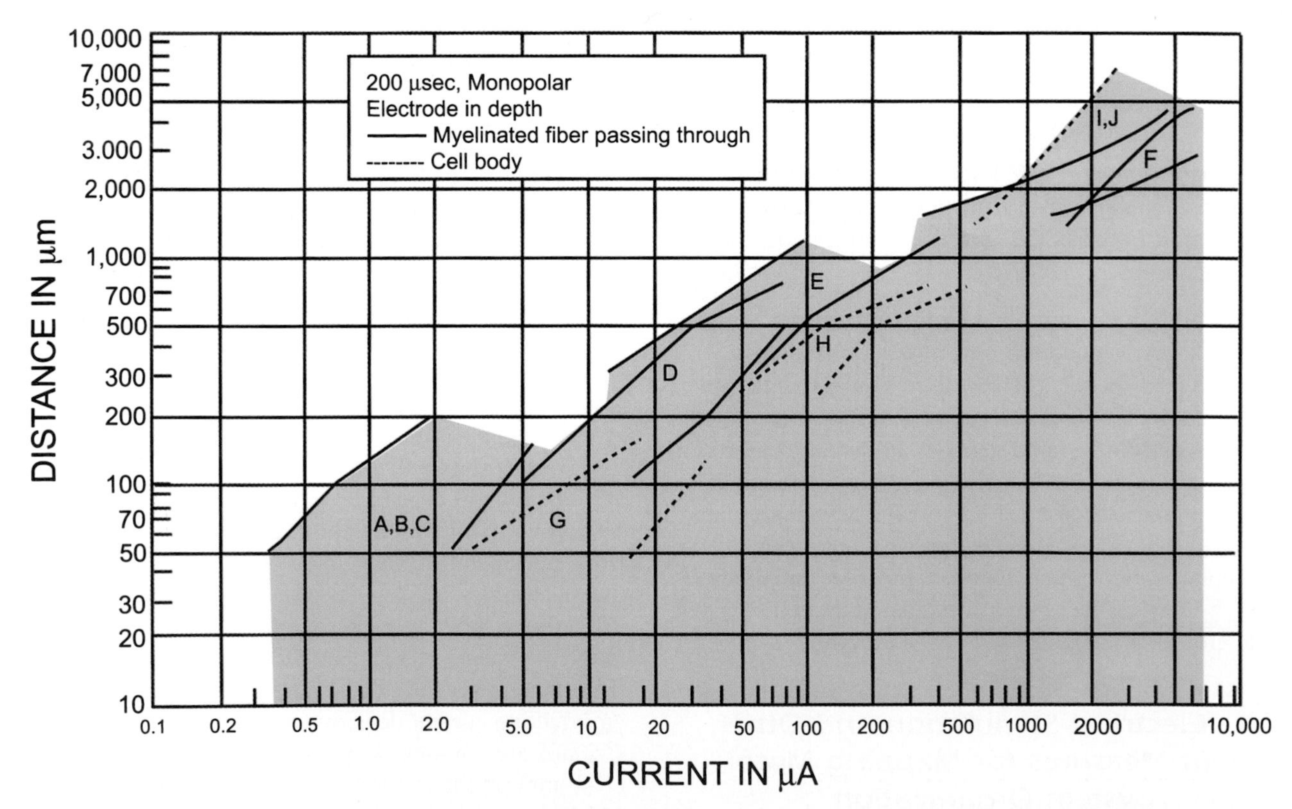

Figure 3 Log–log plot of current distance relations normalized for a 200-μs cathodal monopolar pulse applied in the depth of the brain. Plot is based on data from a variety of studies reported in the literature. See Rank (1975) for references, further explanation, and qualifications that apply.

Using different values of the constant k covering the range of published values, Cheney and Fetz (1985) estimated the number of large and small cortical pyramidal tract neurons that would be activated at different stimulus intensities given estimates of cell density in layer V of the cortex (Andersen *et al.,* 1975). These estimates of current spread suggest that a 10-μA stimulus will directly activate 1–12 large pyramidal tract neurons and 180–2168 small pyramidal tract neurons (Table 3). Asanuma and co-workers (1976) proposed a useful rule of thumb that applies to combined stimulation and unit recording. The rule is that a 10-μA stimulus will directly activate any neuron whose spike can be recorded through the same microelectrode. A 10-μA stimulus would have an effective radius of spread of about 100 μm (intermediate k).

Electrical stimulation is a very useful tool for mapping motor output functions of the brain, but it is essential to try to define what neuronal elements are actually being stimulated for the specific conditions of each experiment. Although this topic has been the source of much debate in the past (Phillips and Porter, 1977), it is safe to assume that extracellular stimulation will virtually always activate a mixture of elements including not only cell bodies but also axons passing through the region stimulated and axon terminals. Interpretation of results, therefore, must be in terms of the total set of potential elements stimulated, not just neuronal cell bodies. For example, stimulating in motor thalamus will not only activate thalamic neurons but also the axon terminals of deep cerebellar nuclear cells. Motor output effects from this stimulation, therefore, could be mediated not only by direct activation of motor cortex neurons that receive cortical input, but also by a so-called axon reflex involving stimulation of cerebellar axon terminals and antidromic conduction to a branch point where the action potential may then conduct orthodromically to the red nucleus, which could also contribute to the motor effects observed.

These same problems also apply to electrical stimulation of the cerebral cortex. It cannot necessarily be assumed that the neurons mediating the effects observed from cortical stimulation at a site are just from the volume of tissue directly activated by physical spread of the stimulus as given by the expression above. In addition to physical spread of the stimulus, there can be physiological spread in which elements directly excited by the stimulus activate other neurons at a distance from the electrode. This could occur through extensive horizontal connections within the cortex, for example, axon collaterals of corticospinal neurons or branches of afferent axons (Huntley and Jones, 1991).

VII. Mapping Motor Output with Transcranial Stimulation of Cortex

The introduction of transcranial electrical stimulation (TES) of the cortex by Merton and Morton (1980) and transcranial magnetic stimulation (TMS) by Barker *et al.* (1985) has provided two new noninvasive techniques by which the output circuitry of the motor system can be investigated and mapped. These techniques are now well established and are widely used both experimentally and in clinical settings to (1) study brain structure–function relationships, (2) map brain motor output, and (3) measure the conduction velocity of central motor pathways (Levy *et al.,* 1991). For motor system mapping with either method, EMG potentials from individual muscles are generally recorded as the output measure. The EMG potentials are usually averaged for several trials and referred to as motor-evoked potentials.

A. Comparison of Electrical and Magnetic Transcranial Stimulation

TES consists of applying a high-voltage, short-duration (1000–2000 V, 10–50 μs) electrical pulse through

Table 3 Number of Pyramidal Tract Neurons Activated by Different Intensities of ICMS[a]

Stimulus	Minimal k (k = 250 μ A/mm²)		Intermediate k (k = 1350 μA/mm²)		Maximal k (k = 3000 μ A/mm²)	
(μ A)	Large	Small	Large	Small	Large	Small
3	3	650	1	120	1	54
5	6	1084	1	201	1	90
10	12	2168	2	402	1	180
15	18	3252	3	603	2	270
20	24	4336	4	804	2	360

Note. Estimates based on evidence of Andersen *et al.* (1975). Large and small refer to size of pyramidal tract neurons.

[a]From Cheney and Fetz (1985).

low-resistance electrodes placed on the skull. The short duration of the stimulus pulse reduces discomfort associated with the stimulus if it is applied in awake subjects. Typical electrodes consist of saline soaked pads about 1 cm in diameter separated by 5–10 cm. The use of large electrodes ensures that current density does not reach tissue-damaging levels. Alternatively, Amassian and Cracco (1987) reported that much of the discomfort is related to muscle contraction produced by the stimulus. They found that effective stimulation could be achieved with longer pulse durations and much lower voltages if the cathode consisted of a semicircular ring placed medially on the skull to the stimulating focal anode. TES is similar to stimulation of the cortical surface in the sense that direct activation of corticospinal neurons occurs at the anode. Therefore, the anode should be placed over the site to be stimulated with the cathode occupying an indifferent site or a ring surrounding the anode.

TMS has provided an important new tool by which to investigate motor system function and organization in humans. This method consists of discharging a large capacitive current (5–10 kA) through a coil placed on the surface of the skull (Fig. 4). TMS current pulses have a duration of 200–300 μs and are produced by discharging a capacitor with a 2–3 kV charge. The current pulse produces a localized magnetic field of up to 2 T. For comparison, the Earth's magnetic field is about 50 μT. The magnetic field flows unimpeded through the skin and skull, decaying as the square of distance from the coil. According to Faraday's law, the magnetic field will induce an electric field. This electric field drives current flow in the brain. The induced current flow is parallel to the plane of the coil and opposite the direction of current flow in the magnetic coil. The rate of change of the magnetic field determines the size of the current induced. Electric fields induced by TMS are strongest near the coil and produce synchronous activation of neurons within a cortical area of 1–2 cm in diameter. The strength and shape of the electric field induced below a circular and figure-8 coil are shown in Fig. 5. Figure-8 coils are now commonly used to

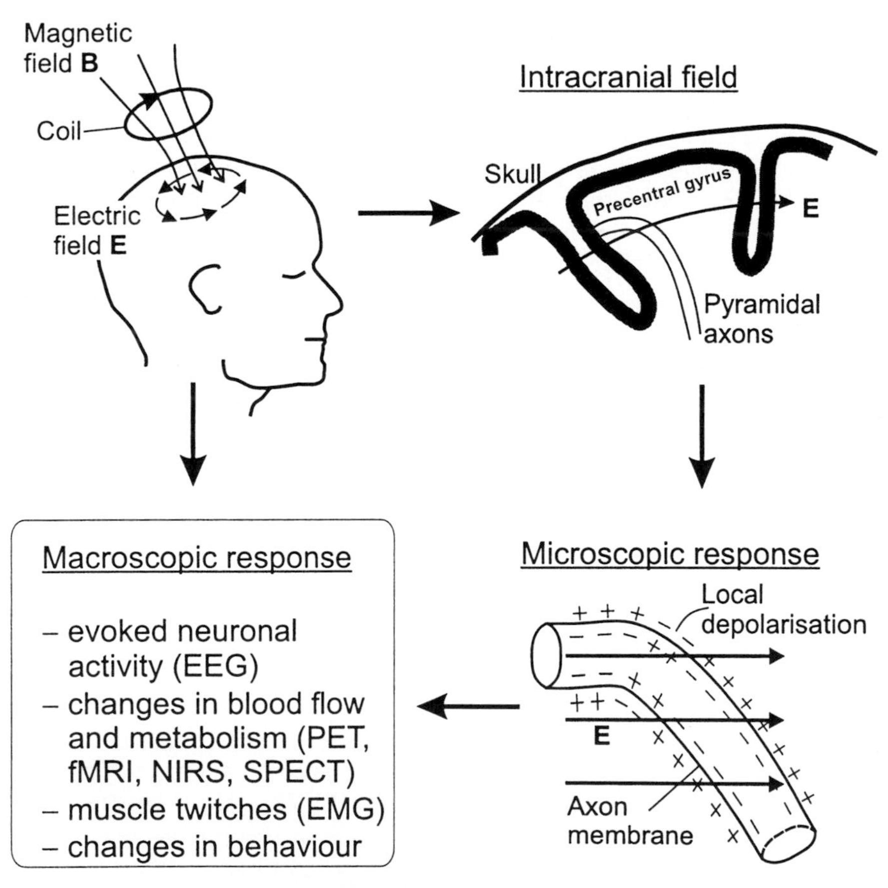

Figure 4 Principles of transcranial magnetic stimulation (TMS). Current applied to the coil generates a magnetic field (B) that induces an electric field (E). The upper right illustrates a section through the precentral gyrus showing the orientation of pyramidal axons in relation to a typical orientation of the intracranial electric field (E). The electric field affects transmembrane potential which will produce depolarization preferentially at points of axon curvature (lower right). At the site of stimulation, TMS will evoke neuronal activity and consequent changes in blood flow and metabolism. Depending on the site stimulated, TMS may evoke muscle contractions, sensory perceptions, or behavioral changes. (Reproduced, with permission, from Ilmoniemi *et al.*, 1999.)

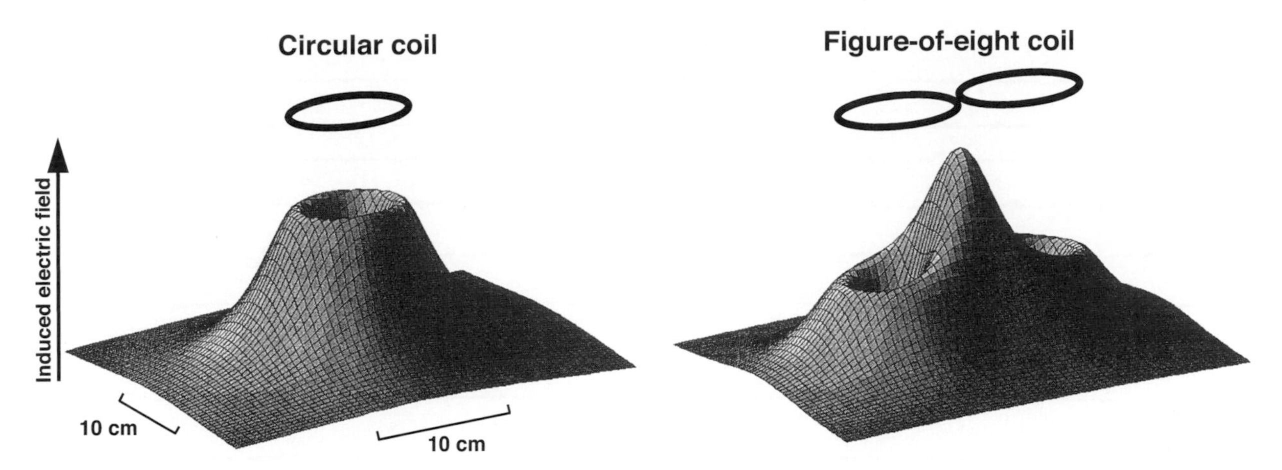

Figure 5 The strength of electric field induced below a circular coil (left) and a figure-8 coil (right). (Reproduced, with permission, from Ilmoniemi *et al.*, 1999.)

focus the electric field and produce a more restricted, punctate area of cortical activation. Using figure-8 coils for mapping, several parameters can be quantified, including the number of excitable scalp positions, location of the optimal position for stimulation, center of gravity (COG; an amplitude-weighted center of the motor map for a muscle or group of muscles), and optimal direction of currents necessary to activate a muscle (Cohen *et al.*, 1998). For the hand muscles, the direction is usually perpendicular to the central sulcus. The areal features of the motor cortical map can easily be demonstrated in humans and other primates with TES. Small movements of the stimulating anode produce corresponding changes in the threshold for different muscles.

As with electrical stimulation, muscles of the distal extremities have the lowest thresholds and are most accessible to magnetic stimulation. However, proximal limb muscles, facial muscles, and muscles of the diaphragm and sphincters can also be activated (Mills, 1991). Once again, the areal features of the motor cortex map are easily demonstrated noninvasively in human subjects with this approach. Individual muscles can also be mapped with TMS although difficulties exist with resolution and the fact that much of motor cortex is buried in the central sulcus. Nevertheless, the best points can be identified and centers of gravity calculated to examine relations between the map representations for different muscles. However, the relationship between the best points on the skull surface for activating particular muscles and the cortical territories mediating those responses is difficult to deduce if maps are referenced only to skull landmarks (Wasserman *et al.*, 1992, 1993, 1994; Wilson *et al.*, 1993). Coregistration of TMS data with MRI data overcomes this limitation (see below). Despite these shortcomings, interesting new data have emerged from this approach. Figure 6 illustrates records obtained in a mapping experiment using TMS to evoke EMG potentials in the left and right deltoid muscle by stimulation of the right and left hemisphere, respectively. Each set of three superimposed records was obtained from scalp sites separated by 1 cm, producing a 1-cm-square grid referenced to CZ. In Fig. 7, these records were used to generate a cortical representation map for four different muscles. Note the extensive overlap in representation for proximal and distal muscles. Interestingly, the right hemisphere representation of deltoid and biceps was larger than the left hemisphere representation. These findings demonstrate that it is possible to generate meaningful motor maps in humans using TMS.

With both magnetic and electrical stimulation, thresholds are lower and latencies shorter if stimulation is applied during voluntary activation of the muscle being tested, although for many muscles this effect may saturate at relatively low levels of muscle contraction (e.g., 10% of maximum voluntary contraction; Han *et al.*, 2001). Reduction in MEP threshold in the presence of voluntary contraction reflects primarily changes in the excitability of spinal motoneurons in the case of TES and a reduction in the time needed to bring the motoneuron to firing threshold. Changes in the excitability of both spinal motoneurons and corticospinal neurons contribute in the case of TMS because excitation of corticospinal neurons is largely transsynaptic (see below).

B. Mode of Activation with Electrical and Magnetic Transcranial Stimulation

Debate about mode of activation with TMS stems from reports that the latencies of TMS-evoked EMG potentials or single-motor-unit potentials are often up to 2 ms longer than for electrical stimulation. Day *et al.* (1989) attributed this latency difference to indirect (transsynaptic) excitation of corticospinal neurons with TMS versus direct excitation with anodal electrical stimulation. While direct activation of cor-

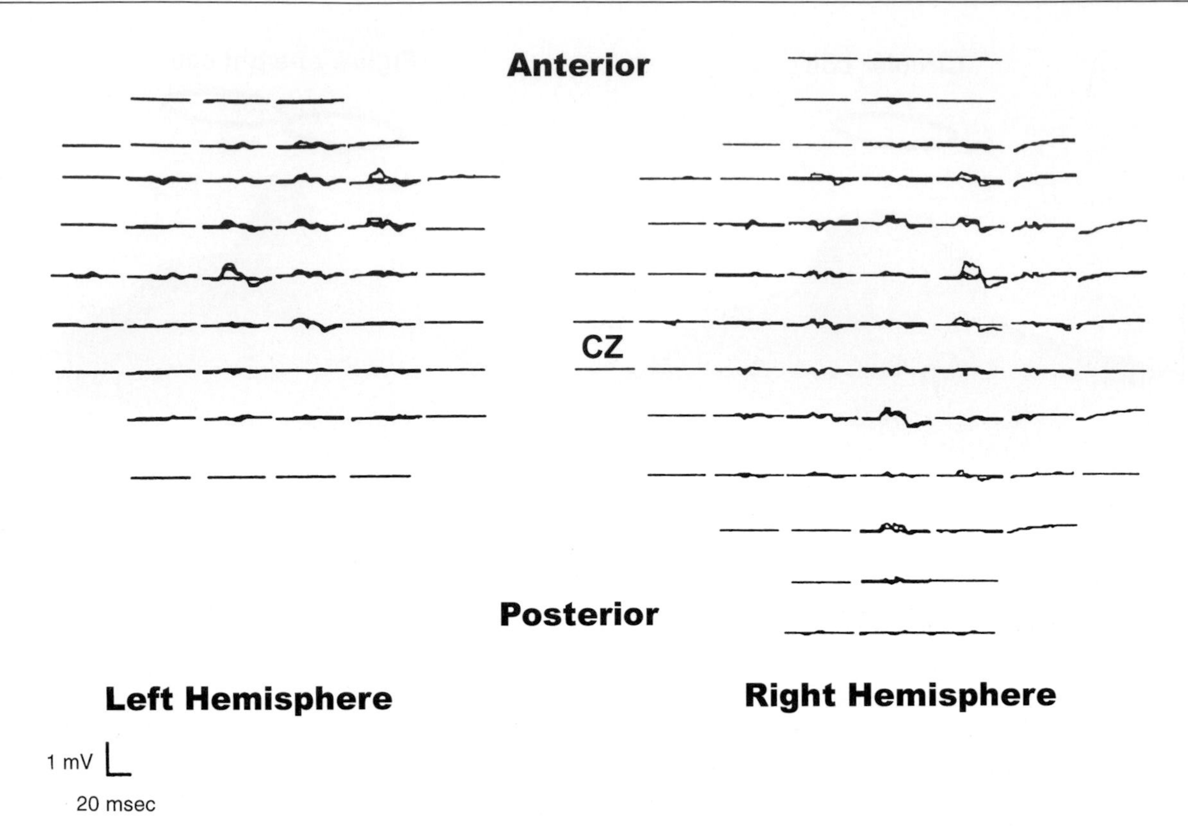

Figure 6 Superimposed motor-evoked potentials (MEPs) in the right and left deltoid from three trials of transcranial magnetic simulation at each scalp position in one subject. The locations of the EMG records correspond to the scalp positions from which they were evoked. Data shown are for stimulation of the hemisphere contralateral to the side of MEP recording. Scalp locations were separated by 1 cm. (Reproduced, with permission, from Wasserman *et al.,* 1992.)

ticospinal neurons with TMS, as illustrated in Fig. 4, occurs under certain conditions, indirect (transsynaptic) activation predominates under other conditions (Day *et al.,* 1989; Edgley *et al.,* 1990; Burke *et al.,* 1993; Di Lazzaro *et al.,* 1998, 1999).

In a comprehensive review on the subject of mode of activation of corticospinal neurons by TES and TMS, Amasian and Deletis (1999) concluded that the site of excitation of neurons with TES depends on the level of depolarization of the neuron at the time the stimulus is delivered. If the neuron is depolarized and near threshold when the stimulus is applied, direct excitation will probably occur at the initial segment or only a few nodes away. On the other hand, for neurons that are more hyperpolarized at the time of stimulation, excitation is transsynaptic, that is, by activation of synaptic inputs to the neuron. Excitation at distant sites (e.g., brain stem) with threshold stimulation was viewed as highly unlikely. However, at intensities substantially above threshold (2×), excitation of corticospinal neurons with TES may shift to sites well below the cortex.

Burke *et al.* (1993) reported that both TMS (coil tangential to the skull) and anodal TES at threshold intensities elicit D waves in anesthetized humans, consistent with direct activation of corticospinal neurons. Descending volleys were recorded with epidural electrodes. As stimulus intensity was increased, I waves were also evoked, corresponding to transsynaptic excitation of corticospinal neurons. Furthermore, increasing the intensity of electrical stimulation above threshold produced step decreases in the latency of the direct descending corticospinal volley. These step decreases in latency with higher intensities of stimulation were attributed to a shift in the site of excitation from the corticospinal neuron soma to an axonal location well below the cortex, perhaps as distant as the medulla (Burke *et al.,* 1995). Using similar methods (circular coil, epidural recordings) in unanesthetized subjects, Houlden *et al.* (1999) reported that the D wave and I waves (one or more) had the same threshold in 4 of 10 subjects, but in 6 subjects I waves had a lower threshold. Similar thresholds for D and I waves from TMS were also reported by Edgely *et al.* (1997) in anesthetized monkeys. In this case, recordings were made from single corticospinal axons at the lumbosacral level. The latency and magnitude of the D wave does not change with anesthesia (Burke *et al.,* 1993; Di Lazzaro *et al.,* 1998; Houlden *et al.,* 1999).

Di Lazzaro *et al.* (1998, 1999) found that at low intensities of stimulation, close to motor threshold, using a figure-8 coil oriented in a posteroanterior direction (inducing a

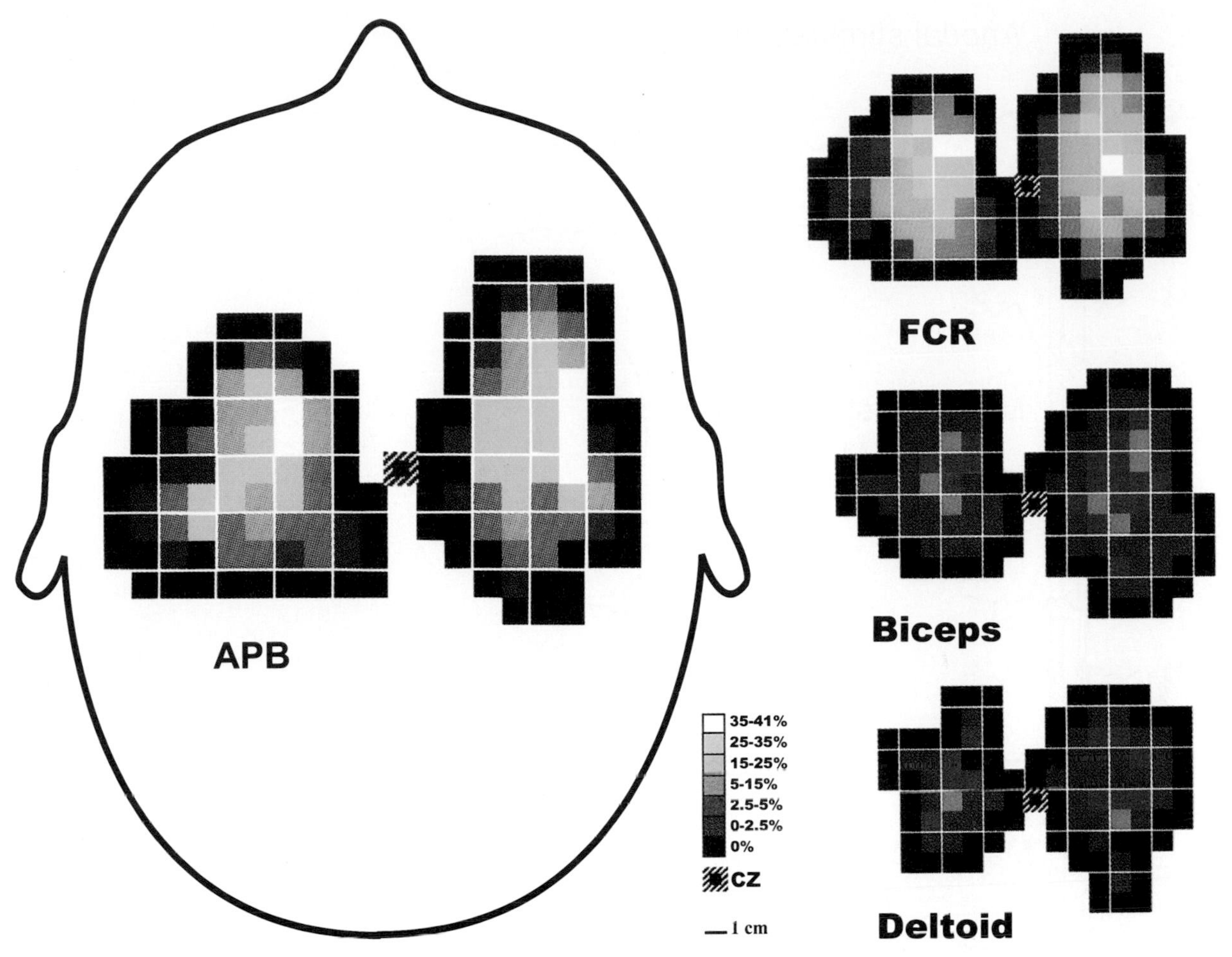

Figure 7 Maps of motor-evoked potential amplitudes for four muscles in one subject. Squares represent stimulation sites located 1 cm apart on the scalp. Shading indicates the magnitude of the motor-evoked potential averaged over the three trials and expressed as a percentage of the maximal compound motor action potential evoked by peripheral nerve stimulation. Maps for proximal muscles show lower amplitudes and fewer excitable positions than those for distal muscles. (Reproduced, with permission, from Wasserman *et al.*, 1992.)

posteroanterior current), the mode of excitation was indirect through excitation of synaptic inputs. Figure 8 illustrates the descending volleys recorded epidurally following anodal TES compared to TMS. Anodal TES elicited a D wave (left dotted line) at all intensities including threshold (active motor threshold). TMS with the coil oriented to induce a posteroanterior current produced only I waves at lower intensities. At an intensity of 21% over threshold, a D wave appeared for the first time. With the coil oriented to induce a lateromedial current, D waves from TMS were present at all intensities tested including threshold. Orienting the coil to produce current in the anteroposterior direction also yielded primarily I waves but with different latencies and different I wave–EMG amplitude relationships, suggesting that different populations of corticospinal neurons were activated (Di Lazzaro *et al.*, 2001). This finding is consistent with the fact that small changes in coil orientation can be used to differentially activate two muscles with similar representation in the cortex (Pascual-Leone *et al.*, 1994). Amassian *et al.* (1989) reported that a sagittal or coronal coil orientation is particularly important for achieving direct excitation of corticospinal neurons with TMS, although the studies of Edgley *et al.* (1990) and Burke *et al.* (1993) show that a transverse coil orientation also produces direct activation of corticospinal neurons. Therefore, coil type, orientation, and stimulus intensity all influence the mode of activation of corticospinal neurons and these factors may have contributed to differences in D and I wave thresholds reported by Burke *et al.* (1993), Burke and Hicks (1999), and Houlden *et al.* (1999) compared to Di Lazzaro *et al.* (1998, 1999). Anesthesia is another factor that may explain differences in thresholds of D and I waves in different studies. Anesthetics could interfere with synaptic transmission to such an extent that the D wave actually becomes lower in threshold than I waves in the descending spinal cord volley. However, even after removal of anesthesia, Burke and Hicks (1999) reported that the D wave had a lower threshold than I waves in their study. In any case,

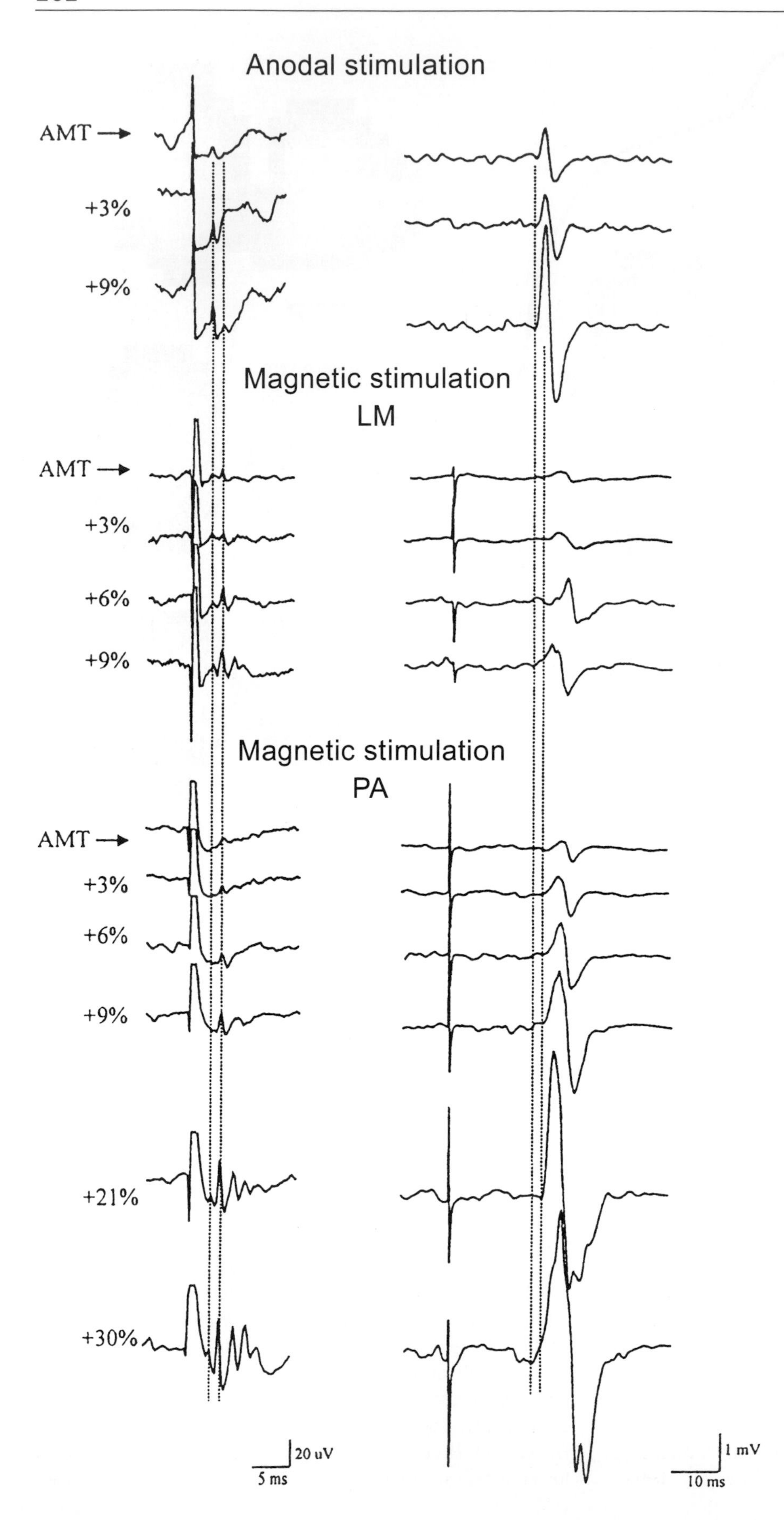

Figure 8 Descending volleys evoked by anodal stimulation (TES) and magnetic stimulation using increasingly strong stimulus intensities beginning at active motor threshold (AMT; threshold in the presence of voluntary muscle activity). Percentages are intensities above threshold for which 100% is maximum stimulator output. On the left, the traces show averaged (5 sweeps for anodal TES and 10 sweeps for magnetic stimulation) epidural recordings from the high cervical cord. The left vertical line indicates the peak latency of the first volley evoked by threshold stimulation (D wave). A longer latency wave appears at higher intensities (I wave). LM is a lateromedially oriented current; PA is a posteroanteriorly mediated current. PA-oriented currents evoke only I waves except at the highest current intensities. LM-oriented currents evoke both D and I waves at threshold. Anodal TES preferentially evokes D waves. Records in the column on the right show the corresponding EMG-evoked responses.

these threshold dependencies emphasize the importance of maintaining, as much as possible, constant stimulation conditions and stable CNS excitability.

At the cellular level, the electric field induced from TMS traverses neuronal membranes containing voltage-sensitive ion channels. The orientation of the electric field relative to the geometry of neuronal and axonal processes influences the extent of depolarization. It is now well accepted that excitation of neurons from TMS is most likely to occur at points where axons bend and at terminations of axons and dendrites (Fig. 4). In this case, activation of corticospinal neurons in the precentral gyrus would most likely occur at the point where cortical axons exit from the gray matter and turn ventrally in the white matter.

C. Coregistration of TMS Maps with Structural and Functional Brain Images

In recent years, it has become common to use frameless coregistration (frameless stereotaxy) procedures to superimpose data from TMS on MR images of brain anatomy from the same subjects. TMS data can also be coregistered with maps from functional imaging studies based on *f*MRI, PET, EEG, and MEG. These methods avoid the use of bulky head clamps and stereotaxic frames that could interfere with TMS mapping. A variety of procedures for achieving coregistration of TMS maps with MR maps of brain anatomy have been described (Krings *et al.*, 1997b; Paus *et al.*, 1997, 1998; Bastings *et al.*, 1998; Classen *et al.*, 1998; Cohen *et al.*, 1998; Ettinger *et al.*, 1998; Karl *et al.*, 2001). A common procedure is to simply digitize the TMS positions on the scalp together with fiducial markers such as the nasion and preauricular points. These fiducial markers are then identified in the subject's MRI so the different coordinate systems can be aligned (Karl *et al.*, 2001). Surface matching algorithms can also be used, although they are computationally more intensive (Wang *et al.*, 1994).

Using a frameless stereotaxic device (Radionics Operating Arm System; Radionics, Burlington, MA), Krings *et al.* (1997b) compared the TMS maps of the first dorsal interosseus muscle (FDI) and flexor carpi radialis (FCR) muscles with the map of cortical activation associated with self-paced hand clenching and index finger to thumb squeezing obtained from *f*MRI. In this procedure, the head was spatially registered to natural landmarks (the tragi and the elateral canthi) which were indicated on the MRI with a cross hair and then referenced to the coordinate space of the frameless stereotaxic device by touching these sites on the subject's head with the tip of the digitizing arm. The MRI was then registered to the head and brain images corresponding to the probe tip could be reconstructed in real time. The probe tip was linked to the middle of the intersection point of a figure-8 coil such that the center of the maximum induced electric field was visualized on an image of the cortical surface. This procedure provided a common coordinate system for localizing TMS sites and *f*MRI data on the same MR map of brain anatomy. Localization accuracy for this frameless stereotaxic procedure defined on the basis of average and maximum error was 1.5 and 2.7 mm, respectively. Figure 9 shows results of

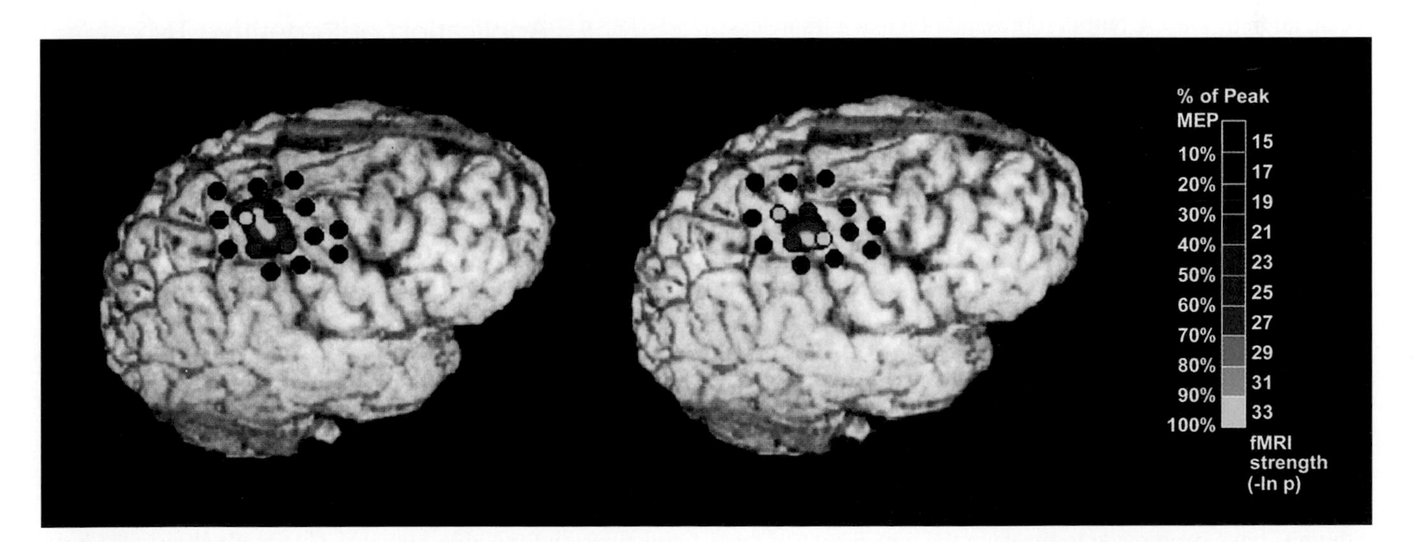

Figure 9 TMS–*f*MRI correlation in a normal subject. During one *f*MRI study the subject repeatedly clenched his right hand into a fist (left), thereby activating mainly hand and forearm muscles. In a second experiment, the subject squeezed his index finger to the thumb (right), thereby activating mainly small hand muscles. The area and statistical strength of *f*MRI activation are indicated. Yellow denotes the strongest hemodynamic responses associated with task performance. The small circles indicate TMS stimulation sites. Their color encodes the amplitude of a forearm muscle MEP (FCR; left) and a small hand muscle MEP (FDI; right) elicited at that site relative to the peak MEP amplitude for the respective muscle in this hemisphere. Note that the peak *f*MRI and peak TMS responses are coincident.

the TMS–*f*MRI correlation in a normal subject. The figure shows *f*MRI activation (color-coded cortical area) for clenching the right hand into a fist (left image) and squeezing the index finger to the thumb (right image). The color-coded dots indicate the amplitude of TMS-evoked potentials in FCR (left image) and FDI (right image) elicited from that cortical site. Note that the peak *f*MRI activation and peak TMS responses are coincident. Also note that with both methods the peak activation of FDI was shifted laterally along the precentral gyrus compared to the peak representation of FCR, consistent with a more lateral somatotopic representation for hand muscles versus forearm muscles. Similar results have been reported by others (Wassermann *et al.* 1996; Terao *et al.*, 1998b; Boroojerdi *et al.*, 1999). For example, Wassermann *et al.* (1996) showed that centers of gravity for muscles mapped with TMS are located within 5–22 mm of coregistered PET activation maxima for the same muscles. More recently, Boroojerdi *et al.* (1999) were able to show that both *f*MRI activation during hand opening and closing and centers of gravity for TMS activation of three intrinsic hand muscles were localized to a knob-like anatomical feature of the precentral gyrus with dimensions of about 10 mm in directions perpendicular and parallel to the central sulcus.

In other studies with patients undergoing surgery for removal of tumors, Krings *et al.* (1997a) demonstrated that the sites of peak activation of muscles (FCR, FDI, and genioglossus) with TMS matched sites of peak activation with direct electrical stimulation of the cortical surface. All TMS responses eliciting more than 75% of the maximum motor-evoked potential fell within 1 cm of the optimal electrical stimulation site. These findings further validate the localization of muscle output representation using frameless stereotaxy to coregister TMS data to MR images of cortical anatomy. Frameless stereotaxic methods for coregistering TMS data with functional and structural MR images have also been applied to studies of function in other cortical areas including the frontal eye fields (Terao *et al.*, 1998a), somatosensory cortex (Karl *et al.*, 2001), and visual cortex (Kosslyn *et al.*, 1999).

D. Mapping the Extent of Brain Activation from TMS

As illustrated in Fig. 5, TMS with a figure-8 coil produces a focused electric field projecting from the point of intersection of the two coils. The current flow induced by this electric field directly excites neurons and axons. Attempts have been made to measure the dimensions of the cortical area directly activated by TMS. Of course this will vary with the procedures used (e.g., coil type and size). Nevertheless, in a study combining stereotaxic TMS with PET imaging of cerebral blood flow coregistered with MRI, Paus *et al.* (1997) showed that a Cadwell figure-8 coil set at 70% of maximum stimulator output activated a cortical tissue volume of 20 × 20 × 10 mm.

In addition to knowing the dimensions of neural tissue directly activated by a TMS pulse, it is important to note that excitation of neural elements by TMS will generate both antidromic and orthodromic volleys of activity which will invade distant regions of the brain possessing anatomical connections to the region stimulated. Therefore, in addition to producing a focus of activated neural elements under the stimulating electrode or coil, TMS and TES will potentially activate brain areas distant from the site of stimulation. Many examples of this type of distant activation by TMS have been reported (Cracco *et al.*, 1989; Ilmoniemi *et al.*, 1997; Paus *et al.*, 1997). In fact, conduction in various central pathways such as the corpus collosum have been investigated by combining TMS with EEG recording (Cracco *et al.*, 1989). A recent study by Brandt *et al.* (2000) provides an excellent example of the extent to which sites distant from the site of stimulation can be activated with transcranial stimulation. They used TES of visual cortex coupled with *f*MRI to identify sites of activation. With the anode located over the right visual cortex, stimulation produced phosphenes in the contralateral lower quadrant of the visual field. Coupled with this perceptual effect, *f*MRI data revealed distant coactivation in the lateral geniculate nucleus, visual cortices (striate and extrastriate), and visuomotor areas including frontal and supplementary eye fields. These findings serve to emphasize that the cortical region underneath the site of stimulation with either TES or TMS is not the only site that might be responsible for the behavioral and/or perceptual effects of stimulation.

E. Application of Repetitive TMS: Reversible Lesions

For many years after its introduction in 1985, TMS was limited to application of single pulses (repetition rate of 1 Hz or less). During this period, TMS was used to investigate a wide range of questions related to motor output organization and brain motor maps. The application of single-pulse TMS in humans can be credited with many important advances in the field of human motor control, including cortical map organization and reorganization (plasticity) associated with injury, developmental disorders, use, and motor learning (Paulus *et al.*, 1999). However, more recently, TMS devices capable of repetitive stimulation at rates up to 50 Hz have been introduced and this has opened a rich new field of possibilities for experimental investigation of cognitive functions and mood in humans. This method has also been tested for effectiveness in treating psychiatric and neurological disorders such as clinical depression, although with somewhat disappointing results to date (Wasserman and Lisanby, 2001). The power of repetitive TMS (rTMS) is its ability to disrupt normal function by producing reversible "virtual brain lesions" (Pascual-Leone *et al.,* 1999). Such reversible lesions provide a powerful method for investigating the functional role of specific cortical areas in various behaviors.

The mechanism of the disruptive effect of rTMS is probably related to both excitatory and inhibitory processes evoked by stimulation (Jahanshahi and Rothwell, 2000). Single TMS pulses produce an initial synchronous activation of neurons under the stimulating coil which will interfere with the normal pattern of activity in the cortical area stimulated. The number of neurons activated will be related to the strength of stimulation. Synchronous activation of large numbers of cortical neurons produces a prolonged GABAergic IPSP that inhibits further neuronal firing for a period of 50–250 ms (Fuhr *et al.,* 1991). With stimulation of motor cortex this is reflected in a period following the motor-evoked potential characterized by an absence of muscle activity (silent period). Stimulating repetitively produces a summation of effects. For example, if one stimulus disrupts activity for 100 ms, then delivery of two pulses at an interval of 100 ms would be expected to disrupt activity for 200 ms.

An elegant example of the use rTMS to produce a virtual brain lesion was a study of the role of visual cortex in Braille reading by subjects who had been blind from an early age (Cohen *et al.,* 1997). Functional imaging studies have shown that primary visual cortex is activated during Braille reading in these subjects (Sadato *et al.,* 1996). To demonstrate that activation of visual cortex was not just an epiphenomenon but critical to the perception of Braille letters, rTMS was performed over the striate (visual) cortex. Repetitive TMS over visual cortex in early or congenitally blind subjects induced reading errors and distorted the tactile perceptions. In contrast, the same stimulation did not affect the performance of sighted control subjects or subjects who had become blind after the age of 14.

In related studies, Kosslyn *et al.* (1999) showed that virtual lesions of visual cortex from rTMS interfere with visual perception and visual imagery (imagining visual scenes). Other studies have applied rTMS to investigate language organization, motor imagery, learning, memory, mood, and emotion (George *et al.,* 1996; Grafman and Wasserman, 1999). Clearly, rTMS has become an important and powerful new tool for the investigation of sensory-, motor-, cognitive-, and mood-related cortical circuits and their function. This method takes imaging studies to a new level by enabling hypothesis testing related to the causal role of specific brain areas in specific functions. However, as pointed out above, interpretation of results from rTMS must take into account not only the areas which are directly excited by the TMS-induced electric field, but also other brain areas whose activity may be altered by virtue of being anatomically connected to the region targeted by TMS.

F. Practical Considerations

TMS has several advantages over TES, including (1) the ability to stimulate brain and deep peripheral nerves without discomfort because no current passes through the skin and high current densities can be avoided, (2) the magnetic fields penetrate high-resistance structures such as the skull without attenuation, and (3) electrical contact with the body is not required so skin preparation is unnecessary (Barker, 1999). The disadvantages are that TMS devices are costly and bulky compared to TES devices.

Safety with TMS is a critical issue and has been investigated extensively. Guidelines for its use have been published (Wasserman, 1998). Single-pulse TMS is a noninvasive and highly useful experimental and clinical tool that poses virtually no risk to the subject. On the other hand, rTMS is more powerful in its effects on the brain and a potentially more dangerous method. It is capable of blocking cortical processes and can have effects on brain function that long outlast the actual period of stimulation. Seizure induction has occurred in patients with epilepsy and in normal subjects. For example, at the National Institute of Neurological Diseases and Stroke, rTMS induced seizures in 4/250 normal subjects, although none of these subjects suffered physical sequelae (e.g., EEG changes, recall deficits) lasting beyond 2 days from the time of the seizure.

VIII. Mapping Motor Output with Electrical Stimulation of the Cortical Surface

Electrical stimulation of the cortical surface has been coupled with several different output measures for the purpose of mapping motor system organization. Most common among these has been evoked movements, intracellular potentials from motoneurons, and evoked EMG activity but other measures have also been studied, including effects on muscle spindle reflex volleys and tension changes in single muscles.

Electrical stimulation of the cortical surface dates back to the early studies of Fritsh and Hitzig in the middle 1800s (Table 1). These early mapping efforts have been detailed in excellent books by Phillips and Porter (1977) and more recently by Porter and Lemon (1993). Although technically somewhat crude by today's standards, these studies were effective in revealing the basic features of somatotopic organization in motor cortex, including the representation of the face and mouth, head, eyes, arm, leg, and trunk. This is referred to as the areal representation. Representation within each of these subdivisions is called the intra-areal representation (Porter and Lemon, 1993). Figure 10 is a summary of the areal and some aspects of the intra-areal features of the motor map based on results of stimulation of the cortical

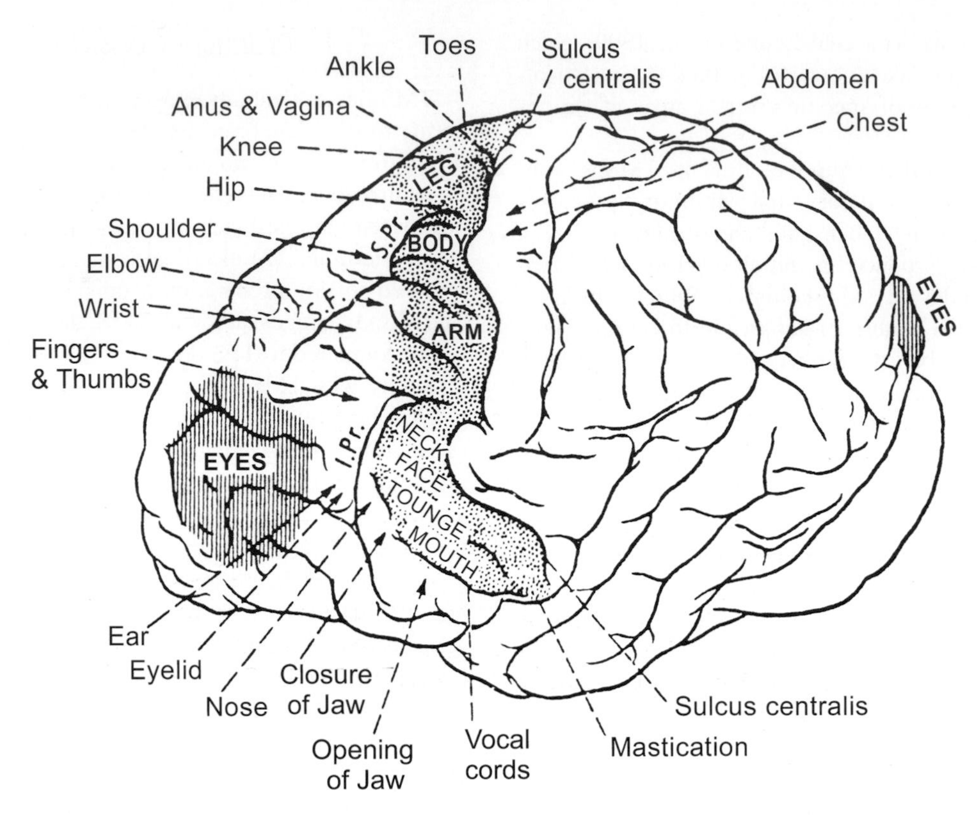

Figure 10 Motor map of the left hemisphere of the chimpanzee. Primary motor cortex is indicated by the fine stippling. Much of it is buried in the central sulcus. Summary based on a large series of experiments. (Reproduced, with permission, from Sherrington, 1906.)

surface in the chimpanzee. In these experiments, primary motor cortex was mapped by applying minimal faradic stimulation to the cortical surface for periods limited to only a few seconds on each trial. The basic pattern of areal representation is clear—the leg is represented most medially followed by the trunk, arm, neck, face, and mouth as one moves laterally along the precentral gyrus. It is interesting to note that the movements observed even with minimal faradic stimulation were not always confined to a single joint.

Stimulation data from several animals can be generalized into a more pictorial map from which the orderly but disproportionate representation of different divisions of the body musculature can be immediately appreciated. An example is Fig. 11, which is taken from Phillips and Porter (1977) and illustrates the elegant motor system mapping work of Woolsey and colleagues (1952, 1964) using stimulation of the cortical surface. The input stimulus in this case was a 60-Hz sinusoidal alternating current delivered as a 2-s train. The movements evoked by stimulation of the precentral cortex at near threshold are illustrated in Fig. 11A. Darkened areas of each figurine indicate the body part that moved. Stippled areas indicate more weakly affected areas. Figure 11B shows that the threshold currents for the movements evoked in Fig. 11A ranged from 0.2 to 2 mA. Data of this type can be used to create a pictorial map of body musculature representation such as that illustrated in Fig. 11C. This map is for a monkey so the figure is referred to as a "simunculus." Similar mapping procedures in humans by Penfield and Rasmussen (1950) yielded the classic motor homunculus that is now commonly reproduced in nearly every textbook of neuroscience and neurology.

This approach to mapping has provided the basic form of the areal map of body musculature representation. The areal features of the motor map have been consistent and reproducible across primate species. What remains less clear is the nature and reproducibility of the intra-areal representation. Woolsey showed that within the arm representation, for example, separate areas exist for the fingers, wrist, elbow, and shoulder. Although often represented as discrete localizations when experimental data are generalized, it should be pointed out that even Woolsey *et al.* (1952) emphasized that cortical foci for different muscles are not discrete patches with sharp boundaries but rather overlapping areas with a best point. This raises a question of interpretation. Are the representations of different muscles or movements overlapping because their respective neurons in the cortex are overlapping or do they overlap because the spread of stimulus current blurs what would otherwise be discrete representations with sharp boundaries? Also, how are the terminals of

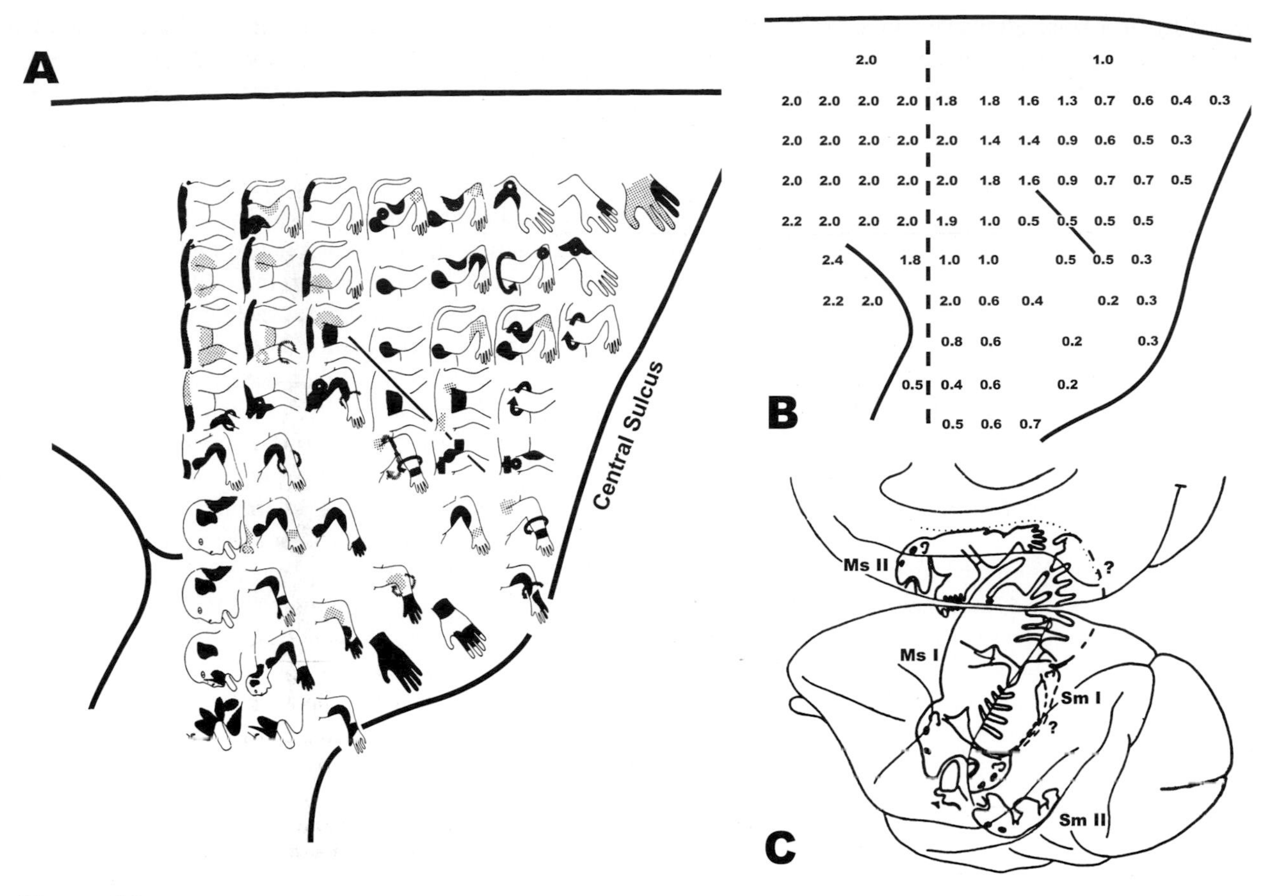

Figure 11 (A) Motor figurines showing responses evoked by near-threshold stimulation of the surface of the monkey's cortex with sinusoidal alternating current in trains of 2-s duration at 2-min intervals. (B) Near-threshold current intensities in milliamperes (RMS) used to evoke the movements illustrated in A. (C) "Simunculus" overlying the dorsolateral and medial surfaces of the monkey brain showing representations in primary motor (Ms I) and supplementary (Ms II) motor areas. Sm I and Sm II indicate somatosensory receiving areas. (Data from Woolsey *et al*, 1952, and Woolsey, 1964, as redrawn in Phillips and Porter, 1977; reproduced with permission of the publisher.)

single cortical cells distributed to different motoneuron pools? Are they confined to a single motoneuron pool or do they diverge to contact multiple motoneuron pools? Clearly, a divergent type of organization at the level of single corticospinal neurons would, in itself, produce an overlapping representation (Fig. 12). These questions arose early and persisted for many years because of the lack of suitable techniques for answering them (Chang *et al.,* 1947). The existence of true overlap of cortical territories containing cells representing different muscles and joint movements

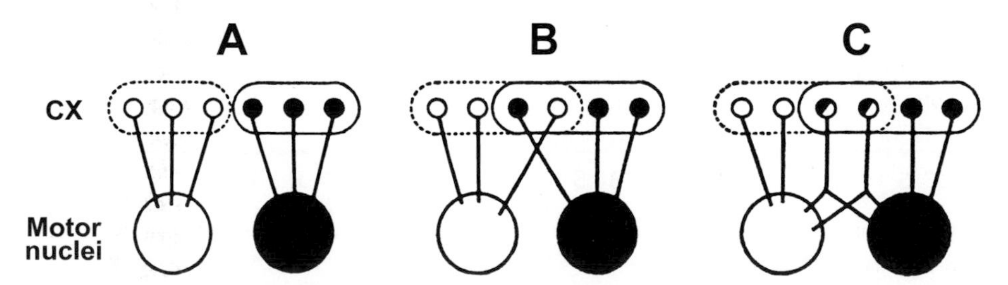

Figure 12 Three variants of corticomotoneuronal organization. Large circles denote two motor nuclei, small circles denote corticospinal neurons and the lines encompassing them denote cortical territories for each muscle. (A) Strict mosaic organization in which cortical cells terminate within a single motor nucleus. (B) Overlapping cortical territories but with terminations of individual corticospinal neurons still confined to a single motor nucleus. (C) Overlapping cortical territories with some cells terminating in multiple motor nuclei. (Modified, with permission, from Jankowska *et al.,* 1975b.)

was generally suspected. However, in more recent years, findings from several newer approaches have provided many answers. These findings will be presented in the section on consistent features of the cortical motor map.

Although stimulation of the cortical surface with large currents to evoke movements in the periphery has been a highly successful and informative mapping method, it does have serious limitations. First, because the surface of the cortex must be exposed, this technique can be used only in animals under anesthesia or in humans undergoing neurosurgical procedures. Another limitation is the fact that subthreshold effects on muscle activity cannot be detected. Muscles that are activated most strongly will dominate the movement that occurs. Yet another limitation is that the electrode dimensions (usually 1–2 mm) are inconsistent with activation of only a small number of neurons near the electrode tip.

Electrical stimulation of the cortical surface can be coupled with other measures for mapping motor output. One very informative and productive approach involves intracellular recording from motoneurons in the spinal cord while stimulating the surface of the cortex with anodal currents. It is well established that surface anodal stimulation is more effective in activating corticospinal cells directly than cathodal stimulation (Landgren *et al.,* 1962). This is especially true for cells deep within the precentral gyrus. Figure 13 shows data from a mapping experiment of this type. The minimum stimulus current for evoking volleys in the lateral corticospinal tract with this method was 0.3 mA and the size of these minimal volleys was comparable to those obtained with minimal intracortical microstimulation. It was estimated that a current of 0.4 mA activated corticospinal neurons in a radius of 0.5 mm, while 0.5-mA stimuli activated cells in a 1-mm radius. In this experiment, Jankowska *et al.* (1975b) defined "total projection area" as the area containing responses that were 30% or greater of the maximum response. In Figs. 13A and 13B, computer-averaged EPSPs are mapped to the cortical surface from which they were elicited. The size of the EPSP elicited from each site indicates the density of corticospinal neurons at that site projecting to the motoneuron being tested. The continuous lines encircle the total projection area (EPSP amplitude 30% of maximum or greater), which is clearly smaller than the total area from which responses could be obtained. The dashed lines encircle optimal projection areas (EPSP amplitude 80% of maximum or greater). In Figs. 13C–13E, these EPSP

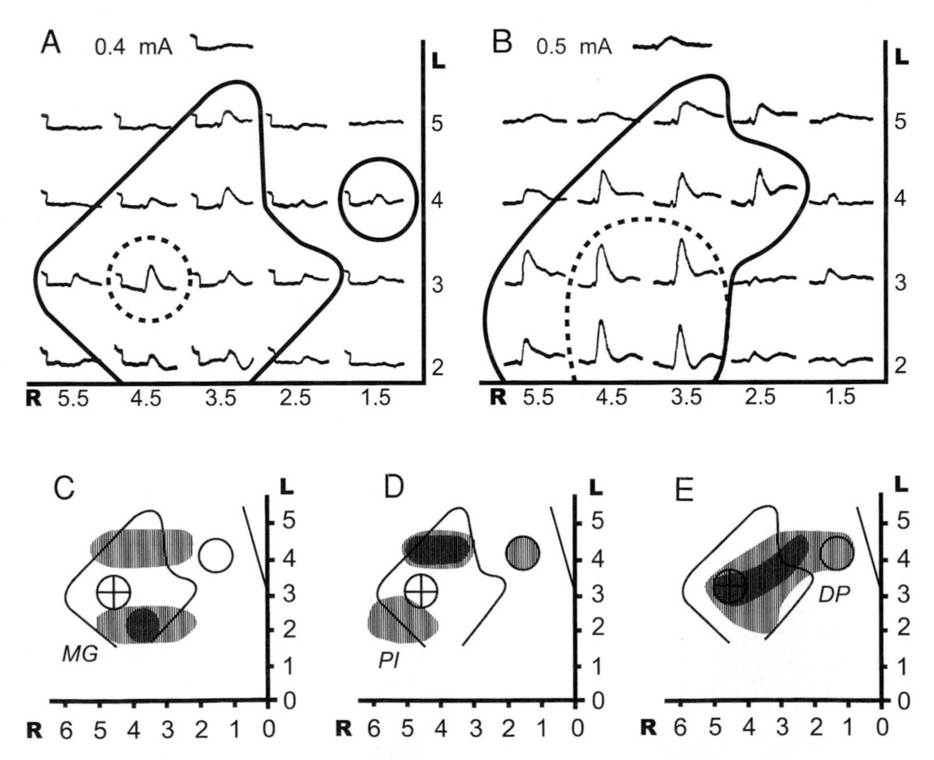

Figure 13 Cortical territories containing corticospinal neurons projecting to a lateral gastrocnemius motoneuron. (**A** and **B**) Averaged records of postsynaptic potentials evoked by 0.4 and 0.5 mA stimuli, respectively, applied to electrode positions spaced at 0.5-mm intervals on the cortical surface. Solid lines enclose total projection areas (EPSP amplitudes 30% or more of the largest ones); dashed lines enclose optimal projection areas (EPSP amplitudes 80% or more of the largest ones). (**C, D,** and **E**) The projection areas for a lateral gastrocnemius motoneuron are compared with the projections areas for motoneurons of two synergist muscles, medial gastrocnemius (MG) and plantaris (Pl) and an antagonist deep peroneal muscle (DP) motoneuron. In (C–E), the total projection area boundary for the lateral gastrocnemius motoneuron and the optimal projection area boundary (circle with cross) are carried over from (A) and (B). (Reproduced, with permission, from Jankowska *et al.,* 1975b.)

data are assembled as a map showing the total projection areas for a lateral gastrocnemius motoneuron (irregularly shaped continuous line from Fig. 13A) and for three additional motoneurons (shaded areas), two of which were from synergist muscles (medial gastrocnemius and plantaris) and one from an antagonist (deep peroneal).

From these data it is clear that the cortical territories representing motoneurons belonging to different muscles are extensively overlapping. This can be conceptualized by defining the collection of all corticospinal cells that project to a single motoneuron as the cortical colony for that motoneuron (Fig. 14). Although cortical colonies for different motoneurons show extensive overlap, the best point on the cortex eliciting the largest EPSP is usually slightly displaced for each motoneuron or muscle. This probably reflects the fact that the geometric centers (COG) of the colonies are not exactly superimposable, and the points containing the greatest density of corticospinal neurons projecting to different motoneurons or muscles are also displaced from each other. It is also noteworthy that the cortical territory from which EPSPs can be evoked in single motoneurons is relatively large (5–12 mm^2). To sum up, these results show that the cortical territories representing different motoneurons and muscles are highly overlapping. However, it is important to note that this method cannot distinguish between the single-muscle and multiple-muscle possibilities for termination of single cells illustrated in Fig. 12. This method reveals the extent of convergence from cortical neurons to motoneurons, but it is not capable of answering questions about the divergence in the projection patterns of single neurons.

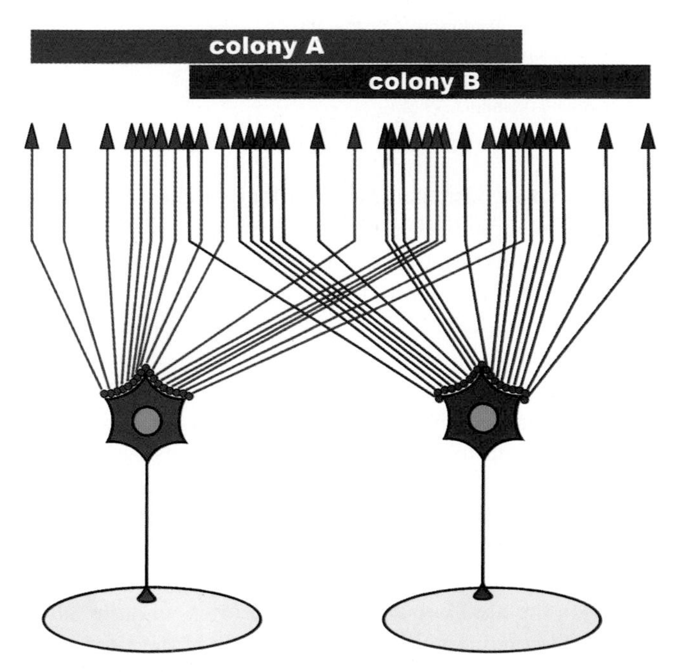

Figure 14 Cortical colonies projecting to two different motoneurons of a single muscle. Note the extensive overlap in cortical territory occupied by the colony and the variations in density of corticospinal neurons within a colony. (Reproduced, with permission, from Andersen *et al.,* 1975.)

IX. Mapping Motor Output with Intracortical Microstimulation

The introduction of intracranial microstimulation (ICMS) by Asanuma and Sakata (1967) provided a substantial improvement in resolution for mapping studies of motor cortex and other CNS output structures. This method consists of applying currents through a microelectrode to activate small clusters of cells in the vicinity of the electrode. Although at a distance from the site of stimulation electrode dimensions do not influence the effective current, dimensions are important for the tissue near the electrode (Rank, 1981). The small size of microelectrodes yields a much higher degree of spatial resolution than is possible with macroelectrodes. With this technique, localized stimulation can be applied directly in the vicinity of cells deep within the precentral gyrus rather than trying to stimulate them by current spread from the surface. A major advantage of microstimulation is not only the spatial resolution it provides, but also the fact that the spike activity of single neurons at the site of stimulation can be recorded. The functional properties of neurons, including relations to active movement and sensory input, can be characterized and correlated with motor output effects from the same site (Asanuma and Rosen, 1972; Rosen and Asanuma, 1972; Wong *et al.,* 1978). This has resulted in major advances in understanding input–output relations of motor cortex and other motor system areas.

ICMS can take several different forms and can be coupled with a variety of output measures, including evoked movements, evoked EMG responses, and averaged EMG responses. Repetitive ICMS generally consists of 10 cathodal or biphasic pulses at 330 Hz in frequency (0.2-ms duration each phase). Excitation occurs with the negative phase of the biphasic pulse so it should come first for timing purposes. Maintaining standard parameters simplifies comparison of results from different labs. Figure 15 illustrates findings of an input–output mapping experiment by Rosen and Asanuma (1972) in which repetitive ICMS at 5 μA was applied to the motor cortex of a monkey to evoke movements in the periphery. Seven electrode penetrations are shown. The movements produced at each site of stimulation are indicated by the symbols, and cutaneous receptive fields of cells are indicated by black areas on drawings of the hand and by notes connected to the map by dotted lines. Two general principles concerning input–output organization emerged from these mapping experiments. First, stimulation within motor cortex generally produces movement of the

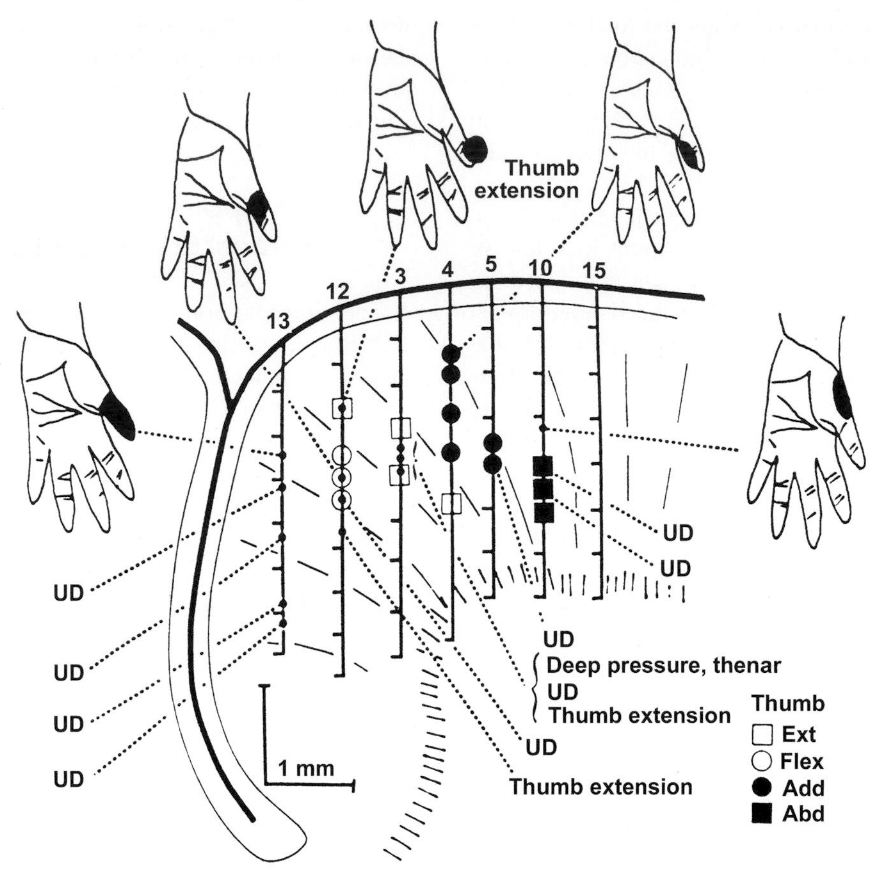

Figure 15 Results of a mapping experiment in which repetitive ICMS at less than 5 μA was used to evoke movements of the thumb. Electrode penetrations are indicated by solid lines; cross bars indicate sites of stimulation. Symbols indicate the specific movements obtained with stimulation. Sites with no symbol denote lack of evoked movement at 5 μA. Sensory receptive fields are connected to specific locations by dotted lines. All evoked movements involved the thumb as did the receptive fields of neurons at the sites of stimulation. UD refers to undriven cells, unresponsive to sensory stimuli. (Reproduced, with permission, from Rosén and Asanuma, 1972.)

body part containing the receptive field of neurons at the site of stimulation. Second, the receptive field is generally located on the body surface in the direction of the movement produced by stimulation at that site. In other words, the activity of a neuron will tend to produce movement in the direction of the neuron's receptive field. This may be a substrate for cutaneous feedback control of movements, particularly manipulative movements involving the hand.

Using repetitive ICMS to evoke EMG activity has the advantage that individual muscle responses can be resolved. Stimulation can also be applied during movement to reveal inhibitory effects. EMG responses from stimulation can be averaged over several trials to reduce variability and provide better records for quantitation and latency measurements. An example comparing these different methods is presented in a later section (Figure 22).

Major advances in understanding the integration of sensory and motor maps of different areas of the brain have resulted from the application of ICMS to brain mapping. ICMS is now considered the technique of choice for many types of mapping experiments. However, ICMS is not without weaknesses that any user of the technique should be aware of. First, it suffers from the same problems that exist with any electrical stimulation technique. Namely, stimulation activates a variety of neuronal elements including not only cell bodies but also axons of passage and afferent terminals. The effects of ICMS can produce excitation at sites distant to the stimulation site by physiological spread of the stimulus resulting from temporal facilitation within the cortical circuitry. This is especially true for repetitive ICMS. Jankowska *et al.* (1975a) demonstrated this by applying ICMS to the site of a pyramidal tract neuron in the hindlimb area of motor cortex while monitoring the descending volley from a fascicle of the lateral corticospinal tract (Fig. 16). Each successive stimulus of a train consisting of three stimuli evoked a descending volley that was larger in amplitude and greater in duration than the preceding stimulus. The most probable explanation of the

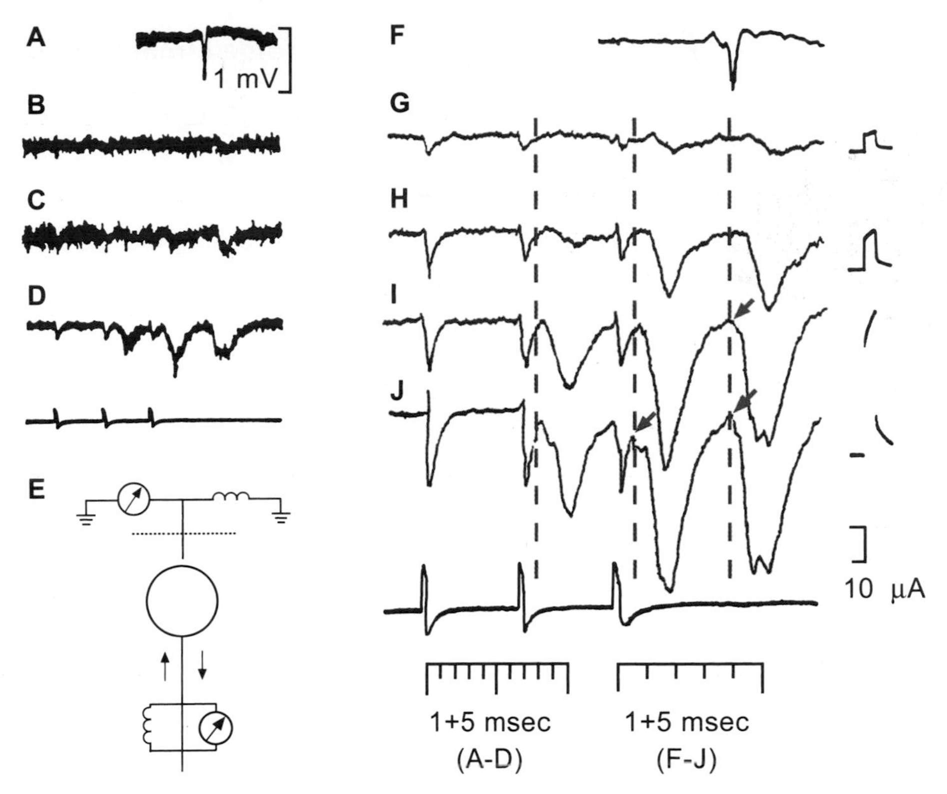

Figure 16 Transsynaptic excitation of corticospinal neurons from repetitive ICMS. (**E**) The experimental paradigm consisting of a microelectrode positioned in the hindlimb area of motor cortex and spinal electrodes placed on a fascicle of the lateral corticospinal tract is illustrated. The cortical microelectrode was positioned to record antidromic spikes of a hindlimb corticospinal neuron (**A** and **F**) evoked from stimuli applied to the spinal cord. The cortical electrode was also used for repetitive ICMS. The spinal cord electrode was used for antidromic stimulation and for recording the descending volleys evoked by repetitive ICMS. (**B–D**) Superimposed unprocessed records and (**G–I**) averaged records of descending volleys evoked by three pulses of repetitive ICMS. (**J**) Descending volleys evoked from a site 300 μm deeper than in (G–I). Current strength is given by the records in the far right column. Arrows in the averaged records indicate descending volleys that occurred with the same latency as the antidromic potential. Note that the three large negative peaks representing most of the descending volley occur at a longer latency than would be expected for direct excitation of corticospinal neurons. Also note the increasing amplitude of the volleys with successive stimuli indicating the presence of temporal summation. (Reproduced, with permission, from Jankowska *et al.*, 1975a.)

increasing descending volley with repetitive ICMS is that corticospinal neurons are excited transsynaptically. Although direct activation of corticospinal neurons with ICMS certainly occurs (Stoney *et al.*, 1968), the effects evoked with repetitive stimulation are predominately from transsynaptic excitation.

X. Mapping Motor Output with High-Density Microelectrode Arrays

Systematic microelectrode recording potentially provides a very high level of spatial resolution for brain mapping because the activity of single neurons can be recorded or small clusters of neurons in the vicinity of the electrode tip can be electrically stimulated (see other sections, this chapter). The traditional single microelectrode method relies upon repeated penetrations into the brain made over a period of many hours in anesthetized monkeys (Nelson *et al.*, 1980; Nudo *et al.*, 1996) or weeks and months in chronically implanted awake monkeys (Park *et al.*, 2000, 2001). In recent years, high-density electrode arrays have become available for simultaneous multichannel recording from the brain and the application of these arrays to mapping is now being explored (Norman *et al.*, 2001). One such array manufactured by Bionic Technologies, Inc., consists of 100 microelectrodes on a square silicon wafer with 400-μm spacing between electrodes (Fig. 17A). These arrays have been implanted in the motor cortex of rhesus monkeys to simultaneously record the activity of populations of neurons in relation to movement (Hatsopoulos *et al.*, 1998; Maynard *et al.*, 1999). One

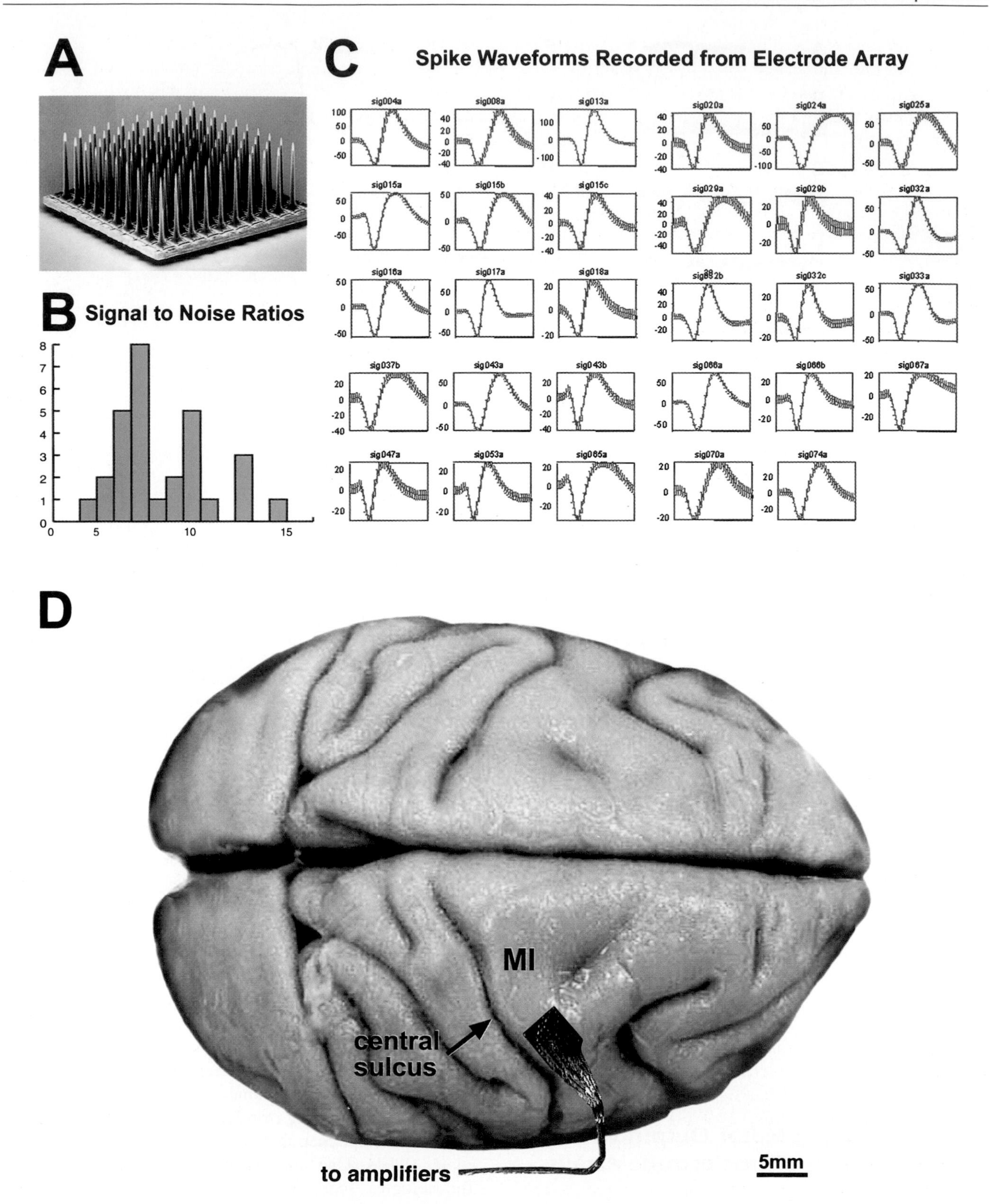

Figure 17 (**A**) "Utah" electrode array manufactured by Bionic Technologies, Inc. These arrays are manufactured from a silicon wafer and consist of a 10×10 matrix of electrodes spaced at 400 μm yielding an array with dimensions of 4×4 mm. Electrode length is limited by wafer thickness to 1.5 mm. (From Bionic Technologies, Inc., website.) (**B**) Distribution of signal-to-noise ratios for spike waveforms shown in C. (**C**) Examples of spike waveforms recorded from the array. (D) The electrode array positioned as it would be for implantation in the hand/arm region of primary motor cortex in the macaque monkey. (Sections B and C courtesy of N. Hatsopoulos and J. Donoghue, unpublished data.)

of the advantages of the "Utah" array is its ease of manipulation and rapid insertion using a pneumatic device. Rapid insertion (200 μs) has the advantage of overcoming tissue viscosity and dimpling that would otherwise occur with slow insertion. The Utah array was also recently applied to mapping the receptive fields of layer 4 neurons in primary visual cortex (area 17) of the anesthetized cat (Norman *et al.*, 2001). Isolated single units occurred on about 25% of the electrodes and multiunit activity on another 20–30%. Although containing fewer electrodes, microwire arrays (NB Labs, Dennison, TX) have been highly successful for multichannel recording of neural activity in rats and New World monkeys with a flat cortex. The yield of single units can be very impressive, for example, 100 units from 96 wires on multiple arrays (Wessberg *et al.*, 2000). These arrays are inserted slowly while monitoring neural activity on all the channels. Multielectrode array methods have the potential to enable visualization of the activity of large numbers of neurons and provide high temporal and spatial resolution. Moreover, using arrays for electrical stimulation also offers the potential for examining motor output maps in awake animals at frequent intervals over long periods of time. As the technology improves, these arrays should provide increased opportunities to investigate new questions concerning brain sensory and motor maps.

XI. Mapping Motor Output with Spike-Triggered Averaging of EMG Activity from Single Neurons

The smallest unit of output from motor cortex and other premotor descending systems is a single neuron. Although extracellular stimulation techniques have been very informative and much of what we know about brain motor maps has been based on stimulation, these techniques are not capable of answering questions about the synaptic targets of single neurons. This is unfortunate because single neurons contain the most detailed information about the nature of the motor map. For example, the alternative modes of output organization illustrated in Fig. 12 can best be answered by examining the distribution of terminals from single neurons to different muscle groups. Spike-triggered averaging of rectified EMG activity was introduced by Fetz and Cheney (1980) and provides a method of identifying the target muscles of corticospinal and other premotoneuronal cells. For the first time, this method has made it possible to investigate in awake animals not only relations between a cell's discharge and movement but also the organization of the cell's synaptic output effects on motoneurons of agonist and antagonist muscles.

The rationale for this method is illustrated in Fig. 18. Premotor neurons with a direct excitatory synaptic linkage to motoneurons will produce individual EPSPs at a fixed latency following discharge of the premotor cell. The magnitude of the EPSPs is too small to reliably discharge the motoneuron with each occurrence. Nevertheless, the EPSPs will depolarize the membrane, bringing the motoneuron closer to firing threshold and transiently increasing its firing probability. Since the neuromuscular junction is normally an obligatory synapse, a one-to-one relationship exists between motoneuron and muscle fiber action potentials. Therefore, motor unit spike trains directly reflect the firing of spinal motoneurons. EMG activity is the sum of the spike trains of a population of motor units within a muscle and provides a measure that can be used to detect changes in firing probability associated with the occurrence of premotor cell spikes.

Further details of the spike-triggered averaging procedure are illustrated in Fig. 19. In this case, the discharge of a single cortical neuron was recorded in relation to the extension phase of a ramp-and-hold wrist movement. EMG activity was full-wave rectified and the 30-ms segment extending from 5 ms before the trigger spikes to 25 ms after the spikes was digitized at 4 kHz and averaged. The EMG signal is normally full-wave rectified to avoid potential cancellation of motor unit potentials with opposite polarities, although Botteron and Cheney (1989) showed that strong effects can also be reliably detected in averages of unrectified EMG activity. The EMG activity associated with each of the first 5 spikes in the record is shown (perispike EMG) along with the cumulative average of these EMG segments. Note that the prominent waveforms occurring about 10 ms after the first cortical spike are largely lost after the EMG segments associated with the first 5 spikes are averaged. However, an average of 2000 spike events shows a clear peak beginning at a latency of 6 ms. Such a transient increase in average EMG activity is referred to as postspike facilitation (PSpF) and is interpreted as evidence of an underlying synaptic linkage between the trigger neuron and the motoneurons. Cortical neurons yielding PSpF are referred to as corticomotoneuronal (CM) cells to emphasize the causal nature of the correlation from which PSpF derives. However, it should be emphasized that these are CM cells in the functional sense and that PSpF does not prove an anatomical connection of the type illustrated in Fig. 2.

The strength of the spike-triggered averaging method is that it can be applied in awake animals, enabling the identification of cells that are causally involved in producing the muscle activity associated with movement. The functional properties of these output neurons are particularly important in understanding the functional role of the descending system to which they belong.

Spike-triggered averaging of rectified EMG activity reveals a cell's correlational linkage with motoneurons. However, compared to spike-triggered averages of intracellularly recorded synaptic potentials (Mendell and

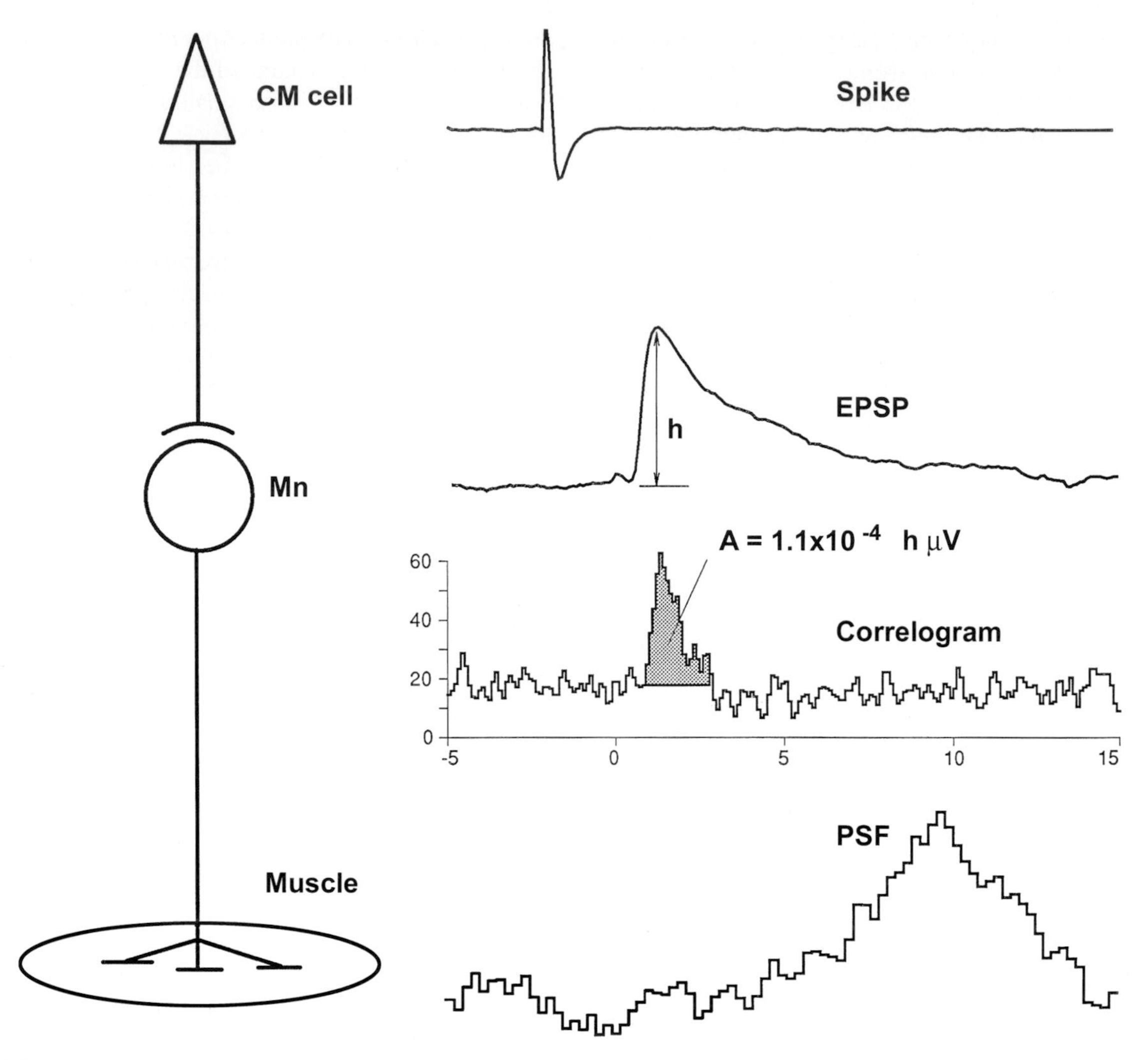

Figure 18 Events mediating the postspike effects from a premotoneuronal cell with monosynaptic connections to motoneurons. Spike discharges are followed by unitary EPSPs and an increase in motor unit firing probability reflecting the derivative of the EPSP. Postspike facilitation (PSpF) in the corresponding average of rectified EMG activity is delayed by conduction time from the spinal cord to the muscle. The area of the correlogram peak was found to be proportional to the height of the EPSP. The time course of PSpF is greater than the correlogram peak for a single motor unit because: (1) PSpF is the sum of facilitation in several motor units with different conduction velocities and (2) the duration of the motor unit potential waveform contributes to PSpF. Data shown are for a muscle spindle Ia afferent (Cope *et al.*, 1987). (Reproduced, with permission, from Fetz *et al.*, 1989.)

Henneman, 1971), it provides less precise information about the timing of synaptic events and, hence, the number of synapses in the linkage. Moreover, although postspike effects can be quantified to provide some measure of the relative strength of an underlying synaptic effect, the magnitude of PSpF must be interpreted with caution. For example, PSpF from a neuron that strongly facilitates a small fraction of recorded motor units might be similar in magnitude to PSpF from a neuron that weakly facilitates a large fraction of the recorded motor units. Cross-correlating premotor cell discharge with single motor units avoids these ambiguities and provides a more exact and interpretable measure of the strength of a synaptic effect (Mantel and Lemon, 1987). The cross-correlogram peaks for individual motor units are weaker and narrower (mean half-width 1.9 ms) than the PSpF of multiunit EMG activity. The results obtained from cross-correlating the discharges of single cortical cells and single motor units suggest that the terminals of individual cortical cells are distributed broadly to most of the motoneurons within a motoneuron pool (Palmer and Fetz, 1985).

Because spike-triggered averaging of EMG activity is a correlational method, other possible explanations of significant postspike events exist. A major requirement of this method is that the discharge of different cells in the population of interest must be independent, that is, not strongly synchronized. Synchronization between cells can occur for many reasons, including common synaptic input, collaterals

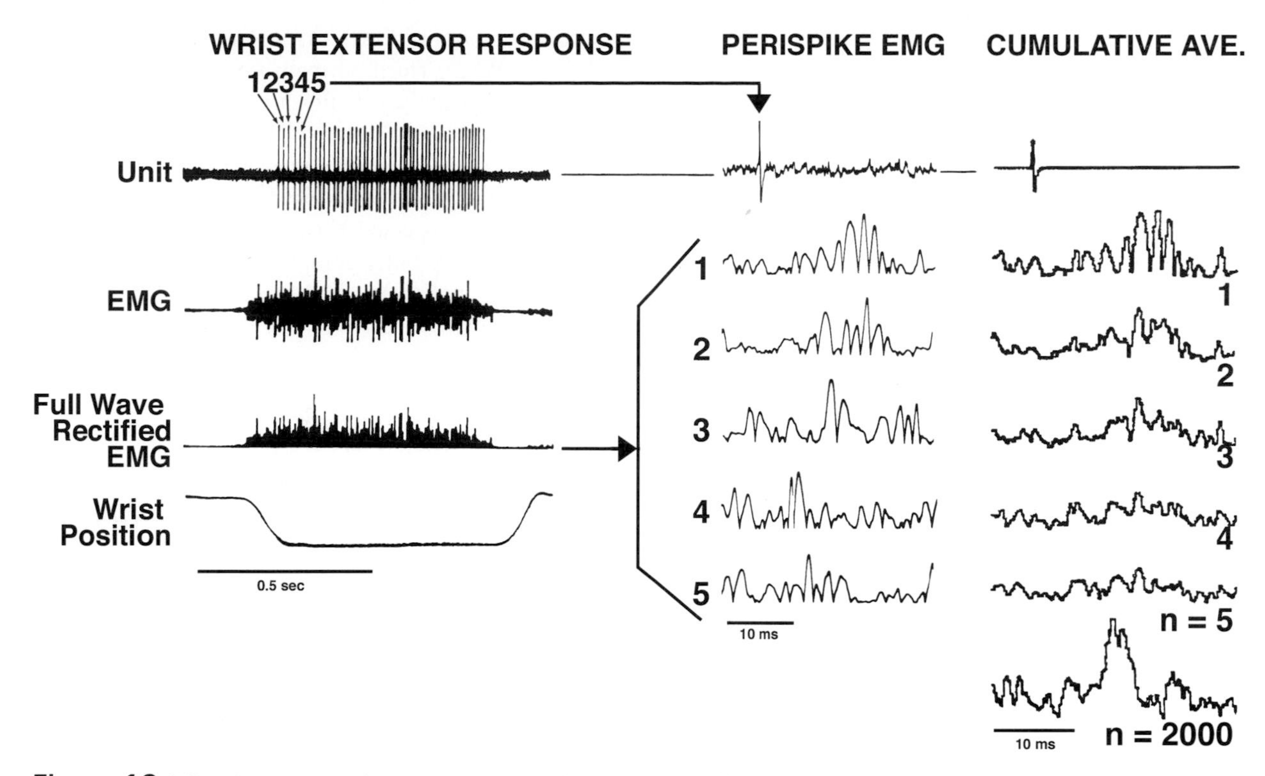

Figure 19 Spike triggered averaging procedure used to detect postspike effects from single premotor neurons. Spikes associated with movement are used to trigger averages of rectified EMG activity. The response at the left illustrates the discharge of a cortical cell and normal and rectified EMG activity of one agonist muscle associated with wrist extension. Thirty-millisecond segments of EMG activity associated with each of the first 5 cortical cell spikes are shown in the middle column. The cumulative average of EMG segments associated with the first 5 cell spikes are shown in the column at the right. Although no clear effects are present after five sweeps, the average of EMG segments associated with 2000 spike events shows a clear, transient postspike facilitation. (Reproduced, with permission, from Fetz and Cheney, 1980.)

of one neuron to another, and synchronous oscillations. Synchrony between cortical cells has been reported in several studies, but it is generally weak and incapable of explaining PSpF. For example, Lemon (1990) showed that a cortical cell that was sharply synchronized with a CM cell still did not produce PSpF despite the presence of synchrony (Fig. 20). Smith and Fetz (1989) obtained similar results and also showed that synchrony spikes contribute a broad peak with an early onset. Flament *et al.* (1992) were able to separate this broad synchrony component from the true PSpF in spike-triggered averages computed from dorsal root ganglion cells (presumed muscle spindle afferents). Figure 21B illustrates an example in which a PSpF was superimposed on a broader synchrony facilitation peak with an early onset latency. In this case, a true PSpF effect could be identified based on a sharp discontinuity in the rising phase of facilitation. Distinguishing synchrony effects from PSpF is essential.

Finally, synchronous oscillations throughout the cerebral cortex have been widely reported (Gray and Singer, 1989). These events are brief oscillations in local field potentials at frequencies ranging from 20 to 70 Hz. Murthy and Fetz (1992) investigated synchronous oscillations in motor and somatosensory cortex. They found that the incidence of synchronous oscillations varied with task conditions from about 0.5/s during a stereotyped movement to 3.5/s during a novel task that required more focused attention. Oscillations sometimes became coherent within the cortex over distances of up to 14 mm including regions of both motor and somatosensory cortex. Others have reported that oscillations are more common during static hold periods of movement tasks than during actual movement (Kilner *et al.,* 1999) and during premovement delay periods (Donoghue *et al.,* 1998). These oscillations may reflect important neural processing events and are certainly of physiological interest for that reason, but for those using spike-triggered averaging and other correlational methods, synchronous oscillations are the source of potential technical problems. Murthy and Fetz (1992) showed that although episodes of synchronous oscillations were not associated with any significant change in the level of average EMG activity, averages compiled by triggering from the same phase of each cycle of the synchronous oscillation revealed correlated oscillations in both flexor and extensor muscle activity. These muscle oscillations seemed

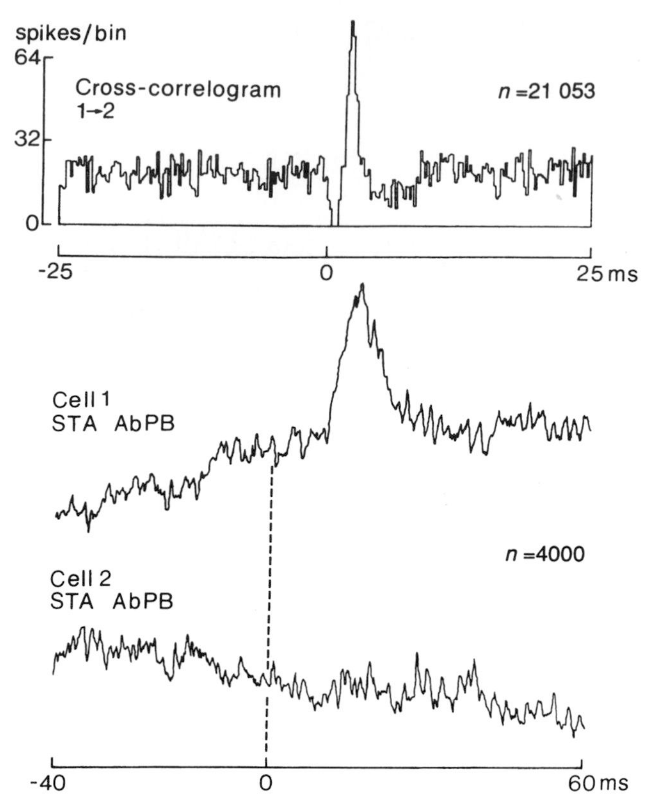

Figure 20 Tests showing that the presence of synchronization with a CM cell is insufficient by itself to produce PSpF. Two cortical cells were recorded through the same microelectrode. Cell 1 was identified as a pyramidal tract neuron. The cross correlogram shows a strong peak in the activity of cell 2 1–2 ms after the discharge of cell 1 at time zero. A spike-triggered average of abductor pollicus brevis from cell 1 revealed a clear PSpF, but no PSpF was observed from cell 2 despite the presence of strong synchronization with cell 1. (Reproduced, with permission, from Lemon, 1990.)

to be strongest during a raisin retrieval task and were negligible during a simple flexion–extension task. The discharges of 49% of single neurons tested were modulated in phase with the cycles of synchronous oscillations.

Synchronous oscillations are of concern for the method of spike-triggered averaging because of their potential to produce significant synchronization in the discharges of corticospinal neurons. The possibility that spike synchronization related to oscillations could compromise results obtained with spike-triggered averaging can be ruled out by specifically excluding cell spikes that occur during synchronous oscillations. In addition, several points argue against synchronous oscillations as a factor contributing to spurious postspike effects. First, the number of spikes synchronized with oscillations is generally small relative to the total number of spikes occurring during a motor task. Furthermore, there is no documented case of a cell producing PSpF solely because of spikes synchronized from oscillations. Second, half of all cortical cells tested did not show discharges correlated with synchronous oscillations (Murthy and Fetz, 1992). Third, of many cells tested in areas 3a, 3b, 1, and 2 of primary somatosensory cortex where synchronous oscillations are coherent with those in motor cortex, only one cell in area 3a produced significant PSpF (Widener and Cheney, 1997). PSpF-producing cells are much rarer than would be expected if widespread synchronous oscillations were significantly compromising spike-triggered averaging data. Finally, the results of spike-triggered averaging were not compromised even in cases in which cells showed synchrony (Fig. 20).

XII. Mapping Motor Output with Stimulus-Triggered Averaging of EMG Activity (Single-Pulse ICMS)

The use of ICMS for motor system mapping was introduced in an earlier section. In a manner analogous to spike-triggered averaging of EMG activity, averages can also be referenced to stimuli applied to the sites of recorded cells. This is referred to as stimulus-triggered averaging of EMG activity (Cheney and Fetz, 1985). In this procedure, averages are computed using the same parameters that were previously used for spike-triggered averaging except that the computer is triggered from microstimuli delivered at a low rate during movements with which the previously recorded cell was related (Fig. 22). The low rate of stimulation (5–15 Hz) avoids spread of excitation by temporal summation. Stimulation must be applied during active movements so EMG activity is present and coupled with averaging because the individual stimuli are generally subthreshold for discharging motoneurons.

Averages resulting from this method can be easily quantified. Onset latencies will reflect conduction time and synaptic transmission in the anatomical pathway from the stimulation site to the muscle. Latency data from stimulus-triggered averages, therefore, are much more meaningful for interpretation purposes than onset latencies obtained with repetitive ICMS. Both excitatory and inhibitory events can be detected. Figure 22 also emphasizes that the results obtained with stimulation reflect the summation of all the stimulated elements, including cell bodies activated directly by the stimulus, axon collaterals, and afferent terminals to corticospinal neurons.

Figure 23 shows an example of poststimulus effects obtained with this method for microstimuli applied to a site in primary motor cortex of a rhesus monkey performing a reach-to-grasp movement task (Park *et al.*, 2001). EMG activity was recorded from 24 muscles of the forelimb, including shoulder, elbow, wrist, digit, and intrinsic hand muscles (Park *et al.*, 2000). At this site, clear poststimulus facilitation effects were observed in both proximal and distal muscles, although the effects in distal muscles were stronger. Systematic electrode

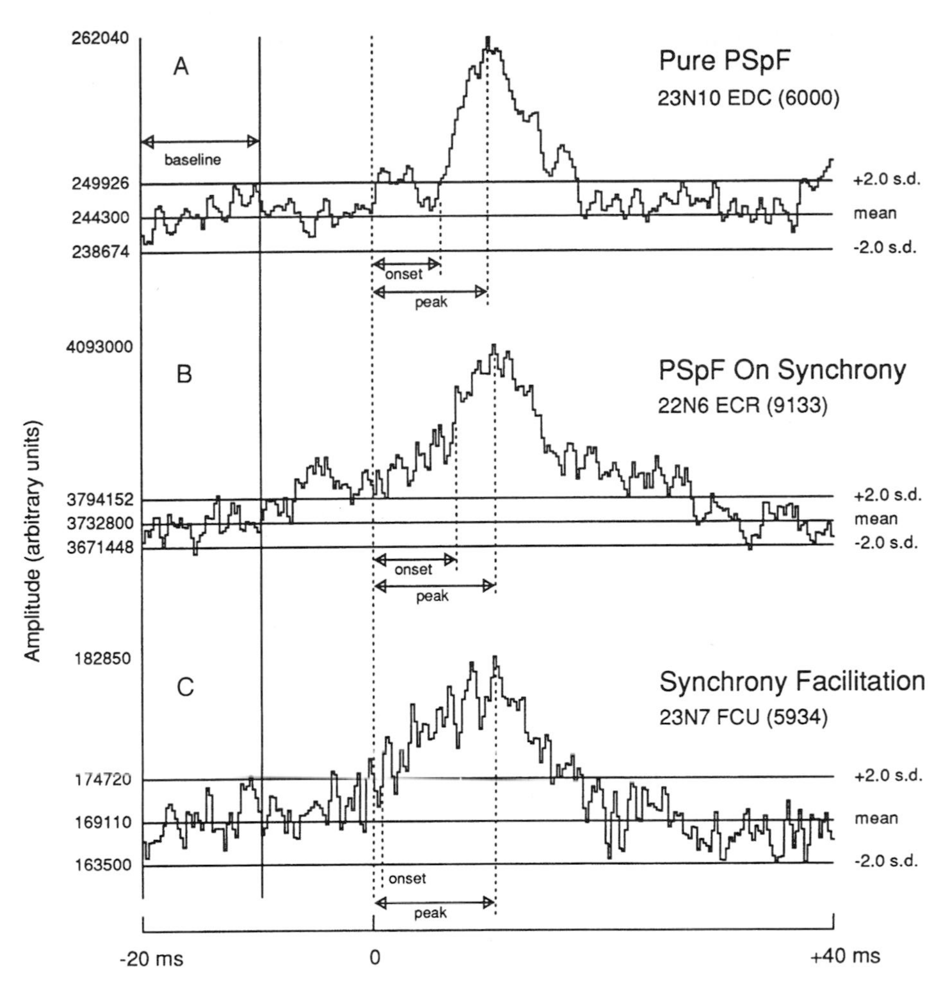

Figure 21 Categorization of PSEs based on presence or absence of synchrony facilitation. (**A**) Pure PSpF with no synchrony facilitation. (**B**) PSpF on synchrony with clear facilitation surrounded by rising and falling synchrony baseline. (**C**) Synchrony facilitation alone. The onset latency for (B) was established by visual examination of the record to identify the point at which the average broke sharply upward from the gradually sloping baseline. Onset latencies are similar for (A) and (B), while the onset of (C) occurs just after the acceptance pulse (indicated by the vertical line at time = 0 ms). The peak latencies for all three types of effects are very similar in this example. (Reproduced, with permission, from McKiernan *et al.*, 1998.)

penetrations were made with 1.0-mm spacing throughout the arm representation of the precentral gyrus to map the representation of each of these 24 muscles. Within each electrode track, stimulation was applied every 0.5 mm. These data were used to map the distribution of proximal (shoulder and elbow) and distal (wrist and digit) muscle representations in primary motor cortex as shown in Figs. 24C and 24D. In these maps, the precentral gyrus has been unfolded and represented in two-dimensional coordinates. While maps for each individual muscle can be plotted from these data, the example in Fig. 24 lumps together distal muscles (wrist, forearm digit, and intrinsic digit) and proximal muscles (shoulder and elbow). The maps represent sites producing facilitation in only distal muscles (blue), in only proximal muscles (red), and in both distal and proximal muscles (purple). The maps of distal and proximal muscle representation are consistent with results from retrograde labeling of corticospinal neurons following injections of tracers in the upper cervical (proximal muscles) and lower cervical (distal muscles) spinal cord (He *et al.*, 1993). The central core of distal muscle representation surrounded by a zone of proximal muscle representation is a consistent feature in the intra-areal forelimb map, which others have also observed in the macaque monkey (Kwan *et al.*, 1978; Wong *et al.*, 1978). In addition, the maps in Fig. 24 revealed for the first time a zone from which effects were obtained in both proximal and distal muscles. The dimensions of this zone are not compatible with simple current spread from pure distal and pure proximal representations. The existence of a specific zone with sites representing proximal and distal muscles in different combinations is consistent with results from spike-triggered averaging studies showing that about half of the corticospinal cells involved in reach-to-grasp

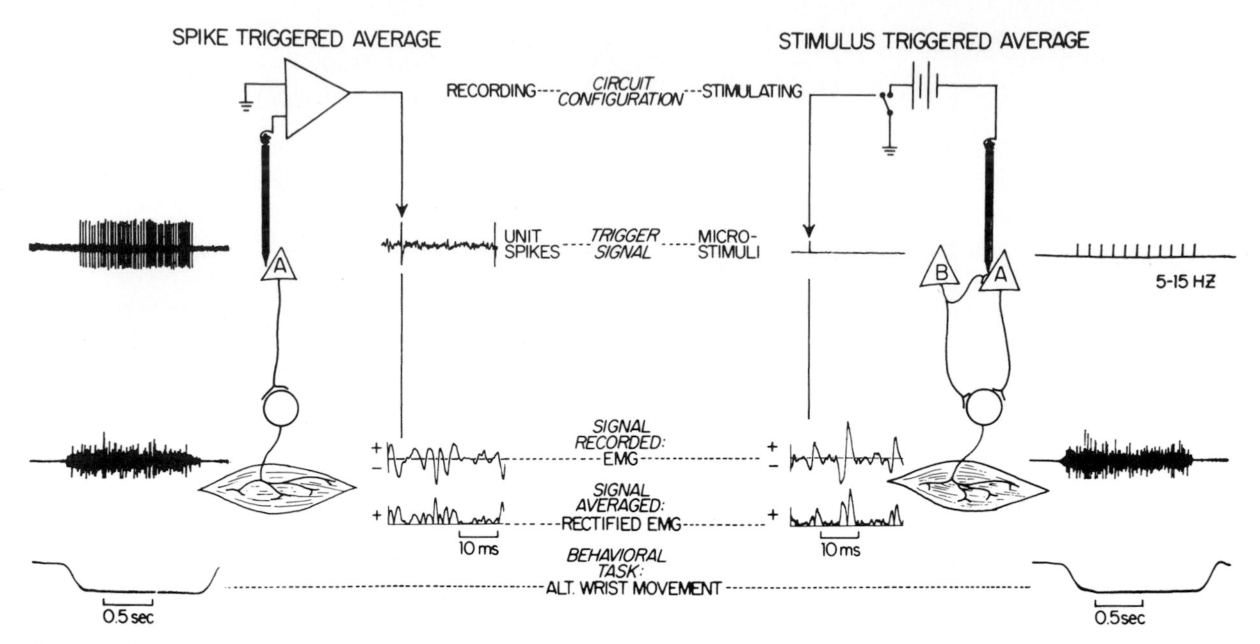

Figure 22 Comparison of spike- and stimulus-triggered averaging methods. For stimulus-triggered averaging, microstimuli are applied at a low rate (15 Hz or less) and under the same behavioral conditions used for computing the spike-triggered averages. Whereas spike-triggered averaging reveals the output organization of a single cell, stimulus-triggered averaging reveals the output organization of all the stimulated neural elements. (Reproduced, with permission, from Cheney and Fetz, 1985.)

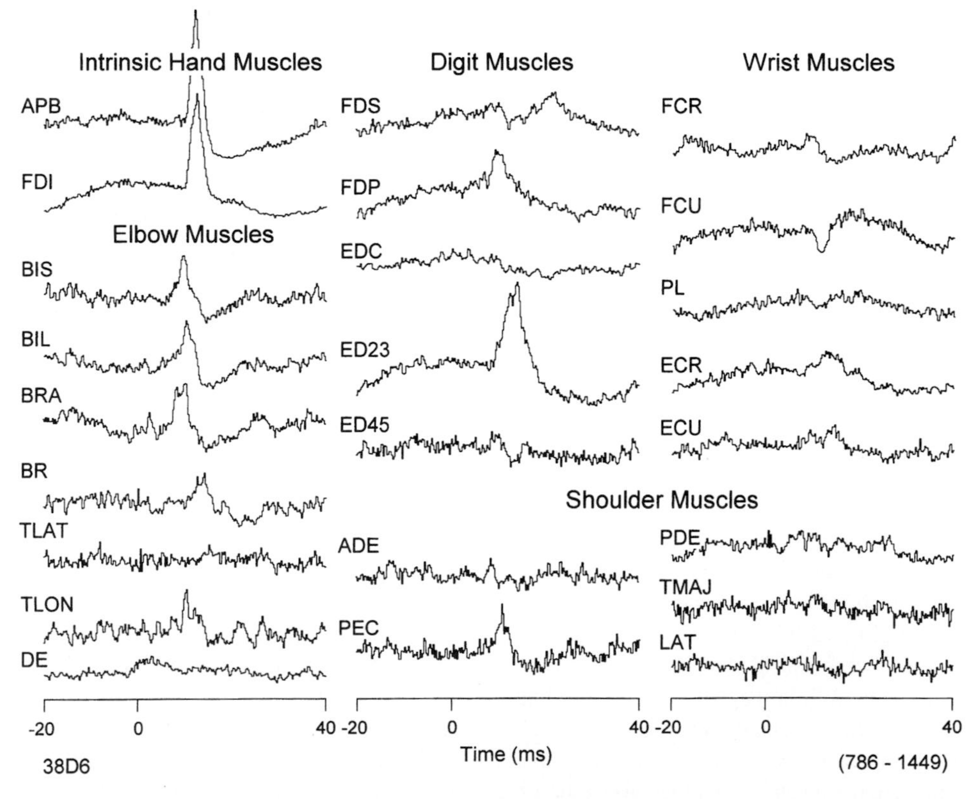

Figure 23 Distribution of poststimulus facilitation (PStF) effects in forelimb muscles from a proximal-distal cofacilitation (PDC) site. Time zero corresponds to the stimulus event used for the average. Stimulation was 15 μA at 15 Hz. Moderate and strong PStF effects were observed in both proximal (BIS, BIL, BRA, BR, TLON, PEC) and distal (APB, FDI, FDP, ED23) forelimb muscles. The range of number of trigger events for different channels is given in parentheses. (Reproduced, with permission, from Park *et al.,* 2001.)

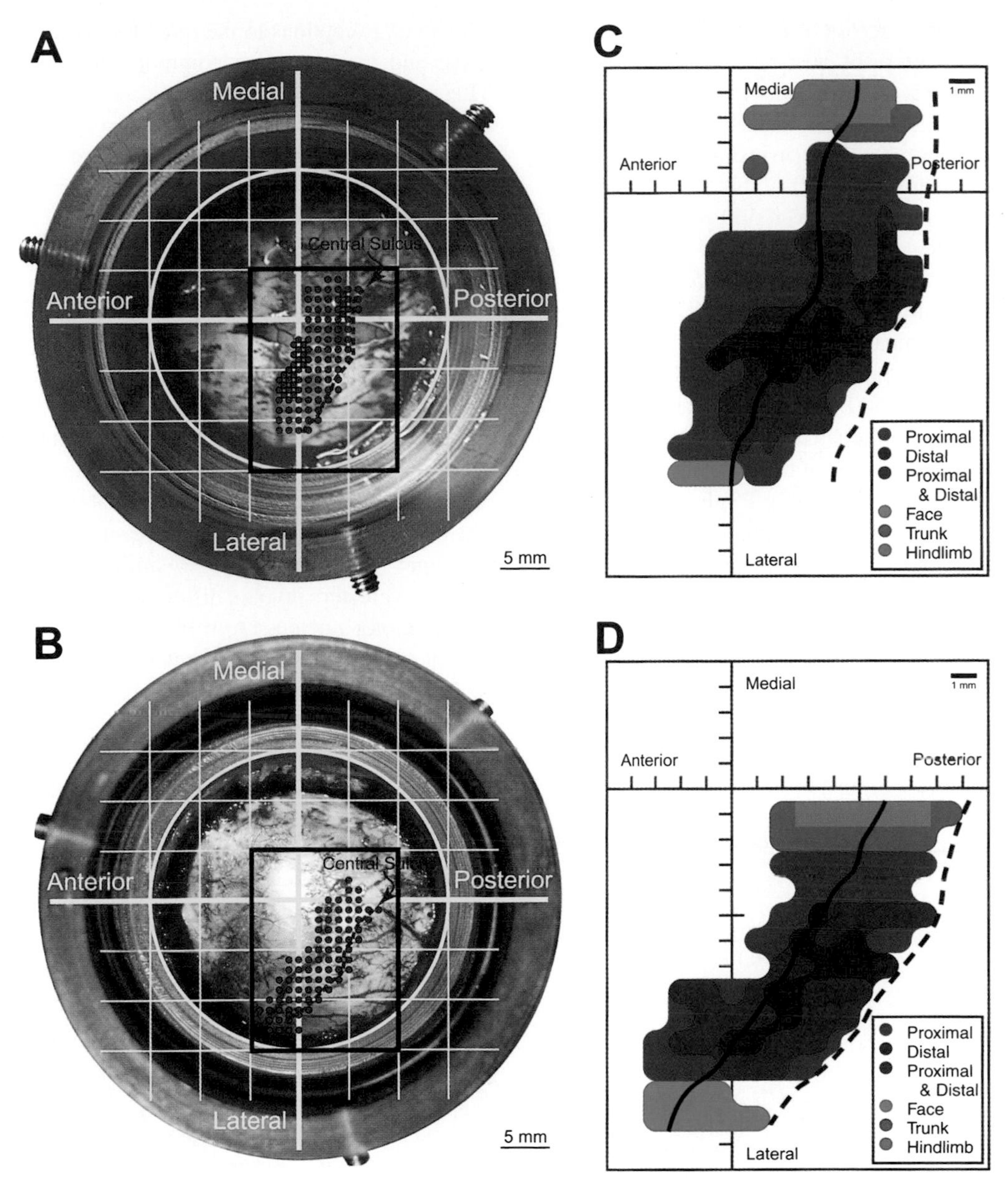

Figure 24 (**A** and **B**) Cortical recording chambers implanted over M1 cortex in two monkeys. The coordinate system (5-mm grid) is overlaid in yellow and locations of electrode tracks are indicated with black-outlined, red dots. The large black rectangle overlying each chamber identifies the cortical area represented in maps (C) and (D). In one monkey (upper chamber), a 15-mm incision was made in the dura for visual identification of the central sulcus. (**C** and **D**) Maps of motor cortex for each monkey represented in two-dimensional coordinates after unfolding the precentral gyrus. Maps were based on strong and moderate PStF effects together with R-ICMS evoked movements. Including weak PStF effects had little effect on map boundaries. (Reproduced, with permission, from Park *et al.,* 2001.)

facilitate at least one proximal and one distal muscle (McKiernan *et al.,* 1998). What might be the function of this zone representing both distal and proximal muscles in a variety of combinations? This question remains to be answered. However, an intriguing hypothesis is that this zone of the forelimb cortical map contains substrates for producing the basic features of coordinated, multijoint movements, for example, extending the limb and withdrawing the limb.

XIII. Comparison of Results from Spike-Triggered Averaging, Stimulus-Triggered Averaging, and Repetitive ICMS

The combined application of spike- and stimulus-triggered averaging has provided significant new insight concerning fundamental features of the motor cortex output map. One of the most significant findings is that the pattern of poststimulus effects across different muscles usually closely matches the pattern of postspike facilitation. Figure 25 is an example in which PSpF was strongest in extensor digitorum 4 and 5 (ED 4,5) and extensor digitorum communus (EDC) but was also clear in extensor carpi ulnaris (ECU) based on an average of 14,000 spike events. A stimulus-triggered average computed from 500 microstimuli at 5 μA applied to the same cortical site showed PStF (poststimulus facilitation) in the same muscles. Moreover, the rank order of PStF by magnitude matched the rank order of PSpF. Although appearing comparable, the absolute magnitude of PStF is actually much greater than that of PSpF because it was obtained with only 500 stimuli compared to 14,000 spikes for PSpF. Because the signal-to-noise ratio increases as the square root of the number of trigger events, the PStF is actually about five times greater than PSpF. Whereas PSpF reflects the output organization of a single cell, PStF reflects the output effects of the population of cells and other neuronal elements that are excited by the stimulus. The fact that PStF involved many CM cells but has the same basic profile across synergist muscles as PSpF from the single CM cell at the same site suggests that neighboring cells activated by the stimulus have similar patterns of synaptic connections with motoneurons. The similarity in target muscle fields of neighboring CM cells has been confirmed by computing spike-triggered averages from adjacent cells simultaneously recorded through the same microelectrode (Cheney and Fetz, 1985). However, exceptions to the rule of similar output patterns do exist and may be more common for muscles of the hand (Lemon, 1990). Nevertheless, it seems reasonable to conclude that just as neighboring sensory neurons in various parts of primary sensory cortex share common receptive fields, CM cells in lamina V of motor cortex may share common muscle fields.

Figure 26 illustrates basic features of the output map postulated by Cheney and Fetz (1985) for motor cortex. The basic module of output is a cluster of CM cells in layer V. The feature shared in common by each cell of a cluster is its muscle field. The similarity in synaptic output from different CM cells of a cluster seems to extend beyond the cell's simple muscle field. The fact that the relative magnitude of PSpF across different target muscles is similar for different cells in a cluster suggests not only that the target muscles are the same, but also that the relative strength of synaptic input to target motor nuclei is also similar. The muscle fields of different clusters involve different muscles; some facilitate a single motor nucleus (A in Fig. 26), but most facilitate different combinations of synergist motor nuclei (for example, B and F). The most common output patterns for CM cells are pure facilitation, in which the cell has no effect on antagonist muscles (clusters A and C), and reciprocal, in which the cells of a cluster not only facilitate agonist muscles but simultaneously suppress antagonists, probably through spinal inhibitory interneurons (B and E).

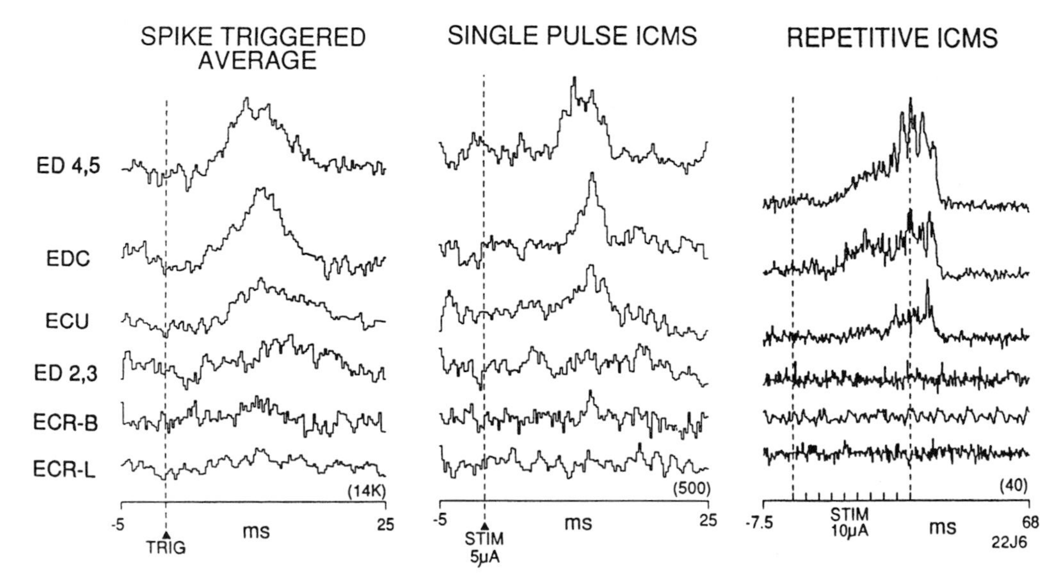

Figure 25 Comparison of results obtained at a CM cell site with spike-triggered averaging of EMG activity, stimulus-triggered averaging of rectified EMG activity (single-pulse ICMS coupled with averaging), and repetitive ICMS coupled with averaging. Repetitive ICMS consisted of 10 pulses at 330 Hz and 10 μA. Responses were averaged over 40 repetitions of the ICMS stimulus train. Effects in stimulus-triggered averaging were based on 500 trigger events and those in spike-triggered averages were based on 14,000 trigger events. Spike-triggered averaging revealed strong PSpFs in ED 4,5 and EDC as well as a clear but weaker response in ECU. No clear effects were present in the other forearm extensor muscles. Poststimulus facilitation in stimulus-triggered averages matched the pattern of postspike facilitation. In this case, the pattern of facilitation obtained with repetitive ICMS also matched the pattern of PSpF. (From Widener and Cheney, unpublished data.)

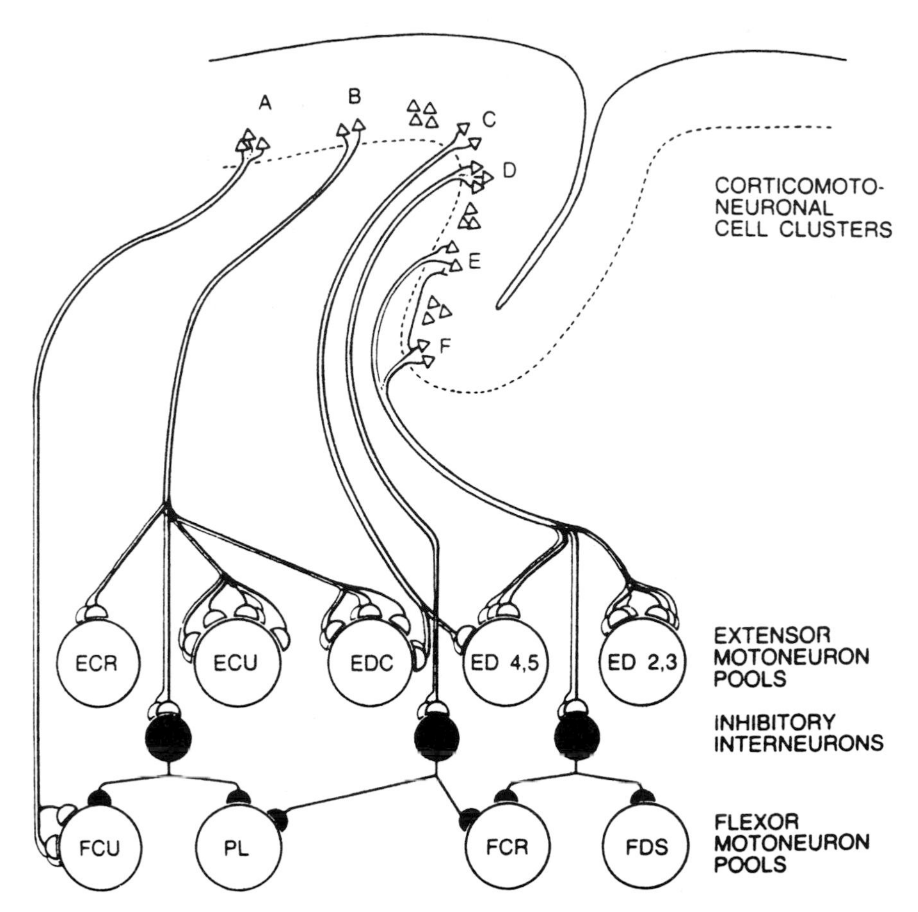

Figure 26 Model illustrating basic features of the motor cortex output map derived from studies with spike- and stimulus-triggered averaging. CM cells occur as clusters or aggregates in which each cell of the aggregate has the same or similar muscles field. (Reproduced, with permission, from Cheney and Fetz, 1985.)

How do the results of repetitive ICMS compare with those from spike- and stimulus-triggered averaging? Although not always the case, in Fig. 25 averages from 40 trains of repetitive ICMS at 10 μA yielded the same profile of muscle effects as observed in spike- and stimulus-triggered averages. Despite the fact that repetitive ICMS involves significant temporal summation and indirect excitation of CM cells, facilitation remained confined to the same muscles. This suggests that in addition to the CM cells directly activated by the stimulus, all stimulated neuronal elements related to the cell cluster may be part of a common functional unit linking the three activated muscles. If this were not true, then spread of excitation from repetitive stimulation would have certainly resulted in effects appearing in other muscles.

Finally, the nature of the evoked movement was obtained for the site illustrated in Fig. 26, completing a full set of output measures. The movement evoked by repetitive stimulation was extension of digit 4. This would have been difficult to predict from the results of spike-triggered averaging, stimulus-triggered averaging, or repetitive ICMS. This illustrates clearly the impossibility of drawing any conclusions about the muscles activated in association with movements evoked by stimulation. Mapping based on EMG recording and averaging is very effective for revealing the distribution of output from the site of stimulation to individual muscles in a form that is reproducible and quantifiable.

XIV. Mapping the Output Terminations of Single Neurons Electrophysiologically

A powerful but tedious method of mapping the output connections of single neurons belonging to descending systems was introduced by Shinoda and Yamaguchi (1978). This method consists of antidromically activating a spinal projecting premotor neuron, for example, a corticospinal neuron, from a stimulating microelectrode in the spinal cord. The threshold for antidromic activation of the axon is determined at each site of stimulation and plotted against depth in the electrode tract. A series of systematic electrode penetrations is made to fully map the path of an axon in the spinal cord and its sites of termination. Figure 27 illustrates the data that are obtained from this type of mapping. In this case, à slowly conducting corticospinal neuron was mapped through the cervical segments of the spinal cord. Multiple points of termination were found based on zones of low threshold. Such findings are consistent with target muscle identification using spike-triggered averaging of EMG activity showing that most CM cells have multiple target muscles rather than just one. These results are also consistent with limited but important anatomical labeling studies discussed at the beginning of this chapter (Fig. 2).

XV. The Future of Electrophysiological Mapping

Over the past decade, noninvasive imaging methods such as functional MRI and PET to map brain functions have generated great interest and excitement. Imaging methods have already produced major advances in localizing function within the brain, including functions that were previously very difficult to investigate in animals, such as thought and emotion. At the same time, new methods for electrophysiological mapping have provided answers to major questions about the structure and function of brain maps. For example,

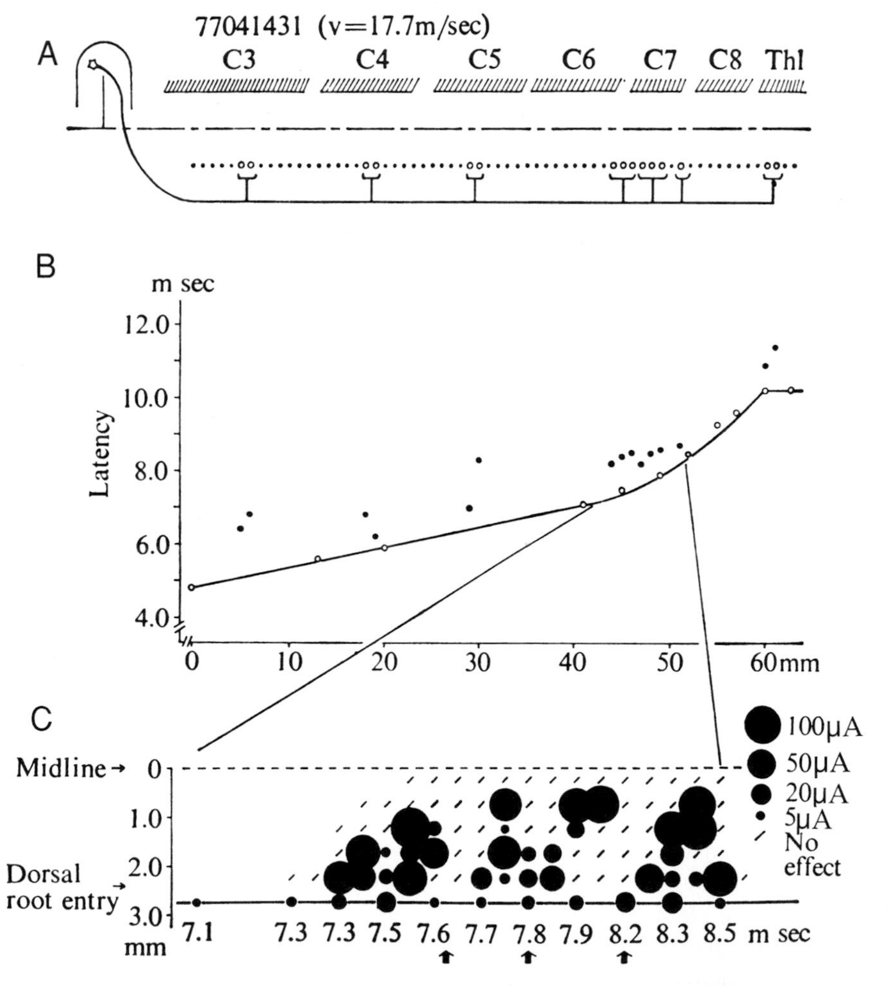

Figure 27 Electrophysiological method for mapping the intraspinal branching pattern of a slowly conducting (17.7 m/s) corticospinal neuron in the cat. (**A**) Schematic diagram of the branching pattern. Dots and open circles indicate the ineffective and effective electrode penetrations, respectively. (**B**) The latency of antidromic spikes relative to distance along the spinal cord. Dots and open circles represent the latencies of spikes evoked from gray matter and the lateral corticospinal tract, respectively. (**C**) Map of the current required for antidromic activation of the corticospinal neuron near the C7 segment. Threshold stimulus currents were measured at 500-μm intervals longitudinally and 100-μm intervals transversely. The lowest threshold values at each penetration are plotted. Dorsal view of the left cervical spinal cord. Three groups of effective stimulating sites were observed at stimulus intensities of 100 μm or less, demonstrating that three independent axon collaterals originated from the stem axon at the level of C7. (Reproduced, with permission, from Shinoda and Yamaguchi, 1978.)

TMS with coregistration to structural and functional MR images can be applied noninvasively in humans and is providing a wealth of new information on the organization of function in the human brain. Repetitive TMS has extended the power of this method and has provided a means of rigorously testing issues of causality arising from functional imaging studies. Methods for averaging EMG activity referenced to the spikes of single cells or individual microstimuli (spike- and stimulus-triggered averaging) have revealed many new features of the output organization of motor cortex and other CNS regions that supply motoneurons. The power of imaging methods lies in the fact that they can provide a global picture of brain function in relation to a particular behavior of interest. In contrast, the power of electrophysiological mapping methods lies in the high degree of spatial and temporal resolution they provide and, in the case of repetitive TMS, the ability to reversibly inactivate cortical regions as a way of testing causal involvement in function. Questions about the detailed microstructure of brain maps, including the organization and function of individual neurons, may always remain the domain of electrophysiological approaches. Given that different techniques have unique strengths and weaknesses and that no one technique for brain mapping has emerged as a methodological panacea, it seems clear that the most effective future strategy for brain mapping will be the continued parallel use of imaging and electrophysiological methods.

Perhaps one of the most important areas for future progress in brain mapping lies in computer modeling and computational neuroscience (Churchland and Sejnowski, 1993). As computer modeling becomes more sophisticated, the potential exists for assimilating the large array of physiological data available from laboratory investigations of brain maps, and the neurons contained within them, into a realistic model of an entire brain system. Such computer models would not only provide a mechanism for quickly exploring the response of a brain system under a broad range of conditions but would also have practical application in prosthetics and robotics. The application of data from electrophysiological studies of the brain together with an increasingly powerful computer technology for modeling brain systems promises to yield a very exciting time of progress over the coming years.

Acknowledgments

This work was supported by NINDS Grants NS39023 and NS38405, NIDA Grant DA12827, and NICHD Center Grant HD02528.

References

Adrian, E. D. (1926). The impulses produced by sensory nerve endings. *J. Physiol. (London)* **61,** 49–72.

Amassian, V. E., and Cracco, R. Q. (1987). Human cerebral cortical responses to contralateral transcranial stimulation. *Neurosurgery* **20,** 148–155.

Amassian, V. E., Cracco, R. Q., and Maccabee, P. J. (1989). Focal stimulation of human cerebral cortex with the magnetic coil: A comparison with electrical stimulation. *Electroencephalogr. Clin. Neurophysiol.* **74,** 401–416.

Amassian, V. E., and Deletis, V. (1999). Relationships between animal and human corticospinal responses. *Electroencephalogr. Clin. Neurophysiol.* **51**(Suppl.), 79–92.

Andersen, P., Hagan, P. J., Phillips, C. G., and Powell, T. P. S. (1975). Mapping by microstimulation of overlapping projections from area 4 to motor units of the baboon's hand. *Proc. R. Soc. London Biol.* **188,** 31–60.

Asanuma, H., and Arnold, A. P. (1975). Noxious effects of excessive currents used for intracortical microstimulation. *Brain Res.* **96,** 103–107.

Asanuma, H., and Sakata, H. (1967). Functional organization of a cortical efferent system examined with focal depth stimulation in cats. *J. Neurophysiol.* **30,** 35–54.

Asanuma, H., Arnold, A., and Zarzecki, P. (1976). Further study on the excitation of pyramidal tract cells by intracortical microstimulation. *Exp. Brain Res.* **26,** 443–461.

Asanuma, H., and Rosen, I. (1972). Topographical organization of cortical efferent zones projecting to distal forelimb muscles in the monkey. *Exp. Brain Res.* **14,** 243–256.

Barker, A. T. (1999). The history and basic principles of magnetic nerve stimulation. *Electroencephalogr. Clin. Neurophysiol.* **51**(Suppl.), 3–21.

Barker, A. T., Jalinous, R., and Freeston, I. L. (1985). Non-invasive magnetic stimulation of human motor cortex. *Lancet* **1,** 1106–1107.

Bastings, E. P., Gage, H. D., Greenberg, J. P., Hammond, G., Hernandez, L., Santago, P., Hamilton, C. A., Moody, D. M., Singh, K. D., Ricci, P. E., Pons, T. P., and Good, D. C. (1998). Co-registration of cortical magnetic stimulation and functional magnetic resonance imaging. *NeuroReport* **22,** 1941–1946.

Boroojerdi, B., Foltys, H., Krings, T., Spetzger, U., Thron, A., and Töpper, R. (1999). Localization of the motor hand area using transcranial magnetic stimulation and functional magnetic resonance imaging. *Clin. Neurophysiol.* **110,** 699–704.

Botteron, G. W., and Cheney, P. D. (1989). Corticomotoneuronal postspike effects in averages of unrectified EMG activity. *J. Neurophysiol.* **62,** 1127–1139.

Brandt, S. A., Brocke, J., Röricht, S., Ploner, C. J., Villringer, A., and Meyer, B.-U. (2000). *In vivo* assessment of human visual system connectivity with transcranial electrical stimulation during functional magnetic resonance imaging. *NeuroImage* **14,** 366–375, 2001.

Breese, C. R., Hampson, R. E., and Deadwyler, S. A. (1989). Hippocampal place cells: Stereotypy and plasticity. *J. Neurosci.* **9,** 1097–1111.

Burke, D., and Hicks, R. (1999). Corticospinal volleys underlying the EMG responses to transcranial stimulation of the human motor cortex. *Electroencephalogr. Clin. Neurophysiol.* **49**(Suppl.), 226–232.

Burke, D., Hicks, R., Gandevia, S. C., Stephen, J., Woodforth, I., and Crawford, M. (1993). Direct comparison of corticospinal volleys in human subjects to transcranial magnetic stimulation and electrical stimulation. *J. Physiol. (London)* **470,** 383–393.

Burke, D., Hicks, R., Stephen, J., Woodforth, I., and Crawford, M. (1995). Trial-to-trial variability of corticospinal volleys in human subjects. *Electroencephalogr. Clin. Neurophysiol.* **97,** 231–237.

Chang, H.-T., Ruch, T. C., and Ward, A. A., Jr. (1947). Topographic representation of muscles in motor cortex in monkeys. *J. Neurophysiol.* **10,** 39–56.

Cheney, P. D., and Fetz, E. E. (1985). Comparable patterns of muscle facilitation evoked by individual corticomotoneuronal (CM) cells and by single intracortical microstimuli in primates: Evidence for functional groups of CM cells. *J. Neurophysiol.* **53,** 786–804.

Churchland, P. S., and Sejnowski, T. J. (1993). "The Computational Brain." MIT Press, Cambridge, MA.

Classen, J., Knorr, U., Werhahn, K. J., Schlaug, G., Kunesch, E., Cohen, L. G., Seitz, R. J., and Benecke, R. (1998). Multimodal output mapping of human central motor representation on different spatial scales. *J. Physiol.* **512,** 163–179.

Cohen, L. G., Celnik, P., Pascual-Leone, A., Corwell, B., Falz, L., Dambrosia, J., Honda, M., Sadato, N., Gerloff, C., Catala, M. D., and Hallett, M. (1997). Functional relevance of cross-modal plasticity in blind humans. *Nature* **389,** 180–183.

Cohen, L. G., Ziemann, U., Chen, R., Classen J., Hallett, M., Gerloff, C., and Butefsch, C. (1998). Studies of neuroplasticity with transcranial magnetic stimulation. *J. Clin. Neurophysiol.* **15,** 305–324.

Cope, T. C., Fetz, E. E., and Matsumura, M. (1987). Cross correlation assessment of the synaptic strength of single Ia fibre connections with triceps surae motoneurons in cats. *J. Physiol.* **390,** 161–188.

Cracco, R. Q., Amassian, V. E., Maccabee, P. J., and Cracco, J. B. (1989). Comparison of human transcallosal responses evoked by magnetic coil and electrical stimulation. *Electroencephalogr. Clin. Neurophysiol.* **74,** 417–424.

Day, B. L., Dressler, D., Maertens De Noordhout, A., Marsden, C. D., Hakashima, K., Rothwell, J. C., and Thompson, P. D. (1989). Electric and magnetic stimulation of human motor cortex: Surface EMG and single motor unit responses. *J. Physiol.* **412,** 449–473.

Di Lazzaro, V., Oliviero, A., Profice, P., Insola, A., Mazzone, P., Tonali, P., and Rothwell, J. C. (1999). Direct recordings of descending volleys after transcranial magnetic and electric motor cortex stimulation in conscious humans. *Electroencephalogr. Clin. Neurophysiol.* **51**(Suppl.), 120–126.

Di Lazzaro, V., Oliviero, A., Profice, P., Saturno, E., Pilato, F., Insola, A., Mazzone, P., Tonali, P., and Rothwell, J. C. (1998). Comparison of descending volleys evoked by transcranial magnetic and electric stimulation in conscious humans. *Electroencephalogr. Clin. Neurophysiol.* **109,** 397–401.

Di Lazzaro, V., Oliviero, A., Saturno, E., Pilato, F., Insola, A., Mazzone, P., Profice, P., Tonali, P., and Rothwell, J. C. (2001). The effect on corticospinal volleys of reversing the direction of current induced in the motor cortex by transcranial magnetic stimulation. *Exp. Brain Res.* **138,** 268–273.

Donoghue, J. P., Sanes, J. N., Hatsopoulos, N. G., and Gaal, G. (1998). Neural discharge and local field potential oscillations in primate motor cortex during voluntary movements. *J. Neurophysiol.* **79,** 159–173.

Edgley, S. A., Eyre, J. A., Lemon, R. N., and Miller, S. (1990). Excitation of the corticospinal tract by electromagnetic and electrical stimulation of the scalp in the macaque monkey. *J. Physiol. (London)* **425,** 301–320.

Edgley, S. A., Eyre, J. A., Lemon, R. N., and Miller, S. (1997). Comparison of activation of corticospinal neurons and spinal motor neurons by magnetic and electrical transcranial stimulation in the lumbosacral cord of the anaesthetized monkey. *Brain* **120,** 839–853.

Ettinger, G. J., Leventon, M. E., Grimson, W. E., Kikinis, R., Gugino, L., Cote, W., Sprung, L., Aglio, L., Shenton, M. E., Potts, G., Hernandez, V. L., and Alexander, E. (1998). Experimentation with a transcranial magnetic stimulation system for functional brain mapping. *Med. Image Anal.* **2,** 133–142.

Fetz, E. E., and Cheney, P. D. (1979). Muscle fields and response properties of primate corticomotoneuronal cells. *Prog. Brain Res.* **50,** 137–146.

Fetz, E. E., and Cheney, P. D. (1980). Postspike facilitation of forelimb muscle activity by primate cortico-motoneuronal cells. *J. Neurophysiol.* **44,** 751–772.

Fetz, E. E., Cheney, P. D., Mewes, K., and Palmer, S. (1989). Control of forelimb muscle activity by populations of corticomotoneuronal and rubromotoneuronal cells. *Prog. in Brain Res.* **80**, 437–449.

Fetz, E. E., Finocchio, D. V., Baker, M. A., and Soso, M. J. (1980). Sensory and motor responses of precentral cortex cells during comparable passive and active joint movements. *J. Neurophysiol.* **43,** 1070–1089.

Flament, D., Fortier, P. A., and Fetz, E. E. (1992). Response patterns and post-spike effects of peripheral afferents in dorsal root ganglia of behaving monkeys. *J. Neurophysiol.* **67,** 875–889.

Fritsch, G., and Hitzig, E. (1870). Uber die elektnsche Erregbarkeit des Grosshirns. *Arch. Anat. Physiol. Wiss. Med.* **37,** 300–332. [Translation by G. von Bonin *in* "The Cerebral Cortex," pp. 73–96, C. C. Thomas, Springfield.]

Fuhr, P., Agostino, R., and Hallett, M. (1991). Spinal motor neuron excitability during the silent period after cortical stimulation. *Electroencephalogr. Clin. Neurophysiol.* **81,** 257–262.

George, M. S., Wassermann, E. M., and Post, R. M. (1996). Transcranial magnetic stimulation: A neuropsychiatric tool for the 21st Century. *J. Neuropsychiatry Clin. Neurosci.* **8,** 373–382.

Grafman, J., and Wassermann, E. (1999). Transcranial magnetic stimulation can measure and modulate learning and memory. *Neuropsychologia* **37,** 159–167.

Gray, C. M., and Singer, W. (1989). Stimulus-specific neuronal oscillations in orientation columns of cat visual cortex. *Proc. Natl. Acad. Sci. USA* **86,** 1689–1702.

Han, T. R., Kim, J. H., and Lim, J. Y. (2001). Optimization of facilitation related to threshold in transcranial magnetic stimulation. *Clin. Neurophysiol.* **112,** 593–599.

Hatsopoulos, N. G, Ojakangas, C. L., and Maynard, E. M. (1998). Detection and identification of ensemble codes in motor cortex. *In* "Neuronal Ensembles: Strategies for Recording and Decoding" (H. E. Eichenbaum and J. L. Davis, eds.), pp. 161–175. Wiley–Liss, New York

He, S., Dum, R. P., and Strick, P. L. (1993). Topographic organization of corticospinal projections from the frontal lobe: Motor areas on the lateral surface of the hemisphere. *J. Neurosci.* **13,** 952–980.

Houlden, D. A., Schwartz, M. L., Tator, C. H., Ashby, P., and MacKay, W. A. (1999). Spinal cord-evoked potentials and muscle responses evoked by transcranial magnetic stimulation in 10 awake human subjects. *J. Neurosci.* **19,** 1855–1862.

Hubel, D. (1957). Tungsten microelectrode for recording from single units. *Science* **125,** 549–550.

Huntley, G. W., and Jones, E. G. (1991). Relationship of intrinsic connections to forelimb movement representations in monkey motor cortex: A correlative anatomic and physiological study. *J. Neurophysiol.* **66,** 390–413.

Ilmoniemi, R. J., Ruohonen, J., and Karhu, J. (1999). Transcranial magnetic stimulation—A new tool for functional imaging of the brain. *Crit. Rev. Biomed. Eng.* **27,** 241–284.

Ilmoniemi, R. J., Virtanen, J., Ruohonen, J., Karhu, J., Aronen, H. J., Näätänen, R., and Katila, T. (1997). Neuronal responses to magnetic stimulation reveal cortical reactivity and connectivity. *NeuroReport* **8,** 3537–3540.

Jahanshahi, M., and Rothwell, J. (2000). Transcranial magnetic stimulation studies of cognition: An emerging field. *Exp. Brain Res.* **131,** 1–9.

Jankowska, E., Padel, Y., and Tanaka, R. (1975a). The mode of activation of pyramidal tract cells by intracortical stimuli. *J. Physiol. (London)* **249,** 617–636.

Jankowska, E., Padel, Y., and Tanaka, R. (1975b). Projections of pyramidal tract cells to α-motoneurones innervating hindlimb muscles in the monkey. *J. Physiol (London)* **249,** 637–667.

Karl, A., Birhaumer, N., Lutzenberger, W., Cohen, L. G., and Flor, H. (2001). Reorganization of motor and somatosensory cortex in upper extremity amputees with phantom limb pain. *J. Neurosci.* **21,** 3609–3618.

Kilner, J. M., Baker, S. N., Salenius, S., Jousmaki, V., Hari, R., and Lemon, R. N. (1999). Task-dependent modulation of 15–30 Hz coherence between rectified EMGs from human hand and forearm muscles. *J. Physiol.* **516,** 559–570.

Krings, T., Buchbinder, B. R., Butler, W. E., Chiappa, K. H., Jiang, H. J., Rosen, B. R., and Cosgrove, G. R. (1997a). Stereotactic transcranial magnetic stimulation: Correlation with direct electrical cortical stimulation. *Neurosurgery* **41,** 1319–1325.

Krings, T., Buchbinder, B. R., Butler, W. E., Chiappa, K. H., Jiang, H. J., Cosgrove, G. R., and Rosen, B. R. (1997b). Functional magnetic resonance imaging and transcranial magnetic stimulation: Complementary approaches in the evaluation of cortical motor function. *Neurology* **48,** 1406–1416.

Kosslyn, S. M., Pascual-Leone, A., Felician, O., Camposano, S., Keenan, J. P., Thompson, W. L., Ganis, G., Sukel, K. E., and Alpert, N. M. (1999).

The role of area 17 in visual imagery: Convergent evidence from PET and rTMS. *Science* **284,** 167–170.

Kwan, H. C., MacKay, W. A., Murphy, J. T., and Wong, Y. C. (1978). Spatial organization of precentral cortex in awake primates. II. Motor outputs. *J. Neurophysiol.* **41,** 1120–1131.

Landgren, S., Phillips, C. G., and Porter, R. (1962). Minimal synaptic actions of pyramidal impulses on some alpha motoneurones of the baboon's hand and forearm. *J. Physiol. (London)* **161,** 91–111.

Lawrence, D. G., Porter, R., and Redman, S. J. (1985). Corticomotoneuronal synapses in the monkey: Light microscopic localization upon motoneurons of intrinsic muscles of the hand. *J. Comp. Neurol.* **232,** 499–510.

Lemon, R. N. (1984). "Methods for Neuronal Recording in Conscious Animals." Wiley, New York.

Lemon, R. N. (1990) Mapping the output functions of the motor cortex. *In* "Signal and Sense: Local and Global Order in Perceptual Maps" (G. Edelman, E. Gall, and W. M. Cowan, eds.), pp. 315–356. Wiley, Chichester, UK.

Lemon, R. N., and Porter, R. (1976). Afferent input to movement-related precentral neurones in conscious monkeys. *Proc. R. Soc. London Biol.* **194,** 313–339.

Levy, W. J., Cracco, R. Q., Barker, A. T., and Rothwell, J. C., eds. (1991). "Magnetic Motor Stimulation: Basic Principles and Clinical Experience." Elsevier, Amsterdam.

Loeb, G. E., and Gans, C. (1986). "Electromyography for Experimentalists." Univ. of Chicago Press, Chicago.

Mantel, G. W. H., and Lemon, R. N. (1987). Cross-correlation reveals facilitation of single motor units in thenar muscles by single corticospinal neurones in the conscious monkey. *Neurosci. Lett.* **77,** 113–118.

Maynard, E. M., Hatsopoulos, N. G., Ojakangas, C. L, Acuna, B. D., Sanes, J. N., Normann, R. A., and Donoghue, J. P. (1999). Neuronal interactions improve cortical population coding of movement direction. *J. Neurosci.* **19,** 8083–8093

McKiernan, B. J., Marcario, J. K., Hill–Karrer, J. and Cheney, P. D. (1998). Corticomotoneuronal (CM) postspike effects on shoulder, elbow, wrist, digit, and intrinsic hand muscles during a reach and prehension task in the monkey. *J. Neurophysiol.* **80**, 1961–1980.

Mendell, L. M., and Henneman, E. (1971). Terminals of single Ia fibers: Location, density and distribution within a pool of 300 homonymous motoneurons. *J. Neurophysiol.* **34,** 171–187.

Merton, P. A., and Morton, H. B. (1980). Stimulation of the cerebral cortex in the intact human subject. *Nature* **285,** 227.

Mills, K. R. (1991). Magnetic brain stimulation: A tool to explore the action of the motor cortex on single human spinal motoneurones. *Trends Neurosci.* **14,** 401–405.

Muller, R. U., and Kubie, J. L. (1989). The firing of hippocampal place cells predicts the future position of freely moving rats. *J. Neurosci.* **9,** 4101–4110.

Murphy, J. T., Kwan, H. C., MacKay, W. A., and Wong, Y. C. (1978). Spatial organization of precentral cortex in awake primates. III. Input–output coupling. *J. Neurophysiol.* **41,** 1132–1139.

Murthy, V. N., and Fetz, E. E. (1992). Coherent 25- to 35-Hz oscillations in the sensorimotor cortex of awake behaving monkeys. *Proc. Natl. Acad. Sci. (USA)* **89,** 5670–5674.

Nelson, R. J., Sur, M., Felleman, D. J., and Kaas, J. H. (1980). Representations of the body surface in postcentral cortex of Macaca fascicularis. *J. Comp. Neurol.* **192,** 611–643.

Normann, R. A., Warren, D. J., Ammermuller, J., Fernandez, E., and Guillory, S. (2001). High-resolution spatio-temporal mapping of visual pathways using multi-electrode arrays. *Vision Res.* **41,** 1261–1275.

Nudo, R. J., Milliken, G. W., Jenkins, W. M., and Merzenich, M. M. (1996). Use-dependent alterations of movement representations in primary motor cortex of adult squirrel monkeys. *J. Neurosci.* **16,** 785–807.

Palmer, S. S., and Fetz, E. E. (1985). Effects of single intracortical microstimuli in motor cortex on activity of identified forearm motor units in behaving monkeys. *J. Neurophysiol.* **54,** 1194–1212.

Park, M. C., Belhaj-Saif, A., and Cheney, P. D. (2000). Chronic recording of EMG activity from large numbers of forelimb muscles in awake macaque monkeys. *J. Neurosci. Methods* **15,** 153–160.

Park, M. C., Belhaj-Saif, A., and Cheney, P. D. (2001). Consistent features in the forelimb representation of primary motor cortex of rhesus macaques. *J. Neurosci.* **21,** 2784–2792.

Pascual-Leone, A., Bartres-Faz, D., and Keenan, J. P. (1999). Transcranial magnetic stimulation: Studying the brain–behavior relationship by induction of "virtual lesions." *Philos. Trans. R. Soc. London B Biol. Sci.* **354,** 1229–1238.

Pascual-Leone, A., Cohen, L. G., Brasil-Neto, J. P., and Hallet, M. (1994). Non-invasive differentiation of motor cortical representation of hand muscle by mapping of optimal current directions. *Electroencephalogr. Clin. Neurophysiol.* **93,** 42–48.

Paulus, W., Hallett, M., Rossini, P. M., and Rothwell, J. C., eds. (1999). Transcranial magnetic stimulation. *Electroencephalogr. Clin. Neurophysiol. Suppl.* **51.**

Paus, T., Jech, R., Thompson, C. J., Comeau, R., Peters, T., and Evans, A. C. (1997). Transcranial magnetic stimulation during positron emission tomography: A new method for studying connectivity of the human cerebral cortex. *J. Neurosci.* **17,** 3178–3184.

Paus, T., Jech, R., Thompson, C. J., Comeau, R., Peters, T., and Evans, A. C. (1998). Dose-dependent reduction of cerebral blood flow during rapid-rate transcranial magnetic stimulation of the human sensorimotor cortex. *J. Neurophysiol.* **79,** 1102–1107.

Penfield, W., and Boldrey, E. (1937). Somatic motor and sensory representation in the cerebral cortex of man as studied by electrical stimulation. *Brain* **60,** 389–443.

Penfield, W., and Rasmussen, T. (1950). "The Cerebral Cortex of Man. A Clinical Study of Localisation of Function." Macmillan Co., New York.

Peters, T., Davey, B., Munger, P., Comeau, R., Evans, A., and Olivier, A. (1996). Three-dimensional multimodal image-guidance for neurosurgery. *IEEE Trans. Med. Imaging* **15,** 121–128.

Phillips, C. G., and Porter, R. (1977). "Corticospinal Neurones." Academic Press, London.

Porter, R., and Lemon, R. N. (1993). "Corticospinal Function and Voluntary Movement." Clarendon Press, Oxford.

Preston, J. B., Shende, M. C., and Uemura, K. (1967). The motor cortex pyramidal system: Patterns of facilitation and inhibition on motoneurons innervating limb musculature of cat and baboon and their possible adaptive significance. *In* "Neurophysiological Basis of Normal and Abnormal Motor Activities" (M. D. Yahr and D. P. Purpura, eds.). Raven Press, New York.

Rank, J. B., Jr. (1975). Which elements are excited in electrical stimulation of mammalian central nervous system: A review. *Brain Res.* **98,** 417–440.

Rank, J. B., Jr. (1981). Extracellular stimulation. *In* "Electrical Stimulation Research Techniques" (M. M. Patterson and R. P. Kesner, eds.). Academic Press, New York.

Rosén, I., and Asanuma, H. (1972). Peripheral afferent inputs to the forelimb area of the monkey motor cortex: Input–output relations. *Exp. Brain Res.* **14,** 257–273.

Sadato, N., Pascual-Leone, A., Grafman, J., Ibanez, V., Deiber, M.-P, Dold, G., and Hallet, M. (1996). Activation of primary visual cortex by Braille reading in blind subjects. *Nature* **380,** 526–528.

Schäfer, E. A. (1900). The cerebral cortex. *In* "Textbook of Physiology" (E. A. Schäfer, ed.), Vol. 2, pp. 697–782. Pentland, Edinburgh/London.

Schieber, M. H., and Hibbard, L. S. (1993). How somatotopic is the motor cortex hand area? *Science* **261,** 489–492.

Sherrington, C. S. (1906). "The Integrative Action of the Nervous System." Scribners, New York. Second Edition, 1947, Cambridge University Press.

Shinoda, Y., and Yamaguchi, T. (1978). The intraspinal branching patterns of fast and slow pyramidal tract neurons in the cat. *J. Physiol. (Paris)* **74,** 237–238.

Shinoda, Y., Yokota, J., and Futami, T. (1981). Divergent projections of individual corticospinal axons to motoneurons of multiple muscles in the monkey. *Neurosci. Lett.* **23,** 7–12.

Smith, W. S., and Fetz, E. E. (1989). Effects of synchrony between primate corticomotoneuronal cells on post-spike facilitation of muscles and motor units. *Neurosci. Lett.* **96,** 76–81.

Soso, M. J. and Fetz, E. E. (1980). Responses of identified cells in postcentral cortex of awake monkeys during comparable active and passive joint movements. *J. Neurophysiol.* **43**, 1090–1110.

Speakman, A. (1987). Place cells in the brain: Evidence for a cognitive map. *Sci. Prog.* **71,** 511–530.

Stoney, S. D., Thompson, W. D., and Asanuma, H. (1968). Excitation of pyramidal tract cells by intracortical microstimulation: Effective extent of stimulating current. *J. Neurophysiol.* **31,** 659–669.

Terao, T., Fukuda, H., Ugawa, Y., Hikosaka, O., Hanajima, R., Furubayashi, T., Sakai, K., Miyauchi, S., Sasaki, Y., and Kanazawa, I. (1998a). Visualization of information flow through human oculomotor cortical regions by transcranial magnetic stimulation. *J. Neurophysiol.* **80,** 936–946.

Terao, T., Ugawa, Y., Sakai, K., Miyauchi, S., Fukuda, H., Sasaki, Y., Takino, R., Hanajima, R., Furubayashi, T., Pütz, B., and Kanazawa, I. (1998b). Localizing the site of magnetic brain stimulation by functional MRI. *Exp. Brain Res.* **121,** 145–152.

Wang, B., Toro, C., Zeffiro, T. A., and Hallett, M. (1994). Head surface digitization and registration: A method for mapping positions on the head onto magnetic resonance images. *Brain Topogr.* **3,** 185–192.

Wassermann, E. M. (1998). Risk and safety of repetitive transcranial magnetic stimulation: report and suggested guidelines from the International Workshop on the Safety of Repetitive Transcranial Magnetic Stimulation, June 5–7, 1996. *Electroencephalogr. Clin. Neurophysiol.* **108,** 1–16.

Wassermann, E. M., and Lisanby, S. H. (2001). Therapeutic application of repetitive transcranial magnetic stimulation: A review. *Clin. Neurophysiol.* **112,** 1367–1377.

Wassermann, E. M., McShane, L. M., Hallett, M., and Cohen, L. G. (1992). Noninvasive mapping of muscle representations in human motor cortex. *Electroencephalogr. Clin. Neurophysiol.* **85,** 1–8.

Wasserman, E. M., Pascual-Leone, A., and Hallet, M. (1994). Cortical motor representation of the ipsilateral hand and arm. *Exp. Brain Res.* **100,** 121–132.

Wasserman, E. M., Pascual-Leone, A., Valls-Solé, J., Toro, C., Cohen, L. G., and Hallett, M. (1993). Topography of the inhibitory and excitatory responses to transcranial magnetic stimulation in a hand muscle. *Electroencephalogr. Clin. Neurophysiol.* **89,** 424–433.

Wassermann, E. M., Wang, B., Zeffiro, T. A., Sadato, N., Pascual-Leone, A., Toro, C., and Hallett, M. (1996). Locating the motor cortex on the MRI with transcranial magnetic stimulation and PET. *NeuroImage* **3,** 1–9.

Wessberg, J., Stambaugh, C. R., Kralik, J. D., Beck, P. D., Laubach, M., Chapin, J. K., Kim, J., Biggs, S. J., Srinivasan, M. A., and Nicolelis, M. A. (2000). Real-time prediction of hand trajectory by ensembles of cortical neurons in primates. *Nature* **408,** 361–365.

Widener, G. L., and Cheney, P. D. (1997). Effects on muscle activity from microstimuli applied to somatosensory and motor cortex during voluntary movement in the monkey. *J. Neurophysiol.* **77,** 2446–2465.

Wilson, S. A., Thickbroom, G. W., and Mastaglia, F. L. (1993). Transcranial magnetic stimulation mapping of the motor cortex in normal subjects. The representation of two intrinsic hand muscles. *J. Neurol. Sci.* **118,** 134–144.

Wong, Y. C., Kwan, H. C., MacKay, W. A., and Murphy, J. T. (1978). Spatial organization of precentral cortex in awake primates. I. Somatosensory inputs. *J. Neurophysiol.* **41,** 1107–1119.

Woolsey, C. N. (1964). Cortical localization as defined by evoked potential and electrical stimulation studies. *In* "Cerebral Localization and Organization" (G. Schaltenbrand and C. N. Woolsey, eds.). Univ. of Wisconsin Press, Madison.

Woolsey, C. N., Settlage, P. H., Meyer, D. R., Spencer, W., Hamuy, T. P., and Travis, A. M. (1952). Patterns of localization in precentral and "supplementary" motor areas and their relation to the concept of a premotor area. *Res. Publ. Assoc. Res. Nerv. Ment. Dis.* **30,** 238–264.

10

Magnetoencephalographic Characterization of Dynamic Brain Activation: Basic Principles and Methods of Data Collection and Source Analysis

Matti Hämäläinen[1] and Riitta Hari

Brain Research Unit, Low Temperature Laboratory, Helsinki University of Technology, 02015-HUT, Espoo, Finland

I. Introduction

Timing is essential for proper brain functioning. Magnetoencephalography (MEG) and electroencephalography (EEG) are at present the only noninvasive human brain imaging tools that provide submillisecond temporal accuracy and thus help to unravel dynamics of cortical function. The advantage of MEG over EEG is that skull and scalp, which affect the electric potential distributions, do not smear the magnetic signals and MEG is thus able to see cortical events "directly through the skull." MEG (and EEG) reflect the electrical currents in neurons directly, rather than the associated hemodynamic or metabolic effects.

In an MEG study (see Fig. 1), one records with sensitive detectors weak magnetic fields generated by currents in the brain. The aim is to pick up the magnetic field at several locations outside the head and then to calculate the most probable source currents in the brain. MEG is a noninvasive

[1]Present address: MGH/MIT/HMS Athinoula A. Martinos Center for Biomedical Imaging, Charlestown, MA.

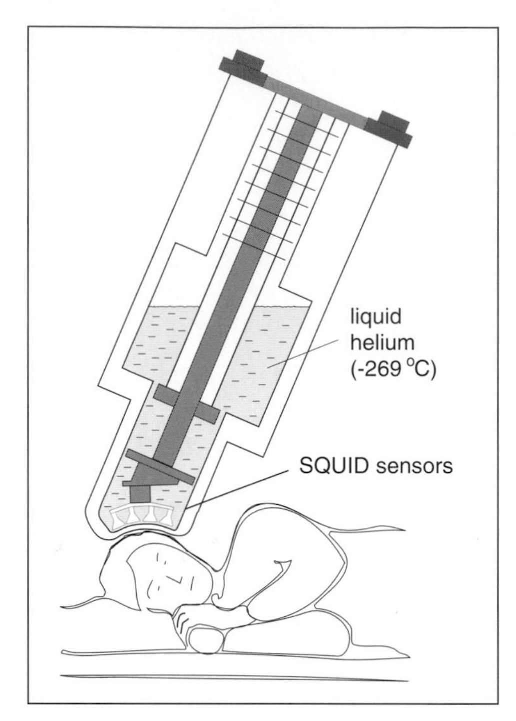

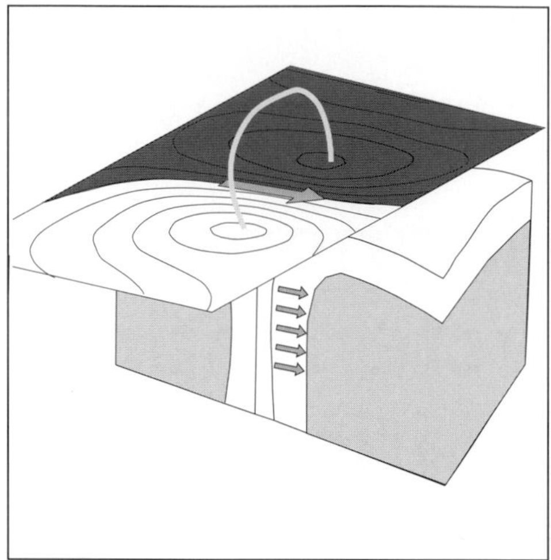

Figure 1 MEG measurement setup. **Left:** The magnetic fields produced by neuronal currents are picked up with an array of superconducting sensors. **Right:** The most probable current configuration in the brain (small arrows) is calculated on the basis of the measured field pattern, in this case a current dipole model (large arrow) was used.

technique well suited to investigation of brain regions embedded within cortical sulci. These cortical areas are poorly accessible even with intracranial recordings but produce an extracranial magnetic field which can be detected with MEG sensors.

The magnetic signals generated by cortical currents are picked up with superconducting coils connected to SQUIDs (superconducting quantum interference devices), the ultrasensitive detectors of magnetic fields. The present state-of-the-art neuromagnetometers contain more than 300 SQUIDs in helmet-shaped arrays so that signals can be recorded simultaneously over the whole neocortex.

Our chapter begins with the discussion of the generation of MEG signals. For this end, a review of the relevant electromagnetic concepts is necessary. Next, we discuss the instrumentation, magnetic shielding, and practical technical aspects of data acquisition and signal processing. We continue with a discussion of several source analysis techniques and conclude with some practical examples of MEG data acquisition and interpretation. More comprehensive information on MEG and its applications in different types of studies can be found in detailed review articles (see, e.g., Hämäläinen *et al.,* 1993; Näätänen *et al.,* 1994; Lounasmaa *et al.,* 1996; Hari, 1999; Hari *et al.,* 2000) and at http://biomag2000.hut.fi/papers_all.html.

II. Generation of Neuromagnetic Fields

A. Sources and Fields

Neuronal currents generate magnetic and electric fields according to principles stated in Maxwell's equations. The neural current distribution can be conveniently described as the primary current, the "battery" in a resistive circuit comprising the head. The postsynaptic currents in the cortical pyramidal cells are the main primary currents giving rise to measurable MEG signals. In many calculations the head can be approximated with a spherically symmetric conductor but more realistic head models for field calculations can be constructed with help of anatomical MR or CT images.

1. Quasistatic Approximation of Maxwell's Equations

The electric field **E** and the magnetic field **B,** induced by the total electric current density, **J,** can be solved from Maxwell's equations. Because **J, E,** and **B** vary in time rather slowly (below 1 kHz) (Plonsey, 1969; Hämäläinen *et al.,* 1993) the sources and fields can be treated in a quasistatic approximation. Thus inductive, capacitive, and displacement effects can be neglected, and the true time-dependent terms in the field equations can be omitted.

In the quasistatic approximation the Maxwell's equations read

$$\begin{aligned} \nabla \cdot \mathbf{E} &= \rho/\varepsilon_o \\ \nabla \times \mathbf{E} &= 0 \\ \nabla \cdot \mathbf{B} &= 0 \\ \nabla \times \mathbf{B} &= \mu_0 \mathbf{J} \end{aligned} \tag{1}$$

where ρ is the charge density and ε_0 and μ_0 are the permittivity and permeability of vacuum, respectively. Particularly, it is seen from the last of the Eqs. (1) that the magnetic field is produced by the total electric current density **J**.

2. Primary Currents

It is useful to divide the total current density, **J**, into two components. The passive volume or return current is proportional to the conductivity (σ) and the electric field (**E**) so that $\mathbf{J}_v = \sigma\mathbf{E}$. $\mathbf{J}_v$ is the result of the effect of the macroscopic electric field on charge carriers in the conducting medium. Everything else is the primary current $\mathbf{J}_p$:

$$\mathbf{J}_p = \mathbf{J} - \sigma\mathbf{E}. \tag{2}$$

Although this equation is true at different scales, it is not possible to include all the microscopic conductivity details in models of MEG activity, and thus σ refers to macroscopic conductivity with at least 1-mm scale. The division of neuronal currents into primary and volume currents is physiologically meaningful. For example, chemical transmitters in a synapse give rise to primary current mainly inside the postsynaptic cell, whereas the volume current flows passively in the medium, with a distribution depending on the conductivity profile. By finding the primary current we can locate the site of the active brain region.

The primary current is the impetus of both EEG and MEG signals. However, as will be shown later in this section, it is easier to calculate the magnetic rather than the electric field due to the symmetries and the particular conductivity distribution of the head. Because EEG electrodes on the scalp or on the cortex are physically in direct contact with the extracellular space, the EEG measures potentials that are associated with the volume currents and driven by the electric field. All currents, both intracellular and extracellular ones, generate magnetic fields. However, as will be shown below, the MEG signals generated in a rather spherical head can be calculated directly from the primary currents without explicit reference to the layered structure of the conductor. Vice versa, solving the inverse problem gives physiologically relevant information on the site, strength, and direction of the intracellular currents at the active brain site.

3. The Current Dipole and Higher Order Current Multipoles

A current dipole with a dipole moment vector **Q** and position $\mathbf{r}_Q$, approximating a localized primary current, is a widely used concept in neuromagnetism and can be thought of as concentration of the primary current $\mathbf{J}_p$ to a single point. In MEG (and EEG) applications, a current dipole is used as an equivalent source for the unidirectional primary current that may extend over relatively wide areas of cortex. Most current distributions in the brain are more complex than those produced by a single current dipole. However, any current distribution can be accurately described by a multipole expansion, by adding higher order terms. In practice, the "dipolarity" of the field patterns depends on, e.g., the viewing distance because the higher order terms decrease more rapidly as a function of distance.

4. Solution to Maxwell's Equations: The Forward Problem

Calculation of the external magnetic field from a given current distribution is called the forward problem, and it requires solution of the Maxwell's Eqs. (1) to a reasonable accuracy. All MEG source modeling approaches are based on the comparison of measured data with signals predicted by the model; thus an accurate forward model is a prerequisite for employing a source model. Understanding the properties of a forward model also provides insight into the generation of MEG signals at a more general level. It may be of interest to note that the mathematics underlying MEG analysis can directly (or better, in reverse order) be applied to understanding effects of transcranial magnetic stimulation on the brain.

In the quasistatic approximation, the electric potential V obeys a Poisson equation,

$$\nabla \cdot (\sigma \nabla V) = \nabla \cdot \mathbf{J}_p \tag{3}$$

while the magnetic field is induced by the total current density **J** and is obtained from the Biot–Savart law,

$$\mathbf{B}(\mathbf{r}) = \frac{\mu}{4\pi}\int_G \mathbf{J}(\mathbf{r}')\times\frac{\mathbf{R}}{R^3}dV, \tag{4}$$

where the integration is performed over a volume G containing all active sources, and $\mathbf{R} = \mathbf{r} - \mathbf{r}'$ is the vector connecting the location of the source at $\mathbf{r}'$ to point $\mathbf{r}$ where the magnetic field is computed.

It can be shown that in an infinite homogeneous volume conductor the volume currents give no contribution to the electric potential or the magnetic field, which are thus solely due to the primary currents, $\mathbf{J}_p$. In general, however, a prerequisite for the calculation of the magnetic field is the knowledge of the electric potential distribution, which gives rise to the volume currents according to

$$\mathbf{J}_v = \sigma\mathbf{E} = -\sigma\nabla V$$

If we assume that the volume at which the currents flow consists of homogeneous compartments, the electric potential and magnetic field can be computed from integral equations involving the values of V on the surfaces bounding the different compartments (Barnard *et al.,* 1967; Geselowitz 1970). This formulation gives rise to the boundary-element method for calculating the electric and magnetic fields, as discussed below.

B. Neural Current Sources

1. Currents Associated with Action Potentials and Postsynaptic Potentials

Electric signals propagate within the brain along nerve fibers (axons) as a series of action potentials (APs). The corresponding primary current can be approximated by a pair of opposite current dipoles, one at the depolarization and another at the repolarization front (see Fig. 2), and this "quadrupolar" source moves along the axon as the activation propagates. The separation of the two dipoles depends on the duration of the AP and on the conduction velocity of the fiber. For a cortical axon with a conduction speed of 5 m/s, the opposite dipoles would be about 5 mm apart. Direct magnetic recordings from isolated frog peripheral nerve (Wikswo *et al.,* 1980; Wikswo and van Egeraat, 1991) and from intact human peripheral nerves (Erné *et al.,* 1988; Hari *et al.,* 1989) support this simple model.

In synapses, the chemical transmitter molecules change the ion permeabilities of the postsynaptic membrane, and a postsynaptic potential (PSP) and current are generated. In contrast to the currents associated with an action potential, the postsynaptic current can be adequately described by a single current dipole oriented along the dendrite. The magnetic field of a current dipole falls off with distance more

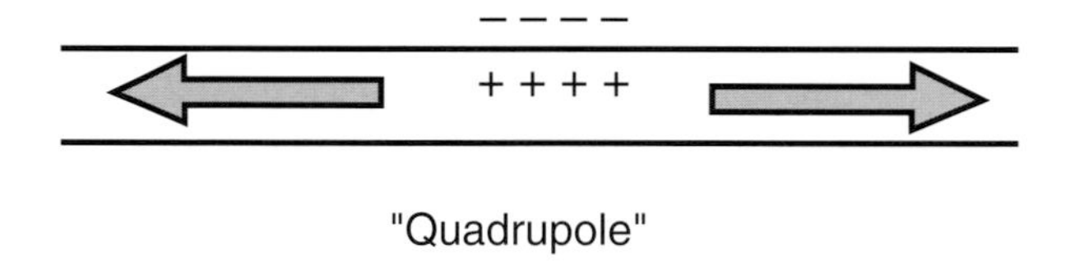

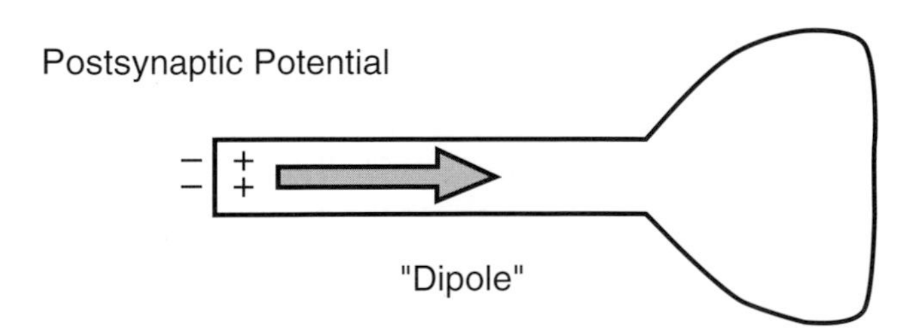

Figure 2 A schematic representation of currents associated with an action potential and a postsynaptic potential. Further details in text.

slowly (in proportion to $1/r^2$) than the field associated with the quadrupolar AP currents (in proportion to $1/r^3$).

Furthermore, temporal summation of currents flowing in neighboring fibers is more effective for synaptic currents, which last up to tens of milliseconds, than for the about 1-ms action potentials. Thus the electromagnetic signals observed outside and on the surface of the head seem to be largely due to the synaptic current flow. In special cases, currents related to action potentials might also significantly contribute to cortical MEG (and EEG) signals, such as high-frequency (about 600 Hz) somatosensory responses (Curio *et al.,* 1994; Hashimoto *et al.,* 1996).

The pyramidal cells are the principal type of neurons in the cortex, with their apical dendrites oriented parallel to each other and perpendicular to the cortical surface. Since neurons guide the current flow, the resultant direction of the electrical current flowing in the dendrites is also perpendicular to the cortical sheet of gray matter.

2. Determinants and Estimates of Dipole Strengths

The strength Q of a current dipole associated with postsynaptic currents depends both on the current I and on the length constant λ of the cellular membrane so that $Q = I\lambda$. When more than one neuron is simultaneously active, the total current I depends on the number of cells and synapses activated, on the synchrony of firing, and on the locations of the currents. Considerable cancellation of magnetic fields due to opposite currents and misalignment of fibers takes place in most brain structures.

The length constant λ of the exponential decay is determined by both the conductance of the membrane and the resistance of the intracellular fluid per unit length. It can also be affected by oscillatory background activity (Bernander *et al.,* 1991) and by back-propagating action potentials. Most likely, large cortical cells contribute to the dipole moment relatively more than small ones because λ increases directly proportional to the diameter of the fiber and because the large cells have more surface area and thereby can obtain more synaptic input.

The magnetic signal per tissue area may increase in some special cases. For example, the spreading depression is associated with strongly enhanced currents due to opening of large ionic channels (Bowyer *et al.,* 1999). In general, calcium- and voltage-sensitive neuronal potassium currents can largely shape the MEG signals (Wu and Okada, 2000), and glial cells (astrocytes), transporting potassium to cerebrospinal fluid, may contribute both as current generators and as amplifiers of synaptic effects. Epileptic discharges are typically associated with strong current densities due to highly synchronous activity. Moreover, the rather unnatural electrical stimulation of peripheral nerves triggers a much more synchronized afferent volley than natural tactile stimulation; accordingly, the early cortical somatosensory responses are stronger for electric than for tactile stimulation (Forss *et al.,* 1994).

The current I through the synapse can be calculated from the change of voltage during a PSP, and for a single PSP, $Q \approx 20$fA. Usually, the current-dipole moments required to explain the evoked magnetic field strengths outside the head are on the order of a few tens of nanoamperes. This would correspond to about a million synapses simultaneously active during a typical evoked response. Since there are approximately 10^5 pyramidal cells per square millimeter of cortex and thousands of synapses per neuron, the simultaneous activation of as few as one synapse in a thousand over an area of 1 mm^2 would suffice to produce a detectable signal.

However, this type of estimation may not have too much relevance in practice because of considerable cancellation of the generated electromagnetic fields in the misaligned neighboring neurons. According to some estimates, microscopic and translaminar cancellation shadows up to 93% of synaptic activity due to spatiotemporal misalignment and asynchronies (Halgren *et al.,* 2000).

Therefore, a better estimate for the activated area might be derived from MEG recordings of guinea pig hippocampal slices which indicate current densities of 4 mA/mm^2, assuming that $\lambda = 0.2$ mm (Okada *et al.,* 1997). Thus a 20-nA dipole moment would correspond to an about 25-mm^2 area of active cortex. The synchronous elements overshadow the effect of the less synchronous ones in the total signal. For example, it may seem counterintuitive that in a population consisting of 10^7 elements (and thereby corresponding to 1 cm^2 of cortex if 1 element is equal to 1 pyramidal cell) 1% synchronous elements could determine more than 96% of the total signal (Hari *et al.,* 1997). Thus MEG (as well as EEG) sees the most synchronous activity only.

Although such a synchronous activity may form only a part of the total activity, it can be functionally highly important. For example, invasive recordings from monkey have shown surprisingly large temporal overlap of neuronal firing in many visual cortical areas classically considered hierarchical processing stages (Schmolesky *et al.,* 1998). Apparently some time markers are required for retaining the temporal order of processing so that the hierarchy will not be violated, and the most synchronous firing could serve such a purpose.

3. Effect of Source Extent

Figure 4 shows that when the activated area is a layer, the dipole is not necessarily located in the center of gravity of the layer but displaced in a characteristic manner. When the layer is extended toward the radial direction, the single equivalent current dipole will be more superficial than the center of gravity of the layer, whereas a layer extended in the orthogonal direction will be mislocated to a larger depth.

C. Conductor Models

1. Spherically Symmetric Conductor

If we approximate the head with a layered spherically symmetric conductor ("sphere model"), the magnetic field of a dipole can be derived from a simple analytic expression (Sarvas, 1987). An important feature of the sphere model is that the result is independent of the conductivities and thicknesses of the layers; it is sufficient to know the center of symmetry. The calculation of the electric potential is more complicated and requires full information on conductivity (Zhang, 1995). Because radial currents do not produce any magnetic field outside a spherically symmetric conductor, MEG is even under realistic conditions to a great extent selectively sensitive to tangential sources. Consequently, EEG data are required for recovering all components of the current distribution. Figure 3 illustrates these essential properties of MEG.

2. More Realistic Volume Conductor Models

The obvious advantage of a simple forward model is that the field can be computed quickly from a simple analytical

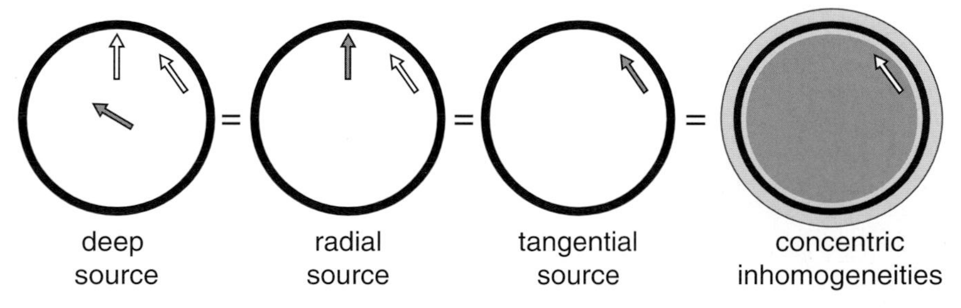

Figure 3 Schematic presentation of the effects of deep, radial, and tangential currents on MEG signals detected outside a spherically symmetric conductor. In all situations the external magnetic field is identical because radial currents anywhere in the sphere do not produce any external magnetic field, sources exactly in the middle of the sphere are always radial, and concentric inhomogeneities do not affect the magnetic field. EEG would see all these currents (tangential, radial, and deep ones) but would be affected by the electric inhomogeneities.

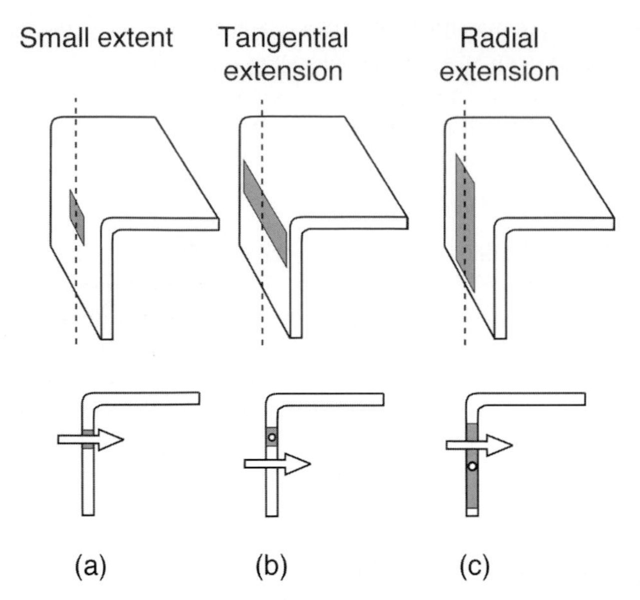

Figure 4 The effect of the size and extension of an active cortical area in the wall of a fissure on the location of the equivalent current dipole (arrow) used to model the layer. For explanation, see text. Adapted, with permission, from Hari (1991).

formula. The sphere model provides accurate enough estimates for many practical purposes but when the source areas are located deep within the brain or in frontal areas it is necessary to use more accurate approaches (Hämäläinen and Sarvas, 1989).

Within a realistic geometry of the head, the Maxwell's equations cannot be solved without numerical techniques. If a boundary-element method (BEM) is applied, electric potential and magnetic field can be calculated numerically starting from integral equations that can be discretized to linear matrix equations (Horacek, 1973; Hämäläinen and Sarvas, 1989).

In most BEM applications to the MEG forward problem, the surfaces are tessellated with triangular elements, assuming either constant or linear variation for the electric potential on each triangle. However, the accuracy of the magnetic-field computation may suffer if a current dipole is located near a triangulated surface. Then some other methods are required to improve the accuracy (Brebbia *et al.,* 1984).

Realistically shaped head geometries of each subject can be extracted from MR images; one example is shown in Fig. 5 The regions of interest, i.e., the brain, the skull, and the scalp, are segmented first. The volumes of the surfaces are then discretized for numerical calculations. The segmentation and tessellation problems are still tedious and nontrivial to solve (Dale *et al.,* 1999; Fischl *et al.,* 1999; Lötjönen *et al.,* 1999a,b).

The conductivity of the skull is low, only 1/80 to 1/100 of the brain's conductivity. Therefore most (about 95%) of the current associated with brain activity is limited to the intracranial space, and a highly accurate model for MEG is obtained by considering only one homogeneous compartment bounded by the skull's inner surface (Hämäläinen and

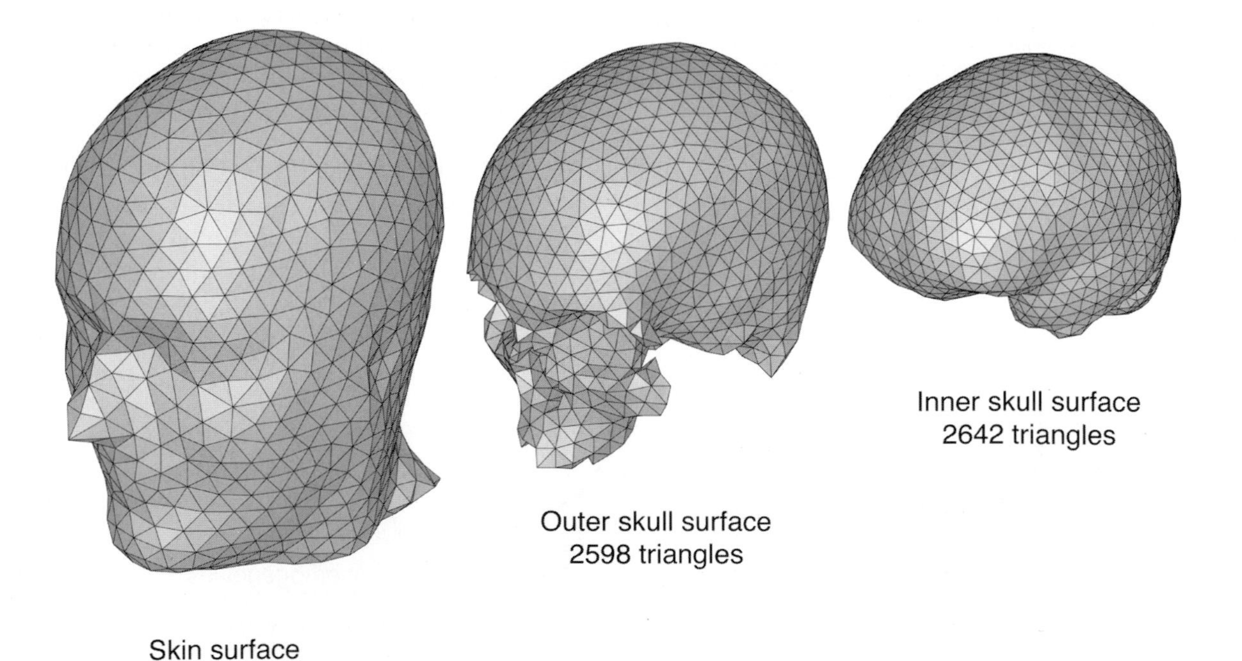

Figure 5 The boundaries between the various compartments of the head employed in a boundary-element model to calculate the potential on the surface of the head and the magnetic field outside. The surfaces have been extracted from MR images of one subject and tessellated automatically with triangles. Courtesy of Mika Seppä.

Sarvas, 1989; Okada *et al.,* 1999). With suitable image processing techniques it is possible to isolate this surface from high-contrast MRI data with virtually no user intervention. The boundary-element model for EEG is more complex because at least three compartments need to be considered: the scalp, the skull, and the brain.

It is also possible to employ the finite-element method (FEM) or the finite difference method (FDM) in the solution of the forward problem. The solution is then based directly on the discretization of the Poisson equation governing the electric potential. In this case any three-dimensional conductivity distribution and even anisotropic conductivity can be incorporated (Buchner *et al.,* 1997). However, the solution is more time consuming than with the boundary-element method. Therefore FEM or FDM have not yet been used in routine source modeling algorithms that require repeated calculation of the magnetic field from different source distributions.

III. Instrumentation and Data Acquisition

A. Instruments

1. Field Strengths, SQUIDs, and SQUID Electronics

Figure 6 shows that magnetic signals from the human brain are extremely weak compared with ambient magnetic field variations and also compared with magnetic signals from other parts of the body. Significant magnetic noise is caused, for example, by fluctuations in the earth's geomagnetic field; by moving vehicles and elevators; by radio, television, and microwave transmitters; and by power-line fields. Thus, rejection of outside disturbances is of utmost importance and is accomplished by avoiding disturbances near the measurement site, by special magnetic shielding, and by designing the sensors to be as insensitive as possible to artifacts. Moreover, various noise cancellation techniques can be applied.

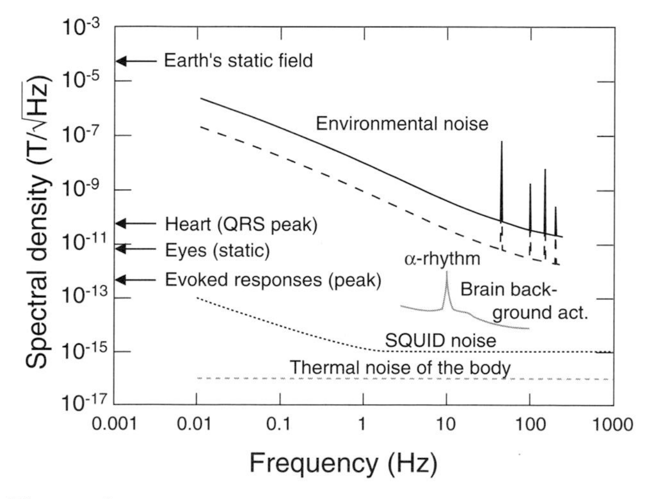

Figure 6 Strengths of various ambient and biological noise signals and of the brain's biomagnetic signals, given as spectral densities as a function of frequency. Courtesy of Jukka Knuutila.

The SQUID, the superconducting quantum interference device, is the only detector that offers sufficient sensitivity for the measurement of the tiny magnetic fields (Lounasmaa, 1974; Ryhänen *et al.,* 1989). The SQUID is a superconducting ring, interrupted by one or two Josephson (1962) junctions. These weak links limit the flow of the supercurrent, which is characterized by the maximum critical current that can be sustained without loss of superconductivity. Dc SQUIDs, with two junctions, are preferred because of lower noise level than in rf SQUIDs (Clarke *et al.,* 1976; Tesche *et al.,* 1985). The SQUIDs operate in liquid helium at a temperature of 4 K (–269°C).

If the current through the SQUID is biased to a suitable value, the voltage over the SQUID becomes a periodic function of the magnetic flux threading the ring. Instead of employing a current bias one can also use voltage biasing whereby the current instead of the voltage over the SQUID is monitored. The period of this characteristic variation is called the flux quantum, $1\Phi_0 = 2.07 \cdot 10^{-15}$ Wb. The SQUID thus acts as an extremely sensitive flux-to-voltage or flux-to-current converter. The voltage over the SQUID or the current passing through it is detected with a sensitive amplifier. The response of the electronics is linearized with a feedback circuit, which keeps the output of the SQUID at a constant value.

2. Flux Transformer Configurations

The sensitivity of the SQUID measuring system to external magnetic noise is greatly reduced by the proper design of the flux transformer, a device used for bringing the magnetic signal to the SQUID. Figure 7 shows some flux transformer configurations.

The simplest flux transformer configuration is the magnetometer, which senses the magnetic field with a single coil loop, thus measuring one component of the magnetic field in one location. A magnetometer is also sensitive to homogeneous fields, which are often caused by distant noise sources.

An axial first-order gradiometer consists of a pickup (lower) coil and a compensation coil, which are identical in area and connected in series but wound in opposition. This transformer is insensitive to spatially uniform changes in the background field but responds to inhomogeneous changes. Therefore, if the signal of interest arises near the lower coil, it will cause a much greater change of field in the pickup loop than in the more remote compensation coil, thus producing a net change in the output. In effect, the lower loop picks up the signal, while the upper coil compensates for variations in the background field.

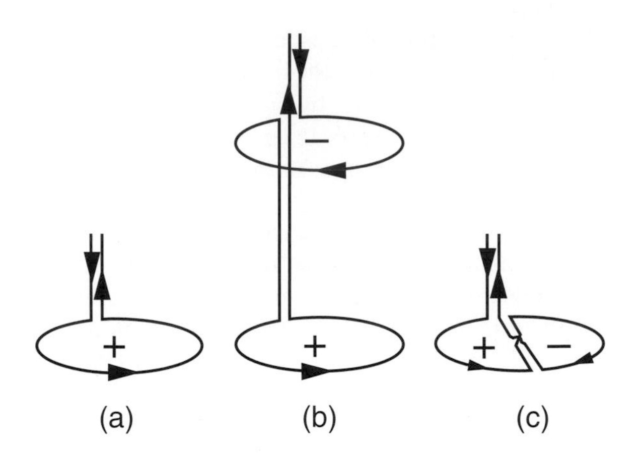

Figure 7 Different flux transformer configurations. (**a**) A magnetometer, (**b**) an axial first-order gradiometer, and (**c**) a first-order planar gradiometer. The plus and minus signs refer to magnetic fluxes of opposite polarities, and the arrows illustrate current directions in the wires.

Traditionally, most MEG measurements have been performed with axial gradiometers. However, the planar configuration measuring horizontal or off-diagonal gradients of the vertical field component (Fig. 7c) has advantages over axial coils: the double-D construction (Cohen, 1979) is compact in size and it can be fabricated easily with thin-film techniques. The locating accuracies of planar and axial gradiometer arrays are essentially the same for typical superficial sources (Carelli and Leoni, 1986; Knuutila *et al.*, 1993). The spatial sensitivity pattern, or "lead field," of off-diagonal gradiometers is narrower and shallower than that of axial gradiometers. These sensors thus collect their signals from a more restricted area near the sources of interest and there is less overlap between lead fields of adjacent sensors in a multichannel array.

Figure 8 shows schematically how the signal distributions picked up by a magnetometer and a planar gradiometer differ. The actual ratio of the maximum signal measured by a planar gradiometer above the source and the peak amplitude of the magnetometer field extrema depends on the depth of the source and the planar gradiometer baseline.

3. State-of-the-Art MEG Systems

The first biomagnetic measurements were performed with single-channel instruments (Cohen, 1968, 1972). However, reliable localization of current sources requires mapping at several locations, which is time consuming with only one channel. In addition, unique spatial features of, for example, ongoing background neuromagnetic signals cannot be studied. Fortunately, during the past 15 years multichannel SQUID systems for biomagnetic measurements have developed into reliable commercial products.

A state-of-the-art multichannel MEG system comprises more than 100 channels in a helmet-shaped array to record the magnetic field distribution all around the brain simultaneously. The dewar containing the sensors is attached to a gantry system, which allows easy positioning of the sensors above the subject's head. It is often also possible to move the bed or the chair horizontally.

Figure 9 shows an example of a modern MEG system in use in our laboratory. This Vectorview (4D Neuroimaging/Neuromag Ltd., Helsinki) device comprises a total of 306 SQUID sensors in 102 three-channel sensor elements, each containing two orthogonal planar gradiometers and one magnetometer. The sensors are arranged into a helmet-shaped array, which covers even the most peripheral areas of the brain, such as frontal and temporal lobes and the cerebellum.

In measurements with the modern whole-scalp neuromagnetometers the subject can be either seated or supine. The former orientation is a natural choice in most experiments on healthy subjects, while the laying position may be preferred for patients and, of course, for measurements during sleep. In addition to the MEG signals, EEG can be recorded simultaneously with nonmagnetic electrodes and leads.

4. Magnetically Shielded Rooms

For additional rejection of external disturbances, MEG measurements are usually performed in a magnetically shielded room. Four different methods can be employed to construct such an enclosure: ferromagnetic shielding, eddy-current shielding, active compensation, and the recently introduced high-T_c superconducting shielding. Combinations of these techniques have been utilized in many experimental rooms (Cohen, 1970; Erné *et al.*, 1981; Kelhä *et al.*, 1982). A typical commercially available room utilized in biomagnetic

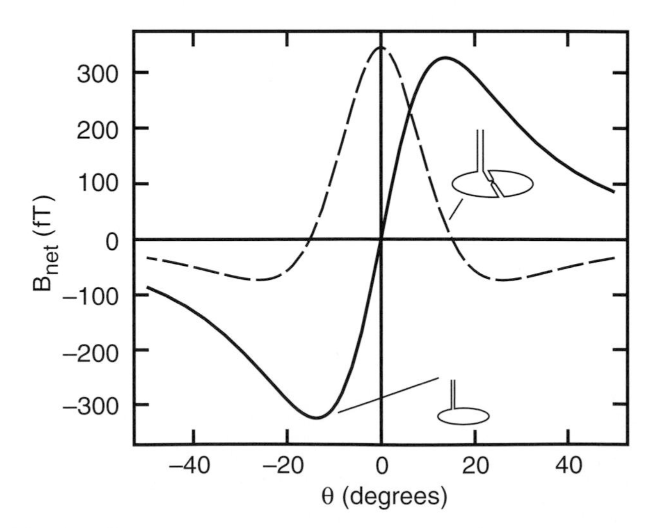

Figure 8 Schematic illustration of the signal strength produced by a current dipole (at angle 0), measured outside a sphere along one line with a magnetometer (solid line) and a planar gradiometer (dashed line). Note that with a magnetometer one detects two field extrema of opposite polarities, whereas the planar gradiometer picks up the maximum signal just above the dipole.

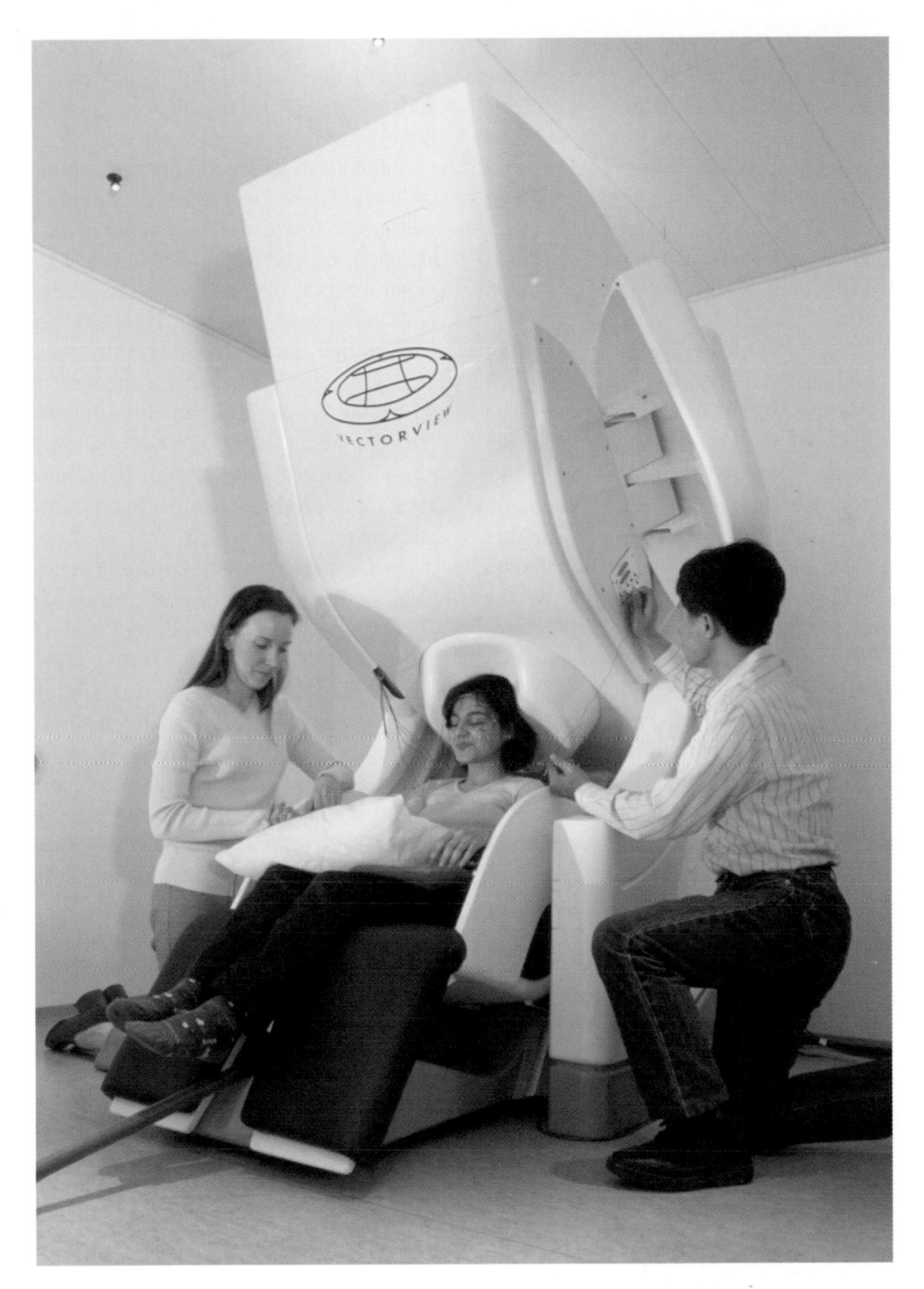

Figure 9 A subject is being prepared for a measurement with the 306-channel Vectorview neuromagnetometer. During the actual measurement the subject's head is covered by the helmet-shaped sensor array.

measurements employs two layers of aluminum and ferromagnetic shielding. The inside floor area is usually 3 by 4 m, and the height is around 2.5 m.

In addition to passive shielding, the external disturbances can also be cancelled using active electronic circuits which are either independent of the actual MEG system or integrated with it. The most traditional active compensation system consists of flux-gate magnetometers and pairs of orthogonal Helmholtz coils positioned around the shielded room. The output of the field sensors is employed to control the current fed to the coils to cancel the detected noise.

5. Noise Cancellation in MEG Systems

In addition to the external active shielding, the MEG system may be equipped with compensation sensors, which detect the background signals relatively far away from the head. A suitable proportion of the outputs of the compensation sensors is added to the outputs of the sensors detecting the brain signals. Commercial multichannel MEG instruments using compensation sensors have been pioneered by CTF Systems, Inc. (Port Coquitlam, BC, Canada). The obvious benefit of this method is that the compensation sensors are almost totally insensitive to brain signals and thus do not affect source identification. On the other hand, the system operation closely depends on the reliability of the noise compensation channels.

It is also possible to compensate for noise without employing separate compensation channels. This approach is utilized in the Vectorview system described above. The compensation, performed in the acquisition and analysis software, is based on the signal-space projection (SSP) method (Uusitalo and Ilmoniemi, 1997). In this method, the data are processed with a spatial filter, which makes suitable linear combinations of the channels. The filter is tuned to produce zero output if the signal pattern measured by the system is a linear combination of a few characteristic patterns accounting for the noise field found in the shielded room. In the terminology of SSP, these field patterns comprise the *noise subspace* to be removed from the data. This subspace is found by making a recording in an empty room (without subject) and by computing an eigenvalue decomposition of this background noise by principal-component analysis. It turns out that three to eight such component patterns, depending on the environment and the properties of the magnetically shielded room, are required to account for the noise fields.

The benefit of the SSP approach is that separate compensation channels are not needed and that the method does not depend on any single measurement channel being operational. However, since the compensation is based on measurements performed by the sensor array close to the brain, the signal patterns ensuing from the brain may also be modified and thus the filter must be applied to the model data calculated from candidate current sources during data analysis.

B. Data Acquisition and Signal Processing

1. Data Acquisition and Sampling

During data acquisition the analog signals are digitized. The commercial MEG systems are designed to provide amplitude sampling fine enough to fully exploit the dynamic range of the sensors.

The temporal sampling interval determines the highest signal frequency that can be collected undistorted. Sampling a signal with a too low frequency will result in aliasing, i.e., the signal may seem to contain components which are not really there. According to the Nyqvist criterion, the sampling frequency has to be at least twice the highest frequency present in the sampled signal. In practice, the data are typically generously oversampled to show smooth waveforms without additional signal reconstruction procedures and to surpass the nonidealities in the analog antialiasing filters.

In addition to the temporal sampling it is also necessary to provide dense enough spatial sampling of the field to be measured. Since the magnetic field is a vector quantity sampled in a three-dimensional space, a multidimensional generalization of the Nyqvist criterion has to be applied. It is of interest to note that if several independent measurements, like those made within the Vectorview system described above, are made at each sampling location, a wider grid spacing can be tolerated without aliasing (Ahonen *et al.,* 1993). The modern whole-head systems provide dense enough spatial sampling of the magnetic field so that the cerebral current sources can be detected without aliasing effects.

In a few cases the signal-to-noise ratio (SNR) is sufficient to allow reliable location of current sources without signal averaging: epileptic discharges and rhythmic background activity have been successfully analyzed from such raw data. When responses to sensory stimuli or voluntary movements are studied one usually resorts to signal averaging. This well-known procedure is based on the assumption that the response of interest is identical after each stimulus and that the noise is additive and uncorrelated across trials. Under these conditions, the SNR improves as the square root of N, where N is the number of averaged responses. It turns out that in many evoked responses the signal starts to emerge from noise after just a few averages, obviously indicating that without averaging the SNR is close to 1.

2. Signal Processing

In many cases, the signal-to-noise ratio for the target phenomena to be studied can be further improved by means of digital signal processing. The most common methods include digital filtering, frequency-domain analysis, coherence and correlation analysis, and spatial filtering. We discuss some essentials of these methods below. Technical treatments of these issues are readily available in textbooks (see, e.g., Jenkins and Watts, 1968; Oppenheim and Schafer, 1975).

It is customary to collect the data with a relatively wide bandwidth to allow selection of the desired final bandpass in postprocessing. In evoked-response studies, the averaged signals are often low-pass filtered digitally to suppress high-frequency noise. If most of the signal energy in spontaneous MEG activity is concentrated to a relatively narrow bandwidth, after bandpass filtering the signal-to-noise ratio may be high enough for source analysis of the unaveraged rhythmic activity.

A time-dependent signal can be presented equivalently in time and frequency domains. However, the Fourier spectra can often reveal some characteristics not easily recognizable in the time-domain analysis. A sophisticated frequency analysis may aim at identifying short time changes in narrow-band rhythmic activity, which requires the calculation of spectrograms and sometimes their averaging time locked to a stimulus. Such calculations can also be performed by employing wavelet transforms (see, e.g., Schiff *et al.,* 1994; Quiroga and Schürmann, 1999). More straightforward analyses have been performed by taking an absolute value (Salmelin and Hari, 1994) or square (Pfurtscheller and Aranibar, 1977; Pfurtscheller, 1992) of the bandpass-filtered spontaneous activity and then averaging the data time-locked to a stimulus, movement, or other event. As a result one can follow the level of the rhythmic activity on the desired frequency band as a function of time, and the activity contributing to the result doesn't need to be time-locked to the triggering event. This approach has been applied in both EEG (Pfurtscheller and Aranibar, 1977) and MEG (Salmelin and Hari, 1994; Salmelin *et al.,* 1995) studies on the temporospatial reactivity of the sensorimotor mu rhythm in relationship to voluntary movements(Salmelin and Hari, 1994; Salmelin *et al.,* 1995).

Cooperative behavior of several brain regions can be studied by calculating cross-correlations or coherence spectra across different signal channels. As described in Section III.A.2, the planar gradiometers employed in some neuromagnetometers are sensitive to currents in the vicinity of each sensor. Therefore, calculation of correlations and coherence between planar gradiometer channels gives useful information about the characteristics of the underlying regions in the brain. Correlation and coherence analysis can be also extended to virtual channels, specifically tuned to detect activity of desired brain areas (Gross *et al.,* 2001).

The rationale for using *spatial* filters instead of traditional filtering in time domain is based on the assumption that the signal of interest has a distribution different from the environmental noise, biological artifacts, or uninteresting brain activity. In Section III.A.5 we already discussed one particular spatial filter, the signal-space projection method. It is evident that the noise subspace to be removed from the data may contain field patterns of, for example, cardiac artifacts, whose contribution can thus be removed, or at least suppressed, by applying a projection operator.

3. Monitoring the Head Position during a Measurement

To obtain the position of the subject's head, an anatomical head-based coordinate system has to be related to a device coordinate system fixed to the MEG sensor array. To this end, three or more small coils can be attached on the surface of the head and their locations can be measured in relation to anatomical landmarks, such as nasion and preauricular points. During the MEG acquisition, excitation currents are fed to the coils and the produced magnetic field is measured (Knuutila *et al.,* 1985; Erné *et al.,* 1987).

The locations of the indicator coils can then be determined by modeling them as magnetic dipoles, i.e., small current loops, and by iteratively adjusting the coil locations and orientations until the best match between the measured signals and those predicted from the coils is achieved. Since the coil locations are thus known both in the device and in the anatomical coordinate systems, their relative locations and orientations can be determined. The head position is measured routinely in the beginning, and sometimes also at the end of each recording, to ensure stable head position.

This routine procedure requires that the subject keeps the head still during the measurements, which may be difficult even for cooperative subjects when the measurement is long or when the subject has a motor task. Individual bite bars can be used to avoid head movements (Singh *et al.,* 1997), but some subjects find them rather uncomfortable; biting may also cause muscle artifacts. Furthermore, bite bars cannot be applied in experiments requiring verbal responses.

Because of head movements, the locations of the inferred current sources are offset from their true values or the results of applying distributed source estimation methods may be blurred more than in the case of an acquisition with a cooperative subject.

Fast methods are being developed to measure head position continuously or intermittently during the whole recording. To speed up the measurement, the coils on the scalp can be activated simultaneously with different temporal patterns. Since the magnetic field is linearly related to the current fed to the coils, the magnetic fields produced by individual coils can be extracted from the measured signal components.

The continuous records of the head movements can then be utilized in different correction methods. If the signal-to-noise ratio is so good that unaveraged signals can be modeled, one can simply use a different coordinate transformation between the anatomical and the device coordinate frames at each time instant. The situation is more complicated if signals are averaged or if continuous data segments containing signals from supposedly identical source distributions are compared. Uutela *et al.* (2001) have recently explored several alternative computational approaches applicable to signal averaging and source modeling in situations with significant head movements during the data acquisition.

C. Artifacts in MEG Recordings

Magnetic artifacts due to fluctuations in the earth's magnetic field, moving vehicles, radio transmitters, or power lines are effectively attenuated in a magnetically shielded room. Noise can also be reduced by higher order gradiometers, but the accompanying decrease in sensitivity and difficulties in interpreting the data discourage this approach. Even within a magnetically shielded environment, disturbances can arise, for example, from stimulators and monitoring devices containing moving magnetic materials. Figure 10 shows some typical MEG artifacts.

1. Eye Blinks and Eye Movements

Eye movements and blinks are important biological sources of MEG artifacts. Blinks may be time-locked to the stimuli, especially if the stimuli are strong and alerting, and the signals can be on the order of 3–4 pT above the lateral aspects of the orbits (Antervo *et al.*, 1985). The corneoretinal potential is the "battery" of both the blink and the eye-movement signals but the generation mechanisms differ. The blink signals are caused by changes in the volume conductor geometry, whereas during eye movements the orientation of the source currents changes in reference to the volume conductor. Because of the large amplitude of these eye-related artifacts, electro-oculogram monitoring is recommended during all MEG recordings to be able to reject epochs coinciding with blinks and eye movements.

2. Cardiac Artifacts

Both magnetocardiogram(MCG)-related and ballistocardiogram-related artifacts may be associated with the cardiac cycle. Jousmäki and Hari (1996) have shown that the cardiac-related MEG artifacts are produced by the electric activity of the heart (MCG), whereas the blood-flow-related susceptibility artifacts are negligible in healthy subjects. The MCG contamination, typically evident as the R peak, is stronger in the left- than in the right-hemisphere MEG channels and depends on the position of the heart with respect to the head. In children, the shorter heart–brain distance may cause more MCG-related artifacts.

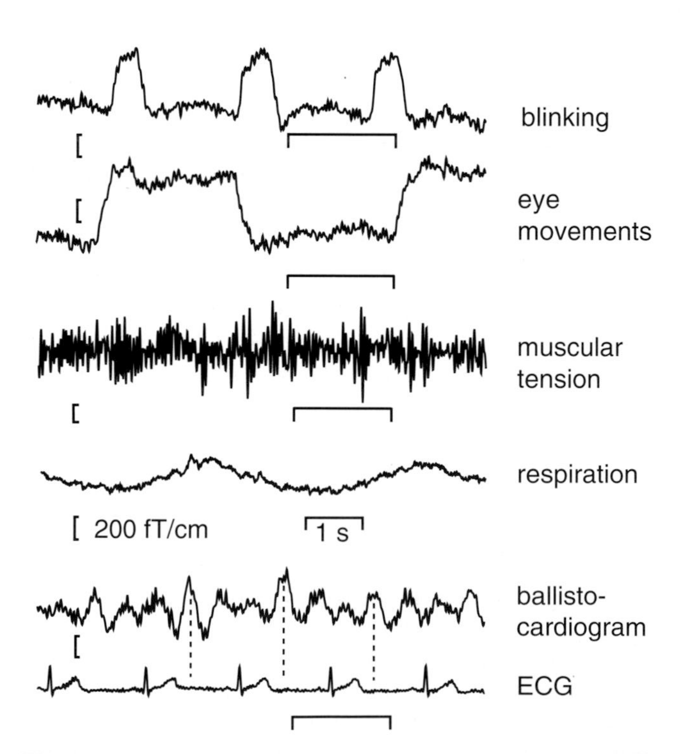

Figure 10 Examples of biological artifacts that may contaminate MEG recordings. The ballistogram time-locked to the cardiac cycle, as indicated by the ECG trace below, was elicited on purpose by putting a piece of magnetic metal on the subject's abdomen.

3. Other Artifacts

Artifacts may also arise, e.g., from respiration if the clothing or the body of the subject contains some magnetic material. Muscle contractions in the neck and face areas can also produce artifacts. However, muscular contamination seems to be weaker in MEG than in EEG measurements, probably because the distance to the sources of the muscle artifacts is significantly larger for MEG sensors than for EEG electrodes.

4. Elimination of Artifacts

Prevention is always the best way to deal with artifacts and it often works for external magnetic noise, stimulation-related artifacts, etc. It is also possible to reject all traces coinciding with some biological artifacts, such as eye blink- and eye movement-related signals. The procedure typically works well but can bias the results in some experiments because certain states of the subject are not included in the analysis.

It is also possible to analyze only signals which follow or precede a known artifact by a fixed time; for example, the analysis can be "gated" by the simultaneously recorded ECG and its large QRS complex (Murayama *et al.*, 2001).

Finally, if artifacts cannot be avoided, their influence on the signals should be minimized. The signal-space projection method may largely suppress some artifacts but the method should be applied with caution when the exact noise pattern and the signal patterns are not known in advance.

IV. Source Analysis

A. The Inverse Problem

The goal of the neuromagnetic inverse problem is to estimate the source current density underlying the MEG signals measured outside the head. Unfortunately, the primary current distribution cannot be recovered uniquely, even if the magnetic field (or the electric potential) were precisely known everywhere outside the head (Helmholtz, 1953). However, it is often possible to use additional physiological information to constrain the problem and to facilitate the solution. One approach is to replace the actual current sources by equivalent generators that are characterized by a few parameters. A unique solution for the parameters may then be obtained from the measured data by, e.g., a least-squares fit. Thus the solution of the forward problem is a prerequisite for most localization studies.

Figure 11 illustrates ambiguities encountered in the solution of the inverse problem. In this simulation, the data were calculated from a single current dipole in the right auditory cortex. When the current dipole model is applied, the correct solution is, naturally, obtained. However, when these same data are analyzed with the minimum-norm estimate (MNE), the result is a smooth, widespread current distribution, because that gives rise to exactly the same field distribution as the current dipole. Vice versa, when a widespread current distribution is analyzed, the result of dipole modeling is, erroneously, a very local source, whereas the MNE

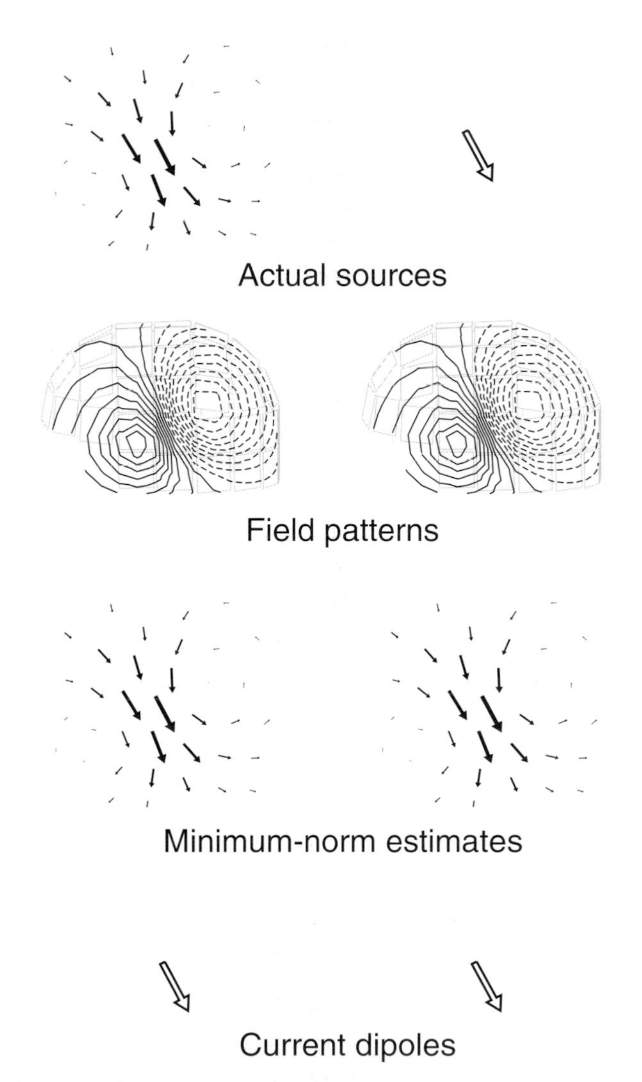

Figure 11 A comparison of minimum-norm estimates and a single current dipole used to represent a distributed source (top left) and a current dipole (top right). Note that the resulting field patterns are identical, as are the minimum-norm and current dipole solutions.

approach gives the correct answer. It is thus evident that the result of the analysis depends crucially on the underlying assumptions of the source modeling method.

B. Statistical Framework: The Bayesian Approach

Most source modeling approaches can be conveniently described within the Bayesian statistical framework ("the statistical inversion theory") (Tarantola, 1987). In this approach physical quantities are modeled as random variables with associated probability distributions. The solution of the inverse problem is then a probability distribution, conditioned on the observation. The estimates for the interesting (target) quantities are calculated from the measurement data on the basis of this conditional a posteriori distribution.

If the observable quantities contained in the measured data are arranged into a vector **y** and the unknowns into **x**, the Bayes theorem states that

$$f_{post}(\mathbf{x}) = Cf_{prior}(\mathbf{x})f(\mathbf{y}|\mathbf{x}), \tag{5}$$

where $f_{post}(\mathbf{x})$ is the a posteriori probability distribution of the unknown quantities, C is a constant, $f_{prior}(\mathbf{x})$ is the a priori distribution reflecting the a priori information we want to employ in the modeling, and $f(\mathbf{y}|\mathbf{x})$ is the distribution of the measured data, in the presence of noise, assuming that the target quantities have the values given in **x**. In MEG analysis, **y** consists of the measured magnetic field values as a function of time and **x** contains the unknown parameters of our current source model.

If the noise is assumed to be Gaussian, the Bayesian approach leads to the standard least-squares fitting method, which actually provides the set of parameters maximizing $f_{post}(\mathbf{x})$, also called the maximum a posteriori estimate (MAP). In addition to the MAP estimate it is customary to calculate confidence limits for the unknown parameters **x**, which describe the "width" of $f_{post}(\mathbf{x})$ around the maximum.

The benefit of resorting to the statistical inversion theory as the initial step of source analysis is that the basic problem of selecting an appropriate model is clearly separated from the technical problems of calculating desired characteristics of $f_{post}(\mathbf{x})$. Furthermore, a priori assumptions, such as anatomical or functional constraints obtained from MRI, *f*MRI, or PET, can be incorporated in a natural fashion. It should be emphasized, though, that computational cost for providing other properties of $f_{post}(\mathbf{x})$ than the MAP estimate can be prohibitive. These properties include, in particular, marginal distributions, which are often needed to visualize $f_{post}(\mathbf{x})$, which is a function of multiple variables. There have been attempts to employ the statistical inversion theory directly to calculate, e.g., a posteriori probabilities of several candidate current source models directly from $f_{post}(\mathbf{x})$ (Schmidt *et al.,* 1999).

C. Parametric Source Models

1. The Current Dipole Model

The simplest physiologically sound model for the neural current distribution consists of one or more point sources, current dipoles. In the simplest case the field distribution measured at one time instant is modeled by that produced by one current dipole (Tuomisto *et al.,* 1983). The best-fitting current dipole, commonly called the equivalent current dipole, can be found reliably by using standard nonlinear least-squares optimization methods (see, e.g., Marquardt, 1963). Some authors have successfully used "dipole density plots," the number of dipoles per unit area, to quantify the strength of activation of a certain brain region (Vieth *et al.,* 1996; Amidzic *et al.,* 2001).

If the same experiment is repeated several times, the locations of the dipoles as well as their amplitudes and orientations will be different because of instrumental noise and ongoing background activity. Therefore, it is useful to compute the dipole parameter confidence limits in addition to the best-fitting parameters (Hämäläinen *et al.,* 1993).

In the time-varying dipole model, first introduced to the analysis of EEG data (Scherg and von Cramon, 1985; Scherg, 1990), an epoch of data is modeled with a set of current dipoles whose orientations and locations are fixed but the amplitudes are allowed to vary with time. This approach corresponds to the idea of small patches of the cerebral cortex or other structures activated simultaneously or in a sequence. The precise details of the current distribution within each patch cannot be revealed by the measurements, performed at a distance in excess of 3 cm from the sources.

If we arrange the epoch of data into a matrix **B,** whose columns contain the measured signals at each time instant, a time-varying multidipole model, consisting of *p* dipoles, can be expressed as

$$\mathbf{B} = \mathbf{G}(\mathbf{r}_1,\ldots,\mathbf{r}_p,\hat{\mathbf{e}}_1,\ldots,\hat{\mathbf{e}}_p)\mathbf{Q} + \mathbf{N}, \quad (6)$$

where **G** is a gain matrix relating the dipole amplitudes to the measurements, **Q** is a matrix whose rows, $\mathbf{q}_1^T,\ldots,\mathbf{q}_p^T$, contain the temporal waveforms of the dipoles, and **N** is a matrix of noise with a Gaussian distribution, independent across different time points, with a known time-independent covariance matrix. The columns of **G**, $\mathbf{g}_1(\mathbf{r}_1,\hat{\mathbf{e}}_1),\ldots,\mathbf{g}_p(\mathbf{r}_p,\hat{\mathbf{e}}_p)$, are the field distributions generated by dipoles located at $\mathbf{r}_1,\ldots,\mathbf{r}_p$ with orientations by $\hat{\mathbf{e}}_1,\ldots,\hat{\mathbf{e}}_p$. The idea of the model is schematically depicted in Fig. 12.

It is evident that an important initial step in dipole modeling is the selection of the number of sources, *p*. This selection can be based on the singular-value decomposition of the data matrix **B.** Ideally, the singular values should first decrease gradually, followed by a plateau corresponding to the additive noise, **N.** In practice, however, such a separation may be far from trivial. One possibility to handle this problem is to build the source model gradually and to require that each new dipole introduced clearly explains some temporal or spatial aspects of the data. In the MUSIC approach described in the next section, the separation of the "noise subspace" corresponding to **N** is not very critical and the model order can be somewhat overestimated without causing a misinterpretation.

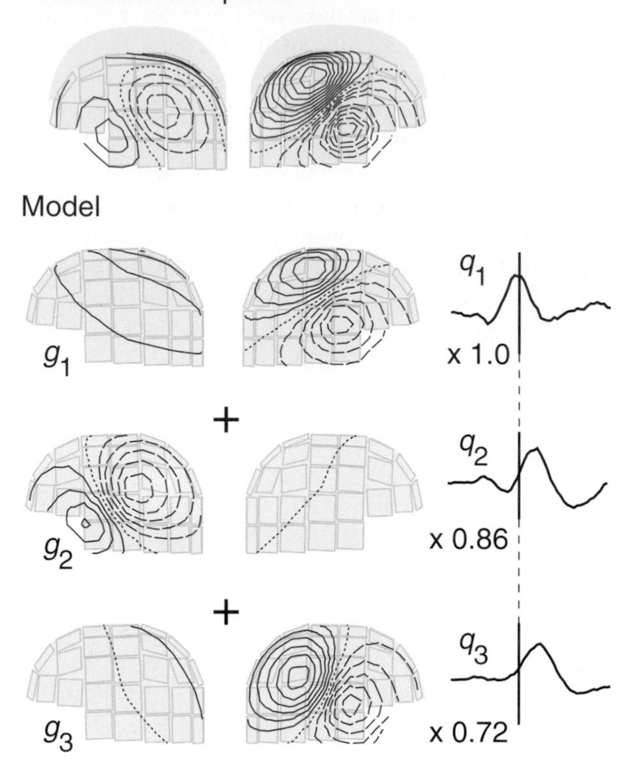

Figure 12 Modeling of MEG data with the time-varying dipole model. The measured field patterns at a given time instant are explained with a linear combination of three field patterns ($\mathbf{g}_1,\ldots,\mathbf{g}_3$) and additive Gaussian noise (not shown). The weights of the component field patterns are determined by the time dependencies of the dipole amplitudes ($\mathbf{q}_1,\ldots,\mathbf{q}_3$) shown by curves in the lower right.

As a result of the modeling one obtains the orientations and locations of the dipoles, as well as the temporal evolutions of the sources' strengths. Again, the optimal source parameters are found by matching the measured data collected over a period of time to those predicted by the model using the least-squares criterion.

From a mathematical point of view, finding the best-fitting parameters for the time-varying multidipole model is a challenging task. Since the measured fields depend nonlinearly on the dipole position parameters, standard least-squares minimization routines may not yield the globally optimal estimates for these parameters. Therefore, more complex optimization algorithms (Uutela *et al.,* 1998; Aine *et al.,* 2000; Huang *et al.,* 2000) and special fitting strategies (Berg and Scherg, 1996) have been suggested to take into account the physiological characteristics related to particular experiments. For each set of dipole position and orientation candidates it is, however, straightforward to calculate the optimal source strength waveforms using linear least-squares optimization methods (see, e.g., Mosher *et al.,* 1992).

Even with advanced numerical methods, the optimization problem of the time-varying dipole model may not be solved adequately. First, there might be competing solutions with different source configurations but indiscernible from each other in the presence of measurement noise. Second, the actual source distribution may not be well approximated by a dipole. Therefore, manually guided fitting strategies are often employed. It is then possible to include "fuzzy" prior information and to build the model gradually in a bottom-up fashion. One can identify the first source at a time when the field pattern reliably suggests a single source or several sources far apart. For example, the primary somatosensory cortex SI can be identified at an earlier time point than the second somatosensory cortex SII after a stimulus to a peripheral nerve (Hari *et al.,* 1993b). When necessary, the effects of known (previously identified) sources can be removed before identifying a new source. The complete model consists of several dipoles with time-varying strengths. An important final step in the model construction is to compare waveforms predicted by the model with the measured signals.

2. MUSIC Approaches

As discussed above, finding the locations of multiple current dipoles is a complex task. If nondipolar sources are present it may be difficult to find the correct solution even if the minimization algorithm would perform optimally for dipoles. Therefore, a less restricted correlation technique known as multiple signal characterization (MUSIC) (Schmidt, 1986) has been introduced for MEG and EEG analysis (Mosher *et al.,* 1992).

The primary assumption in MUSIC approaches is that the time series of the dipoles in Eq. (6) are linearly independent. More recently the method has been extended to synchronous sources (Mosher and Leahy, 1998) and to nondipolar sources (Mosher *et al.,* 1999), corresponding to the higher order terms of the current multipole expansion (Katila, 1983).

The benefit of the MUSIC approach over the traditional least-squares search is that the optimization problem is replaced by searching peaks in a scalar correlation measure calculated over a three-dimensional grid covering the viable dipole locations. In the original "classical" MUSIC approach several local maxima had to be identified at once. The more sophisticated RAP-MUSIC (recursively applied and projected MUSIC) algorithm (Mosher and Leahy, 1999) employs the signal-space projection approach to remove the contributions of the dipoles already identified so that it is sufficient to find the global maximum of the correlation cost function at each iteration of the algorithm.

The RAP-MUSIC algorithm is likely to perform better than a traditional multidipole search in the presence of distributed, nondipolar sources. RAP-MUSIC is also computationally effective and allows the model to be constructed one source at a time.

3. Verification of Dipole Models

The rather complex multidipole models should in many cases be constructed from the bottom up, starting from sources/areas which are already known to be activated by the stimuli used in the study, and then explain the remaining parts of the signals as carefully as possible. It is also important to pay attention to the consistency of the solutions, both within and across subjects. Other types of verification can be obtained by studying patients who have lesions in the assumed source areas or by recording electric signals during surgery. For example, the SII sources first detected by MEG have been later confirmed by intraoperative recordings, which have displayed very similar waveforms from the assumed source areas (Allison *et al.,* 1989).

D. Current Distribution Models

An alternative approach in source modeling is to assume that the sources are distributed within a volume or surface, often called the source space, and then to use various estimation techniques to find out the most plausible source distribution. The source space may be a volume defined by the brain or restricted to the cerebral cortex, determined from MR images. These techniques may provide reasonable estimates of complex source configurations without having to resort to complicated dipole fitting strategies. However, even when the actual source is point-like, its image is typically blurred and can have an extent of a few centimeters in each linear dimension, depending on the method employed (see Fig. 11). Therefore, the size of an activated region in the source images need not relate to the actual dimensions of the source but rather reflects an intrinsic limitation of the imaging method. In fact, without an extremely high signal-to-noise ratio it is unrealistic to claim that it would be possible to determine the extent of a source giving rise to the MEG signals (Nolte and Curio, 2000).

1. Linear Minimum-Norm Solutions

The first current-distribution model applied in MEG analysis was the (unweighted) minimum-norm estimate (Hämäläinen and Ilmoniemi, 1984, 1994), one in a group of linear approaches which can be described in a common framework. Here linearity means that the amplitudes of the currents are obtained by multiplying the data with a (time-independent) matrix. This kind of estimate has been employed by several authors (see, e.g., Dale and Sereno, 1993; Fuchs *et al.,* 1999).

In the Bayesian framework, the measurements are modeled with a distribution of current dipoles whose amplitudes have a Gaussian distribution with a known covariance matrix $\mathbf{C}_q$. The measurements contain noise which is also assumed to be Gaussian with a known covariance matrix $\mathbf{C}_b$. It is common to use a discrete grid of current locations even though a continuous formulation is also possible (Ioannides *et al.,* 1990; Hämäläinen and Ilmoniemi, 1994). In the discrete approach, the MAP estimate for the current distribution is obtained from

$$\mathbf{q}_{\mathrm{MAP}} = \arg\min_{\mathbf{q}} \left\{ (\mathbf{b}-\mathbf{Gq})^T \mathbf{C}_b^{-1}(\mathbf{b}-\mathbf{Gq}) + \mathbf{q}^T \mathbf{C}_q^{-1}\mathbf{q} \right\}, \quad (7)$$

where $\mathbf{q}$ is a vector of the dipole strengths; $\mathbf{G}$ is, as before, the solution of the forward problem, i.e., the gain matrix relating the measured signals $\mathbf{b}$ to the dipole strengths; the superscript T denotes the matrix transpose; and arg min indicates the value at which the function inside the braces is minimized. The first term indicates the difference between the measured data and those predicted by the model, $\mathbf{Gq}$. The second term is the (weighted) size of the current distribution. These estimates are sometimes also called L2-norm solutions because of the quadratic cost function [second term in Eq. (7)] associated with the Gaussian probability distribution assumed for the source strengths. Computationally, this approach has the benefit that the solution is linear:

$$\mathbf{q}_{\mathrm{MAP}} = \mathbf{C}_q \mathbf{G}^{\mathrm{T}}(\mathbf{G}\mathbf{C}_q\mathbf{G}^{\mathrm{T}} + \mathbf{C}_b)^{-1}\mathbf{b}. \quad (8)$$

In the original minimum-norm approach $\mathbf{C}_q = s^2\mathbf{I}$, where $\mathbf{I}$ is an identity matrix and s^2 is the expected variance of the source strength. If $s^2 \to 0$, the estimate vanishes, whereas in the limit $s^2 \to \infty$ we have

$$\mathbf{q}_{\mathrm{MAP}} = \mathbf{G}^{\mathrm{T}}(\mathbf{G}\mathbf{G}^{\mathrm{T}})^{-1}\mathbf{b}. \quad (9)$$

Since the sources are typically close to each other in the source space, some columns of $\mathbf{G}$, i.e., the field distributions of different currents, may be almost similar and the matrix to be inverted above in Eq. (9) is ill-conditioned. Thus, small errors in $\mathbf{b}$ are magnified, and unrealistically large complicated current patterns may appear. According to Eq. (8), the assumption of a finite variance leads to adding a second term inside the matrix inversion, which serves to regularize the problem, i.e., to suppress spurious high-amplitude solutions. When s decreases, the contribution of the second term becomes more significant and larger errors between the measured and the predicted data are thus accepted to avoid explaining the noise. Selection of a reasonable value for the (unknown) variance s^2 thus becomes important.

Several methods can be used to cope with this regularization problem. One possibility is to estimate the signal-to-noise ratio from the data and to use it to define a proper value for the source variance (Dale *et al.,* 2000). Another approach is to consider the relative sizes of the two terms in Eq. (7). The L-curve method (Hansen, 1992) makes this consideration formal by plotting the two terms against each other and by selecting an inflection point at which the rapid decrease of the source power stops and the error in data explanation sets off.

If an equal source variance is assumed throughout the source volume, the MNE is biased toward superficial currents, because the total current necessary to explain the data is smaller if the current elements are closer to the sensors. There have been several attempts to avoid this undesirable tendency. First, the expected source variance can be made depth-dependent. The most popular choice is a lead-field weighting, i.e., $q_{kk} = \beta / \mathbf{g}_k^T \mathbf{g}_k$, where q_{kk} is the variance of the kth source and $\mathbf{g}_k$ is the kth column of $\mathbf{G}$, and β is constant. Deeper currents are thus expected to be stronger to compensate the signal fall-off as a function of distance. Another possibility is to restrict the source space to currents normal to the cortex by using high-resolution MRI information (Dale and Sereno, 1993). This method seems to improve the depth bias to some extent but is not entirely satisfactory either. Finally, it is possible to adjust the expected source variance as a function of source location on the basis of supplementary imaging, e.g., *f*MRI information.

In addition to the depth bias, the MNEs fail to truthfully reflect the actual extent of the underlying current sources. Simulations have shown that the point spread of the estimate, i.e., the image of a focal current source, is a function of the source location (Dale *et al.,* 2000). In addition, the spread depends on the assumed source variance. Thus, any inferences about the size of the estimated areas from MNEs have to be dealt with extreme caution. It is evident that the dipole estimates share a similar problem: the source is always focal even if the true distribution is diffuse (see Fig. 11). In dipole analysis, however, one is usually well aware of this deficiency and thus dipoles are always considered to approximate the activation of a finite-sized patch of cortex. In both cases novice users of MEG may be easily misled when the sources are represented in a visually suggestive and attractive fashion in context with the underlying anatomy obtained from MRI.

Recently, attempts have been made to defeat the problems associated with true vs estimated source extent by presenting statistical parametric maps (SPM) instead of MAP estimates (Dale *et al.,* 2000). These maps present spatial distributions of statistical test variables, which can be used to assign confidence levels in hypothesis testing. For example, by dividing the scalar current strength in the MAP estimate by its variance (Dale *et al.,* 2000) one obtains a test variable to find locations where the estimated current value is significantly different from zero. It turns out that the point-spread functions vary less as a function of source location in the SPMs than in the MAP estimates. However, the strength of a source with fixed extent can vary considerably depending on the synchrony and amount of synaptic activity. Thus it would be highly misleading to trust the source extent estimate of the SPM algorithm under such conditions: apparent changes in the extension of the source area could just reflect changes of the source strength because stronger signals differ statistically significantly from zero over a wider area.

It is also possible to extend the minimum-norm approach to source covariance matrices with nonzero off-diagonal elements. This choice makes currents in different, typically neighboring, nodes of the reconstruction grid behave in an orderly fashion. One particular approach is to define $\mathbf{C}_q$ as the inverse of a discrete spatial Laplacian operator. According to Eq. (7) this means that smooth currents with a smaller Laplacian are preferred. This method, called LORETA (low resolution electromagnetic tomography) (Pascual-Marqui *et al.*, 1995), seems to provide smaller mislocalization as measured by the peak of the current estimate than the linear estimates employing a diagonal $\mathbf{C}_q$ (Fuchs *et al.*, 1999). However, the point-spread functions are wider, i.e., the current estimates are even more blurred than those obtained without the Laplacian constraint.

As an example of a sophisticated application of the minimum-norm estimates, Fig. 13 shows data from a recent paper by Dale *et al.* (2000). These authors employed a cortically constrained source space and presented the current estimates in an anatomical display in which the cortical mantle was inflated to open up the sulci.

2. Minimum-Current Estimates

It is also possible to enter into the source imaging method the assumption that the activated areas have a small spatial extent. For example, the MFT (magnetic field tomography) algorithm obtains the solution as a result of an iteration in which the probability weighting is based on the previous current estimate (Ioannides *et al.*, 1990). Another possibility is to use a probability weighting derived from the MUSIC algorithm, combined with cortical constraints (Dale and Sereno, 1993).

The L1-norm approach employs the sum of the absolute values of the current over the source space as the criterion to select the best current distribution among those compatible with the measurement (Matsuura and Okabe, 1997; Uutela *et al.*, 1999). The amplitudes of the currents are assumed to have an exponential instead of a Gaussian distribution. The MAP estimate is then obtained from

$$\mathbf{q}_{\mathrm{MAP}} = \arg\min_{\mathbf{q}} \left\{ (\mathbf{b}-\mathbf{Gq})^T \mathbf{C}_b^{-1} (\mathbf{b}-\mathbf{Gq}) + \sum_k |q_k| \right\}, \quad (10)$$

where $|q_k|$ is the absolute value of the current at the kth location of the source space. In contrast to the traditional L2-norm cost function, Eq. (7), the L1-norm criterion yields estimates focused to a few small areas within the source space. However, the result of the minimization cannot be any more expressed in a closed form. The L1-norm solutions are sometimes referred to as minimum-current estimates (MCEs) to distinguish them from the L2-norm MNEs.

Figure 14 shows an example of the MCEs in normal-hearing subjects who were viewing sign language (Levänen *et al.*, 2001). The MCEs are from individual left

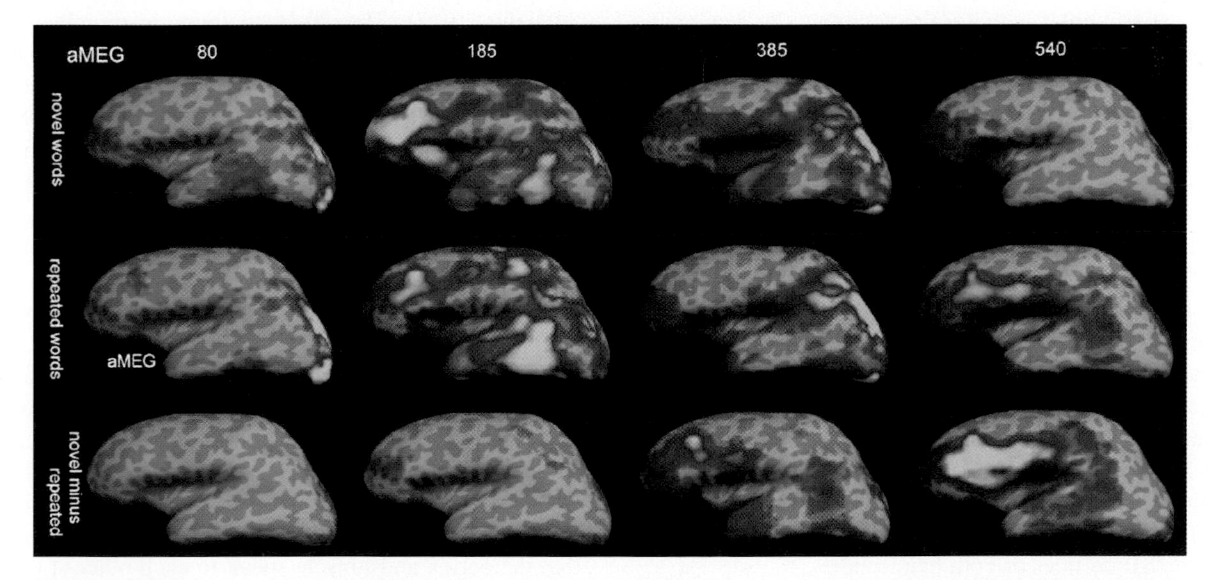

Figure 13 Snapshots of statistical parameter maps calculated from MEG during a verbal size-judgment task in a single subject at selected latencies. The maps are based on test statistics calculated from the anatomically constrained minimum-norm estimates by normalizing them with the expected noise in the current source estimate. Activation spreads rapidly from the visual cortex, around 80 ms, to occipitotemporal, anterior temporal, and prefrontal areas. Activations are displayed on an "inflated" view of the left hemisphere with sulcal and gyral cortices shown in dark and light gray, respectively. The significance threshold for the statistical parametric maps was $P < 0.001$. Adapted, with permission, from Dale *et al.* (2000).

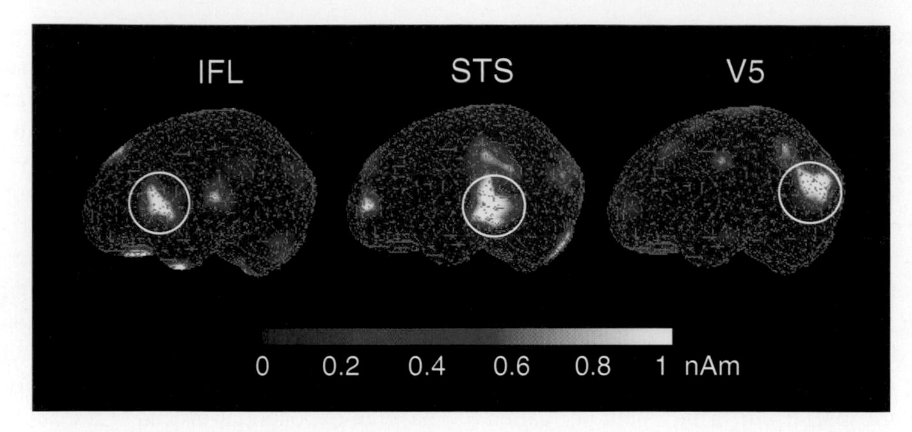

Figure 14 Minimum-current estimates of brain activations in normal-hearing subjects while they viewed sign language (which they did not understand). The activation spots refer to the inferior frontal lobe (IFL; Broca's area), the superior temporal sulcus (STS), and the motion-sensitive visual area V5. Adapted, with permission, from Levänen *et al.* (2001).

hemispheres at different times after the onset of the sign and illustrate activations in the inferior frontal lobe, the superior temporal sulcus (STS), and the visual motion-sensitive area V5.

V. Neuromagnetic Studies

This section briefly discusses examples of neuromagnetic studies, selected to portray how the data analysis methods described in the previous sections are employed in practice. The examples also demonstrate the importance of using a suitable experimental paradigm to facilitate the data analysis and interpretation.

A. Sequence of Evoked Responses

Our first example illustrates the analysis and interpretation of evoked responses to simple somatosensory stimuli; for more details about the characterization of somatosensory-evoked magnetic fields, see, e.g., Hari and Forss (1999). Figure 15 shows a typical evoked response distribution to stimulation of the right median nerve; 124 responses were averaged at all 204 gradiometers of a whole-scalp neuromagnetometer. The main responses peak around 21 ms over the contralateral primary somatosensory cortex SI and later, around 100 ms, over both SII areas at the lateral aspects of the hemispheres. Source analysis with a three-dipole model (see Fig. 16) showed activation first at the left SI hand area and later in the SII cortices of both hemispheres.

The easy distinction between signals from the various somatosensory cortices by MEG recordings allows monitoring of activity changes in association with different tasks and types of sensory inputs, and it has also turned out to be clinically useful in studies of different patient groups. For example, patients with Unverrich–Lundborg-type progressive myoclonus epilepsy show strongly enhanced responses at SI, facilitated callosal transfer to the ipsilateral SI, and absent SII responses (Forss *et al.,* 2001).

B. Cortex–Muscle Coherence

The spontaneous electrical activity of the human brain consists of several rhythmic components, many of which have a characteristic spatial distribution. For example, the primary sensorimotor cortices generate "mu rhythm," which consists of prominent frequencies around 10 and 20 Hz (Hari and Salmelin, 1997). The 20-Hz activity is prominent in the primary motor cortex and, consequently, changes in its level have been utilized as indicators of the functional state of the motor cortex (Schnitzler *et al.,* 1995; Hari *et al.,* 1998).

As an example of the analysis of ongoing spontaneous activity we present some results from the studies on the 20-Hz activity of the motor cortex. These signals are coherent with surface electromyogram during isometric contraction (Conway *et al.,* 1995; Salenius *et al.,* 1997a), reflecting rhythmic drive from the motor cortex to the spinal motor neuron pool. Figure 17 shows that both the EMG signal from an isometrically contracted foot muscle and the simultaneously recorded MEG from the rolandic region are rhythmic but their waveshapes differ. The coherence spectra, also shown in Fig. 17, were calculated between the MEG and the EMG signals. They were clearly above the noise level around 20 Hz, both for hand and foot muscles, suggesting that the cortex and muscle speak to each other at these frequencies (Salenius *et al.,* 1997b). The coherent cortical activity shows a gross somatotopical organization: the maxima of the coherent MEG signals occur laterally along the central sulcus during upper limb contractions and close to the brain midline during lower limb contractions. For

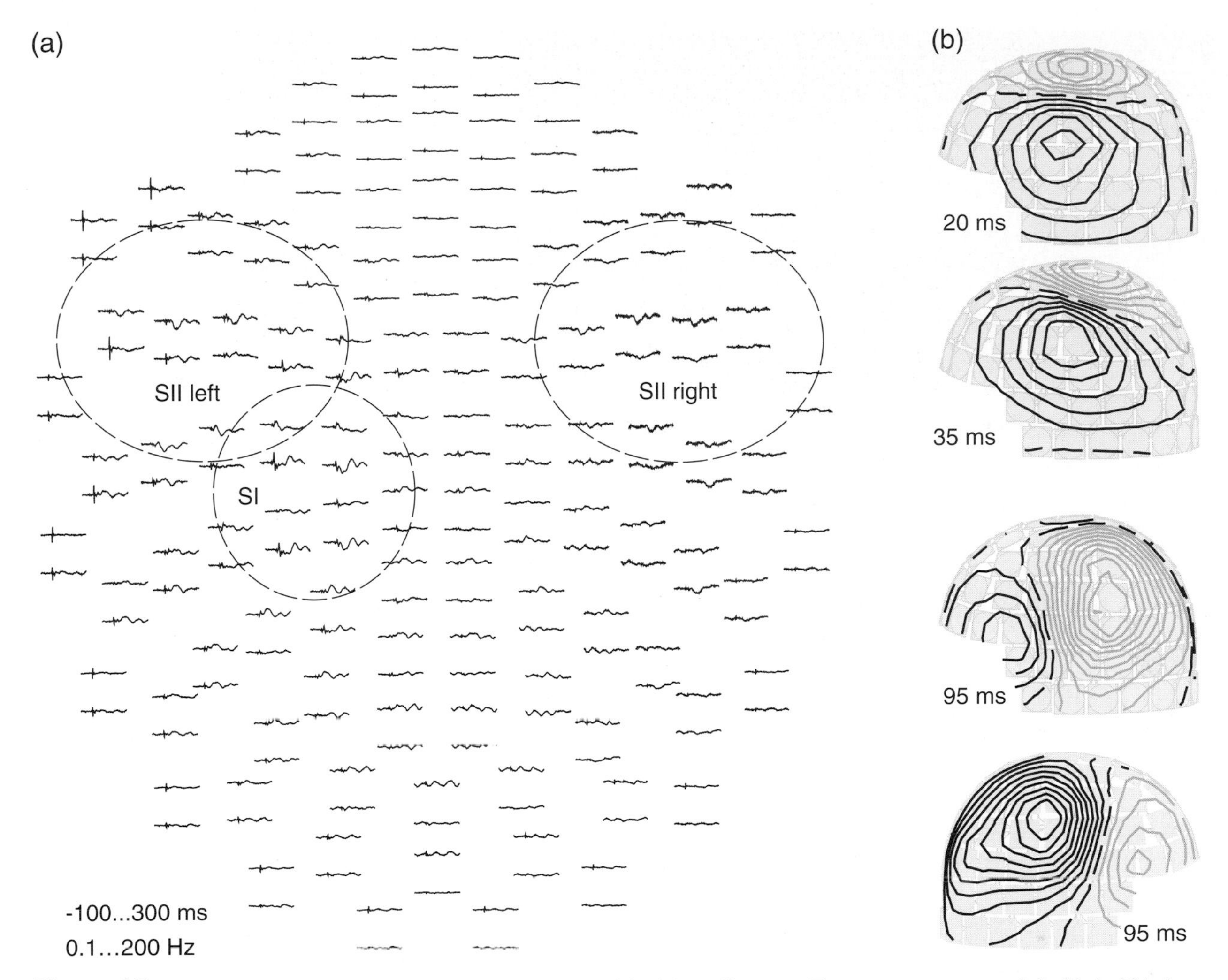

Figure 15 (**a**) Averaged evoked responses to electrical stimulation of the right median nerve. The responses were recorded with the 204 planar gradiometers of the VectorView neuromagnetometer. The upper and lower traces of each response pair depict latitudinal and longitudinal gradients, respectively. The dashed circles point out the areas of largest signals. (**b**) The corresponding field patterns at the peak latencies of the responses superimposed on the sensor array. The solid isocontours illustrate magnetic flux out of the head and the dashed isocontours flux into the head. These patterns agree with activations in the left primary somatosensory cortex SI and in the second somatosensory cortices of both hemispheres. Stimulus artifacts at time 0 are seen on several left-sided channels.

trunk muscles, the motor representation, revealed by means of cortex–muscle coherence, is in-between the foot and the hand muscle representations (Murayama *et al.,* 2001). The coherence between cortex and trunk muscles is technically rather difficult to measure because the signals are small and the surface EMG is easily contaminated by electrocardiographic signals; therefore analysis time-locked to the R peaks of the ECG had to be applied (Murayama *et al.,* 2001).

It is interesting that the MEG signal precedes the EMG with a time lag that systematically increases with the conduction distance from cortex to muscle. Similar results have been obtained from calculations based on cross-correlograms that show phase lags between the MEG and the EMG signals, from phase spectra that show linearities at some frequency ranges, and from MEG signals back-averaged from the EMG onsets (Salenius *et al.,* 1997b; Brown *et al.,* 1998; Gross *et al.,* 2000). The delays between MEG and EMG signals, computed from their phase differences at the best cortex–muscle synchrony, were in excellent agreement with conduction times from the motor cortex to the respective muscle observed in transcranial magnetic stimulation studies (Gross *et al.,* 2000).

The cortex–muscle coherence can also be also used as a tool for identifying the primary motor cortex, and we have routinely applied the MEG–EMG coherence in presurgical evaluation of patients with tumor or epilepsy for such purposes (Mäkelä *et al.,* 2001).

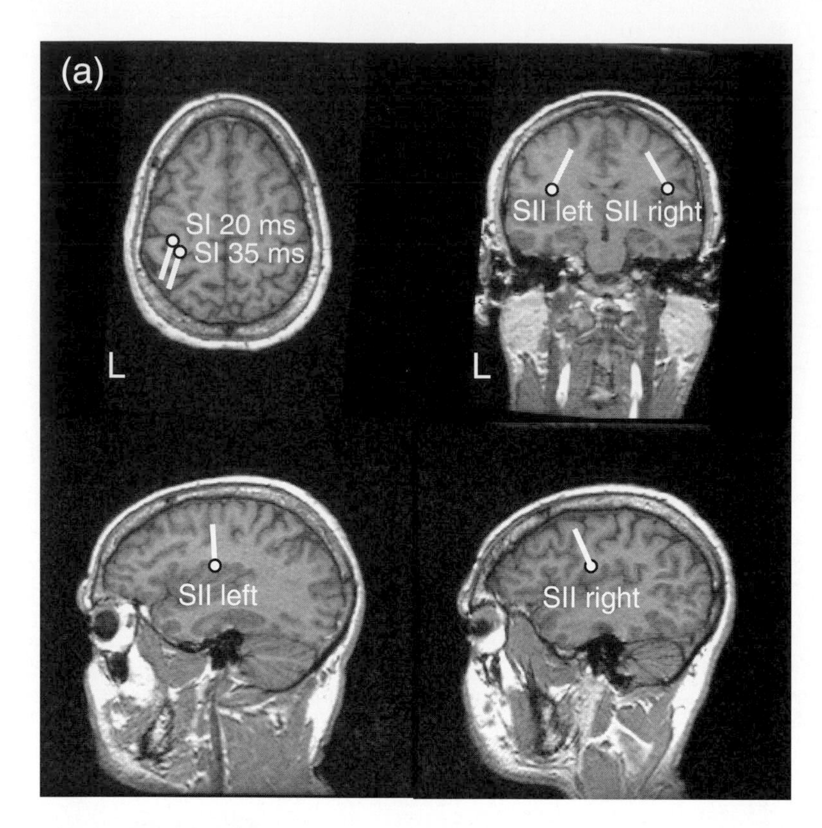

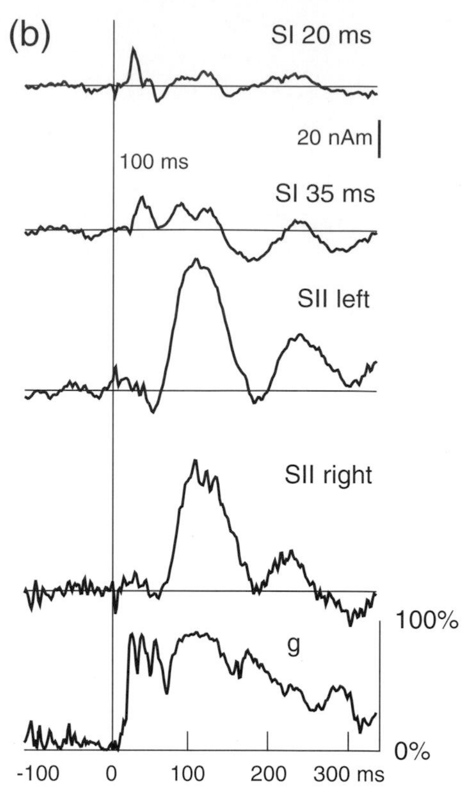

Figure 16 (**a**) The current dipole sources identified for the signals of Fig. 15, superimposed on the MRI slices of the same subject. The dots refer to the source locations (which agree with activations of the left SI and of the bilateral SII cortices) and the bars show the directions of the source currents; note that the intracellular current flow is restricted to the cortex. (**b**) Source strengths as a function of time, derived from a four-dipole model, with two sources in the left SI and one source in the SII cortex of each hemisphere. The lowest trace indicates the goodness of fit of the model in explaining all the 204-channel responses as a function of time.

C. Dynamics of the Human Mirror-Neuron System

Humans copy other persons' actions throughout their whole life, most of the time automatically and effortlessly. Practitioners of sports also know that viewing another person's movements facilitates one's own motor models. This automatic imitation behavior is most likely supported by the mirror-neuron system that was first identified and characterized in the monkey premotor cortical area F5 (Rizzolatti *et al.,* 1996a). This area contains "mirror neurons" that discharge both when a monkey executes hand actions and when he observes the same actions made by another monkey or by the experimenter. The mirror-neuron system matches action observation and execution, and it may play an important role both in action imitation and in understanding the meaning of actions made by other subjects, thereby also having relevance for social interactions.

Several recent brain imaging studies clearly demonstrate the existence of a mirror-neuron system also in the human brain (Fadiga *et al.,* 1995; Rizzolatti *et al.,* 1996b; Hari *et al.,* 1998; Iacoboni *et al.,* 1999; Nishitani and Hari, 2000a; Strafella and Paus, 2000).

Nishtani and Hari (2000a) aimed at identifying the temporal dynamics of the cortical activation sequence within the human mirror-neuron system. They employed a four-dipole model to explain the measured data. Figure 18 shows that when the subjects performed, observed, or imitated right-hand reaching movements which ended with a precision pinch of the top of a manipulandum, activations were found in the left occipital visual cortex, the left posterior inferior frontal area (Broca's area), and the primary motor cortices bilaterally. All these areas are thus involved in the human mirror-neuron system.

During execution, the left Broca's area was activated first (peak ~250 ms before the pinching). This was followed within 100–200 ms by activation in the left primary motor cortex and 150–250 ms later in the right motor cortex. The relative timing of the cortical activation sequence from Broca's area to the left and finally to the right motor cortex was similar also during imitation and observation. However, both Broca's area and motor cortex were activated about twice as strongly during online imitation than during self-paced execution and passive observation. Thus both timing and source strength data suggest that Broca's region, the human counterpart of the monkey

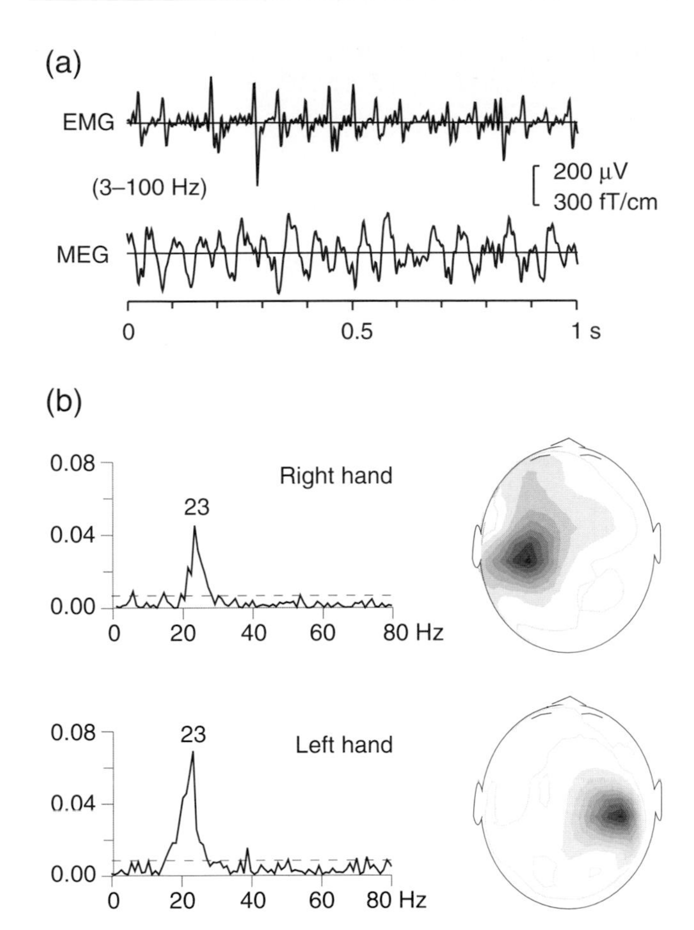

Figure 17 Cortex–muscle coherence. (**a**) Surface electromyogram from isometrically contracted foot muscle and the simultaneously measured MEG signal over the motor foot area. (**b**) Coherence spectra between MEG and rectified EMG from isometrically contracted right and left hand muscles (upper and lower spectra, respectively). The horizontal dashed line indicates the 99% significance level. The schematic heads on the right show the spatial locations of MEG signals corresponding to the strongest peaks in the coherence spectra. Adapted, with permission, from Salenius *et al.* (1997b).

mirror-neuron area F5, plays a key role in the human mirror-neuron system.

More recent studies using photographs of lip forms as stimuli (Nishitani and Hari, 2000b) have shown that activation of Broca's area is often preceded by signals generated in the region of the superior temporal sulcus. The human mirror-neuron system includes at least the STS region, Broca's area, the primary motor cortex, the superior parietal lobe, and the somatosensory cortices.

D. Binaural Hearing Studied in Frequency Domain

In normal hearing, sounds activate our brains through two ears, and the inputs from each ear reach the auditory cortices of both hemispheres (Fig. 19, top). The resulting binaural cortical responses are thus a mixture of inputs from both ears and it has not been possible to find out which part of responses to binaural sounds derives from either ear. Fujiki *et al.* (2001) solved this problem by labeling the auditory inputs from both ears with "frequency-tags": Continuous 1-kHz tones, presented either monaurally to left or right ear or binaurally, were amplitude modulated (left-ear tone at 26.1 Hz and the right-ear tone at 20.1 Hz). An analogous frequency tagging has been previously used to study visual binocular interactions (Brown *et al.,* 1999) and to label different melodies with amplitude modulations (Patel and Balaban, 2000).

Figure 19 shows analysis of the resulting MEG signals in the frequency domain. In the left hemisphere, responses to ipsilateral sounds were significantly suppressed during binaural presentation, whereas responses to contralateral tones were not significantly affected. The left hemisphere's preference for right-ear input was accentuated during binaural hearing, possibly providing the neuronal basis for the well-known "right-ear advantage" in right-handed subjects during dichotic listening. In the right hemisphere, the responses were significantly and similarly suppressed for both contralateral and ipsilateral sounds. This noninvasive analysis of contributions of the two ears to binaural cortical responses indicates that the inputs from the two ears compete strongly in the human auditory cortex but with clear hemispheric differences.

E. Preoperative Functional Localization

MEG has been successfully used in preoperative identification of the somatomotor strip in patients with brain tumors and epilepsy. Figure 20 shows one such example. The functional landmarks used to identify the central sulcus were based on somatosensory responses to hand, foot, and lip stimulation that allowed the identification of the somatosensory gyrus posterior to the central sulcus and on motor cortex–muscle coherence used to pinpoint the precentral motor cortex (Mäkelä *et al.,* 2001). The functional locations are displayed on 3D reconstructions of the individual brain, with the blood vessels, derived from MR angiography, also shown on the exposed brain surface. It has turned out that the vessels are extremely important landmarks for the neurosurgeon, who has to navigate in the operation area with a rather limited field of view. Intracranial recordings and direct cortical stimulations during surgery have given highly concordant results with the noninvasive preoperative MEG evaluation (Mäkelä *et al.,* 2001).

In preoperative evaluation of epileptic patients the main questions are to find out whether the patient has local generators of epileptic discharges, to find where these areas are located with respect to eloquent brain areas, and, in the case of multiple epileptic foci, to determine the temporal relationship between the foci. For example, it is possible to

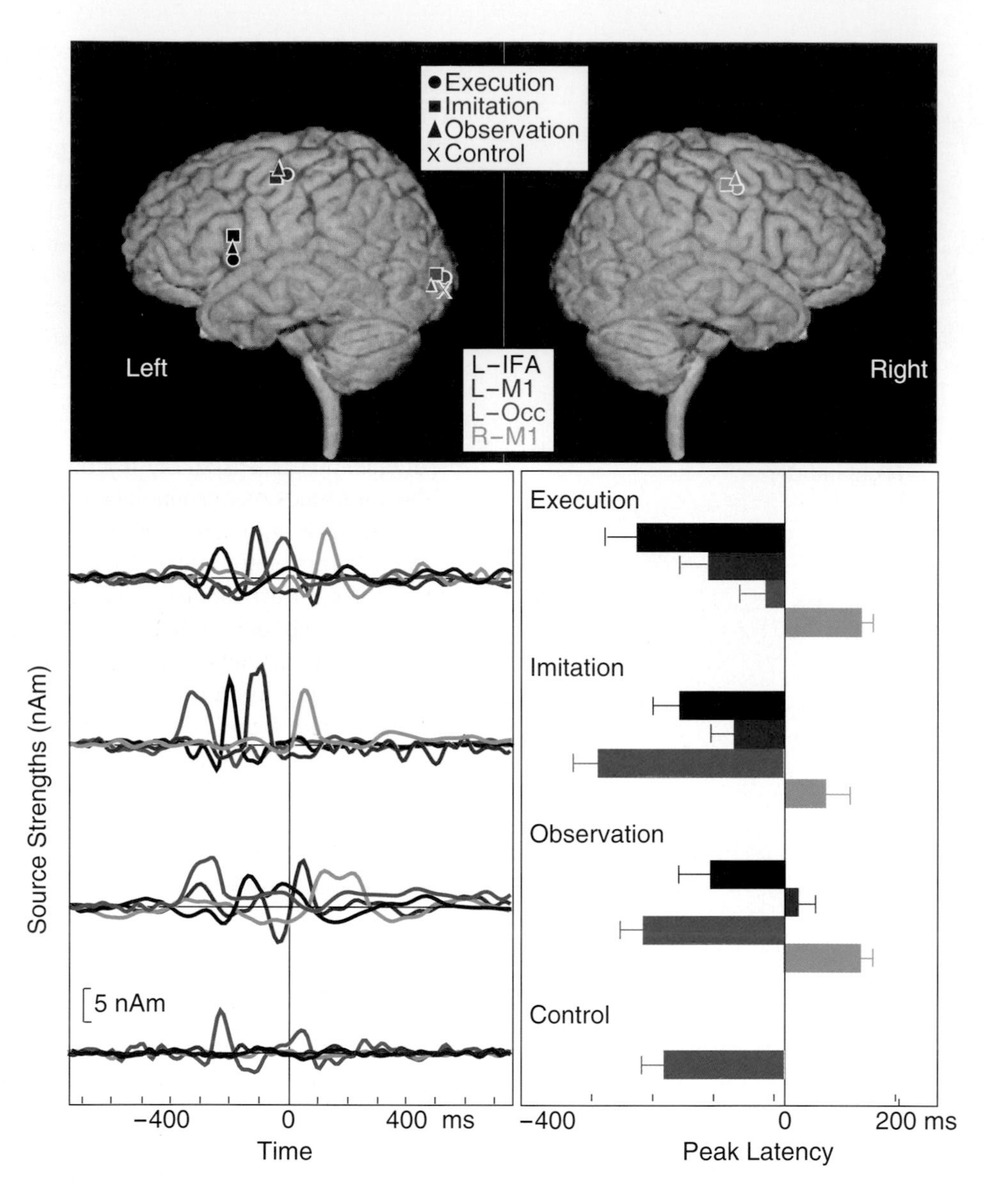

Figure 18 **Top:** The main source locations for one subject while he was executing, imitating (online), and observing reaching hand movements. Each movement ended with pinching the top of a manipulandum, which gave a triggering pulse to signal averaging. In the control condition, the hand approached the manipulandum without pinching. The sources are superimposed on the subject's own three-dimensional MRI brain surface, viewed from left and right sides. **Bottom, left:** Strengths of the main four dipoles in the inferior frontal (IFA) and the occipital (Occ) area of the left hemisphere, and in the motor cortices (M1) of both hemispheres as a function of time; a four-dipole time-varying model was used to explain the data during all conditions. **Bottom, right:** Mean (± SEM) peak latencies of source waveforms in all source areas. Adapted, with permission, from Nishitani and Hari (2000a).

identify a primary and a secondary epileptic focus on the basis of their fixed time delay, typically 20–30 ms (Barth, 1993; Hari *et al.*, 1993a).

VI. Conclusions and Future Directions

With the advent of whole-head neuromagnetometers it has become evident that MEG is a valuable tool for studying both healthy and diseased human brain. The method is totally noninvasive and the measurements can be repeated as desired without risk. In contrast to PET and *f*MRI, MEG and EEG reflect the neural activation directly instead of being indirect measures of blood flow or metabolism. Thus MEG is not hampered, for example, by hemodynamic delays, and it can track brain events at the submillisecond time scale. In contrast to the EEG, the tissues outside the brain do not significantly modify the distribution of the

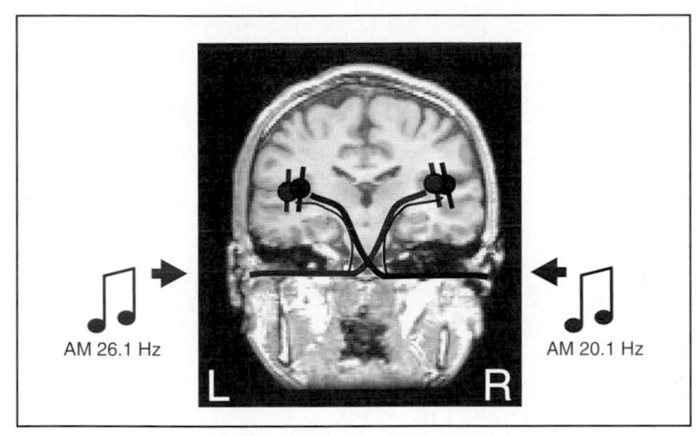

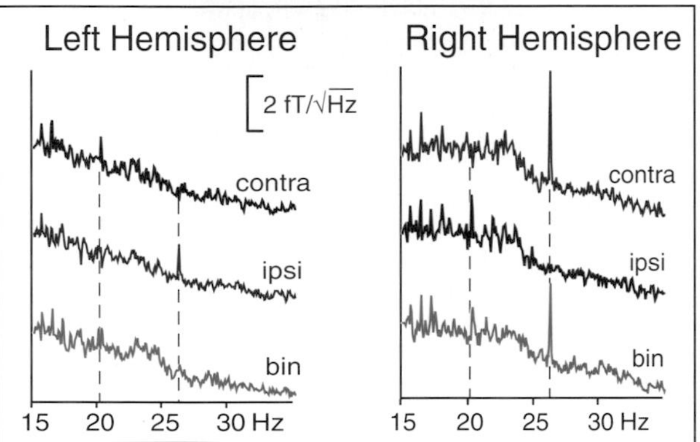

Figure 19 **Top:** Schematic presentation of auditory pathways. The red and blue lines illustrate the left- and right-ear inputs to the auditory cortices; the amplitudes of the corresponding sounds were modulated at 26.1 and 20.1 Hz (AM depth 80%), respectively. The sources of the measured MEG signals are shown as current dipoles in the auditory cortices of both hemispheres. **Bottom:** Frequency spectra (resolution 0.074 Hz) from signals measured over each auditory cortex of one subject. Left-ear sounds elicited spectral peaks at 26.1 Hz and right-ear sounds at 20.1 Hz in both hemispheres; both signals were suppressed during binaural listening. Adapted, with permission, from Fujiki *et al.* (2001).

MEG signals outside the head. Therefore, it is often easier to interpret MEG than EEG data. At best, a source having a small spatial extent can be located with an accuracy of a few millimeters. In addition to source locations and orientations, MEG also provides quantitative information about activation strengths.

The main contribution to MEG signals derives from tangential and relatively superficial currents in the fissural cortex; these areas are difficult to study with other means, including intracranial recordings. EEG is the natural companion of MEG because it provides information about radial currents as well. However, the problems in this combination arising from, e.g., larger systematic errors in EEG than MEG forward modeling are still unsolved.

The signals from deep structures are attenuated both due to the larger distance from the sensors to the sources and due to the effects of symmetry in the almost spherical head. Furthermore, signals from deeper structures are often masked by simultaneous activity of the cortex. Identification of deep sources reported in some MEG studies relies on accurate forward calculations and on the use of the information obtained with whole-head sensor arrays (Tesche and Karhu, 1999).

In contrast to the EEG electrodes, the MEG sensor array is not fixed to the subject's head. Therefore, a head position measurement is necessary to determine the relative location and orientation of the sensor array and the subject. Even if continuous position measurements of head position were available, it might be extremely difficult to study awake young children, and recordings cannot be performed during major epileptic seizures.

Present MEG instruments are designed for adult head size. However, it is conceivable that many epilepsy centers

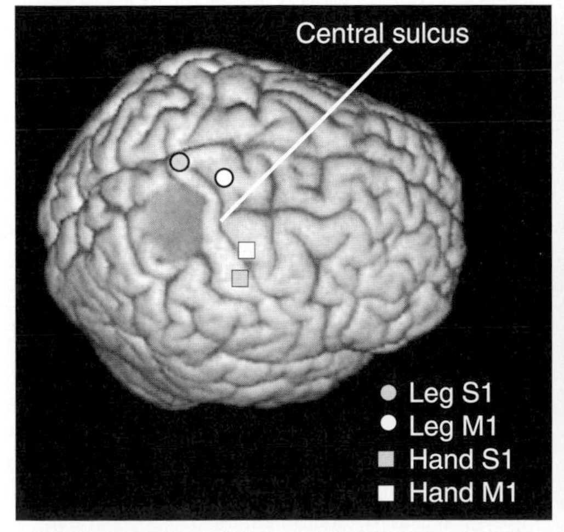

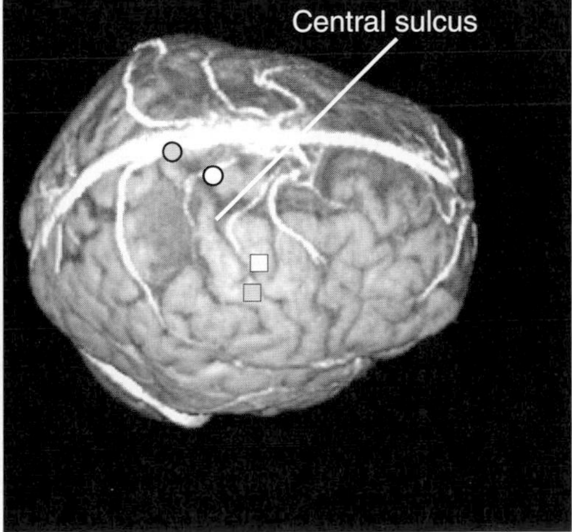

Figure 20 Identification of the central sulcus in a patient who has a brain tumor in the right parietal lobe. The postcentral somatosensory cortex was identified on the basis of source locations for somatosensory responses to electrical stimulation of hand and foot nerves, and the precentral motor cortex was identified on the basis of sources of the cortex–muscle coherence for upper and lower limb muscles. The surface rendering on the right also shows cortical veins which serve as landmarks for the neurosurgeon during operation.

would be willing to invest in a pediatric MEG system once such devices become available.

It is important to note that MEG signals are typically evident without resorting to complicated statistical analysis apart from signal averaging. Thus it is possible to evaluate the signal quality during the data acquisition. Also, conclusions can be made on the basis of single-subject data, which allows studies of individual processing strategies. Thus grand averages, which may often be misleading, are not necessary except as a means to visualize congruent results across subjects. Furthermore, subtractions between conditions are not needed, although possible—again an important difference compared with PET and *f*MRI studies.

The ambiguity of the inverse problem has been often cited as a major drawback of both EEG and MEG. Both methods thus have to rely on a restrictive source model and the analysis is rather difficult for a beginner. It is also perhaps confusing to find that several competing source models are available and sometimes the authors introducing them are not clear enough in stating the underlying assumptions and their consequences for data interpretation. Constraints for the inverse problem can be obtained from other imaging modalities, for example *f*MRI. However, the combination *f*MRI–MEG is not trivial, because the two methods do not necessarily reflect directly the same brain events.

We expect major progress in the development of efficient and automated MEG analysis methods, novel experimental paradigms to fully utilize the benefits of MEG, and reliable routines to combine MEG with other imaging modalities. We anticipate such approaches to significantly increase our understanding of human brain functions, especially their temporal dynamics.

Acknowledgments

This work was supported by the Academy of Finland and the MIND Institute.

References

Ahonen, A. I., Hämäläinen, M. S., Ilmoniemi, R. J., Kajola, M. J., Knuutila, J. E., Simola, J. T., and Vilkman, V. A. (1993). Sampling theory for neuromagnetic detector arrays. *IEEE Trans. Biomed. Eng.* **40,** 859–869.

Aine, C., Huang, M., Stephen, J., and Christner, R. (2000). Multistart algorithms for MEG empirical data analysis reliably characterize locations and time courses of multiple sources. *NeuroImage* **12,** 159–172.

Allison, T., McCarthy, G., Wood, C., Williamson, P., and Spencer, D. (1989). Human cortical potentials evoked by stimulation of the median nerve. II. Cytoarchitectonic areas generating long-latency activity. *J. Neurophysiol.* **62,** 711–722.

Amidzic, O., Riehle, H. J., Fehr, T., Wienbruch, C., and Elbert, T. (2001). Pattern of focal gamma-bursts in chess players. *Nature* **412,** 603.

Antervo, A., Hari, R., Katila, T., Ryhänen, T., and Seppänen, M. (1985). Magnetic fields produced by eye blinking. *Electroencephalogr. Clin. Neurophysiol.* **61,** 247–254.

Barnard, A. C., Duck, I. M., and Lynn, M. S. (1967). The application of electromagnetic theory to electrocardiology. I. Derivation of the integral equations. *Biophys. J.* **7,** 443–462.

Barth, D. S. (1993). The neurophysiological basis of epileptiform magnetic fields and localization of neocortical sources. *J. Clin. Neurophysiol.* **10,** 99–107.

Berg, P., and Scherg, M. (1996). Sequential brain source imaging: Evaluation of localization accuracy. *In* "Recent Advances in Event-Related Brain Potential Research" (C. Ogura, Y. Coga, and M. Shimokochi, eds.). Elsevier, Amsterdam.

Bernander, O., Douglas, R., Martin, K., and Koch, C. (1991). Synaptic background activity determined spatial–temporal integration in pyramidal cells. *Proc. Natl. Acad. Sci. USA* **88,** 11569–11573.

Bowyer, S. M., Tepley, N., Papuashvili, N., Kato, S., Barkley, G. L., Welch, K. M., and Okada, Y. C. (1999). Analysis of MEG signals of spreading cortical depression with propagation constrained to a rectangular cortical strip. II. Gyrencephalic swine model. *Brain Res.* **843,** 79–86.

Brebbia, C., Telles, J., and Wrobel, L. (1984). "Boundary Element Techniques—Theory and Applications in Engineering." Springer-Verlag, Berlin.

Brown, P., Salenius, S., Rothwell, J. C., and Hari, R. (1998). The cortical correlate of the Piper rhythm in man. *J. Neurophysiol.* **80,** 2911–2917.

Brown, R. J., Candy, T. R., and Norcia, A. M. (1999). Development of rivalry and dichoptic masking in human infants. *Invest. Ophthalmol. Visual Sci.* **40,** 3324–3333.

Buchner, H., Knoll, G., Fuchs, M., Rienacker, A., Beckmann, R., Wagner, M., Silny, J., and Pesch, J. (1997). Inverse localization of electric dipole current sources in finite element models of the human head. *Electroencephalogr. Clin. Neurophysiol.* **102,** 267–278.

Carelli, P., and Leoni, R. (1986). Localization of biological sources with arrays of superconducting gradiometers. *J. Appl. Phys.* **59,** 645–650.

Clarke, J., Goubau, W. M., and Ketchen, M. B. (1976). Tunnel junction dc SQUID fabrication, operation, and performance. *J. Low Temp. Phys.* **25,** 99–144.

Cohen, D. (1968). Magnetoencephalography: Evidence of magnetic field produced by alpha-rhythm currents. *Science* **161,** 784–786.

Cohen, D. (1970). Low-field room built at high-field magnet lab. *Phys. Today* **23,** 56–57.

Cohen, D. (1972). Magnetoencephalography: Detection of the brain's electrical activity with a superconducting magnetometer. *Science* **175,** 664–666.

Cohen, D. (1979). Magnetic measurement and display of current generators in the brain. Part I. The 2-D detector. *In* "12th International Conference on Medical and Biological Engineering," p. 15. Petah Tikva, Beilinson Medical Center, Jerusalem.

Conway, B., Halliday, D., Farmer, S., Shahani, U., Maas, P., Weir, A., and Rosenberg, J. (1995). Synchronization between motor cortex and spinal motoneuronal pool during the performance of a maintained motor task in man. *J. Physiol.* **489,** 917–924.

Curio, G., Mackert, B., Burghoff, M., Koetiz, R., Abraham-Fuchs, K., and Harer, W. (1994). Localization of evoked neuromagnetic 600 Hz activity in the cerebral somatosensory system. *Electroencephalogr. Clin. Neurophysiol.* **91,** 483–487.

Dale, A. M., Fischl, B., and Sereno, M. I. (1999). Cortical surface-based analysis. I. Segmentation and surface reconstruction. *NeuroImage* **9,** 179–194.

Dale, A. M., Liu, A. K., Fischl, B. R., Buckner, R. L., Belliveau, J. W., Lewine, J. D., and Halgren, E. (2000). Dynamic statistical parametric mapping: Combining fMRI and MEG for high-resolution imaging of cortical activity. *Neuron* **26,** 55–67.

Dale, A. M., and Sereno, M. I. (1993). Improved localization of cortical activity by combining EEG and MEG with MRI cortical surface reconstruction: A linear approach. *J. Cognit. Neurosci.* **5,** 162–176.

Erné, S., Curio, G., Trahms, L., Trontelj, Z., and Aust, P. (1988). Magnetic activity of a single peripheral nerve in man. *In* "Biomagnetism '87" (K.

Atsumi, M. Kotani, S. Ueno, T. Katila, and S. Williamson, eds.), pp. 166–169. Tokyo Denki Univ. Press, Tokyo.

Erné, S. N., Hahlbohm, H. D., Scheer, H., and Trontelj, Z. (1981). The Berlin magnetically shielded room (BMSR). Section B. Performances. *In* "Biomagnetism" (S. N. Erné, H.-D. Hahlbohm, and H. Lubbig, eds.), pp. 79–87. de Gruyter, Berlin.

Erné, S. N., Narici, L., Pizzella, V., and Romani, G. L. (1987). The positioning problem in biomagnetic measurements: A solution for arrays of superconducting sensors. *IEEE Trans. Magn.* **MAG-23,** 1319–1322.

Fadiga, L., Fogassi, L., Pavesi, G., and Rizzolatti, G. (1995). Motor facilitation during action observation: A magnetic stimulation study. *J. Neurophysiol.* **73,** 2608–2611.

Fischl, B., Sereno, M. I., and Dale, A. M. (1999). Cortical surface-based analysis. II. Inflation, flattening, and a surface-based coordinate system. *NeuroImage* **9,** 195–207.

Forss, N., Salmelin, R., and Hari, R. (1994). Comparison of somatosensory evoked fields to airpuff and electric stimuli. *Electroencephalogr. Clin. Neurophysiol.* **92,** 510–517.

Forss, N., Silén, T., and Karjalainen, T. (2001). Lack of activation of human secondary somatosensory cortex in Unverricht–Lundborg type of progressive myoclonus epilepsy. *Ann. Neurol.* **49,** 90–97.

Fuchs, M., Wagner, M., Köhler, T., and Wischmann, H.-A. (1999). Linear and nonlinear current density reconstructions. *J. Clin. Neurophysiol.* **16,** 267–295.

Fujiki, N., Jousmäki, V., and Hari, R. (2001). Tagging auditory inputs during binaural hearing. Submitted for publication.

Geselowitz, D. (1970). On the magnetic field generated outside an inhomogeneous volume conductor by internal current sources. *IEEE Trans. Magn* **6,** 346–347.

Gross, J., Kujala, J., Hämäläinen, M., Timmermann, L., Schnitzler, A., and Salmelin, R. (2001). Dynamic imaging of coherent sources: Studying neural interactions in the human brain. *Proc. Natl. Acad. Sci. USA* **98,** 694–699.

Gross, J., Tass, P. A., Salenius, S., Hari, R., Freund, H., and Schnitzler, A. (2000). Cortico-muscular synchronization during isometric muscle contraction in humans as revealed by magnetoencephalography. *J. Physiol.* **527,** 623–631.

Halgren, E., Liu, A., Ulbert, I., Klopp, J., Heit, G., and Dale, A. (2000). From synapse to sensor: How much of the EEG/MEG signal is waylaid by spatiotemporal asynchronies? *In* "Book of Abstracts, 12th International Conference on Biomagnetism," p. 36a.

Hämäläinen, M., Hari, R., Ilmoniemi, R., Knuutila, J., and Lounasmaa, O. V. (1993). Magnetoencephalography—Theory, instrumentation, and applications to noninvasive studies of the working human brain. *Rev. Mod. Phys.* **65,** 413–497.

Hämäläinen, M., and Ilmoniemi, R. (1984). Interpreting magnetic fields of the brain: Minimum norm estimates. Report TKK-F-A559, Helsinki Univ. of Technology, Espoo.

Hämäläinen, M., and Ilmoniemi, R. (1994). Interpreting magnetic fields of the brain: Minimum norm estimates. *Med. Biol. Eng. Comput.* **32,** 35–42.

Hämäläinen, M. S., and Sarvas, J. (1989). Realistic conductivity geometry model of the human head for interpretation of neuromagnetic data. *IEEE Trans. Biomed. Eng.* **36,** 165–171.

Hansen, P. C. (1992). Analysis of discrete ill-posed problems by means of the L-curve. *SIAM Rev.* **34,** 561–580.

Hari, R. (1991). On brain's magnetic responses to sensory stimuli. *J. Clin. Neurophysiol.* **8,** 157–169.

Hari, R. (1999). Magnetoencephalography as a tool of clinical neurophysiology. *In* "Electroencephalography: Basic Principles, Clinical Applications and Related Fields" (E. Niedermeyer and F. Lopes da Silva, eds.), 4th ed., Chap. 60, pp. 1107–1134. Williams & Wilkins, Baltimore.

Hari, R., *et al.* (1993a). Parietal epileptic mirror focus detected with a whole-head neuromagnetometer. *NeuroReport* **5,** 45–48.

Hari, R., and Forss, N. (1999). Magnetoencephalography in the study of human somatosensory cortical processing. *Proc. R. Soc. London B* **354,** 1145–1154.

Hari, R., Forss, N., Avikainen, S., Kirveskari, E., Salenius, S., and Rizzolatti, G. (1998). Activation of human primary motor cortex during action observation: A neuromagnetic study. *Proc. Natl. Acad. Sci. USA* **95,** 15061–15065.

Hari, R., Hällström, J., Tiihonen, J., and Joutsiniemi, S. (1989). Multichannel detection of magnetic compound action fields of median and ulnar nerves. *Electroencephalogr. Clin. Neurophysiol.* **72,** 277–280.

Hari, R., Karhu, J., Hämäläinen, M., Knuutila, J., Salonen, O., Sams, M., and Vilkman, V. (1993b). Functional organization of the human first and second somatosensory cortices: A neuromagnetic study. *Eur. J. Neurosci.* **5,** 724–734.

Hari, R., Levänen, S., and Raij, T. (2000). Timing of human cortical activation sequences during cognition: Role of MEG. *Trends Cognit. Sci.* **4,** 455–462.

Hari, R., and Salmelin, R. (1997). Human cortical rhythms: A neuromagnetic view through the skull. *Trends Neurosci.* **20,** 44–49.

Hari, R., Salmelin, R., Mäkelä, J. P., Salenius, S., and Helle, M. (1997). Magnetoencephalographic cortical rhythms. *Int. J. Psychophysiol.* **26,** 51–62.

Hashimoto, I., Mashiko, T., and Imada, T. (1996). Somatic evoked high-frequency magnetic oscillations reflect activity of inhibitory interneurons in the human somatosensory cortex. *Electroencephalogr. Clin. Neurophysiol.,* **100,** 189–203.

Helmholtz, H. (1953). Ueber einige Gesetze der Vertheilung elektrischer Ströme in körperlichen Leitern, mit Anwendung auf die thierisch-elektrischen Versuche. *Ann. Phys. Chem.* **89,** 211.

Horacek, B. M. (1973). Digital model for studies in magnetocardiography. *IEEE Trans. Magn.* **9,** 440–444.

Huang, M. X., Aine, C., Davis, L., Butman, J., Christner, R., Weisend, M., Stephen, J., Meyer, J., Silveri, J., Herman, M., and Lee, R. R. (2000). Sources on the anterior and posterior banks of the central sulcus identified from magnetic somatosensory evoked responses using multistart spatio-temporal localization. *Hum. Brain Mapp.* **11,** 59–76.

Iacoboni, M., Woods, R. P., Brass, M., Bekkering, H., Mazziotta, J. C., and Rizzolatti, G. (1999). Cortical mechanisms of human imitation. *Science* **286,** 2526–2528.

Ioannides, A. A., Bolton, J. P. R., and Clarke, C. J. S. (1990). Continuous probabilistic solutions to the biomagnetic inverse problem. *Inverse Probl.* **6,** 523–542.

Jenkins, G., and Watts, D. (1968). "Spectral Analysis and Its Applications." Holden–Day, San Francisco.

Josephson, B. D. (1962). Possible new effects in superconductive tunnelling. *Phys. Lett.* **1,** 251–253.

Jousmäki, V., and Hari, R. (1996). Cardiac artifacts in magnetoencephalogram. *J. Clin. Neurophysiol.* **13,** 172–176.

Katila, T. E. (1983). On the current multipole presentation of the primary current distributions. *Nuovo Cimento* **2D,** 660–664.

Kelhä, V. O., Pukki, J. M., Peltonen, R. S., Penttinen, A. J., Ilmoniemi, R. J., and Heino, J. J. (1982). Design, construction, and performance of a large volume magnetic shield. *IEEE Trans. Magn.* **18,** 260–270.

Knuutila, J., Ahonen, A. I., Hämäläinen, M. S., Ilmoniemi, R. J., and Kajola, M. J. (1985). Design considerations for multichannel SQUID magnetometers. *In* "SQUID'85: Superconducting Quantum Interference Devices and Their Applications" (H. D. Hahlbohm and H. Lubbig, eds.), pp. 939–944. de Gruyter, Berlin.

Knuutila, J., Ahonen, A., Hämäläinen, M., Kajola, M., Laine, P., Lounasmaa, O., Parkkonen, L., Simola, J., and Tesche, C. (1993). A 122-channel whole-cortex SQUID system for measuring the brain's magnetic fields. *IEEE Trans. Magn.* **29,** 3315–3320.

Levänen, S., Uutela, K., Salenius, S., and Hari, R. (2001). Cortical representation of observed signs: Comparison of deaf signers and hearing non-signers. *Cereb. Cortex* **11,** 506–512.

Lötjönen, J., Magnin, I. E., Nenonen, J., and Katila, T. (1999a). Reconstruction of 3-D geometry using 2-D profiles and a geometric prior model. *IEEE Trans. Med. Imaging* **18,** 992–1002.

Lötjönen, J., Reissman, P. J., Magnin, I. E., and Katila, T. (1999b). Model extraction from magnetic resonance volume data using the deformable pyramid. *Med. Image Anal.* **3,** 387–406.

Lounasmaa, O. V. (1974). "Experimental Principles and Methods below 1K." Academic Press, London.

Lounasmaa, O. V., Hämäläinen, M., Hari, R., and Salmelin, R. (1996). Information processing in the human brain—Magnetoencephalographic approach. *Proc. Natl. Acad. Sci. USA* **93,** 8809–8815.

Mäkelä, J., *et al.* (2001). Three-dimensional integration of brain anatomy and function to facilitate intraoperative navigation around the sensorimotor strip. *Hum. Brain Mapp.* **12,** 180–192.

Marquardt, D. W. (1963). An algorithm for least-squares estimation of nonlinear parameters. *J. Soc. Indust. Appl. Math.* **11,** 431–441.

Matsuura, K., and Okabe, Y. (1997). A robust reconstruction of sparse biomagnetic sources. *IEEE Trans. Biomed. Eng.* **44,** 720–726.

Mosher, J., Lewis, P., and Leahy, R. (1992). Multiple dipole modeling and localization from spatio-temporal MEG data. *IEEE Trans. Biomed. Eng.* **39,** 541–557.

Mosher, J. C., and Leahy, R. M. (1998). Recursive MUSIC: A framework for EEG and MEG source localization. *IEEE Trans. Biomed. Eng.* **45,** 1342–1354.

Mosher, J. C., and Leahy, R. M. (1999). Source localization using recursively applied and projected (RAP) MUSIC. *IEEE Trans. Signal Processing* **47,** 332–340.

Mosher, J. C., Leahy, R. M., Shattuck, D. W., and Baillet, S. (1999). MEG source imaging using multipolar expansions. *In* "IPMI99" (A. Kua, eds.), pp. 98–111. Springer-Verlag, Hungary.

Murayama, N., Lin, Y.-Y., Salenius, S., and Hari, R. (2001). Oscillatory interaction between human motor cortex and trunk muscles during isometric contraction. *NeuroImage,* **14,** 1425–1431.

Näätänen, R., Ilmoniemi, R., and Alho, K. (1994). Magnetoencephalography in studies of human cognitive brain function. *Trends Neurosci.* **17,** 389–395.

Nishitani, N., and Hari, R. (2000a). Temporal dynamics of cortical representation for action. *Proc. Natl. Acad. Sci. USA* **97,** 913–918.

Nishitani, N., and Hari, R. (2000b). Temporal dynamics of human cortical activities related to recognition of lip forms. *In* "Book of Abstracts, 12th International Conference on Biomagnetism," p. 108a.

Nolte, G., and Curio, G. (2000). Current multipole expansion to estimate lateral extent of neuronal activity: A theoretical analysis. *IEEE Trans. Biomed. Eng.* **47,** 1347–1355.

Okada, Y., Wu, J., and Kyuhou, S. (1997). Genesis of MEG signals in a mammalian CNS structure. *Electroencephalogr. Clin. Neurophysiol.* **103,** 474–485.

Okada, Y. C., Lähteenmäki, A., and Xu, C. (1999). Experimental analysis of distortion of magnetoencephalography signals by the skull. *Clin. Neurophysiol.* **110,** 230–238.

Oppenheim, A., and Schafer, R. (1975). "Digital Signal Processing." Prentice Hall International, Englewood Cliffs, NJ.

Pascual-Marqui, R. D., Michel, C. M., and D, L. (1995). Low resolution electromagnetic tomography: A new method for localizing electrical activity in the brain. *Int. J. Psychophysiol.* **18,** 49–65.

Patel, A. D., and Balaban, E. (2000). Temporal patterns of human cortical activity reflect tone sequence structure. *Nature* **404,** 80–84.

Pfurtscheller, G. (1992). Event-related synchronization (ERS): An electrophysiological correlate of cortical areas at rest. *Electroencephalogr. Clin. Neurophysiol.* **83,** 62–69.

Pfurtscheller, G., and Aranibar, A. (1977). Event-related desynchronization detected by power measurements of scalp EEG. *Electroencephalogr. Clin. Neurophysiol.* **42,** 138–146.

Plonsey, R. (1969). "Bioelectric Phenomena." McGraw–Hill, New York.

Quiroga, R. Q., and Schürmann, M. (1999). Functions and sources of event-related EEG alpha oscillations studied with the wavelet transform. *Clin. Neurophysiol.* **110,** 643–654.

Rizzolatti, G., Fadiga, L., Gallese, V., and Fogassi, L. (1996a). Premotor cortex and recognition of motor actions. *Cognit. Brain Res.* **3,** 131–141.

Rizzolatti, G., Fadiga, L., Matelli, M., Bettinardi, V., Paulesu, E., Perani, D., and Fazio, F. (1996b). Localization of grasp representations in humans by PET. Observation versus execution. *Exp. Brain Res.* **111,** 246–252.

Ryhänen, T., Seppä, H., Ilmoniemi, R., and Knuutila, J. (1989). SQUID magnetometers for low-frequency applications. *J. Low Temp. Phys.* **76,** 287–386.

Salenius, S., Forss, N., and Hari, R. (1997a). Rhythmicity of descending motor commands covaries with the amount of motor cortex 20–30 Hz rhythms. *Soc. Neurosci. Abstr.* **23,** 1948.

Salenius, S., Portin, K., Kajola, M., Salmelin, R., and Hari, R. (1997b). Cortical control of human motoneuron firing during isometric contraction. *J. Neurophysiol.* **77,** 3401–3405.

Salmelin, R., Hämäläinen, M., Kajola, M., and Hari, R. (1995). Functional segregation of movement-related rhythmic activity in the human brain. *NeuroImage* **2,** 237–243.

Salmelin, R., and Hari, R. (1994). Spatiotemporal characteristics of rhythmic neuromagnetic activity related to thumb movement. *Neuroscience* **60,** 537–550.

Sarvas, J. (1987). Basic mathematical and electromagnetic concepts of the biomagnetic inverse problem. *Phys. Med. Biol.* **32,** 11–22.

Scherg, M. (1990). Fundamentals of dipole source potential analysis. *In* "Auditory Evoked Magnetic Fields and Potentials" (F. Grandori, M. Hoke, and G. L. Romani, eds.), pp. 40–69. Karger, Basel.

Scherg, M., and von Cramon, D. (1985). Two bilateral sources of the late AEP as identified by a spatio-temporal dipole model. *Electroencephalogr. Clin. Neurophysiol.* **62,** 232–244.

Schiff, S. J., Aldroubi, A., Unser, M., and Sato, S. (1994). Fast wavelet transformation of EEG. *Electroencephalogr. Clin. Neurophysiol.* **91,** 442–455.

Schmidt, D. M., George, J. S., and Wood, C. C. (1999). Bayesian inference applied to the electromagnetic inverse problem. *Hum. Brain. Mapp.* **7,** 195–212.

Schmidt, R. (1986). Multiple emitter location and signal parameter estimation. *IEEE Trans. Ant. Propagat.* **34,** 276–280.

Schmolesky, M. T., Wang, Y., Hanes, D. P., Thompson, K. G., Leutgeb, S., Schall, J. D., and Leventhal, A. G. (1998). Signal timing across the macaque visual system. *J. Neurophysiol.* **79,** 3272–3278.

Schnitzler, A., Salenius, S., Salmelin, R., Jousmäki, V., and Hari, R. (1995). Involvement of primary somatomotor cortex in motor imagery: A neuromagnetic study. *Soc. Neurosci. Abstr.* **21,** 518.

Singh, K. D., Holliday, I. E., Furlong, P. L., and Harding, G. F. A. (1997). Evaluation of MRI-MEG/EEG co-registration strategies using Monte Carlo simulation. *Electroencephalogr. Clin. Neurophysiol.* **102,** 81–85.

Strafella, A. P., and Paus, T. (2000). Modulation of cortical excitability during action observation: A transcranial magnetic stimulation study. *NeuroReport* **11,** 2289–2292.

Tarantola, A. (1987). "Inverse Problem Theory." Elsevier, New York.

Tesche, C., *et al.* (1985). Practical dc SQUIDs with extremely low $1/f$ noise. *IEEE Trans. Magn.* **21,** 1032–1035.

Tesche, C. D., and Karhu, J. (1999). Interactive processing of sensory input and motor output in the human hippocampus. *J. Cognit. Neurosci.* **11,** 424–436.

Tuomisto, T., Hari, R., Katila, T., Poutanen, T., and Varpula, T. (1983). Studies of auditory evoked magnetic and electric responses: Modality specificity and modelling. *Nuovo Cimento* **2D,** 471–494.

Uusitalo, M., and Ilmoniemi, R. (1997). Signal-space projection method for separating MEG and EEG into components. *Med. Biol. Eng. Comp.* **35,** 135–140.

Uutela, K., Hämäläinen, M., and Salmelin, R. (1998). Global optimization in the localization of neuromagnetic sources. *IEEE Trans. Biomed. Eng.* **45,** 716–723.

Uutela, K., Hämäläinen, M., and Somersalo, E. (1999). Visualization of magnetoencephalographic data using minimum current estimates. *NeuroImage* **10,** 173–180.

Uutela, K., Taulu, S., and Hämäläinen, M. (2001). Detecting and correcting for head movements in neuromagnetic measurements. *NeuroImage* **14,** 1424–1431.

Vieth, J. B., Kober, H., and Grummich, P. (1996). Sources of spontaneous slow waves associated with brain lesions, localized by using the MEG. *Brain Topogr.* **8,** 215–221.

Wikswo, J. P., Barach, J. P., and Freeman, J. A. (1980). Magnetic field of a nerve impulse: First measurements. *Science* **208,** 53–55.

Wikswo, J. P., Jr., and van Egeraat, J. M. (1991). Cellular magnetic fields: Fundamental and applied measurements on nerve axons, peripheral nerve bundles, and skeletal muscle. *J. Clin. Neurophysiol.* **8,** 170–188.

Wu, J., and Okada, Y. C. (2000). Roles of calcium- and voltage-sensitive potassium currents in the generation of neuromagnetic signals and field potentials in a CA3 longitudinal slice of the guinea-pig. *Clin. Neurophysiol.* **111,** 150–160.

Zhang, Z. (1995). A fast method to compute surface potentials generated by dipoles within multilayer anisotropic spheres. *Phys. Med. Biol.* **40,** 335–349.

11

Transcranial Magnetic Stimulation

Alvaro Pascual-Leone* and Vincent Walsh†

*Laboratory for Magnetic Brain Stimulation, Beth Israel Deaconess Medical Center, Harvard Medical School, Boston, Massachusetts 02115;

†Experimental Psychology, University of Oxford, Oxford OX1 3UD, United Kingdom

I. Introduction

The term "brain mapping" is not restricted to techniques that can produce a visual or anatomical map of the brain. Several techniques provide other kinds of maps with coordinate systems based in temporal events rather than spatial measurements (ERPs and MEG for example) or behavior (experimental psychology). The task for integrated brain mapping is to find methods, or a language, for making the translations between these different kinds of maps meaningful. The methods available to the neuroscientist can be conceived as occupying different problem spaces according to whether the strength of the technique is based on temporal or spatial resolution (Fig. 1). In this chapter we outline some of the important features and uses of transcranial magnetic stimulation (TMS), a technique that occupies a unique problem space because of the combination of its spatial and temporal resolution and the fact that it is used to stimulate the brain rather than record electrical or metabolic activity.

TMS is a neurophysiologic technique that allows the induction of a current in the brain using a magnetic field to cross the scalp and the skull safely and painlessly. The first example of a physiological effect due to a time-varying magnetic field was reported by d'Arsonval in 1896, who produced phosphenes when a volunteer's head was placed inside a coil driven at 42 Hz. However, it was Anthony Barker and colleagues who in 1984 succeeded in developing a magnetic stimulator that delivered field pulses short enough to allow recording of evoked nerve and muscle action potentials. Since then, there has been a marked increase in the number of magnetic stimulators used clinically and in research worldwide (Pascual-Leone and Meador, 1998). TMS can be used to complement other neuroscience methods in the study of central motor pathways, the evaluation of corticocortical excitability, and the mapping of cortical brain functions. In addition, TMS provides a unique methodology to determine the true functional significance of the results of neuroimaging studies and the causal relationship between focal brain activity and behavior. The origins and development of TMS lie in medical

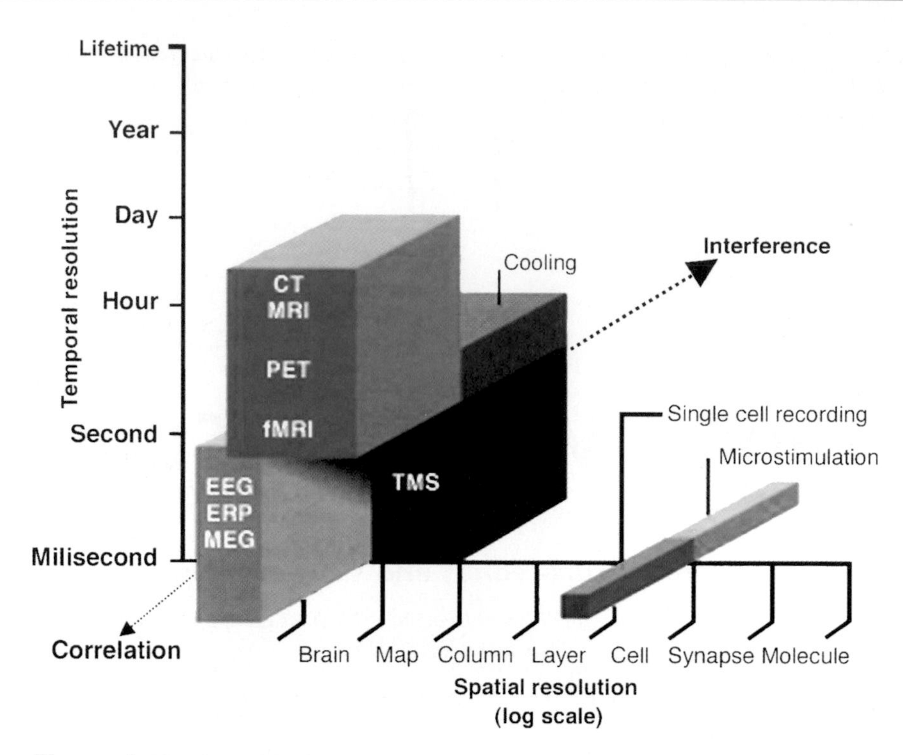

Figure 1 The place of TMS in neuropsychological studies is best thought of in terms of the "problem space" it occupies. The spatial and temporal resolution of TMS compared with other techniques is shown. However, it is not just space and time that make TMS indispensable; it is the ability of TMS transiently to interfere with functions whereas other techniques correlate brain activity with functions. The different volumes occupied by each of the techniques reflect that fact that when one selects a technique, one is also making a selection about the kind of question one can ask within the defined problem space. Reproduced from Walsh and Cowey (2000), with permission.

physics and clinical neurology and these fields have been reviewed in depth by Barker (1999), Mills (1999), George and Belmaker (2000), and Pascual-Leone *et al.* (2002). In the past 5 years in particular there has been rapid development of the use of TMS in cognitive neuroscience, to study psychological questions of vision, attention, memory, or language, and some of these have been studied in the context of plasticity and development. The cognitive neuroscience advances have been reviewed comprehensively elsewhere (Walsh and Cowey, 1998, 2000; Pascual-Leone *et al.*, 1999a, 2000; Walsh and Rushworth, 1999; Rushworth and Walsh, 1999). Our aim in this chapter is to give a brief introduction and overview to the range of different uses to which TMS can be put and in particular to highlight the areas where TMS interfaces with other techniques available (see also Paus' chapter).

II. Basic Principles of Magnetic Brain Stimulation

The basis of magnetic stimulation is electromagnetic induction, which was discovered by Faraday in 1831. A pulse of current flowing through a coil of wire generates a magnetic field. The rate of change of this magnetic field determines the induction of a secondary current in any nearby conductor. In TMS, a current passes through a coil of copper wire that is encased in plastic and held over the subject's head. As a brief pulse of current is passed through the stimulating coil, a magnetic field is generated that passes through the subject's scalp and skull without attenuation (only decaying by the square of the distance). This time-varying magnetic field induces a current in the subject's brain. Therefore, TMS might be best considered a form of "electrodeless, noninvasive electric stimulation."

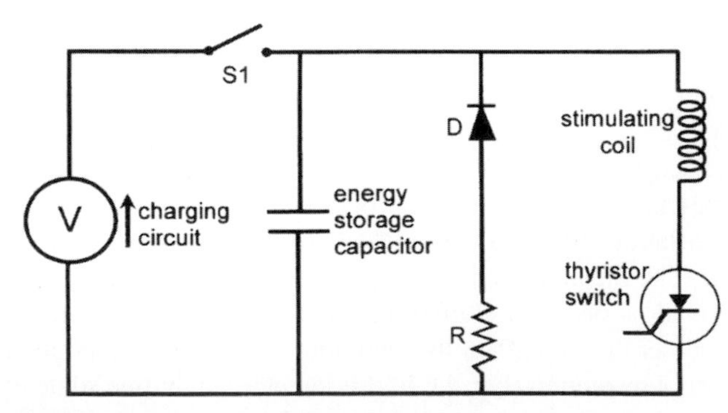

Figure 2 Schematic diagram of a standard (single-pulse) magnetic nerve stimulator. Reproduced from Barker (1999), with permission.

In 1985 Barker *et al.* successfully applied a magnetic pulse over the vertex of the human scalp and elicited hand movements and measured electromyographic (EMG) activity from the first dorsal interosseous (Barker *et al.,* 1985). The basic circuitry of the magnetic stimulator is shown in Fig. 2. A capacitor charged to a high voltage is discharged into the stimulating coil via an electrical switch called a thyristor. This circuitry can be modified to produce rapid, repetitive pulses that are used in repetitive TMS (rTMS). Figure 3 shows the whole sequence of events in TMS from

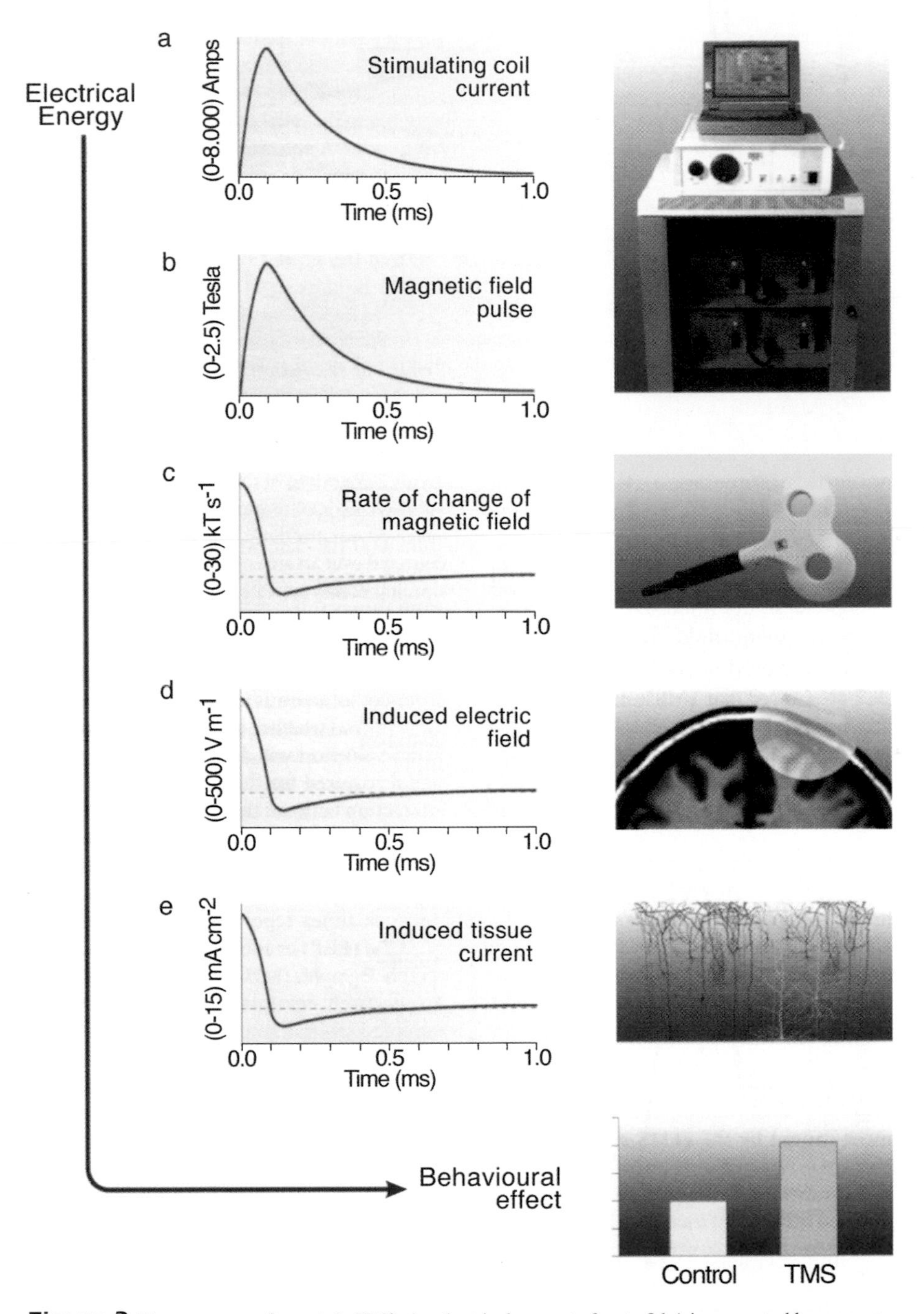

Figure 3 The sequence of events in TMS. An electrical current of up to 8 kA is generated by a capacitor and discharged into a circular, or figure-8-shaped, coil which in turn produces a magnetic pulse of up to 2 T. The pulse has a rise time of approximately 200 μs and a duration of 1 ms and, due to its intensity and brevity, changes at a rapid rate. The changing magnetic field generates an electric field resulting in neural activity or changes in resting potentials. The net change in charge density in the cortex is zero. The pulse shown here is a monophasic pulse but in studies that require rTMS the waveform will be a train of sine-wave pulses, which allow repeated stimulation. Reproduced from Walsh and Cowey (2000), with permission.

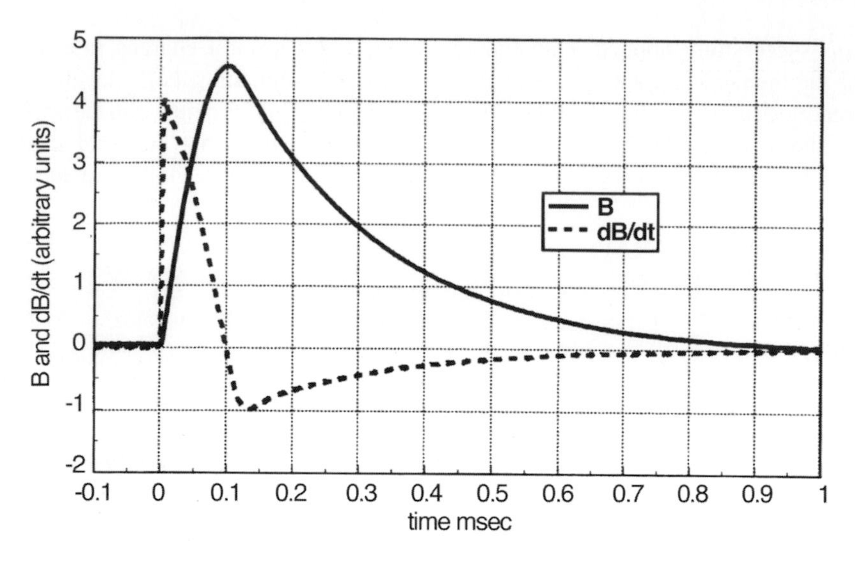

Figure 4 The time course of the magnetic field (*B*) produced by a single-pulse stimulator at the center of a stimulating coil and the resulting electrical field (*dB*/*dt*) waveform (Magstim 200 stimulator). Reproduced from Barker (1999), with permission.

the pulse generation to cortical stimulation. The important points here are that a large current (8 kA in the example shown) is required to generate a magnetic field of sufficient intensity to stimulate the cortex and that the electric field induced in the cortex is dependent upon the rate of change as well as the intensity of the magnetic field. To achieve these requirements the current is delivered to the coil with a very short rise time (approx 200 μs) and the pulse has an overall duration of approximately 1 ms. These demands also require large energy storage capacitors and efficient energy transfer from capacitors to coil, typically in the range of 2000 joules of stored energy and 500 joules transferred to the coil in less than 100 μs. The induced field has two sources (Roth *et al.*, 1991). One is the induction effect from the current in the coil (and this is what is usually meant when discussing TMS); the other is a negligible accumulation of charge on the scalp or between the scalp and the skull. Figures 4 and 5 show the difference between two types of pulses, monophasic and biphasic, that can be produced by magnetic stimulators. The biphasic waveform generally used in rTMS machines differs

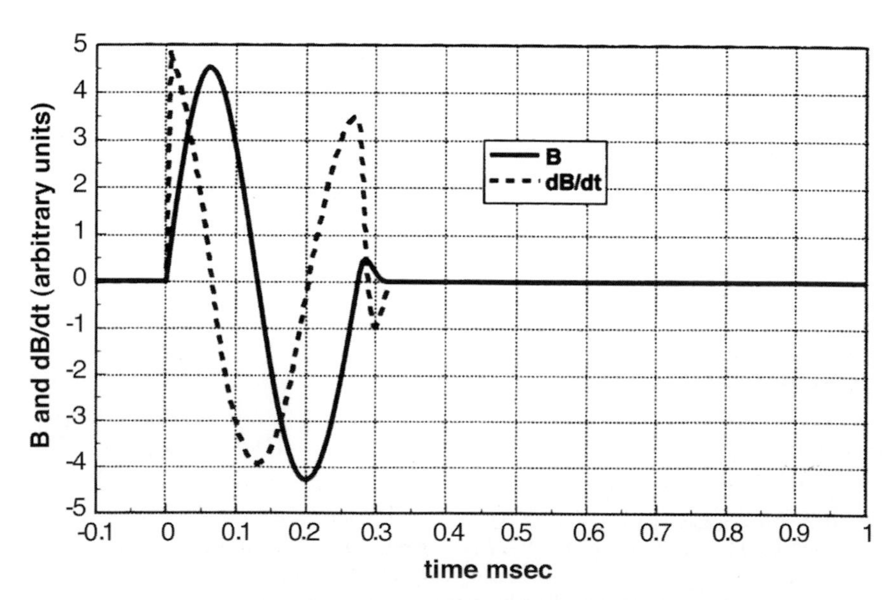

Figure 5 The time course of the magnetic field (*B*) produced by a biphasic repetitive-pulse stimulator at the center of a stimulating coil and the resulting electrical field (*dB*/*dt*) waveform (Magstim Rapid stimulator). Reproduced from Barker (1999), with permission.

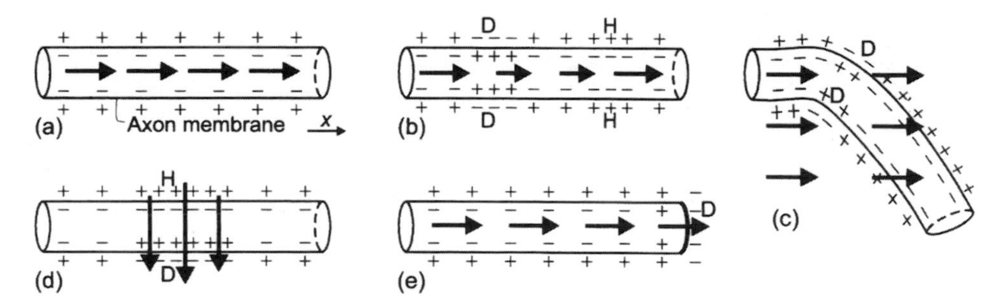

Figure 6 How current flow may activate neurons: schematic illustrations of activation mechanisms. In (a) the current flow in a uniform electric field runs parallel to a neuron and thus causes no change in transmembrane current. In (b) there is a gradient activation due to a nonuniform field along the axon which causes change in transmembrane potentials resulting in action potentials. In (c) the same relationship and end result is seen as in (b), but here the change in transmembrane current is due to spatial variation (bending) of the nerve fiber rather than inhomogeneities in the electric field. In (d) the depolarization is caused by transverse activation of neuron by the induced electric field and (e) represents changes in activation at the axon terminal. Regional depolarization and hyperpolarization are indicated by D and H, respectively. Reproduced from Ruohonen and Ilmoniemi (1999), with permission.

from the monophasic in two ways. First, in the biphasic mode up to 60% of the original energy in the pulse is returned to the capacitor, rendering rTMS more energy efficient and thus enabling the capacitors to recharge more quickly (Jalinous, 1991; Barker, 1999). More importantly for the end user, the biphasic waveform seems to require lower field intensities to induce a current in neural tissue (McRobbie and Foster, 1984). The reasons for the higher sensitivity of neurons to biphasic stimulation have been examined with respect to the properties of the nerve membrane (Reilly, 1992; Wada *et al.,* 1996). The rise time of the magnetic field is important because neurons are not perfect capacitors, they are leaky, and the quicker the rise to peak intensity of the magnetic field, the less time is available for the tissue to lose charge. A fast rise time has other advantages in that it decreases the energy requirements of the stimulator and the heating of the coil (Barker, 1999).

In magnetic stimulation an electric field is induced both inside and outside the axon (Nagarajan *et al.,* 1993). To produce neural activity the induced field must differ *across* the cell membrane. As Fig. 6 shows, if the field is uniform with respect to the cell membrane, no current will be induced; either the axon must be bent across the electric field or the field must traverse an unbent axon. Another way of stating what is visualized in Fig. 6 is that the probability of an induced field activating a neuron is a function of the spatial derivative of the field along the nerve membrane—in Barker's words "the activating function is proportional to the rate of change of the electric field" (Barker, 1999; Reilly, 1992; Abdeen and Stuchley, 1994; Garnham *et al.,* 1995; Maccabee *et al.,* 1993).

The principle of the activating function can be used as a guide in thinking about the site of stimulation. Amassian *et al.* (1992) have modeled the stimulation of bent neurons and calculated that the excitation of straight nerves occurs near the peak electric field, whereas the activation of bent nerves occurs at the positive peak of the spatial derivative. Presumably, where the field and neuron lie in almost the same plane, the spatial derivative is equivalent to the peak field. The different orientations of neurons in the cortex preclude a simple one-to-one mapping from electrical fields in homogeneous conductors to the volume of neural tissue affected. Amassian gives a practical example of how knowledge of the anatomy of the cortical area being stimulated underlies accurate interpretation of the effects of TMS (Fig. 7).

A. Stimulating Coils

Stimulating coils consist of one or more well-insulated coils of copper wire frequently housed in a molded plastic cover. Stimulating coils are available in a variety of shapes and sizes. The two types of coil in most common use are circular and figure of 8 in shape and the regions of effective stimulation produced by these two configurations depend on the geometry of the coil and of the neurons underlying the coil and on local conduction variability. In addition to differences in focality of the induced current, circular and figure-8 coils may differ in the neural structures activated within the brain. Therefore, the stimulating coil employed should be carefully chosen and always considered when interpreting results of TMS studies.

Figure 8 shows the distribution of an induced electric field under a round coil (top) and Fig. 9 (top) the distribution of the spatial derivative of the field with respect to a straight axon that will be hyperpolarized at B and polarized at A ("virtual anode" and "virtual cathode," respectively, in Barker's terminology). Nerves lying tangential to any other

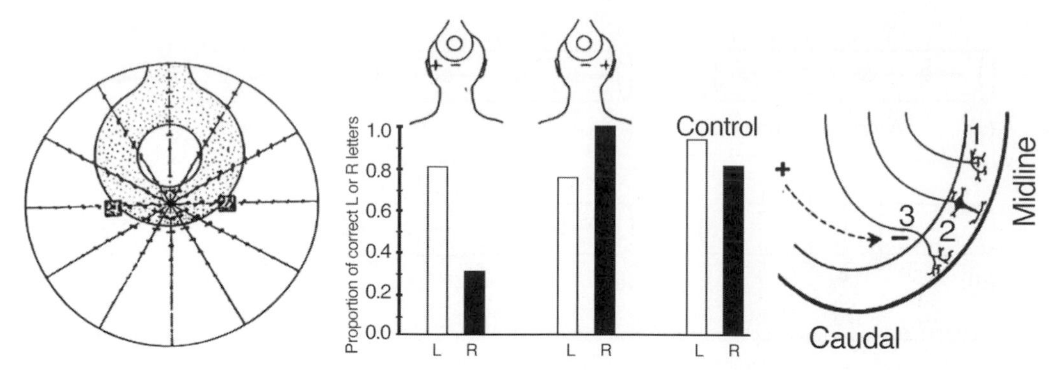

Figure 7 The electric field induced by TMS delivered by a round coil is here modeled (left) in a spherical saline volume conductor. The effect on visual detection of reversing the polarity of the induced electric field is shown (middle) and a schematic of the possible sites of stimulation are shown (right). The clockwise current in the coil (left) induces an anticlockwise electric field and the field intensity diminishes with distance from the peak of stimulation (center of spherical saline bath). The results from one subject show that reversing the direction of the induced field differentially suppresses visual performance in the left or right visual hemifield (Amassian *et al.,* 1994). The most likely point of stimulation is the bend in the axon (3). Excitation of the axonal arborizations (1) is less likely due to relative high resistance, and excitation of the dendritic arbors (2) is less likely due to relative reduced electrical excitability. Reproduced from Amassian *et al.* (1998), with permission.

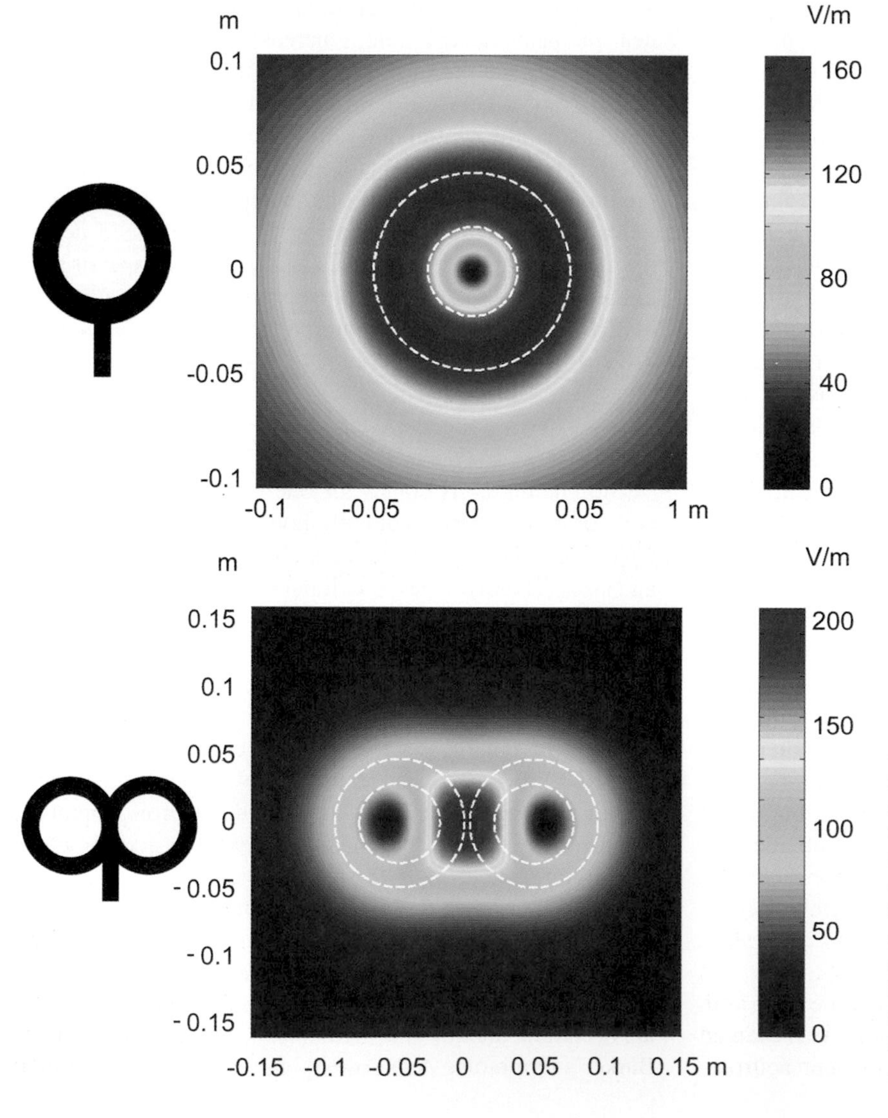

Figure 8 Distribution of the induced electric fields by a circular (top) and figure-8 (bottom) stimulating coil. The circular coil has 41.5-mm inside turn diameter, 91.5-mm outside turn diameter (mean 66.5 mm), and 15 turns of copper wire. The figure-8 coil has 56-mm inside turn diameter, 90-mm outside turn (mean 73 mm), and 9 turns of copper wire on each wing. The outline of each coil is depicted with dashed white lines on the representation of the induced fields. The electric field amplitude is calculated in a plane 20 mm below a realistic model of the coil ($dI/dt = 10^8$ A s^{-1}). Figure created by Anthony Barker.

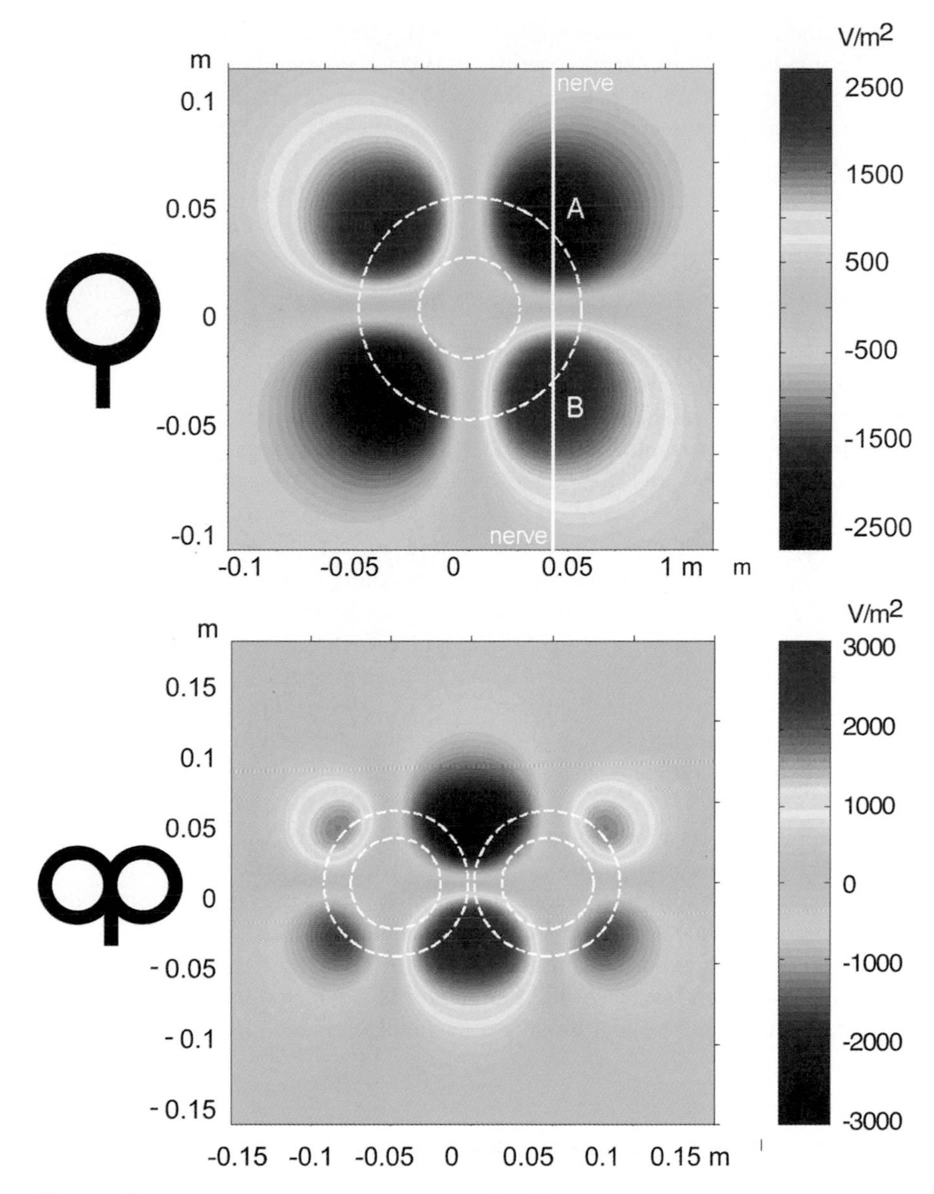

Figure 9 Rate of change of the electric field calculated in the direction of the nerve along the axis AB, measured in the same plane as with coils shown in Fig. 8. Figure created by Anthony Barker.

part of the coil will be similarly stimulated. This does not mean that the effects of TMS are restricted to the cortical area located precisely under the windings of the coil. The neurons receiving stimulation will activate their neighbors and also affect the organization of other interacting pairs of neurons. With this round coil, making contact between only one arc of the coil and the scalp can increase specificity of the area stimulated. The side of the coil with which stimulation is applied will also affect the outcome. With a monophasic pulse, the current travels clockwise with respect to one face of the coil and counterclockwise with respect to the other. This can be used to bias stimulation in one or the other direction and has been used to selectively stimulate one or the other hemisphere while apparently stimulating in the midline (Amassian *et al.,* 1994; Meyer *et al.,* 1991) and to enhance the efficacy of motor cortex stimulation by applying the current direction optimal for stimulation of that region (Brasil-Neto *et al.,* 1992a,b).

Stimulation with a figure-8 coil increases the focality of stimulation (Ueno *et al.,* 1988). This configuration is of two circular coils that carry current in opposite directions and, where the coils meet, there is a summation of the electric field. Figures 8 (bottom) and 9 (bottom) show the induced electric field and the rate of change of the field with respect to a straight neuron. In addition to the new "summated" anode and cathode produced by the figure-8 coil, the two separate windings maintain their ability to induce a field under the outer parts of the windings. However, in experiments in which the center of the figure of 8 is placed over the region of interest, the outer parts of the coil are usually several centimeters away from the scalp and thus unlikely to induce effective fields.

B. Single-Pulse, Paired-Pulse, and Repetitive TMS

TMS can be applied as single pulses, delivering one stimulus every 3 or more seconds to a given cortical region; as pairs of stimuli separated by a variable interstimulus interval of a few milliseconds; or as trains of stimuli at variable frequency delivered to the same brain area for several seconds (Fig. 10).

Paired-pulse TMS can be applied with the two stimuli of the same or different intensities delivered through a single coil to the same brain region. In this manner, paired-pulse TMS can be used to study corticocortical inhibitory and excitatory circuits. Alternatively, paired-pulse TMS can be applied using two coils so that each of the two stimuli affects a different brain region. Using this methodology, paired-pulse TMS can be used to study corticocortical connectivity and interactions.

Repetitive TMS can be applied at relatively slow frequency, delivering one stimulus every second or less. This form of stimulation is referred to as slow (or low frequency) rTMS. Alternatively, rTMS can be applied at higher stimulation frequencies, with stimuli delivered up to 20 times per second. We then speak of rapid or high-frequency rTMS. Slow and rapid rTMS appear to exert differential modulatory effects on cortical excitability (Pascual-Leone *et al.*, 1994a). Furthermore, the differentiation of slow and rapid rTMS is meaningful from the point of view of safety of the technique (Wassermann, 1998).

A single TMS pulse can depolarize a population of neurons and hence evoke a given phenomenon or percept. When applied to the motor cortex, a single TMS pulse of sufficient intensity can induce a movement in a contralateral limb, and when applied to the visual cortex it can induce the perception of a flash of light (phosphene). In addition, a single TMS pulse can transiently disrupt normal brain activity by introducing random neural activity into the stimulated area (see Walsh and Cowey, 2000). If the targeted brain area is *necessary* for the completion of a given task, performance should be impaired. Single TMS pulses disrupt activity for only some tens of milliseconds and provide information on *when* activity contributes essentially to task performance (the "chronometry" of cognition). In this fashion, applied to the motor cortex, single-pulse TMS can investigate the timing of the engagement of the motor cortex in the execution of motor programs (Day *et al.*, 1989a); applied to the somatosensory cortex it can provide insight into the time course of tactile perception (Cohen *et al.*, 1991a); and applied to the occipital cortex it can explore the chronometry of detection and perception of visual stimuli (Amassian *et al.*, 1989).

rTMS offers the advantage of an "offline" paradigm in which magnetic stimulation and task performance are uncoupled in time. Such offline use of rTMS in the study of brain function and cognition is based on studies of motor cortex in which it has been shown that a continuous train of stimulation can modulate cortical excitability beyond the

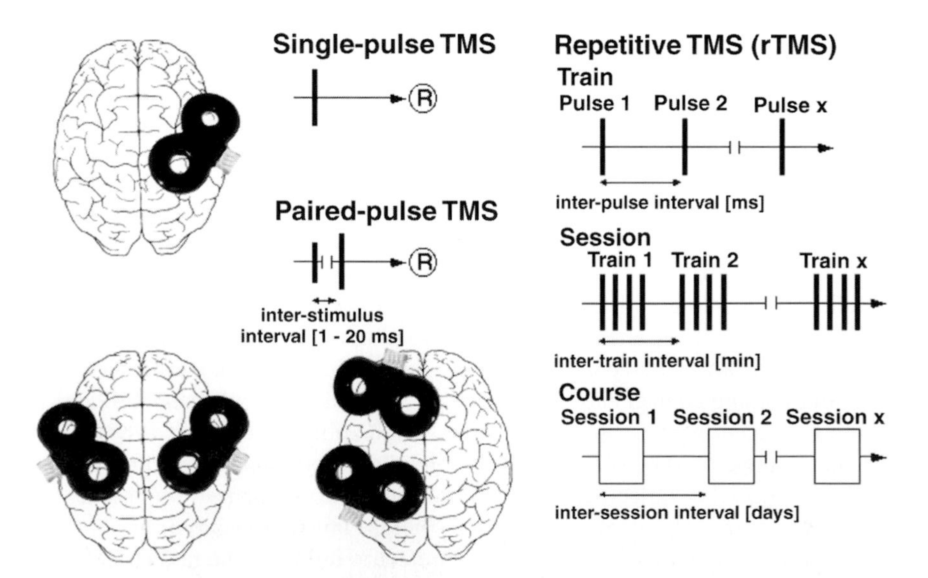

Figure 10 Schematic representation of the different ways of applying TMS: single pulse, paired pulse to a single or to two different brain areas, and repetitive (slow or rapid) TMS.

duration of the rTMS train itself (Chen *et al.,* 1997; Berardelli *et al.,* 1998; Pascual-Leone *et al.,* 1998). Depending on stimulation frequency and intensity, motor cortex excitability can either be enhanced or be reduced as measured with motor-evoked potentials (MEP) (Pascual-Leone *et al.,* 1998; Hallett *et al.,* 1999). Slow rTMS (1 Hz) applied to motor cortex can give rise to a lasting decrease in corticospinal excitability (Chen *et al.,* 1997; Maeda *et al.,* 2000a), while fast rTMS (5, 10, and 20 Hz) can induce an increase in cortical excitability (Pascual-Leone *et al.,* 1994a; Berardelli *et al.,* 1998; Maeda *et al.,* 2000a). It is, however, important to recognize the significant inter- and intraindividual variability of these modulatory effects of rTMS (Maeda *et al.,* 2000a; Fig. 11). Nevertheless, it is hypothesized that application of rTMS to cortical areas other than motor creates similar modulations (decreases or increases) in cortical excitability, which lead to measurable behavioral effects (Pascual-Leone *et al.,* 1999a). This approach has recently been implemented in a number of cognitive studies, including visual perception (Kosslyn *et al.,* 1999), spatial attention (Hilgetag *et al.,* 2001), motor learning (Robertson *et al.,* 2001), working memory (Mottaghy *et al.,* 2001), and language (Shapiro *et al.,* 2001). In addition, this capacity of rTMS of modulating cortical excitability has suggested the possibility of using TMS in therapeutic applications in neuropsychiatric conditions associated with abnormalities in cortical excitability (Pascual-Leone *et al.,* 1998; Wasserman and Lissanby, 2001).

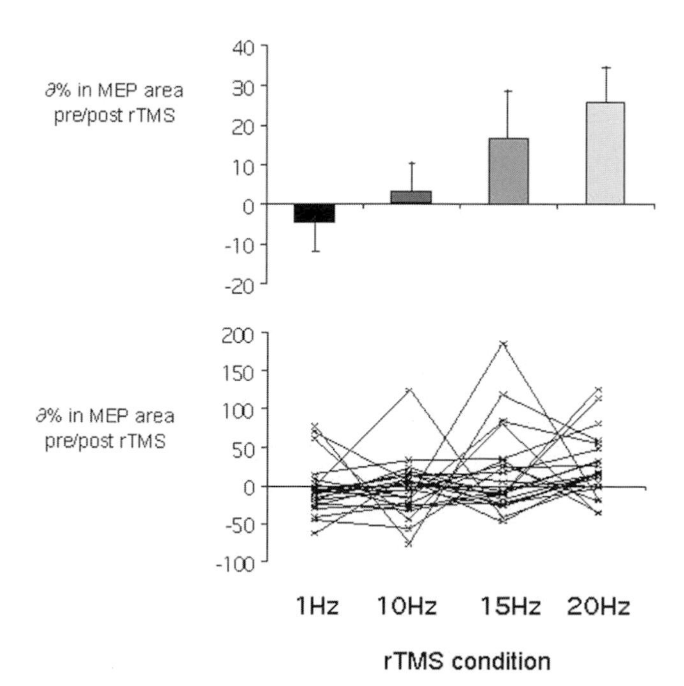

Figure 11 Modified from Maeda *et al.* (2000a,b), with permission. Frequency tuning curves for different subjects are shown. The total number of pulses applied at each rTMS condition was 240. The top shows the mean percentage changes in the averaged MEP area from pre- to post-rTMS. The bars indicate standard error. The bottom shows frequency tuning curves of each individual. Abbreviation: ∂%, percentage change.

C. Positive and Negative Effects of TMS

TMS can have disruptive, "inhibitory" effects on perceptual or motor performance or can sometimes paradoxically improve performance. This leads to the question of whether, in its disruptive or productive modes, TMS stimulates excitatory or inhibitory neurons. If one considers the mechanisms of TMS induction (see above) it becomes readily apparent that TMS cannot be expected to distinguish between excitatory and inhibitory neurons within a region of stimulation, nor can it be expected to distinguish between orthodromic and antidromic direction of stimulation. Delivery of a TMS pulse will randomly excite neurons that lie within the effective induced electrical field. For these reasons it is best to consider TMS as operating in two ways. In its disruptive mode, TMS applied while a subject is trying to perform a task induces neural noise into the signal processing system. Just as the stimulation is likely to be random with respect to inhibition, excitation, and direction of current along any given membrane, so too can it be presumed to be random with respect to the organization of the neural assemblies involved in any particular task. There are some situations in which TMS might be considered to operate in a productive mode and add signal rather than noise, for example, in the functional enhancements produced by TMS (Walsh *et al.,* 1998; Hilgetag *et al.,* 2001) or in the production of phosphenes (Kammer, 1999; Kammer and Nussek, 1998). However, the enhancements reported by Walsh *et al.* and Hilgetag *et al.* were caused by a disruption in one area resulting in disinhibition in a competing region of cortex, and as Kammer has argued cogently, the physiological effects that produce phosphenes are identical with those that produce visual deficits.

III. TMS in Clinical Neurophysiology

When TMS is applied to the motor cortex at appropriate stimulation intensity, it is possible to record MEPs in contralateral extremity muscles (Fig. 12). If the stimulation coil is placed over the spinal column such that nerve roots are stimulated, a radicularly induced MEP can be recorded. The latency difference between the MEP induced by cortical stimulation and the one evoked by radicular activation provides a measure of central motor conduction time (CMCT). Alternatively, the latency of an H reflex of the F wave can be used to obtain more precise measurements of peripheral latency and hence obtain more accurate measures of CMCT. Rossini and Rossi (1998), Rothwell (1997), and Mills (1999) have provided recent reviews on the clinical utility of CMCT determinations.

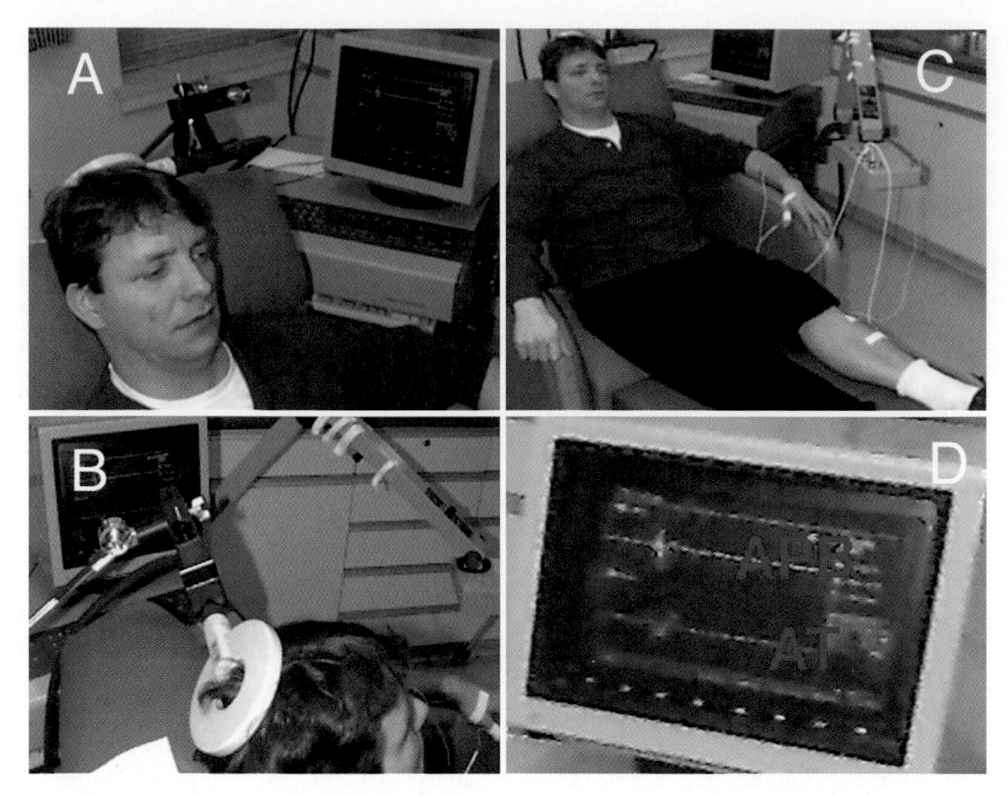

Figure 12 Composite of photographs illustrating the experimental setup for recording of motor evoked potentials to TMS. Julian P. Keenan, Ph.D., demonstrates as a pretend subject for a study. Stimulation is applied transcranially with a circular coil resting on the subject's head and targeting the motor cortex (A and B). The stimulation coil is held in place by a specially designed coil holder (A and B). Surface electrodes are used to record simultaneously the EMG activity in the abductor pollicis brevis (APB) and the anterior tibialis (C and D). EMG activity and evoked motor potentials are recorded and analyzed using a commercially available EMG device (depicted is the Counterpoint by Dantec Medical, Denmark).

A. Measures of Cortical and Corticospinal Excitability

If TMS is applied to the motor cortex, different TMS paradigms can be used to study different components of cortical excitability and provide insight into the function of different neurotransmitter systems. Figure 13 illustrates these different measurements. Single-pulse TMS can be applied to the motor cortex to determine *motor threshold.* Motor threshold refers to the lowest TMS intensity to evoke MEPs in a target muscle in 50% of trials. Motor threshold is felt to represent a measure of membrane excitability in pyramidal neurons. Support for this claim comes from changes in motor threshold induced by antiepileptic medications with prominent sodium and calcium channel-blocking activity but limited or absent neurotransmitter interaction (carbamazepine, phenytoin, or losigamine) (Ziemann *et al.,* 1996c).

Single-pulse TMS can also be applied at suprathreshold intensity to the motor cortex to study the induced *silent period* (Fig. 13). Silent period refers to the suppression of EMG activity in the voluntarily contracted target muscle following the induction of a motor-evoked potential. Studies of segmental spinal excitability during this silent period have established the cortical origin of at least the later part of the evoked EMG silence (Brasil-Neto *et al.,* 1995; Fuhr *et al.,* 1991; Schnitzler and Benecke, 1994; Triggs *et al.,* 1993; Wilson *et al.,* 1993a). This postexcitatory cortically generated inhibition can sometimes be observed in the absence of preceding facilitation (silent period without preceding MEP) (Catano *et al.,* 1997; Triggs *et al.,* 1993; Wassermann *et al.,* 1991) and can be shown to have a cortical origin distinct from the optimal site for activation of a given target muscle (Lewko *et al.,* 1996; Wassermann *et al.,* 1993; Wilson *et al.,* 1993b). The balance of cortical glutamatergic (Faig and Busse, 1996; Prout and Eisen, 1994; Yokota *et al.,* 1996), dopaminergic (Priori *et al.,* 1994; Ziemann *et al.,* 1996a), and GABAergic activity (Inghilleri *et al.,* 1993; Nakamura *et al.,* 1997; Ziemann *et al.,* 1995, 1996b,c) seems to play a critical role in the duration of the silent period to TMS. Indeed, GABA-B activity may be particularly critical for the generation of the silent period. However, there is some debate about precisely when the silent period begins, how one should measure it, and what the underlying physiology is (Mills, 2000; Ziemann and Hallet, 2000). Nevertheless, several uses of the silent period

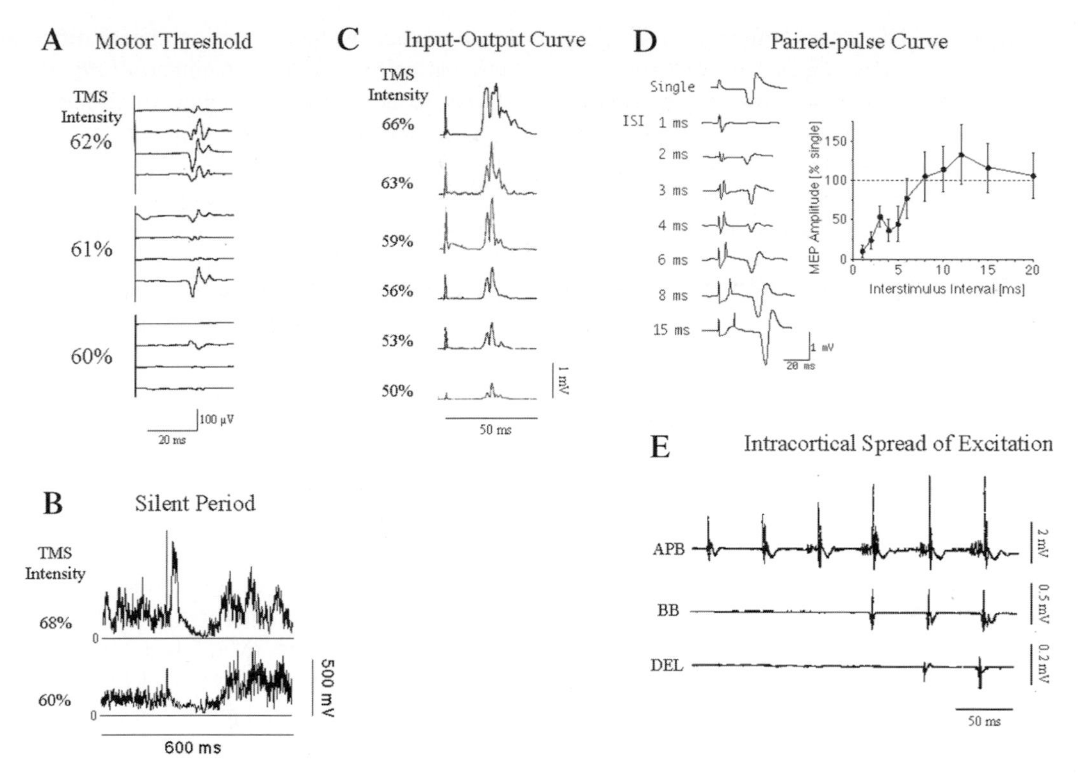

Figure 13 Modified from Pascual-Leone *et al.* (1998), with permission. (A) Representative examples of MEPs induced by TMS at decreasing intensities and recorded from the abductor pollicis brevis muscle in a normal volunteer during the determination of motor threshold. TMS intensity is expressed as a percentage of maximal stimulator output. Note that at an intensity of 62%, MEPs are induced in four of four trials. At an intensity of 61% criterion MEPs (≥50 µV peak-to-peak amplitude) are recorded in only two of four trials. At 60% intensity criterion MEPs are recorded in only one of four trials. In practice, we would rely on 10 consecutively recorded MEPs rather than only on 4. (B) Representative of silent periods evoked by TMS at different stimulator output intensities in a normal volunteer. Reponses are recorded from the first dorsal interosseus muscle. Note the MEP and the postexcitatory silent period at 68% TMS intensity. Note the silent period without preceding MEP at 60% intensity. (C) Examples of MEPs in an input–output curve in a normal volunteer. Rectified and averaged MEPs of a total of 15 single MEPs recorded from the thenar musculature at different stimulus intensities are shown. Note that with increasing TMS intensity there is a progressive increase in MEP amplitude and area, progressive decrease in latency, and progressive prolongation in MEP duration. (D) Paired-pulse curve to TMS in a normal volunteer recorded from the first dorsal interosseus muscle. The conditioning stimulus was applied at 80% of the subject's motor threshold intensity, while the test stimulus was applied at 115%. On the left, representative examples of the recorded MEPs are given for various interstimulus intervals (ISI). On the right, the curve of modulation of MEP amplitude depending on interstimulus interval is displayed for a normal volunteer. MEP amplitude is expressed as percentage of the average amplitude of MEPs evoked by the test stimulus alone (% of single). (E) EMG recording from the abductor pollicis brevis (APB), biceps brachii (BB), and deltoid (DEL) muscles during a train of rTMS (20 Hz, 120% motor threshold intensity) to the optimal scalp position for activation of the hand muscles. Note the progressive increase in amplitude of the MEPs in the APB and the appearance of MEPs in the BB and DEL after four and five stimuli, respectively.

seem particularly relevant. First, the silent period can be used as a marker of cortical modulation and thus changes in length or depth due to learning or disease serve as indicators of the site of damage. Second, because it is independent of previous muscle activity and can be elicited at lower levels of stimulation than MEPs, the silent period is sometimes a more sensitive measure of TMS effects than MEPs.

Single TMS pulses of progressively increasing intensity applied to the motor cortex can be used to generate an *input–output curve* (Fig. 13). The resulting modulation of amplitude of MEPs to increasing intensity of TMS pulses appears to provide a measure of excitatory feedback to corticospinal efferent output (Valls-Sole *et al.,* 1994), which seems glutamatergically mediated (Prout and Eisen, 1994).

Intracortical excitability can be further studied using the *paired-pulse TMS technique* (Fig. 13) (Kujirai *et al.,* 1993). A first, conditioning stimulus is applied, followed, at a variable interval, by a second, test stimulus. The effects obtained depend upon the intensity of the conditioning stimulus, the interval between the stimuli, and the intensity of the test stimulus. The intensity of conditioning and test stimuli influences the effects as different circuits are recruited by different intensities of stimulation. The interstimulus interval (ISI) influences the results as the time constant of each activated circuit may differ. At very short ISIs (<1 ms) it is possible to study neural time constants of the stimulated elements, at ISIs of 1–4 ms it is possible to investigate interactions between I-wave inputs to corticospinal neurons, and at ISIs of 1–20 ms it is possible to investigate corticocortical inhibitory and facilitatory circuits. All these effects appear to be cortically mediated (Kujirai *et al.,* 1993; Valls-Sole *et al.,* 1992; Ziemann *et al.,* 1996d) and intracortical inhibition and facilitation appear to be due to activation of separate circuits (Ziemann *et al.,* 1996d). The effects of different illnesses and medications on the inhibitory and facilitatory phases of the paired-pulse curve suggest that GABAergic and dopaminergic mechanisms are involved. Medications that enhance GABAergic activity have been shown to markedly decrease the degree of corticocortical facilitation evoked by paired TMS stimuli at ISIs of approximately 8–12 ms (Inghilleri *et al.,* 1996; Ziemann *et al.,* 1995, 1996b,c). Conversely, in Parkinson's disease, the dopamine deficiency is associated with reduced corticocortical inhibition at short ISIs (<5 ms) (Berardelli *et al.,* 1996; Ridding *et al.,* 1995), and dopaminergic drugs have been shown to enhance corticocortical inhibition in normal subjects and Parkinsonian patients (Berardelli *et al.,* 1996; Priori *et al.,* 1994; Ridding *et al.,* 1995; Ziemann *et al.,* 1996a). Furthermore, studies suggest that an early phase of facilitation in the paired-pulse curve at approximately 3 ms ISI might be related to glutamatergic, excitatory intracortical modulation (Detsch and Kochs, 1997; Prout and Eisen, 1994; Ziemann *et al.,* 1996d).

Finally, the modulation of the MEPs recorded in contralateral muscles during rTMS trains provides evidence of the pattern of reentry inhibitory and excitatory pathways (Jennum *et al.,* 1995; Pascual-Leone *et al.,* 1994c). Repetitive TMS trains at different intensities and frequencies differentially modulate MEPs. During trains of rTMS at appropriate intensity and frequency, the phenomenon of *intracortical spread of excitation* (ISE) (Fig. 13) has been described (Pascual-Leone *et al.,* 1994c). ISE appears most likely due to the breakdown of GABAergic inhibition. The number of TMS pulses until onset of ISE at a given rTMS frequency and intensity provides a measure of intracortical surround inhibition control which can be shown to be altered, for example, in patients with epilepsy.

These different measures of cortical excitability can be applied to the study of cortical pathophysiology in a variety of neuropsychiatric conditions and may in the future have a profound impact on therapeutic approaches. For example, patients with epilepsy have altered measures of intracortical excitability (Caramia *et al.,* 1996; Jennum and Winkel, 1994; Michelucci *et al.,* 1996; Reutens *et al.,* 1993) that may, in the future, allow differentiation among forms of epilepsy that cannot be predicted on clinical grounds alone. Different antiepileptic drugs, in accordance with their known mechanisms of action, have different effects on intracortical excitability (Ziemann *et al.,* 1996c) and these effects could be used to predict which medication might be best suited to normalize the dysfunction in different patients. Nowadays, in a large number of patients with epilepsy, the choice of an antiepileptic drug for a given patient is made empirically using cost and side-effect profile as principal determinants rather than expected efficacy or mechanisms of action. TMS-derived measures of cortical excitability might in the future guide more pathophysiologically based approaches to neuropharmacology.

B. Strategies for Clinical Applications of TMS

In the present chapter we will concentrate on clinical applications of *transcranial* magnetic stimulation. However, magnetic stimulation can also offer substantial advantages over electric stimulation for the study of nerve root and spinal plexus disorders (Maccabee *et al.,* 1996) or the study of cranial nerves (Benecke *et al.,* 1998).

Most clinical applications of TMS can be performed with a single-pulse magnetic stimulator and a conventional EMG machine. A large circular coil (outer diameter of approximately 10 cm) is sufficient for all routine applications. Smaller circular coils might be required for the stimulation of the facial nerve behind the mandibular angle or the accessory nerve in the posterior fossa. Focal figure-8 coils are needed for reliable hemisphere-selective activation of the motor cortex. The stimulus intensity employed should be expressed in percentage of motor threshold intensity of the recorded muscles at rest. For induction of reliable MEPs in contralateral hand muscles intensities of approximately 120% of motor threshold are sufficient. Induction of MEPs in leg muscles might require maximal stimulator output intensities and the use of specially shaped, double-cone coils.

In general, for most diagnostic applications TMS is applied to the motor cortex and MEPs are recorded using surface electrodes taped over the belly and tendon of the target muscle(s). Frequently, in order to fully interpret the results, motor cortex TMS has to be combined with peripheral nerve, nerve plexus, or spinal root stimulation. As is the case in EMG, the specific sites of stimulation, the recorded muscles, the maneuvers used for facilitation of the motor-evoked potentials, and the evaluation of the different response parameters have to be tailored to the specific questions asked. The following four examples illustrate possible approaches.

(1) In a patient suspected of having a hysterical hemiparesis, TMS can be applied to the motor cortex and MEPs recorded from several arm and leg muscles bilaterally. All target muscles should be relaxed to allow side-to-side comparison of MEPs elicited under similar conditions. A circular TMS coil is most suited to such an application. MEPs of similar amplitude in both hemibodies usually exclude an

organic cause of the hemiparesis, provided that a motor hemineglect and lesions in supplementary or premotor cortical areas are excluded.

(2) In a patient with a suspected spinal cord lesion, TMS is applied to the motor cortex and the MEPs are recorded using surface electrodes from an intrinsic hand muscle (e.g., abductor pollicis brevis or first dorsal interosseus) and a distal leg muscle (e.g., anterior tibial muscle). The comparison of the responses in upper and lower extremities will help define the level of the spinal lesion. The tonic contraction of the target muscles facilitates the spinal motoneurons, thus increasing the likelihood of MEP recording and minimizing conduction velocity. Since it might be difficult to detect small MEPs in contracted muscles and determine their latency, it is useful to rectify and average several responses. In paraplegic patients, electrical stimulation of peripheral nerves to elicit an H reflex or sensory stimuli to elicit a flexor or a Babinski reflex might be used to facilitate spinal leg motoneurons further. MEPs to motor cortex TMS can also be recorded from paravertebral muscles using needle electrodes to define the exact level of the spinal cord lesion more precisely (Meyer *et al.,* 1998a).

(3) In a patient with the clinical picture of motor neuron disease an early diagnosis can be difficult and yet prognostically important as new therapeutic options become available. TMS is applied to the motor cortex and MEPs are recorded from several upper and lower extremity muscles. The target muscles might need to be tonically contracted to facilitate the responses. Rectification and averaging of the EMG responses is useful given the small amplitude of the expected MEPs. H reflexes or magnetic stimulation of the spinal roots can be used for determination of the peripheral nerve conduction. Central motor conduction time is calculated by subtracting this peripheral conduction time from the latency of the TMS response. Central conduction time is prolonged in motor neuron diseases due to the loss of corticospinal cells (Eisen *et al.,* 1990). In addition, the contralateral silent period to motor cortex TMS is significantly briefer than normal. Careful determination of the motor threshold is particularly important in this context as it might help differentiate between amyotrophic and primary lateral sclerosis (Caramia *et al.,* 1997; Eisen *et al.,* 1993). Patients with the former condition have lower than normal motor thresholds, while those with the latter show significantly increased motor thresholds.

(4) In a patient with the suspected diagnosis of multiple sclerosis, TMS is applied to the motor cortex bilaterally using a circular coil and MEPs should be recorded from at least four muscles, ideally bilaterally from two hand and two lower limb muscles. In addition to the mean values of MEP latency and amplitude (Rossini and Rossi, 1998), the variability of both parameters should be carefully considered as a further indicator of impaired corticospinal conduction (Britton *et al.,* 1991).

Measurements of *transcallosal inhibition* might demonstrate a disruption of callosal fibers by periventricular foci even in patients with normal corticospinal responses (Meyer *et al.,* 1995). In addition, such demonstration of transcallosal fiber dysfunction is likely to have prognostic relevance for the cognitive consequences of the multiple sclerosis. For this purpose, TMS is applied to the motor cortex while EMG responses are recorded in tonically contracted ipsilateral hand muscles. In response to the TMS there is a transient suppression of tonic EMG activity due to the transcallosal inhibition of the unstimulated motor cortex (Meyer *et al.,* 1995, 1998a). Several EMG responses are rectified and averaged. Latency, duration, and degree of the transcallosal inhibition can be measured. The latency is defined as the time between the cortical stimulus and the point at which the EMG activity falls under the mean amplitude of the EMG activity before the stimulus. The duration spans between the onset of transcallosal inhibition and the point at which the EMG activity again reaches the mean amplitude of the baseline EMG activity before the stimulus. The degree of EMG suppression is calculated as the percentage decrease of the baseline EMG amplitude before stimulation.

These examples emphasize the notion that diagnostic TMS cannot follow rigid guidelines. Careful thought has to be given to ensure that the number of recorded muscles is sufficient, adequate processing of the responses is performed offline, and suitable facilitation procedures are employed.

There are certainly other important clinical uses of TMS. For example, TMS can help (1) to distinguish between a predominantly demyelinating and an axonal lesion in the descending motor tracts; (2) to detect the level of a hemispheric lesion in relation to the course of transcallosal fibers; (3) to help predict the motor outcome after a vascular cerebral lesions; or (4) to obtain objective data to evaluate the progression of a disease (e.g., myelopathy) or the effects of treatment (e.g., in transverse myelitis). Furthermore, magnetic stimulation can be used for intraoperative monitoring of corticospinal motor tract function during spinal surgery in order to optimize surgical outcomes (Herdman *et al.,* 1993).

C. Mapping Motor Cortical Outputs

TMS can be applied sequentially to different scalp positions as the evoked response from each site of stimulation is recorded, hence generating a spatial map of behavioral manifestations (Hallett, 1996; Walsh, 1998). Most commonly, TMS mapping is applied to motor cortical output and uses the amplitude or the area under the curve of the MEPs as the measure of the motor response (Brasil-Neto *et al.,* 1992a; Thickbroom *et al.,* 1998; Wassermann *et al.,* 1992; Wilson *et al.,* 1993a). However, other neurophysiologic markers or behavioral manifestations, such as the direction of induced finger movements or the performance in a given cognitive task (Wassermann *et al.,* 1998), could be used and "mapped" in the same manner. This type of application of TMS might provide a method for noninvasive, systematic assessment of cortical function in the presurgical planning of neurosurgical procedures. The advantage of TMS over other brain mapping methods, particularly current functional imaging methods, is that it can provide information about true functional significance of the brain area targeted (Krings *et al.,* 1997a). Therefore, the neurosurgeon can be told, not just that a given brain region in some way participates or is associated with a given behavior, but indeed what the consequences will be if that part of the brain is damaged during the surgical procedure. Being able to provide this kind of information would be obviously desirable (Cramer and Bastings, 2000; Krings *et al.,* 1997a). However, work is still needed to fulfill this clinical potential. For example, it is imperative to establish a method of reliably transferring scalp positions (over which TMS is applied) to brain cortical sites (Miranda *et al.,* 1997). Frameless stereotactic methods might be the solution to this problem, but they do not fully address the question of field distribution of the TMS in a real brain (Bohning *et al.,* 1997). Factors like cerebrospinal fluid space, which by virtue of its much greater conductivity compared to brain tissue can significantly distort the electromagnetically induced currents in TMS and hence shift the site of brain stimulation from the strict perpendicular projection of the scalp position of the coil, need to be fully explored. Careful studies correlating the results of TMS and direct cortical stimulation mapping procedures are needed and have only begun to be done for the motor cortex (Krings *et al.,* 1997b, 1998). Similar studies for language representation and other "eloquent" cortical regions are still needed in order to assess the clinical significance of TMS for this purpose.

In research, TMS mapping has proven a valuable tool for exploring issues of plasticity in the adaptation to injury and the acquisition of new skills (Cohen *et al.,* 1998; Pascual-Leone *et al.,* 1999a). Expansion of a specialized cortical area and recruitment of a remote area as a result of learning or brain disease or injury comprise the two most characteristic forms of brain plasticity. The underlying general phenomenon is that neurons in one area assume properties of neurons in an adjacent or remote area. Such remodeling can take place across brain areas within a given modality, for example, within visual, tactile, or motor distributed systems (homotypic or intramodal plasticity), or may bridge across modalities (heterotypic or cross-modal plasticity), as in the case of tactile information processing in the occipital ("visual") cortex in the blind (Cohen *et al.,* 1997; Sadato *et al.,* 1996, 1998). The time course of plasticity is extended, with some changes appearing within seconds of the initial event and continuing many years after an intervention or injury. TMS can be used in this setting to demonstrate plastic changes and to serially track them in time.

Several studies can serve to illustrate this type of application of TMS mapping. Pascual-Leone *et al.* studied how the motor cortical output maps change in normal subjects as they learn to perform with one hand a five-finger exercise on a piano keyboard (Pascual-Leone *et al.,* 1995) and how implicit and explicit knowledge of a sequence influences motor cortical maps during a procedural learning task (Pascual-Leone *et al.,* 1994a). It is important to realize that such serial TMS mapping studies demonstrate a trace or memory of the activation of the motor cortical outputs that took place during the performance of a task rather than the activation during the task itself as would be the case with neuroimaging studies. Therefore, such TMS mapping studies might be revealing the consequences of long-term potentiation or long-term depression on cortical function (Butefisch *et al.,* 2000; Classen *et al.,* 1998) and longer lasting and slower mechanisms of plasticity such as sprouting and establishment of new connections.

In the setting of adjustment of injury, serial TMS mapping following a stroke (Cramer and Bastings, 2000; Rossini *et al.,* 1998) promises to aide our understanding of the mechanisms involved in the recovery of function after a brain lesion and might provide insight into new therapeutic and neurorehabilitation approaches (Liepert *et al.,* 2000). The study of cortical plasticity after amputation provides another clear example of the utility of TMS mapping (Chen *et al.,* 1998; Cohen *et al.,* 1991a). Pascual-Leone *et al.* (1996) tracked the changes in motor cortex excitability from months before to months after a subject lost his right hand and arm and forearm. In the year following the amputation the motor output maps of the amputated biceps and lower facial muscle ipsilateral to the amputated arm expanded over the original representation of the right hand. The expansion was associated with disappearance of phantom sensations and also with the disappearance of the ability of TMS to elicit phantom experience. Figure 14 shows the progressive changes in the area over which EMG responses can be elicited and the gradual diminution of phantom responses.

D. Development and Maturation

Serial studies of cortical excitability (motor threshold, silent period, paired-pulse curves, input–output curves) and TMS mapping can be used for the study of nervous system development and maturation. The development of motor coordination continues throughout childhood and into adolescence and one of the problems the nervous system has to solve is how to maintain motor control over a period of life during which an individual may grow from one-half to two meters and during which the rate of that growth may vary over a 20-fold range. Not only is the child growing in height but the limbs are growing and the area swept by any movements changes as a result. One proposed solution is that the nervous system employs constant conduction times

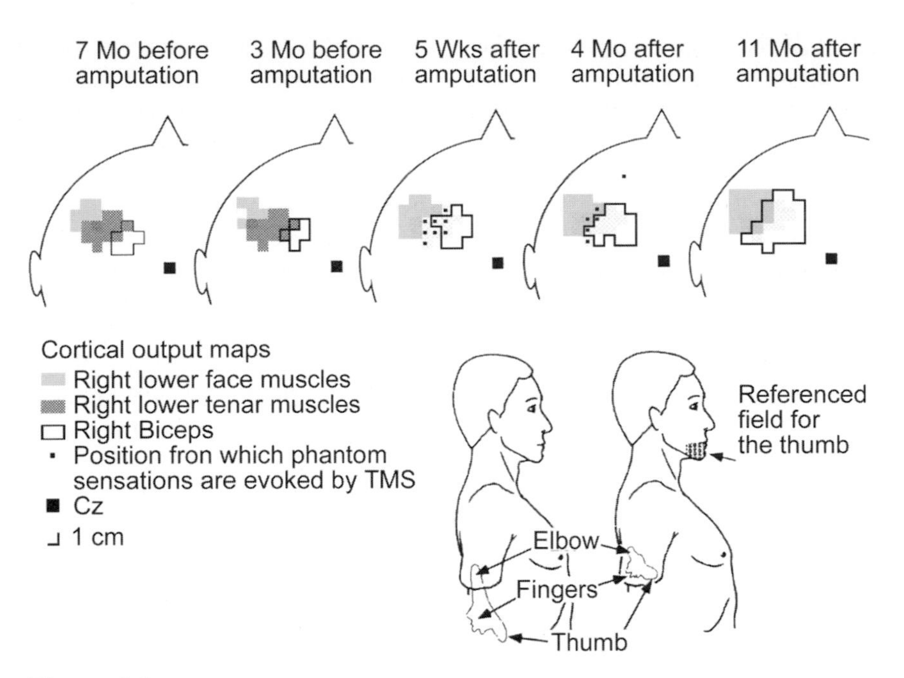

Figure 14 Serial TMS maps in a subject who suffered a traumatic amputation of the right arm. Reproduced from Pascual-Leone *et al.* (1996), with permission.

rather than a more complex mechanism that would be able to track changing timing requirements throughout development. Eyre *et al.* (1991) tested this possibility directly by measuring conduction times and sensitivity to TMS (in the form of MEP threshold) in over 400 subjects between the ages of 32 weeks and 52 years. They applied TMS over the motor cortex and the cervical spine and recorded EMG from the biceps and the hypothenar muscles. They found that cortical-evoked MEPs decreased in latency from 32 weeks until approximately 2 years of age and then plateaued at adult levels. The latencies of response following cervical stimulation were relatively constant until 4–5 years of age and thereafter increased in proportion to arm length across all ages. The motor threshold also decreased markedly over time until approximately 16 years of age. Müller *et al.* (1997) have extended this type of longitudinal, developmental study to the investigation of the ipsilateral corticospinal pathways, and Heinen *et al.* (1998) have explored the question of the relation between motor skills and corticospinal conduction velocity. Future studies should be able to similarly explore developmental questions in nonmotor systems and provide insight into the correlation of longitudinal neurophysiologic changes and the development of symptoms and diseases, such as dyslexia or gait disturbance in cerebral palsy.

E. Potential Future Clinical Uses of TMS

The development of special techniques offers the opportunity of widening the clinical uses of TMS. We have already introduced paired-pulse TMS techniques that can be used to study intracortical excitability. As discussed above, in addition to diagnostic applications in diseases such as dystonia or obsessive–compulsive disorder, paired-pulse TMS could provide neurophysiologic measurements to guide pharmacological interventions. Repetitive TMS can be used in the study of higher cortical functions, and clinically it might be particularly useful for the noninvasive determination of the language-dominant hemisphere (Epstein, 1998). Interference with language comprehension by single-pulse TMS is quite subtle. However, rapid stimulation over the frontotemporal area can produce prominent impairment of speech output. In series reported thus far, success at inducing complete speech arrest has been possible in approximately 75% of the subjects studied with rTMS applied at frequencies ranging from 4 to 30 Hz. Two studies on epileptic patients who had undergone intracarotid amobarbital (Wada) tests found a concordance with the rTMS effects of 100% (Pascual-Leone *et al.*, 1991) and 95% (Jennum *et al.* 1994). However, careful studies in normal subjects have noted that rTMS might overdiagnose atypical language representation (Epstein *et al.*, 1996a, 1996b, 1998). In addition, the speech deficits induced by rTMS may represent anarthria and dysarthria more frequently than aphasia and be primarily due to disruption of the laryngeal or facial motor outputs (Epstein, 1998). Finally, rTMS over the temporalis muscle in order to target lower frontal cortex can be uncomfortable due to associated facial twitching and pain. Therefore, the clinical application of rTMS for determination of the speech-dominant hemisphere requires further investigation. Its sensitivity and specificity for language lateralization require further verification before it can supplant better-established procedures such as the intracarotid amobarbital test.

Image-guided frameless stereotactic techniques provide a method for precise localization of the brain region targeted by TMS applied to the scalp (Paus, 1999). A subject's brain MRI can be used to identify the anatomical substrate for the TMS effects. In this fashion, the location of the motor cortex in relation to a patient's brain lesion can be precisely identified noninvasively (Krings *et al.,* 1997b, 1998). In addition, functional neuroimaging studies can be used to guide the target of TMS, which would then be able to provide causal information about the behavioral significance of the measured brain activity. For example, functional MRI studies might reveal several areas of activity when a patient talks or moves a finger. TMS can sequentially target these different areas of activation of the *f*MRI and selectively and transiently disrupt their function, creating "transient lesions." This approach will provide the neurosurgeon with invaluable information regarding the likely consequences of damage to different brain areas. Therefore, in patients being evaluated for neurosurgical procedures, such information would allow timely planning of the intervention and would likely reduce complications. Nevertheless, at this point, further research is required to test the clinical utility of such TMS applications.

As mentioned above, the capacity of rTMS for modulating cortical excitability has suggested the possibility of using TMS in *therapeutic applications in neuropsychiatric conditions* associated with abnormalities in cortical excitability (Pascual-Leone *et al.,* 1998; Wasserman and Lissanby, 2001). As reviewed by Wasserman and Lissanby, the therapeutic applications of rTMS in diseases such as depression, schizophrenia, acute mania, obsessive–compulsive disorder, focal dystonia, Parkinson's disease, tremor, myoclonus, or epilepsy are potentially exciting, but remain highly preliminary and it would be premature to consider them of established clinical significance at this point.

IV. TMS in Cognitive Neuroscience

A. Creating Virtual Patients

The development of TMS in cognitive neuroscience has been mainly due to the ability of TMS to enhance the lesion analysis approach to psychology by temporarily disrupting sensory or cognitive processes. Single TMS pulses disrupt activity for only some tens of milliseconds and provide information on *when* activity contributes essentially to task performance (the "chronometry" of cognition). Normal cognitive processes can thus be probed with TMS by the creation of "virtual lesions," which offer numerous advantages over the classical neuropsychological approach of inferring brain function from the behavior of brain-lesioned patients. The use of normal subjects in TMS studies removes the possible confounds of size of lesion, general cognitive impairments resulting from the brain injury, and plastic brain reorganization after the insult. In addition, the same subjects can be tested repeatedly while the same paradigm can be applied to multiple participants.

Whereas single-pulse TMS can be viewed as an "online" paradigm (stimulation occurs *during* task performance), rTMS offers the advantage of an "offline" paradigm in which magnetic stimulation and task performance are uncoupled in time. Such offline use of rTMS is based on the fact that a continuous train of stimulation can modulate cortical excitability beyond the duration of the rTMS train itself and increase it or decrease it depending on rTMS frequency and intensity (Pascual-Leone *et al.,* 1998).

B. Single-Pulse Stimulation and Neurochronometry

Amassian and colleagues (1989) were the first to use TMS as a virtual lesion technique in the visual system and also the first to extend this to probe the cortical basis of the well-established psychological phenomenon of visual masking (Amassian *et al.,* 1993a,b). In the first experiment subjects were presented with small, low-contrast trigrams and required to identify the three letters. TMS was applied using a round coil with the lower edge approximately 2 cm above inion. Pulses were given once per trial at a visual stimulus–TMS onset asynchrony of between 0 and 200 ms. Figure 15 shows that TMS was effective in abolishing the subjects' ability to identify the letter if the pulse was delivered between 80 and 100 ms after onset of the visual stimuli. They also demonstrated the retinotopic specificity of the effect by moving the coil slightly to the left, causing a decrease in identifying only letters on the right of the trigram, or to the right, causing a corresponding decrease in

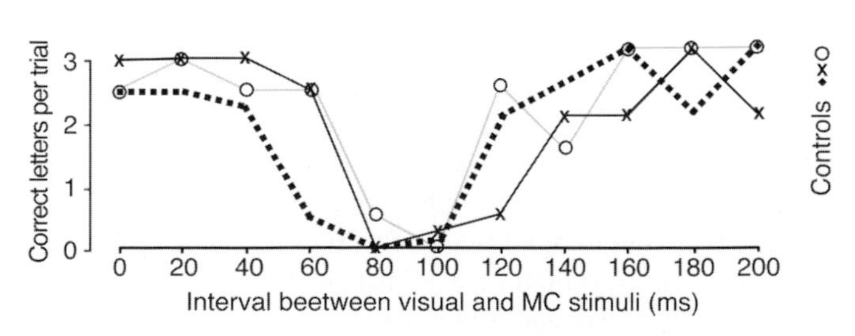

Figure 15 Visual suppression curves of three subjects. The proportion of correct identifications of three briefly flashed dark letters on a bight background is plotted as a function of the delay between stimulus onset and the application of TMS pulse over occipital visual cortex. The magnetic stimulation was delivered with a round coil. Reproduced from Amassian *et al.* (1989), with permission.

identifying letters to the left. They also used vertical trigrams and showed that moving the coil dorsally disrupted perception of the lower letter and moving ventrally interrupted reports of the upper letters.

To make a real test of the specificity of the technique they needed to exclude the possibility that TMS had not made subjects worse on the task because of nonspecific effects on vision. To demonstrate that TMS was having specific effects it should be possible to find an example of two competing stimulus loads and to use TMS to selectively disrupt one in order to unmask the other. They used a classical visual masking paradigm in which subjects were presented with an initial trigram of target letters followed 100 ms later by a second set of masking letters. Following this second set of letters TMS could be applied at a trigram–TMS onset asynchrony of 0–200 ms (Fig. 16). Clearly the presentation of the second set of letters masks the processing of the first, presumably due to some overlapping time period during which initial processing of the second set prevents access to the results of processing the first set. When TMS was applied over the occipital cortex, however, the effects of the second set of stimuli were removed. TMS masked processing of the second set to unmask processing of the first and the time course of the TMS unmasking effect mirrored that of the original masking effect (Fig. 16).

Amassian's work is a good example of how to fuse TMS with psychological models and it laid the foundation for other visual TMS studies, but questions remained. For example, occipital pole stimulation may include several visual areas so other experiments are required to better define the neuroanatomical substrate of the results. The optimal latency of the TMS effect on suppression and masking (80–100 ms) led Amassian *et al.* to suggest that the critical site of stimulation lay beyond the striate cortex because neurons there can respond with shorter latencies. Corthout

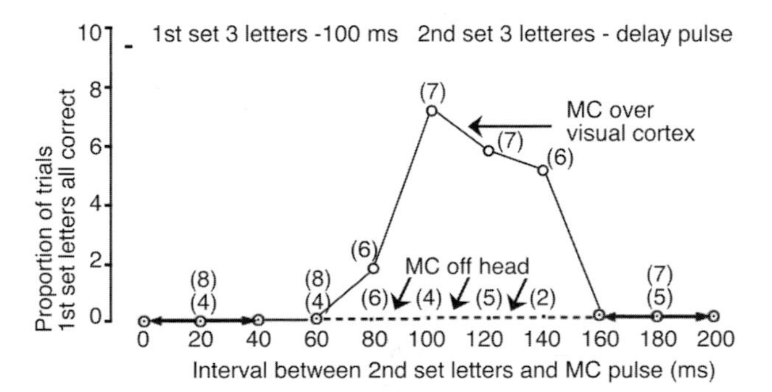

Figure 16 Masking of the first trigram produced by the presentation of a second trigram can be unmasked by TMS suppression of the second trigram. The proportion of trials in which the subjects correctly reported all the letters of the first trigram is presented as a function of the delay between the presentation of the second trigram and the TMS pulse. Numbers in parentheses are the *n* of trials with TMS (top) and with sham TMS (bottom). Reproduced from Amassian *et al.* (1993b), with permission.

et al. (1999a) have disrupted identification of centrally presented letter targets with occipital stimulation as early as 20 ms after stimulus onset, consistent with some reports from single-unit physiology (Wilson *et al.,* 1983; Schmolesky *et al.,* 1998; Celebrini *et al.,* 1993). However, late effective stimulation times may not always mean that higher levels of the visual system are being disrupted and it would not be difficult to launch the counterexplanation that late effects of TMS may be due to disruption of back projections to V1 rather than to disruption of extrastriate areas. Equating TMS time with cortical stage of processing demands corroborating evidence such as supporting single-unit physiology or knowledge of anatomical connections.

C. Virtual Patients: More than "Just" Patients

Modeling neurological patients by transient disruption of focal brain areas with TMS allows the study of brain–behavior relationships avoiding the whim and limitations of natural lesions (Pascual-Leone *et al.,* 1999a). Replicating the effects seen in patients is a good starting point for a TMS study, but it may also be a good end point. Replication is rarely exact and the differences between real and the virtual patients can be important. The work carried out with TMS on visual search and the role of the parietal cortex provides good illustration for this notion. Patients with damage to the right parietal cortex may exhibit a range of deficits that include detection of a conjunction target in a visual search array (Friedman-Hill *et al.,* 1995; Arguin *et al.,* 1990,1993), inability to attend to the left side of visual space (Bisiach *et al.,* 1990, 1994, 1996; Bisiach and Vallar, 1988; Weintraub and Mesulam, 1987), and inaccurate saccadic eye movements. The first two deficits are often linked together and one explanation of these patients' failure to detect conjunction targets is that their spatial attentional problems prevent them from performing what is referred to as "visual binding" (Treisman, 1996). The posterior parietal cortex lies on the dorsolateral surface of the cortex and is easily accessible to TMS. In an attempt to model the effects of right parietal lesions a number of single-pulse studies have been carried out. Ashbridge *et al.* (1997) stimulated right posterior parietal cortex (PPC) while subjects carried out standard "feature" and "conjunction" visual search tasks. Patients with right PPC lesions are impaired on the conjunction tasks but not the feature tasks. TMS over right PPC replicated these two basic findings but with some important differences. Single pulses of TMS were applied at stimulus–TMS onset asynchronies of between 0 and 200 ms and subjects showed two patterns of effect. The reaction time to report "target present" was maximally increased when TMS was applied around 100 ms after visual stimulus onset but to increase the time taken to report "target absent" TMS had to be applied around 160 ms after visual array onset (Fig. 17).

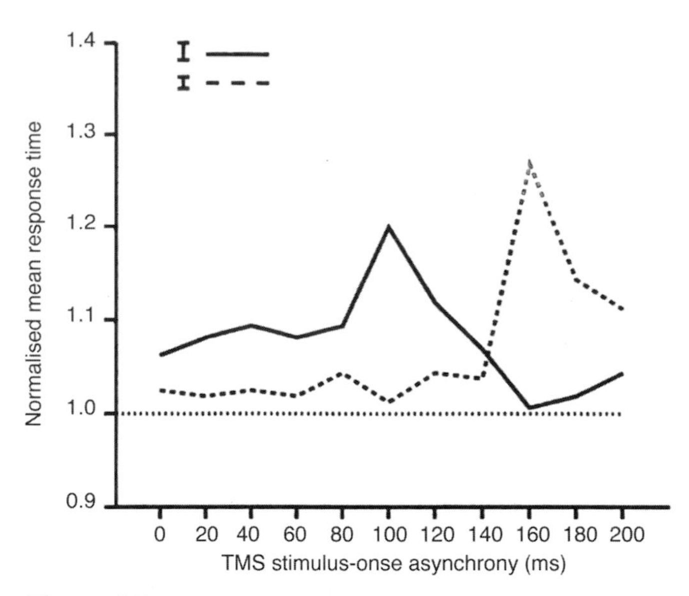

Figure 17 The effects of TMS applied over the right posterior parietal cortex of naïve subjects on a conjunction visual search task (with eight stimuli in the array). Data are normalized to the reaction time on trials when search was performed without TMS. There is a clear effect of TMS on trials when the target was present if the pulse was delivered 100 ms after stimulus onset and also when the target was absent if the pulse was delivered 160 ms after target onset. Solid lines, target present; broken lines, target absent. Vertical bars represent ±1 standard error. Reproduced from Walsh *et al.* (1998), with permission.

Here, then, TMS has replicated the patient data (PPC damage impairs conjunction search) but it adds two further items of information—that the PPC is important for target absent responses and that the mechanisms underlying target present and target absent responses occupy different time windows in PPC. Neuropsychological patients often have an array of problems, which means that standard psychological experimental paradigms have to be modified and reaction time studies might be problematic. TMS provides a means of eliminating such potential confounds. Virtual lesions provide a controlled way to design an experiment such that multiple subjects can be tested on the same paradigm, the contribution of different brain regions to a given behavior can be evaluated, and the role of a given brain area in various behaviors can be assessed. It becomes therefore possible not only to model and reproduce, but also to extend patients' results.

D. Repetitive Pulse Stimulation

The utility of TMS does not reside only in single-pulse neurochronometric studies. If there is no temporal hypothesis under investigation, or a wide temporal window of interest, repetitive-pulse TMS can be used to disrupt behavior. Speech is a subject area in which repetitive-pulse TMS has been particularly valuable to date and illustrates the potential for future development.

One of the most dramatic demonstrations of TMS is magnetically induced speech arrest and several groups have now reported that rTMS over left frontal or either motor cortex can cause subjects to cease speaking or to stutter or repeat segments of words. As far as the neuropsychologist is concerned, this work is preliminary, no more than a calibration experiment in fact, because the emphasis has been on localizing the site of stimulation and/or establishing the most reliable parameters for speech arrest. Similarly, as discussed above, for a clinical application in the presurgical evaluation of patients, more work is required.

Pascual-Leone *et al.* (1991) were the first to induce speech arrest (25-Hz rTMS with a round coil) in a population of epileptic subjects awaiting surgery, and the TMS determination of the dominant hemisphere in all six subjects matched that obtained in the Wada test. The effect was replicated, again in epileptic patients, by Jennum *et al.* (1994; 30-Hz rTMS), whose data also showed a strong concordance with the results of the amobarbital test. The motivation for this and other early experiments on speech was the possibility that TMS could be used to replace the invasive Wada test. In studies which may require hundreds of trials, 25- and 30-Hz frequencies are too high, but a later study by Epstein and colleagues (1996a) identified 4–8 Hz as the optimum range for induction of speech arrest by rTMS in normal subjects. They were also able to distinguish between arrest associated with frontal cortex stimulation, and in the absence of apparent effects on facial muscles, and effects associated with loss of control of the facial muscles. There have been some attempts to examine language functions beyond demonstrations of speech arrest but the best of these have not tested a theoretical prediction and can really be considered as further examples of generalized speech effects. Flitman *et al.* (1998), for example, applied rTMS over frontal and parietal lobes while subjects judged whether a word was congruent with a simultaneously presented picture. Subjects were slower to verify the congruency with TMS but it is not clear whether they were impaired on any particular cognitive aspect of this task or simply that the load on the language system was greater than in the control condition of stating whether the word and picture were surrounded by a rectangular frame.

Three recent studies (Epstein *et al.*, 1999; Bartres-Faz *et al.*, 1999; Stewart *et al.*, 2001a) mark the end of this 10-year period of trying to ascertain the location and reliability of speech arrest effect in normal subjects. All three studies obtained speech arrest lateralized to the left hemisphere with frontal stimulation. Epstein *et al.* suggest that their effects are due to motor cortex stimulation but this is difficult to reconcile with the left unilateral dominance of the effects and also with Bartres-Faz *et al.* and Stewart *et al.*, who provide independent anatomical and physiological evidence of a dissociation between frontal stimulation and pure motor effects (Fig. 18). Bartres-Faz and Stewart's studies both locate the critical site of stimulation to be over

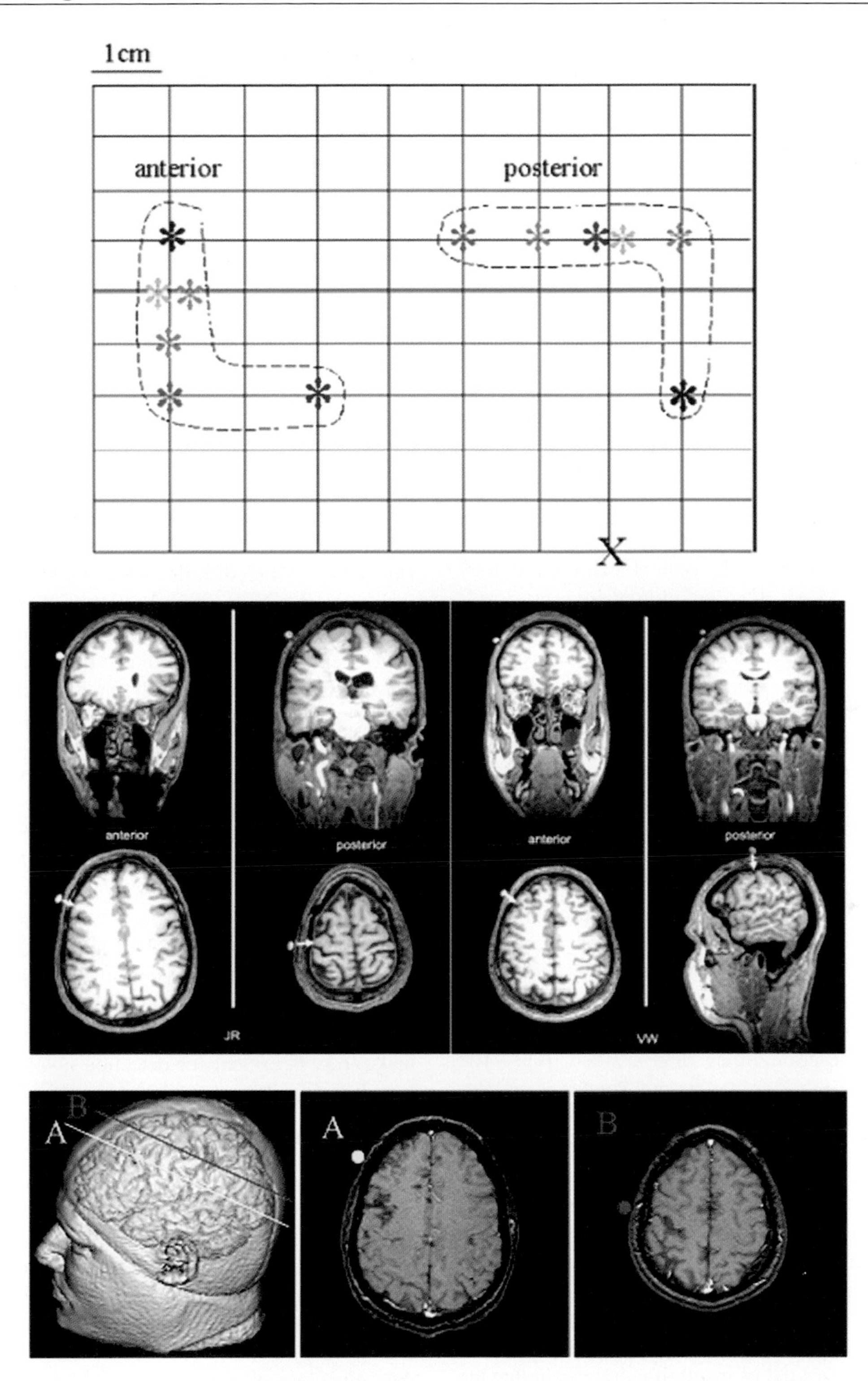

Figure 18 Modified from Stewart *et al.* (2001a), with permission. Top: Asterisks represent the areas that, when stimulated with TMS, produced speech arrest. Stimulation of the anterior site did not produce EMG activity while the posterior site was associated with mentalis muscle activity. Middle: Anatomical MRI showing the anterior and posterior sites that produce speech arrest. Bottom: Modified from Bartres-Faz *et al.* (1999), with permission. The results in a representative subject are presented. The 3D reconstruction of the subject's head MRI (left) demonstrated the sites of TMS application and the level of the axial slices of *f*MRI displayed in the other two panels. The middle illustrates the statistically significant *f*MRI BOLD changes observed during the performance of the verbal fluency task and marks on the scalp the location of the TMS coil for induction of speech arrest. Note that speech arrest is induced by TMS over brain regions that are activated during the verbal fluency task. The right side shows the most representative slice of *f*MRI BOLD activity corresponding to a motor task consisting of opening and closing the right hand, which is shown to be directly under the TMS scalp position that evokes hand movements (but does not lead to speech arrest). Note that the changes corresponding to the motor areas appear more posterior than those responsible for the word generation task. The latter include the areas targeted by rTMS during speech arrest.

the middle frontal gyrus, dorsal to the inferior frontal gyrus and what is usually referred to as Broca's area. These two studies are in agreement with lesion data (e.g., Rostomily *et al.,* 1991), electrical stimulation mapping (Penfield and Roberts, 1959; Ojemann and Mateer,1979; Ojeman, 1983), and PET studies (Ingvar, 1983) that have all shown the several areas, including the middle temporal gyrus, to be important in speech production.

Speech arrest can be obtained from direct electrical stimulation of so many brain regions that it will clearly be very difficult to try to pin down a single area with TMS. The right strategy would seem to be to use TMS to produce language-related dissociations that address theoretical questions. This area is wide open for new approaches using TMS: human lesions that produce language deficits are typically large; animal lesions of course cannot address the question of language. To make use of the localization of speech arrest sites it is not necessary to induce such salient effects on every trial, and we anticipate that the typical neuropsychology experiment will be based on stimulation at intensity levels too low to induce arrest but sufficient to incur reaction time costs in verbal tasks. Stewart *et al.* (2000), for example, have begun to probe parts of the language system by taking the predictions that BA37 has a role in phonological retrieval and object naming. Repetitive-pulse TMS was applied over the posterior region of BA37 of the left and right hemispheres and over the vertex. The rTMS had significant effects on picture naming but no effect on word reading, nonword reading, or color naming. Thus, with respect to object encoding and naming, the posterior region of BA37 would seem to be critical for recognition.

Picture naming was also examined by Topper *et al.* (1998), who applied single-pulse TMS over Wernicke's area and motor cortex. Somewhat paradoxically, TMS over Wernicke's area 500–1000 ms prior to picture presentation resulted in faster reaction times than control trials. The effect was specific to task and area, and Topper *et al.* conclude that TMS "is able to facilitate lexical processes due to a general preactivation of language-related neuronal networks when delivered over Wernicke's area." While these effects are intriguing, they raise several questions about why single-pulse TMS would have facilitatory effects within a system. If generalized arousal within the language system were a tenable explanation, one would have to predict similarly modulated gains whenever TMS was applied over a language-related area. This seems unlikely to be the case. More than in any other kind of result, it is important that the apparently facilitatory effects of TMS are grounded in theoretical frameworks and that the mechanisms proposed in one modality are applicable to others. If, for example, TMS over Wernicke's area facilitates picture naming, then similar facilitations should be obtainable in other modalities. That is to say, if single-pulse TMS over one area facilitates performance in one domain, it should also do so in another. To argue otherwise would go against the physiological similarity of neurons between areas. It is also puzzling that lower intensity TMS produced larger facilitation effects than higher intensity TMS in this study. Further studies of these effects are clearly necessary, but perhaps before basing any further conclusions on a direct facilitation, one should await evidence that an area's primary function can be disabled by TMS.

E. Paired-Pulse TMS: Modulating Intracortical Excitability

The paired-pulse paradigm applied to the motor cortex (Claus *et al.,* 1992; Valls-Sole *et al.,* 1992; Kujirai *et al.,* 1993) can be used to study corticocortical interactions and provides an array of potential clinical applications (see above). A few general findings from standard paired-pulse experiments of potential use in cognitive studies can be stated. Short interstimulus intervals (1–5 ms) can produce intracortical inhibition and slightly longer intervals (7–30 ms) produce facilitation. The mechanisms of these effects have been shown to be mediated by different cortical mechanisms—for example, lower intensity conditioning pulses are required for inhibition than for excitation; coil orientation, and thus direction of current flow, is critical for excitation but not inhibition, and the two phenomena can be independently affected by drugs and neurological disease (see Ziemann, 1999). The ability to potentially increase or decrease sensitivity of a cortical region over a short period of time has clear applications awaiting it in studies of priming, threshold detection, and cortical interactions. The work by Oliveri *et al.* on the role of the parietal lobe in attention employed this strategy for the first time.

Oliveri *et al.* (1999a) used TMS in a tactile stimulus detection task to demonstrate that the right, but not the left, parietal cortex is critical for detection not only of contralateral but also of ipsilateral stimuli. They found that bimanual discrimination is more readily disrupted than unimanual tasks, but only by right parietal TMS. Most importantly, they showed that the contribution of the right parietal cortex takes place around 40 ms after the tactile stimuli are applied, hence suggesting involvement of late cortical events. Fierro *et al.* (2000) extended these results showing that TMS can not only induce extinction to simultaneous visual stimulation of the two hemifields, but can also correct pseudoneglect. The neurophysiology of extinction might in fact be different from that of neglect, the latter being of greater clinical significance (Bisiach *et al.,* 1996; Kinsbourne, 1994; Vallar, 1998). Patients with neglect face tremendous difficulties in rehabilitation as they do not realize the extent of their own limitations. Understanding neglect better will hopefully aid in developing suitable methods for its treatment. Oliveri's and Fierro's results seem to support the widespread notion that the right hemisphere contains representations of both hemi-spaces, while the left hemisphere is concerned with attending only to the contralateral hemispace. However, interhemispheric competition (possibly asymmetrical) of cortical or subcortical structures might be better suited to explain some of these effects. Only interhemispheric competition provides a plausible explanation for the puzzling effects, extensively studied in cats, by which visual hemineglect induced by a lesion of one posterior cortex can be paradoxically reversed by secondary damage to contralateral cortical and subcortical structures (Lomber and Payne, 1996). This notion has been experimentally tested with TMS in humans (see below, figure 23, Hilgetag, *et al.*, 2001). Using exactly the same logic, Oliveri *et al.* (1999b) have used TMS to test this notion in 28 patients with right (n = 14) or left (n = 14) brain lesions. Single-pulse TMS was delivered to frontal and parietal scalp sites of the unaffected hemisphere 40 ms after application of a unimanual or bimanual electric digit stimulus. In patients with right hemispheric damage, left frontal TMS significantly reduced the rate of contralateral extinctions compared with controls. Left parietal TMS did not significantly affect the number of extinctions compared with baseline. Left-brain-damaged patients did not show equivalent results. In them, TMS to the intact, right hemisphere did not alter the recognition of bimanual stimuli. TMS to the left frontal cortex in patients with right hemispheric lesions significantly reduced the rate of contralateral extinctions, even though, as mentioned above, the same type of stimulation did not affect task performance in normal subjects. These results suggest that extinctions produced by right-hemisphere damage may be dependent on a breakdown in the balance of hemispheric rivalry in directing spatial attention to the contralateral hemispace, so that the unaffected hemisphere generates an unopposed orienting response to the side of the lesion (Fig. 19). TMS to the left frontal cortex in patients with right hemisphere damage and contralesional extinction ameliorates their deficit. The mechanism of action of TMS in this setting could involve crossed frontoparietal inhibition. However, interactions at subcortical level cannot be excluded.

Oliveri *et al.* (2000) followed their study of right-brain-damaged patients with neglect, with an experiment in which paired-pulse TMS was used to induce selective intracortical inhibition or facilitation of the unaffected hemisphere depending on the interstimulus interval. The hypothesis was that cortical inhibition would result in an improvement and cortical facilitation in a worsening of contralesional extinction compared with baseline. Paired-pulse TMS with the interstimulus interval set at 1 or 10 ms was applied to the left parietal or frontal cortex at various intervals following bimanual electric digit stimulation. At an interstimulus interval of 1 ms, which leads to intracortical inhibition, paired-pulse TMS led to a greater improvement in extinction than that induced by single-pulse TMS (Oliveri *et al.*, 1999a) (Fig. 20). On the other hand, with paired-pulse TMS at 10 ms, which is believed to increase cortical facilitation, there was a worsening of extinction compared with baseline and a complete reversing of the effects of single-pulse TMS (Fig. 20). These results shed further light on the mechanisms of tactile extinction. In addition, this study illustrates the potential of paired-pulse TMS to selectively modulate intracortical excitability and extend the results of single-pulse TMS.

F. Paired-Pulse TMS: Studying Corticocortical Connectivity

Paired-pulse TMS can consist in pairs of stimuli delivered at the same or different intensity through a single TMS coil, hence targeting a single brain area (see above). Alternatively, paired-pulse TMS can be set up as two stimuli delivered through two different coils targeting separate brain regions with a variable interstimulus interval. Such an application provides a unique opportunity to study corticocortical connectivity and the behavioral role of feedback and feedforward connections between brain regions. Pascual-Leone and Walsh (2001) have provided the first illustration of such a methodology in the study of visual awareness.

The role of primary visual cortex (V1) in awareness is a matter of long-standing and ongoing debate. While some investigators argue that specialized modules of visual cortex are autonomous and can lead to visual awareness on their own (i.e., without V1 contribution; see Zeki and Bartels, 1999; Zeki, 2001), others have maintained that the presence of V1 is necessary for conscious perception (see Stoerig and Cowey, 1997). Indeed, the extent and nature of residual visual capabilities of striate cortex-lesioned patients are still debated today. Destruction of primary visual cortex leads to blindness (hemianopsia) in the visual field contralateral to the lesion. However, numerous reports have detailed some residual capacities (mostly motion detection) in the blind field of destriate patients. One such patient is G.Y., a well-studied subject whose left striate cortex was almost entirely destroyed at the age of 8. The sometimes unconscious (blindsight; see Weiskrantz, 1997) ability of G.Y. to perceive motion is associated with the integrity of the specialized motion area (MT/V5) in both hemispheres (see Zeki and ffytche, 1998). Cowey and Walsh (2000) used the capacity of TMS to induce phosphenes to explore the role V1 plays in visual awareness. When regions corresponding to V1 are stimulated, phosphenes are static and retinotopically organized, whereas MT/V5 stimulation can result in moving phosphenes (Stewart *et al.*, 1999). In G.Y., stimulation of the occipital pole of the intact hemisphere resulted in the perception of static phosphenes, as should be expected in a normal brain. Moreover, stimulating a region corresponding to motion area MT/V5 induced moving phosphenes. When TMS was applied to MT/V5 of the damaged hemisphere, no phosphenes were elicited, despite previous PET data showing MT/V5 activation in response to motion stimuli presented in the blind field (Barbur *et al.*, 1993). The authors

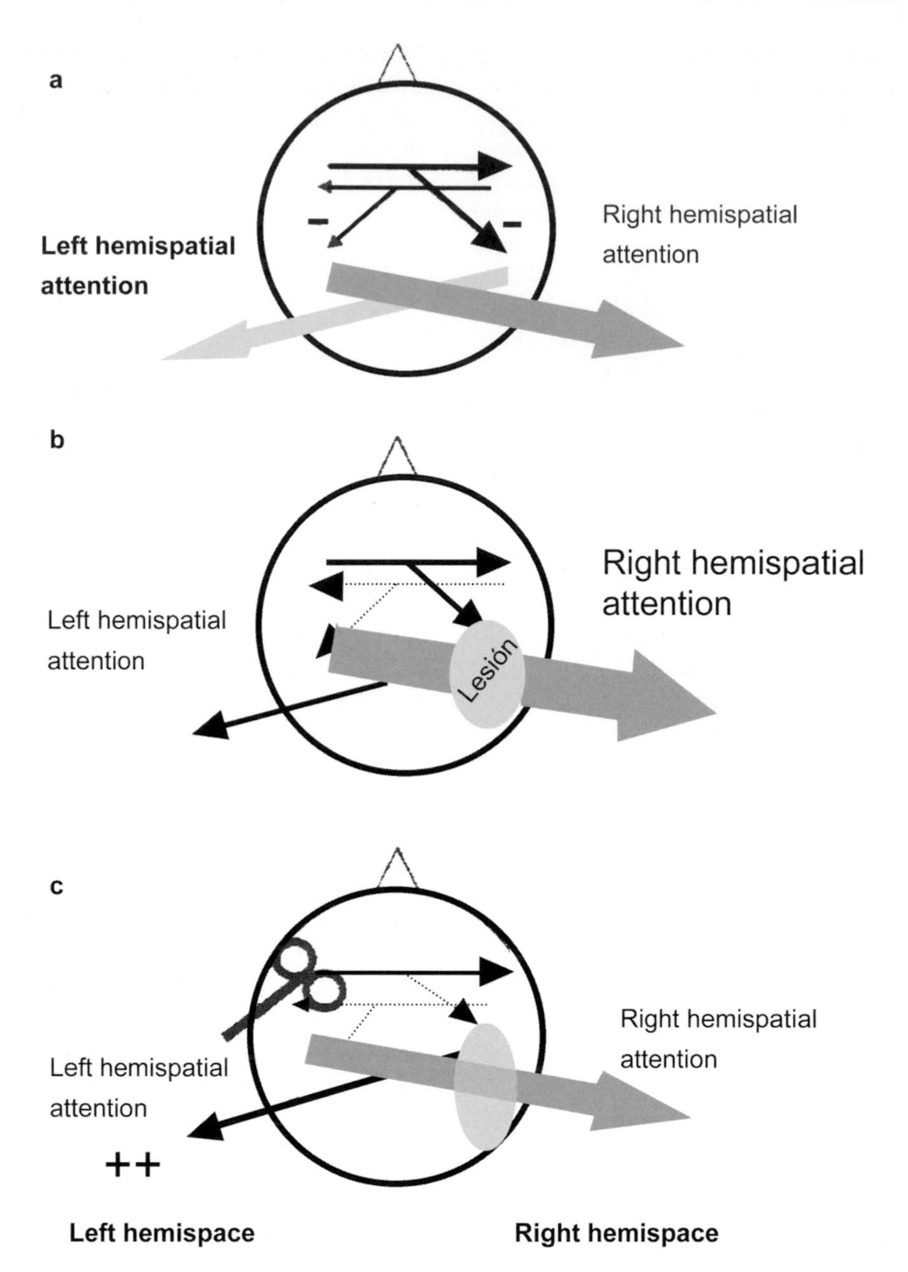

Figure 19 Proposed framework for left- and right-hemispheric contributions to the neural representation of egocentric space. In normal volunteers (A), the mutual inhibitory callosal connections between the two hemispheres are asymmetric with the dominant hemisphere exerting greater inhibition onto the nondominant hemisphere, hence producing a slight hyperorientation to the right side. Following a right-hemisphere stroke (B), the unbalanced effect of the left hemisphere results in excessive attention toward the right (ipsilesional) hemispace (dashed arrow). In right-brain-damaged patients, left frontal TMS interferes with the hypothesized left frontal–right parietal inhibition, thus disinhibiting the right parietal cortex and partially restoring left extinctions (black arrow). Modified from Oliveri *et al.* (1999a, 1999b), with permission.

applied high-intensity TMS at 36 positions over a 5 × 5-cm grid centered over the expected position of V5 without eliciting one phosphene. These observations led to the conclusion that the absence of phosphenes "when TMS was similarly applied to extrastriate visual regions of a patient with hemianopia caused by destruction of V1 and parts of V2 suggests that TMS can only induce conscious visual perception when V1/V2 are intact."

Using paired-pulse TMS delivered to separate brain areas, Pascual-Leone and Walsh (2001) have studied the conscious perception of phosphenes further (Fig. 21). Pascual-Leone and Walsh hypothesized that conscious perception of moving

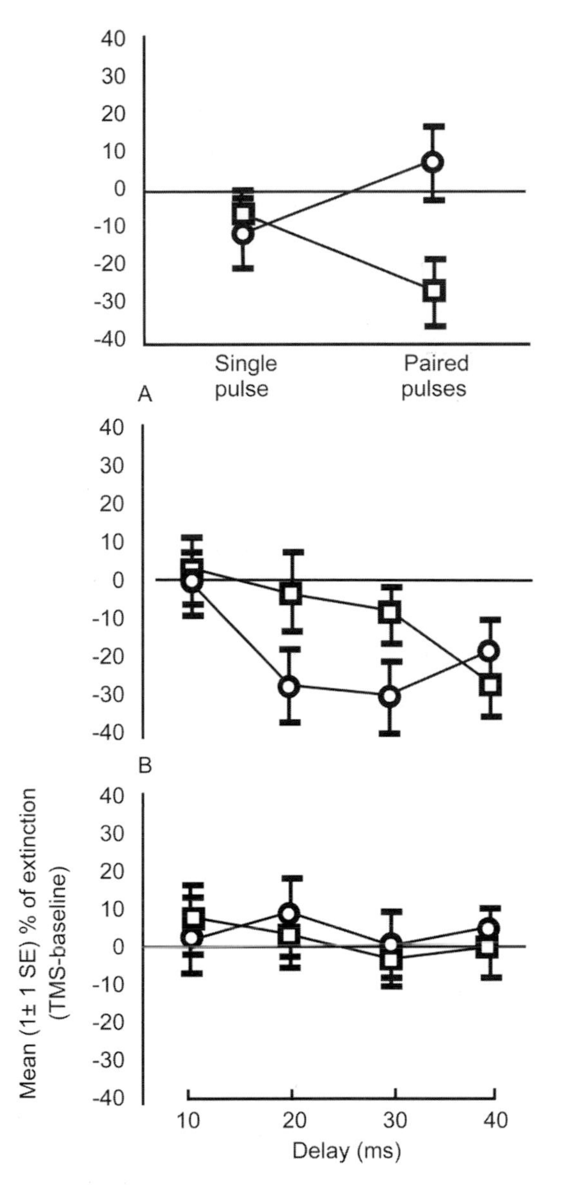

Figure 20 Mean percentage (±standard error) of contralesional extinction for single-pulse and paired-pulse TMS paradigms. Results express the difference between baseline and TMS conditions. Trials with paired-pulse TMS at 1 ms (filled squares) and 10 ms (open circles) show divergent results. See text for details. Modified from Oliveri *et al.* (2000), with permission.

phosphenes, induced by MT/V5 stimulation, may be dependent upon a feedback loop between MT/V5 and V1. A suprathreshold single-pulse TMS over MT/V5 was used to create a moving phosphene in normal subjects. This first pulse was followed by a second, subthreshold pulse over V1 between 0 and 90 ms later. If MT/V5–V1 feedback projections are necessary for the perception of phosphenes, the second pulse should disrupt perception of the attributes encoded by the extrastriate area. Indeed, when the V1 TMS pulse was applied 5 to 45 ms after the MT/V5 pulse, there was a marked decrease in the quality of phosphenes. In five of the eight subjects studied, the double-pulse paradigm completely abolished the perception of phosphenes. These data suggest that magnetic stimulation of V1 disrupts the flow of information going from MT/V5 to primary visual cortex, which usually leads to conscious perception of a moving phosphene. In addition, this study illustrates the potential use of TMS to study the timing of corticocortical interactions and their causal role in behavior and cognition.

G. Online TMS and Offline TMS

The studies discussed above all applied TMS during task performance. TMS can also be used in what has been termed its distal or offline mode. Such offline use of rTMS in the study of cognition is based on studies of motor cortex in which it has been shown that a continuous train of stimulation can modulate cortical excitability beyond the duration of the rTMS train itself (Chen *et al.,* 1997; Berardelli *et al.,* 1998; Pascual-Leone *et al.,* 1998). Slow rTMS (1 Hz) applied to motor cortex can give rise to a lasting decrease in corticospinal excitability (Chen *et al.,* 1997; Maeda *et al.,* 2000) and it seems reasonable to assume a similar suppression of excitability when slow rTMS is applied to nonmotor, cortical areas. This approach has recently been implemented in a number of cognitive studies, including visual perception (Kosslyn *et al.,* 1999), spatial attention (Hilgetag *et al.,* 2001), motor learning (Robertson *et al.,* 2001), working memory (Mottaghy *et al.,* 2001), and language (Shapiro *et al.,* 2001). It is hypothesized that application of slow rTMS to cortical areas other than motor creates similar decreases in cortical excitability which lead to measurable behavioral effects (Pascual-Leone *et al.,* 1999a). In this paradigm, performance on a given task is evaluated before (baseline) and after application of rTMS. This enables study designs in which the potential disruption of ongoing TMS on task performance is eliminated.

In the first application of this "offline" TMS strategy, Kosslyn *et al.* (1999) investigated the role of primary visual cortex in visual imagery using identical task conditions in a rTMS study and in a PET experiment. The PET results revealed activation of V1 during visual imagery and provided the target for the application of TMS. In the TMS experiment subjects received 1-Hz stimulation at 90% of motor threshold for 10 min to the area of activation in striate cortex in the subjects' PET scan. Following rTMS, subjects were required to visualize and compare the properties of memorized images of grating patterns or of real images of the same stimuli. The reaction times of subjects were significantly increased in both real perception and imagery conditions (Fig. 22), showing that area V1 was critical for visual imagery as well as real perception. The effect of TMS was greater for imagery than for real perception, which may reflect the fact that the imagery condition was harder than the perception condition.

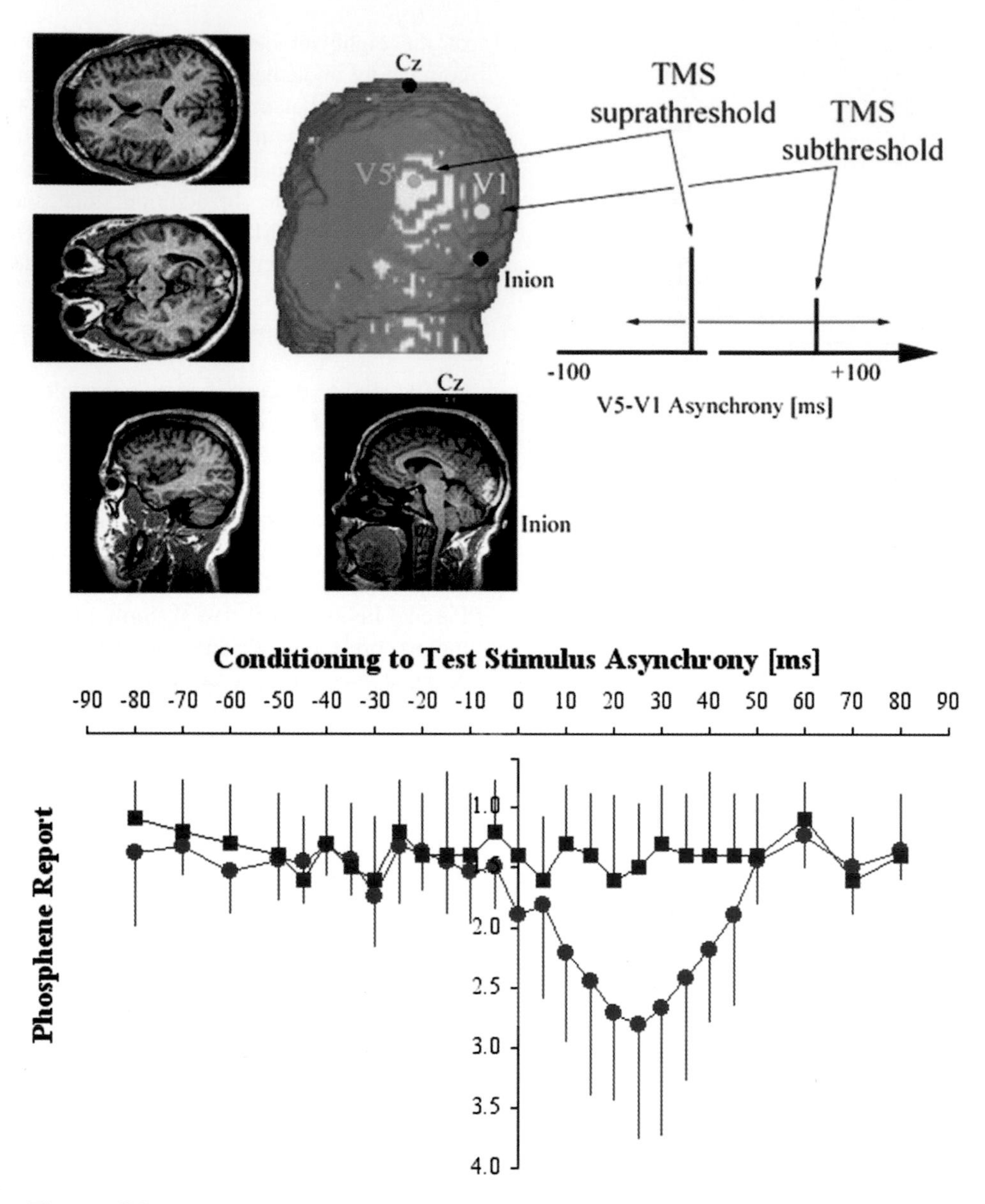

Figure 21 Reproduced from Pascual-Leone and Walsh (2001), with permission. Top: Schematic representation of the experimental design of a V5–V1 interaction study. The brain MR image from one of the study subjects displays a representative example of the site of stimulation for induction of stationary (V1) and moving phosphenes (MT+/V5). The location on the subject's scalp of the center of the intersection of the wings in the 8-shaped TMS coil is projected, perpendicular to the scalp surface, onto the subject's brain as reconstructed from an anatomical MRI. Bottom: Mean responses of all subjects (n = 8) to combined stimulation of V5 and V1. The V5–V1–TMS asynchrony is displayed on the x axis: negative values indicate that V1 received TMS prior to V5, and positive values indicate that V1 was stimulated after V5. The subjects made one of four judgments. The phosphene elicited by V5 TMS was (1) present and moving, (2) present but the subject was not confident to judge whether moving, (3) present but stationary, (4) not observed. TMS over V1 between 10 and 30 ms after TMS over V5 affected the perception of the phosphene.

H. Paradoxical Facilitations

Brain injury sometimes results in functional facilitations (see Kapur, 1996). The two main classes of facilitation have been termed "restorative," wherein a hitherto deficient function has returned (as in the Sprague effect), and "enhancing," in which some damage or loss of function results in the patient performing better than normal subjects at some task. Both classes of facilitation reveal much of interest about the dynamic interactions between different modalities or even components of sensory modalities. Nevertheless, as Kapur notes "such findings have often been ignored or undervalued

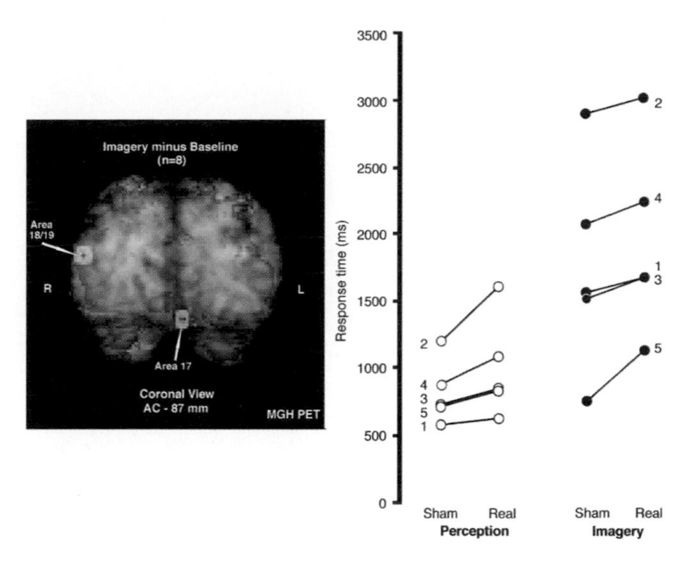

Figure 22 TMS and mental imagery. Results of delivering rTMS over occipital cortex before perception and imagery conditions. "Real" TMS occurred when the magnetic field was directed into area 17 and sham rTMS occurred when the field was diverted away from the head. TMS over visual cortex slowed response times in both perception and imagery conditions in all five subjects. Reproduced from Kosslyn *et al.* (1999), with permission.

in the brain–behavior research literature." Perhaps this is because paradoxical facilitations are less common and less salient than deficits and also more difficult to interpret. Recent neurocomputing work may be useful in imposing some direction and also constraints on the search for and interpretation of facilitatory effects of TMS (Hilgetag *et al.*, 1999; Young *et al.*, 1999, 2000). One simulation, for example, showed that the connectivity of a cortical area was a strong predictor of the effects of lesions on the rest of the network as well as how that area responded to a lesion elsewhere in the network. This may seem like a truism but the kind of connectivity analysis offered by these models is not really taken into account in classical lesion analysis (see also Robertson and Murre, 1999; Rossini and Pauri, 2000) and the modeling work has begun to make these predictions explicit and testable.

In the visual system Walsh *et al.* (1998) stimulated visual area V5 in an attempt to model the "motion-blind" patient L.M. (Zihl *et al.*, 1983) and indeed V5 stimulation did impair performance on visual search tasks that involved scanning complex motion displays. On displays in which motion was absent or irrelevant to task performance, subjects were faster with TMS than in control trials. This can be interpreted as evidence that the separate visual modalities may compete for resources and the disruption of the motion system may have liberated other visual areas from its influence. In this experiment the subjects received blocks of trials of a single type and therefore knew whether the upcoming stimulus array would contain movement or color or form as the important parameter. When the types of trials are interleaved such that the subject does not have advance information the enhancing effects of TMS were not obtained. Thus it seems that a combination of priming (due to the advanced knowledge of the stimuli) and weakening of the V5 system (by TMS) was required to enhance performance on color and form tasks. Conceptually similar is the finding of Seyal *et al.* (1995), who observed improvements in tactile sensitivity as a result of stimulation of the somatosensory cortex ipsilateral to fingers being tested. The interpretation here is also based on disinhibition of the unstimulated hemisphere.

The facilitations reported by Walsh *et al.* were obtained with online TMS but similar effects have been reported using distal TMS. A clear example of interhemispheric rivalry revealed by offline rTMS has recently been reported by Hilgetag *et al.* (2001) in a study designed to address the notion of interhemispheric competition in guiding attention. They found ipsilateral enhancement of visual attention, compared to normal performance (Fig. 23), produced by rTMS of the parietal cortex at stimulation parameters known to reduce cortical excitability. Healthy, right-handed volunteers received rTMS (1 Hz, 10 min) over right or left parietal cortex (at P3, P4 EEG coordinate points, respectively). Subsequently, subjects' attention to ipsilateral visual targets improved significantly while contralateral attention diminished. Additionally, correct detection of bilateral stimuli decreased significantly, coupled with an increase in erroneous responses for ipsilateral unilateral targets. Application of the same rTMS paradigm to motor cortex as well as sham magnetic stimulation indicated that the effect was specific for stimulation of parietal cortex. These results underline the potential of focal brain dysfunction to produce behavioral improvement and provide experimental support for models of visuospatial attention based on the interhemispheric competition of cortical components in a large-scale attentional network.

V. TMS Limitations

A. Safety Considerations

The safety of single-pulse stimulation is well established but further precautions should be taken when using repetitive-pulse TMS. The magnetic field produced by stimulating coils can cause a loud noise and temporary elevations in auditory thresholds have been reported (Pascual-Leone *et al.*, 1993). The use of ear plugs is recommended in all experiments. Some subjects may experience headaches or nausea or may simply find the face twitches and other peripheral effects of TMS too uncomfortable. Such subjects obviously should be released from any obligation to continue the experiments. More serious are the concerns that TMS may induce an epileptic seizure. There are a number of cases of epileptic fits induced by repetitive pulse TMS and caution is necessary. As a guide, any subject with any personal or

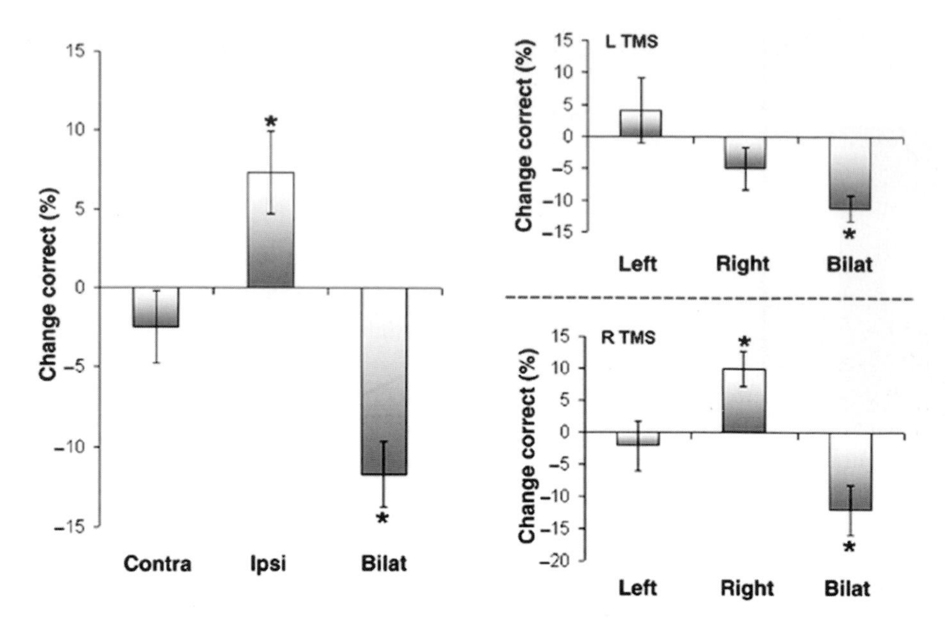

Figure 23 Modified from Hilgetag *et al.* (2001), with permission. Changes in correct stimulus detection after parietal rTMS. The diagrams are based on changes in the number of correctly detected stimuli (relative to the total number of presented stimuli) averaged for both stimulus sizes and all subjects. (a) The pooled data show a significant increase in performance ipsilateral to the parietal rTMS location (increase in relative percentage points 7.3%, SEM 2.6%) and a trend to decreased contralateral performance (reduction by 2.5%, SEM 2.3%). In addition, detection of bilateral stimuli decreased significantly (–11.7%, SEM 2.0%). These trends are also apparent after separating data for (b) left parietal TMS and (c) right parietal rTMS. Significant trends (as determined by *Z* tests, are marked by stars.

family history of epilepsy or other neurological condition should be precluded from taking part in an experiment which does not involve investigation of that condition. Pascual-Leone *et al.* (1993) assessed the safety of rTMS and noted that seizures could be induced in subjects who were not associated with any risk factors. The paper presents some guidelines for the use of rTMS and familiarity with this paper should be a prerequisite of using rTMS. However, the paper is not exhaustive—it is based on only three sites of stimulation and expresses pulse intensity as a percentage of *motor* threshold. It has recently been argued that studies which apply rTMS to areas other than the motor cortex cannot simply lift stimulation parameters and criteria based on motor cortex excitability and assume they transfer to other conditions. There is no necessary relationship between motor cortex excitability and that of other cortical regions (Stewart *et al.*, 2001). It is also recommended that anyone wishing to use rTMS visit the TMS Website (http://pni.unibe.ch/maillist.htm). The TMS community is constantly reviewing safety procedures and this Website is a starting point for access to sound information (although much of it is directed to a clinical audience). A more recent paper (Wassermann, 1998) summarizes the consensus that exists within the TMS community. The adverse effects recorded include seizures, though these are rare, some enhancement effects on motor reaction time and verbal recall, and effects on affect (some subjects have been reported to cry following left prefrontal rTMS and others to laugh). There is little information about potential longer term problems with rTMS but the issue cannot be ducked. If, on the one hand, rTMS is potentially useful in the alleviation of depression (Pascual-Leone, *et al.*, 1996; George *et al.*, 1995, 1996) it must be conceded that rTMS can have longer term effects. It would be disingenuous to suggest that all long-term effects are likely to beneficial rather than deleterious. It should be noted, however, that the improvements in mood as a result of rTMS follow several sessions of magnetic stimulation and the effects were cumulative (Pascual-Leone, *et al.*, 1996; George *et al.*, 1995, 1996). A simple precaution that may be taken is to prevent individual subjects from taking part in repeated experiments over a short period of time. The use of rTMS should follow a close reading of the reports of Pascual-Leone *et al.* (1993) and Wassermann (1998).

Niehaus *et al.* (1999) have approached this question using transcranial Doppler sonography (a noninvasive technique that allows blood flow, as velocities, to be recorded from intracranial arteries; see Bogdahn, 1998) to observe rapid changes in the hemodynamic response to TMS and to compare this with “real” sensory (in this case, visual) stimulation. As Fig. 24 shows, TMS produced changes in blood flow that occurred earlier and were larger in the hemisphere ipsilateral to stimulation over the occipital lobe. There was

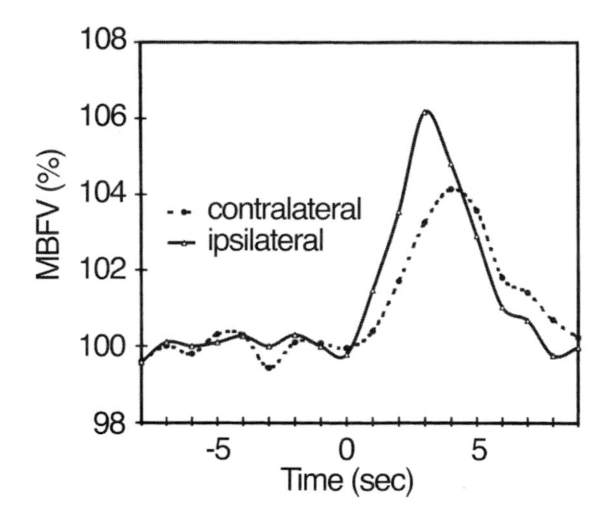

Figure 24 The cerebral hemodynamic response to TMS over the motor cortex in10 subjects. Five trains of 10 Hz were given to each subject. Time-locked average MBFV changes in the middle cerebral artery ipsilateral and contralateral to the stimulation site is shown. Reproduced from Niehaus *et al.* (1999), with permission.

also a close correspondence between blood flow associated with trains of 5-Hz TMS and 5-Hz light flicker (Fig. 25), thus supporting the assumption that blood flow changes evoked by TMS are a reflection of neural activity rather than nonspecific effects on the vascular system. Importantly, there were no long-term changes associated with the rTMS.

B. Spatial Resolution

Several converging lines of evidence now show that there is good reason for confidence in the anatomical, but more

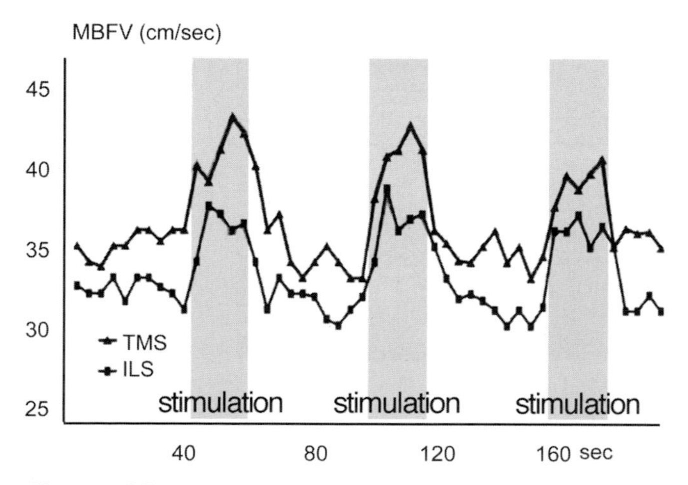

Figure 25 Changes in MBFV in the left posterior cerebral artery during rTMS over the left occipital cortex or visual stimulation with light flicker. Stimulation was performed with rTMS trains of 5 Hz and 20-s duration and intermittent visual stimulation (ILS) with the same frequency and stimulation duration.

importantly in the functional, specificity of TMS. One could simply appeal to the surface validity of TMS—Barker's first demonstration of motor cortex stimulation, for example, was itself strongly suggestive of relatively selective, suprathreshold stimulation of the hand area of the cortex. Perhaps there was some spread of current to arm, shoulder, and face regions of the motor cortex, but in the absence of movements from these parts of the body one must infer that the stimulation was *effectively* precise, i.e., stimulation of the other areas was subthreshold for producing a behavioral effect. There are many other examples of surface validity: phosphenes are more likely if the coil is placed over the visual cortex (Meyer *et al.,* 1991; Kastner *et al.,* 1998; Kammer, 1999; Marg, 1991; Stewart *et al.,* 1999), speech arrest is more likely if stimulation is applied over facial motor or frontal cortex (Pascual-Leone *et al.,* 1991; Epstein *et al.,*1996b; Stewart *et al.,* 2001a), and neglect and extinction-like deficits are more likely if the coil targets the parietal lobe (Pascual-Leone *et al.,* 1994c; Ashbridge *et al.,* 1997; Fierro *et al.,* 2000). Mapping of motor cortex with EMGs also shows precise mapping of the fingers, hand, arm, face, trunk, and legs in a pattern that matches the gross organization of the motor homunculus (Singh *et al.,* 1991), sensitive both to coil location and to intensity (Brasil-Neto *et al.,* 19992a,b). There are also more direct measures of the specificity of TMS. Wassermann *et al.* (1996) mapped the cortical representation of a hand muscle with TMS and coregistered the inferred volumetric fields with anatomical MRIs from each subject, and these were in turn coregistered with PET images obtained while subjects moved the finger that had been mapped with TMS. In all subjects the estimated fields induced by TMS met the surface of the brain at the anterior lip of the central sulcus and extended along the precentral gyrus for a few millimeters anterior to the central sulcus. Compared with the PET activations the MRI locations were all within 5–22 mm—an impressive correspondence across three techniques. A similarly impressive level of correspondence has also been seen in other studies that have correlated TMS with *f*MRI (Terao *et al.,* 1998a,b) and with MEG (Morioka *et al.,* 1995a,b; Ruohonen *et al.,* 1996). There are reasons for caution in interpreting these data (see Wasserman *et al.,* 1996), for example, the hand area activated lies deep in the central sulcus, possibly too deep to be directly activated by TMS and therefore presumably activated transsynaptically. The evidence for transsynaptic activation comes from a comparison of the EMG latencies elicited by electrical or magnetic stimulation (Day *et al.,* 1987, 1989a; Amassian *et al.,* 1990). Magnetically evoked latencies are approximately 1–2 ms longer than electrically evoked ones and this can be explained on the basis of which neurons are most likely to be stimulated by each technique (Rothwell, 1997). TMS is more likely to stimulate neurons that run parallel to the cortical surface, whereas electrical stimulation can directly stimulate pyramidal output

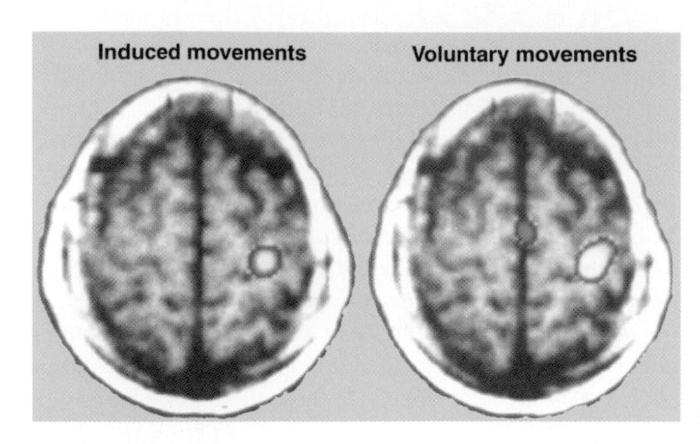

Figure 26 The spatial and functional specificity of TMS is evident in the correspondence between blood flow changes induced by TMS over the motor cortex to produce a finger movement and the activity produced by intentional movement which also produces SMA activity. Reproduced from Siebner *et al.* (1998), with permission.

neurons that run orthogonal to the cortical surface. Thus the 1- to 2-ms delay between electrical and magnetic cortical stimulation may be accounted for by the time taken for the stimulation to be transmitted from the interneurons to the pyramidal cells. Knowledge of which kinds of cells are stimulated based on temporal information can inform the interpretation of functional specificity.

Further evidence of the accuracy of TMS is seen in Fig. 26. Siebner and colleagues compared the changes in regional cerebral blood flow caused by 2-Hz rTMS over the motor cortex, sufficient to elicit an arm movement, with blood flow changes due to the actual movement of the arm. The correspondence was striking. TMS-induced movements and voluntary movements both activated SM1 (area 4) ipsilateral to the site of stimulation. Voluntary movement also activated ipsilateral SMA (area 6) and the motor activity associated with the voluntary movement was more extensive than that elicited by rTMS. This could be because the voluntary arm movement was slightly greater than the TMS movement or because voluntary activity would involve more muscles than TMS activity. Whatever the difference, it is a clear example of the specificity of TMS and the physiological validity of TMS effects. Further evidence comes from studies of TMS effects measured by *f*MRI by George and Bohning and their colleagues (Fig. 27). These studies are important examples of the spatial specificity of TMS—they do not mean that the induced electric field is limited to the functional units stimulated, nor do they suggest that activation of neurons is limited to the areas seen in PET and *f*MRI; but they show unequivocally that the theoretical spread of the induced field is not the determinant of the area of effective stimulation and that the functional localization of TMS is, to a significant degree, under experimenter control.

Studies of EEG responses by Ilmoniemi and colleagues (1997) provide another demonstration of the relative primary and secondary specificity of TMS. As Fig. 28 shows, stimulation over the visual or motor cortex elicits EEG around the site of stimulation in the first few milliseconds after TMS. Within 20–30 ms this activity is mirrored by a secondary area of activity in the homotopic regions of

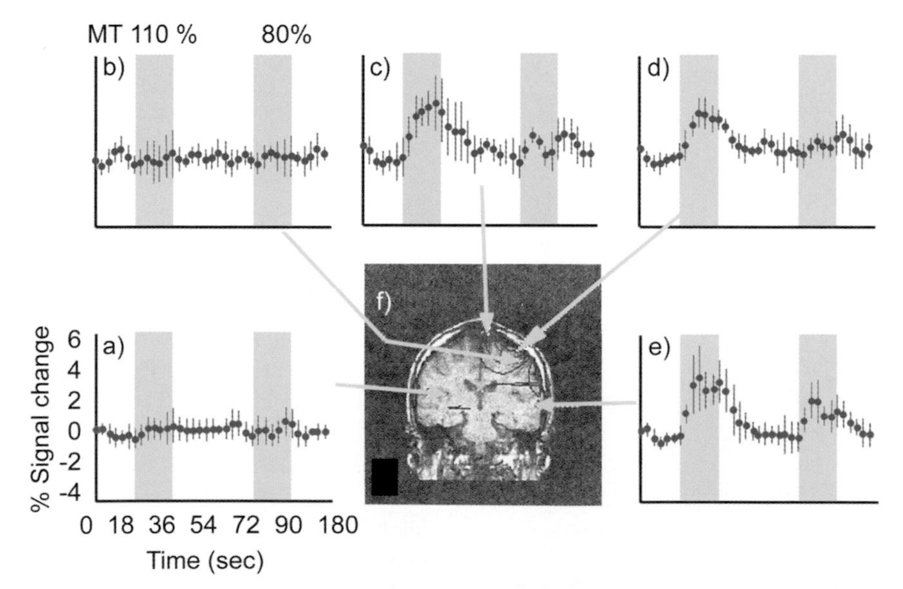

Figure 27 Time activity curves of a subject's brain during rest and with TMS over the thumb representation of the motor cortex. These data were obtained by interleaving BOLD *f*MRI and TMS. The TMS was given at 1 Hz for 8 s The spatial and temporal resolutions of the measurements are approx 2 mm and 3 s. Reproduced from Bohning, *et al.* (1999), with permission.

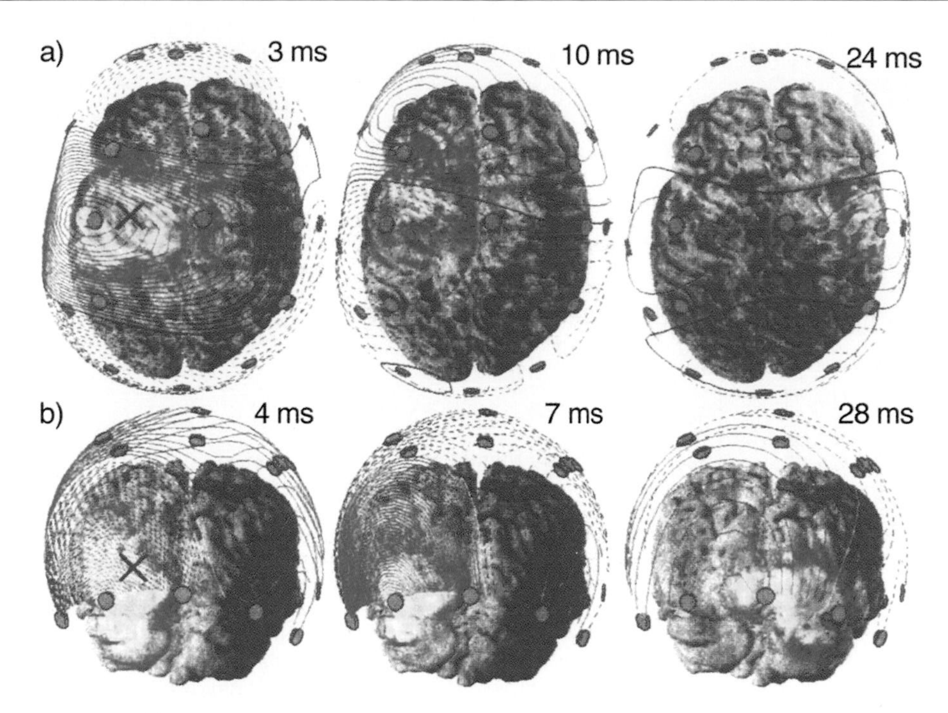

Figure 28 Duration of changes in neural activity induced by TMS. Four milliseconds after TMS over the occipital lobe most of the electrical activity recorded with high-resolution EEG is around the area directly under the TMS stimulation site (marked by the X). By 7 ms this has spread to the midline and by 28 ms there is clearly contralateral activation. Reproduced from Ilmoniemi *et al.* (1997), with permission.

the contralateral hemisphere. These delays in homotopic areas are a rich source of hypotheses regarding the timing of effects in interhemispheric interactions. The utility and specificity of this combination of techniques were further demonstrated by applying TMS to the motor cortex of a patient who had suffered a lesion to the right basal ganglia and had lost fine finger control in his left hand and some control of his left arm. When the intact hemisphere was stimulated EEG responses were seen in both the ipsilateral motor cortices. When the motor cortex ipsilateral to the affected basal ganglia was stimulated, some EEG was seen ipsilaterally but none was transmitted interhemispherically to the intact hemisphere.

Other evidence of the effective focality of TMS is shown in Fig. 29. Pascual-Leone and colleagues compared the correspondence between the sites stimulated in an experiment targeting the motor cortex and the inferior frontal gyrus. The position of the motor cortex across 20 subjects was remarkably consistent with respect to the precentral sulcus, but there was considerable variability in the location of the frontal site relative to the inferior frontal and central sulci.

The depth of penetration of TMS is another important question and as with the question of lateral specificity there is no easy answer, but again there are good reasons to think that the approximations available are meaningful and can be used to guide interpretation of results. Models of the electric field at different depths from the coil suggest that relatively wide areas are stimulated close to the coil, decreasing in surface area as the field is measured at distances farther from the coil. The image offered by these models is of an egg-shaped cone with the apex, which marks the point of the smallest area of stimulation, farthest from the coil. For a standard figure-8 coil, one estimate is that stimulation 5 mm below the coil will cover an area of approximately 7×6 cm. This area decreases to 4×3 cm at 20 mm below the coil, i.e., in the region of the cortical surface (Fig. 30). Calculations of induced electric fields as a function of depth can also be used as a guide to specificity because stimulation at points where the fields overlap allows subtraction of the effects. If a coil at position A disrupts performance on a behavioral task, the effective site of stimulation could be said to be anywhere within, around, or connected to the neurons crossed by the field. If stimulation at neighboring sites B and C fails to disrupt the task, then the overlap in fields between A and B and between A and C can be said to be ineffective regions of the field and the most effective field is the shaded subregion of A. Thus the notion of the effective resolution of TMS can be refined; whereas a single-pulse of TMS cannot be said to have a small, volumetric resolution in the cortex, from a functional point of view it can be shown to have a small scalp resolution and an inferred or subtracted volumetric resolution when multiple sites are compared. A comparison might be made here with *f*MRI and, say, a cortical area such as visual area V5 (Watson *et al.*, 1993); it is clearly

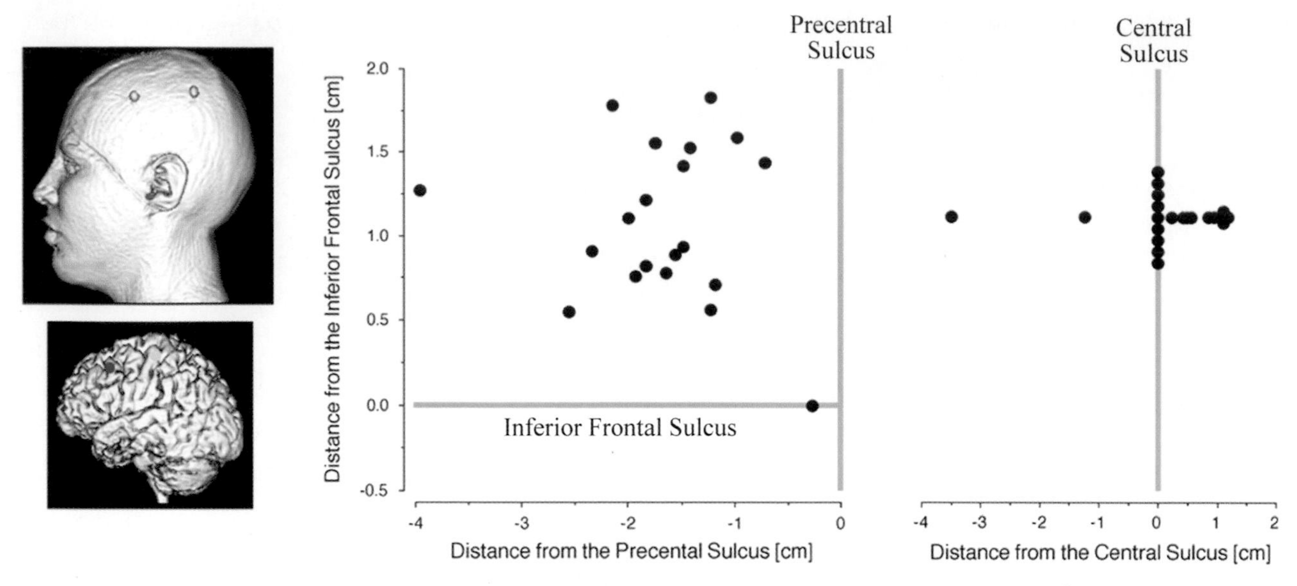

Figure 29 Identification of brain areas targeted by TMS—a comparison across subjects. Optimal sites for activation of abductor pollicis brevis muscle and from a scalp position 5 cm anterior to it and in the same parasagittal plane. Data from 20 subjects are presented. The position of the figure-8 coil on the scalp was marked with a vitamin A capsule and an MRI was obtained and reconstructed in three dimensions. The location of the capsule on the scalp was projected perpendicular to the skull surface onto the brain and the point of intersection with the projection line was marked. The scattergram displays these points (i.e., the brain area targeted by TMS) in relation to the central sulcus and in relation to the precentral and the inferior frontal sulcus. Reproduced from Pascual-Leone *et al.* (1999a), with permission.

not the case that moving visual stimuli activate V5 and V5 alone. Rather, the specificity of this area is, quite properly, inferred by subtracting the activations caused by stationary or colored stimuli or different kinds of visual motion.

C. Temporal Resolution

The cycle of a single pulse of TMS is approximately 1 ms (Fig. 2) and this determines the temporal resolution of the application of TMS. The duration of the effect in the cortex is difficult to determine because the neurons stimulated by the field may take time to recover their normal functional state and normal interactions with other cells. Several TMS studies have applied single-pulse TMS at intervals of 10 ms and obtained effects that suggest TMS can distinguish processes within such a small time window—but the time window is probabilistic rather than fixed and depends on the interaction between the resources the stimulated area is

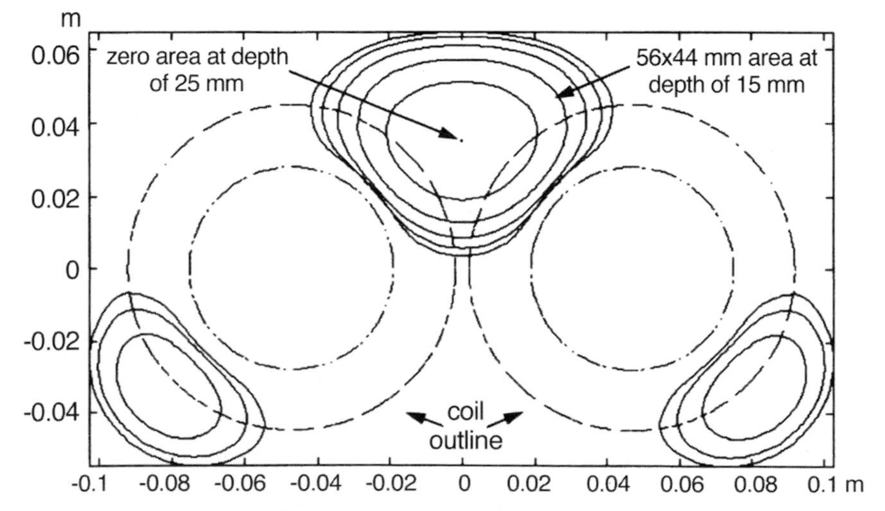

Figure 30 Estimated stimulation areas at depth intervals of 5 mm beginning at the cortical surface. Reproduced from Barker (1999), with permission.

giving to a task and the strength of disruption caused by the TMS pulse.

D. Local and Distant Effects

Other evidence strengthens the correlation between targeted and activated cortical regions. Paus and colleagues (1997, 1998; Paus and Wolforth, 1998; Paus, 1999) have carried out a number of studies in which TMS has been combined with analysis of PET activations using a method of frameless stereotaxy which aligns MRI landmarks and the center of the stimulating coil with an accuracy within 0.4–0.8 cm. The first critical finding of these experiments is that TMS has a major effect approximately under the center of a figure-8 coil, and secondary effects at sites that are known to be anatomically connected. In Chapter 25 Tomás Paus addresses these results and methods further. The finding of distant effects of TMS has obvious relevance in the interpretation of virtual lesion experiments that assume that behavioral consequences of rTMS are due to the disruption of the directly targeted brain region. In some instances, distant effects of rTMS may contribute or even account for behavioral consequences so that careful control experiments targeting different brain areas with TMS are critical.

Acknowledgments

This chapter is dedicated to the memory of Berndt-Ulrich Meyer and Simone Röricht, outstanding neuroscientists and exceptional physicians, but most importantly, in the true sense of the word, good people. We thank Jose M. Tormos for his help with the figures. This work was supported by The Royal Society, The Oxford McDonnell–Pew Center, and the Dr. Hadwen Research Trust (V.W.) and the National Institute of Mental Health (RO1MH60734, RO1MH57980), the National Eye Institute (RO1EY12091), and the Goldberg Family Foundation (A.P.L.).

References

Abdeen, M. A., and Stuchly, M. A. (1994). Modelling of magnetic stimulation of bent neurons. *IEEE Trans. Biomed. Eng.* **41,** 1092–1095.

Amassian, V. E., Cracco, R. Q., Maccabee, P. J., Cracco, J. B., Rudell, A. P., and Eberle, L. (1989). Suppression of visual perception by magnetic coil stimulation of human occipital cortex. *Electroencephalogr. Clin. Neurophysiol.* **74,** 458–462.

Amassian, V. E., Quirk, G. J., and Stewart, A. (1990). A comparison of corticospinal activation by magnetic coil and electrical stimulation of monkey motor cortex. *Electroencephalogr. Clin. Neurophysiol.* **77,** 390–401.

Amassian, V. E., Eberle, L., Maccabee, P. J., and Cracco, R. Q. (1992). Modelling magnetic coil excitation of human cerebral cortex with a peripheral nerve immersed in a brain shaped volume conductor: The significance of fiber-bending in excitation. *Electroencephalogr. Clin. Neurophysiol.* **85,** 291–301.

Amassian, V. E., Maccabee, P. J., Cracco, R. Q., Cracco, J. B., Rudell, A. P., and Eberle, L. (1993a). Measurement of information processing delays in human visual cortex with repetitive magnetic coil stimulation. *Brain Res.* **605,** 317–321.

Amassian, V. E., Cracco, R. Q., Maccabee, P. J., Cracco, J. B., Rudell, A. P., and Eberle, L. (1993b). Unmasking human visual perception with the magnetic coil and its relationship to hemispheric asymmetry. *Brain Res.* **605,** 312–316.

Amassian, V. E., Maccabee, P. J., and Cracco, P. Q. (1994). The polarity of the induced electric field influences magnetic coil inhibition of human visual cortex: Implications for the site of excitation. *Electroencephalogr. Clin. Neurophysiol.* **93,** 21–26.

Amassian, V. E., Cracco, R. Q., Maccabee, P. J., Cracco, J. B., Rudell, A. P., and Eberle, L. (1998). Transcranial magnetic stimulation in study of the visual pathway. *J. Clin. Neurophysiol.* **15,** 288–304.

Arguin, M., Joanette, Y., and Cavanagh, P. (1990). Comparing the cerebral hemispheres on the speed of spatial shifts of visual attention: Evidence from serial search. *Neuropsychologia* **28,** 733–736.

Arguin, M., Joanette, Y., and Cavanagh, P. (1993). Visual search for feature and conjunction targets with an attention deficit. *J. Cognit. Neurosci.* **5,** 436–452.

Ashbridge, E., Walsh, V., and Cowey, A. (1997). Temporal aspects of visual search studied by transcranial magnetic stimulation. *Neuropsychologia* **35,** 1121–1131.

Barbur, J. L., Watson, J. D., Frackowiak, R. S., and Zeki, S. (1993). Conscious visual perception without V1. *Brain* **116,** 1293–1302.

Barker, A. T. (1999). The history and basic principles of magnetic nerve stimulation. *Electroencephalogr. Clin. Neurophysiol. Suppl.* **51,** 3–21.

Barker, A. T., Jalinous, R., and Freeston, I. L. (1985). Non-invasive magnetic stimulation of the human motor cortex. *Lancet* **1,** 1106–1107.

Bartres-Faz, D., Pujol, J., Deus, J., Tormos, J. M., Keenan, J., and Pascual-Leone, A. (1999). Identification of brain areas from which TMS induces speech arrest in normal subjects. *NeuroImage* **9,** S1051.

Beckers, G., and Hömberg, V. (1992). Cerebral visual motion blindness: Transitory akinetopsia induced by transcranial magnetic stimulation of human area V5. *Proc. R. Soc. London Biol. Sci.* **249,** 173–178.

Benecke, R., Meyer, B. U., Schonle, P., and Conrad, B. (1988). Assessment of motor cranial nerve functions using transcranial magnetic stimulation. *Z. EEG-EMG* **19,** 228–233.

Berardelli, A., Inghilleri, M., Rothwell, J. C., Romeo, S., Curra, A., Gilio, F., Modugno, N., and Manfredi, M. (1998). Facilitation of muscle evoked responses after repetitive cortical stimulation in man. *Exp. Brain Res.* **122,** 79–84.

Bisiach, E., Geminiani, G., Berti, A., and Rusconi, M. L. (1990). Perceptual and premotor factors of unilateral neglect. *Neurology* **40,** 8, 1278–1281.

Bisiach, E., Rusconi, M. L., Peretti, V. A., and Vallar, G. (1994). Challenging current accounts of unilateral neglect. *Neuropsychologia* **32,** 1431–1434.

Bisiach, E., Pizzamiglio, L., Nico, D., and Antonucci, G. (1996). Beyond unilateral neglect. *Brain* **119,** 851–817.

Bisiach, E., and Vallar, G. (1988). Hemineglect in humans. *In* "Handbook of Neuropsychology 1" (F. Boller, and J. Grafman, eds.), pp. 195–222. Elsevier, Amsterdam.

Bogdahn, U. (1998). "Transcranial Doppler Sonography." Blackwell Sci., Oxford.

Bohning, D. E. (2000). Introduction and overview of TMS physics. *In* "Transcranial Magnetic Stimulation in Neuropsychiatry" (M. S. George and R. H. Bellmaker, eds.), pp. 13–44. Am. Psychiatric Press, Washington, DC.

Bohning, D. E., Pecheny, A. P., Epstein, C. M., Vincent, D. J., Dannels, W. R., and George, M. S. (1997). Mapping transcranial magnetic stimulation (TMS) fields in vivo with MRI. *NeuroReport* **8,** 2535–2538.

Bohning, D. E., Shastri, A., Nahas, Z., Lorberbaum, J. P., Andersen, S. W., Dannels, W. R., Haxthausen, E. U., Vincent, D. J., and George, M. S. (1998). Echoplanar BOLD fMRI of brain activation induced by concurrent transcranial magnetic stimulation. *Invest. Radiol.* **33,** 6, 336–340.

Bohning, D. E., Shastri, A., Blumenthal, K. M., Nahas, Z., Lorberbaum, J., Roberts, D., Teneback, C., Vincent, D. J., and George, M. S. (1999). A combined TMS/fMRI study of intensity-dependent TMS over motor cortex. *Biol. Psychiatry* **45,** 385–394.

Brasil-Neto, J. P., McShane, L. M., Fuhr, P., Hallett, M., and Cohen, L. G. (1992a). Topographic mapping of the human motor cortex with magnetic stimulation: Factors affecting accuracy and reproducibility. *Electroencephalogr. Clin. Neurophysiol.* **85,** 9–16.

Brasil-Neto, J. P., Coehen, L. G., Panizza, M., Nilsson, J., Roth, B. J., and Hallett, M. (1992b). Optimal focal transcranial magnetic activation of the human motor cortex: Effects of coil orientation, shape of the induced current pulse and stimulus intensity. *J. Clin. Neurophysiol.* **9,** 132–136.

Brasil-Neto, J. P., Cammarota, A., Valls-Sole, J., *et al.* (1995). Role of intracortical mechanisms in the late part of the silent period to transcranial stimulation of the human motor cortex. *Acta Neurol. Scand.* **92,** 383–386.

Britton, T. C., Meyer, B. U., and Benecke, R. (1991). Variability of cortically evoked muscle responses in multiple sclerosis. *Electroencephalogr. Clin. Neurophysiol.* **81,** 186–194.

Butefisch, C. M., Davis, B. C., Wise, S. P., Sawaki, L., Kopylev, L., Classen, J., and Cohen, L. G. (2000). Mechanisms of use dependant plasticity in the human motor cortex. *Proc. Natl. Acad. Sci. USA* **97,** 3661–3665.

Caramia, M. D., Cicinelli, P., Paradiso, C., Mariorenzi, R., Zarola, F., Bernardi, G., and Rossini, P. M. (1991). Excitability changes of muscular responses to magnetic brain stimulation in patients with central motor disorders. *Electroencephalogr. Clin. Neurophysiol.* **81,** 243–250.

Catano, A., Houa, M., and Noel, P. (1997). Magnetic transcranial stimulation: Dissociation of excitatory and inhibitory mechanisms in acute strokes. *Electroencephalogr. Clin. Neurophysiol.* **105,** 29–36.

Celebrini, S., Thorpe, S., Trotter, Y., and Imbert, M. (1993). Dynamics of orientation coding in area V1 of the awake primate. *Visual Neurosci.* **10,** 811–825.

Chen, R., Classen, J., Gerloff, C., Celnik, P., Wassermann, E. M., Hallett, M., and Cohen, L. G. (1997). Depression of motor cortex excitability by low-frequency transcranial magnetic stimulation. *Neurology* **48,** 1398–1403.

Chen, R., Corwell, B., Yaseen, Z., Hallett, M., and Cohen, L. G. (1998). Mechanisms of cortical reorganization in lower-limb amputees. *J. Neurosci.* **18,** 3443–3450.

Classen, J., Liepert, J., Wise, S. P., Hallett, M., and Cohen, L. G. (1998). Rapid plasticity of human cortical movement representation induced by practice. *J. Neurophysiol.* **79,** 1117–1123.

Claus, D., Weis, M., Jahnke, U., Plewe, A., and Brunholzl, C. (1992). Corticospinal conduction studied with magnetic double stimulation in the intact human. *J. Neurol. Sci.* **111,** 180–188.

Corbetta, M., Miezin, F. M., Dobmeyer, S., Shulman, G. L., and Petersen, S. E. (1991). Selective and divided attention during visual discriminations of shape, color, and speed: Functional anatomy by positron emission tomography. *J. Neurosci.* **11,** 2383–2402.

Cohen, L. G., Bandinelli, S., Sato, S., Kufta, C., and Hallett, M. (1991a). Attenuation in detection of somatosensory stimuli by transcranial magnetic stimulation. *Electroencephalogr. Clin. Neurophysiol.* **81,** 366–376.

Cohen, L. G., Bandinelli, S., Findley, T. W., and Hallett, M. (1991b). Motor reorganization after upper limb amputation in man. A study with focal magnetic stimulation. *Brain* **114,** 615–627.

Cohen, L. G., Celnik, P., Pascual-Leone, A., Corwell, B., Falz, L., Dambrosia, J., *et al.* (1997). Functional relevance of cross-modal plasticity in blind humans. *Nature* **389,** 180–183.

Cohen, L. G., Ziemann, U., Chen, R., Classen, J., Hallett, M., Gerloff, C., *et al.* (1998). Studies of neuroplasticity with transcranial magnetic stimulation. *J. Clin. Neurophysiol.* **15,** 305–324.

Corthout, E., Uttl, B., Walsh, V., Hallett, M., and Cowey, A. (1999a). Timing of activity in early visual cortex as revealed by transcranial magnetic stimulation. *NeuroReport* **10,** 1–4.

Corthout, E., Uttl, B., Ziemann, U., Cowey, A., and Hallett, M. (1999b). Two periods of processing in (circum)striate visual cortex as revealed by transcranial magnetic stimulation. *Neuropsychologia* **37,** 137–145.

Cowey, A., and Walsh, V. (2000). Magnetically induced phosphenes in sighted, blind and blindsighted observers. *NeuroReport* **11,** 3269–3273.

Cramer, S. C., and Bastings, E. P. (2000). Mapping clinically relevant plasticity after stroke. *Neuropharmacology* **39,** 842–851.

d'Arsonval, A. (1896). Dispositifs pour la mesure des courants alternatifs de toutes frequences. *C. R. Soc. Biol. (Paris)* **2,** 450–451.

Day, B. J., Dressler, D., Maertens de Noordhout, A., Marsden, C. D., Nakashima, K., Rothwell, J. C., and Thompson, C. D. (1989a). Electric and magnetic stimulation of the human motor cortex: Surface EMG and single motor unit responses. *J. Physiol.* **412,** 449–473.

Day, B. L., Rothwell, J. C., Thompson, P. D., Maertens de Noordhout, A., Nakashima, K., Shannon, K., and Marsden, C. D. (1989b). Delay in the execution of voluntary movement by electrical or magnetic brain stimulation in intact man. Evidence for the storage of motor programs in the brain. *Brain* **112,** 649–663.

Day, B. J., Thompson, P. D., Dick, J. P., Nakashima, K., and Marsden, C. D. (1987). Different sites of action of electrical and magnetic stimulation of the human brain. *Neurosci. Lett.* **75,** 101–106.

Eisen, A., Shytbel, W., Murphy, K., and Hoirch, M. (1990). Cortical magnetic stimulation in amyotrophic lateral sclerosis. *Muscle Nerve* **13,** 146–151.

Eisen, A., Pant, B., and Stewart, H. (1993). Cortical excitability in amyotrophic lateral sclerosis. *Can. J. Neurol. Sci.* **20,** 11–16.

Epstein, C. M. (1998). Transcranial magnetic stimulation: Language function. *J. Clin. Neurophysiol.* **15,** 325–332.

Epstein, C. M., Lah, J. K., Meador, K., Weissman, J. D., Gaitan, L. E., and Dihenia, B. (1996a). Optimum stimulus parameters for lateralized suppression of speech with magnetic brain stimulation. *Neurology* **47,** 1590–1593.

Epstein, C. M., Meador, K., Weissman, J. D., Puhalovich, F., Lah, J. J., Gaitan, L. E., Sheppard, S., and Davey, K. R. (1996b). Localization of speech arrest with transcranial magnetic brain stimulation. *J. Clin. Neurophysiol.* **13,** 387–390.

Epstein, C. M., Meador, K., Loring, D. W., Wright, R. J., Weissman, J. D., Sheppard, S., Lah, J. J., Puhalovich, F., Gaitan, L., and Davey, K. R. (1999). Localization and characterisation of speech arrest during transcranial magnetic stimulation. *Clin. Neurophysiol.* **110,** 1073–1079.

Epstein, C. M., Woodard, J. L., Stringer, A. Y., Bakay, R. A., Henry, T. R., Pennell, P. B., and Litt, B. (2000). Repetitive transcranial magnetic stimulation does not replicate the Wada test. *Neurology* **55,** 1025–1027.

Eyre, J. A., Miller, S., and Ramesh, V. (1991). Constancy of central conduction delays during development in man: Investigation of motor and somatosensory pathways. *J. Physiol.* **434,** 441–452.

Faig, J., and Busse, O. (1996). Silent period evoked by transcranial magnetic stimulation in unilateral thalamic infarcts. *J. Neurol. Sci.* **142,** 85–92.

Fierro, B., Brighina, F., Oliveri, M., Piazza, A., La Bua, V., Buffa, D., and Bisiach, E. (2000). Contralateral neglect induced by right posterior parietal rTMS in healthy subjects. *NeuroReport* **11,** 1519–1521.

Flitman, S. S., Grafman, J., Wassermann, E. M., Cooper, V., O'Grady, J., Pascual-Leone, A., and Hallett, M. (1998). Linguistic processing during repetitive transcranial magnetic stimulation. *Neurology* **50,** 175–181.

Fox, P., Ingham, R., George, M. S., Mayberg, H. S., Ingham, J., Roby, J., Martin, C., and Jerabek, P. (1997). Imaging human intra-cerebral connectivity by PET during TMS. *NeuroReport* **8,** 2787–2791.

Friedman-Hill, S. R., Robertson, L. C., and Treisman, A. (1995). Parietal contributions to visual feature binding: Evidence from a patient with bilateral lesions. *Science* **269,** 853–855.

Fuhr, P., Agostino, R., and Hallett, M. (1991). Spinal motor neuron excitability during the silent period after cortical stimulation. *Electroencephalogr. Clin. Neurophysiol.* **81,** 257–262.

Garnham, C. W., Barker, A. T., and Freeston, I. L. (1995). Measurement of the activating function of magnetic stimulation using combined electrical and magnetic stimuli. *J. Med. Eng. Technol.* **19,** 57–61.

George, M. S., Wassermann, E. M., Williams, W. A., Callahan, A., Ketter, T. A., Basser, P., Hallett, M., and Post, R. M. (1995). Daily repetitive transcranial magnetic stimulation (rTMS) improves mood in depression. *NeuroReport* **6,** 1853–1856.

George, M. S., Wassermann, E. M., Williams, W. A., Steppel, J., Pascual-Leone, A., Basser, P., Hallett, M., and Post, R. M. (1996). Changes in mood and hormone levels after rapid-rate transcranial magnetic stimulation (rTMS) of the prefrontal cortex. *J. Neuropsychiatry Clin. Neurosci.* **8,** 172–180.

Hallett, M. (1996). Transcranial magnetic stimulation: A tool for mapping the central nervous system. *Electroencephalogr. Clin. Neurophysiol. Suppl.* **46,** 43–51.

Hallett, M., Wassermann, E. M., Pascual-Leone, A., and Valls-Sole, J. (1999). Repetitive transcranial magnetic stimulation. The International Federation of Clinical Neurophysiology. *Electroencephalogr. Clin. Neurophysiol. Suppl.* **52,** 105–113.

Heinen, F., Fietzek, U. M., Berweck, S., Hufschmidt, A., Deuschl, G., and Korinthenberg, R. (1998). Fast corticospinal system and motor performance in children: Conduction proceeds skill. *Pediatr. Neurol.* **19,** 217–221.

Herdmann, J., Lumenta, C. B., and Huse, K. O. (1993). Magnetic stimulation for monitoring of motor pathways in spinal procedures. *Spine* **18,** 551–559.

Hess, C. W., Mills, K. R., and Murray, N. M. (1986). Magnetic stimulation of the human brain: Facilitation of motor responses by voluntary contraction of ipsilateral and contralateral muscles with additional observations on an amputee. *Neurosci. Lett.* **71,** 235–240.

Hilgetag, C., Kotter, R., and Young, M. P. (1999). Interhemispheric competition of sub-cortical structures is a crucial mechanism in paradoxical lesion effects and spatial neglect. *Prog. Brain Res.* **121,** 121–141.

Hilgetag, C. C., Théoret, H., and Pascual-Leone, A. (2001). Enhanced visual spatial attention ipsilateral to rTMS-induced 'virtual lesions' of human parietal cortex. *Nat. Neurosci.* **4,** 953–957.

Ilmoniemi, R. J., Virtanen, J., Ruohonen, J., Karhu, J., Aronen, H. J., Naatanen, R., and Katila, T. (1997). Neuronal responses to magnetic stimulation reveal cortical reactivity and connectivity. *NeuroReport* **8,** 3537–3540.

Inghilleri, M., Berardelli, A., Cruccu, G., *et al.* (1993). Silent period evoked by transcranial stimulation of the human cortex and cervicomedullary junction. *J. Physiol.* **466,** 521–534.

Inghilleri, M., Berardelli, A., Marchetti, P., *et al.* (1996). Effects of diazepam, baclofen and thiopental on the silent period evoked by transcranial magnetic stimulation in humans. *Exp. Brain Res.* **109,** 467–472.

Jalinous, R. (1991). Technical and practical aspects of magnetic nerve stimulation. *J. Clin. Neurophysiol.* **8,** 10–25.

Jalinous, R. (1995). "Guide to Magnetic Stimulation." MagStim Co., Whitland, Wales.

Jennum, P., Friberg, L., Fuglsang-Frederiksen, A., and Dam, M. (1994). Speech localization using repetitive transcranial magnetic stimulation. *Neurology* **44,** 269–273.

Jennum, P., and Winkel, H. (1994). Transcranial magnetic stimulation. Its role in the evaluation of patients with partial epilepsy. *Acta Neurol. Scand. Suppl.* **152,** 93–96.

Kammer, T. (1999). Phosphenes and transient scotomas induced by magnetic stimulation of the occipital lobe: Their topographic relationship. *Neuropsychologia* **37,** 191–198.

Kammer, T., and Nusseck, H. G. (1998). Are recognition deficits following occipital lobe TMS explained by raised detection thresholds? *Neuropsychologia* **36,** 1161–1166.

Kapur, N. (1996). Paradoxical functional facilitation in brain-behaviour research. A critical review. *Brain* **119,** 1775–1790.

Kastner, S., Demmer, I., and Ziemann, U. (1998). Transient visual field defects induced by transcranial magnetic stimulation over human occipital pole. *Exp. Brain Res.* **118,** 19–26.

Kosslyn, S. M., Pascual-Leone, A., Felician, O., Camposano, S., Keenan, J. P., Thompson, W. L., Ganis, G., Sukel, K. E., and Alpert, N. M. (1999). The role of area 17 in visual imagery: Convergent evidence from PET and rTMS. *Science* **284,** 167–170. [Published erratum appears in *Science,* 1999, **284,** 197]

Krings, T., Buchbinder, B. R., Butler, W. E., Chiappa, K. H., Jiang, H. J., Cosgrove, G. R., *et al.* (1997a). Functional magnetic resonance imaging and transcranial magnetic stimulation: Complementary approaches in the evaluation of cortical motor function. *Neurology* **48,** 1406–1416.

Krings, T., Buchbinder, B. R., Butler, W. E., Chiappa, K. H., Jiang, H. J., Rosen, B. R., and Cosgrove, G. R. (1997b). Stereotactic transcranial magnetic stimulation: Correlation with direct electrical cortical stimulation. *Neurosurgery* **41,** 1319–1325.

Krings, T., Naujokat, C., and von Keyserlingk, D. G. (1998). Representation of cortical motor function as revealed by stereotactic transcranial magnetic stimulation. *Electroencephalogr. Clin. Neurophysiol.* **109,** 85–93.

Kujirai, T., Caramia, M. D., Rothwell, J. C., Day, B. L., Thompson, B. D., and Ferbert, A. (1993). Cortico-cortical inhibition in human motor cortex. *J. Physiol. (London)* 471, 501–520.

Lamme, V. A., Super, H., and Spekreijse, H. (1998). Feedforward, horizontal, and feedback processing in the visual cortex. *Curr. Opin. Neurobiol.* **8,** 529–535.

Lewko, J. P., Stokic, D. S., and Tarkka, I. M. (1996). Dissociation of cortical areas responsible for evoking excitatory and inhibitory responses in the small hand muscles. *Brain Topogr.* **8,** 397–405.

Liepert, J., Bauder, H., Wolfgang, H. R., Miltner, W. H., Taub, E., and Weiller, C. (2000). Treatment-induced cortical reorganization after stroke in humans. *Stroke* **31,** 1210–1216.

Lomber, S. G., and Payne, B. R. (1996). Removal of 2 halves restores the whole—Reversal of visual hemineglect during bilateral cortical or collicular inactivation in the cat. *Visual Neurosci.* **13,** 1143–1156.

Lomber, S. G., and Payne, B. R. (1999). Assessment of neural function with reversible deactivation methods. *J. Neurosci. Methods* **86,** 105–108.

Maccabee, P. J., Amassian, V. E., Eberle, L. P., and Cracco, R. Q. (1993). Magnetic coil stimulation of straight and bent amphibian and mammalian peripheral nerve in vitro: Locus of excitation. *J. Physiol. (London)* **460,** 201–219.

Maccabee, P. J., Lipitz, M. E., Desudchit, T., Galub, R. W., Cracco, R., and Amassian, V. E. (1996). A new method using neuromagnetic stimulation to measure conduction time within cauda equina. *Electroencephalogr. Clin. Neurophysiol.* **101,** 153–166.

Maeda, F., Keenan, J. P., Tormos, J. M., Topka, H., and Pascual-Leone, A. (2000a). Modulation of corticospinal excitability by repetitive transcranial magnetic stimulation. *Clin. Neurophysiol.* **111,** 800–805.

Maeda, F., Keenan, J. P., Tormos, J. M., Topka, H., and Pascual-Leone, A. (2000b). Interindividual variability of the modulatory effect of repetitive transcranial magnetic stimulation on cortico-spinal excitability. *Exp. Brain Res.* **133,** 425–430.

Mathis, J., de Quervain, D., and Hess, C. W. (1998). Dependence of the transcranially induced silent period on the 'instruction set' and the individual reaction time. *Electroencephalogr. Clin. Neurophysiol.* **109,** 426–435.

Meyer, B. U., Diehl, R., Steinmetz, H., Britten, T. C., and Benecke, R. (1991). Magnetic stimuli applied over motor and visual cortex: Influence of coil position and field polarity on motor responses, phosphenes and eye movements. *Electroencephalogr. Clin. Neurophysiol. Suppl.* **43,** 121–134.

Meyer, B. U., Roricht, S., Grafin von Einsiedel, H., Kruggel, F., and Weindl, A. (1995). Inhibitory and excitatory interhemispheric transfers between motor cortical areas in normal humans and patients with abnormalities of the corpus callosum. *Brain* **118,** 429–440.

Meyer, B. U., Benecke, R., Dressler, D., Haug B., and Conrad, B. (1998a). Fraktionierte Bestimmung zentraler motorischer Leitungszeiten mittels Reizung von Kortex, spinalen Bahnen und Spinalnervenwurzeln: Möglichkeiten und Grenzen. *Z. EEG-EMG* **19,** 234–240.

Meyer, B. U., Roricht, S., and Woiciechowsky, C. (1998b). Topography of fibers in the human corpus callosum mediating interhemispheric inhibition between the motor cortices. *Ann. Neurol.* **43,** 360–369.

Michelucci, R., Passarelli, D., Riguzzi, P., *et al.* (1996). Transcranial magnetic stimulation in partial epilepsy: Drug-induced changes of motor excitability. *Acta Neurol. Scand.* **94,** 24–30.

Mills, K. R. (1999). "Magnetic Stimulation of the Human Nervous System." Oxford Univ. Press, Oxford.

Miranda, P. C., de Carvalho, M., Conceicao, I., Luis, M. L., and Ducla-Soares, E. (1997). A new method for reproducible coil positioning in transcranial magnetic stimulation mapping. *Electroencephalogr. Clin. Neurophysiol.* **105,** 116–123.

Morioka, T., Yamamoto, T., Mizushima, A., Tombimatsu, S., Shigeto, H., Hasuo, K., Nishio, S., Fujii, K., and Fukui, M. (1995a). Comparison of magnetoencephalography, functional MRI, and motor evoked potentials in the localization of the sensory-motor cortex. *Neurol. Res.* **17,** 361–367.

Morioka, T., Mizushima, A., Yamamoto, T., Tobimatsu, S., Matsumoto, S., Hasuo, K., Fujii, K., and Fukui, M. (1995b). Functional mapping of the sensorimotor cortex: Combined use of magnetoencephalography, functional MRI, and motor evoked potentials. *Neuroradiology* **37,** 526–530.

Mottaghy, F., Gangitano, M., Sparing, R., Krause, B., and Pascual-Leone, A. (2002). Segregation of areas related to visual working memory in the prefrontal cortex revealed by rTMS. *Cereb. Cortex,* in press.

Muller, K., Kass-Iliyya, F., and Reitz, M. (1997). Ontogeny of ipsilateral corticospinal projections: A developmental study with transcranial magnetic stimulation. *Ann. Neurol.* **42,** 705–711.

Muri, R. M., Vermersch, A. I., Rivaud, S., Gaymard, B., and Pierrot-Deseilligny, C. (1996). Effects of single-pulse transcranial magnetic stimulation over the prefrontal and posterior parietal cortices during memory-guided saccades in humans. *J. Neurophysiol.* **76,** 2102–2106.

Muri, R. M., Rivaud, S., Gaymard, B., Ploner, C. J., Vermersch, A. I., Hess, C. W., and Pierrot-Deseilligny, C. (1999). Role of the prefrontal cortex in the control of express saccades. A transcranial magnetic stimulation study. *Neuropsychologia* **37,** 199–206.

Nagarajan, S. S., Durand, D. M., and Warman, E. N. (1993). Effects of induced electric fields on finite neuronal structures: A simulation study. *IEEE Trans. Biomed. Eng.* **40,** 1175–1188.

Nakamura, H., Kitagawa, H., Kawaguchi, Y., *et al.* (1997). Intracortical facilitation and inhibition after transcranial magnetic stimulation in conscious humans. *J. Physiol.* **498,** 817–823.

Niehaus, L., Rorricht, S., Scholz, U., and Meyer, B. U. (1999). Hemodynamic response to repetitive magnetic stimulation of the motor and visual cortex. *Electroencephalogr. Clin. Neurophysiol. Suppl.* **51,** 41–47.

Oliveri, M., Rossini, P. M., Pasqualetti, P., Traversa, R., Cicinelli, P., Palmieri, M. G., Tomaiuolo, T., and Caltagirone, C. (1999a). Interhemispheric asymmetries in the perception of unimanual and bimanual cutaneous stimuli. A study using transcranial magnetic stimulation. *Brain* **122,** 1721–1729.

Oliveri, M., Rossini, P. M., Traversa, R., Cicinelli, P., Filippi, M. M., Pasqualetti, P., Tomaiuolo, T., and Caltagirone, C. (1999b). Left frontal transcranial magnetic stimulation reduces contralesional extinction in patients with unilateral right brain damage. *Brain* **122,** 1731–1739.

Oliveri, M., Turriziani, P., Carlesimo, G. A., Koch, G., Tomaiuolo, F., Panella, M., and Caltagirone, C. (2001). Parieto-frontal interactions in visual-object and visual-spatial working memory: Evidence from transcranial magnetic stimulation. *Cereb. Cortex* **11,** 606–618.

Oyachi, H., and Ohtsuka, K. (1995). Transcranial magnetic stimulation of the posterior parietal cortex degrades accuracy of memory-guided saccades in humans. *Invest. Ophthalmol. Visual Sci.* **36,** 1441–1449.

Pascual-Leone, A., Gates, J. R., and Dhuna, A. (1991). Induction of speech arrest and counting errors with rapid-rate transcranial magnetic stimulation. *Neurology* **41,** 697–702.

Pascual-Leone, A., Houser, C. M., Reese, K., Shotland, L. I., Grafman, J., Sato, S., Valls-Sole, J., Brasil-Neto, J. P., Wassermann, E. M., Cohen, L. G., *et al.* (1993). Safety of rapid-rate transcranial magnetic stimulation in normal volunteers. *Electroencephalogr. Clin. Neurophysiol.* **89,** 120–130.

Pascual-Leone, A., Valls-Sole, J., Wassermann, E. M., and Hallett, M. (1994a). Responses to rapid-rate transcranial magnetic stimulation of the human motor cortex. *Brain* **117,** 847–858.

Pascual-Leone, A., Grafman, J., and Hallett, M. (1994b). Modulation of cortical motor output maps during development of implicit and explicit knowledge. *Science* **263,** 1287–1289.

Pascual-Leone, A., Gomez-Tortosa, E., Grafman, J., Alway, D., Nichelli, P., and Hallett, M. (1994c). Induction of visual extinction by rapid-rate transcranial magnetic stimulation of parietal lobe. *Neurology* **44,** 494–498.

Pascual-Leone, A., Nguyet, D., Cohen, L. G., Brasil-Neto, J. P., Cammarota, A., and Hallett, M. (1995). Modulation of muscle responses evoked by transcranial magnetic stimulation during the acquisition of new fine motor skills. *J. Neurophysiol.* **74,** 1037–1045.

Pascual-Leone, A., Peris, M., Tormos, J. M., Pascual, A. P., and Catala, M. D. (1996). Reorganization of human cortical motor output maps following traumatic forearm amputation. *NeuroReport* **7,** 2068–2070.

Pascual-Leone, A., and Meador, K. J. (1998). Is transcranial magnetic stimulation coming of age? *J. Clin. Neurophysiol.* **15,** 300–304.

Pascual-Leone, A., Tormos, J. M., Keenan, J., Tarazona, F., Canete, C., and Catala, M. D. (1998). Study and modulation of human cortical excitability with transcranial magnetic stimulation. *J. Clin. Neurophysiol.* **15,** 333–343.

Pascual-Leone, A., Bartres-Faz, D., and Keenan, J. P. (1999a). Transcranial magnetic stimulation: Studying the brain–behaviour relationship by induction of 'virtual lesions'. *Philos. Trans. R. Soc. London B Biol. Sci.* **354,** 1229–1238.

Pascual-Leone, A., Tarazona, F., Keenan, J. P., Tormos, J. M., Hamilton, R., and Catala, M. D. (1999b). Transcranial magnetic stimulation and neuroplasticity. *Neuropsychologia* **37,** 207–217.

Pascual-Leone, A., Walsh, V., and Rothwell, J. (2000). Transcranial magnetic stimulation in cognitive neuroscience—Virtual lesion, chronometry, and functional connectivity. *Curr. Opin. Neurobiol.* **10,** 232–237.

Pascual-Leone, A., and Walsh, V. (2001). Fast backprojections from the motion area to the primary visual area necessary for visual awareness. *Science* **292,** 510–512.

Pascual-Leone, A., Davey, A., Wassermann, E. M., Rothwell, J., and Puri, B., eds. (2002). "Handbook of Transcranial Magnetic Stimulation." Arnold, London.

Paus, T. (1999). Imaging the brain before, during, and after transcranial magnetic stimulation. *Neuropsychologia* **37,** 219–224.

Paus, T., Jech, R., Thompson, C. J., Comeau, R., Peters, T., and Evans, A. C. (1997). Transcranial magnetic stimulation during positron emission tomography: A new method for studying connectivity of the human cerebral cortex. *J. Neurosci.* **17,** 3178–3184.

Paus, T., and Wolforth, M. (1998). Transcranial magnetic stimulation during PET: Reaching and verifying the target site. *Hum. Brain Mapp.* **6,** 399–402.

Paus, T., Jech, R., Thompson, C. J., Comeau, R., Peters, T., and Evans, A. C. (1998). Dose-dependent reduction of cerebral blood flow during rapid-rate transcranial magnetic stimulation of the human sensorimotor cortex. *J. Neurophysiol.* **79,** 1102–1107.

Priori, A., Berardelli, A., Inghilleri, M., *et al.* (1994). Motor cortical inhibition and the dopaminergic system. Pharmacological changes in the silent period after transcranial brain stimulation in normal subjects, patients with Parkinson's disease and drug-induced Parkinsonism. *Brain* **117,** 317–323.

Prout, A. J., and Eisen, A. A. (1994). The cortical silent period and amyotrophic lateral sclerosis. *Muscle Nerve* **17,** 217–223.

Reilly, J. P. (1992). "Electrical Stimulation and Electropathology." Cambridge Univ. Press, Cambridge.

Reutens, D. C., Puce, A., and Berkovic, S. F. (1993). Cortical hyperexcitability in progressive myoclonus epilepsy: A study with transcranial magnetic stimulation. *Neurology* **43,** 186–192.

Robertson, E. M., Tormos, J. M., Maeda, F., and Pascual-Leone, A. (2001). The role of the dorsolateral prefrontal cortex during sequence learning is specific for spatial information. *Cereb. Cortex* **11,** 628–635.

Robertson, I. H. J. M., and Murre, J. (1999). Rehabilitation of brain damage: Brain plasticity and principles of guided recovery. *Psychol. Bull.* **125,** 544–575.

Rossini, P. M., and Rossi, S. (1998). Clinical applications of motor evoked potentials. *Electroencephalogr. Clin. Neurophysiol.* **106,** 180–194.

Rossini, P. M., Caltagirone, C., Castriota-Scanderbeg, A., Cicinelli, P., Del Gratta, C., Demartin, M., *et al.* (1998). Hand motor cortical area reorganization in stroke: A study with fMRI, MEG and TCS maps. *NeuroReport* **9,** 2141–2146.

Rossini, P. M., and Pauri, F. (2000). Neuromagnetic integrated methods tracking brain mechanisms of sensorimotor areas 'plastic' reorganisation. *Brain Res. Rev.* **33,** 131–154.

Roth, B. J., Saypol, J. M., Hallett, M., *et al.* (1991). A theoretical calculation of the electric field induced in the cortex during magnetic stimulation. *Electroencephalogr. Clin. Neurophysiol.* **81,** 47–56.

Rothwell, J. C. (1997). Techniques and mechanisms of action of transcranial stimulation of the human motor cortex. *J. Neurosci. Methods* **74,** 113–122.

Ruohonen, J., Ravazzani, P., *et al.* (1996). Motor cortex mapping with combined MEG and magnetic stimulation. *Electroencephalogr. Clin. Neurophysiol. Suppl.* **46,** 317–322.

Rushworth, M. F. S., and Walsh, V., eds. (1999). "Transcranial Magnetic Stimulation in Neuropsychology." *Neuropsychologia* **Special Issue 37.**

Sadato, N., Pascual-Leone, A., Grafman, J., Ibañez, V., Deiber, M.-P., Dold, G., and Hallett, M. (1996). Activation of the primary visual cortex by Braille reading in blind subjects. *Nature* **380,** 526–528.

Schnitzler, A., and Benecke, R. (1994). The silent period after transcranial magnetic stimulation is of exclusive cortical origin: Evidence from isolated cortical ischemic lesions in man. *Neurosci. Lett.* **180,** 41–45.

Schmolesky, M. T., Wang, Y., Hanes, D. P., Thompson, K. G., Leutgeb, S., Schall, J. D. and Leventhal, A. G. (1998). Signal timing across the macaque visual system. **79,** 3272–3278.

Seyal, M., Ro, T., and Rafal, R. (1995). Increased sensitivity to ipsilateral cutaneous stimuli following transcranial magnetic stimulation of the parietal lobe. *Ann. Neurol.* **38,** 264–267.

Shapiro, K., Pascual-Leone, A., Mottaghy, F. M., Gangitano, M., and Caramazza, A. (2001). Grammatical distinctions in the left frontal cortex. *J. Cognit. Neurosci.* **13,** 713–720.

Siebner, H. R., Willoch, F., Peller, M., Auer, C., Boecker, H., Conrad, B., and Bartenstein, P. (1998). Imaging brain activation induced by long trains of repetitive transcranial magnetic stimulation. *NeuroReport* **9,** 943–948.

Singh, K. D., Hamdy, S., and Aziz, Q. (1991). Topographic mapping of transcranial magnetic stimulation data on surface rendered MR images of the brain. *Electroencephalogr. Clin. Neurophysiol.* **105,** 345–351.

Stewart, L. M., Battelli, L., Walsh, V., and Cowey, A. (1999). Motion perception and perceptual learning studied by magnetic stimulation. *Electroencephalogr. Clin. Neurophysiol.* **51,** 334–350.

Stewart, L. M., Frith, U., Meyer, B.-U., and Rothwell, J. (2000). TMS over BA37 impairs picture naming. *Neuropsychologia* **39,** 1–6.

Stewart, L. M., Walsh, V., Frith, U., and Rothwell, J. (2001a). TMS can produce two different types of speech disruption. *NeuroImage* **13,** 472–478.

Stewart, L., Walsh, V., and Rothwell, J. (2001b). Motor and phosphene thresholds: A TMS correlation study. *Neuropsychologia* **39,** 415–419.

Stoerig, P., and Cowey, A. (1997). Blindsight in man and monkey. *Brain* **120,** 535–559.

Terao, Y., Fukuda, H., Ugawa, Y., Hikosaka, O., Hanajima, R., Ferubayashi, T., Sakai, K., Miyauchi, S., Sasaki, Y., and Kanazawa, I. (1998a). Visualization of the information flow though the human oculomotor cortical regions by transcranial magnetic stimulation. *J. Neurophysiol.* **80,** 936–946.

Terao, Y., Ugawa, Y., Sakai, K., Miyauchi, S., Fukuda, H., Sasaki, Y., Takino, R., Hanajima, R., Ferubayashi, T., and Kanazawa, I. (1998b). Localizing the site of magnetic brain stimulation by functional MRI. *Exp. Brain Res.* **121,** 145–152.

Thickbroom, G. W., Sammut, R., and Mastaglia, F. L. (1998). Magnetic stimulation mapping of motor cortex: Factors contributing to map area. *Electroencephalogr. Clin. Neurophysiol.* **109,** 79–84.

Thompson, P. D., Day, B. L., Rothwell, J. C., *et al.* (1991). Further observations on the facilitation of muscle responses to cortical stimulation by voluntary contraction. *Electroencephalogr. Clin. Neurophysiol.* **81,** 397–402.

Topper, R., Mottaghy, F. M., Brugmann, M., Noth, J., and Huber, W. (1998). Facilitation of picture naming by focal transcranial magnetic stimulation of Wernicke's area. *Exp. Brain Res.* **121,** 371–378.

Treisman, A. (1996). The binding problem. *Curr. Opin. Neurobiol.* **6,** 171–178.

Triggs, W. J., Cros, D., Macdonell, R. A., *et al.* (1993). Cortical and spinal motor excitability during the transcranial magnetic stimulation silent period in humans. *Brain Res.* **628,** 39–48.

Triggs, W. J., Calvanio, R., and Levine, M. (1997). Transcranial magnetic stimulation reveals a hemispheric asymmetry correlate of intermanual differences in motor performance. *Neuropsychologia* **35,** 1355–1363.

Ueno, S., Tashiro, T., and Harada, K. (1988). Localized stimulation of neural tissues in the brain by means of a paired configuration of time-varying magnetic fields. *J. Appl. Phys.* **64,** 5862–5864.

Valls-Sole, J., Pascual-Leone, A., Wassermann, E. M., *et al.* (1992). Human motor evoked responses to paired transcranial magnetic stimuli. *Electroencephalogr. Clin. Neurophysiol.* **85,** 355–364.

Wada, S., Kubota, H., Maita, S., Yamamoto, I., Yamaguchi, M., Andoh, T., Kawahami, T., Okumura, F., and Takenaka, T. (1996). Effects of stimulus waveform on magnetic nerve stimulation. *Jpn. J. Appl. Physiol.* **35,** 1983–1988.

Walsh, V. (1998). Brain mapping: Faradization of the mind. *Curr. Biol.* **8,** R8–11.

Walsh, V., Ellison, A., Battelli, L., and Cowey, A. (1998). Task-specific impairments and enhancements induced by magnetic stimulation of human visual area V5. *Proc. R. Soc. London B Biol. Sci.* **265,** 537–543.

Walsh, V., and Cowey, A. (1998). Magnetic stimulation studies of visual cognition. *Trends Cognit. Sci.* **2,** 103–110.

Walsh, V., and Rushworth, M. (1999). A primer of magnetic stimulation as a tool for neuropsychology. *Neuropsychologia* **37,** 125–135.

Walsh, V., and Cowey, A. (2000). Transcranial magnetic stimulation and cognitive neuroscience. *Nat. Rev.* **1,** 73–79.

Wassermann, E. M. (1998). Risk and safety of repetitive transcranial magnetic stimulation: Report and suggested guidelines from the International Workshop on the Safety of Repetitive Transcranial Magnetic Stimulation, June 5–7, 1996. *Electroencephalogr. Clin. Neurophysiol.* **108,** 1–16.

Wassermann, E. M., Fuhr, P., Cohen, L. G., *et al.* (1991). Effects of transcranial magnetic stimulation on ipsilateral muscles. *Neurology* **41,** 1795–1799. [Published erratum appears in *Neurology,* 1992, **42,** 1115]

Wassermann, E. M., McShane, L. M., Hallett, M., and Cohen, L. G. (1992). Noninvasive mapping of muscle representations in human motor cortex. *Electroencephalogr. Clin. Neurophysiol.* **85,** 1–8.

Wassermann, E. M., Pascual-Leone, A., Valls-Sole, J., *et al.* (1993). Topography of the inhibitory and excitatory responses to transcranial magnetic stimulation in a hand muscle. *Electroencephalogr. Clin. Neurophysiol.* **89,** 424–433.

Wassermann, E. M., Wang, B., Zeffiro, T. A., Sadato, N., Pascual-Leone, A., Toro, C., and Hallett, M. (1996). Locating the motor cortex on the MRI with transcranial magnetic stimulation and PET. *NeuroImage* **3,** 1–9.

Wassermann, E. M., Tormos, J. M., and Pascual-Leone, A. (1998). Finger movements induced by transcranial magnetic stimulation change with hand posture, but not with coil position. *Hum. Brain Mapp.* **6,** 390–393.

Wassermann, E. M., and Lissanby, S. H. (2001). Therapeutic application of repetitive transcranial magnetic stimulation: A review. *Clin. Neurophysiol.* **112,** 1367–1377.

Watson, J. D. G., Myers, R., Frackowiak, R. S. J., Hajnal, J. V., Woods, R. P., Mazziotta, J. C., Shipp, S., and Zeki, S. (1993). Area V5 of the human brain: Evidence from a combined study using positron emission tomography and magnetic resonance imaging. *Cereb. Cortex* **3,** 79–94.

Weintraub, S., and Mesulam, M. M. (1987). Right cerebral dominance in spatial attention: Further evidence based on ipsilateral neglect. *Arch. Neurol.* **44,** 621–625.

Weiszkrantz, L. (1997). "Consciousness Lost and Found: A Neuropsychological Exploration." Oxford Univ. Press, Oxford.

Wilson, C. L., Babb, T. L., Halgren, E., and Crandall, P. H. (1983). Visual receptive fields and response properties of neurons in human temporal lobe and visual pathways. *Brain* **106,** 473–502.

Wilson, S. A., Lockwood, R. J., Thickbroom, G. W., *et al.* (1993a). The muscle silent period following transcranial magnetic cortical stimulation. *J. Neurol. Sci.* **114,** 216–222.

Wilson, S. A., Thickbroom, G. W., and Mastaglia, F. L. (1993b). Topography of excitatory and inhibitory muscle responses evoked by transcranial magnetic stimulation in the human motor cortex. *Neurosci. Lett.* **154,** 52–56.

Yokota, T., Yoshino, A., Inaba, A., *et al.* (1996). Double cortical stimulation in amyotrophic lateral sclerosis. *J. Neurol. Neurosurg. Psychiatry* **61,** 596–600.

Young, M. P., Hilgetag, C., and Scannell, J. W. (1999). Models of paradoxical lesion effects and rules of inference for imputing function to structure in the brain. *Neurocomputing* **26-27,** 933–938.

Young, M. P., Hilgetag, C., and Scannell, J. W. (2000). On imputing function to structure from the behavioural effects of brain lesions. *Proc. R. Soc. London* **355,** 147–161.

Zangemeister, W. H., Canavan, A. G., and Hoemberg, V. (1995). Frontal and parietal transcranial magnetic stimulation (TMS) disturbs programming of saccadic eye movements. *J. Neurol. Sci.* **133,** 42–52.

Zeki, S. (2001). Localization and globalization in conscious vision. *Annu. Rev. Neurosci.* **24,** 57–86.

Zeki, S., and Bartels, A. (1999). Toward a theory of visual consciousness. *Consciousness Cognit.* **8,** 225–259.

Zeki, S., and ffytche, D. H. (1998). The Riddoch syndrome: Insights into the neurobiology of conscious vision. *Brain* **12,** 25–45.

Ziemann, U., Bruns, D., and Paulus, W. (1996a). Enhancement of human motor cortex inhibition by the dopamine receptor agonist pergolide: Evidence from transcranial magnetic stimulation. *Neurosci. Lett.* **208,** 187–190.

Ziemann, U., Lonnecker, S., and Paulus, W. (1995). Inhibition of human motor cortex by ethanol. A transcranial magnetic stimulation study. *Brain* **118,** 1437–1446.

Ziemann, U., Lonnecker, S., Steinhoff, B. J., *et al.* (1996b). The effect of lorazepam on the motor cortical excitability in man. *Exp. Brain Res.* **109,** 127–135.

Ziemann, U., Lonnecker, S., Steinhoff, B. J., *et al.* (1996c). Effects of antiepileptic drugs on motor cortex excitability in humans: A transcranial magnetic stimulation study. *Ann. Neurol.* **40,** 367–378.

Ziemann, U., Rothwell, J. C., and Ridding, M. C. (1996d). Interaction between intracortical inhibition and facilitation in human motor cortex. *J. Physiol.* **496,** 873–881.

Ziemann, U., and Hallett, M. (2000). Basic neurophysiological studies with TMS. *In* "Transcranial Magnetic Stimulation in Neuropsychiatry" (M. A. George and M. Belmaker, eds.), pp. 45–98. Am. Psychiatric Press, Washington, DC.

Zihl, J., von Cramon, D., *et al.* (1983). Selective disturbance of movement vision after bilateral brain damage. *Brain* **106,** 313–340.

III

Tomagraphic-Based Data Acquisition

12

High-Field Magnetic Resonance

Kamil Ugurbil

Center for Magnetic Resonance Research, University of Minnesota Medical School, Minneapolis, Minnesota 55455

I. Introduction

The history of Nuclear Magnetic Resonance (NMR) is marked by ever increasing number of innovations that have produced novel and surprising uses of this phenomenon for investigating biological processes. No other technique has proven to be so uniquely flexible and dynamic. Since its introduction over fifty years ago [79–81], NMR rapidly evolved to become an indispensable tool in chemical and biochemical research because of its sensitivity to the chemical environment of nuclear spins. In 1973, ability to obtain images with magnetic resonance was introduced [54], leading to the development of magnetic resonance imaging (MRI), which is now solidly established as a research tool and as a non-invasive diagnostic technique in the practice of medicine.

In biochemical and chemical applications of the nuclear magnetic resonance phenomenon, the quest for ever increasing magnetic fields has been and continues to be pursued without hesitation, resulting in a flurry of activity focused at developing instruments operating at higher and higher field strengths. This enthusiasm for higher fields, however, has not been prevalent in the field of human imaging, until perhaps a few years ago. In stark contrast to molecular NMR, it was commonly assumed that higher magnetic fields are detrimental to MR imaging of the human body. This assumption was not based on the existence of any experimental evidence; rather, it followed from concepts and theoretical considerations regarding the interaction of high frequency electromagnetic waves with the conductive human body, leading to the suggestion that human imaging may not be possible beyond 10 MHz (~0.24 Tesla) [40]. Of course, even clinical imaging is now performed mostly at 1.5 T (~63 MHz), and recent work at 3 to 8 Tesla has demonstrated that exquisite anatomical imaging of the human head is achievable at these high fields (e.g. [1, 10, 74, 95, 97] and references therein).

In our laboratory, the effort to pursue high magnetic fields was intricately tied to our interest in developing methods for the acquisition of physiological and biochemical information non-invasively using the nuclear spins of the water molecules and metabolites in the human body. In this effort, a relatively recent and unique accomplishment has been the introduction of the ability to map human brain function non-invasively. The work that lead to the introduction of this methodology from our group [71] (together with concurrent and independent work from MGH [53]) was conducted at 4 Tesla. Today, functional images with sub-centimeter resolution of the entire human brain can be generated in single subjects and in data acquisitions times of several minutes using 1.5 Tesla MRI scanners that are often employed in hospitals for clinical diagnosis. However, there have been accomplishments beyond this type of functional imaging using significantly higher magnetic fields such as 4 Tesla, and recently 7 Tesla in humans, and 9.4 Tesla in animal models. In this chapter, data and concepts relevant to high magnetic fields are reviewed, with the primary focus on efforts related to probing brain function and neurochemistry utilizing imaging and spectroscopy capabilities.

II. Signal-to-Noise Ratio

In all NMR experiments, and especially in *in vivo* applications, gains in signal-to-noise ratio (SNR) are the key to extending the applications of this phenomenon to new frontiers in research. This is well recognized for studies of macromolecules and for *in vivo* spectroscopy where intracellular compounds with submillimolar concentrations are currently observable with difficulty, and the vast majority of intracellular compounds are beyond NMR's reach. SNR limitations, however, are also paramount in demanding functional and physiological studies.

SNR gains can be achieved in going to higher fields. SNR, however, becomes rather complex when high magnetic field (hence high frequencies) are considered with lossy biological samples such as the human body and the human head. The relationship between SNR and resonance frequency, ω, or equivalently field strength has been examined for biological samples in numerous studies [40–42, 102], predicting increases with field strength. At high frequencies such as 170 MHz (^{1}H frequency for 4 Tesla) and above, SNR must be considered as a function of location within a sample and for particular sample geometries. SNR gains with increasing fields have been predicted for the interior of a sphere approximating the size and electromagnetic properties of the human head [41, 98, 99]. Such gains were calculated to be linear up to 200 MHz, and steeper than linear above 200 MHz (^{1}H frequency for 4.7 Tesla) [45].

Field dependence of SNR was experimentally examined by our group in the human head, initially comparing 0.5, 1.5 and 4 Tesla (~21, 64, and 170 MHz, respectively), using a surface coil, documenting that SNR for the ^{1}H nucleus increased at least linearly at the higher frequencies [28]. More recently, with the availability of a 7 Tesla human system in our laboratory, we examined and compared the B_1 field profile and SNR in the human head for 4 and 7 Tesla when using a TEM "volume" head coil [97]. It was expected that a coil with a uniform B_1 field profile when empty will nonetheless lead to a highly non-uniform B_1, SNR, and power deposition over the human head and brain. The B_1 distribution was mapped over the human head using magnetization preparation followed by ultrafast imaging ("turboFlash"). Magnetization preparation was accomplished with a variable-duration hard (square) pulse followed by rapid gradient spoiling to eliminate transverse magnetization. The resultant longitudinal magnetization is directly proportional to $\{\cos(\tau\gamma B_1)\}$ where τ and γ are the pulse duration and gyromagnetic ratio, respectively. In such an experiment, as τ is incremented over a period that exceeds $(1/\gamma B_1)$, the signal intensity oscillates with the frequency γB_1. While the signal amplitude over the image will be affected by variations in B_1 magnitude and the T_1 relaxation affects that may come into play in the ultrafast image acquisition, the *frequency* of signal oscillation will be a function of B_1 only. The results demonstrated that the signal intensity oscillates with higher frequency in the brain center than in the periphery at both field strengths but much more so at 7 Tesla (Figure 1). At 4 T, the B_1 strength in the brain periphery was down 23% from the center value. At 7 T, the peripheral B_1 was 42% lower than the central B_1. The non-uniform B_1 profiles obtained experimentally were both expected and predicted by the Maxwell models of the human head loaded TEM coil [97].

Inhomogeneities in the B_1 distribution within a human head when using a "volume" coil that normally generates a homogeneous RF field implies that SNR must also be measured as a function of location within the brain. Using virtually fully relaxed images when the center of the brain was set to experience a 90 degree pulse (thus, by necessity the periphery undergoes a less than 90 degree rotation), the SNR scaled more than linearly with field magnitude in the center of the brain, and less than linearly in the periphery (Figure 2). When averaged over the entire slice, SNR at 7T increased over 4T by 1.6 fold without considering a difference in the gain of the 7T and 4T coils, 1.76 times when corrected for the gain difference.

The SNR increases measured in going from 4 to 7T, however, are for conditions of same acquisition bandwidth and full relaxation. Such acquisition conditions may be quite appropriate for some imaging applications. However, under many conditions the bandwidth needs to be higher at the higher field strength. For example, in very fast imaging methods such as echo planar imaging (EPI), in order to overcome the deleterious effects of increasing B_O inhomogeneities (and consequently shorter T_2*), the bandwidth must scale with the field as well, to maintain similar distortions. This would decreases the SNR gain to square root of the magnetic field ratio, when comparing two different magnetic fields. Similarly, if signal averaging is necessary with repetitions that do not permit full relaxation, then, SNR gains at the higher magnetic field will also diminish due to the longer T_1 at the higher magnetic field.

For lower gyromagnetic ratio (hence lower resonance frequency) nuclei, SNR gains provided by high magnetic

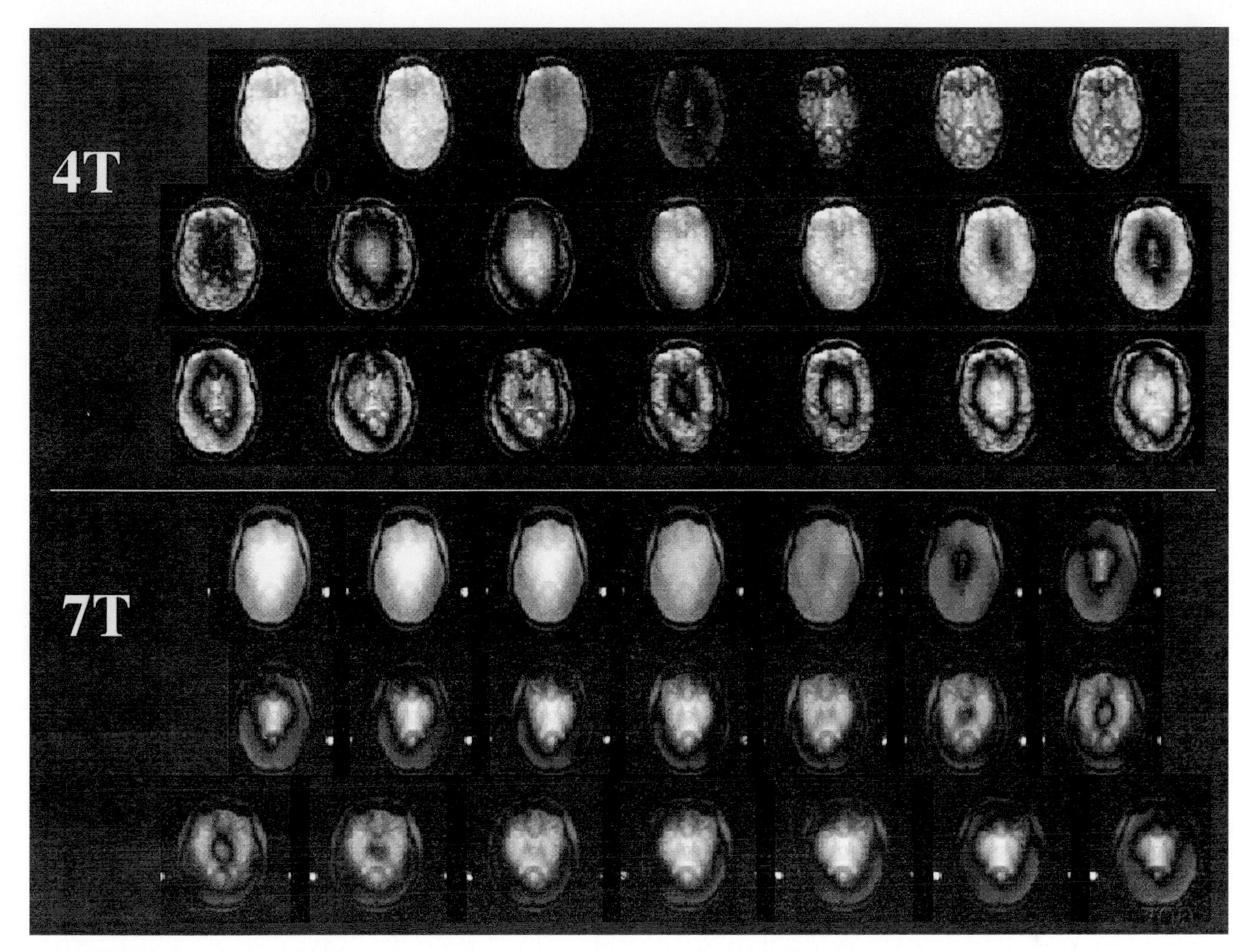

Figure 1A Images of a transverse slice through the human head at 7 Tesla, obtained with magnetization preparation to encode B_1 magnitude as {cos (tgB_1)} and imaging it rapidly using "turboFlash" (center-out k-space sampling, TR/TE = 4.2 ms/2.5 ms, slice thickness = 5 mm, flip angle = 10° at the slice center, matrix size = 128 × 64). The different images correspond to different t while B_1 magnitude was kept constant.

fields can be more dramatic. For example, the ^{17}O nucleus resonates at 27 MHz at 4.7 Tesla as opposed to 200 MHz for ^{1}H nucleus at the same field strength. This is an important nucleus in biological studies because, in principle, it can permit the evaluation of oxygen consumption changes non-invasively. The SNR of ^{17}O detection with MR is elevated ~4 fold with the magnetic field in going from 4.7 Tesla to 9.4 Tesla in the rat brain, within experimental error of expected theoretical maximum of 3.4 [42] to 4 [102] for these low frequencies. In addition, there was no difference for the T_1, T_2, and T_2* between the two fields for ^{17}O incorporated into water, presumably because of the quadrapolar nature of the relaxation in this nucleus. Thus, the SNR gain measured in this case is not diminished by bandwidth and spin relaxation considerations. This means that for this nucleus, a factor of ~3 is expected in going from 4 Tesla to 7 or 8 Tesla, the highest field currently available for human studies. Given the fact that experiments with such nuclei are severely SNR limited, this a significant gain.

III. Functional Brain Imaging

One of the most important accomplishments in MRI research in the last decade is the introduction of methods that can map the areas of altered neuronal activity in the brain, i.e. functional magnetic resonance imaging or fMRI.

The most commonly used method of fMRI is based on Blood Oxygen Level dependent (BOLD) contrast, first described by Ogawa [66–68] in rat brain studies. BOLD contrast is sensitive to the presence of deoxyhemoglobin. Shortly after its introduction, BOLD contrast was successfully used to generate functional images in human brain [2, 53, 71]. BOLD contrast originates from the intravoxel magnetic field inhomogeneity induced by paramagnetic deoxyhemoglobin sequestered in red blood cells, which in turn are compartmentalized within the blood vessels. Magnetic susceptibility differences between the deoxyhemoglobin containing compartments *versus* the surrounding space devoid of this strongly paramagnetic molecule generate magnetic field gra-

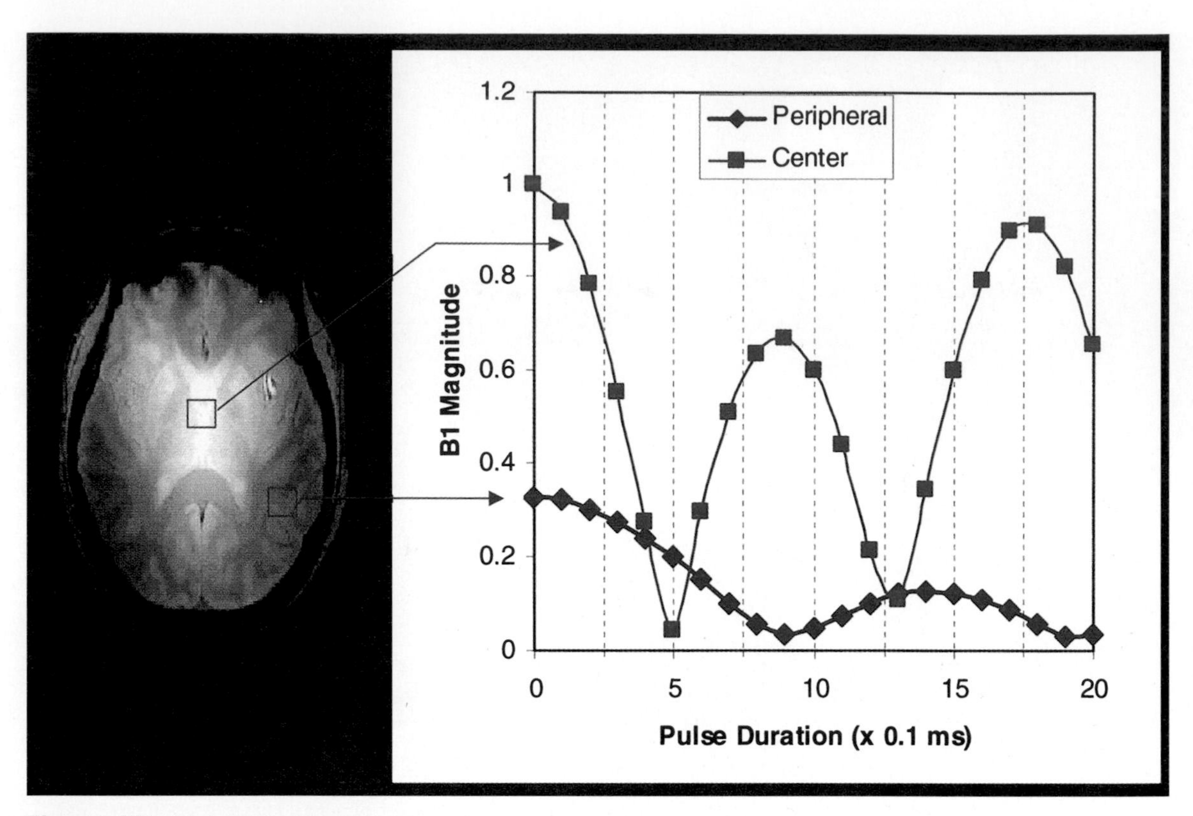

Figure 1B The signal (magnitude) oscillation frequency plotted for two regions of interest in the middle and the periphery of the brain, from the data shown in figure 1A. Oscillation frequency is equal to gB_1 indicating that the B_1 is substantially higher in the center of the brain. From reference [97].

dients across and near the boundaries of these compartments. Therefore, signal intensities in magnetic resonance images sensitized to BOLD contrast are altered if the regional deoxyhemoglobin content is perturbed. This occurs in the brain because of spatially specific metabolic and hemodynamic response to enhanced neuronal activity; it has been suggested that regional blood flow (CBF) increases while oxygen consumption rate (CMR_{O2}) in the same area is not elevated commensurably (e.g. [23, 24, 39, 52, 82]), resulting in decreased extraction fraction and lower deoxyhemoglobin content per unit volume of brain tissue. Consequently, signal intensity in a BOLD sensitive image increases in regions of the brain engaged by a "task" relative to a resting, basal state.

BOLD contrast relies on the interplay between CBF and CMR_{O2} as well as blood volume (CBV). As such, it represents a complex response controlled by several parameters [8, 12, 46, 69, 70, 96, 101]. Recent magnetic resonance imaging techniques, however, can also generate images based on quantitative measures of CBF changes coupled to neuronal activity (e.g. [21, 49, 51–53, 103]). These CBF methods rely on tagging the blood spins differentially within and outside of a well-defined volume. For example, in the FAIR technique [49, 51, 52], frequency-selective inversion pulses are used to invert the longitudinal magnetization within a "slab" along one direction (typically axial); in the absence of blood flow, the spins relax back to thermal equilibrium only by spin-lattice relaxation mechanisms characterized with the time constant T_1. If flow is present, however, the relaxation becomes effectively faster as unperturbed spins outside the inverted slab flow in and replenish the net magnetization within the slab. Consequently, the effective spin-lattice relaxation becomes characterized by a shorter time constant, T_1*, which is related to blood flow. It also follows naturally that if the inversion pulse in FAIR does not define a slab but inverts everything in the whole body (i.e. it is non-selective), blood flow does not enter into to the problem at all provided the arterial blood and tissue T_1 values are not dramatically different. While CBF based fMRI has been consistently improving, it is still not as commonly used as the BOLD techniques because of shortcomings including prolong data acquisition times. Therefore, a large part of the following discussion will still focus on the use of BOLD effect in mapping brain function.

In functional brain imaging, as in many other MR applications, a critical parameter besides SNR is the contrast-to-noise ratio (CNR), associated in this case, with detection of alterations in basal signal intensity during changes in neuronal activity. CNR in functional mapping in the brain, however, is intricately tied with the problem of spatial specificity. For example, increased neuronal activity can induce very large changes in MR signal intensity due to effects associated with macroscopic blood vessels; however, the

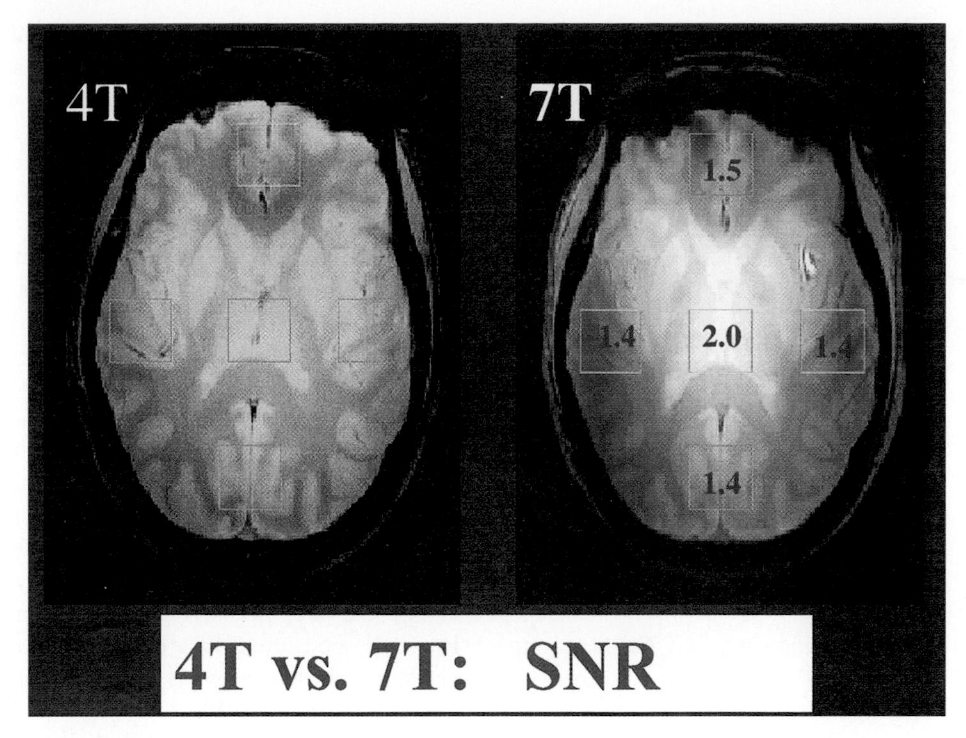

Figure 2 Spin-density images at 4T and 7T of the human head obtained with identical coils and experimental parameters to measure SNR. For regions of interests identified with boxes, SNR increase at the 7T relative to 4T is given on the 7T images. From reference [97].

origin of these large signal changes will not precisely co-localize with the actual site of neuronal activation [62, 83]. Such signal increases, although large in CNR, are undesirable since they are not useful due to their lack of spatial specificity for most except the crudest of functional imaging studies. In functional imaging, ideally only signals associated with microvasculature are desired, and the important problem is to increase the CNR for contributions coming from these vessels while suppressing the CNR associated with large blood vessels; this is exactly what high fields provide. CNR specific to microvasculature is small, but increases supralinearly with the magnetic field. Even at 7 Tesla, however, they do not reach the very large magnitudes that can be seen for macrovascular signals at lower fields such as 1.5T. These macrovascular signals have high CNR, albeit, lack accuracy and as such are undesirable.

Thus, the most important attribute of high magnetic fields in BOLD functional imaging studies is the improvements in spatial accuracy so that it is limited by capillary density. This is an important, in fact an imperative, step if fMRI is to make contributions to brain science beyond its current achievements. While present fMRI capabilities are and may remain useful for a long time in cognitive psychology studies, they are ultimately not adequate given what is known about brain function from single unit studies. For example, it is well recognized from animal experiments that neurons in the visual system with similar functional behavior are clustered together in patches that can be as small as 300 to 500 microns in diameter on the cortical surface (e.g. ocular dominance columns, orientation columns). Such clustering has been proposed for some higher order visual areas (e.g. inferio-temporal (IT) cortex in the monkey with respect to face and object recognition) based on electrophysiology studies [27, 100]. Such clustering may and likely does exist for many other higher order functions of the brain. Ability to examine such fine functional parcellation in cortical space requires high specificity, non-invasive functional imaging methods that can provide high-resolution information over a large volume if not the whole brain. Electrode recordings, while highly accurate, vastly undersample the brain and cannot be practically performed over large areas of the cortex. Optical imaging methods do not provide information deeper than ~1 mm on the cortical surface and cannot access large area of cortex within the multiple folds and sulci in primate brains.

Except in a few cases aimed at examining the BOLD mechanisms BOLD contrast fMRI relies on gradient echo (i.e. T_2* weighted) approaches where, after signal excitation, a delay in tens of milliseconds is introduced before data acquisition, covering full or part of the k-space (Figure 3). The plethora of applications that rely on fMRI for understanding brain function all utilize this approach. However, this BOLD response (the only BOLD effect that is large enough to be routinely usable in fMRI *applications* so far) lacks spatial specificity in the few millimeters to submillimeter scale. This was demonstrated recently using the cat

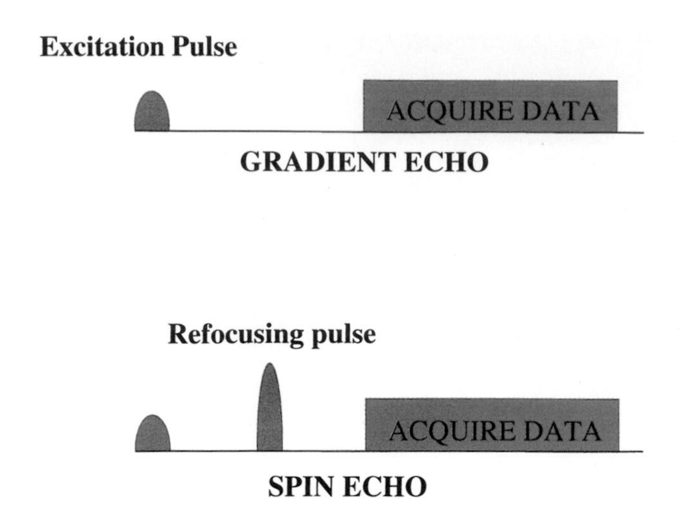

Figure 3 Schematics of gradient versus Hahn spin-echo imaging sequence. The latter differs from the former only with respect to the refocusing pulse.

visual system and the iso-orientation columns. Gradient echo based positive BOLD effect did not provide the necessary specificity to image iso-orientation domains [17, 18, 48] and concurrent recordings of fMRI and single unit recordings from multiple sites, showed that T_2* based positive BOLD effect *could not* depict electrical activity of neurons in a scale smaller than 4 to 5 mm at 4.7 T [47].

This lack of spatial specificity is ascribed to two mechanisms: blood flow increases that accompany neuronal activity alterations exceed the boundaries of active neurons, and large vessels contribute to the T_2* based BOLD signals. The first is a physiological problem related to the point spread function of the blood flow response to increased neuronal activity; it concerns the question of how accurate the boundaries of the blood flow increase are relative to the region of elevated neuronal activity. fMRI does not directly detect neurotransmission or electrical activity of neurons; instead it relies on the secondary and tertiary signals that evolve as a result of the alteration in neuronal activity. Therefore, the spatial accuracy of these secondary and tertiary metabolic and hemodynamic responses that accompany elevated neuronal activity ultimately determine the accuracy of the fMRI maps. The second mechanism that results in the degradation of spatial specificity concerns the nature of the coupling between these metabolic and hemodynamic changes to MR detectable signals. Here, the source of the problem is the macrovascular "inflow" and BOLD effects.

Imaging sequences used for BOLD fMRI can be (but need not be) sensitive to changes in blood flow. Blood flow increases induced by elevated neuronal activity must also result in blood flow increases in large blood vessels that supply blood or drain the activated tissue. If the imaging sequence is flow sensitive, then these vessels appear as "activated", leading to a macrovascular "inflow" artifact [62, 83]. Such an apparent "activation" can be several centimeters away from the area of activation, and depict not the activated tissue in the brain but also provide an angiography-like images of the vessels that supply and drain that tissue [83]. Although this problem probably plagued a majority of early fMRI studies, it is straightforward to suppress by allowing near full relaxation in between excitations, and thus will not be considered further. However, large veins can appear in functional images through the BOLD effect as well [62]. Deoxyhemoglobin changes that originate in the tissue with altered neuronal activity propagate into the draining veins until they are diluted out by vessels that bring in blood from tissues which were not activated. In this case, one is concerned with BOLD effects associated with blood itself (i.e. intravascular effects), and with extravascular BOLD effects associated with large vessels (e.g. > 10 micron diameter) (see discussion below). As will be discussed in greater detail, all intravascular and blood related effects are diminished with increasing magnetic field due to rapidly decreasing T_2 of blood, and are virtually undetectable at 7 or 9.4 Tesla. However, the extravascular effects, of course, persist or even get larger at high magnetic fields in gradient recalled echo (i.e. T_2*) weighted fMRI.

Extravascular BOLD effects associated with large vessels that continue to degrade spatial specificity in the positive T_2* BOLD fMRI at high magnetic fields can be eliminated by the use Hahn spin echo fMRI (Figure 3). As explained in greater detail later on, spin-echoes "refocus" and consequently eliminate extravascular BOLD effects associated with large blood vessels. However, Hahn spin echoes contain information about deoxyhemoglobin through a different mechanism; diffusion induced dynamic averaging of susceptibility gradients surrounding deoxyhemoglobin containing compartments and, within the blood vessels, water exchange between the deoxyhemoglobin-containing red blood cell interior and plasma. Diffusion distance in the time scale characterized by the T_2 of water protons in brain tissue is small (e.g. ~10 microns for 100 ms, based on experimentally measured diffusion constant of 10^{-3} mm^2s^{-1} for water in the rat brain [55]). Consequently, the dynamic averaging is prominent mainly in the tissue space surrounding small blood vessels, and, in blood, around the red blood cells. Thus, Hahn spin echo fMRI signals can originate from extravascular space *around the microvasculature*, as well as from blood itself within the blood vessels. The former provides spatial specificity in the hundred micron spatial scale because capillaries are separated on the average by 50 *mm* [77]. The latter effect, however, is associated with blood containing deoxyhemoglobin irrespective of the blood vessel size containing that blood; thus, this effect can be present both in large and small blood vessels and hence degrades spatial specificity of fMRI. However, the T_2 of venous blood decreases quadratically with B_0 [92] and is diminished from ~90 to ~180 ms at 1.5 Tesla [3, 104] to 7–13 ms at 7 Tesla [105] to ~6 ms at 9.4 Tesla [55] significantly smaller than brain tissue T_2, and TE values that would be used in fMRI studies at such field strengths. Thus, blood effects in fMRI that degrade spatial specificity, and dominate low magnetic field functional images [7, 61, 88], even when Hahn spin echoes are used, are suppressed at high magnetic fields.

When the goal is high specificity functional mapping and this is then pursued with Hahn spin echo BOLD studies at ultra high magnetic fields, the issue of CNR becomes CNR due to capillaries and possibly small post-capillary venules. The microvascular contribution to the BOLD effect is intrinsically small. It is virtually undetectable at 1.5 or even 3T as demonstrated by experiments using Stejskal-Tanner gradients that suppress flowing blood and thus the blood contribution to Hahn spin-echo fMRI signals. But it increases quadratically with magnetic field magnitude and becomes sufficiently large to yield robust, ~10% changes in the rat cortex at 9.4 Tesla [55], and in the human brain at 7 Tesla. Therefore, significant gains are expected in going to 9.4T compared to the 7 Tesla system that is available for human studies at CMRR.

The concepts summarized above regarding physiological point spread function of the fMRI signals, T_2 versus T_2*, and blood versus tissue effects are explained in greater detail below where BOLD mechanisms are reviewed.

A. Specificity of Blood Flow Increases that Accompany Increased Neuronal Activity

It has been well documented that stimulus-induced increases in glucose consumption rate are localized to individual iso-orientation columns of the visual cortex which are ~500 μm in diameter and are 1.0–1.5 mm apart [56]. Optical imaging with intrinsic signals that monitor hemoglobin signals [58, 59] demonstrated that after the onset of stimulation, deoxyhemoglobin time course was biphasic, showing an initial increase in deoxyhemoglobin content, reaching a maximum in about 2 to 3 sec, followed by a decrease culminating in lower deoxyhemoglobin levels compared to the pre-activation basal state. This biphasic response is thought to reflect an early onset of elevated oxygen consumption rate, followed by an increase in blood flow that ultimately reverses the elevation in deoxygenation. It was shown that the initial increase deoxyhemoglobin can be used to resolve functionally active, individual orientation columns. In marked contrast, the stimulus-evoked delayed elevation in total hemoglobin content (i.e. CBV), were reported to be spatially diffuse and less suited for resolving the columnar layouts [58, 59]. Based on the assumption that CBV and CBF responses coincide in spatial extent, the conclusion regarding the specificity of CBV are also extended to the CBF response that accompany alterations in neuronal activity. Thus, it was suggested that "the brain waters the whole garden for the sake of a single thirsty flower".

While optical imaging with intrinsic signals does not provide a direct measure of CBF, MR based methods do. Specificity of the CBF response at the tissue level can be evaluated directly using CBF based fMRI. Though less widely used than the BOLD technique because of its relatively poor temporal resolution (on the order of 1–6 s in comparison to sub-second resolution for BOLD), CBF-based fMRI can be targeted to map flow into the capillaries and the tissue (due to exchange across capillary walls), avoiding large draining vein contributions. Using the flow-sensitive alternating inversion recovery (FAIR) fMRI technique [49] for imaging CBF changes at the tissue level (i.e. imaging tissue perfusion), the point spread function of perfusion increases under a single-stimulus condition was recently investigated in the cat visual cortex [18].

The veracity of the CBF-based single-condition columnar maps was evaluated using the complementarity of orthogonal orientation stimuli. Functional iso-orientation columns responding to two orthogonal stimuli should approximately occupy complementary domains. Figure 4 shows the results of the complementary test. Each patchy image is from a stimulus with a grating (parallel lines) oriented either 45° or 135° from the vertical. These are orthogonal orientations. The patchy clusters from the two orthogonal orientations indeed occupy complementary cortical territories and that the "complementarity" between orthogonal maps was statistically significant [18]. The CBF time courses and percent changes are illustrated both for the correct and orthogonal orientation (Figure 4) in iso-orientation domains responding to 45° or 135° gratings. A marked CBF increase (~55%) following *45° or 135° stimulus* was observed in the complementary regions tuned to these orientations, while the stimulation with the orthogonal orientation lead to a significantly smaller CBF increase (on the average 3.3±0.6 fold smaller).

These results demonstrate that CBF increases that follow neuronal activation are not "perfectly" localized at the iso-orientation column level but there is a sufficient difference between active and neighboring inactive columns in the CBF response that a CBF based iso-orientation map can be generated. Yet, in the same system, when one looks at the positive BOLD changes, iso-orientations domains are indistinguishable [17, 18, 48]. This lack of specificity, then, does not represent a fundamental physiological limitation; rather, it must arise predominantly due to the nature of the BOLD based imaging signals and how they couple to the underlying changes in CBF and metabolism.

The ability to obtain iso-orientation maps by CBF response is not necessarily contradictory to the optical imaging results based on CBV maps. Although CBV changes occur in both active and adjacent inactive columns, CBV increase associated with the active columns may exceed that observed in the inactive columns in the cat iso-orientation domains [91]. In addition, the contrast for the

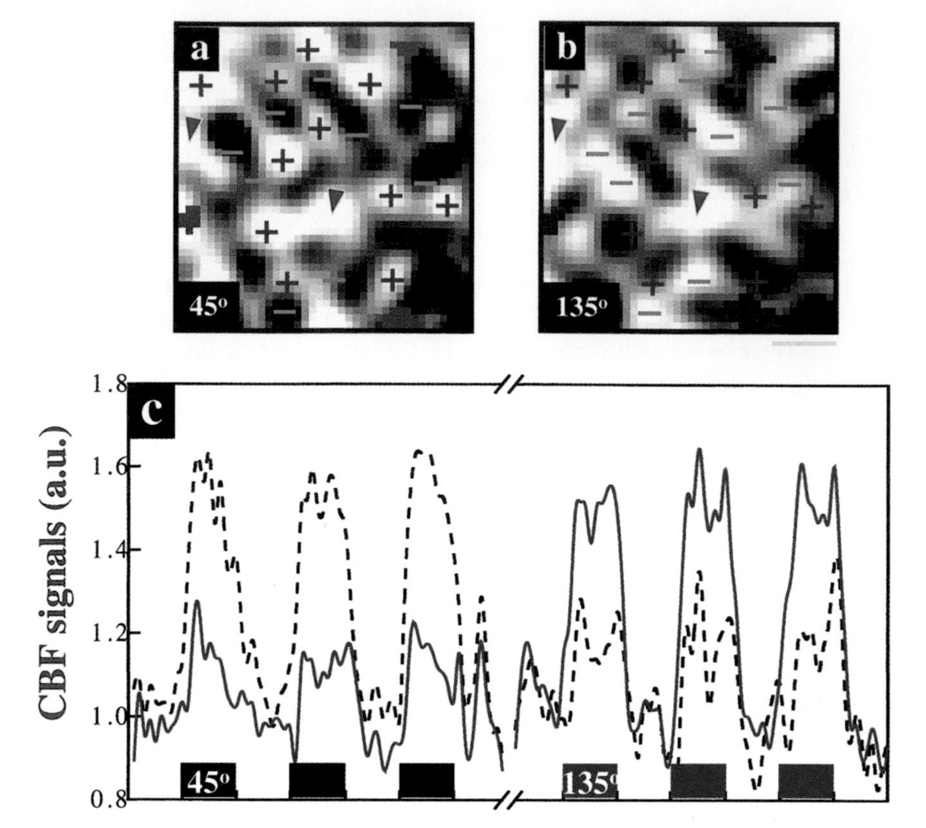

Figure 4 Maps of iso-orientation domains in the cat visual cortex obtained with CBF based mapping for two complimentary orientations (45 and 135 degree gratings). Bottom trace shows the blood flow response to the correct orientation stimuli and the orthogonal orientation stimuli for 45 and 135 degree orientationsactivation in the active and inactive columns. From reference [18]

CBF and CBV based methods is not necessarily the same. While CBV and CBF changes most likely co-localize, it is important to consider that the "contrast" in these two measurements are likely to be different. It has been demonstrated in monkeys using hypercapnia [29] that CBV and CBF alterations are related according to

$$\Delta CBV/CBV + 1 = (\Delta CBF/CBF + 1)^{0.38} \quad (1)$$

where ΔCBV/CBV and ΔCBF/CBF are relative CBV and CBF changes. Assuming this relationship to be applicable to stimulation of iso-orientation domains, it can be used to calculate the contrast for CBV mapping of iso-orientation domains from the MR based CBF data presented above. Taking $\{\Delta CBF_{stim}/CBF_{basal} + 1\}$ to be 1.55 and 1.17, for the domains tuned and orthogonal the stimulation orientation, respectively, the ratio $\{CBF_{stim\text{-}tuned}/CBF_{stim\text{-}othogonal}\}$ is 1.33, where $CBF_{stim\text{-}tuned}$ and $CBF_{stim\text{-}othogonal}$ are the CBF attained in an iso-orientation domain for the tuned and orthogonal orientation stimuli, respectively; the corresponding ratio for CBV, calculated based on Equation 1 would be 1.11. Thus, CBV would have less "contrast" relative to CBF and it would be more difficult to generate CBV based iso-orientation based column maps with CBV compared to CBF.

Specificity of the CBF response with respect to functional clusters will naturally depend on the size of the clusters relative to arteriol density, unless some regulation of CBF at the capillary level exists. Thus, unless we invoke capillary level regulation, iso-orientation columns in the cat *versus*, for example, the ferret where the iso-orientation columns are smaller, or *versus* the ocular dominance columns in the monkey may not be the same. Therefore, care must be exercised in generalizing from the data on CBF mapping of iso-orientation columns in the cat cortex to other functional organizations.

B. Mechanistic Considerations—Extravascular Effects

If BF response is relatively specific to the iso-orientation domains in the cat visual cortex, why does the positive BOLD response lack any sign of this specificity n the same animal preparation? This lack of specificity can be explained by the temporal evolution of the deoxyhemoglobin changes along the post-capillary vascular tree and role played by the vasculature in mediating the coupling between BOLD MR signals and underlying changes in CBF and metabolism.

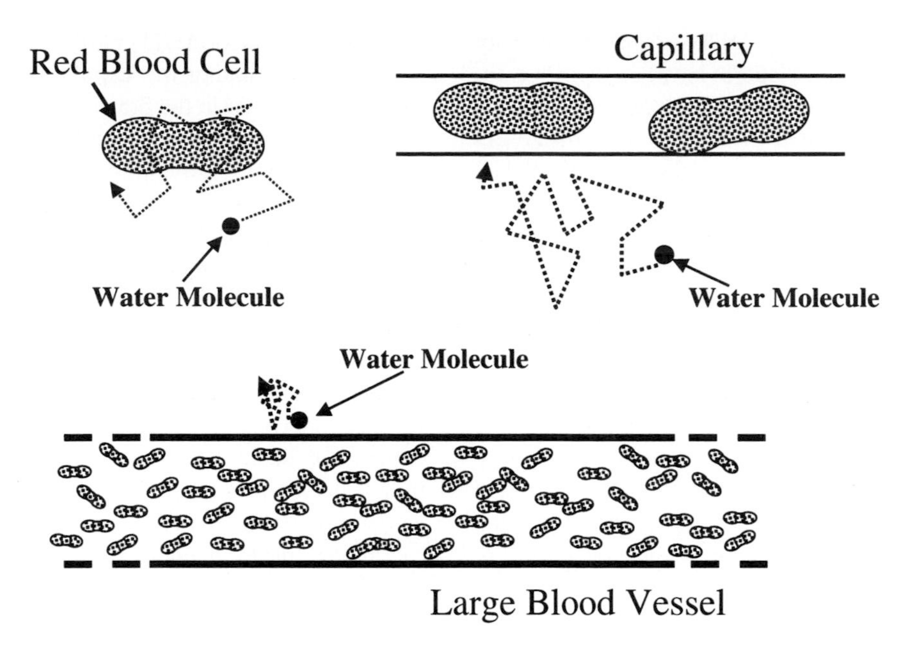

Figure 5 Dynamic and static averaging regimes based on diffusion distances relative to the size of the compartment that differs in magnetic susceptibility compared to the surrounding tissue. The magnetic field gradients are most prominent in the vicinity of the compartments with different susceptibility, i.e. red blood cells, capillaries, and large blood vessels. In case of the large blood vessels, diffusion distances are not large compared to vessel radius, and hence do not lead to dynamic averaging.

The BOLD effect should be divided into two major categories: extravascular and intravascular (i.e. blood related) BOLD effects. If one considers an infinite cylinder as an approximation for a blood vessel with magnetic susceptibility difference Δx, then the magnetic field expressed in angular frequency, at any point in space, will be perturbed from the applied magnetic field ω_o [89]. Inside the cylinder, the perturbation, $\Delta\omega_B$, will be given by the equation

$$\Delta\omega_B^{in} = 2\pi\Delta x_o (1-Y)\omega_o\{\cos^2(\theta)-1/3\} \quad (2)$$

At any point outside the cylinder, the magnetic field will vary depending on the distance and orientation relative to the blood vessel and the external magnetic field direction, according to the equation:

$$\Delta\omega_B^{out} = 2\pi\Delta x_o (1-Y)\omega_o\{r_b/r\}^2 \sin^2(\theta)\cos(2\phi) \quad (3)$$

In these equations, Δx_o is the maximum susceptibility difference expected in the presence of fully deoxygenated blood, Y is the fraction of oxygenated blood present, r_b designates the cylinder radius, r is the distance from the point of interest to the center of the cylinder in the plane normal to the cylinder. Note that outside the cylinder, the magnetic field changes rapidly over a distance comparable to two or three times the cylinder radius; at a distance equal to the diameter of the cylinder from the cylinder center, $\Delta\omega_B^{out}$ is already down to 25% of its value at the cylinder boundary.

If such a blood vessel is present in a given voxel, the magnetic field within this voxel will be inhomogeneous. The effect of this inhomogeneity in tissue can be understood in terms of *dynamic and static averaging*, the former arising as a result of the diffusive motion of the water molecules.

First, let us ignore the blood in the intravascular space (i.e. inside the cylinder) and focus on the *extravascular space* only. If the typical diffusion distances during the delay TE are comparable to the distances spanned by the magnetic field gradients, than during this delay, the magnetic field inhomogeneities will be dynamically time-averaged. Typical TE values used in fMRI experiments depend on the field strength and the specifics of the pulse sequence, but in general range from ~30 to ~100 ms. Thus, blood vessel size compared to the diffusion distances in this ~30 – 100 ms time domain becomes a critical parameter in the BOLD effect (Figure 5). In this time scale, small blood vessels, e.g. capillaries, that contain deoxyhemoglobin will contribute to the dynamic averaging and result in a signal decay that will be characterized with a change in apparent T_2[8, 70, 101]. In a spin-echo experiment with a single refocusing pulse in the middle of the delay period (i.e. a Hahn Spin echo), the dynamic averaging that has taken place during the first half of the echo will not be recovered. Of course, applying many refocusing pulses as in a Carr-Purcell pulse train or applying a large B_1 field (relative to the magnitude of the magnetic field inhomogeneity) for spin-locking during this delay will reduce or even eliminate this signal loss due to dynamic averaging. In a gradient echo measurement, dynamic averaging will also occur during the entire delay TE. If the imaging voxel contains only such small blood vessels at a density such that one-half the average distance between them is comparable to diffusion distances (as is the case in the brain where capillaries are separated on the average by 25 μm [77]), then the entire signal from the voxel will be affected by dynamic averaging.

In considering the movement of water molecules around blood vessels, we need not be concerned with the exchange that ultimately takes place between intra- and extravascular water across capillary walls. Typical lifetime of the water in capillaries exceeds 500 ms [22, 75, 76], significantly longer than the typical T_2 and T_2* values in the brain tissue and longer than the period TE typically employed in fMRI studies.

For larger blood vessels, complete dynamic averaging for the entire voxel will not be possible. Instead, there will be "local" or "partial" dynamic averaging over a subsection of the volume spanned by the magnetic field gradients generated by the blood vessel. However, there will be signal loss from the voxel due to *static averaging* if refocusing pulses are not used or asymmetric spin echoes are employed. Following the excitation and rotation onto the plane transverse to the external magnetic field, the bulk magnetization vector of the nuclear spins will precess about the external magnetic field with the angular frequency ω_B^{out}. A water molecule at a given point in space relative to the blood vessel will see a "locally" time-averaged ω_B^{out}, $\bar{\omega}_B^{out}$, which will vary with proximity to the large blood vessel. Thus, signal in the voxel will then be described according to equation

$$S(t) = \sum_k s_{ok} e^{-TE/T_{2k}} (e^{-i\bar{\omega}_k TE}), \qquad (4)$$

where the summation is performed over the parameter k, which designates small volume elements within the voxel; the time-averaged magnetic field experienced within these small volume elements is $\bar{\omega}_k$ in angular frequency units. The summation over k thus covers the entire voxel. Because $\bar{\omega}_k$ TE varies across the voxel, signal will be "dephased" and lost with increasing echo time TE. This signal loss occurs from "*static averaging*". In this domain, if the variation $\bar{\omega}_k$ over the voxel is relatively large, signal decay can be approximated with a single exponential time constant T_2*. In a spin-echo, the static dephasing will be refocused and thus eliminated.

Figure 6 illustrates R_2* (i.e. $1/T_2$*) as a function of the radius of the cylinder from modeling studies by Ogawa *et al.* [70]. In this calculation, all other relaxation mechanisms that can contribute to transverse relaxation of water protons are ignored; only the effect of the susceptibility difference between the cylinders and their surrounding is considered. The imaging voxel was divided into many cubes ($<10^6$) with an edge dimension L, each of which contained a single cylinder of length L. The volume of the cylinder (i.e. "blood volume" b_v) relative to the volume of the cube is given by $((\pi r_b^2 L)/L^3)$). Figure 6 displays the results as a function of cylinder radius for $\Delta x_O(1\text{-}Y)\omega_O$ equal to 32, 48, and 64 Hz, $b_v = 0.02$ (simulating the capillary blood volume to total tissue volume ratio in the brain [77]), and TE = 40 ms (often used in 4 Tesla studies). Given the maximal susceptibility difference between fully oxygenated and fully deoxygenated blood [68], and taking Y to be 0.6, typical of venous blood in the brain, the frequency difference $\Delta x_o(1\text{-}Y)\omega_o$ is calculated to be 43 Hz at 4 Tesla. The data in Figure 6 demonstrate that at a radius less than ~5 to 10 μm, depending on Δx_o $(1\text{-}Y)\omega_o$ magnitude, R_2* decreases because of dynamic averaging; the cylinder radius where this R_2* decrease becomes apparent is smaller at larger $\Delta x_o(1\text{-}Y)\omega_o$ as expected. Above ~10 μm, R_2* is approximately independent of the radius as the static regime dominates for all $\Delta x_o(1\text{-}Y)\omega_o$ values considered in these calculations. These modeling efforts also suggest that deoxyhemoglobin containing microvessels in the brain, i.e. capillaries (mean diameter 5 μm [77]) and some of the venules (which can range in diameter from 5 to 20 μm [11]) would be in the "dynamic" averaging regime.

The net result of these calculations (again so far not considering the intravascular effects) yield the following terms for contributions to R_2* (i.e. $1/T_2$*)

$$R_2^* = \alpha\{\Delta x_o \omega_o (1-Y)\} b_{vl} \quad \text{(large vessels)} \qquad (5)$$

$$R_2^* = R_2 = \eta\{\Delta x_o \omega_o (1-Y)\}^2 b_{vs} p \quad \text{(small vessels)} \qquad (6)$$

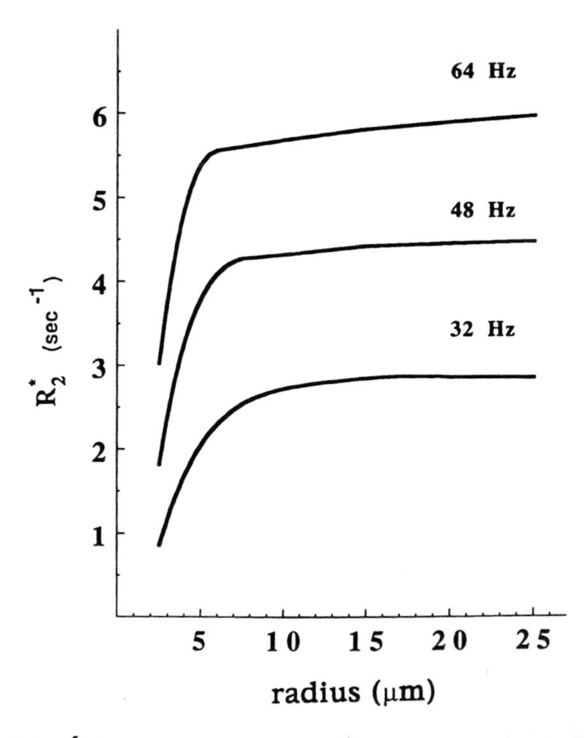

Figure 6 The susceptibility induced R_2^* in the presence of water diffusion plotted as a function of cylinder radius. Plots are shown at three different values of frequency shifts (32, 48, 64 Hz). Echo time = 40 msec, fractional vascular volume (i.e. volume in cylinder relative to voxel volume) = 0.02. From reference [70].

where α and η are constants, ν_o is the external magnetic field in frequency units (rad/sec) (i.e. $\omega_o = \gamma B_o$), $\{\Delta x_o \omega_o(1\text{-}Y)\}$ is the frequency shift due to the susceptibility difference between the cylinder simulating the deoxyhemoglobin containing blood vessel, b_{vl} is the blood volume for large blood vessels (veins and venules with a *radius* greater than ~ 5 *mm* for 4T) and b_{vs} is the *small* vessel blood volume (capillaries and small venules, less than ~5 μm in *radius* that permit dynamic averaging), and p is the fraction of active small vessels (i.e. filled with deoxyhemoglobin containing red blood cells).

An important prediction of the modeling studies is that there are large and small vessel *extravascular* BOLD effects. This has implications with respect to the specificity of functional images generated by the BOLD effect. While capillaries are uniformly distributed in tissue and are sufficiently high in density, large venous vessels are not; consequently BOLD effects associated with large vessels will not be as closely correlated with actual site of neuronal activity. An important feature of equation (6.) is the fact microvascular contributions varies as the square of the external magnetic field for small vessels where the effect is dominated by dynamic averaging. In contrast, the dependence on the external magnetic field is linear for large blood vessels because they are in the static averaging domain.

Another important point that can be surmised from equations 5 and 6 is that the BOLD effect will be proportional to three physiologic parameters: regional cerebral blood flow and oxygen consumption rate (CMR_{o2}) {since (1-Y) = CMR_{o2}/CBF} and regional blood volume. Neuronal activity is coupled to all of these physiologic parameters.

C. Mechanistic Considerations—Blood Contribution (Intravascular Effects) in BOLD

In the blood, hemoglobin is also compartmentalized within red blood cells. Thus, when the deoxy form is present, there are field gradients around the red cells. However, because the dimensions are very small compared to diffusion distances, the effect is dynamically averaged and becomes a T_2 effect only. The dynamic averaging in this case also involves exchange across the red blood cell membrane that is highly permeable to water. The exchange is between two compartments, plasma and the interior of the red blood cell where the magnetic field is significantly different because of the presence of paramagnetic deoxyhemoglobin. Thus, in the presence of deoxyhemoglobin containing red blood cells, apparent T_2 of blood decreases. Therefore, even when we neglect the extravascular effect described before, the T_2 of *blood itself* will change when the content of deoxyhemoglobin is altered by elevated neuronal activity and this will lead to a signal change in the apparent T_2 or T_2* weighted image. This effect will be present wherever the content of deoxyhemoglobin has changed, thus potentially both in large and small blood vessels.

Either from the exchange model between two sites of different frequencies (e.g. due to susceptibility difference between red blood cells and surrounding plasma) formulated by Luz-Meiboom [57] or simply diffusion in the presence of magnetic field gradients generated by the red blood cells, blood T_2 can be written as

$$1/T_2 = A_o + K(1-Y)^2 \quad (7)$$

where Y is the fractional oxy-hemoglobin content, A_o is a field independent term and K is a parameter that would scale quadratically with the magnetic field and would depend also on the the echo time used in a single spin echo measurement. Based on previous studies, it is generally thought that the exchange model dominates [9]. Using the Luz-Meibaum model, Equation 7 can be further elaborated as

$$1/T_2 = A_o + [K_o \omega_o^2 (1 - \frac{2\tau_{ex}}{t} \tanh \frac{t}{2\tau_{ex}})](1-Y)^2, \quad (8)$$

where K_o is a constant, ω_o is the magnetic field in frequency units, and τ is the echo time (i.e TE in imaging terminology) for the spin-echo measurement, and τ_{ex} is the weighted average of the water lifetime in the erythrocyte and plasma compartments. Note that when diffusion or exchange is present, T_2 is itself a TE dependent parameter, and is not just a constant as defined in the classical formulation of Bloch Equations. Hence the reason why we keep referring to it as "apparent" T_2.

For a single spin echo experiment using an echo time of 24 ms, the blood T_2 was reported to be ~94 ms for Y = 0.6 and behaved according to Equation 7 with constants A_O and K equal to 4.0 s^{-1} and and 41.5 s^{-1}, respectively [104]. This is very similar to the in vivo venous blood T_2 reported for the brain [72], but somewhat shorter than the T_2 values reported by Barth and Moser [3] for Y ~72% even when adjusted for the same Y value. Using the data provided by Wright *et al.* [104] would predict apparent T_2 values of 19.5 ms at 4T and 6.7 ms at 7T for echo time of 24 ms. These extrapolated values are in relatively good agreement with apparent blood T_2 (averaged over echo times) measured at 4T and 7T [105], especially in view of confounding differences introduced in T_2 measurements of blood due to variations in the oxygenation level, temperature of blood during measurements, settling of blood cells, and the echo time dependence.

The blood contribution to gradient or Hahn spin-echo weighted images, based on modeling, are illustrated in Figure 7. The figure displays $\Delta S_{blood}/S_{tissue}+S_{blood}$ which is given by

$$\frac{\Delta S}{S} = \frac{\alpha(e^{-TE/T_2(blood,\, stim)} - e^{-TE/T_2(blood,\, ctrl)})}{(1-\alpha)e^{-TE/T_2(tissue,\, stim)} + \alpha e^{-TE/T_2(blood,\, ctrl)}}$$

$$\frac{\Delta S_{blood}}{S_{blood+tissue}} = \frac{\alpha\{e^{-TE\cdot R_2(blood,ctrl)}[-TE\cdot \Delta R_2(blood,stim)]\}}{\alpha e^{-TE\cdot R_2(blood,ctrl)} + (1-\alpha)e^{-TE\cdot R_2(tissue,ctrl)}} \quad (9)$$

where

$$\Delta S_{blood} = S_0[e^{-TE\cdot R_2(blood,stim)} - e^{-TE\cdot R_2(blood,ctrl)}] \approx e^{-TE\cdot R_2(blood,ctrl)}[-TE\cdot \Delta R_2(blood,stim)]. \quad (10)$$

In the above equations, TE is the echo time, S is the signal intensity at TE, S_o is the voxel signal intensity at TE = 0, $R_2 = 1/T_2$ for blood under stimulation and control conditions, and ΔR_2*(blood, stim)* = {R_2*(blood, ctrl)*— R_2*(blood, stim)*}. The approximation in Equation 10 is based on the expectation that the term TE· ΔR_2*(blood, stim)* is expected to be much smaller than 1 for activation studies. For example ΔR_2^* was reported to be 1.5 s^{-1} for gray matter tissue and blood regions at 7 Tesla [105]. Note that for activation studies, ΔR_2*(blood, stim)* is a negative number so that ΔS is positive. The modeling used the data of Wright *et al.* [104] for 24 ms TE, Y = 0.6 at 1.5T and extrapolated for the different field strengths; the data were "normalized" relative to 1.5T by taking ΔR_2*(blood, stim)* as equal to 1 at 1.5T *at echo time* τ *or TE >>* τ_{ex} *where it becomes independent of echo time*, and scaling it up for field magnitude and TE dependence according to Equation (7.) and (8.). τ_{ex} was estimated to be 7 ms from Equation (7.) using data on varying Y at different echo times provided by Wright *et al.* [104] this is in excellent agreement with a value of 7.2 ms that can be calculated from hematocrit or total hemoglobin content data summarized previously [72, 96]. Since blood signals are most important when relatively large blood vessels and/or blood volume exists in the voxel, we have performed the simulations for 10% blood volume.

Figure 7 demonstrates that at around brain tissue T_2's (90, 80, 67 and 55 ms for gray matter at 1.5T, 3T, 4T, 7T and 9.4T, respectively) blood contribution is large at 1.5T but is virtually eliminated at 7 and 9.4 T because of rapidly decreasing apparent T_2 of blood. These predictions were experimentally verified at 4, 7 and 9.4 Tesla, using Stejskal-Tanner "diffusion weighting" gradients (discussed further on).

Blood contribution comes into the BOLD phenomenon in a second special way when blood occupies a large fraction of the volume of the voxel, in other words *when there exists a large blood vessel* in the voxel. When deoxyhemoglobin is present in the blood, the blood water will dynamically average the gradients surrounding the red blood cells and will behave as if it encounters a uniform magnetic field given by Equation 1. This will differ from the magnetic field experienced by the rest of the voxel. In the immediate vicinity of the blood vessel, the magnetic field will vary and approach a constant value in tissue distant from the blood vessel. For simplicity, we can neglect the gradients near the blood vessel and consider the voxel to be composed of two large bulk magnetic moments, one associated with blood and the rest with the extravascular volume. These magnetic moments will precess at slightly different frequencies; therefore, the signal from the voxel will decrease with time as the two moments lose phase coherence. In this scenario, the

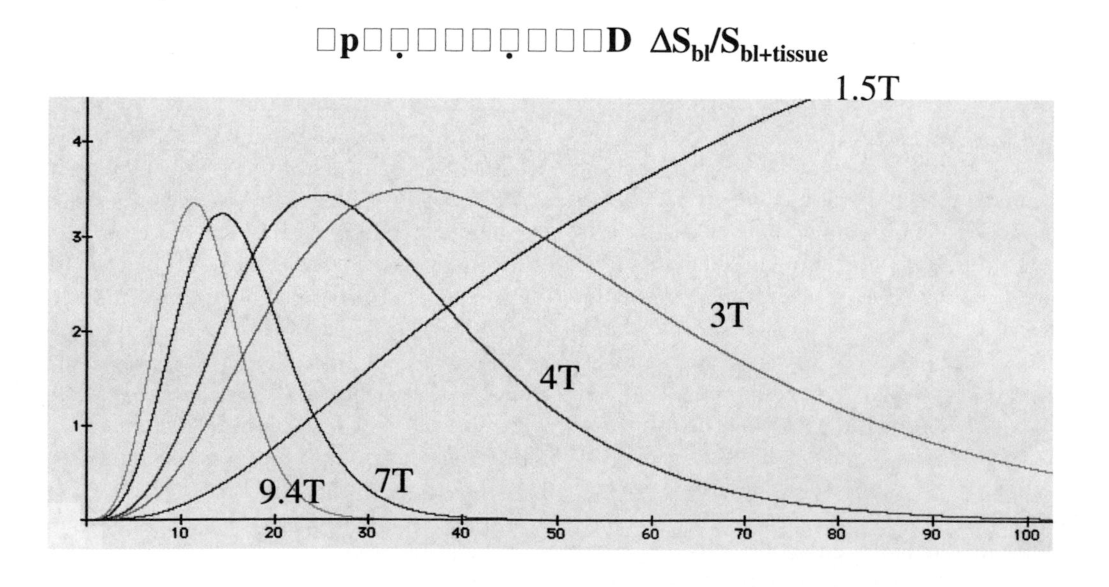

Figure 7 Calculations of $DS_{blood}/S_{blood} + S_{tissue}$ for 10% blood volume for magnetic fields of 1.5T, 3T, 4T, 7T, 9.4T as a function of TE.

signal can even oscillate as the phase between the two magnetic moments increase and then decrease. We refer to this as type-2 blood effect in fMRI.

When a voxel only contains capillaries, the blood volume is ~2% [77]; hence, type-2 blood effect cannot exist for such a voxel. However, when a large blood vessel or vessels are present in the voxel, blood volume can significantly increase and become comparable to or even exceed the tissue volume in the voxel. If, of course, the voxel is smaller than the blood vessel dimensions, and the entire voxel is occupied by blood, then type-2 effect does not come into play either.

Note that the type-2 blood effect and the extravascular BOLD effect in the static averaging regime are similar in nature because they both involve signal modulation due to dephasing of magnetization within a voxel. The main difference is the presence of a large blood component, and the lack of sufficient variation in resonance frequency within the voxel to approximate the signal modulation as an exponential signal decay in the type-2 blood effect. Similar to the extravascular BOLD effect in the static averaging domain, type-2 blood effect will be refocused in a symmetric spin echo and thus nulled. Consequently, it will *not* be present in purely spin-echo BOLD and functional images derived from it.

With respect to high field considerations, the rapidly decaying venous blood signals due to shortening of blood T_2 implies that type-2 blood effects will also diminish with increasing field magnitude. These effects probably are the largest source of BOLD signals in gradient echo fMRI studies at low magnetic fields. However, there contributions would be negligible at 7 Tesla at typical echo times of 25 ms or more.

D. Inflow Effects in BOLD-Based fMRI and the Effect of High Fields

As previously discussed, macrovascular flow change can lead to signal alterations in images intended to report on BOLD contrast (e.g. [20, 26, 50, 62, 83]). This is because a slice-selective image is inherently flow sensitive if the signal within the slice does not attain full relaxation between consecutive signal excitations. In single slice studies, *allowing full relaxation in between RF pulses eliminates this problem completely*; however, this condition often is not satisfied in many studies since it leads to a loss in SNR. Consequently, such studies essentially obtain images of macrovascular flow rather then the much smaller BOLD changes. This problem was demonstrated with clarity by comparing presumably BOLD based "functional" images with vessel images in two and three dimensions [4, 83].

This macrovascular "inflow" artifact can have both a venous and an arterial component. Because of the rapidly decreasing T_2 of venous blood at increasing magnetic fields, the venous component of this "inflow artifact" will also disappear with increasing fields, unless the data acquisition is performed with effectively fast TE's compared to blood T_2. Even the arterial component will be diminished since the arterial T_2 is also very short at high magnetic field, albeit not as short as the venous T_2. The arterial T_2 in anesthetized rat was measured at 9.4 Tesla to be 40 ms as opposed to about 7 ms for venous T_2 at Y ~ 0.6 [55].

E. Experimental Studies of Blood Contribution to fMRI at Different Magnetic Fields

In an fMRI experiment, images are collected subsequent to signal excitation and echo formation, either by a gradient reversal or application of a refocusing RF pulse, as previously discussed. During the delay after excitation and before echo formation, it is possible to apply a pair of gradient pulses with opposing or same polarity depending on whether the experiment is a gradient recalled echo or a spin-echo experiment, respectively. When the water molecules are static in time, then such gradient pulses will ideally have no effect on the image. In the presence of diffusion, such pulses will lead to signal loss since the spatially dependent dephasing during first part of the gradient pulses will not be redone. This pulsed gradient pair have been introduced for diffusion measurements by Stejskal and Tanner [90]; hence often the use of such gradients to alter the image signal intensity is referred to as "diffusion weighting" even though there are additional perturbations that arise from the use of such gradients.

In experiments employing the Stejskal-Tanner gradients, the important parameters are the magnitude and the duration of the gradient pulse and the time separation between them. Frequently, the results are evaluated in terms of a parameter b which is equal to $(\gamma G\delta)^2 (\Delta\text{-}\delta/3)$ where g is the gyromagnetic ratio (rad/sec/Gauss), G is the magnetic field gradient magnitude (Gauss/cm), δ is the duration of the gradient pulse, and Δ is the separation in time of the onset of the two gradient pulses. In simple isotropic diffusion, the MR signal in the presence of Stejskal-Tanner gradients, decays according to exp(-bD) where D is the diffusion constant.

For flowing spins, b does not have an immediately obvious physical meaning. If there is flow, the spins will acquire a phase, Φ. For constant velocity along the direction of the gradient, and gradients that are applied along one direction and back to back but with opposite polarity (i.e. $\Delta = \delta$, and $G = G_o$ for $t = 0$ to δ, and $G = -G_o$ for $t = \delta$ to 2δ), the flowing spins will acquire a phase of $3/2\gamma G_o \nu\delta^2$ (e.g. see [37] for a review). Within blood vessels, however, flow is not uniform especially for large diameter vessels. Furthermore, the blood vessels may change directions within a voxel, and there may be several different blood vessels with different flow rates and/or different orientations relative to the gradent directions. Since the blood signal detected from the voxel will be a sum of all of these, the net result can be signal cancellation due to dephasing of flowing spins. As a result, these

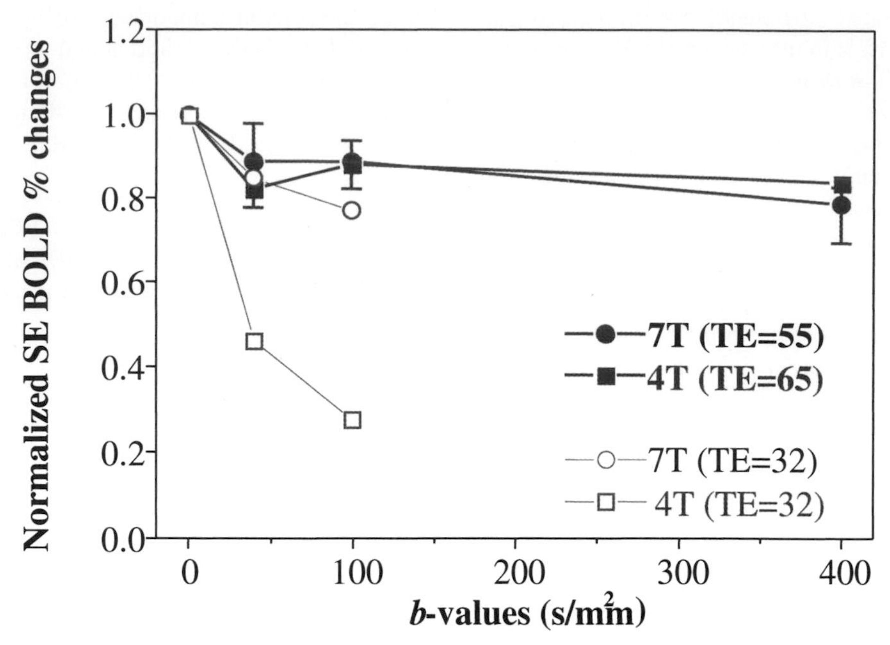

Figure 8 Normalized percent signal intensity increase during visual stimulation in the human brain as a function of b "diffusion weighting" in a spin echo sequence at 4 T and 7 T for two different echo times. From reference [19].

pair of gradients can very effectively suppress flowing spins in blood vessels.

The Stejskal-Tanner pulsed gradients can be used to distinguish between intra-and extra-vascular BOLD effects in functional images. Such experiments conducted at 1.5 Tesla have concluded that most of the BOLD based signal increase during elevated neuronal activity is eliminated by bipolar gradients, leading to the conclusion that most of the fMRI signal at 1.5 T arises from *intravascular* or blood related effects [7, 88].

The effect of the Stejskal-Tanner gradients on brain tissue signal intensity at 4 and 7 Tesla was examined (Figure 8) and found to agree well with the modeling predictions (Figure 7) based on the increasingly shorter T_2 of blood at elevated magnetic fields [19]. The dependence on the b- value for diffusion-weighted spin-echo BOLD data, averaged for all subjects, are illustrated in Figure 8 for an echo time of 32 ms for both at 4 and 7 T, and for echo times of 65 and 55 ms for 4T and 7T, respectively. The 32 ms echo time is shorter than tissue T_2 at both fields. Blood contributions are expected to be echo time dependent and diminish with increasing echo time for TE values exceeding the blood T_2 (Figure 7). At about 32 ms TE, blood contribution to 7 Tesla is expected to be minimal (Figure 7) since T_2 of blood is short [105]. However, for 4T where the T_2 of blood is ~20 ms the blood is still expected to contribute (Figure 7). The 32 ms TE data in Figure 8 emphasizes that at this echo time, signal changes associated with activation which are attributable to blood (and thus can be suppressed with diffusion gradients) are a small fraction of the total signal change at 7 T; in contrast signal changes associated with activation at this echo time arise predominantly from blood at 4 Tesla. This would still be the case in gradient echo experiments where typical echo times employed in these fMRI studies at these fields would be approximately 30 ms for both fields. At the longer echo times, however, as expected, percent changes at both fields were only slightly reduced by the diffusion-weighting gradients.

At 9.4 Tesla the effect of the Stejskal-Tanner gradients are similar to 7T results reported for the human brain above [55]. In a Hahn spin-echo weighted fMRI study conducted in the rat brain (forepaw stimulation, symmetric spin-echo with one 180° pulse), we observed that the activation is not altered at all going from very small to very high b values (Figure 9). Therefore, one can conclude that at this very high magnetic field, there exists a strong and dominant BOLD effect originating from microscopic vessels.

The longer echo times used in the data presented in Figure 8 correspond to gray matter T_2 at these two magnetic fields and represent the optimum TE for spin-echo fMRI assuming that the "noise" in the spin-echo fMRI data (i.e. image to image fluctuations in signal intensity) is TE independent; this certainly would be the case at very high resolution where intrinsic image SNR dominates the "noise" in the fMRI series. At these long and "optimal" echo times, both 7T and 4T are equally devoid of blood related degradations in specificity. The advantage of the 7T in this case would be the supralinear increase in the BOLD effect (i.e. increase in CNR) expected at the higher magnetic field associated with microvasculature. This is important in view of the fact that

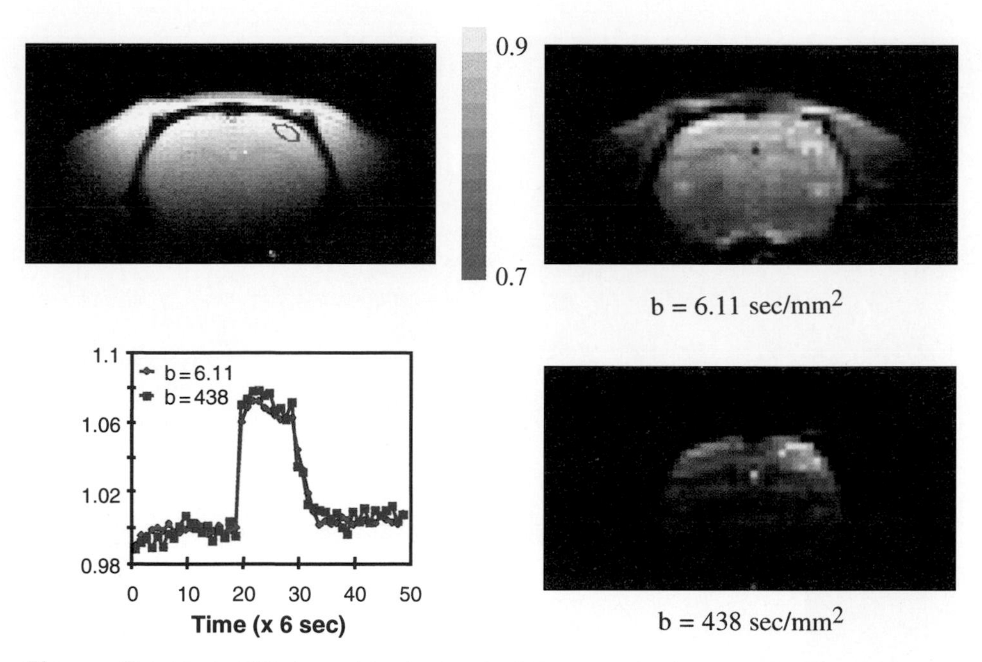

Figure 9 9.4 Tesla Diffusion-weighted spin-echo fMRI maps with b values of 6.1 (top right) and 438 s/mm^2 (bottom right) overlaid on one of the original, consecutively acquired EPI images (BOLD and diffusion weighted) collected during the functional imaging study. Coronal single-slice single-shot spin-echo EPI images of rat brain were acquired with a matrix size of 64 × 32, a FOV of 3.0 × 1.5 cm^2, a slice thickness of 2 mm, and t_e = 30 msec. Somatosensory stimulation was used. Color bar indicates a maximum cross-correlation value from 0.7 to 0.9. Signal intensity (shown in background) was significantly reduced by bipolar gradients, as expected due to diffusion. Localized activation is observed at the somatosensory cortex in the contralateral side of a stimulated forepaw. Foci of activation site (color) agree very well in both fMRI maps. A Turbo FLASH image with a region of interest is shown in the upper left corner and time courses of diffusion-weighted images within the ROI are shown in the bottom left corner. If the macro-vascular contribution were significant, relative BOLD signal changes would decrease when a higher b value is used. However, relative signal changes remained the same in both images, *suggesting that extravascular and /micro-vascular components predominantly contribute to spin echo BOLD at 9.4T*. From reference [55]

this CNR is low even at 4 Tesla. Note, however, that similarly long echo times are not useful for 1.5T because blood T_2 is equal to or exceeds tissue T_2. At that field magnitude, the blood contribution is substantial at TE values in the vicinity of tissue T_2 (~90 ms), and are increasing with TE (Figure 7) until about 150 ms (not shown in Figure 7).

The experiments with the "diffusion" weighted gradients confirm the expectation that blood effects are negligible at high fields. However, experimentally, we are unable to judge at what level of the vascular tree we were able to eliminate the blood signals from the image using the diffusion gradients. Since the apparent T_2 of venous blood is so short, we expect that at all levels of the venous tree (venules to large veins), blood signals must be eliminated provided we use echo times that are about three fold larger than the apparent venous blood T_2. The intra-capillary blood, however, is unlikely to be fully suppressed. In the capillary, red blood cell density is non-uniform and oxygenation level varies from the arterial value at the beginning to the venous value at the end of the capillary.

From the point of view of obtaining accurate functional images, it is *immaterial* as to whether capillary blood still contributes or not at high fields; in either case, spatial resolution is dictated by the capillary density. From the viewpoint of understanding BOLD mechanisms, this question is relevant but at the present unresolved. However, investigators who have realized and documented the importance of the intravascular BOLD effects at 1.5T have suggested that even at high magnetic fields, BOLD is predominantly of intravascular origin. This claim, however, was tested by experiments [64, 65, 106]. Namely, rather than looking at water, one can examine the BOLD effect on metabolites that are sequestered intracellularly. Any such BOLD effect is *entirely* extravascular. The existence of extravascular spin-echo BOLD effect at 4 and 7 Tesla was recently documented in our laboratory [64, 65, 106]. Thus, at high fields of 4 and 7 Tesla, there exists *extravascular spin-echo* BOLD effects in the human brain.

In view of the extensive theoretical considerations backed by extensive data demonstrating that spin-echo BOLD at high (but not low) magnetic fields are specific to microvasculature and is devoid of extravascular and intravascular BOLD contributions from veins and venules

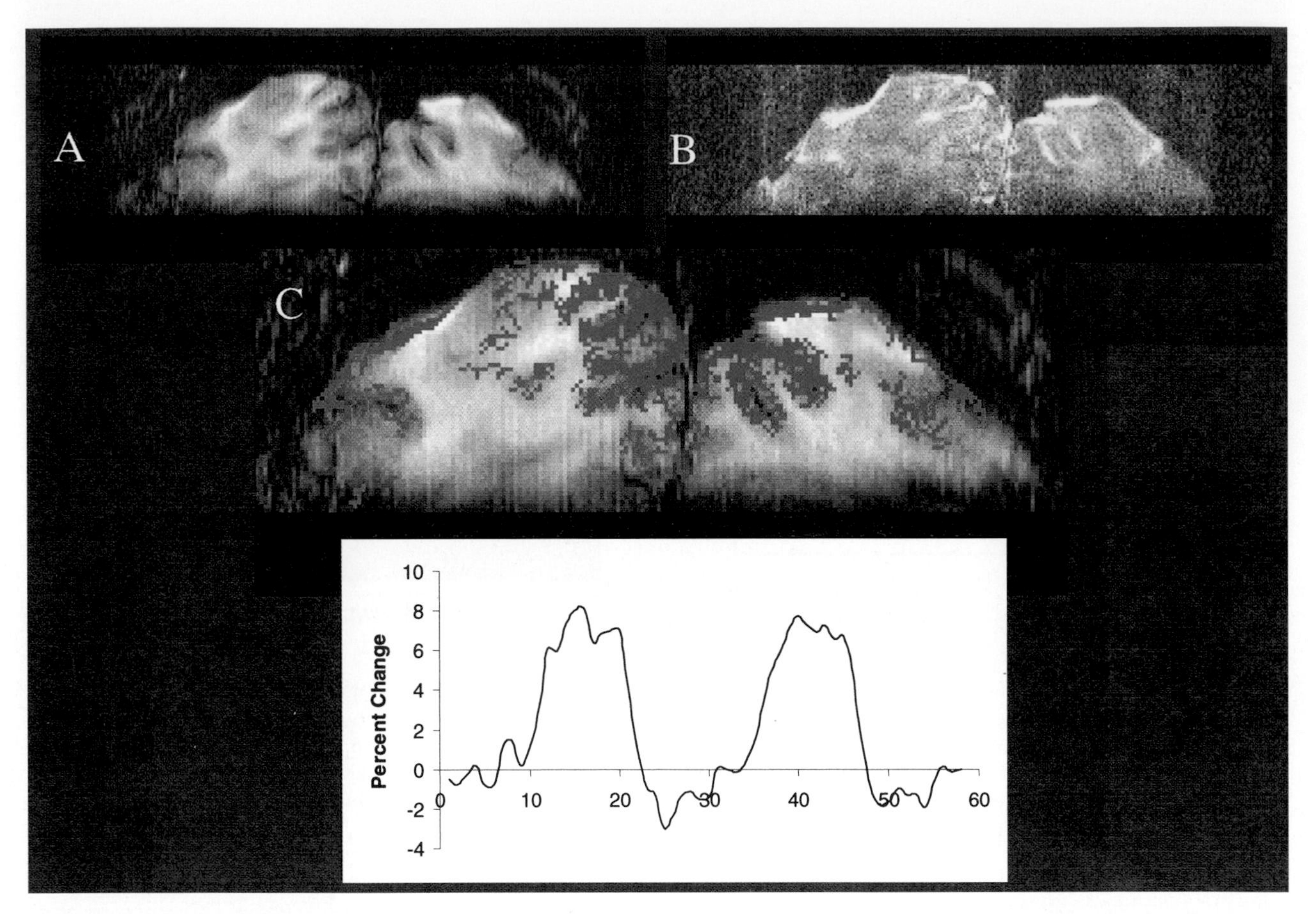

Figure 10 $0.5 \times 0.5 \times 2$ mm^3 Spin-Echo Images of the visual activation in the human visual cortex obtained at 7 Tesla. Single Shot Spin Echo EPI with field of view reduction to 3.2 cm × 12.8 cm (64 × 256) (partial Fourier) TE/TR: 70 msec/2s. From reference [19].

greater than about 10 micron diameter, the method of choice for BOLD fMRI with high spatial accuracy should be the spin-echo approach with a single refocusing pulse. The main question as to whether this can be routinely performed concerns the magnitude of this effect. The spin-echo BOLD response is weaker than the commonly used gradient echo BOLD approaches. But they are expected to increase supralinearly with magnetic field. At 7 Tesla, the effect is large enough to obtain very high resolution images. This is illustrated in Figure 10 which depicts $0.5 \times 0.5 \times 2$ mm^3 spin-echo BOLD images of the human visual cortex, showing about 6 to 8% signal intensity changes. Note how the image depicts accurately the cortical gray matter.

IV. Spectroscopy at High Magnetic Fields

Spectroscopy studies also benefit significantly from unique advantages besides improved SNR at higher field strengths. Higher B_0 fields yield improved resolution due to chemical shift and spectral simplifications of coupled spin systems.

^{31}P NMR studies have been used both in animal models and in human studies to examine tissue bioenergetics, including the bioenergetics of increased neuronal activity. These studies have so far relied on detection of adenosine triphosphate (ATP), inorganic phosphate (Pi), and phosphocreatine (PCr). Significant gains are realized at the higher fields for this nucleus as shown in Figure 11. These data obtained with rats, document better than linear gains in SNR in spatially localized ^{31}P spectra (Figure 11); such gains permit investigations of brain bioenergetics in primate brain with greater accuracy and spatial resolution, and have been utilized in the human brain for evaluating pathology [16, 73] and brain activation [14].

^{1}H spectra also benefit from SNR and resolution gain in going to high magnetic fields. In ^{1}H spectra obtained from the human brain, glutamate is resolved from glutamine partially at 4T [38] but not at 1.5T; these resonances are fully resolved at 9.4 Tesla [31]. Figure 12 displays localized ^{1}H NMR spectra from human volunteers obtained at 1.5 Tesla clinical system (Siemens VISION), our 4 Tesla system and a dog brain spectrum obtained with our 9.4 Tesla/31 cm magnet.

An excellent example of the resolution afforded by high fields in ^{1}H spectroscopy is the detection of 5.23 ppm resonance of α-glucose at 4 T (and higher fields) despite its proximity to H_2O [31, 93], providing for the first time the

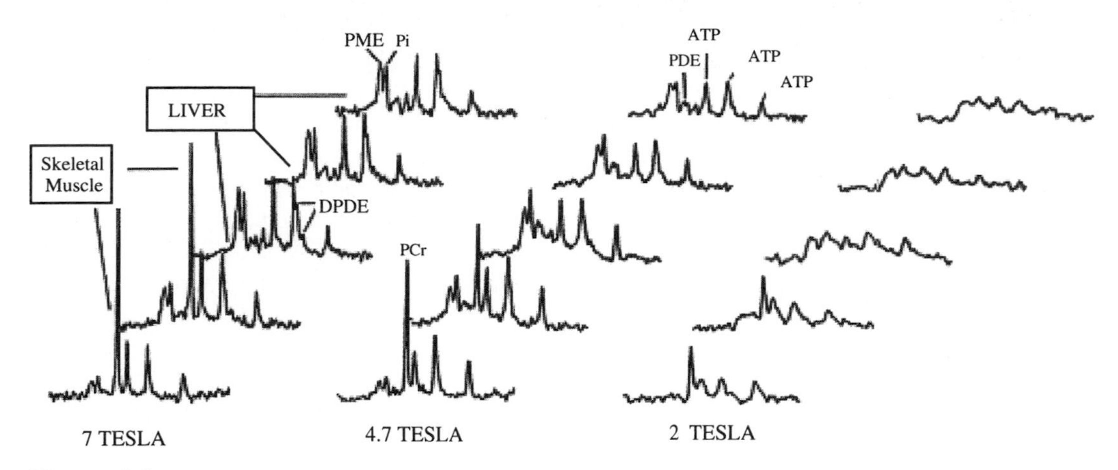

Figure 11 2, 4.7, and 7 Tesla spectra acquired in five voxels over the abdomen of the same intact, and reversibly anesthetized rat. These spectra were obtained using identical techniques, voxel volumes, number of scans, interpulse delay, and *identical total data acquisition time*.

ability to measure cerebral glucose quantitatively by ^{1}H MRS without complications from overlapping resonances. All previous ^{1}H MRS studies concerned with MR detection of glucose at lower fields (e.g. [13, 25, 63]) have relied on the 3.4 ppm peak of glucose which suffers from significant overlap from other resonances such as taurine in this very crowded region of the brain spectrum. As such, the 3.4 ppm resonance cannot and should not be quantitatively used for measuring absolute glucose content in the human brain. The 5.23 ppm peak of glucose, on the other hand, is isolated and does not suffer from overlap. However, this resonance is too close to the very large H_2O peak to be detectable at the lower fields strength because selective suppression of the latter without affecting the former has not been possible at the lower magnetic fields. The increased chemical shift resolution at 4T or higher fields, however, permits selective suppression of the H_2O resonance and detection of the 5.23 ppm 1-H of α-glucose peak.

Recently, we have been able to perform ^{1}H studies in the human brain at 7T [93]. Significant gains relative to 4T was obtained in resolution and SNR, and improved resolution for the coupled spins were observed. However, we have also noted that effective resolution in human proton spectra is expected to increase with field magnitude beyond 7T even if linewidths also scale up linearly with the field magnitude. This is because the spectrum mostly consists of coupled spins and the effective widths of the often overlapping multiplets do not increase as the linewidth of the individual resonances due to the overlap. This is illustrated using simulations in Figure 13. Note the glutamate and glutamine triplet at about 2.3–2.4 ppm, indicated by arrows in the 4, 7 and 9.4T spectra. The resolution between these resonances increase even though the linewidths of the individual resonances scale linearly with the field magnitude in this simulation. This is because the effective width of the triplet is determined by both the scalar coupling and the linewidths and the former is field independent.

Spatially localized MRS has been employed to measure aspects of brain function and neurotransmission directly. Recent efforts have focused on understanding the coupling between cellular bioenergetics and neuronal activity (e.g. [15, 32–34, 43, 44, 84, 85]). The ^{13}C nucleus, the only NMR-detectable, stable carbon isotope that is normally present at 1.1% abundance, has played an important role in these efforts. If ^{13}C-enriched glucose is given to a living organism, the ^{13}C label is incorporated through metabolism into several positions in many different compounds. Of

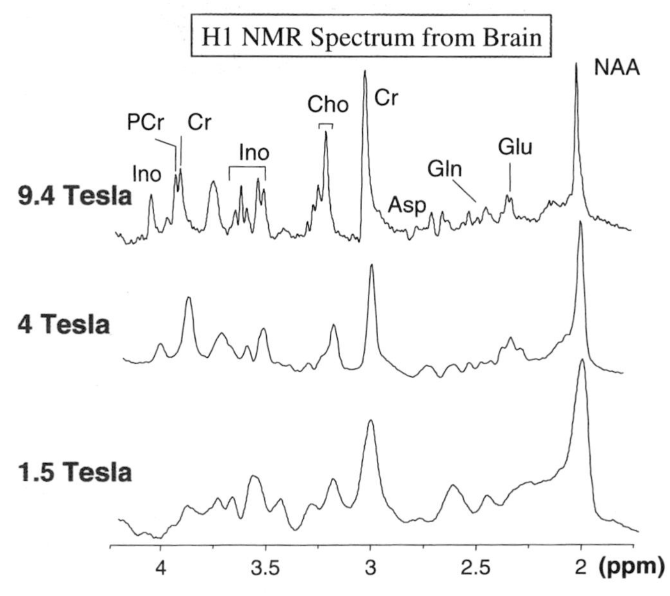

Figure 12 (DRY)STEAM 1H NMR spectroscopy localization (TE = 20 ms, TM = 33–36 ms, TR = 3 s) at 1.5T, and 4 for the human brain and at 9.4 for a dog brain. From reference [36].

interest is the incorporation of label into the intermediates of tricarboxylic acid (TCA) cycle, which generates reducing equivalents utilized in oxygen consumption. The label is then transferred into the amino acids glutamate, aspartate and glutamine. The ability to monitor the labeling of intracellular compounds in intact cells was first demonstrated using E. coli [94]. Today, using advanced shimming techniques [30] and three-dimensional localization methods, such highly specific data can be obtained in human and animal brains from relatively small volume elements.

The human brain ^{13}C spectra in Figure 14 (obtained at 4 Tesla in our laboratory) illustrate the detection of many labeled compounds from a *relatively small, localized region* in the visual cortex, and the first time measurement of neurotransmission rate in the *human* visual cortex [33]. A subsequent effort on humans at 2.1 Tesla [84] was able to report basically the same results, but used at least three-fold larger volumes (144 ml vs. 45 ml) and longer data acquisition times for signal detection due to the lower field employed. In these studies, it is possible to extract information from the compartmentalized metabolic pathways. Most important is the glutamate-glutamine cycling between glia and the neurons, the two major cell types that are present in the brain: Glutamate is the major excitatory neurotransmitter. Once it is released into the synaptic cleft and binds the post-synaptic receptors during neurotransmission, it is scavenged rapidly by nearby glial processes [5, 6]. Glutamate is then converted to glutamine by glutamine-synthase, which is present only in the glia [60]; glutamine is subsequently transported to the neurons, converted to glutamate to replenish vesicular glutamate. Thus, glutamine labeling in ^{13}C experiments can in principle occur predominantly through glutamate release through neurotransmission and hence reflect the kinetics of neurotransmission. Therefore, from such 13C studies, it is possible to calculate rates of glutamatergic neurotransmission, TCA cycle turnover, cerebral oxidative glucose utilization (CMR_{glu}) and even break down the latter into glial and neuronal contributions. Most important finding from our studies is the conclusion that neurotransmission rate equals ~60% of total oxidative glucose consumption rate in the awake human brain and that there is significant oxidative ATP production rate in the glia, apparently in contradiction with earlier conclusions reached from experiments on the anesthetized rat brain data [86].

Such detailed studies have not yet been applied to investigate neuronal activation in the human or animal brains due to limitations in signal-to-noise ratio of the measurements. However, a subset of the data in the direct ^{13}C measurements

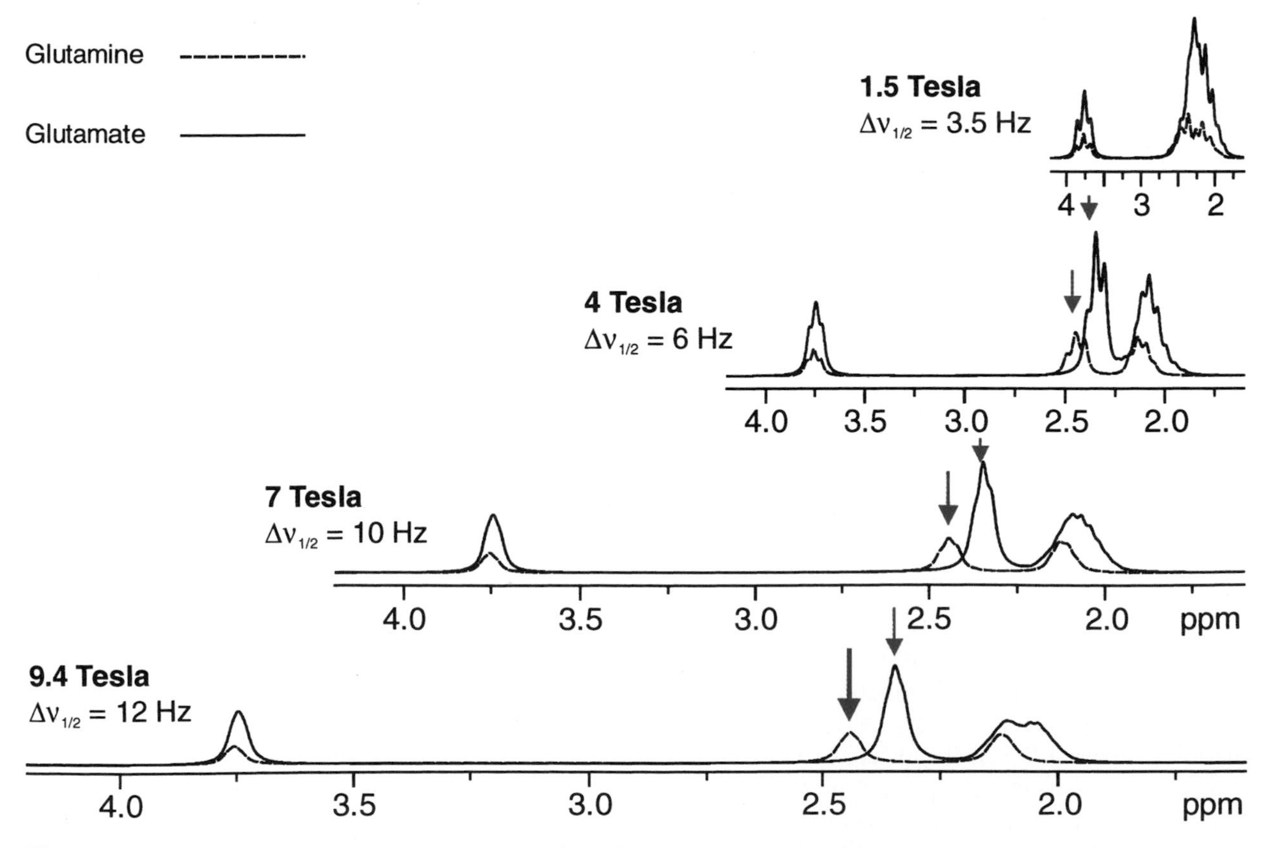

Figure 13 Simulated ^{1}H NMR spectra of glutamine and glutamate at different magnetic field strengths. Linewidths corresponded to values typical for very well shimmed volumes of the human brain. Concentration ratio [Glu]/[Gln] was set to 3. Frequency scale in Hz is identical in all three spectra. From reference [93]

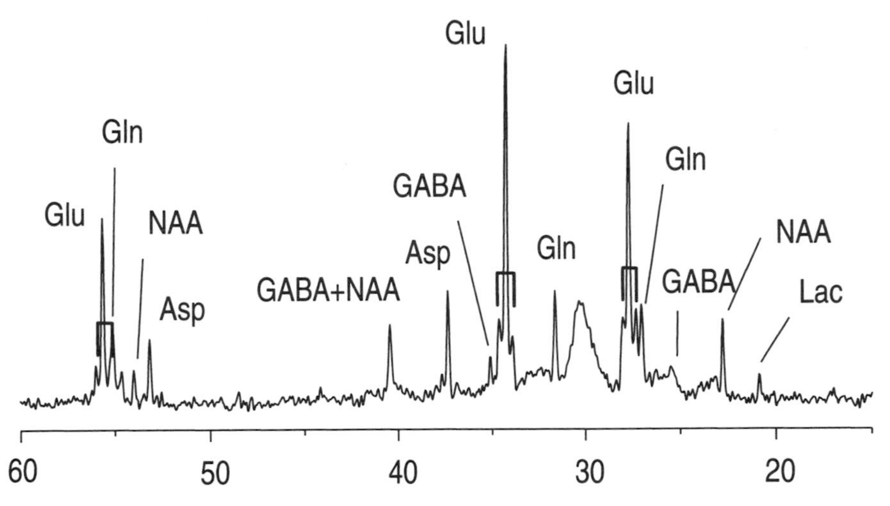

Figure 14 Localized direct detection, ^{13}C spectrum in the human brain at 4 Tesla. 72 mL, 50 min data acquisition. From reference [35].

can be acquired with indirect detection with higher sensitivity using the protons attached to the ^{13}C nuclei, permitting calculations of changes in TCA cycle turnover and oxidative CMR_{glu} during neuronal activation. The experiment demonstrating this is illustrated in Figure 15 where hemifield visual stimulation was used to obtain spectroscopic data (together with CBF and BOLD fMRI) from an activated region in one hemisphere while the other hemisphere served as control [15].

The study illustrated in Figure 15 is concerned with the coupling of oxidative metabolism and neuronal activity. It is

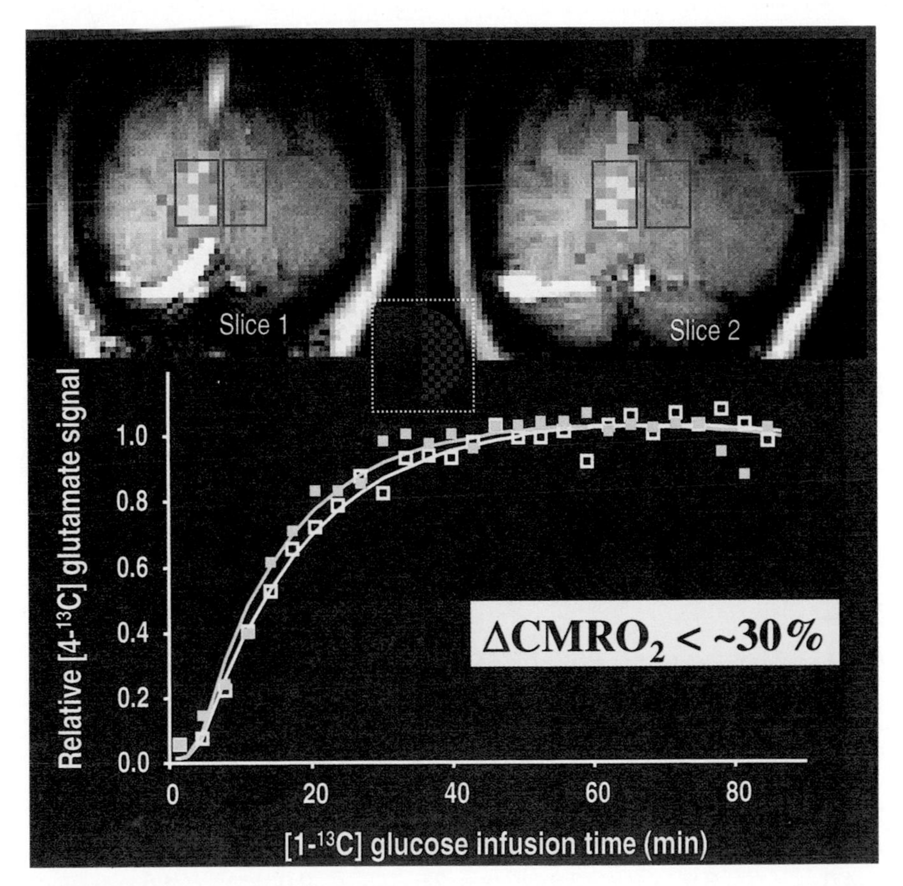

Figure 15 Hemifield visual stimulation, showing activation in only one half of the brain (depicted in coronal images of two slices). Spectra of glutamate C4 was monitered with indirect detection at 4T using two 6 cc voxels each located in one hemisphere (shown as overlayed on the images). Time course of the glutamate C4 signals as they build up in two hemispheres. From reference [15]

well known that under resting conditions, the cerebral metabolic rate of glucose consumption (CMR_{glc}) is well coupled to CMR_{O2} as well as to cerebral blood flow (CBF) in the human brain ([87] and references therein), and the glucose metabolism is almost completely through oxidation. However, this appears not to be the case during increased neuronal activity. Based on PET studies, the increases of CMR_{O2} (0–5%) were found to be much less than the elevation in CBF and CMR_{glc} (40–51%) during visual and somatosensory stimulations [23, 24]. This early PET result remains to this day highly debated, with the intensity of debate having increased recently in view of its significance in understanding the BOLD response that has come to play such a prominent role in neuroscience research. The difficulties and complexities associated with PET measurement of CMR_{O2} have also led to skepticism about the validity of the data that generated this controversial concept.

Resolution of this problem requires new studies, especially using techniques that avoid PET specific errors. Measurements in a single subject within a single experiment are also crucial in order to avoid large variances generated by intersubject averaging. This was recently accomplished in the human brain using isotopic turnover rate of glutamate from infused {1 – ^{13}C} glucose using indirect detection through coupled protons (Figure 15). Glutamate labeling kinetics was followed for "activated" and resting states simultaneously in the same individual during hemifield visual stimulation by acquiring data in two distinct small volumes (6 ml each) that span the primary visual cortex [15]. High field of 4 Tesla employed in these studies enabled the use such a small volume so that most of the volume was activated brain tissue, thus minimizing partial volume effects. Hemifield visual stimulation selectively activates the primary visual cortex of the contralateral hemisphere, and thus approximately half the primary visual cortex volume that normally would be engaged during full field stimulation. Spectra can be acquired from two volumes positioned so that only one of the two covers the "activated" region within one hemisphere while the other covers the analogous but non-activated region within the other hemisphere. Figure 15 demonstrates an image showing the activation in one hemisphere, the location of the spectroscopic voxels and the time courses of glutamate labeling from the activated and control voxels. These data put an upper limit on the increase in CMR_{O2} of 30% as opposed larger increases in CBF, supporting the concept that CMR_{O2} is not stochiometrically coupled to increases in CBF and CMR_{glu}.

At fields of 4 Tesla or lower, poor chemical shift dispersion of the proton nucleus prevents detection of the wealth of metabolites observed in the direct ^{13}C spectrum. Therefore, in the study aimed at measuring $CMRO_2$ changes associated with neuronal activation using indirect detection through ^1H nucleus [15], we were not able to observe all the metabolites detected in the ^{13}C spectrum at 4T. For modeling the very complex set of metabolic reactions involved in extracting neurotransmission rates, and accurate metabolic rates coupled to neurotransmission, it is imperative that the multiple carbons of metabolites and multiple metabolites are detected so that the

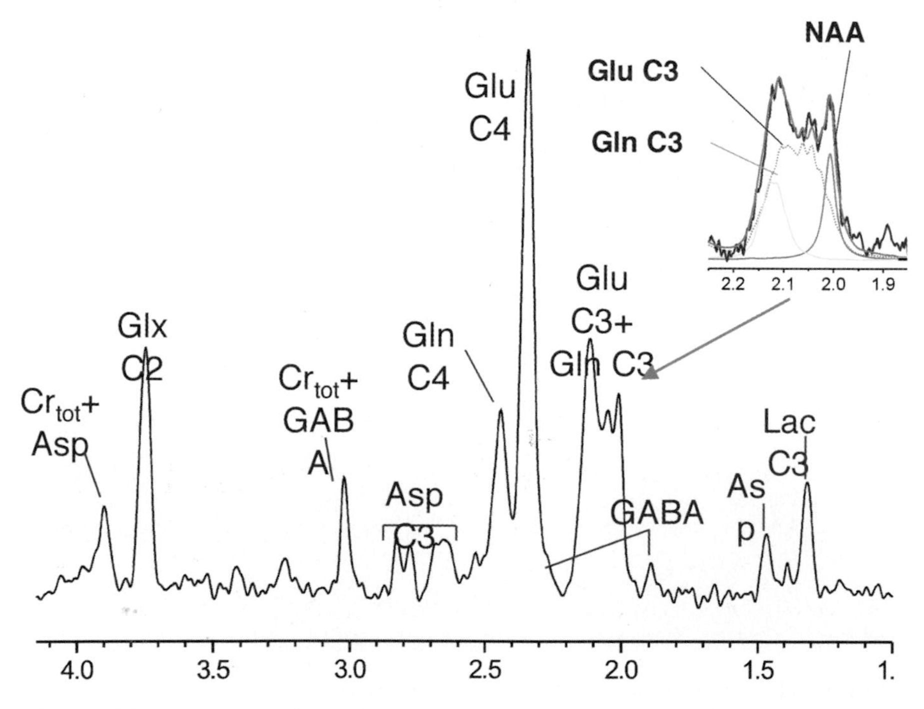

Figure 16 Indirect detection of C13 labled metabolites at 9.4 Tesla in the rat brain. This is an analog of the spectrum illustrated in Figure 14. From reference [78].

model is not underdetermined (see [34]). However, again at 9.4T in the rat brain we were able to illustrate that this was doable [78]. A example of this is illustrated in (Figure 16).

Acknowledgments

The work reported here from the Center for Magnetic Resonance Research, University of Minnesota, was supported by the National Research Resources (NCRR) division of NIH, Grant P41 RR08079.

References

1. Abduljalil, A. M., A. Kangarlu, X. Zhang, R. E. Burgess, and P. M. Robitaille, *Acquisition of human multislice MR images at 8 Tesla*. J Comput Assist Tomogr, 1999. **23**(3): p. 335–40.
2. Bandettini, P. A., E. C. Wong, R. S. Hinks, R. S. Tikofsky, and J. S. Hyde, *Time course EPI of human brain function during task activation*. Magn Reson Med, 1992. **25**(2): p. 390–7.
3. Barth, M. and E. Moser, *Proton NMR relaxation times of human blood samples at 1.5 T and implications for functional MRI*. Cell Mol Biol (Noisy-le-grand), 1997. **43**(5): p. 783–91.
4. Belle, V., C. Delon-Martin, I. R. Massarell, J. Decety, J. Le Bas, A. Benabid, and C. Segebarth, *Intracranial gradient-echo and spin-echo functional MR angiography in humans*. Radiology, 1995. **195**(3): p. 739–46.
5. Bergles, D. E., J. S. Diamond, and C. E. Jahr, *Clearance of glutamate inside the synapse and beyond*. Curr Opin Neurobiol, 1999. **9**(3): p. 293–8.
6. Bergles, D. E., J. A. Dzubay, and C. E. Jahr, *Glutamate transporter currents in bergmann glial cells follow the time course of extrasynaptic glutamate*. Proc Natl Acad Sci USA, 1997. **94**(26): p. 14821–5.
7. Boxerman, J. L., P. A. Bandettini, K. K. Kwong, J. R. Baker, T. L. Davis, B. R. Rosen, and R. M. Weisskoff, *The intravascular contribution to fMRI signal change: Monte Carlo modeling and diffusion-weighted studies in vivo*. Magn Reson Med, 1995. **34**(1): p. 4–10.
8. Boxerman, J. L., L. M. Hamberg, B. R. Rosen, and R. M. Weisskoff, *MR contrast due to intravscular magnetic susceptibility perturbations*. Magn Reson Med, 1995. **34**: p. 555–556.
9. Bryant, R. G., K. Marill, C. Blackmore, and C. Francis, *Magnetic relaxation in blood and blood clots*. Magn Reson Med, 1990. **13**(1): p. 133–44.
10. Burgess, R. E., Y. Yu, A. M. Abduljalil, A. Kangarlu, and P. M. Robitaille, *High signal-to-noise FLASH imaging at 8 Tesla*. Magn Reson Imaging, 1999. **17**(8): p. 1099–103.
11. Burton, A. C., *Role of geometry, of size and shape, in the microcirculation*. Fed Proc, 1966. **25**(6): p. 1753–60.
12. Buxton, R. B. and L. R. Frank, *A model for the coupling between cerebral blood flow and oxygen metabolism during neural stimulation*. Journal of Cerebral Blood Flow & Metabolism., 1997. **17**(1): p. 64–72.
13. Chen, W., E. Novotny, X.-H. Zhu, D. Rothman, and R. G. Shulman, *Localized 1H NMR measurement of glucose consumption in human brain during visual stimulation*. Proc Natl Acad Sci USA, 1993. **90**: p. 9896–9900.
14. Chen, W., X. H. Zhu, G. Adriany, and K. Ugurbil, *Increase of creatine kinase activity in the visual cortex of human brain during visual stimulation: a 31P magnetization transfer study*. Magn Reson Med, 1997. **38**(4): p. 551–7.
15. Chen, W., X. H. Zhu, R. Gruetter, E. R. Seaquist, G. Adriany, and K. Ugurbil, *Study of tricarboxylic acid cycle flux changes in human visual cortex during hemifield visual stimulation using (1)H-[(13)C] MRS and fMRI*. Magn Reson Med, 2001. **45**(3): p. 349–355.
16. Chu, W. J., H. P. Hetherington, R. I. Kuzniecky, T. Simor, G. F. Mason, and G. A. Elgavish, *Lateralization of human temporal lobe epilepsy by 31P NMR spectroscopic imaging at 4.1 T*. Neurology, 1998. **51**(2): p. 472–9.
17. Duong, Q. D., D.-S. Kim, K. Ugurbil, and K. S-G., *Spatio-Temporal Dynamics of BOLD fMRI Signals: Towards mapping Submillimeter Cortical columns Using the early Negative Response*. Magn Reson Med, 2000. **44**(2): p. 231–242.
18. Duong, T. Q., D. S. Kim, K. Ugurbil, and S. G. Kim, *Localized cerebral blood flow response at submillimeter columnar resolution*. Proc Natl Acad Sci USA, 2001. **98**(19): p. 10904–10909.
19. Duong, T. Q., E. Yacoub, G. Adriany, P. Andersen, X. Hu, J. T. Vaughan, K. Ugurbil, and S. G. Kim. *Microvascular BOLD Contribution at 4 and 7 Tesla in the Human Brain: Diffusion-weighted, Gradient-Echo and Spin-Echo fMRI*. in *9th Annual Meeting of the International Society of Magnetci resonance in Medicine (ISMRM)*. 2001. Glasgow, UK.
20. Duyn, J. H., C. T. W. Moonen, G. H. Van Yperen, R. W. De Boer, and P. R. Luyten, *Inflow versus deoxyhemoglobin effects in BOLD functional MRI using gradient echoes at 1.5T*. NMR in Biomed, 1994. **7**(1/2): p. 83–88.
21. Edelman, R. E., B. Siewer, D. G. Darby, V. Thangaraj, A. C. Nobre, M. M. Mesulam, and S. Warach, *Quantitative mapping of cerebral blood flow and functional localization with echo-planar MR imaging and signal targeting with alternating radio frequency*. Radiology, 1994. **192**: p. 513–520.
22. Eichling, J. O., M. E. Raichle, R. L. Grubb, and M. M. Ter-Pogossian, *Evidence of the limitations of water as a freely diffusuble tracer in brain of the Rheusus monkey*. Circ Res, 1974. **35**(3): p. 358–364.
23. Fox, P. T. and M. E. Raichle, *Focal physiological uncoupling of cerebral blood flow and oxidative metabolism during somatosensory stimulation in human subjects*. Proc Natl Acad Sci USA, 1986. **83**(4): p. 1140–4.
24. Fox, P. T., M. E. Raichle, M. A. Mintun, and C. Dence, *Nonoxidative glucose consumption during focal physiologic neural activity*. Science, 1988. **241**(4864): p. 462–4.
25. Frahm, J., K. D. Kruger, K. D. Merboldt, and A. Kleinschmidt, *Dynamic uncoupling and recoupling of perfusion and oxidative metabolism during focal brain activation in man*. Magn Reson Med, 1996. **35**: p. 143–148.
26. Frahm, J., K.-D. Merboldt, W. Hanicke, A. Kleinschmidt, and H. Boecker, *Brain or vein-oxygenation or flow? On signal physiology in functional MRI of human brain activation*. NMR in Biomed, 1994. **7**(1/2): p. 45–53.
27. Fujita, I., K. Tanaka, M. Ito, and K. Cheng, *Columns for visual features of objects in monkey inferotemporal cortex*. Nature, 1992. **360**: p. 343–346.
28. Gati, J. S., R. S. Menon, K. Ugurbil, and B. K. Rutt, *Experimental determination of the BOLD field strength dependence in vessels and tissue*. Magn Reson Med, 1997. **38**(2): p. 296–302.
29. Grubb, R. L., Jr., M. E. Raichle, J. O. Eichling, and M. M. Ter-Pogossian, *The effects of changes in PaCO2 on cerebral blood volume, blood flow, and vascular mean transit time*. Stroke, 1974. **5**(5): p. 630–9.
30. Gruetter, R., *Automatic, localized in vivo adjustment of all first- and second-order shim coils*. Magn Reson in Med, 1993. **29**(6): p. 804–811.
31. Gruetter, R., M. Garwood, K. Ugurbil, and E. R. Seaquist, *Observation of resolved glucose signals in 1H NMR spectra of the human brain at 4 Tesla*. Magn Reson Med, 1996. **36**(1): p. 1–6.
32. Gruetter, R., E. J. Novotny, S. D. Boulware, M. G. F., D. L. Rothman, J. W. Prichard, and S. R. G., *Localized 13C NMR spectroscopy of amino acid labeling from [1–13C] D-glucose in the human brain*. J Neurochem, 1994. **63**: p. 1377–1385.

33. Gruetter, R., E. R. Seaquest, S. W. Kim, and K. Ugurbil, *Localized in vivo 13C NMR of glutamate metabolism in the human brain. Initial results at 4 Tesla*. Dev Neurosci, 1998. **20**(4–5): p. 380–8.
34. Gruetter, R., B. Seaquist, and K. Ugurbil, *A mathematical model of compartmentalized neurotransmitter metabolism in the human brain*. Am J Physiol Endocrinol Metab, 2001. **281**(1): p. E100–E112.
35. Gruetter, R., E. R. Seaquist, S. Kim, and K. Ugurbil, *Localized in vivo 13C-NMR of glutamate metabolism in the human brain: initial results at 4 tesla*. Dev Neurosci, 1998. **20**(4–5): p. 380–8.
36. Gruetter, R., S. A. Weisdorf, V. Rajanayagan, M. Terpstra, H. Merkle, C. L. Truwit, M. Garwood, S. L. Nyberg, and K. Ugurbil, *Resolution improvements in in vivo 1H NMR spectra with increased magnetic field strength*. J Magn Reson, 1998. **135**(1): p. 260–4.
37. Haacke, E. M., A. S. Smith, W. Lin, J. S. Lewin, D. A. Finelli, and J. L. Duerk, *Velocity quantification in magnetic resonance imaging*. Top Magn Reson Imaging, 1991. **3**(3): p. 34–49.
38. Hetherington, H. P., G. F. Mason, J. W. Pan, S. L. Ponder, J. T. Vaughan, D. B. Twieg, and G. M. Pohost, *Evaluation of Cerebral Gray and White Matter Metabolite Differences by Spectroscopic Imaging at 4.1T*. Magn Reson Med, 1994. **32**: p. 565–571.
39. Hoge, R. D., J. Atkinson, B. Gill, G. R. Crelier, S. Marrett, and G. B. Pike, *Linear coupling between cerebral blood flow and oxygen consumption in activated human cortex*. Proc Natl Acad Sci USA, 1999. **96**(16): p. 9403–8.
40. Hoult, D. I. and P. C. Lauterbur, *The sensitivity of the zeugmatographic experiment involving human samples*. J Magn Reson, 1979. **34**: p. 425–433.
41. Hoult, D. I. and D. Phil, *Sensitivity and power deposition in a high-field imaging experiment*. J Magn Reson Imaging, 2000. **12**(1): p. 46–67.
42. Hoult, D. I. and R. E. Richards, *The signal-to-noise ratio of the nuclear magnetic resonance phenomenon*. J. Magn. Reson., 1976. **24**(71): p. 71–85.
43. Hyder, F., J. R. Chase, K. L. Behar, G. F. Mason, M. Siddeek, D. L. Rothman, and R. G. Shulman, *Increased tricarboxylic acid cycle flux in rat brain during forepaw stimulation detected with 1H[13C]NMR*. Proc Natl Acad Sci USA, 1996. **93**(15): p. 7612–7.
44. Hyder, F., D. L. Rothman, G. F. Mason, A. Rangarajan, K. L. Behar, and R. G. Shulman, *Oxidative glucose metabolism in rat brain during single forepaw stimulation: a spatially localized 1H[13C] nuclear magnetic resonance study*. J Cereb Blood Flow Metab, 1997. **17**(10): p. 1040–7.
45. Keltner, J. R., J. W. Carlson, M. S. Roos, S. T. S. Wong, T. L. Wong, and T. F. Budinger, *Electromagnetic Fields of Surface Coil in Vivo NMR at High Frequencies*. Magn Reson Med, 1991. **22**: p. 46–480.
46. Kennan, R. P., J. Zhong, and J. C. Gore, *Intravascular susceptibility contrast mechanisms in tissue*. Magn Reson Med, 1994. **31**: p. 9–31.
47. Kim, D.-S., I. Ronen, C. Olman, H. Merkle, S.-G. Kim, K. Ugurbil, and L. J. Toth, *Neural correlates of Blood Oxygenation Level Dependent Functional MRI*. SUBMITTED, 2001.
48. Kim, D. S., T. Q. Duong, and S. G. Kim, *High-resolution mapping of iso-orientation columns by fMRI*. Nat Neurosci, 2000. **3**(2): p. 164–9.
49. Kim, S.-G., *Quantification of relative cerebral blood flow change by flow-sensitive alternating inversion recovery (FAIR) technique: application to functional mapping*. Magn Reson Med, 1995. **34**: p. 293–301.
50. Kim, S.-G., K. Hendrich, X. Hu, H. Merkle, and K. Ugurbil, *Potential pitfalls of functional MRI using conventional gradient-recalled echo techniques*. NMR in Biomed, 1994. **7**(1/2): p. 69–74.
51. Kim, S.-G. and N. V. Tsekos, *Perfusion imaging by a flow-sensitive alternating inversion recovery (FAIR) technique: application to functional brain imaging*. Mag Reson Med, 1997. **37**(3): p. 425–35.
52. Kim, S. G. and K. Ugurbil, *Comparison of Blood Oxygenation and Cerebral Blood flow Effects in fMRI; Estimation of Relative Oxygen Consumption Change*. Magn Reson Med, 1997. **38**(1): p. 59–65.
53. Kwong, K. K., J. W. Belliveau, D. A. Chesler, I. E. Goldberg, R. M. Weisskoff, B. P. Poncelet, D. N. Kennedy, B. E. Hoppel, M. S. Cohen, R. Turner, and *et al.*, *Dynamic magnetic resonance imaging of human brain activity during primary sensory stimulation*. Proc Natl Acad Sci USA, 1992. **89**(12): p. 5675–9.
54. Lauterbur, P. C., *Image Formation by Induced Local Interaction: Examples Employing Nuclear Magnetic Resonance*. Nature, 1973. **241**: p. 190–191.
55. Lee, S.-P., A. C. Silva, K. Ugurbil, and K. S-G., *Diffusion weighted spin echo fMRI at 9.4 T: Microvascular/Tissue contribution to BOLD signal changes*. Mag Reson Med, 1999. **42**(5): p. 919–28.
56. Lowel, S., H. J. Bischof, B. Leutenecker, and W. Singer, *Topographic relations between ocular dominance and orientation columns in the cat striate cortex*. Exp Brain Res, 1988. **71**(1): p. 33–46.
57. Luz, Z. and M. S., *Nuclear magnetic resonance study of the protolysis of trimethylammonium ion in aqueous solution: order of the reaction with respect to the solvent*. J Chem Phys, 1963. **39**: p. 366–370.
58. Malonek, D., U. Dirnagl, U. Lindauer, K. Yamada, I. Kanno, and A. Grinvald, *Vascular imprints of neuronal activity: relationships between the dynamics of cortical blood flow, oxygenation, and volume changes following sensory stimulation*. Proc Natl Acad Sci USA, 1997. **94**(26): p. 14826–31.
59. Malonek, D. and A. Grinvald, *Interactions between electrical activity and cortical microcirculation revealed by imaging spectroscopy: Implication for functional brain mapping*. Science, 1996. **272**: p. 551–554.
60. Martinez-Hernandez, A., K. P. Bell, and N. M. D., *Glutamine synthetase: Glial localization in Brain*. Science, 1976. **195**: p. 1356–8.
61. Menon, R. S., X. Hu, G. Adriany, P. Andersen, S. Ogawa, and K. Ugurbil. *Comparison of SE-EPI, ASE-EPI and conventional EPI applied to functional neuroimaging: The effect of flow crushing gradients on the BOLD signal*. in *Proceedings of the Society of Magnetic Resononace*. 1994.
62. Menon, R. S., S. Ogawa, D. W. Tank, and K. Ugurbil, *4 Tesla gradient recalled echo characteristics of photic stimulation-induced signal changes in the human primary visual cortex*. Magn Reson Med, 1993. **30**(3): p. 380–6.
63. Merboldt, K. D., H. Bruhn, W. Hanicke, T. Michaelis, and J. Frahm, *Decrease of glucose in the human visual cortex during photic stimulation*. Magn Reson Med, 1992. **25**: p. 187–194.
64. Michaeli S., Garwood M., Zhu X.-H., DelaBarre L., Ugurbil K., and C. W. *Separation of BOLD Contribution Related to Dynamic Dephasing Regime in Waterand Metabolites of Human Visual Cortex at 7T: A comparison Study UsingCarr-Purcell and Hahn Spin Echo MRS*. in *9th Annual Metting of the International society of Magnetic Resonance in Medicine (ISMRM)*. 2001. Glasgow, UK: ISMRM.
65. Michaeli S., Zhu X.-H., Garwood M., Ugurbil K., and C. W. *Probing of Staticand Dynamic BOLD Effect and Non-BOLD Effect of Cerebral Metabolites by Functional MRS in Human Visual Cortex at 4T and 7T*. in *9th Annual Metting of the International Society of Magnetic Resonance in Medicine (ISMRM)*. 2001. Glasgow, UK: ISMRM.
66. Ogawa, S., T.-M. Lee, A. R. Kay, and D. W. Tank, *Brain Magnetic Resonance Imaging with Contrast Dependent on Blood Oxygenation*. Proc Natl Acad Sci USA, 1990. **87**: p. 9868–9872.
67. Ogawa, S., T.-M. Lee, A. S. Nayak, and P. Glynn, *Oxygenation-sensitive contrast in magnetic resonance image of rodent brain at high magnetic fields*. Magn Reson Med, 1990. **14**: p. 68–78.
68. Ogawa, S. and T. M. Lee, *Magnetic Resonance Imaging of Blood Vessels at High Fields: in Vivo and in Vitro Measurments and Image Simulation*. Magn Reson Med, 1990. **16**: p. 9–18.

69. Ogawa, S., R. S. Menon, S. G. Kim, and K. Ugurbil, *On the characteristics of functional magnetic resonance imaging of the brain*. Annu Rev Biophys Biomol Struct, 1998. **27**: p. 447–74.
70. Ogawa, S., R. S. Menon, D. W. Tank, S.-G. Kim, H. Merkle, J. M. Ellermann, and K. Ugurbil, *Functional Brain Mapping by Blood Oxygenation Level-Dependent Contrast Magnetic Resonance Imaging*. Biophys J, 1993. **64**: p. 800–812.
71. Ogawa, S., D. W. Tank, R. Menon, J. M. Ellermann, S. G. Kim, H. Merkle, and K. Ugurbil, *Intrinsic signal changes accompanying sensory stimulation: functional brain mapping with magnetic resonance imaging*. Proc Natl Acad Sci USA, 1992. **89**(13): p. 5951–5.
72. Oja, J. M., J. S. Gillen, R. A. Kauppinen, M. Kraut, and P. C. van Zijl, *Determination of oxygen extraction ratios by magnetic resonance imaging*. J Cereb Blood Flow Metab, 1999. **19**(12): p. 1289–95.
73. Pan, J. W., E. M. Bebin, W. J. Chu, and H. P. Hetherington, *Ketosis and epilepsy: 31P spectroscopic imaging at 4.1 T*. Epilepsia, 1999. **40**(6): p. 703–7.
74. Pan, J. W., J. T. Vaughan, R. I. Kuzniecky, G. M. Pohost, and H. P. Hetherington, *High resolution neuroimaging at 4.1T*. Magn Reson Imaging, 1995. **13**(7): p. 915–21.
75. Paulson, O. B., M. M. Hertz, T. G. Bolwig, and N. A. Lassen, *Filtration and diffusion of water across the blood-brain barrier in man*. Microvasc Res, 1977. **13**(1): p. 113–24.
76. Paulson, O. B., M. M. Hertz, T. G. Bolwig, and N. A. Lassen, *Water filtration and diffusion across the blood brain barrier in man*. Acta Neurol Scand Suppl, 1977. **64**: p. 492–3.
77. Pawlik, G., A. Rackl, and R. J. Bing, *Quantitative capillary topography and blood flow in the cerebral cortex of cats: an in vivo microscopic study*. Brain Res, 1981. **208**(1): p. 35–58.
78. Pfeuffer, J., I. Tkac, I. Y. Choi, H. Merkle, K. Ugurbil, M. Garwood, and R. Gruetter, *Localized in vivo 1H NMR detection of neurotransmitter labeling in rat brain during infusion of [1–13C] D-glucose*. Magn Reson Med, 1999. **41**(6): p. 1077–83.
79. Purcell, E. M., H. C. Torrey, and R. V. Pound, *Resonance absorbtion by nuclear magnetic moments in a solid*. Physical Review, 1945. **69**: p. 37–38.
80. Rabi, I. I., S. Millman, and P. Kusch, *The molecular beam resonance method for measuring nuclear magnetic moments*. Physical Review, 1939. **55**: p. 526–535.
81. Rabi, I. I., J. R. Zacharias, S. Millman, and P. Kusch, *A new method of measuring nuclear magnetic moment*. Physical Review, 1938. **53**: p. 318.
82. Raichle, M. E. *Circulatory and metabolic correlates of braln function in normal humans*, in *Handbook of Physiology-The Nervous System*, F. Plum, Editor. 1987, Am. Phys. Soc., Bethesda. p. 643–674.
83. Segebarth, C., V. Belle, C. Delon, R. Massarelli, J. Decety, J.-F. Le Bas, M. Decorpts, and A. L. Benabid, *Functional MRI of the human brain: Predominance of signals from extracerebral veins*. NeuroReport, 1994. **5**: p. 813–816.
84. Shen, J., K. F. Petersen, K. L. Behar, P. Brown, T. W. Nixon, G. F. Mason, O. A. Petroff, G. I. Shulman, R. G. Shulman, and D. L. Rothman, *Determination of the rate of the glutamate/glutamine cycle in the human brain by in vivo 13C NMR*. Proc Natl Acad Sci USA, 1999. **96**(14): p. 8235–40.
85. Sibson, N. R., A. Dhankhar, G. F. Mason, D. L. Rothman, K. L. Behar, and R. G. Shulman, *In vivo 13C NMR measurements of cerebral glutamine synthesis as evidence for glutamate-glutamine cycling*. Acad Sci U S A, 1997. **94**(6): p. 2699–2704.
86. Sibson, N. R., A. Dhankhar, G. F. Mason, D. L. Rothman, K. L. Behar, and R. G. Shulman, *Stoichiometric coupling of brain glucose metabolism and glutamatergic neuronal activity*. Proc Natl Acad Sci U S A, 1998. **95**(1): p. 316–21.
87. Siesjo, B., *Brain Energy Metabolism*. 1978, Wiley: New York. p. 101–110.
88. Song, A. W., E. C. Wong, S. G. Tan, and J. S. Hyde, *Diffusion weighted fMRI at 1.5 T*. Magn Reson Med, 1996. **35**(2): p. 155–8.
89. Springer, C. S. and Y. Xu, *Aspects of Bulk Magnetic Susceptibility in in vivo MRI and MRS*, in *New Developments in Contrast Agent Research*, P. A. Rink and R. N. Muller, Editors. 1991, European Magnetic Resonance Forum: Blonay, Switzerland. p. 13–25.
90. Stejskal, E. O. and J. E. Tanner, *Spin diffusion measurements: spin-echoes in the presence of a time dependent field gradient*. J. Chem. Phys., 1965. **42**: p. 288–292.
91. Tanafuji, M. in *2001 Minnesota Workshops*. 2001. Minneapolis, MN.
92. Thulborn, K. R., J. C. Waterton, and P. M. Matthews, *Dependence of the transverse relaxation time of water protons in whole blood at high field*. Biochem Biophys Acta, 1992. **714**: p. 265–272.
93. Tkac, I., P. Andersen, G. Adriany, H. Merkle, K. Ugurbil, and R. Gruetter, *In vivo (1)H NMR spectroscopy of the human brain at 7 T*. Magn Reson Med, 2001. **46**(3): p. 451–6.
94. Ugurbil, K., T. R. Brown, J. A. den Hollander, P. Glynn, and R. G. Shulman, *High-resolution 13C nuclear magnetic resonance studies of glucose metabolism in Escherichia coli*. Proc Natl Acad Sci USA, 1978. **75**(8): p. 3742–6.
95. Ugurbil, K., M. Garwood, J. Ellermann, K. Hendrich, R. Hinke, X. Hu, S. G. Kim, R. Menon, H. Merkle, S. Ogawa, and *et al.*, *Imaging at high magnetic fields: initial experiences at 4 T*. Magn Reson Q, 1993. **9**(4): p. 259–77.
96. van Zijl, P. C., S. M. Eleff, J. A. Ulatowski, J. M. Oja, A. M. Ulug, R. J. Traystman, and R. A. Kauppinen, *Quantitative assessment of blood flow, blood volume and blood oxygenation effects in functional magnetic resonance imaging [see comments]*. Nat Med, 1998. **4**(2): p. 159–67.
97. Vaughan, J. T., M. Garwood, C. M. Collins, W. Liu, L. DelaBarre, G. Adriany, P. Andersen, H. Merkle, R. Goebel, M. B. Smith, and K. Ugurbil, *7T vs. 4T: RF power, homogeneity, and signal-to-noise comparison in head images*. Magn Reson Med, 2001. **46**(1): p. 24–30.
98. Vesselle, H. and R. E. Collin, *The Signal-to-Noise Ratio of Nuclear Magnetic Resonance Surface Coils and Application to Lossy Dielectric Cylinder Model-Part I: Theory*. IEEE Transactions on Biomedical Engineering, 1995. **42**(5): p. 497–505.
99. Vesselle, H. and R. E. Collin, *The Signal-to-Noise Ratio of Nuclear Magnetic Resonance Surface Coils and Application to Lossy Dielectric Cylinder Model-Part II: The case of cylinderical window studies*. IEEE Transactions on Biomedical Engineering, 1995. **42**(5): p. 507–520.
100. Wang, G., K. Tanaka, and M. Tanifuji, *Optical imaging of functional organization in the monkey inferotemporal cortex. 1996*. Science, 1996. **272**: p. 1665–1668.
101. Weisskoff, R. M., C. S. Zuo, J. L. Boxerman, and B. R. Rosen, *Microscopic susceptibility variation and transverse relaxation: theory and experiment*. Magn Reson Med, 1994. **31**(6): p. 601–10.
102. Wen, H., A. S. Chesnick, and R. S. Balaban, *The Design and Test of a New Volume Coil for High Field Imaging*. Magn Reson Med, 1994. **32**: p. 492–498.
103. Wong, E. C., R. B. Buxton, and L. R. Frank, *Quantitative imaging of perfusion using a single subtraction (QUIPSS and QUIPSS II)*. Magn Reson Med, 1998. **39**(5): p. 702–8.
104. Wright, G. A., B. S. Hu, and A. Macovski, *Estimating oxygen saturation of blood in vivo with MR imaging at 1.5 T*. J Magn Reson Imaging, 1991. **1**(3): p. 275–83.
105. Yacoub, E., A. Shmuel, J. Pfeuffer, P. F. Van De Moortele, G. Adriany, P. Andersen, J. T. Vaughan, H. Merkle, K. Ugurbil, and X. Hu, *Imaging brain function in humans at 7 Tesla*. Magn Reson Med, 2001. **45**(4): p. 588–94.
106. Zhu, X. H. and W. Chen, *Observed BOLD effects on cerebral metabolite resonances in human visual cortex during visual stimulation: A functional (1)H MRS study at 4 T*. Magn Reson Med, 2001. **46**(5): p. 841–7.

13

Functional MRI

Joseph B. Mandeville and Bruce R. Rosen

Massachusetts General Hospital NMR Center, MGH/MIT/HMS Athinoula A. Martinos Center for Biomedical Imaging Charlestown, Massachusetts 02129

I. Introduction

The safe, noninvasive, and flexible nature of magnetic resonance imaging (MRI) has contributed to the pervasiveness of this technique in research and clinical applications. Following the initial developmental stages in the 1970s and the clinical expansion of MRI for structural imaging during the 1980s, the 1990s were characterized by a rapid evolution of methods for assessing normal and pathological brain function. MRI affords a wide variety of functional contrast mechanisms, including relaxation processes that are intrinsic to tissue, chemical shift imaging or localized spectroscopy (Chapter 14), diffusion-weighted imaging (Chapter 15), and perfusion-sensitive methods based upon paramagnetic contrast agents or endogenous processes. This chapter will discuss the perfusion-weighted techniques that can track changes in brain function continuously with both high temporal resolution (seconds) and high spatial resolution (millimeters).

Although *f*MRI based upon blood oxygen level-dependent (BOLD) signal and blood flow has been available for a decade, advances during the past several years have clarified many aspects of the physical and physiological bases of these contrast mechanisms. A better understanding of the *f*MRI temporal response, together with evolving statistical strategies (Chapter 22), has enabled experimental designs that are less limited by the technology and more suited to the types of rapid, interleaved presentations that have traditionally been employed by cognitive neuroscience. Higher magnetic field strengths and improved data acquisition techniques have increased the sensitivity of endogenous contrast mechanisms, and exogenous contrast agents have changed the *f*MRI landscape in animal models by boosting sensitivity by half an order of magnitude at common field strengths. These advances have facilitated increases in spatial resolution that are beginning to test the biological limits of the vascular response.

In order to provide the necessary background for an understanding of *f*MRI mechanisms and limitations, a brief section on basic MRI precedes a discussion of the physics, physiology, sensitivity, and spatio temporal resolution of *f*MRI, concentrating on dynamic measurements of BOLD signal, cerebral blood flow (CBF), and cerebral blood volume (CBV).

II. MRI: A Brief Primer

A. The Source of Signal

Everyone is familiar with some aspects of magnetism, which causes ferromagnets to stick to refrigerators and compass needles to align with the earth's magnetic poles. Magnetism and electricity, once seen as separate phenomena, are now known to be two manifestations of the electromagnetic force. Charge and spin are both intrinsic characteristics of matter, in the same way as mass. An isolated charge is an electric field source, causing other charges to be attracted or repelled according to their relative sign. In the same way, an isolated spin (or magnetic "dipole moment") is a magnet field source, and such dipoles interact with each other and with external magnetic fields.

Like all common matter in our experience, biological bodies are composed of atoms and molecules, which in turn are assemblies of protons, neutrons, and electrons. For each of these molecular building blocks, charge and spin are *quantized*, meaning that they can assume only certain discrete values. A positive and negative charge form a highly favorable energetic pairing, so that electrically charged species are very short lived in biological samples. In an externally applied magnetic field, spin orientation also causes a splitting of energy levels. However, the magnetic-spin coupling is very weak, even relative to thermal energy within a sample. As a consequence, the preference toward spin orientation (both electronic and nuclear) persists within a biological sample, but not as a dominant energetic factor. Although most spins pair off in the same way as charge, an unpaired spin is ultimately required in order to magnetically separate either electronic or nuclear populations.

The process of exciting electron resonances within an external magnetic field is known as electron spin resonance (ESR). For biological imaging, drawbacks of ESR include the short relaxation times of the resonances, the relatively large energy of splitting, and hence the potential for excessive heating (microwaves are very efficient at boiling water, and people). In contrast, most nuclear resonances have relatively long relaxation times (due to weaker interactions with the environment), and excitation is applied using less powerful radio frequencies (RF), which are much less efficient at heating water and other biological materials. The noninvasive nature and the relatively long relaxation times of *nuclear magnetic resonance* (NMR) are tremendous advantages for a biological imaging modality. Because NMR probes properties of nuclei, whereas electronic interactions determine biochemistry, there is no causal relationship between the nuclei that can be imaged with NMR and the biochemical function of the associated molecules. It is a happy coincidence that the most abundant molecule in the brain is water, carrying two hydrogen nuclei (protons) with unpaired nuclear spins.

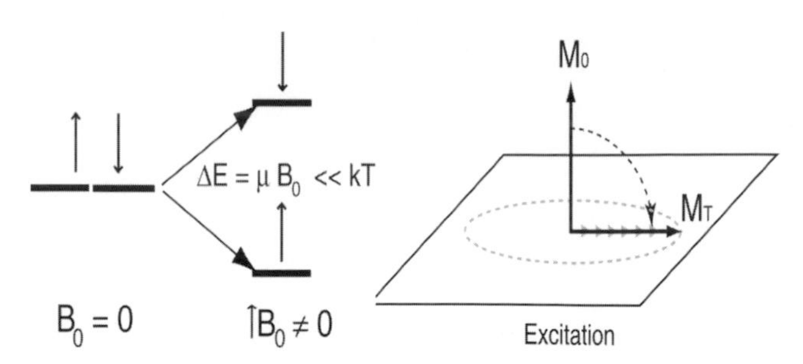

Figure 1 The application of external magnetic field (B_0) produces a small splitting of energy levels that causes a net alignment of spins with the magnetic field and an observable bulk magnetization (M_0). RF excitation promotes aligned spins to the higher energy state, causing the longitudinal magnetization to be tipped into the transverse plane, where it can be detected by induction.

Figure 1 illustrates how spin orientation produces a splitting of energy levels that favors alignment with an externally applied magnetic field (B_0). The net *polarization* of the sample depends upon the small orientation-dependent energy splitting of a magnetic dipole moment (μ), $\Delta E = -\mu \cdot B_0$, relative to the temperature-dependent Brownian motion of the sample (polarization ~$\mu \mathbf{B}_0/kT$). Sample polarization (the net alignment of the total spin population) is only about 1 part in 10^5 at room temperature and at clinical MRI field strengths, meaning that spins have only the slightest tendency to align with an externally applied magnetic field. Fortunately, an average brain volume contains about 2×10^{20} hydrogen nuclei ("water protons") per cubic millimeter. The combination of a small net polarization with a large number of molecules produces a net *bulk magnetization* that is of sufficient size to be detectable by NMR methods.

One of the difficulties of understanding the NMR phenomena is that the origin of the effect is most easily understood on the quantum mechanical level, whereas bulk magnetization is commonly described using notions of classical mechanics. The two viewpoints connect in the transition from one spin to an ensemble of spins. If a single spin can take on orientations aligned or antialigned with a magnetic field ($\pm\frac{1}{2}$), then two spins can be both aligned (+1) or antialigned (−1) with the applied field, or they can cancel each other (0). Three spins can produce a net alignment with values of $+\frac{3}{2}$, $+\frac{1}{2}$, $-\frac{1}{2}$, or $-\frac{3}{2}$. A very large number of spins can assume any (nearly) continuous distribution of orientations with respect to the magnetic field. In this way, the physics governing the quantum mechanical description of the spin state of a large sample, like a brain, is well described by the classical description of the bulk, or net, magnetization.

B. NMR Excitation

NMR excitation involves the promotion of lower energy (aligned) spins to the higher energy (antialigned) state through RF irradiation. Excitation occurs when the RF energy equals the energy splitting of spin states, which is proportional to the local magnetic field (B) with a scaling factor that depends upon the nucleus of interest. Because the energies of photons are proportional to their frequencies, the RF frequency (f) also must be proportional to the field strength:

$$f = (\gamma/2\pi)\,B. \quad (1)$$

This is the concept of *resonance*, wherein a system can be excited only at certain natural energy states and, hence, frequencies. Across a small distribution of magnetic fields within a sample (ΔB), excitation can occur throughout only a small range of frequencies (Δf). For water, the proportionality constant $\gamma/2\pi$ is 42.6 MHz per Tesla, so that the frequency required for water excitation at 1.5 T is about 64 MHz.

In classical terms, the bulk magnetization vector precesses about the main magnetic field at a frequency proportional to the field. As RF radiation promotes spins from the lower energy state to the higher energy state, the bulk magnetization is tipped into a plane transverse to the magnetic field (Fig. 1).

C. NMR Detection

Immediately following excitation, the magnetization that was *longitudinal* to B_0 has been converted into *transverse magnetization*. As the transverse magnetization vector rotates about the main field, it can be detected through the process of *induction*. If a conducting loop is oriented so that the transverse magnetization produces a time-varying flux through the loop, as depicted in Fig. 2, an electrical current will be generated in the loop. The NMR receiver then digitizes the induced current in order to record the magnitude of the transverse magnetization as a function of time. This simple detection system, when combined with certain manipulations of the magnetic field and other experimental parameters, enables spatial encoding across the sample, as well as multiple types of image contrast.

D. Transverse Relaxation (T_2 and T_2*)

As soon as the longitudinal magnetization is tipped into the transverse plane, it begins to decay as the spins precess at slightly different rates, each according to the local value of the magnetic field as described by Eq. (1). Variations in the local magnetic field arise from a variety of factors, including imperfections in the magnet, induced macroscopic magnetic fields within the sample, and the local molecular environments. Relative to the average phase within the sample, the phase of each spin ($\Delta\phi$) will accrue according to the local frequency offset ($\Delta f = \gamma/2\pi\ \Delta B$) and the time ($t$) from the excitation pulse, $\Delta\phi = 2\pi\,\Delta f\,t$. A progressive loss of spin coherence produces decay of transverse magnetization, as illustrated in Fig. 2. When the orientation of the spin population has randomized, the net vector sum in the transverse plane is zero, and no more signal can be detected. The exponential time constant for signal decay following a single RF pulse is referred to as T_2*.

A *spin echo* can be used to reduce the decay rate of transverse magnetization. By inserting a second RF pulse to flip magnetization by 180° at a time halfway between the RF excitation pulse and the collection of signal, the phase accumulation of each spin is reversed ($\Delta\phi \rightarrow -\Delta\phi$). Spins that experience the same field before and after the refocusing pulse will return to a net relative phase of zero at the time of signal collection, regardless of the local value of the magnetic field. *A spin echo refocuses magnetic field heterogeneities that are static in time.* Thus, some processes that contribute to T_2* do not contribute to T_2, which is always

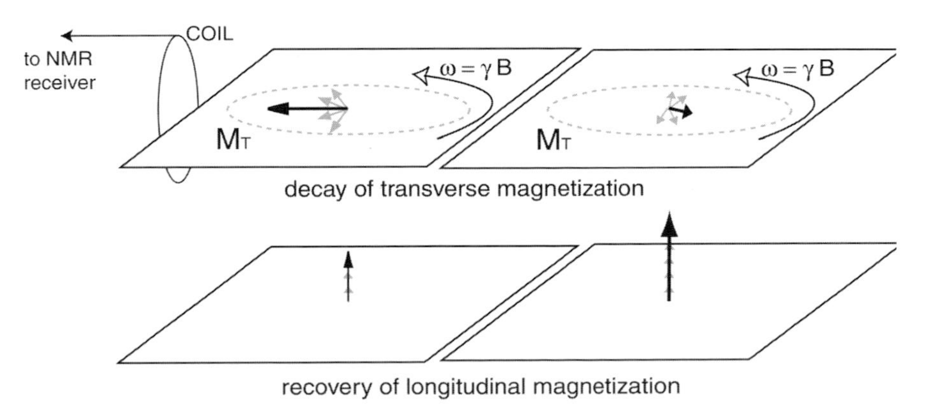

Figure 2 **Top:** Because individual spins each experience slightly different local magnetic fields, they rotate at different rates in the transverse plane, leading to a progressive loss of coherence and signal. **Bottom:** As transverse magnetization decays, longitudinal magnetization recovers along the direction of the main magnetic field.

longer than T_2*. T_2 is not very sensitive to variations in the magnetic field on a large spatial scale, which cause signal dropout on T_2* images. However, spin echoes do not refocus signal from nuclei subject to processes like flow and diffusion that cause spins to sample time-varying magnetic fields.

The rate of decay of transverse magnetization sets the time scale for data collection following an RF excitation; T_2 has values of 65–90 ms in brain tissue at 1.5 T. If we refer to the time between excitation and data collection as the *echo time* (T_E), then transverse magnetization decays at a rate set by T_2 (spin echoes) or T_2* (gradient echoes),

$$M_T(T_E) = M_T(0)\exp(-T_E/T_2), \quad (2)$$

where $M_T(0)$ is the initial value immediately following the RF excitation, and $M_T(T_E)$ is the value at the spin echo time T_E.

E. Longitudinal Recovery (T_1)

Longitudinal magnetization also has a characteristic relaxation time (T_1). Immediately after an excitation, spins begin to reestablish thermal equilibrium as they preferentially repopulate the energetically favored state that is aligned with the magnetic field. Following a 90° excitation from the equilibrium state, longitudinal magnetization will begin an exponential recovery from zero to the maximum value (M_0): $M_L = M_0(1 - \exp(-t/T_1))$. In the more general case, repeated excitations lead to a steady value of longitudinal magnetization just prior to each excitation that depends upon the *repetition time* (T_R) relative to T_1 and the flip angle of the RF pulse (α). The amount of transverse magnetization produced by an excitation depends upon the available longitudinal magnetization and the sine of the flip angle. The net T_1-dependent contribution to the signal is

$$M_L(T_R) = M_0 \frac{1-\exp(-T_R/T_1)}{1-\cos\alpha\exp(-T_R/R_1)}\sin\alpha. \quad (3)$$

For repetition times with T_R much greater than T_1, a 90° flip angle maximizes the signal. At shorter repetition times, the flip angle that produces the maximum signal is called the Ernst angle (Ernst and Anderson, 1966) and is calculated as $\cos(\alpha_E) = \exp(-T_R/T_1)$. Figure 3 shows signal as a function of flip angle for a T_R of 500 ms and T_1 values of 900, 1000, and 1100 ms. At flip angles equal to the Ernst angle and larger, signal is sensitive to the value of T_1. At flip angles less than about one-half of the Ernst angle, sensitivity to T_1 is lost.

The T_1 dependence of signal also has implications for the signal-to-noise ratio per unit time. As T_R decreases below T_1, the maximum signal per excitation (which occurs at the Ernst angle for each T_R) decreases at a rate that scales roughly with the square root of T_R. This means that increasing the rate of data collection is not an effective means of increasing the signal-to-noise ratio per unit time when T_R falls below T_1. For instance, the signal per excitation using a T_R of 100 ms is slightly greater than $1/\sqrt{10}$ of the value that can be obtained at a T_R of 1 s. By averaging 10 such images, the signal-to-noise ratio increases by $\sqrt{10}$, yielding a signal value only about 5% greater than a single image using a T_R of 1 s. Figure 3 depicts the signal-to-noise ratio per unit time as a function of the repetition rate, using the Ernst angle for each value of T_R. Only for T_R values above T_1 is temporal averaging an effective strategy for increasing the signal-to-noise ratio in a given amount time, relative to the choice of a longer T_R value.

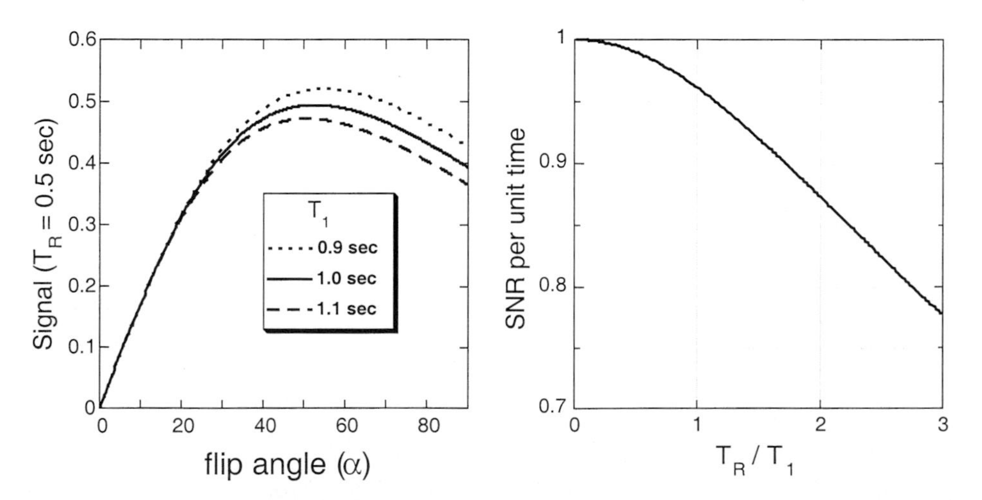

Figure 3 **Left:** For repetition times faster than T_1, a maximum level of signal is obtained by using a flip angle less than 90°. At flip angles less than about one-half of the Ernst angle, T_1 sensitivity is lost. **Right:** The signal-to-noise ratio per unit time has little dependence upon the repetition time for values below T_1.

Finally, it is important to note that the equation above is appropriate for *saturation recovery* acquisitions of signal, which encompass the majority of MRI acquisition strategies. Additional T_1 weighting can be obtained by spin preparation prior to excitation. A common example is *inversion recovery*, in which a 180° pulse initially inverts longitudinal magnetization.

F. Image Contrast

Although we have not yet addressed image formation, the previous discussion of relaxation times leads naturally to the subject of image contrast. MRI signal fundamentally depends upon the number of nuclear spins of interest in a sample volume, the rate of recovery for longitudinal magnetization (T_1), and the rate of decay of transverse magnetization (T_2 or T_2*). The signal in the NMR receiver is proportional to the transverse magnetization [Eq. (2)], which is produced from the longitudinal magnetization [Eq. (3). The density of nuclear spins (ρ) sets the overall scaling factor:

$$S \sim \rho \frac{1-\exp(-T_R / T_1)}{1-\cos\alpha \exp(-T_R / T_1)} \sin\alpha \quad \exp(-T_E / T_2). \quad (4)$$

The choice of T_R and T_E determines image contrast. Choosing T_E much smaller than the cerebral distribution of T_2 values eliminates T_2 weighting in images. Conversely, an echo time within the range of normal T_2 values maximizes T_2 weighting. T_1 weighting is eliminated either by choosing a very long T_R (>>T_1) or by using a shorter T_R in conjunction with a flip angle that is below the Ernst angle. T_1 sensitivity is maximized by choosing T_R within the range of normal T_1 values and by using a flip angle equal to or greater than the Ernst angle.

G. Spatial Encoding

From the discussion of the simple NMR receiver, it is clear that the digitized signal reports only the total magnitude of the transverse magnetization as a function of time. The transverse magnetization vector must carry all the information about spatial location and contrast for every volume element (or voxel) in an image. At any time, the magnitude of the bulk magnetization vector depends upon the spin density and relaxation rates, which provide contrast, and the spin coherence, which enables spatial encoding. The trick of simultaneously encoding both space and contrast is to manipulate spin phase as a function of both space and time. From Eq. (1), the accumulated phase is a time integral of the field at any point in space, $\varphi(t) = 2\pi \int f(t)dt = \gamma \int B(x,y,z,t)dt$. By applying a series of small magnetic field gradients on top of the large scanner field (e.g., $B(x) = B_0 + G_x x$), space can be encoded in all three dimensions, with time providing a mechanism for contrast. For three-dimensional spatial encoding, the number of required phase manipulations equals the total number of image voxels ($N_x \times N_y \times N_z$).

Three-dimensional spatial information can be obtained either through three-dimensional phase encoding or, more commonly for *f*MRI, through two-dimensional spatial encoding plus slice selection. Slices are selected by applying a magnetic field gradient in a desired direction (e.g., z), so that the field (and hence the spin rotation frequency) has a spatial dependence ($B = B_0 + G_z z$). An RF excitation with a central frequency (f_0) and bandwidth (Δf) will excite a resonance across a slice at a specific z_0 location ($f_0 = \gamma/2\pi\, B_0 + \gamma/2\pi\, G_z z_0$) with a known slice thickness ($\Delta f = \gamma/2\pi\, G_z\, \Delta z$). Following slice selection, the problem of spatial encoding has been reduced to two dimensions.

The relationship between physical space and data acquisition methods is simplified by replacing the formulation based upon spin phase, $\varphi(t) = \gamma \int_0^t B(x,y,z,t')\, dt'$, with the abstraction of *spatial frequency* (k), using units like cycles per centimeter. Images are then reconstructed from "raw data" (or "k space") into real space using a Fourier summation of spatial frequencies:

$$\vec{k}(t) = \gamma / 2\pi \int_0^t \vec{G}(t')dt',$$
$$\rho(\vec{r}) = \int \rho(\vec{k}) \exp(i\vec{k} \cdot \vec{r}) d\vec{k}. \quad (5)$$

While the concept of k space may not seem like a conceptual simplification at first, in fact the conjugate nature of k space and real space simplifies the understanding of any pulse sequence. To encode $N_x \times N_y$ points in real space, we need to sample $N_x \times N_y$ spatial frequencies. In order to obtain a spatial resolution of δx in one dimension, we need to span a range of spatial frequencies $\Delta k_x = 1/\delta x$. In order to cover a field of view Δx, we need to obtain complex samples at intervals of $\delta k_x = 1/\Delta x$. These relationships connect real space to the required gradient strengths and sampling rates, using Eq. (5). Any trajectories through k space with the proper sampling density (corresponding to the field of view) and excursion lengths (corresponding to resolution) can produce an image. The most common *f*MRI sampling strategies are discussed in a later section.

H. The Magnitude of Signal

MRI is a sensitivity-limited technique. Optimizing the signal-to-noise ratio (SNR) is important for MRI studies of all types, and this is particularly true for functional MRI. The SNR depends upon a variety of quantities, as summarized in the following equation:

$$\text{SNR} \propto \rho\, B_0\, r^3\, S_c f_1 f_2 \sqrt{T_{ACQ}}. \quad (6)$$

f_1 and f_2 depend upon T_1 and T_2, as described previously [see Eq. (4)] and provide contrast in an image. The other symbols refer to the spin density (ρ), magnetic field strength (B_0), dimension (r) of a voxel, sensitivity of the detection coil (Sc), and total acquisition time (T_{ACQ}). Each of these contributions is discussed below.

1. Spin Density and Magnetic Field

The mechanisms of NMR excitation and detection place limits upon the strength of MRI signal. Unlike scattering methods, such as optical imaging or X-ray CT, we cannot simply increase the flux of photons to probe the sample with higher sensitivity. Because MRI is an emission technique, the intrinsic sensitivity per unit volume depends upon the density of spins and the spin polarization with respect to the externally applied magnetic field.

The means of altering sample polarization are limited, since the small energy splitting produced by the interaction of nuclear spins with the applied magnetic field has to compete with thermal energy. However, it is worthwhile mentioning some polarization strategies from other NMR subfields. In nonbiological settings, such as solid-state physics, extremely low temperatures help significantly to boost the polarization for NMR experiments. Unfortunately, meaningful temperature reductions on the Kelvin scale would leave a human subject frozen and lifeless. For imaging of the lungs, high sample polarization of ^{129}Xe or ^{3}He can be generated by optically polarizing the nuclei before inhalation (Guenther *et al.*, 2000). The long relaxation times of these gases become shorter as they dissolve in the blood, however, so this method does not look particularly promising for *f*MRI of the brain.

The only established method for increasing sample polarization for *f*MRI is the use of higher magnetic fields. Because the energy splitting of parallel and antiparallel spin states is proportional to the strength of the field (Fig. 1), polarization increases in proportion to the magnetic field strength (B_0). For this reason, considerable energy has been devoted to the development of higher field strengths during the past several decades. Although 1.5 T is still the most common field strength for clinical human scanners, 3- and 4-T systems are proliferating, and a few systems with field strengths of 7 T and above are in use (Kangarlu *et al.*, 1999; Yacoub *et al.*, 2001). For animal *f*MRI studies, horizontal bore magnets with field strengths of 9.4 T are not uncommon.

The linear dependence of SNR on the field strength is not completely obvious, because the signal actually increases as the square of the field. The second B_0 factor is due to the electromotive force that produces the electric current in the detection coil by induction. The induced current is proportion to dB/dt, which in turn is proportional to the spin precession frequency, $\omega = \gamma B_0$. However, this extra B_0 factor is canceled by the same effect in the noise, which is dominated by electromagnetic currents in the sample volume (Hoult and Lauterbur, 1978).

2. Coil Sensitivity

Sc is a factor specifying the detection sensitivity of the detection coil(s). Although multiple factors contribute to the sensitivity of the NMR receiver, the dominant factor is the coil size. As the size of a coil is reduced, spatial coverage is sacrificed for stronger signals. For volume coils, like the typical birdcage design (Hayes *et al.*, 1985) provided for head imaging on clinical systems, there is little opportunity to reduce the coil size, since the head must fit completely within the coil. For surface coils, however, smaller designs can produce significantly improved SNR for targeted brain regions. This issue is discussed in more detail later.

3. Acquisition Time

Temporal averaging is always an effective way to increase SNR for any given set of experimental parameters. The acquisition time (T_{ACQ}) in Eq. (6) refers to the cumulative time that the NMR receiver is accepting signal. This is the data acquisition time following one excitation multiplied by all such excitations. In terms of the number of spatial encoding steps in each dimension (e.g., N_x) the number of image averages (N_{EX}), and the time between digital samples of the signal (δt=1/bandwidth), the acquisition time is $T_{ACQ} = N_{EX}\, N_x N_y\, N_z\, \delta t$. Acquisition methods like echo-planar imaging use the available magnetization efficiently by employing long acquisition times per excitation. The practical limit of the acquisition time following a single excitation is set by the decay rate of signal, T_2 or T_2*, and the accumulation of phase errors from field imperfections that lead to image distortions and artifacts.

4. Voxel Size

The final term to address in Eq. (6) is the voxel volume, r^3, which combines with ρ to determine the total number of NMR visible nuclear spins in the sample volume. After all other factors have been optimized, the voxel volume must deliver enough SNR to enable the detection of very small functional signal changes in an appropriate amount of time. SNR requirements thus usually provide the practical limit to *f*MRI spatial resolution.

III. From MRI to fMRI

A. General Considerations

1. Time and Space

Due to the high abundance of water in the brain and the safe, noninvasive nature of MRI, images with exquisite spatial resolution can be generated given sufficient averaging time. However, functional signal changes using MRI techniques are typically a few percent or less, so a few images are not sufficient to achieve sufficient statistical power in maps of brain activation—a lot of images are required. As suggested by Eq. (6), spatial and temporal resolution are intimately related through the signal-to-noise ratio. In fact, the amount of imaging time required to obtain a fixed value of SNR increases in inverse proportion to the sixth power of a voxel dimension! As an example, a typical choice of *f*MRI voxel size at 1.5 T using a standard head coil is 3.5 mm on each side, producing an SNR of 100–200 following a single excitation. In order to reduce the voxel size to 1 mm on a side, a total acquisition time of 10 min would be required in order to obtain the same SNR using the same experimental setup. The upshot is that *f*MRI sacrifices spatial resolution for the sake of temporal resolution. In the usual MRI parlance, there is no such thing as a free lunch.

2. *f*MRI of Water

Given the low spin polarization in biological samples and the requirement of high temporal resolution for functional brain imaging, it is essential for rapid imaging to have as many spins as possible in a sample volume. Protons are by far the most abundant nuclear species in the brain that can be imagined by NMR. Water carries two protons per molecule and accounts for 3/4 of the brain weight (Laiken and Fanestil, 1990), corresponding to a molecular abundance of about 40 moles per liter. In contrast, the other major ^{1}H metabolites like glutamate, glutamine, and glucose are present at levels that are reduced by factors of 1000 or more (Stanley *et al.*, 1995). For this reason, *f*MRI typically refers to imaging of water protons, and this restricted definition will be employed here.

3. Blood-Based Contrast

We might hypothesize that water diffusion, which is a clinically useful MRI marker of neuronal compromise during ischemia (Moseley *et al.*, 1990), could be altered during normal neuronal activation. Further, magnetic fields produced by the electrical activity of neuronal populations can be detected outside the brain by magnetoencephalography using exquisitely sensitive probes (e.g., Dale and Halgren, 2001)—can MRI detect the same fields? Unfortunately, there is little evidence to date of any endogenous MRI methods that are directly sensitive to neuronal activity.

Moreover, no existing MRI contrast agents cross the normal blood–brain barrier, which prevents passage of large polar molecules. This means that the type of receptor-binding studies enabled with PET are precluded by MRI. To circumvent this limitation in rodent studies, mannitol injection has been used to open the blood–brain barrier to enable manganese to penetrate into the brain (Lin and Koretsky, 1997). Since the chemical properties of manganese are similar to those of calcium, this paramagnetic contrast agent enters cells through calcium channels. Increased local brain activity leads to an increase in manganese concentration and a brighter signal on T_1-weighted images.

Though an exciting research method, the requirement for breakdown of the blood–brain barrier and the neurotoxicity of manganese preclude ready extension of this method to humans. Excluding animal studies that eliminate the brain–blood barrier, *f*MRI relies exclusively upon the time-dependent characteristics of the blood supply to infer neuronal activity through flow–metabolism coupling. The *f*MRI methods addressed in the remainder of the chapter include contrast based upon blood oxygenation (T_2 or T_2*), blood volume (T_2 or T_2*), and blood flow (T_1).

B. *f*MRI Origins

1. Background

The basic notion that changes in cerebral activity require concomitant changes in the blood supply was postulated more than a century ago (Roy and Sherrington, 1890; James, 1890). More definitive experiments awaited technical developments in the second half of the past century. The first *in vivo* method for regional measurements of CBF used diffusible tracers in conjunction with invasive autoradiography (Kety, 1960). Sokoloff and colleagues combined the Kety method with a new technique, invasive 2-deoxyglucose autoradiography (Sokoloff *et al.*, 1977), to solidify the relationship between regional changes in CBF and metabolism in animal models (Sokoloff, 1981).

During the 1970s and 1980s, positron emission tomography extended autoradiographic methods to noninvasive measurements in human subjects, and the flexibility of this modality further enabled a wide variety of other means for assessing regional cerebral function. Of particular importance for later MRI methods, PET enabled functional measurements of CBF (Ter-Pogossian *et al.*, 1969), $CMRO_2$ (Ter-Pogossian *et al.*, 1969), and CBV (Grubb *et al.*, 1974) in normal human subjects. An important, and unexpected, piece of physiology emerged from concurrent measurements of flow and metabolism: stimulus-induced functional changes in CBF and glucose uptake exceeded the changes in oxygen utilization (Fox and Raichle, 1986; Fox *et al.*, 1988). The consequences of this observation for *f*MRI would become apparent several years later.

2. Exogenous Contrast Agent

Injected gadolinium-based compounds have a long history as MRI contrast agents (Brady *et al.*, 1982; Weinmann *et al.*, 1984; Lauffer, 1987; Villringer *et al.*, 1988). By dynamically tracking the passage of a bolus of contrast agent through the brain and applying tracer kinetic analysis to the concentration–time profile (Lassen and Perl, 1979), maps of both CBF and CBV can be generated (Rosen *et al.*, 1989; Ostergaard *et al.*, 1996). The combination of rapid imaging and exogenous contrast agent provided a means to track physiological changes in animal models during the 1980s (Belliveau *et al.*, 1986, 1988, 1990; Rosen *et al.*, 1989). This method was applied to human studies of baseline physiology (Belliveau *et al.*, 1989; Rosen *et al.*, 1989) and subsequently led to the first studies of functional brain activation by Belliveau *et al.*, (1991).

An important difference between PET and MRI measurements of CBV is that the physics of compartmentalized intravascular contrast agent leads to an amplification of signal changes, due to magnetic field gradients that extend from the vessels into the extravascular space (Villringer *et al.*, 1988). Although the average blood volume fraction is only 3–5% in the brain, typical values of signal attenuation in the human *f*MRI study by Belliveau were 35–40%. However, in some ways the bolus technique is similar to PET: only a few boluses can be injected into each subject, and the temporal resolution is poor.

As discussed in the original human *f*MRI study (Belliveau *et al.*, 1991), the temporal resolution can be vastly increased by employing a contrast agent with a long blood half-life. Such agents have enabled measurements of CBV with high temporal resolution in animal models (Hamberg *et al.*, 1996), while providing much greater sensitivity for functional brain mapping than endogenous contrast (Mandeville *et al.*, 1998; Kennan *et al.*, van Bruggen *et al.*, 1998).

3. Endogenous Contrast Agent: BOLD Signal

Pauling and Coryell (1936) first measured the magnetic susceptibilities of hemoglobin in the oxygenated and deoxygenated states. Subsequent work (Thulborn *et al.*, 1982) quantified changes in the transverse and longitudinal relaxation rates of blood as a function of deoxyhemoglobin concentration. These early studies demonstrated that deoxyhemoglobin can act as a paramagnetic MRI contrast agent and change T_2 (or T_2*), in the same way as gadolinium. Ogawa and colleagues (1990b) found that signal loss occurred around large veins in high field images and that the effect was dependent upon the oxygenation state of the hemoglobin. They further suggested that the BOLD signal could potentially be used for functional brain imaging in a way similar to that of previous studies based upon exogenous contrast agents, but with better temporal resolution (Ogawa *et al.*, 1990a). The contrast mechanisms proposed for exogenous contrast agent (Villringer *et al.*, 1988) and BOLD signal (Ogawa *et al.*, 1990b) were essentially the same, and invasive optical studies were already mapping brain function in animals based upon the oxygenation state of hemoglobin (Grinvald *et al.*, 1986; Frosting *et al.*, 1990). Shortly after the Ogawa observations, other studies also associated blood oxygenation with T_2* signal changes *in vivo* (Turner *et al.*, 1991a,b; Hoppel *et al.*, 1991). Following the first public presentation of human BOLD *f*MRI results in 1991 (Brady, 1991) (see also Bandettini *et al.*, 1992), a series of studies demonstrated the utility of the BOLD method for noninvasive mapping of human brain function (Kwong *et al.*, 1992; Ogawa *et al.*, 1992; Bandettini *et al.*, 1992; Frahm *et al.*, 1992; Turner *et al.*, 1993). Further refinements in hardware, acquisition methods, and experimental design have contributed to make BOLD signal a powerful tool in the neuroscience armamentarium.

4. Endogenous Blood Flow

Cerebral blood flow can be measured by MRI using a technique that is fundamentally different from those described above, in that the physical mechanism is not dependent upon a paramagnetic species. Following an excitation pulse that modifies the steady-state value of the longitudinal magnetization in an image voxel in a living body, the recovery rate of longitudinal magnetization contains a contribution from magnetically unsaturated blood water that flows into the voxel (Detre *et al.*, 1992). Thus, the degree of magnetic labeling of blood proximal to the image volume can be used to control the flow-dependent recovery rate of longitudinal magnetization and in that way produce images of CBF. This basic method was employed, together with BOLD signal, in the earliest *f*MRI studies (Kwong *et al.*, 1992). Because the functional sensitivity of arterial spin labeling (ASL) is lower than that of the BOLD method, and T_1-weighted acquisition methods are more difficult to apply to large image volumes with the same temporal resolution as T_2 methods, ASL has found more limited application for *f*MRI than BOLD signal. However, the ability to quantitatively measure CBF makes this an important technique in the *f*MRI toolbox.

IV. Physics and Physiology

A. The Physical Basis of *f*MRI

The common *f*MRI methods rely upon the protons of water as the source of signal. In principle, this still leaves a large number of candidate contrast mechanisms for functional brain imaging, including the local water concentration, transverse and longitudinal relaxation rates, and phase of MRI signal. Changes in relaxation rates provide the basis for all common forms of *f*MRI. The mechanisms of NMR relaxation in a biological sample are numerous, but the dominant sources are the interactions between magnetic moments ("dipole–dipole coupling") and between each magnetic moment and the local magnetic field. The local molecular environments within the brain, and the motional processes of water diffusion and exchange between compartments, contribute to the intrinsic MRI relaxation times (T_1, T_2, and T_2*) of brain tissues and fluids. However, various physiological processes associated with local brain activity can also alter relaxation times, and this is the lever that allows *f*MRI.

1. *f*MRI by Transverse Relaxation

Within a heterogeneous volume such as the brain, the magnetic field is not the same at a microscopic level as the large magnetic field applied by the MRI magnet (B_0). The field sampled by a single water molecule depends upon the local environment, due to the *magnetic susceptibilities* of nearby structures and molecules. The presence of a large external field induces small internal magnetic fields within materials. Nearly all molecules in the body are *diamagnetic*, meaning that the orbital motion of the electrons produces a magnetic field that opposes the externally applied field. A few molecules have an unpaired electron in the outer orbital shell; in this case, the intrinsic magnetic moment of the electron, which preferentially aligns with the magnetic field, exceeds the diamagnetic effect produced by the orbital motion. Such materials are *paramagnetic*. Within the brain, paramagnetic species are sparse and can be treated as isolated magnetic field sources that perturb the local field on a microscopic scale.

For T_2 or T_2* contrast mechanisms, the measured signal depends upon the degree of spin coherence within the volume. Because spins precess at frequencies proportional to the local magnetic field ($\omega = \gamma B_0$), nonisotropic field disturbances lead to a loss of coherence and thus to a loss of signal. Paramagnetic molecules are T_2 and T_2* *contrast agents*, since they perturb the magnetic field and alter the signal: more contrast agent leads to less signal. It is simplest to think of this effect in terms of the relaxation rate (e.g., $R_2 = 1/T_2$). Relative to the relaxation rate in the absence of a paramagnetic agent (R_2^0), contrast agent provides additional relaxation that proceeds at a rate proportional to the concentration of agent ([A]) in a single compartment: $R_2 = R_2^0 + R_2^{AGENT} = R_2^0 + K$ [A]. This process is analogous to a concentration-limited chemical reaction, which proceeds at a rate proportional to the less abundant of the two reactants. In the same way, the relaxation rate of water due to agent (R_2^{AGENT}) depends upon how many water molecules sample the magnetic fields produced by the less abundant contrast agent.

One method of perturbing the local field in the brain is injection of *exogenous contrast agent* into the blood stream. Commonly injected contrast agents in human or animal research include gadolinium, dysprosium, and iron oxide. The concentration of contrast agent in an image voxel at any time (t) depends upon the concentration of agent in the blood and the concentration of blood in the voxel, which is CBV (V): $R_2^{AGENT}(t) = K\,[A](t)\,V(t)$. In order to make sense of time-dependent changes in R_2^{AGENT}, it is necessary to vary either the blood concentration of the agent or the blood volume, but not both simultaneously. *Bolus injections* of contrast agents with short blood half-lives are commonly used in clinical settings to measure relative regional CBV by integrating the temporal response of $R_2^{AGENT}(t)$ (Rosen *et al.*, 1989). Such measurements are particularly useful in diagnosing the pathophysiology of stroke (Sorensen *et al.*, 1997; Warach, 2001) and brain tumors (Mathews *et al.*, 1997). In order to measure dynamic changes in CBV, it is necessary to use a *steady-state contrast agent* with a very long blood half-life, such as certain iron oxide agents (e.g., Weissleder *et al.*, 1990; Josephson *et al.*, 1990). In this case, temporal measurements of signal correspond to temporal changes in the blood volume fraction (Hamberg *et al.*, 1996), $R_2^{AGENT}(t) = K\,[A]_0\,V(t)$. By normalizing the changes in relaxation rate to the initial value measured during injection of contrast agent, activation-induced blood volume changes can be converted to percentage CBV: $\Delta V(t)/V(0) = \Delta R_2(t)/R_2^{AGENT}(0)$.

While injection of contrast agent can produce arbitrarily large percentage changes in the signal, BOLD *f*MRI relies upon a weaker *endogenous contrast agent* that is carried in the blood. Like the surrounding blood and tissue, oxygenated hemoglobin is diamagnetic. As hemoglobin traverses the capillaries of the brain, some oxygen dissociates from the hemoglobin to enable the combustion that drives brain metabolism. The deoxygenated form of hemoglobin is a paramagnetic contrast agent (Pauling and Coryell, 1936; Thulborn *et al.*, 1982; Ogawa *et al.*, 1990a). Just like an injected contrast agent, deoxyhemoglobin (Hbr, for reduced hemoglobin) decreases the signal in the brain. Unlike an injected contrast agent, deoxyhemoglobin is always present in the living brain. This means that the MRI signal can either increase or decrease from the basal level depending on whether [Hbr] decreases or increases in response to a stimulus.

a. Compartmentalization. Vascular heterogeneity and the exchange of water between the intravascular and the extravascular spaces complicate a quantitative description of magnetic susceptibility contrast. However, two dominant themes arise out of analytical (Ogawa *et al.*, 1993b; Yablonskiy and Haacke, 1994) and numerical (Fisel *et al.*, 1991; Weisskoff *et al.*, 1994; Kennan *et al.*, 1994; Bandettini and Wong, 1995; Boxerman *et al.*, 1995b) descriptions of magnetic susceptibility contrast in the brain. The first important principle is that intravascular contrast agents produce gradients that extend into the extravascular space, and NMR techniques provide a means to preferentially weight the extravascular signal by vessel size. The second principle is that much of the signal arises from the blood water, rather than the extravascular space, at low levels of blood magnetization, such as low-field BOLD imaging. A large contribution from the intravascular space effectively nullifies vessel selectivity based upon extravascular signal and also leads to prominent vascular artifacts in low-field BOLD imaging.

b. Extravascular signal—spin and gradient echoes. At very high levels of blood magnetization, as produced by injection of an intravascular contrast agent or by deoxygenated hemoglobin at very high magnetic field strengths like 9.4 T (Lee *et al.*, 1999), blood T_2 becomes so short that blood signal disappears at the echo times typically used for *f*MRI. In this case, *f*MRI signal arises solely from water in the extravascular space, due to magnetic field perturbations originating inside vessels. During functional brain activation, changes in blood volume and blood magnetization alter the spatially heterogeneous magnetic field, leading to T_2* (gradient echo) signal changes that have little dependence on the relative distribution of vessel sizes in each parenchymal voxel. Conversely, spin-echo acquisitions are strongly dependent upon the vessel size distribution (Fisel *et al.*, 1991; Weisskoff *et al.*, 1994). In the absence of diffusion, extravascular water molecules would experience spatially dependent but temporally static field distributions, which are refocused by spin echoes. Extravascular T_2 signal changes can be detected only due to water diffusion, which produces an average molecular displacement (L) of about 25 μm during a typical *f*MRI acquisition time ($L = \sqrt{6DT_E}$, $D \sim 3\ \mu m^2/ms$, T_E = 35–40 ms). Within the complicated space surrounding vessels of multiple orientations and sizes, each vessel contributes an induced magnetic field contribution that falls off at a rate inversely proportional to the square of the vessel radius, with a strong angular dependence (Chu *et al.*, 1990). Induced magnetic fields around very large vessels appear almost like temporally static fields for water molecules during a spin-echo acquisition of signal, whereas the rapidly varying field sampled by water around small vessels produces phase variations that cannot be recovered by a spin echo.

Figure 4 illustrates the dependence of extravascular signal upon vessel size, as calculated using a Monte Carlo computer simulation (Boxerman *et al.*, 1995b). Above a vessel radius of about 10 μm, the simulation predicted that gradient-echo signal has a uniform sensitivity to vessel radii, whereas spin-echo signal peaks at radii typical of capillaries. Experimental validation has been demonstrated in a model system (Weisskoff *et al.*, 1994). *In vivo* evidence for MRI vessel size selectivity was provided by a comparison of MRI and histology in normal brain relative to tumors that exhibited angiogenic vessel growth (Dennie *et al.*, 1998).

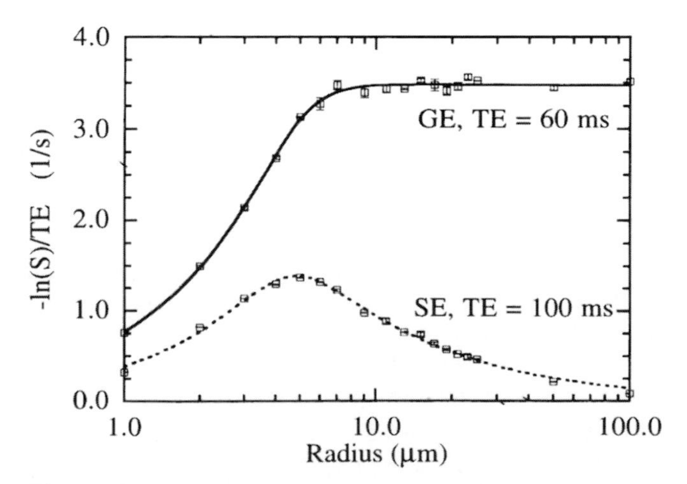

Figure 4 Simulated relaxation rate changes as a function of vessel size due to the extravascular gradients produced by an intravascular contrast agent (Boxerman *et al.*, 1995b). Reprinted by permission of the publisher.

Figure 4 suggests both advantages and disadvantages of spin echoes. By preferentially weighting the signal toward capillaries, spin echoes may improve spatial specificity relative to gradient echoes by eliminating signal contributions from large vessels somewhat distant to the site of activation; this prediction has yet to be verified. On the downside, spin echoes reduce functional sensitivity by eliminating about 75% of the extravascular signal that is seen using gradient echoes (Boxerman *et al.*, 1995b).

c. Intravascular signal. At the low levels of blood magnetization produced by the BOLD effect at field strengths like 1.5 to 4 T, a significant portion of the *f*MRI signal is due to relaxation rate changes in the blood water. By applying external magnetic field gradients to eliminate most of the signal from flowing blood, it has been estimated that blood water produces about two-thirds of the gradient-echo task-induced BOLD signal at 1.5 T (Boxerman *et al.*, 1995a; Song *et al.*, 1996), one-third at 3 T (Buxton *et al.*, 1998b), and no contribution at 9.4 T (Lee *et al.*, 1999). Intravascular signal at 1.5 T produces an even somewhat larger component of spin echo BOLD signal changes (Zhong *et al.*, 1998; Oja *et al.*, 1999). While this intravascular contribution is an important part of BOLD contrast that enables *f*MRI at field strengths like 1.5 T, is also leads to signal amplification in large draining veins, which may reduce the spatial specificity of the method.

2. *f*MRI by Longitudinal Relaxation

a. Exogenous contrast agent. Paramagnetic agents generally affect T_1 as well as T_2. Whereas T_2 shortening leads to a loss of signal by rapid decay of transverse magnetization, T_1 shortening increases signal due to the more rapid recovery of longitudinal magnetization following each excitation. Whereas T_2 or T_2* relaxation can occur at a distance from the intravascular contrast agent, due to the extension of intravascular magnetic field perturbations into the extravascular space, T_1 relaxation requires close contact of agent and water. Although the exchange of water across the blood–brain barrier increases the volume fraction of proton spins that are accessible by the T_1 method (Donahue *et al.*, 1996), all water protons within the brain are accessible by T_2 relaxation processes. Hence, T_1 contrast agents will be ignored in this chapter, in favor of the more sensitive T_2 method.

b. Arterial spin labeling. It is possible to perform *f*MRI without the use of any endogenous or exogenous paramagnetic contrast agents, because the flow (f) of magnetically unsaturated water into a volume contributes to the recovery rate of longitudinal magnetization, $R_1^{APPARENT} = R_1 + f/\lambda$ (Detre *et al.*, 1992), where λ is the blood–tissue water partition coefficient.

In a manner analogous to CBF measurements using tracer-kinetic methods (Lassen and Perl, 1979) employed in autoradiography or PET, *f*MRI uses water as a nearly ideal diffusible tracer. Unlike methods that employ radioactive decay to report the local concentration of tracer, the *f*MRI method provides a small signal ($f \sim 60$ ml/100 mg/min $\sim 0.01\text{s}^{-1}$) riding on top of a large background ($R_1 \sim 1\ \text{s}^{-1}$), so accurate background subtraction and signal averaging are required. Because the decay rate of magnetically labeled water is similar to cerebral blood transit times, accurate CBF measurements require measurements of blood transit times or methods that reduce transit time sensitivity (Alsop and Detre, 1996; Wong *et al.*, 1998).

Figure 5 illustrates how MRI employs magnetic labeling of blood water to measure CBF. The application of RF power at the level of a carotid artery changes the longitudinal relaxation rate in the brain a few seconds later, as the magnetically labeled blood water enters the image volume. By subtracting images acquired with and without prior labeling, the small blood flow contribution to longitudinal relaxation can be isolated, corresponding to the perfusion territory for the single artery (Zaharchuk *et al.* 1999a). This method has numerous variants, which collectively are referred to as either *arterial spin labeling* or arterial spin tagging. The two major subtypes are continuous and pulsed ASL. Continuous ASL applies a long duration of RF power at a fixed spatial location. The use of small labeling coils can prevent saturation of off-resonance macromolecules in the brain (Zhang *et al.*, 1995), an effect that must be controlled when using one coil for RF transmission and reception (Alsop and Detre, 1998). Pulsed ASL applies a spatial label at one point in time within a large proximal tissue volume (e.g., Edelman *et al.*, 1994; Kwong *et al.*, 1995; Kim, 1995; Wong *et al.*, 1998) and avoids significant off-resonance saturation of macromolecules.

The exact relationship of CBF to the signal difference (ΔS) formed between labeled and unlabeled acquisitions depends upon the explicit type of ASL and assumptions about transit times and water exchange across the blood–brain barrier (e.g., Buxton *et al.*, 1998a). A general form is

$$\frac{\Delta S}{S} = 2\alpha\ f/\lambda\ \frac{e^{-t/T_1}}{\varepsilon}. \tag{7}$$

α is the labeling efficiency ($0 \rightarrow 1$). The factor of 2 arises due to the subtraction between inverted and noninverted blood. The exponential factor reflects blood magnetization lost to T_1 decay, due to a combination of blood transit times and delays inserted after labeling in order to reduce blood transit time dependencies (Alsop and Detre, 1996). The factor of ε accounts for the increased level of steady-state magnetization produced by continuous labeling relative to pulsed labeling (ε=1 for continuous ASL, and ε=*e* for pulsed ASL) (Chesler and Kwong, 1995).

B. The Physiological Basis of fMRI

In order to use *f*MRI as a tool, it is essential to have some understanding of the physiology, as well as the physics, of the method. Unfortunately, at the most basic and important level, the physiology is incompletely understood. While it is well known that increased neuronal activity requires metabolic support, and that blood flow provides the substrates for metabolism, holes remain in our descriptions of metabolic requirements and pathways, as well as the subsequent transduction of the hemodynamic response. Of course, these issues are not peculiar to *f*MRI. The link between neuronal activity and blood flow also forms the basis for functional

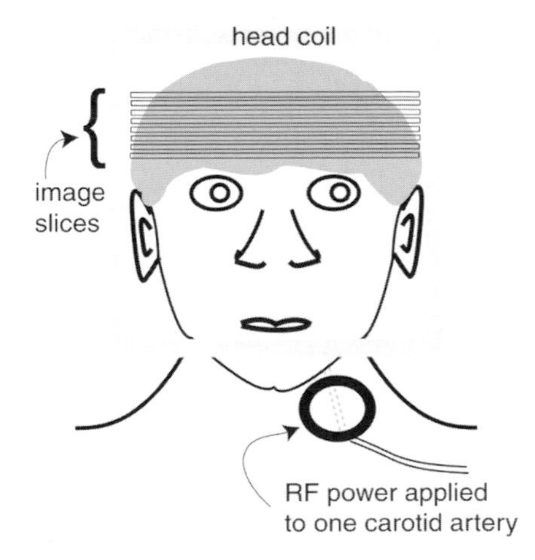

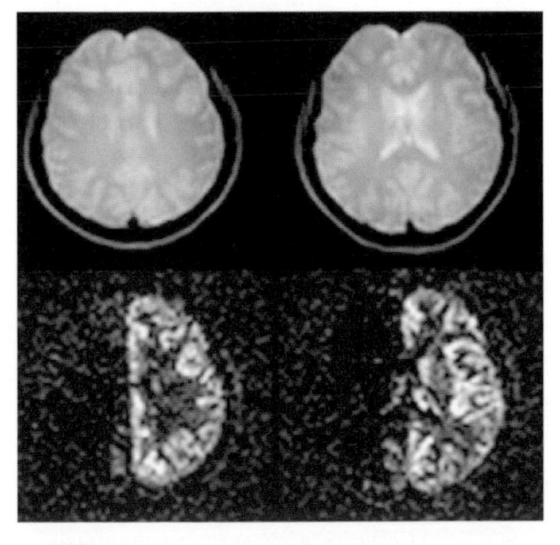

Figure 5 Arterial spin labeling requires the application of RF power proximal to the image volume in order to magnetically label blood water. The effect on the images (right, top row) is subtle, but a subtraction of images with and without labeling reveals a map of blood flow in the region supplied by the labeled artery after signal averaging (adapted from Zaharchuk *et al.*, 1999a).Reprinted by permission of the publisher.

mapping using optical methods and PET. The preponderance of evidence suggests that the neuronally modulated responses of metabolism and blood flow are integral representations of local (pre) synaptic activity (Jueptner and Weiller, 1995).

While a review of the huge body of literature on flow–metabolism coupling is beyond the scope of this chapter, a more modest goal would be a discussion of the mechanistic relationships between blood flow, oxygen utilization, and blood volume as they relate to *f*MRI. From this vantage, as summarized in the right diagram of Fig. 6, neuronal activity is a detached abstraction, and the objective becomes a description of the interrelationships between observable functional quantities that contribute to *f*MRI signal. The discussion will follow the prevalent viewpoint, which is that CBF, CBV, $CMRO_2$, and BOLD signal form an interconnected set of roughly redundant quantities that are coupled through mechanical constraints during normal brain activation.

1. A Framework for BOLD Signal

BOLD signal derives from the local concentration of deoxygenated hemoglobin, which is modulated by several factors. The generator of this paramagnetic contrast agent is metabolism ($CMRO_2$). Blood oxygenation, and hence blood magnetization, depends upon the balance of oxygen flow into and out of a region. The rate of oxygen inflow is proportional to CBF. During functional activation, increased CBF produces a washout of Hbr contrast agent, counteracting the effect of increased $CMRO_2$. A last factor is the local blood volume fraction, which determines the deoxyhemoglobin content of a voxel at any level of blood oxygenation. As blood vessels swell, magnetic fields extend further into the tissue, causing signal loss in the extravascular space.

In order to build a more concrete framework for BOLD contrast, we can employ the approximation that changes in the BOLD relaxation rate scale with changes in deoxyhemoglobin concentrations, $\Delta R_2^{BOLD} = K\,\Delta[\text{Hbr}]$, where K depends upon the magnetic field strength and the sample volume. A BOLD framework is based upon conservation of oxygen mass (Fick's Law), which states that the steady-state unidirectional extraction of oxygen from the blood is the difference between the flow of oxygen into and out of the volume, $F_{O2}^{IN} - F_{O2}^{OUT} = dV_{O2}/dt$. The resulting expression takes a form like the following (Ogawa *et al.*, 1993a):

$$\Delta R_2 = -K[Hbr]_0 \left\{ \frac{\Delta F}{F_0} - \frac{\Delta V}{V_0} - \frac{\Delta M}{M_0} \right\}, \quad (8)$$

F, V, and M refer to CBF, CBV, and $CMRO_2$, respectively, and the subscript "0" indicates baseline values prior to stimulation. BOLD signal changes are positive when the quantity in brackets is positive. The observation of positive stimulus-induced signal changes means that *relative changes in CBF exceed the combined effect of changes in CBV and $CMRO_2$*. The quantity inside the brackets is amplified by the resting-state BOLD relaxation rate, which depends upon the physiology of the baseline state. $[\text{Hbr}]_0$ is proportional to V_0 and M_0 and inversely proportional to F_0. Figure 6, right, summarizes the four variables in Eq. (8) that combine to generate BOLD signal. A few caveats are in order regarding this equation:

1. For simplicity, the equation has been linearized, which is only appropriate for small functional changes.
2. It is not necessary to assume an absolutely linear relationship between ΔR_2 and $\Delta[\text{Hbr}]$. More complete formulations use a linear dependence between ΔR_2 and blood volume, but a slightly superlinear dependence with blood oxygenation, due to intravascular signal contributions (Davis *et al.*, 1998; Mandeville *et al.*, 1999a; Hoge *et al.*, 1999a).
3. The term $\Delta V/V_0$ should actually be replaced by the relative change in total venous hemoglobin. By substituting the

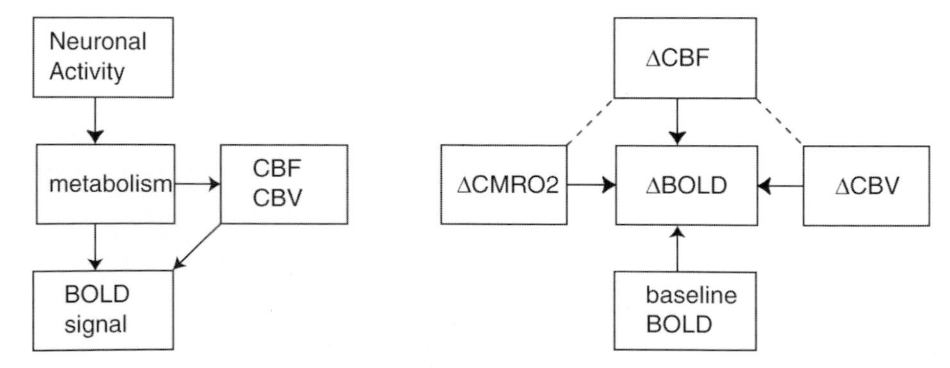

Figure 6 **Left:** A block diagram showing the multistep path to the *f*MRI observables of blood flow, blood volume, and BOLD signal. **Right:** BOLD signal changes result from a combination of changes in CBF, CBV, and $CMRO_2$, together with an amplification factor than depends upon baseline physiology. The dashed lines indicate presumed coupling relationships.

relative change in blood volume, we are assuming that (a) the cerebral hematocrit does not change during a functional response and (b) the relative change in total blood volume serves as an accurate surrogate market for the relative change in venous blood volume.

4. The equation is meant to express steady-state changes, but it is not accurate on the time scale of the cerebral mean transit time (a few seconds). A rapid change in CBF will produce an effect on BOLD signal that is both delayed and dispersed by transit through the vascular bed (refer to Section VI.A, Temporal Resolution).

The framework of Eq. (8) contains no real information about the physiology of the functional response. This is an accountant's view of BOLD signal that does not specify how oxygen is allocated, as long as the total amount is conserved. However, the quantities within the brackets apparently are not free to change arbitrarily during brain activation. The reproducibility of BOLD results across systems and stimuli makes us confront the issue of *coupling*. As indicated by the dashed lines in Fig. 6, right, the common position is that relative changes in CBF, $CMRO_2$, and CBV are all functionally coupled during normal brain activation.

2. CBF and $CMRO_2$

Global (Kety–Schmidt method) and regional (PET) measurements of basal cerebral oxygen and glucose utilization have shown a molar ratio consistently less than 6 ($CMRO_2$/CMRglu ~ 5.5; Mazziotta and Phelps, 1986), suggesting that oxidative metabolism of glucose ($C_6H_{12}O_6$ + $6O_2 \rightarrow 6H_2O + 6CO_2$) is the primary, but not sole, source of basal energy. Tight linear couplings have been shown for regional basal measurements of CBF versus CMRglu (Sokoloff, 1981) and CBF versus $CMRO_2$ (Mazziotta and Phelps, 1986). Such regional relationships are often described as the natural result of long-term developmental adjustments in cerebral physiology enabled by processes like angiogenesis (Villringer, 1999).

During stimulus-induced activation, the brain has more limited means of affecting critical quantities like capillary area and vascular resistance at the local level, so it is not obvious that temporal relationships between such physiological quantities should follow the regional trajectories determined from the baseline (or average) state. The PET measurements of the mid-1980s (Fox and Raichle, 1986; Fox *et al.*, 1988) demonstrated that focal, stimulus-induced changes in oxygen consumption did not, in fact, adhere to previously measured regional relationships with glucose consumption and blood flow. With apparent reference to the regional coupling of CBF and $CMRO_2$, the magnitude of mismatch of those quantities during brain activation was originally labeled as an "uncoupling."

The current viewpoint, based upon both theoretical considerations and experimental data, has evolved to reembrace the idea of CBF–$CMRO_2$ coupling, but along a trajectory different from the regional relationship. Buxton and Frank (1997) provided a compelling description of the essential physiology that leads to a smaller relative change in $CMRO_2$ than in CBF, due to physical constraints on the rate of oxygen delivery during brain activation. Oxygen in the blood is delivered to the brain by diffusion along an oxygen concentration gradient that falls to a few millimeters of mercury in the tissue. Since the brain does not have significant oxygen reserves, blood flow must increase to deliver additional oxygen. An increase in blood flow does not deliver tissue oxygen at the same rate, however, since the reduced blood transit time through the capillaries decreases the oxygen extraction fraction. An increase in capillary surface area would be one possible way of counteracting the reduced extraction fraction. However, all cerebral capillaries are perfused in the baseline state (Gobel *et al.*, 1991; Kuschinsky, 1996); unlike skeletal muscle, the total capillary surface area cannot be significantly altered by capillary recruitment.

Within the Buxton–Frank diffusion-limited model of oxygen delivery, the exact scaling between relative changes in blood flow ($f = F/F_0$) and relative changes in oxygen delivery (m) depends upon the baseline value of the extraction fraction (E_0) and the extent of capillary dilation ($v = V/V_0$): $m = f(1-(1-E_0)^{v/f})/E_0$. Assuming a baseline extraction fraction of 0.4, the net scaling between fractional changes in oxygen delivery and flow is about 1/6 without capillary swelling, as indicated in Fig. 7. The dashed line in the figure assumes a 5% increase in capillary radius for a 60% increase

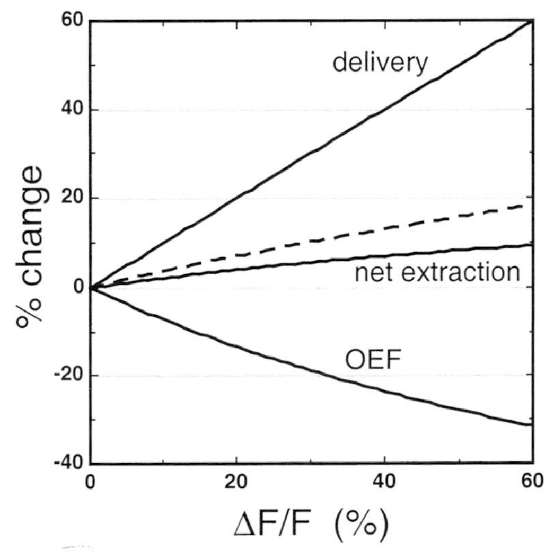

Figure 7 Coupling between CBF and $CMRO_2$ as predicted by a diffusion-limited model of oxygen delivery (Buxton and Frank, 1997). Total oxygen delivery to an area scales with CBF, but the oxygen extraction fraction (OEF) falls due to the decreased resident time of the oxygen in the capillaries. The net effect is an increase in oxygen delivery to the tissue that is smaller on a relative scale than the change in CBF. The dashed line shows the effect of a small increase in capillary diameter.

in blood flow. This small capillary swelling doubles the delivery of oxygen, leading to a mismatch ratio of 1/3.

Empirically, the influence of $CMRO_2$ on BOLD signal can be deduced by comparing the responses of CBF and BOLD signal using stimuli that either increase $CMRO_2$ (e.g., visual stimuli) or have no effect on $CMRO_2$ (Davis *et al.*, 1998; Mandeville, *et al.*, 1996b). Using hypercapnia as a means to modulate CBF without inducing concomitant changes in oxygen utilization (Eklöf *et al.*, 1973), $CMRO_2$ has been shown to increase significantly during focal activation of the human visual cortex (Davis *et al.*, 1998; Hoge *et al.*, 1999a; Kim *et al.*, 1999). By using graded levels of visual stimulus and hypercapnia, a linear coupling was measured between relative changes in CBF and $CMRO_2$ for flow modulations up to 50% (Hoge *et al.*, 1999a). Figure 8 shows empirical measurements in the BOLD–CBF plane for graded levels of hypercapnia and visual stimuli. Assuming CBF and CBV couple identically for these hemodynamic challenges, the reduction of BOLD signal at a given level of CBF is due the influence of $CMRO_2$. Figure 8b demonstrates a linear coupling of $CMRO_2$ to perfusion over this range of CBF.

Within the group of previously mentioned *f*MRI studies, relative changes in flow exceed relative changes in oxygen utilization by a factor of 2 to 3. These empirical data are not inconsistent with the model calculation of Fig. 7 that includes a small degree of capillary dilation, which was not included in the original model.

3. CBF and CBV

Cerebral vascular resistance is defined by its relationship to CBF and the total pressure drop across a vascular bed. In the brain, intravascular pressure drops from mean arterial blood pressure in large arteries to venous pressure in the large veins. The brain increases CBF during normal brain activation by reducing cerebral vascular resistance, corresponding to an increase in CBV.

The physics of fluid flow through a pipe helps us understand the relationship between resistance, volume, and flow. For an ideal Poiseuille (or laminar) fluid, resistance is inversely proportional to the fourth power of the radius. A relative change in blood flow (F) can be produced either by a change in the pressure drop across the pipe (P) or by a change in volume, $F/F_0 = P/P_0\,(V/V_0)^2$. Surprisingly, the simplest possible model—a single pipe—produces a reasonable description of the relative changes in blood flow and volume between hemodynamic steady states. PET data found that hypercapnia-induced changes in blood flow and volume were best fit by the relationship $F/F_0 = (V/V_0)^{2.6}$ (Grubb *et al.*, 1974). Not surprisingly, a single-pipe model ultimately proves to be too simplistic to describe other phenomena. Blood flow and blood volume exhibit different temporal responses (Mandeville *et al.*, 1999b), which cannot occur in a single pipe, and the magnitude of pressure-related changes in blood volume appears to be smaller than predicted by a single pipe (Zaharchuk *et al.*, 1999b)

In order to produce a more realistic (but still simplistic) model, we turn to a series of pipes in the standard resistive model of the brain vasculature, as shown in Fig. 9. Each pipe is a phenomenological representation of the resistance and volume in a portion of the vascular tree. For instance, the resistance in an individual capillary is very high, but the effective resistance of a capillary bed is low due to the large number of parallel capillaries branching from a single arteriole. Small arterioles possess a lion's share of resistance but a mouse's share of volume, and this order is reversed for venules. This allotment of resistance and volume enables local regulation of blood flow.

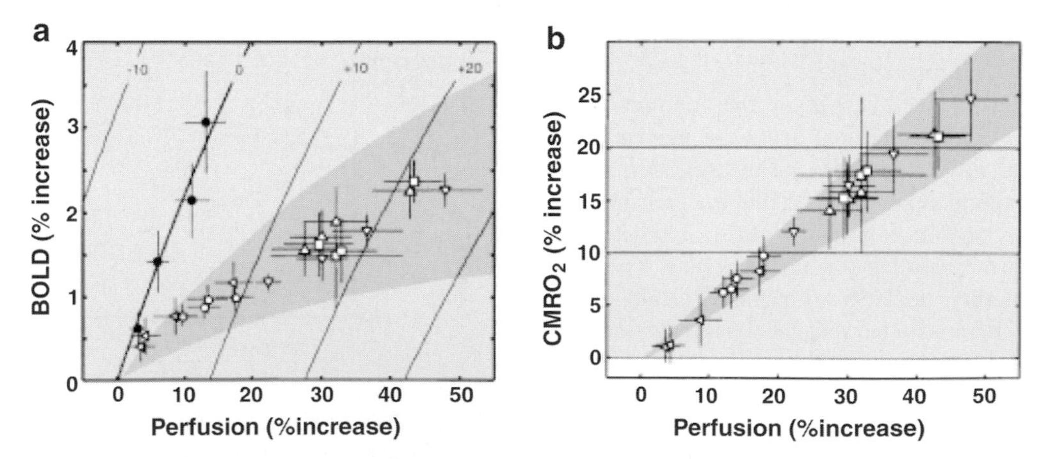

Figure 8 (a) BOLD signal versus perfusion, with solid lines representing iso- $CMRO_2$ contours. A stimulus that does not alter $CMRO_2$ (hypercapnia) traces a trajectory that is clearly different from a variety of visual stimuli (shaded area), which increase oxygen utilization and thereby lower BOLD signal at flow levels matched to hypercapnia. The resulting calculation of changes in oxygen utilization relative to CBF is shown in (**b**). (Hoge *et al.*, 1999a). Reprinted by permission of the publisher.

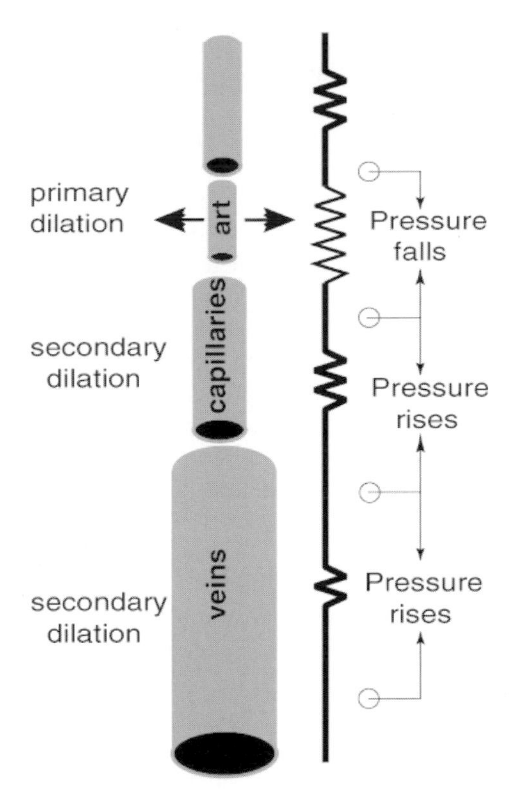

Figure 9 A schematic of vessel dilation assuming that capillaries and veins simply respond to changes in pressure inside the vessels. Because the total pressure drop across the vasculature is fixed, a dilation of arterioles shifts pressure to downstream compartments, causing a pressure-induced dilation in those vessels.

At any point in time, the series of pipes acts like an electrical voltage divider: the pressure drop across any compartment is simply equal to the total pressure drop times the fraction of the total resistance that resides in that compartment (e.g., $P_{art} = P_{TOT}\ R_{art}/R_{TOT}$). While the total pressure drop across the vascular tree is fixed, the pressure drop across any single compartment is not constant during brain activation. As regulating arterioles dilate to satisfy a metabolic demand, the arteriole proportion of total resistance decreases, corresponding to a reduced pressure drop across the arteriole compartment. This extra pressure must be shifted to other compartments, according to the relative resistances in each compartment at any time, in order to maintain a fixed total pressure drop across the entire bed.

Arteriole dilation must be sufficiently large to produce an increase in blood flow through the arteriole compartment even in the face of a reduced driving pressure across that compartment (pressure shifts from arterioles to other compartments). Hence, dilation in the smallest arterioles can be remarkably large (Iadecola *et al.*, 1997)—much greater than predictions of laminar flow that don't account for the pressure change. Conversely, the increase in pressure across capillaries and veins facilitates much of the flow increase in those compartments, so dilations of those compartments are smaller than predictions based upon a pipe model with a fixed pressure drop. Data demonstrate that even capillaries show small changes in radii (Atkinson *et al.*, 1990; Duelli and Kuschinsky, 1993), suggesting that "capillary vasomotion occurs and that capillaries are not rigid tubes" (Atkinson *et al.*, 1990). The change in total blood volume, which is the observable quantity for noninvasive brain mapping, is a sum over compartmental changes, weighted by the respective volume fractions. Large relative arteriole changes occur in a very small blood volume fraction, and modest relative changes in capillary and venous volume occur in a much larger volume fraction.

In this framework, it is possible to write a simple expression for the relationship between relative changes in blood flow and venous blood volume. If capillaries and veins react like an elastic material with a pressure–volume relationship ($V{\sim}P^{1/\beta}$), then it can be shown that relative changes in flow and venous volume are coupled as $F/F_0 = (V/V_0)^{\alpha+\beta}$ (Mandeville *et al.*, 1999b). The exponent reflects contributions from resistance (α=2 for laminar flow) and venous compliance. In an elastic response, flow and volume would change with about the same time constant during activation. In fact, plasma volume reacts more slowly than blood flow for stimuli of relatively long duration (Mandeville *et al.*, 1998). In the absence of a change in hematocrit, it can be shown that the only way to explain such data is by an increase in venous compliance following stimulation (Mandeville *et al.*, 1999b). This suggests that venous vessels do not react like an elastic material, but instead have viscoelastic properties: $F/F_0 = (V/V_0)^{\alpha+\beta(t)}$. Stress relaxation of venous smooth muscle, or "delayed compliance," provides one possible mechanistic explanation for the slow dilation of venous vessels. In excised vessels, a rapid change in pressure leads to an elastic response, followed by a continued slow change in vessel radius (Porciuncula *et al.*, 1964). As an alternative (or secondary) reason for the apparent change in venous compliance, cerebrospinal fluid may slowly redistribute during volume expansion produced by an increase in CBV. Again, this mechanism leads to a time-dependent change in venous compliance.

The preceding discussion follows a point of view developed in the context of Windkessel theory (Frank, 1899; Mandeville *et al.*, 1999b), wherein the temporal evolution of resistance and compliance determine the relationship between venous flow and volume. The modified Balloon model (Buxton *et al.*, 1998c) contains a conceptually similar phenomenological description, in which slow venous volume changes are incorporated by replacing the rate of venous outflow in the original model (Buxton *et al.*, 1998d) with an empirical time constant.

4. Baseline BOLD Relaxation Rate

The final factor that influences the magnitude (but not temporal response) of BOLD signal is the baseline value of the BOLD relaxation rate. At a given field strength, this factor simply reflects the baseline amount of deoxyhemoglobin in an image voxel, which depends upon the local blood oxygenation and volume. The basal blood level of deoxygenated hemoglobin is determined by the ratio of $CMRO_2$ to CBF. The fact that PET experiments have consistently measured a linear coupling between basal metabolism and CBF means that venous blood oxygenation can be considered to be much more constant across the brain than either CBF or $CMRO_2$, which can vary by several factors. Hence, the amplification factor outside the braces in Eq. (8) has a strong regional dependence that primarily reflects CBV. Physiological alterations, such as those produced by anesthesia or hypercapnia, can alter the global scaling of BOLD signal by resetting the ratio of basal $CMRO_2$ to CBF and altering CBV.

5. Caveats

Although the previous description of BOLD signal and related quantities represents a plausible and coherent story of the essential physiology, there are numerous caveats in terms of claiming a quantitative or thorough description of the physiology of MRI signal. As a quantitative description, this formulation has numerous regional dependencies on the blood volume fraction, oxygen extraction fraction, arterial oxygenation, etc. The relative contributions of intravascular and extravascular signal depend upon the pulse sequence, field strength, distribution of vessels, etc. *f*MRI methods that attempt to untangle these complicated dependencies reply upon calibration procedures (Davis *et al.*, 1998; van Zijl *et al.*, 1998) and generally choose parameters in such a way as to minimize unknown or calibrated quantities.

As a thorough description of *f*MRI physiology, there are additional caveats. Clearly, blood flow can be increased without requiring the delivery of additional oxygen for metabolism. Hypercapnia is one such method (Eklöf *et al.*, 1973). Moreover, certain drugs have direct effects on the vasculature, as exemplified by anesthetics. While some anesthetics (e.g., barbiturates) decrease both $CMRO_2$ and CBF, anesthetics like halothane and isoflurane ("vasodilators") decrease $CMRO_2$ but increase CBF at high doses (Michenfelder, 1990). Common to these examples, however, is the delivery of oxygen in excess of demand. A more interesting question is the extent to which the brain has a surplus of oxygen delivery that can buffer against reductions in blood flow. Hypocarbia reduces CBF without an apparent alteration of $CMRO_2$, but task performance and electrical activity are altered in normal subjects (Halgren *et al.*, 1977). Just as a decrease in blood transit time leads to a reduced oxygen extraction fraction during brain activation, the increased transit time during reduced blood flow leads to an increased extraction fraction, a well-known neuroprotective mechanism (Powers, 1991). Although the diffusion-limited model of oxygen delivery appears to describe the essential physiology for normal brain activation, it is not clear that oxygen availability is at the very boundary of hypoxia.

Caveats also surround the relationship between CBF and CBV. If CBF and CBV primarily reflect the arteriole and venous beds, respectively, then it is possible that certain circumstances can lead to dissociation. As an example, cocaine possesses neuroactive and vasoactive properties. Intravenous cocaine administration in rodent models leads to parenchymal blood volume increases in multiple brain regions (Marota *et al.*, 2000), where arteriole dilation and perfusion pressure presumably govern CBV, but large venous structures in the same animals exhibit negative changes in apparent CBV. Thus, dilation of the largest vessels may not be coupled to perfusion in special cases.

6. Flow–Metabolism Coupling

While the basic physics of *f*MRI is in place, as well as an understanding of the interrelationships between BOLD signal and direct contributions like CBF, a significant knowledge gap disconnects neuronal activity from metabolism and thereafter to blood flow. Although these connections have been the focus of basic physiological studies for decades, *f*MRI can serve as an efficient survey tool in order to focus questions on specific brain regions and stimuli, which can then be investigated using the invasive methods of classical neuroscience in animal models. Initial studies of direct comparisons between neuronal activity and *f*MRI signal have been performed using spatially registered acquisition methods (Disbrow *et al.*, 2000; Janz *et al.*, 2001; Logothetis *et al.*, 2001), though much work remains to be done to fully elucidate both the relationships and their underlying mechanisms of action.

V. Sensitivity

*f*MRI is a sensitivity-limited technique. Signal-to-noise ratios are typically 100–200 in BOLD *f*MRI sessions, and stimulus-induced signal changes are a few percent or less. Thus, functional signal changes are often at the level of noise. The issue of detection sensitivity affects almost every aspect of an experiment, including the threshold for identifying subtle changes in local brain activity, the useful spatial resolution for *f*MRI, and the time required for a desired statistical power (hence, the duration of a session or the number of stimuli that will produce useful information in a single session). Defined as the contrast-to-noise ratio per unit time, sensitivity depends upon both the physiology and the physics of a given technique. It is often convenient to think of sensitivity as the product of the relative change in a variable of interest (flow, volume, oxygenation, etc.), which depends upon both the degree of activation and the physiology of the functional variable, times an amplification factor expressing the intrinsic sensitivity per unit change:

$$\text{sensitivity} = (\text{activation}) \times (\text{amplification}). \qquad (9)$$

The amplification factor varies by the functional variable of interest and the measurement technique. For instance, relative changes in CBF exceed relative changes in CBV, but the amplification factor reverses the order of sensitivity for MRI techniques. Although the amplification factor is generally independent of the degree of activation, it can depend strongly upon baseline physiology. Since blood flow, blood volume, and oxygen utilization vary by several factors within the brain, the amplification factor can apply a spatial filter to the pattern of brain activation.

A. SNR and CNR

The SNR and the contrast-to-noise ratio (CNR) each have several definitions in the literature, depending upon whether one is referring to spatial or temporal characteristics of the MRI signal. In order to make nice pictures of anatomy, it is necessary to optimize "anatomical SNR" (aSNR), as described in Eq. (6). aSNR can be defined as the signal in a voxel divided by the standard deviation of the signal across a region without structure, generally selected to be a region outside the brain in air. This is a way of referring to the signal without an arbitrary scaling factor, since noise provides the normalization. Radiologists care strongly about this definition, which determines the quality of a single image. One can also define anatomical CNR as a contrast index between different regions in the brain relative to the noise level. Differentiation of tissue types depends upon the anatomical CNR.

For detecting changes in local brain activity, different definitions come into play. Functional SNR (fSNR) refers to the time-averaged value of signal divided by the temporal standard deviation of the signal: fSNR = $\bar{S}_t/\sigma_t$. Detection sensitivity for *f*MRI is synonymous with the functional contrast-to-noise ratio per unit time (fCNR), or the ratio of time-dependent signal changes (δS_t) to time-dependent noise:

$$\text{fCNR} = \delta S_t / \sigma_t = \text{fSNR} \bullet \delta S_t/\bar{S}_t. \qquad (10)$$

B. Physiological Noise

The anatomic and functional definitions of SNR and CNR can differ significantly due to the temporal nature of the noise, which is defined by the signal structure in the absence of an external stimulus. Some noise is associated with the electronics of the MRI scanner. A significant portion of the noise arises from currents generated in the sample; this noise will be similar for a human subject and a bottle of saline solution and will dominate in low signal settings (such as very high spatial resolution). A last (and more typically dominant) part of the noise arises from physiological processes.

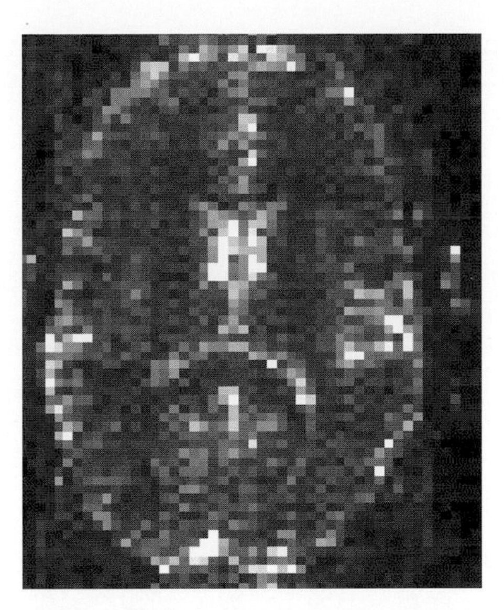

Figure 10 Signal variance without an external stimulus shows structured noise, which is highest at the edge of the brain and around ventricles and major sulci. (Courtesy of Douglas N. Greve and Anders M. Dale).

Sources of physiological noise include bulk motion, microscopic motion from the flow of blood or cerebral spinal fluid, changes in blood oxygenation or volume during the cardiac or respiratory cycles, and changes in the magnetic field distribution associated with body motion or respiration. A typical map of image variance, produced from a subject at rest in the scanner, is shown in Fig. 10.

Physiological noise should include a component that is proportional to the signal. For instance, it is well known that BOLD signal (or noise) is amplified by the resting state blood volume fraction (Ogawa *et al.*, 1993a; Bandettini and Wong, 1997; Buxton *et al.*, 1998d; Mandeville *et al.*, 2001). Since gray matter is roughly twice as vascular as white matter, BOLD signal changes and physiological noise are expected to be about two times larger in gray matter for the same fractional change in blood flow. In fact, resting-state BOLD signal fluctuations are linearly correlated with the magnitude of task-induced signal changes in some experimental settings (Hyde *et al.*, 2001), and a portion of the noise has an echo-time dependence, suggesting a BOLD source (Kruger and Glover, 2001).

To better understand the effect of physiological noise, we can subdivide the noise into the parts that are independent of the signal and correlated with the signal: $\sigma_t = \sqrt{\sigma_0^2 + (\lambda S)^2}$ (Kruger *et al.*, 2001). Physiological noise reduces fSNR relative to aSNR by an amount that depends upon the signal-dependent contribution to the noise:

$$\text{fSNR} = \text{aSNR} / \sqrt{1 + \lambda^2\,\text{aSNR}^2}. \qquad (11)$$

Figure 11 compares measured anatomic and functional signal-to-noise ratios, varied by using two field strengths and changing the flip angle of the RF excitation (Kruger

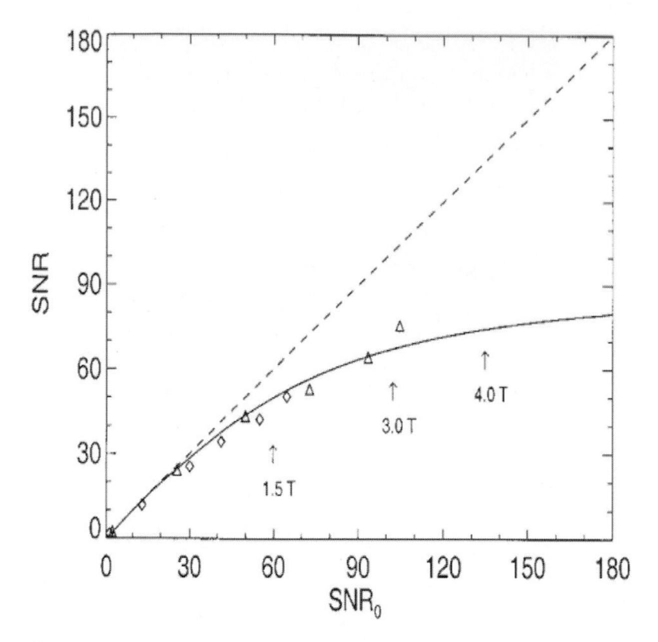

Figure 11 Functional SNR (*y* axis) does not increase linearly with anatomical SNR (*x* axis) due to signal fluctuations associated with physiological processes (Kruger *et al.*, 2001). Reprinted by permission of the publisher.

et al., 2001). The essential point is that functional SNR does not necessarily increase in proportion to the signal strength; the dependence of fSNR changes between regimes of low and high SNR. In the low-SNR regime, fSNR and aSNR increase proportionally, due to increased voxel volume, field strength, coil sensitivity, etc. At a very high level of aSNR, further improvements in the signal will no longer serve to improve the results of functional imaging studies, because physiological noise then dominates other noise contributions. In this high-SNR regime, it is better to trade signal for improved spatial resolution, since the resulting reduction in signal will be partially compensated by a corresponding reduction in the noise.

C. Reducing Physiological Noise

Of course, it is sometimes possible to reject a portion of the noise that finds its way into a functional time series. These strategies can often be quite useful, although it has proven difficult to find a single general strategy that provides significant noise rejection for all applications. The strategies for reducing the noise can be divided into basic types: a) gating the acquisition of images on the cardiac or respiratory cycles and b) retrospective correction based upon recordings of physiological data or temporal subsampling of physiological processes.

A nice example of gating to improve *f*MRI sensitivity was shown in the brain stem, where cardiac gating was found to be essential for detecting sound-related activation of the inferior colliculi (Guimaraes *et al.*, 1998). Gating significantly reduced the variance of signal in the brain stem in this single-slice study, but it produced little effect on the variance of cortical signal. Previous studies demonstrated brain stem velocities of 2–3 mm/s, with a high beat-to-beat reproducibility, that were associated with the cardiac cycle (Feinberg and Mark, 1987; Poncelet *et al.*, 1992).

Physiological gating is an imperfect strategy and may not generalize to all applications. For instance, it is possible in principle to gate on both the cardiac and the respiratory cycles for a single slice, but this method does not generalize to multislice studies using efficient data collection rates. Although gating reduces cardiac-induced motion in some brain regions, it also introduces signal fluctuations from variations in the repetition time (T_R) when the average T_R is not long compared to T_1. In a single-slice study, T_R-induced fluctuations have been retrospectively corrected using the individual values of T_R by fitting for the T_1 of each voxel (Guimaraes *et al.*, 1998). Three-dimension gated acquisitions will be statistically inefficient due to cardiac restrictions on T_R, while multislice studies using contiguous slices should ensure that imperfect slice profiles do not cause cross talk from T_R variations.

Cyclical noise fluctuations related to the cardiac or respiratory cycles can be corrected retrospectively after identifying the phase of each cycle through external monitoring or the use of navigator echoes (Le Bihan *et al.*, 1986), which are low-frequency MRI projections that are collected in addition to the image data. A disadvantage of navigator echoes is that the noise fluctuations need to be temporally oversampled, a criterion that is generally incompatible with multislice coverage of large brain volumes. Thus, external monitoring would seem a preferable general approach for identifying cardiac and respiratory noise.

The effectiveness of reducing physiological noise by such methods is hampered by the fact that the noise spectrum cannot generally be characterized by a few discrete frequencies. A major component of temporal autocorrelation is contained at very low frequencies (~0.1 Hz) unrelated to the cardiac or respiratory cycles (Weisskoff *et al.*, 1993). Furthermore, undersampled cardiac or respiratory cycles produce noise that can alias across the entire frequency spectrum. Noise filters can increase *f*MRI sensitivity only by avoiding frequencies in the stimulus-induced response. Complicating matters further, physiological noise is spatially variant (e.g., Wowk *et al.*, 1997). Ongoing research is attempting to determine methods to characterize the noise, both for improving sensitivity and for improving the accuracy of statistical tests applied to temporally correlated noise (e.g., Purdon *et al.*, 2001).

D. Optimizing the Signal

In terms of detection sensitivity, functional CNR is clearly the quantity of ultimate interest for functional brain imaging. Since percentage changes in signal are generally beyond our control (excluding the choice of imaging tools—BOLD, CBV, or ASL), optimization of sensitivity boils down to an optimization of the signal-to-noise ratio. While anatomic SNR is an imperfect index of functional SNR, it is at least a monotonic index that is independent of bulk motion and easy to measure. In practice, *f*MRI physicists and engineers spend significant effort trying to provide high anatomic SNR and good temporal stability from the MRI scanner. Equation (6) and the related discussion address the quantities that contribute to aSNR.

E. Sensitivity of BOLD fMRI

Since BOLD signal is the most common basis for human *f*MRI studies, it is important to understand some of the practical issues that affect the detection sensitivity of this method. Changes in the concentration of deoxyhemoglobin produce changes in the transverse relaxation rate (ΔR_2 or ΔR_2*) which are small, so we can approximate the fractional signal change ($\exp\{-T_E\,\Delta R_2^*\} - 1$) as simply $-T_E\,\Delta R_2$:

$$\mathrm{CNR}_{\mathrm{BOLD}} = \mathrm{fSNR}\,(-T_E\,\Delta R_2^*). \qquad (12)$$

As an example, a robust BOLD activation at 1.5 T might produce $\Delta R_2^* = -0.5\ \mathrm{s}^{-1}$, which corresponds to a 2% signal increase using an echo time of 40 ms (i.e., $-T_E\,\Delta R_2^* = 0.02$). With a signal-to-noise ratio of 100, the contrast of the BOLD method would be only twice as large as the level of the noise! This dictates the commonly used method to increase statistical power through temporal averaging. Of course, signal changes in large veins in the vicinity of an activated brain region can be much larger. Medium to large veins often exhibit 5 to 10 times larger signal changes than a more representative section of tissue, even at high fields like 4 T (Gati *et al.*, 1997).

The best choice of echo time for the BOLD method occurs around T_2* (or T_2 using spin echoes). From Eq. (12), the percentage signal change increases linearly with echo time, whereas SNR decreases as a mono exponential function of the echo time (SNR $\propto e^{-T_E R_{2^*}}$). The product of these terms is a gamma function in the form xe^{-x}, which is optimized at $x = T_E R_2^* = 1$. The exact choice of echo time is not critical, since an echo time of $T_2^*/2$ decreases sensitivity by only about 20% with respect to the value at T_2*. Assuming physiological noise contains a BOLD contribution that varies with the echo time, one can show that the effect of this noise should be to slightly flatten the gamma function of echo time, so that it falls off more slowly than predicted for $T_E < T_2^*$. However, numerous empirical studies have shown that the echo time dependence roughly follows the expected shape (Menon *et al.*, 1993; Gati *et al.*, 1997; Kruger *et al.*, 2001).

There is generally good reason to set T_E on the low side of T_2*. When using gradient echoes, T_2* can be regionally quite inhomogeneous, due to macroscopic magnetic susceptibility artifacts around air–tissue interfaces (e.g., sinuses) that shorten T_2*, producing signal dropout. The degree of signal incoherence produced by magnetic nonuniformity scales with the echo time, so reducing T_E can have a positive effect on regions with signal dropout. However, the effectiveness of manipulating the echo time is often limited, so more sophisticated strategies are generally required to recover signal in regions with bad magnetic susceptibility artifacts, as described at the end of this section.

Although an entire chapter in this volume is dedicated to high-field imaging, a discussion of BOLD sensitivity would be incomplete without some mention of the advantage of higher fields, which represent a "double win" for BOLD *f*MRI in principle. Relative signal changes (i.e., ΔR_2*) increase in a roughly linear manner between 0.5 and 4 T (Gati *et al.*, 1997), and anatomic SNR also increases linearly with field strength [Eq.(6)]. If not for the increase in physiological noise, as summarized in Fig. 11, BOLD sensitivity might increase quadratically with field. Unfortunately, most *f*MRI experiments operate in an SNR regime in which physiological noise dominates other sources. Nevertheless, even a linear improvement in sensitivity with magnetic field strength makes the pursuit of higher fields worthwhile, and extra signal that is not useful for sensitivity can be invested in spatial resolution. However, a major concern for very high fields like 7 T is that susceptibility artifacts will force compromises on the echo time or the use of multi shot sequences, leading to a loss of some of the high-field benefit. Clever acquisition and postprocessing strategies may obviate this shortcoming, and several groups are pursuing such strategies.

F. Sensitivity of CBV fMRI

Although the BOLD and CBV methods of *f*MRI derive from different physiological sources, the underlying physics of the two methods is essentially the same. In either case, changes in the local concentration of contrast agent (deoxyhemoglobin for BOLD signal or an exogenous agent for the CBV method) cause perturbations in the local magnetic field. However, the use of an exogenous contrast agent presents the experimenter with another degree of freedom—the dose of contrast agent—that alters the characteristics of the signal and changes the amplification factor for T_2 and T_2* *f*MRI.

A simple description of *f*MRI signal as a function of time (t) is $S(t) = S_0 \exp(-T_E R_2^{\mathrm{TOTAL}}(t))$ The present of a contrast agent provides an additional source of signal relaxation, $R_2^{\mathrm{TOTAL}}(t) = R_2^{\mathrm{AGENT}}(t) + R_2^{\mathrm{OTHER}}$. Repeating Eq. (12), and then reformulating the expression, the CNR for endogenous (BOLD) or exogenous contrast agent is

$$\mathrm{CNR}_{\mathrm{T2}}(t) =$$

$$\left\{S_0 e^{-T_E R_2^{\mathrm{OTHER}}}\right\}\left\{T_E R_2^{\mathrm{AGENT}}(0) e^{-T_E R_2^{\mathrm{AGENT}}(0)}\right\}\left\{\frac{\Delta R_2(t)}{R_2^{\mathrm{AGENT}}(0)}\right\}. \qquad (13)$$

The first term on the right side of this equation is an overall scaling factor that is the same for the BOLD and CBV methods. The last term represents the degree of activation, as reflected through the relative change in the transverse relaxation rate. For CBV-weighted signal, the last term reduces to the percentage change in CBV. For BOLD signal, the last term is related to the fractional change in deoxyhemoglobin. In fact, relative changes in relaxation rates associated with CBV and deoxyhemoglobin appear to be quite similar in magnitude, although opposite in sign, during prolonged stimulation (Mandeville *et al.*, 1999a).

The essential difference between BOLD and the CBV methods, in terms of sensitivity, is summarized by the second term on the right side of Eq. (13). This term depends only upon baseline values of blood magnetization and volume. Between the extreme limits of no contrast agent (maximum SNR, but no means of providing functional signal changes) and an infinite amount of contrast agent (SNR = 0) lies an optimal compromise, in which the presence of contrast agent reduces baseline SNR to e^{-1} of the value without contrast agent (i.e., $T_E R_2^{AGENT} = 1$). At common magnetic field strengths, BOLD imaging lies far below the optimal signal attenuation. The resting value of the BOLD relaxation rate can be empirically measured by combining BOLD and CBF (or CBV) measurements during a functional perturbation that does not alter $CMRO_2$ (Davis *et al.*, 1998; Mandeville *et al.*, 1999a). Although the BOLD relaxation rate varies by location within the brain, a typical value at 1.5 T is about $2s^{-1}$ (Davis *et al.*, 1998). Using an echo time of 40 ms, the amplification factor is about a factor of 5 smaller than optimal, as shown in Fig. 12. When using an injected contrast agent, the dose can be adjusted to the optimal value, the peak of the curve in Fig. 12 (Mandeville *et al.*, 1998).

Empirical measurements of the sensitivity of CBV-weighted imaging produce reasonable agreement with the estimates of Fig. 12. In anaesthetized rodent studies, the empirical sensitivity enhancement provided by the use of iron oxide contrast agent is a factor of about 6 at 2 T (Mandeville *et al.*, 1998, 2001) and a factor of about 2.5 at 4.7 T (Chen *et al.*, 2001). Recent comparisons in awake macaques found an increase of about 5 at 1.5 T (Vanduffel *et al.*, 2001) and 3 at 3 T (authors, unpublished data, see Fig. 15). Theoretical estimates, as well as extrapolation of empirical results, suggest that the BOLD method will achieve sensitivity comparable to the use of exogenous contrast agent at a field of 7–10 T.

At magnetic field strengths up to 4.7 T, the *f*MRI landscape in animal models has been completely changed by iron oxide contrast agents, which provide incomparably better results than BOLD signal. Whereas *f*MRI once focused on proof of principle for drug investigations in animal models, pharmacological *f*MRI has now become a valuable tool for assessing whole brain function following drug stimuli. Whereas the drawbacks of BOLD imaging have been apparent in a difficult model like the awake, behaving macaques at low field, iron oxides make *f*MRI in this model quite straightforward. Figure 13 provides an illustration of the significant advantages provided by the use of exogenous contrast agent relative to low-field BOLD imaging. The results in the awake, behaving macaque have demonstrated that a very large number of studies (>20) can be performed in individual animals without apparent adverse health effects (Vanduffel *et al.*, 2001, and authors).

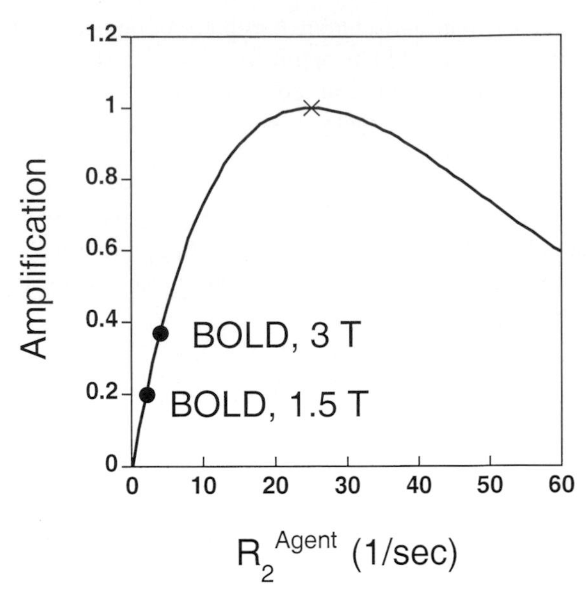

Figure 12 *f*MRI sensitivity using the T_2 or T_2* methods depends upon an amplification factor related to the resting state blood magnetization and volume. The BOLD effect at low magnetic fields is well below the optimal point, which can be reached either by using very high magnetic field strengths or by injecting contrast agent.

While the number of animal *f*MRI studies will continue to expand, the most interesting question is whether iron oxides will become available for human *f*MRI, with clinical applications as the obvious target. The contrast agents that are currently in clinical trials for other applications (Taylor *et al.*, 1999; Sharma *et al.*, 1999) will need to be approved at higher doses in order to realize the compelling advantages for human *f*MRI.

G. Sensitivity of fMRI Using ASL

While ASL has proven to be a useful tool for quantifying CBF and for studying *f*MRI contrast mechanism, it has not yet become a common method for brain mapping, due to reduced sensitivity and volume coverage per unit time relative to T_2*-weighted imaging. A general problem is that ASL is often desired to obtain quantitative CBF, which requires the insertion of time delays in the imaging protocol in order to reduce the dependence on vascular transit times (Alsop and Detre, 1996; Wong *et al.*, 1998). These delays reduce

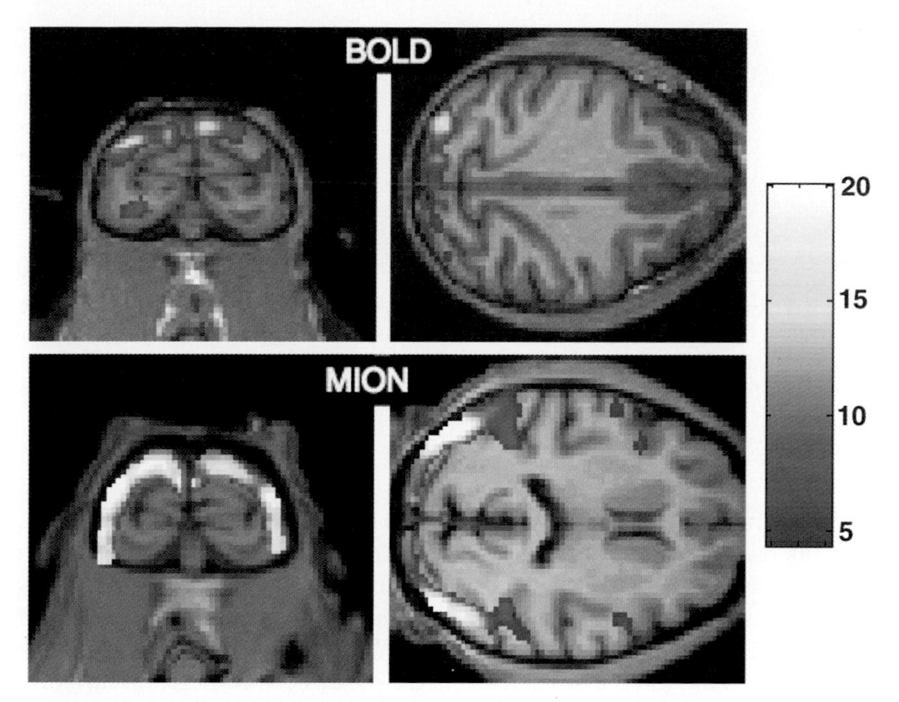

Figure 13 Representative functional maps obtained in awake, behaving macaques during single MRI sessions using either BOLD imaging (**Top**) or iron oxide contrast agent (**Bottom**) at a field strength of 1.5 T. Coronal (**Left**) and horizontal (**Right**) projections through primate V1 shows statistical T scores registered onto a T1-weighted anatomic image. The statistical test compared visual presentation of random dot patterns to a gray uniform field (figure courtesy of Wim Vanduffel and Denis Fize; see Vanduffel *et al.*, 2001).

detection sensitivity, due to T_1 decay of the label. Thus, sensitivity and accuracy produce conflicting requirements for ASL.

In comparing the intrinsic sensitivity of ASL with BOLD signal, it is important to keep in mind that the flow-dependent "ASL signal" that is commonly discussed in the literature is actually a small difference between signals acquired with and without proximal labeling of blood. In terms of the unsubtracted signal (S) and the ASL difference signal (D), the CNR per unit time is

$$\mathrm{CNR}_{\mathrm{ASL}} = \frac{\delta D}{N} = \frac{S}{N}\frac{D}{S}\frac{\delta D}{D} = \mathrm{fSNR}\frac{D}{S}\frac{\delta F}{F}. \quad (14)$$

The ASL-induced signal asymmetry (D/S) is given by Eq. (7). In order to compare the sensitivities of ASL and BOLD signal, imagine the most favorable hypothetical scenario for ASL. Neglecting T_1 decay of the label, assuming perfect labeling efficiency of the blood water ($\alpha = 1$) using the CASL method, and using typical values of T_1 (1 s) and basal CBF for human brain ($f \sim 0.01\ \mathrm{s}^{-1}$), $D/S \approx 2fT_1 = 0.02$. The use of a very short echo time for the ASL method can produce an SNR that is greater than the BOLD SNR by a factor of about 2.5, due to the use of $T_E = T_2^*$ for the BOLD method. From Fig. 6, a 50% change in CBF corresponds to about a 2.5% change in BOLD signal at 1.5 T. Putting all the terms together and comparing Eq. (14) (ASL) with Eq. (12) (BOLD), ASL signal changes can be about as large as BOLD changes at 1.5 T under the most favorable ASL conditions. In fact, BOLD-sized signal changes can be obtained using continuous labeling with a short postlabeling delay in a plane adjacent to a single-image slice obtained by short-echo time spiral imaging, albeit at half the temporal resolution of the BOLD method (Wong *et al.,* 2001). The conclusion is that the sensitivity of ASL falls short of BOLD signal even under the most favorable ASL conditions. Figure 14 shows a typical comparison at 1.5 T using interleaved BOLD and ASL acquisitions. ASL statistical scores are lower than those for the BOLD method, and the volume coverage of brain was limited by the ASL technique.

While ASL loses out to BOLD signal in terms of sensitivity and volume coverage, there are some applications to which it might be better suited, excluding the obvious cases when one actually requires CBF. There is evidence that ASL signal may be better localized than BOLD signal to the primary site of activation (Luh *et al.,* 2000). Moreover, as a T_1 method, ASL can in principle avoid the T_2^* image artifacts associated with magnetic susceptibility interfaces (see next section). This advantage can be expected to become more significant at high fields. ASL sensitivity will increase with field strength, due to increased signal strength [Eq. (6)] and longer T_1.

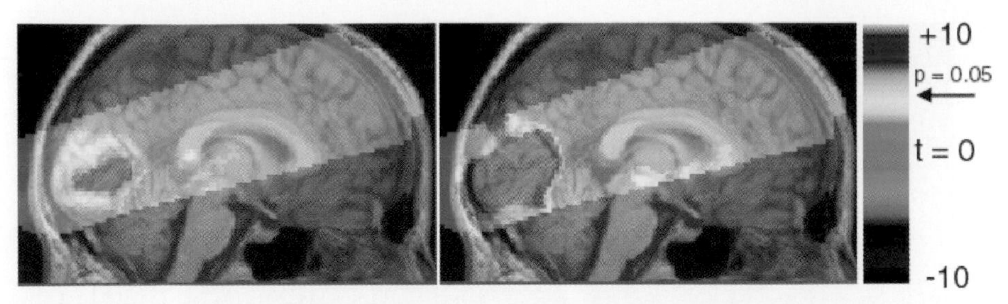

Figure 14 Average flow (left) and BOLD (right) functional maps across four subjects using an interleaved ASL/BOLD acquisition. The colored region designates the overlapping functional volume within the subjects, and the scale shows a *t* statistics. ASL acquisition used the QUIPPS II method (Wong *et al.*, 1998) with a 1200-ms inversion delay, an 800-ms post labeling delay, and a 20-ms gradient echo time. Figure courtesy of Richard D. Hoge.

One avenue that offers significant potential for improved ASL sensitivity is reducing signal from static tissue. Because the ASL signal difference depends only upon T_1 recovery, changes in proton density and transverse relaxation only add noise to the measurements. By collecting signal during inversion recovery as the static tissue passes through the null point, the sensitivity of three-dimensional ASL is significantly enhanced (Ye *et al.,* 2000). Static tissue nulling has been implemented in a two-dimensional single-slice ASL method, providing reduced physiological noise and a two-fold improvement in the *f*MRI sensitivity of ASL at 3 T (Duyn *et al.,* 2001).

H. Susceptibility Artifacts

Magnetic susceptibility image artifacts lead to regionally heterogeneous T_2^* sensitivity, a problem that becomes progressively worse with higher magnetic field strengths. Magnetic field gradients extend into brain tissue near magnetic susceptibility interfaces, particularly air–tissue boundaries around sinuses. Such gradients are generally nonlinear and are too steep to be corrected by static shimming of the magnetic field. When gradients occur in the plane of a two-dimensional image, the effects on spatial encoding primarily lead to distortion or blurring. When gradients occur across an image slice, the result is signal loss. Spin echoes (T_2) provide one means of recovering signal near susceptibility interfaces. However, spin echoes sacrifice functional sensitivity by a factor of 3–4 relative to gradient echoes (Bandettini *et al.,* 1994; Boxerman *et al.,* 1995b). When using gradient echoes (T_2^*), slice oversampling is the only way to recover signal loss while also sampling signal in magnetically well-behaved regions. The most temporally and statistically efficient encoding strategies employ multiple acquisitions with variable slice compensation gradients following each RF excitation (Yang *et al.,* 1997; Song, 2001).

Since ASL relies upon T_1 rather than T_2^*, that method in principle can avoid susceptibility artifacts altogether by using spin echoes or very short echo times. Relative to gradient echo BOLD imaging, the use of ASL is much like the use of the spin-echo BOLD method, in that signal is gained in some regions while sacrificing sensitivity in other regions relative to the gradient-echo BOLD method.

When using exogenous contrast agent in animal models, the dose of contrast agent is an extra degree of freedom not offered by the BOLD method. In order to minimize susceptibility artifacts, gradient echo times can be shortened, or spin echoes can be employed. Either strategy avoids a loss of sensitivity only by increasing the dose. For spin-echo imaging, the three to four times smaller value of ΔR_2 relative to ΔR_2^* can be compensated by using three to four times more contrast agent. For gradient-echo imaging, sensitivity can be maintained by increasing the dose in rough proportion to the inverse of the echo time (Mandeville *et al.,* 1998).

VI. Resolution

A. Temporal Resolution: Data Acquisition and Vascular Response

The temporal response of *f*MRI can be characterized in terms of either the image collection rate or the temporal response of the cerebral blood supply following a stimulus. In the big picture, T_1 recovery of signal and the vascular response time are remarkably compatible and set a natural time scale of a few seconds for *f*MRI sampling. Because the vascular response enforces a low-pass filter on inferences of neuronal activation obtained using *f*MRI, the nature of the response has implications for experimental designs.

1. Rapid Spatial Encoding

As suggested in the brief description of MRI spatial encoding, image data collection can be performed in many ways, using transverse magnetization from a single excitation or multiple excitations. It is generally desirable in most applications to employ methods that encode an entire image following a single excitation. Two-dimensional single-shot image acquisitions are categorized as echo-planar imaging (EPI) (Mansfield , 1977). Because an entire image is collected on a time scale of tens of milliseconds, EPI effectively freezes intra image bulk motion and physiological fluctuations. The long acquisition times provide an intrinsically high SNR per unit time at the price of image artifacts, due to phase errors that accumulate during the long acquisitions and lead to distortion, blurring, and signal dropout near magnetic susceptibility interfaces.

Although images with a given field of view and resolution can be constructed from any *k*-space trajectory with a sufficient sampling density and excursion length, the order of data collection affects the way in which artifacts propagate throughout an image. The direction of the least rapid phase progression is generally called the phase encoding direction. In a rectilinear sampling order, frequency offsets from chemical shifts (e.g., fat) and magnetic field nonuniformity will primarily cause shifts in the low frequency (phase-encoding direction) in single-shot imaging, leading to distortion. In a spiral ordering of data samples, the phase-encode direction is radial, so off-resonance frequencies lead to blurring.

Figure 15 shows *k* space trajectories for a rectilinear EPI acquisition (in this case, using a sinusoidal x gradient with linear sampling in time) and a four-shot spiral acquisition. In principle, the rectilinear EPI method should have a somewhat smaller spatial point spread function but more geometrical distortion, whereas the spiral method oversamples the center of *k* space and thus has somewhat better temporal stability, particularly for multi shot acquisitions. In practice, the single-shot methods offer comparable results for *f*MRI detection sensitivity (Yang *et al.,* 1998). The spiral method enables significantly smaller echo times, since acquisition begins in the center of *k* space. Short echo times are particularly advantageous for ASL in order to increase SNR and eliminate unwanted T_2 sensitivity. Both data acquisition methods resample onto Cartesian coordinates for image reconstruction (Jackson *et al.,* 1991), with spiral reconstruction taking somewhat longer due to the need to resample in two dimensions.

In either of the basic encoding strategies shown in Fig. 15 the required gradient strengths increase proportionally with the field of view at constant resolution, and the gradient slew rates increase as the square of the field of view. Thus, targeted acquisitions using a small field of view (Δy for EPI, $\Delta x = \Delta y$ for spiral) can achieve higher resolution for the same acquisition time, if this can be supported by the SNR. "Variable bandwidth" rectilinear EPI can be implemented using trapezoidal gradients (possibly including sinusoidal ramps, to eliminate slew rate discontinuities) in order to flexibly switch between gradient and slew-rate limitations on the minimum acquisition time. For collection of $N \times N$ voxels in a time T, spiral trajectories often employ the form $k = k_0\ (t/T)^{\beta}\ \exp(i\pi N\ (t/T)^{\beta})$ A uniform sampling density ($\sim (1/k)^{2\beta-1}$) would be obtained by $\beta = 1/2$, but then the slew rate at the center of *k* space is a highly limiting factor for the total acquisition time. At values of β greater than about 3/4, gradient and slew limitations both occur at the edges of *k* space. For a flexible spiral implementation that can interactively calculate gradient trajectories, refer to Glover (1999b).

2. MRI Sampling Rate

Using EPI methods for spatial encoding, the acquisition time for a complete image following a single RF excitation is typically 20–40 ms depending upon the field of view, number of resolution elements, and system gradient capabilities. Depending upon the choice of echo time and other details of the MRI pulse sequence (e.g., additional gradients to crush residual transverse magnetization), image acquisition rates

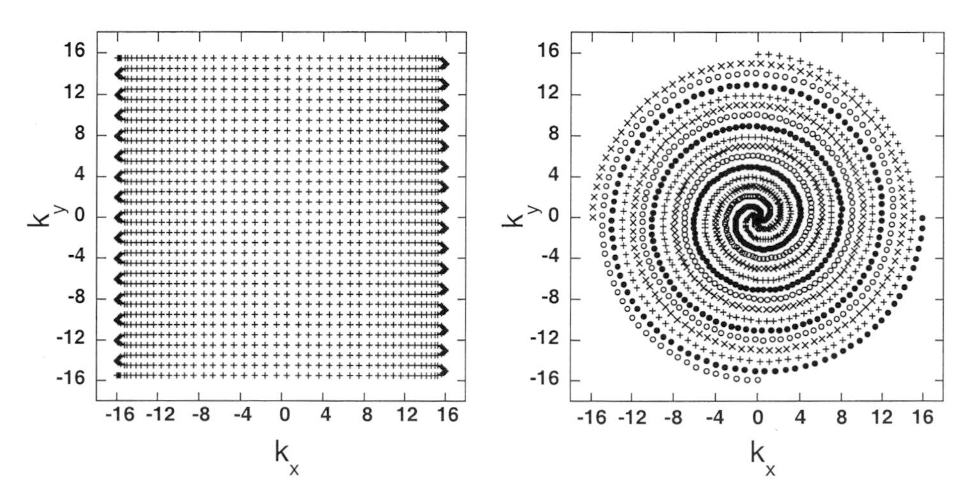

Figure 15 **Left:** k-space trajectories for single-shot EPI with a sinusoidal x gradients and a constant sampling rate. **Right:** Four-shot spiral trajectories using $\beta = 4/5$ (see text).

can approach 20 Hz. In a typical *f*MRI experiment, these rapid acquisition rates are employed for multi slice imaging. For instance, 30–40 slices can be covered in 2 s with an image rate of 15–20 Hz, while providing for about 80% recovery of the longitudinal magnetization between successive excitations of a single slice.

The number of slices (or phase-encode steps in the third dimension) can be reduced to provide more rapid sampling of each slice. This would be desirable, for instance, in studies that attempt to quantify and/or remove sources of physiological noise. However, most routine applications employ multi slice volumetric acquisitions with repetition times somewhat greater than T_1, because there is little penalty in the SNR per unit time for such values of T_R (Fig. 3).

3. Dynamics of the Vascular Response

Following sensory stimulation in animal models, invasive techniques like confocal microscopy and laser-Doppler flowmetry show that arteriole diameters and blood flow begin to respond within the first second (Woolsey *et al.*, 1996). Changes in blood flow and volume are intimately related through conservation of mass (*dV/dt* is a component of flow), and these quantities change concurrently, but with different magnitudes (e.g. Mandeville *et al.*,1999b).

Blood transit through the vascular bed contributes delay and dispersion in *f*MRI signal changes, due to the nature of the MRI reporting techniques. BOLD signal reports the wash-out of deoxyhemoglobin; as such, changes in BOLD signal are delayed with respect to changes in CBF measured using an instantaneous Doppler reporting method (Marota *et al.*, 1999). Arteriole spin labeling reports the wash-in of labeled blood water, and so ASL signal changes are delayed in time with respect to true changes in blood flow (Silva and Kim, 1999). Hemoglobin is deoxygenated in the capillaries, so the delay and dispersion of BOLD signal with respect to the true vascular response will depend upon the capillary and venous transit times (which scale inversely with CBF across species). The extraction fraction for capillary blood water is only slightly below unity at normal flow rates (Eichling *et al.*, 1974), so the delay and dispersion of ASL signal with respect to true changes in CBF will depend upon the arterial and capillary transit times. Because capillary–venous transit times for deoxyhemoglobin are longer than arterial–capillary transit times for labeled blood water, the BOLD impulse response measured in humans is delayed with respect to the impulse response for ASL by about 1 s (Liu *et al.*, 2000). Fits of the BOLD impulse response function in humans generally include a delay of about 2.5 s between onsets of the stimulus and the BOLD response (Boynton *et al.*, 1996); a portion of this delay can be attributed directly to the BOLD technique.

In the few seconds following a stimulus prior to the positive BOLD response, numerous imaging studies have reported the existence of a small "initial dip" in BOLD signal in a subset of activated voxels (Menon *et al.*, *et al.*, 1995; Hu *et al.*, 1997; Kim *et al.*, 2000). This should correspond to an initial increase in deoxyhemoglobin prior to the more robust decrease that provides the typical mapping signal for *f*MRI. Although this response has been reported by optical spectroscopy and interpreted as an increase in oxygen utilization prior to the onset of flow (Malonek and Grinvald, 1996), the details of light propagation through tissue are imperfectly understood, leading to uncertainties in the separation of optical chromophores (Lindauer *et al.*, 2001; Jones *et al.*, 2001). For *f*MRI, this feature of the signal is small, undetectable in some models even at high magnetic field strengths (Marota *et al.*, 1999; Silva *et al.*, 2000), and generally not present in time courses that include all image voxels (e.g., Dale and Buckner, 1997; Vazquez and Noll, 1998; Birn *et al.*, 2001).

During prolonged stimulation, BOLD signal and CBF generally reach plateau values after 5–6 s (De Yoe *et al.*, 1994). Transient features of the neuronal response can produce a variety of temporal features in *f*MRI signals, including postonset overshoots or post offset undershoots in CBF (Hoge *et al.*, 1999b) or habituation during prolonged stimulation. However, sustained responses of CBF and BOLD signal have been demonstrated for up to 20 min for a variety of stimuli (Bandettini *et al.*, 1997), indicating that the magnitude mismatch between changes in CBF and $CMRO_2$ is not a transient uncoupling.

One of the most robust transient features of BOLD signal, noted during the first human *f*MRI experiments (Kwong *et al.*, 1992), is a poststimulus undershoot that is not present in ASL-measured CBF at the same magnitude. The spatio temporal characteristics of the undershoot have been extensively explored in human visual cortex (Kruger *et al.*, 1996; Fransson *et al.*, 1998; Chen *et al.*, 1998). Experimental results in anesthetized rodents have shown that the CBV response contains a component that evolves slowly during and after a stimulus, leading to a flow–volume temporal mismatch with the proper magnitude and time course to explain the poststimulus undershoot (Mandeville *et al.*, 1998, 1999a, b). These same features of the signal have been measured in awake, behaving primates (Fig. 16). Although CBV reacts as rapidly as BOLD signal during the first few seconds following a short stimulus, prolonged stimulation produces a slow evolution of CBV that does not reach plateau by 30 s.

Several interesting questions remain about the temporal evolution of the hemodynamic response during normal brain activation. MRI contrast agents provide a marker of plasma volume, whereas optical imaging experiments have suggested that oxyhemoglobin and total hemoglobin undershoot after stimulation (Royl *et al.*, 2001), in contradiction to the measured MRI response of the plasma. Moreover, the nature of the response just after the onset of stimulation,

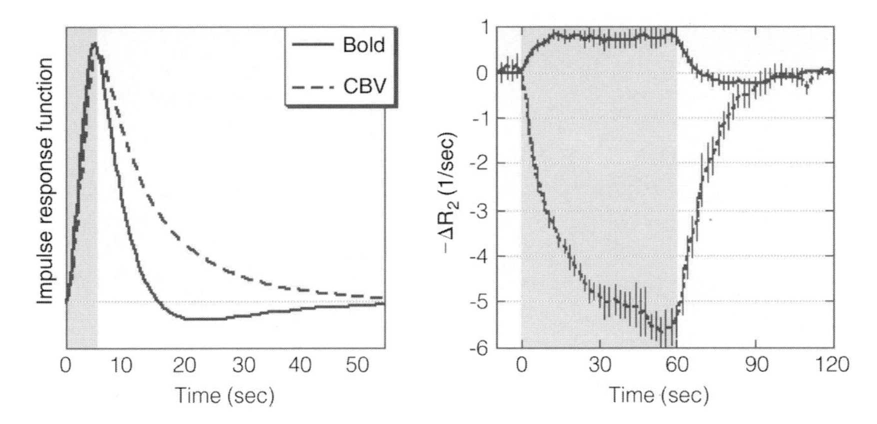

Figure 16 **Left:** The shape of the impulse response functions for BOLD and CBV-weighted *f*MRI signals as fit to data in awake, behaving macaques. **Right:** Relaxation rate changes at 3 T as measured in the same animal model. During prolonged stimulation, temporal evolution of CBV is somewhat slower than that of BOLD signal, and decay of CBV occurs concurrent with a BOLD post stimulus undershoot that is much smaller than the BOLD increase during stimulation.

especially regarding the role of oxygen utilization and the potential spatial specificity of this response relative to the vascular regulation (Buxton, 2001; Vanzetta and Grinvald, 2001), remains an interesting and unresolved issue.

4. Linearity of the Vascular Response

The vascular response is often modeled as a linear, time-invariant transform of the underlying neuronal activity (e.g., Friston *et al.,* 1995). From this point of view, the vascular response is an integral representation of local brain function, and the system output (e.g., BOLD signal) can be predicted for an arbitrarily complicated experimental design by convolution of a BOLD impulse response function with a timing diagram of the stimulus design. The linear model appears to be a good first-order approximation of the relationship between BOLD signal and brain function. BOLD signal changes in the visual cortex scale in a manner roughly linear with stimulus contrast and duration (Boynton *et al.,* 1996), and rapidly presented multiple stimuli produce an overlapping response that is nearly a simple time-shifted summation (Buckner *et al.,* 1996).

In the chain between neuronal activity and *f*MRI signal, nonlinearities can arise at several points. Because BOLD signal measures the wash-out of deoxyhemoglobin, there is finite headroom for signal changes. Thus, BOLD signal changes must become nonlinear with CBF at sufficiently high levels, and data support a saturation effect (Rees *et al.,*1997; Miller *et al.,* 2001). While the temporal evolution of BOLD signal can be predicted from time-shifted stimuli that are 4–6 s or longer, shorter stimuli produce responses that are larger than expected from a linear evolution of signal (Vazquez and Noll, 1998; Robson *et al.,* 1998). Short stimulus separation (< 4 s) may also confound linear-model separation of individual trials (Glover, 1999a). These effects may be due to transient features of the neuronal response or to a vascular source, such as a lack of temporal fidelity on time scales representative of arteriole dilation.

5. Event-Related Experimental Design

Block stimuli of the typed used to generate Fig. 16 replicate PET experiments, rather than taking advantage of the intrinsically higher temporal resolution of *f*MRI. Such designs are statistically efficient, since stimulus durations are generally sufficient to reach a plateau level of signal, and the time spent in hemodynamic transitions is minimal. The high statistical power of block designs is desirable for a host of experiments. Moreover, analyses of block results are relatively insensitive to details of the hemodynamic response. Block designs are amenable to wide variety of analysis strategies (e.g. Bandettini *et al.,* 1993; Friston *et al.,* 1995; Purdon *et al.,* 2001), including simple pooled statistics such as the Student *t* or Komogorov–Smirnov tests, correlation tests, general linear model analyses, or frequency domain analyses.

Other experiments, notably those that probe cognitive processes, must consider confounding issues like neuronal habituation and expectancy during long stimuli in a predictable design (Aguirre and D'Esposito, 1999).These issues can be mitigated by abandoning block paradigms in favour of designs that interleave short stimuli of mixed trial types in randomized orders (Buckner *et al.,* 1996; Zarahn *et al.,* 1997; Friston *et al.,* 1998). Analyses can then employ selective averaging of repeated trials, including posthoc sorting based upon subjects' responses during the experiment. Moreover, such stimulus designs can be compared directly to the results of other modalities, such as MEG and EEG, that employ many short stimuli for temporal averaging.

The statistical power per unit of event-related designs is lower than block designs, which minimize the power lost to hemodynamic transitions by grouping like events into

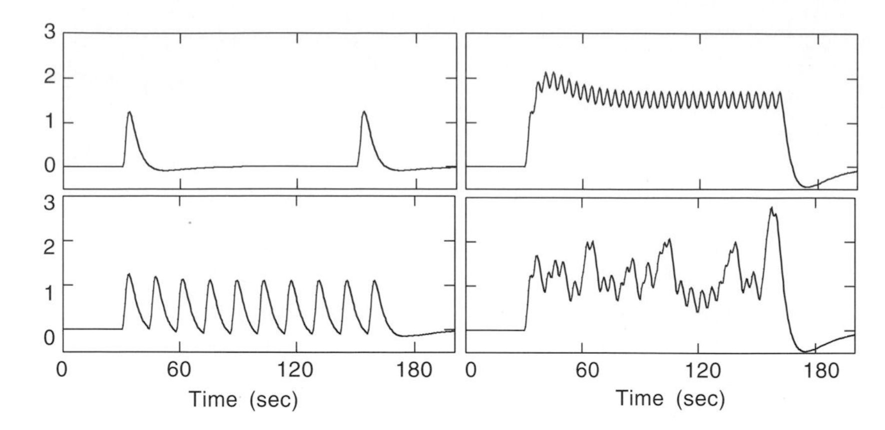

Figure 17 Statistical power versus presentation rate for a fixed stimulus duration (2 s), as illustrated by a general linear model simulation. Short stimuli with a long inter-stimulus interval (ISI) have a low detection power per unit time (**top left**, 2 stimuli). As the presentation rate increases, detection power increases monotonically until an optimal ISI is reached that maximizes the differential area between two conditions (**bottom left**, 10 stimuli, ISI 12 s). The vascular low-pass filter reduces the ability to discriminate conditions at high presentation rates using a fixed ISI (**top right**, 33 stimuli ISI 2 s). Randomizing the ISI recovers contrast and provides high detection power at short average ISI (**bottom right**, 33 stimuli, average ISI 2 s).

unbroken periods of stimulation. Because the vascular response applies a low-pass filter to neuronal activity, detection sensitivity is lost when the presentation rate becomes too rapid and a fixed inter stimulus interval (ISI) is employed, as described and illustrated in Fig. 17. For a fixed ISI, each choice of stimulus duration has an optimal ISI that balances the desire to have many trials against the contrast lost to the vascular filter (Bandettini and Cox, 2000). By randomizing (jittering) the ISI about an average value, this restriction is lifted, and presentation rates can be significantly higher than possible at fixed ISI without suffering a similar degradation of statistical power (Burock *et al.,* 1998).

6. Limits of the Temporal Response

BOLD signal changes in the human visual system have been detected using isolated stimuli as short as 34 ms, but the duration of the response extends over many seconds (Savoy *et al.,* 1995; Rosen *et al.,* 1998). While the slow temporal response of the vasculature is advantageous in terms of extending the detection window for isolated stimuli, there is no evidence that *f*MRI can temporally resolve multiple stimuli on such a time scale. Repeated, short stimuli can produce an overlapping vascular response such that separation of events can be accomplished only by a deconvolution of the measured response with a known temporal point-spread function. Complicating this process, short and rapidly repeated stimuli fall into a temporal regime in which apparent nonlinearities are most visible (Robson *et al.,* 1998; Glover, 1999a). Moreover, vascular heterogeneity leads to a regional distribution of transit delays, with longer onset times generally corresponding to larger vessels and larger signal changes (Lee *et al.,* 1995; Robson *et al.,* 1998). Strategies for improving *f*MRI temporal resolution will need to characterize short impulse responses completely and normalize for vascular delays (Rosen *et al.,* 1998).

7. Multi modal Integration for Improved Temporal Response

While there are no immediate prospects for meaningful improvements in *f*MRI temporal resolution, the combination of *f*MRI spatial localization in conjunction with the temporal resolution of MEG/EEG appears to be a promising avenue (Dale and Halgren, 2001). MEG/EEG measure electromagnetic fields generated by neural activity that can be tracked on a millisecond time scale, but the spatial reconstruction of the sources within the brain is an ill-posed inverse problem. *f*MRI spatial information can be used as a prior constraint to regularize the MEG/EEG spatial reconstruction, in principle enabling real-time movies of electrical activity within distributed neural networks.

B. Spatial Resolution: SNR and Biological Point Spread

Just as for a description of the temporal resolution, a discussion of *f*MRI spatial resolution must address both the probe (NMR) and the sample (the brain). In most experimental settings, the intrinsic sensitivity of MRI is the limiting factor for spatial resolution. As the state of the art continues to advance with the advent of higher magnetic field strengths and better detection hardware (e.g., phased-array coils), spatial resolution is beginning to push up against biological limitations. In order to minimize the biological point-spread function, strategies have evolved that attempt to use the physiology to advantage or instead rely upon the physics of the imaging method to preferentially select a subset of the vessels.

1. *f*MRI (In) sensitivity

At the most basic level, *f*MRI spatial resolution is governed by a fundamental NMR trade-off between space and time, as described in Eq. (6). Spatial resolution can be purchased from the signal-to-noise ratio, which falls as the cube of voxel dimension, but SNR can be replenished only as the square root of time. For a fixed level of SNR, imaging time increases in inverse proportion to the sixth power of the voxel size (time ~ $1/r^6$), as illustrated in Fig. 18.

The basic trade-off between space and time sets a limit for functional MRI—but at what spatial scale? The answer to this question is dependent upon a number of experimental variables, primarily the magnetic field strength and the field of view. The signal strength increases in proportion to the magnetic field [Eq. (6)]. High levels of signal can often be traded for spatial resolution without much loss of sensitivity, which is dominated by physiological noise in the high SNR regime (Fig. 11). For BOLD *f*MRI, relative signal changes also scale with magnetic field strength due to a higher magnetization of the paramagnetic deoxyhemoglobin, and this can be expected to lead to a roughly linear increase in the functional CNR with field strength (refer to Section V, Sensitivity). If the additional CNR is also traded for spatial resolution, then higher field strengths should provide additional signal in proportion to the square of the field, relative to experiments at lower field with equal sensitivity (CNR). In other words, a 7-T experiment using 1.7-mm isotropic voxels should have about the same BOLD functional sensitivity as a 3-T experiment using 3-mm isotropic voxels.

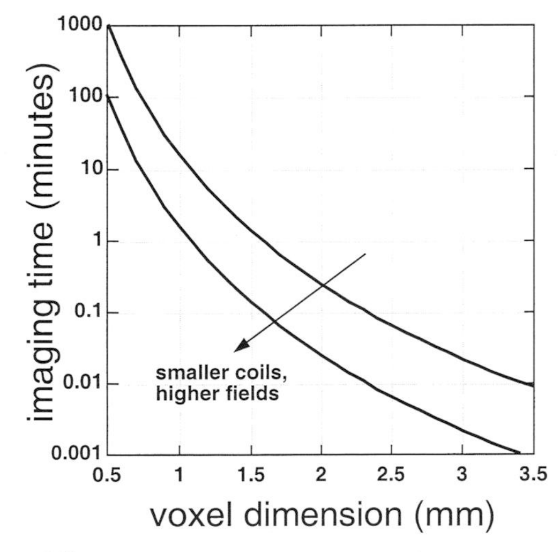

Figure 18 The problem of time versus space at fixed SNR. If a functional image is acquired after a single excitation with a resolution of 3.5 mm on each side (top curve), it will require more than 10 min to obtain 1-mm voxels with the same signal-to-noise ratio and experimental setup. High field strengths and small targeted detection coils lower the curve.

Another strong determinant of the intrinsic SNR per unit time is the size of the field of view that needs to be sampled, because the image volume is intimately related to the sensitivity that can be obtained by detection coils. Smaller coils provide increased sensitivity at the expense of the field of view, so the spatial resolution for targeted applications can exceed the resolution for whole brain imaging. As a rule of thumb, the SNR at the center of a surface coil increases in inverse proportion to the square of the radius of the coil, and the SNR out of the coil plane falls to half-maximum at a distance slightly less than the radius of the coil (Lawry *et al.*, 1990).

The trade-off between spatial coverage and signal strength for a detection coil explains why voxel dimensions used for human functional imaging (typically 3–5 mm) are much larger than those used in small animals like rodents. Small coils for *f*MRI in small animal models easily provide resolution better than 1 mm^3, albeit at relatively coarse anatomical resolution due the smaller brain size.

In order to combine the large spatial coverage associated with large coils and the high SNR provided by smaller designs, a so-called "phased array" of multiple coils can be applied (Roemer *et al.*, 1990). In this design, small coils are placed around the head. Each coil must have a separate RF receiver channel, a feature that hardware vendors have now begun to supply routinely. The image strength at each spatial location is a weighted sum of the information provided by each coil. Phased-array coils should provide increases in signal strength comparable to the use of higher fields, but this technology has been noticeably absent from the *f*MRI field. The four to eight times greater demand on data collection rates, storage, and image reconstruction time has slowed the adoption of this technology for *f*MRI, though more advanced instrument designs should allow such acquisition strategies in the very near future.

For the whole brain functional imaging in humans, the combination of higher field strengths and improved RF coil technology can be expected to provide roughly a 10-fold improvement in the signal-to-noise ratio within a few years relative to the most common *f*MRI experiments today. Even this substantial increase will not allow voxel sizes of less than about 1 mm^3 for most whole brain applications in humans (lower curve in Fig. 18).

2. Biological Specificity: Background

In nonliving biological samples, resolution can be obtained at a few tens of micrometers, the diffusion scale for water, given enough imaging time. In live subjects, bulk motion and cerebral pulsation probably limit the realistically obtainable anatomic resolution to a few hundred micrometers. As described above, the requirement of high temporal resolution further limits spatial resolution for functional studies. An interesting question for *f*MRI, the answer of which is not entirely clear at this time, is whether the spatial point-spread function imposed by the vasculature fixes resolution at a scale even more coarse than the practical *f*MRI limit (> 1–2 mm).

Functionally similar neuronal ensembles form cortical columns in numerous brain regions and species, providing model systems for testing the ultimate resolution of the vascular regulation and *f*MRI. Detailed studies of rodent whisker barrels show that there is no spatial correspondence between individual columns and either arteriole or venular domains, which develop prior to the columns (Woolsey *et al.,* 1996). However, elegant functional maps of cortical columns have been obtained from optical imaging signals based upon the chromophores of hemoglobin (e.g., Frostig *et al.,* 1990; Grinvald *et al.,* 1991). Using differential comparisons of the responses to left and right monocular stimuli, maps with columnar characteristics were generated from optical signals weighted by deoxyhemoglobin, total hemoglobin, and blood plasma (Frostig *et al.,* 1990). The spatiotemporal evolution of optical signals suggested that early components provide better spatial specificity than later signals (Frostig *et al.,* 1990; Malonek and Grinvald, 1996). Although the original optical analyses found an initial increase in deoxyghemoglobin at early time points, recent refinements of the physical description of light propagation through tissue have challenged this interpretation (Lindauer *et al.,* 2001).

Although optical spectroscopic imaging and *f*MRI have many commonalties, there are also numerous differences. While optical signals derive primarily from superficial cortical layers, *f*MRI studies of cortical architecture generally use thick slices through the lamina in order to provide adequate SNR to compensate for high resolution across the columns. While optical and BOLD *f*MRI signals are both sensitive to intravascular signals, *f*MRI methods also have a component of signal that arises from extravascular water. Whereas uncertainties in quantitatively separating the oxygenated and deoxygenated chromophores of hemoglobin currently limit the interpretation of optical signals, the fact that *f*MRI studies have not contributed more to our understanding of vascular regulation is evidence of the extent to which SNR limitations still dominate the *f*MRI field. In contrast to the results of some optical studies that have shown unambiguous maps of cortical columns, the *f*MRI results have been less compelling, and test–retest reliability has been generally lacking.

Despite the SNR hurdle, numerous experiments have begun to test the ultimate biological resolution of *f*MRI. At high spatial resolution, subtle differences can be observed between statistical foci obtained by different *f*MRI methods. Because imaging signals are removed from brain activity by both the physiology of flow–metabolism and the physics connecting physiology to *f*MRI signal, it is important to consider both the physiology and the physics of methods as they pertain to spatial resolution. When attempting to push the state of the art of functional spatial resolution, some *f*MRI studies have emphasized particular advantages of physiology, while others have emphasized advantages of physics.

3. Positive BOLD Signal

It has long been evident that large BOLD signal changes can arise from draining veins that are not well localized to the primary site of activation. Because intravascular contributions to signal are largest at low magnetic fields, and changes in venous oxygenation are progressively diluted with the distance from the site of activation, draining veins are most apparent at low fields near large volumes of activated tissue. Figure 19 provides an illustrative case in a macaque. Relative to the results obtained by the use of exogenous contrast agent, the BOLD method shows vascular artifacts within venous sinuses, as well as signal changes in regions that extend from the sinuses to areas identified using the CBV method.

Although the "brain versus vein" issue is not a major conundrum at the resolutions typically employed for *f*MRI, this issue may present a limitation for very high spatial resolution. In order to amplify contributions from capillaries relative to veins, higher magnetic fields are better. Because changes in venous oxygenation should be slightly delayed relative to changes in capillary oxygenation due to transit delays of blood oxygen, the early component of the BOLD response may provide suppression of venous fields. Using high field and/or short stimuli, the positive part of the BOLD response has produced functional maps with the appearances of columnar structure (Menon *et al.,* 1997; Goodyear and Menon, 2001). A recent study (Cheng *et al.,* 2001)

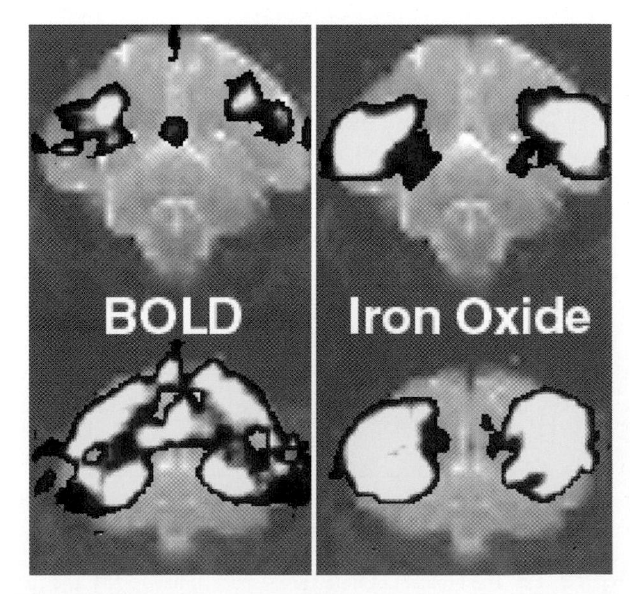

Figure 19 Statistical maps obtained using the BOLD and CBV methods at 3 T show the same basic areas of activition in the visual cortex of the awake macaque. However, the BOLD method also produces signal changes in draining venous sinuses.

showed ocular dominance columnar structure using BOLD signal with a block paradigm at 4 T. This study emphasized geometrical accuracy in the orientation of the thick slices required for adequate SNR.

4. BOLD Initial Dip

The initial dip in BOLD signal, as described in Section VI.A, has been employed in an anesthetized cat model in an attempt to resolve iso-orientation columns in the visual cortex (Kim *et al.,* 2000; Duong *et al.,* 2000). These high-field studies found that the positive BOLD response did not discriminate iso-orientation columns, whereas maps derived from the initial dip, or part of the initial dip (Duong *et al.,* 2000), provided complementary spatial information, in accordance with the known neurophysiology.

The drawbacks of this imaging strategy include the elusive nature of initial dip across studies, as described previously, and the low sensitivity of the method. In the most prominent displays of this feature of the signal, the integrated amplitude of the initial dip suffers an order of magnitude penalty in strength relative to the positive BOLD response. Due perhaps to the low sensitivity of this method, no studies have demonstrated test–retest reliability within the same subject.

5. Arterial Spin Labeling

Following magnetic labeling and arterial transit, approximately 90% of blood water is extracted on each pass through the capillary network at normal levels of blood flow (Eichling *et al.,* 1974). Of the remaining small percentage of blood water that transits directly to veins, the magnetization is suppressed by T_1 relaxation due to the transit time. Thus, the physics of ASL provides suppression of venous fields, even though venous blood flow is identically equal to arterial blood flow at steady-state levels. In an anesthetized cat model, ASL measurements using the FAIR method at 4.7 T found that presentation of orthogonal square-wave gratings produced complementary maps of orientation columns in four of seven data sets (Duong *et al.,* 2001).

Drawbacks of the ASL method include a lower CNR than BOLD signal and reduced volumetric coverage per unit time. Although FAIR ASL can be used effectively on a single slice, multi slice studies can be more easily performed using continuous labeling or pulsed labeling proximal to the image volume. These methods suffer some loss of sensitivity due to transit delays between labeling and image volumes. Due to these delays, volumetric ASL can be expected to have lower sensitivity in humans than in small animal models, in which higher blood flow and shorter distances lead to reduced transit times.

6. CBV Using Contrast Agent

The use of exogenous contrast agent has several nice features in terms of spatial localization. The intravascular part of the signal, which can produce large signal changes in draining veins with the BOLD method, is completely eliminated by the use of contrast agent. Moreover, the dose of contrast agent can be used to target brain parenchyma, providing spatial sensitivity that is roughly uniform across the brain tissue. These qualities have been demonstrated at a coarse resolution like 1 mm^3 in rats (Mandeville and Marota, 1999; Mandeville *et al.,*2001). In terms of the physical description of the signal, the use of contrast agent is analogous to very high field BOLD imaging, but without the corresponding susceptibility artifacts and distortion. In terms of physiology, optical imaging suggests that changes in CBV occur both locally and distant to functional columns, and differential mapping (e.g., right monocular response relative to left monocular response) can isolate columns by canceling volume changes associated with larger vessels (Frostig *et al.,* 1990). This *f*MRI technique has not yet been applied in ultra high resolution studies.

7. Microvascular Weighting Using Spin Echoes

The extravascular contribution to a paramagnetic contrast agent like deoxyhemoglobin can be weighted toward the microvasculature by the use of spin echoes, as shown in Fig. 4 and discussed in that section (Section IV). Although microvascular sensitivity has been demonstrated *in vivo* in studies of tumor angiogenesis (Dennie *et al.,* 1998), similar advantages have not yet been demonstrated for high-resolution *f*MRI.

At low magnetic fields, BOLD spin-echo measurements are strongly weighted by the intravascular signal component (Boxerman *et al.,* 1995a; Song *et al.,* 1996), which has no differential sensitivity based upon vessel size. Higher magnetic fields progressively attentuate the intravascular contribution; diffusion-weighted experiments at 9.4 T show that BOLD signal is exclusively extravascular in origin (Lee *et al.,* 1999). A drawback of the use of spin echoes is that sensitivity is severely reduced by the smaller value of ΔR_2 relative to ΔR_2*. At 1.5 T, the ΔR_2*/ ΔR_2 ratio is about 3.5 (Bandettini *et al.,* 1994). Since progressively larger doses of exogenous agent increasing R_2 and R_2* with a ratio also of about 3.5 (Boxerman *et al.,* 1995b), high-field BOLD imaging can be expected to maintain the relative sensitivity of gradient-and spin-echo imaging. Using exogenous contrast agent, the small size of ΔR_2 relative to ΔR_2* can be compensated in principle by the use of three to four times more contrast agent for experiments in animal models.

VII. Structure–Function Integration

An essential goal of noninvasive brain mapping is the association of function with structure. While this chapter has exclusively addressed *f*MRI methods, rapid advances are occurring in structural MRI, providing information that is

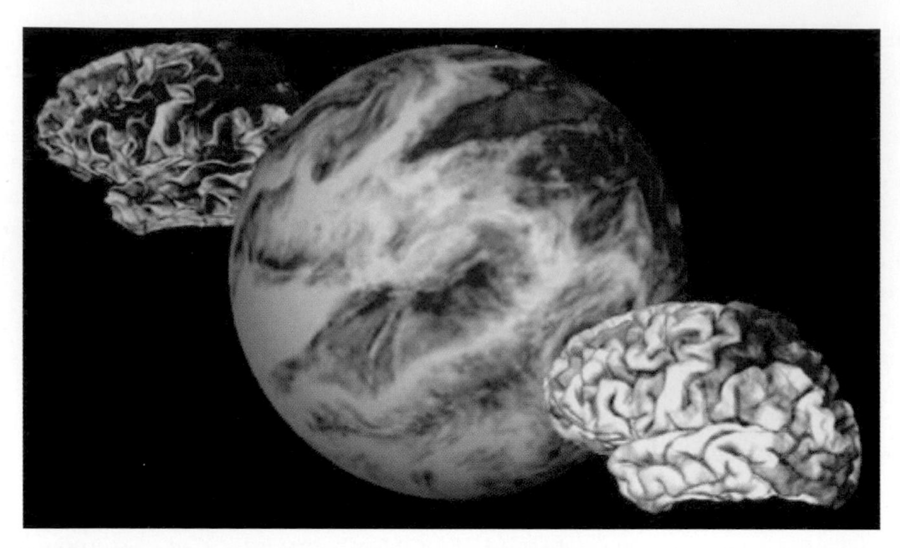

Figure 20 Maps of the average thickness of human cortex, with red indicating thin regions and yellow thick regions, overlaid on the white matter surface (**Left**), the pial surface (**Right**), and a spherical inflated surface (**Middle**), Figure courtesy of Bruce Fischl.

complementary, and often essential, to *f*MRI. Advanced automated or semiautomated algorithms of computational neuroanatomy, together with tissue classification facilitated by optimize pulse sequences, multiple types of contrast, and prior information, are improving spatial registration, tissue segmentation, and morphometry across subjects and within large subpopulations for cross-group comparisons.

It was recently shown that the cortical functional point spread across a large subject population could be substantially reduced, relative to the use of Talairach coordinates for cross-subject registration, by employing an anatomically defined spherical surface-based coordinate system with sulci and gyri used as registration landmarks (Fischl *et al.,* 1999b). The same basic methods—gray/white tissue classification and surface reconstruction (Dale *et al.,* 1999) followed by cortical inflation (Fischl *et al.,* 1999a) and intersubject registration—enabled the maps of average human cortical thickness shown in Fig. 20 (Fischl and Dale, 2000). Similar strategies that employ tissue classification in conjunction with prior anatomical information are being applied to subcortical regions in order to automate morphometric calculations (Fischl *et al.,* 2002) and to improve registration across subjects.

VIII. Future

Although it is impossible to predict new techniques or analysis algorithms that may emerge in coming years to significantly advance the state of the art, it can be seen clearly that remarkable benefits will accrue simply by extending and integrating known avenues for improvement. In terms of *f*MRI sensitivity and spatial resolution, these include (1) the continued proliferation of high-field systems, (2) the application of phased-array coil technology to *f*MRI, (3) modified ASL methods employing static tissue suppression, and (4) the possible application of contrast agents for clinical *f*MRI. The marriage of *f*MRI and MEG/EEG has the potential to revolutionize noninvasive neuroscience by combining millimeter and millisecond resolution. Improved and automated *f*MRI analyses, when integrated with evolving structural analyses and widely disseminated, will facilitate large group comparisons that currently would be prohibitively arduous. These advances, combined with unforeseen developments, should provide the energy to continue the rapid acceleration of *f*MRI applications in basic neuroscience and disease.

References

Aguirre, G. K., and D'Esposito, M. (1999) Experimental design for brain *f*MRI. In "Functional MRI" C. T. W. Moonen and P. A. Bandettini, pp. 369–380. Springer-Verlag, Berlin.

Alsop, D. C., and Detre, J. A. (1996). Reduced transit-time sensitivity in noninvasive magnetic resonance imaging of human cerebral blood flow. *J. Cereb. Blood Flow Metab.* **16**, 1236–1249.

Alsop, D. C. and Detre, J. A. (1998). Multisection cerebral blood flow MR imaging with continuous arterial spin labeling. *Radiology* **208**, 410–416.

Atkinson, J. L., Anderson, R. E. and Sundt, J. T. M. (1990).The effect of carbon dioxide on the diameter of brain capillaries. *Brain Res.* **517**, 333–340.

Bandettini, P. A., and Cox, R. W. (2000). Event-related *f*MRI contrast when using constant interstimulus interval: Theory and experiment. *Magn. Reson. Med.* **43**, 540–548.

Bandettini, P. A., Jesmanowicz, A., Wong, E. C., and Hyde, J. S. 1993. Processing strategies for time-course data sets in functional MRI of the human brain. *Magn. Reson. Med.* **30**, 161–173.

Bandettini, P. A., Kwong, K. K., Davis, T. L., Tootell, R. B., Wong, E. C., Fox, P. T., Belliveau, J. W., Weisskoff, R. M., and Rosen, B. R. (1997).

Characterization of cerebral blood oxygenation and flow changes during prolonged brain activation. *Hum. Brain Mapp.* **5**, 93–109.

Bandettini, P. A., and Wong E. C. (1995). Effects of biophysical and physiological parameters on brain activation-induced R_2* and R_2 changes: Simulations using a deterministic diffusion model. *Int. J. Imaging Syst. Tech.* **6**, 133–152.

Bandettini, P. A., and Wong, E. C. (1997). A hypercapnia-based normalization method for improved spatial localization of human brain activation with *f*MRI. *NMR Biomed.* **10**, 197–203.

Bandettini, P. A., Wong, E. C., Hinks, R. S., Tikofsky, R. S., and Hyde, J. S. (1992). Time course EPI of human brain function during task activation. *Magn. Reson. Med.* **25**, 390–397.

Bandettini, P. A., Wong, E. C., Jesmanowicz, A., Hinks, R. S., and Hyde, J. S. (1994). Spin-echo and gradient-echo EPI of human brain activation using BOLD contrast: A comparative study at 1.5 T. *NMR Biomed.* **7**, 12–20.

Belliveau, J. W., Kantor, H. L., Pykett, I. L., Rzedzian, R. R., Beaulieu, P., Kennedy, D. N., Fisel, C. R., Brady, T. J., and Rosen, B. R. (1988). Real-time proton susceptibility-contrast imaging of hypercapnia induced changes in cerebral physiology. *In* "Seventh Annual Meeting of the Society of Magnetic Resonance in Medicine", San Francisco, p. 1.

Belliveau, J.W., Kennedy, D. N. Jr., McKinstry, R. C., Buchbinder, B. R., Weisskoff, R. M., Cohen, M. S., Vevea, J. M., Brady, T. J., and Rosen, B. R. (1991). Functional mapping of the human visual cortex by magnetic resonance imaging. *Science* **254**, 716–719.

Belliveau, J. W., McKinstry, R. C., Kennedy, D. N., Vevea, J. M., Sorensen, G., Fisel, C. R., Jenkins B., Kantor, H. L., Pykett, I. L., Rzedzian, R. R., Brady, T. J., and Rosen, B. R. (1989). Functional imaging by gamma analysis of contrast enhanced NMR. *In* "Eighth Annual Meeting of the Society of Magnetic Resonance in Medicine", Amtersdam, p. 63.

Belliveau, J. W., Rosen, B. R., Kantor, H. L., Rzedzian, R. R., Kennedy, D. N., McKinstry, R. C., Vevea, J. M., Cohen, M. S., Pykett, I. L., and Brady, T. J. (1990). Functional cerebral imaging by susceptibility-contrast NMR. *Magn. Reson. Med.* **14**, 538–546.

Belliveau, J. W., Villringer, A., Rosen, B. R., Lauffer, R. B., Ackerman, J. L., Buxton, R. B., Frazer, J. C., Johnson, K. A., Moore, J. B., Wedeen, V. J., and Brady, T. J. (1986). Magnetic susceptibility induced signal attenuation changes in rat brain caused by hypercapnia. *In* "Fifth Annual Meeting of the Society of Magnetic Resonance in Medicine", Montreal, p. 273.

Birn, R. M., Saad, Z. S., and Bandettini, P. A. (2001). Spatial heterogeneity of the nonlinear dynamics in the *f*MRI bold response. *NeuroImage* **14**, 817–826.

Boxerman, J. L., Bandettini, P. A., Kwong, K. K., Baker, J. R., Davis, T. L., Rosen, B. R., and Weisskoff, R. M. (1995a). The intervascular contributions to *f*MRI signal change: Monte Carlo modeling and diffusion-weighted studies in vivo. *Magn. Reson. Med.* **34**, 4–10.

Boxerman, J. L., Hamberg, L. M., Rosen, B. R., and Weisskoff, R. M. (1995b). MR contrast due to intravascular magnetic susceptibility perturbations. *Magn. Reson. Med.* **34**, 555–566.

Boynton, G. M., Engel, S. A., Glover, G. H., and Heeger, D. J. (1996). Linear systems analysis of functional magnetic resonance imaging in human VI. *J. Neurosci.* **16**, 4207–4221.

Brady, T. J. (1991). *In* "Tenth Annual Meeting of the Society of Magnetic Resonance in Medicine", San Francisco, p. 2.

Brady, T. J., Goldman, M. R., Pykett, I. L., Buonanno, F. S., Kistler, J. P., Newhouse, J. H., Burt, C. T., Hisnshaw, W. S., and Pohost, G. M. (1982). Proton nuclear magnetic resonance imaging of regionally ischemic canine hearts: Effect of paramagnetic proton signal enhancement. *Radiology* **144**, 343–347.

Buckner, R. L., Bandettini, P. A., O'Craven, K. M., Savoy, R. L., Petersen, S. E., Raichle, M. E., and Rosen, B. R. (1996) Detection of cortical activation during averaged single trials of a cognitive task using functional magnetic resonance imaging. *Proc. Nat. Acad. Sci. USA* **93**, 14878–14883.

Burock, M. A., Buckner, R. L., Woldorff, M. G., Rosen, B. R., and Dale, A. M. (1998). Randomized event-related experimental designs allow for extremely rapid presentation rates using functional MRI. *NeuroReport* **9**, 3735–3739.

Buxton, R. B., (2001). The elusive initial dip. *NeuroImage* **13**, 953–958.

Buxton, R. B., and Frank, L. R. (1997). A model for the coupling between cerebral blood flow and oxygen metabolism during neuronal stimulation. *J.Cereb. Blood Flow Metab.* **17**, 64–72.

Buxton, R. B., Frank, L.R., Wong, E. C., Siewert, B., Warach, S., and Edelman, R. R., (1998a). A general kinetic model for quantitative perfusion imaging with arterial spin labeling. *Magn. Reson. Med.* **40**, 383–396.

Buxton, R. B., Luh, W.-M., Wong, E. C., Frank, L.R., and Bandettini, P. A. (1998b). Diffusion weighting attenuates the BOLD peak signal change but not the post-stimulus undershoot. *In* "International Society of Magnetic Resonance in Medicine Annual Meeting", Sdney, p. 1401.

Buxton, R. B., Wong, E. C., and Frank, L. R. (1998d). Dynamics of blood flow and oxygenation changes during brain activation: The balloon model. *Magn. Reson. Med.* **39**, 855–864.

Chen, W., Zhu, X. H., Kata, T., Andersen, P., and Ugurbil, K. (1998). Spatial and temporal differentiation of *f*MRI BOLD response in primary visual cortex of human brain during sustained visual simulation. *Magn. Reson. Med.* **39**, 520–527.

Chen, Y. I., Mandeville, J. B., Nguyen, T. V., Talele, A., Cavagna, F., and Jenkins, B. G. (2001). Improved mapping of pharmacologically induced neuronal activation using the IRON technique with superparamagnetic iron blood pool agents. *J. Magn. Reson. Imaging* **14**, 517–524.

Cheng, K., Waggoner, R. A., and Tanaka, K. (2001). Human ocular dominance columns as revealed by high-field functional magnetic resonance imaging. *Neuron* **32**, 359–374.

Chesler, D. A., and Kwong, K. K. (1995). An intuitive guide to the T1 based perfusion model. *Int. J. Imaging Syst. Tech.* **6**, 171–174.

Chu, S. C. K., Xu, Y., Balschi, J. A., and Springer, C. S. (1990). Bulk magnetic susceptibility shifts in NMR studies of compartmentalized samples: Use of paramagnetic reagents. *Magn. Reson. Med.* **13**, 239–262.

Dale, A. M., and Buckner, R. L. (1997). Selective averaging of rapidly presented individual trails using *f*MRI. *Hum. Brain Mapp.* **5**, 1–12.

Dale, A. M., Fischl, B., and Sereno, M. I. (1999). Cortical surface-based analysis. I. Segmentation and surface reconstruction. *NeuroImage* **9**, 179–94.

Dale, A. M., and Halgren, E. (2001). Spatiotemporal mapping of brain activity by integration of multiple imaging modalities. *Curr. Opin. Neurobiol.* **11**, 202–208.

Davis, T. L., Kwong, K. K., Weisskoff, R. M., and Rosen, B. R. (1998). Calibrated functional MRI: Mapping the dynamics of oxidative metabolism. *Proc. Natl. Acad. Sci. USA* **95**, 1834–1839.

Dennie, J., Mandeville, J. B., Boxerman, J. L., Packard, S.D., Rosen, B. R., and Weisskoff, R. M. (1998). NMR imaging of changes in vascular mophology due to tumor angiogenesis. *Magn. Reson. Med.* **40**, 793–799.

Detre, J. A., Leigh, J. S., Williams, D. S., and Koretsky, A. P. (1992). Perfusion imaging. *Magn. Reson. Med.* **23**, 37–45.

De Yoe, E. A., Bandettini, P., Neitz, J., Miller, D., and Winans, P. (1994). Functional magnetic resonance imaging (*f*MRI) of the human brain. *J. Neurosci. Methods.* **54**, 171–187.

Disbrow, E. A., Slutsky, D. A., Roberts, T. P., and Krubitzer, L. A. 2000. Functional MRI at 1.5 tesla: A comparison of the blood oxygenation level-dependent signal and electrophysiology. *Proc. Natl. Acad. Sci. USA* **97**, 9718–9723.

Donahue, K. M., Weisskoff, R. M., Chesler, D. A., Kwong, K. K., Bogdanov, A., Jr., Mandeville, J. B., and Rosen, B. R. (1996). Improving MR quantification of regional blood volume with intravascular T1 contrast agents: Accuracy, precision, and water exchange. *Magn. Reson. Med.* **36**, 858–867.

Duelli, R., and Kuschinsky, W. (1993). Changes in brain capillary diameter during hypocapnia and hypercapnia. *J.Cereb. Blood Flow Metab.* **13**, 1025–1028.

Duong, T. Q., Kim, D. S., Ugurbil, K., and Kim, S. G. (2000). Spatiotemporal dynamics of the BOLD *f*MRI signals: Toward mapping submillimeter cortical columns using the early negative response. *Magn. Reson. Med.* **44**, 231–242.

Duong, T. Q., Kim, D. S., Ugurbil, K., and Kim, S. G. (2001). Localized cerebral blood flow response at submillimeter columnar resolution. *Proc. Natl. Acad. Sci. USA* **98**,10904–10909.

Duyn, J. H., Tan C. X., van Gelderen, P., and Yongbi, M. N. (2001). High-sensitivity single-shot perfusion-weighted *f*MRI. *Magn. Reson. Med.* **46**, 88–94.

Edelman, R. R., Siewert, B., Darby, D. G., Thangaraj, V., Nobre, A. C., Mesulam, M. M., and Warach, S. (1994). Qualitative mapping of cerebral blood flow and functional localization with echo planar MR imaging and signal targeting with alternating radiofrequency (EPISTAR). *Radiology* **192**, 1–8.

Eichling, J. O., Raichle, M. E., Grubb, R. L., and Ter-Pogossian, M. M. (1974). Evidence of the limitations of water as a freely diffusible tracer in brain of the rhesus monkey, *Circ.Res.* **35**, 358–364.

Eklöf, B., Lassen, N. A., Nilsson, L., Norberg, K., and Siesjö B. K. (1973). Blood flow and metabolic rate for oxygen in the cerebral cortex of the rat. *Acta Physiol. Scand.* **88**, 587–589.

Ernst, R. R., and Anderson, W. A. (1966). Application of Fourier transform spectroscopy to magnetic resonance. *Rev. Sci. Instrum.* **37**, 93–102.

Feinberg, D. A., and Mark, A. S. (1987). Human brain motion and cerebrospinal fluid circulation demonstrated with MR velocity imaging. *Radiology* **163**, 793–799.

Fischl, B., and Dale, A. M. (2000). Measuring the thickness of the human cerebral cortex from magnetic resonance images. *Proc. Natl. Acad. Sci. USA* **97**, 11050–11055.

Fischl, B., Salat, D.H., Albert, M., Dieterich, M., Haselgrove, C., van der Kouwe, A., Killiany, R., Kenney, D., Klaveness, S., Montillo, A., Makris, N., Rosen, B. R., and Dale, A. M. (2002). Whole brain segmentation: Automated labeling of neuroanatomical structures in the human brain. *Neuron*, **31**, 341–355.

Fischl, B., Sereno, M. I., and Dale, A. M. (1999a). Cortical surface-based analysis. II. Inflation, flattening, and a surface-based coordinate system. *NeuroImage* **9**, 195–207.

Fischl, B., Sereno, M. I., Tootell, R. B., and Dale, A. M. (1999b). High-resolution intersubject averaging and a coordinate system for the cortical surface. *Hum. Brain Mapp.* **8**, 272–284.

Fisel, C. R., Ackerman, J. L., Buxton, R. B., Garrido, L., Belliveau, J.W., Rosen, B. R., and Brady, T. J. (1991). MR contrast due to microscopically heterogeneous magnetic susceptibility: Numerical simulations and applications to cerebral physiology. *Magn. Reson. Med.* **17**, 336–347.

Fox, P. T., and Raichle, M. E. (1986). Focal physiological uncoupling of cerebral blood flow and oxidative metabolism during somatosensory stimulation in human subjects. *Proc. Natl. Acad. Sci. USA* **83**, 1140–1144.

Fox, P. T., Raichle, M. E., Mintun, M. A., and Dence, C. (1988). Nonoxidative glucose consumption during focal physiologic neural activity. *Science* **241**, 462–464.

Frahm, J., Bruhn, H., Merboldt, K., and Hanicke, W. (1992). Dynamic MR imaging of human brain oxygenation during rest and photic stimulation. *J. Magn. Reson. Imaging* **2**, 501–505.

Frank, O. (1899). Die Grundform des arteriellen Pulses. *Z. Biol.* **85**, 91–130.

Fransson, P., Kruger, G., Merboldt, K. D., and Frahm, J. (1998). Temporal characteristics of oxygenation-sensitive MRI responses to visual activation in humans. *Magn. Reson. Med.* **39**, 912–919.

Friston, K. J., Holmes, A. P., Worsley, K. J., Poline, J. P., Frith, C. D., and Frackowiak, R. S. J. (1995). Statistical parametric maps in functional imaging: A general linear approach. *Hum. Brain Mapp.* **2**, 189–210.

Friston, K. J., Josephs, O., Rees, G., and Turner, R. (1998). Nonlinear event-related responses in *f*MRI. *Magn. Reson. Med.* **39**, 41–52.

Frostig, R. D., Lieke, E. E., Tso, D. Y., and Grinvald, A. (1990). Cortical functional architecture and local coupling between neuronal activity and the microcirculation revealed by in vivo high-resolution optical imaging of intrinsic signals. *Proc. Natl. Acad. Sci. USA* **87**, 6082–6086.

Gati, J. S., Menon, R. S., Ugurbil, K., and Rutt, B. K. (1997). Experimental determination of the BOLD field strength dependence in vessels and tissue. *Magn. Reson. Med.* **38**, 296–302.

Glover, G. H. (1999a). Deconvolution of impulse response in event-related BOLD *f*MRI. *NeuroImage* **9**, 416–429.

Glover, G. H. (1999b). Simple analytic spiral K-space algorithm. *Magn. Reson. Med.* **42**, 412–415.

Gobel, U., Theilen, H., Schrock, H., and Kuschinsky, W. (1991). Dynamics of capillary perfusion in the brain. *Blood Vessels,* **28**, 190–196.

Goodyear, B. G., and Menon, R. S. (2001). Brief visual stimulation allows mapping of ocular dominance in visual cortex using *f*MRI. *Hum. Brain Mapp.* **14**, 210–217.

Grinvald, A., Frostig, R. D., Siegel, R. M., and Bartfeld, E. (1991).High-resolution optical imaging of functional brain architecture in the awake monkey. *Proc. Natl. Acad. Sci. USA* **88**, 11559–11563.

Grinvald, A., Lieke, E., Frosig, R. D., Gilbert, C. D., and Wiesel, T. N. (1986). Functional architecture of cortex revealed by optical imaging of intrinsic signals. *Nature*, **324**, 361–364.

Grubb, R. L., Raichle, M. E., Eichling, J. O., and Ter-Pogossian, M. M. (1974). The effects of changes in $PaCO_2$ on cerebral blood volume, blood flow, and vascular mean transit time. *Stroke,* **5**, 630–639.

Guenther, D., Hanisch, G., and Kauczor, H. U. (2000). Functional MR imaging of pulmonary ventilation using hyperpolarized noble gases. *Acta Radiol.* **41**, 519–528.

Guimaraes, A. R., Melcher, J. R., Talavage, T. M., Baker, J. R., Ledden, P., Rosen, B. R., Kiang, N. Y. S., Fullerton, B. C., and Weisskoff, R. M. (1998). Imaging subcortical auditory activity in humans. *Hum. Brain Mapp.* **6**, 33–41.

Halgren, E., Babb, T. L., and Crandall, P.H. (1977). Responses of human limbic neurons to induced changes in blood gases. *Brain Res.* **132**, 43–63.

Hamberg, L. M., Boccalini, P., Stranjalis, G., Hunter, G. J., Huang, Z., Halpern, E., Weisskoff, R. M., Moskowitz, M. A., and Rosen, B. R. (1996). Continuous assessment of relative cerebral blood volume in transient ischemia using steady state susceptibility-contrast MRI. *Magn. Reson. Med.* **35**, 168–173.

Hayes, C. E., Edelstein, W. A., Schenk, J. F., Mueller, O. M., and Eash, M. (1985). An efficient, highly homogeneous radiofrequency coil for whole-body NMR imaging at 1.5 T. *J. Magn. Reson.* **34**, 622–628.

Hoge, R. D., Atkinson, J., Gill, B., Crelier, G. R., Marrett, S., and Pike, G. B. (1999a). Linear coupling between cerebral blood flow and oxygen consumption in activated human cortex. *Proc. Natl. Acad. Sci. USA* **96**, 9403–9408.

Hoge, R. D., Atkinson, J., Gill, B., Crelier, G. R., Marrett, S., and Pike, G. B. (1999b). Stimulus-dependent BOLD and perfusion dynamics in human VI. *NeuroImage* **9**, 573–585.

Hoppel, B. E., Weisskoff, R. M., Thulborn, K. R., Moore, J., and Rosen, B. R. (1991). Measurement of regional brain oxygenation state using echo planar linewidth mapping. *In* "Tenth Annual Meeting of the Society of Magnetic Resonance in Medicine", San Francisco, p. 308.

Hoult, D. I. and Lauterbur, P. C. (1979). The sensitivity of the zeumatographic experiment involving human samples. *J. Magn. Reson.* **34**, 425–433.

Hu, X., Le, T. H., and Ugurbil, K. (1997). Evaluation of the early response in *f*MRI in individual subjects using short stimulus duration. *Magn. Reson. Med.* **37**, 877–884.

Hyde, J. S., Biswal, B. B., and Jesmanowicz, A. (2001). High-resolution *f*MRI using multislice partial k-space GR-EPI with cubic voxels. *Magn. Reson. Med.* **46**, 114–125.

Iadecola, C., Yang, G., Ebner, T. J., and Chen, G. (1997). Local and propagated vascular responses evoked by focal synaptic activity in cerebellar cortex. *J. Neurophysiol.* **78**, 6510.

Jackson, J. I., Meyer, C. H., Nishimura, D. G., and Macovski, A. (1991). Selection of a convolution function for Fourier inversion using gridding. *IEEE Trans. Med. Imaging.* **10**, 473–478.

James, W. (1890). "Principles of Physiology." Holt, New York.

Janz, C., Heinrich, S. P., Kornmayer, J., Bach, M., and Hennig, J. (2001). Coupling of neural activity and BOLD *f*MRI response: New insights by combination of *f*MRI and VEP experiments in transition from single events to continuous stimulation. *Magn. Reson. Med.* **46**, 482–486.

Jones, M., Berwick, J., Johnston, D., and Mayhew, J. (2001). Concurrent optical imaging spectroscopy and laser-Doppler flowmetry: The relationship between blood flow, oxygenation, and volume in rodent barrel cortex. *NeuroImage* **13**, 1002–1015.

Josephson, L., Groman, E. V., Menz, E., Luis, J. M., and Bengele, H. (1990). A functionalized superparamagnetic iron oxide colloid as a receptor directed MR contrast agent. *Magn. Reson. Imaging* **8**, 637.

Jueptner, M., and Weiller, C. (1995). Does measurement of regional cerebral blood flow reflect synaptic activity? Implications for PET and *f*MRI. *NeuroImaging* **2**, 148–156.

Kangarlu, A., Abduljalil, A. M., and Robitaille, P. M. (1999). T1- and T2-weighted imaging at 8 tesla. *J. Comput. Assisted Tomgor.* **23**, 875–878.

Kennan, R. P., Scanley, B. E., Innis, R. B., and Gore, J. C. (1998). Physiological basis for BOLD MR signal changes due to neuronal stimulation: Separation of blood volume and magnetic susceptibility effects. *Magn. Reson. Med.* **40**, 840–846.

Kennan, R. P., Zhong, J., and Gore, J. C. (1994). Intravascular susceptibility contrast mechanisms in tissues. *Magn. Reson. Med.* **31**, 9–21.

Kety, S. (1960). Measurement of local blood flow by the exchange of an inert diffusible substance. *Methods Med. Res.* **8**, 228–236.

Kim, D. S., Duong, T. Q., and Kim, S. G. (2000). High-resolution mapping of iso-orientation columns by *f*MRI. *Nat. Neurosci.* **3**, 164–169.

Kim, S.-G. (1995). Quantification of relative cerebral blood flow change by flow-sensitive alternating inversion recovery (FAIR) technique: Application to functional mapping. *Magn. Reson. Med.* **34**, 293–301.

Kim, S. G., Rostrup, E., Larsson, H. B., Ogawa, S., and Paulson, O.B. (1999). Determination of relative $CMRO_2$ from CBF and BOLD changes: Significant increase of oxygen consumption rate during visual stimulation. *Magn. Reson. Med.***41**, 1152–1161.

Kruger, G., and Glover, G. H. (2001). Physiological noise in oxygenation-sensitive magnetic resonance imaging. *Magn. Reson. Med.* **46**, 631–637.

Kruger, G., Kastrup, A., and Glover, G. H. (2001). Neuroimaging at 1.5 T and 3.0 T: Comparison of oxygenation-sensitive magnetic resonance imaging. *Magn. Reson. Med.* **45**, 595–604.

Kruger, G., Kleinschmidt A., and Frahm, J. (1996). Dynamic MRI sensitized to cerebral blood oxygenation and flow during sustained activation of human visual cortex. *Magn. Reson. Med.* **35**, 797–800.

Kuschinsky, W. (1996). Capillary perfusion in the brain. *Pflugers Arch.* **432**, 42–46.

Kwong, K. K., Belliveau, J. W., Chesler, D. A., Goldberg, I. E., Weisskoff, R. M., Poncelet, B. P., Kennedy, D. N., Hoppel, B. E., Cohen, M. S., Turner, R., Cheng, H.-M., Brady, T. J., and Rosen, B. R. (1992). Dynamic magnetic resonance imaging of human brain activity during primary sensory stimulation. *Proc. Natl. Acad. Sci. USA* **8**, 5675–5679.

Kwong, K. K., Chesler, D. A., Weisskoff, R. M., Donahue, K. M., T. L., D., Ostergaard, L., Campbell, T. A., and Rosen, B. R. (1995). MR perfusion studies with T1-weighted echo planar imaging. *Magn. Reson. Med.* **34**, 878–887.

Laiken, N. D., and Fanestil, D. D. (1990). Body fluids and renal function. *In* "Best and Taylor's Physiological Basis of Medical Practice" (J. B. West, ed.), pp. 406–512. Williams & Wilkins, Baltimore.

Lassen, N. A., and Perl, W. (1979). "Tracer Kinetic Methods in Medical Physiology". Raven Press, New York.

Lauffer, R. B. (1987). Paramagnetic metal complexes as water proton relaxation agents for NMR imaging: Theory and design. *Chem. Rev.* **87**, 901–927.

Lawry, T. J., Weiner, M. W., and Matson, G. B. (1990).Computer modeling of surface coil sensitivity. *Magn. Reson. Med.* **16**, 294–302.

Le Bihan, D., Breton, E., Lallemand, D., Grenier, P., Cabanis, E., and Laval, J. M. (1986). MR imaging of intravoxel incoherent motions: Application to diffusion and perfusion in neurologic disorders. *Radiology* **161**, 401–407.

Lee, A. T., Glover, G. H., and Meyer, C.H. (1995). Discrimination of large venous vessels in time-course spiral blood-oxygen-level-dependent magnetic-resonance functional neuroimaging. *Magn. Reson. Med.* **33**, 745–754.

Lee, S. P., Silva, A. C., Ugurbil, K., and Kim, S. G. (1999). Diffusion-weighted spin-echo *f*MRI at 9.4 T: Microvascular/tissue contribution to BOLD signal changes. *Magn. Reson. Med.* **42**, 919–928.

Lin, Y. J., and Koretsky, A. P. (1997). Manganese ion enhances T1-weighted MRI during brain activation: An approach to direct imaging of brain function. *Magn. Reson. Med.* **38**, 378–388.

Lindauer, U., Royl, G., Leithner, C., Kuhl, M., Gold, L., Gethmann, J., Kohl-Bareis, M., Villringer, A., and Dirnagl, U. (2001). No evidence for early decrease in blood oxygenation in rat whisker cortex in response to functional activation. *NeuroImage* **13**, 988–1001.

Liu, H. L., Pu, Y., Nickerson, L. D., Liu, Y., Fox, P. T., and Gao, J. H. (2000). Comparison of the temporal response in perfusion and BOLD-based event-related functional MRI. *Magn. Reson. Med.* **43**, 768–772.

Logothetis, N. K., Pauls, J., Augath, M., Trinath, T., and Oeltermann, A. (2001). Neurophysiological investigation of the basis of the *f*MRI signal. *Nature* **12**,150–157.

Luh, W. M., Wong, E. C., Bandettini, P. A., Ward, B. D., and Hyde, J. S. (2000). Comparison of simultaneously measured perfusion and BOLD signal increases during brain activation with T(1)-based tissue identification. *Magn. Reson. Med.* **44**, 137–143.

Malonek, D., and Grinvald, A. (1996). Interactions between electrical activity and cortical microcirculation revealed by imaging spectroscopy: Implications for functional brain mapping. *Science* **272**, 551–554.

Mandeville, J. B., Jenkins, B. G., Kosofsky, B. E., Moskowitz, M. A., Rosen, B. R., and Marota, J. J. A. (2001). Regional sensitivity and coupling of BOLD and CBV changes during stimulation of rat brain. *Magn. Reson. Med.* **45**, 443–447.

Mandeville, J. B., and Marota, J. J. A. (1999). Vascular filters and functional MRI: Spatial localization using BOLD and CBV contrast. *Magn. Reson. Med.* **42**, 591–598.

Mandeville, J. B., Marota, J. J. A., Ayata, C., Moskowitz, M. A., Weisskoff, R. M., and Rosen, B. R. (1999a). An MRI measurement of the temporal evolution of relative $CMRO_2$ during rat forepaw stimulation. *Magn. Reson. Med.* **42**, 944–951.

Mandeville, J. B., Marota, J. J. A., Ayata, C., Zaharchuk, G., Moskowitz, M. A., Rosen, B. R., and Weisskoff, R. M. (1999b). Evidence of a cerebrovascular post-arteriole windkessel with delayed compliance. *J.Cereb. Blood Flow Metab.* **19**, 679–689.

Mandeville, J. B., Marota, J. J. A., Kosofsky, B. E., Keltner, J. R., Weissler, R., Rosen, B. R., and Weisskoff, R. M., (1998). Dynamic functional imaging of relative cerebral blood volume during rat forepaw stimulation. *Magn. Reson. Med.* **39**, 615–624.

Mansfield, P. (1977). Multi-planar image formation using NMR spin echos. *J. Phys.* **C10**, L55–L58.

Marota, J. J. A., Ayata, C., Moskowitz, M. A., Weisskoff, R. M., Rosen, B. R., and Mandeville, J. B. (1999). Investigation of the early response to rat forepaw stimulation. *Magn. Reson. Med.* **41**, 247–252.

Marota, J. J. A., Mandeville, J. B., Weisskoff, R. M., Moskowitz, M. A., Rosen, B. R., and Kosofsky, B. E. (2000). Cocaine activation discriminates dopamergic projections by temporal response: An *f*MRI study in rat. *NeuroImage* **11**, 13–23.

Mathews, V. P., Caldemeyers, K. S., Ulmer, J. L., Nguyen, H., and Yuh, W. T. (1997). Effects of contrast dose, delayed imaging, and magnetization transfer saturation on gadolinium-enhanced MR imaging of brain lesions. *J. Magn. Reson. Imaging* **7**, 14–22.

Mazziotta, J. C., and Phelps, M. E. (1986). Positron emission tomography studies of the brain. *In* "Positron Emission Tomography and Autoradiography: Principles and Applications for the Brain and the Heart" (M. E. Phelps, J. C. Mazziotta, and H. R. Schelbert, edns.) pp. 493–597. Raven Press, New York.

Menon, R., Ogawa, S., Tank, D., and Ugurbil, K. (1993). 4 tesla gradient recalled echo characteristics of photic stimulation-induced signal changes in the human primary visual cortex. *Magn. Reson. Med.* **30**, 380–386.

Menon, R. S., Ogawa, S., Hu, X., Strupp, J. S., Andersen, P., and Ugurbil, K. (1995). BOLD based functional MRI at 4 tesla includes a capillary bed contribution: Echo-planar imaging mirrors previous optical imaging using intrinsic signals. *Magn. Reson. Med.* **33**, 453–459.

Menon, R. S., Ogawa, S., Strupp, J. P., and Ugurbil, K. (1997). Ocular dominance in human V1 demonstrated by functional magnetic resonance imaging. *J. Neurophysiol.* **77**, 2780–2727.

Michenfelder, J. D. (1990). Cerebral blood flow and metabolism. *In* "Clinical Neuroanesthesia" (R. F. Cucchiara, and J. D. Michenfelder, eds.). Churchill Livingstone, New York.

Miller, K. L., Luh, W. M., Liu, T. T., Martinez, A., Obata, T., Wong, E. C., Frank, L. R., and Buxton, R. B. (2001). Nonlinear temporal dynamics of the cerebral blood flow response. *Hum. Brain Mapp.* **13**, 1–12.

Moseley, M. E., Cohen, Y., Mintorovitch, J., Chileuitt, L., Shimizu, H., Kucharczyk, J., Wendland, M. F., and Weinstein, P. R. (1990). Early detection of regional cerebral ischemia in cats: Comparison of diffusion and T2 weighted MRI and spectroscopy. *Magn. Reson. Med.* **14**, 330–346.

Ogawa, S., Lee, R. M., and Barrere, B. (1993a). The sensitivity of magnetic resonance image signals of a rat brain to changes in the cerebral venous blood oxygenation. *Magn. Reson. Med.* **29**, 205–210.

Ogawa, S., Lee, T. M., Kay, A. R., and Tank, D. W. (1990a). Brain magnetic resonance imaging with contrast dependent on blood oxygenation. *Proc. Natl. Acad. Sci. USA* **87**, 9868–9872.

Ogawa, S., Lee, T. M., Nayak, A. S., and Glynn, P. (1990b). Oxygenation-sensitive contrast in magnetic resonance image of rodent brain at high magnetic fields. *Magn. Reson. Med.* **14**, 68–78.

Ogawa, S., Menon, R. S., Tank, D.W., Kim, S. G., Merkle, H., Ellermann, J. M., and Ugurbil, K. (1993b). Functional brain mapping by blood oxygenation level-dependent contrast magnetic resonance imaging: A comparison of signal characteristics with a biophysical model. *Biophys. J.* **64**, 803–812.

Ogawa, S., Tank, D. W., Menon, R., Ellermann, J. M., Kim, S. G., Merkle, H., and Ugurbil, K. (1992). Intrinsic signal changes accompanying sensory stimulation: Functional brain mapping with magnetic resonance imaging. *Proc. Natl. Acad. Sci. USA* **89**, 5951–5955.

Oja, J. M., Gillen, J., Kauppinen, R. A., Kraut, M., and van Zijl, P. C. (1999). Venous blood effects in spin-echo *f*MRI of human brain. *Magn. Reson. Med.* **42**, 617–626.

Ostergaard, L., Weisskoff, R. M., Chesler, D. A., Gyldensted, C., and Rosen, B. R. (1996). High resolution measurement of cerebral blood flow using intravascular tracer bolus passages. Part I. Mathematical approach and statistical analysis. *Magn. Reson. Med.* **36**, 715–725.

Pauling, L., and Coryell, C. (1936). The magnetic properties and structure of hemoglobin, oxyhemoglobin and carbon monooxyhemoglobin. *Proc. Natl. Acad. Sci. USA* **22**, 210–216.

Poncelet, B. P., Wedeen, V. J., Weisskoff, R. M., and Cohen, M. S. (1992). Brain parenchyma motion: Measurement with cine echo-planar MR imaging. *Radiology* **185**, 645–651.

Porciuncula, C. I., Armstrong, J., G. G., Guyton, A. C., and Stone, H. L. (1964). Delayed compliance in external jugular vein of the dog. *Am. J. Physiol.* **207**, 728–732.

Powers, W. J. (1991). Cerebral hemodynamics in ischemic cerebrovascular disease. *Ann. Neurol.* **29**, 231–240.

Purdon, P. L., Solo, V., Weisskoff, R. M., and Brown, E. N. (2001). Locally regularized spatiotemporal modeling and model comparison for functional MRI. *NeuroImage* **14**, 912–923.

Rees, G., Howseman, A., Josephs, O., Frith, C. D., Friston, K. J., Frackowiak, R. S., and Turner, R. (1997). Characterizing the relationship between BOLD contrast and regional cerebral blood flow measurements by varying the stimulus presentation rate. *NeuroImage* **6**, 270–278.

Robson, M. D., Dorosz, J. L., and Gore, J. C. (1998). Measurements of the temporal *f*MRI response of the human auditory cortex to trains of tones. *NeuroImage* **7**, 185–198.

Roemer, P. B., Edelstein, W. A., Hayes, C. E., Souza, S. P., and Mueller, O. M. (1990). The NMR phased array. *Magn. Reson. Med.* **16**, 192–225.

Rosen, B. R., Belliveau, J. W., and Chien, D. (1989). Perfusion imaging by nuclear magnetic resonance. *Magn. Reson. Q.* **5**, 263–281.

Rosen, B. R., Buckner, R. L., and Dale, A. M. (1998). Event-related functional MRI: Past, present, and future. *Proc. Natl. Acad. Sci. USA* **95**, 773–780.

Roy, C. S., and Sherrington, C. S. (1890). On the regulation of the blood-supply of the brain. *J. Physiol. (London)* **11**, 85–108.

Royl, G., Leithner, C., Kohl, M., Lindauer, U., Dirnagl, U., Kwong, K., and Mandeville, J. B. (2001). The BOLD post-stimulus undershoot: *f*MRI versus imaging spectroscopy. *In* "International Society of Magnetic Resonance in Medicine Annual Meeting", Glasgow, p. 282.

Savoy, R. L., Bandettini, P. A., O'Craven, K. M., Kwong, K. K., Davis, T. L., Baker, J.R., Weisskoff, R. M., and Rosen, B. R. (1995). Pushing the temporal resolution of *f*MRI: Studies of very brief visual stimuli, onset variability and asynchrony, and stimulus-correlated changes in noise. *In* "International Society of Magnetic Resonance in Medicine Annual Meeting", Nice.

Sharma, R., Saini, S., Ros, P. R., Hahn, P.F., Small, W. C., de Lange, E. E., Stillman, A. E., Edelman, R. R., Runge, V. M., Outwater, E. K., Morris, M., and Lucas, M. (1999). Safety profile of ultrasmall superparamagnetic iron oxide ferumoxtran-10: Phase II clinical trial data. *J. Magn. Reson. Imaging* **9**, 291–294.

Silva, A. C., and Kim, S. G., (1999). Pseudo-continuous arterial spin labeling technique for measuring CBF dynamics with high temporal resolution *Magn. Reson. Med.* **42**, 425–429.

Silva, A. C., Lee, S. P., Iadecola, C., and Kim, S. G. (2000). Early temporal characteristics of cerebral blood flow and deoxyhemoglobin changes during somatosensory stimulation. *J.Cereb. Blood Flow Metab.* **20**, 201–206

Sokoloff, L. (1981). Relationships among local functional activity, energy metabolism, and blood flow in the central nervous system. *Fed. Proc.* **40**, 2311–2316.

Sokoloff, L., Reivich, M., Kennedy, C., DesRosiers, M. H., Patlak, C. S., Pettigrew, K. D., Sakurada, O., and Shinohara, M. (1977). The [14C] deoxyglucose method for the measurement of local cerebral glucose utilization: Theory, procedure, and normal values in the conscious and anesthetized albino rat. *J. Neurochem* **28**, 897–916.

Song, A. W. (2001). Single-shot EPI with signal recovery from the susceptibility-induced losses. *Magn. Reson. Med.* **46**, 407–411.

Song, A. W., Wong, E. C., Tan, S. G., and Hyde, J. S. (1996). Diffusion weighted *f*MRI at 1.5 T. *Magn. Reson. Med.* **35**, 155–158.

Sorensen, A. G., Tievsky, A. L., Ostergaard, L., Weisskoff, R. M., and Rosen, B. R. (1997). Contrast agents in functional MR imaging. *J. Magn. Reson. Imaging* **7**, 47–55.

Stanley, J. A., Drost, D. J., Williamson, P. C., and Thompson, R. T. (1995).The use of a priori knowledge to quantify short echo in vivo ^{1}H MR spectra. *Magn. Reson. Med.* **34**, 17–24.

Taylor, A. M., Panting, J. R., Keegan, J., Gatehouse, P. D., Amin, D., Jhooti, P., Yang, G. Z., McGill, S., Burman, E. D., Francis, J. M., Firmin, D. N., and Pennell, D.J. (1999) Safety and preliminary findings with the intravascular contrast agent NC100150 injection for MR coronary angiography. *J. Magn. Reson. Imaging* **9**, 220–227.

Ter-Pogossian, M. M., Eichling, J. O., Davis, D. O., Welch, M. J., and Metzger, J. M. (1969). The determination of regional cerebral blood flow by means of water labeled with radioactive oxygen 15. *Radiology* **93**, 31–40.

Thulborn, K.R., Waterton, J. C., Matthews, P. M., and Radda, G. K. (1982). Oxygenation dependence of the transverse relaxation time of water protons in whole blood at high field. *Biochim Biophys. Acta* **714**, 265–270.

Turner, R., Jessard, P., Wen, H., Kwong, K. K., Le Bihan, D., Zeffiro, T., and Balaban, R. S. (1993). Functional mapping of the human visual cortex at 4 and 1.5 T using deoxygenation contrast EPI. *Magn. Reson. Med.* **29**, 277–279.

Turner, R., Le Bihan, D., Moonen, C. T., Despres, D., and Frank, J. (1991a). Echo-planar time course MRI of cat brain oxygenation changes. *Magn. Reson. Med.* **22**, 159–166.

Turner, R., Le Bihan, D., Moonen, C. T. W., and Frank, J. (1991b). Echo-planar imaging of deoxygenation episodes in a cat brain at 2 T. *J. Magn. Reson. Imaging* **1**, 227.

Van Bruggen, N., Busch, E., Palmer, J. T., Williams, S. P., and de Crespigny, A. J. (1998).High-resolution functional magnetic resonance imaging of the rat brain: Mapping changes in cerebral blood volume using iron oxide contrast media. *J.Cereb. Blood Flow Metab.* **18**, 1178–1183.

Van Zijl, P. C., Eleff, S. M., Ulatowski, J. A., Oja, J. M., Ulug, A. M., Traystman, R. J., and Kauppinen R. A. (1998). Quantitative assessment of blood flow, blood volume and blood oxygenation effects in functional magnetic resonance imaging, *Nat Med.* **4**, 159–167.

Vanduffel, W., Fize, D., Mandeville, J. B., Nelissen, K., Van Hecke, P., Rosen, B. R., Tootell, R. B. H., and Orban, G. A. (2001). Visual motion processing investigated using contrast-enhanced *f*MRI in awake behaving monkeys. *Neuron* **32**, 565–567.

Vanzetta, I., and Grinvald, A., (2001). Evidence and lack of evidence for the initial dip in the anesthetized rat: Implications for human functional brain imaging. *NeuroImages* **13**, 959–967.

Vazquez, A. L., and Noll, D. C. (1998). Nonlinear aspects of the BOLD response in functional MRI. *NeuroImage* **7**, 108–118.

Villringer, A. (1999) Physiological changes during brain activation. *In* "Functional MRI" (C. T. W. Moonen and P.A. Bandettini edns.), pp. 3–13. Springer-Verlag, Berlin.

Villringer, A., Rosen, B. R., Belliveau, J. W., Ackerman, J. L., Lauffer, R. B., Buxton, R. B., Chao, Y., Wedeen, V. J., and Brady, T. J. (1988). Dynamic imaging with lanthanide chelates in normal brain: Contrast due to magnetic susceptibility effects. *Magn. Reson. Med.* **6**, 164–174.

Warach, S. (2001). New imaging strategies for patient selection for thrombolytic and neuroprotective therapies. *Neurology*, **57**, S48–52.

Weinmann, H. J., Brasch, R. C., Press, W. R., and Wesbey, G. E. (1984). Characteristics of gadolinium–DTPA complex: A potential NMR contrast agent. *Am. J. Roentgenol.* **142**, 619–624.

Weisskoff, R. M., Baker, J. R., Belliveau, J. W., Davis, T. L., Kwong, K. K., Cohen, M. S., and Rosen, B. R. (1993). Power spectrum analysis of functionally-weighted MR data: What's in the noise? *In* "International Society of Magnetic Resonance in Medicine Annual Meeting", New York, p. 7.

Weisskoff, R. M., Zuo, C. S., Boxerman, J. L., and Rosen, B. R. (1994).Microscopic susceptibility variation and transverse relaxation: Theory and experiment. *Magn. Reson. Med.* **31**, 601–610.

Weissleder, R., Elizondo, G., and Wittenberg, K. (1990). Ultrasmall superparamagnetic iron oxide: Characterization of a new class of contrast agents for MR imaging. *Radiology* **175**, 489–493.

Wong, E. C., Buxton, R. B., and Frank, L. R. (1998). Quantitative imaging of perfusion using a single subtraction (QUIPSS and QUIPSS II). *Magn. Reson. Med.* **39**, 702–708.

Wong, E. C., Liu, T., Frank, L., and Buxton, R. (2001). Close tag, short TR continuous ASL for functional brain mapping: High temporal resolution ASL with a BOLD sized signal at 1.5T. *In* "International Society of Magnetic Resonance in Medicine Annual Meeting", Glasgow, p. 1162.

Woolsey, T. A., Rovainen, C. M., Cox, S. B., Henegar, M. H., Liang, G. E., Liu, D., Moskalenko, Y. E., Sui, J., and Wei, L. (1996). Neuronal units linked to microvascular modules in cerebral cortex: Response elements for imaging the brain. *Cereb. Cortex* **6**, 647–660.

Wowk, B., McIntyre, M. C., and Saunders, J. K. (1997). k-space detection and correction of physiological artifacts in *f*MRI. *Magn. Reson. Med.* **38**, 1029–1034.

Yablonskiy, D., and Haacke, E. (1994).Theory of NMR signal behavior in magnetically inhomogeneous tissues: The static dephasing regime. *Magn. Reson. Med.* **32,** 749–763.

Yacoub, E., Shmuel, A., Pfeuffer, J., Van De Moortele, P. F., Adriany, G., Andersen, P., Vaughan, J. T., Merkle, H., Ugurbil, K., and Hu, X. (2001). Imaging brain function in humans at 7 Tesla. *Magn. Reson. Med.* **45**, 588–594.

Yang, Q. W., Dardzinski, B. J., Li, S., Eslinger, P. J., and Smith, M. B. (1997). Multi-gradient echo with susceptibility inhomegeneity compensation (MGESIC): Demonstration of *f*MRI in the olfactory cortex at 3.0 T. *Magn. Reson. Med.* **37**, 331–335.

Yang, Y., Glover, G.H., van Gelderen, P., Patel, A. C., Mattay, V. S., Frank, J. A., and Duyn, J. H. (1998).A comparison of fast MR scan techniques for cerebral activation studies at 1.5 tesla. *Magn. Reson. Med.* **39**,61–67.

Ye, F. Q., Frank, J. A., Weinberger, D. R., and McLaughlin, A. C. (2000). Noise reduction in 3D perfusion imaging by attenuating the statis signal in arterial spin tagging (ASSIST). *Magn. Reson. Med.* **44**, 92–100.

Zaharchuk, G., Ledden, P. J., Kwong, K. K., Reese, T. G., Rosen, B. R., and Wald, L. L. (1999a). Multislice perfusion and perfusion territory imaging in humans with separate label and image coils. *Magn. Reson. Med.* **41**, 1093–1098.

Zaharchuk, G., Mandeville, J. B., Bogdonov, A. A. Jr., Weissleder, R., Rosen, B. R., and Marota, J. J. A. (1999b). Cerebrovascular dynamics of autoregulation and hypotension: An MRI study of CBF and changes in total and microvascular cerebral blood volume during hemorrhagic hypotension. *Stroke* **30**, 197–205.

Zarahn, E., Aguirre, G., and D'Esposito, M. (1997). A trial-based experimental design for *f*MRI. *NeuroImage* **6**, 122–138.

Zhang, W., Silva, A. C., Williams, D. S., and Koretsky, A. P. (1995). NMR measurement of perfusion using arterial spin labeling without saturation of macromolecular spins. *Magn. Reson. Med.* **33**, 370–376.

Zhong, J., Kennan, R. P., Fulbright, R. K., and Gore, J. C. (1998). Quantification of intravascular and extravascular contributions to BOLD effects induced by alteration in oxygenation or intravascular contrast agents. *Magn. Reson. Med.***40**, 526–536.

14

Magnetic Resonance Spectroscopic Imaging

Andrew A. Maudsley

Department of Radiology, University of Miami School of Medicine, Miami, Florida 33136

I. Introduction

Magnetic resonance spectroscopy enables noninvasive analysis of the chemical composition of matter, and when combined with MR imaging techniques, it provides a readily available method for noninvasive mapping of low-molecular-weight metabolites in living systems. These techniques are of interest for biomedical applications since they can monitor tissue metabolism, potentially detecting pathological changes in the absence of structural abnormalities or disease-specific changes where only nonspecific findings may be observed using other imaging modalities.

Nuclear magnetic resonance (NMR) spectroscopy uses the interaction of the magnetic properties of the nucleus with applied static and radio-frequency (RF) magnetic fields to probe the chemical and physical properties of a sample. Discovered in 1946 (Purcell *et al.*, 1946; Bloch, 1946; Bloch *et al.*, 1946), it is now a standard technique for chemical analysis and elucidation of molecular structures and physical properties. The potential use of NMR for chemical analysis of biological tissue was soon realized and NMR was already considered a valuable tool for the biochemist in the 1950s, though these early studies were limited to measurements in solutions. The first studies of mammalian tissue were of objects that were small enough to be placed inside the small (<1 ml) active volumes that were available in the early NMR spectrometer systems, such as a mouse tail (Singer, 1959), excised tissues (Hoult *et al.*, 1974), perfused organs (Gadian *et al.*, 1976), or the head of a mouse (Chance *et al.*, 1978). These early studies demonstrated that NMR spectroscopy could provide valuable information on tissue metabolism and it was widely anticipated that these techniques would be well suited to noninvasive studies of living systems. There was also a particular interest in the potential for clinical applications given the noninvasive nature of the measurement and the requirement for relatively harmless static and radio-frequency magnetic fields (Gadian, 1982). The extension to human studies was then facilitated by the concurrent development, and widespread interest, in NMR imaging methods that occurred in the 1970s (Lauterbur, 1973; Kumar *et al.,* 1975; Mansfield and Maudsley, 1976; Damadian *et al.,* 1977). However, before MR spectroscopy (MRS) would become widely applied to the study of human metabolism it was necessary to await the development of superconducting whole-body imaging magnet systems in the 1980s, which could provide a sufficiently strong and homogeneous magnetic field over the body (Bottomley *et al.,* 1984; Maudsley *et al.,* 1984). As these techniques have reached a broader group of investigators with a greater emphasis on clinical applications, they have become more widely known as MR spectroscopy and MR imaging (MRI).

MRS can detect signals from many tissue metabolites, though for *in vivo* studies the performance is limited by sensitivity considerations. First, the NMR phenomenon itself is relatively insensitive, as the observed signal is dependent on the very small magnetization difference between energy levels of a nucleus possessing angular momentum, or spin (see Chapter 13). Second, the metabolite concentrations are small, typically less than 10 mM, which is considerably smaller than that of tissue water (≈40–45 M), which provides the strong signal used for MRI measurements. Third, while NMR detection sensitivity can be increased by the use of higher magnetic field strengths, the problems of constructing magnet and RF systems suitable for studies in large objects, such as the human body, necessarily involve some loss of sensitivity relative to measurements in small objects. Because of these sensitivity considerations, the majority of *in vivo* MRS studies have concentrated on measurement of the higher concentration metabolites (e.g., >1 mM) and observation using nuclei with high inherent NMR sensitivity, namely ^{1}H or ^{31}P. Even for this selection of metabolites, the spatial resolution for MRS must be significantly reduced in comparison to conventional ^{1}H MRI in order to increase the available signal. Notwithstanding these limitations, MRS offers several advantages, in that it can provide multiparametric information on tissue metabolism in a noninvasive manner, the methods can be readily implemented on conventional MRI scanners, and the limited spatial resolution can be offset by combining the MRS information with higher resolution MRI-based morphologic information conveniently obtained in the same imaging protocol.

The earliest methods for spatially localizing the MRS measurement simply limited the volume from which signals were detected to the immediate proximity of a small detection coil. The obvious limitations to surface structures were soon overcome by the development of a number of single-volume localization methods, whereby spatially selective RF excitation techniques are used to obtain a NMR spectrum from a single volume located at any region within the body (reviewed by Matson and Weiner, 1992). The location and size of the selected volume are typically identified using a MRI scan obtained just prior to the MRS exam. These methods have now become widely available on most commercial MRI instruments; they are relatively easy to implement and obtain good spectral quality in times of a few minutes, though they provide limited spatial information. An alternative approach for spatially localized MRS is to combine the spectroscopic acquisition with the same spatial discrimination techniques used to obtain MRI information (Brown *et al.,* 1982; Haselgrove *et al.,* 1983; Maudsley *et al.,* 1983), using the same principles of encoding multidimensional NMR information described by Ernst *et al.* (Kumar *et al.,* 1975; Aue *et al.,* 1976). Known as magnetic resonance spectroscopic imaging (MRSI), or chemical shift imaging, this technique enables the acquisition of spectroscopic information from multiple contiguous regions distributed over one, two, or three spatial dimensions. For *in vivo* applications, the MRSI data can be processed to create maps of several tissue metabolite distributions and present data in a tomographic format that is frequently preferable for diagnostic imaging and biomedical research studies.

Although single-voxel MRS measurements will continue to play a role for many biomedical applications, it is the spectroscopic imaging methods that provide the extended spatial information required for mapping of metabolite distributions, and this discussion therefore concentrates on these techniques. In addition, only ^{1}H and ^{31}P MRSI studies of human brain are considered, which are widely implemented on commonly available MRI instruments operating at 1.5 T or higher. While several other nuclei are of considerable interest for *in vivo* applications, these are not generally studied using MRSI because of sensitivity limitations. For example, the NMR-observable isotope of carbon, ^{13}C, is present at only 1% abundance, making its measurement insensitive unless high fields and isotopic enrichment are used. Measurement of ^{23}Na is also of interest, which has a relatively good sensitivity, at approximately 1/1000 that for water protons in normal tissue. However, this nucleus does not exhibit a chemical shift (unless shift reagents are applied) and, therefore, can be observed using MRI techniques. Furthermore, the information provided largely reflects extracellular volume in contrast to the metabolic information that will be the focus of this chapter.

II. Basics of in Vivo MR Spectroscopy

A. The NMR Spectrum

An example spectrum obtained from human brain using ^{1}H observation is shown in Fig. 1, together with an illustration of the single-voxel approach to MRS. Here it is

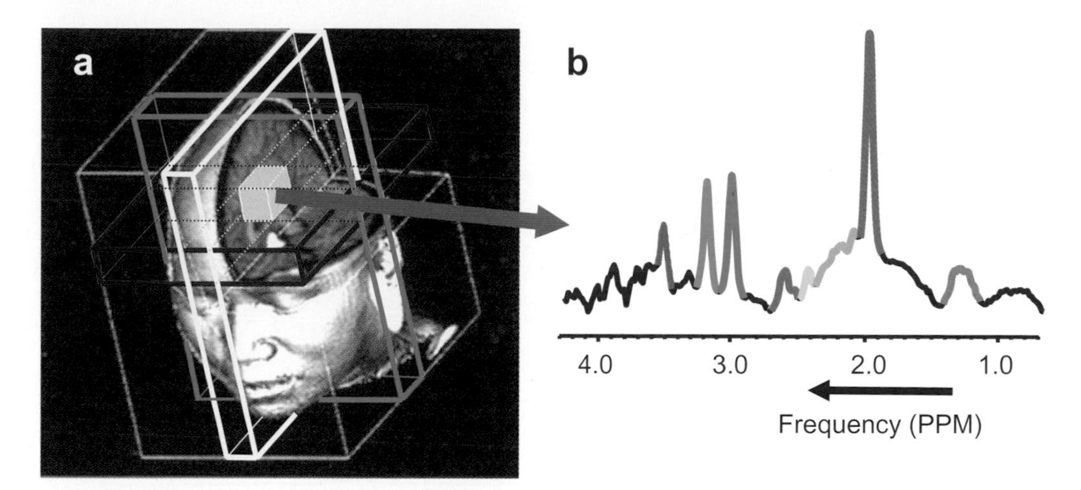

Figure 1 Illustration of the MRS measurement using single-volume localization methods. By using three orthogonal spatially selective RF pulses, a MRS signal is selected from a cubic-shaped volume within the head. The frequency axis of the resultant spectrum then enables discrimination between resonances arising from different metabolites.

assumed that the MRS signal has been localized to a predefined volume by applying spatially selective RF excitation pulses in three orthogonal directions, as for example in the PRESS (point-resolved spectroscopy) (Ordidge *et al.*, 1985; Bottomley, 1987) and STEAM (stimulated echo acquisition mode) (Frahm *et al.*, 1987) sequences. A transient NMR response is elicited from this volume, which is then detected in the absence of any applied magnetic field gradients. This time-domain signal contains multiple overlapping contributions coming from all resonant nuclei within the selected volume, but these can be separated by taking the Fourier transform of the acquired data, which then presents the signal intensity as a function of frequency, and the peaks in this spectrum indicate the resonance frequencies of the sample, as for example in Fig. 1.

The NMR resonance frequency, f (in Hertz), is defined by the strength of the applied magnetic field, B_0 (Tesla), multiplied by the gyromagnetic ratio, γ (Hz/T), which is a constant that differs for each element. However, the magnetic field seen by the individual nuclei in a molecule may differ slightly from the applied B_0 field due to interactions with the surroundings, the primary effect of which is a partial shielding by the effect of the surrounding electrons, i.e.,

$$f = \gamma(B_0 - B_{\text{electron}}) = -\gamma B_0(1 - \sigma), \qquad (1)$$

where the term σ is used to indicate the electron shielding effect, known as the chemical shift. Although this term is dependent on the orientation of the molecule to the B_0 field, only the scalar component is usually observed *in vivo* because of motional averaging effects. This electron shielding is sensitive to the molecular structure; for instance, the proton resonance frequencies of a $-CH_3$ group, $-CH_2$ group, or $-CH$ group are different, and these frequencies may be further modified by neighboring molecular groups. The chemical shift of several resonances is also dependent on physical parameters that may alter the electron shielding effect, including pH, ionic concentrations, and temperature.

The chemical shift provides the primary mechanism for spectroscopic discrimination between different molecular groups and for identification of the individual metabolites from the characteristic patterns of resonances obtained from each metabolite. For example, this enables identification of the resonances labeled *N*-acetylaspartate, creatine, and choline in Fig. 2. Since resonance frequencies scale with the strength of the magnet, and the magnitude of the shielding effect is small, the frequency is normally divided by the spectrometer operating frequency (in MHz) to normalize the scale to parts per million (ppm). This enables direct comparison of spectra obtained at different field strengths. In addition, chemical shifts are quoted relative to the frequency of a reference standard that defines the 0 ppm position for each nucleus. For historical reasons, the ppm scale is plotted from higher to lower frequencies, with higher frequencies corresponding to less electron shielding.

A second effect that influences the appearance of the NMR spectrum is magnetic coupling between neighboring atoms. If neighboring atoms are nuclei with a magnetic spin, their mutual interactions further modify the magnetic field seen by each spin in a manner that depends on which energy levels are populated. Known as the indirect spin–spin coupling effect, because it is mediated by surrounding electrons in contrast to direct coupling between nuclei, the degree of coupling drops off rapidly with distance, but can occur over two or three bond lengths. A single-spin $\frac{1}{2}$ atom can exist in two energy levels ($+\frac{1}{2}$ and $-\frac{1}{2}$); therefore, a coupled spin will exhibit two resonances, reflecting the macroscopic population of spins interacting with the two possible energy states of the coupled atoms. Whereas chemical shifts scale with field strength, spin–spin coupling effects do not, and the strength of the interaction is characterized as a *J*-coupling

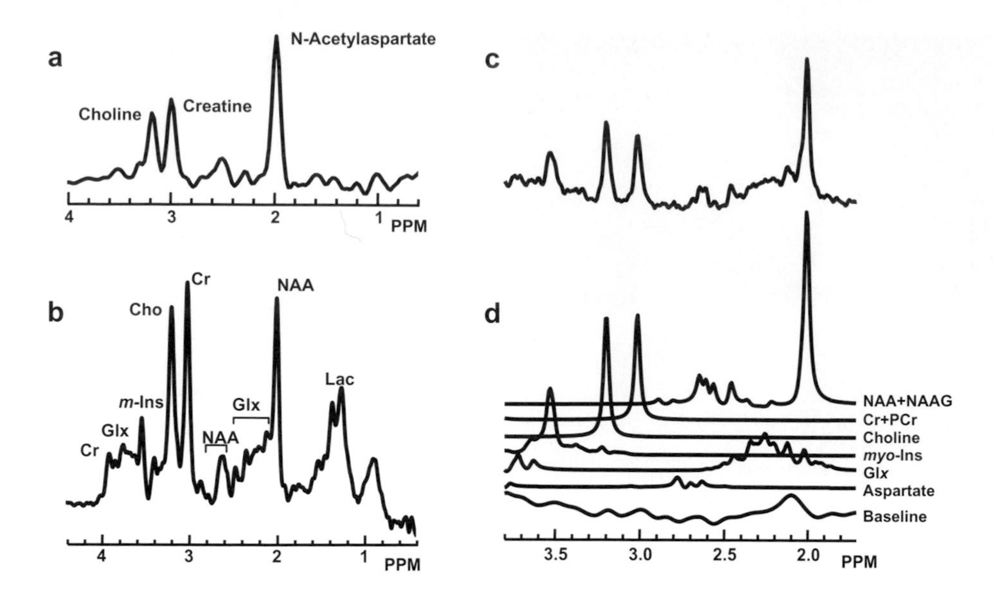

Figure 2 Example of *in vivo* proton spectra of human brain obtained at a field strength of 1.5 T. (**a**) Spectrum from a spin-echo MRSI acquisition for TE = 135 ms. Effective voxel volume is 1.6 ml, obtained with a 30-min acquisition time. (**b**) Single-voxel STEAM acquisition for TE = 30 ms, from an infant with methylmalonic academia that demonstrates increased lactate (Lac) (courtesy of Dr. Peter B. Barker). Data were obtained from an 8-ml volume in 4.3 min. (**c**) Spectrum from a PRESS-localized MRSI for TE = 25 ms (1.6 ml volume, 30-min acquisition). Spectral analysis enables the individual spectral patterns from each metabolite to be separated as shown in (**d**), which has been done for the spectrum shown in (**c**). Abbreviations: creatine (Cr), phosphocreatine (PCr), and inositol (Ins). Glx represents a combination of glutamate and glutamine, whose individual contributions cannot be separated at this field strength.

term, defined in Hertz. The resultant resonance frequency is then given by the sum of the chemical shift and the spin-coupling term.

For protons, the *J*-coupling term is typically smaller than the chemical shift with the result that each of the coupled spins produces a clearly identifiable multiplet resonance group. However, when observed *in vivo,* the linewidths tend to be on the same order, or greater than, the *J*-coupling, so that the coupling patterns are frequently obscured. Another consequence of spin coupling is that the individual resonances in the multiplet group develop different phases as they evolve over time, which is typically not refocused in a spin-echo sequence, so that the resultant overlapping multiplet resonance signals can be of different phase and partially cancel one another. Therefore, the appearance of the spectrum varies as a function of the spin-echo time (TE) and the pulse sequence used, as well as with field strength. Examples of ^{1}H spectra of brain for long-TE and short-TE data acquired at 1.5 T are given in Fig. 2. At long TE (Fig. 2a), the spectrum is dominated by clearly identifiable singlet resonances, while at short TE the spectrum contains several additional signals (Figs. 2b and 2c), though the signals overlap considerably and the individual contributions are difficult to identify. The individual metabolite contributions can be separated using spectral analysis methods, described in a following section, as for example shown in Fig. 2d. Figure 2 also illustrates the improved quality frequently obtained using single-voxel methods (Fig. 2b) over MRSI (Fig. 2c) that is largely due to the relative volumes of tissue sampled, which in this case is a factor of 5 smaller for the MRSI measurement.

It can be seen that on going to longer TE values many resonances are lost due to signal cancellation of *J*-coupled resonance groups, and additional signal losses occur due to transverse relaxation (T_2). Although this also means that valuable metabolite information is lost, long-TE techniques are commonly used to simplify the analysis by removing the complex and overlapping spectral patterns. This can be particularly useful at lower magnetic field strengths at which the limited chemical shift separation results in considerable overlap of the spectra from different metabolites.

One example of spin coupling shown in Fig. 2b is the $^3CH_3$ group of lactate, seen at 1.3 ppm, which appears as two peaks. This signal comes from three protons that are magnetically equivalent, which means that they all have the same resonance frequency. However, all three of these protons are coupled to the single proton bound to the ^2C carbon atom, three bond lengths away, which exists in two energy states and causes the splitting of the CH_3 resonance into two lines, with a separation of 6.9 Hz. The resonance from the proton attached to the ^2C atom is similarly modified by its interaction with the $^3CH_3$ group, which results in a splitting into four lines, with an intensity ratio of 1:3:3:1. However, this multiplet group is centered at 4.1 ppm and cannot generally be seen *in vivo* due to its close proximity to the suppressed water resonance at 4.7 ppm. Other examples of *J*-coupled spectra are given by the NAA CH_2 group (at ≈2.5 ppm) and the spectra for *myo*-inositol and glutamate, though these signals are sufficiently broadened in Figs. 2b and 2c that the underlying multiplet patterns are not discernable.

The shape of each resonance is determined by two factors: the transverse relaxation rate, T_2, of that resonance and the variation of the B_0 field over the sampled volume. If the T_2 dominates, the lineshape will be Lorentzian, though for *in vivo* ^1H MRS the magnetic field inhomogeneity frequently dominates, which causes a spread of resonance frequencies that is commonly approximated by a Gaussian distribution. Typically, a B_0 homogeneity requirement of less than 0.1 ppm over the sampled volume of tissue is required; otherwise distortions of the lineshape complicate the analysis and reduce the peak signal amplitude. Therefore, implementation of MRS typically requires that greater attention be paid to optimizing the B_0 field distribution than is required for MRI. Many of the B_0 field variations in the head are caused by magnetic susceptibility differences between tissue, bone, and air spaces. Over small regions, such as those studied by single-voxel MRS methods, these local B_0 variations can be corrected for by using the available shim systems; however, it becomes a much more difficult problem to fully correct for these field variations over a wide region in the head. Fortunately, for MRSI studies it is possible to relax the homogeneity requirement over the whole head provided that the field variation is small over the region contributing signal to each voxel. Nevertheless, effective shimming is an important consideration for MRSI, and uncorrected field inhomogeneities limit which regions of the brain that can be observed (Spielman *et al.*, 1998; Ebel *et al.*, 2001).

B. The in Vivo Spectrum of Brain

1. The Proton Spectrum

For *in vivo* MRS, observation of protons has several advantages in that it offers the greatest NMR sensitivity, signals can be obtained from many compounds that are present at sufficient concentration, and MRI systems are optimized for observation of this nucleus. However, it also presents several technical challenges. A simple pulse-acquire ^1H MRS measurement of tissue would be dominated by an intense water signal, providing minimal spectroscopic information. Therefore, an essential requirement is to first suppress the water signal. Several methods have been developed (Moonen and van Zijl, 1996), with the most common for *in vivo* studies being frequency-selective excitation of the water resonance using a series of three (or more) RF pulses that have minimal excitation of neighboring metabolite resonances, each followed by a gradient pulse (Haase *et al.*, 1985). This sequence leaves the water magnetization spread out over all phase angles in the transverse plane, resulting in self-cancellation of the resultant signal. The suppression efficiency can be improved by modifying the sequence timing and RF flip angles (Ernst and Hennig, 1995). For MRSI, the quality of water suppression can be adversely affected by inhomogeneity of the B_0 field over the region to be imaged since large B_0 shifts will cause the water resonance to move outside of the suppression region.

Following water suppression, the ^1H MR spectrum of brain is dominated by three singlet resonances, each coming from three protons in the CH_3 groups of *N*-acetylaspartate (NAA) at 2.0 ppm, creatine (Cr) and phosphocreatine (PCr) at 3.03 ppm, and choline (Cho) at 3.2 ppm, which includes contributions from phosphocholine and glycerophosphorylcholine (Fig. 2a). These resonances have relatively long T_2 values, so are conveniently observed at longer TE, and are of considerable interest for many clinical neuroimaging applications. The singlet resonance of NAA is particularly prominent, being both present at relatively high concentration (≈8–16 mM) and relatively well isolated, making it feasible to obtain NAA-only images with higher spatial resolution than is possible for the other metabolites (Guimaraes *et al.*, 1999). Although found in other cell types, NAA has been shown to be an effective marker of neuronal cell integrity, with decreased levels being found in a number of diseases involving neuronal damage (Danielsen and Ross, 1999). Reversible loss and even total absence of NAA have been reported (Barker, 2001), implying that NAA may also reflect neuronal cell metabolism in addition to simply cell density. The creatine resonance is relatively stable in a number of diseases, which has resulted in its frequent use as an internal concentration standard; however, this reference method must be used with caution since its concentration is reduced in stroke and tumors, and it has increased concentrations in gray matter relative to white matter. The third prominent singlet resonance comes from choline that is present in several compounds. The multiple origins of this signal complicate interpretations of altered intensities; however, it is frequently associated with alterations of cell membranes and increased levels are diagnostic indicators of malignancy as well as a number of other diseases. Signals from additional metabolites can be seen at long TE if concentrations are elevated. For example, it is possible to detect the lactate doublet at 1.3 ppm following a stroke or within a tumor, singlet resonances from alanine and acetate are observable with malignancy (Kim *et al.*, 1997), and glycine has been observed with nonketotic hyperglycemia (Heindel *et al.*, 1993).

Since lactate represents a sensitive indicator of decreased oxygenation, its observation has been of interest for many MRS studies. However, the signal at 1.3 ppm can be obscured by overlap with broad lipid resonances that may originate from mobile lipids present in malignant cells or introduced from subcutaneous lipid regions as an artifact due to limitations of the localization procedures. Several methods have been developed to clearly separate the lactate signal from the overlapping resonances. The simplest of these is to make use of the different phase evolution of the two resonances of the lactate doublet due to the *J*-coupling when observed using a spin-echo sequence. For all spin systems, a spin echo is formed at a time TE by applying a refocusing RF pulse at a time TE/2 to reverse the chemical shift evolution. However, if the pulse affects both partners of a coupled spin system, the sign of the mutual *J*-coupling term will also be inverted, resulting in a continuation of the *J* evolution during the second half of the spin-echo sequence. For lactate, the phases of the resultant signals are modulated by the term $\cos(\pi\ J\ \mathrm{TE})$, so that at a time $\mathrm{TE} = 1/J_{\mathrm{lac}} = 144$ ms both of the doublet resonances appear as negative signals. Since any overlapping lipid signals are not inverted, this alone may be sufficient to improve detection of lactate. Improved removal of the background signals can be obtained by subtracting one acquisition, at which the phase of the lactate doublet is inverted, from a second one that is not inverted. This can be done by using an additional frequency-selective pulse on alternating acquisitions to selectively invert the coupled partner, such that the lactate doublet signal is positive on one acquisition and negative on the other.

Several spectral editing methods have been developed for lactate as well as for improving observation of several other metabolites that are otherwise difficult to observe due to spectral overlap with other stronger resonances (de Graaf *et al.,* 2000). Examples of such editing sequences include *in vivo* observation of γ-aminobutyric acid (Petroff and Rothman, 1998), taurine (Hardy and Norwood, 1998), and glutathione (Trabesinger and Boesiger, 2001), though these applications have so far been limited to observing signals from larger volumes typical of single-voxel MRS.

At short TE (e.g., <30 ms) several additional low-molecular-weight metabolites can potentially be observed in brain tissue (Govindaraju *et al.,* 2000), though accurate measurement is limited by sensitivity considerations and the ability to separately quantitate the many strongly overlapping resonances. Although over 100 resonances may be detected from tissue extracts at high field, the majority of these resonances lie within a narrow range of only 3 ppm and it is not possible to detect these *in vivo,* where the resonance linewidth may be on the order of 0.1 ppm and low signal-to-noise ratio further obscures the majority of these signals. For ^{1}H MRSI implemented field strengths below 3 T, measurement of additional metabolites is generally limited to *myo*-inositol, taurine, and a combined signal from glutamate and glutamine. Although glutamate is present at relatively high concentrations (≈6–12 mM/kg), its signal is spread over a wide region, which overlaps strongly with glutamine and other compounds, complicating detection. Some additional lower concentration metabolites can be detected with ^{1}H MRSI at lower field strengths by trading detection sensitivity for larger voxel volumes and longer data acquisition times and perhaps by limiting the imaged region by using surface coil detection. For example, imaging of increased brain histidine following oral loading, estimated at the 1 mM/kg concentration range, has been demonstrated (Büchert *et al.,* 2001).

Short-TE ^{1}H spectra also contain strong signal contributions from macromolecules, including proteins and polypeptides (Behar *et al.,* 1994). Resonances from these relatively immobile compounds have broad linewidths and appear as slowly varying baseline features that further complicate the analysis of these spectra. This background signal has been included in the analysis shown in Fig. 2c and separately plotted. Although lipids represent a large fraction of the brain tissue volume, these high-molecular-weight compounds have relaxation rates that are too short for observation by *in vivo* MRS methods. However, signals from mobile lipids can be observed with necrosis associated with malignancy (Negendank *et al.,* 1996) or multiple sclerosis (Davie *et al.,* 1994). A strong lipid signal is also obtained from subcutaneous regions, which if not reduced can seriously affect the quality of MRSI data over the whole field of view (FOV). Experimental and postprocessing approaches to this problem are described in the following sections.

2. The Phosphorus Spectrum

The sensitivity of ^{31}P observation is approximately 15 times smaller than for protons; however, the *in vivo* spectrum enables monitoring of a small number of compounds that are of considerable importance for energy and lipid metabolism. Therefore, although spatial resolution for ^{31}P MRSI must be further reduced, these studies are of considerable clinical interest. An example of the ^{31}P spectrum of brain is shown in Fig. 3a. It contains three resonance groups from adenosine triphosphate (ATP), a well-defined resonance from phosphocreatine, and less clearly resolved resonances from inorganic phosphate (Pi); phosphodiesters (PDE), chiefly phosphocholine and phosphoethanolamine; and phosphomonoesters (PME). The three ATP resonances are split by *J*-coupling into two doublets (for the γ and α) and a triplet (for β). The Pi peak increases in the presence of anaerobic metabolism and is of additional interest since its resonance frequency is pH dependent within the range of physiological interest. Underlying these resonances is a broad resonance from less mobile molecules such as phosphorylated proteins and the phospholipid bilayer of cell membranes. The resonances of PCr, Pi, and ATP provide

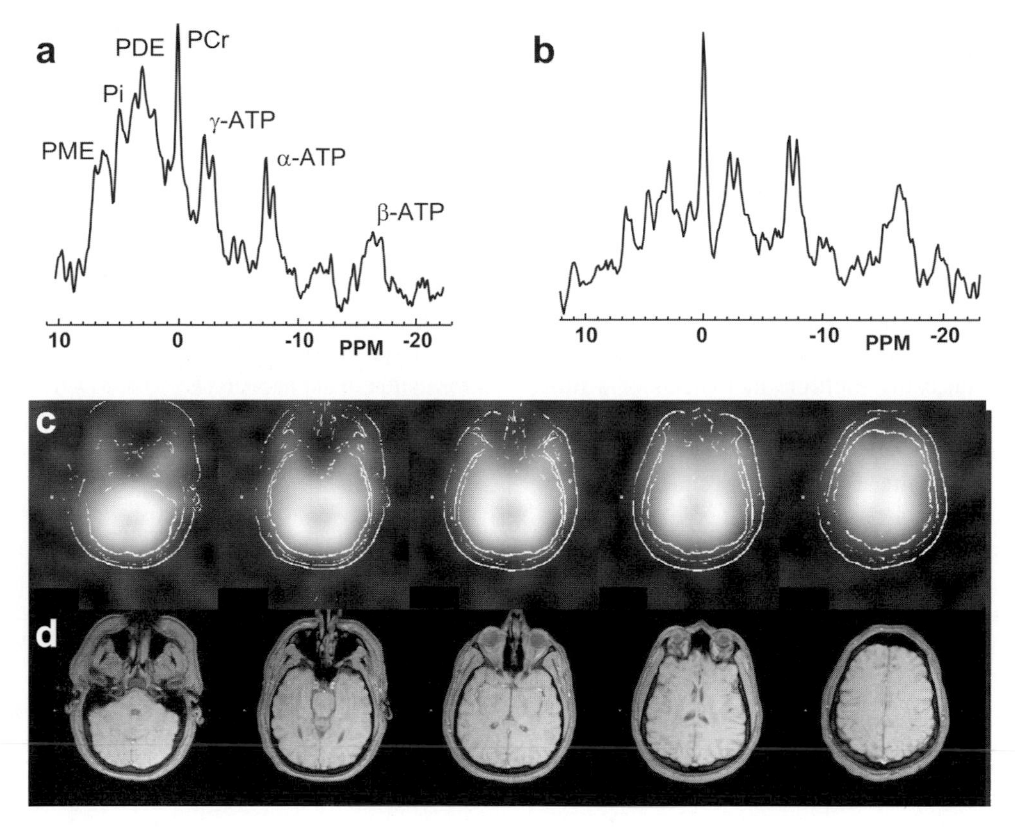

Figure 3 Spectra for phosphorus-31 measurement in brain obtained without (**a**) and with (**b**) proton decoupling and off-resonance excitation. Images of the total ^{31}P distribution are shown in (**c**) together with an overlay derived from the MRI data (**d**) indicating the edges of the head and brain. The 3D MRSI data were obtained at 1.5 T, with an effective spatial resolution of 26 ml.

AU: Please verify part labels

information about high-energy phosphate metabolism, while the PME and PDE metabolites reflect alterations to membrane phospholipids. The underlying broad motion-restricted phospholipid signal can give information on the rigidity of the cell membrane (Estilaei *et al.,* 2001), though observation requires measurement at short TE.

Quantitation of the PDE and PME spectral region can be improved by removing the line broadening due to the ^{1}H–^{31}P coupling using a technique known as proton decoupling, in which RF irradiation is applied at the ^{1}H frequency during the ^{31}P observation. Some additional increase in the ^{31}P signal is also provided through the nuclear Overhauser effect, which transfers energy through direct spin–spin interaction from the ^{1}H to the ^{31}P nucleus. This SNR improvement has also been demonstrated for ^{31}P MRSI of the brain (Arias-Mendoza *et al.,* 1996). An additional simplification of the ^{31}P spectrum can be obtained by using an off-resonance selective irradiation to saturate the broad signal coming from bone and lipid (McNamara *et al.,* 1994). An example of the resultant ^{31}P spectrum of brain using proton decoupling and off-resonance irradiation is shown in Fig. 3b. This provides considerable simplification of the region to the left of PCr and enables clearer identification of Pi and hence more accurate pH measurement.

Also shown in Fig. 3 are images obtained from the total ^{31}P signal in brain, obtained with a 3D MRSI measurement using spherical *k*-space encoding with 12-point diameter, over a 270-mm FOV, for an effective voxel volume of 26 ml. The measurement used a spin-echo excitation, for a TE of 3.2 ms, and a total data acquisition time of 48 min. Since spatial resolution is relatively poor for this type of data, analysis is typically done following selection of spectra from individual voxel locations, rather than viewing images of individual ^{31}P metabolites.

III. MRSI Data Acquisition Methods

Acquisition of MR spectroscopic information in a spatially resolved manner requires the addition of a spectral dimension to conventional MRI data. The acquired data, *s,* therefore includes a time dimension in addition to the usual spatial *k*-space dimensions, and a Fourier transform reconstruction can be used to obtain the final data, *S,* which we can describe by the following relation:

$$s(k_x, k_y, k_z, t) \underset{FT^{-1}}{\overset{FT}{\rightleftarrows}} S(x, y, z, \omega). \tag{2}$$

Therefore, single-slice (2D) MRSI will actually be acquired as a three-dimensional data set, and volumetric (3D) MRSI will be obtained as a four-dimensional data set. Methods for obtaining the spatial information are identical to MRI, and the spectral information can be obtained by similar methods, i.e., either by direct observation or by encoding over multiple measurements. The most common method is to directly sample the spectroscopic time evolution in the absence of any applied magnetic field gradients and to use gradient phase encoding to obtain all spatial k-space information. It is also possible to encode the spectroscopic information using multiple increments of a time evolution period (Sepponen, 1985), followed by direct sampling of spatial information in the presence of an applied readout gradient. This latter method has been less widely used since acquisition of detailed spectroscopic information requires a large number of encoding measurements, though it is feasible for short TR methods or by incorporating alternative spectral analysis procedures that make use of *a priori* information (Ebel *et al.*, 2000).

The MRSI method is illustrated in Fig. 4, for a single-slice excitation through the head. A three-dimensional data set is acquired, which after Fourier transformation produces a set of spectra for all spatial locations, illustrated in Fig. 4b. The simplest method for formation of images corresponding to individual metabolite resonances is to then integrate the spectra over selected spectral regions, for all voxels, producing images illustrated in Fig. 4d. This method of image formation is limited in that contributions from overlapping spectral signals are not distinguished, as for example the NAA and lipid signals that are combined in a single image. Alternatively, individual spectra can be independently analyzed, and improved methods of spectral analysis and image formation can be used, as described in the following section. A coregistered MRI study (Fig. 4c) provides greater structural information to aid in analysis of the lower resolution metabolite images and for selection of spectra from specific locations.

A mathematical model of the 2D spectroscopic imaging acquisition can be described by

$$s(k_x, k_y, t) = \iint_{x\,y} \sum_r \rho(x, y, r) \cdot e^{-i\left((\omega(r)+\gamma\cdot\Delta B_0(x,y))t + k_x x + k_y y + \phi(r) + \phi(x,y)\right) - t/T_2(r)} \cdot \partial x \cdot \partial y \cdot dr. \tag{3}$$

The index r is used to indicate all contributing resonances; $\rho(x,y,r)$ is the signal amplitude distribution for each resonance; $\Delta B_0(x,y)$ is a spatially dependent frequency shift

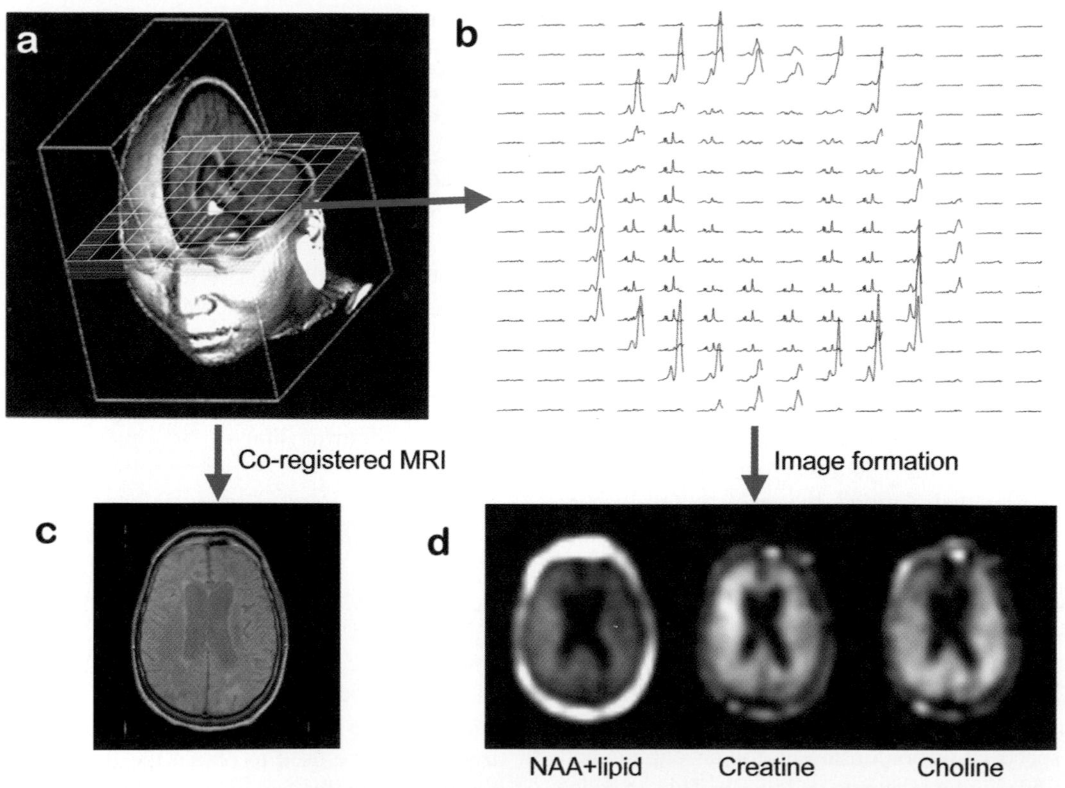

Figure 4 Overview of the 2D MRSI method and data analysis. Signal is obtained from a slice selected within the head (**a**), and spectra are obtained over an array of voxels distributed over the selected plane (**b**). Spectra may be selected for further analysis from specific locations, with the aid of a spatially coregistered MRI (**c**), and metabolite images can be obtained by integrating over different spectral regions, for all voxel locations (**d**).

due to the magnetic field inhomogeneity; γ is the gyromagnetic ratio; $\phi(r)$ is the phase of each resonance, which may be altered by *J*-coupling evolution; $\phi(x,y)$ is a zero-order phase term; $T_2(r)$ is the transverse relaxation rate of each resonance, which may also be considered spatially variant if this information is available; and $i = \sqrt{-1}$.

The definitions of the spatial encoding terms for k_x and k_y in Eq. (3) correspond to those commonly used for MRI, and perhaps the simplest and most commonly implemented method of obtaining these data in the MRSI acquisition is to use gradient phase encoding for all required spatial dimensions, with signal observation in the absence of any applied magnetic field gradients. Pulse sequence diagrams using this method are illustrated in Fig. 5 for 2D and 3D MRSI variants. For these, using k_x as an example, the phase-encoding terms are given as

$$k_x = \gamma \int_0^{t_{enc}} G_x(t) \cdot t \cdot \partial t, \quad (4)$$

where t_{enc} is the time over which the encoding gradient G_x is applied.

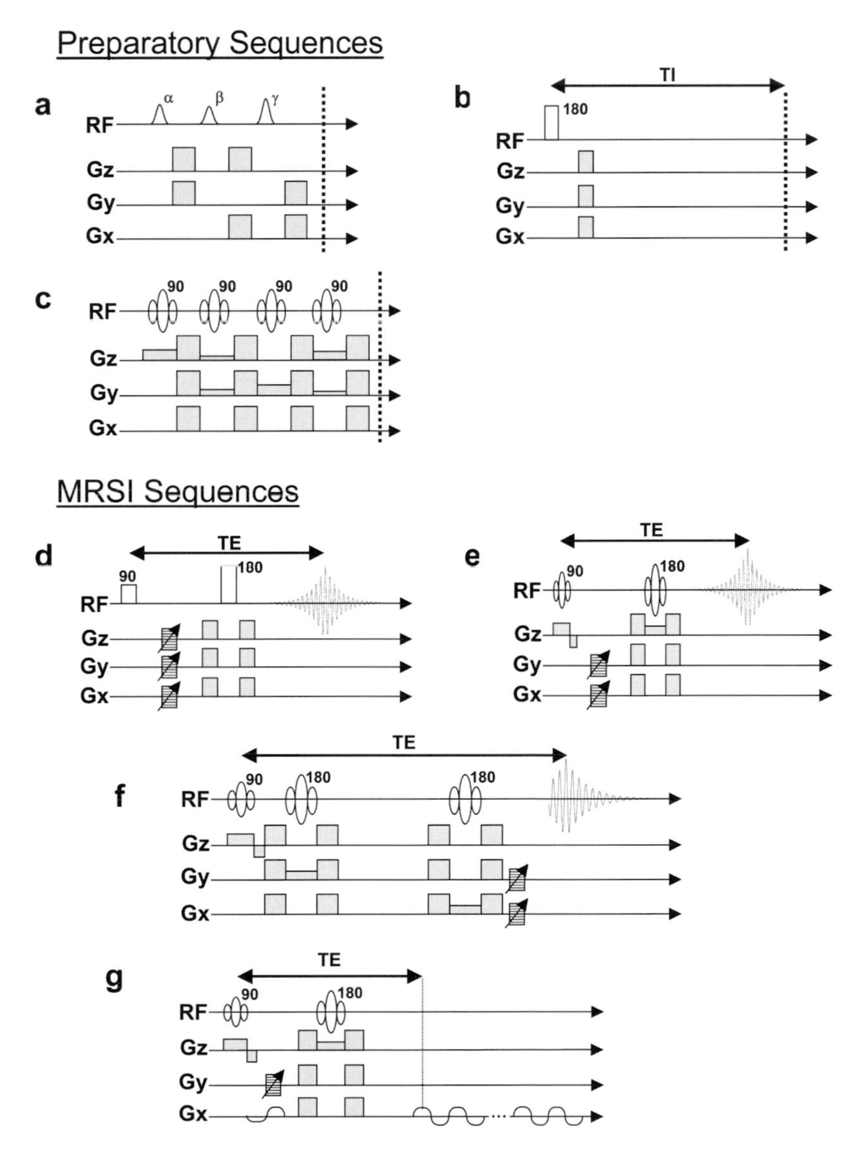

Figure 5 Example components of pulse sequences used for MRSI. Preparatory sequences that commonly precede a ^{1}H MRSI acquisition sequence include (**a**) water suppression, (**b**) inversion-recovery lipid suppression, and (**c**) outer volume saturation. The example MRSI sequences include (**d**) 3D methods, whereby all spatial information is obtained using gradient phase encoding; (**e**) 2D MRSI, which includes slice-selective spin-echo excitation; (**f**) PRESS-selected MRSI, which obtains data only over the selected volume; and (**g**) 2D echo-planar spectroscopic imaging.

Also illustrated in Fig. 5 are several preparatory sequences commonly used in ^{1}H MRSI studies. First is shown the water suppression sequence, described earlier. The next two sequences are commonly used to eliminate signals from subcutaneous lipids. The first of these uses a simple inversion pulse, which inverts all magnetization over the whole object. Since the T_1 of lipid is shorter than most metabolites, if the subsequent signal excitation is carried out at a time, TI, set to the time when the lipid longitudinal magnetization passes through zero, then minimal lipid signal is obtained in the subsequent MRSI sampling period. The third preparatory sequence (Fig. 5c) uses saturation pulses to eliminate all signals from specific spatial locations. The most common implementation is to place four or six slice-selective excitation pulses around the edges of the head, thereby eliminating the subcutaneous lipid signals (Duijn *et al.*, 1992; Posse *et al.*, 1994). A method complementary to the outer volume saturation method is the volume-localized acquisition sequence of Fig. 5f. In this example, the PRESS localization sequence is used to limit the excitation to a region that is fully contained within the brain, though other volume localization methods can also be used.

One consideration associated with volume selection methods, and, indeed, with slice selection in general, is that because the spatial selection pulses are frequency-selective, the position of the selected volume is also dependent on the chemical shift of each resonance. This so-called chemical shift artifact depends on the available gradient strength and RF power, but for most modern MRI systems at 1.5 T this effect means that the localized volume for choline at 3.2 ppm is typically shifted by 1 or 2 mm from that of NAA at 2.0 ppm, and this displacement will increase at higher field strengths. For MRSI, this effect typically means that voxels at the edges of the selected region have slightly different effective volumes for each resonance with a different chemical shift, and these data should therefore be excluded from further analysis. Since phase encoding is independent of the chemical shift, the spatial origins for all remaining voxels within the selected volume are identical for all resonances.

The gradient phase encoding methods of Figs. 5d and 5e become limited for higher spatial resolutions, since a separate measurement is required for each *k*-space point, leading to unacceptably long acquisition times when larger numbers of phase encodings are used. This can be overcome by using high-speed acquisition techniques, one example of which is the echo-planar spectroscopic imaging (EPSI) approach (Guilfoyle *et al.*, 1989) illustrated in Fig. 5g. In this case, the readout period consists of an alternating gradient, causing multiple gradient echoes to be generated as a function of time. The successive echoes provide information on the spectroscopic signal evolution, while the signal sampled for each echo provides one dimension of spatial *k*-space information. If there are M echoes, each of N sample points, then reordering into a $M \times N$ data set and Fourier transformation provides a signal that describes the spectroscopic evolution in one dimension for each of the N points along the spatial readout dimension. Phase encoding is then used to obtain additional spatial dimensions. A further modification of the EPSI method is to use spiral readout gradients (Adalsteinsson *et al.*, 1998), which provide information in two spatial dimensions on each acquisition, potentially providing an N^2 improvement in data collection efficiency. High-speed MRSI methods have been reviewed by Pohmann *et al.* (1997) and Maudsley (1999).

Since *in vivo* MRS is normally limited by sensitivity and multiple acquisitions are required for signal averaging, it may not be immediately obvious that faster acquisition methods are of benefit. However, signal averaging occurs concurrent with the multiple spatial encoding measurements; therefore, EPSI techniques can be effectively used to obtain additional information within the same acquisition time, without loss of sensitivity. One important use is to obtain the additional spatial information for 3D MRSI (Posse *et al.*, 1994; Adalsteinsson *et al.*, 1998), but it also provides opportunities to use weighted *k*-space distributions to reduce truncation artifacts (Adalsteinsson *et al.*, 1999), which are described in a following section (Ebel *et al.*, 2001a), and to obtain additional spectral discrimination using multiple-acquisition editing sequences and 2D NMR methods (Adalsteinsson and Spielman, 1999).

Spatial resolutions for MRSI can be defined in terms of both the "nominal" and the "effective" voxel size or volume. For 2D MRSI the nominal voxel size within the image plane is given by the FOV divided by the number of phase-encoding, or sampled, data points in each dimension, and this is multiplied by the slice thickness to give the nominal voxel volume. However, the final signal at each image voxel represents contributions not only from within the corresponding sample volume, but also from a larger region that includes the effect of smoothing due to the spatial response function (SRF) of the acquisition and data processing, and this region is defined in terms of the effective voxel volume or resolution. For *in vivo* MRSI of whole brain at 1.5 T with acquisition times on the order of 30 min, the typical performance ranges from $\approx$1 ml nominal ($\approx$1.6 ml effective) voxel volumes for ^{1}H observation to $\approx$15 ml ($\approx$25 ml effective) for ^{31}P. However, the spatial and temporal resolution attainable with MRSI can vary considerably from these values. First, surface coil observation can greatly improve reception sensitivity, and nominal resolutions of as small as 0.2 ml have been acquired over small regions close to the surface using this method (Noworolski *et al.*, 1999). The increased sensitivity also enables rapid imaging using EPSI techniques, allowing functional spectroscopy measurements to be performed. For example, 2D ^{1}H MRSI with 1.125-ml nominal resolution has been demonstrated at 1.5 T within 16 s, which also used a phased array reception coil (Posse *et al.*, 1997). Similarly, improved spatial resolution can be obtained by increasing sensitivity by operating at higher field strengths. For example, operation at 4 T provides a theoretical sensitivity improvement of a factor of 2.7 (Hoult and Lauterbur, 1979), and imaging of whole brain has been demonstrated with 0.5-cc nominal (1.15 ml effective) voxel volume for ^{1}H MRSI (Hetherington *et al.,* 1994) and 3.4 ml (0.79 ml effective) for ^{31}P MRSI (Twieg *et al.,* 1994).

IV. Data Processing Methods

A. Basic MRSI Reconstruction

Reconstruction of both spatial and spectral information in MRSI is done using Fourier transformation of the acquired data. Additional preprocessing steps commonly applied include smoothing and interpolation, to improve visual representation of spatial images and spectra, and for ^{1}H MRSI, any residual water signal is further reduced by subtracting an estimation of the water signal obtained using either a low-pass filter convolution or a fit to the data (reviewed by Vanhamme *et al.,* 2001). Following processing, spectra can be examined from any voxel across the imaged region, which is greatly facilitated using a spatially coregistered MRI to enable clearer identification of the region to be examined (Maudsley *et al.,* 1992). Because of the lower spatial resolution, situations will occur in which the MRSI voxel is not centered exactly over a region of interest indicated by the MRI. Interpolation of the spatial dimension, conveniently implemented by zero-filling of the *k*-space data prior to Fourier transform, allows improved registration accuracy for any location over the FOV, and an interpolation factor of 2 is typically sufficient given that the SRF is also relatively broad. Alternatively, the image reconstruction can be repeated with "voxel shifting," which is implemented by multiplying the *k*-space data by a linear phase term in each dimension, to center the voxel over just one specific point of interest (Nelson, 2001).

Images of the metabolite distributions can be created by integrating over the corresponding spectral regions for all points in the spatial dimensions. However, this approach may result in images that contain multiple signal contributions from overlapping resonance groups and baseline signals, as well as variable image intensities coming from spatially dependent frequency shifts, namely the $\Delta B_0(x,y)$ term of Eq. (3). These B_0 shift artifacts can be reduced by correcting the frequency offset of the spectra in each voxel prior to integration, using a map of the B_0 variation across the object. The B_0 map can be obtained from the data itself (Maudsley and Hilal, 1985), from a separate MRSI of the unsuppressed water (Spielman *et al.,* 1989; Maudsley *et al.,* 1994b), or from a coregistered MRI measurement of the B_0 field (Webb *et al.,* 1992). A related procedure using time-domain deconvolution also corrects for time-varying values of $\Delta B_0(x,y,t)$ that may be caused by gradient eddy currents. In this case, the correction function would be derived from a separate MRSI without water suppression (Johnson *et al.,* 1993) or instrument calibration procedure. However, even with B_0 correction, metabolite image formation by signal integration is subject to errors from overlapping signals, and improved identification of individual metabolite contributions is possible using spectral analysis methods described in a following section.

B. Truncation Artifacts

For *in vivo* MRSI, the spatial resolution must be compromised to offset the sensitivity limitations, and optimal sensitivity is obtained when the acquired number of *k*-space data points matches the final spatial resolution. Consequently, a relatively small number of phase-encoding measurements are used. For brain studies at 1.5 T the typical performance ranges from ≈32 spatial sample points in each dimension for ^{1}H observation to as few as 8 for ^{31}P MRSI. However, Fourier transformation of insufficiently sampled data leads to the well-known phenomenon of truncation artifact, also commonly known as Gibbs ringing. Although this effect also occurs in MRI, there the *k*-space sampling is generally sufficiently wide such that there is minimal impact on the image quality; however, this is not the case for MRSI, where it can result in undesirable image intensity artifacts and signal contamination, or leakage, across voxels. These effects can lead to misinterpretation of metabolite images and errors if absolute metabolite quantitation is required. The truncation artifact is illustrated for one spatial dimension in Fig. 6, for sampling of 8 *k*-space points (indicated by those points with value >0 in Fig. 6a) that have been zero-filled to 16 points (the zero-value points) prior to Fourier transformation, which shows the resultant interpolated SRF (Fig. 6b). Ideally, the result should be 100% signal at one point with 50% in the immediate neighbors due to the interpolation, indicating that each voxel in the final reconstructed MRSI data includes signal from the corresponding sample volume only; however, it can be seen that instead the signal is spread over several data points in the form of a sinc function. Each voxel in a MRSI data set therefore includes signal contributions from several neighboring voxels, summed with a value given by the SRF as a function of distance away from that voxel, which is given by the convolution of the actual image intensity distribution with the SRF. In addition, because the *k*-space distribution is asymmetric there is also some signal in the imaginary part of the SRF, as indicated by the dotted line in Fig. 6b. The resultant distortions are most pronounced at sharp changes of intensity such as the edges of the head, resulting in intensity variations that may propagate throughout the image.

Spatial filtering is commonly used to reduce the truncation artifact ringing. This is done by multiplying the *k*-space

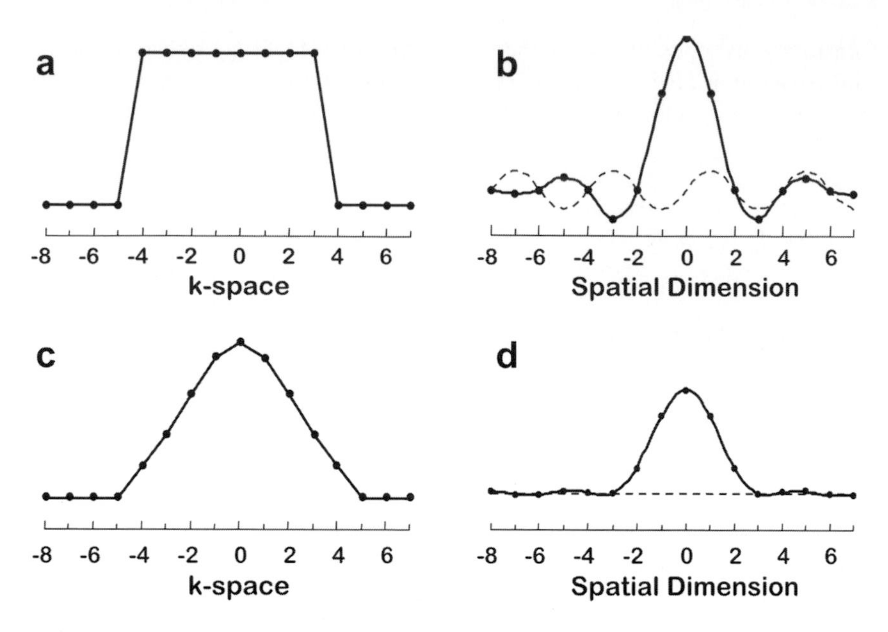

Figure 6 Illustration of the spatial response function typical of MRSI. The relatively small number of *k*-space sample points, indicated by the points with value >0 in plots (**a**) and (**c**), are zero-filled and Fourier transformed to produce the real (solid line) and imaginary (dashed line) function, (**b**) and (**d**), that represents the amplitude with which the corresponding locations in the object contribute to the image voxel, in this case at the center point. Multiplication of the *k*-space data by a weighting function (c) is used to reduce the ringing, with a concomitant loss of spatial resolution (d).

data by a suitable apodization function that reduces the amplitudes of the higher *k*-space data points (Fig. 6c). While this reduces the ringing, there is a concomitant decrease in spatial resolution (Fig. 6d), as indicated by measuring the width of the SRF at 50% of maximum amplitude. An additional modification can be made by taking a symmetric *k*-space sampling distribution, i.e., using an odd number of points, which eliminates the imaginary term of the SRF.

A result similar to spatial filtering can be obtained by weighting the sampled *k*-space data on acquisition, perhaps in conjunction with some additional postprocessing filtering. These two approaches have the same effect on the SRF, though the weighted acquisition will have different noise character and makes better use of the acquired data. Weighted acquisitions can be done by varying the number of signal averages as a function of *k*-space (Parker *et al.,* 1987), which may be possible with ^{31}P studies in which multiple signal averages are required (Hugg *et al.,* 1996) or by varying the *k*-space trajectories used in high-speed MRSI methods, as for example with variable-spiral trajectories (Adalsteinsson *et al.,* 1999). Distributing the *k*-space points over a circular or spherical region (Maudsley *et al.,* 1994a), or a Gaussian-weighted random distribution (Ponder and Twieg, 1994), can also provide some improvements by distributing the ringing more evenly.

The truncation artifact is a property specific to Fourier transformation and can be avoided by using alternative, non-Fourier, image reconstruction methods (Hu *et al.,* 1991; Liang *et al.,* 1992; Plevritis and Macovski, 1995; Wear *et al.,* 1997). While these more computationally intensive techniques are of considerable interest, they remain to be sufficiently developed for general use. A simpler alternative is to improve the spatial reconstruction using postprocessing methods to selectively extrapolate the *k*-space data, based on knowledge of the tissue distributions of the imaged region. This approach is particularly effective for reducing ringing from subcutaneous lipids in ^{1}H MRSI, since this signal comes from a narrow and well-defined region that can be readily identified either from the MRSI data itself or from a coregistered MRI. One such extrapolation procedure is the Papoulis–Gerchberg algorithm, and an example of its application to ^{1}H MRSI data (Haupt *et al.,* 1996) is shown in Fig. 7. This method applies an iterative procedure that progressively estimates additional *k*-space data for the lipid signal only, while leaving the original *k*-space data and the reconstructed data over the brain region unaltered. With this approach, it becomes possible to obtain images of whole brain without requiring volume preselection to eliminate the subcutaneous lipid regions. However, the presence of the high lipid signal also tends to make the MRSI acquisition more sensitive to artifacts caused by subject motion. Limits to the performance of the lipid extrapolation also mean that the method cannot be used for shorter TE acquisitions, where the ratio of the lipid to the metabolite signals becomes extremely large, unless some additional method is used to reduce the lipid signal during data acquisition such as lipid T_1 nulling (Fig. 5b).

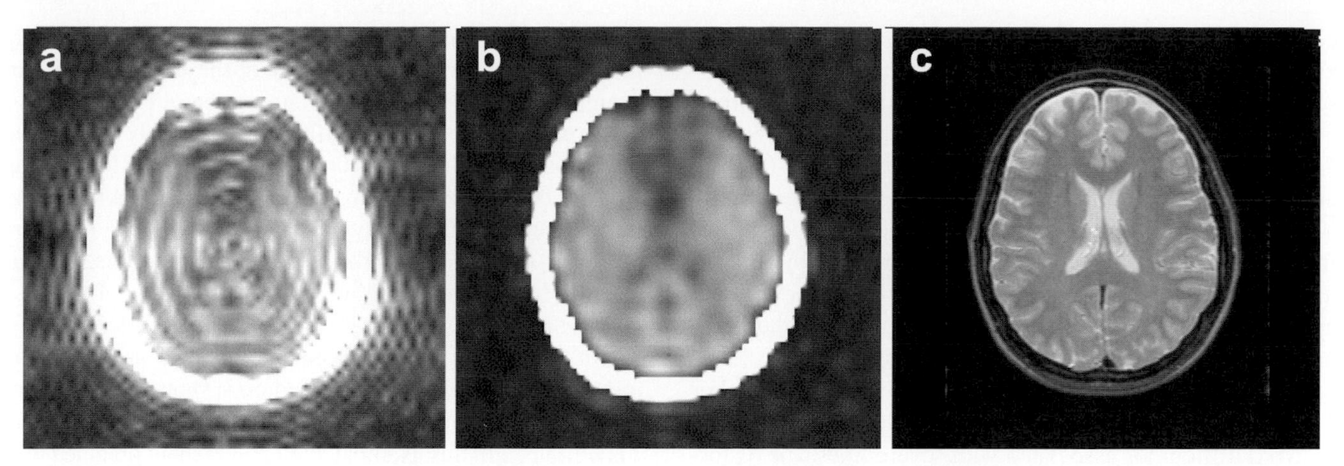

Figure 7 Effect of the truncation artifact for an image obtained by spectral integration of the NAA resonance when a strong subcutaneous lipid signal is present (**a**). Data were obtained for ^{1}H MRSI at 1.5 T with TE = 50 ms and no form of lipid suppression. This artifact can be significantly reduced by applying a postprocessing procedure that extrapolates the *k*-space data for the lipid region only prior to Fourier transform reconstruction (**b**). Also shown are the MRI data for the same slice (**c**). Modified, with permission, from Haupt *et al.* (1996).

An alternative method for accounting for lipid contamination in brain regions is to incorporate the lipid signal into the spectral analysis procedure. Examples using time-domain spectral analysis methods applied to MRSI have been demonstrated (de Beer *et al.,* 1992, 1993) and include an example for a relatively short TE measurement obtained without water suppression (Clayton *et al.,* 2001), though detailed comparison of the applicability of these methods for clinical studies remains to be carried out.

C. Spectral Quantitation Methods

Improved accuracy for measuring individual metabolite signals in a complex spectrum is possible using computer methods that aim to fit a function to the spectral data. If the fitted function is based on a model of the individual metabolite resonances, as well as any variations in the baseline signals, then this approach provides greatly improved separation of overlapping signal contributions over the simple integration approach previously described for metabolite image formation. Many of the spectral fitting methods that have been developed for high-resolution NMR require manual supervision, making them unsuitable for analysis of the large numbers of spectra obtained in MRSI studies as well as being subject to operator bias. However, fully automated spectral analysis methods have also been developed (e.g., see Kay, 1988), and although computationally intensive, these can greatly simplify the task of analyzing large numbers of spectra and their use is considered essential for routine implementation of MRSI. Unfortunately, *in vivo* MRSI data have many undesirable characteristics, including considerable resonance overlap, poor SNR, lineshape distortions, and strong uncharacterized baseline signals, which make many of the automated methods unsuitable for robust analysis. One approach that is widely used for analyzing *in vivo* MRS data is using parametric modeling and optimization methods that make use of *a priori* information on the spectrum (e.g., Barkhuijsen *et al.,* 1985; Provencher, 1993; Stanley *et al.,* 1995; Soher *et al.,* 1998; Slotboom *et al.,* 1998). The success of these procedures depends on an accurate definition of model functions for the metabolites, appropriate accommodation for uncharacterized signals, and frequently a good initial estimate of the model parameters. Fortunately, much is known about the spectral characteristics of *in vivo* MRS signal, and by including this information in the search for the optimal parameters, the procedure becomes faster and more robust in the presence of low SNR. The received signal can be modeled either in the time domain (Vanhamme *et al.,* 2001), as for example by Eq. (3), or in the frequency domain (Mierisová and Ala-Korpela, 2001), and the choice is primarily determined by where the model can be described most accurately or is computationally simplest, given the known and unknown information. There is also some benefit to using mixed-domain signal models, since the metabolite model can be conveniently defined in the time domain, while there is some simplification provided by modeling the baseline signal and doing the least-squares analysis in the frequency domain (Soher *et al.,* 1998; Slotboom *et al.,* 1998).

For *in vivo* MRS, most observable metabolites have been identified and their spectra characterized (e.g., see Robitaille *et al.,* 1991, and Govindaraju *et al.,* 2000, and references therein), which provides considerable *a priori* information to include in the spectral modeling procedure. This prior knowledge can be used to form a set of basis functions, representing the ideal frequencies, amplitudes, and phases of all resonances for each metabolite, that are used within a parametric modeling spectral analysis procedure. Such basis functions can also be generated using NMR measurements of each metabolite in solution (Provencher, 1993), or by computer simulation (Young *et al.,* 1998a), in each case using the same acquisition parameters used for the subsequent *in vivo* measurement. Difficulties due to broad uncharacterized baseline signals can be addressed by adding a smoothly varying signal component in the model (Nelson and Brown, 1987; Provencher, 1993; Soher *et al.,* 1998; Young *et al.,* 1998b), and resonance lineshape distortions from local B_0 inhomogeneity can be addressed by using a variable lineshape model (Provencher, 1993) or by first applying a deconvolution procedure to convert the resonances to a known lineshape (de Graaf *et al.,* 1990; Maudsley *et al.,* 1994b; Morris *et al.,* 1997). However, for MRSI data this latter method must be applied with caution to avoid difficulties associated with severe lineshape distortions (Maudsley *et al.,* 1994b; Wild, 1999).

MRSI data are frequently of lower SNR and contain larger lineshape distortions than are typical for single voxel MRS acquisitions, which increases the need for applying appropriate parameter constraints within any automated analysis procedure. However, the spatial dimensions in MRSI data present an additional opportunity for including prior information to improve the spectral fitting, by incorporating knowledge on the spatial distributions of parameters. For example, the field inhomogeneity distribution [$\Delta B_0(x,y)$ in Eq. (3)] can be separately measured, as previously described, and used to limit the range over which the frequency parameter of the spectral analysis is allowed to vary. Similarly, by using a separate high-speed non-water-suppressed MRSI measurement, the phase term $\phi(x,y)$ can be determined (Spielman *et al.,* 1989). Even if information on spatial distributions of these parameters is not directly available, constraints can be imposed to limit the voxel-to-voxel variations based on the expectation that their spatial dependence varies relatively smoothly. Such spatial constraints have been incorporated in the iterative MRSI reconstruction procedure of Soher *et al.* (1998), by limiting the initial parameter value estimates at the start of each fit iteration based on the statistical properties of the neighboring voxels from the previous fit iteration. This has been found to provide some benefit for voxels at the edges of the brain or at the edges of regions of increased B_0 inhomogeneity.

Any spectral analysis procedure must also consider the uncertainties of the estimation and this is particularly important when clinical decisions are to be based on the result. A calculation of the Cramer–Rao bounds (Cavassila *et al.,* 2001) provides an estimate of the minimum error of the analysis, provided that the model function used is correct, or confidence intervals may be calculated (Young *et al.,* 2000). These measures may be displayed in an image format and viewed alongside images generated following automated analysis of MRSI data to serve as visual and quantitative indicators of how much trust should be placed in interpretation of the images (Young *et al.,* 2000). However, while these methods provide a measure of the quality with which the spectral fit matches the data, they do not necessarily provide information on the quality of the original data. For example, a voxel containing strong baseline variations, perhaps caused by motion artifacts, inadequately suppressed water, or lipid contamination, may be fit well, but should still not be included in any subsequent decisions regarding the outcome of the study. Therefore additional tests are needed to eliminate such data from further analysis, such as testing for unreasonable line widths or low signal to noise (Ebel *et al.,* 2001).

An example of the automated spectral analysis and metabolite image formation procedure is shown in Fig. 8, for ^{1}H MRSI data obtained using 3D EPSI acquisition at TE = 135 ms. In this example, the EPSI readout obtained data for 32 *k*-space sampling points in one spatial dimension and 512 spectral sampling points, while phase encoding was used to obtain an additional 32 × 16 spatial *k*-space points, to obtain a three-dimensional data set. These data have been automatically analyzed to create images for NAA, creatine, and choline, for eight planes, each of 15-mm thickness. Also shown are images representing those voxels that passed a simple quality test based on linewidth of the fit result (Fig. 8d). This test excludes those voxels with excessive B_0 inhomogeneity or with minimal metabolite signal and provides a visual reference to aid in the interpretation of the metabolite images. These images highlight the problems associated with B_0 inhomogeneity, which leads to large lineshape distortions and is the primary cause for rejection of voxels located in lower and frontal brain regions. For comparison with the fitted image results, Fig. 8e includes the image obtained by simple spectral integration over a region including the NAA resonance. These data include a strong additional signal from lipids located around the scalp and the eyes.

D. Signal Normalization

The acquired MRS signal is uncalibrated and, therefore, analysis of MRS data has commonly relied on ratios of the fitted metabolite resonance integrals. This automatically accounts for variations in the reception sensitivity due to inhomogeneity and loading of the RF coil, allowing convenient comparison of data from different regions and between subjects. Analysis of metabolite ratios is appropriate for many applications, particularly if the expected alterations of the two (or more) metabolites occur in different directions, though this also increases the variance of the result. However, for some other applications it has been shown that increased sensitivity can be obtained by analyzing changes of individual metabolites. For these situations, absolute quantitation of metabolite distributions in molar concentration units represents an ideal goal, which would provide direct comparison of metabolite concentrations across multiple MRSI studies as well as with measurements obtained from other methods.

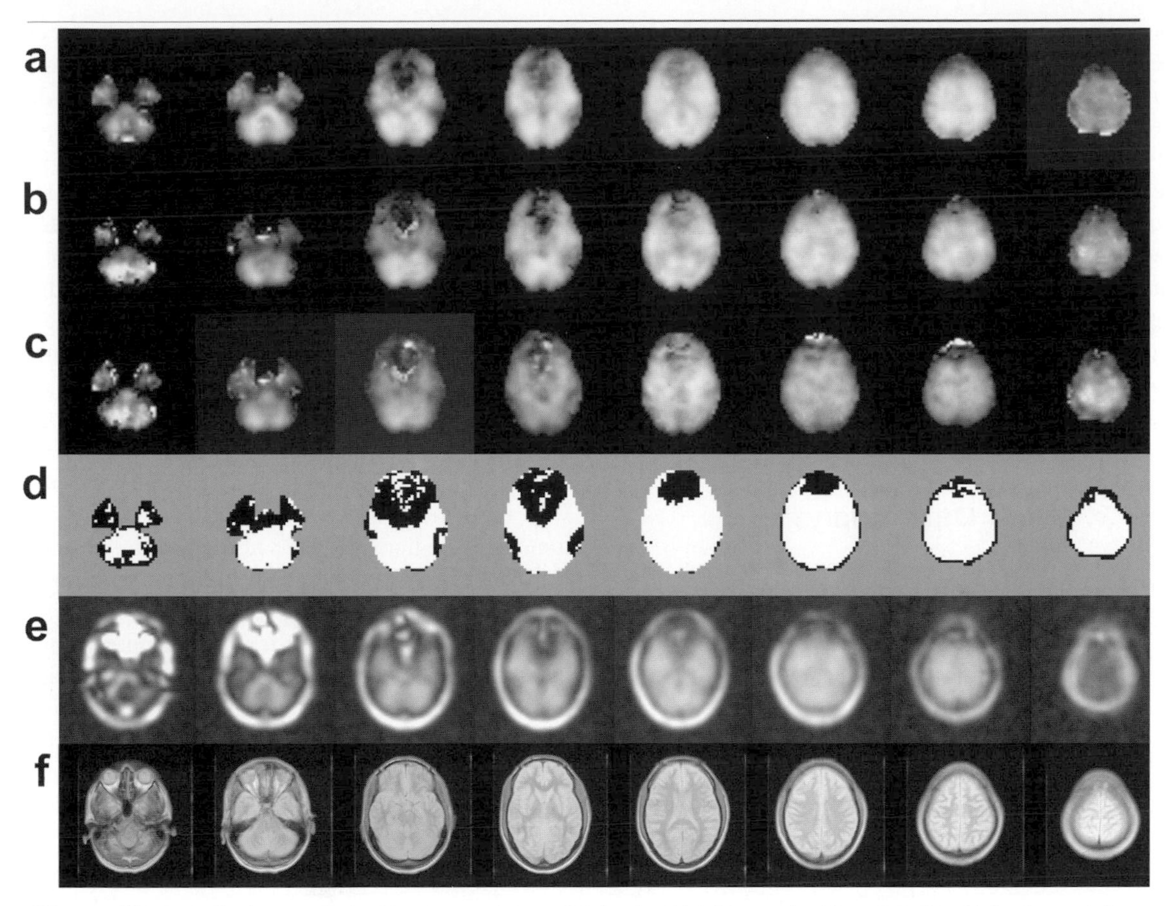

Figure 8 Metabolite images obtained by automated spectral fitting for NAA (**a**), creatine (**b**), and choline (**c**) for data at eight slices through the head obtained using the 3D EPSI method, for TE = 135 ms. In (**d**) are shown are shown results of a spectral quality evaluation, with white indicating accepted voxels and black indicating rejected voxels, which primarily occurred because of B_0 inhomogeneity causing significant line broadening. For comparison, the image obtained by simple spectral integration over the NAA resonance is also shown (**e**) and the MRIs corresponding to the center of each plane of the MRSI (**f**). Modified, with permission, from Ebel *et al.* (2001).

To quantitate MRS signals the data must first be referenced to a concentration standard, while accounting for acquisition parameters, instrumental variables including reception sensitivity distributions, and metabolite relaxation times (Tofts and Wray, 1988; Lara *et al.,* 1993; Danielsen *et al.,* 1995; Soher *et al.,* 1996). Unfortunately, no one metabolite signal can be reliably used as a reference, requiring that other concentration standards be found, and the measurement of relaxation times requires additional data to be acquired, adding significantly to the length of a study and making this impractical for clinical studies. Therefore, it has become common practice to implement a simpler normalization procedure, converting the *in vivo* MRS data to "institutional units," which either omits or uses fixed values to correct for spin relaxation parameters and uses a calibration based on data obtained from a separate measurement. This method allows direct comparison of data acquired at the same field strength and with the same sequence timings, under the assumption of no changes in relaxation rates.

Several methods for obtaining the reference signal measurement have been proposed for MRSI. An external reference can be included within the FOV that includes an observable compound with known volume and concentration (Murphy-Boesch *et al.,* 1998), and calibration of the metabolite signal requires that the B_1 field distribution and the MRSI SRF be taken into account. The phantom replacement method simplifies the calculation by comparing the subject MRSI data with the signal obtained at the corresponding location from a separate MRSI acquisition, using identical parameters, from a phantom containing a solution of known concentration (Lara *et al.,* 1993; Soher *et al.,* 1996). In some situations an internal reference signal can be similarly used, for example using the creatine peak in the subject ^{1}H MRSI data and assuming a concentration of 6.1 mM in white matter (Hetherington *et al.,* 1996), though potential variations of this signal with disease limits application of this method. An alternative reference is available from brain water distributions, including CSF (at 110 M proton concentration). The CSF reference measurement has been implemented using a separate PRESS-localized spectroscopy acquisition (Pan *et al.*, 1998) or from a MRI measurement (Suhy *et al.*, 2000a). The calibration procedure must then account for the volumes of the CSF and metabolite measurements, variations in sensitivity over the imaged region, along with assumed values of the relaxation rates for CSF and the tissue metabolites. Referencing to brain tissue water signal has been recommended for single volume MRS (Christiansen *et al.*, 1992), being readily implemented and having acceptable precision (Keevil *et al.*, 1998), and this method could also be adapted for MRSI by comparison with a proton density MRI.

V. MRSI Data Analysis

A. Standard Analysis Methods

For those studies in which a lesion can be identified from MRI, for example monitoring of malignancy or stroke, or in which the disease is known to affect specific brain regions, MRSI data analysis may be as simple as viewing metabolite ratios or normalized concentration values at one or more operator-selected voxels. This analysis benefits considerably from having a spatially coregistered MRI to identify regions from which to select individual spectra for further analysis and to display alongside the metabolite images to serve as a visual aid. Since the MRI is typically obtained with a much thinner slice thickness than the MRSI data, improved visualization of the metabolite image features is possible by viewing an image obtained by summing all MRI slices that correspond to the MRSI slice thickness (Maudsley *et al.*, 1992). Other capabilities include combined display of multiple metabolite parameters and metabolite and MRI data types using a color display mode. An example of this is shown in Fig. 9, in which metabolite distributions for NAA and choline are separately mapped onto red and green color scales, respectively, and then superimposed on to the MRI data that are displayed in a gray scale (Vigneron *et al.*, 2001). This study shows a region containing a brain tumor, with the result that the green regions clearly identify the tumor, red regions indicate normal ratios of both metabolites, and yellow regions may indicate infiltration into the surrounding normal tissue.

Even if abnormal spectra can be identified by a trained operator, data analysis is commonly done following spectral analysis and comparison of the results with corresponding data from normal subjects, and such operator-independent analysis methods greatly benefit more routine implementation of MRSI for clinical studies. For many other applications, the exact location of interest is not known or cannot

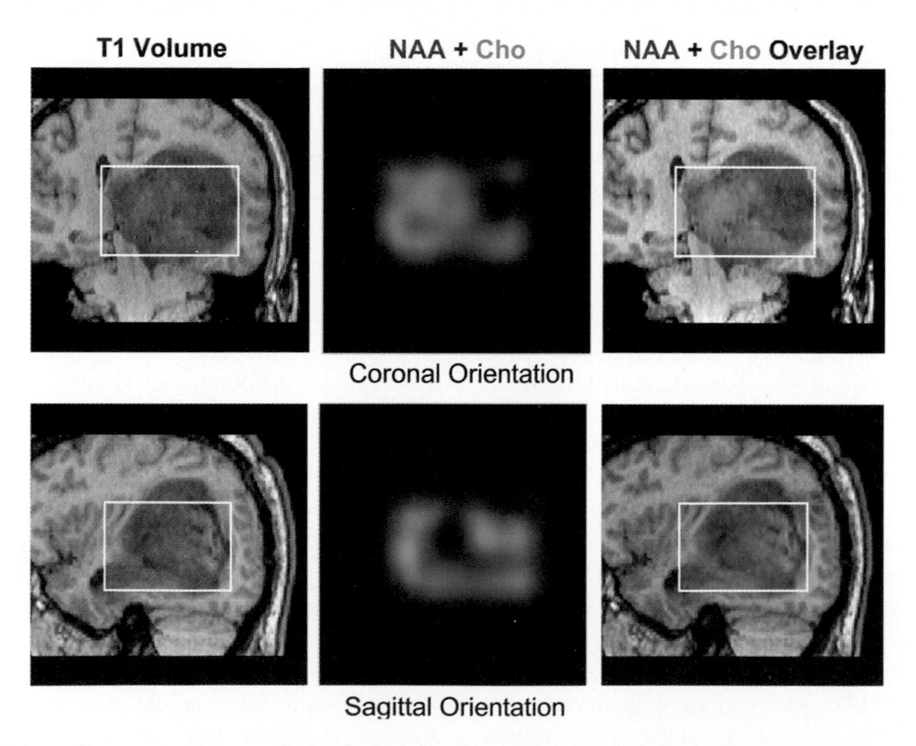

Figure 9 Example of the use of color to display multiparametric metabolite and MRI information. Data are shown for an oligoastrocytoma that demonstrates increased choline (green) and decreased NAA (red), as well as regions with no significant metabolite levels presumed to represent necrosis. This image has then been combined with the MRI to enable visualization of the metabolites in relation to the underlying anatomy. Reproduced, with permission, from Vigneron *et al.* (2001).

be identified from the MRI or MRSI data, and statistical approaches to data analysis are essential. Such comparisons must account for normal biological variability, as well as differences in average metabolite concentrations by brain region, subject variables, and acquisition-dependent variables. Most such analyses have therefore carried out comparisons against results obtained from a matched group of normal subjects, obtained using the same study parameters. The sensitivity of such comparisons, for both individual normalized metabolite values and ratios, can be improved by also accounting for differences in partial volume signal contributions (from CSF) and in the relative gray and white matter tissue fractions, as described in the following section.

Normal variations of metabolites are known to occur as a function of tissue type, brain region, and subject variables. For example, differences with brain region may be on the order of 20% for NAA/Cr and even larger for inositol/Cr, though significant gender or left–right differences have not been found (Komoroski *et al.,* 1999). Intelligence was found to be correlated with NAA and Cho in white matter (Jung *et al.,* 1999), as well as with pH (Rae *et al.,* 1996). Progressive changes of several ^{31}P and ^{1}H metabolite ratios and individual concentrations occur with age (Longo *et al.,* 1993; Chang *et al.,* 1996; Lundbom *et al.,* 1999; Angelie *et al.,* 2001; Schuff *et al.,* 2001a), though the reported findings vary, possibly depending on the measurement technique used and whether tissue contribution was included in the analysis.

With the availability of processing methods that provide fitted metabolite parameters for all MRSI voxel locations, it is possible to implement automatic comparisons of fit results with normal values and, thereby, to automatically identify abnormal tissue regions with a corresponding probability. For subjects with focal abnormalities, the comparison data may be obtained directly from the same subject data provided that normal tissue regions can be identified, and these values then used as an internal control to automatically determine a probability of an abnormality elsewhere in the data. This latter technique has been demonstrated by McKnight *et al.* (2001) to identify voxels indicating malignancy. In this case, the statistical model used assumed a normally linear relationship between Cho and NAA and that the ratio of Cho/NAA is increased with malignancy. Those voxels that fit within this normal distribution could be identified and used as the control data to identify voxels that did not meet this condition.

Given the relatively low SNR of MRSI data, it is not always possible to identify abnormalities directly from individual spectra, and it is frequently unknown which specific tissue regions are likely to be affected. However, an analysis of multiple voxels over a wider region may increase the power of the analysis and enable smaller anomalies to be identified. One such example is the analysis of distributions of metabolite intensities to detect changes in the motor cortex in patients with amyotrophic lateral sclerosis (Schuff *et al.,* 2001b). By viewing histograms of the partial volume-corrected NAA intensity distributions, a distinction between the patient and the normal controls was possible, providing improved sensitivity relative to analysis of a smaller number of operator-selected voxels as well as simplifying the voxel selection procedure. A number of classification techniques can be applied to MRSI data and examples include the use of the principal component analysis for identifying abnormal features (Stoyanova and Brown, 2001) and linear discriminant analysis to assign each voxel to a particular histopathological class on the basis of preestablished classification procedures (Preul *et al.,* 1998; De Edelenyi *et al.,* 2000). In either case, the results can be displayed in an image format.

B. Coanalysis of MRSI and MRI

MRI data are routinely obtained as part of a MRSI study; they provide information that is complementary to MRSI and there are considerable benefits to coanalyzing these two data types to combine metabolic and anatomic information. First, considerable use can be made of information provided by the higher spatial resolution of MRI. The relatively poor spatial resolution of MRSI means that the signal in a specific voxel may originate from multiple tissue types or brain regions, each of which may have different metabolite concentrations. This also means that it may not be possible to select voxels for which the signal corresponds to 100% of one tissue or region, and furthermore, differences in the relative tissue distributions and partial volume contributions may be incorrectly interpreted as alterations of metabolite concentrations. However, these mixed signal effects may at least partially be accounted for by coanalysis of MRSI data with information obtained from coregistered high-resolution tissue segmented MRI data.

An important normalization procedure for MRSI is correction for partial volume signal losses due to CSF (which contributes no observable metabolite signal) to provide metabolite values that represent the concentrations within the tissue, rather than an average value over the voxel volume. The necessary information in the CSF regions can be identified from MRI data, for example using tissue segmentation techniques. In Fig. 10 is shown an example of how significant this contribution can be. Figure 10a shows the CSF regions as identified from segmented MRI data, for each of five 3-mm slices that correspond to the 15-mm slice thickness of a ^{1}H MRSI study. These images have then been integrated (Fig. 10b) and smoothed as described below, to correspond to the spatial resolution of the ^{1}H MRSI data. The resultant image then represents the signal intensity variation in the ^{1}H MRSI data that is due to CSF partial volume contribution. By comparison with the accompanying color scale, it can be seen that in the outer regions of the brain this

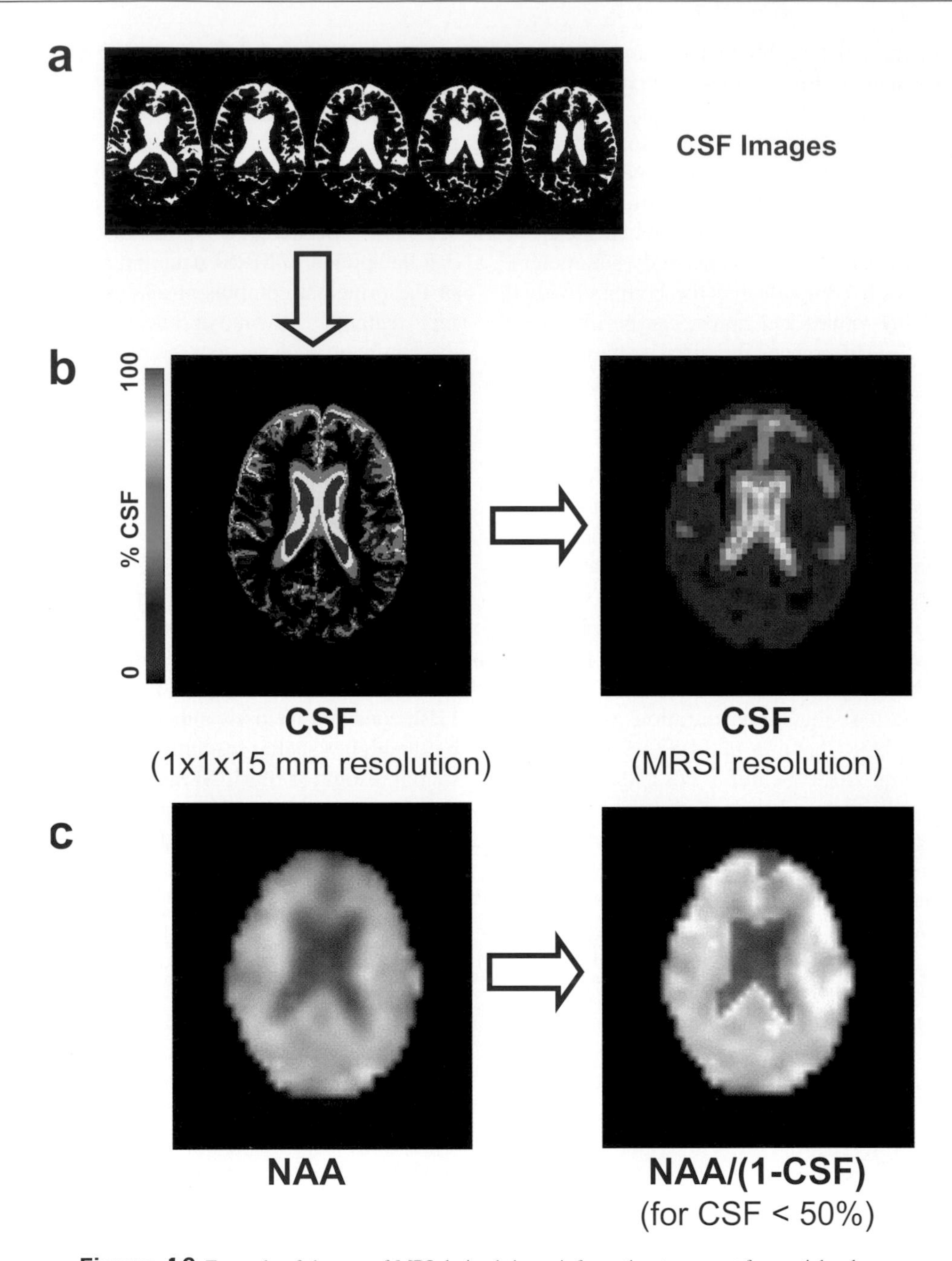

Figure 10 Example of the use of MRI-derived tissue information to correct for partial volume signal losses due to CSF contributions to the MRSI signal. Images of the CSF distribution in a 77-year-old subject (**a**) obtained from tissue segmentation of five 3-mm MRI slices that correspond to the 15-mm slice thickness of the ^{1}H MRSI data are summed and smoothed to correspond to the MRSI spatial resolution (**b**). These data represent the intensity correction factor, which can then be applied to fitted metabolite results (**c**), using an appropriate threshold to limit division by small numbers where CSF concentration is high

correction factor can be up to approximately 40%. By dividing the metabolite images with the relative CSF fraction, the data can then be normalized to account for partial volume signal loss associated with CSF, though appropriate accommodation for regions of low CSF must be taken to avoid errors associated with division by small numbers. If tissue segmentation data are unavailable, the CSF image can also be obtained from a long-TE MRI acquisition using the same slice selection as the MRSI data (Horská *et al.*, 2001).

The smoothing procedure used to calculate the CSF contribution to each MRSI voxel requires knowledge of the MRSI SRF and the distributions of all tissue types, at the higher spatial resolution of the MRI. Additional factors that may alter the metabolite signal intensity must also be taken into account, such as RF coil sensitivity distributions and volume selection profiles. The volume of a specific tissue contributing to the signal at a specific MRSI voxel is then given by the convolution of the MRSI SRF with the tissue distribution function. For two-dimensional data, this must be modified to account for the slice selection profile, which varies spatially as a function of chemical shift offset of each metabolite (Schuff *et al.*, 2001a). Therefore, for tissue, t, contributing to a voxel at position (x,y) for metabolite, m, in a 2D MRSI data set at slice position, z, the effective volume is

$$V_{t,m}(x,y,z) = \sum_{x',y',z'} I_t(x',y',z') \times SRF(x-x', y-y') \times P_m(z-z') \times C_m(x',y',z'), \quad (5)$$

where I_t is the MRI-resolution distribution function for tissue type t; P_m is the slice selection profile, which will vary as a function of the chemical shift of the selected metabolite resonance; and C_m represents additional factors that may alter the intensity distribution, such as RF coil sensitivity and volume selection methods. The indices x' and y' must be chosen to fully cover the SRF within the image plane for the voxel (x,y) and z' must span all MRI slices that cover the MRSI slice selection profile. The input functions for Eq. (5) are illustrated graphically in Fig. 11.

The analysis outlined above for CSF can be extended to separate signal contributions coming from white matter and gray matter by analyzing data from multiple voxels under the assumption that the metabolite concentrations are constant for each tissue type for the selected group of voxels (Hetherington *et al.*, 1996; Mason *et al.*, 1998; Noworolski *et al.*, 1999; Lundbom *et al.*, 1999; McLean *et al.*, 2000; Schuff *et al.*, 2001a). In order to evaluate the signal contributions from each tissue, the resultant signal for each metabolite can be modeled as

$$S_m(x,y,z) = \sum_t V_{t,m}(x,y,z) \times \rho_{t,m} \quad (6)$$

where $\rho_{t,m}$, is the signal for metabolite m and tissue-type t, which has been assumed regionally invariant over the selected voxels. This enables data from a group of voxels to be analyzed using multiparametric linear regression. For brain studies, a three-way linear regression can be performed using the distributions of gray matter, white matter, and CSF, and extrapolation to 100% of each tissue provides the value of $\rho_{t,m}$, which should also result in a value for CSF that is equal to zero. The presence of signal contributions from additional tissue types can also be accommodated by including a bias term in the analysis (Schuff *et al.*, 2001a).

Studies using these methods have demonstrated that for all metabolites observed using ^{1}H or ^{31}P studies the concentrations vary in gray and white matter. Although some differences in the findings have been presented for NAA, these likely reflect differences due to relaxation effects and the methodologies used (reviewed by Schuff *et al.*, 2001a). Additional significance of this data analysis approach is that it can provide increased sensitivity over that possible from analysis of single-voxel spectra for detection of diseases that selectively affect one tissue type, as for example is known to occur for several neurodegenerative diseases. Using these methods it has been shown that the relative gray and white matter metabolite concentrations vary as a function of age (Schuff *et al.*, 2001a) and preferential loss of NAA in gray matter occurs with Alzheimer's disease (Schuff *et al.*, 1998).

For automated analysis of MRSI data, it is necessary to account for tissue volume contributions if comparisons with normal metabolite distributions are required. An example of this is shown in Fig. 12. Here is shown an image generated by highlighting voxels that exhibit significant deviations from normal values of metabolite ratios, while accounting for the relative gray and white matter contribution to each voxel (Hetherington *et al.*, 1996). The distributions of normal metabolite values and ratios were obtained using the linear-regression tissue analysis procedure for MRSI results obtained from 10 control subjects. From these data, the 95% confidence intervals of the metabolite intensity were determined as a function of the tissue distribution contributing to each voxel (Fig. 12a). This information then provided the

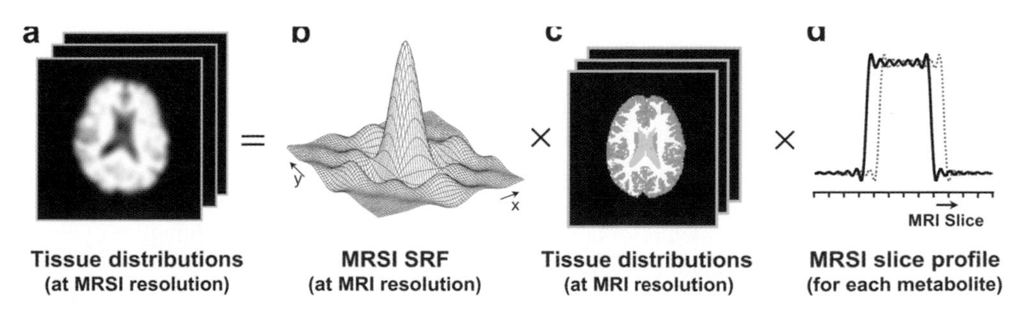

Figure 11 Calculation of the tissue density distributions corresponding to the MRSI spatial resolution (**a**) requires knowledge of the MRSI spatial response function (**b**); the tissue distribution functions, e.g. gray matter, white matter, and CSF (**c**); and the sensitivity profiles in the slice-selection direction, taking into account the chemical shift (**d**).

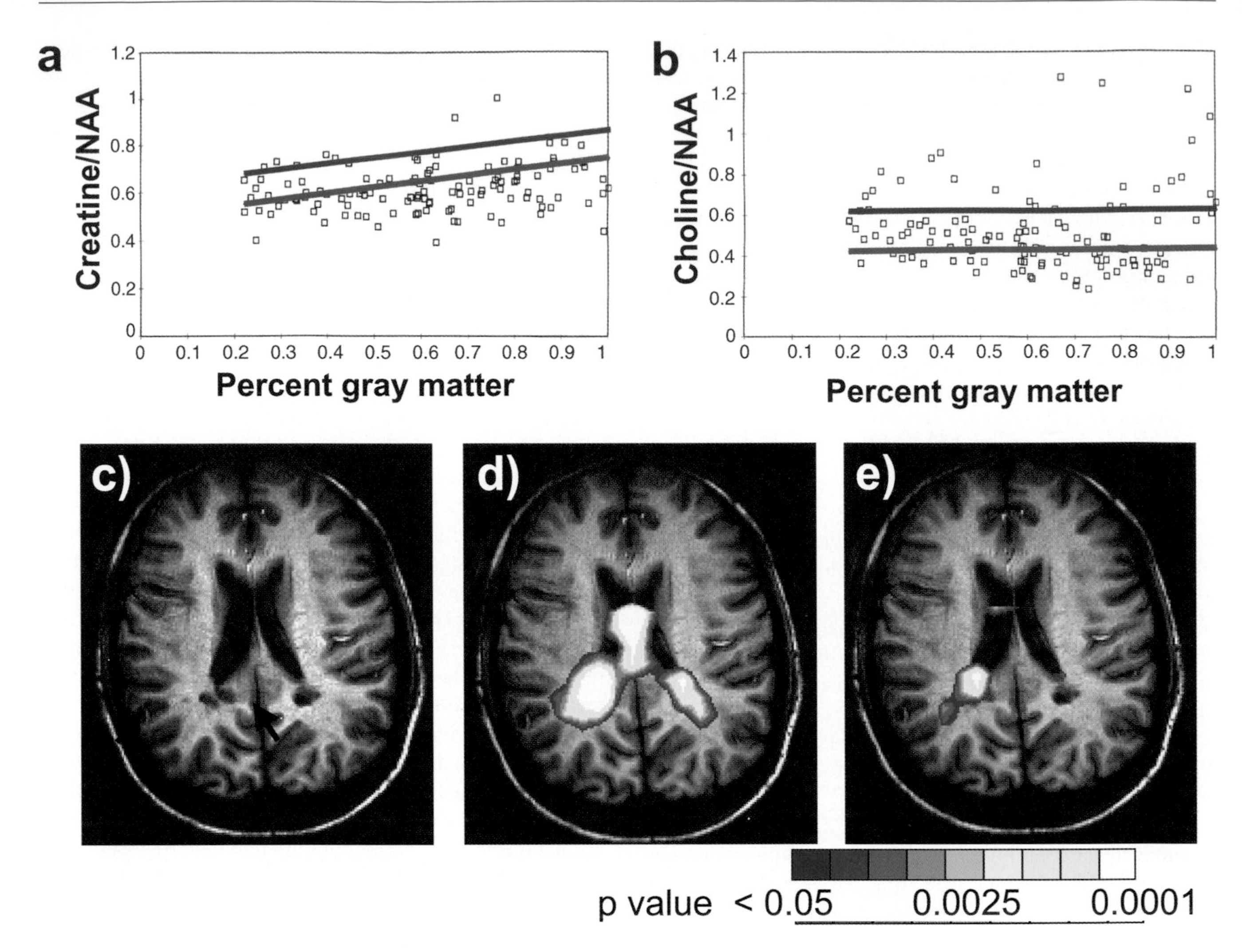

Figure 12 Plots of Cr/NAA (**a**) and Cho/NAA (**b**) against percentage gray matter for all voxels obtained from a ^{1}H MRSI study of a subject with multiple sclerosis. The superimposed lines represent the linear regressions obtained from analysis of normal control subjects (green) and the 95% upper confidence limit of the parameter distributions. Voxels with elevated values lying outside of the 95% confidence limits are then identified and highlighted on the spatially coregistered subject MRI (**c**), to form the corresponding Cr/NAA (**d**) and Cho/NAA (**e**) images. Reproduced, with permission, from Hetherington *et al.* (1996).

comparison data for subsequent MRSI studies of a multiple sclerosis patient, enabling those voxels that lay outside the confidence interval to be identified and subsequently displayed in the image format as shown in Fig. 12b. This same statistically driven voxel identification procedure has been applied to localization of temporal lobe epilepsy and shown to provide increased accuracy and sensitivity in comparison to analysis methods that did not account for tissue distributions (Chu *et al.*, 2000).

MRI and MRSI frequently provide complementary information, and a combined analysis of parameters obtained from both types of data can increase the sensitivity to detect disease. Examples include coanalysis of metabolite levels with volume measurements of brain regions obtained from MRI. Such studies include the combination of metabolite levels and hippocampal volume for lateralization of temporal lobe epilepsy (Cendes *et al.*, 1997), as well as ventricular (Capizzano *et al.*, 2000) and hippocampal volumes (Schuff *et al.*, 1997) for improved discrimination between changes associated with Alzheimer's disease, subcortical ischemic vascular dementia, and aging. Additional studies have included the use of MR image intensities (Fu *et al.*, 1996), magnetization transfer ratios (Pike *et al.*, 1999), and lesion volumetry (Narayana *et al.*, 1998) for assessment of multiple sclerosis; relative MRI intensities as an additional input for discriminant analysis classification of tumors (De Edelenyi *et al.*, 2001); combined analysis of *f*MRI, diffusion tensor imaging, and NAA for epilepsy localization (Krakow *et al.*, 1999); and MRSI and gadolinium contrast enhancement for evaluation of brain tumors (Nelson, 2001).

Comparison of metabolite distributions obtained at different time points enables progression of disease, or changes following therapy, to be assessed. This type of analysis typically requires coregistration of MRSI data obtained at multiple time points, usually done by transforming the later MRSI studies into the coordinate system of the first using a rigid body transformation. The relatively poor SNR and spatial resolution of MRSI make these data unsuitable for calculation of the spatial transformation parameters, though the MRI data are well suited to this purpose. Since the alignment between the MRI and the MRSI data acquired for each study is known, the MRI-determined transformation parameters can then be applied to the MRSI study data. Coregistration for serial evaluation of MRSI has been demonstrated for longitudinal studies of multiple sclerosis (Narayana *et al.,* 1998) and for monitoring tumor progression and response to therapy (Nelson *et al.,* 1999).

Coregistration procedures can be extended to transform both the MRI and the MRSI data into a standardized coordinate system using nonlinear spatial transforms. This allows, for example, measured metabolite parameters to be compared to the normal distribution of values for the same brain region and has been used by Fu *et al.* (1996) for evaluation of multiple sclerosis. It was suggested that this approach allows comparisons of voxels without explicitly taking into account the spatial variations due to tissue distributions, at least for normal brain anatomy. Transformation of MRI data into normalized coordinates also provides a method for automatically identifying larger neuroanatomical structures by comparison with a reference atlas (Toga and Mazziotta, 1996; Ashburner and Friston, 2000). This approach therefore lends itself to the future possibility of automating the selection of voxels from MRSI data over one or more specific brain regions, followed by an automated analysis of those groups.

VI. Applications

The unique value of the information provided by MRS measurements for biomedical applications stems from the fact that different cell types exhibit different metabolic profiles. The added information then provided by MRSI over single-voxel MRS measurements is that the metabolite distributions can be mapped with greater accuracy and correlated with structural information. While MRSI can be applied to the same clinical studies of interest for single-voxel measurements, it is particularly advantageous for situations in which changes of metabolites occur in regions that have no MRI-observable structural abnormality or in which small but widespread metabolite changes occur, perhaps localized to a specific tissue type or brain region. In the first case, the use of MRSI avoids difficulties associated with selection of specific tissue regions for study, and in the second, greater sensitivity is provided by averaging over a wide spatial region.

General reviews of cerebral applications of MRS can be found elsewhere (Vion-Dury *et al.,* 1994; Rudkin and Arnold, 1999; Burlina *et al.,* 2000; Ross and Bluml, 2001). In this section, a brief review is given of only a few selected clinical applications to illustrate situations in which MRSI can provide additional benefit.

A. Brain Tumors

Brain tumors are characterized by changes in several metabolites observed using both ^{1}H and ^{31}P MRS. For ^{1}H MRS studies, the primary metabolite changes are loss of NAA and increased choline, but also include reduced creatine, increased lactate and glycine, and visible lipid resonances associated with necrosis in the tumor (Kuesel *et al.,* 1994); however, correlation of specific metabolite changes with tumor type was found to be poor (Negendank, 1992). Part of the reason for these findings can be attributed to the large degree of spatial heterogeneity associated with many tumors that confounded the results obtained using single-voxel MRS methods. Recent studies indicate that by using MRSI to obtain improved spatial resolutions over a wider region, and by applying pattern recognition analysis, greater accuracy for tumor classification can be achieved (Hagberg, 1998; Preul *et al.,* 1998). The distributions of abnormal metabolite values observed with MRSI may not coincide with those indicated by contrast-enhanced MRI due to the presence of both nonenhancing tumor and enhancing necrosis (Nelson *et al.,* 1999). MRSI is able to identify and distinguish between these regions, as well as distinguishing tumor from radiation effects, and can therefore provide a valuable addition to a MRI examination for treatment planning and monitoring response to therapy.

B. Epilepsy

Many medically refractory temporal lobe epilepsy patients can benefit from neurosurgical removal of the epileptogenic focus; however, no one diagnostic imaging method provides the definitive information needed for localization of the seizure focus. Although neuroimaging techniques such as MRI provide valuable information, the results may be inconclusive. For example, although hippocampal sclerosis may be present with temporal lobe epilepsy, about half of these patients have normal or discordant MRIs (Garcia *et al.,* 1994). While information provided by other neuroimaging techniques, such as PET, may be more conclusive, these diagnostic studies are not widely available. The most successful outcome of surgery is obtained when several localizing studies indicate concordant results (Engel *et al.,* 1990; Laxer and Garcia, 1993; Garcia *et al.,* 1994).

Epileptic seizures are associated with changes in energy metabolism, during both the ictal and the interictal periods, and the seizure focus is associated with long-term changes of metabolite concentrations. MRS measurements therefore provide an additional, and readily available, diagnostic method to provide information for localization of the epileptogenic focus. Findings with ^{1}H MRS indicate decreased levels of NAA and increased creatine and choline, indicating neuronal loss and gliosis in the epileptic foci, and ^{31}P measurements indicate increased Pi and decreased PME (Duncan, 2000). Alterations of pH have also been measured though the results remain inconclusive (Hugg *et al.*, 1992; Chu *et al.*, 1996). Several studies indicate that both ^{1}H and ^{31}P MRS may be more sensitive than MRI in detecting this pathology, as well as detection of bilateral abnormalities (Ende *et al.*, 1997; Chu *et al.*, 2000).

The use of MRSI benefits studies of epilepsy by enabling a wide brain region to be examined and by providing increased sensitivity to detection of small areas of abnormalities. However, measurements in the temporal lobe are more challenging due to strong susceptibility gradients from the petrous temporal bones and the sinuses causing B_0 field inhomogeneity. MRSI studies therefore greatly benefit from the availability of instrumentation with high-order shims and robust shimming methods (McLean *et al.*, 2001).

C. Dementias

The most common cause of dementia in the elderly is Alzheimer's disease (AD), for which no definitive laboratory confirmation is currently available. It is also difficult to distinguish AD from other dementias or, in the early stages, from even normal aging. Brain atrophy may be detected using neuroimaging methods such as MRI, with changes preferentially involving the hippocampus and frontal and parietal cortex, though in the early stage of the disease these changes may be partially offset by gliosis and results show considerable overlap between AD and normal elderly subjects. Proton MRS studies show decreased levels of NAA and increased *myo*-inositol with Alzheimer's disease (Miller *et al.*, 1993), and by combining the MRS information with the MRI results it has been shown that increased discrimination between Alzheimer's disease and control subjects can be obtained (MacKay *et al.*, 1996a; Schuff *et al.*, 1997).

The ability of MRSI to obtain multiple voxel information over a wide field of view offers increased sensitivity for diagnosis of AD since regional- and tissue-specific changes of metabolites can be detected. There is increasing evidence that metabolite changes with dementia are regionally specific. First, it is known that gray matter is preferentially reduced, which can be identified using the regression analysis methods previously described. In addition, recent studies indicate that regional differences of NAA and volume loss may permit distinction between Alzheimer's disease and frontal lobe dementia (Ernst *et al.*, 1997) or subcortical ischemic vascular disease (Capizzano *et al.*, 1999; Fein *et al.*, 2000). Vascular dementia appears to be distinguished by greater involvement in the frontal lobes, indicated by higher NAA/Cho in posterior mesial gray matter, than in AD (MacKay *et al.*, 1996b).

D. Amyotrophic Lateral Sclerosis (ALS)

Diagnosis of ALS is difficult during the early stage of the disease and more accurate evaluation methods would be of considerable importance for the development of potential new therapies (Brooks, 1999). It is characterized by degeneration of motor neurons in the brain stem, spinal cord, and cerebral cortex, and while MRI, PET, and SPECT studies indicate abnormalities in the motor cortex regions these techniques remain limited in terms of specificity and sensitivity (Kalra *et al.*, 1999).

Proton MRS of ALS subjects has shown decreased levels of NAA and increased choline in the motor cortex, as well as increased *myo*-inositol and decreased glutamate with short-TE studies (Bowen *et al.*, 2000). Longitudinal measurements have also demonstrated a correlation of the change in metabolite measures with a decrease of motor function (Suhy *et al.*, 2000b). Although many of these findings have been observed using single-voxel MRS, or single voxels extracted from MRSI data, obtaining data for tissue that is located primarily in the motor cortex, where changes are expected to be the greatest, is difficult. These measurements will therefore include significant contributions from neighboring brain regions as well as containing contributions from both gray and white matter. MRSI offers the opportunity to obtain greater regional specificity by allowing for selection of several voxels, each of relatively small volume, over a wider cortical region, for example along the length of the primary motor cortex (Kalra *et al.*, 1998; Suhy *et al.*, 2000b). Another approach is to select all voxels encompassing the motor cortex and to then perform a histogram analysis of the metabolite concentrations within the selected region. This analysis procedure demonstrates significant differences in the metabolite distributions between ALS subjects and controls (Schuff *et al.*, 1999).

E. Brain Trauma

Traumatic brain injury represents an area in which results of structural neuroimaging measures are poor predictors of the eventual outcome of the injury. While focal lesions can be detected with MRI and CT, the extent of diffuse axonal injury cannot be identified in subjects with mild to moderate injury, who may have poor outcome even without abnormal MRI findings. MRS studies have shown decreased NAA, indicating neuronal injury, and increased choline, suggesting inflammation and demyelination in areas appearing normal on MRI, and good correlation with outcome (Ross *et al.*, 1998; Garnett *et al.*, 2000). Although these findings have been demonstrated using single-volume MRS, the studies remain limited by the inability to identify the most severely injured brain regions. Since diffuse injury can be widespread and remote from the point of injury, greater accuracy can be obtained using MRSI methods that sample a much wider brain region (Wild *et al.*, 1999). Furthermore, since injury may be widespread but relatively mild, integration of data obtained over larger brain regions may enhance the accuracy of the results.

VII. Emerging Technologies

MRSI acquisition techniques have been developed in a number of additional application areas, including mapping of other nuclei, pH imaging (Morikawa *et al.*, 1993; van Sluis *et al.*, 1999; Büchert *et al.*, 2001), temperature mapping (Kuroda *et al.*, 1996), monitoring changes of lactate during brain activation (Richards *et al.*, 1997), and evaluation of tissue oxygenation from the lineshape of water (Al-Hallaq *et al.*, 1998). By using high-speed acquisition techniques a second spectroscopic dimension can be added to obtain spatially localized 2D NMR spectra, which can provide greater spectral discrimination and enable improved separation of overlapping resonances (Adalsteinsson and Spielman, 1999; Mayer *et al.*, 2000; von Kienlin *et al.*, 2000). All of these applications will undoubtedly receive further attention as high-field instruments become more available.

The MRI processing and analysis techniques such as coregistration, atlas matching, and tissue segmentation will play an increasing role in the analysis of MRSI data. First, coanalysis of MRI with MRSI will provide both improved diagnostic sensitivity through coanalysis of multiple parameters. Spatial transformation into normalized coordinates will enable analysis of MRSI data relative to metabolite concentrations for the corresponding region in normal subjects. Atlas matching will enable larger brain regions to be automatically identified so that all MRSI voxels corresponding to that region can be analyzed and the results compared to a database of normal metabolite values. Since it has been shown that brain metabolite distributions vary as a function of tissue type, location, and a number of subject variables, it will be necessary to establish a database of normal metabolite distributions that also accounts for these variables. By including these parameters in the analysis, the diagnostic sensitivity of the MRSI will be increased.

Automatic identification of brain morphology will also facilitate greater automation in the analysis of MRSI data. Currently, data analysis largely relies on a trained operator selecting voxels from a specific brain region for analysis. However, if this selection step can be performed automatically, this would greatly benefit implementation of these methods in the clinical setting. Additional advantages would be removing potential bias from operator selection of specific regions of interest as well as increasing sensitivity for detection of small changes of metabolites distributed over a large region. More routine implementation in the clinical setting will also require faster and more robust automated spectral analysis, with appropriate controls for poor fit results or corrupted data.

VIII. Conclusion

MRSI provides a unique neuroimaging method for evaluation of brain metabolites that can be readily implemented on widely available clinical MRI instruments. These studies can be conveniently integrated with a MRI examination, to provide complementary information that is of interest for many clinical diagnostic applications and biomedical research applications. Although detection sensitivity limits observation to a few of the most abundant metabolites, these compounds have been demonstrated to provide valuable information on oxidative, amino acid, and lipid metabolism. Major clinical application areas include detection of malignancy and neurodegenerative diseases, but also include areas as diverse as psychiatric diseases and encephalopathies.

Despite the fact that numerous studies have demonstrated that MRS offers considerable potential for increasing sensitivity and specificity for detection of disease, implementation of MRSI for routine clinical studies has been slow, and routine clinical protocols have not been established. For those sites that already have the capability, MRSI is in fact a routinely requested diagnostic study with significant impact on patient care; however, currently most of these sites are research institutions that have specialized personnel with expertise in this area, and the techniques remain largely unknown in most clinical diagnostic centers. The reasons for this include the relative complexity of the data processing and analysis methods, which may differ depending on the application, making it difficult to implement the techniques in a routine manner. Additional difficulties are associated with the additional care and time required for the data acquisition, the lack of standardization, and the continuing development of these techniques. As the technology improves and as techniques become more effectively implemented on commercial MR systems, this situation will slowly change. This will require much greater automation of the data acquisition and processing procedures, including improved shimming and calibration methods, and simplified presentation of results in a form suitable for the clinical user. Additional developments must also address lack of standards, unfamiliarity among clinical users, and further analysis of the costs and benefits of this technology.

Acknowledgments

I acknowledge the support of PHS Grants NS35690, AG12119, and NS38029 and thank Drs. A. Ebel, N. Schuff, G. Matson, and M. Weiner for making available much of the data presented in this chapter and for their many suggestions.

References

Adalsteinsson, E., Irarrazabal, P., Topp, S., Meyer, C., Macovski, A., and Spielman, D. M. (1998). Volumetric spectroscopic imaging with spiral-based k-space trajectories. *Magn. Reson. Med.* **39,** 889–898.

Adalsteinsson, E., and Spielman, D. M. (1999). Spatially resolved two-dimensional spectroscopy. *Magn. Reson. Med.* **41,** 8–12.

Adalsteinsson, E., Star-Lack, J., Meyer, C. H., and Spielman, D. M. (1999). Reduced spatial side lobes in chemical-shift imaging. *Magn. Reson. Med.* **42,** 314–323.

Al-Hallaq, H. A., River, J. N., Zamora, M., Oikawa, H., and Karczmar, G. S. (1998). Correlation of magnetic resonance and oxygen microelectrode measurements of carbogen-induced changes in tumor oxygenation. *Int. J. Radiat. Oncol. Biol. Phys.* **41,** 151–159.

Angelie, E., Bonmartin, A., Boudraa, A., Gonnaud, P. M., Mallet, J. J., and Sappey-Marinier, D. (2001). Regional differences and metabolic changes in normal aging of the human brain: Proton MR spectroscopic imaging study. *Am. J. Neuroradiology* **22,** 119–127.

Arias-Mendoza, F., Javaid, T., Stoyanova, R., Brown, T. R., and Gonen, O. (1996). Heteronuclear multivoxel spectroscopy of in vivo human brain: Two-dimensional proton interleaved with three-dimensional ^{1}H-decoupled phosphorus chemical shift imaging. *NMR Biomed.* **9,** 105–113.

Ashburner, J., and Friston, K. (2000). Voxel-based morphometry—The methods. *NeuroImage* **11,** 805–821.

Aue, W. P., Bartholdi, E., and Ernst, R. R. (1976). Two-dimensional spectroscopy: Application to nuclear magnetic resonance. *J. Chem. Phys.* **64,** 2229–2246.

Barker, P. B. (2001). N-acetyl aspartate—A neuronal marker? *Ann. Neurol.* **49,** 423–424.

Barkhuijsen, H., de Beer, R., Bovée, W. M. M. J., and Van Ormondt, D. (1985). Retrieval of frequencies, amplitudes, damping factors, and phases from time domain signals using a least-squares procedure. *J. Magn. Reson.* **61,** 465–481.

Behar, K. L., Rothman, D. L., Spencer, D. D., and Petroff, O. A. C. (1994). Analysis of macromolecule resonances in ^{1}H NMR spectra of human brain. *Magn. Reson. Med.* **32,** 294–302.

Bloch, F. (1946). Nuclear induction. *Phys. Rev.* **70,** 460–474.

Bloch, F., Hansen, W. W., and Packard, M. (1946). The nuclear induction experiment. *Phys. Rev.* **70,** 474–485.

Bottomley, P. A. (1987). Spatial localization in NMR spectroscopy in vivo. *Ann. N. Y. Acad. Sci.* **508,** 333–348.

Bottomley, P. A., Hart, H. A., and Edelstein, W. A. (1984). Anatomy and metabolism of the normal human brain studied by magnetic resonance at 1.5 tesla. *Radiology* **150,** 441–446.

Bowen, B. C., Pattany, P. M., Bradley, W. G., Murdoch, J. B., Rotta, F., Younis, A. A., Duncan, R. C., and Quencer, R. M. (2000). MR imaging and localized proton spectroscopy of the precentral gyrus in amyotrophic lateral sclerosis. *AJNR* **21,** 647–658.

Brooks, B. R. (1999). Diagnostic dilemmas in amyotrophic lateral sclerosis. *J. Neurol. Sci.* **165,** S1–S9.

Brown, T. R., Kincaid, B. M., and Ugurbil, K. (1982). NMR chemical shift imaging in three dimensions. *Proc. Natl. Acad. Sci. USA* **79,** 3523–3526.

Burlina, A. P., Aureli, T., Bracco, F., Conti, F., and Battistin, L. (2000). MR spectroscopy: A powerful tool for investigating brain function and neurological diseases. *Neurochem. Res.* **25,** 1365–1372.

Büchert, M., O'Neill, J., Vermathen, P., and Maudsley, A. A. (2001). In vivo brain pH mapping using 19 × 19 voxel ^{1}H MR spectroscopic imaging of the down field region. *In* "Proceedings of the International Society for Magnetic Resonance in Medicine," p. 1701.

Capizzano, A. A., Schuff, N., Amend, D. L., Tanabe, J. L., Norman, D., Maudsley, A. A., Jagust, W., Chui, H. C., Fein, G., Segal, M. R., and Weiner, M. W. (1999). Subcortical ischemic vascular dementia: Assessment with quantitative MRI and ^{1}H MRSI. *Am. J. Neuroradiology* **20,** 839–844.

Capizzano, A. A., Schuff, N., Amend, D. L., Tanabe, J. L., Norman, D., Maudsley, A. A., Jagust, W., Chui, H. C., Fein, G., Segal, M. R., and Weiner, M. W. (2000). Subcortical ischemic vascular dementia: Assessment with quantitative MR imaging and ^{1}H MR spectroscopy. *AJNR* **21,** 621–630.

Cavassila, S., Deval, S., Huegen, C., Van Ormondt, D., and Graveron-Demilly, D. (2001). Cramer–Rao bounds: An evaluation tool for quantitation. *NMR Biomed.* **14,** 278–283.

Cendes, F., Caramanos, Z., Andermann, F., Dubeau, F., and Arnold, D. L. (1997). Proton magnetic resonance spectroscopic imaging and magnetic resonance imaging volumetry in the lateralization of temporal lobe epilepsy: A series of 100 patients. *Ann. Neurol.* **42,** 737–746.

Chance, B., Nakase, Y., Bond, M., Leigh, J. S., Jr., and McDonald, G. (1978). Detection of ^{31}P nuclear magnetic resonance signals in brain by in vivo and freeze-trapped assays. *Proc. Natl. Acad. Sci. USA* **75,** 4925–4929.

Chang, L., Ernst, T., Poland, R. E., and Jenden, D. J. (1996). In vivo proton magnetic resonance spectroscopy of the normal aging human brain. *Life Sci.* **58,** 2049–2056.

Christiansen, P., Henriksen, O., Stubgaard, M., Gideon, P., and Larsson, H. B. W. (1992). In vivo quantification of brain metabolites by ^{1}H-MRS using water as an internal standard. *Magn. Reson. Imaging* **10,** 107–118.

Chu, W. J., Hetherington, H. P., Kuzniecky, R. J., Vaughan, J. T., Twieg, D. B., Faught, R. E., Gilliam, F. G., Hugg, J. W., and Elgavish, G. A. (1996). Is the intracellular pH different from normal in the epileptic focus of patients with temporal lobe epilepsy? A ^{31}P NMR study. *Neurology* **47,** 756–760.

Chu, W. J., Kuzniecky, R. I., Hugg, J. W., Abou-Khalil, B., Gilliam, F., Faught, E., and Hetherington, H. P. (2000). Statistically driven identification of focal metabolic abnormalities in temporal lobe epilepsy with corrections for tissue heterogeneity using ^{1}H spectroscopic imaging. *Magn. Reson. Med.* **43,** 359–367.

Clayton, D. B., Elliott, M. A., and Lenkinski, R. E. (2001). In vivo proton spectroscopy without solvent suppression. *Concepts Magn. Reson.* **13,** 260–275.

Damadian, R., Goldsmith, M., and Minkoff, L. (1977). Fonar image of the live human body. *Physiol. Chem. Phys. Med. NMR* **9,** 97–100.

Danielsen, E. R., Michaelis, T., and Ross, B. D. (1995). Three methods of calibration in quantitative proton MR spectroscopy. *J. Magn. Reson. B* **106,** 287–291.

Danielsen, E. R., and Ross, B. D. (1999). "Magnetic Resonance Spectroscopy Diagnosis of Neurological Diseases." Dekker, New York.

Davie, C. A., Hawkins, C. P., Barker, G. J., Brennan, A., Tofts, P. S., Miller, D. H., and McDonald, W. I. (1994). Serial proton magnetic resonance spectroscopy in acute multiple sclerosis lesions. *Brain* **117,** 49–58.

de Beer, R., Michels, F., Van Ormondt, D., van Tongeren, B. P. O., Luyten, P. R., and van Vroonhoven, H. (1993). Reduced lipid contamination in in vivo ^{1}H MRSI using time-domain fitting and neural network classification. *Magn. Reson. Imaging* **11,** 1019–1026.

de Beer, R., van den Boogaart, A., Van Ormondt, D., Pijnappel, W. W. F., den Hollander, J. A., Marien, A. J. H., and Luyten, P. R. (1992). Application of time-domain fitting in the quantification of in vivo ^{1}H spectroscopic imaging data sets. *NMR Biomed.* **5,** 171–178.

De Edelenyi, F. S., Estève, F., Grand, S., Segebarth, C., Rubin, C., Décorps, M., Le Bas, J. F., Lefournier, V., *et al.* (2001). Classification of brain tumors using ^{1}H MRSI at an echo time of 272 msec in combination with linear discriminant analysis: Strategies to improve the correct classification rate. *In* "Proceedings of the International Society for Magnetic Resonance in Medicine," p. 1665.

De Edelenyi, F. S., Rubin, C., Estève, F., Grand, S., Décorps, M., Lefournier, V., Le Bas, J. F., and Rémy, C. (2000). A new approach for analyzing proton magnetic resonance spectroscopic images of brain tumors: Nosologic images. *Nat. Med.* **6,** 1287–1289.

de Graaf, A. A., van Dijk, J. E., and Bovee, W. M. M. J. (1990). QUALITY: Quantification improvement by converting lineshapes to the Lorentzian type. *Magn. Reson. Med.* **13,** 343–357.

de Graaf, R. A., and Rothman, D. L. (2000). In vivo detection and quantification of scalar coupled ^{1}H NMR resonances. *Concepts Magn. Reson.* **13,** 32–76.

Duijn, J. H., Matson, G. B., Maudsley, A. A., and Weiner, M. W. (1992). 3D phase encoding ^{1}H spectroscopic imaging of human brain. *Magn. Reson. Imaging* **10,** 315–319.

Duncan J. S. (2000). The epilepsies. *In* "Brain Mapping: The Disorders" (J. C. Mazziotta and A. W. Toga, eds.), pp. 317–355. Academic Press, San Diego.

Ebel, A., Dreher, W., and Leibfritz, D. (2000). A fast variant of ^{1}H spectroscopic U-FLARE imaging using adjusted chemical shift phase encoding. *J. Magn. Reson.* **142,** 241–253.

Ebel, A., and Maudsley, A. A. (2001). Comparison of methods for reduction of lipid contamination for in vivo proton MR spectroscopic imaging of the brain. *Magn. Reson. Med.,* **46,** 706–712.

Ebel, A., Soher, B. J., and Maudsley, A. A. (2001). Assessment of 3D ^{1}H NMR echo-planar spectroscopic imaging using automated spectral analysis. *Magn. Reson. Med.,* **46,** 1072–1078.

Ende, G. R., Laxer, K. D., Knowlton, R. C., Matson, G. B., Schuff, N., Fein, G., and Weiner, M. W. (1997). Temporal lobe epilepsy: Bilateral hippocampal metabolite changes revealed at proton MR spectroscopic imaging. *Radiology* **202,** 809–817.

Engel, J., Jr., Henry, T. R., Risinger, M. W., Mazziotta, J. C., Sutherling, W. W., Levesque, M. F., and Phelps, M. E. (1990). Presurgical evaluation for partial epilepsy: Relative contributions of chronic depth-electrode recordings versus FDG–PET and scalp-sphenoidal ictal EEG. *Neurology* **40,** 1670–1677.

Ernst, T., Chang, L., Melchor, R., and Mehringer, C. M. (1997). Frontotemporal dementia and early Alzheimer disease: Differentiation with frontal lobe H-1 MR spectroscopy. *Radiology* **203,** 829–836.

Ernst, T., and Hennig, J. (1995). Improved water suppression for localized in vivo ^{1}H spectroscopy. *J. Magn. Reson. B* **106,** 181–186.

Estilaei, M. R., Matson, G. B., Payne, G. S., Leach, M. O., Fein, G. F., and Meyerhoff, D. J. (2001). Effects of chronic alcohol consumption on the broad phospholipid signal in human brain: An in vivo ^{31}P MRS study. *Alcohol Clin. Exp. Res.* **25,** 89–97.

Fein, G., Di, S. V., Tanabe, J., Cardenas, V., Weiner, M. W., Jagust, W. J., Reed, B. R., Norman, D., Schuff, N., Kusdra, L., Greenfield, T., and Chui, H. (2000). Hippocampal and cortical atrophy predict dementia in subcortical ischemic vascular disease. *Neurology* **55,** 1626–1635.

Frahm, J., Merboldt, K. D., and Hänicke, W. (1987). Localized proton spectroscopy using stimulated echoes. *J. Magn. Reson.* **72,** 502–508.

Fu, L., Wolfson, C., Worsley, K. J., De Stefano, N., Collins, D. L., Narayanan, S., and Arnold, D. L. (1996). Statistics for investigation of multimodal MR imaging data and an application to multiple sclerosis patients. *NMR Biomed.* **9,** 339–346.

Gadian D. G. (1982). "Nuclear Magnetic Resonance and Its Applications to Living Systems." Clarendon, Oxford.

Gadian, D. G., Hoult, D. I., Radda, G. K., Seeley, P. J., Chance, B., and Barlow, C. (1976). Phosphorus nuclear magnetic resonance studies on normoxic and ischemic cardiac tissue. *Proc. Natl. Acad. Sci. USA* **73,** 4446–4448.

Garcia, P. A., Laxer, K. D., Barbaro, N. M., and Dillon, W. P. (1994). Prognostic value of qualitative magnetic resonance imaging hippocampal abnormalities in patients undergoing temporal lobectomy for medically refractory seizures. *Epilepsia* **35,** 520–524.

Garnett, M. R., Blamire, A. M., Corkill, R. G., Cadoux-Hudson, T. A., Rajagopalan, B., and Styles, P. (2000). Early proton magnetic resonance spectroscopy in normal-appearing brain correlates with outcome in patients following traumatic brain injury. *Brain* **123,** 2046–2054.

Govindaraju, V., Young, K., and Maudsley, A. A. (2000). Proton NMR chemical shifts and coupling constants for brain metabolites. *NMR Biomed.* **13,** 129–153.

Guilfoyle, D. N., Blamire, A., Chapman, B., Ordidge, R. J., and Mansfield, P. (1989). PEEP—A rapid chemical-shift imaging method. *Magn. Reson. Med.* **10,** 282–287.

Guimaraes, A. R., Baker, J. R., Jenkins, B. G., Lee, P. L., Weisskoff, R. M., Rosen, B. R., and Gonzalez, R. G. (1999). Echoplanar chemical shift imaging. *Magn. Reson. Med.* **41,** 877–882.

Haase, A., Frahm, J., Hänicke, W., and Matthaei, D. (1985). ^{1}H NMR chemical shift selective (CHESS) imaging. *Phys. Med. Biol.* **30,** 341–344.

Hagberg, G. (1998). From magnetic resonance spectroscopy to classification of tumors. A review of pattern recognition methods. *NMR Biomed.* **11,** 148–156.

Hardy, D. L., and Norwood, T. J. (1998). Spectral editing techniques for the in vitro and in vivo detection of taurine. *J. Magn. Reson.* **133,** 70–78.

Haselgrove, J. C., Subramanian, V. H., Leigh, J. S., Gyulai, L., and Chance, B. (1983). In vivo one-dimensional imaging of phosphorus metabolites by phosphorus-31 nuclear magnetic resonance. *Science* **220,** 1170–1173.

Haupt, C. I., Schuff, N., Weiner, M. W., and Maudsley, A. A. (1996). Removal of lipid artifacts in ^{1}H spectroscopic imaging by data extrapolation. *Magn. Reson. Med.* **35,** 678–687.

Heindel, W., Kugel, H., and Roth, B. (1993). Noninvasive detection of increased glycine content by proton MR spectroscopy in the brains of two infants with nonketotic hyperglycinemia. *Am. J. Neuroradiology* **14,** 629–635.

Hetherington, H. P., Mason, G. F., Pan, J. W., Ponder, S. L., Vaughan, J. T., Twieg, D. B., and Pohost, G. M. (1994). Evaluation of cerebral gray and white matter metabolite differences by spectroscopic imaging at 4.1 T. *Magn. Reson. Med.* **32,** 565–571.

Hetherington, H. P., Pan, J. W., Mason, G. F., Adams, D., Vaughn, M. J., Twieg, D. B., and Pohost, G. M. (1996). Quantitative ^{1}H spectroscopic imaging of human brain at 4.1 T using image segmentation. *Magn. Reson. Med.* **36,** 21–29.

Horská, A., Calhoun, V., and Barker, P. B. (2001). A rapid method for correction of CSF partial volume in quantitative proton MR spectroscopic imaging. *In* "Proceedings of the International Society for Magnetic Resonance in Medicine," p. 216.

Hoult, D. I., Busby, S. J. W., Gadian, D. G., Radda, G. K., Richards, R. E., and Seeley, P. J. (1974). Observations of tissue metabolites using phosphorus 31 nuclear magnetic resonance. *Nature* **252,** 285–287.

Hoult, D. I., and Lauterbur, P. C. (1979). The sensitivity of the zeugmatographic experiment involving human samples. *J. Magn. Reson.* **34,** 425–433.

Hu, X., Johnson, V., Wong, W. H., and Chen, C.-T. (1991). Bayesian image processing in magnetic resonance imaging. *Magn. Reson. Imaging* **9,** 611–620.

Hugg, J. W., Laxer, K. D., Matson, G. B., Maudsley, A. A., Husted, C. A., and Weiner, M. W. (1992). Lateralization of human focal epilepsy by ^{31}P magnetic resonance spectroscopic imaging. *Neurology* **42,** 2011–2018.

Hugg, J. W., Maudsley, A. A., Weiner, M. W., and Matson, G. B. (1996). Comparison of k-space sampling schemes for multidimensional MR spectroscopic imaging. *Magn. Reson. Med.* **36,** 469–473.

Johnson, G., Jung, K. J., Wu, E. X., and Hilal, S. K. (1993). Self-correction of proton spectroscopic images for gradient eddy current distortions and static field inhomogeneities. *Magn. Reson. Med.* **30,** 255–261.

Jung, R. E., Brooks, W. M., Yeo, R. A., Chiulli, S. J., Weers, D. C., and Sibbitt, W. J. (1999). Biochemical markers of intelligence: A proton MR spectroscopy study of normal human brain. *Proc. R. Soc. London B Biol. Sci.* **266,** 1375–1379.

Kalra, S., Arnold, D. L., and Cashman, N. R. (1999). Biological markers in the diagnosis and treatment of ALS. *J. Neurol. Sci.* **165,** S27–S32.

Kalra, S., Cashman, N. R., Genge, A., and Arnold, D. L. (1998). Recovery of N-acetylaspartate in corticomotor neurons of patients with ALS after riluzole therapy. *NeuroReport* **9,** 1757–1761.

Kay, S. M. (1988). "Modern Spectral Estimation: Theory and Application." Prentice Hall International, Englewood Cliffs, NJ.

Keevil, S. F., Barbiroli, B., Brooks, J. C., Cady, E. B., Canese, R., Carlier, P., Collins, D. J., Gilligan, P., Gobbi, G., Hennig, J., Kügel, H., Leach, M. O., Metzler, D., Mlynárik, V., Moser, E., Newbold, M. C., Payne, G. S., Ring, P., Roberts, J. N., Rowland, I. J., Thiel, T., Tkác, I., Topp, S., Wittsack, H. J., and Podo, F. (1998). Absolute metabolite quantification by in vivo NMR spectroscopy. II. A multicentre trial of protocols for in vivo localised proton studies of human brain. *Magn. Reson. Imaging* **16,** 1093–1106.

Kim, S. H., Chang, K. H., Song, I. C., Han, M. H., Kim, H. C., Kang, H. S., and Han, M. C. (1997). Brain abscess and brain tumor: Discrimination with in vivo H-1 MR spectroscopy. *Radiology* **204,** 239–245.

Komoroski, R. A., Heimberg, C., Cardwell, D., and Karson, C. N. (1999). Effects of gender and region on proton MRS of normal human brain. *Magn. Reson. Imaging* **17,** 427–433.

Krakow, K., Wieshmann, U. C., Woermann, F. G., Symms, M. R., McLean, M. A., Lemieux, L., Allen, P. J., Barker, G. J., Fish, D. R., and Duncan, J. S. (1999). Multimodal MR imaging: Functional, diffusion tensor, and chemical shift imaging in a patient with localization-related epilepsy. *Epilepsia* **40,** 1459–1462.

Kuesel, A. C., Donnelly, S. M., Halliday, W., Sutherland, G. R., and Smith, I. C. (1994). Mobile lipids and metabolic heterogeneity of brain tumours as detectable by ex vivo ^{1}H MR spectroscopy. *NMR Biomed.* **7,** 172–180.

Kumar, A., Welti, D., and Ernst, R. R. (1975). NMR Fourier zeugmatography. *J. Magn. Reson.* **18,** 69–83.

Kuroda, K., Suzuki, Y., Ishihara, Y., Okamoto, K., and Suzuki, Y. (1996). Temperature mapping using water proton chemical shift obtained with 3D-MRSI: Feasibility in vivo. *Magn. Reson. Med.* **35,** 20–29.

Lara, R. S., Matson, G. B., Hugg, J. W., Maudsley, A. A., and Weiner, M. W. (1993). Quantitation of in vivo phosphorus metabolites in human brain with magnetic resonance spectroscopic imaging (MRSI). *Magn. Reson. Imaging* **11,** 273–278.

Lauterbur, P. C. (1973). Image formation by induced local interactions: Examples employing nuclear magnetic resonance. *Nature* **242,** 190–191.

Laxer, K. D., and Garcia, P. A. (1993). Imaging criteria to identify the epileptic focus. Magnetic resonance imaging, magnetic resonance spectroscopy, positron emission tomography scanning, and single photon emission computed tomography. *Neurosurg. Clin. North Am.* **4,** 199–209.

Liang, Z.-P., Boada, F. E., Constable, R. T., Haacke, E. M., Lauterbur, P. C., and Smith, M. R. (1992). Constrained reconstruction methods in MR imaging. *Rev. Magn. Reson. Med.* **4,** 67–186.

Longo, R., Ricci, C., Dalla Palma, L., Vidimari, R., Giorgini, A., den Hollander, J. A., and Segebarth, C. M. (1993). Quantitative ^{31}P MRS of the normal adult human brain. Assessment of interindividual differences and ageing effects. *NMR Biomed.* **6,** 53–57.

Lundbom, N., Barnett, A., Bonavita, S., Patronas, N., Rajapakse, J., Tedeschi, and Di Chiro, G. (1999). MR image segmentation and tissue metabolite contrast in ^{1}H spectroscopic imaging of normal and aging brain. *Magn. Reson. Med.* **41,** 841–845.

MacKay, S., Ezekiel, F., Di Sciafani, V., Meyerhoff, D. J., Gerson, J., Norman, D., Fein, G., and Weiner, M. W. (1996a). Alzheimer disease and subcortical ischemic vascular dementia: Evaluation by combining MR imaging segmentation and H-1 MR spectroscopic imaging. *Radiology* **198,** 537–545.

MacKay, S., Meyerhoff, D. J., Constans, J.-M., Norman, D., Fein, G., and Weiner, M. W. (1996b). Regional gray and white matter metabolite differences in subjects with AD, with subcortical ischemic vascular dementia, and elderly controls with ^{1}H magnetic resonance spectroscopic imaging. *Arch. Neurol.* **53,** 167–174.

Mansfield, P., and Maudsley, A. A. (1976). Line scan proton spin imaging in biological structures by NMR. *Phys. Med. Biol.* **21,** 847–852.

Mason, G. F., Chu, W. J., Vaughan, J. T., Ponder, S. L., Twieg, D. B., Adams, D., and Hetherington, H. P. (1998). Evaluation of ^{31}P metabolite differences in human cerebral gray and white matter. *Magn. Reson. Med.* **39,** 346–353.

Matson, G. B., and Weiner, M. W. (1992). Spectroscopy. *In* "Magnetic Resonance Imaging" (D. D. Stark and W. G. Bradley, eds.), pp. 438–478. Mosby, St. Louis.

Maudsley, A. A. (1999). Methods and applications of high-speed MR spectroscopy. *In* "Ultrafast Magnetic Resonance Imaging in Medicine" (S. Naruse and H. Watari, eds.), pp. 249–254. Elsevier, Tokyo.

Maudsley, A. A., and Hilal, S. K. (1985). Field inhomogeneity correction and data processing for spectroscopic imaging. *Magn. Reson. Med.* **2,** 218–233.

Maudsley, A. A., Hilal, A. K., Perman, W. H., and Simon, H. E. (1983). Spatially resolved high-resolution spectroscopy by "four-dimensional" NMR. *J. Magn. Reson.* **51,** 147–152.

Maudsley, A. A., Hilal, S. K., Simon, H. E., and Wittekoek, S. (1984). In vivo MR spectroscopic imaging with P-31. *Radiology* **153,** 745–750.

Maudsley, A. A., Lin, E., and Weiner, M. W. (1992). Spectroscopic imaging display and analysis. *Magn. Reson. Imaging* **10,** 471–485.

Maudsley, A. A., Matson, G. B., Hugg, J. W., and Weiner, M. W. (1994a). Reduced phase encoding in spectroscopic imaging. *Magn. Reson. Med.* **31,** 645–651.

Maudsley, A. A., Wu, Z., Meyerhoff, D. J., and Weiner, M. W. (1994b). Automated processing for proton spectroscopic imaging using water reference deconvolution. *Magn. Reson. Med.* **31,** 589–595.

Mayer, D., Dreher, W., and Leibfritz, D. (2000). Fast echo planar based correlation-peak imaging: Demonstration on the rat brain in vivo. *Magn. Reson. Med.* **44,** 23–28.

McKnight, T. R., Noworolski, S. M., Vigneron, D. B., and Nelson, S. J. (2001). An automated technique for the quantitative assessment of 3D-MRSI data from patients with glioma. *J. Magn. Reson. Imaging* **13,** 167–177.

McLean, M. A., Woermann, F. G., Barker, G. J., and Duncan, J. S. (2000). Quantitative analysis of short echo time ^{1}H-MRSI of cerebral gray and white matter. *Magn. Reson. Med.* **44,** 401–411.

McLean, M. A., Woermann, F. G., Simister, R. J., Barker, G. J., and Duncan, J. S. (2001). In vivo short echo time ^{1}H-magnetic resonance spectroscopic imaging (MRSI) of the temporal lobes. *NeuroImage* **14,** 501–509.

McNamara, R., Arias-Mendoza, F., and Brown, T. R. (1994). Investigation of broad resonances in ^{31}P NMR spectra of the human brain in vivo. *NMR Biomed.* **7,** 237–242.

Mierisová, S., and Ala-Korpela, M. (2001). MR spectroscopy quantitation: A review of frequency domain methods. *NMR Biomed.* **14,** 247–259.

Miller, B. L., Moats, R. A., Shonk, T., Ernst, T., Woolley, S., and Ross, B. D. (1993). Alzheimer disease: Depiction of increased cerebral myo-inositol with proton MR spectroscopy. *Radiology* **187,** 433–437.

Moonen, C. T., and van Zijl, P. C. (1996). Water suppression in proton spectroscopy of humans and animals. *In* "Encyclopedia of Nuclear Magnetic Resonance" (D. M. Grant and R. K. Harris, eds.), pp. 4943–4955. Wiley, New York.

Morikawa, S., Inubushi, T., Kito, K., and Kido, C. (1993). pH mapping in living tissues: An application of in vivo ^{31}P NMR chemical shift imaging. *Magn. Reson. Med.* **29,** 249–251.

Morris, G. A., Barjat, H., and Horne, T. J. (1997). Reference deconvolution methods. *Prog. Nucl. Magn. Reson. Spectrosc.* **31,** 197–257.

Murphy-Boesch, J., Jiang, H., Stoyanova, R., and Brown, T. R. (1998). Quantification of phosphorus metabolites from chemical shift imaging spectra with corrections for point spread effects and B1 inhomogeneity. *Magn. Reson. Med.* **39,** 429–438.

Narayana, P. A., Doyle, T. J., Lai, D. J., and Wolinsky, J. S. (1998). Serial proton magnetic resonance spectroscopic imaging, contrast-enhanced magnetic resonance imaging, and quantitative lesion volumetry in multiple sclerosis. *Ann. Neurol.* **43,** 56–71.

Negendank, W. (1992). Studies of human tumors by MRS: A review. *NMR Biomed.* **5,** 303–324.

Negendank, W. G., Sauter, R., Brown, T. R., Evelhoch, J. L., Falini, A., Gotsis, E. D., Heerschap, A., Kamada, K., Lee, B. C., Mengeot, M. M., Moser, E., Padavic-Shaller, K. A., Sanders, J. A., Spraggins, T. A., Stillman, A. E., Terwey, B., Vogl, T. J., Wicklow, K., and Zimmerman, R. A. (1996). Proton magnetic resonance spectroscopy in patients with glial tumors: A multicenter study. *J. Neurosurg.* **84,** 449–458.

Nelson, S. J. (2001). Analysis of volume MRI and MR spectroscopic imaging data for the evaluation of patients with brain tumors. *Magn. Reson. Med.* **46,** 228–239.

Nelson, S. J., and Brown, T. R. (1987). A method for automatic quantification of one-dimensional spectra with low signal-to-noise ratio. *J. Magn. Reson.* **75,** 229–243.

Nelson, S. J., Vigneron, D. B., and Dillon, W. P. (1999). Serial evaluation of patients with brain tumors using volume MRI and 3D ^{1}H MRSI. *NMR Biomed.* **12,** 123–138.

Noworolski, S. M., Nelson, S. J., Henry, R. G., Day, M. R., Wald, L. L., Star-Lack, J., and Vigneron, D. B. (1999). High spatial resolution ^{1}H-MRSI and segmented MRI of cortical gray matter and subcortical white matter in three regions of the human brain. *Magn. Reson. Med.* **41,** 21–29.

Ordidge, R. J., Bendall, M. R., Gordon, R. E., and Connelly, A. (1985). Volume selection for in-vivo biological spectroscopy. *In* "Magnetic Resonance in Biology and Medicine" (G. Govil, C. L. Khetrapal, and A. Saran, eds.), pp. 387–397. Tata McGraw–Hill, New Delhi.

Pan, J. W., Twieg, D. B., and Hetherington, H. P. (1998). Quantitative spectroscopic imaging of the human brain. *Magn. Reson. Med.* **40,** 363–369.

Parker, D. L., Gullberg, G. T., and Frederick, P. R. (1987). Gibbs artifact removal in magnetic resonance imaging. *Med. Phys.* **14,** 640–645.

Petroff, O. A., and Rothman, D. L. (1998). Measuring human brain GABA in vivo: Effects of GABA-transaminase inhibition with vigabatrin. *Mol. Neurobiol.* **16,** 97–121.

Pike, G. B., De Stefano, N., Narayanan, S., Francis, G. S., Antel, J. P., and Arnold, D. L. (1999). Combined magnetization transfer and proton spectroscopic imaging in the assessment of pathologic brain lesions in multiple sclerosis. *Am. J. Neuroradiology* **20,** 829–837.

Plevritis, S. K., and Macovski, A. (1995). MRS imaging using anatomically based k-space sampling and extrapolation. *Magn. Reson. Med.* **34,** 686–693.

Pohmann, R., von Kienlin, M., and Haase, A. (1997). Theoretical evaluation and comparison of fast chemical shift imaging methods. *J. Magn. Reson.* **129,** 145–160.

Ponder, S. L., and Twieg, D. B. (1994). A novel sampling method for ^{31}P spectroscopic imaging with improved sensitivity, resolution, and sidelobe suppression. *J. Magn. Reson.* **104,** 85–88.

Posse, S., Dager, S. R., Richards, T. L., Yuan, C., Ogg, R., Artru, A. A., Müller-Gärtner, H. W., and Hayes, C. (1997). In vivo measurement of regional brain metabolic response to hyperventilation using magnetic resonance: Proton echo planar spectroscopic imaging PEPSI. *Magn. Reson. Med.* **37,** 858–865.

Posse, S., DeCarli, C., and Le Bihan, D. (1994). Three-dimensional echo-planar MR spectroscopic imaging at short echo times in the human brain. *Radiology* **192,** 733–738.

Preul, M. C., Caramanos, Z., Leblanc, R., Villemure, J. G., and Arnold, D. L. (1998). Using pattern analysis of in vivo proton MRSI data to improve the diagnosis and surgical management of patients with brain tumors. *NMR Biomed.* **11,** 192–200.

Provencher, S. W. (1993). Estimation of metabolite concentrations from localized in vivo proton NMR spectra. *Magn. Reson. Med.* **30,** 672–679.

Purcell, E. M., Torrey, H. C., and Pound, R. V. (1946). Resonance absorption by nuclear magnetic moments in solids. *Phys. Rev.* **69,** 37–38.

Rae, C., Scott, R. B., Thompson, C. H., Kemp, G. J., Dumughn, I., Styles, P., Tracey, I., and Radda, G. K. (1996). Is pH a biochemical marker of IQ? *Proc. R. Soc. London B Biol. Sci.* **263,** 1061–1064.

Richards, T. L., Dager, S. R., Panagiotides, H. S., Hayes, C. E., Posse, S., Serafini, S., Nelson, J. A., and Maravilla, K. R. (1997). Functional magnetic resonance spectroscopy during language activation. *Int. J. Neuroradiol.* **3,** 490–495.

Robitaille, P. M. L., Robitaille, P. A., Brown, G. G., Jr., and Brown, G. G. (1991). An analysis of the pH-dependent chemical-shift behavior of phosphorus-containing metabolites. *J. Magn. Reson.* **92,** 73–84.

Ross, B., and Bluml, S. (2001). Magnetic resonance spectroscopy of the human brain. *Anat. Rec.* **265,** 54–84.

Ross, B. D., Ernst, T., Kreis, R., Haseler, L. J., Bayer, S., Danielsen, E. R., Blüml, S., Shonk, T., Mandigo, J. C., Caton, W., Clark, C., Jensen, S. W., Lehman, N. L., Arcinue, E., Pudenz, R., and Shelden, C. H. (1998). ^{1}H MRS in acute traumatic brain injury. *J. Magn. Reson. Imaging* **8,** 829–840.

Rudkin, T. M., and Arnold, D. L. (1999). Proton magnetic resonance spectroscopy for the diagnosis and management of cerebral disorders. *Arch. Neurol.* **56,** 919–926.

Schuff, N., Amend, D., Ezekiel, F., Steinman, S., Tanabe, J. L., Norman, D., Jagust, W., Kramer, J. H., Mastrianni, J. A., Fein, G., and Weiner, M. W. (1997). Changes of hippocampal *N*-acetyl aspartate and volume in Alzheimer's disease: A proton MR spectroscopic imaging and MRI study. *Neurology* **49,** 1513–1521.

Schuff, N., Ezekiel, F., Gamst, A., Amend, D., Capizzano, A., Maudsley, A. A., and Weiner, M. W. (2001a). Region and tissue differences of metabolites in normally aged brain using ^{1}H magnetic resonance spectroscopic imaging. *Magn. Reson. Med.* **45,** 899–907.

Schuff, N., Rooney, W. D., Miller, R. G., Gelinas, D. F., Amend, D., Maudsley, A. A., and Weiner, M. W. (2001b). Reanalysis of multislice ^{1}H MRSI in amyotrophic lateral sclerosis. *Magn. Reson. Med.* **45,** 513–516.

Schuff, N., Soher, B. J., Young, K., Maudsley, A. A., Ezekiel, F., Amend, D. L., Fein, G., and Weiner, M. W. (1998). Regression and histogram analysis of ^{1}H MR spectroscopic imaging data. *In* "Proceedings of the International Society for Magnetic Resonance in Medicine," p. 28.

Schuff, N., Vermathen, P., Maudsley, A. A., and Weiner, M. W. (1999). Proton magnetic resonance spectroscopic imaging in neurodegenerative diseases. *Curr. Sci.* **76,** 800–807.

Sepponen, R. E. (1985). A method for imaging of chemical shift or magnetic field distributions. *Magn. Reson. Imaging* **3,** 163–167.

Singer, J. R. (1959). Blood flow rates by nuclear magnetic resonance measurements. *Science* **130,** 1652.

Slotboom, J., Boesch, C., and Kreis, R. (1998). Versatile frequency domain fitting using time domain models and prior knowledge. *Magn. Reson. Med.* **39,** 899–911.

Soher, B. J., van Zijl, P. C., Duyn, J. H., and Barker, P. B. (1996). Quantitative proton MR spectroscopic imaging of the human brain. *Magn. Reson. Med.* **35,** 356–363.

Soher, B. J., Young, K., Govindaraju, V., and Maudsley, A. A. (1998). Automated spectral analysis. III. Application to in vivo proton MR spectroscopy and spectroscopic imaging. *Magn. Reson. Med.* **40,** 822–831.

Spielman, D. M., Adalsteinsson, E., and Lim, K. O. (1998). Quantitative assessment of improved homogeneity using higher-order shims for spectroscopic imaging of the brain. *Magn. Reson. Med.* **40,** 376–382.

Spielman, D. M., Webb, P., and Macovski, A. (1989). Water referencing for spectroscopic imaging. *J. Magn. Reson.* **12,** 38–49.

Stanley, J. A., Drost, D. J., Williamson, P. C., and Thompson, R. T. (1995). The use of a priori knowledge to quantify short echo in vivo ^{1}H MR spectra. *Magn. Reson. Med.* **34,** 17–24.

Stoyanova, R., and Brown, T. R. (2001). NMR spectral quantitation by principal component analysis. *NMR Biomed.* **14,** 271–277.

Suhy, J., Rooney, W. D., Goodkin, D. E., Capizzano, A. A., Soher, B. J., Maudsley, A. A., Waubant, E., Anderson, P. B., and Weiner, M. W. (2000a). ^{1}H MRS comparison of white matter and lesions in primary progressive and relapsing remitting MS. *Multiple Scler.* **6,** 148–155.

Suhy, J., Schuff, N., Miller, R. G., Gatto, N., Maudsley, A. A., and Weiner, M. W. (2000b). ^{1}H MRSI measurement of early detection and longitudinal changes in ALS. *In* "Proceedings of the International Society for Magnetic Resonance in Medicine," p. 630.

Tofts, P. S., and Wray, S. (1988). A critical assessment of methods of measuring metabolite concentrations by NMR spectroscopy. *NMR Biomed.* **1,** 1–10.

Toga, A. W., and Mazziotta, J. C. (1996). "Brain Mapping: The Methods," 1st edition. Academic Press, San Diego.

Trabesinger, A. H., and Boesiger, P. (2001). Improved selectivity of double quantum coherence filtering for the detection of glutathione in the human brain in vivo. *Magn. Reson. Med.* **45,** 708–710.

Twieg, D. B., Hetherington, H. P., Ponder, S. L., den Hollander, J., and Pohost, G. M. (1994). Spatial resolution in ^{31}P metabolite imaging of the human brain at 4.1 T. *J. Magn. Reson. B* **104,** 153–158.

van Sluis, R., Bhujwalla, Z. M., Raghunand, N., Ballesteros, P., Alvarez, J., Cerdán, S., Galons, J. P., and Gillies, R. J. (1999). In vivo imaging of extracellular pH using ^{1}H MRSI. *Magn. Reson. Med.* **41,** 743–750.

Vanhamme, L., Sundin, T., Hecke, P. V., and Huffel, S. V. (2001). MR spectroscopy quantitation: A review of time-domain methods. *NMR Biomed.* **14,** 233–246.

Vigneron, D., Bollen, A., McDermott, M., Wald, L., Day, M., Moyher-Noworolski, S., Henry, R., Chang, S., Berger, M., Dillon, W., and Nelson, S. (2001). Three-dimensional magnetic resonance spectroscopic imaging of histologically confirmed brain tumors. *Magn. Reson. Imaging* **19,** 89–101.

Vion-Dury, J., Meyerhoff, D. J., Cozzone, P. J., and Weiner, M. W. (1994). What might be the impact on neurology of the analysis of brain metabolism by in vivo magnetic resonance spectroscopy? *J. Neurol.* **241,** 354–371.

von Kienlin, M., Ziegler, A., Le Fur, Y., Rubin, C., Decorps, M., and Remy, C. (2000). 2D-spatial/2D-spectral spectroscopic imaging of intracerebral gliomas in rat brain. *Magn. Reson. Med.* **43,** 211–219.

Wear, K. A., Myers, K. J., Rajan, S. S., and Grossman, L. W. (1997). Constrained reconstruction applied to 2-D chemical shift imaging. *IEEE Trans. Med. Imaging* **16,** 591–597.

Webb, P., Spielman, D., and Macovski, A. (1992). Inhomogeneity correction for in vivo spectroscopy by high-resolution water referencing. *Magn. Reson. Med.* **23,** 1–11.

Wild, J. M. (1999). Artifacts introduced by zero order phase correction in proton NMR spectroscopy and a method of elimination by phase filtering. *J. Magn. Reson.* **137,** 430–436.

Wild, J. M., Macmillan, C. S., Wardlaw, J. M., Marshall, I., Cannon, J., Easton, V. J., and Andrews, P. J. (1999). ^{1}H spectroscopic imaging of acute head injury—Evidence of diffuse axonal injury. *MAGMA* **8,** 109–115.

Young, K., Govindaraju, V., Soher, B. J., and Maudsley, A. A. (1998a). Automated spectral analysis. I. Formation of *a priori* information by spectral simulation. *Magn. Reson. Med.* **40,** 812–815.

Young, K., Khetselius, D., Soher, B. J., and Maudsley, A. A. (2000). Confidence images for MR spectroscopic imaging. *Magn. Reson. Med.* **44,** 537–545.

Young, K., Soher, B. J., and Maudsley, A. A. (1998b). Automated spectral analysis. II. Application of wavelet shrinkage for characterization of non-parameterized signals. *Magn. Reson. Med.* **40,** 816–821.

15

Principles, Methods, and Applications of Diffusion Tensor Imaging

Susumu Mori

Department of Radiology, Johns Hopkins University School of Medicine, Baltimore, Maryland 21205

I. Diffusion Measurement by NMR

A. Diffusion and Flow

In this chapter measurements of the molecular diffusion process using magnetic resonance (MR) will be described. Diffusion means incoherent molecular motions by thermal energy, or so-called Brownian motion. For example, if ink is dropped in a cup of water, it starts to spread by diffusion. If there are no obstacles or boundaries in the medium, it is known that the probability of finding a molecule of interest after a certain amount of time follows Gaussian distribution (Fig. 1A). Namely, there is a higher probability at the center of the distribution and the probability gradually decreases in a symmetrical manner. In the following discussion, it is very important not to confuse the diffusion process with flow, which is a coherent molecular motion (Fig. 1B), because these two processes have very different effects on MR signals.

B. Pulsed Magnetic Field Gradient

MRI scanners are always equipped with so-called "pulsed field gradient" instruments. This gradient system is very versatile equipment and, in addition to its main purpose, i.e., imaging, it can be used for many other purposes, one of which is the diffusion measurement.

In both MR spectroscopy and MRI, the strength of magnetic field (called the B_0 field) is kept as homogeneous as possible because the frequency of signal is directly proportional to the B_0. For example, if the B_0 field is inhomogeneous, as shown in Fig. 2B, MR signal from even a simple

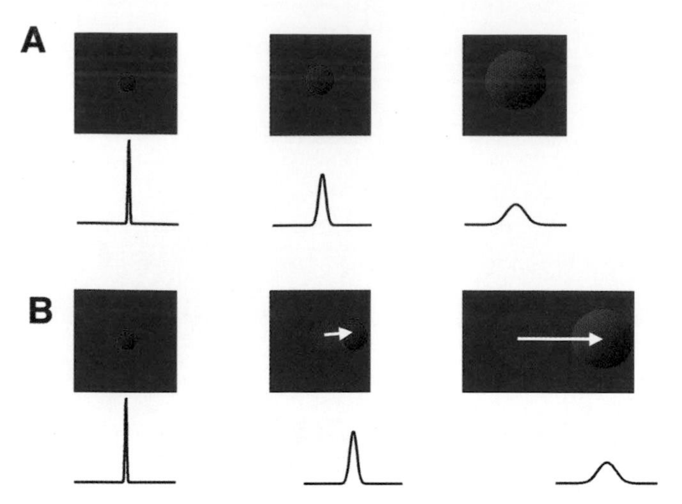

Figure 1 Schematic diagrams of diffusion (**A**) and flow (**B**) processes. In the diffusion process, the probability of finding molecules initially concentrated in one location follows Gaussian distribution, while the center of the distribution stays at the same location. If the center shifts in one direction, it contains flow.

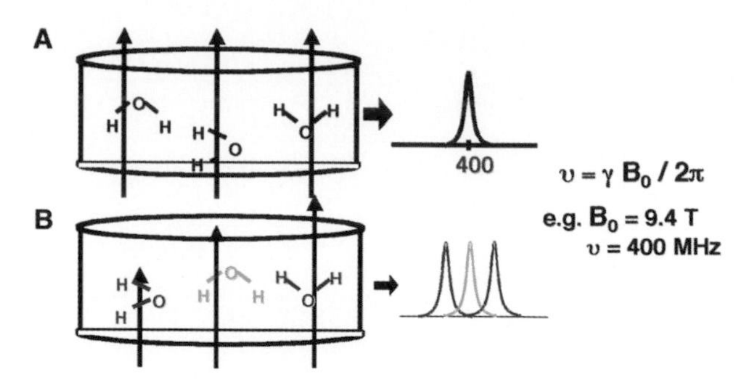

Figure 2 Schematic diagram of field homogeneity and signal frequency. (**A**) Usually, the magnetic field (long arrows) is kept as homogeneous as possible to ensure that every water molecule sees the same field strength and gives the same signal frequency. For example, if the field strength is 9.4 T, all proton MR signal resonates at 400 MHz, giving rise to a single peak. (**B**) If the field is inhomogeneous, each water molecule sees a different field strength. In this example, the red water sees a stronger field and resonates at higher frequency, while the blue water resonates at the lower frequency, resulting in multiple peaks from a tube of water.

sample like a tube of water becomes complicated, which is usually not preferable.

The magnetic field gradient is a technology that introduces a linear magnetic field inhomogeneity (i.e., gradient) on purpose. We can control its duration, orientation, strength, and polarity (Fig. 3).

Figure 4 shows an example of a gradient diagram used to show the time sequence of gradient applications.

C. Effect of Diffusion on NMR Signal

Figure 5 shows a schematic diagram that explains what happens to MR signals if gradients are applied. After a 90° excitation RF pulse, protons at different locations start to give MR signals at the same frequency, which are represented by rotating vectors at the same rate. During a gradient application, protons start to see different B_0 and rotate at different rates depending on their locations. In this example, the red proton sees weaker B_0 and rotates slower and vice *versa* for the blue proton. At the end of the gradient application and when the system regains the homogeneous B_0, the phases (locations of the vectors) of the signals are no longer identical among the protons. Therefore, the first gradient is called the "dephasing" gradient. During the second gradient, because it has the opposite polarity, the red proton starts to rotate faster, catching up with the other protons, and the blue one starts to rotate slower. If the strength and length of the second gradient are identical to those of the first one, the protons should regain the same phase at the end of the second gradient, as if nothing has happened in the past. Therefore the second gradient is called the "rephasing" gradient.

However, this rephasing happens only when protons do not change their locations in between the applications of the two diphase–rephase gradients. As shown in Fig. 6A, what the dephasing gradient does is to "tag" locations of water molecules by their signal phase. If the water moves, it results in breakdown of the nice gradation of the phase across the sample. After the rephasing, the moved protons can be detected because they have phases different from the other unmoved protons. Strictly speaking, MR cannot measure the individual phases but it can detect the imperfect rephasing by the loss of signal because the signal intensity of the sample is proportional to the sum of all the vectors.

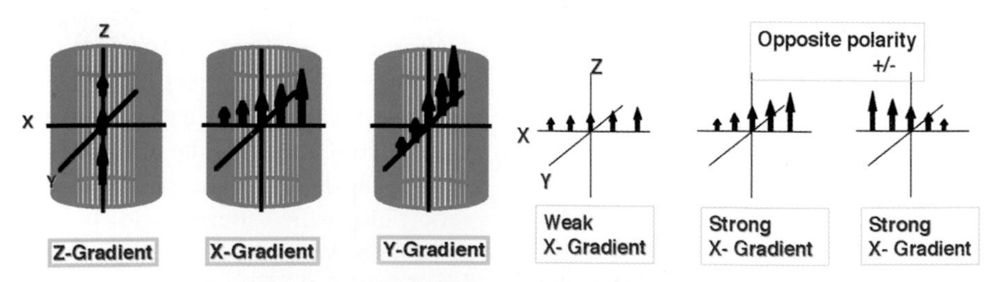

Figure 3 A schematic diagram of gradient orientation, strength, and polarity. The z axis is parallel to the bore of the magnet. Arrows indicate the strength of the magnetic field (B_0).

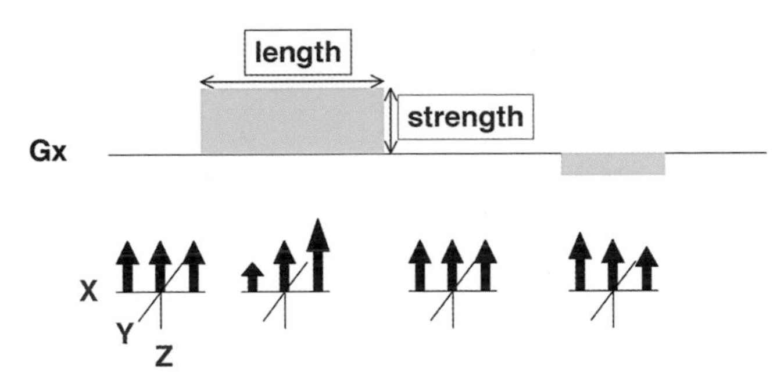

Figure 4 A gradient diagram often used to represent a time sequence of gradient applications. The term *Gx* indicates that the gradient is applied along the x axis. The first gradient was long and strong and the second was short and weak with the opposite polarity.

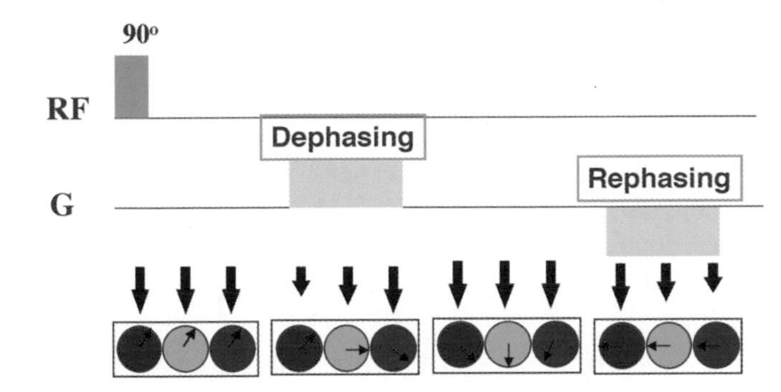

Figure 5 An example of a dephase–rephase experiment by gradient application. Red, green, and blue circles indicate three water molecules located at different positions in a sample tube. Thick arrows indicate the magnetic field strength and thin arrows the "phase" of the MR signals from each molecule.

At this point, I would like to add several important issues about the diffusion measurement by MR. First, it is noninvasive and does not require injection of any chemical tracers. Second, it measures the water motion along a predetermined axis. In this example, water that moved along the horizontal axis (indicated by black boxes) can be detected because it perturbed the phase gradation. However, water motion along the vertical axis (indicated by a green box) does not have any effect and cannot be detected. Third, incoherent (diffusion) and coherent water motions lead to different outcomes in this experiment. As shown in Fig. 6B, coherent motions such as flow or bulk motions of the subject result in perfect refocusing (thus not signal loss) and shift of signal phase. Therefore, the incoherent motion (signal loss) and coherent motion (phase shift) can be, in principle, separated in the MR diffusion measurement. In practice, however, the flow can occur along multiple orientations within a pixel (such as small blood vessels) and bulk motion such as brain pulsation is not rigid and uniform as shown in Fig. 6B. In these cases, the coherent motion often leads to signal loss and interferes with the diffusion measurements. In diffusion measurement, the order of the diffusion process during the measurement is typically 1–20 μm depending on sample, temperature, and pulse sequence, and the experiment is designed in such a way that

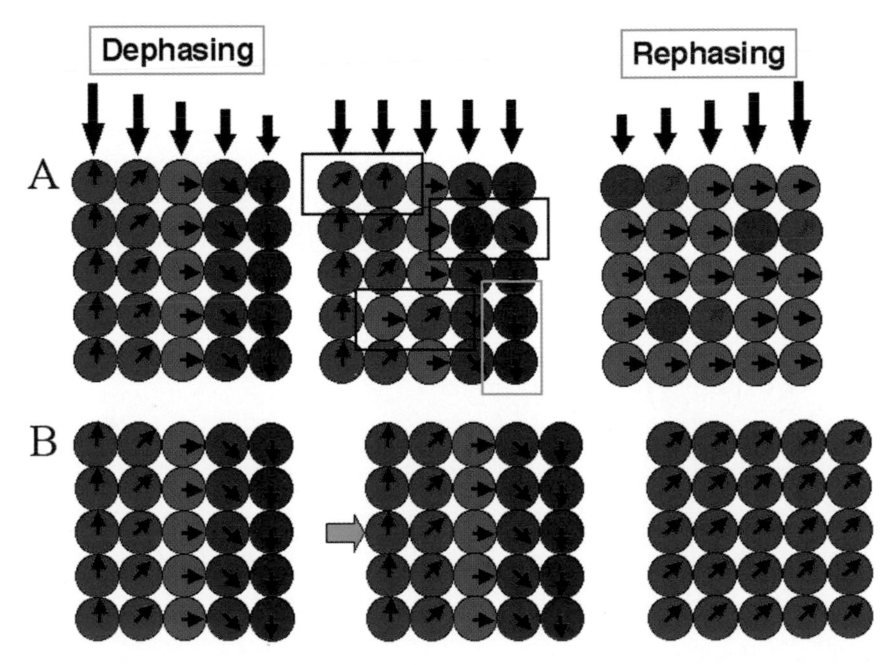

Figure 6 Effects of molecular motions in an experiment with a pair of gradients. Large arrows indicate the strength of magnetic field (B_0) and small arrows the phases of individual protons. The phase is also indicated by gradation of color. (**A**) Diffusion process in which protons move randomly. Protons (circles) that have moved in between the two gradients are indicated by boxes. (**B**) Coherent motion of water such as flow or bulk motion.

the diffusion of this degree leads to the signal loss of typically 10–90% (see next section). In such experiments, water motion due to blood flow is far larger and the blood signal (about 5% of water population in the brain) should be completely dephased. The effect of bulk motions such as brain pulsation and involuntary movement of the subject is more troublesome. This issue will be discussed in more detail under Data Acquisition.

D. Mathematical Description of the Diffusion Effect

From Fig. 6, it should be understood that there are three factors that contribute to the amount of signal loss by the application of a pair of gradients. First, the faster the diffusion is, the more protons change their locations and, thus, the less perfect the rephasing becomes (signal loss). Second the longer the time separation between the two gradient pulses (Δ in Fig. 7), the more time there is for protons to move and this leads to more signal loss. Third, the steeper the phase gradation introduced by the initial dephasing gradient, the more dephasing remains after the rephasing gradient. The amount of the initial phase gradation is proportional to the strength of the gradient (G) and its duration (δ), or in other words the area of the dephasing gradient ($G\delta$). Therefore, the amount of signal loss, S/S_0 (S_0 is the signal intensity without the gradient applications), must be a function of $f(D, \Delta, \delta G)$. If we assume that the diffusion has purely Gaussian probability (Fig. 1), the status of spins at a given moment can be expressed by a function:

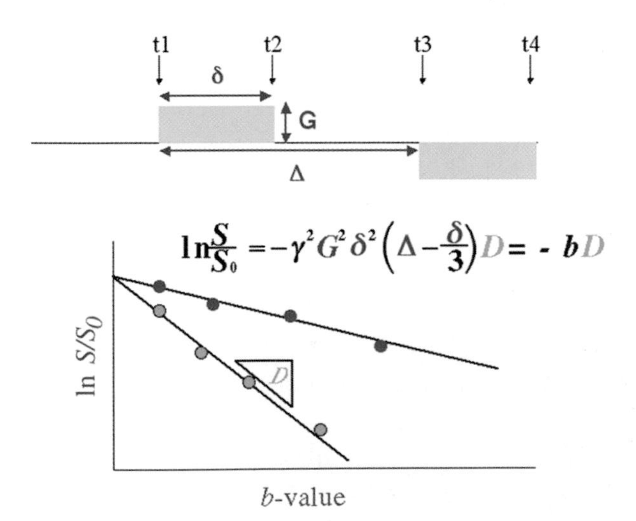

Figure 7 Mathematical description of the signal loss due to the diffusion process in an experiment with a pair of gradients. By performing a series of experiments with different b values, the diffusion constant can be obtained from the slope. Namely, the steeper the slope, the larger the diffusion constant.

$$\left[\frac{S}{S_0}\right] = \left[-\int_0^t D\gamma^2 \int_0^{t'} G(t'')dt''\right]^2 dt'. \tag{1}$$

By integrating the equation from $t1$ to $t4$ in Fig. 7, the signal attenuation in this study can be found to be (Stejskal and Tanner 1965)

$$In\left[\frac{S}{S_0}\right] = -\gamma^2 G^2 \delta^2 (\Delta - \frac{\delta}{3})D. \tag{2}$$

In order to determine the diffusion constant, D, at least two experiments with different b value (= $(\gamma\delta G)^2$ $(\Delta - \delta/3)$) have to be carried out, from which the slope of the signal decay (i.e., diffusion constant) can be calculated (Fig. 7). In actual diffusion experiments, the b value is changed by changing the gradient strength, G. Also, it is common to use a spin-echo sequence (Fig. 8) to avoid T_2* decay during the possibly lengthy gradient application (δ and Δ). Please note that if the two gradients are used in a spin-echo sequence, the polarities of the two gradients have to be the same, because the 180° RF pulse flips the signal phase by 180°.

E. Diffusion Imaging

So far, I have described the diffusion measurement of water inside the entire sample. This technique can be used if one is interested in diffusion of, for example, a homogeneous solution. However, the diffusion constant of water inside a biological sample can be different depending on its locations within the sample. In the late 1980s, the diffusion measurement was combined with imaging and "diffusion imaging" was introduced (Fig. 9) (Le Bihan *et al.*, 1986). In this experiment, a series of images with different b values (called diffusion-weighted images) were obtained from which the diffusion constant at each pixel can be calculated. The calculated image, in which the brightness of each pixel

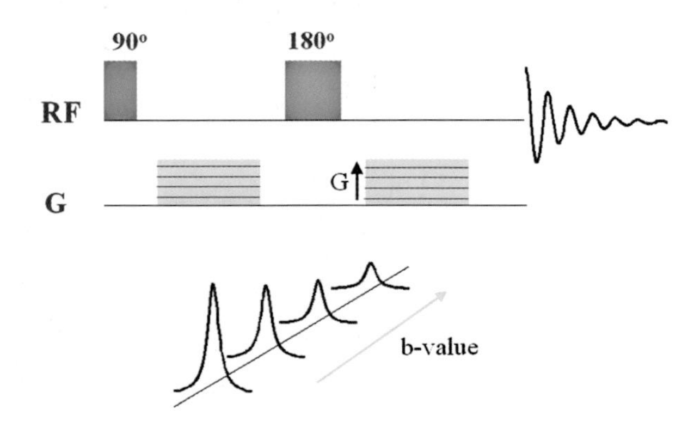

Figure 8 An example of an actual experiment using a spin echo. Four experiments with different gradient strengths are performed. In the spin-echo experiment, the polarity of the two gradients has to be the same.

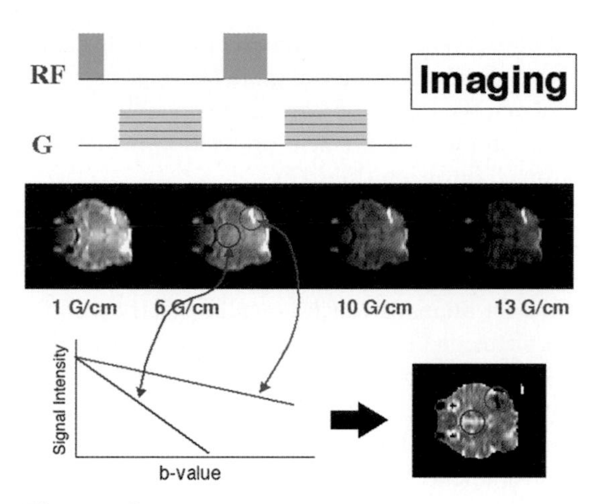

Figure 9 A schematic diagram of a diffusion imaging experiment. Images with different *b* values (middle) are called diffusion-weighted images and a calculated map of the diffusion constant (bottom image) is called the apparent diffusion constant map. Blue color indicates an area with high diffusion constant and red, with low diffusion constant.

shows the diffusion constant, is called an apparent diffusion constant (ADC) map. In the diffusion-weighted images, brain regions with a high diffusion constant (e.g., ventricle shown by the blue circle) tend to be darker and those with a low diffusion constant (red circle), brighter, although the contrast heavily depends on the *b* value. The contrast of the ADC map tends to be the inverse of diffusion-weighted images because the high diffusion constant area becomes brighter and vice versa.

F. Data Acquisition of Diffusion Imaging

Although the principle of diffusion imaging (Fig. 9) may seem straightforward, actual experiments are quite involved mainly due to the phase-shift problem caused by coherent molecular motions, especially bulk motions. In order to better understand this problem, let us go over the process of imaging briefly (Fig. 10). For example, if one wants to obtain an image of a 128 × 128 matrix, a scanner acquires raw data (called time-domain data or *k* space) with 128 × 128 resolution. In conventional imaging, the *k* space is recorded line by line, requiring 128 independent scans. In the *k* space, phase and intensity information from the sample is recorded and after the Fourier transformation (FT), the information is converted to location and intensity (i.e., image). As illustrated in Figs. 6A and 9, the image is sensitized by diffusion by applying a pair of gradients, and the intensity information carries the important information of the diffusion processes. However, if coherent molecular motions occur, phase shifts are introduced into the *k* space. This is equivalent to a shift of the sampling lines in the *k* space. If it happens, it leads to misregistration of proton signals after the FT, which appear as "ghosting" (Fig. 10A) (Turner *et al.*, 1990; Duerk and Simonetti, 1991; Norris, 2001). Our recent study demonstrated that the dominant source of the coherent motion is caused by brain bulk motions due to pulsation (Jiang *et al.*, 2002). The most common way to solve this problem is the use of single-shot echo planner imaging (SS-EPI) (Turner and Le Bihan, 1990; Turner *et al.*, 1990), in which the entire *k* space is obtained with one scan. In this case, even if the diffusion weighting and bulk motions cause the phase shift, the entire *k* space shifts by the same amount (Fig. 10B), which does not have any effect after the FT. The price we have to pay is severe deterioration of image quality (low resolution and image distortion), which is common to SS-EPI. Currently there are two types of techniques being developed to solve these issues. The first approach is based on the multishot imaging technique (Fig. 10A), in which the amount of phase shift is monitored in real time using so-called navigator echoes and the information is used in postprocessing to remove the ghosting (Norris *et al.*, 1992; Anderson and Gore, 1994; Ordidge *et al.*, 1994; Butts *et al.*, 1996). One-dimension, two-dimension (Atkinson *et al.*, 2000),

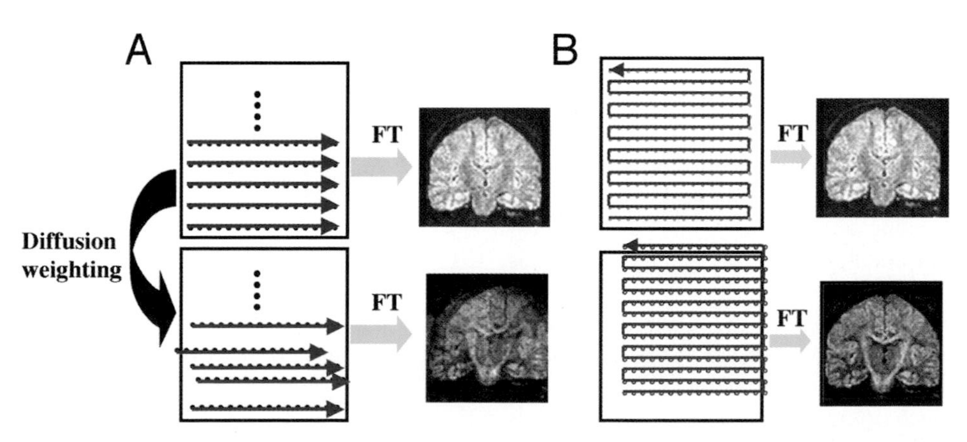

Figure 10 Schematic diagrams explaining the effects of phase errors in multishot imaging (**A**) and single-shot EPI (**B**). Red arrows indicate a sampling line by each scan.

spiral (Butts *et al.*, 1997), or self-navigated approaches (Trouard *et al.*, 1999; Pipe, 2001) have been postulated. The second, technique is based on the SS-EPI (Fig. 10B), in which the image deterioration is ameliorated by reducing the size of the *k* space. New techniques such as SENSE (Pruessmann *et al.*, 1999; Bammer *et al.*, 2001) and SMASH (Sodickson and Manning, 1997) are in this category. These research fields are currently rapidly evolving and in the near future, most of the motion-related problems should be addressed by new hardware, pulse sequences, and postprocessing techniques.

There are several points I would like to stress at this time. First, the phase errors caused by subject motion always exist in MRI. However, their extent is far larger in the diffusion-weighted images due to the application of a pair of gradients. Second, it is very important to understand that SS-EPI is robust against motion-related artifacts not simply because it is rapid. It is because it is insensitive to the phase errors. Third, the motion artifacts discussed above should not be confused with coregistration errors between the diffusion-weighted images. The ADC map or diffusion tensor imaging always requires multiple diffusion-weighted images from which these are calculated on a pixel-by-pixel basis. Therefore, the location of each image has to be exactly coregistered. If the subject moved during the scan, it would cause not only ghosting in the image, but also the coregistration errors during the calculation. This coregistration problem is not a problem specific to diffusion imaging, but it is common to any other MR technique that requires the coregistration of multiple images such as T_2 maps, MT maps, *f*MRI, or perfusion images and should be addressed by postprocessing coregistration schemes.

II. Diffusion Tensor Imaging

A. Isotropic and Anisotropic Diffusion

In the previous section, the principle of diffusion imaging was introduced. One of the unique and important features of the diffusion measurement by MR is that it always detects molecular movement along one predetermined axis (Fig. 6), which is determined by the orientation of an applied gradient. Every MR scanner is equipped with three orthogonal, namely *x*, *y*, and *z*, gradients. By combining these three axis gradients, diffusion along any arbitrary axis can be measured. For example, if equal strength of *x* and *y* gradient is applied simultaneously, diffusion 45° from the *x* and *y* axes can be measured. The orientation of the diffusion measurement is not important if we are interested in freely diffusing water because the results are independent of the measurement orientations. Such orientation-independent diffusion is called "isotropic" diffusion.

In biological systems, the diffusion process can be much more complicated because water molecules see many obstacles and barriers during the diffusion process. If the biological system has an ordered alignment, such as muscle or axonal fibers, the extent of water diffusion may be different depending on the measurement orientation, which is called "anisotropic" diffusion (Fig. 11).

If the sample has anisotropic diffusion, the results of MR diffusion measurements depend on which gradient axis is used. Figure 12 shows an example of diffusion measurements in a rat brain, in which it can be clearly seen that the apparent diffusion constants change considerably if a different gradient orientation is used. For example, a pixel, indicated by pink arrows, has low diffusion constants with *x* and *y* gradients but has a high diffusion constant when it is measured by *z* gradient. In previous studies, there is ample evidence that water tends to diffuse along fibers (Stejskal, 1965; Moseley *et al.*, 1990; Douek *et al.*, 1991; Basser *et al.*, 1994a; van Gelderen *et al.*, 1994a; Scollan *et al.*, 1998), probably because it sees fewer obstacles. From this rat brain study, we can immediately conclude that the pixel indicated by the pink arrow contains axonal fibers that are running perpendicular to the *x* and *y* axes and parallel to *z*. Obviously, the MR diffusion measurement has the capability to provide information on fiber architecture within the sample.

B. How Anisotropy Is Measured

The results shown in Fig. 12 clearly suggest that the ADC measurement along multiple orientations contains important information on the axonal organization of the brain. Then,

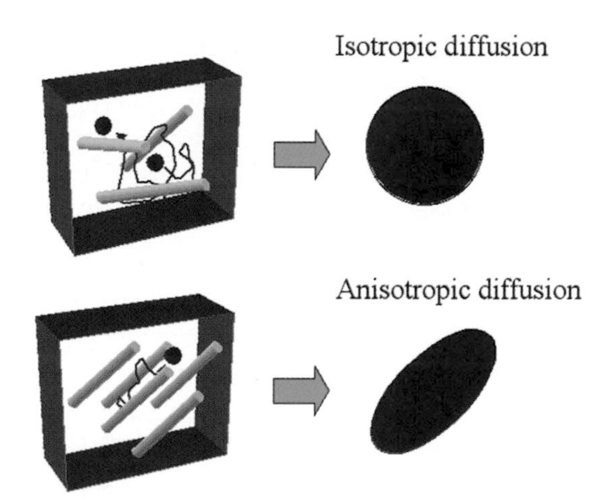

Figure 11 Isotropic and anisotropic diffusion in biological system. When there is coherent alignment, water tends to diffuse along the structure, leading to anisotropic diffusion, in which the extent of diffusion may be different depending on the orientation. On the other hand, if the environment of the water molecules has random structures, diffusion may be isotropic. Probabilities of finding a water molecule after a certain amount of time are spherical for the isotropic diffusion and elliptic for the anisotropic diffusion as shown in the right column. This ellipsoid is called the "diffusion ellipsoid."

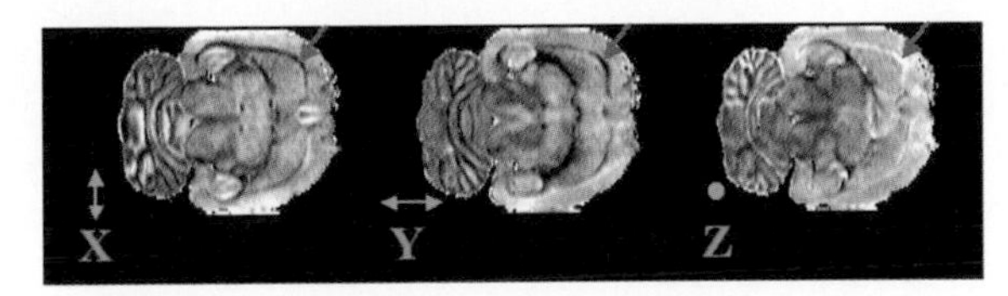

Figure 12 ADC maps of a fixed rat brain measured by x, y, and z gradients. The contrasts heavily depend on the measurement orientations (indicated by green arrows), suggesting water diffusion inside the brain is anisotropic.

the question is how can we fully characterize the anisotropic diffusion and subsequently the fiber architecture. When diffusion is isotropic, the probability of finding a water molecule after a certain amount of time becomes spherical (Fig. 11), which can be described by one parameter (diameter). If water is confined in a homogeneously aligned system, we can assume that the diffusion process leads to an elliptical shape of the probability with the longest axis aligned to the orientation of the fibers (Basser *et al.*, 1994a,b; Basser and Pierpaoli, 1996). Then, our task is to define the shape of the ellipsoid (called the diffusion ellipsoid). The most intuitive way is to measure the ADCs along numerous orientations, from which the shape can be reconstructed. This direct measurement of the diffusion ellipsoid shape is actually becoming popular (Wedeen *et al.*, 2000; Wiegell *et al.*, 2000; Frank, 2001). An alternative way is to measure the ADCs along a smaller number of orientations, from which the shape of the ellipsoid is calculated (Basser *et al.*, 1994a,b; Basser and Pierpaoli, 1996). For this calculation, we need the aide of a mathematical procedure called "tensor" calculation and, thus, this process is called diffusion tensor imaging (DTI).

First of all, I would like to make it clear that the anisotropic diffusion (or the diffusion ellipsoid) cannot be characterized by measurements along the three orthogonal axes, x, y, and z. This can be easily illustrated, as shown in Fig. 13. In this example, the ADC measurements along the three axes lead to the same result (the lengths of the ellipsoid and sphere along the x, y, and z axes are the same), although the two systems have markedly different diffusion properties. Obviously

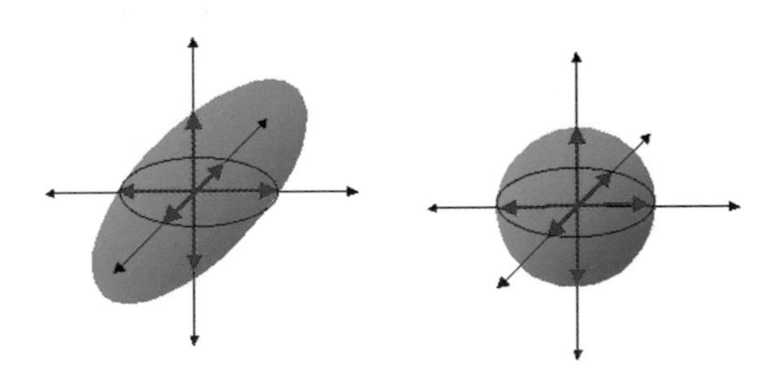

Figure 13 An example of three-axis diffusion measurement in an anisotropic system (**left**) and an isotropic system (**right**), which lead to the same result.

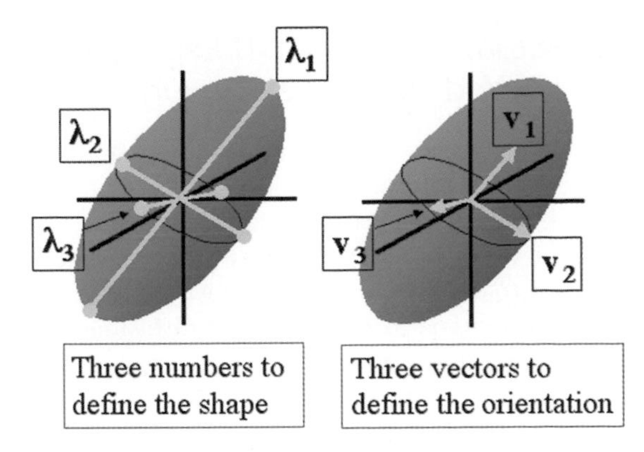

Figure 14 Six parameters needed to define an ellipsoid.

unlike a vector, diffusion cannot be represented by determining its lengths along three orthogonal axes.

In order to fully characterize the diffusion ellipsoid, we need six parameters as shown in Fig. 14. They are three numbers for the length of the longest (λ1), middle (λ2), and shortest (λ3) axes, which define its shape, and three vectors (v1–3) to define the orientations of the axes (called principal axes). In order to keep track of these six parameters, we use a 3 × 3 tensor, called diffusion tensor, $\bar{\bar{\mathbf{D}}}$, which is related to the six parameters by a process called "diagonalization":

$$\bar{\bar{D}} = \begin{bmatrix} D_{xx} & D_{xy} & D_{xz} \\ D_{yz} & D_{yy} & D_{yz} \\ D_{yz} & D_{yz} & D_{zz} \end{bmatrix} \xrightarrow{diagonalization} \lambda_1, \lambda_2, \lambda_3, \nu_1, \nu_2, \nu_3, \tag{3}$$

This diffusion tensor, $\bar{\bar{\mathbf{D}}}$, is a symmetric tensor, which means $D_{ij} = D_{ji}$, and, thus, there are six independent parameters, which makes sense because it intrinsically contains the six parameters of the diffusion ellipsoid.

In order to determine these six elements of $\bar{\bar{\mathbf{D}}}$, not surprisingly, we need to measure at least six diffusion constants along six independent axes (Fig. 15). In the following section, an actual experimental process to determine the $\bar{\bar{\mathbf{D}}}$ will be described.

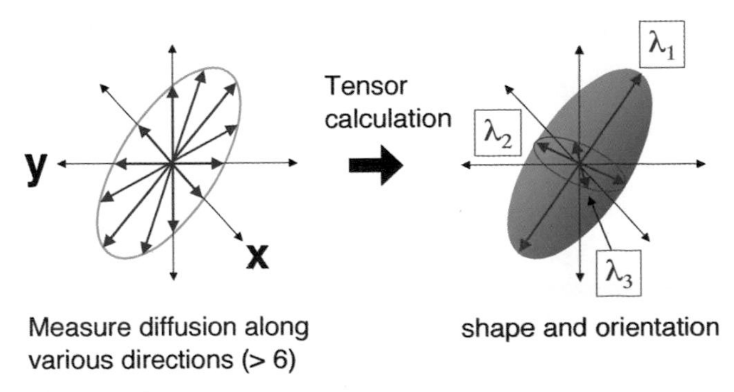

Figure 15 A diffusion ellipsoid can be fully characterized from diffusion measurements along six independent axes.

C. Actual Process of DTI Experiments and Tensor Calculation

In order to measure six diffusion constants along six independent axes, at least seven diffusion-weighted images are needed. This is because to obtain a diffusion constant from a slope of signal attenuation, at least two data points are needed (Fig. 7). The most common way to obtain the six diffusion constants is to obtain a least-diffusion-weighted image, the intensity of which corresponds to S_0 in Eq. (2). Then by obtaining another diffusion-weighted image using, for example, the x gradient, we can calculate a diffusion constant along the x axis (ADCx). The same S_0 data can be also used to obtain ADCy together with a diffusion-weighted image with the y gradient. In this way, six diffusion constants can be obtained using various gradient combinations. Figure 16 shows an example of the measurements, in which diffusion-weighted images by x, y, z, $x + y$, $x + z$, and $y + z$ gradients are used.

In the previous section, it was shown that signal attenuation in the experiment shown in Fig. 7 follows

$$\frac{S}{S_0} = e^{-\gamma^2 G^2 \delta^2 (\Delta - \delta/3) D} = e^{-bD}. \tag{4}$$

This equation is correct only for isotropic diffusion or for diffusion measurement along one axis. For more complete expression for anisotropic media, we have to use equation

$$\ln\left[\frac{S}{S_0}\right] = -\int_0^t \gamma^2 \left[\int_0^{t'} \overline{G(t'')dt''}\right] \cdot \bar{\bar{D}} \cdot \left[\int_0^{t'} \overline{G(t'')dt''}\right] dt'. \tag{5}$$

Again, if we solve this equation for the experiment with a pair of square-shaped gradients (Fig. 7), we obtain

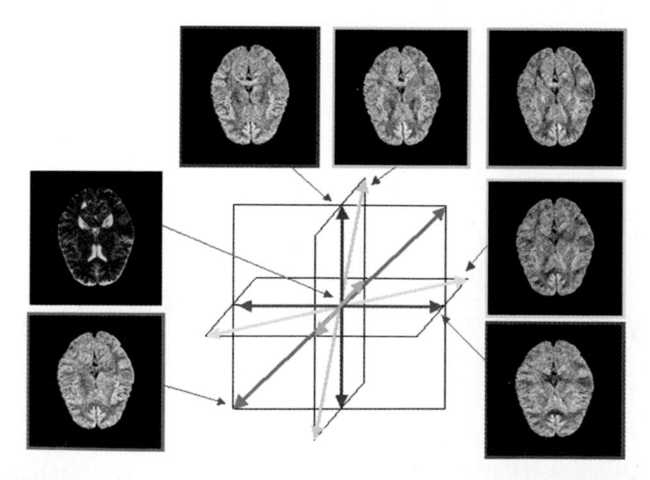

Figure 16 One non-diffusion-weighted image and six diffusion-weighted images along six different axes at least are needed, from which six diffusion constants can be obtained.

$$\frac{S}{S_0} = e^{-\overline{\sqrt{b}}\,\bar{\bar{D}}\,\overline{\sqrt{b}}^{T}}, \tag{6}$$

where $\bar{b}$ is $\gamma^2 \bar{G}^2 \delta^2 (\Delta - \delta/3)$. Here $\bar{G}$(and also $\bar{b}$) is a vector because it contains information on not only gradient strength but also the orientation.

In actual experiments, the six parameters in the $\bar{\bar{\mathbf{D}}}$ are what we want to determine, parameters $\bar{G}$, δ, Δ, and γ are known parameters, and S_0 and S are the experimental results. Please note that this equation has a total of seven unknowns (six in $\bar{\bar{\mathbf{D}}}$ and S_0) and we have seven experimental results (S) with different $\bar{G}$s. Therefore, it should be solvable.

For example, suppose we used the x gradient ($\bar{G}$= [Gx, 0, 0]) and obtained image intensity S_x and plugged this number into Eq. (6); it becomes

$$\frac{S}{S_0} = e^{-\gamma^2 Gx^2 \delta^2 (\Delta - \delta/3) Dxx}, \tag{7}$$

from which we can obtain an element Dxx. Similarly, from experiments using the y or z gradient only, we can obtain Dyy and Dzz. If we apply the same strength of x and y gradient simultaneously (($\bar{G}$= [Gx, Gy, 0]) we obtain image intensity S_{xy}:

$$\frac{S_{xy}}{S_0} = e^{-\gamma^2 \delta^2 (\Delta - \delta/3)(G_x^2 Dxx + 2G_x G_y D_{xy} + G_y^2 D_{yy}}. \tag{8}$$

Because we already know Dxx and Dyy, we can calculate Dxy from this result. Similarly, Dxz and Dyz can be obtained from experiments using the $x + z$ gradient combination and the $y + z$ gradient combination. In this way, from a total of seven diffusion-weighted images, the six elements of the $\bar{\bar{\mathbf{D}}}$ can be determined. Then the diagonalization process gives us $\lambda 1$, $\lambda 2$, $\lambda 3$, $v1$, $v2$, and $v3$ of the diffusion ellipsoid. This calculation has to be repeated for each pixel, from which the diffusion tensor can be obtained for each pixel as shown in Fig. 17.

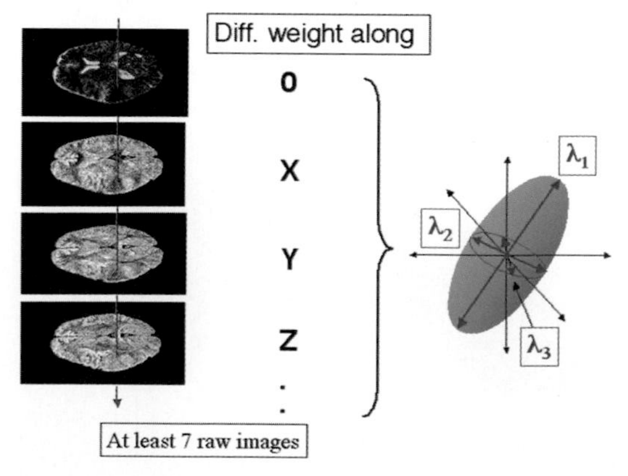

Figure 17 The diffusion tensor is calculated for each pixel from at least seven diffusion-weighted images.

D. Determination of the Tensor Elements from a Fitting Process

As described in the former section, Eq. (6) has seven unknowns, which can be solved from seven measurements (or imaging) with seven different diffusion-weighted images. In practical situations, we often perform more than seven measurements using different gradient strengths, gradient orientations, or signal averaging. In this case, Eq. (6) is overdetermined and the simple calculation process described above may not be appropriate. Furthermore, the signal intensity without any diffusion weighting (S_0) usually cannot be directly obtained because any imaging techniques always have a finite amount of diffusion weighting. Therefore, we should use a fitting technique rather than solve it. For the fitting, we need to expand Eq. (6):

$$\text{In}\frac{S}{S_0} = -\sqrt{\bar{b}}\,\bar{\bar{D}}\sqrt{\bar{b}^T} \rightarrow \text{In } S = \text{In } S_0 - \sqrt{\bar{b}}\,\bar{\bar{D}}\sqrt{\bar{b}^T}$$

$$\sqrt{\bar{b}} = \left[\sqrt{b_x}, \sqrt{b_y}, \sqrt{b_z},\right],$$

$$b_{x,y,z} = \gamma^2 G_{x.y.z}{}^2 \delta^2 (\Delta - \delta/3). \tag{9}$$

Next, we have to expand $\sqrt{\bar{b}}\bar{\bar{\mathbf{D}}}\sqrt{\bar{b}}^T$:

$$\sqrt{\bar{b}}\,\bar{\bar{D}}\sqrt{\bar{b}^T} = \left[\sqrt{b_x}, \sqrt{b_y}, \sqrt{b_z},\right] \begin{bmatrix} D_{xx} & D_{xy} & D_{xz} \\ D_{yx} & D_{yy} & D_{yz} \\ D_{zx} & D_{xy} & D_{zz} \end{bmatrix} \begin{bmatrix} \sqrt{b_x} \\ \sqrt{b_y} \\ \sqrt{b_z} \end{bmatrix}$$

$$= D_{xx}b_x + D_{yy}b_y + D_{zz}b_z + 2D_{xy}\sqrt{b_x}\sqrt{b_y}$$

$$+ 2D_{xz}\sqrt{b_x}\sqrt{b_z} + 2D_{yz}\sqrt{b_y}\sqrt{b_z}$$

$$= \mathbb{\bar{D}}\mathbb{\bar{b}}. \tag{10}$$

Here we defined two new vectors $\mathbb{\bar{D}}$ and $\mathbb{\bar{b}}$, which are defined as

$$\mathbb{\bar{D}} = \left[D_{xx}, D_{yy}, D_{zz}, D_{xy}, D_{xz}, D_{yz}\right],$$

$$\mathbb{\bar{b}} = \left[b_x, b_y, b_z, 2\sqrt{b_x}\sqrt{b_y}, 2\sqrt{b_x}\sqrt{b_z}, 2\sqrt{b_y}\sqrt{b_z}\right] \tag{11}$$

By plugging this back into Eq. (6), it can be rewritten to

$$\text{In S} = \text{In } S_0 - \mathbb{\bar{D}}\mathbb{\bar{b}}. \tag{12}$$

This equation is very similar to a simple linear equation, $y = b - ax$, in which x is an independent variable (equivalent to the **b** vector), y is the result (equivalent to image intensity, S), b is the intercept to the y axis (equivalent to S_0, which is the image intensity with $b = 0$), and a is the slope (equivalent to the diffusion vector). Similar to the equation $y = b - ax$, Eq. (10) can be solved by linear least-square fitting, although the variables are vectors, instead of scalar.

For example, in the experiment shown in Fig. 16, we used the gradient combination [0,0,0], [1,0,0], [0,1,0], [0,0,1], [1/√2, 1/√2,0], [1/√2,0, 1/√2], [0, 1/√2, 1/√2] (defined as $b1, b2, \ldots, b7$), and suppose we get image intensities $S1, S2, \ldots, S7$ from each experiment. Then the entire experiment can be expressed as

$$\begin{bmatrix} S1 \\ S2 \\ S3 \\ S4 \\ S5 \\ S6 \\ S7 \end{bmatrix} = \text{In } S_0 - \mathbb{\bar{D}} \begin{bmatrix} \mathbb{\bar{b}}1 \\ \mathbb{\bar{b}}2 \\ \mathbb{\bar{b}}3 \\ \mathbb{\bar{b}}4 \\ \mathbb{\bar{b}}5 \\ \mathbb{\bar{b}}6 \\ \mathbb{\bar{b}}7 \end{bmatrix}^T = \text{In } S_0 - \mathbb{\bar{D}}$$

$$\begin{bmatrix} 0 & 0 & 0 & 0 & 0 & 0 \\ 1 & 0 & 0 & 0 & 0 & 0 \\ 0 & 1 & 0 & 0 & 0 & 0 \\ 0 & 0 & 1 & 0 & 0 & 0 \\ 1/2 & 1/2 & 0 & 1/2 & 0 & 0 \\ 1/2 & 0 & 1/2 & 0 & 1/2 & 0 \\ 0 & 1/2 & 1/2 & 0 & 0 & 1/2 \end{bmatrix}^T \rightarrow \bar{S}^T = \text{In } S_0 - \mathbb{\bar{D}}\mathbb{\bar{b}}, \tag{(13)}$$

in which $\mathbb{\bar{b}}$ is called the $\mathbb{b}$ matrix. If 30 experiments with 30 different $\bar{bi}$ ($i = 1,2, \ldots, 30$) are performed and 30 images ($S1, \ldots, 30$) are obtained, these numbers can be just appended to the above equation and the best estimated $\bar{D}$ and S_0 can be obtained using a multivariate linear fitting (Basser and Pierpaoli, 1996).

E. Selection of b Matrix

In the above example, the $\mathbb{b}$ matrix was somewhat arbitrarily determined using the combination [0, 0, 0], [1, 0, 0], [0, 1, 0], [0, 0, 1], [1/√2, 1/√2, 0], [1/√2, 0, 1/√2], [0, 1/√2, 1/√2]. In a real situation, we have to consider the optimum $\mathbb{b}$ matrix, which includes the absolute strength of the b value ($\gamma^2|\bar{G}|\delta^2(\Delta - \delta/3)$), the gradient orientation, and the total number of experiments. For example, if acquisition of one diffusion-weighted image takes 1 min and we can afford to spend approximately 20 min for DTI, we can obtain 20 diffusion-weighted images. Then we have several choices for imaging protocols. For example, we can choose b = 1000 s/mm^2 and six gradient orientations (total of 7 diffusion-weighted images, including one image at close to b = 0 s/mm^2). To enhance the signal-to-noise ratio (SNR), the experiments can be repeated three times, giving rise to a total of 21 diffusion-weighted images. In the second option, the 7 diffusion-weighted images can be acquired with three different b values (e.g., $b = 500$, 1000, and 1500 s/mm^2). In the third option, diffusion-weighted images with 20 different gradient orientations with $b = 1000$ s/mm^2 and one repetition can be acquired within a similar amount of time. Recent papers have suggested that the third option may provide the best SNR (Jones *et al.*, 1999a; Skare *et al.*, 2000). The next important question is, "which b value and gradient orientation should we use?"

If the b value is too small, the signal loss by the diffusion process becomes too small and we cannot determine the signal decay slope accurately. If the b value is too strong, we observe a large signal decay and the signal intensity may become lower than the noise level, which we should avoid. Obviously, there is a range of optimum b values, which depends on diffusion constants of the sample and SNR. This is not an easy issue because the range of diffusion constants in brain samples can be substantial due to the large anisotropy, and the SNR heavily depends on imaging parameters. Another important factor is that large b values require longer δ and, thus, longer echo time, which leads to signal loss due to T_2 decay. Practically, $b = 800$–1200 s/mm^2 is the range most often used for clinical studies.

For the measurement orientation (gradient combination), sampling the 3D space uniformly is the most optimal way (Jones *et al.*, 1999a). This is understandable, because our task is to define the shape and orientation of the diffusion ellipsoid, which we do not know *a priori*. The gradient combinations we used in Fig. 16 do not distribute uniformly in the 3D space and, thus, it is not an optimal sampling scheme! Tables for gradient combinations for the uniform distribution for a given number of orientations can be found elsewhere (Jones *et al.*, 1999a; Skare *et al.*, 2000).

F. Technical Notes

In the Section II, I went through application of the tensor theory to describe the anisotropic diffusion. Before I end this section, I would like to mention several important issues that may contradict the tensor theory, which assumes that the environment of water molecules is homogeneous within a pixel. In clinical DTI, the image resolution is typically on the order of 1–5 mm, in which there should be numerous axons with different orientations and curvatures. Within individual axons, there are many water compartments; some may be confined in myelin sheaths and some may be close to the cell membrane, protein filaments, and organelles. Water diffusion properties in each compartment may be markedly different. There are also exchange processes among these multiple compartments. Therefore, our sample is decidedly complicated and inhomogeneous. We should always bear in our mind that our assumption that the diffusion property can be described by a tensor may not hold depending on the situation.

Currently there are many studies under way to investigate the diffusion properties in more detail. For example, it has been shown that the signal decay by diffusion weighting is not linear as shown in Fig. 7 (Niendorf *et al.*, 1996; Stanisz *et al.*, 1997; Buckley *et al.*, 1999; Mulkern *et al.*, 1999; Clark and Le Bihan, 2000). Information about the multicompartment can be obtained by biexponential fitting or by performing q-space analyses. From diffusion measurements along a number of orientation, the shape of the diffusion ellipsoid (or nonelliptic shape) can be directly measured without tensor calculation (Wedeen *et al.*, 2000; Frank, 2001).

Although there is no doubt that these studies provide more precise and important information, it is also true that even with the simple tensor assumption, which gives six parameters at each pixel, we have a tremendous amount of information on white matter anatomy, which is, at present, often beyond our ability to fully appreciate (see Sections III and IV for more detail about the application). Therefore, it is equally important to develop techniques to examine macroscopic brain anatomy based on the tensor theory.

III. Data Visualization and Analysis of DTI

A. 2D Visualization

Once the diffusion tensor is obtained for each pixel, the information consisting of the six elements has to be reduced to a scalar or at least to a vector for visualization or data analysis purposes. The most basic parameters to characterize the ellipsoid are its size, shape, and orientation as shown in Fig. 18.

1. Visualization of the Size of the Ellipsoid

If diffusion is isotropic (spherical), the size of the diffusion can be easily represented by its diameter. However, when diffusion is anisotropic, the most widely used parameter to represent the size of the ellipsoid is called "trace" (van Gelderen *et al.*, 1994a). Trace can be obtained by summing the diagonal terms of the diffusion tensor ($= D_{xx} + D_{yy} + D_{zz}$). In practice, the diagonal term D_{xx} can be obtained by measuring ADC using the x gradient, D_{yy} using the y gradient, and D_{zz} using the z gradient. Therefore, unlike the full tensor calculation, which requires six ADC measurements, the trace can be obtained from just three ADC measurements. An example is shown in

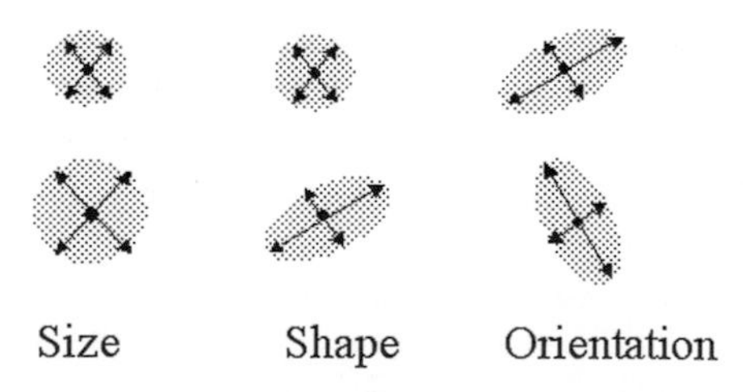

Figure 18 Three parameters to characterize the diffusion ellipsoid.

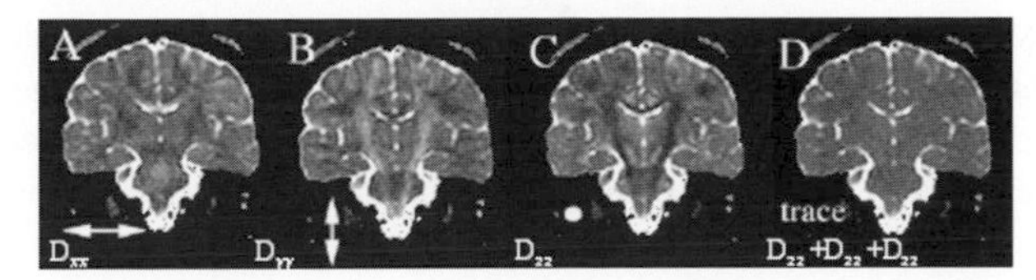

Figure 19 Examples of ADC maps measured using x (**A**), y (**B**), and z (**C**) gradients; each corresponds to D_{xx}, D_{yy}, and D_{zz} components of the diffusion tensor. A trace map can be obtained by adding these three ADC maps and is shown in (**D**).

Fig. 19. From this figure, it can be seen that the trace is quite homogeneous throughout the brain, indicating that regardless of various shapes and orientations of diffusion ellipsoids inside the brain, their size is uniform.

2. Visualization of the Shape of the Ellipsoid

There are many ways to characterize the shape of diffusion ellipsoids which are orientation independent and also are not affected by the size (Basser *et al.*, 1994a,b; Pierpaoli and Basser, 1996). The simplest and most intuitive method is to calculate the ratio of the length of the longest and shortest axes. This method, however, has several unwanted properties. For example, the range of its value is 1 (sphere)–infinity, which is difficult to visualize, and the length of the shortest axis (thus the ratio) is very susceptible to noise. It is preferable to use a parameter that ranges from 0 (isotropy) to 1 (anisotropy) for the visualization purpose. For example,

$$\frac{\lambda_1 - (\lambda_2 + \lambda_3)/2}{(\lambda_1 + \lambda_2 + \lambda_3)} \tag{14}$$

has the preferable property. Namely, it becomes 0 when $\lambda_1 = \lambda_2 = \lambda_3$ (isotropic) and the maximum is 1 when $\lambda_1 >> \lambda_2, \lambda_3$ (extreme anisotropy). The most widely used parameters are

$$FA = \sqrt{\frac{(\lambda_1 - \lambda_2)^2 + (\lambda_2 - \lambda_3)^2 + (\lambda_1 - \lambda_2)^2}{2(\lambda_1{}^2 + \lambda_2{}^2 + \lambda_3{}^2)}},$$

$$RA = \frac{\sqrt{(\lambda_1 - \lambda_2)^2 + (\lambda_2 - \lambda_3)^2 + (\lambda_1 - \lambda_2)^2}}{\lambda_1 + \lambda_2 + \lambda_3},$$

$$VR = \frac{\lambda_1 \lambda_2 \lambda_3}{((\lambda_1 + \lambda_2 + \lambda_3)/3)^3}. \tag{15}$$

Here FA is the fractional anisotropy, RA is the relative anisotropy, and VR is the volume ratio. These parameters all have 0–1 range. Information provided by these parameters is essentially the same. They all indicate how elongated the diffusion ellipsoid is. However, the contrasts they provide are not the same (Ulug and van Zijl, 1999). Among these parameters probably the FA is most widely used. An example of the FA map is shown Fig. 20. It can be seen that the white matter

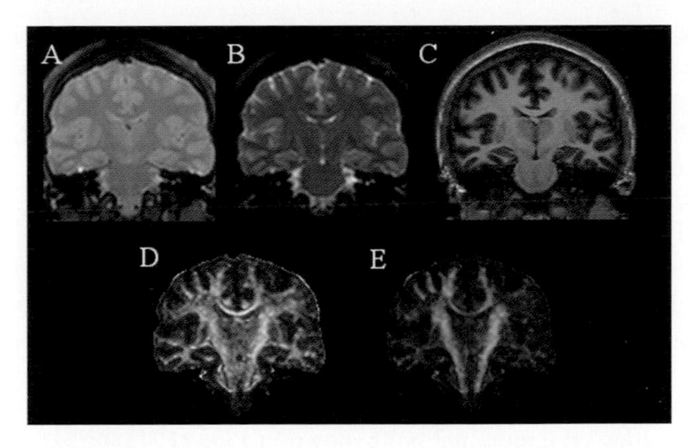

Figure 20 Comparison of proton density (**A**), T_2-weighted (**B**), and T_1-weighted (**C**) images and FA (D) and color map (E) of a healthy subject.

has high FA (Fig. 20D) values, which makes sense because it consists of densely packed axonal fibers. Segmentation of the white matter and gray matter can also be achieved by conventional T_1 and T_2-weighted images (Figs. 20B and 20C). However, detailed inspection of Fig. 20 clearly shows that the gray–white matter contrasts of these images are not identical. Anisotropy, which also shows a very high contrast between white and gray matter, is based on a completely different contrasting mechanism: the directionality of water diffusion given by axonal fibers.

The exact mechanism underlying the anisotropy map is not completely understood (Beaulieu and Allen, 1994; Henkelman *et al.*, 1994; Stanisz *et al.*, 1997). What is known is that the anisotropy drastically increases during early development (Sakuma *et al.*, 1991), and its time course is different from other conventional MRI parameters, such as T_1/T_2 relaxation properties. The changes during development may suggest involvement of the myelin sheath (Sakuma *et al.*, 1991). However, anisotropy has been reported in axonal fibers without myelin sheaths (Beaulieu and Allen, 1994), implying that the increased anisotropy during development may be due to increased fiber density. Our recent studies on coregistered FA and T_2 maps revealed that these parameters are not always correlated (Stieltjes *et al.*, 2001). In any case, it is apparent that the anisotropy provides a new contrasting mechanism that was formally inaccessible and, thus, it is worth pursuing its clinical possibilities as a new diagnostic tool.

3. Visualization of Orientation of the Ellipsoid

The last parameter that can be obtained from DTI is the orientation of diffusion ellipsoids (Fig. 18). The most intuitive way to show the orientation is the vector presentation, in which small lines (vector) indicate the orientations of the longest axis of diffusion ellipsoids. However, unless a small region is magnified, the vector orientation is often difficult to see. To overcome this problem, a color-coded scheme was postulated (Douek *et al.*, 1991; Nakada and Matsuzawa, 1995; Pajevic and Pierpaoli, 1999). An example is also shown in Fig. 20E. In the color map, three orthogonal axes (e.g., right–left, superior–inferior, and anterior–posterior) are assigned to three principal colors (red, green, and blue). If a fiber is running 45° from the red and blue axes, it is assigned magenta, which is a mixture of red and blue.

Compared to conventional images such as T_1- and T_2-weighted images (Figs. 20A–20C), it can be clearly seen that the DTI-based color map carries detailed information about the anatomy of white matter. Using the color maps, some prominent white matter tracts can be immediately identified (Makris *et al.*, 1997; Virta *et al.*, 1999; Stieltjes *et al.*, 2001; Mori *et al.*, 2002b). To further illustrate this point, an atlas-type presentation is shown in **Fig. 21**.

B. 3D-Based Fiber Analyses

1. Technical Descriptions

In the previous section, the usefulness of the color map was emphasized. Namely, it can reveal intra-white matter architectures (white matter tracts) based on pixel-by-pixel fiber orientation information. However, the 2D-based color map can reveal only a cross section of white matter tracts, which often has convoluted structures in the 3D space, and it is often difficult to appreciate their 3D trajectories from the slice-by-slice inspection. Computer-aided 3D tract tracking techniques can be very useful to understand the tract trajectories and their relationships with other white matter tracts and/or gray matter structures.

Roughly speaking, there are two types of tract reconstruction techniques. One is based on "line propagation" (Mori *et al.*, 1998, 1999, 2000; Conturo *et al.*, 1999; Jones *et al.*, 1999a; Xue *et al.*, 1999; Basser *et al.*, 1998, 2000; Lazar *et al.*, 2000;Parker, 2000; Poupon *et al.*, 2000; Tuch *et al.*, 2000; Wedeen *et al.*, 2000; Werring *et al.*, 2000; Stieltjes *et al.*, 2001) and the other is based on "energy minimization" (Parker, 2000; Tuch *et al.*, 2000, 2001). The former strategy is a deterministic approach because it provides only one solution (trajectory) from a given pixel and there is no *a priori* knowledge about the destination of the propagation. In the latter approach, two arbitrary points in the space can be selected and the technique provides the most probable path to connect the two points together with likelihood of the connection. A detailed description of these techniques is beyond the scope of this chapter. In this chapter, the concept of one of the simplest approaches in the line propagation class is described.

In Fig. 22, an example of a vector map is shown. By assuming that the orientation of the largest axis of the diffusion ellipsoid ($\nu 1$) aligns to fiber orientation, this kind of vector map can be obtained. For simplicity, a 2D diagram is used but the actual reconstruction should be performed using a 3D vector field. Note that the raw data, a vector field, comprise discrete information from which we have to reconstruct continuous trajectory coordinates. In this example, tracking is initiated from the center of a pixel at the coordinate (1.5, 1.5) in the continuous coordinate. Then a line is propagated in the continuous coordinate along the direction of the vector of the pixel. Then line the exits the initiating pixel at the coordinate of, e.g., (1.8, 2.0) and enters the next pixel, in which the tracking starts to observe the vector direction of the new pixel.

We applied this simple linear line propagation technique to reconstruct major white matter tracts in the brain stem and cerebral hemisphere and the results qualitatively agree with their known trajectories (Fig. 23).

From Fig. 23, the advantage of performing 3D tract reconstruction can be appreciated. For example, from the color map shown in Fig. 21, the middle cerebellar peduncle (mcp) can be clearly identified; it has a U-shaped trajectory warping around the pons. However, the actual trajectory of the mcp is tilted in the superior–inferior axis as can be seen from the 3D reconstruction as well as postmortem studies (Fig. 23). This kind of information is very difficult to appreciate from the 2D color maps.

Although the 3D tract reconstruction has a clear potential as a tool to visualize and understand 3D configuration of tracts of interest, its limitations should also be empha-

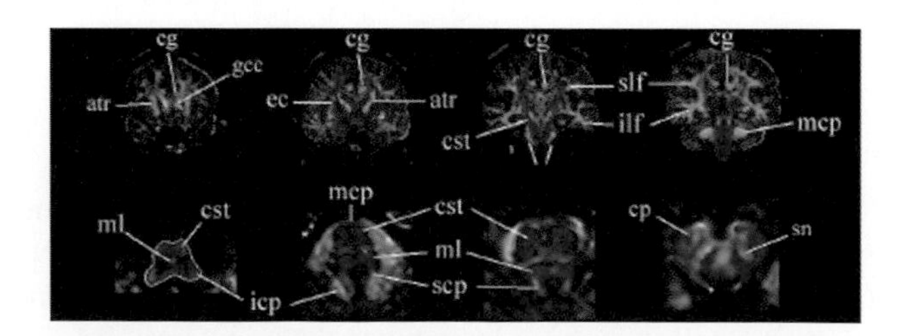

Figure 21 Examples of color maps and identification of prominent white matter tracts. Abbreviations used: cg, cingulum; ec, external capsule; atr, anterior thalamic radiation; cst, corticospinal tract; slf, superior longitudinal fasciculus; ilf, inferior longitudinal fasciculus; ml, medial lemniscus; icp, inferior cerebellar peduncle; mcp, middle cerebellar peduncle; scp, superior cerebellar peduncle; cp, cerebral peduncle; sn, substantia nigra.

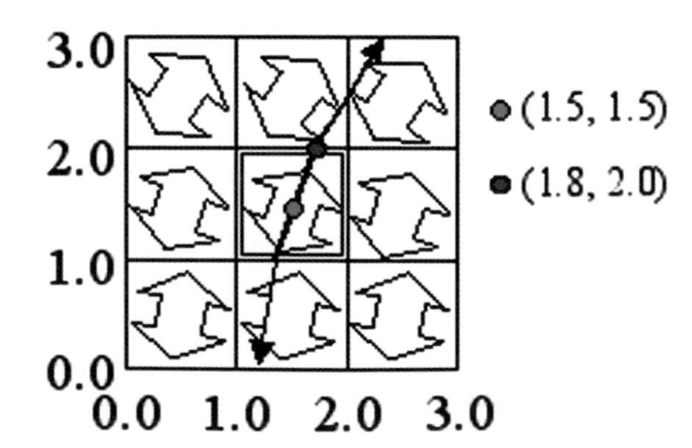

Figure 22 Examples of fiber tracking. Double-headed arrows indicate the orientations of tracts at each pixel, which are obtained from the DTI measurement. Note that the DTI study, which is based on water movement, cannot judge effluent and affluent. Tracking has to be made in both orthograde and retrograde directions.

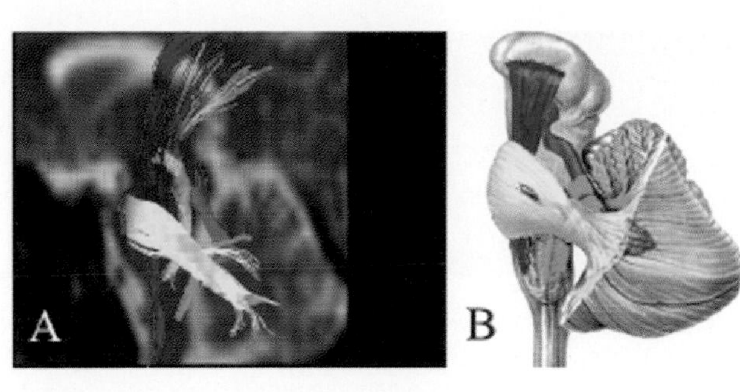

Figure 23 Example of 3D tract reconstruction in the brain-stem region (**A**) and comparison with a postmortem study (**B**). Reconstructed tracts are red, corticospinal tract; blue, medial lemniscus; pink, superior cerebellar peduncle; yellow, middle cerebellar peduncle; and green, inferior cerebellar peduncle. Images are reproduced from Stieltjes *et al.* (2001), with permission.

sized. The first layer of limitation stems from the DTI data acquisition, in which we have only a limited resolution (in the order of 1–5 mm) and measurement errors by noise and artifacts (mostly motion related). The resolution is far larger than individual axons and unless axons form a large bundle with uniform fiber orientation, our pixel-by-pixel information is inevitably "averaged" information of the fiber orientations. This issue immediately leads to two limitations in the DTI-based tract analyses. First, as long as many axons enter into and exit from a pixel, ***it is impossible to obtain a single axonal path and, thus, cellular level neuronal connectivity***. Second, the tensor calculation described in Section II assumes a uniform water diffusion property inside a pixel and, thus, a uniform tract organization, which may not be true. Recent studies by Wedeen's group and by Frank (Wedeen *et al.*, 2000; Wiegell *et al.*, 2000; Frank, 2001) illustrate how this tensor assumption breaks down if a pixel contains inhomogeneous populations of tract, such as tracts with different directions. In certain regions of brain, for example, at pons where the corticospinal tracts and crossing fibers are mixed in a microscopic scale, it has been demonstrated that this breakdown actually occurs in the brain DTI studies. In this case, the vector field obtained from the tensor calculation is not accurate. Inaccuracy caused by the measurement errors further complicated these issues.

Currently, the DTI measurement and data analysis techniques are evolving rapidly and some of these limitations are being addressed. These include data acquisition techniques which are more insensitive to artifacts and have higher signal-to-noise and image resolution, non-tensor-based data analyses, and 3D reconstruction techniques less sensitive to noise. However, we should bear in mind that these fundamental limitations of this MR-based technique cannot be completely elminated.

2. Validation of 3D Reconstruction Techniques

Validation is undoubtedly one of the most important questions, once we get the 3D reconstruction results. However, we may not be able to answer a question like, "is the reconstructed tract true?" in a straightforward manner because the meaning of this question varies depending on the context. First of all, there are multiple layers of validation issues in the 3D reconstruction. These include at least validation of data acquisition, tensor calculation, and 3D reconstruction algorithms. The validation of data acquisition can be easily done by using water and a slow-diffusing phantom. Errors in gradient calibration and waveform can be estimated in this way. The idea that DTI can estimate fiber orientation can be validated using a sample with simple and known fiber orientation such as muscle. For example, we confirmed that DTI can accurately measure rotation of anisotropic samples (Hsu and Mori, 1995; Mori and van Zijl, 1995). Validation by histology using a heart sample was also reported (Scollan *et al.*, 1998). A recent report of DTI of *in vivo* optic nerve and optic tract shows validation of DTI and error analyses due to noise, which contains DTI procedures from data acquisition up to the generation of a tensor-based fiber orientation (vector) map (Lin *et al.*, 2001). Assuming the vector maps are valid (truthfully represent averaged fiber orientation within each pixel), validation and noise analysis of 3D tracking algorithms can be performed using, for example, a Monte Carlo simulation (Lori *et al.*, 1999; Basser *et al.*, 2000; Lazar and Alexander, 2001).

Even if these layers of validation studies have been reported, we still cannot confirm higher level, "biological validation" partly because we are still bounded by the fundamental limitations of image resolution and noise as described above and partly because this validation issue really depends on what kind of biological questions we want to answer. For example, if a tracking result indicates that pixels A and B are connected, can we validate that the two locations are really connected by axons? In other words, can we study cellular-level neuronal connectivity by DTI? We may not be able to validate this level of question because even if some kinds of studies, such as coregistration of DTI and invasive chemical tracer techniques, show that the DTI result is valid, this does not guarantee that the DTI results are always valid because they are prone to, for example, noise and partial volume effects (such as mixed fibers). Therefore, it is extremely important to know what kind of biological questions we want to extract from DTI. In the next section, we consider this issue.

IV. Application Studies

A. Anisotropy Measurement

Analysis of anisotropy is straightforward because it is scalar and can be analyzed in the same way as other MR images. For example, regions of interest can be manually drawn and indicators of the anisotropy, such as fractional anisotropy, can be quantified. Measurements of anisotropy have been performed for various brain diseases, and abnormalities (mostly reduction) have been reported.

Interpretation of the anisotropy change may not be as straightforward as the measurements. This is because we still do not know the exact mechanisms that control the diffusion anisotropy and it is likely that there are multiple factors that lead to the reduction. As a consequence, changes in anisotropy may not pinpoint specific cellular events. On the other hand, as explained in Section III.A.2, it is likely that anisotropy and other conventional contrasts reflect different cellular statuses. Therefore, it is worth examining the anisotropy to judge whether it increases the sensitivity and/or specificity of diagnosing brain diseases. However, it should be stressed that DTI is an inherently time-consuming, low-resolution, and artifact-prone technique and, therefore, it is important to confirm whether anisotropy measurements provide unique and important information that cannot be obtained by conventional MR images and, thus, the DTI studies are justifiable in clinical settings.

B. Study of Fiber Architecture

There is no question that the orientation information obtained from DTI measurements is unique. Especially the elucidation of intra-white matter structures with DTI can greatly enhance the ability of MRI to study normal and abnormal brain neuroanatomy. For example, Fig. 24 shows MR images of a patient who suffered a stroke. Figure 24A shows the infarction in the right frontal lobe. In a posterior slice shown in Fig. 24B, the abnormalities are not clear except for a small T_2 high-intensity region indicated by a yellow arrow. Figure 24C shows a color map at the same slice level as Fig. 24B. In the color map, it can be seen that the superior longitudinal fasciculus (SLF) in the right hemisphere (yellow arrow) has much lower anisotropy (brightness) compared to the contralateral side and the T_2 hyperintensity region actually matches the location of the affected SLF. From these results, the observation made by the conventional T_2-weighted image, namely "small T_2 hyperintense region in the white matter," can be elaborated to "loss of anisotropy and T_2 increase of the SLF."

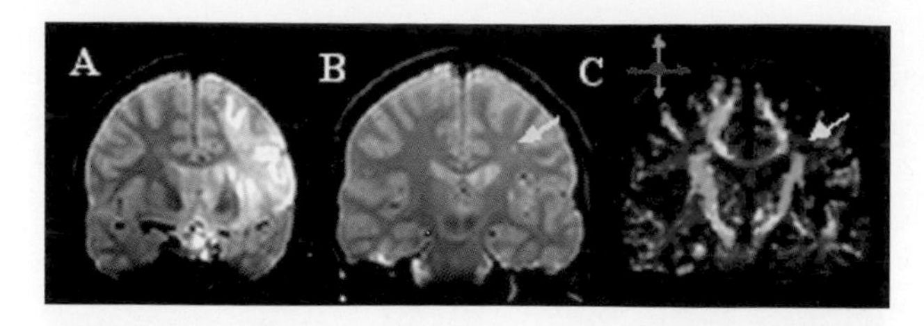

Figure 24 T_2-weighted (**A**, **B**) and color maps (**C**) of a stroke patient. In the color map, red represents fibers running in a right–left orientation, green superior–inferior, and blue anterior–posterior.

Once locations of tracts of interest are identified using the color maps, various MR properties can be studied on a tract-by-tract basis (Virta *et al.*, 1999; Xue *et al.*, 1999; Stieltjes *et al.*, 2001). Examples of tract-specific quantitative studies are shown in Fig. 25. In this way, MRI can study the status of individual white matter tracts.

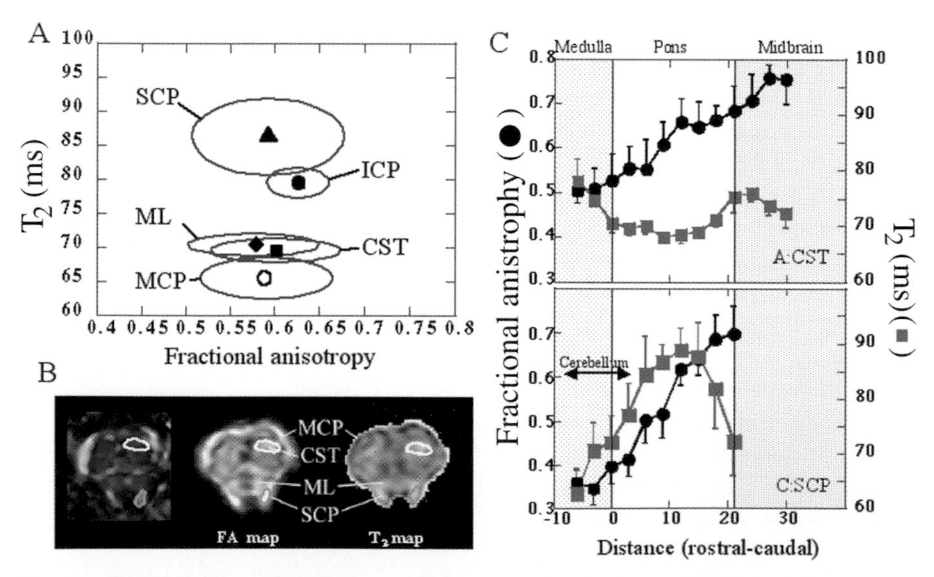

Figure 25 The study of FA and T_2 on a tract-by-tract basis (**A**). Locations of various tracts of interest can be found using the orientation (color) information (color map in **B**), and the coordinates are transferred to FA and T_2 maps to quantify these parameters for each tract. The study shows that each tract has characteristic signatures in these parameters (A). Using the 3D tract reconstruction, these parameters can also be quantified along the tracts (**C**). Note that these parameters are not correlated each other. Abbreviations used: CST, corticospinal tract; SCP, superior cerebellar peduncle; MCP, middle cerebellar peduncle; ICP, inferior cerebellar peduncle; and ML, medial lemniscus. Images are reproduced from Stieltjes *et al.* (2001), with permission.

The capability of identifying individual tracts within the white matter by DTI may allow us to perform white matter parcellation and volumetric studies (Makris *et al.*, 1997; Mori *et al.*, 2002a). This is a very attractive idea because there are numerous neurological conditions in which degeneration or other types of abnormalities of specific white matter are suspected. Figure 26 shows how the parcellation can be improved by incorporating both anisotropy and fiber orientation information. This is still an open field for which parcellation tools need to be developed. For example, a region-growing algorithm based on anisotropy and fiber orientation has been postulated (Mori *et al.*, 2001). In the future, manual or semiautomated white matter parcellation tools based on DTI information and probabilistic map approaches will be important techniques which may profoundly improve brain anatomy studies based on MRI.

C. Application of 3D Tract Reconstruction Techniques

The application studies of the 3D tract reconstruction can be roughly classified into three categories. In the first category, qualitative analysis is applied to study the effects of diseases on the overall architecture of the brain white matter. In these diseases, configuration of white matter tracts or even their existence is the central question. Examples are birth defects, developmental disorders, brain tumors, and chronic stroke (Werring *et al.*, 2000; Wieshmann *et al.*, 2000; Pierpaoli *et al.*, 2001; Mori *et al.*, 2002a). In the second category, abnormalities may or may not be apparent in each patient and thus the study requires statistical analyses to interpret the DTI data. These include psychiatric disorders or neurodegenerative diseases. In this field, statistical data analysis tools are needed. The third category is the study of basic neuroscience of the brain architecture itself, which is uniquely revealed by DTI. In the second and the third categories, it is of central importance to know what kind of "connectivity information" we can obtain from the 3D reconstruction. As was discussed in Section III, it is not appropriate to study cellular-level axonal connectivity using DTI. However, this does not preclude possibilities of obtaining information related to region-to-region connectivity (Tuch *et al.*, 2001) and its alteration due to disease. This area of study has just started and at present many questions are still unanswered.

As an example of a category I study, Fig. 27 shows a brain tumor study which reveals the relationship between the corona radiata and a brain tumor (anaplastic astrocytoma) in the left frontal lobe. It can be clearly seen that the

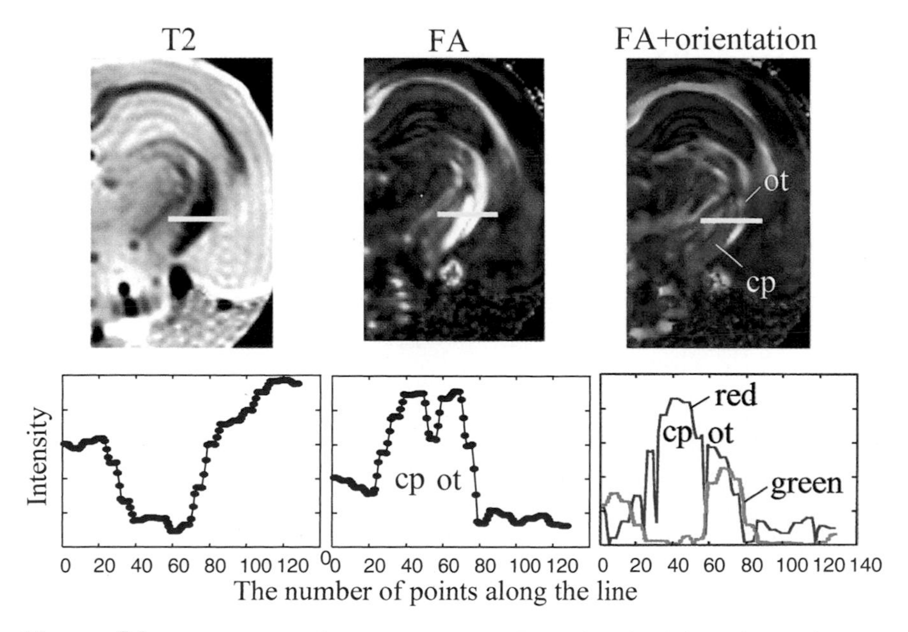

Figure 26 Demonstration of how anisotropy and fiber orientation information can improve parcellation of the white matter. As an example, parcellation of the cerebral peduncle (cp) and the optic tract (ot) of a fixed mouse brain is shown. Plots are intensity profiles of each image along the yellow lines. At this slice level, these two tracts are adjacent to each other and the T_2 contrast could not separate them. The FA map could separate the two due to the low FA regions in between them. By adding the orientation information, the definition of the cp and ot could be much sharper due to their characteristic fiber orientations (red, anterior–posterior; green, right–left; blue component is not shown for clarity).

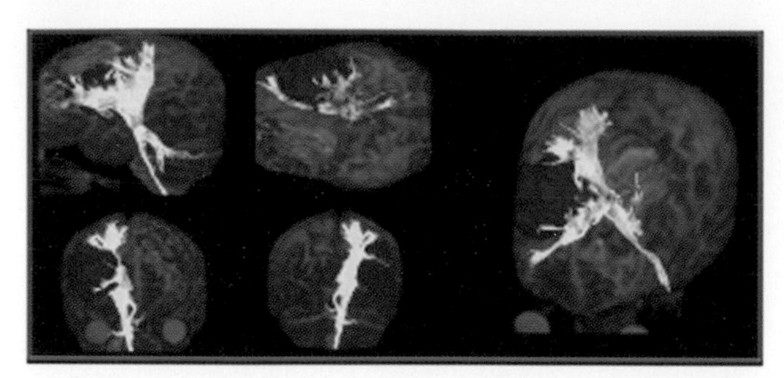

Figure 27 An example of 3D tract reconstruction of the corona radiata in a patient with a brain tumor. Regions with T_2 hyperintensity are shown with red and the corona radiata yellow. Reproduced with permission from Mori, *et al.* (2002a).

corona radiata is displaced medially due to the growth of the tumor. This kind of information may play an important role in preoperative planning to avoid critical white matter pathways in surgeries.

Figure 28 shows an example of the probabilistic map (Mori *et al.*, 2002a), which may prove to be important for category II studies. In this example, 3D trajectories of SLF were obtained from 10 healthy volunteers and the coordinates were normalized to Talairach coordinates using coregistered T_1 anatomical images. After the SLF coordinates from each individual are superimposed in this standard template, we can obtain the pixel-by-pixel probability of finding on the SLF.

D. High-Resolution, High Signal-to-Noise DTI

By using postmortem samples, high "special" resolution, high-SNR images can be obtained. Such images show that characteristic fiber architectures can also be found in many gray matter structures (Fig. 29) (Zhang *et al.*, 2002; Thornton *et al.*, 1997; Mori *et al.*, 2001). For example, layer structures in hippocampus and colliculus, or nuclei in thalamus, can be clearly identified. Similar to the white matter, DTI has a potential to parcellate the gray matter based on its fiber architecture. For animal studies, this technique may be found to be an important tool, including detailed neuroanatomical studies of effects on individual nuclei or layers of gray matter structures due to lesions, pharmacological treatment, developmental defects in transgenic mice, to name a few.

Acquisition of the data set shown in Fig. 29 took 29 h and it is not practical to perform DTI of this resolution in routine clinical studies. However, with the current speed of technical development in data acquisition technologies (both in hardware and in software), in the near future, gray matter DTI will be an interesting research field.

Another interesting research field is the developing brain (Sakuma *et al.*, 1991; Baratti *et al.*, 1997; Huppi *et al.*, 1998; Neil *et al.*, 1998; Jacobs *et al.*, 1999; Mori *et al.*, 2001). It is known that fetal or early neonatal brains have very poor T_1 and T_2 contrasts, probably due to lack of myelination (Garel *et al.*, 1998; Neil *et al.*, 1998). However, these brains already possess complex yet ordered fiber architectures that can be revealed by DTI. Therefore, DTI can be a powerful tool to study internal brain structures in the early developmental phases.

Another interesting example of high-resolution imaging is "high-angular" imaging in which diffusion constants along numerous axes with a number of *b*-value steps are measured (thus high resolution in terms of sampling spaces) (Wedeen *et al.*, 2000). It has been demonstrated that this approach can reveal microscopic fiber architecture of the sample such as structures with mixed fiber orientations.

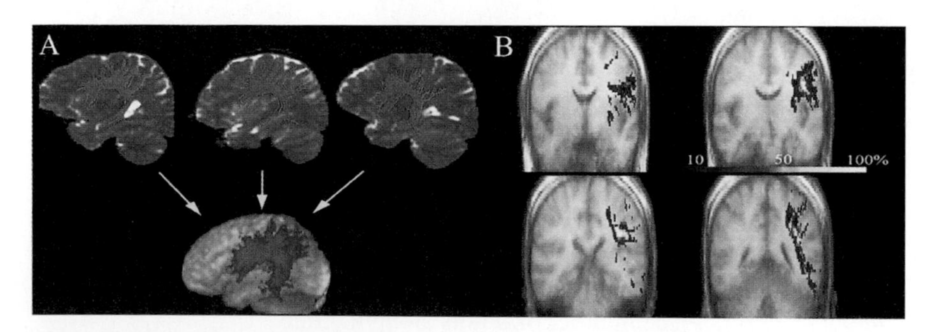

Figure 28 Application of the probabilistic approach to the superior longitudinal fasciculus (SLF) of 10 normal volunteers. The reconstruction result from each individual was standardized into Talairach coordinates (**A**). (**B**) The reconstructed SLF (A) as well as the probability of occurrence over 10 volounteers is shown. It can be seen that some small branches have the lowest probability of 10%, indicating that these pixels had the SLF in only 1 of the 10 subjects. Other branches, such as those in Broca's area and the temporal lobe, have high reproducibility (>50%). Reproduced with permission from Mori, *et al.* (2002b).

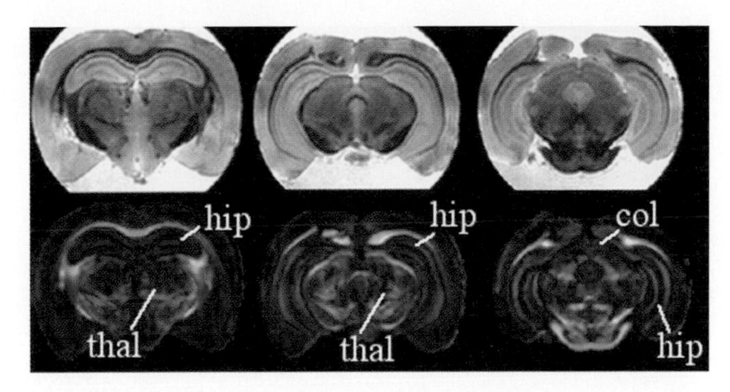

Figure 29 Examples of high-resolution, high-SNR postmortem mouse brain DTI (bottom row) and comparison with coregistered T_2-weighted images (upper row). Abbreviations used: hip, hippocampus; thal, thalamus; and col, colliculus.

V. Summary

In this article, principles of DTI were covered with emphasis on data acquisition and step-by-step description of tensor calculations in Sections I and II. In Section III, various visualization and data analysis techniques were introduced and their applications were demonstrated in Section IV. DTI is still a young research field and is evolving with considerable speed in every aspect. DTI is a unique MRI technique in the sense that it provides orientation information of anatomy. One DTI data set contains a large amount of anatomical information, which often we cannot fully appreciate simply because we do not have appropriate tools to analyze it. Development of computer-aided analysis tools will be, thus, an important research endeavor in the DTI field. On the other hand, it is equally important to understand assumptions and limitations of the technique. I hope that this article gave an overview of the current status of DTI and the future directions of the studies.

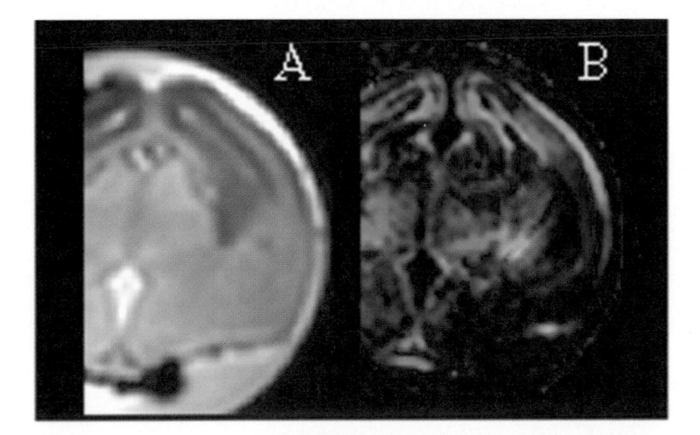

Figure 30 Comparison of a T_2-weighted (A) and a DTI-based color map (B) of E16 embryonic mouse brain (C57BL).

References

Anderson, A. W., and Gore, J. C. (1994). Analysis and correction of motion artifacts in diffusion weighted imaging. *Magn. Reson. Med.* **32**, 379–387.

Atkinson, D., Porter, D. A., Hill, D. L., Calamante, F., and Connelly, A. (2000). Sampling and reconstruction effects due to motion in diffusion-weighted interleaved echo planar imaging. *Magn. Reson. Med.* **44**, 101–109.

Bammer, R., Keeling, S. L., Auer, M., Pruessmann, K. P., Roeschmann, P., *et al.* (2001). "Diffusion Tensor Imaging Using SENSE-Single-Shot EPI." International Society of Magnetic Resonance in Medicine, Glasgow.

Baratti, C., Barnett, A., and Pierpaoli, C. (1997). "Comparative MRI Study of Brain Maturation Using T1, T2, and the Diffusion Tensor." International Society of Magnetic Resonance in Medicine, Vancouver.

Basser, J. B. (1998). "Fiber-Tractography via Diffusion Tensor MRI." Proceedings of the International Society for Magnetic Resonance in Medicine, Sydney.

Basser, P. J., Mattiello, J., and Le Bihan, D. (1994a). MR diffusion tensor spectroscopy and imaging. *Biophys. J.* **66**, 259–267.

Basser, P. J., Mattiello, J., and LeBihan, D. (1994b). Estimation of the effective self-diffusion tensor from the NMR spin echo. *J. Magn. Reson. B* **103**, 247–254.

Basser, P. J., Pajevic, S., Pierpaoli, C., Duda, J., and Aldroubi, A. (2000). In vitro fiber tractography using DT-MRI data. *Magn. Reson. Med.* **44**, 625–632.

Basser, P. J., and Pierpaoli, C. (1996). Microstructural features measured using diffusion tensor imaging. *J. Magn. Reson. B.* **111**, 209–219.

Beaulieu, C., and Allen, P. S. (1994). Determinants of anisotropic water diffusion in nerves. *Magn. Reson. Med.* **31**, 394–400.

Buckley, D. L., Bui, J. D., Phillips, M. I., Zelles, T., Inglis, B. A., *et al.* (1999). The effect of ouabain on water diffusion in the rat hippocampal slice measured by high resolution NMR imaging. *Magn. Reson. Med.* **41**, 137–142.

Butts, K., de Crespigny, A., Pauly, J. M., and Moseley, M. (1996). Diffusion-weighted interleaved echo-planar imaging with a pair of orthogonal navigator echoes. *Magn. Reson. Med.* **35**, 763–770.

Butts, K., Pauly, J., de Crespingy, A., and Moseley, M. (1997). Isotropic diffusion-weighted and spiral-navigated interleaved EPI for routine imaging of acute stroke. *Magn. Reson. Med.* **38**, 741–749.

Clark, C. A., and Le Bihan, D. (2000). Water diffusion compartmentation and anisotropy at high b values in the human brain. *Magn. Reson. Med.* **44**, 852–859.

Conturo, T. E., Lori, N. F., Cull, T. S., Akbudak, E., Snyder, A. Z., *et al.* (1999). Tracking neuronal fiber pathways in the living human brain. *Proc. Natl. Acad. Sci. USA* **96**, 10422–10427.

Douek, P., T. R., Pekar, J., Patronas, N., and Le Bihan, D. (1991). MR color mapping of myelin fiber orientation. *J. Comput. Assisted Tomogr.* **15**, 923–929.

Duerk, J. L., and Simonetti, O. P. (1991). Theoretical aspects of motion sensitivity and compensation in echo-planar imaging. *J. Magn. Reson. Imaging* **1**, 643–650.

Frank, L. R. (2001). Anisotropy in high angular resolution diffusion-weighted MRI. *Magn. Reson. Med.* **45**, 935–939.

Garel, C., Briees, H., Sebag, G., Elmaleh, M., Oury, J.-F., *et al.* (1998). Magnetic resonance imaging of the fetus. *Pediatr. Radiol.* **28**, 201–211.

Henkelman, R., Stanisz, G., Kim, J., and Bronskill, M. (1994). Anisotropy of NMR properties of tissues. *Magn. Reson. Med.* **32**, 592–601.

Hsu, E. W., and Mori, S. (1995). Analytical interpretations of NMR diffusion measurements in an anisotropic medium and a simplified method for determining fiber orientation. *Magn. Reson. Med.* **34**, 194–200.

Huppi, P., Maier, S., Peled, S., Zientara, G., Barnes, P., *et al.* (1998). Microstructural development of human newborn cerebral white matter assessed in vivo by diffusion tensor magnetic resonance imaging. *Pediatr. Res.* **44**, 584–590.

Jacobs, R. E., Ahrens, E. T., Meade, T. J., and Fraser, S. E. (1999). Looking deeper into vertebrate development. *Trends Cell Biol.* **9**, 73–76.

Jiang, H., Golay, X., van Zijl, P. C., and Mori, S. (2002). Origin and minimization of residual motion-related artifacts in navigator-corrected segmented diffusion-weighted EPI of the human brain. *Magn. Reson. Med.*, **47**, 818–822.

Jones, D. K., Horsfield, M. A., and Simmons, A. (1999a). Optimal strategies for measuring diffusion in anisotropic systems by magnetic resonance imaging. *Magn. Reson. Med.* **42**, 515–525.

Jones, D. K., Simmons, A., Williams, S. C., and Horsfield, M. A. (1999b). Non-invasive assessment of axonal fiber connectivity in the human brain via diffusion tensor MRI. *Magn. Reson. Med.* **42**, 37–41.

Lazar, M., and Alexander, A. L. (2001). "Error Analysis of White Matter Tracking Algorithms (Streamlines and Tensorlines) for DT-MRI." Proceedings of the International Society of Magnetic Resonance in Medicine, Glasgow.

Lazar, M., Weinstein, D., Hasan, K., and Alexander, A. L. (2000). "Axon Tractography with Tensorlines." Proceedings of the International Society of Magnetic Resonance in Medicine, Denver, CO.

Le Bihan, D., Breton, E., Lallemand, D., Grenier, P., Cabanis, E., *et al.* (1986). MR imaging of intravoxel incoherent motions: Application to diffusion and perfusion in neurologic disorders. *Radiology* **161**, 401–407.

Lin, C. P., Tseng, W. Y., Cheng, H. C., and Chen, J. H. (2001). Validation of diffusion tensor magnetic resonance axonal fiber imaging with registered manganese-enhanced optic tracts. *NeuroImage* **14**, 1035–1047.

Lori, N. F., Akbuda, E., Snyder, A. Z., Shimony, J. S., and Conturo, T. E. (1999). "Diffusion Tensor Tracking of Human Neuronal Fiber Bundles: Simulation of Effects of Noise, Voxel Size and Data Interpolation." International Society of Magnetic Resonance in Medicine, Denver, CO.

Makris, N., Worth, A. J., Sorensen, A. G., Papadimitriou, G. M., and Reese, T. G., *et al.* (1997). Morphometry of in vivo human white matter association pathways with diffusion weighted magnetic resonance imaging. *Ann. Neurol.* **42**, 951–962.

Mori, S., Crain, B. J., Chacko, V. P., and van Zijl, P. C. M. (1999). Three dimensional tracking of axonal projections in the brain by magnetic resonance imaging. *Ann. Neurol.* **45**, 265–269.

Mori, S., Crain, B. J., and van Zijl, P. C. (1998). "*3D Brain Fiber Reconstruction from Diffusion MRI.*" Proceedings of the International Conference on Functional Mapping of the Human Brain, Montreal.

Mori, S., Fredericksen, K., van Zijl, P. C., Stieltjes, B., Kraut, A. K., *et al.* (2002a). Brain white matter anatomy of tumor patients using diffusion tensor imaging. *Ann. Neurol.*, **51**, 377–380.

Mori, S., Itoh, R., Zhang, J., Kaufmann, W. E., van Zijl, P. C. M., *et al.* (2001). Diffusion tensor imaging of the developing mouse brain. *Magn. Reson. Med.* **46**, 18–23.

Mori, S., Kaufmann, W. E., Davatzikos, C., Stieltjes, B., Amodei, L., *et al.* (2002b). Imaging cortical association tracts in human brain. *Magn. Reson. Imaging*, **47**, 215–223.

Mori, S., Kaufmann, W. K., Pearlson, G. D., Crain, B. J., Stieltjes, B., *et al.* (2000). *In vivo* visualization of human neural pathways by MRI. *Ann. Neurol.* **47**, 412–414.

Mori, S., and van Zijl, P. C. M. (1995). Diffusion weighting by the trace of the diffusion tensor within a single scan. *Magn. Reson. Med.* **33**, 41–52.

Moseley, M. E., Cohen, Y., Kucharczyk, J., Mintorovitch, J., Asgari, H. S., *et al.* (1990). Diffusion-weighted MR imaging of anisotropic water diffusion in cat central nervous system. *Radiology* **176**, 439–445.

Mulkern, R. V., Gudbjartsson, H., Westin, C. F., Zengingonul, H. P., Gartner, W., *et al.* (1999). Multi -component apparent diffusion coefficients in human brain. *NMR Biomed.* **12**, 51–62.

Nakada, T., and Matsuzawa, H. (1995). Three-dimensional anisotropy contrast magnetic resonance imaging of the rat nervous system: MR axonography. *Neurosci. Res.* **22**, 389–398.

Neil, J., Shiran, S., McKinstry, R., Schefft, G., Snyder, A., *et al.* (1998). Normal brain in human newborns: Apparent diffusion coefficient and diffusion anisotropy measured by using diffusion tensor MR imaging. *Radiology* **209**, 57–66.

Niendorf, T., Dijkhuizen, R. M., Norris, D. G., van Lookeren Campagne, M., and Nicolay, K. (1996). Biexponential diffusion attenuation in various states of brain tissue: Implications for diffusion-weighted imaging. *Magn. Reson. Med.* **36**, 847–57.

Norris, D. G. (2001). Implications of bulk motion for diffusion-weighted imaging experiments: Effects, mechanisms, and solutions. *J. Magn. Reson. Imaging* **13**, 486–495.

Norris, D. G., Börnert, P., Reese, T., and Leibfritz, D. (1992). On the application of ultra-fast RARE experiments. *Magn. Reson. Med.* **27**, 142–164.

Ordidge, R. J., Helpern, J. A., Qing, Z. X., Knight, R. A., and Nagesh, V. (1994). Correction of motional artifacts in diffusion-weighted NMR images using navigator echoes. *Magn. Reson. Imaging* **12**, 455–460.

Pajevic, S., and Pierpaoli, C. (1999). Color schemes to represent the orientation of anisotropic tissues from diffusion tensor data: Application to white matter fiber tract mapping in the human brain. *Magn. Reson. Med.* **42**, 526–540.

Parker, G. J. (2000). "Tracing Fiber Tracts Using Fast Marching." International Society of Magnetic Resonance, Denver, CO.

Pierpaoli, C., Barnett, A., Pajevic, S., Chen, R., Penix, L. R., *et al.* (2001). Water diffusion changes in Wallerian degeneration and their dependence on white matter architecture. *NeuroImage* **13**, 1174–1185.

Pierpaoli, C., and Basser, P. J. (1996). Toward a quantitative assessment of diffusion anisotropy. *Magn. Reson. Med.* **36**, 893–906.

Pipe, J. G. (2001). "Multishot Diffusion Weighted FSE with PROPELLAR." International Society of Magnetic Resonance in Medicine, Glasgow.

Poupon, C., Clark, C. A., Frouin, V., Regis, J., Bloch, L., *et al.* (2000). Regularization of diffusion-based direction maps for the tracking of brain white matter fascicules. *NeuroImage* **12**, 184–195.

Pruessmann, K. P., Weiger, M., Scheidegger, M. B., and Boesiger, P. (1999). SENSE: Sensitivity encoding for fast MRI. *Magn. Reson. Med.* **42**, 952–962.

Sakuma, H., Nomura, Y., Takeda, K., Tagami, T., Nakagawa, T., *et al.* (1991). Adult and neonatal human brain: Diffusional anisotropy and myelination with diffusion-weighted MR imaging. *Radiology* **180**, 229–233.

Scollan, D. F., Holmes, A., Winslow, R., and Forder, J. (1998). Histological validation of myocardial microstructure obtained from diffusion tensor magnetic resonance imaging. *Am. J. Physiol.* **275**, H2308–2318.

Skare, S., Hedehus, M., Moseley M. E., and Li, T. Q. (2000). Condition number as a measure of noise performance of diffusion tensor data acquisition schemes with MRI. *J. Magn. Reson.* **147**, 340–352.

Sodickson, D. K., and Manning, W. J. (1997). Simultaneous acquisition of spatial harmonics (SMASH): Fast imaging with radiofrequency coil arrays. *Magn. Reson. Med.* **38**, 591–603.

Stanisz, G. J., Szafer, A., Wright, G. A., and Henkelman, R. M. (1997). An analytical model of restricted diffusion in bovine optic nerve. *Magn. Reson. Med.* **37**, 103–113.

Stejskal, E. (1965). Use of spin echoes in a pulsed magnetic-field gradient to study restricted diffusion and flow. *J. Chem. Phys.* **43**, 3597–3603.

Stejskal, E. O., and Tanner, J. E. (1965). Spin diffusion measurement: Spin echoes in the presence of a time-dependent field gradient. *J. Chem. Phys.* **42**, 288.

Stieltjes, B., Kaufmann, W. E., van Zijl, P. C. M., Fredericksen, K., Pearlson, G. D., *et al.* (2001). Diffusion tensor imaging and axonal tracking in the human brainstem. *NeuroImage* **14**, 723–735.

Thornton, J. S., Ordidge, R. J., Penrice, J., Cady, E. B., Amess, P. N., *et al.* (1997). Anisotropic water diffusion in white and gray matter on the neonatal piglet brain before and after transient hypoxia–ischaemia. *Magn. Res. Imaging* **15**, 433–440.

Trouard, T. P., Theilmann, R. J., Altbach, M. I., and Gmitro, A. F. (1999). High-resolution diffusion imaging with DIFRAD-FSE (diffusion-weighted radial acquisition of data with fast spin-echo) MRI. *Magn. Reson. Med.* **42**, 11–18.

Tuch, D. S., Belliveau, J. W., and Wedeen, V. (2000). "A Path Integral Approach to White Matter Tractography." Proceedings of the International Society of Magnetic Resonance in Medicine, Denver, CO.

Tuch, D. S., Wiegell, M. R., Reese, T. G., Belliveau, J. W., and Wedeen, V. (2001). "Measuring Cortico-cortical Connectivity Matrices with Diffusion Spectrum Imaging." Proceedings of the International Society of Magnetic Resonance in Medicine, Glasgow.

Turner, R., and Le Bihan, D. (1990). Single-shot diffusion imaging at 2.0 tesla. *J. Magn. Reson.* **86**, 445–452.

Turner, R., LeBihan, D., Maier, J., Vavrek, R., Hedges, L. K., *et al.* (1990). Echo-planar imaging of intravoxel incoherent motions. *Radiology* **177**, 407–414.

Ulug, A., and van Zijl, P. C. M. (1999). Orientation-independent diffusion imaging without tensor diagonalization: Anisotropy definitions based on physical attributes of the diffusion ellipsoid. *J. Magn. Reson. Imaging* **9**, 804–813.

van Gelderen, P., de Vleeschouwer, M. H., DesPres, D., Pekar, J., van Zijl, P. C. M., *et al.* (1994a). Water diffusion and acute stroke. *Magn. Reson. Med.* **31**, 154–163.

van Gelderen, P., DesPres, D., van Zijl, P. C. M., and Moonen, C. T. W. (1994b). Evaluation of restricted diffusion in cylinders. Phosphocreatine in rat muscle. *J. Magn. Reson. B* **103**, 247–254.

Virta, A., Barnett, A., and Pierpaoli, C. (1999). Visualizing and characterizing white matter fiber structure and architecture in the human pyramidal tract using diffusion tensor MRI. *Magn. Reson. Imaging* **17**, 1121–1133.

Wedeen, V., Reese, T. G., Tuch, D. S., Weigel, M. R., Dou, J. G., *et al.* (2000). "Mapping Fiber Orientation Spectra in Cerebral White Matter with Fourier-Transform Diffusion MRI." Proceedings of the International Society of Magnetic Resonance in Medicine, Denvor, CO.

Werring, D. J., Toosy, A. T., Clark, C. A., Parker, G. J., Barker, G. J., *et al.* (2000). Diffusion tensor imaging can detect and quantify corticospinal tract degeneration after stroke. *J. Neurol. Neurosurg. Psychiatry* **69**, 269–272.

Wiegell, M., Larsson H., and Wedeen, V. (2000). Fiber crossing in human brain depicted with diffusion tensor MR imaging. *Radiology* **217**, 897–903.

Wieshmann, U. C., Symms, M. R., Parker, G. J., Clark, C. A., Lemieux, L., *et al.* (2000). Diffusion tensor imaging demonstrates deviation of fibres in normal appearing white matter adjacent to a brain tumour. *J. Neurol. Neurosurg. Psychiatry* **68**, 501–503.

Xue, R., van Zijl, P. C. M., Crain, B. J., Solaiyappan, M., and Mori, S. (1999). In vivo three-dimensional reconstruction of rat brain axonal projections by diffusion tensor imaging. *Magn. Reson. Med* **42**, 1123–1127.

Zhang, J., van Zijl, P. C., and Mori, S. (2002). Three dimensional diffusion tensor magnetic resonance micro-imaging of adult mouse brain and hippocumpus. *NeuroImage*, **15**, 892 901.

16

Neuroanatomical Micromagnetic Resonance Imaging

P. T. Narasimhan and Russell E. Jacobs

Beckman Institute, Division of Biology, California Institute of Technology, Pasadena, California 91125

I. Introduction

The correlation of function and structure is one of the overriding goals of magnetic resonance imaging (MRI) of the brain. The assumption underlying much of the research in this area is that we will be able to make detailed models of the brain when we have detailed descriptions of the 3D structures and local physical, chemical, and physiological properties of the brain *and* equally detailed knowledge of how these structures and properties are altered by external stimuli. The models can then be used to assess the presence or absence of disease states, predict the outcome of interventions (surgery, drugs, etc.), and provide insight into how the brain is put together. There are many excellent texts on clinical application of MRI that furnish elaborate looks into the normal and abnormal anatomy of the human brain as visualized by MRI (Stark and Bradley, 1992; Damasio, 1995). In this chapter we focus on the anatomical aspects of magnetic resonance imaging with special emphasis on imaging of the developing nervous system at high spatial resolution.

The nuclear magnetic resonance (NMR) phenomenon in bulk matter was first demonstrated in 1946 by Bloch at Stanford (Bloch *et al.*, 1946) and Purcell at Harvard (Purcell *et al.*, 1946). The fact that the resonance frequency of an individual nucleus is sensitive to its local physicochemical environment (chemical shift phenomenon) has made NMR spectroscopy a common and powerful analytical tool (Roberts, 1959; Derome, 1987; Becker, 2000). Again, because the properties of the NMR signal are exquisitely sensitive to local environment, MRI has become a powerful tool for the clinician, material scientist, biologist, and anyone else interested in the noninvasive examination of optically opaque samples.

The first MR images were produced in 1972 by Lauterbur at SUNY Stony Brook (Lauterbur, 1973), and in 1974 he produced the first images of a live animal. Lauterbur's first MR images employed magnetic field gradients to obtain projections of the proton-spin density from two adjacent water-filled capillary tubes. Using various magnetic field gradient orientations, he was able to reconstruct the image of the tubes from the projections. At about this time, Mansfield's group at the University of Nottingham, England, was pursuing the idea of "NMR diffraction" (Mansfield and Grannell, 1973) as a procedure for imaging. This work subsequently led to the development of the echo-planar imaging (EPI) method (Mansfield, 1977; Mansfield and Pykett, 1978). EPI is one of the fastest MR imaging methods available today. In 1975, Richard Ernst and his co-workers in Zurich developed a very powerful MR imaging technique based on two-dimensional NMR theory and the direct use of Fourier transform techniques to obtain images (Kumar *et al.,* 1975). Almost all modern MRI units now employ this Fourier imaging technique.

MRI is one of the most powerful imaging modalities available for imaging the brain. It is noninvasive in nature. No high-energy radiation is employed, as in X-ray computed tomography (CT). Unlike positron-emission tomography (PET), there is no need to infuse radioactive nuclei into the subject under study. Due to the unique contrast characteristics of the MR image, the investigator can delicately alter the image contrast through adjustment of the many parameters that enter into any MR imaging scheme or by employing different schemes. The MR images can thus provide a wealth of information unavailable through either PET or CT.

Although MR imaging offers many advantages, there are limitations as well: chief among them is the intrinsically low signal level in MR. Due to the physical laws that govern the equilibrium of magnetic moments in a magnetic field, the MR signal obtainable varies as the magnetic field strength; thus, high magnetic fields are favored. However, high magnetic fields of very good uniformity over large volumes are not easily generated, and such magnets are expensive. With current technology, MRI units for humans commonly operate around 1.5 T (1 tesla = 10,000 gauss; the earth's magnetic field at the surface is typically about 0.5 G); however, NMR microscopy is generally carried out at higher field strengths of 7 T or above. As one goes to higher field strengths, the magnet bore size is reduced; thus, at higher field strengths, it is possible to image only smaller animals. Although magnetic field gradients employed for imaging vary, depending on the strength of the steady field and the desired resolution, larger gradient strengths are generally required for higher resolution. The switching of these gradients, demanded by the imaging pulse sequence, can cause eddy currents and lead to image blurring. In order to overcome this problem, specially designed shielded gradients are often employed.

The information content of the MR image depends strongly on how the image is obtained. In this chapter we introduce the basics of the NMR phenomenon and methods of spatial encoding, discuss how contrast arises and how it may be modulated in the image, consider how spatial resolution may be increased, and, finally, show MR brain images from the amphibian, rodent, and primate orders. Our aim is to not provide an exhaustive survey or in-depth analysis, but rather to instill in the reader an appreciation of the variety of kinds and types of information that may be manifest in a magnetic resonance image.

II. Magnetic Resonance Basics

(Morris, 1986; Farrar, 1989)

In the present article, we shall be concerned with protons. Protons are nuclei of hydrogen atoms and have angular momentum $I = \frac{1}{2}(h/2\pi)$ (h is Planck's constant) and, hence, a magnetic moment, μ, that is proportional to I, with proportionality constant γ, the gyromagnetic ratio. Protons have a large γ value in comparison to most other nuclei and are abundantly present in biological samples. In the presence of a magnetic field of strength B_0 along the laboratory z axis, there are two quantum mechanically allowed energy levels for the proton, $-\frac{1}{2}\gamma(h/2\pi)B_0$ and $+\frac{1}{2}\gamma(h/2\pi)B_0$. These levels correspond to the magnetic quantum numbers $-\frac{1}{2}$ and $+\frac{1}{2}$. The distribution of protons between these two energy levels, known as Zeeman levels, depends on temperature and is governed by the Boltzmann equilibrium law. According to this law, the ratio of the population of spins in the two levels is

$$\frac{n\left(-\frac{1}{2}\right)}{n\left(+\frac{1}{2}\right)} = \exp\left(\frac{-\Delta E}{kT}\right), \tag{1}$$

where $n(\pm\frac{1}{2})$ is the population of the $\pm\frac{1}{2}$ Zeeman level, ΔE is the energy difference between the Zeeman levels, T is the absolute temperature, and k is the Boltzmann constant. ΔE in the present case is given by

$$\Delta E = \gamma(h/2\pi)B_0. \tag{2}$$

Equations (1) and (2) lead to the relation

$$\frac{n\left(+\frac{1}{2}\right) - n\left(-\frac{1}{2}\right)}{n} = \frac{\Delta n}{n} \cong \frac{\gamma\left(\frac{h}{2\pi}\right)B_0}{2kT}. \tag{3}$$

$n = n(+\frac{1}{2}) + n(-\frac{1}{2})$ is the total spin population and Δn is the so-called "excess population." It is this excess population in the lower Zeeman level over the upper level that is of importance in magnetic resonance because the MR signal strength is proportional to the net magnetic moment, $\mathbf{M} = \Delta n \mu$, and, hence, Δn. The two factors that really matter, from the point of view of the net magnetization in the sample, are, then, B_0 and T. Due to the nature of biological samples, we usually do not have much flexibility to lower T arbitrarily in order to increase the MR signal. However, the strength of B_0 can be increased to as large a value as practicable. It is here that one faces technological (and financial!) barriers.

Viewed from the laboratory coordinate, the protons in the magnetic field precess around the z axis at an angular frequency $\omega_0 = \gamma B_0$, the Larmor frequency. In the presence of a radio-frequency (RF) field having the frequency ω_0 and appropriate polarization, transitions can be induced between spins in the two Zeeman levels because the precessing magnetic spins can couple to the magnetic component of the RF field. A linearly oscillating RF field in a coil that is oriented perpendicular to the B_0 axis (e.g., along the x axis) can provide a rotating field in the x–y plane; this causes the spins aligned with the main magnetic field to flip between the $+\frac{1}{2}$ and the $-\frac{1}{2}$ states. This oscillating field is termed B_1. We can now imagine a rotating coordinate system (x, y, z), with z coinciding with z of B_0 and the x and y axes rotating around z with an angular velocity ω. In this *rotating frame,* the precessing spins will appear to be stationary when $\omega = \omega_0$. The rotating B_1 field will appear to be stationary at this frequency in the rotating frame. The net effect of this B_1 field is to cause a rotation of the magnetic moment around B_1. With a constant B_1, the angle α, through which the spins are rotated, is given by

$$\alpha = \gamma B_1 \tau, \tag{4}$$

where τ is the duration of application of the RF field. By choosing appropriate values for B_1 and τ, we have the ability to rotate the net magnetic moment through any desired angle! For commonly available B_1 values in the RF range, τ values are on the order of a few micro- or milliseconds for $\alpha = 90°$. We are therefore dealing with RF pulses in MR imaging.

At the end of a 90° pulse applied along the x axis, i.e., $(90°)_x$ pulse, the net magnetic moment is aligned along the y axis in the rotating frame. For an observer in the laboratory frame, the net magnetic moment appears to be rotating in the x–y, or transverse, plane, as a result of which a voltage will be induced in a receiver coil whose axis lies in this plane. The $(90°)_x$ pulse is thus an *excitation pulse* that produces "*transverse magnetization.*" With the help of RF phase-sensitive detectors, we can display the "*in-phase*" and "*out-of-phase*" components of this MR signal corresponding to dispersion and absorption lineshapes in NMR spectroscopy. The reference phase is provided by the phase of the RF transmitter.

The transverse magnetization created in a sample at the end of a 90° excitation pulse lasts for only a short period. The induced signal voltage, which is referred to as the *free induction decay* (FID) signal, decays. The reason for this is that the individual magnetic spins begin rotating synchronously in the x–y plane at the end of the 90° pulse, then lose their coherence and dephase with time (Fig. 1). This dephasing leads to a decrease in **M** and a concomitant decrease in signal intensity.

The decay of the FID signal can be expressed as

$$S(t) = S_0 \exp(-t/\mathrm{T}_2), \tag{5}$$

where S_0 is the initial signal immediately following the 90° pulse, $S(t)$ is the signal as a function of time, t, and T_2 is the characteristic spin dephasing or *transverse relaxation time.* Because this dephasing mechanism is due to the magnetic interaction among the spins, T_2 is also referred to as the *spin–spin relaxation time.* If the ensemble of spins does not dephase too quickly, we can detect the signal by the technique of *spin echoes* (SE), originally discovered by E. L. Hahn (1950). After an interval τ following excitation by a 90° pulse along the x axis, one applies a 180° pulse along the y axis to refocus the spins at a time, 2τ, after the initial 90° pulse. In other words, a signal appears at 2τ = TE, the echo time (Fig. 2). We abbreviate this pulse sequence as $(90°)_x$-τ-$(180°)_y$-τ-echo.

The 180° RF pulse reverses the direction of rotation of the spins (see Fig. 3) and causes the spins to refocus at time 2τ. Such a 180° RF pulse is referred to as a *refocusing pulse.* We can use a succession of 180° refocusing pulses to repeat the echo. The sequence is then referred to as a multiple-echo

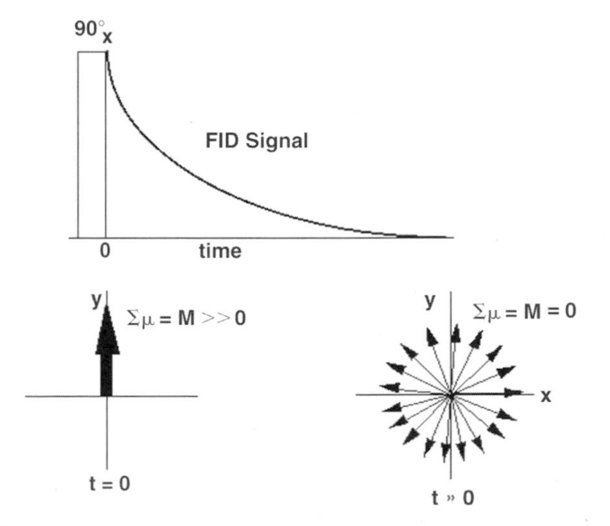

Figure 1 Free induction decay signal following a 90° RF pulse. The spins are in phase at time $t = 0$. After a long time $t \gg 0$ the spins lose their coherence and the net magnetization in the x–y plane is zero, due to the completely random orientation of spins in this plane.

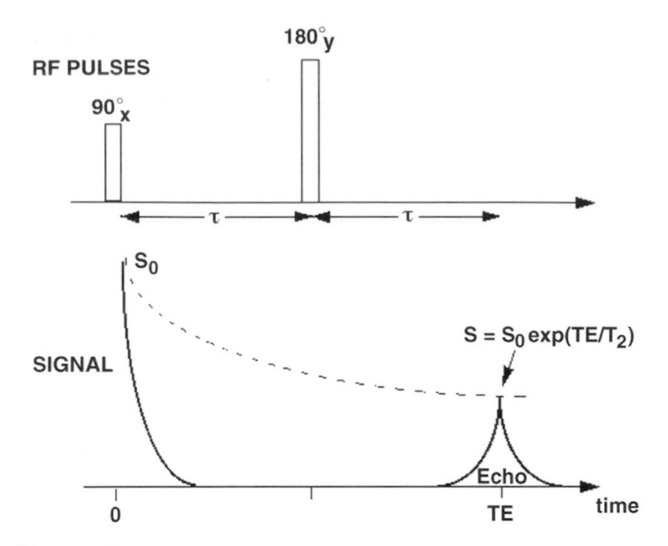

Figure 2 Basic spin echo sequence $(90°)_x$-τ-$(180°)_y$-τ-echo. The dashed line describes signal intensity loss due to T_2 decay.

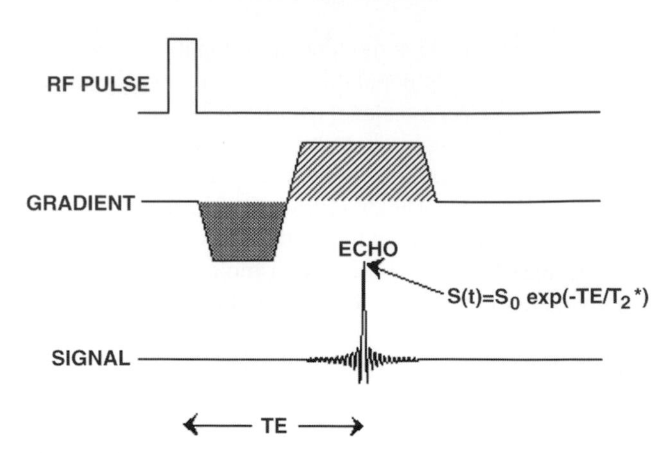

Figure 4 Gradient-echo pulse sequence. The initial negative gradient causes the spin system to fan out or dephase in one direction (say clockwise), while the subsequent positive gradient reverses the direction of motion, causing the spins to rephase completely forming a *gradient echo* at time TE. The echo time is determined by the relative durations and amplitudes of the positive and negative gradients.

sequence. We can also refocus the spins by means of a reversal of a magnetic field gradient (see Fig. 4) to obtain *gradient echoes* (GE). The spin dephasing process is dependent on, among other things, the homogeneity of the magnetic field. T_2 is shortened by inhomogeneity in the magnetic field, which may arise from imperfections in the magnet system or from the sample itself. The measured T_2 value is referred to as T_2^*, and is given by

$$1/T_2^* = 1/T_2 + 1/T_2', \tag{6}$$

where T_2^* is the effective transverse relaxation time, and T_2' is the contribution arising from magnet inhomogeneity, i.e., magnetic field gradients and other factors, such as susceptibility differences within the sample or chemical shift differences. In biological imaging, it is feasible to obtain images that bring out effectively the variations in T_2 in the specimen. Such images are referred to as *T_2-weighted images* and may provide contrast between different regions of the brain, for example, between gray and white matter. Water molecules that are bound to larger molecules exhibit shorter T_2, while mobile protons have larger values.

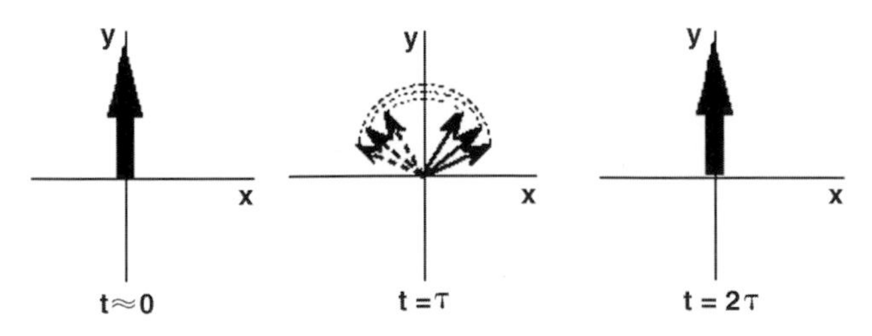

Figure 3 Trajectory of spins during a SE experiment. Spins are initially in phase, all pointing along the *y* axis. At time less than τ they have rotated clockwise and dephased. At time $t = τ$ the spins are flipped about the *y* axis and are now rotating *counter*clockwise. At time $t = 2τ$ the faster rotating spins have caught up with the slower ones and all arrive back at the *y* axis at the same time to form the *spin echo.*

Unlike spin echoes, the intensity of gradient echoes depends on T_2^*. They depend, not simply on T_2, but also on magnetic field inhomogeneity, susceptibility, and chemical shift effects. Thus, the image contrast obtained with a GE scheme can differ from that obtained by SE, and this difference is exploited in brain imaging. Because T_2^* is shorter than T_2, the signal intensity for a given TE is larger with a SE protocol than with a similar GE protocol.

If we apply a 180° pulse along the *x* axis to a sample at equilibrium in a magnetic field, we can invert the net magnetic moment (**M**) so that it now points against the magnetic field. This type of 180° pulse is called an *inversion pulse.* The spins thus inverted tend to go back to their earlier equilibrium condition. In the absence of other perturbations, the recovery to equilibrium is generally an exponential function characterized by a time constant T_1, referred to as the *spin-lattice relaxation time,* because this process entails exchange of energy between the spins and their surroundings (loosely termed the "lattice"). It is also called the *longitudinal relaxation time* to contrast this decay time with T_2. By monitoring the time dependence of the amount of magnetization remaining after a 180° inversion pulse, we can determine T_1. This is done by applying a 90° inspection pulse at variable time, *t,* after the 180° pulse. This scheme is the *inversion recovery* pulse scheme shown in Fig. 5. The signal can be expressed as

$$S(\tau) = S_0[1 - 2\exp(-\tau/T_1)]. \tag{7}$$

If the equilibrium magnetization (M_0) is subjected to a series of 90° pulses, each repeated at an interval TR, the magnetization along the *z* axis (M_z) recovers only partially, and is given by

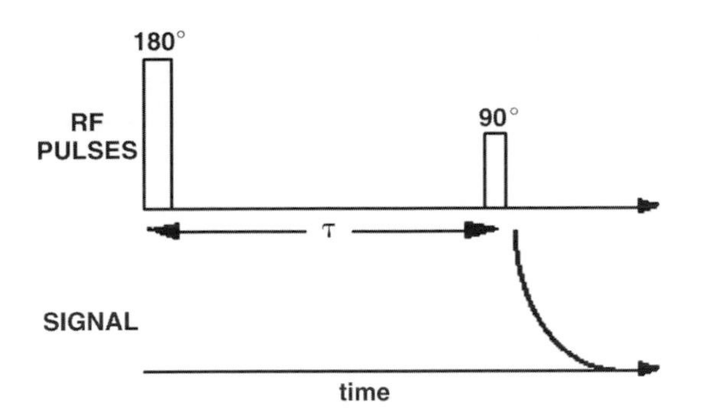

Figure 5 Inversion recovery sequence used to measure T_1 values or to "prepare" the spin system for an imaging sequence.

$$M_z = M_0[1 - \exp(-TR/T_1)]. \quad (8)$$

However, the transverse magnetization and hence the signal-to-noise ratio can be maximized in the case of repeated pulsing by reducing the flip angle. The optimum flip angle, known as the Ernst angle, is $\alpha_E = \cos^{-1}[\exp(-TR/T_1)]$ (Ernst and Anderson, 1966). From the above discussion, it will be clear that T_1 plays an important role in the recovery of the equilibrium magnetization along the z axis and, hence, the signal intensity following repeated pulses. By adjusting the repetition time, TR, it is possible to obtain contrast in MR images between regions differing in T_1. Such images are described as *T_1-weighted images.*

III. Magnetic Resonance Imaging Basics

(Mansfield and Morris, 1982; Morris, 1986; Callaghan, 1991; Vlaardingerbroek and Den Boer, 1996; Haacke *et al.*, 1989; Kuperman, 2000)

In order to obtain information regarding the spatial location of the spins, the MRI technique makes use of magnetic field gradients. A linear gradient along x axis, G_x, applied to a homogeneous sample containing protons will display a distribution of resonance frequencies along x. When such a small linear gradient is applied to a sample placed in a strong field, B_0, the resonance frequency will be a function of its location (x) in G_x:

$$\omega(x) = \omega_0 + \gamma x G_x. \quad (9)$$

Thus, the spatial information is mapped onto a frequency scale. This procedure is referred to as *frequency encoding.* When spins precess in a magnetic field gradient, they tend to dephase. Suppose a group of spins started out in phase at time $t = 0$. In a constant gradient of strength, G_y, the spins acquire a phase ϕ_y during a period of time, t_y, given by:

$$\phi_y = \gamma y G_y t_y. \quad (10)$$

In the case of time-varying gradient we have the relation:

$$\phi_y = \gamma \int y G_y(t) dt. \quad (11)$$

In some special imaging schemes, as for example, spiral scan fast imaging (Ahn *et al.*, 1986; Meyer *et al.*, 1992), both G_x and G_y may vary in a specified manner as a function of time. For the present we consider the case of constant gradients. By using gradients in different orthogonal directions, we can label the positions of the spins by their ϕ values. Usually, the duration of the gradient is fixed and magnitude is varied to obtain different π values. This is the basis of *phase encoding.*

We can excite spins in a slice of the specimen by a combination of a magnetic field gradient and an RF pulse. The RF pulse can be suitably tailored (Mansfield and Morris, 1982; Callaghan, 1991) to excite spins in a chosen bandwidth, say $\Delta\omega$, and by adjusting the value of the gradient G_{slice} and its duration, we excite spins only within this bandwidth, corresponding to a slice in the spatial domain. A typical combination of this type is shown in Fig. 6. The gradient applied during the RF pulse period dephases the spins that one rephases by reversing the gradient, usually for about half the time of the duration of the positive period.

By using a combination of phase and frequency encoding, as described above, we are now in a position to design imaging pulse sequences. A popular three-dimensional (3D) imaging sequence is a combination of two phase and one frequency encoding procedures. 2D imaging sequences usually employ a slice selection scheme for one direction (say, the z axis) and a phase encoding for the y axis, followed by frequency encoding for the x axis. Neglecting relaxation effects, the signal obtained at the end of the sequentially applied phase and frequency encoding in a 2D imaging scheme in which a 90° pulse is used for excitation takes the form

$$S(t_x,t_y) \propto \iint \rho(x,y)\exp[i\gamma(G_x x t_x + G_y y t_y)]dxdy, \quad (12)$$

where $\rho(x,y)$ is the 2D proton spin density distribution and t_x and t_y are the durations of application of gradients G_x and G_y, respectively. Using the relations

$$k_x = \gamma G_x t_x \quad (13)$$

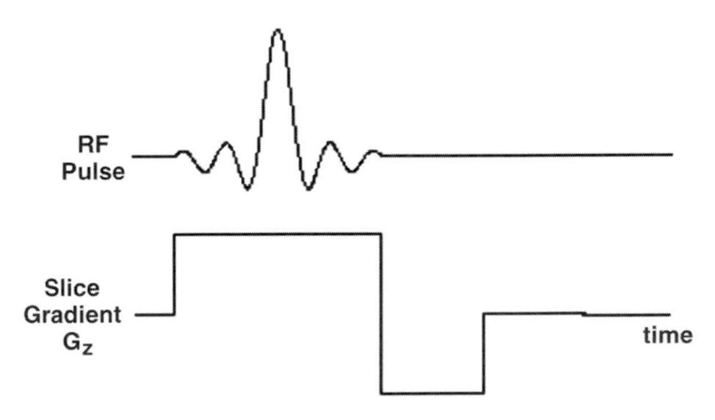

Figure 6 Elementary slice selection. The position, thickness, and intensity profile of the slice are determined by the frequency and shape of the pulse and the gradient intensity.

$$k_y = \gamma G_y t_y, \quad (14)$$

which define the spatial frequencies k_x and k_y, we can recast Eq. (12) as

$$S(t_x,t_y) \propto \iint \rho(x,y)\exp[i(k_x x + k_y y)]dxdy. \quad (15)$$

Fourier transformation (FT) with respect to k_x and k_y yields the 2D image $\rho(x,y)$. In general, for the 3D case:

$$S(\mathbf{k}) \propto \iiint \rho(\mathbf{r})\exp[i\mathbf{k}\cdot\mathbf{r}]d\mathbf{r}. \quad (16)$$

k and **r** form a Fourier pair, and hence we can formally write

$$\rho(\mathbf{r}) \propto \iiint S(\mathbf{k})\exp[-i\mathbf{k}\cdot\mathbf{r}]d\mathbf{k}. \quad (17)$$

Thus, Fourier transformation enables us to go from "k space" to "image space" or "r space." Unlike CT or PET, the data obtained in MRI are initially in k space. A 2D image has data points in the (k_x, k_y) matrix, while the 3D image has its data points in the (k_x, k_y, k_z) array. Before we go into the details of k space and its relation to the image quality, it would be appropriate to outline at least two often-used imaging sequences, namely, spin-echo (Fig. 7) and gradient-echo (Fig. 8) sequences. We show how these sequences are built upon the basic sequences shown in Figs. 2 and 4. For the sake of simplicity, we confine ourselves to 2D imaging.

In both SE and GE sequences an echo is acquired in the presence of the G_x gradient for each particular value of G_y and yields the k_x spatial frequency data points in the "k matrix" for each k_y value. G_y values are usually incremented in equally spaced steps starting with the maximum negative value and proceeding to the maximum positive value. Thus, the data matrix is acquired "line by line," i.e., row after row, in k space (Fig. 9a).

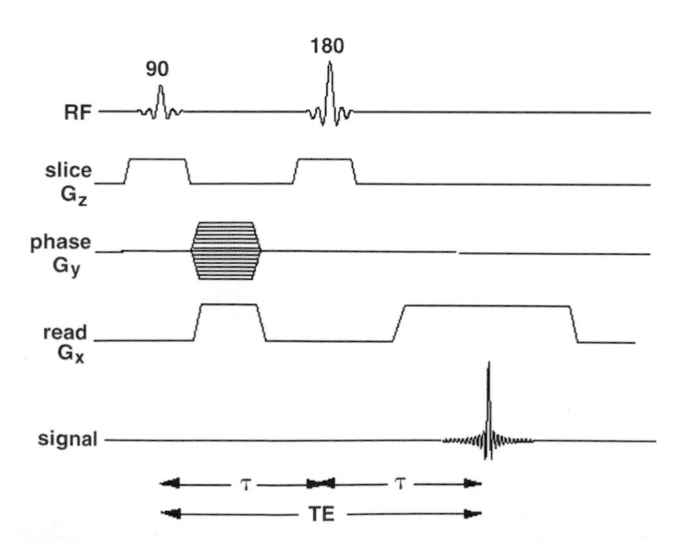

Figure 7 Schematic diagram of a 2D spin-echo imaging sequence. Frequency encoding is done along the x axis, phase encoding is done along the y axis, and a slice is selected perpendicular to the z axis.

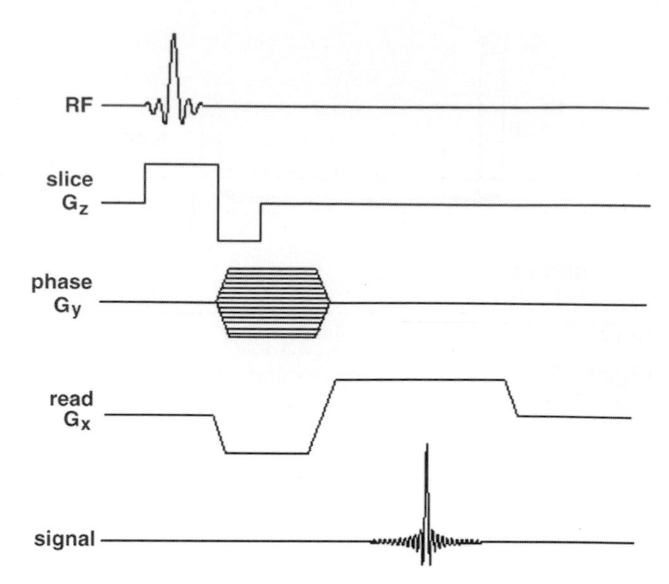

Figure 8 The 2D gradient-echo sequence. Slice selection is performed as in Fig. 6 and the echo produced as in Fig. 4.

The 2D SE sequence shown in Fig. 7 is also known as the 2D FT spin-warp imaging sequence (Edelstein *et al.,* 1980). Figure 8 shows the corresponding gradient-echo version for 2D imaging. Instead of the 180° RF refocusing pulse, the gradient-echo sequence reverses the READ gradient in order to refocus the spins and obtain the echo. The FLASH (fast low angle shot) technique developed by Frahm and co-workers (Frahm *et al.,* 1986a,b; Haase *et al.,* 1986) employs a GE scheme with less rotation angle for the RF excitation pulse so that the longitudinal relaxation does not overshadow the recovery of the magnetization during repeated RF excitation. By keeping the flip angle small and repeating it rapidly, one can maintain a reasonably good value of the transverse magnetization and obtain images in a short time. The FLASH imaging technique has been successful in obtaining brain MR images within the subsecond time regime.

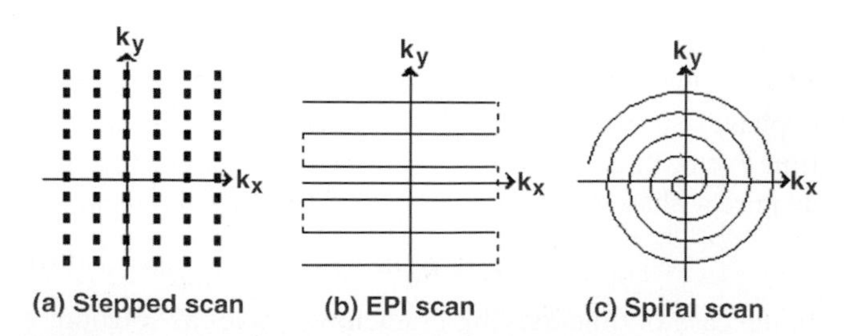

Figure 9 Three different k-space trajectories outlining different strategies for scanning k space.

IV. k Space and MR Images

(Ljunggren, 1983; Twieg, 1983)

It was pointed out earlier that the process of obtaining MRI data consists of acquiring the data in terms of k_x and k_y, in the case of 2D imaging, and k_x, k_y, and k_z, in the case of 3D imaging. Because the dimensions of k are radians/cm, it is common to refer to them as "spatial frequencies." The size of the spatial frequency map is determined by the number of data points along k_x and k_y and the step interval, Δk_i ($i = x,y$). A typical k matrix size is 256×256 and FT converts the k-space data to the image data. The FT process leads to real (R) and imaginary (I) parts of the signal data that can be utilized in constructing different types of images, such as "real," "imaginary," and "modulus" = $(R^2 + I^2)^{1/2}$. The phase angle, ϕ, is defined as $\tan^{-1}(I/R)$, and we can construct "phase" images that are especially helpful in evaluating susceptibility induced inhomogeneity effects in brain images.

In the imaging schemes shown in Figs. 7 and 8 in which phase encoding uses simply *incremented* gradients, we acquire data in k space line by line with one echo for each RF excitation and each phase-encoding step (see Fig. 9a). This type of raster scanning is a rather slow process. Alternate ways of spanning k space offer the advantage of speedier data collection. In the gradient-echo version of the echo-planar imaging method, following a single RF excitation many echoes are obtained by repeated gradient reversals. In the version of EPI known as MBEST-EPI (Stehling *et al.*, 1991), a whole set of data is acquired following "one shot" of RF by scanning all of k space in a zigzag fashion (see Fig. 9b). This mode of data acquisition accounts for the speed of EPI. Another fast imaging method makes use of the spiral scanning of k space (Fig. 9c) (Ahn *et al.*, 1986; Meyer *et al.*, 1992). The SE sequence shown in Fig. 2 can be modified to produce more than one echo from an initial single excitation. By repeating the 180° pulses at intervals as in a Carr–Purcell–Meiboom–Gill sequence (Carr and Purcell, 1954; Meiboom and Gill, 1958) several usable echoes can be generated depending upon the value of T_2. By placing the phase-encode gradients before each echo a multiple-spin-echo (MSE) imaging sequence is obtained. The MSE sequence forms the basis for many fast-spin-echo (FSE) sequences. A popular FSE sequence is the RARE sequence originally outlined by Hennig *et al.* (1986, 1987). RARE is an acronym for rapid acquisition with relaxation enhancement. A single 90° pulse is followed by several 180° refocusing pulses in one train. The phase-encode gradients are varied from echo to echo to traverse different points in the k space with one initial excitation in the spirit of the EPI method. T_2 weighting of the images depends on the order in which data points are obtained. Several modifications of RARE have been proposed (Hennig, 1988; Norris *et al.*, 1992). FSE sequences based on RARE have been examined from the point of view of high-field MR microscopy by Johnson and co-workers (Beaulieu *et al.*, 1993; Zhou *et al.*, 1993).

The Fourier transformation of data points in k space is done using well-established procedures (Brigham, 1974; Bracewell, 1978; Ramirez, 1985). Although Eqs. (12) and (15–17) display the continuous integral form, in practice one uses the discrete FT approach, because the k-space data are discrete rather than continuous. The fast Fourier transform (FFT) algorithm is a very efficient procedure for performing the transformation from k space to r space. Due to the nature of the k-space trajectory, EPI and spiral scanning data need to be reordered before using conventional FFT programs. Although one should acquire the signal data in its entirety, quite often the acquisition is truncated. Consequently, the resulting image may exhibit effects known as truncation artifacts (Henkelman and Bronskill, 1987).

A larger number of data points in the k space leads to better resolution because the theoretical resolution in the ith direction is simply the ratio of the $(\text{FOV})_i$ and the number of data points along $(\text{FOV})_i$. The field of view (FOV) along the phase-encoding axis (y axis) is related to the step size Δk_y, as

$$\Delta k_y = 1/(\text{FOV})_y. \quad (18)$$

The READ gradient G_x and the sampling frequency f_s determine $(\text{FOV})_x$.

$$(\text{FOV})_x = 2\pi f_s/\gamma G_x. \quad (19)$$

Along a horizontal line of k-space data, we have a variation in k_x. In this context, it should be pointed out that "zero-filling" (Bartholdi and Ernst, 1973), or appending zeros to the data (as is done sometimes in the case of limited phase-encode steps), does not improve resolution, but does make the image appear "smoother." Zero-filling is a method of interpolating the original data, increasing the number of calculated pixels (Hedges and Sobering, 1994). In Fig. 9, the top and bottom rows of k_y correspond to high $|G_y|$ values and the signals are, in general, weaker in these rows, compared to the central rows ($k_y = 0$ region). Similarly, the extreme right and left sides of the k map correspond to large $|G_x|$ values, and the signals are weak here. The central portions of the k map correspond to regions of high signal intensity and data in these regions contribute to contrast in the resultant image. Data from the edges of the k map correspond to higher spatial frequencies and contribute to details in the image (Wehrli, 1990; Wood, 1992). Low spatial frequencies are sensitive to motion. Due to the FT relationship between k-space data points and image data points, even one corrupted data point in the k space can affect the quality of the image. Motional effects can thus lead to artifacts in the image. In the context of brain imaging, two factors that cause such artifacts are physical motion and fluid motion. Flow of blood and cerebrospinal fluid (CSF) belong to the latter category. In functional MRI of the brain, physical restraints to minimize head movements are recommended.

Special MR techniques to compensate for motion through corrective algorithms may help further; of these, the "navigator echo" technique makes use of additional, "navigator," echoes interleaved into the imaging sequence. This method uses postprocessing of the navigator echo data with the help of algorithms for adaptive correction of the effects of object motion to suppress artifacts in the final image (Ehman and Felmlee, 1989).

"Keyhole imaging" is a technique that makes novel use of the properties of k-space data in order to produce images more rapidly. In some imaging experiments, such as measurement of the signals in the brain area from a bolus of paramagnetic agent injected into the blood stream, useful data can be obtained by acquiring only selected low-frequency k_y lines in the phase-encoding step. Accordingly, first, an image is obtained prior to the bolus injection, using a full scan of the phase-encode steps. Subsequent images are constructed by updating only the middle regions of the k map and retaining the first set of outer k_y lines of data. New images constructed in this manner by substituting new k-space data into the older set are referred to as keyhole images (Jones *et al.,* 1993; van Vaals *et al.,* 1993). These images can yield valuable information on the rate of transit of the paramagnetic bolus and parameters related to functional imaging. By reducing the field of view temporal resolution can be improved (Hu and Parrish, 1994). Ideally, due to the properties of k space, we should have

$$S^*(-k_x,-k_y) = S(k_x,k_y), \tag{20}$$

where S^* is the complex conjugate of S. We should therefore be able to construct the entire k-space matrix from just one-half of the data. This principle applied to MR imaging is known as "half-Fourier" imaging (Margosian and Schmitt, 1985; Feinberg *et al.,* 1986) and works better with SE than with GE schemes. We note that noise, being random in nature, does not exhibit the conjugate symmetry attributed to the signal in Eq. (20).

V. Signal-to-Noise Ratio (SNR) and Contrast-to-Noise Ratio (CNR)

SNR and CNR are two important quantities that specify the quality of the MR image (Edelstein *et al.,* 1986; Constable *et al.,* 1992). A good SNR value is necessary for the clear delineation of features in a region of interest (ROI) in an image. A lowering of the optimal value of SNR will lead to an image appearing noisy. The entire image is built up of discrete volume elements (voxels) and a reduction in the size of the voxel reduces the signal and thus lowers SNR. The basic challenge in microimaging as contrasted to macroimaging comes from the size of the voxels. The choice of high magnetic fields for microimaging offers a straightforward way to achieve a better SNR.

For two ROIs to be distinguishable, a good CNR value is needed. If the signal from a region, A, is S_A, and that of its neighbor, B, is S_B, then

$$\text{CNR} = (S_A - S_B)/\sigma, \tag{21}$$

where σ is the noise. Noise is generally taken as that of the background. Two kinds of noise may be identified in MRI. One is random noise (σ), arising from receiver electronics (e.g., Johnson noise), from digitization, and from the sample. As the receiver bandwidth increases, noise also increases, because noise is proportional to the square root of the bandwidth. Acquisition time (AT) and bandwidth are related as

$$\text{AT} = N/\text{bandwidth}, \tag{22}$$

where N is the number of data points. Thus, for a given N, a longer acquisition time is favorable from an SNR point of view. The other type of noise is systematic noise (σ_s), arising from sample motion, blood flow, CSF flow, gradient coil movements, truncation effects in signal acquisition, etc. The total noise may be expressed as

$$\sigma = (\sigma_r^2 + \sigma_s^2)^{1/2}. \tag{23}$$

In the presence of random noise, signal averaging improves SNR by adding the signal repetitively and coherently in proportion to the number of scans, n. Due to its statistical nature, noise adds only in proportion to the square root of n, thereby signal averaging leads to an overall increase in SNR. Noise due to periodic motion may be reduced by monitoring the motion and adopting gating techniques for signal acquisition. Thus, effects due to CSF pulsatile motion on MR images may be reduced by peripheral gating, e.g., gating to the pulse in the finger tip. The use of navigator echoes is also helpful in this context as was pointed out above. It is possible to synchronize the RF excitation pulse or receiver pulse to the respiratory or cardiac cycle of the live animal, in the case of *in vivo* studies, with significant suppression of motion artifacts. Commercially manufactured electronic gating units for this purpose are available (see Fig. 10A). A well-designed animal holder that fits securely in the RF probe (see Fig. 10B) can help in reducing motion artifacts.

Two sources of noise are generally distinguished in biological imaging (Hoult and Lauterbur, 1979; Edelstein *et al.,* 1986). These are the coil and its associated probe circuit and the sample itself. Due to the small size of the samples commonly employed sample noise is not as important as coil noise in high-field μMRI. Attempts have been made to design better microcoils (Peck *et al.,* 1995) and improve probe design (Glover *et al.,* 1994; Schoeniger and Blackband, 1994). Thermal cooling of the coil and preamplifier to reduce noise has also been explored (Styles *et al.,* 1989; McFarland and Mortara, 1992; Wright *et al.,* 2000). High-Q coils employing high T_c superconducting materials have also been investigated (Black *et al.,* 1993; Odoj *et al.,* 1998; Hurlston *et al.,* 1999; Miller *et al.,* 1999). All these

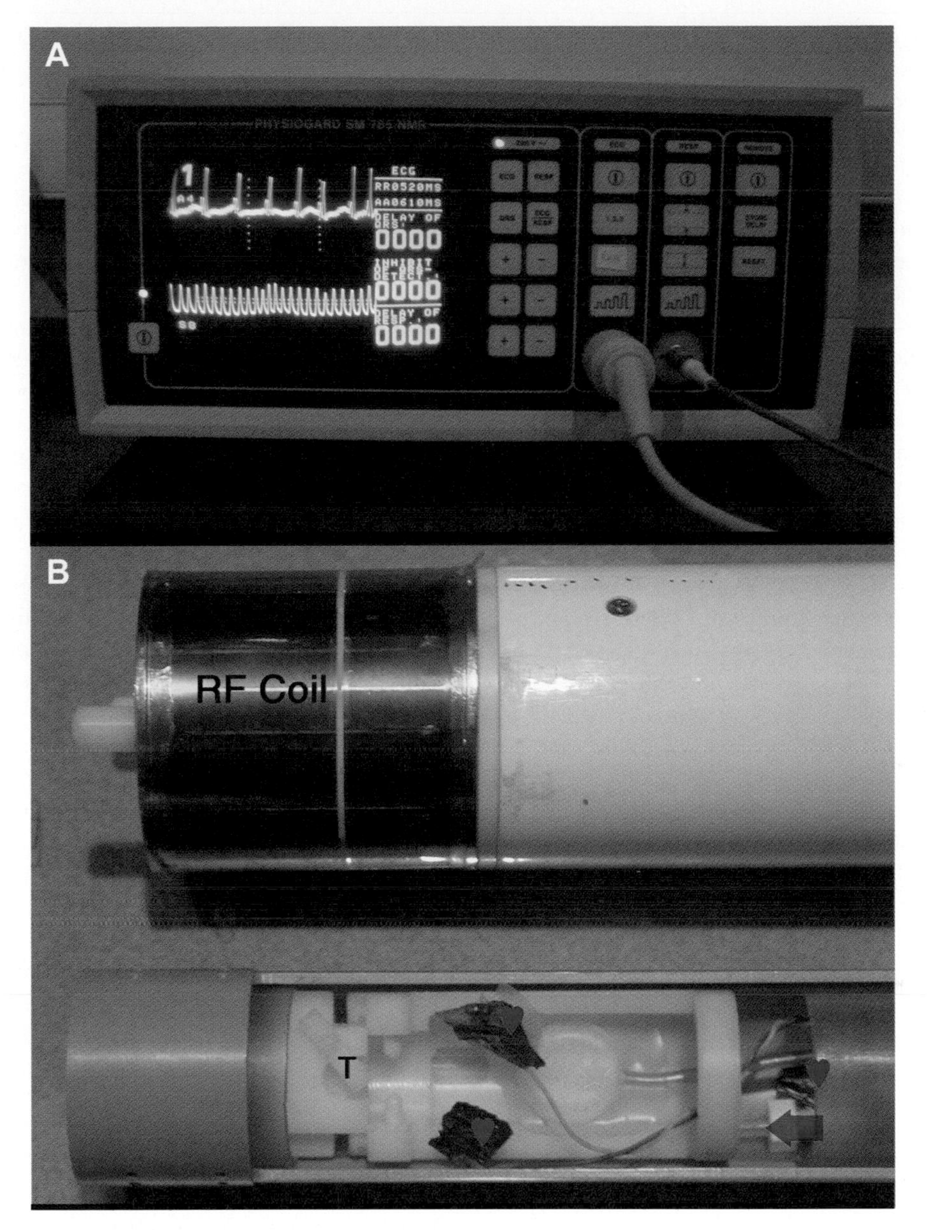

Figure 10 (**A**) Physiogard unit (Bruker Instruments, Inc., Billerica, MA) monitors ECG and respiration of the mouse. This unit connects directly to the Avance console and provides a trigger for respiratory (lower trace) and/or ECG (upper trace) synchronized data collection. (**B**) RF coil and stereotaxic mouse holder for *in vivo* mouse imaging. Only the outer shield of this 30-mm inner diameter birdcage coil is visible, shown in the upper half of the photograph. The mouse to be imaged is placed prone with the upper front teeth in a groove under the adjustable piece labeled "T" with the body to the left. ECG leads of copper foil (red hearts) are wrapped around the forepaws and tail. The small flat balloon (green star) detects respiration via chest movement. Anesthesia is supplied via a line (blue arrow) which carries the gas to just under the animal's mouth. A water jacket surrounds the animal to maintain body temperature, which is monitored with a small rectal probe. Figure courtesy of Drs. Angelique Louie and Robia Pautler.

developments could lead to a substantial improvement in SNR and thus result in improved resolution in μMRI.

VI. T_1- and T_2-Induced Contrast

To span *k* space and for signal averaging purposes, we need to repeat the pulse sequences. If we repeat the SE or GE sequences (Figs. 7 and 8) at a time, TR, that is long in comparison to T_1, we recover the full magnetization along the *z* axis between excitations. From Eq. (8), it is clear that if we repeat the sequence at a faster pace, the longitudinal magnetization will not fully recover between excitations. In fact, for the SE sequences, it can be shown that the signal depends on TR as well as on TE, T_1, and T_2:

$$S(SE) \propto \rho\{1 - 2\exp[-(TR - TE/2)/T_1] + \exp[-TR/T_1]\}\exp[-TE/T_2] = \rho(TR,TE,T_1)\exp[-TE/T_2]. \quad (24)$$

By manipulating the ratio TE/T_2, we get T_2-weighted images. On the other hand, by adjusting the TR/T_1 ratio, we get T_1-weighted images. In brain imaging, a careful choice of TR and TE can thus produce excellent contrast between different areas of the brain. The T_1 value of CSF is longer than most tissue T_1 values (see Table 1) and, by choosing a shorter TR value, we can reduce the intensity from CSF in the SE image. To avoid T_2 weighting, we keep the value of TE short. T_1 values also depend on the main magnetic field strength (B_0). If the molecules bearing protons are mobile and rapidly moving in an isotropic fashion, T_1 equals T_2. If the motion is slow, as in the case of water bound to macromolecules, the T_2 value may be much lower than T_1. Brain MR images can be produced with T_2 weighting to display variations in T_2 values in different areas of the brain. By combining 2D GE techniques with SE, Ordidge *et al.* (1994a) were able to obtain data on the distribution of T_2^* and T_2 values [see Eq. (6)] in the human brain. The difference between these two parameters is claimed to be a better correlate of relative brain iron concentration than T_2^*. An analysis of FLASH versus SE imaging schemes with regard to SNR and CNR is available in the literature (Hendrick *et al.,* 1987). For more detailed considerations of the role of TR, TE, READ gradient strength, and sampling time on SNR and CNR in SE and GE imaging schemes the reader may refer to the work of Haacke, Hendrick, Vinitski, and Seelig (Haacke, 1987; Hendrick, 1987; Vinitski *et al.,* 1987; Link and Seelig, 1990).

In the vicinity of paramagnetic ions, water molecules experience additional spin-lattice relaxation and thus exhibit shorter T_1 in comparison to those distant from the paramagnetic centers. This feature is exploited in imaging using "contrast agents" typically composed of chelated paramagnetic ions. There is a rapidly growing body of literature demonstrating the clinical effectiveness of paramagnetic contrast agents. The capacity to differentiate regions/tissues that may be magnetically similar but histologically distinct is a major impetus for the preparation of these agents (Jackels, 1990). Local differences in relaxation times (T_1 and/or T_2) can be "translated" into image contrast by use of the appropriate MR imaging protocol. For example, regions associated with a gadolinium (Gd^{3+}) ion (nearby water molecules) appear bright in an MR image whereas the normal aqueous solution appears as dark background if T_1-weighted imaging schemes are used. Inversion recovery pulse sequence (Fig. 5) can be used in combination with an SE sequence for this purpose.

Table 1 Proton Relaxation Times in the Adult Human Brain

Tissue	T_1 (ms)	T_2 (ms)
White matter	573 ± 35[a]	
	633 ± 8[b]	
	756[c]	77.4[c]
	939[d]	
Frontal white matter	555 ± 17[e]	
Occipital white matter	588 ± 19[e]	
Gray matter	991 ± 57[a]	
	1148 ± 24[b]	
	1200[c]	79.9[c]
	1354[d]	
	1013 ± 62[e]	
Cerebrospinal fluid	2063 ± 125[a]	
	5127 ± 350[b]	
Ventricle	4282 ± 1562[e]	

[a] Measurements at 1.5 T (Bluml *et al.*, 1993).
[b] Measurements at 1.5 T (Henderson *et al.*, 1999).
[c] Measurements at 1.5 T (Whittall *et al.*, 1997).
[d] Measurements at 4 T (Kim *et al.*, 1994).
[e] Measurements at 1.5 T (Steen *et al.*, 1994).

The lanthanide atom Gd^{3+} has generally been chosen as the metal atom for contrast agents because it has seven unpaired electrons, a high magnetic moment ($\mu^2 = 63\ BM^2$), and a symmetric electronic ground state, (^8S). Transition metals such as high-spin Mn(II) and Fe(III) are also candidates due to their relatively large number of unpaired electrons and high magnetic moments. Once the appropriate metal has been chosen, a suitable ligand or chelate must be found to render the complex nontoxic. Two examples of such chelators are shown in Fig. 11: diethylenetriaminepentaacetic (DTPA) and the macrocyclic 1,4,7,10-tetraazacyclododecane-*N,N′,N″,N‴*-tetraacetic acid (Moi and Meares, 1988; Russell *et al.,* 1989; Meyer *et al.,* 1990; Runge and Gelblumm, 1991). Various features of the ligand can be manipulated to alter the physicochemical properties of the contrast agent—e.g., it may be covalently bonded to a high-molecular-weight polymer to make the agent cell membrane impermeant or attached to a lipid to direct the agent to the cell membrane. Gd-DTPA-dextran (MW 20 kDa) and Gd-DTPA-lipid are two agents currently in use in our laboratory. Contrast agents are finding increased applicability in imaging of the central nervous system and a few applications are illustrated later in this chapter.

Fully oxygenated blood is diamagnetic in contrast to deoxygenated blood, which is paramagnetic (Pauling and Coryell, 1936). Due to this, capillaries carrying deoxygenated

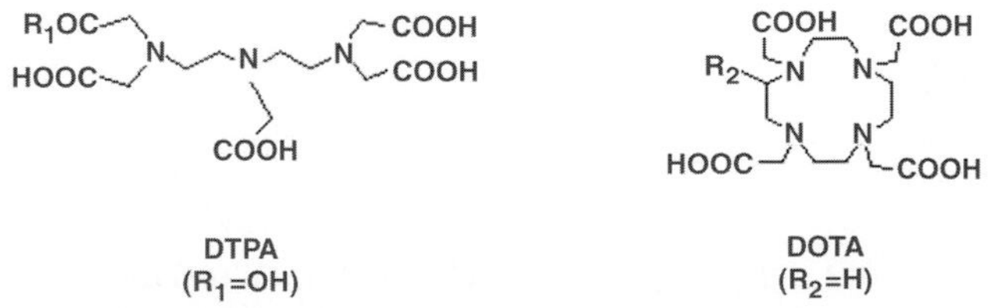

DTPA (R_1=OH)

DOTA (R_2=H)

Figure 11 Metal chelators diethylenetriaminepentaacetic (DTPA) and the macrocyclic 1,4,7,10-tetraazacyclododecane-*N,N′,N″,N‴*-tetraacetic acid used in MR contrast agents.

blood show additional features in brain images obtained with GE sequences. In the neighborhood of capillaries having deoxygenated blood, the induced magnetic field gives rise to magnetic field inhomogeneity. This inhomogeneity depends on oxygenation level, main magnetic field strength, capillary orientation with respect to the magnetic field, and the diameter of the capillary in comparison to the image voxel size. Water protons diffusing in and adjacent to capillaries carrying deoxygenated blood experience a shortening of T_2, and this gives rise to contrast in the GE image. This has been termed BOLD (blood oxygenation level dependent) contrast (Ogawa *et al.*, 1990a,b). The SE image is sensitive to T_2, while the GE image is sensitive to T_2^*, thus the BOLD effect is much more apparent in GE than in SE images. Because blood flow is also a parameter that can contribute to MR signal intensity changes in functional MRI (*f*MRI) experiments, it is common to compare GE images with SE images to separate the BOLD effects from flow effects. The new field of *f*MRI of the brain has been established on the basis of BOLD contrast changes in various regions of the brain following physiological stimulation. The BOLD approach to *f*MRI differs from the earlier dynamic *f*MRI, in which a bolus injection of a contrast agent is used (Belliveau *et al.*, 1991). Because *f*MR images must be obtained in very short periods of time, fast GE methods are employed. In this context, FLASH and GE versions of EPI have been used (Frahm *et al.*, 1993; Turner *et al.*, 1993). Excitation angles in FLASH are kept low, i.e., typically 10–50°. By combining the navigator echo approach with FLASH, it has been shown that motion-related signal fluctuations in *f*MRI can be reduced (Hu and Kim, 1994). The EPI technique is capable of producing images in the millisecond regime and is sensitive to T_2^* in its gradient echo version.

Yet another method for obtaining improved contrast in biological systems is the magnetization transfer (MT) imaging method. In this method the spins associated with less mobile macromolecules which give rise to a broad signal are irradiated from an RF source whose frequency is separated from that of the narrow signal arising from mobile protons. Owing to the presence of dipolar interaction, transfer of magnetization between these two pools of spins can take place, resulting in intensity changes of the mobile spins which are imaged (Balaban and Ceckler, 1992). The quantitative interpretation of MT data involves the various spin–lattice and spin–spin relaxation parameters and has been investigated by several workers (Henkelman *et al.*, 1993). MT contrast imaging has been used to enhance the gray–white matter contrast in brain.

VII. Diffusion-Weighted, Perfusion, and Water Displacement Imaging

Diffusion of nuclear spins in an inhomogeneous magnetic field can cause dephasing and contribute additionally to T_2, making it shorter (Carr and Purcell, 1954). Thus, the echo intensity observed in SE, as well as GE, methods is sensitive to T_1, T_2, and diffusion effects. MR brain images sensitive to diffusion effects, i.e., diffusion-weighted images, can be produced by the use of additional gradients in the image pulse sequence. Diffusion in a constant gradient, G, contributes an additional term, $\exp[-(\gamma^2 DG^2(\mathrm{TE})^3)/12]$, to the description of signal intensity in the SE scheme [Eq. (24)], i.e.,

$$S(\mathrm{SE}) = \rho(\mathrm{TR,TE},T_1)\exp[-\mathrm{TE}/T_2]\exp[-(\gamma^2 DG^2(\mathrm{TE})^3)/12]. \quad (25)$$

In the *pulsed* gradient spin echo (PGSE) MR imaging sequence (Fig. 12A) one introduces two pulsed gradients (one on either side of the 180° refocusing pulse) to obtain diffusion-weighted images (Stejskal, 1965; Stejskal and Tanner, 1965). The contribution to the loss of echo intensity from diffusion is now dependent on the term

$$\exp[-\gamma^2 G\delta^2(\Delta - \delta/3)D] = \exp(-bD^*), \quad (26)$$

instead of the last term in Eq. (25). G is the gradient pulse amplitude, δ its duration, and Δ the time between the leading edges of the two gradient pulses. The sequence shown in Fig. 12B utilizes the stimulated echo and pulsed gradients (PGSTE) to produce diffusion-weighted echo. An FSE sequence for diffusion-weighted imaging has been developed for use in μMRI (Beaulieu *et al.*, 1993). By applying G in different directions, one can probe the anisotropy of D^*. These experiments refer to D^*, the apparent diffusion coefficient, rather than D, the true diffusion constant, because we are dealing in general with restricted diffusion. Restricted diffusion may also be usefully employed to high-

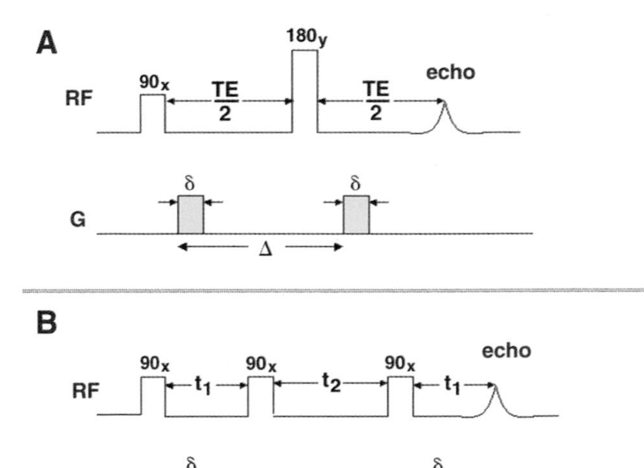

Figure 12 (**A**) Basic pulse gradient spin echo (PGSE) sequence. The width of the gradient pulse is δ and the separation between the leading edges of the gradient pulses is Δ. (**B**) PGSTE sequence is similar to the PGSE sequence, but utilizes stimulated echoes (STE). Phase and frequency encoding gradients are added to these sequences for diffusion-weighted imaging.

light boundaries and interfaces (Hyslop and Lauterbur, 1991; Putz *et al.*, 1992; Callaghan *et al.*, 1993). Reproducible D^* values can be obtained only if motional artifacts are suppressed. In this context, the use of the navigator echo technique has considerably improved the quality of diffusion-weighted brain images (Anderson and Gore, 1994; Ordidge *et al.*, 1994b; de Crespigny *et al.*, 1995; Mori and van Zijl, 1998; Williams *et al.*, 1999).

Referring to Eq. (26) it is clear that large b values have to be employed to obtain significant changes in the echo intensity in the case of spins with low D^* values. By employing higher b values and analyzing the diffusion-weighted images Clark and LeBihan (2000) have recently confirmed the biexponential nature of diffusion attenuation in the human brain. At low b values the diffusion data can be described by a monoexponential decay. However, with higher b values two different D^* values are needed to adequately fit the data. Apparently, one is dealing with two diffusive components—a fast one and a slow one. Studies on rat brain (Niendorf *et al.*, 1996; Assaf and Cohen, 1998) and human brain (Mulkern *et al.*, 1999) earlier pointed to this behavior. "Fast" and "slow" refers to the time scale with reference to the diffusion measurement time. The fast component is thought to be associated with spins in the extracellular region while the slow ones are ascribed to intracellular region.

A number of recent studies have pointed out the importance of diffusion-weighted MR imaging of the brain. Anisotropy in D^* values in the nerve and brain as well as spinal cord has been studied in detail in the hope that it reflects the underlying structural organization. Thus, diffusion along an axon may be expected to be faster than along a perpendicular direction. Diffusion, in general, is characterized by a diffusion tensor with six components. A complete understanding of the diffusion anisotropy requires determination of all the six components and entails application of diffusion gradients in various directions and subsequent processing of the diffusion-weighted image data (Basser and Pierpaoli, 1998). Quantitative and meaningful assessment of diffusion anisotropy is aided by ensuring that the parameters that describe the anisotropy are invariant to the choice of laboratory coordinate system (Basser and Pierpaoli, 1996). Ulug and van Zijl (1999) have described a simple approach that ensures orientational independence of the diffusion tensor anisotropy parameters. There has been much recent interest in this area from the point of view of fiber tract mapping in the brain (Mori *et al.*, 1999, 2001; Xue *et al.*, 1999) and other parts of the nervous system (Beaulieu and Allen, 1994; Pattany *et al.*, 1997; Ahrens *et al.*, 1998; Benveniste *et al.*, 1998; Fenyes and Narayana, 1999).

It has been pointed out that ~5700 capillary vessels traverse a cubic centimeter volume element in the brain cortex (Weiss *et al.*, 1982). Diffusion-sensitized MR images of the working brain are likely to be affected by perfusion defined as the microcirculation of the blood at the capillary network level. A detailed analysis of such perfusion effects on MR brain images has been made and it is possible to separate diffusion and perfusion contributions by examining images obtained from specially designed imaging sequences (Le Bihan *et al.*, 1988). This technique, which has been named the intravoxel incoherent motion method, has been applied to brain imaging. The EPI scheme has also been adopted for this purpose (Turner *et al.*, 1990). NMR methods have been developed for imaging perfusion. These can be broadly classified as steady-state and pulsed methods. Consider the case of a rat brain study of perfusion. In the steady-state or continuous method, inflowing spins in the neck region are continuously saturated and thus "tagged." These saturated spins exchange with bulk water in the brain. When a steady state is reached the regional concentration of saturated spins is governed by regional blood flow and regional T_1. The change in intensity in a slice of the brain image due to saturated spins can be quantitatively related to flow through the Bloch equations (Detre *et al.*, 1992; Zhang *et al.*, 1993). The water protons in the arterial blood are continuously saturated in the arterial spin labeling (ASL) technique and the image intensity studied (Williams *et al.*, 1992). Multislice versions of the ASL method have been studied (Alsop *et al.*, 1996; Alsop and Detre, 1998). In the pulsed ASL method (Edelman *et al.*, 1994) the magnetization in a chosen slab of arterial blood spins is inverted by an hyberbolic secant (sech) pulse (Silver *et al.*, 1985). The tagged spins subsequently cause image intensity change due to perfusion and thus can be quantified. The pulsed ASL technique is suitable for use with fast imaging sequences such as EPI. The pulsed ASL method has been employed in *f*MRI studies (Wong *et al.*, 1997; Silva and Kim, 1999).

Diffusion of water in biological systems can be studied with far superior resolution than normal imaging methods (i.e., k-space methods) by the use of the q-space imaging method. In the q-space method, the echo intensity E_Δ for a fixed Δ in the Stejskal–Tanner procedure is followed as a function of q where

$$q = (\gamma\delta g)/2\pi. \tag{27}$$

δ is the gradient pulse width and g is the gradient strength. The q values used here correspond to high b values [see Eq. (26)]. When water molecules are displaced on an average by a distance r we can relate the probability of displacement $P(r;\Delta)$ during the period Δ to E_Δ using a Fourier transform relation (Cory and Garroway, 1990; Callaghan, 1991):

$$E_\Delta(q) = \int \overline{P}_s(r,\Delta)\exp(i2\pi q \cdot r)dr. \tag{28}$$

By analyzing echo intensity decay as a function of q one can obtain displacement probability profiles. Displacements on the order of even a few micrometers can be detected by q-space techniques. Imaging based on such displacement profiles is referred to as displacement imaging. These techniques are increasingly applied to brain as well as optic nerve and spinal cord (King *et al.,* 1994; Assaf and Cohen, 2000; Assaf *et al.,* 2000). The existence of two types of diffusive components in the brain has been confirmed by *q*-space imaging of rat brain (Assaf and Cohen, 2000) and the slow component has been assigned to intra-axonal water spins. The implication of *q*-space imaging for fiber tract mapping is obvious.

VIII. Microscopic MRI

In this section we outline problems associated with microscopic-resolution MRI and some solutions available at present. Although typical resolutions employed clinically are on the order of a millimeter, the notion of using MRI at microscopic resolutions arose early in the development of this technique (Lauterbur, 1973). Physical limits to the spatial resolution obtainable with MRI were reviewed some years ago by House (1984) and have been discussed in detail recently by Callaghan (1991) and others (Cho *et al.,* 1988; Kuhn, 1990; Blumich and Kuhn, 1992; Zhou and Lauterbur, 1992). Spatial resolution in biological samples is typically limited by linewidth broadening (T_2 effects), diffusion, susceptibility artifacts, and SNR. The MR image resolution is directly related to the number of points across the image and the FOV. From Eq. (19), we see that if the gradient strength is increased, we have a smaller FOV. Likewise, if the sampling frequency is decreased, the FOV decreases. Thus, for microimaging, it would appear that we need large gradients and low sampling frequencies. Decreasing the sampling frequency is equivalent to reducing the bandwidth. Reducing the bandwidth leads to problems, because chemical-shift and susceptibility effects are exaggerated at lower bandwidths. For micro-MR imaging of the brain chemical-shift artifacts do not pose an insurmountable problem because the most often encountered chemically shifted proton signals are from water and fat, and the brain–fat signals are negligible. On the other hand, magnetic susceptibility changes across the brain are significant and often necessitate the use of large bandwidths and correspondingly large gradients. Rofe *et al.* (1995) have experimentally demonstrated the advantage of using large magnetic field gradients in a spin echo-based microimaging sequence to avoid susceptibility and diffusion artifacts. Sharp *et al.* (1993) have developed a spin-echo pulse sequence which they refer to as a line-narrowed 2D FT sequence that avoids distortion associated with the frequency-encoded (READ) direction, thus overcoming susceptibility and chemical-shift artifacts. The basic idea behind the scheme is to sample a series of 180° refocused echoes at their center, when all off-resonant spins are fully refocused. The technique was successfully applied (Bowtell *et al.,* 1995) to micro-MRI of single neurons isolated from abdominal ganglia of adult sea hares (*Aplysia californica*). Rapid sampling rates also allow the use of short echo times and this helps to reduce signal attenuation arising from diffusion and T_2 effects. Deterioration in resolution due to T_2 effects can be compensated for by judicious choice of experimental conditions (e.g., large bandwidth), which in turn influence the SNR of the resulting image (Kuhn, 1990; Callaghan, 1991).

Resolution in T_2*-weighted images can be seriously affected by susceptibility-induced gradients. In 2D imaging the slice selection refocusing gradient has to be properly chosen to compensate for the susceptibility-induced gradient (Frahm *et al.,* 1988; Yang *et al.,* 1998, 1999). Suitably tailored RF pulses (Cho and Ro, 1992; Chen *et al.,* 1998; Glover and Lai, 1998; Chen and Wyrwicz, 1999) can be employed to reduce susceptibility effects in GE imaging. It should be pointed out that several of the methods designed for reducing susceptibility artifacts are, in principle, capable of compensating chemical-shift artifacts also (Weis *et al.,* 1996). 3D GE data acquired in the presence of field inhomogeneity can be processed to yield artifact-free 2D images. (Haacke *et al.,* 1989; Reichenbach *et al.,* 1997). Although time consuming, phase encoding in all three dimensions can be a solution to artifacts arising from frequency shift in the READ direction (Callaghan *et al.,* 1994; Rofe *et al.,* 1995).

Molecular diffusion has a number of effects in the MRI experiment. Diffusion lengths in micro-MRI become comparable to the voxel sizes, and the signals cannot be uniquely ascribed as arising from a given voxel alone. Estimates of the *theoretical* limits of resolution in the MR image arising from these phenomena range from 2 to 0.5 μm (Cho *et al.,* 1988; Kuhn, 1990; Callaghan, 1991). The *practical* spatial resolution is currently determined by SNR that is often limited by the amount of time available to actually acquire the image (i.e., the temporal resolution). For a cubic voxel having side dimension *d,* the imaging time, *t,* is proportional to the square of the desired SNR (Mansfield and Morris, 1982):

$$t \propto (\mathrm{SNR})^2 r^2 (T_1/T_2) f^{-7/2} d^{-6}. \quad (28)$$

f is the spectrometer frequency, and *r* is the radius of the solenoidal RF coil in which the sample is placed. For example, a reasonable SNR clinical MR image can be obtained in about 5 min with a voxel (volume element) size of 1 mm^3. We are interested in voxels on the order of 1–10 μm on a side. Because most of the proton MR signal arises from water in biological samples and water concentration is roughly constant, the SNR change in the image will be proportional to the volume change: a factor of 10^{-8}. Of course, this is true only if all experimental parameters are the same. The challenge in MRI microscopy is to optimize the experimental setup (hardware and software) to overcome the poor intrinsic SNR in order to obtain a respectable image in a reasonable amount of time (Narasimhan *et al.,* 1994).

There are a number of ways of recovering this signal loss. Increasing the main magnetic field strength from 0.2 to 12 T provides a factor of 10^2 improvement. Optimizing the RF coil for small samples can be expected to give another factor of 10^2 to 10^3. Another factor of 10 can be gained by employing three-dimensional volume imaging rather than two-dimensional slice imaging. Application of so-called "fast imaging" pulse sequences (e.g., DEFT, GRASS, FLASH, RARE) can significantly decrease the amount of time necessary to acquire an image, thus providing an equivalent SNR increase on the order of a factor of 10. The smaller sample sizes employed in microscopic scale imaging afford a decreased sample-induced noise in the RF receiver coil that can contribute another factor of 10 to the SNR gain. Combining all of these estimates of SNR enhancements, we arrive at a factor of 10^8 possible improvement in SNR. These estimates are substantiated by MR imaging experiments with spatial resolutions of 10 μm or less that have been achieved by several groups working at field strengths ranging from 4.7 to 12 T (Blumich and Kuhn, 1992). In each of the MRI microscopy experiments reviewed by Blumich and Kuhn some or all of the items leading to SNR enhancement are optimized, employed, and/or modified in order to gain SNR or modulate image contrast. In recent MR microscopy work (Lee *et al.*, 2001) a resolution of 2 μm in a 200-μm voxel volume has been achieved with a geranium leaf stem at 14.1 T. The gradient system employed was capable of a gradient strength of over 1000 G/cm. This work points to submicrometer resolution as a realizable goal for MR microscopy. Submicrometer scale is compatible with biological cell dimensions.

Microscopic MRI techniques are being employed to examine a number of biological systems. Small rodents are often-used model systems for studies in mammalian biochemistry, anatomy, and physiology. Due to their relatively small size they are also prime candidates for high-resolution MR imaging studies. Johnson and co-workers have pioneered the use of *in vivo* MRI microscopy to study the anatomy and pathology of intact individual organs within the larger biological environment (Hedlund *et al.*, 1986a,b; Hollett *et al.*, 1987; Dixon *et al.*, 1988). They obtained detailed images of the rat kidney, thorax, and abdomen using relatively low-field (1.5 T) instrumentation and specially built RF coils. Blurring of the MR image due to motion is a problem that must be addressed in live animal studies. As mentioned earlier, one method is to synchronize the MRI experiment with breathing and/or heartbeat of the animal via physiological monitoring and/or intervention. Recent images of the rat lung show that this can be accomplished quite successfully (Gewalt *et al.*, 1993). In rat brain imaging anesthesia and a cradle to hold the animal's head stationary suffice. Johnson and co-workers have shown that gray–white matter distinctions and many anatomical structures are clearly visible in high-resolution MR images of the rat brain (Johnson *et al.*, 1987). Studies at higher field strengths necessitate miniaturization of some of the physiological monitoring/intervention apparatus because of the much smaller bore size of higher field magnets (e.g., 89-mm diameter for fields greater than ~8.5 T versus 30 cm to 1 m for lower fields). Using an 8.5-T system, Jelinski and co-workers developed a stroboscopic micro-MRI technique that they used to monitor changes in the cross-sectional area of the carotid artery of the rat as a function of the heart cycle and blood pressure changes induced by the vasoconstrictor phenylephrine (Behling *et al.*, 1989). Micro-MR imaging of rodent eyes and xenographs at 9.4 T provides information that correlates well with subsequent histological examinations of the same tissues. Aguayo and co-workers (Aguayo *et al.*, 1987) found that they could resolve cell clusters and structures as small as basement membranes. It is well known that spin relaxation rates and susceptibility effects in imaging change with the strength of the main magnetic field (Callaghan, 1991). Rosen and co-workers (Kennedy *et al.*, 1989) obtained high-precision neuroanatomic information from images of excised rat brains at 9.4 T using a spin-echo imaging protocol. They note that at this high field strength the underlying contrast arises principally from proton density variations rather than relaxation effects. In contrast, local magnetic susceptibility and diffusion differences appear to drive the contrast seen in very high resolution (4.5 μm) images of plant tissue by Mansfield's group using an 11.7-T system (Bowtell *et al.*, 1990). Despite limitations imposed on resolution in MR microscopy by sampling rate, T_2, magnetic susceptibility, and diffusion processes it may still be feasible to identify features in the MR image corresponding to objects whose dimensions are smaller than the MR resolution limit. The basis for this expectation lies in the interplay among the different processes that contribute to the point-spread function of the resonance line. Callaghan and co-workers (Callaghan *et al.*, 1994) observed that small dust particles in a capillary tube of water gave rise to bright spots in the MR image, presumably because of an interaction between local susceptibility-induced gradients and the applied magnetic field gradients.

Contrast in MR images is a valuable key to tissue discrimination and differentiation of pathology. In addition to the well-known contributors to contrast in MR imaging such as T_1, T_2, T_2*, spin density, susceptibility, and diffusion a new form of contrast which depends on intermolecular spin coherence from distant dipoles has been recently identified by Warren and co-workers (1998). Averaging of the dipole–dipole interaction between spins located on different molecules in a fluid is limited to distances which are on the order of diffusion distance. Spins located farther away produce a distant dipolar field which is not averaged out. By suitable pulse sequences and choice of gradients one can detect multiple quantum coherence (MQC) of spins lying within a chosen distance of each other. This important aspect of the distance dependence of MQCs can be exploited for image contrast. MQCs can be converted by suitable RF pulses to observable single-quantum (SQ) coherences and thus employed for imaging. In principle, intermolecular zero-quantum, double-quantum, and MQ coherences (i-ZQCs, i-DQCs, and i-MQCs) can be created. However, higher order coherences lead to signals that are very weak especially from the imaging point of view. High fields and abundance of proton spins facilitate the generation of i-MQCs. For instance, the intensity of images generated from i-ZQCs and i-DQCs is only about 10–15% of that of conventional SQ-generated images in the mouse brain at 11.7 T. i-ZQC images of the rat brain at 9.4 T (Warren *et al.,* 1998) and the human brain at 4 T (Rizi *et al.,* 2000) have been obtained. i-DQC images of the human brain have been obtained at 1.5 T (Zhong *et al.,* 2000a,b). The feasibility of 3D MR microscopy with i-ZQCs and i-DQCs has been demonstrated in the mouse brain at 11.7 T (Narasimhan *et al.,* 2000). The basis for i-MQC contrast being different from those of other sources, i-MQC imaging is expected to play a significant role in areas such as tumor imaging and *f*MRI, among others.

IX. Micromagnetic Resonance Imaging of the Nervous System

A. Overview

One approach to describing how a brain is put together is the creation of atlases of development. These atlases typically take the form of a series of photomicrographs of histologically processed and thin-sectioned specimens taken at a series of time points during development. While a wealth of data is presented, three-dimensional correlations are often obscure and the processing (fixing, staining, sectioning, etc.) introduces variations and possible distortions from the *in vivo* situation (Theiler, 1989; Paxinos *et al.,* 1991; Kaufman, 1992; Schambra *et al.,* 1992). MR imaging offers an alternative approach with a number of advantages. MR imaging can provide three-dimensional images of thick opaque samples in a noninvasive manner, thus obviating the need to sacrifice/fix/stain/section a specimen in order to image it. It is straightforward to analyze the time course of structure and pattern formation by repeated MR imaging of the same animal as it develops (Jacobs and Fraser, 1994). Moreover, all the slices in an MR image are automatically "in register." Any arbitrarily oriented slice can be taken from the 3D images and the images are inherently "stereotactic" in the sense that distances, areas, volumes, angular orientations, etc., are all readily available from the data.

B. Frog Embryo

Figures 13–24 illustrate some MRI microscopy work carried out at high magnetic field strengths. We first describe a cell lineage analysis of frog (*Xenopus laevis*) embryos using a MRI contrast enhancement agent as a cell lineage tracer. Micro-MR imaging is used to follow the first 4 days of development in the frog (Jacobs and Fraser, 1994). An example of a classic host/graft experiment in which two embryos with labeled grafts are followed for 2 days illustrates the great utility of being able to perform longitudinal development studies on the same animal. We then show *in vivo* images of a neonatal mouse pup and *in utero* images of live mouse embryos. Moving up the evolutionary ladder, representative high-resolution images of newborn dwarf and mouse lemur cadavers are presented. These three-dimensional images of small vertebrates have excellent signal to noise and contrast, demonstrating the feasibility of using MRI at high fields to gain neuroanatomical information in a noninvasive manner.

The descendants of individual precursors in the intact embryo can be labeled by microinjection of a stable, nontoxic, membrane-impermeable MRI lineage tracer. Because a complete time series of high-resolution three-dimensional MR images can be analyzed forward or backward in time, one can fully reconstruct the cell divisions and cell movements responsible for any particular descendant(s). Unlike previous methods, in which labeled cells are identified at the termination of the experiment, this technique allows the full kinship relationships of a clone to be determined as the clone expands. To perform cell lineage analyses, a lineage tracer with a large effect on the NMR signal is required. For such an MRI contrast agent to be a good lineage tracer it must meet three criteria: (1) induce a local signal that is characteristically different from that of the rest of the sample, (2) be physiologically inert, and (3) be membrane impermeable, thus remaining within the originally labeled cell and its progeny. DTPA-Gd covalently bound to dextran satisfies these criteria (Gibby, 1988; Gibby *et al.,* 1989). The Gd-DTPA-dextran tracer is an efficient T_1 relaxation agent (Lee, 1991). Thus, our T_1-weighted imaging protocol provides MR images with enhanced intensity in those cells containing the lineage tracer. Because this tracer is a close analog of fluorescent dextran lineage tracers already in use (Bronner and Fraser, 1988), established techniques are utilized for its injection into embryonic cells. Unlike the fluorescent probes that bleach and generate reactive by-products when observed, neither the MRI contrast agent nor the surrounding cytoplasm is perturbed chemically by the imaging experiment.

We show images at 12 μm resolution using a 7-T system with the RF coil, gradient framework, imaging protocol, and sample preparation optimized for *in vivo* micrometer scale imaging (Cho *et al.,* 1988). A single blastomere in the 16-cell embryo was injected with the Gd-DTPA-dextran tracer and the embryo was imaged repeatedly over several days (Jacobs and Fraser, 1994). Figure 13 presents a series of such images in which the label was introduced into a blastomere in the animal hemisphere adjacent to the prospective dorsal midline [blastomere DA in the Wetts and Fraser

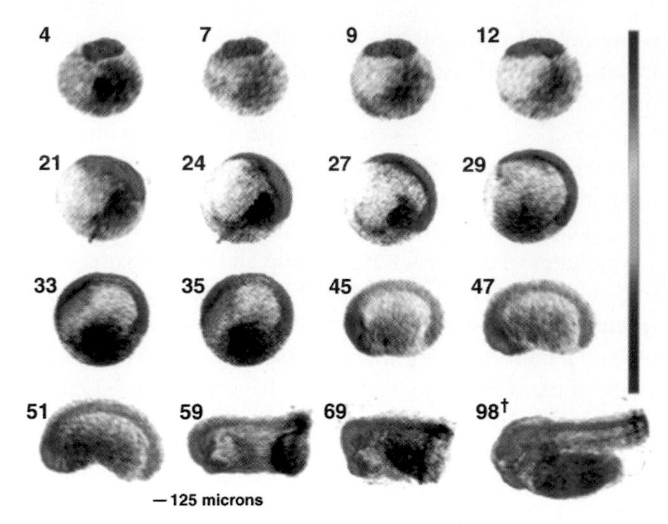

Figure 13 Magnetic resonance images of a single developing frog embryo taken at 16 successive times, noted as hours after fertilization. Descendants of a single 16-cell blastomere, injected with the contrast agent 2 h before the first image was recorded, appear as high-intensity volumes (yellow-red). Gastrulation has begun by hour 24. The embryo passes from an early neurula to the early tailbud stage from hour 45 to hour 98. For visualization, we used VoxelView (Vital Images, Inc., Fairfield, IA). The bottom of the color scale bar indicates low intensity and the top high intensity. To permit the labeled cells to be visualized among the unlabeled cells, the opacity of each voxel was adjusted so that higher voxel intensity values have exponentially higher opacity values. Thus cavities are rendered transparent, the unlabeled cells semitransparent, and labeled cells opaque. This display routine is sufficient to display the majority of the labeled cells except for hours 45–69, at which the thickness of the somite tissue obscures the display of the labeled clone in the neural tube. The bar in the lower left represents 125 μm.

nomenclature (Wetts *et al.,* 1989)] known to give rise to neural tissue in the mature animal. Labeled cells appear as the high-intensity yellow-red volumes in the volume representations. The images show the progression from early cleavage and blastula stages (Fig. 13, top row), through gastrulation (Fig. 13, second row), neurulation (Fig. 13, third row), and tailbud stages (Fig. 13, bottom row).

The Spemann organizer is the prime embryonic region responsible for organizing the body axis of the *X. laevis* embryo (Weinstein and Hemmati-Brivanlou, 1999; Gamse and Sive, 2000; Keller *et al.,* 2000; Lane and Sheets, 2000; Gerhart, 2001; Niehrs, 2001) that occupies the dorsal marginal zone of the embryo at the onset of gastrulation at stage 10 (Nieuwkoop and Faber, 1967; Keller, 1991). Cell labeling experiments have suggested that the region responsible for head induction is initially situated in the center of the blastocoel floor at blastula stages and comes in contact with the ectoderm at the onset of gastrulation. However, none of these morphogenetic events have been observed directly *in vivo*. Because the embryo is almost completely opaque, no direct observation of cell behavior by conventional light microscopy methods is possible and thus most observations rely on interpretations of either tissue explants or the analysis of fixed specimens. Intrinsic image contrast and resolution allow us to identify of a number of structures (blastocoel, archenteron, and blastopore). Even in labeled embryos, animal and vegetal tissue can be distinguished based on their different water and yolk-fat content. In addition, single cells can be labeled by microinjection of either a Gd^{3+}-based or a magnetite-based contrast agent. The dynamics of the clone movement can then be followed in 3D time-lapse series within the context of the whole embryo. Figure 14 shows a comparison of fixed stained x. laevis embryo section and MR images at same stage in which the C1 and C4 blastomeres had been labeled with a T_1 contrast agent at the 32-cell stage. Figure 15 shows a time series of images of development beginning just before gastrulation and progressing to near the end of gastrulation. By labeling the C1 blastomere (fated to form most of the Spemann organizer and axial mesoderm later in development) together with the C4 blastomere, these imaging experiments allow identification of the relationship of the Spemann organizer tissue with ectodermal and endodermal embryonic cells during blastula and gastrula stages.

The images of developing systems shown here point out both the present limitations and the future potential of magnetic resonance imaging microscopy for visualizing the developing CNS. MRI permits structures within the living embryo, usually inaccessible to light microscopy, to be imaged clearly and nondestructively over a period of days. In light microscopy of sectioned embryos, the 3D images must be synthesized from the 2D images, requiring careful alignments of adjacent slices and corrections for distortions from fixation and sectioning. MRI generates a true 3D image with all the data points automatically in register. One may take any arbitrarily oriented 2D "slice" through the 3D image that best allows a detailed examination of the region of interest. These advantages are not without a cost; compared to optical images of histological sections, it is clear that MRI has a smaller signal-to-noise ratio and less resolution. At later stages of development in the frog, the cells become too small to be resolved clearly in the present MR images. Of course, resolution should not be confused with the ability of an imaging technique to detect the presence of an entity smaller than the theoretical spatial resolution of the technique. In fluorescence microscopy, a small highly fluorescent feature will manifest itself as a bright pixel, even when the feature is smaller than the resolution of the light microscope. Therefore, one knows of the existence of the feature and its location to within ±1 pixel. The same is true of MRI as it can readily detect features smaller than the MRI spatial resolution, if the feature(s) of interest is uniquely labeled.

Intrinsic contrast arising from natural variations in local water properties is also readily apparent in MR images of the frog. Figure 16 shows a stage 49 embryo (approximately 1 week after fertilization). All neural tissues appear relatively

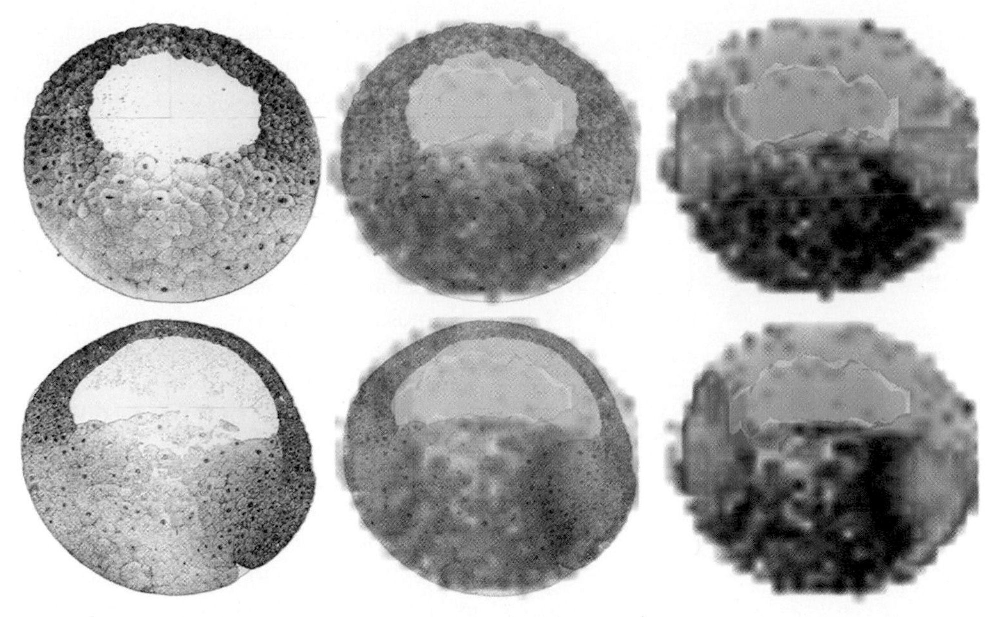

Figure 14 Comparison of histological slices and *in vivo* MR images of early frog embryo. Top series is stage 9 and bottom series is stage 10 (Nieuwkoop and Faber, 1967; Keller, 1991). Left images show histological slices (top is animal pole, bottom is vegetal pole), to the far right side are MR images at the same stage, middle images are overlays of histology and MRI. Yellow cells are progeny of 32-cell stage blastomere C1 (right side of embryo) and C4 (left side of embryo) containing MRI contrast agent introduced at the 32-cell stage (stage 6 after Nieuwkoop and Faber, 1967). Blastomeres C1 and C4 were injected with a high-molecular-weight polyglutamine-based Gd^{2+} contrast agent. Labeled cells were identified using a *seed fill* algorithm and detected regions were mapped to yellow. The threshold level was determined by examining the intensity values of comparable tissue regions in the same embryo that contained no label. Conservative estimates of the threshold ensure that the actual clone region will not be overestimated. The *seed fill* function detects only connected voxels, thus cells that are not part of the main clone are not detected. Experimental parameters: 3D spin-echo protocol, TR/TE = 200/9 ms, 2 averages, data size $128 \times 64 \times 64$ (FOV = $500 \times 250 \times 250$ μm), fat was suppressed. Figure courtesy of Dr. Cyrus Papan.

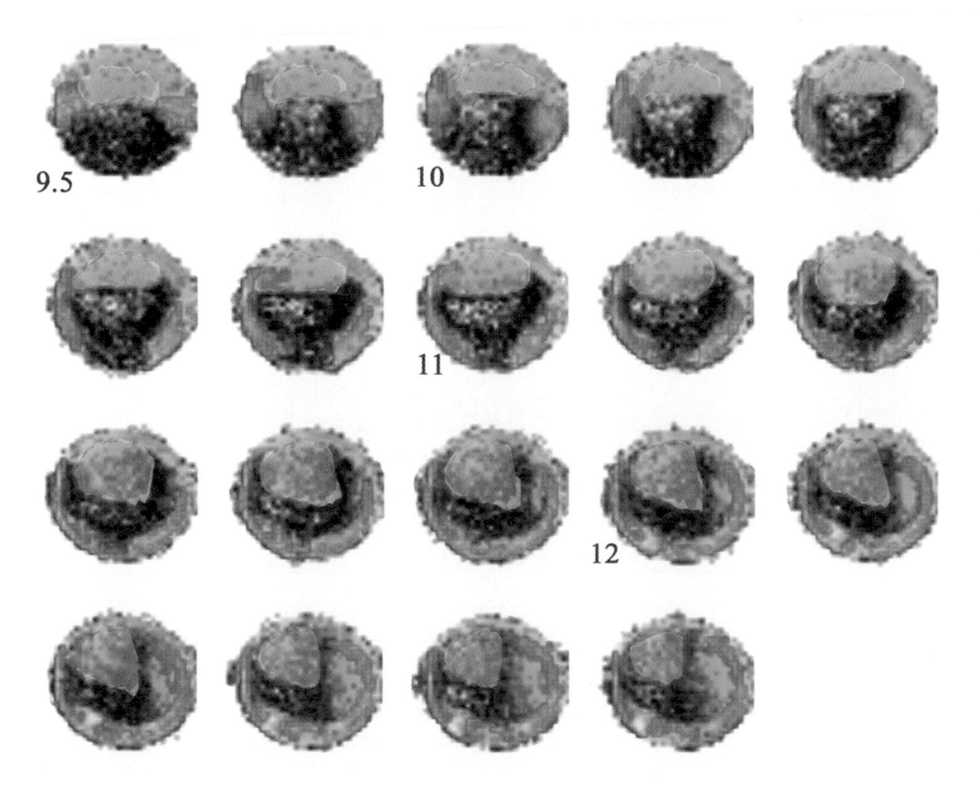

Figure 15 Annotated time series of 3D images showing morphometric movements of two clonal populations (32-cell-stage C1 and C2 blastomeres) during *X. laevis* gastrulation. Stages are noted. Time per image was 55 min and each subsequent image was begun immediately following the previous one. See Fig. 14 for details. Figure courtesy of Dr. Cyrus Papan.

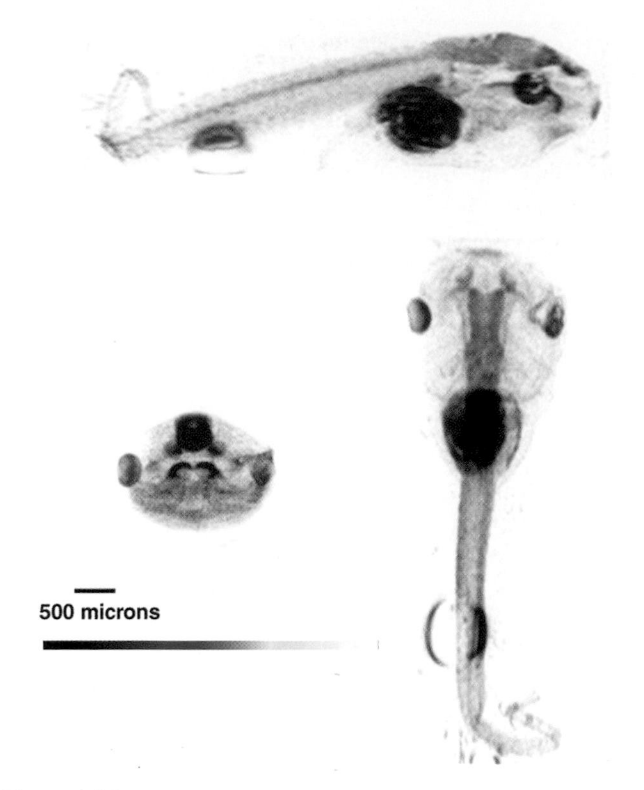

Figure 16 Three views of a MR image of a stage 49 *X. laevis* tadpole showing intrinsic contrast. Susceptibility artifacts from an air bubble near the tail of the animal are quite dramatic.

darker than surrounding tissue. Eyes, nasal cavities, fore/mid/hindbrain segmentation, neural tube, somites, and gut are all apparent. The neural tube is more obvious in the side view as it is thick from that viewpoint, whereas the somites that flank the neural tube are thicker in the dorsal–ventral direction and thus are better revealed in the dorsal view.

C. Mouse

A week-old mouse pup was anesthetized and placed in a 15-mm-i.d. birdcage-type RF coil (Hayes *et al.,* 1985). Temperature was maintained at 30°C by a constant flow of thermostated air passing over the animal. No physiological monitoring (other than visual periodic checking of heartbeat and respiration) was performed, although motion artifacts at the level of the heart and lungs were a constant indication of the continued viability of the preparation. The animal spent a total of ~3 h in the instrument. After the experiments were completed the pup was returned to the mother and monitored over the next 2 days without obvious ill effect. A 3D SE protocol (TR/TE 100/30 ms) was employed with a data set size of 256 × 128 × 128 (field of view 25.6 × 12.8 × 12.8 mm). The data were zero-filled before processing to yield a 3D image with voxels of 50 μm on a side. Figure 17 shows a slice through the 3D image indicating the feasibility of obtaining *in vivo* MR images of mouse neonates to monitor brain development.

A complete atlas of brain development would include 3D images of an individual from conception through adulthood. Gene expression patterns, receptor domains, arrays of innervation in the developing nervous system, cell lineage patterns, and a host of other types of biological processes in embryonic and adult animals take place in three spatial and one temporal dimension. They occur within the context of

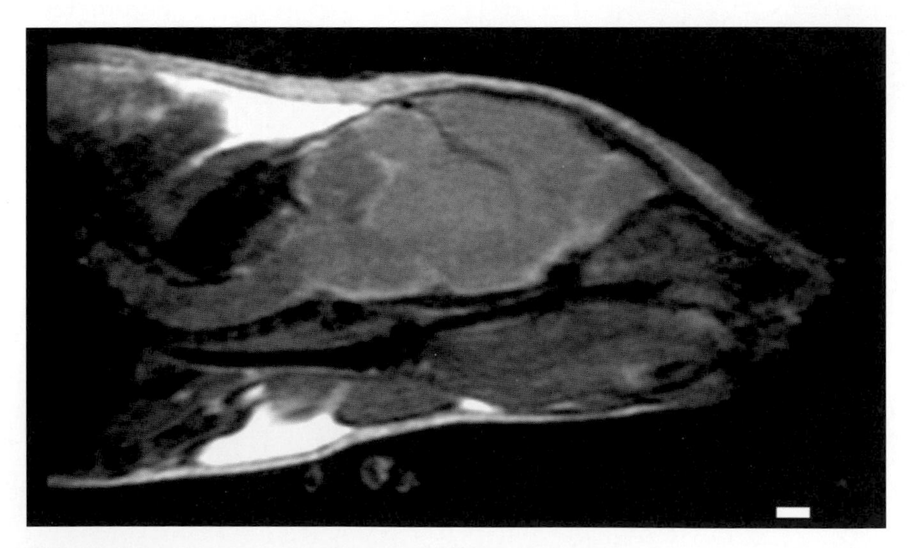

Figure 17 A midsagittal slice through the 3D image of a mouse neonate. Various major subdivisions of the brain as well as dorsal and ventral high-intensity fat deposits are apparent. A 3D SE protocol (TR/TE 100/30 ms) was employed with a data set size of 256 × 128 × 128 (field of view 125.6 × 12.8 × 12.8 mm). The data were zero-filled before processing to yield a 3D image with voxels of 50 μm on a side. Images of this type confirm the feasibility of obtaining *in vivo* MR images of mouse neonates at high resolution. No attempt was made to optimize imaging parameters (i.e., TR or TE).

anatomy of the specific sample being examined. Digital atlases provide a means to put such specific data within the context of normal specimen anatomy, analyze the information in three (or more) dimensions, and examine relationships between different types of information. Initial efforts at using MR imaging to provide the raw data for an anatomical atlas of mouse development are illustrated in Fig. 18 (Dhenain *et al.,* 2001). With this in mind we have initiated attempts at *in utero* MR imaging in the mouse. A 12-day pregnant female is shown in Fig. 19 (top right) alongside a 25-mm Alderman-Grant-type RF coil (Alderman and Grant, 1979) used in the imaging experiments. Figure 19 (top left) shows a portion of one slice taken parallel to the long axis of the female. Four mouse embryos are visible, three in almost coronal sections. Ventricles in the developing brain appear at relatively high intensity. Figure 19 (bottom) shows 19 contiguous slices through one of the embryos. The lower right depicts slice orientation and location in a schematic of the day 12 embryo. The rudimentary brain is visible in frames b–h and convolutions in the snout in frames f–j. Three digits of the right paw are apparent in frame i. In exploring different contrast mechanisms we and many others are employing more esoteric imaging sequences in small-animal microimaging studies. Figure 20 shows an i-DQC slice image of an excised mouse brain taken from a 3D data set. Diffusion tensor imaging can also be used to gain information about the developing nervous system (see Fig. 21). While many of these images are at relatively low resolution and high noise, they have good contrast and illustrate the feasibility of the proposed conception-to-adulthood MRI atlas.

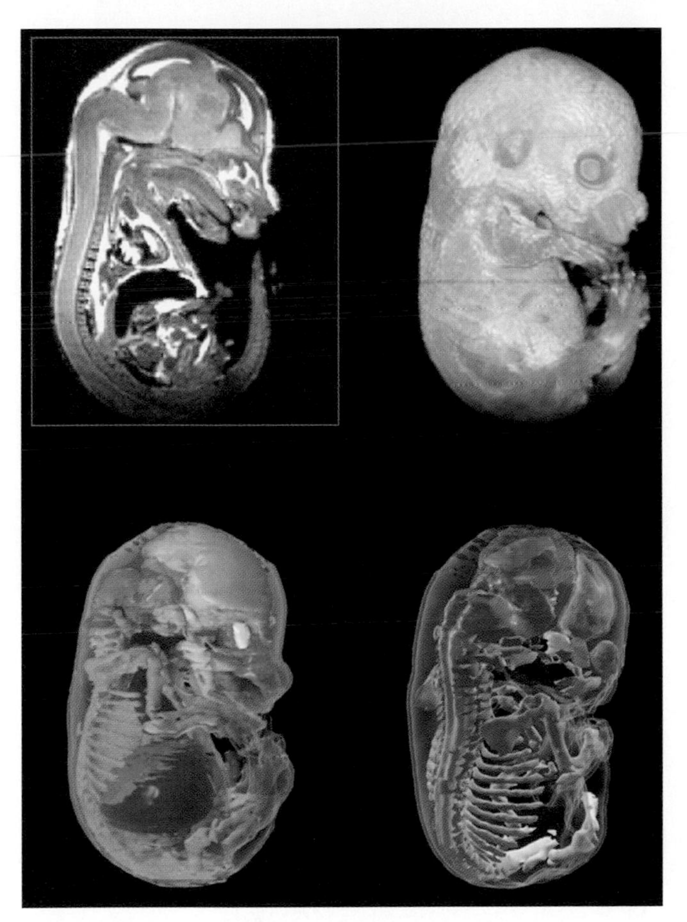

Figure 18 *In vitro* MR images and derived anatomical models of fixed dpc 15 C57/BL6 mouse embryo. A single slice from the 3D spin echo data set is shown in the upper left and surface rendering is shown in the upper right. Anatomical features in the MR slices were outlined by hand, color coded, and converted to surfaces (Dhenain *et al.,* 2001). Lower images show semitransparent volume renderings of anatomical model surfaces. Figure courtesy of Dr. Seth Ruffins.

Recent studies clearly indicate the potential usefulness of Mn^{2+} as a neuronal tract tracer. Three separate studies in two systems (fish and rodent) report that radioactive manganese ion is taken up and transported by neurons (Sloot and Gramsbergen, 1994; Tjalve *et al.,* 1995, 1996). Sloot and Gramsbergen showed that radioactive Mn^{2+} injected into the basal ganglia of rats would migrate to other brain regions (Sloot and Gramsbergen, 1994). Upon administration of colchicine, transport of Mn^{2+} was blocked, indicating that the transport was dependent upon microtubules. Because Mn^{2+} is paramagnetic and an excellent MRI contrast agent its tissue location can be visualized as areas of positive contrast enhancement in T_1-weighted MR images (Geraldes *et al.,* 1986; Fornasiero *et al.,* 1987). Manganese-enhanced MRI (MEMRI) has been used to trace olfactory and visual pathways in mice (Pautler *et al.,* 1998) and the frontal cortical connections in the rhesus macaque (Pautler, 1999; Pautler *et al.,* 1999) and to characterize regional differences in a subset of connections in the olfactory bulbs of Dickie's small eyes (*Sey*) mice and the primary olfactory cortex in the mutant *Reeler* mouse (Pautler, 1999; Pautler *et al.,* 1999). The current model of the mechanism of Mn^{2+} tract tracing is that the Mn^{2+} ions enter neurons through voltage-gated Ca^{2+} channels. Upon entry into the activated neuron, Mn^{2+} is transported along microtubules, most likely in vesicles, where some of it is released at the synaptic terminal. *In vivo* and *in vitro* biochemical data support this model (Narita *et al.,* 1990; Lucaciu *et al.,* 1997; Pautler *et al.,* 1998; Takeda *et al.,* 1998; Pautler, 1999). Furthermore, Mn^{2+} washes out of tissues after several days as evidenced by a return of MRI signal to basal levels (Pautler *et al.,* 1998). This opens up the possibility of using the same animal repeatedly in developmental and time-series studies. Figure 22 shows how focal stereotaxic injection of small amounts of Mn^{2+} in conjunction with MEMRI can highlight active neuronal tracts in the mouse. Comparison of such images in various mice (e.g., strains, mutants, transgenics) and in mice subjected to different experimental protocols (e.g., fear conditioning, administration of drugs of abuse) will illuminate neuronal circuitry associated with these complex differences.

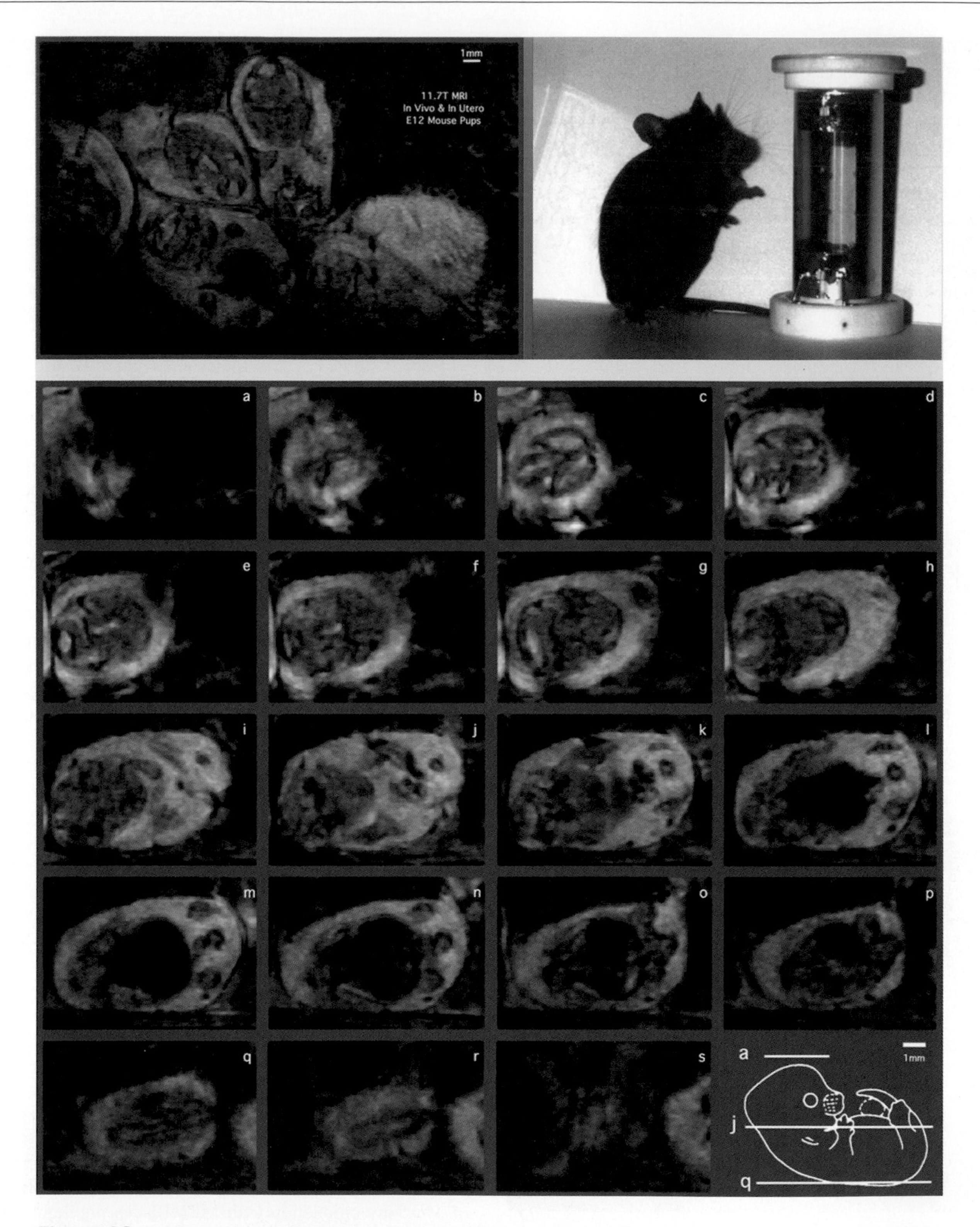

Figure 19 A 12-day pregnant female mouse is shown in the top right alongside the RF coil used to obtain the MR images shown in the bottom. The top left shows a portion of one slice taken parallel to the long axis of the female. Four mouse embryos are visible, three in almost coronal sections. Ventricles in the developing brain appear at relatively high intensity. The bottom shows 19 contiguous slices through another of the embryos. The lower right depicts slice orientation and location in a schematic of the day 12 embryo. The rudimentary brain is visible in frames b–h and convolutions in the snout in frames f–j. Three digits of the right paw are apparent in frame i. While these images are at relatively low resolution and high noise, they have good contrast and illustrate the feasibility of the proposed conception-to-adulthood MRI atlas.

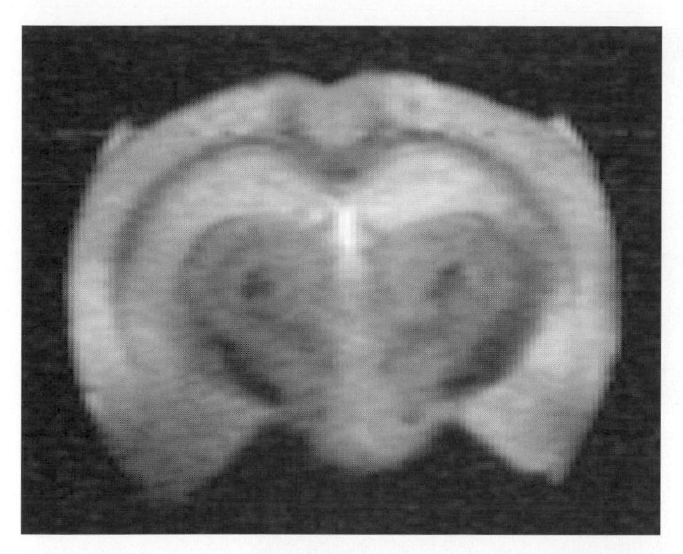

Figure 20 i-DQC image (x–y plane) of an excised mouse brain at 11.7 T. The pulse sequence 90°-$G1$-$t1$-90°-$G2$-$t2$-180°-TE/2-180°-TE/4-ACQ was employed with appropriate phase cycling of RF pulses. The two coherence gradients $G1$ and $G2$ in the ratio 1:2 are applied along the main field direction (z). For this image $G1$ = 2.8 G/cm and $G2$ = 5.6 G/cm. Field of view is 1.5 × 1.5 cm, the in-plane resolution is 98 μm, and the slice thickness along z is 390 μm.

D. Mouse Lemur

The adult gray mouse lemur (*Microcebus murinus*) has a body length of about 12 cm and weight of approximately 60 g. The small size of this prosimian makes it ideal for MRI microscopic studies of the developing brain. The images shown here are images of neonate cadavers and a live anesthetized adult. A 3D SE protocol was used to image the cadaver sample and 3D RARE protocol for live animal imaging. Two slices through the data set shown in Fig. 23 show a wealth of anatomical features, ranging from the lamination of cells in the developing cortical plate to the distinct morphology of the dentate gyrus of the hippocampus. Features can be extracted from this data set, as shown by the volume rendering of the inner ear (white inset). When viewed at full scale, fine structures such as the basilar membrane are clearly rendered.. The complete mouse lemur neonate data set has been published in CD-ROM format (Ghosh *et al.,* 1994). It includes several "movies" through the data, rendered slices in three orthogonal directions, and the "raw" integer 3D data. Data for Figure 24 was recorded in 74 min with data acquisition synchronized to the animal's respiration. Image quality is significantly diminished without respiratory synchronization. As with the *in vitro* images, significant anatomical detail is apparent in these T_2-weighted images of this primitive primate brain.

X. Concluding Remarks

Images are descriptive. They directly address questions of a *what, where,* or *when* nature. Answers to these types of questions can offer insight into the more fundamental *how* types of questions. Understanding the mechanisms that give rise to what is observed in the pictures is the goal of the MR imaging described here. The nervous system is a complex

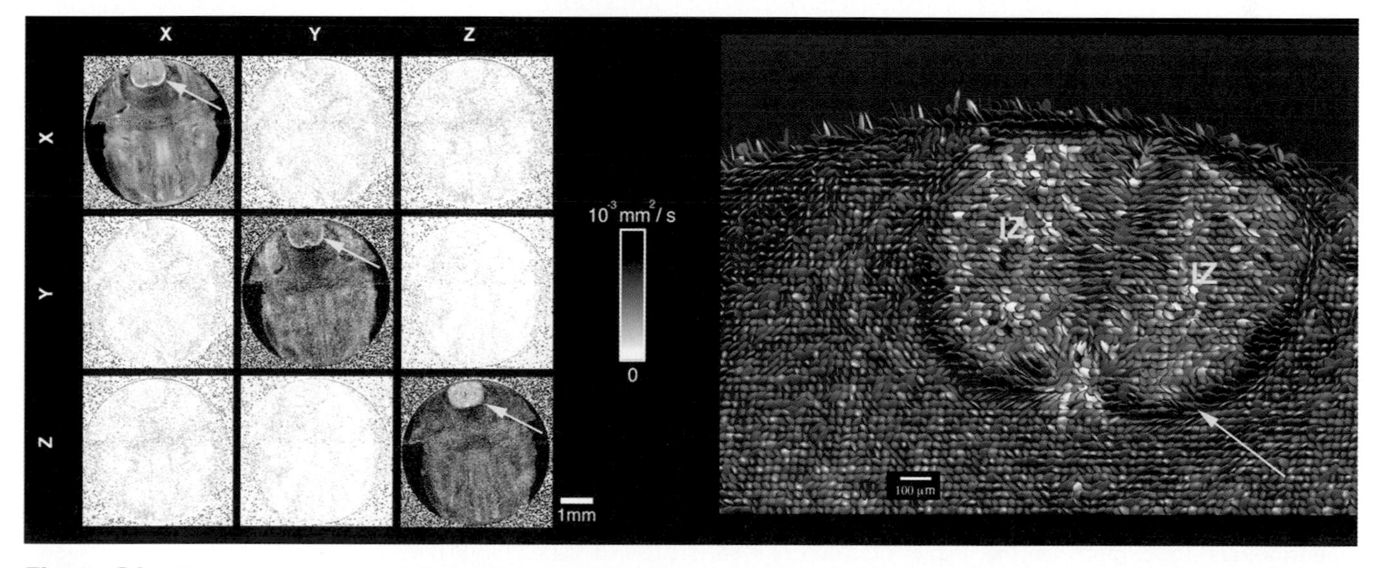

Figure 21 Diffusion tensor images for a single 2D slice through the posterior hindbrain region of a fixed 12.5-day mouse embryo. The left shows elements of the diffusion tensor. In this gray-scale image, the larger the pixel intensity, the larger the diffusion coefficient. *Note.* Among the diagonal elements, there is considerable direction anisotropy in the diffusion coefficient. The right shows the data represented as ellipsoids of diffusion scaled to have the same principal axis length. Arrows point to the marginal zone, and the intermediate zone (IZ) is noted. Maps of nerve tracts can be constructed by computationally "stringing together" ellipsoids in the 3D images. These data were acquired at 11.7 T using a 2D FT-PGSE method with 20 μm in-plane resolution, 300-μm-thick slices, and TE/TR = 16/2000 ms. Figure courtesy of Drs. David Laidlaw and Eric Ahrens.

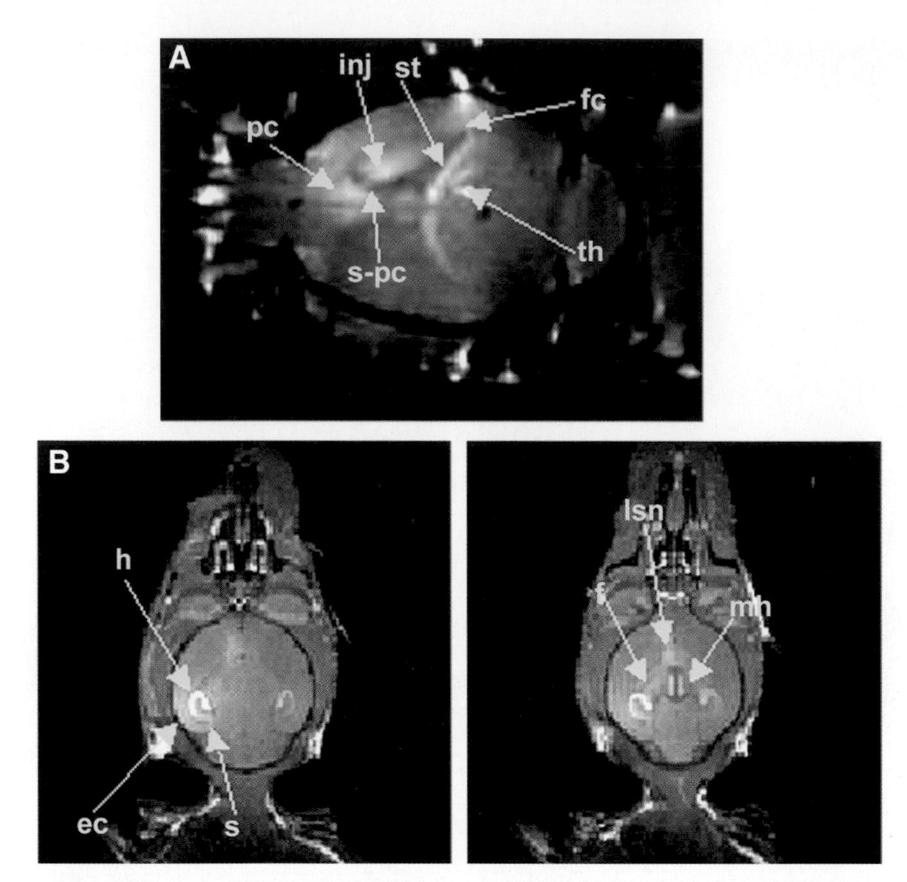

Figure 22 (**A**) One slice from a 3D MEMRI 24-h poststriatal Mn^{2+} injection. inj, injection site in the striatum; pc, prelimbic cortex; s-pc, fine connection from striatum to prelimbic cortex; th, thalamus; st, stria terminalis; fc, fine connection leading to the cortex. Although homotopic labeling was observed with the Mn^{2+}, only structures ipsilateral to the injection are indicated by arrows. (**B**) Two slices from a 3D MEMRI 48 h post-amygdala injection of Mn^{2+}. ec, entorhinal cortex; s, subiculum; h, hippocampus; f, fornix; lsn, lateral septal nucleus; mh, medial habenula. Although homotopic labeling of structures was observed with the Mn^{2+}, only structures ipsilateral to the injection are labeled with arrows with the exception of the medial habenula. Figure courtesy of Dr. Robia Pautler.

interconnected system that arises in vertebrates from an apparently homogeneous neural tube. The description of what happens as this system develops is still at a quite elementary stage. Without a detailed description (what is happening, when it is happening, and where it is happening), it is difficult to formulate and evaluate hypotheses about the underlying mechanisms. In MR imaging in general and microscopic MR imaging in particular, there are many opportunities to explore in the cycle of "observation, hypothesis, model construction, observation, ..." We believe that it is the collaborative efforts of biologists, chemists, physicists, and engineers, along with the overlapping of this cycle in diverse disciplines, that will drive the technology and application of magnetic resonance imaging. Three things drive our efforts in this endeavor: the pretty pictures, hints that we may be gaining a better understanding of how the brain works through these pictures, and having fun.

Acknowledgments

We note the support of the Beckman Institute, the National Institute of Child Health and Human Development (HD25390), the National Center for Research Resources (RR13625), the National Institute of Mental Health (MH61223), the NSF (CCR-0086065), and the Human Brain Project funded jointly by the National Institute of Drug Abuse and the National Institute of Mental Health. We thank David Haring of the Duke Primate Center and John Allman of the Division of Biology at Caltech for generous gifts of the lemur cadaver specimens used in this work.

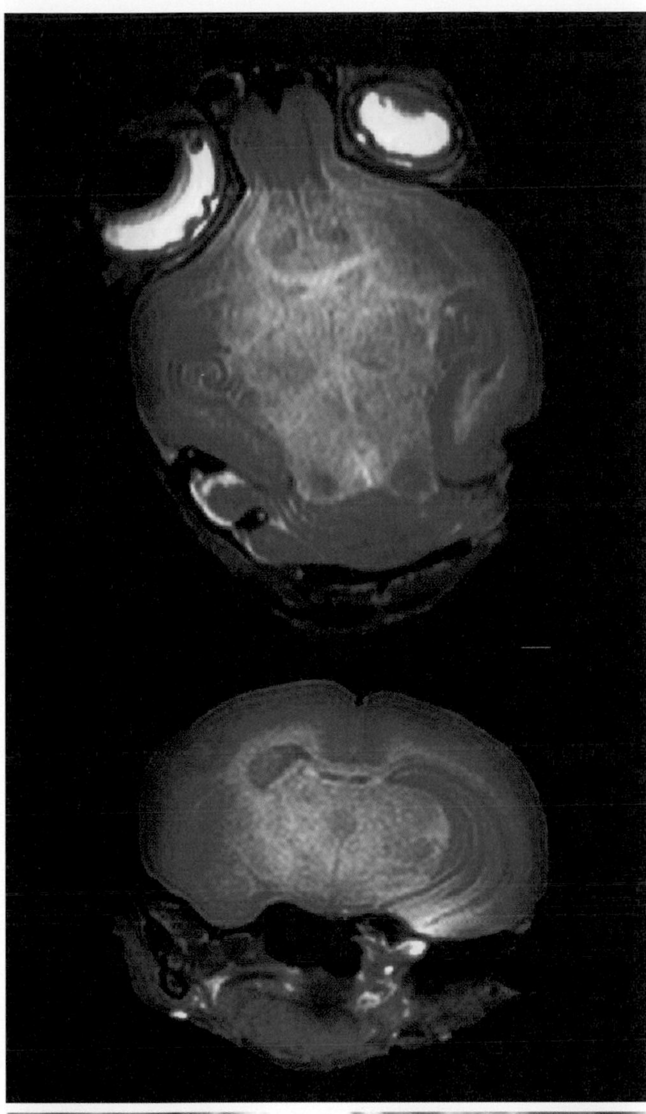

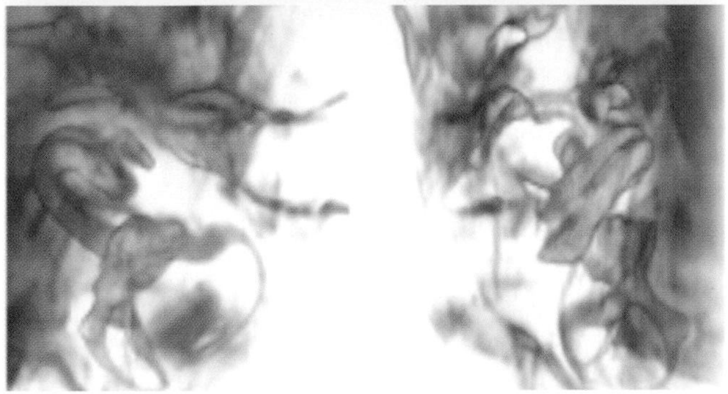

Figure 23 Two slices through a 3D MR image of a dwarf lemur neonate cadaver. A wealth of anatomical features are apparent, ranging from the lamination of cells in the developing cortical plate to the distinct morphology of the dentate gyrus of the hippocampus. Features can be extracted from this data set, as shown by the volume rendering of the inner ear: bottom shows two views of the inner ear rotated around the vertical axis by about 90°. Fine structures such as the basilar membrane and semicircular canals are clearly rendered.

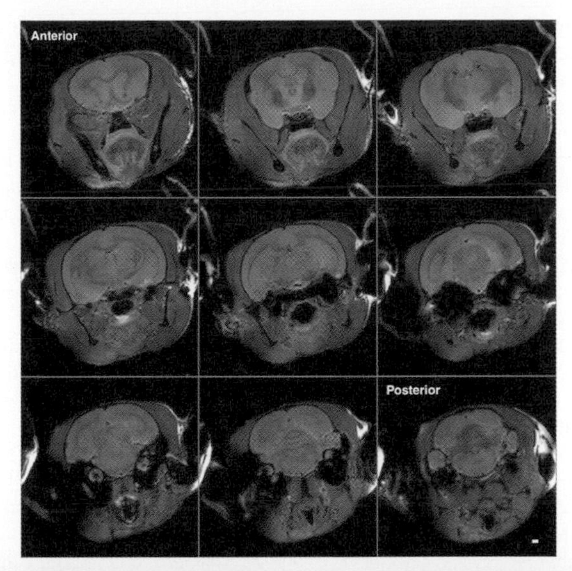

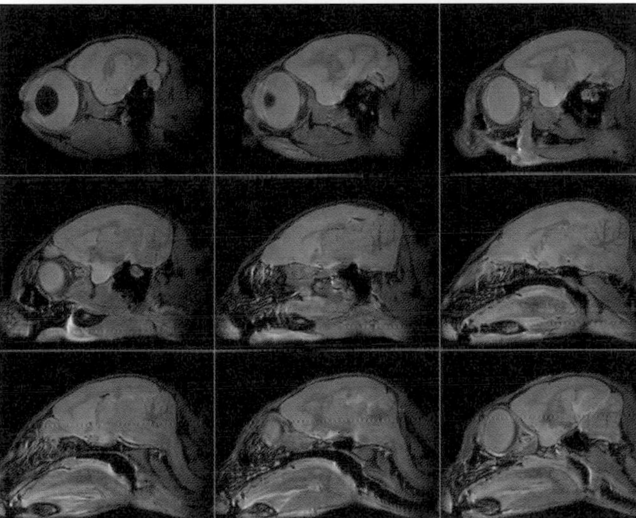

Figure 24 *In vivo* MR images of an 8-year-old mouse lemur taken on a Bruker Avance 500-MHz instrument. The large eyes, as well as a number of anatomical features are apparent in these slices (e.g., cortical layering, hippocampus, fine structure in the cerebellum, corpus callosum, olfactory bulb). Approximately sagittal and coronal slices from a 3D RARE (four echoes) data set are shown. The 256 × 128 × 128 data set was zero-filled to 512 × 256 × 256 before Fourier transformation, TR/TE = 500/12 ms, with a 1 h 14 min experiment time. The animal was anesthetized with isoflurane during the experiment, physiological signs were monitored, and respiratory gating was employed. Scale bar, 1 mm.

References

Aguayo, J. B., Blackband, S. J., Wehrle, J. P., Glickson, J. D., and Mattingly, M. A. (1987). NMR microscopic studies of eyes and tumors with histological correlation. *Ann. N. Y. Acad. Sci.* **508,** 399–413.

Ahn, C. B., Kim, J. H., and Cho, Z. H. (1986). High-speed spiral scan echo planar NMR imaging. *IEEE Trans. Med. Imaging* **5,** 2–7.

Ahrens, E., Laidlaw, D., Readhead, C., Brosnan, C., Fraser, S., *et al.* (1998). MR microscopy of transgenic mice that spontaneously acquire experimental allergic encephalomyelitis. *Magn. Reson. Med.* **40,** 119–132.

Alderman, D. W., and Grant, D. M. (1979). An efficient decoupler coil design which reduces heating in conductive samples in superconducting spectrometers. *J. Magn. Reson.* **36,** 447–451.

Alsop, D. C., and Detre, J. A. (1998). Multisection cerebral blood flow MR imaging with continuous arterial spin labeling. *Radiology* **208,** 410–416.

Alsop, D. C., Murai, H., Detre, J. A., McIntosh, T. K., and Smith, D. H. (1996). Detection of acute pathologic changes following experimental traumatic brain injury using diffusion-weighted magnetic resonance imaging. *J. Neurotrauma* **13,** 515–521.

Anderson, A. W., and Gore, J. C. (1994). Analysis and correction of motion artifacts in diffusion weighted imaging. *Magn. Reson. Med.* **32,** 379–387.

Assaf, Y., and Cohen, Y. (1998). Non-mono-exponential attenuation of water and *N*-acetyl aspartate signals due to diffusion in brain tissue. *J. Magn. Reson.* **131,** 69–85.

Assaf, Y., and Cohen, Y. (2000). Assignment of the water slow-diffusing component in the central nervous system using q-space diffusion MRS: Implications for fiber tract imaging. *Magn. Reson. Med.* **43,** 191–199.

Assaf, Y., Mayk, A., and Cohen, Y. (2000). *Proc. Int. Soc. Magn. Reson. Med.* **8,** 470.

Balaban, R. S., and Ceckler, T. L. (1992). Magnetization transfer contrast in magnetic resonance imaging. *Magn. Reson. Q.* **8,** 116–137.

Bartholdi, E., and Ernst, R. R. (1973). Fourier spectroscopy and the causality principal. *J. Magn. Reson.* **11,** 9–19.

Basser, P. J., and Pierpaoli, C. (1996). Microstructural and physiological features of tissues elucidated by quantitative-diffusion-tensor MRI. *J. Magn. Reson. Ser. B* **111,** 209–219.

Basser, P. J., and Pierpaoli, C. (1998). A simplified method to measure the diffusion tensor from seven MR images. *Magn. Reson. Med.* **39,** 928–934.

Beaulieu, C., and Allen, P. S. (1994). Determinants of anisotropic water diffusion in nerves. *Magn. Reson. Med.* **31,** 394–400.

Beaulieu, C. F., Zhou, X., Cofer, G. P., and Johnson, G. A. (1993). Diffusion-weighted MR microscopy with fast spin-echo. *Magn. Reson. Med.* **30,** 201–206.

Becker, E. D. (2000). "High Resolution NMR: Theory and Chemical Applications." Academic Press, New York.

Behling, R. W., Tubbs, H. K., Cockman, M. D., and Jelinski, L. W. (1989). Stroboscopic NMR microscopy of the carotid artery. *Nature (London)* **341,** 321–323.

Belliveau, J. W., Kennedy, D. N., McKinstry, R. C., Buchbinder, B. R., Weisskoff, R. M., *et al.* (1991). Functional mapping of the human visual cortex by magnetic resonance imaging. *Science* **254,** 716–719.

Benveniste, H., Qui, H., Hedlund, L. W., D'Ercole, F., and Johnson, G. A. (1998). Spinal cord neural anatomy in rats examined by in vivo magnetic resonance microscopy. *Region. Anesth. Pain Med.* **23,** 589–599.

Black, R. D., Early, T. A., Roemer, P. B., Mueller, O. M., Mogro-Campero, A., *et al.* (1993). A high temperature superconducting receiver for nuclear magnetic resonance imaging. *Science* **259,** 793–795.

Bloch, F., Hansen, W. W., and Packard, M. (1946). Nuclear induction. *Phys. Rev.* **69,** 127.

Blumich, B., and Kuhn, W., eds. (1992). "Magnetic Resonance Microscopy." VCH, New York.

Bluml, S., Schad, L. R., Stepanow, B., and Lorenz, W. J. (1993). Spin-lattice relaxation-time measurement by means of a turboflash technique. *Magn. Reson. Med.* **30,** 289–295.

Bowtell, R. W., Brown, G. D., Glover, P. M., McJury, M., and Mansfield, P. (1990). Resolution of cellular structures by NMR microscopy at 11.7 T. *Philos. Trans. R. Soc. A* **333,** 457–467.

Bowtell, R. W., Peters, A., Sharp, J. C., Mansfield, P., Hsu, E. W., *et al.* (1995). NMR microscopy of single neurons using spin-echo and line narrowed 2DFT imaging. *Magn. Reson. Med.* **33,** 790–794.

Bracewell, R. N. (1978). "The Fourier and Its Applications." McGraw–Hill, New York.

Brigham, E. O. (1974). "The Fast Fourier Transform." Prentice Hall International, Englewood Cliffs, NJ.

Bronner, F. M., and Fraser, S. E. (1988). Cell lineage analysis reveals multipotency of some avian neural crest cells. *Nature (London)* **335,** 161–164.

Callaghan, P. T. (1991). "Principles of Nuclear Magnetic Resonance Microscopy." Oxford Univ. Press, New York.

Callaghan, P. T., Coy, A., Forde, L. C., and Rofe, C. J. (1993). Diffusive relaxation and edge enhancement in NMR microscopy. *J. Magn. Reson. A* **101,** 347–350.

Callaghan, P. T., Forde, L. C., and Rofe, C. J. (1994). Correlated susceptibility and diffusion effects in NMR microscopy using both phase-frequency encoding and phase-phase encoding. *J. Magn. Reson.* **B104,** 34–52.

Carr, H. Y., and Purcell, E. M. (1954). Effects of diffusion on free precession in NMR experiments. *Phys. Rev.* **94,** 630–638.

Chen, N., Li, L., and Wyrwixz, A. (1998). Optimized phase preparation and auto-shimming technique for gradient echo MRI. *In* "Proceedings of the Sixth ISMRM," Sydney, p. 127.

Chen, N. K., and Wyrwicz, A. M. (1999). Removal of intravoxel dephasing artifact in gradient-echo images using a field-map based RF refocusing technique. *Magn. Reson. Med.* **42,** 807–812.

Cho, Z. H., Ahn, C. B., Juh, S. C., Lee, H. K., Jacobs, R. E., *et al.* (1988). NMR microscopy with 4-μm resolution theoretical study and experimental results. *Med. Phys.* **15,** 815–824.

Cho, Z. H., and Ro, Y. M. (1992). Reduction of susceptibility artifact in gradient-echo imaging. *Magn. Reson. Med.* **23,** 193–200.

Clark, C. A., and Le Bihan, D. (2000). Water diffusion compartmentation and anisotropy at high b values in the human brain. *Magn. Reson. Med.* **44,** 852–859.

Constable, R. T., Smith, R. C., and Gore, J. C. (1992). Signal-to-noise and contrast in fast spin echo (FSE) and inversion recovery FSE imaging. *J. Comput. Assisted Tomogr.* **16,** 41–47.

Cory, D. G., and Garroway, A. N. (1990). Measurement of translational displacement probabilities by NMR—An indicator of compartmentation. *Magn. Reson. Med.* **14,** 435–444.

Damasio, H. (1995). "Human Brain Anatomy in Computerized Images." Oxford Univ. Press, New York.

de Crespigny, A. J., Marks, M. P., Enzmann, D. R., and Moseley, M. E. (1995). Navigated diffusion imaging of normal and ischemic human brain. *Magn. Reson. Med.* **33,** 720–728.

Derome, A. E. (1987). "Modern NMR Techniques for Chemistry Research." Pergamon, New York.

Detre, J. A., Leigh, J. S., Williams, D. S., and Koretsky, A. P. (1992). Perfusion imaging. *Magn. Reson. Imaging* **23,** 37–45.

Dhenain, M., Ruffins, S., and Jacobs, R. E. (2001). Three dimensional digital mouse atlas using high resolution MRI. *Dev. Biol.* **232,** 458–470.

Dixon, D., Johnson, G. A., Cofer, G. P., Hedlund, L. W., and Maronpot, R. R. (1988). Magnetic resonance imaging pathology. *Toxicol. Pathol.* **16,** 386–391.

Edelman, R. R., Siewert, B., Darby, D. G., Thangaraj, V., Nobre, A. C., *et al.* (1994). Qualitative mapping of cerebral blood-flow and functional localization with echo-planar MR-imaging and signal targeting with alternating radio-frequency. *Radiology* **192,** 513–520.

Edelstein, W. A., Glover, G. H., Hardy, C. J., and Redington, R. W. (1986). The intrinsic signal-to-noise ratio in NMR imaging. *Magn. Reson. Med.* **3,** 604–618.

Edelstein, W. A., Hutchison, J. M. S., Johnson, G., and Redpath, T. (1980). Spin warp NMR imaging and applications to whole body imaging. *Phys. Med. Biol.* **25,** 751–756.

Ehman, R. L., and Felmlee, J. P. (1989). Adaptive technique for high-definition MR imaging of moving structures. *Radiology* **173,** 255–263.

Ernst, R. R., and Anderson, W. A. (1966). Application of Fourier transform spectroscopy to magnetic resonance. *Rev. Sci. Instrum.* **37,** 93–102.

Farrar, T. C. (1989). "Introduction to Pulse NMR Spectroscopy." Farragut, Madison, WI.

Feinberg, D. A., Hale, J. D., Watts, J. C., *et al.* (1986). Halving MR imaging time by conjugation: Demonstration at 3.5 kG. *Radiology* **161,** 527–531.

Fenyes, D., and Narayana, P. (1999). In vivo diffusion tensor imaging of rat spinal cord with echo planar imaging. *Magn. Reson. Med.* **42,** 300–306.

Fornasiero, D., Bellen, J. C., Baker, R. J., and Chatterton, B. E. (1987). Paramagnetic complexes of manganese(II), iron(III), and gadolinium(III) as contrast agents for magnetic resonance imaging: The influence of stability constants on the biodistribution of radioactive aminopolycarboxylate complexes. *Invest. Radiol.* **22,** 322–327.

Frahm, J., Haase, A., and Matthaei, D. (1986a). Rapid 3-dimensional MR imaging using the FLASH technique. *J. Comput. Assisted Tomogr.* **10,** 363–368.

Frahm, J., Haase, A., and Matthaei, D. (1986b). Rapid NMR imaging of dynamic processes using the FLASH technique. *Magn. Reson. Med.* **3,** 321–327.

Frahm, J., Merboldt, K. D., and Hanicke, W. (1988). Direct FLASH MR imaging of magnetic-field inhomogeneities by gradient compensation. *Magn. Reson. Med.* **6,** 474–480.

Frahm, J., Merboldt, K. D., and Hanicke, W. (1993). Functional MRI of human brain activation at high spatial resolution. *Magn. Reson. Med.* **29,** 139–144.

Gamse, D., and Sive, H. (2000). Vertebrate anteroposterior patterning: The Xenopus neurectoderm as a paradigm. *BioEssays* **22,** 976–986.

Geraldes, C. F., Sherry, A. D., Brown, R. D., and Koenig, S. H. (1986). Magnetic field dependence of solvent proton relaxation rates induced by Gd^{3+} and Mn^{2+} complexes of various polyaza macrocyclic ligands: Implications for NMR imaging. *Magn. Reson. Med.* **3,** 242–250.

Gerhart, J. (2001). Evolution of the organizer and the chordate body plan. *Int. J. Dev. Biol.* **45,** 133–153.

Gewalt, S. L., Glover, G. H., Hedlund, L. W., Cofer, G. P., MacFall, J. R., *et al.* (1993). MR microscopy of the rat lung using projection reconstruction. *Magn. Reson. Med.* **29,** 99–109.

Ghosh, P., O'Dell, M., Narasimhan, P. T., Fraser, S. E., and Jacobs, R. E. (1994). Mouse lemur microscopic MRI brain atlas. *NeuroImage* **1,** 345–349. [Available on CD-ROM]

Gibby, W. A. (1988). MR contrast agents: An overview. *Radiol. Clin. North Am.* **26,** 1047–1058.

Gibby, W. A., Bogdan, A., and Ovitt, T. W. (1989). Cross-linked DTPA polysaccharides for magnetic resonance imaging synthesis and relaxation properties. *Invest. Radiol.* **24,** 302–309.

Glover, G., and Lai, S. (1998). Reduction of susceptibility effects in BOLD fMRI using tailored RF pulses. *In* "Proceedings of the Sixth ISMRM," Sydney, p. 298.

Glover, P. M., Bowtell, R. W., Brown, G. D., and Mansfield, P. (1994). A microscope slide probe for high-resolution imaging at 11.7 tesla. *Magn. Reson. Med.* **31,** 423–428.

Haacke, E., Brown, R., Thompson, M., and Venkatesan, R. (1999). "Magnetic Resonance Imaging: Physical Principles and Sequence Design." Wiley, New York.

Haacke, E. M. (1987). The effect of finite sampling in spin echo or field-echo magnetic resonance imaging. *Magn. Reson. Med.* **4,** 407–421.

Haacke, E. M., Tkach, J. A., and Parrish, T. B. (1989). Reduction of T2* dephasing in gradient field echo imaging. *Radiology* **170,** 457–462.

Haase, A., Frahm, J., and Matthaei, D. (1986). FLASH imaging, rapid NMR imaging using low flip-angle pulses. *J. Magn. Reson.* **67,** 258–266.

Hahn, E. L. (1950). Spin echoes. *Phys. Rev.* **80,** 580–594.

Hayes, C. E., Edelstein, W. A., Schnenck, J. F., Mueller, O. M., and Eash, M. (1985). An efficient, highly homogeneous RF coil for whole body imaging at 1.5 T. *J. Magn. Reson.* **66,** 622–628.

Hedges, L. K., and Sobering, G. (1994). Realizing intrinsic resolution in MRI data. *In* "Proceedings of the SMR Second Annual Meeting," San Francisco, p. 839.

Hedlund, L. W., Johnson, G. A., Karis, J. P., and Effmann, E. L. (1986a). MR "microscopy" of the rat thorax. *J. Comput. Assisted Tomogr.* **10,** 948–952.

Hedlund, L. W., Johnson, G. A., and Mills, G. I. (1986b). Magnetic resonance microscopy of the rat thorax and abdomen. *Invest. Radiol.* **21,** 843–846.

Henderson, E., McKinnon, G., Lee, T. Y., and Rutt, B. K. (1999). A fast 3D Look-Locker method for volumetric T-1 mapping. *Magn. Reson. Imaging* **17,** 1163–1171.

Hendrick, R. E. (1987). Sampling time effects on signal-to-noise and contrast-to-noise in spin echo MRI. *Magn. Reson. Imaging* **5,** 31–37.

Hendrick, R. E., Kneeland, J. B., and Stark, D. D. (1987). Maximizing signal-to-noise and contrast-to-noise ratios in FLASH imaging. *Magn. Reson. Imaging* **5,** 117–127.

Henkelman, R. M., and Bronskill, M. J. (1987). Artifacts in magnetic resonance imaging. *Rev. Magn. Reson. Imaging* **2,** 1–126.

Henkelman, R. M., Huang, X. M., Xiang, Q. S., Stanisz, G. J., Swanson, S. D., *et al.* (1993). Quantitative interpretation of magnetization-transfer. *Magn. Reson. Med.* **29,** 759–766.

Hennig, J. (1988). Multiecho imaging sequences with low refocusing flip angles. **78,** 397–407.

Hennig, J., Friedburg, H., and Ott, D. (1987). Fast 3-dimensional imaging of cerebrospinal-fluid. *Magn. Reson. Med.* **5,** 380–383.

Hennig, J., Nauerth, A., and Friedburg, H. (1986). Rare imaging—A fast imaging method for clinical MR. *Magn. Reson. Med.* **3,** 823–833.

Hollett, M. D., Cofer, G. P., and Johnson, G. A. (1987). In situ magnetic resonance microscopy. *Invest. Radiol.* **22,** 965–968.

Hoult, D. I., and Lauterbur, P. C. (1979). The sensitivity of the zeumatographic experiment involving human samples. *J. Magn. Reson.* **34,** 425–433.

House, W. V. (1984). NMR microscopy. *IEEE Trans. Nucl. Sci.* **31,** 570–577.

Hu, X., and Kim, S.-G. (1994). Reduction of signal fluctuation in functional MRI using navigator echoes. *Magn. Reson. Med.* **31,** 495–503.

Hu, X., and Parrish, T. (1994). Reduction of field-of-view for dynamic imaging. *Magn. Reson. Med.* **31,** 691–694.

Hurlston, S. E., Brey, W. W., Suddarth, S. A., and Johnson, G. A. (1999). A high-temperature superconducting Helmholtz probe for microscopy at 9.4 T. *Magn. Reson. Med.* **41,** 1032–1038.

Hyslop, W. B., and Lauterbur, P. C. (1991). Effects of restricted diffusion in microscopic MR imaging. *J. Magn. Reson.* **94,** 501–510.

Jackels, S. C. (1990). Enhancement agents in magnetic resonance and ultrasound. *Pharm. Med. Imaging* **Sect. III, Chap. 20,** 645.

Jacobs, R. E., and Fraser, S. E. (1994). Magnetic resonance microscopy of embryonic cell lineages and movement. *Science* **263,** 681–684.

Johnson, G. A., Thompson, M. B., and Drayer, B. P. (1987). Three-dimensional MRI microscopy of the normal rat brain. *Magn. Reson. Med.* **4,** 351–365.

Jones, R. A., Haraldseth, O., Mueller, T. B., Rinck, P. A., and Oksendal, A. N. (1993). *k*-space substitution: A novel dynamic imaging technique. *Magn. Reson. Med.* **29,** 830–834.

Kaufman, M. H. (1992). "The Atlas of Mouse Development." Academic Press, San Diego.

Keller, R. (1991). Early embryonic development in *Xenopus laevis. In* "Methods in Cell Biology" (B. K. Kay and H. B. Peng, eds.), Vol. 36, pp. 62–113. Academic Press, San Diego.

Keller, R., Davidson, L., Edlund, A., Elul, T., Ezin, M., *et al.* (2000). Mechanisms of convergence and extension by cell intercalation. *Philos. Trans. R. Soc. London Ser. B Biol. Sci.* **355,** 897–922.

Kennedy, D. N., Chang, C., Caviness, V. S., Moore, J., Brady, T. J., *et al.* (1989). Neuroanatomic microscopy of the rodent brain: Imaging strategies and contrast. *In* "Abstracts of the Society of Magnetic Resonance in Medicine, 8th Annual Meeting," p. 979.

Kim, S. G., Hu, X. P., and Ugurbil, K. (1994). Accurate T1 determination from inversion-recovery images—Application to human brain at 4-tesla. *Magn. Reson. Med.* **31,** 445–449.

King, M. D., Houseman, J., Roussel, S. A., van Bruggen, N., Williams, S. R., *et al.* (1994). q-space imaging of the brain. *Magn. Reson. Med.* **32,** 707–713.

Kuhn, W. (1990). NMR microscopy—Fundamentals, limits, and possible applications. *Angew. Chem. Int. Engl.* **29,** 1–112.

Kumar, A., Welti, D., and Ernst, R. R. (1975). NMR Fourier zeumatography. *J. Magn. Reson.* **18,** 69–83.

Kuperman, V. (2000). "Magnetic Resonance Imaging: Physical Principles and Applications." Academic Press, San Diego.

Lane, M. C., and Sheets, M. D. (2000). Designation of the anterior/posterior axis in pregastrula *Xenopus laevis*. *Dev. Biol.* **225,** 37–58.

Lauterbur, P. C. (1973). Image formation by induced local interactions: Examples employing nuclear magnetic resonance. *Nature* **242,** 190–191.

Le Bihan, D., Breton, E., Lallemand, D., Aubon, M. L., Vignaud, J., *et al.* (1988). Separation of diffusion and perfusion in intravoxel incoherent motion MR imaging. *Radiology* **168,** 497–505.

Lee, D. H. (1991). Mechanisms of contrast enhancement in magnetic resonance imaging. *Can. Assoc. Radiol. J.* **42,** 6–12.

Lee, S. C., Kim, K., Kim, J., Lee, S., Yi, J. H., *et al.* (2001). One micrometer resolution NMR microscopy. *J. Magn. Reson.* **150,** 207–213.

Link, J., and Seelig, J. (1990). Comparison of deuterium NMR imaging methods and application to plants. *J. Magn. Reson.* **89,** 310–330.

Ljunggren, S. (1983). A simple graphical representation of Fourier-based imaging methods. *J. Magn. Reson.* **54,** 338–343.

Lucaciu, C. M., Dragu, D., Copaescu, L., and Morariu, V. V. (1997). Manganese transport through human erythrocyte membranes: An EPR study. *Biochim. Biophys. Acta* **1328,** 90–98.

Mansfield, P. (1977). Multiplanar image formation using NMR spin-echoes. *J. Phys. C* **10,** L55–L58.

Mansfield, P., and Grannell, P. K. (1973). NMR 'diffraction' in solids. *J. Phys. C* **6,** L422.

Mansfield, P., and Morris, P. G. (1982). "NMR Imaging in Biomedicine." Academic Press, New York.

Mansfield, P., and Pykett, I. L. (1978). Biological and medical imaging by NMR. *J. Magn. Reson.* **29,** 355–373.

Margosian, P., and Schmitt, F. (1985). Faster MR imaging methods. *Proc. SPIE Med. Image Process.* **593**, 6.

McFarland, E. W., and Mortara, A. (1992). 3-dimensional NMR microscopy—Improving SNR with temperature and microcoils. *Magn. Reson. Imaging* **10,** 279–288.

Meiboom, S., and Gill, D. (1958). Modified spin-echo method for measuring nuclear relaxation times. *Rev. Sci. Instrum.* **29,** 688–691.

Meyer, C. H., Hu, B. S., Nishimura, D. G., and Macovski, A. (1992). Fast spiral coronary artery imaging. *Magn. Reson. Med.* **28,** 202–213.

Meyer, D., Schaefer, M., and Douchet, D. (1990). Advances in macrocyclic gadolinium complexes as magnetic resonance imaging contrast agents. *Invest. Radiol.* **25,** S53.

Miller, J. R., Hurlston, S. E., Ma, Q. Y., Face, D. W., Kountz, D. J., *et al.* (1999). Performance of a high-temperature superconducting probe for in vivo microscopy at 2.0 T. *Magn. Reson. Med.* **41,** 72–79.

Moi, M. K., and Meares, C. F. (1988). The peptide way to the macrocyclic bifunctional chelating agents: Synthesis of *p*-nitobenzyl DOTA. *J. Am. Chem. Soc.* **110,** 6266–6267.

Mori, S., Crain, B. J., Chacko, V. P., and van Zijl, P. C. (1999). Three-dimensional tracking of axonal projections in the brain by magnetic resonance imaging. *Ann. Neurol.* **45,** 265–269.

Mori, S., Itoh, R., Zhang, J. Y., Kaufmann, W. E., van Zijl, P. C. M., *et al.* (2001). Diffusion tensor imaging of the developing mouse brain. *Magn. Reson. Med.* **46,** 18–23.

Mori, S., and van Zijl, P. C. M. (1998). A motion correction scheme by twin-echo navigation for diffusion-weighted magnetic resonance imaging with multiple RF echo acquisition. *Magn. Reson. Med.* **40,** 511–516.

Morris, P. G. (1986). "Nuclear Magnetic Resonance Imaging in Biology and Medicine." Oxford Univ. Press, New York.

Mulkern, R. V., Gudbjartsson, H., Westin, C. F., Zengingonul, H. P., Gartner, W., *et al.* (1999). Multi-component apparent diffusion coefficients in human brain. *NMR Biomed.* **12,** 51–62.

Narasimhan, P. T., Ghosh, P., Fraser, S. E., and Jacobs, R. E. (1994). Magnetic-resonance microscopy—Challenges in biological imaging using a 500 MHz NMR microscope. *Proc. Indian Acad. Sci.–Chem. Sci.* **106,** 1625–1641.

Narasimhan, P. T., Velan, S., and Jacobs, R. (2000). MR microscopic imaging with intermolecular zero- and double-quantum coherences at 11.7 T. *Proc. Int. Soc. Magn. Reson. Med.* **8,** 2074.

Narita, K., Kawasaki, F., and Kita, H. (1990). Mn and Mg influxes through Ca channels of motor nerve terminals are prevented by verapamil in frogs. *Brain Res.* **510,** 289–295.

Niehrs, C. (2001). The Spemann organizer and embryonic head induction. *EMBO J.* **20,** 631–637.

Niendorf, T., Dijkhuizen, R. M., Norris, D. G., Campagne, M. V., and Nicolay, K. (1996). Biexponential diffusion attenuation in various states of brain tissue: Implications for diffusion-weighted imaging. *Magn. Reson. Med.* **36,** 847–857.

Nieuwkoop, P. D., and Faber, J. (1967). "Normal Table of *Xenopus laevis* (Daudin)." North Holland, Amsterdam.

Norris, D., Bornert, P., and Reese, T. (1992). On the application of ultra-fast RARE experiments. *Magn. Reson. Med.* **27,** 142–164.

Odoj, F., Rommel, E., von Kienlin, M., and Haase, A. (1998). A superconducting probehead applicable for nuclear magnetic resonance microscopy at 7 T. *Rev. Sci. Instrum.* **69,** 2708–2712.

Ogawa, S., Lee, T. M., Kay, A. R., and Tank, D. W. (1990a). Brain magnetic resonance imaging with contrast dependent on blood oxygenation. *Proc. Natl. Acad. Sci. USA* **87,** 9868–9872.

Ogawa, S., Lee, T. M., Nayak, A. S., and Glynn, P. (1990b). Oxygenation-sensitive contrast in magnetic resonance image of rodent brain at high magnetic fields. *Magn. Reson. Med.* **14,** 68–78.

Ordidge, R. J., Gorell, J. M., Deniau, J. C., Knight, R. A., and Helpern, J. A. (1994a). Assessment of relative brain iron concentrations using T_2-weighted and T_2*-weighted MRI at 3 tesla. *Magn. Reson. Med.* **32,** 335–341.

Ordidge, R. J., Helpern, J. A., Qing, Z. X., Knight, and Nagash, V. (1994b). Correction of motional artifacts in diffusion-weighted MR images using navigator echoes. *Magn. Reson. Imaging* **12,** 455–460.

Pattany, P. M., Puckett, W. R., Klose, K. J., Quencer, R. M., Bunge, R. P., *et al.* (1997). High-resolution diffusion-weighted MR of fresh and fixed cat spinal cords: Evaluation of diffusion coefficients and anisotropy. *Am. J. Neuroradiol.* **18,** 1049–1056.

Pauling, L., and Coryell, C. D. (1936). The magnetic properties of and structure of hemoglobin and carbon monoxyhemoglobin. *Proc. Natl. Acad. Sci. USA* **22,** 210–216.

Pautler, R. (1999). "In Vivo Neuronal Tract Tracing Utilizing Manganese Enhanced MRI (MEMRI)." Carnegie–Mellon Univ., Pittsburgh. [Ph.D. thesis]

Pautler, R., Olson, C., Williams, D., Ho, C., and Koretsky, A. (1999). In vivo tract tracing using manganese enhanced MRI (MEMRI) in mouse mutants and non-human primates. *Proc. Int. Soc. Magn. Reson. Med.* **7,** 448.

Pautler, R. G., Silva, A. C., and Koretsky, A. P. (1998). In vivo neuronal tract tracing using manganese-enhanced magnetic resonance imaging. *Magn. Reson. Med.* **40,** 740–748.

Paxinos, G., Tork, I., Tecott, L. H., and Valentino, K. L. (1991). "Atlas of the Developing Rat Brain." Academic Press, San Diego.

Peck, T. L., Magin, R. L., and Lauterbur, P. C. (1995). Design and analysis of microcoils for NMR microscopy. *J. Magn. Reson. Ser. B* **108,** 114–124.

Purcell, E. M., Torrey, H. C., and Pound, R. V. (1946). Resonance absorption by nuclear magnetic moments in a solid. *Phys. Rev.* **69,** 37–38.

Putz, B., Barsky, D., and Schulten, K. (1992). Edge enhancement by diffusion in microscopic magnetic resonance imaging. *J. Magn. Reson.* **97,** 27–53.

Ramirez, R. W. (1985). "The FFT Fundamentals and Concepts." Prentice Hall, International., Englewood Cliffs, NJ.

Reichenbach, J. R., Venkatesan, R., Yablonskiy, D. A., Thompson, M. R., Lai, S., *et al.* (1997). Theory and application of static field inhomogeneity effects in gradient-echo imaging. *J. Magn. Reson. Imaging* **7,** 266–279.

Rizi, R. R., Ahn, S., Alsop, D. C., Garrett-Roe, S., Mescher, M., *et al.* (2000). Intermolecular zero-quantum coherence imaging of the human brain. *Magn. Reson. Med.* **43,** 627–632.

Roberts, J. D. (1959). "Nuclear Magnetic Resonance: Applications to Organic Chemistry." McGraw–Hill, New York.

Rofe, C. J., Vannoort, J., Back, P. J., and Callaghan, P. T. (1995). NMR microscopy using large, pulsed magnetic-field gradients. *J. Magn. Reson. Ser. B* **108,** 125–136.

Runge, V. M., and Gelblumm, D. Y. (1991). Future directions in magnetic resonance imaging contrast media. *Top. Magn. Reson. Imaging* **3,** 85.

Russell, E. J., Schaible, T. F., and Dillon, W. (1989). Multicenter double-blind placebo-controlled study of gadopentetate dimeglumine as an MR contrast agent: Evaluation in patients with cerebral lesions. *Am. J. Roentgend.* **152,** 813.

Schambra, U. B., Lauder, J. M., and Silver, J. (1992). "Atlas of the Prenatal Mouse Brain." Academic Press, San Diego.

Schoeniger, J. S., and Blackband, S. J. (1994). The design and construction of a NMR microscopy probe. *J. Magn. Reson. Ser. B* **104,** 127–134.

Sharp, J. C., Bowtell, R. W., and Mansfield, P. (1993). Elimination of susceptibility distortions and reduction of diffusion attenuation in NMR microscopy by line-narrowed 2DFT. *Magn. Reson. Med.* **29,** 407–410.

Silva, A. C., and Kim, S. G. (1999). Pseudo-continuous arterial spin labeling technique for measuring CBF dynamics with high temporal resolution. *Magn. Reson. Med.* **42,** 425–429.

Silver, M. S., Joseph, R. I., and Hoult, D. I. (1985). Selective spin inversion in nuclear magnetic-resonance and coherent optics through an exact solution of the Bloch–Riccati equation. *Phys. Rev. A* **31,** 2753–2755.

Sloot, W. N., and Gramsbergen, J. B. (1994). Axonal transport of manganese and its relevance to selective neurotoxicity in the rat basal ganglia. *Brain Res.* **657,** 124–132.

Stark, D. D., and Bradley, W. G., Jr., eds. (1992). "Magnetic Resonance Imaging." Mosby, St. Louis.

Steen, R. G., Gronemeyer, S. A., Kingsley, P. B., Reddick, W. E., Langston, J. S., *et al.* (1994). Precise and accurate measurement of proton-T1 in human brain in-vivo—Validation and preliminary clinical application. *J. Magn. Reson. Imaging* **4,** 681–691.

Stehling, M. K., Turner, R., and Mansfield, P. (1991). Echo-planar imaging: Magnetic resonance imaging in a fraction of a second. *Science* **254,** 43–50.

Stejskal, E. O. (1965). Use of spin echoes in a pulsed magnetic field gradient to study anisotropic restricted diffusion and flow. *J. Phys. Chem.* **43,** 3597–3602.

Stejskal, E. O., and Tanner, J. E. (1965). Spin diffusion measurements in the presence of time-dependent field gradient. *J. Chem. Phys.* **42,** 288–292.

Styles, P., Soffe, N. F., and Scott, C. A. (1989). An improved cryogenically cooled probe for high-resolution NMR. *J. Magn. Reson.* **84,** 376–378.

Takeda, A., Ishiwatari, S., and Okada, S. (1998). In vivo stimulation-induced release of manganese in rat amygdala. *Brain Res.* **811,** 147–151.

Theiler, K. (1989). "The House Mouse: Atlas of Embryonic Development." Springer-Verlag, New York.

Tjalve, H., Henriksson, J., Tallkvist, J., Larsson, B. S., and Lindquist, N. G. (1996). Uptake of manganese and cadmium from the nasal mucosa into the central nervous system via olfactory pathways in rats. *Pharmacol. Toxicol.* **79,** 347–356.

Tjalve, H., Mejare, C., and Borg-Neczak, K. (1995). Uptake and transport of manganese in primary and secondary olfactory neurones in pike. *Pharmacol. Toxicol.* **77,** 23–31.

Turner, R., Jezzard, P., Wan, H., Kwong, K. K., LeBihan, D., *et al.* (1993). Functional mapping of the human visual cortex at 4 T and 1.5 T using deoxygenation contrast EPI. *Magn. Reson. Med.* **29,** 277–279.

Turner, R., LeBihan, D., Maier, J., Vavrek, R., Hedges, L. K., *et al.* (1990). Echo-planar imaging of intravoxel incoherent motion. *Radiology* **177,** 407–414.

Twieg, D. B. (1983). The k-trajectory formulation of the NMR imaging process with applications in analysis and synthesis of imaging methods. *Med. Phys.* **10,** 610–621.

Ulug, A., and van Zijl, P. (1999). Orientation-independent diffusion imaging without tensor diagonalization: Anisotropy definitions based on physical attributes of the diffusion ellipsoid. *J. Magn. Reson. Imaging* **9,** 804–813.

van Vaals, J. J., Brummer, M. E., Dickson, W. T., Tuithof, H. H., Engels, H., *et al.* (1993). "Keyhole" method for accelerating imaging of contrast agent uptake. *J. Magn. Reson. Imaging* **3,** 671–675.

Vinitski, S., Griffey, R., Fuka, M., Matwiyoff, N., and Prost, R. (1987). Effect of the sampling rate on magnetic resonance imaging. *Magn. Reson. Med.* **5,** 278–285.

Vlaardingerbroek, M., and Den Boer, J. (1996). "Magnetic Resonance Imaging: Theory and Practice." Springer Verlag, Berlin.

Warren, W. S., Ahn, S., Mescher, M., Garwood, M., Ugurbil, K., *et al.* (1998). MR imaging contrast enhancement based on intermolecular zero quantum coherences. *Science* **281,** 247–251.

Wehrli, F. W. (1990). "Fast-Scan Magnetic Resonance: Principles and Applications." Raven Press, New York.

Weinstein, D. C., and Hemmati-Brivanlou, A. (1999). Neural induction. *Annu. Rev. Cell Dev. Biol.* **15,** 411–433.

Weis, J., Gorke, U., and Kimmich, R. (1996). Susceptibility, field inhomogeneity, and chemical shift-corrected NMR microscopy: Application to the human finger in vivo. *Magn. Reson. Imaging* **14,** 1165–1175.

Weiss, H. R., Buchweitz, E., Murtha, T. J., and Auletta, M. (1982). Quantitative regional determination of morphomometric indices of the total and perfused capillary network in the rat brain. *Circ. Res.* **51,** 494–503.

Wetts, R., Serbedzija, G. N., and Fraser, S. F. (1989). Cell lineage analysis reveals multipotent precursors in the ciliary margin of the frog retina. *Dev. Biol.* **136,** 254–263.

Whittall, K. P., MacKay, A. L., Graeb, D. A., Nugent, R. A., Li, D. K. B., *et al.* (1997). In vivo measurement of T2 distributions and water contents in normal human brain. *Magn. Reson. Med.* **37,** 34–43.

Williams, C. F. M., Redpath, T. W., and Norris, D. G. (1999). A novel fast split-echo multi-shot diffusion-weighted MRI method using navigator echoes. *Magn. Reson. Med.* **41,** 734–742.

Williams, D. S., Detre, J. A., Leigh, J. S., and Koretsky, A. P. (1992). Magnetic resonance imaging of perfusion using spin inversion of arterial water. *Proc. Natl. Acad. Sci. USA* **89,** 212–216. [Published erratum appears in *Proc. Natl. Acad. Sci. USA,* 1992, **89,** 4220].

Wong, E. C., Buxton, R. B., and Frank, L. R. (1997). Implementation of quantitative perfusion imaging techniques for functional brain mapping using pulsed arterial spin labeling. *NMR Biomed.* **10,** 237–249.

Wood, M. L. (1992). Fourier imaging. *In* "Magnetic Resonance Imaging" (D. D. Stark and W. G. Bradley, Jr., eds.), Vol. 1, Chap. 2. Mosby, New York.

Wright, A. C., Song, H. K., and Wehrli, F. W. (2000). In vivo MR micro imaging with conventional radiofrequency coils cooled to 77 degrees K. *Magn. Reson. Med.* **43,** 163–169.

Xue, R., van Zijl, P., Crain, B., Solaiyappan, M., and Mori, S. (1999). In vivo three-dimensional reconstruction of rat brain axonal projections by diffusion tensor imaging. *Magn. Reson. Med.* **42,** 1123–1127.

Yang, Q. X., Demeure, R. J., Dardzinski, B. J., Arnold, B. W., and Smith, M. B. (1999). Multiple echo frequency-domain image contrast: Improved signal-to-noise ratio and T2 (T2*) weighting. *Magn. Reson. Med.* **41,** 423–428.

Yang, Q. X., Williams, G. D., Demeure, R. J., Mosher, T. J., and Smith, M. B. (1998). Removal of local field gradient artifacts in T2*-weighted images at high fields by gradient-echo slice excitation profile imaging. *Magn. Reson. Med.* **39,** 402–409.

Zhang, W., Williams, D. S., and Koretsky, A. P. (1993). Measurement of rat brain perfusion by NMR using spin labeling of arterial water: In vivo determination of the degree of spin labeling. *Magn. Reson. Med.* **29,** 416–421.

Zhong, J., Chen, Z., and Kwok, E. (2000a). New image contrast mechanisms in intermolecular double-quantum coherence human MR imaging. *J. Magn. Reson. Imaging* **12,** 311–320.

Zhong, J. H., Chen, Z., and Kwok, E. (2000b). In vivo intermolecular double-quantum imaging on a clinical 1.5 T MR scanner. *Magn. Reson. Med.* **43,** 335–341.

Zhou, X., Cofer, G. P., Suddarth, S. A., and Johnson, G. A. (1993). High-field MR microscopy using fast spin-echoes. *Magn. Reson. Med.* **30,** 60–67.

Zhou, X., and Lauterbur, P. C. (1992). NMR microscopy using projection reconstruction. *In* "Magnetic Resonance Microscopy" (B. Blumich and W. Kuhn, eds.), pp. 3–27. VCH, New York.

17

CT Angiography and CT Perfusion Imaging

M. H. Lev and R. G. Gonzalez

Department of Radiology, Massachusetts General Hospital and Harvard Medical School, Boston, Massachusetts 02114

I. Introduction

A. Head CT Scanning: Brief Historical Background

A revolution in medical imaging occurred in 1971, with the introduction of computed tomography (CT) scanning by Godfrey N. Hounsfield, an engineer from the United Kingdom. For the first time, physicians and neuroscientists could directly visualize the gross structure and cross-sectional anatomy of the brain *in vivo,* without relying on invasive and often painful or dangerous procedures, such as pneumoencephalography or catheter arteriography [1]. Although the earliest scanners, which were manufactured by the company Electrical and Musical Industries (EMI—then the record label for "The Beatles"), took minutes to hours to acquire and reconstruct low spatial and contrast resolution images (13 mm slice thickness, 80 × 80 matrix), they could distinguish the ventricles, sulci, and cisterns of the brain, as well as pathologies including tumor, stroke, and hemorrhage [2, 3]. The original clinical prototype machine was housed at the Atkinson Morely Hospital in Wimbledon in 1971 and was used for 1½ years before a seminal presentation of its utility was made at the April 1972 British Institute of Radiology meeting [2, 4]. Hounsfield was knighted for his contribution to medical science and shared the 1979 Nobel Prize in Medicine with South African physicist Allan Cormack of Tufts University in Boston, who had independently established the mathematical basis for CT image reconstruction a decade earlier. It is noteworthy that the developers of MR imaging, despite MRI's relative complexity, have yet to be awarded a Nobel Prize!

The first clinical CT scanners in the United States, priced at $400,000, were EMI head-only units installed in June 1973 at the Mayo Clinic, in Rochester, Minnesota, and, 1 month later, at the Massachusetts General Hospital, in Boston, Massachusetts [4, 5]. Within 5 years, over 1000 CT scanners had been purchased from any of 17 manufacturers [6]. It was not until 1976 that large-bore, whole-body scanners were produced commercially. The same year also marked the creation of federal legislation mandating scanner owners to obtain certificates of need from their state health departments; this requirement was rescinded during the Reagan era [6]. Since then, there have been numerous hardware and software advances, facilitating the evolution of first- through fourth-generation machines capable of faster scanning times, thinner slices, increased matrix (routinely 512 × 512), fewer artifacts, and faster reconstruction times. With the advent of iodinated CT contrast, it became possible to delineate parenchymal lesions with deficient blood–brain barrier, as well as major intracranial blood vessels [7, 8].

The rapid growth of MRI in the late 1980s and early 1990s threatened to overshadow the popularity of CT imaging for all but a small number of applications, most notably with regard to MRI's undisputed superiority in: (1) delineating long segments of intracranial and extracranial vasculature (MR angiography, MRA) and (2) obtaining high-resolution "functional" images of brain physiology (functional MRI, diffusion-weighted imaging; MR perfusion; and MR spectroscopy). Indeed, CT was eulogized in a 1987 editorial in the journal *Radiology*. CT imaging experienced a renaissance, however, with the development of helical (often incorrectly referred to as "spiral") CT technology in 1991 [9]. Helical CT's speed and resolution made possible rapid, accurate, and convenient noninvasive contrast-enhanced CT angiographic imaging (CTA), although X-ray tube cooling constraints limited the extent of coverage. The annual demand for CT examinations in the United States has increased from approximately 3.6 million in 1980, to 13.3 million in 1990, to 33 million in 1998—although, from the mid-1980s to the 1990s, the sale of new machines was roughly flat [10]. It has been estimated that by 2005, 95% of all CT units in the United States will be helical [11].

A dual-detector Elscint helical CT system was marketed in 1992, but a true second revolution in CT scanning awaited the introduction of multislice CT (MSCT) technology in the late 1990s [12]. The first commercially available four-slice helical scanner was the 1998 General Electric Medical Systems "LightSpeed," capable of performing a complete, high-resolution, neurovascular CTA acquisition—from aortic arch to brain vertex—in under a minute (Fig. 1). More recently, high-quality lower extremity runoff studies, from the supraceliac abdominal aorta to the feet, have also been obtained with MSCT, in only 66 s of scanning time [13]. Other innovations that have prompted the remarkable recent increase in CT usage include parallel advances in networking, picture archival and communication systems (PACS), and 3D workstation technology [14].

Clinically, multislice CT technology has improved patient throughput and increased scanner productivity by as much as 50%, despite the fact that multislice studies are performed using more complicated protocols than are used on single-slice CT scanners [15]. The accessibility of helical scanning has made physiological, "functional" CT imaging of perfusion parameters such as cerebral blood flow (CBF), cerebral blood volume (CBV), and mean transit time (MTT)—first proposed with EMI scanners in 1976—a current clinical reality [16]. The next generation of multislice CT machines will be not only capable of volumetric imaging—with intrinsic slice thickness equal to axial in-plane resolution and 0.5-s gantry rotation speeds—but also integrated into fused imaging systems, together with positron emission tomography (PET) or arteriography units, to create truly coregistered multimodality scanners.

B. Clinical Relevance

1. Anatomy vs Physiology

Conventional MR and CT imaging, despite their ability to sensitively *delineate* tissue anatomy and *detect* head and neck pathology, are limited in their ability to *physiologically characterize* abnormal tissue.

Routine CT scanning is especially advantageous for the accurate detection of acute intracranial hemorrhage and of dystrophic calcification or bony pathology; routine MR imaging is more useful in screening for soft tissue abnormalities. Although the addition of intravenous contrast agents can improve both the sensitivity and the specificity of conventional MR and CT imaging, entities such as tumor, infection, inflammation, and infarction, depending on their age and location, can remain indistinguishable on postcontrast images [17]. Moreover, some disease processes, including psychiatric illnesses, early dementia, and vascular pathologies such as migraine headaches, can present with a grossly normal appearance on conventional imaging. Routine pre- and post-contrast MR and CT alone, for example, cannot reliably discriminate either irreversibly *infarcted* brain tissue from *ischemic* brain tissue at risk for infarction in acute stroke patients or recurrent brain *tumor* from *radiation necrosis* in post-radiation-treated oncology patients.

It is for these reasons that the clinical role of various physiological, "functional" imaging techniques, which could potentially differentiate normal from normal-*appearing* but diseased tissue, is currently of great interest among investigators. Examples of such functional imaging modalities include, but are not limited to, angiographic imaging of arteries and veins, perfusion imaging of tissue blood flow parameters, diffusion imaging of microscopic water motion, and direct imaging of biochemical processes using endogenous (MR spectroscopy) or exogenous (nuclear medicine) tissue tracers.

Perfusion imaging can be accomplished in a variety of ways, using a variety of modalities. MR (with intravenous (IV) contrast or endogenous spin labeling), CT (with IV contrast or inhaled xenon), and nuclear medicine methodologies

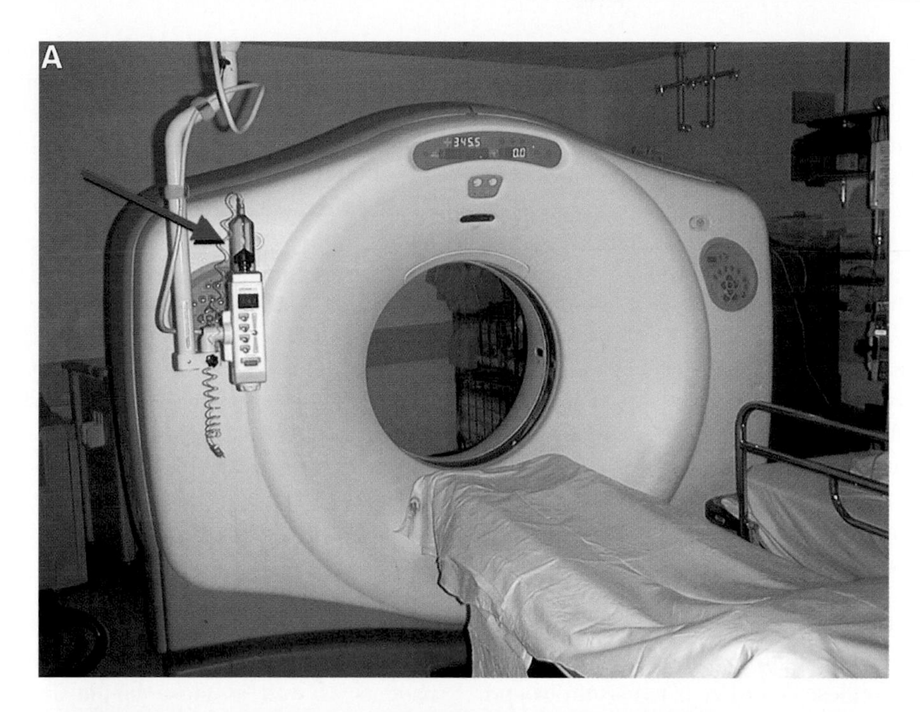

Figure 1 (**A**) A current generation MSCT scanner, the GE LightSpeed. A single-bore IV contrast power injector (Medrad) is also shown (red arrow). (**B**) Control room for the MSCT scanner. The center workstation controls the power injector.

such as PET and thallium-201 single-photon-emission-computed tomography 201 (Tl-SPECT) can all be adapted to measure tissue-level perfusion. Clinical diffusion-weighted imaging is almost exclusively performed using echo-planar MR techniques. Biochemical imaging can be accomplished by MR spectroscopy, as well as by a vast array of nuclear medicine methodologies, including [^{18}F]fluorodeoxyglucose (FDG) PET, methionine PET, and tyrosine PET. Newer contrast agents, such as ultrasmall superparamagnetic iron oxide susceptibility particles, can also be used for blood pool and physiological imaging by virtue of their slow uptake into the reticuloendothelial system (including lymph nodes and liver).

2. CTA Clinical Overview

CT angiography, unlike MR angiographic methods, provides direct anatomic visualization of the vascular lumen. This is because CT pixel intensity is directly proportional to the attenuation, or density, of the contrast agent within a blood vessel. Because MRA techniques, even with the administration of gadolinium, are, at least in part, flow dependent, they can be prone to artifactual signal dropout secondary to *slow* or *in-plane* flow [18]. This difference in sensitivity between CT and MR angiographic methods can be crucial in clinical situations in which the distinction between true vascular occlusion, versus a patent but severely stenotic vessel ("hairline residual lumen"), will influence clinical decision-making. A common example of this situation occurs when a patient presenting with acute stroke or transient ischemic attack (TIA) has complete absence of internal carotid artery flow-related signal enhancement on MRA. In this setting, CTA is typically used to definitively establish the presence or absence of vascular occlusion—potentially obviating the need for more invasive, expensive, and dangerous catheter arteriography.

CTA is typically indicated to expedite the evaluation of any disease process for which visualization of large and intermediate-sized blood vessels is required, including but not limited to aneurysms, carotid artery occlusive disease, intracranial thrombosis (stroke), venous occlusive disease, arterial dissection, vasospasm following subarachnoid hemorrhage, and arterial–venous malformations (AVMs). Our general approach to these disorders is to use ultrasound or MRA as a primary screening exam, with CTA as a "problem-solving" tool for those cases in which disease is likely to be present. In the assessment of acute stroke patients, however—for reasons that will be discussed at length in subsequent sections—CTA is routinely our first-line imaging modality [19]. Indeed, CTA has largely replaced conventional catheter arteriography as the diagnostic test of choice for a variety of vascular disorders, although for preoperative patients, or those with complex multivessel disease, CTA alone may be insufficient. Catheter techniques are in no danger of becoming obsolete, however, as endovascular treatment of numerous neurovascular disorders using glues, coils, and balloons is rapidly gaining acceptance as an alternative to open neurosurgical procedures.

3. CT Perfusion (CTP) Clinical Overview

CT perfusion imaging techniques are relatively new compared to MR-based methods; their clinical applications are therefore less thoroughly reported in the literature [20–22]. Despite this, because the general principles underlying the computation of perfusion parameters such as CBF, CBV, and MTT are the same for both MR and CT, the overall clinical applicability of perfusion imaging using both of these modalities is likely to be similar. A major advantage of CT perfusion imaging is that it is readily available at relatively low cost. This is especially relevant for acute stroke imaging in an emergency department setting, where an unenhanced head CT is already routinely obtained in order to exclude hemorrhage [23]. Additionally, first-pass CTP, unlike MRP, readily provides high-resolution, *quantitative* data using commercially available software. First-pass CTP is currently limited, however, in the extent of coverage that can be obtained during a single bolus injection of contrast. This limitation is sometimes clinically restricting, but is likely to improve as the capability of MSCT scanner software and hardware continues to advance.

MR and CT perfusion techniques are likely to be at least as sensitive and specific as nuclear medicine-based techniques in the accurate grading of intracranial neoplasia and offer the additional advantages of higher intrinsic resolution, convenient coregistration with conventional MR imaging, and time- *and* cost-effective imaging, in patient populations that may already receive frequent conventional MR imaging studies [24, 25]. For stroke imaging, MR also has the advantage of concurrent DWI—a highly specific test for the early detection of irreversibly infarcted tissue [26]. A major disadvantage of MRI methods, however, is their inability to image patients with specific contraindications to MR such as pacemakers, cochlear implants, claustrophobia, and metallic aneurysm clips of uncertain origin. MR studies are also difficult to perform for patients on life-support systems, a group that is likely to include critically ill acute stroke patients as well as those in the immediate postoperative state. In the case of acute stroke imaging, although gradient echo susceptibility imaging has been shown to be highly sensitive for the detection of acute intraparenchymal hemorrhage [27, 28]—an important contraindication to thrombolytic therapy—unenhanced head CT scanning remains the standard of care in excluding a newly hemorrhagic stroke.

4. CTA/CTP Research Overview

In addition to their direct clinical applications, CT angiography and CT perfusion imaging have the potential to play an important role in clinical trials, both as a means of classifying patients into various disease subtypes and as a surrogate marker of treatment response.

In acute stroke disease, for example, despite a multitude of animal studies that have demonstrated a benefit from neuroprotective agents, the only therapy proven in humans to improve outcome has been thrombolysis (both intravenous and intra-arterial) [29–31]. It has been suggested, "if further negative results such as those of recent large neuroprotective trials are to be avoided, trial methodology must be reevaluated" [29]. Specifically, stroke researchers have advocated the use of cerebral imaging techniques to more appropriately select those patients most likely to benefit from a given treatment, thereby increasing the power of clinical trials [29, 32]. In the case of thrombolysis, CTA or CTA combined with CTP could be used to identify patients with proximal large-vessel occlusive thrombus, who are the most appropriate candidates for intra-arterial treatment [19, 33, 34]. The ability of perfusion imaging to quantitatively determine ischemic brain regions that are viable but at risk for infarction if blood flow is not quickly restored—so-called "ischemic penumbra"—might provide a more rational basis for establishing the maximum safe time window for administering thrombolytic agents than the current, arbitrary cutoffs of 3 h postictus for IV and 6 h postictus for intra-arterial (IA) thrombolysis [29, 35]. CT perfusion imaging also has been shown to have prognostic value in predicting both final infarct size and clinical outcome in embolic stroke patients, which may aid in the development of inclusion criteria for clinical trials [36]. Additionally, the detection of early serial changes in CBF and CBV might prove to be a useful surrogate marker for predicting response to varied stroke treatments. Preliminary evidence suggests that perfusion measurements may also be useful in identifying embolic stroke patients with a high risk of hemorrhagic transformation [37–40].

The degree of angiogenesis, or new blood vessel formation, is known to be a critical factor in the growth of brain tumors [41, 42]. Hence, anti-angiogenesis agents, such as the drug Endostatin, are currently the focus of intensive research in the prevention and treatment of cancer. Because the degree of CBV and CBF elevation in cerebral neoplasms has been shown to be proportional to tumor grade, CT perfusion measurements have the potential to serve as surrogate markers of tumor response to experimental therapies [17, 43–45]. Perfusion imaging may also have an important role in studying the physiology of brain tumors. For example, in an MRI study of human cerebral gliomas, dexamethasone was found to cause a dramatic decrease in blood–tumor barrier permeability and regional CBV, but no significant changes in CBF or the degree of edema [46].

Other areas of active perfusion imaging research include dementia and vasospasm. Reductions in hippocampal CBV appear to be present even in early cases of Alzheimer's disease [47, 48]. Symmetrically elevated CBV has been demonstrated in the basal ganglia of patients with AIDS dementia complex [49]. Ischemia due to arterial spasm is the most important cause of morbidity and mortality following subarachnoid hemorrhage from aneurysm rupture; CBF measurements can be used to assess the effectiveness of treatments such as intra-arterial papaverine [50].

5. Summary of CTA/CTP Advantages/Disadvantages

Many of the advantages of CTA/CTP have already been alluded to above. Compared to MRA/MRP methods, CT is typically less expensive, faster, and more readily available in an emergency department setting. Because intravascular contrast enhancement with CTA is not flow dependent, CTA can better distinguish hairline residual lumen from true total occlusion than unenhanced MRA. It can also more accurately discriminate fine increments of vascular stenosis than MRA or ultrasound. CTA has higher spatial resolution, but typically less temporal resolution, than MR, ultrasound, and nuclear medicine techniques. Unlike MRP, CTP can be used to produce truly quantitative CBF, CBV, and MTT maps, but with more limited coverage, or truly quantitative CBV maps covering the whole brain.

Downsides of CTA/CTP include ionizing radiation, the need for intravenous administration of iodinated contrast material, and the potential for long postprocessing and image review times. Therefore, compared to ultrasound and MRA, CTA is less convenient for screening and follow-up studies. Contrast-to-noise ratio is routinely greater for MR and nuclear methods.

II. Technical Background

A. Physical Principles of CT Imaging

All of neuroradiology can be conceptualized as the measurement of the interaction between brain tissue and various forms of energy to produce an image. Image contrast can then be thought of as the difference in signal intensity contributed by tissues of differing composition and structure. Image reconstruction requires the spatial localization of these various contributions to signal intensity. With *ultrasound,* for example, the relative reflection, transmission, and absorption of high-frequency sound waves by different tissues are measured and postprocessed to create a gray-scale view of the relevant anatomy. Spatial localization is based on the time required for reflection of the transmitted sound waves. Similarly, with *MRI,* the absorption and subsequent "echo" of a resonant radiofrequency pulse by *in vivo* hydrogen nuclei (in the presence of a strong, fixed magnetic field) are recorded and postprocessed to create an image. Spatial localization here is accomplished by the clever use and timing of three orthogonal magnetic field gradients, along with a mathematical tool—the Fourier transform—that allows determination of the amplitude corresponding to discrete frequency and phase components of the echoes.

With *CT scanning,* it is the interaction between X-rays and the electron shells of tissue atoms that is measured and postprocessed to create an image. Specifically, a thin, fan-shaped, 80- to 140-kV X-ray beam is projected through the brain onto an array of detectors [51]. The detectors record the degree to which the beam is absorbed, scattered, or transmitted. The degree of beam *attenuation* is proportional to the *density* of the tissue along the ray extending from the X-ray tube to the detector. To a lesser degree, the CT linear attenuation coefficient, m, is also dependent on the effective atomic number of the material comprising the tissue and inversely proportional to the energy of the incoming X-ray beam [52]. Thus, unlike other imaging modalities, the physical property reflected by the gray-scale values displayed on the final CT image is predominantly that of tissue *density.*

In order to collect enough data to construct a full image, the tube–detector pair was originally rotated around the scanner gantry by small increments to produce hundreds of views, the attenuation values of which are all stored [2]. In the newest generation of CT machines, a complete continuous 360° rotation can be accomplished in 0.5 s. Image reconstruction is achieved using a mathematical technique known as *filtered backprojection,* which facilitates calculation of the specific attenuation value at each pixel in the imaged slice, based on all the collected raw data.

1. CT Artifacts

A typical scanning field of view (FOV) for a head CT scan is 20–25 cm, which, together with a 512 by 512 matrix, results in a 0.4- to 0.5-cm pixel size [pixel size = (FOV)/(matrix size)]. Common CT scanning *artifacts* are caused by patient motion, aliasing, and very high attenuation structures (such as metallic aneurysm clips), which produce "streaking" [51]. Detector imbalance can produce circular "rings." Partial volume averaging refers to "contamination" of a voxel's attenuation value by differing structures in an adjacent slice or pixel; in other words, partial volume averaging occurs when a single voxel contains more than one tissue type, each type contributing to that voxel's attenuation value.

Beam hardening artifact, unique to CT scanning, manifests as inappropriately low attenuation values deep to a region of bone or metal. This is a consequence of unequal absorption of different energies from a polychromatic X-ray beam; the beam "hardens" when lower energy X-rays are more preferentially absorbed than are higher energy X-rays. Thus, deep tissues, "behind" a high-density structure, that are disproportionately exposed to "hardened" X-ray beams, appear to have a lower attenuation coefficient, m, than they otherwise would (remember that m is inversely proportional to the energy of the X-ray beam) [52]. Beam hardening results in lower absorption in the center of the object and is sometimes referred to as "cupping" artifact. Superimposed linear streaking can also occur. In modern scanners, beam hardening is minimized by the use of appropriate beam filtration and by applying empirical algorithms during image reconstruction.

2. Hounsfield Scale

CT attenuation values are expressed, according to a linear density scale, as "Hounsfield units" (HU), after Sir Godfrey Newbold Hounsfield, the inventor of CT scanning. In the Hounsfield scale, water is arbitrarily assigned a value of 0 HU. All other CT values are computed according to

$$HU = 1000 \times (\mu_{tissue} - \mu_{H2O})/\mu_{H2O}, \quad (1)$$

in which μ is the CT linear attenuation coefficient.

The HU values for each pixel (which reflect the electron density of the imaged tissue at a given location) are converted into a digital image by assigning a gray-scale intensity to each value—the higher the number, the brighter the pixel intensity. For example, because fat is less dense than water, with an HU value in the –30 to –70 range, fat always appears darker than water in CT images. Some approximate HU values for tissues commonly found on head CT scans are shown in Table 1.

TABLE 1

Hounsfield units	Tissue
>1000	Bone, calcium, metal
100 to 600	Iodinated CT contrast
30 to 500	Punctate calcifications
60 to 100	Intracranial hemorrhage
35	Gray matter
25	White matter
20 to 40	Muscle, soft tissue
0	Water
–30 to –70	Fat
<–1000	Air

The increased attenuation value of parenchymal hematoma, over that of normal intravascular blood, can be attributed to the high density of the "globin" (and not the "heme") component of the retracted clot. Indeed, unenhanced CT scanning remains the primary screening exam for intracranial hemorrhage.

Head CT images are routinely reviewed using standardized "center-level" and "window-width" settings of approximately CL = 30 HU and WW = 80 HU. This means that pixels with attenuation values of 30 HU (the center level) are assigned an intensity at the middle of the gray scale, that pixels with attenuation values of ≤–10 (= 30–40, the center level minus half the window width) are assigned an intensity at the bottom of the gray scale (maximally dark), and that pixels with attenuation values of ≥70 (= 30 + 40, the center level plus half the window width) are assigned an intensity at the top of the gray scale (maximally bright). In other words, using these standardized "head review" windows, fat and air, which both have Hounsfield attenuation values less than –10 HU, would appear equally black on head CT images, whereas a subdural hematoma and the adjacent skull (both greater than 70 HU) would appear equally white (Fig. 2).

The window and level settings used for CT scan image review are known to influence both lesion conspicuity and diagnostic accuracy. For example, in the CT angiographic evaluation of severe carotid artery stenoses, optimal window and level viewing parameters are required for precise lumenal diameter measurement [53–55]. Similarly, the advent of thrombolytic therapy for acute stroke makes the detection of hypodense, ischemic brain parenchyma exceedingly important. The hypodensities associated with cytotoxic edema, however, can be subtle. In animal models of middle cerebral artery stroke, the drop in CT attenuation due

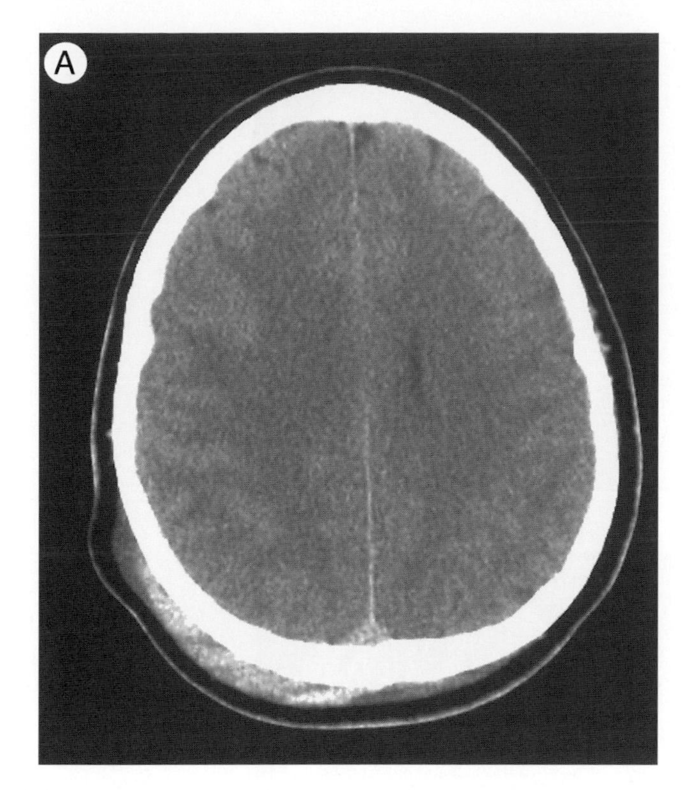

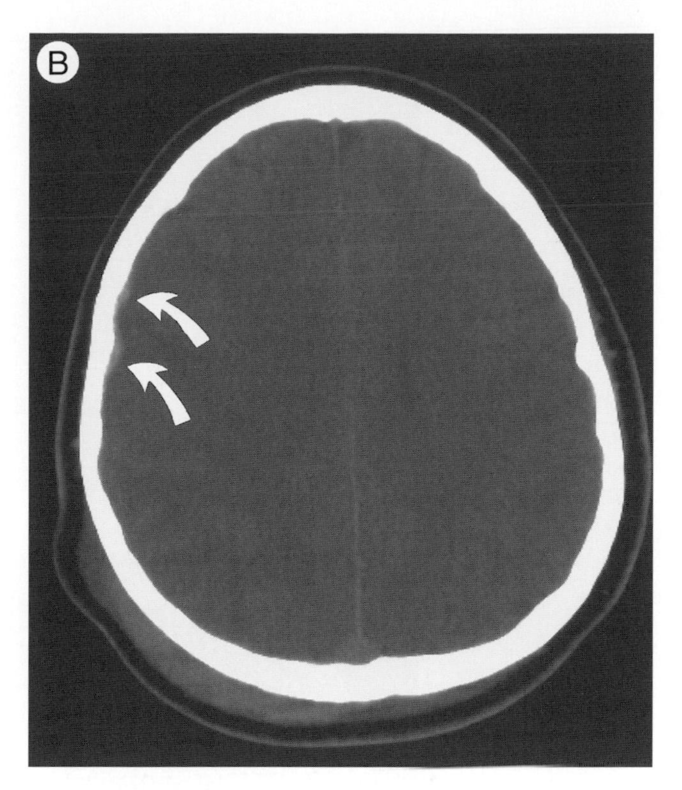

Figure 2 (**A**) Axial head CT from a trauma patient, filmed with "routine head" CT center-level and window-width review settings. A right parietal soft tissue hematoma, but no intracranial hemorrhage, is seen. (**B**) Same image filmed with "subdural hematoma" (see Table 2) center-level and window-width settings. A small crescentic subdural hematoma, adjacent to the inner table of the right frontal bone, can now be visualized (arrows).

to cytotoxic edema is less than 8 HU at 4 h postictus [56]. Because a difference of at least three steps on a 128-step gray-scale image is generally required for lesion detection, such subtle differences in Hounsfield attenuation can be difficult to discern with standard review settings [57]. Not surprisingly, we have shown that the detection of hypodense brain parenchyma is facilitated by soft-copy review using narrowed window-width settings, chosen so as to accentuate the small difference between normal and ischemic gray and white matter attenuation [58]. Specifically, in our series of 21 acute stroke patients, narrowing the mean window width from 80 to 30 HU increased sensitivity for acute stroke detection from 57 to 71% (Fig. 3). Some researchers have proposed taking advantage of these small attenuation changes associated with acute stroke to perform automated image segmentation of acutely infarcted regions [59].

Given the recent marked increased use of freestanding PACS workstations by radiology departments, the potential for increasing diagnostic accuracy through interactive image interpretation, without increasing the time or expense required to film and review hard copy images at multiple different window and level settings, currently exists [60, 61]. Indeed, many major radiology departments are now mostly or completely digital. Table 2 shows some commonly used, "preset" window-width and center-level review settings used during routine, soft-copy head CT interpretation:

The sample values noted in Table 2 should be considered only approximate and often need to be "fine tuned" by the user for optimal viewing. The general principle for determining appropriate soft-copy display settings is to choose the center level to be close to the Hounsfield attenuation of the tissue or tissues of interest, with the window width sufficiently wide such that the bordering tissues fall within the gray scale of the image [53, 58]. For example, as already noted, any tissues with attenuation values greater than 70 HU will appear equally bright using standard routine head settings—hence, a thin, crescentic, subdural hematoma

TABLE 2

Setting name	Center level (HU)	Window width (HU)
Routine head	30	80
Acute stroke	30	30
Skull	250	4000
Subdural hematoma	65	130
CTA	150	450
Soft tissue	0	350
Lung	–500	1500

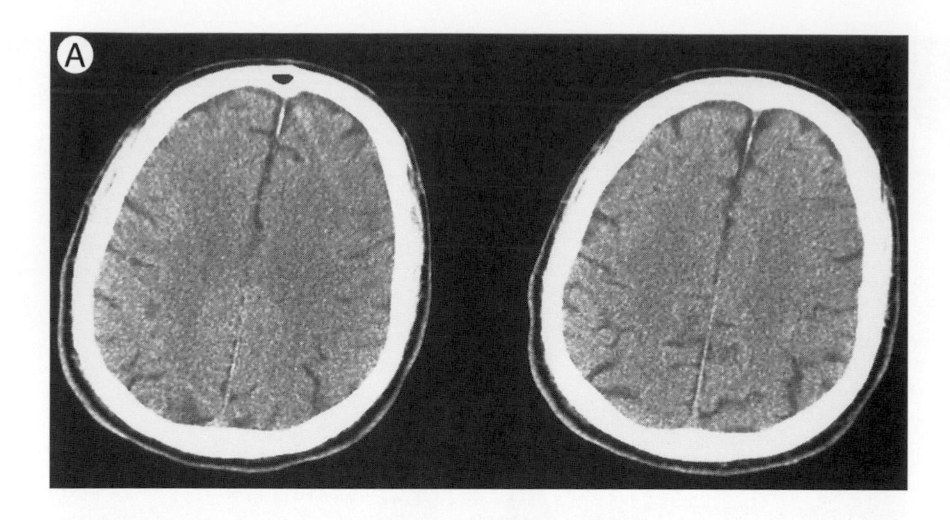

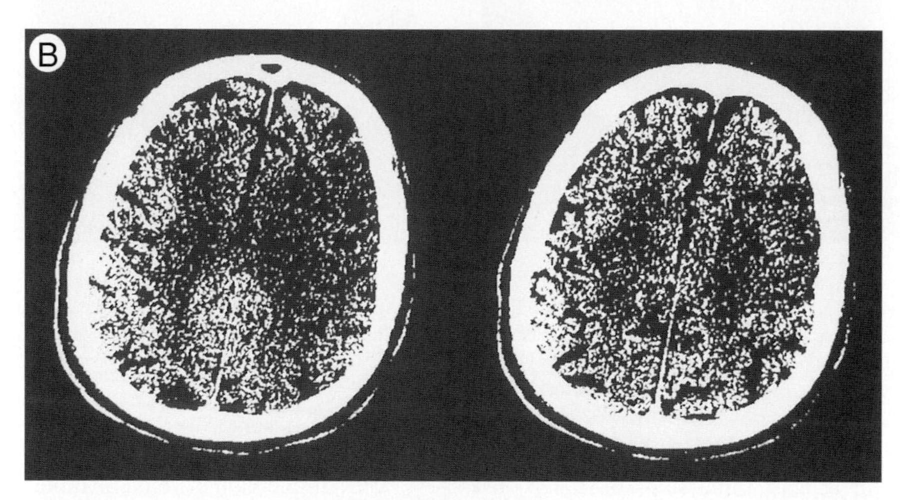

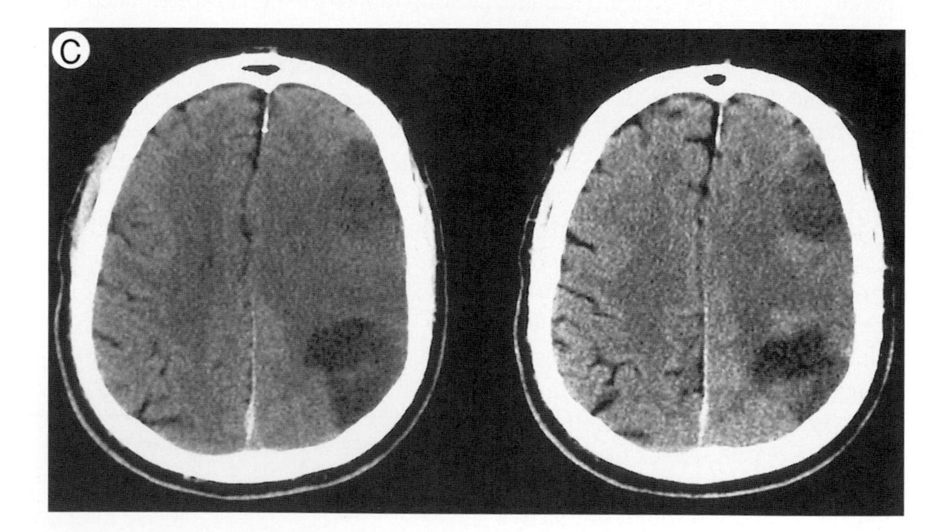

Figure 3 (**A**) Axial head CT from a patient 1.5 h status post onset of new hemiparesis, filmed with routine head CT window and level review settings. (**B**) Same images filmed with acute stroke window and level settings (as per Table 2). Low density of the left frontal lobe is now better visualized, consistent with cytotoxic edema from acute stroke. (**C**) Follow-up scan confirms left frontal infarction.

might be indistinguishable from the overlying skull (Fig. 2). Because the Hounsfield attenuation of blood clot falls within the 60–100 HU range, however, and the dynamic gray scale of images reviewed using subdural hematoma settings ranges from 0 (maximally dark) to 130 (maximally bright) HU, this problem does not arise when scans are viewed using the subdural settings. The dynamic gray scales of such settings, however, are too wide to distinguish subtle changes in gray matter–white matter differentiation and so are not suitable for routine evaluation of the brain parenchyma. Similarly, the window-width and center-level settings used for CTA review are chosen so as to optimize the appearance of intravascular contrast in comparison to surrounding soft tissue structures [53–55].

The precise protocols used for CT scan acquisition depend on both the clinical indication for the study and the generation of CT scanner being used. *Helical scanners,* which can scan more rapidly than conventional "axial mode" scanners, appeared on the scene only during the early 1990s. Even more recent is the development of *multislice helical CT* scanners, which not only have greater speed, but far less strict X-ray tube heating limitations than do single slice helical scanners.

B. Helical CT Background and Definitions

Volumetric CT scanning using helical scanners has revolutionized diagnostic imaging. In addition to neurovascular applications such as CTA and CTP, the current generation of MSCT scanners has made possible the development of screening protocols for lung cancer (high-resolution chest CT), colon cancer (virtual colonoscopy), and cardiac disease (coronary calcium scoring and heart muscle perfusion) [62]. The evaluation of patients with trauma, peripheral vascular disease, renal colic, and liver lesions has also substantially improved. Geoff Rubin of Stanford noted in 1999 that "helical computed tomographic technology has improved over the past eight years with faster gantry rotation, more powerful X-ray tubes, and improved interpolation algorithms, but the greatest advance has been the recent introduction of multislice computed tomography scanners. These scanners provide similar scan quality at a speed gain of 3–6 times greater than single slice computed tomography scanners. This has a profound impact on the performance of computed tomography angiography, resulting in greater anatomic coverage, lower iodinated contrast doses, and higher spatial resolution scans than single slice systems" [9].

1. Single-Slice Helical Scanners

Helical CT scanning permits the rapid acquisition of angiographic type images, with no greater risk to the patient than that of a routine, contrast enhanced CT scan. In helical CT scanning, a *slip-ring* design allows the gantry to freely rotate around a full 360°. CT data are continuously acquired as the scanner table smoothly moves through the rotating gantry, creating a helical "ribbon" of data that can be reconstructed at any slice increment and reformatted in any arbitrary plane (Fig. 4). The term "spiral" CT scanning, which is sometimes used synonymously with helical scanning, is actually a misnomer, because a spiral tapers out to a point, whereas the diameter of a helix remains constant. Moreover, the term "spiral" typically refers to a two-dimensional structure [63].

2. Helical Scanning: Basic Definitions

Optimal single-slice helical scanning requires adequate milliampere second (mAs) and kilovoltage (kV) and therefore a high X-ray tube heat load capacity. The following definitions will facilitate our discussion.

(1) Pitch: The ratio between table travel per complete gantry rotation and X-ray beam collimation thickness, or

$$\text{Pitch} = \frac{\text{table travel per gantry rotation}}{}$$

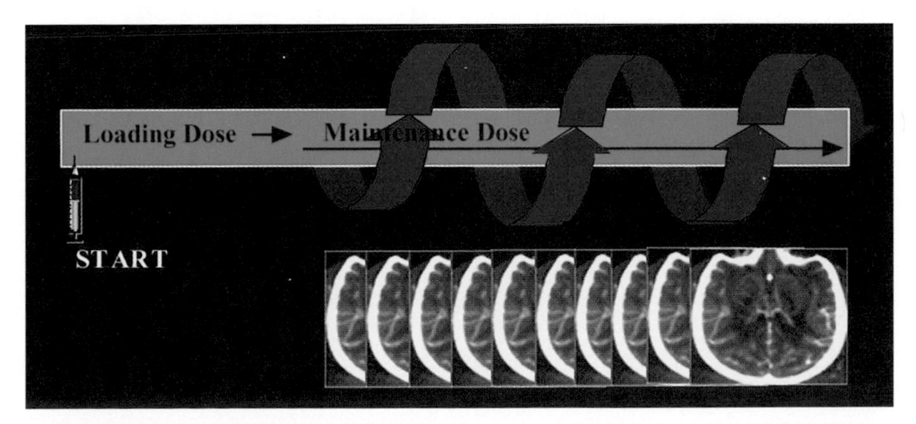

Figure 4 CT image data acquisition. Diagram of the helix created by the continuous rotation of the CT scanner X-ray tube during smooth motion of the CT table through the scanner gantry. (Courtesy of Leena Hamberg, Ph.D., and George J. Hunter, M.D., Boston.)

beam collimation (2)

"Pitch" is therefore unitless [63]. With single-slice scanning, optimal "z direction" (i.e., perpendicular to the scanner gantry, or parallel to the scanner table) and in-plane resolution are achieved using a pitch of 1, in other words, when the table travel in millimeters per rotation equals the X-ray beam thickness in millimeters [14]. At times, however, tube-heating constraints limit the distance that can be covered during a bolus injection of contrast. In such cases, increasing the pitch by a factor of up to the square root of 2 (~1.4) can provide additional coverage without appreciable increase in noise. For multislice scanners, the relationship between pitch and image quality is more complex (Fig. 5). This will be discussed in more detail in the next section.

(2) Vertical distance covered by scan: In designing single-slice scanner protocols, the user typically selects the minimum possible pitch that will permit the desired length of coverage at a desired slice thickness (beam collimation). These parameters are related according to

$$z\text{-axis coverage per scan (mm)} = \text{pitch} \times \text{beam collimation (mm)} \times \text{No. rotations per scan.} \quad (3)$$

In our single-slice CT angiographic protocols, we attempt to adjust our exam parameters such that sufficient intravascular contrast is present throughout the duration of scanning, but that the minimum possible total bolus of contrast is administered. The relationship contrast dose volume = [injection rate (ml/s)] × [total injection time (s)] is used to determine total contrast volume. For reasons that will be discussed at length in subsequent sections, the majority of our CTA protocols employ both a standard 25-s "prep" delay between the start of contrast injection and the initiation of scanning and a 3 ml/s contrast injection rate. At this rate, assuming a normal cardiac output, early arterial enhancement is typically achieved within 17–22 s and early venous enhancement within 20–25 s [21, 34]. Although the determinants of contrast volume redistribution and washout are complex, "useful" arterial enhancement—for the purposes of CT angiographic imaging—typically persists for only 15–20 s after the cessation of contrast injection [64, 65]. Dual-bore CT-compatible power injectors, however, which are currently in development, have the potential to reduce the total required contrast dose by up to 20–40%, with the use of a postbolus saline flush [66, 67]. Such a reduction in dose not only might be safer and cheaper, but also has the added potential to reduce streak artifact caused by pooling of highly concentrated venous contrast at the subclavian, brachiocephalic, and jugular veins (Fig. 6) [66, 68].

(3) Heat capacity: This is a measure of the maximum heat energy that can be tolerated by the anode (the positive terminal) of an X-ray tube during the production of X-rays. Tubes with a high heat capacity are capable of high output for an extended period of time. X-ray tube heat capacity is an important consideration in designing protocols for single-slice, but typically not multislice, scanners, because higher heat capacity tubes are implemented in these systems. The rapid acquisition of the large digital data set required for helical CT places great demands on CT imaging hardware and software. CT *image noise* is inversely proportional to the milliampere-second setting of the X-ray beam through a

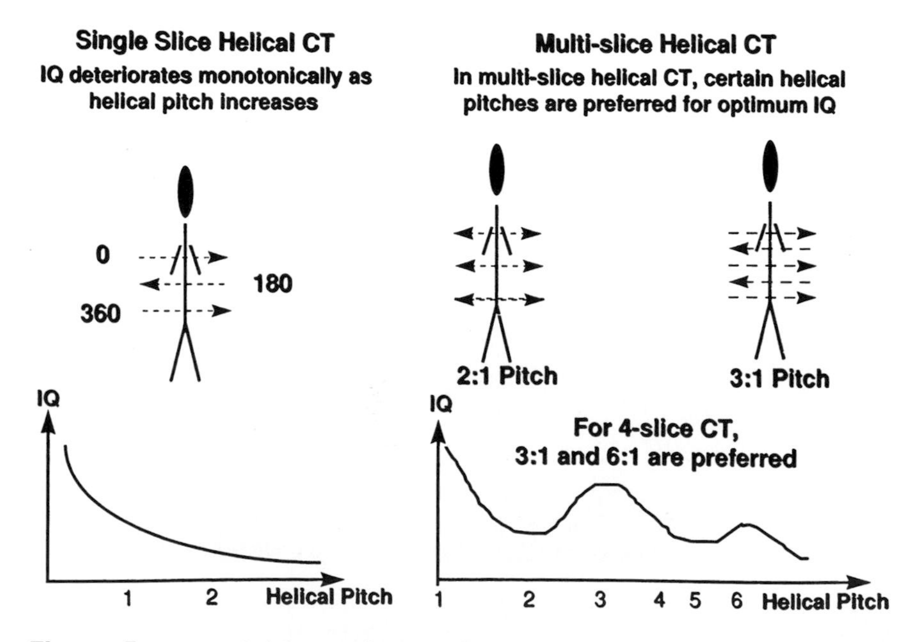

Figure 5 Diagram of pitch versus image quality for single versus multislice scanners. Adapted from Fox *et al.* (1998), Ref. [14], with permission.

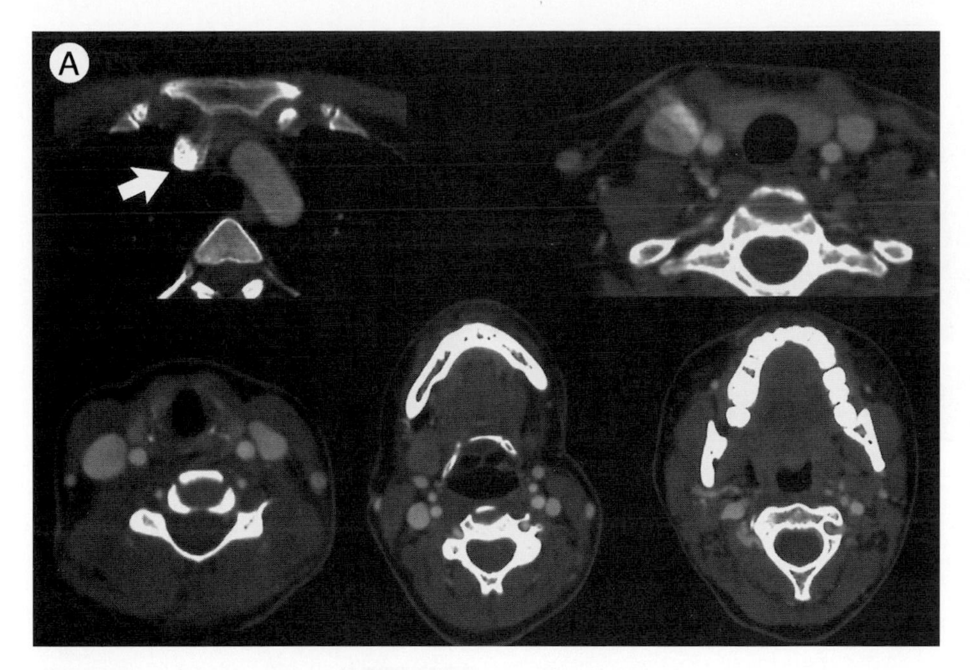

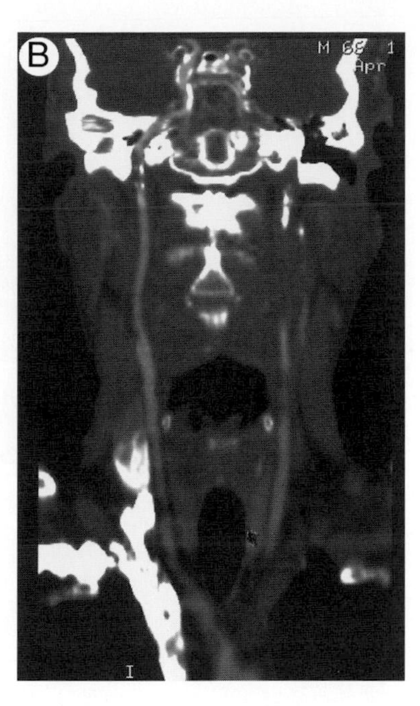

Figure 6 (**A**) Sample axial source images from a CTA acquisition. Note the hyperdense venous contrast at the level of the aortic arch, with adjacent streak artifact (arrow). (**B**) Curved reformat image of the right carotid artery again reveals pooling of hyperdense contrast in the veins of the right thoracic inlet.

given slice, therefore choice of milliampere-second setting involves a trade-off between radiation dose to the patient and image noise. Higher milliampere-second setting also requires a high X-ray tube heat load capacity. Milliampere seconds (the product of the X-ray tube's milliampere setting and the number of seconds required for a complete gantry rotation) is a determinant of photon flux—the number of X-rays passing through the object being imaged [51]. Another user-defined parameter, the kilovoltage setting of the X-ray tube, not only is an additional determinant of photon flux, but also determines the energy distribution of the X-rays produced. Of note, because thinner CT slices have less total photon flux through them than thicker slices, their signal-to-noise ratio is also less (diminished signal relative to fixed background noise); in other words, all else being equal, thinner CT slices are noisier than thick ones if the same technique factors are used.

3. Multislice Scanners

MSCT scanners have the ability to image *entire vascular territories* in under 60 s, well within the time course of the dynamic administration of a single bolus of a contrast agent [14, 69]. The most important features that have facilitated the advancement of MSCT technology have been new detector array designs, as well as higher heat capacity X-ray tubes. The number of detector rows along the scanning axis (*z* direction) of an array can vary from 8 (Siemens) to 34 (Toshiba), and the size of these elements can vary from 0.5 mm centrally (Toshiba) to 5.0 mm peripherally (Siemens) [69]. The total length of the detector array along the *z*-axis direction is typically 2 cm. Marconi, Siemens, and Toshiba MSCT scanners all utilize a mixed array of detector elements, with unequal width (also known as adaptive array detectors) [69]. The "LightSpeed" scanner (GE) utilizes a ceramic-mosaic, equal-width detector matrix comprising 16 rows of detector elements in the *z* direction, each row 1.25 mm thick. The detector matrix consists of approximately 14,000 elements, an over 100-fold increase compared to single-slice scanners. Four independent signals (channels) can be collected from one, two, three, or four of the detector rows, allowing for the acquisition of one, two, or four slices per gantry rotation (Fig. 7). Slice thickness can therefore vary from 1.25 to 10 mm. By the time this chapter appears in print, 8-slice (i.e., 8-channel) MSCT imaging will be a clinical reality. Multislice scanners capable of true isotropic volumetric CT imaging are also not far off; in other words, *z*-axis resolution will equal axial, in-plane resolution [70]. This will be especially advantageous for new applications in cardiac imaging, but will also be important in neurovascular CTA applications such as aneurysm detection. Both 16- and 256-channel MSCT detectors are currently in development, which may also facilitate increased coverage [71].

Unlike older, single-slice helical CT collimators, the new multislice GE collimator consists of two independently controlled tungsten cams. Moreover, a new focal spot geometry, closer to the patient's isocenter, allows for a 20% mAs reduction compared with conventional axial CT scanning. Changes in the slip-ring design, as well as new reconstruction algorithms, also facilitate more rapid scanning.

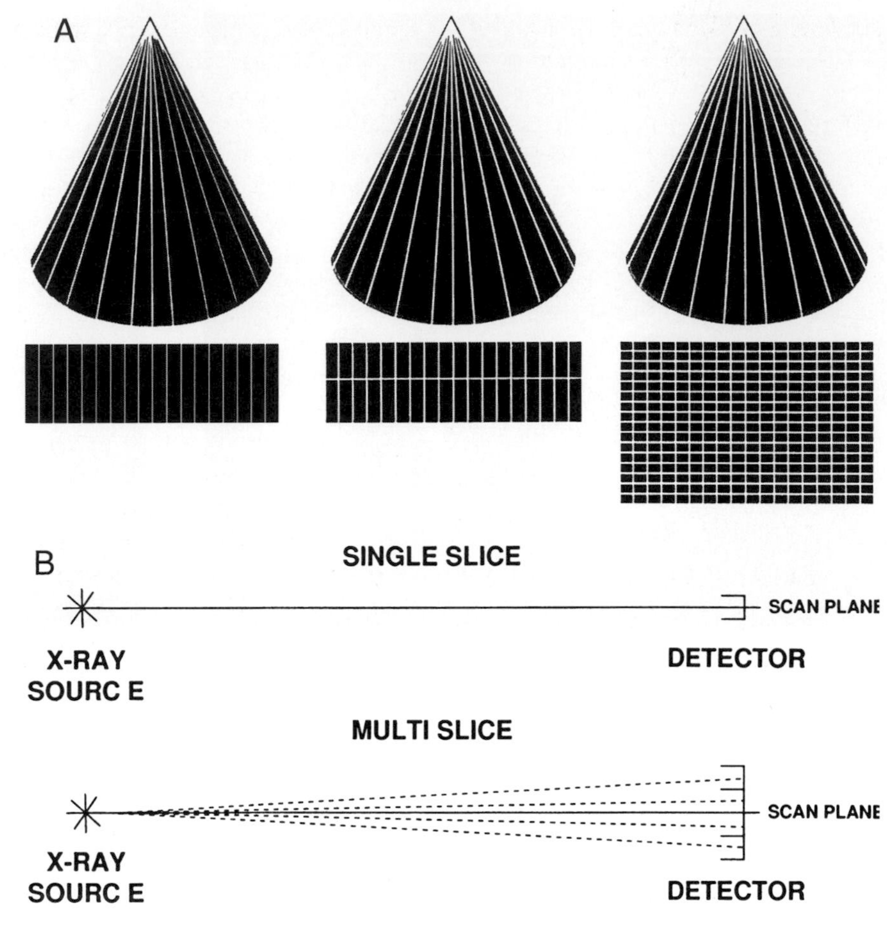

Figure 7 (**A**) Diagram demonstrating three different equal-width detector arrays (left bottom, 1 row; middle bottom, 2 rows; and right bottom, 16 rows). The fan-beam X-ray source is shown for each. (**B**) Diagrammatic representation of beam collimation for single- (top) versus multirow (bottom) detector array. Adapted from Fox *et al.* (1998), Ref. [14], with permission.

Compared to single-slice scanners, there are minimal tube heating constraints with multislice helical scanners. This permits rapid scanning over large vascular territories without interscan delays, using only a single bolus of contrast. As with "conventional" helical CT, initial postprocessing and image reformation can be performed at the scanner console; advanced postprocessing is performed at an independent image processing workstation. Advantages of multislice scanners include reduced total contrast dosage for CTA coverage, shorter scanning duration, and thinner slices versus single-slice scanners [72].

4. MSCT Definitions: Scanning Modes, Pitch, and Detector Configurations

Multislice helical CT scanners can operate in two basic scanning modes—*axial* or *helical.*

Multislice axial scanning is no longer routinely used at our institution for neurological work, due to the speed and improved resolution of MSCT scanning in helical mode. Compared to single-slice helical scanning at 1:1 pitch, however, axial-mode MSCT scanning is advantageous due to its flexibility. On the LightSpeed system, for example, a 4 × 2.5-mm axial scanning configuration uses four channels, each channel composed of input from two 1.25-mm detectors, for a total of eight detector rows used per rotation. This permits X-ray beam coverage of 10 mm in the z direction per rotation; the data are reconstructed into four 2.5-mm-thick slices. Total scanning time is a function of both the time required for actual data acquisition (= time per rotation × number of rotations) and the time required for table movement between rotations. In axial "step and shoot" mode, for example, using a 4 × 5-mm ("4I") configuration, 20 mm of coverage (four 5-mm-thick slices) can be accomplished every 4 to 6 s, typically allowing complete high-resolution brain scanning in under 15 s. Time per rotation is typically 1.0 or 0.8 s, although the newest generation of MSCT scanners is capable of 0.5-s gantry rotations, potentially useful for cardiac imaging.

With multislice helical scanning, because each channel of detector signal generates a *separate* helix of imaging data, the reconstruction algorithms used to merge these data and create MSCT images are complex [73]. Collimation in MSCT is achieved, in part, through the choice of detector configuration. For MSCT, therefore, the definition of pitch is less straightforward than for single detector scanners and can be based on a denominator of either *beam collimation,* as in Eq. (2) or *nominal slice width,* which is based on detector configuration [74]. Indeed, the definition of pitch varies among different CT manufacturers [63]. Some manufacturers have selected the following alternative definition of pitch, different from Eq. (2):

$$\text{Pitch} = \frac{\text{table motion per gantry rotation}}{\text{minimum detector thickness}} \quad (4A)$$

Based on the definition of pitch in Eq. (2), the vertical coverage (z-axis direction) for an MSCT scanner can be computed based on the following relationship, where C is the number of data channels, W is the width of each channel, T is the total scan time, and R is the rotation time:

$$z\text{-axis coverage (mm)} = \text{pitch} \times C \times W \times (T/R). \quad (4B)$$

The original definition of pitch [Eq. (2)], however, is generally preferable to this alternative definition [Eq. (4A)], in that it preserves the basic relationship between radiation dose, overlap of the X-ray beam helices, and pitch already established for single slice CT and so can be unambiguously applied to *either* single- or multislice scanners [74]. For example, a pitch of 3 using the GE MSCT definition of pitch [Eq. (4A)] is equivalent to a pitch of 0.75 using the original definition of pitch [Eq. (2)]; similarly a GE MSCT pitch of 6 is equivalent to a pitch of 1.5 based on the original definition. The relevance of this is that an original pitch value less than 1 implies that radiation overlap is occurring, whereas a pitch greater than (but not equal to) 1.5 implies that image quality may be degraded.

More specifically, whereas in single-slice scanning, image quality is directly proportional to pitch (increased blurring with increased pitch), with multislice scanning—depending on the precise reconstruction algorithms used—there are specific pitches for which image quality is optimized (Fig. 5). For convenience, the GE-compatible definition of pitch [Eq. (4A)] will be used throughout the following discussion. With the LightSpeed scanner, for example, image quality peaks at pitches of 3 and 6; these user-selected, preset pitch options are referred to, respectively, as the *high-quality* (HQ) and *high-speed* (HS) scanning modes [14, 70].

HQ scanning, which uses a pitch of 3:1 (using GE's definition of pitch), is faster than, but at least equal in image quality compared to, single-slice helical CT scanning using a pitch of 1:1 [14, 70, 73, 75]. In HQ scanning, four interleaved (i.e., maximally overlapping) helices are collected and reconstructed per rotation. Interleaved mode permits more data collection, allowing for a reduction in milliampere second and less interpolation, decreasing helical artifact. HS scanning, which uses a pitch of 6:1 (again using GE's definition of pitch), is up to 4× faster than, but at least equal in image quality compared to, single-slice helical CT scanning using a pitch of roughly 1.5–2:1 [75]. In HS mode, four interspaced (i.e., minimally overlapping) helices are collected per rotation. Using HS technique and the scan parameters gantry rotation speed of 0.8 rev/s, 1.25-mm slice thickness, and a 7.5 mm/s table speed, the effective "full width at half-maximum" slice resolution profile is 1.6 mm. At this table speed, the entire distance from the great vessel origins at the aortic arch to the vertex of the skull (approximately 35–40 cm) can be imaged in 45–60 s.

Because of this high speed and resolution, MSCT has become an ideal modality for both CTA imaging of the *neurovascular system*—sometimes obviating the need for conventional catheter X-ray arteriography (XRA)—and CTP imaging of the *brain parenchyma.* Of note, in MSCT scanning, table speed is an *independent* variable. In other words, only the table-speed setting determines total scan time (total scan time = table speed × coverage length), but the pitch setting influences image quality, slice resolution profile, and radiation dose. It is our experience that MSCT artifact at the posterior fossa of the brain causes unacceptable degradation in image quality with a pitch of 6 (HS mode, once again using the GE definition of pitch). The helical, or "star-like" artifacts produced are most severe at bone–soft tissue interfaces. Such artifacts are accentuated at high table speeds or with thick image collimation [70, 73]. The explanation for worsened artifact at helical pitch of 6 is based on the fact than HQ mode (pitch of 3) provides maximal overlap of the helical scanning channels, whereas HS mode (pitch of 6) provides minimal overlap and therefore maximal interpolation—with minimal photon flux through portions of the tissue being scanned along the z axis. For this reason, all of our intracranial MSCT protocols utilize HQ, and not HS, mode.

A drawback, however, to using HQ mode exclusively, is that, although signal-to-noise ratio is increased and therefore image quality is improved with minimal helical artifacts, the minimum obtainable intrinsic slice thickness is also increased. Specifically, the number of detector rows used to form a data acquisition channel—and therefore the minimal achievable intrinsic slice thickness—is related to pitch [using Eq. (4A), the alternative definition of pitch] and table speed in the following way:

$$\text{Minimal intrinsic slice thickness} = \frac{\text{table speed}}{\text{helical pitch}} \quad (5)$$

In other words, all else being held equal, switching from HS mode to HQ mode will double the minimal achievable slice thickness. For example, although HQ at 7.5 mm/rotation through the posterior fossa is superior in overall image quality to HS at 7.5 mm/rotation, without an increased scanning time, the use of HQ technique does result in a higher radiation dose and thicker slices. In this case, the minimal possible slice thickness using HQ is calculated to be 7.5 (table speed)/3 (pitch) = 2.5 mm, whereas the minimal achievable intrinsic slice thickness using HS is 7.5/6 = 1.25 mm. The users can specify, however, the degree of slice overlap in the reformatted images, in order to avoid "zipper" artifact, which is discussed in more detail in a subsequent section (for example, 2.5-mm-thick slices can be reformatted at 1.25-mm intervals). Thus, there is a trade-off between image quality and slice thickness.

5. Radiation Dose Considerations

The topic of CT radiation dose has been very much in the news as of late, especially given the availability of alternative imaging techniques such as MRI and ultrasound, which do not carry the risks of ionizing radiation [10, 76–78]. Nuclear Regulatory Commission guidelines limit occupational radiation exposure to no more than 5 rem/year, a maximal permissible dose with a negligible probability of harmful effects [79]. For comparison purposes, background environmental radiation exposure from natural sources, including radon and cosmic rays, is estimated to be approximately 350 mrem/year (0.35 rem, or 3.5 mSv), and the effective dose from a single chest X-ray is approximately 5–10 mrem [80]. Such federal dose regulations do not apply to diagnostic X-ray examinations of patients, however, because it is assumed that the small per-exam risk from these studies is balanced by the medical necessity, and potential diagnostic benefits, of the test.

Actual patient dose calculations in clinical situations are complex and depend on a multitude of factors including the kilovoltage and milliampere second of the X-ray beam (i.e., beam energy), filtration, collimation, exposure time, irradiated volume, type of tissue being irradiated, patient weight, and geometry of the irradiated region, to name just a few [78, 81, 82].

Also potentially confusing are the units used to measure radiation exposure. Traditionally, the absorbed energy from a radiation source has been measured in "rads," or "radiation absorbed dose"; 1 rad is defined as the absorption of 100 ergs per gram of material. Often a more useful measure is "rem," or "roentgen equivalent man," which relates the absorbed dose in human tissue to the biological damage it causes; rem is calculated as rad multiplied by a correction factor (most typically 1 for gamma rays). One rad is the equivalent of 10 mGy, where gray (Gy) is the SI unit for radiation absorbed dose. Absorbed dose equivalent can be measured using yet another SI unit of radiation exposure, the "sievert" (Sv), where 1 rem = 10 mSv.

Moreover, patient exposure can be calculated in a variety of ways and expressed in quantities such as total dose, surface dose, and organ dose [83]. Surface dose measurements, for example, tend to average about one order of magnitude larger than organ dose measurements resulting from the same X-ray exposure. A useful estimate of radiation risk, especially for comparing different CT protocols, is the "dose-length product," abbreviated DLP, and measured in mGy cm. DLP equals the "CT dose index" ($CTDI_{vol}$), multiplied by the scan length, where $CTDI_{vol}$ is the average dose in the standard head CT dosimetry phantom [84–86]. $CTDI_{vol}$ is measured in milligrays and is often a useful measure by which to compare the radiation output from different CT scanners. $CTDI_{vol}$ is primarily dependent on mAs and kV (i.e., beam energy) [83]. The "effective dose," E, is sometimes employed as a closer estimate of patient risk than DLP, but is much more difficult to determine. Of note, although the definition of pitch varies among different multislice CT scanner manufacturers, because the formula for dose index takes into account the slice thickness and the number of slices per gantry rotation, meaningful comparisons of radiation dose can be made between different MSCT units [63].

MSCT scanner geometry can affect radiation dose in two competing ways. First, the thin septa between detector elements along the z axis absorb radiation, resulting in up to a 4.5% loss of efficiency for many scanners [69]. On the other hand, with MSCT, more of the X-ray beam is used for imaging than in single-slice scanning (in other words, "the 'umbra-to-penumbra' ratio is higher because the ratio of beam collimation to focal spot size is higher") (Fig. 7B) [69]. The overall, net effect of combining these two factors is that the dose efficiency of MSCT is roughly comparable to that of single-slice helical CT [69]. An example of how scanner geometry affects radiation dose occurs with the GE LightSpeed, for which there is a choice of two focal spot sizes—large and small. Use of the smaller focal spot, which can tolerate a tube current of up to only 170 mAs but employs a special focal-spot tracking algorithm designed to reduce dose inefficiencies related to the MSCT cone beam geometry, can result in images of equal quality, but lower total radiation dose, compared to those obtained using higher mAs but with the larger focal spot [74].

There is a trade-off between radiation dose and image noise. High milliampere-second settings result in decreased image noise, or "quantum mottle," due to increased photon flux, but at the expense of undesirably high radiation doses. This principle explains why thinner slices are more grainy, or mottled, in appearance than thicker slices acquired at the same mAs and kV settings—there is less photon flux through a thinner slice and therefore a poorer signal-to-noise ratio. In other words, thinner slices require a higher mAs setting in order to reduce image noise. Multiple contiguous thinner slices can also result in increased dose due to scattered radiation, or focal spot penumbra, extending outside the imaged slice to a proportionately greater degree than is the case with thicker slices [10]. The minimum X-ray dose required for adequate image quality is additionally proportional to tissue depth; it has been calculated that an increase in soft tissue thickness from 4 to 8 cm requires doubling of the milliampere second in order to maintain a constant noise level [84, 85].

In general, our MSCT protocols call for a setting of 140 kV. Our objective is to minimize the radiation dose *absorbed by the patient* by using 140 kV, along with the minimal possible mAs setting for a given application. This is because milliampere second is proportional to the number of photons used for scanning (photon *flux*), whereas kilovoltage reflects both photon number and photon *energy*. At a kV setting of 140, there is less *absorbed* scatter and more through transmission of energetic photons, resulting in—somewhat paradoxically, and again, assuming *minimum possible mAs for a given level of image quality*— an overall *lower absorbed radiation dose* to the patient than would be present using a *lower* kV setting. As already noted above, the smaller distance from focal spot to patient isocenter found in GE MSCT scanners, compared to GE single slice scanners, allows for a 30% reduction in the milliamperes required for scanning, at baseline, compared to single-slice helical scanners. In principle, HS mode has less data overlap, and therefore delivers less radiation dose for a given set of mA and kV settings, than does HQ mode [14, 70].

According to the Radiological Society of North America 2001 Website, "the typical radiation dose from a CT exam of the head and brain is equivalent to the amount of natural background radiation received over a year's time. Among all radiological procedures, radiation exposure from CT of the head is intermediate." Estimates of surface radiation dose for a typical head CT in the literature range from 30 to 70 mGy (3.0–7.0 rad) [82]. Although MSCT scanners provide *increased scanning speed* and *improved scanning resolution* compared to single-slice helical scanners, the trade-off is that of a *possibly increased radiation dose* to the patient, unless the user is careful in the choice of the precise scan mode and scan parameters employed [14, 70]. For example, MSCT at a pitch of 6 may have the same radiation dose as single-slice CT at a pitch of 2, depending on the precise collimation and scan factors chosen. The bottom line here is that radiation doses must be individually computed in the design of any new CTA protocol and that the trade-off between image quality and dose must be carefully considered for a given clinical indication.

Significant decreases in head CT radiation dose may be possible without compromise in image quality simply by a moderate reduction in milliampere second [76, 87]. Indeed, in a study of unenhanced head CTs, our group found that although 90-mAs images were roughly 20% noisier than were 170-mAs images, 90-mAs images were rated by all reviewers as of acceptable diagnostic quality for a variety of indications and that mean gray matter conspicuity was not significantly different for the 170- and 90-mAs groups [88]. These data suggest that reduced-dose CT might be performed for at least some clinical situations—such as monitoring of neuro ICU patients at risk for hydrocephalus—without compromise in diagnostic accuracy.

Ideally, for body applications, mAs settings should be selected based on total body weight—especially for pediatric patients [89]. Studies suggest that, if pediatric patients are imaged using adult scan parameters, the typical radiation dose is at least doubled [10]. It has been estimated that a one-time, uniform body dose of 10 rem, delivered to a 5-year-old child, results in a 1.2–1.5% lifetime increase in fatal cancer risk [10, 79]. Thus, although small, the radiation risks incurred by CT scanning cannot be dismissed as negligible when designing imaging protocols.

Other strategies for CT dose reduction include: (1) adaptive dose modulation—varying the milliampere second dynamically, during scanning, as the patient's anteroposterior and lateral diameters vary, (2) advanced 3D-MSCT interpolation schemes, and (3) longitudinal dose modulation—varying the milliampere second in a preset manner during scanning, according to the body section being imaged. In adaptive dose-modulation studies, agreement between simulations and measured results was better than within 10%, and dose reduction values of 8–56% were found, depending on the phantom geometry and tube current modulation function [84].

Finally, depending on the clinical question being asked, the use of thicker, rather than thinner, slices can permit a reduction in milliampere second [10]. Alternatively, thin slices acquired at a low mAs setting can be reconstructed into thicker, less noisy slices using 3D techniques such as multiplanar reformatting (MPR). Increasing the rate of the table feed can also reduce dose; in single-slice helical scanning, pitch values up to approximately 1.5 do not significantly degrade image quality [90]. Increasing table speed and pitch to the maximum values required to maintain diagnostic quality images will help to lower the total administered radiation dose to the patient [10].

C. CT Angiographic Protocols: General Principles

In designing CTA protocols, the type of scanner (single versus multislice) and its X-ray tube heat capacity are the primary determinants of the maximum coverage length possible during the first-pass circulation of contrast material. Especially for single-slice scanners, protocol design routinely involves a trade-off between a variety of scanning parameters, including pitch, slice thickness, table speed, scan field of view, display field of view, milliampere second, kilovoltage, contrast prep delay, contrast injection rate, and total contrast dose. CT matrix size is generally fixed at 512×512 for a given field of view. The goal of scanning is typically to maximize (1) length of coverage, (2) axial in-plane and z-axis resolution, and (3) signal-to-noise ratio, while minimizing (1) slice thickness, (2) total contrast dose, and (3) total radiation dose to the patient. Clearly, these are conflicting goals.

Some important basic differences between single- and multislice scanners should be kept in mind in approaching protocol design. With single-slice scanners, intrinsic axial slice thickness is best thought of as a function of beam collimation, with image quality a monotonically decreasing function of pitch (i.e., image quality decreases with increasing pitch; Fig. 5). Pitch and table speed are coupled and related to beam collimation—an independent variable—as described by Eqs. (2) and (4B). In other words, for a given beam collimation (and hence a given intrinsic slice thickness), increasing the pitch is equivalent to increasing the table speed, and both pitch and table speed influence not only total coverage length, but also image quality for a given slice.

The relationship between pitch, table speed, intrinsic slice thickness, and image quality for multislice scanners is more complex. Indeed, with MSCT, pitch and table speed are most conveniently thought of as independent variables, with the intrinsic slice thickness determined by the resulting detector configuration as defined by the relationship: "Minimal intrinsic slice thickness = table speed/helical pitch" [Eq. (5)]. Moreover, image quality for a given slice is not simply inversely proportional to pitch, but is variable, depending on the degree of overlap between the acquisition helices. In other words, with MSCT, the user can independently select the table speed and pitch, which, in turn, influence length of coverage, image quality, radiation dose, and intrinsic minimum slice thickness.

Image reformat and reconstruction requirements should also be considered when designing scan protocols.

Reformatting refers to the postprocessing of "raw" image data into "source" images with specified characteristics (Fig. 6). For example, MSCT image data acquired using a detector configuration that produces a minimal intrinsic slice thickness of 2.5 mm can be reformatted into thicker, but not thinner, slices. Such thicker slices are typically even multiples of the intrinsic slice thickness—i.e., 5- or 10-mm-thick slices in the case of 2.5-mm-thick raw image data—and can be created, at the scanner, by the CT technologist after the completion of scanning. Additionally, despite the limitations on minimal obtainable intrinsic slice thickness imposed by the scan acquisition parameters [as demonstrated by Eq. (5)], reformatted slices can be displayed at arbitrary interslice intervals and at arbitrary display field of views. Such smaller slice intervals (for example, 2.5-mm-thick slices reformatted at 1.25-mm interslice spacing) are advantageous for reducing "stair step" or "zipper" artifact in certain two- and three-dimensional reconstructions. Thinner slices, displayed with a smaller field of view, can more accurately depict local vascular anatomy than do thicker, "volume averaged" slices, but, as already discussed, are noisier for a given radiation dose and are therefore more demanding to obtain. With single-slice helical scanning, for clinical applications requiring maximal z-axis resolution, it has been demonstrated that axial slices should be reconstructed with at least a 60% overlap relative to their intrinsic slice thickness [91].

Field of view is defined as the diameter of the circular region scanned or displayed. Field of view determines pixel size, according to the relationship

$$\text{Pixel size} = \text{field of view/matrix size} \tag{6}$$

Scan field of view for the head is typically 20–25 cm. For a 512×512 matrix, this translates into an intrinsic pixel size of 0.4 to 0.5 mm. The physics of CT scanning is such that, when imaging other body parts (such as the aortic arch and great vessel origins through the shoulders) optimal scanning is accomplished when the outer circumference of the scan field of view includes air, i.e., when the scan field of view is larger than the body part being imaged. Intrinsic pixel size, however, is not irrevocably determined by the choice of scan field of view. Raw image data can be reformatted not only into source images with varied slice thicknesses and interslice spacing, but also into source images with varied pixel size, based on user-selected display field of view. Magnification of structures achieved by image reformatting to a smaller display field of view (centered, of course, over the region of interest) is preferable to "geometric" magnification using image display software, because the former, but not the latter, reduces the intrinsic pixel size. In other words, reformatting images to a new field of view, smaller than the scan field of view, results in higher pixel resolution than does geometric magnification.

The term *reconstruction,* as opposed to reformatting, refers to the postprocessing of the reformatted source images into various display projections, in order to create a summary display of vascular or parenchymal anatomy. Common CTA reconstruction techniques, which are discussed at length later in this chapter, include maximum intensity projection (MIP), MPR, shaded surface display (SSD), and volume rendering (VR).

As with MR angiography, patient motion can result in misregistration artifact during CTA image reconstruction. Prior to neurovascular CTA image acquisition, the patient should be instructed to "hold still" and to "breathe slowly and regularly." When the aortic arch and great vessel origins are being scanned, a brief prompt to "hold your breath" is also often required. Moreover, because the carotid bifurcation is located at approximately the level of the hypopharynx and larynx, the patient should be instructed not to swallow. These instructions are important, because although motion correction algorithms for CT imaging exist, most are slow, difficult to implement, and not, therefore, in routine clinical use.

In order to achieve optimal arterial image contrast and resolution, CTA protocols must be tailored to the region being studied, the clinical question being asked, and—for single-slice machines—the heat load capability of the scanner. In general, values for *slice thickness, field of view,* and *pitch* should be as small as possible such that resolution and reconstruction quality of the region of interest (ROI) are optimized, but large enough such that the entire ROI can be covered in a single scan. Conversely, *mAs* values should be large enough to provide sufficient photon flux to limit quantum mottle (image graininess), yet small enough to minimize radiation dose. Image reconstruction interval should also be minimized, in order to reduce step, or zipper, artifact. In general, for any given set of scanning parameters, image quality is better and artifacts are reduced by the use of multislice, rather than single-slice, machines [92].

For example, applying these principles to single-slice scanning of a focal, short-segment, carotid artery plaque, the following scanning parameters are suggested for maximally accurate measurement of the degree of stenosis: collimation 1 mm, pitch 1, reconstruction interval 0.5 mm, scan field of view 18 cm, display field of view 12 cm, kV 140, and mAs 230, with scanning over a narrow slab at the level of maximal stenosis. Screening of the *entire* carotid artery, however, might require collimation 3 mm, pitch 1.2 to 1.5, and, likely, reduced mAs and kV, in order to permit greater *z*-direction coverage [93]. Such limitations are clearly less restrictive with the use of multislice scanners.

Choice of contrast dose and timing in CTA reflects a trade-off between minimizing the quantity of contrast administered, maximizing target *vessel* enhancement, and providing uniform target vessel opacification over a broad area of coverage. Optimizing *parenchymal* enhancement in stroke studies is an additional challenge.

In general, a high bolus injection rate with a brief, 15- to 20-s prep delay is desirable to maximize arterial but minimize venous enhancement over a short *z*-axis distance; this type of injection timing is ideal for aneurysm evaluation [64, 65]. Although the *magnitude* of peak arterial enhancement is a function of injection rate, the *time to peak* arterial enhancement, as well as the *duration* of enhancement (assuming a monophasic injection), is generally not proportional to injection rate per se, but, rather, bears a constant relationship to the time at which the contrast injection is terminated [94]. For studies requiring greater coverage, such as screening for neurovascular stenotic disease of the aortic arch, neck, skull base, and circle of Willis—as well as for stroke studies, which require uniform enhancement through the entire brain parenchyma if "perfused blood volume" measurements are to be obtained—appropriate bolus timing is therefore critical.

A theoretical advantage of a short prep delay is the minimization of venous enhancement, which can interfere with arterial display on 2D and 3D reconstructions. Even with very short prep delays, however, some degree of venous opacification is typically unavoidable, as the transit time from arteries to veins is just on the order of 1 or 2 s. In our experience, with the possible exception of cavernous carotid aneurysm detection, venous opacification *does not* significantly interfere with diagnostic evaluation, so long as careful image reconstruction is performed and, when necessary, axial source images are reviewed.

For the sake of simplicity and speed, we advocate a fixed 25-s prep delay between the onset of contrast administration and the onset of scanning, for all of our intracranial CT arteriography (but not venography) protocols. We have found that, with the exception of patients in atrial fibrillation or with very low cardiac ejection fraction, a 25-s delay is invariably sufficient for adequate arterial opacification. In those patients who do have atrial fibrillation or reduced cardiac output, we therefore employ a 30- to 40-s prep delay as an alternative. The reasoning behind this is that both mathematical and animal models have shown that, when cardiac output decreases—as is the case with atrial fibrillation—the arterial arrival time of an intravenously injected contrast bolus also increases, and there is substantial increase in the degree of peak arterial enhancement [64].

It should be noted that, in the rare event that the bolus is "missed" because scanning commences too soon after the start of contrast administration, the patient should be *immediately rescanned through the same region* in order to obtain a diagnostic study [19]. Of thousands of CTAs performed at our institution, however, this has occurred in only a very small number of cases—most notably in instances of very slow flow within severely stenosed but patent carotid arteries.

Approaches other than using a fixed prep delay time are possible. Scanners equipped with a "smartscan" function can use a variable prep delay in order to minimize venous enhancement; however, this adds complexity to scanning protocols and is seldom required clinically. As an alternative to smartscan, a test bolus can also be used to determine scan delay but will increase the radiation dose. A region of interest is selected in a proximal artery and 10 ml of contrast is injected. This region is scanned continuously using a low milliampere second and kilovoltage technique; the prep delay is chosen as the time corresponding to 50% of maximal test vessel opacification. As with smartscan, however, a test bolus is seldom a clinical necessity.

Logistically, a well-positioned intravenous catheter with a minimum diameter of 20 French (Fr) is required to perform CTA/CTP. If possible, a larger bore, 18-or 16-Fr IV is preferable. Slow contrast delivery through certain types of central venous lines, at maximal rates of 1.5–2.5 ml/s, is also possible using a power injector [95]. Although ionic, iodinated CT contrast agents are potentially neurotoxic when used in the setting of acute stroke, *nonionic* CT contrast agents have been shown to be safe in an animal model of middle cerebral artery (MCA) infarction [96, 97]. The use of nonionic, *iso-osmolar* contrast agents may have additional advantages in high-volume, high-injection-rate CTA/CTP studies—especially in patients at high risk for congestive heart failure. Patient age, renal function, allergy history, history of insulin-dependent diabetes mellitus, and history of previous contrast reactions should all be considered prior to the administration of intravenous contrast.

D. CT Perfusion Protocols: General Principles

Perfusion-weighted CT and MR techniques—in contrast to those of MR and CT *angiography,* which detect bulk vessel flow—are sensitive to capillary, tissue-level blood flow [98]. The idea of contrast-enhanced CT perfusion imaging is certainly not new; as early as 1976, a computerized subtraction technique had been described to measure regional CBV using the EMI scanner. Sodium iothalamate was administered intravenously to increase X-ray absorption in the intracranial circulation, permitting regional differences in CBV to be measured [16]. More recently, but prior to the advent of helical CT scanning, "time-to-peak" analysis of cerebral perfusion was proposed as a means of evaluating stroke patients. Patients with a prolonged, greater than 8 s, time to peak parenchymal enhancement had poor clinical outcomes. This dynamic CT study took 10–15 min longer to perform than a conventional CT exam, and therefore, given the absence of either fast scanning or an approved treatment for acute stroke, never gained clinical acceptance [99].

The generic term "cerebral perfusion" refers to tissue-level blood flow in the brain. This flow can be described using a variety of parameters, which include, but are not limited to, CBF, CBV, and MTT. Working definitions of these parameters are as follows:

Cerebral blood volume is defined as the total volume of blood in a given unit volume of the brain. This definition includes blood in the tissues, as well as blood in the large capacitance vessels including arteries, arterioles, capillaries, venules, and veins. CBV has units of milliliters of blood per 100 g of brain tissue (ml/100 g).

Cerebral blood flow is defined as the volume of blood moving through a given unit volume of brain per unit time (Fig. 8A). CBF has units of milliliters of blood per 100 g of brain tissue per minute (ml/100 g/min).

Mean transit time is a more complex concept. Because the transit time of blood through the brain parenchyma varies depending on the distance traveled between arterial inflow and venous outflow, the mean transit time is defined as the *average* of the transit time of blood through a given brain region (Fig. 8B). Mathematically, mean transit time is related to both CBV and CBF according to the central volume principle, which states that MTT = CBV/CBF [100, 101].

Quantitative noninvasive imaging measurement of these cerebral perfusion parameters is always indirect and is limited by the mathematical model applied to describing the flow of a contrast tracer through the cerebral circulation. Different models vary in their strengths and weaknesses, and the specific model used is in large part dependent on the imaging modality available. With PET and xenon CT imaging, for example, "diffusible tracer" models can be applied, which, generally speaking, involve fewer assumptions regarding steady-state CBF than do the "dynamic, first pass contrast enhanced" models used with MR and CT imaging [102]. Details regarding these models are discussed in a subsequent section of this chapter.

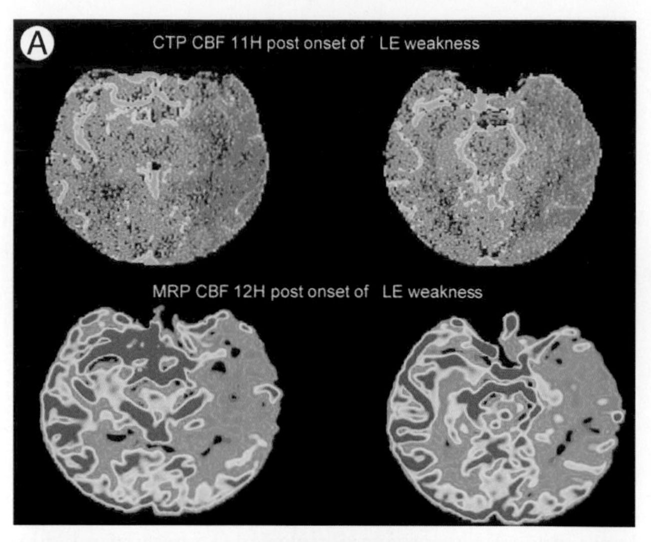

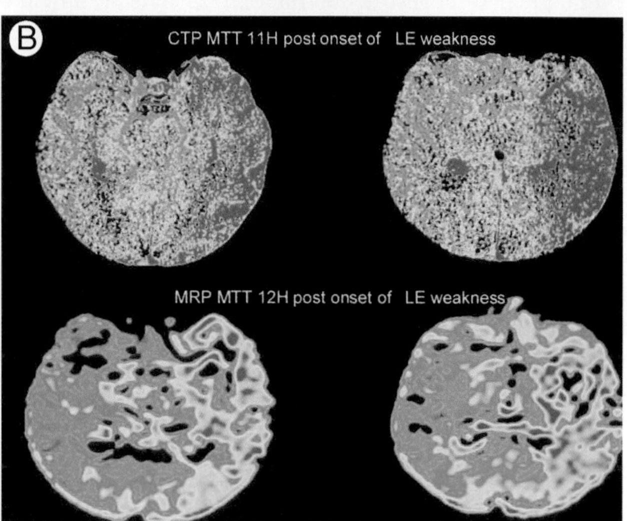

Figure 8 (**A**) CT and MR perfusion CBF maps from a patient with right lower extremity (LE) weakness. There is reduced blood flow in the left hemisphere middle cerebral artery distribution. (**B**) CT and MR perfusion MTT maps from the same patient. There is delayed mean transit time over the left hemisphere middle cerebral artery distribution.

1. Whole-Brain Perfused Blood Volume Maps

Quantitative CT imaging assessment of cerebral perfusion using the "dynamic first pass" model requires cine imaging—the acquisition of multiple consecutive images at a single level or levels over time (Fig. 4). Maps of CBV, CBF, and MTT can then be calculated, on a pixel-by-pixel basis, by analysis of the time-density curves resulting from changes in contrast density over time at each pixel in the imaged section. An alternative technique, ideally suited to many clinical applications because it permits the acquisition of a single set of images of the entire brain and can be performed synchronously with CT angiography using a single contrast bolus, is known as whole-brain perfused blood volume CT imaging or, more generically, as "perfusion-weighted" CT. Although "true" quantitative CBF measurements cannot be obtained using this method, the resulting CT source images can provide a dynamic quantitative measure of CBV—hence the term, "perfused blood volume." Indeed, it has been shown that coregistration and subtraction of conventional, precontrast head CT images from the "whole brain," contrast-enhanced CT source images, and their normalization with the blood value and small to large vessel hematocrit ratio, can provide quantitative maps of perfused blood volume (Fig. 9) [22, 34]. Agreement of this technique with PET-derived measurements of CBV has been good; one study revealed regional values of 4.5 ± 0.6 ml/100 g for cortical gray matter, 2.5 ± 0.6 ml/100 g for white matter, and 3.7 ± 0.4 ml/100 g for the basal ganglia. CBV values in ischemic regions were 1.5 ± 0.4, 0.7 ± 0.7, and 1.8 ± 0.9 ml/100 g, respectively [22].

The perfused blood volume technique, which has been advocated by Hunter and Hamberg, requires the assumption of an approximately steady-state level of contrast during the period of image acquisition [34]. It is for this reason—in order to approach a steady state—that the majority of our neurovascular CT imaging protocols call for a "slow" injection rate of 3 ml/s, with a 25-s prep delay. Theoretically, although they are not yet commercially available, dual-bore CT power injectors, capable of delivering multiphasic contrast injections, could be used to more efficiently reach such a steady state [103, 104]. One study of this topic concluded, "Uniform, prolonged vascular enhancement, which is desirable for CT angiography and essential for steady-state quantification of blood volume in organs, can be achieved with multiphasic injection" [103].

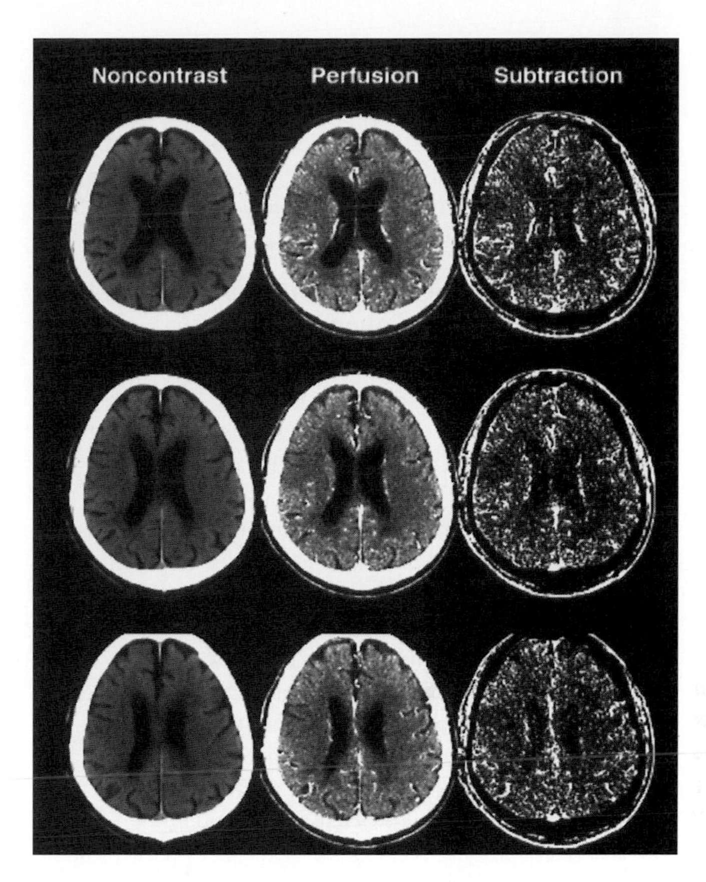

Figure 9 Example of creation of subtraction whole-brain perfused blood volume maps. The unenhanced (**left column**) and enhanced (**middle column**) CT source images are coregistered and subtracted to create pure blood volume images (**right column**). Note the subtle hypodense ischemic lesion with reduced perfused blood volume. Such lesions typically progress to clear-cut infarction. (Courtesy of Nat Alpert, Ph.D., Boston.)

2. Quantitative Single Slab CBV, CBF, and MTT Maps

Although easy to define in theory, the perfusion parameters of CBV, CBF, and MTT can be difficult to quantify in practice. Differing assumptions regarding the inflow and outflow of a tracer substance through a tissue bed lead to different mathematical models for computing perfusion parameters. Such models are known as tracer kinetic models. Dynamic first-pass contrast-enhanced MR and CT perfusion models assume that the tracer (i.e., the contrast) used for perfusion measurement is nondiffusible—that it is neither metabolized nor absorbed by the tissue bed through which it traverses. "Leakage" of contrast material outside of the intravascular space, which can occur in cases of blood–brain barrier breakdown associated with tumor, infection, or inflammation, requires a different model to be used and so adds an additional layer of complexity to the calculations.

As already noted, the first-pass approach to CT and MRI perfusion measurement involves the dynamic IV administration of an intravascular contrast agent, which is tracked with serial imaging during its first-pass circulation through the brain tissue capillary bed. Depending on the assumptions regarding the arterial inflow and the venous outflow of the tracer, the perfusion parameters of CBV, CBF, and MTT can then be computed mathematically. The two major types of mathematical approaches involved in performing these calculations are the deconvolution-based and non-deconvolution-based methods.

In comparison to dynamic MRI techniques, dynamic CT techniques are advantageous due to their convenience, accessibility, high resolution, absence of flow- and susceptibility-related artifact, and, importantly, their ease of quantification.

The MRI techniques, however, are currently advantageous in terms of their multislice, whole brain coverage. Of note, with MRI, endogenous tracers, rather than intravenously administered exogenous contrast agents, can also be used for perfusion imaging. In arterial spin-labeled MR perfusion imaging, for example, the intrinsic T1-weighted MRI signal of red blood cells is used for bolus tracking [17].

3. Theory and Modeling: Deconvolution versus Non-deconvolution Methods

Tracer kinetic techniques for determining cerebral blood flow are well described in the PET literature [105]. These models are based on the single-tissue compartment model for diffusible tracers, originally developed by Kety in 1951 [106].

Non-deconvolution-based perfusion methods rely on the application of the *Fick principle* to a given ROI within the brain parenchyma during the dynamic administration of a contrast bolus. According to this principle, the change in contrast enhancement per unit time within a brain ROI is proportional to the blood flow to that region, multiplied by the difference between the concentrations of contrast within the feeding artery and the draining vein supplying that region. This is simply a "conservation of flow" law. Mathematically, this is expressed by the equation

$$dC_t(t)/dt = \mathrm{CBF} \cdot [C_a(t) - C_v(t)] \quad (7)$$

In the above formula, $C_t(t)$ is the tissue contrast concentration versus time curve (commonly referred to as the time–density curve; TDC) measured within a given brain region. $C_a(t)$ is the time–density curve for the feeding artery (i.e., the arterial contrast concentration versus time curve, also known as the arterial input function, or AIF), and $C_v(t)$ is the time–density curve for the draining vein (i.e., the venous contrast concentration versus time curve). CBF is the local cerebral blood flow in ml blood/100 g brain tissue/min.

The TDC is also known as the *tissue residue function* and reflects the *measured* concentration of contrast agent within a given voxel of brain tissue at a given point in time after the start of contrast administration. In order to create "maps" of cerebral blood flow using cross-sectional imaging techniques, an independent TDC is obtained for each pixel within a slice by acquiring sequential axial images separated by a brief (typically 1 s) time interval. Signal intensity versus time curves can then be generated on a pixel-by-pixel basis for brain tissue [$C_t(t)$], as well as for a feeding artery [$C_a(t)$] and a draining vein [$C_v(t)$]. Because $C_t(t)$, $C_a(t)$, and $C_v(t)$ are known quantities, Eq. (7) can be solved, in principle, on a pixel-by-pixel basis, for CBF. The ease of the mathematical solution to this differential equation, however, is highly dependent on the assumptions made regarding inflow and outflow to the region; some of the more commonly applied models are demonstrated below.

4. No Venous Outflow Assumption

If it can be assumed that imaging is completed prior to contrast reaching the intracranial veins, Eq. (7) reduces to:

$$dC_t(t)/dt = \mathrm{CBF} \cdot C_a(t) \quad (8)$$

Solving for CBF:

$$\mathrm{CBF} = [dC_t(t)/dt]/C_a(t) \quad (9)$$

In practice, therefore, assuming no venous outflow of tracer, the parameter CBF can be *estimated,* for a given pixel, as the ratio of the maximum initial slope of the tissue density concentration curve divided by the peak height of the arterial tissue density curve. Indeed, this method of computing cerebral blood flow is currently used for some CT perfusion protocols [107, 108].

The major advantage of using this method is *simplicity* of calculation. The major disadvantage of using this method, however, is that the assumption of no venous outflow prior to the time of the maximal initial slope of the tissue density curve is generally *not valid* in the brain when contrast is administered at "slow" injection rates of approximately 4–5 ml/s. At higher injection rates, of 10 ml/s or greater, the no venous outflow assumption is more likely to be approximated. Such rates, however, cannot be routinely achieved in the clinics due to practical considerations, limiting the utility of this method when precise quantification of CBF is desired.

5. CBV and MTT Calculations

CBV can be approximated as the area under the "fitted" (smoothed) tissue time–density curve, divided by the area under the fitted arterial time–density curve [109]:

$$\mathrm{CBV} = \int C_t(t)/\int C_a(t) \quad (10)$$

Gamma (variate) "fitting" of the time–density curves is necessary to eliminate the effect of *recirculation* of contrast material after its first pass through the brain capillary bed. This method of CBV calculation is reliable only when the *blood–brain barrier is intact* and the entire bolus of contrast can be assumed to remain in the intravascular compartment. When extensive blood–brain barrier breakdown exists, such as can be seen with tumor, infarction, or abscess, *leakage of contrast material into the extravascular space results in overestimation of CBV*. Methods for correction of contrast leakage when capillary endothelial integrity is compromised have been developed for both MR and CT perfusion imaging [17, 110, 111].

Of note, when it is assumed that contrast concentration in the arteries and capillaries is at a steady state, Eq. (10) forms the basis for the quantitative computation of CBV using the whole-brain perfused blood volume method of Hunter and Hamberg [22, 34]. After soft tissue components have been removed by coregistration and subtraction of the precontrast scan, CBV then simply becomes a function of the density of tissue contrast, normalized by the density of arterial contrast.

If both CBV and CBF are known, MTT can then be calculated by means of the *central volume principle,* which states that

$$\text{MTT} = \text{CBV}/\text{CBF} \qquad (11)$$

Alternatively, the software used to create perfusion maps could be designed so that CBF is calculated, using Eq. (11), from known values of CBV [Eq. (10)] and estimates of transit time. Indeed, when using commercially available or proprietary software to compute perfusion parameters, the user should be aware of precisely how CBF, CBV, and MTT are being calculated and of the advantages and limitations of the underlying model employed.

6. Deconvolution-Based Methods

Direct calculation of CBF, applicable for even relatively slow injection rates, can be accomplished using deconvolution theory [112]. Indeed, the deconvolution model is currently widely accepted for creating *MR* CBF, and hence MTT [with Eq. (11)], perfusion maps. The practical application of the central volume principle to determine cerebral perfusion requires the intravenous administration of an intravascular contrast agent; this tracer is tracked using rapid sequential CT images to construct time–density curves. It is assumed that the signal intensity of contrast material within the arteries, tissue (capillaries), and veins is linearly proportional to the concentration of contrast within these vessels and that blood flow remains a constant during the period of measurement. In order to understand how cerebral blood flow can be calculated from these data using deconvolution, the concept of the *impulse residue function* must be introduced.

The impulse residue function, $R(t)$ (to be distinguished from the "tissue" residue function, already discussed), is defined as the *theoretical, idealized tissue time–density curve* that would result if the entire bolus of contrast being used for the blood flow measurement could be administered *instantaneously* into the artery supplying a given region of brain. The impulse residue function curve is thus characterized by an instantaneous increase in measured signal intensity within the tissue capillary bed, to a plateau value proportional to the blood flow reaching that region, followed by a rapid, continuous decrease toward a zero baseline. The duration of the plateau reflects the length of time during which the injected contrast material remains in the capillary bed and so is proportional to the mean transit time.

Because the precise arterial inlet to a given brain region cannot be localized, and because the arterial input of contrast cannot be made to be truly instantaneous, however, $R(t)$ cannot be directly measured. In clinical practice, the relationship between the measured tissue time–density curve, $C_t(t)$, the arterial time–density curve, $C_a(t)$, CBF, and $R(t)$ is given by the following equation, where "⊗" represents the mathematical "convolution" operator:

$$C_t(t) = \text{CBF} \cdot [C_a(t) \otimes R(t)] \qquad (12)$$

Using an inverse mathematical process to convolution, known as *deconvolution,* and because the tissue and arterial time–density curves, $C_t(t)$ and $C_a(t)$, can be determined directly from the CTP cine images, Eq. (12) can be used to solve for the product CBF · $R(t)$, which is the "scaled" residue function. CBF can now be obtained directly as proportional to the maximum *height* of this scaled residue function curve, whereas CBV is reflected as the *area* under the scaled residue function curve. Once CBF and CBV are known, MTT can be calculated using the central volume principle, Eq. (11).

Mathematically, deconvolution of the arterial (AIF) and tissue curves can be accomplished using a variety of techniques, including the Fourier transform (FT) and the singular value decomposition (SVD) methods. These methods vary in their sensitivity to such factors as: (1) the precise vascular anatomy of the underlying tissue bed being studied and (2) the degree of delay, or dispersal, of the contrast bolus between the measured arterial and tissue time–density curves. The paper by Wirestam *et al.* contains an excellent discussion of the various strengths and weaknesses of these methods [113]. With SVD, the more commonly employed method, dispersal—more so than delay—of the contrast bolus through various collateral channels can invalidate the deconvolution measurement and can be difficult to correct for[1] (see footnote a–d below). Despite this limitation, which can be most marked in acutely ischemic brain regions primarily supplied by collateral blood flow, the deconvolution method of computing CBF has been validated, and used successfully, for MR and CT perfusion imaging in both patient and animal models [110, 114–118].

In the construction of perfusion maps from either CT or MR data sets, voxels comprising the AIF can be selected in a semiautomated manner from an intracranial artery within the imaging volume. Because the deconvolution process can be exquisitely sensitive to image noise, however, dedicated mathematical algorithms are required to prevent oscillations in the computation of the scaled residue function, CBF · $R(t)$,

---

[1] a) Calamante, F., Gadian D. G., and Connelly, A. (2000). Delay and dispersion effects in dynamic susceptibility contrast MRI: simulations using singular value decomposition. *Magn. Reson. Med.*, **44**(3), 466–473.
b) Calamante, F., Gadian D. G., and Connelly, A. (2002). Quantification of perfusion using bolus tracking magnetic resonance imaging in stroke: assumptions, limitations, and potential implications for clinical use. *Stroke*, **33**(4), 1146–1151.
c) Ostergaard, L., Chesler, D. A., Weisskoff, R. M., Sorenson, A. G., and Rosen, B. R. (1999). Modeling cerebral flow and flow heterogeneity from magnetic resonance residue data. *J. Cereb. Blood Flow*, **19**(6), 690--699.
d) Aslop, D. (2002). Perfusion MR imaging. In Atlas, S. (ed.) "Magnetic resonance imaging of the Brain and Spine, Third edition." pp. 215–238. Lippincott, Williams and Wilkens, Philadelphia, Pennsylvania.

in order to preserve the accuracy of the calculated CBF values [119]. As already noted, an advantage of the deconvolution method is that it can provide accurate quantification, despite relatively slow contrast injection rates [112]. In general, deconvolution is also less sensitive to variations in underlying vascular anatomy than are the non-deconvolution-based methods. This is because, for simplicity, the fundamental assumption of most nondeconvolution cerebral perfusion models is that a single feeding artery and a single draining vein support all blood flow to and from a given tissue bed and that the precise arterial, venous, and tissue time–density curves can be uniquely identified by imaging. This assumption is clearly an oversimplification.

Potential imaging *pitfalls* in the computation of cerebral blood flow using the deconvolution method include both (1) patient motion and (2) partial volume averaging, which can cause the arterial input function to be underestimated. The effects of these pitfalls can be minimized by (1) the use of image coregistration software to correct for patient motion as well as by (2) careful choice of ROIs for the AIF (the ROI should be chosen from the center of an artery *orthogonal* to the axial imaging plane, if possible).

7. Optimization of Acquisition Parameters: Contrast, mA, kV, Thickness

Because CT perfusion imaging has only recently gained acceptance as a clinical tool, and because construction of perfusion maps is dependent on the specific mathematical model used to analyze the dynamic, contrast-enhanced data sets, considerable variability exists in the protocols used for CTP scanning. Clearly, for example, algorithm-dependent differences in contrast injection rates exist, with higher injection rates required for the no venous outflow models [107]. Considerably slower injection rates can be used with the deconvolution-based models [112]. Regardless of injection rate, however, as with MR perfusion imaging, higher contrast concentrations are likely to produce maps with improved signal-to-noise ratios [120].

One accepted deconvolution CT perfusion imaging protocol calls for scanning at 80 kV, rather than at a more conventional 120–140 kV. Theoretically—for a constant mAs (in this case 190)—this kV setting might not only reduce the administered radiation dose to the patient, but could also increase the conspicuity of IV contrast, due, in part, to greater importance of the photoelectric effect for 80-kV photons, which are closer to the "*k* edge" of iodine than are 120- to 140-kV photons [121]. This protocol calls for a 4–6 ml/s injection rate of approximately 45–50 ml of a "300" strength (300 mg iodine/ml) nonionic contrast agent. For high-volume, high-injection-rate studies, as has already been discussed, we typically employ iso-osmolar contrast. Scanning begins after a brief, 5-s prep delay, during which time the patency of the IV catheter is quickly checked and the study aborted if contrast extravasation is noted. Images are acquired in cine mode at a rate of approximately one image per second. Improved temporal resolution is possible with some scanners, with acquisition rates as fast as one image per half-second; however, the resulting moderate improvement in tissue-density curve noise may not justify the increased radiation dose.

With most MSCT scanners operating in cine mode, up to 2 cm of *z*-axis direction coverage can be acquired per contrast bolus, typically obtained as either two 10-mm-thick or four 5-mm-thick slices. It is our experience that two 10-mm-thick slices provide sufficient spatial resolution for routine clinical use and, as has already been discussed in the section on radiation dose, have reduced image noise compared with 5-mm-thick slices. The maximum degree of vertical coverage can potentially be doubled, however, using a "toggle table" technique, in which the scanner table moves back and forth, switching between two different cine views, albeit at a reduced temporal resolution of data acquisition [122].

Using the "standard" cine technique, imaging occurs for a total of 45–60 s, sufficient to track the first pass of the contrast bolus through the intracranial vasculature without recirculation effects. The quantitative CBF values obtained using this protocol have been shown to correlate well with those of xenon CT flow measurements in the same set of patients [118].

Perhaps the most important aspect of patient preparation for CTP imaging may be to have an 18- or 20-gauge catheter already placed in an appropriately large vein *prior* to the patient's arrival in the CT suite. It is similarly useful for the power injector to be loaded prior to patient arrival. Total scanning time can be drastically reduced if such details are attended to before the examination. It is important to secure the head with tape or Velcro straps, as motion artifact can severely degrade CTA/CTP image quality. A saline drip will ensure that the IV remains running prior to contrast administration. Gantry angle should be parallel to and above the orbital roof, in order to minimize radiation dose to the lens of the eye. This level is generally appropriate for embolic stroke imaging, because it overlaps portions of the brain supplied by the anterior cerebral arteries, the middle cerebral arteries, and the posterior cerebral arteries.

8. Optimization of Postprocessing: Software Options

Although the precise choice of CTP scanning level is dependent on both the clinical question being asked and other available imaging findings, an *essential caveat* in selecting a CTP slice is that the IMAGED LEVEL MUST CONTAIN A MAJOR INTRACRANIAL ARTERY. This is necessary in order to ensure the availability of an AIF, to be used for the computation of perfusion maps using the deconvolution software.

In urgent clinical cases, perfusion changes can often be observed immediately following scanning by direct visual inspection of the axial source images at the CT scanner console. Soft-copy review at a workstation using "movie" or "cine" mode can reveal relative perfusion changes over time, although advanced postprocessing is required to appreciate subtle changes and to obtain quantification.

Axial source images acquired from a cine CT perfusion study are networked to a freestanding workstation for detailed analysis, including construction of CBF, CBV, and MTT maps. Prior to loading these data into the available software package, the source images should be visually inspected for motion artifact. Images showing significant misregistration with the remaining data set can be deleted or corrected, depending on the sophistication of the existing software.

The computation of quantitative first-pass cine cerebral perfusion maps typically requires some combination of the following user inputs (Fig. 10).

- *Arterial input ROI:* A small ROI (typically 2 × 2 to 4 × 4 pixels in area) is placed over the central portion of a large intracranial artery, preferably an artery orthogonal to the imaging plane in order to minimize "dilutional" effects from volume averaging. An attempt should be made to select an arterial ROI with maximal peak contrast intensity.
- *Venous outflow ROI:* A small venous ROI with similar attributes is selected, most commonly at the superior sagittal sinus. With some software packages, selection of an appropriate venous ROI is critical in producing quantitatively accurate perfusion maps.
- *Baseline:* The baseline is the "flat" portion of the arterial time–density curve, prior to the upward sloping of the curve caused by contrast enhancement. The baseline typically begins to rise after 4–6 s.
- *Postenhancement cutoff:* This refers to the "tail" portion of the time–density curve, which may slope upward toward a second peak value if recirculation effects are present. When such upward sloping at the tail of the TDC is noted, the slice number just prior to the rise in contrast density should be entered as the cutoff value. The perfusion analysis program will subsequently ignore data from slices beyond the cutoff.

Other user defined inputs, such as "threshold" or "resolution" values, are dependent on the specific software package used for image reconstruction. It is worth noting that major variations in the input values described above may not only result in perfusion maps of differing image quality, but, potentially, in perfusion maps with variation in their quantitative values for CBF, CBV, and MTT. As previously noted, special care must often be taken in choosing an optimal venous outflow ROI, because that ROI value may be used to normalize the quantitative parameters.

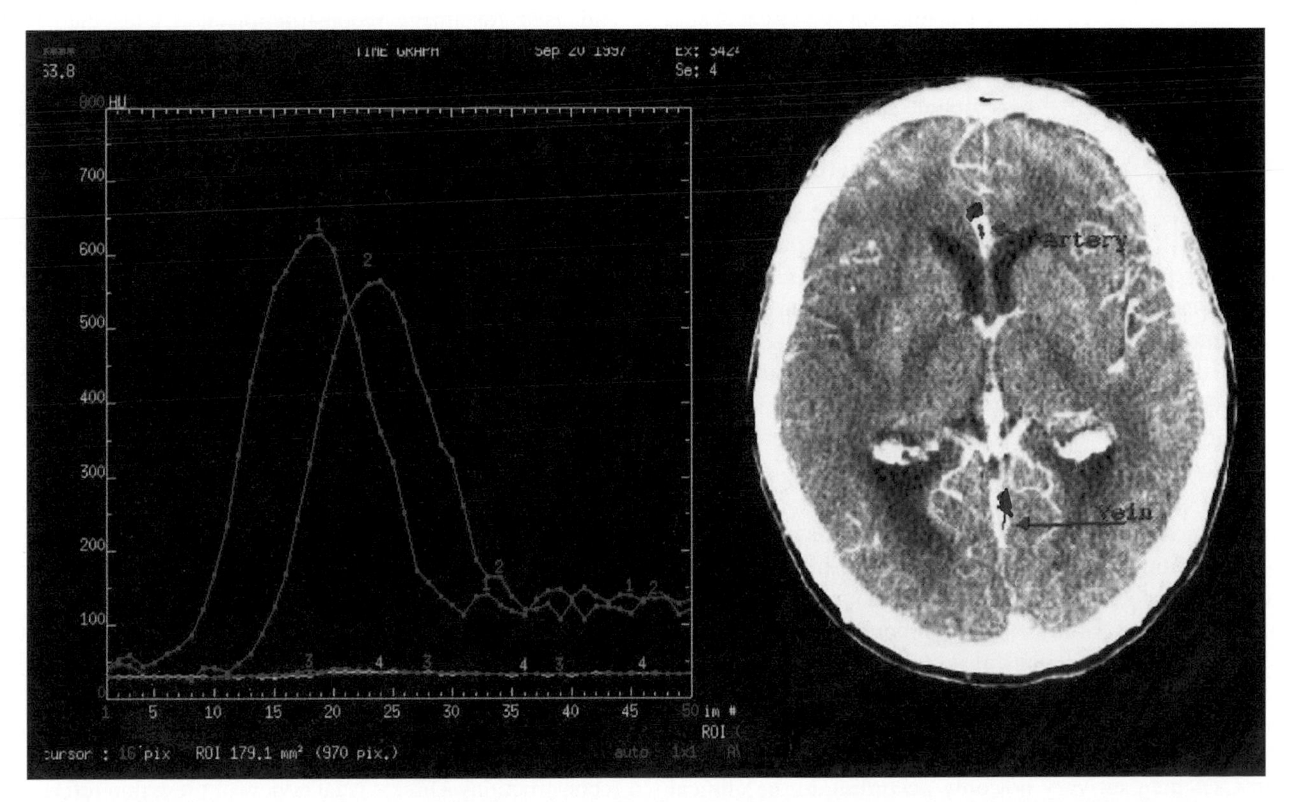

Figure 10 Examples of time–density-curves (TDCs) for the computation of quantitative first-pass cine cerebral perfusion maps. The *y* axis represents Hounsfield units (HU) and the *x* axis represents image number (proportional to time of image acquisition). One image per second was acquired during the dynamic administration of 10 ml/s IV contrast. Curve "1" is the arterial input, curve "2" is the venous outflow, and curves "3" and "4" are unspecified tissue TDCs from gray matter ROIs (not shown). The ROIs corresponding to the arterial and venous TDCs are shown on the source image at right. (Courtesy of Steve Nichols, R.T.R., Boston.) Typically, the torcular or superior sagittal sinus, and not the straight sinus (as shown), is selected as the venous outflow ROI.

E. Postprocessing: General Principles

Postprocessing of CTA images can refer to either image *reformat* or image *reconstruction,* as defined at the start of section II.C. Image reformatting, performed at the CT scanner console, is the recomputation of "raw" CTA image data into "source" images with varying slice thickness, interslice spacing, and display field-of-view.

Reconstruction, which will be emphasized in this section, refers to the manipulation of the resulting, reformatted source images into various two- and three-dimensional projections, in order to create a summary display of vascular or parenchymal anatomy. Although, with current scanners, image reconstruction can be performed at the CT console, this is typically not practical due to patient throughput issues and other logistical considerations. Rather, in routine clinical practice, source images are transferred to a freestanding 3D workstation for analysis.

The major types of two- and three-dimensional CTA reconstruction techniques and their primary clinical applications are outlined below. Reconstructions can be used for either *diagnostic* or display purposes; "display" purposes include *communication* of complex results to referring physicians and *surgical planning*—as is the case for shaded surface displays which delineate the relationship between aneurysm morphology and bony skull base anatomy.

The important 2D-reconstruction techniques include curved reformats (CR), MIP, and MPR. The important 3D-reconstruction techniques include SSD and VR.

1. 2D and 3D Reconstructions: Strengths and Weaknesses

Unlike MRA reconstructions, which are typically completely automated, the time required to create CTA reconstructions can vary from seconds, for simple data sets using preset automated algorithms, to over an hour, for large complex data sets requiring meticulous manual subtraction of bone and venous structures. Indeed, the removal of bone and vein from CTA data sets remains a vexing problem in image reconstruction, often necessitating the use of dedicated "3D" technologists or imaging laboratories in centers with a high volume of CTA studies and limiting the clinical implementation of CTA in centers without these resources. Algorithms for semiautomated bone subtraction currently exist; algorithms for fully automated bone subtraction are in development, but the technical problem of determining the precise boundary between intravascular enhancement and adjacent calcification is nontrivial [123].

The most appropriate reconstruction techniques for a given CTA data set vary not only according to the clinical question being asked, and the presence or absence of surrounding bone or vein, but also according to resource availability. For example, although curved reformats are an ideal screening tool for displaying the entire length of a carotid or vertebral artery (from arch origin to circle of Willis), this reconstruction method can be extremely time and labor intensive and therefore may be unsuitable for routine clinical use in institutions without a dedicated 3D technologist or 3D-imaging lab.

In our 3D lab, we have developed a set of "standard" 2D and 3D views of each major head and neck vessel that should be obtained in every case, depending on the precise clinical indication for the study. The most common daily clinical indications for CTA at our institution include stroke, aneurysm, carotid occlusive disease, dissection, and venous sinus thrombosis. Additional views can, of course, be obtained as needed.

2. Source Images

A potential drawback of CTA reconstruction is the review and storage of the very large data sets created by scanning and reformatting, the so-called "slice pollution" problem. Hundreds of thin axial slices can be generated from a single "arch to vertex" CTA exam. The time required for visual inspection of each of these images, even with state-of-the-art PACS display, can be prohibitive.

Various approaches to this problem have been advanced. With regard to computer storage issues, for example, one group has advocated a technique for compression of raw projection data, which can subsequently be decompressed and used for image reconstruction at a later date [124]. These authors found that projection data files were more compressible than image files. Offline storage of this raw image data, when possible, could also be beneficial.

The ideal strategy for image review of large CTA data sets depends on the body part being imaged and whether plain film or digital PACS display is being used. Researchers at Brigham and Women's Hospital in Boston, for example, have found that hard copy review of high-resolution chest CT images can be accomplished by filming at an "80-on-1" format. Reduction in nodule detection rates at standard viewing distances can be largely compensated for simply by film review at closer distances [125].

Our general approach to neurovascular CTA source image review is to use two different sets of axial images, reformatted as thick and thin slices, respectively, for both visual inspection and 3D reconstruction (Fig. 6). Standard visual review of axial head and neck images can be most efficiently accomplished using the relatively thick slices (typically 5 and 2.5 mm thick, respectively). The thin (1.25 mm thick), maximally overlapping reformatted axial images, on the other hand, are used primarily for image reconstruction. As has already been noted, the use of such thin slices can reduce the appearance of zipper, or stair-step, artifact in most 2D and 3D reconstructions. Direct visual review of these thin reformatted images is typically performed only at selected levels and in order to answer specific questions. Examples of such specific questions include measurement of the precise degree of vascular stenosis detected by, or confirmation of the existence

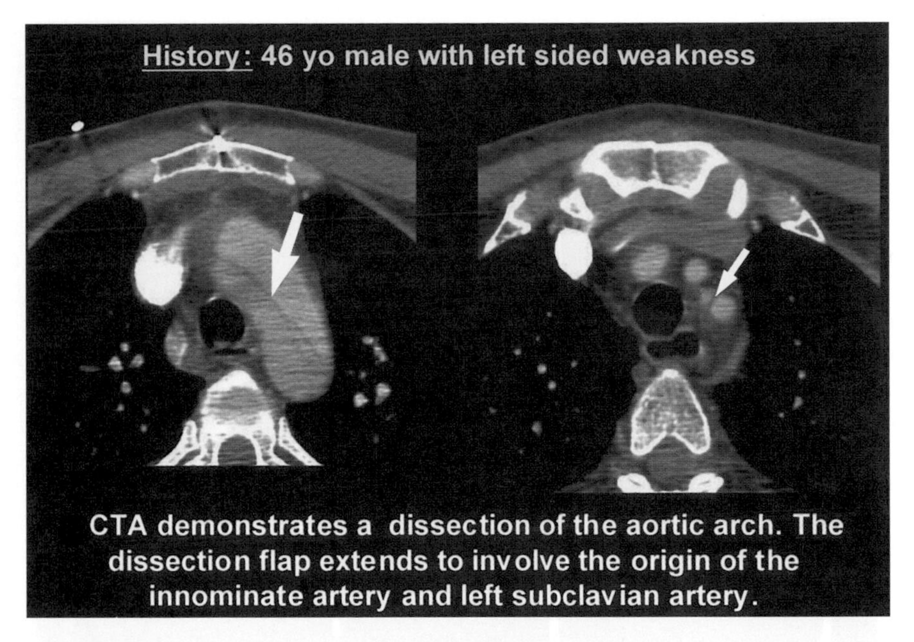

Figure 11 Obviated conventional angiography. Example of axial CTA source images in a case of aortic arch dissection. The dissection flap is a thin hypodense line (arrows) and extends to involve the origin of the innominate artery and left subclavian artery. The patient was a 46-year-old male with left-sided weakness. (Courtesy of Nicholas Petitti, M.D., Albany, NY.)

of an aneurysm suspected based on, the screening 2D and 3D reconstructions [126].

Examination of the thin axial source images is essential for definitive residual lumenal diameter measurement in cases of suspected internal carotid artery stenosis. These measurements are critically dependent on window and level settings, especially for severe stenoses [53–55]. Optimal window and level settings can be chosen "by eye," simply by adjusting image contrast such that the Hounsfield value of the center level is halfway between that of the vascular lumen and the immediately surrounding soft tissue and that both the lumen and the surrounding structures fall within the gray scale of the window width (i.e., such that the regions of interest are neither "too white" nor "too black"). Although studying MRA and not CTA, one group found that, in carotid occlusive disease, review of axial source images—rather than MIP reconstructions—reduced the tendency for overestimation of stenosis and improved the accuracy of detecting 70–99% stenosis [127].

Examination of the thin axial source images is also helpful in confirming the presence of a venous sinus thrombosis or small aneurysm. Small, less than 1- to 3-mm-diameter aneurysms can be especially difficult to detect in the cavernous carotid arteries, due to the confounding effects of venous enhancement in the surrounding cavernous sinus. 3D reconstructions are usually limited in such cases, although once the location of an aneurysm is verified, a limited 2D display may help to better define its relationship to adjacent structures for surgical planning. Direct visual review of the axial source images can also be beneficial in confirming the presence of vascular dissection (Fig. 11).

3. Maximum Intensity Projection

In MIP, those pixels with the highest (*maximum*) Hounsfield unit attenuation along a given ray are projected to create a two-dimensional projected image, as in MRA (Fig. 12). This image can be rotated, so that it is displayed from multiple different angles. Advantages of the MIP technique include speed, ease of use, and the accurate rendering of pixel Hounsfield intensities, which can then be windowed appropriately, allowing for accurate measurements of stenoses. Indeed, Stanford researchers have suggested that MIP is superior to SSD for the accurate CTA assessment of renal artery stenosis [128]. Other groups, including our own, have suggested that the same is true for carotid artery stenosis [129]. Even for gadolinium enhanced MRA, researchers have suggested that the diagnostic utility of MIP and MPR greatly exceeds that of SSD and VR for neurovascular evaluation [130].

Disadvantages of the MIP technique include overlap of bone and venous structures in some projections. As has already been noted, automated methods for bone removal in MIP projections are currently under development. In one feasibility study in which bone pixels are eliminated from CTA images, two observers blinded to the bone elimination method judged the image quality of six MIP studies "to be considerably higher than that of standard subtraction MIP images" and concluded that their method "is an effective

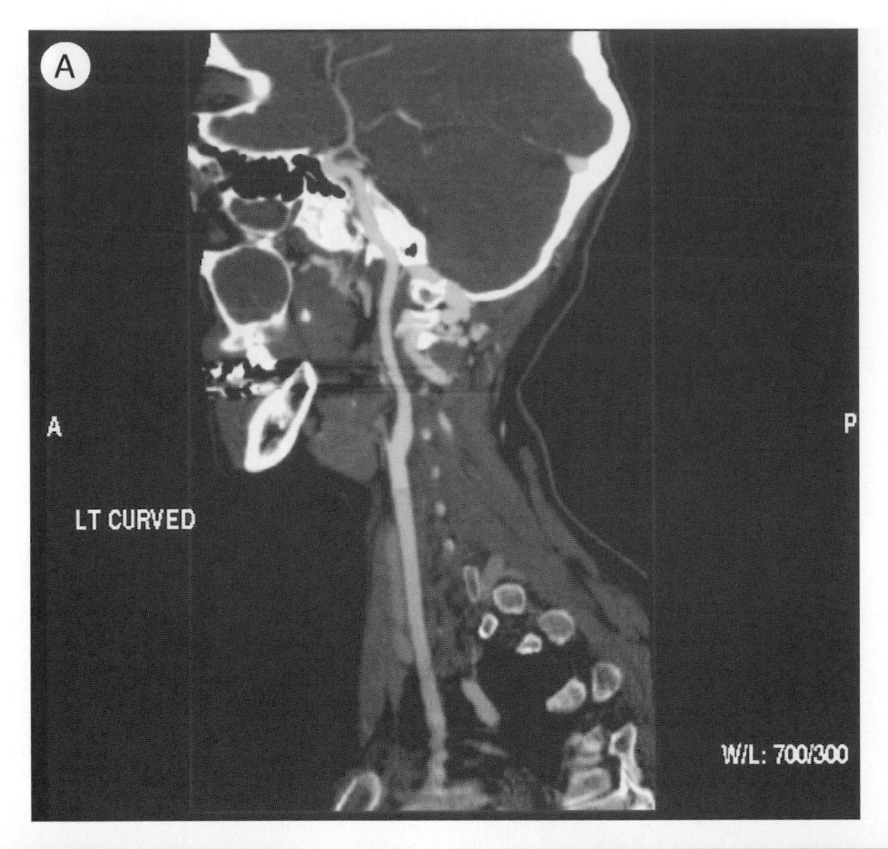

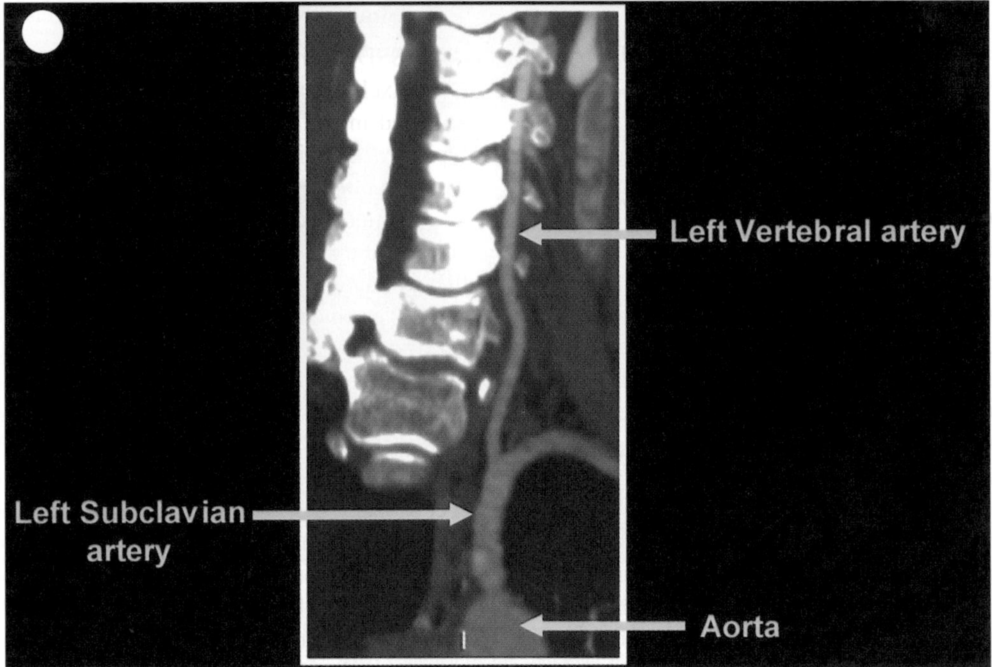

Figure 14 (**A**) Curved reformatted (CR) view of a normal left carotid artery, from origin at the aortic arch to distal MCA branches. (**B**) CR images of extracranial left vertebral artery.

Team members included Robert Drebin, Pat Hanrahan, and Loren Carpenter. VR technology, which requires very fast computers, can also be applied to modeling in aeronautics, seismology, and medical imaging.

VR, which produces images similar to those of SSD, is advantageous in that nonsurface pixels are included in the data set [137]. In volume rendering, various tissue layers can be made transparent and/or colored by thresholding

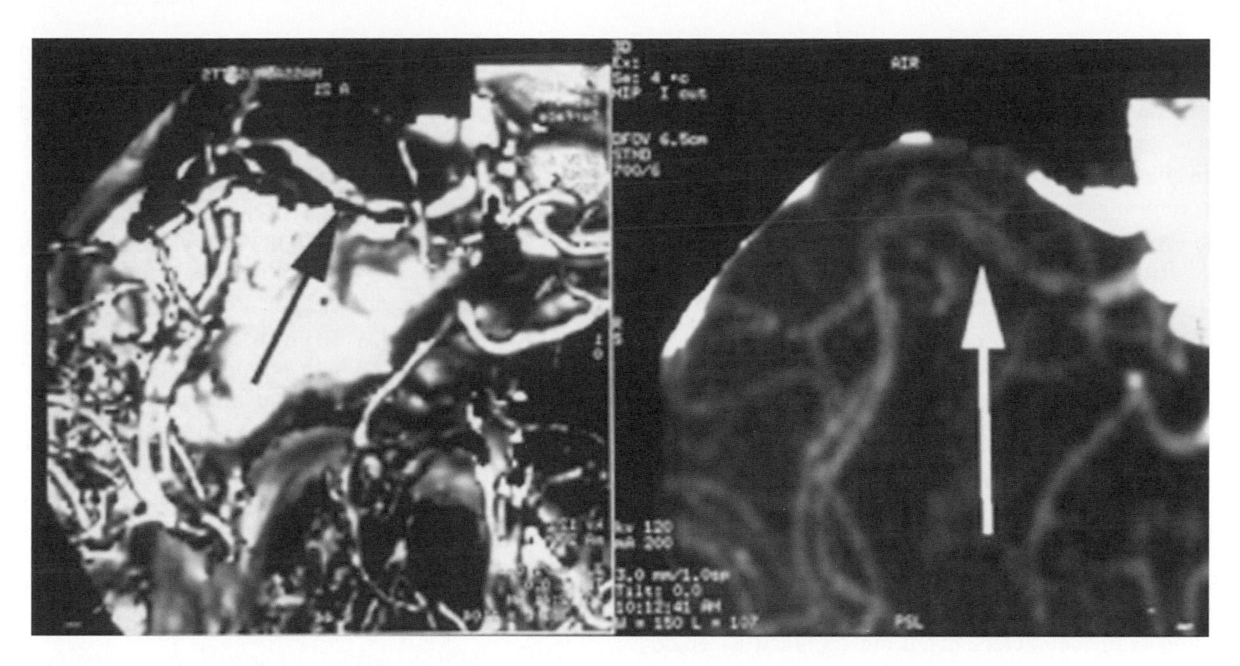

Figure 15 SSD and MIP reconstructions of partially occlusive right MCA thrombus (arrows). Internal lumenal abnormality is better appreciated on the MIP view.

(Fig. 16) [138, 139]. Because, depending on the threshold chosen, there may be overlap between the borders of adjacent structures, VR does not solve the problem of edge detection, which is especially important in measurement of carotid artery lumenal diameters.

Indeed, apparent vessel diameter with VR is a function of contrast concentration, window and level settings, and "opacity" settings (but not brightness). Increasing the opacity setting can make objects appear larger than they really are. With commercial VR software packages, the factory preset opacity settings cannot be assumed to be optimal for the measurement of vascular stenosis, and user interaction is therefore required for accurate measurement of vessel diameters. One group has concluded that CTA with VR can accurately evaluate carotid disease, even when dense calcifications are present, but that "no definite advantage over currently available techniques for CT measurement of stenosis severity was found" [140].

Despite hundreds of published papers comparing the clinical benefits of MIP, MPR, SSD, and VR, few have

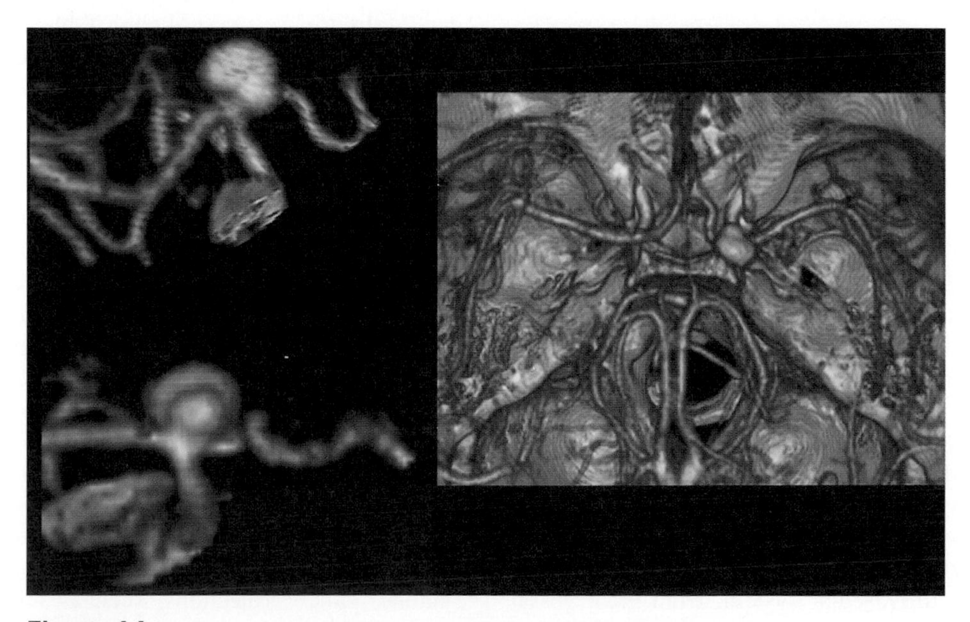

Figure 16 VR views of a multilobed "top of internal carotid artery" aneurysm.

explicitly investigated the relative accuracy of these reconstruction techniques for the most severe degrees of stenosis. Indeed, in one study of vascular phantoms, measurement error was greatest for the smallest lumenal diameters [141]. In that study, the most severely narrowed phantom had an internal lumen diameter of 1.6 mm (9.5-mm phantom with an 83% stenosis). By way of comparison, a "critical" degree of carotid artery stenosis is typically in the 1–1.5 mm range.

III. Scanning Protocols: Acquisition, Postprocessing, Analysis, and Interpretation

In what follows, examples of commonly applied CTA protocols of the intra- and extracranial neurovascular system are surveyed. These are presented as sample protocols only; the specific scan parameters noted are included as guidelines, appropriate to the clinical applications under discussion. These sample protocols were designed for use with the GE scanners in our emergency department, but can be adapted to scanners of any manufacturer.

Due primarily to differences in tube heating constraints, a distinction is made between protocols designed for single slice (Table 3A) and multislice helical scanners (Table 3B). In general, the single-slice protocols are similar to their MSCT counterparts, except for limitations in degree of coverage and minimal obtainable slice thickness. Note, also, in Table 3B, that most of our broad-coverage MSCT protocols call for two discrete scanning acquisitions, or "phases," during the continuous administration of a contrast bolus. The reasons for this are discussed under Section III.A.1, immediately below.

A. Acute Stroke

1. Acquisition

Our sample acute stroke protocol (Tables 3A and 3B) consists of the following components: (1) routine unenhanced head CT, (2) CTA with simultaneous whole-brain perfused blood volume exam, and (3) optional single-slab cine CT perfusion study [23]. The goal of this protocol is to rapidly obtain data regarding intracranial thrombus, vascular stenosis, and parenchymal ischemia in patients with suspected embolic stroke, in order to facilitate their diagnosis and triage to appropriate therapy.

With single-slice scanners, vascular coverage is limited to the circle of Willis and, depending on tube heating constraints, potentially portions of the extracranial internal carotid arteries.

With multislice scanners, high-resolution coverage of the complete neurovascular system, from great vessel origins to the vertex, can be performed in 45–60 s. In order to accomplish this, we employ a two-phase protocol, scanning from skull base to vertex during phase 1 and from arch to skull base during phase 2. The rationale for this protocol design is so that the most critical portion of the exam, the circle of Willis acquisition, is performed first. This not only ensures adequate data collection, but also permits real-time review of the CTA source images at the scanner console during the remainder of the exam. Additionally, by delaying the acquisition through the aortic arch, dense contrast, which can pool in the subclavian vein, is given time to clear, diminishing streak artifact at the great vessel origins.

Single-slab quantitative cine CT perfusion imaging can be performed either independently or as an optional third portion (phase 3) of the stroke exam. When cine perfusion imaging is performed in conjunction with an arch-to-vertex CTA, the MSCT scan parameters can be adjusted so as to minimize total contrast dose, as outlined in Table 3B.

Other MSCT stroke protocols are possible. Some groups have advocated triphasic perfusion CT for diagnosing proximal MCA occlusion, assessing perfusion deficits, and evaluating collateral circulation [142]. With this protocol, sequential images of early, middle, and late phases of enhancement are obtained. Total acquisition time is 5 min.

2. Review of Source Images versus MIP Reconstructions, Speed, and Accuracy

The 5-mm-thick unenhanced head CT source images and the 2.5-mm-thick whole-brain perfused blood volume source images can be reviewed in real time at the scanner console (Fig. 17). After the perfused blood volume images have been reformatted into 1.25-mm-thick slices, they can also be preliminarily inspected at the scanner for evidence of large vessel thrombosis, using appropriate "CTA" window and level settings as described in Table 2.

Proper windowing of both the unenhanced and the enhanced source images is also essential to optimize parenchymal stroke detection [58]. Because the cytotoxic edema of acute stroke causes only minimal reduction in brain Hounsfield attenuation, narrow window and level settings should be chosen so as to exaggerate the subtle difference in density between normal and edematous gray and white matter (Fig. 3). Preset settings of center-level 35 HU, window-width 5–35 HU provide a good first approximation for distinguishing gray from white matter attenuation.

Review of the reformatted axial source images will frequently suffice for the detection of proximal large-vessel circle of Willis occlusions (Fig. 18). For more sensitive evaluation of the middle and distal MCA branches, a superior ("collapsed") MIP view of the circle of Willis is required (Figs. 12 and 17B). This view can be constructed in less than 1 min on most commercially available 3D workstations. Other projections, such as midline sagittal and

TABLE 3A Sample Single-Slice Helical CTA Protocols

Single slice	Contrast	Range	Slice thickness (mm)	Image spacing (mm)	Table feed (mm/s)	Pitch	Mode	kV	mA	Rotation time (s)	Interscan delay (s)	Scan FOV (type)	Display FOV (cm)	Comments
Unenhanced Head CT	None	Foramen magnum to vertex	5	5	N/A	N/A	Axial	140	170	2	2	Head	22	Include bone algorithm for trauma
Neck CTA	3 cc/s, 90–120 cc, 25-s delay	Shoulders (C5/C6) to the base of skull	3	3	3	1:1	Helical	140	≤220 (Max possible)	1	N/A	Head	22	"Head" FOV if above shoulders; "large" FOV if shoulders in field
Head CTA (aneurysm)	4 cc/s, 90–120 cc, 25-s delay	C1 ring to vertex	1	0.5	N/A	1:1	Helical	140	≤220 (Max possible)	1	N/A	Head	12	Delay may be increased up to 40 s for a-fib, low EF
Head CTA (acute stroke)	3 cc/s, 90 cc, 25-s delay	C1ring to vertex	3	3	3	1:1.3	Helical	140	≤220 (Max possible)	1	N/A	Head	22	Reformat 3-mm images to 1-mm thickness for MIP reconstructions

TABLE 3B Sample Multislice Helical CTA Protocols

MSCT	Contrast	Range	Slice thickness (mm)	Image spacing (mm)	Table feed (mm/s)	Mode	kV	mA	Rotation time (s)	Scan FOV (type)	Display FOV (cm)	Comments
1. Unenhanced Head CT	None	Foramen magnum to vertex	5	5	7.5	HQ (Pitch 3)	140	170	1	Head	22	Can be reformatted to 2.5-mm thickness if desired (table feed/pitch = minimum intrinsic slice thickness)
2. Neck only CTA (stenosis, dissection)	3 cc/s, 90–120 cc, 25-s delay	C7/T1 to circle of Willis	2.5	2.5	3.75	HQ (Pitch 3)	140	170	1	Head	22	Provides optimized view of bifurcation. Can be reformatted to 1.25-mm-thick slices for precise lumen measurement
3. Head CTA/CTV (aneurysm/venous thrombosis)	4 cc/s, 90–120 cc, 25–40 s delay	C1 ring to vertex	1.25	0.6	3.75	HQ (Pitch 3)	140	170	1	Head	≤20 (Minimum possible for aneurysm)	40-s delay for CTV; delay (min for may also be increased up to 40 s for a-fib, low ejection fraction
4a. (i) ***Phase 1:*** **Head and neck CTA (acute stroke)**	3 cc/s, 120 cc, 25-s delay	C1 ring to vertex	2.5	2.5	3.75	HQ (Pitch 3)	140	170	1	Head	22	2.5-mm axial slices for visual review, reformat to 1.25-mm thickness for MIP/CR reconstructions
4b. (i) ***Phase 1:*** **Head and neck CTA (acute stroke with single slab profusion)**	2.5 cc/s, 95 cc iso-osmolar, 25-s delay	C1 ring to vertex	2.5	2.5	3.75	HQ (Pitch 3)	140	170	1	Head	22	
4a. (ii) ***Phase 2:*** **Head and neck CTA (acute stroke)**	Continued	Aortic arch to C1 ring	2.5	2.5	7.5	HQ (Pitch 3)	140	250	1	Large	22	Series is combined with phase 1 into single reconstruction. Cannot be reformatted to thinner 1.25 mm slices due to high table speed. Note change in mA, FOV to accomodate shoulders.
4b. (ii) ***Phase 2:*** **Head and neck CTA (acute stroke with single slab profusion)**	Continued	Aortic arch to C1 ring	2.5	2.5	15	HS (Pitch 6)	140	250	1	Large	23	Series is again combined with phase 1 into single reconstruction. Increased table feed and pitch to accommodate smaller contrast dose and maintain thickness.
4c. (iii) ***Phase 3:*** **Head CTA (single slab perfusion)**	4–6 ml/s, 45 cc iso-osmolar, 5-s delay	Superior and parallel to orbital roof, 2 contiguous slices	10	10	0 (table stationary)	Cine 2i	90–140	[illegible]	0.8	Head	22	Can be performed independent of other protocols. See text re: choice of scan mA, kV

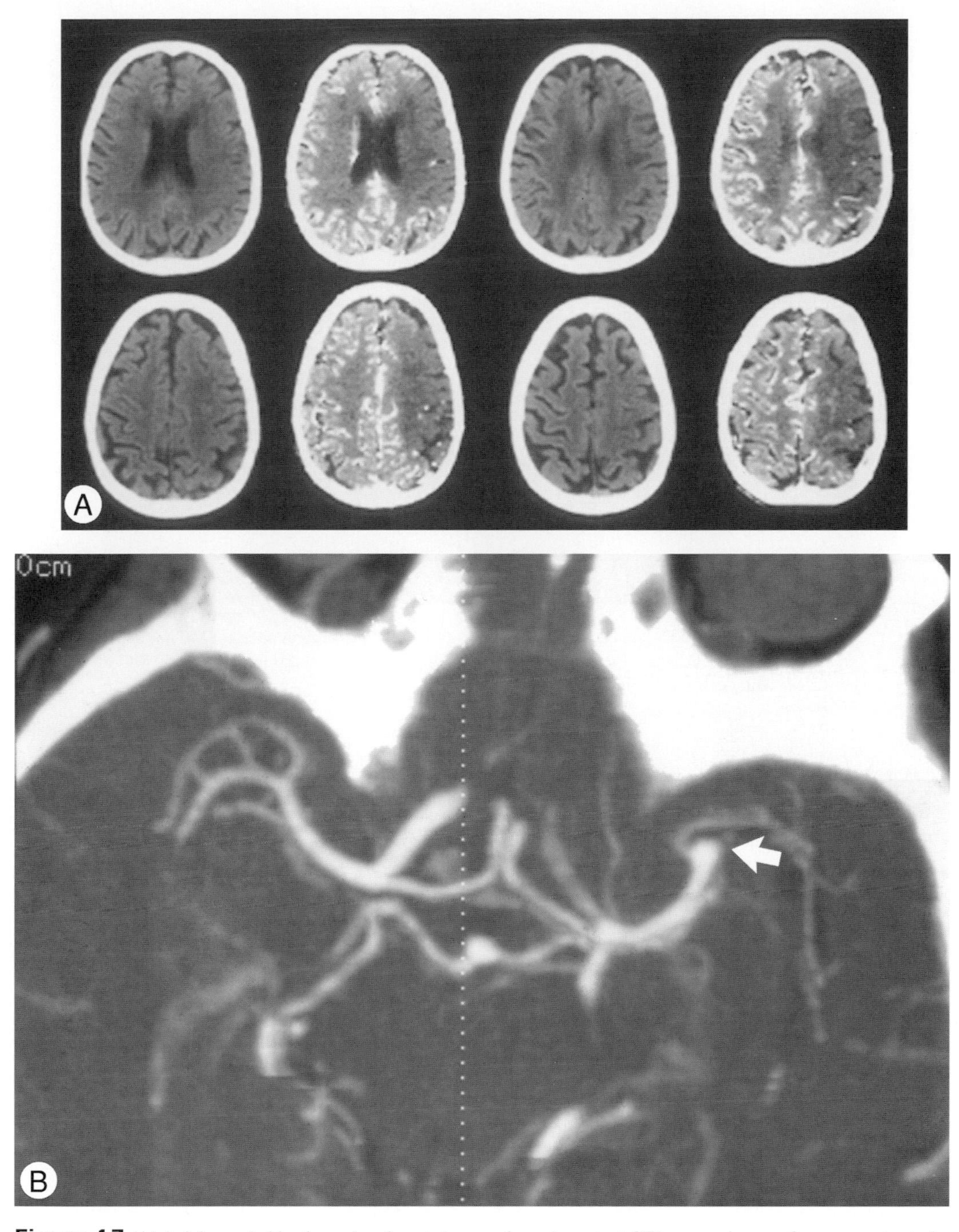

Figure 17 (**A**) Axial pre- (odd columns) and post- (even columns) contrast CTA source images from an acute stroke patient 2 h postictus. There is only subtle left hemispheric hypodensity on the precontrast images. The postcontrast, perfused blood volume-weighted images show more conspicuous, low-density regions. Such blood volume lesions typically progress to complete infarction. (**B**) "Collapsed" MIP circle of Willis view shows occlusive thrombus at left superior division MCA origin (arrow).

coronal MPR reformatted images of the basilar artery, are seldom essential in the acute setting.

Our acute stroke CTA/CTP imaging protocol typically adds a total of 15 min or less—including acquisition, reconstruction, and interpretation time—to the time required to obtain a routine, unenhanced head CT. The accuracy of circle of Willis CTA for the detection of proximal, large vessel thrombus approaches 99% [19].

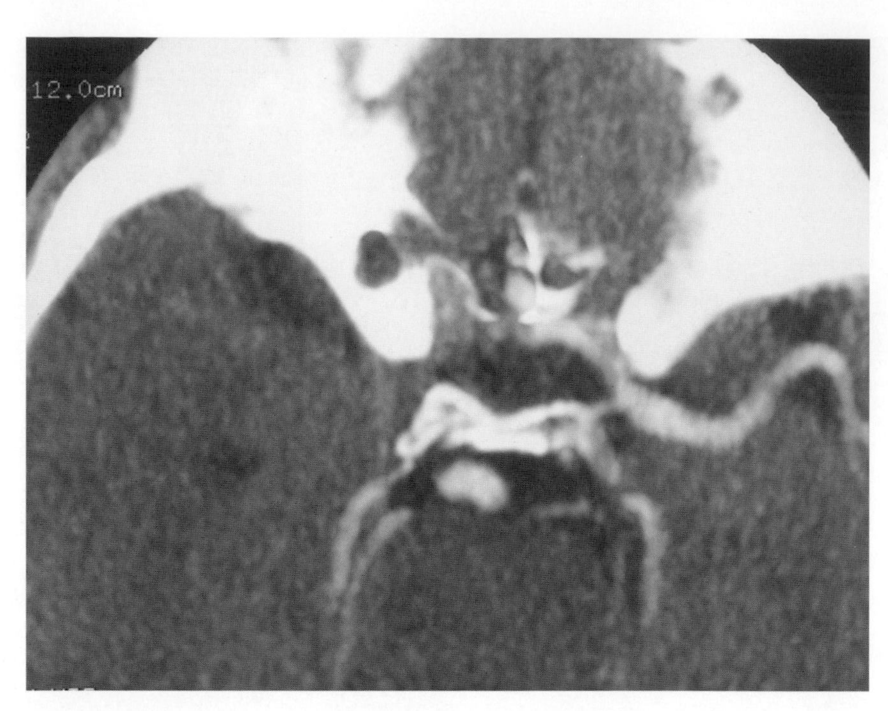

Figure 18 Axial CTA source image at the level of the middle cerebral arteries (MCAs), from an acute stroke patient. There is complete absence of filling of the right supraclinoid ICA and MCA, consistent with occlusive thrombus.

B. Carotid Artery Occlusive Disease: Differences from Stroke Protocol and Optimization of Residual Lumen Measurement

A sample neck CTA protocol is shown in Tables 3A and 3B. It differs from the acute stroke protocol both in its degree of coverage and in its minimal obtainable slice thickness through the extracranial carotid artery. Whereas the stroke protocol is designed to optimize detection of intracerebral thrombus (requiring thin, 1.25-mm-thick slices) but only to *screen* the carotid bifurcation (with 2.5- to 5-mm-thick slices), the neck protocol is designed to optimize precise measurement of lumenal stenoses (with 1.25-mm-thick slices). Indeed, in an early study on the accuracy of carotid stenosis measurement, CTA compared unfavorably with other "gold standard" measurements due to the large (10 mm) slice thickness that was used [143].

Prior to performing CTA of the neck it is important to review prior noninvasive ultrasound, MR, CT, and angiographic studies to know what questions are to be answered. Indications for neck CTA include carotid or basilar stenosis, dissection, or claustrophobia prohibiting MRA.

Screening of large CTA data sets for focal carotid artery stenoses can be achieved quickly by scrolling through the axial source images at a 3D workstation using cine review mode or by review of curved reformatted images through the entire vessel length [53]. Sagittal, coronal, and oblique 2D-MPR images aid in both interpretation and communication of results to referring physicians. MIP or MPR reconstructions can often best demonstrate a tight stenosis or string sign, but are inadequate when heavy circumferential calcification is present. 3D-VR images typically add little relevant diagnostic information, but can be of value in surgical planning. In our lab, we report the degree of vascular stenosis categorically, based on the following "minimal residual lumen diameter" measurement guidelines:

Critical stenosis <1 mm
Severe stenosis <1–1.5 mm
Moderate-to-severe stenosis <1.5–2 mm
Moderate stenosis <2–2.5 mm.

For the vertebrobasilar system, 3D-VR or SSD reconstructions work well for depicting aneurysms and their relation to bony landmarks. 2D-MPR reconstructions, however, are more accurate for depicting subtle internal lumenal abnormalities. The standard projections for an "arch-to-circle of Willis" CTA data set created at our 3D-imaging lab are as follows:

(1) Bilateral coronal and sagittal curved reformats through the entire length of both carotid arteries.
(2) Oblique coronal MPR and MIPs of vertebral and carotid origins.
(3) Oblique MPR and MIPs of bilateral carotid bifurcations.

(4) Oblique coronal and sagittal MIPs of vertebral arteries.
(5) Multiple rotational MIP views through the region of maximal stenosis.

C. Aneurysm/AVM: Differences from Other Protocols and Optimization of Small Aneurysm Detection

In the screening for, or the characterization of, intracranial aneurysms using CTA, the scanning objective is to optimize vascular opacification, slice thickness (as thin as possible), and in-plane resolution (as small a display FOV as possible), at the expense of coverage length (if necessary). Our sample aneurysm CTA protocols (Table 3) reflect these goals, including the use of 0.5- to 0.6-mm interslice spacing in order to improve 3D-reconstruction resolution and an increased, 4 ml/s contrast administration rate. Before performing an aneurysm CTA exam, it can be useful to know the expected location of the aneurysm, based on the pattern of subarachnoid hemorrhage or on prior MRA or catheter arteriographic studies.

Of note, due to the high density of iodinated contrast, aneurysm CTA can be successfully performed even in the setting of subarachnoid hemorrhage. With the proper angling of the patient's head relative to the scanner gantry, CTA can also be performed despite the presence of previously placed aneurysm clips, although streak artifact cannot be completely avoided immediately in the vicinity of the clip.

3D-VR projections are useful for surgical planning, in order to display the relationship of the aneurysm to the skull base; these are especially valuable for carotid–ophthalmic and paraclinoid aneurysms (Fig. 16). MIP and MPR reconstructions can also be of value when there is no overlying bone, especially to demonstrate MCA aneurysms in a "collapsed" circle of Willis projection or to delineate intralumenal thrombus. 2D-CR views are necessary only when overlying bone or veins preclude VR or MIP reconstructions, as is often the case with intracavernous carotid artery aneurysms. The standard projections performed for an aneurysm CTA data set by our 3D-imaging lab are as follows. The objective is to display the origin of every major intracranial vessel (the most common sites of aneurysm formation) in at least two projections:

General: Two views of each vessel, projected at least 30° apart.
(1) "Collapsed" MIP (ACOM, MCA bifurcation, top of ICAs, PCAs).
(2) Oblique axial MPVR or MIP (MCA bifurcation).
(3) "Handlebar" coronal view MIP (anterior circulation).
(4) Oblique coronal MIP (distal VAs, PICA origins).
(5) Bilateral CR coronal and sagittal (ICAs).
(6) VR of aneurysm (for surgical approach).
(7) Coronal oblique MIP (ACOM).
(8) Sagittal oblique MIP (PCOM).
(9) Coronal oblique MPR (VB junction).

(Definitions used: ACOM, anterior communicating artery; MCA, middle cerebral artery; ICA, internal carotid artery; PCA, posterior cerebral artery; VAs, vertebral arteries; PICA, posterior inferior cerebellar artery; VB, vertebrobasilar.)

Similar considerations apply to the scanning of AVMs, although the inability to access direction of flow or definitively locate the AVM nidus is a limitation (Fig. 19) [144].

D. Venous Thrombosis: Differences from Other Protocols and Optimization of Vein versus Bone Separation

CT venography (CTV) of the dural venous sinuses and deep intracranial veins is typically indicated for patients who cannot undergo MR venography (MRV) or as a problem-solving tool in complex cases as an alternative to catheter angiography. Although CTV's inability to conveniently measure flow direction can be a drawback, the flow-related artifacts that so commonly limit MRV do not hinder CTV. Our sample CTV protocol significantly differs from our aneurysm protocol only in that the contrast prep delay is lengthened to 40 s, in order to maximize venous opacification.

An ideal technique for the reconstruction and display of CTV data is the "graded subtraction" method, originally advanced by Casey and Alberico *et al.* [211, 145]. With this method, a series of MIP models of the venous anatomy is created by successive subtraction and dilatation of a bone model of the cranium, facilitating the subtraction of bone with minimum deletion of the underlying vasculature (Fig. 20). The baseline bone model is created by appropriate thresholding of the CTV data set. Initial subtraction of the bone model from the vascular model does not completely remove the bone; a small rind of "bone pixels" remains. These "leftover" pixels are removed from the vascular model by repeated dilatation (by 1-pixel layer) and subtraction of the bone model. Ultimately, the majority of bone has been removed, with the "vascular pixels" intact. After overlying skin has been cut out, the "integral" function of the workstation can be applied, averaging the remaining surface pixels, which smoothes the image.

Helical CTV is a highly accurate technique for assessment of the dural sinuses and typically shows more vascular detail with less technical difficulty than does MRV, especially in children [145]. Our standard 3D-lab CTV views are as follows:

(1) Graded bone subtraction.
(2) MIP (superior view of superior sagittal sinus with inferior cut plane).
(3) MIP (anterior view with anterior cut plane).
(4) MIP (skewed lateral view without cut plane).

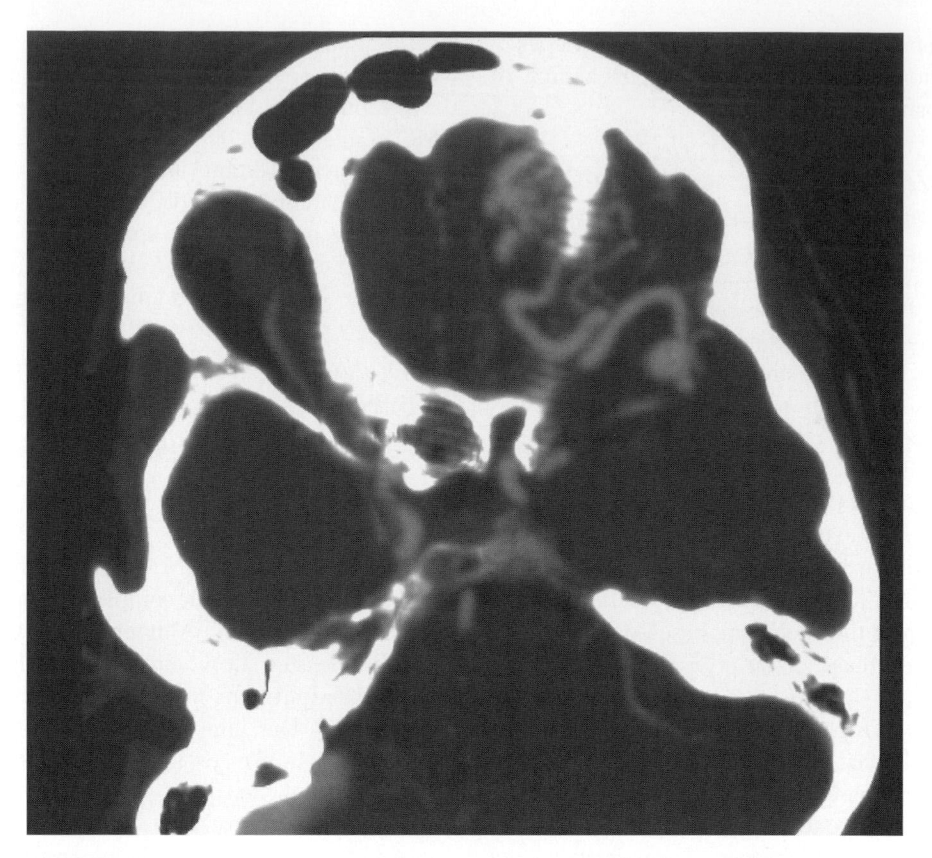

Figure 19 MIP view shows relationship between left frontal AVM and adjacent bone of orbital roof.

(5) Batch oblique MIP slabs, parallel and perpendicular to vein of interest.

IV. Clinical Utility

In this section, potential clinical applications for CTA and CTP imaging will be reviewed.

A. Acute Stroke

1. CTA as a Triage Tool for IA versus IV Thrombolysis, versus Other Management

According to the National Stroke Association Website, "stroke is our nation's third leading cause of death, killing nearly 160,000 Americans every year. Every year approximately 750,000 Americans have a new or recurrent stroke. Over the course of a lifetime, four out of every five American families will be touched by stroke. Approximately one-third of all stroke survivors will have another stroke within five years. Of the 590,000 Americans who survive a stroke each year, approximately 5 to 14 percent will have another stroke within one year. The rate of having another stroke is about 10 percent per year thereafter." Although only approximately 20–25% of strokes are caused by large vessel thrombus amenable to thrombolytic therapy, this group accounts for a majority of the morbidity, mortality, and cost of stroke disease; the application of CTA with CT perfusion imaging as a triage tool for stroke has the potential to reduce these costs [146].

Despite decades of testing promising neuroprotective agents in animal models, the only accepted treatment for acute stroke, to date, of proven benefit in humans, is thrombolytic ("clot-busting") therapy with reverse tissue plasminogen activator. The results of the National Institute of Neurological Disorders and Stroke multicenter trial suggest an absolute 3-h time window for the administration of *intravenous* thrombolytics; beyond 3 h, the risk of severe intraparenchymal hemorrhage exceeds the risk of treatment [31]. The results of the Prolyse in Acute Cerebral Thromboembolism (PROACT) II trial also demonstrated a benefit of *intra-arterial* thrombolytic therapy, administered within 6 h of symptom onset, for the treatment of proximal large-vessel stroke [30]. A major benefit of CTA is its highly accurate detection of proximal, large-vessel, circle of Willis thrombus, useful in the selection of candidates for IA thrombolysis [19].

Typically, patients with acute ischemic stroke symptoms undergo an unenhanced head CT scan as their first imaging

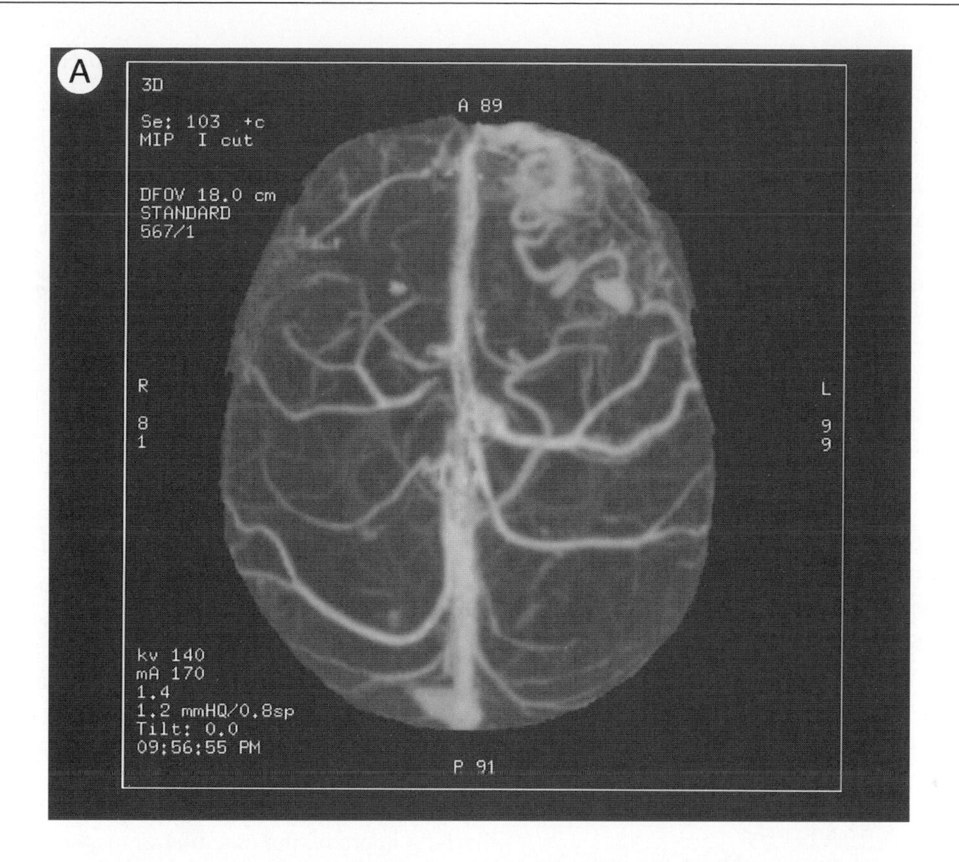

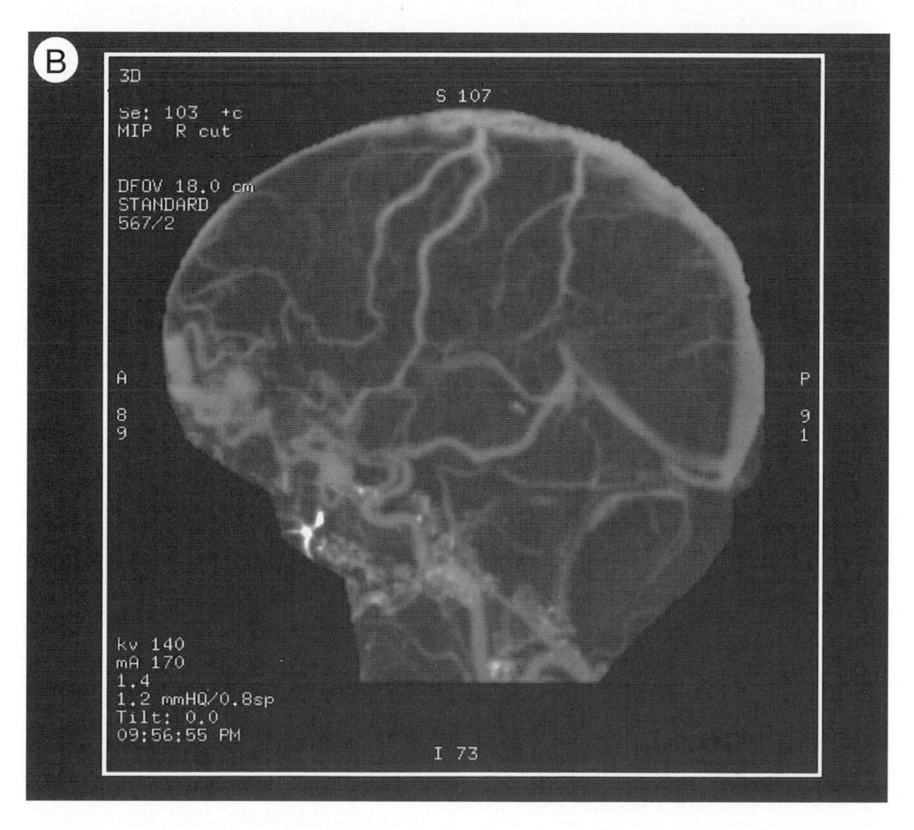

Figure 20 "Graded subtraction" MIP reconstructed superior (**A**) and sagittal (**B**) views from a CTV data set. Note the left frontal AVM with cortical and superior sagittal sinus drainage.

test, to determine if contraindications to thrombolytic treatment exist. Such contraindications include hemorrhage (an absolute contraindication) or a "large" (greater than one-third vascular territory) parenchymal hypodensity (indicating already infarcted brain, a relative contraindication) [31, 147]. In general, early infarction on unenhanced CT is underdetected, even by experienced physicians (Fig. 3) [58]. There is considerable lack of agreement in recognizing and quantifying the early CT changes of stroke [148]. In one study, for example, only 45% of patients were identified correctly, based on CT and clinical criteria, for inclusion in a hypothetical acute stroke treatment trial [149]. Improved methods of recognizing and quantifying early ischemic brain damage are clearly needed [150].

Also, unenhanced CT scanning alone, although of some value in predicting patients most likely to be *harmed by* thrombolysis, is of little value in predicting patients—those with large-vessel vascular occlusions—most likely to *benefit from* thrombolysis. The "ideal" imaging test for hyperacute stroke patients should provide clinically relevant data to assist in the following evaluations:

- Rapid triage to appropriate treatment.
- Assessment of prognosis—including both tissue and clinical outcome.
- Assessment of hemorrhagic risk.
- Rapid stratification of stroke subtype for inclusion in clinical trials.

CTA with CT perfusion imaging has the potential to address each of these issues. Additionally, with further research, CTA with CTP may also help to identify those patients who, by virtue of adequate collateral circulation supplying reversibly ischemic tissue, may be candidates for IV or IA thrombolysis *beyond* the currently accepted time windows [32, 151].

2. Sensitivity and Specificity for Stroke Detection and Delineation

CTA/CTP can be *rapidly* performed, immediately following unenhanced CT, in order to identify intravascular circle of Willis clot in acute stroke patients. Such clot occurs in the middle cerebral artery in over 60% of all embolic strokes. Multiple studies have confirmed the high accuracy of CTA for the detection of large vessel intravascular clot (Figs. 12, 17B, and 18) [19, 152–154]. In studies from our institution, CTA has also been shown to be useful for the evaluation of collateral circulation distal to an occlusion, as well as for improving the conspicuity of acute cerebral ischemia (Fig. 17A) [155, 156]. Furthermore, axial whole-brain perfused blood volume images, which are acquired simultaneously with the CTA data set, can provide quantitative flow information, facilitating the detection of subtle parenchymal ischemic changes associated with distal embolic occlusions [23].

3. Prediction of Final Infarct Size and Clinical Outcome

In addition to facilitating the more accurate *detection* of early stroke, CTP may also be useful in predicting both final infarct size and patient outcome [36]. In one study from our institution, all patients with acute MCA stem occlusions who had on admission whole-brain CT perfusion lesion volumes >100 ml (equal to approximately 1/3 the volume of the MCA territory) had poor clinical outcomes, regardless of recanalization status. Moreover, in those patients from the same cohort who had early complete MCA recanalization, final infarct volume was closely approximated by the size of the initial whole-brain CT perfusion lesion (Figs. 9 and 17A) [36].

Cine single-slab CT perfusion imaging, which can provide quantitative maps of CBF, CBV, and MTT, has the potential to describe regions of so-called "ischemic penumbra"—ischemic but still viable tissue (Figs. 8 and 21). Xenon CT and SPECT studies in acute human stroke following MCA occlusion have revealed that a narrow ischemic penumbra, with CBF values between 7 and 20 ml/100 g/min, surrounds a severely ischemic core [157].[2] Animal studies of CT-CBF in stroke have shown good agreement between infarct core and CBF <10 ml/100 g/min or MTT >6 s [158]. CT-CBV/CT-CBF mismatch may also help in predicting outcome in response to thrombolysis.[3]

The degree of early CBF reduction in acute stroke may also help predict hemorrhagic risk. In a SPECT study of 30 patients who had complete recanalization within 12 h of stoke onset, those with less than 35% of normal cerebellar flow at infarct core were at significantly higher risk for hemorrhage [37]. Preliminary results from our group suggest that severe hypoattenuation, relative to normal tissue, on whole-brain CTP images may also identify ischemic regions more likely to bleed following intra-arterial thrombolysis [40]. Indeed, multiple studies have suggested that severely ischemic regions with early reperfusion are at the highest risk for hemorrhagic transformation [38, 39].

4. Primary Scanning Modality: Availability, Accuracy, Speed, Cost

Advanced imaging of cerebral ischemia can be accomplished with either MRI or CT scanning; MR diffusion and

[2] Iseda, T., Nakano S., Yano, T., Suzuki, Y., and Wakisaka, S. (2002). Time threshold curve determined by single photon emission CT in patients with acute middle cerebral artery occlusion. *Am. J. Neuroradiology*, **23**(4), 572–576.

[3] a) Koroshetz, W. J., and Lev, M. H. (2002). Contrast computed tomography scan in acute stroke: "You can't always get what you want, but you get what you need." *Ann. Neurol.*, **51**(4), 415--416.
b) Wintermark, M., Reichhart, M., Thiran, J. P., Maeder, P., Chalaron, M., Schnyder, P., *et al.* (2002). Prognostic accuracy of cerebral blood flow measurement by perfusion computed tomography, at the time of emergency room admission, in acute stroke patients. *Ann. Neurol.*, **51**(4), 417–432.

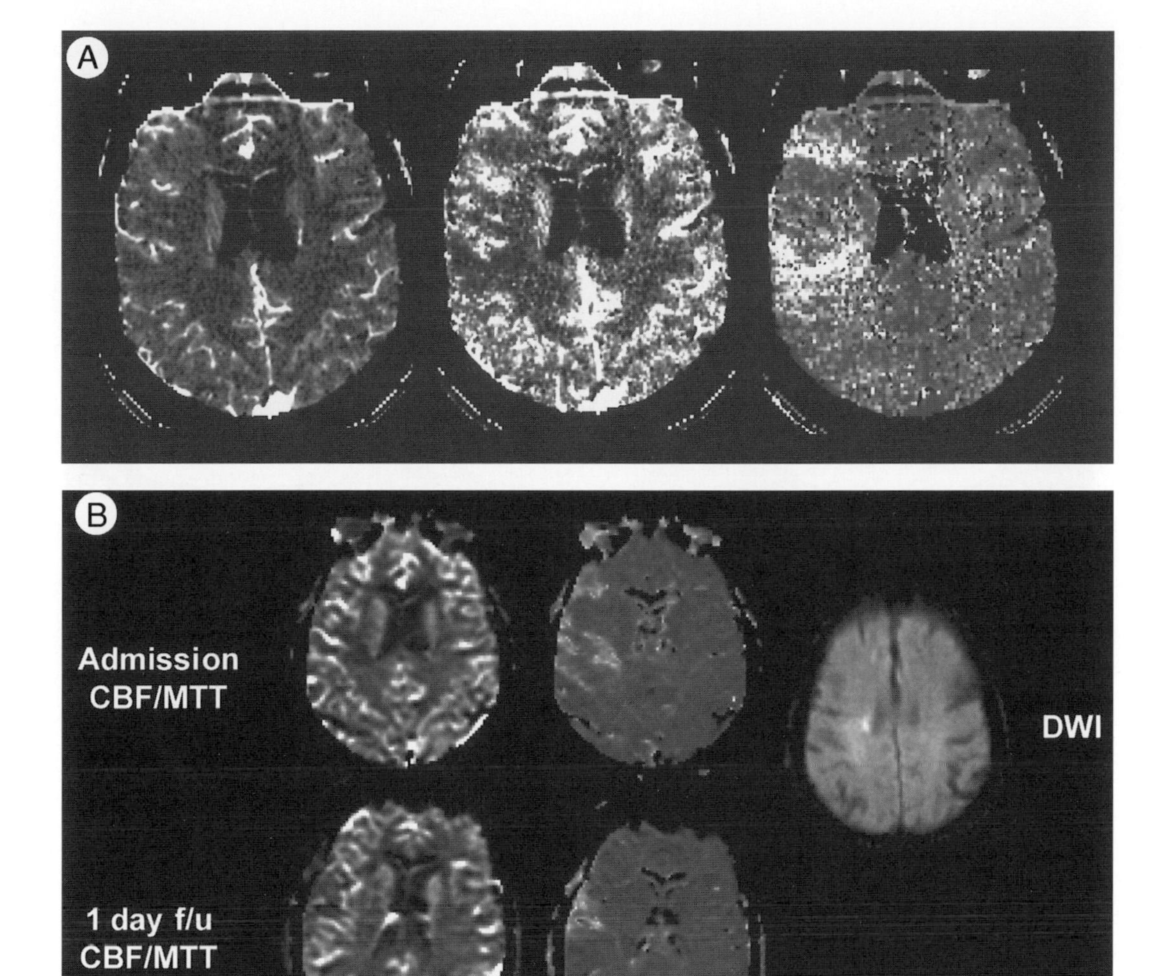

Figure 21 (**A**) First-pass cine CT perfusion maps in a patient with resolving transient ischemic attack and normal CTA. CT-CBV map (left) is normal, but there is a small ring of reduced CBF (middle) and delayed MTT (right) in the right frontal hemisphere, consistent with residual ischemia. This may be due to small distal emboli. (**B**) Follow-up MRI shows continued resolution of the CBF and MTT defects (left), with a small punctate infarct on diffusion-weighted imaging (right).

perfusion imaging have been shown in numerous reports to be highly successful in the evaluation and triage of acute stroke patients[4]. Because helical CT scanners are less expensive and more readily available at most hospital emergency departments than are MRI scanners, however, performing CTA/CTP can be a quick and natural extension of the unenhanced head CT exam—an exam that is routinely obtained as part of the prethrombolysis workup at most institutions [19, 23, 36]. The addition of a CT angiographic study seldom adds more than 10 min of scanning time to that of the conventional CT examination. Required postprocessing could typically be performed in minutes, during which time the patient could be prepared for thrombolysis, should the decision to proceed with treatment be made (Fig. 22) [19, 159].

CTA will become more important as the successful trial of intra-arterial thrombolysis (PROACT) impacts on national stroke treatment trends [30]. In this trial, patients with suspected occlusion of the MCA underwent conventional angiography and, if MCA occlusion was found, they were treated with intra-arterial thrombolysis or placebo. As

[4] a) Sorenson, A. G., Buonanno, F. S., Gonzalez, R. G., Schwamm, L. H., Lev, M. H., Huang-Hellinger, F. R., *et al.* (1996). Hyperacute stroke: evaluation with combined multi-section diffusion-weighted and hemodynamically weighted echo-planar MR imaging. *Radiology*, **199**(2), 391–401.
b) Sunshine, J. L., Bambakidis, N., Tarr, R. W., Lanzieri, C. F., Zaidat, O. O., Suarez, J. I., *et al.* (2001). Benefits of perfusion MR imaging relative to diffusion MR imaging in the diagnosis and treatment of hyperacute stroke. *Am. J. Neuroradiology*, **22**(5), 915–921.
c) Rohl, L., Ostergaard, L., Simonsen, C. Z., Vestergaard-Poulsen, P., Andersen, G., Sakoh, M., *et al.* (2001). Viability thresholds of ischemic penumbra of hyperacute stroke defined by perfusion-weighted MRI and apparent diffusion coefficient. *Stroke*, **32**(5), 1140–1146.

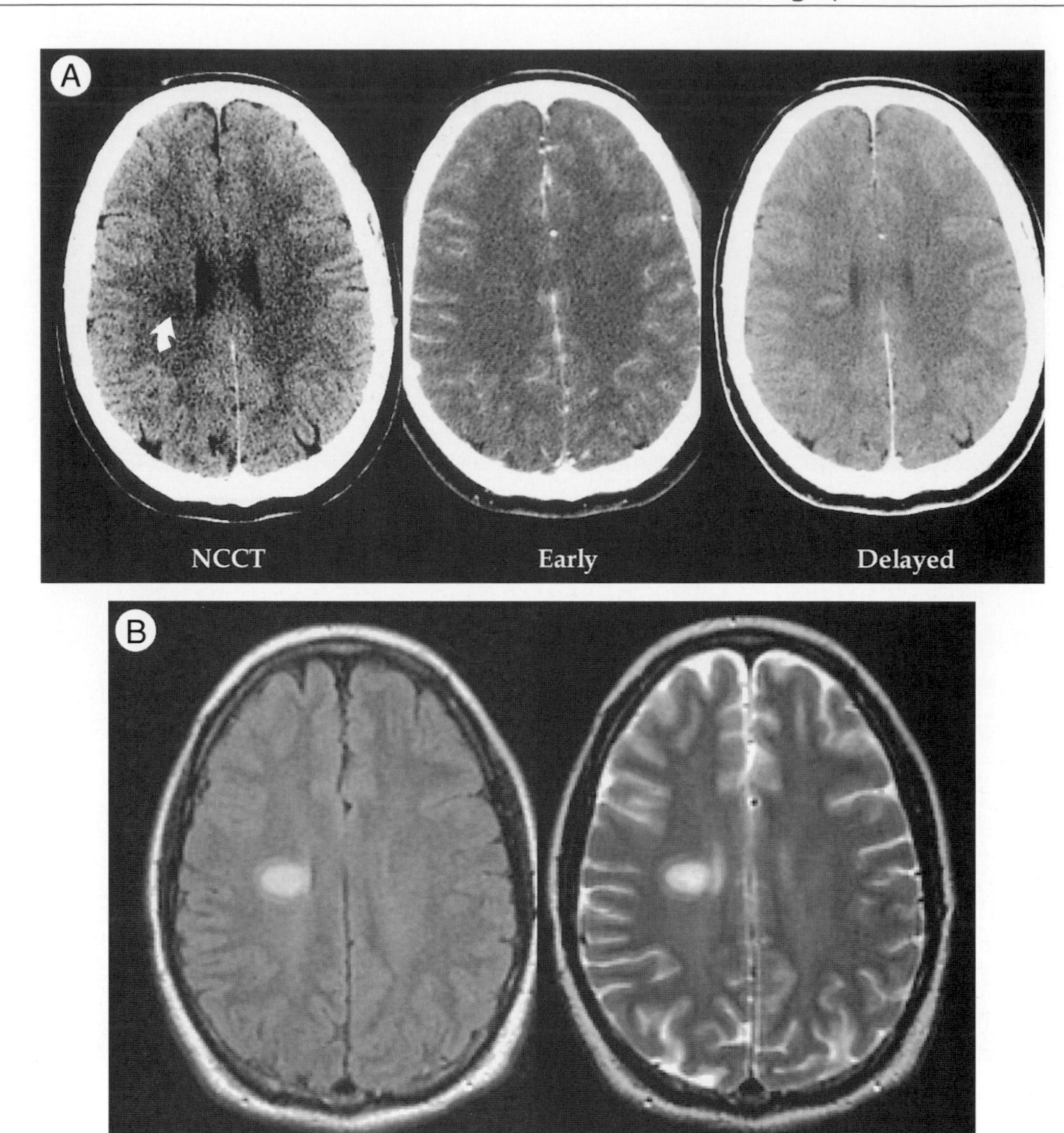

Figure 22 (**A**) Unusual case in which CTA/CTP excluded a stroke mimic. The patient awoke with new-onset left hemiparesis. CTA was normal. Noncontrast CT (NCCT) showed a small right periventricular hypodensity (arrow), which had increasing central enhancement on both early and delayed whole-brain perfused blood volume images. (**B**) Follow-up proton density (left) and T2-weighted (right) MR images confirm the presence of a large multiple sclerosis plaque.

IA thrombolysis use becomes more widespread, CTA could become essential to screen patients for the presence of large-vessel occlusion prior to the initiation of more invasive, risky, and expensive arteriography [159].

B. Carotid Artery Occlusive Disease/Vessel Dissection

1. NASCET (North American Symptomatic Carotid Endarterectomy Trial) Criteria

Dr. C. Miller Fisher of the Massachusetts General Hospital first elucidated the relationship between extracranial carotid artery occlusive disease and embolic stroke in 1951 [160]. Severe atheromatous narrowing of the proximal internal carotid artery predisposes to the formation of platelet-fibrin emboli in regions of turbulent flow; these emboli can be the source of TIA or stroke. By 1954, Eastcott, Pickering, and Rob of Great Britain had reported resection and reconstruction of stenotic carotid arteries, although DeBakey has been credited with the first successful carotid endarterectomy surgery in 1953 [161].

It was not until 1991, however, with the NASCET, that endarterectomy was finally proven to be "highly beneficial to patients with recent hemispheric and retinal transient ischemic attacks or nondisabling strokes and ipsilateral high-grade stenosis (70 to 99 percent) of the internal carotid artery" [162]. Criteria for performing endarterectomy in asymptomatic patients with ≥60% stenoses are more controversial. The 1995 Asymptomatic Carotid Atherosclerosis Study suggested that such patients, who are otherwise good candidates for elective surgery, "will have a reduced 5-year risk of ipsilateral stroke if carotid endarterectomy performed with less than 3% perioperative morbidity and mortality is added to aggressive management of modifiable risk factors" [163].

2. Ultrasound and MRA for Screening; CTA as a Problem-Solving Tool

Ultrasound (US), MRA, CTA, or conventional XRA can all be used to screen for carotid artery occlusive disease [164]. Each of these modalities has strengths and weaknesses. With the recent proliferation of noninvasive modalities, XRA is no longer routinely employed as a first-line diagnostic test for measuring the degree of carotid artery stenosis. Indeed, our general approach to the workup of carotid occlusive disease is to use: (1) US *or* MRA for screening, (2) US *and* MRA together to confirm the presence and degree of a suspected stenosis, and (3) CTA as a *problem-solving* tool when the results of US and MRA are equivocal or when specific anatomic questions are raised [165, 166]. Specifically, carotid artery CTA may be indicated when MRI is contraindicated or cannot be performed or when MRA, together with US, is unable to answer the clinical question, and catheter arteriography is being considered [165–167]. XRA is generally reserved for those few cases with complex, multivessel disease, for which detailed knowledge of collateral flow patterns is necessary in order to make rational treatment decisions [165–168].

An additional advantage of CTA is the possibility of performing "functional" studies, before and after various patient maneuvers. For example, in one of our patients with positional syncope, a small fibrous web was discovered which intermittently obstructed the vertebral artery on extreme head turning (Fig. 23) [169]. In another study helical CT showed "significant narrowing of the costoclavicular space after postural maneuver in symptomatic patients" [170].

3. Accuracy of Lumenal Diameter Measurement Compared with MRA, US, and Arteriography

Many studies have compared the diagnostic accuracy of CTA, MRA, XRA, and US in evaluating carotid lumenal diameter [168, 171–179]. In general, the presence of a flow gap with 2D time-of-flight (TOF) MRA is considered to be a reliable marker of a ≥60% carotid stenosis [180]. Important limitations of TOF MRA, however, include overestimation of the degree of stenosis caused by turbulent flow, motion, and pulsation artifacts and the inability to reliably distinguish very slow flow from true vascular occlusion [18, 172, 176]. Gadolinium-enhanced MRA has the potential to overcome many of these limitations, but it is still relatively early in its development and has not been as thoroughly validated as has CTA in cases with the most severe degrees of stenosis (Fig. 24) [181]. At least one author has recently commented that gadolinium-enhanced MRA, used alone, has not yet been

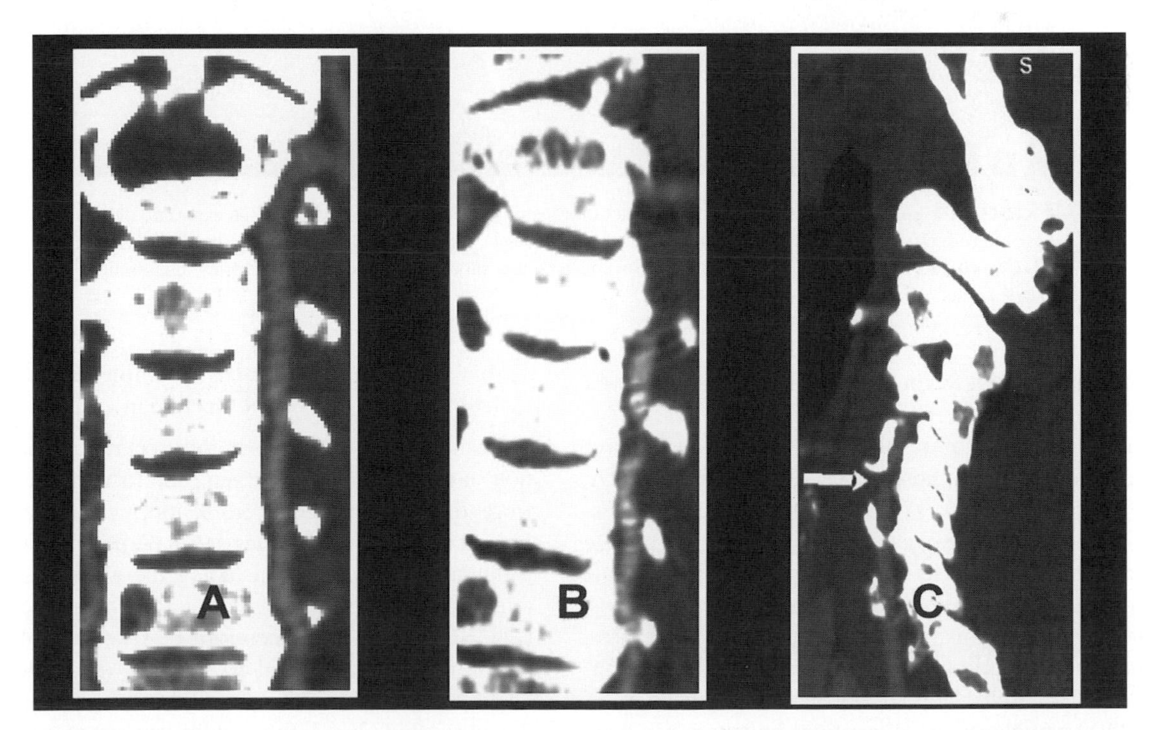

Figure 23 Curved reformat views of the left vertebral artery before (**A**) and after (**B, C**) head turning. The head-turning maneuver replicated the patient's symptoms and revealed a focal transient stenosis at the mid extracranial vertebral artery, which was ascribed to a small fibrous adhesion (arrow).

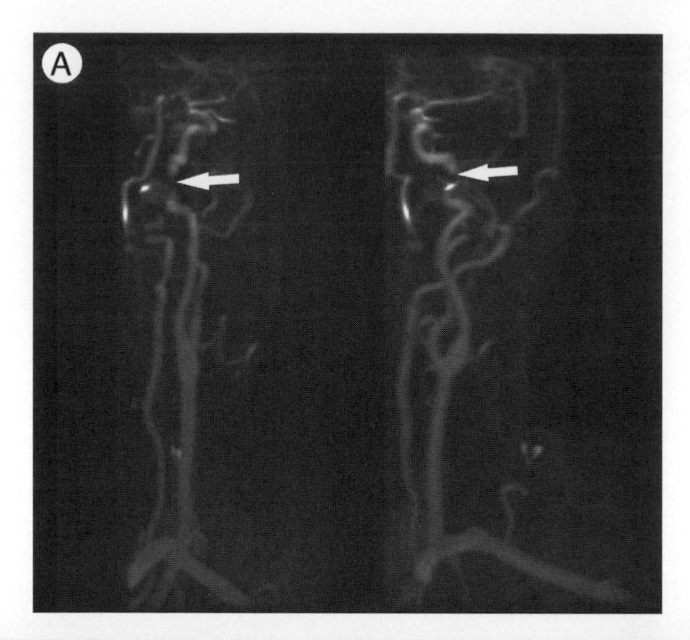

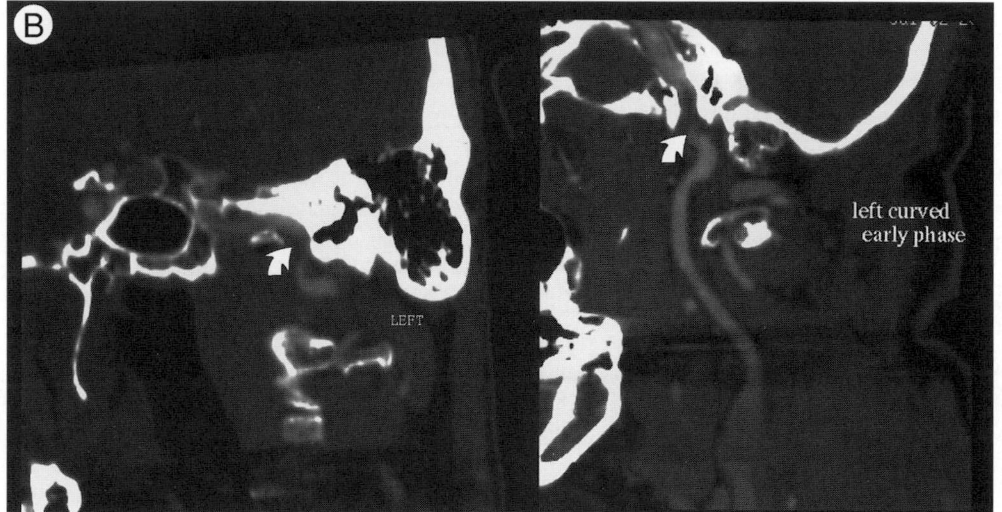

Figure 24 (**A**) Gadolinium-enhanced neck MRA shows a small distal flow gap at the left ICA entry into the skull base, consistent with a severe stenosis or focal occlusion (arrows). Crescentic T1-bright signal suggests dissection. (**B**) Curved reformatted images from a CTA confirm a focal ICA occlusion at the skull base, secondary to dissection.

proven accurate enough to routinely replace XRA in carotid artery evaluation [182].

CTA measurements of residual lumenal diameter have compared favorably with those of XRA, unenhanced MRA, and US [18, 53, 167, 173–178]. In performing such measurements, the user must pay careful attention to appropriate window and level display settings [53–55]. CTA is also able to distinguish important anatomic variants, such as loops, calcifications, aneurysms, and adjacent bony structures, which may have important implications for surgical planning [173, 179]. Only a poor correlation between CTA vessel wall morphology and ulceration, however, has been demonstrated to date [174, 183].

4. Ability to Distinguish Hairline Residual Lumen from True Total Vessel Occlusion

As noted above, CTA is highly accurate in the discrimination of hairline residual lumen (due to very severe stenosis) from true total vascular occlusion, with sensitivities and specificities in the 90% range [18]. A potential pitfall of this technique, however, is mistaking the ascending pharyngeal artery for a hairline patent extracranial internal carotid artery (Fig. 25). Care must therefore be taken to follow the full course of the extracranial ICA on the axial source images, "dot by dot," superior to the skull base, in order to avoid misdiagnosis (Fig. 26). MRA and US are not as reliable as CTA in distinguishing hairline lumen from occlusion, although

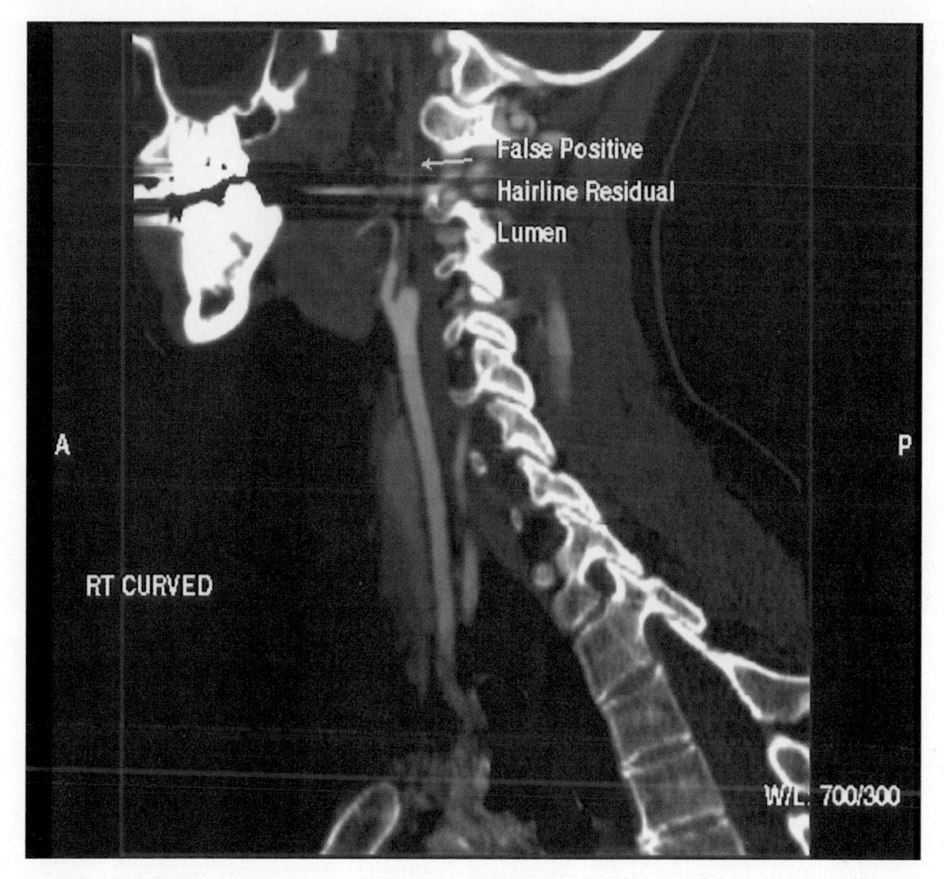

Figure 25 Curved reformat shows a hairline residual lumen along the expected course of the right extracranial ICA. Careful examination of the source images revealed this to be the ascending pharyngeal branch of the *external* carotid artery. The right ICA was completely occluded.

advanced US methods, such as power flow imaging, show promise [184]. XRA remains the "gold standard" for equivocal cases.

5. Chronic Carotid Occlusion and Stroke Risk and Cerebrovascular Reserve

Perfusion imaging is capable of detecting distal hemodynamic change in cases of severe carotid artery occlusive disease. Such hemodynamic changes are primarily of CBF and MTT and not of CBV [185]. Cerebrovascular reserve studies can also be performed using perfusion-imaging techniques. With the administration of diamox, or with a breath hold, the degree to which the cerebral vascular bed is able to compensate for changes in perfusion pressure by autoregulation can be assessed. In PET imaging studies, an increased oxygen extraction fraction has been shown to define a subgroup of patients with symptomatic carotid occlusion who are at high risk for subsequent stroke when treated medically [186]. It remains to be shown, however, whether CTP measurement of such changes can be used to help stratify patients according to stroke risk or to help guide treatment decisions regarding the need for carotid endarterectomy versus medical treatment.

CTP can also monitor flow changes following carotid endarterectomy, including the unpredictable occurrence of reperfusion hyperemia (Fig. 27) [187]. Although such flow changes are complex, in one study early arrival time on the operative side was noted to occur soon after endarterectomy in over half of patients [188].

6. Dissection

Possibly due to increased awareness on the part of clinicians, the diagnosis of carotid and vertebral dissection is increasingly being recognized. Dissections, which are typically treated with many months of coumadin anti-coagulation, are among the more common causes of stroke in young adults. Dissections tend to occur in regions of vascular tethering, such as in the distal extracranial ICA at the skull base (Fig. 24). In one helical CT study of dissection, the vast majority of dissections that presented with occlusion spontaneously resolved during coumadin therapy, although persistent vessel narrowing was typically observed [189]. Pseudoaneurysms associated with dissection were less likely to spontaneously resolve [189].

CTA can also be of value in the evaluation of traumatic vascular injury; one study concluded, "the sensitivity and

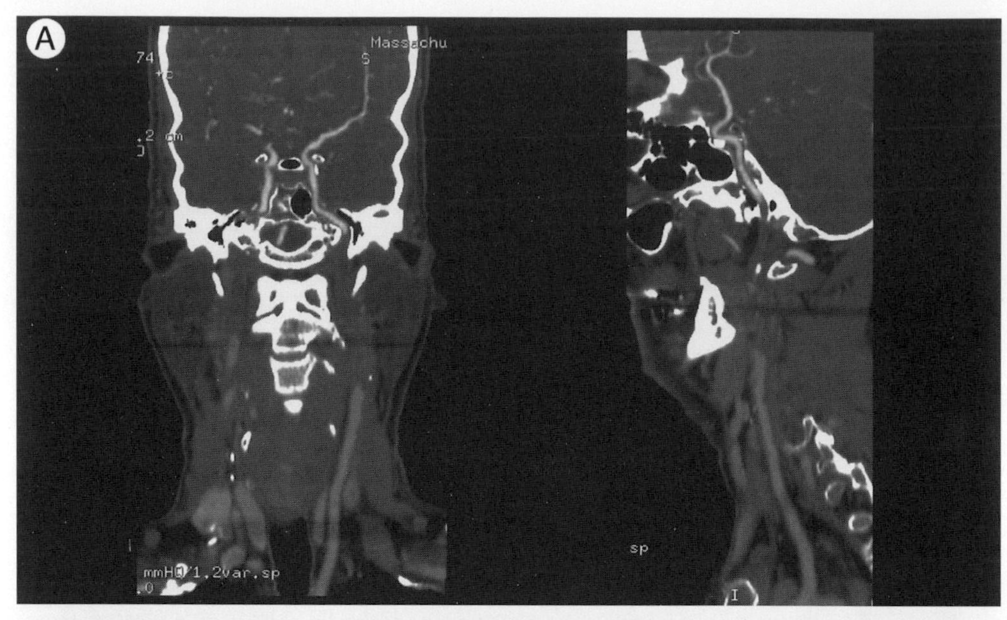

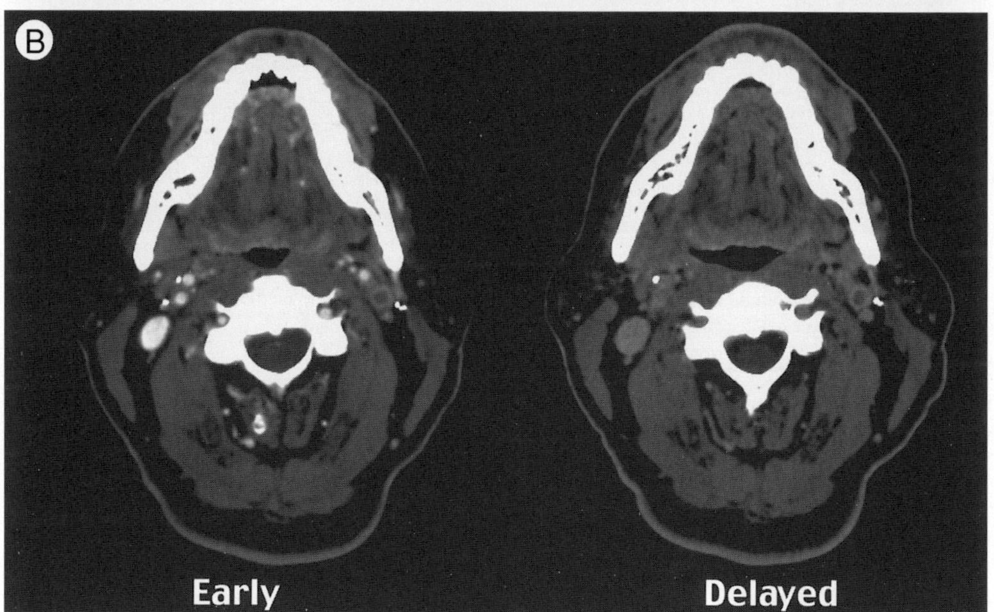

Figure 26 (**A**) Curved reformat views show a possible left extracranial ICA occlusion, although the possibility that the reformatted views were incorrectly constructed must be considered. (**B**) True ICA occlusion is confirmed on early and delayed postcontrast source images.

specificity of helical CT angiography are high for detection of major carotid and vertebral arterial injuries resulting from penetrating trauma" [190].

C. Aneurysm/AVM

Subarachnoid hemorrhage (SAH) from ruptured saccular aneurysms accounts for 6–8% of all strokes; in North America, the incidence of subarachnoid hemorrhage is approximately 11–12/100,000 [191]. Outcome for patients with SAH remains poor, with overall mortality rates of 25% and significant morbidity among approximately 50% of survivors [191].

CTA of aneurysms and AVMs can provide clinically relevant data for both diagnosis and treatment planning (Figs. 16 and 19). Although MRA serves as an adequate noninvasive screening exam for the *detection* of aneurysms, it can be inadequate for the *characterization* of aneurysms [192]. CTA, however, can typically delineate the precise relationship between aneurysms and adjacent bony struc-

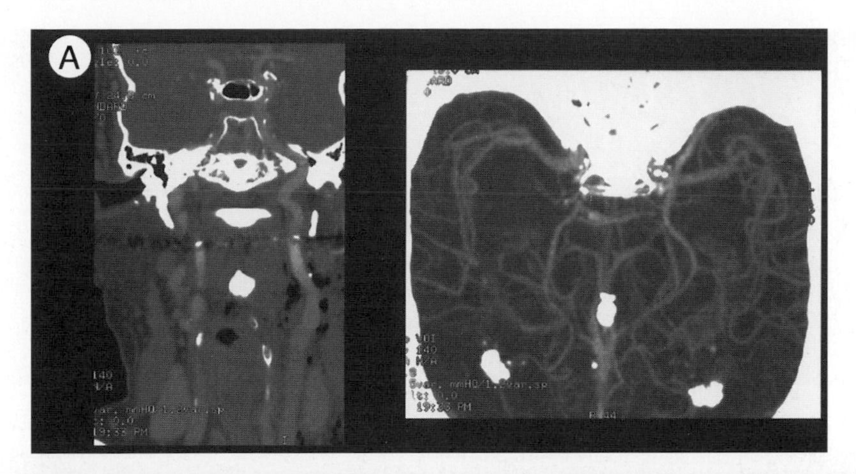

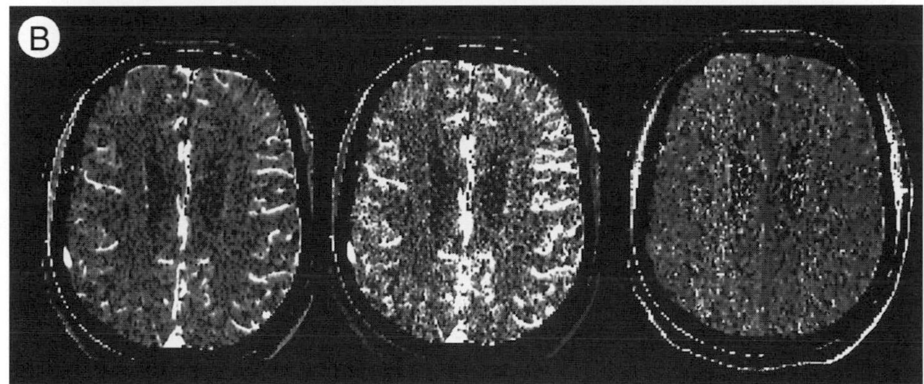

Figure 27 (**A**) Curved (left) and MIP (right) reconstructions of the left ICA circulation, 3 h following carotid endarterectomy. The left ICA is widely patent, and the left MCA shows increased collateral flow with hypervascularity. (**B**) First-pass cine CT perfusion maps show increased blood volume (left), increased blood flow (middle), and decreased transit time (right) over the left MCA territory, consistent with reperfusion hyperemia.

tures Although CTA is not usually performed as an initial screening study for the diagnosis of small, ≤3-mm-diameter aneurysms, CTA might be useful for follow-up of selected low-risk patients with known, unruptured aneurysms. In some instances, CTA can clarify equivocal XRA findings, such as for very large aneurysms whose neck cannot be clearly delineated by standard arteriographic projections.

1. Size and Risk of Rupture: Asymptomatic versus Symptomatic

Aneurysm rupture risk is related to size; for aneurysms <10 mm the risk is estimated at 0.05%/year, and for those >25 mm the risk is as high as 6.00%/year [193]. These risks are up to 11× higher if there has been previous SAH. PCOM and VBJ aneurysms may also be at increased risk of rupture, whereas small intracavernous aneurysms are generally not treated [193]. Multiple aneurysms, or those with daughter sacs, also have a higher risk for rupture [193].

Which unruptured aneurysms should be treated? This is a complex question, the answer to which is dependent on many factors; however, a recent American Heart Association scientific statement suggests that treatment is favored for young patients, with a long life expectancy, previously ruptured aneurysms, large symptomatic aneurysms, aneurysm growth, and low surgical/endovascular risk [194]. Conservative management is favored for those of older age or decreased life expectancy, for asymptomatic or small aneurysms, and for those with comorbid medical conditions [194].

2. Increased Risk with Polycystic Kidney Disease (PCKD) and Positive Family History

Periodic MRA or CTA screening for aneurysms, approximately every 3 years, may be beneficial for patients with PCKD [195]. Familial intracranial aneurysms are a relatively recently identified entity, which may account for up to 25% of all SAH—a 2–4× greater risk than that of isolated SAH [195]. The routine screening of first-degree relatives of aneurysm patients remains controversial and is in part dependent on the expected risk of surgery versus the expected gain in life expectancy [195].

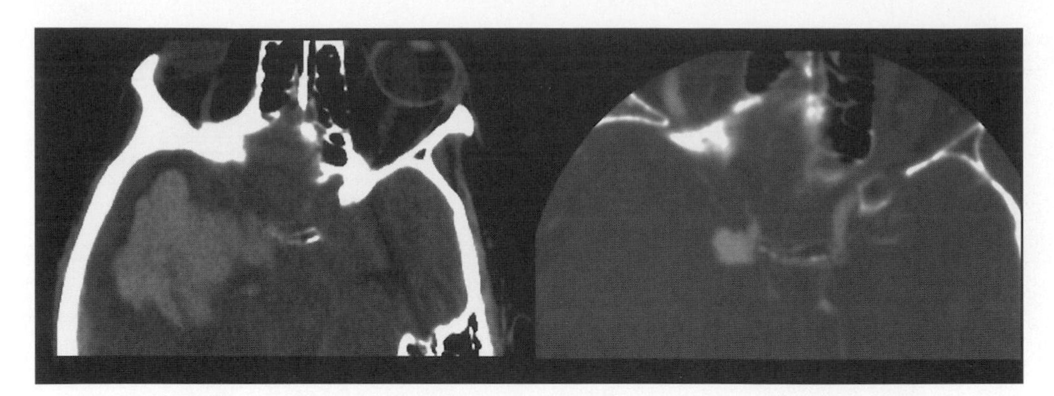

Figure 28 Precontrast axial source image (left) from an aneurysm CTA exam shows a large right subarachnoid and parenchymal high attenuation hemorrhage. Postcontrast source image (right) reveals a densely enhancing supraclinoid ICA aneurysm, not obscured by the hemorrhage.

3. Sensitivity and Specificity versus MRA and Catheter Arteriography for Screening

In 1994, Schwartz *et al.* described 100% sensitivity for the CTA detection of aneurysms >3 mm in size; no aneurysms <3mm were detected [175]. In other early studies, the presence of cisternal SAH did not limit aneurysm detection or reconstruction (Fig. 28) [196, 197].

Subsequent studies have confirmed the excellent sensitivity and specificity of CTA for aneurysm detection in comparison with MRA and XRA [197, 198]. A systematic review of the imaging literature from 1988 to 1998 has suggested that the overall accuracy of both CTA and MRA for aneurysm detection is approximately 90% [199]. Indeed, some authors have advocated CTA as the sole preoperative imaging modality—replacing XRA—for a substantial proportion of patients with ruptured aneurysms [200]. Other authors have also suggested that CTA alone might be sufficient to exclude the presence of aneurysm in cases with a perimesencephalic pattern of SAH and that "only when the complication rate of catheter angiography is <0.2% is arteriography the preferred strategy" [201]. Still others have concluded that CTA, with CR reconstruction, is a reliable diagnostic test for excluding aneurysm in patients presenting with isolated third-nerve palsy and that it can identify that small subgroup requiring definitive XRA [202].

4. 3D Reconstructions as a Primary Modality for Surgical Planning

VR and SSD reconstructions are a favored method for the display of aneurysms for surgical planning, as described at length in Section III.C on scanning protocols. CTA can also be performed in the presence of surgical aneurysm clips. In such cases, the scanner gantry angle should be carefully chosen so as to minimize metallic spray artifact (Fig. 29) [203, 204]. CTA can be used to show residual aneurysm filling adjacent to a clip, vessel lumenal compromise due to recent clip placement, and incorrect clip positioning [205].

5. Vasospasm

Delayed ischemia due to SAH-related vasospasm, and not rebleeding, remains the most significant cause of death and disability following aneurysm rupture [191]. Vasospasm can begin as early as 3 days after SAH, peaks at 6–8 days, and typically resolves within 2 weeks. "Angiographic" spasm, the visually apparent reduction in vessel caliber that can be detected by transcranial Doppler US, CTA, or XRA, should be distinguished from "clinical" spasm, the syndrome of confusion and decreased level of consciousness associated with reduced blood flow to the brain parenchyma. Angiographic spasm occurs in approximately 50% of patients following aneurysmal SAH; clinical spasm occurs in about 30%. Roughly half of clinical spasm cases result in infarction [191].

CTA has the potential to detect angiographic vasospasm, and CTP has the potential to aid in the diagnosis of clinical vasospasm. In a study that compared MIP CTA reconstructions with XRA, it was concluded that CTA was highly accurate for the detection of proximal circle of Willis spasm and only slightly less accurate for the detection of more distal vessel spasm [206]. MR, CT, and SPECT blood flow studies have revealed a correlation between angiographic spasm, reduction in CBF, and clinically symptomatic spasm [207, 208]. In the future, measures of cerebrovascular reserve could prove to be of value in the triage of SAH patients to medical versus endovascular treatment.

In a related study of the cerebral hemodynamics associated with giant aneurysms before and after bypass surgery, the expected autoregulatory responses of increased MTT with increased or preserved CBV were observed. Bypass surgery improved hemodynamics and, in cases with maximal vasodilatation and loss of autoregulation, restored CBF index [209].

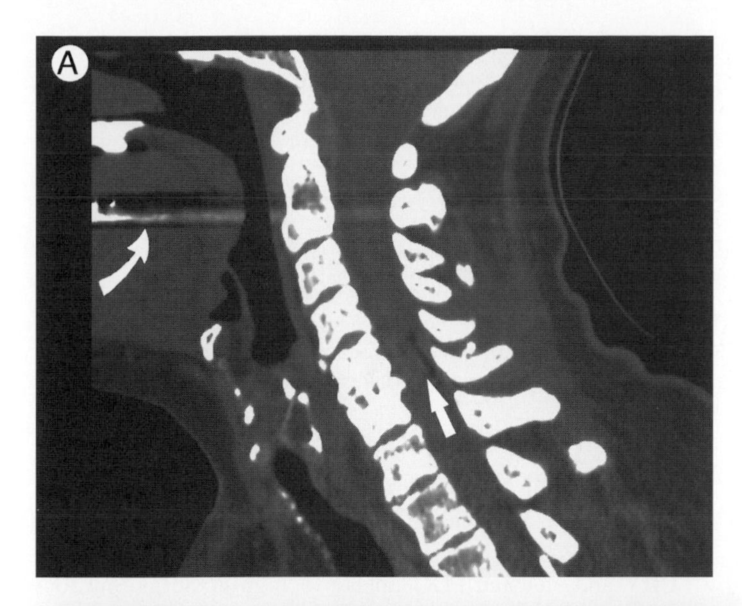

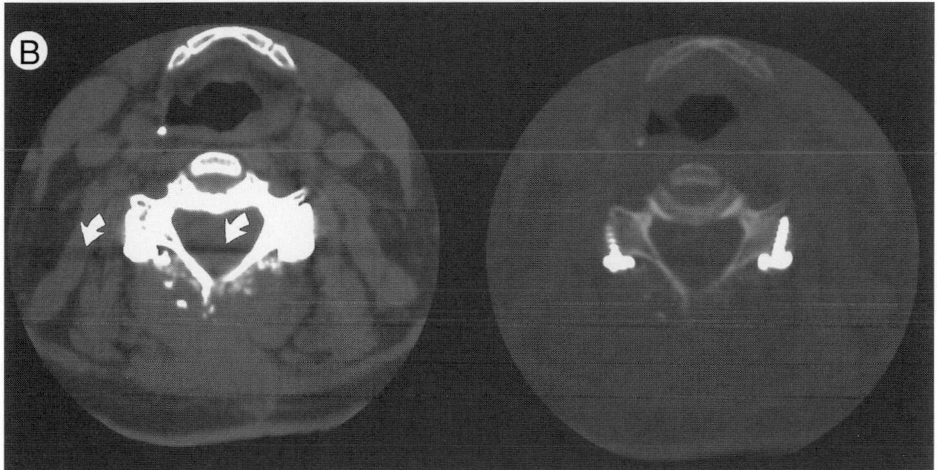

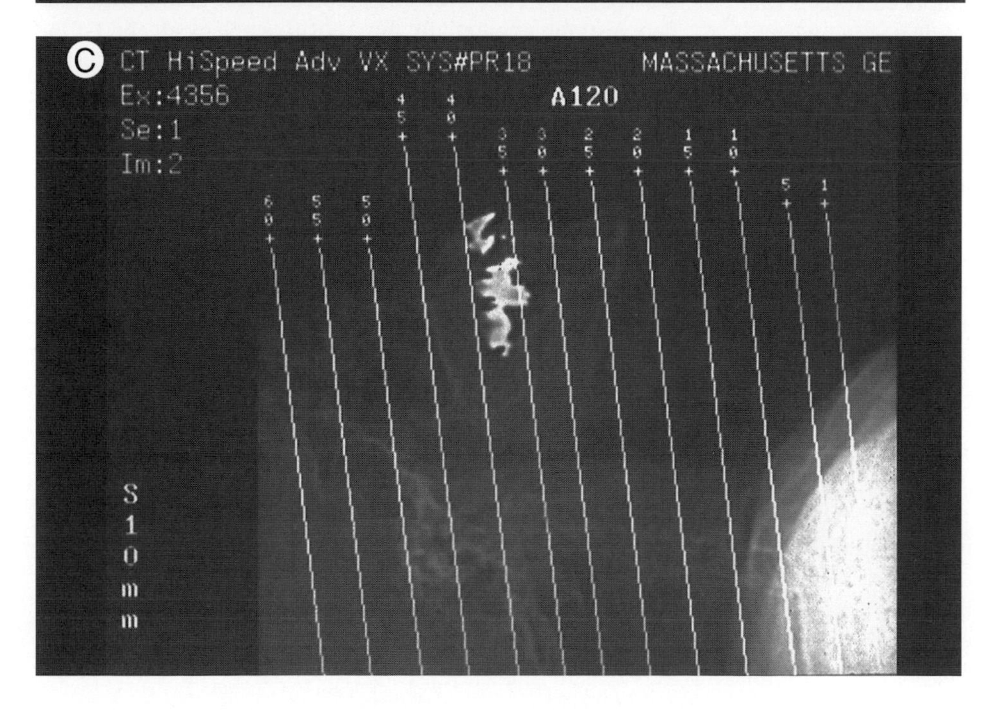

Figure 29 (**A**) Sagittal 2D-MPR of the neck shows metallic streak artifact (curved arrow) and possible air in the spinal canal (short straight arrow). (**B**) Axial source images filmed in soft tissue (left) and bone (right) windows reveal that the apparent air in the spinal canal is streak artifact (arrows), caused by pedicle screws, reconstructed in the sagittal plane. (**C**) The metallic streak artifact caused by dental fillings can be limited to a narrow axial slab by judicious choice of head tilt and gantry angle.

D. Venous Thrombosis: MRV for Screening and CTV as a Problem-Solving Tool

Our general clinical approach to the workup of venous sinus thrombosis, as with carotid occlusive disease, is to use CTV as a problem-solving tool for cases in which the results of screening MR venography are equivocal. CTV not only has the advantages of non-flow-dependent anatomic imaging, but also can more clearly demonstrate adjacent soft tissue and calcific structures, such as meningioma invasion into the dural venous sinuses [210, 211].

CTV has high reliability and can be easier to interpret with fewer artifacts than MRV [211, 212]. In addition, CT venography may more frequently visualize those sinuses or small cerebral veins with low flow, compared with MRV (Fig. 20) [213]. Although MIP reconstructions alone cannot optimize direct visualization of thrombus by either CTV or MRV, in one study, the "graded subtraction" technique with integral display facilitated direct visualization of thrombus [213]. Another study showed that MPR reconstructions were superior to MIP [212].

E. Brain Tumor Angiogenesis

As noted at the Website of The Central Brain Tumor Registry of the United States, brain tumors are the most common solid tumors in children and the second most frequent malignancy of childhood [214, 215]. Brain tumors are the second leading cause of cancer-related deaths in males ages 20–39 and the fifth leading cause of cancer-related deaths in women ages 20–39. About 13,000 people in the United States die of malignant brain tumors each year. Brain tumors represent 23% of cancer-related deaths in male children under the age of 20 and 25% of cancer-related deaths in female children under the age of 20 [214, 215].

1. Background: CBV as a Surrogate Marker for Angiogenesis/Malignant Potential

Much of the current literature regarding perfusion imaging of brain tumors concerns CBV imaging technique. CBF and MTT methods have not, to date, been as well studied, but early results suggest that these, too, have the potential to further the clinical value of existing functional techniques (such as PET and MR spectroscopy). Indeed, in analogy to CBV maps, dynamic first-pass CT-CBF maps of brain tumor regions not only show significantly higher flow than do those of normal regions, but this flow decreases in response to hypocapnia [216]. Arterial spin-labeled MR-CBF techniques, which may be more accurate than contrast-enhanced techniques in regions of severe blood–brain barrier breakdown, have also been shown to correlate well with glioma grade [17]. Newer methods of performing dynamic contrast-enhanced perfusion studies, including measurement of fractional blood volume and microvascular permeability, have additionally shown promise in the clinical evaluation of brain tumors [111].

Possible clinical applications of perfusion imaging of brain tumors include:

- Tumor grading.
- Stereotactic biopsy guidance.
- Distinguishing radiation necrosis from recurrent glioma.
- Determining prognosis and response to treatment.
- As a surrogate marker of response for clinical trials.

2. CBV and CBF for Brain Tumor Grading

The use of both dynamic susceptibility CBV mapping and FDG-PET scanning to localize regions of high-grade glioma has been well studied [43, 217, 218]. The increased CBV and elevated FDG uptake of high-grade glioma foci are functions of the increased vascularity and metabolism of grade III and IV gliomas, compared with lower grade neoplasms [42, 219–221]. In one study of glioma patients who had correlative pathology specimens, foci of maximal tumor CBV were associated with high mitotic activity and vascularity, but not with cellular atypia, endothelial proliferation, necrosis, or cellularity [43].

3. CBV and CBF for Stereotactic Biopsy Guidance

High CBV foci can be found in nonenhancing tumors. In a spin-echo MR-CBV study from our institution, all high-grade astrocytomas were correctly categorized. Of these, almost 25% of the high-grade tumors failed to enhance with contrast. Of the low-grade astrocytomas studied, approximately 75% were correctly classified as low grade, with 10% demonstrating potentially false positive enhancement. Of the oligodendrogliomas studied, however, all of the high grade and 50% of the low grade had elevated CBV values [222]. Thus, in this study, MR-CBV imaging had a high negative predictive value; tumors with uniformly low CBV were unlikely to have high-grade glioma components, regardless of their enhancement characteristics on conventional MR imaging. It is noteworthy that all oligodendrogliomas, independent of grade, had a significant likelihood of containing high blood volume foci. CBV images may therefore offer additional data, not available on conventional MRI, which could help guide stereotactic brain biopsy (Fig. 30) [222]. Researchers using gradient-echo MR-CBV methods have also confirmed strong correlations between tumor grade and degree of CBV elevation [223].

Other highly vascular tumors, such as meningiomas and vascular metastases like renal cell carcinoma and melanoma, also (not surprisingly) demonstrate high CBV compared to normal gray and white matter. Of note, however, care must be taken in imaging extra-axial lesions such as meningiomas, because increased vascular perme-

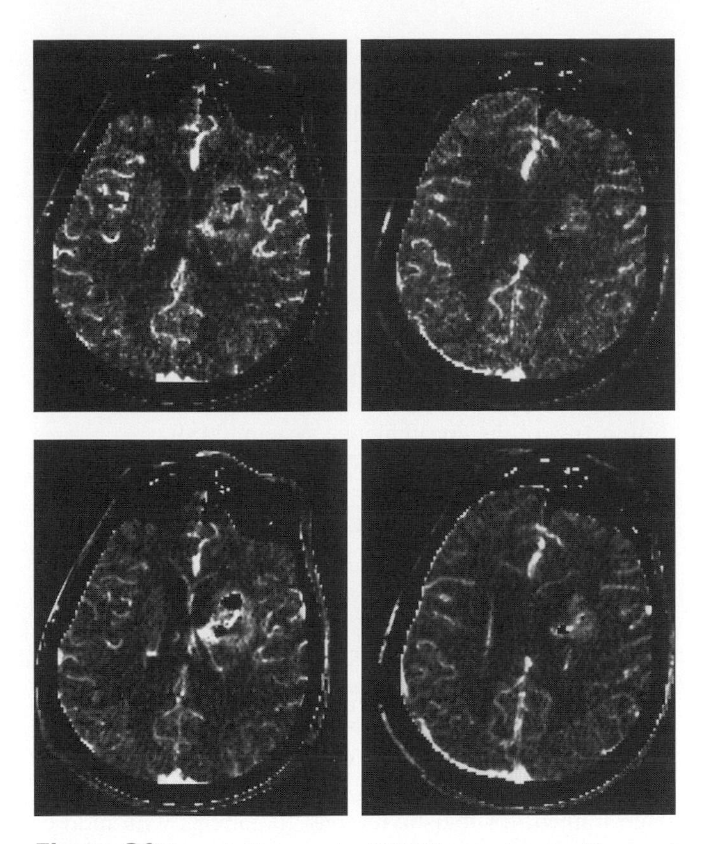

Figure 30 Dynamic first-pass cine CT-CBV maps show focal increased blood volume in the left corona radiata, surrounding an enhancing (not shown) necrotic region that failed to demonstrate high-grade tumor at stereotactic biopsy. Subsequent biopsy of the nonenhancing but elevated CBV regions revealed high-grade, anaplastic astrocytoma.

ability (secondary to the lack of blood–brain barrier) could potentially cause underestimation of the MR-CBV measurements due to gadolinium "leakage" effects [224].

Using conventional CT or MRI alone, AIDS lymphoma cannot be reliably distinguished from toxoplasmosis, as both can show similar patterns of enhancement. In one small series of patients, however, CBV maps were able to better discriminate between these two possibilities—each of the toxoplasmosis lesions had reduced CBV, whereas all of the lymphoma lesions had elevated CBV [225]. Measurement of flow parameters in lymphoma should be approached with caution, however, as steroid treatment can dramatically alter CBV and CBF within these lesions. CBV imaging also has the potential to facilitate the evaluation of diverse head and neck malignancies such as inverted papilloma or lymph node metastases [226, 227].

4. CBV and CBF for Distinguishing Tumor Recurrence from Radiation Necrosis

Although some authors have advocated the use of MR-CBV imaging to help distinguish recurrent tumor from radiation necrosis in post-radiation-treated patients—a distinction not routinely possible using conventional contrast-enhanced MR or CT imaging—the results of studies addressing this issue have been mixed [228, 229]. The theory underlying such imaging is that, because angiogenesis determines blood flow, metabolism, and growth rate of residual or recurrent tumor in irradiated tumor beds, perfusion imaging might permit monitoring of the local microcirculation, thus becoming a surrogate marker for clinical and histological staging [41, 219].

Along these lines, CBV maps *have* been shown to be capable of following changes in the microcirculation of some tumors in response to steroid treatment [46]. In a study of post-high-grade glioma resection patients treated with proton beam irradiation at our institution, however, CBV values did not have a high predictive value in distinguishing radiation necrosis from recurrent tumor, although the predictive value of CBV imaging was considerably greater than that reported for FDG-PET [24, 228]. Also, this result may not be generalizable, as proton-irradiated tumors may behave different from photon-irradiated tumors (the later is the more common treatment). Specifically, the marked blood–brain barrier breakdown in regions of proton beam irradiation can lead to leakage of gadolinium contrast during perfusion imaging, causing underestimation of CBV. As already noted, however, CBF and MTT imaging, as well as arterial spin-labeled MR perfusion imaging, have the potential to improve on these results.

MR cerebral blood volume maps have also been shown to detect tumor progression earlier than do other imaging modalities. In one study which followed tumor evolution in 59 patients, progression was detected by CBV maps an average of 4.5 months earlier than by MRI for 32% of the studies, an average of 4.5 months earlier than by ^{201}Tl-SPECT for 63% of the studies, and an average of 6 months earlier than clinical assessment for 55% of the studies [229]. Size of lesions affected detectability; in 82% of studies with positive MR but negative SPECT, the lesions were smaller than 1.5 cm. A trend toward early detection of *recurrent* tumor has also been reported using CBV imaging [17, 228]. These results suggest that an important application of perfusion imaging may be as a surrogate marker of tumor treatment response in clinical trials—especially given recent interest in cancer therapy with anti-angiogenesis agents.

F. Current Research/Future Indications?

CTA and CTP techniques have the potential to augment current diagnostic approaches in the evaluation of such diverse neurological disorders as ICA occlusion, Alzheimer's type dementia, traumatic brain injury, migraine headache, vasculitis, moya-moya, and brain death [230–232]. For example, multiple studies have estimated that the annual risk of stroke in patients with ischemic symptoms ipsilateral to an occluded ICA is as high as 10%, despite treatment with anti-coagulants or anti-platelet agents

[233]. This underscores the potential benefit of cerebrovascular reserve measurement in stratifying stroke risk in patients with unilateral ICA occlusion [186].

Alzheimer's disease is the most common form of dementia, and the fourth leading cause of death in the United States; nearly half of people over 85 have some symptoms [234]. Approximately 4 million Americans have Alzheimer's disease; this will increase to 14 million in the next 50 years if current trends continue. Alzheimer's disease costs the United States at least $100 billion a year. Therefore, although a "cure" for Alzheimer's disease is unlikely to be available soon, the possibility that early treatment of high-risk patients could delay the onset, or reduce the likelihood, of Alzheimer's disease, even for just a few years, has tremendous public health implications [235]. Such treatments are on the immediate horizon—nonsteroidal anti-inflammatory agents, for example, taken even in moderate doses for more than 2 years prior to symptom onset, have recently been shown to reduce the risk of Alzheimer's-type dementia [235].

A potentially important imaging goal with Alzheimer's disease, therefore, is its early detection and diagnosis—ideally, prior to the onset of symptoms and gross morphological changes. To this end, functional measurements of metabolism, blood volume, and blood flow in the enterorhinal cortex may be of value. Indeed, preliminary investigations have shown early reductions in these parameters [47, 48, 236]. In one study, for example, temporoparietal CBV was found to be reduced up to 17%, even in mildly affected Alzheimer's disease patients [47, 48].

Early subclinical HIV infection can also be associated with dementia; 30–40% of HIV-infected patients show signs of mild cognitive impairment, and 20% of AIDS patients develop frank dementia. FDG-PET and MR-CBV scanning have shown significant agreement in distinguishing normal from abnormal brain regions in these patients [49]. Significant increases in CBV of deep and cortical gray matter have also been discovered in HIV-positive patients; the degree of increase appears to be proportional to the degree of cognitive impairment and may be reversible with zidovudine therapy [237].

Prediction of outcome in traumatic brain injury might also be accomplished using CBF measurement [238]. Changes in CBV, CBF, and blood–brain barrier permeability could potentially be of value in assessing: (1) the development of brain edema following intraparenchymal hemorrhage [239], (2) hemodynamic response following prolonged cardiac arrest [240], (3) the role of ischemia in the development of increased intracranial pressure and cytotoxic edema following severe brain trauma or dissection [241], or (4) the development of vasospasm associated with eclampsia or aneurysmal SAH [242].

Hyperperfusion and vascular dilatation have also been noted to occur during migraine headache aura; these changes resolve after the attack [243]. Moderate decrease in CBF, mild decrease in CBV, and moderate increase in MTT have been observed, without diffusion changes [244].

The imaging and clinical findings of CNS vasculitis are typically nonspecific and may be variable or absent on MRI or XRA. Perfusion and diffusion imaging have been proposed as potential tools for the improved diagnostic assessment of CNS vasculitis [245].

Decreased CBF of the anterior cingulate gyrus, temporal gyrus, and precuneus (brain regions typically associated with working memory and attention), in children with attention deficit hyperactivity disorder, may be correlated with dysfunction of the dopaminergic system [246].

Because the CBV and CBF changes accompanying seizure might occur only during ictus, or in the immediate preictal state, and because EEG leads present technical (but not insurmountable) difficulties for MR scanning, the utility of perfusion imaging in the workup of epilepsy has not yet been established. Increased CBV has been reported during focal status epilepticus [247]. A study using surface CBF monitoring concluded that CBF alterations precede seizure activity, suggesting that vasomotor changes may produce electrical and clinical seizure onset [248].

Although imaging studies of schizophrenic patients are often limited, for logistic reasons, to patients with mild or moderate forms of the disease, perfusion studies of schizophrenics have revealed elevated CBV in the bilateral occipital cortex, basal ganglia, and cerebellum of schizophrenic subjects who had been clinically stable on medications [249].

CTA with CTP, by demonstrating the absence of intracerebral blood flow, may also prove to be of value in confirming the diagnosis of brain death [250].

V. Conclusions

A. CTA/CTP Strengths and Weaknesses

CTA and CTP imaging have a number of advantages and disadvantages compared to MRA, US, XRA, and other functional techniques, such as MR spectroscopy and nuclear medicine methods. CTA provides truly anatomic, non-flow-dependent data that can more accurately determine percentage vessel stenosis, as well as the presence of a hairline vascular lumen, than can standard MRA techniques—although gadolinium-enhanced MRA shows early promise. CTA is convenient—it is available in most hospital emergency departments at low relative cost. CTA is accurate—it can often preclude the need for conventional catheter arteriography. CTA and CTP can also provide high-resolution imaging, with better bony assessment, fewer contraindications, and fewer artifacts from motion and other causes, than

MRI. At present, CT remains the first-line imaging test for the urgent evaluation of intracranial hemorrhage.

On the other hand, MRI uses no ionizing radiation, does not require iodinated contrast, and can produce fully automated 2D and 3D reconstructions. MRA offers physiological imaging of flow direction and velocity and can therefore distinguish arterial from venous flow. MRI also does not have the "slice pollution" issues of CTA. Both modalities can provide "one-stop-shopping," although MRI has more flexibility with regard to additional available techniques, such as MR spectroscopy, arterial spin labeling, and blood oxygen level-dependent imaging.

B. CTA/CTP Clinical Applications

CTA and CTP can be applied to the evaluation of acute stroke, carotid artery occlusive disease, dissection, aneurysm, AVM, venous thrombosis, and brain tumor angiogenesis. Possible future applications include evaluation of cerebrovascular reserve, Alzheimer's-type dementia, traumatic brain injury, and migraine headache, to name a few.

CTA with CTP can be a primary modality for the evaluation of acute stroke or as a problem-solving tool—following screening MR or US—for patients with carotid occlusive or other neurovascular disease. CTA/CTP can also be a surgical planning tool for the treatment of aneurysms and AVMs. Appropriate reconstruction techniques for each of these applications include: (1) real-time thin-slice axial source images and rapid "collapsed" MIP circle of Willis views for defining the site of thrombosis in acute stroke, (2) 2D curved reformats and thin source images for detection and measurement of carotid and vertebral artery stenotic disease, (3) 3D-VR and SSD projections for aneurysm and AVM surgical planning, and (4) MIP and integral "graded subtraction" displays for evaluation of venous sinus thrombosis.

CTA/CTP will likely prove of value in the urgent triage of stroke patients to appropriate therapy, including IV and IA thrombolysis. Advantages of CTA/CTP in the evaluation of stroke are speed, simplicity, accuracy, and convenience. With CTA/CTP, clinically relevant data regarding both vascular patency and parenchymal perfusion can be obtained during a single, brief exam. Such data may have prognostic value with respect to final infarct volume, clinical outcome, and hemorrhagic risk.

In untreated gliomas, the absence of elevated blood volume mitigates against the presence of high-grade tumor, regardless of the enhancement characteristics of the lesion. For posttreatment brain tumors, the situation is more complex, but there is good evidence that perfusion imaging has advantages over both conventional MRI and other, more established, functional imaging techniques [17].

C. Future Developments

As new treatments are developed for stroke, dementia, psychiatric illness, headache, and trauma, the potential clinical applications of CT angiographic and perfusion imaging in the diagnosis, triage, and therapeutic monitoring of these diseases are certain to increase. It is noteworthy that CTA and CTP have also been applied to organs other than the brain. Perfusion imaging may be of value in assessing lymph nodes, breast parenchyma, and other soft tissues for

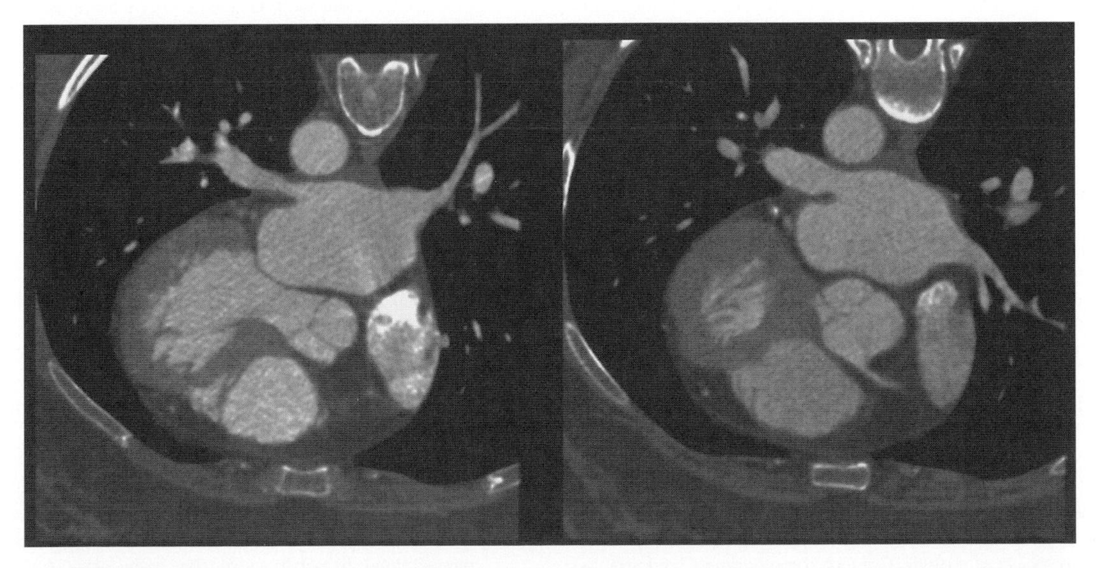

Figure 31 In the future, as scanner technology continues to improve, the "arch-to-vertex" stroke CTA exam has the potential to expand so as to include not only the great vessel origins, but also the left atrium and ventricle of the heart. This might be of value in the assessment of stroke patients with atrial fibrillation.

the presence of malignant disease. Early experiences using perfusion methods to distinguish recurrent tumor from other posttreatment changes in head and neck cancers, as well as to evaluate cardiac ischemia by pharmacological "stress testing," have shown promise [25].

Technical advances in scanner hardware and software will no doubt continue to increase the speed, coverage, and resolution of CTA/CTP imaging (Fig. 31). CTA and CTP offer the promise of improved detection of intracranial disease, which could lead to more efficient utilization of imaging resources and, potentially, to decreased morbidity. CT perfusion techniques may be at least as sensitive and specific as radionuclide-based techniques and offer the added advantages of higher intrinsic resolution, convenient coregistration with anatomic CT images, and time- and cost-effective evaluation of patient cohorts who already receive frequent conventional imaging [25].

Acknowledgments

The authors gratefully acknowledge and deeply appreciate the invaluable editing assistance of George J. Hunter, M.D., and Leena Hamberg, Ph.D. We also acknowledge the insights of, and helpful discussions with, Robert H. Ackerman, M.D., M.P.H., and Director of the MGH Neurovascular Laboratory for many decades, as well as Ting Lee, Ph.D., a driving force behind the development of many of the CT perfusion techniques described herein. We also thank Gordon Harris, Director of the MGH 3D Imaging Lab. Finally, we acknowledge the ongoing efforts of past and present staff and Fellows of the MGH Departments of Neuroradiology, Neurosurgery, and Neurology—most notably Lee Schwamm, M.D., and Walter Koroshetz, M.D., of the MGH Stroke Service—in advancing the clinical applications of CTA and CTP.

References

1. Leeds, N. E., and Kieffer, S. A. (2000). Evolution of diagnostic neuroradiology from 1904 to 1999. *Radiology* **217,** 309–318
2. Ambrose, J., and Hounsfield, G. (1973). Computerized transverse axial tomography. *Br. J. Radiol.* **46,** 148–149.
3. Merino-deVillasante, J., and Taveras, J. M. (1976). Computerized tomography (CT) in acute head trauma. *Am. J. Roentgenol.* **126,** 765–778.
4. Wolpert, S. M. (2000). Neuroradiology classics. *Am. J. Neuroradiol.* **21,** 605–606.
5. New, P. F., Scott, W. R., Schnur, J. A., Davis, K. R., and Taveras, J. M. (1974). Computerized axial tomography with the EMI scanner. *Radiology* **110,** 109–123.
6. Drew. (2001). Historical perspective: CT imaging. *Diagnostic Imaging*, CMP Media, LLC, Manhasset, New York.
7. Sato, O., Kanazawa, I., Kamitani, H., Yamashita, M., and Sano, K. (1976). [Contrast media enhancement in cranial computed tomography—Studies of iodinated contrast infusion (ICI)]. No To Shinkei **28,** 1225–1233.
8. Davis, K. R., New, P. F., Solis, O. J., and Roberson, G. H. (1976). Theoretical considerations in the use of contrast media for computed cranial tomography. *Rev. Interam. Radiol.* **1,** 9–12.
9. Rubin, G. D., Shiau, M. C., Schmidt, A. J., Fleischmann, D., Logan, L., Leung, A. N., *et al.* (1999). Computed tomographic angiography: Historical perspective and new state-of-the-art using multi detector-row helical computed tomography. *J. Comput. Assisted Tomogr.* **23**(Suppl. 1), S83–90
10. Nickoloff, E. L., and Alderson, P. O. (2001). Radiation exposures to patients from CT: Reality, public perception, and policy. *Am. J. Roentgenol.* **177,** 285–287.
11. Amis, E. S., Jr. (1999). Epitaph for the urogram. *Radiology* **213,** 639–640.
12. Berland, L. L., and Smith, J. K. (1998). Multidetector-array CT: Once again, technology creates new opportunities. *Radiology* **209,** 327–329.
13. Rubin, G. D., Schmidt, A. J., Logan, L. J., and Sofilos, M. C. (2001). Multi-detector row CT angiography of lower extremity arterial inflow and runoff: Initial experience. *Radiology* **221,** 146–158.
14. Fox, S. H., Tanenbaum, L. N., Ackelsberg, S., He, H. D., Hsieh, J., and Hu, H. (1998). Future directions in CT technology. *Neuroimaging Clin. North Am.* **8,** 497–513.
15. Jhaveri, K. S., Saini, S., Levine, L. A., Piazzo, D. J., Doncaster, R. J., Halpern, E. F., *et al.* (2001). Effect of multislice CT technology on scanner productivity. *Am. J. Roentgenol.* **177,** 769–772.
16. Zilkha, E., Ladurner, G., Iliff, L. D., Du Boulay, G. H., and Marshall, J. (1976). Computer subtraction in regional cerebral blood-volume measurements using the EMI-scanner. *Br. J. Radiol.* **49,** 330–334.
17. Lev, M., and Rosen, B. (1999). Clinical applications of intracranial perfusion MR imaging. *Neuroimaging Clin. North Am.* **9,** 309–331.
18. Lev, M., Ackerman, R., Rabinov, J., *et al.* (1997). Hairline residual lumen or total occlusion of the internal carotid artery? The clinical utility of spiral CT angiography. *In* "Proceedings of the 35th Annual Meeting of the American Society of Neuroradiology, 1997," Toronto.
19. Lev, M. H., Farkas, J., Rodriguez, V. R., Schwamm, L. H., Hunter, G. J., Putman, C. M., *et al.* (2001). CT angiography in the rapid triage of patients with hyperacute stroke to intraarterial thrombolysis: Accuracy in the detection of large vessel thrombus. *J. Comput. Assisted Tomogr.* **25,** 520–528.
20. Roberts, H. C., Roberts, T. P., and Dillon, W. P. (2001). CT perfusion flow assessment: "Up and coming" or "off and running"? *Am. J. Neuroradiol.* **22,** 1018–1019.
21. Hamberg, L. M., Hunter, G. J., Halpern, E. F., Hoop, B., Gazelle, G. S., and Wolf, G. L. (1996). Quantitative high resolution measurement of cerebrovascular physiology with slip-ring CT. *Am. J. Neuroradiol.* **17,** 639–650.
22. Hamberg, L. M., Hunter, G. J., Kierstead, D., Lo, E. H., Gilberto Gonzalez, R., and Wolf, G. L. (1996). Measurement of cerebral blood volume with subtraction three-dimensional functional CT. *Am. J. Neuroradiol.* **17,** 1861–1869.
23. Lev, M. H., and Nichols, S. J. (2000). Computed tomographic angiography and computed tomographic perfusion imaging of hyperacute stroke. *Top. Magn. Reson. Imaging* **11,** 273–287.
24. Ricci, P., Karis, J., Heiserman, J., Fram, K., *et al.* (1998). Differentiating recurrent tumor from radiation necrosis: Time for re-evaluation of positron emission tomography. *Am. J. Neuroradiol.* **19,** 407–413.
25. Aksoy, F. G., and Lev, M. H. (2000). Dynamic contrast-enhanced brain perfusion imaging: Technique and clinical applications. *Semin. Ultrasound CT MR* **21,** 462–477.
26. Gonzalez, R., Schaefer, P., Buonanno, F., Schwamm, L., Budzik, R., Rordorf, G., *et al.* (1999). Diffusion-weighted MR imaging: Diagnostic accuracy in patients imaged within 6 hours of stroke symptom onset. *Radiology* **210,** 155–162.
27. Atlas, S., Mark, A., Grossman, R., and Gomori, J. (1988). Intracranial hemorrhage: Gradient-echo MR imaging at 1.5T-comparison with spin-echo imaging and clinical applications. *Radiology* **168,** 681–686.
28. Atlas, S., and Thulborn, K. (1998). MR detection of hyperacute parenchymal hemorrhage of the brain. *Am. J. Neuroradiol.* **19,** 1471–1507.
29. Muir, K. W., and Grosset, D. G. (1999). Neuroprotection for acute stroke: Making clinical trials work. *Stroke* **30,** 180–182.

30. Furlan, A., Higashida, R., Wechsler, L., Gent, M., Rowley, H., Kase, C., *et al.* (1999). Intra-arterial prourokinase for acute ischemic stroke. *JAMA* **282,** 2003–2011.
31. National Institute of Neurological Disorders and Stroke rt-PA Stroke Study Group T. (1995). Tissue plasminogen activator for acute ischemic stroke. *N. Engl. J. Med.* **333,** 1581–1587.
32. Warach, S. (2001). New imaging strategies for patient selection for thrombolytic and neuroprotective therapies. *Neurology* **57**(Suppl. 2), S48–52.
33. del Zoppo, G. (1995). Acute stroke—On the threshold of a therapy. *N. Engl. J. Med.* **333,** 1632–1633.
34. Hunter, G. J., Hamberg, L. M., Ponzo, J. A., Huang, H. F., Morris, P. P., Rabinov, J., *et al.* (1998). Assessment of cerebral perfusion and arterial anatomy in hyperacute stroke with three-dimensional functional CT: Early clinical results. *Am. J. Neuroradiol.* **19,** 29–37.
35. Albers, G. (1999). Expanding the window or thrombolytic therapy in acute stroke: The potential role of acute MRI for patient selection. *Stroke* **30,** 2230–2237.
36. Lev, M. H., Segal, A. Z., Farkas, J., Hossain, S. T., Putman, C., Hunter, G. J., *et al.* (2001). Utility of perfusion-weighted CT imaging in acute middle cerebral artery stroke treated with intra-arterial thrombolysis: Prediction of final infarct volume and clinical outcome. *Stroke* **32,** 2021–2028.
37. Ueda, T., Sakaki, S., Yuh, W., Nochide, I., and Ohta, S. (1999). Outcome in acute stroke with successful intra-arterial thrombolysis procedure and predictive value of initial single-photon emission-computed tomography. *J. Cereb. Blood Flow Metab.* **19,** 99–108.
38. Suarez, J., Sunshine, J., Tarr, R., Zaidat, O., Selman, W., Kernich, C., *et al.* (1999). Predictors of clinical improvement, angiographic recanalization, and intracranial hemorrhage after intra-arterial thrombolysis for acute ischemic stroke. *Stroke* **30,** 2094–2100.
39. Ogasawara, K., Ogawa, A., Ezura, M., Konno, H., Suzuki, M., and Yoshimoto, T. (2001). Brain single-photon emission CT studies using 99mTc-HMPAO and 99mTc-ECD early after recanalization by local intraarterial thrombolysis in patients with acute embolic middle cerebral artery occlusion. *Am. J. Neuroradiol.* **22,** 48–53.
40. Swap, C., Lev, M., McDonald, C., Koroshetz, W., Rordorf, G., Buonanno, F., *et al.* (2002). Degree of oligemia by perfusion-weighted CT and risk of hemorrhage after IA thrombolysis. *In* "Stroke—Proceedings of the 27th International Conference on Stroke and Cerebral Circulation, 2002," San Antonio, TX.
41. Folkman, J. (1995). Clinical applications of research on angiogenesis. *N. Engl. J. Med.* **333,** 1757–1763.
42. Folkman, J. (1990). What is the evidence that tumors are angiogenesis dependent? *J. Natl. Cancer Inst.* **82,** 4–6.
43. Aronen, H. J., Gazit, I. E., Louis, D. N., Buchbinder, B. R., Pardo, F. S., Weisskoff, R. M., *et al.* (1994). Cerebral blood volume maps of gliomas: Comparison with tumor grade and histologic findings. *Radiology* **191,** 41–51.
44. Lev, M. H., and Hochberg, F. (1998). Perfusion magnetic resonance imaging to assess brain tumor responses to new therapies. *Cancer Control* **5,** 115–123.
45. Donahue, K. M., Krouwer, H. G., Rand, S. D., Pathak, A. P., Marszalkowski, C. S., Censky, S. C., *et al.* (2000). Utility of simultaneously acquired gradient-echo and spin-echo cerebral blood volume and morphology maps in brain tumor patients. *Magn. Reson. Med.* **43,** 845–853.
46. Ostergaard, L., Hochberg, F. H., Rabinov, J. D., Sorensen, A. G., Lev, M., Kim, L., *et al.* (1999). Early changes measured by magnetic resonance imaging in cerebral blood flow, blood volume, and blood–brain barrier permeability following dexamethasone treatment in patients with brain tumors. *J. Neurosurg.* **90,** 300–305.
47. Harris, G. J., Lewis, R. F., Satlin, A., English, C. D., Scott, T. M., Yurgelun, T. D., *et al.* (1996). Dynamic susceptibility contrast MRI of regional cerebral blood volume in Alzheimer's disease. *Am. J. Psychiatry* **153,** 721–724.
48. Harris, G. J., Lewis, R. F., Satlin, A., English, C. D., Scott, T. M., Yurgelun-Todd, D. A., *et al.* (1998). Dynamic susceptibility contrast MRI of regional cerebral blood volume in Alzheimer disease: A promising alternative to nuclear medicine. *Am. J. Neuroradiol.* **19,** 1727–1732.
49. Gonzalez, R. G., Fischman, A. J., Guimaraes, A. R., Carr, C. A., Stern, C. E., Halpern, E. F., *et al.* (1995). Functional MR in the evaluation of dementia: Correlation of abnormal dynamic cerebral blood volume measurements with changes in cerebral metabolism on positron emission tomography with fluorodeoxyglucose F 18. *Am. J. Neuroradiol.* **16,** 1763–1770.
50. Vajkoczy, P., Horn, P., Bauhuf, C., Munch, E., Hubner, U., Ing, D., *et al.* (2001). Effect of intra-arterial papaverine on regional cerebral blood flow in hemodynamically relevant cerebral vasospasm. *Stroke* **32,** 498–505.
51. Sprawls, P. (1987). "Physical Principles of Medical Imaging." Aspen Publ., Rockville, MD.
52. Denison, C., Carlson, W. D., and Ketcham, R. A. (1997). Three-dimensional quantitative textural analysis of metamorphic rocks using high-resolution computed X-ray tomography. Part I. Methods and technique. *J. Metamorphic Geol.* **15,** 29–44.
53. Lev, M., Ackerman, R., Lustrin, E., and Brown, J. (1995). A procedure for accurate spiral CT angiographic measurement of lumenal diameter. *In* "Proceedings of the 81st Scientific Assembly and Annual Meeting of the Radiological Society of North America, 1995," Chicago.
54. Dix, J., Evans, A., Kallmes, D., Sobel, A., and Phillips, C. (1997). Accuracy and precision of CT angiography in a model of the carotid artery bifurcation. *Am. J. Neuroradiol.* **18,** 409–415.
55. Liu, Y., Hopper, K. D., Mauger, D. T., and Addis, K. A. (2000). CT angiographic measurement of the carotid artery: Optimizing visualization by manipulating window and level settings and contrast material attenuation. *Radiology* **217,** 494–500.
56. Marks, M. (1998). CT in ischemic stroke. *Neuroimaging Clin. North Am.* **8,** 515–523.
57. Constable, R., and Henkelman, R. (1991). Contrast, resolution, and detectability in MR imaging. *J. Comput. Assisted Tomogr.* **15,** 297–303.
58. Lev, M., Farkas, J., Gemmete, J., Hossain, S., Hunter, G., Koroshetz, W., *et al.* (1999). Acute stroke: Improved nonenhanced CT detection—Benefits of soft-copy interpretation by using variable window width and center level settings. *Radiology* **213,** 150–155.
59. Maldjian, J. A., Chalela, J., Kasner, S. E., Liebeskind, D., and Detre, J. A. (2001). Automated CT segmentation and analysis for acute middle cerebral artery stroke. *Am. J. Neuroradiol.* **22,** 1050–1055.
60. Pratt, H., Langlotz, C., Feingold, E., Schwartz, J., and Kundel, H. (1998). Incremental cost of department-wide implementation of a picture archiving and communication system and computed radiography. *Radiology* **206,** 245–252.
61. Bryan, S., Weatherburn, G., Watkins, J., Roddie, M., Keen, J., Muris, N., *et al.* (1998). Radiology report times: Impact of picture archiving and communication systems. *Am. J. Roentgenol.* **170,** 1153–1159.
62. Obuchowski, N. A., Graham, R. J., Baker, M. E., and Powell, K. A. (2001). Ten criteria for effective screening: Their application to multislice CT screening for pulmonary and colorectal cancers. *Am. J. Roentgenol.* **176,** 1357–1362.
63. Silverman, P. M., Kalender, W. A., and Hazle, J. D. (2001). Common terminology for single and multislice helical CT. *Am. J. Roentgenol.* **176,** 1135–1136.
64. Bae, K., Heiken, J., and Brink, J. (1998). Aortic and hepatic contrast medium enhancement at CT. Part II. Effect of reduced cardiac output in a porcine model. *Radiology* **207,** 657–662.
65. Bae, K., Heiken, J., and Brink, J. (1998). Aortic and hepatic contrast medium enhancement at CT. Part I. Prediction with a computer model. *Radiology* **207,** 647–655.

66. Haage, P., Schmitz-Rode, T., Hubner, D., Piroth, W., and Gunther, R. W. (2000). Reduction of contrast material dose and artifacts by a saline flush using a double power injector in helical CT of the thorax. *Am. J. Roentgenol.* **174,** 1049–1053.
67. Hopper, K. D., Mosher, T. J., Kasales, C. J., TenHave, T. R., Tully, D. A., and Weaver, J. S. (1997). Thoracic spiral CT: Delivery of contrast material pushed with injectable saline solution in a power injector. *Radiology* **205,** 269–271.
68. Barmeir, E., Tann, M., Zur, S., and Braun, J. (1998). Improving CT angiography of the carotid artery using the "right" arm. *Am. J. Roentgenol.* **170,** 1657–1658.
69. Rydberg, J., Buckwalter, K. A., Caldemeyer, K. S., Phillips, M. D., Conces, D. J., Jr., Aisen, A. M., *et al.* (2000). Multisection CT: Scanning techniques and clinical applications. *Radiographics* **20,** 1787–1806.
70. Hu, H., He, H. D., Foley, W. D., and Fox, S. H. (2000). Four multi-detector-row helical CT: Image quality and volume coverage speed. *Radiology* **215,** 55–62.
71. Freiherr, G. (2001). CT vendors press quest for more speed, slices. *Diagn. Imaging* **Nov 2001,** 25–26.
72. Rubin, G. D., Shiau, M. C., Leung, A. N., Kee, S. T., Logan, L. J., and Sofilos, M. C. (2000). Aorta and iliac arteries: Single versus multiple detector-row helical CT angiography. *Radiology* **215,** 670–676.
73. Taguchi, K., and Aradate, H. (1998). Algorithm for image reconstruction in multi-slice helical CT. *Med. Phys.* **25,** 550–561.
74. McCollough, C. H., and Zink, F. E. (1999). Performance evaluation of a multi-slice CT system. *Med. Phys.* **26,** 2223–2230.
75. Jones, T. R., Kaplan, R. T., Lane, B., Atlas, S. W., and Rubin, G. D. (2001). Single- versus multi-detector row CT of the brain: Quality assessment. *Radiology* **219,** 750–755.
76. Donnelly, L. F., Emery, K. H., Brody, A. S., Laor, T., Gylys-Morin, V. M., Anton, C. G., *et al.* (2001). Minimizing radiation dose for pediatric body applications of single-detector helical CT: Strategies at a large children's hospital. *Am. J. Roentgenol.* **176,** 303–306.
77. Sternberg, S. (2001). CT scans in children linked to cancer later. *USA Today* Monday, Jan 2, Sect. 1.
78. Hidajat, N., Wolf, M., Nunnemann, A., Liersch, P., Gebauer, B., Teichgraber, U., *et al.* (2001). Survey of conventional and spiral CT doses. *Radiology* **218,** 395–401.
79. Committee on the Biological Effects of Ionizing Radiation BV. (1990). Health effects of exposure to low levels of ionizing radiation: BEIR V. Natl. Research Council, Natl. Acad. Press, Washington, DC.
80. Wall, B. F., and Hart, D. (1997). Revised radiation doses for typical X-ray examinations: Report on a recent review of doses to patients from medical X-ray examinations in the UK by NRPB. National Radiological Protection Board. *Br. J. Radiol.* **70,** 437–439.
81. Hall, E. (1994). "Radiobiology for the Radiologist," 4th edition. Lippincott, Philadelphia.
82. Parry, R. A., Glaze, S. A., and Archer, B. R. (1999). The AAPM/RSNA physics tutorial for residents. Typical patient radiation doses in diagnostic radiology. *Radiographics* **19,** 1289–1302.
83. McCollough, C. H., and Schueler, B. A. (2000). Calculation of effective dose. *Med. Phys.* **27,** 828–837.
84. Kalender, W. A., Wolf, H., and Suess, C. (1999). Dose reduction in CT by anatomically adapted tube current modulation. II. Phantom measurements. *Med. Phys.* **26,** 2248–2253.
85. Kalender, W. A., and Prokop, M. (2001). 3D CT angiography. *Crit. Rev. Diagn. Imaging* **42,** 1–28.
86. Prokop, M. (1999). Protocols and future directions in imaging of renal artery stenosis: CT angiography. *J, Comput, Assisted Tomogr.* **23**(Suppl. 1), S101–110.
87. Cohnen, M., Fischer, H., Hamacher, J., Lins, E., Kotter, R., and Modder, U. (2000). CT of the head by use of reduced current and kilovoltage: Relationship between image quality and dose reduction. *Am. J. Neuroradiol.* **21,** 1654–1660.
88. Bove, P., Lev, M., Hunter, G., Hamberg, L., O'Reilly, C., Saini, S., *et al.* (2001). Gray matter conspicuity and image noise with half dose mAs head CT scanning. *In* "Proceedings of the 87th Scientific Assembly and Annual Meeting of the Radiological Society of North America, 2001," Chicago.
89. Haaga, J. R. (2001). Radiation dose management: Weighing risk versus benefit. *Am. J. Roentgenol.* **177,** 289–291.
90. Wang, G., and Vannier, M. W. (1999). The effect of pitch in multislice spiral/helical CT. *Med. Phys.* **26,** 2648–2653.
91. Brink, J. A., Wang, G., and McFarland, E. G. (2000). Optimal section spacing in single-detector helical CT. *Radiology* **214,** 575–578.
92. Fleischmann, D., Rubin, G. D., Paik, D. S., Yen, S. Y., Hilfiker, P. R., Beaulieu, C. F., *et al.* (2000). Stair-step artifacts with single versus multiple detector-row helical CT. *Radiology* **216,** 185–196.
93. Kallmes, D., Evans, A., Woodcock, R., *et al.* (1996). Optimization of parameters for the detection of cerebral aneurysms: CT angiography of a model. *Radiology* **200,** 403–405.
94. Tello, R., and Seltzer, S. (1999). Effects of injection rates of contrast material on arterial phase hepatic CT. *Am. J. Roentgenol.* **173,** 237–238.
95. Herts, B. R., O'Malley, C. M., Wirth, S. L., Lieber, M. L., and Pohlman, B. (2001). Power injection of contrast media using central venous catheters: Feasibility, safety, and efficacy. *Am. J. Roentgenol.* **176,** 447–453.
96. Kendell, B., and Pullicono, P. (1980). Intravascular contrast injection in ischemic lesions. II. Effect on prognosis. *Neuroradiology* **19,** 241–243.
97. Doerfler, A., Engelhorn, T., von Kommer, R., Weber, J., *et al.* (1998). Are iodinated contrast agents detrimental in acute cerebral ischemia? An experimental study in rats. *Radiology* **206,** 211–217.
98. Villringer, A., Rosen, B. R., Belliveau, J. W., *et al.* (1988). Dynamic imaging with lanthanide chelates in normal brain: Contrast due to magnetic susceptibility effects. *Magn. Reson. Med.* **6,** 164–174.
99. Shih, T. T., and Huang, K. M. (1988). Acute stroke: Detection of changes in cerebral perfusion with dynamic CT scanning. *Radiology* **169,** 469–474.
100. Meier, P., and Zieler, K. (1954). On the theory of the indicator-dilution method for measurement of blood flow and volume. *J. Appl. Physiol.* **6,** 731–744.
101. Roberts, G., and Larson, K. (1973). The interpretation of mean transit time measurements for multi-phase tissue systems. *J. Theor. Biol.* **39,** 447–475.
102. Nambu, K., Suzuki, R., and Hirakawa, K. (1995). Cerebral blood flow: Measurement with xenon-enhanced dynamic helical CT. *Radiology* **195,** 53–57.
103. Bae, K. T., Tran, H. Q., and Heiken, J. P. (2000). Multiphasic injection method for uniform prolonged vascular enhancement at CT angiography: Pharmacokinetic analysis and experimental porcine model. *Radiology* **216,** 872–880.
104. Fleischmann, D., Rubin, G. D., Bankier, A. A., and Hittmair, K. (2000). Improved uniformity of aortic enhancement with customized contrast medium injection protocols at CT angiography. *Radiology* **214,** 363–371.
105. Larson, K. B., Markham, J., and Raichle, M. E. (1987). Tracer-kinetic models for measuring cerebral blood flow using externally detected radiotracers. *J. Cereb. Blood Flow Metab.* **7,** 443–463.
106. Kety, S. (1951). The theory and applications of the exchange of inert gas at the lungs and tissues. *Pharmacol. Rev.* **3,** 1–41.
107. Klotz, E., and Konig, M. (1999). Perfusion measurements of the brain: Using the dynamic CT for the quantitative assessment of cerebral ischemia in acute stroke. *Eur. J. Radiol.* **30,** 170–184.
108. Brix, G., Bahner, M., Hoffman, U., Horvath, A., and Schreiber, W. (1999). Regional blood flow, capillary permeability, and compartmental volumes: Measurements with dynamic CT—Initial experience. *Radiology* **210,** 269–276.
109. Axel, L. (1980). Cerebral blood flow determination by rapid-sequence computed tomography. *Radiology* **137,** 679–686.

110. Nabavi, D., Cenic, A., Craen, R., Gelb, A., Bennet, J., Kozak, R., *et al.* (1999). CT assessment of cerebral perfusion: Experimental validation and initial clinical experience. *Radiology* **213,** 141–149.
111. Roberts, H. C., Roberts, T. P., Brasch, R. C., and Dillon, W. P. (2000). Quantitative measurement of microvascular permeability in human brain tumors achieved using dynamic contrast-enhanced MR imaging: Correlation with histologic grade. *Am. J. Neuroradiol.* **21,** 891–899.
112. Wintermark, M., Maeder, P., Thiran, J. P., Schnyder, P., and Meuli, R. (2001). Quantitative assessment of regional cerebral blood flows by perfusion CT studies at low injection rates: A critical review of the underlying theoretical models. *Eur. Radiol.* **11,** 1220–1230.
113. Wirestam, R., Andersson, L., Ostergaard, L., Bolling, M., Aunola, J. P., Lindgren, A., *et al.* (2000). Assessment of regional cerebral blood flow by dynamic susceptibility contrast MRI using different deconvolution techniques. *Magn. Reson. Med.* **43,** 691–700.
114. Sorensen, A., Copen, W., Ostergaard, L., Buonnano, F., Gonzalez, R., Rordof, G., *et al.* (1999). Hyperacute stroke: Simultaneous measurement of relative cerebral blood volume, relative cerebral blood flow and mean tissue transit time. *Radiology* **210,** 519–527.
115. Ostergaard, L., *et al.* (1996). High resolution of cerebral blood flow using intravascular tracer bolus passages. Part I. Mathematical approach and statistical analysis. *Magn. Reson. Imaging Med.* **36,** 715–725.
116. Nabavi, D. G., Cenic, A., Dool, J., Smith, R. M., Espinosa, F., Craen, R. A., *et al.* (1999). Quantitative assessment of cerebral hemodynamics using CT: Stability, accuracy, and precision studies in dogs. *J. Comput. Assisted Tomogr.* **23,** 506–515.
117. Cenic, A., Nabavi, D. G., Craen, R. A., Gelb, A. W., and Lee, T. Y. (1999). Dynamic CT measurement of cerebral blood flow: A validation study. *Am. J. Neuroradiol.* **20,** 63–73.
118. Wintermark, M., Thiran, J. P., Maeder, P., Schnyder, P., and Meuli, R. (2001). Simultaneous measurement of regional cerebral blood flow by perfusion CT and stable xenon CT: A validation study. *Am. J. Neuroradiol.* **22,** 905–914.
119. Yeung, I., Lee, T., Del Maestro, R., Kozak, R., Bennet, J., and Brown, T. (1992). An absorptiometry method for the determination of arterial blood concentration of injected iodinated contrast agent. *Phys. Med. Biol.* **37,** 1741–1758.
120. Lev, M. H., Kulke, S. F., Weisskoff, R. M., *et al.* (1997). Dose dependence of signal to noise ratio in functional MRI of cerebral blood volume mapping with sprodiamide. *J. Magn. Reson. Imaging* **7,** 523–527.
121. Wintermark, M., Maeder, P., Verdun, F. R., Thiran, J. P., Valley, J. F., Schnyder, P., *et al.* (2000). Using 80 kVp versus 120 kVp in perfusion CT measurement of regional cerebral blood flow. *Am. J. Neuroradiol.* **21,** 1881–1884.
122. Roberts, H. C., Roberts, T. P., Smith, W. S., Lee, T. J., Fischbein, N. J., and Dillon, W. P. (2001). Multisection dynamic CT perfusion for acute cerebral ischemia: The "toggling-table" technique. *Am. J. Neuroradiol.* **22,** 1077–1080.
123. Venema, H. W., Hulsmans, F. J., and den Heeten, G. J. (2001). CT angiography of the circle of Willis and intracranial internal carotid arteries: Maximum intensity projection with matched mask bone elimination-feasibility study. *Radiology* **218,** 893–898.
124. Bae, K. T., and Whiting, B. R. (2001). CT data storage reduction by means of compressing projection data instead of images: Feasibility study. *Radiology* **219,** 850–855.
125. Seltzer, S. E., Judy, P. F., Feldman, U., Scarff, L., and Jacobson, F. L. (1998). Influence of CT image size and format on accuracy of lung nodule detection. *Radiology* **206,** 617–622.
126. Velthuis, B., van Leeuwen, M., Witkamp, T., Boomstra, S., *et al.* (1997). CT angiography: Source images and postprocessing techniques in the detection of cerebral aneurysms. *Am. J. Radiol.* **169,** 1411–1417.
127. Anderson, C. M., Lee, R. E., Levin, D. L., de la Torre Alonso, S., and Saloner, D. (1994). Measurement of internal carotid artery stenosis from source MR angiograms. *Radiology* **193,** 219–226.
128. Rubin, G., Dake, M., Napel, S., Brooke, J., McDonnell, C., Sommer, F., *et al.* (1994). Spiral CT of renal artery stenosis: Comparison of three-dimensional rendering techniques. *Radiology* **190,** 181–189.
129. Alberico, R., Ozsvath, R., Casey, S., and Patel, M. (1996). CT angiography: Shaded surface versus maximum intensity projection. *Am. J. Radiol.* **166,** 1227–1228.
130. Hany, T., Schmidt, M., Davis, C., Gohde, S., and Debatin, J. (1998). Diagnostic impact of four postprocessing techniques in evaluating contrast-enhanced three-dimensional MR angiography. *Am. J. Radiol.* **170,** 907–912.
131. Alpert, N. M., Berdichevsky, D., Levin, Z., Thangaraj, V., Gonzalez, G., and Lev, M. H. (2001). Performance evaluation of an automated system for registration and postprocessing of CT scans. *J. Comput. Assisted Tomogr.* **25,** 747–752.
132. Teasdale, E. (2000). Curved planar reformatted CT angiography: Utility for the evaluation of aneurysms at the carotid siphon. *Am. J. Neuroradiol.* **21,** 985.
133. Vieco, P. (1996). CT angiography: Shaded surface versus maximum intensity projection. *A. J. Radiol.* **166,** 1228–1229.
134. Link, J., Mueller-Huelsbeck, S., Brossmann, J., Grabener, M., Stock, U., and Heller, M. (1995). Prospective assessment of carotid bifurcation disease with spiral CT angiography in surface shaded display (SSD)-technique. *Comput. Med. Imaging Graph.* **19,** 451–456.
135. Papp, Z., Patel, M., Ashatari, M., Takahashi, M., Goldstein, J., Maguire, W., *et al.* (1997). Carotid artery stenosis: Optimization of CT angiography with a combination of shaded surface display and source images. *Am. J. Neuroradiol.* **18,** 759–763.
136. Kuszyk, B. S., and Fishman, E. K. (1998). Technical aspects of CT angiography. *Semin. Ultrasound CT MR* **19,** 383–393.
137. Kuszyk, B. S., Heath, D. G., Ney, D. R., Bluemke, D. A., Urban, B. A., Chambers, T. P., *et al.* (1995). CT angiography with volume rendering: Imaging findings. *Am. J. Roentgenol.* **165,** 445–448.
138. Johnson, P., Heath, D., Bliss, D., Cabral, B., and Fishman, E. (1996). Three dimensional CT: Real-time interactive volume rendering. *Am. J. Roentgenol.* **167,** 581–583.
139. Johnson, P., Fishman, E., Deckwall, J., Calhoun, P., and Heath, D. (1998). Interactive three-dimensional volume rendering of spiral CT data: Current applications in the thorax. *Radiographics* **18,** 165–187.
140. Leclerc, W., Godefrey, O., Lucas, C., Benhaim, J., Michel, T., Leys, D., *et al.* (1999). Internal carotid arterial stenosis: CT angiography with volume rendering. *Radiology* **210,** 673–682.
141. Calhoun, P. S., Kuszyk, B. S., Heath, D. G., Carley, J. C., and Fishman, E. K. (1999). Three-dimensional volume rendering of spiral CT data: Theory and method. *Radiographics* **19,** 745–764.
142. Lee, K. H., Cho, S. J., Byun, H. S., Na, D. G., Choi, N. C., Lee, S. J., *et al.* (2000). Triphasic perfusion computed tomography in acute middle cerebral artery stroke: A correlation with angiographic findings. *Arch. Neurol.* **57,** 990–999.
143. Castillo, M. (1993). Diagnosis of disease of the common carotid artery bifurcation: CT angiography vs catheter angiography. *Am. J. Roentgenol.* **161,** 395–398.
144. Coskun, O., Hamon, M., Catroux, G., Gosme, L., Courtheoux, P., and Theron, J. (2000). Carotid-cavernous fistulas: Diagnosis with spiral CT angiography. *Am. J. Neuroradiol.* **21,** 712–716.
145. Alberico, R., Barnes, P., Robertson, R., and Burrows, P. (1999). Helical CT angiography: Dynamic cerebrovascular imaging in children. *Am. J. Neuroradiol.* **20,** 328–334.
146. Gleason, S., Furie, K. L., Lev, M. H., O'Donnell, J., McMahon, P. M., Beinfeld, M. T., *et al.* (2001). Potential influence of acute CT on inpatient costs in patients with ischemic stroke. *Acad. Radiol.* **8,** 955–964.

147. von Kummer, R., Allen, K. L., Holle, R., Bozzao, L., Bastianello, S., Manelfe, C., *et al.* (1997). Acute stroke: Usefulness of early CT findings before thrombolytic therapy. *Radiology* **205,** 327–333.
148. von Kummer, R., Holle, R., Grzyska, U., Hofmann, E., *et al.* (1996). Interobserver agreement in assessing early CT signs of middle cerebral artery infarction. *Am. J. Neuroradiol.* **17,** 1743–1748.
149. Wardlaw, J., Dorman, P., Lewis, S., and Sandercock, P. (1999). Can stroke physicians and neuroradiologists identify signs of early cerebral infarction on CT? *J. Neurol. Neursosurg. Psychiatry* **67,** 651–653.
150. Grotta, J. C., Chiu, D., Lu, M., Patel, S., Levine, S. R., Tilley, B. C., *et al.* (1999). Agreement and variability in the interpretation of early CT changes in stroke patients qualifying for intravenous rtPA therapy. *Stroke* **30,** 1528–1533.
151. Lee, K. H., Lee, S. J., Cho, S. J., Na, D. G., Byun, H. S., Kim, Y. B., *et al.* (2000). Usefulness of triphasic perfusion computed tomography for intravenous thrombolysis with tissue-type plasminogen activator in acute ischemic stroke. *Arch. Neurol.* **57,** 1000–1008.
152. Knauth, M., R., V., Jansen, O., Hahnel, S., Dorfler, A., and Sartor, K. (1997). Potential of CT angiography in acute ischemic stroke. *Am. J. Neuroradiol.* **18,** 1001–1010.
153. Shrier, D., Tanaka, H., Numaguchi, Y., Konno, S., Patel, U., and Shibata, D. (1997). CT angiography in the evaluation of acute stroke. *Am. J. Neuroradiol.* **18,** 1011–1020.
154. Wildermuth, S., Knauth, M., Brandt, T., Winter, R., Sartor, K., and Hacke, W. (1998). Role of CT angiography in patient selection for thrombolytic therapy in acute hemispheric stroke. *Stroke* **29,** 935–938.
155. Ponzo, J., Hunter, G., Hamburg, L., Farkas, J., Lev, M., Koroshetz, W., *et al.* (1998). Evaluation of collateral circulation in acute stroke patients using CT angiography. *In* "Stroke: Proceedings of the 23rd International Conference on Stroke and Cerebral Circulation, 1998," Orlando, FL.
156. Barest, G., Hunter, G., Hamberg, L., Farkas, J., Ponzo, J., Lev, M., *et al.* (1997). Dynamic contrast enhanced helical CT improves conspicuity of acute cerebral ischemia. *In* "Proceedings of the 83rd Scientific Assembly and Annual Meeting of the Radiological Society of North America, 1997," Chicago.
157. Kaufmann, A. M., Firlik, A. D., Fukui, M. B., Wechsler, L. R., Jungries, C. A., and Yonas, H. (1999). Ischemic core and penumbra in human stroke. *Stroke* **30,** 93–99.
158. Nabavi, D. G., Cenic, A., Henderson, S., Gelb, A. W., and Lee, T. Y. (2001). Perfusion mapping using computed tomography allows accurate prediction of cerebral infarction in experimental brain ischemia. *Stroke* **32,** 175–183.
159. Koroshetz, W. J., and Gonzales, R. G. (1999). Imaging stroke in progress: Magnetic resonance advances but computed tomography is poised for counterattack. *Ann. Neurol.* **46,** 556–558.
160. Fisher, C. M. (1951). Occlusion of the internal carotid artery. *Arch. Neurol. Psychiat.* **65,** 346–377.
161. Eastcott, H., Pickering, G., and Rob, C. (1954). Reconstruction of internal carotid artery in a patient with intermittent attacks of hemiplegia. *Lancet* **267,** 994–996.
162. NASCET Collaborators. (1991). Beneficial effect of carotid endarterectomy in symptomatic patients with high-grade carotid stenosis. North American Symptomatic Carotid Endarterectomy Trial Collaborators. *N. Engl. J. Med.* **325,** 445–453.
163. Executive Committee for the Asymptomatic Carotid Atherosclerosis Study. (1995). Endarterectomy for asymptomatic carotid artery stenosis. Executive Committee for the Asymptomatic Carotid Atherosclerosis Study. *JAMA* **273,** 1421–1428.
164. Brant-Zawadzki, M., and Heiserman, J. E. (1997). The roles of MR angiography, CT angiography, and sonography in vascular imaging of the head and neck. *Am. J. Neuroradiol.* **18,** 1820–1825.
165. Ackerman, R. (1995). Neurovascular non-invasive evaluation. *In* "Radiology: Diagnosis, Imaging, Intervention" (J. Taveras and J. Ferrucci, eds.), Chap. 50. Lippincott, Philadelphia.
166. Ackerman, R., Candia, M., and May, Z. (1999). Technical advances and clinical progress in carotid diagnosis. *Am. J. Neuroradiol.* **20,** 187–189.
167. Lev, M., Ackerman, R., Chehade, R., *et al.* (1996). The clinical utility of spiral computed tomographic angiography in the evaluation of carotid artery disease: Review of our first 50 patients. *In* "Stroke—Proceedings of the 21st International Conference on Stroke and Cerebral Circulation, 1996," San Antonio, TX.
168. Bluemke, D. A., and Chambers, T. P. (1995). Spiral CT angiography: An alternative to conventional angiography. *Radiology* **195,** 317–319.
169. Pettiti, N., Lev, M., Ackerman, R., Nichols, S., and Gonzalez, R. (2001). Rapid evaluation of the complete neurovascular system using multidetector helical CT angiography. *In* "Proceedings of the 39th Annual Meeting of the American Society of Neuroradiology, 2001," Boston.
170. Remy-Jardin, M., Remy, J., Masson, P., Bonnel, F., Debatselier, P., Vinckier, L., *et al.* (2000). Helical CT angiography of thoracic outlet syndrome: Functional anatomy. *Am. J. Roentgenol.* **174,** 1667–1674.
171. Link, J., Brossmann, J., Penselin, V., Gluer, C. C., and Heller, M. (1997). Common carotid artery bifurcation: Preliminary results of CT angiography and color-coded duplex sonography compared with digital subtraction angiography. *Am. J. Roentgenol.* **168,** 361–365.
172. Pan, X. M., Saloner, D., Reilly, L. M., Bowersox, J. C., Murray, S. P., Anderson, C. M., *et al.* (1995). Assessment of carotid artery stenosis by ultrasonography, conventional angiography, and magnetic resonance angiography: Correlation with ex vivo measurement of plaque stenosis. *J. Vasc. Surg.* **21,** 82–88. [Discussion pp. 88–89]
173. Dillon, E., Van Leeuween, M., Fernandez, M. A., Eikelboom, B., and Mali, W. (1993). CT angiography: Application to the evaluation of carotid artery stenosis. *Radiology* **189,** 211–219.
174. Link, J., Brossmann, J., Grabener, M., Mueller-Huelsbeck, S., Steffens, J. C., Brinkmann, G., *et al.* Spiral CT angiography and selective digital subtraction angiography of internal carotid artery stenosis. *Am. J. Neuroradiol.* **17,** 89–94.
175. Schwartz, R., Tice, H., Hooten, S., Hsu, L., and Stieg, P. (1994). Evaluation of cerebral aneurysms with helical CT: Correlation with conventional angiography and MR angiography. *Radiology* **192,** 717–722.
176. Marks, M. P., Napel, S., Jordan, J. E., and Enzmann, D. R. (1993). Diagnosis of carotid artery disease: Preliminary experience with maximum-intensity-projection spiral CT angiography. *Am. J. Roentgenol.* **160,** 1267–1271.
177. Cumming, M. J., and Morrow, I. M. (1994). Carotid artery stenosis: A prospective comparison of CT angiography and conventional angiography. *Am. J. Roentgenol.* **163,** 517–523.
178. Anderson, G. B., Ashforth, R., Steinke, D. E., Ferdinandy, R., and Findlay, J. M. (2000). CT angiography for the detection and characterization of carotid artery bifurcation disease. *Stroke* **31,** 2168–2174.
179. Abdelaziz, O. S., Ogilvy, C. S., and Lev, M. (1999). Is there a potential role for hyoid bone compression in pathogenesis of carotid artery stenosis? *Surg. Neurol.* **51,** 650–653.
180. Heiserman, J. E., Zabramski, J. M., Drayer, B. P., and Keller, P. J. (1996). Clinical significance of the flow gap in carotid magnetic resonance angiography. *J. Neurosurg.* **85,** 384–387.
181. Rothwell, P. M., Pendlebury, S. T., Wardlaw, J., and Warlow, C. P. (2000). Critical appraisal of the design and reporting of studies of imaging and measurement of carotid stenosis. *Stroke* **31,** 1444–1450.
182. Serfaty, J. M., Chirossel, P., Chevallier, J. M., Ecochard, R., Froment, J. C., and Douek, P. C. (2000). Accuracy of three-dimensional gadolinium-enhanced MR angiography in the assessment of extracranial carotid artery disease. *Am. J. Roentgenol.* **175,** 455–463.
183. Oliver, T., Lammie, G., Wright, A., Wardlaw, J., Patel, S., Peek, R., *et al.* (1999). Atherosclerotic plaque at the carotid bifurcation: CT

angiographic appearance with histopathologic correlation. *Am. J. Neuroradiol.* **20,** 897–901.

184. Furst, G., Saleh, A., Wenserski, F., Malms, J., Cohnen, M., Aulich, A., *et al.* (1999). Reliability and validity of noninvasive imaging of internal carotid artery pseudo-occlusion. *Stroke* **30,** 1444–1449.
185. Kluytmans, M., van der Grond, J., and Viergever, M. (1998). Gray matter and white matter perfusion imaging in patients with severe carotid artery lesions. *Radiology* **209,** 675–682.
186. Grubb, R. L., Jr., Derdeyn, C. P., Fritsch, S. M., Carpenter, D. A., Yundt, K. D., Videen, T. O., *et al.* (1998). Importance of hemodynamic factors in the prognosis of symptomatic carotid occlusion. *JAMA* **280,** 1055–1060.
187. Roberts, H. C., Dillon, W. P., and Smith, W. S. (2000). Dynamic CT perfusion to assess the effect of carotid revascularization in chronic cerebral ischemia. *Am. J. Neuroradiol.* **21,** 421–425.
188. Gillard, J., Hardingham, C., Kirkpatrick, P., Antoun, N., Freer, C., and Griffiths, P. (1998). Evaluation of carotid endarterectomy with sequential MR perfusion imaging: A preliminary report. *Am. J. Neuroradiol.* **19,** 1747–1752.
189. Leclerc, X., Lucas, C., Godefroy, O., Tessa, H., Martinat, P., Leys, D., *et al.* (1997). Helical CT for the follow-up of cervical internal carotid artery dissections. *Am. J. Neuroradiol.* **19,** 831–837.
190. Munera, F., Soto, J. A., Palacio, D., Velez, S. M., and Medina, E. (2000). Diagnosis of arterial injuries caused by penetrating trauma to the neck: Comparison of helical CT angiography and conventional angiography. *Radiology* **216,** 356–362.
191. Mayberg, M. R., Batjer, H. H., Dacey, R., Diringer, M., Haley, E. C., Heros, R. C., *et al.* (1994). Guidelines for the management of aneurysmal subarachnoid hemorrhage. A statement for healthcare professionals from a Special Writing Group of the Stroke Council, American Heart Association. *Stroke* **25,** 2315–2328.
192. Atlas, S. W., Sheppard, L., Goldberg, H. I., Hurst, R. W., Listerud, J., and Flamm, E. (1997). Intracranial aneurysms: Detection and characterization with MR angiography with use of an advanced postprocessing technique in a blinded-reader study. *Radiology* **203,** 807–814.
193. International Study of Unruptured Intracranial Aneurysms Investigators. (1998). Unruptured intracranial aneurysms—Risk of rupture and risks of surgical intervention. International Study of Unruptured Intracranial Aneurysms Investigators. *N. Engl. J. Med.* **339,** 1725–1733.
194. Bederson, J. B., Awad, I. A., Wiebers, D. O., Piepgras, D., Haley, E. C., Jr., Brott, T., *et al.* (2000). Recommendations for the management of patients with unruptured intracranial aneurysms: A statement for healthcare professionals from the Stroke Council of the American Heart Association. *Stroke* **31,** 2742–2750.
195. Magnetic Resonance Angiography in Relatives of Patients with Subarachnoid Hemorrhage Study Group. (1999). Risks and benefits of screening for intracranial aneurysms in first-degree relatives of patients with sporadic subarachnoid hemorrhage. *N. Engl. J. Med.* **341,** 1344–1350.
196. Alberico, R. A., Patel, M., Casey, S., Jacobs, B., Maguire, W., and Decker, R. (1995). Evaluation of the circle of Willis with three-dimensional CT angiography in patients with suspected intracranial aneurysms. *Am. J. Neuroradiol.* **16,** 1571–1578. [Discussion pp. 1579–1580]
197. Vieco, P. T., Shuman, W. P., Alsofrom, G. F., and Gross, C. E. (1995). Detection of circle of Willis aneurysms in patients with acute subarachnoid hemorrhage: A comparison of CT angiography and digital subtraction angiography. *Am. J. Roentgenol.* **165,** 425–430.
198. Velthuis, B., Rinkel, G., Ramos, L., Witkamp, T., Berkelbach van der Sprenkel, J., Vandertop, W., *et al.* (1998). Subarachnoid hemorrhage: Aneurysm detection and preoperative evaluation with CT angiography. *Radiology* **208,** 423–430.
199. White, P. M., Wardlaw, J. M., and Easton, V. (2000). Can noninvasive imaging accurately depict intracranial aneurysms? A systematic review. *Radiology* **217,** 361–370.
200. Velthuis, B. K., Van Leeuwen, M. S., Witkamp, T. D., Ramos, L. M., van der Sprenkel, J. W., and Rinkel, G. J. (1999). Computerized tomography angiography in patients with subarachnoid hemorrhage: From aneurysm detection to treatment without conventional angiography. *J. Neurosurg.* **91,** 761–767.
201. Ruigrok, Y. M., Rinkel, G. J., Buskens, E., Velthuis, B. K., and van Gijn, J. (2000). Perimesencephalic hemorrhage and CT angiography: A decision analysis. *Stroke* **31,** 2976–2983.
202. McFadzean, R. M., and Teasdale, E. M. (1998). Computerized tomography angiography in isolated third nerve palsies. *J. Neurosurg.* **88,** 679–684.
203. Brown, J. H., Lustrin, E. S., Lev, M. H., Ogilvy, C. S., and Taveras, J. M. (1999). Reduction of aneurysm clip artifacts on CT angiograms: A technical note. *Am. J. Neuroradiol.* **20,** 694–696.
204. Brown, J. H., Lustrin, E. S., Lev, M. H., Ogilvy, C. S., and Taveras, J. M. (1997). Characterization of intracranial aneurysms using CT angiography. *Am. J. Roentgenol.* **169,** 889–893.
205. Vieco, P. T., Morin, E. E., 3rd, and Gross, C. E. (1996). CT angiography in the examination of patients with aneurysm clips. *Am. J. Neuroradiol.* **17,** 455–457.
206. Anderson, G. B., Ashforth, R., Steinke, D. E., and Findlay, J. M. (2000). CT angiography for the detection of cerebral vasospasm in patients with acute subarachnoid hemorrhage. *Am. J. Neuroradiol.* **21,** 1011–1015.
207. Ohkuma, H., Manabe, H., Tanaka, M., and Suzuki, S. (2000). Impact of cerebral microcirculatory changes on cerebral blood flow during cerebral vasospasm after aneurysmal subarachnoid hemorrhage. *Stroke* **31,** 1621–1627.
208. Nabavi, D. G., LeBlanc, L. M., Baxter, B., Lee, D. H., Fox, A. J., Lownie, S. P., *et al.* (2001). Monitoring cerebral perfusion after subarachnoid hemorrhage using CT. *Neuroradiology* **43,** 7–16.
209. Caramia, F., Santoro, A., Pantano, P., Passacantilli, E., Guidetti, G., Pierallini, A., *et al.* (2001). Cerebral hemodynamics on MR perfusion images before and after bypass surgery in patients with giant intracranial aneurysms. *Am. J. Neuroradiol.* **22,** 1704–1710.
210. Eskey, C. J., Lev, M. H., Tatter, S. B., and Gonzalez, R. G. (1998). Cerebral CT venography in surgical planning for a tentorial meningioma. *J. Comput. Assisted Tomogr.* **22,** 530–532.
211. Casey, S., Alberico, R., Patel, M., *et al.* (1996). Cerebral CT venography. *Radiology* **198,** 163–170.
212. Wetzel, S., Kirsch, E., Stock, K., Klobe, M., Kaim, A., and Radue, E. (1999). Cerebral veins: Comparative study of CT venography with intraarterial digital subtraction angiography. *Am. J. Neuroradiol.* **20,** 249–255.
213. Ozsvath, R. R., Casey, S. O., Lustrin, E. S., Alberico, R. A., Hassankhani, A., and Patel, M. (1997). Cerebral venography: Comparison of CT and MR projection venography. *Am. J. Roentgenol.* **169,** 1699–1707.
214. Surawicz, T. S., McCarthy, B. J., Kupelian, V., Jukich, P. J., Bruner, J. M., and Davis, F. G. (1999). Descriptive epidemiology of primary brain and CNS tumors: Results from the Central Brain Tumor Registry of the United States, 1990–1994. *Neuro-Oncology* **1,** 14–25.
215. Davis, F. G., McCarthy, B., and Jukich, P. (1999). The descriptive epidemiology of brain tumors. *Neuroimaging Clin, North Am.* **9,** 581–594.
216. Cenic, A., Nabavi, D. G., Craen, R. A., Gelb, A. W., and Lee, T. Y. (2000). A CT method to measure hemodynamics in brain tumors: Validation and application of cerebral blood flow maps. *Am. J. Neuroradiol.* **21,** 462–470.
217. Pardo, F. S., Aronen, H. J., Kennedy, D., *et al.* (1994). Functional cerebral imaging in the evaluation and radiotherapeutic treatment

planning of patients with malignant glioma. *Int. J. Radiat. Oncol. Biol. Phys.* **30,** 663–669.
218. DiChiro, G., DeLaPaz, R. L., Brooks, R. A., *et al.* (1982). Glucose utilization of cerebral gliomas measured by FDG and positron emission tomography. *Neurology* **32,** 1323–1329.
219. Okunieff, P., Dols, S., Lee, J., *et al.* (1991). Angiogenesis determines blood flow, metabolism, growth rate and ATPase kinetics of tumors growing in an irradiated bed: ^{31}P and ^{2}H nuclear magnetic resonance studies. *Cancer Res.* **51,** 3289–3295.
220. Brem, S., Cotran, R., and Folkman, J. (1972). Tumor angiogenesis: A quantitative method for histologic grading. *J. Natl. Cancer Inst.* **48,** 347–356.
221. Rozental, J. M., Levine, R. L., Nickles, R. J., and Dobkin, J. A. (1989). Glucose uptake by gliomas after treatment. *Arch. Neurol.* **46,** 1302.
222. Lev, M. H., Barest, G. D., Schaefer, P. W., Csavoy, A. N., Rosen, B. R., Gonzalez, R. G., *et al.* (1997). Is high grade glioma present? The diagnostic value of magnetic resonance relative cerebral blood volume imaging of intracranial neoplasia. *In* "Proceedings of the 83rd Scientific Assembly and Annual Meeting of the Radiological Society of North America, 1997," Chicago. [Supplement]
223. Knopp, E. A., Cha, S., Johnson, G., Mazumdar, A., Golfinos, J. G., Zagzag, D., *et al.* (1999). Glial neoplasms: Dynamic contrast-enhanced T2*-weighted MR imaging. *Radiology* **211,** 791–798.
224. Bruening, R., Wu, R., Yousry, T., Weber, J., Birchtenbreiter, C., Reiser, M. (1996). Regional cerebral blood volume maps of meningiomas before and after partial embolization. *In* "American Society of Neuroradiology, 1996," Seattle, p. 153.
225. Ernst, T., Chang, L., Witt, M., Aronow, H., Cornford, M., Walot, I., *et al.* (1998). Cerebral toxoplasmosis and lymphoma in AIDS: Perfusion MR imaging experience in 13 patients. *Radiology* **208,** 663–669.
226. Lai, P., Yang, C., Pan, H., Wu, M., Chu, S., Ger, L., *et al.* (1999). Recurrent inverted papilloma: Diagnosis with pharmacokinetic dynamic gadolinium-enhanced MR imaging. *Am. J. Neuroradiol.* **20,** 1445–1451.
227. Ahuja, A., Ho, S., Leung, S., Kew, J., and Metreweli, C. (1999). Metastatic adenopathy from nasopharyngeal carcinoma: Successful response to radiation therapy assessed by color duplex sonography. *Am. J. Neuroradiol.* **20,** 151–156.
228. Lev, M. H., Schaefer, P. W., Barest, G. D., Farkas, J., Weisskoff, R. M., Gonzalez, R. G., *et al.* (1997). Radiation necrosis or glioma recurrence? Magnetic resonance relative cerebral blood volume imaging in proton beam treated patients. *In* "Proceedings of the 83rd Scientific Assembly and Annual Meeting of the Radiological Society of North America, 1997," Chicago.
229. Siegal, T., Rubinstein, R., Tzuk-Shina, T., and Gomori, J. M. (1997). Utility of relative cerebral blood volume mapping derived from perfusion magnetic resonance imaging in the routine follow up of brain tumors. *J. Neurosurg.* **86,** 22–27.
230. Petrella, J., DeCarli, C., Dagli, M., Brandin, C., Duyn, J., Frank, J., *et al.* (1998). Age-related vasodilatory response to acetazolamide challenge in healthy adults: A dynamic contrast-enhanced MR study. *Am. J. Neuroradiol.* **19,** 39–44.
231. Kastrup, A., Li, T., Glover, G., and Moseley, M. (1999). Cerebral blood flow-related signal changes during breath-holding. *Am. J. Neuroradiol.* **20,** 1233–1238.
232. Tsuchiya, K., Inaoka, S., Mizutani, Y., and Hachiya, J. (1998). Echo-planar perfusion MR of Moyamoya disease. *Am. J. Neuroradiol.* **19,** 211–216.
233. Adams, H. P., Jr. (1998). Occlusion of the internal carotid artery: Reopening a closed door? *JAMA* **280,** 1093–1094.
234. DeKosky, S. T. (2001). Epidemiology and pathophysiology of Alzheimer's disease. *Clin. Cornerstone* **3,** 15–26.
235. 't Veld, B. A., Ruitenberg, A., Hofman, A., Launer, L. J., van Duijn, C. M., Stijnen, T., *et al.* (2001). Nonsteroidal antiinflammatory drugs and the risk of Alzheimer's disease. *N. Engl. J. Med.* **345,** 1515–1521.
236. Silverman, D. H., Small, G. W., Chang, C. Y., Lu, C. S., Kung De Aburto, M. A., Chen, W., *et al.* (2001). Positron emission tomography in evaluation of dementia: Regional brain metabolism and long-term outcome. *JAMA* **286,** 2120–2127.
237. Tracey, I., Hamberg, L. M., Guimaraes, A. R., Hunter, G., Chang, I., Navia, B. A., *et al.* (1998). Increased cerebral blood volume in HIV-positive patients detected by functional MRI. *Neurology* **50,** 1821–1826.
238. Martin, N. A., Patwardhan, R., Alexander, M., *et al.* (1997). Characterization of cerebral hemodynamic phases following severe head trauma: Hypoperfusion, hyperemia, and vasopasm. *J. Neurosurg.* **87,** 9–19.
239. Yang, G. Y., Betz, A. L., Chenevert, T. L., Brunberg, J. A., and Hoff, J. T. (1994). Experimental intracerebral hemorrhage: Relationship between brain edema, blood flow, and blood–brain barrier permeability in rats. *J. Neurosurg.* **81,** 93–102.
240. Liachenko, S., Tang, P., Yan, B., *et al.* (1997). Cerebral perfusion imaging in an outcome model of prolonged cardiac arrest and resuscitation. *In* "Fifth Scientific Meeting and Exhibition of the Society of Magnetic Resonance in Medicine, 1997," Vancouver.
241. Ito, J., Marmarou, A., Barzo, P., Fatouros, P., and Corwin, F. (1996). Characterization of edema by diffusion-weighted imaging in experimental traumatic brain injury. *J. Neurosurg.* **84,** 97–103.
242. Morriss, M. C., Twickler, D. M., Hatab, M. R., Clarke, G. D., Peshock, R. M., and Cunningham, F. G. (1997). Cerebral blood flow and cranial magnetic resonance imaging in eclampsia and severe preeclampsia. *Obstet. Gynecol.* **89,** 561–568.
243. Masuzaki, M., Utsunomiya, H., Yasumoto, S., and Mitsudome, A. (2001). A case of hemiplegic migraine in childhood: Transient unilateral hyperperfusion revealed by perfusion MR imaging and MR angiography. *Am. J. Neuroradiol.* **22,** 1795–1797.
244. Cutrer, F., Sorensen, G., Weisskoff, R., Ostergaard, L., Sanchez del Rio, M., Lee, E., *et al.* (1998). Perfusion weighted imaging defects during spontaneous migrainous aura. *Ann. Neurol.* **43,** 25–31.
245. Yuh, W., Ueda, T., and Maley, J. (1998). Perfusion and diffusion imaging: A potential tool for improved diagnosis of CNS vasculitis. *Am. J. Neuroradiol.* **20,** 87–89.
246. Imperio, W. (2000). Cerebral blood flow differs in adults with ADHD. *Clin. Psychiatry News* **2000,** 10.
247. Warach, S., Levin, J. M., Schomer, D., Holman, B. L., and Edelman, R. R. (1994). Hyperperfusion of ictal seizure focus demonstrated by magnetic resonance perfusion imaging: SPECT, EEG, and clinical correlation. *Am. J. Neuroradiol.* **15,** 965–968.
248. Weinand, M. E., Carter, L. P., el, S. W., Sioutos, P. J., Labiner, D. M., and Oommen, K. J. (1997). Cerebral blood flow and temporal lobe epileptogenicity. *J. Neurosurg.* **86,** 226–232.
249. Cohen, B. M., Yurgelun, T. D., English, C. D., and Renshaw, P. F. (1995). Abnormalities of regional distribution of cerebral vasculature in schizophrenia detected by dynamic susceptibility contrast MRI. *Am. J. Psychiatry* **152,** 1801–1803.
250. Dupas, B., Gayet-Delacroix, M., Villers, D., Antonioli, D., Veccherini, M. F., and Soulillou, J. P. (1998). Diagnosis of brain death using two-phase spiral CT. *Am. J. Neuroradiol.* **19,** 641–647.

18

Imaging Brain Function with Positron Emission Tomography

Simon R. Cherry* and Michael E. Phelps†

**Department of Biomedical Engineering, University of California at Davis, Davis, California 95616;*

†Department of Molecular and Medical Pharmacology, University of California at Los Angeles School of Medicine, Los Angeles, California 90095

I. Introduction

Positron emission tomography (PET) is a powerful molecular imaging tool that enables regional brain function to be assayed in a fully quantitative and noninvasive manner. PET utilizes radioactively labeled probes that are specific to biochemical pathways or molecular targets to perform *in vivo* assays with exquisite sensitivity. This allows PET to assay biological systems involving concentrations in the nanomolar to picomolar range, without producing significant mass disturbance on the biological system being investigated. In this sense it is an *in vivo* analog of autoradiography and plays a unique role in providing information about brain function which is complementary to the high-resolution anatomical information provided by structural imaging techniques such as magnetic resonance imaging (MRI) and X-ray computed tomography (XCT). PET has been widely applied in both research and clinical environments to provide a window on the living brain, exploring the functional organization of the intact brain during development and periods of neuronal plasticity and its compensatory responses to lesions or surgical resection. PET also has been used to map brain responses to sensory, motor, cognitive, and drug stimulation. In the clinical setting, abnormal patterns of radiotracer uptake have been characterized and are now used to identify disease states with high sensitivity and specificity. The continuing development of novel imaging probes that are labeled with positron-emitting isotopes continues to expand the range and scope of the biological and molecular assays available for investigation by PET imaging, which in turn broadens the possibilities for observing and ultimately understanding the living, functioning human brain.

Since the brain is a complex neural network in which all structural subunits can communicate either directly or indirectly with each other, the capability to image the entire brain *in vivo* with PET provides the means to monitor both local and distributed functions. Combining the functional PET data with the high-resolution anatomical maps produced by MRI using accurate image registration techniques provides powerful data sets which allow structure/function relationships to be identified and provides an efficient means for displaying, analyzing, interpreting, and communicating large amounts of information regarding the interrelated functions of the brain.

Although the focus of this chapter is on the use of PET in the human brain, the applications of PET encompass the entire human body. In particular, total-body metabolic PET surveys for the detection and staging of a wide range of cancers (Hoh *et al.,* 1993; Mankoff and Bellon, 2001; Schiepers and Hoh, 1998) and studies of myocardial viability and heart disease (Camici, 2000; Schelbert, 1991) have direct clinical relevance.

Many research studies also are performed in laboratory animals using PET. By using noninvasive imaging techniques, assays can be performed repeatedly in animal models, without the need to sacrifice the animal. Initially, PET was used primarily in larger animals such as nonhuman primates, because of spatial resolution limitations. But with the arrival of dedicated, high-resolution small-animal PET scanners, studies of the rat and mouse brain are now feasible under certain conditions. The ability to use PET across species provides an important experimental bridge between animal models and man.

Examples of a few of the many different types of assays that can be performed with PET are illustrated in Fig. 1. This chapter describes the basic principles of PET, discusses the production and characteristics of commonly employed PET tracers and probes, examines the design and performance of modern PET scanners, and discusses the process of image reconstruction and the use of tracer kinetic models to turn a time sequence of PET images into quantitative regional biological assays. Finally we review some selected applications of PET in understanding the function of the brain in health and disease (for further recent examples, see Gjedde *et al.,* 2001) and discuss the emerging use of PET in laboratory animals.

II. Basic Overview and Principles of PET

Positron-emitting radionuclides are generally produced by a cyclotron. Atoms from these positron-emitting radionuclides are then used to "tag" molecules of a compound of

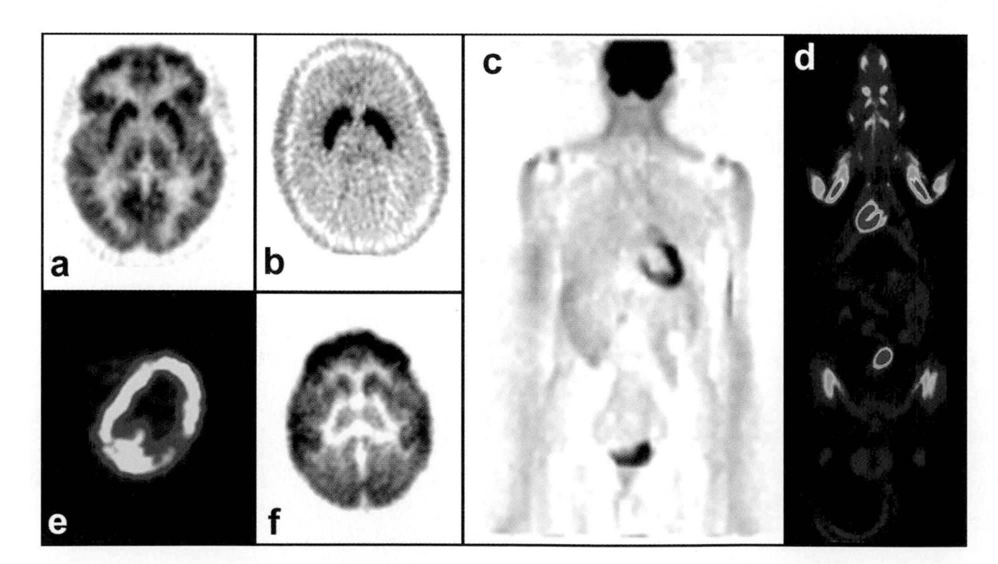

Figure 1 A series of images reflecting the diversity of applications of PET. (**a**) Transaxial section of the brain, human subject, 2-[^{18}F]fluoro-2-deoxy-D-glucose (glucose utilization). (**b**) Transaxial section of the brain, human subject, L-6-[^{18}F]fluoroDOPA (dopamine synthesis). (**c**) Coronal projection of whole body, human subject, 2-[^{18}F]fluoro-2-deoxy-D-glucose (glucose utilization). (**d**) Coronal section through whole body, rat, 2-[^{18}F]fluoro-2-deoxy-D-glucose (glucose utilization). (**e**) Transaxial section through myocardium, human subject, [^{13}N]ammonia (blood flow). (**f**) Transaxial section of the brain, vervet monkey, 2-[^{18}F]fluoro-2-deoxy-D-glucose (glucose utilization).

interest which are introduced into the subject or patient, usually by intravenous injection. These labeled compounds are used to "trace" or "probe" molecular processes and targets and are therefore referred to as molecular tracers or probes. The positron-labeled probe or tracer will distribute in the body according to its delivery, uptake, metabolism, and excretion characteristics. Its spatial distribution and concentration will thus in general vary with time.

At any given time, some of the radioactive atoms that are attached to the probe molecules will decay, emitting a positron and a neutrino. The neutrino passes out of the body without interacting and cannot be detected. The positron rapidly loses energy in collisions with electrons in the tissue and within a very short distance annihilates with one of these electrons. The annihilation converts the mass of the electron and positron into energy, which is liberated in the form of gamma rays. In order to conserve both energy and momentum, the predominant decay mode leads to the back-to-back emission of two gamma rays with an energy of 511 keV, as illustrated in Fig. 2. These gamma rays are sufficiently energetic that they have a reasonable probability of escaping the body without attenuation and can be detected by external detector systems.

A PET scanner consists of circumferential arrays of scintillation detectors which look for "coincidence" events, in which two gamma ray interactions occur almost simultaneously on opposite sides of the head (Fig. 3). If the locations of both gamma rays can be accurately detected, the line along which the annihilation (and therefore the positron emission) took place can be determined. These "lines of response" correspond to projections of the concentration of positron-labeled molecules through the brain. By combining lines of response from many different angles, the data can be reconstructed into cross-sectional images using computed tomography (CT) (Kak and Slaney, 1988). The count density in the resulting images, assuming appropriate data corrections are applied, reflects the concentration of the positron-emitting probe in the tissue. If the kinetics of the probe molecules in the body are known, it is possible to construct tracer kinetic models (Huang and Phelps, 1986). These models can be used in conjunction with a time series of PET images to calculate parameters that are directly related to the rates of biological processes in which the probe is involved.

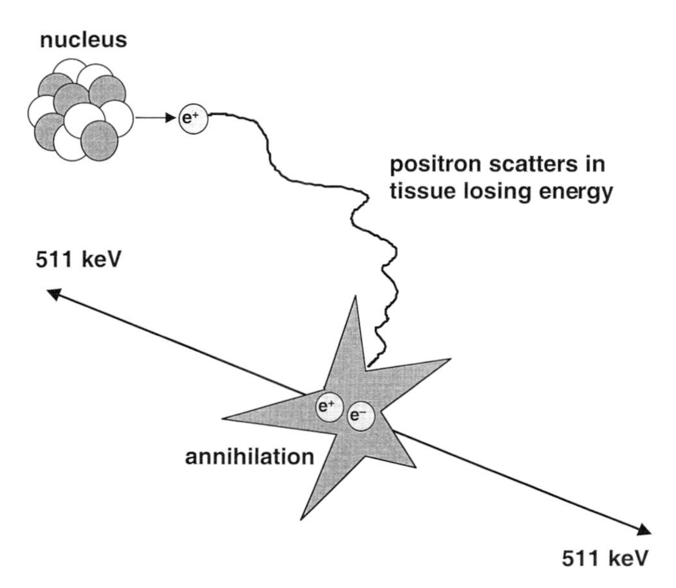

Figure 2 Positron emission and annihilation. A positron is emitted from a labeled probe molecule, losing energy by scattering off atomic electrons before annihilating with an electron to produce two 511-keV gamma rays, which are emitted with an angular separation of 180°. The distance the positron travels before annihilation is small, typically less than 2 mm. Reproduced, with permission, from Cherry and Phelps (1995).

III. Preparation of Positron-Labeled Compounds

One of the unique features of PET is the existence of positron-emitting isotopes of carbon (^{11}C), nitrogen (^{13}N), and oxygen (^{15}O), the major elemental constituents of the body. Fluorine-18 (^{18}F) also is a positron-emitting radioisotope that can be used as a hydrogen or hydroxyl analog. The physical characteristics of these positron-emitting radioisotopes are listed in Table 1. The fact that these isotopes have half-lives in the range of minutes to hours is ideal for medical imaging purposes. If the half-life is too short, there is not sufficient time to label the compound of interest and get the dose to the patient before it decays away. If the half-life is substantially longer than the proposed scanning time, many of the positrons are emitted after the patient has left the scanner, contributing dose to the patient without the benefit of increasing the signal in the images.

The natural ubiquity of carbon, nitrogen, oxygen, and hydrogen (the latter substituted by ^{18}F in PET) in organic molecules gives PET unprecedented flexibility in labeling a

TABLE 1 Physical Properties of Commonly Used Positron-Emitting Isotopes

Isotope	Half-life (min)	Max. energy (keV)	Range (mm) (rms)
Carbon-11	20.4	960	0.69
Nitrogen-13	9.96	1190	0.91
Oxygen-15	2.07	1720	1.44
Fluorine-18	109.8	640	0.38

Note. Energy is maximum possible positron energy (endpoint energy). The root mean square (rms) positron range is measured in water (Derenzo et al., 1982).

wide variety of substrates, substrate analogs, and drugs without disturbing their biological activity. This allows the transfer of tracer assays from the basic biological sciences to man via PET. The extremely high sensitivity of PET also allows the distribution and pharmacokinetics of potentially harmful compounds, such as drugs of abuse and toxins, to be safely examined *in vivo* at a concentration several orders of magnitude below that required to elicit pharmacological or toxic effects. Many hundreds of different compounds have been labeled with positron-emitting isotopes (Fowler and Wolf, 1986; Tewson and Krohn, 1998), and the list continues to grow. Some examples of labeled compounds and their applications in PET are given in Table 2.

The most useful positron emitters, ^{15}O, ^{11}C, ^{13}N, and ^{18}F are produced by an accelerator, usually a cyclotron. In the cyclotron, stable nuclei in a target are bombarded with protons or deuterons (hydrogen with an added neutron) to create the proton-rich state necessary for positron emission. Because of their relatively short half-lives, it is necessary that these isotopes be produced close to the PET scanner. Therefore most PET facilities have their own cyclotrons or receive labeled compounds from regional distribution centers. With ^{18}F, these centers can be located up to about 100 miles from the PET scanner.

TABLE 2 Partial List of Imaging Probes Labeled with Positron Emitters and Their Applications

Labeled compound	Application
Physiological parameters	
[^{15}O]Water, [^{15}O]butanol, $C^{15}O_2$	Blood flow
^{11}CO, $C^{15}O$, ^{68}Ga-EDTA	Blood volume
Metabolism and biosynthesis	
2-[^{18}F]Fluoro-2-deoxy-D-glucose	Glucose metabolism
[^{11}C]Palmitic acid	Free fatty acid metabolism
[1–^{11}C]Acetate	Krebs cycle function
$^{15}O_2$	Oxygen utilization
L-[1–^{11}C]Leucine, L-[*methyl*-^{11}C]methionine	Protein synthesis
3′-Deoxy-3′b6-[^{18}F]fluorothymidine	DNA replication–proliferation
[^{18}F]Fluoromisonidazole, ^{64}Cu-ASTM	Hypoxia
[^{18}F]Fluoride-ion	Bone cell growth
Neuroreceptor/neurotransmitter systems	
[^{11}C]Raclopride, *N*-[^{11}C]methylspiperone	Dopamine D_2 receptor ligand
6-[^{18}F]Fluoro-L-DOPA, 4-[^{18}F]fluoro-*m*-tyrosine	Dopamine synthesis
[^{11}C]CFT, [^{11}C]cocaine	Dopamine reuptake
[^{11}C]Carfentanil, [^{11}C]etorphine	Opiate receptor ligand
Enzyme probes	
5-[^{18}F]Fluoro-2′b6-deoxyuracil	Thymidylate synthetase
[^{11}C]Deprenyl	Monoamine oxidase
Gene expression reporter probes	
8-[^{18}F]Fluoropenciclovir, [^{124}I]FIAU	Transgene expression

The technology for medical cyclotrons continues to advance (McCarthy and Welch, 1998). What started out as large, complicated research machines for nuclear physics groups requiring experienced staff and a large, well-shielded vault have now become compact, automated, self-shielding, and reliable medical devices which can be operated by a single technician. The radioisotope delivery system (RDS) made by Siemens is typical of a modern PET cyclotron (Fig. 4). The RDS accelerates negative hydrogen ions (one proton, two electrons) to around 11 MeV. Once the beam of hydrogen ions has reached the desired energy, it is extracted from the cyclotron by passing it through a thin carbon foil that strips off the two electrons. The resulting proton feels the cyclotron's magnetic field acting in the opposite direction, and the proton leaves the cyclotron and strikes the target (Fig. 5). The great advantage of this system is that it is possible to have multiple ports on the cyclotron, each set up to produce a different radionuclide, and that the beam can be extracted simultaneously into two ports, allowing two radionuclides to be produced simultaneously. These systems are also self-shielded and can be placed in ordinary hospital rooms without any additional shielding requirements.

After bombardment, the target material contains high yields of the positron-emitting isotope in the form of a labeled precursor. This precursor must now be used to label the molecular probe or tracer, and this process must be performed rapidly, before the isotope decays away and without giving undue dose to the personnel responsible for synthesizing the labeled compound. For this reason, biosynthesizers have been developed which automatically take the PET isotopes and, through several steps of unit operations (e.g., solvent and reagent additions, mixing, heating, ion-exchange separation), produce the final labeled compound, in a sterile, pyrogen free form, ready for injection (Satyamurthy *et al.*, 1999). This integrated system, encompassing both cyclotron and biosynthesizer, is referred to as an electronic generator. The whole system is preprogrammed and can be operated by a single technician sitting at a workstation or personal computer. This dramatically simplifies the whole process and leads to rapid and reliable production of positron-labeled isotopes, at the same time significantly reducing the radiation exposure to personnel. As an example of this technology, the popularity of ^{15}O-labeled water for measuring changes in cerebral blood flow (CBF) caused by brain activation has resulted in the development of a simple electronic generator of ^{15}O-labeled H_2O,

Figure 4 Photograph of a compact PET cyclotron. This cyclotron accelerates negative hydrogen ions up to 11 MeV and can produce yields of hundreds of millicuries of ^{11}C, ^{13}N, ^{15}O, and ^{18}F. (Reproduced with permission of CTI PET Systems, Knoxville, TN).

O_2, CO_2, and CO. This device provides a continuous supply of these short-half-life compounds. Similarly, the production of 2-[^{18}F]fluoro-2-deoxy-D-glucose ([^{18}F]FDG), a metabolic tracer widely used in clinical PET, has been automated. It is anticipated that future advances in PET isotope production and automated compound labeling technology will lead to further simplification, further diversity, and lower cost solutions to the production of positron-labeled compounds.

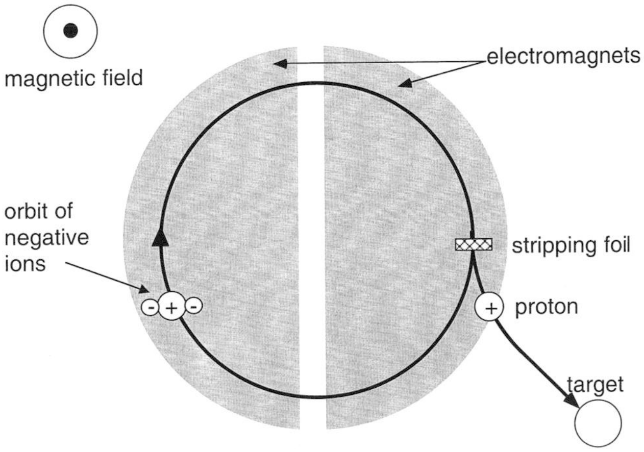

Figure 5 Schematic figure showing how the cyclotron works. Negative hydrogen ions are accelerated to 11 MeV by D-shaped electromagnets. The beam is extracted from the cyclotron by passing it through a thin carbon foil, which strips off the electrons, leaving protons of opposite charge which are then swept out of the beam and strike the target.

IV. PET Scanners

The PET scanner is a sophisticated imaging instrument that provides analytical tomographic measurements of the tissue concentration of positron-labeled compounds. As well as providing spatial information, PET has a temporal resolution of seconds to minutes, which is sufficient to follow many biological processes. Together, the spatial and temporal information obtained from the PET scanner provide the necessary data for converting the recorded time sequence of images into an analytical assay of a biological process.

The tomographic image in PET is formed by recording the two 511-keV gamma rays emitted in positron decay using a ring, or multiple rings, of scintillation detectors. Since the two 511-keV gamma rays are simultaneously emitted 180° apart, electronic coincidence techniques (requiring that two gamma rays are detected within a very short time interval) are used to define the line of emission.

The two key parameters for a PET scanner are spatial resolution (the ability to accurately locate events) and sensitivity (the number of events registered per unit dose injected). Sensitivity is particularly important since signal to noise, and therefore image quality in PET, depends strongly on the number of counts detected. This in turn is determined by the injected dose, the scan length (which is often dictated by the kinetics of the tracer being observed), the efficiency of the detectors, and the solid angle which the detectors subtend to the subject. The spatial resolution of the scanner is dominated by the resolution of the detectors used, although there are additional effects related to positron range and noncollinearity as discussed later. For arrays of single element detectors of width *d*, the spatial resolution is approximately *d*/2 (Hoffman and Phelps, 1986). It is therefore important to use small detector elements to achieve high spatial resolution. Alternatively, continuous, large-area position-sensitive detectors can also be used to achieve relatively high spatial resolution (Karp *et al.,* 1986). Whatever type of detectors are used, the aim in designing a PET scanner is to detect as many of the pairs of 511-keV gamma rays emitted from the body as possible and to determine their interaction position as accurately as possible.

The detectors used in all commercial PET scanners are based around clear, dense, crystalline materials known as scintillators that are coupled to photomultiplier tubes (PMTs) as shown in Fig. 6. These detectors yield relatively good detection efficiencies for high-energy gamma rays relative to other available detectors and have sufficiently fast signals for coincidence timing requirements. When a gamma ray interacts in the scintillating crystal, some of the energy deposited is converted into a flash of visible light. The properties of several scintillators that are appropriate for use in PET are shown in Table 3. The amount of light produced by a scintillator for a 511-keV interaction typically consists of no more than a few thousand photons and is therefore extremely weak. In order to produce a measurable signal, a PMT, optically coupled to one face of the scintillator, is used to detect this scintillation light. Light incident on the photocathode of the PMT generates electrons that are accelerated by a high electric field through several multiplying stages to the output. For each electron liberated at the photocathode, approximately 10^6 electrons are produced at the PMT output, producing a substantial current pulse which is easily detected. To improve the light collection efficiency of the detector, highly reflective materials, such as barium sulfate, are used to surround the faces of the crystal not in contact with the PMT. The physics of gamma ray detection using such detectors is discussed in depth in Knoll (2000).

The first tomographic PET scanner, built in 1975, consisted of 24 large sodium iodide (NaI(Tl)) crystals, each coupled to a PMT and arranged in a hexagonal array (Phelps *et al.,* 1975). This device, referred to as PETT II, for positron emission transaxial tomography, provided a single cross-sectional image plane with an in-plane resolution of about 30 mm and slice thickness of 25 mm. It collected just 12 coincidence lines of response between detectors.

Since this time, the technological advances in PET have been impressive. Bismuth germanate (BGO) has become the scintillator of choice in most PET systems, replacing NaI(Tl) because of its higher stopping power for 511-keV gamma rays (see Table 3). Even though BGO is one of the densest scintillators available, a 3-cm thickness is required to cause 95% of the incident 511-keV gamma rays to interact. The detector unit found in many modern PET scanners consists of a segmented block of BGO scintillator coupled to four PMTs as shown in Fig. 7 (Casey and Nutt, 1986). The BGO is cut such that the crystals are not completely isolated from each other but are optically coupled to varying degrees at the base of the block. These cuts are carefully designed to control the manner in which the scintillation light is shared among the PMTs, allowing the crystal of interaction to be determined by the relative signal observed in the four PMTs. This block detector allows many crystals to be decoded by a relatively small number of PMTs, cutting the cost of the detector and allowing the crystals to be substantially smaller than the smallest PMT available (around 1 cm in diameter). The use of small BGO elements also leads to high spatial resolution.

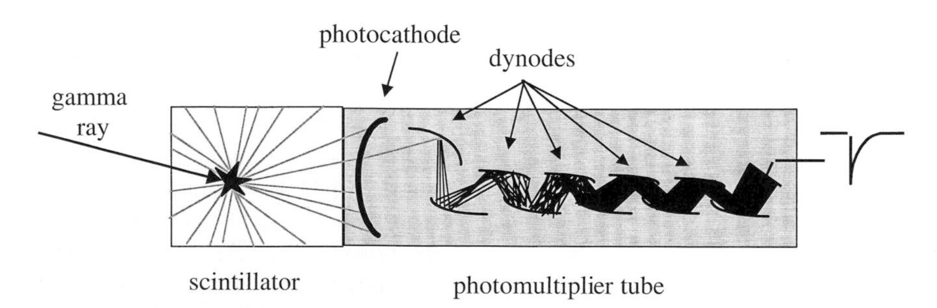

Figure 6 Basic scintillation detector for high-energy gamma rays. Gamma rays interact in the scintillator, causing emission of visible light. Some of this light is converted into electrons at the photocathode of the photomultiplier tube and multiplied by a factor of approximately 3–4 at each of the dynodes to produce a signal of roughly 10^6 electrons for each electron released from the photocathode.

TABLE 3 Properties of Scintillating Materials Suitable for PET

Scintillator	Effective atomic number	Decay time (μ s)	Relative light yield	Peak emission wavelength (nm)
Sodium iodide [NaI(TI)]	50	0.23	100	410
Bismuth germanate (BGO)	74	0.30	15	480
Lutetium oxyorthosilicate (LSO)	66	~0.040	~60–70	420
Gadolinium oxyorthosilicate (GSO)	59	~0.056	~20–25	430

Note. BGO is used in many PET systems because of its high effective atomic number, which provides better stopping power at 511 keV than NaI(TI). LSO is a promising scintillator for PET applications with only slightly lower effective Z than BGO, but with the advantages of a higher scintillation light yield and much shorter decay time. GSO has somewhat lower stopping power, but still retains many of the attractive properties of LSO. The light yield and decay time of LSO and GSO are variable and depend on the exact profile of impurities and the concentration of cerium doping.

A typical modern block detector has 64 individual elements approximately 4 × 4 × 30 mm in size, coupled to four ¾-in. diameter PMT tubes, and is capable of an intrinsic spatial resolution of around 3 mm (Cherry *et al.,* 1995a).

To obtain high sensitivity, a modern PET scanner consists of many of these block detectors, usually arranged in a ring around the subject. For example, the ECAT EXACT HR+ scanner (CTI/Siemens, Knoxville, TN) has 288 block detectors arranged in four rings of 72 blocks per ring around the patient (Fig. 8). There are a total of 18,432 BGO elements and many millions of lines of response are measured. The diameter of the scanner is 82.7 cm and the axial coverage is 15.5 cm. The axial field of view comfortably encompasses the brain, allowing simultaneous collection of data from the entire volume of interest. The ECAT EXACT HR+ produces 63 transaxial image planes with a slice thickness of just over 2.46 mm, which can be resliced into any arbitrary orientation for analysis. Spatial resolution in the reconstructed images can be as high as 4.5 mm depending on the reconstruction filter employed (Brix *et al.,* 1997). The CTI/Siemens ECAT EXACT (Wienhard *et al.,* 1992), EXACT HR (Wienhard *et al.,* 1994), and EXACT 3D (Spinks *et al.,* 2000) scanners, and the GE Medical Systems Advance scanner (DeGrado *et al.,* 1994), have design and performance characteristics similar to those described above for the EXACT HR+, although the dimensions of the detector elements and overall system geometry are somewhat different. A typical set of images from a modern PET scanner, showing the distribution of [^{18}F]FDG in the human brain, is presented in Fig. 9. Note the exquisite delineation of the functional anatomy, in particular the high contrast between gray and white matter and the easy identification of major subcortical structures. The spatial resolution in these images is approximately 5 mm.

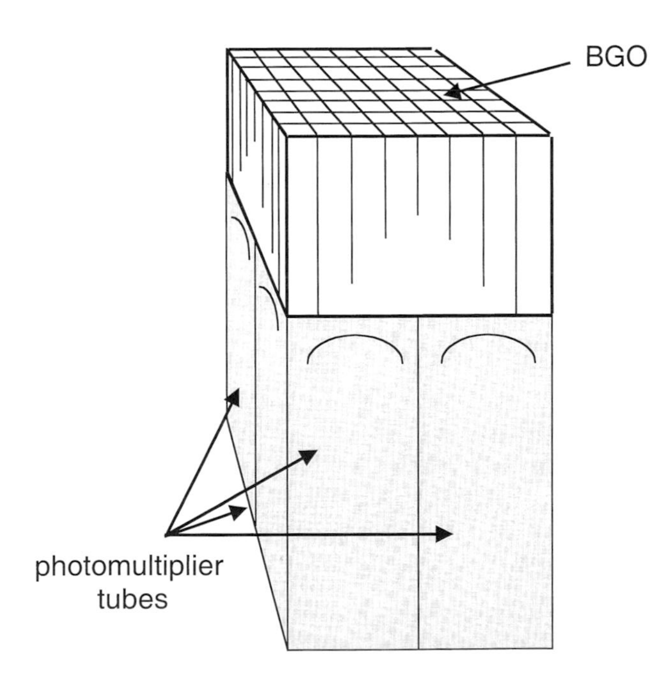

Figure 7 The BGO block detector used in the majority of commercial PET systems. The cuts in the BGO block are carefully designed to control sharing of scintillation light between the four photomultiplier tubes, allowing the crystal of interaction to be determined.

There are also PET systems which are based around large-area NaI(TI) scintillators read out by an array of PMTs. These detectors are analogous to those used in conventional nuclear medicine systems with single gamma-ray-emitting radionuclides, but with the collimator removed. The Phillips/ADAC C-PET system (Smith *et al.,* 1999) consists of six curved 48 (in-plane) × 30 (axial) × 2.5-cm thick NaI(Tl) crystals coupled to 48 PMTs. The position of interaction in the NaI(Tl) crystal is determined by the weighted average of the PMT responses. The HEAD PENN-PET system is based on a single annular NaI(Tl) crystal (42 cm diameter × 30 cm long × 1.9 cm thick) that is decoded in a similar fashion and is designed for brain imaging studies (Freifelder *et al.,* 1994). It can achieve a spatial resolution of around 4 mm. Although the performance of these systems

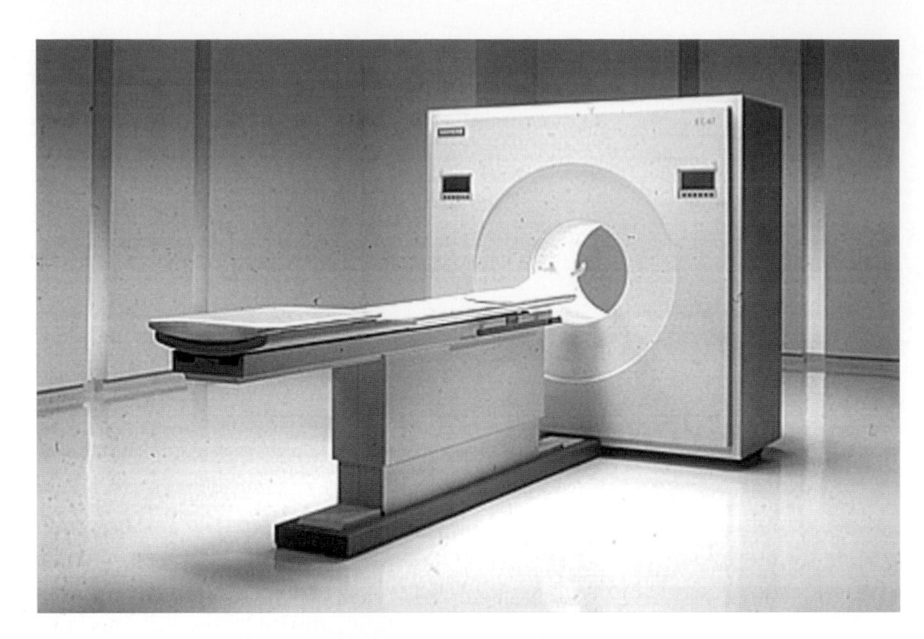

Figure 8 The ECAT EXACT HR+ PET scanner consists of four rings, each made up of 72 block detectors. There are a total of over 18,000 BGO elements. The ring diameter is 82.7 cm and the axial field of view is 15.5 cm. The entire brain can be imaged simultaneously with a spatial resolution of around 4–5 mm. Reproduced with permission of CTI PET Systems.

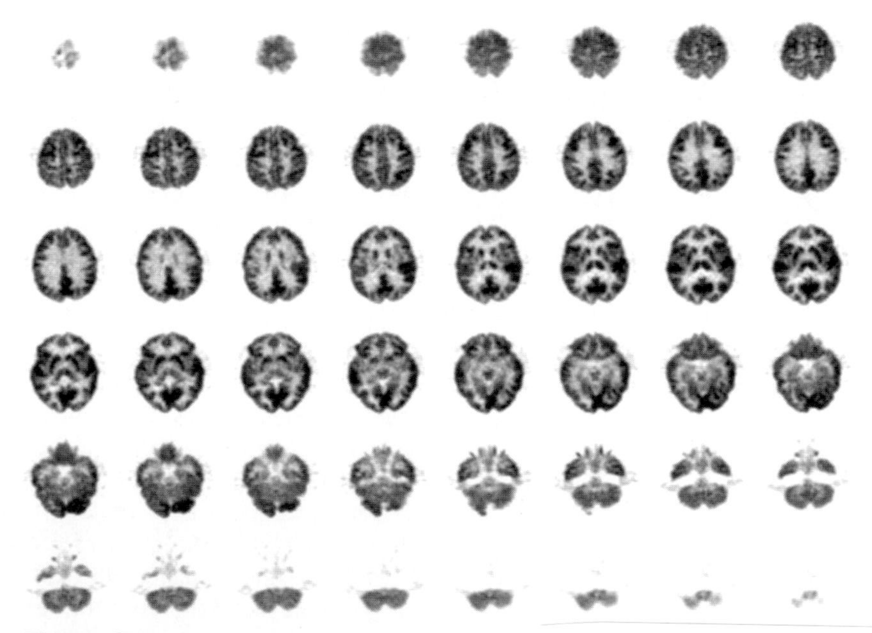

Figure 9 Representative set of brain images from a modern PET scanner. Transaxial sections are shown, running superior (top left) to inferior (bottom right). The tracer is [^{18}F]FDG. Notice the clear delineation of the cortical ribbon and subcortical gray matter structures such as the thalamus and striatum. Image resolution is approximately 5 mm. Data set courtesy of Magnus Dahlbom (UCLA School of Medicine) and Ron Nutt (CTI PET Systems).

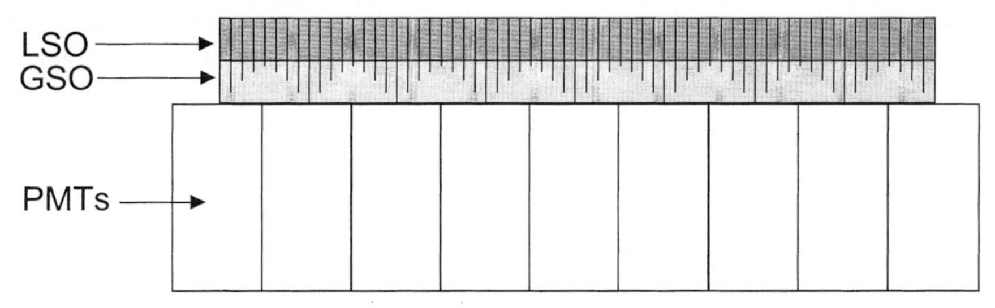

Figure 10 Cross section through detector panel used in the HRRT (high-resolution research tomograph) PET system. Two layers of scintillators with different decay times are used to provide 1-bit depth of interaction information in the detector. The panels are an extension of the block approach (Fig. 7), but each PMT now views four scintillator blocks, reducing the total number of PMTs required to read out the crystals.

tends to be somewhat poorer than that of the BGO block PET systems, due mainly to the lower stopping power of NaI(Tl), high-quality images are obtained for many PET applications.

Two new research PET systems have recently been developed for high-resolution brain imaging applications. Both these systems use fast, cerium-doped scintillators in place of BGO (Table 3). The high-resolution research tomograph system developed by CTI (Wienhard *et al.*, 2001) consists of six detector panels, on which layers of 2 × 2 × 7.5-mm LSO and GSO scintillator crystals are mounted on an array of single-channel PMTs, effectively extending the block design to a large plate (Fig. 10). By using two different scintillators on top of each other, the depth at which the gamma rays interact in the detector can be isolated to the top or bottom scintillator layer by measuring the decay time differences of the pulse (LSO has an ~40-ns decay time, GSO has an ~56-ns decay time). This additional information helps reduce parallax errors that arise from the use of thick detectors and improves spatial resolution while maintaining sensitivity. This system can achieve a spatial resolution as high as 2.5 mm. A dedicated brain imaging system based on GSO has also been developed recently by the University of Pennsylvania (Surti *et al.*, 2000) and consists of arrays of GSO crystals (4 × 4 × 10 mm) coupled to PMTs. Another system based on photodiode readout from the front face of a scintillator array is under development at Berkeley (Moses *et al.*, 1997). By taking the ratio of photodiode signals (front face) and PMT signals (back face) from the scintillator array, continuous depth of interaction information can be obtained. This effectively eliminates parallax errors leading to uniform spatial resolution.

There are two physical factors that ultimately limit the spatial resolution attainable with PET. The 180° emission of 511-keV gamma rays is not exact, because the positron and electron are not completely at rest when they annihilate. The deviation from 180° can be represented by a Gaussian function with a FWHM of 5.7 mrads (Derenzo *et al.*, 1982). The limit that this imposes on the spatial resolution of a PET scanner depends on the diameter of the system and is roughly 1 mm for a 40-cm diameter PET system designed for brain imaging and 2 mm for an 80-cm whole-body PET scanner. The second effect is caused by the fact that the positron travels some distance before combining with an electron and annihilating into the two 511-keV gamma rays (Derenzo, 1979). The positron range, as it is known, is radionuclide dependent and is related to the energy with which the positrons are emitted. The higher their energy, the farther the positrons will travel before annihilating. Positron range effects degrade spatial resolution by between 0.4 and 1.5 mm for the radionuclides most commonly used in PET (see Table 1). Fortunately, the angulation error in 511-keV gamma-ray emission and positron range effects are not additive but rather are convolved together. This yields an overall resolution limit of just over 2 mm for an 80-cm-diameter system using ^{11}C- or ^{18}F-labeled compounds and closer to 1.2 mm for a 40-cm-diameter system designed specifically for brain imaging.

It is an unfortunate fact that most PET studies cannot be reconstructed at the resolution that the scanner is capable of, due to signal-to-noise constraints. Some smoothing must usually be applied to the data during the reconstruction process to improve signal to noise, at the cost of spatial resolution. Therefore, the major limitation in improving spatial resolution in PET is usually not the resolution of the detectors themselves, but the number of coincidence events recorded to form the image. Because of dead-time (the scanner takes a finite time to process each event and is "dead" for this time) and dose considerations, it is usually impractical to increase the injected dose. Longer scan times are also an impractical means for increasing the number of recorded events, as the imaging time is usually dictated by the kinetics of the biological probe or tracer under consideration.

PET scanners initially collected data by recording the coincidence combinations between opposing detectors within the same ring of detectors. This yields a fan beam response from one detector to an array of opposing detectors in the image plane. Modern PET scanners have multiple rings of detectors to allow the simultaneous collection of multiple image planes. As the detector crystals are made smaller, the thickness of the image plane is reduced and more image slices are obtained. However, the number of counts acquired per image plane decreases. To counteract this, data are collected between a number of adjacent detector rings as shown in Fig. 11. This improves sensitivity and, over the central portion of the field of view, does not lead to any degradation in spatial resolution. This is known as 2D acquisition, as the data are collected and reconstructed as a set of independent 2D slices that are then stacked into a 3D volume after image reconstruction. This 2D mode of acquisition is wasteful, as the vast majority of gamma-ray pairs are emitted obliquely with respect to the imaging planes and are rejected. All modern PET scanners now have the capability to collect and reconstruct all gamma-ray pairs that intersect the PET scanner, irrespective of their obliquity. This is referred to as 3D data acquisition and increases the number of gamma rays collected (and therefore the sensitivity of the PET scanner) by roughly a factor of 5 (Cherry *et al.,* 1991; Townsend *et al.,* 1991). This allows many PET data sets to be reconstructed at, or close to, the resolution limit of the PET scanner, while still maintaining adequate signal to noise. The drawbacks of 3D acquisition are the large size of the data sets and the more computationally intensive reconstruction algorithm that is required (Defrise *et al.,* 1990). A further disadvantage is an increased susceptibility to detecting scattered events, that is, gamma rays which scatter in the body before reaching the detectors, leading to mispositioned events. However, the large sensitivity increase is a powerful reason for using 3D acquisition, particularly in highly count-limited situations, such as neuroreceptor studies and [^{15}O]water activation studies. The sensitivity increase can also be used reduce the injected dose, of particular concern in pediatric PET studies, or to reduce the scanning time, which could be useful in a high-throughput clinical environment, while still obtaining the same signal to noise as previously obtained with 2D data acquisition. A direct comparison of standard 2D and the new 3D data acquisition approaches, illustrating the dramatic increase in signal to noise, is shown in Fig. 12.

V. PET Data Correction and Image Reconstruction

With appropriate care, PET images can be fully quantitative, such that there is a linear relationship between the concentration of the labeled imaging probe and the count density measured in the PET images. The absolute concentration can then be determined by cross calibration to a known source. In order to make PET images fully quantitative, a number of correction procedures must be carefully applied. The largest of the correction factors is that required to correct for gamma-ray attenuation by the tissue. Gamma

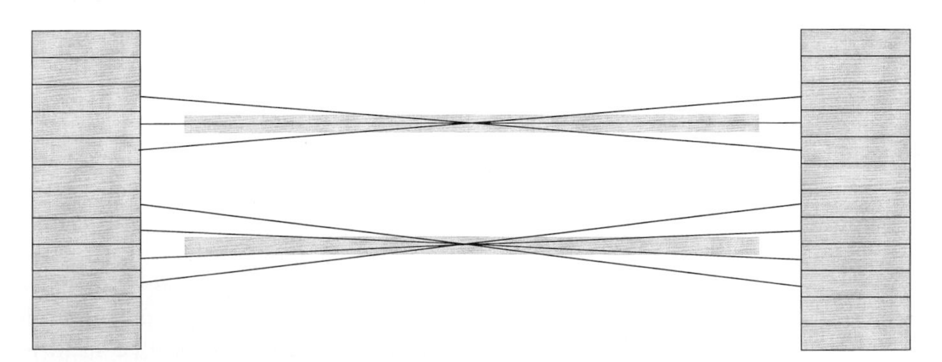

Figure 11 Illustration of 2D data acquisition in PET, in which only lines of response that are parallel, or almost parallel, to the image planes are utilized. The lines of response contributing to two different image planes (shaded areas) are shown. The addition of slightly oblique lines of response improves sensitivity. At the center of the field of view, these oblique lines fall within the image slice and therefore do not degrade spatial resolution for objects close to the center of the scanner. In 3D acquisition (not shown), all possible lines of response connecting detectors in different rings are used, and because of the high obliquity of many lines of response, it is no longer possible to approximate the data as having originated from a set of discrete slices. True 3D approaches are required to reconstruct the data.

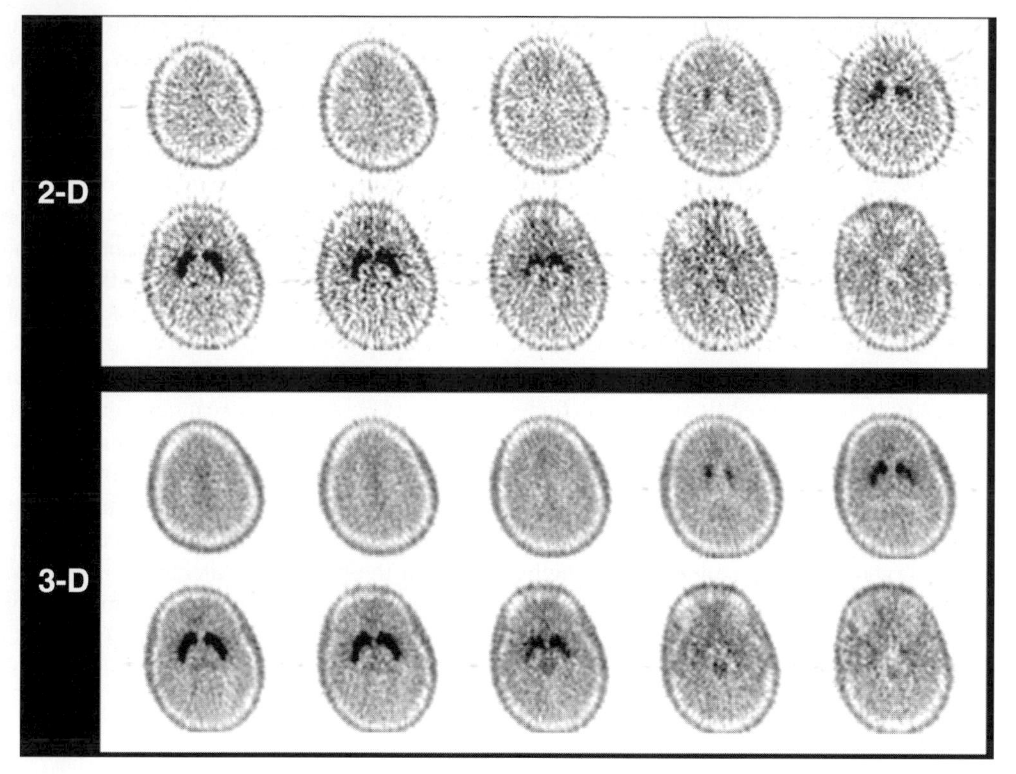

Figure 12 Comparison of 2D and 3D data acquisition and reconstruction in a L-6-[^{18}F]fluoroDOPA study. A 20-min scan was acquired 1 h postinjection using standard 2D acquisition, immediately followed by another 20-min scan using 3D acquisition. There is a substantial increase in signal to noise in the 3D data set for the same scan length and injected dose relative to the 2D data set.

rays emitted from the edge of the brain are more likely to reach the detectors than those emitted from the center of the brain, because they have less tissue to travel through. On average, only one in five gamma-ray pairs emitted from the center of the brain will make it out of the head without being lost due to attenuation. Fortunately, there is a relatively simple way to measure the attenuation correction factors in PET using an external positron-emitting source (Phelps *et al.,* 1975; Ranger *et al.,* 1989).

In addition to the true coincidence events detected by the PET scanner, there are two other classes of events which can contaminate the data (Fig. 13). The first of these, accidental

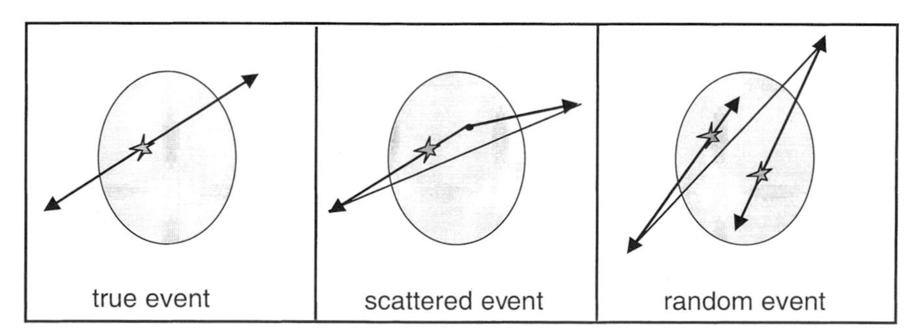

Figure 13 Three classes of coincidence events are detected by PET scanners. In addition to the true coincidence events, scattered and accidental (also known as random) events may be detected. Accidental coincidences randomly add events to the data set, while scatter results in the mispositioning of coincidence events. These types of events must be corrected for in order to obtain fully quantitative images. Reproduced, with permission, from Cherry and Phelps (1995).

or random coincidences, are formed when two unrelated gamma rays happen to strike detectors within the coincidence timing window of the system. This window has to be of a finite temporal width because the gamma rays from an annihilation will never exactly coincide, due to differences in the time taken to arrive at the detectors and differences in electronic processing time. Accidental coincidence events are measured and automatically subtracted in modern PET scanners by using a delayed coincidence window technique (Hoffman *et al.*, 1981). The second class of events, scattered events, are those in which one or both of the 511-keV gamma rays scatter on their way out of the head, but are still detected. Such events will be incorrectly positioned and result in a reduction of contrast in the images. Several techniques for scatter correction have been proposed and implemented (Bergström *et al.*, 1983; Ollinger, 1996; Watson *et al.*, 1997) and this correction is particularly important in 3D PET studies in which the scatter fraction can be as high as 40 or 50% in brain imaging.

There are two further important corrections. Each of the thousands of detectors in a PET system will have a slightly different efficiency. These efficiency differences can be measured, and corrected for, by scanning a source that results in each detector seeing a uniform radiation flux (Badawi *et al.*, 1998; Hoffman *et al.*, 1989). Finally, at higher count rates, the ability of the PET scanner to distinguish each individual event becomes limited due to the electronics. This is known as dead time and results in a count-rate-dependent loss of events. It can be corrected by measuring the count rates on the detector and having knowledge of the dead time in each step of the electronics processing chain (Eriksson *et al.*, 1994; Germano and Hoffman, 1990). Figure 14 shows the count-rate response of the GE Advance scanner for a typical brain study acquired in 3D acquisition mode. In the absence of dead time, the relationship between activity concentration in the brain and the count rate of the scanner would be linear, but the limitations imposed by the decay time of the scintillator and the processing electronics in the scanner limit the maximum count rate to approximately 850,000 cps at an average brain concentration of 2 μCi/ml (Stearns *et al.*, 1995). When corrected for the effects of scatter and acci-

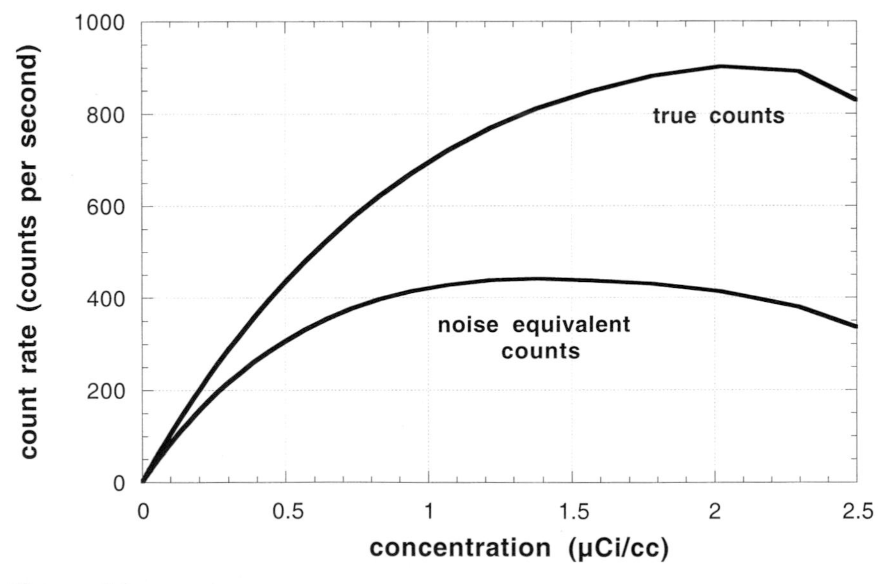

Figure 14 Count-rate performance of the GE Advance scanner for 3D brain studies. The coincidence count rate peaks at 900,000 cps at an activity concentration of 2 μCi/ml. The lower curve results from correcting for the effects of scattered and random coincidence events and is known as the "noise effective count rate." This peaks at just over 400,000 cps at a concentration of 1.25 μCi/cc. In the absence of dead time, the relationship between concentration and counts would be linear. The plot shows that for this particular scanner, doses which lead to concentrations greater than 1 μCi/cc give no improvement in the number of events recorded by the scanner and serve only to increase the dose to the subject. The injected dose for a PET study should therefore be chosen to maximize the noise equivalent count rate, while minimizing the dose to the subject. Data courtesy of Charles Stearns (GE Medical Systems) and Tom Lewellen (University of Washington).

dental coincidences, the noise equivalent count rate (Strother *et al.*, 1990) peaks at 400,000 cps at 1.25 μCi/ml. With increasing dead time, the number of counts detected per unit dose to the subject decreases. For this reason, the dose injected for a PET study should be such that no more than 20–30% of the counts are being lost because of dead time.

After applying all the correction factors and cross-calibrating to a source of known concentration, the pixel intensity in the PET images can be expressed in μCi/ml. It must be noted, however, that for small structures, the activity concentration can still be in error due to resolution effects. Any structure which is smaller than twice the resolution of the PET scanner will suffer from partial volume effects (Hoffman *et al.*, 1979). Because of the limited spatial resolution, activity from small structures is "smeared" over an area which is larger than the structure itself, therefore changing the apparent count density. This effect is shown in Fig. 15 for three objects with the same concentration. As the resolution becomes worse, the concentration in the smallest object is progressively underestimated. Given the resolution of modern PET scanners of around 4–5 mm, any structure less than 1 cm across will suffer from partial volume effects, and this unfortunately includes many structures of interest in the brain. While there is no simple way to correct for partial volume effects for irregularly shaped objects, there are techniques available which make use of coregistered MRI studies, which are blurred to the PET resolution to extract an estimate of the partial volume correction (Müller-Gärtner *et al.*, 1992; Rousset *et al.*, 1998).

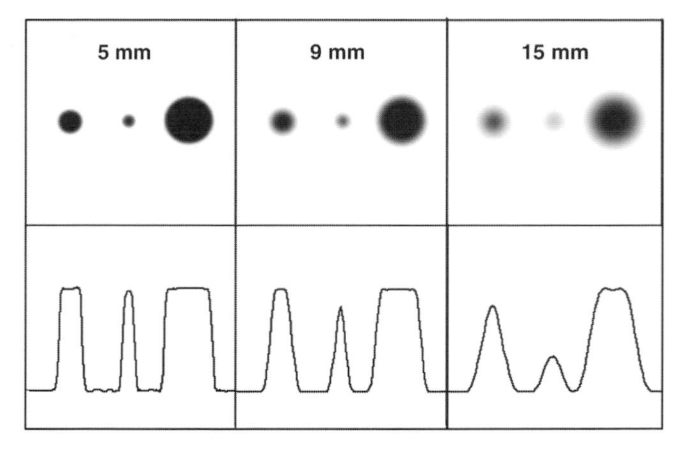

Figure 15 Illustration of the partial volume effect. The three cylinders are 1, 2, and 4 cm in diameter and have identical concentrations as shown in the high-resolution image at left. As the resolution degrades, the smaller cylinders progressively show an underestimation of the concentration. Brain structures are generally small and irregularly shaped and will suffer this same effect to different degrees. Adapted, with permission, from Cherry and Phelps (1995).

After all the correction factors have been applied, the PET projection data are ready for reconstruction. Reconstruction is necessary, because each detector pair sees the sum, or projection, of the radioactivity along the line joining the two detectors, not the activity at any one point. The image reconstruction process seeks to recover the activity distribution from the projection data. The raw projection data for a ring of PET detectors is conveniently represented as a sinogram. The sinogram is a 2D matrix of projection data, in which the vertical axis represents the angle of the projection line and the horizontal axis represents the displacement of the projection line from the center of the field of view. Coincidences between a given detector pair are thus represented by a single element in the sinogram matrix (Fig. 16).

There are essentially two methods for reconstructing the sinogram data into cross-sectional images. The most common, known as filtered backprojection, is an analytic approach which applies a spatially invariant filter to the projection data in the Fourier domain and then backprojects the data into image space (Brooks and Di Chiro, 1976; Kak and Slaney, 1988). The backprojection process involves distributing the counts from a sinogram element along the line defined by the two detectors that contribute to that sinogram element. Modifications of the filtered backprojection method, known as the backprojection/reprojection algorithm and the Fourier rebinning algorithm, are used to reconstruct PET data sets acquired in 3D mode (Bendriem and Townsend, 1998).

The filter that is applied to the projection data prior to backprojection can be modified by a window function to reduce its amplitude at high frequencies. Since the signal-to-noise ratio generally decreases with increasing frequency, this helps to reduce the noise level in the reconstructed images. However, because sharp detail in the images also is represented by high frequencies, this increase in signal to noise comes at the expense of degrading the spatial resolution. The exact choice of window function is always a trade-off between signal to noise and spatial resolution and is often quite subjective.

These analytic methods of image reconstruction have the advantage of computational speed, but also have quite strict sampling criteria and rely on a number of assumptions such as uniform resolution across the field of view. They also do not allow for the incorporation of a priori knowledge into the reconstruction process, nor do they take into account the statistical nature of the data.

The second class of techniques are iterative methods, which try to find the image which best matches the measured projection data using some maximization or minimization criteria (Leahy and Qi, 2000; Shepp and Vardi, 1982). This is more computationally intensive than filtered backprojection, but in many situations may result in somewhat better noise and/or resolution performance. This is because the imaging system and the statistical nature of the data can

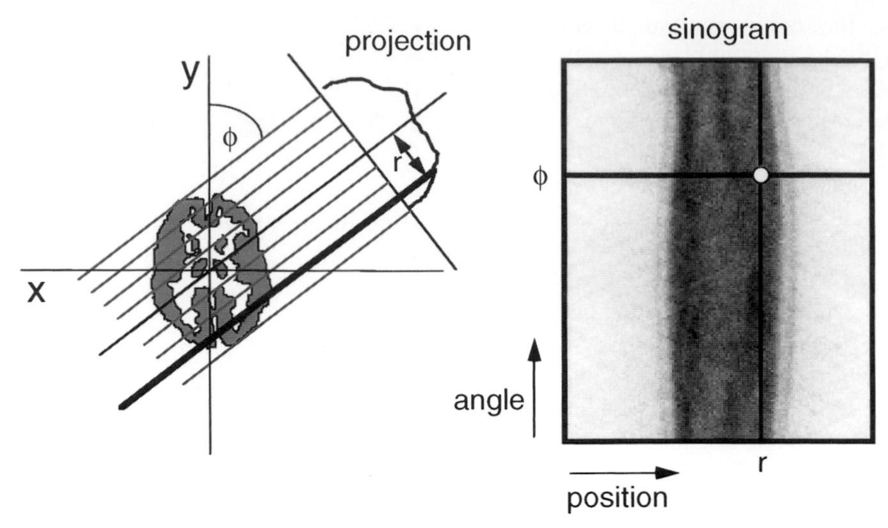

Figure 16 Raw PET data are stored in a 2D matrix known as a sinogram prior to image reconstruction. Each element in the sinogram represents the number of counts detected by a particular detector pair. The convention is such that the vertical axis represents the angle of the line of response and the horizontal axis represents the displacement from the center of the field of view. The events detected from the line of response shown in bold would be stored at the location represented by the white dot. Adapted, with permission, from Cherry and Phelps (1995).

be accurately modeled and incorporated in the reconstruction process (Qi *et al.,* 1998).

VI. Tracer Kinetic Models

An important aspect of PET is the ability to monitor the spatial distribution of the imaging probe or tracer as a function of time. The rate of accumulation or clearance of the probe contains information about the rate of the biological processes or the characteristics of the molecular target (affinity, concentration, etc.) involved. This temporal component to PET is illustrated in Fig. 17, which shows the distribution of L-6-[^{18}F]fluoroDOPA in the human brain at four different times after injection. Tracer kinetic models in PET provide the mathematical framework for calculating the concentration of reactants and products, and the rate of a biological process, based upon the time course of tracer distribution in a dynamic series of PET images and the blood concentration of the biological probe. Compartmental models are the most common tracer kinetic models used in PET. These models are simplifications of biological systems in which a compartment represents the consolidation of microenvironments or processes in which the rate of exchange within a compartment is fast compared to the rate of exchange between compartments. A compartment could therefore represent the positron-emitting isotope in a particular location (e.g., intracellular, extracellular, vascular) or attached to a molecule of a particular chemical species. The models are formulated by differential equations that describe the exchange of the positron-emitting isotope between compartments. For example, Fig. 18 shows the compartmental model and rate constants for exchange between compartments of [^{18}F]FDG.

These compartmental models describe biological systems and, therefore, require extensive biological knowledge to be defined, as well as possible simplifying approximations to aid in their practical formulations. Once a model is properly formulated, a tracer kinetic assay can provide very sensitive and accurate measurements of the rates of the process. Rates of a process involving substrate concentrations in the range of nanomoles/gram and picomoles/gram are routinely achieved. The tracers are used at very high specific activity (radioactivity to mass ratio), so they exert little or no mass effects on the biological systems studied.

PET must model a measurement in the brain from a biological probe administered intravenously. These measurements consist of determining the time course of the supply of the labeled probe in arterial plasma (often by direct blood sampling) and in cerebral tissue (from a dynamic sequence of PET images). The plasma time activity curve represents the input to the tracer kinetic model and is known as the input function. The input function determines how much tracer is available to the biological system as a function of time and can be seen as the driving force behind the observed kinetics. The time course of the labeled probe in the brain will be the convolution of the input function with the kinetics of the biological system. To extract the kinetics of the biological system, the input function must therefore

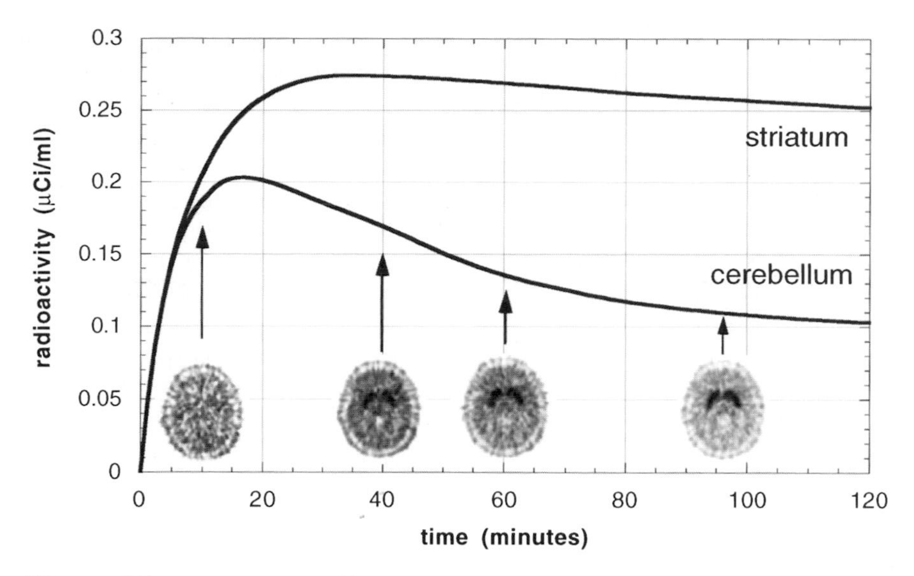

Figure 17 Distribution of L-6-[^{18}F]fluoroDOPA in the human brain as a function of time after injection. Initial image shows a nonspecific distribution which gradually changes over time as L-6-[^{18}F]fluoroDOPA is converted to 6-[^{18}F]fluorodopamine by aromatic amino acid decarboxylase in the striatum. The time–activity curves, as determined from a dynamic sequence of PET scans, are quite different for the striatum and cerebellum, reflecting differences in the kinetics of L-6-[^{18}F]fluoroDOPA in these regions. Reproduced, with permission, from Cherry and Phelps (1995).

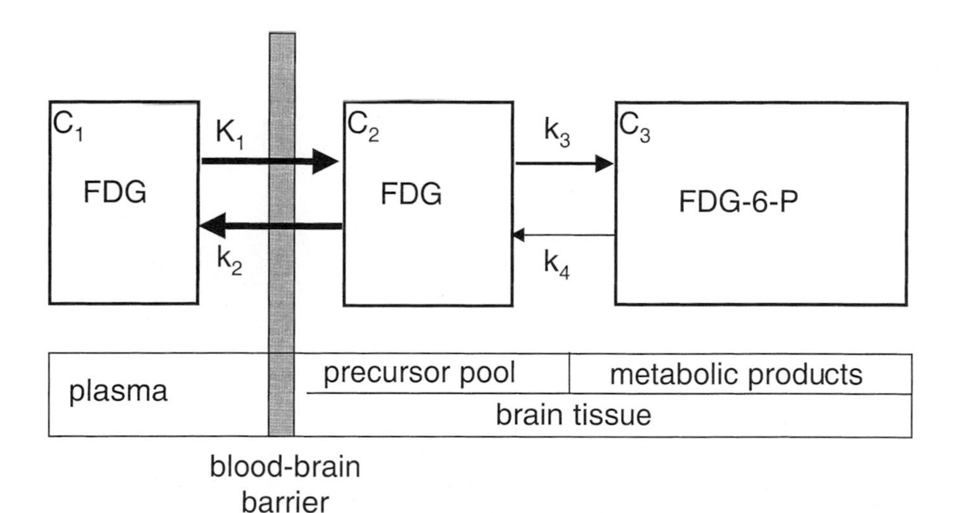

Figure 18 A compartmental model representing the kinetics of [^{18}F]FDG. Delivery is through arterial plasma (C_1) following intravenous injection. FDG is actively transported across the blood–brain barrier by the bidirectional glucose transporter to reach the precursor or free FDG pool (C_2). It can then be phosphorylated by hexokinase, which is represented by rate constant k_3, and enter into the metabolic product compartment (C_3) or be transported back into plasma (k_2). No further metabolism takes place in C_3 and the rate of transfer back to the "free" FDG compartment, k_4, is very small. The rate constants K_1 through k_4 can be determined from a dynamic sequence of PET scans and knowledge of the time course of [^{18}F]FDG in arterial plasma. The thickness of the arrows represent the magnitudes of K_1 through k_4 typically observed in human gray matter.

be known. With the time course of the labeled probe in both arterial plasma and cerebral tissue known, the tracer kinetic model allows the calculation of the rate constants for the exchange of the probe between compartments (i.e., plasma and tissue, reactants and products, ligand and receptor). Thus the model provides a measurement of the flux (e.g., in μmol/min/g) of the process under study.

There are many different types of tracer kinetic assay strategies employed in PET, the majority of which were previously developed and employed in basic biological sciences. Some brief examples will serve to illustrate the different approaches:

A. Cerebral Blood Flow

CBF is measured by using a tracer, such as [^{15}O]water, that freely diffuses through the blood–brain barrier (Table 2). These are referred to as flow-limited tracers. That is, their uptake in brain tissue is determined by blood flow and not by the diffusion rate of the tracer into tissue. These techniques typically use modifications of the Kety–Schmidt technique (Kety and Schmidt, 1948) in which the flow-dependent rate of accumulation or disappearance of a diffusible tracer from the brain is used to measure CBF at the capillary (i.e., nutrient) level in milliliters per minute per gram. A steady-state method for measuring blood flow has also been developed (Jones *et al.,* 1976), which recognizes that the steady-state concentration of [^{15}O]water in the brain is dependent on the short physical half-life of ^{15}O (2 min) and CBF.

B. Natural Substrates and Substrate Analogs

The most commonly used natural substrate is ^{15}O-labeled O_2 for the measurement of the cerebral metabolic rate for oxygen (Frackowiak *et al.,* 1980). Substrate analogs have been used throughout biochemistry to isolate segments of a complex biochemical pathway. The deoxyglucose technique is an example of this approach (Sokoloff *et al.,* 1977). These analogs use the principle of competitive substrate kinetics to formulate the relationship between the analog and the natural substrate for transport carriers, enzymes, and receptors. Common examples of this approach in PET are 2-fluoro-2-deoxy-D-glucose (Phelps *et al.,* 1979; Reivich *et al.,* 1979), as illustrated in Fig. 9, and L-6-fluoroDOPA (Huang *et al.,* 1991), as illustrated in Fig. 16.

C. Ligand–Receptor Binding Assays

The first requirement is that *in vivo* ligand assays of receptors demonstrate the proper anatomical distribution of the receptor as known from *in vitro* studies. Figure 19 shows the distribution of three different receptor systems in the brain using three different PET ligands, illustrating how the appropriate choice of biological probe can isolate the

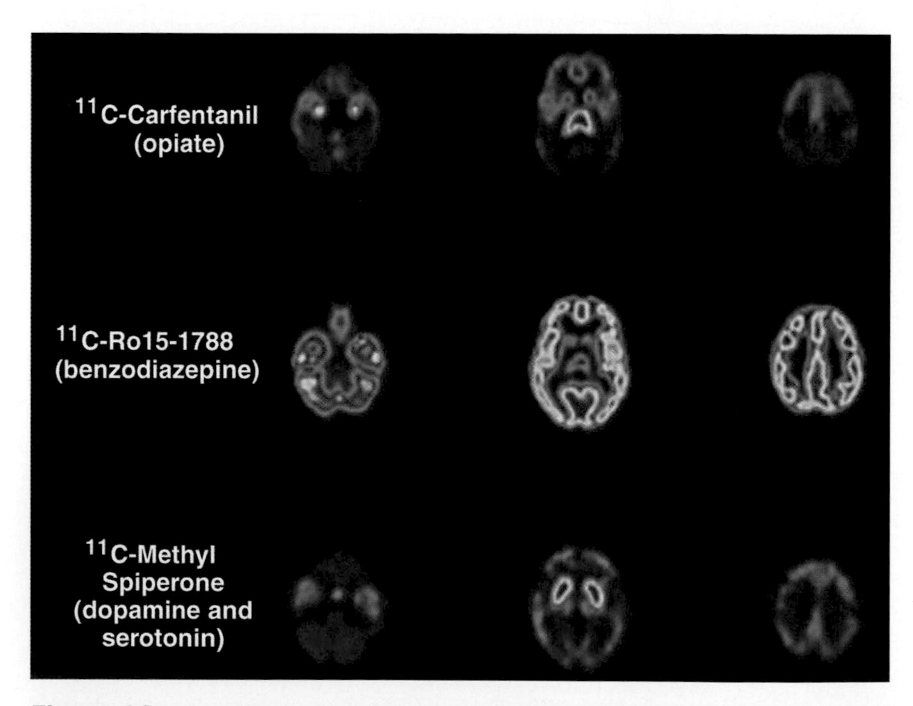

Figure 19 Three different ligand receptor binding assays with PET showing the distribution of opiate, benzodiazepine, dopamine, and serotonin receptors in the human brain. Three different transaxial levels are shown, running inferior (left) to superior (right). Images courtesy of James Frost (Johns Hopkins University).

receptor system of interest. These receptor binding assays have employed classic Scatchard and Hill plots for competitive binding determinations of K_D and B_{max} (Farde *et al.*, 1986). In addition, novel quantitative assay approaches have been developed that take advantage of the temporal measurement capability of PET (Huang *et al.*, 1986; Logan *et al.*, 1990; Patlak and Blasberg, 1985; Perlmutter *et al.*, 1986; Wong *et al.*, 1986) and the use of brain regions that are devoid of the receptor of interest as reference regions (Lammertsma and Hume, 1996).

D. Enzyme Probes

This approach uses a substrate analog or drug that selectively binds to enzymes commonly isolating their active form. This allows enzyme or enzyme activity assays to be performed. These are typically irreversible reactions which require, and indeed employ, high-specific activity tracers. [^{11}C]Deprenyl for the assay of MAO-B is an example of this approach (Fowler *et al.*, 1987).

The possibilities with PET are almost limitless, and this is just a small sampling of some of the biological processes that can be measured *in vivo* with PET. Comprehensive reviews of labeled compounds (Fowler and Wolf, 1986; Tewson and Krohn, 1998), assay strategies (Barrio, 1986), and mathematical formulations of tracer kinetic models (Huang and Phelps, 1986) are provided elsewhere and can be consulted for further details.

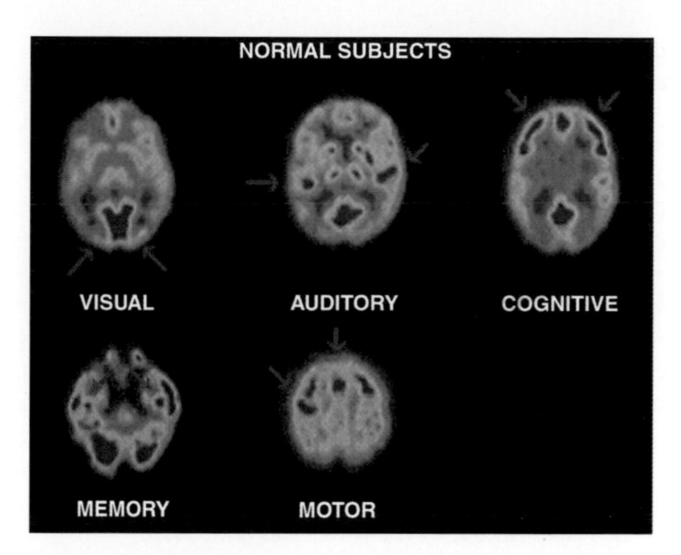

Figure 20 CMRGl increases in response to specific stimuli, as determined with [^{18}F]FDG. Red color denotes highest CMRGl with progressively lower values shown as yellow, green, blue, and purple. Visual: activation of visual cortex (arrows) when subject observes simple visual scene. Auditory: activation of left and right auditory cortices when subject is listening to language and music. Cognitive: activation of frontal cortex when subject is performing a comparison task and planning for a test of comparisons. Memory: activation of hippocampus bilaterally in memory recall task. Motor: activation of left motor cortex and supplementary motor area when subject is touching thumb to the fingers of right hand. Reproduced, with permission, from Phelps (1991).

VII. Task-Specific Mapping of the Human Brain

Although PET has many applications, its use as a noninvasive tool to map function in the human brain has attracted particular attention. The foundation for this work was a series of [^{18}F]FDG studies in normal volunteers which examined the effects of specific stimuli on brain function (Greenberg *et al.*, 1981; Mazziotta *et al.*, 1982; Phelps *et al.*, 1981a,b). Images of the cerebral metabolic rate for glucose (CMRGl) were recorded from subjects who were performing specific tasks during the period (approximately 30 min) when [^{18}F]FDG was being transported into the brain and undergoing metabolism. Since greater than 95% of the brain's ATP for performing work is derived from glucose, it was hypothesized that increased CMRGl, and therefore, increased uptake of [^{18}F]FDG, would occur in regions of the brain that performed the work or the task required of the subject. Some of the first examples of using PET to map the human brain's response to stimuli are shown in Fig. 20. This work demonstrated that robust changes in CMRGl did occur in areas of the brain related to the activating task and laid the foundation for the enormous activity currently seen in brain mapping research.

Not surprisingly, there are also large localized blood flow changes in the brain during task performance. This is because the increased metabolic demand, as measured by [^{18}F]FDG, must be met by increased supply of substrates, which is accomplished by an increase in local blood flow. Measurements of blood flow with PET can be obtained relatively quickly, and, if the tracer is labeled with a short-half-life isotope (such as ^{15}O with a 2-min half-life), repeated studies can be performed in the same individual (Fox *et al.*, 1984). For this reason the majority of PET activation studies now employ blood flow tracers such as [^{15}O]water rather than [^{18}F]FDG. [^{15}O]Water is freely diffusible with delivery to tissue that is flow rather than diffusion limited. It is extracted from plasma into brain tissue on its first pass through the brain and its uptake is highly correlated with regional cerebral blood flow. Blood flow can be quantitatively measured by applying the appropriate compartmental model as described in Section VI. Because of the rapid kinetics in blood flow measurements, the stimulus need be provided for only a minute or two, starting a few seconds before the injection of the tracer, and the PET imaging takes less than 2 min. Within 10 min, the majority of the ^{15}O label has decayed, allowing another study to be performed.

The great advantage of being able to map brain function quickly and repeatedly is that each individual can now serve as their own control. The duration over which the activating task must be performed is also much shorter, reducing concerns over habituation effects and widening the scope of activation paradigms. A typical blood flow activation study would acquire one or more pairs of blood flow images in baseline and activated states (Fig. 21). The resulting images can then be subtracted to reveal areas of the brain in which blood flow differs between the two states. By applying an appropriate statistical analysis, regions of the brain which change significantly and are, therefore, implicated in the activation task can be identified (Fox *et al.,* 1988; Friston, 1994; Friston *et al.,* 1991; Worsley *et al.,* 1992).

There are two schools of thought on the scanning methodology for activation studies. Some centers perform fully quantitative blood flow studies using fast dynamic PET protocols and arterial blood sampling. The PET and blood data are then fit to a simple compartmental model describing the kinetics of [^{15}O]water and the blood flow is extracted as one of the model parameters. Regional blood flow can be determined in units of milliliters per minute per gram (Herscovitch *et al.,* 1983; Huang *et al.,* 1983; Raichle *et al.,* 1983). Many centers, however, employ a semiquantitative or relative approach. For short scan durations of 40–80 s following tracer arrival in the brain, the distribution of [^{15}O]water is almost linearly related to blood flow (Herscovitch *et al.,* 1983). Thus a simple integrated image serves as a map of the blood flow distribution, and multiple studies can be compared by normalizing each separate study to the same total number of counts in the brain (Mazziotta *et al.,* 1985). This approach has the advantage of being much simpler; however, only relative changes in blood flow can be detected. The method of choice depends largely on the question to be answered. In either case, there is a tendency to underestimate blood flow at very high flow rates, because diffusion limitations of the tracers begin to become apparent.

There are a several important issues that must be considered in performing PET activation studies. Careful protocol design is obviously paramount to ensure that the subtracted PET images provide unambiguous information which can be directly related to the activation task. It is particularly important to remove any confounding effects due to issues such as noise and light levels in the scanning room or stress factors related to the injection and/or scanning procedure. In addition, when performing multiple studies over scanning sessions that may last several hours, some movement of the subject is almost inevitable. The majority of this movement is likely to occur between scans during the time required to allow the isotope to decay away. Accurate, automated registration algorithms have been developed to deal with this problem and realign all the PET images prior to averaging and subtraction (Woods *et al.,* 1998). This also allows serial studies from the same subject, but acquired on different days, to be compared.

Because of the limited anatomical information available in the PET blood flow images, it is very useful to superimpose the subtracted blood flow maps onto high-resolution MRI studies of the same subject, thus aiding in relating the functional changes seen in PET with a particular anatomical location. Robust methods to align PET and MRI data sets have been developed with a registration error of less than 2 mm (Woods *et al.,* 1993).

There is one major drawback to the use of blood flow tracers for brain mapping with PET. Because of the short imaging time, combined with dose considerations and count-rate limitations of the PET scanners, the number of events which can be recorded by the PET scanner during a blood flow study is quite limited. Thus, the signal to noise in

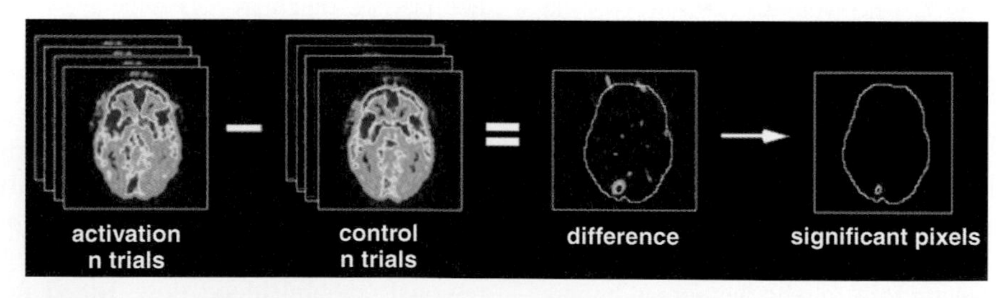

Figure 21 Illustration of the methodology for PET activation studies using blood flow tracers. A series of scans is acquired in activated and control states and is subtracted to produce a difference image. A statistical test is applied to the data to determine which changes in the difference image are statistically significant. This example shows the robust response to a hemifield stimulation of the visual system with a reversing checkerboard pattern using [^{15}O]water as the tracer. The activated area in the visual cortex can clearly be seen, even prior to subtraction.

any one individual scan can be poor. There are also natural biological variations in blood flow from one scan to the next which can be a confounding factor when only one pair of blood flow images is acquired. For this reason, most centers perform between 6 and 18 blood flow scans, with at least 2 or 3 in each condition, averaging them to improve signal to noise (signal to noise is proportional to the square root of the number of counts detected) and reduce the effect of blood flow variations unrelated to the task. This is known as intrasubject averaging. For some activation paradigms, which result in more subtle changes in blood flow, this is still not sufficient to register a significant change. In these situations, it is usually necessary to average across several subjects (Fox *et al.,* 1988). Intersubject averaging requires an accurate algorithm for transforming the images from several different subjects into a standard space (Friston, 1994). It has two drawbacks in that it is susceptible to intersubject variations in gyral anatomy and does not account for functional differences in the way tasks may be handled by different subjects. Despite these difficulties, this approach has been very successful and is now widely used.

In recent years there have been several attempts to optimize the scanning protocol to improve signal to noise in the subtracted images. One advance has been the introduction of 3D PET acquisition and dose fractionation (Cherry *et al.,* 1993). In this scheme many trials are performed in each individual, with relatively low doses. This has the advantage of a fivefold increase in sensitivity from the 3D acquisition, count rates which do not lead to excessive dead-time losses, and several measurements in each condition, which facilitates statistical analysis and helps to average out random biological flow variations unrelated to the task. The signal to noise can be improved by more than a factor of 2 using this approach as illustrated in Fig. 22. Temporal optimization of the protocol, for example, the rate of injecting the radiotracer dose (Silbersweig *et al.,* 1994) and optimization of the scan time (Hurtig *et al.,* 1994), can also result in signal-to-noise improvements. The use of paradigms which switch between activated and baseline states in the middle of a study to exploit the kinetics of the flow tracers has also been evaluated (Cherry *et al.,* 1995b; Volkow *et al.,* 1992). Moderate signal-to-noise

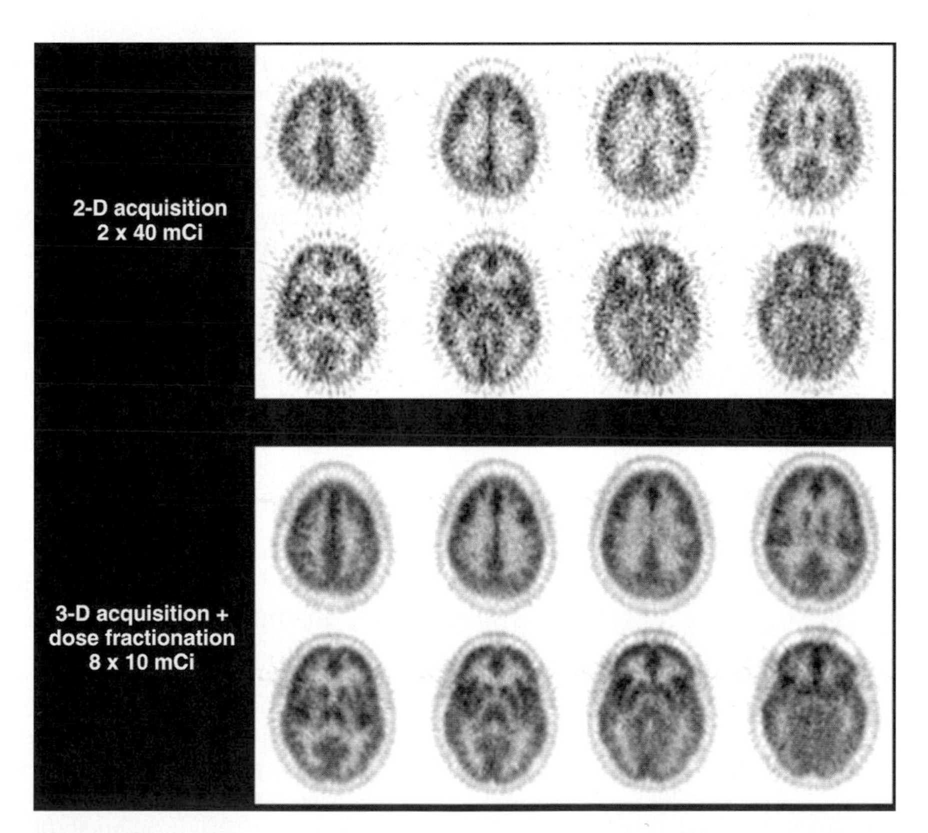

Figure 22 Comparison of 2D activation studies with those obtained using 3D acquisition combined with dose fractionation. A total of 80 mCi of [^{15}O]water was injected for both studies. In 2D, two studies were performed with 40 mCi injected each time. In 3D, the dose was split into 10 mCi fractions and eight studies were performed. The signal to noise is a factor of 2 higher in the 3D study for the same total dose.

improvements can be achieved with these methods if they are carefully implemented.

The introduction of 3D data acquisition, combined with dose fractionation, has widely expanded the possibilities for single-subject activation studies. This is particularly important, as it allows differences between individuals to be examined, it permits longitudinal studies on learning and habituation effects within an individual, and, perhaps most importantly, it allows functional brain mapping studies to be performed in patients. The information that can be derived from such studies is particularly useful for candidates for neurosurgery, in whom the normal functional anatomy is often distorted by pathology. PET activation studies allow the functional anatomy at the surgical site to be mapped, providing important information to the neurosurgeon prior to surgery.

Activation studies are likely to continue to benefit from improved instrumentation (better resolution and sensitivity), optimization of the imaging protocol, sophisticated statistical analysis, and more accurate algorithms for reconstructing, aligning, registering, and transforming PET images. This will permit more subtle activations to be detected, further increasing the utility of this methodology for studying the human brain. These types of studies are, however, not restricted to PET. Similar paradigms are now used with other imaging techniques, particularly functional MRI (*f*MRI) (Belliveau *et al.,* 1991; Kwong *et al.,* 1992; Ogawa *et al.,* 1992), to map the functional organization of the brain in the anticipation, reception, interpretation, and execution of tasks. Indeed, functional MRI has become the technique of choice in many instances, because no ionizing radiation is involved, and therefore very large numbers of trials can be performed in normal subjects, with relatively high spatial and temporal resolution (see Chapter 13). However, PET continues to be used for brain mapping studies at many centers and in some instances has advantages over *f*MRI, for example, studies involving auditory stimulation and for mapping activation of structures deep within the brain.

VIII. Mapping Brain Function in Development and Disease

Activation-related mapping represents just one small area of the application of PET in the human brain. In this section, we briefly review some examples of how PET is used to map development and disease in the living brain and the clinical implications of these findings.

A. Cerebral Development and Neuronal Plasticity

CMRGl studies with PET and [^{18}F]FDG have been used to map the temporal course of "metabolic maturation" of the human brain (Chugani *et al.,* 1987). These studies demonstrated that at birth the highest metabolic activity is in the phylogenetically older structures of the brain, such as sensorimotor cortices, thalamus, brain stem, and cerebellar vermis. As the infant matures in visuospatial and visuosensorimotor integrative functions in the second and third months of life, and primitive reflexes become suppressed, increases in CMRGl occur in parietal, temporal, and primary visual cortex; basal ganglia; and cerebellar hemispheres. The frontal cortex was observed to be the last area to metabolically mature and proceeded from lateral to medial and inferior to superior. These increases in CMRGl of the frontal cortex occurred between 6 and 10 months and were associated with rapid cognitive development. By 1 year, the distribution and absolute values of CMRGl equaled those of the adult. In general, a rise in CMRGl marked the functional entry of a neuroanatomical structure into the expanding behavioral repertoire of the maturing child. The maturation of structures followed an orderly sequence in which ontogeny generally recapitulated phylogeny.

As Fig. 23 shows, cortical CMRGl at birth was about 30% lower than adult values, reached adult values by 1 year, and by 2 to 3 years, exceeded adult values by about a factor of 2. CMRGl remained at these excessive values until about the age of 10 years, when it began to decline, reaching adult values in the latter part of the second decade of life. The high CMRGl observed from ages 2 to 10 corresponds to the period of exuberant connectivity in humans and is probably required to meet the energy demands of the neuronal processes and synapses that are also in excess by about a factor of 2 compared to adults (Huttenlocher, 1979). This period of excessive synapses and nerve terminals, which has been observed in many other species, represents a period of high neuronal plasticity. The ascending portion of the CMRGl development curve corresponds to a time of rapid proliferation of synaptic connections, whereas the descending portion corresponds to a period of selective elimination or "pruning" of excessive connectivity. The selection process is believed to be determined by all forms of learning and stimulus to the child.

B. Epilepsy and Compensatory Reorganization

PET has been used for many years to identify and study epileptogenic foci in human subjects (Cummings *et al.,* 1995). In particular, [^{18}F]FDG studies are used to identify epileptogenic cytoarchitectural disturbances in infants and children who typically have normal MRI and CT scans (Chugani *et al.,* 1988). In addition, PET is used to uniquely assess the functional integrity of the brain outside the epileptogenic region to establish the potential for functional recovery in those children with refractory seizures who are candidates for surgery.

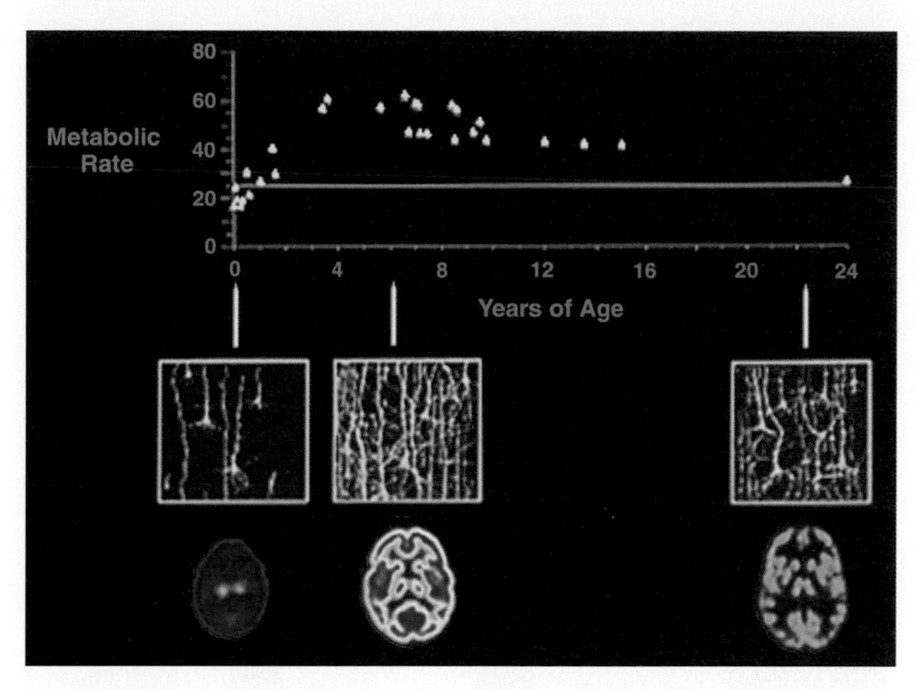

Figure 23 Top: Changes in the metabolic rate for glucose (CMRGl) during development, measured in µmol/min/100 g as measured with PET and [^{18}F]FDG. **Middle:** The density of neuronal processes increases from birth, reaching excessive levels at 6 years and reducing in the adult. **Bottom:** PET studies showing CMRGl at ages of 5 days, 6 years, and adulthood. Adapted, with permission, from Phelps (1991).

PET is now used with [^{18}F]FDG to map the regions for surgical resection (focal cortical resections and hemidecortifications) in refractory seizures. These children have shown remarkable recovery with little functional deficit, in major part due to their unique neuronal plasticity that provides their compensatory recovery. The recovery of these infants and children is both immediate and long term, postsurgery. The activation studies discussed in Section VII are also being used to map the cerebral organization of compensatory recovery, not only in children, but also in adults. These include patients recovering from acute strokes and surgery.

C. Dementia and Movement Disorders

In contrast to end-stage descriptions of disease by autopsy studies, PET studies of CMRGl with [^{18}F]FDG have been used to map the progression of many diseases in living patients throughout the course of their illness. PET has been used to show the progression of Huntington's disease (HD) beginning in the caudate and progressively spreading to the putamen, globus pallidus, and cortical regions, as well as to identify the asymptomatic HD gene carriers (Kuhl *et al.,* 1982; Mazziotta *et al.,* 1987). In Alzheimer's disease, bilateral and often asymmetric CMRGl deficits appear in the superior parietal cortices in early "possible Alzheimer's." With progression of disease, the CMRGl deficit progressively extends throughout the parietal cortex and into the temporal cortex and inferior frontal cortex and, toward the end stage of disease, involves extensive distributions of the association cortex with "relative sparing" of primary cortices and subcortical structures. The extensive literature on this subject has recently been reviewed (Silverman *et al.,* 1999).

L-6-[^{18}F]FluoroDOPA studies of dopamine synthesis have shown dopamine deficiencies of Parkinson's disease in the putamen. This has also been shown with PET in symptomatic patients exposed to MPTP as well as the demonstration of subclinical deficits in asymptomatic MPTP-exposed patients (Calne *et al.,* 1985; Vingerhoets *et al.,* 1994). Figure 24 illustrates the use of PET with two different assays to delineate two different disease processes in the same patient. Images of CMRGl obtained with [^{18}F]FDG are used to demonstrate the characteristic biparietal deficit of Alzheimer's disease, and L-6-[^{18}F]fluoroDOPA images demonstrate the dopamine deficiency of Parkinson's.

D. Brain Tumors

It has been known for many decades that neoplastic degeneration of cells is associated with progressive

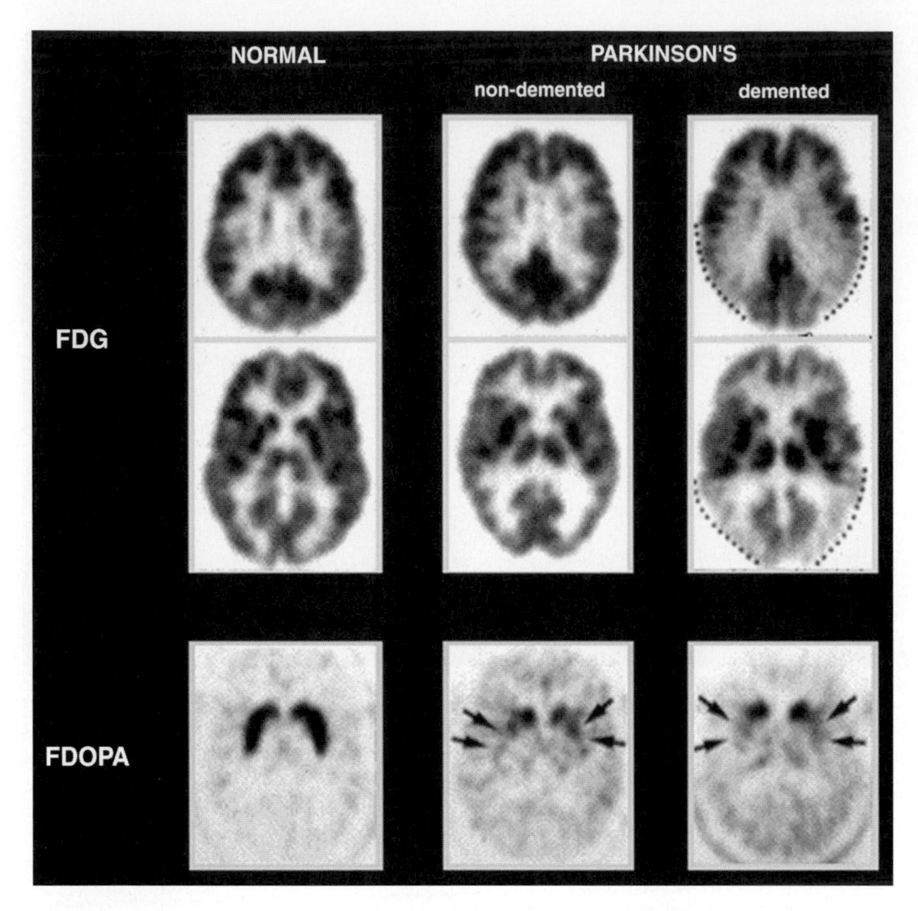

Figure 24 PET studies of CMRGl and dopamine synthesis in Parkinson's disease with and without dementia. Top and middle rows are [^{18}F]FDG CMRGl images just superior to and at the level of the basal ganglia. Bottom row shows images of dopamine synthesis with L-6-[^{18}F]fluoroDOPA. Demented Parkinson's patient has the bilateral metabolic deficit in the parietal and temporal cortices (dotted lines) characteristic of Alzheimer's disease. In addition, the patient has bilateral putaminal dopamine deficits (arrows) of Parkinson's disease. Adapted, with permission, from Phelps (1991).

increases in both aerobic and anaerobic glycolysis. This knowledge has been used to provide a means to identify and grade cerebral tumors with PET *in vivo* (Di Chiro, 1985; Rhodes *et al.,* 1983). In addition, CMRGl studies are used to differentiate recurrence from radiation necrosis and edema, a decision that is difficult, if not impossible, with CT and MRI (Fig. 25). This concept, coupled with the development of whole-body PET scanning (Fig. 1), has been extended to tumors throughout the body (Silverman *et al.,* 1998). The motivation for this has been the critical need for a technique that can effectively screen the whole body for primary and metastatic cancer. Whole-body PET scanning with [^{18}F]FDG has become an important clinical tool in the staging and management of cancer patients.

IX. High-Resolution PET Studies in Animal Models

There is also considerable interest in the use of PET to study laboratory animal models of human disease (Cherry and Gambhir, 2001; Myers, 2001). There is a significant literature on brain imaging in nonhuman primates, in which the brain size is sufficiently large for imaging on clinical PET scanners. Nonhuman primates are often used in the testing of new radiolabeled tracers, to estimate dosimetry for studies in man, and for more invasive studies (for example, imaging combined with microdialysis) that could not be done in man. Several animal PET scanners designed specifically for nonhuman primate brain imaging have been

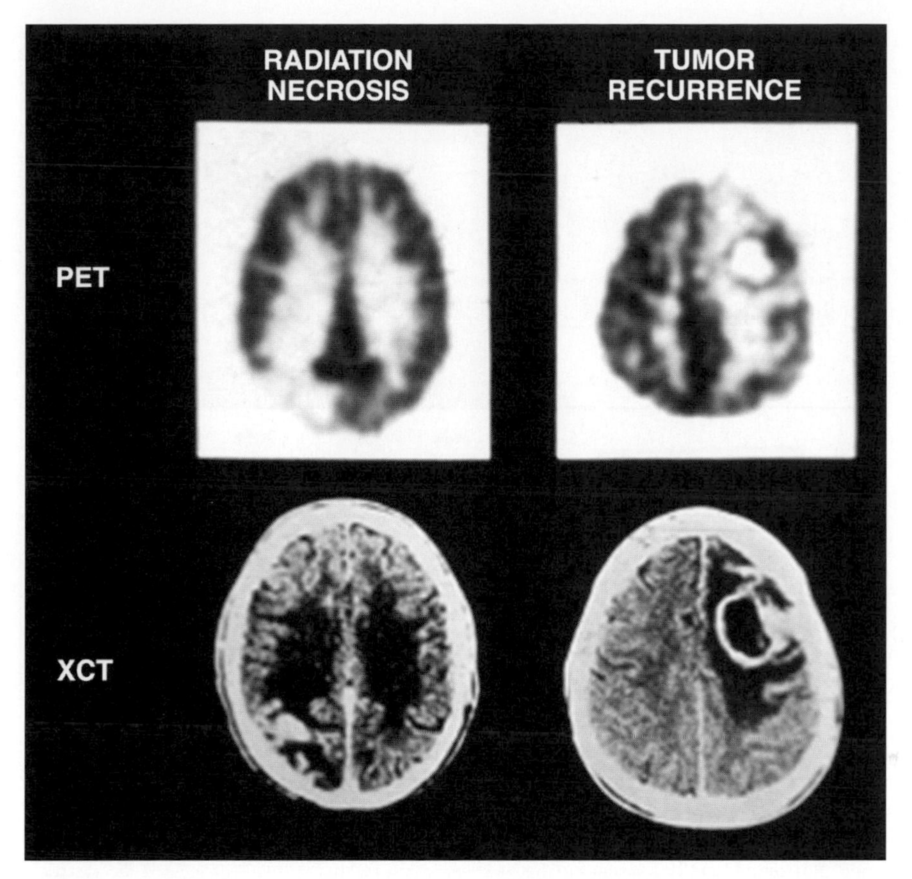

Figure 25 PET and XCT scans of cancer patients following radiotherapy. Abnormalities are apparent on both XCT scans. PET with [^{18}F]FDG is able to differentiate metabolically active recurrent tumor from metabolically inactive radiation necrosis.

developed to provide spatial resolution (typically 2–4 mm) superior to that obtained in human scanners (Cutler *et al.,* 1992; Tai *et al.,* 2001; Watanabe *et al.,* 1992, 1997). The systems developed by Hamamatsu also have the ability to tilt through large angles, so they can image conscious nonhuman primates that have been trained to sit still in a chair and perform certain tasks (Tsukada *et al.,* 1998). These specialized PET scanners have also been used to study development (Moore *et al.,* 2000a), the effects of drugs of abuse (Melega *et al.,* 2000; Tsukada *et al.,* 2000a), and the effects of anesthetics (Tsukada *et al.,* 1999, 2000b) in the nonhuman primate brain.

PET systems have also been developed specifically for imaging smaller laboratory animals such as the mouse and the rat (Bloomfield *et al.,* 1995; Cherry *et al.,* 1997; Jeavons *et al.,* 1999; Lecomte *et al.,* 1996; Tai *et al.,* 2001; Weber *et al.,* 1997; Ziegler *et al.,* 2001). These systems aim for much higher spatial resolution (typically 1–2 mm) in order to be able to visualize major brain structures. These systems have been used extensively in the study of the dopaminergic system in the rat, following a range of surgical and pharmacological interventions (e.g., Fricker *et al.,* 1997; Hume *et al.,* 1996; Unterwald *et al.,* 1997). There have also been studies of brain activation following whisker stimulation in conscious rats using [^{18}F]FDG (Kornblum *et al.,* 2000) and quantitative studies of CMRGl in conscious rats in a traumatic brain injury model (Moore *et al.,* 2000b). Further improvements in resolution and sensitivity of these small animal PET scanners are expected, and using ^{18}F- or ^{11}C-labeled tracers, for which positron range is minimized, submillimeter resolution *in vivo* imaging is almost certainly possible. Figure 26 shows photographs of two specialized animal PET scanners and examples of the images they are capable of generating.

These new high-resolution PET systems are a useful tool in preclinical research designed to evaluate new therapeutic approaches and to monitor therapeutic effects. The ability to follow individual animals longitudinally opens up many new types of studies that were not previously possible. Recent research aimed at merging the methods of molecular biology with imaging are enabling gene expression to be measured by PET (Gambhir *et al.,* 2000), leading to opportunities for mapping the expression of genes of interest in the brain of

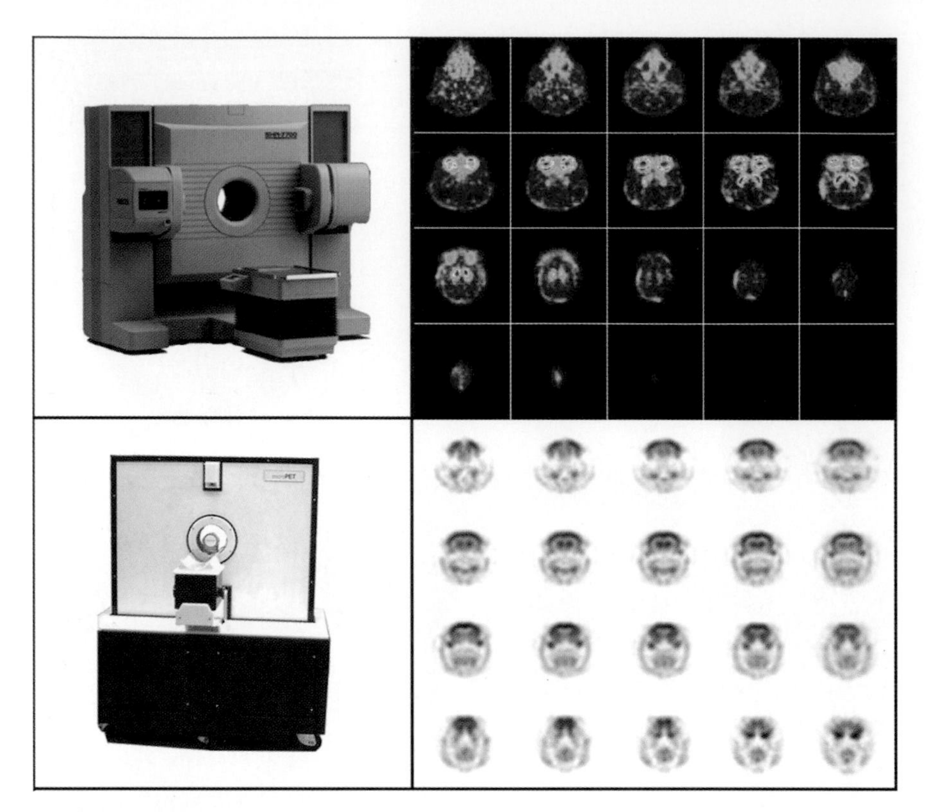

Figure 26 Top: Hamamatsu SHR-7700 animal PET scanner (Watanabe *et al.,* 1997) and transaxial images of rhesus monkey brain obtained with this system following injection of [^{11}C]raclopride. **Bottom:** MicroPET scanner (Cherry *et al.,* 1997) with [^{18}F]FDG images of the rat brain (coronal images).

laboratory animals. The advent of high-resolution small animal PET systems enables PET to be used as an experimental platform across species, from mouse to man. This will be powerful in the study of animal models of human disease and in the development of novel diagnostic and therapeutic approaches, which can then be translated to man.

X. Summary

PET is a powerful noninvasive imaging tool which provides a unique window on the biological processes taking place in the living human brain. It embodies the modern concept of biological imaging, bringing together the highly specific assays of the biological sciences with sophisticated imaging technology. Advances in PET instrumentation and technology continue at a remarkable pace and have brought about an order of magnitude improvement in the resolving power and sensitivity of PET scanners over the past 10 years, resulting in a significantly clearer picture of the brain. The use of PET now extends to studies in small laboratory animals such as the rat and mouse. The scope of biological assays amenable to study with PET continues to grow at a rapid pace, opening up many new and exciting opportunities to image, measure, understand, and communicate information and knowledge about the biological basis of normal function and disease processes of the human brain.

References

Badawi, R. D., Lodge, M. A., and Marsden, P. K. (1998). Algorithms for calculating detector efficiency normalization coefficients for true coincidences in 3D PET. *Phys. Med. Biol.* **43,** 189–205.

Barrio, J. R. (1986). Biochemical principles in radiopharmaceutical design and utilization. *In* "Positron Emission Tomography and Autoradiography" (M. E. Phelps, J. C. Mazziotta, and H. R. Schelbert, eds.), pp. 451–492. Raven Press, New York.

Belliveau, J. W., Kennedy, D. N., McKinstry, R. C., *et al.* (1991). Functional mapping of the human visual cortex by magnetic resonance imaging. *Science* **254,** 716–719.

Bendriem, B., and Townsend, D. W. (1998). "The Theory and Practice of 3D PET." Kluwer Academic, Dordrecht.

Bergström, M., Eriksson, L., Bohm, C., *et al.* (1983). Correction for scattered radiation in a ring detector positron camera by integral transformation of the projections. *J. Comput. Assisted Tomogr.* **7,** 42–50.

Bloomfield, P. M., Rajeswaran, S., Spinks, T. J., *et al.* (1995). The design and physical characteristics of a small animal positron emission tomograph. *Phys. Med. Biol.* **40,** 1105–1126.

Brix, G., Zaers, J., Adam, L.-E., *et al.* (1997). Performance evaluation of a whole-body PET scanner using the NEMA protocol. *J. Nucl. Med.* **38,** 1614–1623.

Brooks, R. A., and Di Chiro, G. (1976). Principles of computer assisted tomography (CAT) in radiographic and radioisotopic imaging. *Phys. Med. Biol.* **21,** 689–732.

Calne, D. B., Langston, J. W., Martin, W. R., *et al.* (1985). Positron emission tomography after MPTP: Observations relating to the cause of Parkinson's disease. *Nature* **317,** 246–248.

Camici, P. G. (2000). Positron emission tomography and myocardial imaging. *Heart* **83,** 475–480.

Casey, M. E., and Nutt, R. (1986). A multicrystal two dimensional BGO detector system for positron emission tomography. *IEEE Trans. Nucl. Sci.* **33,** 460–463.

Cherry, S. R., Dahlbom, M., and Hoffman, E. J. (1991). 3D PET using a conventional multislice tomograph without septa. *J. Comput. Assisted Tomogr.* **15,** 655–668.

Cherry, S. R., and Gambhir, S. S. (2001). Use of positron emission tomography in animal research. *ILAR J.* **42,** 219–232.

Cherry, S. R., and Phelps, M. E. (1995). Positron emission tomography: Methods and instrumentation. *In* "Diagnostic Nuclear Medicine" (M. P. Sandler, R. E. Coleman, F. J. T. Wackers, *et al.*, eds.), pp. 139–159. Williams & Wilkins, Baltimore.

Cherry, S. R., Shao, Y., Silverman, R. W., *et al.* (1997). MicroPET: A high resolution PET scanner for imaging small animals. *IEEE Trans. Nucl. Sci.* **44,** 1161–1166.

Cherry, S. R., Tornai, H. P., Levin, C. S., *et al.* (1995a). A comparison of PET detector modules employing rectangular and round photomultiplier tubes. *IEEE Trans. Nucl. Sci.* **42,** 1064–1068.

Cherry, S. R., Woods, R. P., Doshi, N. K., *et al.* (1995b). Improved signal-to-noise in PET activation studies using switched paradigms. *J. Nucl. Med.* **36,** 307–314.

Cherry, S. R., Woods, R. P., Hoffman, E. J., *et al.* (1993). Improved detection of focal cerebral blood flow changes using three-dimensional positron emission tomography. *J. Cereb. Blood Flow Metab.* **13,** 630–638.

Chugani, H. T., Phelps, M. E., and Mazziotta, J. C. (1987). Positron emission tomography study of human brain functional development. *Ann. Neurol.* **22,** 487–497.

Chugani, H. T., Shewmon, D. A., Peacock, W. J., *et al.* (1988). Surgical treatment of intractable neonatal onset seizures: The role of positron emission tomography. *Neurology* **38,** 1178–1188.

Cummings, T. J., Chugani, D. C., and Chugani, H. T. (1995). Positron emission tomography in pediatric epilepsy. *Neurosurg. Clin.* **6,** 465–472.

Cutler, P. D., Cherry, S. R., Hoffman, E. J., *et al.* (1992). Design features and performance of a PET system for animal research. *J. Nucl. Med.* **33,** 595–604.

Defrise, M., Townsend, D. W., and Geissbuhler, A. (1990). Implementation of three-dimensional image reconstruction for multi-ring tomographs. *Phys. Med. Biol.* **35,** 1361–1372.

DeGrado, T. R., Turkington, T. G., Williams, J. J., *et al.* (1994). Performance characteristics of a whole-body PET scanner. *J. Nucl. Med.* **35,** 1398–1406.

Derenzo, S. E. (1979). Precision measurement of annihilation point spread distributions for medically important positron emitters. *In* "5th International Conference on Positron Annihilation," Sendai, Japan, pp. 819–823.

Derenzo, S. E., Budinger, T. F., Huesman, R. H., *et al.* (1982). Dynamic positron emission tomography in man using small bismuth germanate crystals. *In* "Positron Annihilation" (G. Coleman, S. Sharma, and L. Diana, eds.), pp. 935–941. North Holland, Amsterdam.

Di Chiro, G. (1985). Diagnostic and prognostic value of positron emission tomography using ^{18}F-fluorodeoxyglucose in brain tumors. *In* "Positron Emission Tomography" (M. Reivich and A. Alavi, eds.), pp. 291–309. A. R. Liss, New York.

Eriksson, L., Wienhard, K., and Dahlbom, M. (1994). A simple data loss model for positron camera systems. *IEEE Trans. Nucl. Sci.* **41,** 1566–1570.

Farde, L., Hakan, H., Ehrin, E., *et al.* (1986). Quantitative analysis of D2 dopamine receptor binding in the living human brain by PET. *Science* **231,** 258–261.

Fowler, J. S., MacGregor, R. R., Wolf, A. P., *et al.* (1987). Mapping human brain monoamine oxidase A and B with ^{11}C-labeled suicide inactivators and PET. *Science* **235,** 481–485.

Fowler, J. S., and Wolf, A. P. (1986). Positron emitter-labeled compounds: Priorities and problems. *In* "Positron Emission Tomography and Autoradiography" (M. Phelps, J. Mazziotta, and H. Schelbert, eds.), pp. 391–450. Raven Press, New York.

Fox, P. T., Mintun, M. A., Raichle, M., *et al.* (1984). A non-invasive approach to quantitative functional brain mapping with $H_2{}^{15}O$ and positron emission tomography. *J. Cereb. Blood Flow Metab.* **2,** 89–98.

Fox, P. T., Mintun, M. A., Reiman, E. M., *et al.* (1988). Enhanced detection of focal brain responses using intersubject averaging and change-distribution analysis of subtracted PET images. *J. Cereb. Blood Flow Metab.* **8,** 642–653.

Frackowiak, R. S., Lenzi, G. L., Jones, T., *et al.* (1980). Quantitative measurement of regional cerebral blood flow and oxygen metabolism in man using ^{15}O and positron emission tomography: Theory, procedure, and normal values. *J. Comput. Assisted Tomogr.* **4,** 727–736.

Freifelder, R., Karp, J. S., Geagan, M., *et al.* (1994). Design and performance of the HEAD PENN-PET scanner. *IEEE Trans. Nucl. Sci.* **41,** 1436–1440.

Fricker, R. A., Torres, E. M., Hume, S. P., *et al.* (1997). The effects of donor stage on the survival and function of embryonic striatal grafts in the adult rat brain. II. Correlation between positron emission tomography and reaching behaviour. *Neuroscience* **79,** 711–721.

Friston, K. J. (1994). Statistical parametric mapping. *In* "Functional Neuroimaging" (R. Thatcher, M. Hallett, T. Zeffiro, *et al.*, eds.), pp. 79–93. Academic Press, San Diego.

Friston, K. J., Frith, C. D., Liddle, P. F., *et al.* (1991). Comparing functional (PET) images: The assessment of significant change. *J. Cereb. Blood Flow Metab.* **11,** 690–699.

Gambhir, S. S., Herschman, H. R., Cherry, S. R., *et al.* (2000). Imaging transgene expression with radionuclide imaging technologies. *Neoplasia* **2,** 118–136.

Germano, G., and Hoffman, E. J. (1990). A study of data loss and mispositioning due to pileup in 2-D detectors in PET. *IEEE Trans. Nucl. Sci.* **37,** 671–675.

Gjedde, A., Hansen, S. B., Knudsen, G. M., *et al.* (2001). "Physiological Imaging of the Brain with PET." Academic Press, San Diego.

Greenberg, J. H., Reivich, M., Alavi, A., *et al.* (1981). Metabolic mapping of functional activity in human subjects with the [^{18}F]fluorodeoxyglucose technique. *Science* **211,** 678–680.

Herscovitch, P., Markham, J., and Raichle, M. E. (1983). Brain blood flow measured with intravenous $H_2{}^{15}O$. I. Theory and error analysis. *J. Nucl. Med.* **24,** 782–789.

Hoffman, E. J., Guerrero, T. M., Germano, G., *et al.* (1989). PET system calibration and corrections for quantitative and spatially accurate images. *IEEE Trans. Nucl. Sci.* **36,** 1108–1112.

Hoffman, E. J., Huang, S.-C., and Phelps, M. E. (1979). Quantitation in positron emission computed tomography. 1. Effect of object size. *J. Comput. Assisted Tomogr.* **3,** 299–308.

Hoffman, E. J., Huang, S. C., Phelps, M. E., *et al.* (1981). Quantitation in positron emission computed tomography. 4. Effect of accidental coincidences. *J. Comput. Assisted Tomogr.* **5,** 391–400.

Hoffman, E. J., and Phelps, M. E. (1986). Positron emission tomography: principles and quantitation. *In* "Positron Emission Tomography and Autoradiography" (M. Phelps, J. Mazziotta, and H. Schelbert, eds.), pp. 237–286. Raven Press, New York.

Hoh, C. K., Hawkins, R. A., Glaspy, J. A., *et al.* (1993). Cancer detection with whole-body PET using 2-[^{18}F]fluoro-2-deoxy-D-glucose. *J. Comput. Assisted Tomogr.* **17,** 582–589.

Huang, S.-C., and Phelps, M. E. (1986). Principles of tracer kinetic modeling in positron emission tomography and autoradiography. *In* "Positron Emission Tomography and Autoradiography" (M. Phelps, J. Mazziotta, and H. Schelbert, eds.), pp. 287–346. Raven Press, New York.

Huang, S. C., Barrio, J. R., and Phelps, M. E. (1986). Neuroreceptor assay with positron emission tomography: Equilibrium versus dynamic approaches. *J. Cereb. Blood Flow Metab.* **6,** 515–521.

Huang, S. C., Carson, R. E., Hoffman, E. J., *et al.* (1983). Quantitative measurement of local cerebral blood flow in humans by positron computed tomography and O-15 water. *J. Cereb. Blood Flow Metab.* **3,** 141–153.

Huang, S. C., Yu, D. C., Barrio, J. R., *et al.* (1991). Kinetics and modeling of L-6-[^{18}F]fluoro-dopa in human positron emission tomographic studies. *J. Cereb. Blood Flow Metab.* **11,** 898–913.

Hume, S. P., Lammertsma, A. A., Myers, R., *et al.* (1996). The potential of high-resolution positron emission tomography to monitor striatal dopaminergic function in rat models of disease. *J. Neurosci. Methods* **67,** 103–112.

Hurtig, R. R., Hichwa, R. D., O'Leary, D. S., *et al.* (1994). Effects of timing and duration of cognitive activation in [^{15}O]water PET studies. *J. Cereb. Blood Flow Metab.* **14,** 423–430.

Huttenlocher, P. R. (1979). Synaptic density in human frontal cortex developmental changes and effects of aging. *Brain Res.* **163,** 195–205.

Jeavons, A. P., Chandler, R. A., and Dettmar, C. A. R. (1999). A 3D HIDAC-PET camera with sub-millimetre resolution for imaging small animals. *IEEE Trans. Nucl. Sci.* **46,** 468–473.

Jones, T., Chesler, D. A., and Ter-Pogossian, M. M. (1976). The continuous inhalation of oxygen-15 for assessing regional oxygen extraction in the brain of man. *Br. J. Radiol.* **49,** 339–343.

Kak, A. C., and Slaney, M. (1988). "Principles of Computerized Tomography." IEEE Press, New York.

Karp, J. S., Muehllehner, G., Beerbohm, D., *et al.* (1986). Event localization in a continuous detector using digital processing. *IEEE Trans. Nucl. Sci.* **33,** 550–555.

Kety, S. S., and Schmidt, C. F. (1948). The nitrous oxide method for the quantitative determination of cerebral blood flow in man: Theory, procedure and normal values. *J. Clin. Invest.* **27,** 476–483.

Knoll, G. F. (2000). "Radiation Detection and Measurement." Wiley, New York.

Kornblum, H. I., Araujo, D. M., Annala, A. J., *et al.* (2000). In vivo imaging of neuronal activation and plasticity in the rat brain with microPET, a novel high-resolution positron emission tomograph. *Nat. Biotechnol.* **18,** 655–660.

Kuhl, D. E., Phelps, M. E., Markham, C. H., *et al.* (1982). Cerebral metabolism and atrophy in Huntington's disease determined by FDG and computed tomography scan. *Ann. Neurol.* **12,** 425–434.

Kwong, K. K., Belliveau, J. W., Chesler, D. A., *et al.* (1992). Dynamic magnetic resonance imaging of human brain activity during primary sensory stimulation. *Proc. Natl. Acad. Sci. USA* **89,** 5675–5679.

Lammertsma, A. A., and Hume, S. P. (1996). Simplified reference tissue model for PET receptor studies. *NeuroImage* **4,** 153–158.

Leahy, R. M., and Qi, J. (2000). Statistical approaches in quantitative positron emission tomography. *Stat. Comput.* **10,** 147–165.

Lecomte, R., Cadorette, J., Rodrigue, S., *et al.* (1996). Initial results from the Sherbrooke avalanche photodiode positron tomograph. *IEEE Trans. Nucl. Sci.* **43,** 1952–1957.

Logan, J., Fowler, J. S., Volkow, N. D., *et al.* (1990). Graphical analysis of reversible radioligand binding from time-activity measurements applied to [N-^{11}C-methyl]-(–)-cocaine PET studies in human subjects. *J. Cereb. Blood Flow Metab.* **10,** 740–747.

Mankoff, D. A., and Bellon, J. R. (2001). Positron-emission tomographic imaging of cancer: Glucose metabolism and beyond. *Semin. Radiol. Oncol.* **11,** 16–27.

Mazziotta, J., Phelps, M. E., Huang, S. C., *et al.* (1987). Cerebral glucose utilization reductions in clinically asymptomatic subjects at risk for Huntington's disease. *N. Engl. J. Med.* **316,** 357–362.

Mazziotta, J. C., Huang, S.-C., Phelps, M. E., *et al.* (1985). A noninvasive positron computed tomography technique using oxygen-15-labeled water for the evaluation of neurobehavioral task batteries. *J. Cereb. Blood Flow Metab.* **5,** 70–78.

Mazziotta, J. C., Phelps, M. E., Carson, R. E., *et al.* (1982). Tomographic mapping of human cerebral metabolism: Auditory stimulation. *Neurology* **32,** 921–937.

McCarthy, T. J., and Welch, M. J. (1998). The state of positron emitting radionuclide production in 1997. *Semin. Nucl. Med.* **28,** 235–246.

Melega, W. P., Lacan, G., Desalles, A. A., *et al.* (2000). Long-term methamphetamine-induced decreases of [^{11}C]WIN 35,428 binding in striatum are reduced by GDNF: PET studies in the vervet monkey. *Synapse* **35,** 243–249.

Moore, A. H., Hovda, D. A., Cherry, S. R., *et al.* (2000a). Dynamic changes in cerebral glucose metabolism in conscious infant monkeys during the first year of life as measured by positron emission tomography. *Dev. Brain Res.* **120,** 141–150.

Moore, A. H., Osteen, C. L., Chatziioannou, A. F., *et al.* (2000b). Quantitative assessment of longitudinal metabolic changes in vivo following traumatic brain injury in the adult rat using FDG-microPET. *J. Cereb. Blood Flow Metab.* **20,** 1492–1501.

Moses, W. W., Virador, P. R. G., Derenzo, S. E., *et al.* (1997). Design of a high-resolution, high-sensitivity PET camera for human brains and small animals. *IEEE Trans. Nucl. Sci.* **44,** 1487–1491.

Müller-Gärtner, H. W., Links, J. M., Prince, J. L., *et al.* (1992). Measurement of radiotracer concentration in brain gray matter using positron emission tomography: MRI-based correction for partial volume effects. *J. Cereb. Blood Flow Metab.* **12,** 571–583.

Myers, R. (2001). The biological application of small animal PET imaging. *Nucl. Med. Biol.* **28,** 585–593.

Ogawa, S., Tank, D. W., Menon, R., *et al.* (1992). Intrinsic signal changes accompanying sensory stimulation: Function brain mapping with magnetic resonance imaging. *Proc. Natl. Acad. Sci. USA* **89,** 5951–5955.

Ollinger, J. M. (1996). Model-based scatter correction for fully 3D PET. *Phys. Med. Biol.* **41,** 153–76.

Patlak, C. S., and Blasberg, R. G. (1985). Graphical evaluation of blood-to-brain transfer constants from multiple-time uptake data. *J. Cereb. Blood Flow Metab.* **5,** 584–590.

Perlmutter, J. S., Larson, K. B., Raichle, M. E., *et al.* (1986). Strategies for in vivo measurement of receptor binding using positron emission tomography. *J. Cereb. Blood Flow Metab.* **6,** 154–169.

Phelps, M. E. (1991). PET: A biological imaging technique. *Neurochem. Res.* **16,** 929–940.

Phelps, M. E., Hoffman, E. J., Mullani, N. A., *et al.* (1975). Application of annihilation coincidence detection to transaxial reconstruction tomography. *J. Nucl. Med.* **16,** 210–233.

Phelps, M. E., Huang, S. C., Hoffman, E. J., *et al.* (1979). Tomographic measurement of local cerebral glucose metabolic rate in humans with (F-18) 2-fluoro-2-deoxy-D-glucose: Validation of method. *Ann. Neurol.* **6,** 371–388.

Phelps, M. E., Kuhl, D. E., and Mazziotta, J. C. (1981a). Metabolic mapping of the brain's response to visual stimulation: Studies in humans. *Science* **211,** 1445–1448.

Phelps, M. E., Mazziotta, J. C., Kuhl, D. E., *et al.* (1981b). Tomographic mapping of human cerebral metabolism: Visual stimulation and deprivation. *Neurology* **31,** 517–529.

Qi, J., Leahy, R. M., Hsu, C., *et al.* (1998). Fully 3D Bayesian image reconstruction for the ECAT EXACT HR+. *IEEE Trans. Nucl. Sci.* **45,** 1096–1103.

Raichle, M. E., Martin, W. R. W., Herscovitch, P., *et al.* (1983). Brain blood flow measured with intravenous $H_2{}^{15}O$. II. Implementation and validation. *J. Nucl. Med.* **24,** 790–798.

Ranger, N. T., Thompson, C. J., and Evans, A. C. (1989). The application of a masked orbiting transmission source for attenuation correction in PET. *J. Nucl. Med.* **30,** 1056–1068.

Reivich, M., Kuhl, D. E., Wolf, A., *et al.* (1979). The ^{18}F-fluorodeoxyglucose method for the measurement of local cerebral glucose utilization in man. *Circ. Res.* **44,** 127–137.

Rhodes, C. G., Wise, R. J., Gibbs, J. M., *et al.* (1983). In vivo distribution of the oxidative metabolism of glucose in human cerebral gliomas. *Ann. Neurol.* **14,** 614–626.

Rousset, O. G., Ma, Y., and Evans, A. C. (1998). Correction for partial volume effects in PET: Principle and validation. *J. Nucl. Med.* **39,** 904–911.

Satyamurthy, N., Phelps, M. E., and Barrio, J. R. (1999). Electronic generators for the production of positron-emitter labeled radiopharmaceuticals: Where would PET be without them? *Clin. Positron Imaging* **2,** 233–253.

Schelbert, H. R. (1991). Positron emission tomography for the assessment of myocardial viability. *Circulation* **84**(Suppl. I), 122–131.

Schiepers, C., and Hoh, C. K. (1998). Positron emission tomography as a diagnostic tool in oncology. *Eur. Radiol.* **8,** 1481–1494.

Shepp, L. A., and Vardi, Y. (1982). Maximum likelihood reconstruction for emission tomography. *IEEE Trans. Med. Imaging* **1,** 113–122.

Silbersweig, D. A., Stern, E., Schnorr, L., *et al.* (1994). Imaging transient, randomly occurring neuropsychological events in single subjects with positron emission tomography: An event-related count rate correlational analysis. *J. Cereb. Blood Flow Metab.* **14,** 771–782.

Silverman, D. H., Hoh, C. K., Seltzer, M. A., *et al.* (1998). Evaluating tumor biology and oncological disease with positron-emission tomography. *Semin. Radiol. Oncol.* **8,** 183–196.

Silverman, D. H. S., Small, G. W., and Phelps, M. E. (1999). Clinical value of neuroimaging in the diagnosis of dementia: Sensitivity and specificity of regional cerebral metabolic and other parameters for early identification of Alzheimer's disease. *Clin. Positron Imaging* **2,** 119–130.

Smith, R. J., Adam, L. E., and Karp, J. S. (1999). Methods to optimize whole body surveys with the C-PET camera. *In* "1999 Nuclear Science Symposium and Medical Imaging Conference," Seattle, WA, pp. 1197–1201.

Spinks, T. J., Jones, T., Bloomfield, P. M., *et al.* (2000). Physical characteristics of the ECAT EXACT3D positron tomograph. *Phys. Med. Biol.* **45,** 2601–2618.

Stearns, C. W., Cherry, S. R., and Thompson, C. J. (1995). NECR analysis of 3D brain PET scanner designs. *IEEE Trans. Nucl. Sci.* **42,** 1075–1079.

Strother, S. C., Casey, M. E., and Hoffman, E. J. (1990). Measuring PET scanner sensitivity: Relating count rates to image signal-to-noise ratios using noise equivalent counts. *IEEE Trans. Nucl. Sci.* **37,** 783–788.

Surti, S., Karp, J. S., Freifelder, R., *et al.* (2000). Optimizing the performance of a PET detector using discrete GSO crystals on a continuous lightguide. *IEEE Trans. Nucl. Sci.* **47,** 1030–1036.

Tai, Y. C., Chatziioannou, A., Siegel, S., *et al.* (2001). Performance evaluation of the microPET P4: A PET system dedicated to animal imaging. *Phys. Med. Biol.* **46,** 1845–1862.

Tewson, T. J., and Krohn, K. A. (1998). PET radiopharmaceuticals: State-of-the-art and future prospects. *Semin. Nucl. Med.* **28,** 221–234.

Townsend, D. W., Geissbuhler, A., Defrise, M., *et al.* (1991). Fully 3-dimensional reconstruction for a PET camera with retractable septa. *IEEE Trans. Med. Imaging* **10,** 505–512.

Tsukada, H., Harada, N., Nishiyama, S., *et al.* (2000a). Dose–response and duration effects of acute administrations of cocaine and GBR12909 on dopamine synthesis and transporter in the conscious monkey brain: PET studies combined with microdialysis. *Brain Res.* **860,** 141–148.

Tsukada, H., Harada, N., Nishiyama, S., *et al.* (2000b). Ketamine decreased striatal [^{11}C]raclopride binding with no alterations in static dopamine concentrations in the striatal extracellular fluid in the monkey brain: Multiparametric PET studies combined with microdialysis analysis. *Synapse* **37,** 95–103.

Tsukada, H., Kakiuchi, T., Shizuno, H., *et al.* (1998). Interactions of cholinergic and glutamatergic neuronal systems in the functional activation of cerebral blood flow response: A PET study in unanesthetized monkeys. *Brain Res.* **796,** 82–90.

Tsukada, H., Nishiyama, S., Kakiuchi, T., *et al.* (1999). Isoflurane anesthesia enhances the inhibitory effects of cocaine and GBR12909 on dopamine transporter: PET studies in combination with microdialysis in the monkey brain. *Brain Res.* **849,** 85–96.

Unterwald, E. M., Tsukada, H., Kakiuchi, T., *et al.* (1997). Use of positron emission tomography to measure the effects of nalmefene on D1 and D2 dopamine receptors in rat brain. *Brain Res.* **775,** 183–188.

Vingerhoets, F. J., Snow, B. J., Tetrud, J. W., *et al.* (1994). Positron emission tomographic evidence for progression of human MPTP-induced dopaminergic lesions. *Ann. Neurol.* **36,** 765–770.

Volkow, N. D., Fowler, J. S., Wolf, A. P., *et al.* (1992). Distribution and kinetics of carbon-11-cocaine in the human body measured with PET. *J. Nucl. Med.* **33,** 521–525.

Watanabe, M., Okada, H., Shimizu, K., *et al.* (1997). A high resolution animal PET scanner using compact PS-PMT detectors. *IEEE Trans. Nucl. Sci.* **47,** 1277–1282.

Watanabe, M., Uchida, H., Okada, H., *et al.* (1992). A high resolution PET for animal studies. *IEEE Trans. Med. Imaging* **11,** 577–580.

Watson, C. C., Newport, D., Casey, M. E., *et al.* (1997). Evaluation of simulation-based scatter correction for 3-D PET cardiac imaging. *IEEE Trans. Nucl. Sci.* **44,** 90–97.

Weber, S., Terstegge, A., Herzog, H., *et al.* (1997). The design of an animal PET: Flexible geometry for achieving optimal spatial resolution or high sensitivity. *IEEE Trans. Med. Imaging* **16,** 684–689.

Wienhard, K., Dahlbom, M., Eriksson, L., *et al.* (1994). The ECAT EXACT HR: Performance of a new high resolution positron scanner. *J. Comput. Assisted Tomogr.* **18,** 110–118.

Wienhard, K., Eriksson, L., Grootoonk, S., *et al.* (1992). Performance eval uation of the positron scanner ECAT EXACT. *J. Comput. Assisted Tomogr.* **16,** 804–813.

Wienhard, K., Schmand, M., Casey, M. E., Baker, K., Bao, J., Eriksson, L., Jones, W. F., Knoess, C., Lenox, M., Lercher, M., Lulk, P., Michel, C., Reed, J. H., Richerzhagen, N., Treffort, J., Vollmar, S., Young, J. W., Heiss, W. D., and Nutt, R. (2000). The ECAT HRRT: Performance and first clinical application of the new high resolution research tomograph. *In* 2000 IEEE Nuclear Science Symposium. Conference Record, Lyon, France, Oct. 15–20, 2000. IEEE Press, Piscataway, NJ, U.S.A., pp. 1712–1716, vol. 3.

Wong, D. F., Gjedde, A., and Wagner, H. N. J. (1986). Quantification of neuroreceptors in the living human brain. I. Irreversible binding of ligands. *J. Cereb. Blood Flow Metab.* **6,** 137–146.

Woods, R. P., Grafton, S. T., Holmes, C. J., *et al.* (1998). Automated image registration. I. General methods and intrasubject, intramodality validation. *J. Comput. Assisted Tomogr.* **22,** 139–152.

Woods, R. P., Mazziotta, J. C., and Cherry, S. R. (1993). MRI–PET registration with automated algorithm. *J. Comput. Assisted Tomogr.* **17,** 536–546.

Worsley, K. J., Evans, A. C., Marrett, S., *et al.* (1992). A three-dimensional statistical analysis for CBF activation studies in human brain. *J. Cereb. Blood Flow Metab.* **12,** 900–918.

Ziegler, S. I., Pichler, B. J., Boening, G., *et al.* (2001). A prototype high-resolution animal positron tomograph with avalanche photodiode arrays and LSO crystals. *Eur. J. Nucl. Med.* **28,** 136–143.

19

SPECT Functional Brain Imaging

Michael D. Devous, Sr.

Nuclear Medicine Center and Department of Radiology, The University of Texas Southwestern Medical Center, Dallas, Texas 75390-9061

I. Introduction

Functional brain imaging refers to that set of techniques used to derive images reflecting biochemical, physiologic, or electrical properties of the CNS. The most developed of these techniques for human use are single-photon-emission computed tomography (SPECT), positron emission tomography (PET), and topographic electroencephalography. Of these, SPECT may offer the most widely available tomographic measure of neuronal behavior (there are about 100 PET installations and about 12,000 SPECT installations in the United States). Several recent reviews describe the use of SPECT alone or in combination with PET and/or functional magnetic resonance imaging (*f*MRI) in studies of human cognition, in imaging of neuroreceptor systems, in aiding diagnosis or assessment of progression or treatment response in various psychiatric and neurologic disorders, in neuropharmacologic challenge studies, and in the new field of molecular imaging, including imaging of transgene expression [1–9]. Though PET still provides the highest resolution tomographic images of brain function, modern SPECT images have similar resolution (making any differences relatively inconsequential in clinical application), and ongoing developments in instrumentation are now producing both SPECT and PET devices with 2-mm or better resolution. Also, while the breadth of radiopharmaceuticals available for brain SPECT is not as great as that for PET, the variety of SPECT tracers is expanding rapidly. SPECT perfusion tracers are FDA approved and available for wide distribution, while FDA approval for nationally distributable PET tracers for brain function remains elusive. SPECT is limited by the lack of a direct measure of metabolism. Since cerebral perfusion and metabolism are tightly coupled under most normal and pathologic circumstances, this difference may also be of limited relevance.

This chapter provides an overview of the technical aspects of SPECT functional brain imaging, referring primarily to the most common SPECT brain function measure, regional cerebral blood flow (rCBF). SPECT images of rCBF are influenced by a number of factors separate from pathology, including (1) the quality of the tomographic device, (2) the radiopharmaceutical employed, (3) environmental conditions at the time of radiotracer administration, (4) characteristics of the subject (e.g., age, gender, handedness), (5) the format used for image presentation, and (6) image processing techniques. All but the last aspect (6) are reviewed in this chapter. Image processing per se is covered only modestly since the considerable variety in the details of various methods (e.g., filter choices, methods of reconstruction, attenuation and scatter correction schemes) is beyond the scope of this chapter. The reader is referred to available reviews [4, 10–14]. However, a brief overview of the essential components of image processing necessary to the achievement of high-quality SPECT brain images has been included. In addition, an intercomparison of available techniques for the quantitative measurement of rCBF and a brief description of relevant radiation safety issues are provided.

II. Instrumentation

Tremendous growth in instrumentation over the past 2 decades has resulted in commercially available, very high quality tomographs for SPECT brain imaging and very sophisticated image processing hardware and software. SPECT instruments developed by university-based research paved the way for current devices, but industry has led the computational hardware development process. SPECT brain imaging is now a mature clinical entity. SPECT instruments fall into two categories: non-camera-based and camera-based systems, although the latter dominate both academic and commercial development.

A. Non-Camera-Based Systems

Non-camera-based systems include rotating detector arrays, multidetector scanners, and fixed rings. Rotating detector array devices include the Tomomatic two-, three-, and five-slice machines (Medimatic, Inc.) and the Hitachi 4-head system. The Tomomatic's most characteristic attribute is the capacity for ^{133}Xe SPECT imaging, which requires very high sensitivity and rapid dynamic sampling (i.e., complete tomographic studies every 10 s) [11, 15]. The advantage of ^{133}Xe SPECT is that it yields absolute quantitation of rCBF (ml/min/100 g) without arterial blood sampling. Unfortunately the high sensitivity that this technique requires limits spatial resolution, which for this system is about 16 mm. The Tomomatic, by changing collimators, can also produce moderate-resolution (9–10 mm) rCBF images using ^{123}I- or ^{99m}Tc-labeled tracers. The Hitachi rotating detector array system is also capable of both ^{133}Xe and good-resolution (8–10 mm) static imaging using ^{123}I or ^{99m}Tc tracers [16].

The original fixed-detector research systems were the SPRINT [17], the HEADTOME [18], and the MUMPI [19]. They were designed with fixed detectors or a circular annulus of sodium iodide with an internally rotating collimator. The Shimadzu (HEADTOME) system, available only in Japan, is capable of high-sensitivity ^{133}Xe studies and moderate-resolution (10–12 mm) imaging using ^{123}I or ^{99m}Tc. The most widely available fixed-ring system is the CERASPECT [20], first of the fixed sodium iodide annulus/rotating collimator machines to come to commercial production. It yields 8- to 10-mm resolution images and is the only fixed-ring system still commercially available in the United States.

The original multidetector scanner was developed by Stoddart and colleagues [21] and became commercially available from Strichman, Inc. This is a slice-based tomograph, as are the Hitachi, Shimadzu, and Tomomatic, but it is built with very thick crystals that operate much like pinhole cameras as they traverse through space to obtain tomographic data. Hill and colleagues [22] have demonstrated that this device can image ^{18}F in a single-photon (not PET) mode, as well as ^{99m}Tc and ^{123}I. It cannot perform ^{133}Xe SPECT. In a very recent reincarnation as the NeuroFOCUS system it has demonstrated 3-mm spatial resolution with ^{99m}Tc-labeled rCBF tracers in man, the highest spatial resolution SPECT or PET system that is commercially available (Fig. 1). While resolution is excellent, image quality requires further development.

B. Gamma-Camera-Based Systems

Both single-head and multihead gamma-camera-based systems are vastly more prevalent than dedicated tomographs, primarily because they can do both head and body SPECT. Modern single-head tomographs have overcome many of the limitations of the original systems, such as poor head alignment, magnetic field aberrations, and inadequate uniformity and linearity for tomography. A few systems have also been designed to circumvent shoulders so that minimal radius scanning is possible. Most of these systems provide 7- to 10-mm resolution images with static tracers. Unfortunately, single-head systems suffer from poor sensitivity and prolonged imaging times.

A collaborative team from the University of Texas Southwestern Medical Center at Dallas and the nuclear engineering division of Technicare developed the first three-head gamma-camera-based SPECT system to address the limited sensitivity of single-head systems [23]. This collaboration yielded a system capable of both head and body SPECT at high resolution with static tracers and with adequate sensitivity and rotation speed for dynamic tomography with ^{133}Xe. The first three-head system (PRISM) was installed in

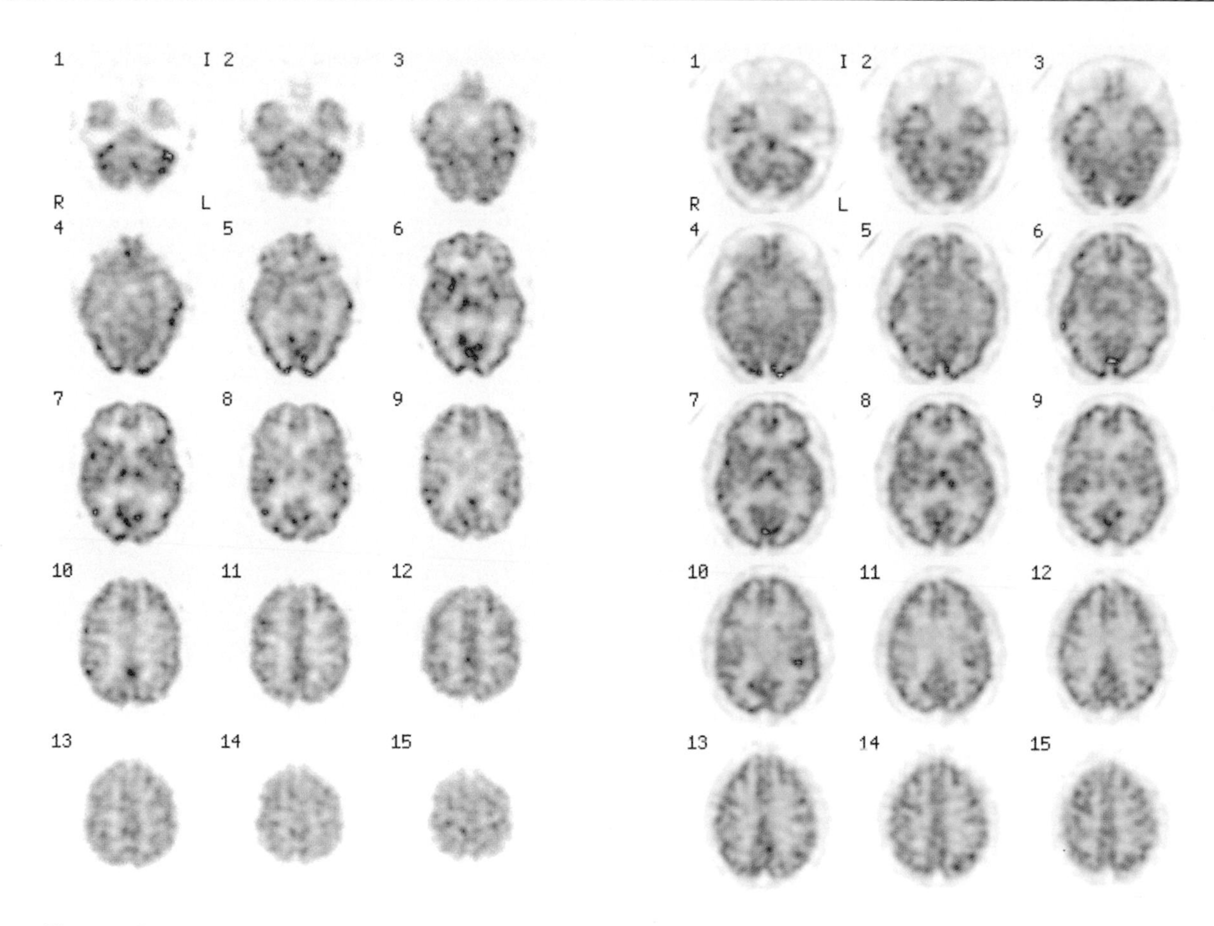

Figure 1 High-resolution rCBF images obtained in two different patients with the NeuroFOCUS high-definition focusing emission tomographic scanner (HDFET) using ^{99m}Tc HMPAO. (Images provided courtesy of NeuroPhysics Corp.)

Dallas in late 1987 under the sponsorship of Ohio Imaging, now a division of Picker/Marconi. Additional three-head SPECT instruments have been produced by Trionix (also as a by-product of the collaboration mentioned above), Toshiba, General Electric, and Siemens. Three-head SPECT systems currently are the most sophisticated instruments for brain SPECT. There are more than 1500 such units installed, indicating wide acceptance of this technology. In addition to good-resolution (6–8 mm) imaging with static tracers, the Picker and the Toshiba systems were designed to permit dynamic, fully quantitative rCBF imaging with ^{133}Xe. The first ^{133}Xe SPECT images in humans from a three-head system (PRISM) were produced in late 1992 by the Dallas group (Fig. 2) [24, 25].

C. Image Processing Essentials

State-of-the-art SPECT systems can be expected to provide high-resolution imaging of statically distributed brain radiopharmaceuticals with patient imaging times of 10–20 min (Fig 3). All of the currently available three-head systems offer excellent spatial resolution: 6-mm resolution in the cortex and about 7 mm at the center of the brain, with appropriate collimators (usually ultrahigh-resolution fan-beam collimators) and ^{99m}Tc-HMPAO (exametazime or Ceretec; Amersham International) or ^{99m}Tc-ECD (bicisate or Neurolite; Dupont).

To achieve such resolution a few simple principles should be followed. First, it is necessary to reconstruct nearly motion-free data. A key instrumentation feature to facilitate collection of motion-free data is the capability of sequential image acquisition. That is, it should be possible to acquire multiple short studies back to back and subsequently discard segments degraded by patient motion (e.g., collecting five 4-min studies to achieve a 20-min total acquisition time). While a 20-min acquisition on a three-head system provides completely adequate data density, studies with less acquisition time may be acceptable for some clinical purposes (Fig. 4).

Next, data must be filtered in three dimensions simultaneously. This can be accomplished by either filtering projection data prior to transverse reconstruction (which is automatically 3D) or filtering reconstructed data if the manufacturer offers true 3D postfiltering. The postreconstruction

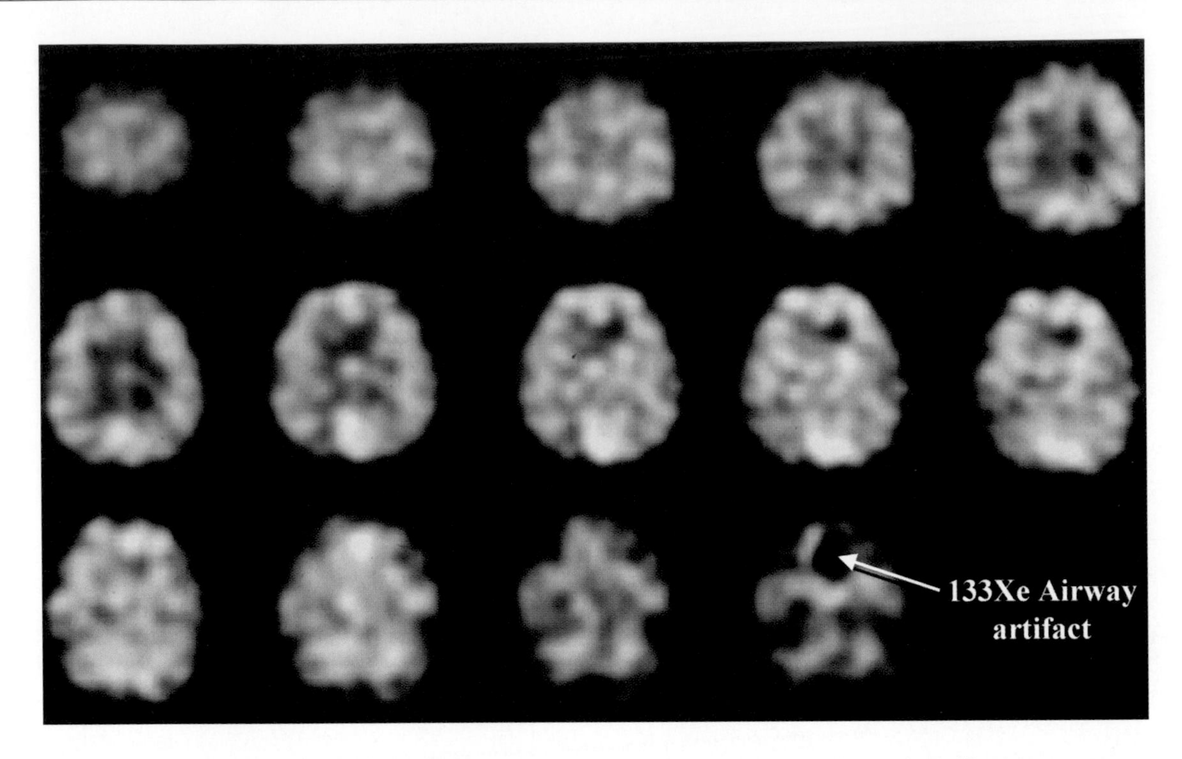

Figure 2 Quantitative (ml/min/100 g) dynamic rCBF images obtained in a normal volunteer using ^{133}Xe and the PRISM 3000S. Images were obtained in 4 min, including 1 min of wash in and 3 min of wash out of the inert gas tracer.

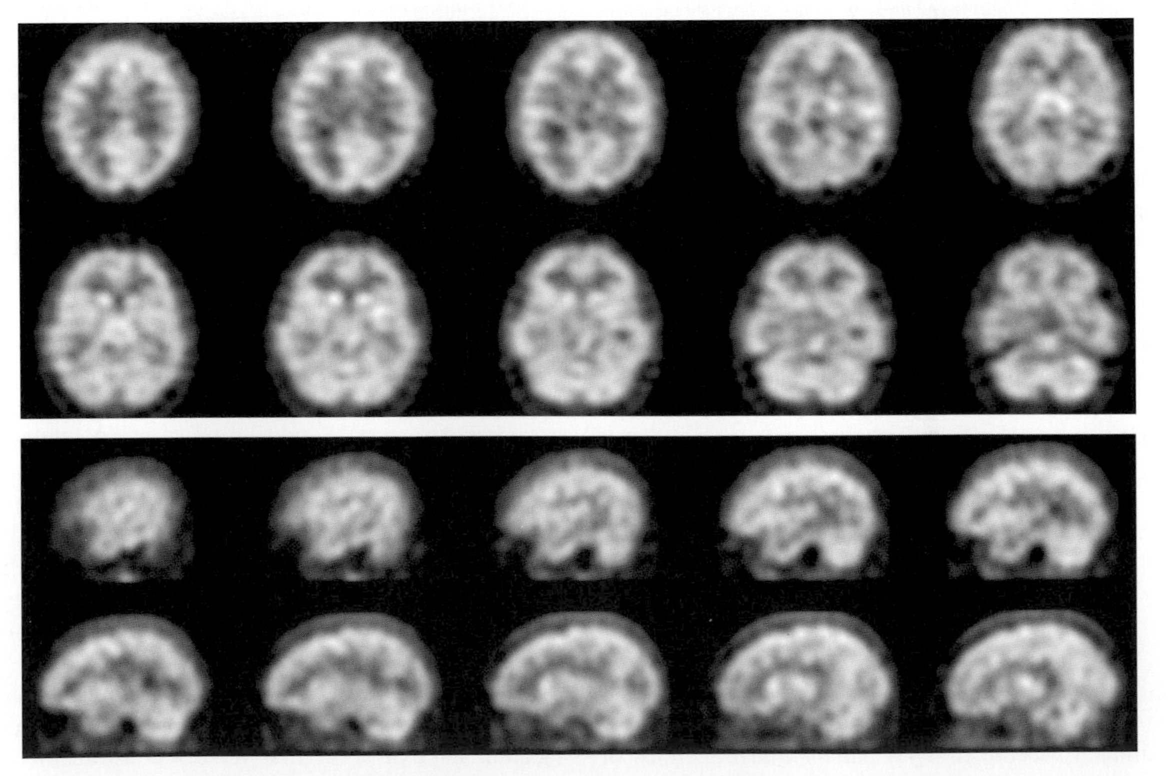

Figure 3 Typical high-resolution rCBF SPECT images obtained using ^{99m}Tc HMPAO and the PRISM 3000S tomograph in a normal volunteer. **Top** shows transverse and **bottom** shows sagittal images.

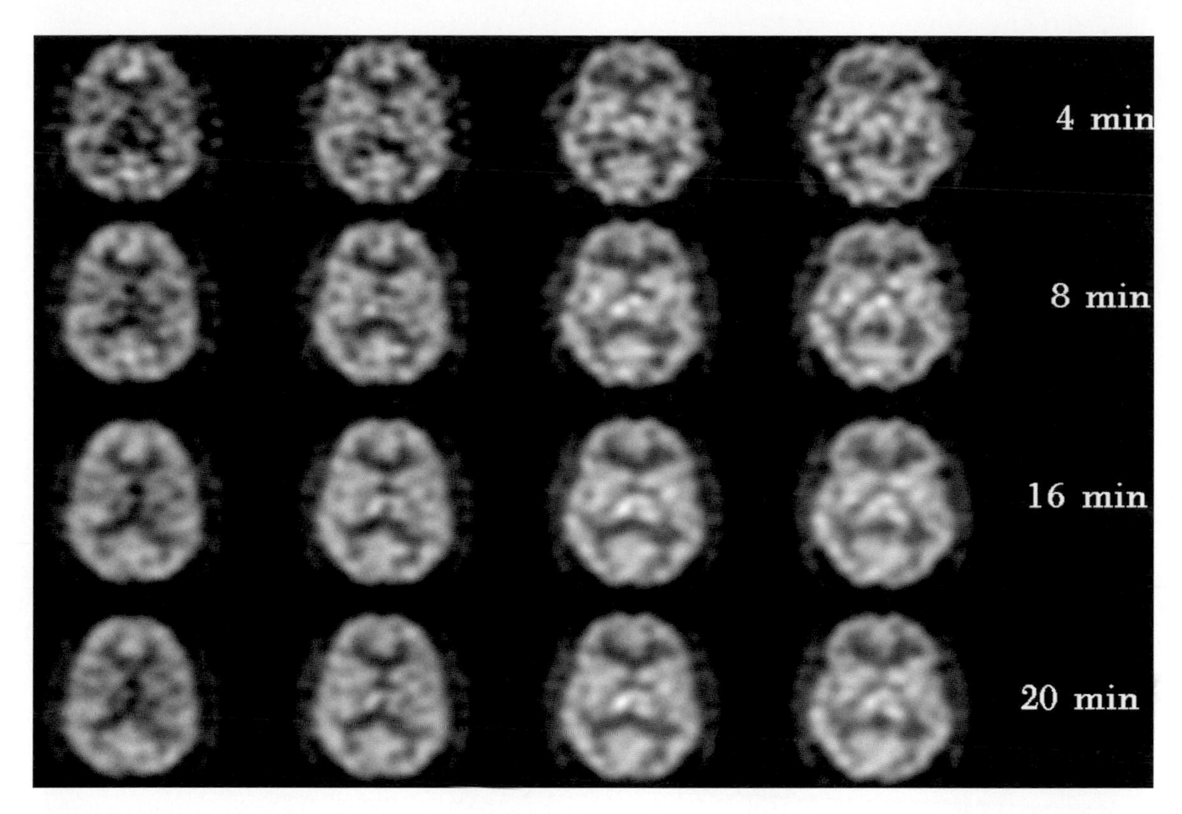

Figure 4 Effect of acquisition time on image quality. Subject was imaged for 20 min in five 4-min sequential blocks using a three-headed gamma camera (Picker 3000S). Either a single projection set was used (t = 4 min) or several sets were summed (time on the right side of each row indicates total summed acquisition time for that image set). Each row was reconstructed and filtered in exactly the same manner. Note that ideal image quality is obtained at 20 min, but acceptable quality is obtained at 16 min, and for rough clinical assessments (such as the presence of a severe acute stroke) even a 4-min acquisition may be adequate.

filtering option has the advantage of being able to view the effects of filter choices on final images and is therefore generally preferred. It is also important to use spatially invariant filters, such as Butterworth, exponential, or similar low-pass filters—optimizing cutoff frequency and order (slope) to maximize resolution while minimizing noise (graininess). Spatially varying filters (e.g., Metz or Weiner) can be used only if they are carefully calibrated for brain, which is seldom done. Without such calibration they can artificially enhance the already substantial contrast present in SPECT brain images (such as that found between gray and white matter or between either of these and CSF) and lead to incorrectly reconstructed rCBF distributions (Fig. 5) and even to creation of "false" lesions.

Transverse reconstruction should be done at single-pixel thickness so that later processing of oblique angle views (sagittal and coronal) will be unhindered by limited sampling errors. Further, no filtering should be conducted during reconstruction as this is by definition a 2D (within slice) filtering process ("no filtering" is often referred to as a "ramp reconstruction").

After reconstruction and filtering the next step should be attenuation correction. Here several basic principles need to be applied. First, use an attenuation correction coefficient that has been calibrated for your system. This can be done by simply imaging a phantom that is an unstructured uniform container of ^{99m}Tc and adjusting the attenuation correction coefficient until it leads to a flat profile through a cross section of the phantom. Second, be sure that the brain is aligned so that its long axis (anterior to posterior) is parallel to the long axis of the correction ellipse. Third, it is important to apply an ellipse shape that conforms to the changing shape of the skull (assuming that you use an approximation for attenuation correction such as the Chang method instead of a direct measurement). Thus, each transverse slice requires a unique ellipse. This process is now automated on most systems through the use of threshold-based edge detection techniques and only takes a few seconds. Direct measurement of attenuation (as is commonly done in PET) is becoming more common with SPECT systems, particularly for cardiac studies. It has not become commonplace for brain studies at least in part due to the very effective ellipse-based uniform attenuation correctors, such as the Chang first-order method. It is not yet clear whether modeling attenuation with an assumed absorber profile or direct measurement with its attendant

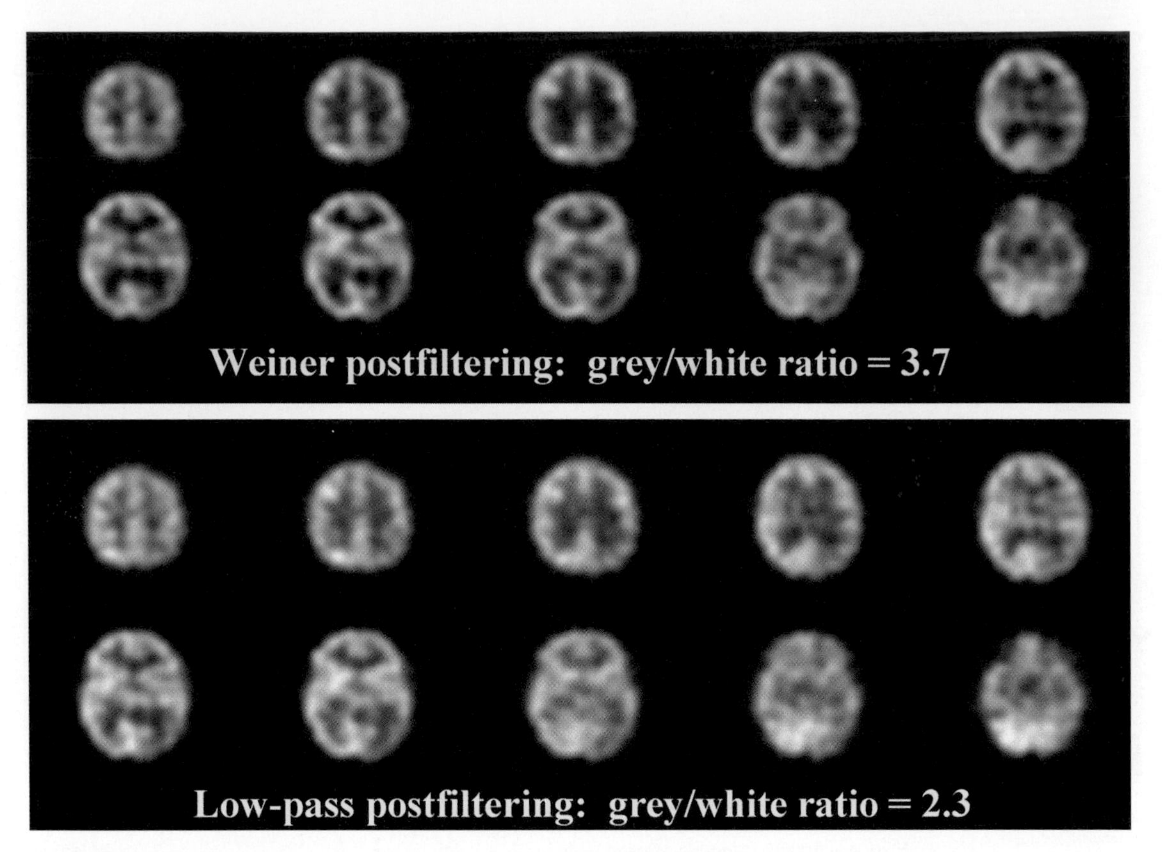

Figure 5 Effect of spatially varying filters on image contrast. **Upper images,** while appearing to have better definition, have had midfrequency data boosted and high-frequency data removed using a Wiener filter. **Lower images** have had high-frequency noise removed using a low-pass filter (Butterworth), but the midfrequency data have not been altered. Wiener filtering produced a gray/white ratio of 3.7, while low-pass filtering produced a gray/white ratio of 2.3. The true gray/white ratio of HMPAO (used here) is approximately 2.8, suggesting that the Weiner filter produced a 32% false contrast enhancement.

introduced noise is the best method to correct for attenuation in SPECT brain studies.

Following reconstruction, filtering, and attenuation correction, the next step depends on the intended application. If the data are to be viewed for clinical interpretation, then the next step is to create oblique angle views. The brain has a complex and convoluted anatomy, and it is not possible to fully evaluate all components from a single (classically transverse) view. At a minimum, transverse, coronal, and sagittal images should be created. Most modern software will allow you to dynamically alter all three angles, correcting for tilt and yaw as well as choosing the transverse angle of interest. For example, you can orient the brain to the canthomeatal line, or to some angle relative to it such as might be used for CT or MRI, without prepositioning the subject to a specific anatomic orientation (see discussion below). In fact, oblique reformatting is now so fast that you can make various sets of images for the same patient to facilitate interpretation relative to several axes if you wish (Fig. 6). Finally, some special orientations can be of interest. One particular view, cut parallel to the long axis of the temporal lobe, is especially useful for evaluating medial temporal lobe structures such as the hippocampus, often the site of abnormalities in epilepsy and memory disorders (Figs. 6 and 7).

However, for most research purposes the reconstructed, filtered, attenuation-corrected image data are submitted to some analysis software. More often than not such analysis depends on voxel-based statistics such as SPM [26]. These analyses first morph image data into a common stereotaxic space, such as that of Talairach. In this regard, SPECT data are no different from PET data, and these techniques are described in detail in other chapters. For lower resolution SPECT systems the software parameters for edge detection may need to be adjusted relative to those used for PET rCBF studies with ^{15}O-labeled H_2O. In our experience, the same parameters work well for both SPECT and PET studies as long as the SPECT studies are acquired and processed as described above [27].

In summary, image processing software should support dynamic filtering, surface-variable attenuation correction, multiple angle (oblique) reconstructions, and three-dimensional as well as conventional cross-sectional displays (Fig. 8). Although numerous permutations of reconstruc-

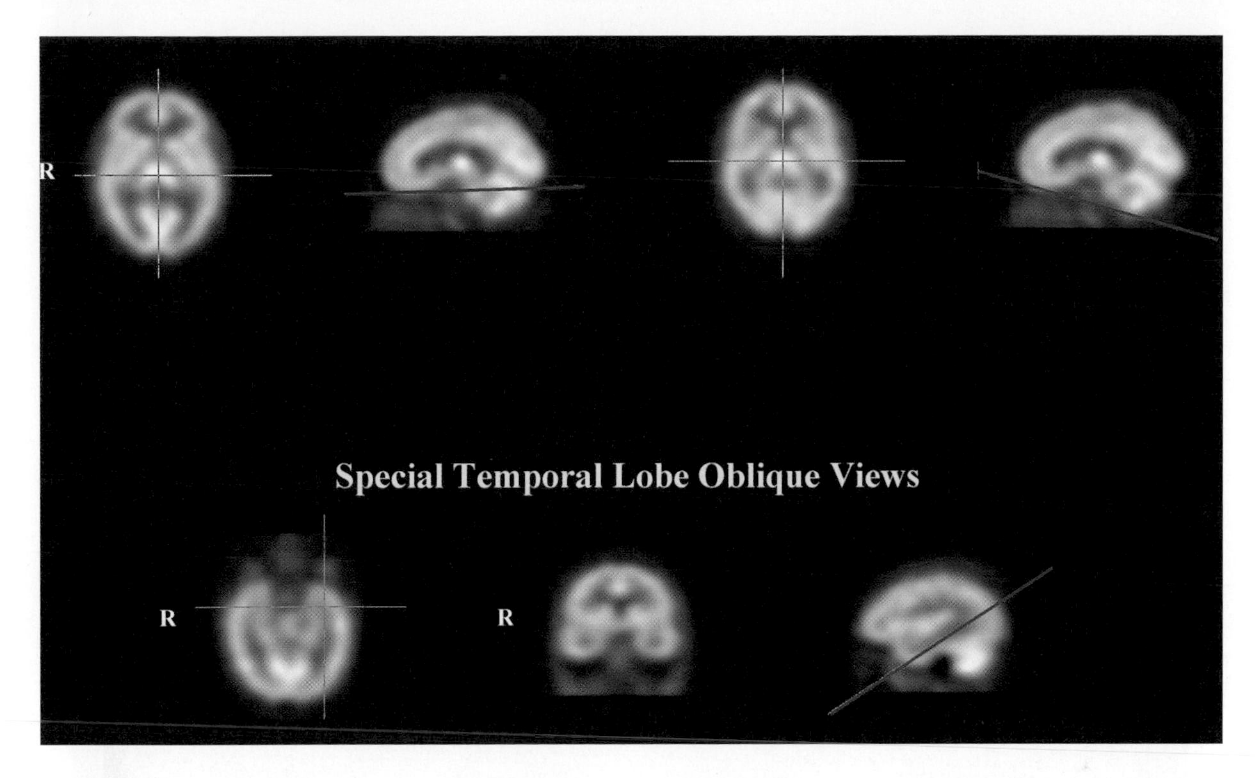

Figure 6 Modern SPECT image processing software can reformat images at any angle. Oblique angle selection: **Top left quadrant** shows a sagittal section for reference with an angle selected that is approximately that of the AC-PC line and the resulting transverse image at the level of the red marker in the sagittal image. **Right upper quadrant** shows an angle approximating the canthomeatal line (red) and the associated transverse slice. Special temporal lobe oblique views: **Bottom row** indicates a cross section obtained parallel to the long axis of the temporal lobe (red), providing in "transverse" views a complete image of the mesial temporal wall, including hippocampus and amygdala. These latter images are particularly useful for the evaluation of temporal lobe epilepsy and memory disorder patients.

tion algorithms, filter choices, and attenuation methods can be pursued, the general guidelines offered above should result in optimal image generation.

D. Dual-Isotope Imaging

Dual-isotope imaging permits simultaneous imaging of ^{99m}Tc- and ^{123}I (or ^{201}Tl)-labeled brain radiopharmaceuticals administered to a single subject. Modern SPECT systems should have adequate energy resolution (≤9%) and multiple-energy-window capability in order to separate ^{99m}Tc and ^{123}I radiotracers in the same patient. We examined the effect of dual-isotope imaging on isotope discrimination as a function of window width and position and on quantitative count recovery [28]. Simultaneous dual-isotope imaging of phantoms separated isotope distributions for 10% asymmetric and for 15 or 10% centered ^{99m}Tc windows when combined with a 10% asymmetric ^{123}I window. Isotope concentrations were recovered as accurately from asymmetric dual-isotope windows as from conventional or asymmetric single-isotope windows.

We applied this technique to the simultaneous measurement of resting rCBF and changes induced by vasodilation (1 g acetazolamide) in 10 subjects with cerebrovascular disease [29]. Resting and vasodilated ^{133}Xe SPECT images were compared to resting (^{99m}Tc-HMPAO) and postacetazolamide (^{123}I-IMP or HIPDM) dual-isotope images (Fig. 9). There was a linear relationship between ^{133}Xe SPECT and dual-isotope SPECT measurements of lesion/cerebellum ratios in baseline, vasodilated, and rest-minus-vasodilated data. A further advantage of the dual-isotope technique is that ^{99m}Tc and ^{123}I images obtained through dual-isotope imaging are by definition in perfect anatomic registration. Unfortunately, rCBF tracers labeled with ^{123}I are not currently commercially available. Additional applications for dual-isotope rCBF imaging included ictal/interictal seizure imaging, monitoring acute therapeutic interventions, and single-session evaluations of cognitive or pharmaceutical challenge tests. Of most relevance today is that this technique can be used in receptor modeling studies by directly measuring rCBF (normally deduced by assumption) with a ^{99m}Tc-labeled flow tracer, while simultaneously using an ^{123}I-labeled receptor ligand.

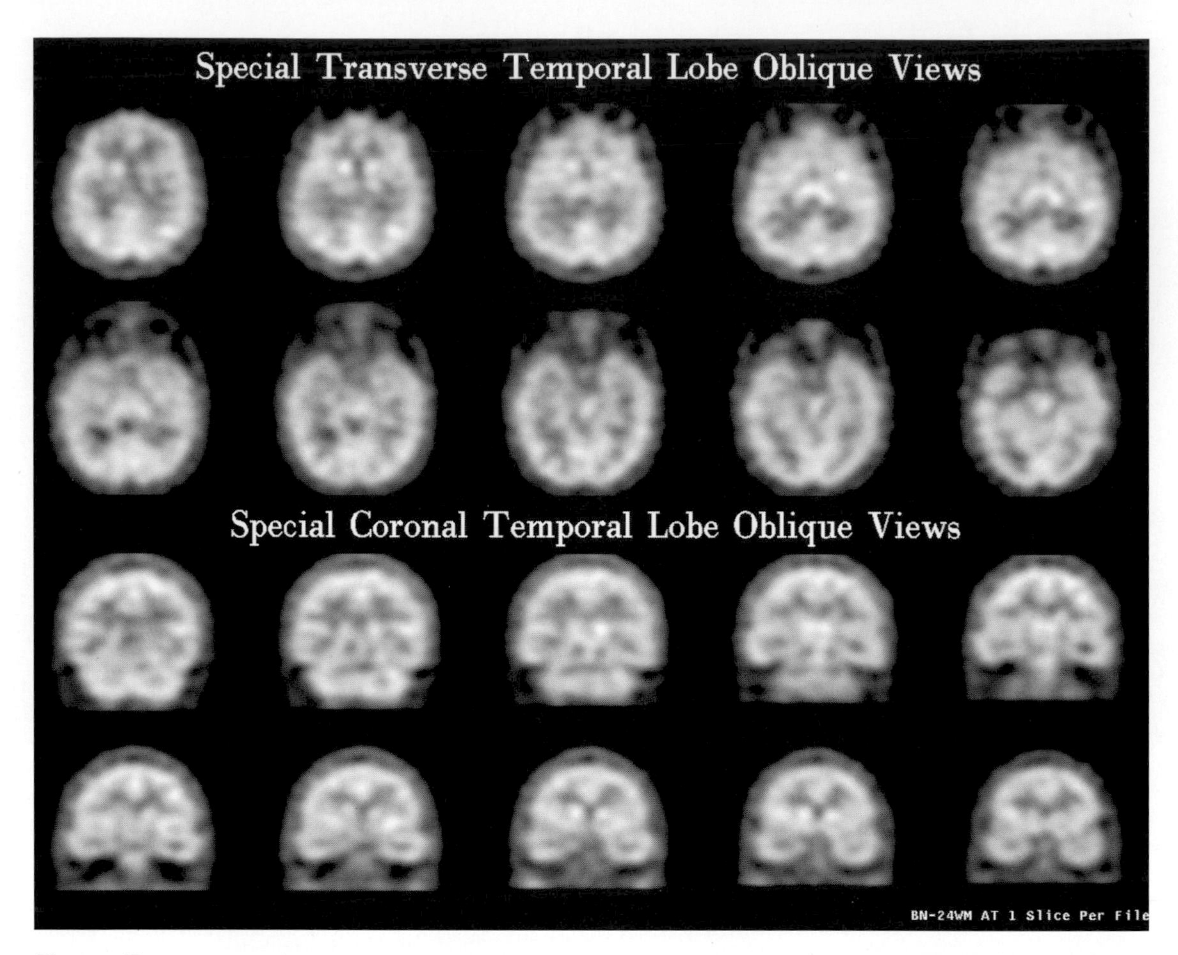

Figure 7 Images derived from orienting along the long axis of the temporal lobe in order to more clearly assess mesial temporal (hippocampus and amygdala) pathology. Special transverse temporal lobe oblique views: **Top two rows** are normal "transverse" sections cut parallel to the long axis of the temporal lobe. Special coronal temporal lobe oblique views: **Bottom two rows** are "coronal" sections cut perpendicular to this same axis.

III. Radiopharmaceuticals

Several ^{99m}Tc-labeled and ^{123}I-labeled radiopharmaceuticals for the SPECT measurement of rCBF have been developed. It is also possible to measure regional cerebral blood volume (rCBV) using SPECT techniques. Receptor imaging with SPECT is still primarily a research tool, although at least one agent for D2 receptor studies (^{123}I-IBZM) and one agent for dopamine and serotonin transporter imaging (^{123}I-β-CIT) are commercially available in Europe. SPECT agents for dopaminergic, serotonergic, noradrenergic, cholinergic, and GABAergic receptor systems are in various stages of clinical trial or preclinical testing. At this moment there is no tracer for the measurement of cerebral metabolism by SPECT. However, two metabolism-related SPECT measurements can be made. The rCBF/rCBV ratio can be measured directly and is related to regional oxygen extraction. Also, several groups have tested a class of ^{99m}Tc- and ^{123}I-labeled agents that permit detection of hypoxic cerebral tissues.

→

Figure 9 Dual-isotope rCBF images in a patient with transient ischemic attacks demonstrating failed vasodilator reserve. Baseline study (^{99m}Tc-HMPAO, **top row**) shows only mild left frontal hypoperfusion, while extensive reserve failure (^{123}I-IMP, **middle row**) is seen after vasodilation with acetazolamide (Diamox). The distribution of failed reserve is seen in the subtraction images (**bottom row**), which are easily obtained since the dual-isotope technique produces image sets that are in perfect anatomic registration.

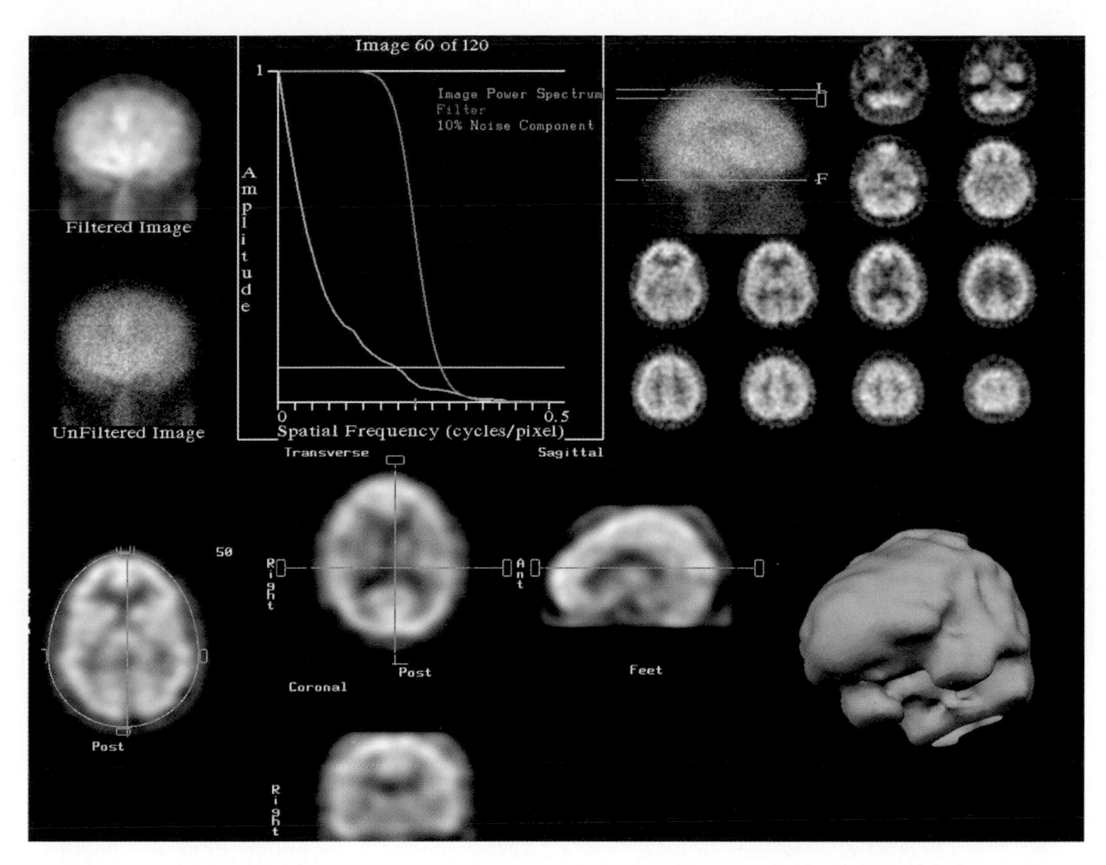

Figure 8 Typical image processing applications that should be available on all SPECT instruments. **Upper row:** Left—Fourier-space filtering; center—patient-specific filter design; right—backprojection reconstruction. **Lower row:** Left—adjustable attenuation correction; center—oblique angle reconstruction; right—three-dimensional display.

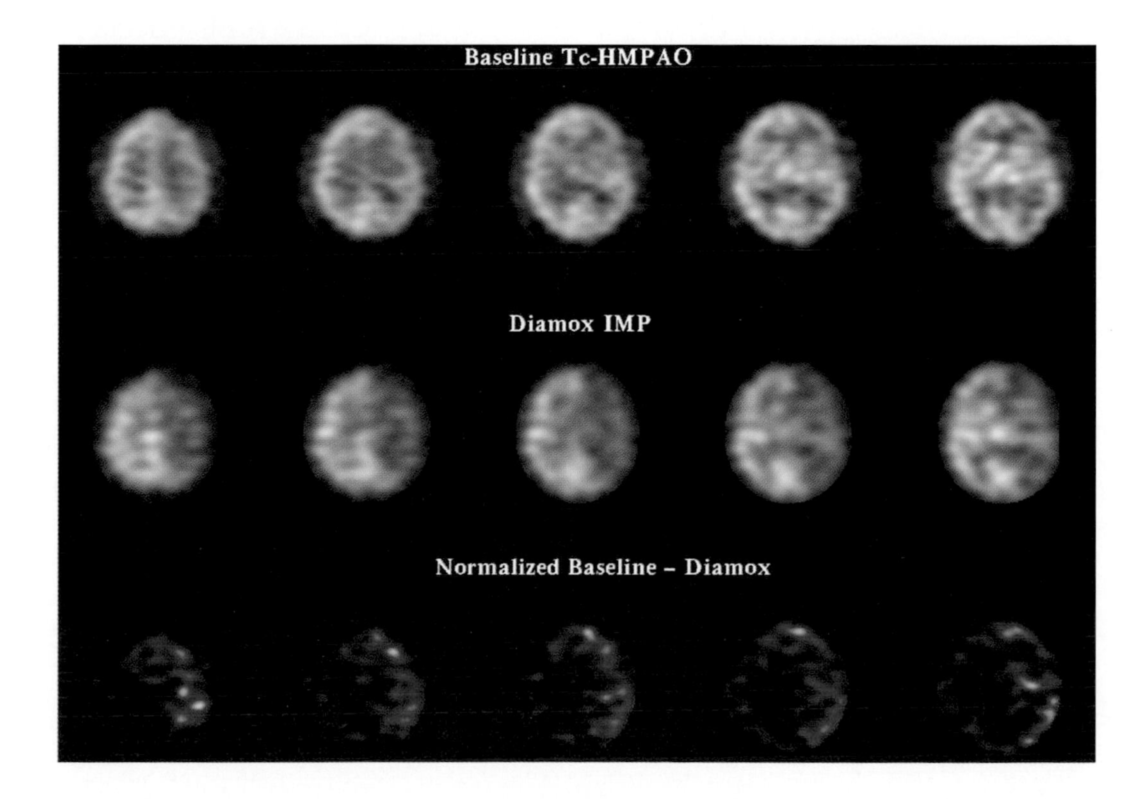

A. Diffusible Tracers

A diffusible tracer is one that passes through the circulation without engaging in metabolism or catabolism. Its concentration in tissue (brain) is therefore dependent only on the concentration gradient between arterial supply and tissue and on the rate of delivery (perfusion). By measuring (directly or indirectly) the arterial concentration, usually known as the input function, and by imaging the rate of brain uptake and clearance, reasonably standard models of diffusible tracer clearance can be used to provide a quantitative measure of rCBF [10, 11, 15]. ^{133}Xe is the original noninvasive brain blood flow marker [30]. It has been in clinical use for several decades and still has significant value [10, 11]. The cerebral transit of ^{133}Xe is very rapid, requiring complete tomographic scans every 10 s to obtain accurate reconstructions [15]. It undergoes no chemical interaction in the brain because it is an inert gas. Therefore its kinetics provide very high fidelity to true rCBF across a broad range of perfusion values and consequently yields superb lesion contrast. The input function can be easily measured by placing a scintillation probe over the lungs. SPECT studies of the transit of ^{133}Xe, when combined with a measure of the input function, can be fit to a mathematical model yielding quantitative estimates of brain blood flow [11, 31, 32]. Unfortunately, the high sensitivity required for dynamic scanning is usually obtained by sacrificing spatial resolution. For the Medimatic devices this is about 16 mm, while for the PRISM (Picker/Marconi) it has been improved to about 12 mm [24, 25]. Since ^{133}Xe has a short *biological* half-life, you can repeat examinations about every 15 min. Short imaging times and rapid isotope clearance greatly facilitate rest and stress brain imaging. In addition, the entire SPECT acquisition process is accomplished in 4 min, an important feature for difficult, uncooperative patients.

^{127}Xe is not currently commercially available in the United States and therefore has not enjoyed the extensive use as an rCBF tracer that has occurred for ^{133}Xe. However, it has several potential advantages (see below), including more optimal photon energy (204 keV vs 81 keV for ^{133}Xe) and reduced radiation dose since its decay does not include the emission of a soft β, in contrast to ^{133}Xe. These factors should combine to provide improved spatial resolution, likely exceeding that available for the ^{99m}Tc-labeled static tracers at a reduced radiation burden. ^{81m}Kr has also been proposed as an rCBF tracer, though its extremely short half life (13 s) would require the use of a steady-state model rather than the clearance models used for the Xe tracers for quantitation. It has been successfully used for cardiac blood flow measurements.

B. Static Tracers

All SPECT rCBF agents other than ^{133}Xe or ^{127}Xe (e.g., IMP, HIPDM, HMPAO, and ECD) were designed for use with rotating gamma cameras, which have low sensitivity. Consequently, rCBF tracers for use with such systems must be relatively stable *in vivo* (at least 60 min). Unlike the diffusible tracers, these radiopharmaceuticals are extracted by the brain on first arterial pass after iv injection and are then retained for several hours. Brain retention (or at least hindered diffusion from brain) is effected by some trapping mechanism, such as metabolic degradation or conformational alteration. Such agents are commonly referred to as "chemical microspheres." Stable distribution permits prolonged imaging times (as long as the patient does not move) so that specialized collimators can be used to produce high resolution images. While count ratios among brain regions correctly represent relative rCBF, most retention mechanisms do not lend themselves to simple mathematical models to provide absolute quantitation.

The original tracer microsphere model works reasonably well for IMP and HIPDM but not for ECD or HMPAO. More sophisticated models have been proposed. Unfortunately, these models depend on knowing the input function, which requires arterial blood sampling (not a routine practice in most nuclear medicine laboratories). If a simple method of measuring the input function is devised, then the microsphere-like compounds can be used to measure absolute rCBF as quantitatively as any other noninvasive modality, including PET (see additional comments under Quantification below). Several models have been proposed, including those of Isaka *et al.* [33] and Matsuda *et al.* [34], each of which have had some measure of success.

C. ^{123}I rCBF Tracers

The first rCBF agent for use on rotating gamma cameras was ^{123}I-IMP, followed almost immediately by ^{123}I-HIPDM. IMP (Spectamine) was developed by Winchell *et al.* [35] at MediPhysics and HIPDM was developed by Kung *et al.* [36] at SUNY. Both are iodinated amines with fairly rapid brain uptake and good extraction. HIPDM has a 6-h brain retention half-life. IMP has a much shorter brain retention half-life (on the order of 60–90 min). Both IMP [37–39] and HIPDM [40] follow higher cerebral blood flow levels more accurately than either HMPAO or ECD. Unfortunately, neither agent is commercially available in the United States, though IMP remains the most widely used rCBF tracer in Japan.

D. ^{99m}Tc rCBF Tracers

Investigators at Amersham [40, 41] and Volkert *et al.* [42] at the University of Missouri developed the first of the technetium agents approved by the FDA for use in humans, ^{99m}Tc-HMPAO (Ceretec). ^{99m}Tc-ECD (Neurolite) is of a class suggested by Kung *et al.* [43] and was developed and commercialized by Dupont [44, 45]. Both agents have good brain uptake. The extraction of ECD is slightly lower, but the contrast (the gray-to-white matter ratio) is higher than for HMPAO primarily because ECD has more rapid background clearance. Both tracers image defects similarly, although there are suggestions that during luxury perfusion following stroke HMPAO will follow perfusion, while ECD will continue to show a defect at the site of injury. There is also a growing body of literature (mostly from Japan) that HMPAO and IMP mark somewhat different territories of damage in cerebrovascular disease (HMPAO underestimates IMP lesion size). There are few data directly comparing ECD and IMP.

ECD initially proved easier to use because HMPAO was unstable *in vitro* and required freshly eluted $^{99m}TcO_4^-$. However, Amersham refined a stabilized version of HMPAO (Fig. 10) that subsequently achieved FDA approval. ECD may be used for up to 6 h after reconstitution, and HMPAO may be used for up to 4 h. ECD clears from the blood and the body faster, consequently producing less radiation exposure per millicurie administered than HMPAO. In general, image contrast with ECD is superior to that with HMPAO, but recent studies with HMPAO obtained 90 min after injection demonstrate the potential for equivalent contrast (see Fig. 3).

In separate studies in human subjects we compared "derived" rCBF for either ^{99m}Tc-ECD [46] or ^{99m}Tc-HMPAO [47] to absolute rCBF. "Derived" rCBF refers to whole-brain-normalized ROI data converted to absolute rCBF values by equating ECD or HMPAO whole-brain counts to whole-brain rCBF (^{133}Xe) for each subject. Excellent correlations were obtained for both agents (Fig. 11).

E. Regional Cerebral Blood Volume

Under most circumstances, rCBF is tightly coupled to tissue metabolism. However, rCBF is not the only hemodynamic parameter that affects tissue metabolism. rCBV and the extraction of oxygen can be important determinants of nutrient availability. Although direct measures of oxygen metabolism or extraction are not available by SPECT, the ratio of rCBF to rCBV is related to the regional oxygen extraction ratio (rOER). Thus, an estimate of the rOER can be obtained from the rCBF/rCBV ratio [48, 49]. rCBV imaging is conducted in a manner analogous to cardiac blood pool imaging: red cells are labeled with ^{99m}Tc, followed by static SPECT of the head. Quantitative values are obtained by SPECT imaging of a reference blood sample drawn from the subject at the time of rCBV imaging [50, 51]. The dual-isotope technique can be used to simultaneously image rCBF and rCBV (Fig. 12).

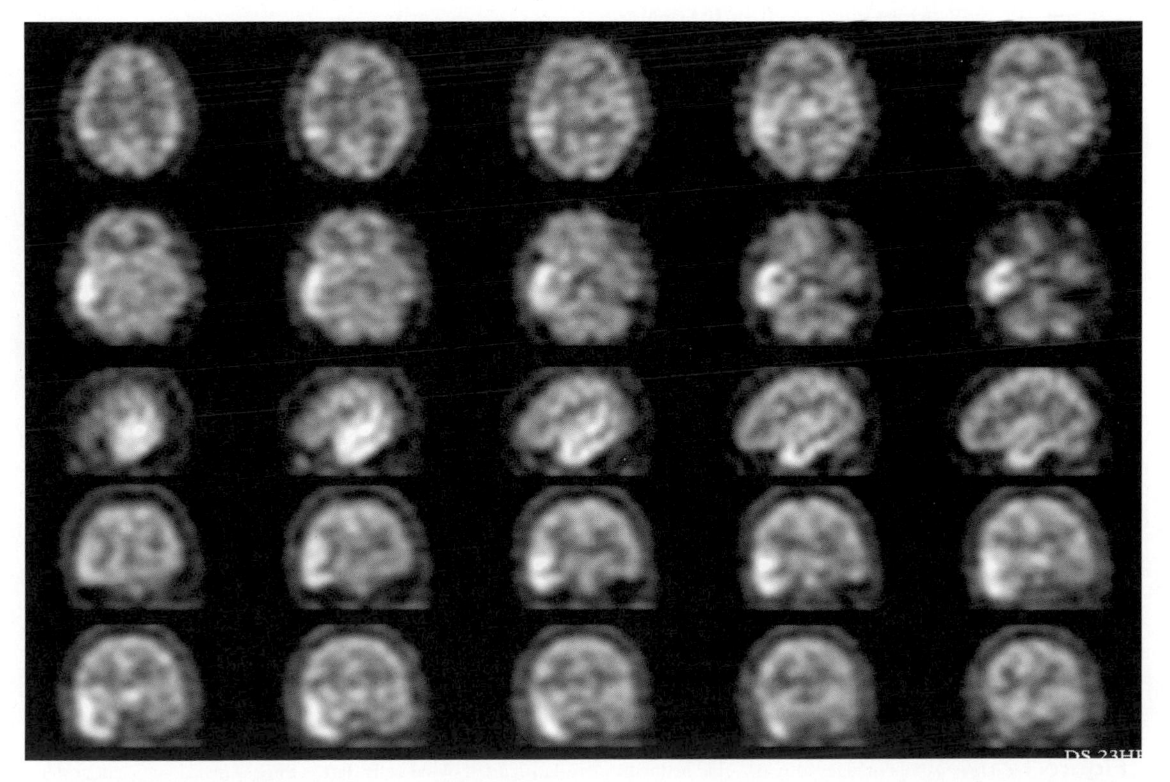

Figure 10 Ictal rCBF SPECT images obtained in a epilepsy patient with complex partial seizures of temporal lobe origin using the stabilized form of ^{99m}Tc-HMPAO. Note dramatically increased tracer uptake at the seizure focus visible in transverse (top), sagittal (middle), and coronal (bottom) views.

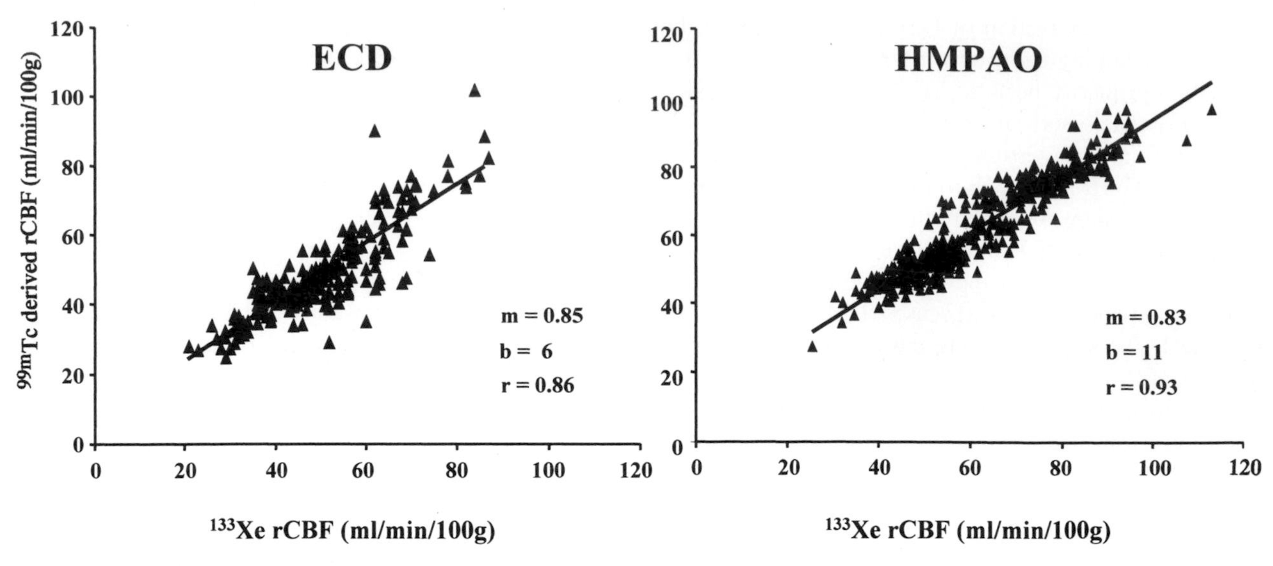

Figure 11 Comparison of the "derived" rCBF for ^{99m}Tc-ECD (left) or ^{99m}Tc-HMPAO (right) relative to rCBF determined by quantitative ^{133}Xe SPECT. "Derived" rCBF refers to whole-brain-normalized ROI data converted to absolute rCBF by equating ECD or HMPAO whole-brain counts to whole-brain rCBF (^{133}Xe) for each subject. **Left:** Regression analysis of "derived" ECD rCBF vs true rCBF. There is a strong linear relationship and ECD follows rCBF up to at least 80 ml/min/100 g. **Right:** Data comparing ^{99m}Tc-HMPAO to ^{133}Xe SPECT illustrating similar results. ECD and HMPAO mildly underestimated high rCBF and mildly overestimated low rCBF.

F. Receptor Imaging

There are currently no FDA-approved radiopharmaceuticals for neuroreceptor imaging in the United States and its clinical role is not yet well established even in Europe and Japan, where there are approved D2 and dopamine transporter tracers. However, early clinical trials and extensive PET experience suggest that SPECT imaging using specific receptor binding agents may soon find major clinical application. Radioligands are currently under study in human

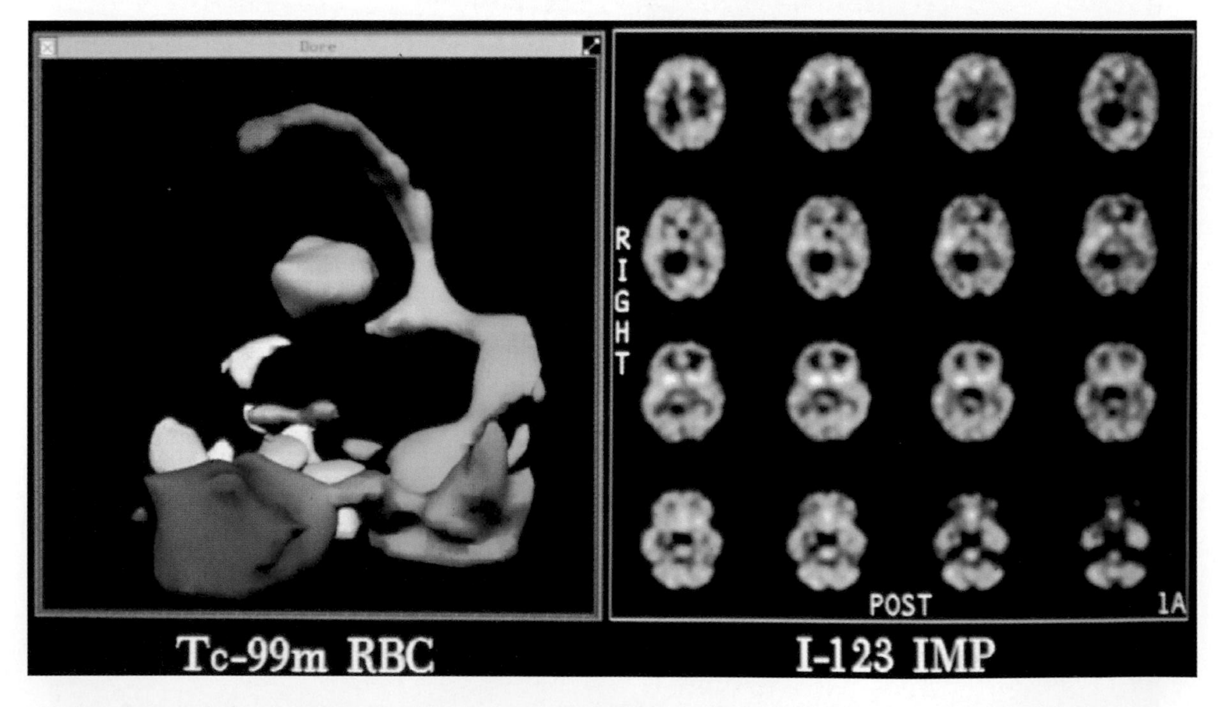

Figure 12 Dual-isotope images of rCBV obtained with ^{99m}Tc-labeled red blood cells (**left,** a three-dimensional right posterior oblique view) and rCBF obtained using ^{123}I-IMP (**right,** transverse cross sections) in a patient with an arteriovenous malformation (AVM). The AVM appears as a large mass just anterior to the descending sagittal sinus in the rCBV image, while it appears as a defect in the rCBF images since it does not retain the perfusion tracer.

trials for quantitating [1] adrenergic, dopaminergic (transporter, D1 and D2) [5, 6, 8, 52–54], serotonergic (transporter, 5-HT_2 and 5-HT_{1A}) [55], benzodiazepine and GABA [53], muscarinic cholinergic [56], and opioid systems.

Neurotransmitter values obtained include distribution volume, receptor density, receptor or transporter occupancy, and binding constants. Most such agents rely on ^{123}I as the radiolabel, although a few ^{99m}Tc-labeled ligands are under investigation. While these compounds differ structurally from their native analogs, their affinity for the specific receptor site often exceeds that of the native compound. In addition to their high affinity for the receptor site, many of these agents also have high total brain uptake (on the order of 10%). Such uptake is comparable to that seen with blood flow agents.

Potential clinical applications for receptor imaging would include the diagnosis of specific neurodegenerative diseases, quantitative assessment of therapeutic interventions designed to alter receptor function, assessment of toxic effects of substances of abuse, and evaluation of interventions capable of producing prophylaxis. Such applications, in combination with or separate from perfusion imaging, may be particularly of interest in the study of psychiatric disorders which are so commonly responsive to neurotransmitter-active pharmaceuticals. Areas for further investigation necessary to make receptor imaging a practical clinical tool include the establishment of correlations between the binding site concentration and the disease process, as well as the development of more accurate methods for absolute quantitation of the distribution of brain radioactivity [1]. Combined use of receptor, perfusion, and structural imaging in a coregistration paradigm may also greatly enhance quantitation. Fortunately, Innis *et al.* have demonstrated that at least receptor affinity can be determined from data reflecting only relative count density [57].

G. Metabolism

Metabolic aspects of neuronal function cannot be directly imaged with SPECT. We do not have, even on the horizon, an oxygen analog. Glucose metabolism has been monitored with great success by PET with [^{18}F]FDG; it has been suggested that it might be possible to iodinate glucose analogs for SPECT that would behave like [^{18}F]FDG. While there have been two promising preliminary reports, neither has come to fruition at this time.

In summary, SPECT measures of rCBF are well developed, perhaps even better than for PET. Both dynamic and static techniques are effective. SPECT can also be used to image rCBV, though clinical application of this technique has been minimal. There is great potential for receptor imaging, which has now moved out of the basic science laboratory and into clinical research. Interest in ^{201}Tl as a SPECT brain agent has also developed because it has been shown to be useful in distinguishing recurrent brain tumor from radiation necrosis after radiation therapy [58, 59] and in staging brain tumors [60–62]. Finally, progress is being made with glucose metabolic imaging at the animal level.

IV. Factors That Affect Image Appearance

A. Environmental Conditions

These include conditions experienced by subjects during radiotracer administration that play a significant role in determining the observed rCBF distribution. The coupling of rCBF to regional metabolism has been frequently demonstrated not only under resting conditions, but also during cognitive or motoric activation [63–69]. Thus, visual, auditory, and somatosensory stimuli can all be expected to impact the regional level of neuronal activity and thus rCBF (Fig. 13). Unfortunately, there are no clearly established standards describing the ideal conditions for any of these environmental parameters. Most investigators use an "eyes and ears open" imaging environment in which subjects are seated during radiotracer injection. Supportive of this choice are data from our group and others that quantitative flow or metabolism values are less variable under such conditions than in the eyes and ears closed setting [11, 70]. Room lights are often dimmed to provide a minimum and relatively standard visual stimulus. The degree to which surrounding personnel provide both auditory and visual stimuli should be (though seldom is) carefully controlled.

The duration of steady-state conditions necessary to ensure minimal variability induced by environmental conditions is also not well established. To some degree this is a radiotracer-dependent issue. For example, ^{99m}Tc-HMPAO has very rapid first-pass extraction, while a component of ^{123}I-IMP is retained by lung and likely affects brain tracer distribution for 5–10 min after injection. ^{133}Xe is a dynamic tracer that must be imaged during administration. In each case, the requirements vary for how long environmental conditions must be constant both before and after tracer administration. As a general rule, we require steady-state environmental conditions for 10 min prior to and after tracer administration for injectables. No requirement after administration is needed for ^{133}Xe.

B. Subject Characteristics

Age, gender, handedness, anxiety, time of day (diurnal variations), blood pressure, arterial carbon dioxide levels, cognitive involvement (attention), and other factors are subject-specific determinants of rCBF. While it is clear that there are both age [65, 71–76] and gender [71, 77–79] effects on whole-brain blood flow, regional effects are somewhat less well characterized [71, 73, 74, 77, 78]. Further, recent

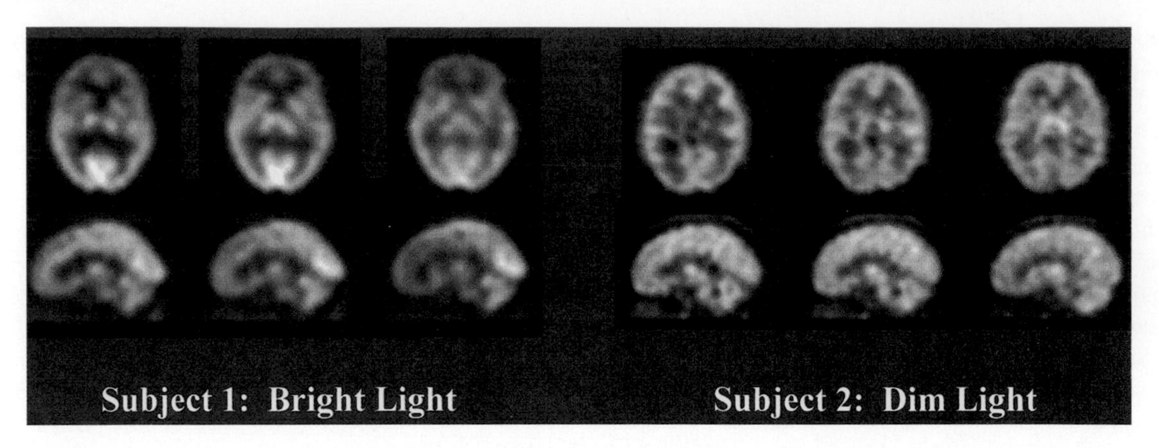

Figure 13 Effect of visual stimulation on rCBF in normal volunteers. **Left** images were obtained from a subject during bright white light stimulation. **Right** images were obtained in a subject with eyes and ears open sitting in a dimly lit room. Note that perfusion decreases in the "dim light" condition relative to the " bright light" condition not only in primary visual cortex, but also in associative cortices.

studies suggest that these two factors affect rCBF in complex ways. For example, the decline in rCBF seen with age is not as marked in active elderly individuals as it is in age-matched inactive subjects [76]. Also, gender effects seem dependent even on gender dominance within a particular gender (e.g., males with more feminine characteristics have higher flow than males with more masculine characteristics) [78]. Age-by-gender interactions are also commonly observed; that is, age changes in rCBF differ by gender and vice versa. Pediatric studies are rare, but also support age and gender effects on both global and regional cerebral blood flow [80].

C. Challenge Studies

Pharmacologic challenges induce alterations in rCBF which can provide useful tools for the discrimination of disease or the elucidation of brain functions as well as conundrums for image interpretation. For example, acetazolamide is a cerebral vasodilator commonly used in the determination of vasodilatory reserve (Fig. 9) [81–84]. However, the consistency with which it alters or preserves regional patterns from a "resting" state is not well known. Caffeine, which may be present in subjects in various quantities, can lead to reductions in both global and regional CBF [85]. The impact of other pharmacologic factors (e.g., antidepressants, antiepileptics) is only just now being elucidated. It is advisable to eliminate all such complicating factors for as long as possible preceding a study. For caffeine, 24 h may be effective. For antidepressants, drug clearance may not be complete for up to several weeks [86].

Cognitive challenges are also sources of valuable discrimination and unwanted variability. For example, mild visual stimulation (dimly lit room) seems to minimize variability among normals relative to deprivation of visual stimuli [11, 70]. Auditory stimuli must similarly be carefully controlled and evaluated [66, 87]. However, specific sensory stimuli can induce asymmetry (Fig. 14) and lead to activation of not only primary and associative sensory cortices,

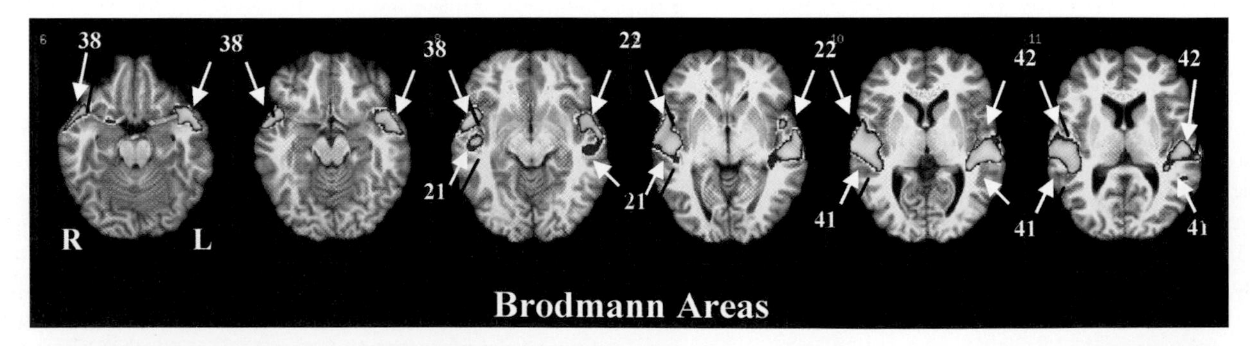

Figure 14 Left-ear speech stimulation: Parametric images derived from an SPM analysis representing areas of significant change (red) in seven normal controls overlaid on an MRI Talairach atlas-based registration model and labeled according to the responding Brodmann areas. Images are arranged in order to correspond (from left to right) to Talairach levels ranging from –16 to +12 mm below/above the AC-PC line. Cognitive activation from speech perception results in mild contralateral rCBF asymmetry in primary auditory cortices and symmetric responses in associative cortices.

but even of remote cortical and subcortical sites [67, 88–90]. In general, consistency is the greatest asset. That is, study all subjects under as identical a set of conditions as possible.

D. Image Presentation for Clinical Interpretation

Recent advances in image processing instrumentation and SPECT tomographs have afforded the opportunity to present image data in a wide variety of formats. The degree to which conventional transverse cross-sectional images provide adequate information for image interpretation is being appropriately challenged. Sagittal, coronal, and other oblique angle reconstructions are readily produced by most computer systems. Experienced observers recognize that certain brain structures are more readily appreciated in nonconventional display orientations. For example, evaluation of the medial aspect of the orbital frontal cortex is more readily performed from sagittal cross sections than from transverse. Evaluation of mesial temporal lobe hypoperfusion in epilepsy is easiest in special oblique views designed to highlight the mesial temporal wall (Fig. 7) than in classical coronal, sagittal, or transverse views. It is wise to review studies in at least three orthogonal orientations.

Similarly, the modality employed for image review has significant impact on interpretation. Conventional film-and-lightbox formats require the reviewer to consider all image data over a similar contrast range and often with a fixed degree of background subtraction. In contrast, direct viewing from a video display affords the opportunity for gray-scale manipulation and dynamic background subtraction. Choosing between gray-scale image displays and color image displays presents the viewer with yet another poorly defined dilemma. In general, most reviewers of high-resolution images prefer gray scales. The human visual system is better suited to gray-scale discrimination across structural boundaries than it is to the same discrimination using a color-based scheme. However, with poorer resolution systems, or foreshortened image data sets (minimum pixel density), color scales can provide enhanced interpretation of abnormalities. Similarly, parametric displays in which color can be used to portray functional information may provide enhanced opportunities for lesion detection (Figs. 9 and 14).

Parametric image displays are moving from the realm of research applications to clinical utility. There is widespread use of both two- and three-dimensional displays to provide an anatomical reference (SPECT, PET, or MRI image) on which to display the results of a statistical analysis (Fig. 14). While this is commonplace in research applications, the use of statistical renderings of any kind in clinical applications has been minimal (in sharp contrast to cardiac SPECT). This has been in part due to the complexity of statistical analyses for functional brain imaging. In general such techniques are not easily transferred from the research environment where they are used by investigators with substantial experience and statistical expertise to the clinical environment for use by nuclear medicine technologists with more general training. Further, there has not been a widely available source of normative data for use in clinical interpretation. Two recent innovations have altered this landscape at least in part. SPM has become so widespread in its utility, and so well validated, that some degree of automation is now possible. Certainly the first step in such an approach, image registration or image fusion, has indeed become automated and widely available on commercial SPECT systems. Further, the Brain Imaging Council of the Society of Nuclear Medicine has established a freely available database of normal SPECT brain images (http://www.snm.org/about/new_councils_1.html). The searchable database allows one to access images for viewing or download raw projection data and reconstructed SPECT scans in various file formats. Demographic and scan parameter information are also provided with the scans. It is hoped that the combination of automated software and a large database will lead to statistical assistance in the interpretation of SPECT brain images as has occurred with the "bull's-eye" and other cardiac analyses.

Recently, three-dimensional surface-rendered displays have become commonplace (Fig. 15) while more sophisticated three-dimensional displays of "see-through" or "smoked-glass" images are not as available (Fig. 16). The "dial-a-lesion" format of most surface-rendered images currently limits their routine applicability. Established standards for count-density threshold settings have not been published. Cinemagraphic displays of projection data also provide a certain three-dimensional quality. Their primary use is in monitoring patient motion as a source of image degradation.

Six conclusions can be drawn regarding image presentation. (1) SPECT rCBF studies should be presented in at least transverse, coronal, and sagittal cross sections. (2) The conditions under which subjects are studied should be established as a laboratory standard and maintained for all clinical and research studies. The most common standard is eyes and ears open in a dimly lit environment with minimal auditory "white noise." (3) The normal distribution of rCBF in such a setting shows symmetric flow distribution between homologous regions. (4) Both age and gender effects can be noted. They are well characterized only for whole-brain blood flow, which is not readily measured by conventional SPECT techniques. (5) Among brain stress tests that could be employed to enhance discrimination of disease, the only well-established technique is the acetazolamide vasodilator test. Under most circumstances it reveals striking asymmetries in disease states of vascular origin. Its regional effects (independent of global flow changes) have not yet been established in normal controls. (6) Future research should focus on the establishment of statistical techniques

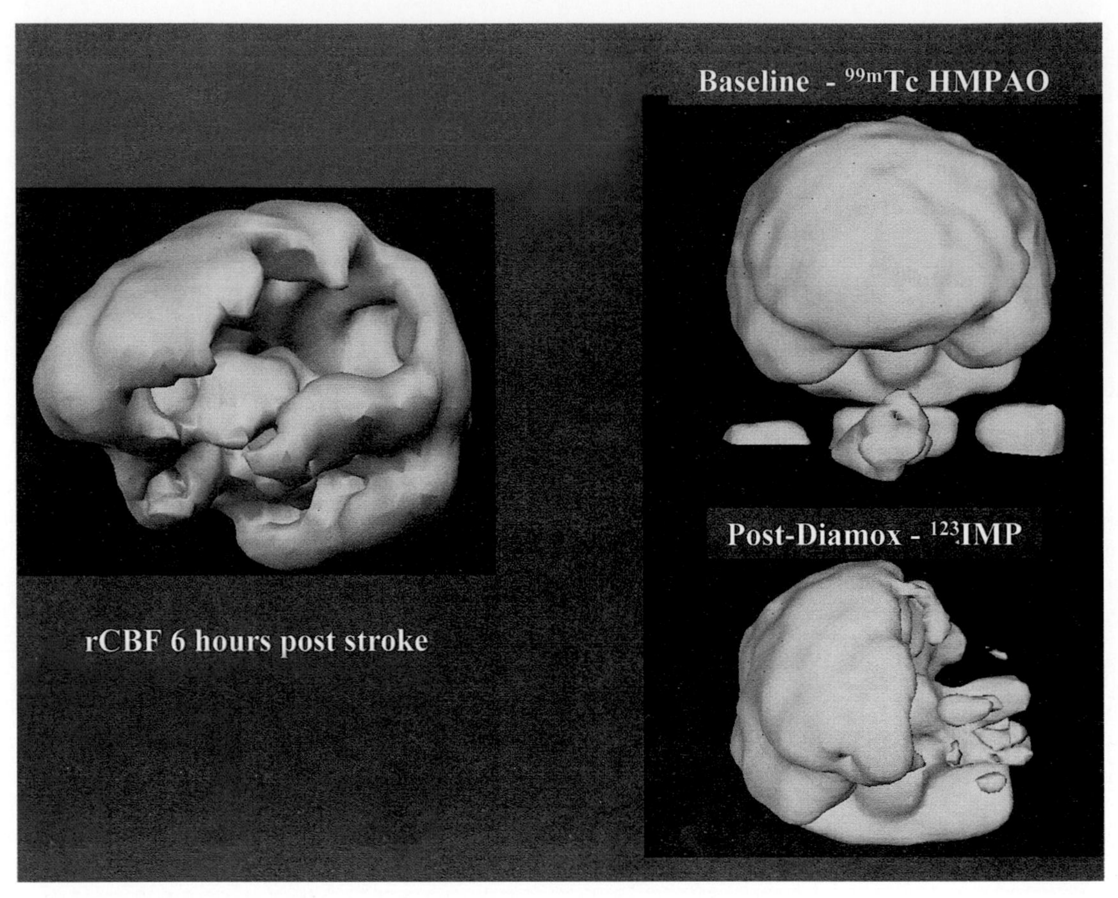

Figure 15 Various three-dimensional surface-rendered rCBF SPECT images **Left:** Acute rCBF image 6 h after a left hemisphere stroke. Surface-rendered images are useful to assess the distribution (vascular territory) of cortical defects, but do not allow visualization of either defect magnitude (count density) or subcortical abnormalities. **Right:** Baseline (top) and post-Diamox (bottom) surface-rendered images in a patient with TIAs. Note extensive vasodilatory reserve failure illustrated in the post-Diamox images.

and normative databases and on the effects of environmental variations on regional cerebral blood flow.

In summary, factors reviewed above that significantly influence the final quality of your SPECT brain images fall broadly into three categories: patient preparation, radiopharmaceuticals, and imaging systems. These key factors are summarized in Table 1.

V. Intercomparison of Neuroimaging Techniques for the Quantification of rCBF

The two most commonly reported imaging procedures for measuring rCBF are SPECT and PET. In addition, functional measures that are related to rCBF (though indirectly) can be obtained with either *f*MRI using the blood oxygenation level-dependent technique or by obtaining a signal that is heavily dependent on water diffusion characteristics (diffusion-weighted MRI). The latter two techniques are still under development, while SPECT and PET are well established for both research and clinical applications.

The current literature suggests that both global and focal effects on neuronal metabolism and cerebral blood flow are observed in a variety of neurologic and psychiatric disorders and it may therefore be important to use imaging procedures that can quantify both global and regional cerebral blood flow. This is possible with both SPECT and PET, but is not currently possible with either *f*MRI or diffusion-weighted MRI. Key characteristics of the most commonly used functional brain imaging procedures are outlined in Table 2. Many aspects of SPECT and PET techniques are similar, but important differences should be carefully considered when choosing the appropriate imaging tool for a particular investigation. For this discussion we restrict ourselves to a description of those characteristics of interest to the measurement of rCBF.

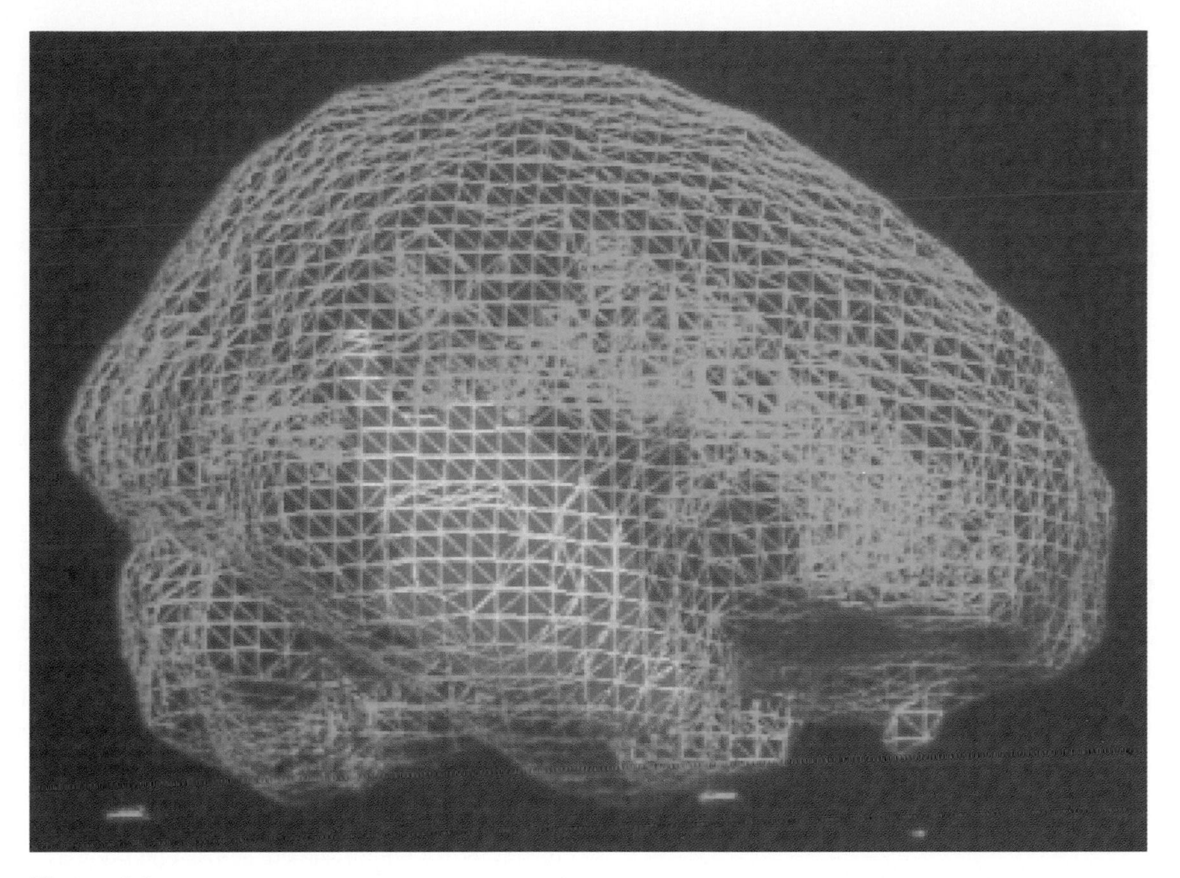

Figure 16 Some "see-through" three-dimensional rCBF SPECT images combine multiple surfaces. In this case, a "wire-cage" surface is used to outline the brain, while a solid body is used to define the location of an area of high flow in an ictal study of rCBF in a seizure patient.

A. PET rCBF Imaging

1. Advantages

A principal advantage of PET for rCBF measurement lies in the tracer, $H_2{}^{15}O$, which functions as a diffusible tracer. Because ^{15}O has a short (2 min) physical half-life, the tracer is rapidly removed from the subject by decay and so repeat imaging can be performed as quickly as 10 min after the initial scan. Also, the rapid image-acquisition process permits detection of rather transient phenomena (rCBF changes as brief as 30 s in duration can be detected, though 2 min is more typical). This is in contrast to regional cerebral glucose metabolism (rCGM) PET imaging with [^{18}F]FDG for which tracer uptake occurs over about 30 min and the resulting image represents the integral of brain activity over that time period (Table 2). Radiation dose is quite low and many scans can safely be conducted in one session (see radiation safety discussion below). Further, subject-specific attenuation correction maps are routinely obtained for PET images, and these enhance the data available to quantitative models.

2. Disadvantages

Unfortunately, the 2-min half-life of ^{15}O requires that one use an on-site cyclotron to produce the tracer and perform rapid chemical procedures to make the chemical entity of interest (in this case water). Further, the subject to be imaged must be fully prepared and in the scanner when the tracer is compounded for use. Therefore, imaging results can be confounded by cognitive effects of anxiety in the scanner or by visual or auditory stimuli present therein. Also, as a consequence of the short half-life of ^{15}O and the rapid transit of $H_2{}^{15}O$, the PET scanner must be operated in a data acquisition mode that maximizes sensitivity to emitted radiation and minimizes the duration of the image acquisition. So while the spatial resolution of PET scanners can be as good as 4 mm when long imaging times can be used (such as in imaging rCGM), when used for rCBF imaging this resolution is typically degraded to about 8 mm (Table 2). Next, while ^{15}O offers minimal radiation exposure per dose to the subject being imaged, its short half-life requires the cyclotron production of substantial quantities so that enough is left after radioactive decay to provide a

TABLE 1

Brain SPECT—Imaging Principles
Patient preparation
Quiet reproducible environment;
eyes/ears open; dimly lit room
IV line placed ~10 min before injection;
quiet ~10 min after injection
Radiopharmaceuticals
rCBF tracers
^{99m}Tc HMPAO Ceretec (Amersham)
^{99m}Tc ECD Neurolite (DuPont)
Delay imaging for best results
90 min for Ceretec
45 min for Neurolite
Imaging
High-resolution collimators;
UHR and fan beam when possible
Minimum radius (heads in close)
"Motion-free" positioning
Patient comfort first
Multiple sequential acquisitions
Single-pixel reconstruction;
no filtering during reconstruction
3D low-pass filtering
Attenuation correction;
shape-conforming ellipse
Oblique reformatting;
transverse, sagittal, coronal, special views

reasonable dose to the subject. This means that radiation safety for PET technologists, chemists, and physicists can be challenging. Finally, PET tomographs and cyclotrons are expensive, and this technique requires a considerable capital investment. Though this is a complex process, these details are well developed and PET rCBF imaging can readily be performed in properly equipped facilities with experienced personnel.

3. PET Quantitation

PET rCBF imaging with $H_2{}^{15}O$ can be fully quantified using a mathematical model that relies principally on the diffusible tracer technique. It is worth noting that water is not completely freely diffusible across the blood–brain barrier, and so $H_2{}^{15}O$ is not a perfect rCBF tracer, yielding values that somewhat underestimate true rCBF especially at higher perfusion rates. Full quantitation requires arterial catheterization and blood sampling during the scan, subsequent counting of the acquired samples, and cross-calibration of the scanner and counting equipment. This process, once established in a laboratory, is straightforward, though time consuming, and permits precise intercomparison of global CBF values within subjects across conditions and between subjects. The value of absolute rCBF determination varies with the desired study. In most of the reported literature the substantial intersubject variance in *normal* absolute global CBF has served primarily as a confound and so it is removed by calculating relative rCBF values either through normalization to global CBF or rCBF in a reference brain region such as cerebellum or occipital lobe or by the use of analysis of variance or a related statistical procedure. However, in the circumstance in which a task, pharmacologic challenge, disease process, or treatment is expected to impact absolute CBF values then full quantitation may be important for within-subject across-condition assessments.

B. SPECT rCBF Imaging

As described above, SPECT rCBF imaging can be accomplished with either of two classes of radiopharmaceuticals: the diffusible tracers ^{133}Xe or ^{127}Xe or the static tracers ^{99m}Tc-HMPAO or ^{99m}Tc-ECD. General characteristics common to both classes and specific characteristics unique to each class are outlined below.

1. General Advantages

SPECT is widely available and the tracers used are FDA approved (no PET tracer for the imaging of neuronal function has received FDA approval for national distribution). Thus, SPECT is the only functional brain imaging technique in common clinical use. SPECT is generally considered to be

TABLE 2 Key Parameters of the Most Common Functional Brain Imagining Procedures

Imaging procedure	Parameter measured	Spatial resolution	Temporal resolution	Test environment
PET ([^{18}F] FDG)	rCGM	4 mm	30 min	Scan room
PET ($H_2{}^{15}O$)	rCBF	8 mm	2 min	Scanner
SPECT (^{99m}Tc)	rCBF	7 mm	20 sec	Anywhere
SPECT (^{133}Xe)	rCBF	12 mm	2 min	Scanner
*f*MRI	α rCBF and rCBV	3–8 mm	100 ms	Loud scanner

less complicated than PET, often only because less rigorous approaches to its use are employed. The radiopharmaceuticals are commercially available and relatively inexpensive (though there is a substantial difference between the classes, ^{133}Xe being about one-fourth the cost of the ^{99m}Tc-labeled tracers). As also previously described, the instrumentation has matured so that distinctions between SPECT and PET spatial resolution are not substantial (Table 1).

2. ^{133}Xe and ^{127}Xe Advantages

Xe SPECT tracers are freely diffusible tracers and are in a sense the SPECT equivalent of ^{15}O for PET. Both ^{133}Xe and ^{127}Xe have short biological half-lives because they are inert gases and are exhaled from the body within a few minutes of the completion of their administration. Therefore, as with ^{15}O, repeat imaging can be performed as quickly as 15 min after the initial scan. The inert gas chemical form makes these the purest of the diffusible tracers, yielding the best fidelity to true rCBF; it also permits administration either by inhalation or by iv injection of dissolved gas, which can be helpful with patients for whom iv placement is challenging. Also, the rapid image acquisition process (about 4 min compared to about 2 min for $H_2{}^{15}O$) also permits detection of transient phenomena (Table 1). Radiation dose is also low (^{127}Xe is similar to ^{15}O, but ^{133}Xe has higher dose per scan due to the emission of a low-energy β particle) and several scans can safely be conducted in one session (see radiation safety discussion below). Only ^{133}Xe is commercially available in the United States at this time and can be obtained in bulk for frequent use (one vial will last about a week since ^{133}Xe has a 5.3-day half-life). A commercial system for the distribution and recycling of ^{127}Xe (which has a 36.4-day half-life) has been proposed but is not yet in production. This would be an important development as ^{127}Xe affords much higher spatial resolution than ^{133}Xe (approximately 6 mm compared to 12 mm).

3. ^{99m}Tc Static Tracer Advantages

^{99m}Tc-HMPAO and ^{99m}Tc-ECD function as "static" rCBF tracers, or chemical microspheres. The fact that they are retained in the brain for several hours provides the unique opportunity to inject subjects away from the scanner and obtain a "snap shot" of rCBF at the time of injection in an environment completely under investigator control; imaging can then occur up to 6 h later. Because the tracer distribution is stable *in vivo* for such a long time, advanced high-resolution scanning techniques can be used to acquire data. Thus these tracers yield the best spatial resolution rCBF images of any technique. In addition, for some pharmacologic and cognitive challenge tests administering the test outside of a scanner environment is critical. The ^{99m}Tc SPECT tracers are the only method currently available for sampling brain activation over a relatively short time period (seconds to minutes) in a sensory-controlled environment.

4. General Disadvantages

It is generally agreed that the spatial resolution for SPECT scanners is inferior to that of PET scanners for comparable tracers. For example, for diffusible tracers the resolution of ^{133}Xe is a little worse than for $H_2{}^{15}O$ and the same would be true for a comparison with ^{99m}Tc tracers if there were a chemical microsphere PET radiopharmaceutical (the closest comparator for this purpose would be [^{18}F]FDG). This difference is typically only a few millimeters and for specific instrument comparisons differences may be nonexistent. If ^{127}Xe becomes available, then it would have the best resolution of the diffusible rCBF tracers for either SPECT or PET and would be comparable to or slightly better than that for the ^{99m}Tc agents. In addition to inferior resolution, SPECT is generally regarded to have poorer quantitative properties. For example, attenuation correction is usually accounted for by direct measurement with PET and only by mathematical approximation for SPECT. This is of importance only for static tracers since the models for quantifying diffusible tracers with SPECT do not rely on attenuation correction (this is not the case for the bolus $H_2{}^{15}O$ PET method). Further, subject-specific attenuation maps are becoming more commonly available on multihead SPECT systems (largely driven by the demands of nuclear cardiology procedures) and thus are likely to become more commonplace in SPECT brain imaging.

5. ^{133}Xe and ^{127}Xe Disadvantages

As with $H_2{}^{15}O$ the subject to be imaged with the Xe tracers must be in the scanner when the tracer is administered. Again, imaging results can be confounded by cognitive effects of anxiety in the scanner or by visual or auditory stimuli present therein. Further, as a consequence of the short biological half-life of the Xe tracers, as with the short physical half-life of ^{15}O, the SPECT scanner must be operated in a data acquisition mode that maximizes sensitivity to emitted radiation and minimizes the duration of the image acquisition, leading to significant reductions in spatial resolution. ^{133}Xe has a relatively low gamma-ray energy (80 keV), making it difficult to obtain good count statistics and further degrading image resolution. While the 5.3-day half-life is useful for "off-the-shelf" use, it does present a radiation safety complication for both storage and disposal. Even though ^{127}Xe would have many advantages over ^{133}Xe, including superior spatial resolution, minimal attenuation artifact, and reduced radiation dose to the subject, its longer half-life further complicates the storage and disposal issues and these are not yet thoroughly solved.

6. ^{99m}Tc Static Tracer Disadvantages

The principal disadvantage of the static tracers is a consequence of one of the advantages—long *in vivo* residence time. In addition to the multihour biochemical half-life, the 6-h physical half-life of ^{99m}Tc means that one must wait four to six half-lives before the radioactive background of an initial scan is below detectable limits. While this feature affords superb control over the imaging environment, it also means that one must wait at least a day, and preferably 2 days, before repeating the imaging sequence. Several means to address this problem have been explored, including split-dose techniques and dual-isotope imaging, but in most settings simply waiting for the decay of the first injection is the preferred solution. Additionally neither ^{99m}Tc-HMPAO nor ^{99m}Tc-ECD is a perfect chemical microsphere and consequently underestimates rCBF values above about 60 ml/min/100 g, with the error increasing in proportion to rCBF (Fig. 11).

7. SPECT Quantitation

As with $H_2{}^{15}O$, SPECT rCBF imaging with the Xe tracers can be fully quantified using a mathematical model that relies principally on the diffusible tracer technique. Fortunately, full quantitation does not requires arterial catheterization but only measurement of the Xe concentration curve over time in the lungs (i.e. the input function), which can be accomplished with a simple radiation detector placed above the chest. This process is also straightforward and can readily be performed in most nuclear medicine departments. Full quantification of ^{133}Xe rCBF actually preceded that by PET since this procedure was developed for two-dimensional nonimaging measurement techniques as early as the 1950s.

Models used to quantitate rCBF for static tracers differ from those for diffusible tracers and are not as well developed. In fact, earlier studies using these tools to examine rCBF used only the semiquantitative relative measures mentioned above. While this method provides a useful index of regional rCBF relationships, it cannot be used to measure absolute change in CBF. Several fully quantitative approaches have been explored, including complete modeling with arterial sampling. These approaches are effective, but have not been popular since they remove from SPECT its advantage of relative simplicity. It is also possible to measure a quantity known as the brain perfusion index that is closely related to global CBF [34]. This process is not substantially more complicated than that required for Xe tracer rCBF modeling but does require that the subject be physically in the scanner at the time of tracer administration. Thus, both classes of SPECT rCBF tracers can be used either for relative or for absolute rCBF measurement with differing degrees of complexity and accuracy, neither of which exceeds that required for PET modeling with $H_2{}^{15}O$.

VI. Radiation Risk Issues

There exists a general view that radiation is harmful at any level and that consequently medical imaging procedures that involve exposure to any radiation whatsoever are associated with some level of risk [91, 92]. Great efforts (and costs) are extended to minimize radiation exposure for patients, research volunteers, and workers. With the advent of functional measures based on MR imaging sequences, it is sometimes put forward that these techniques should be used in favor of PET or SPECT imaging whenever possible to avoid the perceived radiation risk of the radioactive tracer techniques. Much of this impression has been fostered by the widespread regulatory use of the Linear No-Threshold Theory regarding the assessment of radiation risk [93–95]. The purpose of this section is to briefly describe the data regarding low-level radiation and risk and to clarify that in fact SPECT and PET procedures have no more risk than MRI-based procedures.

The Linear No-Threshold Theory assumes that measures of mortality, disease induction, or tissue injury caused by very high radiation exposure levels in nuclear accidents, in atomic bomb exposures, or via intentional radiation therapy treatments can be extrapolated over many orders of magnitude to the much lower levels of radiation exposure incurred in diagnostic imaging procedures by a simple linear interpolation. That is, the injury caused by high level radiation is lowered in linear proportion based on the assumption that the only "zero-risk" state is in fact zero exposure. A single gamma ray stopping in the human body would therefore be associated with at least some risk. However, most data indicate that this is not the case, but actually that risk asymptotes to zero at radiation exposure levels well above those incurred in diagnostic procedures for both adults and children [96–98].

In 1996 the Health Physics Society [99] issued a policy statement indicating that while there is substantial and convincing scientific evidence for health risk at high dose, for exposures below 10 rem "health risks are either too small to be observed or are nonexistent." The whole-body dose of a typical SPECT brain imaging procedure is about 0.1 rem. In its assessment of risk to children for diagnostic imaging procedures used in clinical and research investigations, the Office of the Clinical Director of the NIH stated: "The risk of increased rates of cancer after low-level radiation exposure is not supported by population studies of health hazards from exposure to background radiation, radon in homes, radiation in the workplace or radiotherapy. Compared to the frequency of daily spontaneous genetic mutations, the biologic effect of low-level radiation at the cellular level seems extremely low. Furthermore, the potentiation of cellular repair mechanisms by low-level radiation may result in protective effect from subsequent high-level radiation" [100]. They concluded their risk review by saying: "Health risks from low-level radiation could not be detected above the 'noise' of adverse events of everyday life. In addition, no data were found that demonstrated higher risks with younger age at low-level radiation exposure."

Indeed, there are no data that have ever demonstrated any harm to humans by radiation exposure at diagnostic imaging levels. In fact, current data support the presence of radiation hormesis: that low levels of radiation exposure induce beneficial effects of cellular repair and immune system enhancement [101–103]. This certainly makes sense since we evolved as a race in a background radiation environment many times higher than is currently present. Further, current background radiation levels vary by an order of magnitude across geographic regions without any indication that those living in lower background regions have less cancer prevalence than those in higher background areas. Data from studies in the United States, China, India, Austria, and the United Kingdom show that populations in higher background areas have *decreased* cancer death rates and *increased* longevity [98, 101]. Therefore it should be concluded that neither PET nor SPECT brain imaging procedures are associated with any particular risk over activities of daily living and certainly should not be considered to be any more "risky" than MRI or any of its associated functional imaging derivatives.

VII. Conclusions

SPECT functional brain imaging is a powerful clinical and research tool. There are several clinical applications now documented, a substantial number under active investigation, and an even larger number yet to be studied. Instrumentation continues to improve, although current SPECT tomographs yield excellent image quality. There is a rapidly expanding armamentarium of radiopharmaceuticals. Challenge tests, well developed only in cerebrovascular disease (the acetazolamide test for vasodilatory reserve), offer great promise in elucidating the extent and nature of disease, as well as predicting therapeutic responses. Some standards regarding patient imaging environment and image presentation are emerging. However, much is yet to be learned about the ideal circumstances for the performance and evaluation of SPECT functional brain imaging. Finally, we must keep in mind that SPECT will achieve its full potential as a clinical tool for the management of patients with cerebral pathology only through close cooperation between the nuclear medicine community and our colleagues in neurology, psychiatry, or neurosurgery.

References

1. Ichise, M., Meyer, J. H., and Yonekura, Y. (2001). An introduction to PET and SPECT neuroreceptor quantification models. *J. Nucl. Med.* **42,** 755–763.
2. Iacoboni, M., Baron, J. C., Frackowiak, R. S. J., Mazziotta, J. C., and Lenzi, G. L. (1999). Emission tomography contribution to clinical neurology. *Clin. Neurophysiol.* **110,** 2–23.
3. Phelps, M. E. (2000). The merging of biology and imaging into molecular imaging. *J. Nucl. Med.* **41,** 661–681.
4. Devous, M. D., Sr. (1998). SPECT brain imaging in cerebrovascular disease. *In* "Nuclear Medicine in Clinical Diagnosis and Treatment," 2nd edition (I. P. C. Murray and P. J. Ell, eds.), pp. 631–649. Churchill–Livingstone, London.
5. Brücke, T., Djamshidian, S., Bencsits, G., Pirker, W., Asenbaum, S., and Podreka, I. (2000). SPECT and PET imaging of the dopaminergic system in Parkinson's disease. *J. Neurol.* **247,** IV/2–IV/7.
6. Brooks, D. J. (2000). Morphological and functional imaging studies on the diagnosis and progression of Parkinson's disease. *J. Neurol.* **247,** II/11–II/18.
7. Holman, B. L., and Devous, M. D., Sr. (1992). Functional brain SPECT: The emergence of a powerful clinical method. *J. Nucl. Med.* **33,** 1888–1904.
8. Heinz, A., Jones, D. W., Raedler, T., Coppola, R., Knable, M. B., and Weinberger, D. R. (2000). Neuropharmacological studies with SPECT in neuropsychiatric disorders. *Nucl. Med. Biol.* **27,** 677–682.
9. MacLaren, D. C., Toyokuni, T., Cherry, S. R., Barrio, J. R., Phelps, M. E., Herschman, H. R., and Gambhir, S. S. (2000). PET imaging of transgene expression. *Biol. Psychiatry* **48,** 337–348.
10. Devous, M. D., Sr. (1995). SPECT functional brain imaging. *In* "Clinical SPECT Imaging" (E. L. Kramer and J. Sanger, eds.), pp. 97–128. Raven Press, New York.
11. Devous, M. D., Sr. (2000). SPECT functional brain imaging: Instrumentation, radiopharmaceuticals and technical factors. *In* "Cerebral SPECT Imaging," 3rd edition (R. L. Van Heertum and R. S. Tikofsky, eds.), pp. 1–22. Lippincott, Williams & Wilkins, Philadelphia.
12. Juni, J. E., Waxman, A. D., Devous, M. D., Sr., Tikofsky, R. S., Ichise, M., Van Heertum, R. L., Holman, B. L., Carretta, R. F., and Chen, C. C. (1998). Procedure guideline for brain perfusion SPECT using technetium-99m radiopharmaceuticals. *J. Nucl. Med.* **39,** 923–926.
13. Links, J. M., and Devous, M. D., Sr. (1995). Three-dimensional display in nuclear medicine: A more useful depiction, or only a superficial rendering? *J. Nucl. Med.* **36,** 703–704.
14. Trivedi, M. H., Husain, M. M., and Devous, M. D., Sr. (1997). Functional brain imaging: SPECT. Basic and technical considerations. *In* "Brain Imaging in Clinical Psychiatry" (K. R. Krishnan and P. M. Doraiswamy, eds.), pp. 47–62. Dekker, New York.
15. Stokely, E. M., Sveinsdottir, E., Lassen, N. A., and Rommer, P. (1980). A single photon dynamic computer assisted tomograph (DCAT) for imaging brain function in multiple cross-sections. *J. Comput. Assisted Tomogr.* **4,** 230–240.
16. Kimura, K., Hashikawa, K., Etani, H., *et al.* (1990). A new apparatus for brain imaging: Four-head rotating gamma camera single-photon emission computed tomograph. *J. Nucl. Med.* **31,** 603–609.
17. Rogers, W. L., Clinthorne, N. H., Stamos, J., *et al.* (1984). Performance evaluation of SPRINT, a single-photon ring tomograph for brain imaging. *J. Nucl. Med.* **25,** 1013–1018.
18. Kanno, I., Uemura, K., Miyura, S., and Miyura, Y. (1981). HEADTOME: A hybrid emission tomograph for single-photon and positron emission imaging of the brain. *J. Comput. Assisted Tomogr.* **5,** 216–226.
19. Logan, K. W., and Holmes, R. A. (1984). Missouri University multiplane imager (MUMPI): A high-sensitivity rapid dynamic ECT brain imager. *J. Nucl. Med.* **25,** PI05.
20. Smith, A. P., and Genna, S. (1989). Imaging characteristics of ASPECT, a single-crystal ring camera for dedicated brain SPECT. *J. Nucl. Med.* **30,** 796.
21. Stoddart, H. F., and Stoddart, H. A. (1979). A new development in single-gamma transaxial tomography: Union Carbide focused collimator scanner. *IEEE Trans. Nucl. Sci.* **26,** 2710–2712.
22. Hill, T. C., Stoddart, H. F., Doherty, M. D., Alpert, N. M., and Wolfe, A. P. (1988). Simultaneous SPECT acquisition of CBF and metabolism. *J. Nucl. Med.* **29,** 876.
23. Devous, M. D., Sr., and Bonte, F. J. (1988). Initial evaluation of cerebral blood flow imaging with a high-resolution, high-sensitivity three-headed SPECT system (PRISM). *J. Nucl. Med.* **29,** 912.

24. Devous, M. D., Sr., Gong, W., Payne, J. K., and Harris, T. S. (1993). Dynamic quantitative Xe-133 rCBF SPECT on the PRISM 3-headed tomograph: Human studies. *J. Nucl. Med.* **34,** 68P.
25. Devous, M. D., Sr., Gong, W., and Payne, J. K. (1993). Dynamic quantitative Xe-133 rCBF SPECT on the PRISM 3-headed tomograph: Simulation and phantom studies. *J. Nucl. Med.* **34,** 91P.
26. Friston, K. J., Worsley, K. J., Frackowiak, R. S. J., Mazziotta, J. C., and Evans, A. C. (1994). Assessing the significance of focal activations using their spatial extent. *Hum. Brain Mapp.* **1,** 210–220.
27. Devous, M. D., Sr., Trivedi, M. H., and Rush, A. J. (2001). Regional cerebral blood flow response to oral amphetamine challenge in normal subjects. *J. Nucl. Med.* **42,** 535–542.
28. Devous, M. D., Sr., Lowe, J. L., and Payne, J. K. (1992). Dual-isotope brain SPECT imaging with ^{99m}Tc and ^{123}I: Validation by phantom studies. *J. Nucl. Med.* **33,** 2030–2035.
29. Devous, M. D., Sr., Payne, J. K., and Lowe, J. L. (1992). Dual-isotope brain SPECT imaging with ^{99m}Tc and ^{123}I: Clinical validation using ^{133}Xe SPECT. *J. Nucl. Med.* **33,** 1919–1924.
30. Obrist, W. D., Thompson, H. K., King, C. H., and Wang, H. S. (1967). Determination of regional cerebral blood flow by inhalation of xenon-133. *Circ. Res.* **20,** 124–135.
31. Kanno, I., and Lassen, N. A. (1979). Two methods for calculating regional cerebral blood flow from emission computed tomography of inert gas concentrations. *J. Comput. Assisted Tomogr.* **3,** 71–76.
32. Celsis, P., Goldman, T., Henriksen, L., and Lassen, N. A. (1981). A method for calculating regional cerebral blood flow from emission computerized tomography of inert gas concentrations. *J. Comput. Assisted Tomogr.* **5,** 641–645.
33. Isaka, Y., Furukawa, S., Etani, H., Nakanishi, E., Ooe, Y., and Imaizumi, M. (2000). Noninvasive measurement of cerebral blood flow with ^{99m}Tc-hexamethylpropylene amine oxime single-photon emission computed tomography and 1-point venous blood sampling. *Stroke* **31,** 2203–2207.
34. Matsuda, H., Tsuji, S., Shuke, N., Sumiya, H., Tonami, N., and Hisada, K. (1993). Noninvasive measurements of regional cerebral blood flow using technetium-99m hexamethylpropylene amine oxime. *Eur. J. Nucl. Med.* **20,** 391–401.
35. Winchell, H. S., Baldwin, R. M., and Lin, T. H. (1980). Development of ^{123}I-labeled amines for brain studies: Localization of ^{123}I iodophenylalkylamines in rat brain. *J. Nucl. Med.* **21,** 940–202.
36. Kung, H. F., Tramposh, K., and Blau, M. (1983). A new brain imaging agent: (I123) HIPDM: N,N,N¢-trimethyl-N′-(2-hydroxy-3-methyl-5-iodobenzyl)-1,3-propanediamine. J. Nucl. Med. 24, 66–72.
37. Kuhl, D. E., Barrio, J. R., Huang, S. C., *et al.* (1982). Quantifying local cerebral blood flow by *N*-isopropyl-*p*-[^{123}I]iodoamphetamine (IMP) tomography. *J. Nucl. Med.* **23,** 196–203.
38. Nishizawa, S., Tanada, S., Yonekura, Y., *et al.* (1989). Regional dynamics of *N*-isopropyl-(^{123}I)*p*-iodo-amphetamine in human brain. *J. Nucl. Med.* **30,** 150–156.
39. Nakano, S., Kinoshita, K., Jinnouchi, S., Hiroaki, H., and Watanabe, K. (1989). Comparative study of regional cerebral blood flow images by SPECT using xenon-133, iodine-123 IMP, and technetium-99m HMPAO. *J. Nucl. Med.* **30,** 157–164.
40. Leonard, J.-P., Nowotnik, D. P., and Neirinckx, R. D. (1986). Technetium-99m-*d,l*-HM-PAO: A new radiopharmaceutical for imaging regional brain perfusion using SPECT—A comparison with iodine-123 HIPDM. *J. Nucl. Med.* **27,** 1819–1823.
41. Neirinckx, R. D., Canning, L. R., Piper, I. M., *et al.* (1987). Technetium-99m *d,l*-HM-PAO: A new radiopharmaceutical for SPECT imaging of regional cerebral blood perfusion. *J. Nucl. Med.* **28,** 191–202.
42. Volkert, W. A., Hoffman, T. J., Seger, R. M., Trounter, D. E., and Holmes, R. A. (1984). Tc^{99m}-propylene amine oxime (Tc^{99m}-PnAO): A potential brain radiopharmaceutical. *Eur. J. Nucl. Med.* **9,** 511–516.
43. Kung, H. F., Guo, Y. H., Yu, C.-C., Billings, J., Subramanyam, V., and Calabrese, J. (1989). New brain perfusion imaging agents based on ^{99m}Tc-bis(aminoethanethiol) complexes: Stereoisomers and biodistribution. *J. Med. Chem.* **32,** 437–444.
44. Walovitch, R. C., Hill, T. C., Garrity, S. T., *et al.* (1989). Characterization of technetium-99m-*l,l*-ECD for brain perfusion imaging. Part 1. Pharmacology of technetium-99m ECD in nonhuman primates. *J. Nucl. Med.* **30,** 1892–1901.
45. Leveille, J., Demonceau, G., De Roo, M., *et al.* (1989). Characterization of technetium-99m-*l,l*-ECD for brain perfusion imaging. Part 2. Biodistribution and brain imaging in humans. *J. Nucl. Med.* **30,** 1902–1910.
46. Devous, M. D., Sr., Payne, J. K., and Lowe, J. L. (1993). Comparison of ^{99m}Tc-ECD to ^{133}Xe SPECT in normal controls and in patients with mild to moderate rCBF abnormalities. *J. Nucl. Med.* **34,** 754–761.
47. Payne, J. K., Trivedi, M. H., and Devous, M. D., Sr. (1996). Comparison of 99mTc HM-PAO to 133Xe for the measurement of regional cerebral blood flow by SPECT. *J. Nucl. Med.* **37,** 1735–1740.
48. Gibbs, J. M., Wise, R. J. S., Leendersbs, K. L., Herold, S., Frackowiak, R. S. J., and Jones, T. (1985). Cerebral hemodynamics in occlusive carotid artery disease. *Lancet* **1,** 933–934.
49. Devous, M. D., Sr., and Arora, G. D. (1991). Direct measurement of rOER, rCBF, and rCBV define a relationship between rOER and the rCBF/rCBV ratio in acute stroke. *J. Nucl. Med.* **32,** 960.
50. Knapp, W. H., Kummer, R. V., and Kubler, W. (1986). Imaging of cerebral blood flow-to-volume distribution using SPECT. *J. Nucl. Med.* **27,** 465–470.
51. Toyama, H., Takeshita, G., Takeuchi, A., Anno, H., Ejiri, K., Maeda, H., Katada, K., Koga, S., Ishiyama, N., Kanno, T., and Yamaoka, N. (1990). Cerebral hemodynamics in patients with chronic obstructive carotid disease by rCBF, rCBV, and rCBV/rCBF ratio using SPECT. *J. Nucl. Med.* **31,** 55–60.
52. Kung, H. F., Alavi, A., Chang, W., *et al.* (1990). In vivo SPECT imaging of CNS D-2 dopamine receptors: Initial studies with iodine-123-IBZM in humans. *J. Nucl. Med.* **31,** 573–579.
53. Ichise, M., Kim, Y. J., Ballinger, J. R., *et al.* (1999). SPECT imaging of pre- and postsynaptic dopaminergic alterations in L-dopa-untreated PD. *Neurology* **52,** 1206–1214.
54. Verhoeff, N. P. L. G. (1999). Radiotracer imaging of dopaminergic transmission in neuropsychiatric disorders. *Psychopharmacology* **147,** 217–249.
55. Kasper, S., Tauscher, J., Küfferle, B., Barnas, C., Pezawas, L., and Quiner, S. (1999). Dopamine and serotonin receptors in schizophrenia: Results of imaging-studies and implications for pharmacotherapy in schizophrenia. *Eur. Arch. Psychiatry Clin. Neurosci.* **249,**(Suppl. 4), IV/83–IV/89.
56. Holman, B. L., Gibson, R. E., Hill, T. C., Eckelman, W. C., Albert, M., and Reba, R. C. (1985). Muscarinic acetylcholine receptors in Alzheimer's disease: In vivo imaging with iodine-123-labeled 3-quinuclidinyl-4-iodobenzilate and emission tomography. *JAMA* **254,** 3063.
57. Innis, R. B., Al-Tikriti, M. S., Zoghbi, S. S., *et al.* (1991). SPECT imaging of the benzodiazepine receptor, feasibility of in vivo potency measurements from stepwise displacement curves. *J. Nucl. Med.* **32,** 1754–1761.
58. Kim, K. T., Black, K. L., Marciano, D., *et al.* (1990). Thallium SPECT imaging of brain tumors: Methods and results. *J. Nucl. Med.* **31,** 965–969.
59. Schwartz, R. B., Carvalho, P. A., Alexander, E., III, Loeffler, J. S., Folkerth, R., and Holman, B. L. (1991). Radiation necrosis vs high-grade recurrent glioma: Differentiation by using dual-isotope SPECT with ^{201}Tl and ^{99m}TC-HMPAO. *Am. J. Neuroradiol.* **12,** 1187–1192.
60. Kaplan, W. D., Takvorian, T., Morris, J. H., Rumbaugh, C. L., Connolly, B. T., and Atkins, H. L. (1987). Thallium-201 brain tumor

imaging: A comparative study with pathological correlation. *J. Nucl. Med.* **28,** 47–52.
61. Brismar, T., Collins, V. P., and Kesselberg, M. (1989). Thallium-201 uptake relates to membrane potential and potassium permeability in human glioma cells. *Brain Res.* **500,** 30–36.
62. Black, K. L., Hawkins, R. A., Kim, K. T., Becker, D. P., Lerner, C., and Marciano, D. (1989). Use of thallium-201 SPECT to quantitate malignancy grade of gliomas. *J. Neurosurg.* **71,** 342–346.
63. Baron, G. C., Lebrun-Grandie, P., Collard, P., Crouzel, C., Mestelan, G., and Bousser, M. G. (1982). Noninvasive measurement of blood flow, oxygen consumption and glucose utilization in the same brain regions in man by positron emission tomography. *J. Nucl. Med.* **23,** 391–399.
64. Ingvar, D. H., and Risberg, J. (1967). Increase of regional cerebral blood flow during mental effort in normals and in patients with focal brain disorders. *Exp. Brain Res.* **3,** 195–211.
65. Kety, S. S., and Schmidt, C. F. (1948). The nitrous oxide method for quantitative determination of cerebral blood flow in man: Theory, procedure and normal values. *J. Clin. Invest.* **27,** 476–483.
66. Mazziotta, J. C., Phelps, M. E., Carson, R. E., and Kuhl, D. E. (1982). Tomographic mapping of human cerebral metabolism: Auditory stimulation. *Neurology* **32,** 921–937.
67. Phelps, M. E., Kuhl, D. E., and Mazziotta, J. C. (1981). Metabolic mapping of the brain's response to visual stimulation: Studies in humans. *Science* **211,** 1445–1448.
68. Raichle, M. E., Grubb, R. L., Gado, M. H., Eichling, J. O., and Ter-Pogossian, M. M. (1976). Correlation between regional cerebral blood flow and oxidative metabolism. *Arch. Neurol.* **33,** 523–526.
69. Roland, P. E., Eriksson, L., Stone-Elander, S., and Widen, L. (1987). Does mental activity change the oxidative metabolism of the brain? *J. Neurosci.* **7,** 2373–2389.
70. Mazziotta, J. C., Phelps, M. E., Carson, R. E., and Kuhl, D. E. (1982). Tomographic mapping of human cerebral metabolism: Sensory deprivation. *Ann. Neurol.* **12,** 435–444.
71. Devous, M. D., Sr., Stokely, E. M., Chehabi, H. H., and Bonte, F. J. (1986). Normal distribution of regional cerebral blood flow measured by dynamic single-photon emission tomography. *J. Cereb. Blood Flow Metab.* **6,** 95–104.
72. Gur, R. C., Gur, R. E., Obrist, W. D., Skolnick, B. E., and Reivich, M. (1987). Age and regional cerebral blood flow at rest and during cognitive activity. *Arch. Gen. Psychiatry* **44,** 617–621.
73. Hagstadius, S., and Risberg, J. (1989). Regional cerebral blood flow characteristics and variations with age in resting normal subjects. *Brain Cognit.* **10,** 28–43.
74. Kuhl, D. E., Metter, E. J., Riege, W. H., and Phelps, M. E. (1982). Effects of human aging on patterns of local cerebral glucose utilization determined by the ^{18}F fluorodeoxyglucose method. *J. Cereb. Blood Flow Metab.* **2,** 163–171.
75. Mathew, R. J., Wilson, W. H., and Tant, S. R. (1986). Determinants of resting regional cerebral blood flow in normal subjects. *Biol. Psychiatry* **21,** 907–914.
76. Rogers, R. L., Meyer, J. S., and Mortel, K. F. (1990). After reaching retirement age physical activity sustains cerebral perfusion and cognition. *J. Am. Geriatr. Soc.* **38,** 123–128.
77. Baxter, L. R., Mazziotta, J. C., Phelps, M. E., Selin, C. E., Guze, B. H., and Fairbanks, L. (1987). Cerebral glucose metabolic rates in normal human females vs. normal males. *Psychiatry Res.* **21,** 237–245.
78. Daniel, D. G., Mathew, R. J., and Wilson, W. H. (1988). Sex roles and regional cerebral blood flow. *Psychiatry Res.* **27,** 55–64.
79. Gur, R. C., Gur, R. E., Obrist, W. D., *et al.* (1982). Sex and handedness differences in cerebral blood flow during rest and cognitive activity. *Science* **217,** 659–661.
80. Devous, M. D., Sr., Altuna, D., Furl, N., Gabbert, G., Ngai, W. T., Cooper, W., Chiu, S., Scott, J., Harris, T. S., Payne, J. K., and Tobey, E. A. (2001). The functional neuroanatomy of speech and language development in pediatric normal controls. *J. Nucl. Med.* **42,** 146P.
81. Bonte, F. J., Devous, M. D., Sr., and Reisch, J. S. (1988). The effect of acetazolamide on regional cerebral blood flow in normal human subjects as measured by single photon emission computed tomography. *Invest. Radiol.* **23,** 564–568.
82. Rogg, J., Rutigliano, M., Yonas, H., Johnson, D. W., Pentheny, S., and Latchaw, R. E. (1989). The acetazolamide challenge: Imaging techniques designed to evaluate cerebral blood flow reserve. *Am. J. Neuroradiol.* **10,** 803–810.
83. Sullivan, H. G., Kingsbury, T. B., Morgan, M. E., *et al.* (1987). The rCBF response to diamox in normal subjects and cerebrovascular disease patients. *J. Neurosurg.* **67,** 525–534.
84. Vorstrup, S., Brun, B., and Lassen, N. A. (1986). Evaluation of the cerebral vasodilatory capacity by the acetazolamide test before EC-IC bypass surgery in patients with occlusion of the internal carotid artery. *Stroke* **17,** 1291–1298.
85. Mathew, R. J., Barr, D. L., and Weinman, M. L. (1983). Caffeine and cerebral blood flow. *Br. J. Psychiat.* **143,** 604–608.
86. Rush, A. J., Cain, J. W., Raese, J., Stewart, R. S., Waller, D. A., and Debus, J. R. (1991). The neurobiological bases for psychiatric disorders. *In* "Comprehensive Neurology" (R. N. Rosenberg, ed.), pp. 555–603. Raven Press, New York.
87. Petersen, S. E., Fox, P. T., Posner, M. I., Mintun, M., and Raichle, M. E. (1988). Positron emission tomographic studies of the cortical anatomy of single-word processing. *Nature* **331,** 585–589.
88. Fox, P. T., Mintun, M. A., Raichle, M. E., Miezen, F. M., Allman, J. M., and Van Essen, D. C. (1986). Mapping human visual cortex with positron emission tomography. *Nature* **323,** 806–809.
89. Tobey, E. A., and Devous, M. D., Sr. (2000). Functional brain imaging in cochlear implant users. *ASHA* **5,** 90.
90. Roland, P. S., Tobey, E. A., and Devous, M. D., Sr. (2001). Pre-operative functional assessment of auditory cortex in adult cochlear implant users. *Laryngoscope* **111,** 77–83.
91. Nussbaum, R. H. (1998). The linear no-threshold dose–effect relation: Is it relevant to radiation protection regulation? *Med. Phys.* **25,** 291–299.
92. Bond, V. P., Wielopolski, L., and Shani, G. (1996). Current misinterpretations of the linear no-threshold hypothesis. *Health Phys.* **70,** 877–882.
93. Fry, R. J., Grosovsky, A., Hanawalt, P. C., Jostes, R. F., Little, J. B., Morgan, W. F., Oleinick, N. L., and Ullrich, R. L. (1998). The impact of biology on risk assessment—Workshop of the National Research Council's Board on Radiation Effects Research, July 21–22, 1997, National Academy of Sciences, Washington, DC. *Radiat. Res.* **150,** 695–705.
94. Cohen, B. L. (1995). Test of the linear-no threshold theory of radiation carcinogenesis for inhaled radon decay products. *Health Phys.* **68,** 157–174.
95. Jaworowski, Z. (1997). Beneficial effects of radiation and regulatory policy. *Aust. Phys. Eng. Sci. Med.* **20,** 125–138.
96. Little, M. P., and Muirhead, C. R. (1998). Curvature in the cancer mortality dose response in Japanese atomic bomb survivors: Absence of evidence of threshold. *Int. J. Radiat. Biol.* **74,** 471–480.
97. Ron, E. (1998). Ionizing radiation and cancer risk: Evidence from epidemiology. *Radiat. Res.* **150,** S30–41.
98. Pollycove, M. (1998). Nonlinearity of radiation health effects. *Environ. Health Perspect.* **106**(Suppl. 1), 363–368.
99. Mossman, K. L., Goldman, M., Masse, F., Mills, W. A., Schlager, K. J., and Vetter, R. J. (1996). Radiation risk in perspective. Health Physics Society Position Statement. *Health Phys. Soc. Newsl.* **24,** 2–3.
100. Ernst, M., Freed, M. E., and Zametkin, A. J. (1998). Health hazards of radiation exposure in the context of brain imaging research: Special consideration for children. *J. Nucl. Med.* **39,** 689–698.

101. Pollycove, M. (1995). The issue of the decade: Hormesis. *Eur. J. Nucl. Med.* **22,** 399–401.
102. Bogen, K. T., and Layton, D. W. (1998). Risk management for plausibly hormetic environmental carcinogens: The case of radon. *Hum. Exp. Toxicol.* **17,** 463–467.
103. Azzam, E. I., de Toledo, S. M., Raaphorst, G. P., and Mitchel, R. E. (1996). Low-dose ionizing radiation decreases the frequency of neoplastic transformation to a level below the spontaneous rate in C3H 10T1/2 cells. *Radiat. Res.* **146,** 369–373.

IV
Postmortem

20

Postmortem Anatomy

Jacopo Annese and Arthur W. Toga

Laboratory of NeuroImaging, Department of Neurology, University of California at Los Angeles School of Medicine, Los Angeles, California 90095-1769

I. Introduction

Recent advances in the digital technologies, functional neuroimaging, and brain mapping represent revolutionary approaches to basic and applied human brain neuroscience. These approaches to understanding or therapeutically manipulating the human brain are intrinsically three-dimensional (3D) and require both comprehensive and detailed structural concepts. Full exploitation of these new forms of data and the clinical potential of new neuroimaging techniques requires, or indeed mandates, the use of three-dimensional data processing and complex digital tools in order to capture, understand, and manipulate a neuroanatomical framework in three dimensions. One of the main disadvantages of noninvasive imaging techniques is low resolution. There is an increasingly urgent need to develop very high resolution digital atlases or maps of the human brain which can be used to parcel out and to complement lower resolution noninvasive modalities now utilized in both research and clinical neuroimaging.

A complete brain mapping survey should integrate the largest number of complete individual microstructural brain maps with MRI structural templates and functional localization models such as PET, *f*MRI, EEG, and optical imaging. Technically this means going beyond the information given by imaging *in vivo* to resolve macroscopic and microscopic neural structures, making them accessible to morphometric and statistical analyses at the cellular and even molecular level. This aim is dependent on a methodical survey of cerebral anatomy based on serial reconstruction and complex three-dimensional analysis of structure. This chapter describes the procedures involved in brain mapping in the wet lab and the strategies for visualization and analysis of the structures resolved by histological methods.

The first section deals with technicalities associated with the preparation of a brain specimen. We then explain those histological procedures that are intended to gather as much anatomical information as possible from the specimen. In addition to the craftsmanship of making histological slides is the computational effort to resolve and acquire anatomical data at different levels of resolution. We describe the computational and algorithmic requirements for analyzing, integrating, and visualizing those data. Finally, we shall suggest a *modus operandi* to improve the efficiency and the usefulness of neuroanatomical models in the form of detailed statistical and individual microstructural 3D reconstruction. Ultimately, these microstructural digital maps or data spaces may be used as a framework that can provide the underpinning for a new form of comprehensive atlas capable of incorporating three-dimensional data obtained from biochemical, functional, and neurogenetic studies.

II. The Representation of Anatomy

The medical writings of Homer provide vague insights into the anatomical works of the classic era (Kevorkian, 1959). The earliest accounted public dissections were first practiced at Bologna among medieval Universities whose main motive was the gathering of evidence for legal processes (Singer, 1962). There was no "research" in postmortem examination, even of the most elementary kind. Practitioners of medicine relied unquestioningly on the writings of ancient scholars, those of Galen (121–201 A.D.) and Avicenna (Abu Ali Sina, 980–1037) in particular. The authoritative weight of Galen in early medicine transcended the fact that his own investigations were limited to animal dissection. Nevertheless, human dissection before Vesalius (1514–1564) was mainly conducted in a ceremonial fashion to illustrate the concepts explained in the classic texts and eventually to help memorize anatomical detail.

Anatomical illustration stems from this particular need and has since developed into two basic forms: a very schematic representation to aid medical studies and an artistic, yet very precise, portrait of anatomic structures. Anatomical dissection was used by prominent artists like Michelangelo (1475–1564) for their own search of human form, without any intention to teach others, or in the case of Leonardo da Vinci (1452–1519) in combination with scientific purposes (Mathé, 1978). Incidentally, the considerations faced today in digital neuroimaging already might be recognized in nascent form in the work of Leonardo da Vinci, who realized that a full understanding of anatomic structures could be achieved only through multiple-view representations (Fig. 1).

The popularization of anatomic concepts in the Renaissance and Baroque periods can be attributed to book-printing and wood-cutting techniques applied to anatomic illustration and, more essentially, to the combination of careful dissection and figurative skill. Vesalian illustrations (Vesalius, 1543) are attributed to a pupil of Titian (Choulant, 1945). Sir Christopher Wren, the designer of St. Paul's Cathedral in London, was a faithful auditor during Thomas Willis' dissections and authored the renowned engravings for the *Cerebri Anatome* published in Latin in 1664 (Willis, 1664). The precision of Wren's representations of the brain is exemplified by the faithfulness of his engravings to the surface morphology of the cortical mantle (Clarke and Dewhurst, 1972).

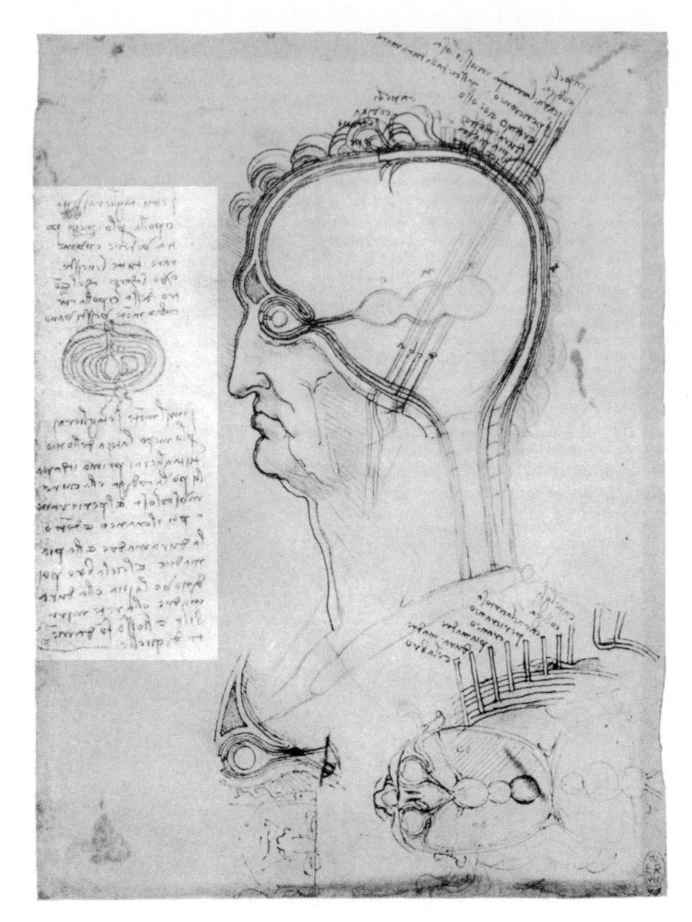

Figure 1 An early drawing (1490–1500) of Leonardo da Vinci that illustrates the descriptions of the brain given by Avicenna (980–1037). He appreciated the idea that cross sections can provide insight into the organization of a three-dimensional object (O' Malley and Saunders, 1952). The highlighted sketch of the onion was intended to illustrate what would be expected when cutting sections through the object. "If you could cut an onion through the middle, you could see and enumerate all the coats or skins which circularly clothe the center of this onion. Likewise if you should cut the human head through the middle, you would first cut the hair, then the scalp, the muscular flesh (galea aponeurotica) and the pericranium, then the cranium and in the interior the dura mater, the pia mater and the brain, then again the pia, the dura mater, the rete mirabile and their foundation, the bone." (Translation from O'Malley and Saunders, 1952; reproduced with permission.) In the 19th century, anatomists used band-saw techniques to produce atlases of the brain based on serial cross sections of different human anatomic specimens through the three orthogonal axes (Gavoy, 1882; MacEwen, 1893). These surveys were more recent efforts at describing the whole volume of the human brain in a comprehensive way.

In the 18th century, lithography, steel engraving, and daguerreotype were popular reproductive techniques, and the precision of anatomical depiction was firmly established, at least for the structures of interest within the brain. This is shown clearly in Sömmerring's own illustrations (Sömmerring, 1796), which were supervised by an artist and an engraver.

By the end of the 19th century major nuclei and even many minor structures of the brain had been mapped, and major pathways of the nervous system were known. This knowledge was represented in comprehensive textbooks of anatomy like the famous "*Big Cunningham*" (Barker, 1899; Cunningham, 1902). Concurrently, the development of histological techniques (Bracegirdle, 1978; Clark and Kasten, 1983) opened the new field of microscopic anatomy, and photography consolidated the absolute accuracy of anatomical illustration.

Modern neuroanatomical studies still rely on time-honored preparation methods and histological techniques. The main advances have been in the ability to localize molecular and genetic markers in the brain and of course in the methods of visualization and analysis of anatomical information acquired from postmortem specimens. Digital imaging has revolutionized anatomical representation by means of image enhancement, segmentation, and 3D reconstruction. Indeed the modern anatomist cannot lack knowledge of the theory and practice of digital image processing (Gonzales and Woods, 1992; Russ, 1999) that has, in a certain sense, taken the place of artistic talent.

III. The Specimen

A. Institutional Procedures

The ideal specimen for postmortem neuroanatomical and histological techniques in the context of brain mapping is the whole unfixed, yet unaltered, cerebrum and cerebellum with the cortical surface intact. Currently, postmortem anatomical sample populations are traditionally minute compared to those of noninvasive imaging programs due to the limitations imposed by institutional policies and methodological constraints. Nowadays a postmortem sample of several hundred human brain specimens such as that surveyed by Sir Grafton Elliot Smith (1871–1937) during his appointment at the University of Cairo from 1900 to 1909 may seem impossible (Smith, 1907; Elkin and Macintosh, 1974).

Particularly in recent years, in most medical institutions the number of autopsies has declined considerably, mainly because of a decreased interest. Interestingly enough, the diagnostic potentials of *in vivo* brain imaging techniques seem to have depreciated neuropathological evaluation at autopsy. Autopsies conducted at the hospital are usually requested by the attending physician; if the patient was recruited into a donors' program the procedure can be potentially immediate, otherwise, the consent of the patient's family is usually required by law. Autopsies are typically performed only by certified technicians or the resident pathologist during working hours on weekdays. The result of these legal and institutional policies is that the postmortem interval, generally taken as the time period from death to the immersion of the specimen in the fixative solution (or freezing), is prolonged severely, often to the point that an otherwise suitable specimen cannot be utilized for research (Fig. 2).

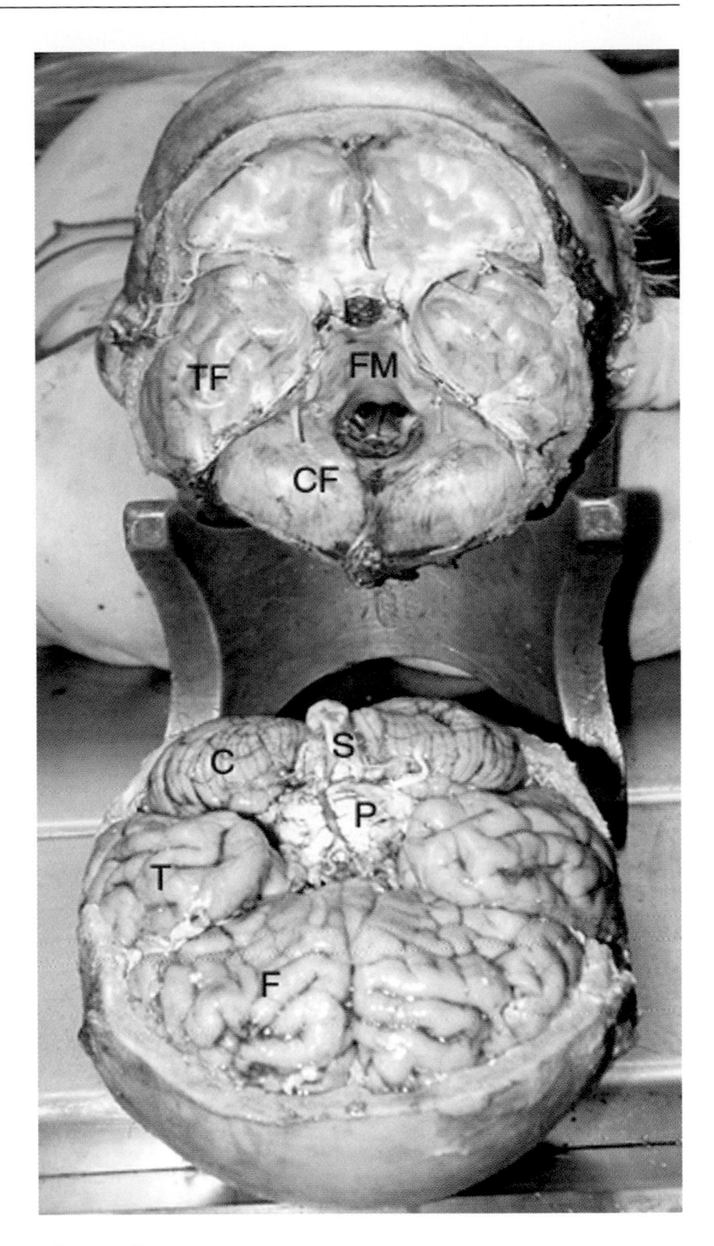

Figure 2 The autopsy procedure. A deep incision is made from behind one ear to the other and the scalp is resected. The top hemisphere of the cranium is then cropped with a vibration saw. Due to the highly inconsistent thickness of the walls of the cranium, deep damage to the surface of the brain is quite common in routine procedures. MRI of the head, if available, can guide the cut through the walls of the skull. The brain and cerebellum are typically dislodged following the resection of the spinal medulla at the level of the foramen magnum (F). The optic and cranial nerves are cut with a scalpel, while the frontal lobes are held away and above the floor of the cranium; the carotids and deep cerebral veins are also severed to free the forebrain. The cerebrum is literally scooped out of the cranium as the tentorium cerebelli is cut in order to release the cerebellum. Alternatively, the brain can be dislodged encased in the calvarium and removed subsequently (Annese, 2002). Because fresh brain tissue is very soft and easily distorted it is quite difficult to handle and must be supported at all times. Archival brain specimens removed from the skull during regular autopsy frequently present damage in the form of deep scars or tearing of tissue. This obviously concerns any systematic topographical study of the cortical mantle. C, cerebellum; CF, cerebellar fossa; F, frontal lobe; FM, foramen magnum; P, pons; S, spinal medulla; T, temporal lobe; TF, temporal fossa.

Following the extraction from the cranium, the brain is ordinarily weighed and subsequently immersed in 10% formaldehyde (a diluted solution of commercial 40% formaldehyde). Because the specimen is heavier than the fixative solution the former is suspended by the carotid arteries to prevent distortion.

Routine pathological examination follows a very conservative protocol. Fixation in formalin lasts generally 2 to 4 weeks at room temperature; the brain is then thoroughly rinsed and cut into thick coronal slabs to inspect deeper structures. In many cases this tissue becomes available for research at a later stage, after small standard blocks are harvested for pathohistological diagnosis. The harvest of these standard sections necessarily injures the complete morphology of the brain. Occasionally, if the fixed brain is presumed normal based on the patient's clinical history and does look normal on the surface it is not sliced, but preserved whole for didactic purposes.

B. Brain Banking

Brain-specimen banking can provide access to postmortem brain tissue that has been prepared and stored to satisfy most research requirements. If the brain is received fresh by the bank it is split midsagittally by a cut through the corpus callosum and other transhemispheric commissures. The cut also severs the brain stem and the cerebellum into two symmetrical halves. One side is prepared for research while the other is sunk in 10% formalin and kept for classic neuropathological evaluation.

Most brain banks follow the latter approach (Tourtellotte and Berman, 1987); however, the design of preservation protocols include aldehyde fixation and cryogenic techniques and may be elaborate enough to accommodate the prerequisites of most neurological studies (Tourtellotte *et al.*, 1993; Vonsattel *et al.*, 1995). Indeed, archival tissue is an invaluable resource by virtue of its consistency, its availability, and the wealth of clinical information that complement each specimen. This strategy of preparation and preservation parcels out the brain into samples for different investigators and research applications. It is therefore unsuitable for a complete and topographically organized anatomical survey of the whole specimen. Another constraint is that most banks have been funded to sustain studies into the causes of neurodegenerative disorders (Jendroska *et al.*, 1993). Normal brains are required and collected for control studies, but their provision is much more problematic.

C. Postmortem Population Statistics

Neurological disease, brain injury, or cerebral infarcts may be the cause of an individual's death, excluding them from a normal sample to be surveyed histologically. Normal aging also can introduce macroscopic (Davies *et al.*, 1994) and microscopic (Peinado, 1998; Peters *et al.*, 2000) changes to the brain. Macroscopic cortical atrophy and widening of ventricular and sulcal spaces can be easily visualized and qualitatively appreciated, while subtle but significant molecular and cellular changes are revealed quantitatively only under microscopic examination (Schulz *et al.*, 1980). This concern is made more complex by indications that specimens of different ages may be differently affected by the histological procedures. For example, the shrinkage that follows the dehydration steps before staining is apparently greater for young adolescent individuals than for older specimens (Sass, 1982; Haug, 1980; Eggers *et al.*, 1984), but only if certain solvents are used (Haug, 1986). It is not known how specimens with a specific pathology, such as Alzheimer's disease, react to histological processing compared to age-matched normal brains.

It is also well known that many biochemical assays on postmortem material are sensitive to the mode of death (Perry and Perry, 1983). Long terminal illness is associated with acidosis of the brain (Perry *et al.*, 1982; Hardy *et al.*, 1985), and the levels of pH in the terminal phase match those that have been found to be neurotoxic in animal models (Siesjo, 1981). Biochemical events in the brain during the agonal state, hypoxia in particular, strongly affect neurochemical parameters such as mRNA preservation (Barton *et al*, 1993; Kingsbury *et al.*, 1995) and enzyme activities (Perry *et al.*, 1982; Yates *et al.*, 1990). Even examined macroscopically, the brain of a patient that was artificially sustained before death will appear severely swollen and edemic. These perturbations are inevitably carried through the histological and analytic procedures and need to be weighed carefully before embarking on a thorough histological investigation that is costly and time consuming.

Aside from issues of specimen "quality," the question of what is a normal human brain specimen is crucial to the development of a representative postmortem anatomic atlas. It is difficult to balance histological sample populations for handedness, gender, age, or other demographically descriptive criteria. Hospital patients or bodies willed to a medical school will most often be of an aged, socially and ethnically homogeneous population, while the average subjects of *f*MRI studies tend to be young students enrolled in the same institution where the scans are conducted. The potential range of brain conditions emphasizes the need to examine carefully individual specimens and their clinical histories before combining the neuroanatomy of different individuals with data from other imaging modalities.

While it is important to appreciate the limitations imposed by the use of postmortem specimens obtained without the luxury of *a priori* selection criteria, it is still necessary to create the largest number of histological maps possible. Also, a consistent set of sampling standards is essential in order to add (not average) individual maps into a common anatomical database.

IV. Preservation of Anatomical Information

Autolysis is the process by which dead tissue digests itself. The specimen has to be fixed as soon as possible to preserve detailed morphology and intact molecular composition. However, differential rates of morbid changes for tissue components of different molecular nature (Barton *et al.,* 1993) mean that specific staining methods will be affected incomparably by the postmortem interval time. For instance, myelinated nerve fibers are quite resilient to decomposition, meaning that the accuracy of topographic studies based on silver impregnation of myelin will be less compromised by longer postmortem intervals compared to other methods. Different neuroanatomical techniques may require mutually exclusive fixation procedures, and careful consideration is often required in choosing a fixation method in order to target the widest spectrum of anatomical information possible.

A. Aldehyde Fixation

The prolonged fixation in 10% formalin of archival tissue, while retaining excellent morphology for classical tinctorial staining, compromises sensitive molecular and neurogenetic applications. Indeed, there are ways to restore the status of morphological components of a tissue that was fixed according to conventional protocols. Microwave technology, among other methods, has been used effectively to retrieve antigenicity and enzymatic activity and to enhance RNA hybridization signal in ethanol- or formalin-archived tissue (Boon and Kok, 1987; Kok *et al.,* 1987; Kok and Boon, 1989; Evers and Uylings, 1994; Newman and Gentleman, 1997; Shiurba *et al.,* 1998; Mueller *et al.,* 2000).

Another relevant yet overlooked consideration is the fact that conventional fixation by immersion of a whole brain wrapped in its own outer membranes is hardly homogeneous at any given time across the volume of the brain. Assuming that very little fixative can infiltrate efficiently between cortical gyri that are tightly packed by the membranes, the agent will have to permeate gradually through the tissue along the total radius of the specimen. This is a gradual process that for a whole human brain can last several weeks. Consequently the deep structures of the brain (e.g., the basal ganglia) and the cortex lining the floors of deep sulci of the mantle (e.g., the fundus of the central sulcus) may still be unfixed after 2 weeks or more of immersion in formalin. One month is a sufficient time for the deeper structures to "catch up" with the surface of the brain. At this point it is generally assumed that the effect of fixation on the tissue will be consistent within the brain and from one archival specimen to another. However, autolysis times will have varied enormously across anatomical structures located at different depths in the brain and even for different regions of the cortical mantle. Therefore, it is likely that the results of quantitative and comparative studies will be affected by the inconsistent and successively delayed fixation. This process might also explain poor and inconsistent yields of RNA and the low specificity of hybridization in archival tissue (but see Mizuno *et al.,* 1998).

All things considered, for the purpose of brain mapping, a practical and consistent method of fixing the whole cerebrum is the ablation of the meninges and the blood vessels before immersing each separate hemisphere in frequent changes of 2–4% paraformaldehyde solution. Because the convolutions of the mantle are no longer held tight by the membranes, the fixative will affect all regions of the cortex simultaneously and will reach deeper structures more quickly. It is much more difficult to remove interstitial meninges after fixation. Removing the pia mater will result in pulling and tearing the cortical gray matter, plus the fixed blood vessels are very resilient and easily cut through cortical tissue.

This method obviously precludes the study of the superficial vascularization of the brain by injection of India ink. However, because the deeper microvasculature would not be affected, angioarchitectonics studies of the cerebrum are still possible (Pfeifer, 1930; Duvernoy *et al.,* 1981). A drawback of this method in a highly convoluted specimen like the human brain is, of course, that the *in vivo* 3D profile of the mantle will be severely distorted, making the integration of postmortem maps with multimodality atlases more problematic. The MRI volume of the same specimen, acquired before histological processing, may serve to register the 3D reconstruction from histological sections to the original shape of the brain (Schormann and Zilles, 1998; Annese *et al.,* 2001). By this method, anatomical data that have been preserved consistently throughout the specimen can be restored to their original topology (Fig. 3).

The brain can also be perfused *in situ* by pumping the fixative solution into the carotid arteries shortly after death (Miller, 1998). Notwithstanding the collapse of the ventricles that is the consequence of the morbid lack of cerebrospinal fluid pressure, the brain is hardened into a fairly natural shape. It is possible that local blood clots or lesions may prevent the fixative from diffusing into certain parts of the tissue. Accordingly, postfixation by immersion is often required after the autopsy nonetheless. The removal of the brain after perfusion is simple mainly because the brain is harder and it has shrunk considerably in the cranium. On the other hand, problems may arise removing the membranes from the cortical mantle. Once fixed, the meninges do not detach easily and can pull and damage the cortex.

Highly susceptible methods, like immunocytochemistry with monoclonal antibodies (MAB) and *in situ* hybridization (ISH), are not compromised by a balanced fixation in 2–4% paraformaldehyde solution. It is therefore possible to couple the optimal preservation of the "molecular ecology" of the tissue (mRNA, enzymes, and structural proteins) with good histologic structural detail. In practice, tinctorial,

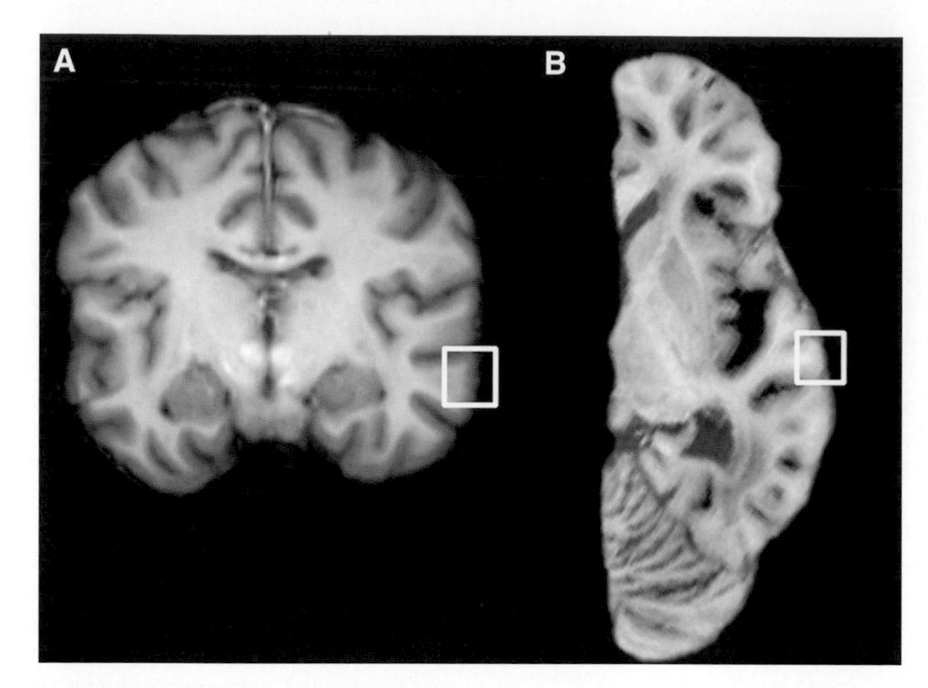

Figure 3 (**A**) The fresh specimen can be imaged much like the brain of a living subject, using a T1-weighted 3D SPGR sequence (GE systems) or T1-weighted 3D FLASH scan (Siemens systems). The scan shown is the average of two consecutive scans, which were acquired using a 3D fast SPGR sequence (TR = 12; minimum TE; flip angle 60°, 4 NEX) in a 1.5-T GE Signa LX (dimensions 192 × 256; minimum TE; slice thickness 1.0 mm). Decent contrast can be obtained in fixed tissue using a fast spin echo (FSE) sequence. The fixed specimen in (**B**) was imaged in a 3-T GE Signa 5.3 with a T2-weighted FSE (dimensions 256 × 256; TR = 5.5; TE = 0.03; flip angle 90°). Voxel values in (B) have been inverted to better compare the contrast with the T1-weighted scan (A). In both cases, during the scans the specimen was immersed in phosphate-buffered saline solution. The fresh brain was housed in a life-size plastic replica of the skull while the fixed brain was rigid enough to be simply resting at the bottom of the container. The higher contrast between gray and white matter in the fresh brain T1-weighted image (A) may justify the difficulties in arranging the scan immediately after the autopsy, yet, MRI of the fixed tissue can be integrated into the histological protocols much more efficiently. The theoretical limit of the resolution in NMR images is determined by the point-spread function of the imaging device (Rossmann, 1969; Robson *et al.*, 1997; Roberts *et al.*, 2000). In practice, in addition to shifts of physical parameters, the postural and physiological movements of the living subject decrease the theoretical precision of the acquisition. This is obviously not a concern in the case of postmortem brains. Accuracy can be enhanced by increasing the number of acquisitions during the same scan, or by averaging different scans, provided no movement occurred between sessions. The autolytic process in the fresh specimen would continue undisturbed during prolonged scans, damaging the anatomical information. On the other hand, fixed tissue could be scanned indefinitely at room temperature. Kruggel *et al.* (2001) have prolonged the scan of a brain fixed in formalin for 12 h. The scan yielded a high enough signal-to-noise ratio to be able to parcel out the cortical mantle into regions that may correspond to architectonic fields, based on voxel intensity profiles.

histochemical, and molecular techniques can be applied to interleaved microtome sections from the same specimen.

B. Cryopreservation

Fresh tissue may also be frozen and stored indefinitely at –70°C without fixation. Fresh-frozen samples are preferred for the isolation of intact genomic DNA or mRNA (Naber *et al.,* 1992). However, freezing can disrupt the morphology of the tissue. The severity of freezing artifacts is less if the tissue freezes very rapidly. Artifacts are formed as cavities in the tissue, caused by the expansion of ice crystals in the neuropil. Relatively thin slabs (1–2 cm thick) cut from the whole brain can be frozen quickly enough in between conducting metallic plates that are precooled to –70°C, bringing the slice to a core temperature of –70°C in approximately 10 min (Nochlin *et al.,* 1993). The tissue can also be snap-frozen by immersion in liquid nitrogen or in isopentane (2-methylbutane) chilled with dry ice to –75°C. Samples larger than a few cubic centimeters, or bigger, such as a whole human hemisphere, cannot be fresh-frozen effectively without risk of damage. Most commonly, deep crevasses occur due to opposing dynamic temperature fronts in the tissue that solidifies. Optimal freezing methods have been evaluated by implanting thermocouples in the tissue or in the embedding medium to measure freezing rates empirically (Rosene and Rhodes, 1990; Nochlin *et al.,* 1993).

There are applications for which freezing is the only procedure that can be considered for tissue preservation. Receptor autoradiography *in vitro* is a very powerful method for topographical brain mapping at the microscopic level (Zilles *et al.,* 1991, and see Chapter 21 in this book). The preparation of the specimen in this case can follow only a definite routine which does not permit any fixation.

V. Preparing the Specimen for Cutting

Even though the main purpose of fixation and freezing is to preserve the morphology and the molecular ecology of the tissue, a consequence of both is hardening of the tissue. Indeed several compounds with embalming properties were used for hardening alone. Notably, Vicq D'Azyr (1786) used alcohol in his work on the brain essentially to toughen the specimen before dissection. Basically, in order to cut slices of even thickness consistently with the microtome it is necessary for the tissue block to be resilient to the distortion caused by the knife blade. The process of embedding was developed to meet these requirements more efficiently, starting from organic materials (originally waxy liver from the dissection table) that supported the delicate biological specimen against the cutting edge of manual microtomes (Bracegirdle, 1978).

A. Infiltration with Embedding Compounds

One has to make a distinction between simple embedding, which is surrounding the surface of the sample, and true infiltration. The latter is a gradual process by which the medium actually diffuses into the tissue. Infiltration with paraffin wax and Celloidin (a proprietary brand of collodion; Schering, Berlin) displaced other embedding media introduced in the second half of the 19th century. The former continues to be used widely in laboratories for routine staining. The tissue is dehydrated in alcohol and immersed in the medium, preferably under vacuum conditions to remove air bubbles and speed up the diffusion process. Celloidin has been used in combination with paraffin (Tyrer, 1999) and without infiltration to prepare whole human hemispheres for thick (>100 μm) serial sections (Heinsen *et al.,* 2000).

The use of plastics or resins (Gerrits, 1988; Kingsley and King, 1989; Hand, 1995; Hand and Phil, 1999), compared to other methods of infiltration, produces minimal intrinsic and perimetric distortion of the histological section. In other words, the relative geometrical alignment between cellular components (neuronal and glial elements) and structural assemblies (nuclei and fiber systems) is maintained within one slice and between sections. One further benefit of this method is that large plastic sections can be cut as thin as 1–2 μm and is therefore exceptional for high-resolution microscopy. The most common plastic compounds used in histology are hydrophobic [methyl methacrylate (MMA)] or hydrophilic (glycol methacrylate, employed in electron microscopy, especially) esters of acrylic acid.

Before the infiltration with paraffin wax, Celloidin, and MMA, the specimen will have to be dehydrated completely with acetone or through graded solutions of alcohol. Both processes take a considerable amount of time to complete, which is a major drawback of these procedures. A whole hemisphere may take as long as several weeks to dehydrate completely and just as long to be infiltrated *in toto*. Another consideration is that MMA plastic sections tend to curl when they are cut and they need to be flattened in a water bath above 50°C in order to be mounted on glass. These steps may preclude certain staining techniques, mainly as a consequence of exposure of the tissue to high temperatures and organic solvents. Nevertheless, the possibility of using MMA for most neuroanatomical techniques has been demonstrated (Hand and Phil, 1999; Mueller *et al.,* 2000). Plastic infiltration is theoretically more advantageous, but it has not been widely tested empirically on very large blocks of brain tissue.

B. Frozen Sections

Large brain sections can be cut frozen from a block of tissue that has been adequately fixed. This method, introduced by Lewis (1877, 1879a,b; Le Brun-Kemper and Galabaruda, 1991), avoids the long delays imposed by the embedding process and produces sections that are not adulterated by exposure to high temperature or organic solvents, as we have mentioned.

Large tissue blocks are frozen with pulverized dry ice or with cold fluids like liquid nitrogen or isopentane as described earlier. The same issues that concern cryopreservation are relevant to the preparation of a frozen block for cutting. The objective is the same: the specimen is to be frozen as quickly as possible to control for ice crystal formation within the tissue. The freezing rate by immersion in –75°C isopentane is roughly twice as fast as that obtained with dry ice (Rosene *et al.,* 1986), but liquid cryogens boil in contact with the warmer tissue, creating an insulating gaseous gap that drastically reduces the rate of cooling.

However, fixed tissue lends itself to treatment with cryoprotective agents, thereby reducing the preoccupation with very rapid freezing. Laboratory-grade sucrose is the most popular compound used to prepare fixed brain tissue for freezing. Even though a combination of glycerol and dimethyl sulfoxide (DMSO) has been shown to be more effective than sucrose (Meryman, 1966, 1971; Rosene *et al.,* 1986), these cryoprotecting agents are not used commonly. This is possibly because the effect of DMSO on histological procedures has not been assessed quantitatively; however, no deleterious repercussions have been reported (Rosene and Rhodes, 1990). Cryoprotective methods are obviously precluded to fresh unfixed tissue because autolysis would continue unopposed while the long infiltration occurs.

For frozen blocks, in addition to cryoprotection, some mechanism is needed to maintain the spatial coherence of structures that would naturally dissociate in the sections, such as the apical portions of the convolutions of the mantle. Gelatin embedding—without infiltration—is a simple and effective procedure (Heinsen and Heinsen, 1991; Hine and

Figure 4 The manufacture of manual rotary microtomes and modern motorized versions of the latter has been traditionally tailored to section small specimen embedded in shallow paraffin blocks. Two large lever-operated sledge designs, both from 1910, are notable exceptions and remain perfectly practical and operational nowadays. One model by Jung was called the "Tetrander" (Mayer, 1910). A long lever pulls the object holder under the fixed blade, and a conveyer belt carries the continuous paraffin ribbon away from the cutting edge. In the same year Cambridge Scientific Instrument Company produced a remarkably similar design (Cambridge Scientific Instrument Co., 1910). Both machines allow one to section very large surfaces (up to 120 × 160 mm). In addition, because of the sturdy lever system, these microtomes are very suitable for cutting large sections from a whole frozen human brain.

Rodriguez, 1992; but see Fig. 7). In practice, the hemispheres can be coated with multiple layers of warm gelatin that congeal at room temperature, after which a brief immersion in cold paraformaldehyde solution hardens the gelatin and makes it insoluble.

VI. Histological Slides

The introduction of true serial sectioning was largely dependent on the newly discovered properties of embedding materials. The observation that paraffin blocks could be cut into uninterrupted ribbons of sections led to the design of the first automatic microtome (Threlfall, 1930) and consequentially to the rapid and economical production of sections of known constants (Bracegirdle, 1978).

A. Blocking

The way to approach a whole human brain specimen for cutting depends critically on the equipment at hand. If a sufficiently large microtome is not available, the decision to sample smaller blocks will be inevitable. Notably, the custom-made microtome that Korbinain Brodmann had designed for his cytoarchitectonic work (Garey, 1994) could accommodate only blocks that measured 3 × in., so that for his human studies he was obliged to cut the brain into smaller chunks (Le Brun-Kemper and Galaburda, 1991). Traditional practice has had a strong influence both on the design of equipment and on the choice of a sampling strategy. In most cases, large microtomes that have been commercially available are limited in the size of the specimen that can be sectioned; this is because they were built to cut standard shallow slabs of tissue that were embedded in paraffin according to conventional clinical histological routines (Fig. 4). In order to cut an intact hemisphere it is often necessary to implement drastic modifications to the microtome (Annese, 2002).

If only a small well-defined area is to be evaluated it is convenient to target the smallest possible volume, because sampling smaller blocks enables fast fixation and embedding processes; but the aim of brain mapping is to conduct a large-scale survey on the entire brain specimen. In this case, rebuilding the whole from smaller blocks is not only cumbersome, it also leaves room for a high degree of imprecision. Whole brain sections (Yakovlev, 1970) allow the direct and clear visualization of structures of interest and their exact spatial relationship. This approach, even though it necessarily involves the implementation of custom-made and prototypical hardware, is indeed the gold standard for brain mapping at histological resolution.

It should be mentioned that a very elaborate method of sampling was used by the scholars of cytoarchitectonics to map the cerebral cortex. Several hundred blocks were cut from the mantle at a right angle to the surface and in order to study the organization of cortical layers (Bolton, 1900; Campbell, 1905; Vogt and Vogt, 1912; Economo and Koskinas, 1925, 1929; Bok, 1959).

B. The Cutting Procedure

Cutting large sections from a paraffin, plastic, or Celloidin block is straightforward on rotary or sledge microtomes, depending on the hardness that the embedding material has achieved (Voogd and Feirabend, 1981; Hand and Phil, 1999). The paraffin film or plastic sections can be retrieved with tweezers and stored dry until staining and mounting.

Sectioning a frozen block poses different problems and is best accomplished on a large sliding microtome equipped with a freezing stage for the specimen. Alternatively, an aluminum stage large enough to accommodate sufficient dry ice can maintain low enough temperatures (see Fig. 5).

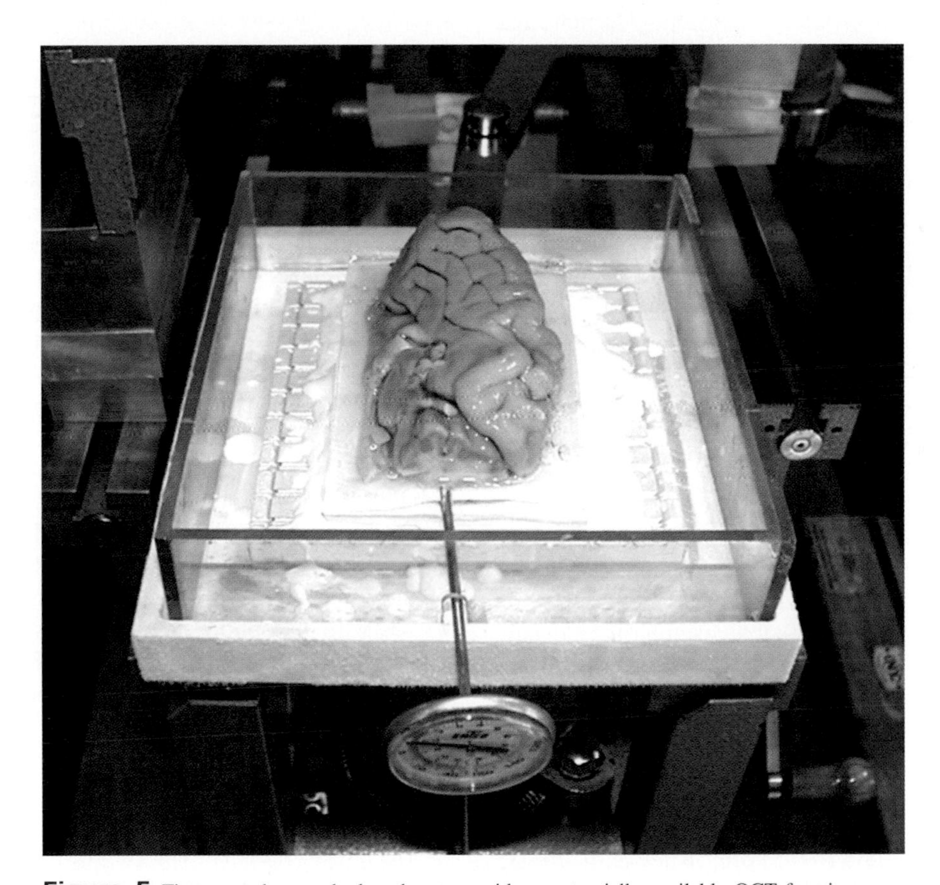

Figure 5 Tissue can be attached to the stage with commercially available OCT freezing compounds (Tissue-Tek, Sakura Finetek U.S.A., Inc.). In the case of large blocks, however, it is preferable to attach the *unfrozen* tissue block to the base of the stage that had been cooled to –30 or to –75°C in a freezer. A layer of frozen distilled water, or better, a diluted sucrose solution (10%), can act as an intermediate pedestal that adheres to the grooved floor of the stage. The brain can, at this point, be covered by pulverized dry ice or lowered in a bath of cooling fluid. If the temperature of the stage rises above –30°C, the block may be pulled off completely. To thaw and reposition the tissue means losing a significant number of sections before the plane of cut is restored. Frequent changes of dry ice are needed to maintain the temperature of both the block and the stage sufficiently low. However, immediately after each change of dry ice the tissue will be too cold and some time is needed to reach the most suitable temperature for cutting (24–30°C). Frequent interruptions and delicate adjustments make the overall cutting procedure a time-consuming effort. Temperature can be monitored with a thermometer implanted in a gelatin base that is attached to the specimen and the stage.

It is extremely important, in fact it is possibly the most crucial step, to affix the block firmly on the stage before cutting. If the frozen block does not remain solidly attached to the base of the stage, the sections will have unpredictable thickness as the knife blade sinks into or pulls through the tissue surface. Unfixed specimens need to be cut on a microtome that is housed in a refrigerated unit (cryomicrotome) or the sections would "dissolve" when they thaw on the blade, at room temperature.

The speed at which a section is cut is crucial and depends on three factors: its thickness, the temperature of the tissue, and the dimensions of the block face. Sucrose or glycerol infiltration facilitates cutting by making the frozen brain more malleable, while fresh-frozen tissue tends to shatter as the blade chatters on the surface. In both cases, the only true way to obtain smooth and intact sections is to calibrate carefully the speed of cut. Cutting the block too fast results in sections that are brittle; cutting too slow means altering the thickness of the section as the surface thaws in contact with the knife. Indeed one common concern regarding frozen sections is that their thickness can be quite variable in a capricious fashion, but this can be effectively minimized by a temperature-controlled freezing stage. At room temperature, the knife cuts and thaws the sections that ripple on its upper surface, and then these are picked up with a wet brush and deposited in sequence in wells containing the appropriate buffered storage solution.

The time to cut a frozen section from a fixed cryoprotected hemisphere at the level of thalamus along its dorsoventral axis may take up to 10 s. The overall process of cutting a hemisphere coronally will last approximately 20 h. By the end of the cutting procedure, a complete series of sections is stored sequentially in individual wells containing storage buffer solution until staining. It should be mentioned that every solution that is in contact with the tissue should always be buffered by salts (e.g., phosphate or sodium acetate) to maintain suitable osmolarity and balanced pH. These are parameters that affect the state of the tissue and the processes of fixation and staining.

Needless to say those techniques that target more labile elements or processes in the tissue require priority. The intensity of enzymatic reaction (Rosene and Mesulam, 1978; Horton and Hedley-Whyte, 1984; Silverman and Tootell, 1987; Wong-Riley, 1979; Wong-Riley *et al.*, 1993) in the tissue deteriorates quickly with time even in sections from tissue fixed with paraformaldehyde. ISH and MAB protocols also will have priority over structural stains like Nissl and silver-impregnation methods (see below), as the latter are much more resilient to delays and storage conditions. Storing sections in buffers at 4°C for long periods of time can ruin the material. Formalin is often added to the buffer in order to preserve the sections, but it does not prevent contamination of the wells, especially against fungal spores (sodium azide can be used more effectively for this purpose). Sections can also be frozen in separate wells in a cryoprotective solution (glycerol 20% and 5% DMSO) and stored at –70°C; they can be otherwise stored frozen already mounted on glass. Sensitive histochemical, immunological, and molecular techniques may be resumed successfully at a later stage provided due care has been taken to prevent freeze-drying.

VII. Histological Methods

If the cutting process was not perturbed by one or several possible mishaps mentioned above, the series of sections may be assumed coplanar, of even thickness, and most importantly, uninterrupted. Perhaps the most significant advantage of postmortem material is the ability to employ a variety of histological recipes to the tissue to resolve different anatomical data. In practice this means assigning a particular interval in the series to each particular stain so that adjacent sections will be processed for different purposes.

Obviously the larger the distance between two sections, the greater the perimetric variability (see Fig. 7). This is a factor to consider in sampling for the stain and the final 3D reconstruction. In the following sections the terms histology and histological will be used in their widest sense, including those techniques that fall more appropriately into the category of enzyme histochemistry, immunohistochemistry, and nucleotide hybridization (ISH) methods.

A. Topographic Stains

Classical histological topographical localization is the subdivision of the brain into regions that have a homogeneous intrinsic organization and discrete borders (Brodmann, 1909; Garey, 1994). Major structural domains of the brain are best visualized by staining neuronal cell bodies with basophilic dyes like cresyl violet or thionine (Nissl, 1894, 1910; but see Merker, 1983, and Uylings *et al.*, 1999). The dyes interact with nucleic acids in the cell (DNA and RNA) and therefore stain both the nucleus and the cytoplasm. As a result, individual neurons clearly stand out colored blue or purple against the background (Simmons and Swanson, 1993; see Fig. 6A).

The Nissl stain is, indeed, the common currency for exchanging neuroanatomical topographical information and is also used to *counterstain* sections that were processed histochemically or with monoclonal antibodies for an orthodox anatomical reference.

Silver-impregnation techniques (Voogd and Feirabend, 1981) may complement Nissl stains when they are used to differentiate neural structures on the basis of the density and disposition of myelinated fibers. Most clinical laboratories employ a relatively simple method of staining fibers with Luxol fast blue, a dye that binds to the porphirines present in the myelin (Klüver and Barrera, 1953). The most sensi-

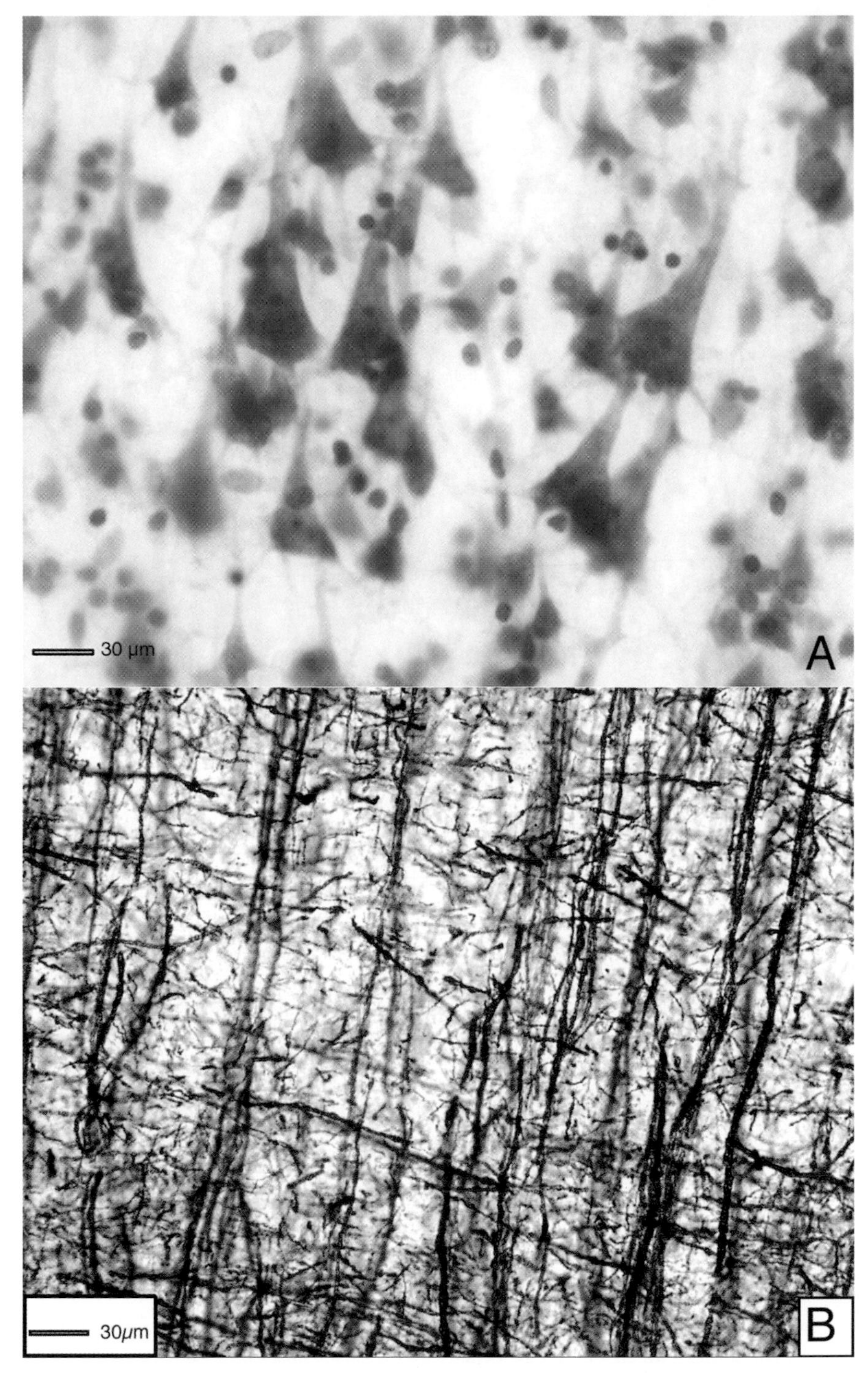

Figure 6 (**A**) Nissl-stained pyramidal neurons in layer III of the human visual cortex. (**B**) The same region stained with the reducing silver technique for myelinated fibers (Gallyas, 1979). Both images were acquired at 40× magnification, at a resolution of 1024 × 1024 dots per inch.

tive method to date, and one that provides greatest structural detail, employs the selective reducing property of myelin at a very specific pH. Colloidal silver is deposited on myelinated axons (Gallyas, 1971, 1979) and the tissue is then developed with the same physical process used in photography. The tissue eventually can be "bleached" (in potassium dichromate) to enhance the contrast of the fibers against the background (see Fig. 6B).

In order to resolve topographical variations, and therefore to segment discrete structures of interest, the section

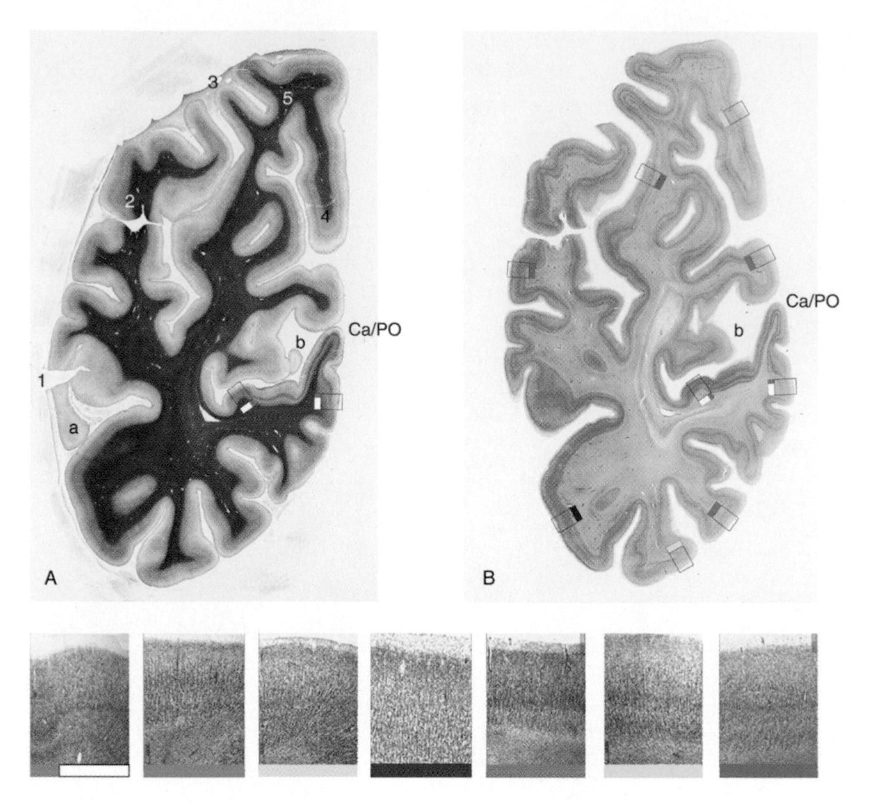

Figure 7 Two coronal sections (60 μm) through the occipital pole at the level of the junction between the calcarine sulcus and the parieto-occipital sulcus (Ca/PO) (Ono, 1990). Section (**A**) is stained for myelinated fibers with a modified reduced silver method (Gallyas, 1979). Section (**B**) is stained with cresyl violet (Nissl stain). In most regions of the brain, especially in the cortex, the differences in the tangential and radial distribution of myelinated fibers are far more conspicuous than cellular inhomogeneities. In other words, at very low power magnification, the borders between two myeloarchitectural types are in every instance sharper than transitions in cytoarchitecture [compare the cortical mantle in (A) and (B)]. This is especially true for the boundary between the primary visual cortex and its bordering area (gray–white border in the flag). The region that is coded gray was seen on fresh tissue more than two centuries ago (Gennari, 1782; Vic D'Azyr, 1876). This "striate cortex" is the exact cortical representation of the entire visual field (Kuljis, 1994). This field is visible also in Fig. 11, in which the two above-mentioned stains are combined. The color-coded boxes below the sections contain the cytoarchitecture of individual cortical areas at 25× magnification. Also shown are several macroscopic artifacts that can occur in processing the tissue. The numbers refer to markers: (1) The gelatin that embedded the whole block shrank when the sections were air dried before staining, pulling and tearing a portion of the tissue (a). (2) Damage that occurred during the autopsy procedure. (3) Gelatin staining. Silver in the gelatin was not removed completely in the differentiation process that was optimized for contrast in the cortex. This staining would interfere with any automatic contour generation to represent the perimeter of the section. (4) Although it is less noticeable macroscopically, a string of dust was attached to the mounted section and prevented silver from depositing on the tissue. (5) Minor knife chatter. This side of the block was too cold although it did not offer any resistance to the knife. A careful comparison of both sections reveals several differences in shape (more noticeably those labeled a and b). The sections are 120 μm apart [(B) is the second section cut after (A)], enough distance to produce a reasonable morphological shift. A comparison between different stains in the same tissue that involves the contour-based registration of homologous sections (see for example Cohen *et al.*, 1998) must take into consideration this effect and the presence of damage that can vary from one section to the next.

must be viewed at low-power magnification so as to have an "impressionistic" view (Le Brun-Kemper and Galaburda, 1991). The borders of a cortical area or a deep cerebral nucleus are identified in the sharp (or gradual) zones of transition of fiber distribution or cellular density, respectively.

Human and comparative histologic collections are composed of hundreds of large-format Nissl- and myelin-stained histological glass slides. The largest collections were the Yakovlev collection in the United States and the Vogt collection in Europe (Vogt and Vogt, 1942; Payne, 1967; Yakovlev, 1970; Kretschmann, *et al.*, 1979). The material has been used as a valuable resource for morphometric studies (see, for example, Kretschmann *et al.*, 1979; Kaufman and Galaburda, 1989; Arnold *et al.*, 1991).

B. Morphological Stains

A particular anatomical territory can also be characterized via the localization of individual histological elements using methods that label neurons in their entirety. These methods are used to visualize neurons clearly for detailed morphological and comparative studies (Valverde, 1970). The Golgi technique, the *reazione nera* (Mazzarello, 1996), is a rather elegant procedure to reveal the individual morphology of brain cellular elements with a clarity and faithfulness unmatched by other histological methods. After immersion in osmium dichromate the tissue is impregnated with a weak solution of silver nitrate, and staining occurs as the result of the deposition of silver particles (Golgi, 1878–1879) in a final reducing solution. Several modifications have been implemented to allow this technique to coexist with other histological methods and also for frozen sections (Ebbesson and Cheek, 1988). It is possible, therefore, to combine this accurate description of neuronal elements with topographic and other morphological stains.

The intriguing aspect of the Golgi stain is that only a fraction of the cells in the sample of brain tissue are stained in an "all or nothing" fashion. This mechanism allows the clear visualization of entire neurons and their processes, as there is not any background interference from adjacent cells. The phenomenon is not yet understood. One theory relates the capriciousness of staining to the level of activity of different neurons before death (Bertram and Sheppard, 1964, Spacek, 1992). If this were true the Golgi stain would possess functional significance (Fig. 8).

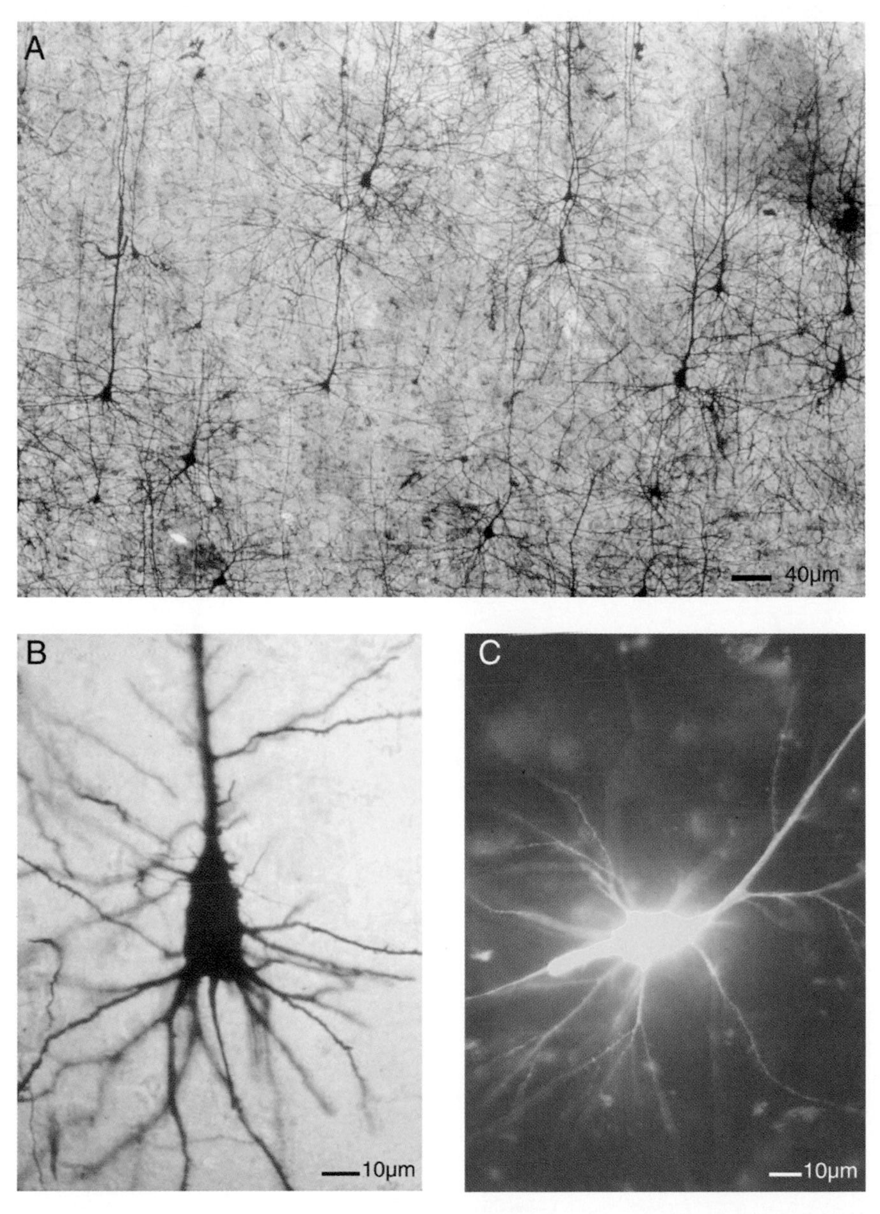

Figure 8 (**A** and **B**) Preparation of the human cortex stained by the "rapid Golgi method" (Scheibel and Scheibel, 1970; Valverde, 1970) showing the "black reaction" in a population of pyramidal neurons (Arnold B. Scheibel, unpublished data). The block of nervous tissue is first immersed in an osmium dichromate solution. It is then impregnated for several days in a weak silver solution. This process is usually repeated three or four times. The Golgi method can be applied to tissue that has been fixed with aldehyde compounds (paraformaldehyde). It can therefore be combined with most other methods. (**C**) Fluorescence microscopy photomicrographs of Lucifer yellow-filled classical pyramidal neuron in the human entorhinal cortex. (Modified from Mikkonen *et al.*, 2000, with permission). The scale is the same as in (A). Compare (B) with (C).

Intracellular injection of fluorescent Lucifer yellow in fixed tissue also provides a Golgi-like stain for morphological analysis of cellular elements (Belichenko, 1991; Buhl, 1987; Belichenko and Dahlstrom, 1995; Elston and Rosa, 1997; Mikkonen *et al.,* 2000).

These methods yield very high contrast so that edge detection and thresholding algorithms to select features of interest automatically are relatively effective.

C. Immunohistochemistry

Histochemical techniques localize neurotransmitters and enzymatic reactions in the brain based on the formation of an insoluble precipitate. This precipitate can be the product of an enzymatic reaction that can indicate a particular metabolic state (Horton and Hedley-Whyte, 1984; Silverman and Tootell, 1987; Wong-Riley *et al.,* 1993) or the presence of neurotransmitter such as acetylcholine (Tago *et al.,* 1986). Histochemical techniques have been corroborated and superceded by immunocytological methods by virtue of their more sensitive and specific localization. In fact, immunization protocols and the hybridoma technique (Kohler and Milstein, 1975; Goding, 1996) have effectively allowed a more direct bridge between structural and functional characterization at the microscopic level. These methods are based on the concept of antigen–antibody interaction and have succeeded in creating very specific and consistent staining agents demonstrating the topographic distribution of distinct classes of molecules (Barnstable, 1980; Zipser and McKay, 1981; Trisler *et al.,* 1981; Hawkes *et al.,* 1982a,b; Hockfield and McKay, 1983). It is assumed that the molecular phenotype that is localized with antibodies may account, in part, for the physiological properties that differentiate neuronal populations. Notably, this methodology has identified structural pockets in the cortex that relate to specific cortical functions in primates (DeYoe *et al.,* 1990; Hof and Morrison, 1995; Hof *et al.,* 1995) and man (Hockfield *et al.,* 1990; Tootell and Taylor, 1995).

Lectins are sugar-binding proteins that interact with glycoconjugates in a manner similar to antigen–antibody interactions. Binding is very specific because each lectin has an affinity for a particular subunit in the carbohydrate chain. All biological membranes, including the excitable membrane of neurons, contain glycoconjugates, and therefore it has been possible to stain specific neuronal populations with lectins derived from a variety of plant sources (Seeger *et al.,* 1996; Hilbig *et al.,* 2001). A great number of lectin species are available commercially and only the staining properties of a few have been examined to date (Fig. 9).

Topographic and morphological stains are not mutually exclusive. In fact it is important to use the former to define territories where neural elements are localized and modeled in detail (Fig. 10).

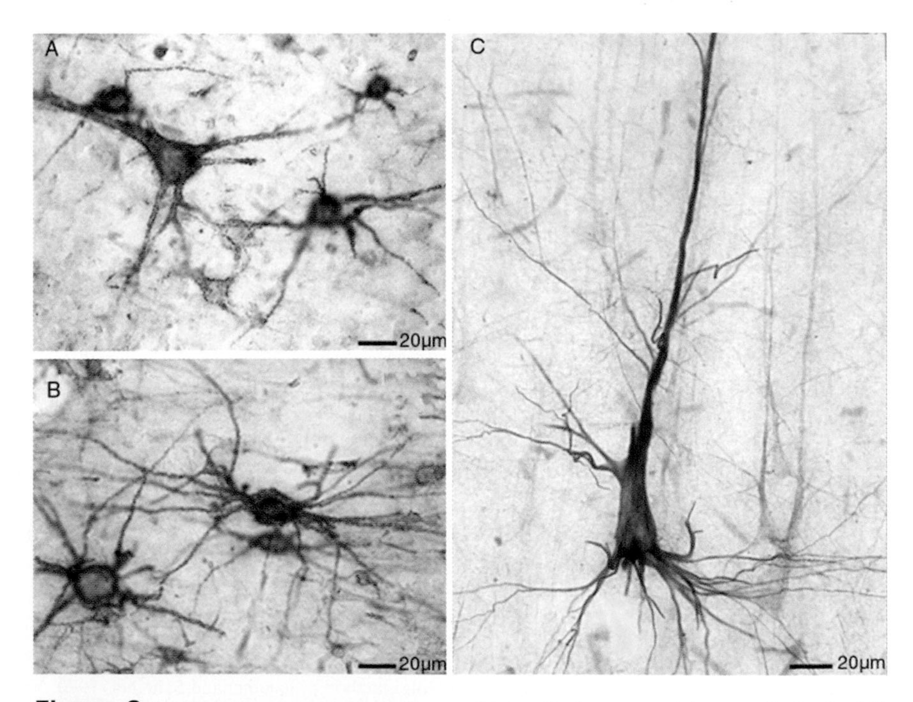

Figure 9 (**A**) Staining of multipolar and pyramidal cortical neurons with the antibody Cat-301 (Hockfield and McKay, 1983; Hendry *et al.,* 1988; DeYoe *et al.,* 1990). A very similar tinctorial pattern (**B**) is expressed by *Wisteria floribunda* agglutinin, a plant lectin that labels residues of glycoproteins within the extracellular matrix. (**C**) The monoclonal antibody SMI32 (Hof and Morrison, 1995) labels only discrete subclasses of pyramidal neurons.

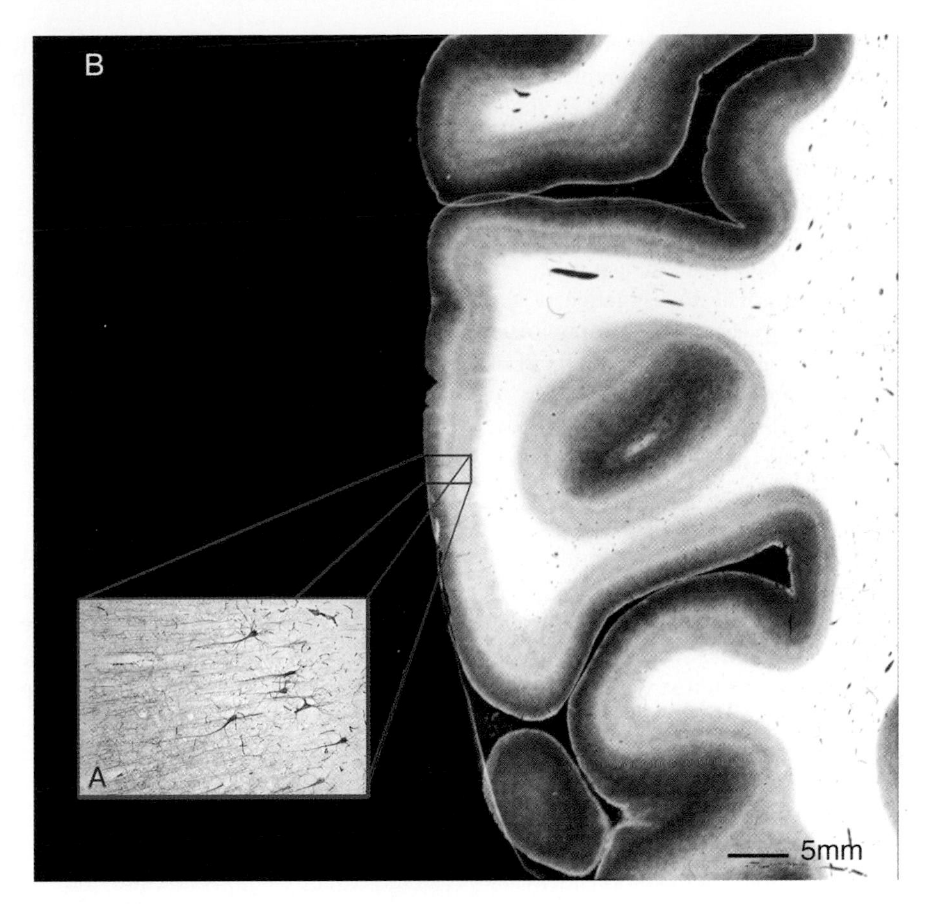

Figure 10 Localization of a subpopulation of pyramidal neurons stained with the monoclonal antibody Cat-301 (Hockfield *et al.,* 1990) within a well-defined myeloarchitectonic area in the visual cortex. Both the specific staining pattern with the MAB Cat-301 and a conspicuous double-striate fiber architecture help to identify this area in the visual cortex that has been shown to be involved in the perception of motion (Watson *et al.,* 1993). Topographical and elemental morphological analysis can be effectively combined by applying multiple histological techniques to the same section or to adjacent sections in a series. This not only means mapping the individual but also provides valuable insight on the interaction of structural—and functional—modules that form the complex architecture of the brain.

D. Fiber Tracts

Even a detailed knowledge of neural structures of the brain is not sufficient in the understanding of brain function if their connectivity is not established. However, in the adult, it is not possible to distinguish between different white matter pathways with histological methods, although it is possible to obtain a certain degree of contrast between bundles of fibers of different myelin content (Bürgel *et al.,* 1997). The use of anterograde and retrograde tracers, such as radioactively labeled amino acids and horseradish peroxidase, respectively, is restricted to animal research. Those few tracers that migrate along axons in fixed postmortem tissue are very slow and short-ranged (Bunt and Hubbard, 1988; Bulkharter and Bernardo, 1989; Tardif and Clarke, 2001).

Flechsig (1901, 1920) identified several pathways in fetal or neonatal brains with silver impregnation techniques. In fact, during development, several of the tracts are not myelinated, allowing one to recognize those that have already developed a myelin sheath, but clear separation between fiber tracts in the forebrain disappears when the process of myelination in the developing brain is completed.

Much of the knowledge on connectivity in the adult postmortem brain is based on the occurrence of histopathologic changes. Brain lesions can alter the composition of distal axons that originate from or pass through the injured regions. The breakdown products of myelin are visible in polarized light, with fluorescence microscopy (Miklossy and Van der Loos, 1991), and with MRI (Danek *et al.,* 1990; Savoiardo *et al.,* 1990; Rademacher *et al.,* 1999). Silver impregnation methods can be modified to stain degenerating axons selectively (Nauta and Gygax, 1954; Mesulam, 1979; Miklossy *et al.,* 1991). Clarke and Miklossy (1990) proposed a subdivision of the human visual cortex based on a pattern of degenerating callosal afferents that derived from cortical lesions in contralateral occipital lobes. It is important to note that nerve fiber degeneration can be resolved within only a limited time window after the lesion and that optimal post-injury survival time seems to vary between fiber tracts (Brodal, 1982).

E. Neurogenetic Methods

It is estimated that more than half of all human genes are expressed in the nervous system. The investigation of the genetic bases of brain complexity and of those nonlinear interactions with environmental factors is necessarily part of the comprehensive effort of brain mapping. One way to interrogate the genomic representation in the brain is to localize the proteins that are expressed by the DNA's encoding sequences. Monoclonal antibodies bind to specific epitopes that do not necessarily correspond to the proteinic expression of particular genes. Instead, *in situ* hybridization (ISH) targets mRNA transcripts directly within the cells (Valentino *et al.,* 1987; Wilkinson, 1992; Swiger and Tucker, 1996). ISH further allows the quantification of the amounts of transcripts detected in different neural structures and the comparison of homologous structures in different specimens or in different developmental or pathologic states (Fig. 11).

Purely quantitative studies do not involve any fixation procedures to preserve the tissue morphology. On the other hand gene expression analysis in a meaningful topographical and morphological context can be done in paraformaldehyde-fixed material so that neural structures remain intact. In addition, this technique can be successfully combined with other histological investigations (Westlake *et al.,* 1994).

Currently there are other methods to examine gene expression in brain tissue. A class of "subtraction" techniques such as differential display PCR (Liang and Pardee, 1992; Tochitani *et al.,* 2001), suppressive subtraction hybridization (Wang and Feuerstein, 2000), and representational difference analysis (Welford *et al.,* 1998; Geschwind *et al.,* 2001) allow the comparison of genotypic expression in different samples. Recent studies have utilized the potential and practicality of "DNA chips" (Lockhart and Barlow, 2001). Oligonucleotide arrays (Duggan *et al.,* 1999) and cDNA microarrays (Schena *et al.,* 1995; DeRisi, 1996) are used to characterize gene expression at a genomic level. Because of the large number of genes that can be screened in one experiment (clone sets of 40,000 human genes are available commercially) microarray technology is a high-throughput method to discover novel genes (Schena, 1996; Geshwind *et al.,* 2001) and to compare gene expression profiles in different tissue samples.

"Forward" genetic approaches (Takahashi *et al.,* 1994) quantify those traits that are expressed in graded fashion in the brain (variations in the size, complexity, and susceptibility to disease of the nervous system) and localize the gene loci involved in their phenotypic expression (Lander and Schork, 1994; Falconer and Mackay, 1996; Kearsey and Pooni, 1996). This strategy is essentially subordinate to the accurate quantification of phenotypic traits (Williams, 1998; Williams *et al.,* 1998; Risch and Merikangas, 1996; Bartley *et al,* 1997; Williams, 2000) and benefits considerably from stereological methods that provide "unbiased" estimates of volume, shape, and neuronal population number (see below; Gundersen *et al.,* 1988).

The publication of the complete human genome will likely have a drastic impact on the neurosciences (Walsh, 2001) and on brain mapping in particular. Powerful profiling

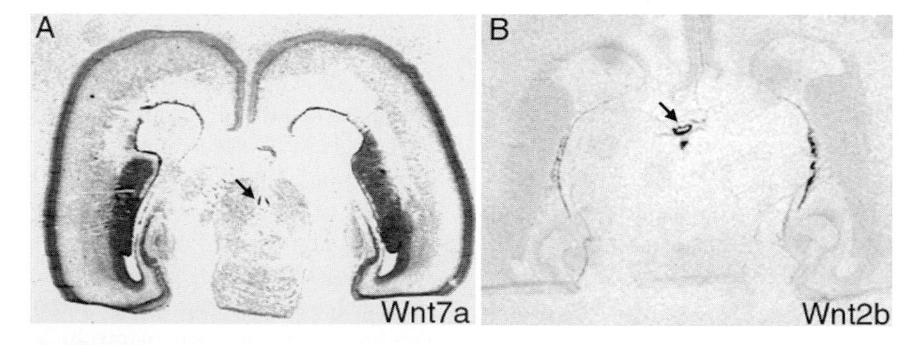

Figure 11 *In situ* hybridization of candidate genes that influence left–right axis formation using ^{35}S-labeled riboprobes (mRNA) hybridized onto 20-μm frozen sections from human brain at 19 to 22 weeks' gestation. Hybridization techniques are based on the principles of nucleic acid hybridization. The brain section is incubated in a supernatant solution containing labeled DNA or mRNA probes (Valentino *et al.,* 1987; Young, 1992). These interact with the RNA transcripts in the tissue, and complementary hybridizations are localized by amplifying the signal from the labels. Multiple transcripts can be detected by combining different species of radioactive, fluorescent, or conjugated labels (Albertson *et al.,* 1995; Morey, 1995; Swiger and Tucker, 1996). Genes of the *Wnt* family act as morphogens that are typically expressed at developmental boundaries and are involved in the axis formation and regionalization of the vertebrate body and CNS. The arrows indicate midline expression. (Modified from Geschwind and Miller, 2001, with permission).

of gene expression in normal and diseased populations achieved with microarray technology (Walsh, 2001; Mirnics, 2001) can be combined with the topographic resolution of ISH.

F. Histological Artifact

Shrinkage is inevitable and possibly the most conspicuous artifact of any histological procedure. The whole brain can undergo significant changes in dimensions when subjected to fixation (Sadowski *et al.,* 1995; Quester and Schroder, 1997), cryoprotection (Rosene *et al.,* 1986), freezing (Pech *et al.,* 1987), or embedding routines (Haug, 1986; Iwadare *et al.,* 1984). Dehydration is a compulsory treatment for histological tissue either before staining or only just before the stained section is sealed between the slide and a thin coverglass. In the case of frozen sections, in which the whole specimen was not dehydrated before cutting, the tissue may shrink on the glass up to 1/5 of its original thickness following dehydration through alcohol and xylene. Plus there is no guarantee that this shrinkage will be uniform throughout the thickness of the section, either (Hatton and Von Barthel, 1999). Different neural structures may well shrink to different extents. It is reasonable to assume, for example, that during deacetylation, a step of fiber silver stain that removes lipids from the tissue, the degree of shrinkage of medullary myelin-rich regions will be greater than that of the densely packed cellular gray nuclei or the cortex. Section to section and batch to batch variations are also expected.

Histological artifacts are extremely difficult to prevent and account for. While changes in volume and shape of the tissue are beyond the control of the experimenter, the damage that can occur at autopsy or during cutting on the microtome can be checked effectively by paying meticulous attention to every procedure involved. Staining will be largely homogeneous and consistent, if sections have been cut appropriately. Histological adulteration obviously affects the visualization and measurement of anatomical data on the section. Methods of quantitative analysis cannot rely on the assumption that a section is artifact-free, but rather they should be designed to overcome the inconsistencies of histological processing as will be discussed below.

VIII. Anatomical Visualization

In the process of making the histological slides the investigator has to consider the tools and methodologies that will be available for the analysis of the data. Section thickness and the intensity of staining are variables that affect the performance of optical systems and image analysis tools. Furthermore, each specific staining procedure is repeated in the series of sections at an arbitrary interval. This interval has to be chosen considering the requirements for a 3D reconstruction that will represent that particular stained series. In short, the anatomist needs to be very well acquainted with the methods to visualize and analyze histological images in order to design, in the wet lab, the appropriate protocols.

A. Acquisition

MRI of the brain *in vivo* at a resolution of 256^2 does not provide the anatomic detail of high-resolution digital imaging of histological material. Indeed, as discussed earlier, one of the most important advances in the development of the histological technique was the ability to produce serial and consistently thin sections from the tissue. These sections can be thin enough for transmitted light (or electrons) to be projected through the tissue in a microscope system, allowing for the analysis of structure at different levels of resolution. Digital imaging devices can be directly mounted on the microscope. With the implementation of macro optics offered by newer research microscopes it is possible to acquire images at microscopic resolution over larger fields of view. These high-resolution image tiles can be recomposed into an image mosaic of the entire histological section (see Fig. 12).

The digitization of analog images is the basis for the enhancement and segmentation of neural structures in histological material. Digital images were initially acquired by computer "frame grabber cards" that converted the analog signal from the video camera into 512 × 512 arrays of values. Digitization is now commonly accomplished by arrays of sensors (charge-coupled devices or CCDs) that translate photon density into gray-scale values for each picture element (pixel). The range of gray levels represents the bit depth of the image. For example, an 8-bit image has 2^8 or 256 brightness levels, while a 16-bit image contains 2^{16} or 65,536 brightness levels.

The resolution of the image is directly related to the number of elements in the CCD array, although some cameras now use smaller arrays to scan the field of view systematically. Chips with up to 4000 × 4000 sensors are commercially available for research or industrial applications, providing very high resolution images with a large number of pixels per unit area.

Important information in stained histological material is also conveyed by color especially in fluorescence microscopy if different structures are labeled with different fluorochromes. Digital color cameras typically acquire images in the form of 24-bit RGB, meaning that 8 bits or 256 levels of brightness are stored for each red, green, and blue channel. Single-chip color cameras employ filters to deliver different wavelengths to each individual transistor. Three-chip cameras that use separate monochrome CCDs for each wavelength component in the incoming light provide higher resolution.

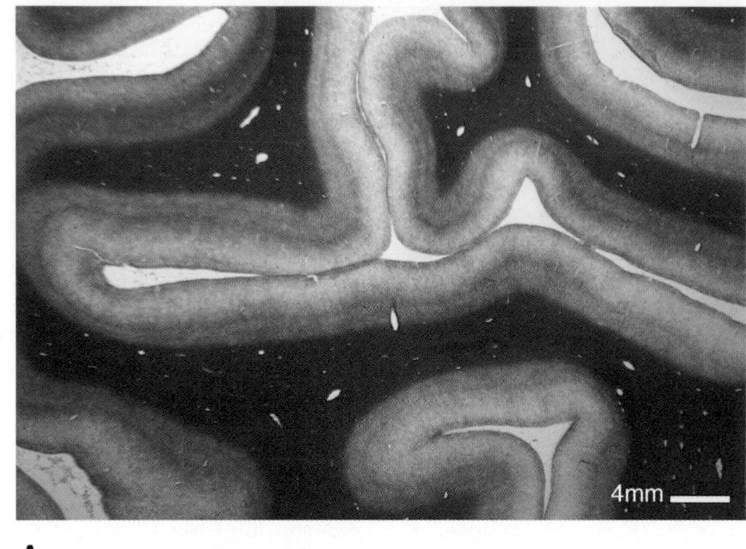

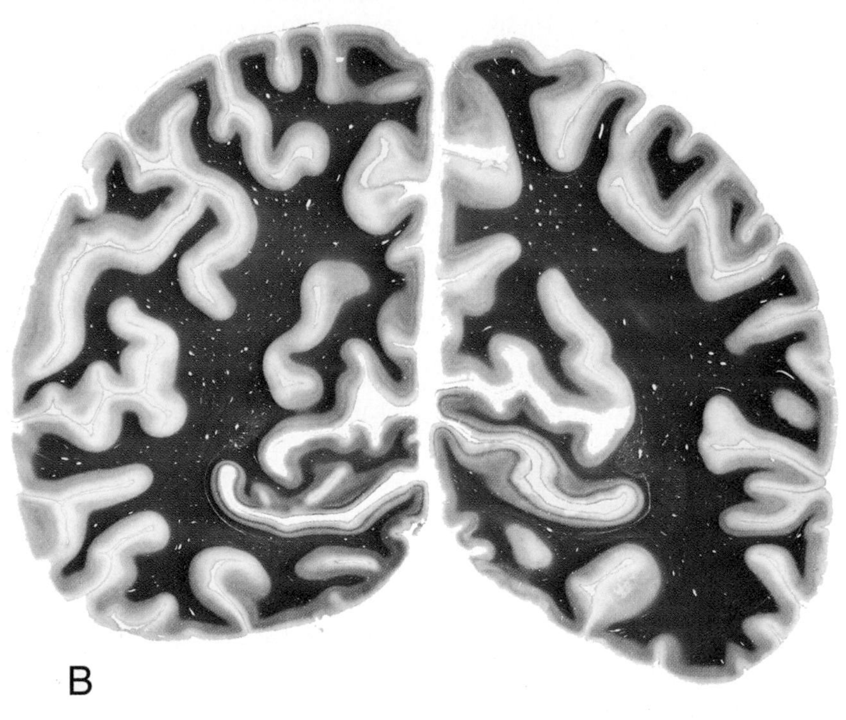

Figure 12 Techniques of image mosaic construction permit the collection of microscopic data over large fields of view. These methods create image "tiles" (**A**) at high imaging resolution from the entire section. The size of the field of view of each tile, and accordingly the number of tiles, can be determined by the user; when reassembled digitally, the result is a mosaic set of microscopic images that describe the entire section. A computer-controlled motorized stage is needed to drive portions of the section under the objective systematically, recording the positions of the boundaries of each tile so that the edges of the tiles will line up appropriately. Motorized stages can be accurate enough for a correct tessellation, but illumination may not be homogeneous throughout the sample. However, macro optics is a feature of new research microscopes that provides even illumination at a very low power magnification (with 0.5× objectives). In addition, shading effects can be corrected (Inoué and Spring, 1997) and changes in illumination can be compensated for by voltage-stabilized alternating-current transformers. This system offers the advantage of acquiring large images with the clarity of microscope objectives and obviously produces extraordinary amounts of data. (**B**) Flat-bed scanning (Schmitt and Eggers, 1999) is an alternative digitizing method which solves *a priori* the technical problems discussed above, but the possibility of multiple magnifications is obviously precluded, limiting the analysis to one level of structural resolution. The whole section was scanned on the transparency module of a flat-bed scanner (Agfa DuoScan) at a resolution of 1024 × 1024 dots per inch.

High resolution and wide dynamic ranges are essential to capture important texture data and boundaries for the segmentation of neural structures.

B. Blockface Imaging

By mounting the digital camera on the microtome in line with the stage (see Fig. 13) it is possible to acquire images directly from the surface of the specimen every time the block surface is eroded by the knife or at regular intervals during cutting. This record, referred to as "blockface imaging" represents a first level of postmortem anatomical visualization.

Specimen blockface photography was first utilized in the 1950s as a means of maintaining a permanent and accurate photographic record of the cut specimen (Ullberg, 1977).

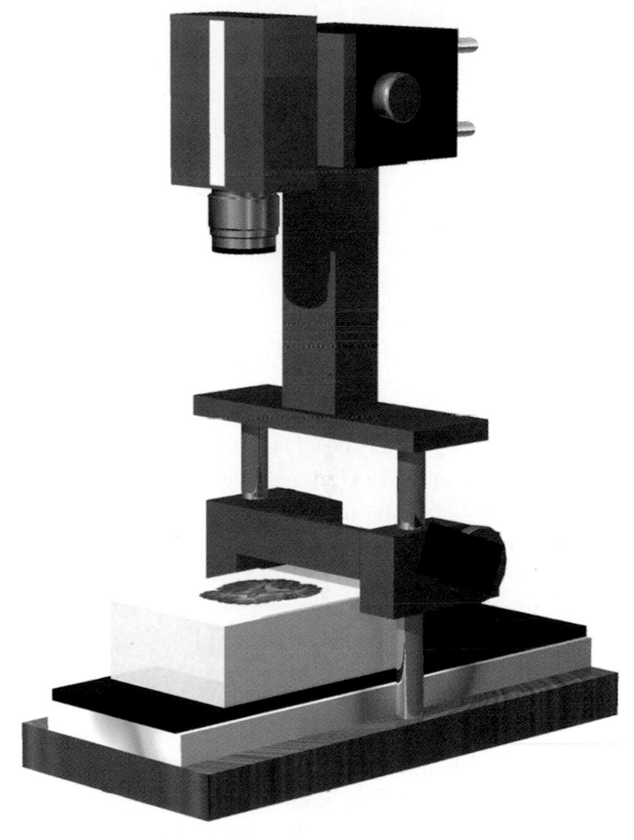

Figure 13 This schematic describes the digital imaging system mounted on a motorized Polycut microtome (Leica Microsystems) used for histological tissue and digital image data collection from large brain specimens. The camera is integrated with the knife descending apparatus so that magnification is held constant as the specimen block is cryoplaned. A fiber optic lighting system, not shown, is also integrated with the camera and knife apparatus so that consistent illumination is maintained throughout the sectioning process. Specimens are embedded in a gelatin block and rigidly fixed to the stage. The sledge includes an automated stop feature that places the specimen block directly under the camera prior to the capture of each digital image. In this manner serial digital images are captured in precise register.

This technique derives from a block surface staining procedure introduced by Hegre and Brashear (1946, 1947) and serial section cinematography (Hegre, 1952) by which the stained surface of the specimen block was photographed by a cine camera capable of single-frame exposure each time a section was cut.

In fresh nonfixed specimens, the contrast between nuclear or cortical gray matter and the myelinated fiber tracts that separates them is sufficient to provide morphologically detailed anatomic image data that surpass the resolution offered by noninvasive imaging techniques. In terms of image contrast, one potential technical obstacle for blockface image capture is show-through of underlying structures through overlying ice or through translucent or clear specimen areas. This effect can reduce the spatial accuracy of subsequential anatomic delineation especially for structures whose boundaries change rapidly along the z axis (perpendicular to the cutting plane). In order to reduce the amount of information introduced into the image from structures lying deep to the blockface surface, flat field high-quality optical lenses at maximum aperture have been used in conjunction with fiber optic assembly that lit the specimen surface at a 45° angle. This method effectively improved the accuracy of structure boundary information (Toga *et al.*, 1994a).

Toga *et al.* (1994b) performed a systematic study to determine the most appropriate methods for high-resolution anatomic cryosectioning of fresh-frozen or whole-fixed head specimens in order to render the topography of 3D data which conformed to the position of brain structures in life. Brains that were fresh-frozen *in situ* were found to be preferable to immersion-fixed brains such as those routinely obtained from autopsies, since immersion fixation of brain can result in global spatial deformations, as discussed above. Further mechanical and digital solutions are aimed at obtaining registered, serial digital images of constant magnification and illumination and produced 3D solid digital images of the specimen (Fig. 14).

Serial images from the blockface can provide a tomographic reference to register histological sections (Hirsch, *et al.*, 1989; Courchesne *et al.*, 1989; see below) and provide high-resolution images that can be combined into detailed anatomic volumes (Quinn *et al.*, 1993; Toga *et al.*, 1994b). Cryoplaning of specimens has also been used previously in the production of digital atlases along with conventional film photography taken from the surface of the preparations (Gerke *et al.*, 1992). Similar methods have been used to yield high-resolution, morphologically detailed images in correlation with other modalities such as MR and computed tomography (CT) (Ho *et al.*, 1988; Katzberg *et al.*, 1988).

This imaging and atlasing technique is potentially limited only by the intrinsic contrast between unstained different classes of tissue in the brain, unless of course the surface is stained before each image is acquired (Hegre and Brashear,

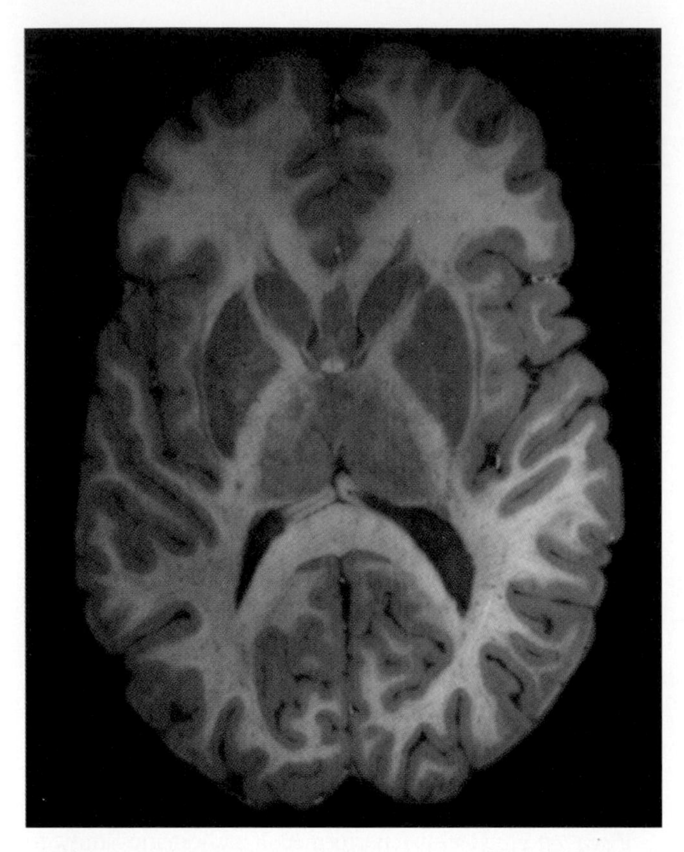

Figure 14 Digital blockface imagery from whole human brain. This image is from the original horizontal plane of section. Specimen was frozen in liquid nitrogen before sectioning. The specimen was not fixed and so gray–white differences are easy to appreciate. Note the coloration of the tissue. The gelatin matrix in which the specimen was embedded was edited away digitally. The slight asymmetry in orientation is not problematic as the images, once reconstructed, are repositioned into coordinate systems.

1947). Subsequent images are intrinsically aligned; therefore, although this technique would never quite match the wealth of morphological information contained in histologically processed slices, it has been extremely successful in providing high-resolution digital images for the accurate segmentation of most brain structures (Toga *et al.,* 1994b).

C. Optical Sectioning

A sequence of images can also be obtained by optically sectioning through a single slice of tissue (the physical section) either with a confocal microscope or with a normal microscope using low numerical aperture objectives. The optics of the confocal microscope have been designed to enhance the axial resolution, and it is particularly advantageous for 3D viewing of a thick section because of its optical sectioning properties (Sheppard, 1994). Confocal microscopy is especially efficient in removing out-of-focus data that are gathered from above or below the image plane. On the other hand, one limitation of the light microscope is that there is no information regarding the depth of an image. The solution is to attach a step drive motor to the microscope stage and a depth encoder to control precisely for the position of the image plane in the z axis. Serial virtual optical sections can be acquired by focusing through the depth (the z axis) of the section, and images from different focal planes can be combined into a 3D reconstruction. Thus, it is possible to obtain stacks of microscopic images from a histological slide on a light microscope. Obviously the images that are derived from this application are aligned to each other. Osborn (1967) published the first detailed optical sectioning technique applied to reconstructing teeth; the method has been incorporated into 3D stereometric techniques and has been used to reconstruct the morphology of microscopic neural structures (Agard, 1984; Shaw and Rawlins, 1991; Peterson, 1999).

The removal of out-of-focus data from an image plane can also be performed by digital deconvolution. The software removes out-of-focus haze from microscope images by applying "nearest neighbor"-type algorithms over at least three optical slices. A calculation of the point-spread function of the microscope system corrects for the amount of excess light in the image. The objects contained in the sharp images from different levels in the sections can be selected and reconstructed in 3D. This method provides accurate models of microscopic neural elements such as neurons and their dendritic trees (Turner *et al.,* 1991).

D. Image Analysis

Computer-based image capture allows the direct interactive or automatic analysis of anatomy, enabling the segmentation of structures of interest and their morphological and quantitative analysis.

The simplest automatic tracing and counting methods are based on thresholding algorithms. Thresholding produces a binary image from the monochrome or color images. Basically two pixel values are assigned relative to the presence or the absence of a particular feature (Moss, 1992). Once the image is converted into binary form it is easier to extract contours and to highlight object features. Naturally, the correct threshold depends on the type of stain and on the overall level of staining, but usually simple threshold methods do not compensate for the noise introduced by uneven staining.

More complex segmentation algorithms can be implemented on the basis of Fourier analysis (Pratt, 1978) and deformable templates (Cohen and Cohen, 1992). Other methods have been proposed that recognize neural structures based on an internal model of what a neuron (Ahrens *et al.,* 1990) or a myelinated fiber (Romero, 2000) "should" look like (Tucker *et al.,* 1984; Merkle, 1989). These methods based on *a priori* knowledge of the object's characteristics are also used for segmentation and labeling of structures in MR images (Szekely *et al.,* 1996; Taylor *et al.,* 1997; Duta and Sonka, 1998).

Many stains, especially at a lower magnification, produce granular images that can be analyzed only with texture-based segmentation (Baba *et al.,* 1996). The borders of a whole Nissl-stained section can be extracted by simple threshold methods or more complex "edge tracking" approaches (Menhardt *et al.,* 1986). However, the segmentation of cell clusters or cortical areas within the image depends on the analysis of the spatial granular distribution of stained elements.

The human visual system is extremely efficient at discriminating subtle and gradual spatial texture variations in an image. In fact, one of the main criticisms of classical histological characterization of structure is the fact that subjective perceptual grouping can bias interpretation of images of stained sections (Lashley and Clark, 1946; Schleicher *et al.,* 1999). Because quantification that transcends simple—or complex—descriptions is a prerequisite for comparing anatomical data, "observer independent" methods were designed to collect an objective analysis of the nervous system (Schleicher *et al.,* 2000). Automatic and semiautomatic image analysis was conducted in the past on sections stained with cresyl violet (Sauer, 1983; Sauer *et al.,* 1986) and the Gallyas silver method (Gallyas *et al.,* 1993). To resolve the granular spatial distribution of staining, the digitization of the image can be based on the density values (Hudspeth *et al.,* 1976) in the image, or a gray-level index that is related to the volume density of cell bodies (Wree *et al.,* 1982). Texture-based approaches (Weszka *et al.,* 1976; Baba *et al.,* 1996) and image-filtering methods (Randen and Husoy, 1999) are very useful for the extraction of the area of anatomical regions of interest in the image. Automatic segmentation of MR brain images is an active field of brain mapping (Lundervold and Storvik, 1995; McIrney and Terzopoulos, 1998; Neumann and Lorenz, 1998; Sato *et al.,* 1998; Kelemen *et al.,* 1999) and the same principles and results can be applied at the microscopic level on histological material.

The key factor in the analysis of histological images is contrast that is, in turn, defined by the pattern of staining. Staining levels that optimally resolve the objects of interest from the background may obscure details of the structure that could be analyzed at a higher resolution. Therefore, the objectives, in terms of feature analysis, should be considered carefully before processing the sample histologically. Section thickness and staining parameters need to be chosen appropriately for specific image analysis applications.

IX. Quantification

A. Stereology

One approach to the representation of anatomy is essentially statistical. Measuring or counting objects in histological material requires the extrapolation from each two-dimensional section to the total volume. The most simple and common method to count neural structures is to extract their profiles in the histological sections manually or by automatic image segmentation. This method overestimates the "true" object number. The error is introduced when a structure, say a neuron, appears in more than one section, because it will be counted more than once. The greater the neuron size relative to the section thickness the greater the error.

An early mathematical solution to this problem is the Abercrombie correction (Abercrombie, 1946) that provides an estimate of the "true" cell number in a series of sections. This type of correction entails specific assumptions regarding the shape and orientation of the objects, for example, that neurons are spherical. Another requirement is the assumption of homogeneity of the structure in which the samples are taken. However, because the brain is highly organized and certainly not homogeneous, such conditions can hardly be met. Miles and Davy (1976) showed that a "representative" sample of a biological tissue, even if heterogeneous, can provide valid estimates about the properties of the volume as long as there is random sampling in the tissue (Baddely, 2001). Sampling is the key issue in stereology. Random sampling also eliminates the bias caused by artifacts introduced in the preparation of the tissue (Weibel, 1979). In other words, for a measurement to be *unbiased* in the statistical sense, every location on the section should have an equal probability of being sampled.

A new generation of stereological (from Greek: *stereos,* solid, and *logos,* knowledge) tools eliminated the restrictions of the former "model-based" methods (Sterio, 1984; Cruz-Orive, 1997; Howard and Reed, 1998). "Design-based" stereology relies on systematic sampling procedures that are independent of the tissue properties (Glaser and Glaser, 2000). The optical dissector is the most frequently implemented of the stereometric methods to estimate population numbers. It consists of a 3D counting box that is optically created in the section (Williams and Rakic, 1988; Gundersen *et al.,* 1988; Guillery and Herrup, 1997). Creating a sample volume optically within the section also reduces the concern for uneven section thickness (Pakkenberg and Gundersen, 1988).

It is still a matter of debate whether design-based counting methods are necessarily more useful than model-based approaches, such as the Abercrombie correction (West, 1999; Benes and Lange, 2001). The traditional focus of stereology has been the estimation of object number and density in space (Haug, 1986; Abercrombie, 1946; Sterio, 1984; Gundersen, 1986; West, 1999; Benes and Lange, 2001). However, the quantification of the structural parameters of an object, such as length, surface, and volume, can be approximated by implementing the same sampling principles (Gundersen *et al.,* 1988; Cruz-Orive, 1997; Peterson and Jones, 1993).

B. Map-Integrated Stereology

The principal limitation of conventional stereology is that the estimates are not linked in any way to larger images of the tissue that are under examination (Glaser and Glaser, 2000). The application of 3D probes requires the use of high-power objective lenses with a low numerical aperture. This means that the field of view will be extremely small and not necessarily representative of the morphology of the whole section. Semiautomated computer-based stereology systems have recently been designed to sample a distribution of fields systematically, while maintaining a record of the position and area of each sampled field relative to the entire section. The experimenter has control over the sampling procedure and determines the boundaries of the structure under examination. The microtome stage is equipped with an *x–y* step motor that is controlled by a computer.

The travel range of the motorized stage has to be chosen in order to cover the entire section. Large computer-controlled stages are commercially available. Although this equipment has been designed for industrial applications, it is the most suitable to survey whole sections of the human brain microscopically. Low-resolution digitization of whole sections provides the map for the navigation of the stereological probes. This not only is practical in the analysis of one individual specimen, but also establishes the basis for a common histological framework for independent studies.

The principles of stereology have been mainly applied to light microscopy, but the method is essentially valid for the analysis of data acquired by other imaging modalities such as confocal microscopy (Peterson, 1999), CT (Pakkenberg, *et al.*, 1989), and magnetic resonance imaging (Roberts *et al.,* 1993, 2000).

X. 3D Reconstruction

Serial section reconstruction is a procedure by which the form of a tissue component is assembled directly or indirectly from serial sections of that tissue cut in a specific plane (Gaunt and Gaunt, 1978). Graphical reconstructions were 2D

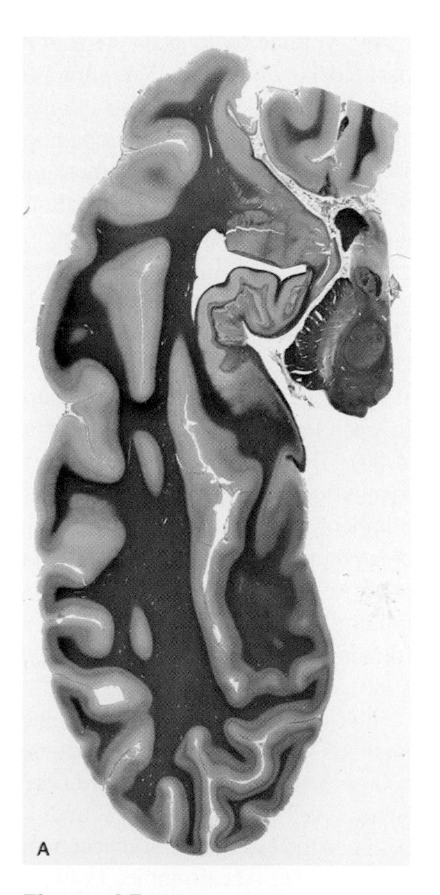

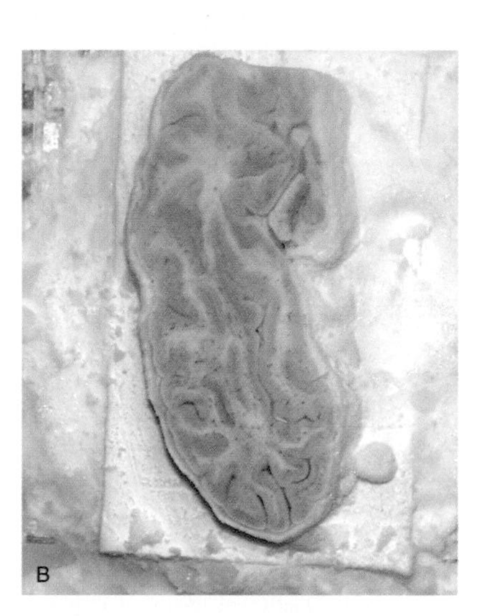

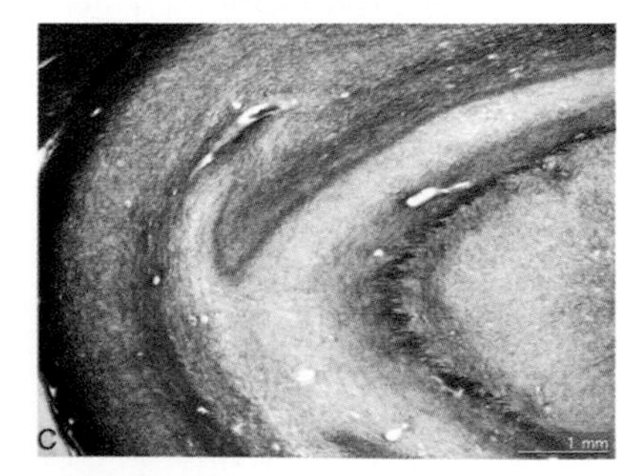

Figure 15 High-resolution images from the cryoplaned blockface (**B**) can be correlated with stained slices for the 3D mapping of microstructural territories. The topological correspondence between the two sets of data is used primarily for the registration of each histological section into a tomographically aligned set. The horizontal section through the temporal lobe (**A**) is stained with a combination of silver impregnation and cresyl violet for the visualization of fiber architecture and of neuronal cell bodies. (**C**) Optically magnified region of the hippocampus.

charts representing the overlap of serial tracings of the histological material. The first application of these techniques implied the examination in translumination of stacks of outlines, drawn purposefully on transparent material (His, 1868). Krieg's (1966, 1967) topographical work on the brain of the monkey and man epitomizes the usefulness of reconstructing series of traced outlines drawn from the histological material. The obvious complement to this technique was the making of 3D solid models from plates of wax (Born, 1883) and celluloid (Vossmaer, 1899; Lebedkin, 1930–1931) of proportionate thickness (Gaunt and Gaunt, 1978).

Virtual reconstruction generated by highly interactive computer graphics systems substituted for graphical and physical laminate models. Virtual reconstructions are not merely representations of objects, but yield accurate measurements and can be manipulated, duplicated, and transferred in the form of readily accessible databases (Haas and Fisher, 1997). The number of computer-based methods that address the specific problems of reconstruction from physical sections (Huijsmans *et al.,* 1986; Drushel, 1993) is extremely small compared to commercial or public domain software that allows 3D reconstruction from tomographic images, produced by CT, MRI, and confocal microscopy. The latter do not need any retrospective rectification or alignment. In any case, there are two basic representation methods for serial image reconstruction: the first is based on the extraction of contour lines and polygon-based surface models and the second is voxel-based and produces a solid body representation of the image data.

A. Alignment

Physical sectioning procedures, unlike various tomographic imaging techniques that are intrinsically aligned, must utilize superimposing schemes to register serial sections because their relationship is lost during histological processing. Therefore they must be aligned and corrected for distortion before 3D reconstruction.

A common approach registers the series by superimposing external fiducial markings that are introduced orthogonally to the cutting plane in the tissue (Gaunt and Gaunt, 1978; Toga and Arnicar, 1985; Goldszal *et al.,* 1995) or in the embedding media (Streicher *et al.,* 1997). This method is suitable only for reconstruction from large sections and cannot be implemented efficiently at the microscopic level. Principal axes techniques have been used to perform the registration of a series of coronally sectioned autoradiograms for 3D reconstructions (Hibbard and Hawkins, 1984). This approach uses calculations to align a vertical and horizontal meridian of successive images but its precision remains quite limited (Schormann and Zilles, 1997). Cross-correlative techniques can be used to align two corresponding images in the spatial domain (Banerjee and Toga, 1994) or the frequency domain (Hibbard and Hawkins, 1988), taking advantage of the correspondence of image density features and performing correlations of pixel intensity. The displacement of any pair of images is calculated in the form of a cross-correlation function and the images are aligned accordingly. Several other procedures utilize information obtained about the shape and form of the image or the presumed consistency from one slice to its immediate neighbor. Local image features can be used to align sets of corresponding points in two images (Rangarajan *et al.,* 1997). If this iconic method (Andreasen *et al.,* 1992; Kim *et al.,* 1997) is applied to subblocks of an image it is possible to calculate local displacement fields that contribute independently to the general rigid transformation of the image (Ourselin *et al.,* 1998). This method is particularly useful in histology where the staining is more or less consistent from one section to the next, and therefore contrast can be different for corresponding structures. Furthermore, the edges of a histological section can be deformed or torn by cutting and handling, making the outlines of the edges inadequate as alignment landmarks.

Manual registration based upon the investigator's perception of best overlap can also be used (Marko *et al.,* 1988). However, even if guided by a comprehensive photographic reference of the whole intact specimen it is very difficult not to introduce a gradual drift of the sections in the *x* or *y* direction. This drift, which can occur translationally or rotationally, would obviously compromise the faithfulness of the reconstruction to the original shape. By this token, it is possible to envisage a "best fit" serial reconstruction of the leaning tower of Pisa that would result in a perfectly straight construction.

For the macroscopic reconstruction of the whole hemisphere the reference to blockface imaging can provide validation to successive transformations of histologically stained slides. The implementation of photographic reference marks (Heard, 1931) for alignment follows the same routines of serial section cinematography mentioned above and is the progenitor of blockface imaging registration methods (Fig. 15).

Blockface techniques, including cryosectioning, often must utilize registration schemes as well, due to either error introduced following knife blade changes or inadvertent loss of the relationship between camera and specimen.

The undistorted MRI volume of the brain has been used as a reference for the alignment of the large whole histological sections (Schormann and Zilles, 1998; Annese *et al.,* 2001). It should be noted, however, that MRI images are susceptible to gradient nonlinearities of the magnetic field in the scanner (Wald *et al.,* 2001). There is, therefore, a certain degree of distortion in the MRI image reference coordinate system, comparable to the extent of brain shape change after death *in situ*.

Histology to MRI registration can be performed two-dimensionally section by section, provided the cutting plane of the microtome blade matches the image plane of the MRI

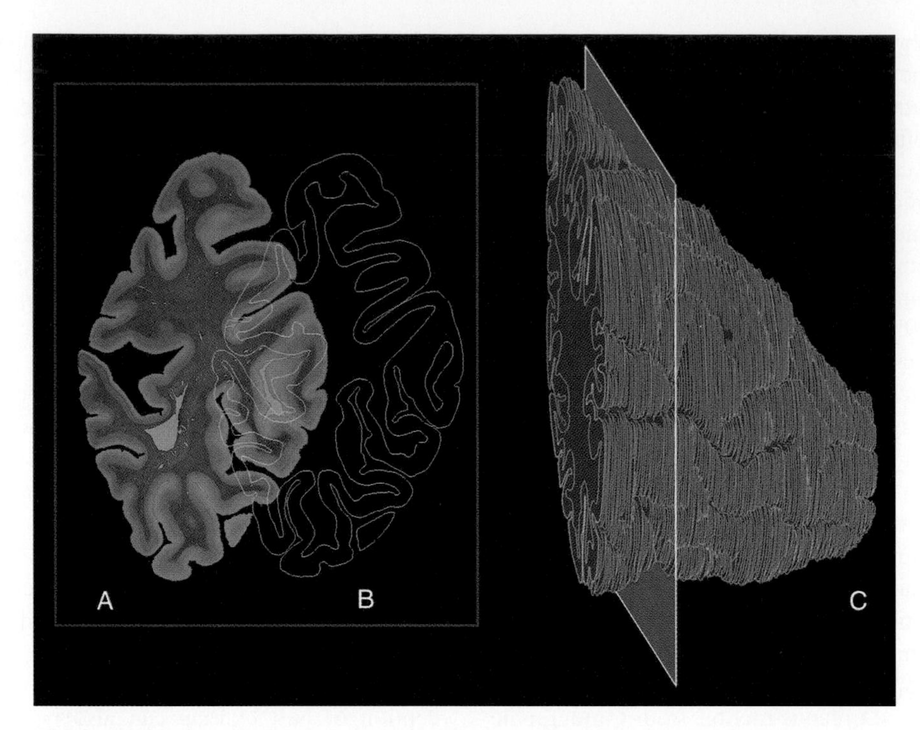

Figure 16 Contour-based reconstruction. Glass-mounted myelin-stained sections from a human left hemisphere (**A**) are optically projected without distortion at a 5–8× constant magnification onto a large-format digitizer (Calcomp USA; Drawing Board III/0.005-in. accuracy). The outer cortical surface and the boundary between gray and white matter are traced and labeled independently (**B**) in the same reference space using custom software (Leonardo 1.0) developed at the Laboratory of NeuroImaging (UCLA School of Medicine), running on a NT workstation. The contours derived from the cortical surface in serial sections are initially aligned manually using a rigid external marker method included in digitizing software and submitted to a 3D reconstruction routine for visualization and editing (**C**). The whole reconstruction can be registered to the volumes obtained from blockface imaging and MRI, respectively, using a surface matching method (Thompson and Toga, 1996). This subsequent registration step normalizes the histological reconstruction to the original shape and eventually to a stereotaxic orientation.

scan. Alternatively, the reconstructed histological "volume" can be aligned to the MRI volume via principal axis transformations (Alpert *et al.,* 1990; Rusinek *et al.,* 1993), voxel intensity matching (Woods *et al.,* 1998), surface matching (Pellizzari *et al.,* 1989; Thompson and Toga, 1996), and thin plate splines (Bookstein, 1991; Jacobs *et al.,* 1999) methods. In addition to correcting alignment, the correlation between histology and MRI is a prerequisite for the integration of microstructural data into stereotaxic atlases and its comparison to large populations of macroanatomical information acquired by structural and functional noninvasive imaging (Toga and Banerjee, 1993; Roland and Zilles, 1996). Histological methods impose a considerable degree of section distortion. Shrinkage could be considered linear to a certain extent, although as noted earlier, different tissues may react differently to fixatives and organic solvents. But the process of mounting the sections onto the glass slides creates unpredictable nonlinear deformations that especially concern the cortical mantle; therefore linear coregistration alone does not match accurately the MRI and histological data sets together. Nonlinear warping can be imposed to histological images to fit the MRI counterparts (Schormann and Zilles, 1998) and "homologous" sections from other individuals (Cohen *et al.,* 1998). The local displacement of an image pair can be approximated by the deformation of a 2D grid (Durr *et al.,* 1989). However, it is important to stress the fact that any image warping deforms the intrinsic architecture of the histological material, and in order to avoid severe microstructural digital "smearing" it is necessary to apply transformations to the contours that represent histological territories (Annese, 2000).

B. Contour-Based Reconstruction

Contour profiles can be traced manually or by automatic feature extraction methods from the image as noted above. The whole histological slide can be projected onto a digitizing tablet. Alternatively sections can be drawn on tracing paper and then traced on the digitizing tablet. Optical projection at a constant magnification provides the resolution necessary to define and trace structural borders accurately. The error in tracing is effectively minimized by the magnification of the image. The larger the image, the greater the number of points that comprise each contour and the greater the accuracy of the reconstruction. This rule also applies in the case of digital images of the sections acquired by a flat-bed scanner or a digital camera.

In addition to the alignment of the contours, in order to reconstruct a model that is geometrically correct, it is necessary to know which pairs of contour lines should be connected between neighboring sections. Homology between contours can be defined by calculating the overlapping rate between each pair of adjacent contours (Baba, 2001) or by matching the contours between two consecutive slices using a heuristic function that takes into account the position of multiple contours centroids (Annese *et al.,* 2001, 2002).

Stacks of contours can be displayed as wire-frame models (see, for example, Herman and Liu, 1977; Macagno *et al.,* 1979; Johnson and Capowski, 1985; Romaya and Zeki, 1985; Braverman and Braverman, 1986) using hidden line removal algorithms to aid the visualization of the reconstruction (Wong *et al.,* 1983; Moss, 1992; Kvasnicka and Thiele, 1995).

Several methods of triangulation have been designed to provide surfaces for clearer visualization (Boissonnat, 1988; Löhner, 1997). The success of these algorithms is inversely proportional to the topological complexity of the object, and as a result the highly convoluted surface of the human brain has presented severe challenges to the implementation of surface tessellation (Carman *et al.,* 1995). A difficult problem arises when the number of contours is different from one section to the next, which means that splits and merges will somehow perturb the surface. Automatic tessellation algorithms that try to locate corresponding points on adjacent contours (Yaegashi, 1987) may not always find the right solution. This limitation creates topological "handles" that must be corrected manually. Finally, surface roughness may occur when the number of sections selected for the reconstruction is inadequate to account for the complexity of the structure. Contour reconstructions are "empty shells" that enclose surfaces or volumes of interest that can be labeled for visualization and measured quantitatively (Fig. 16).

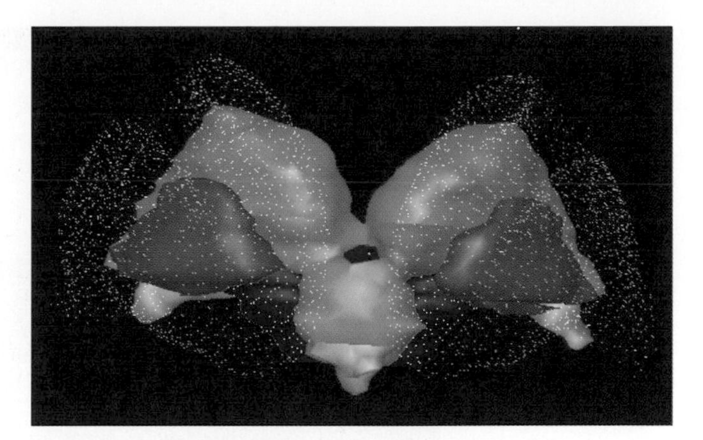

Figure 17 Surface representations of segmented anatomy are constructed from line contours of structure boundaries using wire mesh triangulation. Surface models of structures illustrate the spatial relationship between individual anatomic structures and in the context of the whole brain. Various graphic rendering cues (color, points surfaces, etc.) may be used to help differentiate structures. This rostral-to-caudal view illustrates the globus pallidus, striatum, substantia nigra, and diencephalon rendered as solid objects and points. Anatomic structures that retain spatial coordinates can be displayed individually or nested within the cortex. Surface models are amenable to any orientation and resampling in any viewer-specified plane, as well as application of shading and texture to enhance spatial relationships. Since they are generated directly from anatomic data and retain real-world coordinates, 3D surface models also form the basis for several morphometric measurements such as volume, principal axes, center of mass, and surface area.

C. Voxel-Based Reconstruction

Volume rendering techniques (Drebin *et al.,* 1988; Russ, 1992, 1999) produce similar results. The voxel array is shaped by the alignment procedures into a 3D model based on the image data. A certain degree of image correction is needed before the reconstruction, mainly in the form of equalization of the image intensity values. In fact, the staining of each section can be different, and luminance can vary considerably during the cutting procedure (in the case of cryo blockface images) or the acquisition of the images at the microscope. In addition, the image data between the section images must be interpolated relative to the estimation of the section thickness (Baba, 2001). The voxel-based approach does not explicitly define the surface geometry but can produce 3D models of structure projecting volume elements directly for shading and visualization (Goldwasser and Reynolds, 1987) (Fig. 17). Ray-tracing techniques (Baba *et al.,* 1993) allow the visualization of internal structures in the model by "filtering" the amount of light that passes through the array according to each voxel value. Furthermore the object models can be resampled along the principal axes or along arbitrary planes using cutaways to expose different planes of section (Fig. 18).

Contour tracing and reconstruction produce clear 3D images. If different labels are used to define specific structures, it is possible to toggle on or off their display, which is useful in the visualization of nested anatomical objects. On the other hand, it is very difficult to trace and render in 3D very fine and complicated structures, like a capillary bed. In this case voxel-based representation maybe more useful.

Finally, one of the reasons to create 3D models is to obtain quantitative data. The volume fraction for a specific region can be calculated simply by counting the voxels and dividing by the total number of voxels in the array, assuming the region can be precisely selected by thresholding. For other global parameters, such as surface area or the length

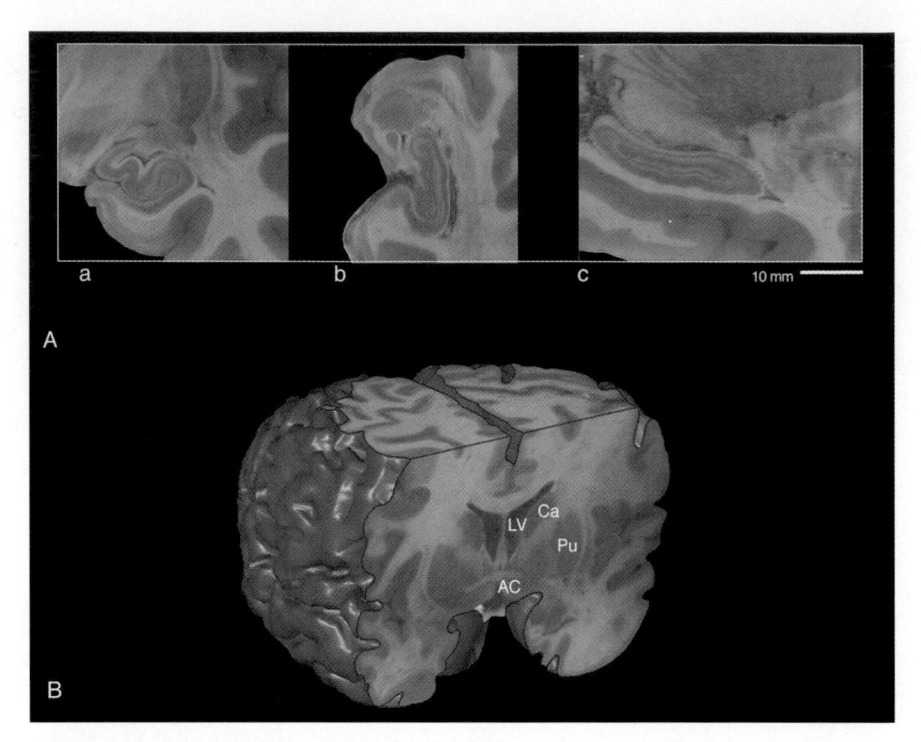

Figure 18 Volume rendering. (**A**) Details of orthogonal resampled planes through the volume of the whole human brain created from unstained blockface images. The coronal (a), horizontal (b), and sagittal (c) images were all derived from the data set at the level of the hippocampus. Note the color and spatial continuity of cryosectioned anatomy. This is, in part, due to the frequency and isotropy of the resampling. This level of resolution is both practical in terms of digital manipulation and efficient in terms of anatomical illustration. It is theoretically possible to create a similarly large-scale volumetric reconstruction out of stained histological data. In practice, because of nonlinear distortions that different slices can undergo independently to correct for the distortion and misalignment between whole histological sections at higher structural resolution could prove extremely problematic. (**B**) A coronal plane through the same 3D volume cuts through the anterior commissure (AC), and exposes the caudate nucleus (Ca), putamen (Pu), and lateral ventricles (LV). The high resolution of the source data permits subsequent resampling along different planes without appreciable loss of anatomic detail. The orientation of the model and intersecting planes can be chosen arbitrarily and interactively. Surface reconstruction techniques enable visualization of spatial relationships and serve to define explicitly the geometric properties of the structures of interest as a prerequisite for 3D morphometric measurements.

and curvature of structural boundaries, direct measurements in 3D are not necessarily superior to stereological estimates from 2D images. If shape is simple enough (close to spherical) 2D measurement may suffice, otherwise complex information on the distribution of shape and size can be measured only in 3D.

Topological properties are closely related to shape parameters. They may include number of nodes or branches of an object (Aigeltinger *et al.,* 1972), and they are obviously extremely relevant to the analysis of neuronal morphology or measuring the complexity of cortical areas that follow the folding of the mantle (Annese *et al.,* 2001). It is impossible to derive topological properties of 3D structures from 2D images, therefore the measurements must be made directly on the 3D data. Digitized points in the 2D approximation of the surfaces can be converted into parametric grids of uniformly spaced points as a regular rectangular mesh stretched over the selected structure. The parametric grid imposed on the objects of interest provides a computational structure that supports measurement of geometric shape parameters, average models, and statistical maps (Thompson *et al.,* 1996). This method has been useful to detect complex changes of macroscopic brain structures during development and in normal and diseased conditions. Likewise, it can be applied effectively to the study of the topology of individual neurons and other microscopic structural elements.

XI. Epilogue

An axiom of anatomy is that there exists a relationship between the form and the function of any biological struc-

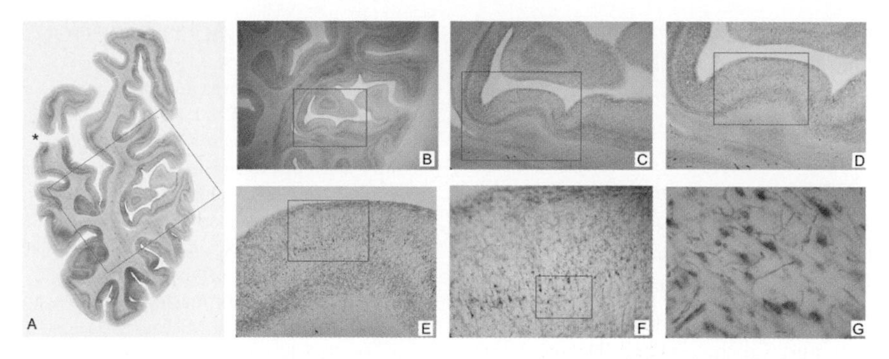

Figure 19 Coronal section (50 μm thickness) stained for acetylcholinesterase histochemistry with a modified Kanovski–Roots method (Tago *et al.*, 1986). Higher magnification images are confined to the primary visual cortex. Cholinergic pathways to the cortex originate from magnocellular neurons in the basal forebrain (Mesulam and Guela, 1991). This input affects most aspects of cortical function, notably attention, memory, and emotion (Mesulam, 1996). Compare (A) with Figs. 7A and 7B (the asterisk indicates damage that occurred during the removal of the brain at autopsy). Table 1 illustrates the computational storage requirements for histological data contained in the section at different levels of magnification. (**A**) The entire histological section was acquired with an Agfa DuoScan flat-bed scanner at a resolution of 1024 × 1024 dots per inch (dpi) (size of the image file 70.16 MB). Images (**B–G**) were acquired with a Polaroid digital camera mounted on with an Olympus AX 70 Provis research microscope at resolution of 1600 × 1200 dpi. The size of the volume data (in the last column of Table 1) was estimated for the entire hemisphere, assuming the same size for every section along the horizontal axis and that only one optical plane was imaged in each section.

ture. It follows that a model of the microscopic structural architecture of the brain will yield invaluable insights into its functions.

The nervous system is an exquisitely differentiated and intricate object, which gives rise to the question of how much anatomical detail is necessary in order to understand the principles underlying brain function. In this context, the smallest significant structural elements are the neurons and the axons and dendrites that constitute their connections. Merkle (1989) and Montgomery (1996) describe the issues involved in large-scale 3D reconstructions from transmission electron microscopy images. The resolving power of light microscopy is determined by the wavelength of the light employed and is theoretically 0.2 μm; the size of neurons (the length of the major axis) varies from approximately 4 (small granule cells) to 100 μm (large cortical motor neurons). It follows that by combining light microscopy with image analysis algorithms "one can imagine identifying every cellular component and encoding the x–y–z co-ordinates of each going through the entire length and depth of a section" (Blackstad, 1970) (Fig. 19 and Table 1).

Reconstructions of nervous tissue comprising up to hundreds of neurons have been accomplished either with light or with electron microscopy (Macagno *et al.*, 1979; Stevens *et al.*, 1980; Kropf *et al.*, 1985). Most dramatic was the reconstruction of all 959 cells of the nematode including 338 cells in the nervous system (Sulston *et al.*, 1983). These undertakings provide a fascinating glance into the possibility of recreating the neuronal landscape of the human brain.

The goal of building such a model from 3D reconstructed data is not totally unrealistic, given that optical sectioning and digital image mosaicking techniques can, theoretically, provide a complete reconstruction of a single section at any optical resolution. The main difficulty lies in the precise alignment of the 3D data set of one section to the adjacent

Table 1

Magnification	Field of view	Pixel size	Slice bytes	Volume bytes
0.5 × (B)	40 × 30 mm	25 × 25 μm	3.66 MB	12.87 GB
1.25× (C)	16 × 12 mm	10 × 10 μm	9.16 MB	32.73 GB
2.0× (D)	10 × 7.5 mm	6.25 × 6.25 μm	14.69 MB	51.50 GB
4× (E)	5.3 × 3.8 mm	3.13 × 3.13 μm	29.28 MB	103.37 GB
10× (F)	2 × 1.5 mm	1.25 × 1.25 μm	73.24 MB	257.05 GB
40× (G)	0.49 × 0.37 mm	0.31 × 0.31 μm	292.78 MB	1.01 TB

one. Histological preparation can be optimized to facilitate this process. In addition, linear and nonlinear registration techniques could provide sufficient alignment for meaningful interpretation of the reconstruction over a number of histological slices.

Even though the global structure might not be deducible from the examination of isolated components, local 3D reconstruction at high magnification provides essential descriptions of the organization of different brain structures. However, it is estimated that in the neocortex alone 13.8 billion neurons (Blinkov and Glezer, 1968; Haug, 1986) share some 164 trillion synaptic connections (Tang *et al.*, 2001). It follows that a large-scale analysis of the brain at microscopic resolution cannot be realistically approached by 3D reconstruction alone; local stereological quantitative statistics must complement 3D reconstruction. In addition, stereological methods can also effectively overcome most limitations imposed by the physical inconsistencies of histological sections that, as we discussed, cannot be avoided.

In practice, image analysis methods applied to "topographic" histological stains define the exact boundaries of anatomical regions of interest (AROI) where "map-based" (Glaser and Glaser, 2000) stereological analysis and 3D reconstruction are undertaken.

Issues of population variability are of course ineluctable, and in fact at higher levels of resolution and magnification generalizations are extremely difficult because of the impressive variation across individuals. It is difficult to predict and correct for the differences in the total number of neurons in different individuals, not to mention the almost infinite variation in 3D arborization pattern of cellular connections. This limitation is exacerbated if one considers disease states, such as Alzheimer's disease, and the dynamic variability imposed by the effect of aging and plasticity. Indeed, any structure– function relationship is not represented sufficiently at a single moment in time. Developmental and neurobehavioral influences introduce a fourth chronological dimension to microstructural maps, making generalizations even more difficult.

A present-day solution is to project individual stereological neurostatistics into a common map-integrated database, provided anatomical homologies are established correctly. Regarding the precise 3D organization of brain structure, one objective is to relate particular neural architecture to specific neurobehavior in the same individual. Second, it may be possible to establish a canonical representation of the human brain, at very high resolution, by mapping the microscopic peculiarities that are visualized and modeled in homologous AROI. Recent advances in digital technologies have drastically reduced concerns for the acquisition and storage of microanatomical information in individual and statistical form. Indeed, much of the "backwardness of human neuroanatomy" (Crick and Jones, 1993) can be made up for by the capture and interpretation of large high-resolution data spaces obtained from postmortem human brains.

Acknowledgments

Part of this work was supported by research grants from the National Library of Medicine (LM05639); a Human Brain Project grant known as the International Consortium for Brain Mapping, which is supported jointly by NIMH and NIDA; and a P41 grant (RR13642). The authors thank the Laboratory of NeuroImaging for scientific support and Cammy Babaie and Grace Park for assistance in the lab and with the manuscript. The authors also thank Dr. D. Geschwind and Dr. M. Mikkonen for making their original figures available. Also, we thank Professor Arnold E. Scheibel (UCLA School of Medicine) for making his own histological material available for Figs. 8A and 8B and Dr. Michael S. Mega (UCLA School of Medicine) for Fig. 3B. The histological material presented in this chapter is based on studies conducted by J.A. at the Center for Cognitive Neuroscience and Dartmouth-Hitchcock Medical Center. J.A. acknowledges the partial support of the Hitchcock Foundation (Dartmouth Medical School) and of the Albert Cass Foundation (Rockerfeller University). J.A. also thanks Dr. Michael S. Gazzaniga (Director Center for Cognitive Neuroscience, Dartmouth College), Dr. W. F. Hickey (Chair, Dept. of Pathology, Dartmouth-Hitchcock Medical Center), John Hutchins for assistance with digital photography, the staff of the Department of Pathology, and the staff of the Department of Radiology of the Dartmouth–Hitchcock Medical Center for the invaluable experience relevant to the topics discussed in this chapter.

References

Abercrombie, M. (1946). Estimation of nuclear population from microtome sections. *Anat. Rec.* **94,** 239–247.

Agard, D. A. (1984). Optical sectioning microscopy: Cellular architecture in three dimensions. *Annu. Rev. Biophys. Bioeng.* **13,** 191–219.

Ahrens, P., Schleicher, A., Zilles, K., and Werner, L. (1990), Image analysis of Nissl-stained neuronal perikarya in the primary visual cortex of the rat: Automatic detection and segmentation of neuronal profiles with nuclei and nucleoli. *J. Microsc.* **157,** 349–365.

Aigeltinger, E. H., Craig, K. R., and DeHoff, R. T. (1972). Experimental determination of the topological properties of three dimensional microstructures. *J. Microsc.* **95,** 69–81.

Albertson, D. G., Fishpool, R. M., and Birchall, P. S. (1995). Fluorescence in situ hybridization for the detection of DNA and RNA. *Methods Cell Biol.* **48,** 339–364.

Alpert, N. M., Bradshaw, J. F., Kennedy, D., and Correia, J. A. (1990). The principal axes transformation—A method for image registration. *J. Nucl. Med.* **31,** 1717–1722.

Andreasen, A., Drewes, A. M., Assentoft, J. E., and Larsen, N. E. (1992). Computer-assisted alignment of standard serial sections without use of artificial landmarks: A practical approach to the utilization of incomplete information in 3-D reconstruction of the hippocampal region. *J. Neurosci. Methods* **45,** 199–207.

Annese, J., Pitiot, A., and Toga, A. W. (2001). Complex topological analysis of the human striate cortex. *NeuroImage* **13,** S858.

Annese, J. (2002). A new histological survey of the visual cortex of man. Dartmouth College (Ph.D. Thesis).

Arnold, S. E., Hyman, B. T., Van Hoesen, G. W., and Damasio, A. R. (1991). Some cytoarchitectural abnormalities of the entorhinal cortex in schizophrenia. *Arch. Gen. Psychiatry* **48,** 625–632.

Baba, N. (2001). Computer-aided three-dimensional reconstruction from serial section images. *In* "Image Analysis: Methods and Applications" (D. P. Hader, ed.), pp. 329–354. CRC Press, Boca Raton, FL.

Baba, N., Ichise, N., and Tanaka, T. (1996). Image area extraction of biological objects from a thin section image by statistical texture analysis. *J. Electron Microsc. (Tokyo)* **45,** 298–306.

Baba, N., Satoh, H., and Nakamura, S. (1993). Serial section image reconstruction by voxel processing. *Bioimages* **1,** 105.

Baddeley, A. (2001). Is stereology 'unbiased'? *Trends Neurosci.* **24,** 375–376. [Discussion pp. 378–380]

Banerjee, P., and Toga, A. W. (1994). Image alignment by integrated rotational and translational transformation matrix. *Phys, Med. Biol.* **39,** 1969–1988.

Barker, L. F. (1899). "The Nervous System and Its Constituent Neurons." Appleton. New York.

Barnstable, C. J. (1980). Monoclonal antibodies which recognize different cell types in the rat retina. *Nature* **286,** 231–235.

Bartley, A. J., Jones, D. W., and Weinberger, D. R. (1997). Genetic variability of human brain size and cortical gyral patterns. *Brain* **120,** 257–269.

Barton, A. J., Pearson, R. C., Najlerahim, A., and Harrison, P. J. (1993). Pre- and post mortem influences on brain RNA. *J. Neurochem.* **61,** 1–11.

Belichenko, P. V. (1991). [A method of intracellular administration of Lucifer yellow for the study of human brain using autopsy specimens.] *Arkh. Anat. Gistol. Embriol.* **100,** 81–84.

Belichenko, P. V., and Dahlstrom, A. (1995). Studies on the 3-dimensional architecture of dendritic spines and varicosities in human cortex by confocal laser scanning microscopy and Lucifer yellow microinjections. *J. Neurosci. Methods* **57,** 55–61.

Benes, F. M., and Lange, N. (2001). Two-dimensional versus three-dimensional cell counting: A practical perspective. *Trends Neurosci.* **24,** 11–17.

Bertram, E. G., and Sheppard, C. J. R. (1964). A possible explanation for the Golgi impregnation of neurons. *Anat. Rec.* **148,** 413.

Blackstad, T. W. (1970). Electron microscopy of Golgi preparations for the study of neuronal relations. *In* "Contemporary Research Methods in Neuroanatomy" (W. J. H. Nauta and S. O. E. Ebbesson, eds.), pp. 187–215.

Blinkov, S. M , and Glezer, I. I. (1968). "The Human Brain in Figures and Tables: A Quantitative Handbook." Basic Books, New York.

Boissonat, J. D. (1988). Shape reconstruction from planar cross sections. *Comp. Vision Graph. Image Process.* **44,** 1–29.

Bok, S. T. (1959). "Histonomy of the Cerebral Cortex." Elsevier, Amsterdam/New York.

Bolton, J. S. (1900). The exact histological localisation of the visual area of the human cerebral cortex. *Philos. Trans. R. Soc. London* **193,** 165–222.

Bookstein, F. L. (1991). "Morphometric Tools for Landmark Data" Cambridge Univ. Press, Port Chester, NY.

Boon, M. E., and Kok, L. P. (1987). "Microwave Cookbook of Pathology." Coulomb Press Leyden, Leiden.

Born, G. (1883). Die Plattenmodellirmethode. *Arch. Mikrosc. Anat. Entwicklungsmech.* **22,** 584–599.

Bracegirdle, B. (1978). "A History of Microtechnique" Heinemann. London.

Braverman, M. S., and Braverman, I. M. (1986). Three-dimensional reconstructions of objects from serial sections using a microcomputer graphics system. *J. Invest. Dermatol.* **86,** 290–294.

Brodal, A. (1982). Anterograde and retrograde degeneration of nerve cells in the central nervous system. *In* "Histology and Histopathology of the Nervous System" (W. Haymaker and R. D. Adams, eds.), pp. 276–362. Springfield, IL.

Brodmann, K. (1909). "Vergleichende Localisationslehre der Grosshirnrinde in ihren Prinzipien dargestellt auf Grund des Zellebaus." Barth, Leipzig.

Brodmann, K. (1912). Neue ergibnisse uber die vergleichende Lokalisation der Grosshirnrinde mit besonderer Berucksichtigung des Stirnhirns. *Anat. Anz. Suppl.* **41,** 157.

Buhl, E. H., and Lubke, J. (1989). Intracellular Lucifer yellow injection in fixed brain slices combined with retrograde tracing, light and electron microscopy. *Neuroscience* **28,** 3–16.

Bunt, S. M., and Hubbard, B. M. (1988). Fibre tracing with carbocyanine dyes in the *post mortem* human brain. *Eur. J. Neurosci. (Suppl.)* **1,** 131.

Burgel, U., Mecklenburg, I., Blohm, U., and Zilles, K. (1997). Histological visualization of long fiber tracts in the white matter of adult human brains. *J. Hirnforsch.* **38,** 397–404.

Burkhalter, A., and Bernardo, K. L. (1989). Organization of cortico-cortical connections in human visual cortex. *Proc. Natl. Acad. Sci. USA* **86,** 1071–1075.

Campbell, A. W. (1905). "Histological Studies on the Localisation of Cerebral Function." Cambridge Univ. Press, Cambridge, UK.

Cannestra, A. F., Santori, E. M., Holmes, C. J., and Toga, A. W. (1997). A three-dimensional multimodality brain map of the nemestrina monkey. *Brain Res. Bull.* **43,** 141–148.

Carman, G. J., Drury, H. A., and Van Essen, D. C. (1995). Computational methods for reconstructing and unfolding the cerebral cortex. *Cereb. Cortex* **5,** 506–517.

Choulant, L. (1945). "History and Bibliography of Anatomic Illustration." Hafner, New York.

Clark, G., and Kasten, F. (1983). "History of Staining." Williams & Wilkins, Baltimore.

Clarke, E., and Dewhurst, K. (1972). "An Illustrated History of Brain Function." Sandford, Oxford.

Clarke, S., and Miklossy, J. (1990). Occipital cortex in man: Organization of callosal connections, related myelo- and cytoarchitecture, and putative boundaries of functional visual areas. *J. Comp. Neurol.* **298,** 188–214.

Cohen, F. S., Yang, Z., Huang, Z., and Nissanov, J. (1998). Automatic matching of homologous histological sections. *IEEE Trans. Biomed. Eng.* **45,** 642–649.

Cohen, L. D., and Cohen, I. (1992). Deformable models for 3D medical images using finite elements and balloons. *In* "IEEE International Conference on Computer Vision and Pattern Recognition." Urbana, IL.

Courchesne, E., Press, G., Murakami, J., Berthoty, D., Grafe, M., Wiley, C., and Hesselink, J. (1989). The cerebellum in sagittal planes—Anatomic-MR correlation. 1. The vermis. *Am. J. Neuroradiol.* **10,** 659–665.

Crick, F., and Jones, E. (1993). Backwardness of human neuroanatomy. *Nature* **361,** 109–110.

Cruz-Orive, L. M. (1997). Stereology of single objects. *J. Microsc.* **186,** 93–107.

Cunningham, D. J. (1902). "Textbook of Anatomy." Pentland, Edinburgh/London.

Danek, A., Bauer, N., and Fries, W. (1990). Tracing of neuronal connections in the human brain by magnetic resonance imaging in vivo. *Eur. J. Neurosci.* **2,** 112–115.

Davies, P. C., Mirra, S., and Alazraki, N. (1994). The brain in older persons with and without dementia: Findings on MR, PET, and SPECT images. *Am. J. Roentg.* **162,** 1267–1278.

DeRisi, J., Penland, L., Brown, P. O., Bittner, M. L., Meltzer, P. S., Ray, M., Chen, Y., Su, Y. A., and Trent, J. M. (1996). Use of a cDNA microarray to analyse gene expression patterns in human cancer. *Nat. Genet.* **14,** 457–460.

DeYoe, E. A., Hockfield, S., Garren, H., and Van Essen, D. C. (1990). Antibody labeling of functional subdivisions in visual cortex: Cat-301 immunoreactivity in striate and extrastriate cortex of the macaque monkey. *Visual Neurosci.* **5,** 67–81.

Drebin, R., Carpenter, L., and Hanrahan, P. (1988). Volume rendering. *Comp. Graph.* **22,** 65–74.

Drushel, R. F. (1993). Survey of methods for 3-dimensional reconstruction: 1991–1993. Comp. graphics. visualization: Usenet, 1993.

Duggan, D. J., Bittner, M., Chen, Y., Meltzer, P., and Trent, J. M. (1999). Expression profiling using cDNA microarrays. *Nat. Genet.* **21,** 10–14.

Durr, R., Peterhans, E., and von der Heydt, R. (1989) Correction of distorted images. *Eur. J. Cell Biol.* **48**(Suppl. 25), 85.

Duta, N., and Sonka, M. (1998). Segmentation and interpretation of MR brain images: An improved active shape model. *IEEE Trans. Med. Imaging*, **17,** 1049–1062..

Duvernoy, H. M., Delon, S., and Vannson, J. L. (1981). Cortical blood vessels of the human brain. *Brain Res. Bull.* **7,** 519–579.

Ebbesson, S. O., and Cheek, M. (1988). The use of cryostat microtomy in a simplified Golgi method for staining vertebrate neurons. *Neurosci. Lett.* **88,** 135–138.

Economo, C. v., and Koskinas, G. N. (1925). "Die Cytoarchitectonik der Hirnrinde des erwachsenen Menschen." Springer-Verlag, Vienna.

Economo, C. v., and Koskinas, G. N. (1929). "The Cytoarchitectonics of the Human Cerebral Cortex." Oxford Univ. Press, Oxford.

Eggers, R., Haug, H., and Fischer, D. (1984). Preliminary report on macroscopic age changes in the human prosencephalon: A stereologic investigation. *J. Hirnforsch.* **25,** 129–139.

Elkin, A. P., and Macintosh, N. W. G. (1974). "Grafton Elliot Smith: The Man and His Work." Sydney Univ. Press, Sydney.

Elston, G. N., and Rosa, M. G. (1997). The occipitoparietal pathway of the macaque monkey: Comparison of pyramidal cell morphology in layer III of functionally related cortical visual areas. *Cereb. Cortex* **7,** 432–452.

Evers, P., and Uylings, H. B. (1994). Microwave-stimulated antigen retrieval is pH and temperature dependent. *J. Histochem. Cytochem.* **42,** 1555–1563.

Falconer, D. S., and Mackay, T. F. C. (1996). "Introduction to Quantitative Genetics." Longman, Harlow.

Felleman, D. J., and Van Essen, D. C. (1991). Distributed hierarchical processing in the primate cerebral cortex. *Cereb. Cortex* **1,** 1–47.

Flechsig, P. (1901). Developmental (myelogenetic) localisation of the cerebral cortex in the human subject. *Lancet* **2,** 1027–1029.

Flechsig, P. (1920). "Anatomie des menschichen Gehirns und Ruckenmarks auf Myelogenetischer Grundlage." Thieme, Leipzig.

Gallyas, F. (1971). A principle for silver staining of tissue elements by physical development. *Acta Morphol. Acad. Sci. Hung.* **19,** 57–71.

Gallyas, F. (1979). Silver staining of myelin by means of physical development. *Neurol. Res.* **1,** 203–209.

Gallyas, F., Hsu, M., and Buzsaki, G. (1993). Four modified silver methods for thick sections of formaldehyde-fixed mammalian central nervous tissue: 'Dark' neurons, perikarya of all neurons, microglial cells and capillaries. *J. Neurosci. Methods* **50,** 159–164.

Garey, L. J. (1994). "Brodmann's Localisation in the Cerebral Cortex." Smith–Gordon, London.

Gaunt, W. A., and Gaunt, P. N. (1978). "Three Dimensional Reconstruction in Biology." Pitman, Tunbridge Wells, UK.

Gavoy, E. (1882). "Atlas d'Anatomie Topographique du Cerveau et des Localisation Cerebrales." Actave Doin, Paris.

Gennari, F. (1782). "De Peculiari Structura Cerebri: Nonnullisque Ejus Morbis." Ex Regio Typographeo, Parma, Italy.

Gerke, M., Schutz, T., Vogt, H., and Kretschmann, H. J. (1992). Computer-assisted 3D-reconstruction and statistics of the limbic system. 2. Spatial statistics of the hippocampal formation, the fornix, and the mammillary bodies. *Anat. Embryol. (Berlin)* **186,** 137–143.

Gerritts, P. O. (1988). Immunohistochemistry on glycol methacrylate embedded tissues: Possibilities and limitations. *J. Histotechnol.* **11,** 243–246.

Geschwind, D. H., and Miller, B. L. (2001). Molecular approaches to cerebral laterality: Development and neurodegeneration. *Am. J. Med. Genet.* **101,** 370–381.

Geschwind, D. H., Ou, J., Easterday, M. C., Dougherty, J. D., Jackson, R. L., Chen, Z., Antoine, H., Terskikh, A., Weissman, I. L., Nelson, S. F., and Kornblum, H. I. (2001). A genetic analysis of neural progenitor differentiation. *Neuron* **29,** 325–339.

Glaser, J. R., and Glaser, E. M. (2000). Stereology, morphometry, and mapping: The whole is greater than the sum of its parts. *J. Chem. Neuroanat.* **20,** 115–126.

Goding, W. J. (1996). "Monoclonal Antibodies: Principles and Practice." Academic Press, London.

Goldszal, A. F., Tretiak, O. J., Hand, P. J., Bhasin, S., and McEachron, D. L. (1995). Three-dimensional reconstruction of activated columns from 2-[^{14}C]deoxy-<E5>d</E5>-glucose data. *NeuroImage* **2,** 9–20.

Goldwasser, S. M., and Reynolds, R. A. (1987). Real-time display and manipulation of 3-D medical objects: The voxel processor architecture. *Comput. Vision Graph. Image Process.* **39,** 1.

Golgi, C. (1878–1879). Di una nuova reazione apparentemente nera delle cellule nervose cerebrali ottenuta col bicloruro di mercurio. *Arch. Sci. Med.* **3,** 1–7.

Gonzales, R. C., and Woods, R. E. (1992). "Digital Image Processing." Addison–Wesley, Reading, MA.

Guillery, R. W., and Herrup, K. (1997). Quantification without pontification: Choosing a method for counting objects in sectioned tissues. *J. Comp. Neurol.* **386,** 2–7.

Gundersen, H. J. (1986). Stereology of arbitrary particles. A review of unbiased number and size estimators and the presentation of some new ones, in memory of William R. Thompson. *J. Microsc.* **143,** 3–45.

Gundersen, H. J. (1988). The nucleator. *J. Microsc.* **151,** 3–21.

Gundersen, H. J., Bagger, P., Bendtsen, T. F., Evans, S. M., Korbo, L., Marcussen, N., Moller, A., Nielsen, K., Nyengaard, J. R., Pakkenberg, B., *et al.* (1988). The new stereological tools: Disector, fractionator, nucleator and point sampled intercepts and their use in pathological research and diagnosis. *APMIS* **96,** 857–881.

Haas, A., and Fischer, M. S. (1997). Three-dimensional reconstruction of histological sections using modern product-design software. *Anat. Rec.* **249,** 510–516.

Hand, N., and Phil, M. (1999). Plastic embedding for light microscopy in 2000: A guide to developments and techniques for semi-thin sections. *In* "National Society for Histotechnology 1999 Symposium/Convention," Providence, RI.

Hand, N. M. (1995). The naming and types of acrylic resins. *UK NEQAS Newsl.* **6,** 15.

Hardy, J. A., Wester, P., Winblad, B., Gezelius, C., Bring, G., and Eriksson, A. (1985). The patients dying after long terminal phase have acidotic brains: Implications for biochemical measurements on autopsy tissue. *J. Neural Transm.* **61,** 253–264.

Hatton, W. J., and von Bartheld, C. S. (1999). Analysis of cell death in the trochlear nucleus of the chick embryo: Calibration of the optical disector counting method reveals systematic bias. *J. Comp. Neurol.* **409,** 169–186.

Haug, H. (1980). The significance of quantitative stereologic experimental procedures in pathology. *Pathol. Res. Pract.* **166,** 144–164.

Haug, H. (1986). History of neuromorphometry. *J. Neurosci. Methods* **18,** 1–17.

Hawkes, R., Niday, E., and Matus, A. (1982a). Monoclonal antibodies identify novel neural antigens. *Proc. Natl. Acad. Sci. USA* **79,** 2410–2414.

Hawkes, R., Ng, M., Niday, E., and Matus, A. (1982b). Immunocytochemical localization of identified proteins in brain by monoclonal antibodies. *Prog. Brain Res.* **56,** 77–86.

Heard, O. O. (1931). A photography method of orienting serial sections for reconstruction. *Anat. Rec.* **49,** 59–69.

Hegre, E. S. (1952). A new research tool and technique for the biologist. *Virginia J. Sci.* **2,** 10–12.

Hegre, E. S., and Brashear, A. D. (1946). Block surface staining. *Stain Technol.* **21,** 161–164.

Hegre, E. S., and Brashear, A. D. (1947). The block surface method of staining, as applied to the study of embryology. *Anat. Rec.* **97,** 21–28.

Heinsen, H., Arzberger, T., and Schmitz, C. (2000). Celloidin mounting (embedding without infiltration): A new, simple and reliable method for producing serial sections of high thickness through complete human brains and its application to stereological and immunohistochemical investigations. *J. Chem. Neuroanat.* **20,** 49–59.

Heinsen, H., and Heinsen, Y. L. (1991). Serial thick, frozen, gallocyanin stained sections of human central nervous system. *J. Histotech.* **14,** 167–173.

Hendry, S. H., Jones, E. G., Hockfield, S., and McKay, R. D. (1988). Neuronal populations stained with the monoclonal antibody Cat-301 in the mammalian cerebral cortex and thalamus. *J. Neurosci.* **8,** 518–542.

Herman, G. T., and Liu, H. K. (1977). Display of three-dimensional information in computed tomography. *J. Comput. Assisted Tomogr.* **1,** 155–160.

Hibbard, L. S., and Hawkins, R. A. (1984). Three-dimensional reconstruction of metabolic data from quantitative autoradiography of rat brain. *Am. J. Physiol.* **247,** E412–419.

Hibbard, L. S., and Hawkins, R. A. (1988). Objective image alignment for three-dimensional reconstruction of digital autoradiograms. *J. Neurosci. Methods* **26,** 55–74.

Hilbig, H., Bidmon, H. J., Blohm, U., and Zilles, K. (2001). *Wisteria floribunda* agglutinin labeling patterns in the human cortex: A tool for revealing areal borders and subdivisions in parallel with immunocytochemistry. *Anat. Embryol. (Berlin)* **203,** 45–52.

Hine, B., and Rodriguez, R. (1992). Rapid gelatin embedding procedure for frozen brain tissue sectioning. *J. Histol.* **15,** 121–122.

Hirsch, W. L., Kemp, S. S., Martinez, A. J., Curtin, H., Latchaw, R. E., and Wolf, G. (1989). Anatomy of the brainstem: Correlation of in vitro MR images with histologic sections. *Am. J. Neuroradiol.* **10,** 923–928.

His, W. (1868). "Untersuchungen uber die Erste Anlage des Wirbeltierleibes." Vogel, Leipzig.

Ho, P. S., Yu, S. W., Sether, L., Wagner, M., and Haughton, V. M. (1988). MR and cryomicrotomy of C1 and C2 roots. *Am. J. Neuroradiol.* **9,** 829–831.

Hockfield, S., and McKay, R. D. (1983). A surface antigen expressed by a subset of neurons in the vertebrate central nervous system. *Proc. Natl. Acad. Sci. USA* **80,** 5758–5761.

Hockfield, S., Tootell, R. B., and Zaremba, S. (1990). Molecular differences among neurons reveal an organization of human visual cortex. *Proc. Natl. Acad. Sci. USA* **87,** 3027–3031.

Hof, P. R., and Morrison, J. H. (1995). Neurofilament protein defines regional patterns of cortical organization in the macaque monkey visual system: A quantitative immunohistochemical analysis. *J. Comp. Neurol.* **352,** 161–186.

Hof, P. R., Mufson, E. J., and Morrison, J. H. (1995). Human orbitofrontal cortex: Cytoarchitecture and quantitative immunohistochemical parcellation. *J. Comp. Neurol.* **359,** 48–68.

Horton, J. C., and Hedley-Whyte, E. T. (1984). Mapping of cytochrome oxidase patches and ocular dominance columns in human visual cortex. *Philos. Trans. R. Soc. London B Biol. Sci.* **304,** 255–272.

Howard, C. V., and Reed, M. G. (1998). "Unbiased Stereology: Three-Dimensional Measurement in Microscopy." Springer-Verlag, Berlin.

Hudspeth, A. J., Ruark, J. E., and Kelly, J. P. (1976). Cytoarchitectonic mapping by microdensitometry. *Proc. Natl. Acad. Sci. USA* **73,** 2928–2931.

Huijsmans, D. P., Lamers, W. H., Los, J. A., and Strackee, J. (1986). Toward computerized morphometric facilities: A review of 58 software packages for computer-aided three-dimensional reconstruction, quantification, and picture generation from parallel serial sections. *Anat. Rec.* **216,** 449–470.

Inoue, S., and Spring, K. R. (1997). "Videomicroscopy," Plenum, New York.

Iwadare, T., Mori, H., Ishiguro, K., and Takeishi, M. (1984). Dimensional changes of tissues in the course of processing. *J. Microsc.* **136,** 323–327.

Jacobs, M. A., Windham, J. P., Soltanian-Zadeh, H., Peck, D. J., and Knight, R. A. (1999). Registration and warping of magnetic resonance images to histological sections. *Med. Phys.* **26,** 1568–1578.

Jendroska, K., Patt, S., Janisch, W., Cervos-Navarro, J., and Poewe, W. (1993). How to run a "brain bank": Clinical and institutional requirements for "brain banking." *J. Neural Transm. Suppl.* **39,** 71–75.

Johnson, E. M., and Capowski, J. J. (1985). Principles of reconstruction and three-dimensional display of serial sections using a computer. *In* "The Microcomputer in Cell and Neurobiology Research."

Katzberg, R. W., Westesson, P. L., Tallents, R. H., Anderson, R., Kurita, K., Manzione, J. V., Jr., and Totterman, S. (1988). Temporomandibular joint: MR assessment of rotational and sideways disk displacements. *Radiology* **169,** 741–748.

Kaufmann, W. E., and Galaburda, A. M. (1989). Cerebrocortical microdysgenesis in neurologically normal subjects: A histopathologic study. *Neurology* **39,** 238–244.

Kearsey, M. J., and Pooni, H. S. (1996). "The Genetical Analysis of Quantitative Traits." Chapman & Hall, London.

Kelemen, A., Szekely, G., and Gerig, G. (1999). Elastic model-based segmentation of 3-D neuroradiological data sets. *IEEE Trans. Med. Imaging* **18,** 828–8239.

Kevorkian, J. (1959). "The History of Dissection." Philosophical Library, New York.

Kim, B., Boes, J. L., Frey, K. A., and Meyer, C. R. (1997). Mutual information for automated unwarping of rat brain autoradiographs. *NeuroImage* **5,** 31–40.

Kingsbury, A. E., Foster, O. J., Nisbet, A. P., Cairns, N., Bray, L., Eve, D. J., Lees, A. J., and Marsden, C. D. (1995). Tissue pH as an indicator of mRNA preservation in human post-mortem brain. *Brain Res. Mol. Brain Res.* **28,** 311–318.

Kingsley, T. C., and King, R. (1989). Plastic or paraffin? *Lancet* **1,** 563.

Kluver, H., and Barrera, E. (1953). A method for the combined staining of cells and fibers in the nervous system. *J. Neuropathol. Exp. Neurol.* **12,** 400.

Kohler, G., and Milstein, C. (1975). Continuous cultures of fused cells secreting antibody of predefined specificity. *Nature* **256,** 495–497.

Kok, L. P., and Boon, M. E. (1989). Microwaves for microscopy. *J. Microsc.* **158,** 291–322.

Kok, L. P., Boon, M. E., and Suurmeijer, A. J. (1987). Major improvement in microscopic-image quality of cryostat sections. Combining freezing and microwave-stimulated fixation. *Am. J. Clin. Pathol.* **88,** 620–623.

Kretschmann, H. J., Schleicher, A., Grottschreiber, J. F., and Kullmann, W. (1979). The Yakovlev Collection: A pilot study of its suitability for the morphometric documentation of the human brain. *J. Neurol. Sci.* **43,** 111–126.

Krieg, V. J. S. (1966). "Functional Neuroanatomy." Evanston, IL.

Krieg, V. J. S. (1967). Reconstruction from serial sections. *In* "Stereology" (H. Elias, ed.). New York.

Kropf, N., Sobel, I., and Levinthal, C. (1985). "Serial Section Reconstruction Using CARTOS: The Microcomputer in Cell and Neurobiology Research." Elsevier, New York.

Kruggel, F., Bruckner, M. K., Arendt, T., Wiggins, C. J., and von Cramon, D. Y. (2001). Analyzing the neocortical fine-structure. *In* "Lecture Notes in Computer Science (IPMI 2001)" (M. F. Insana and R. M. Leahy, eds.). Berlin/Heidelberg.

Kuljis, R. O. (1994). The human primary visual cortex. *In* "Cerebral Cortex" (A. Peters and E. G. Jones, eds.), pp. 469–497. New York/London.

Kvasnicka, H. M., and Thiele, J. (1995). [3-Dimensional reconstruction of serial sections in light microscopy.] *Pathologe* **16,** 128–138.

Lander, E. S., and Schork, N. J. (1994). Genetic dissection of complex traits. *Science* **265,** 2037–2048.

Lashley, K. S., and Clark, G. (1946). The cytoarchitecture of the cerebral cortex of *Ateles:* A critical examination of architectonic studies. *J. Comp. Neurol.* **85,** 223.

Le Brun-Kemper, T., and Galaburda, A. M. (1991). Principles of architectonics. *In* "Cerebral Cortex. Cellular Component of the Cerebral Cortex" (A. Peters and E. Jones, eds.), pp. 35–57. New York.

Lebedkin, S. (1930–1931). Die rationelle Technik der Herstellung von plastichen Rekonstruktionen und die Zelluloidmodelle. *Z. Wiss. Mikrosk. Mikrosk. Tech.* **47,** 294–317.

Lewis, B. (1877). A new freezing microtome for the preparation of sections of brain and spinal chord. *J. Anat. Physiol.* **11,** 537.

Lewis, B. (1879a). On the comparative structure of the cortex cerebri. *Brain* **1,** 77.

Lewis, B. (1879b). Application of freezing methods to the microscopic examination of the brain. *Brain* **1,** 348.

Liang, P., and Pardee, A. B. (1992). Differential display of eukaryotic messenger RNA by means of the polymerase chain reaction. *Science* **257,** 967–971.

Lockhart, D. J., and Barlow, C. (2001). Expressing what's on your mind: DNA arrays and the brain. *Nat. Rev. Neurosci.* **2,** 63–68.

Lohner, R. (1997). Automatic unstructured grid generators. *Finite Elem. Anal. Des.* **25,** 111–134.

Lundervold, A., and Storvik, G. (1995). Segmentation of brain parenchyma and cerebrospinal fluid in multispectral magnetic resonance images. *IEEE Trans. Med. Imaging* **14,** 339–349.

Macagno, E. R., Levinthal, C., and Sobel, I. (1979). Three-dimensional computer reconstruction of neurons and neuronal assemblies. *Annu. Rev. Biophys. Bioeng.* **8,** 323–351.

MacEwen, W. (1893). "Atlas of Head Sections," Macmillan Co., New York.

Marko, M., Leith, A., and Parsons, D. (1988). Three-dimensional reconstruction of cells from serial sections and whole-cell mounts using multilevel contouring of stereo micrographs. *J. Electron Microsc. Tech.* **9,** 395–411.

Mathé, J. (1978). "Leonardo da Vinci: Anatomical Drawings." Miller Graphics, distributed by Crown.

Mayer, P. (1910). Ein neues Mikrotom: Das Tetrander. *Z. Wiss. Mikrosk.* **27,** 52–62.

Mazzarello, P. (1996). "La Struttura Nascosta: La Vita di Camillo Golgi." Litosei, Rastignano.

McIrney, T., and Terzopoulos, D. (1998). Deformable models in medical image analysis: A survey. *Med. Image Anal.* **1,** 91–108.

McKay, R. D., Hockfield, S., Johansen, J., and Frederiksen, K. (1983). The molecular organization of the leech nervous system. *Cold Spring Harbor Symp. Quant. Biol.* **48**(Pt. 2), 599–610.

Menhardt, W., Lockhausen, J., Dallas, W. J., and Kristen, U. (1986). An environment for three-dimensional shaded perspective display of cell components. *Micron Microsc. Acta* **17,** 349.

Merker, B. (1983). Silver staining of cell bodies by means of physical development. *J. Neurosci. Methods* **9,** 235–241.

Merkle. (1989). Large scale analysis of neural structures. Report No. P89-00173, 11/10/1989. Xerox Corp., Palo Alto, CA.

Meryman, H. T. (1966). The interpretation of freezing rates in biological materials. *Cryobiology* **2,** 165–170.

Meryman, H. T. (1971). Cryoprotective agents. *Cryobiology* **8,** 173–183.

Mesulam, M. M. (1979). Tracing neural connections of human brain with selective silver impregnation: Observations on geniculocalcarine, spinothalamic, and entorhinal pathways. *Arch. Neurol.* **36,** 814–818.

Mesulam, M. M. (1996). The systems-level organization of cholinergic innervation in the human cerebral cortex and its alterations in Alzheimer's disease. *Prog. Brain Res.* **109,** 285–297.

Mesulam, M. M., and Geula, C. (1991). Acetylcholinesterase-rich neurons of the human cerebral cortex: Cytoarchitectonic and ontogenetic patterns of distribution. *J. Comp. Neurol.* **306,** 193–220.

Mikkonen, M., Pitkanen, A., Soininen, H., Alafuzoff, I., and Miettinen, R. (2000). Morphology of spiny neurons in the human entorhinal cortex: Intracellular filling with Lucifer yellow. *Neuroscience* **96,** 515–522.

Miklossy, J., Clarke, S., and Van der Loos, H. (1991). The long distance effects of brain lesions: Visualization of axonal pathways and their terminations in the human brain by the Nauta method. *J. Neuropathol. Exp. Neurol.* **50,** 595–614.

Miklossy, J., and Van der Loos, H. (1991). The long-distance effects of brain lesions: Visualization of myelinated pathways in the human brain using polarizing and fluorescence microscopy. *J. Neuropathol. Exp. Neurol.* **50,** 1–15.

Miles, R. E., and Davy, P. J. (1976). Precise and general conditions for the validity of a comprehensive set of stereological fundamental formulas. *J. Microsc.* **107,** 211–226.

Miller, D. C. (1998). Use of perfusion fixation for improved neuropathologic examination. *Arch. Pathol. Lab. Med.* **122,** 949.

Mirnics, K. (2001). Microarrays in brain research: The good, the bad and the ugly. *Nat. Rev. Neurosci.* **2,** 444–447.

Mizuno, T., Nagamura, H., Iwamoto, K. S., Ito, T., Fukuhara, T., Tokunaga, M., Tokuoka, S., Mabuchi, K., and Seyama, T. (1998). RNA from decades-old archival tissue blocks for retrospective studies. *Diagn. Mol. Pathol.* **7,** 202–208.

Montgomery, K. (1996). "Automated Reconstruction of Neural Elements from Transmission Electron Microscope Images." Univ. of California, Santa Cruz. [Ph.D. thesis]

Morey, A. L. (1995). Non-isotopic in situ hybridization at the ultrastructural level. *J. Pathol.* **176,** 113–121.

Moss, V. A. (1992). Fundamentals of 3-D reconstruction of serial sections on a microcomputer. *In* "Visualization in Biomedical Microscopies" (A. Kreite, ed.), pp. 19–43. New York/Weinheim.

Mueller, M., Wacker, K., Hickey, W. F., Ringelstein, E. B., and Kiefer, R. (2000). Co-localization of multiple antigens and specific DNA: A novel method using methyl methacrylate-embedded semithin serial sections and catalyzed reporter deposition. *Am. J. Pathol.* **157,** 1829–1838.

Naber, S. P., Smith, L. L., Jr., and Wolfe, H. J. (1992). Role of the frozen tissue bank in molecular pathology. *Diagn. Mol. Pathol.* **1,** 73–79.

Nauta, W. J. H., and Gygax, P. A. (1954). Silver impregnation of degenerating axons in the central nervous system: A modified technique. *Stain Technol.* **29,** 91–93.

Neumann, A., and Lorenz, C. (1998). Statistical shape model based segmentation of medical images. *Comput. Med. Imaging Graph.* **22,** 133–143.

Newman, S. J., and Gentleman, S. M. (1997). Microwave antigen retrieval in formaldehyde-fixed human brain tissue. *Methods Mol. Biol.* **72,** 145–152.

Nissl, F. (1894). Mitteilungen zur Anatomie der Nervenzelle. *Z. Psychiatr.* **370.**

Nissl, F. (1910). Nervensystem. *In* "Enzyklopadie der Mikroskopschen Technik" (P. Ehrlich, R. Krause, M. Mosse, and K. Weigert, eds.), pp. 243–285. Berlin/Vienna.

Nochlin, D., Mackenzie, A. P., Bryant, E. M., Norwood, T. H., and Sumi, S. M. (1993). A simple method of rapid freezing adequately preserves brain tissue for immunocytochemistry, light and electron microscopic examination. *Acta Neuropathol.* **86,** 645–650.

O'Malley, C. D., and Saunders, C. M. (1952). "Leonoardo da Vinci on the Human Body: The Anatomical, Physiological, and Embryological Drawings of Leonardo da Vinci, with Translations, Emendations and a Biographical Introduction." Schuman, New York.

Ono, M., Kubic, S., and Abernathey, C. D. (1990). "Atlas of the Cerebral Sulci." Thieme, Stuttgart.

Osborn, J. W. (1967). A mechanistic view of dentinogenesis and its relation to the curvatures of the processes of the odontoblasts. *Arch. Oral Biol.* **12,** 275–280.

Ourselin, S., Roche, A., Subsol, G., Pennec, X., and Sattonnet, C. (1998). Automatic alignment of histological sections for 3D reconstruction and analysis, pp. 27. Sophia Antinopolis.

Pakkenberg, B., Boesen, J., Albeck, M., and Gjerris, F. (1989). Unbiased and efficient estimation of total ventricular volume of the brain obtained from CT-scans by a stereological method. *Neuroradiology* **31,** 413–417.

Pakkenberg, B., and Gundersen, H. J. (1988). Total number of neurons and glial cells in human brain nuclei estimated by the disector and the fractionator. *J. Microsc.* **150,** 1–20.

Pakkenberg, B., and Gundersen, H. J. (1989). New stereological method for obtaining unbiased and efficient estimates of total nerve cell number in human brain areas exemplified by the mediodorsal thalamic nucleus in schizophrenics. *APMIS* **97,** 677–681.

Payne, E. E. (1967). A technique for the examination of the brain: Large thin sections mounted on paper. *J. Pathol. Bacteriol.* **93,** 721–723.

Pech, P., Bergstrom, K., Rauschning, W., and Haughton, V. M. (1987). Attenuation values, volume changes and artifacts in tissue due to freezing. *Acta Radiol.* **28,** 779–782.

Peinado, M. A. (1998). Histology and histochemistry of the aging cerebral cortex: An overview. *Microsc. Res. Tech.* **43,** 1–7.

Pelizzari, C. A., Chen, G. T., Spelbring, D. R., Weichselbaum, R. R., and Chen, C. T. (1989). Accurate three-dimensional registration of CT, PET, and/or MR images of the brain. *J. Comput. Assisted Tomogr.* **13,** 20–26.

Perry, E. K., and Perry, R. H. (1983). Human brain neurochemistry—Some *post mortem* problems. *Life Sci.* **33,** 1733–1743.

Perry, E. K., Perry, R. H., and Tomlinson, B. E. (1982). The influence of agonal status on some neurochemical activities of *post mortem* human brain tissue. *Neurosci. Lett.* **29,** 303–307.

Peters, A., Moss, M. B., and Sethares, C. (2000). Effects of aging on myelinated nerve fibers in monkey primary visual cortex. *J. Comp. Neurol.* **419,** 364–376.

Peterson, D. A. (1999). Quantitative histology using confocal microscopy: Implementation of unbiased stereology procedures. *Methods* **18,** 493–507.

Peterson, D. A., and Jones, D. G. (1993). Determination of neuronal number and process surface area in organotypic cultures: A stereological approach. *J. Neurosci. Methods* **46,** 107–120.

Pfeifer, R. A. (1930). "Grundlegende Untersuchungen fur die Angioarchitektonik des Menschelichen Gerhins." Springer-Verlag, Berlin.

Pratt, W. (1978). "Digital Image Processing." Wiley, New York.

Quester, R., and Schroder, R. (1997). The shrinkage of the human brain stem during formalin fixation and embedding in paraffin. *J. Neurosci. Methods* **75,** 81–89.

Quinn, B., Ambach, K. A., and Toga, A. W. (1993). Three-dimensional cryomacrotomy with integrated computer-based technology in neuropathology. *Lab. Invest.* **68,** 121A.

Rademacher, J., Engelbrecht, V., Burgel, U., Freund, H., and Zilles, K. (1999). Measuring in vivo myelination of human white matter fiber tracts with magnetization transfer MR. *Neuroimage* **9**, 393–406.

Randen, T., and Husøy, J. H. (1999). Filtering for texture classification: A comparative study. *IEEE Trans. PAMI* **21,** 291–310.

Rangarajan, A., Chui, H., Mjolsness, E., Pappu, S., Davachi, L., Goldman-Rakic, P., and Duncan, J. (1997). A robust point-matching algorithm for autoradiograph alignment. *Med. Image Anal.* **1,** 379–398.

Risch, N., and Merikangas, K. (1996). The future of genetic studies of complex human diseases. *Science* **273,** 1516–1517.

Roberts, N., Cruz-Orive, L. M., Reid, N. M., Brodie, D. A., Bourne, M., and Edwards, R. H. (1993). Unbiased estimation of human body composition by the Cavalieri method using magnetic resonance imaging. *J. Microsc.* **171,** 239–253.

Roberts, N., Puddephat, M. J., and McNulty, V. (2000). The benefit of stereology for quantitative radiology. *Br. J. Radiol.* **73,** 679–697.

Robson, M. D., Gore, J. C., and Constable, R. T. (1997). Measurement of the point spread function in MRI using constant time imaging. *Magn. Reson. Med.* **38,** 733–740.

Roland, P. E., and Zilles, K. (1996). Functions and structures of the motor cortices in humans. *Curr. Opin. Neurobiol.* **6,** 773–781.

Romaya, J., and Zeki, S. (1985). Rotable three-dimensional computer reconstructions of the macaque monkey brain. *J. Physiol.* **371,** 25.

Romero, E., Cuisenaire, O., Denef, J. F., Delbeke, J., Macq, B., and Veraart, C. (2000). Automatic morphometry of nerve histological sections. *J. Neurosci. Methods* **97,** 111–122.

Rosen, G. D., and Harry, J. D. (1990). Brain volume estimation from serial section measurements: A comparison of methodologies. *J. Neurosci. Methods* **35,** 115–124.

Rosene, D. L., and Mesulam, M. M. (1978). Fixation variables in horseradish peroxidase neurohistochemistry. I. The effect of fixation time and perfusion procedures upon enzyme activity. *J. Histochem. Cytochem.* **26,** 28–39.

Rosene, D. L., and Rhodes, K. J. (1990). Cryoprotection and . *In* "Quantitative and Qualitative Microscopy" (P. M. Conn, ed.), pp. 360–390. New York.

Rosene, D. L., Roy, N. J., and Davis, B. J. (1986). A cryoprotection method that facilitates cutting frozen sections of whole monkey brains for histological and histochemical processing without freezing artifact. *J. Histochem. Cytochem.* **34,** 1301–1315.

Rossman, K. P. (1969). Point spread-function, line spread function, and modulation transfer function: Tools for the study of imaging systems. *Radiology* **93,** 257–272.

Rusinek, H., Tsui, W. H., Levy, A. V., Noz, M. E., and de Leon, M. J. (1993). Principal axes and surface fitting methods for three-dimensional image registration. *J. Nucl. Med.* **34,** 2019–2024.

Russ, J. C. (1992). Automatic vs. computer-assisted 3-D reconstruction. *In* "Proceedings of the 50th EMSA Meeting," San Francisco, p. 1050.

Russ, J. C. (1999). "The Image Processing Handbook." CRC Press, Boca Raton, FL; IEEE Press, New York.

Sadowski, M., Morys, J., Berdel, B., and Maciejewska, B. (1995). Influence of fixation and histological procedure on the morphometric parameters of neuronal cells. *Folia Morphol.* **54,** 219–226.

Sass, N. L. (1982). The age-dependent variation of the embedding-shrinkage of neurohistological sections. *Mikroskopie* **39,** 278–281.

Sato, Y., Nakajima, S., Shiraga, N., Atsumi, H., Yoshida, S., Koller, T., Gerig, G., and Kikinis, R. (1998). Three-dimensional multi-scale line filter for segmentation and visualization of curvilinear structures in medical images. *Med. Image Anal.* **2,** 143–168.

Sauer, B. (1983). Semi-automatic analysis of microscopic images of the human cerebral cortex using the grey level index. *J. Microsc.* **129**(Pt. 1), 75–87.

Sauer, B., Dietl, H. W., Kretschmann, H. J., and Mehraein, P. (1986). A quantitative study of the cerebral cortex in Alzheimer's disease and senile dementia using an automatic image analyzing system. *J. Hirnforsch.* **27,** 695–702.

Savoiardo, M., Strada, L., Girotti, F., Zimmerman, R. A., Grisoli, M., Testa, D., and Petrillo, R. (1990). Olivopontocerebellar atrophy: MR diagnosis and relationship to multisystem atrophy. *Radiology* **174**, 693–696.

Scheibel, M. E., and Scheibel, A. B. (1970). The rapid Golgi method: Indian Summer or Renaissance? *In* "Contemporary Research Methods in Neuroanatomy" (W. J. H. Nauta and S. O. E. Ebbesson, eds.), pp. 1–11. New York/Heidelberg/Berlin.

Schena, M. (1996). Genome analysis with gene expression microarrays. *BioEssays* **18,** 427–431.

Schena, M., Shalon, D., Davis, R. W., and Brown, P. O. (1995). Quantitative monitoring of gene expression patterns with a complementary DNA microarray. *Science* **270,** 467–470.

Schleicher, A., Amunts, K., Geyer, S., Kowalski, T., Schormann, T., Palomero-Gallagher, N., and Zilles, K. (2000). A stereological approach to human cortical architecture: Identification and delineation of cortical areas. *J. Chem. Neuroanat.* **20,** 31–47.

Schleicher, A., Amunts, K., Geyer, S., Morosan, P., and Zilles, K. (1999). Observer-independent method for microstructural parcellation of cerebral cortex: A quantitative approach to cytoarchitectonics. *NeuroImage* **9,** 165–177.

Schmitt, O., and Eggers, R. (1999). Flat-bed scanning as a tool for quantitative neuroimaging. *J. Microsc.* **196,** 337–346.

Schormann, T., and Zilles, K. (1997). Limitations of the principal-axes theory. *IEEE Trans. Med. Imaging* **16,** 942–947.

Schormann, T., and Zilles, K. (1998). Three-dimensional linear and nonlinear transformations: An integration of light microscopical and MRI data. *Hum. Brain Mapp.* **6,** 339–347.

Schulz, U., Hunziker, O., Frey, H., and Schweizer, A. (1980). *Post mortem* changes in stereological parameters of cerebral neurons. *Pathol. Res. Pract.* **166,** 260–270.

Seeger, G., Luth, H. J., Winkelmann, E., and Brauer, K. (1996). Distribution patterns of *Wisteria floribunda* agglutinin binding sites and parvalbumin-immunoreactive neurons in the human visual cortex: A double-labelling study. *J. Hirnforsch.* **37,** 351–366.

Shaw, P. J., and Rawlins, D. J. (1991). Three-dimensional fluorescence microscopy. *Prog. Biophys. Mol. Biol.* **56,** 187–213.

Sheppard, C. J. R. (1994). Confocal microscopy: Basic principles and system performance. *In* "Multidimensional Microscopy" (P. C. Cheng, T. H. Lin, W. L. Wu, and J. L. Wu, eds.), pp. 1–32. New York/Berlin.

Shiurba, R. A., Spooner, E. T., Ishiguro, K., Takahashi, M., Yoshida, R., Wheelock, T. R., Imahori, K., Cataldo, A. M., and Nixon, R. A. (1998). Immunocytochemistry of formalin-fixed human brain tissues: Microwave irradiation of free-floating sections. *Brain Res. Brain Res. Protocols* **2,** 109–119.

Siesjo, B. K. (1981). Cell damage in the brain: A speculative synthesis. *J. Cereb. Blood Flow Metab.* **1,** 155–185.

Silverman, M. S., and Tootell, R. B. (1987). Modified technique for cytochrome oxidase histochemistry: Increased staining intensity and compatibility with 2-deoxyglucose autoradiography. *J. Neurosci. Methods* **19,** 1–10.

Simmons, D. M., and Swanson, L. W. (1993). The Nissl stain. *In* "Neuroscience Protocols."

Singer, C. (1962). Beginnings of academic practical anatomy. *In* "History and Bibliography of Anatomic Illustration" (L. Choulant, ed.), pp. 21A–21R. New York/London.

Smith, C. G. (1981). "Serial Dissections of the Human Brain." Urban & Schwarzberg, Baltimore.

Smith, G. E. (1907). A new topographical survey of the human cerebral cortex, being an account of the anatomically distinct cortical areas and their relationship to the cerebral sulci. *J. Anat.* **41,** 237–254.

Sömmerring, S. T. (1796). "Uber das Organ der Seele." Nicolovius, Konigsberg.

Spacek, J. (1992). Dynamics of Golgi impregnation in neurons. *Microsc. Res. Tech.* **23,** 264–274.

Sterio, D. C. (1984). The unbiased estimation of number and sizes of arbitrary particles using the disector. *J. Microsc.* **134,** 127–136.

Stevens, J. K., McGuire, B. A., and Sterling, P. (1980). Toward a functional architecture of the retina: Serial reconstruction of adjacent ganglion cells. *Science* **207,** 317–319.

Streicher, J., Weninger, W. J., and Muller, G. B. (1997). External marker-based automatic congruencing: A new method of 3D reconstruction from serial sections. *Anat. Rec.* **248,** 583–602.

Sulston, J. E., Schierenberg, E., White, J. G., and Thomson, J. N. (1983). The embryonic cell lineage of the nematode Caenorhabditis elegans. *Dev. Biol.* **100,** 64–119.

Swiger, R. R., and Tucker, J. D. (1996). Fluorescence in situ hybridization: A brief review. *Environ. Mol. Mutagen* **27,** 245–254.

Szekely, G., Kelemen, A., Brechbuhler, C., and Gerig, G. (1996). Segmentation of 2-D and 3-D objects from MRI volume data using constrained elastic deformations of flexible Fourier contour and surface models. *Med. Image Anal.* **1,** 19–34.

Tago, H., Kimura, H., and Maeda, T. (1986). Visualization of detailed acetylcholinesterase fiber and neuron staining in rat brain by a sensitive histochemical procedure. *J. Histochem. Cytochem.* **34,** 1431–1438.

Takahashi, J. S., Pinto, L. H., and Vitaterna, M. H. (1994). Forward and reverse genetic approaches to behavior in the mouse. *Science* **264,** 1724–1733.

Tang, Y., Nyengaard, J. R., De Groot, D. M., and Gundersen, H. J. (2001). Total regional and global number of synapses in the human brain neocortex. *Synapse* **41,** 258–273.

Tardif, E., and Clarke, S. (2001). Intrinsic connectivity of human auditory areas: A tracing study with DiI. *Eur. J. Neurosci.* **13,** 1045–1050.

Taylor, C. J., Cootes, T. F., Lanitis, A., Edwards, G., Smyth, P., and Kotcheff, A. C. (1997). Model-based interpretation of complex and variable images. *Philos. Trans. R. Soc. London B Biol. Sci.* **352,** 1267–1274.

Thompson, P. M., Schwartz, C., and Toga, A. W. (1996). High-resolution random mesh algorithms for creating a probabilistic 3D surface atlas of the human brain. *NeuroImage* **3,** 19–34.

Thompson, P. M., and Toga, A. W. (1996). Surface-based technique for warping three-dimensional images of the brain. *IEEE Trans. Med. Imaging* **15,** 402–417.

Threlfall, R. (1930). The origin of the automatic microtome. *Biol. Rev.* **5,** 357–336.

Tochitani, S., Liang, F., Watakabe, A., Hashikawa, T., and Yamamori, T. (2001). The occ1 gene is preferentially expressed in the primary visual cortex in an activity-dependent manner: A pattern of gene expression related to the cytoarchitectonic area in adult macaque neocortex. *Eur. J. Neurosci.* **13,** 297–307.

Toga, A. W., Ambach, K., Quinn, B., Hutchin, M., and Burton, J. S. (1994a). *Post mortem* anatomy from cryosectioned whole human brain. *J. Neurosci. Methods* **54,** 239–252.

Toga, A. W., Ambach, K. L., and Schluender, S. (1994b). High-resolution anatomy from in situ human brain. *NeuroImage* **1,** 334–344.

Toga, A. W., and Arnicar, T. L. (1985). Image analysis of brain physiology. *Comp. Graph. Appl.* **5,** 20–25.

Toga, A. W., and Banerjee, P. K. (1993). Registration revisited. *J. Neurosci. Methods* **48,** 1–13.

Tootell, R. B., and Taylor, J. B. (1995). Anatomical evidence for MT and additional cortical visual areas in humans. *Cereb. Cortex* **5,** 39–55.

Tourtellotte, W. W., and Berman, K. (1987). Brain banking. *In* "Encyclopedia of Neuroscience" (G. Adelman, ed.), pp. 156–158. Boston/Basel/Stuttgart.

Tourtellotte, W. W., Rosario, I. P., Conrad, A., and Syndulko, K. (1993). Human neuro-specimen banking 1961–1992: The National Neurological Research Specimen Bank (a donor program of pre- and post-mortem tissues and cerebrospinal fluid/blood; and a collection of cryopreserved human neurological specimens for neuroscientists). *J. Neural Transm. Suppl.* **39,** 5–15.

Trisler, G. D., Schneider, M. D., and Nirenberg, M. (1981). A topographic gradient of molecules can be used to identify neuron position. *Proc. Natl. Acad. Sci. USA* **78,** 2145–2149.

Tucker, L. W., Cornejo, H. J., and Capowski, J. J. (1984). Image understanding and the cell world model. *In* "Quantitative Neuroanatomy in Neurotransmitter Research" (L. Agnati and K. Fuxe, eds.). Wenner-Gren Center, Stockholm.

Turner, J. N., Szarowski, D. H., Smith, K. L., Marko, M., Leith, A., and Swann, J. W. (1991). Confocal microscopy and three-dimensional reconstruction of electrophysiologically identified neurons in thick brain slices. *J. Electron Microsc. Tech.* **18,** 11–23.

Tyrer, N. M. (1999). Celloidin–wax sandwich microtomy: A novel and rapid method for producing serial semithin sections. *J. Microsc.* **196,** 273–278.

Uylings, H. B., Zilles, K., and Rajkowska, G. (1999). Optimal staining methods for delineation of cortical areas and neuron counts in human brains. *NeuroImage* **9,** 439–445.

Valentino, K. L., Eberwine, J. H., and Barchas, J. D. (1987). "In Situ Hybridization: Applications to Neurobiology." Oxford Univ. Press, New York.

Valverde, F. (1970). The Golgi method: A tool for comparative analyses. *In* "Contemporary Research Methods in Neuroanatomy" (W. J. H. Nauta and S. O. E. Ebbesson. eds.), pp. 12–31. New York/Heidelberg/Berlin.

Vesalius, A. (1543). "De Humani Corpora Fabrica." Ex Officina Joannis Operini, Basel.

Vicq d'Azyr, F. (1786). "Traite d'Anatomie et de Physiologie." F. A. Didot, Paris.

Vogt, C., and Vogt, O. (1919). Allgemeinere ergebnisse unserer Hirnforschung. *J. Psychol. Neurol.* **25,** 399–462.

Vogt, C., and Vogt, O. (1942). Morphologische gestaltungen unter normalen und pathologen bedingugen: Ein hirnanatomischer beitrag zu ihrer kenntnis. *J. Psychol. Neurol.* **50**, 11–310.

Vonsattel, J. P., Aizawa, H., Ge, P., DiFiglia, M., McKee, A. C., MacDonald, M., Gusella, J. F., Landwehrmeyer, G. B., Bird, E. D., Richardson, E. P., Jr., *et al.* (1995). An improved approach to prepare human brains for research. *J. Neuropathol. Exp. Neurol.* **54,** 42–56.

Voogd, J., and Feirabend, H. K. P. (1981). Classic methods in neuroanatomy. *In* "Methods in Neurobiology" (R. Lahue, ed.), pp. 301–364. New York/London.

Vossmaer, G. C. J. (1899). Einfache Modification zur Herstellung von Plattendiagrammen. *Anat. Anz.* **16,** 269–271.

Wald, L., Schmitt, F., and Dale, A. (2001). Systematic spatial distortion in MRI due to gradient non-linearities. *NeuroImage* **13,** S50.

Walsh, C. A. (2001). Neuroscience in the post-genome era: An overview. *Trends Neurosci.* **24,** 363–364.

Wang, X., and Feuerstein, G. Z. (2000). Suppression subtractive hybridisation: Application in the discovery of novel pharmacological targets. *Pharmacogenomics* **1,** 101–108.

Watson, J. D., Myers, R., Frackowiak, R. S., Hajnal, J. V., Woods, R. P., Mazziotta, J. C., Shipp, S., and Zeki, S. (1993). Area V5 of the human brain: Evidence from a combined study using positron emission tomography and magnetic resonance imaging. *Cereb. Cortex* **3,** 79–94.

Weibel, E. R. (1979). "Stereological Methods." Academic Press, London/New York.

Welford, S. M., Gregg, J., Chen, E., Garrison, D., Sorensen, P. H., Denny, C. T., and Nelson, S. F. (1998). Detection of differentially expressed genes in primary tumor tissues using representational differences analysis coupled to microarray hybridization. *Nucleic Acids Res.* **26,** 3059–3065.

West, M. J. (1999). Stereological methods for estimating the total number of neurons and synapses: Issues of precision and bias. *Trends Neurosci.* **22,** 51–61.

Westlake, T. M., Howlett, A. C., Bonner, T. I., Matsuda, L. A., and Herkenham, M. (1994). Cannabinoid receptor binding and messenger RNA expression in human brain: An in vitro receptor autoradiography and in situ hybridization histochemistry study of normal aged and Alzheimer's brains. *Neuroscience* **63,** 637–652.

Weszka, J. S., Dyer, C. R., and Rosenfeld, A. (1976). A comparative study of texture measures or terrain classification. *IEEE Trans. Syst. Man Cybernet.* **269.**

Wilkinson, D. G. E. (1992). "In Situ Hybridization: A Practical Approach." Oxford Univ. Press, New York.

Williams, R. W. (1998). Neuroscience meets quantitative genetics: Using morphometric data to map genes that modulate CNS architecture. *In* "Society for Neuroscience Education Committee." Los Angeles.

Williams, R. W. (2000). Mapping genes that modulate mouse brain development: A quantitative genetic approach. *In* "Mouse Brain Development" (A. F. Goffinet and P. Rakic, eds.), pp. 21–49. New York.

Williams, R. W., and Rakic, P. (1988). Three-dimensional counting: An accurate and direct method to estimate numbers of cells in sectioned material. *J. Comp. Neurol.* **278,** 344–352.

Williams, R. W., Strom, R. C., and Goldowitz, D. (1998). Natural variation in neuron number in mice is linked to a major quantitative trait locus on Chr 11. *J. Neurosci.* **18,** 138–46.

Willis, T. (1664). "Cerebri Anatome." Martin & Allestry, London.

Wong, Y. M., Thompson, R. P., Cobb, L., and Fitzharris, T. P. (1983). Computer reconstruction of serial sections. *Comput. Biomed. Res.* **16,** 580–586.

Wong-Riley, M. (1979). Changes in the visual system of monocularly sutured or enucleated cats demonstrable with cytochrome oxidase histochemistry. *Brain Res.* **171,** 11–28.

Wong-Riley, M. T., Hevner, R. F., Cutlan, R., Earnest, M., Egan, R., Frost, J., and Nguyen, T. (1993). Cytochrome oxidase in the human visual cortex: Distribution in the developing and the adult brain. *Visual Neurosci.* **10,** 41–58.

Woods, R. P., Grafton, S. T., Watson, J. D., Sicotte, N. L., and Mazziotta, J. C. (1998). Automated image registration. II. Intersubject validation of linear and nonlinear models. *J. Comput. Assisted Tomogr.* **22,** 153–165.

Wree, A., Schleicher, A., and Zilles, K. (1982). Estimation of volume fractions in nervous tissue with an image analyzer. *J. Neurosci. Methods* **6,** 29–43.

Yaegashi, H., Takahashi, T., and Kawasaki, M. (1987). Microcomputer-aided reconstruction: A system designed for the study of 3-D microstructure in histology and histopathology. *J. Microsc.* **146,** 55–65.

Yakovlev, P. I. (1970). Whole brain serial histological sections. *In* "Neuropathology Methods and Diagnosis" (C. G. Tedeschi, eds.), pp. 371–378. Little Brown, Boston.

Yates, C. M., Butterworth, J., Tennant, M. C., and Gordon, A. (1990). Enzyme activities in relation to pH and lactate in *post mortem* brain in Alzheimer-type and other dementias. *J. Neurochem.* **55,** 1624–1630.

Young, W. S. (1992). In situ hybridization with oligodeoxyribonucleotide probes. *In* "In Situ Hybridization: A Practical Approach" (D. G. Wilkinson, ed.), pp. 33–44. New York.

Zilles, K., Qu, M. S., Schroder, H., and Schleicher, A. (1991). Neurotransmitter receptors and cortical architecture. *J. Hirnforsch.* **32,** 343–356.

Zipser, B., and McKay, R. (1981). Monoclonal antibodies distinguish identifiable neurones in the leech. *Nature* **289,** 549–554.

21

Quantitative Analysis of Cyto- and Receptor Architecture of the Human Brain

Karl Zilles,[*,†] **Axel Schleicher,**[†] **Nicola Palomero-Gallagher,**[*] **and Katrin Amunts**[*]

[*]*Institute of Medicine, Research Center Juelich, D-52425 Juelich, Germany;*

[†]*C. and O. Vogt Institute of Brain Research, Heinrich–Heine University Duesseldorf, Duesseldorf, Germany*

I. Introduction

Many architectonic, e.g., cyto- and myeloarchitectonic, maps of the human cerebral cortex have been published over the past hundred years (Fig. 1) since the existence of discrete variations in the laminar structure of the cerebral cortex were first observed and described. Gennari (1782), Vicq d'Azyr (1786), and Baillarger (1840) reported the presence of white stripes of myelinated fibers in unstained brain tissue in a part of the cerebral cortex which was later identified as the primary visual cortex. Betz (1874) described the presence of giant pyramidal cells in a region anterior to the central sulcus, in the primary motor cortex, thus providing the first cytoarchitectonic analysis of a defined brain region. Campbell's study presented the first complete map of the cerebral cortex (Campbell, 1905). More detailed maps of the cerebral cortex were published by Elliot Smith (1907), Brodmann (1909), the Vogts (1910, 1911, 1919), Flechsig (1920), von Economo and Koskinas (1925), the Russian school (Sarkisov *et al.,* 1949), Bailey and von Bonin (1951), and Braak (1980). The most influential map is that of Korbinian Brodmann (1909), a co-worker of the Vogts, who published several detailed descriptions of the cytoarchitecture of cortical regions (Brodmann, 1903, 1905, 1908, 1919). Brodmann's map still dominates present concepts of the microstructural organization of the human cerebral cortex

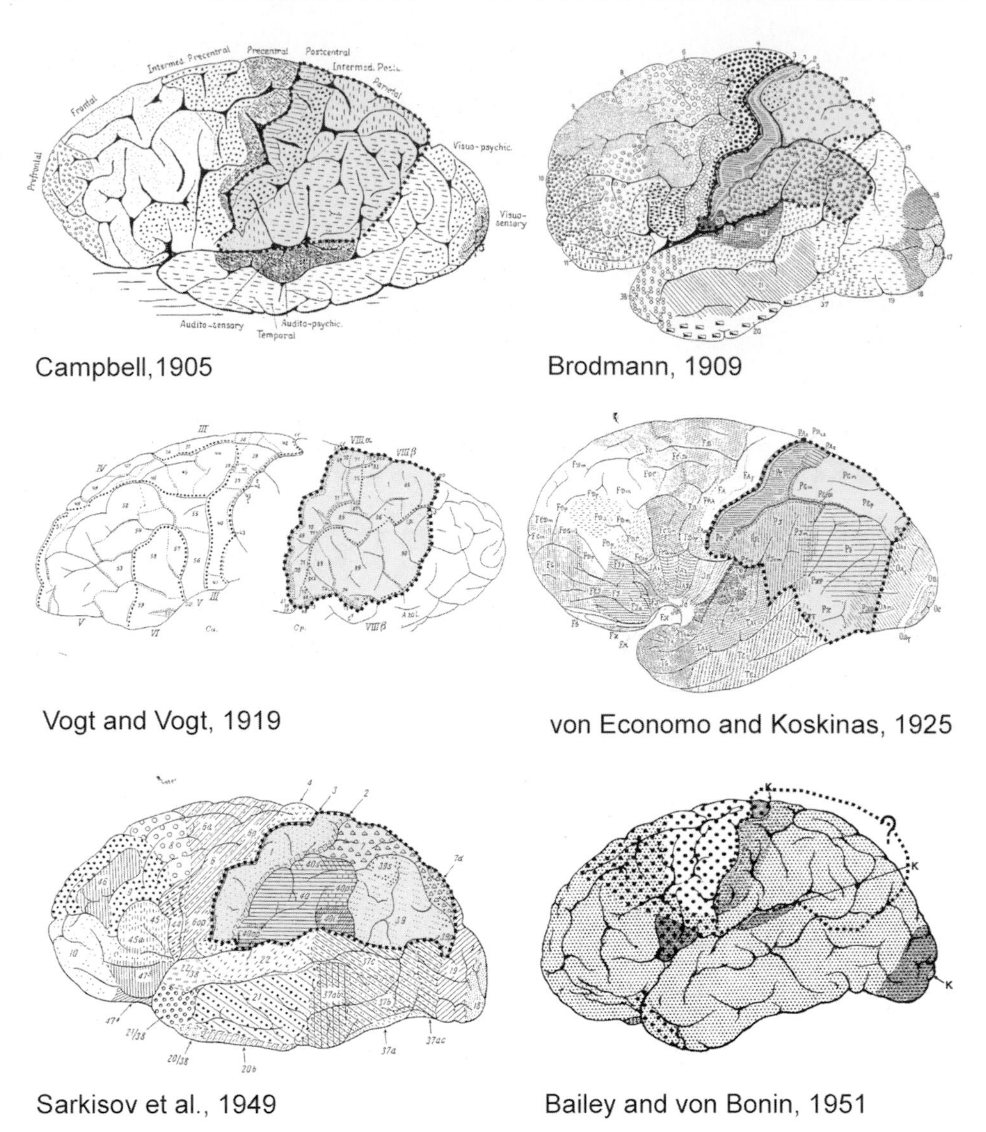

Figure 1 Classical architectonic maps adapted from Campbell (1905), Brodmann (1909), C. and O. Vogt (1919), von Economo and Koskinas (1925), the Russian school (Sarkisov *et al.*, 1949), and Bailey and von Bonin (1951). The map of Bailey and von Bonin is shown in its black-and-white version as published by Chusid (1964). Only the lateral views are shown. The approximate extent of the parietal cortex is marked by a black dotted line in each of the maps. In Bailey and von Bonins's map, however, this marking is arbitrary, since these authors did not map parietal areas as separate entities. Hence, they classified almost the whole temporo-occipitoparietal complex as one cytoarchitectonic entity. Note that the maps of Brodmann, Vogt and Vogt, von Economo and Koskinas, and Sarkisov also differ, e.g., in the number of parietal areas and subareas, in the ventral extent of parietal areas (see von Economo and Koskinas' map compared to others), and in their caudal extent (see Brodmann's map compared to others). Reproduced with permission.

and, via the atlas of Talairach and Tournoux (1988), the topographical interpretation of functional imaging data.

Brodmann's research was based on the working hypothesis that the cerebral cortex is composed of multiple cortical areas, each of them characterized by a distinct cytoarchitecture and function. Cytoarchitecture should be more or less constant within a cortical area and changes abruptly at its border. For example, Brodmann's area (BA) 4 was conceptualized as the anatomical equivalent of the primary motor cortex, which guides voluntary movements (Fritsch and Hitzig, 1870). Although for the vast majority of cortical areas such a microstructure–function relationship could not be rigorously tested at the time, Brodmann (and also Campbell) took architectonic localization for granted. Old (Vogt and Vogt, 1919) and more recent studies in nonhuman primates have demonstrated by means of combined electrophysiological–neuroanatomical studies, that Brodmann's basic idea was true. Accordingly, response properties of neurons change across cytoarchitectonic borders (Tanji and Kurata, 1989; Matelli *et al.,* 1991; Luppino *et al.,* 1991). In contrast to some studies (Kleist, 1934), Brodmann did not argue for an extreme localizational concept. That is, he did not try to relate each complex function to one distinct cytoarchitectonic area. In order to avoid confusion between histological data and unproven evolutionary and functional speculations, he created a "neutral" nomenclature by numbering different cytoarchitectonic areas mainly according to their dorsoventral sequence. Brodmann's map and cytoarchitectonic analyses constitute an impressive scientific achievement and they have influenced research on the structural and functional organization of the human cerebral cortex for many decades.

In the context of increasing spatial and functional resolution of data provided by recent imaging techniques, however, several drawbacks of these early architectonic maps became critical for their practical use as anatomical references. These disadvantages concern (i) their presentation in two-dimensional schematic drawings, (ii) a subjective, i.e., observer-dependent, definition of the criteria which describe a cortical area and the precise location of its borders, and (iii) the neglect of intersubject differences both in brain macroscopy and in architecture. In addition, mapping techniques have been developed during the past years which reflect functionally highly relevant information based on, e.g., immunohistochemistry of transmitters and cytoskeletal elements and receptor autoradiography. The mapping of different receptor binding sites has recently become a powerful tool to reveal the (receptor) architectonic organization of the cerebral cortex (Zilles and Schleicher, 1995; Zilles *et al.,* 1996; Geyer *et al.,* 1996, 1997, 1998; Zilles and Clarke, 1997; Zilles and Palomero-Gallagher, 2001).

Although a complete map of the human cerebral cortex, which overcomes all of the above-mentioned drawbacks and which considers more recent mapping techniques (e.g., receptor architectonics), remains a project for future research; such maps have already been elaborated for several cortical regions, e.g., parts of the visual, the sensorimotor, and the auditory cortex. The present chapter will:

▲ critically discuss the principles of classical architectonic mapping in the context of recent imaging techniques;

▲ present an observer-independent approach for a quantitative analysis of cortical areas and their borders, which is based on a multivariate statistical analysis of the cytoarchitecture;

▲ illustrate the application of this approach for the cytoarchitectonic mapping of the human visual cortex;

▲ introduce the principles and methodological basis of the quantitative mapping of neurotransmitter receptors; and

▲ discuss some future developments of architectonic mapping.

II. Principles of Cytoarchitectonic Analysis

A. Structural Organization of the Human Cerebral Cortex (Layers and Columns)

The cerebral cortex can be subdivided into isocortex (neocortex) and allocortex (paleo- and archicortex). The isocortex has a thickness of 2 (e.g., visual cortex) to 5 mm (e.g., primary motor cortex). It shows a laminar organization which consists of six *horizontal layers* (parallel to the cortical surface) and *vertical cell columns.* The allocortex contains fewer than six layers in most regions, but can also present more than six layers in some areas (e.g., entorhinal cortex).

The layers of the isocortex are defined according to the packing density of neuronal cell bodies and the proportion and spatial arrangement of different neuronal cell types, as well as cell sizes and shapes (Fig. 2). The layers of the isocortex are:

▲ Lamina I (molecular layer). It occupies the most superficial parts of the cortex and contains only a few scattered, small neuronal cell bodies.

▲ Lamina II (external granular layer). It shows many densely packed, small granular cell bodies which look like granules when Nissl-stained sections are viewed at a low magnification.

▲ Lamina III (external pyramidal layer). It is characterized by small- and medium-sized pyramidal cells. Their apical dendrites reach Lamina I. In some cortical regions, very large pyramids are found in the lower third of lamina III. The axons of layer III pyramids project mainly to other cortical regions of the ipsi- and contralateral (via corpus callosum) hemisphere.

▲ Lamina IV (internal granular layer). Numerous, densely packed small granular cells can be found in this layer. Its thickness varies considerably between the different isocortical regions, reaching the largest width in primary sensory areas. Whereas layer IV is not recognizable in the adult motor cortex (BA 4 and BA 6), it consists of three sublayers (which can be even further subdivided) in the primary visual cortex (BA 17 = area V1). Layer IV is the major target of the thalamocortical input (afferent fibers from the dorsal thalamus and the metathalamus).

▲ Lamina V (internal pyramidal layer). It consists mainly of pyramidal cells of large but variable size. In BA 4, extremely large pyramidal cells (= Betz cells, giant pyramids, cell body height reaching up to 130 μm) are present in this layer. The axons of the pyramidal cells reach other cortical areas in the ipsi- or contralateral hemisphere via short and long fiber tracts of the white matter, or they descend to the basal ganglia, brain stem, and spinal cord in the corticostriatal, corticonuclear, and corticospinal tracts, respectively.

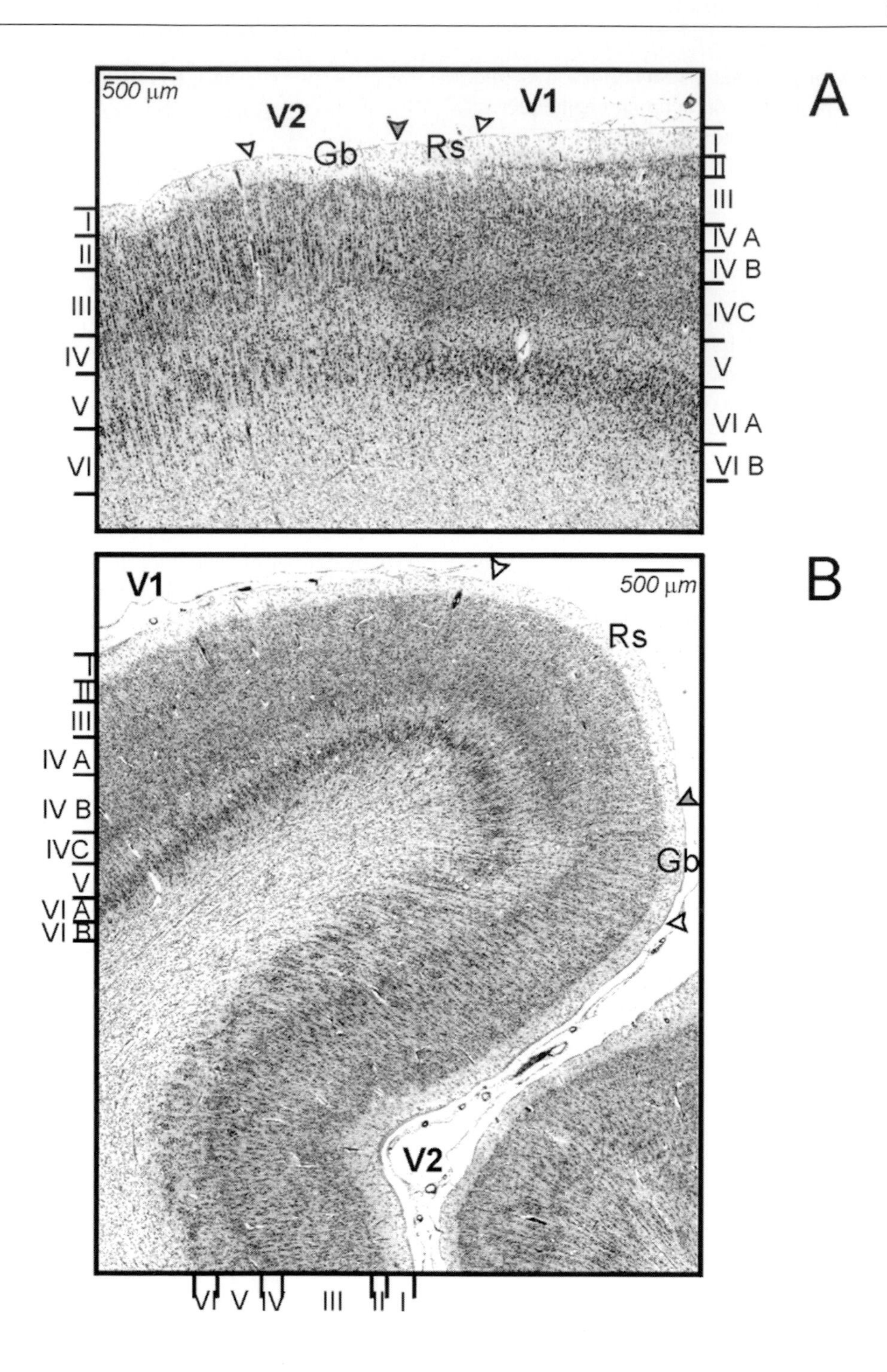

Figure 2 Cytoarchitecture and borders of areas V1 and V2 of the human visual cortex. The border between both areas can easily be recognized irrespective of its location on the crown of a gyrus (**A**) or in the wall of a sulcus (**B**). A silver technique for staining cell bodies was applied, which gives results similar to those of Nissl staining, but with a higher contrast (Merker, 1983). Cytoarchitectonic layers are marked by Roman numerals. The cytoarchitecture as indicated by the thickness and composition of layer IV changes abruptly at the border between areas V1 and V2 (gray arrowheads). Note that near the border between areas V1 and V2, but within V1, the cytoarchitecture changes gradually (= fringe area, Rs, "Randsaum"). In addition, subarea Gb (border tuft, "Grenzbüschel") can be delineated. These subdivisions in myelm stained sections were described by Sanides and Vitzthum (1965a,b). They can be verified using an observer-independent approach for the definition of cortical borders (Amunts *et al.*, 2000).

▲ Lamina VI (multiform layer). It contains cell bodies of different shapes, e.g., spindle-like, polygonal. Their axons project mainly into subcortical brain regions.

The *vertical columns* can be viewed as vertically oriented aggregates of cell bodies separated from each other by neuropil-rich space. The radius of a column is 30–50 μm. Columns interact with each other and seem to be derived from the ontogenetic cell columns migrating along the radial glia of the developing cortex (for an overview see Buxhoeveden *et al.,* 2000).

Since these columns are not clearly visible across all cortical areas, and since their visibility is strongly influenced by the sectioning plane of the histological sections, they have not been used as a major criterion for architectonic mapping. Therefore, the cytoarchitectonic method, as described in this chapter, is based mainly on the horizontal, laminar characteristics of the isocortex.

B. Important Cytoarchitectonic Criteria

Important criteria for cytoarchitectonic mapping are (i) the absolute thickness of cortical layers, (ii) the proportionate thickness of a layer relative to the other cortical layers and to the total cortical depth, (iii) the presence of clearly recognizable laminar borders and vertical columns, (iv) the packing density and size of neuronal cell bodies, (v) the homogeneous or clustered distribution of cell bodies throughout the layers, and (vi) the presence of special cell types such as Betz cells (Betz, 1874). The latter cell type is a characteristic unique to BA 4. Betz cells cannot be found in any other cortical area. Another highly specific criterion is the subdivision of the internal granular layer (layer IV) of the primary visual cortex (area V1 or BA 17) into three sublayers, of which sublayer IVb corresponds to the macroscopically visible Gennari stripe (Fig. 2). Such distinctive architectonic features are, unfortunately, extremely rare. For example, relatively large pyramidal cells can also be found in the Area postcentralis gigantopyramidalis of von Economo and Koskinas (1925), in layers III and V of BA 44 and 45, in the extrastriate visual cortex, and in the temporal cortex (Bailey and von Bonin, 1951).

Verbal cytoarchitectonic descriptions of different cortical areas are very often redundant. This can be illustrated by the remarkably similar descriptions of layer III of three completely different cortical areas which were made in the atlas of the Russian school (Sarkisov *et al.,* 1949). The first description concerns BA 45: "Layer III is relatively thick. It contains relatively large pyramidal cells, the sizes of which increase from the surface to the white matter. Based on this increase in cell size, layer III can be subdivided into three sublayers. Sublayer III^3 contains many, very large pyramidal cells. These cells form either clusters, or are located separately. Sometimes, pyramidal cells intermingle with cells of layer IV" [translation by one of the authors (K.A.)]. Large pyramidal cells of layer III are also the most prominent feature in the description of motor BA 6: "Lamina III can be subdivided into three sublayers. The lower part of III^3 contains many pyramidal cells. Some of them reach very large sizes. These pyramidal cells often form clusters of 8–10 cells." This description of layer III of BA 6 also sounds similar to that of visual BA 18: "...Lamina III can be subdivided into three sublayers....[S]ublayer III^3 contains considerably larger cells than III^2. The large, well-shaped pyramidal cells of layer III^3 are closely related to layer IV and, intermingle partly with cells of layer IV."

Obviously, the purely verbal description of layer III does not enable unambiguous classification of these three cortical areas. Consequently, a reliable definition of these areas requires *multiple* criteria including, as a rule, more than one cortical layer as well as additional topographical information.

It has to be noted that multiple criteria were often weighted by the observer with respect to their significance, reliability, visibility, etc. For example, the primary visual area BA 17 (or V1) was defined by the appearance and disappearance of the highly complex layer IV (Fig. 2). The characteristic structure of layer IV is, therefore, the most important cytoarchitectonic criterion for the definition of V1. However, not all cytoarchitectonic features change simultaneously with this feature. For example, when moving from area V1 to V2, the cell density of layer VI shows the most prominent change within the first few millimeters of the adjoining area V2 (= BA 18) (Grenzbüschel, i.e., "border tuft" of area V2; Sanides and Vitzthum, 1965a,b) and not at the border between V1 and V2 (Fig. 2B). Only after the border tuft region has been passed does lamina VI become less cell dense, reaching the low cell density typical of area V2. Thus, the precise localization of the V1/V2 border is based on a "leading" criterion (i.e., layer IV) and on the neglect of other cytoarchitectonic features (e.g., cell density of layer VI).

Another example of the experience-based weighting of cytoarchitectonic criteria in classical mapping studies is the definition of the border between BA 6 and BA 4. Possible criteria for the detection of this border are the existence of the giant pyramidal cells (Betz cells) solely in BA 4 or the appearance in BA 6 of a layer III, with relatively larger pyramidal cells than those present in layer V. Since distances between Betz cells differ depending on their location within the precentral gyrus (e.g., in its most ventral part Betz cells are relatively rare compared to the more dorsal parts), the position of the BA 4/BA 6 border may shift up to several millimeters depending on which "leading" criterion is selected (Betz cells of BA 4 vs large pyramidal cells of BA 6).

Most of the cytoarchitectonic criteria used in the classical studies were qualitative statements, not quantitative. For example, the authors did not precisely define what a "large" or a "giant" cell was. Von Economo and Koskinas (1925) were among the first who tried to make such descriptions less dependent from the observer by measuring the thickness of layers, the height and width of cells, and the number of cells per millimeter field of view. They supported their verbal description of each area by adding quantitative data. These measurements, however, disclosed another aspect of cytoarchitectonic features: variability. For example, the height of pyramidal cells of layer III_c of area F*B* (corresponds approximately to BA 6) varied between 35 and 80 μm (von Economo and Koskinas, 1925). If the stem of the apical dendrite was also included into this measurement, the height of the cell body increased up to 130 μm, the size of a Betz cell. Thus, cell measures may vary due to methodological reasons, but intersubject variability of cell sizes also contributes greatly to this problem and does not permit the use of such criteria for the "differential" diagnosis of large or giant pyramidal cells.

Moreover, not only do single morphometric parameters vary considerably between different subjects, but also the whole laminar pattern is highly variable. This was shown in a recent study (Amunts *et al.,* 1999) in which intersubject differences in cytoarchitecture were quantitatively analyzed by measuring cell packing density profiles in the same cortical area of 10 postmortem brains. The intersubject differences in the laminar pattern were then compared to differences between various cortical areas of one brain. Comparison was performed by calculating a multivariate distance measure (for methods see Sections III.B and III.F) between feature vectors of the profiles. This study showed that interareal differences within one brain may have the same magnitude as intersubject differences of one and the same area. It clearly demonstrates that cytoarchitectonic analysis should be based on measurements and not on verbal descriptions.

C. Differences between Classical Cortical Maps

The highly complex nature of cytoarchitectonic criteria, the considerable intersubject differences, and the observer-dependent selection and definition of criteria caused several inconsistencies between the classical maps. As examples, see the maps of Campbell (1905), Brodmann (1909), Vogt and Vogt (1919), von Economo and Koskinas (1925), Sarkisov *et al.* (1949), and Bailey and von Bonin (1951) (Fig. 1).

▲ The maps differ in the number of identified cortical areas. Whereas Campbell's map distinguishes three parietal areas on the dorsal surface, that of von Economo and Koskinas shows nine areas with further subdivisions. The map of Bailey and von Bonin represents an extreme point of view. These authors denied the existence of clearly recognizable architectonic differences between the posterior part of the parietal lobe, the major part of the temporal lobe, and the occipital cortex outside the koniocortical BA 17. They questioned whether there is any objective basis for a detailed cytoarchitectonic map at all and came to the final conclusion that "... vast areas are so closely similar in structure as to make any attempt at subdivisions unprofitable, if not impossible" (Bailey and von Bonin, 1951). As a consequence, their cytoarchitectonic map is based on a division into four main types: with granules (koniocortex), without granules (agranular cortex), with large pyramids in layer III, and the allocortex, as well as four combinations of these main types. In contrast to other maps, they do not show sharp borders for most of the areas, but gradual transitions.

▲ The association of cortical areas with lobes differs between authors. For example, von Economo and Koskinas defined an area P*E*γ as part of the parietal cortex. Although this area has no analogue in Brodmann's map, it might be part of BA 19, located on the occipital lobe (von Economo and Koskinas, 1925).

▲ The maps differ in their classification principles and nomenclature. In order to avoid confusion of histological data and unproven evolutionary and functional concepts, Brodmann created his system of a "neutral" nomenclature by numbering different cytoarchitectonic areas mainly according to their appearance during his investigation of a series of horizontal sections in a dorsoventral sequence. Von Economo and Koskinas introduced a more complex subdivision of the human cortex into cortical areas which show regional peculiarities (subareas). Their nomenclature is less "neutral," since it considered the location of an area within a certain lobe (e.g., "P" for the parietal lobe), areal characteristics (e.g., P*E,* whereby "*E*" indicates the Regio parietalis superior), and subareal peculiarities (P*E*γ, γ corresponds to the magnopyramidal part of P*E*), as well as transitional areas which share cytoarchitectonic peculiarities of two neighboring areas (e.g., P*ED* with cytoarchitectonic features of P*E* and P*D*). Von Economo and Koskinas weighted cytoarchitectonic criteria for the differential definition of areas and subareas. Such a subtle classification, of course, is problematic, if based on a subjective judgment of the "weight" of a certain criterion.

▲ Differences between the maps concern the size and the shape of cortical areas. Compare, e.g., the size of BA 7 on the free surface between Brodmann's and Sarkisov's map. The latter shows a much smaller extent than Brodmann's map.

▲ The locations of the areas with respect to macroscopical landmarks differ between the maps. For example, area 39 of Brodmann's and Sarkisov's maps are ventrally limited by the extension of the superior temporal sulcus. Von Economo's and Koskinas' corresponding area P*G* reaches a more ventral level, that of the inferior temporal sulcus.

▲ The sulcal pattern itself and, finally, the surrounding areas differ between the maps. Whereas, e.g., Brodmann described that areas 46 and 9 of the frontal cortex are located in the dorsal neighborhood of areas 44 and 45, the Russian school reported that areas 8 and 9, but not area 46, constitute the neighborhood of areas of 44 and 45.

Some of the mentioned differences, obviously, might be caused by observer-dependent criteria for the definition of a cortical area, others seem to be influenced by intersubject variability in microscopical and macroscopical brain structure. Last but not least, these maps cannot be compared with each other since they do not show borders within the sulci. The part of the cortex, however, which is buried in the sulci constitutes approximately two-thirds of the whole cortical surface (Zilles *et al.,* 1988a) and, therefore, cannot be neglected.

D. Application of Classical Architectonic Maps for the Anatomical Interpretation of Functional Imaging Data

Differences between these maps may lead to different conclusions concerning the anatomical location of functional imaging data. Moreover, this application in functional imaging needs three-dimensional maps and not maps in which only the superficially exposed part of the cortex is shown. One possible solution was proposed by Talairach and Tournoux (1988), who used Brodmann's schematic surface drawing as the basis of an architectonic parcellation in a stereotaxic atlas. Talairach and Tournoux simply transferred Brodmann's areas to their own atlas brain by trying to identify corresponding sulcal patterns in both brains, assuming a strong association between the sulcal pattern and the borders of cortical areas. This procedure has several disadvantages:

▲ A strong association between sulci and the location of cortical borders was already doubted by Brodmann. He mentioned that "... a schematic drawing can reflect only the major spatial relationships, and therefore, precise topographical associations cannot be considered in general or only in a distorted manner; this is true in particular for all those cortical regions which have borders in the neighborhood of sulci and those regions which are located in the depth of such a cortical region" (Brodmann, 1908; own translation). The lack of a precise association of areal borders and macroscopical landmarks such as gyri and sulci has been demonstrated frequently in the past years (Geyer *et al.,* 1996, 1999; Amunts *et al.,* 1999, 2000; Morosan *et al.,* 2001; Grefkes *et al.,* 2001). It represents, in particular, an important issue in higher associative areas (Rademacher *et al.,* 1993).

▲ Intersubject and interhemispheric differences were neglected. Talairach and Tournoux used a single brain, different from that examined by Brodmann. The transformation of cortical areas of Brodmann's map to the Talairach atlas is based on the assumption that both brains possess a comparable micro- and macroanatomy. In contrast to the brain examined by Brodmann, the Talairach/Tournoux brain was never the subject of a cytoarchitectonic analysis. In addition, the brains depicted in both Brodmann's and Talairach's maps differ in shape, size, and sulcal pattern from other individual brains. Furthermore, important interhemispheric differences, e.g., with respect to the frequency and to the course of sulci and gyri (Ono *et al.,* 1990; Duvernoy, 1991; Kretschmann and Weinrich, 1996) and with respect to the location and size of cytoarchitectonic areas such as differences in the volumes of BA 44 (Amunts *et al.,* 1999) and topographic differences of visual areas V1 and V2 (Amunts *et al.,* 2000), were not considered in either map.

▲ Brodmann's map does not contain any information concerning the lateromedial extent of cytoarchitectonic areas. This information was more or less arbitrarily introduced by Talairach and Tournoux.

The critical analysis of the classical cortical maps led to the development of new approaches to cytoarchitectonic and receptor architectonic analysis of the cerebral cortex, some of which will be described in the next sections.

III. Observer-Independent Mapping of the Human Cerebral Cortex

A. Postmortem Brains and Histological Processing

Serial sections (20 μm thick) through formalin-fixed and paraffin-embedded complete human brains were mounted on glass slides and processed with a silver staining technique for demonstrating all cell bodies (Merker, 1983). This staining procedure leads to a high contrast between background and cellular compartments (Fig. 3). It enables a reliable segmentation of the darkly stained structures using automated gray-value thresholding with an image analyzer. Since each brain was completely sectioned in only one of the three possible sectioning planes, mapping of cortical areas in a distinct hemisphere was based on one plane.

B. Hard- and Software

Image acquisition and processing as well as the analysis of cytoarchitecture by means of profiles (see next paragraph) were performed using a KS400 image analyzing system (Zeiss, Germany). Processing of profiles, feature extraction, border detection, and multivariate statistics were applications of MATLAB for Windows including the Statistics Toolbox (Version 6.1; MathWorks, USA). The human brain atlas (Roland and Zilles, 1994) and the software used for the creation of population maps were run on UNIX-based Workstations.

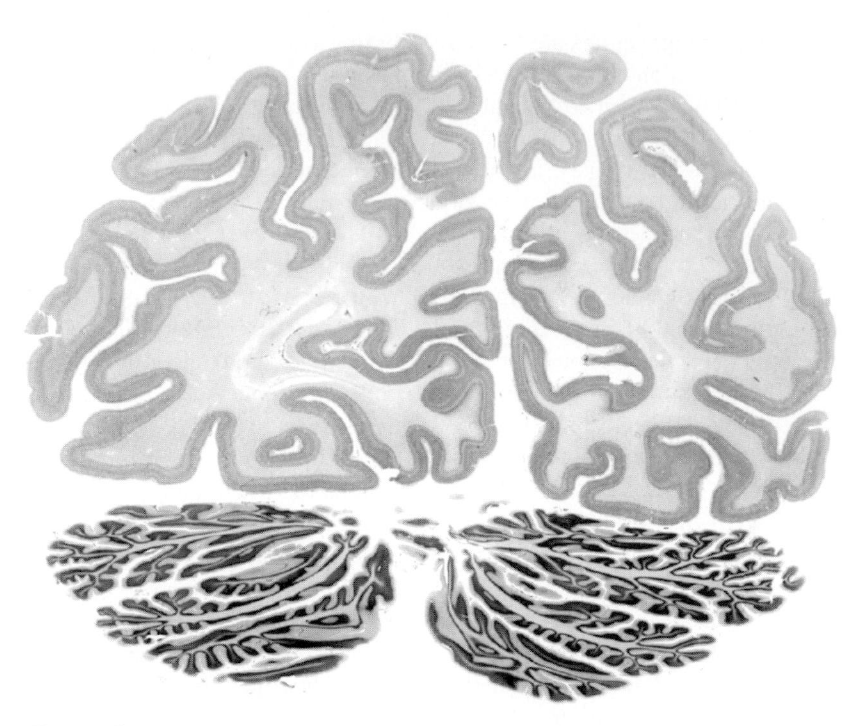

Figure 3 Coronal histological section through a complete adult human brain such as is used for cytoarchitectonic mapping viewed from occipital. Staining for cell bodies by Merker (1983).

C. Image Processing and Profile Extraction of the Histological Sections

Data of quantitative cytoarchitectonic studies were acquired from interactively defined, rectangular regions of interest (ROI; Fig. 4) at low microscopic resolution (Planapo 6.3 × 1.25; Zeiss, Germany). Using an automatic scanning procedure, each ROI was completely covered by a continuous, mosaic-like pattern and scanned using TV frames 524 × 524 μm in size, 512 × 512 pixels, and 8-bit gray resolution (Schleicher *et al.,* 1999). From each TV frame, a binary image was generated by adaptive thresholding (e.g., Castleman, 1979). This segmentation was found to be superior to thresholding using a fixed gray value threshold, since the threshold is derived from the local gray value difference between bright background and dark cellular profiles and is adjusted to variations of background staining within each image. The binary image is subdivided by a grid of square measuring fields. The spacing of this grid defines the spatial resolution of the scanning procedure, which should result in a representation of the total cortical width by not less than 64 pixels. This guarantees that none of the six cortical layers will be sampled with substantially fewer than 10 pixels, and differences in the cortical widths between various cortical areas will not affect shape analysis of the laminar pattern. For most of the hitherto analyzed human cortical areas, this constraint resulted in a spatial resolution of 16 to 30 μm. The areal fraction of cellular profiles (Schleicher and Zilles, 1990) was determined in each measuring field (Fig. 5). The resulting data matrix [= gray-level index (GLI) image] was stored for profile extraction.

GLI data are not affected by local variations in staining intensity, which cannot be avoided even when using highly standardized histological procedures. This makes the GLI superior to optical density measurements, which depend on the staining intensities of the background and of the cellular structures. In addition, GLI data are closely correlated with the volume density of neurons, since the contribution of glial and endothelial cells does not vary significantly in their laminar distribution (Tower and Young, 1973; Wree *et al.,* 1982).

In order to register GLI profiles, two lines defining the starting and end points of the traverses were interactively traced on the GLI image (Fig. 6). The first line was set at the layer I/layer II border. The second line was traced along the layer VI/white matter border. Attempts to define these lines by an automatic procedure were not successful in all cortical regions analyzed. The two lines were used by a GLI profile-extracting algorithm to calculate equidistant traverses vertically oriented to the cortical layers and as parallel as possible to the vertically oriented cell columns. For a detailed description of the algorithm, see Schleicher *et al.* (2000). GLI profiles were extracted from the GLI image along the course of each traverse. The signal-to-noise ratio of the GLI profiles was improved by smoothing each GLI image with two passes of a 3 × 3-pixel average filter prior to extracting the profiles (Fig. 7). Starting and end points of the traverses and all GLI profiles of a ROI were stored in a database.

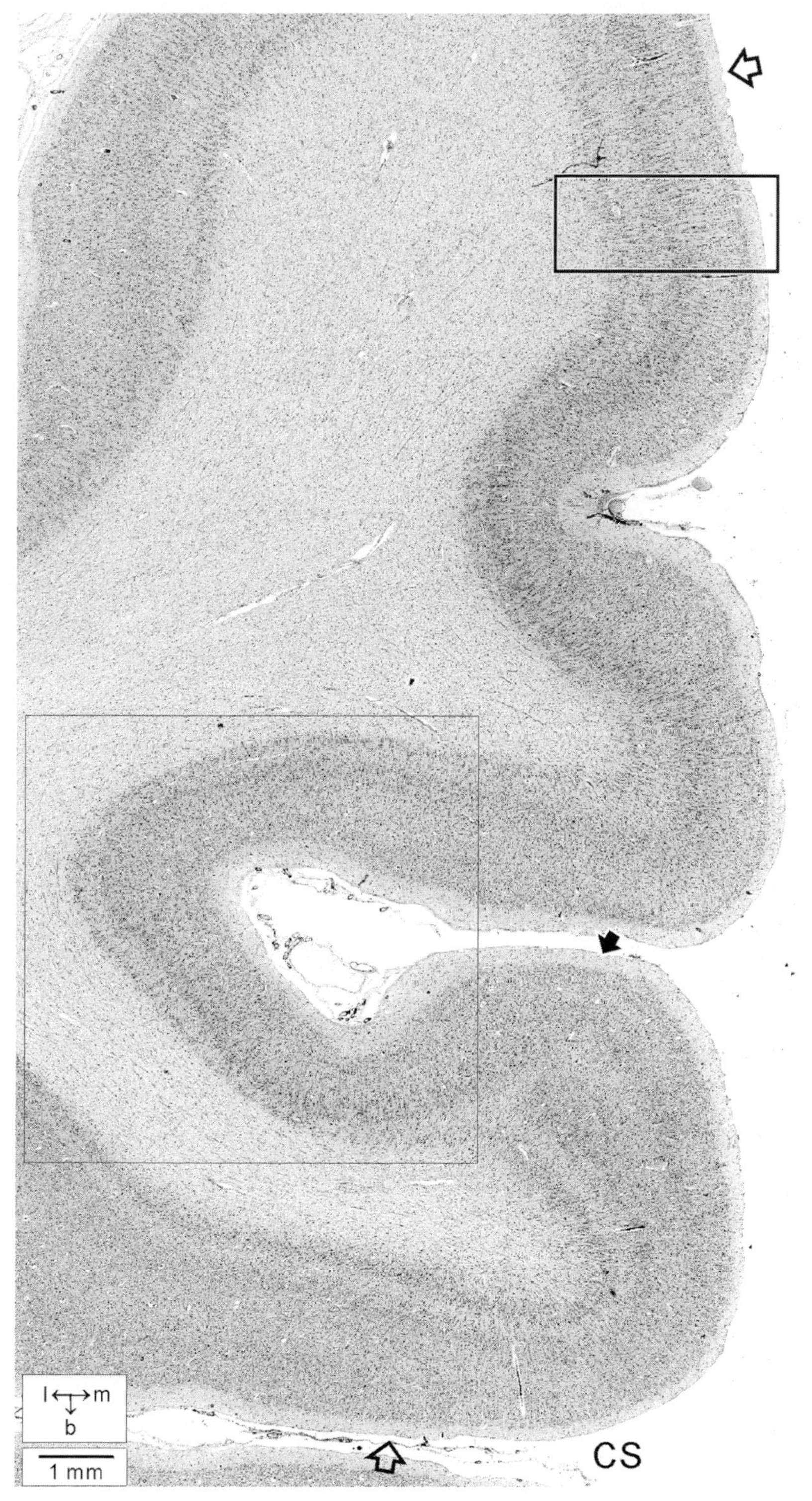

Figure 4 High-resolution image of a region of interest (ROI) in a histological section (20 μm) of an adult human brain, occipital cortex, cut in the coronal plane and stained for cell bodies (Merker, 1983). The parcellation pattern of the cortical areas covered by this ROI (V1, V2d, V2d) is not completely obvious to pure visual inspection. Only the V1/V2d border can easily be detected and is indicated by a black arrow. The region in the small rectangular frame is depicted in Fig. 5 to demonstrate details of cytoarchitecture and the corresponding gray-level index image. The larger rectangular frame delineates a subregion used to demonstrate the analysis of dissimilarity between cortical sectors (distance analysis) in Fig. 8. The two open arrows define a cortical ROI, the subsequent analysis of which is shown in Fig. 6. The cross indicates orientation: l, lateral; m, medial; b, basal. CS, calcarine sulcus.

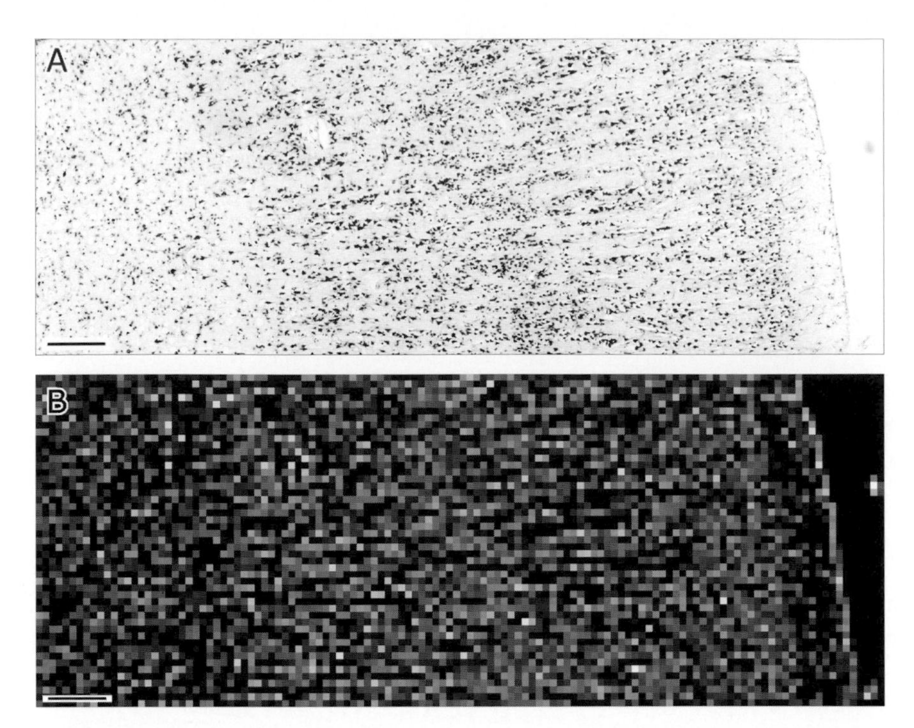

Figure 5 Cytoarchitecture and GLI image of the human visual cortex. (**A**) High-resolution image (1417 × 530 pixels) of the cortical sector highlighted within the small rectangle of Fig. 4. (**B**) GLI image (124 × 49 pixels; pixel values are GLI values 0% ≤ GLI ≤ 100%) of the cortical sector shown in (A) at a spatial resolution of 20 × 20 µm;, i.e., each pixel represents an estimate of the local volume density of neurons in a square measuring field of this size. Values in this GLI image are scaled to gray values for printing (0% black; 100% white). Scale bars, 200 µm. The whole cortical depth is covered by approximately 100 pixels.

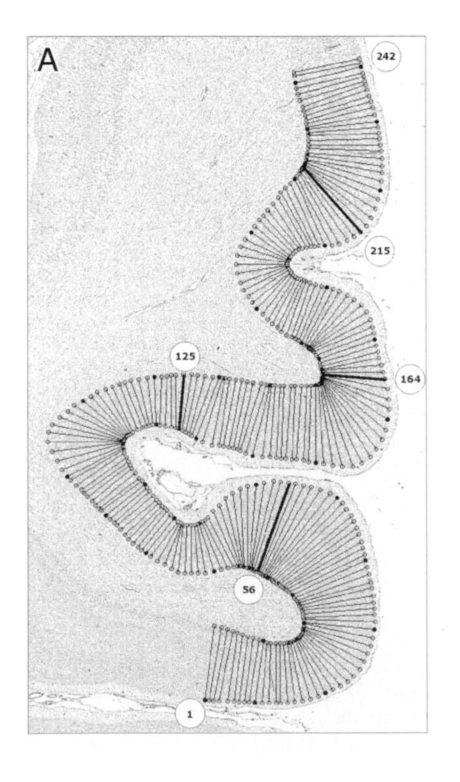

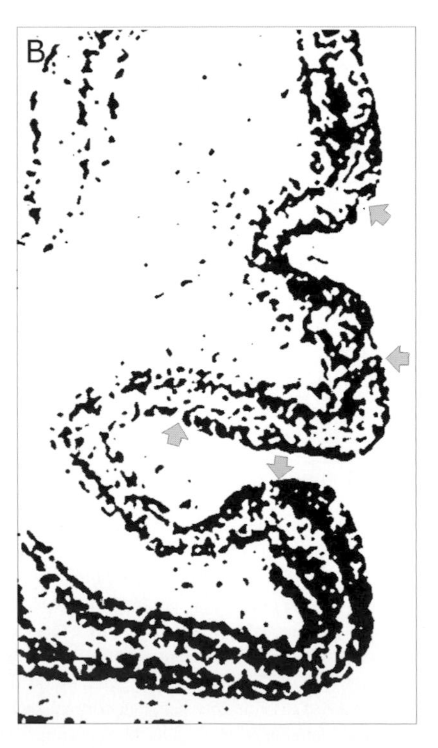

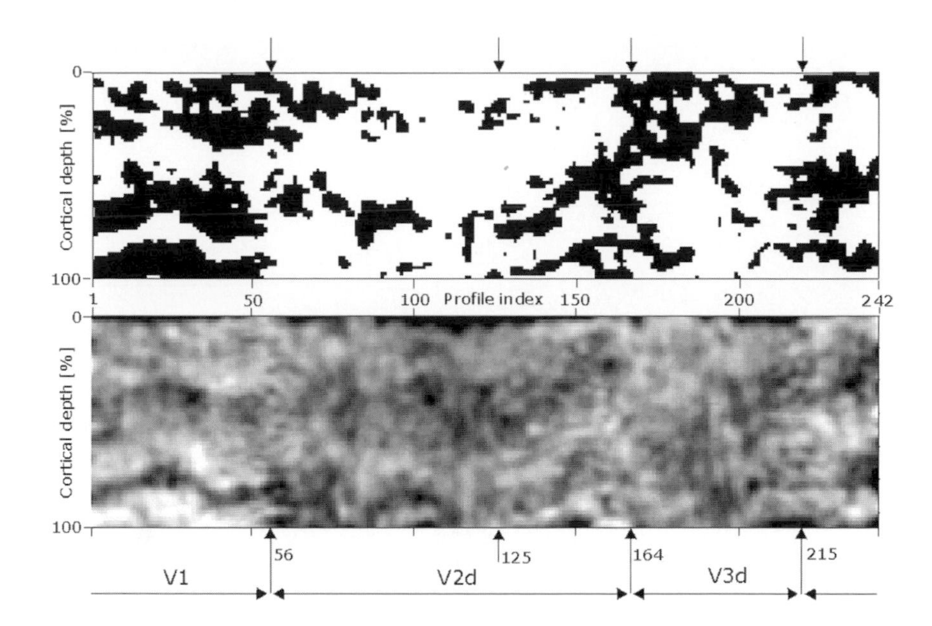

Figure 7 "Flattened" cortex constructed from the GLI profiles indicated in Fig. 6. The 242 GLI profiles were standardized to a cortical depth of 100% (ordinate) and arranged in sequential order (abscissa) from left to right. 0%, layer I/II border; 100%, layer VI/white matter border. Signal-to-noise ratio was improved by averaging the corresponding GLI values across three neighboring profiles and replacing the central profile by the average (moving average). **Bottom:** GLI image after linear contrast enhancement: black, 0%; white, 20%. **Top:** Binary image of the GLI image shown below created by thresholding GLI values. Black, GLI ≥ 16%. The result is a flattened reconstruction of the cortex from the binary image shown in Fig. 6. The locations of areal borders of areas V1, V2d, and V3d are indicated by arrows. V1 is characterized by high GLI values in supra- and in infragranular layers; V2d shows a generally reduced GLI level compared to V1 and V3d. In V3d, the GLI values are increased in supragranular layers compared to V2d.

D. Feature Extraction and Shape Analysis

The procedure for locating areal borders is based on the concept that each cortical area has a typical cytoarchitecture, i.e., a typical laminar pattern, which differs from that of neighboring areas. The laminar pattern is represented by GLI profiles as a quantitative measure. Comparing pairs of neighboring GLI profiles, maximum differences in profile shape can be expected at the interface between two neighboring cortical areas, i.e., the two GLI profiles are extracted from two different cortical areas.

A detailed description of the border-locating procedure was given in Schleicher *et al.* (1998, 1999). Profile shape was quantified by extracting a feature vector of ten elements from each profile. Five features were extracted from the original profile (Fig. 8). For instance, meany.o is the mean GLI across cortical layers (16.28% in this example). Four features were calculated by treating the profile as a frequency distribution. The cortical depth is the x value, and the GLI value is the frequency at that x value. The first four central moments about the mean (meanx.o to kurt.o) were calculated (Dixon *et al.,* 1988): meanx.o is the mean cortical depth and indicates the x coordinate of the center of gravity of the area beneath the profile curve; std.o is the standard deviation, skew.o the skewness, and kurt.o the kurtosis of the frequency distribution. Additional five features, meany.d to kurt.d, were

Figure 6 (**A**) Sampling of GLI profiles obtained from the ROI depicted in Fig. 4. Each line indicates the position of a GLI profile extracted from the GLI image along the course of this traverse from the layer I/layer II transition to the layer VI/white matter border (Schleicher *et al.,* 2000). The two curved lines which mark the starting and end points of the traverses were defined interactively using a pointing device prior to the automated positioning of traverses. Profile index runs from 1 to 242. Indices 56, 125, 164, and 215 are highlighted with bold lines and indicate areal borders established using distance analysis (see Fig. 10). Spacing between traverses is 120 μm (center-to-center distance). GLI image was inverted for visualization. (**B**) Image enhancement in order to visualize the laminar pattern (Zilles *et al.,* 1978). The GLI image of the ROI of Fig. 4 was smoothed by median filtering (5 ×; 5 kernel, three passes) and thresholded at a GLI value of 15%. Pixel values above this threshold were set to black. Thus, layers with high volume densities of neurons form black stripes. Arrows mark the interareal borders indicated in (A). For scaling see Fig. 4.

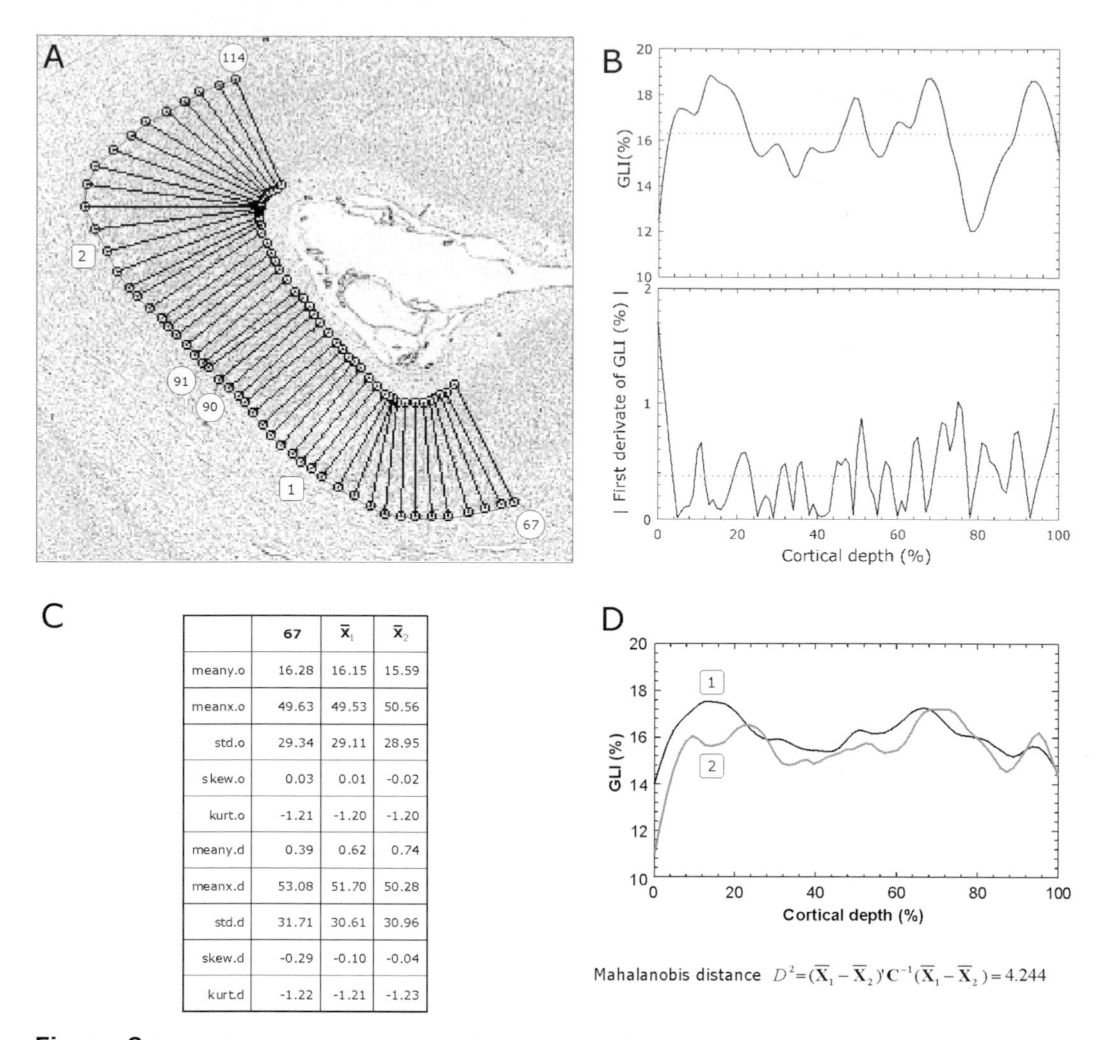

	67	$\bar{X}_1$	$\bar{X}_2$
meany.o	16.28	16.15	15.59
meanx.o	49.63	49.53	50.56
std.o	29.34	29.11	28.95
skew.o	0.03	0.01	-0.02
kurt.o	-1.21	-1.20	-1.20
meany.d	0.39	0.62	0.74
meanx.d	53.08	51.70	50.28
std.d	31.71	30.61	30.96
skew.d	-0.29	-0.10	-0.04
kurt.d	-1.22	-1.21	-1.23

Figure 8 GLI profile extraction and analysis of dissimilarity in the laminar pattern of two cortical sectors (distance analysis). (A) Region of interest marked in Fig. 4 by the large rectangle. Profile traverses 67 to 114 from the set of traverses indicated in Fig. 6A were superimposed to the inverted GLI image of that region. The profiles define two blocks of profiles, each of which encompasses 24 profiles: block (1) comprises profiles 67 to 90, block (2) profiles 91 to 114. (B) Graphs of an individual profile (GLI profile at position 67). Top: Original profile. Bottom: Absolute of the first derivative of the original profile. Cortical depth was standardized to 100%. Dotted horizontal lines indicate means [meany.o (i.e., the mean GLI) 16.3%, meany.d 0.387). The feature vector which was extracted from this profile is shown in (C). (C) Names of features (first column); second column, feature vector of profile 67 shown in (B), third and fourth columns, mean feature vectors of the profiles in block (1) and in block (2), calculated by averaging each of the 10 features across the 24 individual feature values in each block. (D) Mean profiles of block (1) and block (2). Profiles differ in shape, i.e., the regions from which they were obtained, differ in cytoarchitecture. Mean profiles were calculated by averaging the corresponding profile values across the profiles in each block. Note that feature vectors describing the mean profiles would be different from the mean feature vectors shown in (C), since mean profiles were smoothed by averaging and their shape characteristics are thus different from those of individual profiles. The equation used to calculate the Mahalanobis distance D^2 as a measure of dissimilarity between the laminar pattern in the two blocks and the resulting distance value are indicated. This distance value is assigned to position 90, see Fig. 10A. C^{-1} is the pooled variance–covariance matrix calculated from the 48 individual feature vectors from both blocks (Bartels, 1979; Dixon et al., 1988).

extracted from the absolute value of the differential quotient of the same profile (Figs. 8B and 8C). The differential quotient is a numerical approximation of the first derivative of the profile and quantifies its local slope. Both sets of five features each were combined into one feature vector. Prior to calculating the Euclidean distance function (see below), feature values from all profiles in a ROI were standardized to z scores in order to assign equal weight to each of the features (Dixon *et al.,* 1988).

The use of central moments offered an essential advantage to the shape analysis of laminar patterns since central moments can be interpreted in neurobiological terms. As an example, a mean cortical depth (mean*x*.o) below 50% and a positive skewness (skew.o) indicate a profile skewed to the right. In terms of cytoarchitecture, this corresponds to a laminar pattern with a higher cell density in the supragranular compared to the infragranular cortical layers (Fig. 9A). Conversely, a laminar pattern with dominant infragranular layers is characterized by a mean cortical depth above 50% and a negative skewness (Fig. 9). For a more detailed interpretation of shape-describing features see Figs. 9C–9F.

E. Location of Areal Borders

Various measures for dissimilarity between groups of multivariate data exist (Gower, 1985). We calculated the Euclidean distance *E* and the Mahalanobis distance *D* (Bartels, 1979) between the feature vectors of two blocks of profiles which represent two neighboring cortical sectors (Fig. 8). Both distances have different characteristics. *E* is the distance between the two centroids (means or mean feature vectors) of the two 10-dimensional clusters. It is independent of the degree of dispersion within each cluster. *D* accounts for this dispersion and decreases with increasing dispersion, even when the distance between centroids remains constant. In addition, it considers the mutual correlations between the features forming the feature vector.

Distance functions were established by calculating the distances between all pairs of neighboring cortical sectors and plotting the values as a function of the sectors position (Fig. 10). This was accomplished by shifting the two blocks of profiles across the cortical region with an increment equal to the distance between two neighboring profiles and by calculating the distance values *E* and *D* at each position.

Maxima in distance functions are hints to areal borders, but they have to be checked for their statistical and biological significance before they can be accepted as architectonically relevant borders. A major advantage of the Mahalanobis distance is the possibility of carrying out the Hotelling's T^2 test to prove the significance of a difference between the mean feature vectors of two multivariate distributions (Bartels, 1981). We applied this test in combination with a Bonferroni adjustment of the *P* values for multiple comparisons (Dixon *et al.,* 1988). Maxima which were found to be significant and not exceeded by another significant maximum within a neighborhood of the size of one block of profiles on both sides were accepted as main maxima and interpreted as putative areal borders. Main maxima found in one histological section were biologically validated by comparing spatially corresponding maxima from adjoining sections. Since the probability of locating such maxima at comparable sites in a stack of sections by reasons other than changes in cortical organization is exceedingly small, this procedure excluded biologically meaningless maxima which may be caused by artifacts (e.g., ruptures, folds) or local inhomogeneities in microstructure such as blood vessels or atypical cell clusters. The consistency of the parcellation pattern in the human somatosensory cortex across stacks of neighboring sections was illustrated in Geyer *et al.* (1999).

Distance functions depend on the spatial resolution of the distance analyzing procedure, which is defined by the width

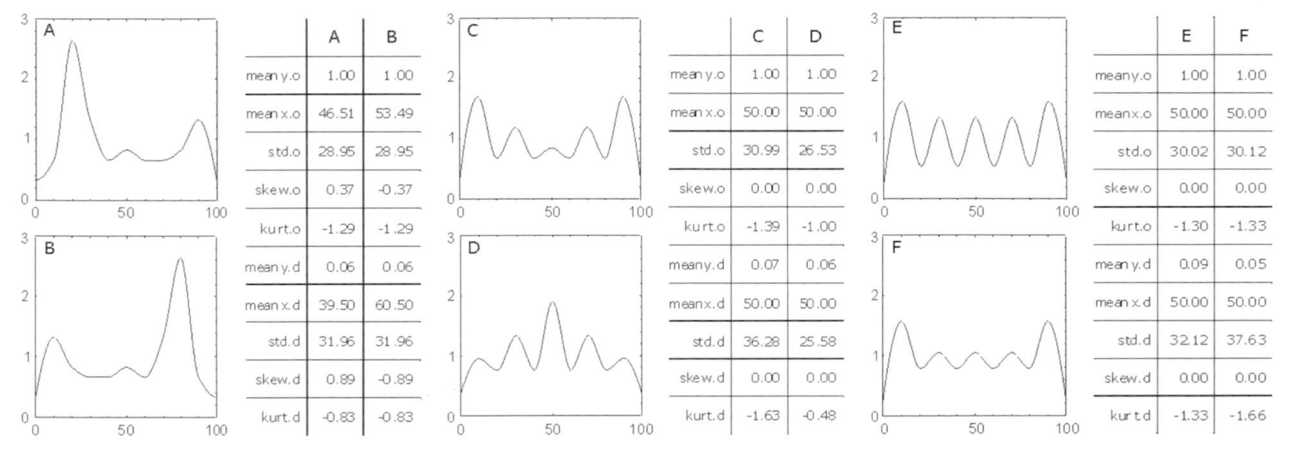

	A	B
meany.o	1.00	1.00
meanx.o	46.51	53.49
std.o	28.95	28.95
skew.o	0.37	-0.37
kurt.o	-1.29	-1.29
meany.d	0.06	0.06
meanx.d	39.50	60.50
std.d	31.96	31.96
skew.d	0.89	-0.89
kurt.d	-0.83	-0.83

	C	D
meany.o	1.00	1.00
meanx.o	50.00	50.00
std.o	30.99	26.53
skew.o	0.00	0.00
kurt.o	-1.39	-1.00
meany.d	0.07	0.06
meanx.d	50.00	50.00
std.d	36.28	25.58
skew.d	0.00	0.00
kurt.d	-1.63	-0.48

	E	F
meany.o	1.00	1.00
meanx.o	50.00	50.00
std.o	30.02	30.12
skew.o	0.00	0.00
kurt.o	-1.30	-1.33
meany.d	0.09	0.05
meanx.d	50.00	50.00
std.d	32.12	37.63
skew.d	0.00	0.00
kurt.d	-1.33	-1.66

Figure 9 Shape analysis of GLI profiles by feature extraction. Three pairs of model profiles (A/B, C/D, and E/F) were generated to demonstrate typical differences between the profiles in each pair. Ten shape-describing features (meany.o to kurt.d) were extracted from each profile (Schleicher et al., 1999) and are listed to the right of each pair. Features meany.o to kurt.o were calculated from the original profiles, and features meany.d to kurt.d from the first derivative of the profiles (derivative not shown). Cortical depth (abscissa) 0 to 100%, mean GLI across layers (ordinate, meany.o) is standardized to 1.0. (A, B) Profile (B) is profile (A) mirrored at a cortical depth of 50%. The difference in shape between both profiles is reflected by the features meanx.o, skew.o, meanx.d, and skew.d. (C, D) Both profiles are symmetrical to the cortical depth of 50% (meanx.o, meanx.d = 50%, skew.o, skew.d = 0), but differ with respect to the distribution of laminar differentiation: in (C), main lamination is located near the layer I/layer II transition and the layer VI/white matter border, in (D), main lamination occurs around the profile center (cortical depth 50%). The features differ mainly in std.o, std.d, kurt.o, kurt.d. (E, F) Both profiles differ with respect the overall degree of lamination. In (F), the lamination is less pronounced than in (E). This is mainly reflected by meany.d.

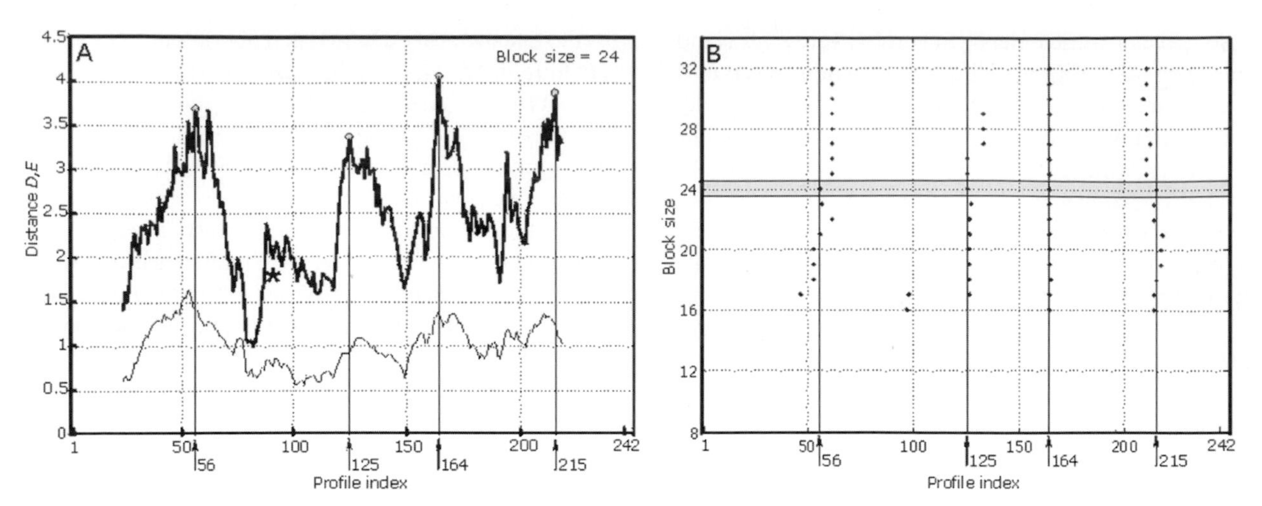

Figure 10 Distance analysis based on the set of 242 profiles indicated in Fig. 6A. (A) Mahalanobis distance function D (upper bold curve) and Euclidean distance function E (lower thin curve, not drawn to scale) established using cortical sectors with 24 profiles (block size 24). Main maxima in the Mahalanobis distance function indicating putative interareal borders were found at positions 56, 125, 164, and 215 and are marked by drop lines on the abscissa. Main maxima of the Mahalanobis distance function show significant P values as established using the Hotelling's T^2 statistic and are not allowed to include higher distance values within a neighborhood of one block size to either side (Schleicher et al., 1999). The corresponding positions are indicated by bold lines in Fig. 6A and arrows in Fig. 6B. The Mahalanobis distance value found in the example described in Fig. 8 at position 90 is indicated by an asterisk. Distance functions are restricted to a range from index 24 to 218 set by the interface of the two adjoining blocks of 24 profiles each at the first and the last position within the ROI. (B) Dependency of the location of main maxima on the width of cortical sectors (block size) demonstrated for the set of traverses shown in Fig. 6A. Each dot represents one main maximum with coordinates defined by (i) the profile index which marks the location of the putative interareal border and (ii) the block size used to establish the corresponding Mahalanobis distance function. Abscissa, profile index which indicates the distance from the first traverse in 120-μm increments, identical to (A). Ordinate, width of the cortical sector (block size) used to establish the corresponding distance function. Each row of points indicates all main maxima of that distance function and thus represents a distinct parcellation pattern. The main maxima originating from the distance function with block size 24 shown in (A) are highlighted by a shaded rectangle and by vertical lines which correspond to the drop lines in (A). All 17 parcellation patterns in this graph describe the ROI shown in Fig. 6A. Block sizes ranging from 8 to 32 were used. No main maxima were found for block sizes smaller than 16. In this graph, most areal borders were represented by main maxima which showed a constant regular vertical arrangement and which were therefore independent of a distinct setting of the block size. These borders showed only minor variations in location and indicate prominent interareal borders (e.g., index 164 marks the V2d/V3d border, see also Fig. 7). The two main maxima at profiles indices 97 and 98 could not be confirmed in neighboring sections and were therefore not accepted as interareal borders.

of the cortical sectors (number of profiles in each sector, block size). In order to avoid an *a priori* selection of a distinct value of the block size, we analyzed each ROI repeatedly with the same set of profiles, but with systematically increasing block sizes. Figure 10B illustrates the distance functions originating from block sizes in the range of 8 to 32 profiles per block.

Another important factor which may influence the shape of the profiles and, thus, the distance function, is cortical folding (Bok and van Kip, 1939). In order to get an insight into the properties of the Mahalanobis and the Euclidean distance functions with regard to the effect of cortical folding on border detection, a lamination pattern model, which was characterized by a shift in width of the cortex and its layers similar to that caused by the natural cortical folding, was analyzed (Schleicher *et al.,* 1999). The application of the distance functions to the laminar model clearly demonstrated the main advantage of the Mahalanobis distance function: it is insensitive to gradual changes in the laminar pattern. Only relatively abrupt changes within a narrow cortical sector such as those occurring at areal borders were detected. The Mahalanobis distance analysis can, therefore, be efficiently applied to cortices with gyri and sulci, where shifts in the laminar pattern are frequently induced by cortical folding (Armstrong *et al.,* 1995) as occurs in the human brain. Cortical regions where the typical laminar pattern is impaired by very oblique or completely tangential sectioning of the curved cortical surface, however, cannot be analyzed, since in these cases one or more cortical layers are not represented in the GLI image. Thus, the application of the Mahalanobis distance for detecting abrupt changes in cytoarchitecture requires sectioning, which results in a presentation of all cortical layers in the ROI. This problem, however, cannot be solved by pure visual inspection either and is therefore not a specific condition of the quantitative cytoarchitectonic approach.

F. Dissimilarities in Cytoarchitecture between Cortical Areas

After having established the borders of cortical areas, differences between the cytoarchitecture of two or more areas (= cytoarchitectonic dissimilarity between areas) can be statistically analyzed. This approach is based on the hypothesis that the feature vectors of profiles measured in different cortical areas constitute different and distinct clusters in a 10-dimensional feature space (10 dimensional, because a profile is defined by 10 features; see above). Vice versa, if profiles come from the same area, they should be members of the same cluster. This approach is, therefore, similar in some aspects to that used for border detection. It differs with respect to the sampling of the profiles: whereas two sets of profiles from two neighboring cortical sectors of equal size have to be compared for the detection of borders, the analysis of dissimilarities between cortical areas relies on a comparison of representative samples of profiles of two or more cortical areas (Fig. 11). A canonical analysis is used for visualization of the results. For statistical background and comparison to the principal component analysis see Uylings *et al.* (1989). An example, including statistical testing, with data obtained from a sample of 20 brains is illustrated in Amunts *et al.* (1997).

The above-described tools, which were originally developed to analyze the cytoarchitecture of the cerebral cortex, can also be applied to examine other kinds of cortical architecture, such as myelo- or receptor architecture. The latter is of special interest, since it can be assumed to provide information concerning cortical organization at a molecular level, particularly at the level of signal transduction. In this context, however, some additional methodical aspects have to be taken into account. Whereas the classical histological techniques (e.g., cresyl violet, silver impregnation) for the demonstration of cytoarchitecture can easily be applied to large series of brains (important for subsequent statistical analysis), such methodical standards are more difficult to achieve in receptor architectonics. Methodical problems in receptor architectonic mapping concern, e.g., the preparation of large cryostat sections of unfixed, frozen human brains and the large amounts of expensive, labeled receptor ligands necessary for visualization of numerous receptors of the diverse neurotransmitter systems in the brain.

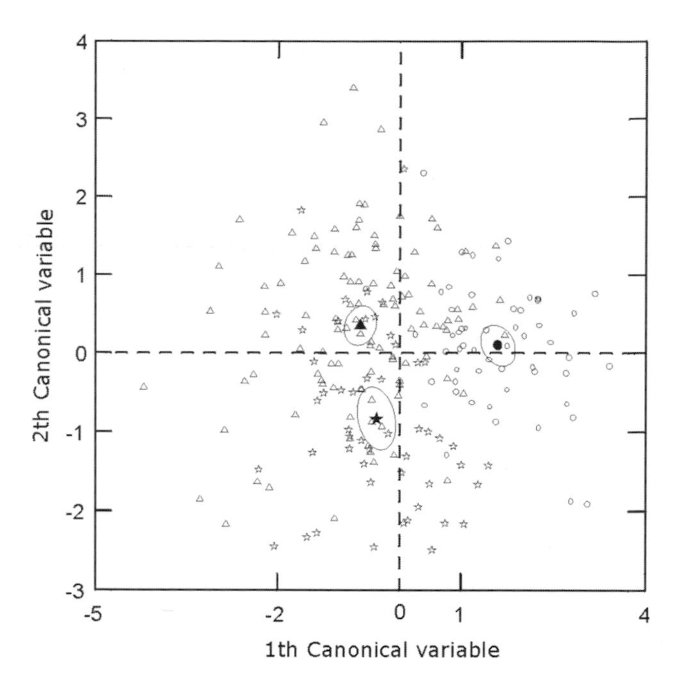

Figure 11 Multivariate distances between profiles of areas V1 (*circles* = 56), V2d (*triangles* = 108), and V3d (*stars* = 51). The profiles were extracted from the ROI as shown in Fig. 5A. In order to determine whether the entire set of profiles differs between the areas, canonical variables were calculated from the feature vectors. These are linear combinations of the raw features which were chosen in such a way that they spaced out the mean feature vectors of the given cortical area to the greatest possible extent (Krzanowski, 1988). The first two canonical variables were used as coordinates to plot the shape characteristics of each profile. Open symbols represent individual feature vectors, filled symbols the mean feature vector (centroid) of each area. Centroids are surrounded by 95% confidence ellipses. The three clusters originating from the three areas partially overlap, indicating a high degree of variation of profile shape within each cluster. Such variation is influenced by local inhomogeneities of cytoarchitecture which correspond to blood vessels, cell clusters, or vertically oriented structures such as cell columns (Mountcastle, 1978). Separation between clusters (areas) is indicated by the distances between centroids; confidence ellipses do not overlap. The first canonical variable separates V1 from V2d and V3d, the second canonical variable V2d from V3d. The distances of V1 from either V2d and V3d clearly exceed the distance between V2d and V3d, indicating a more similar laminar pattern across V2d and V3d, which is different from the laminar pattern of V1. This relationship reflects the well-known fact that it is much easier to distinguish area V1 from area V2 than to cytoarchitectonically delineate areas V2d and V3d. The latter two areas seem to be much more similar to each other than area V1 is similar to any other cortical area.

IV. Quantitative Autoradiography of Different Receptor Binding Sites

Quantitative *in vitro* receptor autoradiography is a powerful tool which reveals the chemoarchitectonic organization of the cerebral cortex (Zilles, 1991, 1992; Zilles *et al.*, 1995; Zilles and Schleicher, 1995). Additionally, measurements of up- or down-regulation of receptors under different physiological and pathological conditions, local differences in receptor affinity, and competition studies with different ligands can be performed with high spatial resolution (Qü *et al.*, 1998; Zilles *et al.*, 1998, 1999; Que *et al.*, 1999).

Autoradiography combines a considerable degree of spatial resolution with a high sensitivity, therefore permitting the anatomical identification of receptor localization as well as the visualization of low receptor concentrations. Furthermore, in combination with computerized image analysis, it enables accurate and reproducible *quantification* of receptor densities (Schleicher and Zilles, 1988; Zilles *et al.*, 1988b).

The degree of spatial resolution obtainable with *in vitro* receptor autoradiography is based mainly on the type of isotope with which the ligands are labeled. The results discussed in the present chapter were obtained using tritiated ligands, since this isotope permits a better local resolution compared with other isotopes such as ^{125}I and ^{14}C. The structures of interest should exceed 50 μm in their smallest dimension to be resolvable with receptor autoradiography. This method has been used repeatedly to study receptor distribution patterns in the cerebral cortex (Geyer *et al.*, 1996, 1997, 1998; Zilles *et al.*, 1995; Zilles and Palomero-Gallagher, 2001; Mazziotta *et al.*, 2001).

Diverse methodological aspects deserve special attention when using quantitative receptor autoradiography to analyze the distribution and densities of neurotransmitter binding sites in the cerebral cortex: pre- and postmortem conditions, tissue processing, labeling procedure, and automated image analysis.

A. Pre- and Postmortem Conditions

Autoradiographical analysis of postmortem human brain tissue or biopsies raises a series of methodological problems, some of which are also relevant for studies in animal models; others, however, are unique in experiments carried out with human brain sections and include both pre- and postmortem conditions.

The most important premortem factor, which has been described in numerous reports, is the effect that neurological diseases can have on the distribution, density, and affinity of specific neurotransmitter receptors (Tedroff, 1999; Blows, 2000; Mihailescu and Drucker-Colin, 2000; Mann *et al.*, 2001). Therefore, only brains obtained from patients who died without a history of neurological or psychiatric disorders were used for the chemoarchitectonic mapping of the human cerebral cortex. Furthermore, binding site density and affinity can also be affected by aging depending on the receptor type under consideration. A consistent finding is the age-related decrease in the density of glutamatergic NMDA receptors, which seems to be accompanied by regionally specific changes in the interaction between glutamate and other neurotransmitters such as dopamine and GABA (Wenk and Barnes, 2000; Adams *et al.*, 2001; Segovia *et al.*, 2001).

The effect of postmortem delay in the freezing of brain tissue as well as of prolonged storage of the frozen tissue prior to analysis on receptor binding assays are potential artifacts that may limit interpretation of the effects of disease on receptor populations. However, only a relatively small number of reports discuss the problems caused by these circumstances. Some receptors show a surprisingly high stability, with binding site densities remaining constant up to 70–80 h postmortem [e.g., NMDA (Kornhuber *et al.*, 1988), GABA (Lloyd and Dreksler, 1979), M_1 (Burke and Greenbaum, 1987), D_2 (Kontur *et al.*, 1994), and 5-HT_2 receptors (Gross-Isseroff *et al.*, 1990; Kontur *et al.*, 1994)]. Decreases in densities and affinities were described for D_1 and 5-HT_{1A} receptors (Kontur *et al.*, 1994), as well as for [^{3}H]N-methylscopolamine binding sites (Rodriguez-Puertas *et al.*, 1996). Conversely, the density of benzodiazepine binding sites increased with increasing postmortem delay of freezing (Whitehouse *et al.*, 1984).

Prolonged storage of deep frozen tissue is inevitable when analyzing a statistically significant sample of human brains. Storage of tissue for up to 3 years resulted in stable [^{3}H]N-methylscopolamine binding site densities (Whitehouse *et al.*, 1984; Rodriguez-Puertas *et al.*, 1996). Furthermore, no significant changes were observed when [^{3}H]GABA and [^{3}H]prazosin binding properties were examined after a storage period of at least 6 years (Lloyd and Dreksler, 1979; Faull *et al.*, 1988). Although it cannot be ruled out that other receptors may be vulnerable, current knowledge provides evidence that a storage time of up to 6 years has no influence on the stability of most receptors.

B. Tissue Processing

Both the quality of receptor autoradiographs and the preservation of histological stainings are highly dependent on the handling of the brain immediately after autopsy. Although fixation of the brain before deep freezing and cutting clearly improves the quality of histological stainings, it also impairs the structure of receptor proteins and, consequently, leads to changes in specific and nonspecific binding, altering the ratio between both parameters to different degrees (Rotter *et al.*, 1979; Herkenham, 1988; Zilles *et al.*, 1988b; Zilles and Schleicher, 1995). Therefore we use only unfixed, deep frozen brains for receptor autoradiography.

Immediately after autopsy, the brains were photographed and the hemispheres and brain stem were separated and stored in plastic bags on crushed ice before further dissection. The meninges and blood vessels were not removed, since this process, however carefully it is carried out, causes damage to the brain surface and leads to a partial loss of cortical layer I. Each hemisphere was cut into coronal, sagittal, or horizontal slabs (1.5–3.0 cm thick), which were placed on a sheet of strong aluminum foil to preserve a flat sectioning surface and to avoid distortions. Each slab comprises parts, or the complete circumference, of a hemisphere. The foil with the tissue was slowly immersed in *N*-methylbutane at –50°C for 10–15 min. This method enabled fast freezing of the brain tissue, avoiding freeze artifacts such as the appearance of ice crystals, which would destroy cellular morphology. The tissue was then stored in a deep freezer at –70°C in air-tight plastic bags to protect it from freeze-artifacts.

C. Labeling Procedure

The brain tissue was serially sectioned in a cryostat microtome for large sections into 20-μm sections at –20°C. The sections were thaw-mounted on gelatin-coated glass slides and freeze-dried overnight. Alternating sections were incubated with tritiated ligands alone (total binding) or with the tritiated ligands and a receptor type-specific displacing agent (nonspecific binding) or were stained with modified silver methods that produce Nissl-like images (Merker, 1983) or visualize myelinated fibers (Gallyas, 1979). The latter histologically stained sections enabled a precise microscopical identification of architectonically defined areas and layers.

The autoradiographical labeling method is carried out following standardized protocols previously described (Monaghan and Cotman, 1982; Billard *et al.*, 1984; Monaghan *et al.*, 1984; Wong *et al.*, 1988; Vilaró *et al.*, 1992; Zilles *et al.*, 1993; Nénonéné *et al.*, 1994; Rabow *et al.*, 1995; Marks *et al.*, 1998). In short, it consists of three steps: a preincubation, a main incubation, and a rinsing step. Incubation protocols of the receptors discussed in the present chapter are summarized in Table 1.

The aim of the *preincubation step* is to rehydrate the sections and to wash out endogenous substances which bind to the examined receptor and thus block the binding site for the tritiated ligand. In the *main incubation step,* adjacent sections are incubated in a buffer solution containing the tritiated ligand (in nM concentrations) or the tritiated ligand (in nM concentrations) plus an unlabeled specific displacer (in μM concentrations). Since the incubation of a brain section with a labeled ligand demonstrates the total binding of this ligand, the incubation with the tritiated ligand in the presence of a specific displacer is necessary to determine what proportion of the total binding sites is occupied by nonspecific, and thus nondisplaceable, binding. Specific binding is the difference between total and nonspecific binding. Nonspecific binding is taken into consideration only when it amounts to more than 10% of the total binding sites marked by the ligand. Finally, the *rinsing step* stops the binding procedure and eliminates surplus tritiated ligand as well as buffer salts, thus preventing artifacts on the film emulsion during exposure.

The radioactively marked sections were coexposed with plastic tritiated standards of varying but known concentrations of radioactivity against β-sensitive films for 4 to 10 weeks. After development of the film, the spatial distribution of optical densities in the autoradiograph indicates the local concentration of radioactivity present in the brain tissue and thus represents a measure of the local binding site concentrations.

D. Automated Image Analysis

These autoradiographs are further processed using densitometry with a video-based image analyzing technique (Schleicher and Zilles, 1988; Loats and Links, 1991; Zilles and Schleicher, 1991, 1995). The first step in the evaluation of autoradiographs is the image acquisition. Autoradiographs are digitized up to 768× with a spatial resolution of 512 pixels and 8-bit gray-value resolution (shades of gray ranging from 0, black, to 255, white). Digitization is carried out by means of a KS400® image analyzing system (Zeiss, Germany) connected to a CCD camera (Sony, Tokyo) with an S-Orthoplanar 60-mm macro lens (Zeiss, Germany), which is corrected for geometric distortions. For each exposed film, a blank area (= reference field) is placed on an illumination box, and the intensity of the light source and aperture of the macro lens are adjusted to obtain a mean gray value of 220, which is calculated as the mean gray value of all the pixels in the reference image, which is stored for shading correction. A reference gray value well below 255 is chosen in order to avoid saturation effects in the camera. Furthermore, gray values below 20, which cannot be reliably resolved by a CCD camera, should not occur at any place on the film. Therefore, exposure time of the film is adjusted for each ligand according to this constraint. Additional prerequisites for correct densitometry are reduction of stray light, sufficient warm-up of the light source as well as of the camera to avoid shifts in the system, and a homogeneous light intensity (Schleicher and Zilles, 1988; Zilles and Schleicher, 1995), which is obtained by means of a light box fitted with a double opal glass diffuser. To permit full-contrast densitometry, the smallest structures of interest in an image must be covered by more than 1 pixel in the x and y directions (Ramm *et al.*, 1984) in order to avoid biasing by the point-spread function of neighboring pixels, which may belong to other anatomical structures. Since image acquisition with a video system involves an inevitable noise component caused by non-spatially correlated discrete isolated pixel variations (Pratt, 1978), eight images of the same autoradiograph are optionally averaged during acquisition in order to improve the signal-to-noise ratio. During image acquisition, a shading correction is also carried out. Shading is the variation of gray values within an image of a homogeneous object and is caused by the video target, camera electronics, illumination source, and camera lens (Schleicher and Zilles, 1988; Zilles and Schleicher, 1995). Shading causes a dependency of the gray values of an object on its position in the measuring field and leads, therefore, to a decrease in the number of resolvable gray values, and, if neglected, induces considerable artifacts in the measurements. Shading correction is achieved by using the abovementioned reference image [$R(x,y)$, x and y are pixel coordinates], which contains a homogeneous and empty (without any brain tissue) film area, and its mean gray value (C), to transform each pixel of the digitized autoradiographs [$A(x,y)$] into corrected values [$SA(x,y)$] by means of Eq. (1):

$$SA(x,y) = A(x,y)\frac{C}{R(x,y)}(1 \leq x, y \leq 512). \tag{1}$$

Following these preparatory steps, an autoradiograph can be visualized as an intensity gray value image. These images, are now converted into images with pixel values representing concentrations of radioactivity or fmol binding sites/mg protein (Schleicher and Zilles, 1988; Zilles and Schleicher,

TABLE 1 Summary of Incubation Conditions for Receptor Autoradiography.

^{3}H-Ligand	Displacer	Incubation buffer	Preincubation	Main incubation	Rinsing
Prazosin [0.2 nM]	Phentolamine [10 μM]	50 mM Tris–HCI (pH 7.4)	30 min at 37 °C	45 min at 30 °C	2 × 5 min at 4 °C 2 × ⇵ in distilled H_2O
UK-14, 304 [1.4 nM]	Noradrenalin [100 μM]	50 mM Tris–HCI (pH 7.7) +100 μM $MnCl_2$	15 min at 22 °C 22 °C	90 min at 22 °C	5 min at 4 °C 2 × ⇵ in distilled H_2O
AMPA [10 nM]	Quisqualate [10 μM]	50 mM Tris–acetate (pH 7.2) +100 mM KSCN*	3 × 10 min at 4 °C	45 min at 4 °C	4 × 4 s in buffer at 4 °C 2 × 2 s in acetone/glutaraldehyde at 4 °C [100 ml/2.5 ml]
Kainate [8 nM]	Kainate [100 μM]	50 mM Tris–citrate (pH 7.1) +10 mM Ca–acetate*	3 × 10 min at 4 °C	45 min at 4 °C	4 × 4 s in buffer at 4 °C 2 × 2 s in acetone/glutaraldehyde at 4 °C [100 ml/2.5 ml]
MK-801 [5 nM]	(+) MK-801 [100 μM]	50 mM Tris–HCI (pH 7.2) +30 μM glycine* +50 μM spermidine*	15 min at 22 °C	60 min at 22 °C	2 × 5 min at 4 °C 2 × ⇵ in distilled H_2O
Oxotremorine-M [0.8 nM]	Carbachol [1 μM]	20 mM Hepes–Tris (pH 7.5) +10 mM $MgCl_2$	20 min at 22 °C	60 min at 22 °C	2 × 2 min at 4 °C 2 × ⇵ in distilled H_2O
Epibatidine [0.5 nM]	Nicotine [10 μM]	15 mM Hepes–Tris (pH 7.5) +120 mM NaCl +5.4 mM KCl +0.8 mM $MgCl_2$ +1.8 mM $CaCl_2$	20 min at 22 °C	90 min at 4 °C	1 × 5 min at 4 °C 2 × ⇵ in distilled H_2O
Muscimol [3 nM]	GABA [10 μM]	50 mM Tris–citrate (pH 7.0)	3 × 5 min at 4 °C	40 min at 4 °C	3 × 3 s at 4 °C 2 × ⇵ in distilled H_2O
8-OH-DPAT [1 nM]	Serotonin [10 μM]	170 mM Tris–HCl (pH 7.6) +4 mM $CaCl_2$ +0.01% ascorbic acid	30 min at 22 °C	60 min at 22 °C	1 × 5 min at 4 °C 2 × ⇵ in distilled H_2O
SCH-23390 [0.5 nM]	SKF 83566 [1 μM]	50 mM Tris–HCl (pH 7.4) +120 mM NaCl +5 mM KCl +2 mM $CaCl_2$ +1 mM $MgCl_2$ +1 μM mianserin*	20 min at 22 °C	90 min at 22 °C	2 × 10 min at 4 °C 2 × ⇵ in distilled H_2O

Note: [^{3}H] Prazosin binds mainly to the noradrenergic α_1 receptor, and [^{3}H]UK-14,304 labels the high-affinity α_2 (α_{2h}) binding sites (Zilles *et al.*, 1993). The glutamatergic α-amino-3-hydroxy-5-methyl-4-isoxalone propionic acid (AMPA), kainate, and *N*-methyl-D-aspartate binding sites were labeled with [^{3}H]AMPA, [^{3}H]kainate, and [^{3}H]MK-801, respectively (Monaghan and Cotman, 1982; Monaghan *et al.*, 1984; Wong *et al.*, 1988). Muscarinic M_2 cholinergic binding sites were visualized with [^{3}H]oxotremorine-M, which labels mainly the M_2 receptor (Vilaró *et al.*, 1992), and nicotinic cholinergic binding sites with [^{3}H]epibatidine (Marks *et al.*, 1998). [^{3}H]Muscimol labels the γ-aminobutyric acid (GABA) GABA a receptors (Rabow *et al.*, 1995), [^{3}H]8-OH-DPAT the serotoninergic 5-HT_{1A} binding sites (Nénonéné et al., 1994), and [^{3}H]SCH-23390 the dopaminergic D1 receptors (Billard *et al.*, 1984). *Substances added to buffer only during the main incubation.

1995). This scaling is performed in two stages: first, the gray value images of coexposed tritium standards are used to compute a non-linear calibration curve, thus defining the relationship between the gray values of the autoradiographs and concentrations of radioactivity (Fig. 12). Each of the standards has a known amount of radioactivity, which, in the case of brain homogenate standards, is determined in an adjacent standard section by liquid scintillation counting. For each of the standards, the amount of radioactivity (R) is converted to the concentration of binding sites (C_b) using Eq. (2),

$$C_b = \frac{R}{E \cdot B \cdot W_b \cdot S_a} \cdot \frac{K_D + L}{L}, \tag{2}$$

where E is the efficiency of the scintillation counter, B is the number of decays per unit of time and radioactivity, W_b is the protein weight of a standard, S_a is the specific activity of the ligand, K_D is the dissociation constant of the ligand, and L is the free concentration of the ligand during incubation. The nonlinear correlation between gray values and increasing concentrations of binding sites must be emphasized. Due to this nonlinear dependency, it is imperative that the range of gray values in the standards covers the gray value range found in the digitized autoradiograph, since no valid extrapolation can be carried out. Second, the gray value of each pixel in an autoradiographic image is converted into a corresponding concentration of radioactivity by interpolation into the calibration curve and subsequently linearly transformed into new gray values in order to create an image in which the gray values are a linear function of the concentration of radioactivity (Fig. 12).

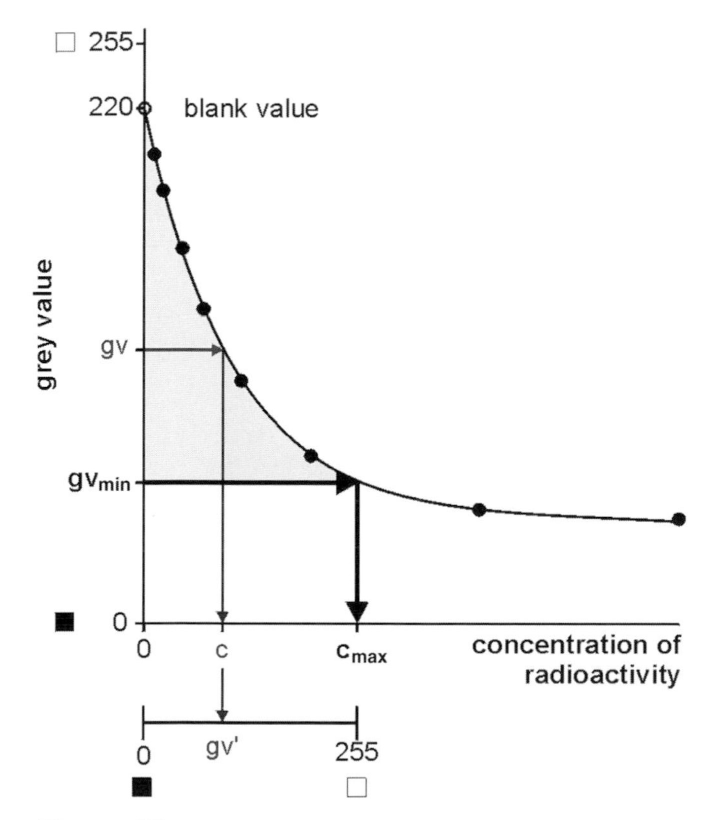

Figure 12 Calibration curve demonstrating the nonlinear dependency between gray values and concentrations of radioactivity of tissue standards. The lowest gray value (gv_{min}) of all autoradiographs and the calibration curve are used to calculate the maximum concentration of radioactivity (c_{max}). The gray value of each pixel in an image is converted into a corresponding receptor density by interpolation into the calibration curve and subsequently linearly transformed into new gray values (ranging from 0, black, to 255, white) in order to create an image in which the gray values are a linear function of the concentration of radioactivity.

Now the mean ligand concentration of an anatomically defined brain region can be quantified. Two different strategies can be applied depending on whether mean laminar or regional densities are to be extracted. The laminar distribution of neurotransmitter receptors within a given area can be characterized by means of density profiles such as introduced in Section III.C of this chapter (Schleicher *et al.*, 2000).

In order to extract numerical values for the mean regional receptor densities in an anatomically defined area, comparison with an adjacent cell-body-stained histological section is necessary, since not all anatomical structures are regularly associated with clear-cut differences in receptor densities. Therefore, in order to precisely identify and delineate the regions of interest in an autoradiograph, a print of the digitized autoradiograph and an adjacent cell-body-stained section are superimposed by means of a microscope equipped with a drawing tube (Schleicher and Zilles, 1988; Zilles and Schleicher, 1995). Both the histological section and the hard copy of the autoradiograph are visible simultaneously in the microscope, and the cytoarchitectonic borders are ink-traced on the hard copy. This traced contour of a brain region is then used as a template on a graphics tablet connected with a PC in which the digitized autoradiograph is stored. All pixel values within the contour line delineating the ROI are selected from the stored image, and the mean receptor concentration per unit protein (fmol/mg protein) contained in a specific region over a series of three to five sections is calculated by weighting the means from single sections by their areal size.

Pseudo color coding of autoradiographs is carried out solely to provide a clear visual impression of regional and laminar receptor distribution patterns. Since the complete range of available gray values is not necessarily used by the frequency distribution of actually occurring receptor densities, images are linearly contrast enhanced, thus preserving the absolute linear scaling between gray values and receptor densities, while improving the optical presentation of the images. The full range of 256 gray values (0–255) obtained after contrast enhancement is then pseudo-color-coded (Figs. 13, 15, and 16). The assignment of colors to the density ranges can be done in an arbitrary fashion, but the spectral arrangement of 11 colors to equally spaced density ranges results in an optimized visualization of the density pattern of the autoradiographs (Zilles and Schleicher, 1995).

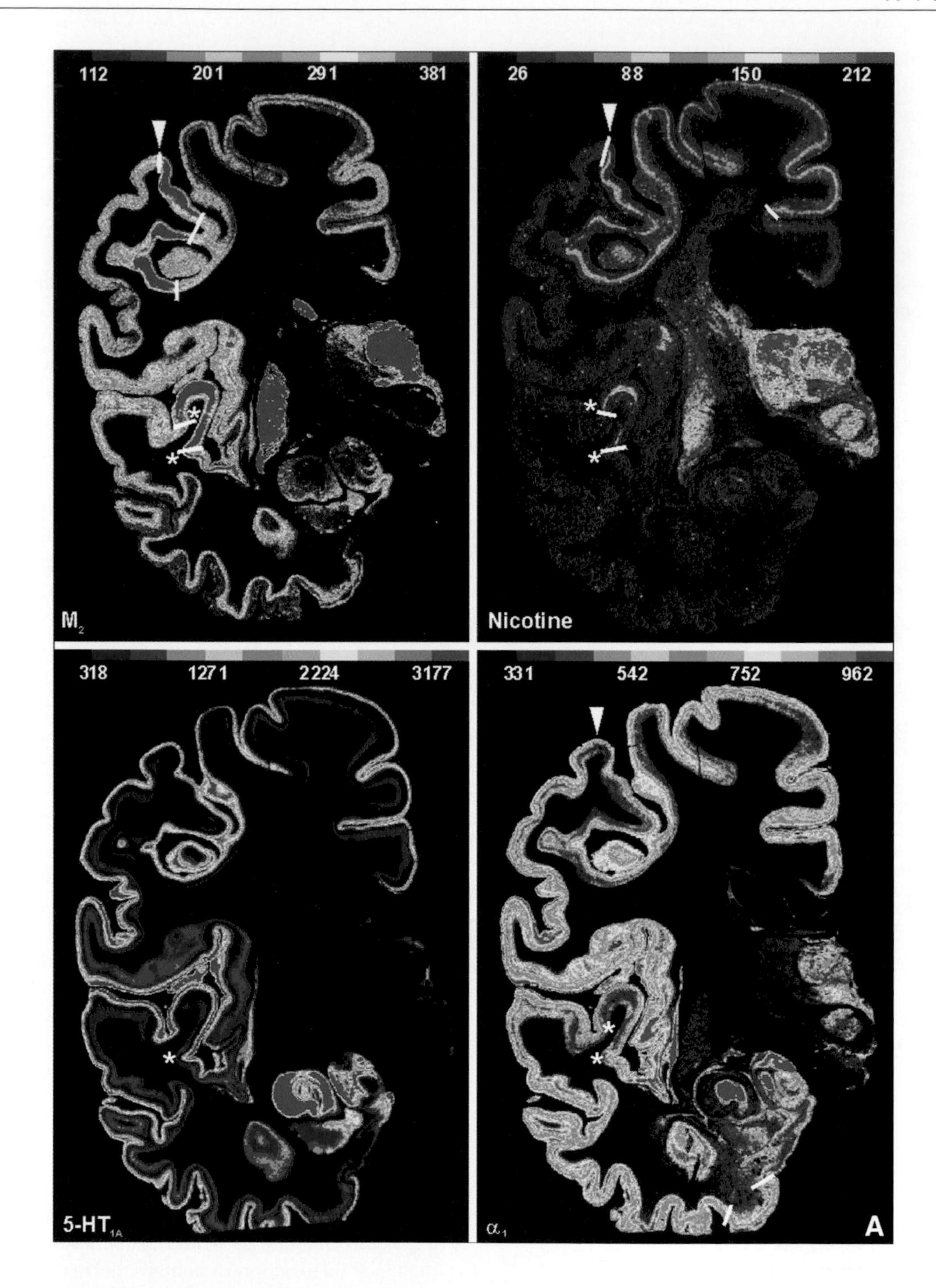

Figure 13 The regional and laminar distribution patterns of receptors for acetylcholine (M2 and nicotine), glutamate (AMPA and NMDA), GABA ($GABA_A$), dopamine (D_1), noradrenaline (α1) and serotonin (5-$HT1_A$) were visualized in neighboring coronal sections through a human hemisphere. Incubation protocols are summarized in Table 1. The scale above each image indicates the color coding of receptor densities in fmol/mg protein for that specific section. Note that, although each of the depicted receptor types does not indicate all areal borders, there is a perfect agreement in the location of those borders which are displayed by several receptors, i.e., the border between BA 1 and 3 (marked with an arrowhead) or the borders of the primary auditory cortex (BA 41, marked with asterisks). Unimodal primary sensory regions stand out due to their conspicuously high M_2 receptor densities. Primary sensorimotor areas are highlighted from the rest of the cerebral cortex by the laminar distribution pattern of nicotinic receptors. Anatomical landmarks are labeled on the schematic drawing in Fig. 14.

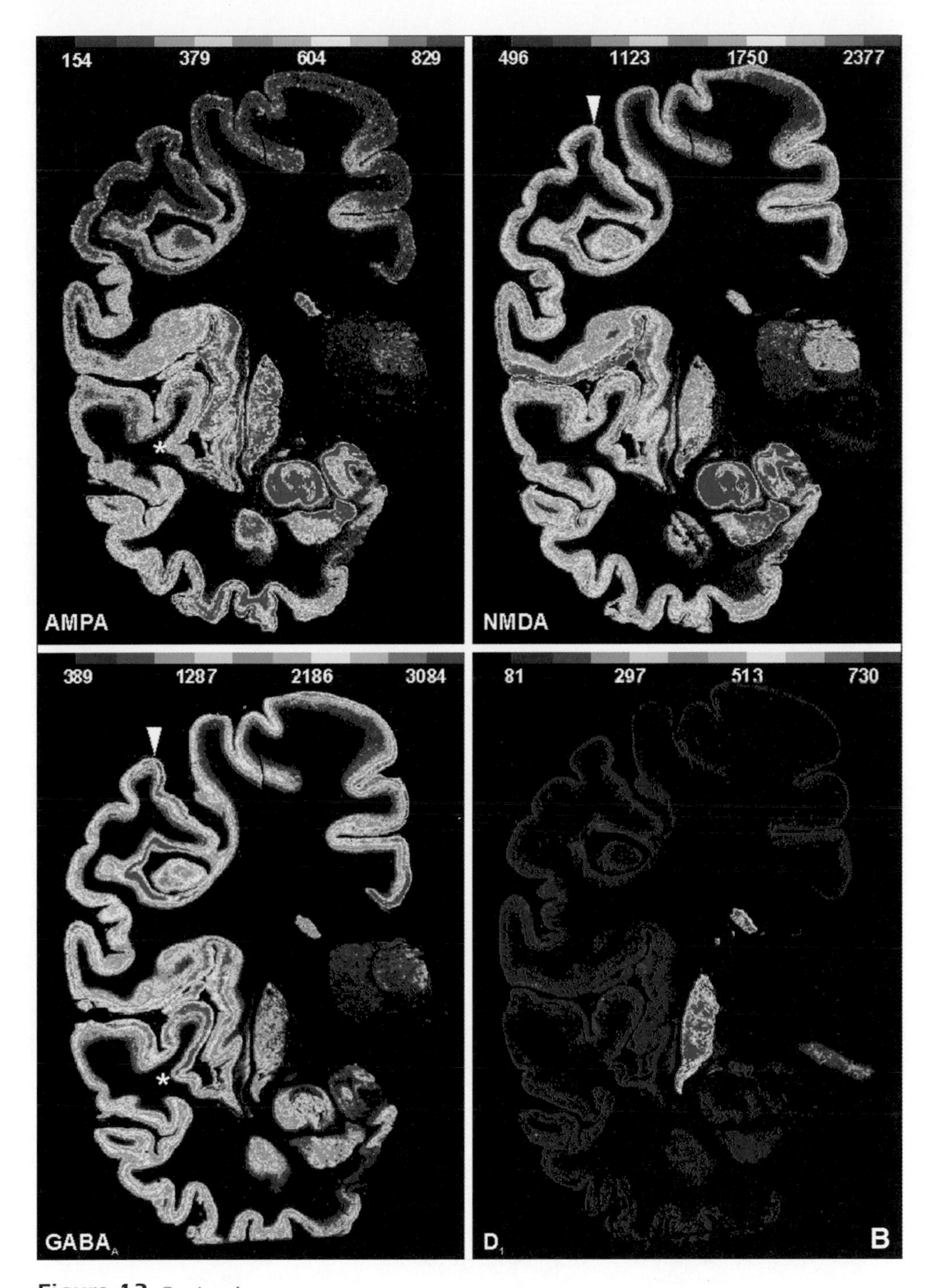

Figure 13 *Continued*

E. Receptor Autoradiography in Human Brain Mapping

Receptors for glutamate, GABA, acetylcholine, noradrenaline, serotonin, and dopamine are heterogeneously distributed throughout the human cerebral cortex (Figs. 13, 15, and 16). They show clear regional differences both in their mean densities and in their laminar distribution patterns. Furthermore, these variations are also present between different receptor types for a single neurotransmitter, i.e., glutamatergic AMPA and NMDA receptors or muscarinic cholinergic M_2 and nicotinic cholinergic receptors (Fig. 13). Although each receptor does not indicate all areal borders, there is a perfect agreement in the location of those borders which are displayed by several receptors (Figs. 13, 15, and 16).

The M_2 receptor antagonist [^{3}H]oxotremorine-M selectively emphasizes primary sensory areas (Figs. 13, 14, and 15). The primary somatosensory, auditory (Fig. 13), and visual cortices (Fig. 15) contain conspicuously higher M_2 receptor densities than any other cortical region of the human brain. Contrary to the M_2 receptors, which by all means clearly visualize further cortical parcellations, i.e., auditory and inferior temporal association cortices, the nicotinic receptors exclusively accentuate the family of sensorimotor areas (Fig. 13).

Receptors for classical neurotransmitters faithfully reflect the complex laminar structure of the primary visual cortex (Fig. 15, inset). Furthermore, their regional heterogeneity also enables the parcellation of extrastriate visual areas (Zilles and Clarke, 1997).

The hippocampus has a clear laminar structure (Fig. 16A), which is associated with segregated input, output, and intrinsic fiber systems and is therefore a favorable model for comparisons of anatomical structures with functionally and neurochemically identified neuronal systems. The regional and laminar distribution patterns of receptors for classical neurotransmitters in the hippocampus are highly correlated with its anatomical structure (Zilles *et al.*, 1993). For example, the main targets of the perforant pathway are the dendrites of the granule cells, located in the molecular layer of the dentate gyrus. The AMPA receptors show a preferential localization in this layer (Fig. 16B). The granule cells send their axons, the mossy fibers, to the proximal dendrites of CA3 pyramidal cells (lucidum layer). Glutamate neurotransmission at the level of the lucidum layer is associated with kainate receptors, which were present at extremely high concentrations in this hippocampal layer (Fig. 16C). The CA3 pyramidal cells project in turn to the dendritic region of CA1 pyramidal cells via the glutamatergic Schaffer collaterals. Consequently, high densities of AMPA receptors are visible in the target region of the Schaffer collaterals. The noradrenergic α_{2h} receptors are also found in high densities in the lacunosum–molecular layer of the CA1 region (Fig. 16D), where dendrites of CA1 pyramids are present. This spatial association of receptors of one neurotransmitter with a neuronal projection system using a different transmitter points to complex interactions between transmitter systems in an anatomically defined structure (Zilles *et al.*, 1991).

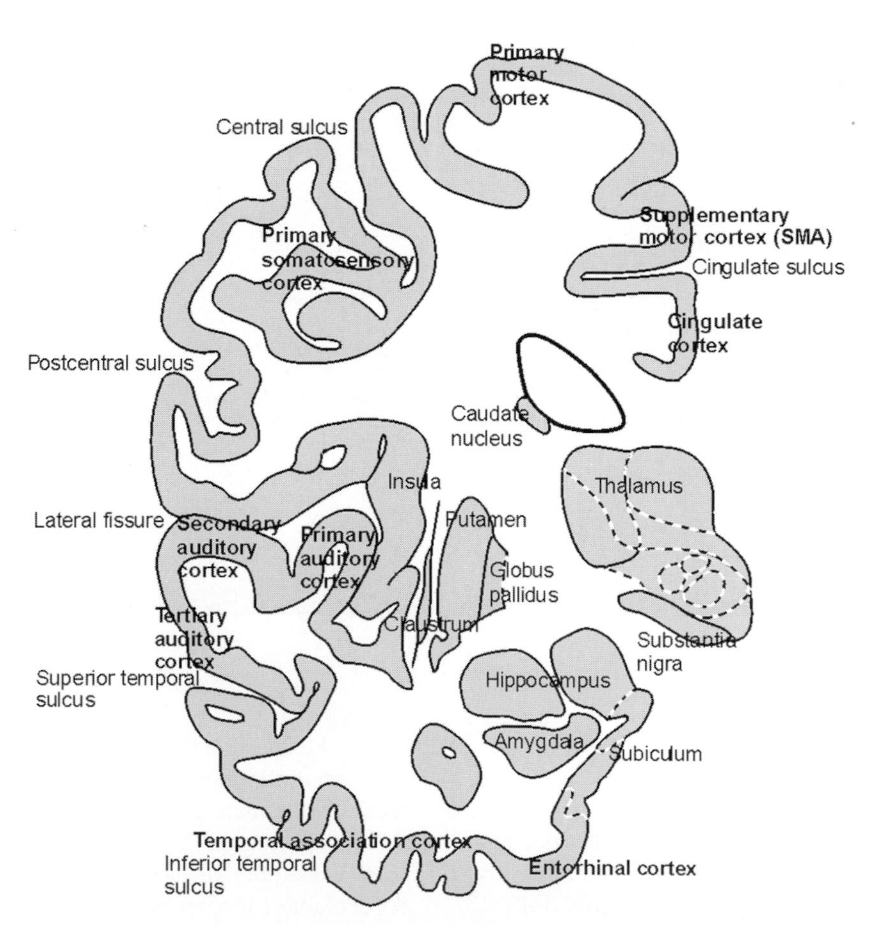

Figure 14 Schematic drawing of the section from Fig. 13 in which M_2 receptors were visualized. Functional areas (boldface) and anatomical landmarks have been labeled.

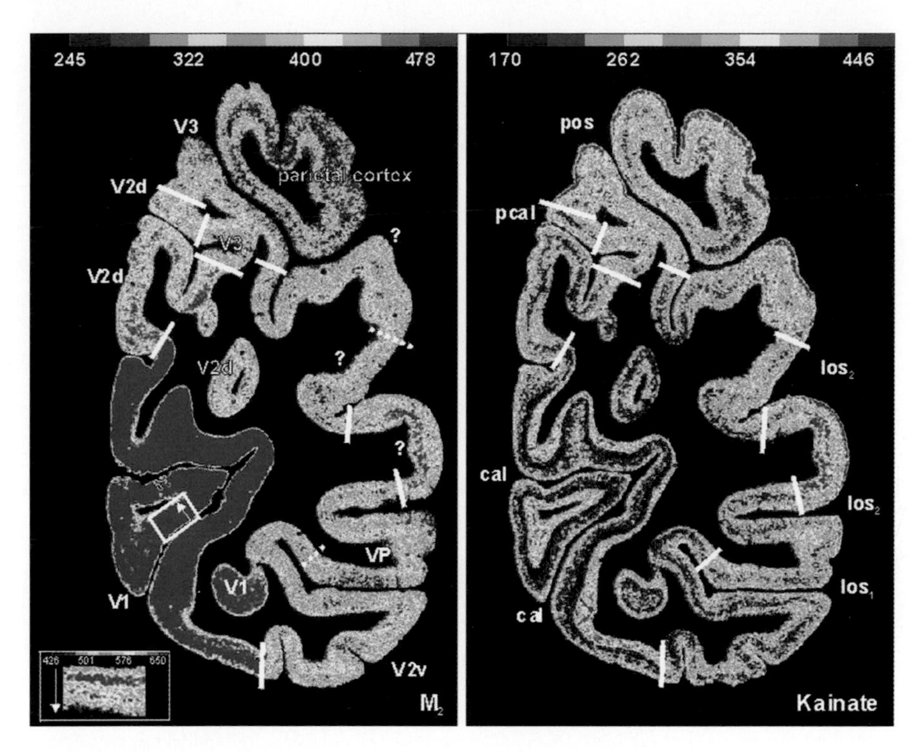

Figure 15 The regional and laminar distribution patterns of cholinergic M_2 and glutamatergic kainate receptors were visualized in neighboring coronal sections through a human occipital pole. Incubation protocols are summarized in Table 1. Inset: Detail of area V1, in which contrast enhancement was optimized to demonstrate the laminar distribution pattern of the M_2 receptors. The inset is rotated compared to the whole section in order to depict layer I up and the cortex/white matter border down. The scale above each image indicates the color coding of receptor densities in fmol/mg protein for that specific section. The areas marked by question marks have not yet been systematically mapped. Incubation protocols are summarized in Table 1. cal, calcarine sulcus; pcal, paracalcarine sulcus; pos, parieto-occipital sulcus; los_2, superior lateral occipital sulcus; los1, inferior lateral occipital sulcus.

Receptors for classical neurotransmitters are also differentially distributed throughout the basal ganglia (Fig. 13). The diverse thalamic nuclei are characterized by varying receptor densities. Cholinergic nicotinic and M_2 receptors show the highest binding site densities within the thalamus, whereas serotoninergic 5-HT_{1A} receptors are practically nonexistent in this region. A very detailed parcellation of the thalamus is provided by the cholinergic nicotinic receptors, which clearly differentiate the medial and ventrolateral thalamic nuclei. In the striatum, differences are not only visible between the putamen and the caudate nucleus (higher $GABA_A$ receptor densities in the caudate nucleus than in the putamen), but also within these nuclei. That is, the putamen shows an inhomogeneous dorsoventral distribution of the kainate and NMDA receptors. Furthermore, receptors for classical neurotransmitters are differentially distributed between the striosomes and the matrix of both the striatum and the caudate nucleus (Bauer, 1999). In the substantia nigra, the highest concentration of binding sites visualized is that of the dopaminergic D_1 receptors (Fig. 13).

It is important to stress the fact that changes in receptor densities should not be interpreted as being a mere reflection of variations in the degree of cell packing density in a given region or cortical layer. That is, a high receptor density does not necessarily imply a high cell packing density, and vice versa. One and the same cytoarchitectonically defined region may contain highest densities of one receptor and lowest of another. Clear examples are, among others, the primary auditory cortex and the hippocampus (Figs. 13 and 16). The primary auditory cortex shows extremely high M_2 receptor densities, whereas it contains one of the lowest α_1 binding site concentrations measured in the human brain. Conversely, the hippocampus is characterized by highest α_1 and lowest M_2 receptor densities. Furthermore, within the hippocampus, the lucidum layer, target of the mossy fibers, contains highest kainate receptor densities but only low AMPA receptor concentrations (Figs. 16B and 16C). This lack of correlation between receptor density and cell packing density is plausible, since by far the majority of transmitter receptors demonstrated by receptor autoradiography are located on dendrites, which represent a major proportion of the cell-body-free neuropil compartment. Thus, receptor concentration is not correlated with cell packing density, which is defined as volume density of cell bodies. It

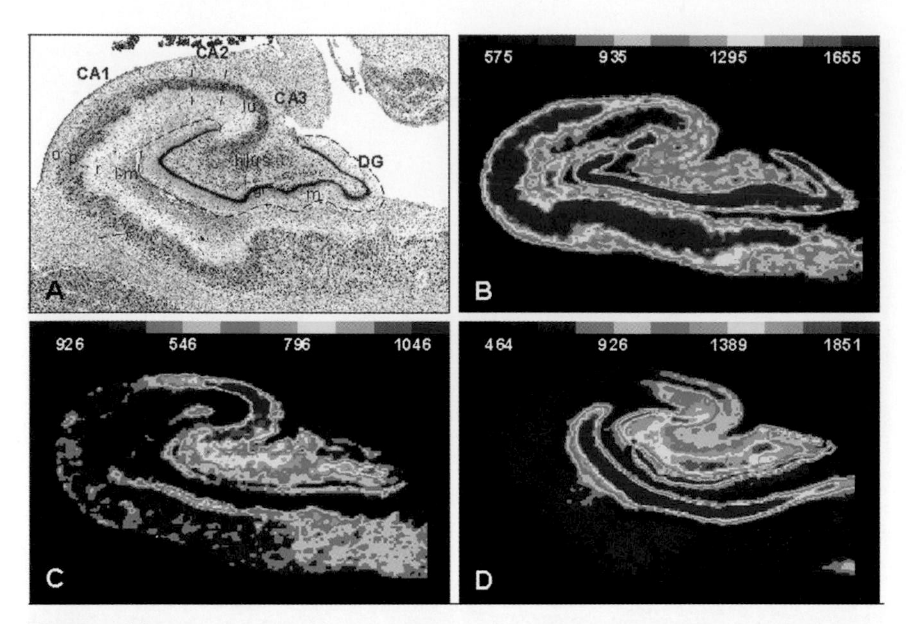

Figure 16 Coronal sections through the human hippocampus showing a histological cell body staining (Merker, 1983) (A) as well as the distribution patterns of noradrenergic α_{2h} receptors (B), glutamatergic AMPA (C), and kainate (D) receptors. The scale above each image indicates the color coding of receptor densities in fmol/mg protein. Incubation protocols are summarized in Table 1. CA1–CA3, Ammon's horn subfields; DG, dentate gyrus; o, oriens layer; p, pyramidal layer; r, radiatum layer; l-m, lacunosum–molecular layer; lu, lucidum layer; m, molecular layer; g, granular layer; hilus, CA4 subfield and polymorphic layer of the fascia dentata.

is interesting to note that this situation is in stark contrast to immunohistochemical receptor studies, which demonstrate single protein subunits of a receptor and not the native receptor complex. These subunits are frequently accumulated in the cell body, and thus their local concentrations are clearly associated with cell packing density.

The complex codistribution patterns of various receptors in architectonically defined brain regions stimulated the introduction of a new analytical procedure, the so-called receptor fingerprint (Fig. 17). Receptor fingerprints of cortical areas are polar coordinate plots of the mean regional densities of several different receptors over all cortical layers in a single, architectonically defined brain region. They demonstrate the site-specific balance between different receptor types and transmitter systems. These fingerprints may differ between regions by their shapes and/or sizes, thus representing the locally specific neurochemical organization at the receptor level. The shapes of the fingerprints differ between the motor, the unimodal sensory (Fig. 17A), and the associative (Fig. 17B) isocortices, as well as the allocortex (Fig. 17E), indicating the functionally specific balances between the different receptors in these different areas. Fingerprints may also define families of several cortical areas which are similar regarding the balance between different receptors. Differences in areal size of fingerprints may represent different hierarchical levels within a functional system, i.e., a larger receptor fingerprint for the primary visual and auditory areas than for their respective secondary cortices (Figs. 17C and 17D). An analysis of receptor fingerprints in the mesial motor areas (primary, supplementary, and presupplementary motor areas) of macaque monkeys clearly demonstrates identical shapes (i.e., all three areas belong to the "motor" family), but increasing sizes (i.e., proportionally increasing receptor densities from the primary motor through the supplementary to the presupplementary motor area) of their receptor fingerprints (Geyer *et al.,* 1998).

The distribution patterns of receptors for classical neurotransmitters reveal a more detailed cortical parcellation than that described by classical brain maps, i.e., the cytoarchitectonic map of Brodmann (1909). This lack of congruence does not imply a total lack of correspondence between classical parcellation schemes and the areas revealed by receptor autoradiography. In some cases, a total correspondence exists, i.e., in BA 17 (Fig. 15). In other instances, receptor autoradiography leads to a further parcellation of a region originally described by Brodmann. Such is the case of BA 4, which, according to Brodmann, is not further subdivided based on cytoarchitectonic features. A recent combined receptor autoradiographical and functional imaging study demonstrated that BA 4 can be subdivided at least into an anterior (4a) and a posterior (4b) component based both on differential receptor distribution patterns and quantitative cytoarchitecture functional activations (Geyer *et al.,* 1996).

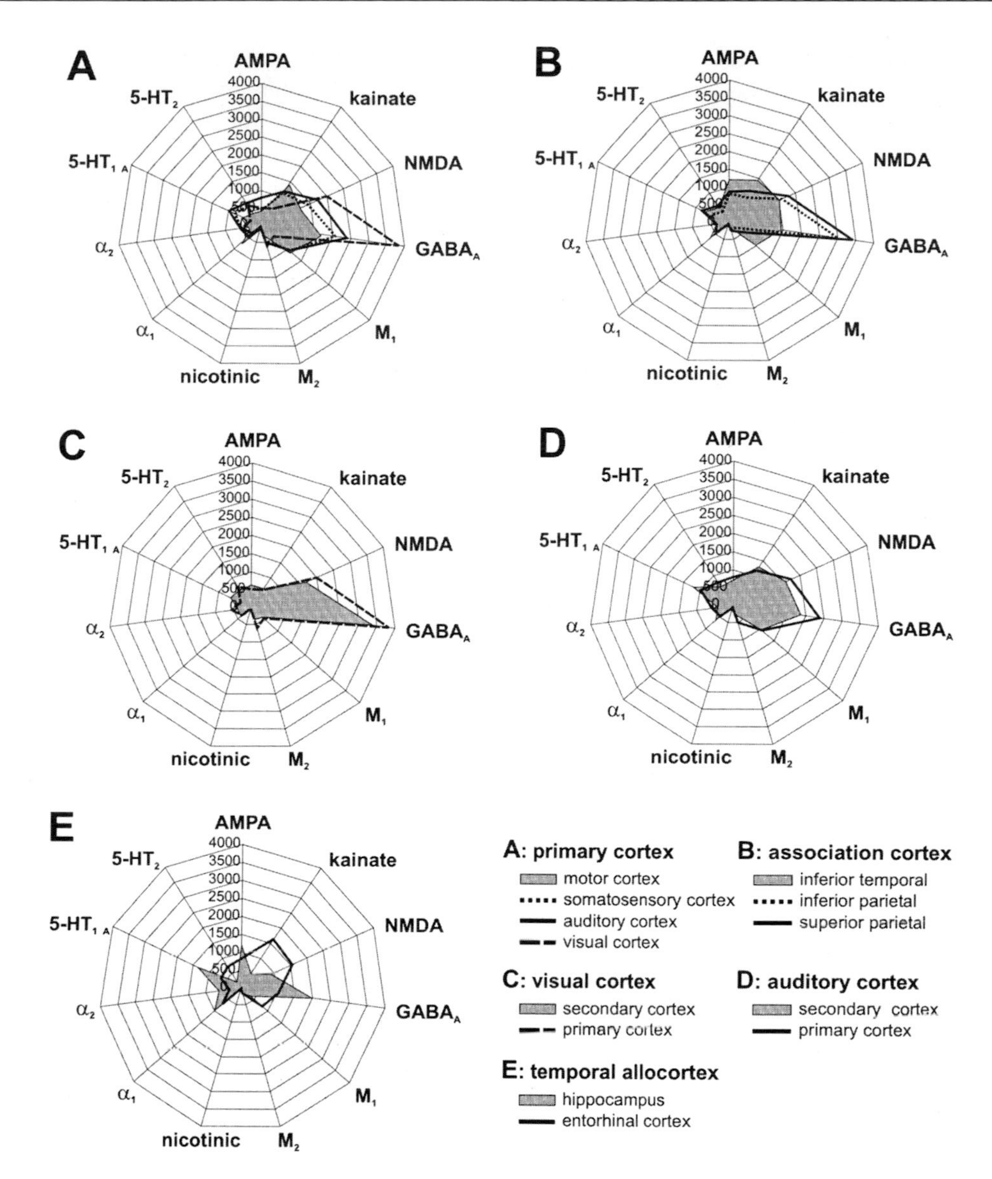

Figure 17 Receptor fingerprints of glutamatergic AMPA, kainate, and NMDA receptors; GABAergic GABAA receptors; muscarinic cholinergic M_1 and M_2 receptors; a nicotinic cholinergic receptor; adrenergic α_1 and α_2 receptors; and serotoninergic 5-HT$_{1A}$ and 5-HT$_2$ receptors in motor, somatosensory, visual, auditory, and higher association isocortices as well as in the allocortex. The polar coordinate plots (binding site densities in 0–4000 fmol/mg protein) indicate the mean receptor density in each cortical area averaged over all cortical layers.

Such increasingly detailed parcellations can also be expected in other cortical regions, e.g., the parietal cortex (Zilles and Palomero-Gallagher, 2001) and the visual cortex. Compare, e.g., the detailed maps of the macaque visual cortex (Fellemann and v. Essen, 1991; Sereno *et al.,* 1995; Tootell *et al.,* 1997) with the rough tripartition of the visual cortex as described by Brodmann (1909), von Economo and Koskinas (1925), and the Russian school (Sarkisov *et al.,* 1949) (Fig. 1). Although detailed cyto- and myeloarchitectonic maps of the parietal lobe were published half a century ago (Vogt, 1911; Gerhardt, 1940; Batsch, 1956), these maps were forgotten during the following years because, at the time of their publication, the proposed parcellations could not be functionally interpreted. However, recent studies of the chemoarchitecture of this region suggest a detailed parcellation into numerous areas based on regional and laminar receptor distribution patterns. At least some aspects of this ongoing parcellation support the results of the early maps of Vogt (1911), Batsch (1956), and Gerhard (1940) and add further subdivisions (Zilles and Palomero-Gallagher, 2001) compatible with recent functional imaging studies (Bremmer *et al.,* 2001; Weiss *et al.,* 2000; Fink *et al.,* 2000a,b, 2001a,b).

V. Perspectives of Architectonic Mapping

A more detailed parcellation scheme of the human cerebral cortex than that described by most of the classical architectonic maps is in accordance with the growing evidence for functionally segregated brain regions. A major challenge lies in understanding how this functional heterogeneity may be represented in structural terms. Understanding the regional distributions of neurotransmitter receptors is likely to provide a crucial intermediary level of description between function and structure, since different cytoarchitectonic and functional areas have obviously different mean receptor densities as well as distinct laminar distribution patterns.

Another way for a better understanding of brain function and the underlying anatomy is to compare architectonic maps obtained in postmortem brains with activation maps obtained in functional imaging studies in a common spatial reference system. Since these two kinds of maps stem from different subsets of brains, such a comparison must be performed on a probabilistic basis.

A. Probabilistic Architectonic Maps

Such three-dimensional cytoarchitectonic maps were proposed by us during past years. As a prerequisite, this approach requires some special preparation steps of the postmortem brains: MR imaging of postmortem brains was carried out as previously described (Zilles *et al.,* 1995; Amunts *et al.,* 1996; Geyer *et al.,* 1996; Roland and Zilles, 1998). Imaging was performed with a Siemens 1.5-T magnet (Erlangen, Germany) with a T_1-weighted 3D FLASH sequence covering the entire brain (flip angle 40°, repetition time TR = 40 ms, echo time TE = 5 ms for each image). Each volume consisted of 128 sagittal sections, the spatial resolution was $1 \times 1 \times 1.17$ mm (Steinmetz *et al.,* 1990). The image space was discretized into voxels in the form of rectangular parallelepipeds of uniform size, each with a resolution of 8 bits corresponding to 256 gray values. Image sequences of the postmortem brains were 3D reconstructed.

The whole brains were then embedded in paraffin and serially sectioned. 5000–7500 sections (20 μm) were acquired per brain in the coronal, horizontal, or sagittal plane. Each 60th section of an entire series was used for further analysis. Sections were mounted on glass slides and silver stained (Merker, 1983), resulting in a Nissl-stain-like appearance. Images of the histological sections were digitized and 3D reconstructed. The 3D reconstructions were warped to the MR sequences of the same brain, which had been obtained prior to embedding and sectioning. This enabled exclusion of inevitable distortions caused by histological techniques such as, for instance, sectioning and mounting on glass slides. Warping was performed by using both linear (i.e., scaling, translating, rotating) and nonlinear transformations. A modified principle axes theory and a movement model for large deformations were applied for the warping of the 3D data sets of postmortem brains to the standard brain in order to compensate for intersubject variability in size, shape, and sulcal pattern of the brains and to achieve a maximal anatomical overlap between the different postmortem brains and a reference brain (Schormann *et al.,* 1995, 1996; Schormann and Zilles, 1998). Brains were aligned according to the AC-PC plane (Talairach and Tournoux, 1988).

Cortical maps of the sensorimotor (Geyer *et al.,* 1996, 1999; Grefkes *et al.,* 2001), auditory (Morosan *et al.,* 2001; Rademacher *et al.,* 2001), and visual cortices (Amunts *et al.,* 2000) as well as of Broca's region (Amunts *et al.,* 1999) have already been defined in 10 human postmortem brains using the above-described observer-independent approach for the definition of areal borders (Schleicher *et al.,* 1998, see Section III.E). The borders of the cortical areas of the 10 individual brains were traced in the corresponding digitized histological sections of each brain and warped to the reference brain. This enabled the calculation of so-called probability (= population) maps which contain, for each voxel of the standard reference brain, the probability with which a certain cortical area is present in it (Fig. 18).

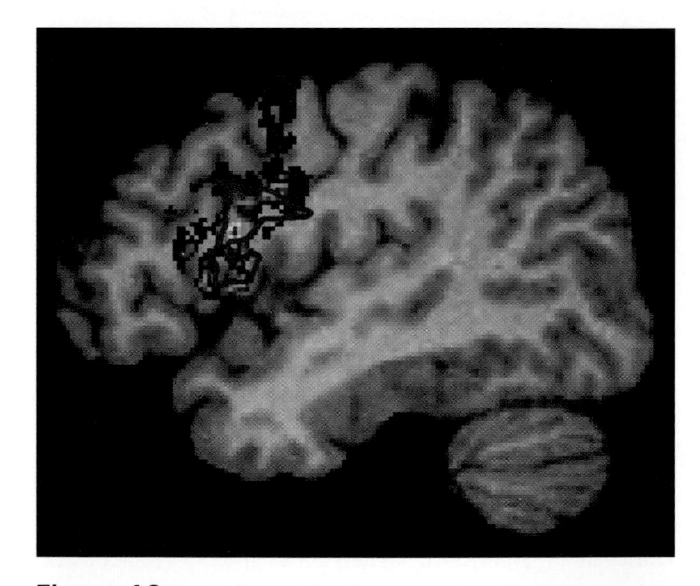

Figure 18 Comparison of significant activation of a fMRI study in a dual somatosensory reaction time task with the probability map of BA 44. When the task interval was short (<300 ms), the first task interfered with the second, and an activation occurred in the right inferior frontal gyrus. The functional activation (purple contour) was overlaid on the probability map of BA 44 at x = 44 of the human brain atlas (Roland and Zilles, 1994). The arbitrary color scale of the probability map, which ranged from blue (P = 0.1) to red (P > 0.6), represents the likelihood of those voxels representing BA 44. The activation overlapped approximately 20% of the cytoarchitectonically defined BA 44 (Herath et al., 2001).

B. Analysis of the Relationship of Brain Structure and Function

The functional relevance of the architectonic parcellation of the cortex can be tested by superimposing probabilistic maps with the results of recent PET, *f*MRI, and MEG studies in a common spatial reference system. In contrast to traditional architectonic maps and to the atlas of Talairach and Tournoux (1988), probabilistic architectonic maps (i) enable a 3D representation of cortical areas, (ii) are based on an observer-independent architectonic definition of cortical areas, and (iii) provide quantitative information concerning intersubject variability of the topography of each cortical area.

Using a common spatial reference system for combined analysis of PET and cytoarchitectonic data, it has been shown that the processing of both real and illusory contours activates architectonically defined area V2 (Larsson *et al.,* 1999). Another example is the cytoarchitectonic subdivision of area 4 into an anterior and a posterior part, which was first observed in receptor architectonic sections (Zilles and Schleicher, 1995) and then defined in cytoarchitectonic sections and superimposed with PET data, which enabled correlation of these areas with functional specificities (Geyer *et al.,* 1996). Further combined cytoarchitectonic/functional imaging studies have been performed in the sensorimotor and visual cortices (Naito *et al.,* 1999, 2000; Ehrsson *et al.,* 2000; Bodegard *et al.,* 2000a,b; Binkofski *et al.,* 2000b).

The application of probabilistic maps of BA 44 and 45 as the putative correlates of Broca's region have shown that the functional involvement of this region may be more complex than originally thought. Whereas its involvement in language (Indefrey *et al.,* 2001) is widely accepted, investigation into its role in movement (Binkofski *et al.,* 2000a) and higher cognitive function such as in dual task interference (Herath *et al.,* 2001) is only commencing.

In conclusion, architectonic brain mapping has become more objective and less observer-dependent. Furthermore, recent techniques, e.g., neurotransmitter receptor autoradiography and immuno- and enzyme histochemical methods, have added functional meaningful information. Multimodal mapping and quantitative analysis of interareal differences promote a new and more complex concept of a cortical area. Finally, the combined analysis of architectonic maps and functional imaging studies by applying recent 3D probabilistic atlas systems enables testing of the functional significance of architectonic parcellations and a systematic search for new, functionally relevant cortical areas. Thus, architectonic brain mapping has changed considerably during the past 100 years and opens exciting perspectives for future research.

Acknowledgments

The authors thank U. Blohm, R. Dohm, B. Machus, and N. Ivens for excellent histological and autoradiographic processing of brain tissue. Published and unpublished work for this chapter was supported by the Deutsche Forschungsgemeinschaft (SFB 194/A6). The Human Brain Project/Neuroinformatics research was jointly funded by the National Institute of Mental Health, the National Institute of Neurological Disorders and Stroke, the National Institute on Drug Abuse, and the National Cancer Institute. We are particularly grateful to P. E. Roland and his co-workers, for a long-lasting collaboration in the development and testing of the functional aspects of cytoarchitectonic probability maps, and to J. Mazziotta and A. Toga for their support of receptor-autoradiographic mapping as shown in Fig. 13.

References

Adams, M. M., Smith, T. D., Moga, D., Gallagher, M., Wang, Y., Wolfe, B. B., Rapp, P. R., and Morrison, J. H. (2001). Hippocampal dependent learning ability correlates with *N*-methyl-D-aspartate (NMDA) receptor levels in CA3 neurons of young and aged rats. *J. Comp. Neurol.* **432,** 230–243.

Amunts, K., Malikovic, A., Mohlberg, H., Schormann, T., and Zilles, K. (2000). Brodmann's areas 17 and 18 brought into stereotaxic space—Where and how variable? *NeuroImage* **11,** 66–84.

Amunts, K., Schlaug, G., Schleicher, A., Steinmetz, H., Dabringhaus, A., Roland, P. E., and Zilles, K. (1996). Asymmetry in the human motor cortex and handedness. *NeuroImage* **4,** 216–222.

Amunts, K., Schleicher, A., Bürgel, U., Mohlberg, H., Uylings, H. B. M., and Zilles, K. (1999). Broca's region revisited: Cytoarchitecture and intersubject variability. *J. Comp. Neurol.* **412,** 319–341.

Amunts, K., Schmidt-Passos, F., Schleicher, A., and Zilles, K. (1997). Postnatal development of interhemispheric asymmetry in the cytoarchitecture of human area 4. *Anat. Embryol.* **196,** 393–402.

Armstrong, E., Schleicher, A., Omran, H., Curtis, M., and Zilles, K. (1995). The ontogeny of human gyrification. *Cereb. Cortex* **1,** 56–63.

Bailey, P., and von Bonin, G. (1951). "The Isocortex of Man." Univ. of Illinois Press, Urbana.

Bartels, P. (1979). Numerical evaluation of cytologic data. II. Comparison of profiles. *Anal. Quant. Cytol.* **1,** 77–83.

Bartels, P. (1981). Numerical evaluation of cytologic data. VII. Multivariate significance tests. *Anal. Quant. Cytol.* **3,** 1–8.

Batsch, E.-G. (1956). Die myeloarchitektonische Untergliederung des Isocortex parietalis beim Menschen. *J. Hirnforsch.* **2,** 226–258.

Bauer, A. (1999). "Verteilung von Neurotransmitterrezeptoren in den Stammganglien des Menschen. Eine Quantitative Autoradiographische Untersuchung." Berichte des Forschungszentrums Jülich, Jülich.

Betz, W. (1874). Anatomischer Nachweis zweier Gehirncentra. *Zentralbl. Med. Wiss.* **37/38.**

Billard, W., Ruperto, V., Crosby, G., Iorio, L. C., and Barnett, A. (1984). Characterization of the binding of ^{3}H-SCH 23390, a selective D-1 receptor antagonist ligand, in rat striatum. *Life Sci.* **35,** 1885–1893.

Binkofski, F., Amunts, K., Stephan, K. M., Posse, S., Schormann, T., Freund, H.-J., Zilles, K., and Seitz, R. J. (2000a). Broca's region subserves imagery of motion: A combined cytoarchitectonic and fMRI study. *HBM* **11,** 273–285.

Binkofski, F., Geyer, S., Fink, G. R., Buccino, G., Seitz, R. J., Zilles, K., and Freund, H.-J. (2000b). Differential activation of area 4p and 4a in motor-related attention. *NeuroImage* **11,** S7.

Blows, W. T. (2000). Neurotransmitters of the brain: Serotonin, noradrenaline (norepinephrine), and dopamine. *J. Neurosci. Nurs.* **32,** 234–238.

Bodegard, A., Geyer, S., Naito, E., Zilles, K., and Roland, P. E. (2000a). Somatosensory areas in men activated by moving stimuli: Cytoarchitectonic mapping and PET. *NeuroReport* **11,** 187–191.

Bodegard, A., Ledberg, A., Geyer, S., Naito, E., Zilles, K., and Roland, P. E. (2000b). Object shape differences reflected by somatosensory cortical activation in human. *J. Neurosci.* **20,** 1–5.

Bok, S., and van Kip, M. (1939). The size of the body and the size and the number of the nerve cells in the cerebral cortex. *Acta Nederl. Morphol.* **3,** 1–22.

Braak, H. (1980). "Architectonics of the Human Telencephalic Cortex." Springer-Verlag, Berlin/Heidelberg/New York.

Bremmer, F., Schlack, A., Shah, N. J., Zafiris, O., Zilles, K., and Fink, G. R. (2001). Polymodal motion processing in posterior parietal and premotor cortex: A human fMRI study strongly implies equivalencies between humans and monkeys. *Neuron* **29,** 287–296.

Brodmann, K. (1903). Beiträge zur histologischen Lokalisation der Grosshirnrinde. II. Der Calcarinustyp. *J. Psychol. Neurol.* **II,** 133–159.

Brodmann, K. (1905). Beiträge zur histologischen Lokalisation der Grosshirnrinde. V. Über den allgemeinen Bauplan des Cortex pallii bei den Mammaliern und zwei homologe Rindenfelder im besonderen. Zugleich ein Beitrag zur Furchenlehre. *J. Psychol. Neurol.* **VI,** 275–303.

Brodmann, K. (1908). Beiträge zur histologischen Lokalisation der Grosshirnrinde. VI. Die Cortexgliederung des Menschen. *J. Psychol. Neurol.* **X,** 231–246.

Brodmann, K. (1909). "Vergleichende Lokalisationslehre der Grosshirnrinde in ihren Prinzipien dargestellt auf Grund des Zellenbaues." Barth, Leipzig.

Brodmann, K. (1919). Beiträge zur histologischen Lokalisation der Grosshirnrinde. I. Die Regio Rolandica. *Beiträge zur Lokalisation der Grosshirnrinde* **II,** 79–107.

Burke, R. E., and Greenbaum, D. (1987). Effect of postmortem factors on muscarinic receptor subtypes in rat brain. *J. Neurochem.* **49,** 592–596.

Buxhoeveden, D. P., Switala, A. E., Roy, E., and Casanova, M. F. (2000). Quantitative analysis of cell columns in the cerebral cortex. *J. Neurosci. Methods* **97,** 7–17.

Campbell, A. W. (1905). "Histological Studies on the Localisation of Cerebral Function." Cambridge Univ. Press, Cambridge, UK.

Castleman, K. (1979). "Digital Image Processing." Prentice Hall International, Englewood Cliffs, NJ.

Chusid, J. G. (1964). Black-and-white supplement for the color brain map of Bailey and von Bonin. *Neurology* **14,** 154–157.

Dixon, W. J., Brown, M. B., Engelman, L., Hill, M. A., and Jennrich, R. I. (1988). "BMDP: Statistical Software Manual." Univ. of California Press, Berkley.

Duvernoy, H. (1991). "The Human Brain: Surface, Three-Dimensional Sectional Anatomy and MRI." Springer-Verlag, Vienna/New York.

Ehrsson, H. H., Naito, E., Geyer, S., Amunts, K., Zilles, K., Forssberg, H., and Roland, P. E. (2000). Simultaneous movements of upper and lower limbs are coordinated by motor representations that are shared by both limbs: A PET study. *Eur. J. Neurosci.* **12,** 3385–3398.

Elliot Smith, G. (1907). A new topographical survey of the human cerebral cortex, being an account of the distribution of the anatomically distinct cortical areas and their relationship to the cerebral sulci. *J. Anat.* **41,** 237–254.

Faull, K. F., Bowersox, S. S., Zeller-DeAmicis, L., Maddaluno, J. F., Ciaranello, R. D., and Dement, W. C. (1988). Influence of freezer storage time on cerebral biogenic amine and metabolite concentrations and receptor ligand binding characteristics. *Brain Res.* **450,** 225–230.

Fellemann, D. J., and v. Essen, D. C. (1991). Distributed hierarchical processing in the primate cerebral cortex. *Cereb. Cortex* **1,** 1–47.

Fink, G. R., Marshall, J. C., Gurd, J., Weiss, P. H., Zafiris, O., Shah, N. J., and Zilles, K. (2001a). Deriving numerosity and shape from identical visual displays. *NeuroImage* **13,** 46–55.

Fink, G. R., Marshall, J. C., Shah, N. J., Weiss, P. H., Halligan, P. W., Grosse-Ruyken, M., Ziemons, K., Zilles, K., and Freund, H.-J. (2000a). Line bisection judgements implicate right parietal cortex and cerebellum as assessed by fMRI. *Neurology* **54,** 1324–1331.

Fink, G. R., Marshall, J. C., Weiss, P. H., Shah, N. J., Toni, I., Halligan, P. W., and Zilles, K. (2000b). "Where" depends on "What": A differential functional anatomy for position discrimination in one- versus two-dimensions. *Neuropsychologia* **38,** 1741–1748.

Fink, G. R., Marshall, J. C., Weiss, P. H., and Zilles, K. (2001b). The neural basis of vertical and horizontal line bisection judgements: An fMRI study of normal volunteers. *NeuroImage* **14,** 59–67.

Flechsig, P. (1920). "Anatomie des Menschlichen Gehirns und Rückenmarks auf Myelogenetischer Grundlage." Thieme, Leipzig.

Fritsch, G., and Hitzig, E. (1870). Über die elektrische Erregbarkeit des Grosshirns. *Arch. Anat. Physiol. Wiss. Med.* **Jg. 1870,** 300–332.

Gallyas, F. (1979). Silver staining of myelin by means of physical development. *Neurol. Res.* **1,** 203–209.

Gerhardt, E. (1940). Die Cytoarchitektonik des Isocortex parietalis beim Menschen. *J. Psychol. Neurol.* **49,** 367–419.

Geyer, S., Ledberg, A., Schleicher, A., Kinomura, S., Schormann, T., Bürgel, U., Klingberg, T., Larsson, J., Zilles, K., and Roland, P. E. (1996). Two different areas within the primary motor cortex of man. *Nature* **382,** 805–807.

Geyer, S., Matelli, M., Luppino, G., Schleicher, A., Jansen, Y., Palomero-Gallagher, N., and Zilles, K. (1998). Receptor autoradiographic mapping of the mesial and premotor cortex of the macaque monkey. *J. Comp. Neurol.* **397,** 231–250.

Geyer, S., Schleicher, A., and Zilles, K. (1997). The somatosensory cortex of human: Cytoarchitecture and regional distributions of receptor-binding sites. *NeuroImage* **6,** 27–45.

Geyer, S., Schleicher, A., and Zilles, K. (1999). Areas 3a, 3b, and 1 of human primary somatosensory cortex. I. Microstructural organisation and interindividual variability. *NeuroImage* **10,** 63–83.

Gower, J. (1985). Measures of similarity, dissimilarity, and distance. In "Encyclopedia of Statistical Sciences" (S. Kotz and N. L. Johnson, eds.). Wiley, New York.

Grefkes, C., Geyer, S., Schormann, T., Roland, P., and Zilles, K. (2001). Human somatosensory area 2: Observer-independent cytoarchitectonic mapping, interindividual variability, and population map. *NeuroImage* **14,** 617–632.

Gross-Isseroff, R., Salama, D., Israeli, M., and Biegon, A. (1990). Autoradiographic analysis of age-dependent changes in serotonin 5-HT_2 receptors of the human brain postmortem. *Brain Res.* **519,** 223–227.

Herath, P., Klingberg, T., Young, Y., Amunts, K., and Roland, P. (2001). Neural correlates of dual task interference can be dissociated from those of divided attention: An fMRI study. *Cereb. Cortex* **11,** 796–805.

Herkenham, M. (1988). Influence of tissue treatment on quantitative receptor autoradiography. *In* "Molecular Neuroanatomy" (F. W. van Leeuwen, R. M. Buijs, C. W. Pool, and O. Pach, eds.), pp. 111–120. Elsevier, Amsterdam.

Indefrey, P., Brown, C. M., Hellwig, F., Amunts, K., Herzog, H., Seitz, R. J., and Hagoort, P. (2001). A neural correlate of syntactic encoding during speech production. *Proc. Natl. Acad. Sci. USA* **98,** 5933–5936.

Kleist, K. (1934). "Gehirnpathologie." Barth, Leipzig.

Kontur, P. J., al-Tikriti, M., Innis, R. B., and Roth, R. H. (1994). Postmortem stability of monoamines, their metabolites and receptor binding in rat brain regions. *J. Neurochem.* **62,** 282–290.

Kornhuber, J., Retz, W., Riederer, P., Heinsen, H., and Fritze, J. (1988). Effect of antemortem and postmortem factors on [^{3}H]glutamate binding in the human brain. *Neurosci. Lett.* **93,** 312–317.

Kretschmann, H.-J., and Weinrich, W. (1996). "Dreidimensionale Computergraphik Neurofunktionaler Systeme." Thieme, Stuttgart.

Krzanowski, W. (1988). "Principles of Multivariate Analysis." Oxford Univ. Press, Oxford.

Larsson, J., Amunts, K., Gulyas, B., Malikovic, A., Zilles, K., and Roland, P. E. (1999). Neuronal correlates of real and illusory contour perception: Functional anatomy with PET. *Eur. J. Neurosci.* **11,** 4024–4036.

Lloyd, K. G., and Dreksler, S. (1979). An analysis of [^{3}H]gamma-aminobutyric acid (GABA) binding in the human brain. *Brain Res.* **163,** 77–87.

Loats, H. L., and Links, J. M. (1991). Digital image processing in autoradiography. *In* "Autoradiography and Correlative Imaging" (W. E. Stumpf and H. F. Solomon, eds.), pp. 467–484. Academic Press, San Diego/New York/Boston/London/Sydney/Tokyo/Toronto.

Luppino, G., Matelli, M., Camarda, R. M., Gallese, V., and Rizzolatti, G. (1991). Multiple representations of body movements in mesial area 6 and the adjacent cingulate cortex: An intracortical microstimulation study in the macaque monkey. *J. Comp. Neurol.* **311,** 463–482.

Mann, J. J., Brent, D. A., and Arango, V. (2001). The neurobiology and genetics of suicide and attempted suicide: A focus on the serotonergic system. *Neuropsychopharmacology* **24,** 467–477.

Marks, M. J., Smith, K. W., and Collins, A. C. (1998). Differential agonist inhibition identifies multiple epibatidine binding sites in mouse brain. *J. Pharmacol. Exp. Theor.* **285,** 377–386.

Matelli, M., Luppino, G., and Rizzolatti, G. (1991). Architecture of superior and mesial area 6 and the adjacent cingulate cortex in the macaque monkey. *J. Comp. Neurol.* **311,** 445–462.

Mazziotta, J., Toga, A., Evans, A., Fox, P., Lancaster, J., Zilles, K., Woods, R., Paus, T., Simpson, G., Pike, B., Holmes, C., Collins, L., Thompson, P., Macdonald, D., Iacoboni, M., Schormann, T., Amunts, K., Palomero-Gallagher, N., Geyer, S., Parsons, L., Narr, K., Kabani, N., Le Goualher, G., Feidler, J., Smith, K., Boomsma, D., Pol, H. H., Cannon, T., Kawashima, R., and Mazoyer, B. (2001). A four-dimensional probabilistic atlas of the human brain. *J. Am. Med. Inform. Assoc.* **8,** 401–430.

Merker, B. (1983). Silver staining of cell bodies by means of physical development. *J. Neurosci.* **9,** 235–241.

Mihailescu, S., and Drucker-Colin, R. (2000). Nicotine and brain disorders. *Acta Pharmacol. Sin.* **21,** 97–104.

Monaghan, D. T., and Cotman, C. W. (1982). The distribution of [^{3}H]kainic acid binding sites in cat CNS as determined by autoradiography. *Brain Res.* **252,** 91–100.

Monaghan, D. T., Yao, D., and Cotman, C. W. (1984). Distribution of [^{3}H]AMPA binding sites in rat brain as determined by quantitative autoradiography. *Brain Res.* **324,** 160–164.

Morosan, P., Rademacher, J., Schleicher, A., Amunts, K., Schormann, T., and Zilles, K. (2001). Human primary auditory cortex: Cytoarchitectonic subdivisions and mapping into a spatial reference system. *NeuroImage* **13,** 684–701.

Mountcastle, V. (1978). An organizing principle for cerebral function: The unit module and the distributed system. *In* "The Mindful Brain: Cortical Organization and the Group-Selective Theory of Higher Brain Function" (G. Edelmann and V. Mountcastle, eds.), pp. 7–51. MIT Press, Cambridge, MA.

Naito, E., Ehrsson, H. H., Geyer, S., Zilles, K., and Roland, P. E. (1999). Illusory arm movements activate cortical motor areas: A positron emission tomography study. *J. Neurosci.* **19,** 6134–6144.

Naito, E., Kinomura, S., Geyer, S., Kawashima, R., Roland, P. E., and Zilles, K. (2000). Fast reaction to different sensory modalities activates common fields in the motor areas, but the anterior cingulate cortex is involved in the speed of reaction. *J. Neurophsysiol.* **83,** 1701–1709.

Nénonéné, E. K., Radja, F., Carli, M., Grondin, L., and Reader, T. A. (1994). Heterogeneity of cortical and hippocampal 5-HT$_{1A}$ receptors: A reappraisal of homogenate binding with 8-[^{3}H]hydroxydipropylaminotetralin. *J. Neurochem.* **62,** 1822–1834.

Ono, M., Kubik, S., and Abernathey, C. D. (1990). "Atlas of the Cerebral Sulci." Thieme, Stuttgart/New York.

Pratt, W. K. (1978). "Digital Image Processing." Wiley, New York.

Que, M., Witte, O. W., Neumann-Haefelin, T., Schiene, K., Schroeter, M., and Zilles, K. (1999). Changes of GABA$_A$ and GABA$_B$ receptor binding following cortical photothrombosis: A quantitative receptor autoradiographic study. *Neuroscience* **93,** 1233–1240.

Qü, M. S., Mittmann, T., Luhmann, H., Schleicher, A., and Zilles, K. (1998). Long-term changes of ionotropic glutamate and GABA receptors after unilateral permanent focal cerebral ischemia in the mouse brain. *Neuroscience* **85,** 29–43.

Rabow, L. E., Russek, S. J., and Farb, D. H. (1995). From ion currents to genomic analysis: Receptor advances in GABA(A) receptor research. *Synapse* **21,** 189–274.

Rademacher, J., Caviness, J., Steinmetz, H., and Galaburda, A. M. (1993). Topographical variation of the human primary cortices: Implications for neuroimaging, brain mapping, and neurobiology. *Cereb. Cortex* **3,** 313–329.

Rademacher, J., Morosan, P., Schormann, T., Schleicher, A., Werner, C., Freund, H.-J., and Zilles, K. (2001). Probabilistic mapping and volume measurement of human primary auditory cortex. *NeuroImage* **13,** 669–683.

Ramm, P., Kulick, J. H., and Farb, D. H. (1984). Video and scanning microdensitometer-based imaging systems in autoradiographic densitometry. *J. Neurosci. Methods* **11,** 89–100.

Rodriguez-Puertas, R., Pascual, J., and Pazos, A. (1996). Effects of freezing storage time on the density of muscarinic receptors in the human brain postmortem: An autoradiographic study in control and Alzheimer's disease brain tissues. *Brain Res.* **728,** 65–71.

Roland, P. E., and Zilles, K. (1994). Brain atlases—A new research tool. *Trends Neurosci.* **17,** 458–467.

Roland, P. E., and Zilles, K. (1998). Structural divisions and functional fields in the human cerebral cortex. *Brain Res. Rev.* **26,** 87–105.

Rotter, A., Birdsall, N. J. M., Burgen, A. S. V., Field, P. M., Hulme, E. C., and Raisman, G. (1979). Muscarinic receptors in the central nervous system of the rat. I. Technique for autoradiographic localization of the binding of [^{3}H]propylbenzilylcholine mustard and its distribution in the forebrain. *Brain Res. Rev.* **1,** 141–166.

Sanides, F., and Vitzthum, H. G. (1965a). Die Grenzerscheinungen am Rande der menschlichen Sehrinde. *Dtsch. Ztsch. Nervenheilk.* **187,** 708–719.

Sanides, F., and Vitzthum, H. G. (1965b). Zur Architektonik der menschlichen Sehrinde und den Prinzipien ihrer Entwicklung. *Dtsch. Ztsch. Nervenheilk.* **187,** 680–707.

Sarkisov, S. A., Filimonoff, I. N., and Preobrashenskaya, N. S. (1949). "Cytoarchitecture of the Human Cortex Cerebra." Medgiz, Moscow. [In Russian]

Schleicher, A., Amunts, K., Geyer, S., Kowalski, T., Schormann, T., Palomero-Gallagher, N., and Zilles, K. (2000). A stereological approach to human cortical architecture: Identification and delineation of cortical areas. *J. Chem. Neuroanat.* **20,** 31–47.

Schleicher, A., Amunts, K., Geyer, S., Kowalski, T., and Zilles, K. (1998). An observer-independent cytoarchitectonic mapping of the human cortex using a stereological approach. *Acta Stereol.* **17,** 75–82.

Schleicher, A., Amunts, K., Geyer, S., Morosan, P., and Zilles, K. (1999). Observer-independent method for microstructural parcellation of cerebral cortex: A quantitative approach to cytoarchitectonics. *NeuroImage* **9,** 165–177.

Schleicher, A., and Zilles, K. (1988). The use of automated image analysis for quantitative receptor autoradiography. *In* "Molecular Neuroanatomy" (F. W. van Leeuwen, R. M. Buijs, C. W. Pool, and O. Pach, eds.), pp. 147–157. Elsevier, Amsterdam.

Schleicher, A., and Zilles, K. (1990). A quantitative approach to cytoarchitectonics: Analysis of structural inhomogeneities in nervous tissue using an image analyser. *J. Microsc.* **157,** 367–381.

Schormann, T., Dabringhaus, A., and Zilles, K. (1995). Statistics of deformations in histology and improved alignment with MRI. *IEEE Trans. Med. Imaging* **14,** 25–35.

Schormann, T., Henn, S., and Zilles, K. (1996). A new approach to fast elastic alignment with application to human brains. *Lect. Notes Comp. Sci.* **1131,** 437–442.

Schormann, T., and Zilles, K. (1998). Three-dimensional linear and nonlinear transformations: An integration of light microscopical and MRI data. *HBM* **6,** 339–347.

Segovia, G., Porras, A., Del Arco, A., and Mora, F. (2001). Glutamatergic neurotransmission in aging: A critical perspective. *Mech. Ageing Dev.* **122,** 1–29.

Sereno, M. I., Dale, A. M., Reppas, J. B., Kwong, K. K., Belliveau, J. W., Brady, T. I., Rosen, B. R., and Tootell, R. B. H. (1995). Borders of multiple visual areas in humans revealed by functional magnetic resonance imaging. *Science* **268,** 889–893.

Steinmetz, H., Rademacher, J., Jäncke, L., Huan, Y., Thron, A., and Zilles, K. (1990). Total surface of temporoparietal intrasylvian cortex: Diverging left–right asymmetries. *Brain Lang.* **39,** 357–372.

Talairach, J., and Tournoux, P. (1988). "Coplanar Stereotaxic Atlas of the Human Brain." Thieme, Stuttgart.

Tanji, J., and Kurata, K. (1989). Changing concepts of motor areas of the cerebral cortex. *Brain Dev.* **11,** 374–377.

Tedroff, J. M. (1999). Functional consequences of dopaminergic degeneration in Parkinson's disease. *Adv. Neurol.* **80,** 67–70.

Tootell, R. B. H., Mendola, J. D., Hadjikhani, N. K., Ledden, P. J., Liu, A. K., Reppas, J. B., Sereno, M. I., and Dale, A. M. (1997). Functional analysis of V3A and related areas in human visual cortex. *J. Neurosci.* **17,** 7060–7078.

Tower, D., and Young, O. (1973). The activities of butyryl cholinesterase and carbonic anhydrase, the rate of anaerobic glycolysis, and the question of a constant density of glial cells in cerebral cortices of various mammalian species from mouse to whale. *J. Neurochem.* **20,** 269–278.

Uylings, H, van Pelt, J., Verwer, R., and McConnell, P. (1989). Metric analysis of neuronal tree patterns. *In* "Computer Techniques in Neuroanatomy" (J. Capowsky, ed.), pp. 241–264. Plenum, New York.

Vilaró, M. T., Wiederhold, K. H., Palacios, J. M., and Mengod, G. (1992). Muscarinic M_2 receptor mRNA expression and receptor binding in cholinergic and non-cholinergic cells in the rat brain: A correlative study using *in situ* hybridization, histochemistry and receptor autoradiography. *Neuroscience* **47,** 367–393.

Vogt, C. (1910). Die myeloarchitektonische Felderung des menschlichen Stirnhirns. *J. Psychol. Neurol.* **15,** 221–238.

Vogt, C., and Vogt, O. (1919). Allgemeinere Ergebnisse unserer Hirnforschung. *J. Psychol. Neurol.* **25,** 292–398.

Vogt, O. (1911). Die Myeloarchitektonik des Isocortex parietalis. *J. Psychol. Neurol.* **18,** 107–118.

von Economo, C., and Koskinas, G. N. (1925). "Die Cytoarchitektonik der Hirnrinde des Erwachsenen Menschen." Springer-Verlag, Berlin.

Weiss, P. H., Marshall, J. C., Tellmann, L., Halligan, P. W., Freund, H.-J., Zilles, K., and Fink, G. R. (2000). Neural consequences of acting in near versus far space: A physiological basis for clinical dissociations. *Brain* **123,** 2531–2541.

Wenk, G. L., and Barnes, C. A. (2000). Regional changes in the hippocampal density of AMPA and NMDA receptors across the lifespan of the rat. *Brain Res.* **885,** 1–5.

Whitehouse, P. J., Lynch, D., and Kuhar, M. J. (1984). Effects of postmortem delay and temperature on neurotransmitter receptor binding in a rat model of the human autopsy process. *J. Neurochem.* **43,** 553–559.

Wong, E. H., Knight, A. R., and Woodruff, G. N. (1988). [^{3}H]MK-801 labels a site on the *N*-methyl-D-aspartate receptor channel complex in rat brain membranes. *J. Neurochem.* **50,** 274–281.

Wree, A., Schleicher, A., and Zilles, K. (1982). Estimation of volume fractions in nervous tissue with an image analyzer. *J. Neurosci. Methods* **6,** 29–43.

Zilles, K (1991). Codistribution of receptors in the human cerebral cortex. *In* "Receptors in the Human Nervous System" (F. A. O. Mendelsohn and G. E. Paxinos, eds.), pp. 165–206. Academic Press, San Diego.

Zilles, K. (1992). Neurotransmitter receptors in the forebrain: Regional and laminar distribution. *Prog. Histochem. Cytochem.* **26,** 229–240.

Zilles, K., Armstrong, E., Schleicher, A., and Kretschmann, H. J. (1988a). The human pattern of gyrification in the cerebral cortex. *Anat. Embryol.* **179,** 173–179.

Zilles, K., and Clarke, S. (1997). Architecture, connectivity and transmitter receptors of human extrastriate visual cortex. Comparison with non-human primates. *In* "Cerebral Cortex" (Rockland *et al.,* eds.), Vol. 12, pp. 673–742. Plenum, New York.

Zilles, K., Dabringhaus, A., Geyer, S., Amunts, K., Qü, M., Schleicher, A., Gilissen, E., Schlaug, G., Seitz, R., and Steinmetz, H. (1996). Structural asymmetries in the human forebrain and the forebrain of non-human primates and rats. *Neurosci. Behav. Rev.* **20,** 593–605.

Zilles, K., and Palomero-Gallagher, N. (2001). Cyto-, myelo-, and receptor architectonics of the human parietal cortex. *NeuroImage* **14,** 58–520.

Zilles, K., Qü, M., and Schleicher, A. (1993). Regional distribution and heterogeneity of alpha-adrenoreceptors in the rat and human central nervous system. *J. Hirnforsch.* **34,** 123–132.

Zilles, K., Qü, M., Schleicher, A., and Luhmann, H. J. (1998). Characterization of neuronal migration disorders in neocortical structures: Quantitative receptor autoradiography of ionotropic glutamate, $GABA_A$ and $GABA_B$ receptors. *Eur. J. Neurosci.* **10,** 3095–3106.

Zilles, K., Qü, M., Schröder, H., and Schleicher, A. (1991). Neurotransmitter receptors and cortical architecture. *J. Hirnforsch.* **32,** 343–356.

Zilles, K., Qü, M. S., Köhling, R., and Speckmann, E.-J. (1999). Ionotropic glutamate and γ-aminobutyric acid receptors in human epileptic neocortical tissue: Quantitative in vitro receptor autoradiography. *Neuroscience* **94,** 1051–1061.

Zilles, K., Schlaug, G., Matelli, M., Luppino, G., Schleicher, A., Qü, M., Dabringhaus, A., Seitz, R., and Roland, P. E. (1995). Mapping of human and macaque sensorimotor areas by integrating architectonic, transmitter receptor, MRI and PET data. *J. Anat.* **187,** 515–537.

Zilles, K., and Schleicher, A. (1991). Quantitative receptor autoradiography and image analysis. *Bull. Assoc. Anat.* **75,** 117–121.

Zilles, K., and Schleicher, A. (1995). Correlative imaging of transmitter receptor distributions in human cortex. *In* "Autoradiography and Correlative Imaging" (W. Stumpf and H. Solomon, eds.), pp. 277–307. Academic Press, San Diego.

Zilles, K., Schleicher, A., and Kretschmann, H.-J. (1978). A quantitative approach to cytoarchitectonics. I. The areal pattern of the cortex of Tupaia belangeri. *Anat. Embryol.* **153,** 195–212.

Zilles, K., Schleicher, A., Rath, M., and Bauer, A. (1988b). Quantitative receptor autoradiography in the human brain. Methodical aspects. *Histochemistry* **90,** 129–137.

V

Analysis

22

Statistics I: Experimental Design and Statistical Parametric Mapping

Karl J. Friston

The Wellcome Department of Cognitive Neurology, University College London, London, United Kingdom WC1N 3BG

I. Introduction

This chapter is about making regionally specific inferences in neuroimaging. These inferences may be about differences expressed when comparing one group of subjects to another or, within subjects, over a sequence of observations. They may pertain to structural differences (e.g., in voxel-based morphometry; Ashburner and Friston, 2000) or neurophysiological indices of brain functions (e.g., *f*MRI). The principles of data analysis are very similar for all of these applications and constitute the subject of this chapter. We will focus on the analysis of *f*MRI time series because this covers most of the issues that are likely to be encountered in other modalities. Generally, the analysis of structural images and PET scans is simpler because they do not have to deal with correlated errors, from one scan to the next.

A general issue, in data analysis, is the relationship between the neurobiological hypothesis one posits and the statistical models adopted to test that hypothesis. This chapter begins by reviewing the distinction between functional *specialization* and *integration* and how these principles serve as the motivation for most analyses of neuroimaging data. We will address the design and analysis of neuroimaging studies from both these perspectives but note that both have to be integrated for a full understanding of brain mapping results.

Statistical parametric mapping is generally used to identify functionally specialized brain regions and is the most prevalent approach to characterizing functional anatomy and disease-related changes. The alternative perspective, namely that provided by functional integration, requires a different set of (multivariate) approaches that examine the relationship between changes in activity in one brain area and another. Statistical parametric mapping is a voxel-based approach, employing classical inference, to make some comment about regionally specific responses to experimental factors. In order to assign an observed response to a particular brain structure, or cortical area, the data must conform to a known anatomical space. Before considering statistical modeling, this chapter deals briefly with how a time series of images is realigned and mapped into some standard anatomical space (e.g., a stereotactic space). The general ideas behind statistical parametric mapping are then described and illustrated with attention to the different sorts of inferences that can be made with different experimental designs. *f*MRI is special, in the sense that the data lend themselves to a signal processing perspective. This can be exploited to ensure that both the design and the analysis are as efficient as possible. Linear time invariant models provide the bridge between inferential models employed by statistical mapping and conventional signal processing approaches. Temporal autocorrelations in noise processes represent another important issue, specific to *f*MRI, and approaches to maximizing efficiency in the context of serially correlated error terms will be discussed. Nonlinear models of evoked hemodynamics will be considered here because they can be used to indicate when the assumptions behind linear models are violated. *f*MRI can capture data very fast (in relation to other imaging techniques), engendering the opportunity to measure event-related responses. The distinction between event- and epoch-related designs will be discussed from the point of view of efficiency and the constraints provided by nonlinear characterizations. Before considering multivariate analyses we will close the discussion of inferences about regionally specific effects, by looking at the distinction between fixed- and random-effect analyses and how this relates to inferences about the subjects studied or the population from which these subjects came. The final section will deal with functional integration using models of effective connectivity and other multivariate approaches.

II. Functional Specialization and Integration

The brain appears to adhere to two fundamental principles of functional organization, *functional integration* and *functional specialization*, through which the integration within and among specialized areas is mediated by effective connectivity. The distinction relates to that between *localizationism* and *(dis)connectionism* that dominated thinking about cortical function in the 19th century. Since the early anatomic theories of Gall, the identification of a particular brain region with a specific function has become a central theme in neuroscience. However, functional localization per se was not easy to demonstrate: For example, a meeting that took place on August 4, 1881, addressed the difficulties of attributing function to a cortical area, given the dependence of cerebral activity on underlying connections (Phillips *et al.*, 1984). This meeting was entitled "Localization of Function in the Cortex Cerebri." Goltz (1881), although accepting the results of electrical stimulation in dog and monkey cortex, considered that the excitation method was inconclusive, in that movements elicited might have originated in related pathways, or current could have spread to distant centers. In short, the excitation method could not be used to infer functional localization because localizationism discounted interactions, or functional integration among different brain areas. It was proposed that lesion studies could supplement excitation experiments. Ironically, it was observations on patients with brain lesions some years later (see Absher and Benson, 1993) that led to the concept of *disconnection syndromes* and the refutation of localizationism as a complete or sufficient explanation of cortical organization. Functional localization implies that a function can be localized in a cortical area, whereas specialization suggests that a cortical area is specialized for some aspects of perceptual or motor processing and that this specialization is anatomically segregated within the cortex. The cortical infrastructure supporting a single function may then involve many specialized areas whose union is mediated by the functional integration among them. In this view functional specialization is meaningful only in the context of functional integration and vice versa.

The functional role played by any component (e.g., cortical area, subarea, or neuronal population) of the brain is largely defined by its connections. Certain patterns of cortical projections are so common that they could amount to rules of cortical connectivity. "These rules revolve around one, apparently, overriding strategy that the cerebral cortex uses—that of functional segregation" (Zeki, 1990). Functional segregation demands that cells with common functional properties be grouped together. This architectural constraint necessitates both convergence and divergence of cortical connections. Extrinsic connections among cortical regions are not continuous but occur in patches or clusters. This patchiness has, in some instances, a clear relationship to functional segregation. For example, V2 has a distinctive cytochrome oxidase architecture, consisting of thick stripes, thin stripes, and interstripes. When recordings are made in V2, directionally selective (but not wavelength or color selective) cells are found exclusively in the thick stripes. Retrograde (i.e., backward) labeling of cells in V5 is limited to these thick stripes. All the available physiological evidence suggests that V5 is a functionally homogeneous area that is specialized for visual motion. Evidence of this nature supports the notion that patchy connectivity is the anatomical infrastructure that mediates functional segregation and specialization. If it is the case that neurons in a given cortical area share a common responsiveness (by virtue of their

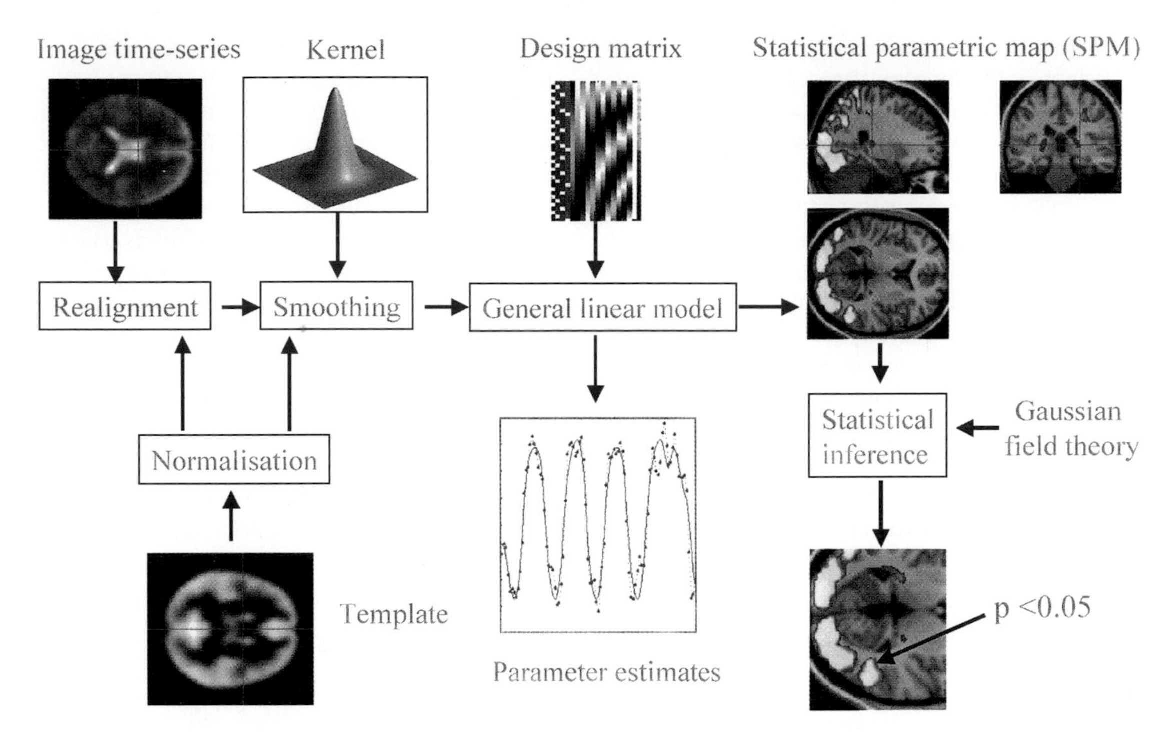

Figure 1 This schematic depicts the data transformations that start with an imaging data sequence and end with a statistical parametric map (SPM). SPMs can be thought of as "X-rays" of the significance of an effect. Voxel-based analyses require the data to be in the same anatomical space: This is effected by realigning the data (and removing movement-related signal components that persist after realignment). After realignment the images are subject to nonlinear warping so that they match a template that already conforms to a standard anatomical space. After smoothing, the general linear model is employed to (i) estimate the parameters of the model and (ii) derive the appropriate univariate test statistic at every voxel (see Fig. 3). The test statistics that ensue (usually *t* or *F* statistics) constitute the SPM. The final stage is to make statistical inferences on the basis of the SPM and Gaussian random field theory (see Fig. 6) and characterize the responses observed using the fitted responses or parameter estimates.

extrinsic connectivity) to some sensorimotor or cognitive attribute, then this functional segregation is also an anatomical one. Challenging a subject with the appropriate sensorimotor attribute or cognitive process should lead to activity changes in, and only in, the area of interest. This is the anatomical and physiological model upon which the search for regionally specific effects is based.

The analysis of functional neuroimaging data involves many steps that can be broadly divided into (i) spatial processing, (ii) estimating the parameters of a statistical model, and (iii) making inferences about those parameter estimates with their associated statistics (see Fig. 1). We will deal first with spatial transformations: In order to combine data from different scans from the same subject, or data from different subjects, it is necessary that they conform to the same anatomical frame of reference. This is the subject of the next section.

III. Spatial Realignment and Normalization

The analysis of neuroimaging data generally starts with a series of spatial transformations. These transformations aim to reduce artifactual variance components in the voxel time series that are induced by movement or shape differences among a series of scans. Voxel-based analyses assume that the data from a particular voxel all derive from the same part of the brain. Violations of this assumption will introduce artifactual changes in the voxel values that may obscure changes, or differences, of interest.. Even single-subject analyses proceed in a standard anatomical space, simply to enable reporting of regionally specific effects in a frame of reference that can be related to other studies.

The first step is to realign the data in order to "undo" the effects of subject movement during the scanning session. After realignment the data are then transformed using linear or nonlinear warps into a standard anatomical space. Finally, the data are usually spatially smoothed before entering the analysis proper.

A. Realignment

Changes in signal intensity over time, from any one voxel, can arise from head motion and this represents a serious confound, particularly in *f*MRI studies. Despite restraints on head movement, cooperative subjects still show displacements of up to a millimeter or so. Realignment involves (i) estimating the six parameters of an affine "rigid-body" transformation that minimizes the (sum of squared) differences between each successive scan and a reference scan (usually the first or the average of all scans in the time series) and (ii) applying the transformation by resampling the data using trilinear, sinc, or cubic spline interpolation. Estimation of the affine transformation is usually effected with a first-order approximation of the Taylor expansion of the effect of movement on signal intensity using the spatial derivatives of the images (see below). This allows for a simple iterative least-squares solution (that corresponds to a Gauss–Newton search) (Friston *et al.*, 1995a). For most imaging modalities this procedure is sufficient to realign scans to, in some instances, a hundred micrometers or so (Friston *et al.*, 1996a). However, in *f*MRI, even after perfect realignment, movement-related signals can still persist. This calls for a final step in which the data are adjusted for residual movement-related effects.

B. Adjusting for Movement-Related Effects in *f*MRI

In extreme cases as much as 90% of the variance, in a *f*MRI time series, can be accounted for by the effects of movement *after* realignment (Friston et al., 1996a). Causes of these movement-related components are due to movement effects that cannot be modeled using a *linear* affine model. These nonlinear effects include (i) subject movement between slice acquisition, (ii) interpolation artifacts (Grootoonk *et al.*, 2000), (iii) nonlinear distortion due to magnetic field inhomogeneities (Andersson *et al.*, 2001), and (iv) spin-excitation history effects (Friston *et al.*, 1996a). The latter can be pronounced if the TR (repetition time) approaches T_1, making the current signal a function of movement history. These multiple effects render the movement-related signal (y) a nonlinear function of displacement (x) in the nth and previous scans, $y_n = f(x_n, x_{n-1}, \ldots)$. By assuming a sensible form for this function, its parameters can be estimated using the observed time series and the estimated movement parameters x from the realignment procedure. The estimated movement-related signal is then simply subtracted from the original data. This adjustment can be carried out as a preprocessing step or embodied in model estimation during the analysis proper. The form for $f(x)$, proposed in Friston *et al.* (1996a), was a nonlinear autoregression model that used polynomial expansions to second order. This model was motivated by spin-excitation history effects and allowed displacement in previous scans to explain the current movement-related signal. However, it is also a reasonable model for many other sources of movement-related confounds. Generally, for TRs of several seconds, interpolation artifacts supersede (Grootoonk *et al.*, 2000) and first-order terms, comprising an expansion of the current displacement in terms of periodic basis functions, appear to be sufficient.

This subsection has considered *spatial* realignment. In multislice acquisition different slices are acquired at slightly different times. This raises the possibility of *temporal* realignment to ensure that the data from any given volume were sampled at the same time. This is usually performed using sinc interpolation over time and only when (i) the temporal dynamics of evoked responses are important and (ii) the TR is sufficiently small to permit interpolation. Generally timing effects of this sort are not considered problematic because they manifest as artifactual latency differences in evoked responses from region to region. Given that biophysical latency differences may be in the order of a few seconds, inferences about these differences are made only when comparing different trial types at the *same* voxel. Provided the effects of latency differences are modeled, this renders temporal realignment unnecessary in most instances.

C. Spatial Normalization

After realigning the data, a mean image of the series, or some other coregistered (e.g., a T_1-weighted) image, is used to estimate the warping parameters that map it onto a template that already conforms to some standard anatomical space (e.g., Talairach and Tournoux, 1988). This estimation can use a variety of models for the mapping, including (i) a 12-parameter affine transformation, in which the parameters constitute a spatial transformation matrix, (ii) low-frequency basis spatial functions (usually a discrete cosine set or polynomials), in which the parameters are the coefficients of the basis functions employed, and (ii) a vector field specifying the mapping for each control point (e.g., voxel). In the latter case, the parameters are vast in number and constitute a vector field that is bigger than the image itself. Estimation of the parameters of all these models can be accommodated in a simple Bayesian framework, in which one is trying to find the deformation parameters θ that have the maximum posterior probability $P(\theta|y)$ given the data y, where $P(\theta|y) = P(y|\theta)P(\theta)$. Put simply, one wants to find the deformation that is most likely given the data. This deformation can be found by maximizing the probability of getting the data, assuming the current estimate of the deformation is true, times the probability of that estimate being true. In practice the deformation is updated iteratively using a Gauss–Newton scheme to maximize $P(\theta|y)$. This involves jointly minimizing the likelihood and prior potentials $H(y|\theta) = -\ln P(y|\theta)$ and $H(\theta) = -\ln P(\theta)$. The likelihood potential is generally taken to be the sum of squared differences between the template and deformed image and reflects the probability of actually getting that image if the transformation was correct. The prior potential can be used to incorporate prior information about the likelihood of a given warp. Priors can be determined empirically or motivated by constraints on the mappings. Priors play a more essential role as the number of parameters specifying the mapping increases and are central to high-dimensional warping schemes (Ashburner *et al.*, 1997).

In practice most people use an affine or spatial basis function warps and use iterative least squares to minimize the posterior potential. A nice extension of this approach is that the likelihood potential can be refined and taken as difference between the index image and the best (linear) combination of templates (e.g., depicting gray, white, CSF, and skull tissue partitions). This models intensity differences that are unrelated to registration differences and allows different modalities to be coregistered (see Fig. 2).

A special consideration is the spatial normalization of brains that have gross anatomical pathology. This pathology can be of two sorts: (i) quantitative changes in the amount of a particular tissue compartment (e.g., cortical atrophy) or (ii) qualitative changes in anatomy involving the insertion or deletion of normal tissue compartments (e.g., ischemic tissue in stroke or cortical dysplasia). The former case is, generally, not problematic in the sense that changes in the amount of cortical tissue will not affect its optimum spatial location in reference to some template (and, even if it does, a disease-specific template is easily constructed). The second sort of pathology can introduce substantial "errors" in the normalization unless special precautions are taken. These usually involve imposing constraints on the warping to ensure that the pathology does not bias the deformation of undamaged tissue. This involves "hard" constraints implicit in using a small number of basis functions or "soft" constraints implemented by increasing the role of priors in Bayesian estimation. An alternative strategy is to use another modality that is less sensitive to the pathology as the basis of the spatial normalization procedure or to simply remove the damaged region from the estimation by masking it out.

D. Coregistration of Functional and Anatomical Data

It is sometimes useful to coregister functional and anatomical images. However, with echo-planar imaging, geometric distortions of T_2* images, relative to anatomical T_1-weighted data, are a particularly serious problem because of the very low frequency per point in the phase-encoding direction. Typically for echo-planar *f*MRI, magnetic field inhomogeneity, sufficient to cause dephasing of 2p through the slice, corresponds to an in-plane distortion of a voxel. "Unwarping" schemes have been proposed to correct for the distortion effects (Jezzard and Balaban, 1995). However, this distortion is not an issue if one spatially normalizes the functional data.

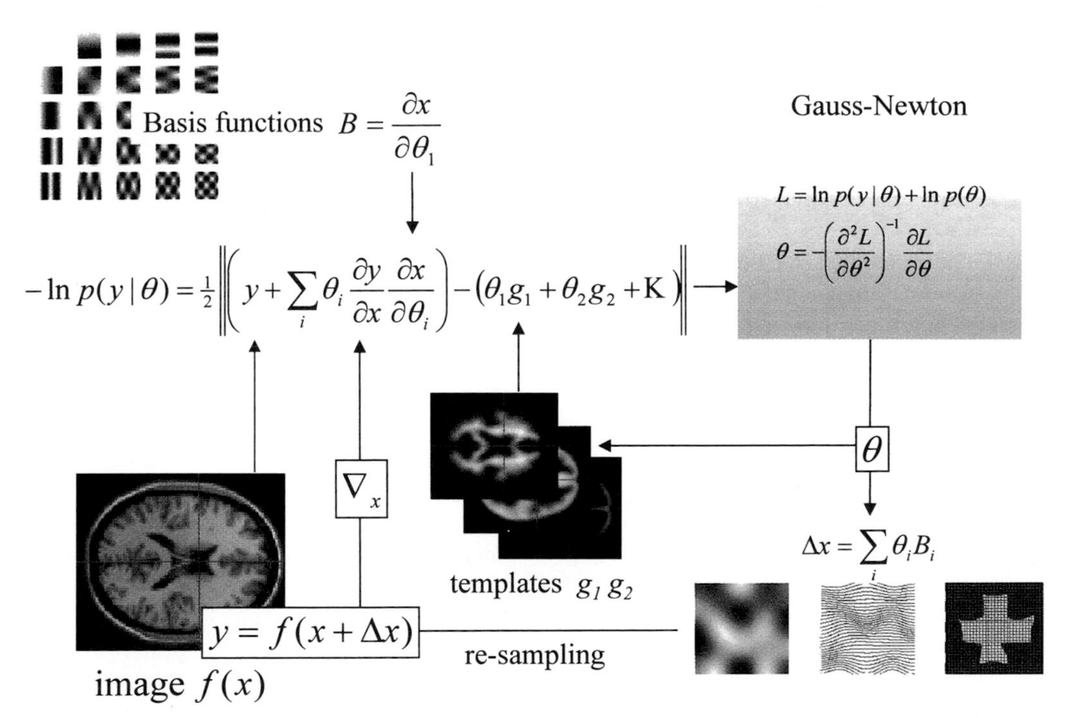

Figure 2 Schematic illustrating a Gauss–Newton scheme for maximizing the posterior probability P(θ|y) of the parameters for spatially normalizing an image. This scheme is iterative. At each step the conditional estimate of the parameters is obtained by jointly minimizing the likelihood and the prior potentials. The former is the difference between a resampled (i.e., warped) version *y* of the image *f* and the best linear combination of some template *g*. These parameters are used to mix the templates and resample the image to progressively reduce both the spatial and the intensity differences. After convergence the resampled image can be considered normalized.

E. Spatial Smoothing

The motivations for smoothing the data are fourfold: (i) By the matched-filter theorem, the optimum smoothing kernel corresponds to the size of the effect that one anticipates. The spatial scale of hemodynamic responses is, according to high-resolution optical imaging experiments, about 2 to 5 mm. Despite the potentially high resolution afforded by *f*MRI an equivalent smoothing is suggested for most applications. (ii) By the central limit theorem, smoothing the data will render the errors more normal in their distribution and ensure the validity of inferences based on parametric tests. (iii) When making inferences about regional effects using Gaussian random field theory (see Section IV) one of the assumptions is that the error terms are a reasonable lattice representation of an underlying and smooth Gaussian field. This necessitates smoothness to be substantially greater than voxel size. If the voxels are large, then they can be reduced by subsampling the data and smoothing (with the original point-spread function) with little loss of intrinsic resolution. (iv) In the context of intersubject averaging it is often necessary to smooth more (e.g., 8 mm in *f*MRI or 16 mm in PET) to project the data onto a spatial scale in which homologies in functional anatomy are expressed among subjects.

IV. Statistical Parametric Mapping

Functional mapping studies are usually analyzed with some form of statistical parametric mapping. Statistical parametric mapping refers to the construction of spatially extended statistical processes to test hypotheses about regionally specific effects (Friston *et al.*, 1991). Statistical parametric maps (SPMs) are image processes with voxel values that are, under the null hypothesis, distributed according to a known probability density function, usually the Student t or F distributions. These are known colloquially as t or F maps. The success of statistical parametric mapping is due largely to the simplicity of the idea. Namely, one analyzes each and every voxel using any standard (univariate) statistical test. The resulting statistical parameters are assembled into an image—the SPM. SPMs are interpreted as spatially extended statistical processes by referring to the probabilistic behavior of Gaussian fields (Adler, 1981; Worsley *et al.*, 1992, 1996; Friston *et al.*, 1994a). Gaussian random fields model both the univariate probabilistic characteristics of a SPM and any nonstationary spatial covariance structure. "Unlikely" excursions of the SPM are interpreted as regionally specific effects, attributable to the sensorimotor or cognitive process that has been manipulated experimentally.

Over the years statistical parametric mapping has come to refer to the conjoint use of the *general linear model* (GLM) and *Gaussian random field* (GRF) theory to analyze and make classical inferences about spatially extended data through SPMs. The GLM is used to estimate some parameters that could explain the data in exactly the same way as in conventional analysis of discrete data. GRF theory is used to resolve the multiple comparison problem that ensues when making inferences over a volume of the brain. GRF theory provides a method for correcting P values for the search volume of a SPM and plays the same role for *continuous* data (i.e., images) as the Bonferroni correction for the number of discontinuous or *discrete* statistical tests.

The approach was called SPM for three reasons: (i) to acknowledge *significance probability mapping*, the use of interpolated pseudo-maps of P values used to summarize the analysis of multichannel ERP studies, (ii) for consistency with the nomenclature of parametric maps of physiological or physical parameters (e.g., regional cerebral blood flow, rCBF, or volume, rCBV, parametric maps), and (iii) in reference to the *parametric* statistics that comprise the maps. Despite its simplicity there are some fairly subtle motivations for the approach that deserve mention. Usually, given a response or dependent variable comprising many thousands of voxels one would use *multivariate* analyses as opposed to the *mass-univariate* approach that SPM represents. The problems with multivariate approaches are that (i) they do not support inferences about regionally specific effects, (ii) they require more observations than the dimension of the response variable (i.e., number of voxels), and (iii) even in the context of dimension reduction, they are usually less sensitive to focal effects than mass-univariate approaches. A heuristic argument, for their relative lack of power, is that multivariate approaches estimate the model's error covariances using lots of parameters (e.g., the covariance between the errors at all pairs of voxels). In general, the more parameters (and hyperparameters) an estimation procedure has to deal with, the more variable the estimate of any one parameter becomes. This renders inferences about any single estimate less efficient.

An alternative approach would be to consider different voxels as different levels of an experimental or treatment factor and use classical analysis of variance, not at each voxel (cf. SPM), but by considering the data sequences from all voxels together, as replications over voxels. The problem here is that regional changes in error variance, and spatial correlations in the data, induce profound nonsphericity[1] in

[1]Sphericity refers to the assumption of identically and independently distributed error terms (iid). Under iid the probability density function of the errors, from all observations, has spherical isocontours, hence *sphericity*. Deviations from either of the iid criteria constitute nonsphericity. If the error terms are not identically distributed then different observations have different error variances. Correlations among error terms reflect dependencies among the error terms (e.g., serial correlation in *f*MRI time series) and constitute the second component of nonsphericity.

the error terms. This nonsphericity would again require large numbers of (hyper)parameters to be estimated for each voxel using conventional techniques. In SPM the nonsphericity is parameterized in the most parsimonious way with just two (hyper)parameters for each voxel. These are the error variance and smoothness estimators (see Section IV.B and Fig. 2). This minimal parameterization lends SPM a sensitivity that usually surpasses other approaches. SPM can do this because GRF theory implicitly imposes constraints on the nonsphericity implied by the continuous and (spatially) extended nature of the data. This is the only constraint on the behavior of the error terms implied by the use of GRF theory and is something that conventional multivariate and equivalent univariate approaches are unable to accommodate, to their cost.

Some analyses use statistical maps based on nonparametric tests that eschew distributional assumptions about the data. These approaches may, in some instances, be useful but are generally less powerful (i.e., less sensitive) than parametric approaches (see Aguirre *et al.*, 1998). Their original motivation in *f*MRI was based on the (specious) assumption that the residuals were not normally distributed. Next we consider parameter estimation in the context of the GLM. This is followed by an introduction to the role of GRF theory when making classical inferences about continuous data.

A. The General Linear Model

Statistical analysis of imaging data corresponds to (i) modeling the data to partition observed neurophysiological responses into components of interest, confounds, and error and (ii) making inferences about the interesting effects in relation to the error variance. This inference can be regarded as a direct comparison of the variance due to an interesting experimental manipulation with the error variance (cf. the F statistic and other likelihood ratios). Alternatively, one can view the statistic as an estimate of the response, or difference of interest, divided by an estimate of its standard deviation. This is a useful way to think about the t statistic.

A brief review of the literature may give the impression that there are numerous ways to analyze PET and *f*MRI time series with a diversity of statistical and conceptual approaches. This is not the case. With very few exceptions, every analysis is a variant of the general linear model. This includes (i) simple t tests on scans assigned to one condition or another, (ii) correlation coefficients between observed responses and boxcar stimulus functions in *f*MRI, (iii) inferences made using multiple linear regression, (iv) evoked responses estimated using linear time invariant models, and (v) selective averaging to estimate event-related responses in *f*MRI. Mathematically they are all identical. The use of the correlation coefficient deserves special mention because of its popularity in *f*MRI (Bandettini *et al.*, 1993). The significance of a correlation is identical to the significance of the equivalent t statistic testing for a regression of the data on the stimulus function. The correlation coefficient approach is useful but the inference is effectively based on a limiting case of multiple linear regression that obtains when there is only one regressor. In *f*MRI many regressors usually enter into a statistical model. Therefore, the t statistic provides a more versatile and generic way of assessing the significance of regional effects and is preferred over the correlation coefficient.

The general linear model is an equation, $Y = X\beta + \varepsilon$, that expresses the observed response variable Y in terms of a linear combination of explanatory variables X plus a well-behaved error term (see Fig. 3 and Friston *et al.*, 1995b). The general linear model is variously known as "analysis of covariance" or "multiple regression analysis" and subsumes simpler variants, like the "t test" for a difference in means, to more elaborate linear convolution models such as finite impulse response (FIR) models. The matrix X that contains the explanatory variables (e.g., designed effects or confounds) is called the *design matrix*. Each column of the design matrix corresponds to some effect one has built into the experiment or that may confound the results. These are referred to as explanatory variables, covariates, or regressors. The example in Fig. 1 relates to a *f*MRI study of visual stimulation under four conditions. The effects on the response variable are modeled in terms of functions of the presence of these conditions (i.e., boxcars smoothed with a hemodynamic response function) and constitute the first four columns of the design matrix. There then follows a series of terms that are designed to remove or model low-frequency variations in signal due to artifacts such as aliased biorhythms and other drift terms. The final column is whole-brain activity. The relative contribution of each of these columns is assessed using standard least squares, and inferences about these contributions are made using t or F statistics, depending upon whether one is looking at a particular linear combination (e.g., a subtraction) or all of them together. The operational equations are depicted schematically in Fig. 3. In this scheme the general linear model has been extended (Worsley and Friston, 1995) to incorporate intrinsic nonsphericity, or correlations among the error terms, and to allow for some specified temporal filtering of the data. This generalization brings with it the notion of *effective degrees of freedom*, which are less than the conventional degrees of freedom under identically and independently distributed assumptions (see footnote 1). They are smaller because the temporal correlations reduce the effective number of independent observations. The t and F statistics are constructed using Satterthwaite's approximation. This is the same approximation used in classical nonsphericity corrections such as the Geisser–Greenhouse correction. However, in the Worsley and Friston (1995) scheme, Satterthwaite's approximation is used to construct the statistics and appropriate degrees of freedom, not simply to provide a post hoc correction to the degrees of freedom.

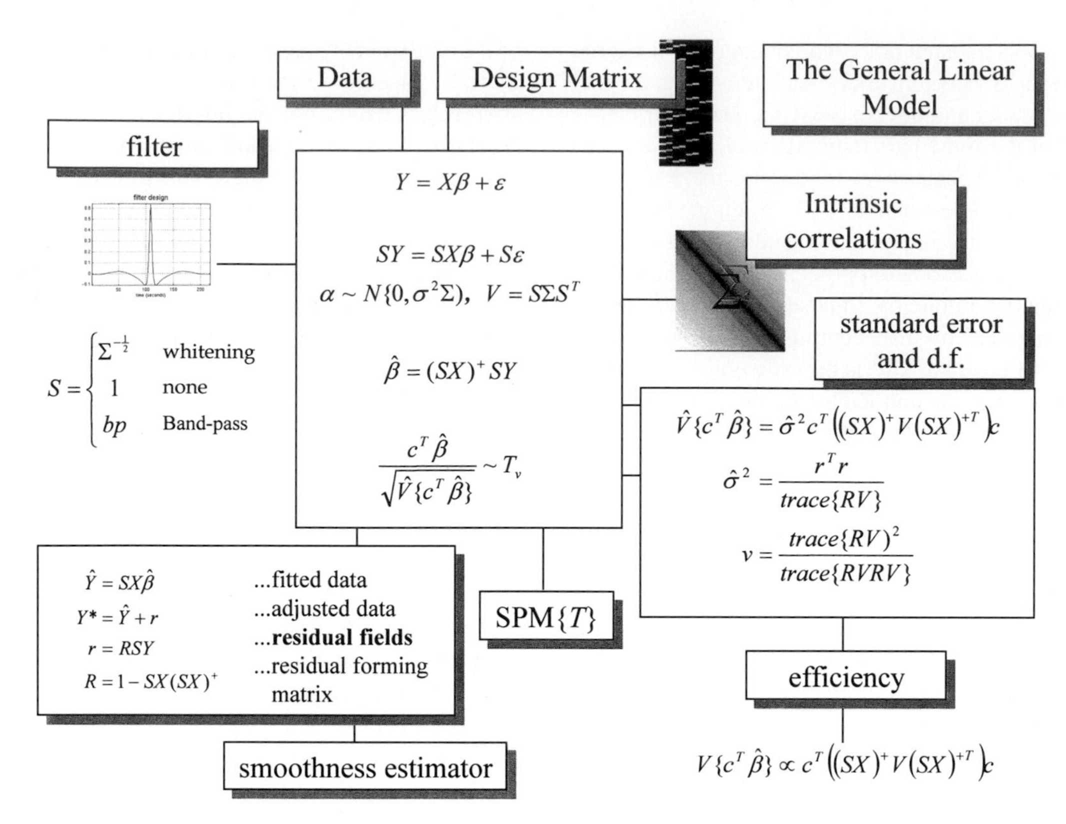

Figure 3 The general linear model. The general linear model is an equation expressing the response variable Y in terms of a linear combination of explanatory variables in a design matrix X and an error term with assumed or known autocorrelation Σ. In *f*MRI the data can be filtered with a convolution matrix S, leading to a generalized linear model that includes (intrinsic) serial correlations and applied (extrinsic) filtering. Different choices of S correspond to different (de)convolution schema as indicated on the upper left. The parameter estimates obtain in a least-squares sense using the pseudo-inverse (denoted by $^+$) of the filtered design matrix. Generally an effect of interest is specified by a vector of contrast weights c that give a weighted sum or compound of parameter estimates referred to as a *contrast*. The t statistic is simply this contrast divided by its estimated standard error (i.e., square root of its estimated variance). The ensuing t statistic is distributed with v degrees of freedom. The equations for estimating the variance of the contrast and the degrees of freedom associated with the error variance are provided on the right. Efficiency is simply the inverse of the variance of the contrast. These expressions are useful when assessing the relative efficiency of an experimental design. The parameter estimates can either be examined directly or used to compute the fitted responses (see lower left). "Adjusted data" refers to data from which estimated confounds have been removed. The residuals r obtain from applying the residual-forming matrix R to the data. These residual fields are used to estimate the smoothness of the component fields of the SPM used in Gaussian random field theory (see Fig. 6).

The equations summarized in Fig. 3 can be used to implement a vast range of statistical analyses. The issue is therefore not so much the mathematics but the formulation of a design matrix X appropriate to the study design and inferences that are sought. The design matrix can contain both covariates and indicator variables. Each column of X has an associated unknown parameter. Some of these parameters will be of interest (e.g., the effect of particular sensorimotor or cognitive condition or the regression coefficient of hemodynamic responses on reaction time). The remaining parameters will be of no interest and pertain to confounding effects (e.g., the effect of being a particular subject or the regression slope of voxel activity on global activity). Inferences about the parameter estimates are made using their estimated variance. This allows one to test the null hypothesis that all the estimates are zero using the F statistic to give an SPM$\{F\}$ or that some particular linear combination (e.g., a subtraction) of the estimates is zero using a SPM$\{t\}$. The t statistic obtains by dividing a contrast or compound (specified by contrast weights) of the ensuing parameter estimates by the standard error of that compound. The latter is estimated using the variance of the residuals about the least-squares fit. An example of a contrast weight *vector* would be [−1 1 0 0 ...] to compare the difference in responses evoked by two conditions, modeled by the first two condition-specific regressors in the design matrix. Sometimes several contrasts of parameter estimates are jointly interesting, for example, when using polynomial (Büchel *et al.*, 1996) or basis function expansions of some experimental factor. In these instances, the SPM$\{F\}$ is used and is specified with a matrix of contrast weights that can be thought of as a collection of "*t* contrasts" that one wants to test together. An "*F*-contrast" may look like

$$\begin{bmatrix} -1 & 0 & 0 & 0 & K \\ 0 & 1 & 0 & 0 & K \end{bmatrix},$$

which would test for the significance of the first or second parameter estimates. The fact that the first weight is –1 as opposed to 1 has no effect on the test because the *F* statistic is based on sums of squares.

In most analyses the design matrix contains indicator variables or parametric variables encoding the experimental manipulations. These are formally identical to classical analysis of (co)variance (i.e., AnCova) models. An important instance of the GLM, from the perspective of *f*MRI, is the linear time-invariant (LTI) model. Mathematically this is no different from any other GLM. However, it explicitly treats the data sequence as an ordered time series and enables a signal processing perspective that can be very useful.

In Friston *et al.* (1994b) the form of the hemodynamic impulse response function (HRF) was estimated using a least-squares deconvolution and a time-invariant model, in which evoked neuronal responses are convolved with the HRF to give the measured hemodynamic response (see also Boynton *et al.*, 1996). This simple linear framework is the cornerstone for making statistical inferences about activations in *f*MRI with the GLM. An impulse-response function is the response to a single impulse, measured at a series of times after the input. It characterizes the input–output behavior of the system (i.e., voxel) and places important constraints on the sorts of inputs that will excite a response. The HRFs estimated in Friston *et al.* (1994b) resembled a Poisson or gamma function, peaking at about 5 s. Our understanding of the biophysical and physiological mechanisms that underpin the HRF has grown considerably in the past few years (e.g., Buxton and Frank, 1997). Figure 4 shows some simulations based on the hemodynamic model described in Friston *et al.* (2000a). Here, neuronal activity induces some autoregulated signal that causes transient increases in rCBF. The resulting flow increases dilate the venous balloon, increasing its volume (v) and diluting venous blood to decrease deoxyhemoglobin content (q). The BOLD signal is roughly proportional to the concentration of deoxyhemoglobin (q/v) and follows the rCBF response with about a second's delay.

Knowing the forms that the HRF can take is important for several reasons, not least because it allows for better statistical models of the data. The HRF may vary from voxel to voxel and this has to be accommodated in the GLM. To allow for different HRFs in different brain regions the notion of temporal basis functions, to model evoked responses in *f*MRI, was introduced (Friston *et al.*, 1995c) and applied to event-related responses in Josephs *et al.* (1997) (see also Lange and Zeger, 1997). The basic idea behind temporal basis functions is that the hemodynamic response induced by any given trial type can be expressed as the linear combination of several (basis) functions of peristimulus time. The convolution model for *f*MRI responses takes a stimulus function encoding the supposed neuronal responses and convolves it with a HRF to give a regressor that enters into the design matrix. When using basis functions the stimulus function is convolved with all the basis functions to give a series of regressors. The associated parameter estimates are the coefficients or weights that determine the mixture of basis functions that best models the HRF for the trial type and voxel in question. We find the most useful basis set to be a canonical HRF and its derivatives with respect to the key parameters that determine its form (e.g., latency and dispersion). The nice thing about this

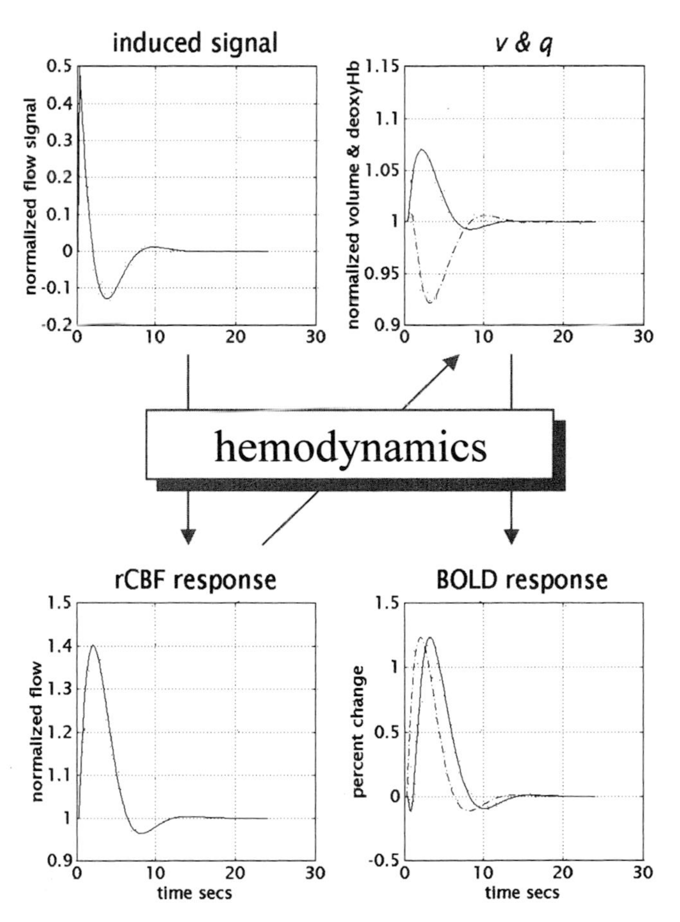

Figure 4 Hemodynamics elicited by an impulse of neuronal activity as predicted by a dynamic biophysical model (see Friston *et al.*, 2000a, for details). A burst of neuronal activity causes an increase in flow-inducing signal that decays with first-order kinetics and is down regulated by local flow. This signal increases rCBF, which dilates the venous capillaries, increasing its volume (v). Concurrently, venous blood is expelled from the venous pool, decreasing deoxyhemoglobin content (q). The resulting fall in deoxyhemoglobin concentration leads to a transient increase in BOLD (blood oxygenation level-dependent) signal and a subsequent undershoot.

approach is that it can partition differences among evoked responses into differences in magnitude, latency, or dispersion that can be tested for using specific contrasts and the SPM{t} (see Friston *et al.*, 1998b, for details).

Temporal basis functions are important because they enable a graceful transition between conventional multilinear regression models with one stimulus function per condition and FIR models with a parameter for each time point following the onset of a condition or trial type. Figure 5 illustrates this graphically (see figure legend). In summary, temporal basis functions offer useful constraints on the form of the estimated response that retain (i) the flexibility of FIR models and (ii) the efficiency of single regressor models. The advantage of using several temporal basis functions (as opposed to an assumed form for the HRF) is that one can model voxel-specific forms for hemodynamic responses and formal differences (e.g., onset latencies) among responses to different sorts of events. The advantages of using basis functions over FIR models are that (i) the parameters are estimated more efficiently and (ii) stimuli can be presented at any point in the interstimulus interval. The latter is important because time-locking stimulus presentation and data acquisition give a biased sampling over peristimulus time and can lead to differential sensitivities, in multislice acquisition, over the brain.

B. Statistical Inference and the Theory of Gaussian Fields

Inferences using SPMs can be of two sorts depending on whether one knows where to look in advance. With an anatomically constrained hypothesis, about effects in a particular brain region, the uncorrected P value associated with the height or extent of that region in the SPM can be used to test the hypothesis. With an anatomically open hypothesis (i.e., a null hypothesis that there is no effect anywhere in a specified volume of the brain) a correction for multiple dependent comparisons is necessary. The theory of Gaussian random fields provides a way of correcting the P value that takes into account the fact that neighboring voxels are not independent by virtue of continuity in the original data. Provided the data are sufficiently smooth the GRF correction is less severe (i.e., is more sensitive) than a Bonferroni correction for the number of voxels. As noted above GRF

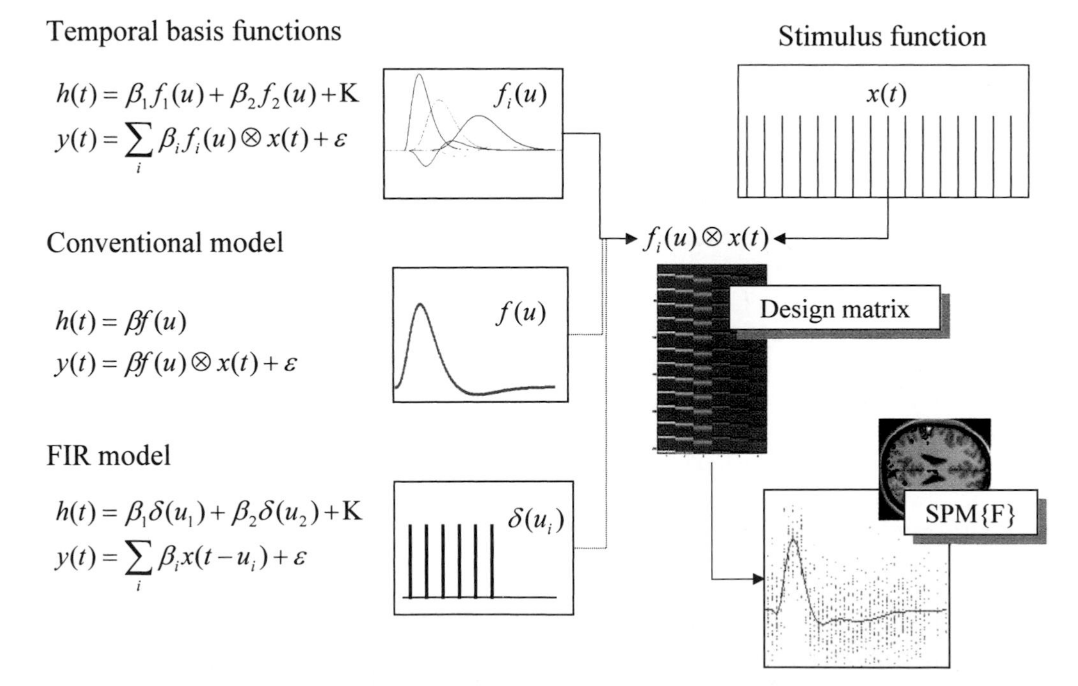

Figure 5 Temporal basis functions offer useful constraints on the form of the estimated response that retain (i) the flexibility of FIR models and (ii) the efficiency of single-regressor models. The specification of these constrained FIR models involves setting up stimulus functions $x(t)$ that model expected neuronal changes [e.g., boxcars of epoch-related responses or spikes (δ functions) at the onset of specific events or trials]. These stimulus functions are then convolved with a set of basis functions $f_i(u)$ of peristimulus time u, that model the HRF, in some linear combination. The ensuing regressors are assembled into the design matrix. The basis functions can be as simple as a single canonical HRF (middle), through to a series of delayed δ functions (bottom). The latter case corresponds to a FIR model and the coefficients constitute estimates of the impulse response function at a finite number of discrete sampling times, for the event or epoch in question. Selective averaging in event-related *f*MRI (Dale and Buckner, 1997) is mathematically equivalent to this limiting case.

theory deals with the multiple-comparisons problem in the context of continuous, spatially extended statistical fields, in a way that is analogous to the Bonferroni procedure for families of discrete statistical tests. There are many ways to appreciate the difference between GRF and Bonferroni corrections. Perhaps the most intuitive is to consider the fundamental difference between a SPM and a collection of discrete t values. When declaring a connected volume or region of the SPM to be significant, we refer collectively to all the voxels that comprise that volume. The false-positive rate is expressed in terms of connected (excursion) sets of voxels above some threshold, under the null hypothesis of no activation. This is not the expected number of false-positive voxels. One false-positive volume may contain hundreds of voxels, if the SPM is very smooth. A Bonferroni correction would control the expected number of false-positive voxels, whereas GRF theory controls the expected number of false-positive *regions*. Because a false-positive region can contain many voxels the corrected threshold under a GRF correction is much lower, rendering it much more sensitive. In fact the number of voxels in a region is somewhat irrelevant because it is a function of smoothness. The GRF correction discounts voxel size by expressing the search volume in terms of smoothness or resolution elements (*resels*). See Fig. 6. This intuitive perspective is expressed formally in terms of differential topology using the *Euler characteristic* (Worsley *et al.*, 1992). At high thresholds the Euler characteristic corresponds to the number of regions exceeding the threshold.

There are only two assumptions underlying the use of the GRF correction: (i) The error fields (but not necessarily the data) are a reasonable lattice approximation to an underlying random field with a multivariate Gaussian distribution. (ii) These fields are continuous, with a twice-differentiable autocorrelation function. A common misconception is that the autocorrelation function has to be Gaussian. It does not. The only way in which these assumptions can be violated is if (i) the data are not smoothed (with or without subsampling of the data to preserve resolution), violating the reasonable lattice assumption, or (ii) the statistical model is misspecified so that the errors are not normally distributed. Early formulations of the GRF correction were based on the assumption that the spatial correlation structure was wide-sense stationary. This assumption can now be relaxed due to a revision of the way in which the smoothness estimator enters the correction procedure (Kiebel *et al.*, 1999). In other words, the corrections retain their validity, even if the smoothness varies from voxel to voxel.

1. Anatomically Closed Hypotheses

When making inferences about regional effects (e.g., activations) in SPMs, one often has some idea about where the activation should be. In this instance a correction for the entire search volume is inappropriate. However, a problem remains in the sense that one would like to consider activations that are "near" the predicted location, even if they are not exactly coincident. There are two approaches one can adopt: (i) prespecify a small search volume and make the appropriate GRF correction (Worsley *et al.*, 1996) or (ii) used the uncorrected P value based on spatial extent of the nearest cluster (Friston, 1997). This probability is based on getting the observed number of voxels, or more, in a given cluster (conditional on that cluster existing). Both these procedures are based on distributional approximations from GRF theory.

2. Anatomically Open Hypotheses and Levels of Inference

To make inferences about regionally specific effects the SPM is thresholded, using some height and spatial extent thresholds that are specified by the user. Corrected P values can then be derived that pertain to (i) the number of activated regions (i.e., number of clusters above the height and volume threshold)—set level inferences, (ii) the number of activated voxels (i.e., volume) comprising a particular region—cluster level inferences, and (iii) the P value for each voxel within that cluster—voxel level inferences. These P values are corrected for the multiple dependent comparisons and are based on the probability of obtaining c, or more, clusters with k, or more, voxels, above a threshold u in an SPM of known or estimated smoothness. This probability has a reasonably simple form (see Fig. 6 for details).

Set level refers to the inference that the number of clusters comprising an observed activation profile is highly unlikely to have occurred by chance and is a statement about the activation profile, as characterized by its constituent regions. Cluster-level inferences are a special case of set-level inferences that obtain when the number of clusters $c = 1$. Similarly voxel-level inferences are special cases of cluster-level inferences that result when the cluster can be small (i.e., $k = 0$). Using a theoretical power analysis (Friston *et al.*, 1996b) of distributed activations, one observes that set-level inferences are generally more powerful than cluster-level inferences and that cluster-level inferences are generally more powerful than voxel-level inferences. The price paid for this increased sensitivity is reduced localizing power. Voxel-level tests permit individual voxels to be identified as significant, whereas cluster- and set-level inferences allow clusters or sets of clusters to be only declared significant. It should be remembered that these conclusions, about the relative power of different inference levels, are based on distributed activations. Focal activation may well be detected with greater sensitivity using voxel-level tests based on peak height. Typically, people use voxel-level inferences and a spatial extent threshold of zero. This reflects the fact that characterizations of functional anatomy are generally more useful when specified with a high degree of anatomical precision.

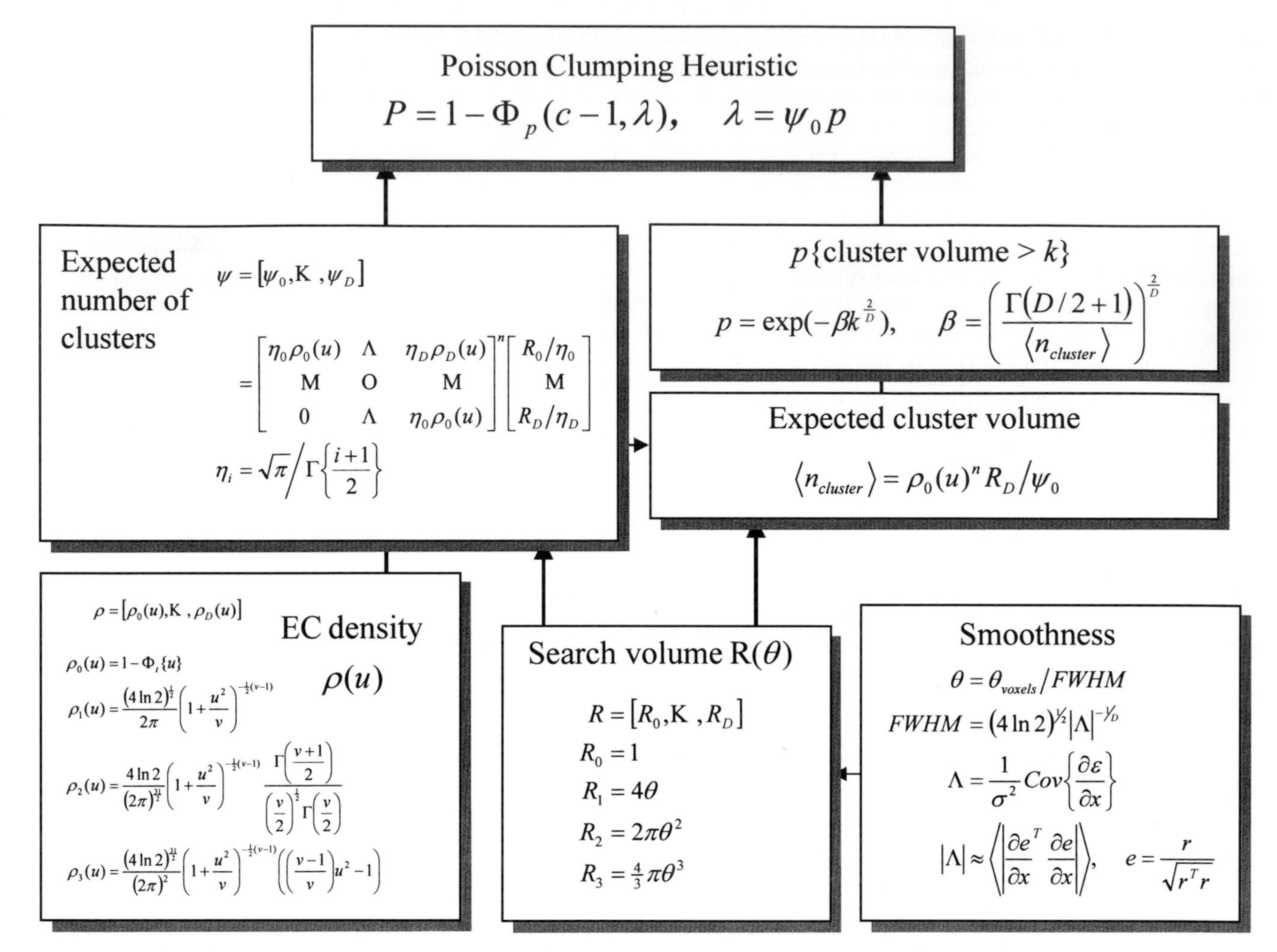

Figure 6 Schematic illustrating the use of Gaussian random field theory in making inferences about activations in SPMs. If one knew where to look exactly, then inference could be based on the value of the statistic at a specified location in the SPM, without correction. However, if one did not have an anatomical constraint *a priori*, then a correction for multiple dependent comparisons has to be made. These corrections are usually made using distributional approximations from GRF theory. This schematic deals with a general case of n SPM$\{t\}$'s whose voxels all survive a common threshold u (i.e., a conjunction of n component SPMs). The central probability, upon which all voxel, cluster, or set-level inferences are made, is the probability P of getting c or more clusters with k or more resels (resolution elements) above this threshold. By assuming that clusters behave like a multidimensional Poisson point process (i.e., the Poisson clumping heuristic) P is simply determined. The distribution of c is Poisson with an expectation that corresponds to the product of the expected number of clusters, of any size, and the probability that any cluster will be bigger than k resels. The latter probability is shown using a form for a single Z-variate field constrained by the expected number of resels per cluster (< > denotes expectation or average). The expected number of resels per cluster is simply the expected number of resels in total divided by the expected number of clusters. The expected number of clusters is estimated with the Euler characteristic (EC) (effectively the number of blobs minus the number of holes). This estimate is in turn a function of the EC density for the statistic in question (with degrees of freedom v) and the resel counts. The EC density is the expected EC per unit of D-dimensional volume of the SPM where the D-dimensional volume of the search space is given by the corresponding element in the vector of resel counts. Resel counts can be thought of as a volume metric that has been normalized by the smoothness of the SPM's component fields expressed in terms of the full width at half-maximum (FWHM). This is estimated from the determinant of the variance–covariance matrix of the first spatial derivatives of e, the normalized residual fields r (from Fig. 3). In this example equations for a sphere of radius θ are given. Φ denotes the cumulative density function for the subscripted statistic in question.

V. Experimental Design

This section considers the different sorts of designs that can be employed in neuroimaging studies. Experimental designs can be classified as *single-factor* or *multifactorial* designs; within this classification the levels of each factor can be *categorical* or *parametric*. We will start by discussing categorical and parametric designs and then deal with multifactorial designs.

A. Categorical Designs, Cognitive Subtraction, and Conjunctions

The tenet of cognitive subtraction is that the difference between two tasks can be formulated as a separable cognitive or sensorimotor component and that regionally specific differences in hemodynamic responses, evoked by the two tasks, identify the corresponding functionally specialized area. Early applications of subtraction range from the functional anatomy of word processing (Petersen *et al.*, 1989) to functional specialization in extrastriate cortex (Lueck *et al.*, 1989). The latter studies involved presenting visual stimuli with and without some sensory attribute (e.g., color, motion). The areas highlighted by subtraction were identified with homologous areas in monkeys that showed selective electrophysiological responses to equivalent visual stimuli.

Cognitive conjunctions (Price and Friston, 1997) can be thought of as an extension of the subtraction technique, in the sense that they combine a series of subtractions. In subtraction one tests a *single* hypothesis pertaining to the activation in one task relative to another. In conjunction analyses *several* hypotheses are tested, asking whether all the activations, in a series of task pairs, are jointly significant. Consider the problem of identifying regionally specific activations due to a particular cognitive component (e.g., object recognition). If one can identify a series of task pairs whose differences have only that component in common, then the region which activates, in all the corresponding subtractions, can be associated with the common component. Conjunction analyses allow one to demonstrate the context-invariant nature of regional responses. One important application of conjunction analyses is in multisubject *f*MRI studies, in which generic effects are identified as those that are conjointly significant in all the subjects studied (see Section VII).

B. Parametric Designs

The premise behind parametric designs is that regional physiology will vary systematically with the degree of cognitive or sensorimotor processing or deficits thereof. Examples of this approach include the PET experiments of Grafton *et al.* (1992) that demonstrated significant correlations between hemodynamic responses and the performance of a visually guided motor tracking task. On the sensory side Price *et al.* (1992) demonstrated a remarkable linear relationship between perfusion in periauditory regions and frequency of aural word presentation. This correlation was not observed in Wernicke's area, where perfusion appeared to correlate, not with the discriminative attributes of the stimulus, but with the presence or absence of semantic content. These relationships or *neurometric functions* may be linear or nonlinear. Using polynomial regression, in the context of the GLM, one can identify nonlinear relationships between stimulus parameters (e.g., stimulus duration or presentation rate) and evoked responses. To do this one usually uses a SPM{F} (see Büchel *et al.*, 1996).

The example provided in Fig. 7 illustrates both categorical and parametric aspects of design and analysis. These data were obtained from a *f*MRI study of visual motion processing using radially moving dots. The stimuli were presented over a range of speeds using *isoluminant* and *isochromatic* stimuli. To identify areas involved in visual

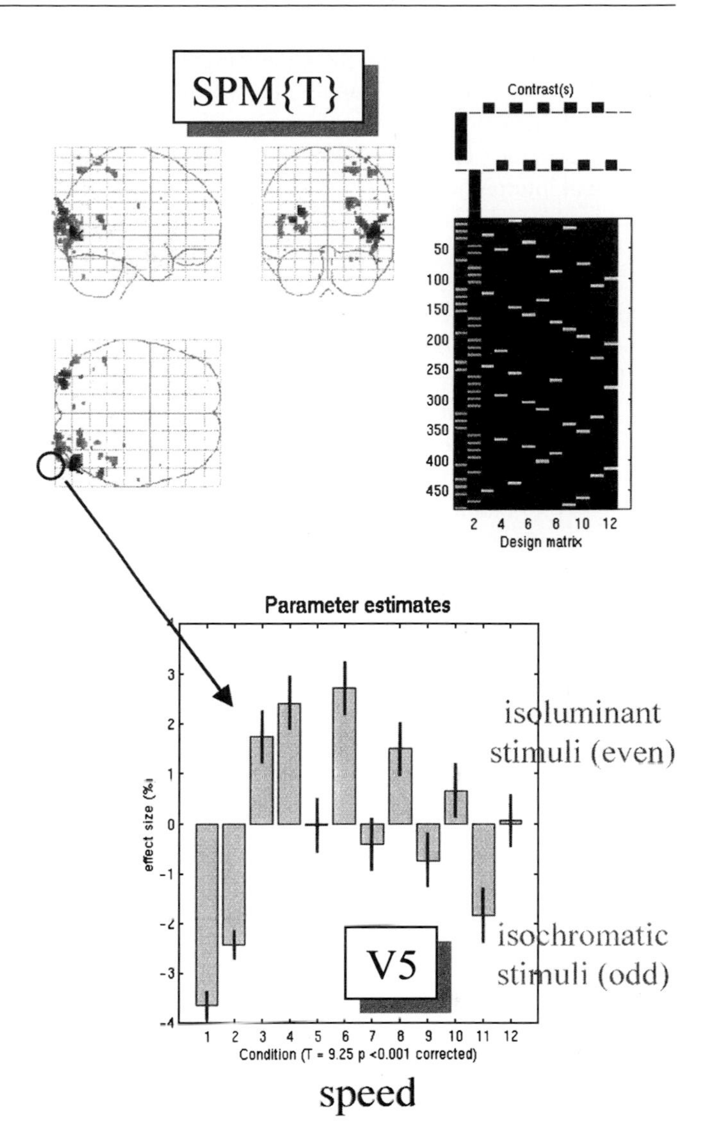

Figure 7 **Top right**: Design matrix. This is an image representation of the design matrix. Contrasts are the vectors of contrast weights defining the linear compounds of parameters tested. The contrast weights are displayed over the column of the design matrix that corresponds to the effects in question. The design matrix here includes condition-specific effects (boxcars convolved with a hemodynamic response function). Odd columns correspond to stimuli shown under isochromatic conditions and even columns model responses to isoluminant stimuli. The first two columns are for stationary stimuli and the remaining columns are for conditions of increasing speed. The final column is a constant term. **Top left**: SPM{t}. This is a maximum intensity projection of the SPM{t} conforming to the standard anatomical space of Talairach and Tournoux (1988). The t values here are the minimum t values from both contrasts, thresholded at P = 0.001 uncorrected. The most significant conjunction is seen in left V5. Bottom: Plot of the condition-specific parameter estimates for this voxel. The t value was 9.25 (P < 0.001 corrected—see Fig. 6).

motion a stationary dots condition was subtracted from the moving dots conditions (see the contrast weights on the upper right). To ensure significant motion-sensitive responses, using both color and luminance cues, a conjunction of the equivalent subtractions was assessed under both viewing contexts. Areas V5 and V3a are seen in the ensuing

SPM$\{t\}$. The t values in this SPM are simply the minimum of the t values for each subtraction. Thresholding this SPM$\{t_{\text{min}}\}$ ensures that all voxels survive the threshold u in each subtraction separately. This *conjunction* SPM has an equivalent interpretation; it represents the intersection of the excursion sets, defined by the threshold u, of each *component* SPM. This intersection is the essence of a conjunction. The expressions in Fig. 6 pertain to the general case of the minimum of n t values. The special case where $n = 1$ corresponds to a conventional SPM$\{t\}$.

The responses in left V5 are shown at the bottom of Fig. 7 and speak to a compelling inverted "U" relationship between speed and evoked response that peaks at around 8° per second. It is this sort of relationship that parametric designs try to characterize. Interestingly the form of these speed-dependent responses was similar using both stimulus types, although luminance cues are seen to elicit a greater response. From the point of view of a factorial design there is a main effect of cue (isoluminant vs isochromatic), a main (nonlinear) effect of speed, but no speed by cue interaction.

Clinical neuroscience studies can use parametric designs by looking for the neuronal correlates of clinical (e.g., symptom) ratings over subjects. In many cases multiple clinical scores are available for each subject and the statistical design can usually be seen as a multilinear regression. In situations in which the clinical scores are correlated principal component analysis or factor analysis is sometimes applied to generate a new, and smaller, set of explanatory variables that are orthogonal to each other. This has proved particularly useful in psychiatric studies in which syndromes can be expressed over a number of different dimensions (e.g., the degree of psychomotor poverty, disorganization, and reality distortion in schizophrenia; see Liddle *et al.*, 1992). In this way, regionally specific correlates of various symptoms may point to their distinct pathogenesis in a way that transcends the syndrome itself. For example psychomotor poverty may be associated with left dorsolateral prefrontal dysfunction irrespective of whether the patient is suffering from schizophrenia or depression.

C. Multifactorial Designs

Factorial designs are becoming more prevalent than single-factor designs because they enable inferences about interactions. At its simplest an interaction represents a change in a change. Interactions are associated with factorial designs in which two or more factors are combined in the same experiment. The effect of one factor on the effect of the other is assessed by the interaction term. Factorial designs have a wide range of applications. An early application, in neuroimaging, examined physiological adaptation and plasticity during motor performance, by assessing time by condition interactions (Friston *et al.*, 1992a). Psychopharmacological activation studies are further examples of factorial designs (Friston *et al.*, 1992b). In these studies cognitively evoked responses are assessed before and after being given a drug. The interaction term reflects the pharmacological modulation of task-dependent activations. Factorial designs have an important role in the context of cognitive subtraction and additive factors logic by virtue of being able to test for interactions, or context-sensitive activations (i.e., to demonstrate the fallacy of "pure insertion"; see Friston *et al.*, 1996c). These interaction effects can sometimes be interpreted as (i) the integration of the two or more (cognitive) processes or (ii) the modulation of one (perceptual) process by another being manipulated. See Fig. 8 for an example. From the point of view of clinical studies interactions are central. The effect of a disease process on sensorimotor or cognitive activation is simply an interaction and involves replicating a subtraction experiment in subjects with and without the pathophysiology being studied. Factorial designs can also embody parametric factors. If one of the factors has a number of parametric levels, the interaction can be expressed as a difference in regression slope of regional activity on the parameter, under both levels of the other (categorical) factor. An important example of factorial designs that mix categorical and parameter factors are those looking for *psychophysiological interactions*. Here the parametric factor is brain activity measured in a particular brain region. These designs have proven useful in looking at the interaction between bottom-up and top-down influences within processing hierarchies in the brain (Friston *et al.*, 1997). This issue will be addressed below in Section VIII from the point of view of effective connectivity.

VI. Designing fMRI Studies

In this section we consider *f*MRI time series from a signal-processing perspective with particular reference to optimal experimental design and efficiency. *f*MRI time series can be viewed as a linear admixture of signal and noise. Signal corresponds to neuronally mediated hemodynamic changes that can be modeled as a (non)linear convolution of some underlying neuronal process, responding to changes in experimental factors, by a HRF. Noise has many contributions that render it rather complicated in relation to other neurophysiological measurements. These include neuronal and nonneuronal sources. Neuronal noise refers to neurogenic signal not modeled by the explanatory variables and has the same frequency structure as the signal itself. Nonneuronal components have both white [e.g., RF (Johnson) noise] and colored components (e.g., pulsatile motion of the brain caused by cardiac cycles and local modulation of the static magnetic field B_0 by respiratory movement). These effects are typically low frequency or wide-band (e.g., aliased cardiac-locked pulsatile motion). The superposition of all these components induces temporal correlations among the error terms (denoted by Σ in Fig. 3)

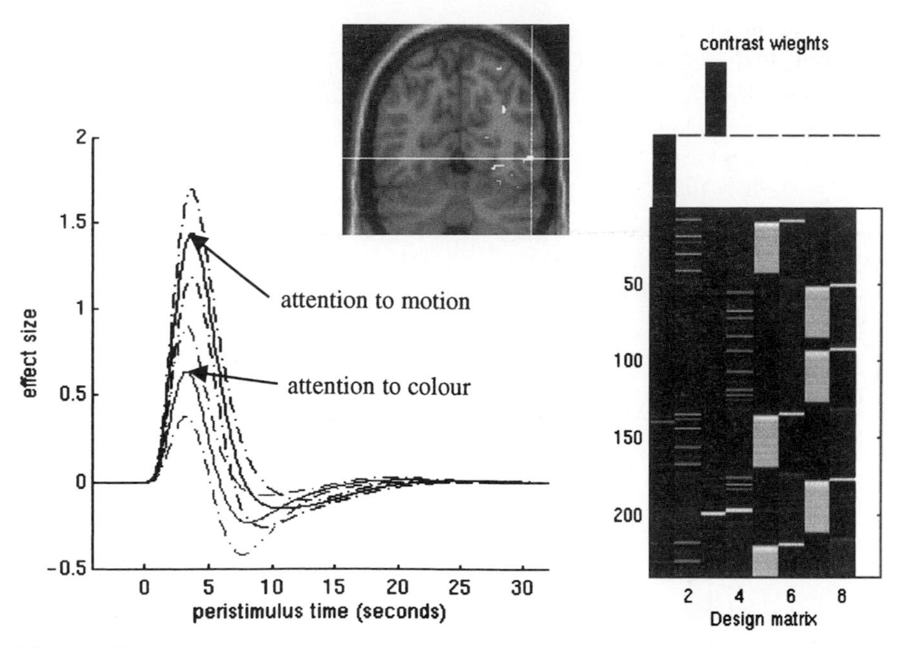

Figure 8 Interactions between set- and event-related responses: attentional modulation of V5 responses. Results showing how to assess an interaction using an event-related design. Subjects viewed stationary monochromatic stimuli that occasionally changed color and moved at the same time. These compound events were presented under two levels of attentional set (attention to color and attention to motion). The event-related responses are modeled, in an attention-specific fashion, by the first four regressors (δ functions convolved with a hemodynamic response function and its derivative) in the design matrix on the right. The simple main effects of attention are modeled as similarly convolved boxcars. The interaction between attentional set and visually evoked responses is simply the difference in evoked responses under both levels of attention and is tested for with the appropriate contrast weights (upper right). Only the first 256 rows of the design matrix are shown. The most significant modulation of evoked responses under attention to motion was seen in left V5 (inset). The fitted responses and their standard errors are shown on the left as functions of peristimulus time.

that can affect sensitivity to experimental effects. Sensitivity depends upon (i) the relative amounts of signal and noise and (ii) the efficiency of the experimental design. Efficiency is simply a measure of how reliable the parameter estimates are and can be defined as the inverse of the variance of a contrast of parameter estimates (see Fig. 3). There are two important considerations that arise from this perspective on *f*MRI time series: The first pertains to optimal experimental design and the second to optimum (de)convolution of the time series to obtain the most efficient parameter estimates.

A. The Hemodynamic Response Function and Optimum Design

As noted above, a LTI model of neuronally mediated signals in *f*MRI suggests that, whatever the frequency structure of experimental variables, only those that survive convolution with the HRF can be estimated with any efficiency. By convolution theorem the experimental variance should therefore be designed to match the transfer function of the HRF. The corresponding frequency profile of this transfer function is shown in Fig. 9—solid line). It is clear that frequencies around 0.03 Hz are optimal, corresponding to periodic designs with 32-s periods (i.e., 16-s epochs). Generally, the first objective of experimental design is to comply with the natural constraints imposed by the HRF and ensure that experimental variance occupies these intermediate frequencies.

B. Serial Correlations and Filtering

This is quite a complicated but important area. Conventional signal processing approaches dictate that whitening the data engenders the most efficient parameter estimation. This corresponds to filtering with a convolution matrix S (see Fig. 3) that is the inverse of the intrinsic convolution matrix K ($KK^T = \Sigma$). This *whitening* strategy renders the generalized least-squares estimator in Fig. 3 equivalent to the Gauss–Markov estimator or minimum variance estimator. However, one generally does not know the form of the intrinsic correlations, which means it has to be estimated. This estimation usually proceeds using a restricted maximum likelihood (ReML) estimate of the serial correlations, among the

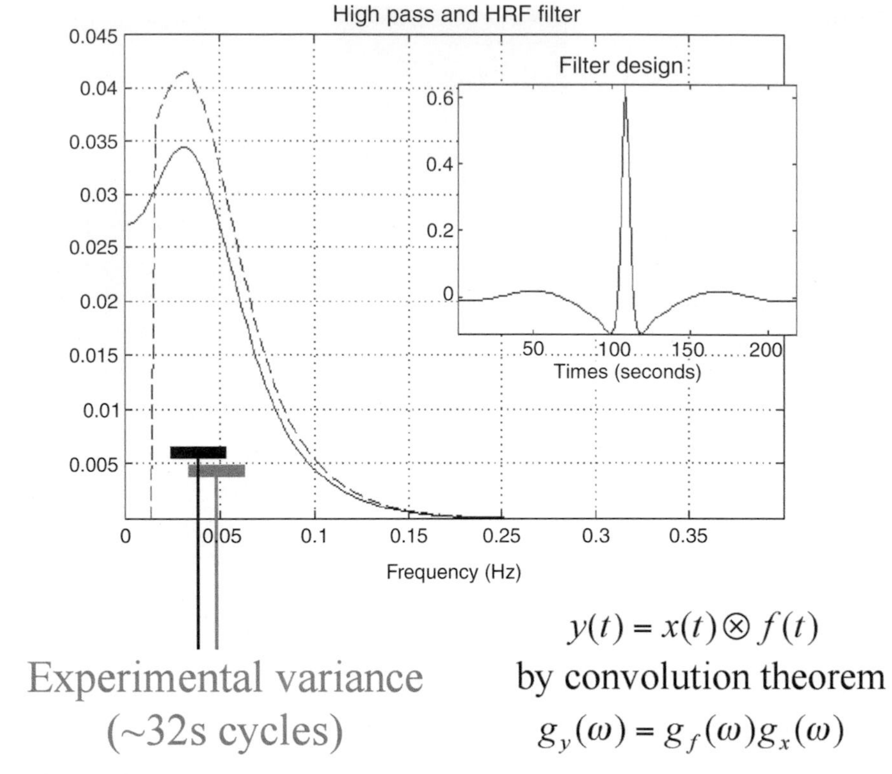

Figure 9 A signal-processing perspective. Modulation transfer function of a canonical hemodynamic response function (HRF), with (broken line) and without (solid line) the addition of a high-pass filter. This transfer function corresponds to the spectral density of a white noise process after convolution with the HRF and places constraints on the frequencies that survive convolution with the HRF. This follows from the convolution theorem (summarized in the equations). The inset is the filter expressed in time, corresponding to the spectral density that obtains after convolution with the HRF and high-pass filtering.

residuals, that properly accommodates the effects of the residual-forming matrix and associated loss of degrees of freedom. However, using this estimate of the intrinsic nonsphericity to form a Gauss–Markov estimator at each voxel has several problems. (i) First the estimate of nonsphericity can itself be very inefficient, leading to bias in the standard error (Friston *et al.*, 2000b). (ii) ReML estimation requires an iterative procedure at every voxel and this is computationally prohibitive. (iii) Adopting a different form for the serial correlations at each voxel means the effective degrees of freedom and the null distribution of the statistic will change from voxel to voxel. This violates the assumptions of GRF results for *t* and *F* fields (although not very seriously). There are a number of different approaches to these problems that aim to increase the efficiency of the estimation and reduce the computational burden. The approach we have chosen is to forgo the efficiency of the Gauss–Markov estimator and use a generalized least-square GLS estimator, after approximately whitening the data with a high-pass filter. The GLS estimator is unbiased and, luckily, is identical to the Gauss–Markov estimator if the regressors in the design matrix are periodic.[2] After GLS estimation, Σ is estimated using ReML and the resulting estimate of $V = S\Sigma S^T$ entered into the expression for the standard error and degrees of freedom provided in Fig. 3. To ensure this nonsphericity estimate is robust, we assume it is the same at all voxels. Clearly this is an approximation but can be motivated by the fact we have applied the same high-pass temporal convolution matrix S to all voxels. This ameliorates any voxel-to-voxel variations in $V = S\Sigma S^T$ (see Fig. 3).

The reason that high-pass filtering approximates a whitening filter is that there is a preponderance of low frequencies in the noise. *f*MRI noise has been variously characterized as a 1/*f* process (Zarahn *et al.*, 1997) or an autoregressive process (Bullmore *et al.*, 1996) with white noise (Purdon and Weisskoff, 1998). Irrespective of the exact form these serial correlations take, high-pass filtering suppresses low-frequency components in the same way that whitening would. An example of a band-pass filter with a

[2]More exactly, the GLS and ML estimators are the same if X lies within the space spanned by the eigenvectors of Toeplitz autocorrelation matrix Σ.

high-pass cut-off of 1/64 Hz is shown in the inset of Fig. 7. This filter's transfer function (the broken line in the main panel) illustrates the frequency structure of neurogenic signals after high-pass filtering.

C. Spatially Coherent Confounds and Global Normalization

Implicit in the use of high-pass filtering is the removal of low-frequency components that can be regarded as confounds. Other important confounds are signal components that are artifactual or have no regional specificity. These are referred to as *global confounds* and have a number of causes. These can be divided into physiological (e.g., global perfusion changes in PET, mediated by changes in pCO_2) and nonphysiological (e.g., transmitter power calibration, B_1 coil profile, and receiver gain in *f*MRI). The latter generally scale the signal before the MRI sampling process. Other nonphysiological effects may have a nonscaling effect (e.g., Nyquist ghosting, movement-related effects). In PET it is generally accepted that regional changes in rCBF, evoked neuronally, mix additively with global changes to give the measured signal. This calls for a global normalization procedure in which the global estimator enters into the statistical model as a confound. In *f*MRI, instrumentation effects that scale the data motivate a global normalization by proportional scaling, using the whole brain mean, before the data enter into the statistical model.

It is important to differentiate between global confounds and their estimators. By definition the global mean over intracranial voxels will subsume regionally specific effects. This means that the global estimator may be partially collinear with effects of interest, especially if the evoked responses are substantial and widespread. In these situations global normalization may induce apparent deactivations in regions *not* expressing a physiological response. These are not artifacts in the sense they are real, relative to global changes, but they have little face validity in terms of the underlying neurophysiology. In instances in which regionally specific effects bias the global estimator, some investigators prefer to omit global normalization. Provided drift terms are removed from the time series, this is generally acceptable because most global effects have slow time constants. However, the issue of normalization-induced deactivations is better circumnavigated with experimental designs that use well-controlled conditions, which elicit differential responses in restricted brain systems.

D. Nonlinear System Identification Approaches

So far we have considered only LTI models and first-order HRFs. Another signal-processing perspective is provided by nonlinear system identification (Vazquez and Noll, 1998). This section considers nonlinear models as a prelude to the next subsection on event-related *f*MRI, in which nonlinear interactions among evoked responses provide constraints for experimental design and analysis. We have described an approach to characterizing evoked hemodynamic responses in *f*MRI based on nonlinear system identification, in particular the use of *Volterra series* (Friston *et al.*, 1998a). This approach enables one to estimate Volterra kernels that describe the relationship between stimulus presentation and the hemodynamic responses that ensue. Volterra series are essentially high-order extensions of linear convolution models. These kernels therefore represent a nonlinear characterization of the HRF that can model the responses to stimuli in different contexts and interactions among stimuli. In *f*MRI, the kernel coefficients can be estimated by (i) using a second-order approximation to the Volterra series to formulate the problem in terms of a general linear model and (ii) expanding the kernels in terms of temporal basis functions. This allows the use of the standard techniques described above to estimate the kernels and to make inferences about their significance on a voxel-specific basis using SPMs.

One important manifestation of the nonlinear effects, captured by the second-order kernels, is a modulation of stimulus-specific responses by preceding stimuli that are proximate in time. This means that responses at high stimulus presentation rates saturate and, in some instances, show an inverted U behavior. This behavior appears to be specific to BOLD effects (as distinct from evoked changes in cerebral blood flow) and may represent a hemodynamic refractoriness. This effect has important implications for event-related *f*MRI, in which one may want to present trials in quick succession (see below).

The results of a typical nonlinear analysis are given in Fig. 10. The results on the right represent the average response, integrated over a 32-s train of stimuli as a function of stimulus onset asynchrony (SOA) within that train. These responses were based on the kernel estimates (Fig. 10, left) using data from a voxel in the left posterior temporal region of a subject obtained during the presentation of single words at different rates. The solid line represents the estimated response and shows a clear maximum at just less than 1 s. The dots are responses based on empirical data from the same experiment. The broken line shows the expected response in the absence of nonlinear effects (i.e., that predicted by setting the second-order kernel to zero). It is clear that nonlinearities become important at around 2s, leading to an actual diminution of the integrated response at subsecond SOAs. The implication of this sort of result is that (i) SOAs should not really fall much below 1 s and (ii) at short SOAs the assumptions of linearity are violated. It should be noted that these data pertain to single-word processing in auditory association cortex. More linear behaviors may be expressed in primary sensory cortex where the feasibility of using minimum SOAs as low as 500 ms has been demonstrated

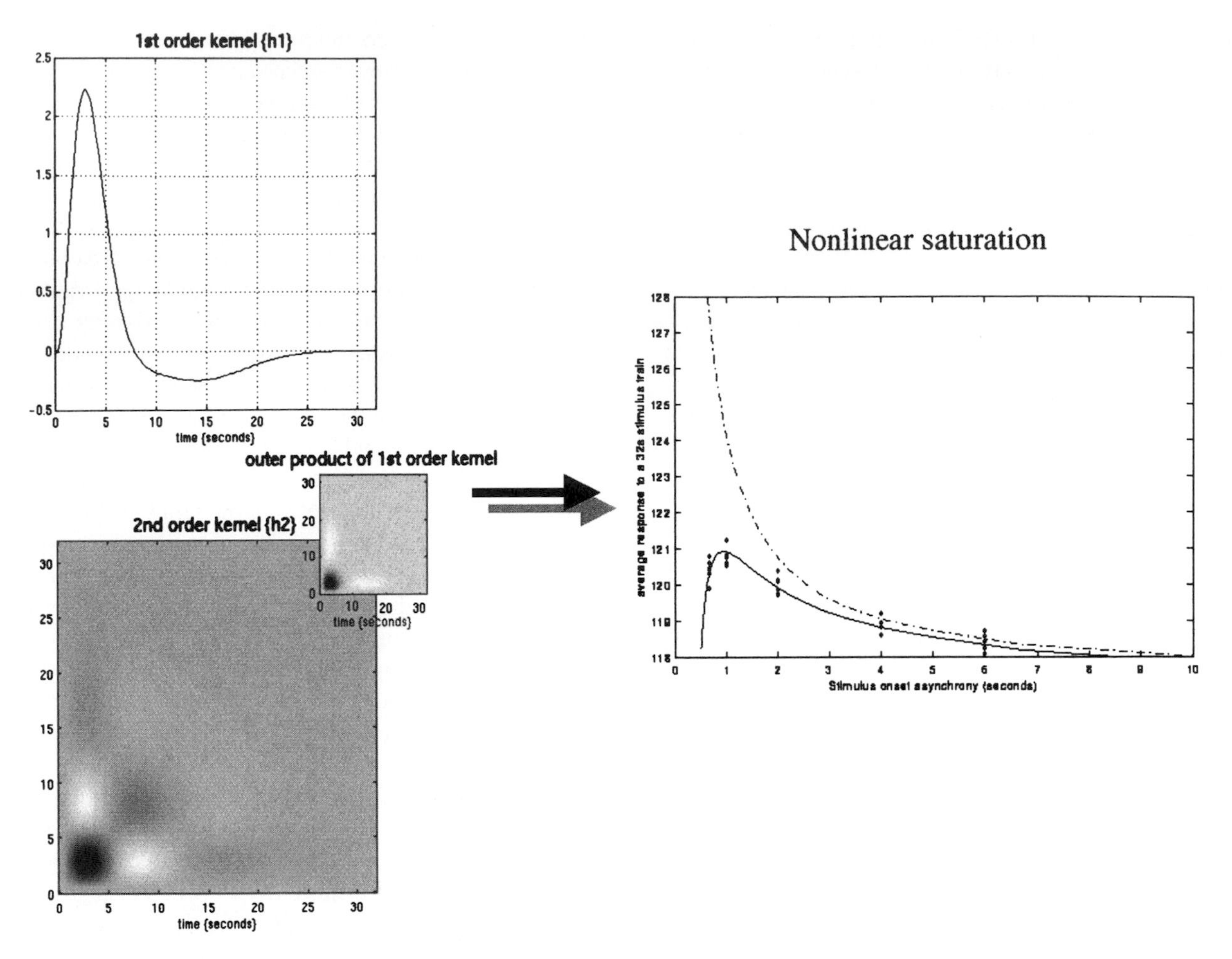

Figure 10 Left: Volterra kernels from a voxel in the left superior temporal gyrus at –56, –28, 12 mm. These kernel estimates were based on a single-subject study of aural word presentation at different rates (from 0 to 90 words per minute) using a second-order approximation to a Volterra series expansion modeling the observed hemodynamic response to stimulus input (a δ function for each word). These kernels can be thought of as a characterization of the second-order hemodynamic response function. The first-order kernel κ_1 (top) represents the (first-order) component usually presented in linear analyses. The second-order kernel (bottom) is presented in image format. The color scale is arbitrary; white is positive and black is negative. The inset on the right represents $\kappa_1\kappa_1^T$, the second-order kernel that would be predicted by a simple model that involved linear convolution with κ_1 followed by some static nonlinearly. **Right:** Integrated responses over a 32-s stimulus train as a function of SOA. Solid line, estimates based on the nonlinear convolution model parameterized by the kernels on the left. Broken line, the responses expected in the absence of second-order effects (i.e., in a truly linear system). Dots, empirical averages based on the presentation of actual stimulus trains.

(Burock *et al.*, 1998). This lower bound on SOA is important because some effects are detected more efficiently with high presentation rates. We now consider this from the point of view of event-related designs.

E. Event- and Epoch-Related Designs

A crucial distinction, in experimental design for *f*MRI, is that between *epoch*- and *event*-related designs. In SPECT and PET only epoch-related responses can be assessed because of the relatively long half-life of the radiotracers used. However, in *f*MRI there is an opportunity to measure event-related responses that may be important in some cognitive and clinical contexts. An important issue, in event-related *f*MRI, is the choice of interstimulus interval or more precisely SOA. The SOA, or the distribution of SOAs, is a critical factor in experimental design and is chosen, subject to psychological or psychophysical constraints, to maximize the efficiency of response estimation. The constraints on the SOA clearly depend upon the nature of the experiment but are generally satisfied when the SOA is small and derives from a random distribution. Rapid presentation rates allow for the maintenance of a particular cognitive or attentional set, decrease the latitude that the subject has for engaging alternative strategies, or incidental processing, and allows the integration of event-related paradigms using *f*MRI and electrophysiology. Random SOAs ensure that preparatory or anticipatory factors do not confound event-related responses and ensure a uniform context in which events are presented. These constraints speak to the well-documented advantages of event-related *f*MRI over conventional blocked designs (Buckner *et al.*, 1996, Clark *et al.*, 1998).

In order to compare the efficiency of different designs it is useful to have some common framework that accommodates them all. The efficiency can then be examined in relation to the parameters of the designs. Designs can be *stochastic* or *deterministic* depending on whether there is a random element to their specification. In stochastic designs (Heid *et al.*, 1997) one needs to specify the probabilities of an event occurring at all times those events could occur. In deterministic designs the occurrence probability is unity and the design is completely specified by the times of stimulus presentation or trials. The distinction between stochastic and deterministic designs pertains to how a particular realization or stimulus sequence is created. The efficiency afforded by a particular event sequence is a function of the event sequence itself and not of the process generating the sequence (i.e., deterministic or stochastic). However, within stochastic designs, the design matrix X, and associated efficiency, are random variables, and the *expected* or average efficiency over realizations of X is easily computed.

$$X = SB$$

$$1/\text{Efficiency} \propto c^T \left\langle X^T X \right\rangle^{-1} c$$

$$\left\langle X^T X \right\rangle = \left\langle B^T S^T S B \right\rangle = p^T (S^T S - 1) p + diag(p)$$

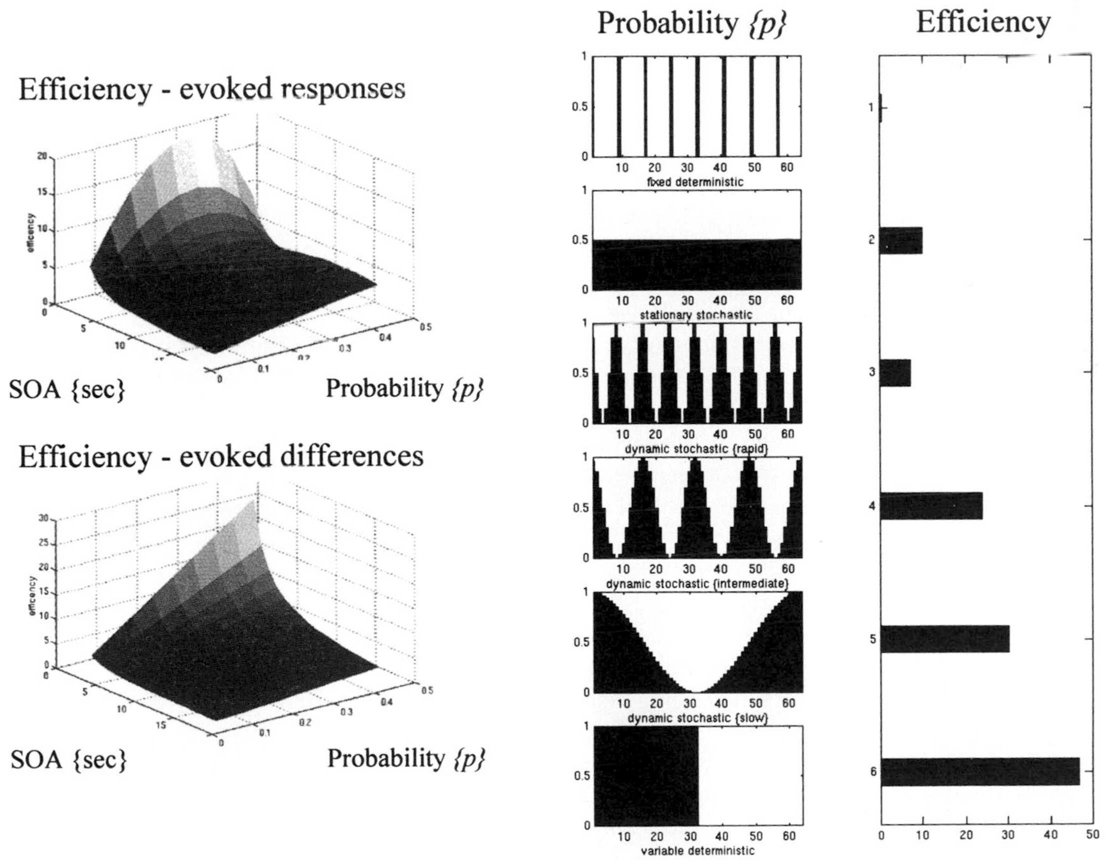

Figure 11 Efficiency as a function of occurrence probabilities P for a model X formed by postmultiplying S (a matrix containing n columns, modeling n possible event-related responses every SOA) by B. B is a random binary vector that determines whether the nth response is included in X, where $\langle B \rangle = P$. Right: A comparison of some common designs. A graphical representation of the occurrence probabilities P expressed as a function of time (seconds) is shown on the left and the corresponding efficiency is shown on the right. These results assume a minimum SOA of 1 s, a time-series of 64 s, and a single trial type. The expected number of events was 32 in all cases (apart from the first). Left: Efficiency in a stationary stochastic design with two event types, both presented with probability P every SOA. The upper graph is for a contrast testing for the response evoked by one trial type and the lower graph is for a contrast testing for differential responses.

In the framework considered here (Friston *et al.*, 1999a) the occurrence probability P of any event occurring is specified at each time that it could occur (i.e., every SOA). Here P is a vector with an element for every SOA. This formulation engenders the distinction between *stationary* stochastic designs, in which the occurrence probabilities are constant, and *non-stationary* stochastic designs, in which they change over time. For deterministic designs the elements of P are 0 or 1, the presence of a 1 denoting the occurrence of an event. An example of P might be the boxcars used in conventional block designs. Stochastic designs correspond to a vector of identical values and are therefore stationary in nature. Stochastic designs with temporal modulation of occurrence probability have time-dependent probabilities varying between 0 and 1. With these probabilities the expected design matrices and expected efficiencies can be computed. A useful thing about this formulation is that by setting the mean of the probabilities P to a constant, one can compare different deterministic and stochastic designs given the same number of events. Some common examples are given in Fig. 11 (right) for an SOA of 1 s and 32 expected events or trials over a 64-s period (except for the first deterministic example with 4 events and an SOA of 16 s). It can be seen that the least efficient is the sparse deterministic design (despite the fact that the SOA is roughly optimal for this class), whereas the most efficient is a block design. A slow modulation of occurrence probabilities gives high efficiency while retaining the advantages of stochastic designs and may represent a useful compromise between the high efficiency of block designs and the psychological benefits and latitude afforded by stochastic designs. However, it is important not to generalize these conclusions too far. An efficient design for one effect may not be the optimum for another, even within the same experiment. This can be illustrated by comparing the efficiency with which evoked responses are detected and the efficiency of detecting the difference in evoked responses elicited by two sorts of trials.

Consider a stationary stochastic design with two trial types. Because the design is stationary the vector of occurrence probabilities, for each trial type, is specified by a single probability. Let us assume that the two trial types occur with the same probability $\mathbf{P}$. By varying $\mathbf{P}$ and SOA one can find the most efficient design depending upon whether one is looking for evoked responses per se or differences among evoked responses. These two situations are depicted on the left of Fig. 11. It is immediately apparent that, for both sorts of effects, very small SOAs are optimal. However, the optimal occurrence probabilities are not the same. More infrequent events (corresponding to a smaller $\mathbf{P} = 1/3$) are required to estimate the responses themselves efficiently. This is equivalent to treating the baseline or control condition as any other condition (i.e., by including null events, with equal probability, as further event types). Conversely, if we are interested only in making inferences about the differences, one of the events plays the role of a null event and the most efficient design ensues when one or the other event occurs (i.e., $\mathbf{P} = ½$). In short, the most efficient designs obtain when the events subtending the differences of interest occur with equal probability.

Another example of how the efficiency is sensitive to the effect of interest is apparent when we consider different parameterizations of the HRF. This issue is sometimes addressed through distinguishing between the efficiency of response *detection* and response *estimation*. However, the principles are identical and the distinction reduces to how many parameters one uses to model the HRF for each trial type (one basis function is used for detection and a number are required to estimate the shape of the HRF). Here the contrasts may be the same but the shape of the regressors will change depending on the temporal basis set employed. The conclusions above were based on a single canonical HRF. Had we used a more refined parameterization of the HRF, say using three-basis functions, the most efficient design to estimate one basis function coefficient would not be the most efficient for another. This is most easily seen from the signal-processing perspective in which basis functions with high-frequency structure (e.g., temporal derivatives) require the experimental variance to contain high-frequency components. For these basis functions a randomized stochastic design may be more efficient than a deterministic block design, simply because the former embodies higher frequencies. In the limiting case of FIR estimation the regressors become a series of stick functions (see Fig. 5), all of which have high frequencies. This parameterization of the HRF calls for high frequencies in the experimental variance. However, the use of FIR models is contraindicated by model selection procedures (Henson *et al.*, manuscript in preparation) that suggest only two or three HRF parameters can be estimated with any efficiency. Results that are reported in terms of FIRs should be treated with caution because the inferences about evoked responses are seldom based on the FIR parameter estimates. This is precisely because they are estimated inefficiently and contain little useful information.

VII. Inferences about Subjects and Populations

In this section we consider some issues that are generic to brain mapping studies that have repeated measures or replications over subjects. The critical issue is whether we want to make an inference about the effect in relation to the *within-subject variability* or with respect to the *between-subject variability*. For a given group of subjects, there is a fundamental distinction between saying that the response is significant relative to the variability with which that response in measured and saying that it is significant in relation to the intersubject variability. This distinction relates directly to the difference between *fixed-* and *random*-effect analyses. The following example tries to make this clear. Consider what would happen if we scanned six subjects during the performance of a task and baseline. We then constructed a statistical model, in which task-specific effects were modeled separately for each subject. Unknown to us, only one of the subjects activated a particular brain region. When we examine the contrast of parameter estimates, assessing the mean activation over all the subjects, we see that it is greater than zero by virtue of this subject's activation. Furthermore, because that model fits the data extremely well (modeling no activation in five subjects and a substantial activation in the sixth) the error variance, on a scan-to-scan basis, is small and the t statistic is very significant. Can we then say that the group shows an activation? On the one hand we can say, quite properly, that the mean group response embodies an activation but clearly this does not constitute an inference that the group's response is significant (i.e., that this sample of subjects shows a consistent activation). The problem here is that we are using the *scan-to-scan* error variance and this is not necessarily appropriate for an inference about group responses. In order to make the inference that the group showed a significant activation one would have to assess the variability in activation effects from *subject to subject* (using the contrast of parameter estimates for each subject). This variability now constitutes the proper error variance. In this example the variance of these six measurements would be large relative to their mean, and the corresponding t statistic would not be significant.

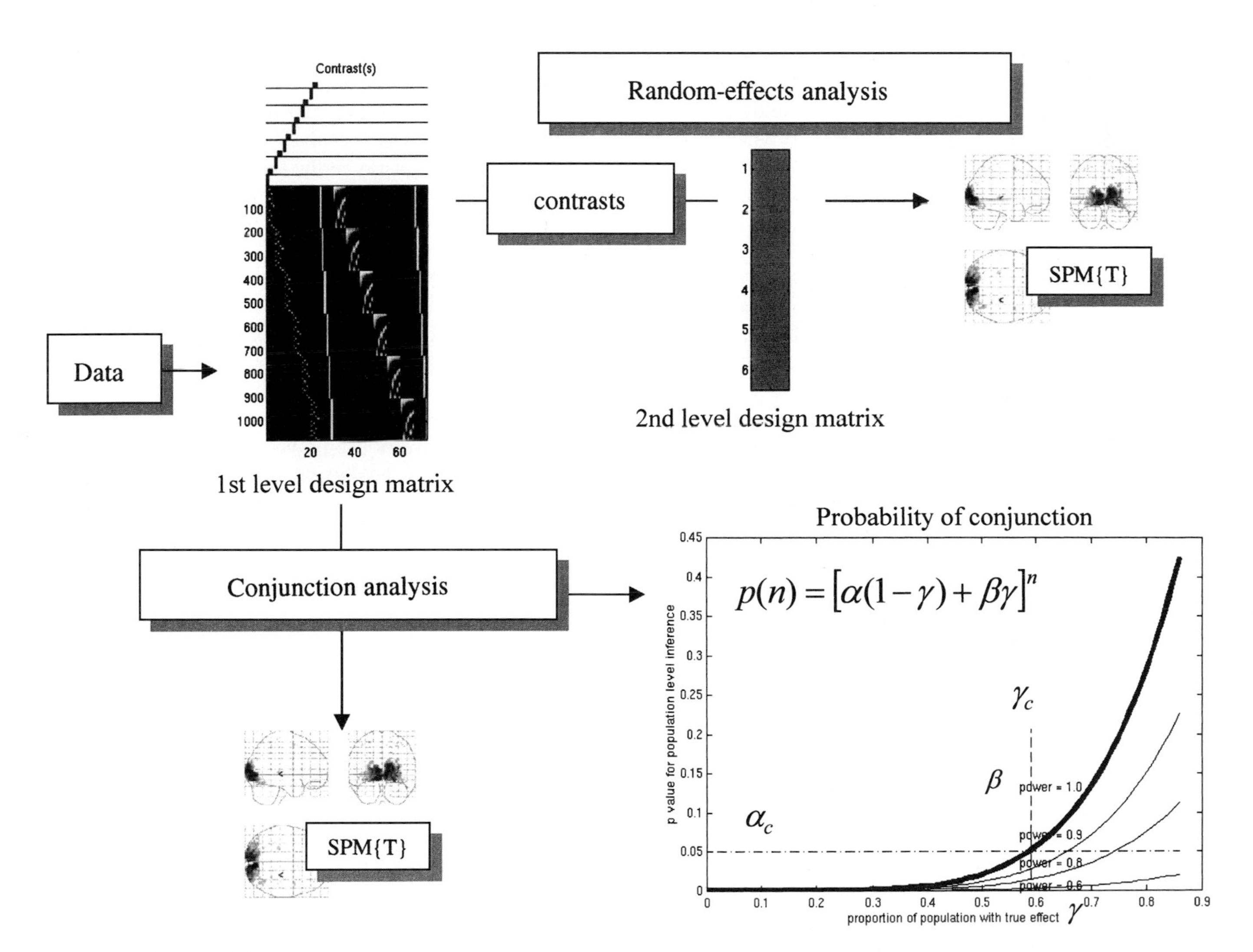

Figure 12 Schematic illustrating the implementation of random-effect and conjunction analyses for population inference. The lower right graph shows the probability $P(n)$ of obtaining a conjunction over n subjects, conditional on a certain proportion γ of the population expressing the effect, for a test with specificity of $\alpha = 0.05$, at several sensitivities ($\beta = 1$, 0.9, 0.8, and 0.6). The critical specificity for population inference α_c and the associated proportions of the population γ_c are denoted by the broken lines.

The distinction, between the two approaches above, relates to how one computes the appropriate error variance. The first represents a fixed-effect analysis and the second a random-effect analysis (or more exactly a mixed-effects analysis). In the former the error variance is estimated on a scan-to-scan basis, assuming that each scan represents an independent observation (ignoring serial correlations). Here the degrees of freedom are essentially the number of scans (minus the rank of the design matrix). Conversely, in random-effect analyses, the appropriate error variance is based on the activation from subject to subject, in which the effect per se constitutes an independent observation and the degrees of freedom fall dramatically to the number of subjects. The term "random effect" indicates that we have accommodated the randomness of differential responses by comparing the mean activation to the variability in activations from subject to subject. Both analyses are perfectly valid but only in relation to the inferences that are being made: Inferences based on fixed-effects analyses are about the particular subject(s) studied. Random-effects analyses are usually more conservative but allow the inference to be generalized to the population from which the subjects were selected.

A. Random-Effects Analyses

The implementation of random-effect analyses in SPM is fairly straightforward and involves taking the contrasts of parameters estimated from a *first-level* (fixed-effect) analysis and entering them into a *second-level* (random-effect) analysis. This ensures that there is only one observation (i.e., contrast) per subject in the second-level analysis and that the error variance is computed using the subject-to-subject variability of estimates from the first level. The nature of the inference made is determined by the contrasts entered into the second level (see Fig. 12). The second-level design matrix simply tests the null hypothesis that the contrasts are zero (and is usually a column of 1's, implementing a single-sample t test).

The reason this multistage procedure emulates a full mixed-effects analyses, using a hierarchical observation model, rests upon the fact that the design matrices for each subject are the same (or sufficiently similar). In this special case the estimator of the variance at the second level contains the right mixture of variance induced by observation error at the first level and between-subject error at the second. It is important to appreciate this because the efficiency of the design at the first level percolates up to higher levels. It is therefore important to use efficient strategies at all levels in a hierarchical design.

B. Conjunction Analyses and Population Inferences

In some instances a fixed-effects analysis is more appropriate, particularly to facilitate the reporting of a series of single-case studies. Among these single cases it is natural to ask what are common features of functional anatomy (e.g., the location of V5) and what aspects are subject-specific (e.g., the location of ocular dominance columns). One way to address commonalties is to use a conjunction analysis over subjects. It is important to understand the nature of the inference provided by conjunction analyses of this sort. Imagine that in 16 subjects the activation in V5, elicited by a motion stimulus, was greater than zero. The probability of this occurring by chance, in the same area, is extremely small and is the P value returned by a conjunction analysis using a threshold of $P = 0.5$ ($t = 0$) for each subject. This result constitutes evidence that V5 is involved in motion processing. However, note that this is not an assertion that each subject activated significantly (we require the t value only to be greater than zero for each subject). In other words, a significant conjunction of activations is not synonymous with a conjunction of significant activations.

The motivations for conjunction analyses, in the context of multisubject studies, are twofold. (i) They provide an inference, in a fixed-effect analysis testing the null hypotheses of no activation in any of the subjects, that can be much more sensitive than testing for the average activation. (ii) They can be extended to make inferences about the population as described next.

If, for any given contrast, one can establish a conjunction of effects over n subjects using a test with a specificity of $\propto$ and sensitivity β, the probability of this occurring by chance can be expressed as a function of γ, the proportion of the population that would have activated (see the equation in Fig. 12 lower right). This probability has an upper bound α_c corresponding to a critical proportion γ_c that is realized when (the generally unknown) sensitivity is 1. In other words, under the null hypothesis that the proportion of the population evidencing this effect is less than or equal to γ_c, the probability of getting a conjunction over n subjects is equal to, or less than, α_c. In short a conjunction allows one to say, with a specificity of α_c, that more than γ_c of the population show the effect in question. Formally, we can view this analysis as a conservative $100(1 - \alpha_c)\%$ confidence region for the unknown parameter γ. These inferences can be construed as statements about how typical the effect is, without saying that it is necessarily present in every subject.

In practice, a conjunction analysis of a multisubject study comprises the following steps: (i) A design matrix is constructed in which the explanatory variables pertaining to each experimental condition are replicated for each subject. This subject-separable design matrix implicitly models subject-by-condition interactions (i.e., different condition-specific responses among sessions). (ii) Contrasts are then specified that test for the effect of interest in each subject to give a series of SPM$\{t\}$ that can be reported as a series of "single-case" studies in the usual way. (iii) These SPM$\{t\}$'s are combined at a threshold u (corresponding to the specificity $\propto$ in

Fig. 12) to give a SPM{t_{min}} (i.e., conjunction SPM). The corrected P values associated with each voxel are computed as described in Fig. 6. These P values provide for inferences about effects that are common to the particular subjects studied. Because we have demonstrated regionally specific conjunctions, one can also proceed to make an inference about the population from which these subjects came using the confidence region approach described above (see Friston *et al.*, 1999b, for a fuller discussion).

VIII. Effective Connectivity

A. Functional and Effective Connectivity

Imaging neuroscience has firmly established functional specialization as a principle of brain organization in man. The functional integration of specialized areas has proven more difficult to assess. Functional integration is usually inferred on the basis of correlations among measurements of neuronal activity. Functional connectivity has been defined as *correlations between remote neurophysiological events*. However, correlations can arise in a variety of ways: For example, in multiunit electrode recordings they can result from stimulus locked transients evoked by a common input or reflect stimulus-induced oscillations mediated by synaptic connections (Gerstein and Perkel, 1969). Integration within a distributed system is usually better understood in terms of effective connectivity: Effective connectivity refers explicitly to *the influence that one neural system exerts over another*, either at a synaptic (i.e., synaptic efficacy) or at a population level. It has been proposed that "the [electrophysiological] notion of effective connectivity should be understood as the experiment- and time-dependent, simplest possible circuit diagram that would replicate the observed timing relationships between the recorded neurons" (Aertsen and Preißl, 1991). This speaks to two important points: (i) Effective connectivity is dynamic, i.e., activity- and time-dependent, and (ii) it depends upon a model of the interactions. The estimation procedures employed in functional neuroimaging can be classified as (i) those based directly on regression (Friston *et al.*, 1995d) or (ii) structural equation modeling (McIntosh and Gonzalez-Lima, 1994) (i.e., path analysis).

There is a necessary relationship between approaches to characterizing functional integration and multivariate analyses because the latter are necessary to model interactions among brain regions. Multivariate approaches can be divided into those that are inferential in nature and those that are data led or exploratory. We will first consider multivariate approaches that are universally based on functional connectivity or covariance patterns (and are generally exploratory) and then turn to models of effective connectivity (that usually allow for some form of inference)

B. Eigenimage Analysis and Related Approaches

Most analyses of covariances among brain regions are based on the singular value decomposition (SVD) of the between-voxel covariances in a neuroimaging time series In Friston *et al.* (1993) we introduced voxel-based principal component analysis (PCA) of neuroimaging time series to characterize distributed brain systems implicated in sensorimotor, perceptual, or cognitive processes. These distributed systems are identified with principal components or *eigenimages* that correspond to spatial modes of coherent brain activity. This approach represents one of the simplest multivariate characterizations of functional neuroimaging time series and falls into the class of exploratory analyses. Principal component or eigenimage analysis generally uses SVD to identify a set of orthogonal spatial modes that capture the greatest amount of variance, expressed over time. As such the ensuing modes embody the most prominent aspects of the variance–covariance structure of a given time series. Noting that the covariances among brain regions are equivalent to functional connectivity renders eigenimage analysis particularly interesting because it was among the first ways of addressing functional integration (i.e., connectivity) with neuroimaging data. Subsequently, eigenimage analysis has been elaborated in a number of ways. Notable among these are canonical variate analysis (CVA) and multidimensional scaling (Friston *et al.*, 1996d,e). Canonical variate analysis was introduced in the context of ManCova (multiple analysis of covariance) and uses the generalized eigenvector solution to maximize the variance that can be explained by some explanatory variables relative to error. CVA can be thought of as an extension of eigenimage analysis that refers explicitly to some explanatory variables and allows for statistical inference.

In *f*MRI, eigenimage analysis (Sychra *et al.*, 1994) is generally used as an exploratory device to characterize coherent brain activity. These variance components may, or may not, be related to experimental design and endogenous coherent dynamics have been observed in the motor system (Biswal *et al.*, 1995). Despite its exploratory power eigenimage analysis is fundamentally limited for two reasons. First, it offers only a linear decomposition of any set of neurophysiological measurements and second the particular set of eigenimages or spatial modes obtained is uniquely determined by constraints that are biologically implausible. These aspects of PCA confer inherent limitations on the interpretability and usefulness of eigenimage analysis of biological time series and have motivated the exploration of nonlinear PCA and neural network approaches (e.g., Mørch *et al.*, 1995).

Two other important approaches deserve mention here. The first is independent component analysis (ICA). ICA uses entropy maximization to find, using iterative schemes,

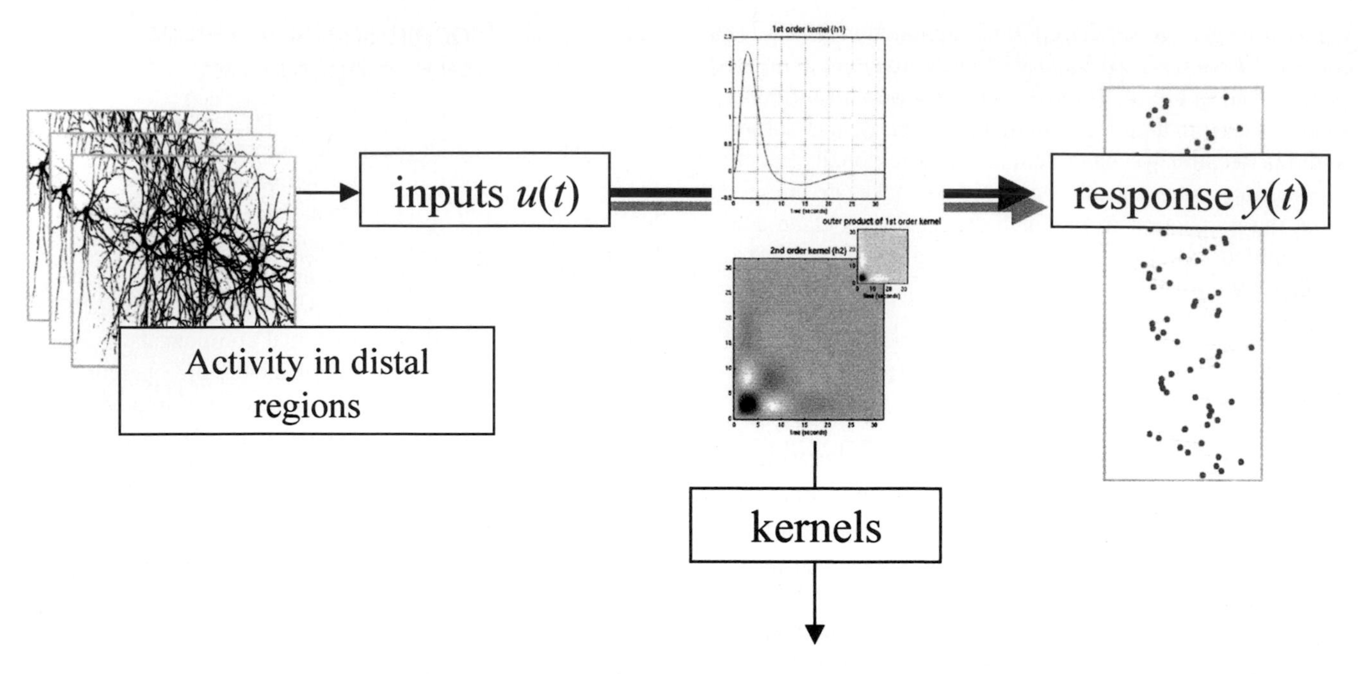

Volterra series a general nonlinear input-state-output characterization

$$y(t) = \kappa_0 + \sum_{i=1}^{\infty} \int_0^t \mathrm{K} \int_0^t \kappa_i(\sigma_1, \mathrm{K}\ \sigma_i) u(\sigma_1) \mathrm{K}\ u(\sigma_i) d\sigma_1 \mathrm{K}\ d\sigma_i$$

V5

$$\kappa_i(\sigma_1) = \frac{\partial^i y(t)}{\partial u(\sigma_1)}, \qquad \kappa_i(\sigma_1, \sigma_2) = \frac{\partial^2 y(t)}{\partial u(\sigma_1) \partial u(\sigma_2)}, \qquad \Lambda$$

PPC V2

n.b. Volterra kernels are synonymous with effective connectivity

Figure 13 Schematic depicting the causal relationship between the outputs and the recent history of the inputs to a nonlinear dynamical system, in this instance a brain region or voxel. This relationship can be expressed as a Volterra series, which expresses the response or output $y(t)$ as a nonlinear convolution of the inputs $u(t)$, critically without reference to any (hidden) state variables. This series is simply a functional Taylor expansion of $y(t)$ as a function of the inputs over the recent past. κ_i is the ith order kernel. Volterra series have been described as a "power series with memory" and are generally thought of as a high-order or "nonlinear convolution" of the inputs to provide an output. Volterra kernels are useful in characterizing the effective connectivity or influences that one neuronal system exerts over another because they represent the causal characteristics of the system in question. Neurobiologically they have a simple and compelling interpretation—*they are synonymous with effective connectivity*. It is evident that the first-order kernel embodies the response evoked by a change in input at $t - \sigma_1$. In other words it is a time-dependant measure of *driving* efficacy. Similarly the second-order kernel reflects the *modulatory* influence of the input at $t - \sigma_1$ on the evoked response at $t - \sigma_2$, and so on for higher orders. Note. Volterra kernels are synonymous with effective connectivity.

spatial modes or their dynamics that are approximately *independent*. This is a stronger requirement than *orthogonality* in PCA and involves removing high-order correlations among the modes (or dynamics). It was initially introduced as *spatial* ICA (McKeown *et al.*, 1998) in which the independence constraint was applied to the modes (with no constraints on their temporal expression). More recent approaches use, by analogy with magneto- and electrophysiological time series analysis, *temporal* ICA in which the dynamics are enforced to be independent. This requires an initial dimension reduction (usually using conventional eigenimage analysis). Finally, there has been an interest in cluster analysis (Baumgartner *et al.*, 1997). Conceptually, this can be related to eigenimage analysis through multidimensional scaling and principal coordinates analysis. In cluster analysis voxels in a multidimensional scaling space are assigned belonging probabilities to a small number of clusters, thereby characterizing the temporal dynamics (in terms of the cluster centroids) and spatial modes (defined by the belonging probability for each cluster). These approaches eschew many of the unnatural constraints imposed by eigenimage analysis and can be a useful exploratory device.

C. Characterizing Nonlinear Coupling among Brain Areas

Linear models of effective connectivity assume that the multiple inputs to a brain region are linearly separable. This assumption precludes activity-dependent connections that are expressed in one context and not in another. The resolution of this problem lies in adopting nonlinear models like the Volterra formulation that include interactions among inputs. These interactions can be construed as a context- or activity-dependent modulation of the influence that one region exerts over another, where that context is instantiated by activity in further brain regions exerting modulatory effects. These nonlinearities can be introduced into structural equation modeling using so-called "moderator" variables that represent the interaction between two regions in causing activity in a third (Büchel and Friston, 1997). From the point of view of regression models modulatory effects can be modeled with nonlinear input–output models and in particular the Volterra formulation described above. In this instance the inputs are not stimuli but activities from other regions. Because the kernels are high order they embody interactions over time and among inputs and can be thought of as explicit measures of effective connectivity (see Fig. 13). An important thing about the Volterra formulation is that it has a high face validity and biological plausibility. The only thing it assumes is that the response of a region is some analytic nonlinear function of the inputs over the recent past. This function exists even for complicated dynamical systems with many (unobservable) state variables. Within these models, the influence of one region on another has two components: (i) the direct or *driving* influence of input from the first (e.g., hierarchically lower) region, irrespective of the activities elsewhere and (ii) an activity-dependent, *modulatory* component that represents an interaction with inputs from the remaining (e.g., hierarchically higher) regions. These are mediated by the first- and second-order kernels, respectively. The example provided in Fig. 14 addresses the modulation

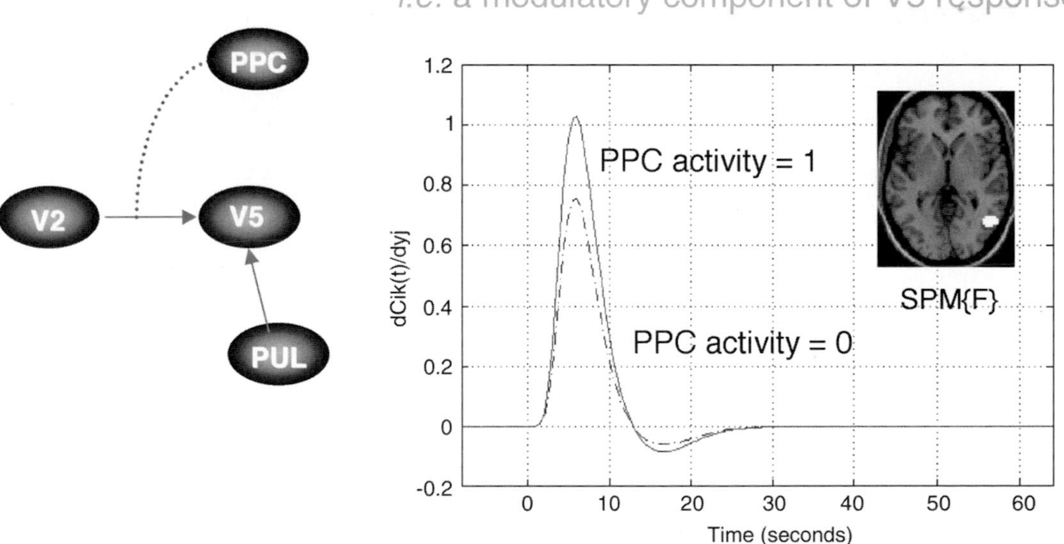

Figure 14 **Left:** Brain regions and connections comprising the effective connectivity model formulated in terms of a Volterra series (see Fig. 13). **Right:** Characterization of the effects of V2 inputs on V5 and their modulation by posterior parietal cortex (PPC). The broken lines represent estimates of V5 responses when PPC activity is zero, according to a second-order Volterra model of effective connectivity with inputs to V5 from V2, PPC, and the pulvinar (PUL). The solid curves represent the same response when PPC activity is 1 standard deviation of its variation over conditions. It is evident that V2 has an activating effect on V5 and that PPC increases the responsiveness of V5 to these inputs. The inset shows all the voxels in V5 that evidenced a modulatory effect ($P < 0.05$ uncorrected). These voxels were identified by thresholding a SPM{F} testing for the contribution of second-order kernels involving V2 and PPC (treating all other terms as nuisance variables). The data were obtained with *f*MRI under identical stimulus conditions (visual motion subtended by radially moving dots) while manipulating the attentional component of the task (detection of velocity changes).

of visual cortical responses by attentional mechanisms (e.g., Treue and Maunsell, 1996) and the mediating role of activity-dependent changes in effective connectivity:

The right side of Fig. 14 shows a characterization of this modulatory effect in terms of the increase in V5 responses, to a simulated V2 input, when posterior parietal activity is zero (broken line) and when it is high (solid lines). In this study subjects were studied with *f*MRI under identical stimulus conditions (visual motion subtended by radially moving dots) while manipulating the attentional component of the task (detection of velocity changes). The brain regions and connections comprising the model are shown in the top. The bottom shows a characterization of the effects of V2 inputs on V5 and their modulation by posterior parietal cortex (PPC) using simulated inputs at different levels of PPC activity. It is evident that V2 has an activating effect on V5 and that PPC increases the responsiveness of V5 to these inputs. The inset shows all the voxels in V5 that evidenced a modulatory effect ($P < 0.05$ uncorrected). These voxels were identified by thresholding a SPM$\{F\}$ testing for the contribution of second-order kernels involving V2 and PPC while treating all other components as nuisance variables. The estimation of the Volterra kernels and statistical inference procedure are described in Friston and Büchel (2000c).

Acknowledgment

The Wellcome Trust funded this work.

References

Absher, J. R., and Benson, D. F. (1993). Disconnection syndromes: An overview of Geschwind's contributions. *Neurology* **43**, 862–867.

Adler, R. J. (1981). *In* "The Geometry of Random Fields." Wiley, New York.

Aertsen, A., and Preißl, H. (1991). Dynamics of activity and connectivity in physiological neuronal networks. *In* "Non Linear Dynamics and Neuronal Networks" (H. G. Schuster, ed.), pp. 281–302. VCH, New York.

Aguirre, G. K., Zarahn, E., and D'Esposito, M. (1998). A critique of the use of the Kolmogorov–Smirnov (KS) statistic for the analysis of BOLD *f*MRI data. *Magn. Reson. Med.* **39**, 500–505.

Andersson, J. L., Hutton, C., Ashburner, J., Turner, R., and Friston, K. (2001). Modeling geometric deformations in EPI time series. *NeuroImage* **13**, 903–919.

Ashburner, J., Neelin, P., Collins, D. L., Evans, A., and Friston, K. (1997). Incorporating prior knowledge into image registration. *NeuroImage* **6**, 344–352.

Ashburner, J., and Friston, K. J. (1999). Nonlinear spatial normalization using basis functions. *Hum. Brain Mapp.* **7**, 254–266.

Ashburner, J., and Friston, K. J. (2000). Voxel-based morphometry—The methods. *NeuroImage* **11**, 805–821.

Bandettini, P. A., Jesmanowicz, A., Wong, E. C., and Hyde, J. S. (1993). Processing strategies for time course data sets in functional MRI of the human brain. *Magn. Reson. Med.* **30,** 161–173.

Baumgartner, R., Scarth, G., Teichtmeister, C., Somorjai, R., and Moser, E. (1997). Fuzzy clustering of gradient-echo functional MRI in the human visual cortex. Part 1. Reproducibility. *J. Magn. Reson. Imaging* 7, 1094–1101.

Biswal, B., Yetkin, F. Z., Haughton, V. M., and Hyde, J. S. (1995). Functional connectivity in the motor cortex of resting human brain using echo-planar MRI. *Magn. Reson. Med.* **34,** 537–541.

Boynton, G. M., Engel, S. A., Glover, G. H., and Heeger, D. J. (1996). Linear systems analysis of functional magnetic resonance imaging in human V1. *J. Neurosci.* **16**, 4207–4221.

Büchel, C., Wise, R. J. S., Mummery, C. J., Poline, J.-B., and Friston, K. J. (1996). Nonlinear regression in parametric activation studies. *NeuroImage* **4,** 60–66.

Büchel, C., and Friston, K. J. (1997). Modulation of connectivity in visual pathways by attention: Cortical interactions evaluated with structural equation modeling and *f*MRI. *Cereb. Cortex* **7**, 768–778.

Buckner, R., Bandettini, P., O'Craven, K., Savoy, R., Petersen, S., Raichle, M., and Rosen, B. (1996). Detection of cortical activation during averaged single trials of a cognitive task using functional magnetic resonance imaging. Proc. Natl. Acad. Sci. USA **93**, 14878–14883.

Bullmore, E. T., Brammer, M. J., Williams, S. C. R., Rabe-Hesketh, S., Janot, N., David, A., Mellers, J., Howard, R., and Sham, P. (1996). Statistical methods of estimation and inference for functional MR images. *Magn. Reson. Med.* **35,** 261–277.

Burock, M. A., Buckner, R. L., Woldorff, M. G., Rosen, B. R., and Dale, A. M. (1998). Randomized event-related experimental designs allow for extremely rapid presentation rates using functional MRI. *NeuroReport* **9**, 3735–3739.

Buxton, R. B., and Frank, L. R. (1997). A model for the coupling between cerebral blood flow and oxygen metabolism during neural stimulation. *J. Cereb. Blood Flow Metab.* **17,** 64–72.

Clark, V. P., Maisog, J. M., and Haxby, J. V. (1998). *f*MRI study of face perception and memory using random stimulus sequences. *J. Neurophysiol.* **76**, 3257–3265.

Dale, A., and Buckner, R. (1997). Selective averaging of rapidly presented individual trials using *f*MRI. *Hum. Brain Mapp.* **5**, 329–340.

Friston, K. J. (1997). Testing for anatomical specified regional effects. *Hum. Brain Mapp.* **5**, 133–136.

Friston, K. J., Frith, C. D., Liddle, P. F., and Frackowiak, R. S. J. (1991). Comparing functional (PET) images: The assessment of significant change. *J. Cereb. Blood Flow Metab.* **11,** 690–699.

Friston, K. J., Frith, C., Passingham, R. E., Liddle, P. F., and Frackowiak, R. S. J. (1992a). Motor practice and neurophysiological adaptation in the cerebellum: A positron tomography study. *Proc. R. Soc. London Ser. B* **248,** 223–228.

Friston, K. J., Grasby, P., Bench, C., Frith, C. D., Cowen, P. J., Little, P., Frackowiak, R. S. J., and Dolan, R. (1992b). Measuring the neuromodulatory effects of drugs in man with positron tomography. *Neurosci. Lett.* **141,** 106–110.

Friston, K. J., Frith, C., Liddle, P., and Frackowiak, R. S. J. (1993). Functional connectivity: The principal component analysis of large data sets. *J. Cereb. Blood Flow Metab.* **13,** 5–14.

Friston, K. J., Worsley, K. J., Frackowiak, R. S. J., Mazziotta, J. C., and Evans, A. C. (1994a). Assessing the significance of focal activations using their spatial extent. *Hum. Brain Mapp.* **1**, 214–220.

Friston, K. J., Jezzard, P. J., and Turner, R. (1994b). Analysis of functional MRI time-series. *Hum. Brain Mapp.* **1**, 153–171.

Friston, K. J., Ashburner, J., Frith, C. D., Poline, J.-B., Heather, J. D., and Frackowiak, R. S. J. (1995a). Spatial registration and normalization of images. *Hum. Brain Mapp.* **2**, 165–189.

Friston, K. J., Holmes, A. P., Worsley, K. J., Poline, J. B., Frith, C. D., and Frackowiak, R. S. J. (1995b). Statistical parametric maps in functional imaging: A general linear approach. *Hum. Brain Mapp.* **2,** 189–210.

Friston, K. J., Frith, C. D., Turner, R., and Frackowiak, R. S. J. (1995c). Characterizing evoked hemodynamics with *f*MRI. *NeuroImage* **2**, 157–165.

Friston, K. J., Ungerleider, L. G., Jezzard, P., and Turner, R. (1995d). Characterizing modulatory interactions between V1 and V2 in human cortex with *f*MRI. *Hum. Brain Mapp.* **2,** 211–224.

Friston, K. J., Williams, S., Howard, R., Frackowiak, R. S. J., and Turner, R. (1996a). Movement related effects in *f*MRI time series. *Magn. Reson. Med.* **35,** 346–355.

Friston, K. J., Holmes, A., Poline, J.-B., Price, C. J., and Frith, C. D. (1996b). Detecting activations in PET and *f*MRI: Levels of inference and power. *NeuroImage* **4**, 223–235.

Friston, K. J., Price, C. J., Fletcher, P., Moore, C., Frackowiak, R. S. J., and Dolan, R. J. (1996c). The trouble with cognitive subtraction. *NeuroImage* **4**, 97–104.

Friston, K. J., Poline, J.-B., Holmes, A. P., Frith, C. D., and Frackowiak, R. S. J. (1996d). A multivariate analysis of PET activation studies. *Hum. Brain Mapp.* **4**, 140–151.

Friston, K. J., Frith, C. D., Fletcher, P., Liddle, P. F., and Frackowiak, R. S. J. (1996e). Functional topography: Multidimensional scaling and functional connectivity in the brain. *Cereb. Cortex* **6,** 156–164.

Friston, K. J., Büchel, C., Fink, G. R., Morris, J., Rolls, E., and Dolan, R. J. (1997). Psychophysiological and modulatory interactions in neuroimaging. *NeuroImage* **6,** 218–229.

Friston, K. J., Josephs, O., Rees, G., and Turner, R. (1998a). Non-linear event-related responses in *f*MRI. *Magn. Reson. Med.* **39,** 41–52.

Friston, K. J., Fletcher, P., Josephs, O., Holmes, A., Rugg, M. D., and Turner, R. (1998b). Event-related *f*MRI: Characterizing differential responses. *NeuroImage* **7**, 30–40.

Friston, K. J., Zarahn, E., Josephs, O., Henson, R. N., and Dale, A. M. (1999a). Stochastic designs in event-related *f*MRI. *NeuroImage*. **10,** 607–619.

Friston, K. J., Holmes, A. P., Price, C. J., Buchel, C., and Worsley, K. J. (1999b). Multisubject *f*MRI studies and conjunction analyses. *NeuroImage* **10**, 385–396.

Friston, K. J., Mechelli, A., Turner, R., and Price, C. J. (2000a). Nonlinear responses in *f*MRI: The Balloon model, Volterra kernels, and other hemodynamics. *NeuroImage* **12,** 466–477.

Friston, K. J., Josephs, O., Zarahn, E., Holmes, A. P., Rouquette, S., and Poline, J. (2000b). To smooth or not to smooth? Bias and efficiency in *f*MRI time-series analysis. *NeuroImage* **12,** 196–208.

Friston, K. J., and Buchel, C. (2000c). Attentional modulation of effective connectivity from V2 to V5/MT in humans. *Proc. Natl. Acad. Sci. USA* **97**, 7591–7596.

Gerstein, G. L., and Perkel, D. H. (1969). Simultaneously recorded trains of action potentials: Analysis and functional interpretation. *Science* **164,** 828–830.

Girard, P., and Bullier, J. (1989). Visual activity in area V2 during reversible inactivation of area 17 in the macaque monkey. *J. Neurophysiol.* **62**, 1287–1301.

Goltz, F. (1881). In "Transactions of the 7th International Medical Congress" (W. MacCormac, ed.), Vol. I, pp. 218–228.Kolkmann, London.

Grafton, S., Mazziotta, J., Presty, S., Friston, K. J., Frackowiak, R. S. J., and Phelps, M. (1992). Functional anatomy of human procedural learning determined with regional cerebral blood flow and PET. *J. Neurosci.* **12,** 2542–2548.

Grootoonk, S., Hutton, C., Ashburner, J., Howseman, A. M., Josephs, O., Rees, G., Friston, K. J., and Turner, R. (2000). Characterization and correction of interpolation effects in the realignment of *f*MRI time series. *NeuroImage* **11,** 49–57.

Heid, O., Gönner, F., and Schroth, G. (1997). Stochastic functional MRI. *NeuroImage* **5,** S476.

Hirsch, J. A., and Gilbert, C. D. (1991). Synaptic physiology of horizontal connections in the cat's visual cortex. *J. Neurosci.* **11,** 1800–1809.

Jezzard, P., and Balaban, R. S. (1995). Correction for geometric distortion in echo-planar images from B0 field variations. *Magn. Reson. Med.* **34**, 65–73.

Josephs, O., Turner, R., and Friston, K. J. (1997). Event-related *f*MRI. *Hum. Brain Mapp.* **5,** 243–248.

Kiebel, S. J., Poline, J. B., Friston, K. J., Holmes, A. P., and Worsley, K. J. (1999). Robust smoothness estimation in statistical parametric maps using standardized residuals from the general linear model. *NeuroImage* **10**, 756–766.

Lange, N., and Zeger, S. L. (1997). Non-linear Fourier time series analysis for human brain mapping by functional magnetic resonance imaging (with discussion). *J. R. Stat. Soc. Ser. C* **46,** 1–29.

Liddle, P. F., Friston, K. J., Frith, C. D., and Frackowiak, R. S. J. (1992). Cerebral blood-flow and mental processes in schizophrenia. *J. R. Soc. Med.* **85,** 224–227.

Lueck, C. J., Zeki, S., Friston, K. J., Deiber, M. P., Cope, N. O., Cunningham, V. J., Lammertsma, A. A., Kennard, C., and Frackowiak, R. S. J. (1989). The color centre in the cerebral cortex of man. *Nature* **340,** 386–389.

McIntosh, A. R., and Gonzalez-Lima, F. (1994). Structural equation modeling and its application to network analysis in functional brain imaging. *Hum. Brain Mapp.* **2,** 2–22.

McKeown, M., Jung, T.-P., Makeig, S., Brown, G., Kinderman, S., Lee, T.-W., and Sejnowski, T. (1998). Spatially independent activity patterns in functional MRI data during the Stroop color naming task. *Proc. Natl. Acad. Sci. USA* **95,** 803–810.

Mørch, N., Kjems, U., Hansen, L. K., Svarer, C., Law, I., Lautrup, B., and Strother, S. C. (1995). Visualization of neural networks using saliency maps. *In* "IEEE International Conference on Neural Networks," Perth, pp. 2085–2090.

Petersen, S. E., Fox, P. T., Posner, M. I., Mintun, M., and Raichle, M. E. (1989). Positron emission tomographic studies of the processing of single words. *J. Cognit. Neurosci.* **1**, 153–170.

Phillips, C. G., Zeki, S., and Barlow, H. B. (1984). Localisation of function in the cerebral cortex: Past present and future. *Brain* **107,** 327–361.

Purdon, P. L., and Weisskoff, R. M. (1998). Effect of temporal autocorrelation due to physiological noise and stimulus paradigm on voxel-level false-positive rates in *f*MRI. *Hum. Brain Mapp.* **6,** 239–495.

Price, C. J., Wise, R. J. S., Ramsay, S., Friston, K. J., Howard, D., Patterson, K., and Frackowiak, R. S. J. (1992). Regional response differences within the human auditory cortex when listening to words. *Neurosci. Lett.* **146,** 179–182.

Price, C. J., and Friston, K. J. (1997). Cognitive conjunction: A new approach to brain activation experiments. *NeuroImage* **5**, 261–270.

Sychra, J. J., Bandettini, P. A., Bhattacharya, N., and Lin, Q. (1994). Synthetic images by subspace transforms. I. Principal component images and related filters. *Med. Phys.* **21,** 193–201.

Talairach, P., and Tournoux, J. (1988). "A Stereotactic Coplanar Atlas of the Human Brain." Thieme, Stuttgart.

Treue, S., and Maunsell, H. R. (1996). Attentional modulation of visual motion processing in cortical areas MT and MST. *Nature* **382,** 539–541.

Vazquez, A. L., and Noll, C. D. (1998). Nonlinear aspects of the BOLD response in functional MRI. *NeuroImage* **7,** 108–118.

Worsley, K. J., Evans, A. C., Marrett, S., and Neelin, P. (1992). A three-dimensional statistical analysis for rCBF activation studies in human brain. *J. Cereb. Blood Flow Metab.* **12**, 900–918.

Worsley, K. J., and Friston, K. J. (1995). Analysis of fMRI time-series revisited—Again. *NeuroImage* **2,** 173–181.

Worsley, K. J., Marrett, S., Neelin, P., Vandal, A. C., Friston, K. J., and Evans, A. C. (1996). A unified statistical approach for determining significant signals in images of cerebral activation. *Hum. Brain Mapp.* **4,** 58–73.

Zarahn, E., Aguirre, G. K., and D'Esposito, M. (1997). Empirical analyses of BOLD *f*MRI statistics. I. Spatially unsmoothed data collected under null-hypothesis conditions. *NeuroImage* **5,** 179–197.

Zeki, S. (1990). The motion pathways of the visual cortex. In "Vision: Coding and Efficiency" (C. Blakemore, ed.), pp. 321–345. Cambridge Univ. Press, Cambridge, UK.

23

Statistics II: Correlation of Brain Structure and Function

Roger P. Woods

Department of Neurology and Division of Brain Mapping, UCLA School of Medicine, Los Angeles, California 90024

I. Introduction

The modern era of correlating brain structure and function dates back well over a century. The earliest work in this regard was based on functional observations of patients made during life and then correlated with structural abnormalities that could be determined only postmortem. The elegant work of Broca (1865), demonstrating that language deficits result from damage to the left frontal lobe, is an excellent example of the power of this simple observational methodology. With the advent of invasive neurosurgical procedures, correlation of brain structure and function in living humans became possible. The pioneering work of Penfield and others (Penfield and Roberts, 1959; Ojemann, 1991) performing intraoperative cortical mapping allowed the correlation of structure and function at a much higher level of detail than can easily be achieved based on pathologic observation. These techniques also introduced the ability to manipulate an experimental variable since the neurosurgeon is able to choose the site of electrical stimulation. However, such techniques are obviously quite invasive and are therefore limited to humans with brain lesions that justify neurosurgical intervention. Furthermore, these lesions themselves may alter the relationship between brain structure and function (Woods *et al.*, 1988), making it difficult to generalize the findings to a normal population.

The modern era of noninvasive brain imaging has added new dimensions to the correlation of structure and function in humans. First, noninvasive structural brain imaging has made it possible to localize brain pathology while the patient is still alive. As a result, modern neurobehaviorists, following in the tradition of Broca, are now able to examine brain function both before and after localizing the relevant area of pathology (Damasio and Frank, 1992). This approach allows for prediction of structure based on function and prediction of function based on structure, a situation well suited to framing and testing new hypotheses.

In addition to providing data on structure, modern brain imaging is also able to provide functional information Neuronal firing leads to localized physiologic changes that can be measured using imaging techniques. Some of the earliest work in this regard utilized the relationship between neuronal firing and glucose metabolism using positron emission tomography (PET). By manipulating the environment of the subject as an experimental variable, it was possible to induce anatomically discrete increases in uptake of the radiolabeled tracer [^{18}F]fluorodeoxyglucose (FDG) in those regions of the brain that were predicted *a priori* to respond to such stimuli (Mazziotta *et al.*, 1982a,b,c). For a variety of technical reasons, most of the more recent work with functional imaging of the brain has focused on the coupling between neuronal firing and the vascular system rather than glucose metabolism. It has been known since the time of Sherrington (1906) that increases in neuronal firing are associated with local increases in cerebral blood flow. Pursuing work in experimental animals, Herscovitch *et al.* (1983) and Raichle *et al.* (1983) were the first to apply blood flow measurement techniques to humans for brain imaging using the radioactive tracer $H_2{}^{15}O$ with PET. Early work in this field required pooling of data across multiple subjects in order to improve signal to noise while minimizing radiation exposure (Fox *et al.*, 1985). Advances in PET methodology now make it possible to perform such experiments on individual normal subjects or patients (Cherry *et al.*, 1993; Grafton *et al.*, 1993). The coupling between neuronal firing and vascular changes has also been exploited to perform functional brain mapping using magnetic resonance imaging (MRI) (Kwong *et al.*, 1992). Unlike the PET technique, which measures blood flow directly, the MRI technique exploits the fact that the increase in blood flow in activated areas of brain exceeds metabolic oxygen demands, leading to an increase in blood oxygenation in activated brain areas that can be detected as a MRI signal difference between oxygenated and deoxygenated hemoglobin. Finally, reminiscent of Sherrington's work in the last century, the optical intrinsic signal technique (Grinvald *et al.*, 1986) utilizes direct observation of the brain surface using modern optical methods to detect subtle changes associated with brain activity. Previously confined to animal research, this technique is now being applied intraoperatively in human subjects (Haglund *et al.*, 1992). The coupling between neuronal firing and vascular changes is presumably responsible for some of the signal measured by optical intrinsic imaging, although other factors such as metabolism may also contribute.

While the progress in functional brain mapping using imaging techniques since the 1970s has been impressive, the future challenges are substantial. The magnitude of this challenge is well illustrated by the map of cortical visual areas in the macaques monkey prepared by van Essen *et al.* (1992) (Fig. 1). The visual cortex of the macaque is composed of a multitude of discrete functional units that are interconnected in a hierarchical but complex fashion. The topographic organization of these units is sufficiently stereotyped to allow an anatomic map of this topography to be prepared. While some of these areas and their boundaries can be defined on the basis of anatomic criteria, many of the boundaries can currently be located only on the basis of *in vivo* electrical recording. Consequently, a situation exists where important subdivisions of brain anatomy can currently be defined only on the basis of functional patterns of neuronal firing in response to behaviorally relevant stimuli. Sir Francis Crick has recently pointed out that our knowledge of anatomic cortical connectivity in humans lags far behind our knowledge of such connectivity in laboratory animals (Crick and Jones, 1993). A very similar situation prevails with regard to our knowledge of the functional anatomy of visual and other systems. Different primate species differ from one another even in the organization of the most basic visual areas (Sereno *et al.*, 1994), so any conclusions drawn from the macaque should be extrapolated to the human only with great caution. To the extent that abilities and attributes unique to humans (e.g., language) motivate our interest to explore brain function, no animal model is likely to provide sufficient information. A detailed map of the entire human cortical surface analogous to the map of the macaque visual system is an ambitious goal to pursue for human structure-function correlation.

Requirements for safety and noninvasiveness make studies of the human brain particularly challenging. Direct electrical recording of the sort used for mapping macaque cortex is possible only under extraordinary circumstances (i.e., brain pathology justifying neurosurgery). These circumstances may themselves limit the generalizeability of the resulting conclusions. Consequently, detailed maps of human cortex need to be built on observations made using a variety of techniques and data drawn from a large number of subjects, both normal and abnormal. In order to cross-validate techniques and to refine hypotheses, it will be necessary to integrate and compare different types of functional information with in the same individual. As an extreme example, a patient with a brain tumor causing a neurobehavioral deficit might undergo preoperative and post-operative functional mapping using PET and functional MRI, preoperative and postoperative source localization using electroencephalography (EEG) and magnetoencephalography (MEG), intraoperative optical intrinsic signal measurements, and intraoperative electrocorticography. The combination of many technique in a single individual provides the opportunity to develop a more complete understanding of the underlying physiology that can then be used to better understand and characterize the strengths and weaknesses of each individual technique.

In addition to using different techniques to study a single individual, it will also be important to apply any particular

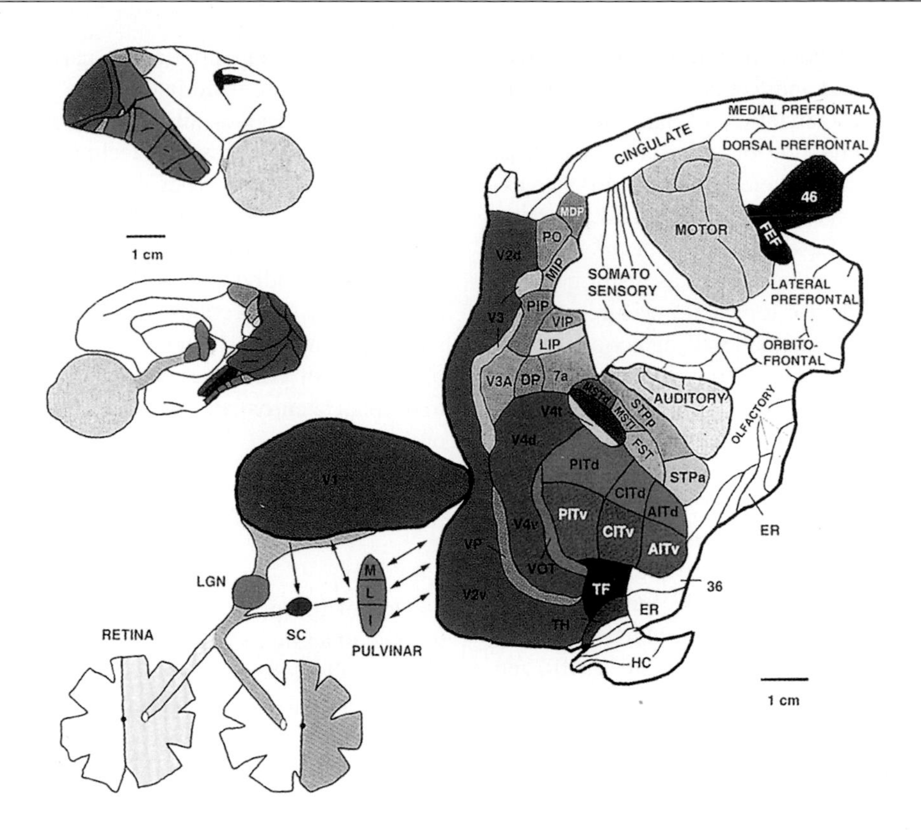

Figure 1 Functional visual cortical areas of the macaque brain. The smaller insets show the relationship between the flattened map and the sulci and gyri of the macaque brain. Each separate area has a distinct physiologic function. Many of the functional areas are quite small and are more precisely localized by virtue of their relationships to one another than by their global relationship to gross brain morphology. The human brain is much larger than the macaque brain and presumably has additional functional areas that subserve functional abilities that are unique to humans. An analogous functional map of the human cortex is not yet available at this level of detail, but the production of such a map for the entire human cortex is one of the objectives of ongoing human brain mapping investigations in laboratories worldwide.

technique to many different individuals. The ability to integrate and compare information across subjects is a difficult challenge. While it is clear that there are many anatomical consistencies shared by all humans (excluding those with certain types of brain pathology), it is also clear that there is considerable anatomic variability. Furthermore, interactions between variability in brain structure and variability in brain function may complicate efforts to look at either of these factors in isolation. However, an understanding of the genetic and environmental factors that make individuals different from one another should prove to be a powerful tool for understanding the principles of brain organization. Furthermore, an understanding of the uniqueness of each human brain is a topic of considerable scientific as well as general interest.

The material in this chapter is divided into three main sections. The first section is devoted to the tools and issues related to the multimodality integration of information from a single individual. The second section provides a series of case vignettes that illustrate the use of these intrasubject techniques for human patients. The third section focuses on issues of intersubject registration and comparison, with particular emphasis on the selection of tools and models.

II. Intrasubject Multimodal Integration

Experimental approaches to correlate function and structure in humans can be subdivided into two basic categories. In the first category of techniques, a particular region of brain anatomy is selected and subjected to some experimental manipulation and the functional consequences are then observed. In some cases (e.g., classical neurobehavorial methods) the choice of anatomic site is an "experiment of nature" and there may be little or no opportunity to define that individual's premorbid function and behavior. In such cases, the functional assessment must be made in comparison to norms established from other individuals. Other techniques in this first category, including intraoperative cortical stimulation (Penfield and Roberts, 1959; Ojemann, 1991), Wada testing [anesthetization of part or all of a hemisphere by selective injection of a barbiturate such as sodium amobarbital though a catheter threaded into the vasculature of the brain (Rauch *et al.* 1992a,b)], and magnetic transcranial stimulation (Pascual-Leone *et al.*, 1992), provide the opportunity to use the subject as his or her own control and potentially allow for repeated measurements of both baseline and experimentally altered behavior. For all of these techniques, it is desirable to limit the experimental manipulation to a region that is as anatomically discrete as possible in hopes of producing a very focal functional deficit or response that will allow a conclusive relationship between structure and function to be inferred. When a positive result is obtained, these techniques can be uniquely powerful, but negative results are more difficult to interpret because anatomically distributed networks with redundant parallel pathways are known to be involved in many aspects of higher cortical functioning (Mesulam, 1990), and such redundancy may minimize the functional consequences of a focal manipulation.

For the second category of techniques, it is the behavioral task that is manipulated experimentally. Signals produced by the brain during the performance of an experimental task are compared to signals produced during performance of a control task using a detection system capable of determining both the magnitude and the spatial origin of the signal within the brain, EEG, MEG, PET, single photon emission computed tomography (SPECT), MRI, and optical intrinsic signal imaging are all techniques that utilize this approach. These techniques vary in terms of the accuracy and resolution with which the three-dimensional spatial origin of a signal can be determined (see Chapter 2). The techniques also vary considerably in terms of the strengths of the signals measured and the number of repeated measurements required to overcome signal-to-noise limitations. Just as it is important to use anatomically discrete experimental manipulations for the techniques described in the preceding paragraph, for these techniques, it is important that the stimulation and control tasks differ from one another in a way that defines a particular function as precisely as possible. The advantage of these techniques is that they are able to simultaneously demonstrate multiple anatomic areas that participate in a particular brain function, even in settings where redundant parallel pathways do not demand obligatory participation of any given single anatomic region. However, the weakness of these techniques is that, in isolation, they cannot demonstrate that a particular brain region is critically involved in a particular function.

The goal of intrasubject multimodal integration is to allow information from all types of techniques that provide information about the correlation between structure and function to be merged together into a single framework. The individual's neuroanatomy provides a particularly good common frame of reference for integrating the various techniques. Thanks to the ready availability of high resolution structural magnetic resonance imaging techniques, the problem of integrating two different types of functional/structural information can usually be reduced to the problem of mapping the two different functional measures onto the appropriate voxels of the individual's MRI scan. Consequently, the following three subsections will focus on the problems of mapping functional imaging data onto MRI data, mapping other types of functional data onto MRI data, and assigning traditional neuroanatomic labels to structures identified on MRI.

A. Mapping Functional Imaging Data to MRI

One of the primary advantages of working with tomographic imaging techniques is that tomographic instruments are carefully designed to maintain an orderly and mathematically straightforward relationship between the three-dimensional object being imaged and the representation of that object that is produced. With modern scanners, the anatomic distortions are generally sufficiently small that an accurate mapping from one tomographic image to another can generally be made by performing a set of rigid body rotations and translations after correcting for any differences in voxel sizes. Consequently, in contrast to the intersubject registration problems considered in the third section of this chapter, the choice of the appropriate mathematical model for intrasubject registration is straightforward. The rigid body model incorporates a total of six parameters that can be conceptualized as rotations around and translations along each of the three major coordinate axes. A number of excellent methods for determining these six parameters have been devised. This subsection is not intended to serve as a comprehensive review of all of these methods, but rather to provide an illustrative overview of the types of approaches that are available. Several recent reviews of various image registration techniques are available.

Although the distinction between intramodality and intermodality registration is unimportant for some of these methods, these two problems are fundamentally different. Consequently, the intramodality and intermodality registration problems will be considered separately.

1. Intramodality Registration

All functional imaging techniques are dependent on a comparison of at least two different independent images. One of these images is acquired during a control state and the other image is acquired during a stimulation state with the difference between the stimulation and the control tasks representing the function of interest. In the ideal situation, the subject being imaged is completely motionless both during and between the acquisition of the two different images. However, in reality, head movements between acquisitions are extremely common, even with the use of fairly restrictive head restraint devices. This problem even occurs with very short interacquisition times such as those used in functional magnetic resonance imaging. Because the head is able to move in three dimensions, even simple visual comparison of two images by someone with neuroanatomic expertise can be difficult to perform accurately. Quantitative analysis of functional data, based on the subtraction of images or analogous statistical procedures, demands that the images being compared be aligned as precisely as possible. Even small degrees of misalignment can result in substantial artifacts giving the illusion of changes where there are none and vice versa. Intramodality registration of the various functional images is therefore an essential antecedent to the subtraction method of analysis. Three fundamentally different approaches to the problem of intramodality registration are described here to illustrate the similarities and differences among various registration techniques.

a. Point landmarks and the Procrustes problem. An intuitively straightforward way to achieve intramodality registration is to rely on an expert user to identify a series of point landmarks that represent identical locations in each of the images. For a rigid body model, only three points would need to be identified in each image, assuming that the points could be identified with perfect accuracy. Such accuracy can rarely, if ever, be achieved. Even a slight variation in the measured distance between any homologous pair of landmarks represents an error in landmark localization that cannot be reconciled with a rigid body model without further assumptions. The solution to this problem that is most commonly used is to minimize the mean squared distance between homologous landmarks when computing the six rigid body model parameters. This minimization, which has an exact, noniterative solution (Schonemann, 1966; Sibson, 1978), is known as the "Procrustes" problem in reference to the character in Greek mythology who tied travelers to a bed and stretched them or cut off body parts so that they would fit the size of the bed (Hurley and Cattell, 1962). The strength of the Procrustes approach is that it is computationally very simple. The appropriate rigid body transformation parameters can be derived almost instantaneously by computer. The accuracy of the derived transformation improves with the number of homologous point landmarks identified, and Evans et al. (1989) have shown that the identification of 15 landmarks on PET images is sufficient to give registration accuracy on the order of 0.5 mm if each individual landmark can be identified with an accuracy of at least 5 mm. However, the amount of time required to identify the landmarks in the first place is not trivial, and expert neuroanatomic guidance is required.

b. Surface matching. As an alternative to point-based registration techniques, Pellizzari et al. (1989) have developed a technique that matches surfaces to one other. The outer surface of the skull or of the brain is used most commonly for this technique. The surfaces of interest are identified as a series of contours on the two images being registered. Unlike the Procrustes problem, there is no noniterative method for minimizing the distance between the two surfaces, so a directed iterative exploration of the rigid body parameter space is required. User guidance to initialize the algorithm at a reasonable location in parameter space decreases the likelihood of identifying an incorrect local minimum. Registration accuracy on the order of 2 to 3 mm can be attained (Pelizzari et al., 1989). Compared to the point landmark Procrustes approach, this technique has the advantage of not requiring expert neuroanatomic guidance. Accurate identification of the appropriate surfaces is a critical prerequisite to this approach. The use of images that include the entire head within the field of view is also very helpful in ensuring that the surfaces have a sufficiently complex shape to define a robust and unique minimum distance (Turkington et al., 1995).

c. Automated image registration. An algorithm for intramodality registration that does not require identification of any point or surface landmarks has been developed (Woods et al., 1992). Instead, this algorithm depends on an overall measure of the similarity between the images to guide the registration (Fig. 2). Conceptually, the algorithm depends on the notion that if two images acquired using the same technique are perfectly aligned to one another, then the ratio of one image to the other on a voxel-by-voxel basis should be fairly uniform across voxels. Even a slight degree of misregistration should lead to a substantial degree of nonuniformity that can be quantitated by computing the standard deviation of the voxel-by-voxel ratio. The algorithm that is used iteratively seeks to minimize this standard deviation by adjusting the parameters of the rigid body model. This technique is accurate to within 1–2 mm even for very noisy PET data. The technique is also not dependent on a large field of view and can even be used successfully on single slice functional MRI data. The algorithm is relatively insensitive to a wide range of starting positions and can be initialized at any arbitrary site in parameter space if desired. Although computationally intensive, this algorithm converges within a few minutes on standard workstations available in most imaging centers. This technique has performed favorably in terms of both accuracy and speed compared to a number of alternative registration techniques (Strother et al., 1994).

2. Intermodality Registration

Whereas intramodality registration is primarily motivated by the need to subtract or compare images, the development of intermodality registration has mostly been driven by the desire to provide better anatomic localization. PET, SPECT, and sometimes even CT imaging techniques have

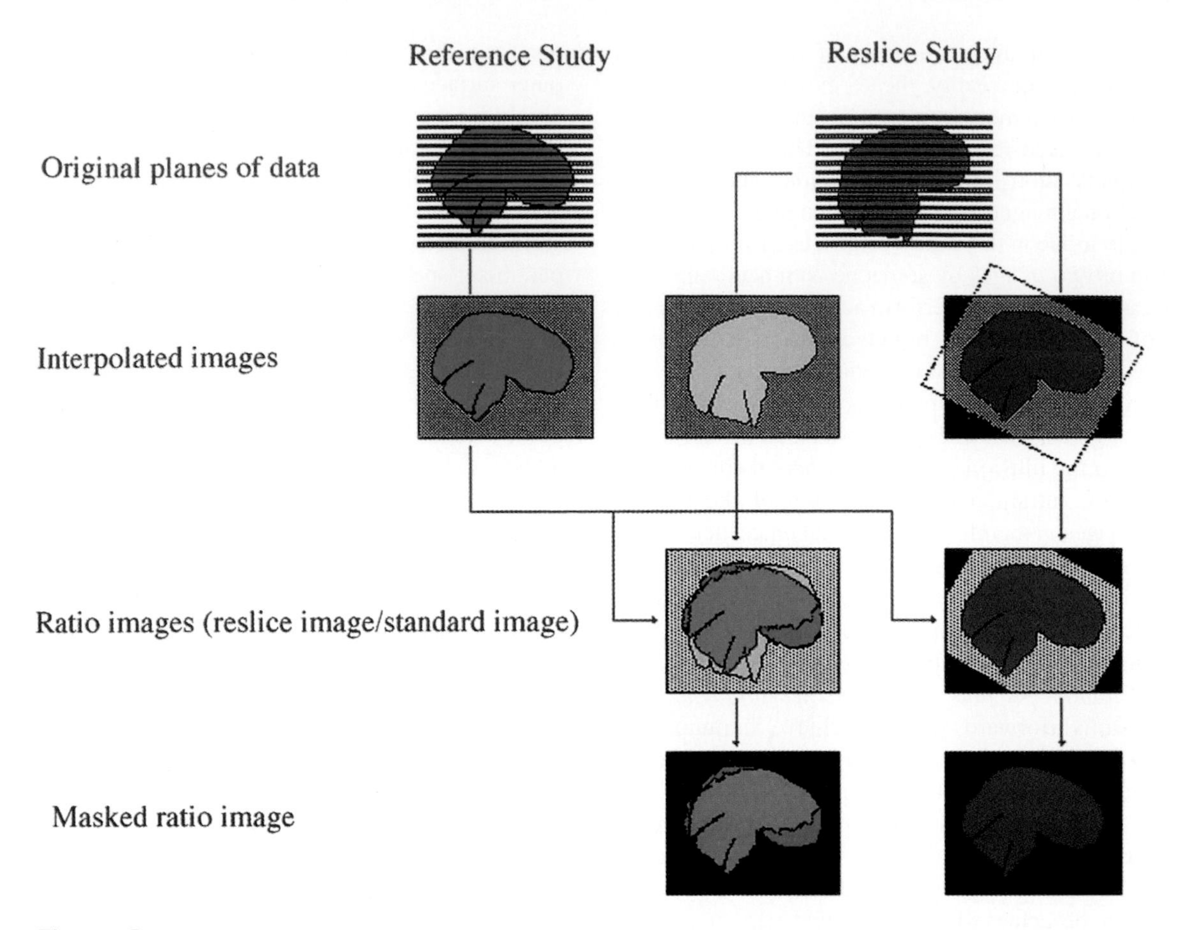

Figure 2 Overview of the strategy used by the automated image registration algorithms. The two image sets to be registered are represented schematically in blue and orange on the first row. The blue study has been designated as the "reference study" and the objective is to reorient the orange "reslice study" to match it as closely as possible. Tomographic data are frequently acquired with voxels that are not perfectly cubic, so both data sets must be interpolated to cubic voxels (second row) before rigid body rotation can be applied. The brain shown in yellow shows the orange image after interpolation with no rigid body rotation or translation. The brain in red shows the orange images after interpolation and rigid body rotation and translation to the position required for registration with the blue image set. The yellow and red images represent only two of an infinite number of possible orientations of the orange data set. The automated registration algorithm attempts to iteratively converage upon the position shown in red. It does this by computing the ratio of the reoriented "reslice study" to the "reference study" on a voxel-by-voxel basis. This is shown schematically in the third row. After masking areas outside the brain in the "reference study," the resulting masked ratio image (fourth row) has heterogeneous values when the images are poorly registered (yellow position) and homogeneous values when the images are well registered (red position). The heterogeneity of the masked ratio image can be quantitated by computing the standard deviation of the voxel values across all unmasked voxels. Calculus is used to minimize this standard deviation by adjusting the parameters that dictate the repositioning of the orange data set. The algorithm concludes that the images are registered when the standard deviation has been minimized. This technique is described in detail in Woods *et al.* (1992) and has been modified for intermodality registration (e.g., registration of MRI and PET images) as described in Woods *et al.* (1993). The same general strategy is also employed by the 7, 9, 12, and 15 parameter intersubject registration models described in the third section of this chapter, but in these cases, image distortion is not limited to rigid body rotation and translation.

sufficiently poor resolution or discrimination between tissue types that it is not possible to identify landmarks such as sulci and gyri with precision. For example, although resolution has steadily improved, current PET imaging methodology used for functional imaging allows for unequivocal identification of only the most prominent gyral and sulcal landmarks. Given that functional organization in primary cortical areas is known to be fairly tightly linked to gyral patterns, it is desirable to place any functional imaging activation site in an anatomic context. In the past, anatomic localization in PET data has been estimated by mapping the data directly into Talairach space (see the third section of this chapter) without the benefit of MRI registration (Fox *et al.*, 1985; Friston *et al.*, 1989). This approach does not allow gyral locations to be estimated with sufficient certainty to fully resolve even such prominent structures as the precentral and postcentral gyri in any given individual (Talairach *et al.*, 1967) (see Fig. 8).

Registration of a lower resolution PET, SPECT, or CT image to a MR image from the same individual offers a straightforward solution to the problem of anatomic localization. The Procrustes and surface matching approaches described in the previous section can also be applied in this context without modification. Because the two images being registered are likely to have very different image characteristics and resolution, considerably more neuroanatomic experience and expertise are required to identify homologous point landmarks across imaging modalities. Likewise, for surface matching approaches, it is important to ensure that the surface identification algorithm performs similarly across imaging modalities or else to introduce additional scaling parameters into the rigid body rotation model to allow adjustments for any differences that may be present.

At first glance, the automated registration technique described in the previous section would not appear applicable to the intermodality problem since the ratio of a PET image to a coregistered MRI image should be nonuniform. This nonuniformity is predominantly due to the fact that different tissue types will show markedly different ratios. However, within a given tissue type, a much more uniform ratio is expected. We have taken advantage of this fact to generate a modified version of the automated algorithm that maximizes the uniformity of the ratio within each tissue type, where tissue typing is based on MRI voxel intensity (Woods *et al.*, 1993). In order to use voxel intensity successfully for segmenting into tissue types, it is necessary to manually edit scalp, skull, and meninges from the MRI images. This approach has been validated using fiducial markers rigidly attached to the skull of patients prepared for epilepsy surgery and generates registration results with an accuracy of 2–3 mm. As with the intramodality algorithm, computation times are practical on standard workstations and the accuracy of the technique compares favorably with that of other available registration methods (Strother *et al.*, 1994).

3. Registration of Nontomographic Functional Imaging Data

For tomographic data, the solution of registration problems is greatly facilitated by the fact that the images provide a full three-dimensional set of coordinate locations. This is not the case with electrophysiologic or optical techniques. For functional imaging using optical intrinsic signals, it is currently necessary to ensure that each of the images is obtained with the camera and lighting oriented in exactly the same way with respect to the cortical surface. Intramodality registration to account for differences in head position relative to the camera has not been developed and would clearly require some means of assigning three-dimensional coordinate locations to each point on the surface. This could either be achieved by using multiple cameras to provide stereoscopic depth perception analogous to that in the human visual system or the information might be derived by registering two-dimensional images from a single viewing point with a three-dimensional MRI data set.

B. Mapping Other Types of Functional Data to MRI

1. Noninvasive EEG and MEG

For noninvasive EEG and MEG the problem of registration to MRI data can conceptually be broken down into two components. The first component involves relating the locations of the signal detecting elements to the MRI data. This is most easily achieved by digitizing the location of the scalp surface and the location of the detector elements in the same frame of reference (Wang *et al.*, 1994). The scalp surface coordinate locations can then be matched to the same surface on MRI using the surface matching approach described earlier. This effectively allows the EEG electrodes or the MEG detectors to be mapped directly onto the MRI scan.

Unlike tomographic noninvasive imaging techniques, MEG and EEG do not allow for unambiguous three-dimensional localization of the source of signals without further constraints. Consequently, the second phase of registering MRI to functional MEG or EEG data is not straightforward. This phase involves localizing the signal relative to the electrodes or detector elements. In order to facilitate solution of this "inverse problem" (Balish and Muratore, 1990), anatomic information from the MRI scan itself may be helpful (Wieringa *et al.*, 1994). Likewise, there is considerable interest in utilizing the results of functional imaging studies to constrain the anatomic localization of activation sites while using the electrical data to provide temporal information on a time scale much shorter than that currently accessible to functional imaging techniques. This sort of approach is an excellent example of how integration of different functional imaging techniques into anatomic space defined by MRI scanning can provide information beyond that available when using the techniques in isolation.

2. Mapping Intraoperative Data onto MRI

In order to map intraoperative data onto a MRI or other imaging study, it is necessary to establish a fixed coordinate system within the operating room environment. Traditionally, stereotactic head frames have been used for this purpose (Zhang *et al.*, 1990). The stereotactic frame is rigidly and invasively anchored to the patient's skull in a standard orientation with respect to external landmarks. The degree of standardization across subjects is sufficiently good that stereotactic coordinates were once used to relate anatomic structures to relatively crude anatomic landmarks for surgical purposes long before tomographic imaging techniques were available (Talairach *et al.*, 1967). The Talairach coordinate system described in the third section of this chapter was developed specifically for this purpose. If the stereotactic frame is present during magnetic resonance or other tomographic imaging, the frame itself provides a reliable external fiducial system of landmarks that can be used for registration purposes.

More recently, interest has grown in the use of intraoperative "frameless stereotaxy" (Barnett *et al.*, 1993; Olivier *et al.*, 1994; Sandeman *et al.*, 1994). Using this technique, a probe is used to digitize the three-dimensional coordinates of various anatomic landmarks on the patient and these landmarks are registered to homologous landmarks on a MRI scan in the operating room environment. So long as the patient (and more importantly, the brain) does not move, this technique provides the same information that could be obtained from a stereotactic head frame without the need for the frame itself. Other imaging modalities can then be registered to the MRI scan to provide integration of all types of information into the operating room frame of reference.

Developments are currently under way to provide actual MRI scanning intraoperatively (Crelier *et al.*, 1994; Jolesz and Blumenfeld, 1994). This technique promises to unify the operating room and the MR images into a single frame of reference without the need for registration. This approach will allow the imaging information to be updated instantaneously as tissue is resected or as the brain swells. It may someday be possible to perform intraoperative functional imaging and to integrate this information on line with all of the other relevant functional and structural information. Advances in virtual reality technology may even eventually make it possible for a neurosurgeon to visualize preoperative and intraoperative functional information projected directly onto the surgeon's view of the operative field.

C. Anatomic Labeling of MRI Data

MRI scanning provides an enormous amount of anatomic detail at high resolution. However, when formatted as two-dimensional slices, it is extremely difficult to identify the three-dimensional pattern of gyration that is needed to name or label the various sulci or gyri. Numerous commercial packages are available for converting the two-dimensional images into three-dimensional surface images that are much better suited to the identification of such surface landmarks. The combination of high resolution MRI and volume rendering can produce images that rival direct visual inspection of the brain surface. At the present time, expert neuroanatomic opinion is still required to apply the proper neuroanatomic names to the various sulci and gyri. At some point in the future, it is hoped that this task will be relegated to a computer algorithm that incorporates the necessary neuroanatomic expertise. Some of the intersubject registration techniques discussed in the third section of this chapter may serve to simplify this process by putting brains into a relatively standard orientation and shape. However, gyral patterns are sufficiently variable across subjects that a relatively sophisticated pattern recognition system will almost certainly be required (Ono *et al.*, 1990). While major landmarks are generally easily identified, some sulci and gyri are so variable that it is difficult to apply a universal unambiguous nomenclature. A better understanding of the genetic, developmental, and functional significance of these variations will be important in determining how best to apply an appropriate systematic nomenclature to these areas. Ongoing projects to collect MRI data from representative samples of various human populations are likely to contribute substantially to the database necessary to make these decisions. Cytoarchitectonic and chemoarchitectonic anatomic boundaries may be even more relevant to function than simple gyral and sulcal patterns. Surprisingly little is known about the relationship between cytoarchitectonic boundaries and gyral landmarks. In addition to cytoarchitectonics, biochemical markers may be useful to delineate important functional boundaries. The study of cryomacrotome sections of human brains allows for the integration of microscopic and macroscopic anatomy with the precision necessary to determine the extent to which there is a consistent relationship between the macroscopic and the microscopic landmarks. Precise quantitation of the relationship among gross anatomic landmarks, microscopic anatomic or biochemical landmarks, and functional landmarks will be essential for developing a probabilistic atlas of function–structure relationships. Such an atlas is a first step toward developing a complete functional anatomic map of the human brain in a generic sense and a good starting point for developing such a map of any given individual's brain for clinical purposes.

III. Clinical Examples of Intrasubject Multimodal Registration

The following case histories illustrate the use of intrasubject multimodal registration in a clinical context. All of the images described here were registered using the automated image registration techniques (Woods *et al.*, 1992, 1993) described in the preceding section.

A. Case Number 1

Clinicopathophysiologic correlation in a patient with new onset deafness. A 50-year-old businessman suffered a left subcortical hemorrhage in 1990, with no permanent clinical deficits. In 1991, he suffered a second contralateral subcortical hemorrhage that resulted in permanent deafness. As shown in Figs. 3A and 3B, both hemorrhages were located ventral and lateral to the posterior half of the putamen.

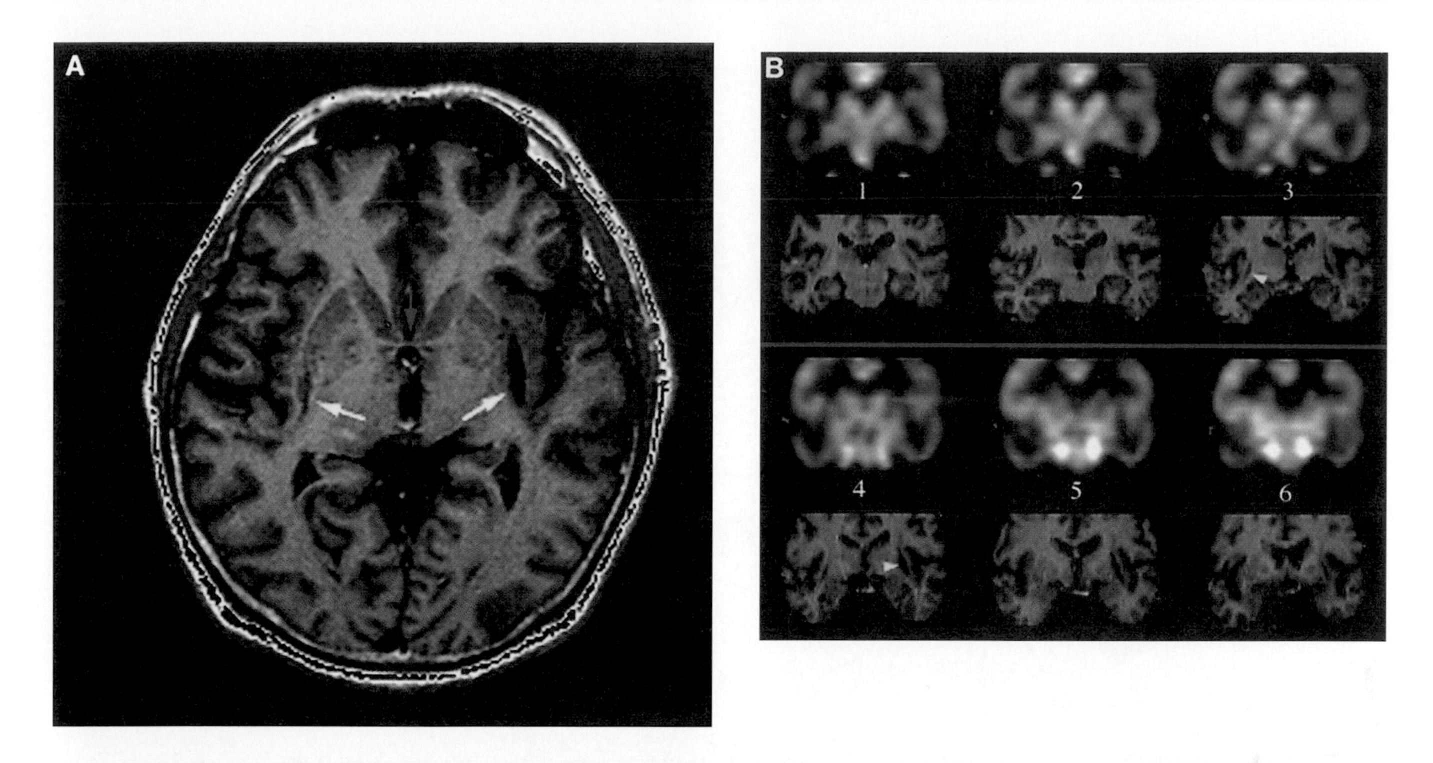

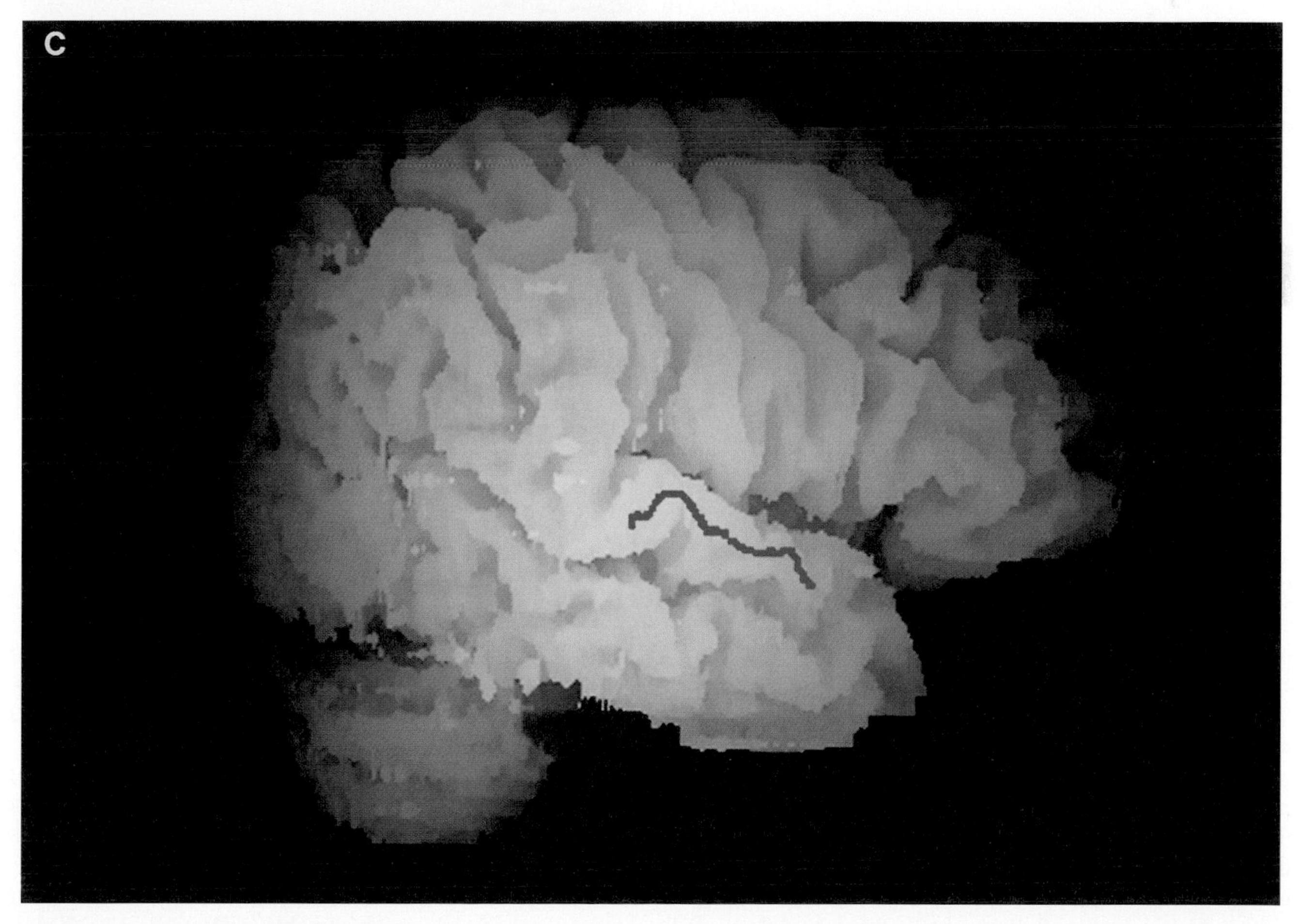

Figure 3 *For captions, see overleaf.*

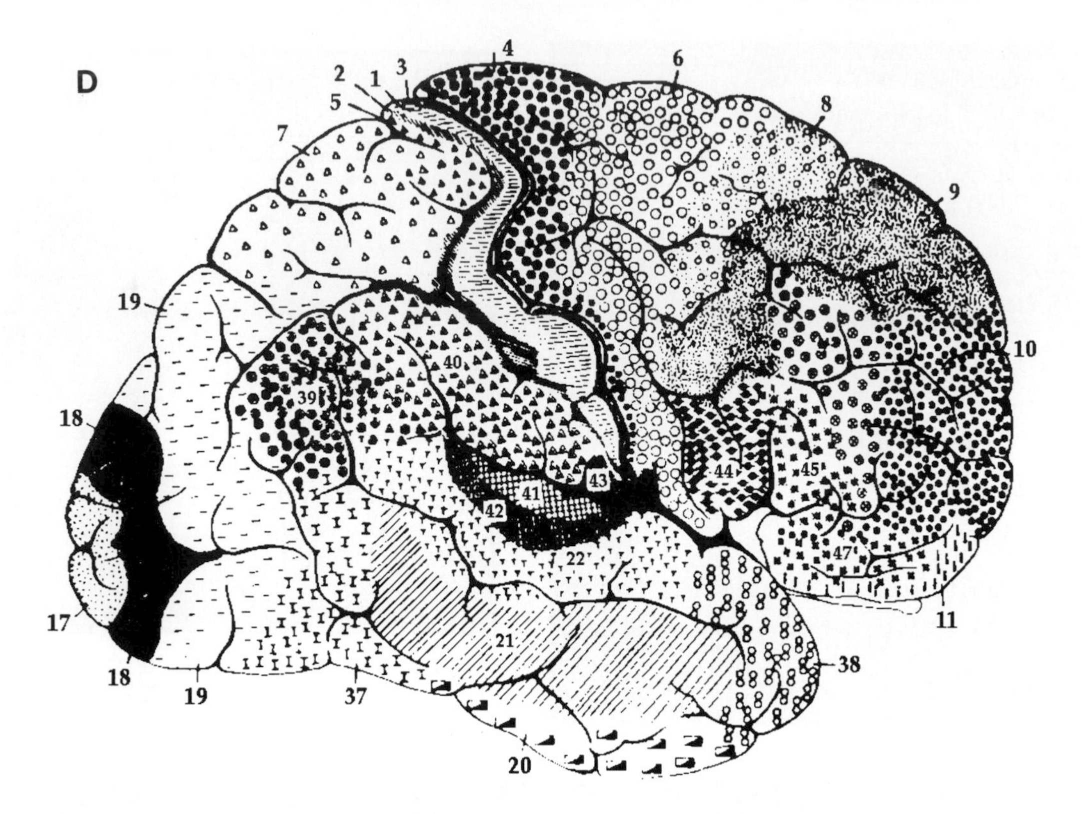

Figure 3 A patient with deafness following subcortical hemorrhages (see Case Number 1). (**A**) A transverse section of a MRI scan passing through the right- and left-sided hemorrhages (yellow arrows) and the anterior commissure (red arrow). (**B**) Coronal sections of coregistered PET blood flow and MRI scans of the patient. The six sections start at the posterior end of the Sylvian fissure and move forward at 5.2-mm intervals. The points shown in pink correspond to the crest of the right superior temporal gyrus as traced on the 3D rendering of the cortical surface shown in **C**. The subcortical hemorrhages are again shown in yellow on the coronal MRI sections. Note the profound hypoperfusion of the right superior temporal gyrus on the PET study despite normal anatomy on the MRI scan. (**D**) Cytoarchitectonic map of Brodmann showing the close correspondence between Brodmann areas 41, 42, and 22 and the right hemisphere brain region hypoperfused as a result of subcortical deafferentation by the most recent subcortical hemorrhage. The Brodmann map was derived by electronically mirroring the published Brodmann map of the left hemisphere to create an approximate "right hemisphere" for comparison purposes. Right-left brain asymmetries are prominent around the Sylvian fissure, and this may account for some of the anatomic discrepancies between the brain of the patient and the Brodmann brain. Other anatomic differences (e.g., the overall shape of the brain) may result from fixation artifacts in Brodmann's pathological specimen, population differences (the patient is from the Philippines), or simple individual variability. Note also that the top of the brain was out of the field of view of the PET scanner and, to a lesser extent, out of the field of view of the MRI scanner, accounting for the missing data at the top of the brain in the patient images.

Several other patients with deafness secondary to bilateral strokes in this same location have been reported (Tanaka *et al.*, 1991). This case illustrates the classical neurobehavioral approach of correlating a functional deficit, in this case deafness, with a particular anatomic lesion. This is a rather unusual example of use of the classical technique in the sense that the requirement for two anatomic lesions actually proves the existence of a redundant parallel pathway. The initial hemorrhage presumably interrupted projections from the left medial geniculate nucleus to the left auditory cortex in the superior temporal gyrus. No clinical deficit resulted at that time because the parallel pathway from the right medial geniculate nucleus to the right auditory cortex remained intact, allowing the patient to hear (both medial geniculate nuclei receive auditory information from both ears and project this binaural information onward to the auditory cortex). The functional damage caused by the initial hemorrhage was unmasked when the second hemorrhage destroyed the redundant parallel pathway.

The patient underwent PET blood flow scanning using the tracer $H_2{}^{15}O$. Data from this study were registered to the patient's MRI scan. Figures 3B and 3C demonstrate that the anatomically normal right superior temporal gyrus was markedly hypoperfused. This hypoperfusion at a site distant from the hemorrhage presumably reflects decreased neuronal firing in this region due to deafferentation, a process known as "diaschisis" (Meyer *et al.*, 1993). Comparison of the blood flow images, the three-dimensional renderings of the cortical surface, and the cytoarchitectonic maps of Brodmann (1909) (Fig. 3D) shows that the area of diaschisis corresponds to Brodmann's areas 41, 42, and 22. Areas 41 and 42 are known to receive projections from the medial geniculate and project in turn to area 22 (Carpenter, 1976). The presence of relatively normal blood flow in the left superior temporal gyrus may indicate some degree of reorganization since the time of the earlier left-sided injury or may indicate that the injury itself was less extensive. Blood flow measurements were made both at rest and with the presentation of auditory stimulation with repeated bursts of white noise. The patient was unable to hear these bursts, and no change in blood flow was produced in the superior temporal gyrus of either hemisphere. This case illustrates the value of anatomic data from MRI scanning when interpreting the lower resolution PET images and the value of combining information from different imaging modalities. It also illustrates how three-dimensional rendering of the cortical surface identified by MRI can be used to assign proper neuroanatomic labels to features seen on lower resolution PET studies and how this allows for comparison with published data from other sources such as cytoarchitectonic data. The combination of the clinical history, MRI scanning after each of the clinical events, blood flow measurements, functional clinical assessment, functional imaging, and knowledge of the neuroanatomic organization of the auditory system allows cross validation of the many different sources of information and a unified assessment of the patient's clinical situation.

B. Case Number 2

Preoperative motor mapping in a patient with a motor cortex lesion. A 36-year-old man with epilepsy since age 4 was found to have a cystic lesion in the right prefrontal gyrus on MRI scanning. The patient's seizures had become refractory to treatment with anticonvulsant medications and surgery was planned for both diagnostic and therapeutic purposes. Functional mapping of the primary motor cortex can be easily and reliably performed with PET in normal subjects (Grafton *et al.*, 1991) and in patients (Grafton *et al.*, 1994), and such preoperative mapping was requested to assist with surgical planning. Figure 4A shows a three-dimensional rendering of the cortical surface based on the patient's MRI scan, illustrating the value of rendering techniques in identifying cortical landmarks. The patient underwent a series of $H_2{}^{15}O$ blood flow studies both at rest and while tracking a moving object on a computer screen by moving his wrist, his shoulder, or his foot. Figure 4A shows the relationship between the sites identified as hand, shoulder, and foot motor cortex and the surface anatomy. The cystic lesion has been digitally resected revealing extensive representation of motor cortex controlling the distal left upper extremity in the floor of the cyst. The neurosurgeon was advised that resection of the cyst would probably result in postoperative motor deficits involving the left hand. These concerns were discussed with the patient and a decision was made to proceed with the surgery. Intraoperative photographs are shown in Figs. 4B and 4C. Frameless stereotaxy (Fig. 4D) was used to interconvert MRI coordinates and locations in the operative field. Intraoperatively, electrical stimulation immediately posterior to the cystic lesion produced movements of the shoulder. Stimulation over the lesion itself produced movement of the thumb and the second and third digits. An intraoperative biopsy demonstrated that the cystic lesion was a mixed glioma with predominantly ependymoma features so the lesion was resected. After the resection, electrical stimulation of the remaining tissue produced shoulder movements, but no movements of the hand could be elicited. Postoperatively the patient had weakness of the left hand which showed some modest improvement over the following week.

This case illustrates the integration of preoperative noninvasive functional assessment with intraoperative functional evaluation. For the functional imaging study, a motor task was manipulated as the experimental variable and the relevant anatomic sites were revealed by the data analysis. In contrast, for the intraoperative functional assessment, an anatomic site of electrical stimulation was the experimental variable, and the function was revealed by observing the patient's motor response. Both mapping techniques produced similar results, and the ability to interconvert coordinate locations on the MRI and in the operating room facilitated rapid assessment and integration of all of the available information. Finally, the resection of the cystic lesion, together with the postoperative deficits of the left hand, provided a classical neurobehavioral assessment of the relationship between function and structure, with results identical to those predicted by functional imaging and by intraoperative cortical mapping.

C. Case Number 3

A 30-year-old woman first began to experience seizures as a child after a bout of meningitis. The seizures were often associated with a visual aura. On examination, a left visual field deficit was noted by her physician. Formal field visual testing demonstrated sparing of the central 18° of vision (see Fig. 5C), accounting for the fact that the patient herself had previously been unaware of this deficit. Despite medication, the patient experienced as many as 20 seizures per day. Electroencephalography suggested that the seizures were arising in the right occipital lobe, and a surgical workup for possible right occipital lobe resection to control the seizures was initiated. MRI scanning demonstrated multiple right occipital abnormalities interspersed with more

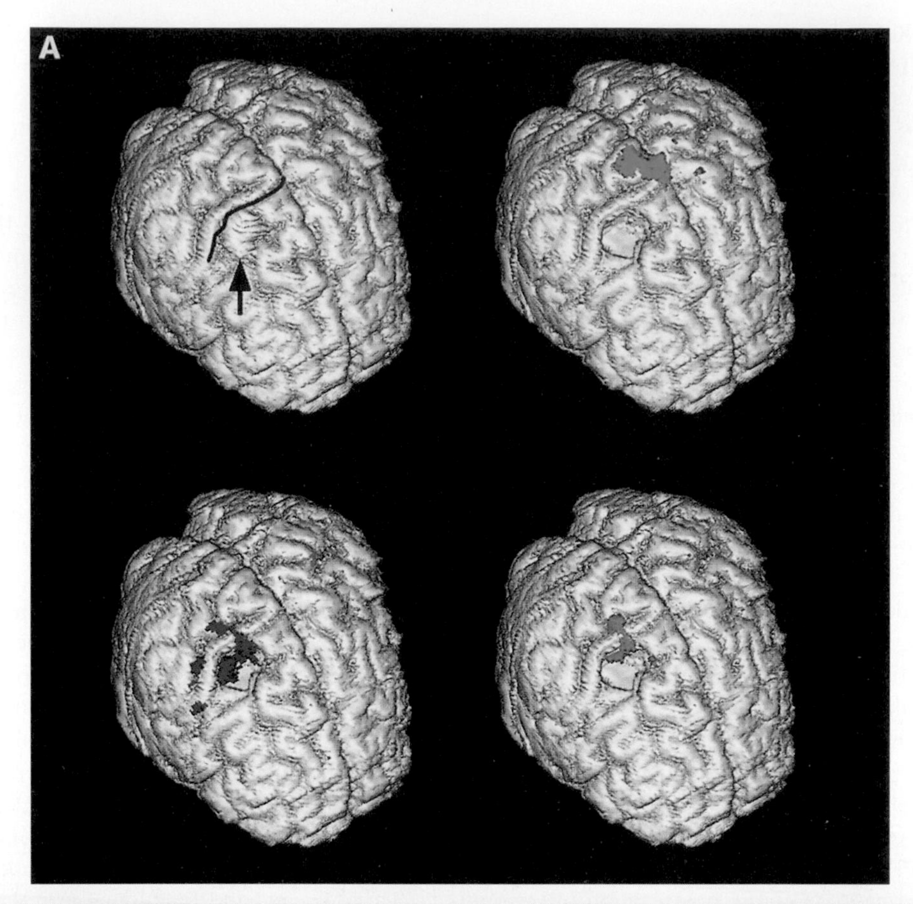

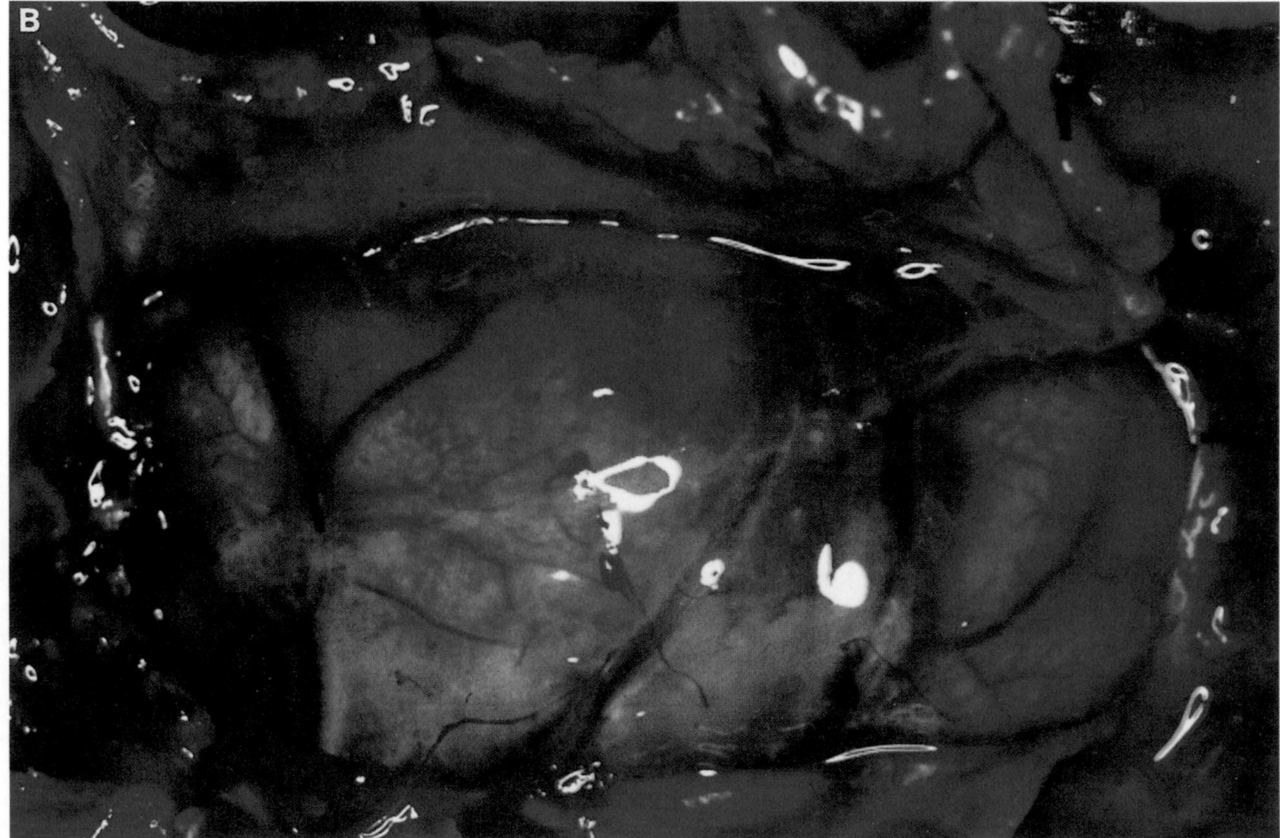

Figure 4 *For captions, see opposite.*

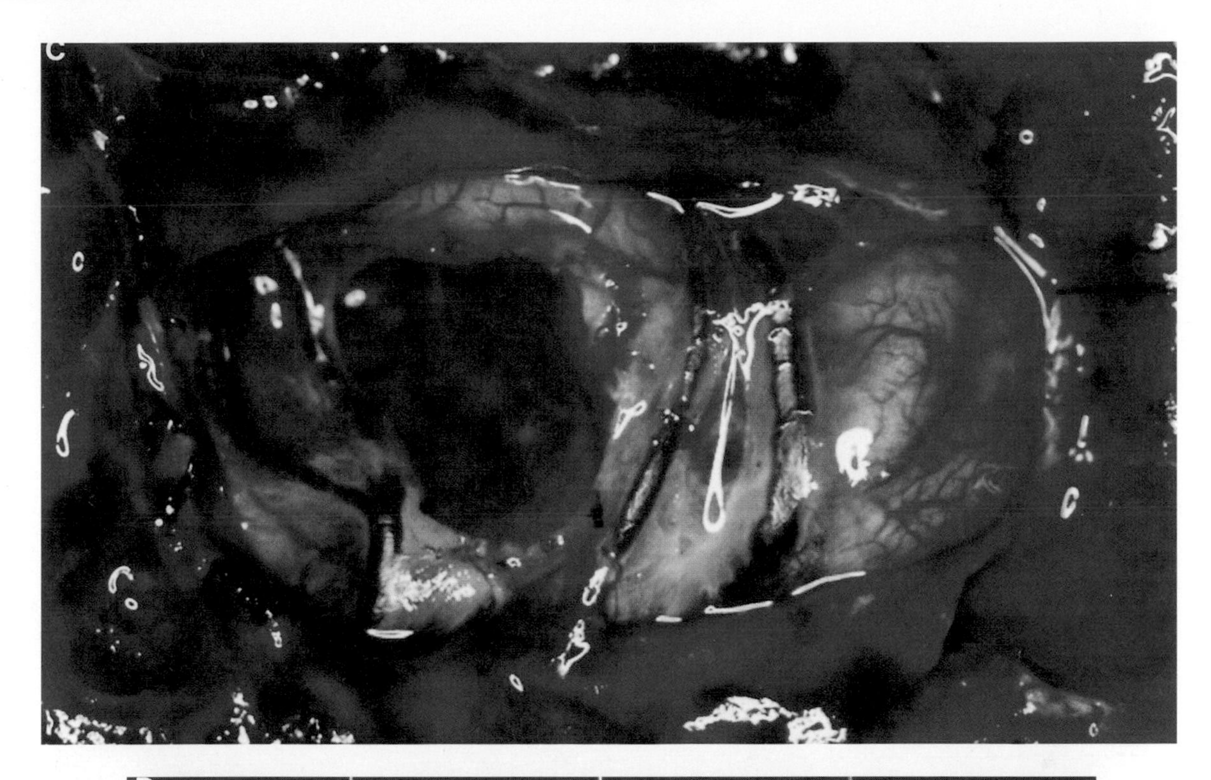

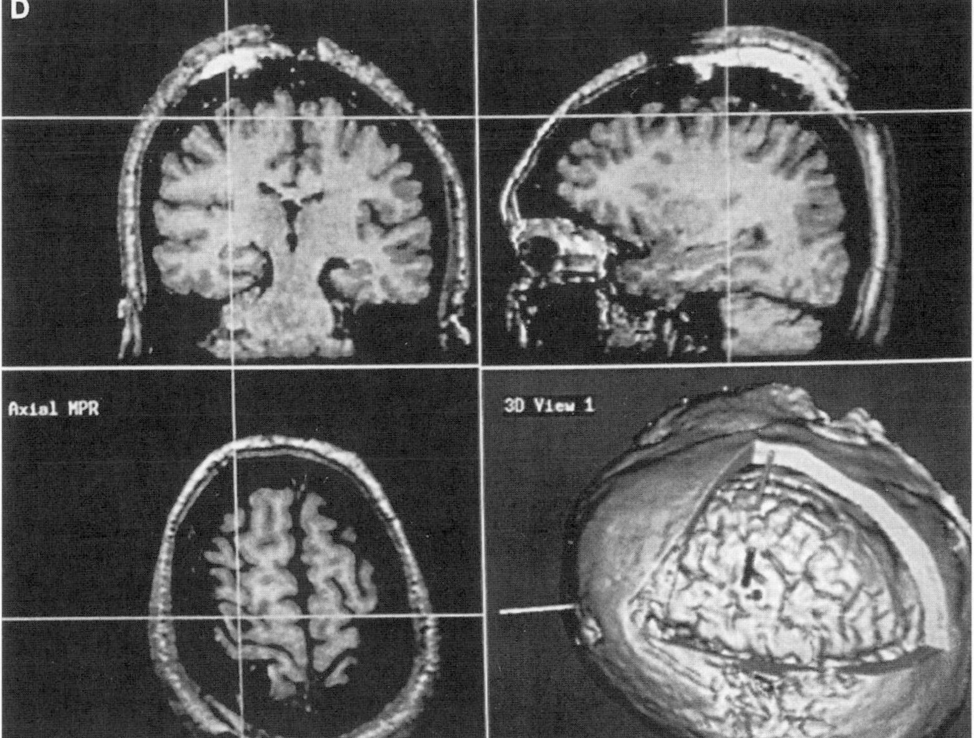

Figure 4 Motor function in a patient with epilepsy and a right frontal cystic lesion (see Case Number 2). (**A**) Three-dimensional reconstructions of the patient's cortical surface based on MRI scanning. The black arrow points to the cystic lesion, the roof of which has been digitally resected in the other three images. The lesion extends posteriorly to the central sulcus which is identified (based on anatomic considerations) and traced in black. Areas where a motor tracking task produced increased blood flow are shown on the other three images. The areas shown in green were produced by motor tracking with the right foot, the areas in blue by arm tracking involving shoulder movements, and the areas in red by hand tracking involving wrist and finger movements. (**B**) Intraoperative photograph of the cystic lesion before it was unroofed. (**C**) Intraoperative photograph of the cystic lesion after it was unroofed. Note the relatively small amount of cortex actually exposed in the operative field and how difficult it is to orient the intraoperative photographs with respect to the three-dimensional renderings of the MRI scans. (**D**) The frameless stereotaxy system display used intraoperatively for this patient. By establishing common landmarks on the MRI scan and on the patient's head, a probe within the operative field can be localized on the MRI scan during the surgery. The cystic lesion can be seen at the marked locations on coronal, sagittal, and transverse images.

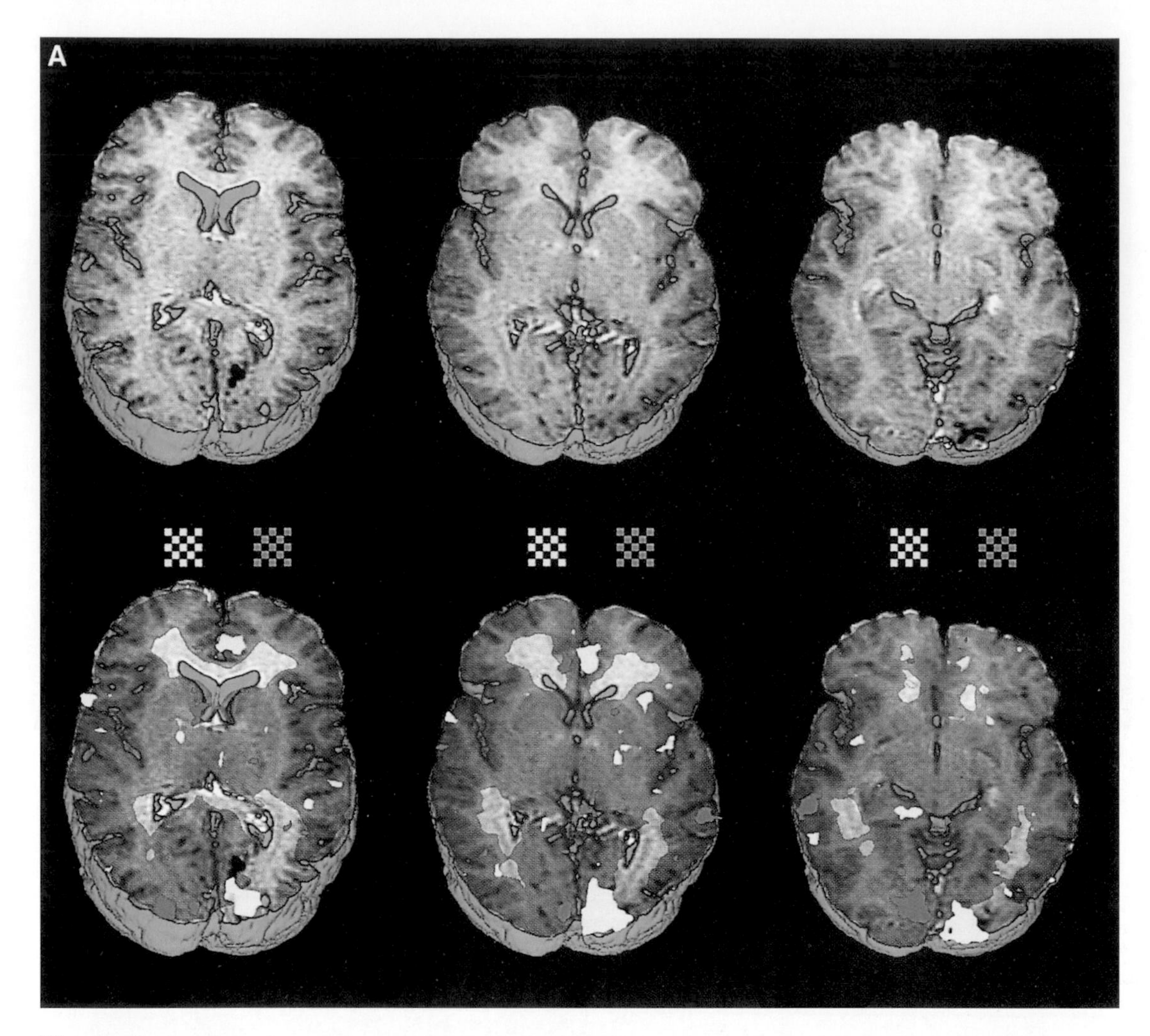

Figure 5 Functional assessment of vision in a woman with occipital epilepsy (see Case Number 3). (**A**) Three-dimensional rendering of the brain surface in these images is shown in blue. The top row shows the MRI scan on the cut surface of the brain. Note the structurally abnormal dark areas in the right occipital region. The bottom row shows PET data superimposed onto MRI data. Glucose metabolism as assessed by [^{18}F]fluorodeoxyglucose uptake is shown in red with a display threshold chosen such that all cortical areas should be above the threshold. Note that the right occipital cortex is hypometabolic with no red seen in this region. The areas in yellow are regions where increased blood flow was seen in response to an alternating checkerboard pattern in the left (abnormal) visual field. The areas in orange show blood flow increases in response to similar functional stimulation of the right visual field. Note that the right occipital region does respond strongly to left visual field stimulation. (**B**) Preoperative (left) and postoperative (right) MRI scans demonstrating the area of surgical resection. These images are displayed with the right hemisphere on the right to facilitate comparison with the three-dimensional renderings. The right occipital area that responded to left hemifield stimulation during the PET blood flow study is shown in pink. Note that the surgical resection spared most of this area. The preoperative and postoperative MRI scans were done using different pulse sequences, accounting for the differences in gray–white contrast between the two image sets. Because of these differences, the intermodality automated image registration algorithm was used for coregistration. (**C**) Preoperative (top) and postoperative (bottom) visual fields. The fields for the right eye are shown on the right and the left eye on the left. The preoperative fields shown in pink were tested with a 3-mm test object. The postoperative fields shown in green were tested with a 2-mm test object and are more restricted as a result. The patient was able to count fingers in postoperative areas of the visual fields shown in orange and could detect hand motion in postoperative areas shown in yellow.

normal appearing tissue (Figs. 5A and 5B). Visual-evoked potential studies were difficult to interpret, suggesting a markedly abnormal organization of the visual cortex. In order to gain a more detailed anatomic understanding of the visual cortex organization, $H_2{}^{15}O$ PET studies were performed. Previous studies in normals have demonstrated that an 8-Hz alternating checkerboard pattern produces marked blow flow response in primary visual cortex (Fox and Raichle, 1985). Consequently, 8-Hz right or left hemifield alternating checkerboard patterns were used during the PET

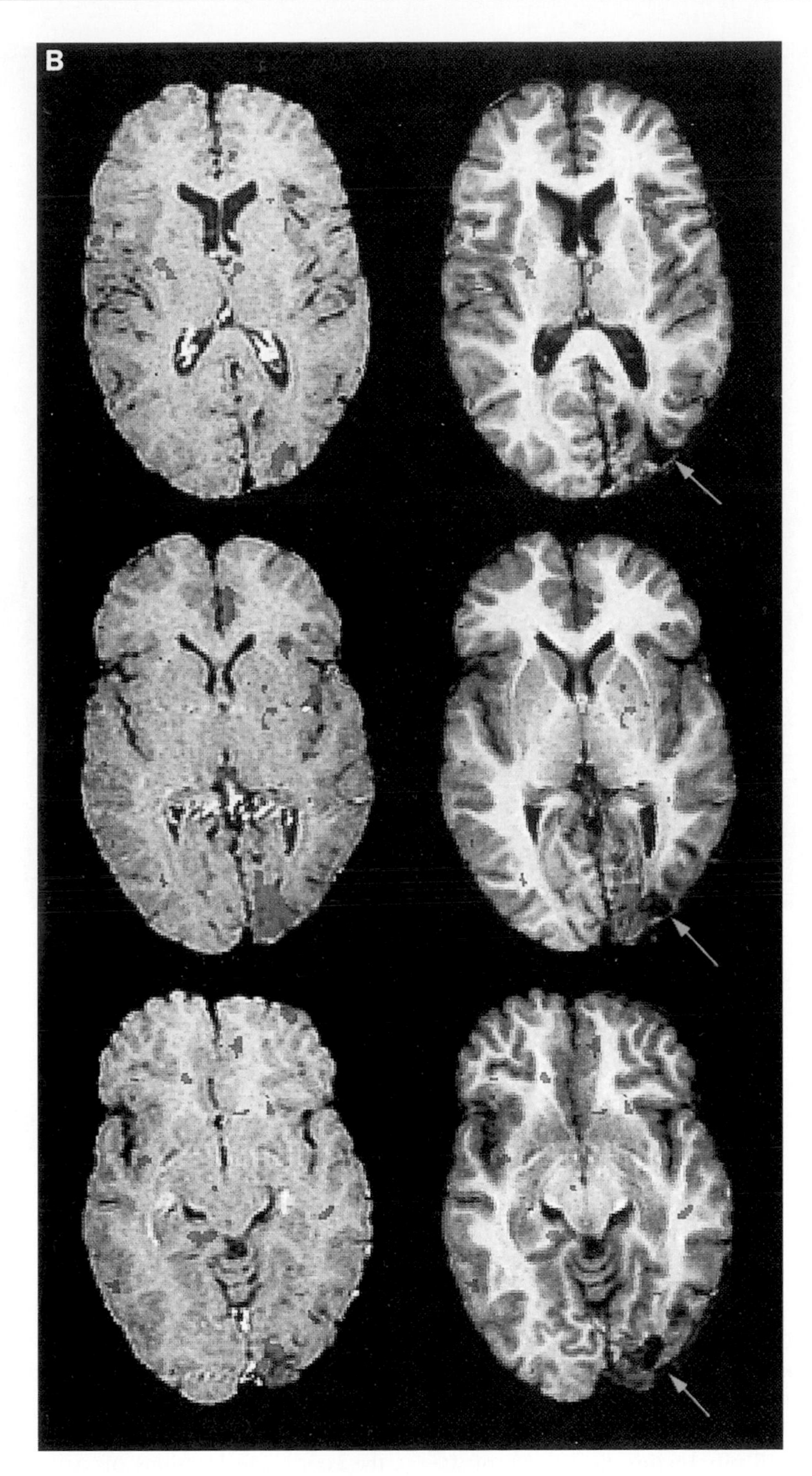

Figure 5 *(continued)*

study and compared to a control task without checkerboard stimulation.

Figure 5A shows the results of the PET functional mapping coregistered with the patient's preoperative MRI scan. An area in the right occipital lobe responded to left visual field stimulation (yellow area in Fig. 5A) and is presumably responsible for the sparing of central vision on the left. The area of activation is shifted posteriorly in the right

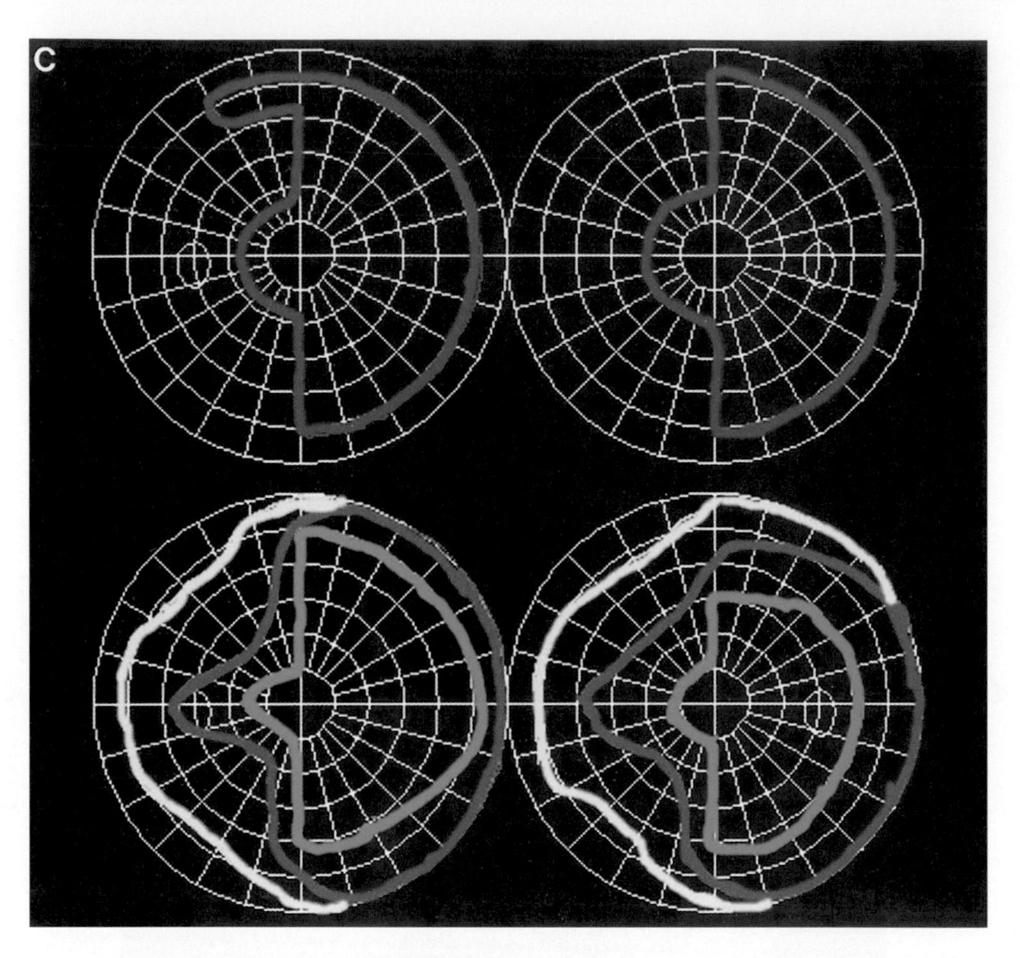

Figure 5 *(continued)*

hemisphere relative to that in the left hemisphere. The location in the left hemisphere (orange area in Fig. 5A) is more typical of that seen in normal subjects. The atypical positioning in the right hemisphere either may indicate reorganization of visual cortex or could simply represent structural displacement by the abnormalities seen on MRI.

The patient's physicians were notified that the activation PET study indicated that the right hemisphere was functioning to provide vision in the left visual field and that this function would be lost if a right occipital lobectomy was performed for seizure control. A detailed review of the EEG recordings suggested that a more limited resection of the occipital lobe might spare the functional area demonstrated by the PET scan, while still eliminating the source of the seizures. The notion that the seizures were arising more laterally was also supported by a metabolic PET study using [^{18}F]fluorodeoxyglucose. Between seizures, FDG PET studies commonly demonstrate decreased metabolism in the area that gives rise to the seizures (Engel *et al.*, 1990). Registration of the FDG PET study to the MRI scan (see areas shown in red in Fig. 5A) showed relatively normal metabolism in the region thought to be mediating vision in the right occipital lobe with marked hypometabolism more laterally.

After discussion with the patient, a decision was made to proceed with the more focal surgical plan. MRI scanning was performed with the patient wearing a stereotactic head frame, the PET $H_2{}^{15}O$ data were registered to this MRI scan, and the coordinates of the tissue in the right occipital lobe thought to mediate vision were determined in the stereotactic frame of reference. With the head frame in place, the patient was taken to the operating room and the relevant areas of occipital cortex were exposed. Special goggles were used to produce visual stimulation during the surgery, making it possible to confirm with electrical measurements of evoked potentials the coordinate locations of visual cortex predicted by PET. The epileptogenic region was resected. The area of resection can be seen in relation to the preoperative PET study results in Fig. 5B where postoperative and preoperative MRI scans have been registered to one another.

Pathologic examination of the resected tissue indicated that the structural lesions seen on MRI scanning were not due to tumor or infection and most likely represented changes secondary to the episode of meningitis in childhood. In

3 years of postoperative follow-up, the patient has had a single visual aura and only one seizure. Postoperatively, she retains vision centrally in the left visual field (Fig. 5C).

This case illustrates the integration of multiple modalities of functional assessment. Preoperative, intraoperative, and postoperative functional measures were all unified into a neuroanatomic context. In this case, it was important not only to localize normally functioning tissue but also to localize abnormally functioning tissue that gave rise to the seizures. Without the benefit of precise localization of functioning visual cortex, it would not have been feasible to perform such a limited resection, and a right occipital lobectomy, sacrificing all left visual field vision, would have been necessary. This case also incorporates an element of classical neurobehavioral assessment in the sense that visual fields were assessed preoperatively and postoperatively. Fortunately, the surgery did not create any new deficits, so the only conclusion that can be drawn by comparing the behavioral evaluations before and after surgery is that the resected tissue was not critically involved in visual field processing. One of the major goals of functional imaging of surgical patients is to minimize the conclusions that can be drawn using classical neurobehavioral techniques by preventing resection of brain areas that would result in clinical deficits.

IV. Intersubject Multimodal Integration

Since most information relevant to the relationship between structure and function within individuals can be placed in an anatomic context based on each individual's particular anatomy, the problem of intersubject multimodal integration can be reduced to the problem of accounting for the anatomic variability between subjects. Most commonly, this anatomic information will be provided by MRI scanning although in some cases lower resolution PET data or higher resolution cryomacrotome data (see Chapter 20) may also be utilized. Before considering the tools available for registering anatomic images across subjects, it is useful to review the various reasons for wanting to do so in the first place. The following practical applications are all dependent upon intersubject registration to account for individual anatomic variability:

1. Signal-to-noise enhancement by pooling data across subjects. Averaging of data across subjects is primarily used in situations where it is not possible to collect sufficient information from a single subject. This approach is frequently utilized in functional PET studies is order to minimize individual radiation exposure (Fox *et al.*, 1988). This topic will be discussed further later in relation to the Talairach coordinate system. Pooling of data across subjects can also be used in combination with classical neurobehavioral techniques by mapping brain lesions associated with a particular functional deficit into some common space and looking for an area of overlap.

2. Intersubject comparison of individuals or populations. In order to make statistically valid comparisons across individuals or groups of some functional signal arising from the brain, it is obviously important to ensure that anatomically homologous areas are being compared.

3. Coordinate-based communication of brain locations. As illustrated by the visual system of the macaque, the number of distinct functional anatomic areas that can ultimately be delineated may be larger than the number of convenient conventional neuroanatomic labels based on gyral patterns. Precise communication of neuroanatomic sites can be facilitated by the use of an accurate coordinatebased brain atlas.

4. Morphometric studies of anatomic variability. Here the object is not to eliminate or to control for anatomic variability but rather to quantitate it in order to better understand the genetic and developmental factors that affect brain size and shape. In addition, such studies can be helpful in choosing better models to use in controlling for such variability.

5. Intersubject registration as a starting point for computerized landmark identification and labeling. Even fairly simple methods of controlling for brain size and shape are likely to simplify the strategies required for automated definition of landmarks. Once such landmarks are identified, it may be possible to use them as a basis for even more accurate registration across subjects. Equally important, this approach should allow for user independent segmentation of neuroanatomic structures.

A. Mathematical Models for Intersubject Registration

Unlike the intrasubject registration problem, there is no *a priori* mathematical model of choice for intersubject registration. The biological rules of genetics and development are certain to be complex, and simple mathematical models with finite numbers of parameters should be viewed as only an approximation of this complexity. The very notion of registering data across subjects with continuous transformations may be ill posed at some level. For example, it is already known that some subjects may have three parallel gyri or sulci in areas where other individuals only have one or two (Ono *et al.*, 1990). Even more concerning is the possibility that the organization of functional areas on the cortical mantle could potentially be topologically nonequivalent across subjects. Thus, a particular functional landmark might border on partially or even completely different sets of functional landmarks in different subjects. Finally, different criteria may yield conflicting definitions of homology across subjects, with anatomic and functional criteria in disagreement.

These concerns are not raised to discourage attempts at intersubject registration, but rather to point out that the mathematical model used is *only* a model. Consequently, the choice of model depends on the nature of the data being analyzed and the question that is being asked. Practical considerations may predicate the use of a simpler model in preference to a more complex one. For example, averaging of PET data to improve signal to noise is an application where minimal user interaction is desirable and where the traditional use of heavy smoothing has made precise superimposition of cortical landmarks relatively unimportant. In contrast, for studies of brain morphometrics, precise identification of cortical and subcortical landmarks by a user with neuroanatomic expertise is essential to study design. For any particular problem, it is important to decide what degree of accuracy (using neuroanatomic expertise as a "gold standard") is required and to validate both the choice of model and the choice of tool to ensure that this degree of accuracy is met.

Simple visual inspection of a group of human brains provides compelling evidence that a simple rigid body model with six parameters is almost invariably inadequate to account for intersubject differences (Fig. 6). The only possible exception is the case of genetically identical twins without brain disease. Although a seven parameter model with global rescaling as the seventh parameter is a possibility, a nine parameter model (see discussion of the Talairach space) is the simplest model commonly used for human data. Increasing genetic distance between subjects may demand the use of increasingly sophisticated models to maintain a given level of accuracy. For example, attempts to reconcile the anatomy of Japanese brains with those of European brains suggest that a significant nonlinear distortion may be required (Thurfjell *et al.*, 1994). While such variability may be viewed initially as an unwanted nuisance, its presence is ultimately likely to enhance the depth of our understanding of the relationship between brain structure and function.

B. Tools for Intersubject Registration

The description of a tool for intersubject registration can be broken down into three subcomponents. The first subcomponent is the mathematical model, discussed earlier, that defines and limits the results that the tool can derive. The second subcomponent is a series of rules that govern the behavior of the tool and define the criteria for concluding that registration of the images is complete. Finally, if designed to serve as an atlas, the tool must provide some means of assigning an unambiguous (hopefully unique) coordinate for every location in the image being registered to the atlas. This section is not intended to provide a comprehensive review of all of the techniques used for intersubject registration, but rather to give an overview of the sorts of approaches that are currently in use. Attention will focus primarily on methods that have been implemented fully in three dimensions. While most two-dimensional techniques can be generalized to three dimensions, addition of a third dimension adds sufficient computational complexity to make many two-dimensional approaches impractical in three dimensions at the present time. Furthermore, proper validation of a tool for intersubject registration demands the use of a fully three-dimensional approach. For a number of reasons, it is appropriate to subdivide the discussion of intersubject registration tools into a discussion of those tools that utilize linear models and those tools that do not.

1. Linear Registration Tools

In the most general sense, a model can be considered to be linear if all points that were colinear prior to registration remain colinear after registration. This constraint is surprisingly nonrestrictive and allows for as many 15 independent parameters in the model. These linear models have a number of advantages, including conceptual and mathematical simplicity. The transformations can be formatted in terms of straightforward linear algebra, and the ready availability of algorithms and software for performing linear algebraic computations simplifies the implementation of such models. In addition, if linear transformations are combined or inverted, the result is still a simple linear transformation. This allows for a very general treatment of all linear models. Finally, linear models that do not incorporate perspective transformation have the advantage of being nonlocal in their effect. As a result, properties such as image resolution remain constant throughout the image after transformation. The primary disadvantage of linear models is that they cannot account for nonlinear distortions, and such distortions are clearly prevalent in biology. This issue has been discussed in detail by Bookstein (1991).

Depending on the particular constraints applied, an infinite number of different linear models are possible. Only five of these models are discussed in this chapter and for convenience they will be defined here. The models are also illustrated in Fig. 7.

1. Six parameter model. This is the rigid body rotation model described in the first section of this chapter. It will not be considered further in relation to intersubject registration.

2. Seven parameter model. This model adds a global rescaling factor to the six parameter model. This model is rarely used in practice and will not be discussed further.

3. Nine parameter model. In this model, a 6 parameter rigid body rotation and translation is followed by rescaling along the x, y, and z axes. The fact that these three axes receive preferential treatment makes the 9 parameter model less general than the other four models described here. Whereas the other four models generate parameters belonging to the same model when inverted or combined, inversion or combination of the results of 9 parameter models will generally require 12 parameters for their description. For all discussions of 9 parameter models in this chapter, the y–z plane will be assumed to coincide with the midsagittal plane

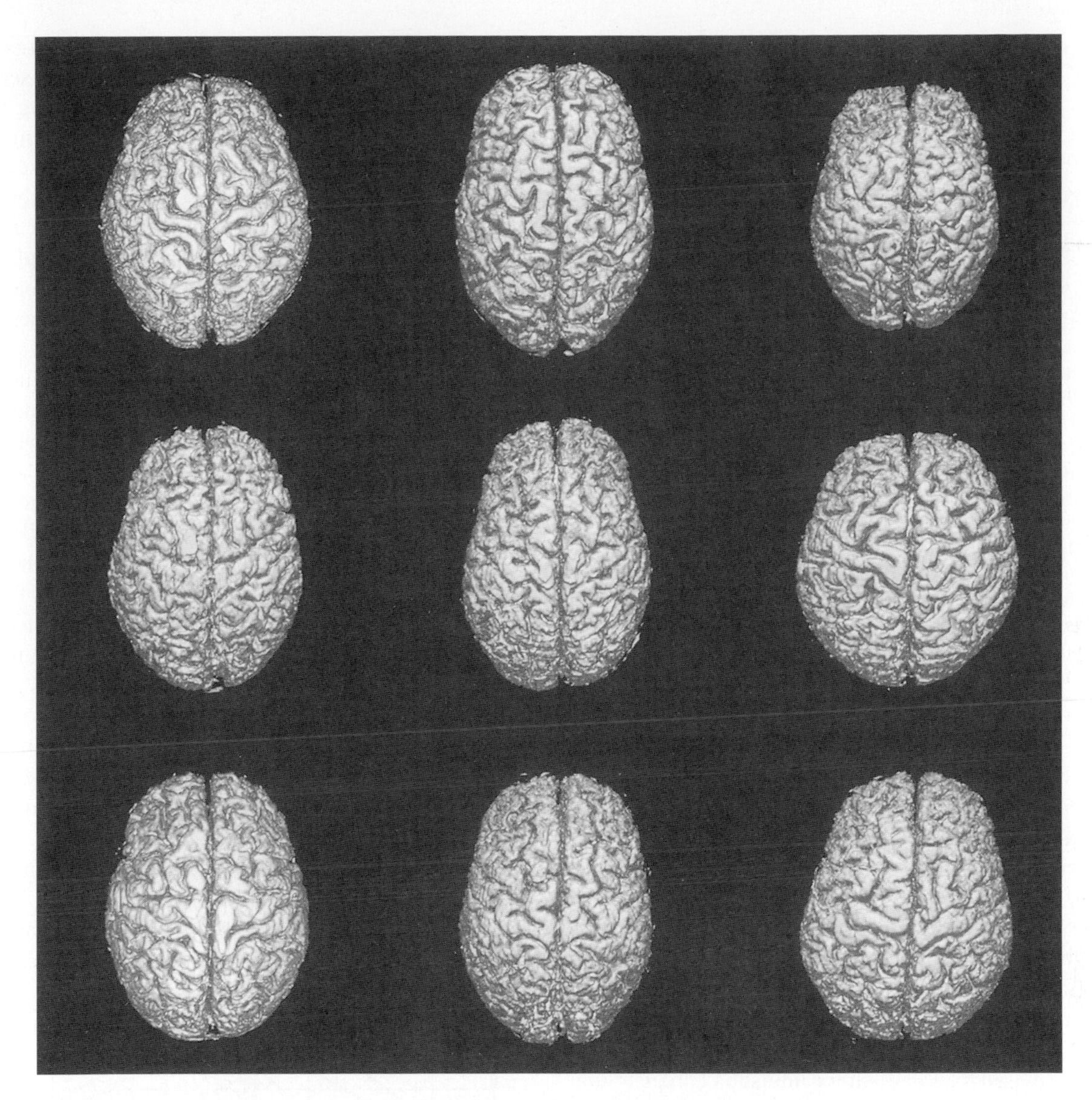

Figure 6 Three-dimensional renderings of the brains of nine healthy subjects. All of the brains are shown at the same scale and from the same orientation. Three of the subjects were female and six were male. Two subjects were black, one was Hispanic, and the other six were Caucasian. All subjects were right handed. Note the marked variability in brain size and shape.

of the brain and the y axis will be assumed to pass through the anterior and posterior commissures.

4. Twelve parameter model. This model, also known as the affine model, requires that parallel lines remain parallel after transformation. In addition to rigid body rotations and translations, this model can be viewed as allowing independent rescaling along any three arbitrarily chosen coordinate axes. In a sense, the 12 parameter model can be viewed as a superset containing all of the 9 parameter models involving rescaling along prespecified axes. Parallel lines will remain parallel after 12 parameter affine transformation.

5. Fifteen parameter model. This is the most general linear model and incorporates all of the features present in the 12 parameter model as well as allowing parallel lines to converge after transformation. This effectively introduces an element of perspective. The additional 3 parameters in this model specify the reference point in space that determines the three-dimensional perspective distortion. If this reference point is placed at infinity, a 12 parameter affine model results. Consequently, perspective distortions are always linear but are not generally affine.

a. The manual Talairach intersubject registration technique. Newcomers to the field of functional neuroimaging, particularly those reading the PET activation literature for the first time, are often bewildered by the abandonment of traditional neuroanatomic nomenclature in favor of a set of coordinates in "Talairach space." This bewilderment may be transiently alleviated by obtaining a copy of the 1988

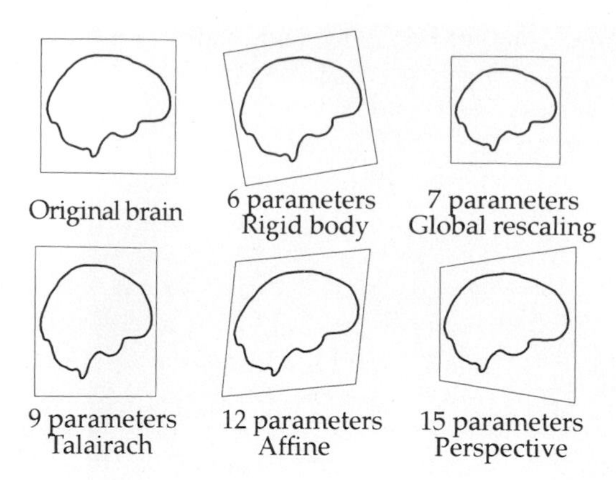

Figure 7 Linear transformation models. Two-dimensional representations of the types of distortions allowed by linear models with various numbers of parameters. Each model includes all of the degrees of freedom available to models with fewer parameters. The 6 parameter model allows for rigid body rotations and translations. The 7 parameter model allows global rescaling. The 9 parameter model allows independent rescaling along the three primary axes of the Talairach model. The 12 parameter model allows independent rescaling along any arbitrary set of axes (i.e., shears). The 15 parameter model allows perspective distortion and is the only one of these linear models that is not affine (i.e., parallel lines do not necessarily remain parallel after transformation with the 15 parameter model).

Talairach atlas (Talairach and Tournoux, 1988) where these three-dimensional coordinates can be referenced against the tracings of a single human brain mapped into the Talairach coordinate system using the original rules and methods described by Talairach and co-workers. However, the bewilderment is likely to return when trying to understand how the coordinates were actually derived from a given set of MRI or PET data. This section will first describe the original method devised by Talairach for intersubject registration (Talairach *et al.*, 1967; Talairach and Tournoux, 1988). An understanding of the original method is important in clarifying subsequent modifications and in interpreting the 1988 Talairach atlas. However, the reader is forewarned that although many groups indicate that they map data into "Talairach space," neither the mathematical model nor the specific rules defined by Talairach are in common usage. A number of different variations of the Talairach method are in use and will be described at the end of this section.

The Talairach method was originally devised for stereotactic surgery on structures deep within the brain. The goal of the method was to utilize neuroimaging studies available in the 1950s to identify landmarks from which the position of deep brain structures could be interpolated with reasonable accuracy. Registration of cortical structures was not a primary goal of this work. The adoption of the Talairach methodology for functional imaging studies can be directly attributed to the methodical thoroughness of Talairach's work which included validation of the performance of his method for a large number of cortical structures as published in the 1967 atlas (Talairach *et al.*, 1967) (see Fig. 8). Since Talairach's interest focused on identifying the location of deep brain structures, it is not surprising that the primary landmarks that he identified as being useful are far from the cortical surface. In particular, two white matter tracts that cross the midline of the brain anterior and posterior to the thalamus, known as the anterior and posterior commissures, were chosen as the primary landmarks. After identifying the anterior and posterior commissures and the midsagittal plane, the first step in the Talairach transformation is to perform a rigid body rotation of the brain such that the midsagittal plane is oriented vertically and the line through the anterior and posterior commissures is oriented horizontally.

With the brain oriented in this position, the next step in the Talairach transformation is to define the most extreme x, y, and z coordinate locations for any point in the brain. In defining the z coordinate extreme at the bottom of the brain, the cerebellum is not included. It should be noted that the exact locations of these most extreme positions are dependent on the definition of the locations of the anterior and posterior commissures and are therefore secondary rather than primary landmarks. In Talairach's original method, all points are mapped by linear interpolation between the nearest of the eight identified landmarks (the anterior commissure, the posterior commissure, and the two most extreme x, y, and z coordinate locations) (Talairach *et al.*, 1967; Talairach and Tournoux, 1988). This effectively results in 12 piecemeal linear transformations that are continuous with one another where they meet along four planes

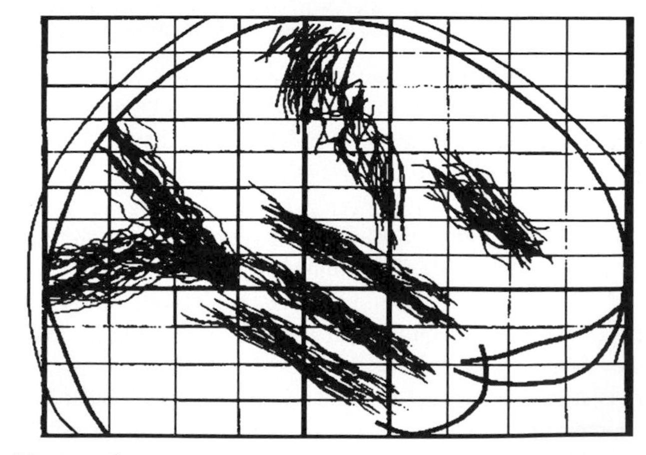

Figure 8 Residual intersubject variability after Talairach transformation. The image shown here is modified from the 1967 Talairach atlas and was mapped into "Talairach space" using the approach originally described by Talairach *et al.* The landmarks shown are the occipital sulcus, the parietooccipital sulcus, the central sulcus, the inferior frontal sulcus, and the superior, middle, and inferior temporal sulci. Compare the amount of residual variability to the sizes of the functional areas shown for the macaque brain in Fig. 1.

oriented parallel to the major coordinate axes and passing through the anterior or posterior commissure. Each of these piecemeal linear transformations effectively utilizes a 9 parameter mathematical model since a rigid body rotation is followed by independent rescaling along each of the three coordinate axes. The 12 piecemeal transformations share a number of common constraints, and the total number of independent parameters that must be defined in a Talairach transformation using Talairach's original methodology is 13.

Talairach's original validation work for cortical landmarks has been extended to modern MRI data by Steinmetz *et al.* (1989, 1990) and by Talairach and Tournoux (1993) in the form of a new 1993 atlas. As with the earlier work based on pathologic specimens and lower resolution imaging techniques, the results are good but somewhat disappointing to anyone hoping to assign reliable coordinates to small functional brain areas. The transformation does a reasonably good job at mapping homologous anatomic structures to similar coordinate locations, but the amount of residual variability is still large enough to allow most gyri to be confused with their nearest neighbor. This variability also serves to illustrate that the brain of the single individual in the 1988 Talairach atlas can be used only as a rough guide for interconverting coordinate locations and anatomic structures. Reliance on population-based locations such as those shown in the 1967 atlas (Fig. 8) are more likely to be reliable, an issue that will be discussed further below in relation to the "average brain."

The 1967 Talairach atlas did not define numerical coordinate locations. When Fox *et al.* (1985) first adopted a modified version of the Talairach approach for functional imaging, they chose the midpoint between the anterior and the posterior commissures as the origin for the coordinate system. Unfortunately, when the 1988 Talairach atlas was published, it utilized the anterior commissure as the origin of a numerical coordinate system, and this origin is now frequently used in the functional imaging literature. This results in a y axis discrepancy of a few millimeters between coordinate locations reported using the two different origins. A similar possibility for confusion exists with regard to the coordinate locations of the extreme points of the brain. Some groups have chosen to adopt the brain dimensions of the single individual in the 1988 atlas, whereas others have used one of the brains in the 1967 atlas or even the average dimensions of the group of individuals under study. It is also common to abandon the use of 12 piecemeal linear transformations in favor of a single global nine parameter transformation. In general, a single nine parameter transformation will not be able to precisely match the anterior commissure, the posterior commissure, and all six brain extrema simultaneously. Various solutions to this dilemma (usually by ignoring some landmarks in favor of others) can be used and will result in coordinates that differ slightly from those derived using the original Talairach approach (Senda *et al.*, 1993). It is important to be aware of these methodologic differences and the variability that they can introduce when interpreting Talairach coordinates reported in the literature and when deriving Talairach coordinates for publication (see Fig. 9).

b. Semiautomated and automated linear intersubject registration tools. Friston *et al.* (1989) and Minoshima *et al.* (1993) have developed automated methods for recognizing the AC-PC line in PET data, allowing for less user-intensive approaches to this first component of a nine parameter linear transformation. Both of these techniques retain the sequential characteristics of the original Talairach approach in the sense that the rigid body transformation is completed and not further modified at the time that the scaling factors are computed. Friston *et al.* (1991a,b) have also implemented a fully automated mapping into "Talairach space" that involves a number of nonlinear transformations, many of which are applied on a plane-by-plane basis, to match PET data to a series of templates. The continued use of the term "Talairach space" in relation to this highly nonlinear transformation model relates to the fact that the template itself was originally created by averaging together a group of brains that had all been registered using a linear nine parameter model, but the overall approach should clearly be classified as a nonlinear intersubject registration tool.

Collins *et al.* (1994) have reported a fully automated implementation of a 9 parameter linear model for MRI data. Unlike the work of Friston *et al.* (1991a,b) and Minoshima *et al.* (1993), this model does not attempt to identify the rigid body rotation in isolation from the rescaling parameters, but rather modifies both of these sets of parameters iteratively while searching for the best possible fit. A correlation function is used to quantitate the goodness of fit at each stage of the algorithm. A similar 9 parameter fully automated algorithm has been successfully implemented and applied to MRI and PET data. The same ratio criteria described in the first section of this chapter have been used as a measure of the goodness of fit, minimizing the standard deviation of this ratio across all voxels. An anatomic validation of this model has been completed, analogous to the anatomic validation of the original Talairach model (Fig. 10). For MRI data, the results are quite similar to those obtained using the traditional (13 parameter piecemeal) Talairach approach and are better than those achieved with a single, manually defined 9 parameter transformation. Seven, 12, and 15 parameter automated linear models have also been implemented in order to compare the different mathematical models to one another (Fig. 10). Improvements are evident as the number of parameters are increased from 7 to 9 to 12. Differences between the 12 and the 15 parameter models are minimal despite considerable residual intersubject variability, suggesting that perspective

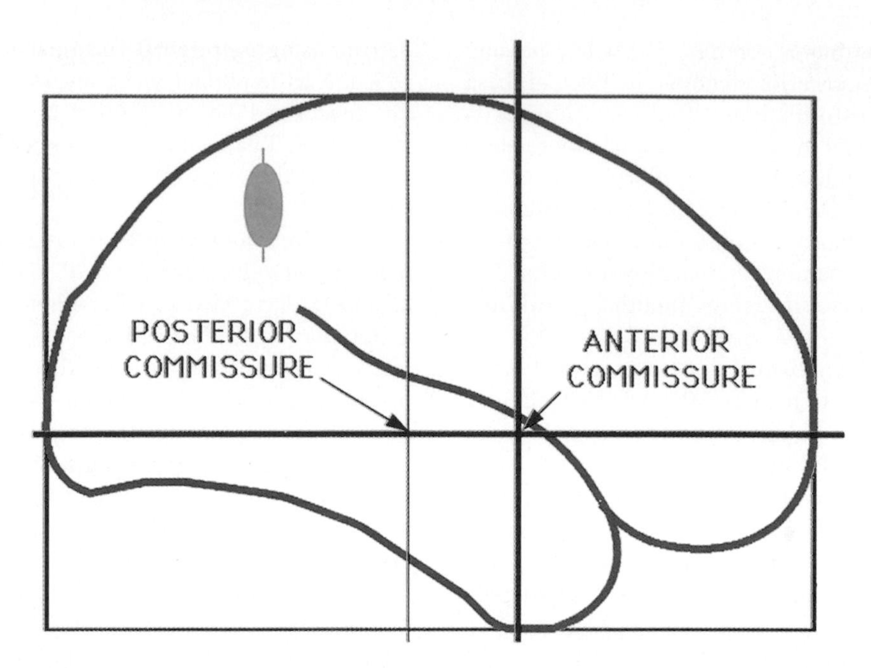

Figure 9 What is the Talairach y coordinate of the area shown in green? The brain depicted has already been oriented with the line between the commissures horizontal and the midsagittal plane vertical as required by all techniques for mapping into "Talairach space." The anterior and posterior commissures are labeled. By definition, the y axis coincides with the line between the commissures, the AC-PC line. In the 1988 Talairach atlas, the anterior commissure is defined as the origin of the coordinate system, but some published PET studies have used the midpoint between the anterior and the posterior commissures as the origin. Linear interpolation can be used to map the area shown in green into the coordinate space of the 1988 Talairach atlas, but no universal standard dictates which landmarks should be used for interpolation. In the method originally described by Talairach *et al.*, the back of the brain and the posterior commissure are used as interpolation landmarks. Alternatively, the coordinate can be interpolated between the back of the brain and the anterior commissure. In some cases, the anterior commissure may be poorly localized, in which case the coordinate can be interpolated between the back of the brain and the front of the brain. Other variations are also possible such as computing the distance from the landmark to the (anterior commissure as a fraction of the distance from the front of the brain to the back of the brain. All of these methods should give similar results in this particular case since the brain shown here was traced from the 1988 Talairach atlas itself. In real cases, the proportionate distances between the four landmarks anterior commissure, posterior commissure, front of brain, and back of brain) are variable, so each of the possibilities just outlined will give a different result. Different results will also be obtained if the 1967 Talairach atlas is used as a reference instead of the 1988 atlas since the brains in the 1967 atlas have different overall dimensions. Finally, it is possible to simply compute the real distance from the anterior commissure to the site of interest and to report this absolute value (without reference to any published atlas) as a coordinate location in real space with the brain in a "Talairach orientation." Problems similar to those encountered along the y axis also occur along the x (right-left) and z (top-bottom) axes, but only three landmarks are identified along these axes instead of four, making the situation somewhat less complicated. The answer to the question in the caption title is that the question itself is ill posed unless the mathematical model, the rules for using the model, and the conventions for assigning coordinates are all fully specified.

distortions may be nonbiological in nature and not worth including as part of a registration model is biological images. The 12 parameter affine model is currently the model of choice in this author's laboratory for intersubject registration and analysis of functional PET data.

The limitations of linear intersubject registration tools are made evident by the residual anatomic variability shown in Fig. 10. Dissatisfaction with this degree of variability is the driving force behind the development of nonlinear registration techniques. Unless some nonlinear parameter with special biological validity can be identified, a large number of nonlinear parameters may be required to significantly improve upon the results that can be obtained with these linear approaches. In any case, the best results obtainable by

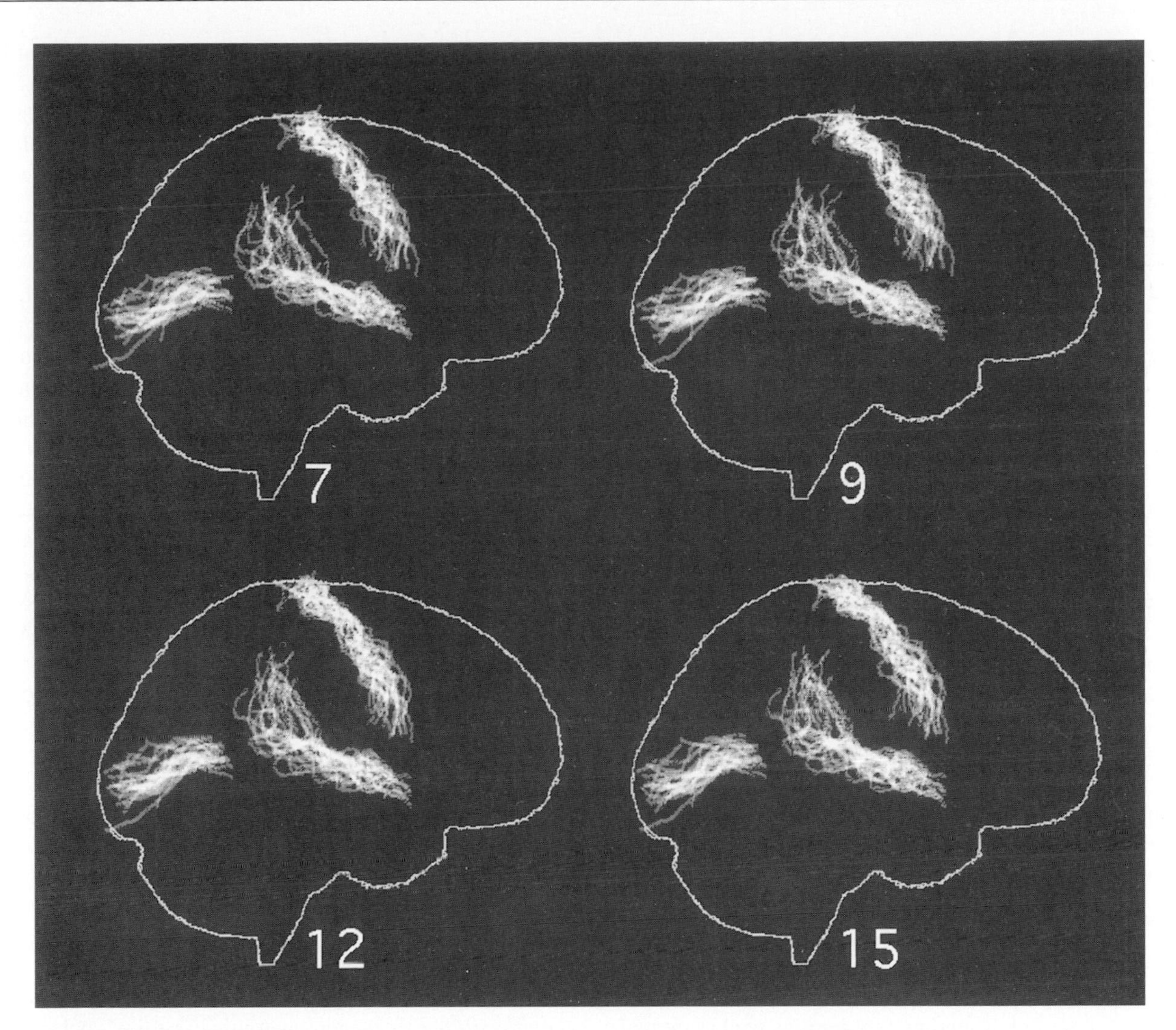

Figure 10 Automated intersubject registration with linear models. Three cortical landmarks (right calcarine sulcus, right precentral gyrus, and right superior temporal gyrus and its posterior extension) from 22 subjects have been mapped into a common frame of reference using a linear automated image registration algorithm incorporating 7, 9, 12, or 15 parameters. A better superimposition of landmarks results as the number of parameters is increased from 7 to 9 to 12. The addition of the three additional perspective parameters does not lead to better results for the 15 parameter model than was seen for the 12 parameter model. Note the marked variability of the posterior aspect of the superior temporal gyrus. Considerable right-left asymmetry was also seen in this region (left hemisphere results are not shown).

a linear method should be the standard against which nonlinear techniques should be judged.

c. The "average brain". Despite the amount of intersubject variability of cortical landmarks in the Talairach frame of reference and even variability in the way that these landmarks are placed into the coordinate system, the coordinate locations of PET activation foci have been found to be reasonably reproducible across laboratories. This reproducibility can be largely attributed to the fact that PET data analysis has typically used data pooled from a large number of subjects. Identification of an activation focus involves identifying either the maximal or the most central point of blood flow change in the pooled data analysis (Fox, 1991). This effectively results in spatial averaging of the coordinates that would have been seen in any given individual. As with any averaging procedure, this results in a decrease in variability. The use of data averaging can make the performance of any model at a population level superior to its performance on an individual level. The better that a model performs on an individual level, the less need there will be to average together multiple subjects to achieve reproducible results. As attention focuses increasingly on analysis of data from individual subjects, the need for better mathematical models will increase. Likewise, higher resolution data will demand more accurate registration than has been required in the past for low resolution data.

The notion of an "average brain" is a useful one, particularly if the amount of variability within the population under study is well defined. A brain that minimizes the average amount of distortion required to register it to any individual in the population is an ideal choice to serve as a target atlas for that population. Since overall orientation and position are arbitrary, an "average brain" can easily be oriented to make the "average" AC-PC line horizontal and the "average" midsagittal plane vertical. The "average" extrema of the brain can be defined to generate an "average brain Talairach atlas." The averaged target brain used for the automated registrations shown in Fig. 10 was generated in this way, using the shape information derived by the 12 parameter linear algorithm to define a shape for the target brain that would optimize the performance of the 7 and 9 parameter models. Distortions of the target brain (e.g., affine distortion to match the brain of a single individual) lead to poorer registration and less accurate coordinate locations from the 7 and 9 parameter models.

Finally, it is worth keeping in mind that the average brain of one population may be different from the average brain of another population and that accurate reconciliation of landmark locations across populations may not be possible with any linear model. The discrepancy between Japanese and European brains mentioned earlier in this chapter illustrates this concern (Thurfjell *et al.*, 1994).

2. Nonlinear Intersubject Registration Tools

Nonlinear tools for intersubject registration are still in a relatively early stage of development. Nonlinearity introduces a number of confounding problems as well as tremendous computational complexity. For example, nonaffine transformation will generally lead to variable image resolution and may invalidate the assumptions implicit in statistical methods of accounting for the large number of independent hypotheses tested when comparing images to one another (Friston *et al.*, 1991a,b; Worsley *et al.*, 1992). At the present time, these nonlinear tools should be used with caution unless a nonlinear approach is dictated on theoretical grounds by the nature of the question being asked or unless the nonlinear approach has been shown to be superior to an affine linear approach with regard to the landmarks of interest.

The study of morphometrics is one case where theoretical considerations dictate the use of a nonlinear approach. Bookstein (1989) has developed morphometric analysis techniques such as the use of thin-plate splines whereby landmarks identified using neuroanatomic expertise can be used to define and quantitate differences between brains. Bookstein's approach has also been used on a limited basis for registering PET data for analysis (Evans *et al.*, 1991), but it is limited in this context because of the need for labor intensive definitions of multiple point landmarks.

Another context that has motivated the use of nonlinear models is in the matching of the cortical and ventricular surfaces. For example, the nonlinear components of the "Talairach" methodology developed by Friston *et al.* (1991a,b) were designed to maximize the overlap of cortical surfaces as a prelude to image analysis. Similarly, work of Gee *et al.* (1993) has focused on the matching of both cortical and ventricular surfaces. Like many nonlinear models, these approaches begin with a linear transformation and then apply additional nonlinear distortions. Several other nonlinear approaches have been implemented in two dimensions, but have not been extended to three dimensions. Since brains are three-dimensional structures, two-dimensional nonlinear registration techniques are of questionable utility.

For many of these nonlinear tools, the nonlinearity is manifest by an extremely large number of parameters. In the most extreme case, each voxel in an image may be independently remapped to some new coordinate location specified by three parameters applicable only to that particular voxel. There is a danger with models incorporating so many parameters that the results may look extremely good in terms of making the registered images appear to be similar to one another, but anatomic considerations may show very poor results. Ideally, part of the validation procedure for such models should involve verification that homologous landmarks identified in the original images using neuroanatomic expertise are mapped to homologous locations after registration. Anatomic validation is not currently available for most of the nonlinear tools discussed here.

Although most of the nonlinear tools are driven by computation rather than neuroanatomic expertise, such expertise can be incorporated. The Karolinska computerized brain atlas is a good example of such a tool (Bohm *et al.*, 1991; Thurfjell *et al.*, 1994). This atlas is based on cryomacrotome data from postmortem brains, and the user applies a series of linear and nonlinear transformations of the atlas in order to optimally match the atlas to the particular brain under consideration. These transformations can then be reversed to map the various brains of interest into a common space defined by the original cryomacrotome data. This approach has the appeal of allowing neuroanatomic expertise to be applied without the need for identification of explicit point landmarks. The major drawbacks of the technique are that it is labor intensive and user dependent. A quantitative analysis of the use of the various parameters available with this tool might provide some biological guidance with respect to the particular nonlinear distortions that should be preferentially incorporated by future, more automated techniques.

C. Future Directions

Intersubject registration is essentially a biological problem and further directions will be dictated primarily by experimentation and observation, not by theoretical exposition. The approach to intersubject registration will evolve with changing needs and with added information. If the goal of intersubject registration is to superimpose functionally equivalent landmarks using structural information, then understanding of how tightly function is coupled to macroscopic structure evident on MRI will be an essential first step. A clear understanding of genetic factors that control morphology and population differences is also a high priority item at present. Finally, new noninvasive means of identifying structural landmarks beyond those currently evident on MRI scanning could provide an important source of additional information to be exploited in generating better superimposition of homologous structures. This information could come in the form of novel biochemical markers or in the form of a higher resolution noninvasive definition of white matter tracts. For example, it has been suggested that visual area V5 in the visual cortex holds a consistent relationship to the gyral pattern of the region and that its precise location can be identified by the presence of particular heavy myelination (Clarke and Miklossy, 1990; Watson *et al.*, 1993). If this observation is valid throughout and across populations, identification of the heavy myelination using MRI scanning might provide a reliable landmark with both structural and functional validity. Likewise, the ability to image cortical columns noninvasively would provide a wealth of functionally relevant structural information that could dramatically alter approaches to intersubject registration. The notion of defining global transformations that will approximately map homologous functional areas onto one another may ultimately be replaced by the identification of validated, possibly microscopic, landmarks identified noninvasively on the basis of anatomic, biochemical, or perhaps even functional criteria.

V. Conclusion

While much of the discussion in this chapter has treated function and structure as if they were totally independent, this is clearly not the case in the brain. A great deal of a function of a particular brain area is dictated by its anatomic connections to other brain areas. Thus at the microscopic level, structure and function are intimately linked. At a higher level of organization, it is clear that parcellation of specific functions into anatomically distinct locations is a general principle of brain organization. This functional parcellation is so well characterized in some brain areas that "functional landmarks" can be identified to complement or even substitute for traditional structural landmarks. Interestingly, recent developmental work suggests that the functional activation of brain pathways can even be responsible for molding brain structure and anatomic connectivity. The eventual goal of brain mapping is to understand the rules that govern both the structural and the functional aspects of the brain. In the short term, this goal will be directed toward developing functional maps of the entire cortex analogous to van Essen's map of the macaque visual system, but in the long term a more profound picture is likely to emerge that will unite genetics, development, structure, function, and the representation of information in the brain into a unified whole.

Acknowledgments

The participation of the healthy volunteers, the patients, and the many UCLA physicians (Bruce Enos, Michel Lesvesque, Itzhak Fried. Mark Morrow, and Bruce Dobkin) involved in the clinical assessment and treatment of the patients is gratefully acknowledged. John C. Mazziotta, M. D., Ph.D., Simon Cherry, Ph.D., Scott T. Grafton, M. D., and John D. G. Watson, M. D., were valued collaborators on much of the original work summarized here. This work was supported by NINDS (1 K08 NS01646-01). Department of Energy Contract DE-FC03-87ER60615, generous gifts from the Pierson–Lovelace Foundation, the Ahmanson Foundation, and the Charles A. Dana Foundation and a grant from the Brain Mapping Medical Research Organization.

References

Balish, M., and Muratore, R. (1990). The inverse problem in electroencephalography and magnetoencephalography. *Adv. Neurol.* **54**, 79–88.

Barnett, G. H., Kormos, D. W., *et al.* (1993). Use of a frameless, armless stereotactic wand for brain tumor localization with two-dimensional and three-dimensional neuroimaging. *Neurosurgery* **33**(4), 674–678.

Bohm, C., Greitz, T., *et al.* (1991). Specification and selection of regions of interest (ROIs) in a computerized brain atlas. *J. Cereb. Blood Flow Metab.* **11**(2), A64–A68.

Bookstein, F. L. (1989). Principal warps: Thin-plate splines and the decomposition of deformations. *IEEE Trans. Pattern Anal. Mach. Intell.* **11**, 567–585.

Bookstein, F. L. (1991). "Morphometric Tools for I apdmark Data." Cambridge University Press. New York.

Broca, P. (1865). Sur la faculté du langage articulé. *Bull. Soc. Anthropol. (Paris)* **6**, 337–393.

Brodmann, K. (1909). "Vergleichende Lokalisationlehre der Gross-hirnrinde in ihren Prinzipien dargestellt auf Grund des Zellenbaues." Barth, Leipzig.

Carpenter, M. B. (1976). "Human Neuroanatomy." Williams & Wilkins, Baltimore.

Cherry, S. R., Woods, R. P., *et al.* (1993). Improved detection of focal cerebral blood flow changes using three-dimensional positron emission tomography. *J. Cereb. Blood Flow Metab.* **13**(4), 630–638.

Clarke, S., and Miklossy, J. (1990). Occipital cortex in man: Organization of callosal connections, related myelo- and cytoarchitecture, and putative boundaries of functional visual areas. *J. Comp. Neurol.* **298**(2), 188–214.

Collins, D. L., Neelin, P., *et al.* (1994). Automatic 3D intersubject registration of MR volumetric data in standardized Talairach space. *J. Comput. Assisted Tomogr.* **18**(2), 192–205.

Crelier, G. R., Fischer, S. E., *et al.* (1994). Real-time image reconstruction system for interventional magnetic resonance surgery. *Technol. Health Care* **2**(4), 267–273.

Crick, F., and Jones, E. (1993). Backwardness of human neuroanatomy. *Nature* **361**(6408), 109–110.

Damasio, H., and Frank, R. (1992). Three-dimensional in vivo mapping of brain lesions in humans. *Arch. Neurol.* **49**(2), 137–143.

Engel, J. J., Henry, T. R., *et al.* (1990). Presurgical evaluation for partial epilepsy: Relative contributions of chronic depth-electrode recordings versus FDG-PET and scalp-sphenoidal ictal EEG. *Neurology* **40**(11), 1670–1677.

Evans, A. C., Dai, W., *et al.* (1991). Warping of a computerized 3-D atlas to match brain image volumes for quantitative neuroanatomical and functional analysis. *SPIE Image Process.* **1445**, 236–246.

Evans, A. C., Marrett, S., *et al.* (1989). Anatomical-functional correlative analysis of the human brain using three dimensional imaging systems. *SPIE Med. Imag. III: Image Process.* **1092**, 264–274.

Fox, P. T. (1991). Physiological ROI definition by image subtraction. *J. Cereb. Blood Flow Metab.* **11**(2), A79–A82.

Fox, P. T., Mintun, M. A., *et al.* (1988). Enhanced detection of focal brain responses using intersubject averaging and change-distribution analysis of subtracted PET images. *J. Cereb. Blood Flow Metab.* **8**(5), 642–653.

Fox, P. T., Perlmutter, J. S., *et al.* (1985). A stereotactic method of anatomical localization for positron emission tomography. *J. Comput. Assisted Tomogr.* **9**(1), 141–153.

Fox, P. T. and Raichle, M. E. (1985). Stimulus rate determines regional brain blood flow in striate cortex. *Ann. Neurol.* **17**(3), 303–305.

Friston, K. J., Frith, C. D., *et al.* (1991a). Comparing functional (PET) images: The assessment of significant change. *J. Cereb. Blood Flow Metab.* **11**(4), 690–699.

Friston, K. J., Frith, C. D. *et al.* (1991b). Plastic transformation of PET images. *J. Comput. Assisted Tomogr.* **15**(4), 634–639.

Friston, K. J., Passingham, R. E., *et al.* (1989). Localisation in PET images: Direct fitting of the intercommissural (AC-PC) line. *J. Cereb. Blood Flow Metab.* **9**(5), 690–695.

Gee, J. C., Reivich, M., *et al.* (1993). Elastically deforming 3D atlas to match anatomical brain images. *J. Comput. Assisted Tomogr.* **17**(2), 225–236.

Grafton, S. T., Martin, N. A., *et al.* (1994). Localization of motor areas adjacent to arteriovenous malformations. A positron emission tomographic study. *J. Neuroimag.* **4**(2), 97–103.

Grafton, S. T., Woods, R. P., *et al.* (1991). Somatotopic mapping of the primary motor cortex in humans: Activation studies with cerebral blood flow and positron emission tomography. *J. Neurophysiol.* **66**(3), 735–743.

Grafton, S. T., Woods, R. P., *et al.* (1993). Within-arm somatotopy in human motor areas determined by positron emission tomography imaging of cerebral blood flow. *Exp. Brain. Res.* **95**(1), 172–176.

Grinvald, A., Lieke, E., *et al.* (1986). Functional architecture of cortex revealed by optical imaging of intrinsic signals. *Nature* **324**(6095), 361–364.

Haglund, M. M., Ojemann, G. A., *et al.* (1992). Optical imaging of epileptiform and functional activity in human cerebral cortex. *Nature* **358**(6388), 668–671.

Herscovitch, P., Markham, J., *et al.* (1983). Brain blood flow measured with intravenous H2(15)O. I. Theory and error analysis. *J. Nucl. Med.* **24**(9), 782–789.

Hurley, J. R. and Cattell, R. B. (1962). The PROCRUSTES program: Producing direct rotation to test a hypothesized factor structure. *Behav. Sci.* **7**, 258–262.

Jolesz, F. A., and Blumenfeld, S. M. (1994). Interventional use of magnetic resonance imaging. *Magn. Reson Q.* **10**(2), 85–96.

Kwong, K. K., Belliveau, J. W., *et al.* (1992). Dynamic magnetic resonance imaging of human brain activity during primary sensory stimulation. *Proc. Natl. Acad. Sci. U.S.A.* **89**(12), 5675–5679.

Mazziotta, J. C., Phelps, M. E., *et al.* (1982a). Tomographic mapping of human cerebral metabolism: Auditory stimulation. *Neurology* **32**(9), 921–937.

Mazziotta, J. C., Phelps, M. E., *et al.* (1982b). Tomographic mapping of human cerebral metabolism: Sensory deprivation. *Ann. Neurol.* **12**(5), 435–444.

Mazziotta, J. C., Phelps, M. E., *et al.* (1982c). Anatomical localization schemes for use in positron computed tomography using a specially designed headholder. *J. Comput. Assisted Tomogr.* **6**(4), 848–853.

Mesulam, M. M. (1990). Large-scale neurocognitive networks and distributed processing for attention, language, and memory. *Ann. Neurol.* **28**(5), 597–613.

Meyer, J. S., Obara, K., *et al.* (1993). Diaschisis. *Neurol. Res.* **15**(6), 362–366.

Minoshima, S., Koeppe, R. A., *et al.* (1993). Automated detection of the intercommissural line for stereotactic localization of functional brain images. *J. Nucl. Med.* **34**(2), 322–329.

Ojemann, G. A. (1991). Cortical organization of language. *J. Neurosci.* **11**(8), 2281–7.

Olivier, A., Germano, I. M., *et al.* (1994). Frameless stereotaxy for surgery of the epilepsies: Preliminary experience. *J. Neurosurg.* **81**(4), 629–633.

Ono, M., Kubik, S. *et al.* (1990). "Atlas of the Cerebral Sulci." Thieme, Stuttgart.

Pascual-Leone, A., Valls-Sol, J., *et al.* (1992). Effects of focal transcranial magnetic stimulation on simple reaction time to acoustic, visual and somatosensory stimuli. *Brain* **115**(Pt 4), 1045–1059.

Pelizzari, C. A., Chen, G. T., *et al.* (1989). Accurate three-dimensional registration of CT, PET, and/or MR images of the brain. *J. Comput. Assisted Tomogr.* **13**(1), 20–26.

Penfield, W., and Roberts, L. (1959). "Speech and Brain Mechanisms." Princeton University Press, Princeton.

Raichle, M. E., Martin, W. R., *et al.* (1983). Brain blood flow measured with intravenous $H_2{}^{15}O$. II. Implementation and validation. *J. Nucl. Med.* **24**(9), 790–798.

Rauch, R. A., Vinuela, F., *et al.* (1992a). Preembolization functional evaluation in brain arteriovenous malformations: The ability of superselective Amytal test to predict neurologic dysfunction before embolization. *Am. J. Neuroradiol.* **13**(1), 309–314.

Rauch, R. A., Vinuela, F., *et al.* (1992b). Preembolization functional evaluation in brain arteriovenous malformations: The superselective Amytal test. *Am. J. Neuroradiol.* **13**(1), 303–308.

Sandeman, D. R., Patel, N., *et al.* (1994). Advances in image-directed neurosurgery: Preliminary experience with the ISG Viewing Wand compared with the Leksell G frame. *Br. J. Neurosurg.* **8**(5), 529–544.

Schonemann, P. H. (1966). A generalized solution of the orthogonal Procrustes problem. *Psychometrika* **31**(1), 1–10.

Senda, M., Kanno, I., *et al.* (1993). Comparison of three anatomical standardization methods regarding foci localization and its between subject variation in the sensorimotor activation. *In* "Quantification of Brain Function: Tracer Kinetics and Image Analysis in Brain PET" (K. Uemura, N. A. Lassen, T. Jones, and I. Kanno, eds.), pp. 439–445. Excerpta Medica. Amsterdam.

Sereno, M. I., McDonald, C. T. *et al.* (1994). Analysis of retinotopic maps in extrastriate cortex. *Cereb. Cortex* **6**, 601–620.

Sherrington, C. (1906). "The Integrative Action of the Nervous System." Yale University Press. New Haven, CT.

Sibson, R. (1978). Studies in the robustness of multidimensional scaling: Procrustes statistics. *J. R. Stat. Soc. B* **40**(2), 234–238.

Steinmetz, H., Furst, G., *et al.* (1989). Cerebral cortical localization: Application and validation of the proportional grid system in MR imaging. *J. Comput. Assisted Tomogr.* **13**(1), 10–9.

Steinmetz, H., Furst, G., *et al.* (1990). Variation of perisylvian and calcarine anatomic landmarks within stereotaxic proportional coordinates. *Am. J. Neuroradiol.* **11**(6), 1123–1130.

Strother, S. C., Anderson, J. R., *et al.* (1994). Quantitative comparisons of image registration techniques based on high-resolution MRI of the brain. *J. Comput. Assisted Tomogr.* **18**(6), 954–962.

Talairach, J., Szikla, G., *et al.* (1967). "Atlas d'anatomie stéréotaxique du télencéphale." Masson, Paris.

Talairach, J., and Tournoux, P. (1988). "Co-planar Stereotaxic Atlas of the Human Brain." Thieme, Stuttgart.

Talairach, J., and Tournoux P., (1993). "Referentially Oriented Cerebral MRI Anatomy: An Atlas of Stereotaxic Anatomical Correlations for Gray and White Matter." Thieme, Stuttgart.

Tanaka, Y., Kamo, T., *et al.* (1991). So-called cortical deafness: Clinical, neurophysiological and radiological observations. *Brain*. 2385–2401

Thurfjell, L., Bohm, C., *et al.* (1994). Accuracy and precision in image standardization in intra-and intersubject comparisons. *In* "Functional Neuroimaging: Technical Foundations" (R. W. Thatcher, M., Hallett, T., Zeffiro, E. R. John, and M. Huerta. eds.) Academic Press, San Diego.

Turkington, T. G., Hoffman, J. M., *et al.* (1995). Accuracy of surface fit registration for PET and MR brain images using full and incomplete brain surfaces. *J. Comput. Assisted Tomogr.* **19**(1), 117–124.

van Essen, D. C., Anderson, C. H. *et al.* (1992). Information processing in the primate visual system: An integrated systems perspective. *Science* **255**(5043), 419–423.

Wang, B., Toro, C., *et al.* (1994). Head surface digitization and registration: A method for mapping positions on the head onto magnetic resonance images. *Brain Topogr.* **6**(3), 185–192.

Watson, J. D. G., Myers, R., *et al.* (1993). Area V5 of the human brain: Evidence from a combined study using positron emission tomography and magnetic resonance imaging. *Cereb. Cortex* **3**(2), 79–94.

Wieringa, H. J., Peters, M. J., *et al.* (1994). Integration of MEG. EEG, and MRI. *In* "Functional Neuroimaging: Technical Foundations". (R. W. Thatcher, M. Hallett, T. Zeffiro, E. R. John, and M. Huerta, eds.). Academic Press, San Diego.

Woods, R. P., Cherry, S. R., *et al.* (1992). Rapid automated algorithm for aligning and reslicing PET images. *J. Comput. Assisted Tomogr.* **16**(4), 620–633.

Woods, R. P., Dodrill, C. B. *et al.* (1988). Brain injury, handedness, and speech lateralization in a series of amobarbital studies. *Ann. Neurol.* **23**(5), 510–518.

Woods, R. P., Mazziotta, J. C., *et al.* (1993). MRI-PET registration with automated algorithm. *J. Comput. Assisted Tomogr.* **17**(4), 536–546.

Worsley, K. J., Evans, A. C., *et al.* (1992). A three-dimensional statistical analysis for CBF activation studies in human brain. *J. Cereb. Blood Flow Metab.* **12**(6), 900–918.

Zhang, J., Levesque, M. F., *et al.* (1990). Multimodality imaging of brain structures for stereotactic surgery. *Radiology* **175**(2), 435–441.

24

Advanced Nonrigid Registration Algorithms for Image Fusion

Simon K. Warfield,* Alexandre Guimond, Alexis Roche, Aditya Bharatha, Alida Tei, Florin Talos, Jan Rexilius, Juan Ruiz-Alzola, Carl-Fredrik Westin, Steven Haker, Sigurd Angenent, Allen Tannenbaum, Ferenc Jolesz, and Ron Kikinis

* *Department of Radiology, Brigham and Women's Hospital and Harvard Medical School, Boston, Massachusetts 02115*

I. Introduction

Medical images are brought into spatial correspondence, or *aligned*, by the use of registration algorithms. Nonrigid registration refers to the set of techniques that allow the alignment of data sets that are mismatched in a nonrigid, or nonuniform, manner. Such misalignments can result from physical deformation processes or can be a result of morphological variability. For example, physical deformation in the brain can occur during neurosurgery as a result of such factors as swelling, cerebrospinal fluid (CSF) loss, hemorrhage, and the intervention itself. Nonrigid deformation is also characteristic of the organs and soft tissues of the abdomen and pelvis. In addition, nonrigid morphological differences can arise when comparisons are made among image data sets acquired from different individuals. These changes can be a result of normal anatomical variability or the product of pathological processes. Because the gross structure of the brain is essentially similar among humans (and even among related species), the factors described above tend to produce *local* nonrigid shape differences.

Nonrigid brain registration techniques have numerous applications. They have been used to align scans of different brains, permitting the characterization of normal and pathological morphological variation (brain mapping). They have also been used to align anatomical templates with specific data sets, thus facilitating segmentation (i.e., segmentation by registration). More recently, these techniques have been used to capture changes which occur during neurosurgery.

With the ongoing development of robust algorithms and advanced hardware platforms, further applications in surgical visualization and enhanced functional image analysis are inevitable.

One exciting application of nonrigid registration algorithms is in the automatic registration of multimodal image data. Rigid registration of multimodal data has been greatly facilitated by the framework provided by mutual information (MI). However, MI-based strategies to effectively capture large nonrigid shape differences are still being explored. An alternate approach is to normalize multimodality images and thus reduce the problem to a monomodality match. In the first section, we present a nonrigid registration method which uses an intensity transform which allows a single intensity in one modality to be mapped onto (up to) two intensities. The method is iterative, combining in each iteration an intensity correction and a geometric transform using intensity-similarity criteria. The method is applied in two cases with promising results.

In the next section, we turn our attention to the issue of image registration and fusion during neurosurgery. It is common to desire to align preoperative data with images of the patient acquired during neurosurgery. It is now widely acknowledged that during neurosurgical operations, nonrigid changes in the shape of the brain occur as a result of the intervention itself and due to reactive physiological changes. These deformations ("brain shift") make it difficult to relate preoperative image data to the intraoperative anatomy of the patient. Since preoperative imaging is not subject to the same time constraints and limitations in tissue contrast selection methods as intraoperative imaging, a major goal has been to develop robust nonrigid registration algorithms for matching of preoperative image data onto intraoperative image data. We present our biomechanical modeling algorithm, which can capture nonrigid deformations based on surface changes and infer volumetric deformation using a finite element discretization. We also describe our early prospective experience using the method during neurosurgical cases and provide examples of the enhanced visualizations which are produced.

In the third section, we build upon the theme of physics-based models by presenting a novel inhomogeneous elasticity model which uses a local similarity measure to obtain an initial sparse estimate of the deformation field. The method includes automatic feature-point extraction using a nonlinear diffusion filter. Correspondence detection is achieved by maximizing a local normalized cross-correlation. The sparse estimates of the deformation field calculated at the feature points are then introduced as external forces, restricting the registration process so that the deformation field is fixed at those points. An advantage of the method is that feature points and correspondences are established automatically. Thus neither segmentation nor the manual identification of correspondences is required.

In the fourth section we discuss registration of diffusion tensor MRI data and introduce a framework for nonrigid registration of tensor data (including the special case of vector data). The approach is based on a multiresolution template matching scheme followed by interpolation of the sparse displacement field using a Kriging interpolator. After the data are warped, the tensors are locally realigned based on information from the deformation gradient of the displacement.

In the fifth section, we present a novel method for producing area-preserving surface deformations, and more general mass-preserving area and volume deformations, based on the minimization of a functional of Monge–Kantorovich type. The theory is based around the *optimal mass transport problem* of minimizing the cost of redistributing a certain amount of mass between two distributions given *a priori*. Here the cost is a function of the distance each bit of material is moved, weighted by its mass. The problem of optimal transport is classical and has appeared in econometrics, fluid dynamics, automatic control, transportation, statistical physics, shape optimization, expert systems, and meteorology. We show how the resulting low-order differential equations may be used for image registration.

The challenge of nonrigid registration remains one of the outstanding open problems in medical image analysis. New algorithm developments, often targeted toward specific clinical applications, have helped to identify further unsolved issues. This chapter provides an overview of the nonrigid registration algorithms being pursued today at the Surgical Planning Laboratory.

II. Intermodality and Multicontrast Images

A. Introduction

Automatic registration techniques of brain images have been developed following two main trends: (1) registration of multimodal images using low- to intermediate-degree transformations (fewer than a few hundred parameters) and (2) registration of monomodal images using high-dimensional volumetric maps (elastic or fluid deformations with hundreds of thousands of parameters). The first category mainly addresses the fusion of complementary information obtained from different imaging modalities. The second category's predominant purpose is the evaluation of either the anatomical evolution process present in a particular subject or the anatomical variations between different subjects. Despite promising early work such as that of Hata (1998), dense transformation field multimodal registration has, so far, remained relatively unexplored.

Research on multimodal registration culminated with the concept of MI (Viola and Wells, 1995, 1997; Wells *et al.*,

1996b; Hata *et al.*, 1996; Maes *et al.*, 1997), leading to a new class of rigid/affine registration algorithms. In this framework, the registration of two images is performed by maximizing their MI with respect to the transformation space. A significant reason for the success of MI as a similarity measure resides in its generality, as it does not use any prior information about the relationship between the intensities of the images. For instance, MI does not assume a linear relationship as is typically the case in standard optical flow techniques. Also, unlike some earlier approaches, MI does not require the identification of corresponding features in the images to be registered.

Significant work has been done in establishing the applicability of MI for nonrigid registration (Gaens *et al.*, 1998; Maintz *et al.*, 1998; Meyer *et al.*, 1999; Likar and Pernus, 2000; Hellier and Barillot, 2000; Rueckert *et al.*, 2000; Hermosillo *et al.*, 2001). Some authors have further improved the robustness of the approach by modifying the original MI measure, either by including some prior information on the joint intensity distribution (Maintz *et al.*, 1998; Likar and Pernus, 2000) or by using higher order definitions of MI which incorporate spatial information (Rueckert *et al.*, 2000).

Our approach described here stems from the observation that a number of multimodal rigid registration problems can be solved in practice using similarity measures other than MI, one of which is the correlation ratio (CR) (Roche *et al.*, 1998). The CR is much more constrained than MI as it assumes a functional, though nonlinear, relationship between the image intensities. In other words, it assumes that one image could be made similar to the other by a simple intensity remapping. Thus, the CR method amounts to an adaptive estimation strategy in which one image is alternately corrected in intensity and in geometry to progressively match the other.

For most combinations of medical images, the functional dependence assumption is generally valid for a majority of anatomical structures, but not for all of them. Although this problem does not turn out to be critical in a rigid/affine registration context, we observe that it may seriously hamper the estimation of a high-dimensional transformation. We propose here an extension of the functional dependence model, which we call the bifunctional model, to achieve better intensity corrections. While the bifunctional model is more realistic than the functional one, it remains strongly constrained and thus enables a good conditioning of the multimodal nonrigid registration problem.

B. Method

The registration algorithm described here is iterative and each iteration consists of two parts. The first part transforms the intensities of anatomical structures of a source image S so that they match the intensities of the corresponding structures of a target image T. The second part is concerned with the registration of S (after intensity transformation) with T using an optical flow algorithm.

1. Intensity Transformation

The intensity correction process starts by defining the set C of intensity pairs from corresponding voxels of T and S. Hence, the set C is defined as

$$C = \{(S(x), T(x)); 1 \leq x \leq N\}, \tag{1}$$

where N is the number of voxels in the images. $S(x)$ and $T(x)$ correspond to the intensity value of the xth voxel of S and T, respectively, when adopting the customary convention of considering images as one-dimensional arrays. We shall now show how to perform intensity correction if we can assume that a single intensity value in S has either (1) exactly one corresponding intensity value in T (monofunctional dependence) or (2) at least one and at most two corresponding intensity values in T (bifunctional dependence).

a. Monofunctional Dependence Assumption. Our goal is to characterize the mapping from voxel intensities in S to those in T, knowing that some elements of C are erroneous, i.e., they would not be present in C if S and T were perfectly matched. Let us assume here that the intensity in T is a function of the intensity in S corrupted with an additive stationary Gaussian white noise η,

$$\mathrm{T}(x) = f(S(x)) + \eta(x), \tag{2}$$

where f is an unknown function to be estimated. This is exactly the model employed by Roche *et al.* (2000), which leads to the correlation ratio as the measure to be maximized for registration. In that approach, for a given transformation, one seeks the function that best describes T in terms of S. In a maximum-likelihood context, this function is actually a least-squares (LS) fit of T in terms of S.

The major difference between our respective problems is that we seek a high-dimensional geometrical transformation. As opposed to affine registration in which the transformation is governed by the majority of good matches, elastic deformations may be computed using mainly *local* information (i.e., gradients, local averages, etc.). Hence, we cannot expect good displacements in one structure to correct for bad ones in another; we have to make certain each voxel is moved properly at each iteration. For this, since the geometrical transformation is found using intensity similarity, the most precise intensity transformation is required. Consequently, instead of performing a standard LS regression, we have opted for a robust linear regression estimator which will remove outlying elements of C during the estimation of the intensity transformation. To estimate f we use the least-trimmed-squares (LTS) method followed by a binary reweighted least-squares (RLS) estimation (Rousseeuw and Leroy, 1987). The combination of these two methods provides a very robust regression technique with outlier detection, while ensuring that a maximum of pertinent voxel pairs are taken into account.

Different types of functions can be used to model f. In Guimond *et al.* (2001) we made use of polynomial functions. The intensity correspondence between T and S is then defined as

$$T(x) = \theta_0+\theta_1 S(x)+\theta_2 S(x)^2+\ldots+\theta_p S(x)^p, \quad (3)$$

where $\theta = [\theta_0, \ldots \theta_p]$ needs to be estimated and p is the degree of the polynomial function. This model is adequate to register images that have a vast range of intensities; the restricted polynomial degree imposes intensity space constraints on the correspondences, mapping similar intensities in S to similar intensities in T.

In the case in which S is a labseled image, neighboring intensities in S will usually correspond to different structures. Hence the intensity space constraint is no longer required. f is then modeled as a piecewise constant function, such that each label of S is mapped to the LTS/RLS estimate of intensities corresponding to that label in T.

b. Bifunctional Dependence Assumption. Functional dependence as expressed in Eqs. (2) and (3) implies that two structures having similar intensity ranges in S should also have similar intensity ranges in T. With some combinations of images, this is a crude approximation. For example, CSF and bones generally give similar intensity values in T1-weighted images, while they appear with very distinct values in PD-weighted scans. Conversely, CSF and gray matter are well contrasted in T1-weighted images, while they correspond to similar intensities in PD-weighted scans.

To circumvent this difficulty, we have developed a strategy that enables the mapping of an intensity value in S to not only one, but two possible intensity values in T. This method is a natural extension of the previous section. Instead of computing a single function that maps the intensities of S to those of T, two functions are estimated and the mapping becomes a weighted sum of these two functions.

We start with the assumption that if a point has an intensity s in S, the corresponding point in T has an intensity t that is normally distributed around two possible values depending on s, $f_1(s)$ and $f_2(s)$. In statistical terms, this means that given s, t is drawn from a mixture of Gaussian distribution.

$$P(t|s) = \pi_1(s)N(f_1(s), \sigma^2) + \pi_2(s)N(f_2(s), \sigma^2), \quad (4)$$

where $\pi_1(s)$ and $\pi_2(s) = 1 - \pi_1(s)$ are mixing proportions that depend on the intensity in the source image, and σ^2 represents the variance of the noise in the target image. Consistent with the functional relationship, we will restrict ourselves to polynomial intensity functions, i.e., $f_1(s) = \theta_0 + \theta_1 s + \theta_2 s^2 + \ldots + \theta_p s^p$, and $f_2(s) = \psi_0 + \psi_1 s + \psi_2 s^2 + \ldots + \psi_p s^p$.

An intuitive way to interpret this modeling is to state that for any voxel, there is a binary 'selector' variable $\varepsilon = \{1, 2\}$ that would tell us, if it was observed, which of the two functions f_1 or f_2 actually serves to map s to t. Without knowledge of ε, the best intensity correction to apply to S (in the minimum variance sense) is a weighted sum of the two functions,

$$f(s,t) = P(\varepsilon=1|\, s,t)\, f_1(s)+P(\varepsilon=2|\, s,t)\, f_2(s), \quad (5)$$

in which the weights correspond to the probability that the point be mapped according to either the first or the second function. To estimate the functions, we employ a sequential strategy that performs two successive LTS/RLS regressions as in the monofunctional case. Details on how the other parameters are determined can be found in Guimond *et al.* (2001).

2. Geometrical Transformation

Having completed the intensity transformation stage, we end up with an intensity-corrected version of the source image, which will be denoted S^*. In the monofunctional case $S^*(x) = f(S(x))$ and in the bifunctional case $S^*(x) = f(S(x), T(x))$. We may assume that S^* is roughly of the same modality as T in the sense that corresponding anatomical structures have similar intensities in S^* and T. The geometrical transformation problem may then be treated in a monomodal registration context.

Many algorithms that deform one brain so its shape matches that of another have been developed (Maintz and Viergever, 1998; Toga, 1999). The procedure used here was influenced by a variety of optical flow methods, primarily the Demons algorithm (Thirion, 1995, 1998). At a given iteration n, each voxel x of T is displaced according to a vector $v_n(x)$ so as to match its corresponding anatomical location in S^*_n. We use the scheme

$$v_{n+1}(x) = G_\sigma \otimes \left(v_n + \frac{S^*_n \circ h_n(x) - T(x)}{\| \nabla S^*_n \circ h_n(x) \|^2 + [S^*_n \circ h_n(x) - T(x)]^2} \nabla S^*_n \circ h_n(x) \right) \quad (6)$$

where G_σ is a 3D Gaussian filter with isotropic variance σ^2, $\otimes$ denotes the convolution, $\circ$ denotes the composition, ∇ is the gradient operator, and the transformation $h_n(x)$ is related to the displacement by $h_n(x) = x + v_{n(x)}$. As is common with registration methods, we also make use of multilevel techniques to accelerate convergence. Details about the number of levels and iterations as well as filter implementation issues are addressed in Section II. C. We show here how our method can be related to three other registration methods: the minimization of the sum of squared difference (SSD) criterion, optical flow, and the Demons algorithm.

3. Relation to SSD Minimization

In the SSD minimization framework, one searches for the transformation h that minimizes the sum of squared differences between the transformed source image and the target image. The SSD is then defined as

$$SSD(h) = \frac{1}{2} \sum_{x=1}^{N} \left[S^* \circ h(x) - T(x) \right]^2. \quad (7)$$

The minimalization of Eq. (7) may be performed using a gradient descent algorithm. By differentiating the above equation, we get for a given x: $\nabla SSD(h) = -[S^* \circ h(x) - T(x)]\nabla S^* \circ h(x)$. Thus, the gradient descent consists of an iterative scheme of the form

$$v_{n+1} = v_n + \alpha[S_n^* \circ h_n(x) - T(x)\nabla S_n^* \circ h_n(x)\,, \quad (8)$$

where α is the step length. If we set α to a constant value, this method corresponds to a first-order gradient descent algorithm. Comparing Eq. (8) to Eq. (6), we see that our method sets

$$\alpha = \frac{1}{\| \nabla(S \circ h_n)(x) \|^2 + [T(x) - S \circ h_n(x)]^2} \quad (9)$$

and applies a Gaussian filter to provide a smooth displacement field. Cachier *et al.* (1999; Pennec *et al.*, 1999) have shown that using Eq. (9) closely relates Eq. (6) with a second-order gradient descent of the SSD criterion, in which each iteration n sets h_{n+1} to the minimum of the SSD quadratic approximation at h_n. We refer the reader to these articles for a more technical discussion on this subject.

4. Relation to Optical Flow

T and S are considered as successive time samples of an image sequence represented by $I(x,t)$ where $x = (x_1, x_2, x_3)$ is a voxel position in the image and computed by constraining the brightness of brain structures to be constant in time so that

$$\frac{dI(x,t)}{dt} = 0. \quad (10)$$

It is well known that Eq. (10) is not sufficient to provide a unique displacement for each voxel. In fact, this constraint leads to

$$f(x) = -\frac{\partial I(x,t)/\partial t}{\| \nabla_x I(x,t) \|^2} \nabla_x I(x,t), \quad (11)$$

which is the component of the displacement in the direction of the brightness gradient (Horn and Schunck, 1981).

Other constraints need to be added to Eq. (10) to obtain the displacement components in other directions. Many methods have been proposed to fulfill this purpose and thus regularize the resulting vector field (Barron *et al.*, 1994). One that can be computed very efficiently was proposed by Thirion (1998) in his description of the Demons registration method, using a complete grid of demons. It consists of smoothing each dimension of the vector field with a Gaussian filter G_σ. He also proposed to add $[\partial I(x,t)/\partial t]^2$ to the denominator of Eq. (11) for numerical stability when $\nabla_x I(x,t)$ is close to zero, a term which serves the same purpose as α^2 in the original optical flow formulation of Horn and Schunck (1981). As is presented by Bro-Nielsen and Gramkow (1996), this kind of regularization approximates a linear elasticity transformation model.

With this in mind, the displacement that maps a voxel position in T to its position in S is found using an iterative method,

$$v_{n+1}(x) = G_\sigma \otimes \left(v_n + \frac{\partial I(x,t)/\partial t}{\| \nabla_x I(x,t) \|^2 + [\partial I(x,t)/\partial t]^2} \nabla_x I(x,t) \right). \quad (12)$$

Spatial derivatives may be computed in several ways (Horn and Schunck, 1981; Brandt, 1997; Simoncelli, 1994). We have observed from practical experience that our method performs best when they are computed from the resampled source image of the current iteration. As shown in Section II.B.3, this is in agreement with the SSD minimization. Temporal derivatives are obtained by subtracting the target images from the resampled source image of the current iteration. These considerations relate Eq. (12) to Eq. (6). The reader should note that the major difference between this method and other optical flow strategies is that regularization is performed *after* the calculation of the displacements in the gradient direction instead of using an explicit regularization term in a minimization framework.

5. Relation to the Demons Algorithm

Our algorithm is actually a small variation of the Demons method (Thirion, 1995, 1998) using a complete grid of demons, itself closely related to optical flow as described in the previous section. The Demons algorithm finds the displacements using the following formula:

$$v_{n+1}(x) = G_\sigma \otimes \left(v_n + \frac{S \circ h_n(x) - T(x)}{\| \nabla T(x) \|^2 + [S \circ h_n(x) - T(x)]^2} \nabla T(x) \right). \quad (13)$$

In comparing Eqs. (13) and (6), it is apparent that the only difference between our formulation and the Demon method is that derivatives are computed on the resampled source image of the current iteration. This modification was performed following the observations on the minimization of the SSD criterion.

C. Results and Discussion

In the following section we present registration results involving images obtained from multiple modalities. First, we show a typical example in which monofunctional dependence can be assumed: the registration of an atlas (Collins *et al.*, 1998b) with a T1-weighted MR image. We next present an example in which bifunctional dependence may be assumed: the registration of a PD-weighted image with the same T1-weighted image.

All of the images used in this section have a resolution of $1 \times 1 \times 1$ mm^3 and respect the neurological convention, i.e., on coronal slices, the patient's left is on the left side of the image. Before registration, images are affinely registered using the correlation ratio method (Roche *et al.*, 1998).

The multilevel process was performed at three resolution levels, namely 4, 2, and 1 mm per voxel. Displacement fields at one level are initialized from the result of the previous level. The initial displacement field v_0 is set to zero. One hundred twenty-eight iterations are performed at 4 mm/voxel, 32 at 2 mm/voxel, and 8 at 1 mm/voxel. These are twice the numbers of iterations used for registration of monomodal images using the conventional Demons algorithm. We believe that making use of a better stopping criterion, such as the difference of the SSD values between iterations, would probably improve the results shown below. This aspect is currently under investigation. The Gaussian filter G_σ used to smooth the displacement field has a standard deviation of 1 voxel regardless of the resolution. This models stronger constraints on the deformation field at the beginning of the registration process to correct for gross displacements and weaker constraints near the end when fine displacements are sought. The resampling process makes use of trilinear interpolation, except in the case of the atlas in which nearest-neighbor interpolation is used.

Computation time to obtain the following results is around 60 min on a 450-MHz PC with 500 MB of RAM (10 min at 4 mm, 20 min at 2 mm, and 30 min at 1 mm). Most of this time ($\approx 85\%$) is devoted to the intensity correction part, which has not been optimized in this first version of our program. The other 15% is taken by the standard registration code, which is stable and well optimized.

1. Monofunctional Dependence

We present here the result of registering the atlas with a T1-weighted image. This is a typical example of monofunctional dependence between the intensities of the images to register: since the atlas can be used to generate realistic MR images, it is safe to assume a functional dependence between the intensity of the atlas and those of the T1-weighted image. Also, since the source image S is a labeled image, the function f is modeled as a piecewise constant function. In this case, each intensity level (10 in all) corresponds to a region from which to estimate the constant function.

The result of registration is presented in Fig. 1. The first image (Fig. 1a) shows one slice of the atlas. The second (Fig. 1b) is the corresponding slice of the T1-weighted image. The third and fourth images (Fig. 1c and 1d) present the result of registering the atlas with the T1-weighted image using our algorithm. Figure 1c shows the result without the intensity transformation; we have simply applied to the atlas the geometrical transformation resulting from the registration procedure. Figure 1d shows the image resulting from the registration process. It has the same shape as the T1-weighted image (Fig. 1b) and intensities have been transformed using the intensity correction. To facilitate the visual assessment of registration accuracy, contours obtained using a Canny–Deriche edge detector (on the T1-weighted image) have been overlaid over each image in Fig. 1.

Figure 1e shows the joint histogram of intensities after registration. Values have been compressed logarithmically and normalized as is depicted in the color scale. The histogram is colorcoded and ranges from dark red representing high point densities to light yellow depicting low point densities. The values of the piecewise constant function f are overlaid as white dots.

2. Bifunctional Dependence

When registering images from different modalities, monofunctional dependence may not necessarily be assumed. We presented in Section II.B.1 such an example: the registration of PD- and T1-weighted images. The main problem in this case is that the CSF/GM intensity of the PD image needs to be mapped to two different intensities in the T1-weighted scan.

To solve this problem, we applied the method described in Section II.B.1 to register PD- and T1-weighted images for which two polynomial functions of degree 12 are estimated. This polynomial degree was set arbitrarily to a relatively high value to allow significant intensity transformations.

As shown in Figs. 1f–1j, the CSF, which is white in the PD-weighted image (Fig. 1f) and black in the T1-weighted image (Fig. 1g), is well registered. Also, the intensity transformation is adequate (Fig. 1i). The same comments apply to the GM, which is white in the PD-weighed image (Fig. 1f) and gray in the T1-weighed image (Fig. 1g).

Figure 1j presents the joint histogram of the two images after registration. The functions f_1 and f_2 found during the registration process are superimposed, the blue line corresponds to f_1 and the green to f_2. The line width for a given intensity s is proportional to the value of the corresponding $\pi_\varepsilon(s)$.

As can be observed in Fig. 1j, the polynomial functions f_1 and f_2 fit well with the high-density clusters of the joint histogram. In particular, we see that the CSF/GM intensity values from the PD-weighted image (with values around 220) get mapped to two different intensity values in the T1-weighted scan: 75 and 45. The mapping to 75 represents the GM (red polynomial) while the mapping to 45 represents CSF (blue polynomial).

Note that in the registration of the atlas with the T1-weighted image and the PD- with the T1-weighted image, we have in both cases deformed the image with the most information (i.e., the one with the greater number of perceivable structures). This is simply because our algorithm permits many structures of the deformed image to be mapped to a single intensity. But a single intensity in the deformed image can be mapped to at most two intensities in the target image.

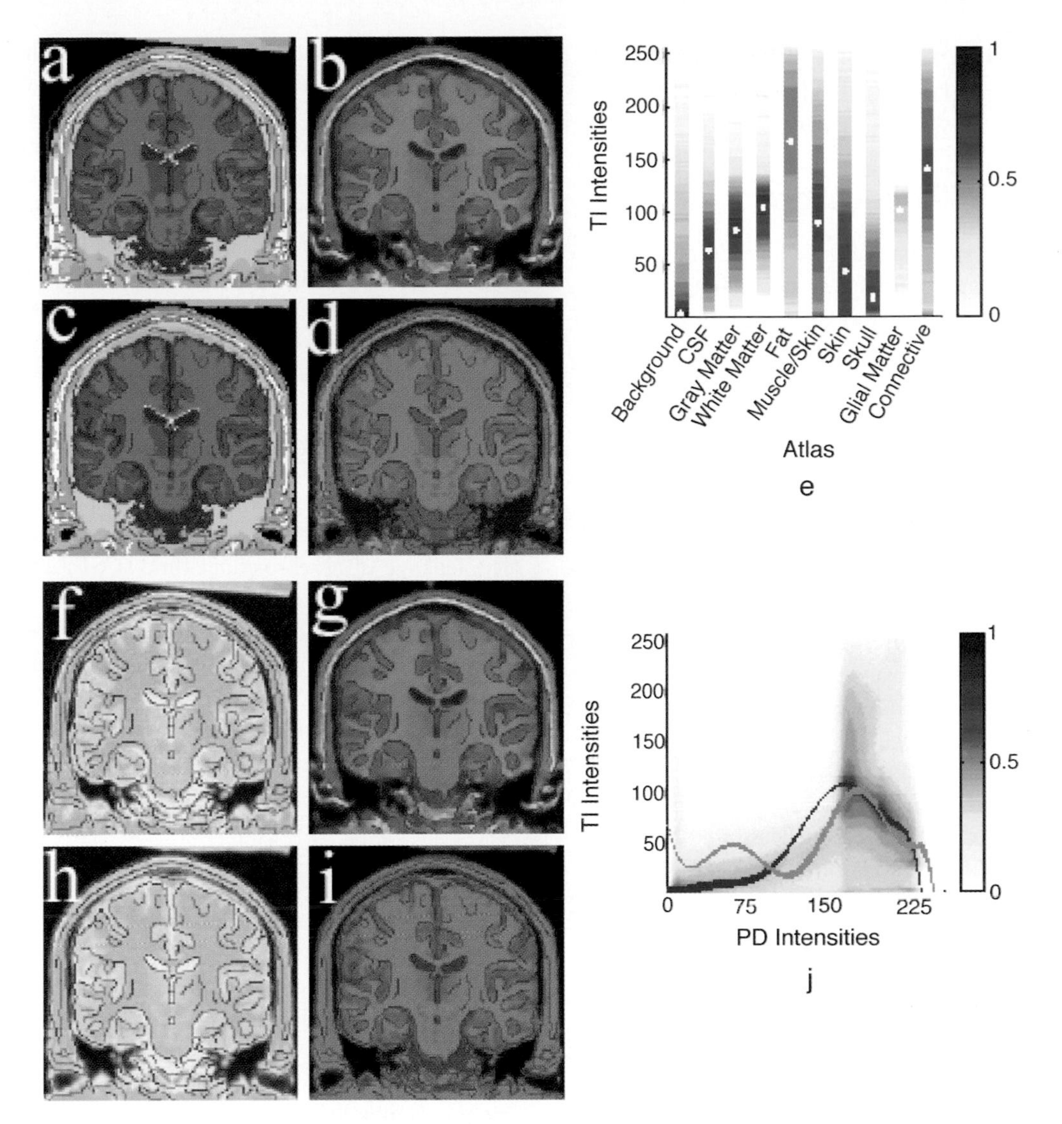

Figure 1 Results of 3D registration. **Registration of atlas with T1-weighted image.** (**a**) Atlas (source). (**b**) T1-weighted image (target). (**c**) Atlas without intensity correction, after registration with T1. (**d**) Atlas with intensity correction, after registration with T1. (**e**) The joint histogram of the atlas and T1-weighted image after registration; values range from dark red representing high point densities to light yellow depicting low point densities; the white dots correspond to the intensity transformation found by registering the atlas with the T1-weighted image and assuming monofunctional dependence (piecewise constant function). **Registration of PD-weighted with T1-weighted image.** (**f**) PD-weighted image (source). (**g**) T1-weighted image (target). (**h**) PD-weighted image without intensity correction, after registration with T1-weighted image. (**i**) PD-weighted image with intensity correction after registration with T1-weighted image. (**j**) The joint histogram of PD-weighted image and T1-weighted image, after registration; the blue line corresponds to f_1 and the green to f_2; the line width for a given intensity value s in T1 corresponds to the value of the corresponding $\pi_\varepsilon(s)$. Contours were obtained using a Canny–Deriche edge detector on the targets (b and g) and overlaid on the other images to better assess the quality of registration. The joint histogram values have been compressed logarithmically and normalized as is depicted in the color scale.

Hence, it is always better to use the image with the greater number of structures visible as the source image.

D. Conclusion

We have presented an original method to perform nonrigid registration of multimodal images. This iterative algorithm is composed of two steps: the intensity transformation and the geometrical transformation. Two intensity transformation models were described, which assume either monofunctional or bifunctional dependence between the intensity values in the images being matched. Both of these models were built using robust estimators to enable precise and accurate transformation solutions.

III. Image Fusion during Neurosurgery with a Biomechanical Model of Brain Deformation

A. Introduction

A critical goal of neurosurgery is to accurately locate, access, and remove intracranial lesions without damaging healthy brain tissue. The overriding goal is to preserve neurological function. This requires the precise delineation of the functional anatomy and morphology of the patient's brain, as well as lesion margins. The similar visual appearance of healthy and diseased brain tissue (e.g., as with infiltrating tumors) and the inability of the surgeon to see critical structures beneath the brain surface can pose difficulties during the operation. Some critical structures, such as white matter fiber tracts, may not be visible at all. Moreover, the difficulty in perceiving lesion (e.g., tumor) boundaries makes complete resection extremely difficult (Jolesz, 1997).

Over the past decade, advances in image-guided neurosurgery (IGNS) techniques have contributed to the growth of minimally invasive neurosurgery. These procedures must be carried out in operating rooms which are specially equipped with imaging systems. These systems are used to acquire images intraoperatively, as necessitated by the procedure. The improved visualization of deep structures and the improved contrast between the lesion and the healthy tissue (depending on the modality) allow the surgeon to plan and execute the procedure with greater precision.

IGNS has largely been a visualization-driven task. In the past, it had not been possible to make quantitative assessments of intraoperative imaging data, and instead physicians relied on qualitative judgments. In order to create a rich visualization environment which maximizes the information available to the surgeon, previous work has been concerned with image acquisition, registration, and display. Algorithm development for computer-aided IGNS has focused on improving imaging quality and speed. Another major focus has been to develop sophisticated multimodality image registration and fusion techniques, to enable fusion of preoperative and intraoperative images. However, clinical experience with IGNS involving deep brain structures has revealed the limitations of existing rigid registration approaches. This motivates the search for nonrigid techniques that can rapidly and faithfully capture the morphological changes that occur during surgery. In the future, the use of computer-aided surgical planning will include not only three-dimensional (3D) models but also information from multiple imaging modalities, registered into the patient's reference frame. Intraoperative imaging and navigation will thus be fully integrated.

Various imaging modalities have been used for image guidance. These include, among others, digital subtraction angiography, computed tomography, ultrasound, and intraoperative magnetic resonance imaging (IMRI). IMRI represents a significant advantage over other modalities because of its high spatial resolution and superior soft tissue contrast. However, even the most advanced intraoperative imaging systems cannot provide the same image resolution or tissue contrast selection features as preoperative imaging systems. More over, intraoperative imaging systems are by necessity limited in the amount of time available for imaging. Multimodality registration can allow preoperative data that cannot be acquired intraoperatively [such as nuclear medicine scans (SPECT/PET), functional MRI, MR angiography] to be visualized alongside intraoperative data.

B. Nonrigid Registration for IGNS

During neurosurgical operations, changes occur in the anatomical position of brain structures and adjacent lesions. The leakage of cerebrospinal fluid (CSF) after opening the dura, hyperventilation, administration of anesthetic and osmotic agents, and retraction and resection of tissue all contribute to shifting of the brain parenchyma. This makes information based upon preoperatively acquired images unreliable. The loss of correspondence between pre- and intraoperative images increases substantially as the procedure continues. These changes in the shape of the brain have been widely recognized as nonrigid deformations called "brain shift" (see Nabavi *et al.*, 2001).

Suitable approaches to capture these shape changes and to create integrated visualizations of preoperative data in the configuration of the deformed brain are currently in active development. Previous work aimed at capturing brain deformations for neurosurgery can be grouped into two categories. In the first category are those approaches that use some form of biomechanical model (recent examples include Hagemann *et al.*, 1999; Skrinjar and Duncan, 1999; Miga *et al.*, 1999; Skrinjar *et al.*, 2001; Ferrant *et al.*, 2000b). In the second category are those approaches that use phenomenological methods, relying upon image-related criteria (recent examples include Hill *et al.*, 1998; Hata, 1998; Ferrant *et al.*, 1999b; Hata *et al.*, 2000).

Purely image-based matching may be able to achieve a visually pleasing alignment, once issues of noise and intensity artifact are accounted for. However, in our work on intraoperative matching we favor physics-based models which ultimately may be expanded to incorporate important material properties (such as inhomogeneity, anisotropy) of the brain, once these are determined. However, even among physics-based models, there exists a spectrum of approaches, usually involving a trade-off between physical plausibility and speed.

A fast surgery simulation method has been described by Bro-Nielsen (1996). Here, high computational speeds were obtained by converting a volumetric finite element model into a model with only surface nodes. The goal of this work was to achieve very high graphics speeds consistent with interactive computation and display. This is achieved at the cost of simulation accuracy. This type of model is best suited to computer graphics-oriented display, for which high frame rates are needed.

A sophisticated finite element-based biomechanical model for two-dimensional brain deformation simulation has been proposed in Hagemann *et al.*, (1999). In this work, correspondences were established by manual interaction. The elements of the finite element model were the pixels of the two-dimensional image. The manual determination of correspondences can be time consuming and is subject to human error. Moreover, when methods are generalized to three dimensions, the number of points which must be identified can be very large. Due to the realities of clinical practice, two-dimensional results are not practical. A (three-dimensional) voxelwise discretization approach, while theoretically possible, is extremely expensive from a computational standpoint (even considering a parallel implementation) because of the number of voxels in a typical intraoperative MRI data set, leading to a large number of equations to solve ($256 \times 256 \times 60 = 3{,}932{,}160$ voxels, which corresponds to 11,796,480 displacements to determine). Downsampling can lead to fewer voxels, but leads to a loss of detail.

Edwards *et al.,* (1997) presented a two-dimensional three-component model for tracking intraoperative deformation. This work used a simplified material model. However, the initial 2D multigrid implementation required 120–180 min when run on a Sun Microsystems Sparc 20, which may limit its feasibility for routine use.

Skrinjar and Duncan (1999) have presented a very interesting system for capturing real-time intraoperative brain shift during epilepsy surgery. In this context, brain shift occurs slowly. A very simplified homogeneous brain tissue material model was adopted. Following the description of surface-based tracking from intraoperative MRI driving a linear elastic biomechanical model in Ferrant *et al.*, (2000b), Skrinjar *et al.* presented a new implementation (Skrinjar *et al.*, 2001) of their system using a linear elastic model and surface-based tracking from IMRI with the goal of eventually using stereoscopic cameras to obtain intraoperative surface data and hence to capture intraoperative brain deformation.

Paulsen *et al.* (1999) and Miga *et al.* (1999, 2001) have developed a sophisticated finite element model to simulate brain deformation. Their model is interesting because it incorporates simulations of forces associated with tumor tissue, as well as those resulting from retraction and resection. A limitation of the existing approach is that the preoperative segmentation and tetrahedral finite element mesh generation currently require around 5 h of operator time. Nevertheless, this approach holds promise in actually predicting brain deformation.

The real-time needs of surgery dictate that any algorithm used for prospective image matching must rapidly, reliably, and accurately capture nonrigid shape changes in the brain which occur during surgery. Our approach is to construct an unstructured grid representing the geometry of the key structures in the image data set. This technique allows us to use a finite element discretization that faithfully models key characteristics in important regions while reducing the number of equations to solve by using mesh elements that span multiple voxels in other regions. The algorithm allows the projection of preoperative images onto intraoperative images, thereby allowing the fusion of images from multiple modalities and spanning different contrast mechanisms. We have used parallel hardware, parallel algorithm design, and efficient implementations to achieve rapid execution times compatible with neurosurgery.

C. Method

Figure 2 is an overview, illustrating the image analysis steps used during intraoperative image registration. The critical image processing tasks include segmentation (the identification of anatomical structures) and registration. Segmentation data are used both for preoperative planning and to create intraoperative segmentations. The segmentation data are used to calculate an initial affine transformation (rotation, translation, scaling) which rigidly registers the images, thus initializing the data for nonrigid matching using our biomechanical simulation. Using the biomechanical model, the volumetric deformation is inferred through a mechanical simulation with boundary conditions established via surface matching. This sophisticated deformation model can be solved during neurosurgery, providing enhanced intraoperative visualization.

1. Preoperative Data Acquisition and Processing

The time available for image processing during surgery is extremely limited compared to that available preoperatively. Consequently, preoperative data acquisition can be more comprehensive, and more extensive image analysis (for example segmentation) can be performed.

A variety of manual (Gering *et al.*, 1999), semiautomated (Kikinis *et al.*, 1992; Yezzi *et al.*, 2000), and automated (Warfield *et al.*, 2000a,b; Kaus *et al.*, 1999)

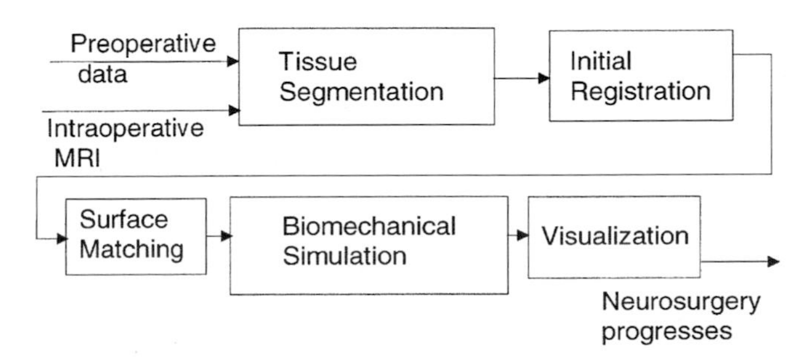

Figure 2 Schematic illustrating image analysis tasks carried out during neurosurgery.

segmentation approaches are available. We select the most accurate, robust approach based upon the preoperative data and the particular critical structures. For the matching experiments which will be described below, we have used an anatomical atlas, although other preoperative data such as magnetic resonance angiography or diffusion tensor images could ultimately also be used. The atlas was constructed from a high-resolution scan of a single patient, in which over 200 structures were segmented (Kikinis *et al.*, 1996) using a combination of automated and interactive techniques. During surgery, we are especially interested in the corticospinal tract, a region of white matter which can be difficult or impossible to directly observe with conventional MRI and which must be preserved. We have previously shown that we can project the corticospinal tract from the atlas onto patient scans for preoperative surgical planning (Kaus *et al.*, 2000).

2. Intraoperative Image Processing

Intraoperative image processing consists of: (1) acquiring one or more intraoperative volumetric data sets, (2) constructing a segmentation of the intraoperative acquisition, (3) computing an affine registration of the preoperative data onto the new acquisition, (4) identifying the correspondences between key surfaces of the preoperative and intraoperative data, and (5) solving a biomechanical model to infer a volumetric deformation field, applying the deformation to the preoperative data and constructing a new visualization merging critical structures from the preoperative data with the intraoperative data.

a. Segmentation of Intraoperative Volumetric Images. In the experiments conducted below, a rapid segmentation of the brain and ventricles was obtained using a binary curvature-driven evolution algorithm (Yezzi et al., 2000). The region identified as brain or ventricle was then interactively corrected to remove misclassified tissue using the software described by Gering et al. (2001). This approach allows the surgeon to inspect and interactively edit the segmentation data, increasing its accuracy.

Alternatively, we have experimented with automated intraoperative segmentation (Warfield *et al.*, 1998b, 2000a) utilizing tissue classification in a multichannel feature space using a model of expected anatomy as an initial template. Automated approaches will likely be preferable once they have been validated.

b. Unstructured Mesh Generation and Surface Representation. We have implemented a mesh generator which is optimized for use with biomedical structures, building upon previously described techniques (Schroeder et al., 1996; Geiger, 1993). During mesh generation, we extract an explicit representation of the surface of the brain and ventricles based on the preoperative segmentation. We also create a volumetric unstructured mesh using a multiresolution version of the marching tetrahedra algorithm. The mesher begins by subdividing the image into cubes, which are then divided into five tetrahedra using an alternating split pattern which prevents diagonal crossings on the shared faces. The mesh is iteratively refined in the region of complex boundaries, and then a marching tetrahedra-like approach is applied to this multiresolution mesh. For each cell of the final mesh, the label value of each vertex is checked and, if different, the tetrahedron is divided along the edge having different node labels. A detailed description can be found in Ferrant et al. (1999b, 2000a).

The meshing process is extremely robust, allowing us to generate tetrahedral meshes of the brain and ventricles from rapid segmentations of each volumetric intraoperative acquisition carried out during the surgery. This facilitates intraoperative matching from one volumetric acquisition to the next.

c. Affine Registration of Preoperative to Intraoperative Image Datasets. For affine registration (rotation, translation, scaling), we use a fast parallel implementation of a robust algorithm which is based upon aligning segmented image data, using a rapidly converging multiresolution search strategy (Warfield et al., 1998a). Applying the resulting transform, segmentations and gray-scale data from the preoperative and intraoperative scans are rigidly registered.

d. Volumetric Biomechanical Simulation of Brain Deformation. During the procedure, the brain undergoes nonrigid shape changes for the reasons described above. During IGNS the surgeon is able to acquire a new volumetric MRI when he or she wishes to review the current configuration of the entire brain. A volumetric deformation field relating earlier acquisitions to this new scan is computed by first matching surfaces from the earlier acquisition to the current acquisition and then calculating the volumetric displacements by using the surface displacements as boundary conditions. The critical concept is that forces are applied to the volumetric model that will produce the same surface displacements as were obtained by the surface matching. The biomechanical model can then be used to compute a volumetric deformation map.

di. Establishing Surface Correspondences. The surfaces of the brain and lateral ventricles are iteratively deformed using a dual active surface algorithm. Image-derived forces are applied iteratively to an elastic membrane surface model of the early scan, thereby deforming it so as to match the boundary of the current acquisition. The derived forces are a decreasing function of the image intensity gradients, so as to be minimized at the edges of objects in the volume. We have included prior knowledge about the expected gray level and gradients of the objects being matched to increase the convergence rate of the process. This algorithm is fully described by Ferrant *et al.* (1999a).

dii. Biomechanical Simulation of Volumetric Brain Deformation. We treat the brain as a homogeneous linearly elastic material. The deformation energy of an elastic body Ω, under no initial stresses or strains and subject to externally applied forces, can be described by the model (Zienkiewicz and Taylor, 1994)

$$E(u)=\frac{1}{2}\int_{\Omega}\sigma^{T}\varepsilon\, d\Omega-\int_{\Omega}u^{T}F\, d\Omega, \tag{14}$$

where the variables are given in terms of the stress vector, σ; the strain vector, ε; the forces $F = F(x,y,z)$ applied to the elastic body (forces per unit volume, surface forces, or forces concentrated at the nodes of the mesh); and $\mathbf{u} = (u(x,y,z), v(x,y,z), w(x,y,z))^{T,}$ the displacement vector field we wish to compute. Since we are using a linear elasticity framework, we assume small deformations. Hence the strain vector ε is given by

$$\varepsilon=\left(\frac{\partial u}{\partial x},\frac{\partial v}{\partial y},\frac{\partial w}{\partial z},\frac{\partial u}{\partial y}+\frac{\partial v}{\partial x},\frac{\partial v}{\partial z}+\frac{\partial w}{\partial y},\frac{\partial w}{\partial x}+\frac{\partial u}{\partial z}\right)^{T}, \tag{15}$$

which can be written as $\varepsilon = \mathbf{L}\,\mathbf{u}$ where $\mathbf{L}$ is a linear operator. The elastomechanical relation between stresses and strains can be expressed by the generalized Hooke's law as

$$\sigma=\left(\sigma_{x},\sigma_{y},\sigma_{z},\tau_{xy},\tau_{yz},\tau_{zx},\right)^{T}=D\varepsilon. \tag{16}$$

Assuming isotropic material properties for each point, we obtain a symmetric elasticity matrix $\mathbf{D}$ in the form

$$D=\frac{E}{(1+\nu)(1-2\nu)}\begin{bmatrix}1-\nu & \nu & \nu & 0 & 0 & 0\\ \nu & 1-\nu & \nu & 0 & 0 & 0\\ \nu & \nu & 1-\nu & 0 & 0 & 0\\ 0 & 0 & 0 & \frac{1-2\nu}{2} & 0 & 0\\ 0 & 0 & 0 & 0 & \frac{1-2\nu}{2} & 0\\ 0 & 0 & 0 & 0 & 0 & \frac{1-2\nu}{2}\end{bmatrix}, \tag{17}$$

with physical parameters E (Young's modulus) and ν (Poisson's ratio). See Zienkiewicz and Taylor, (1994) for the full details.

For the discretization, we use the finite element method applied over the volumetric image domain so that the total potential energy can be written as a sum of potential energies for each element: $E(\mathbf{u}) = \Sigma_{e=1}^{Nnodes} E^e(\mathbf{u}^e)$. The mesh is composed of tetrahedral elements and thus each element is defined by four mesh nodes. The continuous displacement field $\mathbf{u}$ everywhere within element e of the mesh is defined as a function of the displacement at the element's nodes $\mathbf{u}_i^e$ weighted by the element's interpolating functions $N_i^e(\mathbf{x})$,

$$\mathbf{u}(\mathbf{x})=\sum_{i=1}^{N_{nodes}}\mathbf{I}N_i^e(\mathbf{x})\mathbf{u}_i^e. \tag{18}$$

Linear interpolating functions are used to define the displacement field inside each element. The interpolating function of node i of tetrahedral element e is defined as

$$N_i^e(\mathbf{x})=\frac{1}{6V^e}\left(a_i^e+b_i^e x+c_i^e y+d_i^e z\right). \tag{19}$$

The computation of the volume of the element V^e and the interpolation coefficients are detailed in Zienkiewicz and Taylor (1994; pp. 91–92).

The volumetric deformation of the brain is found by solving for the displacement field that minimizes the deformation energy described by Eq. (14). For our finite element approach this is described by

$$\delta E(\mathbf{u})=\sum_{e=1}^{M}\delta E^e(\mathbf{u}^e)=0, \tag{20}$$

where

$$\delta E^e(\mathbf{u}^e)=\sum_{i=1}^{N_{nodes}}\frac{\partial}{\partial u_i^e}E^e(\mathbf{u}^e)\delta u_i^e+\sum_{i=1}^{N_{nodes}}\frac{\partial}{\partial v_i^e}E^e(\mathbf{u}^e)\delta v_i^e+\sum_{i=1}^{N_{nodes}}\frac{\partial}{\partial w_i^e}E^e(\mathbf{u}^e)\delta w_i^e, \tag{21}$$

Since δu_i^e, and δu_i^e and δw_i^e are independent, defining matrix $\mathbf{B}^e = (\mathbf{B}_i^e)_{i=1}^{Nnodes}$ with $\mathbf{B}_i^e = \mathbf{L}N_i^e$ for every node i of each element e yields the equation

$$0=\int_{\Omega}\mathbf{B}^{eT}\mathbf{D}\mathbf{B}^e\mathbf{u}^e d\boldsymbol{\Omega}-\int_{\Omega}\mathbf{N}^{eT}\mathbf{F}^e d\boldsymbol{\Omega}, \tag{22}$$

with the element stiffness matrix $\mathbf{K}^e = \int_{\Omega}\mathbf{B}^{eT}\mathbf{D}\mathbf{B}^e\, d\Omega$. An assembly of the equations for all elements finally leads to a global linear system of equations, which can be solved for the displacements resulting from the forces applied to the body:

$$\mathbf{Ku} = \mathbf{F}. \tag{23}$$

The displacements at the boundary surface nodes are fixed to match those generated by the active surface model. Let $\tilde{\mathbf{u}}$ be the vector representing the displacement to be imposed at the boundary nodes. The elements of the rows of the stiffness matrix $\mathbf{K}$ corresponding to the nodes for which a displacement is to be imposed are set to zero and the diagonal elements of these rows are set to one. The force vector $\mathbf{F}$ is set to equal the displacement vector for the boundary nodes: $\mathbf{F} = \tilde{\mathbf{u}}$ (Zienkiewicz and Taylor, 1994). In this way solving Eq. (23) for the unknown displacements will produce a deformation field over the entire volumetric mesh that matches the prescribed displacements at the boundary surfaces.

e. Hardware and Implementation. The volumetric deformation of the brain is computed by solving for the displacement field that minimizes the energy described by Eq. (14), after fixing the displacements at the surface to match those generated by the active surface model.

Three variables, representing the x, y, and z displacements, must be determined for each element of the finite element mesh. Each variable gives rise to one row and one column in the global K matrix. The rows of the matrix are divided equally among the CPUs available for computation and the global matrix is assembled in parallel. Each CPU assembles the local $\mathbf{K}^e$ matrix for each element in its subdomain. Although each CPU has an equal number of rows to process, because the connectivity of the mesh is irregular, some CPUs may do more work than others.

Following matrix assembly, the boundary conditions determined by the surface matching are applied. The global K matrix is adjusted such that rows associated with variables that are determined consist of a single nonzero entry of unit magnitude on the diagonal.

The volumetric biomechanical brain model system of equations (and the active surface membrane model equations) are solved using the Portable, Extensible Toolkit for Scientific Computation (PETSc) package (Balay *et al.*, 1997, 2000a) using the Generalized Minimal Residual solver with block Jacobi preconditioning. During neurosurgery, the system of equations was solved on a Sun Microsystems SunFire 6800 symmetric multiprocessor machine with 12 750-MHz UltraSPARC-III (8 MB Ecache) CPUs and 12 GB of RAM. This architecture gives us sufficient compute capacity to execute the intraoperative image processing prospectively during neurosurgery.

f. Intraoperative visualization. Once the volumetric deformation field has been computed, it can be applied to earlier data to warp it into the current configuration of the patient anatomy. The imaging data can then be displayed by texture mapping onto flat planes to facilitate comparisons with current intraoperative data as well as prior scans. Triangle models of segmented structures (i.e., based on registered volumetric data) can be used to display surface renderings of critical anatomical structures, overlaid on intraoperative image data. This allows ready appreciation of the 3D anatomy of these segmented structures together with the imaging data in the form of planes passing through or over the 3D triangle models (Gering et al., 2001). This augments the surgeon's ability to see critical structures which must be preserved (such as the corticospinal tract) and to better appreciate the lesion and its relationship to other structures.

Figure 3 shows the open-configuration magnetic resonance scanner optimized for imaging during surgical procedures (Jolesz, 1997; Black *et al.*, 1997). The image we constructed was presented on the LCD and increased the information available to the surgeon as the operation progressed.

D. Results and Discussion

The image registration strategy described here has been applied prospectively during several neurosurgical cases.

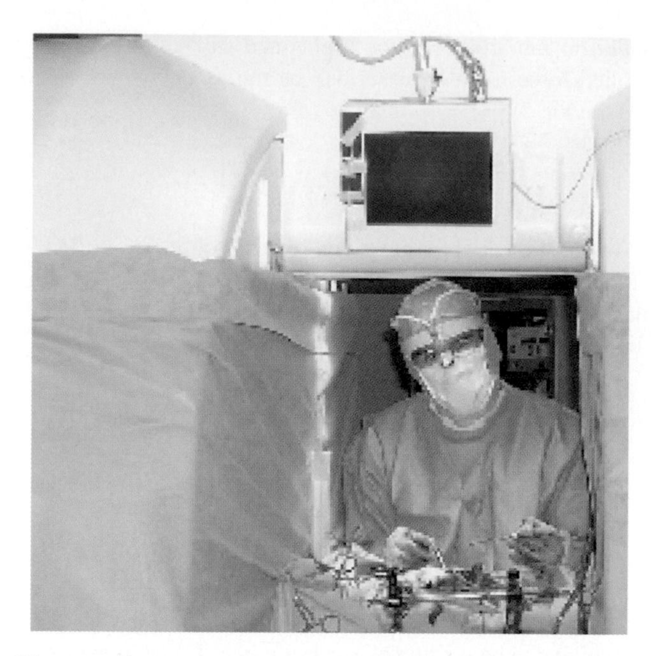

Figure 3 Open-configuration magnetic resonance scanner during neurosurgery.

We present here illustrative results which demonstrate the ability of our algorithm to capture intraoperative brain deformations.

The enhancement provided by intraoperative nonrigid registration to the surgical visualization environment is shown by our matching the corticospinal tract of a preoperatively prepared anatomical atlas to the initial and subsequent intraoperative scans of a subject. This matching was carried out prospectively during the neurosurgery, demon-

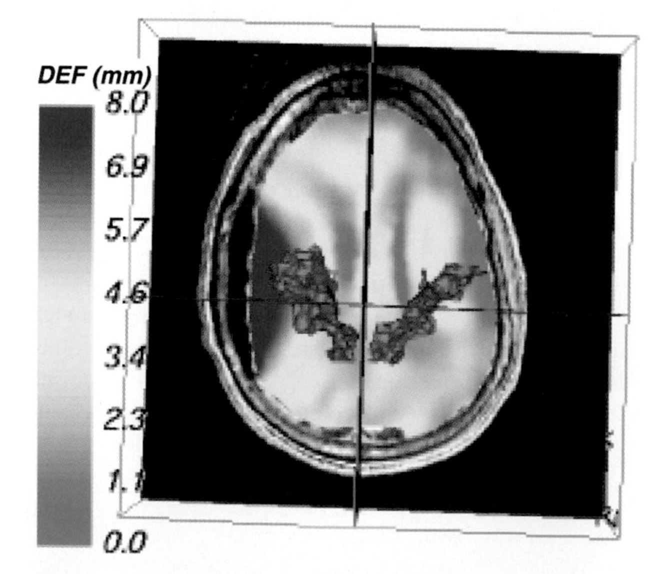

Figure 4 The corticospinal tract from our anatomical atlas is shown in blue, projected into the shape of the brain of the subject shown in Fig. 3.

strating the practical value of the approach and its ability to meet the real-time constraints of surgery. We have also conducted parallel scaling experiments which have yielded very encouraging results. The entire image analysis process can be completed in less than 10 min, which has been adequate to display the information to the surgeon. Interestingly, the most computationally intensive task (the biomechanical simulation) has also been optimized the most and is now the fastest step. We anticipate that segmentation techniques requiring less user interaction will result in significant improvements in speed.

Figure **4** shows the corticospinal tract from our anatomical atlas projected into the shape of the brain of the subject. This visualization helps the surgeon to better appreciate the 3D relationship of this essential structure to the lesion and other regions of the brain. The corticospinal tract cannot be readily observed in IMRI acquisitions.

Figure 5 is a typical case illustrating the amount of brain deformation which can occur during neurosurgery, as well as the effectiveness of our algorithm in capturing this shift during neurosurgery. As shown, the quality of the match is significantly better than can be obtained through rigid registration alone.

Our early experience has shown that our intraoperative biomechanical simulation of brain deformation is a robust and reliable method for capturing the changes in brain shape that occur during neurosurgery. The registration algorithm requires no user interaction and the parallel imple-

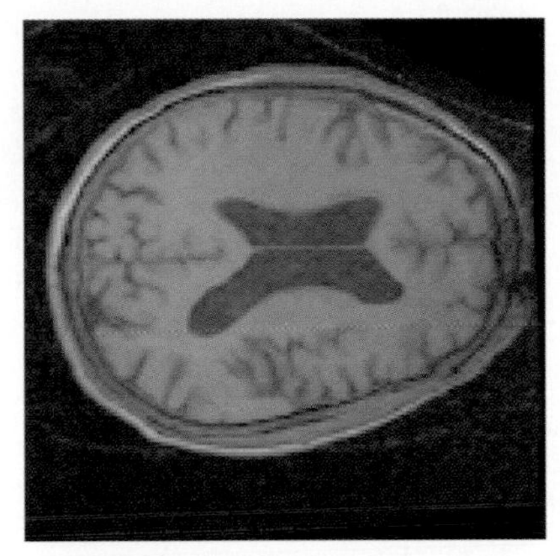

(a) A single slice from an early 3D IMRI scan.

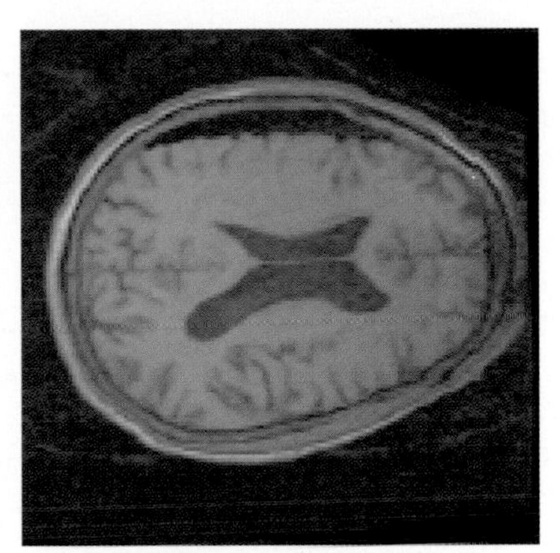

(b) The corresponding slice in a later 3D IMRI scan, showing significant brain shift has occurred.

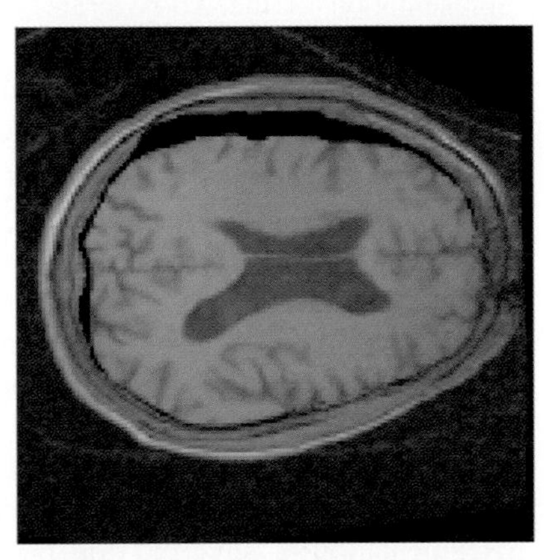

(c) The matched slice of the first volume after simulation of the brain deformation.

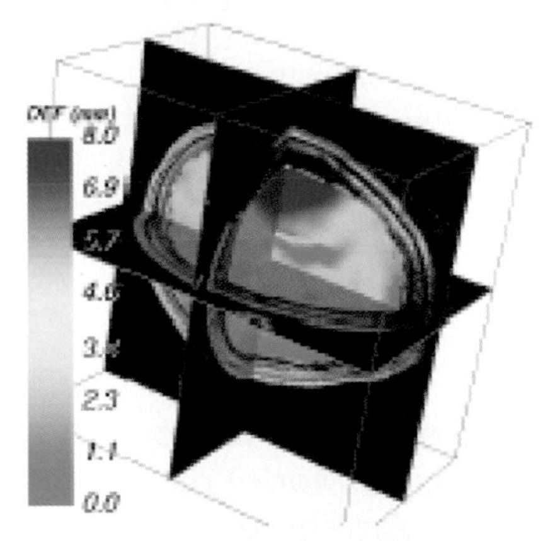

(d) Visualization of the magnitude of the deformation field computed in matching image (a) to image (b).

Figure 5 Two-dimensional slices through three-dimensional data, showing the match of the simulated deformation of the initial brain onto the actual deformed brain. The quality of the match is significantly better than can be obtained through rigid registration alone. (a) A single slice from an early 3D IMRI scan. (b) The corresponding slice in a later 3D IMRI scan, showing that significant brain shift has occurred. (c) The matched slice of the first volume after simulation of the brain deformation. (d) Visualization of the magnitude of the deformation field computed in matching image (a) to image (b).

mentation is sufficiently fast to be used intraoperatively. We intend to incorporate patient-specific preoperative data in place of the anatomical atlas to increase the surgical value of the intraoperative updates.

As we refine our approach, we expect to appreciate performance gains based on more automated segmentation methods and further optimized parallel implementations which address load imbalances. Improvements in the accuracy of the match could result from a more sophisticated model of the material properties of the brain (such as more accurate modeling of the cerebral falx and the lateral ventricles). Sophisticated MR imaging methods such as diffusion tensor MRI now enable the preoperative imaging of inhomogeneous anistropic white matter structure, which could be incorporated into the material model. Ultimately, the prediction of brain deformation, as opposed to the capture of observed deformation described here, will most likely require a nonlinear material model together with extensive monitoring of physiological data. Such prediction could be used to indicate when new intraoperative imaging is necessary to appropriately update both the simulation model and the surgeon's understanding of the brain shape.

E. Conclusion

Nonrigid changes in brain morphology occur during neurosurgery and limit the usefulness of preoperative imaging for intratreatment planning and surgical navigation. Intraoperative nonrigid registration can add significantly to the value of intraoperative imaging. It provides for quantitative monitoring of therapy application, including the ability to make quantitative comparisons with a preoperatively defined treatment plan and enables preoperative image data to be aligned with the current configuration of the brain of the patient. We have shown that even a relatively complex biomechanical model can be initialized and solved during neurosurgery, providing enhanced surgical visualization. Ultimately, such approaches may provide a truly integrated, multimodality environment for surgical navigation and planning.

IV. Physics-Based Regularization with an Empirical Model of Anatomical Variability

An important issue in nonrigid registration for computer-assisted neurology and neurosurgery is the generation of deformation fields that reflect the transformation of an image in a realistic way with respect to the given anatomy. Due to lack of image structure, noise, intensity artifacts, computational complexity, and a restricted time frame (e.g., during surgery), it is not feasible to measure directly the deformation occuring at each voxel. This leads to estimates of the deformation field only at sparse locations, which have to be interpolated throughout the image.

Recently, physics-based elastic and viscous fluid models for nonrigid registration have become popular (Bajcsy and Kovacic, 1989), since they have the potential to constrain the underlying deformation in a plausible manner. However, viscous fluid models (Lester *et al.*, 1999; Wang and Staib, 2000) have to be chosen carefully, since they allow large deformations. This is not always suitable for medical applications concerning the brain. Furthermore, viscous fluid models driven by alignment of similar gray values may allow anatomically incorrect matches of different but adjacent structures through the same mechanism by which large-deformation matches are permitted. For example, one gyrus may flow from the source brain to match two or more different gyri in a target brain, producing an anatomically incorrect match.

In terms of physics-based elastic models, recent work (Davatzikos, 1997; Ferrant *et al.*, 2000b) has proposed an active surface algorithm computed at the boundary of a regarded structure as an initial estimate of the deformation field which was then introduced into a volumetric elastic model to infer the deformation inside and outside the surface. A drawback of this method is that although it has been shown to be accurate close to the object's boundary, away from the boundaries the solution could be less accurate. The work by Wang and Staib (2000) represents an improvement in that statistical shape information (based on a set of images with manually identified boundary points) was included as an additional matching criterion. Even though such methods are promising for specific brain structures, a robust 3D shape representation of the whole brain still remains difficult to achieve.

In Collins (1994) another nonrigid registration algorithm was proposed, based on an iterative refinement of a local similarity measure using a simplex optimization. In this approach the deformation field was constrained only by smoothing after correspondence estimation and thus can be accurate only for specific regions of the brain. To achieve better results, the method was improved by introducing various gyri and sulci of the brain as geometric landmarks (Collins *et al.*, 1998a).

In order to obtain realistic deformations, we propose here a physics-based elastic model. The method does not require a segmentation or have the drawback that initial estimates of the deformation are generated only for the boundary of a considered structure. Instead, these estimates are calculated based on a template matching approach with a local similarity measure. Furthermore we have incorporated a model for inhomogeneous elasticities into our algorithm. The discretization of the underlying equation is done by a finite element technique, which has become a popular method for medical imaging applications (e.g., Bro-Nielsen, 1998), and (Ferrant *et al.*, 2000b).

A. Method

The process of registration can be described as an optimization problem that minimizes the deformation energy between a template and a reference image. Assuming that both images represent the same physical object, the deformation that aligns them is therefore related to the theorem of minimum potential energy. The idea of our registration process can now be described as follows: based on a set of points extracted out of an image as described in Section IV.B, an initial sparse estimate of the deformation field is found by a local normalized cross-correlation (Section IV.C). In a next step nonrigid registration is performed using an elastic model (Section IV.D), which is constrained by the sparse estimates computed in the previous step.

B. Feature Point Extraction

Let V denote the domain of a volume $S: \Omega \rightarrow \mathbb{R}$ with voxel positions $\mathbf{x} = (x,y,z)^T$, $\mathbf{x} \in \Omega$. In a first step a set of feature points is extracted out of the reference image. For that purpose we calculate the gradient magnitude out of blurred image intensities. In order to obtain suitable feature points for an initial sparse estimate of the deformation field, only voxels higher than 2 standard deviations above the mean of the magnitude of the gradient are used for the correspondence detection (Section IV.C). Figure 6 shows this process for one slice of a MR scan of the brain.

To overcome the poor edge-preserving properties of linear low-pass filters, we use a nonlinear diffusion filter. A filtered version p of volume S can be described as a solution of the partial differential equation

$$\partial_t p = \text{div}(g(|\nabla p_\sigma|^2)\nabla p), \tag{24}$$

with Neumann boundary conditions and the original image as initial state (Weickert, 1997). The diffusion function $g: \mathbb{R} \rightarrow \mathbb{R}$ is used to reduce the noise sensitivity and thus depends on the magnitude of the gradient of smoothed image intensities, computed by convolving p with a Gaussian kernel of standard deviation σ. The idea of the diffusion function is to stop the filtering process at regions with high gradients (i.e., at edges in an image) and to provide a value close to zero there. In our method, we use a diffusion function proposed by Weickert (1997):

$$g(x^2) = \begin{cases} 1 & \text{for } x^2 = 0 \\ 1 - \exp\left(\frac{-C}{(x/\lambda)^8}\right) & \text{for } x^2 > 0 \end{cases} \tag{25}$$

The parameter λ separates regions of low contrast from those of high contrast. For values greater than λ, the filtering is reduced, while for values less than or equal to λ stronger smoothing is applied. For the constant C, Weickert proposes $C = 3.31448$, which gives visually good results and gives the flux $f(x) = x \cdot g(x^2)$ the expected behavior (i.e., f is increasing for values $|x| \leq \lambda$ and decreasing for values $|x| > \lambda$). As an *a priori* determination of λ is very difficult, the contrast parameter was set interactively for each volume in our approach. Furthermore a parallel additive operator splitting scheme was used for computational efficiency. See Weickert *et al.* (1998) for details.

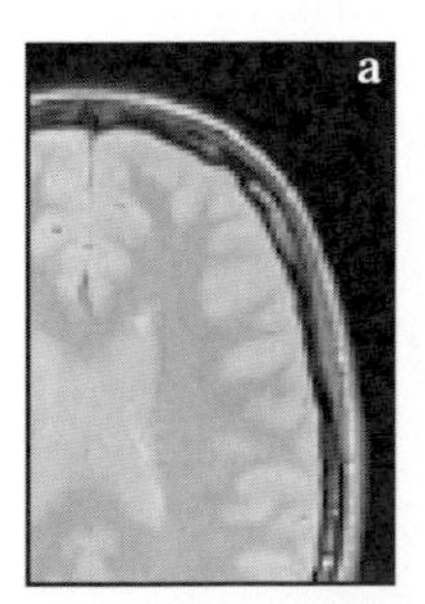

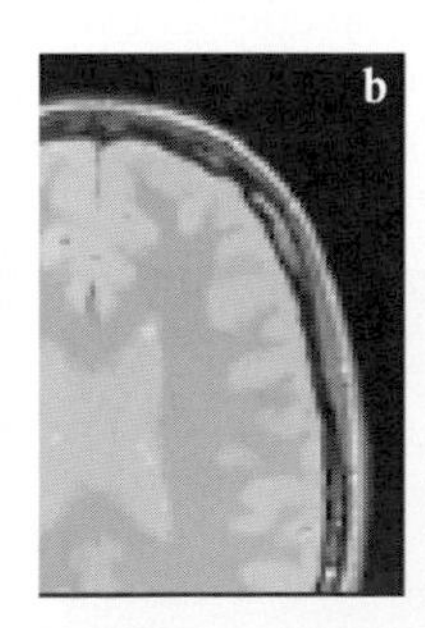

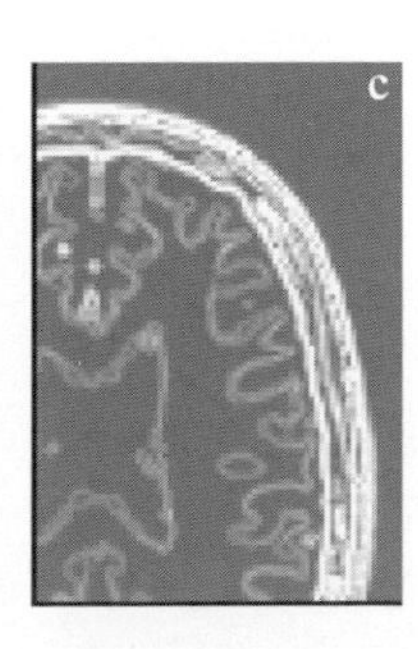

Figure 6 Illustration of feature point extraction. For a better visual impression only a detail of the image is shown. (**a**) Slice of an MR scan. (**b**) Slice after applying a nonlinear diffusion filter. (**c**) Magnitude of the gradient of the blurred image after thresholding.

C. Correspondence Detection

After feature points are extracted, the correspondences between the reference image R and the template image T is computed for these points. A common way to minimize the distance between regions of two volumes consists of finding the global optimum of a function which measures their similarity. This can be expressed as a cost function $C: \mathbb{R}^n \rightarrow \mathbb{R}$, which is optimized for a voxel $\mathbf{x}$ between two regions of R and T in terms of a given transformation T_ϑ. The search space is restricted by a set of parameter vectors $\vartheta \in \mathbb{R}^n$

Our approach uses the local normalized cross-correlation (NCC)

$$C(\vartheta) = \frac{\sum_{k \in \mathcal{N}(x)} f(R,k) \cdot f(T, T_\vartheta^{-1}(k))}{\sqrt{\sum_{k \in \mathcal{N}(x)} f^2(R,k) \cdot \sum_{k \in \mathcal{N}(x)} f^2(T, T_\vartheta^{-1}(k))}}, \quad \forall x \in \Omega, \tag{26}$$

which is maximized at a given voxel by a brute force search. Therefore we assume a window of size $(w \times w \times w)$ around a voxel $\mathbf{x}$ in the reference image and compute the maximal NCC by shifting a window of similar size in the template image. In Eq. (26), this window is described by a local neighborhood of a voxel $\mathbf{x}$ defined as $\mathcal{N}(\mathbf{x}) = \{(x - w, y - w, z - w)^T, \ldots, (x + w, y + w, z + w)^T\}$. The search space in our method is restricted to translations because other transformations like rotations or scaling would be of higher computational complexity. Furthermore the NCC is computed only for voxels with high gradient magnitudes calculated out of blurred image intensities, as described in Section (IV.B). For a better performance for large data sets the optimization is solved in parallel.

D. Interpolation from Sparse Displacement Estimates

The sparse deformation estimates obtained at the feature points computed by a local normalized cross-correlation are now introduced as external forces into an elastic model. We use a similar energy term as described in Section III.C.2 using the finite element method for the discretization. Hence we seek the deformation **u** that minimizes Eq. (14)—repeated here for convenience:

$$E(\mathbf{u}) = \frac{1}{2}\int_{\Omega} \sigma^{\mathrm{T}} \varepsilon d\Omega - \int_{\Omega} \mathbf{u}^{\mathrm{T}} \mathbf{F} d\Omega. \qquad (26)$$

The underlying idea is again to restrict the registration process so that the resulting deformation field is *a priori* fixed by the estimates at these points.

For a volume of 256 × 256 × 124 voxels, the linear system of equations we obtain has approximately 532,000 unknowns, which is also solved in parallel with the PETSc package (Balay *et al.*, 1997, 2000a,b). The execution time for the whole registration process is usually about 5 min on a cluster with 12 CPUs (see Section III.C.2 for details).

1. Inferring Empirically Observed Anatomical Variability

In order to describe the mechanical behavior of tissue undergoing a deformation, the relation between stress and strain is expressed by an elasticity matrix D, generated during the matrix assembly. For isotropic material two parameters are needed: Young's modulus E as a measure of stiffness and Poisson's ratio ν as a measure of incompressibility.

Typically elasticity parameters have been set arbitrarily and homogeneously (Bajcsy and Kovacic, 1989; Ferrant *et al.*, 2000b), which is only a rough approximation of the underlying tissue. Recently Lester *et al.* (1999) applied an inhomogeneous viscous fluid model to brain and neck registration. Manual segmentations of the bone were used as a region of high stiffness. Davatzikos (1997) applied inhomogeneities to brain warping, setting the elasticity parameters of the brain four times higher than their value in the ventricles.

Our approach differs in that inhomogeneous elasticity parameters are derived from an empirical estimate of anatomical variability, so that each discrete element can obtain its own material properties during the matrix assembly. We used a set of 154 MR scans of the brain, first segmented into white matter, gray matter, CSF, and background using an EM-based statistical classification algorithm (Wells *et al.*, 1996a). In the next step, the head of each scan was aligned to an arbitrarily selected scan out of this database, using global affine transformations (Warfield *et al.*, 1998a) and our nonrigid registration method. Figure 7 shows the result for the tissue classes after nonrigid registration, averaged over all scans. In order to generate a model for inhomogeneous elasticities, we use an entropy measure for each voxel.

Therefore we define the joint voxelwise entropy as

$$h(s1, s2, s3, s4) = -\sum_{i-1}^{4} p(si)\ln(p(si)), \qquad (27)$$

where each S_i represents the sum over all scans for one of the four different segmented tissue classes at a certain voxel. According to these results, the elasticity parameters are computed for every voxel. We choose a linear mapping for the computed joint voxelwise entropy of the identified brain tissues where the Poisson ratio ν was scaled in the range of $\nu \in [0.1, 0.4]$ while Young's elasticity modulus E had a range of $E \in [2kPa, 10\,kPa]$. The background was set to a low stiffness ($E = 1kPa$) and incompressibility parameter ($\nu = 0.05$). Figure 8 shows a slice of the computed model and the associated values for ν.

E. Illustration of Nonrigid Registration with Homogeneous and Inhomogeneous Elasticities

In order to demonstrate the behavior of our deformation model with homogeneous and inhomogeneous elasticities, the algorithm was applied to register 159 MR scans of the brain of young adults. Each scan was first globally registered to an arbitrarily chosen data set by an affine transformation (Warfield *et al.*, 1998a). The nonrigid registration with homogeneous and inhomogeneous elasticities was then applied to the aligned data. Figure 9 shows the results of the matching process after averaging over all scans. Because we are performing registration among different subjects, a global affine transformation normally will not be able to

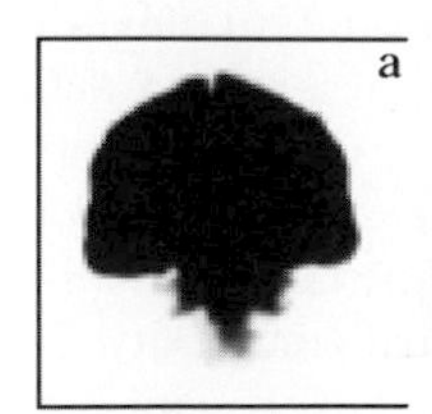

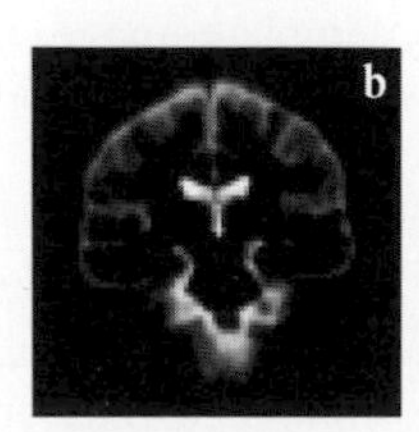

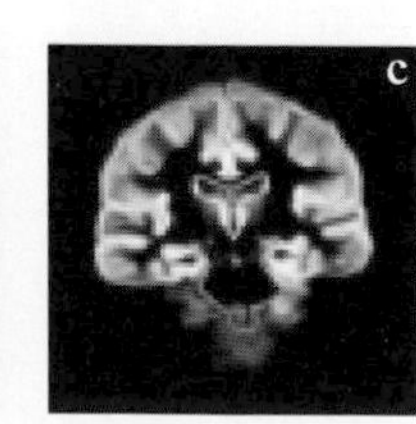

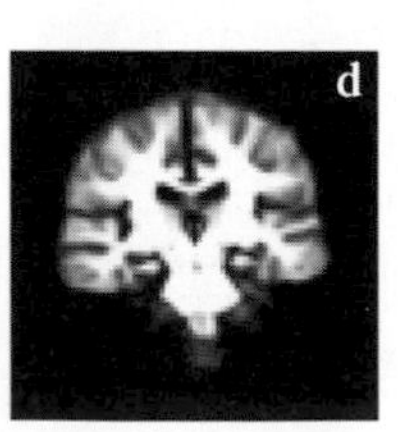

Figure 7 Images from the averaged volume for different tissue classes after nonrigid registration. Dark regions imply a slight overlapping. (**a**) Background, (**b**) CSF, (**c**) Gray matter, (**d**) White matter.

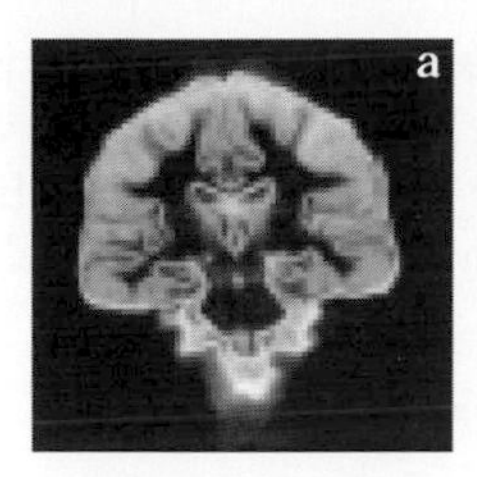
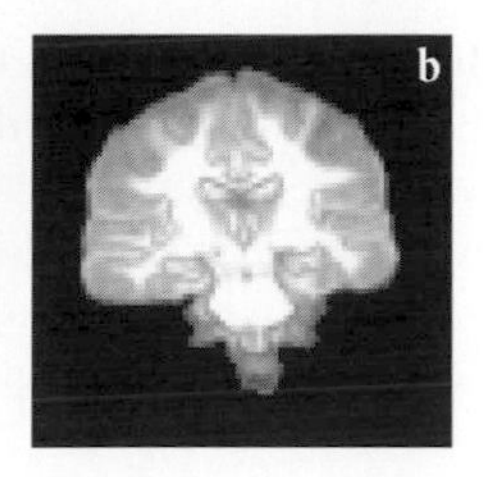

Figure 8 Model of empirically observed variability. (**a**) Slice of the model after voxelwise entropy computation. Dark regions imply a low entropy value, (**b**) Computed incompressibility parameter (Poisson ratio ν) for each voxel of the same slice. Dark regions imply a low value for ν.

align reference and template image properly. This leads to a blurred average image (Fig. 9b). The alignment for the elastic model is shown in Fig. 9c for homogeneous and in Fig. 9d for inhomogeneous elasticities. In the case of homogeneous elasticities we use $E = 3kPa$ for the Young's elasticity modulus and $\nu = 0.4$ for the Poisson ratio, as used by Ferrant *et al.* (2000b).

An analysis of the summed squared differences showed an improvement of 2% using inhomogeneous elasticities. This rather small effect is due to the setting of feature points in our experiments. As can be seen in Fig. 8, large regions of white matter have only a small range of anatomical variability. In other words, the large number of fixed deformation estimates constrains the interpolation done by the elastic model. Further research will investigate new approximation schemes to address this.

V. Registration of Diffusion Tensor Images

A. Introduction

A large amount of research has been done over the past 2 decades on the registration of medical images provided by different imaging modalities, resulting in a proliferation of algorithms with a solid theoretical background. Nonscalar imaging modalities are emerging in radiology. For example phase contrast angiography MRI (Dumoulin *et al.*, 1989) provides a description of speed and direction of blood flow, and diffusion tensor MRI (DT-MRI) (LeBihan *et al.*, 1986; Basser *et al.*, 1994; Pierpaoli *et al.*, 1996) provides diffusion

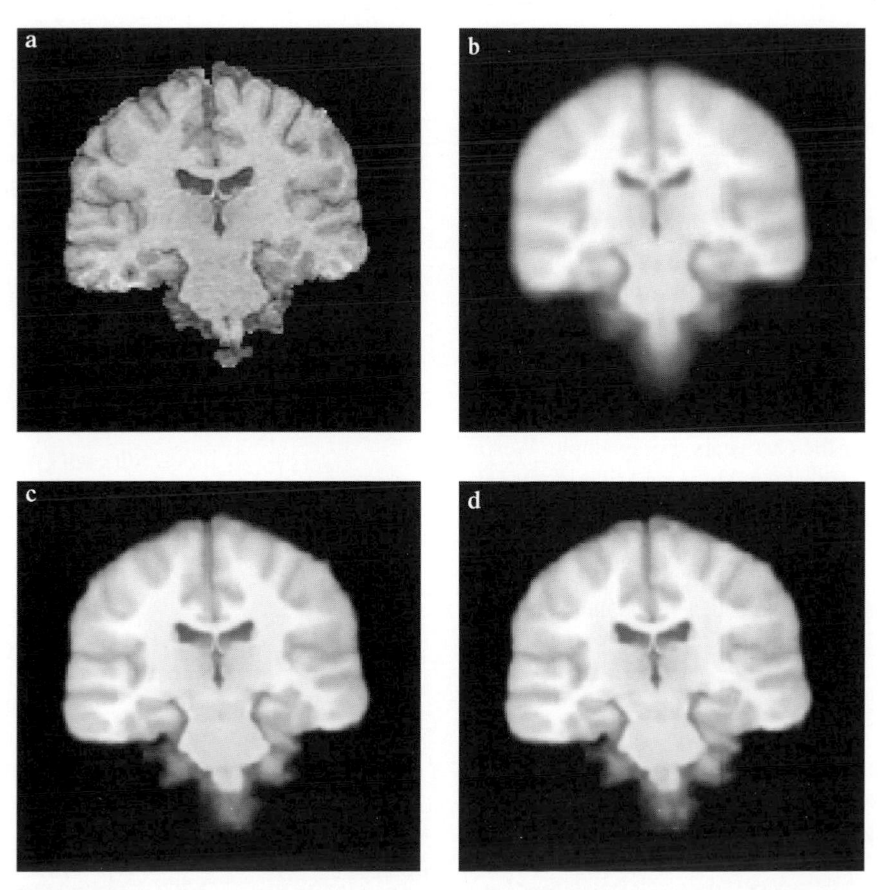

Figure 9 Results of global affine and nonrigid registration applied to 159 subjects. (**a**) Slice of reference volume, (**b–d**) Results after registration and averaging over all scans using: (b) global affine registration, (c) nonrigid registration with homogeneous elasticities, (d) nonrigid registration with inhomogeneous elasticities.

tensors describing local mobility of water molecules in tissue. The increasing clinical relevance of such image modalities has prompted research focused on registration methods supporting them.

Although the theory that will be presented in this chapter is general and valid for any data dimensions and arbitrary tensor data (including the special case of vectors) the driving example throughout this section will be registration of DT-MRI data. DT-MRI is a relatively recent MR imaging modality used for relating image intensities to the relative mobility of endogeneous tissue water molecules. In DT-MRI, a tensor describing local water diffusion is calculated for each voxel from measurements of diffusion in several directions. To measure diffusion, the Stejskal–Tanner imaging sequence is used (Stejskal and Tanner, 1965). This sequence uses two strong gradient pulses, symmetrically positioned around a 180° refocusing pulse, allowing for controlled diffusion weighting. DT-MRI has shown its clinical value in early assessment of brain ischemia and stroke by showing the decreased ability of the affected tissues to diffuse water (Hajnal and Bydder, 1997; Provenzale and Sorensen, 1999). Since MRI methods, in general, obtain a macroscopic measure of a microscopic quantity (which necessarily entails intravoxel averaging), the voxel dimensions influence the measured diffusion tensor at any given location in the brain. Factors affecting the shape of the apparent diffusion tensor (shape of the diffusion ellipsoid) in the white matter include the density of fibers, the degree of myelination, the average fiber diameter, and the directional similarity of the fibers in the voxel. The direction of maximum diffusivity is described by the eigenvector corresponding to the largest eigenvalue. This is descriptive of the orientation of white matter fiber tracts in the central nervous system. This is due to the restricted diffusion caused by the presence of a tightly packed sheath of myelin surrounding the axons (Basser *et al.*, 1994; Peled *et al.*, 1998). Some postprocessing algorithms suited to DT-MRI have arisen over the past years. For example, Westin *et al.* (2001) describe anisotropy analysis and filtering of DT-MRI data, and Ruiz *et al.* (2001a) describe an approach to point landmark detection in tensor data. The ability to visualize and automatically trace white matter tracts is expected to play a major role in basic neurosciences, in the understanding of neurological disorders (especially those associated with white matter demyelination), aging, and brain development (Poupon *et al.*, 1998, 1999; Weinstein *et al.*, 1999; Westin *et al.*, 2001).

The approach presented here stems from our work presented in Ruiz-Alzola *et al.*, (2000, 2001a, b) and it is based on template matching by locally optimizing a similarity function (Section V.C). A local structure detector for generic tensor fields (Section V.D) allows us to constrain the matching to highly structured areas. In order to obtain a dense deformation field, the sparse estimates from the template matching are interpolated. The whole approach is embedded in a multiresolution scheme using a Gaussian pyramid in order to deal with moderate deformations and decrease the influence of false optima. We also present (Section V.E) some illustrative results carried out on synthetic and clinical data.

B. Registration of DT-MRI Data

In addition to our own work (Ruiz-Alzola *et al.*, 2000, 2001, 2002), previous work in diffusion tensor registration includes the efforts of Alexander and co-workers (Alexander *et al.*, 1999; Alexander and Gee, 2000). They extend the multiresolution elastic matching paradigm in Bajcsy and Kovacic (1989) and Gee and Bajcsy (1999) to tensor data. Tensor reorientation is not included in the regularization term, but tensors are reoriented in each iteration according to the estimated displacement field. Several strategies to estimate the tensor reorientation from the displacement field are also investigated.

We state the problem of registration as a mapping of a reference anatomy, depicted by the signal $S_r(\mathbf{x})$, to a deformed one, represented by the signal $S_d(\mathbf{x})$. Equation (28) describes a model to characterize the relationship between both signals, where D models the deformation applied to the reference signal and both H and the noise v model the interscan differences:

$$S_d(\mathbf{x}) = H[D[S_r(\mathbf{x})];\mathrm{x}]+\nu(\mathbf{x}). \tag{28}$$

The deformation D represents a space-variant shift system and, hence, its response to a signal $S(\mathbf{x})$ is $D[S(\mathbf{x})] = S(\mathbf{x} + \mathbf{d}(\mathbf{x}))$, where $\mathbf{d}(\mathbf{x})$ is a displacement field. With regard to the differences between the system generating the images (signals), we consider H to be a nonmemory, possibly space-variant, system depending on a set $\mathbf{h}(\mathbf{x}) = (h_1(\mathbf{x}) \ldots h_p(\mathbf{x}))^t$ of unknown parameters and the noise to be spatially white and with zero mean. With these simplifications and defining $S_r^H(\mathbf{x}) = H[S_r(\mathbf{x});\mathbf{x}]$, the model Eq. (28) reduces to

$$S_d(\mathbf{x}) = S_r^H(\mathbf{x}+\mathbf{d}(\mathbf{x}))+\nu(\mathbf{x}). \tag{29}$$

The goal of registration is to find the displacement field $\mathbf{d}(\mathbf{x})$ that makes the best match between $S_r(\mathbf{x})$ and $S_d(\mathbf{x})$ according to Eq. (29).

C. Template Matching

Several schemes can be used to estimate the displacement field in Eq. (29). When there is no *a priori* probabilistic information about the signal and noise characterization, a *least-squares* (Moon and Stirling, 2000) approach is a natural choice. For this, all that is required is a suitable definition of an inner product and, thereafter, an induced norm. Note that scalar, vector, and tensor fields are applications of a real domain onto Euclidean vector spaces and this allows us to define the inner product between fields by means of the integral over the whole domain of the inner products between their values. Let us consider the functional set $\mathcal{F} = \{f: D \rightarrow V\}$, where D is a real domain and V is a Euclidean space. Then an inner product can be defined on $\mathcal{F}$ as $< f_1, f_2 > = \int_D w(\mathbf{x}) < f_1(\mathbf{x}), f_2(\mathbf{x}) > d\mathbf{x}$, where $w(\mathbf{x})$ is a weighting function for the inner product. Note that the inner product in the left-hand side is defined between fields and in the right-hand side, inside the integral, is defined between values of the field.

The least squares estimator is obtained by minimizing a cost function [Eq. (30)] that consists of the squared norm of the estimation error:

$$C^*(\mathbf{d}(\mathbf{x});\mathbf{h}(\mathbf{x})) = \| S_d(\mathbf{x}) - S_r^H(\mathbf{x}+\mathbf{d}(\mathbf{x})) \|^2 . \quad (30)$$

The dependency on the unknown parameters $\mathbf{h}(\mathbf{x})$ can be removed by estimating them using constrained least-squares schemes. For example, if the parameters are assumed to be constant all over the spatial domain, a least-squares estimation can be obtained, $\hat{\mathbf{h}}(\mathbf{d}(\mathbf{x})) = \hat{\mathbf{h}}(S_d(\mathbf{x}), S_r(\mathbf{x} + \mathbf{d}(\mathbf{x})))$, and substituted in C^* to obtain a new cost function [Eq. (31)] that depends only on $\mathbf{d}$ (see Ruiz-Alzola *et al.*, 2001b, for further details):

$$C(\mathbf{d}(\mathbf{x})) - C^*(\mathbf{d}(\mathbf{x}); \hat{\mathbf{h}}(\mathbf{d}(\mathbf{x}))). \quad (31)$$

The optimization of $C(\mathbf{d}(\mathbf{x}))$ in order to obtain the displacement field $\mathbf{d}(\mathbf{x})$ is a daunting task that requires additional constraints to make it feasible. *Template matching* trades off accuracy and computational burden to approximate a solution for this optimization problem. It essentially consists of defining a template from the neighborhood of every point of the deformed data set. Each of these templates is then compared or *matched* against the neighborhoods of tentatively correspondent points in the reference data set and a similarity measure is obtained for each of them. The tentative point whose neighborhood provides the biggest similarity is selected as corresponding to the current point in the deformed data set and the displacement between both points is obtained. There is a fundamental trade-off to be considered in the design of the neighborhoods: they must be nonlocal, and hence large in size, in terms of the $S_d(\mathbf{x})$ space frequencies to avoid the *ill-posedness* arising from the lack of discriminant structure [aperture problem (Poggio *et al.*, 1985)], and they must be local, and hence small in size, in terms of the unknown displacement field spatial frequencies to guarantee the validity of the local deformation model. Adaptive templates with different sizes and weights can help to deal with this problem.

Let $T(\mathbf{x} - \mathbf{x}_0)$ be a window function centered in a generic point $\mathbf{x}_0$ in the deformed data set and designed following the previous remarks. The template matching assumptions transform Eq. (29) into Eq. (32), which holds for every point $\mathbf{x}_0$ in the deformed data set:

$$T(\mathbf{x}-\mathbf{x}_0)S_d(\mathbf{x}) = T(\mathbf{x}-\mathbf{x}_0)S_r^H(\mathbf{x}+\mathbf{d}(\mathbf{x})) + \mathrm{v}(\mathbf{x}). \quad (32)$$

Equation (32) has an intuitive interpretation: any neighborhood in the deformed data set around a point $\mathbf{x}_0$, defined by the window function $T(\mathbf{x} - \mathbf{x}_0)$, corresponds to a neighborhood in the reference data set defined by the window function $T(\mathbf{x} - \mathbf{x}_0 - \mathbf{d}(\mathbf{x}))$, which has been warped by the deformation field. Template matching assumes that a model is chosen for the displacement field and for the parameters of the transformation $\mathbf{h}(\mathbf{x})$ in a neighborhood of the point $\mathbf{x}_0$ to be registered. For example the deformation field model may constrain the template just to shift along the coordinate axes (translation) or to undergo rigid motions, hence allowing also rotations or possibly even allowing stretch and twist. In any case the model for the local deformation must be such that it depends only on a few parameters, in order to make the search computationally feasible. With respect to the parameters, the common choice is to assume them constant in the neighborhood.

The template matching splits a complex global optimization problem, i.e., coupled searching for all the displacements, into many simple local ones, i.e., searching independently for the displacement of each point using template matching in each case. For example, for the common case in which the displacement field is assumed to be constant inside the template, the cost function [Eq. (30)] reduces to a set of cost functions

$$C^*(\mathbf{d}(\mathbf{x}_0);\mathbf{h}(\mathbf{x})) = \| T(\mathbf{x}-\mathbf{x}_0)(S_d(\mathbf{x}) - S_r^H(\mathbf{x}+\mathrm{d}(\mathbf{x}_0))) \|^2, (33)$$

where $\mathbf{x}_0$ refers to every point in the deformed data set. One of the main characteristics of template matching is the absence of any global regularization that constrains the local variability of the estimated deformation field. While this prevents getting trapped in false optima that are far from the absolute optimum, as global optimization methods are prone to, noise can produce high-frequency artifacts on the estimated deformation. Hence a further refinement of the solution may be advisable depending on the application, either postfiltering the estimated deformation or using it as an initial solution for a global optimization scheme.

1. Similarity Functions

A *similarity function* is a monotonic function of the cost [Eq. (30)], $SF(\mathbf{d}(\mathbf{x})) = F[C(\mathbf{d})]$, which leaves the locations of the optima unchanged and remains invariant with respect to the unknown parameters. The local nature of the template matching method makes it necessary to define a similarity function $SF(\mathbf{d}(\mathbf{x}_0))$ for every point in the deformed data set which is to be matched onto the reference one, i.e., the monotonic function is applied to Eq. (33). In this section the least-squares method referred to above is used to obtain suitable local similarity functions for the template matching of generic tensor fields.

Let us first consider that H is the identity mapping and that the displacement field is constant inside the template. Direct use of Eq. (33) leads to

$$SF_{SSD}(\mathbf{d}(\mathbf{x}_0)) = \|T(\mathbf{x}-\mathbf{x}_0)(S_d(\mathbf{x})-S_r(\mathbf{x}+\mathbf{d}(\mathbf{x}_0)))\|^2, \quad (34)$$

which corresponds to the well-known *sum of squared differences* similarity function. Extending it by using inner products and assuming that $\|T(\mathbf{x} - \mathbf{x}_0)\ S_r(\mathbf{x} + \mathbf{d}(\mathbf{x}_0))\|^2$ is almost constant for all possible $\mathbf{d}(\mathbf{x}_0)$ leads to an alternative similarity function that corresponds to the *correlation* measure:

$$SF_C(\mathbf{d}(\mathbf{x}_0)) = \langle T^2(\mathbf{x}-\mathbf{x}_0)S_d(\mathbf{x}),S_r(\mathbf{x}+\mathbf{d}(\mathbf{x}_0))\rangle. \quad (35)$$

Let us now consider that H is a space-invariant affine transformation of the intensity. In this case

$$\mathrm{T}(\mathbf{x}-\mathbf{x}_0)S_d(\mathbf{x}) = aT(\mathbf{x}-\mathbf{x}_0)S_r(\mathbf{x}+\mathbf{d}(\mathbf{x}_0))+b\mathrm{T}(\mathbf{x}-\mathbf{x}_0)\,1(\mathbf{x})+\nu(\mathbf{x}) \quad (36)$$

where $1(\mathbf{x})$ refers to the one tensor function (all the components are equal to 1 everywhere). The cost of Eq. (33) turns out to be

$$C^*(\mathbf{d}(\mathbf{x}_0);a,b) = \|T(\mathbf{x}-\mathbf{x}_0)(S_d(\mathbf{x})-aS_r(\mathbf{x}+\mathbf{d}(\mathbf{x}_0))-b1(\mathbf{x}))\|^2 \quad (37)$$

A similarity function invariant to a and b can be obtained by replacing these coefficients with their least-squares estimation and minimizing the resulting cost. Details can be found in Ruiz-Alzola *et al.* (2001b). The resulting similarity function is the absolute value of a generalized version of the well-known *correlation coefficient*,

where

$$SF_\rho(\mathbf{d}(\mathbf{x}_0)) = \left\| \left\langle \frac{\mathbf{s} - \frac{1}{\|\mathbf{t}\|^2}\langle \mathbf{s},\mathbf{t}\rangle \mathbf{t}}{\left\|\mathbf{s} - \frac{1}{\|\mathbf{t}\|^2}\langle \mathbf{s},\mathbf{t}\rangle \mathbf{t}\right\|}, \frac{\mathbf{p}(\mathbf{d}(\mathbf{x}_0)) - \frac{1}{\|\mathbf{t}\|^2}\langle \mathbf{p}(\mathbf{d}(\mathbf{x}_0)),\mathbf{t}\rangle \mathbf{t}}{\left\|\mathbf{p}(\mathbf{d}(\mathbf{x}_0)) - \frac{1}{\|\mathbf{t}\|^2}\langle \mathbf{p}(\mathbf{d}(\mathbf{x}_0)),\mathbf{t}\rangle \mathbf{t}\right\|} \right\rangle \right\|, \quad (38)$$

$$\mathbf{s} = T(\mathbf{x}-\mathbf{x}_0)S_d(\mathbf{x}), \quad (39)$$

$$\mathbf{p} = T(\mathbf{x}-\mathbf{x}_0)S_r(\mathbf{x}+\mathbf{d}(\mathbf{x}_0)), \quad (40)$$

$$\mathbf{t} = T(\mathbf{x}-\mathbf{x}_0)1(\mathbf{x}). \quad (41)$$

The application of the equations above requires a proper definition of the inner product,

$$\langle S_1(\cdot),S_2(\cdot)\rangle = \int_D S_{1_{i1\cdots in}}(\mathbf{x})S_{2_{i1\cdots in}}(\mathbf{x})d\mathrm{x}, \quad (42)$$

and its induced norm,

$$\| S(\cdot)\|^2 = \int_D S_{i1\cdots in}(\mathbf{x})S_{i1\cdots in}(\mathbf{x})d\mathbf{x}. \quad (43)$$

We assume that the tensors are Cartesian (defined with respect to an orthonormal basis) and we are using the Einstein notation for sums (any repetition of an index entails a summing over this index). Note that any implementation relies on sampled data and therefore the integrals above become sums.

2. Warped Vectors and Tensors

Vector and tensor data are linked to the body under inspection and, thereafter, any warping of the supporting tissue will lead to a consequent warping or reorientation of these data. The warping of the domain can be expressed by the transformation

$$\mathbf{x} = \mathbf{T}(\mathbf{x}') = \mathbf{x}' + \mathbf{d}(\mathbf{x}'), \quad (44)$$

where $\mathbf{x}$ stands for points in the reference data set and $\mathbf{x}'$ for points in the deformed one. Moreover, the transformation is assumed to be differentiable and hence the neighborhoods of the correspondent points $\mathbf{x}$ and $\mathbf{x}'$ are related through

$$d\mathbf{x} = \left[\nabla \otimes \mathbf{T}(\mathbf{x}')\right]d\mathbf{x}', \quad (45)$$

where the *deformation gradient* $[\nabla\otimes\mathbf{T}(\mathbf{x}')]$ can be easily recognized as the Jacobian matrix $\mathbf{J}(\mathbf{x}')$ of the transformation $\mathbf{T}(\mathbf{x}')$:

$$\left[\nabla \otimes \mathbf{T}(\mathbf{x}')\right] \equiv \mathbf{J}(\mathbf{x}') = \frac{\delta\mathbf{T}(\mathbf{x}')}{\delta\mathbf{x}'}. \quad (46)$$

Equation (45) simply states that, as far as the transformation is differentiable, a linear mapping relates the local neighborhoods of both points. For finite-size neighborhoods, the deformation gradient corresponds to a linear approximation, as a Taylor's expansion clearly shows:

$$\mathbf{x} + \Delta\mathbf{x} \approx \mathbf{T}(\mathbf{x}' + \Delta\mathbf{x}') = \mathbf{T}(\mathbf{x}') + \frac{1}{1!}\frac{\delta\mathbf{T}(\mathbf{x}')}{\delta\mathbf{x}'}\Delta\mathbf{x}' \quad (47)$$

$$\Delta\mathbf{x} \approx \frac{\delta\mathbf{T}(\mathbf{x}')}{\delta\mathbf{x}'}\Delta\mathbf{x}'. \quad (48)$$

In this work it will be assumed that the linear approximation is valid since the function data, vectors or tensors, are related to infinitesimal properties of the tissue. Consequently, two vectors $\mathbf{v}$ and $\mathbf{v}'$ are locally related as

$$\mathbf{v} = \mathbf{J}(\mathbf{x}')\mathbf{v}' \quad (49)$$

Since two second-order diffusion tensors $\mathbf{P}$ and $\mathbf{P}'$ can be considered associated to quadratic forms, they are related by

$$\mathbf{P}' = \mathbf{J}^t(\mathbf{x}')\mathbf{P}\mathbf{J}(\mathbf{x}'). \quad (50)$$

Equation (51) provides a theoretical way to estimate the alteration of diffusion tensors due to a deformation field. Nevertheless it is not clear that DT-MRI data actually are modified according to this model especially in areas of high anisotropy, i.e., the white matter fiber tracts, where these data are most relevant. The idea here is that the shape of the diffusion tensor should be preserved through the transformation and hence it must be reoriented only as an effect of local rotation and shear. This essentially means that the deformation field affects only the directional properties of diffusion and not its strength along its principal axes. For example, in a reference frame intrinsic to a fiber tract diffusion should remain invariant with respect to the deformation. This has motivated a search for tensor transformations that maintain the shape and include both the effect of local rotation and the shear. Early experiments on this topic have begun (Sierra, 2001). An ad hoc solution to this problem is to scale the resulting tensor after Eq. (51) is applied so as, for example, to preserve the ellipsoid volume or normalize the largest eigenvalue. Another possibility is to modify the deformation gradient so as to avoid undesirable effects such as the scaling (Alexander *et al.*, 1999). Nevertheless much research is still needed to clarify the appropriate tensor transformation to be used.

A mathematical tool to deal with this problem is the *Polar Decomposition Theorem* (see for example Segel, 1987, from the theory of nonlinear elasticity). It allows us to deal not only with infinitesimal but also with finite deformations. The theorem states that for any nonsingular square matrix, such as the *deformation gradient* $\mathbf{J}(\mathbf{x}')$, there are unique symmetric positive definite matrices $\mathbf{U}(\mathbf{x}')$ and $\mathbf{V}(\mathbf{x})'$ and also a unique orthonormal matrix $\mathbf{R}(\mathbf{x}')$ such that

$$\mathbf{J}(\mathbf{x}') = \mathbf{R}(\mathbf{x}')\mathbf{U}(\mathbf{x}') = \mathbf{V}(\mathbf{x}')\mathbf{R}(\mathbf{x}'). \tag{51}$$

This leads to important geometric interpretations of the geometric mapping. For example, notice that a sphere is first stretched by the mapping in the directions of the eigenvectors of $\mathbf{U}(\mathbf{x}')$ and then rotated by $\mathbf{R}(\mathbf{x}')$. A transformation such that $\mathbf{R}(\mathbf{x}') = \mathbf{I}$ is said to be a *pure strain* at $\mathbf{x}'$, while if $\mathbf{U}(\mathbf{x}') = \mathbf{V}(\mathbf{x}') = \mathbf{I}$ it is said to be a *rigid rotation* at that point.

As mentioned above, the matching approach to data registration relies on a model for the local displacement field inside the template. In order to perform the matching, the vectors or tensors inside the template must be reoriented according to the hypothesized model. Note that if a simple model such as just shifting the template along the coordinate axes is adopted, i.e., assuming a constant displacement field for all the points inside, the vectors or tensors would not be reoriented. Similarly, if the model is rigid motion no stretching of the vectors or tensors should be considered. From a practical point of view, no reorientation is performed during matching and therefore a constant displacement field is assumed inside the template. This is not a limitation as long as the local rotation is small and in fact it is accepted in conventional template matching of scalar data. The reorientation is then calculated once the displacement field—and its gradient—has been estimated.

D. Structure Measures

Matching must be constrained to areas with local high discriminant structure. Depending on the data set, this approach will lead to very sparse control points to estimate the deformation field. The applicability of the method ultimately depends on the characteristics of the deformation field—it being obvious that if the deformation field has a large spatial variability, the sparse displacement field estimated from the detected points will suffer from spectral aliasing. Incrementing the sampling density by accepting low structure points is possible depending on the noise characteristics of the data, since it is unacceptable to allow noise to provide the discriminant information that drives the template matching. When it is not possible to provide enough points according to the frequency properties of the deformation field, it might be necessary to resort to regularized schemes, such as elastic models, that use the whole data set. Alternatively, in some applications, an additional channel of data is customarily provided. This is the case, for example, in DT-MRI using EPI sequences in which additional T2-weighted images are provided. Therefore it is possible to estimate different sparse displacement fields from T2 and DT-MRI and combine them in order to estimate the whole displacement field, providing constraints both from structural T2 images and from the diffusion tensors (the white matter fiber tracts).

In order to identify the areas of high structure, we use a measure of cornerness (Ruiz-Alzola *et al.*, 2000, 2001, 2002) which generalizes the locally averaged outer product (i.e., the correlation matrix) of the gradient field (Rohr, 1997) frequently used in the scalar case.

E. Results

In this section results on both synthetic and real data are presented. The real imaging data were acquired on a GE Signa 1.5-T Horizon Echospeed 5.6 scanner, using a line scan diffusion technique, with the following parameters: no averaging, effective TR = 2.4 s, TE = 65 ms, b_{high} = 750 s/mm^2, FOV 22 cm, voxel size 4.8 × 1.6 × 1.5 mm^3, 6- kHz readout bandwidth, acquisition matrix 128 × 128.

Figure 10a shows the clusters of points that have been detected as highly structured in a DT-MRI sagittal image overlaid on the corresponding T2-weighted MRI image and Fig. 10b shows these clusters in the portion corresponding to the highlighted square. Recall that the matching is performed using these clusters, not the isolated landmarks. Notice that the diagonal components of the tensor field provide stronger and more structured signals than the off-diagonal ones and how the structure detector finds the thin structures in these images. It must be recalled that the cooperation of all the components is what provides this result. In order to obtain the clusters, we have normalized the tensor field components to fit into the interval [−1, 1] (weaker com-

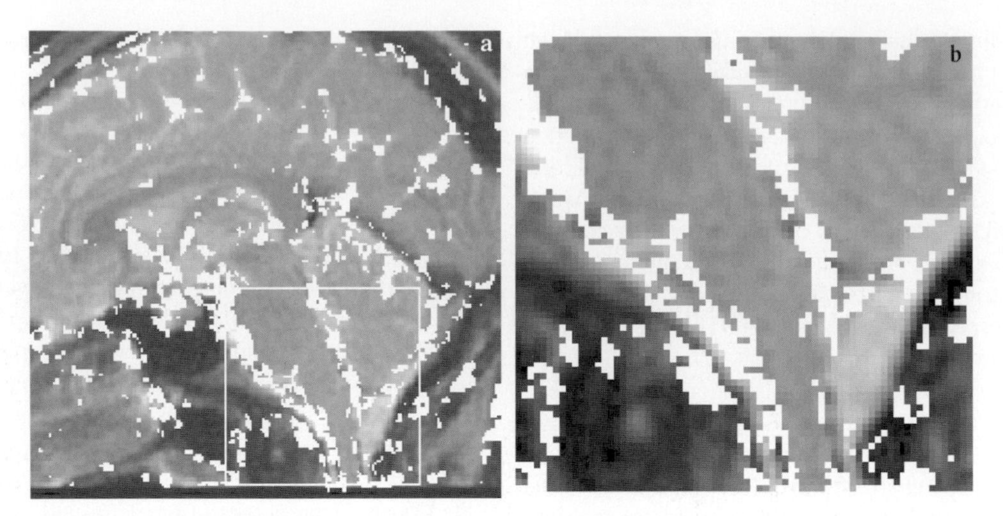

Figure 10 (**a**) High-structure clusters overlaid on T2W MRI. (**b**) Detail of the clusters inside the square.

ponents do not reach the extrema values). The estimations of the gradient and the generalized correlation matrices have been made using 3×3 neighborhoods. The difficulty of presenting illustrative results from volume data using 2D figures has motivated us to report this experiment using a single DT-MRI slice (the tensors in it are 3D). Nevertheless, the method is essentially N-dimensional and it can be directly applied to volumes of data using the same parameters, just adding one more dimension in the definition of the neighborhoods.

In order to assess the overall performance of our nonrigid registration method, Fig. 11a shows a sagittal MRI slice of the corpus callosum that is deformed by a synthetic Gaussian field as depicted in Fig. 11b. In order to estimate the deformation, a Gaussian pyramid decomposition is obtained, performing the template matching on structure areas in each level and interpolating using Kriging (Ruiz-Alzola *et al.*, 2000, 2001). Figure 11c shows the result of reconstructing the original image with the deformation field estimated with our approach, using the Kriging estimator with an exponential variogram.

Figure 12a shows a slice of a DT-MRI data set of the corpus callosum on which the principal eigenvector directions have been represented using a color coding ranging from blue (in-plane projection) to red (orthogonal to plane) (Peled *et al.*, 1998). The whole approach has been applied to warp this data set into another corresponding to a different individual, shown in Fig. 12b, using three levels of a Gaussian pyramid and an exponential variogram for the Kriging interpolator that is limited to take into account the eight closest samples. Figure 12c shows a T2W zoomed version of the right-hand side of the former, corresponding to the posterior corpus callosum and the estimated deformation field.

F. Conclusions

We have described a framework for nonrigid registration of scalar, vector, and tensor medical data. The approach is local, since it is based on template matching, and resorts to a multiresolution implementation using a Gaussian pyramid in order to provide a coarse-to-fine approximation to the solution. This allows the method to handle moderate defor-

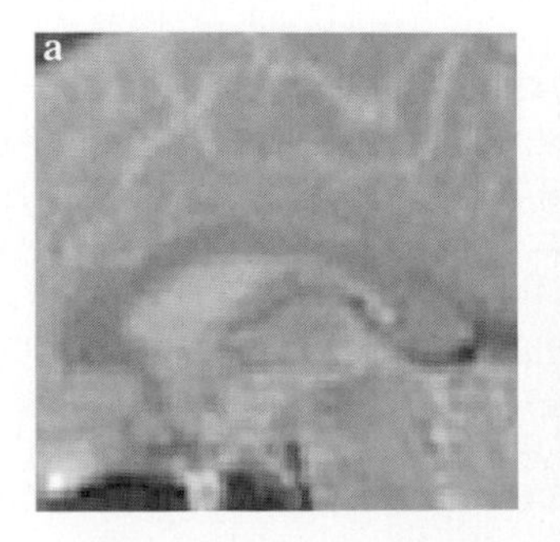

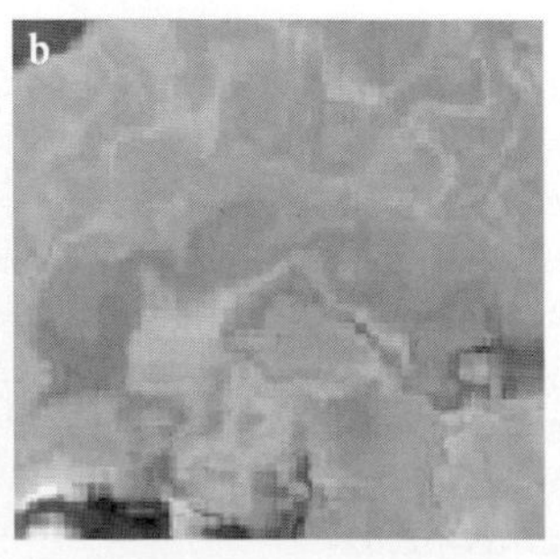

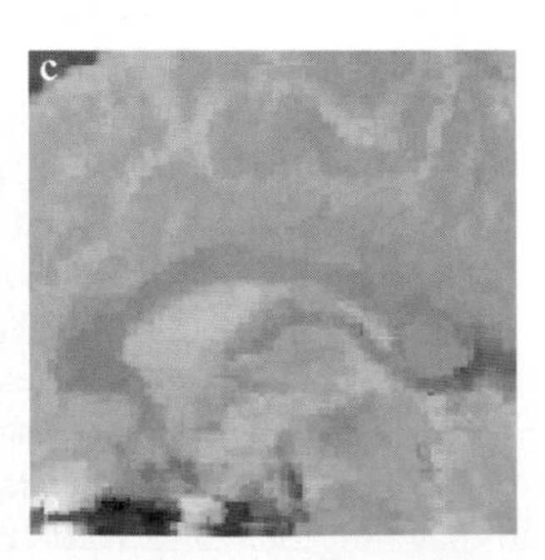

Figure 11 Synthetic deformation. (**a**) Original MRI, (**b**) deformed, and (**c**) reconstructed original using estimated deformation field.

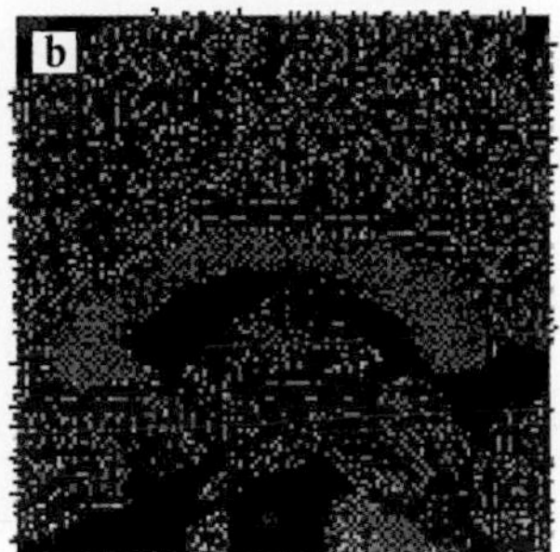

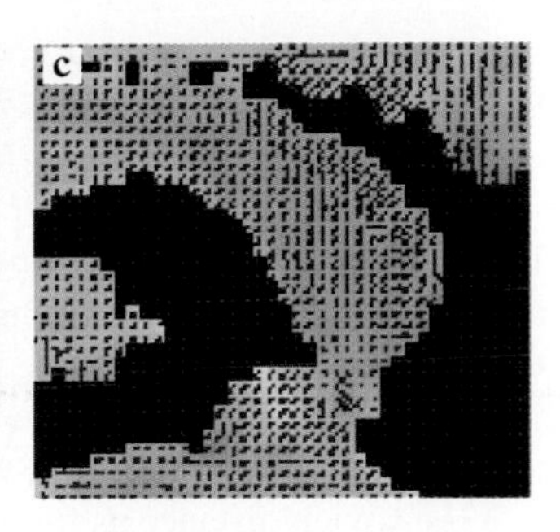

Figure 12 DT-MRI interpatient warping. (**a, b**) DT-MRI of different individuals (**c**) Zoomed T2W of the posterior corpus of (a) and estimated deformation.

mations and avoids false local solutions. The method does not assume any global *a priori* regularization and therefore avoids the computational burden associated with those approaches.

VI. The Monge–Kantorovich Problem and Image Registration

In this section, we present a method for producing area-preserving surface deformations, and more general mass-preserving area and volume deformations, based on the minimization of a functional of Monge–Kantorovich type. The theory is based on the problem of minimizing the cost of redistributing a certain amount of mass between two distributions given *a priori*. Here the cost is a function of the distance each bit of material is moved, weighted by its mass. We show how the resulting low-order differential equations may be used for registration.

A. The Monge–Kantorovich Problem

Here we present a method for image warping and elastic registration based on the classical problem of optimal mass transport. The mass transport problem was first formulated by Gaspar Monge in 1781 and concerned finding the optimal way, in the sense of minimal transportation cost, of moving a pile of soil from one site to another. This problem was given a modern formulation in the work of Kantorovich (1948) and so is now known as the *Monge–Kantorovich problem*. This type of problem has appeared in econometrics, fluid dynamics, automatic control, transportation, statistical physics, shape optimization, expert systems, and meteorology (Rachev and Rüschendorf, 1998).

The method we introduce in this section is designed for elastic registration and is based on an optimization problem built around the L^2 Monge–Kantorovich distance taken as a similarity measure. The constraint that we put on the transformations considered is that they obey a mass preservation property. We will assume that a rigid (nonelastic) registration process has already been applied before applying our scheme.

Our method is based on differential equations and in this sense may be thought of as being in the same class with optical flow and elastic deformation model approaches to registration. See Hata *et al.*, (2000), Davatzikos (1997), and Lester *et al.* (1999) for representative examples of these methods. Our method also has a strong connection to computational fluid dynamics, an area which has previously been successfully applied to brain registration problems (Christensen *et al.*, 1996).

Our method, however, has a number of distinguishing characteristics. It is parameter free. It utilizes all of the gray-scale data in both images and places the two images on equal footing. It is thus symmetrical, the optimal mapping from image A to image B being the inverse of the optimal mapping from B to A. It does not require that landmarks be specified. The minimizer of the distance functional involved is unique; there are no other local minimizers. The functional at the heart of the method is such that the corresponding differential equations governing its minimization are of low order. Finally, it is specifically designed to take into account changes in density that result from changes in area or volume.

We believe that this type of elastic warping methodology is quite natural in the medical context, in which density can be a key measure of similarity. It also occurs in functional imaging, in which one may want to compare the degree of activity in various features deforming over time and obtain a corresponding elastic registration map. A special case of this problem occurs in any application in which volume or area-preserving mappings are considered.

1. Formulation of the Problem

We now give a modern formulation of the Monge–Kantorovich problem. We assume we are given, *a priori*, two subdomains Ω_0 and Ω_1 of $\mathbf{R}^d$, with smooth boundaries, and a pair of positive density functions, μ_0 and μ_1 defined on Ω_0 and Ω_1, respectively. We assume $\int_{\Omega_0} \mu_0 = \int_{\Omega_1} \mu_1$ so that the same total mass is associated with Ω_0 and Ω_1. We consider diffeomorphisms $\tilde{u}$ from Ω_0 to Ω_1 which map one density to the other in the sense that

$$\mu_0 = |D\tilde{u}|\mu_1 \circ \tilde{u}, \tag{53}$$

which we will call the *mass preservation* (MP) property, and write $\tilde{u} \in MP$. Equation (53) is called the *Jacobian equation*. Here $|D\tilde{u}|$ denotes the determinant of the Jacobian map $D\tilde{u}$, and $\circ$ denotes composition of functions. In particular, Eq. (53) implies that if a small region in Ω_0 is mapped to a larger region in Ω_1, then there must be a corresponding decrease in density in order for the mass to be preserved.

There may be many such mappings, and we want to pick out an optimal one in some sense. Accordingly, we define the squared L^2 Monge–Kantorovich distance as follows:

$$d_2^2(\mu_0,\mu_1) = \inf_{\tilde{u} \in MP} \int \| \tilde{u}(x) - x \|^2 \mu_0(x)dx. \quad (54)$$

An *optimal MP map* is a map which minimizes this integral while satisfying the constraint (53). The Monge–Kantorovich functional [Eq. (54)] is seen to place a penalty on the distance the map $\tilde{u}$ moves each bit of material, weighted by the material's mass. A fundamental theoretical result (Brenier, 1991; Gangbo and McCann, 1996) is that there is a unique optimal $\tilde{u} \in MP$ transporting μ_0 to μ_1 and that this $\tilde{u}$ is characterized as the gradient of a function w, i.e., $\tilde{u} = \nabla w$. This theory translates into a practical advantage, since it means that there are no nonglobal minima to stall our solution process.

2. Computing the Transport Map

There have been a number of algorithms considered for computing an optimal transport map. For example, methods have been proposed based on linear programming (Rachev and Rüschendorf, 1998) and on Lagrangian mechanics closely related to ideas from the study of fluid dynamics (Benamou and Brenier, 2000). An interesting geometric method has been formulated by Cullen and Purser (1984). Here, we follow closely the work in Haker and Tannenbaum (2001a,b).

Let u: $\Omega_0 \rightarrow \Omega_1$ be an initial mapping with the MP property. Inspired by Brenier (1991) and Gangbo (1994), we consider the family of MP mappings of the form $\tilde{u} = u \circ s^{-1}$, as s varies over MP mappings from Ω_0 to itself, and try find an s which yields a $\tilde{u}$ without any curl, that is, such that $\tilde{u} = \nabla w$. Once such an s is found, we will have the Monge–Kantorovich mapping $\tilde{u}$. We will also have $u = \tilde{u} \circ s = (\nabla w) \circ s$, known as the *polar factorization* of u with respect to μ_0 (Brenier, 1991).

3. Removing the Curl

Our method assumes that we have found an initial MP mapping u. This can be done for general domains using a method of Moser (1965; Dacorogna and Moser, 1990) or, for simpler domains, using a type of histogram specification. Once an initial MP u is found, we need to apply the process which will remove its curl. It is easy to show that the composition of two MP mappings is an MP mapping, and the inverse of an MP mapping is an MP mapping. Thus, since u is an MP mapping, we have that $\tilde{u} = u \text{ o } s^{-1}$ is an MP mapping if

$$\mu_0 = | Ds | \mu_0 \circ s. \quad (55)$$

In particular, when μ_0 is constant, this equation requires that s be area or volume preserving.

Next, we will assume that s is a function of time and determine what s_t should be to decrease the L^2 Monge–Kantorovich functional. This will give us an evolution equation for s and in turn an equation for $\tilde{u}_t$ as well, the latter being the most important for implementation. By differentiating $\tilde{u} \circ s = u$ with respect to time, we find

$$\tilde{u}_t = -D\tilde{u}s_t \circ s^{-1}. \quad (56)$$

Differentiating Eq. (55) with respect to time yields

$$\text{div}(\mu_0 s_t \circ s^{-1}) = 0, \quad (57)$$

from which we see that s_t and $\tilde{u}_t$ should have the following forms:

$$\text{div}(\mu_0 s_t \circ s^{-1}) = 0, \; \tilde{u}_t = \frac{-1}{\mu_0} D\tilde{u}\zeta \quad (58)$$

for some vector field ζ on Ω_0, with div(ζ) = 0 and $\langle \zeta, \vec{n} \rangle = 0$ on $\partial\Omega_0$. Here $\vec{n}$ denotes the normal to the boundary of Ω_0. This last condition ensures that s remains a mapping from Ω_0 to itself, by preventing the flow of s, given by $s_t = (1/\mu_0\, \zeta) \circ s$, from crossing the boundary of Ω_0. This also means that the range of $\tilde{u} = u \circ s^{-1}$ is always $u(\Omega_0) = \Omega_1$.

Consider now the problem of minimizing the Monge–Kantorovich functional:

$$M = \int \| \tilde{u}(x) - x \|^2 \mu_0(x)dx. \quad (59)$$

Taking the derivative with respect to time, and using the Helmholtz decomposition $\tilde{u} = \nabla w + \chi$ with div(χ) = 0, we find from Eq. (58) that

$$-\frac{1}{2} M_t = \int \langle \tilde{u}, \zeta \rangle = \int \langle \chi, \zeta \rangle. \quad (60)$$

Thus, in order to decrease M, we can take $z = \chi$ with corresponding formulas (58) for s_t and $_{\tilde{u}t}$, provided that we have div(χ) = 0 and $\langle \chi, \vec{n} \rangle = 0$ on $\partial\Omega_0$. Thus it remains to show that we can decompose $\tilde{u}$ as $\tilde{u} = \nabla w + \chi$ for such a χ.

4. Gradient Descent: R^d

We let w be a solution of the Neumann-type boundary problem

$$\Delta w = \text{div}(\tilde{u}), \; \langle \nabla w, \vec{n} \rangle = \langle \tilde{u}, \vec{n} \rangle \; on \; \partial\Omega_0, \quad (61)$$

and set $\chi = \tilde{u} - \nabla w$. It is then easily seen that χ satisfies the necessary requirements. Thus, by Eq. (58), we have the following evolution equation for $\tilde{u}$:

$$\tilde{u}_t = -\frac{1}{\mu_0} D\tilde{u}(\tilde{u} - \nabla\Delta^{-1}\mathrm{div}(\tilde{u})). \quad (62)$$

This is a first-order nonlocal scheme for $\tilde{u}_t$ if we count Δ^{-1} as minus 2 derivatives. Note that this flow is consistent with respect to the Monge–Kantorovich theory in the following sense. If $\tilde{u}$ is optimal, then it is given as $\tilde{u} = \nabla w$, in which case $\tilde{u} - \nabla\Delta^{-1}\mathrm{div}(\tilde{u}) = \nabla w - \nabla\Delta^{-1}\mathrm{div}(\nabla w) = 0$ so that by Eq. (62), $\tilde{u}_t = 0$.

5. Gradient Descent: $\mathbf{R}^2$

The situation is somewhat simpler in the $\mathbf{R}^2$ case, due to the fact that a divergence-free vector field χ can in general be written as $\chi = \nabla^{\perp}h$ for some scalar function h, where $\perp$ represents rotation by 90°, so that $\nabla^{\perp}h = (-h_y, h_x)$. In this case, we solve Laplace's equation with a Dirichlet boundary condition and derive the evolution equation

$$\tilde{u}_t = -\frac{1}{\mu_0} D\tilde{u}\nabla^{\perp}\Delta^{-1}\mathrm{div}(\tilde{u}^{\perp}). \quad (63)$$

6. Generalizations

We note that we can define a generalized Monge–Kantorovich functional as

$$M = \int \Phi \circ (\tilde{u} - i)\mu_0, \quad (64)$$

where $\Phi : \mathbf{R}^d \to \mathbf{R}$ is a positive strictly convex C^1 cost function and i is the identity map $i(x) = x$. In particular, the L^2 Monge–Kantorovich problem described above corresponds to the cost function $\Phi(x) = \|x\|^2$. If we define

$$\Psi := (\nabla\Phi) \circ (\tilde{u} - i), \quad (65)$$

then analysis similar to that above shows that M_t must be of the form

$$M_t = -\int \langle \Psi, \zeta \rangle, \quad (66)$$

where, as before,

$$\zeta = \mu_0 s_t \circ s^{-1} \quad (67)$$

is a divergence-free vector field. This analysis yields an evolution equation of the form

$$\tilde{u}_t = -\frac{1}{\mu_0} D\tilde{u}(\Psi - \nabla\Delta^{-1}\mathrm{div}\Psi), \quad (68)$$

where it is understood that the Laplacian is inverted with respect to appropriate boundary conditions.

Further, a purely local flow equation for the minimization of the Monge–Kantorovich functional may be obtained by setting

$$\zeta = \nabla\mathrm{div}\Psi - \Delta\Psi. \quad (69)$$

It is straightforward to check that in this case $\mathrm{div}(\zeta) = 0$ and

$$M_t = -\frac{1}{2}\int \|\mathrm{curl}\Psi\|^2 \le 0. \quad (70)$$

The corresponding second-order local evolution equation for $\tilde{u}$ is

$$\tilde{u}_t = -\frac{1}{\mu_0} D\tilde{u}(\nabla\mathrm{div}\Psi - \Delta\Psi), \quad (71)$$

and Eq. (70) shows that at optimality we must have curl $\Psi = 0$, so $\Psi = \nabla w$ for some function w.

7. Defining the Warping Map

Typically in elastic registration, one wants to see an explicit warping which smoothly deforms one image into the other. This can easily be done using the solution of the Monge–Kantorovich problem. Thus, we assume now that we have applied our gradient descent process as described above and that it has converged to the Monge–Kantorovich optimal mapping $\tilde{u}^*$. It is shown in Benamou and Brenier (2000) that the flow $X(x, t)$ defined by

$$X(x,t) = x + t(\tilde{u}^*(x) - x) \quad (72)$$

is the solution to a closely related minimization problem in fluid mechanics and provides appropriate justification for using Eq. (72) to define our continuous warping map X between the densities μ_0 and μ_1. See McCann (1997) for applications and a detailed analysis of the properties of this *displacement interpolation*.

B. Implementation and Examples

We note that even though our non-local method requires that the Laplacian be inverted during each iteration, the problem has been set up specifically to allow for the use of standard fast numerical solvers which use FFT-type methods and operate on rectangular grids (Press *et al.*, 1992).

We illustrate our methods with a pair of examples. In Fig. 13 we show a brain deformation sequence. One slice each from two MR data sets, acquired at the Brigham and Women's Hospital, were used. The first data set was preoperative, the second was acquired during surgery, after craniotomy and opening of the dura. Both were preprocessed to remove the skull. The Monge–Kantorovich mapping was found using the evolution Eq. (62) with intensity values as densities, scaled slightly so that the sum of the intensities was the same for both images. This process took roughly 10 min on a single processor Sun Ultra 10. A full 3D volumetric data set can take several hours to process. The displacement interpolation [Eq. (72)] together with Eq. (53) for the intensities was then used to find the continuous deformation through time. The first image, in the upper left, shows a planar axial slice at time $t = 0.00$. The bottom right is an axial slice at time

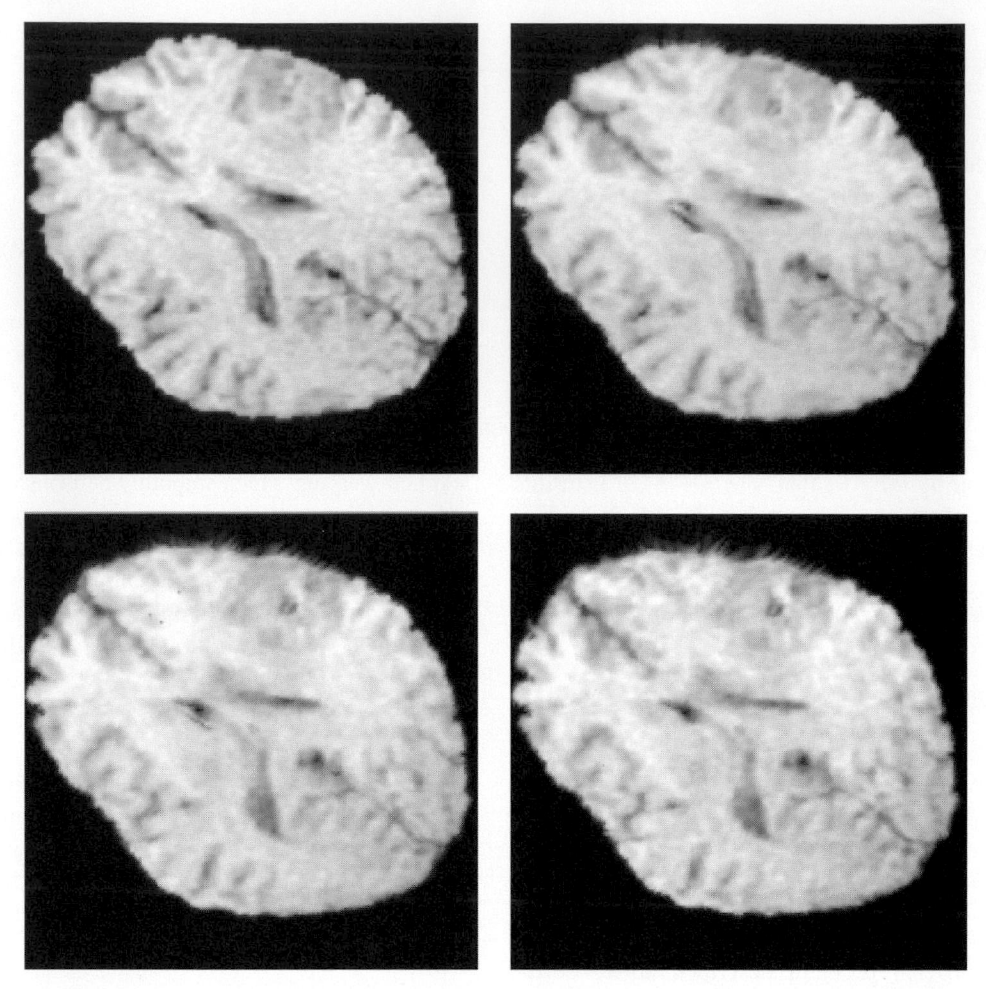

Figure 13 A brain deformation sequence. The upper left (time $t = 0.00$) and lower right ($t = 1.00$) are the input into the algorithm. The upper right ($t = 0.33$) and lower left ($t = 0.67$) represent the interpolation using the Monge–Kantorovich displacement map and enforcing the preservation of mass.

$t = 1.00$. Together, these images represent the input to our algorithm. The upper right and lower left images represent the computed interpolation at time $t = 0.33$ and $t = 0.66$, respectively.

The second example shows an application of our method to surface warping. Figure 14 shows a portion of the white matter surface obtained by segmenting an MRI scan. We cut the surface end to end and flattened it into the plane using a conformal mapping technique (Angenent *et al.*, 1999a,b), as shown in the left of Fig. 15. It is well known that a surface of nonzero Gaussian curvature cannot be flattened by any means without some distortion. The conformal mapping is an attempt to preserve the appearance of the surface through the preservation of angles. The conformal mapping is a "similarity in the small" and so features on the surface appear similar in the flattened representation, up to a scaling factor. Further, the conformal flattening maps can be calculated simply by solving systems of linear equations. For a triangulated surface of a few hundred thousand triangles, this takes only a few minutes on a single processor computer. Parallelization can be achieved using freely available numerical linear algebra software.

However, in some applications it is desirable to be able to preserve areas instead of angles, so that the sizes of surface structures are accurately represented in the plane. The Monge–Kantorovich approach allows us to find such an area-correct flattening. The idea here is that the conformal flattening should be altered by moving points around as little as possible. Once we have conformally flattened the surface, we define a density μ_0 to be the Jacobian of the inverse of the flattening map and set μ_1 to a constant. The Monge–Kantorovich optimal mapping is then area-correcting by Eq. (53). The resulting map took just a few minutes to calculate. Detail of the conformal surface flattening and the area-corrected flattening are shown in Fig. 15. Although corrected for area, surface structures are still clearly discernible. The curl-free nature of the Monge–Kantorovich mapping avoids distortion effects often associated with area-preserving maps.

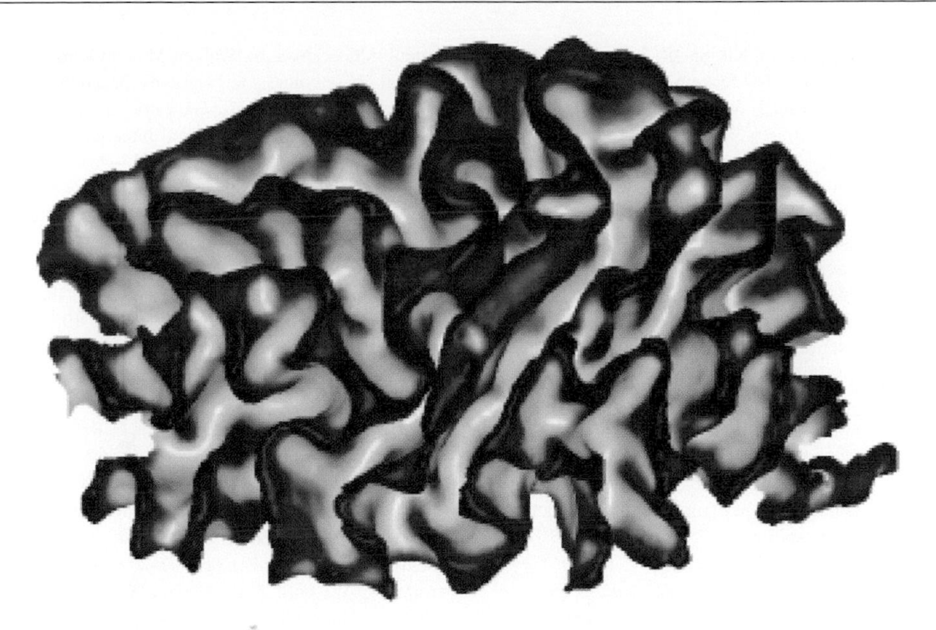

Figure 14 White matter surface.

C. Conclusions

In this section, we presented a natural method for image registration and surface warping based on the classical problem of optimal mass transportation. Although applied here to the L^2 Monge–Kantorovich problem, the method used to enforce the mass preservation constraint is general, as shown in Section VI.A.6 and will have other applications. For example, any weighted linear combination of the Monge–Kantorovich functional and a standard L^2 energy functional or other matching functional can be used. These ideas are a current area of research.

Acknowledgments

This work was funded in part by the Spanish Government (Ministerio de Educación y Culture) with a visiting research fellowship (FPU PRI1999–0175) (J.R.A.); the European Commission and the Spanish Government (CICYT), with the joint research grant 1FD97–0881–C02–01; a New Concept Award from the Center for Integration of Medicine and Innovative Technology (S.K.W.); and NIH P41 RR13218, NIH P01 CA67165, R01 CA86879, and NIH R01 RR11747.

References

Alexander, D., and Gee, J. (2000). Elastic matching of diffusion tensor images. *Comput. Vision Image Understand.* **77**, 233–250.

Alexander, D., Gee, J., and Bajcsy, R. (1999). Strategies for data reorientation during non-rigid transformations of diffusion tensor images. *In* "Lecture Notes in Computer Science: Medical Image Computing and Computer Assisted Intervention—MICCAI '98" (C. Taylor and A. Colchester, eds.), Vol. 1496, pp. 463–472. Springer-Verlag, Cambridge, UK.

Angenent, S., Haker, S., Tannenbaum, A., and Kikinis, R. (1999a). Laplace–Beltrami operator and brain surface flattening. *IEEE Trans. Med. Imaging* **18**, 700–711.

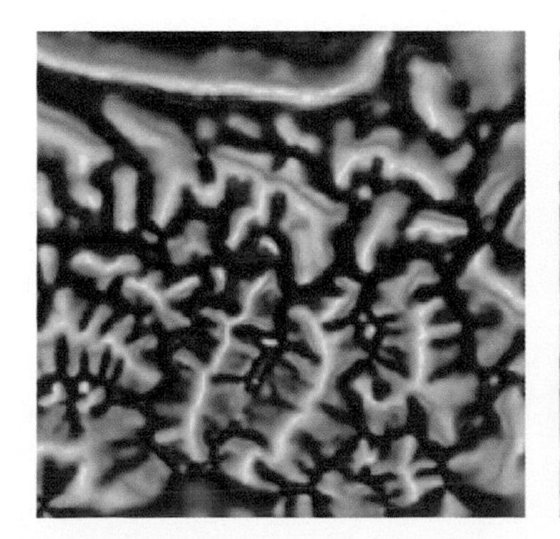

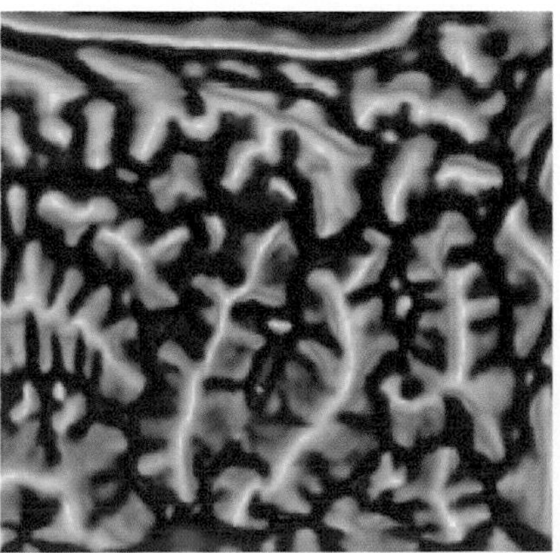

Figure 15 Conformal (left) and area-preserving (right) mappings.

Angenent, S., Haker, S., Tannenbaum, A., and Kikinis, R. (1999b). On area preserving maps of minimal distortion. *In* "System Theory: Modeling, Analysis, and Control" (T. Djaferis and I. Schick, eds.), pp. 275–287. Kluwer Academic, Dordrecht.

Bajcsy, R., and Kovacic, S. (1989). Multiresolution elastic matching. *Comput. Vision Graph. Image Process.* **46**, 1–21.

Balay, S., Gropp, W. D., McInnes, L. C., and Smith, B. F. (1997). Efficient management of parallelism in object oriented numerical software libraries. *In* "Modern Software Tools in Scientific Computing" (E. Arge, A. M. Bruaset, and H. P. Langtangen, eds.), pp. 163–202. Birkhauser, Basel.

Balay, S., Gropp, W. D., McInnes, L. C., and Smith, B. F. (2000a). "PETSc 2.0 User's Manual." Argonne National Laboratory. [Technical Report ANL-95/11—Revision 2.0.28]

Balay, S., Gropp, W. D., McInnes, L. C., and Smith, B. F. (2000b). PETSc home page. http://www.mcs.anl.gov/petsc.

Barron, J. L., Fleet, D. J., and Beauchemin, S. S. (1994). Performance of optical flow techniques. *Int. J. Comput. Vision* **12**, 43–77.

Basser, P., Mattiello, J., and LeBihan, D. (1994). Estimation of the effective self-diffusion tensor from the NMR spin-echo. *J. Magn. Reson. B* **103**, 247–254.

Benamou, J.-D., and Brenier, Y. (2000). A computational fluid mechanics solution to the Monge–Kantorovich mass transfer problem. *Numerische Math.* **84**, 375–393.

Black, P. M., Moriarty, T., Alexander, E., Stieg, P., Woodard, E. J., Gleason, P. L., Martin, C. H., Kikinis, R., Schwartz, R. B., and Jolesz, F. A. (1997). The development and implementation of intraoperative MRI and its neurosurgical applications. *Neurosurgery* **41**, 831–842.

Brandt, J. W. (1997). Improved accuracy in gradient-based optical flow estimation. *Int. J. Comput. Vision* **25**, 5–22.

Brenier, Y. (1991). Polar factorization and monotone rearrangement of vector-valued functions. *Comp. Pure Appl. Math.* **64**, 375–417.

Bro-Nielsen, M. (1996). Surgery simulation using fast finite elements. *In* "Visualization in Biomedical Computing," pp. 529–534. Springer-Verlag, Heidelberg, Germany.

Bro-Nielsen, M. (1998). Finite element modeling in medical VR. *Proc. IEEE Spec. Issue Virtual Augmented Reality Med.* **86**, 490–503.

Bro-Nielsen, M, and Gramkow, C. (1996). Fast fluid registration of medical images. *In* "Lecture Notes in Computer Science," Vol. 1131, "Proceedings of the 4th International Conference on Visualisation in Biomedical Computing," in Hamburg, Germany, September 22–25, 1996 (K. H. Höhne and R. Kikinis, eds.), pp. 267–276. Springer-Verlag, Berlin.

Cachier, P., Pennec, X., and Ayache, N. (1999). Fast non rigid matching by gradient descent: Study and improvements of the "demons" algorithm. Institut National de Recherche en Informatique et en Automatique. http://www.inria.fr/rrrt/rr-3706.html. [Technical Report 3706]

Christensen, G. E., Rabbit, R. D., and Miller, M. I. (1996). Deformable templates using large deformation kinematics. *IEEE Trans. Med. Imaging* **5**, 1435–1447.

Collins, D. (1994). "3D Model-Based Segmentation of Individual Brain Structures for Magnetic Resonance Imaging Data." Montreal Neurological Institute, Montreal. [Ph.D. thesis].

Collins, D., Goualher, G. L., and Evans, A. (1998a). Non-linear cerebral registration with sulcal constraints. *In* "MICCAI 1998", pp. 974–984. Cambridge, MA.

Collins, D. L., Zijdenbos, A. P., Kollokian, V., Sled, J. G., Kabani, N. J., Holmes, C. J., and Evans, A. C. (1998b). Design and construction of a realistic digital brain phantom. *IEEE Trans. Med. Imaging* **17**, 463–468.

Cullen, M., and Purser, R. (1984). An extended Lagrangian theory of semi-geostrophic frontogenesis. *J. Atmos. Sci.* **41**, 1477–1497.

Dacorogna, B., and Moser, J. (1990). On a partial differential equation involving the Jacobian determinant. *Ann. Inst. H. Poincaré Anal. Non Linéaire* **7**, 1–26.

Davatzikos, C. (1997). Spatial transformation and registration of brain images using elastically deformable models. *Comp. Vision Image Understand. Spect. Issue Med. Imaging* **66**, 207–222.

Dumoulin, C., Souza, S., Walker, M., and Wagle, W. (1989). Three-dimensional phase contrast angiography. *Magn. Reson. Med.* **9**, 139–149.

Edwards, P. J., Hill, D. L. G., Little, J. A., and Hawkes, D. J. (1997). Deformation for image guided interventions using a three component tissue model. *In* "IPMI'97," pp. 218–231.

Ferrant, M., Cuisenaire, O., and Macq, B. (1999a). Multi-object segmentation of brain structures in 3D MRI using a computerized atlas. *SPIE Med. Imaging '99* **3661–2**, 986–995.

Ferrant, M., Nabavi, A., Macq, B., and Warfield, S. K. (2000a). Deformable modeling for characterizing biomedical shape changes. *In* "Lecture Notes in Computer Science", Vol. 1953, "DGCI2000: Discrete Geometry for Computer Imagery," December 13–15, 2000, Uppsala, Sweden (G. Borgefors, I. Nyström, and G. Sanniti di Baja, eds.), pp. 235–248. Springer-Verlag, Heidelberg.

Ferrant, M., Warfield, S. K., Guttmann, C. R. G. Mulkern, R. V., Jolesz, F. A., and Kikinis, R. (1999b). 3D image matching using a finite element based elastic deformation model. *In* "MICCAI99: Second International Conference on Medical Image Computing and Computer-Assisted Intervention," September 1999, 19–22, Campbridge, England (C. Taylor and A. Colchester, eds.), pp. 202–209. Springer-Verlag, Heidelberg.

Ferrant, M., Warfield, S. K., Nabavi, A., Macq, B., and Kikinis, R. (2000b). Registration of 3D intraoperative MR images of the brain using a finite element biomechanical model. *In* "MICCAI 2000: Third International Conference on Medical Robotics, Imaging and Computer Assisted Surgery," October 11–14, 2000, Pittsburgh (A. M. DiGioia and S. Delp, eds.), pp. 19–28. Springer-Verlag, Heidelberg.

Gaens, T., Maes, F., Vandermeulen, D., and Suetens, P. (1998). Non-rigid multimodal image registration using mutual information. *In* "Lecture Notes in Computer Science," Vol. 1496, "Proceedings of the First International Conference on Medical Image Computing and Computer-Assisted Intervention (MICCAI'98)," October 11–13, 1998, Cambridge, MA, pp. 1099–1106. Springer Verlag, Berlin.

Gangbo, W. (1994). An elementary proof of the polar factorization of vector-valued functions. *Arch. Rational Mech. Anal.* **128**, 381–399.

Gangbo, W., and McCann, R. (1996). The geometry of optimal transportation. *Acta Math.* **177**, 113–161.

Gee, J., and Bajcsy, R. (1999). Elastic matching: Continuum mechanical and probabilistic analysis. *In* "Brain Warping," pp. 193–198. Academic Press, San Diego.

Geiger, B. (1993). "Three Dimensional Modeling of Human Organs and Its Application to Diagnosis and Surgical Planning." INRIA, Rocquencourt. [Technical Report 2105]

Gering, D., Nabavi, A., Kikinis, R., Grimson, W., Hata, N., Everett, P., Jolesz, F., and Wells, W. (1999). An integrated visualization system for surgical planning and guidance using image fusion and interventional imaging. *In* "MICCAI 99: Proceedings of the Second International Conference on Medical Image Computing and Computer Assisted Intervention," pp. 809–819. Springer Verlag, Berlin.

Gering, D., Nabavi, A., Kikinis, R., Hata, N., O'Donnell, L., Grimson, W., Jolesz, F., Black, P., and Wells, W., III. (2001). An integrated visualization system for surgical planning and guidance using image fusion and an open MR. *J. Magn. Reson. Imaging* **13**, 967–975.

Guimond, A., Roche, A., Ayache, N., and Meunier, J. (2001). Three-dimensional multimodal brain warping using the Demons algorithm and adaptive intensity corrections. *IEEE Trans. Med. Imaging* **20**, 58–69.

Hagemann, A., Rohr, K., Stiel, H., Spetzger, U., and Gilsbach, J. (1999). Biomechanical modeling of the human head for physically based, non-rigid image registration. *IEEE Trans. Med. Imaging* **18**, 875–884.

Hajnal, J., and Bydder, G. (1997). Clinical uses of diffusion weighted imaging. *In* "Advanced MR Imaging Techniques." Dunitz, London, UK.

Haker, S., and Tannenbaum, A. (2001a). Optimal transport and image registration. Submitted for publication.

Haker, S., and Tannenbaum, A. (2001b). Optimal transport and image registration. *In* "IEEE Workshop on Variational and Level Set Methods in Computer Vision, ICCV 2001."

Hata, N. (1998). "Rigid and Deformable Medical Image Registration for Image-Guided Surgery." University of Tokyo, Tokyo. [Ph.D. thesis]

Hata, N., Nabavi, A., Wells, W. M., Warfield, S. K., Kikinis, R., Black, P. M., and Jolesz, F. A. (2000). Three-dimensional optical flow method for measurement of volumetric brain deformation from intraoperative MR images. *J. Comput. Assisted Tomogr.* **24**, 531–538.

Hata, N., Wells, W. M., Halle, M., Nakajima, S., Viola, P., Kikinis, R., and Jolesz, F. A. (1996). Image guided microscopic surgery system using mutual-information based registration. *In* "Visualization in Biomedical Computing," pp. 307–316. Springer-Verlag, Heidelberg, Germany.

Hellier, P., and Barillot, C. (2000). Multimodal non-rigid warping for correction of distortions in functional MRI. *In* "Lecture Notes in Computer Science," Vol. 1935, "Third International Conference on Medical Image Computing and Computer-Assisted Intervention (MICCAI'00)," pp. 512–520. Springer-Verlag, Pittsburgh.

Hermosillo, G., Chefd'Hotel, C., and Faugeras, O. (2001). A variational approach to multi-modal image matching. Institut National de Recherche en Informatique et en Automatique. [Technical Report 4117]

Hill, D., Maurer, C., Maciunas, R., Barwise, J., Fitzpatrick, J., and Wang, M. (1998). Measurement of intraoperative brain surface deformation under a craniotomy. *Neurosurgery* **43**, 514–526.

Horn, B. K. P., and Schunck, B. G. (1981). Determining optical flow. *Artif. Intell.* **17**, 185–203.

Jolesz, F. (1997). Image-guided procedures and the operating room of the future. *Radiology* **204**, 601–612.

Kantorovich, L. V. (1948). On a problem of Monge. *Uspekhi Mat. Nauk.* **3**, 225–226.

Kaus, M. R., Nabavi, A., Mamisch, C. T., Wells, W. M., Jolesz, F. A., Kikinis, R., and Warfield, S. K. (2000). Simulation of corticospinal tract displacement in patients with brain tumors. *In* "MICCAI 2000: Third International Conference on Medical Robotics, Imaging and Computer Assisted Surgery," October 11–14, 2000, *Pittsburgh* (A. M. DiGioia and S. Delp, eds.), pp. 9–18. Springer-Verlag, Heidelberg.

Kaus, M. R., Warfield, S. K., Nabavi, A., Chatzidakis, E., Black, P. M., Jolesz, F. A., and Kikinis, R. (1999). Segmentation of MRI of meningiomas and low grade gliomas. *In* "MICCAI 99: Second International Conference on Medical Image Computing and Computer-Assisted Intervention," September 1999, 19–22, *Cambridge, England* (C. Taylor and A. Colchester, eds.), pp. 1–10. Springer-Verlag, Heidelberg.

Kikinis, R., Shenton, M. E., Gerig, G., Martin, J., Anderson, M., Metcalf, D., Guttmann, C. R. G., McCarley, R. W., Lorenson, W. E., Cline, H., and Jolesz, F. (1992). Routine quantitative analysis of brain and cerebrospinal fluid spaces with MR imaging. *J. Magn. Reson. Imaging* **2**, 619–629.

Kikinis, R., Shenton, M. E., Iosifescu, D. V., McCarley, R. W., Saiviroonporn, P., Hokama, H. H., Robatino, A., Metcalf, D., Wible, C. G., Portas, C. M., Donnino, R. M., and Jolesz, F. A. (1996). A digital brain atlas for surgical planning, model driven segmentation, and teaching. *IEEE Trans. Visualizat. Comput. Graph.* **2**, 232–241.

LeBihan, D., Breton, E., Lallemand, D., Grenier, P., Cabanis, E., and Laval-Jeantet, M. (1986). MR imaging of intravoxel incoherent motions: Applications to diffusion and perfusion in neurologic disorders. *Radiology* **161**, 401–407.

Lester, H., Arridge, S., Jansons, K., Lemieux, L., Hajnal, J., and Oatridge, A. (1999). Nonlinear registration with the variable viscosity fluid algorithm. *In* "IPMI 1999," pp. 238–251.

Likar, B., and Pernus, F. (2000). A hierarchical approach to elastic registration based on mutual information. *Image Vision Comput.* **19**, 33–44.

Maes, F., Collignon, A., Vandermeulen, D., Marchal, G., and Suetens, P. (1997). Multimodality image registration by maximization of mutual information. *IEEE Trans. Med. Imaging* **16**, 187–198.

Maintz, J. B. A., Meijering, E. H. W., and Viergever, M. A. (1998). General multimodal elastic registration based on mutual information. *In* "SPIE Proceedings," Vol. 3338, "Medical Imaging 1998; Image Processing (MI'98)", Febuary 23–26, 1998, San Diego (K. M. Hanson, ed.), pp. 144–154. Int. Soc. Opt. Eng., Bellingham, WA.

Maintz, J. B. A., and Viergever, M. A. (1998). A survey of medical image registration. *Med. Image Anal.* **2**, 1–36.

McCann, R. J. (1997). A convexity principle for interacting gases. *Adv. Math.* **128**, 153–179.

Meyer, C. R., Boes, J. L., Kim, B., and Bland, P. H. (1999). Probabilistic brain atlas construction: Thin plate spline warping via maximization of mutual information. *In* "Lecture Notes in Computer Science," Vol. 1679, "Proceedings of the Second International Conference on Medical Image Computing and Computer-Assisted Intervention (MICCAI'99)", September 19–22, 1999, Cambridge, UK, pp. 631–637. Springer-Verlag, Cambridge, UK.

Miga, M., Paulsen, K., Lemery, J., Hartov, A., and Roberts, D. (1999). In vivo quantification of a homogeneous brain deformation model for updating preoperative images during surgery. *IEEE Trans. Med. Imaging* **47**, 266–273.

Miga, M. I., Roberts, D. W., Kennedy, F. E., Platenik, L. A., Hartov, A., Lunn, K. E., and Paulsen, K. D. (2001). Modeling of retraction and resection for intraoperative updating of images. *Neurosurgery* **49**, 75–85.

Moon, T. K., and Stirling, W. C. (2000). "Mathematical Methods and Algorithms for Signal Processing." Prentice Hall International, Englewood Cliffs, NJ.

Moser, J. (1965). On the volume elements on a manifold. *Trans. Am. Math. Soc.* **120**, 286–294.

Nabavi, A., Black, P., Gering, D., Westin, C., Mehta, V., Jr., R. P., Ferrant, M., Warfield, S., Hata, N., Schwartz, R., 3rd, W. W., Kikinis, R., and Jolesz, F. (2001). Serial intraoperative magnetic resonance imaging of brain shift. *Neurosurgery* **48**, 787–797.

Paulsen, K., Miga, M., Kennedy, F., Hoopes, P., Hartov, A., and Roberts, D. (1999). A computational model for tracking subsurface tissue deformation during stereotactic neurosurgery. *IEEE Trans. Med. Imaging* **47**, 213–225.

Peled, S., Gudbjartsson, H., Westin, C.-F., Kikinis, R., and Jolesz, F. A. (1998). Magnetic resonance imaging shows orientation and asymmetry of white matter fiber tracts. *Brain Res.* **780**, 27–33.

Pennec, X., Cachier, P., and Ayache, N. (1999). Understanding the "Demon's algorithm": 3D non-rigid registration by gradient descent. *In* "Lecture Notes in Computer Science, Proceedings of the Second International Conference on Medical Image Computing and Computer-Assisted Intervention (MICCAI'99)", September 19–22, 1999, Cambridge, UK. Springer-Verlag, Berlin.

Pierpaoli, C., Jezzard, P., Basser, P., Barnett, A., and Chiro, G. D. (1996). Diffusion tensor MR imaging of the human brain. *Radiology* **201**, 637–648.

Poggio, T., Torre, V., and Koch, C. (1985). Computational vision and regularization theory. *Nature* **317**, 314–319.

Poupon, C., Clark, C., Frouin, V., LeBihan, D., Bloch, I., and Mangin, J.-F. (1999). Inferring the brain connectivity from MR diffusion tensor data. *In* "Lecture Notes in Computer Science," Vol. 1679, "Medical Image Computing and Computer-Assisted Intervention (MICCAI'98)" (W. M. Wells, A. Colchester, and S. Delp, eds.), pp. 453–462. Springer-Verlag, Cambridge, UK.

Poupon, C., Mangin, J.-F., Frouin, V., Regis, J., Poupon, F., Pachot-Clouard, M., LeBihan, D., and Bloch, I. (1998). Regularization of MR diffusion tensor maps for tracking brain white matter bundles. *In* "Lecture Notes in Computer Science," Vol. 1996, "Medical Image Computing and Computer-Assisted Intervention (MICCAI'98)" (W. M. Wells, A. Colchester, and S. Delp, eds.), pp. 489–498. Springer-Verlag, Cambridge, UK.

Press, W., Teukolsky, S., Vetterling, W., and Flannery, B. (1992). "Numerical Recipes in C: The Art of Scientific Computing." Cambridge Univ. Press, Cambridge, UK.

Provenzale, J., and Sorensen, A. (1999). Diffusion-weighted MR imaging in acute stroke: The oretical considerations and clinical applications. *Am. J. Roentgenol.* **173**, 1459–1467.

Rachev, S., and Rüschendorf, L. (1998). "Mass Transportation Problems," Vols. I and II. Springer-Verlag, New York.

Roche, A., Malandain, G., and Ayache, N. (2000). Unifying maximum likelihood approaches in medical image registration. *Int. J. Imaging Syst. Technol.* **11**, 71–80.

Roche, A., Malandain, G., Pennec, X., and Ayache, N. (1998). The correlation ratio as a new similarity measure for multimodal image registration. *In* "Lecture Notes in Computer Science," Vol. 1996, "Proceedings of the First International Conference on Medical Image Computing and Computer-Assisted Intervention (MICCAI'98), October 11–13, 1998. Cambridge, MA, (W. M. Wells, A. Colchester, and S. Delp, eds.), pp. 1115–1124. Springer-Verlag, Berlin. http://www.inria.fr/rrrt/rr-3378.html.

Rohr, K. (1997). Differential operators for detecting point landmarks. *Image Vision Comput.* **15**, 219–233.

Rousseeuw, P. J., and Leroy, A. M. (1987). "Robust Regression and Outlier Detection." Wiley Series in Probability and Mathematical Statistics. Wiley, New York.

Rueckert, D., Clarkson, M., Hill, D., and Hawkes, D. (2000). Non-rigid registration using higher-order mutual information. *In* "SPIE Proceedings," Vol. 3979, "Medical Imaging," pp. 438–447. Int. Soc. Opt. Eng., San Diego.

Ruiz-Alzola, J., Kikinis, R., and Westin, C.-F. (2001a). Detection of point landmarks in multi-dimensional tensor data. *Signal Process.*, **81**, 2243–2247.

Ruiz-Alzola, J., Westin, C.-F., Warfield, S. K., Alberola, C., Maier, S., and Kikinis, R. (2002). Nonrigid registration of 3D tensor medical data. *Med. Image Anal.*, **6**(2), 143–161.

Ruiz-Alzola, J., Westin, C.-F., Warfield, S. K., Nabavi, A., and Kikinis, R. (2000). Nonrigid registration of 3D scalar, vector and tensor medical data. *In* "Lecture Notes in Computer Science," Vol. 1935, "Medical Image Computing and Computer-Assisted Intervention (MICCAI '00)" (S. L. Delp, A. M. DiGioia, and B. Jaramaz, eds.), pp. 541–550. Springer-Verlag, Berlin.

Schroeder, W., Martin, K., and Lorensen, B. (1996). "The Visualization Toolkit: An Object-Oriented Approach to 3D Graphics." Prentice Hall International, Englewood Cliffs, NJ.

Segel, L. (1987). "Mathematics Applied to Continuum Mechanics." Dover, New York.

Sierra, R. (2001). Nonrigid registration of diffusion tensor images. Swiss Federal Institute of Technology, Zurich. [Master's thesis]

Simoncelli, E. P. (1994). Design of multi-dimensional derivative filters. *In* "International Conference on Image Processing", Austin, TX. IEEE Press, New York.

Skrinjar, O., and Duncan, J. S. (1999). Real time 3D brain shift compensation. *In* "IPMI'99," pp. 641–649.

Skrinjar, O., Studholme, C., Nabavi, A., and Duncan, J. (2001). Steps toward a stereo-camera-guided biomechanical model for brain shift compensation. *In* "Proceedings of International Conference of Information Processing in Medical Imaging," pp. 183–189.

Stejskal, E., and Tanner, J. (1965). Spin diffusion measurements: Spin echoes in the presence of a time-dependent field gradient. *J. Chem. Phys.* **42**, 288–292.

Thirion, J.-P. (1995). Fast non-rigid matching of 3D medical images. Institut National de Recherche en Informatique et en Automatique, Sophia-Antipolis. http://www.inria.fr/rrrt/rr-2547.html. [Technical Report 2547]

Thirion, J.-P. (1998). Image matching as a diffusion process: An analogy with Maxwell's demons. *Med. Image Anal.* **2**, 243–260.

Toga, A. W. (1999). "Brain Warping." Academic Press, San Diego.

Viola, P. and Wells, W. M. (1995). Alignment by maximization of mutual information. *In* "Fifth International Conference on Computer Vision (ICCV)", pp. 16–23. IEEE Press, New York.

Viola, P., and Wells, W. M. (1997). Alignment by maximization of mutual information. *Int. J. Comput. Vision* **24**, 137–154.

Wang, Y., and Staib, L. (2000). Physical model-based non-rigid registration incorporating statistical shape information. *Med. Image Anal.* **4**: 7–20.

Warfield, S. K., Jolesz, F., and Kikinis, R. (1998a). A high performance computing approach to the registration of medical imaging data. *Parallel Comput.* **24**, 1345–1368.

Warfield, S. K., Jolesz, F. A., and Kikinis, R. (1998b). Real-time image segmentation for image-guided surgery. *In* "SC 1998: High Performance Networking and Computing Conference," November 1998, 7–13, 1998, Orlando, Vol. 1114, pp. 1–14. IEEE, Press, New York.

Warfield, S. K., Kaus, M., Jolesz, F. A., and Kikinis, R. (2000a). Adaptive, template moderated, spatially varying statistical classification. *Med. Image Anal.* **4**, 43–55.

Warfield, S. K., Mulkern, R. V., Winalski, C. S., Jolesz, F. A., and Kikinis, R. (2000b). An image processing strategy for the quantification and visulization of exercise induced muscle MRI signal enhancement. *J. Magn. Reson. Imaging* **11**, 525–531.

Weickert, J. (1997). "Anisotropic Diffusion in Image Processing." Teubner Verlag, Stuttgart.

Weickert, J., ter Haar Romeny, B., and Viergever, M. (1998). Efficient and reliable schemes for nonlinear diffusion filtering. *IEEE Trans. Image Process.* **7**, 398–410.

Weinstein, D. M., Kindlmann, G. L., and Lundberg, E. C. (1999). Tensorlines: Advection–diffusion based propagation through diffusion tensor fields. *In* "IEEE Visualization '99" (D. Ebert, M. Gross, and B. Hamann, eds.), pp. 249–254. IEEE Press, San Francisco.

Wells, W., Kikinis, R., Grimson, W., and Jolesz, F. (1996a). Adaptive segmentation of MRI data. *IEEE Trans. Med. Imaging* **15**, 429–442.

Wells, W. M., Viola, P., Atsumi, H., Nakajima, S., and Kikinis, R. (1996b). Multi-modal volume registration by maximization of mutual information. *Med. Image Anal.* **1**, 35–51.

Westin, C.-F., Maier, S. E., Mamata, H., Nabavi, A., Jolesz, F. A., and Kikinis, R. (2001). Processing and visualization of diffusion tensor MRI. *Med. Image Anal.*, **6**(2), 93–108.

Yezzi, A., Tsai, A., and Willsky, A. (2000). Medical image segmentation via coupled curve evolution equations with global constraints. *In* "Mathematical Methods in Biomedical Image Analysis," pp. 12–19. IEEE, New York.

Zienkiewicz, O. C., and Taylor, R. L. (1994). "The Finite Element Method: Basic Formulation and Linear Problems," 4th edition. McGraw–Hill, New York.

25

Combination of Transcranial Magnetic Stimulation and Brain Mapping

Tomáš Paus

Montreal Neurological Institute, McGill University, Montreal, Canada H3A 2B4

I. Introduction

Three principal reasons have motivated recent advances in combining transcranial magnetic stimulation (TMS) with brain mapping and its use to study brain–behavior relationships in health and disease. First, TMS allows the investigator to manipulate neural activity in space and time and, as such, provides a tool for testing the *causality* of structure–function correlations revealed by functional imaging techniques. Second, TMS combined with a concurrent measurement of neural activity with PET, *f*MRI, or EEG serves as a behavior-independent assay of *cortical excitability and connectivity*. Third, the measurement of TMS-induced changes in neural activity in general, and specific neurotransmitter systems in particular, furthers our understanding of the potential *treatment effects* of brain stimulation and the pathophysiology of certain brain disorders. This chapter provides an overview of several ways in which TMS has been combined with brain imaging to achieve the above goals. In order to facilitate the reader's understanding of TMS-induced effects on brain activity, I shall also briefly review several key issues related to the origin of TMS-induced excitatory and inhibitory phenomena and to the role of excitatory neurotransmission in driving the hemodynamic signal, the most common index of brain activity.

II. Neurophysiological Underpinnings of the Signal

A. Modulation of Neural Activity by TMS

TMS uses a time-varying magnetic field to induce electrical current in spatially restricted regions of the cerebral cortex; the induced current flows in the direction opposite to that of electrical current in the stimulating coil. Most of the commercially available stimulators generate a brief (<200 μs) mono-phasic or biphasic current, a distinction worth considering vis-à-vis possible differences in the induced charge accumulation and associated behavioral effects (Corthout *et al.,* 2001). TMS can be used in three different modes: (1) single-pulse TMS (> ~4-s interval between two pulses); (2) paired-pulse TMS (1- to 30-ms intervals between two pulses and >4 s between pairs); and (3) repetitive TMS (1- to 20-Hz trains of pulses). At the behavioral level, TMS elicits a variety of phenomena falling into two broad categories. Positive phenomena are most often observed after stimulation of primary motor (i.e., muscle twitch) and sensory (e.g., phosphene) cortices. Negative phenomena (or "virtual lesions") are typically manifested by a decrement in behavioral performance, such as an increase in detection threshold or a rise in reaction time. The neurophysiological mechanisms underlying such TMS-induced behavioral effects are not well understood, however. Below, I review some relevant data obtained in TMS studies of motor physiology; although limited to one particular neural system, these studies illustrate the nature of the interaction between the stimulating coil and the underlying brain tissue.

Single-pulse TMS applied over the primary motor cortex (M1) elicits motor-evoked potentials (MEPs) in contralateral muscles. TMS exerts its influence on corticospinal neurons primarily through activation of their afferents; direct excitation of corticospinal axons is small (Rothwell, 1997; Nakamura *et al.,* 1996). Transcranial electrical stimulation (TES), on the other hand, elicits a muscle response largely through direct activation of the corticospinal axons. This TMS–TES difference illustrates an important principle: TMS-induced effects depend on the geometry of the coil relative to that of the stimulated neural elements. Cortical currents induced by a coil positioned tangentially to the skull would be maximal in dendrites and axons located in the plane parallel to the plane of the coil and oriented along the axis passing through the virtual anode (+) and cathode (–). TES and TMS induce most of the currents perpendicular and tangential to the skull/brain surface, respectively. This leads to predominant activation of vertical (TES) and horizontal (TMS) fibers and, therefore, direct (D-wave) and indirect (I-wave) activation of corticospinal neurons, respectively.

TMS appears to activate both excitatory and inhibitory interneurons. This is best illustrated in studies of the silent period and those of intracortical inhibition and excitation (for a minireview, see Chen, 2000). A single TMS pulse applied during a tonic contraction of the muscle suppresses ongoing muscle activity. This suppression, or silent period, lasts for about 200 ms when induced by TMS but only about 100 ms when elicited by TES (Inghilleri *et al.,* 1993); it is believed that the late part of a TMS-induced silent period is of cortical origin (Fuhr *et al.,* 1991; Hallett, 1995; Chen *et al.,* 1999). Intracortical inhibition and excitation are assessed using the paired-pulse paradigm whereby two TMS pulses are applied in rapid succession (1- to 30-ms between-pulse interval); the first pulse is of subthreshold intensity (conditioning stimulus), while the second pulse is suprathreshold (test stimulus). Paired-pulse TMS with short (1–5 ms) and long (8–30 ms) between-pulse intervals elicits, respectively, suppression and facilitation of the muscle response (Kujirai *et al.,* 1993; Nakamura *et al.,* 1997; Ziemann *et al.,* 1996a). Early suppression and later facilitation suggest somewhat different dynamics of inhibitory and excitatory processes activated by TMS in the motor cortex under these conditions. Pharmacological (Ziemann *et al.,* 1996b; Werhahn *et al.,* 1999) and *in vitro* (Avoli *et al.,* 1997) studies suggest that the short- (paired-pulse) and long- (silent period) lasting cortical inhibition is mediated, respectively, by GABA-A and GABA-B receptors. It remains to be seen, however, whether similar dynamics of TMS-induced excitation and inhibition exist in other cortical regions.

Long-lasting effects of repetitive TMS (rTMS) on motor excitability may provide useful information regarding stimulation parameters and duration of rTMS-induced changes in cortical excitability in general. Repetitive TMS of the primary motor cortex typically induces decreases and increases in the MEP amplitude immediately following, respectively, low- (1 Hz) and high- (15–20 Hz) frequency stimulation applied for 4–15 min at different intensities (Chen *et al.,* 1997; Maeda *et al.,* 2000a,b). Intermediate frequency of rTMS (5 Hz) has been shown to reduce intracortical inhibition when tested with the paired-pulse paradigm (Wu *et al.,* 2000; Peineman *et al.,* 2000). The above effects appear to last up to 30 min following the cessation of rTMS (Chen *et al.,* 1997; Peineman *et al.,* 2000; Muellbacher *et al.,* 2000; Gerschlager *et al.,* 2001). Using the combined TMS–PET method, we have now shown that similar effects can be induced in the prefrontal cortex (Paus *et al.,* 2001a; see below).

B. Neural Correlates of the Hemodynamic Response

The most common parameters measured with PET and *f*MRI in studies of brain–behavior relationships are cerebral blood flow (CBF) and blood oxygenation-level-dependent (BOLD) signal, respectively. In both cases, regional changes in brain activity are inferred from the measured changes in local hemodynamics. Although there is no dispute about the existence of a relationship between local hemodynamics and brain activity, there is still little agreement about the roles of various neural events in driving the hemodynamic signal. At least two issues need to be considered when interpreting the functional significance of CBF/BOLD responses to TMS: (1) the relative importance of neuronal firing vs synaptic activity occurring in the sampled tissue and (2) the relative contributions of excitatory and inhibitory neurotransmission.

Intuitively, a brain region requires more energy and, hence, more blood flow when neurons located in that region increase their firing rate. Several recent experiments suggest, however, that firing rate is not the best predictor of local changes in hemodynamics. Using simultaneous recordings of single-unit activity, field potentials and CBF in the rat cerebellar cortex, Mathiesen *et al.* (1998) demonstrated that electrical stimulation of parallel fibers inhibited spontaneous firing of Purkinje cells located in the sampled cortex while, at the same time, it *increased* CBF and field potentials in the same tissue sample. A strong correlation (r = 0.985) was observed between postsynaptic activity, indicated by the summed field potentials, and activity-dependent increase in CBF. Similar findings were recently obtained in the monkey visual cortex, where the local field potentials provided a better estimate of BOLD responses to visual stimulation than did multiunit activity; while multiunit activity increased only briefly at the onset of stimulation, the field-potential increase was sustained throughout the presence of the stimulus (Logothetis *et al.,* 2001). In addition, Moore and colleagues showed that even subthreshold synaptic activity in the rat primary somatosensory cortex is accompanied by significant variations in hemodynamic signal (Moore *et al.,* 1996, 1999).

Although it is likely that the exact relationship between neuronal firing, input synaptic activity, and CBF might vary across different brain regions, it is safe to assume that postsynaptic activity is the primary driving force of the signal measured with PET and *f*MRI. The Mathiesen *et al.* study also addresses the second issue of interest, namely the relative contribution of excitatory and inhibitory neurotransmission to changes in local hemodynamics. In their study, blockage of GABAergic transmission during electrical stimulation of parallel fibers did not attenuate stimulation-induced increase in local blood flow, suggesting that the increase was primarily due to the excitatory component of the postsynaptic input (Mathiesen *et al.,* 1998). As suggested previously (e.g., Paus *et al.,* 1995; Akgören *et al.,* 1996), a direct link between excitatory neurotransmission and local blood flow may be related to the role of nitric oxide in coupling blood flow to synaptic activity. Nitric oxide (NO) is one of the signals leading to dilation of small vessels in the vicinity of "active" synapses (Northington *et al.,* 1992; Iadecola, 1993). It is known that glutamate activates NO synthase through an increase in the intracellular level of calcium and that, under physiological conditions, entry of calcium into a cell is almost exclusively linked to excitatory neurotransmission. Such a relationship between excitatory transmission and regional CBF would also hold for other coupling models, including that of the activity-dependent astrocyte–neuron system (Magistretti and Pellerin, 1999).

Overall, it is likely that changes in local hemodynamics reflect a sum of excitatory postsynaptic inputs in the sample of scanned tissue. The firing rate of "output" neurons may be related to local blood flow (Heeger *et al.,* 2000; Rees *et al.,* 2000) but only inasmuch as it is linked in a linear fashion to excitatory postsynaptic input. Inhibitory neurotransmission may lead to decreases in CBF indirectly, through its presynaptic effects on postsynaptic excitation.

III. Combination of TMS and Brain Mapping

A. Brain Mapping before TMS: Structural MRI and Frameless Stereotaxy

TMS allows the investigator to manipulate neural activity in space and time and, as such, provides a tool for testing *causality* of structure–function correlations revealed by functional imaging techniques. The complementarity of TMS and functional neuroimaging is best exemplified by the following "two-stage" approach: (1) in Stage I, PET or *f*MRI is used to identify a cortical region with statistically significant task-related changes in local hemodynamics and (2) in Stage II, TMS is used to interfere with neural activity in this "target" region during the performance of the same task.

Precise localization of the TMS coil relative to the brain is the prerequisite for such two-stage brain-mapping studies. Frameless stereotaxy is well suited to this purpose; this method allows one to position the stimulating coil over the target region under the guidance of an MR image of the subject's brain. Frameless stereotaxy uses a 3D position sensor to mark a set of identifiable anatomical landmarks of the subject's head, such as the bridge of the nose and the tragus of the ear, in order to coregister the subject's MRI and his/her head (Fig. 1). Following the coregistration, the coil is positioned over the target location under real-time MR guidance. In addition, the exact position and orientation of the coil at the time of stimulation can be recorded and plotted online on the cortical surface (Fig. 2). Accuracy of the frameless stereotaxy is slightly inferior to that based on a fiducial frame and varies between 4 and 8 mm (Zinreich *et al.,* 1993). A frameless-stereotaxy system comprises two major components: position sensor and tracking software. A commercial TMS-dedicated system has been developed that combines the hardware and software necessary for coil positioning and tracking in real time (BrainSight Frameless; Rogue Research, Inc., Montreal, Canada).

Identification of the target region can be based on a single-subject imaging study or on group averages of PET/*f*MRI data. In the former case, the same individual participates in both the imaging and the TMS stages. For example, in one of our recent studies (Fung *et al.,* 2001) we used *f*MRI to identify, in each individual, a region located in the inferior temporal cortex (ITC) in which the BOLD signal increases with the difficulty of a picture-naming task

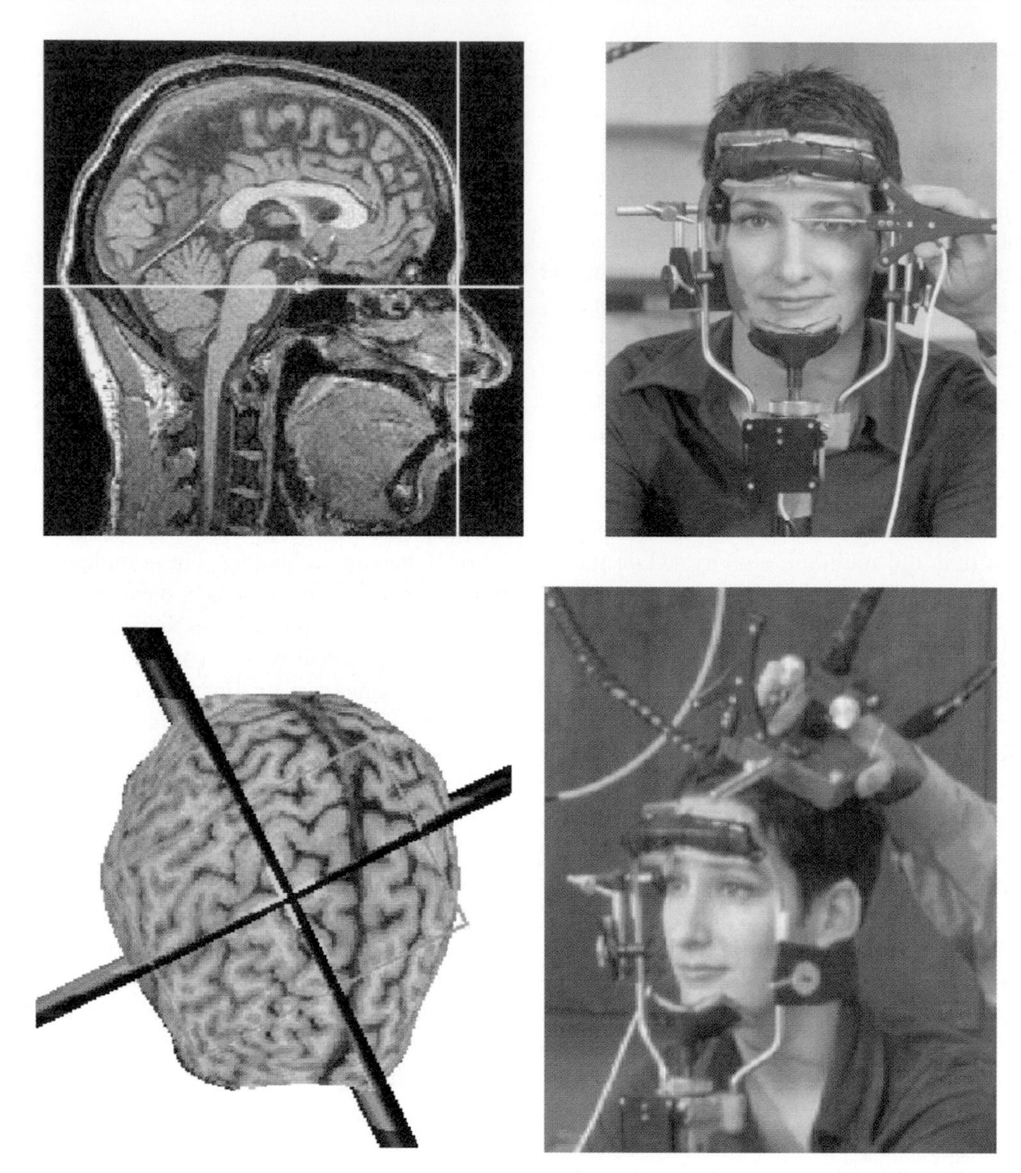

Figure 1 Frameless stereotaxy. **Top:** Registration of the subject's head with the corresponding magnetic resonance (MR) image. A computer-linked probe is touching the bridge of the nose (right); the matching location is highlighted by a cross hair on the MR image (left). **Bottom:** A location targeted by TMS (left) and the probe–coil interface used to position the coil over this location (right).

(Fig. 3, bottom left). Subsequently, each subject participated in an rTMS experiment in which the coil was positioned, with frameless stereotaxy, over the "activated" region and a brief train of rTMS was applied during the picture naming (Fig. 3, top); rTMS was applied to the left and right ITC, as well as to a control region in the left superior parietal cortex. The results showed a significant increase in naming latency during the stimulation of the left ITC; no changes were observed when applying rTMS over the right ITC or the parietal cortex (Fig. 3, bottom right). Although both the left and the right ITC showed task-specific changes in BOLD signal, the rTMS results suggest that only the left ITC is necessary for picture naming.

Considering a relatively high consistency in the location of task-related "activations" across individuals, single-subject imaging studies may not always be necessary to identify the target region. An alternative approach takes advantage of standardized stereotaxic space and uses *x, y,* and *z* coordinates of "activation" peaks identified in previous group-based PET or *f*MRI studies (Paus *et al.,* 1997). For example, we have used a probabilistic location of the frontal eye field (FEF) in a TMS study of visuospatial attention (Grosbras and Paus, 2002); the FEF was identified through a meta-analysis of previous oculomotor blood-flow activation studies (Paus, 1996), its *x, y,* and *z* coordinates were transformed from the standardized stereotaxic space

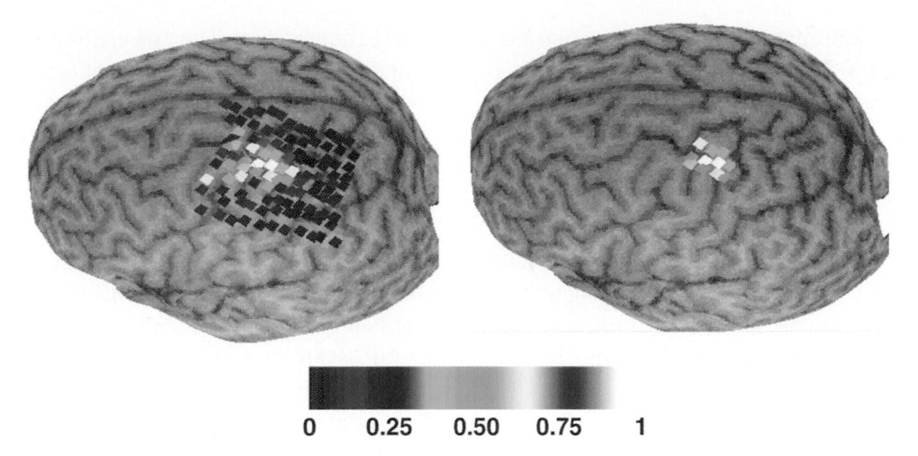

Figure 2 Motor-evoked potentials (MEPs) and coil position. Normalized amplitude of MEPs acquired with the Magstim figure-8 coil and plotted over the cortical surface. The coil locations were sampled with frameless stereotaxy on a 5-mm grid at the time of each TMS pulse; each point indicates the location of the presumed virtual "maximum" of the coil, estimated to be 10 mm in front of its physical center. The arrow indicates the central sulcus at the level of the presumed hand representation ("the knob").

(Talairach and Tournoux, 1988) to the subject's brain coordinate ("native") space, and the coil was positioned over this location with frameless stereotaxy. The transformation from the "Talairach" to the "native" space simply uses the inverse version of the MRI_{native}–$MRI_{Talairach}$ transformation matrix. Single-pulse TMS applied 50 ms before the expected onset

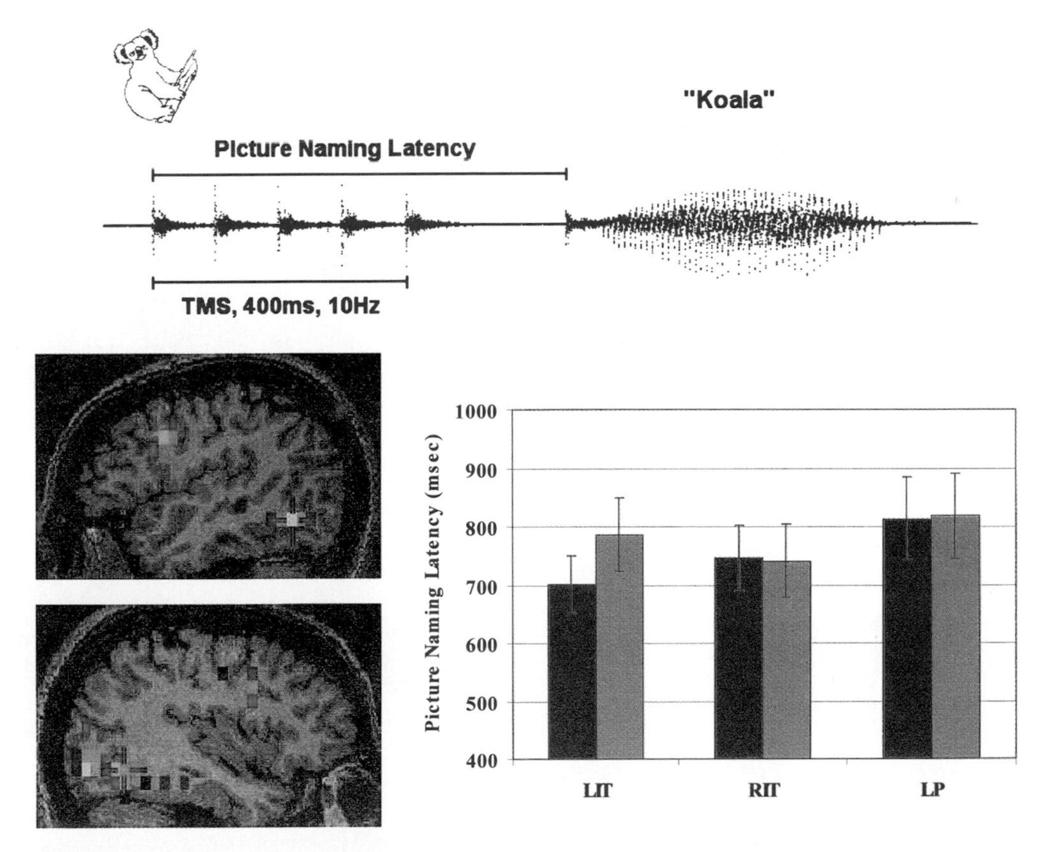

Figure 3 A two-phase brain-mapping experiment. In Phase I, functional MRI was used to identify, in each individual, a region located in the inferior temporal cortex (ITC) in which the BOLD signal increased with the difficulty of a picture-naming task (**bottom left**). In Phase II, such "activated" regions were stimulated with a brief train of repetitive TMS during the picture naming (**top**); the naming latency is shown for the left and right ITC and a control region in the left parietal (LP) lobe (**bottom right**). TMS increased significantly the latency when applied over the left ITC only.

of a saccadic eye movement increased latency of contralateral saccades in half of the subjects included in the study. To validate further this "probabilistic" approach, we used average coordinates of the hand representation in the primary motor cortex and positioned a figure-8 coil over this location; Fig. 4 provides a comparison of the MEP values obtained when stimulating over such probabilistic location with those obtained in the same subject by sampling many locations over the putative M1. No statistically significant differences were found between the two approaches, thus confirming the usefulness of probabilistic locations for targeting cortical regions of interest.

B. Brain Mapping during TMS

TMS combined with a concurrent measurement of neural activity with PET, *f*MRI, or EEG serves as a behavior-independent assay of *cortical excitability and connectivity*. Functional techniques based on the correlational analysis of EEG (Gevins *et al.,* 1981; Thatcher *et al.,* 1986; Tucker *et al.,* 1986) or CBF (Friston 1994; Friston *et al.,* 1996; McIntosh and Gonzales-Lima, 1994; Stroether *et al.,* 1995; Paus *et al.,* 1996) are indirect and may not reveal actual neural connectivity. The correlational studies suffer a major limitation in that the engagement of a subject in performing a task confounds the data being acquired: the observed "coactivations" may reflect relationships between different components of behavior rather than connectivity. The use of TMS to change brain activity directly overcomes the necessity to engage the subject in the performance of a behavioral task. This approach represents a major advantage mainly in studies that are aimed at evaluating the effects of various treatments on the excitability and connectivity of specific neural systems. For example, rehabilitation-induced changes in the motor system are often studied by requiring the patient to perform a similar motor task before and after the intervention. But if successful, the intervention would inevitably affect the movement to be performed during scanning and, in turn, make it difficult to interpret the comparison of the two (before, after) functional data sets. TMS circumvents this problem by giving the investigator full control over the stimulation in a manner independent of the intervention.

In the following three sections, I shall provide an overview of the key methodological aspects involved in the concurrent use of TMS during PET, functional MRI, and

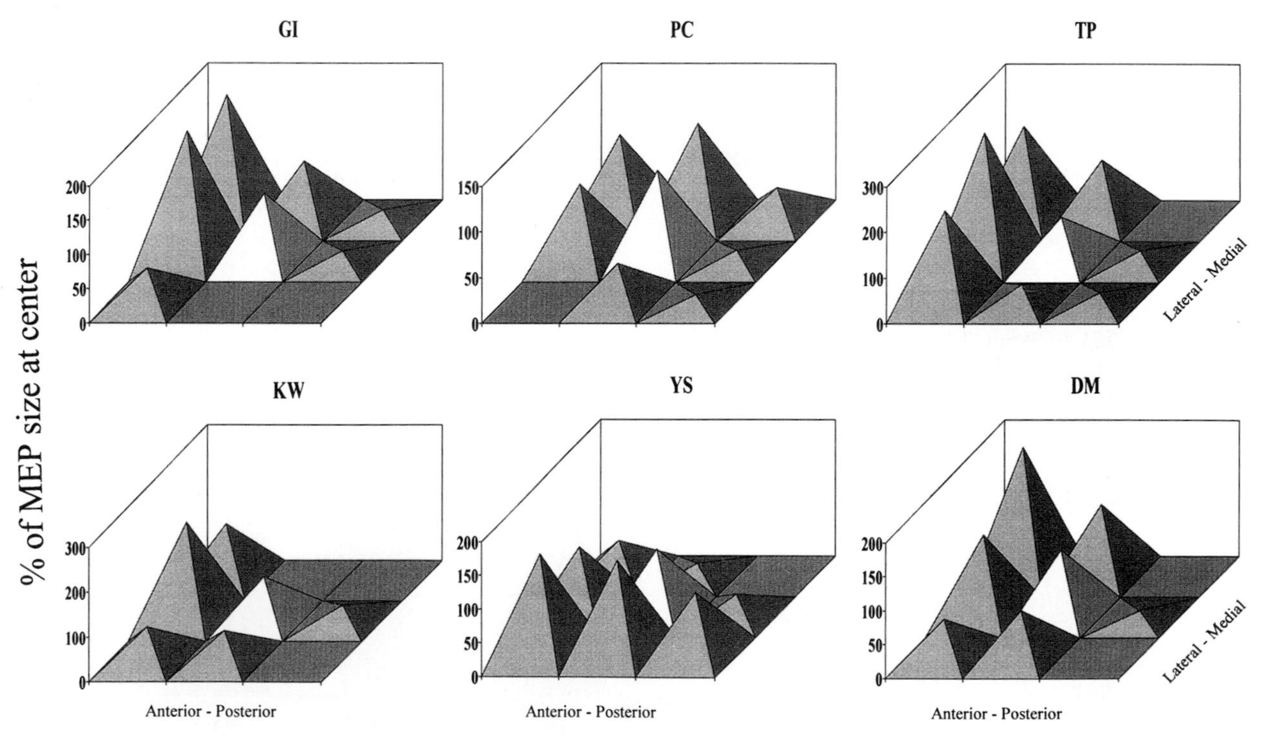

Figure 4 MEPs obtained at the probabilistic location of the left sensorimotor cortex and its vicinity. Each plot shows the amplitude of MEPs obtained in one of six subjects by stimulating at the probabilistic location of the left primary sensorimotor cortex (x = –31, y = –22, z = 52; Paus et al., 1998) and at eight other sites on a 10-mm grid around it. Average amplitudes were calculated from MEPs obtained in response to five single pulses applied at each location and expressed as a percentage of the average obtained at the probabilistic location (center of graph, stippled pyramid). The stimulation was carried out with the Magstim figure-8 coil, the virtual "maximum" of which (see legend to Fig. 2) was positioned over each of the sites using frameless stereotaxy. Note that in most subjects, the largest MEP amplitude was obtained anterior to the probabilistic location of the primary sensorimotor cortex, which was located on the central sulcus in all subjects.

multichannel EEG, illustrating two of these approaches (PET, EEG) with several examples from my laboratory.

1. TMS during PET

Several technical issues need to be considered in TMS/PET experiments. First of all, the strong, albeit brief, coil-generated magnetic field can affect photomultipliers and the related electronic circuits housed about 20 cm away from the subject's head in the gantry of the PET scanner. Even though the field falls off quickly with distance, photomultipliers are sensitive to the interfering effects of magnetic fields as small as 10^{-4} T. Using a single-detector assembly and a figure-8 coil, positioned 19 cm from the photomultipliers, we examined such possible effects and showed that operating the coil even at 40% of the maximum output of the stimulator causes serious distortions in the crystal identification matrix (Paus *et al.,* 1997; Thompson *et al.,* 1998). We were able to prevent these distortions by placing four sheets of well-grounded mu-metal between the coil and the PET detector; the mu-metal causes, however, attenuation of gamma rays and, in turn, a decrease in the number of detected coincidence counts. The second important issue is that of possible movement of the coil relative to the subject's head and the effect of such a misalignment on attenuation corrections. Each PET session begins with a transmission scan that provides information about the location and density of various objects located between the brain, i.e., the "source" of gamma rays in subsequent emission scans, and the PET detectors. In the combined TMS/PET studies, the coil is the single most attenuating object. Any movement of the coil or the head after the transmission scan will result in an incorrect application of the attenuation corrections when calculating the distribution of counts measured in emission scans (Turkington *et al.,* 1994). Thus, false-positive areas of significant differences in rCBF could emerge in cases in which a movement occurred between the TMS-on and TMS-off scans. Lee *et al.* (2000) used a phantom to evaluate the effect of coil movement that had occurred between the acquisition of transmission and emission scans, respectively, and found that a 2-cm movement of the coil in the tangential plane caused up to 50% change in the number of counts recorded from beneath the coil. To minimize the possibility of such movement-related artifacts, we use a bite bar in all TMS/PET studies. The bite bar is placed in the subject's mouth during the coil positioning, transmission scan, and all emission scans, but it is retracted during each 10-min break between successive emission scans; return of the bite bar to the same position is achieved with a special lock-in mechanism.

The PET-based measurements of local and distal effects of TMS on blood flow allow one to assess, respectively, cortical excitability and connectivity and their changes brought about by various interventions. By way of example, I shall describe the design and some of the results of a TMS/PET study aimed at studying corticocortical connectivity of the middorsolateral frontal cortex (MDL-FC) and its modulation by brief periods of repetitive TMS (Paus *et al.,* 2001a). In this study, the target region was chosen based on a probabilistic location of the left MDL-FC ($x = -40$, $y = 32$, $z = 30$), as revealed by a previous PET study of verbal working memory (Fig. 5B). Using the inverse MRI_{native}–$MRI_{Talairach}$ transformation matrix, x, y, and z coordinates of this location were calculated for each individual's "native" brain space and a figure-8 coil was positioned over this location using frameless stereotaxy; during the study, the coil was held with a rigid arm mounted at the back of the scanner's gantry. The transmission scan was carried out and used not only for the attenuation corrections but also for the verification of coil positioning (Fig. 5C; Paus and Wolforth, 1998). Six 60-s water-bolus ($H_2{}^{15}O$) emission scans were acquired afterward: two baseline scans with no TMS applied and four TMS scans during which 30 pairs of pulses were administered with intensity at the individual's motor threshold (Fig. 5A). White noise (90 dB) was played over insert earphones during all scans to attenuate coil-generated clicks. The double-pulse TMS was applied during the scans at 0.5-Hz frequency to provide us with a measure of cortical excitability and connectivity of the left MDL-FC. In addition, we wished to examine the putative modulatory effect of high-frequency repetitive TMS on the MDL-FC excitability/connectivity. To this effect, we applied two series of rTMS between the first and the last TMS scans; the following TMS parameters were used for each series: 15 1-s trains, 10 pulses in each train (i.e., 10 Hz), 10-s between-train intervals, intensity at motor threshold (see Fig. 5A). The stimulation site was identical for the double-pulse and repetitive TMS. In response to the double-pulse TMS applied before rTMS, CBF decreased both at the stimulation site and in several distal regions presumably connected to the site, including the anterior cingulate cortex (Figs. 5D and E). Based on the rationale outlined in Section II, such CBF decreases most likely reflect a net decrease in excitatory synaptic activity mediated by TMS-induced release of GABA. Following the two series of rTMS, this "suppression" response was reversed, resulting in double-pulse-induced increases in CBF that were maximal during the last TMS scan (Fig. 5F). Using correlational analysis, a network of cortical regions was revealed in which the blood-flow response to double-pulse TMS covaried with that at the stimulation site, including the contralateral MDL-FC and the anterior cingulate cortex (Fig. 5G). Overall, this study demonstrated that a mere 30 trains (300 pulses) of 10-Hz rTMS can induce subtle changes in cortical excitability and connectivity of the stimulated region. Such a putative reversal of the initial "inhibitory" response to low-frequency TMS is akin the phenomenon of "long-term transformation," i.e., the transformation of the hyperpolarizing GABA-mediated inhibitory postsynaptic potentials into depolarizing responses as observed *in vitro* (e.g., Sun *et al.,* 2000, 2001).

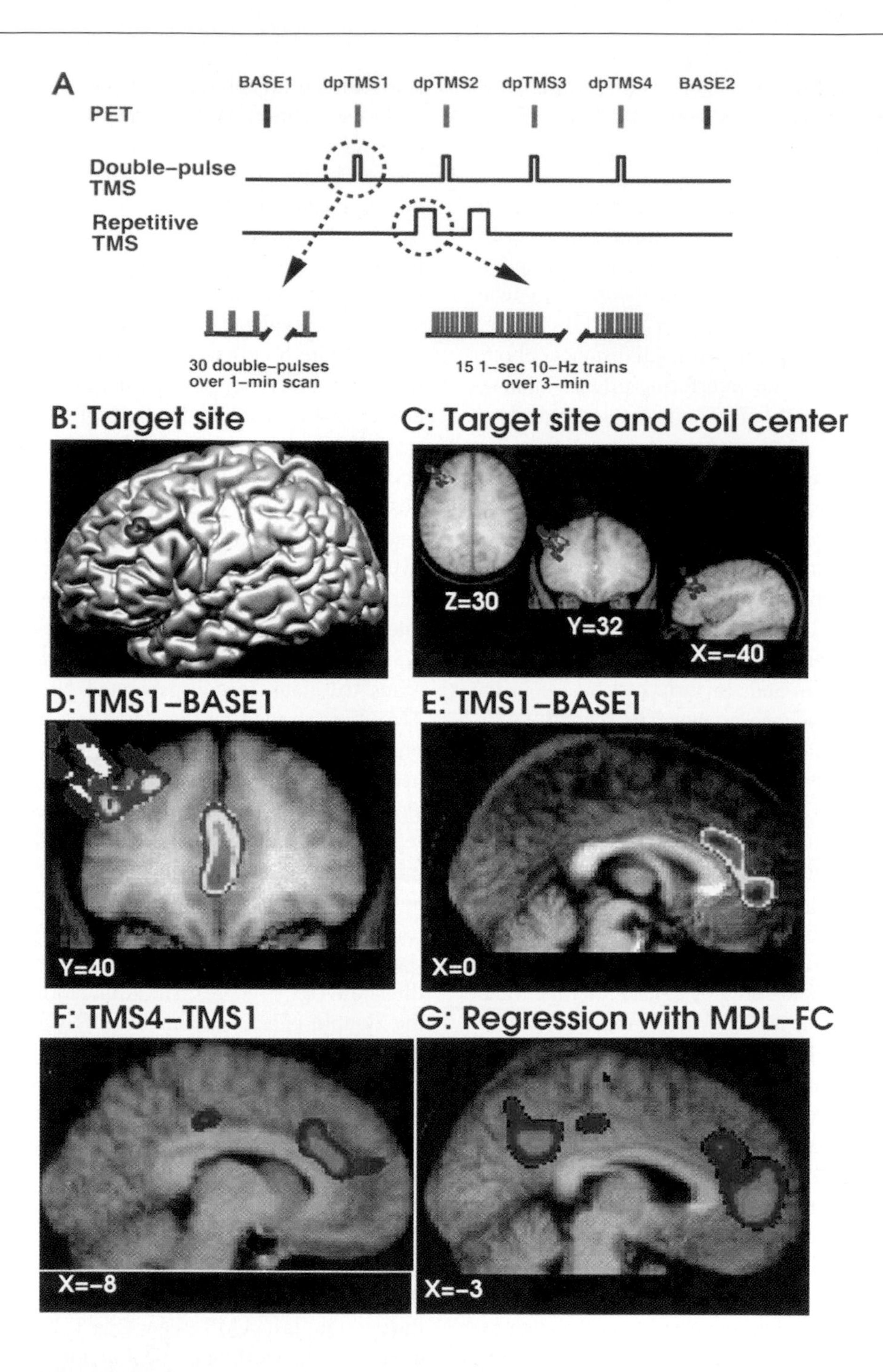

Figure 5 Modulation of corticocortical connectivity by repetitive TMS. The flowchart (**A**) indicates the sequence of events during the TMS/PET study; the PET scans were repeated every 10 min (base, no TMS applied; dpTMS, double-pulse TMS). The target site (**B**) within the left MDL-FC was selected from a previous blood-flow activation study by Petrides *et al.* (1993); the "peak" is located just above the left inferior frontal sulcus. The proximity of the target site (cross hair) and the coil center derived from the transmission scans in the eight subjects (color "bars") demonstrate the successful positioning of the coil with frameless stereotaxy (**C**). The results of subtraction (**D–F**) and regression (**G**) analyses of blood-flow data are shown; the images depict the exact locations that showed statistically significant decreases (D, E) and increases (F) in blood flow and significant positive correlation with blood flow at the stimulation site (G). The thresholded maps of *t*-statistic values ($t > 3.0$ or $t < -3.0$) are superimposed on coronal (D) and sagittal (E–G) sections through the average magnetic resonance image of the eight subjects. All images are aligned within the standardized stereotaxic space. Reproduced from Paus *et al.* (2001a), with permission.

The combination of TMS and $H_2^{15}O$ PET has now been used in several other studies, most of which focused on TMS-induced changes in the excitability and connectivity of the primary motor cortex (Fox *et al.*, 1997; Paus *et al.*, 1998; Siebner *et al.*, 2001; Strafella and Paus, 2001). The only overt behavior involved in some of these studies was that of a TMS-elicited muscle twitch. It is important to emphasize, however, that careful considerations should be given to other potential behavioral confounds, including the coil-generated click and the knocking sensations on the scalp, discomfort often associated with TMS (and its fluctuation over time), as well as possible direct effects of TMS on mood and cognition. Some investigators have begun using TMS/PET during the performance of cognitive tasks (Mottaghy *et al.*, 2000). Interpretation of functional data obtained during such TMS/cognition studies represents, however, a great challenge because of the mix of direct effects of TMS and the task performance, respectively, as well as their interaction. Perhaps, future studies of this type will take advantage of long-lasting effects of repetitive TMS and will be carried out with *f*MRI acquired before and after rTMS (see Section III.C).

In addition to the above-described studies of brain activity with the water-bolus technique, PET offers the opportunity to investigate TMS effects on brain metabolism (e.g., Siebner *et al.*, 1998, 2001) and on neural transmission. The measurement of TMS-induced release of specific neurotransmitters is particularly attractive, for it will allow us to delineate the neurochemical pathways involved in mediating behavioral and treatment effects of TMS. We have begun this research by studying TMS-induced release of dopamine; [^{11}C]raclopride was employed to measure release of dopamine in the human striatum in response to rTMS of the left MDL-FC, with the left occipital cortex used as a control site (Strafella *et al.*, 2001). On two successive days, three series of 15 10-Hz trains of rTMS were applied 10 min apart with a circular coil at either of the two sites. [^{11}C]raclopride was injected immediately after the end of the last rTMS series and the tracer uptake in the brain was measured over the next 60 min. Voxel-wise [^{11}C]raclopride binding potential (BP) was calculated using a simplified reference tissue method (Lammertsma and Hume, 1996; Gunn *et al.*, 1997) to generate statistical parametric images of change in BP (Aston *et al.*, 2000). This analysis revealed a significant decrease in BP in the left caudate nucleus following rTMS of the left MDL-FC, compared with rTMS of the occipital cortex (Fig. 6). Such a reduction in [^{11}C]raclopride BP is indicative of an increase in extracellular dopamine concentration (Laruelle *et al.*, 1997; Endres *et al.*, 1997). It is likely that this TMS-induced focal release of dopamine in the ipsilateral caudate nucleus is mediated by excitatory corticostriatal projections known to originate in high density in the primate prefrontal cortex and to synapse at the vicinity of nigrostriatal dopaminergic nerve terminals. Overall, this study confirms the feasibility of using PET to investigate TMS-induced changes in specific neurotransmitter systems of the human brain and opens up new avenues for studies of the pathophysiology of neurological and psychiatric disorders.

2. TMS during Functional MRI

Compared with PET, functional MRI offers several advantages, including higher spatial and temporal resolution, as well as no ionizing radiation. Although the feasibility of this approach has already been demonstrated (Bohning *et al.*, 1998, 1999, 2000a,b; Baudewig *et al.*, 2001), technical

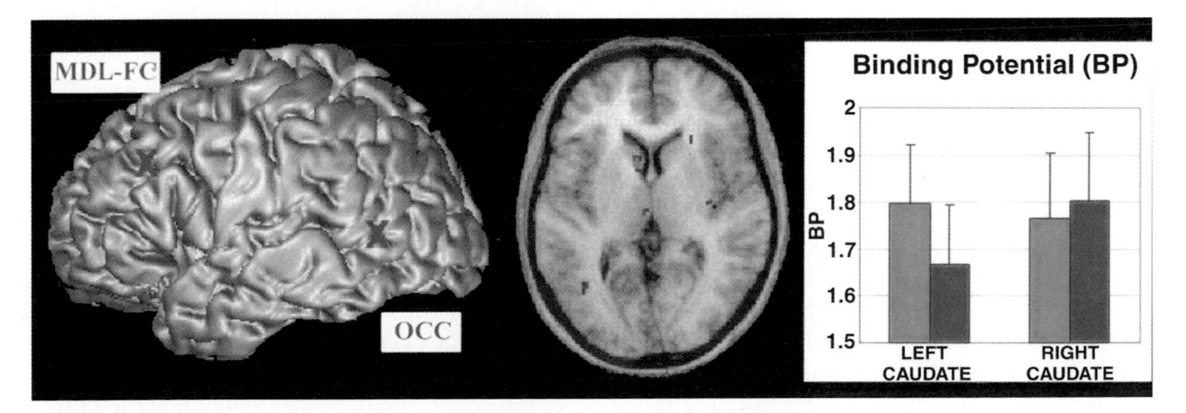

Figure 6 Dopamine release induced by repetitive TMS of the prefrontal cortex. **Left:** Location (red markers) of the two stimulation sites, the left middorsolateral prefrontal cortex (MDL-FC) and the left occipital (OCC) cortex, on the MRI of one subject in stereotaxic space. **Middle:** Transverse (z = 6) section of the statistical parametric map of the change in [^{11}C]raclopride binding potential (BP) overlaid upon the average MRI of all subjects in stereotaxic space. The peak in the left caudate nucleus shows the location where [^{11}C]raclopride BP changed significantly following rTMS of the left middorsolateral prefrontal cortex. **Right:** Mean ± SEM values of the binding potential in the left and right caudate nucleus in response to repetitive TMS of the left MDL-FC (purple) and the left occipital cortex (blue). From Strafella *et al.* (2001).

challenges associated with the use of TMS inside an MR scanner remain significant. Bohning and George (2001) reviewed recently their experience with this approach and identified the following issues requiring the close attention of those interested in the use of TMS inside an MR scanner. The application of brief TMS pulses (~2 T) in a static (~1–4 T) magnetic field increases the forces and torques exerted on the coil; these are minimized by the use of biphasic TMS pulses and a figure-8 coil with its two wings positioned within the same plane and rigidly encapsulated. The presence of the TMS coil generates a significant susceptibility artifact that can be reduced by the appropriate orientation of the EPI slice relative to the coil plane (Baudewig *et al.,* 2000). Another significant source of artifact is eddy currents induced by the TMS pulse. Eddy currents persist for up to 150 ms after the pulse, thus requiring precise synchronization of the image acquisition and TMS (Bohning *et al.,* 1998). The stimulator and the cable connecting it with the coil are potential sources of radiofrequency (RF) interference; this can be minimized with the help of appropriate RF filters placed between the scanner room and the adjacent control area housing the stimulator and by routing the cable away from the subject (Shastri *et al.,* 1999). Finally, the rather closed space of most MR scanners and the use of a headcoil restrict significantly the accessibility of different cortical regions to the stimulating coil. The combined effect of the mechanical forces and the closed space also increases the intensity of the coil-generated acoustic artifact.

The majority of previous TMS/*f*MRI studies targeted the primary motor cortex (M1), either with single-pulse TMS (Bohning *et al.,* 1998, 2000a) or with repetitive TMS (1 Hz, Bohning *et al.,* 1999, 2000b; 10 Hz, Baudewig *et al.,* 2001). Suprathreshold TMS invariably resulted in significant increases in the BOLD signal under the coil (Bohning *et al.,* 1998, 1999, 2000; Baudewig *et al.,* 2001) and, in some cases, in the contralateral M1 (Bohning *et al.,* 2000a) and in the supplementary motor area, SMA (Baudewig *et al.,* 2001). The location and magnitude of BOLD changes in the stimulated M1 were comparable to those observed during voluntary movement. For example, Bohning used a 21-s block design during which 21 pulses were delivered; they observed 2–3% increase in BOLD signal, the center of which was located about 4 mm from the center of "activation" related to voluntary hand movements. Baudewig *et al.* (2001) used event-related *f*MRI and observed significant increase in BOLD signal in response to 1-s trains of 10-Hz rTMS applied with the intensity 10% above the motor threshold; these effects were observed both under the coil and in the vicinity of SMA. When applied over the premotor cortex, however, the same stimulation failed to induce any changes in the BOLD signal. This lack of BOLD increases induced by TMS of a non-M1 region raises an important question regarding the sensitivity of the *f*MRI vis-à-vis TMS; it is also possible that some of the above positive findings obtained with suprathreshold TMS of M1 are related to reafference from the activated muscles rather than to the direct stimulation of the tissue. Overall, it remains to be seen whether *f*MRI proves suitable for routine studies of cortical excitability and connectivity, despite the many technical challenges inherent in this particular combination of TMS and brain imaging.

3. TMS during Multichannel EEG

The high temporal resolution of EEG affords unique insights into the dynamics of TMS-induced changes in cortical excitability and connectivity. Cracco, Amassian, and colleagues were the first to combine TMS with EEG in their studies of *trans*-callosal and frontocerebellar responses (Cracco *et al.,* 1989, Amassian *et al.,* 1992). Ilmoniemi *et al.* (1997) perfected this technique by combining TMS with a 60-channel EEG system; they observed clear *trans*-callosal EEG responses to magnetic stimulation of the primary motor and visual cortex. To prevent the saturation of EEG amplifiers by the magnetic pulse, these authors designed a sample-and-hold circuit that pins the amplifier output to a constant level during the pulse. The amplifiers recover within 3 ms after the pulse; if there are no other sources of artifacts (see below), such a fast-recovering amplification system allows investigators to measure immediate TMS-induced changes in the EEG signal. The overheating of the electrodes placed close to the coil (Roth *et al.,* 1992) is prevented by making the electrodes from low-conductivity material (purified silver with Ag/AgCl coating; Virtanen *et al.,* 1996) and by cutting a slit in the electrode to interrupt eddy currents. As pointed out above, TMS often induces peripheral effects, including activation of scalp muscles. In our experience, local muscle artifacts as well as widespread artifacts due to the relative movement of the scalp against the electrode cap persist for up to 20 ms after the pulse. This effectively prevents one from evaluating evoked potentials elicited by TMS within this time window and, hence, to measure the speed of neural transmission between and within the hemispheres. Until satisfactory technical solutions are found to deal with this problem, we have focused our TMS/EEG studies on the assessment of TMS-induced cortical oscillations.

Event-related synchronization (ERS) and desynchronization allow one to study changes in the spontaneous rhythmic activity of the brain elicited by different events. In the case of a TMS/EEG study, an application of the TMS pulse becomes the event of interest; instead of "natural" activation of a given cortical region by sensory, motor, or cognitive events, one stimulates the region directly. In the first study of this type, we have applied single suprathreshold TMS pulses to the left M1 and recorded EEG with the TMS-compatible 60-channel system (Paus *et al.,* 2001b). The waveform elicited by single-pulse TMS consisted of a positive peak at 30 ms (P30) after the pulse, followed by two negative peaks at 45 (N45) and 100 (N100) ms (Fig. 7). The potential maps revealed that the P30 component was distributed centrally, the N45 component formed a dipole centered over the stimulation site (i.e., the left M1), and the N100 component had a wide distribution with slight predominance over the left central region. Using the ERS analytical approach described in detail elsewhere (Pfurtscheller, 1999), we have observed a striking burst of EEG synchronization in the β frequency range (Fig. 8); this synchronization was time-locked to the onset of the TMS pulse. Several control experiments were carried out to rule out the possibility that the observed changes were related to different TMS-induced peripheral effects, including somatosensory stimulation of the scalp, peripheral feedback from the activated muscle, and auditory stimulation due to the coil-generated click. Altogether, our findings are consistent with the notion that TMS applied to M1 induces transient synchronization of spontaneous activity of cortical neurons within the 15–30 Hz frequency range. As such, they corroborate previous studies of cortical oscillations in the motor cortex and point to the potential of the combined TMS/EEG approach for further investigations of cortical rhythms in the human brain.

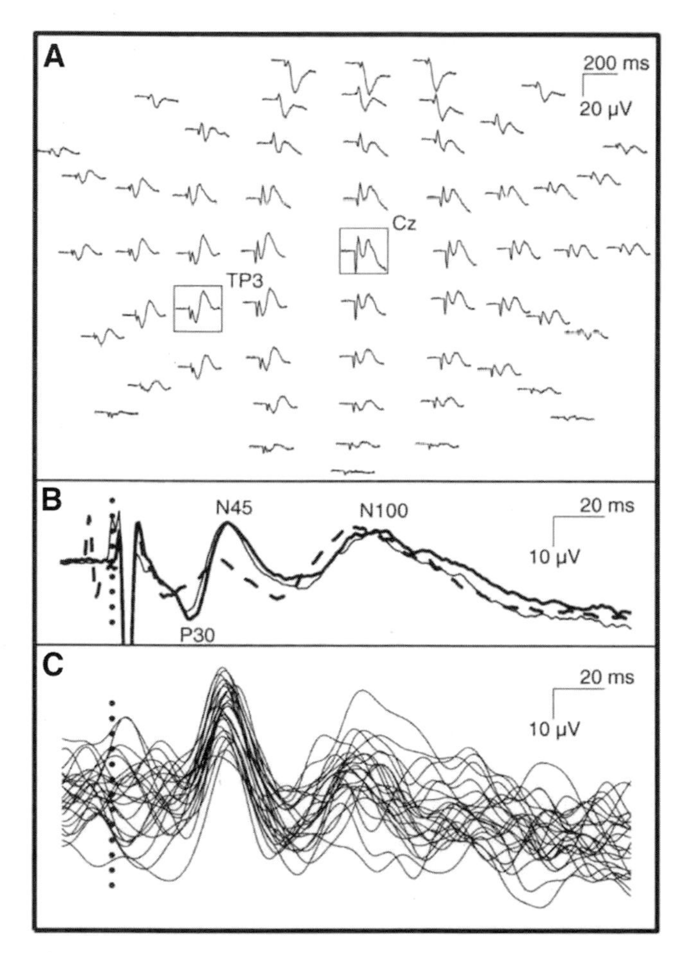

Figure 7 EEG potentials elicited by single- and paired-pulse TMS applied over the left motor cortex. (A) Grand average (seven subjects, 100–120 trials/subject) of the EEG response to single-pulse TMS at all scalp locations. (B) Grand average of the EEG response recorded at Cz in single-pulse (thick solid line), 3-ms paired-pulse (thin solid line), and 12-ms paired-pulse (dashed line) trials. (C) 20 single traces of EEG randomly selected from 120 traces recorded during single-pulse TMS in one subject (PS). The dotted line indicates TMS onset. Reproduced from Paus *et al.* (2001b), with permission.

C. Brain Mapping after TMS

All techniques described in the preceding section are well suited to evaluating short- and long-term effects of repetitive TMS. Several studies of the short-term (minutes, hours) effects of rTMS have already been carried out, either with the water-bolus PET (Paus *et al.*, 2001a; see above) or with [^{18}F]fluorodeoxy-D-glucose (Siebner *et al.*, 2000). Such studies will continue to provide important insights into the neural events underlying plasticity of the human cerebral cortex both at "baseline" and in relation to various cognitive processes. It is likely that *f*MRI will become the main imaging modality in this endeavor; it is ideally suited to the assessment of the minute-by-minute dynamics of TMS-induced changes in brain activity, as well as to the investigation of short-term consequences of brief periods of rTMS (e.g., 10-min 1-Hz stimulation) on task-related activity.

Studies of long-term (days, weeks) effects of rTMS are crucial for our understanding of the neural mechanisms mediating some of the treatment effects of rTMS. Several reports have been published in which SPECT (Peschina *et al.,* 2001; Zheng, 2000; Tenebach *et al.,* 1999) or PET (Speer *et al.,* 2000; Kimbrell *et al.,* 1999) was used to measure before and after rTMS treatment of major depression; the treatment typically consisted of 10 days of high-frequency (10 to 20 Hz) rTMS applied daily over the left prefrontal cortex. In the majority of these studies, such treatment resulted in the increase of perfusion/metabolism in the prefrontal cortex. Kimbrell *et al.* (1999) and Speer *et al.* (2000) employed both high- (20 Hz) and low-frequency (1 Hz) stimulation and observed, respectively, an increase and decrease in CBF and/or glucose metabolism in the frontal cortex and other brain regions. Interpretation of the previous studies is somewhat limited, however, due to the lack of detailed information about the exact coil position during treatment and, in many cases, also because of the absence of an adequate control stimulation and/or comparison group. Furthermore, in most of the published studies, the blood-flow or metabolic measurements were acquired during a resting baseline. In order to facilitate interpretation of the long-term effects of rTMS, it may be useful to evaluate these by comparing, for example, the level of cortical excitability and connectivity probed by TMS applied during the scanning (see above). Overall, it is likely that future imaging studies will provide useful information vis-à-vis potential therapeutic effects of rTMS in depression, as well as in other psychiatric

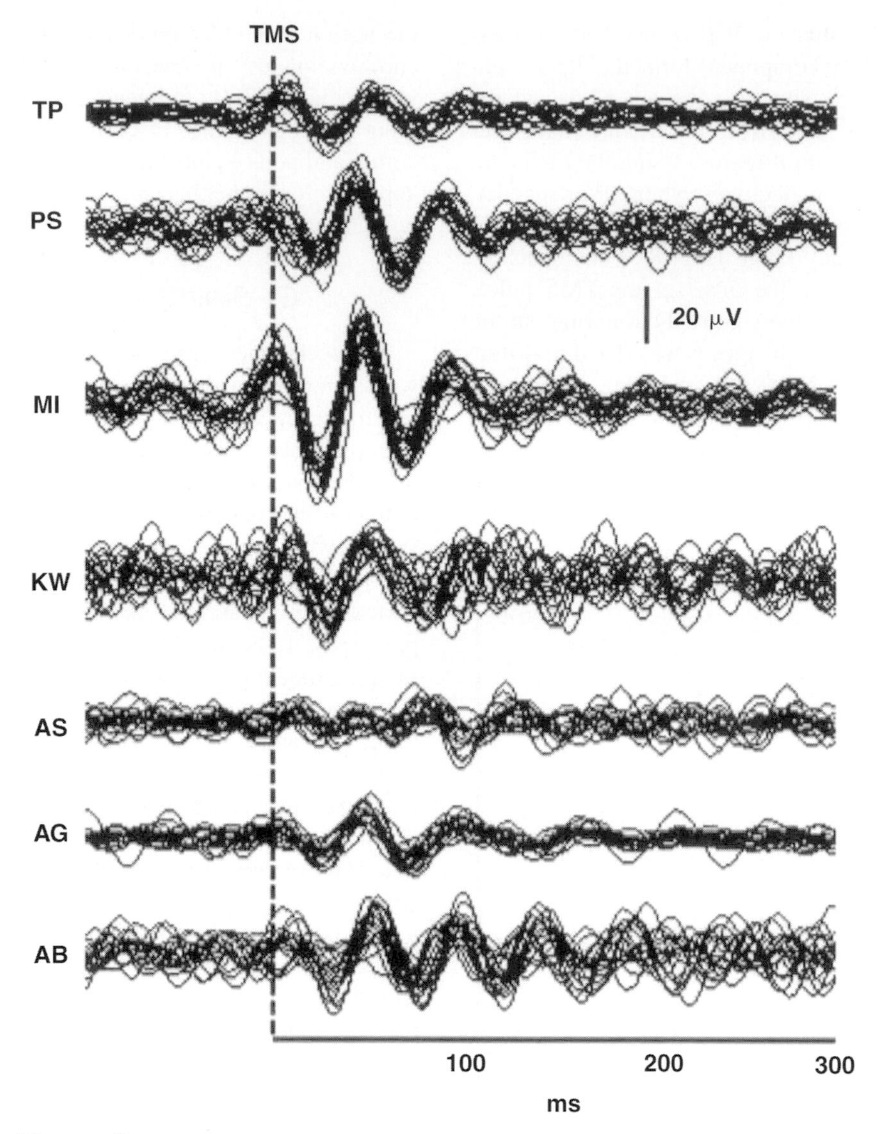

Figure 8 Cortical oscillations elicited by single-pulse TMS applied over the left motor cortex. Temporal evolution of EEG activity filtered in the β range (15–30 Hz) using single traces recorded during single-pulse TMS in seven subjects; 20 traces are superimposed for each subject. Note that the onset of the oscillation appears to precede the TMS pulse; but this time difference is a technical one, resulting from the filtering. Reproduced from Paus *et al.* (2001b), with permission.

and neurological disorders (reviewed by Post and Keck, 2001; Wassermann and Lisanby, 2001). The combination of TMS and imaging may, in this context, lead to improvement of the treatment protocols, as well as an increased understanding of the pathophysiology of these brain disorders.

IV. Conclusion

The combination of brain stimulation and brain imaging is a powerful tool for mapping the human brain. Not only do the two approaches complement each other, one measuring and the other manipulating brain activity, but their combination also gives neuroscientists a unique tool for the assessment of cortical excitability and connectivity. The latter approach is also proving valuable for investigating the neural mechanisms underlying treatment effects of repetitive TMS.

Acknowledgments

I thank Drs. Bruce Pike, Antonio Strafella, Kate Watkins, and Gabriel Leonard for comments on the manuscript. I also thank Dr. Daryl Bohning for useful discussion of some of the technical aspects of TMS during *f*MRI. Data shown in Figs. 2 and 4 were collected and analyzed by Philippe Chouinard, Kate Watkins, Ysbrand van der Werf, and Roch Comeau. The author's research is supported by the Canadian Institutes of Health Research, the Canadian Foundation for Innovation, and the National

Science and Engineering Research Council of Canada. Cadwell Laboratories, Inc., and Rogue Research, Inc., donated equipment and services to the TMS laboratory.

References

Akgoren, N., Dalgaard, P., and Lauritzen, M. (1996). Cerebral blood flow increases evoked by electrical stimulation of rat cerebellar cortex: Relation to excitatory synaptic activity and nitric oxide synthesis. *Brain Res.* **710,** 204–214.

Amassian, V. E., Cracco, R. Q., Maccabee, P. J., and Cracco, J. B. (1992). Cerebello-frontal cortical projections in humans studied with the magnetic coil. *EEG Clin. Neurophysiol.* **85,** 265–272.

Amassian, V. E., Cracco, R. Q., Maccabee, P. J., Cracco, J. B., Rudell, A. P., and Eberle, L. (1998). Transcranial magnetic stimulation in study of the visual pathway. *J. Clin. Neurophysiol.* **15,** 288–304.

Aston, J. A., Gunn, R. N., Worsley, K. J., Ma, Y., Evans, A. C., and Dagher, A. (2000). A statistical method for the analysis of positron emission tomography neuroreceptor ligand data. *NeuroImage* **12,** 245–256.

Avoli, M., Hwa, G., Louvel, J., Kurcewicz, I., Pumain, R., and Lacaille, J. C. (1997). Functional and pharmacological properties of GABA-mediated inhibition in the human neocortex. *Can. J. Physiol. Pharmacol.* **75,** 526–534.

Basar, E., and Bullock, T. H., eds. (1992). "Brain Dynamics: Progress and Perspectives." Birkhauser, Boston.

Baudewig, J., Paulus, W., and Frahm, J. (2000). Artifacts caused by transcranial magnetic stimulation coils and EEG electrodes in T(2)*-weighted echo-planar imaging. *Magn. Reson. Imaging* **18,** 479–484.

Baudewig, J., Siebner, H. R., Bestmann, S., Tergau, F., Tings, T., Paulus, W., and Frahm, J. (2001). Functional MRI of cortical activations induced by transcranial magnetic stimulation (TMS). *NeuroReport,* **12,** 3543–3548.

Bohning, D. E., and George, M. S. (2001). Interleaved transcranial magnetic stimulation (TMS) and BOLD fMRI as a new tool to image and test brain circuits and behavior. Education Workshop held at the 7th Annual Meeting of the Organization for Human Brain Mapping, Brighton.

Bohning, D. E., Shastri, A., McConnell, K. A., Nahas, Z., Lorberbaum, J. P., Roberts, D. R., Teneback, C., Vincent, D. J., and George, M. S. (1999). A combined TMS/fMRI study of intensity-dependent TMS over motor cortex. *Biol. Psychiatry* **45,** 385–394.

Bohning, D. E., Shastri, A., McGavin, L., McConnell, K. A., Nahas, Z., Lorberbaum, J. P., Roberts, D. R., and George, M. S. (2000b). Motor cortex brain activity induced by 1-Hz transcranial magnetic stimulation is similar in location and level to that for volitional movement. *Invest. Radiol.* **35,** 676–683.

Bohning, D. E., Shastri, A., Nahas, Z., Lorberbaum, J. P., Andersen, S. W., Dannels, W. R., Haxthausen, E. U., Vincent, D. J., and George, M. S. (1998). Echoplanar BOLD fMRI of brain activation induced by concurrent transcranial magnetic stimulation. *Invest. Radiol.* **33,** 336–340.

Bohning, D. E., Shastri, A., Wassermann, E. M., Ziemann, U., Lorberbaum, J. P., Nahas, Z., Lomarev, M. P., and George, M. S. (2000a). BOLD-fMRI response to single-pulse transcranial magnetic stimulation (TMS). *J. Magn. Reson. Imaging* **11,** 569–574.

Chen, R. (2000). Studies of human motor physiology with transcranial magnetic stimulation. *Muscle Nerve, Suppl. g,* S26–S32.

Chen, R., Classen, J., Gerloff, C., Celnik, P., Wassermann, E. M., Hallett, M., and Cohen, L. G. (1997). Depression of motor cortex excitability by low-frequency transcranial magnetic stimulation. *Neurology* **48,** 1398–1403.

Chen, R., Lozano, A. M., and Ashby, P. (1999). Mechanism of the silent period following transcranial magnetic stimulation. Evidence from epidural recordings. *Exp. Brain Res.* **128,** 539–542.

Collins, D. L., Neelin, P., Peters, T. M., and Evans, A. C. (1994). Automatic 3D intersubject registration of MR volumetric data in standardized Talairach space. *J. Comput. Assisted Tomogr.* **18,** 192.

Corthout, E., Barker, A. T., and Cowey, A. (2001). Transcranial magnetic stimulation: Which part of the current waveform causes the stimulation. *Exp. Brain Res.,* **141,** 128–132.

Cracco, R. Q., Amassian, V. E., Maccabee, P. J., and Cracco, J. B. (1989). Comparison of human transcallosal responses evoked by magnetic coil and electrical stimulation. *EEG Clin. Neurophysiol.* **74,** 417–424.

Endres, C. J., Kolachana, B. S., Saunders, R. C., Su, T., Weinberger, D., Breier, A., Eckelman, W. C., and Carson, R. E. (1997), Kinetic modelling of [^{11}C] raclopride: Combined PET–microdialysis studies. *J. Cereb. Blood Flow Metab.* **9,** 932–942.

Fox, P., Ingham, R., George, M. S., Mayberg, H., Ingham, J., Roby, J., Martin, C., and Jerabek, P. (1997). Imaging human intra-cerebral connectivity by PET during TMS. *NeuroReport* **8,** 2787–2791.

Friston, K. J. (1994). Functional and effective connectivity in neuroimaging: A synthesis. *Hum. Brain Mapping* **2,** 56–78.

Friston, K. J., Frith, C. D., Fletcher, P., *et al.* (1996). Functional topography: Multidimensional scaling and functional connectivity in the brain. *Cereb. Cortex* **6,** 156–164.

Fuhr, P., Agostino, R., and Hallett, M. (1991). Spinal motor neuron excitability during the silent period after cortical stimulation. *Electroencephalogr. Clin. Neurophysiol.* **81,** 257–262.

Fung, T. D., Chertkow, H., Paus, T., and Whatmough, C. (2002). Transcranial magnetic stimulation of the left inferior temporal cortex slows down picture naming. *J. Int. Neuropsychol. Soc.,* **8,** 206.

Gerschlager, W., Siebner, H. R., and Rothwell, J. C. (2001). Decreased corticospinal excitability after subthreshold 1 Hz rTMS over lateral premotor cortex. *Neurology* **57,** 449–455.

Gevins, A., Doyle, J. C., Cuttilo, B. A., *et al.* (1981). Electrical potentials in human brain during cognition: New method reveals dynamic patterns of correlation. *Science* **213,** 918–922.

Gevins, A., Le, J., Brickett, P., Reutter, B., and Desmond, J. (1991). Seeing through the skull: Advanced EEGs use MRIs to accurately measure cortical activity from the scalp. *Brain Topogr.* **4,** 125–131.

Grosbras, M. H., and Paus, T. (2002). Transcranial magnetic stimulation of frontal eye-field: effects on visual perception and attention. *J. Cog. Neurosci.* (in press).

Gunn, R. N., Lammertsma, A., Hume, S. P., and Cunningham, V. J. (1997). Parametric imaging of ligand–receptor binding in PET using a simplified reference region model. *NeuroImage* **6,** 279–287.

Hallett, M. (1995). Transcranial magnetic stimulation: Negative effects. *Adv. Neurol.* **67,** 107–113.

Heeger, D. J., Huk, A. C., Geisler, W. S., and Albrecht, D. G. (2000). Spikes versus BOLD: What does neuroimaging tell us about neuronal activity? *Nat. Neurosci.* **3,** 631–633.

Iadecola, C. (1993). Regulation of the cerebral microcirculation during neural activity: Is nitric oxide the missing link? *Trends Neurosci.* **16,** 206–214.

Ilmoniemi, R. J., Virtanen, J., Ruohonen, J., Karhu, J., Aronen, H. J., Naatanen, R., and Katila, T. (1997). Neuronal responses to magnetic stimulation reveal cortical reactivity and connectivity. *NeuroReport* **8,** 3537–3540.

Inghilleri, M., Berardelli, A., Cruccu, G., and Manfredi, M. (1993). Silent period evoked by transcranial stimulation of the human cortex and cervicomedullary junction. *J. Physiol.* **466,** 521–534.

Kimbrell, T. A., Little, J. T., Dunn, R. T., Frye, M. A., Greenberg, B. D., Wassermann, E. M., Repella, J. D., Danielson, A. L., Willis, M. W., Benson, B. E., Speer, A. M., Osuch, E., George, M. S., and Post, R. M. (1999). Frequency dependence of antidepressant response to left prefrontal repetitive transcranial magnetic stimulation (rTMS) as a function of baseline cerebral glucose metabolism. *Biol. Psychiatry* **46,** 1603–1613.

Knowles, R. G., Palacios, M., Palmer, R. M. J., and Moncada, S. (1989). Formation of nitric oxide from *l*-arginine in the central nervous system: A transduction mechanism for stimulation of the soluble guanylate cyclase. *Proc. Natl. Acad. Sci. USA* **86,** 5159–5162.

Kujirai, T., Caramia, M. D., Rothwell, J. C., Day, B. L., Thompson, P. D., Ferbert, A., Wroe, S., Asselman, P., and Marsden, C. D. (1993). Corticocortical inhibition in human motor cortex. *J. Physiol. (London.* **471,** 501–519.

Lammertsma, A. A., and Hume, S. P. (1996). Simplified reference tissue model for PET receptor studies. *NeuroImage* **4,** 153–158.

Laruelle, M., Iyer, R. N., Al-Tikriti, M. S., Zea-Ponce, Y., Malison, R., Zoghbi, S. S., Baldwin, R. M., Kung, H. F., Charney, D. S., Hoffer, P. B., Innis, R. B., and Bradberry, C. V. (1997). Microdialysis and SPECT measurements of amphetamine-induced dopamine release in non human primates. *Synapse* **25,** 1–14.

Lee, J. S., Kim, K. M., Lee, D. S., Paek, M. Y., Ahn, J. Y., Paus, T., Park, K. S., Chung, J. K., and Lee, M. S. (2000). Significant effect of subject movement during brain PET imaging with transcranial magnetic stimulation. *J. Nucl. Med.* **41,** 845.

Logothetis, N. K., Pauls, J., Augath, M., Trinath, T., and Oeltermann, A. (2001). Neurophysiological investigation of the basis of the fMRI signal. *Nature* **412,** 150–157.

Maeda, F., Keenan, J. P., Tormos, J. M., Topka, H., and Pascual-Leone, A. (2000a). Modulation of corticospinal excitability by repetitive transcranial magnetic stimulation. *Clin. Neurophysiol.* **111,** 800–805.

Maeda, F., Keenan, J. P., Tormos, J. M., Topka, H., and Pascual-Leone, A. (2000b). Interindividual variability of the modulatory effects of repetitive transcranial magnetic stimulation on cortical excitability. *Exp. Brain Res.* **133,** 425–430.

Magistretti, P. J. and Pellerin, L. (1999). Cellular mechanisms of brain energy metabolism and their relevance to functional brain imaging. *Philos. Trans. R. Soc. Lond. B. Biol. Sci.*, **354,** 1155–1163.

Mathiesen, C., Caesar, K., Akgoren, N., and Lauritzen, M. (1998). Modification of activity-dependent increases of cerebral blood flow by excitatory synaptic activity and spikes in rat cerebellar cortex. *J. Physiol.* **512,** 555–566.

McIntosh, A. R., and Gonzalez-Lima, F. (1994). Structural equation modelling and its application to network analysis in functional brain imaging. *Hum. Brain Mapping* **2,** 2–22.

Moore, C. I., Nelson, S. B., and Sur, M. (1999). Dynamics of neuronal integration in rat somatosensory cortex. *Trends Neurosci.* **22,** 513–520.

Moore, C. I., Sheth, B., Basu, A, Nelson, S., and Sur, M. (1996). What is the neural correlate of the optical imaging signal? Intracellular receptive field maps and optical imaging in rat barrel cortex. *Soc. Neurosci. Abstr.* **22.**

Mottaghy, F. M., Krause, B. J., Kemna, L. J., Topper, R., Tellmann, L., Beu, M., Pascual-Leone, A., and Muller-Gartner, H. W. (2000). Modulation of the neuronal circuitry subserving working memory in healthy human subjects by repetitive transcranial magnetic stimulation. *Neurosci. Lett.* **280,** 167–170.

Muellbacher, W., Ziemann, U., Boroojerdi, B., and Hallett, M. (2000). Effects of low-frequency transcranial magnetic stimulation on motor excitability and basic motor behavior. *Clin. Neurophysiol.* **111,** 1002–1007.

Nakamura, H., Kitagawa, H., Kawaguchi, Y., and Tsuji, H. (1996). Direct and indirect activation of human corticospinal neurons by transcranial magnetic and electrical stimulation. *Neurosci. Lett.* **210,** 45–48.

Nakamura, H., Kitagawa, H., Kawaguchi, Y., and Tsuji, H. (1997). Intracortical facilitation and inhibition after transcranial magnetic stimulation in conscious humans. *J. Physiol. (London)* **498,** 871–823.

Northington, F. J., Matherne, G. P., and Berne, R. M. (1992). Competitive inhibition of nitric oxide synthase prevents the cortical hyperaemia associated with peripheral nerve stimulation. *Proc. Natl. Acad. Sci. USA* **89,** 6649–6652.

Paus, T. (1996). Location and function of the human frontal eye-field: A selective review. *Neuropsychologia* **34,** 475–483.

Paus, T. (1999). Imaging the brain before, during, and after transcranial magnetic stimulation. *Neuropsychologia* **37,** 219–224.

Paus, T., Castro-Alamancos, M., and Petrides, M. (2001a). Cortico-cortical connectivity of the human mid-dorsolateral frontal cortex and its modulation by repetitive transcranial magnetic stimulation. *Eur. J. Neurosci.* **14,** 1405–1411.

Paus, T., Jech, R., Thompson, C. J., Comeau, R., Peters, T., and Evans, A. (1997). Transcranial magnetic stimulation during positron emission tomography: A new method for studying connectivity of the human cerebral cortex. *J. Neurosci.* **17,** 3178–3184.

Paus, T., Jech, R., Thompson, C. J., Comeau, R., Peters, T., and Evans, A. C. (1998). Dose-dependent reduction of cerebral blood flow during rapid-rate transcranial magnetic stimulation of the human sensorimotor cortex. *J. Neurophysiol.* **79,** 1102–1107.

Paus, T., Marrett, S., Worsley K. J., and Evans, A. C. (1995). Extra-retinal modulation of cerebral blood-flow in the human visual cortex: Implications for saccadic suppression. *J. Neurophysiol.* **74,** 2179–2183.

Paus, T., Marrett, S., Worsley, K., and Evans, A. C. (1996). Imaging motor-to-sensory discharges in the human brain: An experimental tool for the assessment of functional connectivity. *NeuroImage* **4,** 78–86.

Paus, T., Sipila, P. K., and Strafella, A. P. (2001b). Synchronization of neuronal activity in the human sensori-motor cortex by transcranial magnetic stimulation: A combined TMS/EEG study. *J. Neurophysiol.* **86,** 1983–1990.

Paus, T., and Wolforth, M. (1998). Transcranial magnetic stimulation during PET: Reaching and verifying the target site. *Hum. Brain Mapping* **6,** 399–402.

Peinemann, A., Lehner, C., Mentschel, C., Munchau, A., Conrad, B., and Siebner, H. R. (2000). Subthreshold 5-Hz repetitive transcranial magnetic stimulation of the human primary motor cortex reduces intracortical paired-pulse inhibition. *Neurosci. Lett.* **296,** 21–24.

Peschina, W., Conca, A., Konig, P., Fritzsche, H., and Beraus, W. (2001). Low frequency rTMS as an add-on antidepressive strategy: Heterogeneous impact on ^{99m}Tc-HMPAO and ^{18}F-FDG uptake as measured simultaneously with the double isotope SPECT technique. Pilot study. *Nucl. Med. Commun.* **22,** 867–873.

Petrides, M., Alivisatos, B., Meyer, E., and Evans, A. C. (1993). Functional activation of the human frontal cortex during the performance of verbal working memory tasks. *Proc. Natl. Acad. Sci. USA* **90,** 878.

Pfurtscheller, G. (1999). EEG event-related desynchronization (EDR) and event-related synchronization (ERS). *In* "Electroencephalography: Basic Principles, Clinical Applications, and Related Fields," 4th edition (E. Niedermeyer and F. Lopes da Silva, eds.), pp. 958–967. Williams & Wilkins, Baltimore.

Post, A., and Keck, M. E. (2001). Transcranial magnetic stimulation as a therapeutic tool in psychiatry: What do we know about the neurobiological mechanisms? *J. Psychiatr. Res.* **35,** 193–215.

Rees, G., Friston, K., and Koch, C. (2000). A direct quantitative relationship between the functional properties of human and macaque V5. *Nat. Neurosci.* **3,** 716–723.

Roth, B. J., Pascual-Leone, A., Cohen, L. G., and Hallett, M. (1992). The heating of metal electrodes during rapid-rate magnetic stimulation: A possible safety hazard. *Electroencephalogr. Clin. Neurophysiol.* **85,** 116–123.

Rothwell, J. C. (1997). Techniques and mechanisms of action of transcranial stimulation of the human motor cortex. *J. Neurosci. Methods* **74,** 113–122.

Rothwell, J. C. (1999). Paired-pulse investigations of short-latency intracortical facilitation using TMS in humans. *Electroencephalogr. Clin. Neurophysiol. Suppl.* **51,** 113–119.

Shajahan, P. M., Glabus, M. F., Gooding, P. A., Shah, P. J., and Ebmeier, K. P. (1999). Reduced cortical excitability in depression. Impaired post-exercise motor facilitation with transcranial magnetic stimulation. *Br. J. Psychiatry* **174,** 449–454.

Shastri, A., George, M. S., and Bohning, D. E. (1999). Performance of a system for interleaving transcranial magnetic stimulation with steady-state magnetic resonance imaging. *Electroencephalogr. Clin. Neurophysiol. Suppl.* **51,** 55–64.

Siebner, H. R., Peller, M., Willoch, F., Minoshima, S., Boecker, H., Auer, C., Drzezga, A., Conrad, B., and Bartenstein, P. (2000). Lasting cortical activation after repetitive TMS of the motor cortex: A glucose metabolic study. *Neurology* **54,** 956–963.

Siebner, H. R., Takano, B., Peinemann, A., Schwaiger, M., Conrad, B., and Drzezga, A. (2001). Continuous transcranial magnetic stimulation during positron emission tomography: A suitable tool for imaging regional excitability of the human cortex. *NeuroImage* **14,** 883–890.

Siebner, H. R., Willoch, F., Peller, M., Auer, C., Boecker, H., Conrad, B., and Bartenstein, P. (1998). Imaging brain activation induced by long trains of repetitive transcranial magnetic stimulation. *NeuroReport* **9,** 943–948.

Speer, A. M., Kimbrell, T. A., Wassermann, E. M. D., Repella, J., Willis, M. W., Herscovitch, P., and Post, R. M. (2000). Opposite effects of high and low frequency rTMS on regional brain activity in depressed patients. *Biol. Psychiatry* **48,** 1133–1141.

Strafella, A., and Paus, T. (2001). Cerebral blood-flow changes induced by paired-pulse transcranial magnetic stimulation of the primary motor cortex. *J. Neurophysiol.* **85**, 2624–2629.

Strafella, A., Paus, T., Barrett, J., and Dagher, A. (2001). Repetitive transcranial magnetic stimulation of the human prefrontal cortex induces dopamine release in the caudate nucleus. *J. Neurosci.* **21,** RC157 (1–4).

Strother, S. C., Kanno, I., and Rottenberg, D. A. (1995). Principal component analysis, variance partitioning, and "functional connectivity." *J. Cereb. Blood Flow Metab.* **15,** 353–360.

Sun, M. K., Nelson, T. J., and Alkon, D. L. (2000). Functional switching of GABAergic synapses by ryanodine receptor activation. *Proc. Natl. Acad. Sci. USA* **97**, 12300–12305.

Sun, M., Dahl, D., and Alkon, D. L. (2001). Heterosynaptic transformation of GABAergic gating in the hippocampus and effects of carbonic anhydrase inhibition. *J. Pharmacol. Exp. Ther.* **296,** 811–817.

Talairach, J., and Tournoux, P. (1988). "Co-planar Stereotaxic Atlas of the Human Brain." Thieme, New York.

Teneback, C. C., Nahas, Z., Speer, A. M., Molloy, M., Stallings, L. E., Spicer, K. M., Risch, S. C., and George, M. S. (1999). Changes in prefrontal cortex and paralimbic activity in depression following two weeks of daily left prefrontal TMS. *J. Neuropsychiatry Clin. Neurosci.* **11,** 426–435.

Thatcher, R. W., Krause, P. J., and Hrybyk, M. (1986). Cortico-cortical associations and EEG coherence: A two-compartmental model. *EEG Clin. Neurophysiol.* **64,** 123–143.

Thompson, C. J., Paus, T., and Clancy, R. (1998). Magnetic shielding requirements for PET detectors during transcranial magnetic stimulation. *IEEE Trans. Nucl. Sci.* **45,** 1303–1307.

Tucker, D. M., Roth, D. L., and Bair, T. B. (1986). Functional connections among cortical regions: Topography of EEG coherence. *EEG Clin. Neurophysiol.* **63,** 242–250.

Turkington, T. G., Coleman, R. E., Schubert, S. F., and Ganin, A. (1994). Evaluation of post-injection transmission measurement in PET. *IEEE Trans. Nucl. Sci.* **41,** 1538–1544.

Virtanen, J., Rinne, T., Ilmoniemi, R. J., and Naatanen, R. (1996). MEG-compatible multichannel EEG electrode array. *Electroencephalogr. Clin. Neurophysiol.* **99,** 568–570.

Wassermann, E. M. (1998). Risk and safety of repetitive transcranial magnetic stimulation: Report and suggested guidelines from the International Workshop on the Safety of Repetitive Transcranial Magnetic Stimulation, June 5–7, 1996. *Electroencephalogr. Clin. Neurophysiol.* **108,** 1–16.

Wassermann, E. M., and Lisanby, S. H. (2001). Therapeutic application of repetitive transcranial magnetic stimulation: A review. *Clin. Neurophysiol.* **112,** 1367–1377.

Werhahn, K. J., Kunesch, E., Noachtar, S., Benecke, R., and Classen, J. (1999). Differential effects on motor cortical inhibition induced by blockade of GABA uptake in humans. *J. Physiol.* **517,** 591–597.

Wu, T., Sommer, M., Tergau, F., and Paulus, W. (2000). Lasting influence of repetitive transcranial magnetic stimulation on intracortical excitability in human subjects. *Neurosci. Lett.* **287,** 37–40.

Zheng, X. M. (2000). Regional cerebral blood flow changes in drug-resistant depressed patients following treatment with transcranial magnetic stimulation: A statistical parametric mapping analysis. *Psychiatry Res.* **100,** 75–80.

Ziemann, U., Rothwell, J. C., and Ridding, M. C. (1996a). Interaction between intracortical inhibition and facilitation in human motor cortex. *J. Physiol. (London)* **496,** 873–881.

Ziemann, U., Lonnecker, S., Steinhoff, B. J., and Paulus, W. (1996b). Effects of antiepileptic drugs on motor cortex excitability in humans: A transcranial magnetic stimulation study. *Ann. Neurol.* **40,** 367–378.

Zinreich, S. J., Tebo, S., Long, D. M., Brem, H., Mattox, D., Loury, M. E., Vander Volk, C., Kotch, W., Kennedy, D. W., and Bryan, R. N. (1993). Frameless stereotactic integration of CT imaging data. *Radiology* **188,** 735–742.

26

Volume Visualization

Andreas Pommert, Ulf Tiede, and Karl Heinz Höhne

Institute of Mathematics and Computer Science in Medicine (IMDM), University Hospital Hamburg–Eppendorf, 20251 Hamburg, Germany

I. Introduction

Medical imaging technology has experienced a dramatic change over the past three decades. Previously, only X-ray radiographs were available, which showed the depicted organs as superimposed shadows on photographic film. With the advent of modern computers, new *tomographic* imaging modalities like computed tomography (CT), magnetic resonance imaging (MRI), and positron emission tomography (PET), which deliver cross-sectional images of a patient's anatomy and physiology, were developed. These images show different organs free from overlays with unprecedented precision. Even the three-dimensional (3D) structure of organs can be recorded if a sequence of parallel cross sections is taken.

For many clinical tasks like surgical planning, it is necessary to understand and communicate complex and often malformed 3D structures. Experience has shown that the "mental reconstruction" of objects from cross-sectional images is extremely difficult and strongly depends on the observer's training and imagination. For these cases, it is advantageous to present the human body as a surgeon or anatomist would see it.

The aim of *volume visualization* in medicine, also known as *3D imaging,* is to create precise and realistic perspective views of objects from tomographic volume data. The resulting images, even though they are of course two-dimensional, are often called *3D images* or *3D reconstructions,* to distinguish them from 2D cross sections or conventional radiographs. The first attempts date back to the late 1970s, with the first clinical applications reported on the visualization of bone from CT in craniofacial surgery and orthopedics. The first 3D images of the complete human brain from MRI as shown in Fig. 1 were presented about one decade later (Bomans *et al.,* 1987; Höhne *et al.,* 1987).

An overview of the volume visualization pipeline as presented in this chapter is shown in Fig. 2. After the

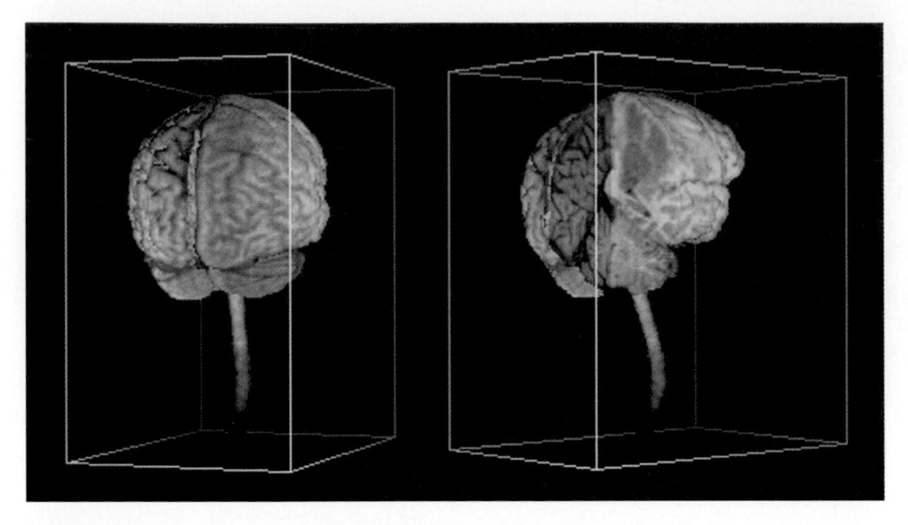

Figure 1 The first ever 3D images of the brain of a living person from MRI (1987). The flattening of the sulci of the right hemisphere (left) was induced by an occipital metastasis and an edema, revealed by a cut (right).

acquisition of one or more series of tomographic images, the data usually undergo some preprocessing such as image filtering, interpolation, and image fusion, if data from several sources are to be used. From this point, one of several paths may be followed.

In *contour extraction,* an object is reconstructed from its contours on adjacent cross sections. This early approach is hardly used nowadays and will not be covered here. All other approaches start from a contiguous image volume. The more traditional surface extraction methods first create an interme-

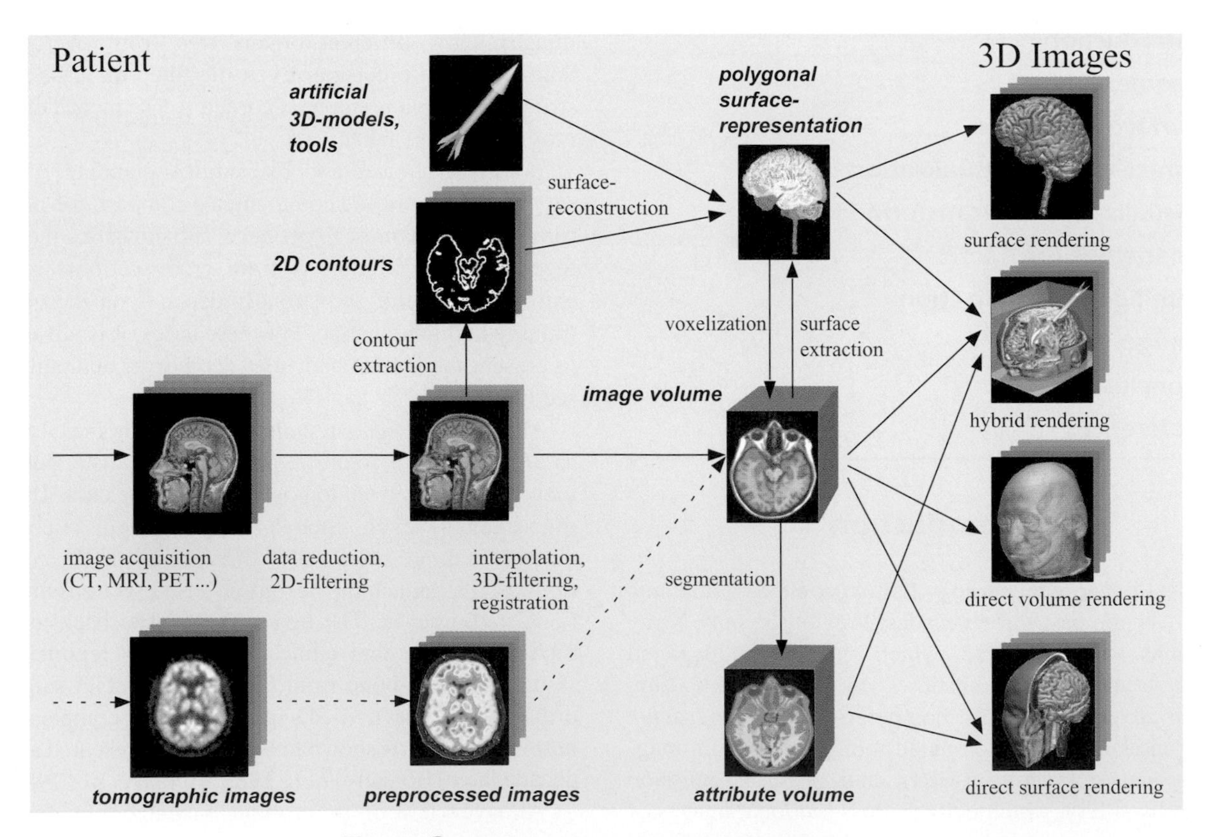

Figure 2 Overview of the volume visualization pipeline.

diate surface representation of the objects to be shown. It may then be rendered with any standard computer graphics utility. More recently, direct volume visualization methods which create 3D views directly from the volume data have been developed. These methods use the full image intensity information to render surfaces, cuts, or transparent and semi-transparent volumes. They may or may not include an explicit segmentation step for the identification and labeling of the objects to be rendered.

Extensions to the volume visualization pipeline not shown in Fig. 2, but also covered here, include the visualization of transformed data and intelligent visualization. Furthermore, aspects of image quality are discussed.

A. Related Fields

Volume visualization has its roots in three other fields of computer science, which are image processing, computer vision, and computer graphics. *Image processing* deals with any image-to-image transformation, such as filters or geometric transformations (Jähne, 1997; Pratt, 1991; Seul *et al.,* 2000). Most steps in volume visualization can therefore be considered as applying special image processing methods.

The aim of *computer vision,* also known as *image understanding,* is to create symbolic descriptions (in terms of names, relationships, etc.) of the contents of an image (Duda *et al.,* 2000; Parker, 1996). In volume visualization, the more low-level functions of image segmentation are used in order to identify different parts of a volume which may be either displayed or removed.

Computer graphics provides methods to synthesize images from numerical descriptions (Foley *et al.,* 1995; Watt, 2000). These techniques were originally developed for the realistic display of human-defined objects, such as technical models from *computer-aided design.* Objects in 3D space are usually represented by small surface patches such as triangles or higher order curves. Contributions of computer graphics to volume visualization include data structures, projection techniques, and illumination models.

B. Preprocessing

The data we consider usually come as a spatial sequence of 2D cross-sectional images. When they are put on top of each other, a contiguous *image volume* is obtained. The resulting data structure is an orthogonal 3D array of volume elements or *voxels,* each representing an intensity value, equivalent to picture elements or *pixels* in 2D. This data structure is called the *voxel model.* In addition to intensity information, each voxel may also contain *labels,* describing its membership to various objects, and/or data from different sources (*generalized voxel model;* Höhne *et al.,* 1990).

Many algorithms for volume visualization work on *isotropic* volumes in which the voxel spacing is equal in all three dimensions. In practice, however, only very few data sets have this property, especially for CT. In these cases, the missing information has to be approximated in an *interpolation* step. Quite a simple method is linear interpolation of the intensities between adjacent images. Higher order functions such as splines usually yield better results for fine details (Marschner and Lobb, 1994; Möller *et al.,* 1997).

With respect to later processing steps, especially intensity-based segmentation, it is often desirable to improve the signal-to-noise ratio of the data, using image or volume filtering. Well-known *noise filters* are average, median, and Gaussian filters. These methods, however, tend to smooth out small details as well. Better results are obtained with *anisotropic diffusion* filters, which largely preserve object boundaries (Gerig *et al.,* 1992a). In MRI, another obstacle may be low-frequency intensity inhomogeneities, which can be corrected to some extent (Arnold *et al.,* 2001).

II. Segmentation

An image volume usually represents a large number of different structures obscuring each other. To display a particular one, we thus have to decide which parts of the volume we want to use or ignore. A first step is to partition the image volume into different regions which are homogeneous with respect to some formal criteria and corresponding to real (anatomical) objects. This process is called *segmentation.* In a subsequent *interpretation* step, the regions may be identified and labeled with meaningful terms such as "white matter" or "ventricle." While segmentation is rather easy for a human expert, it has turned out to be extremely difficult for a computer.

All segmentation methods can be characterized as being either binary or fuzzy, corresponding to the principles of binary and fuzzy logic, respectively (Winston, 1992). In *binary segmentation,* the question whether a voxel belongs to a certain region is always answered yes or no. This information is a prerequisite, e.g., for creating surface representations from volume data. As a drawback, uncertainty or cases in which an object takes up only a fraction of a voxel (*partial volume effect*) cannot be handled properly. Strict yes–no decisions are avoided in *fuzzy segmentation,* in which a set of probabilities is assigned to every voxel, indicating the evidence for different materials. Fuzzy segmentation is closely related to the direct volume rendering methods (see Section IV.E).

Following is a choice of segmentation methods used for volume visualization, ranging from classification and edge detection to recent approaches such as deformable models, atlas registration, and scale-space and interactive segmentation. In practice, these basic approaches are often combined. For further reading, the excellent survey on medical image analysis (Duncan and Ayache, 2000) is recommended.

A. Classification

A straightforward approach to segmentation is to *classify* a voxel depending on its intensity, no matter where it is located. A very simple but nevertheless important example is *thresholding:* a certain intensity range is specified with lower and upper threshold values. A voxel belongs to the selected class if and only if its intensity level is within the specified range. Thresholding is the method of choice for selecting bone or soft tissue in CT. In direct volume visualization, it is often performed during the rendering process itself so that no explicit segmentation step is required.

Instead of a binary decision based on a threshold, Drebin *et al.* use a fuzzy maximum likelihood classifier which estimates the percentages of the different materials represented in a voxel, according to Bayes' rule (Drebin *et al.,* 1988). This method requires that the gray level distributions of different materials are different from each other and known in advance.

Simple classification schemes are not suitable if the structures in question have mostly overlapping or even identical gray-level distributions, such as different soft tissues from CT or MRI. Segmentation becomes easier if multispectral images are available, such as T_1- and T_2-weighted images in MRI, emphasizing fat and water, respectively. In this case, individual threshold values can be specified for every parameter. To generalize this concept, voxels in an n-parameter data set can be considered as n-dimensional vectors in an n-dimensional *feature space*. This feature space is partitioned into subspaces, representing different tissue classes or organs. This is called the *training phase:* in supervised training, the partition is derived from feature vectors which are known to represent particular tissues (Cline *et al.,* 1990; Pommert *et al.,* 2001). In unsupervised training, the partition is generated automatically (Gerig *et al.,* 1992b). In the subsequent *test phase,* a voxel is classified according to the position of its feature vector in the partitioned feature space.

With especially adapted image acquisition procedures, classification methods have successfully been applied to considerable numbers of two- or three-parametric MRI data volumes (Cline *et al.,* 1990; Gerig *et al.,* 1992b). Quite frequently, however, isolated voxels or small regions are classified incorrectly, such as subcutaneous fat in the same class as white matter. To eliminate these errors, a *connected components analysis* may be carried out to determine whether the voxels which have been classified as belonging to the same class are part of the same (connected) region. If not, some of the regions may be discarded.

Classification and connected components analysis may also be combined into a single *region growing* algorithm (Cline *et al.,* 1987). Starting from a user-selected seed voxel, all neighboring voxels are added to the region if they satisfy certain intensity-based criteria. This process is continued recursively from the newly added voxels until no further suitable neighbors can be found. As a major problem, it is often difficult to strike a good balance between *undersegmentation* (several anatomical objects present in a region) and *oversegmentation* (an anatomical object contains several segmented regions) of the data.

Instead of intensity values alone, tissue *textures* may be considered, which are determined using local intensity distributions (Saeed *et al.,* 1997). A survey of intensity-based classification methods is presented in Clarke *et al.* (1995).

B. Edge Detection

Another classic approach to segmentation is the detection of edges, using first or second derivatives of the 3D intensity function. These edges (in 3D, they are actually surfaces; it is, however, common to speak about edges) are assumed to represent the borders between different tissues or organs.

There has been much debate over what operator is most suitable for this purpose. The Canny operator locates the maxima of the first derivative (Canny, 1986). While the edges found with this operator are very accurately placed, all operators using the first derivative share the drawback that the detected contours are usually not closed, i.e., they do not separate different regions properly. An alternative approach is to detect the zero-crossings of the second derivative. With a 3D extension of the Marr–Hildreth operator, the complete human brain as shown in Fig. 1 was segmented and visualized from MRI for the first time (Bomans *et al.,* 1987; Höhne *et al.,* 1987). A free parameter of the Marr–Hildreth operator has to be adjusted to find a good balance between under- and oversegmentation.

C. Deformable Models

All segmentation methods discussed so far share the drawback that there are no constraints with respect to the spatial properties of the segmented objects. One way to represent general knowledge about possible shapes is the use of *deformable models,* based on parameterized curves or surfaces.

As an important class of deformable models, *active contours,* iteratively evolve toward selected image features, thereby minimizing an energy function based on external and internal forces (Blake and Isard, 1998). External forces describe how well the contour is fitted to features detected in the image data, such as high gradient magnitudes. Internal forces describe the tension of the contour itself. In many cases, segmentation with active contours is implemented as a 2D operation, taking a contour from one slice as an initial template for the next, but true 3D algorithms have also been developed (Székely *et al.,* 1996). Instead of minimizing an energy function, an evolving shape can also be described as a propagating wavefront, with the local speed depending on image features or even the evolution of other shapes, using the so-called level set formalism (Teo *et al.,* 1997; Zeng *et al.,* 1999).

Automatic segmentation using deformable models is generally considered superior to other methods discussed so far (Duncan and Ayache, 2000). However, deformable models are often very sensitive to the initial placement of a contour and can handle only more or less smooth shapes. Surveys may be found in Baillard *et al.* (2001) and McInerney and Terzopoulos (1996).

D. Atlas Registration

A more explicit representation of prior knowledge about object shape are is the anatomical atlas. Segmentation is based on the registration of the image volume under consideration with a prelabeled volume that serves as a target atlas. Once the registration parameters are estimated, the inverse transformation is used to map the anatomical labels back onto the image volume, thus achieving the segmentation (see Section VI).

In general, these atlases do not represent one individual, but "normal" anatomy and its variability in terms of a probabilistic spatial distribution, obtained from numerous cases. Methods based on atlas registration were reported suitable for the automatic segmentation of various brain structures, including lesions and objects poorly defined in the image data (Arata *et al.,* 1995; Collins *et al.,* 1999; Kikinis *et al.,* 1996). However, registration may not be very accurate. For improved results, atlas registration may be complemented with intensity-based classification (Collins *et al.,* 1999).

E. Scale-Space Segmentation

Another interesting idea is to investigate object features in *scale space,* i.e., at different levels of image resolution. This approach allows us to ignore irrelevant image detail. One such method developed by Pizer *et al.* considers the symmetry of previously determined shapes, described by medial axes (Pizer *et al.,* 1998b). The resulting ridge function in scale space is called the *core* of an object. It may be used, e.g., for interactive segmentation, in which the user can select, add, or subtract regions, or move to larger "parent" or smaller "child" regions in the hierarchy. Other applications, such as automatic segmentation or registration, are currently being investigated (Pizer *et al.,* 1998a).

F. Interactive Segmentation

Even though there is a great variety of promising approaches for automatic segmentation, "...no one algorithm can robustly segment a variety of relevant structure in medical images over a range of datasets" (Duncan and Ayache, 2000). In particular, the underlying model assumptions may not be flexible enough to handle various pathologies. Therefore, there is currently a strong tendency to combine simple, but fast, operations carried out by the computer with the unsurpassed recognition capabilities of the human observer (Olabarriaga and Smeulders, 2001).

A practical interactive segmentation system was developed by Höhne and Hanson (1992) and later extended to handle multiparametric data (Pommert *et al.,* 2001; Schiemann *et al.,* 1997). Regions are initially defined with thresholds; the user can subsequently apply connected components analysis, volume editing tools, or operators from mathematical morphology. Segmentation results are immediately visualized on orthogonal cross sections and 3D images, in such a way that they may be corrected or further refined in the next step. With this system, segmentation of gross structures is usually a matter of minutes (Fig. 3).

III. Surface Extraction

Surface extraction bridges the gap between volume visualization and more traditional computer graphics (Foley *et al.,* 1995; Watt, 2000). The key idea is to create intermediate surface descriptions of the relevant objects from the volume data. Only this information is then used for rendering images. If triangles are used as surface elements, this process is called *triangulation.*

An apparent advantage of surface extraction is the potentially very high data reduction from volume to surface representations. Resulting computing times can be further reduced if standard data structures such as polygon meshes are used which are supported by standard computer graphics hard- and software.

On the other hand, the extraction step eliminates most of the valuable information on the cross-sectional images. Even simple cuts are meaningless because there is no information about the interior of an object, unless the image volume is also available at rendering time. Furthermore, every change of surface definition criteria such as adjusting a threshold requires a recalculation of the whole data structure.

The classic method for surface extraction is the Marching Cubes algorithm, developed by Lorensen and Cline (1987). It creates an *isosurface,* approximating the location of a certain intensity value in the data volume. This algorithm basically considers cubes of $2 \times 2 \times 2$ contiguous voxels. Depending on whether one or more of these voxels are inside the object (i.e., above a threshold value), a surface representation of up to four triangles is placed within the cube. The exact location of the triangles is found by linear interpolation of the intensities at the voxel vertices. The result is a highly detailed surface representation with subvoxel resolution (Fig. 4).

Various modifications of the Marching Cubes algorithm have been developed. These include the correction of topological inconsistencies (Natarajan, 1994) and improved

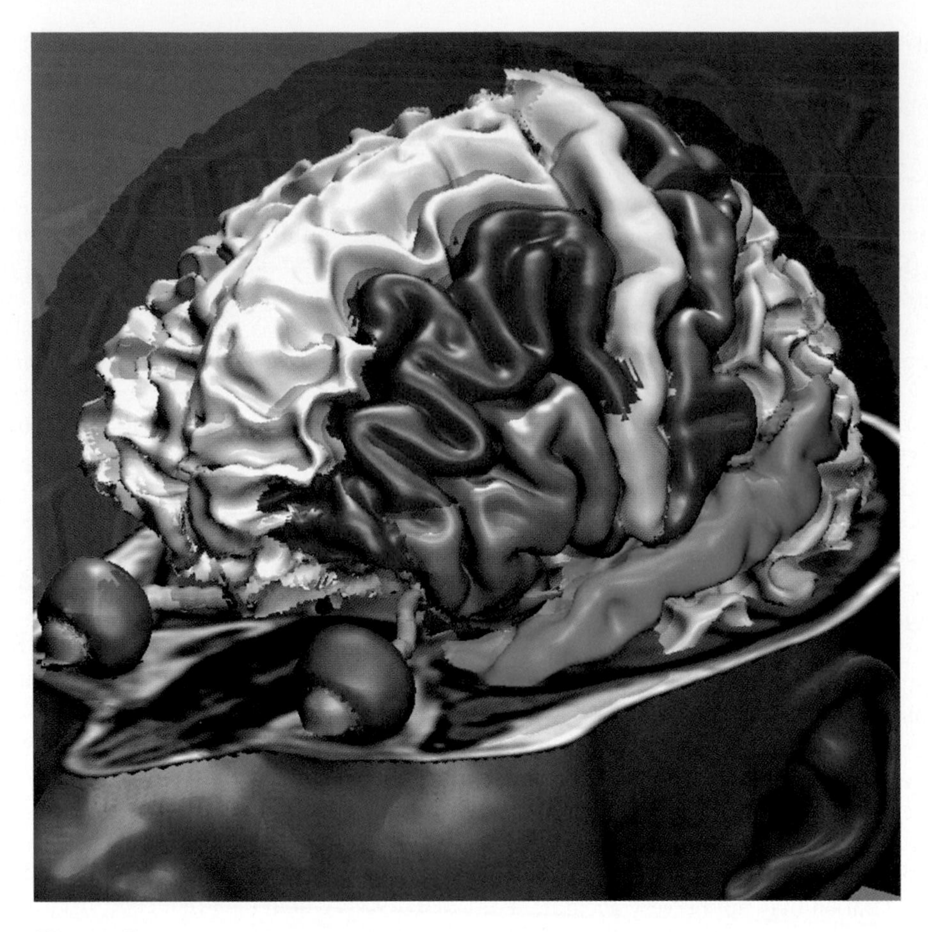

Figure 3 Results of an interactive segmentation of a brain from MRI. While white matter and gray matter were separated using a threshold, the gyri shown here in different colors were segmented using volume editing tools.

accuracy by better approximating the true isosurface in the volume data, using higher order curves (Hamann *et al.,* 1997) or an adaptive refinement (Cignoni *et al.,* 2000).

As a major practical problem, the Marching Cubes algorithm typically creates hundreds of thousands of triangles when applied to clinical data. As has been shown, these numbers can be reduced considerably by a subsequent simplification of the triangle meshes (Cignoni *et al.,* 1998; Schroeder *et al.,* 1992; Wilmer *et al.,* 1992).

IV. Direct Volume Visualization

In *direct volume visualization,* images are created directly from the volume data. Compared to surface extraction, the major advantage is that all image information which has originally been acquired is kept during the rendering process. As shown by Höhne *et al.* (1990), this makes it an ideal technique for interactive data exploration. Threshold values and other parameters which are not clear from the beginning can be changed interactively. Furthermore, direct volume visualization allows a combined display of different aspects such as opaque and semitransparent surfaces, cuts, and maximum intensity projections. A current drawback of direct volume visualization is that the large amount of data which has to be handled requires a high-performance computer for realtime applications (Parker *et al.,* 1999). However, special volume graphics acceleration boards for PCs with limited functionality are already available (Pfister *et al.,* 1999).

A. Scanning the Volume

In direct volume visualization, we basically have the choice between two scanning strategies: pixel by pixel (image order) or voxel by voxel (volume order). These strategies correspond to the image and object order rasterization algorithms used in computer graphics (Foley *et al.,* 1995).

In *image order* scanning, the data volume is sampled on rays along the viewing direction. This method is commonly known as *ray casting:*

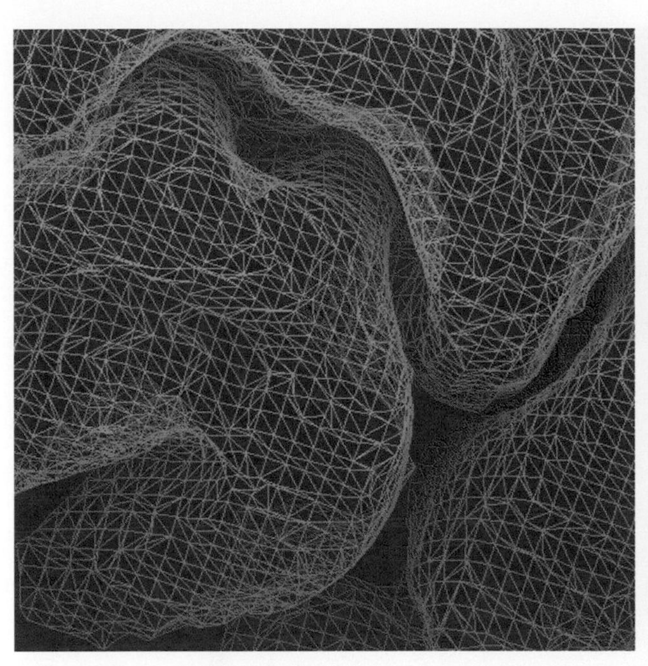
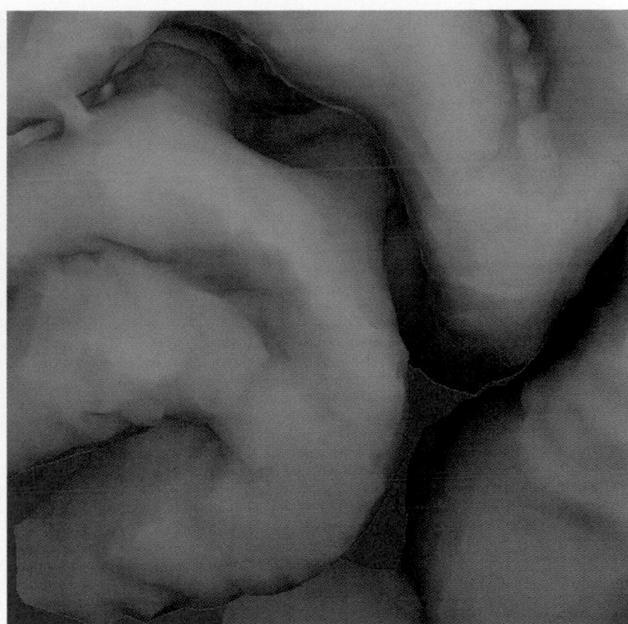

Figure 4 Triangulated (left) and shaded (right) gyri from MRI, created with the Marching Cubes algorithm.

```
FOR each pixel on image plane DO
    FOR each sampling point on associated viewing ray DO
        compute contribution to pixel
```

The principle is illustrated in Fig. 5. At the sampling points, the intensity values are interpolated from the neighboring voxels, using trilinear interpolation or higher order curves (Marschner and Lobb, 1994; Möller *et al.,* 1997). Along the ray, visibility of surfaces and objects is easily determined. The ray can stop when it meets an opaque surface. Yagel *et al.* extended this approach to a full *ray tracing* system which follows the viewing rays as they are reflected on various surfaces (Yagel *et al.,* 1992). Multiple light reflections between specular objects can thus be handled.

Image order scanning can be used to render both voxel and polygon data at the same time, known as *hybrid rendering* (Levoy, 1990). Image quality can be adjusted by choosing smaller (oversampling) or wider (undersampling) sampling intervals. Unless stated otherwise, all 3D images shown in this chapter were rendered with a ray casting algorithm.

As a drawback, the whole input volume must be available for random access to allow arbitrary viewing directions. Furthermore, interpolation of the intensities at the sampling points requires a high computing power. A strategy to reduce computation times is based on the observation that most of the time is spent traversing empty space, far away from the objects to be shown. If the rays are limited to scanning the data only within a predefined *bounding volume* around these objects, scanning times are greatly reduced (Šrámek and Kaufman, 2000; Tiede, 1999; Wan *et al.,* 1999).

In *volume order* scanning, the input volume is sampled along the lines and columns of the 3D array, projecting a chosen aspect onto the image plane in the direction of view:

```
FOR each sampling point in volume DO
    FOR each pixel projected onto DO
        compute contribution to pixel
```

The volume can either be traversed in back-to-front order from the voxel with maximal distance to the voxel with minimal distance to the image plane or vice versa in front-to-back order. Scanning the input data as they are stored, these techniques are reasonably fast even on computers with small main memories. However, implementation of display algorithms is usually much more straightforward using the ray casting approach.

B. Direct Surface Rendering

Using one of the scanning techniques described, the visible surface of an object can be rendered directly from the volume data. This approach is called *direct surface rendering*. To determine the surface position, a threshold or an object membership label may be used, or both may be combined to obtain a highly accurate isosurface (Pommert *et al.,* 2001; Tiede *et al.,* 1998; Tiede, 1999). Examples are shown in Figs. 3 and 6. Note that the position of the observer may also be inside the object, thus creating a *virtual endoscopy*.

For realistic display of the surface, one of the *illumination models* developed in computer graphics may be used.

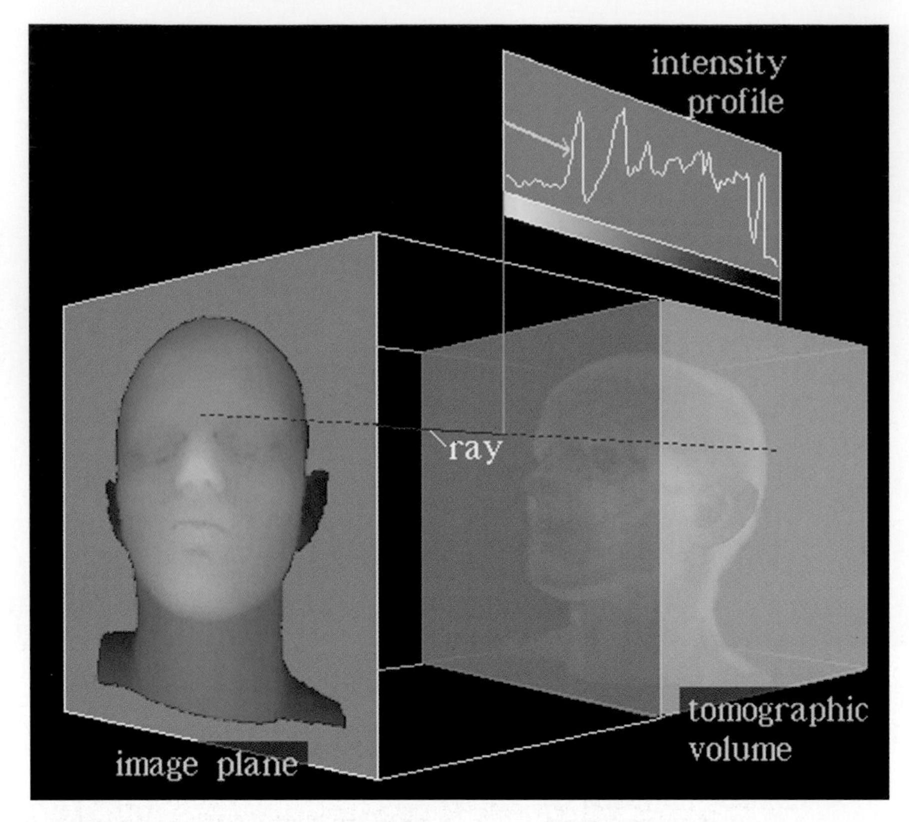

Figure 5 Principle of ray casting. In this case, an intensity threshold is used to find the object surface along each viewing ray.

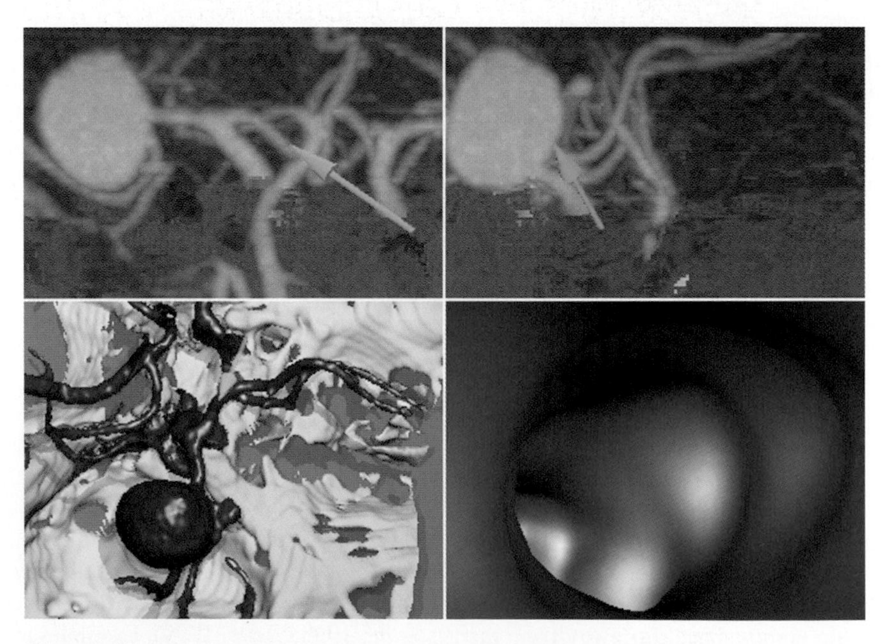

Figure 6 Virtual catheter examination of an aneurysm of the right middle cerebral artery from CT angiography, illustrating various rendering techniques. **Top left:** Maximum intensity projection (MIP) in a coronal view. **Top right:** MIP, sagittal view. **Bottom left:** Overview of the cranial base from CT in conjunction with blood vessels. **Bottom right:** Endoscopic view of a blood vessel. The current position and view direction of the camera are indicated by arrows.

These models, such as the *Phong* shading model, take into account both the position and the type of simulated light sources, as well as the reflection properties of the surface (Foley *et al.,* 1995; Watt, 2000). A key input into these models is the local surface inclination, described by a *normal vector* perpendicular to the surface.

As shown by Höhne and Bernstein, a very accurate estimate of the local surface normal vectors can be obtained from the image volume (Höhne and Bernstein, 1986). Due to the partial volume effect, the intensities in the 3D neighborhood of a surface voxel represent the relative proportions of different materials inside these voxels. The surface inclination is thus described by the local *gray-level gradient,* i.e., a 3D vector of the partial derivatives. A number of methods to calculate the gray-level gradient are presented and discussed in Marschner and Lobb (1994), Tiede *et al.* (1990), and Tiede (1999).

C. Cut Planes

Once a surface view is available, a very simple and effective method of visualizing interior structures is cutting (Fig. 1). When the original intensity values are mapped onto the cut plane, they can be better understood in their anatomical context (Höhne *et al.,* 1990). A special case is selective cutting, in which certain objects are left untouched (Fig. 7).

D. Integral and Maximum Intensity Projection

A different way to look into an object is to integrate the intensity values along the viewing ray. If applied to the whole data volume, this is a step back to the old X-ray projection technique. If applied in a selective way, this *integral projection* is nevertheless helpful in certain cases (Höhne *et al.,* 1990; Tiede *et al.,* 1990).

For small bright objects such as vessels from CT or MR angiography, *maximum intensity projection* (MIP) is a suitable display technique (Fig. 6). Along each ray through the data volume, the maximum gray level is determined and projected onto the image plane. The advantage of this method is that neither segmentation nor shading is needed, which may fail for very small vessels. But there are also

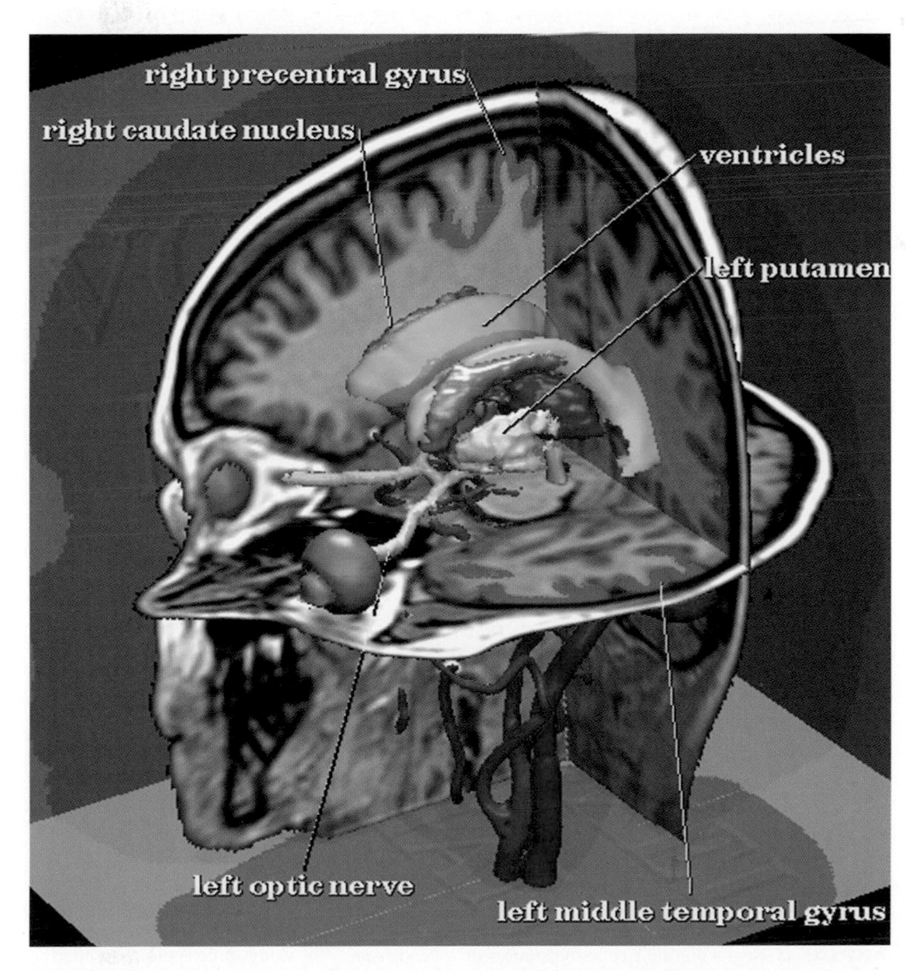

Figure 7 Direct volume visualization allows a convenient exploration of interior structures by cutting. The original MR intensity values have been mapped onto the cut planes.

some drawbacks: as light reflection is totally ignored, maximum intensity projection does not give a realistic 3D impression. Spatial perception can be improved by real-time rotation (Mroz *et al.*, 2000) or by a combined presentation with other surfaces or cut planes (Höhne *et al.*, 1990).

E. Direct Volume Rendering

Direct volume rendering is the visualization equivalent of fuzzy segmentation (see Section II). For medical applications, these methods were first described by Drebin *et al.* (1988) and Levoy (1988). A commonly assumed underlying model is that of a colored, semitransparent gel with suspended reflective particles. Illumination rays are partly reflected and change color while traveling through the volume.

Each voxel is assigned a color and an opacity. This opacity is the product of an object-weighting function and a gradient-weighting function. The object-weighting function is usually dependent on the intensity, but it can also be the result of a more sophisticated fuzzy segmentation algorithm. The gradient-weighting function emphasizes surfaces for 3D display. All voxels are shaded, using the gray-level gradient method (see Section IV.B). The shaded values along a viewing ray are weighted and summed up. A somewhat simplified recursive equation which models frontal illumination with a ray casting system is given as follows:

Variable		**Range**
I	Intensity of reflected light	
p	Index of sampling point on ray	[0, … , maximum depth of scene]
L	Incoming light	
α	Local opacity	[0.0, … , 1.0]
s	Local shading component	[0.0, … , 1.0]

$$I(p,L) = \alpha(p) \cdot L \cdot s(p) + (1.0 - \alpha(p)) \cdot I(p + 1,(1.0 - \alpha(p)) \cdot \mathrm{L})$$

Since binary decisions are avoided in volume rendering, the resulting images are very smooth and show a lot of fine details (Fig. 8). Another advantage is that even coarsely

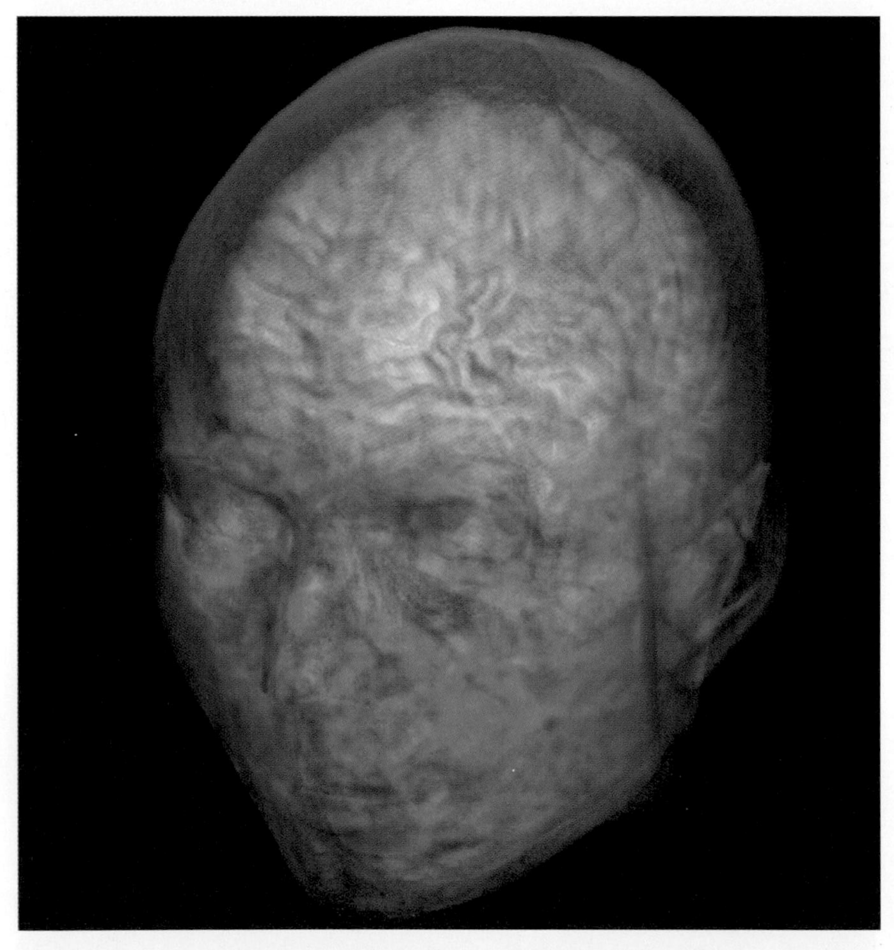

Figure 8 Segmentation-free visualization of a head from MRI using direct volume rendering. Since many objects have similar gray values, it is very difficult to emphasize particular surfaces such as that of the brain.

defined objects can be rendered (Tiede *et al.,* 1990). On the other hand, the more or less transparent images produced with volume rendering are often hard to understand so that their value is often questionable (Fig. 8). To some extent, spatial perception can be improved by rotating the object.

Another problem is the very large number of parameters which have to be specified to define the weighting functions. Attempts to generate the weighting functions automatically from the image volume have been published only recently (Kindlmann and Durkin, 1998). Furthermore, volume rendering is comparatively slow because weighting and shading operations are performed for many voxels on each ray. Various techniques to speed up image generation have been developed, including *shear–warp,* a combination of methods for faster access to volume data used for ray casting (Lacroute and Levoy, 1994); *splatting,* a volume order scanning algorithm for volume rendering (Westover, 1990); and a method to utilize the 3D texture mapping buffer of graphics workstations (Cabral *et al.,* 1994). A comparison of various volume-rendering methods is presented in Meißner *et al.* (2000). Current research issues also include methods for more accurate numerical approximations of the volume-rendering equations, thus reducing some artifacts (de Boer *et al.,* 1997; Williams *et al.,* 1998).

V. Visualization of Transformed Data

While both surface extraction and direct volume visualization operate in a 3D space, 3D images may be created from other data representations as well. One such method is *frequency domain volume rendering,* which creates 3D images in Fourier space, based on the projection-slice theorem (Totsuka and Levoy, 1993). This method is very fast, but the resulting images are limited to rather simple integral projections (see Section IV.D).

A more general approach is *wavelet* transformation (Muraki, 1995). These transformations provide a multiscale representation of 3D objects, with the size of represented detail locally adjustable. Wavelets are especially suitable for the visualization of very large data sets, such as the Visible Human (Ihm and Park, 1999). Amounts of data and rendering times may thus be dramatically reduced.

VI. Image Fusion

For many applications, it is desirable to combine or *fuse* information from different imaging modalities. For example, functional imaging techniques such as magnetoencephalography (see Chapter 10), functional MRI (see Chapter 13), or PET (see Chapter 18) show various physiological aspects, but give little or no indication for the localization of the observed phenomena. For their interpretation, a closely matched description of the patient's morphology is required, as obtained in MRI.

In general, image volumes obtained from different sources do not match geometrically. It is therefore required to transform one volume with respect to the other. This process is known as *image registration.* The transformation may be defined using corresponding *landmarks* in both data sets. In a simple case, artificial markers attached to the patient, which are visible on different modalities, are available (Bromm and Scharein, 1996). Otherwise, pairs of preferably stable matching points such as the AC-PC line may be used. A more robust approach is to interactively match larger features such as surfaces (Schiemann *et al.,* 1994). Figure 9 shows the result of the registration of a PET and an MRI data set. All these techniques may also be applied in scale space at different levels of resolution (Pizer *et al.,* 1998a).

In a fundamentally different approach, the results of a registration step are evaluated at every point of the

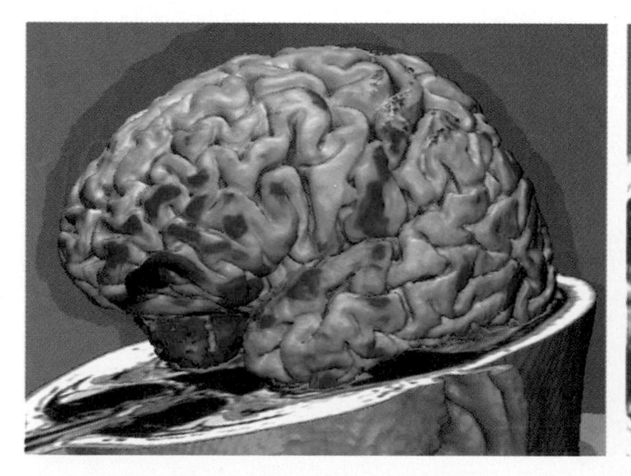

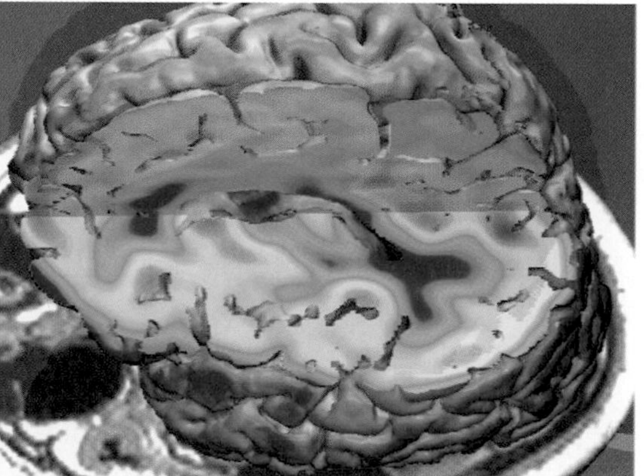

Figure 9 Fusion of different imaging modalities for therapy control in a clinical study of obsessive compulsive disorder. MRI showing morphology is combined with a PET scan, showing glucose metabolism. Since the entire volume is mapped, the activity can be explored at any location of the brain.

combined volume, based on intensity values (Studholme *et al.,* 1996; Wells *et al.,* 1996). Starting from a coarse match, registration is achieved by adjusting position and orientation until the *mutual information* ("similarity") between both data sets is maximized. These methods are fully automatic, do not rely on a possibly erroneous definition of landmarks, and seem to be more accurate than others (West *et al.,* 1997). A more detailed presentation of image fusion may be found in Chapter 24 of this volume.

VII. Intelligent Visualization

While in classical medicine, knowledge about the human body is represented in books and atlases, present-day computer science allows for new, more powerful and versatile computer-based representations of knowledge. The most straightforward examples are multimedia CD-ROMs containing collections of classical pictures and text, which may be browsed arbitrarily. Although computerized, such media still follow the old paradigm of text printed on pages, accompanied by pictures.

Using methods of volume visualization, spatial knowledge about the human body may be much more efficiently represented by computerized three-dimensional models. If such models are connected to a knowledge base of descriptive information, they can even be interrogated or disassembled by addressing names of organs (Brinkley *et al.,* 1999; Golland *et al.,* 1999; Höhne *et al.,* 1996; Pommert *et al.,* 2001).

A suitable data structure for this purpose is the *intelligent volume* (Höhne *et al.,* 1995), which combines a detailed spatial model enabling realistic visualization with a symbolic description of human anatomy (Fig. 10). The spatial model is represented as a 3D volume as described above. The membership of voxels to an object is indicated by labels which are stored in *attribute volumes* congruent to the image volume. Different attribute volumes may be generated, e.g., for structure or function. Further attribute volumes may be added which contain, e.g., the incidence of a tumor type or a time tag for blood propagation on a per-voxel basis.

The objects themselves bear attributes as well. These may be divided into two groups: first, attributes indicating meaning such as names, pointers to text or pictorial explanations, or even features like vulnerability or mechanical properties, which might be important, e.g., for surgical simulation, and second, attributes defining their visual appearance, such as color, texture, and reflectivity. In addition, the model describes the interrelations of the objects with a semantic network. Examples of relations are *part of* or *supplied by*.

Once an intelligent volume is established, it can be explored by freely navigating in both the pictorial and the descriptive world. A viewer can compose arbitrary views from the semantic description or query semantic information for any visible voxel of a picture. An atlas of the human brain which contains spatial and semantic descriptions of basic structure, function, and blood supply is shown in Fig. 11. Apart from educational purposes, such atlases are also a powerful aid for the interpretation of clinical images (Kikinis *et al.,* 1996; Nowinski and Thirunavuukarasuu, 2001; Schiemann *et al.,* 1994; Schmahmann *et al.,* 1999).

Because of the high computational needs, three-dimensional anatomical atlases are not yet suitable for

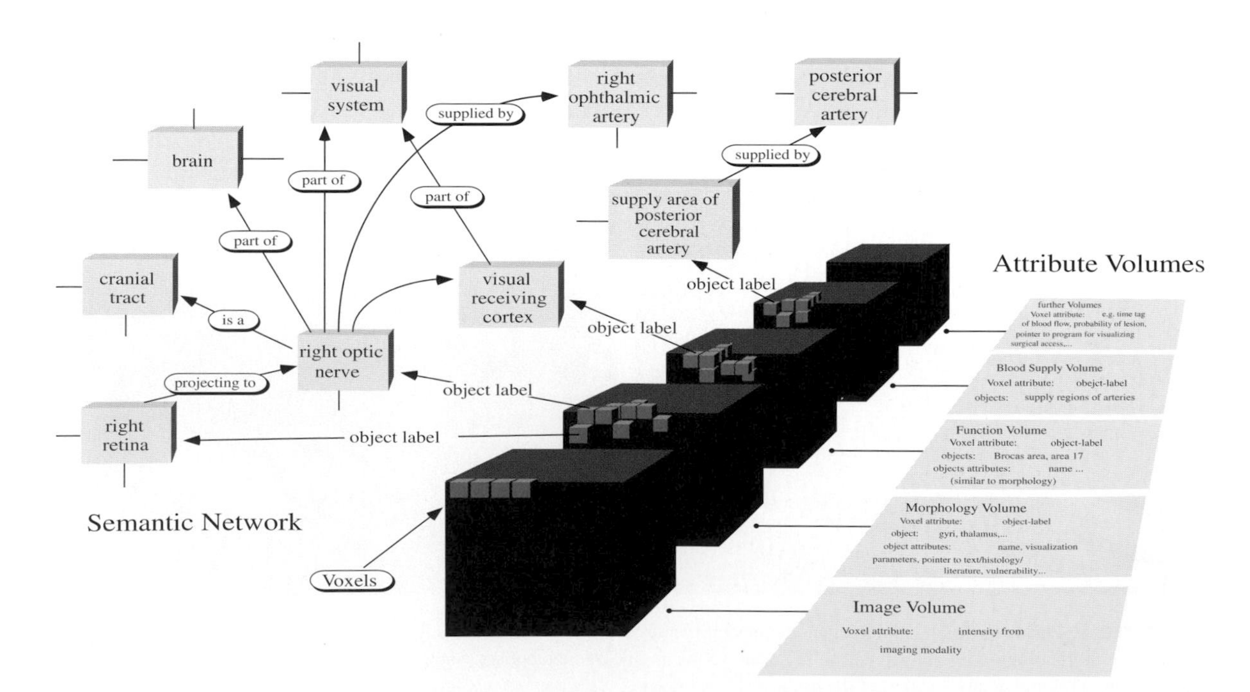

Figure 10 Basic structure of the intelligent volume, integrating spatial and symbolic descriptions of anatomy.

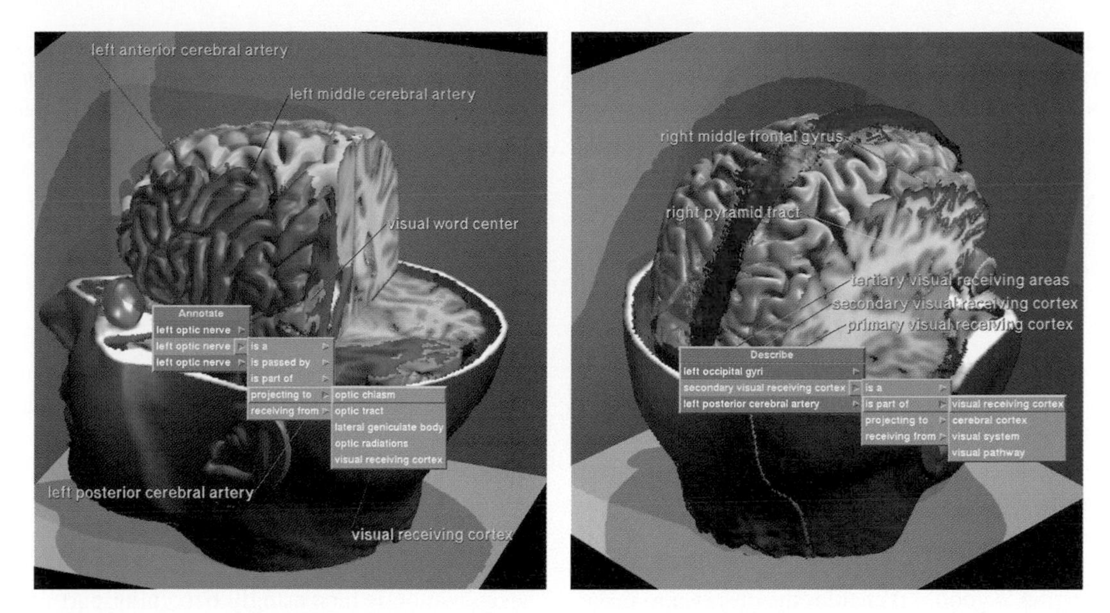

Figure 11 Exploration of the brain by mouse click, using a three-dimensional anatomical atlas. **Left:** Information available for the optic nerve appears as a cascade of pop-up menus. **Right:** The pop-up menu indicates regional, functional, and blood supply areas to which a voxel belongs.

present-day PCs. Schubert *et al.* make such a model available for interactive exploration via precomputed *Intelligent QuickTime VR Movies* (Schubert *et al.*, 1999), which may be viewed on any personal computer. A three-dimensional atlas of regional, functional, and radiological anatomy of the brain based on this technique has been published recently (Höhne *et al.*, 2001). A screenshot is shown in Fig. 12.

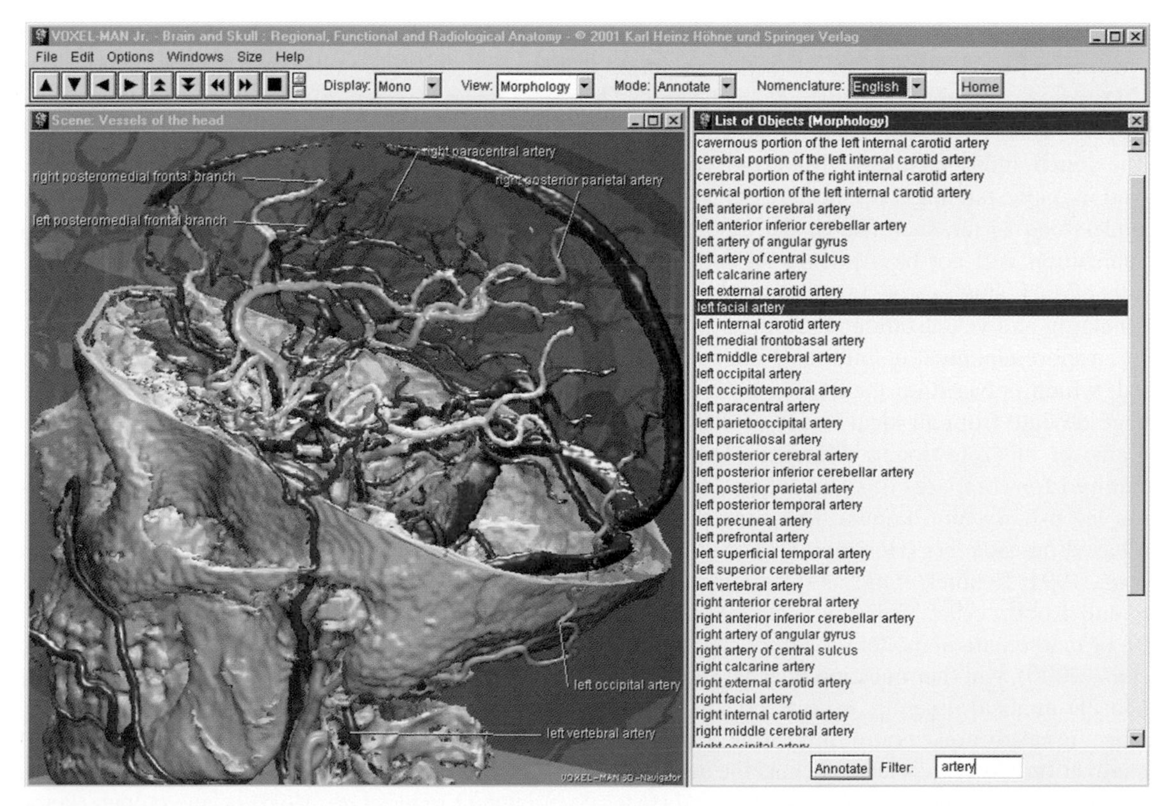

Figure 12 User interface of VOXEL-MAN 3D-Navigator: Brain and Skull, a PC-based three-dimensional atlas of regional, functional, and radiological anatomy. The user may navigate freely in both the pictorial (**left**) and the descriptive (**right**) context.

To date, most three-dimensional anatomical atlases are based on the data derived from one individual only. The interindividual variability of organ shape and topology in space and time is thus not yet part of the model. Methods for measuring and modeling variability are currently being developed (Mazziotta *et al.,* 1995; Styner and Gerig, 2001; Thompson *et al.,* 2000).

VIII. Image Quality

For clinical applications, it is of course important to ensure that the 3D images really show the true anatomical situation or at least to know about their limitations. But when is an image good or bad? A straightforward definition of image quality is given as follows: An image is good if it suits the observer's needs. For example, an image used in diagnostic imaging is good if it enables the observer to make the right diagnosis. Image quality can thus be measured in terms of sensitivity, specificity, diagnostic accuracy, or ROC index (Alder *et al.,* 1995; Fox *et al.,* 1995; Pilgram *et al.,* 1994; Vannier *et al.,* 1994).

However, this definition has some problems from a practical point of view. First, results strongly depend on factors outside the image, such as the observer's experience and the task. They may thus be different if these factors are changed. Second, no measures are at hand for application areas other than diagnostics, such as therapy planning or even brain research. Third, observer studies are extremely costly. Therefore, for certain tasks such as the detection of small signals in 2D images in nuclear medicine, mathematical model observers were developed (Barrett *et al.,* 1993). However, the much more complex visual and cognitive tasks involved in understanding a perspective 3D image are only little understood so far, such that model observers for volume visualization will not be available for some time. Fourth, results of such studies give few or no clues on how to improve imaging and visualization procedures.

Therefore, a more technical definition of image quality is often used, which is based on the question: How much does an image deviate from an ideal image of that scene? In order to answer this question, a reference ("gold standard") is required for comparison, since the true anatomical situation is usually not known for clinical cases. Studies are based on cadavers (Hemmy and Tessier, 1985; Pommert *et al.,* 1991; Rusinek *et al.,* 1991), simulated data (Marschner and Lobb, 1994; Collins *et al.,* 1998; Tiede *et al.,* 1990), or mathematical models (Bentum *et al.,* 1996; Pommert *et al.,* 2000). Another question is, what should be measured? In 2D medical imaging, aspects such as image resolution or signal-to-noise ratio are often used. In volume visualization, other measures such as the accuracy of surface position (Pommert *et al.,* 2000) or surface normal vectors (Bentum *et al.,* 1996; Marschner and Lobb, 1994; Tiede *et al.,* 1990) seem to be more appropriate, since these are more critical for the visual impression. This way, typical ranges of error could be estimated. However, an integrated description of all steps and parameters of the volume visualization pipeline with respect to the resulting image quality is not yet available.

IX. Conclusions

Medical volume visualization has come a long way from the first experiments to the current, highly detailed renderings. As the rendering algorithms are improved and the quality of the resulting images is investigated, 3D images are not just pretty pictures, but a powerful source of information for research, education, and patient care. In certain areas such as craniofacial surgery or traumatology, volume visualization is increasingly becoming part of the standard preoperative procedures. As we have shown, it is almost certain that the techniques described will revolutionize large parts of medical education. Software packages for volume visualization such as the Visualization Toolkit VTK (Visualization Toolkit, 2001) or the VolPack rendering library (VolPack Volume Rendering Library, 1995) are available over the Internet.

A number of problems still impair an even broader use of volume visualization in medicine. First, and most important, the segmentation problem is still unresolved. It is no coincidence that volume visualization is most accepted in areas where clinicians are interested in bone from CT, which can be segmented using a simple threshold. Especially for MRI, however, automatic segmentation methods are still far from being generally applicable, while interactive procedures are too time consuming. As has been shown, there is research in different directions going on; in many cases, methods have already proven valuable for specific applications.

The second major problem is the design of a user interface which is suitable in a clinical environment. Currently, there are still a large number of rather technical parameters for controlling segmentation, registration, shading, and so on. Acceptance in the medical community will certainly depend heavily on progress in this field.

Third, most current workstations are not yet able to deliver 3D images fast enough. For the future, it is certainly desirable to interact with the workstation in real time. However, with computing power further increasing, this problem will be overcome in a few years.

Currently, a number of applications based on volume visualization are becoming operational, such as surgical simulation systems and three-dimensional atlases. Another intriguing idea is to combine volume visualization with *virtual reality* systems, which enable the clinician to fly through (Krüger *et al.,* 1995) or even touch (Petersik *et al.,* 2002) a virtual patient. In *augmented reality,* images from the real and virtual worlds are superimposed to guide the surgeon during an intervention (Colchester *et al.,* 1996). Integration of volume visualization with virtual reality and robotics toward *computer-assisted surgery* will certainly be a major topic in the coming decade.

Acknowledgments

All images presented in this chapter were created at the IMDM, with contributions from Bernhard Pflesser, Martin Riemer, Thomas Schiemann, Rainer Schubert, Frank Wilmer, and our much missed late colleague Markus Urban. Research in interventional neuroradiology (Fig. 6) is in cooperation with Christoph Koch and Hermann Zeumer (Department of Neuroradiology). The study on obsessive compulsive disorder (Fig. 9) was done in cooperation with Ralph Buchert, Malte Clausen (Department of Nuclear Medicine), and Jost Obrocki (Department of Psychiatry). The raw image volumes used for Figs. 1, 3, 7, 8, and 11 are courtesy of Siemens Medical Systems, Erlangen, Germany. Our thanks also go to Bernhard Pflesser for his comments, Renate Reche for her help in preparing the manuscript, and Kristine Pommert (BBC London) for checking our English.

References

Alder, M. E., Deahl, S. T., and Matteson, S. R. (1995). Clinical usefulness of two-dimensional reformatted and three-dimensionally rendered computerized images: Literature review and a survey of surgeons' opinions. *J. Oral Maxillofac. Surg.* **53,** 375–386.

Arata, L. K., Dhawan, A. P., Broderick, J. P., Gaskil-Shipley, M. F., Levy, A. V., and Volkow, N. D. (1995). Three-dimensional anatomical model-based segmentation of MR brain images through principal axes registration. *IEEE Trans. Biomed. Eng.* **42,** 1069–1078.

Arnold, J. B., Liow, J.-S., Schaper, K. A., Stern, J. J., Sled, J. G., Shattuck, D. W., Worth, A. J., Cohen, M. S., Leahy, R. M., Mazziotta, J. C., and Rottenberg, D. A. (2001). Qualitative and quantitative evaluation of six algorithms for correcting intensity nonuniformity effects. *NeuroImage* **13,** 931–943.

Baillard, C., Hellier, P., and Barillot, C. (2001). Segmentation of brain 3D MR images using level sets and dense registration. *Med. Image Anal.* **5,** 185–194.

Barrett, H. H., Yao, J., Rolland, J. P., and Myers, K. J. (1993). Model observers for assessment of image quality. *Proc. Natl. Acad. Sci. USA* **90,** 9758–9765.

Bentum, M. J., Malzbender, T., and Lichtenbelt, B. B. (1996). Frequency analysis of gradient estimators in volume rendering. *IEEE Trans. Visualizat. Comput. Graph.* **2,** 242–254.

Blake, A., and Isard, M. (1998). "Active Contours." Springer-Verlag, London.

Bomans, M., Riemer, M., Tiede, U., and Höhne, K. H. (1987). 3D-Segmentation von Kernspin-Tomogrammen. *In* "Informatik-Fachberichte," Vol. 149, "Mustererkennung 1987, Proceedings of the 9th DAGM Symposium" (E. Paulus, ed.), pp. 231–235. Springer-Verlag, Berlin.

Brinkley, J. F., Wong, B. A., Hinshaw, K. P., and Rosse, C. (1999). Design of an anatomy information system. *IEEE Comput. Graph. Appl.* **19,** 38–48.

Bromm, B., and Scharein, E. (1996). Visualisation of pain by magnetoencephalography in humans. *In* "Lecture Notes in Computer Science," Vol. 1131, "Visualization in Biomedical Computing: Proceedings of VBC '96" (K. H. Höhne and R. Kikinis, eds.), pp. 477–481. Springer-Verlag, Berlin.

Cabral, B., Cam, N., and Foran, J. (1994). Accelerated volume rendering and tomographic reconstruction using texture mapping hardware. *In* "Proceedings of the 1994 Symposium on Volume Visualization" (A. Kaufman and W. Krueger, eds.), pp. 91–98. ACM Press, New York.

Canny, J. (1986). A computational approach to edge detection. *IEEE Trans. Pattern Anal. Mach. Intell.* **8,** 679–698.

Cignoni, P., Montani, C., and Scopigno, R. (1998). A comparison of mesh simplification algorithms. *Comput. Graph.* **22,** 37–54.

Cignoni, P., Ganovelli, F., Montani, C., and Scopigno, R. (2000). Reconstruction of topologically correct and adaptive trilinear isosurfaces. *Comput. Graph.* **24,** 399–418.

Clarke, L. P., Velthuizen, R. P., Camacho, M. A., Heine, J. J., Vaidyanathan, M., Hall, L. O., Thatcher, R. W., and Silbiger, M. L. (1995). MRI segmentation: Methods and applications. *Magn. Reson. Imaging* **13,** 343–368.

Cline, H. E., Dumoulin, C. L., Hart, H. R., Lorensen, W. E., and Ludke, S. (1987). 3D reconstruction of the brain from magnetic resonance images using a connectivity algorithm. *Magn. Reson. Imaging* **5,** 345–352.

Cline, H. E., Lorensen, W. E., Kikinis, R., and Jolesz, F. (1990). Three-dimensional segmentation of MR images of the head using probability and connectivity. *J. Comput. Assisted Tomogr.* **14,** 1037–1045.

Colchester, A. C. F., Zhao, J., Holton-Tainter, K. S., Henri, C. J., Maitland, N., Roberts, P. T. E., Harris, C. G., and Evans, R. J. (1996). Development and preliminary evaluation of VISLAN, a surgical planning and guidance system using intra-operative video imaging. *Med. Image Anal.* **1,** 73–90.

Collins, D. L., Zijdenbos, A. P., Kollokian, V., Sled, J. G., Kabani, N. J., Holmes, C. J., and Evans, A. C. (1998). Design and construction of a realistic digital brain phantom. *IEEE Trans. Med. Imaging* **17,** 463–468.

Collins, D. L., Zijdenbos, A. P., Barré, W. F. C., and Evans, A. C. (1999). ANIMAL+INSECT: Improved cortical structure segmentation. *In* "Lecture Notes in Computer Science," Vol. 1613, "Information Processing in Medical Imaging: Proceedings of IPMI '99" (A. Kuba, M. Samal, and A. Todd-Pokropek, eds.), pp. 210–223. Springer-Verlag, Berlin.

de Boer, M. W., Gröpl, A., Hesser, J., and Männer, R. (1997). Reducing artifacts in volume rendering by higher order integration. *In* "Proceedings of IEEE Visualization '97." IEEE Comput. Soc., Los Alamitos, CA.

Drebin, R. A., Carpenter, L., and Hanrahan, P. (1988). Volume rendering. *Comput. Graph.* **22,** 65–74.

Duda, R. O., Hart, P. E., and Stork, D. G. (2000). "Pattern Classification," 2nd edition. Wiley, New York.

Duncan, J. S., and Ayache, N. (2000). Medical image analysis: Progress over two decades and the challenges ahead. *IEEE Trans. Pattern Anal. Mach. Intell.* **22,** 85–105.

Foley, J. D., van Dam, A., Feiner, S. K., and Hughes, J. F. (1995). "Computer Graphics: Principles and Practice," 2nd edition. Addison–Wesley, Reading, MA.

Fox, L. A., Vannier, M. W., West, O. C., Wilson, A. J., Baran, G. A., and Pilgram, T. K. (1995). Diagnostic performance of CT, MPR, and 3DCT imaging in maxillofacial trauma. *Comput. Med. Imaging Graph.* **19,** 385–395.

Gerig, G., Kübler, O., Kikinis, R., and Jolesz, F. A. (1992a). Nonlinear anisotropic filtering of MRI data. *IEEE Trans. Med. Imaging* **11,** 221–232.

Gerig, G., Martin, J., Kikinis, R., Kübler, O., Shenton, M., and Jolesz, F. A. (1992b). Unsupervised tissue type segmentation of 3D dual-echo MR head data. *Image Vision Comput.* **10,** 349–360.

Golland, P., Kikinis, R., Halle, M., Umans, C., Grimson, W. E. L., Shenton, M. E., and Richolt, J. A. (1999). Anatomy Browser: A novel approach to visualization and integration of medical information. *Comput. Aided Surg.* **4,** 129–143.

Hamann, B., Trotts, I., and Farin, G. (1997). On approximating contours of the piecewise trilinear interpolant using triangular rational-quadratic Bézier patches. *IEEE Trans. Visualizat. Comput. Graph.* **3**, 215–227.

Hemmy, D. C., and Tessier, P. L. (1985). CT of dry skulls with craniofacial deformities: Accuracy of three-dimensional reconstruction. *Radiology* **157,** 113–116.

Höhne, K. H., and Bernstein, R. (1986). Shading 3D-images from CT using gray level gradients. *IEEE Trans. Med. Imaging* **MI-5,** 45–47.

Höhne, K. H., and Hanson, W. A. (1992). Interactive 3D-segmentation of MRI and CT volumes using morphological operations. *J. Comput. Assisted Tomogr.* **16,** 285–294.

Höhne, K. H., Tiede, U., Riemer, M., Bomans, M., Heller, M., and Witte, G. (1987). Static and dynamic three-dimensional display of tissue structures from volume scans. *Radiology* **165,** 420.

Höhne, K. H., Bomans, M., Pommert, A., Riemer, M., Schiers, C., Tiede, U., and Wiebecke, G. (1990). 3D-visualization of tomographic volume data using the generalized voxel-model. *Visual Comput.* **6,** 28–36.

Höhne, K. H., Pflesser, B., Pommert, A., Riemer, M., Schiemann, T., Schubert, R., and Tiede, U. (1995). A new representation of knowledge concerning human anatomy and function. *Nat. Med.* **1,** 506–511.

Höhne, K. H., Pflesser, B., Pommert, A., Riemer, M., Schiemann, T., Schubert, R., and Tiede, U. (1996). A virtual body model for surgical education and rehearsal. *IEEE Comput.* **29,** 25–31.

Höhne, K. H., Petersik, A., Pflesser, B., Pommert, A., Priesmeyer, K., Riemer, M., Schiemann, T., Schubert, R., Tiede, U., Urban, M., Frederking, H., Lowndes, M., and Morris, J. (2001). "VOXEL-MAN 3D Navigator: Brain and Skull. Regional, Functional and Radiological Anatomy." Springer-Verlag Electronic Media, Heidelberg. [2 CD-ROMs, ISBN 3-540-14910-4]

Ihm, I., and Park, S. (1999). Wavelet-based 3D compression scheme for interactive visualization of very large volume data. *Comput. Graph. Forum* **18,** 3–15.

Jähne, B. (1997). "Digital Image Processing: Concepts, Algorithms, and Scientific Applications," 4th edition. Springer-Verlag, Berlin.

Kikinis, R., Shenton, M. E., Iosifescu, D. V., McCarley, R. W., Saiviroonporn, P., Hokama, H. H., Robatino, A., Metcalf, D., Wible, C. G., Portas, C. M., Donnino, R. M., and Jolesz, F. A. (1996). A digital brain atlas for surgical planning, model driven segmentation, and teaching. *IEEE Trans. Visualizat. Comput. Graph.* **2,** 232–241.

Kindlmann, G., and Durkin, J. W. (1998). Semi-automatic generation of transfer functions for direct volume rendering. *In* "1998 ACM/IEEE Symposium on Volume Visualization" (W. E. Lorensen and R. Yagel, eds.), pp. 79–86. Assoc. Comput. Mach., New York.

Krüger, W., Bohn, C.-A., Fröhlich, B., Schüth, H., Strauss, W., and Wesche, G. (1995). The responsive workbench: A virtual work environment. *IEEE Comput.* **28,** 42–48.

Lacroute, P., and Levoy, M. (1994). Fast volume rendering using a shear-warp factorization of the viewing transformation. *In* "Proceedings of SIGGRAPH '94," pp. 451–458, Orlando, FL.

Levoy, M. (1988). Display of surfaces from volume data. *IEEE Comput. Graph. Appl.* **8,** 29–37.

Levoy, M. (1990). A hybrid ray tracer for rendering polygon and volume data. *IEEE Comput. Graph. Appl.* **10,** 33–40.

Lorensen, W. E., and Cline, H. E. (1987). Marching Cubes: A high resolution 3D surface construction algorithm. *Comput. Graph.* **21,** 163–169.

Marschner, S. R., and Lobb, R. J. (1994). An evaluation of reconstruction filters for volume rendering. *In* "Proceedings of IEEE Visualization '94" (R. D. Bergeron and A. E. Kaufman, eds.), pp. 100–107. IEEE Comput. Soc., Los Alamitos, CA.

Mazziotta, J. C., Toga, A. W., Evans, A. C., Fox, P., and Lancaster, J. (1995). A probabilistic atlas of the human brain: Theory and rationale for its development. *NeuroImage* **2,** 89–101.

McInerney, T., and Terzopoulos, D. (1996). Deformable models in medical image analysis: A survey. *Med. Image Anal.* **1,** 91–108.

Meißner, M., Huang, J., Bartz, D., Mueller, K., and Crawfis, R. (2000). A practical evaluation of popular volume rendering algorithms. *In* "Proceedings of the 2000 IEEE Symposium on Volume Visualization and Graphics," pp. 81–90, Salt Lake City, UT.

Möller, T., Machiraju, R., Mueller, K., and Yagel, R. (1997). Evaluation and design of filters using a Taylor series expansion. *IEEE Trans. Visualizat. Comput. Graph.* **3,** 184–199.

Mroz, L., Hauser, H., and Gröller, E. (2000). Interactive high-quality maximum intensity projection. *Comput. Graph. Forum* **19,** 341–350.

Muraki, S. (1995). Multiscale volume representation by a DOG wavelet. *IEEE Trans. Visualizat. Comput. Graph.* **1,** 109–116.

Natarajan, B. K. (1994). On generating topologically consistent isosurfaces from uniform samples. *Visual Comput.* **11,** 52–62.

Nowinski, W. L., and Thirunavuukarasuu, A. (2001). Atlas-assisted localization analysis of functional images. *Med. Image Anal.* **5,** 207–220.

Olabarriaga, S. D., and Smeulders, A. W. M. (2001). Interaction in the segmentation of medical images: A survey. *Med. Image Anal.* **5,** 127–142.

Parker, J. R. (1996). "Algorithms for Image Processing and Computer Vision." Wiley, New York.

Parker, S., Parker, M., Livnat, Y., Sloan, P.-P., Hansen, C., and Shirley, P. (1999). Interactive ray tracing for volume visualization. *IEEE Trans. Visualizat. Comput. Graph.* **5,** 238–250.

Petersik, A., Pflesser, B., Tiede, U., Höhne K. H., and Leuwer, R. (2002). Haptic volume interaction with anatomic models at sub-voxel resolution. *In* "10th International Symposium on Haptic Interfaces for Virtual Environment and Teleoperator Systems, Proc. Haptics 2002," pp. 66–72, Orlando, FL.

Pfister, H., Hardenbergh, J., Knittel, J., Lauer, H., and Seiler, L. (1999). The VolumePro real-time ray-casting system. *In* :Proceedings of SIGGRAPH 1999," pp. 251–260, Los Angeles, CA.

Pilgram, T. K., Vannier, M. W., Marsh, J. L., Kraemer, B. B., Rayne, S. C., Gado, M. H., Moran, C. J., McAlister, W. H., Shackelford, G. D., and Hardesty, R. A. (1994). Binary nature and radiographic identifiability of craniosynostosis. *Invest. Radiol.* **29,** 890–896.

Pizer, S. M., Eberly, D., Fritsch, D. S., and Morse, B. S. (1998a). Segmentation, registration, and measurement of shape variation via image object shape. *IEEE Trans. Med. Imaging* **18,** 851–865.

Pizer, S. M., Eberly, D., Fritsch, D. S., and Morse, B. S. (1998b). Zoom-invariant vision of figural shape: The mathematics of cores. *Comput. Vision Image Understanding* **69,** 55–71.

Pommert, A., Höltje, W.-J., Holzknecht, N., Tiede, U., and Höhne, K. H. (1991). Accuracy of images and measurements in 3D bone imaging. *In* "Computer Assisted Radiology, Proceedings of CAR '91" (H. U. Lemke, M. L. Rhodes, C. C. Jaffe, and R. Felix, eds.), pp. 209–215. Springer-Verlag, Berlin.

Pommert, A., Tiede, U., and Höhne, K. H. (2000). Accuracy of isosurfaces in volume visualization. *In* "Vision, Modeling, and Visualization, Proceedings of VMV 2000" (B. Girod, G. Greiner, H. Niemann, and H.-P. Seidel, eds.), pp. 365–371. IOS Press, Amsterdam.

Pommert, A., Höhne, K. H., Pflesser, B., Richter, E., Riemer, M., Schiemann, T., Schubert, R., Schumacher, U., and Tiede, U. (2001). Creating a high-resolution spatial/symbolic model of the inner organs based on the Visible Human. *Med. Image Anal.* **5,** 221–228.

Pratt, W. K. (1991). "Digital Image Processing," 2nd edition. Wiley, New York.

Rusinek, H., Noz, M. E., Maguire, G. Q., Kalvin, A., Haddad, B., Dean, D., and Cutting, C. (1991). Quantitative and qualitative comparison of volumetric and surface rendering techniques. *IEEE Trans. Nucl. Sci.* **38,** 659–662.

Saeed, N., Hajnal, J. V., and Oatridge, A. (1997). Automated brain segmentation from single slice, multislice, or whole-volume MR scans using prior knowledge. *J. Comput. Assist. Tomogr.* **21,** 192–201.

Schiemann, T., Höhne, K. H., Koch, C., Pommert, A., Riemer, M., Schubert, R., and Tiede, U. (1994). Interpretation of tomographic images using automatic atlas lookup. *In* "Visualization in Biomedical Computing 1994, Proceedings of SPIE 2359" (R. A. Robb, ed.), pp. 457–465, Rochester, MN.

Schiemann, T., Tiede, U., and Höhne, K. H. (1997). Segmentation of the Visible Human for high quality volume based visualization. *Med. Image Anal.* **1,** 263–271.

Schmahmann, J. D., Doyon, J., McDonald, D., Holmes, C., Lavoie, K., Hurwitz, A. S., Kabani, N., Toga, A., Evans, A., and Petrides, M. (1999).

Three-dimensional MRI atlas of the human cerebellum in proportional stereotaxic space. *NeuroImage* **10,** 233–260.

Schroeder, W. J., Zarge, J. A., and Lorensen, W. E. (1992). Decimation of triangle meshes. *Comput. Graph.* **26,** 65–70.

Schubert, R., Pflesser, B., Pommert, A., Priesmeyer, K., Riemer, M., Schiemann, T., Tiede, U., Steiner, P., and Höhne, K. H. (1999). Interactive volume visualization using "intelligent movies." *In* "Health Technology and Informatics," Vol. 62, "Medicine Meets Virtual Reality, Proceedings of MMVR '99" (J. D. Westwood, H. M. Hoffman, R. A. Robb, and D. Stredney, eds.), pp. 321–327. IOS Press, Amsterdam.

Seul, M., O'Gorman, L., and Sammon, M. J. (2000). "Practical Algorithms for Image Analysis: Description, Examples, and Code." Cambridge Univ. Press, Cambridge, UK.

Šrámek, M., and Kaufman, A. (2000). Fast ray-tracing of rectilinear volume data using distance transforms. *IEEE Trans. Visualizat. Comput. Graph.* **6,** 236–251.

Studholme, C., Hill, D. L. G., and Hawkes, D. J. (1996). Automated 3-D registration of MR and CT images of the head. *Med. Image Anal.* **1,** 163–175.

Styner, M., and Gerig, G. (2001). Medial models incorporating object variability for 3D shape analysis. *In* "Lecture Notes in Computer Science," Vol. 2082, "Information Processing in Medical Imaging, Proceedings of IPMI 2001" (M. F. Insana and R. M. Leahy, eds.), pp. 502–516. Springer-Verlag, Berlin.

Székely, G., Kelemen, A., Brechbühler, C., and Gerig, G. (1996). Segmentation of 2-D and 3-D objects from MRI volume data using constrained elastic deformations of flexible Fourier contour and surface models. *Med. Image Anal.* **1,** 19–34.

Teo, P. C., Sapiro, G., and Wandell, B. (1997). Creating connected representations of cortical gray matter for functional MRI visualization. *IEEE Trans. Med. Imaging* **16,** 852–863.

Thompson, P. M., Woods, R. P., Mega, M. S., and Toga, A. W. (2000). Mathematical/computational challenges in creating deformable and probabilistic atlases of the human brain. *Hum. Brain Mapping* **9,** 81–92.

Tiede, U. (1999). "Realistische 3D-Visualisierung Multiattributierter und Multiparametrischer Volumendaten." Fachbereich Informatik, Universität Hamburg. [Ph.D. thesis]

Tiede, U., Höhne, K. H., Bomans, M., Pommert, A., Riemer, M., and Wiebecke, G. (1990). Investigation of medical 3D-rendering algorithms. *IEEE Comput. Graph. Appl.* **10,** 41–53.

Tiede, U., Schiemann, T., and Höhne, K. H. (1998). High quality rendering of attributed volume data. *In* "Proceedings of IEEE Visualization '98" (D. Ebert, H. Hagen, and H. Rushmeier, eds.), pp. 255–262. IEEE Comput. Soc., Los Alamitos, CA.

Totsuka, T., and Levoy, M. (1993). Frequency domain volume rendering. *Comput. Graph.* **27,** 271–278.

Vannier, M. W., Pilgram, T. K., Marsh, J. L., Kraemer, B. B., Rayne, S. C., Gado, M. H., Moran, C. J., McAlister, W. H., Shackelford, G. D., and Hardesty, R. A. (1994). Craniosynostosis: Diagnostic imaging with three-dimensional CT presentation. *Am. J. Neuroradiol.* **15,** 1861–1869.

Visualization Toolkit (2001). The Visualization Toolkit. http://public.kitware.com/.

VolPack Volume Rendering Library (1995). The VolPack Volume Rendering Library. Stanford University, Palo Alto, CA. http://graphics.stanford.edu/software/volpack/.

Wan, M., Kaufman, A.,, and Bryson, S. (1999). High performance presence-accelerated ray casting. *In* "Proceedings of IEEE Visualization '99," pp. 379–386, San Francisco.

Watt, A. (2000). "3D Computer Graphics," 3rd edition. Addison–Wesley, Reading, MA.

Wells, W. M., III, Viola, P., Atsumi, H., Nakajima, S., and Kikinis, R. (1996). Multi-modal volume registration by maximization of mutual information. *Med. Image Anal.* **1,** 35–51.

West, J., Fitzpatrick, J. M., Wang, M. Y., Dawant, B. M., Maurer, C. R., Kessler, R. M., Maciunas, R. J., Barillot, C., Lemoine, D., Collignon, A., Maes, F., Suetens, P., Vandermeulen, D., van den Elsen, P. A., Napel, S., Sumanaweera, T., Harkness, B., Hemler, P. F., Hill, D. L. G., Hawkes, D. J., Studholme, C., Maintz, J. B. A., Viergever, M. A., Malandain, G., Pennec, X., Noz, M. E., Maguire, G. Q., Pollack, M., Pelizzari, C. A., Robb, R. A., Hanson, D., and Woods, R. P. (1997). Comparison and evaluation of retrospective intermodality brain image registration techniques. *J. Comput. Assisted Tomogr.* **21,** 554–566.

Westover, L. (1990). Footprint evaluation for volume rendering. *Comput. Graph.* **24,** 367–376.

Williams, P. L., Max, N. L., and Stein, C. M. (1998). A high accuracy volume renderer for unstructured data. *IEEE Trans. Visualizat. Comput. Graph.* **4,** 37–54.

Wilmer, F., Tiede, U., and Höhne, K. H. (1992). Reduktion der Oberflächenbeschreibung triangulierter Oberflächen durch Anpassung an die Objektform. *In* "Mustererkennung 1992, Proceedings of the 14th DAGM-Symposium" (S. Fuchs and R. Hoffmann, eds.), pp. 430–436. Springer-Verlag, Berlin.

Winston, P. H. (1992). "Artificial Intelligence," 3rd edition. Addison–Wesley, Reading, MA.

Yagel, R., Cohen, D., and Kaufman, A. (1992). Discrete ray tracing. *IEEE Comput. Graph. Appl.* **12,** 19–28.

Zeng, X., Staib, L. H., Schultz, R. T., and Duncan, J. S. (1999). Segmentation and measurement of the cortex from 3D MR images using coupled surfaces propagation. *IEEE Trans. Med. Imaging* **18,** 148–157.

VI

Databases and Atlases

27

The International Consortium for Brain Mapping: A Probabilistic Atlas and Reference System for the Human Brain

John Mazziotta for the International Consortium for Brain Mapping

Ahmanson–Lovelace Brain Mapping Center, Department of Neurology Radiological Sciences, and Medical and Molecular Pharmacology, UCLA School of Medicine, Los Angeles, California 90024

I. Introduction

Classic atlases of the human brain or other species have been derived from a single brain, or brains from a very small number of subjects, and have employed simple scaling factors to stretch or constrict a given subject's brain to match the atlas. The result is a rigid and often inflexible system that disregards useful information about morphometric (i.e., dimensionality) and densitometric (i.e., intensity) variability among subjects.

This chapter reviews the rationale for and development of a probabilistic atlas and reference system of the human brain derived from a large series of subjects, representative of the entire species, with retention of information about variability. The atlas includes structural as well as functional information. Such a project must take on the problems inherent in dealing with a variable biological structure and function but, when successful, provides a system that is realistic in its complexity, that has defined accuracy and errors, and that, as a benefit, contributes new neurobiological information. Such a strategy will also spawn atlases of other species as well as human atlases of pathologic conditions such as Alzheimer's disease, autism, schizophrenia, multiple sclerosis, and many others (see Chapter 28). These disease-specific atlases can then be used to demonstrate the natural history of disease progression and will find utility in clinical trials in which experimental therapies can be examined for their impact on disease progression using an automated, objective, and quantifiable reference atlas of the natural history of the disease state.

II. Motivation for Developing a Probabilistic Human Brain Atlas

The relationship between structure and function in the human brain, at either a macro- or a microscopic level, is complex and poorly understood. We are not proposing to unravel this complexity with the data collected in the context of building this atlas. Rather, we will continue to develop a probabilistic framework in which appropriate data sets can be entered, across an ever-increasing number of modalities, between subjects, laboratories, and experiments such that, in time, the aggregate data from populations will provide even greater (in both quality and quantity) insights into this important relationship. Our perspective on brain function is typically equated with the methods available to measure it. For the tomographic brain imaging techniques, the results produce macroscopic estimates of *where* gross functional changes (typically of a hemodynamic nature) are occurring. Electromagnetic techniques can provide direct information about *when* these events occur and indirect information about where. The development of a probabilistic reference system and atlas for the human brain simply provides the framework in which to place these ever-accumulating data sets in a fashion that allows them to be related to one another and that begins to provide insights into the relationship between micro- and macroscopic structure and function.

A. Growth of Neuroscience and Lost Opportunities

The growth of neuroscience in the past 25 years has been extraordinary. Annually, over 20,000 individuals attend the meeting of the Society for Neuroscience in the United States. At that meeting, over 40,000 papers have been presented in the past 3 years. Brain mapping and neuroimaging have witnessed a similarly exponential rise in interest, output, and productivity, although on a smaller scale. Throughout the neuroscience community, there is a general frustration with the volume of data that is generated and its relative inaccessibility in forms other than narrative text. Consider, for example, that over 13,000 Society for Neuroscience abstracts are published in hard copy and electronically each year. Faced with such a staggering volume of information, the individual neuroscientist typically retreats to his or her small scientific niche, resulting in ever-increasing specialization and isolation within the field.

At the same time, funding for neuroscience research has a limited return on its investment in that only a small fraction of raw data that are collected through such funds is analyzed fully, far less is interpreted and published. Even when published, their narrative format requires arduous comparisons across experiments, methods, and species.

If there were a system that provided a logical and organized means by which to maintain data from meetings, individual experiments, or the field as a whole, referenced to the anatomy of the brain, the species, and the stage in development or duration of a pathologic process, highly automated, content-based queries would vastly improve access, allow immediate comparisons among experiments and laboratories, provide a manageable format to assess new data at meetings or through periodic publications, provide for electronic experiments and hypothesis generation using the data of others to test theories, and greatly increase the value of every dollar spent on neuroscience research. Such an outcome requires sophisticated neuroinformatics tools, dedicated scientists committed to the successful completion of such a project in a practical fashion, and a paradigm shift in the sociology of neuroscientists with regard to information sharing (Koslow, 2000). Nevertheless, the benefits of such an approach are enormous on their own and of even greater value if one extrapolates the current situation to even higher numbers of neuroscientists and data sets in the future.

B. Data Suffocation

As the quality of neuroscientific data improves, so too does its magnitude. As spatial resolution in imaging data changes by one order of magnitude in one dimension, the volume of data points increases by a factor of 1000. *In vivo* imaging instruments are now routinely capable of producing 1-mm^3 resolution elements, whereas microscopic and ultrastructural studies achieve spatial resolutions 1000–100,000 times better. If one considers that 50–75,000 genes code for proteins of relevance to the human nervous system at some point during the life span, the impact of assaying and storing such information across a range of spatial resolutions is staggering. Current genomic technology, and future advances in it, makes feasible the ability to generate vast amounts of genetic information. All of these data are in search of an organizational home referenced to the location of the sample in neuroanatomical terms and the time frame of the sample as a function of the age of the organism. Once again, the brain's architecture becomes the most appropriate and intuitively sensitive structure in which to organize such data so as to optimize correlations between biologically related data sets. While the remainder of this article focuses on imaging data of the human brain, it is important to note the magnitude of the information management problem for neuroscience as a whole.

C. Data Integration

To demonstrate the practical uses of the probabilistic reference system, an example is taken from actual experience, namely, an experiment performed by Watson and colleagues (1993) to identify the visual motion area of the human brain (i.e., V5 or MT) (Ungerleider and Desimone, 1986) using relative cerebral blood flow (CBF) (Mazziotta *et al.,* 1985; Fox and Mintun, 1989) measured with PET (Fig. 1). In the experiment, each subject had multiple PET-CBF studies in two states, the first viewing a stationary pattern of targets and the second, with the targets moving. The significant difference between the data sets collected in these two states (Friston *et al.,* 1991) was then superimposed on MRI data using the AIR alignment and registration algorithm (Woods *et al.,* 1992, 1993). The result of this experiment demonstrated consistent bilateral activations of the dorsolateral, inferior occipital cortex in each subject. Further, a consistent relationship between the site of increased blood flow and the frequently observed ascending limb of the inferior temporal sulcus (Ono *et al.,* 1990) was found (Fig. 1A). Because the investigators were knowledgeable about temporo-occipital anatomy and physiology, they recognized that this location had also been identified by Flechsig (1920) as a portion of the human cerebral cortex (Flechsig Feld 16) that is myelinated at birth (Bailey and von Bonin, 1951) (Fig. 1B). These observations have been repeatedly confirmed by independent laboratories demonstrating the human V5 (Zeki *et al.,* 1991) area as a frequently detectable and robust functional landmark at the temporo-occipital junction (Dumoulin *et al.,* 2000).

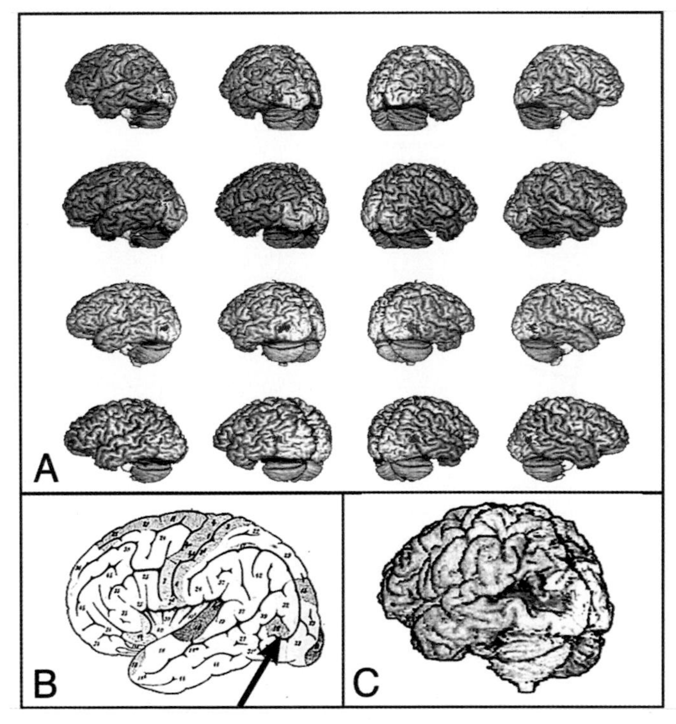

Figure 1 Illustrative example of human visual area V5. (**A**) Four separate subjects demonstrating bilateral CBF-PET activation of V5 superimposed on their respective structural MRI studies (Watson *et al.,* 1993). Note the consistent relationship seen between the activated site (red) and the ascending limb of the inferior temporal sulcus that also coincides with the cortical region [seen in (**B**), arrow) identified by Flechsig (1920) as being myelinated at birth. (**C**) Patient studied by Zihl *et al.* (1983, 1991) with damage to the V5 area resulting in a selective disturbance of visual motion perception.

Now envision this experiment performed using neuroinformatics tools previously developed or proposed for the probabilistic reference system. Prior to performing the V5 PET experiment, each subject would perform a Functional Reference Battery of tasks (see below in Section III F), thereby providing functional landmarks throughout the brain. Following the experiment, anatomical warping and segmentation tools would be used to automatically segment and label the anatomical regions of the brain for each subject. Alignment and registration by functional landmarks would show the effects of functional registration on macroscopic anatomy. Functional alignment and registration using the V5 activation sites would automatically demonstrate the consistent relationship between that functional region and the ascending limb of the inferior temporal gyrus. Further, it would quantitate, in probabilistic terms, the spatial relationships between the sulcal/gyral anatomy and the functionally activated zone across subjects. Differences in responses could be related to demographic, clinical, and genotypic (Bartley *et al.,* 1997; Zilles *et al.,* 1997) information, if these were collected as part of the experiment, and related to population data already available in the four-dimensional database. Cyto- and chemoarchitectural data, as they begin to populate the database, would be available for automated reference with regard to this cortical zone (Clark and Miklossy, 1990; Rademacher *et al.,* 1993). Time-series data from EEG or MEG would show the temporal relationships of this region to others (Dale *et al.,* 1999; Ahlfors *et al.,* 1999). Lesion data could also be accessed if such data sets had been added as an attribute (Zihl *et al.,* 1991) (Fig. 1C). This is in contrast to the current situation in which activated cortical regions are identified and one must laboriously search the literature to try and identify, qualitatively in experiments with different characteristics, qualities, and attributes, the regions of the brain that are of experimental interest for a given neuroscientific question.

III. Strategy and Rationale

A. Overall Concept

The goal of the International Consortium for Brain Mapping (ICBM) is to develop a voxel-based, probabilistic atlas of the human brain from a large sample of normal individuals, ages 18 to 90, with a wide ethnic and racial distribution. The data set is designed to contain a substantial amount of demographic information describing the subjects' background, family history, habits, diet, and many other features. In addition, clinical and behavioral evaluations include neurological examinations, psychiatric screening, handedness scores, and neuropsychological tasks. One cubic millimeter multispectral MRI studies including T1-, T2-, and proton density-weighted pulse sequences are obtained consistently. A subset of subjects also have functional imaging using a standardized battery of tasks and employing functional MRI, positron emission tomography, and event-related potentials. DNA samples will be acquired from 5800 of the subjects and made available for genotyping.

From an organizational point of view, eight laboratories in seven countries on three continents participate in the core data collection and analysis. These sites were selected because of their expertise in brain imaging, their capacity to perform a large number of studies in a consistent fashion, and the fact that most sites had different imaging devices and computer platforms, thereby requiring the consortium to solve problems of interoperability and data differences from different acquisition devices.

It was decided early in the planning for the program that in situations in which the optimal solution to a given problem (e.g., data analysis pathway, visualization scheme) was not known, each laboratory would independently try to solve these problems. Once a laboratory-specific solution was obtained, appropriate algorithms would be distributed to consortium participants and evaluated. Ultimately, these algorithms were sent to outside laboratories for independent evaluation and comparison with methods developed by non-consortium groups. In each case, the optimal strategy was then incorporated into the final approach used by the consortium. This was a "real-world" situation designed to produce the optimal result through competition. As each successful component of these competitions emerged, it was incorporated into the overall ICBM strategy for data analysis, visualization, and distribution. Thus, while each laboratory developed an independent strategy for processing data, the consortium as a whole made the commitment to a unified, centralized strategy for the pooled results, thereby resulting in a single atlas rather than a federation of atlases. The latter would result in inconsistencies in data analysis and confounding factors for users of the atlas in the long run.

The principles, practices, and tools developed through the ICBM have also spawned a series of other atlas projects on different populations (Fig. 2). Probabilistic atlases for children (i.e., birth to age 18 years) and disease states (e.g., Alzheimer's disease, traumatic brain injury, multiple sclerosis, autism, schizophrenia, stuttering, cerebral infarction) are under development. These population- and disease-specific

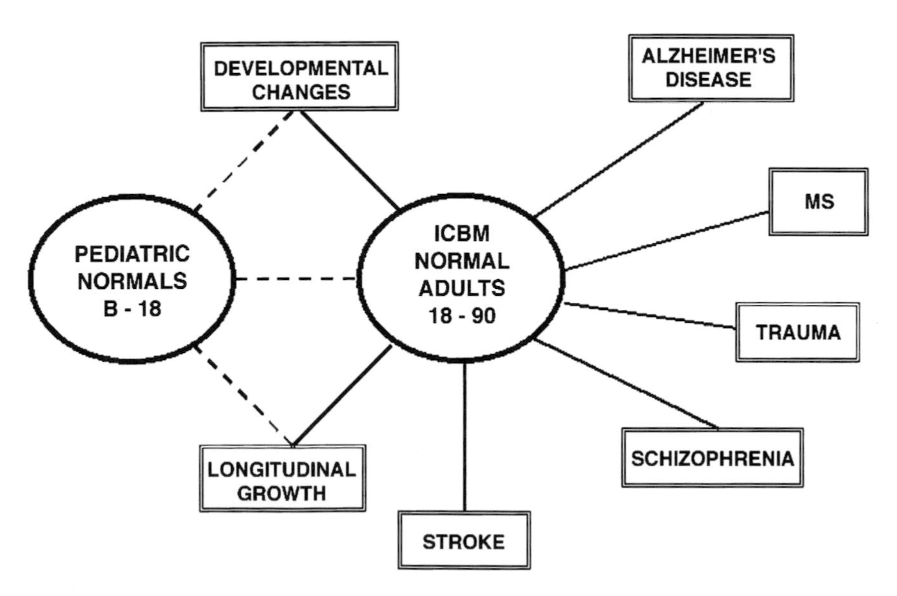

Figure 2 The normal adult ICBM atlas includes individuals between the ages of 18 and 90, as indicated in the center. As a result of the practices, principles, and methods developed through this core project, a number of other atlases and databases have been spawned. Most logical was the development of a pediatric database for structural and functional brain anatomy. This is indicated on the left side. Through the use of data sets in both of these atlases as well as the methods developed in the consortium program, it has been and will be possible to perform interesting cross-sectional and longitudinal studies across age ranges. A number of disease-specific atlases have been developed and are discussed in the text. A number of these are listed here. It is anticipated that there will be a continuing growth in the number of different disease-specific atlases and that these atlases will all become related to the normal ICBM probabilistic atlas through a common focus on neuroanatomy and a four-dimensional data structure.

atlases have been developed for different reasons but employ similar principles and many of the same tools used for the normal adult brain atlas described here.

We also consider a part of this project to be the development of a reference system. The atlas will describe brain structure and function in three spatial domains and a temporal one referenced to the age of the subjects. Attributes (e.g., blood flow, receptor density, behaviors inducing blood flow changes at specific sites, signs and symptoms associated with lesions at specific sites, literature references) are then superimposed on the basic atlas. As such, the atlas becomes the architectural framework for the reference system, the former being grounded in the four physical dimensions and the latter being extensible, based on the interests and data sets available by consortium participants and future users.

B. Probabilistic

Since there is no single, unique representation for the human brain that is representative of the entire species, its variance must be captured in an appropriate framework. The framework that we have chosen is a probabilistic one in which the intersubject variability is captured as a multidimensional distribution. These probabilities can change if subpopulations are sampled because of the shifting distributions. The probabilistic approach was relatively new to neuroanatomical thinking when we first proposed it in 1992. The only previous related strategies had to do with postmortem analyses that reported distributions for structure sizes and dimensions for certain select regions of the brain (Filimonoff, 1932). In recent years, the probabilistic strategy has been more widely used (Roland and Zilles, 1994, 1996, 1998; Mazziotta *et al.,* 1995b) and many probabilistic atlases are now being developed for such species as the monkey and the mouse.

For many psychiatric and behavioral problems, the relationship between structural abnormalities and disease is not straightforward. For example, while many studies have demonstrated that schizophrenics are more likely to have certain structural abnormalities on MRI (e.g., reduced brain size, ventricular enlargement, altered gray matter density in association cortex), none of these abnormalities is sufficiently distinctive or specific to make MRI scanning a useful procedure in the routine clinical diagnosis of schizophrenia. Similar situations prevail for behavioral disorders such as dyslexia and autism.

C. Neuroanatomy Is the Language of Neuroscience

1. Many Nomenclatures

The basic language of neuroscience is neuroanatomy. However, as in any global topic, many languages and dialects exist. Analogous to air traffic control systems, the ultimate solution to the development of a useable brain atlas requires location references expressed as coordinates and a common language to express them (for air traffic control it is the English language). In developing the probabilistic atlas, it was our intention to be able to accommodate multiple languages and meanings. As such, it was important to build a hierarchical nomenclature system in which aliases could be referenced and the boundaries to which they referred adjusted, based on the language selected. This resulted in the requirement for a nomenclature editing system (BrainTree, see below) and an approach that ultimately allows translation from one neuroanatomical language to another without the requirement to force all investigators to use a single, arbitrarily chosen language. It remains clear that the final solution does require a coordinate-based approach devoid of many of the ambiguities associated with qualitative naming of structures.

2. BrainTree

We have developed a system that provides a graphical relationship between anatomic nomenclature and its relationship with the structure or system to which it belongs. To link this nomenclature to a three-dimensional space from the atlas, BrainTree relies on a two-coordinate bounding box for each of the nodes, producing a defined region of three-dimensional space that entirely encompasses the named structure. The user can select a structure on the basis of its standard nomenclature and have its coordinates passed on to standard display or measurement tools. Hence, the BrainTree program provides a facile interface between an editable hierarchical nomenclature system and the indexable three-dimensional coordinate space. Furthermore, the nomenclature can easily be extended to include the myriad of aliases that are common in neuroanatomy or even relate the structural names that provide an association between species (Toga *et al.,* 1996).

D. Use of a Large Population

It is clear that the use of large populations in the development of an atlas that is intended to capture the variance in structure and function of the human brain is an essential requirement. Such a large population can be newly acquired, as was done in this project, or could be the result of pooling smaller studies to produce a metadatabase. The latter approach was rejected because, after examining reports in the literature of smaller sample size projects, it was clear that there was such a wide range of methodological and strategic differences among these studies as to make their pooling difficult, if not impossible. Technical issues such as voxel size, slice thickness, scanning parameters, and many others would cause difficulties in any attempt to produce a homogeneous final product. The same can be said of subject selection and description. As might be expected, a wide range of criteria were used in selecting subject populations, including the definition of normality. Screening tests and demographic and background information, as well as neurological and psychiatric examinations, vary from study to study, adding to the incompatibility of the pooled results. If one also includes functional information, the situation is far worse. Since brain function is obtained by having subjects perform tasks, any slight variation in the task presentation, the psychophysics, or the strategy employed by the subject in performing the task will cause unpredictable differences among experiments thereby adding methodological variance in the pooled data and confounding the final product. Thus, it was decided to prospectively collect a sample of a large number of subjects for which these confounding factors could be controlled.

Given the need to have much larger populations of subjects than had previously been available, the current program is now intended to include 7000 normal subjects obtained from geographical locations as disparate as Japan and Scandinavia and spanning the age range from 18 to 90 years. Special efforts have been made to obtain a wide range of racial and ethnic diversity. In addition, 342 twin pairs (half mono- and half dizygotic) are also part of this sample. The data set for each subject includes a detailed historical description of medical, developmental, psychological, educational, and other demographic features. In addition, behavioral data including neurological, neuropsychological, and neuropsychiatric examinations are part of the data set. In 5800 subjects DNA samples are being collected, stored, and made available for genotyping. This large sample size allows the opportunity to provide realistic estimates about the variance of structure and function for brain regions, the relationships between structure and function at macro- and microscopic levels, and true phenotype–genotype–behavioral comparisons. The large sample size also increases statistical power in making such inferences about the population or when the atlas is used as a comparison sample for investigations involving other groups, be they normal or pathologic. Finally, as the sample size increases, the opportunity to select subpopulations of meaningful size also increases.

E. Target and Reference Brains

A fundamental concept of our consortium's project was to distinguish between target and reference brains (Fig. 3). We have defined the target brain to be the data set, derived from one or, at best, a few individuals, that has the richest collection of data available. Theoretically, this would be the brain of a normal individual studied with *in vivo,* high-resolution, structural, and functional imaging and then, after death, having detailed postmortem analysis including cyto- and chemoarchitecture. If a series of such brains could be studied, then a probabilistic target brain would emerge. Given the high resolution of the postmortem data, target brains would be the most informative with regard to anatomical and chemical localizations. While we have studied a few individuals (all elderly) who had both *in vivo* macroscopic brain imaging and, through the UCLA Willed Body Program, postmortem cryosectioning, we typically do not have both *in vivo* and postmortem data sets of the same individual. As such, a synthesis of this information into an optimized target brain has been the practical solution to date.

In contradistinction to the target brain, reference brains are derived from large populations of subjects typically through *in vivo* imaging of structure and function. These data sets provide information about variance in the population for both structure and function but at a three-dimensional spatial resolution that is three orders of magnitude lower than that of the target brains.

Target and reference brains are used for different purposes. Target brains are, as the name implies, the target to which an unlabeled data set can be warped. The unlabeled data set then picks up the anatomical, functional, or other attributes of each voxel. Once it is back-transformed to its original shape, the new data set will have the appropriate anatomical and functional labels for all brain regions. A certain percentage of these labels will be erroneous based on imperfections of the warping system, an incomplete understanding of the anatomy of homologous brain regions between subjects, and errors in the primary labeling of the target brain. Reference brains provide data about distributions of brain regions and can be divided into subpopulations for specific purposes. Reference brains give estimates of anatomical and functional regions in a population of individuals and, as such, can be used to determine confidence limits when a new data set falls outside the range of normality or expected variance for a given population. Taken together, these two tools provide important but very different vehicles for analyzing existing or new data sets with regard to brain structure and function.

F. Function

It is important to emphasize from the outset that our motivation for studying functional landmarks in this project is completely analogous to the motivation for studying structural anatomic landmarks. Specifically, the ICBM atlas will use functional landmarks to augment atlasing methods that are currently based primarily on macroscopic structural anatomy in the same way that these anatomic methods now augment atlasing methods that were previously based on stereotaxis with simple proportional scaling. An important distinction must be made between functional imaging to answer neuroscience questions, which is not proposed here, and functional imaging to serve as a neuroinformatics tool, which is our intent. Whereas neuroscience functional imaging studies currently use individual macroscopic

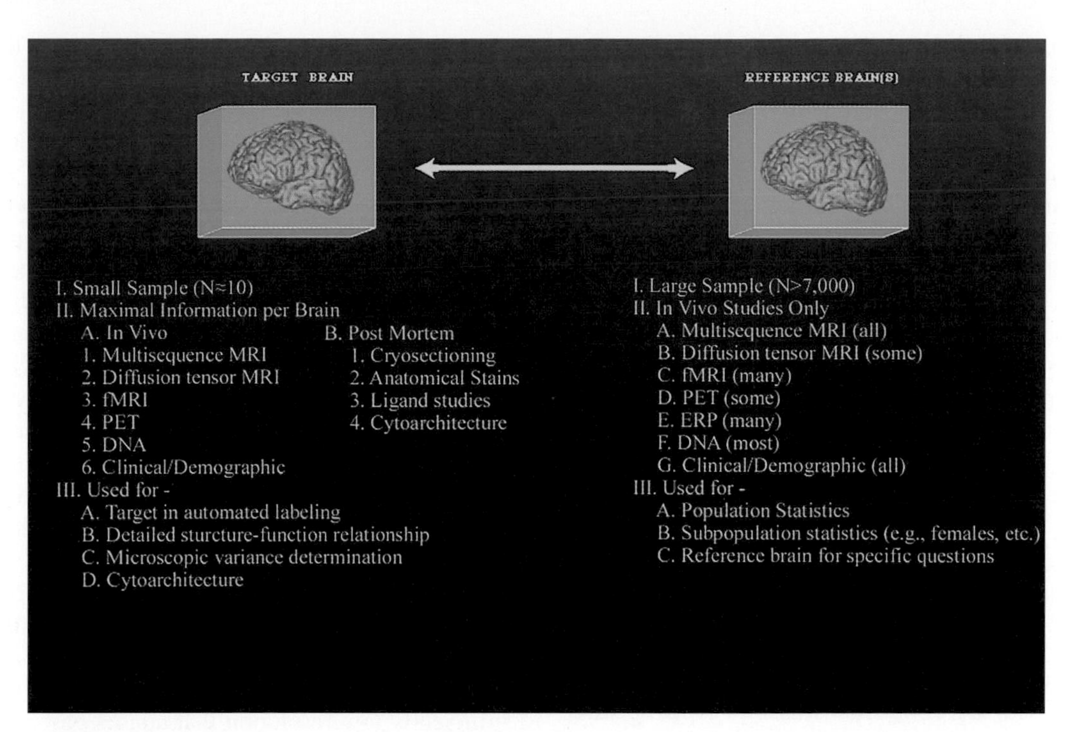

Figure 3 Target versus reference brains. Fundamental to the development of the probabilistic atlas is the concept that there are two types of data sets that are required for a comprehensive system. The first is a target brain or brains. The criteria for this resource are listed. Ideally, single individuals who are normal and studied during life will have the complete complement of *in vivo* imaging studies performed. After death, these same individuals would be studied again using appropriate imaging studies such as MRI, and then detailed cytoarchitectural and chemoarchitectural analyses of their brains would result in a very information-rich ultimate data set. Because it is unlikely that such a situation will occur (although we have studied a few individuals ante- and postmortem), typically, the postmortem and *in vivo* studies are obtained on separate individuals. These target brains are used to better understand microscopic and macroscopic structure–function relationships but also to take a newly acquired data set and warp it to match the target, thereby picking up the anatomical labels for each structure on a voxel-by-voxel basis. When the new study is back transformed to its original shape and configuration, all structures in the brain will have the appropriate labels with an error rate defined by the confidence limits of the warping algorithm and other factors discussed in the text. Reference brains are also described. These represent large populations of *in vivo* studies in which population statistics about variance for structure and function can be obtained. Reference brains can also be sampled to provide subpopulations appropriate to a unique set of descriptors. The size of the sample set for the reference brains will dictate whether subpopulations of sufficient size will be available for specific projects.

anatomic landmarks to define a common neuroinformatics framework for comparing and combining data from different subjects, we anticipate that future neuroscience functional imaging studies will complement this macroscopic anatomy with functional anatomic landmarks, identified in each individual through a selected battery of neuroinformatics tasks. In general, we expect that the neuroinformatics tasks and the functional landmarks that they produce may be completely unrelated to the tasks constituting the primary focus of the neuroscientific investigation. It is our objective to develop and validate tasks that are well suited to producing functional landmarks. These tasks will be the first of what we expect to be an ever-growing library of tasks that will constitute a Functional Reference Battery (FRB). The FRB will be used to develop a new generation of brain atlases through novel warping techniques that move beyond macroscopic anatomy and into the realm of functional and cytoarchitectonic similarities as the fundamental basis for homologous mapping of one brain to another.

The major theoretical and practical issue in identifying homologous brain structures, and in warping strategies designed to compare brains within a population of subjects, is a critical issue with regard to both three-dimensional and surface geometries and representations. In this project we have based all aspects of the atlas development on three-dimensional, voxel-based strategies. Nevertheless, this neither obviates nor limits one's capacity to address special issues related to cortical surface topology. In fact, a significant fraction of the program has been focused on development of appropriate cortical surface extraction and cortical interface (e.g., gray–white interface) identifications. It is important to understand the appropriate constraints that must be imposed to preserve cortical surface topology for both the cerebrum and the cerebellum (Van Essen and Feldman, 1991; Van Essen and Drury, 1997; Van Essen *et al.,* 1998; Fischl *et al.,* 1999).

To understand the motivation to identify functional landmarks, it is important to understand the differences and similarities between functional landmarks and anatomic landmarks with respect to meeting the objectives of neuroinformatics research. Neuroanatomy and, specifically, neuroanatomic landmarks have been the basis that formed the framework for indexing neuroscience information from a number of specific sources collected across spatial scales. Explicit in the plan was the notion that the atlas system would need to continue to adapt in an iterative fashion to accommodate improvements in spatial scale and in the models used to map data into a single neuroanatomic framework. A self-critical evaluation of the methodologies used for structural atlasing alone reveals areas in need of extension:

1. Macroscopic Landmarks from Structural MRI Studies Provide a Suboptimal Basis for Appropriate Mapping of Individual Anatomy into a Unified Neuroinformatics Framework

Three independent lines of research serve to demonstrate the difficulties of relying exclusively on macroscopic anatomic landmarks as a neuroinformatics framework. The first evidence comes from the significant progress made in warping three-dimensional anatomic data to match a target template. In the absence of brain pathology, it is computationally feasible to use high order, nonlinear warps to generate a one-to-one correspondence between brains, even while requiring perfect alignment of unambiguous cortical anatomic features such as the crests of gyri and the depths of sulci. However, even with the inclusion of such anatomic constraints, these mappings are not unique. Mechanical properties such as viscosity or elasticity must be ascribed to the brain tissues to find a solution that is optimal from the standpoint of those presumed mechanical properties (Christensen *et al.,* 1993; Davatzikos, 1997; Schormann *et al.,* 1996). In the absence of more restrictive constraints or independent external standards, different solutions can all lead to equally good (from the standpoint of visual inspection or image similarity criteria) but mutually inconsistent answers, indicating that with macroscopic anatomic data alone, the mapping problem is substantially underconstrained. While it might be a valid computer science goal to identify the transformation that perfectly maps one brain onto another, while minimizing some intuitively appealing quantity, this is not necessarily the best neuroinformatics goal. A more appropriate goal from a neuroinformatics standpoint is *to maximize the genuine homology of points that are brought into correspondence by the transformation.* Functional landmarks will provide additional constraints on intersubject warping that will help to meet this important neuroinformatics goal.

Various criteria can be used to define homology, and conflicts between macroscopic homologies and microscopic cytoarchitectonic homologies (Rademacher *et al.,* 1993) constitute the second line of research demonstrating the problems of relying solely on macroscopic structural anatomy. Recent postmortem cyto- and chemoarchitectonic studies have shown that even some sulcal and gyral features that were once thought to be almost perfectly correlated with nearby cytoarchitectonic boundaries are in fact only approximately correlated (Zilles *et al.,* 1997; Geyer *et al.,* 1997, 1999, 2000; Amunts *et al.,* 1999, 2000). It is our explicit bias that homologies based on function and cytoarchitectonics are more fundamental to neuroscience, and hence to its informatics, than homologies based on sulcal and gyral anatomy.

The third line of research that highlights the difficulties of an informatics framework that is based solely on structural anatomy comes from the rapidly expanding field of *f*MRI. A decade ago, functional imaging with PET was of sufficiently low resolution that atlases based on the simple proportionality of the original Talairach system were adequate to ensure that homologous activation sites would overlap from subject to subject and that the resulting group results would be interpreted as consistent across laboratories. Subsequent improvements in PET image resolution have justified the adoption of the more sophisticated techniques based on structural MRI scanning and MRI–PET coregistration that are in widespread use today (Woods *et al.,* 1993). The high resolution possible with *f*MRI, and the fact that statistically significant responses are readily identified in *f*MRI data from a single subject, demands much more accurate mapping of homologous landmarks from every individual subject and threatens to make methods that rely solely on macroscopic anatomy obsolete. Ideally, a neuroinformatics framework should seek to stay a step ahead of such developments. Functional links may also be of particular value for patient populations in which normal function may persist even in the presence of substantial anatomic distortions. Providing the necessary link between global and local anatomy is a problem that will require new tools, new approaches, and new population data acquired specifically for that purpose.

2. Cytoarchitectonic Studies Provide an Insufficient Basis for Quantifying Relevant Intersubject Variability in the Population

As mentioned above, cytoarchitectonic studies in a small number of subjects can be extremely powerful in demonstrating the potential range of intersubject variability, hence, our motivation to begin to incorporate such data into the ICBM atlas. However, the collection of cytoarchitectonic data is extremely demanding in terms of time and resources.

These realities make it unlikely that reliable population estimates of the variability between structural and functional anatomy will be quantified for many brain regions any time soon using these techniques. Since cytoarchitectonics cannot be identified *in vivo,* such data may help to define general rules (e.g., a given cytoarchitectonic field is most likely to be located at position X in women and at position Y in men), but will not help to identify the individualized exceptions to such rules. In contrast, functional imaging is well suited to population-based studies and functional imaging can be applied routinely to living individuals. To a first approximation, functional landmarks can be viewed as an *in vivo* proxy for cytoarchitectonic landmarks. It should be explicitly stated that it is not our primary intent to equate a given functional landmark with a given cytoarchitectonic region. Indeed, it is clear that a one-to-one relationship will sometimes not exist, since functional subdivisions are present as maps within some cytoarchitectonic areas (e.g., M1 and V1) and since adjacent, functionally correlated areas can be distinguished cytoarchitectonically. Rather, we view cytoarchitectonic anatomy and functional anatomy as intrinsically intertwined features that reveal an underlying pattern of brain organization that provides an optimal framework for neuroscience research and neuroinformatics challenges. By warping brains in a way that brings homologous functional landmarks into concordance, we expect to simultaneously bring nearby cytoarchitectonic regions into better superimposition, even if we do not explicitly know the identities of the cytoarchitectonic regions or even the locations of their boundaries (Fig. 4). Capturing the unique spatial information represented by functional landmarks is an important front for neuroinformatics research, one that will provide routine, direct access to this fundamentally important level of brain organization.

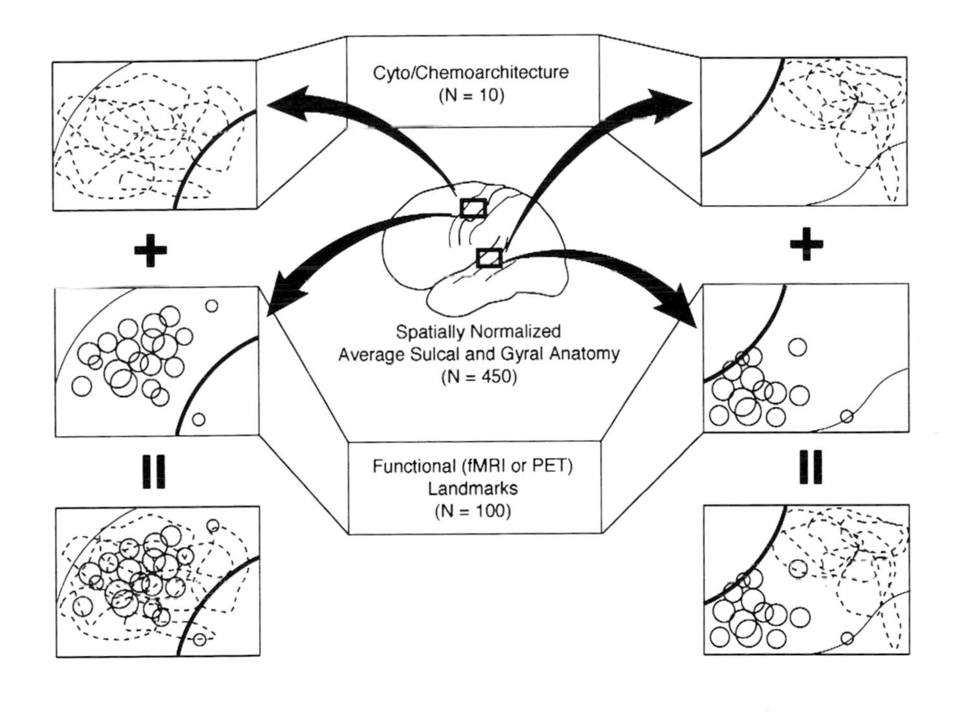

Figure 4 Hypothetical example of concordant and discordant relationships between macroscopic function, derived from *f*MRI or PET, and microscopic cyto- or chemoarchitecture from postmortem specimens. First, all data sets are spatially normalized in probabilistic space at a macroscopic scale (e.g., gyri/sulci) and then the relative positions of functional landmarks (center row) are examined relative to the sites of cyto- or chemoarchitectural zones (top row). There is superimposition of the sites on the left (concordance) but not on the right (discordance). With systems and sets of data such as those proposed in this renewal, it will be possible to add an ever-increasing body of this type of information, leading to structure–function relationship insights throughout the cortex, deep nuclei, cerebellum, and brain stem.

3. Properties of Good and Informative Functional Landmarks (table I)

The minimal attributes of a good functional landmark are that it be unambiguously detectable in individuals and that the variability in its location within individuals be small. "Small" is a relative term, and contexts for making this judgment will be explicitly defined below, along with specific consideration of how "landmarks" can be extracted from functional images. Additional desirable attributes of functional landmarks and the tasks that produce them are listed in Table 1. We make an important conceptual distinction here between a "good" functional landmark and an "informative" functional landmark. In order to also be considered informative, a good functional landmark should provide unique information that could not have been determined purely on macroscopic anatomic grounds. For example, a functional task for identifying primary visual cortex might not prove to be particularly informative since human cytoarchitectonic data indicate that striate cortex consistently maps to the calcarine fissure (Polyak, 1957) (though some variability is present, as reviewed by Aine *et al.* in 1996). In contrast, a good functional landmark in a frontal region where gyral anatomy is quite variable might be highly informative. Caution is generally indicated in trying to predict in advance which functional landmarks will ultimately prove informative since detailed studies of the relationship between cytoarchitectonic and macroscopic anatomy are still relatively rare, as are data comparing functional and structural anatomy. Indeed, with careful study, some traditional assignments of functional areas to specific sulcal or gyral locations are proving to be less reliable than previously expected (Aine *et al.,* 1996; Zilles *et al.,* 1997). Likewise, data from some functional studies have identified previously unsuspected function–structure correlations. A good example of this latter situation comes from the previously discussed studies of putative human area V5, where substantial variability in Talairach coordinate location across subjects turned out to be largely explained by a highly consistent relationship between V5 (defined functionally) and the intersection of the ascending limb of the inferior temporal sulcus and the lateral occipital sulcus (Fig. 1A) (Watson *et al.,* 1993). Because of the difficulty in predicting which landmarks will be informative, we have primarily focused our attention on identifying tasks that produce good functional landmarks and identifying these landmarks in a representative population. These data are being evaluated to determine how informative the landmarks actually are and to look for currently unrecognized structure–function correlations.

Those who are primarily involved in functional imaging neuroscience (as opposed to neuroinformatics) research may be surprised that our criteria for a good functional landmark do not include that the landmark should have a consistent location across subjects. When trying to answer neuroscience questions, there are situations in which variability across subjects is undesirable—one hopes that a functional task will produce responses at a highly consistent anatomically standardized location across subjects so that overlapping regions of response will increase statistical significance and so that the consistency of location will increase confidence that the areas seen in each individual are truly homologous. It is, therefore, perhaps counterintuitive that the exact opposite situation applies to functional landmarks to be used for neuroinformatics. A functional landmark that is always present in the exact same location in every subject, when using current methods of anatomic standardization, is sure to be uninformative, provid-

Essential criteria for good functional landmarks
1. Universally (or nearly universally) identifiable in individual subjects without ambiguity
2. Location in individuals stable with repeated testing

Pragmatic criteria for good functional landmarks
1. Location insensitive to environmental variation (e.g., background noise levels, room lighting)
2. Location insensitive to educational background, native language, gender, etc.
3. Subject performance verifiable or irrelevant
4. Minimal opportunity for diverse cognitive strategies
5. Tasks simple enough to be applicable to cognitively impaired patient populations or children

Desirable features for good functional landmarks
1. Location independent of imaging modality
2. Underlying physiology understood
3. Identifiable simultaneously with many other landmarks produced by a single task and control

Criteria for informative functional landmarks
1. Must meet essential criteria for good functional landmarks
2. Must provide unique spatial information not predictable from macroscopic anatomy

Table I. Functional Landmarks

ing only redundant information that could have been derived from the anatomic data alone.

The neuroinformatics goal here is to use functional landmarks to provide a new source of valid, independent anatomic information that cannot be detected using macroscopic anatomy and to use this information to improve the homologous mapping of different subjects to one another or to an atlas. The result should be better mapping from one subject to another that will serve to improve local homology, a goal that should prove advantageous when subsequently analyzing neuroscience functional imaging data in these same subjects. Two major and one minor assumption are implicit in this line of reasoning and need to be explicitly stated: (1) despite the variation in location, it is critical that the functional landmarks that are identified in each subject are truly homologous, (2) methodological variability in establishing the location of the functional landmark *within* each subject must be small compared to the true anatomical variability in the standardized location of the landmark *across* subjects, and notable, though less important, (3) some preservation of local topology is assumed, so that establishing the location of a functional landmark will indeed improve the homologous mapping of nearby brain regions.

An important implication of the last two assumptions is that the value of a functional landmark will vary: (1) depending on the amount of within-subject variability (more variability decreases its value), (2) depending on the amount of local intersubject variability (more variability increases its value), and (3) depending on its proximity to the nearby regions where better mapping is desired (greater proximity increases its value). If the goal is to improve mapping throughout the brain, numerous functional landmarks may be needed, whereas local mapping may be improved with just one strategically placed functional landmark. The value of proximity raises an important consideration: in functional neuroscientific imaging experiments, why bother to use the locations of established functional landmarks that may be unrelated to the task of interest rather than simply using the locations produced by the primary task itself? There are at least two good answers to this question: (1) Unless landmarks produced by the primary task have been determined to be good landmarks (implying considerable prior investigation), the resulting mapping may actually lead to less reliable homologous mappings than anatomic data alone, and (2) statistical models for evaluating group significance would be invalidated by such a procedure unless separate trials were used for mapping and for addressing the primary neuroscience question. Consequently, appropriate use of the landmarks produced by the primary task being investigated as functional landmarks would require that these landmarks be validated and used in exactly the same way as any other nearby functional landmark. The use of landmarks will also depend, in part, on the brain region(s) of interest for a given experiment and the interests of the investigator.

G. Analysis Strategy

At the outset of this project, it was unclear what the optimal analysis strategy would be for both the structural and the functional aspect of the program. Given the large number of subjects, each with multispectral MRI data sets and many with functional imaging studies as well, it was clear that the tools to be developed would have to function in an automated, or at least semiautomated, fashion to be feasible. Further, reliable automaticity would be a general benefit to the brain imaging field, given the labor-intensive aspects of manual image editing. It was also clear that certain steps would be required to process data in what we have called an ICBM "analysis pipeline." These steps include:

- Screening data for obviously incomplete or artifact-laden studies and rejecting them,
- Intensity normalization in three dimensions for each pulse sequence,
- Alignment and registration across pulse sequences and studies within a given subject,
- Tissue classification (i.e., gray and white matter, cerebrospinal fluid, other),
- "Scalping" whereby extracranial structures are removed,
- Spatial normalization of each subject to a target where anatomical labels can be obtained automatically,
- Surface feature extraction, and
- Visualization

Given this sequence of tasks, it was unclear, in most cases, what the optimal solution for each would be. Rather than making an *a priori* decision and have all consortium members work to achieve it, an alternate approach was chosen. It was decided that each of the primary laboratories in the consortium would work to solve each step in the analysis pipeline independently and in parallel. These laboratory-specific algorithms would then be locally optimized. Once a given laboratory was satisfied with the performance and documentation of their approach, it would be distributed to the other participating laboratories for alpha testing. If an algorithm failed to perform adequately or was awkward to use because of hardware platform incompatibilities or other factors, it was rejected. Those algorithms that performed well across consortium laboratories were ultimately sent to an independent group (David Rottenberg, M.D., Stephen Strother, Ph.D., and colleagues at the University of Minnesota) for beta testing. This independent testing included not only the ICBM algorithms for a given module in the analysis pipeline but also any other algorithms that could be identified worldwide that purported to perform the same functions. During beta testing algorithms were evaluated with simulated as well as real data sets selected by the beta test laboratory and evaluated for documentation, ease of installation, computation time, accuracy, and precision. The results of these evaluations were then published

(Strother *et al.,* 1994; Arnold *et al.,* 2001). The winners of this competition were then selected for the ICBM analysis pipeline (Fig. 5) and will be the basis for the mass data analysis of all data sets. While it was decided that it was important to analyze all 7000 studies in a consistent manner so that users would know the methodology, algorithms, and versions of the algorithms from which the results were derived, this in no way precluded individual laboratories in the ICBM or elsewhere from using their own strategies for data analysis on the original data sets which are provided through digital libraries (see below). This strategy has been successful in that it established an internal competition whereby the best solution emerged rather than an *a priori* and hypothetical prediction that might have fallen far short of the optimal outcome.

H. Visualization

Similar to the approach chosen for analysis, it was decided to keep an open mind as to how to present the data developed by the consortium. Given the probabilistic nature of the resultant data, the decision is not straightforward and has not yet been fully resolved. It may well be that the optimal solution is to select many avenues and that users of this system choose for themselves. Approaches that flatten (Carman *et al.,* 1995; Felleman and Van Essen, 1991; Van Essen and Drury, 1997; Van Essen *et al.,* 1998) or inflate (Dale *et al.,* 1999; Fischl *et al.,* 1999) the cortex have been proposed and well described. As a visualization tool, these strategies allow cortical anatomy to be seen in its entirety at the expense of the more familiar, three-dimensional appearance of the brain. It is important to note here that visualization, simply as a tool to view the data, must be distinguished from the use of these tools to identify homologies between regions in different brains or different species. In this context we consider these strategies only as a visualization tool as our approach to homology identification was described earlier with regard to macro- and microscopic, structure–function considerations. Each visualization strategy has its benefits and limitations. The traditional three-dimensional view of the brain in its natural state obviates the ability to see brain regions hidden in folded cortex or deep structures without providing tools for translucency or sectioning. Flattening or inflating the surfaces will produce areas of compression and expansion that alter the data from their original state but make all surface regions visible. Providing all of these avenues will allow the user to choose among them given a specific purpose. The user can choose

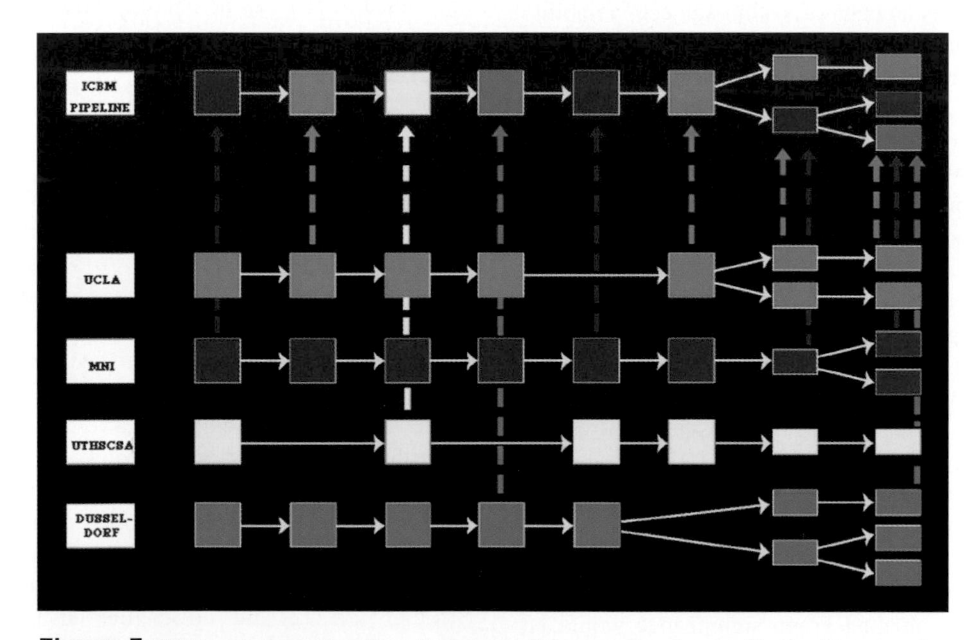

Figure 5 ICBM analysis pipeline. Since it was impossible to predict what specific algorithm or mathematical strategy would be optimal for each step in the analysis of structural and functional data collected for subjects in this consortium project, the original core laboratories elected to each develop independent strategies for each step. Once complete and tested within a given laboratory, they were distributed among consortium participants for alpha testing. After the consortium members were satisfied with the performance at this phase, all consortium-developed algorithms were delivered to an independent laboratory, not part of the consortium, for beta testing. The beta testing included not only the ICBM-developed algorithms but also any other algorithms identifiable worldwide that purported to perform the same function. The best (see text) algorithm was then selected for incorporation into the ICBM pipeline. All data in the final atlas will be processed through this unified, single pathway. The bottom four boxes (white) in the left column represent consortium sites contributing algorithms to the pipeline.

whether the benefits and insights provided by a given visualization strategy outweigh the disadvantages or artifacts induced by the visualization scheme.

I. Database

1. Digital Libraries

In addition to the derived data organized in the databases described above, digital libraries and data warehouses (Fig. 6) of complete data sets will also be provided through the ICBM project to the neuroimaging community. These data sets include those with "raw" data (i.e., complete, three-dimensional, multispectral MRI structural studies of individual subjects), "scalped" (i.e., extracranial structures removed) data sets, and intensity-normalized, "scalped" data sets. Access to such information may allow investigators to obtain normal control data for neuroimaging experiments or to test various methods for image analysis and display without the requirement to acquire original data on their own. Most problematic will be the distribution of raw data sets, as the potential for compromising subject confidentiality is an issue. Since the experimental subject's face could be reconstructed from the raw data sets, one strategy would be to alter or eliminate facial structures from the data set prior to distribution.

2. Four Dimensional

There currently exists no comprehensive database for the storage of complete, individual subject, neuroimaging data sets for the human brain that is both electronically accessible and efficient in its interactions with neuroscientists. This reduces the value of both clinical and research dollars spent on the acquisition of these important and interesting studies. The physical world is organized in four dimensions and, thus, forms a logical and comprehensive organizational framework for the ICBM database. Plans anticipate the future inclusion of time-series data from dynamic, functional

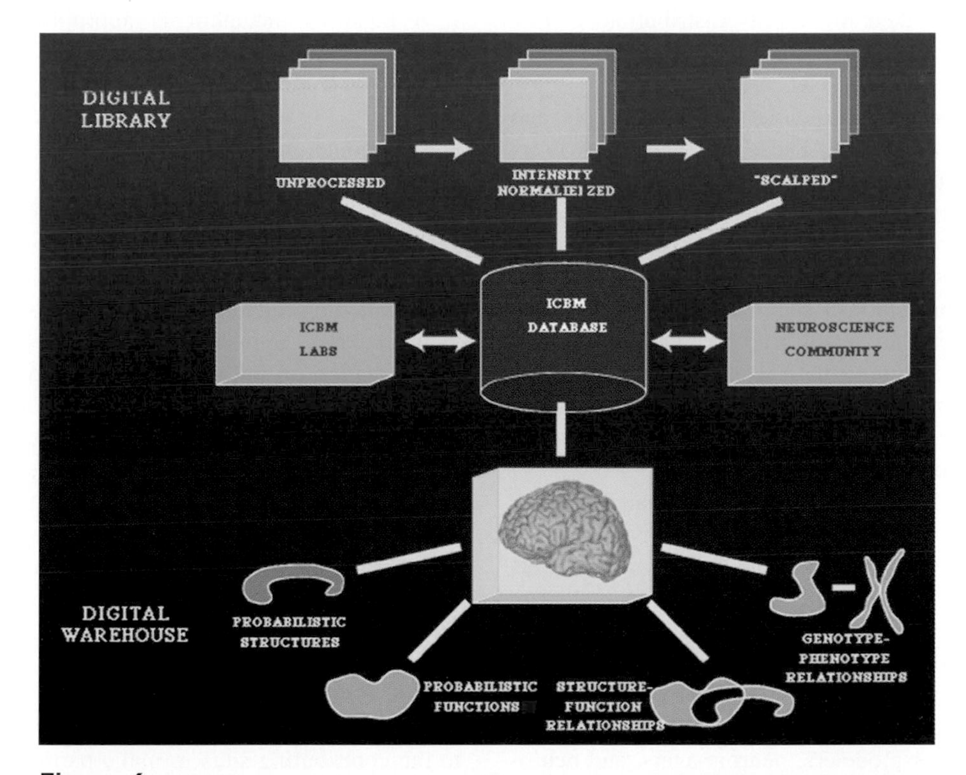

Figure 6 Digital libraries and data warehouses. We envision two outputs of the probabilistic atlas to the neuroimaging and neuroscientific communities. The first is termed "digital warehouse." In this warehouse we will catalog complete MRI studies from large numbers of subjects. They will be held in three forms: raw data (with the face corrupted for confidentiality reasons), intensity-normalized images, and "scalped," intensity-normalized images from which extracranial tissues have been removed. A user would be able to select a certain list of descriptors (e.g., females, ages 25–30, right-handed) and will be told how many subjects and data sets fulfill this requirement. These data sets can then be transferred (either electronically or through magnetic media, depending on the size) to the requesting investigator. Data warehouses provide database interactions where probabilistic structures and high-level queries can be obtained by interacting with the entire population of 7000 studies or user-selected subpopulations. Such queries can be representational or symbolic and may ultimately be object oriented. These queries may involve the imaging, clinical, behavioral, or genetic data.

data acquisition methods such as *f*MRI, EEG, and MEG, requiring the fourth dimension. It is expected that spatiotemporal and purely temporal patterns of brain activity will constitute functional entities and markers of their own. These can be used for the following purposes:

(i) Since function will be defined in the future by brain locations and timing of activity, the probabilistic reference will incorporate temporal and spatiotemporal brain activity information.

(ii) Spatiotemporal and temporal functional markers will be used for most of the same purposes described for the spatial functional markers (*f*MRI, PET) in this paper, warping, correlations across subjects, an additional source of information in calculating population distributions—these can be used, for example, to inform studies with small populations.

(iii) Temporal and spatiotemporal information will be used to correlate brain activity across subjects in the temporal dimension.

(iv) They can also be used as priors for brain source estimation methods, and their probability distributions can be used for Bayesian procedures in EEG and MEG brain source localization (Schmidt *et al.,* 1999).

With this data structure, queries-by-content tools and strategies are being developed. These tools will allow users of the database to submit a query in the form of actual data (e.g., a two-dimensional image of a portion of the brain or a three-dimensional block of data) and ask the database to search for matches using wavelet-based techniques that have previously been demonstrated to be successful for two-dimensional Internet searches of graphic material (Wang *et al.,* 1997). The expansion of these approaches to three and, eventually, four dimensions will be an important neuroinformatics milestone that will find uses far beyond the applications in this consortium.

Further, a system organized in this fashion, and the tools associated with it, will allow for efficient, convenient, and comprehensive access by neuroscience clients to the ever-growing data in the ICBM probabilistic reference system. The goal is not to develop physiologic models of brain function, neural connectivity, and other important neurobiological questions. But these exciting opportunities will be more easily achieved by providing a system of database interactions and structure for modelers, neuroimagers, and neuroscientists, in general. We envision that, once established and populated with data, the probabilistic reference system, organized in this fashion, will allow for "electronic" hypothesis generation and experimentation using previously collected, well-described, and effectively organized data.

3. Attributes

It is conceptually important to understand that the database architecture, while organized in four dimensions, to match the organization of the nervous system, can have a very high number of attributes all referenced to these basic four dimensions. These additional attributes need not be specified at the time of establishing the data sample or the data set. Some can be derived and others can be added at a later point through further examination of the original subjects (e.g., longitudinal studies, other methodologies) or by further analysis of existing data (e.g., genotyping of stored DNA samples).

The most difficult challenge to the actual organization of such a database is the scaling and referencing of data across major spatial or temporal domains. While originally developed to have a fundamental spatial unit of resolution of 1 mm^3, there is no reason why microscopic and ultrastructural information cannot appropriately populate the individual 1-mm^3 voxels of the macroscopic data set. The same can be said of temporal information but the exact manner of binning of time-series information will require judicious attention to the types of queries anticipated of such data sets.

4. Central vs Distributed

A business metaphor is appropriate here. Fledgling industries rarely do well when trying to establish standards, means of communication, and interoperability methods that are designed to result in a reliable and durable outcome for a given community. Examples abound, including telecommunications, aviation, electronics, meteorology, and others. In most of these cases, a well-designed, centralized approach established both the problems and the solutions that later led to deregulated, decentralized systems that were linked by regulatory groups, industrial standards, and metadatabases. Similarly, in the burgeoning field of neuroinformatics, an initial centralized approach appears both desirable and manageable. It allows for a straightforward and easily monitored means of distributing data sets on a continuous basis. A centralized approach can also monitor the required submission of attributes derived from the data sets back into the database as a measure of successful, reciprocal sharing of data and results (Bloom, 1996; Pennisi, 1999). Finally, in order to even attempt such a project, there must exist a critical mass of data analysis tools, organization, and reputation to make participation attractive psychologically and sociologically. The common goal and ultimate result must also be sufficiently valuable to the contributing sites to make participation compelling. The results of participation must be worth more than the sum of the individual parts.

J. Real World

The ICBM has always maintained a "real-world" environment such that the participating sites use different equipment, software, and protocols reflecting a microcosm of the larger neuroscience, neuroimaging, and neuroinformatics communities and forcing us to develop solutions to problems through flexible, compatible systems rather than rigid standards, protocols, and equipment requirements. The significance of this feature is that the products are not platform-, institution-, or protocol-specific.

1. Interoperability

Interoperability was an important concern early in the development of the ICBM atlas. So important was the requirement to develop interoperable tools and data sets that a conscious decision was made to deliberately utilize imaging instruments, computing hardware, and file formats that differed among the participating sites. This forced certain principles and rules to be utilized in the development of software and the exchange of data, the goal being accessibility of any ultimate end-user to all of these products. The psychology and sociology of any advanced research field is to develop homemade tools and to maintain intralaboratory file structures. The experience in the ICBM was no different. As such, we developed translators that would allow data sets to be transferred among sites with an agreed-upon file format (MINC; Neelin *et al.,* 1998) but that was translated into the "home" file format upon receipt at any of the participating sites. A similar strategy was used for algorithms. This simplistic approach has worked quite well, allowing a relatively seamless exchange of information.

2. Quality Control

If the ICBM atlas is to be a growing resource, tools that have been developed, thus far, will ultimately be open to the entire neuroscientific community for the future additions of data sets. How then will we ensure the quality of data from investigators? Having pondered and debated this question for many years and having examined the approaches used by other fields, the simple answer is that we cannot ensure a certain level of quality control in a completely open data exchange program. Not only is this impractical but it may also lead to the erroneous exclusion of data that might someday be deemed valuable. If there were some filter on the input of data, what would the review process be? How can we predict how tomorrow's observations will be judged by today's standards? We cannot. Further, in a practical sense, such an approach would immediately become backlogged with data sets awaiting "review" by some "panel of experts" whose opinions might change as time and experience progresses. What we can provide, however, is a system by which users of such data sets can select their own level of confidence about the populations or results that they sample. For example, a user might request all information about a certain region of the brain for a given demographic population of subjects. Most of these data would be of high quality and reliably collected but some of them would undoubtedly include experimental, methodological, and other errors. Nevertheless, it would give the user a complete picture of all of the information available about their query. At the other end of the spectrum, consider a user who is interested in only the most accurate information about a given site in the brain for a certain population. That user could request data that was obtained only from the results of peer-reviewed, published, and independently reproduced data collections. Thus, just as the data sets can be filtered using demographic, anatomical, or clinical criteria, they can also be filtered and queried by confidence level. "Let the user beware" is the only rational approach to developing such a system.

IV. Methods and Results

A. MRI

1. Basic Principles

Multispectral anatomical MRI data for the ICBM project were acquired using optimized protocols matched as closely as possible across the different scanner manufacturers and field strengths (3.0 T, GE; 1.5 T, Philips; and 2 T, Elscint). The protocol design goals were to achieve whole-head 1-mm isotropic T1-weighted image volumes and whole-head 1 × 1 × 2-mm T2- and proton density (PD)-weighted volumes.

2. Averaging

The accuracy of brain atlases is constrained by the resolution and signal-gathering powers of available imaging equipment. In an attempt to circumvent these limitations, and to produce a high-resolution *in vivo* human neuroanatomy, we investigated the usefulness of intrasubject registration for post hoc magnetic resonance signal averaging (Holmes *et al.,* 1998). Twenty-seven high-resolution (7 × 0.78 and 20 × 1.0 mm^3) T1-weighted MRI volumes were acquired from a single subject, along with 12 double-echo T2/PD-weighted volumes. These volumes were automatically registered to a common stereotaxic space in which they were subsampled and intensity averaged. The resulting images were examined for anatomical quality and usefulness for other analytical techniques.

The quality of the resulting images from the combination of as few as 5 T1 volumes was visibly enhanced. The signal-to-noise ratio was expected to increase as the root of the number of contributing scans, to 5.2 for an N of 27. The improvement in the $N = 27$ average was great enough that fine anatomical details, such as thalamic subnuclei and the gray bridges between the caudate and the putamen, became sharply defined. The gray–white matter boundaries were also enhanced, as was the visibility of any finer structure that was surrounded by tissue of varying T1 intensity. The T2 and PD average images were also of higher quality than single scans but the improvement was not as dramatic as that of the T1 volumes. Overall, the enhanced signal in the averaged images resulted in higher quality anatomical images with improved results for other postprocessing techniques. The high quality of the enhanced images permits novel uses of the data and extends the possibilities for *in vivo* human neuroanatomical explorations. Post hoc registration and averaging of MRI scans is a robust method for the enhancement of MR images. There is a significant reduction in noise in averaged images that reveals previously unobservable structure. The high quality of the resulting images opens the door to other forms of postprocessing and suggests even further applications. The method itself is very straightforward and can easily be employed.

B. Postmortem Cryosectioned Material

1. Data Acquisition

Mapping the human brain and its functions requires a comprehensive anatomic framework. This reasoning dictated the need in our consortium to obtain high-resolution, digital, whole-brain, postmortem data sets. The fact that recent advances in anatomic digital imaging techniques now permit unrestricted visualization in multiple cut planes and three-dimensional regional or subregional analyses when appropriate primary data sets are available (Spitzer and Whitlock, 1992; Wertheim, 1989) made this approach feasible. Digital representations also offer the opportunity for morphometric comparisons and sophisticated mapping between anatomic and metabolic imaging modalities (Payne and Toga, 1990; Toga and Arnicar-Sulze, 1987). The primary source data for human brain atlasing must include not only very fine spatial detail but also image color and texture to convey the subtle characteristics that make it possible to distinguish subnuclear and laminar differences. Further, the incorporation of an appropriate spatial coordinate system is critical as a framework for intersubject morphometrics. High-resolution anatomic data sets serve as references for the accurate interpretation of clinical data from the PET, CT, and MRI modalities as well as the mapping of transmitters, their receptors (Fig. 7), and other regional biological characteristics.

Thus, we designed a system of histologic and digital processing protocols for the acquisition of high-resolution, digital imagery from postmortem cryosectioned whole human brain and head for computer-based, three-dimensional representation and visualization (Cannestra *et al.,* 1997; Toga *et al.,* 1997). High-resolution (1024^2 pixel) serial images can be captured directly from a cryoplaned blockface using an integrated color digital camera and fiber optic illumination system mounted over a modified cryomacrotome. The system can process tissue treated in a variety of ways, including fixed, fresh, frozen, or otherwise prepared for sectioning at micrometer increments. Sometimes it is desirable to section the tissue while still *in situ*. Specimens frozen and sectioned with the cranium intact preserve brain spatial relationships

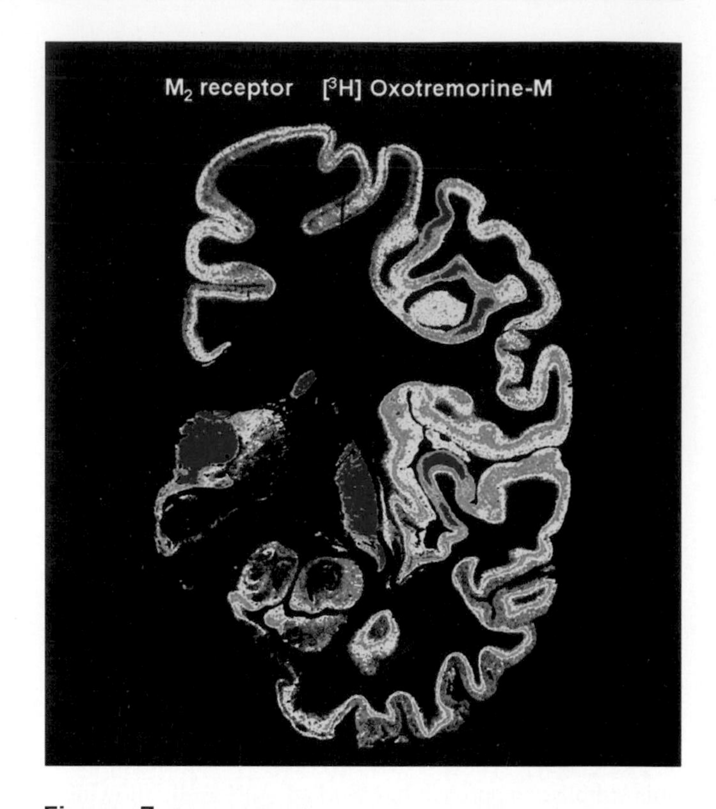

Figure 7 Coronal image demonstrating tritiated muscarinic receptors from one hemisphere of a cryosectioned brain and demonstrating the anatomical detail that such chemoarchitectural maps can provide. When serial sections are obtained and stained for a wide range of receptors, anatomical features, and gene expression maps, a tremendous wealth of information is available for comparison with sites of functional activation obtained using *in vivo* techniques and macroscopic brain structure (gyri, sulci, deep nuclei, white matter tracts). Having a probabilistic strategy for relating these different types of anatomies will provide previously unavailable insights about the relationship of structure and function on both microscopic and macroscopic levels for the human brain and, by analogy, for the brains of other species (see Figs. 4 and 8). The analysis of the regional and laminar distribution patterns of transmitter receptors is a powerful tool for revealing the architectonic organization of the human cerebral cortex. We succeeded in preparing extra-large serial cryostat sections through an unfixed and deep-frozen human hemisphere. Neighboring sections were incubated with tritiated ligands for the demonstration of 15 different receptors of all classical transmitter systems. The distribution of [^{3}H]oxotremorine-M binding to cholinergic muscarinic M2 receptors is shown here as an example. Even a cursory inspection of a color-coded receptor autoradiograph permits the distinction of numerous borders of cortical areas and subcortical nuclei by localized changes in receptor density and regional/laminar patterns. For example, the M2 receptor subtype clearly labels the primary sensory cortices (at the level of the section shown, e.g., the primary somatosensory area BA3b and the primary auditory area BA41) by very high receptor densities sharply restricted to both areas. The different receptors allow the multimodal molecular characterization of each area or nucleus by the so-called receptor fingerprint typing. A receptor fingerprint of a brain region consists of a polar plot based on the mean density of each receptor in the same architectonical unit (area, nucleus, layer, module, striosome, etc.). The following areas and nuclei could be delineated in the present example: (1) cingulate cortex, (2) motor cortex, (3) primary somatosensory cortex (BA3b), (4) inferior parietal cortex, (5) insular cortex, (6) primary auditory cortex (BA41), (7) nonprimary auditory cortex, (8) inferior temporal association cortex, (9) entorhinal cortex, (10) mediodorsal thalamic nucleus, and (11) putamen (K. Zilles, A. Toga, N. Palomero-Gallagher, and J. Mazziotta, unpublished observation).

and anatomic bony landmarks. Color preservation is superior in unfixed tissue but unfixed heads were incompatible with decalcification and cryoprotection procedures. Thus, section collection from such specimens was complicated by bone fragmentation. Collection of 1024^2 images from whole brains results in a spatial resolution of 200 μm/pixel in a 1- to 3-Gbyte data space. Even higher three-dimensional spatial resolution is possible by primary image capture of selected regions such as hippocampus or brain stem or by using higher resolution cameras. Discrete registration errors can be corrected using image processing strategies such as cross-correlative and other algorithmic approaches. Data sets are amenable to resampling in multiple planes as well as scaling and transpositioning into standard coordinate systems. These methods enable quantitative measurements for comparison between subjects or to atlas data. These techniques allow visualization and measurement at resolutions far higher than those available through other *in vivo* imaging technologies and provide greatly enhanced contrast for delineation of neuroanatomic structures, pathways, and subregions.

The use of cryosectioned anatomic images as a gold standard for mapping the human brain requires a complete understanding of the assumptions and errors introduced by this method. While there are several obvious advantages to using these data as a reference for other tomographic and *in vivo* mappings, their collection requires sophisticated instrumentation and representative *postmortem* material. Spatial resolution, the inclusion of bony anatomy, full color, block-face reference for histologically stained sections, and the resulting registered three-dimensional volumetric data sets are important aspects of this method. Neverthcless, cryosectioning approaches, like all others, introduce distortion during acquisition and processing. Sources of errors include postmortem brain changes and artifacts associated with tissue handling. A major source of error is related to specimen preparation prior to sectioning. Removal of the cranium and subsequent brain deformation, perfusion protocols, or freezing alter the spatial configuration of the data set.

While three-dimensional data at this resolution are difficult to acquire, they are necessary for careful studies of morphometric variability and the generation of digital comprehensive neuroanatomic atlases. Ultimately, what is needed is the combined use of cryosectioned data as the source of higher resolution raw and stained anatomy spatially referenced to an *in vivo* electronically acquired data set such as MRI.

2. Cyto- and Chemoarchitecture

A major effort in this project is to obtain cyto- and chemoarchitectural data from postmortem brains to enter into the probabilistic database for comparison with *in vivo* studies. An example of this approach is described for Broca's area. The putative anatomical correlates of Broca's speech region, i.e., Brodmann's areas 44 and 45 (Brodmann, 1909), are of considerable interest in functional imaging studies of language. It is a long-standing matter of discussion whether anatomical features are associated with the functional lateralization of speech (Galaburda, 1980; Hayes and Lewis, 1995, 1996; Jacobs *et al.,* 1993; Simonds and Scheibel, 1989; Scheibel *et al.,* 1985). Furthermore, the precise position and extent of both areas in stereotaxic space and their intersubject variability still remain to be analyzed, since Brodmann's delineation is highly schematic, is not documented in sufficient detail, and does not contain any statement about intersubject variability.

We studied the cytoarchitecture of Brodmann's areas 44 and 45 in 10 human postmortem brains using cell-body-stained (Merker, 1983), 20-μm-thick serial sections through complete brains (Amunts *et al.,* 1999). Cytoarchitectonic borders of both areas were defined using an observer-independent approach, which is based on the automated high-resolution analysis of the packing density of cell bodies (gray-level index or GLI) from the border between layers I and II to the cortex/white matter border (Schleicher *et al.,* 1999). These profiles are perpendicular to the cortical surface and define the laminar pattern of cell bodies. Thus, the profiles are a quantitative expression of the most important cytoarchitectonic feature. Multivariate statistical analysis was used for locating significant differences between the shapes of adjacent GLI profiles along the cortical extent. Those locations represent cytoarchitectonic borders. GLI profiles were also used for investigating interhemispheric differences in cytoarchitecture. Significant interhemispheric differences in cytoarchitecture (i.e., differences in GLI profiles between right and left areas) were found in both areas 44 and 45. Profiles obtained as internal controls from the neighboring ventral premotor cortex did not show any lateralization.

The positions of the borders of areas 44 and 45 with respect to sulci and gyri showed a high degree of intersubject variability (Fig. 8). This concerned the sulcal pattern, i.e., the presence, course, and depths of sulci, as well as the spatial relation of areal borders with these sulci. The position of a cytoarchitectonic border could vary up to 1.5 cm with respect to the bottom of one and the same sulcus in different brains. Thus, sulci and gyri are not reliable and precise markers of cytoarchitectonic borders. Although there was a considerable intersubject variability in volume of areas 44 and 45 (N = 10), area 44 was larger on the left than on the right side in all cases of our sample. We could not find any significant left–right differences in the volume of area 45.

The extent and position of areas 44 and 45 were analyzed in the three-dimensional space of the standard reference brain of the European Computerized Human Brain Database (Roland and Zilles, 1996) after the above-described microstructural definition of the areal borders. MR imaging (3D FLASH-scan; Siemens 1.5-T magnet) was performed

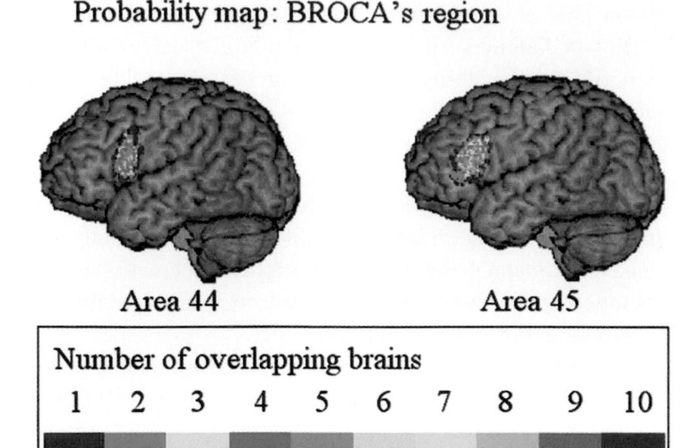

Figure 8 Location and extent of Broca's region (Brodmann areas 44 and 45). Areas 44 and 45 as defined in serial coronal sections of an individual brain after three-dimensional reconstruction; lateral views of the left hemisphere. Probability maps of Broca's region, based on microscopic analysis of 10 human brains can be referenced, also in a probabilistic fashion, to functional activation sites associated with the functions of Broca's area using the multimodality probabilistic atlas strategy. The overlap of individual postmortem brains is color coded for each voxel of the reference brain (color bar); for example, 7 of 10 brains overlapped in the yellow-marked voxels. Left in the image is left in the brain.

on postmortem brains prior to histology. Corrections of deformations inevitably caused by the histological technique were performed by matching MRI and corresponding histological volumes (Schormann and Zilles, 1997; Schormann *et al.,* 1995). Brain volumes were finally transformed to the spatial format of the reference brain. For both steps, a movement model for large deformations was applied (Schormann *et al.,* 1996, 1997; Schormann and Zilles, 1998). The superimposition of individual cytoarchitectonic areas in the standard reference format resulted in probability maps (Fig. 8). These maps quantitatively describe the degree of intersubject variability in extent and position of both areas. They serve as a basis for topographical interpretations of functional imaging data obtained in PET and *f*MRI experiments (Amunts *et al.,* 1998).

The observed intersubject variability in the extent and cytoarchitecture of Broca's region has to be considered when correlating data of functional imaging studies with the underlying cortical structures. Interhemispheric differences in the volume of area 44, and in the cytoarchitecture of both areas, may contribute to functional lateralization which is associated with Broca's region.

C. Warping and Segmentation Strategies

1. Segmentation

a. Manual voxel segmentation and labeling. We developed a general image analysis package, DISPLAY, which provides a wide range of capabilities for (i) interactive three-dimensional exploration of image volumes using simultaneous orthogonal planes and surface-rendered representations; (ii) manual labeling of image voxels; (iii) archival/recall of labeled three-dimensional objects such as brain regions, pathological masses, tissue class maps; and (iv) morphological operations such as the dilate/erode/open/close primitives. DISPLAY has become a standard utility within the ICBM, and elsewhere, for labeling brain regions (Evans *et al.,* 1996; Paus *et al.,* 1996a,b; Penhune *et al.,* 1996). However, the use of manual tools for labeling large numbers of MRI data sets is prohibitively time consuming and subject to interrater variability. We have therefore developed a series of algorithms for automated image segmentation.

b. Correction for three-dimensional intensity nonuniformity—N3. A major problem for automated MR image segmentation is the slowly varying change in signal intensity over the image, caused principally by nonuniformities in the radiofrequency field. Apparent signal from any one tissue type is therefore different from one brain area to another, confusing automated-segmentation algorithms that assume constant signal for one tissue type. We have developed a fully automated three-dimensional technique for inhomogeneity correction. The method maximizes the entropy of the intensity histogram to maximize its structure. The effect of inhomogeneity is modeled as a convolution histogram by a blurring kernel and the effective kernel can be estimated and deconvolved by iterative entropy maximization. The method is applicable to any pulse sequence, field strength, and scanner (Sled *et al.,* 1998, 1997). In the previously described competition among algorithms, the N3 approach of Sled *et al.* (1997, 1998) proved superior (Arnold *et al.,* 2001) and has been selected for the ICBM data analysis pipeline (Fig. 5).

c. Tissue classification—INSECT. We have developed a series of algorithms for tissue classification (Kamber *et al.,* 1992, 1995; Zijdenbos *et al.,* 1996). They are used for automatically processing multispectral (T1-, T2-, PD-weighted) data sets from large numbers of subjects, known as INSECT (Intensity-Normalized Stereotaxic Environment for Classification of Tissues). All data are corrected for field inhomogeneity (Sled *et al.,* 1998), interslice normalization, and intersubject intensity normalization. Stereotaxic transformation is then performed (Collins *et al.,* 1994) and an artificial neural network classifier identifies gray/white/CSF tissue types (Zijdenbos *et al.,* 1996; Evans *et al.,* 1997).

d. Regional parcellation—ANIMAL. Manual labeling of brain voxels is both time consuming and subjective. We have developed an automated algorithm to perform this labeling in three dimensions (Collins *et al.,* 1995). The ANIMAL algorithm (Automated Nonlinear Image Matching and Anatomical Labeling) deforms one MRI volume to match another, previously labeled, MRI volume. It builds up the three-dimensional, nonlinear deformation field in a piecewise linear fashion, fitting cubical neighborhoods in sequence. The algorithm is applied iteratively in a multiscale hierarchy. At each step, image volumes are convolved with a three-dimensional Gaussian blurring kernel of successively smaller width (32-, 16-, 8-, 4-, and 2-mm full width at half-maximum). Anatomical labels are defined in the new volume by interpolation from the original labels, via the spatial mapping of the three-dimensional deformation field.

2. Warping Strategies

Atlases can be greatly improved if they are elastically deformable and can fit new image sets from incoming subjects. Local warping transformations (including local dilations, contractions, and shearing) can adapt the shape of a digital atlas to reflect the anatomy of an individual subject, producing an *individualized* brain atlas. Introduced by Bajcsy and colleagues at the University of Pennsylvania (Broit, 1981; Bajcsy and Kovacic, 1989; Gee *et al.,* 1993, 1995), this approach was adopted by the Karolinska Brain Atlas Program (Seitz *et al.,* 1990; Thurfjell *et al.,* 1993; Ingvar *et al.,* 1994), where warping transformations were applied to a digital cryosectioned atlas to adapt it to individual CT or MR data and coregistered functional scans.

Image-warping algorithms, specifically designed to handle three-dimensional neuroanatomic data (Christensen *et al.,* 1993; 1996; Collins *et al.,* 1994, 1995; Thirion, 1995; Rabbitt *et al.,* 1995; Davatzikos, 1996; Thompson and Toga, 1996; Bro-Nielsen and Gramkow, 1996; Schormann *et al.,* 1996, 1997; Schormann and Zilles, 1998; Ashburner *et al.,* 1997; Woods *et al.,* 1998), can transfer all the information in a three-dimensional digital brain atlas onto the scan of any given subject, while respecting the intricate patterns of structural variation in their anatomy. These transformations must allow any segment of the atlas anatomy to grow, shrink, twist, and rotate, to produce a transformation that encodes local differences in topography from one individual to another. Deformable atlases (Seitz *et al.,* 1990; Evans *et al.,* 1991; Miller *et al.,* 1993; Gee *et al.,* 1993; Christensen *et al.,* 1993; Sandor and Leahy, 1994; 1995; Rizzo *et al.,* 1995) resulting from these transformations can carry three-dimensional maps of functional and vascular territories into the coordinate system of different subjects. The transformations also can be used to equate information on different tissue types, boundaries of cytoarchitectonic fields, and their neurochemical composition (Amunts *et al.,* 1998, 1999, 2000; Geyer *et al.,* 1996, 1997, 1999, 2000).

Warping algorithms calculate a three-dimensional deformation field which can be used to nonlinearly register one brain with another (or with a neuroanatomic atlas). The resultant deformation fields can subsequently be used to transfer physiologic data from different individuals to a single anatomic template (Geyer *et al.,* 1996; Larsson *et al.,* 1999; Naito *et al.,* 1999, 2000; Bodegard *et al.,* 2000a,b). This enables functional data from different subjects to be compared and integrated in a context from which confounding effects of anatomical shape differences are factored out. Nonlinear registration algorithms, therefore, support the integration of multisubject brain data in a stereotaxic framework and are increasingly used in functional image analysis packages (Seitz *et al.,* 1990; Friston *et al.,* 1995).

Any successful warping transform for cross-subject registration of brain data must be high-dimensional, in order to accommodate fine anatomic variations (Christensen *et al.,* 1996; Thompson and Toga, 1998). This warping is required to bring the atlas anatomy into structural correspondence with the target scan at a very local level. Another difficulty arises from the fact that the topology and connectivity of the deforming atlas have to be maintained under these complex transforms. This is difficult to achieve in traditional image warping manipulations (Christensen *et al.,* 1995). Physical continuum models of the deformation address these difficulties by considering the deforming atlas image to be embedded in a three-dimensional deformable medium which can be either an elastic material or a viscous fluid (Schormann *et al.,* 1996). The medium is subjected to certain distributed internal forces, which reconfigure the medium and eventually lead the image to match the target. These forces can be based mathematically on the local intensity patterns in the data sets, with local forces designed to match image regions of similar intensity.

3. Automated Methods

The intersubject differences in the anatomy of the brain can be large, even after alignment, making anatomical segmentation inaccurate (Galaburda *et al.,* 1978; Geschwind and Levitsky, 1968; Gur *et al.,* 1980; Steinmetz *et al.,* 1991; Zilles *et al.,* 1995, 1997). Without any perceived pathology, structures in the brain can differ in shape and size, as well as in relative orientation to each other (Roland and Zilles, 1994; Mazziotta *et al.,* 1995a,b). Affine transformation is often insufficient for the labeling and segmentation of structures. Automated image registration (AIR) algorithms can be used to align MR data with previously labeled and segmented brains by maximizing a measure of intensity similarity, such as three-dimensional cross-correlation (Collins *et al.,* 1994), ratio image uniformity (Woods *et al.,* 1992), or mutual information (Viola and Wells, 1995; Wells *et al.,* 1997). These techniques can be used in a nonlinear fashion to obtain better results, but they still develop errors with small structures and in the borders of larger structures. The following steps have been used:

• Each pulse sequence for each subject is intensity normalized within and between slices.

• Pulse sequences are aligned and registered within subjects (between pulse sequences) using AIR (Woods *et al.,* 1992).

• The intensity-normalized and aligned data sets from each subject are spatially normalized to our labeled target.

• Skull and scalp stripping is accomplished using the Leahy algorithm (Sandor and Leahy, 1997).

• Manual editing and segmentation are performed with SEG (UCLA) or DISPLAY (MNI).

To ensure accuracy, a previously labeled atlas is registered to the target brain via a nonlinear technique that captures the desired structures in the region of interest (ROI) defined by a probability density of where the structure of interest lies. An iterative procedure is used by which the ROI in the atlas is registered to the ROI projected into the target brain via a high-dimensional warping technique that allows all segments of the anatomy to grow, shrink, twist, and rotate. The ROI can then be refined to include greater detail and a closer approximation to the desired structure. The refined ROI in the atlas is again registered with the high-dimensional techniques to the target brain. This is repeated until the ROI is equal to the desired anatomical structure, within some allowed error estimate. In this way, a successive approximation from lobar, to gyral, to subgyral, to nuclear resolution labels can be achieved.

The registration techniques for the high-dimensional warps do not have to be limited to the previously mentioned intensity-based techniques. Edges and surfaces can be automatically computed and used to determine boundaries of structures (Sandor and Leahy, 1997; Lohmann, 1998; Duta *et al.,* 1999; LeGoualher *et al.,* 1999, Zhou *et al.,* 1998; Zhou and Toga, 1999). These boundaries can be aligned and used to align the tissue that surrounds them via continuum mechanical techniques guiding the tissue flow (Thompson *et al.,* 1996a,b).

Ultimately, it is the combination of different registration techniques in the proper order that archive a registration accurate enough to transfer the boundaries of one segmentation in the atlas to that of the target volume. When combined with the automated selection of a given atlas from a database of populations and the use of probabilistic information from the template associated with the class of interest, it will be possible to accurately label and segment any digital brain volume.

D. Surface Methods

1. Surface Extraction

Vast numbers of anatomical models can be stored in a population-based atlas (Thompson and Toga, 1997, 2000). These models provide detailed information on the three-dimensional geometry of the brain and how it varies in a population. By averaging models across multiple subjects, subtle features of brain structure emerge that are obscured in an individual due to wide cross-subject differences in anatomy (Thompson *et al.,* 2000b,c). These modeling approaches have recently uncovered striking patterns of disease-specific structural differences in Alzheimer's disease (Thompson *et al.,* 1997, 1998, 2000b), schizophrenia (Narr *et al.,* 2000), and fetal alcohol syndrome (Sowell *et al.,* 2000), as well as strong linkages between patterns of cortical organization and age (Thompson *et al.,* 2000a), gender (Thompson *et al.,* 2000b), cognitive scores (Mega *et al.,* 1997), and genotype (Le Goualher *et al.,* 2000) (see Chapter 28 for details). To illustrate the approach, Fig. 9 shows a model of the lateral ventricles, in which each element is represented by a three-dimensional surface mesh. These surface models can often be extracted automatically from image data, using recently developed algorithms based on deformable parametric surfaces (Thompson *et al.,* 1996a, 1996b; MacDonald 1998) or voxel coding (Zhou and Toga, 1999). Once an identical computational grid (or surface mesh) is imposed on the same structure in different subjects, an average anatomical model can be created for a group. This is done by averaging the three-dimensional coordinate locations of boundary points that correspond across subjects.

2. Cortical Surface Analysis Algorithms

a. Cortical surface segmentation. Multiple surface deformation (MSD) is a fully automated procedure for fitting and unfolding the entire human cortex, using an algorithm which automatically fits a three-dimensional mesh model to the cortical surface extracted from MRI. MSD uses an iterative minimization of a cost function that balances the distance of the deforming surface from (i) the target surface and (ii) the previous iteration surface. Specification of the relative weight of these competing forces allows MSD to range from unconstrained (data-driven) deformation to tightly constrained (model-preserving) deformation. Further shape-preserving constraints are also employed. The initial mesh surface can be chosen arbitrarily to be a simple geometric object, such as a sphere, an ellipsoid, or two independently fitted hemispheres (MacDonald *et al.,* 1994). Recently, MSD has been extended to allow simultaneous extraction of both inner and outer surfaces of the cortical mantle, using linked concentric mesh models (MacDonald *et al.,* 2000). Corresponding vertices in each surface are elastically linked using distance range constraints. Intersurface cross intersection and intrasurface self-intersection constraints prevent impossible topologies. These two factors allow for a deeper penetration of the deforming surfaces into the cortical sulci since areas where infolding of the outer (gray–CSF) boundary is indistinct due to partial volume effects are areas where the inner (gray–white) boundary is usually well distinguished. MSD can operate upon raw image intensity or upon fuzzy-classified tissue maps. Extraction of both surfaces yields a measurement of cortical thickness at each surface vertex. The thickness measurement can be defined in a variety of ways: (i) distance between corresponding vertices, (ii) closest approach of one surface to each vertex of the other surface, or (iii) distance between surfaces along the surface normal at each vertex of one surface. These

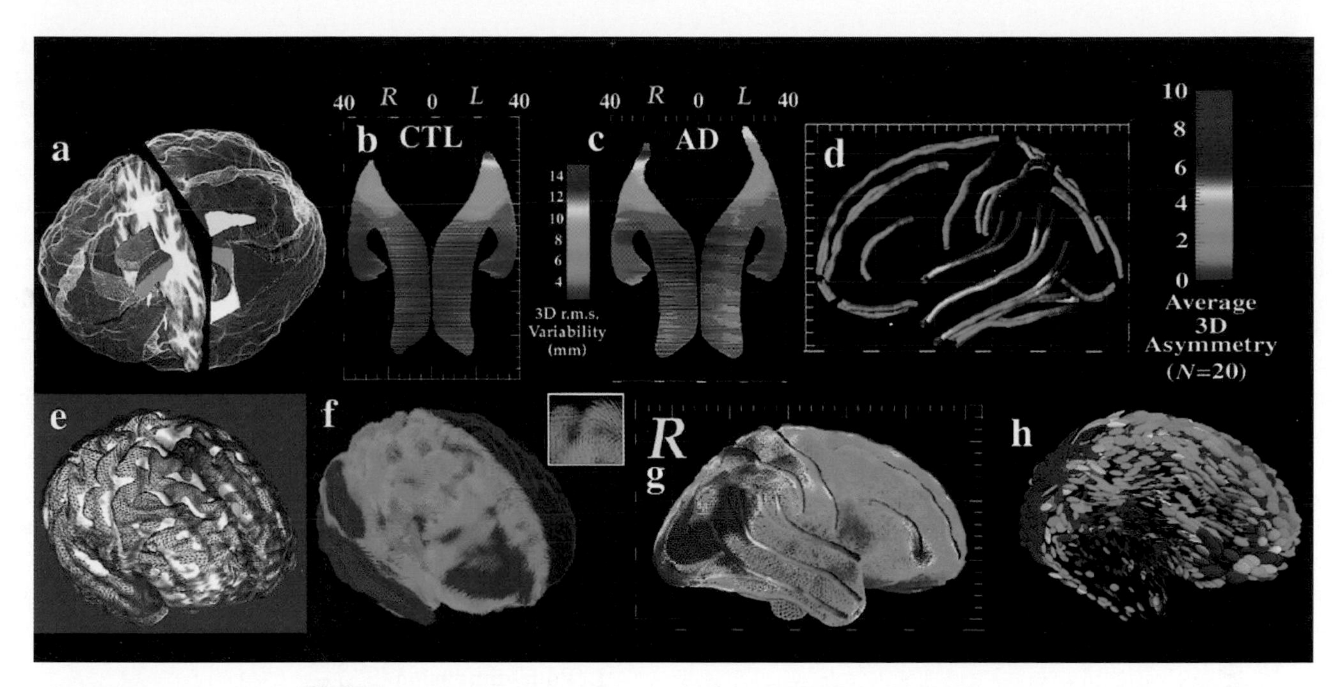

Figure 9 Surface models. Three-dimensional models can be created to represent major structural and functional interfaces in the brain. (**a**) A model of the lateral ventricles, in which each element is a three-dimensional parametric surface mesh. (**b** and **c**) Average ventricular models from a group of patients with Alzheimer's disease ($N = 10$) and matched elderly controls ($N = 10$). Note the larger ventricles in the patients and a prominent ventricular asymmetry (left larger than right). These features emerge only after averaging models for groups of subjects. Average population maps of cortical anatomy (**d**) reveal a clear asymmetry of the perisylvian cortex. (**e**) An individual's cortex (brown mesh) overlaid on an average cortical model for a group. Differences in cortical patterns are encoded by computing a three-dimensional elastic deformation (**f**) (pink colors, large deformation) that reconfigures the average cortex into the shape of the individual, matching elements of the gyral pattern exactly. These deformation fields (f) provide detailed information on individual deviations and can be averaged across subjects to create three-dimensional variability maps, demonstrating fundamental patterns of anatomical variability in the brain (**g**) (Thompson *et al.*, 2000). Tensor maps using color ellipsoids (h) reveal the directions in which anatomical variation is greatest. The ellipsoids are more elongated in the directions in which structures tend to vary the most. Pink colors denote the largest variation while blue colors show the least. These statistical data can be used to detect patterns of abnormal anatomy in new subjects. Severe abnormality is detected (red colors), while corresponding regions in a matched elderly control subject are signaled as normal. (Courtesy of P. Thompson, UCLA School of Medicine, Los Angeles, CA.)

definitions give rise to different absolute values for cortical thickness (closest approach must yield the smallest value, by definition) but the variation in thickness over the whole cortex is generally very similar among the distance measures (MacDonald *et al.*, 2000). The method has been applied to a set of 102 MRI volumes from the ICBM database which have been previously mapped automatically into stereotaxic space (Collins *et al.*, 1994) and used to generate various group results by averaging the three-dimensional location of corresponding vertices across subjects. The average outer cortical surface obtained when simultaneously fitting both surfaces exhibits a dramatic increase detail compared with that obtained when fitting only the outer surface, a consequence of the deeper penetration into individual sulci. Since the average cortex can be used as the starting point for mesh modeling of any individual surface, this is likely to lead to faster and more accurate extraction of individual cortical surfaces in future. Moreover, the average cortical surface is used by some groups to constrain electrophysiological inverse solutions (e.g., Harmony *et al.*, 1999) and an improved specification of this surface can be expected to improve that process. The cortical thickness maps exhibit the expected variation in cortical thickness, the temporal poles having the thickest cortex (4–6 mm) and the posterior bank of the central sulcus having the thinnest (1.8–2.5 mm).

This approach has been tested against manual estimates for 20 regions (10 per hemisphere) using 40 brain MRI studies. Validity was determined by an anatomist labeling the CSF–gray and gray–white borders of selected gyri and by allowing the algorithm to determine the CSF–gray and gray–white borders for the same region. The distance between the CSF–gray and gray–white tags determined the cortical thickness at that point. The manual and automatic methods were in agreement for all but 4 of 20 regions tested. The 4 regions where the results were statistically different between the two methods were the insula in both hemispheres, the cuneus, and the parahippocampus in the right hemisphere. Thus, the automatic algorithm is valid for most of the cortex and provides a reasonable alternative to manual *in vivo* measurement except in regions where cortex is adjacent to other gray matter structures.

b. Sulcal extraction and labeling. We have implemented an automated sulcal extraction and labeling algorithm (SEAL) (LeGoualher *et al.*, 1999, 2000). At every voxel on the MSD isosurface, SEAL calculates the two principal curvatures: the mean curvature and the Gaussian curvature. Voxels with negative mean curvature, belonging to sulci, are extracted and pruned to obtain a set of superficial sulcal traces. SEAL extracts the buried sulcus with an "active ribbon" that evolves in three dimensions from a superficial trace to the bottom of a sulcus by optimizing an energy function based on (i) maximizing distance between starting and current trace position (i.e., for increased penetration), (ii) maximizing distance to any other sulcal voxel (i.e., stay within sulcus), and (iii) minimizing distance from the median sulcal locus, as defined by the mLvv "ridge" operator. To encode the extracted information, we defined a relational graph structure composed of two main features, arcs and vertices. Arcs contain a surface representing the interior of a sulcus. Points on this surface are expressed in stereotaxic coordinates. For each arc, we store length, depth, and orientation, as well as attributes, e.g., hemisphere, lobe, sulcus type. Each vertex stores its three-dimensional location and its connecting arcs. We have written functions to access this data structure that allow a systematic description of the sulci themselves and their interconnections. Sulcal labeling is performed semiautomatically within DISPLAY by tagging a sulcal trace in the three-dimensional graph and selecting from a menu of candidate labels. The menu is restricted to most likely candidates by the use of spatial priors for sulcal distribution. Given these SPAMs, the user is provided with the probability that the selected arc belongs to a particular sulcus.

E. Database

Several approaches can be used in the creation of databases to accommodate the diversity of data types and structures needed to adequately represent brain structure and function in four dimensions. Whereas a map is a collection of information, a representation of our understanding of the brain, a database is designed with more interactions in mind. Its function is to organize and archive data records *and* provide an efficient and comprehensive query mechanism. Modern digital maps have only begun to incorporate database functionality.

One of the first database brain maps was developed by Bloom *et al.* (1990). They created an electronic version of atlas delineations from the Paxinos and Watson (1986) neuroanatomic atlas of the rat brain. These outlines were equated with coordinates and nomenclature so that the user could request information regarding structural groupings and systems. Since the system was based upon a hypercard (Apple Computer Corp.) database, the user could add information to this anatomic framework as an anatomy laboratory organizer. In a similar vein, Swanson (1992) provided a digital version of his anatomic delineations to his atlas of the rat brain. Cortical connectivity in the macaque monkey also has been organized as a database (Felleman and Van Essen, 1991). The most sophisticated attempt in the human brain mapping literature is BrainMap (Fox *et al.*, 1994; Fox and Lancaster, 1994). This database incorporates a true relational database structure intended to encapsulate data from diverse studies of brain structure and function. The anatomic framework is based upon the Talairach system and the database relates information about the activation task, the methods, the bibliography, and other pertinent data. Usually the image data are excluded and only boundary information is retained. We have chosen to include source image data in the database in a manner that supports both visualization and exploratory query. The query modes include both spatial query of the source image data and query based on a reference anatomical coordinate system.

1. Four-dimensional

A four-dimensional database allows for the intuitive referencing of information by time (age) and place in the nervous system. This works well within a species but requires separate atlases for any given species with links between them to be established only when sufficient information is available to identify true anatomical or functional homologues between the two populations. Once established, the attributes associated with each four-dimensional point could similarly be linked between species-specific atlases and their associated probability estimates.

2. Daemon and BrainMap

An automated coordinate-based system to retrieve brain labels from the 1988 Talairach atlas, called the Talairach Daemon (TD), was previously introduced (Lancaster *et al.*, 1997). The TD system and its three-dimensional database of labels for the 1988 Talairach atlas were tested for labeling of functional activation foci. The TD system labels were compared with author-designated labels of activation coordinates from over 250 published functional brain mapping studies and with manual atlas-derived labels from an expert group using a subset of these activation coordinates. Automated labeling by the TD system compared well with authors' labels, with a 70% or greater label match averaged over all locations. Author-label matching improved to greater than 90% within a search range of ±5 mm for most sites. An adaptive gray matter (GM) range-search utility was evaluated using individual activations from the M1 mouth region (30 subjects, 52 sites). An 87% label match to Brodmann area labels (BA 4 and BA 6) was achieved within a search range of ±5 mm. Using the adaptive GM range search, the TD system's overall match with authors' labels (90%) was better than that of an expert group (80%). When used in concert with authors' deeper knowledge of an experiment, the TD system provides consistent and comprehensive labels for brain activation foci. Additional suggested applications of the TD system include interactive labeling, anatomical grouping of activation foci, lesion-deficit analysis, and neuroanatomy education.

V. Other Issues

A. Isolated Brain Regions

The more difficult problem, compared to working with whole brain, three-dimensional data sets, is that of entering microscopic data from brain sites that are analyzed on a regional basis (e.g., the study of the isolated hippocampus). Nevertheless, such data can also be incorporated into the probabilistic reference system and atlas. Such a problem will require landmarks to appropriately localize regional data in the global atlas brain.

Consider a series of postmortem cryomacrotome human brains that are stained with a series of conventional and commonly used neuroanatomical "landmark" stains (e.g., Nissl, acetylcholinesterase). Using state-of-the-art imaging devices, these sections would be digitized and sampled at a 20-μm resolution. The resultant data sets would be warped and entered into the probabilistic atlas as an additional feature. Then consider an investigator who studies GABA receptors in the human hippocampus. This investigator would like to see where the receptors from the hippocampi of a given epileptic patient population fall with regard to other data in the probabilistic reference system. In preparing the tissue, this investigator would process every *N*th section using one of the landmark stains that are part of the probabilistic atlas. The investigator would then digitize the information from both the GABA receptor sections and the landmark-stained sections. Using alignment, registration, and warping tools that are part of the atlas system, the investigator would register the landmark-stained sections with the atlas and then use the same mathematical transformations to enter the GABA receptor information into the hippocampal region of the atlas. Once referenced, database queries and visualization of these new data could be performed in the atlas system. A similar approach allows referencing between newly acquired *in vivo* data and stored postmortem specimens that should aid in relating functional localization with macroscopic and microscopic anatomy (Rademacher *et al.,* 1992; Larsson *et al.,* 1999; Naito *et al.,* 1999, 2000; Bodegard *et al.,* 2000a,b).

B. EEG/MEG

The ICBM atlas is based on neuroanatomy. This is the most fundamental language of communication in neuroscience. As such, it allows appropriate reference and localization to any structure in the brain from any signal source. In the development of the reference system, cross-sectional and tomographic data have been the initial data sets. Once established, however, appropriate vehicles for entering nontomographic data will be developed. For EEG data, for example, systems already exist to localize scalp electrode placement three-dimensionally, either through the use of a paired tomographic image set or by nontomographic localization methods (Gevins *et al.,* 1994).

C. Sociology

Any endeavor to organize information across laboratories, or especially across an entire field, requires attention to the sociology involved (Koslow, 2000). Frustration with existing methods must be high enough and the solutions good enough (in terms of practicality, economics, and implementation) that it will be adopted. Such a transition is made easier if rigid new standards are not imposed on the structure or organization of data generated in a given laboratory but rather the tools are available to translate such data into the framework and form required for interaction with the database and atlas. This is the strategy we have employed. Perhaps the most important, if not critical, step is the willingness on the part of the community to share data in all its forms (including raw data) to allow for the full implementation of such a system. Such strategies will require participation of traditional final end products of research (e.g., journals) as well as academic recognition for data provided to such systems. Finally, it is always important to have a consensus from the community before embarking on the construction of a complex system such as this one. Wide participation, frequent requests for input, and distributed testing of products are all helpful in establishing a successful system that is accepted by the community for which it was intended.

VI. Limitations and Deliverables

A. Deliverables

A project of this size and scope has a very large overhead at the front end. This results in frustrations for the participants as well as for the community which it is intended to serve. Nevertheless, prior to the release of any individual products (e.g., algorithms, data sets) or the atlas itself, sufficient documentation, validation, and a critical mass of test subjects must be acquired in order to be confident of the outcome. We have described our strategy for developing a competitive approach to algorithm development that has at least five phases: theory, initial development, alpha testing, beta testing, and general release. A number of such algorithms have already completed this lengthy process. These include AIR (Woods *et al.,* 1992,

1993, 1998) (now distributed to over 1300 laboratories worldwide), N3 (Sled *et al.*, 1997, 1998) (now available at Website http://www.bic.mni.mcgill.ca/software/N3/), an MRI environment simulator (available at Website http://www.bic.mni.mcgill.ca/brainweb/), and algorithms of other components of the ICBM analysis pipeline that are well on their way through this competitive process and will be released when complete. The same can be said of data sets. Both human, *in vivo* MRI studies and cryosection data sets and examples can be found at the ICBM Website.

In an attempt to allow the general neuroimaging and neuroscientific communities to have access to some of the more basic data that have been collected thus far, we are in the process of developing digital libraries. These libraries were described above and will contain raw images (with facial features corrupted) and intensity-normalized images from multiple pulse sequences, as well as normalized and "scalped," multiple pulse sequences for each subject. This will allow investigators to search for selected subpopulations and use the resultant data for normal controls, methodological developments, and many other, currently unforeseen, uses.

B. Limitations

Every project has its limitations. This one is no different. When faced with the opportunity to evaluate 7000 normal individuals, there is a tendency to be all-inclusive and attempt to collect every potential type of information available. At the onset of this study the contributing investigators met and discussed all of the possible data sets that could be collected from a human subject. The list was long and will not be reiterated here. We opted to start with those data sets that would provide structural imaging of the highest resolution in the largest number of subjects for the best price. We felt that in later years, and in subsequent iterations, it might be possible to add other data sets. In fact, this was done with the addition of functional imaging using *f*MRI, PET, and event-related potentials. Nevertheless, it was not possible to add information about the vasculature from MR angiography, neurotransmitter systems through PET or SPECT ligand studies, cerebral perfusion through perfusion MRI or PET, chemical information about the brain from MR spectroscopy, or data sets that describe major white matter tracts in the brain using diffusion tensor imaging or connectivity using combinations of transcranial magnetic stimulation and PET or *f*MRI. These are all issues that would be extremely important and valuable to add in the future. Those that have been selected and implemented reflect the basic criteria list noted above as well as the realistic constraints associated with finances, subject risk, time burdens and institutional review board (IRB) criteria. In fact, the reason that only 5800 of the 7000 subjects have DNA samples relates to IRB rules in certain countries with regard to the collection and distribution of genetic materials and information about subjects.

We believe that having neuroanatomy as the basis for building the ICBM Probabilistic Atlas and Reference System was the logical and correct starting point. Other factors that can be added as attributes will be a function of practicality, finances and interests, of the field. They will also be dictated by advances and developments in methodologies.

VII. Conclusions

There is no question that the development of systems and tools such as the probabilistic atlas will have a specific and not insignificant cost associated with them. Also true is the fact that increments in neuroscientific research funding have not kept pace with the growth of the field in terms of numbers of investigators or the magnitude of their projects.

The thoughtful response to this statement requires, however, an honest appraisal of the ultimate goals of neuroscientific research. If such research is designed to produce the most accurate understanding of normal brain function and diseases that affect it, then tools that will enhance the accuracy of results and the comparison of results between subjects and laboratories, make more rigorous the confirmation or refutation of data, and guard against its loss should have a high priority. A system such as a probabilistic atlas for a given species, or potentially across species, provides a means by which to rigorously store, compare, and analyze data over time and between laboratories. Such a system does not currently exist. Further, by virtue of data exchange and comparison, integration within the broad field of neuroscience will begin.

One could rephrase the above statement into a question and ask, "What would it cost not to develop such integrated systems?" The costs, in our opinion, would be the progressive and continued reduction in the value of every dollar spent on future neuroscientific research because of the progressively unmanageable amounts and types of data that are generated by neuroscientists. Lacking the tools to manage, compare, and analyze these data sets will make funds spent for their acquisition of lesser impact than if such data could be preserved and referenced in an ever-evolving and integrated approach.

Clearly, it would be optimal if funding for systems and approaches to integrate data across not only neuroscience, but also computer science, informatics, and potentially other related fields, could come from new sources. In fact, this is already happening. Contributions to the funding of the initial round of the Human Brain Project (Huerta *et al.*, 1993) in the United States came from sources that are both traditional and novel for funding neuroscientific research.

By having small contributions from many countries and agencies, the burden on any one country or agency is small but the impact for the neuroscience community is significant. It is hoped that an expanded participation by agencies and contributors outside of traditional pathways as well as the potential for generating new appropriations based on interest in this international effort will result in the creation of systems such as the one described in this report as well as others contemplated or funded through the auspices of the Human Brain Project in the United States without detracting from traditional neuroscientific funding.

The development of a probabilistic atlas and reference system for the human brain is a formidable goal and one that involves participation from many sites around the world and investigators committed to the end product. The creation of a probabilistic atlas of the human brain is not an exercise in library science. It is a series of fundamental, hypothesis-driven experiments in merging mathematical and statistical approaches with morphological and physiological problems posed with regard to the nervous system. It will create new data and insights into the organization of the human nervous system in health and disease, its development, and its evolution. When successful, it will provide previously unprecedented tools for organizing, storing, and communicating information about the human brain throughout development, maturation, adult life, and old age. It will be a natural prelude to studies of patients with cerebral disorders and provide the first mechanism by which phenotype–genotype–behavioral comparisons can be made on a macroscopic and microscopic level. These results will provide the first insights into the structure–function organization of the human brain across all structures and a wide range of ages. Its design anticipates the continuing evolution in the quality, resolution, and magnitude of data generated by existing technologies that are used to map the human brain and even anticipates that many future technologies, unknown today, will be applicable because the entire system is organized using the architecture of the brain as its guiding principle. The result will allow electronic experimentation and hypothesis generation, facilitated communication among investigators, and an objective way of assessing new information gleaned either at scientific meetings or through publications. Developing such a system is an open-ended project with constant evolution, improvement, and expansion both in the numbers of subjects included and in the range of attributes associated with each. The results should be far more than a data structure and organizational system. Rather, the system should provide new insights and new opportunities for neuroscientists to utilize data in their own laboratories as well as others to more rapidly, effectively, and efficiently make progress in understanding human brain function in health and disease.

Acknowledgments

This work was supported by a grant from the Human Brain Project (P20-MHDA52176), funded by the National Institute of Mental Health, National Institute for Drug Abuse, National Cancer Institute, and National Institute for Neurological Disease and Stroke. For generous support, the author also thanks the Brain Mapping Medical Research Organization, the Pierson–Lovelace Foundation, The Ahmanson Foundation, the Tampkin Foundation, the Jennifer Jones Simon Foundation, the Robson Family, and National Center for Research Resources Grants RR12169 and RRO8655. Special thanks to Laurie Carr for the preparation of the manuscript and Andrew Lee for preparation of illustrations. The author thanks the faculties and staffs of the participating organizations for their dedication and participation in this program. Finally, we thank the subjects who participated in these investigations for their time, interest, and commitment.

References

Ahlfors, S. P., Simpson, G. V., Dale, A. M., Belliveau, J. W., Liu, A., Korvenoja, A., Virtanen, J., Huotilanen, M., Tootell, R. B. H., Aronen, H. J., and Ilmoniemi, R. J. (1999). Spatiotemporal activity of a cortical network for processing visual motion revealed by MEG and *f*MRI. *J. Neurophysiol.* **82,** 2545–2555.

Aine, C. J., Supek, S., George, J. S., Ranken, D., Lewine, J., Sanders, J., Best, E., Tiee, W., Flynn, E. R., and Wood, C. C. (1996). Retinotopic organization of human visual cortex: Departures from the classical model. *Cereb. Cortex* **6,** 354–361.

Amunts, K., Klingberg, T., Binkofski, F., Schormann, T., Seitz, R. J., Roland, P. E., *et al.* (1998). Cytoarchitectonic definition of Broca's region and its role in functions different from speech. *NeuroImage* **7,** 8.

Amunts, K., Malikovic, A., Mohlberg, H., Schormann, T., and Zilles, K. (2000). Brodmann's areas 17 and 18 brought into stereotaxic space—Where and how variable? *NeuroImage* **11,** 66–84.

Amunts, K., Schleicher, A., Bürgel, U., Mohlberg, H., Uylings, H. B. M., and Zilles, K. (1999). Broca's region revisited: Cytoarchitecture and intersubject variability. *J. Comp. Neurol.* **412,** 319–341.

Arnold, J. B., Liow, J.-S., Schaper, K. A., *et al.* (2001). Quantitative and qualitative evaluation of six algorithms for correcting intensity non-uniformity effects. *NeuroImage,* in press.

Ashburner, J., Neelin, P., Collins, D. L., Evans, A., and Friston, K. (1997). Incorporating prior knowledge into image registration. *NeuroImage* **6,** 344–352.

Bailey, P., and von Bonin, G. (1951). "The Isocortex of Man." University Press, Urbana, IL.

Bajcsy, R., and Kovacic, S. (1989). Multi-resolution elastic matching. *Comput. Vision Graph. Image Process.* **46,** 1–21.

Bartley, A. J., Jones, D. W., and Weinberger, D. R. (1997). Genetic variability of human brain size and cortical gyral patterns. *Brain* **120,** 257–269.

Bloom, F. E. (1996). The multidimensional database and neuroinformatics requirements for molecular and cellular neuroscience. *NeuroImage* **4,** S12–S13.

Bloom, F. E., Young, W. G., and Kim, Y. M. (1990). "Brain Browser: Hypercard Application for the Macintosh." Academic Press, San Diego.

Bodegård, A., Geyer, S., Naito, E., Zilles, K., and Roland, P. E. (2000a). Somatosensory areas in man activated by moving stimuli: Cytoarchitectonic mapping and PET. *NeuroReport* **11,** 187–191.

Bodegård, A., Ledberg, A., Geyer, S., Naito, E., Larsson, J., Zilles, K., and Roland, P. (2000b). Object shape differences reflected by somatosensory cortical activation in human. *J. Neurosci.* **20,** 1–5.

Brodmann, K. (1909). "Vergleichende Lokalisationslehre der Grosshirnrinde in Ihren Prinzipien Dargestellt auf Grund des Zellenbaues." Barth, Leipzig.

Broit, C. (1981). "Optimal Registration of Deformed Images." Univ. of Pennsylvania, Philadelphia. [Ph.D. dissertation]

Bro-Nielsen, M., and Gramkow, C. (1996). Fast fluid registration of medical images. *In* "Proceedings of the Fourth International Conference on Visualization in Biomedical Computing (VBC '96)," Hamburg, September 22–25 (K. H. Hohne and R. Kikinis, eds.), pp. 267–76. Springer-Verlag, Berlin.

Cannestra, A. F., Santori, E. M., Holmes, C. J., and Toga, A. W. (1997). A three-dimensional multimodality map of the nemistrina monkey. *Brain Res. Bull.* **5,** 147–153.

Carman, G. J., Drury, H. A., and Van Essen, D. C. (1995). Computational methods for reconstructing and unfolding the cerebral cortex. *Cereb. Cortex* **5,** 506–517.

Christensen, G., Rabbitt, R. D., and Miller, M. I. (1993). A deformable neuroanatomy textbook based on viscous fluid mechanics. *In* "Proceedings of the 1993 Conference on Information Science Systems" (Prince and Runolfsson, eds.), pp. 211–216. Johns Hopkins Univ., Baltimore.

Christensen, G. E., Rabbitt, R. D., and Miller, M. I. (1996). Deformable templates using large deformation kinematics. *IEEE Trans. Image Process.* **5,** 1435–1447.

Christensen, G. E., Rabbitt, R. D., Miller, M. I., Joshi, S. C., Grenander, U., Coogan, T. A., and Van Essen, D. C. (1995). Topological properties of smooth anatomic maps. *In* "Information Processing in Medical Imaging" (Y. Bizais, C. Barillot, and R. Di Paola, eds.), pp. 101–112. Springer-Verlag, Berlin.

Clark, S., and Miklossy, J. (1990). Occipital cortex in man: Organization of callosal connections, related myelo- and cytoarchitecture, and putative boundaries of functional visual areas. *J. Comp. Neurol.* **298,** 188–214.

Collins, D. L., Holmes, C. J., Peters, T. M., and Evans, A. C. (1995). Automatic 3D model-based neuroanatomical segmentation. *Hum. Brain Mapp.* **3,** 190–208.

Collins, D. L., Neelin, P., Peters, T. M., and Evans, A. C. (1994). Automatic 3D registration of MR volumetric data in standardized Talairach space. *J. Comput. Assisted Tomogr.* **18,** 192–205.

Dale, A. M., Fischl, B., and Sereno, M. I. (1999). Cortical surface-based analysis. I. Segmentation and surface reconstruction. *NeuroImage* **9,** 179–194.

Davatzikos, C. (1996). Spatial normalization of 3D brain images using deformable models. *J. Comput. Assisted Tomogr.* **20,** 656–665.

Davatzikos, C. (1997). Spatial transformation and registration of brain images using elastically deformable models. *Comput. Vision Image Understand.* **66,** 207–222.

Dumoulin, S. O., Bittar, R. G., Kabani, N. J., Baker, C. L., LeGoualher, G., Pike, G. B., and Evans, A. C. (2000). A new neuroanatomical landmark for the reliable identification of human area V5/MT: A quantitative analysis of sulcal patterning. *Cereb. Cortex* **10,** 454–463.

Duta, N., Sonka, M., and Jain, A. K. (1999). Learning shape models from examples using automatic shape clustering and Procrustes analysis. *In* "Lecture Notes in Computer Science," Vol. 1613, "Information Processing in Medical Imaging. 16th International Conference, IPMI'99: Proceedings" (A. Kuba, M. Samal, and Todd-Pokropek, eds.), pp. 370–375. Springer-Verlag, Berlin.

Evans, A. C., Collins, D. L., and Holmes, C. J. (1996). Automatic 3D regional MRI segmentation and statistical probability anatomy maps. *In* "Quantification of Brain Function Using PET" (R. Myers, V. Cunningham, D. Bailey, and T. Jones, eds.), Chap. 25, pp. 123–130. Academic Press, San Diego.

Evans, A. C., Dai, W., Collins, D. L., Neelin, P., and Marrett, S. (1991). Warping of a computerized 3D atlas to match brain image volumes for quantitative neuroanatomical and functional analysis. *SPIE Med. Imaging* **1445,** 236–247.

Evans, A. C., Frank, J. A., Antel, J., and Miller, D. H. (1997). The role of MRI in clinical trials of multiple sclerosis: Comparison of image processing techniques. *Ann. Neurol.* **41,** 125–132.

Felleman, D., and Van Essen, D. (1991). Distributed hierarchical processing in the primate cerebral cortex. *Cereb. Cortex* **1,** 1–47.

Filimonoff, I. N. (1932). Über die Variabilität der Grosshirnrindenstruktur. Mitteilung II. Regio occipitalis beim erwachsenen Menschen. *J. Psychol. Neurol.* **44,** 2–96.

Fischl, B., Sereno, M. I., Tootell, R. B., and Dale, A. M. (1999). High-resolution intersubject averaging and a coordinate system for the cortical surface. *Hum. Brain Mapp.* **8,** 272–284.

Flechsig, P. (1920). "Anatomie des Menschlichen Gehirns und Ruckenmarks." Thieme, Leipzig.

Fox, P. T., and Lancaster, J. L. (1994). Neuroscience on the net. *Science* **266,** 994–996.

Fox, P. T., and Mintun, M. A. (1989). Noninvasive functional brain mapping by change-distribution analysis of averaged PET images of $H_2^{15}O$ tissue activity. *J. Neurosci.* **7,** 913–922.

Fox, P. T., Mikiten, S., Davis, G., and Lancaster, J. (1994). Brain Map: A database of human functional brain mapping. *In* "Functional Neuroimaging" (R. W. Thatcher, M. Hallett, T. Zeffiro, E. R. John, and M. Huerta, eds.), pp. 95–105. Academic Press, San Diego.

Friston, K. J., Frith, C. D., Liddle, P. F., and Frackowiak, R. S. J. (1991). Comparing functional (PET) images: The assessment of significant change. *J. Cereb. Blood Flow Metab.* **11,** 690–699.

Friston, K. J., Holmes, A. P., Worsley, K. J., Poline, J. P., Frith, C. D., and Frackowiak, R. S. J. (1995). Statistical parametric maps in functional imaging: A general linear approach. *Hum. Brain Mapp.* **2,** 189–210.

Galaburda, A. M. (1980). La region de Broca: Observations anatomiques faites un siecle apres la mort de son decoveur. *Rev. Neurol.* **136,** 609–616.

Galaburda, A. M., LeMay, M., Kemper, T. L., and Geschwind, N. (1978). Right–left asymmetries in the brain. *Science* **199,** 852–856.

Gee, J. C., Reivich, M., and Bajcsy, R. (1993). Elastically deforming an atlas to match anatomical brain images. *J. Comput. Assisted Tomogr.* **17,** 225–236.

Gee, J. C., LeBriquer, L., Barillot, C., Haynor, D. R., and Bajcsy, R. (1995). Bayesian approach to the brain image matching problem. Institute for Research in Cognitive Science Technical Report 95-08, April 1995.

Geschwind, N., and Levitsky, W. (1968). Human brain: Left–right asymmetries in temporal speech region. *Science* **161,** 186–187.

Gevins, A. S., Lee, J., Martin, N., Reutter, R., Desmond, J., and Brickett, P. (1994). High resolution EEG: 124-channel recording, spatial deblurring and MRI integration methods. *EEG Clin. Neurophysiol.* **90,** 337–358.

Geyer, S., Ledberg, A., Schleicher, A., Kinomura, S., Schormann, T., Bürgel, U., Klingberg, T., Larsson, J., Zilles, K., and Roland, P. E. (1996). Two different areas within the primary motor cortex of man. *Nature* **382,** 805–807.

Geyer, S., Schleicher, A., and Zilles, K. (1997). The somatosensory cortex of human: Cytoarchitecture and regional distributions of receptor-binding sites. *NeuroImage* **6,** 27–45.

Geyer, S., Schleicher, A., and Zilles, K. (1999). Areas 3a, 3b, and 1 of human primary somatosensory cortex. 1. Microstructural organization and interindividual variability. *NeuroImage* **10,** 63–83.

Geyer, S., Schormann, T., Mohlberg, H., and Zilles, K. (2000). Areas 3a, 3b, and 1 of human primary somatosensory cortex. 2. Spatial normalization to standard anatomical space. *NeuroImage,* in press.

Gur, R. C., Packer, I. K., Hungerbuhler, J. P., Reivich, M., Obrist, W., Amarnek, W., and Sackeim, H. (1980). Differences in the distribution of gray and white matter in human cerebral hemispheres. *Science* **207,** 1226–1228.

Harmony, T., Fernandez, T., Silva, J., Bosch, J., Valdes, P., Fernandez-Bouzas, A., Galan, L., Aubert, E., and Rodriguez, D. (1999). Do specific EEG frequencies indicate different processes during mental calculation? *Neurosci. Lett.* **266,** 25–28.

Hayes, T. L., and Lewis, D. A. (1995). Anatomical specialization of the anterior motor speech area: Hemispheric differences in magnopyramidal neurons. *Brain Lang.* **49,** 289–308.

Hayes, T. L., and Lewis, D. A. (1996). Magnopyramidal neurons in the anterior motor speech region. *Arch. Neurol.* **53,** 1277–1283.

Holmes, C. J., Hoge, R., Collins, L., Woods, R., Toga, A. W., and Evans, A. C. (1998). Enhancement of magnetic resonance images using registration for signal averaging. *J. Comput. Assisted Tomogr.* **22,** 324–333.

Huerta, M., Koslow, S., and Leshner, A. (1993). The Human Brain Project: An international resource. *Trends Neurosci.* **16,** 436–438.

Ingvar, M., Eriksson, L., Greitz, T., Stone-Elander, S., Dahlbom, M., Rosenqvist, G., af Trampe, P., and von Euler, C. (1994). Methodological aspects of brain activation studies: Cerebral blood flow determined with [^{15}O]butanol and positron emission tomography. *J. Cereb. Blood Flow Metab.* **14,** 628–638.

Jacobs, B., Batal, H. A., Lynch, B., Ojemann, G., Ojemann, L. M., and Scheibel, A. B. (1993). Quantitative dendritic and spine analysis of speech cortices: A case study. *Brain Lang.* **44,** 239–253.

Kamber, M., Collins, D. L., Shinghal, R., Francis, G. S., and Evans, A. C. (1992). Model-based 3D segmentation of multiple sclerosis lesions in dual-echo MRI data. *In* "Proceedings of the 2nd Conference on Visualization in Biomedical Computing," Chapel Hill, NC, Vol. 1808, pp. 590–600.

Kamber, M., Shinghal, R., Collins, D. L., Francis, G. S., and Evans, A. C. (1995). Model-based 3D segmentation of multiple sclerosis lesions in magnetic resonance brain images. *IEEE Trans. Med. Imaging* **14,** 442–453.

Koslow, S. H. (2000). Should the neuroscience community make a paradigm shift to sharing primary data? *Nat. Neurosci.* **3,** 863–865.

Lancaster, J. L., Rainey, L. H., Summerlin, J. L., Freitas, C. S., Fox, P. T., Evans, A. C., Toga, A. W., and Mazziotta, J. C. (1997). Automated labeling of the human brain: A preliminary report on the development and evaluation of a forward-transform method. *Hum. Brain Mapp.* **5,** 238–242.

Larsson, J., Amunts, K., Gulyás, B., Malikovic, A., Zilles, K., and Roland, P. E. (1999). Neuronal correlates of real and illusory contour perception: Functional anatomy with PET. *Eur. J. Neurosci.* **11,** 4024–4036.

LeGoualher, G., Argenti, A., Duyme, M., Baare, W., Hulshoff Pol, H., Barillot, C., and Evans, A. (2000). Statistical sulcal shape comparisons: Application to the detection of genetic encoding of the central sulcus shape. *NeuroImage,* in press.

LeGoualher, G., Procyk, E., Collins, D. L., Venugopal, R., Barillot, C., and Evans, A. C. (1999). Automated extraction and variability analysis of sulcal neuroanatomy. *IEEE Trans. Med. Imaging* **18,** 206–217.

Lohmann, G. (1998). Extracting line representations of sulcal and gyral patterns in MR images of the human brain. *IEEE Trans. Med. Imaging* **17,** 1040–1048.

MacDonald, D. (1998). "A Method for Identifying Geometrically Simple Surfaces from Three-Dimensional Images." McGill Univ., Montreal. [Ph.D. dissertation]

MacDonald, D., Avis, D., and Evans, A. C. (1994). Multiple surface identification and matching in magnetic resonance imaging. *Proc. SPIE* **2359,** 160–169.

MacDonald, D., Kabani, N., Avis, D., and Evans, A. C. (2000). Automated 3D extraction of inner and outer surfaces of cerebral cortex from MRI. *NeuroImage,* in press.

Mazziotta, J. C., Huang, S. C., Phelps, M. E., Carson, R. E., MacDonald, N. S., and Mahoney, K. (1985). A non-invasive positron computed tomography technique using oxygen-15 labeled water for the evaluation of neurobehavioral task batteries. *J. Cereb. Blood Flow Metab.* **5,** 70–78.

Mazziotta, J. C., Toga, A. W., Evans, A., Fox, P., and Lancaster, J. (1995a). Digital brain atlases. *Trends Neurosci.* **18,** 210–211.

Mazziotta, J. C., Toga, A. W., Evans, A. C., Fox, P., and Lancaster, J. (1995b). A probabilistic atlas of the human brain: Theory and rationale for its development. *NeuroImage* **2,** 89–101.

Mega, M. S., Chen, S., Thompson, P. M., Woods, R. P., Karaca, T. J., Tiwari, A., Vinters, H., Small, G. W., and Toga, A. W. (1997). Mapping pathology to metabolism: Coregistration of stained whole brain sections to PET in Alzheimer's disease. *NeuroImage* **5,** 147–153.

Merker, B. (1983). Silver staining of cell bodies by means of physical development. *J. Neurosci.* **9,** 235–241.

Miller, M. I., Christensen, G. E., Amit, Y., and Grenander, U. (1993). Mathematical textbook of deformable neuroanatomies. *Proc. Natl. Acad. Sci. USA* **90,** 11944–11948.

Naito, E., Ehrsson, H. H., Geyer, S., Zilles, K., and Roland, P. E. (1999). Illusory arm movements activate cortical motor areas: A positron emission tomography study. *J. Neurosci.* **19,** 6134–6144.

Naito, E., Kinomura, S., Geyer, S., Kawashima, R., Roland, P. E., and Zilles, K. (2000). Fast reaction to different sensory modalities activates common fields in the motor areas, but the anterior cingulate cortex is involved in the speed of reaction. *J. Neurophysiol.* **83,** 1701–1709.

Narr, K. L., Thompson, P. M., Sharma, T., Moussai, J., Cannestra, A. F., and Toga, A. W. (2000). Mapping morphology of the corpus callosum in schizophrenia. *Cereb. Cortex* **10,** 40–49.

Neelin, P. D., MacDonald, D., Collins, D. L., and Evans, A. C. (1998). The MINC file format: From bytes to brains. *NeuroImage* **7,** S786.

Ono, M., Kubik, S., and Abernathy, C. (1990). "Atlas of the Cerebral Sulci." Thieme, Stuttgart.

Paus, T., Otaky, N., Caramanos, Z., MacDonald, D., Zijdenbos, A., D'Avirro, D., Gutmans, D., Holmes, C. J., Tomaiuolo, F., and Evans, A. C. (1996a). In-vivo morphometry of the intrasulcal gray-matter in the human cingulate, paracingulate and superior-rostral sulci: Hemispheric asymmetries, gender differences, and probability maps. *J. Comp. Neurol.* **376,** 664–673.

Paus, T., Tomaiuolo, F., Otaky, N., MacDonald, D., Petrides, M., Atlas, J., Morris, R., and Evans, A. C. (1996b). Human cingulate and paracingulate sulci: Pattern, variability, asymmetry, and probabilistic map. *Cereb. Cortex* **6,** 207–214.

Paxinos, G., and Watson, C. (1986). "The Rat Brain in Stereotaxic Coordinates." Academic Press, Sydney.

Payne, B. A., and Toga, A. W. (1990). Surface mapping brain function on 3D models. *IEEE Comput. Graph. Appl.* **10,** 33–41.

Penhune, V. B., Zatorre, R. J., MacDonald, J. D., and Evans, A. C. (1996). Interhemispheric anatomical differences in human primary auditory cortex: Probabilistic mapping and volume measurement from MR scans. *Cereb. Cortex* **6,** 617–672.

Pennisi, E. (1999). Keeping genome databases clean and up to date. *Science* **286,** 447–450.

Polyak, S. L. (1957). "The Vertebrate Visual System." Univ. of Chicago Press, Chicago.

Rabbitt, R. D., Weiss, J. A., Christensen, G. E., and Miller, M. I. (1995). Mapping of hyperelastic deformable templates using the finite element method. *Proc. SPIE* **2573,** 252–265.

Rademacher, J., Galaburda, A., Kennedy, D., Filipek, P., and Caviness, V. (1992). Human cerebral cortex: Localization, parcellation, and morphometry with magnetic resonance imaging. *J. Cognit. Neurosci.* **4,** 352–374.

Rademacher, J., Caviness, V. S., Steinmetz, H., and Galaburda, A. M. (1993). Topographical variation of the human primary cortices: Implications for neuroimaging, brain mapping and neurobiology. *Cereb. Cortex* **3,** 313–329.

Rizzo, G., Gilardi, M. C., Prinster, A., Grassi, F., Scotti, G., Cerutti, S., and Fazio, F. (1995). An elastic computerized brain atlas for the analysis of clinical PET/SPET data. *Eur. J. Nucl. Med.* **22,** 1313–1318.

Roland, P. E., and Zilles, K. (1994). Brain atlases: A new research tool. *Trends Neurosci.* **17,** 458–467.

Roland, P. E., and Zilles, K. (1996). The developing European Computerized Human Brain Database for all imaging modalities. *NeuroImage* **4,** 39–47.

Roland, P. E., and Zilles, K. (1998). Structural divisions and functional fields in the human cerebral cortex. *Brain Res. Rev.* **26,** 87–105.

Sandor, S. R., and Leahy, R. M. (1994). Matching deformable atlas models to pre-processed magnetic resonance brain images. *In* "Proceedings of the IEEE Conference on Image Processing," Vol. 3, pp. 686–690.

Sandor, S. R., and Leahy, R. M. (1995). Towards automated labeling of the cerebral cortex using a deformable atlas. *In* "Information Processing in Medical Imaging" (Y. Bizais, C. Barillot, and R. Di Paola, eds.), pp.127–138. Springer-Verlag, Berlin.

Sandor, S., and Leahy, R. (1997). Surface-based labeling of cortical anatomy using a deformable atlas. *IEEE Trans. Med. Imaging* **16,** 41–54.

Scheibel, A. B., Paul, L. A., Fried, I., Forsythe, A. B., Tomiyasu, U., Wechsler, A., *et al.* (1985). Dendritic organization of the anterior speech area. *Exp. Neurol.* **87,** 109–117.

Schleicher, A., Amunts, K., Geyer, S., Morosan, P., and Zilles, K. (1999). Observer-independent method for microstructural parcellation of cerebral cortex: A quantitative approach to cytoarchitectonics. *NeuroImage* **9,** 165–177.

Schmidt, D. M., George, J. S., and Wood, C. C. (1999). Bayesian inference applied to the electromagnetic inverse problem. *Hum. Brain Mapp.* **7,** 195–212.

Schormann, T., and Zilles, K. (1997). Limitations of the principal axes theory. *IEEE Trans. Med. Imaging* **16,** 942–947.

Schormann, T., and Zilles, K. (1998). Three-dimensional linear and nonlinear transformations: An integration of light microscopical and MRI data. *Hum. Brain Mapp.* **6,** 339–347.

Schormann, T., Dabringhaus, A., and Zilles, K. (1995). Statistics of deformations in histology and improved alignment with MRI. *IEEE Trans. Med. Imaging* **14,** 25–35.

Schormann, T., Dabringhaus, A., and Zilles, K. (1997). Extension of the principal axis theory for the determination of affine transformations. *In* "Informatik Aktuell," pp. 384–391. Springer-Verlag, Berlin.

Schormann, T., Henn, S., and Zilles, K. (1996). A new approach to fast elastic alignment with application to human brains. *Lect. Notes Comput. Sci.* **1131,** 437–442.

Seitz, R. J., Bohm, C., Greitz, T., Roland, P. E., Eriksson, L., Blomqvist, G., Rosenqvist, G., and Nordell, B. (1990). Accuracy and precision of the computerized brain atlas programme for localization and quantification in positron emission tomography. *J. Cereb. Blood Flow Metab.* **10,** 443–457.

Simonds, R. J., and Scheibel, A. B. (1989). The postnatal development of the motor speech area: A preliminary study. *Brain Lang.* **37,** 43–58.

Sled, J., Zijdenbos, A., and Evans, A. (1997). A comparison of retrospective intensity non-uniformity correction methods for MRI. *In* "Information Processing in Medical Imaging," pp. 459–464. Springer-Verlag, Berlin.

Sled, J., Zijdenbos, A., and Evans, A. (1998). A non-parametric method for automatic correction of intensity non-uniformity in MRI data. *IEEE Trans. Med. Imaging* **17,** 87–97.

Sowell, E. R., Mattson, S. N., Thompson, P. M., Jernigan, T. L., Riley, E. P., and Toga, A. W. (2000). Mapping corpus callosum morphology and its neurocognitive correlates: The effects of prenatal alcohol exposure. Submitted for publication.

Spitzer, V., and Whitlock, D. G. (1992). High resolution imaging of the human body. *Biol. Photogr.* **60,** 167–172.

Steinmetz, H., Volkman, J., Jancke, L., and Freund, H. (1991). Anatomical left–right asymmetry of language-related temporal cortex is different in left and right handers. *Ann. Neurol.* **29,** 315–319.

Strother, S. C., Anderson, J. R., Xu, X.-L., Low, J.-S., Boar, D. C,. and Rottenberg, D. A. (1994). Quantitative comparisons of image registration techniques based on high-resolution MRI of the brain. *J. Comput. Assisted Tomogr.* **18,** 954–962.

Swanson, L. W. (1992). "Computer Graphics File," Version 1.0. Elsevier, Amsterdam.

Talairach, J., and Tournoux, P. (1988). Principe et technique des etudes anatomiques. *In* "Co-planar Stereotaxic Atlas of the Human Brain—3-Dimensional Proportional System: An Approach to Cerebral Imaging." Thieme, New York.

Thirion, J.-P. (1995). "Fast Non-rigid Matching of Medical Images." INRIA Internal Report 2547, Projet Epidaure, INRIA, France.

Thompson, P. M., and Toga, A. W. (1996). A surface-based technique for warping 3-dimensional images of the brain. *IEEE Trans. Med. Imaging* **15,** 1–16.

Thompson, P. M., and Toga, A. W. (1997). Detection, visualization and animation of abnormal anatomic structure with a deformable probabilistic brain atlas based on random vector field transformations. *Med. Image Anal.* **1,** 271–294.

Thompson, P. M., and Toga, A. W. (1998). Surface-based strategies for high-dimensional brain image registration. *In* "Brain Warping" (A. W. Toga, ed.). Academic Press, San Diego, in press.

Thompson, P. M., and Toga, A. W. (2000). Elastic image registration and pathology detection. *In* "Handbook of Medical Image Processing" (I. Bankman, ed.). Academic Press, in press.

Thompson, P. M., Schwartz, C., Lin, R. T., Khan, A. A., Toga, A. W., and Collins, R. C. (1995). 3D statistical analysis of sulcal variability in the human brain using high-resolution cryosection images. *Proc. Soc. Neurosci.* **21,** 154.

Thompson, P. M., Schwartz, C., and Toga, A. W. (1996a). High-resolution random mesh algorithms for creating a probabilistic 3D surface atlas of the human brain. *NeuroImage* **3,** 19–34.

Thompson, P., Schwartz, C., Lin, R. T., Khan, A. A., and Toga, A. W. (1996b). 3D statistical analysis of sulcal variability in the human brain. *J. Neurosci.* **16,** 4261–4274.

Thompson, P. M., MacDonald, D., Mega, M. S., Holmes, C. J., Evans, A. C., and Toga, A. W. (1997). Detection and mapping of abnormal brain structure with a probabilistic atlas of cortical surfaces. *J. Comput. Assisted Tomogr.* **21,** 567–581.

Thompson, P. M., Moussai, J., Khan, A. A., Zohoori, S., Goldkorn, A., Mega, M. S., Small, G. W., Cummings, J. L., and Toga, A. W. (1998). Cortical variability and asymmetry in normal aging and Alzheimer's disease. *Cereb. Cortex* **8,** 492–509.

Thompson, P. M., Giedd, J. N., Woods, R. P., MacDonald, D., Evans, A. C., and Toga, A. W. (2000a). Growth patterns in the developing brain detected by using continuum mechanical tensor maps. *Nature* **404,** 190–193.

Thompson, P. M., Mega, M. S., and Toga, A. W. (2000b). Disease-specific brain atlases. *In* "Brain Mapping: The Disorders" (A. W. Toga, J. C. Mazziotta, and R. S. J. Frackowiak, eds.), pp. 131–177. Academic Press, San Diego.

Thompson, P. M., Woods, R. P., Mega, M. S., and Toga, A. W. (2000c). Mathematical/computational challenges in creating deformable and probabilistic atlases of the human brain. *Hum. Brain Mapp.* **9,** 81–92.

Thurfjell, L., Bohm, C., Greitz, T., and Eriksson, L. (1993). Transformations and algorithms in a computerized brain atlas. *IEEE Trans. Nucl. Sci.* **40,** 1167–1191.

Toga, A. W., and Arnicar-Sulze, T. L. (1987). Digital image reconstruction for the study of brain structure and function. *J. Neurosci. Methods* **20,** 7–21.

Toga, A. W., Goldkorn, A., Ambach, K., Chao, K., Quinn, B. C., and Yao, P. (1997). Postmortem cryosectioning as an anatomic reference for human brain mapping. *Comput. Med. Image Graph.* **21,** 131–141.

Toga, A. W., Thompson, P. M., Holmes, C. J., and Payne, B. A. (1996). Informatics and computational neuroanatomy. *J. Am. Med. Inf. Assoc.* **6,** 299–303.

Ungerleider, L. G., and Desimone, R. (1986). Cortical connections of visual area MT in the macaque. *J. Comp. Neurol.* **248,** 190–222.

Van Essen, D. C., and Drury, H. A. (1997). Structural and functional analyses of human cerebral cortex using a surface-based atlas. *J. Neurosci.* **17,** 7079–7102.

Van Essen, D. C., Drury, H. A., Joshi, S., and Miller, M. I. (1998). Functional and structural mapping of human cerebral cortex: Solutions are in the surfaces. *Proc. Natl. Acad. Sci. USA* **95,** 788–795.

Viola, P. A., and Wells, W. M. (1995). Alignment by maximization of mutual information. *In* "5th IEEE International Conference on Computer Vision," Cambridge, MA, pp. 16–23..

Wang, J. Z., Wiederhold, G., and Firschein, O. (1997). System for screening objectionable images using Daubechies' wavelets and color histograms, interactive distributed multimedia systems and telecommunication services. *In* "Proceedings of the 4th European Workshop (IDMS'97)" (R. Steinmetz and L. C. Wolf, eds.). Springer-Verlag, Darmstadt.

Watson, J. D., Myers, R., Frackowiak, R. S., Hajnal, J. V., Mazziotta, J. C., Shipp, S., and Zeki, S. (1993). Area V5 of the human brain from a combined study using positron emission tomography and magnetic resonance imaging. *Cereb. Cortex* **3,** 79–94.

Wells, W. M., Viola, P., Atsumi, H., Nakajima, S., and Kikinis, R. (1997). Multi-modal volume registration by maximization of mutual information. *Med. Image Anal.* **1,** 35–51.

Wertheim, S. L. (1989). The brain database: A multimedia neuroscience database for research and teaching. *In* "Proceedings of the 13th Annual Symposium on Computer Applications in Medical Care" (L. C. Kingsland, III, ed.). IEEE Comput. Soc., Los Alamitos, CA.

Woods, R. P., Cherry, S. R., and Mazziotta, J. C. (1992). A rapid automated algorithm for accurately aligning and reslicing positron emission tomography images. *J. Comput. Assisted Tomogr.* **16,** 620–633.

Woods, R. P., Grafton, S. T., Holmes, C. J., Cherry, S. R., and Mazziotta, J. C. (1998). Automated image registration. I. General methods and intrasubject, intramodality validation. *J. Comput. Assisted Tomogr.* **22,** 139–52.

Woods, R. P., Mazziotta, J. C., and Cherry, S. R. (1993). MRI-PET registration with automated algorithm. *J. Comput. Assisted Tomogr.* **17,** 536–46.

Zeki, S., Watson, J. D. G., Lueck, C. J., Friston, K. J., Kennard, C., and Frackowiak, R. S. J. (1991). A direct demonstration of functional specialization in human visual cortex. *J. Neurosci.* **11,** 641–649.

Zhou, Y., Kaufman, A., and Toga, A. W. (1998). Three-dimensional skeleton and centerline generation based on an approximate minimum distance field. *Visual Comput.* **14,** 303–314.

Zhou, Y., and Toga, A. W. (1999). Efficient skeletonization of volumetric objects. *IEEE Trans. Vision Comput. Graph.* **5,** 196–209.

Zihl, J., von Cramon, D., Mai, D., and Schmid, C. (1991). Disturbance of movement vision after bilateral posterior brain damage. Further evidence and follow up observations. *Brain* **114,** 2235–2252.

Zijdenbos, A. P., Evans, A. C., Riahi, F., Sled, J. G., Chui, H.-C., and Kollokian, V. (1996). Automatic quantification of multiple sclerosis lesion volume using stereotaxic space. *In* "Proceedings of the 4th International Conference on Visualization in Biomedical Computing, VBC '96, Hamburg, pp. 439–448..

Zilles, K., Schlaug, G., Matelli, M., Luppino, G., Schleicher, A., Qu, M., Dabringhaus, A., Seitz, R., and Roland, P. (1995). Mapping of human and macaque sensorimotor areas by integrating architectonic, transmitter receptor, MRI and PET data. *J. Anat.* **187,** 515–537.

Zilles, K., Schleicher, A., Langemann, C., Amunts, K., Morosan, P., Palomero-Gallagher, N., Schormann, T., Mohlberg, H., Bürgel, U., Steinmetz, H., Schlaug, G., and Roland, P. E. (1997). Quantitative analysis of sulci in the human cerebral cortex: Development, regional heterogeneity, gender difference, asymmetry, intersubject variability and cortical architecture. *Hum. Brain Mapp.* **5,** 218–221.

28

Subpopulation Brain Atlases

Paul M. Thompson, Michael S. Mega, and Arthur W. Toga

Laboratory of Neuro Imaging, Brain Mapping Division, and Alzheimer's Disease Center, Department of Neurology, UCLA School of Medicine, Los Angeles, California 90095-1769

I. Population-Based Brain Imaging

Recent developments in brain imaging have revolutionized medicine and neuroscience. The ability to image the living brain has greatly accelerated the collection and databasing of brain maps (Mazziotta *et al.,* 1995; Huerta and Koslow, 1996; Fox, 1997; Toga and Thompson, 1998a; Letovsky *et al.,* 1998; Van Horn *et al.,* 2001). These maps store information on anatomy and physiology, from whole-brain to molecular scales. Some capture functional changes that occur over milliseconds and others anatomical changes occurring over entire lifetimes (see, e.g., Toga and Mazziotta, 1996; Frackowiak *et al.,* 1997; for recent reviews).

This rapid collection of brain images from healthy and diseased subjects has stimulated the development of mathematical algorithms that compare, pool, and average brain data across whole populations. Brain structure is complex and varies widely from one individual to another. New approaches in computer vision (Bankman, 1999; Fitzpatrick and Sonka, 2000; MacDonald *et al.,* 2000; Xu *et al.,* 1999; Shattuck and Leahy, 2001; Kriegeskorte and Goebel, 2001), anatomical modeling (Thompson *et al.,* 2001a; Fischl *et al.,* 2001), differential geometry (Grenander and Miller, 1998; Haker *et al.,* 1999; Hurdal *et al.,* 1999), and statistical field theory (Friston *et al.,* 1995; Worsley *et al.,* 1999; Cao and Worsley, 1999; Taylor and Adler, 2000) are being formulated to capture this variation, encode it, and detect disease-specific patterns (Thompson *et al.,* 1997, 2001a). Statistics that describe how brain structure and function vary in a population can greatly empower the analysis of new images (Ashburner *et al.,* 1997; Gee and Bajcsy, 1998; Dinov *et al.,* 2000;

Kang *et al.*, 2001). Population statistics can help algorithms find brain structures in new scans and detect abnormalities (Thompson *et al.*, 1997, 2001a; Csernansky *et al.*, 1998; Ashburner and Friston, 2000; Narr *et al.*, 2001a; Fischl and Dale, 2001). They can also identify systematic patterns of anatomy and function (e.g., Giedd *et al.*, 1999a; Sowell *et al.*, 1999, 2001; Paus *et al.*, 1999; Good *et al.*, 2001a), and uncover surprising relationships between genotype and phenotype (Styner and Gerig, 2001; Thompson *et al.*, 2001a; Cannon *et al.*, 2001a).

A. Increased Automation

Brain mapping analyses often draw upon hundreds or even thousands of images (Evans *et al.*, 1994; Good *et al.*, 2001a). Computational approaches must therefore distil information from these images in a highly automated way. This challenge has driven developments in automated image registration and warping methods (Woods *et al.*, 1998a; Toga, 1998; Ashburner *et al.*, 1999; Guimond *et al.*, 1999; Thompson *et al.*, 2000a; Shattuck and Leahy, 2001), as well as techniques for rapid image segmentation and labeling (Collins *et al.*, 1995). As brain databases grow at a near-exponential pace, very large image analyses can now be run through client-server software pipelines. These intensive analyses may be performed on a remote server, aided by supercomputing resources to mine data for patterns and population trends (Toga *et al.*, 2001a; cf. Zijdenbos *et al.*, 1996; Warfield *et al.*, 1998; Megalooikonomou *et al.*, 2000).

In this chapter, we review recent advances in image analysis that have enabled the creation of *population-based brain atlases* (Mazziotta *et al.*, 1995, 2001; Thompson *et al.*, 2000a). These atlases combine imaging data from healthy and diseased populations and have a range of exciting applications in neuroscience. They describe how the brain varies with age, gender, and demographics. They also provide a comprehensive approach for studying a particular population subgroup, with a specific disease or neuropsychiatric disorder, for instance.

B. What Do Population-Based Atlases Contain?

Traditional single-subject atlases represent anatomy in a 3D coordinate system. Population-based atlases do so as well, but they contain anatomical models from many individuals. They store population averages, templates, and statistical maps, to summarize features of the population.

Central to the concept of all atlases is the idea of a common 3D reference space. The anatomy of the atlas, and new datasets that are aligned to it, can then be referred to using 3D spatial coordinates. Image registration techniques are typically used to align new data sets with the anatomic atlas (see Toga and Thompson, 2001a, for a review). Multiple data sets can then be compared in a common coordinate space. Multimodality atlases (Fig. 1) bring together and correlate brain maps from diverse imaging devices. Changes in anatomical morphology can then be related to differences in underlying neurochemistry and molecular content (Poxton *et al.*, 1998; Mega *et al.*, 1999a). Atlases may also contain derived computational maps of various types. For instance they may contain normative statistics on anatomic variability or on rates of brain change in development or disease. Dynamic maps of growth patterns in development are of particular interest (Lange *et al.*, 1997; Giedd *et al.*, 1999a; Sowell *et al.*, 1999; Thompson *et al.*, 2001a).

Relating all these structural and dynamic maps to genetic, therapeutic, and cognitive factors presents specific mathematical challenges. Among the armory of mathematical tools in building an atlas are cortical flattening approaches (Section IV), warping algorithms (Section VIII), modeling approaches from population genetics (Section IX), and new results in random field theory. These mathematical tools give a multisubject atlas its power to reveal generic patterns of brain structure and function not observable in an individual (Thompson *et al.*, 2000a).

Disease-specific brain atlases are a particular type of population atlas. They provide a unique perspective on the anatomy and physiology of a particular disease (Mega *et al.*, 1999a; Thompson *et al.*, 2000a,b,c, 2001a; Narr *et al.*, 2001a,b; Cannon *et al.*, 2001a). These atlases store multimodality imaging data from a specific clinical subpopulation, such as patients with Alzheimer's disease (Mega *et al.*, 1997, 1999a), schizophrenia (Narr *et al.*, 2001a; Thompson *et al.*, 2001a), or a psychiatric disorder such as fetal alcohol syndrome (Sowell *et al.*, 2001). The populations may be stratified to reflect a particular clinical subgroup, including those at familial risk for a disease (Cannon *et al.*, 2001a), those receiving different medications (Thompson *et al.*, 2001a), or those with a specific symptom profile or genotype.

C. Genetic Atlases

Inclusion of genetic data in these atlases makes it possible to go beyond describing a disease to investigating its causes. Novel mathematics can be used to mine imaging data and to identify genetic sources of variation. This allows the direct mapping of genetic influences on brain structure and lets us quantify heritability for different features of the brain (Thompson *et al.*, 2001a; see Section IX). Familial, twin, and genetic linkage studies have recently begun to expand the atlas concept to tie together genetic and imaging studies of disease. Atlases that contain genetic brain maps, and a means to analyze them, can help screen relatives for inherited disease (Cannon *et al.*, 2001a). They also offer a framework to mine large imaging databases for risk genes and quantitative trait loci (Gottesman, 1997), as well as genetic and environmental triggers of disease.

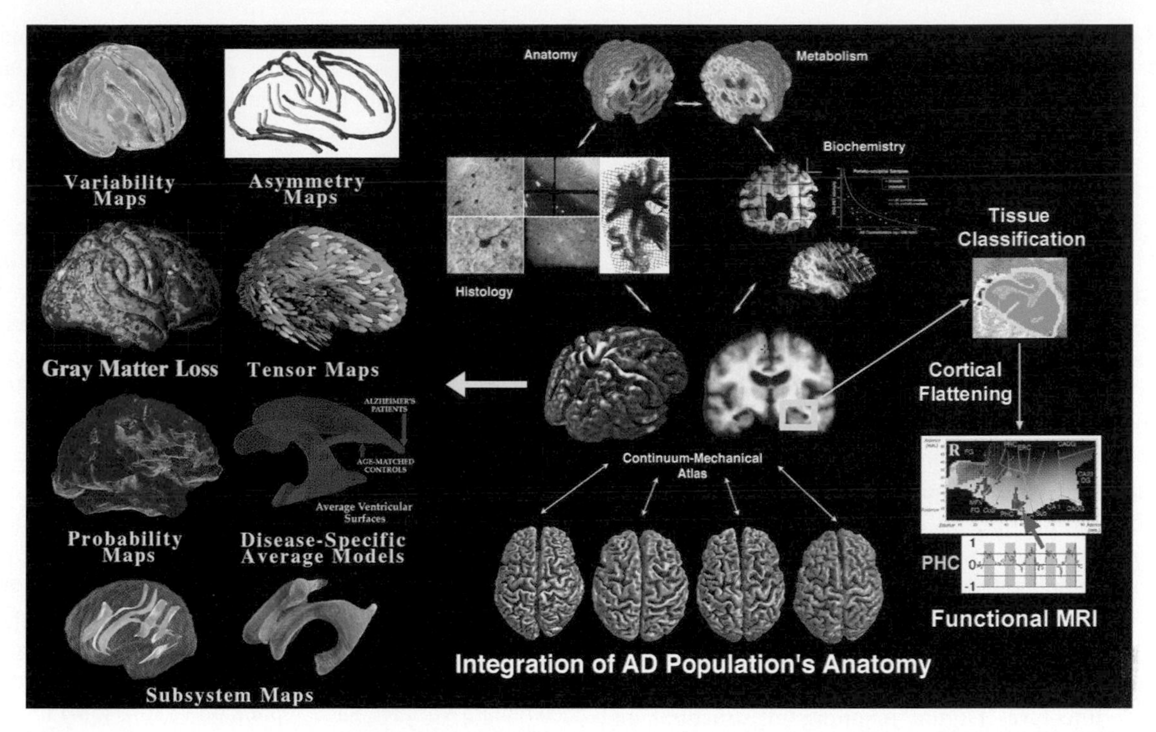

Figure 1 Elements of a disease-specific atlas. This schematic shows the types of maps and models contained in a disease-specific brain atlas (Thompson *et al.*, 2000a; Mega *et al.*, 2000). A diverse range of computational anatomical tools is required to generate these average brain image templates (continuum-mechanical atlas), models, and maps. Disease-specific brain atlases, such as this one based on patients with Alzheimer's disease (AD), allow imaging data from diverse modalities to be compared and correlated in a common 3D coordinate space. 3D anatomical models (e.g. cortical surfaces, **bottom row**) were extracted from a database of structural MRI data from AD patients. Models of these and other structures were digitally averaged and used to synthesize an average brain template (continuum-mechanical atlas, **middle**) with well-resolved anatomical features in the mean shape and size for the population (see Section V for details). By rotating and scaling new images to occupy the same space as this template, models of subcortical, ventricular, and deep nuclear structures can be built (**lower left**). Average models for patients and controls then be used to compute average patterns and statistics of cortical variability and asymmetry (**top left**), to chart average profiles of gray matter loss in a group, and to detect atrophy in a group or individual (probability maps, **left column**). Mega *et al.* (1997, 1999a) also fused histologic maps of postmortem neurofibrillary tangle staining density, biochemical maps of β-amyloid distribution, and 3D metabolic FDG-PET data obtained 8 h before death, in the same patient with AD (**top middle**). By classifying gray and white matter (tissue classification) and unfolding the topography of the hippocampus (**right**), Zeineh *et al.* (2001) revealed the fine-scale anatomy and dynamics of brain activation during memory tasks, using high-resolution functional MRI (time course shown for activation in right parahippocampal cortex; PHC). Atlasing techniques can represent and compare these diverse data sets in a common coordinate space, enabling novel multisubject and cross-modality comparisons.

II. Atlases in Brain Mapping

A. Brain Maps and Atlases

Since the development of computerized tomography (CT; Hounsfield, 1973) and magnetic resonance imaging techniques (Lauterbur, 1973), maps of brain structure have typically been based upon 3D tomographic images (Damasio, 1995). Angiographic or spiral CT techniques can also visualize vascular anatomy (Fishman, 1997), while diffusion tensor images can even reveal fiber topography *in vivo* (Turner *et al.,* 1991; Jacobs and Fraser, 1994; Mori *et al.,* 2001). These brain maps can be supplemented with high-resolution information from anatomic specimens (Talairach and Tournoux, 1988; Ono *et al.,* 1990; Duvernoy, 1991) and a variety of histologic preparations which reveal regional cytoarchitecture (Brodmann, 1909) and regional molecular content such as myelination patterns (Smith, 1907; Mai *et al.,* 1997), receptor binding sites (Geyer *et al.,* 1997), protein densities, and mRNA distributions. Other brain maps have concentrated on function, quantified by positron emission tomography (PET; Minoshima *et al.,* 1994), functional MRI (Le Bihan, 1996), electrophysiology (Avoli *et al.,* 1991; Palovcik *et al.,* 1992), or optical imaging (Cannestra *et al.,* 1996). Additional maps have been developed to represent neuronal connectivity and circuitry (Van Essen and Maunsell, 1983), based on compilations of empirical evidence (Brodmann, 1909; Berger, 1929).

Each of these brain mapping approaches produces data with a different scale and resolution, and none is inherently comparable with any other. Their correlative potential is therefore underexploited unless algorithms are applied to align them into a common reference space. This common coordinate system can be provided by a *brain atlas*.

B. Aligning Multiple Subjects' Data to an Atlas

To address difficulties in comparing brain maps, brain atlases (e.g., Talairach and Tournoux, 1988; Swanson, 1992; Evans *et al.*, 1994; Mazziotta *et al.*, 1995; Spitzer *et al.*, 1996; Kikinis *et al.*, 1996; Drury and Van Essen, 1997; Schmahmann *et al.*, 1999) provide a structural framework in which individual brain maps can be integrated. Most brain atlases are based on a detailed representation of a single subject's anatomy in a standardized 3D coordinate system, or stereotaxic space. The chosen data set acts as a template on which other brain maps (such as functional images) can be overlaid. The anatomic data provide the additional detail necessary to accurately localize activation sites, as well as providing other structural perspectives such as chemoarchitecture. Digital mapping of structural and functional image data into a common 3D coordinate space is a prerequisite for many types of brain imaging research, as it supplies a quantitative spatial reference system in which brain data from multiple subjects and modalities can be compared and correlated.

C. Talairach Reference System

The first brain atlas used widely by the brain mapping community was that defined by the neurosurgeon Jean Talairach (Talairach and Tournoux, 1988). Using a stereotaxic device anchored to a patient's skull, neurosurgeons can accurately position surgical apparatus within a patient's brain to target biopsy locations, epileptic foci, and vascular lesions identified in 3D reference coordinates. The Talairach atlas was developed before intraoperative imaging, to make it easier to identify deep nuclei in stereotaxic coordinates. At the time, these structures were imaged with very limited resolution using pneumoencephalography.

D. Alignment to an Atlas

In addition to a series of labeled anatomical plates, reconstructed from histologic material, Talairach defined a mechanism to transfer new images onto the atlas. In the Talairach stereotaxic system, piecewise affine transformations are applied to 12 rectangular regions of the brain, defined by vectors from the anterior and posterior commissures to the extrema of the cortex. These transformations reposition the anterior commissure of the subject's scan at the origin of the 3D coordinate space, vertically align the interhemispheric plane, and horizontally orient the line connecting the two commissures. Each point in the incoming brain image, after it is "warped" into the atlas space, is labeled by an (x,y,z) address referable to the atlas brain. Although originally developed for surgical purposes, the Talairach stereotaxic system rapidly became an international standard for reporting functional activation sites in PET studies, allowing researchers to compare and contrast results from different laboratories (Fox *et al.*, 1985, 1988; Friston *et al.*, 1989, 1991). A recent tool for meta-analysis of functional imaging data, the *Talairach Daemon,* contains a digital version of the Talairach atlas. Using the atlas as a reference coordinate system, it provides spatially indexed links to Brodmann areas, probabilistic maps of activation foci, and related scientific literature (Lancaster *et al.*, 2000).

E. MRI Atlas Templates

The Talairach templates were based on post mortem sections of the brain of a 60-year-old female subject, which clearly did not reflect the *in vivo* anatomy of subjects in activation studies. The atlas plates were also compromised by having a variable slice separation (3 to 4 mm) and inconsistent data from orthogonal planes. To address these limitations, a composite T1-weighted MRI data set was constructed from 305 young normal subjects (239 males, 66 females; age 23.4 ± 4.1 years) whose scans were individually mapped into the Talairach system by a nine-parameter linear transformation, intensity normalized, and averaged on a voxel-by-voxel basis (Evans *et al.*, 1994). The resulting average brain made it easier to develop automated image alignment methods to map new MRI and PET data into a common space. The International Consortium for Brain Mapping (ICBM; Mazziotta *et al.*, 1995; 2001) subsequently applied the same image-averaging procedure to a subset of 152 brains. This produced a template that is widely used as part of the Statistical Parametric Mapping image analysis package (SPM99b; Friston *et al.*, 1995).

F. Aligning New Data to an Atlas

New MR data are typically aligned with an atlas template by defining a measure of intensity similarity between the overlapping data set and atlas. This measure of fit is optimized by tuning the parameters of the alignment transformation until the similarity is maximized (Woods *et al.*, 1998a; Ashburner, 2001). Intensity-based registration measures include 3D cross-correlation (Collins *et al.*, 1994a, 1995), ratio image uniformity (Woods *et al.*, 1992, 1993), or mutual information (Viola *et al.*, 1995; Wells *et al.*, 1997), or the summed squared differences in intensity between the scans (Christensen *et al.*, 1993, 1996; Ashburner *et al.*, 1998; Woods *et al.*, 1998a). Both linear (global) transforms and nonlinear transforms may be used, and these registration methods are reviewed in detail elsewhere (Toga, 1998; Thompson *et al.*, 2001a). Any alignment transformation defined for one modality, such as MRI, can be identically applied to another modality, such as PET, if a previous cross-modality intrasubject registration has been performed (Woods *et al.,* 1993). For the first time then, PET data could be mapped into stereotaxic space via a correlated MR data set (Woods *et al.,* 1993; Evans *et al.,* 1994). Registration algorithms therefore made it feasible to automatically map data from a variety of modalities into an atlas coordinate space based directly on the Talairach reference system.

G. Other Standardized Brain Templates

Individual brains differ substantially in structure. If MRI scans are averaged after only a linear alignment into stereotaxic space, the resulting data set is blurred in the more variable anatomic regions (Evans *et al.,* 1994). Many structures are washed away rather than reinforced. This problem can be circumvented using nonlinear registration to define a mean anatomic template for a population, with well-resolved cortical features (see Section V; Collins *et al.,* 1994a; Thompson *et al.,* 2000a). Alternatively, an atlas may be based on a single individual, as in the Karolinska Brain Atlas project (Greitz *et al.,* 1991; Roland and Zilles, 1996) and the Digital Anatomist project in Seattle (Sundsten *et al.,* 1991; Brinkley *et al.,* 1997). Other atlasing projects have preferred multimodality data from a single subject, including the Harvard Brain Atlas (Kikinis *et al.,* 1996), the VOXEL-Man atlas (Höhne *et al.,* 1992; Tiede *et al.,* 1993; Pommert *et al.,* 1996), and the Visible Human Project (Spitzer, 1996).

H. High-Resolution Atlas Templates

A recent innovation in the collection of atlas quality MRI uses multiple scans ($N = 27$) and a registered average of a single individual to overcome the lack of contrast and relatively poor signal to noise of tomographic data (Holmes *et al.,* 1998; Kochunov *et al.,* 2001a). This high-resolution brain template is used in the BrainWeb simulator (Collins *et al.,* 1998) and the SPM functional imaging package (Friston *et al.,* 1995) and serves as a template in a recent cerebellar atlas (Schmahmann *et al.,* 1999). Unlike with a population-based atlas, some bias is implicit in selecting an individual brain as an atlas target. These biases may be minimized by (1) selecting a brain that deviates minimally from a population (Lancaster *et al.,* 2000) or (2) explicitly transforming an individual brain, using image warping methods, to reflect the mean geometry of a group (Kochunov *et al.,* 2001a).

I. Interoperability

A vexing issue with the proliferation of atlas templates is the need to relate imaging data that have been aligned to different atlases, sometimes with different registration methods (Haller and Vannier, 1998). Efforts to mutually register the atlases themselves (Letovsky *et al.,* 1998) have resulted in the fusion of the Talairach atlas with the Ono sulcal atlas (Ono *et al.,* 1990). The Talairach templates have recently been fused (Nowinski *et al.,* 2000) with the Schaltenbrand and Wahren brain atlas (1977), which represents deep nuclear anatomy and is often consulted in the preoperative planning of thalamic surgery (Niemann and van Nieuwenhofen, 1999). Because of differences in brain shape, the ICBM average brain templates are slightly larger (in particular higher, deeper, and longer) than the original Talairach brain. Formulas have been estimated to reconcile data expressed in each system (Brett *et al.,* 2001; Rorden and Brett, 2001). These formulas refer activation foci in ICBM space to Brodmann areas labeled on the Talairach atlas.

J. Population Maps of Cytoarchitecture

Because Brodmann areas defined on the Talairach atlas are only approximate, probabilistic maps of cytoarchitecture are currently being compiled in standardized atlas spaces, such as that defined by the ICBM templates (Geyer *et al.,* 2001; Rademacher *et al.,* 2001; Grefkes *et al.,* 2001; Amunts and Zilles, 2001). Brodmann areas may also be defined on unfolded cortical templates (Carman *et al.,* 1995; Hurdal *et al.,* 1999; Van Essen *et al.,* 2001; Zeineh *et al.,* 2001; Fischl and Dale, 2001; Thompson *et al.,* 2001a; see Section IV). These templates can, in some cases, be projected back into individual data sets in atlas space to help localize activation foci (Drury and Van Essen, 1997; Rorden and Brett, 2001). Finally, research is under way to transfer brain mapping data between anatomic templates with nonlinear warping approaches (Fig. 2; Toga, 1998). For different applications, different templates may have different merits and biases (Thompson *et al.,* 2001a; Kochunov *et al.,* 2001a,b). A reasonable goal is therefore to develop mathematics to compute consistent (e.g., symmetric and transitive) mappings between the atlas templates that are most widely used (Christensen *et al.,* 1999).

III. Anatomical Modeling

The development of group anatomic atlases is motivated by understanding anatomical variations in human populations. Even after aligning brain data into a common reference space, such as that defined by an atlas, individual anatomy still varies widely. These variations are so great that disease-specific differences are hard to identify. Computational techniques have therefore been designed to encode these statistical variations, which are greatest at the cortex. Population atlases can then identify systematic effects on brain structure, such as differences due to age, gender, disease processes, therapy, or genetic risk.

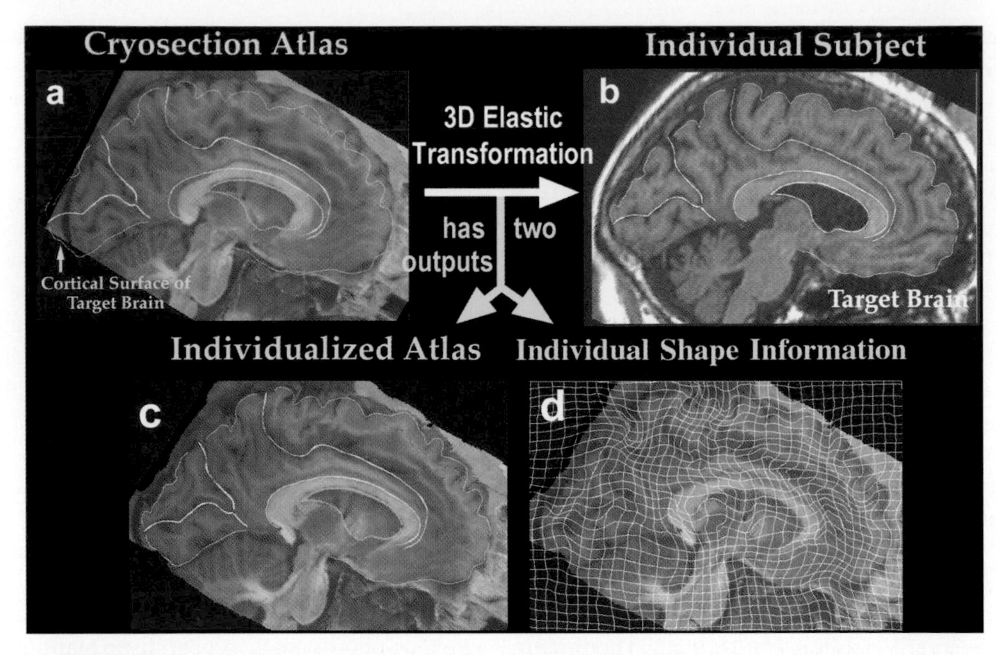

Figure 2 Computing anatomical differences with a deformable brain atlas. When a cryosection atlas of the brain (**a**) is deformed to match the anatomy of an individual patient (**b**), here imaged with 3D MRI, there are two useful products. The first is a high-resolution anatomical template that is customized to reflect the individual's anatomy (**c**), and the second is a mathematical record of the shape differences between the atlas and the individual (warped grid) (**d**). These fields can be analyzed statistically to quantify differences in brain structure and detect abnormal anatomy. The transformation of the atlas onto the target MRI is here performed by constraining functionally important surfaces to match, while extending the deformation to the full 3D volume (Thompson and Toga, 1996).

A. Computational Anatomy

Efforts to uncover new patterns of altered structure and function in individuals and clinical populations have led to the new field of *computational anatomy* (Grenander and Miller, 1998; Thompson *et al.*, 2001a; Fischl *et al.*, 1999; Davatzikos *et al.*, 1996; Guimond *et al.*, 2001; Worsley *et al.*, 2000; Ashburner, 2001; see Thompson and Toga, 2001, for a review). This growing field has powerful applications in neuroscience, revealing, for example, how the brain grows in childhood (Thompson *et al.*, 2000a), how genes affect brain structure (Thompson *et al.*, 2001a; Cannon *et al.*, 2001a; Styner and Gerig, 2001), and how diseases such as Alzheimer's, schizophrenia, or multiple sclerosis evolve over time or respond to therapy (Zijdenbos *et al.*, 1996; Subsol *et al.*, 1997; Freeborough and Fox, 1998; Thompson *et al.*, 2001a; Haney *et al.*, 2001a). Fundamental to the goals of computational anatomy is the ability to create *average* maps of brain structure. Average templates are under rapid development for the Macaque brain (Grenander and Miller, 1998) and for individual structures such as the corpus callosum (Davatzikos, 1996; Gee *et al.*, 1998), central sulcus (Manceaux-Demiau *et al.*, 1998), cingulate and paracingulate sulci (Paus *et al.*, 1996; Thompson *et al.*, 1997), and hippocampus (Haller *et al.*, 1997; Joshi *et al.*, 1998; Csernansky *et al.*, 1998; Thompson *et al.*, 1999) and for transformed representations of the human and Macaque cortex (Van Essen *et al.*, 1997; Grenander and Miller, 1998; Thompson *et al.*, 1999; Fischl and Dale, 2001).

One approach to creating average anatomical maps is based on the concept of *parametric mesh* modeling. Anatomical models based on parametric meshes can be valuable components of population atlases; they can identify and localize systematic anatomic differences in patients with diseases such as Alzheimer's, schizophrenia, and developmental disorders. We review their construction and uses next.

B. Parametric Meshes

The modeling of structures in the brain using parametric curves and surfaces (Fig. 3) offers exciting advantages for detecting generic patterns in anatomy (Thompson *et al.*, 1996a,b, 1998; Joshi *et al.*, 1998; Vaillant *et al.*, 1998; MacDonald *et al.*, 2000; Fischl *et al.*, 2001; Styner and Gerig, 2001; Gerig *et al.*, 2001). The idea is to overlay a computational grid on the boundary of an anatomical structure, such as the ventricles (Fig. 4) or the cortex (see Section IV). The surface grid structure helps in building average templates of anatomy and statistics that capture its variation (Fig. 3). In simple cases, surface models may represent a

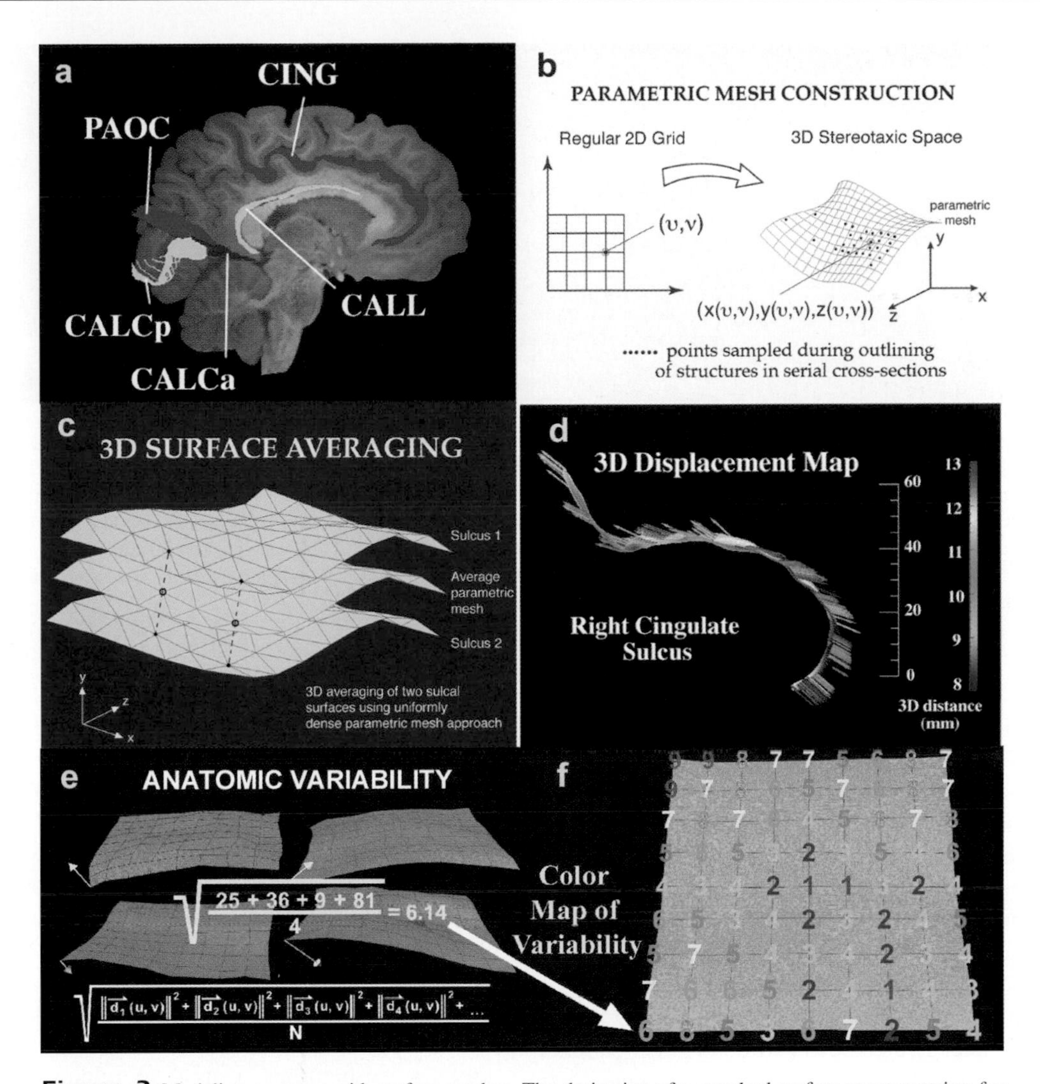

Figure 3 Modeling anatomy with surface meshes. The derivation of a standard surface representation for each structure makes it easier to compare anatomical models from multiple subjects. An algorithm converts a set of digitized points on an anatomical structure boundary [e.g., deep sulci (**a**)] into a parametric grid of uniformly spaced points in a regular rectangular mesh stretched over the surface [(**b**); Thompson *et al.,* 1996a]. By averaging nodes with the same grid coordinates across subjects (**c**), an average surface is produced for the group. However, information on each subject's individual differences is retained as a vector-valued displacement map (**d, e**). This map indicates how that subject deviates locally from the average anatomy. The root mean square magnitude (e) of these deviations provides a variability measure whose values can be visualized using a color code (**f**). These maps can be stored to measure variability in different anatomic systems, including ventricular and deep sulcal surfaces (Thompson *et al.,* 1998). A more complex method measures cross-subject variations in gyral patterns, with a surface matching procedure that better reflects anatomical variations at the cortex (see Section IV). These maps can be stored to measure variability (f) and detect typical (or abnormal) patterns of brain structure in different anatomic systems.

single structure such as the corpus callosum (Thompson *et al.,* 1998; Fig. 5), deep sulcal surfaces (Thompson *et al.,* 1996a,b, 1997, 1998; Le Goualher *et al.,* 2000), deep motor nuclei (Blanton *et al.,* 2000), or the hippocampus and amygdala (Narr *et al.,* 2001a; Thompson *et al.,* 2001a). The resulting computational models can represent anatomical structures in geometric detail, including 3D surface boundaries (e.g., basal ganglia, cortex), as well as 2D and 3D curves (e.g., the corpus callosum at midline, sulcal fundi lying in the cortex). As well as serving as key building blocks for brain atlases, surface models can also drive algorithms for nonlinear warping (Thompson and Toga, 1996; Davatzikos *et al.,* 1996) and can assist in computing growth patterns and other dynamic changes in anatomy (Thompson *et al.,* 2000a; Section VIII).

Parametric mesh modeling approaches define a mapping of a regular 2D grid onto a complex 3D surface (Fig. 3). The concept is similar to stretching a regular net over an object.

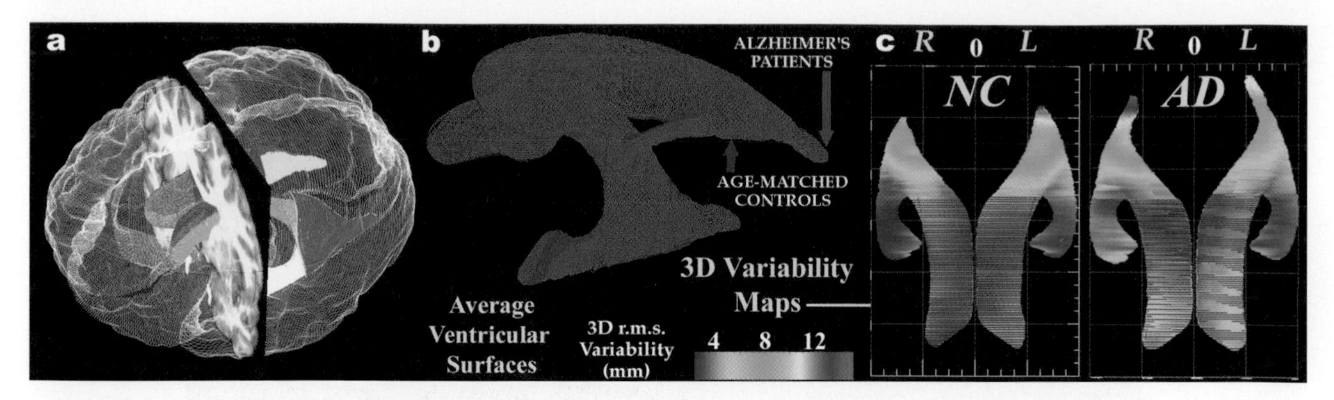

Figure 4 Population-based maps of average ventricular anatomy in normal aging and Alzheimer's disease. In patients and controls, 3D parametric surface meshes (Thompson *et al.*, 1996a) were used to model 14 ventricular elements (**a**), and meshes representing each surface element were averaged by hemisphere in each group. (**b**) An average model for Alzheimer's patients (red; AD) is superimposed on an average model for matched normal controls (blue; NC). Mesh averaging reveals enlarged occipital horns in the Alzheimer's patients and high stereotaxic variability (**c**) in both groups. Extreme variability at the occipital horn tips also contrasts sharply with the stability of the septal and temporal ventricular regions. A top view of these averaged surface meshes reveals localized asymmetry, variability, and displacement within and between groups. These subcortical asymmetries emerge only after averaging of anatomical maps in large groups of subjects.

The explicit geometry provided by this approach makes it easy to derive shape descriptors such as surface curvature, extent, area, fractal dimension, and geometric complexity (Thompson *et al.*, 1996a; see Blanton *et al.*, 2000; Narr *et al.*, 2000 for applications). These morphometric statistics may be altered in disease, even if structure volumes are not statistically different. Narr *et al.* (2000) observed that callosal area did not discriminate schizophrenic groups from healthy controls, while shape measures provided a distinct group separation. Wang *et al.* (2001) found that hippocampal shape descriptors had greater power to distinguish patients from controls than volumetry, while others have argued that each approach provides complementary information (Gerig *et al.*, 2001; Thompson *et al.*, 2001a).

C. Advantages of Surface Models

Surface modeling also provides advantages for data visualization. The surface format may be triangulated and rendered graphically or even animated (Toga, 2001). Texture maps, statistics, and dynamic patterns of functional activations may also be superimposed on graphical renderings of anatomy (e.g., Höhne *et al.*, 1996; Kikinis *et al.*, 1996; Thompson *et al.*, 1997; Zeineh *et al.*, 2001). Imposition of an identical regular structure on surfaces from different subjects makes them easy to compare and average together. Points on each surface with the same mesh coordinate occupy similar positions in relation to the geometry of the surface they belong to and are therefore regarded as homologous. More complex methods can enforce additional correspondences when averaging surfaces, such as sulcal curves lying in the cortex; these issues are covered in Section IV.

D. Generating Surfaces

Some surface models are easy to extract automatically. Examples include the cortex, cerebellar surface, corpus callosum, and hippocampus (Haller *et al.*, 1997; Joshi *et al.*, 1998; MacDonald *et al.*, 2000; Shattuck and Leahy, 2001; Fischl and Dale, 2001; Pitiot *et al.*, 2002). Structures that are more difficult to extract automatically can be traced manually in serial sections, or in 3D, using a formal anatomical protocol with quantified reliability (e.g., Sowell *et al.*, 2001; Zhou *et al.*, 2001). Manual traces are subsequently converted to uniform mesh format using a regridding algorithm, which makes the sampled points spatially uniform (Thompson *et al.*, 1996a,b). Typically in building a brain atlas, a mixture of manual and automated methods is used, to ensure greatest accuracy. For example, the cortex may be extracted automatically, but gyral landmarks may be traced on it manually using a protocol with known, quantified reliability.

E. Anatomical Averaging

An average anatomical surface is generated for a group of subjects by averaging the vector locations of corresponding surface points across the subject group. This process is repeated for each point on the surface. Suppose we have a family, F, of N different surface meshes representing a particular structure in N separate individuals, drawn from a specific population. If $\mathbf{r}_i(u,v)$ is the 3D position in stereotaxic space of the point with parametric coordinates (u,v) on the ith mesh, then the average sulcal surface is given by another mesh of the form

$$\mathbf{r}_\mu(\mathrm{u,v}) = (1/N)\Sigma_{i=1 \text{ to } N}\, \mathbf{r}_i(\mathrm{u,v}),\ \forall(\mathrm{u,v}). \quad (1)$$

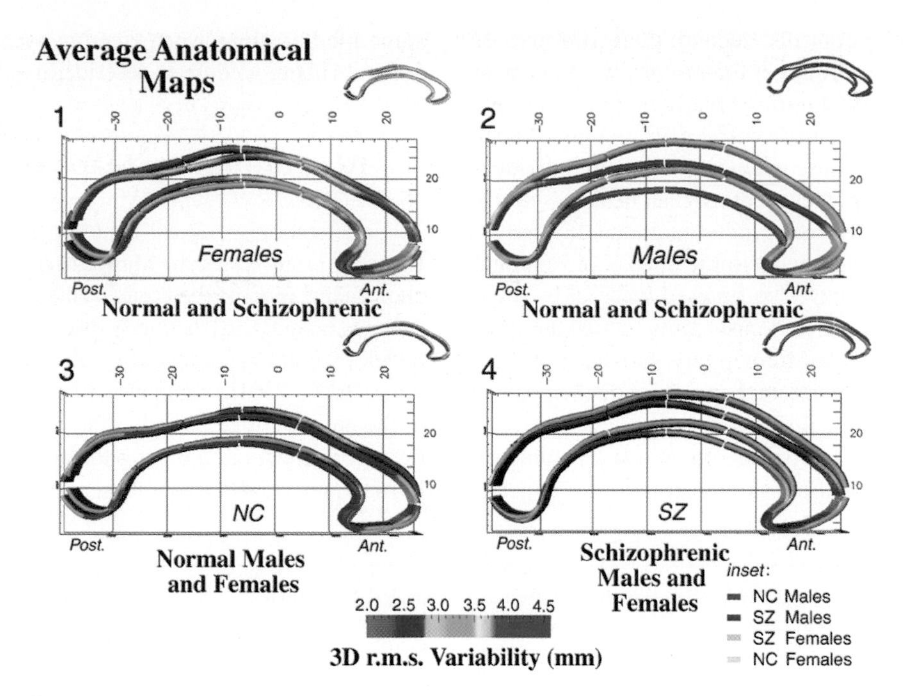

Figure 5 Corpus callosum in schizophrenia. Data from Narr *et al.*, 2000, courtesy of Katherine Narr. Midsagittal corpus callosum boundaries were averaged from 25 patients with chronic schizophrenia (DSM-III-R criteria; 15 males, 10 females; age 31.1 ± 5.6 years) and from 28 control subjects matched for age (30.5 ± 8.7 years), gender (15 males, 13 females), and handedness (1 left-handed subject per group). Profiles of anatomic variability around the group averages are also shown (in color) as an rms deviation from the mean. Anatomical averaging reveals a pronounced and significant bowing effect in the schizophrenic patients relative to normal controls. Male patients show a significant increase in curvature for superior and inferior callosal boundaries ($P < 0.001$), with a highly significant sex-by-diagnosis interaction ($P < 0.004$). The sample was stratified by sex and diagnosis and separate group averages show that the disease induces less bowing in females (**1**) than in males (**2**). While gender differences are not apparent in controls (**3**), a clear gender difference is seen in the schizophrenic patients (**4**). Abnormalities localized in a disease-specific atlas can therefore be analyzed to reveal interactions between disease and demographic parameters.

Information on anatomic variability can also be retained and shown as a *variability map* (Thompson *et al.,* 1996a, 1998). To map variability, individual deviations from the average surface are measured by computing 3D displacement maps (Fig. 3). These are patterns of 3D displacement vectors that would be required to reshape the average surface into the shape of a specific individual. If surface locations $\mathbf{r}_i(u,v)$ in subject *i* are indexed by parametric coordinates (u,v), then deviations of these locations in individual *i* from the mean anatomical surface are given by the set of displacement vectors

$$\mathbf{d}_i(u,v) = \mathbf{r}_i(u,v) - \mathbf{r}_\mu(u,v), \tag{2}$$

for all pairs of corresponding grid points $\mathbf{r}_i(u,v)$ and $\mathbf{r}_\mu(u,v)$. For each (u,v), the associated displacement maps $\mathbf{d}_i(u,v)$ represent a sample of (vector) observations from a zero-mean, spatially anisotropic probability distribution (Thompson *et al.,* 1996a). The variability of the surface points can be encoded using the covariance matrix, or tensor, of the deformation maps:

$$\Psi_F(u,v) = [1/(N-1)] \sum_{i=1 \text{ to } N} \mathbf{d}_i(u,v)\mathbf{d}_i(u,v)^T. \tag{3}$$

This matrix stores the shape, or preferred directions, of anatomic variability in the brain (Thompson *et al.,* 1996a; Cao and Worsley, 1999; Section VI). Perhaps the simplest measure of anatomic variability is the root mean square magnitude of the 3D displacement vectors, assigned to each point, in the surface maps from individual to average. This variability pattern can be visualized as a color-coded map. This map helps pick out regions that are highly variable across subjects, such as the occipital horns of the ventricles (Fig. 4) and the gyral patterns of the perisylvian cortices.

F. Emerging Features

An example application is the analysis of ventricular anatomy in Alzheimer's disease. Two features are observed in the average anatomical maps (Fig. 4) that may not be apparent, or would be difficult to localize, with conventional volumetry. First, the ventricles are larger in Alzheimer's disease than in healthy controls; second, there is a marked ventricular asymmetry (left larger than right), which is most pronounced in the occipital horns. Finally there is considerable anatomic variation (red colors) in occipital horn regions, which obscures these average patterns in individual data sets. Once identified, these features can be assessed statistically in new data sets, or an individual anatomy can be compared with the average maps in the atlas. Conventional morphometric statistics may also be used to describe these differences, such as lengths and curvatures, which are typically easiest to understand. Alternatively the shapes can be compared directly, using mathematics to encode the population variability in anatomical shape.[1]

In building average brain models in an atlas, parametric mesh averaging works well for simple structures, but needs some modifications when creating average templates of the cortex. This procedure is described next.

IV. Population Maps of the Cortex

Understanding cortical anatomy and function is a major focus in brain research. Many diseases affect the anatomy and organization of the cortex. The cortex also changes over time, as in aging, Alzheimer's disease (Mega *et al.,* 2000), or developmental disorders (Sowell *et al.,* 1999; Thompson *et al.,* 1998, 2001a; Blanton *et al.,* 2000). The gyral patterns of the human cortex provide a fairly reliable guide to its functional organization, although the congruence is not absolute (Brodmann, 1909; Rademacher *et al.,* 1993; Amunts and

---

[1]*How can anatomical shapes be compared?* So far (e.g., in Fig. 3) we have considered variations of surface points individually (Thompson *et al.,* 1999; Figs. 3 and 4). The notion of overall (global) shape similarity and variability can also be encoded using a *point distribution model* or *active shape model* (Cootes *et al.,* 1995; Davies *et al.,* 2001; Frangi *et al.,* 2001; Poxton *et al.,* 1998). These statistical models are popular in the computer vision community for face recognition (Sclaroff and Pentland, 1994) and statistically guided image segmentation (Duta and Sonka, 1998). They use principal component analysis to describe shape variation in a population using a small number of parameters. By capturing the notion of allowable anatomical shapes, they make it easier to find structures in new images and implicitly quantify the abnormality in an anatomic structure. An anatomical shape is considered as a single vector of coordinates. The statistical distribution of different shapes is estimated from a training set, or a reference atlas, of examples. For N shapes described as vectors of k different 3D landmark points $\mathbf{s}_i = (x_{11}, x_{12}, x_{13}, \ldots, x_{k1}, x_{k2}, x_{k3})$, the individual anatomies form a distribution in a $3k$-dimensional space, which is approximated by a linear model of the form

$$\mathbf{s} = \mathbf{s}_\mu + \mathbf{\Phi b}. \tag{4}$$

Here $\mathbf{s}_\mu = [1/N]\ (\Sigma_{i=1 \text{ to } N}\ \mathbf{s}_i)$ is the mean shape vector, $\mathbf{b}$ is the shape parameter vector of the anatomical model, and Φ is a matrix whose columns are the principal components (unit eigenvectors) of the covariance matrix:

$$\mathbf{S} = [1/(N-1)]\ (\Sigma_{i=1 \text{ to } N}\ (\mathbf{s}_i - \mathbf{s}_\mu)\ (\mathbf{s}_i - \mathbf{s}_\mu)^T). \tag{5}$$

The shape information in the population is then compacted in a new matrix $\mathbf{\Phi^*}$, whose columns are the first t eigenvectors of $\mathbf{\Phi}$, corresponding to the largest eigenvalues, $\lambda_1 \geq \ldots \geq \lambda_t$. These are the *principal modes* of shape variation. They describe the most significant types of variation, in terms of the proportion of variance explained. Individually they describe how boundary points tend to move together as the shape varies. Then we can approximate any anatomical shape by adding a linear combination of the eigenvectors to the mean vector, or center point, which describes the average shape: $\mathbf{s^*} = \mathbf{s}_\mu + \mathbf{\Phi^* b^*}$. Here $\mathbf{b^*} = (b_1 \ldots b_t)$ is a t-dimensional vector of weights, or shape parameters, which can be computed for a given shape, from the formula

$$\mathbf{b}_i^* = \mathbf{\Phi}^{*\mathrm{T}}(\mathbf{s}_i - \mathbf{s}_\mu). \tag{6}$$

If Gaussian distributions are assumed, the abnormality of an anatomical shape can be measured by its Mahalanobis distance from the mean (cf. Thompson *et al.,* 1996) :

$$D^2_t = \Sigma_{i=1 \text{ to } t}\ b_i^2/\lambda_i^2. \tag{7}$$

For large N, this measure follows a χ^2 distribution with t degrees of freedom and is proportional to the logarithm of the probability of obtaining the shape from the measured distribution of shapes. Multivariate analysis can then assess effects of diagnosis, gender, or other covariates, on the shape vector $\mathbf{b}_i^*$ (Joshi *et al.,* 1998; Gerig *et al.,* 2001).

Anatomical shape classification. The above approach represents anatomical shapes in terms of the eigenvectors of statistical variation, learned from a sample or training set. Wang and Staib (2000) apply this shape encoding in a Bayesian framework to locate the corpus callosum and basal ganglia in new images, based on a training set. A more complex variant of this approach is Grenander's shape modeling approach (Grenander and Miller, 1998). This models anatomical shapes in terms of a basis of mathematical functions, such as spherical harmonics (Thompson and Toga, 1996; Csernansky *et al.,* 1998; Gerig *et al.,* 2001), Oboukhov expansions (Joshi *et al.,* 1998; Gerig *et al.,* 2001), or elliptical Fourier series (Staib and Duncan, 1992). These functions are in fact eigenfunctions of self-adjoint differential operators (such as the Laplacian and Cauchy–Navier operator), and they can be thought of as the modes of vibration of physical systems described by partial differential equations (Grenander and Miller, 1998; Thompson *et al.,* 2001a). Sclaroff and Pentland (1994) also represent shape differences in terms of the physical modes of vibration of the original shape. Both of these statistical and physical approaches are designed to create efficient representations of complex shapes in terms of a small number of numerical parameters. The resulting parameter set can then be subjected to multivariate analysis or linear discriminant analysis to classify and characterize anatomical shape abnormalities in disease (Joshi *et al.,* 1998). Golland *et al.* (2001) extended this approach to classify 3D models of the hippocampus/amygdala complex in schizophrenia relative to healthy controls. Rather than use surfaces directly, they used a 3D distance map as a shape descriptor, whose value at any point is the nearest distance to the surface. A classifier was developed based on support vector machines, a popular classification technique in machine learning theory. A measure based on the Vapnik–Chervonenkis dimension (Vapnik, 1995) was employed to estimate the smallest sample size necessary for statistical significance, an important feature in atlas construction (Golland *et al.,* 2000).

Zilles, 2001). Since most imaging studies of brain function focus on the cortex, it is especially important to be able to pool cortical brain mapping data from subjects whose anatomy is different (Zeineh *et al.*, 2001). Despite interest in analyzing patterns of cortical variation for interesting effects, general patterns of organization are hard to discern, as are systematic alterations in disease. Sulcal pattern variation (Fig. 6; Steinmetz *et al.*, 1990; Ono *et al.*, 1990; Thompson *et al.*, 1996a; Le Goualher *et al.*, 1999, 2000; MacDonald *et al.*, 2000) also complicates attempts to define statistical criteria for abnormal cortical anatomy.

In the following section we outline a framework for analyzing cortical anatomy. Specialized algorithms compare and average cortical anatomy across subjects and groups,

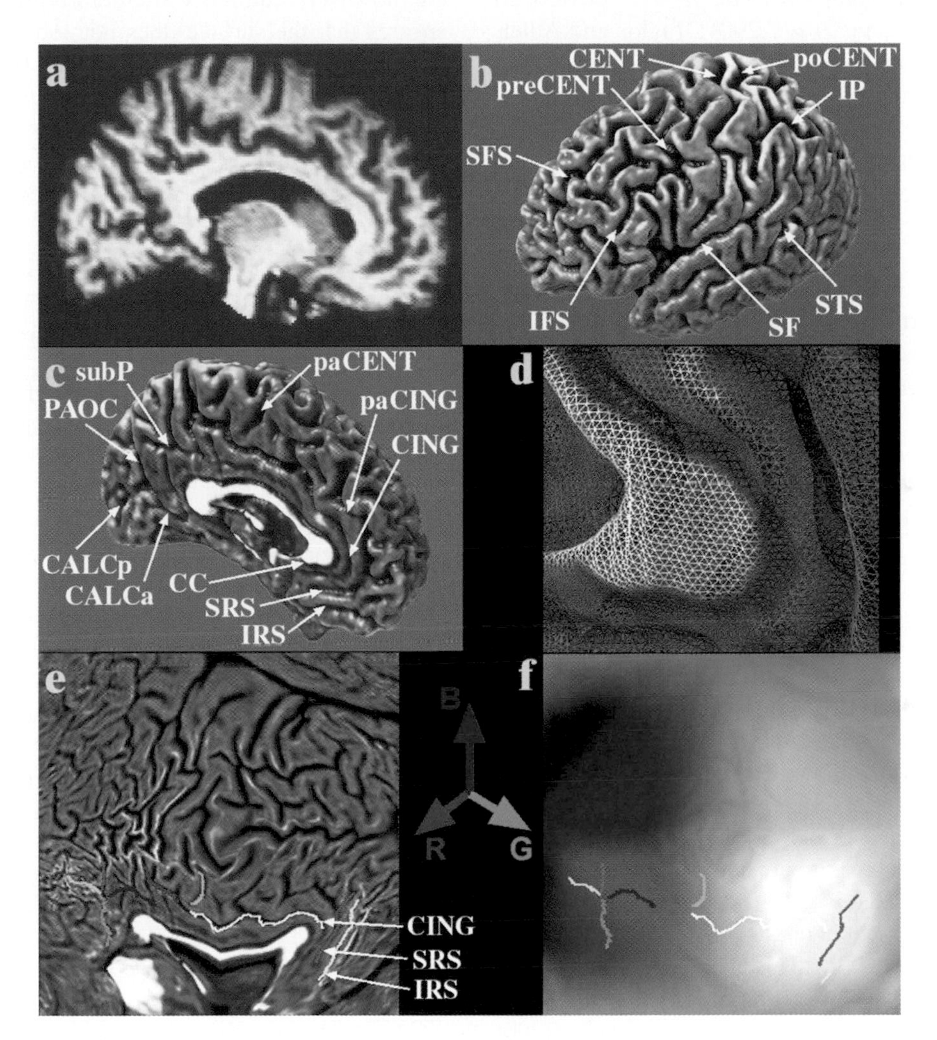

Figure 6 Measuring differences in cortical anatomy. Based on an individual's 3D MRI scan (**a**), detailed surface models of the cerebral cortex can be generated (**b** and **c**). A template of 3D curved lines is delineated on these surfaces, capturing the morphology of the sulcal pattern. On the lateral brain surface, important functional landmarks include the central (CENT), pre- and postcentral (preCENT, poCENT) sulci; superior and inferior frontal sulci (SFS, IFS); intraparietal sulcus (IP); Sylvian fissure (SF); and superior temporal sulcus (STS). Medial surface landmarks include the corpus callosum (CC); anterior and posterior calcarine (CALCa/p), parieto-occipital, subparietal, paracentral, paracingulate, and cingulate sulci; and superior and inferior rostral sulci. A spherically parameterized, triangulated 3D mesh represents the cortical surface; (**d**) shows the grid structure around the anterior corpus callosum. When the parameter space of the surface is flattened out (**e**), landmarks in the folded brain surface can be reidentified (e.g., IRS, SRS, etc.). (The white patch by the corpus callosum is where the surface model cuts across the white matter of the brain stem). To avoid loss of 3D information in the flattening, a color code is used to store where each flat map location came from in 3D, with red colors brighter where the lateral (x) coordinate is larger, green colors brighter where the posterior-to-anterior coordinate (y) is larger, etc. The warping of these color maps (Figs. 7 and 8), and the averaging of the resulting images, provides a surprising strategy for creating average cortical models for a group of subjects and for exploring cortical pattern variation.

map its variation and asymmetry, and chart patterns of abnormality or brain change. Rapid developments are being made in the mathematics of cortical mapping (Angenent *et al.*, 1999; Hurdal *et al.*, 1999; Joshi *et al.*, 1995; Haker *et al.*, 2000; Bertalmio *et al.*, 2000; Fischl *et al.*, 2001). Signal detection methods based on random field theory, used widely in brain mapping, are also being optimized to handle cortical surface data (Thompson *et al.*, 1997, 2001a; Cao and Worsley, 1999; Chung *et al.*, 2001; Taylor and Adler, 2000). These approaches draw upon parametric surface methods, but supplement them with additional warping, gyral pattern matching, and diffusion smoothing approaches that make comparison of cortical data tractable.

A. Cortical Modeling

A major challenge in investigations of disease is to determine (1) whether cortical organization is altered and, if so, which cortical systems are implicated and (2) whether normal features of cortical organization are lost, such as sulcal pattern asymmetries (Kikinis *et al.*, 1994; Bilder *et al.*, 2000; Narr *et al.*, 2001; Sowell *et al.*, 2001). These questions motivate methods to create a disease-specific average models of the cortex and a statistical framework to compare individual and group average models with normative data.

B. Cortical Parameterization

Several methods exist to generate surface models of the cortex from 3D MRI scans. Some of these impose a tiled, parametric grid structure on the anatomy, which is used as a coordinate framework to support subsequent computations. In "bottom-up" approaches (e.g., Fischl *et al.*, 1999; Haker *et al.*, 1999; Shattuck and Leahy, 2001), a voxel-based segmentation of white matter is generated first, using a tissue classifier or level set methods (Sapiro, 2000). Its topology is then corrected using graph theoretic methods (Shattuck and Leahy, 2001; Han *et al.*, 2001). This creates a single, closed, simply connected surface homeomorphic to a sphere (Fischl *et al.*, 1999; Hurdal *et al.*, 1999; Rettman *et al.*, 2000; Shattuck and Leahy, 2001). The surface is tiled using triangulation methods such the Marching Cubes algorithm (Lorensen and Kline, 1987). The gridded surface is then inflated, using iterative smoothing, to a spherical shape. Inverting this inflation mapping allows a spherical coordinate system to be projected back onto the 3D model, for subsequent computations. Alternatively the 3D surface may be flattened to a 2D plane (Fig. 7; Drury and Van Essen, 1997; Van Essen *et al.*, 1997; Thompson *et al.*, 1997; Angenent *et al.*, 1999; Hurdal *et al.*, 1999), inducing an alternative 2D parameterization onto the original 3D surface.

A second ("top-down") type of surface extraction method (Davatzikos, 1996; MacDonald, 1998; Kabani *et al.*, 2001) begins with a spherical or ellipsoidal surface that is already tiled. This parametric surface is successively moved, under image-dependent forces, reshaping it into the complex geometry of the cortical boundary (see Xu *et al.*, 1999, for work on *gradient vector flow*). This avoids the need for topology correction, as a single, fixed, grid structure is established at the start and mapped with a continuous deformation onto each anatomy. Complex constraints are, however, required while deforming the surface. These ensures that the surface does not self-intersect and adapts fully to the target geometry. The first (bottom-up) strategy turns the cortex into a sphere, while the latter approach deforms a sphere onto the cortex. Both approaches allow one to project a coordinate system onto the anatomy, so that cortical locations can be referred to in surface-based coordinates.

C. Mapping Gyral Pattern Differences in a Population

Once cortical models are available for a large number of subjects, in a common 3D coordinate space, patterns of cortical variability can be calculated. The major gyri and sulci of the cortical surface have a similar spatial layout across subjects (Ono *et al.*, 1990; Regis, 1994), even though their geometry varies substantially. In one approach (Thompson *et al.*, 2000a), a maximal set, or template, is specified, containing all primary sulci that consistently occur in normal subjects[2] (Figs. 6b and 6c show some of these). This set of sulcal curves can be reliably identified manually by trained raters, so long as a formalized protocol and detailed anatomical criteria are followed (Sowell *et al.*, 2001). Automated labeling of sulci is also the focus of intense study (Mangin *et al.*, 1994; Vaillant and Davatzikos, 1997; MacDonald, 1998; Lohmann *et al.*, 1999; Royackkers *et al.*, 1999; Le Goualher *et al.*, 1999; Zhou *et al.*, 1999; Rettman *et al.*, 2000; Tao *et al.*, 2001; Caunce and Taylor, 2001).

[2]Several complications arise in identifying corresponding sulci across subjects. These can be only partially resolved using information on which sulci border known architectonic fields (Brodmann, 1909; Rademacher *et al.*, 1993). Approximately a quarter of normal brain hemispheres have two cingulate gyri (the "double parallel" conformation; Ono *et al.*, 1990; Regis, 1994; Paus *et al.*, 1999), and some individuals have two Heschl's gyri (Leonard, 1996), while others have only one. When there are two cingulate sulci, the outer (paracingulate) sulcus arguably matches the single sulcus in an individual with only one, as it bounds the Brodmann areas belonging to the limbic system. Interrupted sulci, in which a sulcal curve is broken into several segments, may also need to be connected and modeled as a single curve to facilitate matching (cf. Thompson *et al.*, 1999; Sebastian *et al.*, 2000). In rare cases, some pairs of sulci, such as the postcentral sulcus and the marginal ramus of the cingulate, meet the superior margin of the interhemispheric fissure in a different anterior-to-posterior order. Modeling of the graph-theoretic structure and connectivity of the sulci may also be necessary for a fuller understanding of cortical variation (Mangin *et al.*, 1994).

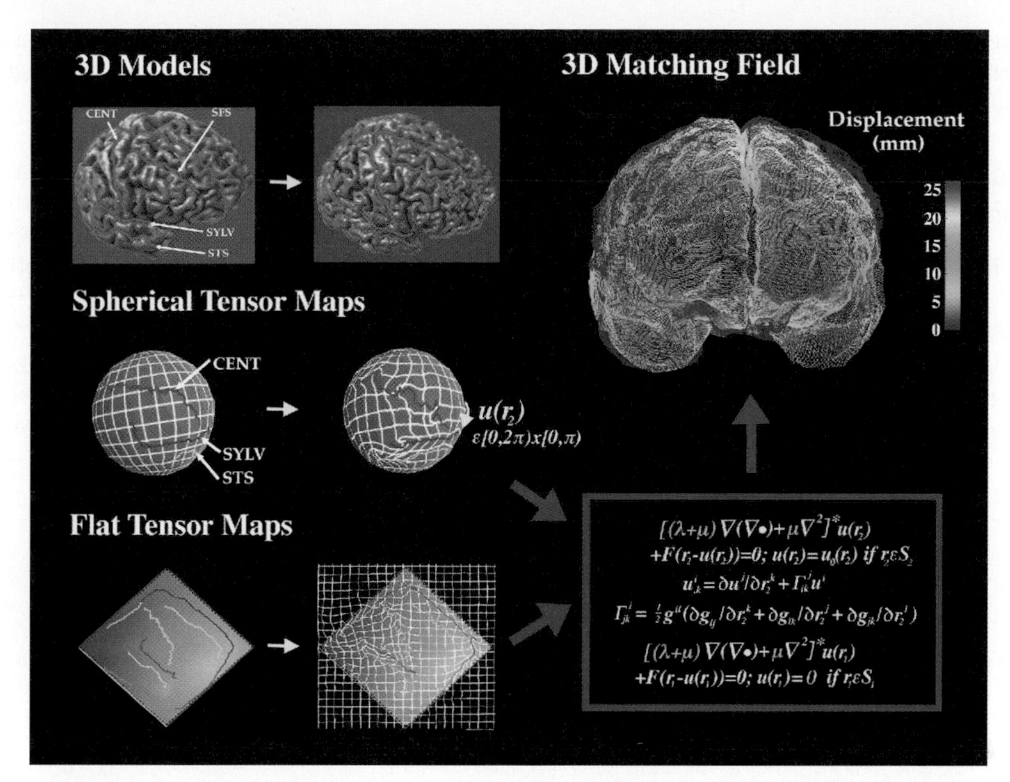

Figure 7 Cortical pattern matching. Cortical anatomy can be compared, for any pair of subjects (3D models; **top left**), by computing a 3D deformation field that reconfigures one subject's cortex onto the other (3D matching field; **top right**). In this mapping, gyral patterns can be constrained to match their counterparts in the target brain. To do this, flattening or inflation of the extracted cortical surface provides a continuous inverse mapping from each subject's cortex to a sphere or plane. A vector field $\mathbf{u}(\mathbf{r})$ in the parameter space can then drive the gyral pattern elements into register on the sphere (see spherical flow). The full mapping (**top middle**) is recovered in 3D space as a displacement vector field matching cortical regions in one brain into precise structural registration with their counterparts in the other brain. Tensor maps (**middle and lower left**). Different amounts of local dilation and contraction [encoded in the metric tensor of the mapping, $g_{jk}(\mathbf{r})$] are required to transform the cortex into a simpler two-parameter surface. These variations complicate the direct application of 2D warping equations for matching their features. Using a covariant tensor approach (red box) the regularization operator L is replaced by its covariant form $L^{\ddagger}$. Correction terms (Christoffel symbols, Γ^{i}_{jk}) compensate for fluctuations in the metric tensor of the inflation and flattening procedures. This (1) makes the matching field independent of the underlying gridding of the surface (spherical or planar) and (2) eliminates effects of metric distortions that occur in the inflation or flattening procedure.

D. Matching Cortical Patterns

Cortical anatomy can be compared, between any pair of subjects, by computing the warped mapping that elastically transforms one cortex into the shape of the other. Due to variations in gyral patterning, cortical differences among subjects will be severely underestimated unless elements of the gyral pattern are matched from one subject to another. This matching is also required for cortical averaging; otherwise, corresponding gyral features will not be averaged together. Transformations can therefore be developed that match large networks of gyral and sulcal features with their counterparts in the target brain (Thompson and Toga, 1996, 1997; Davatzikos, 1996; Van Essen *et al.*, 1997; Fischl *et al.*, 1999). In a recent anatomically based approach, we matched 38 elements of the gyral pattern, including the major features that are consistent in their incidence and topology across subjects (see Thompson *et al.*, 2001a, for details; Sowell *et al.*, 2000; cf. Ono *et al.*, 1990; Leonard, 1996; Kennedy *et al.*, 1998).

To find good matches among cortical regions we performed the matching process in the cortical surface's parametric space, which permits more tractable mathematics (Fig. 7). This vector flow field in the parametric space indirectly specifies a correspondence field in 3D, which drives one cortical surface into the shape of another. This mapping not only matches overall cortical geometry, but also matches the entire network of the 38 landmark curves with their counterparts in the target brain and thus is a valid encoding of cortical variation. Details of this procedure are given in footnote 3.[3]

E. Making a Well-Resolved Average Cortical Model for a Group

The intersubject variability of the cortex is computed by first creating an average cortex for each subject group and measuring individual differences from the deformation mappings that drive the average model onto each individual. By defining probability distributions on the space of deformation transformations applied to the average template (as in Section III), statistical parameters of these distributions are estimated from the databased anatomic data. This determines the magnitude and directional biases of anatomic variation. To do this, all 38 gyral curves for all subjects are first transferred to the parameter space (Fig. 7e). Next, each curve is uniformly reparameterized to produce a regular curve of 100 points whose corresponding 3D locations are uniformly spaced. A set of 38 average gyral curves for the group is created by vector averaging all point locations on each curve. This *average curve template* (*curves* in Fig. 8a) serves as the target for alignment of individual cortical patterns (Thompson *et al.,* 2000a; Zeineh *et al.,* 2001). Each individual cortical pattern is

---

[3]*Mathematics of spherical and planar maps of cortex.* We recently applied this matching approach to measure anatomic variability in a database of 96 cortical models (extracted from an MRI database with the algorithm of MacDonald, 1998; Thompson *et al.,* 2001a). Since cortical models were created by driving a tiled, spherical mesh into the configuration of each subject's cortex, any point on the cortical surface maps to exactly one point on the sphere and vice versa. Each cortical surface is parameterized with an invertible mapping, D_p,D_q: $(r,s) \rightarrow (x,y,z)$, so sulcal curves and landmarks in the folded brain surface can be reidentified in a spherical map (cf. Fischl *et al.,* 1999). To retain relevant 3D information, cortical surface point position vectors (x,y,z) in 3D stereotaxic space were color-coded using a unique RGB color triplet, to form an image of the parameter space in color image format (Fig. 6f). These spherical locations, indexed by two parameters, can also be mapped to a plane (Fig. 6e; Thompson *et al.,* 1997). Cortical differences between any pair of subjects were calculated as follows. A flow field was first calculated that elastically warps one flat map onto another from the other subject (Fig. 7; or equivalently, one spherical map onto the other). On the sphere, the parameter shift function $\mathbf{u}(\mathbf{r}):\Omega \rightarrow \Omega$, is given by the solution $F_{pq}:\mathbf{r} \rightarrow \mathbf{r} - \mathbf{u}(\mathbf{r})$ to a curve-driven warp in the spherical parametric space $\Omega = [0,2\pi) \times [0,\pi)$ of the cortex (Fig. 6; Thompson *et al.,* 1997). For points $\mathbf{r} = (r,s)$ in the parameter space, a system of simultaneous partial differential equations can be written for the flow field $\mathbf{u}(\mathbf{r})$:

$$L^{\ddagger}(\mathbf{u}(\mathbf{r})) + \mathbf{F}(\mathbf{r} - \mathbf{u}(\mathbf{r})) = \mathbf{0},\ \forall\ \mathbf{r} \in \Omega, \text{ with } \mathbf{u}(\mathbf{r}) = \mathbf{u}_0(\mathbf{r}),\ \forall\ \mathbf{r} \in M_0 \cup M_1. \tag{8}$$

Here M_0, M_1 are sets of points and (sulcal or gyral) curves for which displacement vectors $\mathbf{u}(\mathbf{r}) = \mathbf{u}_0(\mathbf{r})$ matching corresponding anatomy across subjects are known. The flow behavior is modeled using equations derived from continuum mechanics, and these equations are governed by the Cauchy–Navier differential operator

$$L = \mu\nabla^2 + (\lambda + \mu)\ \nabla\ (\nabla^T\cdot), \tag{9}$$

with body force $\mathbf{F}$ (Gee *et al.,* 1998; Gramkow, 1996). The only difference is that $L^{\ddagger}$ is the *covariant* form of the differential operator L (for reasons explained in footnote 2). This approach not only guarantees precise matching of cortical landmarks across subjects, but also creates mappings that are independent of the surface metrics and therefore independent of how the surfaces are gridded (parameterized).

Warping one cortex to another. Since the cortex is not a *developable* surface, it cannot be given a parameterization whose metric tensor is uniform (Van Essen and Maunsell, 1983). As in fluid dynamics or general relativity applications, the intrinsic curvature of the solution domain can be taken into account when computing flow vector fields in the cortical parameter space and mapping one mesh surface onto another. In the *covariant tensor* approach (Thompson *et al.,* 2001a), correction terms (Christoffel symbols, $\mathbf{\Gamma}^{\mathbf{i}}_{\mathbf{jk}}$) make the necessary adjustments for fluctuations in the metric tensor of the mapping procedure. In the partial differential equation (8), we replace L by the covariant differential operator $L^{\ddagger}$. In $L^{\ddagger}$, all L's partial derivatives are replaced with *covariant* derivatives (Burke, 1985). These covariant derivatives are defined with respect to the metric tensor of the surface domain where calculations are performed. The covariant derivative of a (contravariant) vector field, $u^i(\mathbf{x})$, is defined as

$$u^i_{,k} = \partial u^j/\partial x^k + \Gamma^j_{ik}\ u^i, \tag{10}$$

where the *Christoffel symbols of the second kind* (Einstein, 1914), $\mathbf{\Gamma}^{\mathbf{j}}_{\mathbf{ik}}$, are computed from derivatives of the metric tensor components $g_{jk}(\mathbf{x})$:

$$\Gamma^{\mathbf{i}}_{\mathbf{jk}} = (1/2)g^{\mathrm{il}}\ (\partial g_{\mathrm{lj}}/\partial \mathrm{x}^{\mathrm{k}} + \partial g_{\mathrm{lk}}/\partial \mathrm{x}^{\mathrm{j}} - \partial g_{\mathrm{jk}}/\partial \mathrm{x}^{\mathrm{l}}). \tag{11}$$

These correction terms are then used in the parameter space flow that matches one cortex with another. Note that a parameterization-invariant variational formulation could also be used to minimize metric distortion when mapping one surface to another. If P and Q are cortical surfaces with metric tensors $g_{jk}(u^i)$ and $h_{jk}(\xi^\alpha)$ in local coordinates u^i and ξ^α (i, α = 1,2), the *Dirichlet energy* of the mapping $\xi(u)$ is defined as $E(\xi) = \int_P e(\xi)\ (u)\ dP$, where $e(\mathrm{x})\ (u) = g^{ij}(u)\ \partial\xi^\alpha(u)/\partial u^i\ \partial\xi^\beta(u)/\partial u^j\ h_{\alpha\beta}(\xi(u))$ and $dP = (\sqrt{\det[g_{ij}]})du^1du^2$. The Euler equations, whose solution $\mathrm{x}^a(u)$ minimizes the mapping energy, are (Liseikin, 1991)

$$0 = L(\xi^i) = \Sigma_{\mathrm{m}=1 \text{ to } 2}\ \partial/\partial \mathrm{u}^{\mathrm{m}}\ [(\sqrt{\det[\mathrm{g}^{\mathrm{ru}}]})\ \Sigma_{\mathrm{l}=1 \text{ to } 2}\ g^{ml}{}_{ur}\ \partial\xi^i/\partial \mathrm{u}^{\mathrm{l}}]\ (i = 1,2). \tag{12}$$

The resulting (harmonic) map (1) minimizes the change in metric from one surface to the other and (2) is again independent of the parameterizations (spherical or planar) used for each surface. The harmonic energy is therefore a functional defined on a quotient space, being invariant to the action of the reparameterization group on each surface [related algorithms for minimizing harmonic energies, invariant under reparameterization, have been developed in level set methods for image restoration (Bertalmio *et al.,* 2000), for signal detection and smoothing on surfaces (Chung *et al.,* 2001), in modeling liquid crystals (Alouges, 1997), and in Polyakov's formulation of string theory (Polyakov, 1987)].

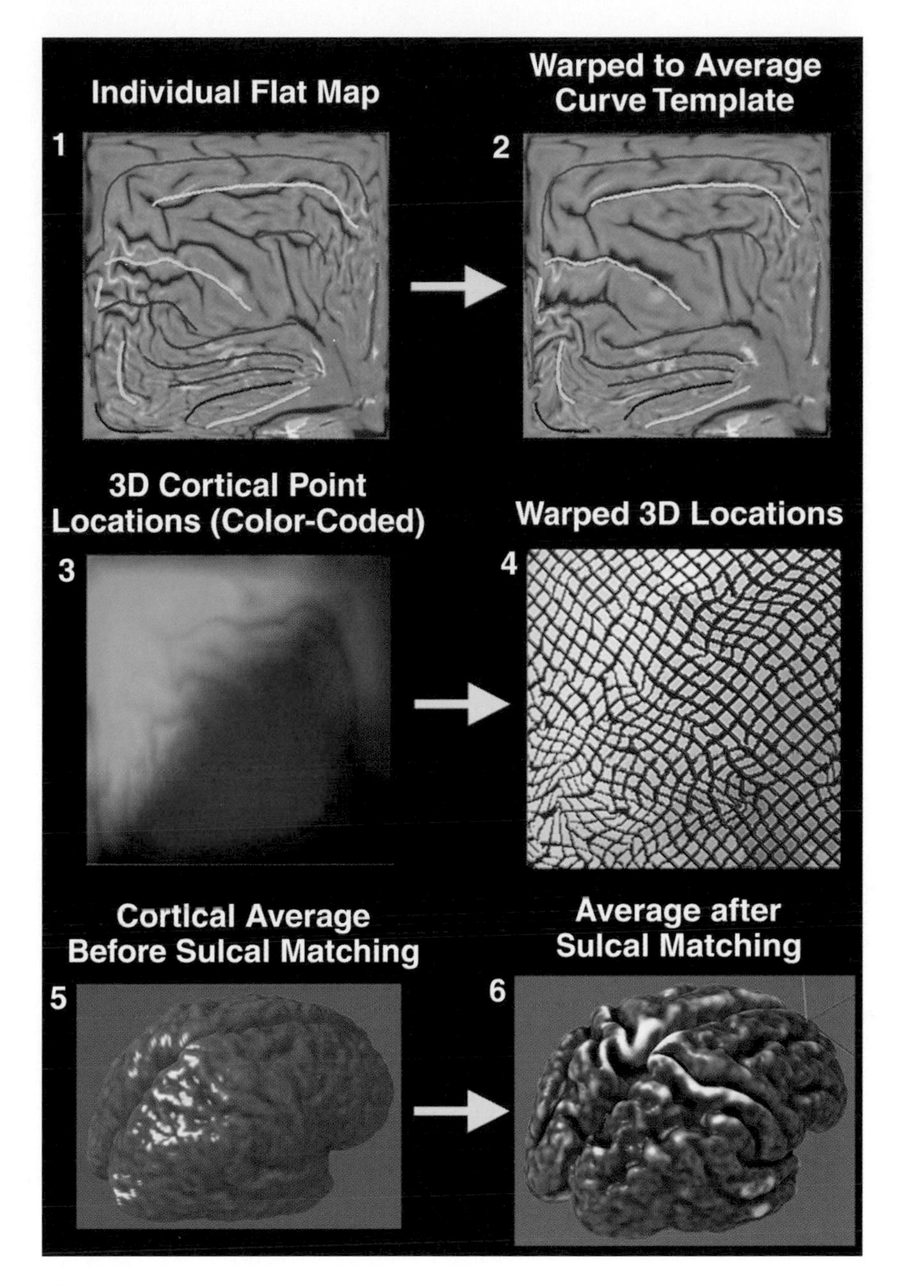

Figure 8 Cortical pattern matching and averaging. A well-resolved average cortical model (**6**) for a group of subjects can be created by first flattening each subject's cortical model to a 2D square (**1;** see also Figs. 6 and 7). A color-coded map (**3**) stores a unique color triplet (RGB) at each location in the 2D parameter space encoding the (x,y,z) coordinate of the 3D cortical point mapped to that 2D location. By averaging these color maps pixel by pixel across subjects, and then decoding the 3D colors into a surface model, a smooth cortical model (**5**) is produced. However, a well-resolved average model (**6**) is produced, with cortical features in their group mean location, if each subject's color map is first flowed (**4**) so that sulcal features are driven into the configuration of a 2D average sulcal template (**2**). The average curve set is defined by 2D vector averaging of many subjects' flattened curves. In this flow (4), codes indexing similar 3D anatomical features are placed at corresponding locations in the parameter space and are thus reinforced in the group average (**6**).

transformed into the average curve configuration using a flow field in the parameter space (Fig. 8, panel 2; cf. Bakircioglu *et al.*, 1999). By carrying a color code (that indexes 3D locations; Fig. 8, panel 3) along with the vector flow that aligns each individual with the average folding pattern, information can be recovered at a particular location in the average

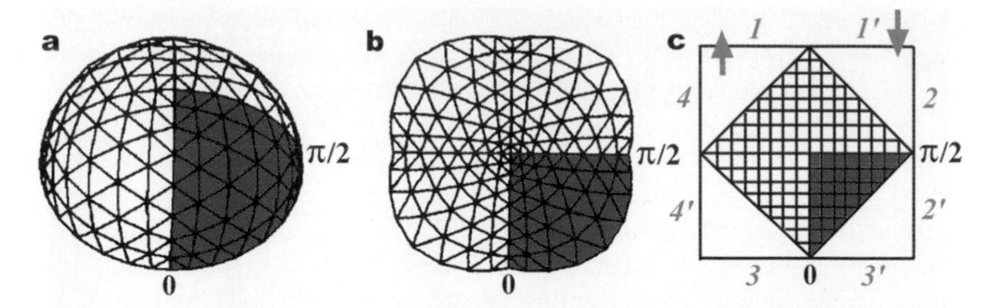

Figure 9 Computational grids used for cortical parameterization and matching. Cortical data can be compared from one individual to another by first imposing a spherical grid structure onto the anatomy of the cortex (Thompson *et al.,* 1997; MacDonald, 1998; Fischl and Dale, 2001). A specialized geodesic grid structure (**a**) is applied to each hemisphere, and the resulting discrete, triangulated mesh can be further transformed to a plane (**b**). One octant of the sphere is highlighted in red. (b) is adapted, with permission, from the work of Zhang and Hebert (1999), who employ the theory of harmonic mappings (Pinkall and Polthier, 1993) to project 3D surface data into the plane, with minimal metric distortion (cf. Thompson *et al.,* 2001a). In our approach (Thompson *et al.,* 2000a; Fig. 7), we match cortical features from one individual to another by computing flows that align features in spherical space (cf. Bakircioglu *et al.,* 1999; Fischl *et al.,* 1999; for similar approaches). Partial differential equations are used to perform the matching. The periodic (spherical) parameter space can also be represented using a gridded square (**c**), in which boundary elements *1* and *1'*, *2* and *2'*, etc., are regarded as topologically adjacent (green arrows). All these grid structures are designed to allow efficient calculations on cortical data, including flow fields that match cortical anatomy across subjects (Fig. 8).

folding pattern (Fig. 8) specifying the 3D cortical points mapping each subject to the average.[4]

This produces a new coordinate grid (Fig. 8, panel 4) on a given subject's cortex in which particular grid points appear in the same location across subjects relative to the mean gyral pattern. By averaging these 3D positions across subjects, an average 3D cortical model can be constructed for the group. An example of this type of cortical average, based on nine subjects with Alzheimer's disease, is shown in Fig. 8, panel 6. The resulting mapping is guaranteed to average together all points falling on the same cortical locations across the set of brains and ensures that corresponding features are averaged together.

F. Advances in Cortical Surface Matching: Related Approaches

The above mathematics is motivated by the need to (1) enforce sulcal correspondences when cortical data are averaged together and (2) create mappings that do not depend on the way the surfaces are gridded (the covariant approach). Recent advances in level set theory and implicit partial differential equations (PDEs) promise to simplify and accelerate this process (Sapiro, 2001; Bertalmio *et al.,* 2000; Chung *et al.,* 2001). To solve equations that describe complex flows in a spherical parameter space, a peculiar grid structure is used, with circular wraparound (Fig. 9), so that variables describing the flow can be stored at each node in the grid. Fast solution of the flow equations takes advantage of a class of spatial data structures, based on the Platonic solids (Fig. 9). These "geodesic spheres" allow hierarchical gridding of the parameter space (MacDonald, 1998) and minimize distortions in grid cell area and shape toward the sphere's poles (Sahr and White, 1998).

[4]In the language of Lie algebras, corresponding 3D cortical points across the subject database are defined as the *pull-back* $D_p{}^*(\mathbf{r})$ (Burke, 1985) of the parameterization mappings D_p: $(r,s) \rightarrow (x,y,z)$ under the covariant vector flow $\mathbf{u}(\mathbf{r})$ that maps each subject to the average curve template. [For any smooth function D_p: $\Omega \rightarrow R^n$ and any diffeomorphic map $\mathbf{u}(\mathbf{r})$: $\Omega \rightarrow N$, there is a function on N, $D_p{}^*$: $N \rightarrow R^n$ called the pull-back of D_p by $\mathbf{u}(\mathbf{r})$ and defined by $D_p \circ \mathbf{u}$ (Burke, 1987)].

G. Cortical Flattening

Minimizing distortion when flattening the cortex is a topic of intense research (Carman *et al.,* 1995; Drury and Van Essen, 1997; Chow, 1998; Hurdal *et al.,* 1999). This work is related to texture mapping methods in the computer graphics literature (Kanai *et al.,* 1997; Zhang and Hebert, 1999) and theoretical work on minimal surfaces and soap films (Oprea, 1998). Pinkall and Polthier (1993) pioneered the idea of using harmonic mappings to flatten surfaces, as mathematical results exist that guarantee that grid cells are minimally distorted and do not fold over in the process (Eells and Sampson, 1964). Ingenious extensions to this method have generated conformal (angle-preserving) cortical maps using the Laplace–Beltrami operator (Angenent *et al.,* 1999; Haker *et al.,* 1999) or circle-packing methods (Hurdal *et al.,* 1999). For some applications (e.g., cortical matching), covariant approaches avoid the need for isometric flattening of the cortex, by incorporating the surface metrics in the flows (Thompson *et al.*, 2000a; Chung *et al.*, 2001). Related work by Fischl and Dale (2001) projects cortical surfaces into a spherical coordinate system while minimizing a distortion metric. They then compute a warp between two cortices by minimizing a penalty function. This consists of a curvature matching to improve the registration, an areal term to prevent folding, and a metric distortion term to favor the preservation of surface distances. All these methods aim to fulfil the goal of averaging data from corresponding cortical regions, when constructing population-based maps of the cortex. The result is an ability to analyze structural and functional data with greater statistical power (Thompson *et al.*, 1998; Rex *et al.*, 2000; Zeineh *et al.*, 2001) and a method to build average cortical models whose geometry is well-resolved and consistent with other data in a population-based atlas.

V. Brain Averaging

In making population-based brain templates, a key challenge is to average individual images together so that common features of the population are reinforced. If MR images are only globally aligned and averaged together, cortical features are washed away (Fig. 10).

Maps deforming individual cortical patterns to a group average shape can also help generate a brain template with the mean shape for a group and with sharply defined geometry. We recently used high-dimensional transformations to create a mean image template for a group of patients with Alzheimer's disease (AD), whose anatomy is not well accommodated by existing brain atlases or imaging templates (Thompson *et al.*, 2001a). We introduce this idea now, as in later sections we will typically use an average brain coordinate space as the space in which anatomical variability is quantified.

A. Average Brain Templates

To make a mean image template for a group, several approaches are possible (Evans *et al.*, 1994; Collins *et al.*, 1994a; Subsol, 1995; Grenander and Miller, 1998; Guimond *et al.*, 1999; Thompson *et al.*, 2000a; Woods *et al.*, 2000; Miller *et al.*, 2002). If scans are mutually aligned using only a linear transformation (Fig. 10), the resulting average brain is blurred in the more variable

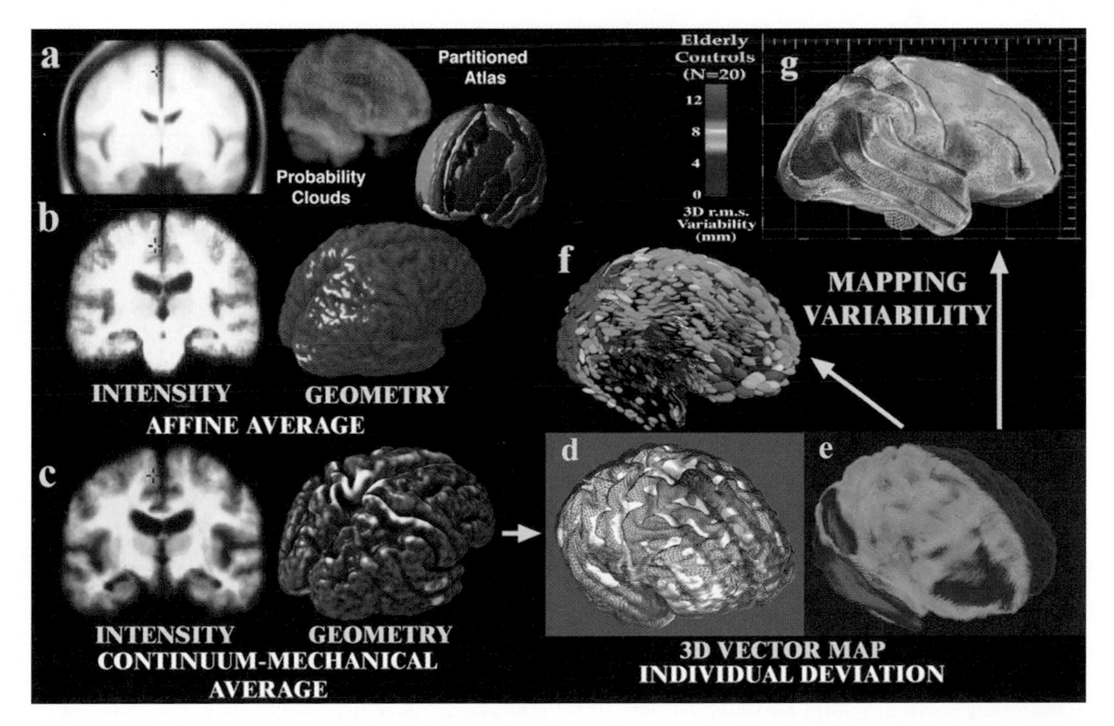

Figure 10 Average and probabilistic brain templates. Direct averaging of imaging data after a simple affine transform into stereotaxic space washes cortical features away [(**a**) Evans *et al.*, 1994; $N = 305$ normals; (**b**) shows a similar approach with $N = 9$ Alzheimer's patients]. By first averaging a set of vector-based 3D geometric models, and warping each subject's scan into the average configuration (as in Fig. 2), a well-resolved average brain template is produced (**c**). Deformation vector maps (**e**) store individual deviations (brown mesh) from a group average [white surface (**d**)], and their covariance fields (**f**) store information on the preferred directions and magnitude (**g**) of anatomic variability (pink colors, large variation; blue colors, less).

anatomic regions, and cortical features are washed away. The resulting average brain also tends to exceed the average dimensions of the component brain images. By averaging geometric and intensity features separately (cf. Ge *et al.,* 1995; Bookstein, 1997; Grenander and Miller, 1998; Christensen *et al.,* 1999; Thompson *et al.,* 2000a), a template can be made with the mean intensity and geometry for a patient population. To illustrate this, we generated an initial image template for a group of Alzheimer's patients by (1) using automated linear transformations (Woods *et al.,* 1993) to align the MRI data with a randomly selected image, (2) intensity-averaging the aligned scans, and then (3) recursively reregistering the scans to the resulting average affine image. The resulting average image was adjusted to have the mean affine shape for the group using matrix exponentiation to define average transformations (Woods *et al.,* 1998a). Images and a large set of anatomical surface models (84 per subject) were then linearly aligned to this template, and an average surface set was created for the group. Displacement maps driving the surface anatomy of each subject into correspondence with the average surface set were then computed and were extended to the full volume with surface-based elastic warping (Thompson *et al.,* 2000a). These warping fields reconfigured each subject's 3D image into the average anatomic configuration for the group. By averaging the reconfigured images (after intensity normalization), a crisp image template was created to represent the group (Fig. 10). Note the better resolved cortical features and sharper definition of tissue boundaries in the average images after high-dimensional cortical registration. If desired, this AD-specific atlas can retain the coordinate matrix of the Talairach system [with the anterior commissure at (0,0,0)] while refining the gyral map of the Talairach atlas to encode the unique anatomy of the AD population. By explicitly computing matching fields that relate gyral patterns across subjects, a well-resolved and spatially consistent set of probabilistic anatomical models and average images can be made to represent the average anatomy and its variation in a subpopulation.

B. Uses of Average Templates

Average brain templates have a variety of uses. If functional imaging data from Alzheimer's patients are warped into an atlas template based on young normals, signals in regions with selective atrophy in disease are artificially expanded to match their scale in young normals. Biases can therefore result. If the atlas has the average geometry for the diseased group, which may include atrophy, least distortion is applied by warping data into the atlas. Since the template (in Fig. 10) also has the average affine shape for the group (Woods *et al.,* 1998a), least distortion is applied when either linear or nonlinear approaches are used. The notion of least distortion can be formulated precisely using (1) mean vector fields (Thompson *et al.,* 2000a; Kochunov *et al.,* 2001a), (2) matrix and deformation tensor metrics (Woods *et al.,* 2000), or (3) the L^2 norm on the Hilbert space of deformation field coefficients (Grenander and Miller, 1998; cf. Martin *et al.,* 1994) or indirectly through a continuum-mechanical operator or regularization functional that defines what it means for a distortion to be irregular (Christensen *et al.,* 1999; Miller and Younes, 2001). These different definitions of average result from the fact that the idea of how "far away" one anatomy is from another can be formulated with different mathematical metrics[5] (Miller *et al.,* 2002). The resulting notion of population average (a template that is least far from the individuals) is determined accordingly.

As a result, average brain image templates may be defined in various different ways. The one in Fig. 10 has the mean geometry *and mean intensity* for a group (Thompson *et al.,* 2000a). Kochunov *et al.* (2001a) extended this idea to optimize the geometry of an individual image (the average of 27 MRI scans of the same subject mentioned in Section II; Holmes *et al.,* 1998). The individuality of brain shape was removed by first deforming this high-resolution template to 30 brains and applying the mean deformation field to the template. Interestingly, automated registration approaches were able to reduce anatomic variability to a statistically greater degree if this specially prepared image template was used as a registration target (Kochunov *et al.,* 2001a). With smaller deformations, local minima of the

[5]*Defining the average brain.* For a given nonlinear registration algorithm, and after affine components of deformation are factored out, a "mean-field average brain template" is one for which

$$\Sigma_{i=1 \text{ to } N} \int_\Omega \| u_i(x) \|^p \, dx , \tag{13}$$

is minimal, when $u_i(x)$ are the deformations mapping it onto a large set of other brains (p = 1 or 2 correspond to different norms). This template is closest to all brains; on average, brains have to be warped the least "distance" to match it. Alternatively, a "least-distortion average brain template" is one for which

$$\Sigma_{i=1 \text{ to } N} \int_\Omega \| L^{\ddagger} \mathbf{u}_i(\mathbf{x}) \|^p \, d\mathbf{x} , \tag{14}$$

is minimal. Here L is a (possibly covariant, see above) differential operator that measures the irregularity of the deformation field and ‡ denotes covariant differentiation with respect to the metric of the base manifold (this has no effect unless we are averaging nonflat manifolds, such as cortical surfaces, for which the Christoffel symbols do not vanish). Extending these ideas to registration algorithms that use velocity fields to ensure diffeomorphic mappings (e.g., Christensen *et al.,* 1996; see above), Miller and Younes (2001) show that

$$argmin \; \mathbf{v} \int_{\Omega \times [0,1]} \| L\mathbf{v}(\mathbf{x}, t) \|^2 \, d\mathbf{x} \, dt , \tag{15}$$

defines a metric on the space of diffeomorphisms, where V is the space of all velocity fields (paths) that deform the reference anatomy at $t = 0$ onto a target anatomy at time $t = 1$. In their formulation, a mean brain template would be one for which the following average energy is minimized:

$$\Sigma_{i=1 \text{ to } N} \int \| L(\mathbf{v}_i(\mathbf{x}, t)) \|^2 \, d\mathbf{x} \, dt. \tag{16}$$

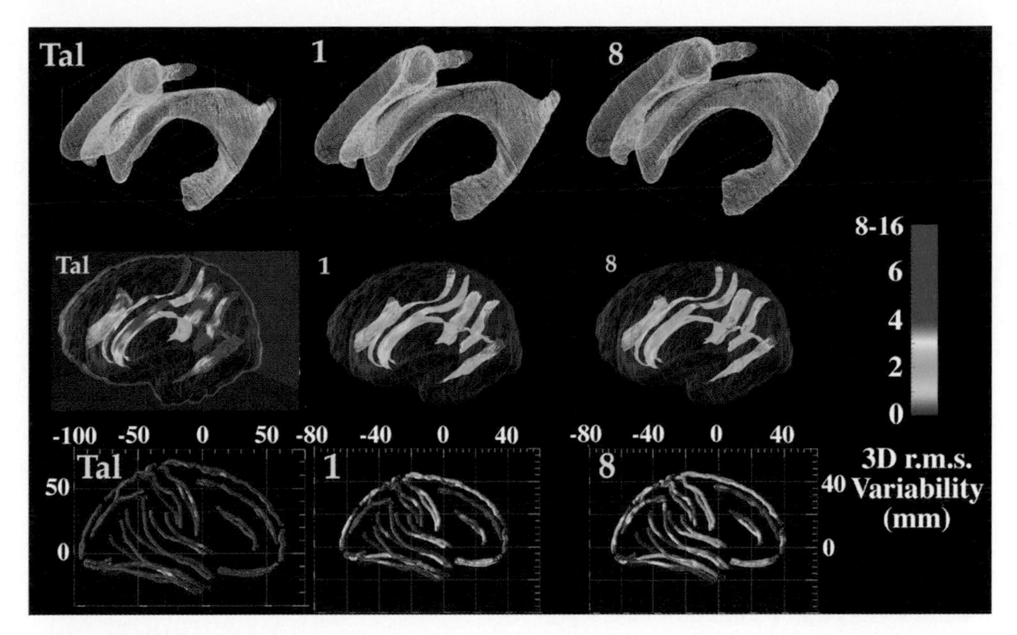

Figure 11 Comparing different registration approaches. The ability of different registration algorithms to reduce anatomic variability in a group of subjects ($N = 9$, Alzheimer's patients) is shown here. Digital anatomic models for each subject were mapped into common coordinate spaces using the transformations specified in the Talairach atlas (Tal; Talairach and Tournoux, 1988), as well as automated affine (or first-order) and eighth-order polynomial mappings as implemented in the Automated Image Registration package (Woods *et al.*, 1993, 1998a, 1999). After applying each type of mapping to the models from all subjects, the residual variability of ventricular (**top row**) and deep cortical surfaces (**middle row**) and superficial sulci (**bottom row**) is shown as a color coded map across each structure. The color represents the 3D root mean square distance from individual models to an average model for each structure, in which distance is measured to the near-point on the average mesh model (Thompson *et al.*, 1999). As expected, polynomial transformations reduce variability more effectively than affine transformations, and both outperform the Talairach system. At the cortex, model-driven registration can be used, if desired, to explicitly match gyral patterns (Thompson *et al.*, 2000a), improving registration still further.

registration measure may be avoided. Convergence may also be faster, as the parameter space is searched for an optimal match. This optimality of average brain templates may help in normalizing individual brains (Fig. 11) and may also be advantageous when databases are mined for information using nonlinear registration as an information source (Thompson *et al.*, 2000a).

VI. Atlas Statistics: Probabilistic Atlases

Once anatomic data are aligned with an average brain template, or atlas, a variety of statistics and maps can be computed. Some of these reflect the variability of different anatomic features in the atlas coordinate system. Features of interest include variations in gray matter distribution, structural asymmetry, incidence of sulci, and gyral patterning; they may also include rates of brain change in developing and diseased subpopulations. The result is a *probabilistic atlas* (Mazziotta *et al.*, 1995, 2001; Thompson *et al.*, 1997, 2000a, 2001a; Chiavaras *et al.*, 2001) that provides normative criteria to detect abnormal anatomy in an individual or group. Disease-related differences, such as altered asymmetries, or gray matter deficits, can also be mapped, using the atlas as a reference standard. With appropriate statistical models, the effects of aging, medication, and genetics on brain structure and brain change can be gauged in whole populations.

A. Mapping Anatomic Variability

As noted in Section III, maps of anatomic variability can help assess structural abnormalities in an individual or group, by encoding what normal variations are likely to be. By using cortical pattern matching to identify corresponding cortical locations in 3D space, rather than simple image averaging (Figs. 10a and 10b), deformation maps can be recovered mapping many individual subjects into gyrus-by-gyrus correspondence with the average cortex (Fig. 10e). Anatomic variability can thus be defined at each point on the average cortical mesh as the root mean square magnitude of the 3D displacement vectors, assigned to each point, in the surface maps from individual to average. This variability pattern is visualized as a color-coded map (Fig. 10g). This map shows the anatomic differences, due to gyral pattern variation, that remain after affine alignment of MR data into a brain template with the mean shape and intensity for the group.

B. Deformation-Based Morphometry

Deformations that align individual anatomies with an atlas standard can be analyzed to detect systematic differences in anatomy (a technique called "deformation-based morphometry"; Thompson *et al.,* 1997; Ashburner *et al.,* 1999; Gaser *et al.,* 1999; Ashburner, 2001; Good *et al.,* 2001a). The deformation fields contain detailed information on individual morphometry, and their statistics can be stored in an atlas. Depending on whether these fields are stored as 3D deformation *vectors* (Thompson *et al.,* 1997; Cao and Worsley, 1999), as *tensors* (Davatzikos *et al.,* 1996; Thompson *et al.,* 2000a; Pettey and Gee, 2001; Section VIII), or as a set of *basis function coefficients* that parameterize the nonlinear warp (Csernansky *et al.,* 1998; Ashburner, 2001), the analysis of structural differences proceeds a little differently (see also Good *et al.,* 2001a). Because of their utility in applying atlases to understand disease, we describe these approaches next (see Thompson *et al.,* 2001a, for a more detailed review).

C. Random Vector Fields

In a *random vector field* approach (Thompson *et al.,* 1997; Cao and Worsley, 1999), affine components of the deformation fields are first factored out. After this, the deformation vector required to match the structure at position **x** in the average cortex with its counterpart in subject *i* can be modeled as

$$\mathbf{W}_i(\mathbf{x}) = \boldsymbol{\mu}(\mathbf{x}) + \boldsymbol{\Sigma}(\mathbf{x})^{1/2}\boldsymbol{\varepsilon}_i(\mathbf{x}). \quad (17)$$

Here $\boldsymbol{\mu}(\mathbf{x})$ is the mean deformation vector for the population (which approaches the zero vector for large N), $\boldsymbol{\Sigma}(\mathbf{x})$ is a nonstationary, anisotropic covariance tensor field estimated from the mappings, $\boldsymbol{\Sigma}(\mathbf{x})^{1/2}$ is the upper triangular Cholesky factor tensor field, and $\boldsymbol{\varepsilon}_i(\mathbf{x})$ can be modeled as a trivariate random vector field whose components are independent zero-mean, unit variance, stationary random fields. This 3D probability distribution makes it possible to visualize the principal directions (eigenvectors), as well as the magnitude of gyral pattern variability, as a "tensor map" (Fig. 12b). These characteristics are highly heterogeneous across the cortex. For any desired confidence threshold α, $100(1 - \alpha)\%$ *confidence regions* for possible locations of points corresponding to **x** on the average cortex are given by nested ellipsoids $\mathbf{E}_{\lambda(\alpha)}(\mathbf{x})$ in displacement space (Fig. 12b; Thompson *et al.,* 1996a; Thirion *et al.,* 1998; Cao and Worsley, 1999). The shape of these ellipsoids is

$$\mathbf{E}_{\lambda}(\mathbf{x}) = \{\boldsymbol{\mu}(\mathbf{x}) + \boldsymbol{\lambda}[\boldsymbol{\Sigma}(\mathbf{x})]^{-1/2}\mathbf{p} | \forall \mathbf{p} \in \mathbf{pB}(\mathbf{0};1)\}, \quad (18)$$

where $\mathbf{B}(\mathbf{0};1)$ is the unit ball in R^3, and

$$\lambda(\alpha) = [[N(N-3)/3(N^2-1)]^{-1} F_{\alpha,3,N-3}]^{1/2}, \quad (19)$$

where $F_{\alpha,3,N-3}$ is the critical value of the F distribution such that $\Pr\{F_{3,N-3} = F_{\alpha,3,N-3}\} = \alpha$ and N is the number of subjects.

D. Detecting Group Anatomic Differences with Random Fields

This confidence limit representation of anatomy can also detect group differences in brain structure. The significance of a difference in brain structure between two subject groups (e.g., patients and controls) of N_1 and N_2 subjects is assessed by calculating the sample mean and variance of the deformation fields ($j = 1,2$),

$$\mathbf{W}^{\mu}_j(\mathbf{x}) = \boldsymbol{\Sigma}_{i=1 \text{ to } Nj}\, \mathbf{W}_{ij}(\mathbf{x})/N_j$$

$$\boldsymbol{\Psi}(\mathbf{x}) = (1/[N_1 + N_2 - 2])\{\boldsymbol{\Sigma}_{j=1 \text{ to } 2}\, \boldsymbol{\Sigma}_{i=1 \text{ to } Nj} [\mathbf{W}_{ij}(\mathbf{x}) - \mathbf{W}^{\mu}_j(\mathbf{x})][\mathbf{W}_{ij}(\mathbf{x}) - \mathbf{W}^{\mu}_j(\mathbf{x})]^{\mathrm{T}}\}, \quad (20)$$

and computing the following statistical map (Thompson *et al.,* 1997; Cao and Worsley, 2001):

$$\mathrm{T}^2(\mathbf{x}) = \{\mathrm{N}_1\mathrm{N}_2/(\mathrm{N}_1 + \mathrm{N}_2)(\mathrm{N}_1 + \mathrm{N}_2 - 2)\}[\mathbf{W}^{\mu}_2(\mathbf{x}) - \mathbf{W}^{\mu}_1(\mathbf{x})]^{\mathrm{T}} [\boldsymbol{\Psi}(\mathbf{x})]^{-1}[\mathbf{W}^{\mu}_2(\mathbf{x}) - \mathbf{W}^{\mu}_1(\mathbf{x})]. \quad (21)$$

Under the null hypothesis, $(N_1 + N_2 - 2)\mathrm{T}^2(\mathbf{x})$ is a stationary Hotelling's T^2-distributed random field. At each point, if we let $\nu = (N_1 + N_2 - 2)$ and we let the dimension of the search space be $d = 3$, then

$$F(\mathbf{x}) = ((\nu - d + 1)/d)\mathrm{T}^2(\mathbf{x}) \sim F_{d,(\nu - d + 1)}. \quad (22)$$

In other words, the field can be transformed point-wise to a Fisher–Snedecor F distribution (Thompson *et al.,* 1997), and these statistics of abnormality can be plotted in color across the cortex (Fig. 12c). To obtain a P value for the effect that is adjusted for the multiple comparisons involved in assessing a whole field of statistics (cf. Friston *et al.,* 1995; Holmes *et al.,* 1996), Cao and Worsley (1999) examined the distribution of the global maximum T^2_{max} of the resulting T^2-distributed random field under the null hypothesis. Alternatively a significance value for the whole experiment can be assigned by estimating their fraction of the statistical map that exceeds any threshold by permutation (Holmes *et al.,* 1996; Bullmore *et al.,* 1999; Sowell *et al.,* 1999). This nonparametric approach avoids assumptions about the spatial autocorrelation of the process and has been widely used in functional imaging analyses (e.g., Holmes *et al.,* 1996). Subjects are randomly assigned to groups and the distribution of accidental clusters is tabulated empirically. We have recently used this approach to detect developmental changes in brain asymmetry and gray matter distribution, as well as gray matter loss in Alzheimer's disease and schizophrenia (see Sections VII and VIII).

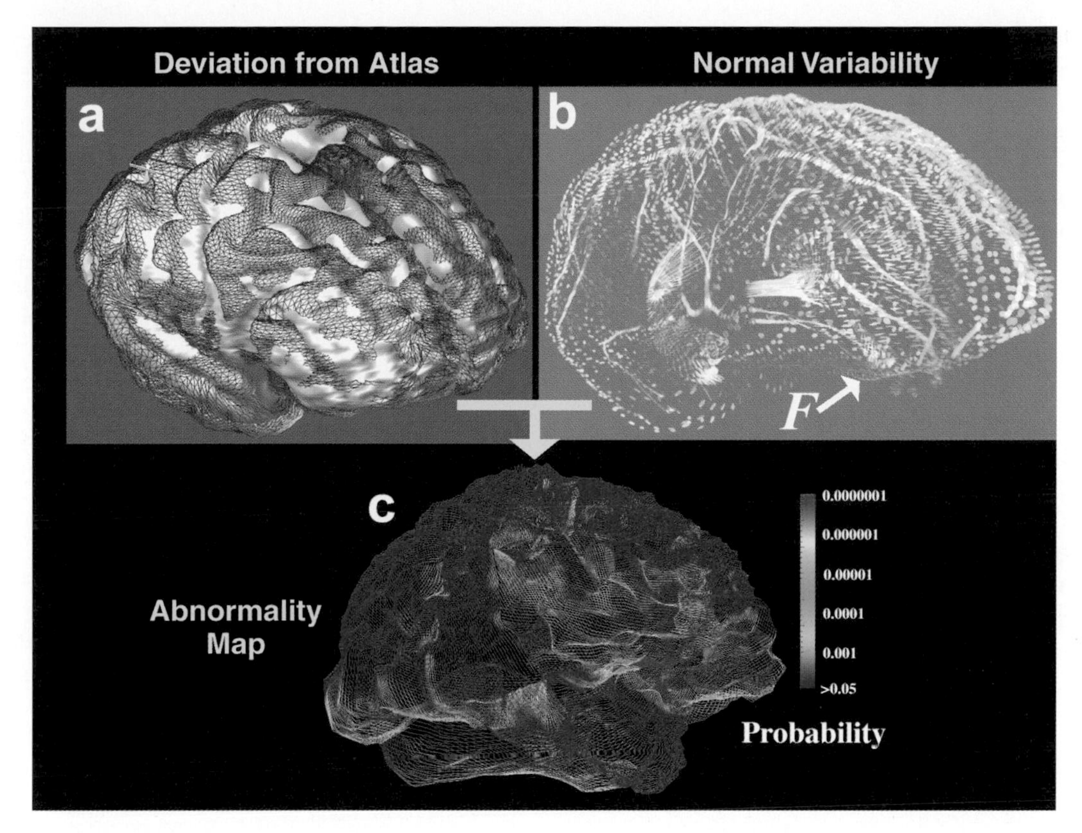

Figure 12 Mapping cortical shape anomalies. Due to individual anatomical differences, an individual subject's brain may not match the shape of an anatomical atlas and also may deviate from an average cortical model prepared for a group [(**a**) white mesh]. However, elastic warping algorithms can apply local dilations and contractions to the average brain model, deforming its shape to match the individual anatomy so that key surfaces and landmarks correspond. These deformations also store detailed information on how specific individuals [e.g., brown mesh (**a**)] deviate from the atlas. Mean anatomical shapes and confidence limits on normal variation (**b**) can be computed. If individual deviations (**a**) are calibrated against the probability distributions that capture normal variation, abnormality maps (**c**) may be generated indicating the probability of finding the anatomy in its observed configuration in a normal population. Here, in a patient with mild Alzheimer's disease, atrophic changes are easiest to detect in orbitofrontal regions where normal variation is least [labeled F in (b); red colors in (c); data from Thompson *et al.,* 1997, 1998].

E. Mapping Brain Asymmetry

By analysis of variance in 3D deformation fields that match different subjects' anatomies, it is also possible to map patterns of brain asymmetry. To do this, we differentiate intrasubject (between hemisphere), intersubject, and intergroup contributions to brain variation in human populations, and again we detect significant differences using null distributions for features in Hotelling's T^2 random fields. Mapping brain asymmetry in a group is an interesting application (Watkins *et al.,* 2001), as asymmetry is linked with functional lateralization (Strauss *et al.,* 1983), handedness (Witelson, 1989), and language function (Davidson and Hugdahl, 1994) and is thought to be diminished in some diseases (cf. Kikinis *et al.,* 1994, Narr *et al.,* 2001a). Although the mappings computed so far specify the set of cortical points that correspond across subjects, an average map of asymmetry cannot be computed without an additional set of mappings to define the points that correspond across hemispheres. To do this, all left-hemisphere sulcal curves are projected into the cortical parameter space, reflected in the vertical axis, and averaged with their flattened counterparts in the right hemisphere, to produce a second *average curve template*. Color maps (as in Fig. 8c) representing point locations in the left and right hemispheres are then subjected to a second warp, or flow, that transforms corresponding features in each hemisphere to the same location in parameter space. 3D deformation fields can then be recovered matching each brain hemisphere with a reflected version of the opposite hemisphere (cf. Thompson *et al.,* 1998; Thirion *et al.,* 1998; Gerig *et al.,* 2001; Wang *et al.,* 2001). These parameter flows are advantageous in that the asymmetry fields are also *registered;* in other words, asymmetry measures can be averaged across corresponding anatomy at the cortex. This is not necessarily the case if warping fields are averaged at the same coordinate locations in stereotaxic space (cf. Fig. 10a). The pattern of mean brain asymmetry for a group of 20 subjects is shown in Fig. 13. The resulting asymmetry fields $\mathbf{a}_i(\mathbf{r})$ (at parameter space location $\mathbf{r}$ in subject i) were treated as observations from a spatially parameterized random vector field, with mean $\boldsymbol{\mu}_a(\mathbf{r})$ and a nonstationary covariance tensor $\boldsymbol{\Sigma}_a(\mathbf{r})$ (Fig. 13). The significance α of deviations from symmetry can be assessed using a T^2 or F statistic that indicates evidence of significant asymmetry in cortical patterns between hemispheres,

$$\alpha(\mathbf{r}) = F_{3,N-3}^{-1}\,([(N-3)/3(N-1)]T^2(\mathbf{r})), \tag{23}$$

where $T^2(\mathbf{r}) = N[\boldsymbol{\mu}_a(\mathbf{r})^T\,\boldsymbol{\Sigma}_a^{-1}(\mathbf{r})\,\boldsymbol{\mu}_a(\mathbf{r})]$.

Using this asymmetry mapping technique, we recently observed that brain asymmetry appears to increase during childhood and adolescence (Sowell *et al.,* 2001). There may also be significant asymmetries in the degree to which genes affect brain structure (Thompson *et al.,* 2001a). Encoded knowledge on the statistics of brain asymmetry can also help detect departures from normal asymmetry and even the emergence of lesions (sometimes termed *dissymmetry:* see Thirion *et al.,* 2000; Joshi *et al.,* 2001).

F. Tensor-Based Morphometry

Tensor-based morphometry is a related approach for detecting morphometric differences in a population, based on the deformation maps, or warping fields, that align each individual data set with an atlas standard (Davatzikos *et al.,* 1996; Machado and Gee, 1998). The local deformation tensor is a matrix containing the local derivatives of the warping field, and it can be used to tell whether a structure is larger or smaller relative to an atlas. Although the deformation tensor also contains information on the principal directions in which a structure is compressed or enlarged relative to an atlas, for simplicity, its determinant (also called the Jacobian) is often used instead. This is a single number reflecting whether a structure is smaller or larger than the atlas and by how much. Essentially it is a dilation factor, expressing the ratio of a structure's size in a specific individual to its counterpart in the atlas. Gaser *et al.* (1999, 2001) have employed these dilation maps to localize morphometric differences in the ventricles, as well as thalamic and temporal regions, in a cohort of patients with schizophrenia (N = 160; cf. Andreasen *et al.,* 1994). Studholme *et al.* (2001) located points where voxel-level differences in

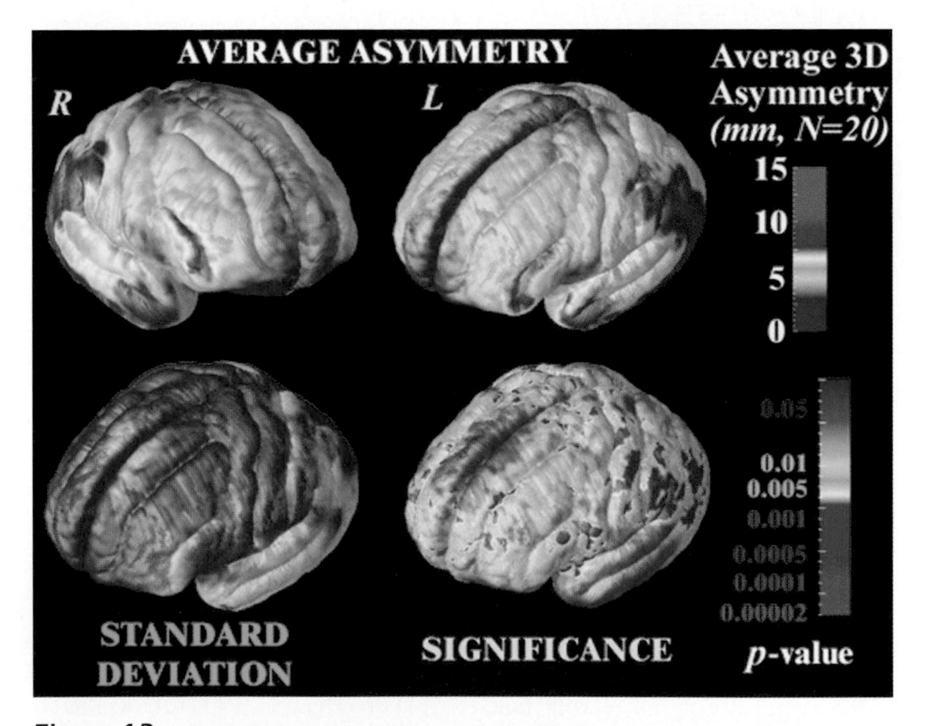

Figure 13 Mapping brain asymmetry in a population. The average magnitude of brain asymmetry in a group (N = 20, elderly normals) can be assessed based on warping fields that map the cortical pattern of one hemisphere onto a reflected version of the other and then flow the observations again so that corresponding measures can be averaged across subjects. Variations in asymmetry are also nonstationary across the cortex (**lower left**), and a Hotelling's T^2 statistical field can be computed to map the significance of the asymmetry (**lower right**) relative to normal anatomic variations (see text for mathematical details).

the Jacobian occurred between groups of patients with frontotemporal dementia, Alzheimer's disease, and semantic dementia. This indicated spatially consistent shape differences induced by each particular disease. Davatzikos *et al.* (1996) and Pettey and Gee (2001) employed a similar approach to localize differences between males and females in the anatomy of the corpus callosum (a controversial topic reviewed in Thompson *et al.*, 2001a; see also Bermudez and Zatorre, 2001). The use of tensor maps to localize brain growth in children and anatomical change in dementia is discussed in Section VIII.

G. Specialized Approaches for Detecting Cortical Differences

Several conventional approaches for signal detection (and enhancement) in functional images are based on mathematics that describes the distribution of features in random fields derived from images (e.g., SPM; Worsley *et al.*, 1994a,b, 2000; Friston *et al.*, 1995; Lange, 1996). These statistical equations become even more complicated when the data lie on curved manifolds such as the cortex (Goebel, 2000; Jones *et al.*, 2000; Taylor and Adler, 2000; Andrade, 2001). Specifically, if cortical data are flattened, the spatial autocorrelation of the data in flat space will depend to some degree on the local parameterization tensor (or distortion) involved in flattening the surface.

This apparent barrier, however, can be turned into an advantage: if a surface model is available, it is possible to selectively enhance signals in the cortex, by searching and filtering data within the surface only (Jones *et al.*, 2000; Chung *et al.*, 2001). For functional activations, as well as for structural attributes of thickness, asymmetry, and shape, the surface null distributions can be estimated for clusters exceeding a given height and spatial extent and the total spatial extent of these clusters, as these help in detecting subtle effects. To estimate significance values for these maps, the roughness tensor, Λ (or its inverse, the smoothness tensor, $S = \Lambda^{-1}$), is crucial (Poline *et al.*, 1995; Kiebel *et al.*, 1999); it is defined as the covariance matrix of the partial derivatives of the process along each of the D coordinate axes, with variances $\mathrm{Var}[\partial X/\partial x_i]$ on the diagonal and off-diagonal elements $\mathrm{Cov}[\partial X/\partial x_i, \partial X/\partial x_j]$. Usually the smoothness is calculated not from the data themselves, which may contain a physiological signal, but from the residuals after fitting a linear statistical model which removes linear effects of the experimental parameters. The roughness tensor of the process is not likely to be stationary within the cortical surface, which can complicate the application of standard results in Gaussian field theory (Worsley *et al.*, 1999). To alleviate this problem, a partial differential equation,

$$g^{ij}(\partial^2\mathbf{u}/\partial r^i\partial r^j) + \partial/\partial u^j(S^{ij})\mathbf{u}_r i = 0, \tag{24}$$

can be run in the parameter space of the group average cortex. This generates a deformed grid $\mathbf{u}(\mathbf{r})$ whose deformation gradient tensor approximates the smoothness tensor S^{ij} of the normalized residuals of the data on the surface (here g^{ij} is the contravariant metric tensor of the grid). If the smoothness tensor has nonzero curvature (and is therefore not realizable as a deformation tensor), the deformation gradient approximates it in the Frobenius matrix norm. Relative to this new computational grid, the residuals become stationary and isotropic, and P values for effects such as gray matter reductions can be evaluated. This approach for detecting statistical effects in images is known as *statistical flattening* (which is to be distinguished from cortical flattening); it can be achieved using the Nash Embedding theorem (Worsley *et al.*, 1999) or by running the above PDE on the data (Thompson *et al.*, 2001a). In a related approach, cortical signals can be detected more powerfully by smoothing data using covariant filters on the cortical surface. This technique is analogous to 3D data smoothing in volumetric PET and *f*MRI studies and is under investigation in the computer vision and image restoration literature (Sapiro, 2001; Bertalmio *et al.*, 1999; see also Thompson *et al.*, 2000a). To increase the detection sensitivity to interesting functional and structural effects, a Laplace–Beltrami flow, induced by the Laplace–Beltrami operator $\nabla^2_{LB}I$, can be run in the cortical parameter space. This produces a scale space of diffused data $I(\mathbf{x}, t)$ (Nielsen *et al.*, 1994; Worsley *et al.*, 1996; Sochen *et al.*, 2000; Chung *et al.*, 2001) on the cortex, which acts as a prefilter to enhance detection of effects at different scales (Huiskamp, 1991):

$$I(\mathbf{x}, t_{n+1}) = I(\mathbf{x}, t_n) + \Delta t \cdot \nabla^2_{LB} I(\mathbf{x}, t_{n+1}). \tag{25}$$

In this process, covariant derivatives on the cortical manifold are computed from the gradients of the base vectors on a logical grid in parameter space (cf. Section IV). The resulting ability to adaptively filter data on the cortical sheet maximizes the statistical power to detect changes and differences in cortical structure and function in a population atlas, including diffuse effects across the cortex that may emerge in a population.

VII. Applications to Development and Disease

The population-based atlasing approaches introduced so far have been applied to study brain structure in Alzheimer's disease (Thompson *et al.*, 2000a,b; Mega *et al.*, 1999a); chronic, first-episode, and childhood-onset schizophrenia (Narr *et al.*, 2000, 2001a,b; Cannon *et al.*, 2000); fetal alcohol syndrome (Sowell *et al.*, 2001); and brain changes during childhood and adolescence (Thompson *et al.*, 2000a, 2001a; Sowell *et al.*, 2001a,b; Blanton *et al.*, 2000).

An interesting application is in visualizing the average profile of gray matter loss across the cortex in Alzheimer's disease, based on a large number of subjects at a specific stage in the disease. First we describe a cross-sectional study in which each subject is imaged once; longitudinal data are described in the next section. Gyral pattern variation makes it difficult to make inferences about where exactly gray matter is lost in a group. If gray matter maps are directly averaged together in stereotaxic space (e.g., Fig. 10a), it is difficult to localize results to specific cortical regions. To address this, we used cortical pattern matching to help compute group averages and statistics. First, we segmented all images in the database with a previously validated Gaussian mixture classifier. Maps of gray matter, white matter, cerebrospinal fluid, and a background class were created for each subject (Fig. 14). The proportion of gray matter lying within 15 mm of each cortical point was then plotted as an attribute on each cortex and aligned across subjects by projecting it into flat space and warping the resulting attribute field with the elastic matching technique (as in Fig. 8). [Again, the gray matter proportion can be thought of as a scalar attribute $G(\mathbf{r})$ defined in the cortical parameter space, which can be subjected to a flow field $\mathbf{u}(\mathbf{r})$ to compensate for gyral pattern differences.] By averaging the aligned maps, and texturing them back onto a group average model of the cortex, the average magnitude of gray matter loss was computed for the Alzheimer's disease population (Fig. 14; Thompson *et al.*, 2000a). Regions with up to 30% reduction in gray matter were sharply demarcated from adjacent regions with little or no loss. The group effect size was measured by attaching a field of t statistics, $t(\mathbf{r})$, to the cortical parameter space and computing the area of the t field on the group average cortex above a fixed threshold ($P < 0.01$, uncorrected). For groups that are not demographically matched, more sophisticated regression models could be applied, resulting in F fields that indicate the significance of the overall fit and of how individual model parameters help explain the loss. In a multiple comparisons correction, the significance of the overall effect was confirmed to be $P < 0.01$, by permuting the assignment of subjects to groups 1,000,000 times.

VIII. Dynamic Brain Maps

A. Dynamic Brain Change

Everyone's brain shrinks with age, and not in a uniform way. Diseases such as Alzheimer's cause changes in the overall rates and patterns of brain change. Population-based atlases can store key statistics on the rates of these brain changes. These are especially relevant to the understanding of development (Lange *et al.,* 1997) as well as relapsing-remitting diseases such as multiple sclerosis (Guttmann *et al.,* 1995; Thirion *et al.,* 1997; Rey *et al.,* 1999; Collins *et al.,* 2001; Welti *et al.,* 2001) and tumor growth (Haney *et al.,* 2001a,b). They can provide normative criteria for early brain change in patients with dementia (Jernigan *et al.,* 1991;

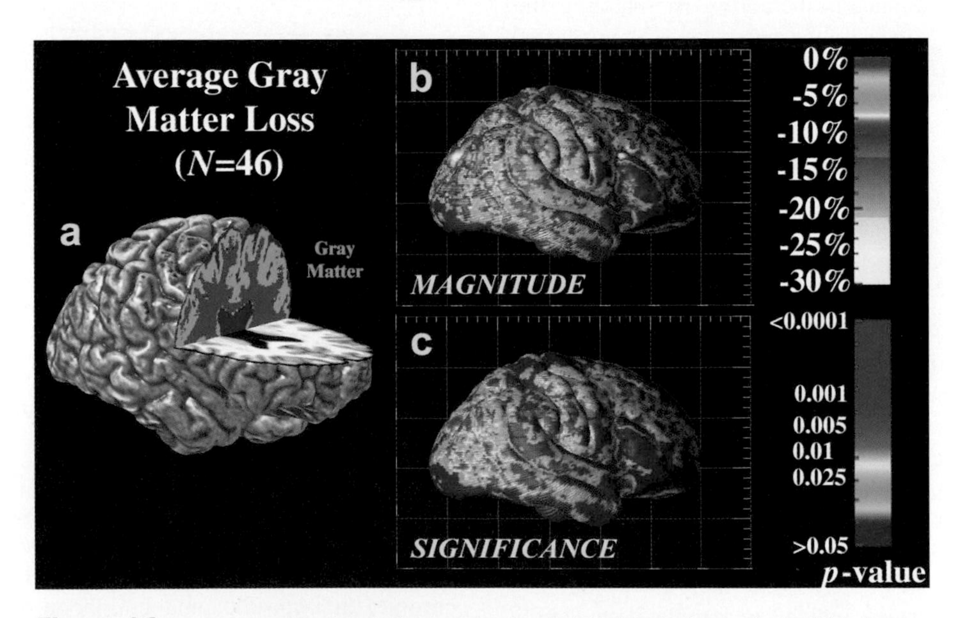

Figure 14 Average patterns of cortical gray matter loss in Alzheimer's disease. Scalar fields that represent the density of gray matter across the cortex (**a**) can be aligned using elastic matching of cortical patterns. A localized and highly significant loss of gray matter (**b** and **c**) is revealed in temporoparietal cortices of Alzheimer's patients relative to matched elderly controls, in a pattern similar to the metabolic and perfusion deficits seen early in the disease. If longitudinal data are available, scalar fields representing the rates of gray matter loss can also be compared (see Fig. 15).

DeCarli *et al.,* 1992; Janke *et al.,* 2001; Thompson *et al.,* 2001a), in those with mild cognitive impairment (Studholme *et al.,* 2001), or in those at genetic risk for Alzheimer's disease (Small *et al.,* 2000). An interesting application is the compilation of dynamic maps to characterize diseases with childhood or adolescent onset, which we illustrate next.

In a developmental application (Thompson *et al.,* 2001a; Fig. 15), the gray matter mapping procedure, described above, was applied to longitudinal MRI data from 12 schizophrenic patients and 12 adolescent controls scanned at both the beginning and the end of a 5-year interval. The goal was to estimate the average rate of gray matter loss throughout the cortex, by matching cortical patterns and comparing changes in disease with normal changes in controls. Cortical models and gray matter measures were elastically matched first within each subject across time, to compute individual rates of loss, and then flowed into an average configuration using flat space warping (Fig. 8). The resulting maps (Fig. 15) showed dynamic loss of gray matter in superior parietal, sensorimotor, and some frontal brain regions (up to 5% annually). Group differences were highly significant ($P < 0.01$, permutation test; Fig. 15), relative to healthy controls and nonschizophrenic controls matched for medication and IQ, and were linked with psychotic symptom severity (for details, see Thompson *et al.,* 2001a).

B. Tensor Maps of Brain Change

Maps of brain change over time can also be based on a deformation mapping concept. In this approach, a 3D elastic deformation is calculated. This deformation, or warping field, drives an image of a subject's anatomy at a baseline time point to match its shape in a later scan. Image-warping algorithms have evolved over many years (Toga, 1998). Now mappings can be calculated in a very exact way that matches a large number of the key functional and anatomic elements in the scans to be matched. This results in a very complex transformation, often with up to a billion parameters, from which local volume changes in tissues can be calculated (Miller *et al.,* 1993; Christensen *et al.,* 1993, 1995, 1996; Collins *et al.,* 1994a, 1995; Davatzikos, 1996; Thompson and Toga, 1996; Davis *et al.,* 1996, 1997; Bro-Nielsen and Gramkow, 1996; Dupuis *et al.,* 1998; Gee *et al.,* 1993, 1995, 1998; Freeborough and Fox, 1998; Thompson *et al.,* 1999, 2001a; Cachier *et al.,* 1999; Janke *et al.,* 2001). In capturing brain change, deformation-based methods can be complementary to voxel-based morphometric methods (Ashburner and Friston, 2000; Good *et al.,* 2001a) and methods that estimate whole brain atrophic rates (Subsol *et al.,* 1997; Calmon and Roberts, 2000; Collins *et al.,* 2001; Smith *et al.,* 2001). Voxel-based methods typically use a simple pixel-by-pixel subtraction of scan intensities registered rigidly across time.

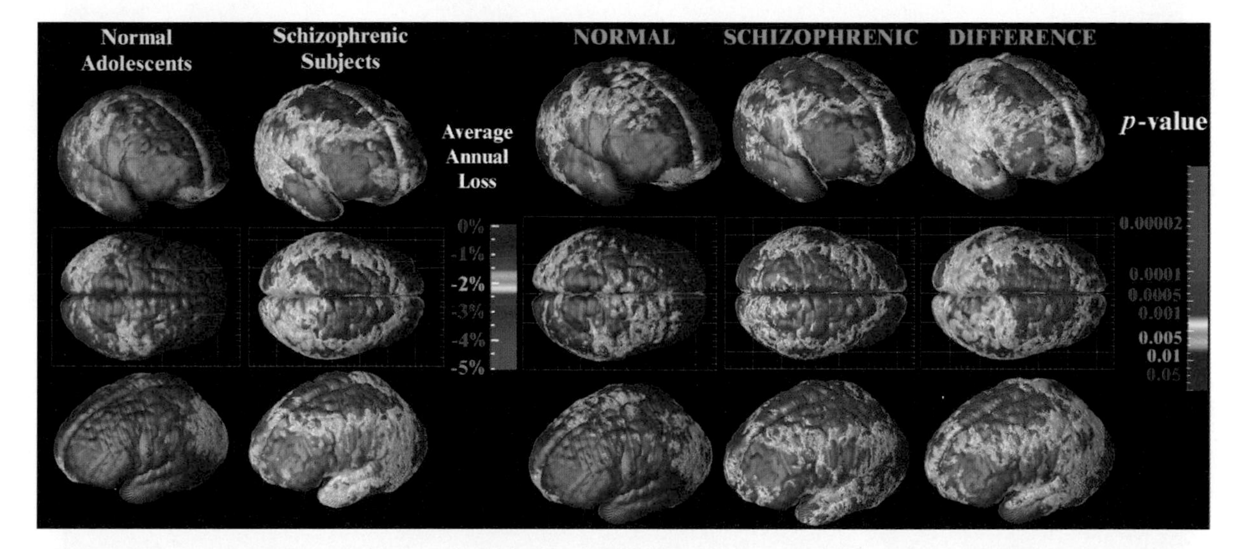

Figure 15 Average rates of gray matter loss in normal adolescents and in schizophrenia. (**Left two columns**) 3D maps of brain changes, derived from high-resolution magnetic resonance images (MRI scans), acquired repeatedly from the same subjects over a 5-year time span, reveal profound, progressive gray matter loss in schizophrenia (right column). Average rates of gray matter loss during the 5-year period from 13 to 18 years of age are displayed on average cortical models for the group. Severe loss is observed (red and pink colors; up to 5% annually) in parietal, motor, and temporal cortices, while inferior frontal cortices remain stable (blue colors; 0–1% loss). Dynamic loss is also observed in the parietal cortices of normal adolescents, but at a much slower rate. (**Right three columns**) These maps show the local significance of the dynamic brain change in normal adolescents and in schizophrenic subjects. By comparing the average rates of loss in disease (middle column) with the loss pattern in normal adolescents (first column), the normal variability in these changes can also be taken into account and the significance of disease-specific change can be established (last column). Data from Thompson *et al.,* 2001a.

Deformation methods, however, can distinguish local from global effects and true tissue loss from shifts in anatomy. These can confound image subtraction methods.

C. Mapping Growth Patterns

We recently developed a deformation-based approach to detect an anterior-to-posterior wave of growth in the brains of young children scanned repeatedly between the ages of 3 and 15 (Thompson *et al.,* 2000a). Parametric surface meshes were built to represent anatomical structures in a series of scans over time, and these were matched using a fully volumetric deformation. Dilation and contraction rates, and even the principal directions of growth, can be derived by examining the eigenvectors of the deformation gradient tensor, or the local Jacobian matrix of the transform that maps the earlier anatomy onto the later one (see Fig. 16). Applications include measuring the statistics of brain growth (Thompson *et al.,* 2000a) and measuring tumor response to novel chemotherapy agents (Haney *et al.,* 2001a). By building probability densities on registered tensor fields (e.g., Thompson *et al.,* 2000a; Chung *et al.,* 2001), a quantitative framework can be established to detect normal and aberrant brain change and its modulation by medication in clinical trials.

D. Mathematical Details

Deformation-based methods to track brain change have often been based on continuum mechanics, which describes physical models of elastic or fluid bodies (reviewed in Toga, 1998; Thompson and Toga, 2000; cf. Freeborough and Fox, 2000; Haney *et al.,* 2001a,b; Chung *et al.,* 2001). The 3D shape of one brain, imaged with MRI at one time point, is reconfigured to match its shape in a later image (intensity changes over time may also be modeled; q.v. Joshi *et al.,* 2001). A complex 3D deformation field is computed that matches large numbers of surface, curve, and point landmarks in the two brains (see Thompson *et al.,* 2000a for details and a review of similar methods by other groups). Important anatomic and functional interfaces are matched up when one scan is deformed into the shape of the other.

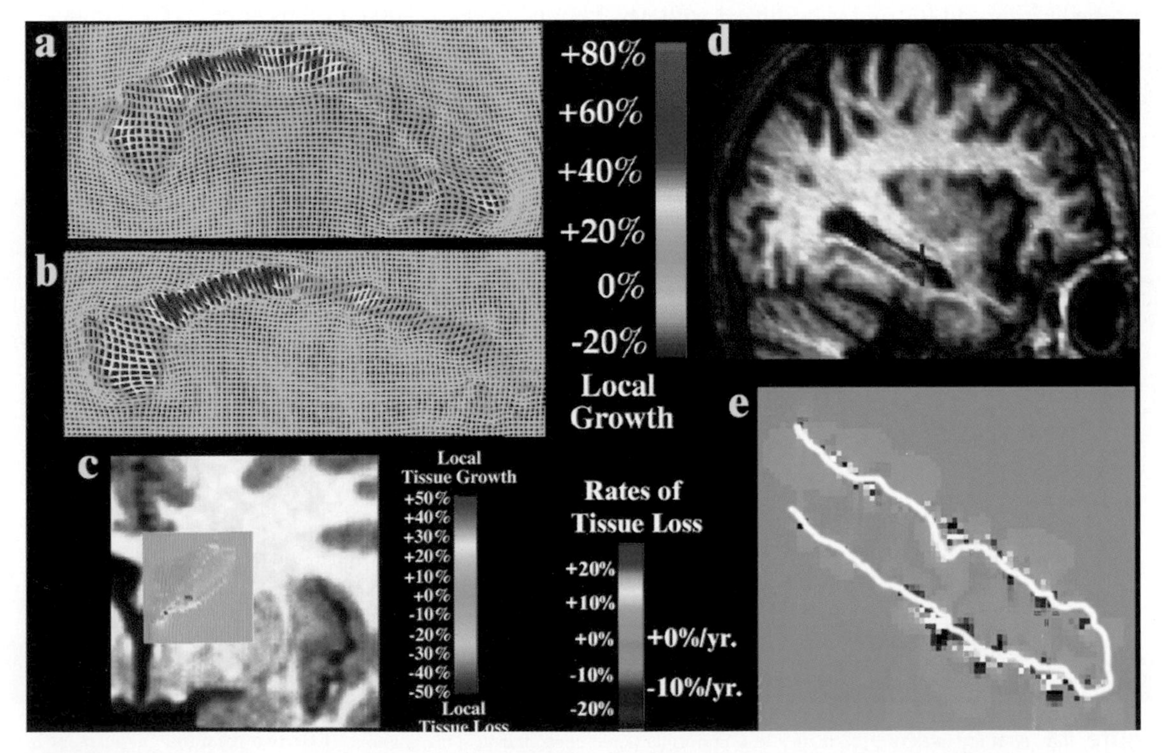

Figure 16 Tensor maps of brain change: visualizing growth and atrophy. If follow-up (longitudinal) images are available, the dynamics of brain change can be measured with *tensor mapping* approaches (Thompson *et al.,* 2000a). These map volumetric change at a local level and show local rates of tissue growth or loss. Fastest growth is detected in the isthmus of the corpus callosum in two young girls identically scanned at ages 6 and 7 (**a**) and at ages 9 and 13 (**b**). Maps of loss rates in tissue can be generated for the developing caudate [(**c**), here in a 7- to 11-year-old child) and for the degenerating hippocampus (**d** and **e**). In (e), a female patient with mild Alzheimer's disease was imaged at the beginning and end of a 19-month interval with high-resolution MRI. The patient, age 74.5 years at first scan, exhibits faster tissue loss rates in the hippocampal head (10% per year, during this interval) than in the fornix. These maps may ultimately help elucidate the dynamics of therapeutic response in an individual or a population (Thompson *et al.,* 2000a, 2001a; Haney *et al.,* 2001a).

Parametric mesh models of brain structures are used to drive a 3D deformation vector map U:$\mathbf{x} \rightarrow \mathbf{u}(\mathbf{x})$, which is derived from the Navier equilibrium equations for linear elasticity:

$$\mu\nabla^2\mathbf{u} + (\lambda + \mu)\ \nabla\ (\nabla \cdot \mathbf{u}(\mathbf{x})) + \mathbf{F}(\mathbf{x} - \mathbf{u}(\mathbf{x})) = \mathbf{0},\ \forall\ \mathbf{x} \in \mathrm{R}. \quad (26)$$

All the terms in this equation just describe forces and distortions in a 3D material, in which the image is considered to be embedded. R is a discrete lattice representation of the scan to be transformed, $\nabla \cdot \mathbf{u}(\mathbf{x}) = \sum \partial u_j / \partial x_j$ is the divergence, or cubical dilation of the medium, ∇^2 is the Laplacian operator, which measures the irregularity of the deformation, $\mathbf{F}(\mathbf{x})$ is the internal force vector, and Lamé's coefficients λ and μ refer to the elastic properties of the medium. Matching of cortical surfaces, across time and subsequently across subjects (for data averaging) can also be enforced (Section IV). Mappings based on high-dimensional elastic and fluid models can recover extremely complex patterns of change (Fig. 17; see also Freeborough and Fox, 1998; Christensen *et al.,* 1996; Rey *et al.,* 1999); their mathematics is reviewed elsewhere (Thompson *et al.,* 2000a).

E. Population-Based Atlasing of Brain Change

In different individuals, growth processes occur in anatomies that are geometrically different. Additional warping techniques are needed to compare growth profiles across subjects. This additional warping is needed to compute average profiles and to define statistical differences in rates of growth or loss. Mathematically, if $U^i(\mathbf{x},t_i)$ is the 3D displacement vector required to deform the anatomy at position $\mathbf{x}$ in subject i at reference time 0 to its corresponding homologous position at time t_i, then a linear approximation of the local rate of volumetric growth (Chung *et al.,* 2001) can be written in terms of the identity tensor and displacement gradient tensor as

$$\Lambda^i(\mathbf{x}) = \partial J^i / \partial t = \det\ (I + \nabla U^i)/t_i. \quad (27)$$

If A^i is the secondary deformation mapping transforming the baseline anatomy of individual i onto the atlas (Warping Field in Fig. 18), then the set of registered growth maps $\Lambda^i(A^i(\mathbf{x}))$ (shown in the final panel of Fig. 18) can be treated as observations from a spatially parameterized random field, whose mean and variance can be estimated. Statistical effects of age, gender, genotype, or medication can then be detected using random field theory to produce statistical maps (Thompson *et al.,* 2001a; Ashburner, 2001).

F. Improved Dynamic Models

In developing dynamic atlases for clinical applications, there is a particular interest in modeling developmental

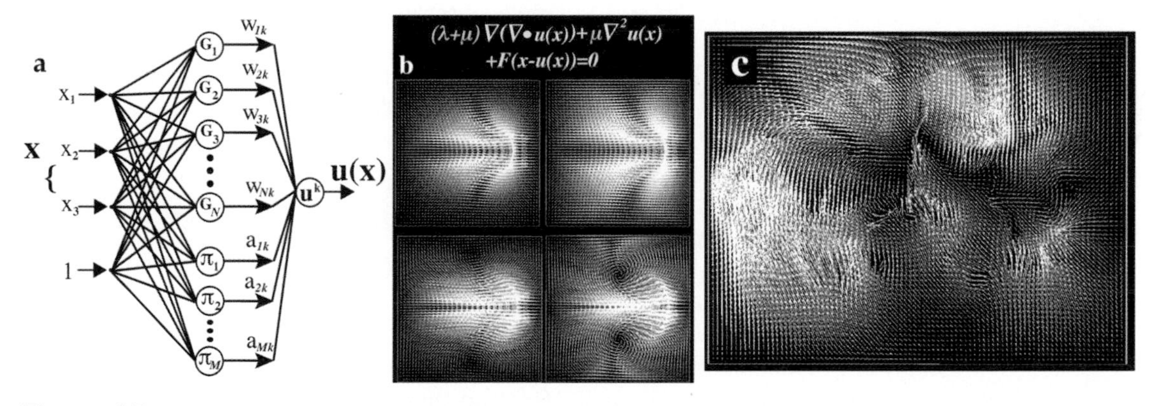

Figure 17 Deforming anatomical templates with neural nets and continuum mechanical flows. The complex transformation required to reconfigure one brain into the shape of another (Fig. 2) can be determined using neural networks (**a**) or continuum-mechanical models (**b** and **c**), which describe how real physical materials deform. In Davis *et al.* (1997), each of the three deformation vector components, $\mathbf{u}^k(\mathbf{x})$, is the output of the neural net when the position in the image to be deformed, $\mathbf{x}$, is input to the net. Outputs of the hidden units (G_i, π_m) are weighted using synaptic weights, w_{ik}. If landmarks constrain the mapping, the weights are found by solving a linear system. Otherwise, the weights can be tuned so that a measure of similarity between the deforming image and the target image is optimized. Continuum-mechanical models (b) can also be used to compute these deformation fields (Davatzikos *et al.,* 1996; Christensen *et al.,* 1996; Gee *et al.,* 1998; Thompson *et al.,* 2000a; Miller *et al.,* 2002). These models describe how real physical materials deform. Different choices of the Lamé elasticity coefficients, λ and μ, in the Cauchy–Navier equations [shown in continuous form (b)] result in different deformations, even if the applied internal displacements are the same. For brain image transformations, values of elasticity coefficients can be chosen to limit the amount of curl (middle right) in the deformation field. (*Note.* To help visualize differences, displacement vector fields have been multiplied by a factor of 10, but the elasticity equations are valid only for small deformations). (c) shows the complexity of a typical deformation field, in this case one used to reconfigure a histologic section stained for molecular content. Curve and surface anatomic landmarks are used to constrain the mapping, and the Cauchy–Navier equations are solved to estimate how the rest of the 3D volume deforms. (a) is adapted, with permission, from Davis *et al.* (1997).

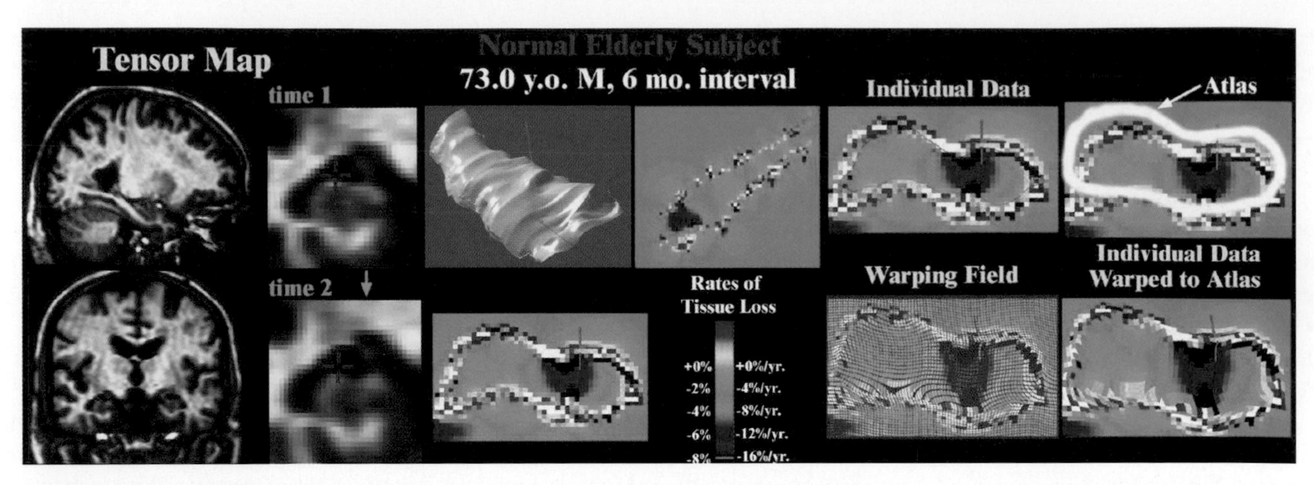

Figure 18 Tensor maps of local volumetric loss in normal elderly individuals. Local volume loss patterns in the hippocampus of an elderly subject (here, over a 6-month interval) are hard to appreciate from raw MRI data (**left**). They can be localized by using 3D surface models to drive a 3D continuum-mechanical PDE (see Fig. 17) from which dynamic statistics of loss are derived. Comparison and averaging of this loss rate data across subjects requires a second PDE to convect the attribute data onto an average neuroanatomical atlas (final four images; see Thompson *et al.,* 1997, 2000a, 2001a for methods and applications).

processes that speed up or slow down. Diseases may accelerate, or they may be attenuated by therapy. If individuals are scanned more than twice over large time spans, this presents the opportunity for more accurate detection of brain change and encoding of these changes in a group atlas. To compare growth patterns in different groups of subjects, the general linear model can be used to analyze the registered growth profiles (or degenerative profiles). For the *i*th individual's *j*th measure we have

$$Y_{ij} = f(\text{Age}_{ij}, \underline{\beta}) + e_{ij}. \tag{28}$$

Here Y_{ij} signifies the outcome measure at a voxel or surface point, such as growth or tissue loss, $f()$ denotes a constant, linear, quadratic, cubic, or other function of the individual's age for that scan, and $\underline{\beta}$ denotes the regression/ANOVA coefficients to be estimated. Age (Age_{ij}) may be replaced by time from the onset of disease, the start of medication, or from the onset of puberty (Giedd *et al.,* 1999a). This flexibility will allow one to temporally register dynamic patterns using criteria that are expected to bring into line temporal features of interest that appear systematically in a group (Janke *et al.,* 2001). In the above model, the coefficient vector, $\underline{\beta}$, is assumed to be constant, i.e., a fixed effect. The e_{ij} are assumed normally distributed and uncorrelated both between and within individuals. If multiple scans are available over time, a random-effects model can also model brain changes in a population:

$$Y_{ij} = a_i + f(\text{Age}_{ij}, \underline{\beta}) + e_{ij}. \tag{29}$$

Here the model is the same as the General Linear Model except for the a_i term, which is called a *random effect* (Pinheiro and Bates, 2000; Ripley, 2000). It describes the correlation between an individual's multiple scans. Random-effects models may also be fitted with *correlated* errors (Davidian and Giltinan, 1995; Verbeke and Molenberghs, 1997). If this is done, e_{ij} and e_{ik} (k not equal to j) are assumed correlated with the correlation a function of the time elapsed between the two measurements (Giedd *et al.,* 1999a). In models whose fit is confirmed as significant, e.g., by permutation, loadings on nonlinear parameters may be visualized as attribute maps $\underline{\beta}(\mathbf{x})$. This reveals the topography of accelerated or decelerated brain change (Thompson *et al.,* 2001a). The result is a formal approach to assess whether, and where, brain change is speeding up or slowing down. This is key feature in developmental or medication studies and a key element of developmental atlases that are currently being built.

IX. Genetic Brain Maps

One of the most exciting frontiers of brain imaging is its linkage with genetic data in large human populations. Linking brain structure and genotype is important to understand:

(1) the normal heritability of brain structure (Oppenheim *et al.,* 1989; Bartley *et al.,* 1997; Biondi *et al.,* 1998; Pfefferbaum *et al.,* 2000; LeGoualher *et al.,* 2000; Baare *et al.,* 2001; Thompson *et al.,* 2001a) and

(2) how deficits are inherited in diseases for which there are known genetic risks (e.g., Alzheimer's, schizophrenia).

These genetic studies can be set up in several ways. If a candidate marker, or risk gene, is known (e.g., apolipoprotein E, or ApoE, in Alzheimer's disease; Roses, 1997), an individual's genetic status can be used as a covariate to mine for effects of the risk gene on brain structure or function (Small *et al.,* 2000; Laakso *et al.,* 2000; Reiman *et al.,* 2001). For other diseases that are polygenic, candidate markers may be elusive (e.g., schizophrenia, autism). In these cases, genetic effects on brain structure may be tested using twin, familial, or discordance designs (see Lohmann *et al.*, 1999; Thompson *et al.*, 2001a; Cannon *et al.*, 2001a; Narr *et al.*, 2001a; Styner and Gerig, 2001; Molloy *et al.*, 2001; for examples).

A. Genetic Influences on Brain Structure

The few existing studies of brain structure in twins suggest that the overall volume of the brain itself (Tramo *et al.*, 1998) and some brain structures, including the corpus callosum (Oppenheim *et al.*, 1989; Pfefferbaum *et al.*, 2000) and ventricles, are somewhat genetically influenced, while gyral patterns, observed qualitatively or by comparing their 2D projections, are much less heritable (Bartley *et al.*, 1997; Biondi *et al.*, 1998; cf. Le Goualher *et al.*, 2000).

B. Genetic Brain Maps

In a recent approach, we developed a brain atlas based on twins to determine genetic influences on brain structure (Thompson *et al.*, 2001a; Plomin and Kosslyn, 2001). We compared the average differences in cortical gray matter density (Wright *et al.*, 1995; Sowell *et al.*, 1999; Thompson *et al.*, 2000a; Ashburner and Friston, 2000; Good *et al.*, 2001a) in groups of unrelated subjects, as well as in dizygotic (DZ) and monozygotic (MZ) twins. This approach is known as the *classical twin design*. Although both types of twins share gestational and postgestational rearing environments, DZ twins share, on average, half their segregating genes, while MZ twins are normally genetically identical (with rare exceptions due to somatic mutations). Maps of intrapair gray matter differences, generated within each MZ and DZ pair, were elastically realigned for averaging across the pairs within each group, prior to intergroup comparisons. First, maps of intrapair variance and broad-sense heritability were computed using Falconer's method (Falconer, 1989) to determine all genic influences on the phenotype, at each cortical point. Heritability, which is usually denoted by h^2, is a useful statistical construct that estimates the amount of variation in a structural attribute, or a behavioral trait, that is attributable to genetic factors. It can be estimated in different ways (Feldman and Otto, 1997), but is usually defined as twice the difference between MZ and DZ intraclass correlation coefficients (see Fig. 19).

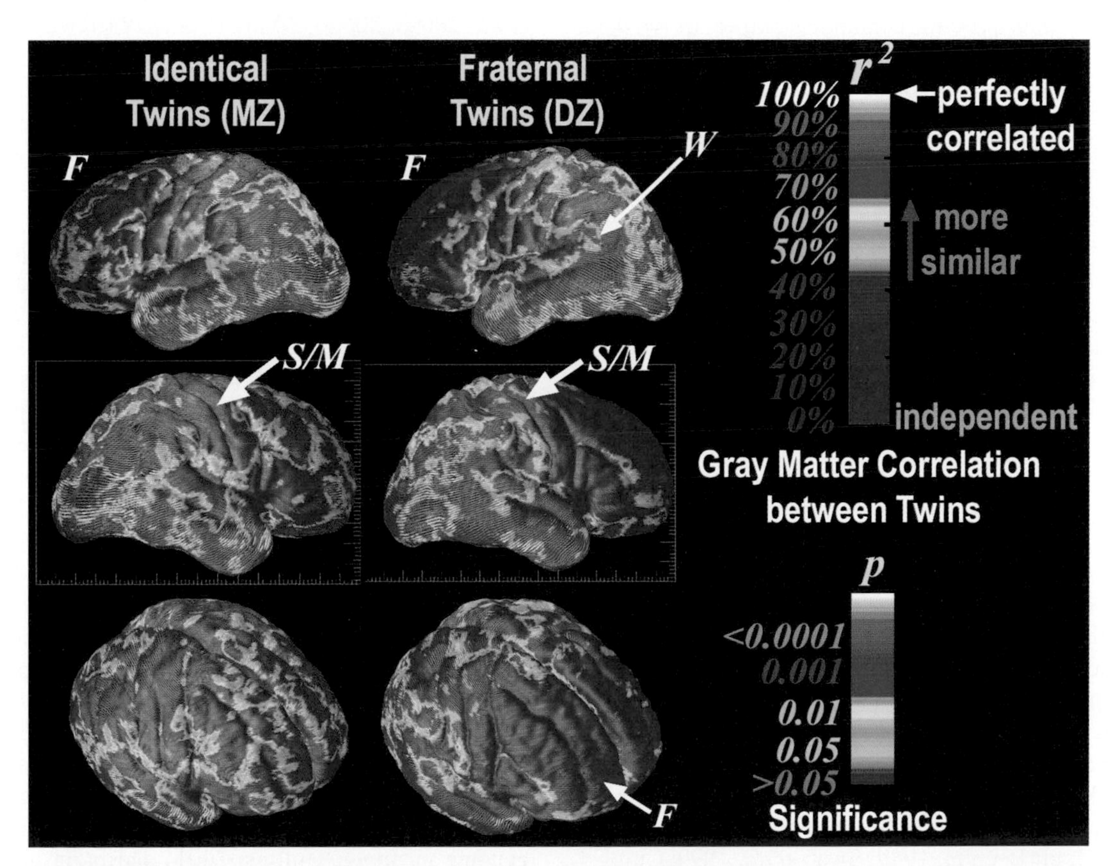

Figure 19 Correlation between twins in gray matter distribution. Genetically identical twins are almost perfectly correlated in their gray matter distribution, with near-identity in frontal (F), sensorimotor (S/M), and perisylvian language cortices. Fraternal twins are significantly less alike in frontal cortices, but are 90–100% correlated for gray matter in perisylvian language-related cortex, including supramarginal and angular territories and Wernicke's language area (W). The significance of these increased similarities, visualized in color, is related to the local intraclass correlation coefficients (r) and can be transformed into maps of heritability and genetic influences on brain structure (Thompson *et al.*, 2001a; Plomin and Kosslyn, 2001).

By treating the loss of variance with increasing genetic affinity as an observation from an *F*-distributed random field, we identified a genetic cascade in which within-pair correlations were highest for MZ twins, lower for DZ twin pairs, and lowest of all for unrelated subjects. Specific regions of cortex were more heritable than others. We plotted these correlations across the cortex and assessed their statistical significance. The resulting maps indicate a successively increasing influence of common genetics.

C. Cognitive Linkages

Genetically identical twins displayed only 10–30% of normal differences (Fig. 19; red and pink colors) in a large anatomical band spanning frontal, sensorimotor, and Wernicke's language cortices. This suggests strong genetic control of brain structure in these regions. Intriguingly, the highly heritable frontal gray matter differences were also linked to Spearman's *g*, which measures successful test performance across multiple cognitive domains ($P < 0.018$). Like IQ, this widely used measure isolates a component of intellectual function common to multiple cognitive tests and has been shown to be highly heritable across many studies, even more so than specific cognitive abilities ($h^2 = 0.62$, McClearn *et al.*, 1997; cf. Feldman and Otto, 1997; $h^2 = 0.48$, Devlin *et al.*, 1997; $h^2 = 0.6$–0.8, Finkel *et al.*, 1998; cf. Swan *et al.*, 1990; Loehlin *et al.*, 1989; Chipuer *et al.*, 1990; Plomin and Petrill, 1997). Genetic factors may therefore contribute to structural differences in the brain that are statistically linked with cognitive differences. The resulting genetic brain maps reveal a strong relationship between genes, brain structure, and behavior, suggesting that highly heritable aspects of brain structure may also play a fundamental role in determining individual differences in cognition (Thompson *et al.*, 2001a; Plomin and Kosslyn, 2001).

D. Discordance Studies

Parallel studies of heritability are also under way, mapping genetic components of deficits in schizophrenia (Cannon *et al.*, 2001a; cf. Noga, 1999; Styner and Gerig, 2001). These correlational models of genetic determination (Fig. 19) are among the simplest. They can be extended to path analyses, or structural equation models widely used in population genetics, in cases in which sample sizes are large enough to reliably estimate their parameters. More advanced models have also been proposed to estimate gene–environment interaction and covariance terms, when gauging genetic influences on phenotype (Neale and Cardon, 1992; cf. Loehlin, 1989; Swan *et al.*, 1990; Chipuer *et al.*, 1990; Plomin and Petrill, 1997; Finkel *et al.*, 1998; Molloy *et al.*, 2001).

The resulting techniques for genetic brain mapping represent an exciting new dimension in computational anatomy. When genetic brain maps are included in population-based atlases, they begin to shed light on familial liability and potential genetic triggers for human brain diseases (Cannon *et al.*, 2001a).

X. Subpopulation Selections

A key advantage of a population-based brain atlas is that it can be stratified, according to genetic, demographic, or therapeutic criteria, to reflect a more constrained subset of the population. Differences in a diseased population, or one with known genetic risk, can be visualized by reference to a normative standard. Normative atlases based on young normals can store a rich variety of structural and functional data (Mazziotta *et al.*, 1995, 2001). These atlases have recently been expanded to incorporate data from elderly and developing populations (Paus *et al.*, 1999; Thompson *et al.*, 2000a, 2001a; Chung *et al.*, 2001), as well as data from the whole gamut of imaging devices (Toga and Mazziotta, 1996; Roland and Zilles, 1996). Additional disease-specific atlases are under development for populations with Alzheimer's disease (Mega *et al.*, 1997, 1999a; Thompson *et al.*, 2000a, 2000b; Dinov *et al.*, 2000), whose brain morphology is not well accommodated by existing templates. The ability to stratify these atlases by demographic criteria shows promise in uncovering gender effects and interactions in disease (Fig. 5; Narr *et al.*, 2000). The ability to select subpopulations from the database makes it possible to compare different groups with similar or overlapping symptom profiles (Thompson *et al.*, 2001a). This is of particular interest in populations of patients with complex disorders such as schizophrenia. As cohort studies are performed, the growing atlas provides a basis to contrast chronically medicated and first-episode patients (Narr *et al.*, 2001a), childhood-onset and adult patients (Rapoport *et al.*, 1999; Thompson *et al.*, 2001a), schizoaffective and bipolar patients, patients receiving different medications, and even family members with difference genetic risks (Cannon *et al.*, 2001). The incorporation of discordant twins (Noga, 1999; Cannon *et al.*, 2001) and familial pedigree data (Kaprio *et al.*, 1990) into population atlases shows enormous promise for understanding disease transmission in human populations. The associated tools for mapping genetic influences on brain structure (Lohmann *et al.*, 1999; Thompson *et al.*, 2001a; Styner and Gerig, 2001) are also beginning to offer the potential to map inheritance patterns and identify brain regions with specific genetic risk. Informatics projects that link data from international genome mapping (Collins, 2001) and brain mapping projects (Huerta and Koslow, 1996) will also allow atlases to be more easily mined for genotype–phenotype relationships. This will facilitate the quest for specific genetic markers and quantitative trait loci that confer liability for a particular disease, personality trait, or intellectual ability (Gottesman *et al.*, 2000; Plomin and Kosslyn, 2001).

The growth of brain imaging databases also presents opportunities to determine unforeseen patterns in large data sets with exploratory data mining or data profiling techniques (Megalooikonomou *et al.*, 2000). Exploratory techniques such as unsupervised learning or independent components analysis (Bell and Sejnowski, 1995; Hyvärinen *et al.*, 2001) have been applied fruitfully for source identification in EEG and *f*MRI time series (Makeig *et al.*, 1996; McKeown *et al.*, 1998a). These techniques use information theory (Shannon, 1948; Hyvärinen *et al.*, 2001) and neural networks to uncover fundamental factors that govern variation in data sets. They often attempt to find a projection of the entire data set, using a "demixing" matrix, or even a nonlinear mapping, that maximizes the entropy of the data or minimizes the mutual information (Bell and Sejnowski, 1995). The output is a clustering of the data, or representation in terms of latent variables or factors, which explain the structure of the observations and may reveal their underlying

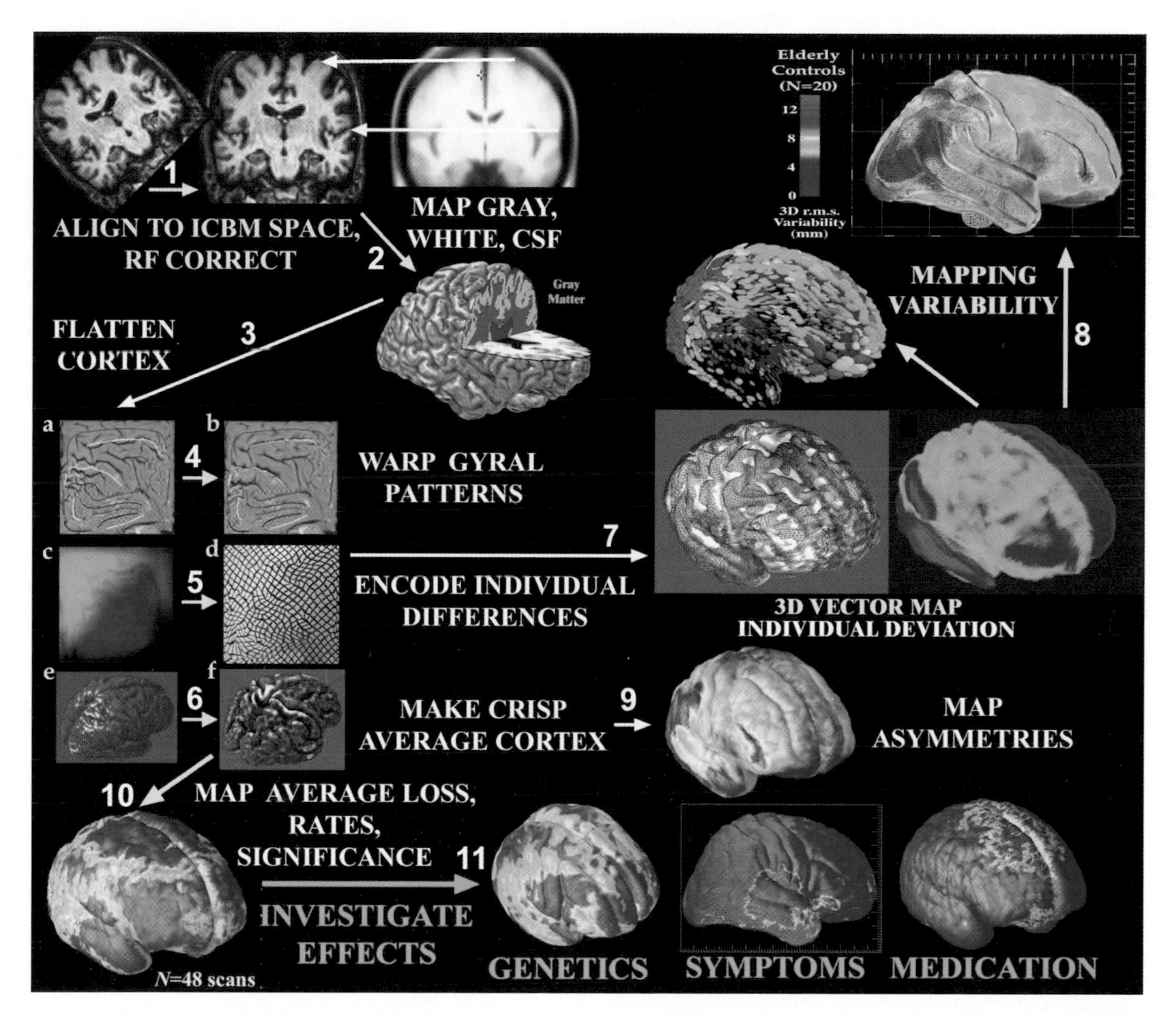

Figure 20 Creating brain maps and anatomical models. An image analysis pipeline (Thompson *et al.*, 2001a) is shown here. It can be used to create maps that reveal how brain structure varies in large populations, differs in disease, and is modulated by genetic or therapeutic factors. This approach aligns new 3D MRI scans from patients and controls (**1**) with an average brain template based on a population (here the ICBM template is used, developed by the International Consortium for Brain Mapping). Tissue classification algorithms then generate maps of gray matter, white matter, and CSF (**2**). To help compare cortical features from subjects whose anatomy differs, individual gyral patterns are flattened (**3**) and aligned with a group average gyral pattern (**4**). If a color code indexing 3D cortical locations is flowed along with the same deformation field (**5**), a crisp group average model of the cortex can be made (**6**), relative to which individual gyral pattern differences (**7**), group variability (**8**), and cortical asymmetry (**9**) can be computed. Once individual gyral patterns are aligned to the mean template, differences in gray matter distribution or thickness (**10**) can be mapped, pooling data from homologous regions of cortex. Correlations can be mapped between disease-related deficits and genetic risk factors (**11**). Maps may also be generated visualizing linkages between deficits and clinical symptoms, cognitive scores, and medication effects.

causes. In the future, data mining techniques may be applied to isolate novel patterns in population-based atlases, such as diagnostic subtypes, developmental stages, or unsuspected groupings in the data. As atlases are populated with thousands of data sets, a resurgence is likely in information-theoretic mathematical tools that extract information from images, compare disparate data sets, and detect patterns with statistical and visual power.

XI. Conclusion

In this chapter we presented a mathematical framework to create population-based brain atlases. This emerging field in medical imaging is already uncovering fundamental features of brain structure and function in health and disease. Brain data are so complex and variable that it is essential to rely on brain atlases, templates, and anatomical models in large-scale investigations. Deformable and probabilistic atlases can warehouse population-based data in a common 3D reference frame. They capture anatomic variability using a variety of mathematical approaches. The interest in cortical anatomy, in particular, has motivated specialized approaches to analyze its structure (Fig. 20). Finally, we suggested several new directions for future atlas development. Dynamic and genetic brain maps, among other new techniques, are beginning to reveal how the brain develops, how diseases progress, and how genes affect complex patterns of brain structure. The resulting armory of tools shows enormous promise in shedding light on the complex structural and functional organization of the human brain.

Acknowledgments

This work was supported by research grants from the National Center for Research Resources (P41 RR13642), the National Institute of Neurological Disorders and Stroke, and the National Institute of Mental Health (NINDS/NIMH NS38753) and by a Human Brain Project grant to the International Consortium for Brain Mapping, funded jointly by NIMH and NIDA (P20 MH/DA52176). Additional support was provided by the National Library of Medicine (LM/MH05639) and the National Science Foundation (BIR 93-22434). Special thanks go to Neal Jeffries for his advice on longitudinal random effects models and to our colleagues Katherine Narr, Jay Giedd, Judith Rapoport, Tyrone Cannon, Elizabeth Sowell, Christine Vidal, David MacDonald, Alan Evans, Roger Woods, Colin Holmes, and John Mazziotta and many others whose support has been invaluable in these investigations.

References

Alouges, F. (1997). A new algorithm for computing liquid crystal stable configurations: The harmonic mapping case. *SIAM J. Number Anal.* **34,** 1708–1726.

Amunts, K., and Zilles, K. (2001). Advances in cytoarchitectonic mapping of the human cerebral cortex. *Neuroimaging Clin. North Am.* **11,** 151–169.

Andrade, A. (2001). Scale space searches in cortical surface analysis of fMRI data. *In* "Proceedings of the 7th International Conference on Human Brain Mapping," Brighton.

Andreasen, N. C., Arndt, S., Swayze, V., Cizadlo, T., Flaum, M., O'Leary, D., Ehrhardt, J. C., and Yuh, W. T. C. (1994). Thalamic abnormalities in schizophrenia visualized through magnetic resonance image averaging. *Science* **266,** 294–298.

Angenent, S., Haker, S., Tannenbaum, A., and Kikinis, R. (1999). Conformal geometry and brain flattening. *Proc. MICCAI* **1999,** 271–278.

Ashburner, J. (2001). "Computational Neuroanatomy." Univ. of London. [Ph.D. thesis].

Ashburner, J., and Friston, K. J. (1999). Nonlinear spatial normalization using basis functions. *Hum. Brain Mapp.* **7,** 254–266.

Ashburner, J., and Friston, K. J. (2000). Voxel-based morphometry—The methods. *NeuroImage* **11,** 805–821.

Ashburner, J., Hutton, C., Frackowiak, R., Johnsrude, I., Price, C., and Friston, K. (1998). Identifying global anatomical differences: Deformation-based morphometry. *Hum. Brain Mapp.* **6,** 348–357.

Ashburner, J., Neelin, P., Collins, D. L., Evans, A. C., and Friston, K. J. (1997). Incorporating prior knowledge into image registration. *NeuroImage* **6,** 344–352.

Avoli, M., Hwa, G. C., Kostopoulos, G., Oliver, A., and Villemure, J. G. (1991). Electrophysiological analysis of human neocortex in vitro: Experimental techniques and methodological approaches. *Can. J. Neurol. Sci.* **18,** 636–639.

Baare, W. F., Hulshoff Pol, H. E., Boomsma, D. I., Posthuma, D., de Geus, E. J., Schnack, H. G., van Haren, N. E., van Oel, C. J., and Kahn, R. S. (2001). Quantitative genetic modeling of variation in human brain morphology. *Cereb. Cortex* **11,** 816–824.

Bakircioglu, M., Joshi, S., and Miller, M. I. (1999). Landmark matching on brain surfaces via large deformation diffeomorphisms on the sphere. *Proc. SPIE Med. Imaging.*

Bankman, I. N. (1999). "Handbook of Medical Imaging: Processing and Analysis." Academic Press, San Diego.

Bartley, A. J., Jones, D. W., and Weinberger, D. R. (1997). Genetic variability of human brain size and cortical gyral patterns. *Brain* **120,** 257–269.

Bell, A. J., and Sejnowski, T. J. (1995). An information-maximization approach to blind separation and blind deconvolution. *Neural Comput.* **7,** 1129–1159.

Berger, H. (1929). Uber das Elektrenkephalogramm des Menschen. *Arch. Psychiatr. Nervenkr.* **87,** 527–580.

Bermudez, P., and Zatorre, R. J. (2001). Sexual dimorphism in the corpus callosum: Methodological considerations in MRI morphometry. *NeuroImage* **13,** 1121–1130.

Bertalmio, M., Osher, S. J., Cheng, L. T., and Sapiro, G. (2000). Variational problems and partial differential equations on implicit surfaces: The framework and examples in image processing and pattern formation. UCLA CAM Rep., 00(23).

Bilder, R. M., Goldman, R. S., Robinson, D., Reiter, G., Bell, L., Bates, J. A., Pappadopulos, E., Willson, D. F., Alvir, J. M., Woerner, M. G., Geisler, S., Kane, J. M., and Lieberman, J. A. (2000). Neuropsychology of first-episode schizophrenia: Initial characterization and clinical correlates. *Am. J. Psychiatry* **157,** 549–559.

Biondi, A., *et al.* (1998). Are the brains of monozygotic twins similar? A three-dimensional MR study. *Am. J. Neuroradiol.* **19,** 1361–1367.

Blanton, R. E., Levitt, J. L., Thompson, P. M., Capetillo-Cunliffe, L. F., Sadoun, T., Williams, T., McCracken, J. T., and Toga, A. W. (2000). Mapping cortical variability and complexity patterns in the developing human brain. *Psychiatry Res.* **107,** 29–43.

Bookstein, F. L. (1997). Landmark methods for forms without landmarks: Morphometrics of group differences in outline shape. *Med. Image Anal.* **1,** 225–243.

Brett, M. (1999). The MNI brain and the Talairach atlas. Internet course notes, URL: http://www.mrc-cbu.cam.ac.uk/Imaging/mnispace.html.

Brinkley, J. F., Bradley, S. W., Sundsten, J. W., and Rosse, C. (1997). The Digital Anatomist information system and its use in the generation and delivery of Web-based anatomy atlases. *Comput. Biomed. Res.* **30,** 472–503.

Brodmann, K. (1909). Vergleichende Lokalisationslehre der Grosshirnrinde in ihren Prinzipien dargestellt auf Grund des Zellenbaues, Barth, Leipzig. *In* "Some Papers on the Cerebral Cortex," pp. 201–230. Thomas, Springfield, IL, 1960. [Translated as "On the comparative localization of the cortex."]

Bro-Nielsen, M., and Gramkow, C. (1996). Fast fluid registration of medical images. *In* "Lecture Notes in Computer Science," Vol. 1131, "Visualization in Biomedical Computing, Hamburg, Germany" (K. H. Höhne and R. Kikinis, eds.), pp. 267–276. Springer-Verlag, Berlin.

Bullmore, E. T., Suckling, J., Overmeyer, S., Rabe-Hesketh, S., Taylor, E., and Brammer, M. J. (1999). Global, voxel, and cluster tests, by theory and permutation, for a difference between two groups of structural MR images of the brain. *IEEE Trans. Med. Imaging* **18,** 32–42.

Burke, W. L. (1985). "Applied Differential Geometry." Cambridge Univ. Press, Cambridge, UK.

Cachier, P., Pennec, X., and Ayache, N. (1999). Fast non rigid matching by gradient descent: Study and improvements of the "Demons" algorithm. INRIA Technical Rep. RR-3706, June 1999.

Calmon, G., and Roberts, N. (2000). Automatic measurement of changes in brain volume on consecutive 3D MR images by segmentation propagation. *Magn. Reson. Imaging* **18,** 439–453.

Cannestra, A. F., Blood, A. J., Black, K. L., and Toga, A. W. (1996). The evolution of optical signals in human and rodent cortex. *NeuroImage* **3,** 202–208.

Cannon, T. D., Thompson, P. M., van Erp, T., Toga, A. W., Huttunen, M., Lönnqvist, J., and Standertskjöld-Nordenstam, C.-G. (2001a). A probabilistic atlas of cortical gray matter changes in monozygotic twins discordant for schizophrenia. Presented at the International Congress on Schizophrenia Research, April 28–May 2, 2001.

Cannon, T. D., Thompson, P. M., van Erp, T., Toga, A. W., Poutanen, V.-P., Huttunen, M., Lönnqvist, J., Standertskjöld-Nordenstam, C.-G., Narr, K. L., Khaledy, M., Zoumalan, C. I., Dail, R., and Kaprio, J. (2001b). Cortex mapping reveals heteromodal gray matter deficits in monozygotic twins discordant for schizophrenia. In press.

Cannon, T. D., Mednick, S. A., and Parnas, J. (1989). Genetic and perinatal determinants of structural brain deficits in schizophrenia. *Arch. Gen. Psychiatry* **46,** 883–889.

Cao, J., and Worsley, K. J. (1999). The geometry of the Hotelling's T-squared random field with applications to the detection of shape changes. *Ann. Stat.* **27,** 925–942.

Carman, G. J., Drury, H. A., and Van Essen, D. C. (1995). Computational methods for reconstructing and unfolding the cerebral cortex. *Cereb. Cortex* **5,** 506–517.

Caunce, A., and Taylor, C. J. (2001). Building 3D sulcal models using local geometry. *Med. Image Anal.* **5,** 69–80.

Chiavaras, M. M., LeGoualher, G., Evans, A., and Petrides, M. (2001). Three-dimensional probabilistic atlas of the human orbitofrontal sulci in standardized stereotaxic space. *NeuroImage* **13,** 479–496.

Chipuer, H. M., Rovine, M. J., and Plomin, R. (1990). LISREL modeling: Genetic and environmental influences on IQ revisited. *Br. J. Dev. Psychol.* **10,** 110.

Chow, S. (1998). "Finite Element Decomposition of the Human Neocortex." Texas A&M Univ., College Station. [Ph.D. thesis]

Christensen, G. E., Johnson, H. J., Haller, J. W., Melloy, J., Vannier, M. W., and Marsh, J. L. (1999). Synthesizing average 3D anatomical shapes using deformable templates. *Proc. SPIE Med. Imaging* **3661.**

Christensen, G. E., Miller, M. I., Marsh, J. L., and Vannier, M. W. (1995). Automatic analysis of medical images using a deformable textbook. *In* "Proceedings of the Society for Computer Assisted Radiology," pp. 152–157. Springer-Verlag, Berlin.

Christensen, G. E., Rabbitt, R. D., and Miller, M. I. (1993). A deformable neuroanatomy textbook based on viscous fluid mechanics. *In* "27th Annual Conference on Information Sciences and Systems," pp. 211–216.

Christensen, G. E., Rabbitt, R. D., and Miller, M. I. (1996). Deformable templates using large deformation kinematics. *IEEE Trans. Image Process.* **5,** 1435–1447.

Chung, M. K., Worsley, K. J., Paus, T., Cherif, C., Collins, D. L., Giedd, J. N., Rapoport, J. L., and Evans, A. C. (2001). A unified statistical approach to deformation-based morphometry. *NeuroImage*, in press.

Chung, M. K., Worsley, K. J., Taylor, J., Ramsay, J., Robbins, S., and Evans, A. C. (2000). Diffusion smoothing on the cortical surface. *Hum. Brain Mapp.*.

Collins, D. L., Holmes, C. J., Peters, T. M., and Evans, A. C. (1995). Automatic 3D model-based neuroanatomical segmentation. *Hum. Brain Mapp.* **3,** 190–208.

Collins, D. L., Le Goualher, G., Venugopal, R., Caramanos, A., Evans, A. C., and Barillot, C. (1996). Cortical constraints for non-linear cortical registration. *In* "Lecture Notes in Computer Science," Vol. 1131, "Visualization in Biomedical Computing, Hamburg, Germany" (K. H. Höhne and R. Kikinis, eds.), pp. 307–316. Springer-Verlag, Berlin.

Collins, D. L., Montagnat, J., Zijdenbos, A. P., Evans, A. C., and Arnold, D. L. (2001). Automated estimation of brain volume in multiple sclerosis with BICCR. *In* "Lecture Notes in Computer Science," Vol. 2082, "Proceedings of the Annual Symposium on Information Processing in Medical Imaging" (M. F. Insana and R. M. Leahy, eds.), pp. 141–147, Springer-Verlag, Berlin.

Collins, D. L., Neelin, P., Peters, T. M., and Evans, A. C. (1994a). Automatic 3D intersubject registration of MR volumetric data into standardized Talairach space. *J. Comput. Assisted Tomogr.* **18,** 192–205.

Collins, D. L., Peters, T. M., and Evans, A. C. (1994b). An automated 3D non-linear image deformation procedure for determination of gross morphometric variability in the human brain. *Proc. SPIE Visualization in Biomed. Comp.* **3,** 180–190.

Collins, D. L., Zijdenbos, A. P., Kollokian, V., Sled, J. G., Kabani, N. J., Holmes, C. J., and Evans, A. C. (1998). Design and construction of a realistic digital brain phantom. *IEEE Trans. Med. Imaging* **17,** 463–468.

Collins, F. S. (2001). Contemplating the end of the beginning. *Genome Res.* **11(5),** 641–643. [Review]

Cootes, T. F., Taylor, C. J., Cooper, D. H., and Graham, J. (1995). Active shape models—Their training and application. *Comput. Vision. Image Understand.* **61,** 38–59.

Csernansky, J. G., Joshi, S., Wang, L., Haller, J. W., Gado, M., Miller, J. P., Grenander, U., and Miller, M. I. (1998). Hippocampal morphometry in schizophrenia by high dimensional brain mapping. *Proc. Natl. Acad. Sci. USA* **95,** 11406–11411.

Csernansky, J. G., Wang, L., Joshi, S., Miller, J. P., Gado, M., Kido, D., McKeel, D., Morris, J. C., and Miller, M. I. (2000). Early DAT is distinguished from aging by high-dimensional mapping of the hippocampus: Dementia of the Alzheimer type. *Neurology* **55,** 1636–1643.

Damasio, H. (1995). "Human Brain Anatomy in Computerized Images." Oxford Univ. Press, Oxford/New York.

Davatzikos, C. (1996). Spatial normalization of 3D brain images using deformable models. *J. Comput. Assisted Tomogr.* **20,** 656–665.

Davatzikos, C., Vaillant, M., Resnick, S. M., Prince, J. L., Letovsky, S., and Bryan, R. N. (1996). A computerized approach for morphological analysis of the corpus callosum. *J. Comput. Assisted Tomogr.* **20,** 88–97.

Davidian, M., and Giltinan, D. (1995). "Nonlinear Models for Repeated Measurement Data." Chapman & Hall, London/New York.

Davidson, R. J., and Hugdahl, K. (1994). "Brain Asymmetry." MIT Press, Cambridge, MA.

Davies, R. H., Cootes, T. F., and Taylor, C. J. (2001). A minimum description length approach to statistical shape modelling. *In* "Proceedings of Information Modeling in Medical Imaging," Davis, CA (M. Insana and R. Leahy, eds.), pp. 50–63.

Davis, M. H., Khotanzad, A., and Flamig, D. P. (1996). 3D image matching using a radial basis function neural network. *In* "Proceedings of the 1996 World Congress on Neural Networks," San Diego, CA, Sept. 15–18, 1996, pp. 1174–1179.

Davis, M. H., Khotanzad, A., Flamig, D. P., and Harms, S. E. (1997). A physics based coordinate transformation for 3D image matching. *IEEE Trans. Med. Imaging* **16,** 317–328.

DeCarli, C., Haxby, J. V., Gillette, J. A., Teichberg, D., Rapoport, S. I., and Schapiro, M. B. (1992). Longitudinal changes in lateral ventricular volume in patients with dementia of the Alzheimer type. *Neurology* **42,** 2029–2036.

Devlin, B., Daniels, M., and Roeder, K. (1997). The heritability of IQ. *Nature* **388,** 468–471

Dinov, I. D., Mega, M. S., Thompson, P. M., Lee, L., Woods, R. P., Holmes, C. J., Sumners, D. L., and Toga, A. W. (2000). Analyzing functional brain images in a probabilistic atlas: A validation of sub-volume thresholding. *J. Comput. Assisted Tomogr.* **24,** 128–138.

Drury, H. A., and Van Essen, D. C. (1997). Analysis of functional specialization in human cerebral cortex using the Visible Man surface based atlas. *Hum. Brain Mapp.* **5,** 233–237.

Drury, H. A., Van Essen, D. C., Corbetta, M., and Snyder, A. Z. (2000). Surface-based analyses of the human cerebral cortex. *In* "Brain Warping" (A. W. Toga, ed.), Chap. 19. Academic Press, San Diego.

Dupuis, P., Grenander, U., and Miller, M. I. (1998). Variational problems on flows of diffeomorphisms for image matching. *Q. Appl. Math.* **56,** 587–600.

Duta, N., and Sonka, M. (1998). Segmentation and interpretation of MR brain images: An improved active shape model. *IEEE Trans. Med. Imaging* **17,** 1049–1062.

Duvernoy, H. M. (1991). "The Human Brain." Springer-Verlag, New York.

Eells, J., and Sampson, J. H. (1964). Harmonic mappings of Riemannian manifolds. *Am. J. Math.* **86,** 109–160.

Einstein, A. (1914). Covariance properties of the field equations of the theory of gravitation based on the generalized Theory of Relativity. *Z. Math. Phys.* **63,** 215–225. [Published as Kovarianzeigenschaften der Feldgleichungen der auf die verallgemeinerte Relativitätstheorie gegründeten Gravitationstheorie.]

Evans, A. C., Collins, D. L., Neelin, P., MacDonald, D., Kamber, M., and Marrett, T. S. (1994). Three-dimensional correlative imaging: Applications in human brain mapping. *In* "Functional Neuroimaging: Technical Foundations" (R. W. Thatcher, M. Hallett, T. Zeffiro, E. R. John, and M. Huerta, eds.), pp. 145–162.

Falconer, D. S. (1989). "Introduction to Quantitative Genetics, 3rd edition." Longman, Essex.

Feldman, M. W., and Otto, S. P. (1997). Twin studies, heritability, and intelligence. *Science* **278,** 1383–1384. [Discussion pp. 1386–1387]

Finkel, D., Pedersen, N. L., Plomin, R., and McClearn, G. E. (1998). Longitudinal and cross-sectional twin data on cognitive abilities in adulthood: The Swedish Adoption/Twin Study of Aging. *Dev. Psychol.* **34,** 1400–1413.

Fischl, B., and Dale, A. M. (2001). Measuring the thickness of the human cerebral cortex from magnetic resonance images. *Proc. Natl. Acad. Sci. USA* **97,** 11050–11055.

Fischl, B., Sereno, M. I., Tootell, R. B. H., and Dale, A. M. (1999). High-resolution inter-subject averaging and a coordinate system for the cortical surface. *Hum. Brain Mapp.* **8,** 272–284.

Fishman, E. K. (1997). High resolution three-dimensional imaging from subsecond helical CT datasets: Applications in vascular imaging. *Am. J. Radiol.* **169,** 441–443.

Fitzpatrick, J. M., and Sonka, M. (2000). "Handbook of Medical Imaging," Vol. II, "Medical Image Processing and Analysis." SPIE–Int. Soc. Opt. Eng., Bellingham, WA.

Fox, P. T. (1997). The growth of human brain mapping. *Hum. Brain Mapp.* **5,** 1–2.

Fox, P. T., Mintun, M. A., Reiman, E. M., and Raichle, M. E. (1988). Enhanced detection of focal brain responses using inter-subject averaging and change distribution analysis of subtracted PET images. *J. Cereb. Blood Flow Metab.* **8,** 642–653.

Fox, P. T., Perlmutter, J. S., and Raichle, M. (1985). A stereotactic method of localization for positron emission tomography. *J. Comput. Assisted Tomogr.* **9,** 141–153.

Frackowiak, R. S. J., Friston, K. J., Frith, C. D., Dolan, R. J., and Mazziotta, J. C. (1997). "Human Brain Function," Academic Press, San Diego.

Frangi, A. F., Rueckert, D., Schnabel, J. A., and Niessen, W. (2001). Automatic 3D ASM construction via atlas-based landmarking and volumetric elastic registration. *In* "Proceedings of the Information Processing in Medical Imaging Annual Meeting," 2001, pp. 78–91

Freeborough, P. A., and Fox, N. C. (1998). Modeling brain deformations in Alzheimer disease by fluid registration of serial 3D MR images. *J. Comput. Assisted Tomogr.* **22,** 838–843.

Friston, K. J. (1997). Testing for anatomically specified regional effects. *Hum. Brain Mapp.* **5,** 133–136.

Friston, K. J., Frith, C. D., Liddle, P. F., and Frackowiak, R. S. J. (1991). Plastic transformation of PET images. *J. Comput. Assisted Tomogr.* **9,** 141–153.

Friston, K. J., Holmes, A. P., Worsley, K. J., Poline, J. P., Frith, C. D., and Frackowiak, R. S. J. (1995). Statistical parametric maps in functional imaging: A general linear approach. *Hum. Brain Mapp.* **2,** 189–210.

Friston, K. J., Passingham, R. E., Nutt, J. G., Heather, J. D., Sawle, G. V., and Frackowiak, R. S. J. (1989). Localization in PET images: Direct fitting of the intercommissural (AC-PC) line. *J. Cereb. Blood Flow Metab.* **9,** 690–695.

Gaser, C., Nenadic, I., Buchsbaum, B. R., Hazlett, E. A., and Buchsbaum, M. S. (2001). Deformation-based morphometry and its relation to conventional volumetry of brain lateral ventricles in MRI. *NeuroImage* **13,** 1140–1145.

Ge, Y., Fitzpatrick, J. M., Kessler, R. M., and Jeske-Janicka, M. (1995). Intersubject brain image registration using both cortical and subcortical landmarks. *SPIE Image Proc.* **2434,** 81–95.

Gee, J. C., and Bajcsy, R. K. (1998). Elastic matching: Continuum-mechanical and probabilistic analysis. *In* "Brain Warping" (A. W. Toga, ed.). Academic Press, San Diego.

Gee, J. C., LeBriquer, L., Barillot, C., Haynor, D. R., and Bajcsy, R. (1995). Bayesian approach to the brain image matching problem. Inst. Res. Cognit. Sci. Technical Rep. 95-08, April 1995.

Gee, J. C., Reivich, M., and Bajcsy, R. (1993). Elastically deforming an atlas to match anatomical brain images. *J. Comput. Assisted Tomogr.* **17,** 225–236.

Gerig, G., Styner, M., Shenton, M. E., and Lieberman, J. A. (2001). Shape versus size: Improved understanding of the morphology of brain structures. *In* "Lecture Notes in Computer Science," Vol. 2208, "Proceedings of MICCAI 2001," pp. 24–32, Springer-Verlag, Berlin.

Geyer, S., Schleicher, A., Schormann, T., Mohlberg, H., Bodegard, A., Roland, P. E., and Zilles, K. (2001). Integration of microstructural and functional aspects of human somatosensory areas 3a, 3b, and 1 on the basis of a computerized brain atlas. *Anat. Embryol. (Berlin)* **204,** 351–366.

Geyer, S., Schleicher, A., and Zilles, K. (1997). The somatosensory cortex of man: Cytoarchitecture and regional distributions of receptor binding sites. *NeuroImage* **6,** 27–45.

Giedd, J. N., Jeffries, N. O., Blumenthal, J., Castellanos, F. X., Vaituzis, A. C., Fernandez, T., Hamburger, S. D., Liu, H., Nelson, J., Bedwell, J., Tran, L., Lenane, M., Nicolson, R., and Rapoport, J. L. (1999a). Childhood-onset schizophrenia: Progressive brain changes during adolescence. *Biol. Psychiatry* **46,** 892–898.

Giedd, J. N., Blumenthal, J., Jeffries, N. O., Castellanos, F. X., Liu, H., Zijdenbos, A., Paus, T., Evans, A. C., and Rapoport, J. L. (1999b). Brain development during childhood and adolescence: A longitudinal MRI study. *Nat. Neurosci.* **2,** 861–863.

Goebel, R. (2000). A fast automated method for flattening cortical surfaces. *NeuroImage* **11,** S680.

Golland, P., Grimson, W. E. L., Shenton, M. E., and Kikinis, R. (2000). Small sample size learning for shape analysis of anatomical structures, *In* "Lecture Notes in Computer Science," Vol. 1935, "Proceedings of MICCAI 2000: Third International Conference on Medical Robotics, Imaging and Computer Assisted Intervention," pp. 72–82.

Golland, P., Grimson, W. E. L., Shenton, M. E., and Kikinis, R. (2001). Deformation analysis for shaped based classification. *In* "Lecture Notes in Computer Science," Vol. 2082, "Proceedings of the 17th International Conference on Information Processing and Medical Imaging," 2001, pp. 517–530.

Good, C. D., Ashburner, J., and Frackowiak, R. S. J. (2001a). Computational neuroanatomy: New perspectives for neuroradiology. *Rev. Neurol. (Paris)* **157,** 797–806.

Good, C. D., Johnsrude, I. S., Ashburner, J., Henson, R. N., Friston, K. J., and Frackowiak, R. S. J. (2001b). A voxel-based morphometric study of ageing in 465 normal adult human brains. *NeuroImage* **14,** 21–36.

Gottesman, I. I. (1997). Twins: En route to QTLs for cognition. *Science* **276,** 1522–1523.

Gramkow, C. (1996). "Registration of 2D and 3D Medical Images." Denmark Tech. Univ. [M.Sc. thesis]

Grefkes, C., Geyer, S., Schormann, T., Roland, P., and Zilles, K. (2001). Human somatosensory area 2: Observer-independent cytoarchitectonic mapping, interindividual variability, and population map. *NeuroImage* **14,** 617–631.

Greitz, T., Bohm, C., Holte, S., and Eriksson, L. (1991). A computerized brain atlas: Construction, anatomical content, and some applications. *J. Comput. Assisted Tomogr.* **15,** 26–38.

Grenander, U., and Miller, M. I. (1998). Computational anatomy: An emerging discipline. Department of Mathematics, Brown Univ., New York. [Technical Rep.]

Guimond, A., Meunier, J., and Thirion, J.-P. (1999). Average brain models: A convergence study. INRIA Technical Rep. RR-3731.

Guttmann, C. R., Ahn, S. S., Hsu, L., Kikinis, R., and Jolesz, F. A. (1995). The evolution of multiple sclerosis lesions on serial MR. *Am. J. Neuroradiol.* **16,** 1481–1491.

Haker, S., Angenent, S., Tannenbaum, A., and Kikinis, R. (1999). Nondistorting flattening maps and the 3-D visualization of colon CT images. *IEEE Trans. Med. Imaging* **19,** 665–670.

Haller, J. W., Banerjee, A., Christensen, G. E., Gado, M., Joshi, S., Miller, M. I., Sheline, Y., Vannier, M. W., and Csernansky, J. G. (1997). Three-dimensional hippocampal MR morphometry with high-dimensional transformation of a neuroanatomic atlas. *Radiology* **202,** 504–510.

Haller, J. W., and Vannier, M. W. (1998). Mapping the visible human brain to the Talairach atlas. Presented at the Human Brain Mapping Conference, Montreal.

Han, X., Xu, C., Braga-Neto, U., and Prince, J. L. (2001). Graph-based topology correction for brain cortex segmentation. *In* "Information Processing in Medical Imaging, 2001," pp. 395–401.

Haney, S., Thompson, P. M., Cloughesy, T. F., Alger, J. R., Frew, A., Torres-Trejo, A., Mazziotta, J. C., and Toga, A. W. (2001a). Mapping response in a patient with malignant glioma. *J. Comput. Assisted Tomogr.*

Haney, S., Thompson, P. M., Cloughesy, T. F., Alger, J. R., and Toga, A. W. (2001b). Tracking tumor growth rates in patients with malignant gliomas: A test of two algorithms. *Am. J. Neuroradiol.* **22,** 73–82.

Hohne, K. H., Pflesser, B., Pommert, A., Riemer, M., Schubert, R., Schiemann, T., Tiede, U., and Schumacher, U. (2001). A realistic model of human structure from the Visible Human data. *Methods Inf. Med.* **40,** 83–89.

Holmes, A. P., Blair, R. C., Watson, J. D. G., and Ford, I. (1996). Nonparametric analysis of statistic images from functional mapping experiments. *J. Cereb. Blood Flow Metab.* **16,** 7–22.

Holmes, C. J., Hoge, R., Collins, L., Woods, R., Toga, A. W., and Evans, A. C. (1998). Enhancement of MR images using registration for signal averaging. *J. Comput. Assisted Tomogr.* **22,** 324–333.

Hounsfield, G. N. (1973). Computerized transverse axial scanning (tomography). I. Description of system. *Br. J. Radiol.* **46,** 1016–1022.

Huerta, M. F., and Koslow, S. H. (1996). Neuroinformatics: Opportunities across disciplinary and national borders. *NeuroImage* **4,** S4–6.

Huiskamp, G. (1991). Difference formulas for the surface Laplacian on a triangulated surface. *J. Comput. Phys.* **95,** 477–496.

Hurdal, M. K., Sumners, D. L., Stephenson, K., Bowers, P. L., and Rottenberg, D. A. (1999). Circlepack: Software for creating quasi-conformal flat maps of the brain. *In* "Fifth International Conference on Functional Mapping of the Human Brain," p. S250.

Hyvärinen, Karhunen, and Oja (2001). "Independent Component Analysis." Wiley, New York.

Jacobs, R. E., and Fraser, S. E. (1994). Magnetic resonance microscopy of embryonic cell lineages and movements. *Science* **263,** 681–684.

Janke, A. L., Zubicaray, G. D., Rose, S. E., Griffin, M., Chalk, J. B., and Galloway, G. J. (2001). 4D deformation modeling of cortical disease progression in Alzheimer's dementia. *Magn. Reson. Med.* **46,** 661–666.

Jernigan, T. L., Archibald, S. L., Fennema-Notestine, C., Gamst, A. C., Stout, J. C., Bonner, J., and Hesselink, J. R. (2001). Effects of age on tissues and regions of the cerebrum and cerebellum. *Neurobiol. Aging* **22,** 581–594.

Jones, S. E., Buchbinder, B. R., and Aharon, I. (2000). Three-dimensional mapping of cortical thickness using Laplace's equation. *Hum. Brain Mapp.* **11,** 12–32.

Joshi, S., Grenander, U., and Miller, M. I. (1997). On the geometry and shape of brain sub-manifolds. *IEEE Trans. Pattern Anal. Mach. Intell.* **11,** 1317–1343.

Joshi, S., Lorenzen, P., Gerig, G., and Bullitt, E. (2001). Tumor-induced structural and radiometric asymmetry in brain images. *Int. J. Comput. Vision.*

Joshi, S., and Miller, M. I. (2000). Landmark matching via large deformation diffeomorphisms. *IEEE Trans. Image Process.* **9,** 1357–1370.

Joshi, S. C., Miller, M. I., Christensen, G. E., Banerjee, A., Coogan, T. A., and Grenander, U. (1995). Hierarchical brain mapping via a generalized Dirichlet solution for mapping brain manifolds. *Proc. SPIE Conf. Opt. Sci. Eng. Instrum.* **2573,** 278–289.

Kabani, N., Le Goualher, G., MacDonald, D., and Evans, A. C. (2001). Measurement of cortical thickness using an automated 3-D algorithm: A validation study. *NeuroImage* **13,** 375–380.

Kanai, T., Suzuki, H., and Kimura, F. (1997). 3D geometric metamorphosis based on harmonic maps. *In* "Proceedings of Pacific Graphics '97," October 1997, pp. 97–104.

Kang, K. W., Lee, D. S., Cho, J. H., Lee, J. S., Yeo, J. S., Lee, S. K., Chung, J. K., and Lee, M. C. (2001). Quantification of F-18 FDG PET images in temporal lobe epilepsy patients using probabilistic brain atlas. *NeuroImage* **14,** 1–6.

Kaprio, J., Koskenvuo, M., and Rose, R. J. (1990). Change in cohabitation and intrapair similarity of monozygotic (MZ) cotwins for alcohol use, extraversion, and neuroticism. *Behav. Genet.* **20,** 265–276.

Kennedy, D. N., Lange, N., Makris, N., Bates, J., Meyer, J., and Caviness, V. S., Jr. (1998). Gyri of the human neocortex: An MRI-based analysis of volume and variance. *Cereb. Cortex* **8,** 372–384.

Kiebel, S. J., Poline, J. B., Friston, K. J., Holmes, A. P., and Worsley, K. J. (1999). Robust smoothness estimation in statistical parametric maps using standardized residuals from the general linear model. *NeuroImage* **10,** 756–766.

Kikinis, R., Shenton, M. E., Gerig, G., Hokama, H., Haimson, J., O'Donnell, B. F., Wible, C. G., McCarley, R. W., and Jolesz, F. A. (1994). Temporal lobe sulco-gyral pattern anomalies in schizophrenia: An in vivo MR three-dimensional surface rendering study. *Neurosci. Lett.* **182,** 7–12.

Kikinis, R., Shenton, M. E., Iosifescu, D. V., McCarley, R. W., Saiviroonporn, P., Hokama, H. H., Robatino, A., Metcalf, D., Wible, C. G., Portas, C. M., Donnino, R., and Jolesz, F. (1996). A digital brain atlas for surgical planning, model-driven segmentation, and teaching. *IEEE Trans. Visualizat. Comput. Graph.* **2,** 232–241.

Kochunov, P., Lancaster, J., Thompson, P. M., Boyer, A., Hardies, J., and Fox, P. T. (2000). Validation of an octree regional spatial normalization method for regional anatomical matching. *Hum. Brain Mapp.* **11,** 193–206.

Kochunov, P., Lancaster, J., Thompson, P. M., Toga, A. W., Brewer, P., Hardies, J., and Fox, P. T. (2001a). An optimized individual target brain in the Talairach coordinate system. In press.

Kochunov, P., Lancaster, J., Thompson, P. M., Woods, R. P., Hardies, J., and Fox, P. T. (2001b). Regional spatial normalization: Towards an optimal target. *J. Comput. Assisted Tomogr.,* in press.

Kriegeskorte, N., and Goebel, R. (2001). An efficient algorithm for topologically correct segmentation of the cortical sheet in anatomical MR volumes. *NeuroImage* **14,** 329–346.

Laakso, M. P., Frisoni, G. B., Kononen, M., Mikkonen, M., Beltramello, A., Geroldi, C., Bianchetti, A., Trabucchi, M., Soininen, H., and Aronen, H. J. (2000). Hippocampus and entorhinal cortex in frontotemporal dementia and Alzheimer's disease: A morphometric MRI study. *Biol. Psychiatry* **47,** 1056–1063.

Lancaster, J. L., Woldorff, M. G., Parsons, L. M., Liotti, M., Freitas, C. S., Rainey, L., Kochunov, P. V., Nickerson, D., Mikiten, S. A., and Fox, P. T. (2000). Automated Talairach atlas labels for functional brain mapping. *Hum. Brain Mapp.* **10,** 120–131.

Lange, N. (1996). Statistical approaches to human brain mapping by functional magnetic resonance imaging. *Stat. Med.* **15,** 389–428.

Lange, N., Giedd, J. N., Castellanos, F. X., Vaituzis, A. C., and Rapoport, J. L. (1997). Variability of human brain structure size: Ages 4–20 years.

Lauterbur, P. (1973). Image formation by induced local interactions: Examples employing nuclear magnetic resonance. *Nature* **242,** 190–191.

Lawrie, S. M., and Abukmeil, S. S. (1998). Brain abnormality in schizophrenia: A systematic and quantitative review of volumetric magnetic resonance imaging studies. *Br. J. Psychiatry* **172,** 110–120.

Le Bihan, D. (1996). Functional MRI of the brain: Principles, applications and limitations. *Neuroradiology* **23,** 1–5.

Le Goualher, G., Argenti, A. M., Duyme, M., Baare, W. F., Hulshoff Pol, H. E., Boomsma, D. I., Zouaoui, A., Barillot, C., and Evans, A. C. (2000). Statistical sulcal shape comparisons: Application to the detection of genetic encoding of the central sulcus shape. *NeuroImage* **11,** 564–574.

Le Goualher, G., Procyk, E., Collins, D. L., Venugopal, R., Barillot, C., and Evans, A. C. (1999). Automated extraction and variability analysis of sulcal neuroanatomy. *IEEE Trans. Med. Imaging* **18,** 206–217.

Leonard, C. M. (1996). Structural variation in the developing and mature cerebral cortex: Noise or signal? *In* "Developmental Neuroimaging: Mapping the Development of Brain and Behavior" (R. W. Thatcher, G. Reid Lyon, J. Rumsey, and N. Krasnegor, eds.), pp. 207–231. Academic Press, San Diego.

Letovsky, S. I., Whitehead, S. H., Paik, C. H., Miller, G. A., Gerber, J., Herskovits, E. H., Fulton, T. K., and Bryan, R. N. (1998). A brain image database for structure/function analysis. *Am. J. Neuroradiol.* **19,** 1869–1877.

Liseikin, V. D. (1991). On a variational method for generating adaptive grids on N-dimensional surfaces. *Dokl. Akad. Nauk. CCCP* **319,** 546–549.

Loehlin, J. C. (1989). Partitioning environmental and genetic contributions to behavioral development. *Am. Psychol.* **44,** 1285.

Lohmann, G., and von Cramon, D. Y. (2000). Automatic labelling of the human cortical surface using sulcal basins. *Med. Image Anal.* **4,** 179–188.

Lohmann, G., von Cramon, D. Y., and Steinmetz, H. (1999). Sulcal variability of twins. *Cereb. Cortex* **9,** 754–763.

Lorensen, W. E., and Cline, H. E. (1987). Marching cubes: A high resolution 3D surface construction algorithm. *Comp. Graph.* **21,** 163–169.

MacDonald, D. (1998). "A Method for Identifying Geometrically Simple Surfaces from Three Dimensional Images." McGill Univ., Montreal. [Ph.D. thesis]

MacDonald, D., Kabani, N., Avis, D., and Evans, A. C. (2000). Automated 3-D extraction of inner and outer surfaces of cerebral cortex from MRI. *NeuroImage* **12,** 340–356.

Machado, A., and Gee, J. (1998). "Atlas Warping for Brain Morphometry." *In* "Proceedings of SPIE Medical Imaging 1998: Image Processing, San Diego and Bellingham, WA.

Mai, J., Assheuer, J., and Paxinos, G. (1997). "Atlas of the Human Brain." Academic Press, San Diego.

Makeig, S., Bell, A. J., Jung, T. P., and Sejnowski, T. J. (1996). Independent component analysis of electroencephalographic data. *In* "Advances in Neural Information Processing Systems" (D. Touretzky, M. Mozer, and M. Hasselmo, eds.), Vol. 8, pp. 145–151.

Manceaux-Demiau, A., Bryan, R. N., and Davatzikos, C. (1998). A probabilistic ribbon model for shape analysis of the cerebral sulci: Application to the central sulcus. *J. Comput. Assisted Tomogr.,* in press.

Mangin, J.-F., Frouin, V., Bloch, I., Regis, J., and Lopez-Krahe, J. (1994). Automatic construction of an attributed relational graph representing the cortex topography using homotopic transformations. *SPIE* **2299,** 110–121.

Martin, J. (1995). "Characterization of Neuropathological Shape Deformations." Radiological Sciences Program, MIT Department of Nuclear Engineering, Cambridge, MA. [Ph.D. thesis]

Mazziotta, J. C., Toga, A. W., Evans, A. C., Fox, P., and Lancaster, J. (1995). A probabilistic atlas of the human brain: Theory and rationale for its development. *NeuroImage* **2,** 89–101.

Mazziotta, J. C., Toga, A. W., Evans, A. C., Fox, P. T., Lancaster, J., Zilles, K., Woods, R. P., Paus, T., Simpson, G., Pike, B., Holmes, C. J., Collins, D. L., Thompson, P. M., MacDonald, D., Schormann, T., Amunts, K., Palomero-Gallagher, N., Parsons, L., Narr, K. L., Kabani, N., Le Goualher, G., Boomsma, D., Cannon, T., Kawashima, R., and Mazoyer, B. (2001). A probabilistic atlas and reference system for the human brain. *J. R. Soc.,* in press.

McClearn, G. E., Johansson, B., Berg, S., Pedersen, N. L., Ahern, F., Petrill, S. A., and Plomin, R. (1997). Substantial genetic influence on cognitive abilities in twins 80 or more years old. *Science* **276,** 1560–1563.

McKeown, M. J., Jung, T. P., Makeig, S., Brown, G. G., Kindermann, S. S., and Sejnowski, T. J. (1998a). Spatially independent activity patterns in functional magnetic resonance imaging data during the Stroop color-naming task. *Proc. Natl. Acad. Sci. USA* **95,** 803–810.

McKeown, M. J., Makeig, S., Brown, G. G., Jung, T.-P., Kindermann, S. S., Bell, A. J., and Sejnowski, T. J. (1998b). Analysis of fMRI data by blind separation into independent components. *Hum. Brain Mapp.* **6,** 1–31.

Mega, M. S., Chen, S., Thompson, P. M., Woods, R. P., Karaca, T. J., Tiwari, A., Vinters, H., Small, G. W., and Toga, A. W. (1997). Mapping pathology to metabolism: Coregistration of stained whole brain sections to PET in Alzheimer's disease. *NeuroImage* **5,** 147–153.

Mega, M. S., Chu, T., Mazziotta, J. C., Trivedi, K. H., Thompson, P. M., Shah, A., Cole, G., Frautschy, S. A., and Toga, A. W. (1999a). Mapping biochemistry to metabolism: FDG-PET and beta-amyloid burden in Alzheimer's disease. *NeuroReport* **10,** 2911–2917.

Mega, M. S., Fennema-Notestine, C., Dinov, I. D., Thompson, P. M., Archibald, S. L., Lindshield, C. J., Felix, J., Toga, A. W., and Jernigan, T. L. (2000). Construction, testing, and validation of a sub-volume probabilistic human brain atlas for the elderly and demented populations. Presented at the 6th International Conference on Functional Mapping of the Human Brain, San Antonio, TX, June 2000.

Mega, M. S., Thompson, P. M., Cummings, J. L., Back, C. L., Xu, L. Q., Zohoori, S., Goldkorn, A., Moussai, J., Fairbanks, L., Small, G. W., and Toga, A. W. (1998). Sulcal variability in the Alzheimer's brain: Correlations with cognition. *Neurology* **50,** 145–151.

Mega, M. S., Thompson, P. M., Toga, A. W., and Cummings, J. L. (1999b). Brain mapping in dementia. *In* "Brain Mapping: The Disorders" (A. W. Toga and J. C. Mazziotta, eds.). Academic Press, San Diego.

Megalooikonomou, V. M., Ford, J., Shen, L., Makedon, F., and Saykin, A. (2000). Data mining in brain imaging. *Stat. Methods Med. Res.* **9,** 359–394.

Miller, M. I., Christensen, G. E., Amit, Y., and Grenander, U. (1993). Mathematical textbook of deformable neuroanatomies. *Proc. Natl. Acad. Sci. USA* **90,** 11944–11948.

Miller, M. I., and Younes, L. (2001). Group actions, homeomorphisms, and matching: A general framework. Technical report available at http://www.cmla.ens-cachan.fr/Utilisateurs/younes/.

Miller, M. I., Younes, L., and Trouve, A. (2002). On the metrics and Euler–Lagrange equations of computational anatomy. In press.

Minoshima, S., Koeppe, R. A., Frey, K. A., Ishihara, M., and Kuhl, D. E. (1994). Stereotactic PET atlas of the human brain: Aid for visual interpretation of functional brain images. *J. Nucl. Med.* **35,** 949–954.

Molloy, E., *et al.* (2001). The relationship between brain morphometry and cognitive abilities in healthy pediatric monozygotic twins. Presented at the 7th Annual Meeting of the Organization for Human Brain Mapping (OHBM), Brighton, England, June 10–14, 2001; available at http://www.academicpress.com/www/journal/hbm2001/10228.html.

Mori, S., Itoh, R., Zhang, J., Kaufmann, W. E., van Zijl, P. C., Solaiyappan, M., and Yarowsky, P. (2001). Diffusion tensor imaging of the developing mouse brain. *Magn. Reson. Med.* **46,** 18–23.

Narr, K. L., Thompson, P. M., Sharma, T., Moussai, J., Blanton, R., Anvar, B., Edris, A., Krupp, R., Rayman, J., Khaledy, M., and Toga, A. W. (2001a). Three-dimensional mapping of temporo-limbic regions and the lateral ventricles in schizophrenia: Gender effects. *Biol. Psychiatry* **50,** 84–97.

Narr, K. L., Thompson, P. M., Sharma, T., Moussai, J., Cannestra, A. F., and Toga, A. W. (2000). Mapping corpus callosum morphology in schizophrenia. *Cereb. Cortex* **10,** 40–49.

Narr, K. L., Thompson, P. M., Sharma, T., Moussai, J., Zoumalan, C. I., Rayman, J., and Toga, A. W. (2001b). 3D mapping of gyral shape and cortical surface asymmetries in schizophrenia: Gender effects. *Am. J. Psychiatry* **158,** 244–255.

Narr, K. L., van Erp, T., Cannon, T. D., Woods, R. P., Thompson, P. M., Jang, S., Poutanen, V.-P., Huttunen, M., Lönnqvist, J., Standertskjöld-Nordenstam, C.-G., Mazziotta, J. C., and Toga, A. W. (2001c). A twin study of genetic contributions to hippocampal morphology in schizophrenia. Presented at the 31st International Meeting of the Society for Neuroscience, San Diego, CA, November 10–15, 2001.

Neale, M. C., and Cardon, L. R. (1992). "Methodology for Genetic Studies of Twins and Families." Kluwer Academic, Boston.

Nielsen, M., Florack, L., and Deriche, R. (1994). Regularization and scale space. INRIA Technical Rep. RR-2352, Sept. 1994.

Niemann, K., and van Nieuwenhofen, I. (1999). One atlas—Three anatomies: Relationships of the Schaltenbrand and Wahren microscopic data. *Acta Neurochir. (Vienna)* **141,** 1025–1038.

Noga, J. T. (1999). Schizophrenia as a brain disease: What have we learned from neuroimaging studies of twins? *Medscape Mental Health* **4.**

Nowinski, W. L., Yang, G. L., and Yeo, T. T. (2000). Computer-aided stereotactic functional neurosurgery enhanced by the use of the multiple brain atlas database. *IEEE Trans. Med. Imaging* **19,** 62–69.

Ono, M., Kubik, S., and Abernathey, C. D. (1990). "Atlas of the Cerebral Sulci." Thieme, Stuttgart.

Oppenheim, J. S., Skerry, J. E., Tramo, M. J., and Gazzaniga, M. S. (1989). Magnetic resonance imaging morphology of the corpus callosum in monozygotic twins. *Ann. Neurol.* **26,** 100–104.

Oprea, J. (1998). "The Mathematics of Soap Films: Explorations with Maple." Student Mathematical Library, Vol. 10.

Palovcik, R. A., Reid, S. A., Principe, J. C., and Albuquerque, A. (1992). 3D computer animation of electrophysiological responses. *J. Neurosci. Methods* **41,** 1–9.

Paus, T., Tomaioulo, F., Otaky, N., MacDonald, D., Petrides, M., Atlas, J., Morris, R., and Evans, A. C. (1996). Human cingulate and paracingulate sulci: Pattern, variability, asymmetry and probabilistic map. *Cereb. Cortex* **6,** 207–214.

Paus, T., Zijdenbos, A., Worsley, K., Collins, D. L., Blumenthal, J., Giedd, J. N., Rapoport, J. L., and Evans, A. C. (1999). Structural maturation of neural pathways in children and adolescents: in vivo study. *Science* **283,** 1908–1911.

Pettey, D. J., and Gee, J. C. (2001). Using a linear diagnostic function and non-rigid registration to search for morphological differences between populations: An example involving the male and female corpus callosum. *In* "Information Processing in Medical Imaging," 2001, pp. 372–379.

Pfefferbaum, A., Sullivan, E. V., Swan, G. E., and Carmelli, D. (2000). Brain structure in men remains highly heritable in the seventh and eighth decades of life. *Neurobiol. Aging* **21,** 63–74.

Pinheiro, J. C., and Bates, D. M. (2000). "Mixed-Effects Models in S and S-PLUS." Statistics and Computing Series. Springer-Verlag, New York.

Pinkall, U., and Polthier, K. (1993). Computing discrete minimal surfaces and their conjugates. *Exp. Math.* **2,** 15–36.

Pitiot, A., Thompson, P. M., and Toga, A. W. (2002). Spatially and temporally adaptive elastic template matching. *IEEE Trans. Med. Imaging*.

Plomin, R., and Kosslyn, S. M. (2001). Genes, brain and cognition. *Nat. Neurosci.* **4,** 1153–1154.

Plomin, R., and Petrill, S. A. (1997). Genetics and intelligence: What's new? *Intelligence* **24,** 53.

Poline, J. B., Worsley, K. J., Holmes, A. P., Frackowiak, R. S., and Friston, K. J. (1995). Estimating smoothness in statistical parametric maps: Variability of p values. *J. Comput. Assisted Tomogr.* **19,** 788–796.

Polyakov, A. M. (1987). "Gauge Fields and Strings." Harwood, New York.

Pommert, A., Riemer, M., Schiemann, T., Schubert, R., Tiede, U., and Hohne, K. H. (1996). Three-dimensional imaging in medicine: Methods and applications. *In* "Computer-Integrated Surgery: Technology and Clinical Applications" (R. H. Taylor, S. Lavallee, G. C. Burdea, and R. Moesges, eds.). MIT Press, Cambridge, MA.

Poxton, D., Graham, J., and Deakin, J. F. W. (1998). Detecting asymmetries in hippocampal shape and receptor distribution using statistical appearance models and linear discriminant analysis. Presented at Medical Image Understanding and Analysis, July 6–7, 1998, Leeds Univ., England.

Rademacher, J., Burgel, U., Geyer, S., Schormann, T., Schleicher, A., Freund, H. J., and Zilles, K. (2001). Variability and asymmetry in the human precentral motor system: A cytoarchitectonic and myeloarchitectonic brain mapping study. *Brain* **124,** 2232–2258.

Rademacher, J., Caviness, V. S., Jr., Steinmetz, H., and Galaburda, A. M. (1993). Topographical variation of the human primary cortices: Implications for neuroimaging, brain mapping and neurobiology. *Cereb. Cortex* **3,** 313–329.

Rapoport, J. L., Giedd, J., Kumra, S., Jacobsen, L., Smith, A., Lee, P., Nelson, J., and Hamburger, S. (1997). Childhood-onset schizophrenia: Progressive ventricular change during adolescence. *Arch. Gen. Psychiatry* **54,** 897–903.

Rapoport, J. L., Giedd, J. N., Blumenthal, J., Hamburger, S., Jeffries, N., Fernandez, T., Nicolson, R., Bedwell, J., Lenane, M., Zijdenbos, A., Paus, T., and Evans, A. (1999). Progressive cortical change during adolescence in childhood-onset schizophrenia: A longitudinal magnetic resonance imaging study. *Arch. Gen. Psychiatry* **56,** 649–654.

Régis, J. (1994). "Anatomie Sulcale Profonde et Cartographie Fonctionnelle du Cortex Cérébral." Univ. of Marseille. [Ph.D. thesis, in French]

Reiman, E. M., Caselli, R. J., Chen, K., Alexander, G. E., Bandy, D., and Frost, J. (2001). Declining brain activity in cognitively normal apolipoprotein E epsilon 4 heterozygotes: A foundation for using positron emission tomography to efficiently test treatments to prevent Alzheimer's disease. *Proc. Natl. Acad. Sci. USA* **98,** 3334–3339.

Rettmann, M. E., Han, X., and Prince, J. L. (2000). Watersheds on the cortical surface for automated sulcal segmentation. *In* "IEEE Workshop on Mathematical Methods in Biomedical Image Analysis," pp. 20–27.

Rex, D. E., Pouratian, N., Thompson, P. M., Cunanan, C. C., Sicotte, N. L., Collins, R. C., and Toga, A. W. (2000). Cortical surface warping applied to group analysis of fMRI of tongue movement in the left hemisphere. *Proc. Soc. Neurosci.*.

Rey, D., Subsol, G., Delingette, H., and Ayache, N. (1999). Automatic detection and segmentation of evolving processes in 3D medical images: Application to multiple sclerosis. *In* "Lecture Notes in Computer Science, Proceedings of the Information Processing in Medical Imaging '99 Annual Meeting," Visegrad, Hungary. Springer-Verlag, Berlin.

Roland, P. E., and Zilles, K. (1996). The developing European computerized human brain database for all imaging modalities. *NeuroImage* **4,** S39–47.

Rorden, C., and Brett, M. (2001). Stereotaxic display of brain lesions. *Behav. Neurol.* **12,** 191–200.

Roses, A. D. (1997). A model for susceptibility polymorphisms for complex diseases: Apolipoprotein E and Alzheimer disease. *Neurogenetics* **1,** 3–11. [Review]

Royackkers, N., Desvignes, M., Fawal, H., and Revenu, M. (1999). Detection and statistical analysis of human cortical sulci. *NeuroImage* **10,** 625–641.

Sahr, K., and White, D. (1998). Discrete global grid systems. Proceedings of the 30th Symposium on the Interface. *Comput. Sci. Stat.* **30,** 269–278.

Sapiro, G. (2001). "Geometric Partial Differential Equations and Image Processing." Cambridge Univ. Press, Cambridge, UK.

Schaltenbrand, G., and Wahren, W. (1977). "Atlas for Stereotaxy of the Human Brain," 2nd edition. Thieme, Stuttgart.

Schmahmann, J. D., Doyon, J., McDonald, D., Holmes, C., Lavoie, K., Hurwitz, A. S., Kabani, N., Toga, A., Evans, A., and Petrides, M. (1999). 3D MRI atlas of the human cerebellum in proportional stereotaxic space. *NeuroImage* **10,** 233–260.

Sclaroff, S., and Pentland, A. (1994). On modal modeling for medical data: Underconstrained shape description and data compression. *In* "Proceedings of the IEEE Workshop on Biomedical Image Analysis, Seattle, June 1994.

Sebastian, T. B., Klein, P. N., Kimia, B. B., and Crisco, J. J. (2000). Constructing 2D curve atlases. *In* Proceedings of the IEEE Workshop on Mathematical Methods in Biomedical Image Analysis (MMBIA '00).

Shannon, C. E. (1948). A mathematical theory of communication. *Bell Syst. Technical J.* **27,** 379–423, 623–656.

Shattuck, D., and Leahy, R. M. (1999). Topological refinement of volumetric data. *Proc. SPIE Med. Imaging* **3661,** 204–213.

Shattuck, D. W., and Leahy, R. M. (2001). Automated graph-based analysis and correction of cortical volume topology. *IEEE Trans. Med. Imaging* **20,** 1167–1177.

Small, G. W., Ercoli, L. M., Silverman, D. H., Huang, S. C., Komo, S., *et al.* (2000). Cerebral metabolic and cognitive decline in persons at genetic risk for Alzheimer's disease. *Proc. Natl. Acad. Sci. USA* **97,** 6037–6042.

Smith, G. E. (1907). A new topographical survey of the human cerebral cortex, being an account of the distribution of the anatomically distinct cortical areas and their relationship to the cerebral sulci. *J. Anat.* **41,** 237–254.

Smith, S. M., De Stefano, N., Jenkinson, M., and Matthews, P. M. (2001). Normalized accurate measurement of longitudinal brain change. *J. Comput. Assisted Tomogr.* **25,** 466–475.

Sochen, N., Kimmel, R., and Malladi, R. (1998). A general framework for low level vision. *IEEE Trans. Image Process.* **7,** 310–318.

Sowell, E. R., Thompson, P. M., Holmes, C. J., Jernigan, T. L., and Toga, A. W. (1999). Progression of structural changes in the human brain during the first three decades of life: In vivo evidence for post-adolescent frontal and striatal maturation. *Nat. Neurosci.* **2,** 859–861.

Sowell, E. R., Thompson, P. M., Mega, M. S., Zoumalan, C. I., Lindshield, C., Rex, D. E., and Toga, A. W. (2000). Gyral pattern delineation in 3D: Surface curve protocol. Available via the Internet: http://www.loni.ucla.edu/~esowell/new_sulcvar.html.

Sowell, E. R., Thompson, P. M., Tessner, K. D., and Toga, A. W. (2001). Accelerated brain growth and cortical gray matter thinning are inversely related during post-adolescent frontal lobe maturation. *J. Neurosci.* **21,** 8819–8829.

Spitzer, V., Ackerman, M. J., Scherzinger, A. L., and Whitlock, D. (1996). The visible human male: A technical report. *J. Am. Med. Inf. Assoc.* **3,** 118–130. [http://www.nlm.nih.gov/extramural_research.dir/visible_human.html]

Staib, L. H., and Duncan, J. S. (1992). Boundary finding with parametrically deformable models. *IEEE Trans. PAMI* **14,** 1061–1075.

Steinmetz, H., Furst, G., and Freund, H.-J. (1990). Variation of perisylvian and calcarine anatomic landmarks within stereotaxic proportional coordinates. *Am. J. Neuroradiol.* **11,** 1123–1130.

Strauss, E., Kosaka, B., and Wada, J. (1983). The neurobiological basis of lateralized cerebral function: A review. *Hum. Neurobiol.* **2,** 115–127.

Studholme, C., Cardenas, V., Schuff, N., Rosen, H., Miller, B., and Weiner, M. W. (2001). Detecting spatially consistent structural differences in Alzheimer's and fronto temporal dementia using deformation morphometry. *MICCAI* **2001,** 41–48.

Styner, M., and Gerig, G. (2001). Medial models incorporating object variability for 3D shape analysis. *In* "Proceedings of the Information Processing in Medical Imaging Annual Meeting," Davis, CA, pp. 502–516.

Subsol, G., Roberts, N., Doran, M., Thirion, J. P., and Whitehouse, G. H. (1997). Automatic analysis of cerebral atrophy. *Magn. Reson. Imaging* **15,** 917–927.

Sundsten, J. W., Kastella, K. G., and Conley, D. M. (1991). Videodisc animation of 3D computer reconstructions of the human brain. *J. Biocommun.* **18,** 45–49.

Swan, G. E., Carmelli, D., Reed, T., Harshfield, G. A., Fabsitz, R. R., and Eslinger, P. J. (1990). Heritability of cognitive performance in aging twins: The National Heart, Lung, and Blood Institute Twin Study. *Arch. Neurol.* **47,** 259–262.

Swanson, L. W. (1992). "Brain Maps: Structure of the Rat Brain." Elsevier, Amsterdam.

Talairach, J., and Tournoux, P. (1988). "Co-planar Stereotaxic Atlas of the Human Brain." Thieme, New York.

Tao, X., Han, X., Rettmann, M. E., Prince, J. L., and Davatzikos, C. (2001). Statistical study on cortical sulci of human brains. *In* "Proceedings of the 17th Information Processing in Medical Imaging Annual Meeting," June 2001.

Taylor, J. E., and Adler, R. J. (2000). Euler characteristics for Gaussian fields on manifolds. *Ann. Probability,* in press.

Thirion, J.-P., and Calmon, G. (1997). Deformation analysis to detect and quantify active lesions in 3D medical image sequences. INRIA Technical Rep. 3101, Feb. 1997.

Thirion, J.-P., Prima, S., and Subsol, S. (1998). Statistical analysis of dissymmetry in volumetric medical images. *Med. Image Anal.,* in press.

Thompson, P. M., Cannon, T. D., Narr, K. L., van Erp, T., Khaledy, M., Poutanen, V.-P., Huttunen, M., Lönnqvist, J., Standertskjöld-Nordenstam, C.-G., Kaprio, J., Dail, R., Zoumalan, C. I., and Toga, A. W. (2001a). Genetic influences on brain structure. *Nat. Neurosci.* **4,** 1253–1258.

Thompson, P. M., de Zubicaray, G., Janke, A. L., Rose, S. E., Dittmer, S., Semple, J., Gravano, D., Han, S., Herman, D., Hong, M. S., Mega, M. S., Cummings, J. L., Doddrell, D. M., and Toga, A. W. (2001b). Detecting dynamic (4D) profiles of degenerative rates in Alzheimer's disease patients, using high-resolution tensor mapping and a brain atlas encoding atrophic rates in a population. *In* "7th Annual Meeting of the Organization for Human Brain Mapping," Brighton, June 2001.

Thompson, P. M., Giedd, J. N., Woods, R. P., MacDonald, D., Evans, A. C., and Toga, A. W. (2000a). Growth patterns in the developing brain detected by using continuum-mechanical tensor maps. *Nature* **404,,** 190–193.

Thompson, P. M., MacDonald, D., Mega, M. S., Holmes, C. J., Evans, A. C., and Toga, A. W. (1997). Detection and mapping of abnormal brain structure with a probabilistic atlas of cortical surfaces. *J. Comput. Assisted Tomogr.* **21,** 567–581.

Thompson, P. M., Mega, M. S., Narr, K. L., Sowell, E. R., Blanton, R. E., and Toga, A. W. (2000b). Brain image analysis and atlas construction. *In* "SPIE Handbook on Medical Image Analysis" (M. Fitzpatrick, ed.). SPIE—Int. Soc. Opt. Eng., Bellingham, WA.

Thompson, P. M., Mega, M. S., and Toga, A. W. (2000c). Disease-specific brain atlases. *In* "Brain Mapping: The Disorders" (A. W. Toga and J. C. Mazziotta, eds.). Academic Press, San Diego.

Thompson, P. M., Mega, M. S., Vidal, C., Rapoport, J. L., and Toga, A. W. (2001c). Detecting disease-specific patterns of brain structure using cortical pattern matching and a population-based probabilistic brain atlas. *In* "Lecture Notes in Computer Science," Vol. 2082, "IEEE Conference on Information Processing in Medical Imaging," Davis, CA (M. Insana and R. Leahy, eds.), pp. 488–501. Springer-Verlag, Berlin.

Thompson, P. M., Mega, M. S., Woods, R. P., Blanton, R. E., Moussai, J., Zoumalan, C. I., Aron, J., Cummings, J. L., and Toga, A. W. (2001d). Early cortical change in Alzheimer's disease detected with a disease-specific population-based brain atlas. *Cereb. Cortex* **11,** 1–16.

Thompson, P. M., Moussai, J., Khan, A. A., Zohoori, S., Goldkorn, A., Mega, M. S., Small, G. W., Cummings, J. L., and Toga, A. W. (1998). Cortical variability and asymmetry in normal aging and Alzheimer's disease. *Cereb. Cortex* **8,** 492–509.

Thompson, P. M., Narr, K. L., Blanton, R. E., and Toga, A. W. (2002). Mapping structural alterations of the corpus callosum during brain development and degeneration. *In* "The Corpus Callosum" (M. Iacoboni and E. Zaidel, eds.). Kluwer Academic, Dordrecht/Norwell, MA, in press.

Thompson, P. M., Schwartz, C., Lin, R. T., Khan, A. A., and Toga, A. W. (1996a). 3D statistical analysis of sulcal variability in the human brain. *J. Neurosci.* **16,** 4261–4274.

Thompson, P. M., Schwartz, C., and Toga, A. W. (1996b). High-resolution random mesh algorithms for creating a probabilistic 3D surface atlas of the human brain. *NeuroImage* **3,** 19–34.

Thompson, P. M., and Toga, A. W. (1996). A surface-based technique for warping 3-dimensional images of the brain. *IEEE Trans. Med. Imaging* **15,** 1–16.

Thompson, P. M., and Toga, A. W. (1997). Detection, visualization and animation of abnormal anatomic structure with a deformable probabilistic brain atlas based on random vector field transformations. *Med. Image Anal.* **1,** 271–294.

Thompson, P. M., and Toga, A. W. (2000). Elastic image registration and pathology detection. *In* "Handbook of Medical Image Processing" (I. Bankman, R. Rangayyan, A. C. Evans, R. P. Woods, E. Fishman, and H. K. Huang, eds.). Academic Press, San Diego.

Thompson, P. M., and Toga, A. W. (2002). A framework for computational anatomy. In press.

Thompson, P. M., Vidal, C. N., Giedd, J. N., Gochman, P., Blumenthal, J., Nicolson, R., Toga, A. W., and Rapoport, J. L. (2001e). Mapping adolescent brain change reveals dynamic wave of accelerated gray matter loss in very early-onset schizophrenia. *Proc. Natl. Acad. Sci. USA* **98,** 11650–11655.

Thompson, P. M., Woods, R. P., Mega, M. S., and Toga, A. W. (2000e). Mathematical/computational challenges in creating population-based brain atlases. *Hum. Brain Mapp.* **9,** 81–92.

Tiede, U., Bomans, M., Hohne, K. H., Pommert, A., Riemer, M., Schiemann, T., Schubert, R., and Lierse, W. (1993). A computerized three-dimensional atlas of the human skull and brain. *Am. J. Neuroradiol.* **14,** 551–559. [Discussion pp. 560–561]

Toga, A. W. (1998). "Brain Warping." Academic Press, San Diego.

Toga, A. W., and Mazziotta, J. C. (1996). "Brain Mapping: The Methods," 1st edition. Academic Press, San Diego.

Toga, A. W., Rex, D. E., and Ma, J. (2001a). A graphical interoperable processing pipeline. *In* "International Conference on Human Brain Mapping," 2001, Brighton, Abstract 266.

Toga, A. W., and Thompson, P. M. (1998a). An introduction to brain warping. *In* "Brain Warping" (A. W. Toga, ed.). Academic Press, San Diego.

Toga, A. W., and Thompson, P. M. (1998b). Multimodal brain atlases. *In* "Medical Image Databases" (S. T. C. Wong, ed.), pp. 53–88. Kluwer Academic, Dordrecht/Norwell, MA.

Toga, A. W., and Thompson, P. M. (1999). An introduction to maps and atlases of the brain. *In* "Brain Mapping: The Applications" (A. W. Toga and J. C. Mazziotta, eds.). Academic Press, San Diego.

Toga, A. W., and Thompson, P. M. (2000). Brain atlases and image registration. *In* "Handbook of Medical Image Processing" (I. Bankman, R. Rangayyan, A. C. Evans, R. P. Woods, E. Fishman, and H. K. Huang, eds.). Academic Press, San Diego.

Toga, A. W., and Thompson, P. M. (2001a). Maps of the brain. *Anat. Rec.* **265,** 37–53. [Review]

Toga, A. W., and Thompson, P. M. (2001b). The role of image registration in brain mapping. *Image Vision Comput. J.* **19,** 3–24.

Toga, A. W., and Thompson, P. M. (2002). New approaches in brain morphometry. *J. Gerontol.,* in press.

Toga, A. W., Thompson, P. M., Mega, M. S., Narr, K. L., and Blanton, R. E. (2001b). Probabilistic approaches for atlasing normal and disease-specific brain variability. *Anat. Embryol. (Berlin)* **204,** 267–282.

Turner, R., Le Bihan, D., and Chesnick, A. S. (1991). Echo-planar imaging of diffusion and perfusion. *Magn. Reson. Med.* **19,** 247–253. [Review]

Vaillant, M., and Davatzikos, C. (1997). Finding parametric representations of the cortical sulci using an active contour model. *Med. Image Anal.* **1,** 295–315.

Van Essen, D. C., Drury, H. A., Joshi, S. C., and Miller, M. I. (1997). Comparisons between human and macaque using shape-based deformation algorithms applied to cortical flat maps. 3rd International Conference on Functional Mapping of the Human Brain, Copenhagen, May 19–23, 1997. *NeuroImage* **5,** S41.

Van Essen, D. C., Lewis, J. W., Drury, H. A., Hadjikhani, N., Tootell, R. B., Bakircioglu, M., and Miller, M. I. (2001). Mapping visual cortex in monkeys and humans using surface-based atlases. *Vision Res.* **41,** 1359–1378.

Van Essen, D. C., and Maunsell, J. H. R. (1983). Hierarchical organization and functional streams in the visual cortex. *Trends Neurol. Sci.* **6,** 370–375.

Van Horn, J. D., Grethe, J. S., Kostelec, P., Woodward, J. B., Aslam, J. A., Rus, D., Rockmore, D., and Gazzaniga, M. S. (2001). The Functional Magnetic Resonance Imaging Data Center (fMRIDC): The challenges and rewards of large-scale databasing of neuroimaging studies. *Philos. Trans. R. Soc. London B Biol. Sci.* **356,** 1323–1339.

Vapnik, V. N. (1995). "The Nature of Statistical Learning Theory." Springer-Verlag, New York.

Verbeke, G., and Molenberghs, G. (2000). "Linear Mixed Models for Longitudinal Data." Springer-Verlag, New York.

Viola, P. A., and Wells, W. M. (1995). Alignment by maximization of mutual information *In* "5th IEEE International Conference on Computer Vision," Cambridge, MA, pp. 16–23.

Wang, L., Joshi, S. C., Miller, M. I., and Csernansky, J. G. (2001). Statistical analysis of hippocampal asymmetry in schizophrenia. *NeuroImage* **14,** 531–545.

Wang, Y., and Staib, L. (1998). Boundary finding with correspondence using statistical shape models. *In* "Proceedings of the Conference on Computer Vision and Pattern Recognition," Santa Barbara, CA, pp. 338–345.

Warfield, S., Robatino, A., Dengler, J., Jolesz, F., and Kikinis, R. (1998). Nonlinear registration and template driven segmentation. *In* "Brain Warping" (A. W. Toga, ed.), Chap. 4, pp. 67–84. Academic Press, San Diego.

Watkins, K. E., Paus, T., Lerch, J. P., Zijdenbos, A., Collins, D. L., Neelin, P., Taylor, J., Worsley, K. J., and Evans, A. C. (2001). Structural asymmetries in the human brain: A voxel-based statistical analysis of 142 MRI scans. *Cereb. Cortex* **11,** 868–877.

Weinberger, D. R. (1995). Schizophrenia as a neurodevelopmental disorder: A review of the concept. *In* "Schizophrenia" (S. R. Hirsch and D. R. Weinberger, eds.), pp. 294–323. Blackwood Press, London.

Wells, W. M., Viola, P., Atsumi, H., Nakajima, S., and Kikinis, R. (1997). Multi-modal volume registration by maximization of mutual information. *Med. Image Anal.* **1,** 35–51.

Welti, D., Gerig, G., Radü, E. W., Kappos, L., and Székely, G. (2001). Spatio-temporal segmentation of active multiple sclerosis lesions in serial MRI data. *In* "Information Processing in Medical Imaging," 2001, pp. 438–445.

Witelson, S. F. (1989). Hand and sex differences in the isthmus and genu of the human corpus callosum: A postmortem morphological study. *Brain* **112,** 799–835.

Woods, R. P., Cherry, S. R., and Mazziotta, J. C. (1992). Rapid automated algorithm for aligning and reslicing PET images. *J. Comput. Assisted Tomogr.* **16,** 620–633.

Woods, R. P., Dapretto, M., Sicotte, N. L., Toga, A. W., and Mazziotta, J. C. (1999). Creation and use of a Talairach-compatible atlas for accurate, automated, nonlinear intersubject registration, and analysis of functional imaging data. *Hum. Brain Mapp.* **8,** 73–79.

Woods, R. P., Grafton, S. T., Watson, J. D. G., Sicotte, N. L., and Mazziotta, J. C. (1998a). Automated image registration. II. Intersubject validation of linear and nonlinear models. *J. Comput. Assisted Tomogr.,* in press.

Woods, R. P., Mazziotta, J. C., and Cherry, S. R. (1993). MRI–PET registration with automated algorithm. *J. Comput. Assisted Tomogr.* **17,** 536–546.

Woods, R. P., Mega, M. S., and Thompson, P. M. (1998b). Use of automated polynomial warping to create an MRI atlas specific for the study of Alzheimer's disease. *Ann. Neurol.* **44(3),** 449.

Woods, R. P., Thompson, P. M., Mazziotta, J. C., and Toga, A. W. (2000). A definition of average brain size, orientation and shape. Presented at the 6th International Conference on Functional Mapping of the Human Brain, San Antonio, TX, June 2000.

Worsley, K. J. (1994a). Quadratic tests for local changes in random fields with applications to medical images. Department of Mathematics and Statistics, McGill Univ., Montreal. [Technical Rep. 94-08]

Worsley, K. J. (1994b). Local maxima and the expected Euler characteristic of excursion sets of chi-squared, F and t fields. *Adv. Appl. Probability* **26,** 13–42.

Worsley, K. J., Andermann, M., Koulis, T., MacDonald, D., and Evans, A. C. (1999). Detecting changes in nonisotropic images. *Hum. Brain Mapp.* **8,** 98–101.

Worsley, K. J., Marrett, S., Neelin, P., and Evans, A. C. (1996). Searching scale space for activation in PET images. *Hum. Brain Mapp.* **4,** 74–90.

Wright, I. C., McGuire, P. K., Poline, J. B., Travere, J. M., Murray, R. M., Frith, C. D., Frackowiak, R. S. J., and Friston, K. J. (1995). A voxel-based method for the statistical analysis of gray and white matter density applied to schizophrenia. *NeuroImage* **2,** 244–252.

Xu, C., Pham, D. L., Rettmann, M. E., Yu, D. N., and Prince, J. L. (1999). Reconstruction of the human cerebral cortex from magnetic resonance images. *IEEE Trans. Med. Imaging* **18,** 467–480.

Zeineh, M. M., Engel, S. A., Thompson, P. M., and Bookheimer, S. (2001). Unfolding the human hippocampus with high-resolution structural and functional MRI. *New Anat. (Anat. Rec.)* **265,** 111–120.

Zhang, D., and Hebert, M. (1999). Harmonic shape images: A representation for 3D free-form surfaces. *In* "Proceedings of the 2nd International Workshop on Energy Minimization Methods in Computer Vision and Pattern Recognition (EMMCVPR '99)," York, England, July, 1999.

Zhou, Y., Thompson, P. M., and Toga, A. W. (1999). Automatic extraction and parametric representations of cortical sulci. *Comput. Graph. Appl.* **19,** 49–55.

Zijdenbos, A., Evans, A., Riahi, F., Sled, J., Chui, H.-C., and Kollokian, V. (1996). Automatic quantification of multiple sclerosis lesion volume using stereotactic space. *In* "Proceedings of the 4th Visualization in Biomedical Computing," Hamburg, Germany.

VII

Emerging Concepts

29

Radionuclide Imaging of Reporter Gene Expression

Gobalakrishnan Sundaresan and Sanjiv S. Gambhir

Crump Institute for Molecular Imaging, Department of Molecular and Medical Pharmacology, School of Medicine, University of California at Los Angeles, Los Angeles, California 90095-1770

I. Overview of Molecular Imaging

Insights into cell and organism function can, in part, be gained from studies of the patterns of gene expression that encode for normal biological processes such as replication, migration, and signal transduction. Many disease conditions occur due to altered patterns of gene expression. Due to advances in the fields of molecular biology and biological imaging, various molecular imaging assays are now available to noninvasively and repetitively study gene expression in living subjects. *In vivo* imaging of gene expression is very useful as one can not only determine the location(s) of cells expressing a particular gene of interest, but also determine the magnitude and time variation of gene expression. The gene of interest may be an endogenous or exogenous gene introduced into the organism/tissue of interest. In the case of imaging exogenous gene expression one can also determine the efficacy of the gene-delivery system that was used to introduce the gene of interest. By the use of appropriate imaging probes, various modern imaging instruments such as single photon emission computed tomography (SPECT), positron emission tomography (PET), magnetic resonance imaging (MRI), and recently developed optical methods can be employed to study molecular events by imaging living subjects.

In this chapter we first review the instrumentation for molecular imaging, followed by background material on reporter genes. We then describe approaches to using reporter gene technology with radionuclide imaging technologies such as PET and SPECT. We characterize in detail the herpes simplex virus type 1 thymidine kinase (HSV1-tk) PET reporter gene and the dopamine-type 2 receptor (D_2R) PET reporter gene, as well as other reporter genes. We detail applications of reporter genes for gene therapy and for indirectly imaging endogenous gene expression. We briefly describe nonradionuclide approaches to reporter gene imaging (also see Chapter 30) and discuss specific issues of relevance to the neuroscience community. Finally we describe issues for human gene therapy and the role imaging is playing and will play in the future in clinical applications.

II. Instrumentation for Molecular Imaging

Molecular imaging requires high-sensitivity instrumentation to detect relatively low levels of reporter probes that are designed to accumulate in cells expressing a specific gene. Although no imaging technology allows for quantitative imaging of individual cells at any depth within living subjects, including small animals, large groups of cells (10^5–10^6) contained within a resolution element can be imaged with various imaging technologies (Cherry and Gambhir, 2001). Mice are particularly well suited to various imaging applications because of their rapid breeding time, relatively low cost, and ease of genetic modification to make transgenic knock-ins or knockouts. Of the various imaging technologies available, the radionuclide-based approaches of PET and SPECT offer unique advantages for molecular imaging assay development. PET in particular is highly suitable because of its high sensitivity (~10^{-11} M), quantitative features, and tomographic capability.

PET is an analytical imaging tool with which compounds labeled with positron-emitting radioisotopes are used to measure biological processes with only trace amounts of the radiolabeled probes. Molecules labeled with positron-emitting radionuclides, injected intravenously in trace quantities, are retained in tissues as a result of either binding to a receptor or conversion due to action by an enzyme. Relatively small amounts of PET imaging probes are used so as to avoid any pharmacological response. Levels of PET imaging probes are typically orders of magnitude lower than the concentrations of the corresponding endogenous or pharmacologic agents required to elicit a biological response.

Small-animal PET systems, such as the microPET (Cherry *et al.,* 1997; Chatziioannou *et al.,* 1999), have been developed to specifically image rodents with a spatial resolution of ~1.8 mm^3 (see Chapter 18). These systems are now being further refined to approach spatial resolutions of ~1 mm^3. Mice can typically be volumetrically imaged in 10–40 min, and if only a limited region (e.g., brain) needs to be scanned, even 5 min may be sufficient. Dynamic imaging is also possible to produce time–activity curves of a specific region of interest. Brain imaging of mice is currently without adequate resolution for most applications, but for rats can be sufficient (Kornblum *et al.,* 2000). Other small-animal PET systems as well as small-animal SPECT also exist and are reviewed elsewhere (Chatziioannou, 2002; Cherry and Gambhir, 2001).

III. Reporter Genes

Reporter genes are nucleic acid sequences encoding for easily assayed proteins. They are very efficient tools to monitor the efficacy of gene delivery vehicles and of gene expression. Reporter genes can also "report" on different properties and events such as the strength of promoters, the intracellular fate of a gene product, a result of protein trafficking, and the efficiency of translation initiation signals. The introduced reporter gene driven by a promoter of choice is sometimes referred to as a "transgene." The reporter gene is cloned downstream of a regulatory region (e.g., promoter/enhancer) that is usually responsible for the controlled expression of a specific gene. These regulatory regions can be tissue or event specific in their actions. These regulatory regions drive the transcription (the process of converting DNA to mRNA) of the cloned reporter gene. Thus, by introducing a reporter gene driven by a promoter of choice into target tissue(s), one can *indirectly* monitor expression of the gene whose promoter has been cloned. The promoter can be constitutive, leading to continuous transcription, or inducible, leading to controlled expression. The promoter can also be cell specific, allowing expression of the reporter gene to be restricted to certain cells. Various aspects of gene expression, including promoter/regulatory elements, inducible promoters, and endogenous gene expression (Grandaliano *et al.,* 1995; Ikenaka and Kagawa, 1995), can be studied using the reporter gene. Reporter gene analysis provides an inexpensive, rapid, and sensitive assay that can be used to study gene delivery and gene expression. Use of reporter genes avoids having to develop a specific probe to evaluate the expression of every new gene of interest. For monitoring endogenous genes, the reporter gene approach is indirect and does not in general provide identical information if the endogenous protein is directly assayed. Nevertheless, the use of reporter genes can be an important approach because of their ability to study numerous processes without developing numerous reporter probes.

Among the more commonly used conventional reporter genes are those that encode for the following proteins: (a) chloramphenicol acetyltransferase, which transfers radioactive acetyl groups to chloramphenicol; detection by thin-layer chromatography and autoradiography (Nomura *et al.,* 1997; Zhou *et al.,* 1997); (b) β-galactosidase (GAL), which hydrolyzes colorless galactosides to yield colored products (Lee *et al.,* 1999; Misslitz *et al.,* 2000); (c) β-glucuronidase, which hydrolyzes colorless glucuronides to yield colored products; (d) β-lactamase, which catalyzes hydrolysis of a cephalosporin monitored by a change in fluorescence emission of a substrate (Sauvonnet and Pugsley, 1996; Zlokarnik *et al.,* 1998); (e) firefly luciferase, which oxidizes luciferin, emitting photons (Gnant *et al.,* 1999; Karp and Oker-Blom, 1999), and *Renilla* luciferase, which catalyzes oxidation of coelentrazine, leading to bioluminescence (Lorenz *et al.,* 1996; Inouye and Shimomura, 1997); and (f) green fluorescent protein (GFP), which fluoresces, due to energy transfer, on irradiation (Kunert *et al.,* 2000; Leffel *et al.,* 1997; Seul and Beyer, 2000). Most of these conventional reporter gene methods are easy to use for living cells, but for living animals require tissue sampling or animal sacrifice.

IV. Adapting the Reporter Gene Concept for Radionuclide Imaging

The ideal reporter gene for use with living animals would have the following characteristics: (i) When the reporter gene is expressed, the reporter protein should produce specific reporter probe accumulation only in the cells in which it is expressed. (ii) When the reporter gene is not expressed, there should be no accumulation of the reporter probe in cells. (iii) There should be no immune response to the reporter protein and the reporter probe should not significantly perturb the cell. (iv) The reporter probe should be stable and its magnitude of accumulation should not depend on other endogenous substrates. (v) The reporter probe should rapidly be cleared from blood and other tissues in which the reporter gene is not expressed in order to lead to good imaging contrast at the sites of reporter gene expression. Other desirable characteristics are also important and have been reviewed earlier (Gambhir *et al.,* 2000b). Various approaches are possible to adapt the reporter gene concept for imaging with PET or SPECT. The reporter gene can be chosen so that it encodes an enzyme that is capable of trapping a tracer (a radiolabeled substrate or reporter probe) within the cell (Fig. 1A). A second approach uses a reporter gene that encodes an intracellular and/or extracellular receptor capable of binding a tracer (radioactive ligand) (Fig. 1B). A cell surface receptor can also potentially recycle and leave ligand inside the cell prior to returning to the surface of the cell to bind another ligand. The accumulation of the tracer in both approaches is dependent on the magnitude of the expression of reporter gene. Through the choice of the right tracer and optimization of its pharmacokinetics, it is possible to achieve imaging signal predominantly only in those areas in which the reporter gene is expressed. This reporter gene-imaging paradigm is independent of a particular delivery vector; it can be used with any of the several currently available vectors (e.g., retrovirus, adenovirus, adeno-associated virus, lentivirus, liposomes) as well as in transgenic animal models.

A. Herpes Simplex Virus Type 1 Thymidine Kinase as a PET Reporter Gene

One of the most widely used reporter gene systems for use with radionuclide approaches is the HSV1-tk reporter gene. Thymidine kinases are present in all mammalian cells; they phosphorylate thymidine for incorporation into DNA. Acycloguanosines (e.g., acyclovir, ACV) and guanosines (e.g., ganciclovir, GCV; penciclovir, PCV) are much more effectively phosphorylated by the viral TK (HSV1-TK) than by the mammalian TKs (*Note.* HSV1-tk refers to the gene, HSV1-TK refers to the enzyme). Unlike mammalian thymidine kinases, HSV1-TK has relaxed substrate specificity and is able to phosphorylate acycloguanosine, guanosine, and thymidine derivatives, which get "trapped" in their phosphorylated charged state inside the cell. If phosphorylated acycloguanosines are present in high concentrations, they may lead to cell death, either due to chain termination or by inhibition of DNA polymerase (Boehme, 1984; Furman *et al.,* 1984; Miller and Miller, 1980; Reardon, 1989). If HSV1-tk expression can be restricted to only tumor cells, then HSV1-tk will function as a "suicide" gene, because its expression can be used to trap prodrugs and thereby kill the cells expressing HSV1-tk. When radiolabeled versions of these prodrugs are used in nonpharmacological (trace) doses they can serve as SPECT or PET reporter probes. Thus trace quantities of radiolabeled acycloguanosines/guanosines are used to produce tomographic images after their retention, as a consequence of phosphorylation by HSV1-TK. It must be emphasized that the PET imaging of reporter gene assay uses relatively small amounts of high-specific-activity probes.

Two main categories of substrates have been investigated as reporter probes for imaging HSV1-tk reporter gene expression: derivatives of thymidine/uracil [e.g., 2′-fluoro-2′-deoxy-1-β-D-arabinofuranosyl-5-iodo-uracil (FIAU) labeled with radioactive iodine] and derivatives of guanosine (e.g., PCV radiolabeled with fluorine-18). These two major classes of reporter probes share in common the ability to be phosphorylated by HSV1-TK, leading to their accumulation in cells expressing HSV1-tk (Gambhir *et al.,* 1999a).

Positron labeled 8-[^{18}F]fluoroganciclovir ([^{18}F]FGCV) and 8-[^{18}F]fluoropenciclovir ([^{18}F]FPCV) have been synthesized (Namavari *et al.,* 2000). Differences in substrate accumulation by cells expressing HSV1-tk was studied and it was found that [^{18}F]FPCV was two- to threefold more effective than [^{18}F]FGCV in detecting hepatic HSV1-tk expression in mice tail vein injected with adenovirus carrying the

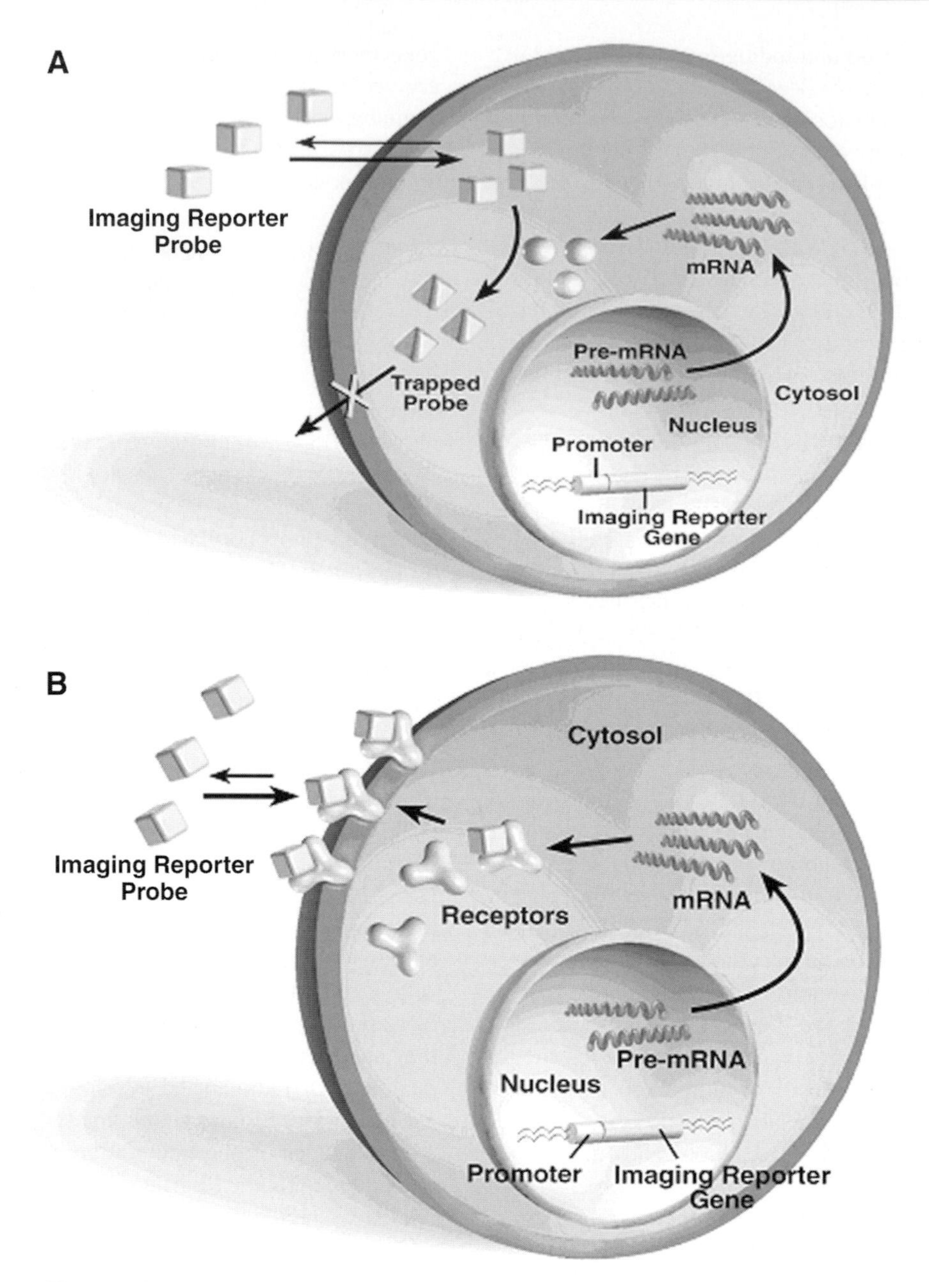

Figure 1 A schematic diagram representing two different approaches for imaging of reporter gene expression using PET/SPECT. A reporter gene introduced into the cell can encode for (**A**) an enzyme (e.g., HSV1-TK) that leads to trapping of a radiolabeled probe or (**B**) an intracellular and/or extracellular receptor (e.g., D_2R), which would lead to trapping of a radiolabeled ligand. In both cases significant trapping of the radiolabeled probe will occur only if the imaging reporter gene is expressed.

HSV1-tk gene and in stably transfected C6 rat glioma xenografts (Iyer *et al.,* 2001a). [^{18}F]FPCV is, therefore, a better substrate for HSV1-TK than is [^{18}F]FGCV.

Alauddin and colleagues (1996) first described the labeling of the side chain of ganciclovir instead of labeling the C8 position. They first synthesized 9-[(3-[^{18}F]fluoro-1-hydroxy-2-propoxy)methyl]guanine ([^{18}F]FHPG) and further tested this *in vivo* as an imaging probe for HSV1-tk (Alauddin *et al.,* 1999). Their experiments in tumor-bearing nude mice demonstrated that the tumor uptake of the radiotracer is three- and sixfold higher at 2 and 5 h, respectively, in transduced cells compared with the control cells. Alauddin and Conti (1998) also synthesized 9-(4-[^{18}F]fluoro-3-hydroxymethylbutyl)guanine ([^{18}F]FHBG), the side-chain-fluorinated analogue of penciclovir, and showed that it may be a better imaging probe for HSV1-tk than [^{18}F]FHPG. We recently showed that [^{18}F]FHBG is a better substrate than [^{18}F]FPCV for imaging HSV1-tk reporter gene expression (Iyer *et al.,* 2000). In summary, [^{18}F]FHBG appears to be the best acycloguanosine derivative for PET imaging studied to date.

The reporter probe FIAU radiolabeled with various forms of radioactive iodine is also useful for imaging HSV1-tk reporter gene expression and has been validated in a number of experimental models (reviewed in Sadelain and Blasberg, 1999; Gambhir *et al.,* 2000b). SPECT imaging of HSV1-tk expression was performed with ^{131}I-labeled FIAU. These early studies found that clinically relevant levels of HSV1-tk gene expression in transfected tissue can be imaged with [^{131}I]FIAU and a gamma camera or SPECT (Tjuvajev *et al.,* 1996). Later, Tjuvajev and coworkers (1998) imaged HSV1-tk reporter gene expression with [^{124}I]FIAU, in rats bearing subcutaneous tumors. The longer half-life of iodine-124 (4.18 days) relative to fluorine-18 (110 min) can be exploited to let background signal from the bladder and gastrointestinal tract clear prior to imaging. A recent report compared [^{124}I]FIAU and [^{18}F]FHPG as substrates for the wild type HSV1-TK enzyme and concluded that [^{124}I]FIAU may be the preferred substrate for imaging the HSV1-tk reporter gene expression (Brust *et al.,* 2001). Comparisons between [^{18}F]FHBG and [^{124}I]FIAU as imaging substrates for HSV1-tk are also in progress (Iyer *et al.,* 2000, Tjuvajev *et al.,* 2000). These initial studies still do not provide consistent evidence as to whether [^{124}I]FIAU or [^{18}F]FHBG is the preferred reporter probe for imaging HSV1-tk reporter gene expression and this issue will require further investigation. Further details of the HSV1-tk reporter gene approach, including a comparison of various tracers, are detailed in review articles (Gambhir *et al.,* 2000b; Ray *et al.,* 2001).

B. Mutant HSV1-tk Reporter Gene

The HSV1-tk gene was initially proposed as a therapeutic gene in suicide gene therapy as an approach for treating cancer. Expression of HSV1-tk can convert pharmacologic levels of prodrugs (e.g., ACV, GCV, PCV) into toxic forms leading to cell death. To improve its therapeutic efficacy, mutant HSV1-TK enzymes were created by site-directed mutagenesis (Black *et al.,* 1996) that are more effective than wild-type HSV1-TK at phosphorylating ACV/GCV and simultaneously less effective than the wild-type at utilizing endogenous thymidine. Further, a replication-deficient adenovirus (AdCMV-HSV1-sr39tk) capable of expressing HSV1-sr39tk was constructed and its effectiveness in utilizing FPCV was compared with that of another virus expressing the wild-type HSV1-tk (AdCMV-HSV1-tk) (Gambhir *et al.,* 2000a). The mutant HSV1-TK enzyme (HSV1-sr39TK) was successfully applied in PET imaging with [^{18}F]FPCV (Gambhir *et al.,* 2000a) and was found to be more effective in utilizing ganciclovir substrates than the wild-type HSV1-TK enzyme.

The development and use of the mutant HSV1-sr39TK illustrate the ability to engineer the reporter protein and reporter probe to be optimized for each other, leading to a marked gain in imaging signal. The sensitivity of the HSV1-sr39TK/[^{18}F]FHBG system is severalfold greater than the sensitivity of the HSV1-TK/[^{18}F]FGCV system (Iyer *et al.,* 2000) and is the current reporter gene/reporter probe of choice in our laboratories. Future approaches in which the HSV1-tk reporter gene can be mutated to produce reporter proteins even further optimized for substrate conversion should produce even more sensitive assays.

C. Dopamine Type 2 Receptor Reporter Gene

A radionuclide reporter gene system that is receptor based uses the D_2R, which binds spiperone [3-(2′[^{18}F]fluoroethyl)spiperone ([^{18}F]FESP)] intra- and extracellularly and thus results in probe accumulation in D_2R-expressing cells/tissue (MacLaren *et al.,* 1998, 1999). The D_2R receptor is a transmembrane protein expressed predominantly in striatum and pituitary and important for mediating the effects of dopamine to control movements. This receptor exerts its response through a G-protein-mediated signaling cascade involving adenyl cyclase as a second messenger. Spiperone, a D_2R antagonist, can be labeled with fluorine-18 to yield radiolabeled [^{18}F]FESP and, thus, has been applied for PET imaging of the D_2R reporter gene (MacLaren *et al.,* 1999). Recently, we have investigated the potential for PET imaging of two mutant D_2R receptors (D_2R80A and D_2R194A) that can uncouple the downstream cAMP-dependent signaling cascade while retaining the property of binding with [^{18}F]FESP (Liang *et al.,* 2001). These two mutants have an advantage of overcoming the undesirable effects of ectopic expression of D_2R receptors on cellular biochemistry by uncoupling signal transduction. The expression of HSV1-sr39tk and D_2R reporter genes can be imaged simultaneously in living mice, with microPET, using two different tracers (e.g., [^{18}F]FPCV and [^{18}F]FESP, respectively) (Fig. 2) (Iyer *et al.,* 2001a). Formal sensitivity comparisons between the HSV1-sr39tk and the D_2R reporter genes have yet to be performed.

D. Other Reporter Gene Systems

Somatostatin receptor subtype II (SSTr2) is another reporter gene that has been extensively studied. There already exist various somatostatin analogues (e.g., radiolabeled octreotide) for imaging cells expressing somatostatin receptor. SSTr2 is normally expressed primarily in the pituitary

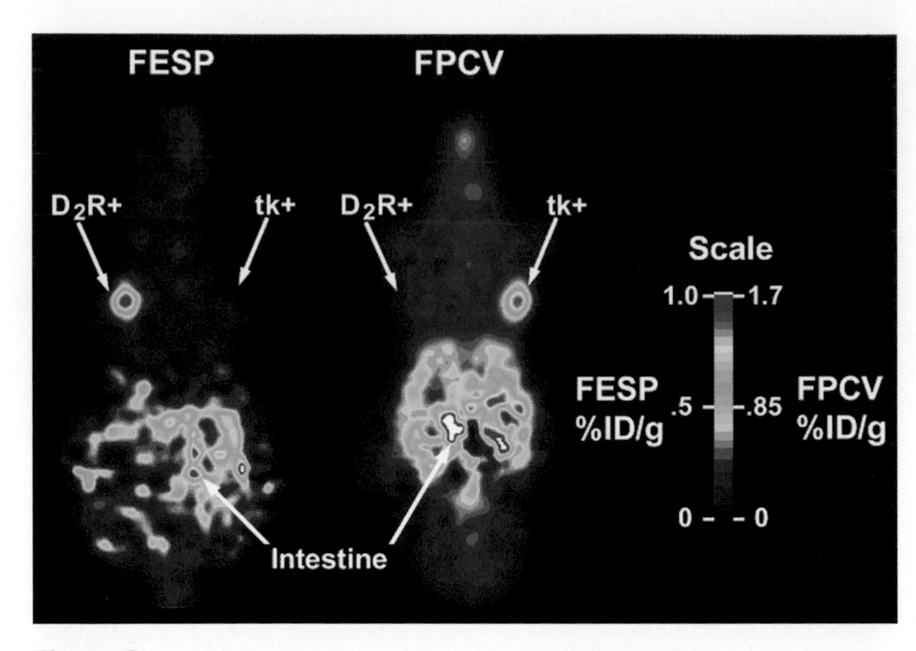

Figure 2 MicroPET imaging of two PET reporter genes (HSV1-tk and D_2R) in the same mouse using two different reporter probes (^{18}F-labeled FPCV for HSV1-tk and ^{18}F-labeled FESP for D_2R). These images show specific accumulation of probes in a mouse carrying a tumor stably expressing HSV1-tk on the right shoulder and a separate tumor stably expressing D_2R on the left shoulder. The accumulation of [^{18}F]FPCV and [^{18}F]FESP in each tumor reflects trapping due to HSV1-tk and D_2R expression, respectively. Background signal in the intestinal tract is seen due to clearance of the probes via the hepatobiliary system. The percentage injected dose per gram (%ID/g) is shown in separate scales for [^{18}F]FPCV and [^{18}F]FESP. Reproduced from Iyer *et al.* (2001a), by permission.

gland and also in other tissues like the thyroid, pancreas, gastrointestinal tract, kidney, and lung. When SSTr2 is used as a reporter gene, the receptor can be expressed on the surface of cells that would not normally express this gene and, therefore, the use of various tracers can image SSTr2 reporter gene expression. This approach has been applied with small-animal models and SPECT imaging and is discussed in further detail in a review article (Rogers *et al.*, 2000).

One of the earliest reporter gene approaches investigated cytosine deaminase (CD) as a reporter gene with 5-[^{3}H]fluorocytosine (5-FC) as the reporter probe (Haberkorn *et al.*, 1996). Cytosine deaminase, which is expressed in yeasts and bacteria but not in mammalian cells, converts the antifungal agent 5-fluorocytosine to the highly toxic 5-fluorouracil (5-FU). The 5-FC does not incorporate into the mammalian DNA synthesis pathway; however, its by-product 5-FU can block DNA and protein synthesis due to substitution of uracil by 5-FU in RNA and inhibition of thymidilate synthetase by 5-fluorodeoxyuridine monophosphate, resulting in impaired DNA biosynthesis. However, lack of sufficient accumulation of the reporter probe in cytosine deaminase-expressing cells (due to efflux of 5-FU) limits its use as an imaging reporter gene (Haberkorn *et al.*, 1996). More recently a magnetic resonance spectroscopy (MRS)-based approach using some kinetic modeling has been described using the CD gene (Stegman *et al.*, 1999) and shows promise. The conversion of ^{19}F-labeled 5-FC to 5-FU was followed by MRS in subcutaneous human colorectal carcinoma xenografts in nude mice by using H29 cell lines stably transfected with the yeast cd (ycd) gene and a three-compartment model was described for noninvasive estimation of ycd transgene expression. Earlier, Aboagye and associates (1998) demonstrated the feasibility of noninvasive magnetic resonance spectroscopy and spectroscopic imaging (chemical shift imaging) to detect activation of the prodrug 5-FC to the cytotoxic species 5-FU by monoclonal antibody–CD conjugates. They have used monoclonal antibodies conjugated with CD to deliver the reporter enzyme to the cells of interest instead of incorporating the reporter gene into the cells. The utility of an MRS-based approach warrants further investigation.

Another candidate for a reporter gene is the sodium/iodide symporter (NIS). The NIS is an intrinsic membrane protein. Mandell and associates (1999) demonstrated a method to concentrate radiation for tumor imaging or killing. They cloned the rat sodium/iodide symporter gene into a retroviral vector for transfer into cancer cells. Their results show that the transformed cancer cells can be selectively killed by the induced accumulation of ^{131}I by the sodium/iodide symporter. Similar experiments using ^{125}I with the human sodium/iodide symporter also showed promising results (Haberkorn *et al.*, 2001). To develop human NIS gene transfer for radioiodide therapy for patients with brain tumors, Cho and co-workers (2000) have constructed recombinant adenoviruses that can express exogenous hNIS in human glioma cells. Further, NIS-expressing adenovirus has also been shown to function in tumor cells of various origins and to be very efficient in triggering significant iodide uptake by a tumor, outlining the potential of this novel cancer gene therapy approach for a targeted radiotherapy (Boland *et al.*, 2000). This sodium/iodide symporter gene may prove to be a useful PET reporter gene if its expression is used to transiently accumulate ^{124}I, which is a positron emitter (half-life of 4.18 days). Several other reporter gene approaches have been studied and are listed with appropriate references in Table 1 (adapted from Ray *et al.*, 2001).

TABLE 1 Summary of Reporter Gene/Probe Systems (Adapted from Ray et al., 2001)

Reporter gene	Mechanism	Imaging agents	Imaging	References
Cytosine deaminase	Deamination	5-[^{3}H] Fluorocytosine 5-[^{19}F] Fluorocytosine	Cell culture study MRS	Haberkorn *et al.*, 1996; Stegman *et al.*, 1999
Herpes simplex virus type 1 thymidine kinase (HSV1-tk)	Phosphorylation	[^{131}I]FIAU, [^{14}C]FIAU	SPECT, gamma camera	Tjuvajev *et al.*, 1996, 1999b
		[^{131}I]FIAU	SPECT, gamma camera	
		[^{124}I]FIAU	PET	Tjuvajev *et al.*, 1999a
		[$^{123/125}$I]FIAU	Gamma camera	Haubner *et al.*, 2000
		[^{125}I]IVDU, [^{125}I]IVFRU, [^{125}I]IVFIAU, [^{125}I]IVAU	Cell culture	Morin *et al.*, 1997
		[^{125}I]FIAU, [^{125}I]FIRU	Cell culture	Wiebe *et al.*, 1999
		[^{3}H]FFUdR	Cell culture	Germann *et al.*, 1998,
		[^{14}C]GCV, [^{3}H]GCV	Autoradiography	Gambhir *et al.*, 1998; Haberkorn *et al.*, 1997, 1998
		[^{18}F]GCV	PET	Gambhir *et al.*, 1999b, 1998
		[^{18}F]PCV	PET	Iyer *et al.*, 2001a
		[^{18}F]FHPG	PET	Alauddin *et al.*, 1996, 1999; de Vries *et al.*, 2000; Hospers *et al.*, 2000; Hustinx *et al.*, 2001
		[^{18}F]FHBG	PET	Alauddin and Conti, 1998; Yaghoubi *et al.*, 2001a
Mutant herpes simplex virus type 1 thymidine kinase (HSV1-sr39-tk)	Phosphorylation	[^{18}F]PCV [^{18}F]FHBG	Cell culture, PET PET	Gambhir *et al.*, 2000a; Sun *et al.*, 2001; Yaghoubi *et al.*, 2001a; Yu *et al.*, 2000
Dopamine-2 receptor	Receptor–ligand	[^{18}F]FESP	PET	MacLaren *et al.*, 1999; Sun *et al.*, 2001; Yaghoubi *et al.*, 2001b; Yu *et al.*, 2000
Mutant dopamine-2 receptor	Receptor–ligand	[^{18}F]FESP	PET	Liang *et al.*, 2001
Somatostatin receptor	Affinity binding	[^{111}In]DTPA-D-Phe^{1}-octreotide	Gamma camera	Rogers *et al.*, 1999
		[^{64}Cu]TETA-octreotide	Tumor uptake study	Buchsbaum *et al.*, 1999
		[^{188}Re]Somatostatin analogue, ^{99m}Tc-somatostatin analogue	Gamma camera	Rogers *et al.*, 2000; Zinn *et al.*, 2000
Oxotechnetate-binding fusion proteins	Binding via transchelation	[^{99m}Tc] Oxotechnetate gamma camera	Autoradiography,	Bogdanov *et al.*, 1997, 1998
Gastrin-releasing peptide receptor	Affinity binding	[^{125}I]mIP-Des-Met^{14}-bombesin (7–13)NH_2	Cell culture	Baidoo *et al.*, 1998; Rogers *et al.*, 1997a,b; Rosenfeld *et al.*, 1997
		[^{125}I]Bombesin	Cell culture	
		[^{99m}Tc]Bombesin analogue	Cell culture	
Sodium/iodine symporter (NIS)	Active symport	^{131}I	Gamma camera	Boland *et al.*, 2000; Haberkorn, 2001
Tyrosinase	Metal binding to melanin	Synthetic metallomelanins, ^{111}In, Fe	Cell culture/MRI	Enochs *et al.*, 1997; Weissleder *et al.*, 1997
Green fluorescent protein (GFP)	GFP gene expression resulting in fluorescence	Fluorescence	Fluorescence microscopy	Hasegawa *et al.*, 2000; Pfeifer *et al.*, 2001; Yang *et al.*, 2000a,b, 2001, 1998, 1999

Continued

TABLE 1—Continued

Reporter gene	Mechanism	Imaging agents	Imaging	References
Luciferase (firefly)	Luciferase–luciferin reaction in presence of oxygen, Mg^{2+}, and ATP	Bioluminescence	Charge-coupled device (CCD) camera	Contage *et al.*, 1997, 1998
Luciferase (*Renilla*)	Luciferase–luciferin reaction in presence of oxygen, no other cofactors required	Bioluminescence	CCD camera	Bhaumik and Gambhir, 2001
Cathepsin D	Quenched NIRF fluorochromes	Fluorescence activation	CCD camera	Tung *et al.*, 1999, 2000; Weissleder *et al.*, 1999
β-galactosidase	Hydrolysis of β-glycoside bond	1-(2-(β-galactopyranosyloxy) propy1)-4,7,10-tris (carboxymethyl)-1,4,7, 1-(tetraazacyclododecane)ga dolinium(III) or EgadMe	MRI	Louie *et al.*, 2000
Engineered transferrin receptor (TfR)	Receptor–ligand, internalization	Superparamagnetic iron, Tf-MION, Oxide-tran	MRI	Weissleder *et al.*, 2000

V. Application of in Vivo Reporter Gene Imaging to Monitor Gene Therapy Regimens

Precise localization and quantitative assessment of the location(s), magnitude, and time variation of gene expression are highly desirable for the evaluation of gene therapy trials. It would be ideal if the expression of each therapeutic gene could be analyzed directly by an *in vivo* imaging technique, using a gene-specific imaging probe. However, this is not feasible as most protein products or mRNA of therapeutic transgenes lack appropriate reporter probes that can be radiolabeled. Therefore, indirect or inferential imaging strategies are being explored to indirectly measure the level of therapeutic gene expression. The basic concept adopted in these strategies is to coexpress a therapeutic gene product and a reporter gene product in a coordinately regulated fashion. The goal of these approaches is to quantitatively image reporter gene expression and from that infer levels of therapeutic gene expression. Several approaches (Fig. 3) are currently being developed and are discussed next.

A. Bicistronic Approach

The discovery of cap-independent translation and existence of internal ribosomal entry sites (IRES) in polioviruses and encephalomyocarditis virus opened a new gateway to construct bicistronic vectors for gene therapy (Jang *et al.*, 1988, 1989; Sonenberg and Pelletier, 1989). In a bicistronic expression cassette, one coding sequence (e.g., encoding therapeutic protein) is placed in a cap-dependent position proximal to the IRES. A second coding sequence (e.g., encoding a reporter protein) placed distal to the IRES can be translated by a cap-independent mechanism. Both genes are under the control of the same promoter and transcribed into a single mRNA, which is then translated into two different proteins due to the presence of the IRES sequence. Since the two proteins are translated from a common message, expression of the two proteins should be proportional to one another. Thus, by measuring changes in the level of one of the protein products, the changes in the relative level of the second protein product can be inferred. The validity of this concept was tested by taking advantage of the availability of two different PET reporter genes. A bicistronic vector from which both D_2R and HSV1-sr39tk reporter genes were coexpressed from a common cytomegalovirus (CMV) promoter with the aid of the endomyocarditis virus IRES was used to transfect C6 glioma cells. Clones stably expressing differing levels of the D_2R–IRES–HSV1-sr39tk reporter gene were grown as xenograft tumors on nude mice and imaged in a microPET (Fig. 4) (Yu *et al.*, 2000). The levels of D_2R and HSV1-sr39tk reporter gene expressions were found to be well correlated, using [^{18}F]FESP and [^{18}F]FHBG as imaging probes, respectively. Tjuvajev and colleagues (1999b) showed in a mouse xenograft model using FIAU SPECT imaging of HSV1-tk reporter gene expression good correlation with a second reporter gene, β-galactosidase. The analysis of β-galactosidase was done following sacrifice of the animal. Therefore, for both PET and SPECT imaging, the IRES-based approach has been preliminarily validated as a method to couple the expression of two genes.

Although the IRES sequence leads to proper translation of the downstream cistron from a bicistronic vector, translation from the IRES can be cell type specific and the magni-

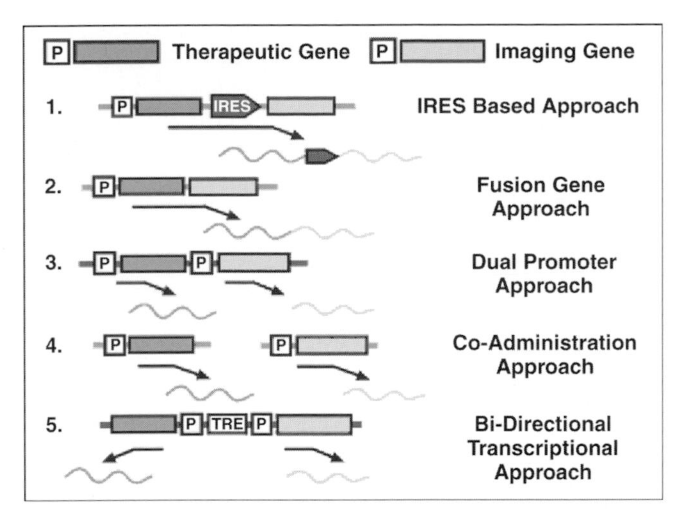

Figure 3 Different approaches and strategies that are used to coexpress the therapeutic gene and reporter gene in a coordinately regulated fashion are shown along with the resulting mRNA. The promoter (P) used to drive the therapeutic gene and/or the reporter gene can be constitutive or inducible. (**1**) *IRES-Based Approach.* An internal ribosomal entry site (IRES) links two genes and hence they are transcribed from a single common promoter leading to one mRNA. Two proteins are then translated from this single mRNA. See also Fig. 4. (**2**) *Fusion Gene Approach.* Two genes are fused together with a spacer sequence in between them. A single mRNA is transcribed under a single promoter, which upon translation yields a single fusion protein retaining the properties of both of the original individual proteins and their expression is absolutely coupled. (**3**) *Dual Promoter Approach.* Two identical promoters are used to drive expression of two different genes leading to two mRNAs. (**4**) *Coadministration Approach.* Estimation of the location and expression levels of a therapeutic gene delivered through a vector can be assessed by coadministering a second vector, which is similar to the first vector, having the therapeutic gene in all respects except that a reporter gene replaces the therapeutic gene. (**5**) *Bidirectional Transcriptional Approach.* Bidirectional promoters that are regulated by a tetracycline-responsive element can be made to induce the expression of two different genes in a highly correlated manner. See also Fig. 5.

tude of expression of the gene placed distal to the IRES is often attenuated (Kamoshita *et al.,* 1997; Yu *et al.,* 2000). This can lead to a lower imaging sensitivity, and methods to improve this approach are currently under investigation. The experiments with "super IRES" sequences (Chappell *et al.,* 2000; Wang *et al.,* 2001) may help in addressing these problems and will require further investigation.

B. Fusion Approach

In recent years, fusion gene/protein technology has become a powerful tool in molecular/cell biology, biochemistry, and gene therapy. A fusion gene construct contains two or more different genes joined so that their coding sequences are in the same reading frame and thus a single protein with properties of both of the original proteins is produced. Examples include HSV1-TK–GFP (Loimas *et al.,* 1998; Jacobs *et al.,* 1999; Steffens *et al.,* 2000) and HSV1-TK–luciferase–Neo (Strathdee *et al.,* 2000). An advantage of a fusion approach is that expression of both genes is stoichiometrically coupled. This approach, however, cannot be generalized, as many fusion proteins do not yield functional activity for both of the individual proteins or may not localize in an appropriate subcellular compartment. Every new therapeutic gene has to be fused to a reporter gene, and often the reporter protein and/or therapeutic protein activity is likely to be partially compromised. This approach is therefore not as generalizable as the other approaches that are available.

C. Dual-Promoter Approach

Two different genes expressed from distinct promoters within a single vector (e.g., pCMV–D_2R–pCMV–HSV1-sr39tk) can potentially be a useful way to couple the expression of two genes. This approach may avoid some of the attenuation and tissue variation problems of an IRES-based approach and is currently under active investigation (Wang *et al.,* 2001). However, in this approach the expression of the two genes may become uncoupled if the transcriptional activity of the two identical promoters is altered due to the inherent nature of the region where the vector integrates into the host genome. Also mutation in one or both promoters can alter the transcriptional activity. More work will, therefore, be needed to carefully study the potential utility of this approach in specific applications.

D. Co-vector Administration Approach

Another relatively simple approach to analyze the expression pattern of a therapeutic gene is to coadminister another reporter gene that is identical to the vector having the therapeutic gene in all respects except that the therapeutic gene is replaced with the reporter gene. In other words, the therapeutic and reporter genes are cloned in two different vectors but driven by same promoter. This approach has recently been validated (Yaghoubi *et al.,* 2001b) and shows that the expression of two PET reporter transgenes, HSV1-sr39tk and D_2R, driven by same CMV promoter but cloned in separate adenoviral vectors is well correlated at the multicell (macroscopic) level when delivered simultaneously. Although on an individual-cell basis substantial differences in infection, selection, etc., might occur, at a macroscopic/organ level such individual cell variation is not measurable. This very simple approach to indirectly imaging ectopic gene expression might be a useful alternative in various experimental situations.

E. Bidirectional Transcriptional Approach

For many gene therapy applications it would be desirable not only to deliver the therapeutic gene to the desired target(s), but also to be able to regulate levels of therapeutic

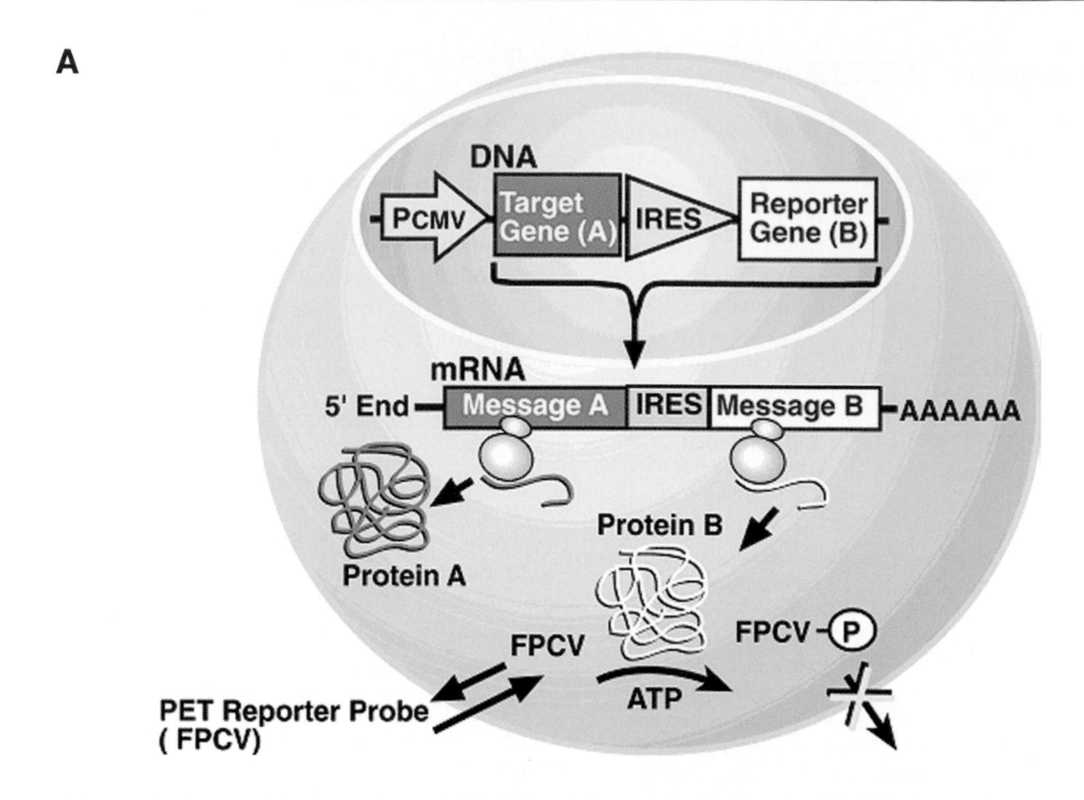

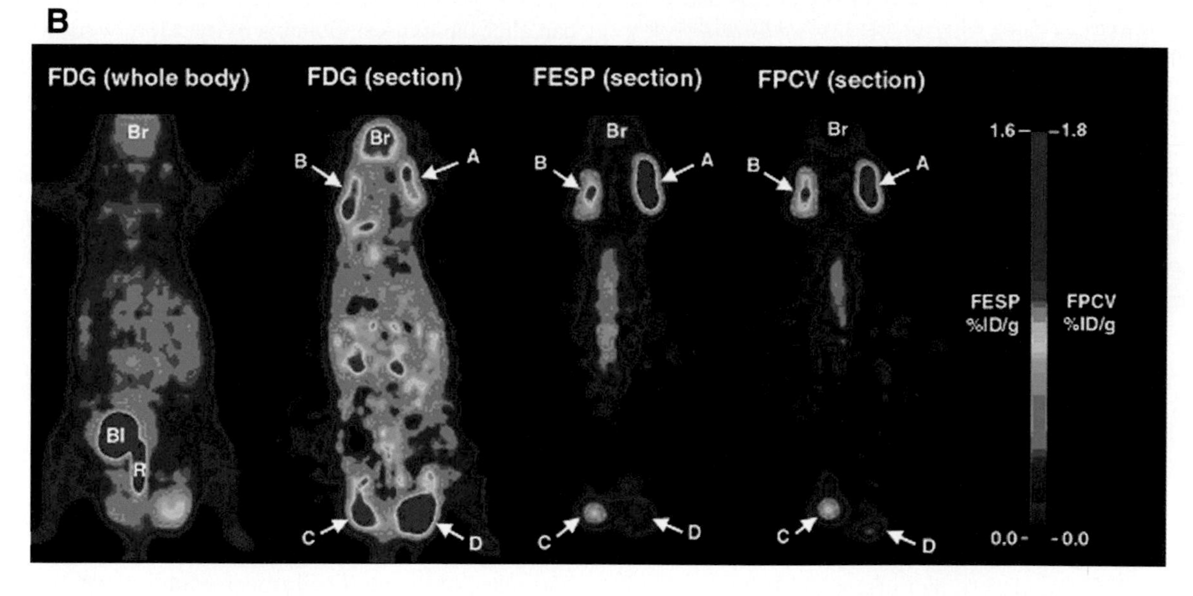

Figure 4 (**A**) Schematic representation of imaging reporter gene expression by PET from a bicistronic vector containing an IRES. Diagram shows how two proteins are coded and expressed from a bicistronic vector having an IRES site. Both gene A and gene B are coexpressed from the same vector and expression of gene B (a PET reporter) can be imaged quantitatively by trapping of a PET reporter probe, which will give an indirect measurement of the expression of gene A. (**B**) MicroPET imaging of bicistronic gene expression (pCMV–D_2R–IRES–sr39tk) in which both genes were imaged by using two different PET reporter probes ([^{18}F]FESP and [^{18}F]FPCV) in the same animal. Three C6 cell lines stably transfected with pCMV–D_2R–IRES–sr39tk (A, B, and C) and the parental C6 control cell line (D) were injected at four different sites in a single mouse. Sequential imaging of the tumors with [^{18}F]FDG, [^{18}F]FESP, and [^{18}F]FPCV after 10 days showed specific accumulation of [^{18}F]FESP and [^{18}F]FPCV in the tumors with stably transfected cells, while accumulation of [^{18}F]FDG is seen in all four tumors in the FDG section image. The [^{18}F]FDG (whole body) image shows the mouse outline and is provided for reference. The [^{18}F]FESP and [^{18}F]FPCV tumor images are highly correlated, illustrating that when one monitors expression of one of the two genes, one can infer levels of expression of the second gene. Reproduced from Yu *et al.* (2000), with permission.

gene expression. This can be achieved by the use of bidirectional expression vectors in which expression of two genes can be coregulated. We have recently validated this approach in a tumor xenograft model by microPET imaging of living mice using a bidirectional vector, which expresses the D_2R and HSV1-sr39tk reporter genes (Sun *et al.,* 2001). A minimal CMV promoter, which drives these two genes in this bidirectional vector, is induced by a tetracycline-responsive element (TRE). Transcription is significantly enhanced when a fusion protein, rtetR–VP16, binds to the TRE (Figs. 5A and 5B). The fusion protein binding can be titrated by using tetracycline and its analogs (e.g., doxycycline). Good correlations between expressions of the two PET reporter genes used were found. In our experiments HeLa cells that specifically express the rtetR–VP16 transactivator protein were used. The gene coding for this protein can also be cotransfected from a separate vector or can potentially be incorporated into the same bidirectional vector. This approach has the unique feature that levels of doxycycline can regulate levels of transcription. This approach avoids the attenuation and tissue variation problems of an IRES-based approach and may prove to be one of the most robust approaches developed to date. This dual gene expression cassette experiment demonstrates the ability of PET reporter gene analyses to monitor pharmacologically regulated alterations in gene expression and also proves to be another approach whereby correlated expression of therapeutic and reporter genes can be achieved.

VI. Indirect Imaging of Endogenous Gene Expression through Coupling Endogenous Promoters with Reporter Genes

One of the key limitations for many of the imaging assays developed to date with the various imaging modalities is the lack of generalizable approaches. By generalizability it is meant that minimal new chemistry need be involved for each new cellular/molecular target, so that one can easily target a whole array of potential cellular targets. For each protein target, one usually has to develop a new imaging probe in order to image the relative levels of that particular protein target. One potentially generalizable approach is to develop a tracer that can interact with pre-mRNA or mRNA of the gene of interest to image transcription directly, such as modified radiolabeled antisense oligodeoxynucleotide probes targeted toward a specific mRNA (reviewed in Gambhir, 2000; Tavitian, 2000). This approach would be a very powerful approach, but has yet to be fully worked out in living subjects. Another is to build a reporter probe for the endogenous protein of interest such that the reporter probe binds specifically to the protein or is modified by the protein and hence trapped. Developing reporter probes for each protein is a difficult problem and not easily generalizable. One potential solution to the issue of a generalizable approach is to couple endogenous promoters to a reporter gene and then to use the same reporter probe regardless of the choice of promoter. This approach is described in detail next.

In many of the imaging applications to date, reporter genes have been driven by a constitutive exogenous promoter such as the CMV promoter (Tjuvajev *et al.,* 1998; Gambhir *et al.,* 1999a), which allows for continuous transcription of the reporter gene, but does not allow for tracking of endogenous gene expression. A reporter gene can be linked to any endogenous promoter (e.g., albumin promoter) of choice and, therefore, the endogenous gene promoter can modulate expression of the reporter gene.

The feasibility of indirectly imaging endogenous gene expression by PET reporter gene imaging in transgenic animals and assessment of the transcriptional regulation of such an endogenous gene through imaging the suppression and induction of reporter gene expression was recently reported (Green *et al.,* 2002). In this study a transgenic mouse model in which the expression of a wild-type HSV1-tk PET reporter gene driven by the albumin promoter was modulated by dietary alterations, and comparisons of HSV1-tk reporter gene expression with endogenous albumin gene expression were made. Protein deprivation resulted in the suppression of endogenous albumin gene expression, indirectly observed as lower [^{18}F]FHBG accumulation in the liver when the transgenic mice were imaged using microPET. Restoration of protein in the diet led to induction and an increase in the expression of albumin, demonstrated by microPET imaging as significant accumulation of [^{18}F]FHBG in the livers of transgenic mice. The transcriptional regulation brought about by the manipulation of dietary protein content was investigated and correlated in control nontransgenic mice, in which the levels of GAPDH-normalized albumin mRNA correlated with the absence or presence of dietary protein. Hence, despite the unavailability of a tracer to directly monitor albumin mRNA or protein, successful imaging and correlations were made of endogenous albumin gene expression. This work demonstrates that the *in vivo* expression of endogenous genes can be indirectly monitored in transgenic animals expressing a PET reporter gene. Many reasons might uncouple the endogenous gene expression with reporter gene expression, so these indirect approaches must be validated for each specific endogenous gene and an application of interest.

Other investigators have also shown that DNA-damage-induced up-regulation of p53 transcriptional activity can be imaged and correlated with the expression of p53-dependent downstream genes (Doubrovin *et al.,* 2001). Mutation in the p53 gene is one of most common causes of human cancer (Hollstein *et al.,* 1991, Levine *et al.,* 1994). Doubrovin and

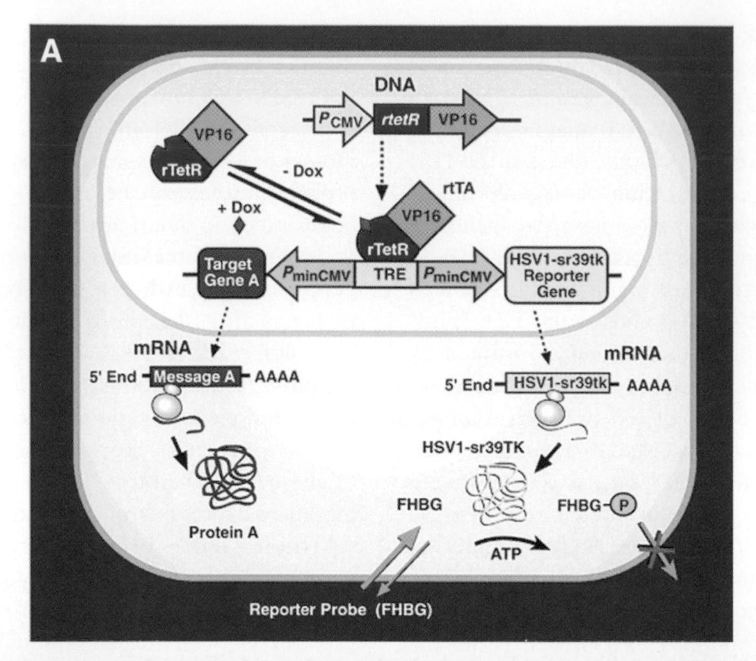

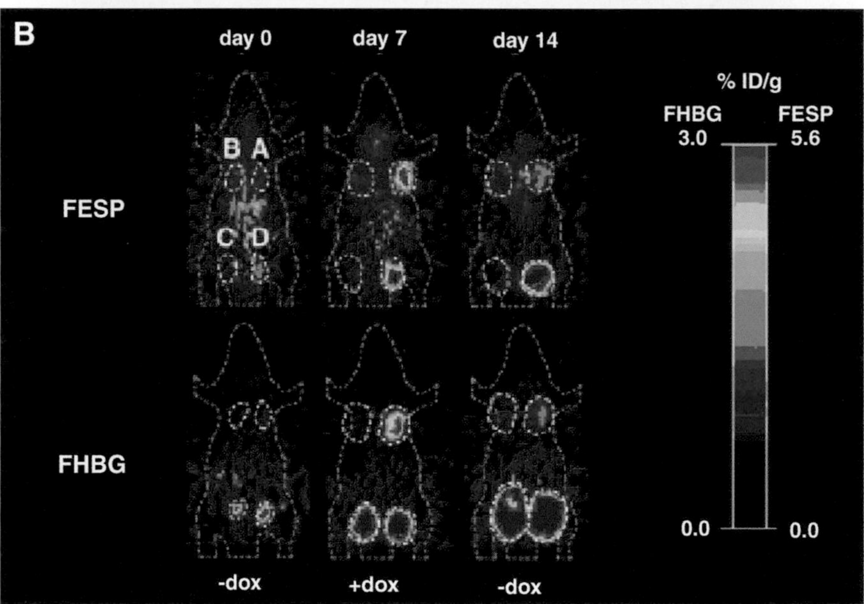

Figure 5 (**A**) Schematic diagram showing the basis by which induction of two different genes can be simultaneously achieved with the use of a single bidirectional tetracycline-responsive element (TRE). In this example, the fusion protein rtetR–VP16, which is constitutively expressed, binds to the TRE site and transactivates the two minimal CMV promoters leading to the simultaneous expression of the two genes. This binding will occur only in the presence of tetracycline or an analog such as doxycycline (Dox). Varying the levels of Dox administration can simultaneously control gene induction and the magnitude of expression of the two different proteins coded by those genes. (**B**) Sequential microPET imaging of tumors in a nude mouse with or without gene induction by Dox. Tumor A has a bidirectional promoter driving the expression of the two PET reporter genes, D_2R and HSV1-sr39tk. Tumor B is a negative control. Tumor C is a positive control that has been stably transfected with a HSV1-sr39tk gene under the control of a constitutive (CMV) promoter. Tumor D is another positive control having cells that are stably transfected with an IRES-based gene expression system, expressing the D_2R and HSV1-sr39tk PET reporter genes. When the tumors (dotted circles) were 5 mm in diameter, microPET imaging was done with [^{18}F]FESP and [^{18}F]FHBG on sequential days. The top row shows the expression of the D_2R gene and the bottom row that of the HSV1-sr39tk gene. Doxycycline (500 μg/ml) added to the drinking water led to gene induction in tumor A, shown here at day 7. The induced expression of both the genes was turned off if Dox administration was withdrawn (day 14). Also note the absence of any tracer accumulation in tumor B. Tracer accumulation is seen in tumors C and D irrespective of the presence or absence of doxycycline. Note the accumulation of both tracers in tumor D due to the expression of two PET reporter genes, unlike tumor C, which encodes only HSV1-sr39tk. Reproduced from *Sun et al.* (2001), with permission.

colleagues (2001) have used a HSV1-tk/GFP (TKGFP) dual reporter gene placed under the control of an artificial *cis*-acting p53-specific enhancer in a retroviral construct that was transduced into U87 glioma and SaOS-2 osteosarcoma (p53–/–) cells. In this study with the *cis*-p53/TKGFP reporter system, PET imaging and independent measurements of p53 activity and the expression levels of downstream genes demonstrated that it is possible to sufficiently detect the transcriptional up-regulation of genes in the p53 signal transduction pathway.

One of the key problems in using endogenous promoters to drive reporter gene expression can be the relatively weak transcriptional activity of some promoters. This leads to insufficient levels of mRNA and therefore insufficient levels of reporter protein. In order to address this problem, we have validated a two-step transcriptional activation/amplification (TSTA) in which the weak promoter drives expression of an intermediate protein (GAL4–VP16), which in turn is used to amplify expression of the reporter gene (Iyer *et al.,* 2001b) (Fig. 6). This approach was tested using human prostate cancer cells, the prostate-specific enhancer (PSE), the firefly luciferase reporter gene, and optical charged-coupled device (CCD) imaging. It was shown that the TSTA approach was able to produce an imaging signal, whereas directly trying to drive reporter gene expression from the relatively weak PSE did not lead to detectable imaging signal. The work also showed, in cell culture, a gain in expression when HSV1-sr39tk was used as the reporter gene. This work demonstrates the potential to eventually apply a TSTA approach for gene therapy vectors and transgenic applications. Further, we demonstrated that a single plasmid bearing the TSTA components expressed firefly luciferase at 20-fold higher levels than the CMV enhancer (Zhang *et al.*, 2002). Recently, we modified the yeast two-hybrid system to be inducible (IY2H) and used it to detect protein-protein

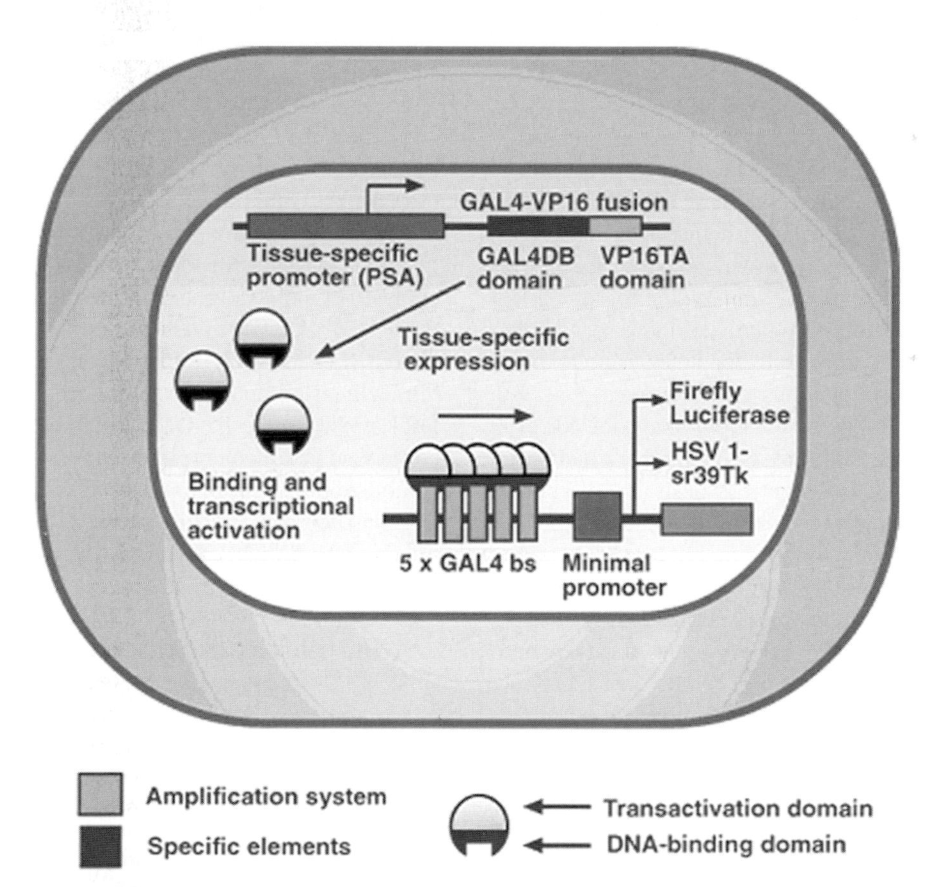

Figure 6 The two-step transcriptional activation (TSTA) system by which enhancement of gene expression can be achieved due to gene induction of the GAL4–VP16 fusion protein expressed from a tissue-specific but relatively weak promoter. Shown is a tissue-specific prostate-specific antigen (PSA) promoter, which is relatively weak in gene induction. In the first step this weak promoter is used to express the fusion protein GAL4–VP16, which performs the second step by binding to specific GAL4 binding sites (GAL4 bs) proximal to the minimal promoter driving either a PET (HSV1-sr39tk) or an optical (firefly luciferase) reporter gene. This binding of the transactivator protein in the second step leads to an enhancement of gene expression from a minimal promoter. Many orders of enhancement in gene expression with tissue specificity can be achieved by this TSTA system. Reproduced from Iyer *et al.,* 2001b, with permission.

interactions in living mice using bioluminescent optical imaging (Ray *et al.*, 2002). Protein-protein interactions modulate many intracellular events including gene expression and signal transduction. The ability to image this interaction in living animals should have important implications for the *in vivo* evaluation of protein-protein interactions in cells maintained in their natural environment and for the *in vivo* evaluation of new pharmaceuticals targeted to modulate protein-protein interactions.

VII. Antisense Reporter Probes for Imaging Endogenous Gene Expression in Vivo

Antisense oligonucleotides are small chains of nucleic acids capable of binding cellular RNA by a hybridization mechanism. Paterson and colleagues (1977) were the first to report that gene expression can be modified with exogenous nucleic acids by the use of single-stranded DNA to inhibit translation of a complementary RNA, and later Zamecnik and Stephenson (1978) showed that a synthetic oligonucleotide, complementary to the 3′ end of the Rous sarcoma virus, could inhibit the formation of new virus. Much progress has subsequently been made and currently there are a number of ongoing therapeutic clinical trials in hematology and oncology representing a growing interest in antisense technology (reviewed in Koller *et al.*, 2000; Tamm *et al.*, 2001).

Imaging mRNA levels with antisense reporter probes may be a very useful approach for many applications. Transcription of any endogenous or exogenous gene will lead to mRNA. Messenger RNA levels may not correlate well with protein levels, but in cases in which a good correlation exists or in cases in which it is desirable to know if transcription has occurred, it may be useful to image mRNA levels. The use of antisense reporter probes would be particularly useful for endogenous genes, because unlike the reporter gene imaging approach it is not necessary to introduce any reporter vectors. Several key issues for using antisense reporter probes as imaging agents exist and are reviewed elsewhere (Gambhir, 2000; Tavitian, 2000).

The antisense approach has been successfully used to image firefly luciferase gene expression in experimental brain tumors generated by implantation of cells permanently transfected with the firefly luciferase gene while using a peptide nucleic acid (PNA) as an antisense reporter probe (Shi *et al.*, 2000). Degradation and delivery of the antisense probe within the body are major issues while designing the probe for therapy or *in vivo* imaging. The PNAs are a useful class of antisense agents as these molecules are metabolically stable and are not significantly bound by proteins and do not activate RNase H. In this study (Shi *et al.*, 2000), a PNA that was antisense around the methionine initiation codon of the firefly luciferase mRNA was designed with a carboxyl-terminal tyrosine to enable radiolabeling with iodine-125. It also had an amino-terminal biotin residue to enable conjugation to a blood–brain barrier (BBB) drug targeting system, which is required for transportation as PNAs cross the BBB poorly. The drug targeting system comprised a conjugate of streptavidin (to which the biotin residue will bind) and a monoclonal antibody (MAb) against the transferrin receptor (TfR). This MAb undergoes receptor-mediated transcytosis across the BBB via the brain capillary endothelial TfR. This MAb drug targeting system enabled transport of the PNA antisense radiopharmaceutical into the tumor cell owing to abundant expression of the TfR on the tumor cell membrane. Film autoradiography of the brain and luciferase enzyme activity measurements in the tumor extracts confirmed the specific hybridization of the ^{125}I-antiluciferase PNA to the target luciferase mRNA. This method reflects *in vivo* gene expression because the radiolabeled antisense imaging agent was administered *in vivo* and not applied to tissue sections *ex vivo*. If other laboratories are able to reproduce these initial results, this will be an exciting approach to imaging with antisense reporter probes.

In another study, the expression of the gene encoding glial fibrillary acidic protein was imaged in experimental brain tumors *in vivo* with a 25-mer phosphorothioate oligodeoxynucleotide (PS-ODN) (Kobori *et al.*, 1999). This PS-ODN contained a 5′ amino group that was labeled with ^{11}C for imaging with PET. This PS-ODN also contained a cholesterol moiety at the 3′ terminus to facilitate transport of the PS-ODN across cellular membranes. This lipidization approach enabled the targeting of gene expression in brain intracellular compartments. But the addition of the cholesterol moiety to the PS-ODN eliminates the solubility of the compound in aqueous solution. It then becomes necessary to solubilize the agent in an organic solvent, dichloromethane, which is neurotoxic. This approach will require more investigation and further validation.

VIII. Nonradionuclide Approaches to Reporter Gene Imaging

Radionuclide approaches offer a highly sensitive approach to imaging reporter gene expression that can easily be extended from animal studies to human. Nevertheless, other modalities, when used in the appropriate setting, may have desirable characteristics. MRI techniques recently have obtained encouraging initial results and are reviewed in detail in a separate chapter. MRI approaches lack intrinsic sensitivity compared to PET/SPECT, and therefore, significant limitations may exist for the routine use of MRI, especially in clinical settings (see Chapter 13).

Optical reporter genes such as firefly luciferase, *Renilla* luciferase, or GFP are currently being used to monitor reporter gene expression *in vivo*. The expression of optically assayed reporter genes can be monitored *in vivo* in small living animals, using digital cameras, video cameras, or sensitive cooled CCD cameras (Contag *et al.,* 1997; Wu *et al.,* 2001). Firefly and *Renilla* luciferase reporter gene expression can be imaged using a high-sensitivity CCD camera. Bioluminescence is produced through the interaction of firefly luciferase with its substrate D-luciferin (injected peritoneally or via tail vein) in the presence of magnesium, oxygen, and ATP (Contag *et al.,* 1998, 2000). *Renilla* luciferase has the advantage of not requiring ATP or cofactors and has also recently been validated in living mice in our laboratories while injecting mice with the substrate coelenterazine (Bhaumik and Gambhir, 2001). Imaging both firefly and *Renilla* luciferase in the same living animal has also recently been validated by our laboratories (Bhaumik and Gambhir, 2001). For fluorescence approaches with GFP and its mutants, an input wavelength must be provided and an output wavelength of light is produced (Yang *et al.,* 2000a). The bioluminescent and fluorescent optical approaches have the distinct advantage of ease of use and low cost; however, light scatter and absorption primarily limit their use to small-animal models with very limited potential human applications. The fluorescence approaches with GFP and its mutants also suffer from a relatively high background signal because of autofluorescence of tissues. The D-luciferin/firefly luciferase approach allows potential imaging in the brain because of the ability of D-luciferin to cross the blood–brain barrier (Benaron *et al.,* 1997). This may also be the case for *Renilla* luciferase and coelenterazine, but needs to be verified. It is likely that continued refinement of these optical approaches would lead to rapid throughput and multiplexing methods for imaging reporter gene expression in living rodents.

IX. Specific Issues for Neuroscience Applications

Imaging reporter or endogenous gene expression in the brain is especially difficult because targeting of reporter probes to intracellular spaces in brain cells is a "two-barrier" drug targeting problem. These probes must first cross the BBB and then must reach the target cells in the brain. Therapeutic options for the treatment of malignant brain tumors have also been limited, in part, because of the presence of the BBB. Techniques for imaging gene expression in the brain can utilize various strategies to circumvent this problem. There is a blood–brain barrier disruption (BBBD) in various forms of brain tumors, which can be exploited by techniques using the suicide gene HSV1-tk and various prodrugs and imaging probes.

The ability of a number of substances to alter the BBB has also been exploited. Mannitol is one of the most widely used BBBD agents (Kroll and Neuwelt, 1998). In addition to mannitol various other hypertonic solutions like arabinose, lactamide, saline, urea, and several radiographic contrast agents are also used to cause BBBD. However, it is important that the disruption must be transient and reversible and the disrupting agent must not interfere with the therapeutic regimen. Another alternative is to administer the vector and reporter probe intracranially by stereotactic infusion. These approaches are undergoing active investigation by several laboratories.

Both the HSV1-tk/HSV1-sr39tk and the D_2R PET reporter genes are not optimal for most central nervous system applications. All of the reporter probes for HSV1-tk/HSV1-sr39tk developed to date show very poor penetration across the BBB. Therefore, imaging with any of these reporter probes is useful only if some BBBD is present. The BBBD must allow for the reporter probe both to enter and to exit from regions in which the reporter probe is not trapped. The D_2R reporter gene with [^{18}F]FESP can work with brain-specific applications but only if reporter gene expression is away from normal D_2R expression in the striatum. More work will be needed to develop other PET reporter genes/reporter probes for use within the brain. The use of optical reporter genes such as firefly luciferase looks attractive for the studies of gene expression in the brain of small animals. This is especially important because D-luciferin, which is the substrate for firefly luciferase, crosses the BBB, and hence the firefly luciferase could be used as a reporter gene for gene imaging studies in the brain of small animals.

X. Human Gene Therapy of Brain Tumors and Imaging Studies

Gene therapy offers the promise of augmenting traditional cancer therapies (drugs, radiation, and surgery). Malignant gliomas have been a primary target for gene therapy partly because of their dismal prognosis despite advances in neurosurgical techniques, radiation, and drug therapies. Some of the difficulties encountered include inaccessibility to resective surgery because of anatomical location and single-cell invasion of surrounding brain tissue. Most of the clinical trials for glioma treatment involving suicide genes have investigated the therapeutic potential of the transfer of the HSV1-tk followed by the administration of the prodrug GCV. In the first clinical trial (Ram *et al.,* 1997), 15 patients (12 with recurrent malignant gliomas, 2 with metastatic melanomas, and 1 with metastatic breast carcinoma) were treated with stereotactic injections of murine fibroblast virus-producing cells, which produce retroviruses carrying the HSV1-tk gene. This pretreatment was followed by the administration of GCV. Despite the low transduction efficiency (<0.17% of the tumor cells), this study showed a reduction of the tumor volume in five tumors. In another study (Klatzmann *et al.,* 1998), 12 patients with recurrent glioblastoma were treated by direct injection of HSV1-tk retroviral-producing cells into the surgical cavity margins after tumor debulking. Overall median survival was 206 days, with 25% of the patients surviving longer than 12 months. In HSV1-tk suicide gene therapy as well as other gene therapy approaches imaging can help to optimize the therapy. Current transgene/vector strategies, including novel genes, combinational therapies, and new delivery modalities, have been reviewed (Engelhard, 2000; Lam and Breakefield, 2001).

Studies to image reporter gene expression in human subjects are now beginning. We reported a study to measure the kinetics, biodistribution, stability, dosimetry, and safety of [^{18}F]FHBG in healthy human volunteers, prior to eventually imaging patients undergoing HSV1-tk gene therapy (Yaghoubi *et al.,* 2001a). This study indicated that due to the properties of stability, rapid blood clearance, low background signal, biosafety, and acceptable dosimetry [^{18}F]FHBG should be a good reporter probe for HSV1-tk imaging in humans, although [^{18}F]FHBG does not significantly cross the BBB. This may eventually allow imaging of gene therapy performed not only with HSV1-tk, but with any therapeutic gene coupled to HSV1-tk for indirect imaging of the therapeutic gene. In another recent study (Jacobs *et al.,* 2001a) showed that [^{124}I]FIAU, another PET reporter probe for HSV1-tk, cannot penetrate the BBB and thus is not a good marker probe for noninvasive localization of HSV1-tk in the central nervous system. However, this probe is useful for areas where the BBB is disrupted (e.g., glioblastoma).

Recently, Jacobs *et al.* (2001b) used [^{124}I]FIAU-PET imaging of humans in a prospective gene-therapy trial to investigate the safety of intratumorally infused liposome–gene complex followed by ganciclovir administration. In this study, vector-mediated HSV1-tk gene expression was followed by [^{124}I]FIAU-PET in five patients (age range 49–67 years) with recurrent glioblastomas. Intratumoral infusion through stereotactically placed catheters was done in these patients. After vector administration, in one of five patients specific [^{124}I]FIAU-associated radioactivity was observed within the infused tumor (Fig. 7). In the four patients in whom no specific [^{124}I]FIAU accumulation was observed, the histology showed a significantly lower number of proliferating tumor cells per voxel. The authors hypothesized that a certain critical number of tk-gene-transduced tumor cells per voxel (threshold) have to be present so that tk-gene-associated accumulation of [^{124}I]FIAU can be measured and detected by PET. These preliminary findings in a small group show that [^{124}I]FIAU-PET imaging of HSV1-tk expression in patients may be feasible and that vector-mediated gene expression may predict therapeutic effect. Much more work is still needed to better characterize HSV1-tk reporter gene imaging in the brain as well as in other regions.

The other PET/SPECT reporter genes also have various tracers that have been previously used in human subjects

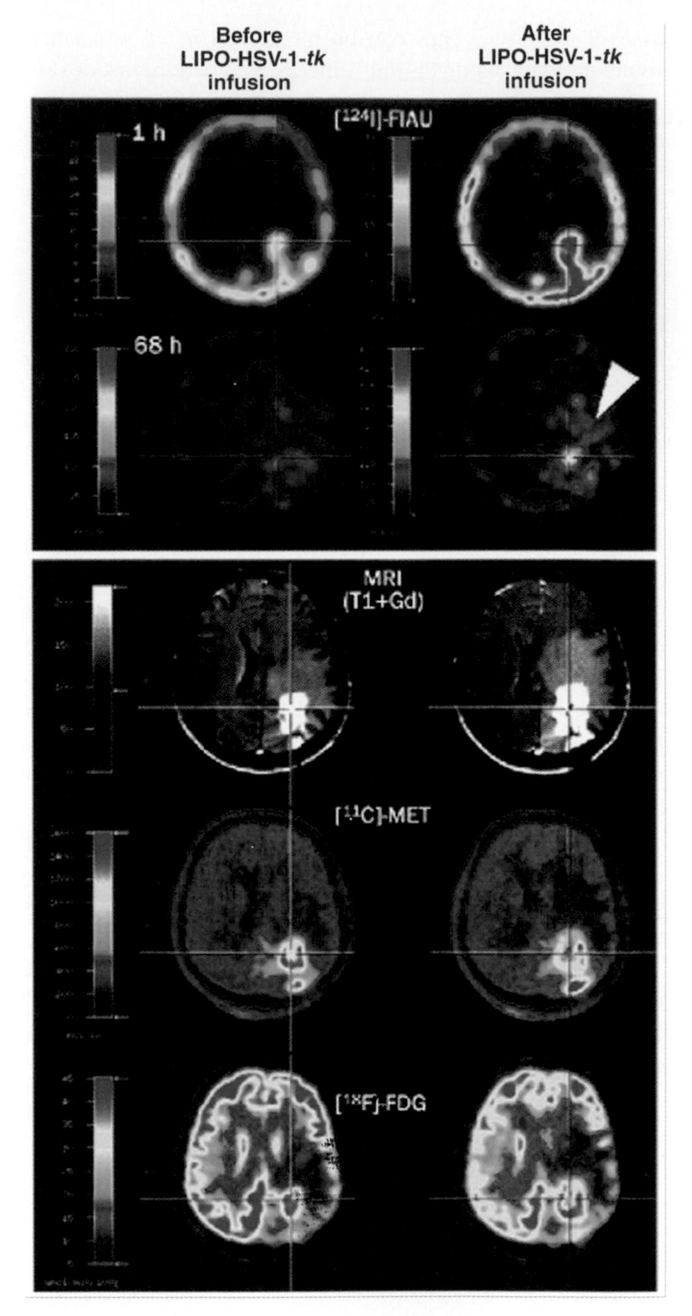

Figure 7 PET brain imaging of HSV1-tk suicide gene therapy in a patient. Coregistration of FIAU-PET, MRI, methionine (MET)-PET, and FDG-PET before and after intratumorally infused liposome–gene complex containing HSV1-tk (LIPO–HSV-1-tk). The white arrowhead marks the region within the tumor where specific [^{124}I]FIAU retention was imaged after LIPO–HSV-1-tk transduction. The cross hairs in the right column indicate signs of necrosis after ganciclovir treatment (5 mg/kg a day over 14 days) that was started 4 days after vector application. Reproduced from Jacobs *et al.* (2001b), with permission.

and, therefore, can likely be translated to human applications in gene therapy. With the rapid progress in human gene therapy and molecular imaging it is likely that radionuclide approaches will play a major role in optimizing gene therapy.

XI. Conclusion

The use of reporter genes and reporter probes will have important applications for both animal research and human gene therapy. Imaging of tumor models with reporter genes has been the primary focus to date. Although much has been accomplished with PET/SPECT reporter genes to date, additional work is needed for neuroscience applications. With continued refinement, the reporter gene approaches should become a powerful tool for basic and clinical research of normal function and neurological disorders.

References

Aboagye, E. O., Artemov, D., Senter, P. D., and Bhujwalla, Z. M. (1998). Intratumoral conversion of 5-fluorocytosine to 5-fluorouracil by monoclonal antibody–cytosine deaminase conjugates: Noninvasive detection of prodrug activation by magnetic resonance spectroscopy and spectroscopic imaging. *Cancer Res.* **58,** 4075–4078.

Alauddin, M. M., Conti, P. S., Mazza, S. M., Hamzeh, F. M., and Lever, J. R. (1996). 9-[(3-[^{18}F]-Fluoro-1-hydroxy-2-propoxy)methyl]guanine ([^{18}F]-FHPG): A potential imaging agent of viral infection and gene therapy using PET. *Nucl. Med. Biol.* **23,** 787–792.

Alauddin, M. M., and Conti, P. S. (1998). Synthesis and preliminary evaluation of 9-(4-[^{18}F]-fluoro-3-hydroxymethylbutyl)guanine ([^{18}F]FHBG): A new potential imaging agent for viral infection and gene therapy using PET. *Nucl. Med. Biol.* **25,** 175–180.

Alauddin, M. M., Shahinian, A., Kundu, R. K., Gordon, E. M., and Conti, P. S. (1999). Evaluation of 9-[(3-^{18}F-fluoro-1-hydroxy-2-propoxy) methyl]guanine ([^{18}F]-FHPG) in vitro and in vivo as a probe for PET imaging of gene incorporation and expression in tumors. *Nucl. Med. Biol.* **26,** 371–376.

Baidoo, K. E., Scheffel, U., Stathis, M., Finley, P., Lever, S. Z., Zhan, Y., and Wagner, H. N., Jr. (1998). High-affinity no-carrier-added 99mTc-labeled chemotactic peptides for studies of inflammation in vivo. *Bioconjugate Chem.* **9,** 208–217.

Benaron, D. A., Contag, P. R., and Contag, C. H. (1997). Imaging brain structure and function, infection and gene expression in the body using light. *Philos. Trans. R. Soc. London B Biol. Sci.* **352,** 755–761.

Bhaumik, S., and Gambhir, S. S. (2001). Optical imaging of Renilla luciferase reporter gene expression in living mice. *Proc. Natl. Acad. Sci. USA,* **99,** 377–382.

Black, M. E., Newcomb, T. G., Wilson, H. M., and Loeb, L. A. (1996). Creation of drug-specific herpes simplex virus type 1 thymidine kinase mutants for gene therapy. *Proc. Natl. Acad. Sci. USA* **93,** 3525–3529.

Boehme, R. E. (1984). Phosphorylation of the antiviral precursor 9-(1,3-dihydroxy-2-propoxymethyl)guanine monophosphate by guanylate kinase isozymes. *J. Biol. Chem.* **259,** 12346–12349.

Bogdanov, A., Petherick, P., Marecos, E., and Weissleder, R. (1997). In vivo localization of diglycylcysteine-bearing synthetic peptides by nuclear imaging of oxotechnetate transchelation. *Nucl. Med. Biol.* **24,** 739–742.

Bogdanov, A., Jr., Simonova, M., and Weissleder, R. (1998). Design of metal-binding green fluorescent protein variants. *Biochim. Biophys. Acta* **1397,** 56–64.

Boland, A., Ricard, M., Opolon, P., Bidart, J. M., Yeh, P., Filetti, S., Schlumberger, M., and Perricaudet, M. (2000). Adenovirus-mediated transfer of the thyroid sodium/iodide symporter gene into tumors for a targeted radiotherapy. *Cancer Res.* **60,** 3484–3492.

Brust, P., Haubner, R., Friedrich, A., Scheunemann, M., Anton, M., Koufaki, O. N., Hauses, M., Noll, S., Noll, B., Haberkorn, U., Schackert, G., Schackert, H. K., Avril, N., and Johannsen, B. (2001). Comparison of [^{18}F]FHPG and [$^{124/125}$I]FIAU for imaging herpes simplex virus type thymidine kinase gene expression. *Eur. J. Nucl. Med.* **28,** 721–729.

Buchsbaum, D. J., Rogers, B. E., Khazaeli, M. B., Mayo, M. S., Milenic, D. E., Kashmiri, S. V., Anderson, C. J., Chappell, L. L., Brechbiel, M. W., and Curiel, D. T. (1999). Targeting strategies for cancer radiotherapy. *Clin. Cancer Res.* **5,** 3048s–3055s.

Chappell, S. A., Edelman, G. M., and Mauro, V. P. (2000). A 9-nt segment of a cellular mRNA can function as an internal ribosome entry site (IRES) and when present in linked multiple copies greatly enhances IRES activity. *Proc. Natl. Acad. Sci. USA.* **97,** 1536–1541.

Chatziioannou, A. F. (2002). Molecular imaging of small animals with dedicated PET tomographs. *Eur. J. Nucl. Med. Mol. Imaging* **29,** 98–114.

Chatziioannou, A. F., Cherry, S. R., Shao, Y., Silverman, R. W., Meadors, K., Farquhar, T. H., Pedarsani, M., and Phelps, M. E. (1999). Performance evaluation of microPET: A high-resolution lutetium oxyorthosilicate PET scanner for animal imaging. *J. Nucl. Med.* **40,** 1164–1175.

Cherry, S. R., Shao, Y., Silverman, R. W., Meadors, K., Siegel, S., Chatziioannou, A., Young, J. W., Jones, W. F., Moyers, J. C., Newport, D., Boutefnouchet, A., Farquhar, T. H., Andreaco, M., Paulus, M. J., Binkley, D. M., Nutt, R., and Phelps, M. E. (1997). MicroPET: A high resolution PET scanner for imaging small animals. *IEEE Trans. Nucl. Sci.* **44,** 1161–1166.

Cherry, S. R., and Gambhir, S. S. (2001). Use of positron emission tomography in animal research. *ILAR J.* **42,** 219–232.

Cho, J. Y., Xing, S., Liu, X., Buckwalter, T. L., Hwa, L., Sferra, T. J., Chiu, I. M., and Jhiang, S. M. (2000). Expression and activity of human Na^+/I^- symporter in human glioma cells by adenovirus-mediated gene delivery. *Gene Ther.* **7,** 740–749.

Contag, C. H., Spilman, S. D., Contag, P. R., Oshiro, M., Eames, B., Dennery, P., Stevenson, D. K., and Benaron, D. A. (1997). Visualizing gene expression in living mammals using a bioluminescent reporter. *Photochem. Photobiol.* **66,** 523–531.

Contag, P. R., Olomu, I. N., Stevenson, D. K., and Contag, C. H. (1998). Bioluminescent indicators in living mammals. *Nat. Med.* **4,** 245–247.

Contag, C. H., Jenkins, D., Contag, P. R., and Negrin, R. S. (2000). Use of reporter genes for optical measurements of neoplastic disease in vivo. *Neoplasia* **2,** 41–52.

de Vries, E. F. J., van Waarde, A., Harmsen, M. C., Mulder, N. H., Vaalburg, W., and Hospers, G. A. P. (2000). [C-11]FMAU and [F-18]FHPG as PET tracers for herpes simplex virus thymidine kinase enzyme activity and human cytomegalovirus infections]. *Nucl. Med. Biol.* **27,** 113–119.

Doubrovin, M., Ponomarev, V., Beresten, T., Balatoni, J., Bornmann, W., Finn, R., Humm, J., Larson, S., Sadelain, M., Blasberg, R., and Gelovani Tjuvajev, J. (2001). Imaging transcriptional regulation of p53-dependent genes with positron emission tomography in vivo. *Proc. Natl. Acad. Sci. USA* **98,** 9300–9305.

Engelhard, H. H. (2000). Gene therapy for brain tumors: The fundamentals. *Surg. Neurol.* **54,** 3–9.

Enochs, W. S., Petherick, P., Bogdanova, A., Mohr, U., and Weissleder, R. (1997). Paramagnetic metal scavenging by melanin: MR imaging. *Radiology* **204,** 417–423.

Furman, P. A., St Clair, M. H., and Spector, T. (1984). Acyclovir triphosphate is a suicide inactivator of the herpes simplex virus DNA polymerase. *J. Biol. Chem.* **259,** 9575–9579.

Gambhir, S. S. (2000). Imaging gene expression: Concepts and future outlook. *In* "Diagnostic Nuclear Medicine" (C. Schiepers, ed.), pp. 253–271. Springer-Verlag, Berlin.

Gambhir, S. S., Barrio, J. R., Wu, L., Iyer, M., Namavari, M., Satyamurthy, N., Bauer, E., Parrish, C., MacLaren, D. C., Borghei, A. R., Green, L. A., Sharfstein, S., Berk, A. J., Cherry, S. R., Phelps, M. E., and Herschman, H. R. (1998). Imaging of adenoviral-directed herpes simplex virus type 1 thymidine kinase reporter gene expression in mice with radiolabeled ganciclovir. *J. Nucl. Med.* **39,** 2003–2011.

Gambhir, S. S., Barrio, J. R., Phelps, M. E., Iyer, M., Namavari, M., Satyamurthy, N., Wu, L., Green, L. A., Bauer, E., MacLaren, D. C.,

Nguyen, K., Berk, A. J., Cherry, S. R., and Herschman, H. R. (1999a). Imaging adenoviral-directed reporter gene expression in living animals with positron emission tomography. *Proc. Natl. Acad. Sci. USA* **96,** 2333–2338.

Gambhir, S. S., Barrio, J. R., Herschman, H. R., and Phelps, M. E. (1999b). Imaging gene expression: principles and assays. *J. Nucl. Cardiol.* **6,** 219–233.

Gambhir, S. S., Bauer, E., Black, M. E., Liang, Q., Kokoris, M. S., Barrio, J. R., Iyer, M., Namavari, M., Phelps, M. E., and Herschman, H. R. (2000a). A mutant herpes simplex virus type 1 thymidine kinase reporter gene shows improved sensitivity for imaging reporter gene expression with positron emission tomography. *Proc. Natl. Acad. Sci. USA* **97,** 2785–2790.

Gambhir, S. S., Herschman, H. R., Cherry, S. R., Barrio, J. R., Satyamurthy, N., Toyokuni, T., Phelps, M. E., Larson, S. M., Balatoni, J., Finn, R., Sadelain, M., Tjuvajev, J., and Blasberg, R. (2000b). Imaging transgene expression with radionuclide imaging technologies. *Neoplasia* **2,** 118–138.

Germann, C., Shields, A. F., Grierson, J. R., Morr, I., and Haberkorn, U. (1998). 5-Fluoro-1-(2′-deoxy-2′-fluoro-beta-D-ribofuranosyl) uracil trapping in Morris hepatoma cells expressing the herpes simplex virus thymidine kinase gene. *J. Nucl. Med.* **39,** 1418–1423.

Gnant, M. F., Noll, L. A., Irvine, K. R., Puhlmann, M., Terrill, R. E., Alexander, H. R., Jr., and Bartlett, D. L. (1999). Tumor-specific gene delivery using recombinant vaccinia virus in a rabbit model of liver metastases. *J. Natl. Cancer Inst.* **91,** 1744–1750.

Grandaliano, G., Choudhury, G. G., and Abboud, H. E. (1995). Transgenic animal models as a tool in the diagnosis of kidney diseases. *Semin. Nephrol.* **15,** 43–49.

Green, L., Yap, C., Nguyen, K., Barrio, J., Namavari, M., Satyamurthy, N., Phelps, M., Sandgren, E., Herschman, H., and Gambhir, S. S. (2002). Indirect monitoring of endogenous gene expression by positron emission tomography (PET) imaging of reporter gene expression in transgenic mice. *Mol. Imaging Biol.,* **4,** 71–81.

Haberkorn, U., Oberdorfer, F., Gebert, J., Morr, I., Haack, K., Weber, K., Lindauer, M., van Kaick, G., and Schackert, H. K. (1996). Monitoring gene therapy with cytosine deaminase: In vitro studies using tritiated-5-fluorocytosine. *J. Nucl. Med.* **37,** 87–94.

Haberkorn, U., Altmann, A., Morr, I., Knopf, K. W., Germann, C., Haeckel, R., Oberdorfer, F., and van Kaick, G. (1997). Monitoring gene therapy with herpes simplex virus thymidine kinase in hepatoma cells: Uptake of specific substrates. *J. Nucl. Med.* **38,** 287–294.

Haberkorn, U., Khazaie, K., Morr, I., Altmann, A., Muller, M., and van Kaick, G. (1998). Ganciclovir uptake in human mammary carcinoma cells expressing herpes simplex virus thymidine kinase. *Nucl. Med. Biol.* **25,** 1998.

Haberkorn, U., Henze, M., Altmann, A., Jiang, S., Morr, I., Mahmut, M., Peschke, P., Kubler, W., Debus, J., and Eisenhut, M. (2001). Transfer of the human NaI symporter gene enhances iodide uptake in hepatoma cells. *J. Nucl. Med.* **42,** 317–325.

Hasegawa, S., Yang, M., Chishima, T., Miyagi, Y., Shimada, H., Moossa, A. R., and Hoffman, R. M. (2000). In vivo tumor delivery of the green fluorescent protein gene to report future occurrence of metastasis. *Cancer Gene Ther.* **7,** 1336–1340.

Haubner, R., Avril, N., Hantzopoulos, P. A., Gansbacher, B., and Schwaiger, M. (2000). In vivo imaging of herpes simplex virus type 1 thymidine kinase gene expression: Early kinetics of radiolabelled FIAU. *Eur. J. Nucl. Med.* **27,** 283–291.

Hollstein, M., Sidransky, D., Vogelstein, B., and Harris, C. C. (1991). p53 mutations in human cancers. *Science* **253,** 49–53.

Hospers, G. A. P., Calogero, A., van Waarde, A., Doze, P., Vaalburg, W., Mulder, N. H., and de Vries, E. F. J. (2000). Monitoring of herpes simplex virus thymidine kinase enzyme activity using positron emission tomography. *Cancer Res.* **60,** 1488–1491.

Hustinx, R., Shiue, C. Y., Alavi, A., McDonald, D., Shiue, G. G., Zhuang, H., Lanuti, M., Lambright, E., Karp, J. S., and Eck, S. L. (2001). Imaging in vivo herpes simplex virus thymidine kinase gene transfer to tumour-bearing rodents using positron emission tomography and [^{18}F]FHPG. *Eur. J. Nucl. Med.* **28,** 5–12.

Ikenaka, K., and Kagawa, T. (1995). Transgenic systems in studying myelin gene expression. *Dev. Neurosci.* **17,** 127–136.

Inouye, S., and Shimomura, O. (1997). The use of *Renilla* luciferase, *Oplophorus* luciferase, and apoaequorin as bioluminescent reporter protein in the presence of coelenterazine analogues as substrate. *Biochem. Biophys. Res. Commun.* **233,** 349–353.

Iyer, M., Bauer, E., Barrio, J. R., Nguyen, K., Namavari, M., Satyamurthy, N., Toyokuni, T., Phelps, M. E., Herschman, H. R., and Gambhir, S. S. (2000). Comparison of FPCV, FHBG and FIAU as reporter probes for imaging herpes simplex virus type 1 thymidine kinase gene expression. *J. Nucl. Med.* **41,** 80P.

Iyer, M., Barrio, J. R., Namavari, M., Bauer, E., Satyamurthy, N., Nguyen, K., Toyokuni, T., Phelps, M. E., Herschman, H. R., and Gambhir, S. S. (2001a). 8-[^{18}F]Fluoropenciclovir: An improved reporter probe for imaging HSV1-tk reporter gene expression in vivo using PET. *J. Nucl. Med.* **42,** 96–105.

Iyer, M., Wu, L., Carey, M., Wang, Y., Smallwood, A., and Gambhir, S. S. (2001b). Two-step transcriptional amplification as a method for imaging reporter gene expression using weak promoters. *Proc. Natl. Acad. Sci. USA,* **98,** 14595–14600.

Jacobs, A., Dubrovin, M., Hewett, J., Sena-Esteves, M., Tan, C. W., Slack, M., Sadelain, M., Breakefield, X. O., and Tjuvajev, J. G. (1999). Functional coexpression of HSV-1 thymidine kinase and green fluorescent protein: Implications for noninvasive imaging of transgene expression. *Neoplasia* **1,** 154–161.

Jacobs, A., Braunlich, I., Graf, R., Lercher, M., Sakaki, T., Voges, J., Hesselmann, V., Brandau, W., Wienhard, K., and Heiss, W. D. (2001a). Quantitative kinetics of [I-124]FIAU in cat and man. *J. Nucl. Med.* **42,** 467–475.

Jacobs, A., Voges, J., Reszka, R., Lercher, M., Gossmann, A., Kracht, L., Kaestle, C., Wagner, R., Wienhard, K., and Heiss, W. D. (2001b). Positron-emission tomography of vector-mediated gene expression in gene therapy for gliomas. *Lancet* **358,** 727–729.

Jang, S. K., Kräusslich, H. G., Nicklin, M. J., Duke, G. M., Palmenberg, A. C., and Wimmer, E. (1988). A segment of the 5′ nontranslated region of encephalomyocarditis virus RNA directs internal entry of ribosomes during in vitro translation. *J. Virol.* **62,** 2636–2643.

Jang, S. K., Davies, M. V., Kaufman, R. J., and Wimmer, E. (1989). Initiation of protein synthesis by internal entry of ribosomes into the 5′ nontranslated region of encephalomyocarditis virus RNA in vivo. *J. Virol.* **63,** 1651–1660.

Kamoshita, N., Tsukiyama-Kohara, K., Kohara, M., and Nomoto, A. (1997). Genetic analysis of internal ribosomal entry site on hepatitis C virus RNA: Implication for involvement of the highly ordered structure and cell type-specific transacting factors. *Virology* **233,** 9–18.

Karp, M., and Oker-Blom, C. (1999). A streptavidin–luciferase fusion protein: Comparisons and applications. *Biomol. Eng.* **16,** 101–104.

Klatzmann, D., Valery, C. A., Bensimon, G., Marro, B., Boyer, O., Mokhtari, K., Diquet, B., Salzmann, J. L., and Philippon, J. (1998). A phase I/II study of herpes simplex virus type 1 thymidine kinase "suicide" gene therapy for recurrent glioblastoma. Study Group on Gene Therapy for Glioblastoma. *Hum. Gene Ther.* **9,** 2595–2604.

Kobori, N., Imahori, Y., Mineura, K., Ueda, S., and Fujii, R. (1999). Visualization of mRNA expression in CNS using ^{11}C-labeled phosphorothioate oligodeoxynucleotide. *NeuroReport* **10,** 2971–2974.

Koller, E., Gaarde, W. A., and Monia, B. P. (2000). Elucidating cell signaling mechanisms using antisense technology. *Trends Pharmacol. Sci.* **21,** 142–148.

Kornblum, H. I., Araujo, D. M., Annala, A. J., Tatsukawa, K. J., Phelps, M. E., and Cherry, S. R. (2000). In vivo imaging of neuronal activation and plasticity in the rat brain by high resolution positron emission tomography (microPET). *Nat. Biotechnol.* **18,** 655–660.

Kroll, R. A., and Neuwelt, E. A. (1998). Outwitting the blood–brain barrier for therapeutic purposes: Osmotic opening and other means. *Neurosurgery* **42,** 1083–1099. [Discussion pp. 1099–1100]

Kunert, A., Hagemann, M., and Erdmann, N. (2000). Construction of promoter probe vectors for Synechocystis sp. PCC 6803 using the light-emitting reporter systems Gfp and LuxAB. *J. Microbiol. Methods* **41,** 185–194.

Lam, P. Y., and Breakefield, X. O. (2001). Potential of gene therapy for brain tumors. *Hum. Mol. Genet.* **10,** 777–787.

Lee, J. H., Federoff, H. J., and Schoeniger, L. O. (1999). G207, modified herpes simplex virus type 1, kills human pancreatic cancer cells in vitro. *J. Gastrointest. Surg.* **3,** 127–131. [Discussion pp. 132–123]

Leffel, S. M., Mabon, S. A., and Stewart, C. N., Jr. (1997). Applications of green fluorescent protein in plants. *Biotechniques* **23,** 912–918.

Levine, A. J., Perry, M. E., Chang, A., Silver, A., Dittmer, D., Wu, M., and Welsh, D. (1994). The 1993 Walter Hubert Lecture: The role of the p53 tumour-suppressor gene in tumorigenesis. *Br. J. Cancer* **69,** 409–416.

Liang, Q., Satyamurthy, N., Barrio, J., Toyokuni, T., Phelps, M., Gambhir, S. S., and Herschman, H. (2001). Noninvasive, quantitative imaging, in living animals, of a mutant dopamine D2 receptor reporter gene in which ligand binding is uncoupled from signal transduction. *Gene Ther.* **8,** 1490–1498.

Loimas, S., Wahlfors, J., and Jänne, J. (1998). Herpes simplex virus thymidine kinase–green fluorescent protein fusion gene: New tool for gene transfer studies and gene therapy. *Biotechniques* **24,** 614–618.

Lorenz, W. W., Cormier, M. J., O'Kane, D. J., Hua, D., Escher, A. A., and Szalay, A. A. (1996). Expression of the Renilla reniformis luciferase gene in mammalian cells. *J. Biolumin. Chemilumin.* **11,** 31–37.

Louie, A. Y., Huber, M. M., Ahrens, E. T., Rothbacher, U., Moats, R., Jacobs, R. E., Fraser, S. E., and Meade, T. J. (2000). In vivo visualization of gene expression using magnetic resonance imaging. *Nat. Biotechnol.* **18,** 321–325.

MacLaren, D. C., Gambhir, S. S., Cherry, S. R., Barrio, J. R., Satyamurthy, N., Toyokuni, T., Berk, A., Wu, L., Phelps, M. E., and Herschman, H. R. (1998). Repetitive and non-invasive in vivo imaging of reporter gene expression using adenovirus delivered dopamine D2 receptor as a PET reporter gene and FESP as a PET reporter probe. *J. Nucl. Med.* **39,** 35P.

MacLaren, D. C., Gambhir, S. S., Satyamurthy, N., Barrio, J. R., Sharfstein, S., Toyokuni, T., Wu, L., Berk, A. J., Cherry, S. R., Phelps, M. E., and Herschman, H. R. (1999). Repetitive, non-invasive imaging of the dopamine D2 receptor as a reporter gene in living animals. *Gene Ther.* **6,** 785–791.

Mandell, R. B., Mandell, L. Z., and Link, C. J., Jr. (1999). Radioisotope concentrator gene therapy using the sodium/iodide symporter gene. *Cancer Res.* **59,** 661–668.

Miller, W. H., and Miller, R. L. (1980). Phosphorylation of acyclovir (acycloguanosine) monophosphate by GMP kinase. *J. Biol. Chem.* **255,** 7204–7207.

Misslitz, A., Mottram, J. C., Overath, P., and Aebischer, T. (2000). Targeted integration into a rRNA locus results in uniform and high level expression of transgenes in Leishmania amastigotes. *Mol. Biochem. Parasitol.* **107,** 251–261.

Morin, K. W., Atrazheva, E. D., Knaus, E. E., and Wiebe, L. I. (1997). Synthesis and cellular uptake of 2′-substituted analogues of (E)-5-(2-[^{125}I]iodovinyl-2′-deoxyuridine in tumor cells transduced with the herpes simplex type-1 thymidine kinase gene. Evaluation as probes for monitoring gene therapy. *J. Med. Chem.* **40,** 2184–2190.

Namavari, M., Barrio, J. R., Toyokuni, T., Gambhir, S. S., Cherry, S. R., Herschman, H. R., Phelps, M. E., and Satyamurthy, N. (2000). Synthesis of 8-[(18)F]fluoroguanine derivatives: In vivo probes for imaging gene expression with positron emission tomography. *Nucl. Med. Biol.* **27,** 157–162.

Nomura, T., Takakura, Y., and Hashida, M. (1997). [Cancer gene therapy by direct intratumoral injection: Gene expression and intratumoral pharmacokinetics of plasmid DNA.] *Gan To Kagaku Ryoho* **24,** 483–488.

Paterson, B. M., Roberts, B. E., and Kuff, E. L. (1977). Structural gene identification and mapping by DNA–mRNA hybrid-arrested cell-free translation. *Proc. Natl. Acad. Sci. USA* **74,** 4370–4374.

Pfeifer, A., Kessler, T., Yang, M., Baranov, E., Kootstra, N., Cheresh, D., Hoffman, R., and Verma, I. (2001). Transduction of liver cells by lentiviral vectors: Analysis in living animals by fluorescence imaging. *Mol. Ther.* **3,** 319–322.

Ram, Z., Culver, K. W., Oshiro, E. M., Viola, J. J., DeVroom, H. L., Otto, E., Long, Z., Chiang, Y., McGarrity, G. J., Muul, L. M., Katz, D., Blaese, R. M., and Oldfield, E. H. (1997). Therapy of malignant brain tumors by intratumoral implantation of retroviral vector-producing cells. *Nat. Med.* **3,** 1354–1361.

Ray, P., Bauer, E., Iyer, M., Barrio, J. R., Namavari, M., Phelps, M. E., Herschman, H. R., and Gambhir, S. S. (2001). Monitoring gene therapy with reporter gene imaging. *Semin. Nucl. Med.* **31,** 312–320.

Ray, P., Pimenta, H., Paulmurugan, R., Berger, F., Phelps, M. E., Iyer, M., and Gambhir, S. S. (2002). Noninvasive quantitative imaging of protein-protein interactions in living subjects. *Proc. Natl. Acad. Sci. USA* **99,** 3105–3110.

Reardon, J. E. (1989). Herpes simplex virus type 1 and human DNA polymerase interactions with 2′-deoxyguanosine 5′-triphosphate analogues. Kinetics of incorporation into DNA and induction of inhibition. *J. Biol. Chem.* **264,** 19039–19044.

Rogers, B. E., Curiel, D. T., Mayo, M. S., Laffoon, K. K., Bright, S. J., and Buchsbaum, D. J. (1997a). Tumor localization of a radiolabeled bombesin analogue in mice bearing human ovarian tumors induced to express the gastrin-releasing peptide receptor by an adenoviral vector. *Cancer* **80,** 2419–2424.

Rogers, B. E., Rosenfeld, M. E., Khazaeli, M. B., Mikheeva, G., Stackhouse, M. A., Liu, T., Curiel, D. T., and Buchsbaum, D. J. (1997b). Localization of iodine-125-mIP-Des-Met14-bombesin (7-13)NH_2 in ovarian carcinoma induced to express the gastrin releasing peptide receptor by adenoviral vector-mediated gene transfer. *J. Nucl. Med.* **38,** 1221–1229.

Rogers, B. E., McLean, S. F., Kirkman, R. L., Della Manna, D., Bright, S. J., Olsen, C. C., Myracle, A. D., Mayo, M. S., Curiel, D. T., and Buchsbaum, D. J. (1999). In vivo localization of [(111)In]-DTPA-D-Phe1-octreotide to human ovarian tumor xenografts induced to express the somatostatin receptor subtype 2 using an adenoviral vector. *Clin. Cancer Res* **5,** 383–393.

Rogers, B. E., Zinn, K. R., and Buchsbaum, D. J. (2000). Gene transfer strategies for improving radiolabeled peptide imaging and therapy. *Q. J. Nucl. Med.* **44,** 208–223.

Rosenfeld, M. E., Rogers, B. E., Khazaeli, M. B., Mikheeva, G., Raben, D., Mayo, M. S., Curiel, D. T., and Buchsbaum, D. J. (1997). Adenoviral-mediated delivery of gastrin-releasing peptide receptor results in specific tumor localization of a bombesin analogue in vivo. *Clin. Cancer Res.* **3,** 1187–1194.

Sadelain, M., and Blasberg, R. G. (1999). Imaging transgene expression for gene therapy. *J. Clin. Pharmacol.* **Suppl.,** 34S–39S.

Sauvonnet, N., and Pugsley, A. P. (1996). Identification of two regions of Klebsiella oxytoca pullulanase that together are capable of promoting beta-lactamase secretion by the general secretory pathway. *Mol. Microbiol.* **22,** 1–7.

Seul, K. H., and Beyer, E. C. (2000). Mouse connexin37: Gene structure and promoter analysis. *Biochim. Biophys. Acta* **1492,** 499–504.

Shi, N., Boado, R. J., and Pardridge, W. M. (2000). Antisense imaging of gene expression in the brain in vivo. *Proc. Natl. Acad. Sci. USA* **97,** 14709–14714.

Sonenberg, N., and Pelletier, J. (1989). Poliovirus translation—A paradigm for a novel initiation mechanism. *BioEssays* **11,** 128–132.

Steffens, S., Frank, S., Fischer, U., Heuser, C., Meyer, K. L., Dobberstein, K. U., Rainov, N. G., and Kramm, C. M. (2000). Enhanced green fluorescent protein fusion proteins of herpes simplex virus type 1 thymidine kinase and cytochrome P450 4B1: Applications for prodrug-activating gene therapy. *Cancer Gene Ther.* **7,** 806–812.

Stegman, L. D., Rehemtulla, A., Beattie, B., Kievit, E., Lawrence, T. S., Blasberg, R. G., Tjuvajev, J. G., and Ross, B. D. (1999). Noninvasive quantitation of cytosine deaminase transgene expression in human tumor xenografts with in vivo magnetic resonance spectroscopy. *Proc. Natl. Acad. Sci. USA.* **96,** 9821–9826.

Strathdee, C. A., McLeod, M. R., and Underhill, T. M. (2000). Dominant positive and negative selection using luciferase, green fluorescent protein and beta-galactosidase reporter gene fusions. *Biotechniques* **28,** 210–212, 214.

Sun, X., Annala, A., Yaghoubi, S., Barrio, J. R., Nguyen, K., Toyokuni, T., Satyamurthy, N. M., Phelps, M. E., Herschman, H., and Gambhir, S. S. (2001). Quantitative imaging of gene induction in living animals. *Gene Ther.* **8,** 1572–1579.

Tamm, I., Dorken, B., and Hartmann, G. (2001). Antisense therapy in oncology: New hope for an old idea? *Lancet* **358,** 489–497.

Tavitian, B. (2000). In vivo antisense imaging. *Q. J. Nucl. Med.* **44,** 236–255.

Tjuvajev, J. G., Finn, R., Watanabe, K., Joshi, R., Oku, T., Kennedy, J., Beattie, B., Koutcher, J., Larson, S., and Blasberg, R. G. (1996). Noninvasive imaging of herpes virus thymidine kinase gene transfer and expression: A potential method for monitoring clinical gene therapy. *Cancer Res.* **56,** 4087–4095.

Tjuvajev, J. G., Avril, N., Oku, T., Sasajima, T., Miyagawa, T., Joshi, R., Safer, M., Beattie, B., DiResta, G., Daghighian, F., Augensen, F., Koutcher, J., Zweit, J., Humm, J., Larson, S. M., Finn, R., and Blasberg, R. (1998). Imaging herpes virus thymidine kinase gene transfer and expression by positron emission tomography. *Cancer Res.* **58,** 4333–4341.

Tjuvajev, J. G., Joshi, A., Callegari, J., Lindsley, L., Joshi, R., Balatoni, J., Finn, R., Larson, S. M., Sadelain, M., and Blasberg, R. G. (1999a). A general approach to the non-invasive imaging of transgenes using cis-linked herpes simplex virus thymidine kinase. *Neoplasia* **1,** 315–320.

Tjuvajev, J. G., Chen, S. H., Joshi, A., Joshi, R., Guo, Z. S., Balatoni, J., Ballon, D., Koutcher, J., Finn, R., Woo, S. L., and Blasberg, R. G. (1999b). Imaging adenoviral-mediated herpes virus thymidine kinase gene transfer and expression in vivo. *Cancer Res.* **59,** 5186–5193.

Tjuvajev, J. G., Doubrovin, M., Akhurst, T., Balatoni, J., Alauddin, M., Beattie, B., Larson, S. M., Conti, P. S., and Blasberg, R. G. (2000). Direct comparison of HSV1-tk PET imaging probes: FIAU, FHPG, FHBG. *J. Nucl. Med.* **41,** 277P.

Tung, C. H., Bredow, S., Mahmood, U., and Weissleder, R. (1999). A cathepsin D sensitive near infrared fluorescence probe for in vivo imaging of enzyme activity. *Bioconjugate Chem.* **10,** 892–896.

Tung, C. H., Mahmood, U., Bredow, S., and Weissleder, R. (2000). In vivo imaging of proteolytic enzyme activity using a novel molecular reporter. *Cancer Res.* **60,** 4953–4958.

Wang, Y. L., Iyer, M., Annala, A. J., Chappell, S. A., Nguyen, K., Herschman, H. R., Mauro, V. P., and Gambhir, S. S. (2001). New approaches for linking PET and therapeutic reporter gene expression for imaging gene therapy with increased sensitivity. *J. Nucl. Med.* **42,** 75P.

Wiebe, L. I., Knaus, E. E., and Morin, K. W. (1999). Radiolabelled pyrimidine nucleosides to monitor the expression of HSV-1 thymidine kinase in gene therapy. *Nucleosides Nucleotides* **18,** 1065–1066.

Weissleder, R., Simonova, M., Bogdanova, A., Bredow, S., Enochs, W. S., and Bogdanov, A., Jr. (1997). MR imaging and scintigraphy of gene expression through melanin induction. *Radiology* **204,** 425–429.

Weissleder, R., Tung, C. H., Mahmood, U., and Bogdanov, A. (1999). In vivo imaging of tumors with protease-activated near-infrared fluorescent probes. *Nat. Biotechnol.* **17,** 375–378

Weissleder, R., Moore, A., Mahmood, U., Bhorade, R., Benveniste, H., Chiocca, E. A., and Basilion, J. P. (2000). In vivo magnetic resonance imaging of transgene expression. *Nat. Med.* **6,** 351–355.

Wu, J. C., Sundaresan, G., Iyer, M., and Gambhir, S. S. (2001). Noninvasive optical imaging of firefly luciferase reporter gene expression in skeletal muscles of living mice. *Mol. Ther.* **4,** 297–306.

Yaghoubi, S., Barrio, J. R., Dahlbom, M., Iyer, M., Namavari, M., Satyamurthy, N., Goldman, R., Herschman, H. R., Phelps, M. E., and Gambhir, S. S. (2001a). Human pharmacokinetic and dosimetry studies of [(18)F]FHBG: A reporter probe for imaging herpes simplex virus type-1 thymidine kinase reporter gene expression. *J. Nucl. Med.* **42,** 1225–1234.

Yaghoubi, S. S., Wu, L., Liang, Q., Toyokuni, T., Barrio, J. R., Namavari, N., Satyamurthy, N., Phelps, M. E., Herschman, H. R., and Gambhir, S. S. (2001b). Direct correlation between positron emission tomographic images of two reporter genes delivered by two distinct adenoviral vectors. *Gene Ther.* **8,** 1072–1080.

Yang, M., Hasegawa, S., Jiang, P., Wang, X. O., Tan, Y. Y., Chishima, T., Shimada, H., Moossa, A. R., and Hoffman, R. M. (1998). Widespread skeletal metastatic potential of human lung cancer revealed by green fluorescent protein expression. *Cancer Res.* **58,** 4217–4221.

Yang, M., Jiang, P., Sun, F. X., Hasegawa, S., Baranov, E., Chishima, T., Shimada, H., Moossa, A. R., and Hoffman, R. M. (1999). A fluorescent orthotopic bone metastasis model of human prostate cancer. *Cancer Res.* **59,** 781–786.

Yang, M., Baranov, E., Jiang, P., Sun, F. X., Li, X. M., Li, L. N., Hasegawa, S., Bouvet, M., Al-Tuwaijri, M., Chishima, T., Shimada, H., Moossa, A. R., Penman, S., and Hoffman, R. M. (2000a). Whole-body optical imaging of green fluorescent protein-expressing tumors and metastases. *Proc. Natl. Acad. Sci. USA* **97,** 1206–1211.

Yang, M., Baranov, E., Moossa, A. R., Penman, S., and Hoffman, R. M. (2000b). Visualizing gene expression by whole-body fluorescence imaging. *Proc. Natl. Acad. Sci. USA* **97,** 12278–12282.

Yang, M., Baranov, E., Li, X. M., Wang, J. W., Jiang, P., Li, L., Moossa, A. R., Penman, S., and Hoffman, R. M. (2001). Whole-body and intravital optical imaging of angiogenesis in orthotopically implanted tumors. *Proc. Natl. Acad. Sci. USA* **98,** 2616–2621.

Yu, Y., Annala, A. J., Barrio, J. R., Toyokuni, T., Satyamurthy, N., Namavari, M., Cherry, S. R., Phelps, M. E., Herschman, H. R., and Gambhir, S. S. (2000). Quantification of target gene expression by imaging reporter gene expression in living animals. *Nat. Med.* **6,** 933–937.

Zamecnik, P. C., and Stephenson, M. L. (1978). Inhibition of Rous sarcoma virus replication and cell transformation by a specific oligodeoxynucleotide. *Proc. Natl. Acad. Sci. USA* **75,** 280–284.

Zhang, L., Adams, J. Y., Billick, E., Ilagan, R., Iyer, M., Le, K., Smallwood, A., Gambhir, S. S. , Carey, M., and Wu, L. (2002). Molecular engineering of a two-step transcription amplification (TSTA) system for transgene delivery in prostate cancer. *Mol Ther.* **5,** 223–232.

Zhou, D., Zhou, C., and Chen, S. (1997). Gene regulation studies of aromatase expression in breast cancer and adipose stromal cells. *J. Steroid Biochem. Mol. Biol.* **61,** 273–280.

Zinn, K. R., Buchsbaum, D. J., Chaudhuri, T. R., Mountz, J. M., Grizzle, W. E., and Rogers, B. E. (2000). Noninvasive monitoring of gene transfer using a reporter receptor imaged with a high-affinity peptide radiolabeled with 99mTc or 188Re. *J. Nucl. Med.* **41,** 887–895

Zlokarnik, G., Negulescu, P. A., Knapp, T. E., Mere, L., Burres, N., Feng, L., Whitney, M., Roemer, K., and Tsien, R. Y. (1998). Quantitation of transcription and clonal selection of single living cells with beta-lactamase as reporter. *Science* **279,** 84–88.

30

Mapping Gene Expression by MRI

Angelique Y. Louie, Joseph A. Duimstra, and Thomas J. Meade

Division of Biology and The Beckman Institute, California Institute of Technology, Pasadena, California 91125

I. Introduction

From the mapping of the human genome to the identification of disease-causing genetic defects, developments in molecular biology have advanced to a stage at which gene therapy is a viable consideration for treatment. There have been mixed successes with the known examples of gene therapy and the reasons for failure are often obscured by the inability to adequately monitor the location and extent of expression of the introduced genes. In assessing the efficacy of treatment it would be invaluable to be able to characterize the distribution and amount of gene expression and to correlate these results with the therapeutic effects that are observed. In this chapter we will cover the use of magnetic resonance techniques for imaging gene expression.

Historically, the absence of appropriate probes has prevented the use of magnetic resonance imaging (MRI) for detecting biochemical events such as gene expression. However, in recent years there has been a surge of development in new contrast agents that are indicators for biological activity, such as pH (Helpern *et al.*, 1987; Mikawa *et al.*, 1998; Zhang *et al.*, 1999), ion concentration (Li *et al.*, 1999), enzyme cleavage (Louie *et al.*, 2000; Moats *et al.*, 1997), and protein binding (Curtet *et al.*, 1998; Konda *et al.*, 2000; Kresse *et al.*, 1998; Sipkins *et al.*, 1998). The aim of this chapter is to provide an overview of recent progress in MRI contrast agent development that allows imaging of gene expression in living systems. This is a new field with relatively few examples in the literature. But the advances described here are at the cutting edge and efforts are under way to develop new and better probes. These types of agents may pave the way for expanding the applications of MRI in the study of biology and as a tool in the clinic. The ability to image living specimens noninvasively by MRI would be an invaluable tool for tracking the progress of gene therapy.

II. MRI Contrast Agents

The basics of magnetic resonance imaging and the application of MRI contrast agents to improve image quality are covered in the chapter by Narasimhan and Jacobs. Here we will elaborate upon the mechanism of action of contrast agents.

Magnetic resonance contrast agents are typically composed of chelated paramagnetic ions. A paramagnetic ion in solution can act as a transceiver of radiation. The unpaired electrons residing on the ion form a permanent magnetic dipole with a magnitude dependent on the number of unpaired spins. This dipole is constantly reorienting itself due to inherent electronic relaxation and the tumbling of the ion in solution. The reorientation of the electronic spin dipole can result in radiation just as reorientation of the precessing nuclear magnetization **M** does. The interplay between the paramagnetic ion's electronic spin moment and the magnetization vector of the nuclear spin can give rise to an increased relaxation rate (a shorter T_1) of the nucleus under investigation, typically water protons.

The fluctuating electronic dipole generates a broad band of frequencies from zero to τ_c^{-1}, where τ_c is the correlation time of the fluctuations. If ω_H is within this range, then it is affected by the paramagnetic ion. This distribution of frequencies is known as the spectral density function $J(\omega_H, \tau_c)$ (Banci *et al.*, 1991; Clementi and Luchinat, 1998):

$$J(\omega_H, \tau_c) \propto \frac{\tau_c}{1+\omega_H^2 \tau_c^2}$$

It is Lorentzian in shape and simple differentiation shows that J is maximal when $w_H = 1/\tau_c$.

The relationship between the spectral density function, J, and T_1 can be seen by factoring T_1 into its components and applying the Solomon–Bloembergen–Morgan (SBM) equations. First, the observed T_1 can be separated into the inherent relaxation due to the diamagnetic environment, T_{1dia}, and the effect induced by the presence of a paramagnetic ion, T_{1para}. T_{1para} may be further broken down into inner sphere (i.s.) and outer sphere (o.s.) contributions (Caravan *et al.*, 1999):

$$\left(\frac{1}{T_1}\right)_{obs} = \left(\frac{1}{T_1}\right)_{dia} + \left(\frac{1}{T_1}\right)_{para}$$

$$\left(\frac{1}{T_1}\right)_{para} = \left(\frac{1}{T_1}\right)_{i.s.} + \left(\frac{1}{T_1}\right)_{o.s.}$$

Typically the o.s. effects are calculated using a hard-sphere approximation for the paramagnetic ion. Here, we will concentrate on the i.s. factors due to their applicability to physiologically activated contrast agents.

Further description of $T_{1i.s.}$ depends upon the contrast agent under study. The following equation shows that the i.s. contribution to the relaxation rate, $1/T_1$, depends directly on both the concentration of the contrast agent, P_m, and the number of inner-sphere water molecules, q. Furthermore, it is inversely dependent upon the residency lifetime of the water molecule, τ_m:

$$\left(\frac{1}{\mathbf{T}_1}\right)_{i.s} = \frac{q\mathbf{P}_m}{\mathbf{T}_{1m} + \tau_m}$$

It is convenient to define a concentration-independent quantity known as the relaxivity, R, of a contrast agent. Relaxivity has the units of $mM^{-1}\ s^{-1}$ and is equal to $(1/T_1)$/(contrast agent concentration).

The effect of T_{1m} on $T_{1i.s.}$ is less obvious. T_{1m} can be factored further into the sum of dipole–dipole (DD) and scalar coupling (SC) contributions. The SC mechanism is a through-bond process whereby the unpaired electron spin density is delocalized directly onto the nucleus in question. This occurs through some linear combination of the s orbitals of the diamagnetic species with the orbitals containing the unpaired electrons (Drago, 1992). In the case of gadolinium, the SC mechanism may be neglected due to the poor covalency of the f orbitals in the lanthanides. Therefore, T_{1m} closely approximates T_1^{DD} for gadolinium-based contrast agents. From the SBM equations, we see that $1/T_1^{DD}$ is given by (Caravan *et al.*, 1999)

$$\frac{1}{T_1^{DD}} = \frac{2}{15}\frac{\gamma_H^2 g^2 \mu_B^2 S(S+1)}{r^6}\left[\frac{3\tau_{c1}}{1+\omega_H^2\tau_{c1}^2} + \frac{7\tau_{c2}}{(1+\omega_S^2\tau_{c2}^2)}\right]$$

where

$$\frac{1}{\tau_{ci}} = \frac{1}{T_{ie}} + \frac{1}{\tau_m} + \frac{1}{\tau_R} \quad i = 1,2$$

Inspection of the above equation shows how T_1^{DD} depends on the various dynamic processes occurring at the molecular level. First, $1/T_1^{DD}$ increases as the square of the spin quantum number, S. So the use of Gd(III) with seven unpaired electrons will give an almost twofold increase in relaxation rate compared to Mn(II) or Fe(III), which have five unpaired electrons. Next, the rate falls off rapidly with distance, r, since the interaction between the paramagnetic ion and water protons is a dipole–dipole process. Finally, the actual relaxation enhancement is a convolution of three factors contained in the spectral density function. If any of these three factors is much faster than the others, it will determine the relaxation enhancement. Recall that the spectral density function is maximal when $\tau_c = 1/\omega_H$. Thus, efficient water proton relaxation results when the factors influencing τ_c are balanced. Small-molecule contrast agents are typically governed by the rotational correlation time τ_R, which is on the order of 10^{-10} s, while T_{ie}, the electronic relaxation time for gadolinium, is magnetic field dependent and is around 10^{-9} s for the fields employed, and τ_m can vary between 10^{-6} and 10^{-7} s (Caravan *et al.*, 1999). Examination of the required T_{ie} time shows why other lanthanides, having larger angular momentum values due to spin-orbit coupling, fail to give better relaxation enhancement. The spin-orbit coupling gives rise to efficient electronic relaxation pathways, with T_{ie}'s in the 10^{-13} to 10^{-14} s range (Clementi and Luchinat, 1998). This gives a frequency distribution which does not overlap well with the proton Larmor frequency and hence the relaxation gain is minimal even though the magnitude of the dipole is large.

There are numerous reports of MR contrast agents developed over the past 10 years (Caravan *et al.,* 1999). Virtually all of these compounds have been developed in an attempt to discover an agent that maximizes T_1 relaxation rate, and emphasis has been on Gd(III)-based agents. The reduction in T_1 of water protons caused by these agents reduces the saturation effects and hence allows more scans to be taken in a given time. The similarity in size between Gd(III) and Ca(II) stipulates the confinement of the lanthanide to a sequestering ligand in order to prevent toxicity. Since gadolinium, like all rare earths, lacks strong bonding orbitals and is coordinately labile, it must be bound in a chelate structure that can prevent its uptake by the organism under study. This requirement reduces the number of coordination sites accessible to water molecules, thus reducing the effectiveness of the relaxation enhancement. Nonetheless, significant contrast enhancement is still realized and several contrast agents based upon highly kinetically and thermodynamically stable cyclic and acyclic amine–carboxylate chelates have been approved for clinical usage (Fig. 1) (Caravan *et al.,* 1999). These agents enhance contrast wherever they are present and contrast enhancement is derived from differential localization of the agents. Imaging of particular biochemical events has been more the realm of MRS. New probes needed to be developed before MRI could be used to assess biological activity.

III. Biochemically Activated MR Contrast Agents

Recently researchers have developed contrast agents that are sensitive to a number of biologically relevant targets such as pH, Ca^{2+} concentration, and enzyme activity (Louie *et al.,* 2000; Louie and Meade, 2000; Li *et al.,* 1999; Helpern *et al.,* 1987; Mikawa *et al.,* 1998; Moats *et al.,* 1997; Zhang *et al.,* 1999). These examples have provided proof of the principle that contrast agents can act as indicators of biological activity and could open MRI to a wide range of new applications. Particularly relevant to the study of gene expression has been the development of contrast agents that are sensitive to the activity of β-galactosidase, the product of the classic marker gene lacZ.

Unlike existing contrast agents that constantly enhance signal, these agents behave in a conditional fashion and do not act until cleaved by the product of the lacZ gene, the

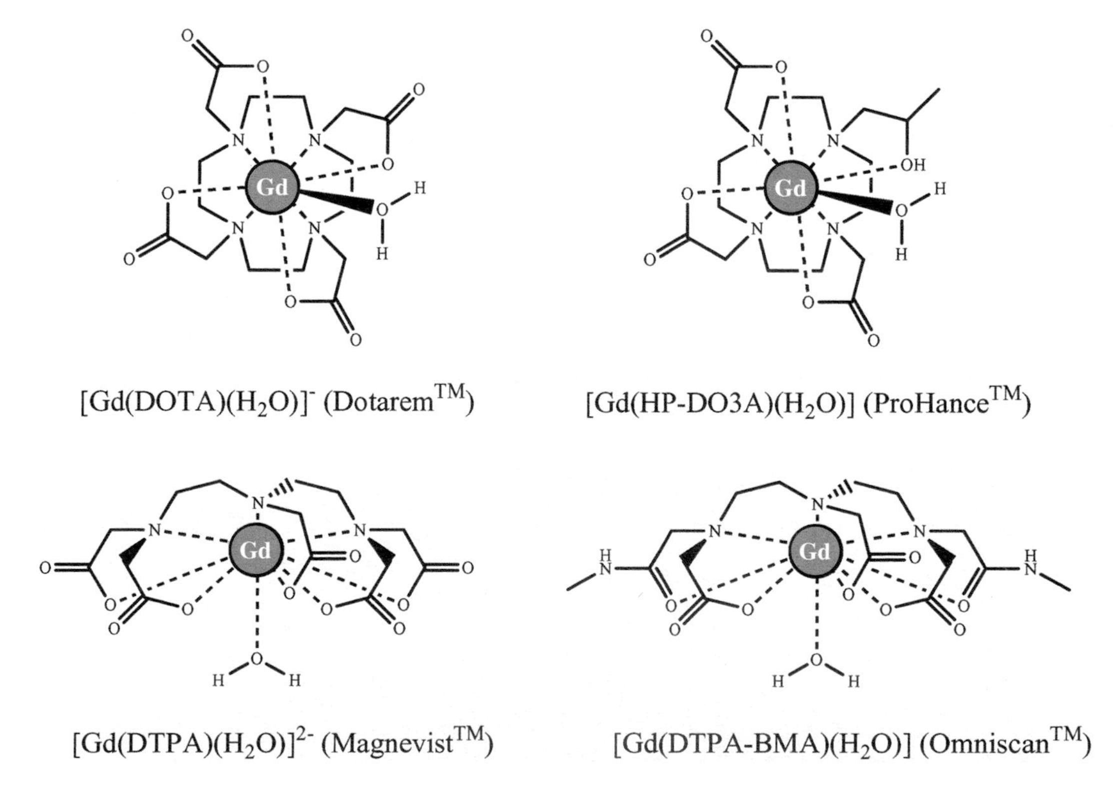

Figure 1 Commercially available contrast agents. These gadolinium-based contrast agents are approved for clinical use (figure by Matt Allen, California Institute of Technology).

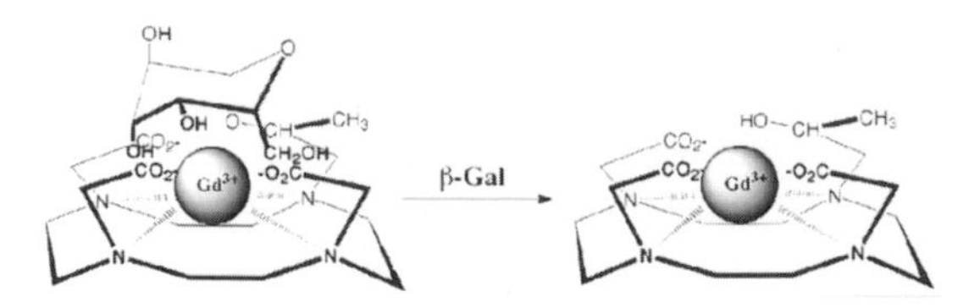

Figure 2 β-Galactosidase-activated contrast agent. Gadolinium has nine coordination sites, seven of which are occupied in EgadMe by the ligand "cage." In addition, this agent possesses a galactopyranose cap (left) that shields a free coordination site of gadolinium from water. This cap is attached via a β-galactosidase-cleavable linker (red bond). Upon cleavage (right), the sugar cap is removed and access of water to the gadolinium ion (magenta) is unblocked. Reproduced from Louie *et al.* (2000), with permission.

enzyme β-galactosidase (β-gal). As discussed earlier, the interaction between water protons and paramagnetic metal centers of contrast agents results in signal enhancement (for T_1-weighted images). Examination of the relaxation equations detailed above shows that changing the number of coordinated water molecules, q, and the distance between the paramagnetic ion and the water protons, r, will affect the inner-sphere relaxation rate. Thus, the synthesis of a contrast agent in which Gd(III) has variable water access would provide a contrast agent of variable relaxivity. If an external physiological process regulates water access, then the contrast agent would be a reporter of the physiological process itself. This principle was applied to synthesize the β-gal-sensitive contrast agent.

These MR contrast agents are based on the framework of a clinical contrast agent, Gd(HP-DO3A), that has been modified with a carbohydrate "cap" that blocks access of water to the gadolinium. Figure 2 presents a schematic of one of these complexes illustrating the exposure of the gadolinium ion (magenta) as the sugar cap is removed (left to right). When access of water to gadolinium is blocked, signal enhancement by the contrast agent is turned "off." The cap is attached to the contrast agent through a β-gal-cleavable linker. Enzyme cleavage releases the cap and opens water access to the gadolinium ion, turning the contrast agent "on." *In vitro* characterization of this agent demonstrated that measured relaxivity changes by a factor of 3 upon cleavage. Interestingly, the change in relaxivity for a similar agent, which lacked the methyl group on the linker between the macrocycle base and the sugar cap, was only 20%. This indicates that structural features can cause large variations in the ability of the sugar cap to prevent water access, suggesting that the ability of these agents to enhance contrast can be tuned via structural modifications.

Experiments in a *Xenopus* embryo system demonstrated the ability of MRI to detect gene expression in living animals (Louie *et al.*, 2000). In these studies contrast agents were introduced into animals in the presence or absence of lacZ expression. Contrast agent was injected into both cells of *Xenopus* embryos at the two-cell stage and then one of these cells was injected with mRNA for or linearized plasmids carrying lacZ. The first cleavage approximates halving the embryo so the two cells represent the future right and left sides of the animal. The injections created an internal control in which one side of the embryo contained contrast agent only, while the other side contained contrast agent *and* either the mRNA or the gene encoding β-gal. Twenty-four hours after injection the animals were anesthetized and imaged by MRI. Following MR imaging the animals were treated to detect β-gal. Figure 3 presents a top view of an embryo by three imaging techniques: fluorescence microscopy, MRI, and bright-field microscopy. In these experiments the mRNA for green fluorescent protein was

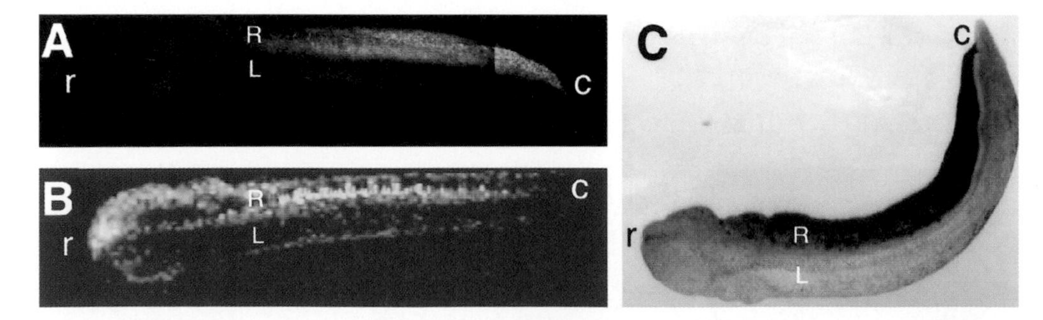

Figure 3 EgadMe detects expression of mRNA in *Xenopus* embryos. *Xenopus* embryos at the two-cell stage were injected with EgadMe, a β-galactosidase-activated contrast agent, into both cells. One of these cells was also injected with the mRNA for lacZ, the gene expressing β-galactosidase, and the mRNA for green fluorescent protein (GFP). GFP was injected as a visual marker for success of the injection. Top views of a *Xenopus* embryo (head to left) are given by three methods. (**A**) Fluorescence image. Top view shows that the injection was localized primarily to the right side of the embryo. (**B**) MR image. High signal intensity by MRI is found on the right side of the embryo. (**C**) Bright-field image of embryo after X-gal staining. Staining for the presence of β-galactosidase shows that enzyme is located primarily on the right side. Scale bar, 1 mm. Reproduced from Louie *et al.* (2000), with permission.

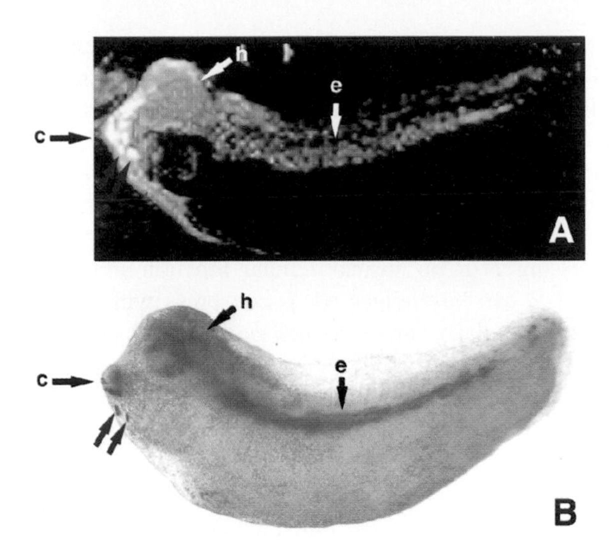

Figure 4 EgadMe detects gene expression in *Xenopus* embryos. In these experiments, *Xenopus* embryos at the two-cell stage were injected with EgadMe into both cells and one of these cells was then injected with linearized plasmid carrying the lacZ gene. (**A**) MR image. Bright signal is found in a stripe of endoderm (e) along the length of the embryo and the head (h), as well as in two distinct spots (arrows) ventral to the cement gland (c). (**B**) Bright-field image after X-gal staining. Staining for β-galactosidase shows that the distribution pattern for the enzyme correlates well with the regions of higher intensity in the MR image. Scale bar, 1 mm. Reproduced from Louie *et al.* (2000), with permission.

co-injected with that for lacZ as a visual marker for injection efficiency. The fluorescence image of Fig. 3A shows that the mRNA injection went to the right side of the embryo (head to the left in all views). The MR image shows distinct enhancement of signal on the side containing the mRNA (Fig. 3B). This correlates well with the location of enzyme expression as detected by traditional staining techniques (Fig. 3C).

The ability of this agent to detect gene expression was determined by injecting linearized plasmids carrying lacZ rather than mRNA (otherwise keeping the same procedure as above). In Fig. 4 we see that both the MR image of the live embryo (Fig. 4A) and bright-field image of the fixed and stained embryo (Fig. 4B) show labeling of a stripe of endoderm along the length of the embryo (e). Labeling is also found in the head (h) and in two distinct spots (red arrows) just ventral to the cement gland (c). The location of enzyme expression correlates well with regions of high signal intensity in the MR image, demonstrating that the contrast agent accurately detects *in vivo* lacZ expression.

The results from these studies are promising and inspire tantalizing possibilities for mapping gene expression in a variety of systems. Such applications will require further refinement of the contrast agents in order to facilitate their delivery to tissues and to maximize their detectability *in vivo*. The currently described generation of enzyme-cleavable agents does not freely cross cell membranes, so intracellular accumulation is limited in most tissues (does not cross the blood–brain barrier). MR imaging requires contrast agent concentrations of 100 μM and up for detection. Furthermore, the kinetics of cleavage for these compounds is slow, a few orders of magnitude slower than for the free substrate. These limitations must be overcome before we can begin using this method to map gene expression in the brain. However, the current contrast agents are useful for specific questions in tissues that readily accumulate the agent or which are amenable to injection. The ability to non-invasively detect gene expression is a major advancement for MR imaging research.

IV. Targeted MR Contrast Agents

An alternative to generating activatable agents is to couple the agents to targeting moieties. These moieties direct the agents to accumulate in specific sites or to bind to particular proteins. The expression of endogenous genes then can be followed by tracking their protein products. Typical methods to accomplish this include binding the contrast agents to proteins through antibody recognition or through protein–protein interactions, such as between a receptor and its ligand (Kresse *et al.,* 1998; Sipkins *et al.,* 2000). The main limitation of this approach is the need to accumulate sufficient contrast agent concentration to produce measurable signal. The study of genes with low levels of expression would be contraindicated. The introduction of exogenous genes (transgenes) for gene therapy or for generating transgenic animals requires expression at high levels and is a better candidate for detection by MRI.

A few recent reports investigate the use of new expression markers and evaluate their effectiveness as markers for MRI of gene expression. One study highlights the use of the transferrin receptor (Tf-R) as a potential transporter for contrast agent uptake (Weissleder *et al.,* 2000). The transferrin receptor is found in most cell types and is part of the iron regulatory system. The receptor binds to transferrin, an iron-carrying protein, and brings it into the cell. Normal expression of the receptor is under feedback regulation to prevent excessive uptake of iron by the cell. However, cells can be engineered to overexpress transferrin. In a recent study, investigators transfected gliosarcoma cells with the transferrin receptor gene (ETR+) to create cells that overexpress the receptor. These cells, and their control counterparts (ETR–), were implanted in the flanks of nude mice to produce tumors that were ETR+ and ETR–. Endogenous sources of diferric iron were insufficient to produce differences in the tumors' image intensities. Therefore, mice were injected intravenously with a contrast agent that is designed to be recognized by the receptor.

These agents consist of human holotransferrin covalently conjugated to low molecular-weight dextrans coating 3-nm

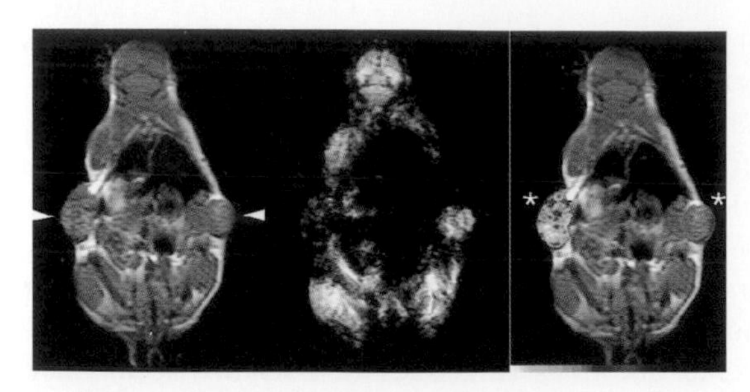

Figure 5 *In vivo* imaging of a single mouse with ETR+ (left arrowheads) and ETR– (right arrowheads) flank tumors. (**A**) T1-weighted coronal SE image (imaging time 3.5 min; voxel resolution 300 × 300 × 3000 μm). ETR+ and ETR– have similar signal intensities. (**B**) T2-weighted gradient-echo image corresponding to the image in A showing substantial differences between ETR+ and ETR– tumors (imaging time 8 min; voxel resolution 330 × 300 × 3000 μm). These differences were most pronounced using T2- and T2*-weighted imaging pulse sequences. (**C**) Composite image of a T1-weighted spin-echo image obtained for anatomic detail with superimposed R_2 changes after Tf-MION administration, as a color map. *Difference in R_2 changes between the ETR+ and the ETR– tumors. Scale bar (bottom left) represents 10 mm, $n = 1$. Reproduced from Weissleder *et al.* (2000), with permission.

monocrystalline iron oxide nanoparticles (Fe Tf-MION). The transferrin is recognized by the receptor and the entire particle is endocytosed by the cell, bringing in Fe, a paramagnetic ion that acts as a contrast agent by affecting the T_2 rate. In Fig. 5 we see three coronal views of a mouse that has been treated as described. The ETR+ tumor lies on the left side (left arrowheads) while the ETR– tumor rests on the right (right arrowheads). Figure 5A shows a T_1-weighted spin-echo image of the animal, illustrating that the tumors have similar intensity. Figure 5B gives a T_2-weighted gradient-echo image of the same animal showing that the ETR+ tumor is considerably darker than the ETR– tumor. Figure 5C presents a composite of the T_1-weighted image with the R_2 difference image, highlighting the regions with largest changes in relaxivity as brightest. These results demonstrate that overexpression of the transferrin receptor in conjunction with the Fe Tf-MION successfully increases the iron content in the ETR+ cells such that measurable MRI contrast can be achieved.

It remains to be determined whether the transferrin receptor can serve as a viable genetic marker. The size of the receptor, coupled with its presence as part of normal cellular homeostasis, is a negative attribute for a reporter gene. Further studies are necessary to evaluate the effect of overexpressing Tf-R on normal cellular function and to verify that Tf-R can be engineered to coexpress with a gene of interest. Similarly, more studies need to be performed to assess how well normal systems tolerate increased levels of iron, both freely circulating and after cellular uptake. An additional limitation for studies of the brain is that transferrin does not cross the blood–brain barrier and it is likely that the large Tf-labeled MION also will not.

Other reports in the literature attempt to directly track the delivery of genes. These couple MRI contrast agents to the gene delivery vehicle to monitor the transport of DNA to specific locations. While these methods do not report on gene expression per se they could prove useful for mapping sites of delivery for gene therapy vectors.

Early studies adapted a system that was targeted to the transferrin receptor pathway (Wagner *et al.,* 1991). In that system positively charged poly-L-lysine chains bound electrostatically to negatively charged DNA, thus neutralizing the overall charges and causing the molecules to condense. When transferrin molecules are conjugated to the poly-L-lysine before condensation, the resulting DNA/polylysine particles are taken up through the transferrin receptor pathway and DNA is successfully expressed (Zenke *et al.,* 1990). Kayyem *et al.* modified this system by attaching gadolinium diethylenetriaminepentaacetic acid (Gd-DPTA) to poly-D-lysine and using a mixture of the Gd-DTPA-modified chains and transferrin-modified chains to form complexes with DNA (Kayyem *et al.,* 1995). Poly-D-lysine with an average length of 180 residues was modified with an average of 22 Gd-DTPA per molecule (Gd-DPTA-PDL). Transferrin polylysine with an average chain length of 220 residues and molar ratio of 2.2–2.5, transferrins per polylysine, was synthesized (Tf-PL).

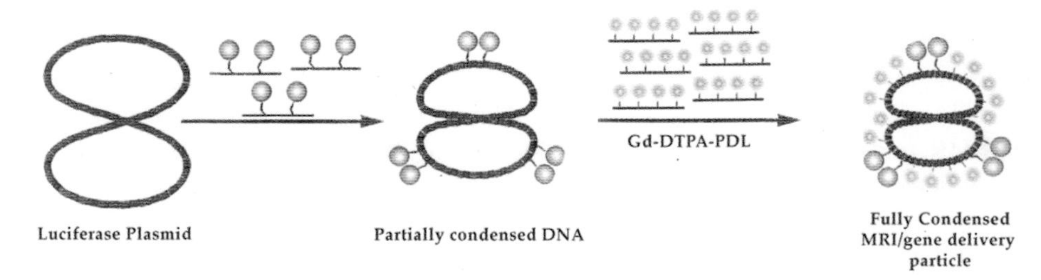

Figure 6 Summary of the formation of ternary complexes for receptor-mediated uptake of DNA and MRI contrast agents. DNA is partially condensed by the addition of suboptimal quantities of transferrin–polylysine. Full condensation to electroneutrality follows the addition of Gd-DTPA-PDL, producing particles with high transfection efficiency and MRI contrast enhancement. Reproduced from Kayyem *et al.* (1995), with permission.

These derivatized polylysine chains were mixed with DNA, carrying the firefly luciferase gene, at a ratio of 6:3:4 by weight of DNA:Tf-PL:Gd-DPTA-PDL (Fig. 6). The resulting particles were tested on K562 human leukemia cells in culture.

Uptake of the particles was shown to be competitively inhibited by free transferrin. Transfection efficiency was measured by the firefly luciferase assay and found to be dramatically enhanced by the presence of the particles. This may be due to the change in net charge for particle-bound DNA versus free DNA. Assays using DNA without the delivery vehicle or in the presence of inhibitor produced background levels of expression, while in the presence of the delivery vehicles production of luciferase was in the 0.4–0.8 ng range. MRI demonstrated that the cells treated with the condensed particles had significantly enhanced intensity over control cells treated with particles containing PDL instead of Gd-DTPA-PDL or control cells treated in the presence of excess free transferrin as an inhibitor of uptake. These studies demonstrated that Gd-DTPA could be introduced as an MRI tracer of gene delivery. The applicability *in vivo* for this system requires further investigation.

More recently, investigators have studied MR tracking of polylysine–DNA gene delivery vehicles that contain polylysine modified dextran chains surrounding a superparamagnetic core (deMarco *et al.,* 1998). These particles are not targeted for uptake by any particular receptor system, but do enhance delivery of genes into cells. Particles were formed by coprecipitating iron chloride in the presence of dextran and modifying the resulting particles with polylysine. The polylysine-modified particles were then condensed with DNA carrying the gene for green fluorescent protein (GFP). The particles were 20 nm in diameter and contained an average of 25 dextran molecules. Extent of polylysine modification and amount of iron in the core were not reported but 4.2–10 linkage sites for polylysine were introduced into dextran and initial dextran/iron coprecipitation took place with a ratio of 39.6:1:2.7, dextran:Fe^{2+}:Fe^{3+} (by weight). Complete particles were introduced into human 293 embryonal kidney cells and transfection efficiency was determined by fluorescence microscopy as the percentage of fluorescent cells. Cells transfected with particles showed measurable expression of GFP in the range of 0.3–4.1%. This is considerably lower than the rate using the $CaPO_4$ method, 52.1%, but higher than for DNA or dextran particle alone (no transfection). *In vivo* tests of the ability of the particles to enhance MR images were performed in rats in parallel with autoradiographic studies using [^{33}P]dCTP-labeled constructs. Particles were injected into the corpus callosum of rats and appeared as a dark spot centered at the site of injection in spin-echo MR images, which corresponded with localization of the particles in autoradiographic images. These studies suggest that there is sufficient contrast to detect gene delivery using these particles; however, the very low transfection efficiency of the system must be improved before this method can be considered as a viable means of gene delivery for therapy purposes.

For both of these systems of gene delivery with MR tracking capability, further characterization *in vivo* is required to evaluate the potential immunogenicity of the delivery vehicles as well as their clearance times from blood. It is also important to bear in mind that identifying the site of delivery for a gene is an indirect method to measure gene expression in that it does not always correlate with the sites or levels of expression of the delivered gene.

V. Magnetic Resonance Spectroscopy and Gene Expression

Magnetic resonance spectroscopy is the subject of another chapter in this book, but we will highlight a few recent reports that are relevant to the study of gene expression. While MRS does not produce three-dimensional images, the technique does provide accurate measurement of gene expression in short time frames and may eventually be harnessed to produce true spatial "images." MRS is capable of identifying cellular metabolites by their chemical shifts. This technique has been successfully applied to study the metabolic effects of overexpressing certain genes in transgenic mice; these genes include creatine kinase, phosphofructokinase, ornithine decarboxylase, and utrophin (Budinger *et al.,* 1999; Goudemant *et al.,* 1998; Kauppinen *et al.,* 1992; Nicolay *et al.,* 1998; Peled-Kamar *et al.,* 1998). More recently investigators have explored the use of MRS-detectable enzymes as reporter genes. We will discuss a few examples of these here.

Two recent reports center on the use of MRS to detect virally mediated gene transfer (Auricchio *et al.,* 2001; Walter *et al.,* 2000). Investigators utilized the MRS-visible markers creatine kinase and arginine kinase to detect gene expression in the liver and muscle, respectively. Both reports use virally mediated methods to introduce the genes with the ultimate goal of employing MRS as a tool to monitor the effectiveness of gene therapy. The liver studies employed creatine kinase, an enzyme that catalyzes the reaction

$$H^+ + PCr + ADP \leftrightarrow ATP + Cr,$$

where PCr is phosphocreatine and Cr is creatine. This enzyme is found in high levels in muscle and brain, but not commonly in the liver. Introduction of creatine kinase into the liver in transgenic animals appears to have no ill effects on the liver (Koretsky *et al.,* 1990), so the enzyme may be useful as a marker in this organ. Its utility for brain mapping is arguable, given that it is already found in high concentration there. But it is instructive to examine the development of this marker for liver as an indication of the applicability of the technique to gene expression in general. The activity of creatine kinase is detected by MRS by measuring the presence of a peak at the chemical shift corresponding to phosphocreatine. In the liver experiments mouse creatine kinase was engineered into an adenovirus vector as described (Auricchio *et al.,* 2001); these vectors tend to target the liver. Adenoviral vectors carrying the creatine kinase gene were then introduced into mice by tail-vein injection. Seven days after injection the mice received an intraperitoneal injection of 2% creatine, then 1 h later MRS spectra were obtained. The investigators used a surface coil placed over exposed liver and isolated the tissues surrounding the liver with copper shielding to eliminate MR signals they might contribute. Results are shown in Fig. 7.

Figure 7A gives a representative spectrum from six control livers showing the lack of a peak where phosphocreatine typically appears compared to Fig. 7B, illustrating the appearance of the phosphocreatine peak in transduced livers (representative sample from six). These results demonstrate that transgenic creatine kinase activity produces a measurable MRS signal above control and hints at the ability of this entity to serve as a marker for gene expression. While the surgical excision of the liver and shielding

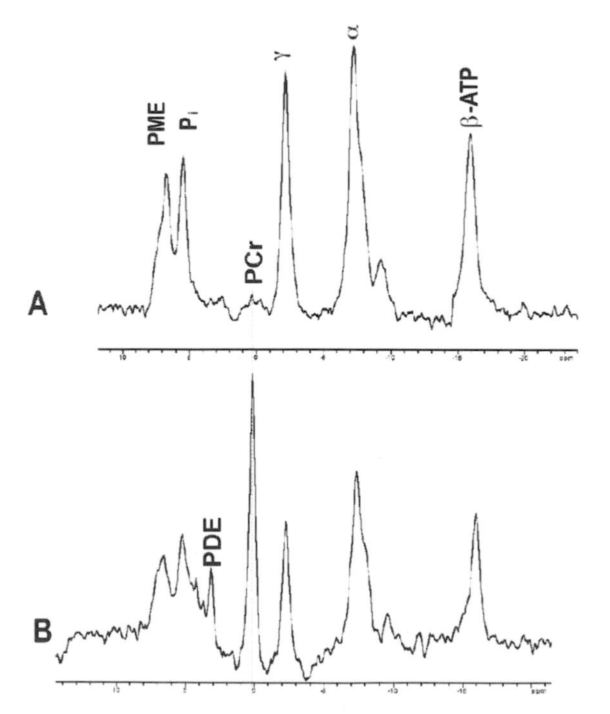

Figure 7 *In vivo* ^{31}P detection of murine livers transduced with the creatine kinase gene. (**A**) Representative spectrum from control livers (n = 6). (**B**) Spectrum from transduced livers (n = 6). PME, phosphomonoesters; P_i, inorganic phosphate; γ, α, and β-ATP, γ,α,β resonance of adenosine triphosphate. Reproduced from Auricchio *et al.* [copyright (2001) National Academy of Sciences, U.S.A.], with permission.

from surrounding tissues to exclude confounding signals from muscle hardly qualify as noninvasive imaging, the authors propose that in the future these types of measurements can be made noninvasively by using localized MRS techniques (such as ^{31}P 3D spectroscopy). These techniques are currently available for use in humans but are difficult to implement on mice. Problems with probe design, spatial resolution, and signal-to-noise ratio are cited as limitations. The use of an endogenous protein as a marker also limits the technique due to high background in some tissues and potential metabolic effects of the gene product in tissues.

A related study measures gene expression in muscle using a different marker system, arginine kinase; this is the invertebrate analog of creatine kinase (Walter *et al.,* 2000). In this system the ^{31}P MRS signal is supplied by phosphoarginine, a unique phosphorus signal in mammalian muscle. Similarly, these studies employ an adenovirus construct to introduce the arginine kinase into mice. The arginine kinase-carrying vectors were injected into the interstitial space of the anterior and posterior muscle compartment of the right hindlimb of neonatal mice. Animals were imaged 12–13 days after injection using a 5-mm surface coil positioned over the hindlimb. The contralateral limb was imaged as a control. Peaks for phosphoarginine and phosphocreatine are both present in the injected limb but phosphoarginine is absent in the control, uninjected limb (refer to paper). Given these results showing that arginine kinase produced MRS-visible phosphoarginine signal in muscle this system has potential for applications in the brain, a tissue with creatine kinase levels similar to those in muscle. Arginine kinase, while of invertebrate origin, still can act as an ATP buffer in mammalian systems, and the consequences of this and increased arginine levels caused by the presence of the introduced gene must be considered before wide use of the enzyme as a marker for gene therapy applications is possible.

A third study used ^{19}F MRS to quantitate transgene expression in tumor xenografts for purposes somewhat different from those of the two studies above. These investigators introduced cytosine deaminase (CD), a microbial enzyme, into cells and used the enzyme to activate a prodrug for chemotherapy of solid tumors (Stegman *et al.,* 1999). Cytosine deaminase catalyzes the conversion of low-toxicity 5-fluorocytosine (5-FC) to the chemotherapeutic agent 5-fluorouracil (5-FU). Cells stably transfected to express yeast cytosine deaminase or their untransfected counterparts were injected subcutaneously into the hindlimbs of nude mice. Tumors were allowed to develop to 250 mm^3, and then the animals were injected intraperitoneally with 5-fluorocytosine. Animals were imaged by MRS every 20 min after injection of 5-FC using a surface coil over the hindlimb. In these experiments, the researchers looked for production of a peak corresponding to 5-FU. Representative spectra are shown in Fig. 8. The upper trace

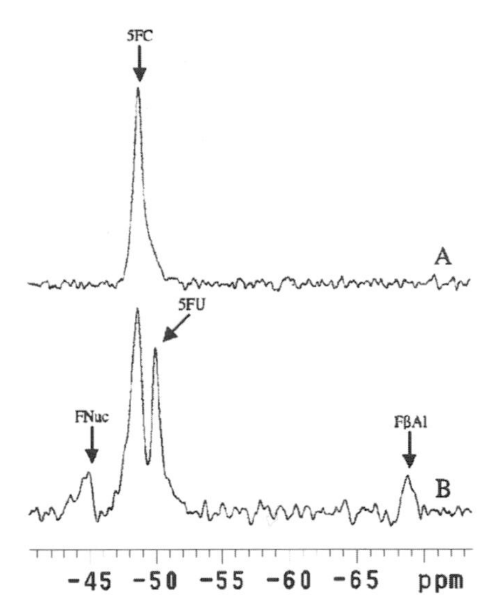

Figure 8 Representative ^{19}F spectra from subcutaneous tumors obtained 120–150 min after ip injection of 1 g/kg 5-FC. (**A**) Untransduced HT29 carcinoma. (**B**) HT29/yCD carcinoma. The chemical shift of FβA1 relative to NaF is –68.2 ppm. Reproduced from Stegman *et al.* [copyright (1999) National Academy of Sciences, U.S.A.], with permission.

is for a tumor developing from untransfected cells, the lower trace is for tumors from CD-expressing cells. These spectra were obtained 120–150 min after injection of 5-FC. These spectra demonstrate that 5-FU is produced only in CD-expressing cells. This is a specialized system in which gene expression is harnessed to produce an active, MRS-visible chemotherapeutic *in situ* and, thus, is not directly applicable to brain mapping. However, the method shows promise for MRS as a measurement tool for gene expression and its future in brain mapping rests in the development of appropriate marker genes/gene products.

^{31}P and ^{19}F spectroscopy suffer from low and limited spatial resolution sensitivity compared to ^{1}H spectroscopy. However, these results show exciting promise for using MRS to *quantify* expression of introduced genes while the ^{1}H MR imaging techniques outlined earlier in the chapter provide merely a qualitative assessment of expression. ^{1}H MRS has been applied in many studies of brain function and neurochemistry (Burlina *et al.,* 2000). It is intriguing to consider the possibility of adapting one of those metabolic markers for use in measuring gene expression in the brain much as the above-described systems were modified from metabolic MRS applications to track gene expression in other tissues.

VI. Conclusion

The promising groundwork reviewed here hints at the power of MRI for *in vivo* imaging of gene expression. MRI has traditionally been employed primarily for imaging of anatomy. With the development of new probes the applications for MRI can extend to map functionally active areas in tissues. But, further innovations are required before the techniques presented here will be applicable to mapping gene expression in the brain. The issue of delivery of molecules into the brain and the development of appropriate markers in the brain will be the challenge for both techniques. With further investigation it is hoped that the array of MR contrast agents available will grow to extend the applications for MRI, just as optical probes expanded the realm of applications of optical imaging.

Acknowledgments

A.Y.L. acknowledges support from the National Institute of General Medical Sciences (F32 GM202526). The authors thank the Biological Imaging Center of the Beckman Institute and the National Institutes of Health [Grants NIAID (AI47003), the Human Brain Project NIMH/NIDA (DA08944), and NCRR (RR13625)] for support.

References

Auricchio, A., Zhou, R., Wilson, J., and Glickson, J. (2001). In vivo detection of gene expression in liver by ^{31}P nuclear magnetic resonance spectroscopy employing creatine kinase as a marker gene. *Proc. Natl. Acad. Sci. USA* **98,** 5205–5210.

Banci, L., Bertini, I., and Luchinat, C. (1991). Nuclear and electron relaxation: The magnetic nucleus-unpaired electron coupling in solution. New York.

Budinger, T., Benaron, D., and Koretsky, A. (1999). Imaging transgenic animals. *Annu. Rev. Biomed. Eng.* **1,** 611–648.

Burlina, A., Aureli, T., Bracco, F., Conti, F., and Battistin, L. (2000). MR spectroscopy: A powerful tool for investigating brain function and neurological diseases. *Neurochem. Res.* **25,** 1365–1372.

Caravan, P., Ellison, J., McMurry, T., and Lauffer, R. (1999). Gadolinium(III) chelates as MRI contrast agents: Structure, dynamics, and applications. *Chem. Rev.* **99,** 2293–2352.

Clementi, V., and Luchinat, C. (1998). NMR and spin relaxation in dimers. *Acc. Chem. Res.* **31,** 351–361.

Curtet, C., Maton, F., Havet, T., Slinkin, M., Mishra, A., Chatal, J., and Muller, R. (1998). Polylysine-Gd-DTPA(n) and polylysine-Gd-DOTA(n) coupled to anti-CEA F(ab′) (2) fragments as potential immunocontrast agents—Relaxometry, biodistribution, and magnetic resonance imaging in nude mice grafted with human colorectal carcinoma. *Invest. Radiol.* **33,** 752–761.

deMarco, G., Bogdanov, A., Marecos, E., Moore, A., Simonova, M., and Weissleder, R. (1998). MR imaging of gene delivery to the central nervous system with an artificial vector. *Radiology* **66,** 65–71.

Drago, R. S. (1992). "Physical Methods for Chemists," 2nd edition. Saunders College Publ., Fort Worth, TX.

Goudemant, J., Deconnick, N., Tinsley, J., Demeure, R., and Robert, A. (1998). Expression of truncated utrophin improves pH recovery in exercising muscles of dystrophic mdx mice: A ^{31}P NMR study. *Neuromusc. Disord.* **8,** 371–379.

Helpern, J., Curtis, J., Hearchen, D., Smith, M., and Welch, K. (1987). The development of a pH-sensitive contrast agent for NMR ^{1}H-NMR. *Magn. Reson. Med.* **5,** 302–305.

Kauppinen, R. A., Halmekyto, M., Alhonen, L., and Janne, J. (1992). Nuclear magnetic resonance spectroscopy study on energy metabolism, intracellular pH, and free Mg^{2+} concentration in the brain of transgenic mice overexpression in human ornithine decarboxylase gene. *J. Neurochem.* **58,** 831–836.

Kayyem, J., Kumar, R., Fraser, S., and Meade, T. (1995). Receptor-targeted cotransport of DNA and magnetic-resonance contrast agents. *Chem. Biol.* **2,** 615–620.

Konda, S., Aref, M., Brechbiel, M., and Wiener, E. (2000). Development of a tumor-targeting MR contrast agent using the high-affinity folate receptor—Work in progress. *Invest. Radiol.* **35,** 50–57.

Koretsky, A., Brosnan, M., Chen, L., Chen, J., and Dyke, T. V. (1990). NMR detection of creatine kinase expressed in liver of transgenic mice: Determination of free ADP levels. *Proc. Natl. Acad. Sci. USA* **87,** 3112–3116.

Kresse, M., Wagner, S., Pfefferer, D., Lawaczeck, R., Elste, V., and Semmler, W. (1998). Targeting of ultrasmall superparamagnetic iron oxide (USPIO) particles to tumor cells in vivo by using transferrin receptor pathways. *Magn. Reson. Med.* **40,** 236–242.

Li, W., Fraser, S., and Meade, T. (1999). A calcium-sensitive magnetic resonance imaging contrast agent. *J. Am. Chem. Soc.* **121,** 1413–1414.

Louie, A., Huber, M., Ahrens, E., Rothbacher, U., Moats, R., Jacobs, R., Fraser, S., and Meade, T. (2000). In vivo visualization of gene expression using magnetic resonance imaging. *Nat. Biotechnol.* **18,** 321–325.

Louie, A., and Meade, T. (2000). Recent advances in MRI: Novel contrast agents shed light on *in vivo* biochemistry. *New Technol. Life Sci. Trends Guide* **December,** 7–11.

Mikawa, M., Miwa, N., Brautigam, M., Akaike, T., and Maruyama, A. (1998). A pH-sensitive contrast agent for functional magnetic resonance imaging (MRI). *Chem. Lett.* **7,** 693–694.

Moats, R., Fraser, S., and Meade, T. (1997). A 'smart' magnetic resonance imaging agent that reports on specific enzymatic activity. *Angew. Chem., Int. Ed.* **36,** 726–728.

Nicolay, K., VanDorsen, F. A., Reese, T., Kruiskamp, M. J., Gellerich, J. F., *et al.* (1998). In situ measurements of creatine kinase flux by NMR. The lessons from bioengineered mice. *Mol. Cell Biol.* **184,** 195–208.

Peled-Kamar, M., Degani, H., Bendel, P., Margalit, R., and Groner, Y. (1998). Altered brain glucose metabolism in transgenic-PFKL mice with elevated L-phosphofructokinase:in vivo NMR studies. *Brain Res.* **810,** 138–145.

Sipkins, D., Cheresh, D., Kazemi, M., Nevin, L., Bednarski, M., and Li, K. (1998). Detection of tumor angiogenesis in vivo by alpha(v)beta(3)-targeted magnetic resonance imaging. *Nat. Med.* **4,** 623–626.

Sipkins, D., Gijbels, K., Tropper, F., Bednarski, M., Li, K., and Steinman, L. (2000). ICAM-1 expression in autoimmune encephalitis visualized using magnetic resonance imaging. *J. Neuroimmunol.* **104,** 1–9.

Stegman, L., Rehemtulla, A., Beattie, B., Kievit, E., Lawrence, T., Blasberg, R., Tjuvajev, J., and Ross, B. (1999). Noninvasive quantitation of cytosine deaminase transgene expression in human tumor xenografts with in vivo magnetic resonance spectroscopy. *Proc. Natl. Acad. Sci. USA* **96,** 9821–9826.

Wagner, E., Cotton, M., Foisner, R., and Birnstiel, M. L. (1991). Transferrin polycation DNA complexes—The effect of polycations on the structure of the complex and DNA delivery to cells. *Proc. Natl. Acad. Sci. USA* **88,** 4255–4259.

Walter, G., Barton, E., and Sweeney, H. (2000). Noninvasive measurement of gene expression in skeletal muscle. *Proc. Natl. Acad. Sci. USA* **97,** 5151–5155.

Weissleder, R., Moore, A., Mahmood, U., Bhorade, R., Beneviste, H., Chiocca, E. A., and Basilion, J. P. (2000). In vivo magnetic resonance imaging of transgene expression. *Nat. Med.* **6,** 351–354.

Zenke, M., Steinlein, P., Wagner, E., Cotton, M., Beug, H., and Birnstiel, M. L. (1990). Receptor-mediated endocytosis of transferrin polycation conjugates: An efficient way to introduce DNA into hematopoietic cells. *Proc. Natl. Acad. Sci. USA* **87,** 3655–3659.

Zhang, S., Wu, K., and Sherry, A. (1999). A novel pH-sensitive MRI contrast agent. *Angew. Chem., Int. Ed.* **38,** 3192–3194.

31

Speculations about the Future

John C. Mazziotta* **and Arthur W. Toga**[†]

**Ahmanson–Lovelace Brain Mapping Center, Department of Neurology, Radiological Sciences, and Medical and Molecular Pharmacology and* [†]*Laboratory of Neuro Imaging, Division of Brain Mapping, Department of Neurology, UCLA School of Medicine, Los Angeles, California 90024*

I. Introduction

Those who predict the future often have the luxury of the fact that few remember their predictions long enough to determine whether they were accurate or not. The second edition of a volume, however, provides just such an opportunity. In the first edition of *Brain Mapping: The Methods* (Toga and Mazziotta, 1996) we speculated on advances that would occur in the methods of brain mapping in the years immediately following its publication. Not surprisingly, those predictions that were accompanied by early examples of the specific applications all matured as methods or applications in the field. We thought it would be useful to begin by reviewing those past predictions to rate our abilities as fortune tellers and to provide you, the reader, with some basis for determining the credibility of our predictions in this edition.

II. Previous Predictions and Their Outcomes

A. Successful Predictions

1. Improved Spatial and Temporal Resolution and Sampling

While some predictions about the future are highly speculative, others are almost inevitable. In most every scientific discipline, from astronomy to physics, there is always the motivation to increase resolution and sampling frequency. This applies to both the spatial and the temporal domains. This is certainly true in the field of brain mapping. The demand for ever-improving spatial and temporal resolution in both structural and functional imaging has been a constant pressure driven by the need to acquire higher quality data sets about the human brain from all methods. This motivation is satisfied only when the method itself matches the anatomy or physiology to be studied. An example of this is the electrophysiological measurements obtained with electroencephalographic (EEG) techniques. Here, temporal resolution can equal or exceed the physiological rate of neuronal firing. At the same time, spatial resolution of EEG methods is far below that which would be required for an accurate map that localized the source of the signals. Thus, while no advances were made or needed in terms of temporal resolution, the need for improved spatial resolution with EEG techniques has motivated a trend toward increasing electrodes applied to the scalp for the collection of electrophysiological source information.

The other situation in which the drive for methodological improvements in the spatial and temporal domain is satisfied is when there is an absolute physical limit to the resolution of a given method. An example of this principle would be PET imaging. With PET, absolute limitations on spatial resolution are determined by the physical characteristics of positron behavior, specifically the positron range (the distance a positron must travel in tissue before it encounters an electron and the two annihilate) in tissue and the angulation of annihilation radiation (180 ± 0.126°) (Evins *et al.,* 1999). For the first limitation, positron range causes mislocation of the event of interest (positron emission by the radionuclide) in proportion to the range effect. The range effect is, in turn, proportional to positron energy. The angulation error also causes mislocation of the event of interest, reaching 2.2 mm for systems with 100-cm detector diameters. It is proportionally small with systems having smaller diameters (Evins *et al.,* 1999). Had PET instruments achieved spatial resolutions that matched these physical constraints, further development toward improving spatial resolution would have been futile. This actually is not the case and, thus, there are still theoretical opportunities to improve spatial resolution with this technique. For PET instruments that are capable of imaging human subjects, practical and financial constraints have curtailed net increments in spatial resolution improvements since the time of our predictions in the first edition of this text. Nevertheless, very high resolution, small-animal PET tomographs (Cherry *et al.,* 1997; Cherry and Gambhir, 2001; Kornblum *et al.,* 2000) have continued to improve in spatial resolution for the reasons stated at the beginning of this section.

We predicted that significant improvements in spatial and temporal resolution as well as sampling would occur with magnetic resonance imaging (MRI) techniques. This has certainly been realized in the 5 years since the first edition was published. For structural MRI, spatial resolution in the range of 1 mm^3 is now routinely achieved and associated with sampling frequencies that are in the same range. Thus, data sets which sample the entire brain structurally frequently use voxel sizes that are $1 \times 1 \times 1$ mm. Through the use of high-field MRI devices, submillimeter spatial resolutions have been obtained. Using 3-, 4-, 7-, and 8-T instruments, it has been possible to produce images in living human subjects with spatial resolutions under 1 mm. Similar advances have been achieved with improvements in spatial resolution for MR angiography and functional MRI (*f*MRI). With this latter technique, published reports have demonstrated submillimeter spatial resolutions. Since the average diameter of a human cerebral cortical column is on the order of 300 μm, these advanced methods are capable of resolving events that occur at this important spatial frequency. All these advances in increased spatial resolution with MRI refer to its use in the human brain using large-bore instruments.

Specialized MRI devices designed to image small animals at ultrahigh fields (e.g., ≥11 T) continue to produce images that have spatial resolutions far in excess of that described for large-bore human devices (e.g., 10–100 μm) (Hyder *et al.,* 2000; Kida *et al.,* 2000). With these instruments, it has been possible to examine cellular groups in living specimens. This achievement was not only predicted 5 years ago but initial demonstrations were illustrated at that time. The ability to perform *f*MRI experiments with these ultrahigh-field instruments (Chapter 12) has also been realized (Hsu *et al.,* 1998; Jezzard *et al.,* 1997; Kamada *et al.,* 1999; Kida *et al.,* 2000; Lahti *et al.,* 1998; Mandeville *et al.,* 1999; Peeters *et al.,* 1999; Reese *et al.,* 2000; Yang *et al.,* 1998). Thus, *f*MRI studies of rodents and small primates are now a reality and will extend the range of experimentation using these devices to laboratory animals. The significance of this achievement should not be underestimated. Currently, ultra high-field *f*MRI in the human brain has been achieved with a spatial resolution of $0.5 \times 0.5 \times 2.0$ mm^3 (see Chapter 12, Fig. 10). The ability to perform the identical functional imaging experiment in a human subject and in an animal allows for validation of animal models and for a continuum between human and animal experimentation. This situation has existed for many decades using PET technology and autoradiography as an animal analogue. The advent of small-animal PET devices (Cherry *et al.,* 1997; Cherry and Gambhir, 2001; Kornblum *et al.,* 2000) produced the same type of continuum of experimentation as has now been achieved with MRI but without the necessity to sacrifice the animal, thereby allowing longitudinal experiments to be undertaken. The net effect of this parallel experimental design, between humans and animals, and the identification of validated animal models of human conditions will result in the ability to test, in specially designed animal models (e.g., transgenic rodents), paradigms, drugs, or manipulations that would be impractical or unethical in human subjects. The results of such experiments may demonstrate the feasibility of a technique which could then be refined or changed with regard to its delivery, so as to make human experimentation possible.

Techniques such as magnetoencephalography have also increased their spatial resolution by increasing the number of superconducting quantum interference devices (SQUIDs) in a given instrument. This increased density of sensors results in improved spatial resolution for detecting electrophysiological events. The same can be said of optical intrinsic signal (OIS) (Toga *et al.,* 1995; Grinvald *et al.,* 1986) and near-infrared-OIS techniques. Here, improved camera designs, light frequency filtering, and other methods (e.g., motion artifact correction) have led to significant improvements in spatial resolution using these optical techniques.

Temporal resolution has also improved, with the most substantial gains being realized using MRI methodologies. Sampling frequency has increased not only due to improved data transfer and bandwidth characteristics of the instruments themselves, but also with regard to the advent of event-related MRI experimental paradigms. Using this approach, strategies similar to those pioneered using event-related potentials (ERPs) and EEG techniques have been applied to the data collection schemes and signal averaging strategies applied to *f*MRI data sets (Goldman *et al.,* 2001; Lemieux *et al.,* 2001). This results in the ability to resolve events that are at or near the temporal time frame dictated by hemodynamic responses (i.e., 1–10 s). Such strategies should go a long way to expand the range of experiments that can be contemplated using *f*MRI methods in both animals and human subjects.

2. The Development and Refinement of Less Invasive Techniques

Just as there is constant pressure to improve the spatial and temporal resolving power of brain imaging techniques, so too there is pressure to reduce their invasiveness. Invasiveness may be defined as risk exposure to the subject and can range from potential toxin or radiation exposure to the setting in which the procedure must be performed. In this latter category, one may include catheterization of arteries in the body in order to produce conventional angiograms or techniques that can be applied directly to the brain only during operative procedures or in animal experiments. An example in the former category would be the considerable replacement of conventional angiography with computed tomographic angiography (CTA) or MR angiography (Lev, 2002). In fact, as improvements in resolution continue to occur, it seems likely that diagnostic angiography will become entirely noninvasive, using either CTA or MR angiography, and that conventional angiography will be utilized only as part of a therapeutic regimen (e.g., the placement of coils, balloons, stents, or drugs for the treatment of arteriovenous malformations, aneurysms, arterial stenoses, or tumors).

An example of a recent modification of an existing technique to make it less invasive can be found in the method of optical intrinsic signal imaging (Grinvald *et al.,* 1986; Malonek *et al.,* 1995; Gratton *et al.,* 2001; Toga *et al.,* 1995). Originally developed for animal experimentation (Grinvald *et al.,* 1986) and requiring the exposed surface of the cortex for data gathering, recent methods using near-infrared portions of the light spectrum have allowed the technique to be used in a noninvasive fashion without the requirement for the removal of the scalp, skull, and dura. This approach was first applied in infants, in whom openings in the skull (i.e., fontanelles) or thin portions of the bone (i.e., infantile temporal bones) allowed the relatively low electromagnetic radiation of near-infrared frequencies to be transmitted to the surface of the brain and their reflected light measured using the OIS method (Kusaka *et al.,* 2001). Refinements in this technique now make it possible to utilize such measurements in normal adult subjects (Benaron *et al.,* 2000; Boas *et al.,* 2001; Dunn *et al.,* 2001; Kusaka *et al.,* 2001; Szapiel, 2001; Zourabian *et al.,* 2000). While light scattering and absorption reduce the sensitivity of the technique and, therefore, its spatial resolution, it nevertheless is a demonstration of the concept that invasive techniques will, when possible, progressively evolve to less invasive applications.

In a similar vein, in the 5 years since the publication of our initial predictions, noninvasive *f*MRI studies, which utilize endogenous signals from hemoglobin in the cerebral vasculature, have largely replaced blood flow studies performed in normal subjects using MRI with exogenous contrast agents, PET, or single-photon-emission computed tomography (SPECT). A trade-off in this evolution is the fact that the purely noninvasive *f*MRI studies produced only relative measurements of cerebral blood flow and cannot provide absolute values which, at present, still require exogenous agents with potential toxic risks.

3. The Application of Techniques from Other Fields

We predicted that methodological strategies developed in other disciplines would be applied to brain mapping problems. There is an abundance of examples that have occurred in the past 6 years. Most notable are statistical techniques that have been transferred from disciplines such as applied mathematics and statistics for utilization with brain mapping data. In addition, *in vitro* techniques for monitoring gene expression have been modified for *in vivo* use, either combined with radioisotope techniques and applied to PET (Phelps, 2000) or combined with gadolinium-based compounds for utilization with MRI (Louie *et al.,* 2000). More direct and obvious technology transfers have occurred between allied fields such as those for measuring chemical constituents in a sample using nuclear magnetic resonance assay techniques as applied to *in vivo* magnetic resonance spectroscopy (Prichard and Rosen, 1994; Prichard *et al.,* 1991) in living subjects.

4. The Merger of Multiple Modalities

At the time of our first predictions, the alignment and registration of data sets from multiple tomographic modalities was already a reality (Woods *et al.,* 1992, 1993, 1998). These approaches made it possible to integrate information in the same subject from data sets acquired with different tomographic techniques (e.g., CT, PET, MRI, *f*MRI). There was also considerable evidence that these same strategies would be useful for averaging data in the same subject over time from the same modality (e.g., serial MRI studies in patients with degenerative diseases or brain tumors to measure rates of change in atrophy or tumor growth). We predicted, and have seen realized in the interval 6 years, that different types of data sets have been integrated with one another.

One example of this has been the integration of *f*MRI, EEG, and magnetoencephalographic (MEG) data sets (Ahlfors *et al.,* 1999; Bonmassar *et al.,* 2001). When applied in the same subject using the same behavioral paradigm, *f*MRI provides useful information on functional localization with good spatial resolution (~3–4 mm) but relatively low temporal resolution. MEG provides the opposite, that is, high temporal resolution with relatively low spatial resolution. The integration of the two techniques produces a combined data set that has both high spatial and high temporal resolution. Nevertheless, this type of strategy is not without its pitfalls. The source of the *f*MRI signal results from hemodynamic changes in perfusion associated with increased neuronal activity resulting from the net increase (or decrease) in cell firing in large numbers of cells over an integrated time frame. The MEG data result from synchronous firing of cells of sufficient number to be detected by surface SQUID devices in a reliable fashion. Thus, the two signal sources are not identical, and using one (i.e., MEG) to provide temporal information for the other (i.e., *f*MRI), with which spatial resolution is excellent, may lead to erroneous results. Various strategies are now under way to determine the weighting factors that each should use for selecting the sites and timing of events, for example, the specific relationship between cellular electrophysiology and *f*MRI responses (Logothetis *et al.,* 2001). If successful, the integrated use of both techniques may result in composite data sets that have improved quality over either of the individual two individually.

*f*MRI has also been integrated with EEG data for two purposes. In the first, the two methods have been used in a fashion analogous to the *f*MRI–MEG data sets discussed above, namely, to provide composite high spatial and temporal resolution data sets in which timing of events is obtained from the EEG data and spatial localization from the *f*MRI data. Clinically, *f*MRI–EEG integration has been to identify pathologic electrophysiologic events and their site of origin. In patients with epilepsy, focal "spikes" occur in the EEG record as subclinical manifestations of the propensity to have clinical seizures. Spike localization is an important aspect of epilepsy diagnostics and the selection of appropriate therapy. The poor spatial resolution of scalp EEG, however, can often make such localization quite difficult. At the time of publication of the first edition of *Brain Mapping: The Methods,* there had already been the demonstration of the ability, at 1.5 T (Ives *et al.,* 1993; Detre *et al.,* 1995), to monitor the timing and location of epileptic spikes in *f*MRI–EEG data sets. Nevertheless, artifacts produced by the radiofrequency signals from the *f*MRI data acquisition contaminated the EEG and movement of electrode wires in the MRI's magnetic field, caused by voluntary subject movement or the ballistocardiogram, resulted in *f*MRI artifacts. In the interval since our last prediction, many of these problems have been solved or minimized. In fact, it is now possible to do "spike mapping" using combined *f*MRI and EEG in MRI units that function at higher fields (e.g., 3 T) (Goldman *et al.,* 2001; Lemieux *et al.,* 2001). This type of integrated strategy should be very useful in diagnostic applications relevant to patients with epilepsy or in studies of normal subjects in which *f*MRI ERPs are of interest (Bonmassar *et al.,* 2001).

5. Improvements and Enhancements in Automated Analyses, Statistical Strategies, and Image Display

Similar to the predicted improvements in spatial and temporal resolution of brain mapping techniques, our predictions on data analysis and display were almost inevitable. The pressure to pool data among subjects within an experiment or from groups of subjects among experiments was recognized as being important for the field and also critical for determining the validity of experimental hypotheses tested across different laboratories and in the performance of meta-analyses across many brain mapping data sets (Fox *et al.,* 1994; Fox and Lancaster, 1994). The application of probability theory and probabilistic approaches to data organization have made such predictions a reality in the past 5 years. Given that no two individuals have identical brain anatomies, for either structure or function, a probabilistic approach was proposed (Mazziotta *et al.,* 1995a,b) and has proven to be one of the most practical strategies for describing and quantifying this type of structural and functional variance. The use of classical probability theories as well as the use of conditional probabilities to describe these relationships has resulted in probabilistic atlases produced at a macroscopic (Mazziotta *et al.,* 1995a,b, 2001a,b; Roland *et al.,* 1994, 1996; Zilles *et al.,* 1997) level from *in vivo* human imaging as well as for microscopic data including cyto- (Amunts *et al.,* 1998, 1999, 2000; Geyer *et al.,* 1999, 2000) and chemoarchitecture (Geyer *et al.,* 1997; Zilles *et al.,* 1995) derived from postmortem human specimens. Similar approaches can and will be applied in nonhuman species as well.

Anyone working in a scientific discipline that is relatively new and rapidly advancing recognizes that the tools for data analysis are typically developed within a given laboratory and utilized locally. They are often developed by individual faculty members or trainees and, as a result, often have limited documentation and poor interoperability when attempts are made to port such software packages and algorithms to other users. This has certainly been the case in the brain mapping field. We predicted that the exchange of analysis algorithms would be refined and advanced in the late 1990s. In addition, we predicted that automated image analysis would become much more commonplace and largely supplant the requirement to manually segment data sets into anatomical or functional regions. This prediction has proven to be quite accurate. Many groups have distributed software packages for general use, following laboratory-specific efforts to improve software interoperability and documentation. Examples of this type include the automated image registration package of Roger Woods, M.D. (Woods *et al.,* 1992, 1993, 1998) (now distributed to over 1655 laboratories) and statistical parametric mapping (Friston *et al.,* 1991, 1995) (now distributed to over 3039 users worldwide). The ability to obtain software packages or modules (containing individual algorithms for a specific step in an analysis pathway) has resulted in the strategy of combining such modules into "analysis pipelines." Such pipelines (Toga *et al.,* 2001) allow the selection of various modules depending on their performance (Strother *et al.,* 1994; Arnold *et al.,* 2001) and on the specific analysis question at hand. The result of this process is the exchange of software, which is labor-intense to develop, among laboratories, thereby obviating the need for each laboratory to develop its own analysis strategies and algorithms for every possible application.

In the area of image display, we predicted new directions would come from those which were available in 1996. At that point, the brain was displayed either in multiple (typically orthographic) two-dimensional planes, to allow the visualization of all of the three-dimensional components of brain structure and function, or as three-dimensional renderings of the entire brain or portions of it, often using transparent or translucent surfaces to visualize deeper structures. Since that time, strategies to flatten (Felleman and Van Essen, 1991; Van Essen and Drury, 1997; Van Essen *et al.,* 1998) or inflate (Dale *et al.,* 1999; Carman *et al.,* 1995; Sereno *et al.,* 1994) the brain have provided important additional visualization tools (Fig. 1). These strategies were designed for visualization of outer cortical surfaces for either the cerebrum or the cerebellum where important structural or functional information is hidden in sulci and, therefore, not apparent in three-dimensional renderings. Flattening or inflating allows one to visualize the entire cortical surface in a single view. The benefits of such approaches, however, are not without a price. Even early cartographers recognized that flattening the three-dimensional surface of the earth into a two-dimensional format required distortions (local expansions or compressions of the surface map) or the cutting of the surface in places in order to flatten it completely. In addition, traditional and intuitive landmarks (for example, major fissures) are no longer intuitively visualized in flattened or inflated renderings. Nevertheless, if the scientific question is better served by being able to see the entire surface of the cortex in one view, this advantage may outweigh artifacts produced in the creation of such a visualization display. In either case, these tools have been and are continuing to be developed. They provide an option for brain mapping investigators to explore and determine the relative merits of each.

6. Increasing Translational Research from Theory to Basic Science Research to Clinical Applications

Another safe prediction was that ideas that were first formulated in a theoretical context would be tested in a research environment and that those that proved successful would ultimately find their way into the clinical arena. A number of examples exist that demonstrate this process. One was noted above for "spike mapping," in that the theoretical idea of imaging spikes identified in EEG data sets could be utilized as a probe to analyze *f*MRI data and enhance spatial localization of these events. Such strategies are now being applied in patients with epilepsy to improve diagnostic accuracy and, potentially, to help guide therapeutic interventions. Advanced analysis strategies for morphometric evaluation of brain structure have resulted in the probabilistic atlases noted above, which quantify variance within the human population (Mazziotta *et al.,* 1995b, 2001a,b). Subpopulations drawn from these larger data sets can provide comparison populations for smaller groups or individual subjects (Thompson *et al.,* 1998, 2000a,b). Using such an approach, combined with advanced image segmentation analysis strategies, it is possible to find subtle structural abnormalities in the brain. This too has been applied in clinical environments. For example, small and qualitatively subtle heterotopias that produce focal epilepsy are frequently missed when conventional MRI studies of brain structure are qualitatively analyzed by neuroradiologists. This is because the variance in cortical thickness is large among individuals in the population. Nevertheless, when that variance can be quantified through probabilistic strategies and compared to single individuals, such heterotopias have been identified (Bastos *et al.,* 1999) because they fall outside of the confidence limits for the normal variance of the brain in a given cortical region. In some cases patients have had as many as three or four qualitative MRI studies demonstrating no abnormality even when the EEG focus of seizures had been identified and could be used by the radiologist to bias their qualitative examination of structural MRI studies for abnormalities. The quantitative techniques have been able identify these "missed" heterotopias (Bastos *et al.,* 1999), in some cases leading to surgical resection of these cortical zones and improvement or resolution of the patient's seizure disorder.

The use of brain mapping techniques for both pre- and intraoperative evaluation of patients also demonstrates the delivery of these methods, developed for research purposes, into clinical practice. Now patients that have lesions that require surgical resection and that are near critical cortical areas (e.g., language, motor, visual) have preoperative brain mapping to identify the sites of these functionally important cortical zones so that the resection strategy can avoid them

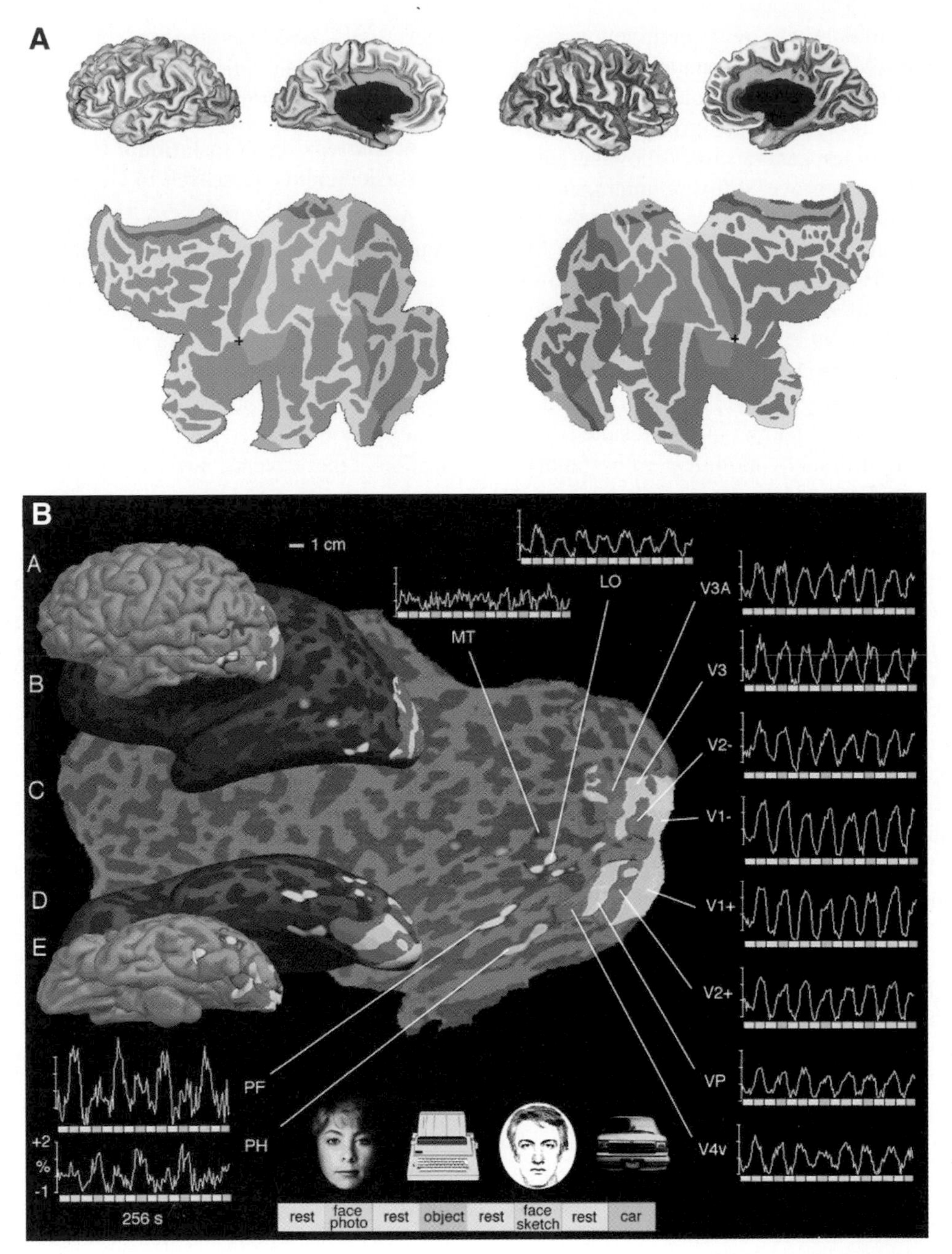

Figure 1 Visualization strategies for the human brain. Techniques to flatten the cerebral and cerebellar cortices of the brain have been advocated by some groups because they provide a means by which to see the entire cortical surface as opposed to the conventional three-dimensional view in which cortical regions that are hidden in sulci cannot be conveniently seen. (**A**) Flattening strategies have been applied to both the cerebral and the cerebellar cortices. Analogous in some ways to the flattening strategies used by cartographers to produce flat maps of the earth (see Fig. 5), such techniques require cutting the surface to produce this type of projection. In addition, certain cortical regions must be stretched or compressed on the planar surface. Flat maps of the left and right cerebral hemispheres are shown here. Each map was aligned by making the mean orientation of the fundus of the central sulcus on the flat map match the visually estimated average orientation of the lips of the central sulcus in the 3D reconstruction. For a region that is folded but not intrinsically curved, a mean curvature of ±0.5 mm^{-1} (maximum on the scale) is equivalent to a cylinder of 1-mm radius. **Top:** Medial and lateral views of the intact hemispheres, with lobes identified according to landmarks delineated by Ono *et al.* (1990) and suitably colored (occipital lobe in pink, parietal lobe in green, temporal lobe in blue, frontal lobe in beige, and the limbic lobe in lavender). **Bottom:** The same flat maps with lobes colored and with darker shading applied to all regions of buried cortex, i.e., cortex not externally visible in the intact hemisphere, as determined from the original image slices and from the 3D surface and volume reconstructions. Artificial cuts were introduced to reduce distortion in the flat maps. Courtesy of David van Essen, Ph.D., Washington University, St. Louis, Missouri (van Essen and Drury, 1997). (**B**) "Inflating" the brain is another strategy that has been advocated to provide visualization of the entire cortical surface. This approach also provides a means of visualizing cortical regions hidden in sulci. While this approach does not require cutting of the cortical surface, it does require regional compression or expansion of cortical regions to produce a continuous sheet. The use of either flattening or inflation techniques can provide a valuable strategy for visualizing data despite the fact that artifactual compression or expansion of certain regions is required. In general, if the visualization strategy provides insights in excess of the induced artifacts, then they will prove useful for certain applications. Reproduced from Dale *et al.* (1999), with permission.

(Kettenbach *et al.,* 1999). At UCLA alone, over 400 such patients have been evaluated in this manner. Concomitantly, the use of intraoperative brain mapping techniques such as OIS can provide an ongoing means for updating the location of functionally important cortical areas, as the brain becomes deformed during the surgical procedure due to therapeutic osmotic dehydration of the brain, ventricular drainage, or local swelling associated with the resection itself (Toga *et al.,* 1995). Refinements in the delivery of such techniques in patients as well as advanced strategies for displaying such information for the neurosurgeon are likely to continue in the years to come.

7. Transcranial Magnetic Stimulation

While transcranial magnetic stimulation (TMS) has been used in its modern form for nearly 15 years, its application on a widespread basis to questions of cerebral physiology and for the mapping of brain function in cortical regions was in its infancy at the time of the writing of the first edition of this text. We accurately predicted that this technique would be an area of considerable growth in the years to come. TMS is now in much more widespread use, in isolation as well as in combination with imaging techniques such as MRI, *f*MRI, and PET (see below) (Paus *et al.,* 1997; Fox *et al.,* 1997) (Fig. 2).

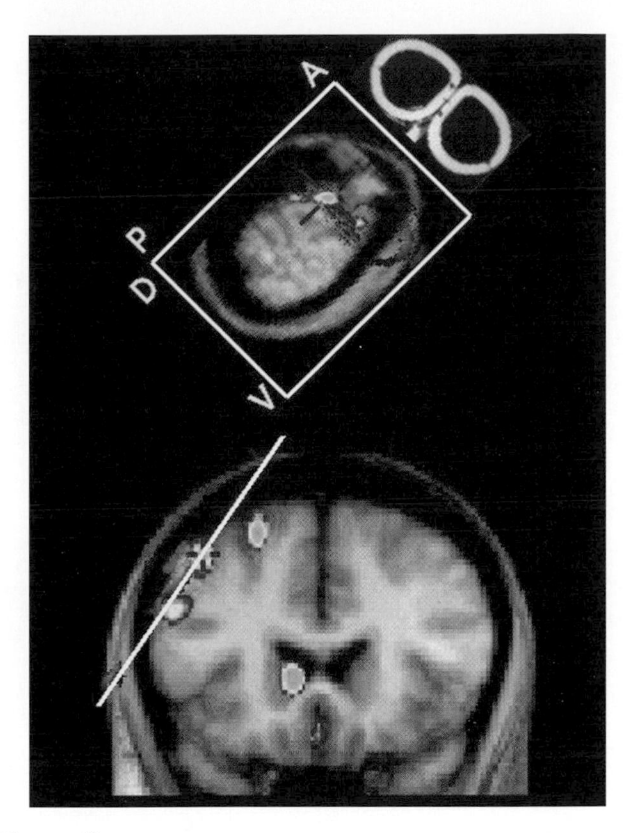

Figure 2 Transcranial magnetic stimulation (TMS) and positron emission tomography (PET). The combined use of these two modalities provides a means by which to evaluate regional connections in the human brain. Cerebral sites of interest can be identified by having subjects perform behavioral tasks using either PET or *f*MRI to acquire relative blood flow measurements. These sites are then targeted with the TMS stimulator while the subject is positioned in a PET device. In this example, repetitive TMS stimulation is delivered to a region of the left midportion of the dorsal lateral prefrontal cortex. **Bottom:** Significant TMS-induced decreases in blood flow obtained in the vicinity of the stimulation site (cross hair) and in the ipsilateral caudate nucleus. **Top:** Tangential section parallel with the coil plane and positioned 20 mm below its surface. The average location of the stimulating coil was derived from transmission scans obtained in each subject; the coil X-ray is shown for comparison. Reproduced from Paus *et al.* (2002), with permission.

8. Neurofeedback

We predicted that brain mapping methods would be used in a closed-loop system in which a subject, performing a task while undergoing physiologic measurement of brain activity, would view the actual changes in their brain function in real time as the task proceeded. Specifically, we suggested that this form of "neurofeedback" could be used to guide and enhance performance during the learning of a new task. No such studies have yet been performed. Nevertheless, there have been strategies whereby *f*MRI has been used to evaluate the interaction of separate subjects in separate tomographs performing competitive or cooperative tasks (Berns *et al.,* 2001). While not a within-subject learning paradigm, this experimental design demonstrates the ability to link interactions between physiologic data and behavioral strategies performed by subjects during the course of a given experiment. We anticipate that similar approaches will be employed in the future and that attempts at neurofeedback, as originally defined and predicted, will ultimately be made.

9. Four-Dimensional Rate-of-Change Studies

The physical world is organized in four dimensions, three in space and one in time. Tomographic brain mapping techniques produce true three-dimensional data while surface techniques such as EEG and OIS produce two-dimensional surface maps. The ability to combine temporal information with two- or three-dimensional spatial information can result in rate-of-change maps. Most commonly used with three-dimensional structural MRI data, four-dimensional rate-of-change maps have demonstrated cross-sectional and longitudinal changes of the corpus callosum during natural development (Thompson *et al.,* 2000a) (Fig. 3). Similarly, these same strategies have been applied in pathologic conditions to observe the four-dimensional growth of tumors, either as a measure of their natural course or to evaluate selective treatment strategies (Haney *et al.,* 2001a,b). The same approach has been used to monitor tissue loss (i.e., atrophy) in neurodegenerative disorders such as Alzheimer's disease (Thompson *et al.,* 1998, 2000a,b). It is anticipated that four-dimensional rate-of-change maps will be applied to functional information measured during the processes of learning, functional loss in degenerative disease, and recovery of function following acute injury such as trauma and stroke, with and without the applications of neurorehabilitation strategies.

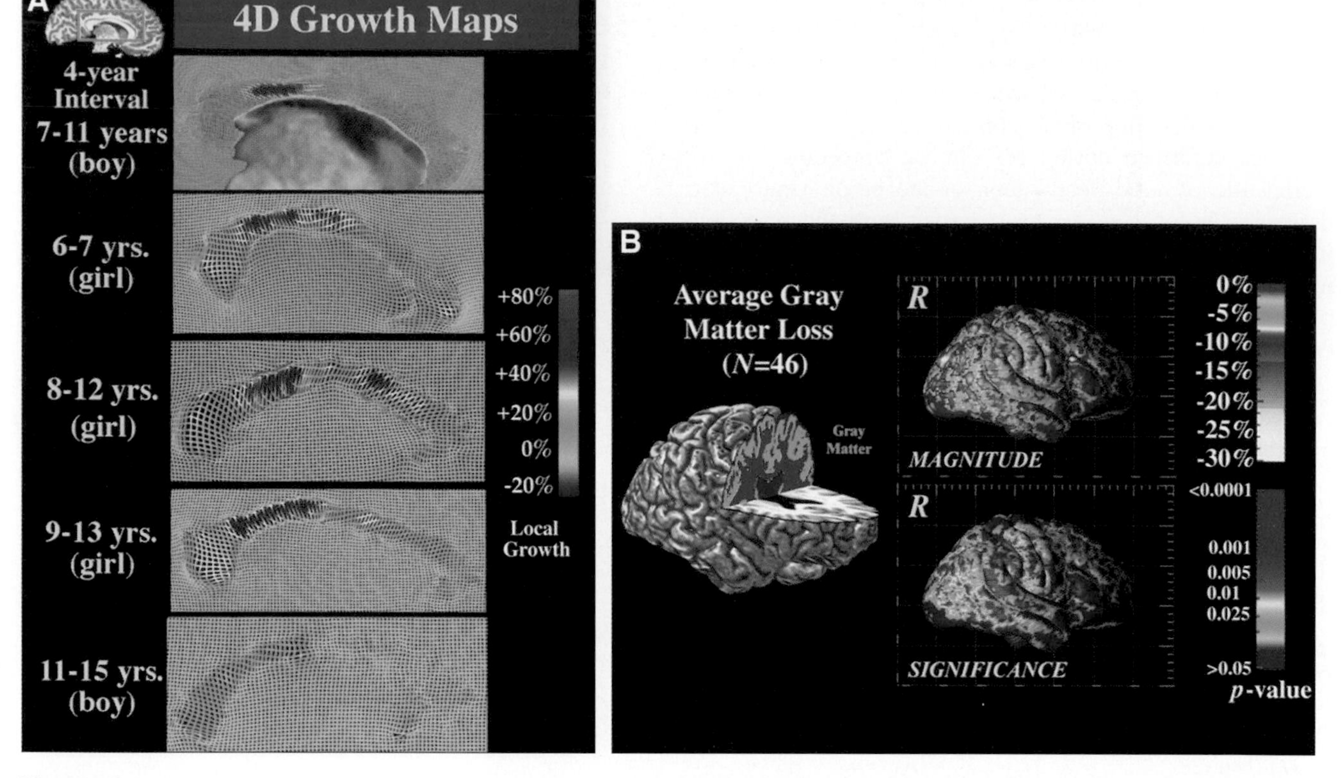

Figure 3 Rate-of-change maps in four dimensions. Four-dimensional maps (three dimensions in space and one in time) provide the means by which to evaluate changes in structure (as seen here) and function. Many applications are of interest in both basic neuroscience and for clinical use. (**A**) Tensor maps of brain growth. Growth rates are revealed for a range of individual children, based on MRI scans acquired over intervals of several years. Tensor maps allow growth rates to be mapped in an individual child. They uncover regions in which growth rates are fastest and visualize changes in these patterns across time. Red colors indicate fastest growth in fiber regions that innervate temporoparietal cortices, a brain region which is functionally specialized to support language function. Note how drastically reduced growth is in the 11–15 year age range, which also marks the end of a critical period when children are most efficient at learning new languages. (**B**) Mapping gray matter loss in a population. Maps of gray matter density across the cortex can be aligned using elastic matching of cortical patterns. A localized and highly significant loss of gray matter is revealed in temporoparietal cortices of patients with mild to moderate Alzheimer's disease ($N = 26$) relative to matched elderly controls ($N = 20$), in a pattern similar to the metabolic and perfusion deficits seen early in the disease. Other clinical applications include the demonstration of the expansion of brain regions such as the ventricular system with increasing hydrocephalus or tumor growth. Reproduced from Thompson *et al.* (2000b, 2001), with permission.

10. Probabilistic Atlases

In the early 1990s the first references to the use of probability strategies to design methods for the establishment of human brain mapping atlases were described (Mazziotta *et al.,* 1995b; Roland *et al.,* 1996). In the 5 years since the first edition, probabilistic strategies have been applied to the data sets acquired *in vivo* from MRI devices for structural imaging studies by the International Consortium for Brain Mapping (Mazziotta *et al.,* 1995b, 2001a,b) (Fig. 4). Concomitantly, similar strategies have been applied to postmortem microscopic data (Toga *et al.,* 1997) with regard to both probabilistic cytoarchitectural and chemoarchitectural data sets (Zilles *et al.,* 1995, 1997). The value of using probabilistic approaches was discussed above in Section II.A.5 as a means of capturing, organizing, and quantifying variance in structural and functional imaging set derived from human subjects. The fact that such strategies would be applied to brain mapping data has been realized in the interval since our initial prediction. It seems likely that this same approach will only continue in the years to come. The use of more advanced probabilistic theories and the application of conditional probabilities to segmented human brain data sets will, most likely, also be seen.

11. International and Global Projects

The field of brain mapping is truly international. This is reflected in the annual meeting of the Organization for Human Brain Mapping at which representation from laboratories located on continents throughout the world has routinely occurred. Brain mapping equipment is quite expensive and the resultant experiments costly to conduct. As a result, we anticipated that international collaborative efforts would be established for both expediency and economic reasons.

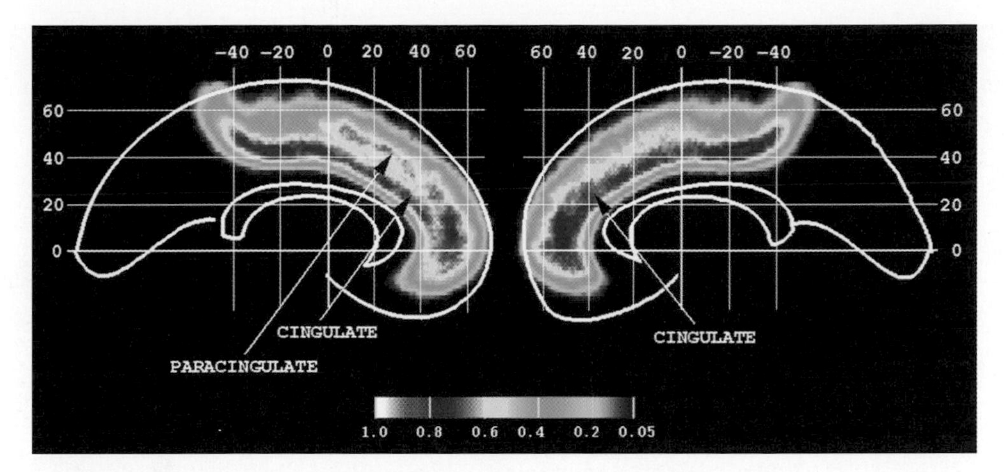

Figure 4 Probabilistic human brain regions. Because of the variance in human brain structure and function, a rational way of expressing anatomy in groups is by the use of probabilistic strategies. Here the cingulate cortex is seen as a probabilistic average of 247 subjects. By combining information from a large number of individuals, the variance in human brain regions can be quantified and described in terms of both magnitude and spatial extent. When local regions are combined for all portions of the brain, a probabilistic atlas emerges that can quantify the structural differences in a large population of human subjects and, when the same approach is used to evaluate functional data from PET, *f*MRI, or other techniques, functional variance can also be quantified in this probabilistic strategy. Reproduced from Paus *et al.* (1996), with permission.

The first approach to developing global systems for sharing data and for structuring meta-analyses of functional brain information sets was provided by BrainMap (Fox *et al.,* 1994; Fox and Lancaster, 1994). This initial effort led the way for future techniques that actually require the acquisition of data as part of large-scale global or international projects.

The International Consortium for Brain Mapping (Mazziotta *et al.,* 1995b, 2001a,b) is a vivid example of a global project. In this project, laboratories on four continents and in eight countries contribute to data to develop a probabilistic atlas and reference system for the human brain (Fig. 5). Additional collaborators participate in other phases of the project as well as commercial companies to evaluate potential products and the rational distribution and maintenance of such a system over longer time durations. Similarly, the European Computerized Data Atlas (Roland *et al.,* 1996) was an attempt to develop collaborative efforts within the continent of Europe for similar purposes. Most recently, an approach to sharing functional MRI data by a group in the United States represents a clearinghouse strategy for sharing costly *f*MRI data that has been acquired in the course of neuroscientific research (Van Horn *et al.,* 2002).

Global collaboration requires both the proper funding structure to support such large-scale endeavors and a shift in the sociology of individual investigators and laboratories (Koslow, 2000). Rather than remaining isolated and publishing individual data sets, laboratories and investigators must pool their efforts in order to realize results that are of a larger scale, more economical, and more expedient than they could accomplish on their own.

B. Unsuccessful Predictions

1. Neurofabrication

In the first edition, we anticipated that integrated sensors and actuators would be created to meet the demands of biological instrumentation (Wise and Najafi, 1991). We felt that these advances would result in instruments that could be guided into the fluid-filled spaces that surround the brain, the parenchyma itself, or the blood vessels. Such methods could provide continuous recording and/or the dispensing of chemical agents, some of which may be visible by external imaging devices. The resulting data could locally measure vessel diameter, intravascular pressure, and even chemical and electrical properties of neural and vascular tissue. Practically none of these concepts have been employed since our original prediction was made. While endoscopic procedures have been used neurosurgically and dialysis probes have been implanted into brain parenchyma for chemical assays, none of the anticipated instrumentation advances have been realized. Nevertheless, these concepts are still viable and if such devices are developed and coupled with transceivers, they offer the future possibility of remote monitoring and control of indwelling nanodevices to effect brain perfusion and neuronal conduction.

2. Clinical Brain Mapping Databases

We predicted that a vast amount of clinical information about the human brain would begin to be organized into databases that would allow access and research applications to be applied to this costly, unique, and massive collection

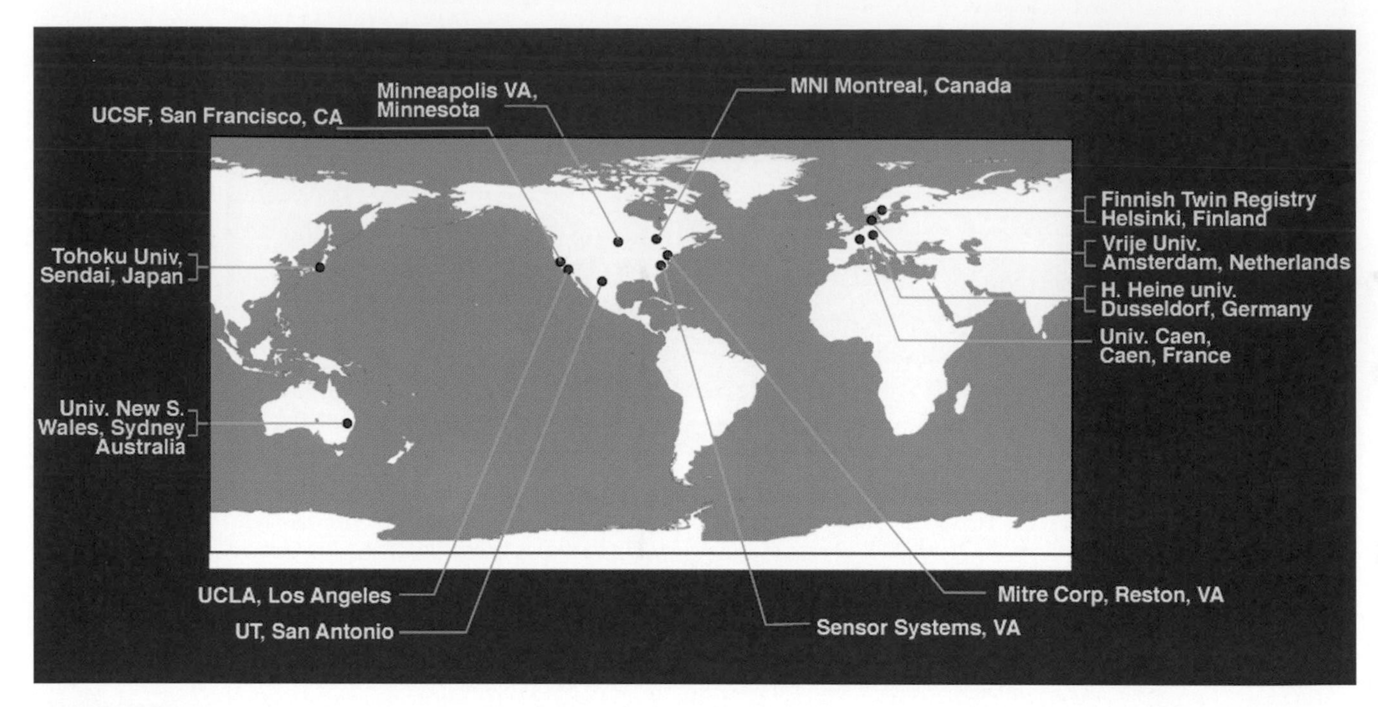

Figure 5 Global collaborations in brain mapping projects. The International Consortium for Brain Mapping (Mazziotta *et al.*, 1995b, 2001b) is an example of an international collaboration to develop a probabilistic atlas and reference system for the human brain. Both academic centers and commercial interests are involved in this venture involving many groups and disciplines. Seven thousand subjects will be studied in this program obtained at eight sites, in seven countries, on four continents. Such global efforts will expedite the pace of neuroscientific discoveries in the brain mapping field as well as foster the exchange and communication of information in all forms among participating groups.

of information. We felt that data such as intraoperative cortical stimulation, electrode recordings, selective intracranial barbiturate administration (Wada testing) (Desmond *et al.*, 1995; Hunter *et al.*, 1999), and other clinical data sets would no longer be isolated in patient charts but rather shared in large databases accessible by the neuroscientific and brain mapping community. Almost none of these predictions have come to fruition. We believe that this is due, in part, to the monumental effort required to accomplish such a task as well as issues related to patient confidentiality and recent ethics committee rules about sharing such personal and proprietary data. Nevertheless, the concept remains a viable one and we continue to anticipate such an eventuality.

III. New Predictions

A. Technical

1. Microimaging Techniques

As was discussed at the beginning of this chapter, it is a safe, if not inevitable, prediction that progressive increases in spatial resolution will lead to "microscopic" *in vivo* imaging techniques in the coming years. Evidence for this can already be found in the use of high-field (>3 T) MRI devices applied to studies in the human brain. Imaging studies at 4, 7, and 8 T have already demonstrated their safety (Robitaille *et al.*, 1998) and ability to produce extremely detailed information about the human brain (Fig. 6). The fact that resolution in the range of 10 μm can be achieve in animals using small-bore instruments at 11 T or greater demonstrates the feasibility of these strategies. With these very high-field techniques, it will be possible, in the future, to investigate very detailed fine structure of the *in vivo* human brain including subnuclear and regionally specific portions of deep brain structures and cortical regions. A critical discovery in this regard was the observation that the energy required for proton excitation at very high magnetic fields (e.g., 8 T) is much lower than predicted by nuclear magnetic resonance theory (Robitaille *et al.*, 1998). This results in fewer concerns related to power absorption and specific absorption rates at ultrahigh fields in human MRI devices. With regard to functional imaging, *f*MRI techniques at these very high fields should result in reliable *f*MRI experiments in animals such as small rodents, leading the way to examination of transgenic differences in response patterns and providing controllable genotype–phenotype–behavioral experiments. In human subjects, these ultra high field resolution studies should unveil the columnar structure of the cortex as the unit of functional organization and control of distributed cortical networks.

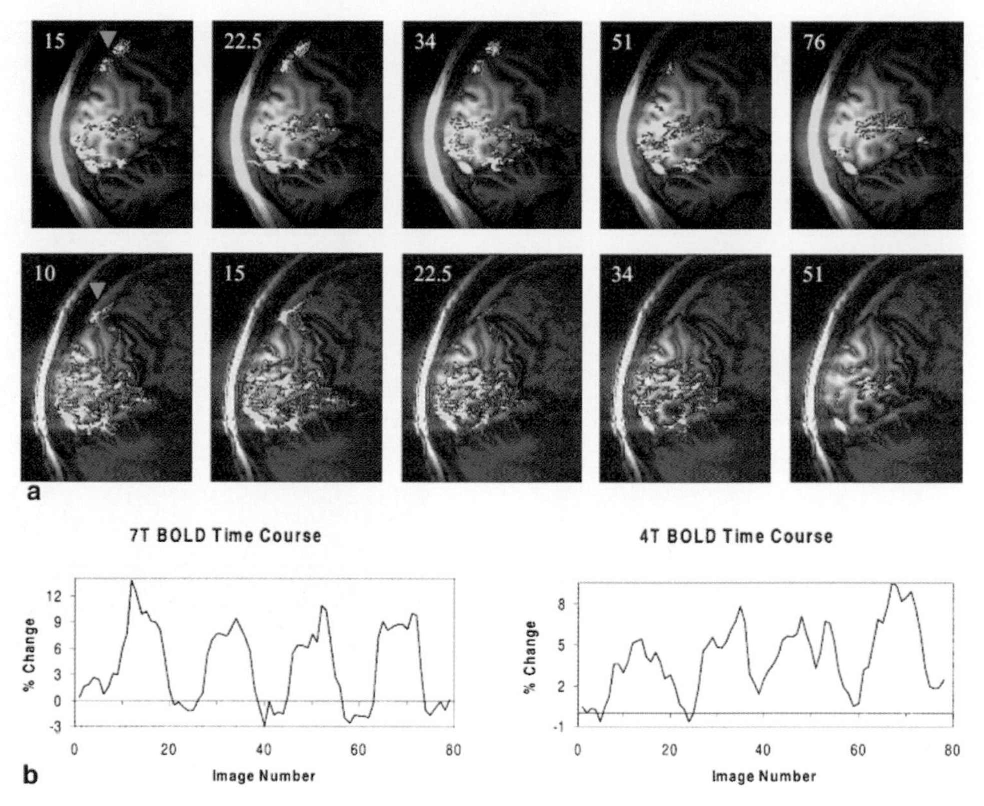

Figure 6 Human cerebral anatomy demonstrated by ultra high field (7 T) MR imaging. With high magnetic fields and appropriate gradients, MR images can demonstrate the fine structure of the human brain *in vivo*. The only limitations to this type of strategy are the inherent physiologic movements in the tissue generated by the ballistocardiogram and voluntary movements of subjects who are required to remain still for extended periods of time. Reproduced from Ugurbil *et al.* (2001), with permission.

Such strategies will undoubtedly be linked to new approaches to apply MR techniques in animals or detailed anatomical regions of the human brain. One approach has been to use the resting T_2* for signal detection. This method is based on the resting, instead of dynamic, changes in the oxygen-dependent signal and, therefore, allows high spatial resolution that can identify small structures in passive human subjects or animals (Small *et al.,* 2000). Changes in resting neuronal function are correlated with resting oxyhemoglobin concentrations as shown in a study in which EEG measurements were coupled with near-infrared spectroscopy (Hoshi *et al.,* 1998). Similar studies have shown a disproportionate change in oxyhemoglobin in the setting of brain disease (Falligatter *et al.,* 1997). Because the resting T_2* signal is known to be sensitive to resting oxyhemoglobin levels in blood vessels as well as in brain (Ogawa *et al.,* 1990) and cardiac tissue (Atalay *et al.,* 1993), the resting T_2* signal map should provide physiologic evidence of neuronal function with improved spatial resolution as demonstrated by Small and colleagues (2000). Their technique known as ROXY (resting oxyhemoglobin)-dependent signal measurement has been used to identify hippocampal subregion responses in human subjects (Fig. 7) as well as mice. They demonstrated that the hippocampal signal was significantly diminished in elderly subjects with memory decline compared to age-matched controls and that this difference was regionally specific for different portions of the hippocampus. Specifically, signal intensity from the subiculum was correlated with memory performance. Since this approach does not require an activation task, it can be used in anesthetized normal and genetically modified mice. In fact, this has already been demonstrated (Small *et al.,* 2000) and is sufficiently sensitive to detect functional changes in the absence of underlying anatomical abnormalities. It is predicted that studies such as these and others like it will advance the application of very high resolution functional techniques and obviate the need for subject participation and the performance of behavioral tasks.

In the area of postmortem analysis, we anticipate a continued resurgence of pure anatomical studies. Such work will include the extension of probabilistic maps of both cyto- and chemoarchitectural regions of the brain (Zilles *et al.,* 1995; Geyer *et al.,* 1997) (Figs. 8 and 9), as well as deep structures including the basal ganglia, thalamus, and brain stem. Further, it is anticipated that agents capable of examining trans-synaptic relationships, such as those

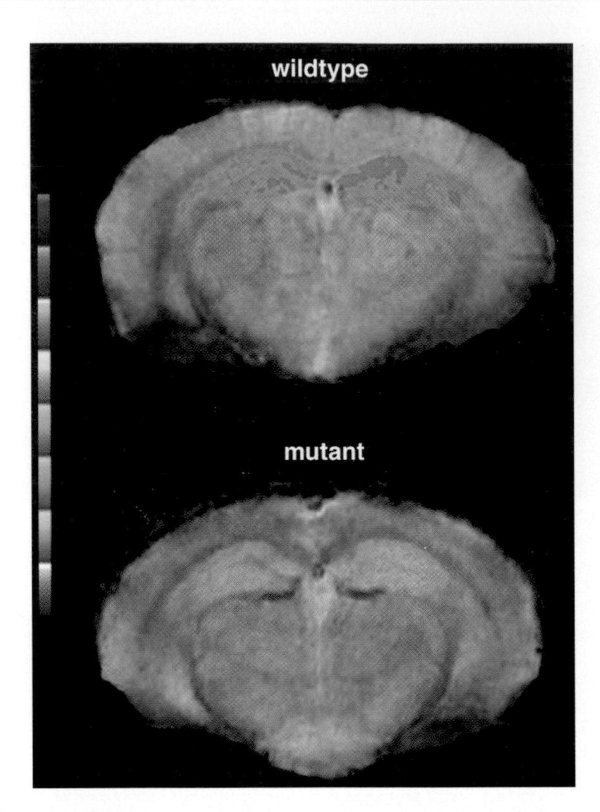

Figure 7 Resting oxygen level-dependent (ROXY) imaging obtained with MRI to show the basal state of perfusion in the hippocampus. The strategy utilizes the approach of measuring T^*_2 values from brain regions without the requirement for subjects to perform tasks. This approach has been demonstrated in human subjects and also in animals at a microscopic level. For example, this approach has been used to evaluate hippocampal changes in transgenic mice, as illustrated here. Courtesy of Scott Small, Ph.D., Columbia University, New York, New York (Small *et al.,* 2000); reproduced by permission of the publisher.

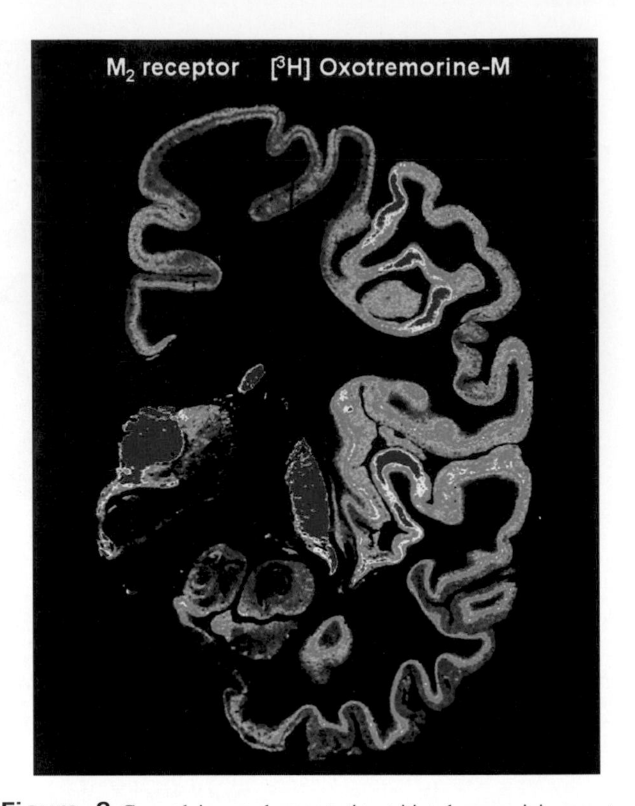

Figure 8 Coronal image demonstrating tritiated muscarinic receptors from one hemisphere of a cryosectioned brain and demonstrating the anatomical detail that such chemoarchitectural maps can provide. When serial sections are obtained and stained for a wide range of receptors, anatomical features, and gene expression maps, a tremendous wealth of information is available for comparison with sites of functional activation obtained using *in vivo* techniques and macroscopic brain structure (gyri, sulci, deep nuclei, white matter tracts). Having a probabilistic strategy for relating these different types of anatomies will provide previously unavailable insights about the relationship of structure and function on both microscopic and macroscopic levels for the human brain and, by analogy, for the brains of other species (see Fig. 9). The analysis of the regional and laminar distribution patterns of transmitter receptors is a powerful tool for revealing the architectonic organization of the human cerebral cortex. This approach requires the preparation of extra-large serial cryostat sections through an unfixed and deep-frozen human hemisphere. Neighboring sections were incubated with tritiated ligands for the demonstration of 15 different receptors of all classical transmitter systems. The distribution of [^{3}H]oxotremorine-M binding to cholinergic muscarinic M2 receptors is shown here as an example. Even a cursory inspection of a color-coded receptor autoradiograph permits the distinction of numerous borders of cortical areas and subcortical nuclei by localized changes in receptor density and regional/laminar patterns. For example, the M2 receptor subtype clearly labels the primary sensory cortices (at the level of the section shown, e.g., the primary somatosensory area BA3b and the primary auditory area BA41) by very high receptor densities sharply restricted to both areas. The different receptors allow the multimodal molecular characterization of each area or nucleus by the so-called receptor fingerprint typing. A receptor fingerprint of a brain region consists of a polar plot based on the mean density of each receptor in the same architectonical unit (area, nucleus, layer, module, striosome, etc.) (K. Zilles, A. Toga, N. Palomero-Gallagher, and J. Mazziotta, unpublished observation).

identified in animals using tract tracing strategies, will also become available. Automated strategies (Zilles *et al.,* 1997) for cytoarchitectural differentiation of cortical regions will reduce the onerous manual labor required to perform these measurements today. The same will be true of chemoarchitectural analyses in which neurotransmitter-specific regions can be separated, one from another, and their concordance or discordance with cytoarchitectural zones, functional activation sites, and macroscopic (e.g., gyral and sulcal) anatomy can be determined (Mazziotta *et al.,* 2001a,b).

2. Tracer Techniques

A key area for the utilization of tracer techniques in human brain mapping experiments will be in the area of monitoring gene expression in human subjects. The ability to image gene expression *in vivo* in humans would provide an important link between the disciplines of molecular biology and phenotype assessment in human subjects. These two rapidly expanding fields would then be joined with a common purpose, namely, the visualization of molecular events that cause disease and interventions, at a molecular level, that can prevent, delay, arrest, or reverse disease progression. All current approaches require tracer techniques.

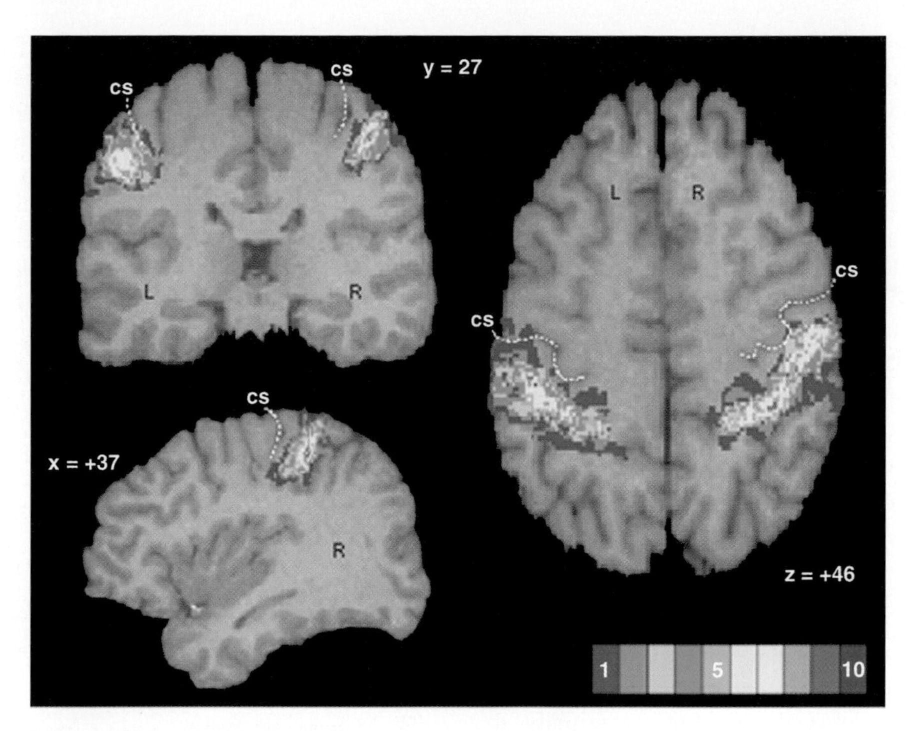

Figure 9 Location and extent of Brodmann area 2 defined from postmortem brain anatomy. The area is defined in serial sections of individual brains after three-dimensional reconstruction. Probability maps of this region, based on microscopic analysis of *N* human brains can be referenced, also in a probabilistic fashion, to functional activation sites associated with the functions of this somatosensory area using the multimodality probabilistic atlas strategy. The overlap of individual postmortem brains is color coded for each voxel of the reference brain. Coronal, sagittal, and horizontal sections through the reference brain at Talairach coordinates $y = -27$, $x = +37$, and $z = +46$, respectively. Population map of area 2 is superimposed. Representation of area 2 in *N* brains is color coded from dark blue ($N = 1$) to dark red ($N = 10$) (cf. color bar). L, left hemisphere; R, right hemisphere; cs, central sulcus. Reproduced from Grefkes *et al.* (2001), with permission.

These can use radioisotope tracers and imaging methods such as PET or SPECT or nonradioactive contrast agents visible with MRI. The more general the solution to the problem of tracing such events, the more generalizable and useful the result. The approach is similar to the standard molecular biology experiment, in which a reporter gene and a reporter probe produce a signal that can be visualized in the cell, typically with some type of optical (e.g., fluorescence) change. Two promising approaches have been described.

To illustrate the strategy that can be used with PET, consider the situation in which the protein product of an expressed PET reporter gene is an enzyme, such as herpes simplex virus thymidine kinase, and the PET reporter probe is a fluorine-18-labeled analog of gancyclovir, [^{18}F]fluorogancyclovir (Phelps, 2000). The strategy works as follows. The PET reporter gene is incorporated into the genome of an adenovirus that is administered to the test subject or animal. Subsequently, [^{18}F]fluorogancyclovir is injected intravenously. It diffuses into cells and, if there has been no gene expression of the reporter gene, it will diffuse back out of cells and will be cleared from circulation by the kidneys. If gene expression has occurred, the [^{18}F]fluorogancyclovir will be phosphorylated by the reporter gene enzyme and trapped in the cell. One molecule of the enzyme can phosphorylate many molecules of the reporter probe so that there is an amplification effect in this enzyme-mediated approach (Phelps, 2000).

Two MRI strategies are under consideration for demonstrating gene expression. In the first, a molecule is created with a gadolinium ion at its center, positioned in such a way that the ion has limited access to water protons and, as such, minimally perturbs the MRI signal. Site-specific placement of a portion of the molecule that can be cleaved by the product of a reporter gene then exposes the gadolinium ion, making it more accessible to water protons and changing the MRI signal at local sites of gene expression. For example, the site-specific placement of a galactopyranocyl ring in the macromolecule can be cleaved by the commonly used reporter gene β-galactosidase (Louie *et al.*, 2000). An alternate approach is to create particles composed of DNA and a polylysine molecule modified with a paramagnetic contrast agent such as gadolinium (Louie *et al.*, 2000). The cotransport of the DNA and MRI contrast agents demonstrates areas of gene expression that can target specific cells *in vivo* as a marker of this process (Kayyem, 1995). These tools will be extremely useful in both research and clinical settings as gene therapy becomes a reality in the years to come.

Finally, it may be possible to produce viral-assisted mapping of neuronal inputs into various cerebral nuclei for the purpose of imaging. This strategy has already been demonstrated using fluorescent markers. DeFalco and colleagues (2001) reported on the development of a pseudorabies virus that can be used for retrograde tracing from selected neurons. This virus encodes a green fluorescent protein marker and replicates only in neurons that express certain agents (e.g., Cre-recombinase) and in neurons in synaptic contact with the originally infected cells. These authors injected virus into the arcuate nucleus of mice that express Cre only in those neurons that also express neuropeptide-Y or the leptin receptor. It was demonstrated that these arcuate neurons receive inputs from the hypothalamus, amygdala, and specific cortical regions (Fig. 10). The strategy of neuronal tracing may prove useful in studies of other complex circuits and may be of great value in trying to identify neuronal connectivity, initially in animal models and possibly in humans, if viral delivery can be made noninvasive and safe.

3. Optical, Magnetophysiological, and Electrophysiological Tomography

Computed tomography, as the name implies, represents a method of calculating a cross section from a three-dimensional volume based on the mathematical analysis of projection data passing through that section at many angles. By changing the energy source and detection system for the data acquisition at all these angles, various anatomical and physiological data sets can be acquired. With MRI, the energy source is radiofrequency electromagnetic radiation and the detector is a radiofrequency antenna positioned inside a strong static magnetic field. For computed tomography, the energy source is X-rays and detectors measure photon fluxes that vary with the tissue attenuation of the X-rays as the X-ray beam passes through

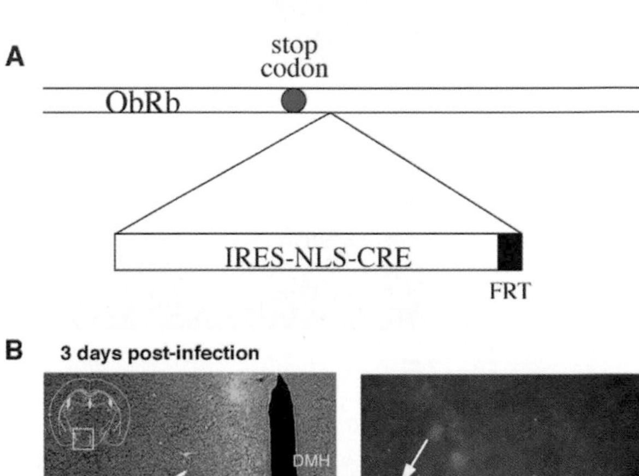

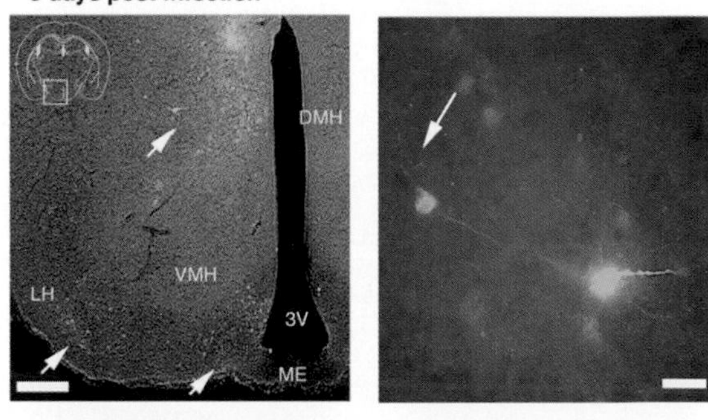

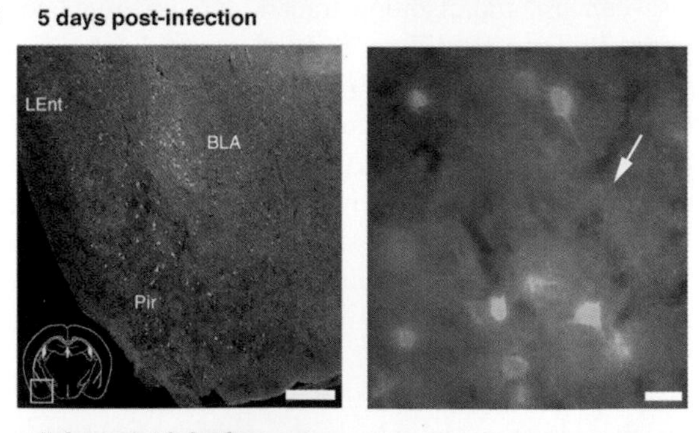

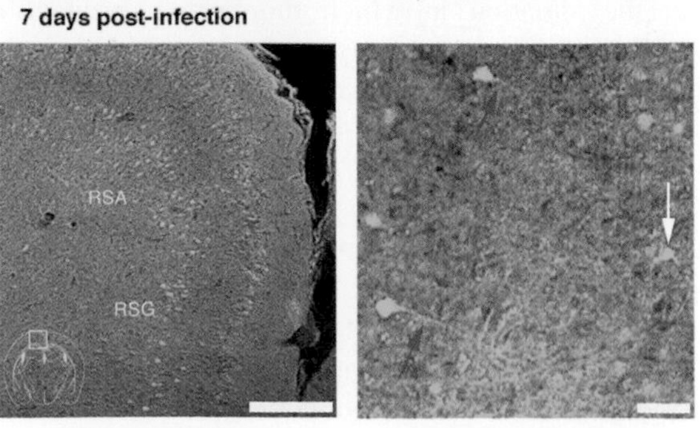

Figure 10 Viral-assisted mapping of neuronal imprints in cerebral nuclei of the mouse using a pseudorabies virus. Ba2001 infection of ObRb-IRES-Cre recombinant mice was performed. A total of 1.3×10^6 pfu of Ba2001 was injected into the arcuate nucleus of ES-cell-derived mice that express Cre from an IRES element inserted into the 3′ untranslated region of the ObRb gene. Animals were sacrificed postinfection at the times indicated. Brains were sectioned and GFP expression was visualized by anti-GFP immunofluorescence (see B–D). Low0power (10×) and high-power (40×) images are shown on the left and right, respectively. (**A**) ObRb IRES-Cre targeting construct. A cassette containing an IRES-NLS-Cre and *neo* gene, flanked by frt sites, was inserted immediately 3′ of the stop codon in the last exon of ObRb. This cassette was then used for recombination in ES cells and positive recombinant ES clones were used to generate ObRb "knock-in" mice. (**B**) Expression of GFP in the hypothalamus at 3 days postinfection. At 3 days, GFP was expressed in the arcuate, DMH, LH, and VMH. Higher magnification of the DMH (right) revealed a possible connection between two DMH neurons (white arrow). Scale bars: left, 200 μm; right, 25 μm. (**C**) GFP expression in the limbic and cortical brain regions. GFP expression in the BLA, Pir, and LEnt at 5 days postinfection (left). Population of GFP-expressing neurons in the basolateral amygdala (BLA) (right). Scale bars: left, 200 μm; right, 25 μm. (**D**) GFP expression in the retrosplenial cortex at 7 days postinfection. GFP-expressing neurons in the granular restrosplenial and agranular cortex. Right illustrates cells with the morphology of pyramidal cells (red arrows) and interneurons (white arrow). Scale bars: left, 200 μm; right, 50 μm. Reproduced from DeFalco *et al.* (2001), with permission.

the tissue. With nuclear techniques, such as PET and SPECT, the energy sources are the radionuclides that emit the products of radioactive decay within tissue and are detected by scintillation crystals linked to photomultiplier tubes surrounding the subject. By making such measurements at many angles, the positions of these radioactive sources within a slice of tissue, imaged tomographically, can be determined.

The same strategy can be applied to optical, magnetophysiological, or electrophysiological information. Here, the energy sources are optical reflectance, electrical fields, or magnetic fields that are induced within the brain itself. By having sufficient sensors (e.g., optical sensors, EEG electrodes, or SQUIDs for MEG studies) around the head, it is possible, theoretically, to reconstruct three-dimensional slice information analogous to that acquired by traditional tomographic techniques. Such a strategy would result in optical or electrophysiologic information that retains its high temporal resolution, already achievable with conventional techniques, with the enhanced spatial resolution and localization power that tomographic strategies provide. While only in its infancy, such strategies have been discussed for some time. It seems feasible that the technical and practical aspects of these problems will be resolved in the years to come and true optical, EEG, and MEG tomographic studies will become a reality.

B. Neuroscience

1. Neuronal Events

Neuronal events take place in the millisecond time frame. Most functional imaging strategies rely on changes in cerebral perfusion or blood volume that typically take seconds to develop. Combining this relatively long physiologic "distance" from the event of importance (i.e., one wants to measure neuronal firing but is really measuring hemodynamic changes) with the methodological requirements to acquire enough data about events to produce a statistically reliable image, one finds that current functional imaging approaches are measurements of integrated events during time frames that are one to two orders of magnitude longer than the events of interest. Nevertheless, one can use signal averaging techniques to improve on this situation. EEG recordings obtained in MR imaging devices capable of producing *f*MRI data have been used to identify the site of epileptic spikes (see above). We predict that this same strategy will be used in healthy subjects who repeatedly perform a task and where EEG data are collected in the standard ERP strategy (Bonmassar *et al.*, 2001) and with animals during depth electrode recordings (Logothetis *et al.*, 2001). With such an approach, the electrophysiologic data from the EEG could be combined with the functional data from MRI and a composite map having both high spatial and temporal resolution will be achieved.

An alternate strategy is to measure changes in ionic concentrations that are associated with neuronal events. Here again, two strategies can be employed. The first is to use MRI and specific natural elements, such as sodium (e.g., intracellular versus extracellular), which can be imaged directly in high-field scanners. The second approach is to use molecules that have "caged" structures and that open at specific ionic concentrations to expose a portion of the molecule that changes the magnetic resonance signal (Li *et al.*, 1999) (Fig. 11). Caged gadolinium molecules have been developed that open at specific calcium concentrations, potentially giving a local map of calcium concentration in the brain. Similar strategies can be used with different ionic probes. By signal averaging to improve temporal resolution, an image that more closely estimates neuronal activity could be developed. As yet, none of these "ionic imaging" strategies have been used in humans but we predict that they will be in the years to come.

2. Connectivity

One may use a combination of imaging techniques to investigate neuronal connectivity (Fox *et al.*, 1997; Paus *et al.*, 1997). Three techniques are typically combined. Consider the following example. A healthy subject performs a motor task during an *f*MRI study to identify the hand area of the motor cortex. Three-dimensional images demonstrating the location of that functional area in that subject's brain are reconstructed. The subject is then positioned in a PET device and a TMS stimulator is targeted, using frameless stereotaxy, to the coordinates of the subject's hand area of the motor cortex. At the time of injection of the PET blood flow tracer, the TMS device is fired, artificially activating the hand area of the motor cortex, thereby causing it to appear on the functional image. Since this activation will also be propagated orthodromically and antidromically, both inputs and outputs from that subject's motor cortex will also increase their relative perfusion and be visualizable in the PET image. The strategy allows for the identification of regions in the brain that are functionally connected through the combination of these three brain mapping methods. By performing such studies in large numbers of subjects and in many brain locations, considerable data can be added to probabilistic atlases, providing the important attribute of connectivity to maps of brain structure and function. Such methods can be used in conjunction with anatomical connection data derived from diffusion MRI techniques (Conturo *et al.*, 1999; Stieltjes *et al.*, 2001). Further, these methods can be compared with functional connectivity maps (Friston *et al.*, 1993; Hyde *et al.*, 1995) to determine the relationship between anatomical and functional connectivity within the human brain.

Connectivity may prove to be one of the most valuable, interesting, and currently least addressed issues of functional neuroanatomy. Connectivity will be critical to explorations of

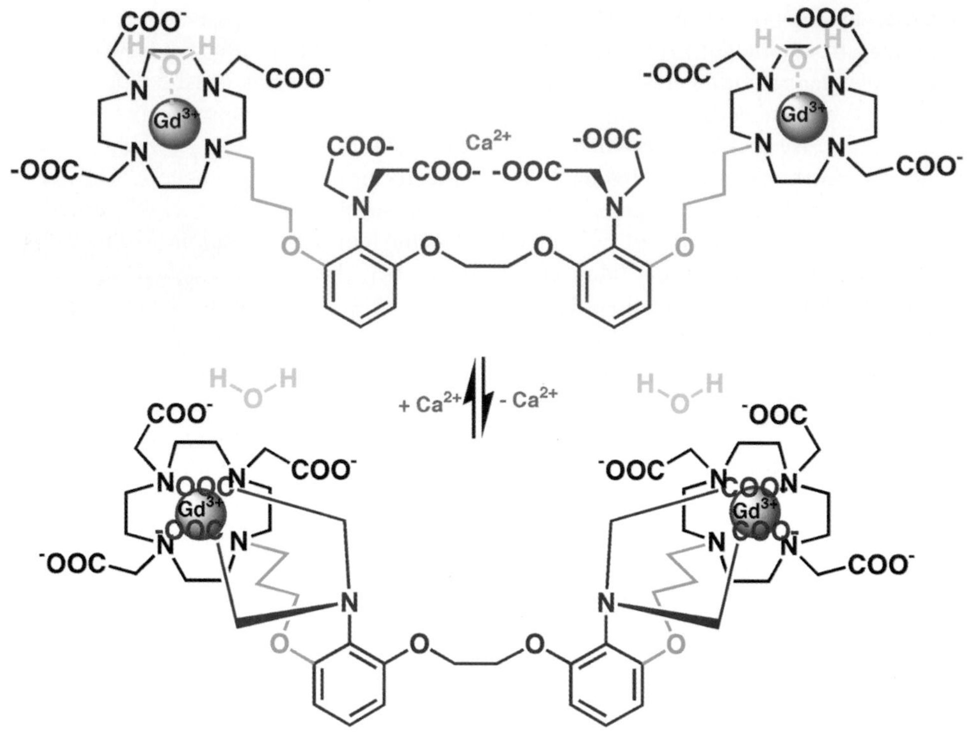

Figure 11 The MRI contrast agent (DOPTA-GD) shows how the relaxivity of the complex is controlled by the presence or absence of the divalent ion Ca^{2+}. By structurally modulating inner-sphere access of water to a chelated Gd^{3+} ion a substantial change in T_1 upon the addition of Ca^{2+} is observed. DOPTA-GD has a Ca^{2+} dissociation constant of 0.96 μM and the relaxivity of the complex increases approximately 80% when Ca^{2+} is titrated to a Ca^{2+}-free solution. Importantly, the agent is selective for binding Ca^{2+} ions versus Mg^{2+} and H^{+}. This new class of MRI contrast agents should provide the ability to image *in vivo* processes of biological and clinical significance. Reproduced from Li *et al.* (1999), with permission.

topics as important and profound as consciousness. In discussions of this biological phenomenon, the concept of signal reentry has been often suggested. "Binding" is the term used to describe the phenomenon whereby information from numerous functional channels is "bound" together to form a conscious precept, either from external sensory inputs or from mental reconstruction of previous memories. Taken together, information derived from effective, anatomical, and other forms of connectivity (e.g., microscopic techniques in postmortem specimens) will be valuable, if not critical, components in developing better theoretical frameworks for understanding cerebral organization and primary or high-level mental processes.

3. Structure–Function Relationships

Combining information about structure and function will be a mainstay of brain mapping strategies in the years to come. We predicted this will occur at both microscopic and macroscopic levels. Population studies from large numbers of subjects will provide information about the anatomical locations and their variance for macroscopic cerebral structures, including gyri, sulci, and the size, shape, and position of deep nuclei as well as white matter tracts (Mazziotta *et al.,* 2001a,b; Steinmetz *et al.,* 1991; Paus *et al.,* 1996a,b; Penhume *et al.,* 1996; Geschwind and Levitsky, 1968; Ono *et al.,* 1990; Talairach and Tournoux, 1988; Thompson *et al.,* 1998). Similar large population data sets will be obtained using functional techniques to provide estimates of functional anatomy in large segments of the population (Dumoulin *et al.,* 2000). Previously mentioned strategies for the automated identification of microscopic cyto- and chemoarchitectural data from significant numbers of postmortem human specimens will provide probability estimates for these variables at a microscopic level (Clark and Miklossy, 1990; Amunts *et al.,* 1998, 2000; Rademacher *et al.,* 1992, 1993; Geyer *et al.,* 2000; Hayes and Lewis, 1995), an issue that has been discussed for the past century (Brodmann, 1909; Bailey and von Bonin, 1951). By combining structural and functional data sets at microscopic and macroscopic levels (Bodegård, 2000), it will be possible to understand, for the first time in large populations in which information about variance in these parameters is quantified, what the important structure–function relationships in the brain actually are. For example, is there concordance or discordance between cytoarchitectural zones and major gyri and sulci in the human brain? Do chemoarchitectural zones, populated by transmitter-specific neurons, correspond to cytoarchitectural regions? Do activation sites in the brain correspond more closely to macroscopic anatomy such as gyri and sulci, or are they better aligned with cyto- or chemoarchitectural regions? If it turns out that functional brain responses are best correlated with cyto- or chemoarchitectural regions, standardized functional tasks may be used as surrogate markers of these more difficult to obtain and costly microscopic assays (Mazziotta *et al.,* 2001a,b).

This information will also have practical importance with regard to clinical questions. For example, if one can develop a probabilistic distribution of functional activation for key sites in the brain, for the targeting of therapeutic lesions, or for locating the placement site of stimulation electrodes (e.g., subthalamic electrodes in patients with Parkinson's disease) will be greatly facilitated and potentially more therapeutically efficacious. The same type of information would be useful in presurgical planning for patients who are under consideration for surgical therapy of resectable cerebral lesions such as tumors or epileptic foci.

4. White Matter Anatomy

We have already discussed improving our understanding of connectivity in the human brain using various methods and agents. Understanding the gross anatomical distribution of white matter tracts will be similarly important. The advent of tensor diffusion imaging (Basser *et al.,* 1996; Conturo *et al.,* 1999; Pierpaoli *et al.,* 1996; Stieltjes *et al.,* 2001) has already provided the opportunity to see the position and direction of large bundles of nerve fibers and their myelin sheaths in human subjects, a topic that has languished for most of the past century (Flechsig, 1920). By collecting such data in large populations, probabilistic estimates of the location and variance of fiber tracts can be developed. By evaluating changes in these tracts during maturation, development (Paus *et al.,* 2001b), and aging, it may be possible to get a better sense of how the brain responds to these natural events and the implications associated with them. This will further enhance the interpretation of activation studies performed to monitor changes in the responses of gray matter regions throughout the human age range.

While less attention has been paid to white matter anatomy and function by the brain mapping community, a resurgence of interest in this area will certainly emerge as the new techniques discussed above mature and because questions will arise relevant to brain maturation and aging that rely heavily on information of this sort. Already, MRI has provided an indirect measure of myelinization (Fig. 12) (Paus *et al.,* 2001b) that can provide important new insights into the relationship between gray and white matter maturation during the human life span.

Finally, a detailed understanding of white matter anatomy will be critical in pre- and intraoperative surgical planning (Kettenbach *et al.,* 1999) in which disruption of a critical

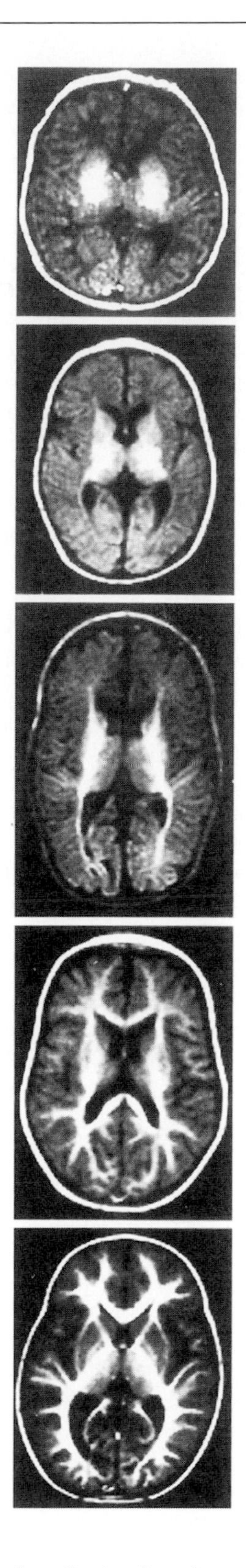

Figure 12 Stages of myelination. Inversion–recovery images (TR = 2800 or 2400 ms, TI = 600 ms, B_0 = 0.6 T) of infant brains acquired at the following stages of myelination (**left to right**): I (1st month), II (2nd month), III (3rd–6th month), IV (7th–9th month), and V (>9th month). Reproduced from Paus *et al.* (2001b) (modified from van der Knaap and Valk, 1990), with permission.

white matter connection is undoubtedly as detrimental to the patient as the resection of the gray matter sites of origin or targets for specific functional systems.

5. "Tissue Clocks"

The combination of EEG and *f*MRI techniques makes possible the spatial identification of particular electrophysiologic events. A critical component of this type of analysis will be in the identification of certain oscillating circuits in the brain that may serve as timing devices or basic principles for cerebral organization (Goldman *et al.,* 2001; Cohen *et al.,* 2001). Such oscillators may play an important role in attentional systems, consciousness, and other global neuronal events. Already, the correlation of power spectra analyses from EEG combined with *f*MRI have produced some initial estimates of localization for sites in the brain associated with certain frequency bands in the EEG signal (Goldman *et al.,* 2001; Backes and van Dijk, 2001). Additional effort on this front is likely to produce important and interesting results in the years to come.

A similar, though more clinical, strategy would involve the understanding of "tissue clocks" for ischemic brain injury (Koroshetz *et al.,* 1997). MRI and PET studies have already provided new insights into the hemodynamic and metabolic events that result in cerebral ischemia and infarction (Derdeyn *et al.,* 1999; Firlik *et al.,* 1998; Lutsep *et al.,* 1997; Marchal *et al.,* 1993; Moseley *et al.,* 1990; Gilman, 1998; Schlaug *et al.,* 1999). However, patients who arrive at hospitals with new ischemic events must be screened for the duration of time between the onset of the event and potential therapy (e.g., tPA). Currently, a 3-h window is allowable and predictive of efficacious intervention using this approach (Hacke *et al.,* 1995; Katzan *et al.,* 2000; National Institute of Neurological Disorders and Stroke rt-PA Stroke Study Group, 1995). While this seems straightforward, the determination of this time interval is often difficult and fraught with errors. For example, a patient may awaken with a new deficit and be unable to determine when it actually occurred. The patient may be unable to report the information personally (e.g., secondary to aphasia) and family members may not have a clear or accurate estimate of when a given patient's symptoms began. Finally, the therapeutic window for the use of agents such as tPA has been determined through clinical trials for which an arbitrary time frame was selected and efficacy determined in a binary fashion (i.e., it worked or it didn't). It may be that there are selected patients for which a time window determined by these strategies is inappropriate, at either end of the temporal spectrum. For all these reasons, it is highly desirable that a tissue clock (Baird and Warach, 1998; Kidwell *et al.,* 2000; Koroshetz *et al.,* 1997; Staroselskaya *et al.,* 2001) be developed using objective and noninvasive tools. It is conceivable that some combination of perfusion, diffusion, and other MRI-based techniques may, in aggregate, provide such an estimate of tissue viability and a window of therapeutic intervention. It is predicted that rigorous, multipulse sequence evaluations of such patients with MRI methods will result in just such a tissue clock, thereby enhancing the application of such therapies to a wider range of patients and providing a better understanding of why and when such methods are efficacious.

6. Large-Scale Databases

In the initial section of this chapter we discussed the development of large-scale databases for the organization of basic science and clinical information about the human brain. We predict that this process will continue and expand in scope. By having large databases of basic neuroscientific information, derived from brain mapping techniques, along with clinical data sets, derived from individual patients but accumulated into accessible and conveniently queried database systems, the pace, scope, and range of neuroscientific progress will be greatly enhanced. The case for this strategy has been made by many over the past 10 years (Bloom, 1996; Fox *et al.,* 1994; Mazziotta *et al.,* 1995b, 2001a,b; Huerta *et al.,* 1993; Koslow, 2000; Roland *et al.,* 1996, Van Horn *et al.,* 2002). In addition, data that are currently lost will be retained. That is, for a given funded study, most investigators reject a certain portion of data because they feel they are contaminated by artifacts. Other data, though valid, are not among the hypotheses proposed for evaluation in a given study. Other data, while potentially part of the hypothesis for a given experiment, are beyond the rigorous interpretation of a given investigator and will not be interpreted because they are too speculative. Still fewer data are actually published. As a result there is a tremendous data reduction from what is actually collected to what is delivered to the neuroscientific community in the form of narrative publications. Those data that are delivered are typically presented in narrative form, requiring arduous and time-consuming investigations by readers to determine supportive or conflicting results among similar experiments. As such, the value of every dollar invested in this type of research is greatly diluted by the process outlined above. Accumulating all data in databases, while it requires a shift in technology, sociology, and investigative behavior, retains all the information for interpretation, at some point, and for easy comparison with other experiments. The same strategy can also lead to electronic hypothesis generation that does not require the costly expenditure of new dollars on pilot experiments. For all these reasons, we predict and, in fact, believe it is inevitable that large-scale, multilaboratory, international databases will emerge and become a standard as they have in the fields of genomics and proteomics (Koslow, 2000; Mazziotta *et al.,* 2001a,b). The examples discussed above in Section II.A.11 illustrate that this process has already begun.

7. Phenotype–Genotype Correlations

The simultaneous collection of phenotypic information from brain mapping experiments in the human nervous system with genotype information derived from the analysis of subjects' DNA will allow for true phenotype–genotype experimentation to flourish in the years to come. If such studies are performed in large enough populations and include twins (both mono- and dizygotic), estimates of heritability for certain traits or disorders should emerge (Mazziotta *et al.,* 2001a,b). There is evidence for this type of evaluation already present in the literature (Bartley *et al.,* 1997; Bookheimer *et al.,* 2000; Egan *et al.,* 2001; Thompson *et al.,* 2001). Studies that seek to determine the relationship between phenotype and genotype can use cohorts of twins to understand the contribution of genetics to human behavior while controlling for environmental factors with twins that are reared together or apart. Thus, the full gamut of genetic and environmental factors can be put in a framework that is manageable and, potentially, scientifically rigorous. While the uses of such information are obvious in studies of patients with specific cerebral disorders, particularly those that are inherited, a wide range of previously unanswered or even unconsidered questions will emerge in the realm of basic neuroscience and for the normal human brain. We predict that such observations will become commonplace in the years to come.

8. Autonomic Nervous System

Most human brain mapping studies have focused on somatic aspects of human sensation and motor output. Few studies have been devoted to autonomic function in the nervous system (Blok *et al.,* 1998; Fowler, 1999; Reiman *et al.,* 1996). We anticipate that this will change. Studies focused on such autonomic functions as the regulation of respiration, heart rate, micturition, defecation, and sweating will be linked with behavioral paradigms associated with the generation of pain, fear, and other cognitive and emotional factors. The lack of information on these autonomic systems is currently a defect in our experimental design planning and should be factored into studies that focus on nonautonomic systems since they are always a contributing factor to the net sum of functional responses associated with these processes. Consider the extreme example of the subject who has volunteered to participate in an experiment on face recognition using visual stimuli. Unfortunately, the subject has a full bladder and did not report this to the investigator prior to being placed in the brain mapping device to perform the proposed experiment. With time, the conscious effort of this subject will progressively shift in focus to his or her autonomic nervous system, providing less and less attentive focus to the task at hand. While this is an extreme example, it nevertheless demonstrates that fact that autonomic function is a constant part of neuronal activity, despite the fact that it is typically monitored in the "background" and is ignored by both the participating subject and the investigator who develops a given experimental design. Understanding the functions of this component of the central nervous system will be important in our future understanding of global brain–behavior relationships and will undoubtedly be part of future experimentation.

9. Consciousness

The understanding of consciousness has been said to be one of the greatest scientific questions we can solve (Atkinson *et al.,* 2000; Georgieff *et al.,* 1998; Tononi and Edelman, 1998; Zeman, 2001). Elusive both in its definition and in strategies to explore its basis, brain mapping techniques do provide a mechanism by which to begin to explore this vexing scientific question. The key to success in this area, in our opinion, is to establish unique and innovative experimental designs that capture "conscious moments." Examples of experiments that do just that include those that have been reported for binocular rivalry (Lumer *et al.,* 1998) and the perception of ambiguous figures (Kleinschmidt *et al.,* 1998) (Fig. 13). In these situations, the stimulus is constant and the perceptions of the subject are constant but there are brief moments in time at which the perception switches from one interpretation to another. Event-related MRI studies that selectively image those areas responsible for the perceptual switch provide clues to the conscious process. Designs such as these combined with previously discussed strategies for measuring neuronal events, cerebral connectivity, and functional anatomy may provide an effective means by which to explore hypotheses such as "binding" and others that have been put forth as a basis for the conscious experience. Additional data on cerebral cyto- and chemoarchitecture are critical in assessing theories of consciousness (Jones, 1998). The development of methods to observe and localize oscillating circuits in the brain using *f*MRI and EEG (Buckow *et al.,* 2001; Gevins *et al.,* 1994; Cohen *et al.,* 2001; Bonmassar *et al.,* 2001; Goldman *et al.,* 2001; Harmony *et al.,* 1999; Mayhew *et al.,* 1996; Seghier *et al.,* 2000) and their correlation with lesion studies (Barbur *et al.,* 1993; Rees *et al.,* 2000; Silbersweig *et al.,* 1995) can further contribute to insights into this important problem.

10. Imaging Emotions

Like autonomic function, emotional interpretation and experience are part of everyday life. Nevertheless, this aspect of human behavior is often neglected in the experimental design of brain mapping studies. Specific experimentation devoted to the understanding of emotional (Adolphs *et al.,* 2000; Bocher *et al.,* 2001; Damasio *et al.,* 2000; George *et al.,* 1996; Nelson *et al.,* 2001) and moral (Dolan, 1999; Heekeren *et al.,* 2001) states is now accelerating in its pace. Such studies can be performed in normal subjects to explore such experiences as humor, lying (Spence *et al.,* 2001), fear, love, depression, sexual desire, anger, empathy (Meltzkoff, 1999), and others or in patient populations in which affective states have reached a pathological magnitude (e.g., clinical depression, mania, phobias). We predict that this will be an

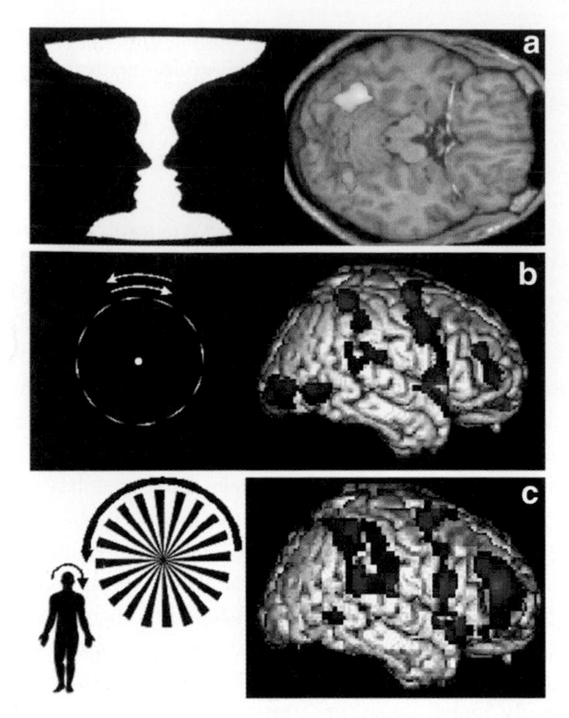

Figure 13 Imaging aspects of consciousness. Ambiguous or bistable images provide insights into conscious processes. Since the contrast in BOLD *f*MRI is endogenous and, thus, constantly present, studies of brain activity changes occurring at unpredictable time points can be imaged and registered by other means (e.g., behavioral reports) and analyzed by post hoc correlations. Examples of this approach are provided in experiments addressing event-related activations during spontaneous switches of perception when confronted with ambiguous images, in which visual perception can become bistable and spontaneously switch between one perceptual interpretation of the visual stimulus and another. As no physical stimulus change is involved in such paradigms, event-related activations occurring during perceptual switches can be attributed to brain processes involved in perceptual synthesis, i.e., the constructive side of vision. In a series of experiments, three different visual ambiguities were studied: (**a**) ambiguous figures such as the "vase or face" picture, (**b**) directional ambiguity of apparent motion (the spinning wheel illusion), and (**c**) the ambiguity of object or illusory self-motion (circular vection) when looking at wide-field coherent motion. Common to all experiments, perceptual switches involved activation of a predominantly right-sided frontoparietal network. The type of perceptual change was reflected by attribute-specific activations in specialized extrastriate cortex, i.e., (a) the fusiform gyrus for object category, (b) the occipitotemporal junction (putative MT homologue) and kinetic occipital area for direction of motion, and (c) the anterior portion of the occipitotemporal junction (putative MSTd homologue) for object vs self-motion (circular vection). Courtesy of Andreas Kleinschmidt, Frankfurt; Richard Frackowiak, Kai Thilo, Adolfo Bronstein, London; Christian Büchel, Hamburg; and Philipp Sterzer, Frankfurt.

important and ever-growing aspect of brain mapping studies in the years to come.

In addition to monitoring, through brain mapping techniques, emotional states in normal subjects and patients, there is also the opportunity to induce such states. One vivid example comes from the study of a patient with Parkinson's disease who had a subthalamic stimulator placed (Bejjani *et al.,* 1999). When a specific electrode in the stimulator was activated, the patient developed profound depression, the functional anatomy of which was identified through a PET study performed concomitantly with the activation of this electrode. Similar examples can be anticipated in patients who undergo stereotactic radiosurgery cingulatomies. We believe that such observations will be more commonplace in the future due to the increasing use of intracerebral stimulators and therapeutic lesion placement and will be extremely instructive in the understanding the functional anatomy of human emotion in both normal and pathologic states.

11. Studies of Infants and Children

There has already been considerable interest in the study of infants and children using brain mapping methods. Unfortunately, many brain mapping methods require subject immobilization or invasive techniques (such as the administration of radioisotopes), thereby making such studies impractical or unethical in immature subjects. Nevertheless, it has been possible, under selective circumstances and with behavioral conditioning, to evaluate children and even infants, using MRI strategies. Electrode caps have made possible the study of EEG and ERP data from infants and young children. Recent advances in the use of OIS and near-infrared optical imaging may also make it possible to obtain data with high spatial and temporal resolution from infants and children. Subjects in this spectrum of the human age range provide critically important information about functional organization of the brain and, because of the magnitude and rate of plastic changes in this age group, the ability to understand such processes and their natural course. When performed longitudinally (Thompson *et al.,* 2000a) or in large-population cross-sectional studies (Sowell *et al.,* 1999), critical insights about brain function will emerge. We predict that such studies will be performed and applied to such important questions such as learning theory, human development, education, and neuronal plasticity.

12. Plasticity

The study of human learning will dominate a large component of brain mapping studies in the years to come. This is true not only because of its importance, as an observation unto itself, but also because of its valuable therapeutic insights for patients with damage to the central nervous system. We predict that brain mapping experiments will be used to evaluate not only the natural course of learning, but also strategies designed to accelerate this process. In addition, studies will be performed to understand the relearning of tasks that were learned in an inappropriate way for a variety of reasons. For example, it has been determined that patients with dyslexia have difficulty in auditory discrimination of certain strings of sounds (Merzenich *et al.,* 1996; Tallal *et al.,* 1996). By slowing the rate of delivery of such sounds, affected individuals can begin to appreciate differences and better understand the total sound patterns. Coincidentally, success in this auditory domain results in improved reading ability. By using an interactive strategy and intense relearning of this process (i.e., FastForword) sustained improvements in reading ability have been achieved. While this is but a single example, similar strategies will undoubtedly be tested as a way of improving education for both normal children and those with learning disabilities.

In patients, the use of brain mapping techniques in individuals with acute brain injury, such as those that follow trauma or cerebral infarction, or chronic brain injury, such as those that occur in the course of neurodegenerative disorders, will be vitally important for the neurorehabilitation of such individuals. Given the fact that it has been estimated that there are over 54 million neurologically disabled people in the United States alone, the magnitude and importance of this problem dictate that it will be an area targeted for research with brain mapping techniques. In such patient groups plastic changes could be observed during the natural course of recovery from injury or during active neurorehabilitation strategies using behavioral, pharmacological, or combined approaches. For a recent review of this topic the reader is referred to Chapters 23–26 of *Brain Mapping: The Disorders* (Mazziotta *et al.,* 2000b) and other reviews (Mazziotta, 2000).

Occasionally, plastic reorganization of the brain leads to additional pathologic states. Common examples include phantom limb pain, thalamic pain syndromes, and certain types of dystonia. Brain mapping techniques have and will continue to explore these pathologic reorganizations in order to determine a better understanding of how they occur and the means to minimize or ameliorate the symptoms that they produce.

C. Clinical

1. Experimental Drug Trials

Experimental drug trials are both important and expensive. Any approach that can be used to make such trials more economical and shorter in duration will certainly be adopted. Through the combined use of probabilistic normal and disease-specific atlases, it may be possible to use data from brain mapping methods to serve as a surrogate marker in clinical trials designed to evaluate a pharmacological or behavioral therapy.

Consider the following example. A new experimental drug has been developed for the treatment of relapsing and remitting multiple sclerosis. It is clinically important to compare the efficacy of this drug relative to conventional treatment and to the natural course of the disease. Three groups of subjects are selected, one receiving the experimental drug, one on conventional therapy, and one receiving a placebo. Subjects are serially studied using gadolinium-enhancing lesions as a surrogate marker of disease burden (Weiner *et al.,* 2000). A duration is determined for the trial, with equally spaced MRI evaluations prior to and after the onset of the proposed interventions. Probabilistic atlases are developed from the longitudinal studies in each subject for all three groups. The placebo group provides an estimate of the natural history of disease burden during the interval. The conventional therapy atlas can be used to show the impact of traditional approaches on the natural course of the disease. Last, the experimental therapeutic group is compared with the other two to determine, in an objective, noninvasive, and quantifiable manner, the relative efficacy of the proposed new therapy relative to the natural course of the disease and conventional treatments.

This approach should be extremely appealing because it has the potential to make clinical trials less expensive and more objective. Further, since the data can be continually reanalyzed and used in comparisons with other experimental drug trials, the results of such work are additive rather than performed in isolation. This further enhances the value of the funds spent to perform such studies. Already examples of this type of work have been obtained in patients with multiple sclerosis (Evans *et al.,* 1997; Kamber *et al.,* 1992, 1995; Zijdenbos *et al.,* 1994, 1996) and have been proposed for patients with Alzheimer's disease (Bookheimer *et al.,* 2000). A critical factor in the utilization of such a strategy is that there is a biologically significant and important marker of disease burden that can be estimated using the brain mapping method at hand. Without a strong correlation between the surrogate marker and the biological significance of the disease process, it would be unwise and ineffective to base the efficacy of a given therapeutic strategy on the surrogate marker derived from brain mapping methods (Weiner *et al.,* 2000). We believe that such surrogate markers will be validated and can serve as an important means by which to improve the efficacy and economics of clinical trials on experimental therapeutics.

2. The Use of Brain Mapping Techniques in the Acute Clinical Setting

We have already discussed the possibility of using brain mapping methods, particularly MRI strategies, to develop a tissue clock for the evaluation of patients with acute ischemic injury of the brain. The same approach may be useful in other acute settings such as trauma. In patients with acute head trauma, particularly if it is severe, it is difficult or impossible to determine the ultimate outcome and quality of the patient during the first few days following injury. Extraordinary costs and emotional anguish by the family are expended for patients who might otherwise be judged as nonviable. Concomitantly, it is conceivable that treatment is withheld, delayed, or terminated in some individuals who might ultimately have a better-than-expected clinical outcome. If it were possible to reliably determine the prognosis for such patients in the acute setting, using one or more brain mapping methods, the net result will be a reduced economic burden to families and health care plans as well as the important and substantial reduction in family anguish over the ambiguity of this type of clinical situation. We predict that brain mapping methods will be used to determine prognosis and will be an important part of the evaluations of such patients.

It is quite likely, and in fact inevitable, that brain mapping techniques will continue to be of great value in the pre- and intraoperative evaluations of patients undergoing neurosurgical procedures (Lee *et al.,* 1999; Kettenbach *et al.,* 1999).

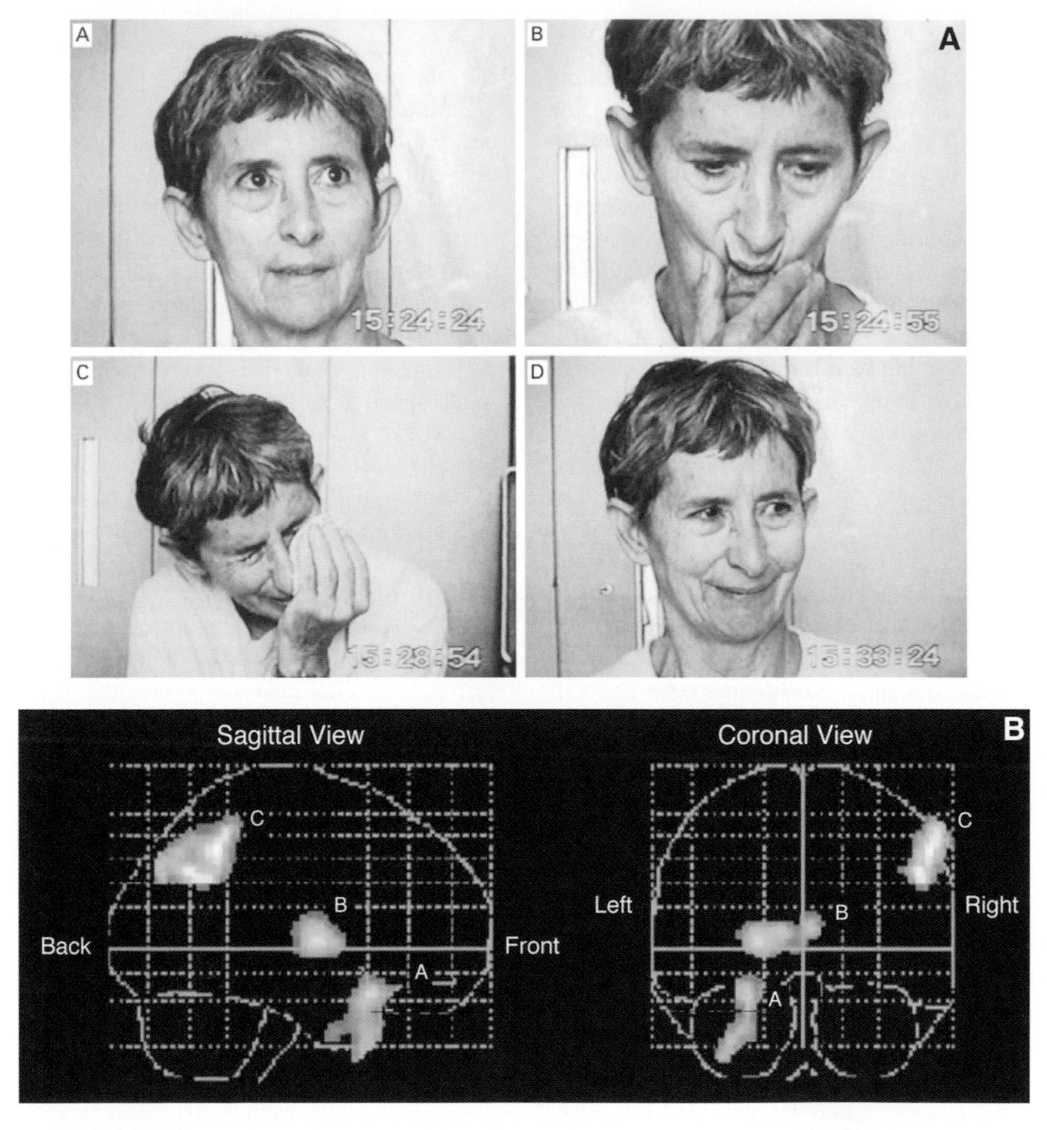

Figure 14 Depression induced by deep brain electrical stimulation in a patient with Parkinson's disease. (**A**) Videotaped images of the patient's facial expressions during depression elicited by high-frequency stimulation through an electrode implanted in the left subthalamic region for treatment of Parkinson's disease. Stimulation only through a single contact of the electrode placed on the left side caused depression. The actual time of each recording is indicated on the photograph. Image A shows the patient's usual expression while receiving levodopa. Image B shows a change in the facial expression 17 s after stimulation began. Image C shows the patient crying and expressing despair 4 min and 16 s after the start of stimulation. Image D shows the patient laughing 1 min and 20 s after the stimulator was turned off. (**B**) Brain regions activated during an acute episode of depression induced by high-frequency stimulation of the left substantia nigra. Regional cerebral blood flow was measured by positron-emission tomography with [^{15}O]water. There were foci of activation in the left orbitofrontal cortex (Brodmann's area 47), spreading to the left amygdala (A); in the left globus pallidus, spreading to the anterior thalamus (B); and in the right parietal lobe (Brodmann's area 40) (C). Reproduced, with permission, from Bejjani *et al., N. Engl. J. Med.* **340**(19), 1476–1480,

This will include the identification of critical gray matter structures to be avoided in the planning of resections as well as white matter pathways and tracts that connect such areas. Brain mapping data will be combined with surgical navigation systems and probabilistic atlases to provide neurosurgeons with a virtual environment during the course of operations (Kettenbach *et al.,* 1999). In many cases, this will make neurosurgical procedures safer, of shorter duration, and, potentially, less invasive if stereotactic or endoscopic strategies are employed.

3. Therapeutic Brain Mapping Techniques

We tend to think of brain mapping methods as passive and observational. They provide information about the structure and function of the brain with varying degrees of invasiveness. Nevertheless, some brain mapping techniques can also serve a therapeutic role. A good example is repeated stimulation with TMS and its effects on brain function (Epstein *et al.,* 1996; Pascual-Leone *et al.,* 1994; Paus *et al.,* 1997). Already this technique has been used as a potential treatment for depression in patients proven to be intractable to pharmacological and behavioral therapy (George *et al.,* 1995, 1999; Levkovitz, 2001; Maeda *et al.,* 2002; Pascual-Leone *et al.,* 1996; Post *et al.,* 1999). We predict that such strategies of focal, repetitive transcranial magnetic stimulation will prove successful in the treatments of certain disorders, including depression and some types of focal epilepsy. In addition, the use of other energy sources, such as focused ultrasound, may find a place in the treatment of lesions of the brain such as tumors or arteriovenous malformations. A by-product of implanted electrodes (Fig. 14) is that they may produce unanticipated (Bejjani *et al.,* 1999), but potentially therapeutic, responses in given subjects (see for example, Section III.B.10, in which there is a description of a subthalamic electrode that produced depression in a Parkinson's patient). We believe that these types of patient observations combined with theoretical strategies for brain stimulation will result in new therapeutic avenues for a wide range of disorders, including mental illness, addictive syndromes, and neurodegenerative diseases.

IV. Conclusion

Brain mapping methods and neuroimaging techniques have transformed the way we understand the structure and function of the human brain as well as the way we practice clinical neurology and neurosurgery today. The years to come will see only further improvements and opportunities. As a large array of new therapeutic agents enter the field of clinical neuroscience, imaging will provide the means to visualize the success or failure of these strategies, quantitatively using modern, objective, and automated strategies that will increase the speed, efficacy, and cost savings of clinical trials in the transition of potential therapies from the bench to the bedside. Ever-improving functional techniques will demonstrate the brain's chemical subsystems, integrated networks, and complex temporal choreography that result in the wide range of human behaviors, such as learning (Iacoboni *et al.,* 1999), and their aberration in disease states (Merzenich *et al.,* 1996; Tallal *et al.,* 1996).

Based on the predictions in this chapter and from observations of brain mapping techniques that are already known as well as others, yet to be discovered, brain imaging will provide the bridge among the phenotypes, the genotypes, and the behaviors of the human species. By monitoring gene expression, gene therapy, and other molecular events, the continuum between molecular biology and clinical neuroscience will become linked through brain mapping, allowing us to understand how genetic and cellular events operate in the complex and integrated networks of the human brain.

As we stated in the first edition of this text, almost every great leap in neuroscience has been preceded by the development of instrumentation or methodology. The rate of methodological progress has been steadily increasing, resulting in more comprehensive and useful maps of brain structure and function. There also has been an ever-closer relationship between neuroscience and those disciplines that can help accelerate the development of tools to map the brain. Just as importantly, international cooperation among laboratories is now a reality through electronic networks and programmatic funding mechanisms. These forces, combined with the dedication of the investigators themselves, will ultimately lead to a complete map of the human brain.

Acknowledgments

Partial support for this work was provided by a grant from the Human Brain Project (P20-MHDA52176), the National Institute of Mental Health, the National Institute for Drug Abuse, the National Cancer Institute, and the National Institute for Neurologic Disease and Stroke. For generous support the authors also thank the Brain Mapping Medical Research Organization, the Brain Mapping Support Foundation, the Pierson–Lovelace Foundation, The Ahmanson Foundation, the Tamkin Foundation, the Jennifer Jones-Simon Foundation, the Capital Group Companies Charitable Foundation, the Robson Family, the Northstar Fund, and the National Center for Research Resources Grants RR12169, RR13642, and RR08655. The authors also thank Laurie Carr, Kami Longson, and Palma Piccioni for preparation of the manuscript and Andrew Lee for preparation of the graphic materials.

References

Adolphs, R., Damasio, H., Tranel, D., Cooper, G., and Damasio, A. (2000). A role for somatosensory cortices in the visual recognition of emotion as revealed by three-dimensional lesion mapping. *J. Neurosci.* **20,** 2683–2690.

Ahlfors, S. P., Simpson, G. V., Dale, A. M., *et al.* (1999). Spatiotemporal activity of a cortical network for processing visual motion revealed by MEG and fMRI. *J. Neurophysiol.* **82,** 2545–2555.

Amunts, K., Klingberg, T., Binkofski, F., Schormann, T., Seitz, R. J., Roland, P. E., and Zilles, K. (1998). Cytoarchitectonic definition of Broca's region and its role in functions different from speech. *NeuroImage* **7,** 8.

Amunts, K., Malikovic, A., Mohlberg, H., Schormann, T., and Zilles, K. (2000). Brodmann's areas 17 and 18 brought into stereotaxic space—Where and how variable? *NeuroImage* **11,** 66–84.

Amunts, K., Schleicher, A., Bürgel, U., Mohlberg, H., Uylings, H. B. M., and Zilles, K. (1999). Broca's region revisited: Cytoarchitecture and intersubject variability. *J. Comp. Neurol.* **412,** 319–341.

Arnold, J., Liow, J.-S., Schaper, K., Stern, J., Sled, J., Shattack, D., Woth, A., Cohen, M., Leahy, R., Mazziotta, J., and Rottenberg, D. (2001). Qualitative and quantitative evaluation of six algorithms for correcting intensity non-uniformity effects. *NeuroImage* **13,** 931–943.

Atalay, M. K., Forder, J. R., Chacko, V. P., Kawamoto, S., and Zerhouni, E. A. (1993). Oxygenation in the rabbit myocardium: Assessment with susceptibility-dependent MR imaging. *Radiology* **189,** 759–764.

Atkinson, A. P., Thomas, M. S. C., and Cleeremans, A. (2000). Consciousness: Mapping the theoretical landscape. *Trends Cognit. Sci.* **4,** 372–382.

Backes, W., and van Dijk, P. (2001). Simultaneous sampling of event-related BOLD responses. *NeuroImage* **13,** S2.

Bailey, P., and von Bonin, G. (1951). "The Isocortex of Man." University Press, Urbana, IL.

Baird, A. E., and Warach, S. (1998). Magnetic resonance imaging of acute stroke. *J. Cereb. Blood Flow Metab.* **18,** 583–609.

Barbur, J. L., Watson, J. D. G., Frackowiak, R. S. J., and Zeki, S. (1993). Conscious visual perception without V1. *Brain* **116,** 1293–1302.

Bartley, A. J., Jones, D. W., and Weinberger, D. R. (1997). Genetic variability of human brain size and cortical gyral patterns. *Brain* **120,** 257–269.

Basser, P. J., and Pierpaoli, C. (1996). Microstructural and physiological features of tissues elucidated by quantitative-diffusion-tensor MRI. *J. Magn. Reson.* **3,** 209–219.

Bastos, A. C., Comeau, R., Andermann, F., Melanson, D., Cendes, F., Dubeau, F., Fontaine, S., Tampieri, D., and Olivier, A. (1999). Diagnosis of subtle focal dysplastic lesions: Curvilinear reformatting from three-dimensional magnetic resonance imaging. *Ann. Neurol.* **46,** 88–94.

Bejjani, B. P., Damier, P., Arnulf, I., Thivard, L., Bonnet, A. M., Dormont, D., Cornu, P., Pidoux, B., Samson, Y., and Agid, Y. (1999). Transient acute depression induced by high-frequency deep-brain stimulation. *N. Engl. J. Med.* **340,** 1476–1480.

Benaron, D. A., Hintz, S. R., Villringer, A., Boas, D., Kleinschmidt, A., Frahm, J., Hirth, C., Obrig, H., van Houten, J. C., Kermit, E. L., Cheong, W.-F., and Stevenson, D. K. (2000). Noninvasive functional imaging of the human brain using light. *J. Cereb. Blood Flow Metab.* 20, 469–477.

Berns, G. S., McClure, S. M., Wiest, M., Karpov, I., Pagnoni, G., Dhamala, M., and Montague, P. R. (2001). Hyperscan: Functional coupling of human brains across the internet. *Neurosci. Abstr.*

Blok, B. F., Sturms, L. M., and Holstege, G. (1998). Brain activation during micturition in women. *Brain* **121,** 2033–2042.

Bloom, F. E. (1996). The multidimensional database and neuroinformatics requirements for molecular and cellular neuroscience. *NeuroImage* **4,** S12–S13.

Boas, D. A., Gaudette, T., Strangman, G., Cheng, X., Marota, J. J. A., and Mandeville, J. B. (2001). The accuracy of near infrared spectroscopy and imaging during focal changes in cerebral hemodynamics. *NeuroImage* **13,** 76–90.

Bocher, M., Chisin, R., Parag, Y., Freedman, N., Weil, Y. M., Lester, H., Mishani, E., and Bonne, O. (2001). Cerebral activation associated with sexual arousal in response to a pornographic clip: A ^{15}O-H_2O PET study in heterosexual men. *NeuroImage* **14,** 105–117.

Bodegård, A., Geyer, S., Naito, E., Zilles, K., and Roland, P. E. (2000). Somatosensory areas in man activated by moving stimuli: Cytoarchitectonic mapping and PET. *NeuroReport* **11,** 187–191.

Bonmassar, G., Van De Moortele, P.-F., Purdon, P., Jaaskelainen, I., Ives, J. R., Vaughan, T. J., Ugurbil, K., and Belliveau, J. (2001). 7 tesla interleaved EEG and fMRI recordings: BOLD measurements. *NeuroImage* **13,** S4.

Brodmann, K. (1909). "Vergleichende Lokalisationslehre der Grosshirnrinde in Ihren Prinzipien Dargestellt auf Grund des Zellenbaues." Barth, Leipzig.

Buckow, C., Kohl, M., Zank, H., Hellmuth, O., Uludag, Steinbrink, J., Israel, H., and Villringer, A. (2001). Assessment of cortical activation by multi channel topography. *NeuroImage* **13,** S5.

Carman, G. J., Drury, H. A., and Van Essen, D. C. (1995). Computational methods for reconstructing and unfolding the cerebral cortex. *Cereb. Cortex* **5,** 506–517.

Cherry, S. R., and Gambhir, S. (2001). Use of positron emission tomography in animal research. *ILAR J.* **42,** 219–232.

Cherry, S. R., Shao, Y., Silverman, R. W., Meadors, K., Siegel, S., Chatziioannou, A., Young, J. W., Jones, W. F., Moyers, J. C., Newport, D., Boutefnouchet, A., Farquhar, T. H., Andreaco, M., Paulus, M. J., Binkley, D. M., Nutt, R., and Phelps, M. E. (1997). MicroPET: A high resolution PET scanner for imaging small animals. *IEEE Trans. Nucl. Sci.* **44,** 1161–1166.

Clark, S., and Miklossy, J. (1990). Occipital cortex in man: Organization of callosal connections, related myelo- and cytoarchitecture, and putative boundaries of functional visual areas. *J. Comp. Neurol.* **298,** 188–214.

Cohen, M. S., Goldman, R. I., Stern, J., and Engel, J. (2001). Simultaneous EEG and fMRI made easy. *NeuroImage* **13,** S6.

Conturo, T., Lori, N. F., Cull, T. S., Akbudak, E., Snyder, A. Z., Shimony, J. S., McKinstry, R. C., Burton, H., and Raichle, M. E. (1999). Tracking neuronal fiber pathways in the living human brain. *Proc. Natl. Acad. Sci. USA* **96,** 10422–10427.

Dale, A. M., Fischl, B., and Sereno, M. I. (1999). Cortical surface-based analysis. I. Segmentation and surface reconstruction. *NeuroImage* **9,** 179–194.

Damasio, A. R., Grabowski, T. J., Bechara, A., Damasio, H., Ponto, L. L. B., Parvizi, J., and Hichwa, R. D. (2000). Subcortical and cortical brain activity during the feeling of self-generated emotions. *Nat. Neurosci.* **3,** 1049–1056.

DeFalco, J., Tomishima, M., Liu, H., Zhao, C., Cai, X.-L., Marth, J. D., Enquist, L., and Friedman, J. M. (2001). Virus-assisted mapping of neural inputs to a feeding center in the hypothalamus. *Science* **291,** 2608–2613.

Derdeyn, C. P., Grubb, R. L., Jr., and Powers, W. J. (1999). Cerebral hemodynamic impairment: Methods of measurement and association with stroke risk. *Neurology* **53,** 251–259.

Desmond, J. E., Sum, J. M., Wagner, A. D., Demb, J. B., Shear, P. K., Glover, G. H., Gabrieli, J. D., and Morrell, M. J. (1995). Functional MRI measurement of language lateralization in Wada-tested patients. *Brain* **118,** 1411–1419.

Detre, J. A., Sirven, J. I., Alsop, D. C., O'Connor, J. J., and French, J. A. (1995). Localization of subclinical ictal activity by functional magnetic resonance imaging: Correlation with invasive monitoring. *Ann. Neurol.* **38,** 618–624.

Dolan, R. J. (1999). On the neurology of morals. *Nat. Neurosci.* **2,** 927–929.

Dumoulin, S. O., Bittar, R. G., Kabani, N. J., Baker, C. L., LeGoualher, G., Pike, G. B., and Evans, A. C. (2000). A new neuroanatomical landmark for the reliable identification of human area V5/MT: A quantitative analysis of sulcal patterning. *Cereb. Cortex* **10,** 454–463.

Dunn, A. K., Boaly, H., Moskowitz, M. A., and Boas, D. A. (2001). Dynamic imaging of cerebral blood flow using laser speckle. *J. Cereb. Blood Flow Metab.* **21,** 195–201.

Egan, M. F., Goldberg, T. E., Kolachana, B. S., Callicott, J. H., Mazzanti, C. M., Straub, R. E., Goldman, D., and Weinberger, D. R. (2001). Effect of COMT Val108/158 Met genotype on frontal lobe function and risk for schizophrenia. *Proc. Natl. Acad. Sci. USA* **98,** 6917–6922.

Epstein, C. M., Lah, J. K., Meador, K., Weissman, J. D., Gaitan, L. E., and Dihenia, B. (1996). Optimum stimulus parameters for lateralized suppression of speech with magnetic brain stimulation. *Neurology* **47,** 1590–1593.

Evans, A. C., Frank, J. A., Antel, J., and Miller, D. H. (1997). The role of MRI in clinical trials of multiple sclerosis: Comparison of image processing techniques. *Ann. Neurol.* **41,** 125–132.

Evins, C. S., Hoffman, E. J., and Cherry, S. R. (1999). Positron range calculation and its effect on positron emission tomography spatial resolution. *Phys. Med. Biol.* **44,** 781–799.

Falligatter, A. J., Roesler, M., Sitzmann, L., Heidrich, A., Mueller, T. J., and Strik, W. K. (1997). Loss of functional hemispheric asymmetry in Alzheimer's dementia assessed with near-infrared spectroscopy. *Brain Res. Cognit. Brain Res.* **6,** 67–72.

Felleman, D., and Van Essen, D. (1991). Distributed hierarchical processing in the primate cerebral cortex. *Cereb. Cortex* **1,** 1–47.

Firlik, A. D., Rubin, G., Yonas, H., and Wechsler, L. R. (1998). Relation between cerebral blood flow and neurologic deficit resolution in acute ischemic stroke. *Neurology* **51,** 177–182.

Flechsig, P. (1920). "Anatomie des Menschlichen Gehirns und Ruckenmarks." Thieme, Leipzig.

Fowler, C. J. (1999). Neurological disorders of micturition and their treatment. *Brain* **122,** 1213–1231.

Fox, P. T., and Lancaster, J. L. (1994). Neuroscience on the net. *Science* **266,** 994–996.

Fox, P. T., Ingham, R. J., George, M. S., Mayberg, H. S., Ingham, J. C., Roby, J., Martin, C., and Jerabek, P. (1997). Imaging human intra-cerebral connectivity by PET during TMS. *NeuroReport* **8,** 2787–2791

Fox, P. T., Mikiten, S., Davis, G., and Lancaster, J. (1994). Brain Map: A database of human functional brain mapping. *In* "Functional Neuroimaging" (R. W. Thatcher, M. Hallett, T. Zeffiro, E. R. John,and M. Huerta, eds.), pp. 95–105.Academic Press, San Diego.

Friston, K. J., Frith, C. D., Liddle, P. F., and Frackowiak, R. S. J. (1991). Comparing functional (PET) images: The assessment of significant change. *J. Cereb. Blood Flow Metab.* **11,** 690–699.

Friston, K. J., Frith, C. D., and Frackowiak, R. S. J. (1993). Time-dependent changes in effective connectivity measured with PET. *Hum. Brain Mapp.* **1,** 69–79.

Friston, K. J., Holmes, A. P., Worsley, K. J., Poline, J. P., Frith, C. D., and Frackowiak, R. S. J. (1995). Statistical parametric maps in functional imaging: A general linear approach. *Hum. Brain Mapp.* **2,** 189–210.

George, M. S., Lisanby, S. H., and Sackeim, H. A. (1999). Transcranial magnetic stimulation: Applications in neuropsychiatry. *Arch. Gen. Psychiatry* **56,** 300–311.

George, M. S., Wasserman, E. M., Williams, W. A., Callahan, A., Ketter, T. A., Basser, P., Hallett, M., and Post, R. M. (1995). Daily repetitive transcranial magnetic stimulation (rTMS) improves mood in depression. *NeuroReport* **6,** 1853–1856.

George, M. S., Wassermann, E. M., Williams, W. A., Steppel, J., Pascual-Leone, A., Basser, P., Hallett, M., and Post, R. M. (1996). Changes in mood and hormone levels after rapid-rate transcranial magnetic stimulation (rTMS) of the prefrontal cortex. *J. Neuropsychiatry Clin. Neurosci.* **8,** 172–180.

Georgieff, N., and Jeannerod, M. (1998). Beyond consciousness of external reality: A "who" system for consciousness of action and self-consciousness. *Consciousness Cognit.* **7,** 465–477.

Geschwind, N., and Levitsky, W. (1968). Human brain: Left–right asymmetries in temporal speech region. *Science* **161,** 186–187.

Gevins, A. S., Lee, J., Martin, N., Reutter, R., Desmond, J., and Brickett, P. (1994). High resolution EEG: 124-channel recording, spatial deblurring and MRI integration methods. *Electroencephalogr. Clin. Neurophysiol.* **90,** 337–358.

Geyer, S., Schleicher, A., and Zilles, K. (1997). The somatosensory cortex of human: Cytoarchitecture and regional distributions of receptor-binding sites. *NeuroImage* **6,** 27–45.

Geyer, S., Schleicher, A., and Zilles, K. (1999). Areas 3a, 3b, and 1 of human primary somatosensory cortex. 1. Microstructural organization and interindividual variability. *NeuroImage* **10,** 63–83.

Geyer, S., Schormann, T., Mohlberg, H., and Zilles, K. (2000). Areas 3a, 3b, and 1 of human primary somatosensory cortex. 2. Spatial normalization to standard anatomical space. *NeuroImage* **11,** 684–696.

Gilman, S. (1998). Medical progress: Imaging the brain: Second of two parts. *N. Engl. J. Med.* **338,** 889–898.

Goldman, R., Stern, J., Engel, J., and Cohen, M. (2001). Tomographic mapping of alpha rhythm using simultaneous EEG/fMRI. *NeuroImage* **13,** S1291.

Gratton, G., Goodman-Wood, M. R., and Fabiani, M. (2001). Comparison of neuronal and hemodynamic measure of the brain response to visual stimulation: An optical imaging study. *Hum. Brain Mapp.* **13,** 13–25.

Grefkes, C., Geyer, S., Schormann, T., Roland, P., and Zilles, K. (2001). Human somatosensory area 2: Observer-independent cytoarchitectonic mapping., interindividual variability., and population map. *NeuroImage* **14,** 617–631.

Grinvald, A., Lieke, E., Frostig, R. D., Gilbert, C. D., and Wiesel, T. N. (1986). Functional architecture of cortex revealed by optical imaging of intrinsic signals. *Nature* **324,** 361–364.

Hacke, W., Kaste, M., Fieschi, C., Toni, D., Lesaffre, E., von Kummer, R., Boysen, G., Bluhmki, E., Höxter, G., and Mahagne, M. H. (1995). Intravenous thrombolysis with recombinant tissue plasminogen activator for acute hemispheric stroke. *JAMA* **274,** 1017–1025.

Haney, S., Thompson, P. M., Cloughesy, T. F., Alger, J. R., Frew, A., Torres-Trejo, A., Mazziotta, J. C., and Toga, A. W. (2001a). Mapping therapeutic responses in a patient with malignant glioma. *J. Comput. Assisted Tomogr.* **25,** 529–536.

Haney, S., Thompson, P. M., Cloughesy, T. F., Alger, J. R., and Toga, A. W. (2001b). Tracking tumor growth rates in patients with malignant gliomas: A test of two algorithms. *Am. J. Neuroradiol.* **22,** 73–82.

Harmony, T., Fernandez, T., Silva, J., Bosch, J., Valdes, P., Fernandez-Bouzas, A., Galan, L., Aubert, E., and Rodriguez, D. (1999). Do specific EEG frequencies indicate different processes during mental calculation? *Neurosci. Lett.* **266,** 25–28.

Hayes, T. L., and Lewis, D. A. (1995). Anatomical specialization of the anterior motor speech area: Hemispheric differences in magnopyramidal neurons. *Brain Lang.* **49,** 289–308.

Heekeren, H. R., Wartenburger, I., Schmidt, H., Denkler, C., Schwintowski, H. P., and Villringer, A. (2001). The functional anatomy of moral judgment—An fMRI-study. *NeuroImage* **13,** S417.

Hoogerwerf, A. C., and Wise, K. D. (1994). A three-dimensional microelectrode array for chronic neural recording. *IEEE Trans. Biomed. Eng.* **41,** 1136–1146.

Hoshi, Y., Kosaka, S., Xie, Y., Kohri, S., and Tamura, M. (1998). Relationship between fluctuations in the cerebral hemoglobin oxygenation state and neuronal activity under resting conditions in man. *Neurosci. Lett.* **245,** 147–150.

Hsu, E. W., Hedlund, L. W., and MacFall, J. R. (1998). Functional MRI of the rat somatosensory cortex: Effects of hyperventilation. *Magn. Reson. Med.* **40,** 421–426.

Huerta, M., Koslow, S., and Leshner, A. (1993). The Human Brain Project: An international resource. *Trends Neurosci.* **16,** 436–438.

Hunter, K. E., Blaxton, T. A., Bookheimer, S. Y., Figlozzi, C., Gaillard, W. D., Grandin, C., Anyanwu, A., and Theodore, W. H. (1999). ^{15}O water positron emission tomography in language localization: A study comparing positron emission tomography visual and computerized region of interest analysis with the Wada test. *Ann. Neurol.* **45,** 662–665.

Hyde, J. S., Biswal, B., Yetkin, F. Z., and Haughton, V. M. (1995). Functional connectivity determined from analysis of physiological fluctuations in a series of echo-planar images. *Hum. Brain Mapp.* **S1,** 287.

Hyder, F., Kennan, R. P., Kida, I., Mason, G. F., Behar, K. L., and Rothman, D. (2000). Dependence of oxygen delivery on blood flow in rat brain: A 7 tesla nuclear magnetic resonance study. *J. Cereb. Blood Flow Metab.* **20,** 485–498.

Iacoboni, M., Woods, R. P., Brass, M., Bekkering, H., Mazziotta, J. C., and Rizzolatti, G. (1999). Cortical mechanisms of human imitation. *Science* **286,** 2526–2528.

Ives, J., Warach, S., Schmitt, F., Edelman, R. R., and Schomer, D. L. (1993). Monitoring the patient's EEG during echo planar MRI. *Electroencephalogr. Clin. Neurophysiol.* **87:**417–20.

Jezzard, P., Rauschecker, J. P., and Malonek, D. (1997). An in vivo model for functional MRI in cat visual cortex. *Magn. Res. Med.* **38,** 699–705.

Jones, E. G. (1998). Viewpoint: The core and matrix of thalamic organization. *Neuroscience* **85.** 331–345.

Kamada, K., Pekar, J. J., and Kanwal, J. S. (1999). Anatomical and functional imaging of the auditory cortex in awake mustached bats using magnetic resonance technology. *Brain Res.* **4,** 351–359.

Kamber, M., Collins, D. L., Shinghal, R., Francis, G. S., and Evans, A. C. (1992). Model-based 3D segmentation of multiple sclerosis lesions in dual-echo MRI data. *Proc. SPIE Visualizat. Biomed. Comp.* **1808,** 590–600.

Kamber, M., Shinghal, R., Collins, D. L., Francis, G. S., and Evans, A. C. (1995). Model-based 3D segmentation of multiple sclerosis lesions in magnetic resonance brain images. *IEEE Trans. Med. Imaging* **14,** 442–453.

Katzan, I. L., Furlan, A. J., Lloyd, L. E., Frank, J. I., Harper, D. L., Hinchey, J. A., Hammel, J. P., Qu, A., and Sila, C. A. (2000). Use of tissue-type plasminogen activator for acute ischemic stroke. *JAMA* **283,** 1151–1158.

Kayyem, J. F., Kumar, R. M., Fraser, S. E., and Meade, T. J. (1995). Receptor-targeted co-transport of DNA and magnetic resonance contrast agents. *Chem. Biol.* **2,** 615–620.

Kettenbach, J., Wong, T., Kacher, D., Hata, N., Schwartz, R. B., Black, P. M., Kikinis, R., and Jolesz, F. A. (1999). Computer-based imaging and interventional MRI: Applications for neurosurgery. *Comput. Med. Imaging Graph.* **23,** 245–258.

Kida, I., Kennan, R. P., Rothman, D. L., Behar, K. L., and Hyder, F. (2000). High-resolution CMR(O2) mapping in rat cortex: A multiparametric approach to calibration of BOLD image contrast at 7 tesla. *J. Cereb. Blood Flow Metab.* **20,** 847–860.

Kidwell, C. S., Saver, J. L., Mattiello, J., Starkman, S., Vinuela, F., Duckwiler, G., Gobin, Y. P., Jahan, R., Vespa, P., Kalafut, M., and Alger, J. R. (2000). Thrombolytic reversal of acute human cerebral ischemic injury shown by diffusion/perfusion magnetic resonance imaging. *Ann. Neurol.* **47,** 462–469.

Kleinschmidt, A., Büchel, C., Zeki, S., and Frackowiak, R. S. J. (1998). Human brain activity during spontaneously reversing perception of ambiguous figures. *Proc. R. Soc. London B* **265,** 2427–2433.

Kornblum, H. I., Araujo, D. M., Annala, A. J., Tatsukawa, K. J., Phelps, M. E., and Cherry, S. R. (2000). In vivo imaging of neuronal activation and plasticity in the rat brain by high resolution positron emission tomography (microPET). *Nat. Biotech.* **18,** 655–660.

Koroshetz, W. J., and Gonzalez, G. (1997). Diffusion-weighted MRI: An ECG for "brain attack?" *Ann. Neurol.* **41,** 565–566.

Koslow, S. H. (2000). Should the neuroscience community make a paradigm shift to sharing primary data? *Nat. Neurosci.* **3,** 863–865.

Kusaka, T., Isobe, K., Nagano, K., Okubo, K., Yasuda, S., Kondo, M., Itoh, S., and Onishi, S. (2001). Estimation of regional cerebral blood flow distribution in infants by near-infrared topography using indocyanine green. *NeuroImage* **13,** 944–952.

Lahti, K. M., Ferris, C. F., Li, F., Sotak, C. H., and King, J. A. (1998). Imaging brain activity in conscious animals using functional MRI. *J. Neurosci. Methods* **82,** 75–83.

Lee, C. C., Ward, H. A., Sharbrough, F. W., Meyer, F. B., Marsh, W. R., Raffel, C., So, E. L., Cascino, G. D., Shin, C., Xu, Y., Riederer, S. J., and Jack, C. R., Jr. (1999). Assessment of functional MR imaging in neurosurgical planning. *Am. J. Neuroradiol.* **20,** 1511–1519.

Lemieux, L., Salek-Haddadi, A., Josephs, O., Allen, P., Toms, N., Scott, C., Krakow, K., Turner, R., and Fish, D. R. (2001). Event-related fMRI with simultaneous and continuous EEG: Description of the method and initial case report. *NeuroImage* **14,** 780–787

Lev, M. (2002). CT angiography and CT perfusion imaging. *In* "Brain Mapping: The Methods" (A. W. Toga and J. C. Mazziotta, eds.), 2nd edition. Academic Press, San Diego, in press.

Levkovitz, Y. (2001). Transcranial magnetic stimulation and antidepressive drugs share similar cellular effects in rat hippocampus. *Neuropsychopharmacology* **24,** 608–616.

Li, W.-H., Fraser, S. E., and Meade, T. J. (1999). A calcium-sensitive magnetic resonance imaging contrast agent. *J. Am. Chem. Soc.* **121,** 1413–1414.

Logothetis, N. K., Pauls, J., Augath, M., Trinath, T., and Oeltermann, A. (2001). Neurophysiological investigation of the basis of the fMRI signal. *Nature* **412,** 150–157.

Louie, A., Ahrens, E., Jacobs, R., Fraser, S., and Meade, T. (2000). In vivo visualization of gene expression by magnetic resonance imaging. *Nat. Biotech.,* in press.

Lumer, E. D., Friston, K. J., and Rees, G. (1998). Neural correlates of perceptual rivalry in the human brain. *Science* **280,** 1930–1934.

Lutsep, H. L., Albers, G. W., DeCrespigny, A., Kamat, G. N., Marks, M. P., and Moseley, M. E. (1997). Clinical utility of diffusion-weighted magnetic resonance imaging in the assessment of ischemic stroke. *Ann. Neurol.* **41,** 574–80.

Maeda, F., and Pascual-Leone, A. (2002). A transcranial magnetic stimulation: Studying motor neurophysiology of psychiatric disorders and their treatment. *Psychopharmacology,* in press.

Malonek, D., and Grinvald, G. (1995). Local autoregulation of cerebral blood flow and blood volume following natural stimulation revealed by optical imaging and spectroscopy of HbO_2 and rHb. *J. Cereb. Blood Flow Metab.* **15,** S79.

Mandeville, J. B., Marota, J. J., Ayata, C., Moskowitz, M. A., Weisskoff, R. M., and Rosen, B. R. (1999). MRI measurement of the temporal evolution of relative CMRO(2) during rat fore paw stimulation. *Magn. Reson. Med.* **42,** 944–951.

Marchal, G., Serrati, C., Rioux, P., Petit-Taboué, M. C., Viader, F., de la Sayette, V., Le Doze, F., Lochon, P., Derlon, J. M., and Orgogozo, J. M. (1993). PET imaging of cerebral perfusion and oxygen consumption in acute ischemic stroke: Relation to outcome. *Lancet* **341,** 925–927.

Mayhew, J. E., Askew, S., Zheng, Y., Porrill, J., Westby, G. W. M., Redgrave, P., Rector, D. M., and Harper, R. M. (1996). Cerebral vasomotion: A 0.1-Hz oscillation in reflected light imaging of neural activity. *NeuroImage* **4,** 183–193.

Mazziotta, J. C. (2000). Imaging: Window on the brain. *Arch. Neurol.* **37,** 1413–1421.

Mazziotta, J. C., Toga, A., Evans, A., Fox, P., Lancaster, J., Zilles, K., Simpson, G., Woods, R., Paus, T., Pike, B., Holmes, C., Collins, L., Thompson, P., MacDonald, D., Schormann, T., Amunts, K., Palomero-Gallagher, N., Parsons, L., Narr, K., Kabani, N., LeGoualher, G., Boomsma, D., Cannon, T., Kawashima, R., and Mazoyer, B. (2001a). A probabilistic atlas and reference system for the human brain. *Philos. Trans. R. Soc. London B,* in press.

Mazziotta, J. C., Toga, A., Evans, A., Fox, P., Lancaster, J., Zilles, K., Simpson, G., Woods, R., Paus, T., Pike, B., Holmes, C., Collins, L., Thompson, P., MacDonald, D., Schormann, T., Amunts, K., Palomero-Gallagher, N., Parsons, L., Narr, K., Kabani, N., LeGoualher, G., Feidler, J., Smith, K., Boomsma, D., Hulshoff Pol, H., Cannon, T., Kawashima, R., and Mazoyer, B. (2001b). A four-dimensional probabilistic atlas of the human brain. *J. Am. Med. Inf. Assoc.* **8,** 401–430.

Mazziotta, J. C., Toga, A. W., Evans, A., Fox, P., and Lancaster, J. (1995a). Digital brain atlases. *Trends Neurosci.* **18,** 210–211.

Mazziotta, J. C., Toga, A. W., Evans, A., Fox, P., and Lancaster, J. (1995b). A probabilistic atlas of the human brain: Theory and rationale for its development. *NeuroImage* **2,** 89–101.

Mazziotta, J. C., Toga, A. W., and Frackowiak, R. S. J. (2000). "Brain Mapping: The Disorders." Academic Press, San Diego.

Meltzoff, A. N. (1999). Origins of theory of mind, cognition and communication. *J. Commun. Disord.* **32,** 251–269.

Merzenich, M., Jenkins, W., Johnston, P. S., Schreiner, C., Miller, S. L., and Tallal, P. (1996). Temporal processing deficits of language-learning impaired children ameliorated by training. *Science* **271,** 77–81.

Moseley, M., Kucharczyk, J., Mintorovitch, Cohen, Y., Kurhanewicz, J., Derugin, N., Asgari, H., and Norman, D. (1990). Diffusion-weighted MR imaging of acute stroke: Correlation of T2 weighted and magnetic susceptibility-enhanced MR imaging in cats. *Am. J. Neuroradiol.* **11,** 423–429.

National Institute of Neurological Disorders and Stroke rt-PA Stroke Study Group (1995). Tissue plasminogen activator for acute ischemic stroke. *N. Engl. J. Med.* **333,** 1581–1587.

Nelson, E. E., Nitschke, J. B., Rusch, B. D., Oakes, T. R., Anderle, M. J., Ferber, K. L., Pederson, A. J. C., and Davidson, R. J. (2001). Motherly love: An fMRI study of mothers viewing pictures of their infants. *NeuroImage* **13,** S450.

Ogawa, S., Lee, T. M., Nayak, A. S., and Glynn, P. (1990). Oxygenation-sensitive contrast in magnetic-resonance image of rodent brain at high magnetic-fields. *Magn. Reson. Med.* **14,** 68–78.

Ono, M., Kubik, S., and Abernathy, C. (1990). "Atlas of the Cerebral Sulci." Thieme, Stuttgart.

Pascual-Leone, A., Grafman, J., Cohen, L. G., Roth, B. J., and Hallett, M. (1994). Transcranial magnetic simulation: A new tool for the study of higher cognitive functions in humans. *In* "Handbook of Neuropsychology" (F. Boller and J. Graffman, eds.). Elsevier, New York.

Pascual-Leone, A., Rubio, B., Pallardo, F., and Catala, M. D. (1996). Rapid-rate transcranial magnetic stimulation of left dorsolateral prefrontal cortex in drug-resistant depression. *Lancet* **358,** 234–237.

Paus, T., Castro-Alamancos, M., and Petrides, M. (2002). Cortico-cortical connectivity of the human mid-dorsolateral frontal cortex and its modulation by repetitive transcranial magnetic stimulation. *Eur. J. Neurosci.,* in press.

Paus, T., Collins, D. L., Evans, A. C., Leonard, G., Pike, B., and Zijdenbos, A. (2001a). Maturation of white matter in the human brain: A review of magnetic-resonance studies. *Brain Res. Bull.* **54,** 255–266.

Paus, T., Collins, D. L., Evans, A. C., Leonard, G., Pike, B., and Zijdenbos, A. (2001b). Maturation of white matter in the human brain: A review of magnetic resonance studies. *Brain Res. Bull.* **54,** 255–266.

Paus, T., Jech, R., Thompson, C. J., Comeau, R., and Evans, A. C. (1997). Transcranial magnetic stimulation during positron emission tomography: A new method for studying connectivity of the human cerebral cortex. *J. Neurosci.* **17,** 3178–3184.

Paus, T., Otaky, N., Caramanos, Z., MacDonald, D., Zijdenbos, A., D'Avirro, D., Gutmans, D., Holmes, C. J., Tomaiuolo, F., and Evans, A. C. (1996a). In-vivo morphometry of the intrasulcal gray-matter in the human cingulate, paracingulate and superior-rostral sulci: Hemispheric asymmetries, gender differences, and probability maps. *J. Comp. Neurol.* **376,** 664–673.

Paus, T., Tomaiuolo, F., Otaky, N., MacDonald, D., Petrides, M., Atlas, J., Morris, R., and Evans, A. C. (1996b). Human cingulate and paracingulate sulci: Pattern, variability, asymmetry, and probabilistic map. *Cereb. Cortex* **6,** 207–214.

Peeters, R. R., Verhoye, M., Vos, B. P., Van Dyck, D., Van Der Linden, A., and De Schutter, E. (1999). A patchy horizontal organization of the somatosensory activation of the rat cerebellum demonstrated by functional MRI. *Eur. J. Neurosci.* **11,** 2720–2730.

Penhune, V. B., Zatorre, R. J., MacDonald, J. D., and Evans, A. C. (1996). Interhemispheric anatomical differences in human primary auditory cortex: Probabilistic mapping and volume measurement from MR scans. *Cereb. Cortex* **6,** 617–672.

Phelps, M. E. (2000). PET: The merging of biology and imaging into molecular imaging. *J. Nucl. Med.,* in press.

Pierpaoli, C., Jezzard, P., Basser, P. J., Barnett, A., and Di Chiro, G. (1996). Diffusion tensor MR imaging of the human brain. *Radiology* **3,** 637–648.

Post, R. M., Kimbrell, T. A., McCann, U. D., Dunn, R. T., Osuch, E. A., Speer, A. M., and Weiss, S. R. (1999). Repetitive transcranial magnetic stimulation as a neuropsychiatric tool: Present status and future potential. *J. ECT* **15,** 39–59.

Prichard, J. W., and Rosen, B. R. (1994). Functional study of the brain by NMR. *J. Cereb. Blood Flow Metab.* **14,** 365–372.

Prichard, J. W., Rothman, D., Novony, E., Petroff, O., Kuwabara, T., Avison, M., Howseman, A., Hanstock, C., and Shulman, R. (1991). Lactate rise detected by ^{1}H NMR in human visual cortex during physiologic stimulation. *Proc. Natl. Acad. Sci. USA* **88,** 5829–5831.

Rademacher, J., Galaburda, A., Kennedy, D., Filipek, P., and Caviness, V. (1992). Human cerebral cortex: Localization, parcellation, and morphometry with magnetic resonance imaging. *J. Cognit. Neurosci.* **4,** 352–374.

Rademacher, J., Caviness, V. S., Steinmetz, H., and Galaburda, A. M. (1993). Topographical variation of the human primary cortices: Implications for neuroimaging, brain mapping and neurobiology. *Cereb. Cortex* **3,** 313–329.

Rees, G., Wojciullik, E., Clarke, K., Husain, M., Frith, C., and Driver, J. (2000). Unconscious activation of visual cortex in the damaged right hemisphere of a parietal patient with extinction. *Brain* **123,** 1624–1633.

Reese, T., Bjelke, B., Porszasz, R., Baumann, D., Bochelen, D., Sauter, A., and Rudin, M. (2000). Regional brain activation by bicuculline visualized by functional magnetic resonance imaging. Time-resolved assessment of bicuculline-induced changes in local cerebral blood volume using an intravascular contrast agent. *NMR Biomed.* **13,** 43–49.

Reiman, E. M., Armstrong, S. M., Matt, K. S., and Mattox, J. H. (1996). The application of positron emission tomography to the study of the normal menstrual cycle. *Hum. Reprod.* **11,** 2799–2805.

Robitaille, P.-M. L., Abduljalil, A. M., Kangarlu, A., Zhang, X., Yu, Y., Burgess, R., Bair, S., Noa, P., Yang, L., Zhu, H., Palmer, B., Jiang, Z., Chakeres, D. M., and Spigos, D. (1998). Human magnetic resonance imaging at 8T. *NMR Biomed.* **11,** 263–265.

Roland, P. E., and Zilles, K. (1994). Brain atlases: A new research tool. *Trends Neurosci.* **17,** 458–467.

Roland, P. E., and Zilles, K. (1996). The developing European Computerized Human Brain Database for all imaging modalities. *NeuroImage* **4,** 39–47.

Schlaug, G., Benfield, A., Baird, A. E., Siewert, B., Lövblad, K. O., Parker, R. A., Edelman, R. R., and Warach, S. (1999). The ischemic penumbra: Operationally defined by diffusion and perfusion MRI. *Neurology* **53,** 1528–1537.

Seghier, M., Dojat, M., Delon-Martin, C., Rubin, J., Warnking, C., Segebarth, C., and Bullier, J. (2000). Moving illusory contours activate primary visual cortex: An fMRI study. *Cereb. Cortex* **10,** 663–670.

Sereno, M. I., McDonald, C. T., and Allman, J. M. (1994). Analysis of retinotopic maps in extrastriate cortex. *Cereb. Cortex* **4,** 601–620.

Silbersweig, D. A., Stern, E., Frith, C. D., Seward, J., Holmes, A., Schnorr, L., Seward, J., Holmes, A., Schnorr, L., Cahill, C., McKenna, P., Chua, S., Jones, T., and Frackowiak, R. S. J. (1995). Mapping the neural correlates of auditory hallucinations in schizophrenia: Involuntary perceptions in the absence of external stimuli. *Hum. Brain Mapp. Suppl.* **1,** 422.

Sled, J., Zijdenbos, A., and Evans, A. (1998). A non-parametric method for automatic correction of intensity non-uniformity in MRI data. *IEEE Trans. Med. Imaging* **17,** 87–97.

Small, S. A., Wu, E. X., Bartsch, D., Perera, G. M., Lacefield, C. O., DeLaPaz, R., Mayeux, R., Stern, Y., and Kandel, E. R. (2000). Imaging physiologic dysfunction of individual hippocampal subregions in humans and genetically modified mice. *Neuron* **28,** 653–664.

Sowell, E. R., Thompson, P. M., Holmes, C. J., Jernigan, T. L., and Toga, A. W. (1999). In vivo evidence for post adolescent brain maturation in frontal and striated regions. *Nat. Neurosci.* **2,** 859–861.

Spence, S. A., Farrow, T., Herford, A., Zheng, Y., Wilkinson, I. D., Brook, M. L., and Woodruff, P. W. R. (2001). A preliminary description of the behavioral and functional anatomical correlates of lying. *NeuroImage* **13,** S477.

Staroselskaya, I., Chaves, C., Silver, B., Linfante, I., Edelman, R., Caplan, L., Warach, S., and Baird, A. (2001). Relationship between magnetic resonance arterial patency and perfusion–diffusion mismatch in acute ischemic stroke and its potential clinical use. *Arch. Neurol.* **58,** 1069–1074.

Steinmetz, H., Volkman, J., Jancke, L., and Freund, H. (1991). Anatomical left–right asymmetry of language-related temporal cortex is different in left and right handers. *Ann. Neurol.* **29,** 315–319.

Stieltjes, B., Kaufmann, W. E., van Ziljl, P. C. M., Fredericksen, K., Pearlson, G. D., Solaiyappan, M., and Mori, S. (2001). Diffusion tensor imaging and axonal tracking in the human brainstem. *NeuroImage* **14,** 723–735.

Strother, S. C., Anderson, J. R., Xu, X.-L., Low, J.-S., Boar, D. C., and Rottenberg, D. A. (1994). Quantitative comparisons of image registration techniques based on high-resolution MRI of the brain. *J. Comput. Assisted Tomogr.* **18,** 954–962.

Szapiel, S. (2001). Optical imaging and its role in clinical neurology. *Arch. Neurol.* **58,** 1061–1065.

Talairach, J., and Tournoux, P. (1988). Principe et technique des études anatomiques. *In* "Co-planar Stereotaxic Atlas of the Human Brain—3-Dimensional Proportional System: An Approach to Cerebral Imaging." Thieme, New York.

Tallal, P., Miller, S. L., Bedi, G., Byma, G., Wang, X., Nagarajan, S. S., Schreiner, C., Jenkins, W. M., and Merzenich, M. M. (1996). Language comprehension in language-learning impaired children improved with acoustically modified speech. *Science* **271,** 81–84.

Thompson, P. M., Moussai, J., Khan, A. A., Zohoori, S., Goldkorn, A., Mega, M. S., Small, G. W., Cummings, J. L., and Toga, A. W. (1998). Cortical variability and asymmetry in normal aging and Alzheimer's disease. *Cereb. Cortex* **8,** 492–509.

Thompson, P. M., Giedd, J. N., Woods, R. P., MacDonald, D., Evans, A. C., and Toga, A. W. (2000a). Growth patterns in the developing brain detected by using continuum mechanical tensor maps. *Nature* **404,** 190–193.

Thompson, P. M., Mega, M. S., and Toga, A. W. (2000b). Disease-specific brain atlases. *In* "Brain Mapping: The Disorders" (A. W. Toga, J. C. Mazziotta, and R. S. J. Frackowiak, eds.), pp. 131–177. Academic Press, San Diego.

Thompson, P., Cannon, T. D., Narr, K. L., van Erp, T., Poutanen, V. P., Hutteunen, M., Lonnqvist, J., Standertskjold-Nordenstam, C. G., Kaprio, J., Khaledy, M., Dail, R., Zoumalen, C., and Toga, A. W. (2001). Genetic influences on brain structure. *Nat. Neurosci.,* in press.

Toga, A., Cannestra, A., and Black, K. (1995). The temporal/spatial evolution of optical signals in human cortex. *Cereb. Cortex* **5,** 561–565.

Toga, A. W., Goldkorn, A., Ambach, K., Chao, K., Quinn, B. C., and Yao, P. (1997). Postmortem cryosectioning as an anatomic reference for human brain mapping. *Comp. Med. Image Graph.* **21,** 131–141.

Toga, A., and Mazziotta, J. (1996). "Brain Mapping: The Methods," 1st edition. Academic Press, San Diego.

Toga, A. W., Rex, D. E., and Ma, J. (2001). A graphical interoperable processing pipeline. *NeuroImage* **13,** S266.

Tononi, G., and Edelman, G. M. (1998). Consciousness and complexity. *Science* **282,** 1846–1851.

van der Knaap, M. S., and Valk, J. (1990). MR imaging of the various stages of normal myelination during the first year of life. *Neuroradiology* **31,** 459–470.

Ugurbil, K. (2001). *Magn. Reson. Med.* **45,** 588–594.

Van Essen, D. C., and Drury, H. A. (1997). Structural and functional analyses of human cerebral cortex using a surface-based atlas. *J. Neurosci.* **17,** 7079–7102.

Van Essen, D. C., Drury, H. A., Joshi, S., and Miller, M. I. (1998). Functional and structural mapping of human cerebral cortex: Solutions are in the surfaces. *Proc. Natl. Acad. Sci. USA* **95,** 788–795.

Van Horn, J. D., Grethe, J. S., Kostelec, P., Woodward, J. B., Aslam, J. A., Rus, D., Rockmore, D., and Gazzaniga, M. S. (2001). The Functional Magnetic Resonance Imaging Data Center (fMRIDC): The challenges and rewards of large-scale databasing of neuroimaging studies. *Philos. Trans. R. Soc. London B* **356,** 1–17.

Weiner, H. L., Guttmann, C. R., Khoury, S. J., Orav, E. J., Hohol, M. J., Kikinis, R., and Jolesz, F. A. (2000). Serial magnetic resonance imaging in multiple sclerosis: Correlation with attacks, disability, and disease stage. *J. Neuroimmunol.* **104,** 164–173.

Wise, K. D., and Najafi, K. (1991). Microfabrication techniques for integrated sensors and microsystems. *Science* **254,** 1335–1342.

Woods, R. P., Cherry, S. R., and Mazziotta, J. C. (1992). A rapid automated algorithm for accurately aligning and reslicing positron emission tomography images. *J. Comput. Assisted Tomogr.* **16,** 620–633.

Woods, R. P., Grafton, S. T., Holmes, C. J., Cherry, S. R., and Mazziotta, J. C. (1998). Automated image registration. I. General methods and intrasubject, intramodality validation. *J. Comput. Assisted Tomogr.* **22,** 139–152.

Woods, R. P., Mazziotta, J. C., and Cherry, S. R. (1993). MRI–PET registration with automated algorithm. *J. Comput. Assisted Tomogr.* **17,** 536–546.

Yang, X., Renken, R., Hyder, F., Siddeek, M., Greer, C. A., Shepherd, G. M., and Shulman, R. G. (1998). Dynamic mapping at the laminar level of odor-elicited responses in rat olfactory bulb by functional MRI. *Proc. Natl. Acad. Sci. USA* **95,** 7715–7720.

Zeman, A. (2001). Consciousness. *Brain* **124,** 1263–1289.

Zijdenbos, A. P., and Dawant, B. M. (1994). Brain segmentation and white matter lesion detection in MR images. *Crit. Rev. Biomed. Eng.* **22,** 401–465.

Zijdenbos, A. P., Evans, A. C., Riahi, F., Sled, J. G., Chui, H.-C., and Kollokian, V. (1996). Automatic quantification of multiple sclerosis lesion volume using stereotaxic space. *In* "Proceedings of the 4th International Conference on Visualization in Biomedical Computing VBC '96, Hamburg, pp. 439–448.

Zilles, K., Schlaug, G., Matelli, M., Luppino, G., Schleicher, A., Qu, M., Dabringhaus, A., Seitz, R., and Roland, P. (1995). Mapping of human and macaque sensorimotor areas by integrating architectonic, transmitter receptor, MRI and PET data. *J. Anat.* **187,** 515–537.

Zilles, K., Schleicher, A., Langemann, C., Amunts, K., Morosan, P., Palomero-Gallagher, N., Schormann, T., Mohlberg, H., Bürgel, U., Steinmetz, H., Schlaug, G., and Roland, P. E. (1997). Quantitative analysis of sulci in the human cerebral cortex: Development., regional heterogeneity, gender difference, asymmetry, intersubject variability and cortical architecture. *Hum. Brain Mapp.* **5,** 218–221.

Zourabian, A., Siegel, A., Chance, B., Ramanujan, N., Rhode, M., and Boas, D. A. (2000). Trans-abdominal monitoring of fetal arterial blood oxygenation using pulse oximetry. *J. Biomed. Opt.* **5,** 391–405.

Index

A

B

C

D

E

F

G

H

I

L

M

N

O

R

S

T

V

W

Y

Z